**Energy Technology
Handbook**

OTHER McGRAW-HILL HANDBOOKS OF INTEREST

Baumeister and Marks • Standard Handbook for Mechanical Engineers
Beeman • Industrial Power Systems Handbook
Brady • Materials Handbook
Callender • Time-Saver Standards for Architectural Design Data
Carrier Air Conditioning Company • Handbook of Air Conditioning System Design
Considine • Chemical and Process Technology Encyclopedia
Considine • Encyclopedia of Instrumentation and Control
Considine • Process Instruments and Controls Handbook
Crocker and King • Piping Handbook
Croft • American Electrician's Handbook
De Chiara • Time-Saver Standards for Building Types
Dudley • Gear Handbook
Emerick • Handbook of Mechanical Specifications for Buildings and Plants
Emerick • Heating Handbook
Factory Mutual Engineering Division • Handbook of Industrial Loss Prevention
Fink • Electronics Engineer's Handbook
Fink and Carroll • Standard Handbook for Electrical Engineers
Gartmann • De Laval Engineering Handbook
Harris • Handbook of Noise Control
Harris and Crede • Shock and Vibration Handbook
Heyel • The Foreman's Handbook
Hicks • Standard Handbook of Engineering Calculations
King and Brater • Handbook of Hydraulics
Korn and Korn • Mathematical Handbook for Scientists and Engineers
LeGrand • The New American Machinists' Handbook
Lewis • Facilities and Plant Engineering Handbook
Lund • Industrial Pollution Control Handbook
Machol • System Engineering Handbook
Mantell • Engineering Materials Handbook
Maynard • Industrial Engineering Handbook
Merritt • Building Construction Handbook
Morrow • Maintenance Engineering Handbook
Perry • Chemical Engineers' Handbook
Perry • Engineering Manual
Rossnagel • Handbook of Rigging
Rothbart • Mechanical Design and Systems Handbook
Shand • Glass Engineering Handbook
Society of Manufacturing Engineers • Tool and Manufacturing Engineers' Handbook
Streeter • Handbook of Fluid Dynamics
Truxal • Control Engineers' Handbook

Energy Technology Handbook

Prepared by 142 Specialists

Douglas M. Considine, P.E., *editor-in-chief*

Consulting Engineer
Los Angeles, California

McGRAW-HILL BOOK COMPANY

New York   St. Louis   San Francisco   Auckland   Bogotá
Düsseldorf   Johannesburg   London   Madrid   Mexico
Montreal   New Delhi   Panama   Paris   São Paulo
Singapore   Sydney   Tokyo   Toronto

Library of Congress Cataloging in Publication Data

Main entry under title:
Energy technology handbook.

Includes index.
1. Power (Mechanics)—Handbooks, manuals, etc.
2. Power resources—Handbooks, manuals, etc. I. Considine, Douglas Maxwell.
TJ163.9.E54   621   76-17653
ISBN 0-07-012430-2

The editors for this book were Harold B. Crawford, Winifred Eisler, and Lester Strong,
the designer was Naomi Auerbach, and the production supervisor
was Teresa F. Leaden. It was set in Caledonia
by University Graphics, Inc.

Printed and bound by The Kingsport Press.

12163 40 - 210

Overview of Contents

For the detailed contents of any subsection, consult the title page of that subsection. The alphabetical index is at the end of the book.

Contents

v

GAS TECHNOLOGY*

PETROLEUM TECHNOLOGY

*NOTE: Several SNG processes are described in Handbook Section on *Gas Technology*. SNG
from waste materials is described in Handbook Section on *Chemical Fuels Technology*.

CHEMICAL FUELS TECHNOLOGY

NUCLEAR ENERGY TECHNOLOGY

SOLAR ENERGY TECHNOLOGY

Contributors

A. N. ADDIE, B.S., M.S.M.E. *Manager, Advance Engineering, Electro-Motive Division, General Motors Corporation, La Grange, Illinois.* (RAIL POWER SYSTEMS)

M. M. ADIBI. *Industry Consultant, International Business Machines Corporation, Houston, Texas.* (TRENDS IN TECHNICAL USE OF COMPUTER SYSTEMS IN THE POWER INDUSTRY)

O. J. ADLHART, Ph.D. *Systems Department, Engelhard Minerals & Chemicals Corporation, East Newark, New Jersey.* (FUEL CELLS)

SAYEED AKHTAR, Ph.D. *Project Leader,* Synthoil *Exploratory Engineering, Pittsburgh Energy Research Center, Bureau of Mines, U. S. Department of the Interior, Pittsburgh, Pennsylvania.* (SYNTHOIL PROCESS FOR CONVERTING COAL TO NONPOLLUTING FUEL OIL)

MARK T. ATWOOD. *Manager of Laboratories, The Oil Shale Corporation, Golden, Colorado.* (OIL SHALE RETORTING)

J. D. BAILIE, B.S.M.E. *Director of Technical Sales, Petroleum Chemicals Division, Ethyl Corporation, Houston, Texas.* (METHYLCYCLOPENTADIENYL MANGANESE TRICARBONYL COMBUSTION IMPROVER)

JOHN BAKOSS, B.S. *Staff Engineer, J. H. Fletcher & Co., Huntington, West Virginia.* (ROOF CONTROL FOR UNDERGROUND MINING)

M. S. BALDWIN, B.S.E.E. *Generation Consultant, Power Generation Systems, Westinghouse Electric Corporation, Pittsburgh, Pennsylvania.* (SUPERCONDUCTING TURBINE GENERATORS)

P. BARRY BALDWIN, B.A., M.A. *Assistant to Vice-President, Marketing, Marine Transport Lines, Inc., a subsidiary of General American Transportation Corporation, New York.* (SEAGOING VESSELS FOR TRANSPORTING LIQUID AND GASEOUS FUELS)

CLAYTON G. BALL, Ph.D. *Consultant, Paul Weir Company Incorporated, Chicago, Illinois.* (COAL RESERVES—ESTIMATES) (COAL EXPLORATION)

WAYNE T. BARRETT, Ph.D. *President, Foote Mineral Company, Exton, Pennsylvania.* (LITHIUM FOR THERMONUCLEAR FUSION REACTORS)

RALPH BLOOM, JR., B.S.Ch.E. *Manager, Economics and Planning, COGAS Development Company, Princeton, New Jersey.* (COGAS PROCESS FOR CONVERSION OF COAL TO GAS AND OIL PRODUCTS)

R. S. BOLTON, B.E., *Diploma Imperial College. Chief Geothermal Engineer, Ministry of Works New Zealand, Wellington North, New Zealand.* (GEOTHERMAL ENERGY IN NEW ZEALAND)

N. S. BOODMAN, B.S., M.S. *Section Supervisor, Research Laboratory, U.S. Steel Corporation, Monroeville, Pennsylvania.* (CARBONIZING/HYDROGENATION METHOD FOR PRODUCING METALLURGICAL COKE)

R. G. BOWEN. *Consulting Geologist, Bend, Oregon.* (GEOTHERMAL ENERGY)

T. C. BUECHLEY, B.S., M.S.Ch.E. *Development Engineer, Halliburton Services, Duncan, Oklahoma.* (DRILLING AND PRODUCTION)

R. J. BUIST, B.S., M.S. *Engineering Manager, Thermoelectrics Department, Borg-Warner Corporation, Des Plaines, Illinois.* (THERMOELECTRIC COOLING)

JOHN O. BURCKLE. *Research Chemical Engineer, Office of Research and Development, U.S. Environmental Protection Agency, National Environmental Research Center, Cincinnati, Ohio.* (PARTICULATE EMISSION MEASUREMENT METHODS)

JAMES C. BURKE, B.S., M.S. *Arthur D. Little, Inc., Cambridge, Massachusetts.* (SOLAR CLIMATE CONTROL)

FRANKLIN B. CARLSON, B.S.Ch.E. *Manager, Pilot Plant and Engineering, The Oil Shale Corporation, Rocky Flats Research Center, Golden, Colorado.* (THE TOSCOAL PROCESS FOR PYROLYSIS OF COAL)

MAURICE CARRIGY, M.Sc. *Senior Advisor on Oil Sands, Alberta Mines and Minerals, and Senior Research Officer, Alberta Research Council, Edmonton, Canada.* (FUEL FROM TAR SANDS)

RICHARD A. CHAPMAN, B.S., M.S.C.E. *Senior Civil Engineer, Solid and Hazardous Waste Research Laboratory, National Environmental Research Center, U.S. Environmental Protection Agency, Cincinnati, Ohio.* (SOLID WASTE SYSTEMS FOR ENERGY GENERATION)

L. T. CLERE, B.S.Ch.E. *Associate Environmental Engineer, Sun Oil Company, Toledo, Ohio.* (BIO-OXIDATION PROCESS REDUCES WATER USE)

J. H. CRONIN, B.A., B.S. *Advisory Engineer, Transmission and Distribution Systems Engineering, Westinghouse Electric Corporation, East Pittsburgh, Pennsylvania. Member, Institute of Electrical and Electronics Engineers.* (TRENDS IN ELECTRIC POWER TRANSMISSION)

A. B. CROSSMAN, B.A., B.S.C.E. *Vice President, Brown & Root International Ltd., Houston, Texas.* (MARINE PIPELINES)

GEORGE P. CURRAN, B.S.Ch.E. *Manager of Gasification Research, Conoco Coal Development Company, Library, Pennsylvania.* (THE CO_2 ACCEPTOR COAL GASIFICATION PROCESS)

W. V. DAILEY. *Technical Products Division, Mine Safety Appliances Company, Pittsburgh, Pennsylvania.* (MONITORING OF HYDROCARBON EMISSIONS) (MONITORING OF CARBON MONOXIDE EMISSIONS)

J. L. S. DALEY, B.S. *Deceased, formerly Development Associate, Battery Engineering Department, Union Carbide Corporation, Cleveland, Ohio.* (BATTERIES)

BILL DICKIE, Q.C. *Formerly Minister of Mines and Minerals, Alberta, Edmonton, Canada.* (FUEL FROM TAR SANDS)

RICHARD K. DORAN, M.E., M.S. *Associate Director, Gas Requirements Agency, Denver, Colorado.* (FUTURE GAS CONSUMPTION OF THE UNITED STATES)

ERNEST A. DRAEGER, B.S. *Project Engineer, McNally Pittsburg Manufacturing Corporation, Pittsburg, Kansas.* (COAL PREPARATION PLANTS)

HENRY E. DUCKHAM, JR., B.S.Ch.E. *Process Manager in Cryogenices, The M. W. Kellogg Company, a division of Pullman Incorporated, Houston, Texas.* (CHEMICAL FEEDSTOCKS FROM NATURAL GAS)

GORDON L. DUGGER, Ph.D. *Applied Physics Laboratory, The Johns Hopkins University, Laurel, Maryland.* (OCEAN THERMAL ENERGY CONVERSION)

S. G. DUKELOW. *Manager, Power Generation Marketing, Bailey Meter Company, a subsidiary of The Babcock & Wilcox Company, Wickliffe, Ohio.* (CONTROLS FOR FOSSIL FUEL/STEAM ENERGY CONVERSION)

HERMAN F. FELDMAN, B.S., M.S.Ch.E. *Manager, Coal Research Programs, Battelle Columbus Laboratories, Columbus, Ohio.* (PRODUCTION OF HIGH-BTU GAS BY THE HYDRANE PROCESS)

J. FILE, Ph.D. *Plasma Physics Laboratory, Princeton University, Princeton, New Jersey.* (ENERGY) (SUPERCONDUCTING MAGNETS AS APPLIED TO FUSION REACTORS)

CARL E. FINK, B.S.Ch.E. *Project Manager of Pilot Plant, Conoco Coal Development Company, Rapid City, South Dakota.* (THE CO_2 ACCEPTOR COAL GASIFICATION PROCESS)

E. J. FISCH, B.S. *Staff Engineer, Process Engineering-Licensing Department, Shell Development Company, Houston, Texas.* (ABSORPTION OF ACIDIC GASES FROM NATURAL GAS AND SNG)

J. ROBERT FLETCHER, B.S.M.E. *President, J. H. Fletcher & Co., Huntington, West Virginia.* (ROOF CONTROL FOR UNDERGROUND MINING)

S. FRIEDMAN, B.Ch.E. *Assistant Research Supervisor and Supervisory Chemical Engineer, Pittsburgh Energy Research Center, Bureau of Mines, U.S. Department of the Interior, Pittsburgh, Pennsylvania. Member, American Chemical Society, American Institute of Chemical Engineers.* (WOOD-TO-OIL PROCESS)

JOHN F. S. FRITH, Ph.D. *Assistant Manager, Chemical Engineering Department, The*

Lummus Company, a subsidiary of Combustion Engineering, Inc., Bloomfield, New Jersey. (THE SYNTHANE COAL GASIFICATION PROCESS)

JOSEPH GELZER. *Manager of Product Development in the Software and Computer Systems Engineering Organization, The Boeing Company, Seattle, Washington.* (SOLAR CELLS AND ARRAYS)

A. S. GIBSON, B.S.M.E. *Manager, Product Engineering, Fast Breeder Reactor Department, Energy Systems and Technology Division, General Electric Company, Sunnyvale, California.* (THE LIQUID METAL FAST BREEDER REACTOR)

P. N. GLOVER. *Chairman, Editorial Committee of Potential Gas Committee, Mineral Resources Institute, Colorado School of Mines Foundation, Inc., Golden, Colorado.* (POTENTIAL SUPPLY OF NATURAL GAS IN THE UNITED STATES)

ROBERT J. GRACE, B.S.Ch.E. *Process Engineering Manager, Bituminous Coal Research, Inc., Monroeville, Pennsylvania.* (BI-GAS PROCESS FOR PRODUCTION OF HIGH-BTU PIPELINE GAS FROM COAL)

JAMES A. GRAY, B.E., M.E., Ph.D. *Supervisor Chemical Engineer, Coal Gasification – Hydrane Process, Pittsburgh Energy Research Center, Bureau of Mines, U.S. Department of the Interior, Pittsburgh, Pennsylvania.* (PRODUCTION OF HIGH-BTU GAS BY THE HYDRANE PROCESS)

DANIEL L. GREGORY, B.S. *Senior Specialist Engineer, Space-Based Power Study, Research and Engineering Division, Boeing Aerospace Company, Seattle, Washington.* (HEAT ENGINES FOR SOLAR POWER) (SATELLITE ENERGY SYSTEMS FOR EARTH POWER)

WILLEM GROENENDAAL. *Shell Internationale Petroleum Maatschappij B.V., The Hague, The Netherlands.* (CLAUS SULFUR RECOVERY UNITS AND CLAUS OFF-GAS TREATING)

E. A. GROH. *Consulting Geologist, Portland, Oregon.* (GEOTHERMAL ENERGY)

ALFRED L. HABOUSH, B.S. *Senior Project Manager, General Atomic Company, San Diego, California.* (NUCLEAR POWER PLANT—CONCEPT TO STARTUP)

PIET J. HALBMEYER. *Head of the Oil Gasification and Technical Back-Up, Licensing Division, Shell Internationale Petroleum Maatschappij B.V., The Hague, The Netherlands.* (GASIFICATION AS BASIS FOR COMBINED-CYCLE POWER GENERATION)

KENNETH E. HASTINGS, Ph.D. *Manager, Synthetic Gas Projects, Cities Service Research and Development Company, Cranbury, New Jersey.* (HYDRODESULFURIZATION OF PETROLEUM CRUDE OIL FRACTIONS AND PETROLEUM PRODUCTS)

VICTOR G. HERMAN, Ph.D. *New York, N.Y.* (DYNAMOMETERS FOR MEASUREMENT OF RELATIVELY SMALL FORCES)

E. WENDELL HEWSON, Ph.D. *Professor and Chairman, Department of Atmospheric Sciences, Oregon State University, Corvallis, Oregon.* (ELECTRICAL ENERGY FROM THE WIND)

A. F. HILDEBRANDT, Ph.D. *Director, Solar Energy Laboratory, University of Houston, Houston, Texas.* (SOLAR TOWER ENERGY COLLECTOR)

E. S. HILL, Ph.D. *Consultant, Houston, Texas.* (ABSORPTION OF ACIDIC GASES FROM NATURAL GAS AND SNG)

JAN C. HOOGENDOORN. *Manager, Process Division, South African Coal, Oil and Gas Corporation Limited, Sasolburg, Republic of South Africa.* (COAL GASIFICATION AT SASOL)

J. HUEBLER, B.S. *Senior Vice-President, Institute of Gas Technology, Chicago, Illinois.* (THE HYGAS COAL GASIFICATION PROCESS)

G. C. HUMPHREYS, Bach. Tech. Sci. *Head of Methanol Group, Process Engineering Department, Davy Powergas Ltd., London, England.* (METHANOL PRODUCTION AND FUEL METHANOL VS. LNG)

J. C. JANKA, B.S., M.S. *Associate Chemical Engineer, Insitute of Gas Technology, Chicago, Illinois.* (HYDROCRACKING-HYDROGASIFICATION PROCESS FOR PRODUCING PIPELINE GAS FROM CRUDE OIL)

R. B. JAQUES, Engineer of Mines. *Formerly Systems Engineer, Black Mesa Pipeline, Inc., Flagstaff, Arizona.* (COAL TRANSPORTATION BY SLURRY PIPELINE)

STEPHEN A. JOHNSON, B.S.Ch.E. *Senior Research Engineer, Riley Stoker Corporation, Worcester, Massachusetts.* (NO_x CONTROL BY FURNACE AND BURNER DESIGN)

W. G. JOHNSON. *Manager, Chemical Process Department, The Rust Engineering Company, a subsidiary of Wheelabrator-Frye Inc., Birmingham, Alabama.* (COAL PURIFICATION BY SOLVENT REFINING)

JOHN F. JONES, Ph.D. *Technical Director of Coal Research, FMC Corporation, Princeton, New Jersey.* (CHAR-OIL-ENERGY DEVELOPMENT—PROJECT COED)

J. KEVIN JONES, First Class Honors Degree (Ch.E.). *Manager of Technical Operations, Petrocarbon Developments Limited, Houston, Texas. Formerly, Process Design Superintendent, Petrocarbon Developments Limited, Manchester, England.* (CRYOGENIC UPGRADING OF LOW-BTU GASES)

JOHN D. KEENAN, Ph.D. *Assistant Professor of Civil and Urban Engineering, University of Pennsylvania, Philadelphia, Pennsylvania.* (BIOCHEMICAL SOURCES OF FUELS)

DONALD B. KENDALL, B.S.E.E. *Formerly Chief Scale Engineer, Toledo Scale Company, Division of Reliance Electric Company, Toledo, Ohio. Member, American Railway Engineering Association, Institute of Electrical and Electronics Engineers.* (ELECTRIC DYNAMOMETERS)

H. J. KLOOSTER, B.S.Ch.E. *Manager, Process Engineering, Fluor Engineers and Constructors, Inc., Los Angeles, California.* (THE TOTAL ENERGY CONCEPT IN REFINERY AND PROCESS PLANT PLANNING)

JOHN A. KNOX, B.S., M.S. *Chemical Service Coordinator, Environmental Control Department, Halliburton Services, Duncan, Oklahoma.* (ACIDIZING)

C. J. KUHRE, B.S., M.S.Ch.E. *Process Engineer, Shell Oil Company, Houston, Texas.* [NONCATALYTIC, PARTIAL-OXIDATION GASIFICATION PROCESS (SHELL)]

IHOR. A. KUNASZ, Ph.D. *Chief Geologist, Foote Mineral Company, Exton, Pennsylvania.* (LITHIUM FOR THERMONUCLEAR FUSION REACTORS)

R. F. LAWRENCE, B.S.E.E. *Manager, Transmission and Distribution Systems Engineering, Westinghouse Electric Corporation, East Pittsburgh, Pennsylvania. Fellow, Institute of Electrical and Electronics Engineers.* (TRENDS IN ELECTRIC POWER TRANSMISSION)

B. S. LEE, Ph.D. *Director, Coal Gasification Research, Institute of Gas Technology, Chicago, Illinois.* (THE HYGAS COAL GASIFICATION PROCESS)

LAWRENCE M. LIDSKY, Ph.D. *Associate Professor of Nuclear Engineering, Massachusetts Institute of Technology, Cambridge, Massachusetts,* (FUSION POWER)

BALDUR LINDAL. *Consulting Engineer, Reykjavik, Iceland.* (GEOTHERMAL ENERGY FOR SPACE AND PROCESS HEATING)

DONALD R. LOHSE, B.S.C.E. *Product Specialist, Reciprocating Compressors, Air Power Division, Joy Manufacturing Company, Michigan City, Indiana.* (UTILIZING WASTE HEAT FROM AIR COMPRESSORS)

DAVID L. LORD, B.S.Ch.E., M.S. Pet. Eng. *Group Leader, Halliburton Services, Duncan, Oklahoma.* (DRILLING AND PRODUCTION)

J. L. LUMMUS, B.S. *Special Research Group Supervisor, Research Department, Amoco Production Company, Tulsa, Oklahoma.* (DRILLING OPTIMIZATION)

HOWARD L. MALAKOFF, B.S., M.S., D.Sc. *General Manager, COGAS Development Company, Princeton, New Jersey.* (COGAS PROCESS FOR CONVERSION OF COAL TO GAS AND OIL PRODUCTS)

H. C. McLAUGHLIN, B.S., M.S. *Senior Research Chemist, Halliburton Services, Duncan, Oklahoma.* (DRILLING TECHNIQUES—WATER PROBLEMS) (WELL COMPLETION—WATER CONTROL)

JOSEPH F. McMAHON, B.S., M.S., Ch.E. *Manager, Process Technology Department, Process Plants Division, Foster Wheeler Energy Corporation, Livingston, New Jersey.*

(THE FLUIDIZED BED HYDROGENATION PROCESS FOR SUBSTITUTE NATURAL GAS PRODUCTION)

JAMES E. McNULTY, B.S.   *Vice-President, Paul Weir Company Incorporated, Chicago, Illinois.* (COAL RESERVES—ESTIMATES) (COAL EXPLORATION)

A. S. MEHTA, B. S., M.S.   *Formerly Staff Engineer, Chemical Process Department, The Rust Engineering Company, a subsidiary of Wheelabrator-Frye Inc., Birmingham, Alabama.* (WOOD-TO-OIL PROCESS)

E. F. MOHLER, JR.   *Senior Environmental Engineer, Sun Oil Company, Toledo, Ohio.* (BIO-OXIDATION PROCESS REDUCES WATER USE)

RALPH W. MOIR, Sc.D.   *Research Physicist, Lawrence Livermore Laboratory, University of California, Livermore, California.* (DIRECT ENERGY CONVERSION IN FUSION REACTORS)

J. G. MONTFORT, B.S.C.E.   *Vice-President and General Manager, Black Mesa Pipeline, Inc., Flagstaff, Arizona.* (COAL TRANSPORTATION BY SLURRY PIPELINE)

JAAP E. NABER.   *Koninklijke/Shell–Laboratorium, Amsterdam, The Netherlands.* (CLAUS SULFUR RECOVERY UNITS AND CLAUS OFF-GAS TREATING)

ROY G. NEVILLE, M.Sc., Ph.D., D.Sc., F.R.I.C.   *Chemical and Environmental Consultant, San Carlos, California.* (WET SCRUBBING PROCESSES—SO_x AND NO_x REMOVAL CHEMISTRY)

SHOGO NOJIMA.   *Director, Japan Gasoline Co., Ltd., Tokyo, Japan.* (METHANE-RICH GAS PROCESS FOR SUBSTITUTE NATURAL GAS)

H. OMAN.   *Supervisor of Electrical Power Research, Boeing Aerospace Company, Seattle, Washington.* (SOLAR CELLS AND ARRAYS)

W. JEFF OSBORNE, B.S.   *Formerly Supervising Process Engineer, Davy Powergas Inc., Lakeland, Florida.* [SULFUR DIOXIDE RECOVERY (WELLMAN-LORD PROCESS)]

ALAN PASTERNAK.   *Formerly Lawrence Livermore Laboratory, University of California, Livermore, California.* (METHYL ALCOHOL—A POTENTIAL FUEL FOR TRANSPORTATION)

C. O. PEINADO, B.S., M.S.   *Director, Product Applications Division, General Atomic Company, San Diego, California. Registered Professional Engineer (California).* (HIGH-TEMPERATURE GAS-COOLED REACTORS)

JOHN A. PHINNEY, Ph.D.   *Consultant, Conoco Coal Development Company, Library, Pennsylvania.* [THE CONSOL SYNTHETIC-FUEL PROCESS (CSF PROCESS)]

WILLIAM T. PLASS.   *Principal Plant Ecologist, United States Department of Agriculture, Forest Service, Princeton, West Virginia.* (SEEDING AND PLANTING TO ACHIEVE LAND-MANAGEMENT OBJECTIVES)

F. POSTHUMA.   *Executive Director (Retired), Rotterdam-Europoort. Port Consultant, Rotterdam, The Netherlands.* (SUPERPORTS)

WILLIAM L. PRICE, B.S.M.E.   *Product Development Manager, Engineering Works Division, Dravo Corporation, Pittsburgh, Pennsylvania.* (BARGE MOVEMENT OF COAL)

E. A. RASSINIER, Ph.D.   *Chairman, Potential Gas Committee, Mineral Resources Institute, Colorado School of Mines Foundation, Inc., Golden, Colorado.* (POTENTIAL SUPPLY OF NATURAL GAS IN THE UNITED STATES)

ALBERT H. RAWDON, B.S., M.S.M.E.   *Director of Research and Development, Riley Stoker Corporation, Worcester, Massachusetts.* (NO_x CONTROL BY FURNACE AND BURNER DESIGN)

ROBERT M. REED, Ph.D.   *Consulting Chemical Engineer, Louisville, Kentucky.* (HYDROGEN)

ROGER F. RENSVOLD, M.A.   *Development Chemist, Halliburton Services, Duncan, Oklahoma.* (WORK-OVERS WATER CONTROL)

DONALD M. ROSS.   *Consulting Engineer, Lancaster, California. Formerly Deputy Director, U.S. Air Force Rocket Propulsion Laboratory, Edwards, California.* (PROPELLANTS)

PAUL F. H. RUDOLPH, Dipl. Ingenieur.   *Director, Lurgi Mineralötechnik GmbH,*

Frankfurt/Main, Federal Republic of Germany. (THE LURGI PROCESS FOR COAL GASIFICATION)

KENNETH I. SAVAGE, Ph.D. *Assistant Technical Director, Commercial Testing & Engineering Company, Chicago, Illinois.* (COAL TESTING)

F. C. SCHORA, JR., B.S.Ch.E. *Vice-President, Process Research, Institute of Gas Technology, Chicago, Illinois.* (THE HYGAS COAL GASIFICATION PROCESS)

K. A. SCHOWALTER, Ph.D. *Chief Staff Engineer, Chemicals and Plastics Research Laboratory, U.S. Steel Corporation, Monroeville, Pennsylvania.* (CARBONIZING/HYDROGENATION METHOD FOR PRODUCING METALLURGICAL COKE)

J. G. SEAY, B.S., M.S.Ch.E. *Senior Advisor, Management Science, Institute of Gas Technology, Chicago, Illinois.* (HYDROCRACKING-HYDROGASIFICATION PROCESS FOR PRODUCING PIPELINE GAS FROM CRUDE OIL)

S. SENDEROFF, Ph.D. *Senior Research Associate, Battery Products Division, Parma Technical Center, Union Carbide Corporation, Cleveland, Ohio.* (BATTERIES)

ROBERT H. SIMON. *Director, Gas-Cooled Fast Breeder Reactor Program, General Atomic Company, San Diego, California.* (GAS-COOLED FAST BREEDER REACTOR)

JAMES R. SMALL, A.B., M.S. *Product Manager, Process Instruments, Instrument Products Division, E. I. DuPont DeNemours & Co., Inc., Wilmington, Delaware.* (SOURCE MONITORING OF NO_x AND SO_2)

LAWRENCE F. SMALL, Ph.D. *Associate Professor, School of Oceanography, Oregon State University, Corvallis, Oregon.* (PHOTOSYNTHESIS)

PAUL SMARDO. *Assistant Manager, Coal Handling Systems, McNally Pittsburg Manufacturing Corporation, Pittsburg, Kansas.* (COAL HANDLING AND STORAGE SYSTEMS)

EDWIN J. STAHL, JR. B.S.Ch.E. *Section Supervisor, Fracturing Section, Chemical Research and Development Department, Halliburton Services, Duncan, Oklahoma.* (HYDRAULIC FRACTURING) (FRACTURING IN WORK-OVERS)

M. R. STEFFENSON, B.S.Ch.E. *Vice-President, Parr Instrument Company, Moline, Illinois.* (BOMB CALORIMETRY FOR TESTING SOLID AND LIQUID FUELS)

D. STEINER, Ph.D. *Member of the Thermonuclear Division and Coordinator of the Fusion Reactor Technology Program, Oak Ridge National Laboratory, Oak Ridge, Tennessee.* (TRITIUM BREEDING REQUIREMENTS IN FUSION REACTOR BLANKETS)

D. R. STEPHENS, Ph.D. *Project Leader, In-Situ Coal Gasification, Lawrence Livermore Laboratory, University of California, Livermore, California.* (IN-SITU COAL GASIFICATION)

C. S. STERRETT, B.S.M.E. *Program Manager, Superconducting Electrical Machinery Department, Westinghouse Electric Corporation, Pittsburgh, Pennsylvania.* (SUPERCONDUCTING TURBINE GENERATORS)

JOHN D. SUDBURY, Ph.D. *General Manager, Research Division, Conoco Coal Development Company, Library, Pennsylvania.* (THE CO_2 ACCEPTOR COAL GASIFICATION PROCESS)

JOHN D. SWARD, B.B.A. *The Atchison, Topeka and Santa Fe Railway Company, Chicago, Illinois.* (RAILROAD TRANSPORTATION OF COAL)

J. A. SYKES, JR., Ph.D. *Formerly with Shell Development Company, Houston, Texas.* [NONCATALYTIC, PARTIAL-OXIDATION GASIFICATION PROCESS (SHELL)]

DONALD J. SZOSTAK, B.S. *Vice-President, Engineering, Marine Transport Lines, Inc., a subsidiary of General American Transportation Corporation.* (SEAGOING VESSELS FOR TRANSPORTING LIQUID AND GASEOUS FUELS)

GORDON L. TAFT, B.S. *Product Specialist, Rotary Screw Compressors, Air Power Division, Joy Manufacturing Company, Michigan City, Indiana.* (UTILIZING WASTE HEAT FROM AIR COMPRESSORS)

P. B. TARMAN, B.S., M.S.Ch.E. *Director, Chemical Processing Research, Institute of Gas Technology, Chicago, Illinois.* (HYDROCRACKING–HYDROGASIFICATION PROCESS FOR PRODUCING PIPELINE GAS FROM CRUDE OIL)

COLIN H. TAYLOR, B.Sc. *Chief Process Engineer (Gas), Woodall-Duckham Limited,*

Crawley, Sussex, England. [THE CATALYTIC RICH GAS (CRG) PROCESS FOR GASIFICA-TION OF LIGHT HYDROCARBONS]

P. L. THIGPEN, B.S.Ch.E. *Principal Process Engineer, Chemical Process Department, The Rust Engineering Company, a subsidiary of Wheelabrator-Frye Inc., Birmingham, Alabama.* (COAL PURIFICATION BY SOLVENT REFINING) (WOOD-TO-OIL PROCESS)

DAVID P. THORNTON, JR., B.S.Ch.E. *Manager of Technical Publications, Universal Oil Products Company, Des Plaines, Illinois.* (PETROLEUM)

WALTER A. TROEGER, B.S., M.S. *Engineering Department, Weston Instruments, Inc., Newark, New Jersey.* (ELECTRIC POWER AND ENERGY MEASUREMENT)

GODFRIED J. VAN DEN BERG, Ph.D. *Head of the Conversion Processes and Project Technology Division, Shell Internationale Petroleum Maatschappi B.V., The Hague, The Netherlands.* (GASIFICATION AS BASIS FOR COMBINED-CYCLE POWER GENERATION)

PIET J. J. VAN DOORN. *Specialist, Licensing Division, Shell Internationale Petroleum Maatschappi, B.V., The Hague, The Netherlands.* (GASIFICATION AS BASIS FOR COM-BINED-CYCLE POWER GENERATION)

ROGER P. VAN DRIESEN, B.S.Ch.E. *Vice President, Cities Service Research and Development Company, Cranbury, New Jersey.* (HYDRODESULFURIZATION OF PETROLEUM CRUDE OIL FRACTIONS AND PETROLEUM PRODUCTS)

L. L. VANT-HULL, Ph.D. *Solar Thermal Program Manager, Solar Energy Laboratory, University of Houston, Houston, Texas.* (SOLAR TOWER ENERGY COLLECTOR)

R. W. VAN SCOY, Ph.D. *Formerly Process Engineer, Process Engineering-Licensing Department, Shell Oil Company, Houston, Texas.* (ABSORPTION OF ACIDIC GASES FROM NATURAL GAS AND SNG)

L. W. VERMEULEN. *Metallurgical Engineer, Rio Algom Mines Limited, Toronto, Canada.* (URANIUM RESOURCES AND PROCESSING)

JESSE D. WALTON, JR., B.S. *Principal Research Engineer, Head, Special Projects Office, Applied Sciences Laboratory, Engineering Experiment Station, Georgia Institute of Technology, Atlanta, Georgia.* (HIGH-TEMPERATURE SOLAR ENERGY)

C. W. WARNER, B.S., M.S.E.E. *Development Consultant, Cutler-Hammer, Inc., Mil-waukee, Wisconsin.* (CALORIMETRIC MEASUREMENT OF GASEOUS FUELS) (GAS-MIX-ING CONTROL SYSTEMS)

JOHANNES A. WESSELINGH. *Koninklijke/Shell—Laboratorium, Amsterdam, The Netherlands.* (CLAUS SULFUR RECOVERY UNITS AND CLAUS OFF-GAS TREATING)

JAMES A. WILLIAMSON, B.S., M.S. *Senior Applications Chemist, Instrument Products Division, E. I. DuPont deNemours & Co., Inc., Wilmington, Delaware.* (SOURCE MON-ITORING OF NO_x AND SO_2)

GORDON R. WOODCOCK, B.S., M.S. *Engineering Supervisor, Future Space Transpor-tation Systems Analysis, Research and Engineering Division, Boeing Aerospace Com-pany, Seattle, Washington.* (HEAT ENGINES FOR SOLAR POWER) (SATELLITE ENERGY SYSTEMS FOR EARTH POWER)

PAUL M. YAVORSKY, Ph.D. *Research Supervisor of the Exploratory Engineering Section, Pittsburgh Energy Research Center, Bureau of Mines, U.S. Department of the Interior, Pittsburgh, Pennsylvania.* (SYNTHOIL PROCESS FOR CONVERTING COAL TO NONPOLLUTING FUEL OIL) (PRODUCTION OF HIGH-BTU GAS BY THE HYDRANE PROCESS)

BILL M. YOUNG, B.S., M.S.Ch.E. *Group Leader, Water and Sand Control Section, Chemical Research and Development, Halliburton Services, Duncan, Oklahoma.* (SAND CONTROL—WELL COMPLETIONS) (SAND CONTROL WORK-OVERS)

MARIO ZAMORA, P.E. *Manager, Staff Engineering, IMCO Services, a division of Halliburton Company, Houston, Texas.* (GEOPRESSURED DRILLING) (WELL CONTROL)

A. H. ZEITZ, JR., B.S.M.E. *Manager, Product Application, Petroleum Chemicals Divi-sion, Ethyl Corporation, Ferndale, Detroit, Michigan.* (METHYLCYCLOPENTADIENYL MANGANESE TRICARBONYL COMBUSTION IMPROVER)

Preface

This Handbook concentrates on those fundamental technologies which relate to energy sources, energy reserves, energy conversion, energy transportation and transmission, and to an extent limited both by time and space in the preparation of this volume, energy distribution, utilization, and the energy/environmental interface. Obviously, a single volume of reasonable length cannot accommodate all related technology. It is obvious from reading the inputs of over 140 scientists and engineers represented in this book that energy is indeed one of the most interdisciplinarian of scientific topics. Hopefully, the users of this book will find that it represents a working example of the concept of technology and information transfer. Numerous difficult decisions had to be made by the authors and editor pertaining to the scope and depth of nearly seven score separate articles. The extent of information in other permanent reference sources, particularly pertaining to the more traditional and conventional aspects of energy technology, was always given careful consideration so that maximum utilization of the space available could be in the interest of the more promising, relatively new concepts in the energy sciences. Nevertheless, in all instances, an attempt was made to include overviews of well-established practices, effectively coupled with terse reviews of solid progress made during the past decade or two.

First conceived at a time prior to the present sense of urgency in matters dealing with energy, this volume was not put together hastily, even though all persons involved in the project believe that it will serve an urgent need. The passing needs for information on energy have been and are continuing to be served by periodicals, proceedings, and other works of a semipermanent nature. Although undoubtedly future editions of this Handbook will appear as dictated by the progress of energy technology, a serious and dedicated effort has been made to concentrate in this volume on fundamentals and data of a basic and reasonably permanent nature that will auger well for at least the next few years.

Energy as a topic is now uppermost in the minds of diplomats and politicians, and, needless to say, these factors will continue to have a bearing on technological incentives and accomplishments. Consequently, the temptation to editorialize even in technical discussions is high. But such dialogue is not the charter of an engineering Handbook, and much effort has been exerted to confine descriptions to scientific and engineering aspects sans political connotations. The editors have followed a path of middle ground in terms of energy conservation on the one hand and of full concentration on increased energy production on the other hand. There seems to be no denying that significant quantities of energy have been wasted and that reasonable conservation measures are indicated. But also, in contrast, there seems to be no denying that a past abundance of relatively inexpensive energy has underpinned the industrial, economic, and technological leadership of the developed nations.

If all the seriously proposed new or more appropriately termed nonconventional energy sources were a reality today, the problem would be one of finding ways to use energy instead of conserving it. Needless to say, many years will be required to achieve practical, economic ways to perfect a number of the nonconventional energy approaches. During the next decade or so of further development of nonconventional means, considerable shrinkage will occur, some proposals simply not being able to meet practical operating tests. Further, it is hoped that some breakthroughs will occur which will hasten the "weeding out" process. But the time of this first edition does not seem appropriate for such judgments. Consequently, a very wide variety of concepts is presented here. Considering these numerous approaches to energy solutions, the editors have attempted to remove exaggerated claims or biases on the parts of the naturally and thankfully zealous experts, while still permitting some enthusiasm and persuasion to shine through.

Throughout this project, it became clear that there will be increasing stress on the regional aspects of energy, in terms of both sources and consumption, so that instead of a universal approach to energy emerging, there will be regionally customized approaches, such regional borders not necessarily following national, state, or provincial boundaries. The editors have kept this factor in mind and attempted to schedule topics, some of which may be largely of regional interest, while others are of worldwide interest.

In terms of organization, all major sections of this Handbook except the last section are oriented to the major sources of energy. At this stage in the development of energy technology it appeared that this extension of prior editorial approaches, namely, concentrating on "fuels," seems in order. The last section of the book, on power technology trends, stresses energy utilization and the energy/environment interface. For obvious reasons, the major energy sources cannot be described in a fully parallel manner. The conventional fuels—coal, gas, and petroleum—are described from the standpoints of properties, statistics, notably reserves, exploration for new sources, testing, mining and gathering, preparation for consumption, transportation, and physical and chemical treatment and conversion in the interest of consuming efficiency and reduction of environmental impact. Much emphasis is given to new conversion processes, that is, the superimposition of new technology on established energy sources. As the less conventional energy sources are approached, there is even greater emphasis on newer concepts—some proved, others unproved on a large scale. Thus, under the chemical fuels, there is extensive coverage of biochemical sources, the conversion of solid wastes, the possibility of a future hydrogen economy, the pros and cons of shipping LNG versus methyl alcohol, new concepts in batteries and fuel cells. Since this Handbook is devoted to energy technology on a broad base, the entire emphasis is not confined to ultimately the large-scale production of electricity. The discussion of propellants, for example, is indicative of this broad approach.

Under nuclear energy technology, emphasis is given to current and next-generation reactors, but very significant attention is also given to fusion power. Nuclear raw materials, such as uranium and lithium, also are described in terms of resources and economic availability.

As one reviews the information on solar power, the variety of concepts is striking, some of them only having in common the concept of capturing energy from the sun. The variety of arrangements, ranging from solar tower energy collectors and solar absorption-heat-pipe systems to solar cells, satellite approaches, ocean thermal energy conversion, photosynthesis, and electrical energy from the wind, comprises a vast new technology in itself.

After a comprehensive review of the fundamentals of geothermal energy, meaning in depth is given by descriptions of specific geothermal fields in New Zealand, California and the western United States, and Iceland. The differences in the characteristics of fields found in different parts of the world are described, and indicate the greater suitability in some cases for direct use of the thermal energy in production of electric power; and in other cases, the better use of such energy for space and process heating.

Conventional hydropower approaches are described with particular attention to pumped storage concepts. Detailed attention is also given to tidal power which for some parts of the world offers interesting potential.

In the section on power technology trends, considerable attention is given to steam generation, particularly to the advancements in this basic technology, and to combined-cycle power generation. The use of gas and expansion turbines and magnetohydrodynamic generators for augmentation of conventional approaches is stressed. Electric power production and requirements are reviewed, notably from a statistical base. Advances in superconducting turbine generator design, and trends in electric power transmission and in the greater and more effective use of computers in the electric utility field are described. Essential measurements of electric power and energy are detailed. The general organizational approach to the energy/environment interface is by way of specific pollutants, such as the sulfur and nitrogen oxides and the particulates. In addition to ways of combating pollution after the fact so to speak, with scrubbers, absorbers, and the like, attention is also given to reduction of pollution through improved furnace and burner design, and the use of combustion improving chemicals. Particular attention is given to the monitoring by instrumental means of the sulfur and nitrogen oxides, carbon monoxide, hydrocarbon, and particulate emissions. Better use of water in energy production, recovery of heat from processes and air compressors, and the total energy concept in process plant planning are described.

Although this volume essentially has been conceived as a tool for scientists and engineers and technologists already in or new to the energy field, hopefully it will also provide a further service toward the enlightenment of legislators, government leaders worldwide, economists, planners, social reformers, politicians, and the public information media and other persons and organizations that make and influence vital decisions which will shape the energy situation of the next several decades. Not only is a keener understanding of energy technology required on the part of decision makers, but also the basic starting point is to convince numerous decision makers that by and large the solutions to a presently crisis-oriented energy program, whether motivated politically, economically, conservationally, or environmentally, will come from technology.

DOUGLAS M. CONSIDINE
Editor-in-Chief

Acknowledgments

Many hundreds of persons have made this volume possible. Inputs to the editor are represented not only by the distinguished authors present, but by numerous advisors and counselors, some of whom are formally recognized here. There are still others who helped the project at some juncture during its preparation. The thousands of references cited throughout the volume attest to the large backdrop of information—an energy data bank so to speak that has been created over the years, and of which this single volume represents a select distillation.

R. C. Bell, *Nuclear Energy Division, General Electric Company, San Jose, California.*

V. M. Brown and **J. F. Walsh,** *National Petroleum Council, Washington, D.C.*

W. E. Caine, *Houston, Texas.*

D. J. Carney and **M. J. Warzbach,** *United States Steel Corp., Pittsburgh, Pennsylvania.*

R. Cohen, *National Science Foundation, Washington, D.C.*

T. L. Cramer, *Institute of Gas Technology, Chicago, Illinois.*

J. F. Ebdon, *Texas Eastern Transmission Corp., Houston, Texas.*

D. M. Evans, *The Potential Gas Agency, Colorado School of Mines, Goldon, Colorado.*

H. F. Feldman and **R. J. Anderson,** *Battelle Memorial Institute, Columbus, Ohio.*

E. G. Foster, *Shell Oil Company, Houston, Texas.*

J. R. Garvey, *Bituminous Coal Research, Inc., Monroeville, Pennsylvania.*

J. E. Hodges and **V. M. Smyth,** *American Petroleum Institute, Washington, D.C.*

L. Hoover, *American Geological Institute, Falls Church, Virginia.*

M. K. Hubbert and **R. F. Meyer,** *U.S. Geological Survey, Reston, Virginia.*

A. E. Humphrey, *College of Engineering and Applied Science, University of Pennsylvania, Philadelphia, Pennsylvania.*

E. V. Larson, *United States Department of Agriculture, Forest Service, Northeastern Forest Experiment Station, Upper Darby, Pennsylvania.*

M. Lebarbier, *Électricité de France, Paris, France.*

S. W. Lombard, *South Africa Coal, Oil and Gas Corporation Limited, Sasolburg, Republic of South Africa.*

J. J. Miller, *Bailey Meter Company, subsidiary of Babcock & Wilcox, Wickliffe, Ohio.*

R. G. Mills, *Princeton Plasma Physics Laboratory, Princeton University, Princeton, New Jersey.*

C. C. Newton, *Pacific Gas and Electric Company, San Francsico, California.*

R. C. Reid, *Department of Chemical Engineering, Massachusetts Institute of Technology, Cambridge, Massachusetts.*

V. J. Rinaldi, *G. V. Barkman, and M. M. Matthews, Westinghouse Electric Corporation, Pittsburgh, Pennsylvania.*

J. D. Ryan, *Combustion Engineering, Inc., Windsor, Connecticut.*

T. Sharp, *Davy Powergas Ltd., London, England.*

J. W. Shelton, *Department of Geology, Oklahoma State University, Stillwater, Oklahoma.*

R. A. Siskin, *American Gas Association, Arlington, Virginia.*

K. E. Swan, *General Atomic Company, San Diego, California.*

J. C. Thompson, *General American Transportation Corp., Chicago, Illinois.*

W. T. Tierney and **W. J. Coppoc,** *Texaco Inc., Beacon, New York.*

P. A. Wagner III, *Edison Electric Institute, New York, N.Y.*

R. Wegehoft and **S. Anderson,** *Aeromotor Div., Braden Industries, Inc., Broken Arrow, Oklahoma.*

C. C. Williams, III, *Shell Oil Company, Houston, Texas.*

Special thanks are also in order for D. L. Gregory, *Boeing Aerospace Company, Seattle, Washington,* for assisting in assembling information sources on certain aspects of solar energy technology; Prof. L. M. Lidsky, *Massachusetts Institute of Technology, Cambridge, Massachusetts,* for outlining the editorial coverage of fusion power; D. L. Lord and several of his associates at *Halliburton Services, Duncan, Oklahoma,* for outlining and providing much of the editorial coverage in the area of oil drilling and production; and A. S. Wonstolen and C. W. Hasek, *Babcock & Wilcox, Barberton, Ohio* for coordinating the inputs on steam generation technology.

Energy

By Joseph File* and Douglas M. Considine†

In most contemporary texts and those of the last several decades, energy generally has been defined simply as "the ability or capacity to do work." This is a broadening of the earlier definition in terms of Newtonian mechanics, which was "a property of moving masses."

The concept of energy is central to thermodynamics, quantitative chemistry, and electromagnetism. Consider Einstein's mass-energy equation, $E = mc^2$ for the interconversion of mass and energy, where E = energy in ergs; m = mass in grams; and c is the velocity in centimeters per second. Or, Planck's equation, which expresses the fundamental law of quantum theory, stating that the energy transfers associated with radiation are made up of definite quanta of energy proportional to the frequency of the radiation: $E = h\nu$, where E = the value of the quantum units of energy; ν = the frequency of radiation; and h is the elementary quantum of action, more commonly known as Planck's constant (6.6255×10^{-27} erg-second—the proportionality factor that, when multiplied by the frequency of a photon, gives the energy of the photon).

Although the fundamental definition of energy can be stated briefly, it immediately calls for an explanation of work, and of power. In the strict physical sense, work is performed only when a force is exerted on a body while the body moves at the same time in such a way that the force has a component in the direction of motion. The amount of work done during motion from point a to point b can be expressed by

$$W = \int_a^b F \cos\theta \, ds$$

where F is the total force exerted and θ is the angle between the direction of F and the direction of the elemental displacement, ds. In the cgs system, the unit of work is the dyne-centimeter or erg; in the mks system, the newton-meter or joule; and in the English system, the foot-pound.

Power is defined as the rate at which work is performed. The average

*Plasma Physics Laboratory, Princeton University, Princeton, New Jersey.
†Editor-in-Chief, *Energy Technology Handbook*.

power accomplished by an agent during a given period of time is equal to the total work performed by the agent during the period, divided by the length of the time interval. The instantaneous power can be expressed simply as

$$P = dW/dt$$

In the cgs system, power has the units of ergs per second; in the mks system, units of joules per second (or watts); and in the English system, units of foot-pounds per second. A common engineering unit is the horsepower, defined as 550 foot-pounds per second; or 33,000 foot-pounds per minute.

Now, returning to the basic definition of energy as the capacity for performing work. This definition may be better understood when stated as: "The energy is that which is diminished when work is done by an amount equal to the work so done." The units of energy are identical with the units of work previously given.

Energy can exist in a variety of forms, some forms more immediately recognizable as being capable of performing work than other more abstruse forms. Forms in which energy is not dependent upon mechanical motion are generally referred to as forms of potential energy. The most common example in this category is gravitational potential energy. A body near the earth's surface, for example, undergoes a change in potential energy when it is changed in elevation, the amount being equal to the product of the weight of the body and the change in elevation.

Potential energy also may be stored in an elastic body, such as a spring or a container of compressed gas. It may exist in the form of chemical potential energy, as measured by the amount of energy made available when given substances react chemically. Potential energy also exists in the nuclei of atoms and can be released by certain nuclear rearrangements.

Kinetic energy is the energy associated with mechanical motion of bodies. It is quantitatively equal to $\frac{1}{2}mv^2$, where m is the mass of a body moving with velocity v. In the case of rotational motion, the kinetic energy is more easily calculated, using the expression $\frac{1}{2}I\omega^2$, where I is the moment of inertia of the body about its axis of rotation and ω is the angular velocity. Kinetic energy, like all forms of energy, is a scalar quantity (having magnitude but not direction). In a system made up of an assembly of particles, such as a given volume of gas, the total kinetic energy is equal to the sum of the kinetic energies of all the molecules contained in the volume. Calculation of the energy of such systems is successfully treated theoretically on the basis of statistical averages.

Within a given system, energy may be transformed back and forth from one form to another, without changing the total energy of the system. A simple example is the pendulum, in which the energy is periodically converted from gravitational potential energy to kinetic energy and then back to gravitational potential energy. A similar situation, but on a submicroscopic scale, occurs in solid materials where the atoms are vibrating under the effect of interatomic rather than gravitational forces. As the temperature of a solid increases, the energy associated with the vibration of the atoms increases. This example illustrates, on a macroscopic scale, how heat can be considered a form of energy. Regardless of the material involved, any amount of heat absorbed or released may be quantitatively expressed as an amount of energy. A gram-calorie of heat is equivalent to 4.19 joules; and in the English system, a British thermal unit (Btu) is equivalent to 778 foot-pounds.

Potential energy is also present in electrical and magnetic fields. The energy available in a region of electric field is equal to $E^2/8\pi$ per unit volume, where E is the electric field strength. Within a given volume, the total energy represented by the electric field is the integral of $E^2/8\pi$ over the volume. Similarly, the energy represented by a magnetic field may be independently calculated by integrating $H^2/8\pi$ over any given volume, where H represents the magnetic field strength. In the case of an electrically charged capacitor, the total energy in the electric field, and hence in the capacitor, can be shown to be $\frac{1}{2}CV^2$. Here C is the capacitance and V the electric potential to which the capacitor is charged. Similarly, the total energy in the magnetic field associated with an inductor carrying an electric current is $\frac{1}{2}LI^2$, where L is the inductance and I is the current.

Electromagnetic radiation is a combination of rapidly alternating electric and magnetic fields. Energy is associated with these fields and is exchanged between the electric and magnetic forms. This energy in a quantum of electromagnetic radiation, such as light or gamma radiation, can be expressed in different ways, but is commonly given by $E = h\nu$, as previously mentioned.

The expression previously given for the kinetic energy, $\frac{1}{2}mv^2$, is not accurate when the velocity approaches that of the velocity of light. The theory of relativity requires a correction to be made, and the exact kinetic energy, T, may be calculated in terms of the mass, m_0, of light in vacuum, c, as follows:

$$T = m_0c^2\left[\left(1 - \frac{v^2}{c^2}\right)^{-1/2} - 1\right]$$

Notice that this formula may also be written

$$T = (m - m_0)c^2$$

where m is the variable quantity, $m_0\left(1 - \dfrac{v^2}{c^2}\right)^{-1/2}$

This latter quantity represents the mass of the body, reducing to m_0 when v is zero, and approaching infinity as v approaches the speed of light.

This example illustrates another result of the theory of relativity, namely, the equivalence of mass and energy. Rewriting the last equation:

$$m = m_0 + \frac{T}{c^2}$$

The mass is seen to increase linearly with the kinetic energy of the body, the proportionality factor being c^2. It should be noted that even the rest mass, m_0, represents an amount of energy equal to $m_0 c^2$. The total energy of a body of mass m can be generally given as

$$E = mc^2; \text{ or } E = m_0 c^2 + T$$

In dealing with radiation, it is customary to express energy in terms of electron volts. An electron volt is equal to the amount of work done when an electron moves through an electric field produced by a potential difference of 1 volt. One electron volt is equivalent to 1.60×10^{-12} erg. When charged particles, such as electrons or protons, are accelerated, their kinetic energy is stated in terms of electron volts (eV) or million electron volts (MeV).

A basic principle of physics known as the conservation of energy requires that, within any closed system, the total energy must remain constant. Energy can be changed from one form to another, but the total, so long as no energy is added to or lost from the system, must be constant. In the case of the swinging pendulum, decreases in kinetic energy reappear as increases in potential energy and vice versa. Eventually, of course, the pendulum will stop due to the effect of frictional forces. At that time, all of the kinetic energy and gravitational potential energy will have been converted to heat.

In another example involving a radioactive atom, the total energy represented by the atom and the emitted radiation must be constant. If a gamma ray is emitted, the rest mass of the atom will be decreased by an amount equivalent to the sum of the energy of the gamma ray and the recoil kinetic energy of the atom, which will be very small.

AVAILABILITY OF ENERGY

It has long been established that there is a general direct correlation of the gross national product of a nation with its energy consumption. In the past decade or so this notion has been brought to a fine focus. It is not at all surprising that almost universal interest has been generated in this most important subject.

In the early days of inhabitance of this planet, people depended solely on the sun and wood-burning forces for their energy. In about the

12th century, after finding coal deposits, the energy needs of people were satisfied mainly by burning fossil fuels. It has taken perhaps half a billion years or more for the remains of animal and plant life under certain chemical conditions to decompose into the earth's present stores of fossil fuels. Only negligible amounts of fossil fuels were used during the million or so years since humans evolved on this planet, and it is interesting to note that it was in the 12th century that the first large scale use of coal began, when mines were worked in northern England. Coal at first was used solely as a fuel to produce heat. Later, with the invention of the steam engine in the early part of the 18th century, coal and petroleum products (found for the first time in the middle of the 19th century) began to be used to convert heat to mechanical and electrical energy.

Since the invention of the steam engine, fossil fuels have been used at increasingly higher rates. As people progressed through the Industrial Revolution, the need and desire for better lives led to more and more use of energy. One finds many estimates of world fossil fuel reserves, most of which seem to be consistent within an order of magnitude or so. All of these estimates lead one to conclude that the earth's fossil fuel reserves are indeed finite and that, at the present rate of consumption, will soon become very scarce. For example, let us take the estimate made by M. King Hubbert, in a paper presented to the Fourth Symposium on Engineering Problems of Fusion Research at Washington, D.C., April, 1971, that up to 1970, about 2 percent of the earth's coal resources and 14 percent of the earth's oil resources had been used. More than ½ billion years were required to produce these fossil fuels. In about 3½ centuries, the human population has been able to deplete (at ever increasing rates) about 2 percent of the world's coal and in about 1 century about 14 percent of the earth's petroleum products. These statistics should and do cause concern.

It is now universally conceded that fossil fuels are finite. It is only a matter of time before fossil fuel reserves will essentially be depleted. Estimates of reserves of fossil fuels all reach the same conclusion, differing only in detail. Extended use of these reserves in the current manner will continue for no more than several generations, or at most a few centuries. People, therefore, now recognize the need for alternate fuel sources, economically convertible to electrical and thermal energy. If possible, use should be made of existing sources which are not yet developed to economic practicality.

Some of those alternate and existing sources, all of which have major advantages and disadvantages are: *solar energy,* which has an estimated average power potential of 24 watts per square meter of earth's surface (assuming 10 percent efficiency); *water power,* major sources of which are still underdeveloped, has an estimated potential of 10^6 megawatts, but is only available in certain areas of the world; also available only to limited areas where tidal changes are significant, *tidal energy* has an estimated power potential of 10^6 megawatts. *Geothermal energy*, a relatively small source of energy (about 10^{20} joules, or about 3×10^6 megawatt years) is subject to the same restrictions as the two previously mentioned sources; *nuclear fission*, a source of energy now in limited use, but under attack by environmental groups in the United States and other parts of the world, is slowly being developed. With the exclusive use of breeder reactors, nuclear fission is a source which can easily offer 100 times more energy than all of the fossil fuels combined; *nuclear fusion*, a source now entering the scientific feasibility stage, has a great deal of promise and availability. The deuterium used as a fuel is plentiful in water, and lithium used to breed tritium is relatively abundant. For example, in the United States and its surrounding oceans alone, there is enough available low-cost lithium, that if it is utilized for all the United States energy uses, it would correspond to about a 90-century supply at 1970 consumption rates.

These and other sources of energy are discussed in detail in later portions of this book, but it is hoped that this short introduction serves to focus on the importance of energy sources to future generations of people on this planet.

Breadth in the "Packaging" of Energy. Various energy packets, ranging from 1 electron volt (1.6×10^{-19} joule) to the daily energy output of the sun (total, in all directions, of 10^{32} joules), are plotted here for comparative purposes. There is a vast range in magnitude of energy packets. For materials, energy equivalents are given.

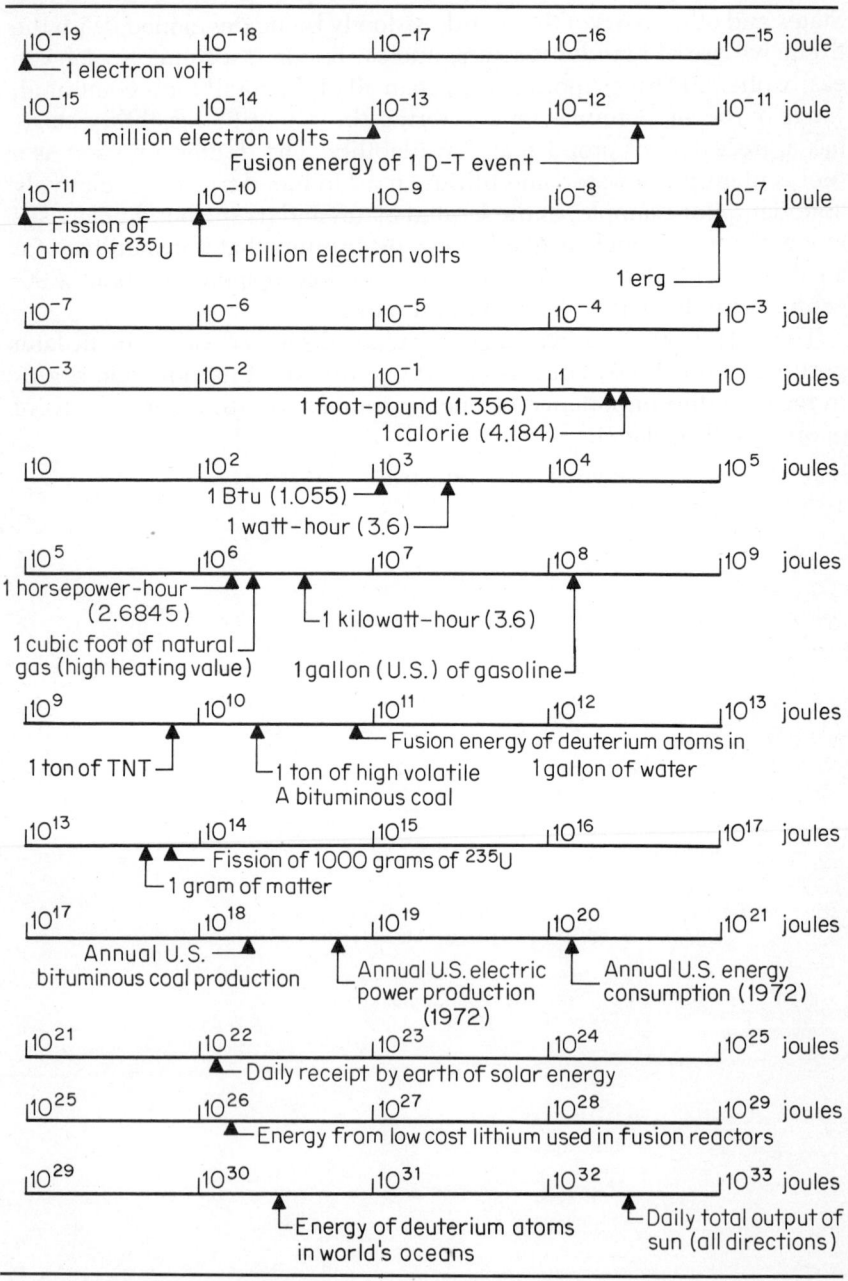

Coal Technology

SAYEED AKHTAR, B.Sc., M.Sc. Ph.D. *Project Leader, Synthoil Exploratory Engineering, Pittsburgh Energy Research Center, Energy Research and Development Administration, Pittsburgh, Pennsylvania. Member, American Chemical Society. (Synthoil Process for Converting Coal to Nonpolluting Fuel Oil)*

JOHN BAKOSS, B.S. *Staff Engineer, J. H. Fletcher & Co., Huntington, West Virginia. (Roof Control for Underground Mining)*

CLAYTON G. BALL, Ph.D. *Consultant, Paul Weir Company Incorporated, Chicago, Illinois. Fellow, The Geological Society of America. Member, American Institute of Consulting Engineers; American Institute of Mining, Metallurgical and Petroleum Engineers; American Institute of Professional Geologists; Illinois Mining Institute; Mining and Metallurgical Society of America; Society of Economic Geologists. Registered Professional Engineer and Certified Professional Geologist. (Coal Reserves— Estimates) (Coal Exploration)*

RALPH BLOOM, JR., B.S.Ch.E. *Manager, Economics and Planning, COGAS Development Company, Princeton, New Jersey. Member, American Chemical Society. Licensed Professional Engineer (New York). (Cogas Process for Conversion of Coal to Gas and Oil Products)*

N.S. BOODMAN, B.S., M.S. *Section Supervisor, Research Laboratory, United States Steel Corporation, Monroeville, Pennsylvania. Member, American Chemical Society, Pittsburgh Catalysis Society. (Carbonization/ Hydrogenation Method for Producing Metallurgical Coke)*

FRANKLIN B. CARLSON, B.S.Ch.E. *Manager, Pilot Plant and Engineering. The Oil Shale Corporation, Rocky Flats Research Center, Golden, Colorado. Member, American Institute of Chemical Engineers. Registered Professional Engineer (Texas). (The Toscoal Process for Pyrolysis of Coal)*

GEORGE P. CURRAN, B.S.Ch.E.   *Manager of Gasification Research, Conoco Coal Development Company, Library, Pennsylvania. (The CO_2 Acceptor Coal Gasification Process)*

ERNEST A. DRAEGER, B.S.   *Project Engineer, McNally Pittsburg Manufacturing Corporation, Pittsburg, Kansas. Member, Society of Mining Engineers. (Coal Preparation Plants)*

HERMAN F. FELDMAN, B.S., M.S.Ch.E.   *Manager, Coal Research Programs, Battelle Columbus Laboratories, Columbus, Ohio. Member, American Chemical Society, American Institute of Chemical Engineers. (Production of High-Btu Gas by the Hydrane Process)*

CARL E. FINK, B.S.Ch.E.   *Project Manager of Pilot Plant, Conoco Coal Development Company, Rapid City, South Dakota. (The CO_2 Acceptor Coal Gasification Process)*

J. ROBERT FLETCHER, B.S.M.E.   *President, J. H. Fletcher & Co., Huntington, West Virginia. Member, American Institute of Mining, Metallurgical and Petroleum Engineers; Illinois Mining Institute; West Virginia Mining Institute. Registered Professional Engineer (Illinois, West Virginia). (Roof Control for Underground Mining)*

JOHN F. S. FRITH, Ph.D.   *Assistant Manager, Chemical Engineering Department, The Lummus Company, a subsidiary of Combustion Engineering, Inc., Bloomfield, New Jersey. Member, British Institute of Petroleum, Institute of Petroleum. (The Synthane Coal-Gasification Process)*

ROBERT J. GRACE, B.S.Ch.E.   *Process Engineering Manager, Bituminous Coal Research, Inc., Monroeville, Pennsylvania. Member, American Institute of Mining, Metallurgical and Petroleum Engineers; Society of Mining Engineers. Registered Professional Engineer (Pennsylvania). (Bi-Gas Process for Production of High-Btu Pipeline Gas from Coal)*

JAMES A. GRAY, B.E., M.E., Ph.D.   *Supervisory Chemical Engineer, Coal Gasification—Hydrane Process, Pittsburgh Energy Research Center, Energy Research and Development Administration, Pittsburgh, Pennsylvania, Member, American Institute of Chemical Engineers. (Production of High-Btu Gas by the Hydrane Process)*

JAN C. HOOGENDOORN. *Manager, Research and Development, South African Coal, Oil and Gas Corporation Limited, Sasolburg, Republic of South Africa. (Coal Gasification at Sasol)*

J. HUEBLER, B.S. *Senior Vice-president, Institute of Gas Technology, Chicago, Illinois, Fellow, American Institute of Chemists. Member, American Gas Association, American Standards Association. Registered Professional Engineer (Ohio). (The Hygas Coal-Gasification Process)*

R. B. JAQUES. *System Engineer, Black Mesa Pipeline, Inc., Flagstaff, Arizona. Member, American Institute of Mining, Metallurgical and Petroleum Engineers. (Coal Transportation by Slurry Pipeline)*

W. C. JOHNSON. *Formerly Manager, Chemical Process Department, The Rust Engineering Company, a subsidiary of Wheelabrator-Frye Inc., Birmingham, Alabama. (Coal Purification by Solvent Refining)*

JOHN F. JONES, Ph.D. *Technical Director of Coal Research, FMC Corporation, Princeton, New Jersey. Member, American Chemical Society, American Institute of Chemical Engineers. (Char-Oil-Energy Development—Project Coed)*

B. S. LEE, Ph.D. *Assistant Vice-President, Process Research, Institute of Gas Technology, Chicago, Illinois. Member, American Association for the Advancement of Science; American Chemical Society; American Institute of Chemical Engineers; American Institute of Mining, Metallurgical and Petroleum Engineers. Registered Professional Engineer (Illinois, Massachusetts, New York). (The Hygas Coal Gasification Process)*

HOWARD L. MALAKOFF, B.S., M.S., D.Sc. *General Manager, COGAS Development Company, Princeton, New Jersey. Member, American Association for the Advancement of Science, American Chemical Society, American Institute of Chemists, American Institute of Chemical Engineers, American Petroleum Institute, Association of Research Directors, Commercial Development Association, Inc., Industrial Research Institute, National Association of Manufacturers, Society of Automotive Engineers, Society of Chemical Industry. (Cogas Process for Conversion of Coal to Gas and Oil Products)*

JAMES E. McNULTY, B.S. *Vice-president, Paul Weir Company Incorporated, Chicago, Illinois. Member, American Institute of Mining, Metallurgical and Petroleum Engineers; Illinois Mining Institute; Rocky Mountain Coal Mining Institute. (Coal Reserves—Estimates) (Coal Exploration)*

J. G. MONTFORT, B.S.C.E. *Vice-president and General Manager, Black Mesa Pipeline, Inc., Flagstaff, Arizona. (Coal Transportation by Slurry Pipeline)*

JOHN A. PHINNEY, Ph.D. *Consultant, Conoco Coal Development Company, Library, Pennsylvania. [The Consol Synthetic-Fuel Process (CSF Process)]*

WILLIAM T. PLASS, B.S., M.S. *Principal Plant Ecologist, Northeastern Forest Experiment Station, Forest Service, U.S. Department of Agriculture, Princeton, West Virginia. (Seeding and Planting to Achieve Land-Management Objectives)*

WILLIAM L. PRICE, B.S.M.E. *Products Development Manager, Engineering Works Division, Dravo Corporation, Pittsburgh, Pennsylvania. Member, American Institute of Iron and Steel Engineers; American Institute of Mining, Metallurgical and Petroleum Engineers. (Barge Movement of Coal)*

PAUL F. H. RUDOLPH, Dipl. Ingenieur. *Director, Lurgi Mineralötechnik GmbH, Frankfurt/Main, Federal Republic of Germany. (The Lurgi Process for Coal Gasification)*

KENNETH I. SAVAGE, Ph.D. *Assistant Technical Director, Commercial Testing & Engineering Company, Chicago, Illinois. Member, American Chemical Society; American Institute of Chemical Engineers; American Institute of Mining, Metallurgical and Petroleum Engineers; American Society for Testing and Materials; Pennsylvania Society of Professional Engineers. Registered Professional Engineer (Illinois). (Coal Testing)*

F. C. SCHORA, JR., B.S.Ch.E. *Vice-president, Process Research, Institute of Gas Technology, Chicago, Illinois. Member, American Chemical Society, American Gas Association, American Institute of Chemical Engineers. Registered Professional Engineer (Massachusetts). (The Hygas Coal-Gasification Process)*

K. A. SCHOWALTER, Ph.D. *Chief Staff Engineer, Chemicals and Plastics, Research Laboratory, United States Steel Corporation, Monroeville, Pennsylvania. Member, American Chemical Society, American Institute of Chemical Engineers, Society of Plastics Engineers. (Carbonization/Hydrogenation Method for Producing Metallurgical Coke)*

PAUL SMARDO. *Assistant Manager, Coal Handling Systems, McNally Pittsburg Manufacturing Corporation, Pittsburg, Kansas. (Coal Handling and Storage Systems)*

M. R. STEFFENSON, B.S.Ch.E. *President, Parr Instrument Company, Moline, Illinois. Member, American Chemical Society, Scientific Apparatus Makers Association. (Bomb Calorimetry for Testing Solid and Liquid Fuels)*

D. R. STEPHENS, Ph.D. *Project Leader, In-Situ Coal Gasification, Lawrence Livermore Laboratory, University of California, Livermore, California. Member, American Association for the Advancement of Science, American Chemical Society, American Geophysical Union. (In-Situ Coal Gasification)*

JOHN D. SUDBURY, Ph.D. *General Manager, Research Division, Conoco Coal Development Company, Library, Pennsylvania. (The CO_2 Acceptor Coal Gasification Process)*

JOHN D. SWARD, B.B.A. *The Atchison, Topeka and Santa Fe Railway Company, Chicago, Illinois. Member, American Society of Traffic and Transportation. Association of Interstate Commerce Commission Practitioners. (Railroad Transportation of Coal)*

P.L. THIGPEN, B.S.Ch.E. *Principal Process Engineer, Chemical Process Department, The Rust Engineering Company, a subsidiary of Wheelabrator-Frye Inc., Birmingham, Alabama (Coal Purification by Solvent Refining)*

PAUL M. YAVORSKY, B.S., M.S.,Ph.D. *Research Supervisor of the Exploratory Engineering Section, Pittsburgh Energy Research Center, Energy Research and Development Administration, Pittsburgh, Pennsylvania. Member, American Chemical Society. (Synthoil Process for Converting Coal to Nonpolluting Fuel Oil) (Production of High-Btu Gas by the Hydrane Process)*

Coal—Properties and Statistics

Coal is a readily combustible rock containing more than 50 percent by weight and more than 70 percent by volume of carbonaceous material including inherent moisture, formed from compaction and induration of variously altered plant remains similar to those in peat. Differences in the kinds of plant materials account for the *type of coal*. Differences in the degree of metamorphism account for the *rank of coal*. Differences in the range of impurity account for the *grade of coal*. These characteristics are used in the classification of coals and are summarized for major coals in the tables of this section.

The fermentation of vegetable matter under conditions of no air and abundant moisture where volatiles are retained, resulting in the formation of bitumens, such as peat and coal, is known as *bituminous fermentation*. The metamorphic transformation of bituminous coal into anthracite is known as *anthracitization. Coalification* is the alteration or metamorphism of plant material into coal: the biochemical process of diagenesis and the geochemical process of metamorphism in the formation of coal. The peat-to-anthracite theory of coal formation is described as a process in which the progressive ranks of coal are indicative of the degree of coalification and, by inference, of the relative geological age of the deposit. Peat, as the initial stage of coalification, is of recent geological age. Lignite, as an intermediate stage, is usually Tertiary or Mesozoic, and bituminous coal and anthracite, as the more advanced stages of coalification, are usually Carboniferous. More detail on the geology of coal will be found in the section on *Coal Exploration.*

The major coals may be defined* as follows:

Anthracite Coal. Coal of the highest metamorphic rank, in which the fixed carbon content is between 92 and 98 percent. It is hard and black, and has a semimetallic luster and semiconchoidal fracture. Anthracite ignites with difficulty and burns with a short, blue flame and without smoke. Anthracite coal is also known as *hard coal, stone coal, kilkenny coal,* and *black coal.*

Semianthracite Coal. Coal having a fixed-carbon content of between 86 and 92 percent. It is between bituminous coal and anthracite coal in metamorphic rank, although its physical properties more closely resemble those of anthracite.

Semibituminous Coal. Coal that ranks between bituminous coal and semianthracite. It is harder and more brittle than bituminous coal, has a high fuel ratio, and burns without smoke. Semibituminous coal is also known as *metabituminous coal,* which is defined as containing 89 to 91.2 percent carbon, analyzed on a dry, ash-free basis. The term *smokeless coal* also is used.

Bituminous Coal. Coal that ranks between subbituminous and semibituminous coal and that contains 15 to 20 percent volatile matter. It is dark brown-to-black in color and burns with a smoky flame. Bituminous coal is the most abundant rank of coal and is commonly Carboniferous in age. The most common synonym is *soft coal.*

Subbituminous Coal. A black coal intermediate in rank between lignite and bituminous coals, or in some classifications the equivalent of *black lignite.* It is distinguished from lignite by higher carbon and lower moisture content. Further classification of subbituminous coal is made on the basis of calorific value:

Subbituminous A Coal. A type of subbituminous coal having 10,500 or more, but less than 13,000 Btu per pound.

Subbituminous B Coal. A type of subbituminous coal having 9500 or more, but less than 10,500 Btu per pound.

Subbituminous C Coal. A type of subbituminous coal having 8300 or more, but less than 9500 Btu per pound.

Lignite Coal. A brownish-black coal that is intermediate in coalification between peat

---

*Excerpted by permission from the "Glossary of Geology," published and copyrighted by the American Geological Institute, Washington, D.C., 1972.

and subbituminous coal; consolidated coal with a calorific value less than 8300 Btu per pound, on a moist, mineral-matter-free basis. Synonyms include *brown lignite* and *brown coal*. Further classification of lignite is made on the basis of calorific value:

Lignite A.   A lignite that contains 6300 or more, but less than 8300 Btu per pound. Also known as *black lignite*.

Lignite B.   A lignite that contains less than 6300 Btu per pound. Also known as brown lignite or brown coal.

Peat is an unconsolidated deposit of semicarbonized plant remains of a water-saturated environment, such as a bog or fen, and of persistently high moisture content (at least 75 percent). It is considered an early stage or rank in the development of coal. The carbon content is about 60 percent; oxygen content about 30 percent. Structures of the vegetal matter can be seen. When dried, peat burns freely.

Peat Coal refers to two materials: (1) a coal transitional between peat and brown coal or lignite, and (2) an artificially carbonized peat used as a fuel.

Cannel Coal is a compact, tough *sapropelic coal* that contains spores and is characterized by dull to waxy luster, conchoidal fracture, and massiveness. It is attrital and high in volatiles. By American standards, it must contain less than 5 percent anthraxylon. Synonyms include *candle coal, kennel coal, cannel, cannelite, parrot coal,* and *curley cannel*. A sapropelic coal is derived from organic residues (finely divided plant material, spores, algae, etc.) in stagnant or standing bodies of water. Putrefaction is under anaerobic conditions instead of by peatification.

CLASSIFICATION OF COAL BY RANK

Coals are classified in order to identify end use and also to provide data useful in specifying and selecting burning and handling equipment and in the design and arrangement of heat-transfer surfaces.* One classification of coal is by rank, i.e., according to the degree of metamorphism, or progressive alteration, in the natural series from lignite to anthracite. Volatile matter, fixed carbon, inherent or bed moisture (equilibrated moisture at 30°C and 97 percent humidity), and oxygen are all indicative of rank, but no one item completely defines it. In the ASTM† classification, the basic criteria are the fixed carbon and the calorific values calculated on a mineral-matter-free basis.

In establishing the rank of coals, it is necessary to use information showing an appreciable and systematic variation with age. For the older coals, a good criterion is the "dry, mineral-matter-free fixed carbon or volatile." However, this value is not suitable for designating the rank of the younger coals. A dependable means of classifying the latter is the "moist, mineral-matter-free Btu," which varies little for the older coals, but appreciably and systematically for younger coals.

Major coals are classifed according to rank or age in Table 1. The criteria given in the prior paragraph are used in classifying the older and younger coals. In Table 2, seventeen United States coals are arranged in the order of the classification given in Table 1. The basis for the two ASTM criteria (the fixed carbon and the calorific values calculated on a moist, mineral-matter-free basis) is shown in Fig. 1 for over 300 typical coals of the United States. The classes and groups of Table 1 are indicated in Fig. 1. For the anthracitic and low- and medium-volatile bituminous coals, the moist, mineral-matter-free calorific value changes very little; hence, the fixed-carbon criterion is used. Conversely, for the high-volatile bituminous, subbituminous and lignitic coals, the moist, mineral-matter-free calorific value is used, since the fixed carbon is almost the same for all classifications.

There are classifications of coal by rank (or type) which are currently in limited use on the European Continent. These are the *International Classification of Hard Coals by Type*, and the *International Classification of Brown Coals*. The systems were developed by a Classification Working Party established in 1949 by the Coal Committee of the Economic Commission for Europe.

The classification system of hard coals, which also includes a simplified statistical grouping combining coals having the same general characteristics, is given in Table 3.

---

*Some of this information is excerpted by permission from "Steam, Its Generation and Use." 38th ed., published and copyrighted by The Babcock and Wilcox Company, New York, 1972.
†American Society for Testing and Materials.

The term *hard coal* is based on European terminology and is defined as coal with a calorific value of more than 10,260 Btu per pound on the moist, ash-free basis. The term *type* is equivalent to rank in American coal classification terminology, and the term *class* approximates the ASTM rank.

The nine classes of coal, based on dry, ash-free volatile matter content, and moist, ash-free calorific value, are divided into groups according to their caking properties. Either the

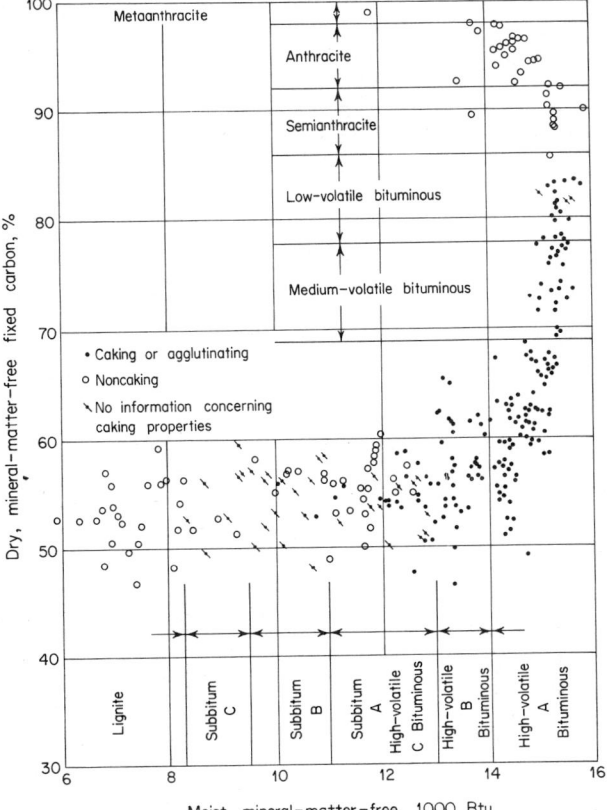

Fig. 1. Distribution plot for over 300 coals of the United States, illustrating ASTM classification by rank as defined in Table 1.

free-swelling test or the Roga test may be used to determine caking properties (a measure of the behavior of coal when heated rapidly).

The coal groups are further subdivided into subgroups according to their coking properties, which are a measure of the behavior of the coal when heated slowly. Either the Audibert-Arnu test or the Gray-King coking test may be used to determine these coking properties. See also sections in this handbook on *Coal Testing;* and *Bomb Calorimetry for Testing Solid and Liquid Fuels.*

A three-figure code number is used to express coal classification. The first figure indicates the class of the coal, the second figure the group, and the third figure the subgroup. For example, a 635 coal would be a class 6 coal with a free-swelling index greater than 4 and expansion (dilatation) greater than 140.

The rank classification is about the same as the ASTM classification except for determination of calorific-value correction. The International system uses the ash-free basis, whereas the ASTM system uses the mineral-matter-free basis. American coals fit in the

TABLE 1. Classification of Coals by Rank* (ASTM D 388)

Class	Group	Fixed carbon limits, % (dry mineral-matter-free basis)		Volatile matter limits, % (dry mineral-matter-free basis)		Calorific value limits, Btu/lb (moist† mineral-matter-free basis)		Agglomerating character
		Equal or greater than	Less than	Greater than	Equal or less than	Equal or greater than	Less than	
I Anthracitic	1. Meta-anthracite	98	—	—	2	—	—	Nonagglomerating
	2. Anthracite	92	98	2	8	—	—	
	3. Semianthracite†	86	92	8	14	—	—	
II Bituminous	1. Low volatile bituminous coal	78	86	14	22	—	—	Commonly agglomerating¶
	2. Medium volatile bituminous coal	69	78	22	31	—	—	
	3. High volatile A bituminous coal	—	—	31	—	14,000§	—	
	4. High volatile B bituminous coal	—	—	—	—	13,000§	14,000	
	5. High volatile C bituminous coal	—	—	—	—	11,500	13,000	
						10,500¶	11,500	Agglomerating
III Subbituminous	1. Subbituminous A coal	—	—	—	—	10,500	11,500	Nonagglomerating
	2. Subbituminous B coal	—	—	—	—	9,500	10,500	
	3. Subbituminous C coal	—	—	—	—	8,300	9,500	
IV Lignitic	1. Lignite A	—	—	—	—	6,300	8,300	
	2. Lignite B	—	—	—	—	—	6,300	

*This classification does not include a few coals, principally nonbanded varieties, which have unusual physical and chemical properties and which come within the fixed carbon or calorific value of the high-volatile bituminous and subbituminous ranks. All these coals either contain less than 48% dry mineral-matter-free fixed carbon or have more than 15,500 moist mineral-matter-free Btu per pound.

†Moist refers to coal containing its natural inherent moisture, but not including visible water on the surface of the coal.

‡If agglomerating, classify in low-volatile group of the bituminous class.

§Coals having 69% or more fixed carbon on the dry, mineral-matter-free basis shall be classified according to fixed carbon, regardless of calorific value.

¶It is recognized that there may be nonagglomerating varieties in these groups of the bituminous class, and there are notable exceptions in the high-volatile C bituminous group.

NOTE: The terms *mineral-matter-free fixed carbon* and *mineral-matter-free Btu* are defined by the following formulas:

Parr formulas:

$$\text{Dry Mm-free FC} = \frac{FC - 0.15S}{100 - (M + 1.08A + 0.55S)} \times 100, \%$$

$$\text{Dry Mm-free VM} = 100 - \text{dry Mm-free FC}, \%$$

$$\text{Moist Mm-free Btu} = \frac{Btu - 50S}{100 - (1.08A + 0.55S)} \times 100, \text{ per lb}$$

Approximation formulas:

$$\text{Dry Mm-free FC} = \frac{FC}{100 - (M + 1.1A + 0.1S)} \times 100, \%$$

$$\text{Dry Mm-free VM} = 100 - \text{dry Mm-free FC}, \%$$

$$\text{Moist Mm-free Btu} = \frac{Btu}{100 - (1.1A + 0.1S)} \times 100, \text{ per lb}$$

where Mm = mineral matter; Btu = heating value per lb; FC = fixed carbon, %; VM = volatile matter, %; M = bed moisture, %; A = ash, %; S = sulfur, %; all for coal on a moist basis.

TABLE 2. Seventeen Selected United States Coals Arranged in Order of ASTM Classification

No.	Coal rank Class	Coal rank Group	State	County	Coal analysis bed moisture basis M	VM	FC	A	S	Btu	Rank FC	Rank Btu
1	I	1	Pa.	Schuylkill	4.5	1.7	84.1	9.7	0.77	12,745	99.2	14,280
2	I	2	Pa.	Lackawanna	2.5	6.2	79.4	11.9	0.60	12,925	94.1	14,880
3	I	3	Va.	Montgomery	2.0	10.6	67.2	20.2	0.62	11,925	88.7	15,340
4	II	1	W.Va.	McDowell	1.0	16.6	77.3	5.1	0.74	14,715	82.8	15,600
5	II	1	Pa.	Cambria	1.3	17.5	70.9	10.3	1.68	13,800	81.3	15,595
6	II	2	Pa.	Somerset	1.5	20.8	67.5	10.2	1.68	13,720	77.5	15,485
7	II	2	Pa.	Indiana	1.5	23.4	64.9	10.2	2.20	13,800	74.5	15,580
8	II	3	Pa.	Westmoreland	1.5	30.7	56.6	11.2	1.82	13,325	65.8	15,230
9	II	3	Ky.	Pike	2.5	36.7	57.5	3.3	0.70	14,480	61.3	15,040
10	II	3	Ohio	Belmont	3.6	40.0	47.3	9.1	4.00	12,850	55.4	14,380
11	II	4	Ill.	Williamson	5.8	36.2	46.3	11.7	2.70	11,910	57.3	13,710
12	II	4	Utah	Emery	5.2	38.2	50.2	6.4	0.90	12,600	57.3	13,560
13	II	5	Ill.	Vermilion	12.2	38.8	40.0	9.0	3.20	11,340	51.8	12,630
14	III	1	Mont.	Musselshell	14.1	32.2	46.7	7.0	0.43	11,140	59.0	12,075
15	III	2	Wyo.	Sheridan	25.0	30.5	40.8	3.7	0.30	9,345	57.5	9,745
16	III	3	Wyo.	Campbell	31.0	31.4	32.8	4.8	0.55	8,320	51.5	8,790
17	IV	1	N.D.	Mercer	37.0	26.6	32.2	4.2	0.40	7,255	55.2	7,610

NOTE: For definition of coal rank class, see Table 1.

Data on Coal (Bed Moisture Basis):
M = equilibrium moisture, %; VM = volatile matter, %; FC = fixed carbon, %; A = ash, %; S = sulfur, %; Btu = high heating value, Btu per lb; Rank FC = dry mineral-matter-free fixed carbon, %; rank Btu = moist mineral-matter-free Btu per lb; all calculations by Parr formulas (see Table 1).

TABLE 3. International Classification of Hard Coals by Type and Their Statistical Grouping*

The first figure of the code number indicates the class of the coal, determined by volatile-matter content up to 33% volatile matter and by calorific parameter above 33% volatile matter. The second figure indicates the group of coal, determined by caking properties. The third figure indicates the subgroup, determined by coking properties.

Groups (determined by caking properties)			Code numbers (Class number)										Subgroups (determined by coking properties)		
Group number	Free-swelling index (crucible-swelling number)	Roga index	0	1	2	3	4	5	6	7	8	9	Sub-group number	Audibert-Arnu dilatometer	Gray-King
3	>4	>45					435	535	635				5	>140	>G8
						334	434	534	634				4	>50–140	G5–G8
						333	433	533	633	733			3 (VA/VB/VC/VD)	>0–50	G1–G4
						332a 332b	432	532	632	732	832		2	≧0	E–G
2	2½–4	>20–45				323	423	523	623	723	823		3 (IV/VIA)	≧0–50	G1–G4
						322	422	522	622	722	822		2	≧0	E–G
						321	421	521	621	721	821		1 (VIB)	Contraction only	B–D
1	1–2	>5–20			212	312	412	512	612	712	812		2 (VII)	≧0	E–G
					211	311	411	511	611	711	811		1	Contraction only	B–D
0	0–½	0–5		100	200	300	400	500	600	700	800	900	0	Nonsoftening	A

Statistical group labels: I (100); II (200); III (300, 311, 321…); IV (333, 323, 432, 422…); VA, VB, VC, VD; VIA, VIB; VII.

Class parameters	0	1	2	3	4	5	6	7	8	9
Class number	0	1	2	3	4	5	6	7	8	9
Volatile matter (dry, ash-free) →	0–3	>3–10 (A >3–6.5 / 6.5; B >6.5–10 / 10)	>10–14	>14–20	>20–28	>28–33	>33	>33	>33	>33
Calorific parameter† →	—	—	—	—	—	—	>13,950	>12,960–13,950	>10,980–12,960	>10,260–10,980

Classes: Determined by volatile matter up to 33% volatile matter and by calorific parameter above 33% volatile matter.

As an indication, the following classes have an approximate volatile-matter content of:
Class 6: 33–41% volatile matter
Class 7: 33–44% volatile matter
Class 8: 35–50% volatile matter
Class 9: 42–50% volatile matter

*SOURCE: U.S. Bureau of Mines.
†Gross calorific value on moist ash-free basis (86°F/96% relative humidity) Btu per pound.

NOTES:
1. Where the ash content of coal is too high to allow classification according to the present systems, it must be reduced by laboratory float-and-sink method (or any other appropriate means). The specific gravity selected for flotation should allow a maximum yield of coal with 5–10% of ash.
2. 332a > 14–16% volatile matter. 332b > 16–20% volatile matter.

International classification as shown by the open-faced blocks in Table 4. A general comparison of the systems with several American coals is shown in Table 5.

The International Classification of Brown Coals is shown in Table 6, where the term *brown coal* refers to coal containing less than 10,260 Btu per pound on the moist, ash-free basis. The six classes of coal, based on the ash-free equilibrium-moisture content, are divided into groups according to their tar yield on a dry, ash-free basis. This grouping indicates the values of these low-rank coals as fuel and as raw material for chemical purposes. Brown coals with high tar content are generally used as raw material in the chemical industry rather than as fuel.

Other criteria for the classification of coal by rank have been proposed by various authorities. A method of classifying, by computing the heating value of the residual coal with moisture M, ash A, and sulfur S removed, has been used by Lord. This is called the H value.

$$H = \frac{Btu - 4050S}{100 - (M + A + S)} \times 100 \qquad (1)$$

Although the H values as listed in Table 7 are of interest, they do not arrange the 17 coals in the same order as the ASTM classification.

Another classification method, reported by Perch and Russell, is based on the following ratio:

$$\text{Ratio} = \frac{\text{Moist, } Mm\text{-free Btu}}{\text{Dry, } Mm\text{-free VM}} \qquad (2)$$

where Mm = mineral matter, and VM = volatile matter, %

Values of this ratio for the 17 coals listed in Table 2 arrange these coals in Table 7 in exactly the same order as the ASTM method. The very high ratio for coal No. 1 is not shown, since it is of little importance. It is said that this Perch and Russell ratio is closely related to the coking properties of coals used in coke-oven practice.

It is of interest to consider the validity of other criteria in the classification of coal by rank based on the dry, ash-free analyses appearing in Table 7. Carbon (C) comes close to lining up the coals in the same order as the ASTM rank. As a criterion, the carbon content would be quite simple, but it cannot be used as a suitable basis for classification. Oxygen (O_2) generally decreases with the age of coals, but it is not consistent. The volatile matter (VM), fixed carbon (FC), hydrogen (H_2), and the Btu do not follow the same order as the ASTM method. Nitrogen (N_2) varies little for all ranks.

Since the quality of volatile matter in coal is an index of the extent of the conversion of the original carbohydrates to hydrocarbons, studies have been made suggesting that the heating value of the volatile matter will serve as an accurate criterion for classification by rank. This criterion, developed on a "pure coal" basis as calculated in Eq. (6), is listed in Table 7 under Btu per pound VM_{pc}.

Composition of the fixed carbon in all types of coal is substantially all carbon. The variable constituents of coals can, therefore, be considered as concentrated in the volatile matter. One index of the quality of the volatile matter, its heating value, is perhaps the most important property insofar as combustion is concerned and bears a direct relation to the properties of the pure coal (dry, mineral-matter-free). The volatile matter in coals of lower rank, where the conversion of carbohydrates to hydrocarbons has not progressed far, is relatively high in water and CO_2 and, consequently, low in heating value. The volatile matter in coals of higher rank is relatively high in hydrocarbons, such as methane (CH_4), and thus relatively high in heating value.

The analyses and heating values of the coal must be converted to the mineral-matter-free (pure coal) basis in order to establish reasonably accurate heating values for volatile matter. The only difference in the conversion used for this method and the conversion used in the ASTM Standard D 388 procedure is that half the sulfur* is assumed to be

---

*Sulfur is present in coal in three forms: (1) combined with iron as pyrite (pyritic) or marcasite, (2) organic sulfur, and (3) sulfate sulfur or sulfur combined with iron or calcium together with oxygen as $FeSO_4$ or $CaSO_4$. Sulfur seldom occurs in the free state in coal. Sulfate sulfur usually is found in weathered coal.

TABLE 4. Grouping of United States Coals in International System of Classification*

The first figure of the code number indicates the class of the coal, determined by volatile-matter content up to 33% volatile matter and by calorific parameter above 33% volatile matter. The second figure indicates the group of coal, determined by caking properties. The third figure indicates the subgroup, determined by coking properties.

Group number	Alternative group parameters: Free-swelling index (crucible-swelling number)	Roga index	Code numbers — Class → 0	1	2	3	4	5	6	7	8	9	Subgroup number	Alternative subgroup parameters: Audibert-Arnu dilatometer	Gray-King
3	>4	>45					435	535	635				5	>140	>G₈
3	>4	>45				334	434	534	634	734			4	>50–140	G₅–G₈
3	>4	>45				333	433	533	633	733			3	>0–50	G₁–G₄
3	>4	>45				332 a b	432	532	632	732	832		2	≦0	E–G
2	2½–4	>20–45				323	423	523	623	723	823		3	>0–50	G₁–G₄
2	2½–4	>20–45				322	422	522	622	722	822		2	≦0	E–G
2	2½–4	>20–45				321	421	521	621	721	821		1	Contraction only	B–D
1	1–2	>5–20			212	312	412	512	612	712	812		2	≦0	E–G
1	1–2	>5–20			211	311	411	511	611	711	811		1	Contraction only	B–D
0	0–½	0–5	000 A | B	100	200	300	400	500	600	700	800	900	0	Nonsoftening	A

Class parameters:

	0	1	2	3	4	5	6	7	8	9
Class number	0	1	2	3	4	5	6	7	8	9
Volatile matter (dry, ash-free) →	0–3	>3–10 (>3–6.5 | >6.5–10)	>10–14	>14–20	>20–28	>28–33	>33	>33	>33	>33
Calorific parameter† →	—	—	—	—	—	—	>13,950	>12,960–13,950	>10,980–12,960	>10,260–10,980

Classes: Determined by volatile matter up to 33% volatile matter and by calorific parameter above 33% volatile matter.

As an indication, the following classes have an approximate volatile-matter content of:
Class 6: 33–41% volatile matter
Class 7: 33–44% volatile matter
Class 8: 35–50% volatile matter
Class 9: 42–50% volatile matter

*SOURCE: U.S. Bureau of Mines.
†Gross calorific value on moist ash-free basis (86°F/96% relative humidity) Btu per pound.
NOTES:
1. Where the ash content of coal is too high to allow classification according to the present systems, it must be reduced by laboratory float-and-sink method (or any other appropriate means). The specific gravity selected for flotation should allow a maximum yield of coal with 5–10% of ash.
2. 332a > 14–16% volatile matter. 332b > 16–20% volatile matter.

TABLE 5. System Comparison of Selected United States Coals*

State	County	Seam	ASTM	International
			Classification	Classification
West Virginia	McDowell	Pocahontas No. 3	*lvb*	333
Pennsylvania	Clearfield	Lower Kittaning	*mvb*	435
West Virginia	Monongalia	Pittsburgh	*hvAb*	635
Illinois	Williamson	No. 6.	*hvBb*	734
Illinois	Knox	No. 6	*hvCb*	821

*SOURCE: U.S. Bureau of Mines.
NOTES *lvb* = low-volatile bituminous; *mvb* = medium-volatile bituminous;
       hvAb = high-volatile A bituminous; *hvBb* = high-volatile B bituminous;
       hvCb = high-volatile C bituminous.

pyritic on the pure-coal basis. The assumption that only half the sulfur is pyritic and the remainder organic is in closer agreement with the average for a large number of United States coals. The formulas for converting the analyses and coal heating values to the pure coal basis and for calculating the heating value of the volatile matter are:

$$VM_{pc} = \frac{VM - (0.08A + 0.2S)}{100 - (1.08A + 0.2625S + M)} \times 100, \% \tag{3}$$

$$FC_{pc} = \frac{FC - 0.0625S}{100 - (1.08A + 0.2625S + M)} \times 100, \% \tag{4}$$

$$Btu_{pc} = \frac{Btu - 26.2S}{100 - (1.08A + 0.2625S + M)} \times 100, \ Btu/lb \tag{5}$$

$$\frac{Btu}{lb \ VM_{pc}} = \frac{Btu_{pc} - \left(\dfrac{14,460}{100} \ FC_{pc}\right)}{VM_{pc}} \times 100, \ Btu/lb \tag{6}$$

where VM = volatile matter, %; FC = fixed carbon, %; A = ash, %; S = sulfur, %; M = bed moisture, %; Btu = heating value per pound; and subscript *pc* designates pure coal basis

The Btu per pound volatile matter, calculated from the foregoing formulas, is given in Table 7 for 16 of the 17 coals listed, covering the entire range of rank. The order of these values follows generally the ASTM classification of the coals, although the correlation is

TABLE 6. International Classification of Brown Coals*

Group no.	Group parameter tar yield (dry, ash free), %	Code number					
40	>25	1040	1140	1240	1340	1440	1540
30	>20–25	1030	1130	1230	1330	1430	1530
20	>15–20	1020	1120	1220	1320	1420	1520
10	>10–15	1010	1110	1210	1310	1410	1510
00	>10 and less	1000	1100	1200	1300	1400	1500
Class number		10	11	12	13	14	15
Class parameter, i.e., total moisture, ash free, %		20 and less	20 . to 30	30 to 40	40 to 50	50 to 60	60 to 70

SOURCE: U.S. Bureau of Mines
NOTES:
*Gross calorific value below 10,260 Btu per pound. Moist ash-free basis(86°F/96% relative humidity).
The total moisture content refers to freshly mined coal. For internal purposes, coals with a gross calorific value over 10,260 Btu per pound (moist ash-free basis), considered in the country of origin as brown coals, but classified under this system, to ascertain, in particular, their suitability for processing. When the total moisture content is over 30%, the gross calorific value is always below 10,260 Btu per pound.

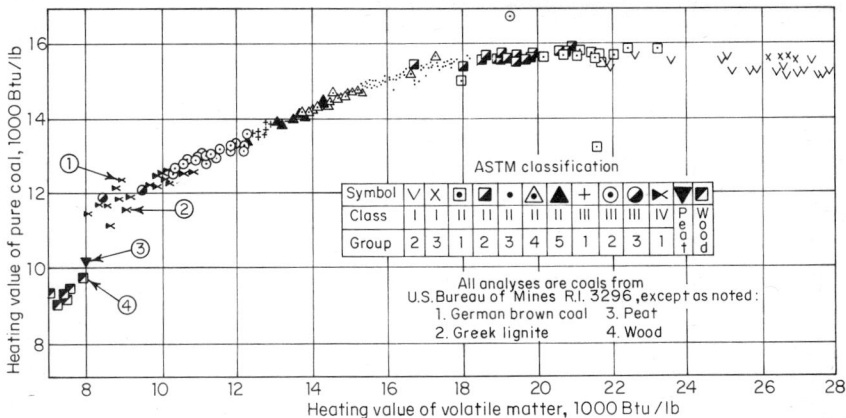

Fig. 2. Illustration of a suggested coal classification, using the relationship of the respective heating values of "pure coal" and the volatile matter.

not as good as with the Perch and Russell ratio. The range of values of Btu per pound volatile matter, from about 8000 to 28,000 (Fig. 2), is large and may serve as a useful classification.

The relation of the heating value of the volatile matter and the heating value of the pure coal is shown in Fig. 2 for a large number of coals. It is evident that a fair line could be drawn, without serious error, to indicate the path of relationship.

COMMERCIAL SIZES OF COAL

Bituminous Coal. The sizes of bituminous coal are not well standardized, but the following sizings are common:

Run of Mine. This is coal shipped as it comes from the mine without screening. It is used for both domestic heating and steam production.

Run of Mine (8 in.). This is run-of-mine coal with oversize lumps broken up.

Lump (5 in.). This size will not go through a 5-inch round hole. It is used for hand firing and domestic purposes.

Egg (5 in. × 2 in.). This size goes through 5-inch and is retained on 2-inch round-hole screens. It is used for hand firing, gas producers, and domestic firing.

Nut (2 in. × 1¼ in.). This size is used for small industrial stokers, gas producers, and hand firing.

Stoker coal (1¼ in. × ¾ in.). This is used largely for small industrial stokers and domestic firing.

Slack (¾ in. and under). This is used for pulverizers, cyclone furnaces, and industrial stokers.

Anthracite Coal. Definite sizes of anthracite are standardized and indicated in Table 8. The broken, egg, stove, nut, and pea sizes are largely used for hand-fired domestic units and gas producers. Buckwheat and rice are used in mechanical types of firing equipment.

COAL PRODUCTION AND CONSUMPTION

As indicated by Table 9, only six nations account for nearly all coal production. Bituminous production in the United States by states is given in Table 10. It is interesting to note the relatively high rates of coal production in some states in earlier years. See the "Highest year" column at the left of the table. Bituminous enjoyed particularly high production during the World War I years. Declines in production during the postwar period may be attributed to several factors: (1) the generally depressed economy of the 1930s, (2) the swing of the railroads from steam to Diesel locomotives, (3) changes in processes by certain coal-consuming industries, notably the steel industry, (4) the change-over to oil and gas heating for commercial buildings and residences, initiated by factors of

TABLE 7. Study of the Suitability of Other Criteria in the Classification of Coals

Coals (Table 2 no.)	Btu Lord's H value	Perch & Russell ratio	Btu per pound, VM_{pc}	Coal analysis, dry, ash-free basis						Btu	Grindability
				VM	FC	C	H_2	O_2	N_2		
1	14,950	—	—	2.0	98.0	93.9	2.1	2.3	0.3	14,850	37
2	15,180	2520	25,685	7.3	92.7	93.5	2.6	2.3	0.9	15,100	26
3	15,410	1358	23,330	13.6	86.4	90.7	4.2	3.3	1.0	15,325	83
4	15,765	907	21,750	17.7	82.3	90.4	4.8	2.7	1.3	15,670	100
5	15,840	834	21,155	19.8	80.2	89.4	4.8	2.4	1.5	15,615	112
6	15,765	688	19,785	23.5	76.5	88.6	4.8	3.1	1.6	15,540	105
7	15,930	611	19,570	26.5	73.5	87.6	5.2	3.3	1.4	15,630	95
8	15,500	445	17,230	35.2	64.8	85.0	5.4	5.8	1.7	15,265	88
9	15,460	389	16,930	39.0	61.0	85.5	5.5	6.7	1.6	15,370	56
10	15,230	322	15,430	45.8	54.2	80.9	5.7	7.4	1.4	14,730	57
11	14,800	320	14,875	43.8	56.2	80.5	5.5	9.1	1.6	14,430	60
12	14,359	318	14,200	43.2	56.8	79.8	5.6	11.8	1.7	14,260	50
13	14,830	300	14,690	49.3	50.7	79.2	5.7	9.5	1.5	14,400	61
14	14,170	295	13,885	40.8	59.2	80.9	5.1	12.2	1.3	14,110	55
15	13,145	229	11,435	42.8	57.2	75.9	5.1	17.0	1.6	13,100	43
16	13,055	181	11,570	49.0	51.0	74.0	5.6	18.6	0.9	12,970	52
17	12,400	170	9,945	45.3	54.7	72.7	4.9	20.8	0.9	12,330	45

NOTES: For calculations of Lord's H Value and the Perch & Russell ratio, and the heating value of the volatile matter (pure coal basis), refer to Eqs. (1) through (6) in accompanying text. The subscript pc stands for a "pure coal" basis. The dry ash-free VM may be used, instead of the VM_{pc}, without appreciable error. Grindability is determined by ASTM Method D 409.

TABLE 8.   Commercial Sizes of Anthracite*

	Diameter of holes, inches	
Trade name	Passes through	Retained on
Broken	4⅜	3¼–3
Egg	3¼–3	2⁷⁄₁₆
Stove	2⁷⁄₁₆	1⅝
Nut	1⅝	1³⁄₁₆
Pea	1³⁄₁₆	⁹⁄₁₆
Buckwheat	⁹⁄₁₆	⁵⁄₁₆
Rice	⁵⁄₁₆	³⁄₁₆

*ASTM D 3-10 testing procedure. Graded on round-hole screens.

TABLE 9.   Principal Worldwide Producers of Coal

Nation	Estimated tonnage (annual), millions of tons	Percent of major producers
United States	556	28.1
U.S.S.R.	529	26.7
People's Republic of China	452	22.8
United Kingdom	162	8.2
Poland	160	8.1
Federal Republic of Germany (West)	122	6.1

NOTE: Because of the absence of a central and reliable collection agency for worldwide coal production data, figures of this type are, at best, educated estimates. The last reliable worldwide data available are for 1971—hence these figures are given here. See also table at end of later article on *Coal Reserves—Estimates.*

TABLE 10.   Bituminous Coal Production in the United States

State	Highest year before 1971 Year	Amount	Production, million tons 1966	1968	1970	1972
Alabama	1926	21.001	14.219	16.440	20.560	20.814
Arkansas	1907	2.670	0.236	0.211	0.268	0.428
Colorado	1917	12.483	5.222	5.558	6.025	5.522
Illinois	1918	89.291	63.571	62.441	65.119	65.523
Indiana	1918	30.679	17.326	18.486	22.263	25.949
Iowa	1917	8.966	1.025	0.876	0.987	0.851
Kansas	1918	7.562	1.122	1.268	1.627	1.227
Kentucky	1970	125.305	93.156	101.156	125.305	121.187
Maryland	1907	5.533	1.222	1.447	1.615	1.640
Missouri	1917	5.671	3.582	3.205	4.447	4.551
Montana	1944	4.844	0.419	0.519	3.447	8.221
New Mexico	1970	7.361	2.755	3.429	7.361	8.248
N. Dakota	1970	5.639	3.543	4.487	5.639	6.632
Ohio	1970	55.351	43.341	48.323	55.351	50.967
Oklahoma	1920	4.849	0.843	1.089	2.427	2.624
Pennsylvania	1918	178.551	81.443	76.200	80.491	75.939
Tennessee	1956	8.848	6.309	8.148	8.237	11.260
Utah	1947	7.429	4.635	4.316	4.733	4.802
Virginia	1968	36.966	35.565	36.966	35.016	34.028
Washington	1918	4.082	0.059	0.178	0.037	2.634
W. Virginia	1947	176.157	149.681	145.921	144.072	123.743
Wyoming	1945	9.847	3.670	3.829	7.222	10.928
Other States*	—	—	0.937	0.752	0.683	7.668
Total U.S.	1947	630.624	533.881	545.245	602.932	595.386

*Production includes Alaska, Arizona, California, Georgia, Idaho, Michigan, N. Carolina, Oregon, S. Dakota, and Texas.

NOTE: Total production for United States to end of 1970 from earliest records available is set at 35,004,344,000 tons.

TABLE 11. Bituminous Coal Consumption and Exports (United States)

million net tons

Year	Electric power utilities	Rail-roads (Class 1)	Coking coal	Steel & rolling mills	Cement mills	Other manu-facturing*	Retail sales	Total U.S. con-sumption	Total exports
1940	49.126	85.130	81.386	14.169	5.559	110.853	84.687	430.910	16.466
1945	71.603	125.120	95.349	14.241	4.203	129.754	119.297	559.567	27.956
1950	88.262	60.969	103.845	10.877	7.923	97.904	84.422	454.202	25.468
1955	140.550	15.473	107.377	7.353	8.529	91.110	53.020	423.412	51.277
1960	173.882	2.101	81.015	7.378	8.216	77.432	30.405	380.429	36.541
1965	242.729	‡	94.779	7.466	8.873	86.269	19.048	459.164	50.181
1966	264.202	‡	95.892	7.117	9.149	89.941	19.965	486.266	49.302
1967	271.784	‡	92.272	6.330	8.922	84.009	17.099	480.416	49.528
1968	294.739	‡	90.765	5.657	9.391	83.054	15.224	498.830	50.637
1969	308.461	‡	92.901	5.560	†	85.687	14.666	507.275	56.234
1970	318.921	‡	96.009	5.410	†	83.207	12.072	515.619	70.944
1971	326.280	‡	82.809	5.560	†	68.862	11.351	494.862	56.632
1972	348.612	‡	87.272	4.850	†	67.294	8.748	516.776	
1973	386.879	‡	93.634	6.356	†	60.953	8.200	556.022	

SOURCE: U.S. Bureau of Mines
NOTES: *Includes bunker fuel. ‡Included in Other manufacturing.

TABLE 12.   Bituminous Coal Mines in the United States Exceeding One Million Tons of Annual Production

Name of mine	State*	Production, million tons 1972	Type	Operator
Navajo	New Mexico	6.898	S	1
River King	Illinois	6.776	S	2
Colstrip	Montana	5.501	S	3
Sinclair	Kentucky (W)	5.477	S	2
No. 10	Illinois	4.693	D	2
River Queen	Kentucky (W)	4.661	D & S	2
Moss No. 3	Virginia	4.582	D	4
Captain	Illinois	4.481	S	5
Muskingum	Ohio	4.310	S	6
Lynnville	Indiana	4.174	S	2
Egypt Valley	Ohio	3.822	S	7
Ayrgem	Kentucky (W)	3.183	S	8
Robena	Pennsylvania	3.008	D	9
Humphrey No. 7	W. Virginia	3.008	D	7
Black Mesa	Arizona	2.954	S	2
Vogue	Kentucky (W)	2.811	S	2
Burning Star No. 2	Illinois	2.802	D	7
Ken	Kentucky (W)	2.771	D & S	2
Dave Johnston	Wyoming	2.618	D	10
Centralia	Washington	2.597	S	11
Kopperston Nos. 1 & 2	W. Virginia	2.521	D	12
No. 24	Illinois	2.507	D	13
Homestead	Kentucky (W)	2.469	S	2
Wright	Indiana	2.444	S	8
Orient No. 3	Illinois	2.444	D	14
Loveridge	W. Virginia	2.439	D	7
Leahy	Illinois	2.356	S	8
No. 1	Pennsylvania	2.327	D	15
Ireland	W. Virginia	2.300	D	7
Fox	Pennsylvania	2,300	S	16
Universal	Indiana	2.253	S	2
Mathies	Pennsylvania	2.205	D	17
Sunnyhill	Ohio	2.171	D	2
Enos	Indiana	2.137	S	13
Osage No. 3	W. Virginia	2.128	D	7
Inland	Illinois	2.126	D	18
Squaw Creek	Indiana	2.087	S	2
Maple Creek	Pennsylvania	2.062	D	9
No. 26	Illinois	2.041	D	13
Jane	Pennsylvania	1.999	D	19
Colonial	Kentucky (W)	1.991	S	20
Hillsboro	Illinois	1.991	D	7
Monterey No. 1	Illinois	1.974	D	21
Arkwright No. 1	W. Virginia	1.936	D	7
Montour No. 4	Pennsylvania	1.935	D	7
Hamilton	Kentucky (W)	1.911	D	22
Itmann	W. Virginia	1.896	D	23
Sorensen	Wyoming	1.844	D	24
Orient No. 6	Illinois	1.844	D	14
Robinson Run No. 95	W. Virginia	1.837	D	7
Minnehaha	Indiana	1.808	S	8
McElroy	W. Virginia	1.755	D	7
No. 21	Illinois	1.733	D	13
Fidelity No. 11	Illinois	1.726	S	25
Harris No. 1 & 2	W. Virginia	1.718	D	12
Warwick	Pennsylvania	1.714	D	26
Gary No. 14	W. Virginia	1.682	D	9
Concord	Alabama	1.663	D	9
Shoemaker	W. Virginia	1.643	D	7
Streamline	Illinois	1.628	S	5
Buckheart No. 17	Illinois	1.620	S	25
Big Sky	Montana	1.601	S	2

TABLE 12.   Bituminous Coal Mines in the United States Exceeding One Million Tons of Annual Production   (Continued)

Name of mine	State*	Production, million tons 1972	Type	Operator
Gateway	Pennsylvania	1.599	D	27
Norris	Illinois	1.589	S	7
Beaulah	N. Dakota	1.530	S	28
Blacksville No. 2	W. Virginia	1.526	D	7
Mecco	Illinois	1.519	S	29
Matthews	Tennessee	1.518	D	7
Ayrcoe	Indiana	1.510	S	8
Powhatan No. 3	Ohio	1.509	D	30
Volunteer	Kentucky (W)	1.507	S	31
Powhatan No. 1	Ohio	1.506	D	30
Star	Kentucky (W)	1.494	D	2
Wharton	W. Virginia	1.492	D	12
No. 9	Kentucky (W)	1.479	D	32
Bishop	W. Virginia	1.471	D	33
Franklin No. 25	Ohio	1.470	D	7
Power	Maryland	1.459	S	2
Blackfoot No. 5	Indiana	1.450	S	13
Big Horn	Wyoming	1.433	S	34
Federal No. 1	W. Virginia	1.426	D	12
Glenharold	N. Dakota	1.418	S	7
Hawthorn	Indiana	1.414	S	2
Burning Star No. 3	Illinois	1.412	S	7
Dugger	Indiana	1.390	S	2
Gary No. 2 & 10	W. Virginia	1.372	D	9
Lucerne No. 6	Pennsylvania	1.365	D	35
Marianna 58	Pennsylvania	1.354	D	36
Dotiki	Kentucky (W)	1.336	D	37
Somerset 60	Pennsylvania	1.332	D	36
Providence No. 1	Kentucky (W)	1.330	D	22
Gilbraltar	Kentucky (W)	1.323	S	38
Lynch Winifrede	Kentucky (E)	1.297	D	9
Georgetown No. 24	Ohio	1,284	S	7
Latta	Indiana	1.279	S	2
Simco-Peabody	Ohio	1.278	D & S	2
Olga	W. Virginia	1.260	D	39
Fabius	Alabama	1.254	S	40
Murdock	Illinois	1.241	D	32
Chinook	Indiana	1.225	S	8
V. C. No. 1	W. Virginia	1.223	D	41
Morris Creek	W. Virginia	1.207	D	42
Rosebud	Wyoming	1.200	S	43
Beatrice	Virginia	1.195	D	44
Virginia Pocahontas No. 1	Virginia	1.184	D	22
Maxine	Alabama	1.166	D	45
Paradise	Kentucky	1.154	S	20
Keystone No. 1	W. Virginia	1.140	D	12
Warrior	Alabama	1.139	S & A	2
Segco No. 1	Alabama	1.133	D	46
Stone	Kentucky (E)	1.117	D	12
Leatherwood No. 1	Kentucky (E)	1.107	D	47
Eagle No. 2	Illinois	1.101	D	2
Marty	Kentucky (E)	1.100	S & A	48
Orient No. 4	Illinois	1.092	D	14
Ellsworth No. 51	Pennsylvania	1.088	D	36
Sunspot	Illinois	1.070	S	8
Powhatan No. 5	Ohio	1.062	D	30
Pursglove No. 15	W. Virginia	1.054	D	7
Delta	Illinois	1.046	S	8
Williams No. 98	W. Virginia	1.045	D	7
Indian Head	N. Dakota	1.037	S	30
Bolt	W. Virginia	1.034	D	49
Rose Valley No. 6	Ohio	1.011	D	7

Name of mine	State*	Production, million tons 1972	Type	Operator
Bee Veer	Missouri	1.010	S	2
Federal No. 2	W. Virginia	1.009	D	12
Lynch No. 32	Kentucky (E)	1.006	D	9
Kellerman	Alabama	1.000	S	50

*(W) = western Kentucky; (E) = eastern Kentucky.
S = strip mine; D = deep mine; A = auger mine.

Operators

1 Utah International Inc.	26 Duquesne Light Co.
2 Peabody Coal Co.	27 Gateway Coal Co.
3 Western Energy Co.	28 Knife River Coal Co.
4 Pittston Co.	29 Midland Coal Co.
5 Southwestern Illinois Coal Corp.	30 North American Coal Corp.
6 Central Ohio Coal Co.	31 Cimarron Coal Corp.
7 Consolidation Coal Co.	32 Zeigler Coal Co.
8 Amax Coal Co.	33 Bishop Coal Co.
9 United States Steel Corp.	34 Big Horn Coal Co.
10 Pacific Power & Light Co.	35 Helvetia Coal Co.
11 Washington Irrigation District	36 Bethlehem Mines Corp.
12 Eastern Associated Coal Corp.	37 Webster County Coal Corp.
13 Old Ben Coal Corp.	38 Gibraltar Coal Corp.
14 Freeman Coal Mining Corp.	39 Olga Coal Co.
15 Florence Mining Co.	40 Arch Coal Co.
16 C & K Coal Co.	41 Valley Camp Coal Co.
17 Mathies Coal Co.	42 Central Appalachian Coal Co.
18 Inland Steel Co.	43 Rosebud Coal Sales Co.
19 Rochester & Pittsburgh Coal Co.	44 Beatrice Pocahontas Co.
20 Pittsburg & Midway Coal Mining Co.	45 Alabama By-Products Corp.
21 Monterey Coal Co.	46 Southern Electric Generating Co
22 Island Creek Coal Co.	47 Blue Diamond Coal Co.
23 Itmann Coal Co.	48 Marty Corp.
24 Kemmerer Coal Co.	49 Ranger Fuel Corp.
25 United Electric Coal Co.	50 Kellerman Mining Co.

TABLE 13.   Bituminous Coal Production by Type of Mining (United States)

Year	Type of mining (Percent of total bituminous mined)		
	Strip mining	Underground mining	Auger mining
1955	25	74	1
1956	25	73	2
1957	25	73	2
1958	28	70	2
1959	29	69	2
1960	29	69	2
1961	30	68	2
1962	31	67	2
1963	31	66	3
1964	31	66	3
1965	32	65	3
1966	34	63	3
1967	34	63	3
1968	34	63	3
1969	35	62	3
1970	40	56	4
1971	47	50	3

TABLE 14. Types of Bituminous Coal Mines in the United States

State	1966 Underground	1966 Strip	1966 Auger	1968 Underground	1968 Strip	1968 Auger	1970 Underground	1970 Strip	1970 Auger
Alabama	116	61	6	80	60	2	35	91	3
Alaska		4			3			3	
Arizona								1	
Arkansas	4	4		2	6		2	6	
Colorado	65	6	1	49	5	1	40	8	
Illinois	34	49		33	36		28	31	
Indiana	13	38		9	35		6	32	
Iowa	8	15		3	12		3	10	
Kansas		5			4			5	
Kentucky	1465	111	128	1121	123	151	1104	410	207
Maryland	23	27	2	21	33	7	16	28	6
Missouri	1	13			12			9	
Montana	8	4		7	4		4	4	
New Mexico	4	3		3	3		1	3	
N. Dakota		25			22			20	
Ohio	89	274	63	55	263	54	44	217	45
Oklahoma	3	10	1	1	6	1	2	8	1
Pennsylvania	410	618	60	261	489	55	198	555	54
S. Dakota		1							
Tennessee	144	49	10	114	59	9	116	80	7
Utah	25			23			20		
Virginia	1002	66	65	658	86	65	566	154	83
Washington	4	1		3	2		1	1	
W. Virginia	1318	179	100	934	220	109	748	418	153
Wyoming	5	9		4	9		4	9	
Total U.S.	4741	1572	436	3381	1492	454	2939	2103	559
All types		6749			5327			5601	

convenience and cleanliness and later by environmental concerns, and (5) a declining cost advantage of coal over other more convenient fuels. Although electric power utilities switched to oil and gas in large numbers, it is interesting to note that while most other major coal-consuming areas were declining, the tonnage delivered to the utilities maintained a modest growth. This reflects, of course, the tremendous increase in electric power generation, particularly since World War II. Coal exports have remained reasonably steady. These observations are documented by the figures given in Table 11.

The largest bituminous coal mines in the United States are identified by type and operator and location in Table 12. The steady swing from underground mining in proportion to strip mining is depicted by the percentage figures of Table 13. The types of mining used are further broken down by states in Table 14.

The impact of gasification, liquefaction, and other processes for converting coal into cleaner, more easily transportable fuels will be large. Additional statistics appear throughout this section on *Coal Technology*. For economic appraisals of coal technology, reference to "U.S. Energy Outlook—Coal Availability," and subsequent reports which may issue is suggested. The foregoing booklet (approximately 300 pages), prepared for the U.S. Department of the Interior, is obtainable from the National Petroleum Council, Washington, D.C.

Coal Reserves—Estimates

BY JAMES E. MCNULTY* AND CLAYTON G. BALL†

Published estimates and quotations of estimates of amounts of coal reserves in the United States vary substantially according to sources and to bases which are not always clearly defined. The widely cited estimates of coal resources published by the U.S. Geological Survey‡ establish a resource base of 3.2 trillion tons, about half of which (1.6 trillion tons)

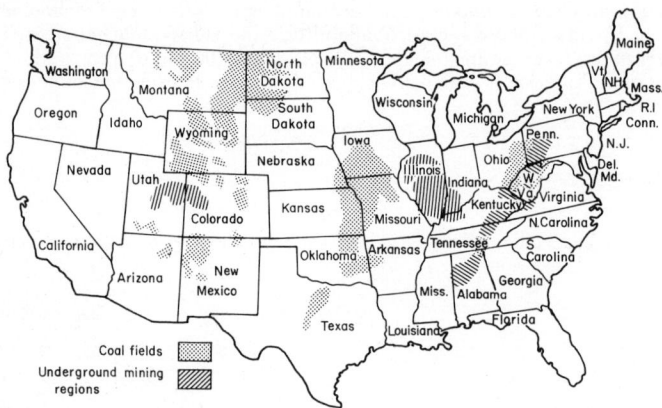

Fig. 1. Underground coal reserves and production (minable by underground mining methods). *(Reproduced by permission of source: National Petroleum Council.)*

	Billions of tons				Life of recoverable reserves at % growth rate, years		
Region	Remaining measured and indicated reserves*	Economically available reserves†	Recoverable reserves‡	1970 production, millions of tons	0%	3%	5%
1. W. Virginia, Pennsylvania	92.7	67.1	33.5	145.8	230	69	50
2. Mercer, McDowell, and Wyoming Counties, W. Virginia	9.1	9.1	4.6	N.A.	—	—	—
3. Illinois, Indiana, Ohio	83.1	59.5	29.7	52.3	568	96	68
4. Kentucky, Tennessee, Virginia	34.5	24.4	12.2	95.0	129	52	40
5. Utah, Colorado	21.9	13.3	6.7	8.6	774	106	74
6. Alabama	1.6	0.6	0.3	9.1	35	23	20
Other	106.3	35.2	17.6	N.A.	—	—	—
Total§	349.1	209.2	104.6	338.8	309	80	58

Notes:

*Bituminous coal in seams more than 28 inches thick; subbituminous coal and lignite in seams more than 5 feet thick; less than 1000 feet of overburden.

†Excludes lignite, subbituminous coal less than 10 feet thick and bituminous coal less than 42 inches thick.

‡Based on 50 percent recovery of economically available reserves.

§May not add correctly due to rounding.

*Vice-president, †Consultant, Paul Weir Company Incorporated, Chicago, Illinois.

‡Paul Averitt, Coal Resources of the United States, *U.S. Geol. Surv. Bull.* 1275, Washington, D.C.

is usually considered as recoverable, the other half necessarily being left in the ground for support during eventual mining operation or otherwise lost because of location below cities, towns, or lakes; between depleted areas; etc. Of this 1.6 trillion recoverable tons, approximately half, or 780 billion tons, were determined from actual mapping and exploration to depths of 3000 feet below ground surface, with the remaining 820 trillion

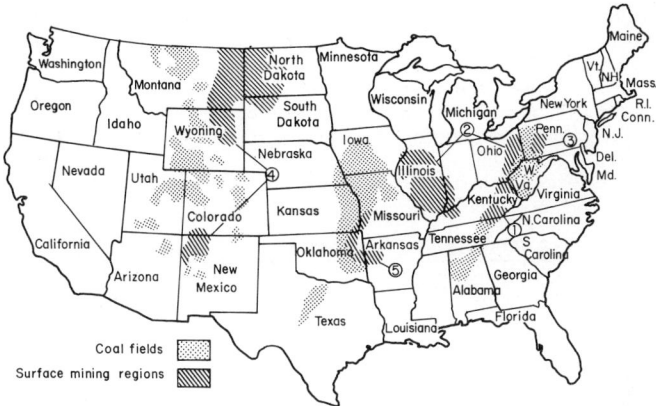

Fig. 2. Surface coal reserves and production (minable by surface mining methods). *(Reproduced by permission of source: National Petroleum Council.)*

Region	Recoverable reserves, billions of tons	1970 production, millions of tons	Life of reserves at % growth rate, years		
			0%	3%	5%
1. Kentucky, W. Virginia, Virginia, Tennessee	4.2	101.2	42	27	23
2. Illinois, Indiana, Iowa, Ohio	5.6	91.0	62	36	29
3. Pennsylvania	0.8	25.1	32	23	19
4. Colorado, Montana, New Mexico, Wyoming	23.8	19.1	1,246	122	85
5. Oklahoma, Kansas, Missouri	1.6	8.3	193	65	48
6. North Dakota	2.1	5.6	375	85	62
Other	6.9	13.8	500	95	67
Total	45.0	264.1	170	61	46

tons considered as occurring in unmapped or unexplored areas to depths of 6000 feet below the surface.

As determined by mapping and exploration in a number of representative states, approximately 200 billion of the 780 billion recoverable tons are projected as occurring in bituminous coal and anthracite beds 42 inches or more in thickness and in subbituminous coal and lignite beds of 10 feet or more in thickness, generally less than 1000 feet below the surface.§ The remaining 580 billion tons are hence in thinner beds less than 1000 feet below the surface and in beds of all thicknesses from 1000 to 3000 feet below the surface. Of the 200 billion recoverable tons of thicker coal, approximately one-half, or 100 billion tons, is based on a relative abundancy of information, with the remaining 100 billion tons being projected with a reasonable degree of assurance into areas having lesser available data.

Although these 100 billion tons by no means represent a finite amount of reserves for future recovery, they do represent those reserves as known today that have the most favorable characteristics for acquisition of individual areas of reserves of sufficient magnitude to support large-scale mining operations. Coal is being and will be mined at depths below 1000 feet from the surface, in thicknesses less than those cited above, and from the

§*Ibid.*, table 5, p. 33.

areas not yet fully known, which will undoubtedly be developed into minable categories by additional programs of exploration as the better-known areas become depleted.

In 1973 the National Petroleum Council prepared estimates of "economically available reserves" in the principal coal-producing regions of the United States as presented in Figs. 1 and 2 for underground and surface mining, respectively.¶

Later studies indicate that the amounts of recoverable surface reserves shown in Fig. 2 should be materially increased, essentially in the Western states (Regions 4 and 6).

It can be concluded from these data that known coal reserves susceptible to today's mining practices and economics are sufficient to last well into the next century. Further projections, however, even though coal resources as such will still be abundant, will require substantial changes in mining technology and will be greatly influenced, of course, by intervening changes in comparative patterns of fuel utilization and demand.

EDITOR'S NOTE: The following listing is based on information published by the National Coal Association, Washington, D.C., 1972.

Location of Worldwide Coal Reserves

Country	% of world total	
ASIA		
U.S.S.R.	19.9	
People's Republic of China	16.7	
India	1.0	
Japan	0.2	
Others	0.1	
	37.9	37.9
NORTH AMERICA		
U.S.A.	48.2	
Canada	1.4	
Mexico	0.1	
	49.7	49.7
EUROPE		
Germany	4.8	
U.K.	2.8	
Poland	1.3	
Czechoslovakia	0.3	
France	0.2	
Belgium	0.1	
Netherlands	0.1	
Others	0.4	
	10.0	10.0
AFRICA		
Republic of South Africa	1.1	
Others	—	
	1.1	1.1
AUSTRALASIA		
Australia	1.0	
Others	—	
	1.0	1.0
SOUTH AND CENTRAL AMERICA		
Colombia	0.2	
Venezuela	0.1	
Others	—	
	0.3	0.3
		100.0

¶"U.S. Energy Outlook: Coal Availability," National Petroleum Council, Washington, D.C., 1973.

Coal Exploration

BY JAMES E. MCNULTY* AND CLAYTON G. BALL†

This description includes a discussion of the geology of coal, the characteristics of coal occurrence to be investigated, and exploration methods, including associated data compilation and interpretation, and postexploration decision making.

GEOLOGY OF COAL

Although coal is composed of organic (plant) derivatives, it is nevertheless a distinctive type of rock which is interspersed as individual beds within other types of sedimentary rock beds such as sandstones, shales, clays, limestones, and various mixtures thereof. Geological studies indicate that the plant material which eventually became the coal deposits as known today was originally accumulated in upland bogs, coastal or near-coastal swamps, or delta plains, under conditions very similar to those existing today in areas such as the Okefenokee Swamp in Georgia and the Everglades of Florida. The original coal-forming bogs, swamps, or delta accumulations may have varied in size from but a few acres to hundreds of square miles.

It was necessary, after periods of plant accumulation, that these swamps became submerged, either through land subsidence or sea-level rises, to the extent that they were buried by muds or sands washed down from adjacent highlands. As subsidence continued, calcareous forms may also have been deposited in deeper water within or above the muds and sands. In certain suitable geological and climatic periods, such intervals of plant growth and interrupted but continuing submersion were repeated many times before final cessation of each such period.

During subsequent geological eras and depending on varying increases in combinations of depth of burial, heat, and pressure over differing lengths of time, the buried plant accumulations became compacted and progressed successively through one or more stages of coalification, either from peat to lignite, to subbituminous coal, to bituminous coal, or to anthracite. The intervening or covering muds, sands, and calcareous debris, meanwhile, became the shales, sandstones, and limestones that form the stratigraphic series in which minable coal beds may be contained.

There have been three major geological periods when conditions within the United States were especially suitable over large areas for persistent and repeated plant growth with subsequent preservation and coalification. Essentially all the predominately bituminous coal beds in the Appalachian Province extending from Pennsylvania (also including the anthracite beds in central Pennsylvania) into northeastern Alabama; in the contiguous Eastern Interior Region of Illinois, southwestern Indiana, and western Kentucky; in the contiguous Western Interior Region of Iowa, Kansas, Missouri, northeastern Oklahoma, and northwestern Arkansas; and in the separated central portion of Texas (exclusive of Texas lignite) were formed in the "Pennsylvanian" (Carboniferous) period of geologic time, which is dated as approximately 300 million years ago.

The predominately bituminous and subbituminous coal beds in the Rocky Mountain Province, extending in large, separated regions from central Montana into northeastern Arizona and northwestern New Mexico, were formed in the Cretaceous period of geologic time, dated as approximately 100 million years ago. The subbituminous coal and lignite beds in the Great Plains Province, which includes northeastern Wyoming, eastern Montana, western North Dakota, and northwestern South Dakota, were formed in the early Tertiary period of geologic time, approximately 65 million years ago.

Each of these three major periods of coal-forming time lasted for many millions of years. Generally speaking, the geological and climatic conditions conducive to periodic accumulations of plant material were relatively more stable during the Pennsylvanian period than

*Vice-president, †Consultant, Paul Weir Company Incorporated, Chicago, Illinois.

in the other two periods. The coal beds of the Western Interior Region and those east of the Mississippi River hence tend to be more persistent in extent and more uniform in coal-bed characteristics than those in the Rocky Mountain and Great Plains Provinces where shorter and more irregular periods suitable for plant accumulation resulted in generally more-scattered and relatively more-lenticular coal deposits.

CHARACTERISTICS OF COAL OCCURRENCE TO BE INVESTIGATED

Aside from purely geological resource mapping, the primary purpose of coal exploration, of course, is to ascertain the commercial potentials of mining and marketing or otherwise utilizing the coal either from a specific and hitherto unmined area or in unexplored areas bordering areas of current mining. Depending on the desired or feasible rate of annual production, the amounts of coal reserves required to support a new mine designed for an economic life of up to 20 years or more range from a comparatively few to 300 million tons or more. Such amounts, depending on the coal-bed thicknesses and the suitable or necessary mining methods, whether surface or underground, might require exploration of areas for a single mine from as little as 1000 acres to as much as 50 square miles or more in extent.

Despite their seeming abundance in the broad Provinces and Regions described above, it should be pointed out that the coal beds in themselves form only a very small percentage of the total thicknesses of the overall sedimentary strata comprising the "Coal Measures" in the same coal-bearing areas; the thicknesses of individual coal beds within the United States generally range from a few millimeters (horizon markers) to as much as 100 feet or more in some instances. The numbers of individual coal beds of present-day commercial significance may vary from less than 10 feet in some coal-bearing areas to over 100 feet in others. These numbers, however, are rarely, if ever, found in full vertical sequence at any one particular spot, but are usually distributed unevenly in single beds or small groups of beds around the margin or within the interior of the generally basin-shaped areas of coal-bearing strata.

In addition to the simple determination of ranges in depth and of the apparent lateral extent and thickness of coal in an area undergoing coal-bed exploration, from which the character of mine opening and the amount of coal in place can be readily determined, the most appropriate coal-exploration program in any area must be designed to assess the minability and quality of the coal to the fullest extent necessary or possible. Many of the factors affecting the minability and quality of coal, some of which may be relatively local while others may be regional, are related not only to the original deposition of the coal-bed material but also to the subsequent geological history of the area, both short term and long term. The nature and degree of effect of any such depositional or geological factors, as predetermined by exploration, contribute greatly to establishing the comparative ease or difficulty of mining and the comparative value of the potential mine product for commercial utilization.

A variety of detrimental irregularities may accompany or interrupt an otherwise orderly accumulation of plant material either during swamp growth or very shortly thereafter. While many coal beds or portions of beds are relatively low in ash content, other beds or portions of beds may contain depositional admixtures of particles of mud or silt which were washed or blown into the swamp during plant growth. Where relatively abundant, these particles serve to increase the ash content of the eventual coal bed and to decrease, correspondingly, its quality.

During periods of prolonged swamp flooding, layers of mud or silt may have been deposited on the preexisting plant accumulations, such deposition then having been followed by additional plant accumulation. Such layers of impurities between underlying and overlying accumulations of plant material, eventually hardening into shales or silty shales, are designated as "partings" and may range from knife-edge thickness to thicknesses of up to 12 inches or more. Such thicknesses of partings and the number of such partings within a single coal bed, both decrease the quality of the coal as mined and also impair the mining procedures, especially if such partings eventually become pyritized.

The partings described above are not always evenly deposited over large portions or the entire extent of a coal-forming swamp, but may become progressively thicker toward the source of the deposited material. In such cases the partings are wedge-shaped, with the overlying plant material occurring increasingly higher above the underlying plant material and frequently becoming increasingly thinner to the point of eventual disappearance. Such types of deposition of impurities result in "splitting" the total thickness of the coal bed into two or more diverging "benches," with resulting difficulties in mining and possible isolation of the lower bench into a single bed too thin to be economically recoverable.

In some coal beds in some areas, relatively flat, lenticular masses ranging up to several feet in diameter and composed of pyrite, calcite, or siderite, or combinations thereof (geologically described as "concretions") were formed either during plant growth or immediately after cessation of such growth during deposition of the next-overlying material representing the eventual immediate "roof" of the coal bed. Where present in the coal bed, such concretions may impede operation of mining equipment and, when present immediately above the coal bed, may represent a serious hazard because of their tendency to drop out of the roof unexpectedly during mining operations.

Although not widely prevalent, there are some coal beds which contain areas of relatively narrow, vertical, or steeply slanting veins of clay, sometimes partially pyritized. These irregularly shaped veins, usually extending down into or through the coal bed from the overlying strata, may range from but a few inches to several feet in thickness, but may extend for hundreds of feet in length. They frequently occur in intersecting patterns, with the distances between the more-or-less parallel veins of either segment of the pattern varying up to several hundred feet. Their origin is not clearly understood, but they seem to have been formed by some sort of shrinkage and in-filling shortly after the cessation of plant material accumulation in that particular deposit, and after initial stages of compaction, loss of original moisture, and incipient coalification had begun. Since occurrences of such veins usually tend to impair the stability of the mine roof, they represent inferior mining conditions and decrease the quality of the coal as mined.

Even as in present times, the streams flowing within the coal-forming areas during coal-forming geological periods originated and deepened their valleys by downward erosion. Such drainage channels might often cut close to, into, or through any lower bed or beds of accumulated plant material, even after such beds had been previously covered with later sedimentary materials. During subsequent stages of subsidence or burial by later depositions of the overall coal-bearing regions, these stream channels were filled with mud, silt, or frequently with coarse sand, which later became a water-bearing sandstone. Where such channels fillings penetrated a coal bed, they obviously represent barriers to systematic mine layouts. Where their bases approach the top of a minable coal bed, serious roof-support problems may be created.

From initial deposition and burial under overlying sedimentary materials through all succeeding geological periods into modern times, and regardless of the particular stage of coalification attained, today's coal beds have been continually subjected to the action of groundwater. In many cases such action has been negligible. In others, under certain conditions substantial amounts of inorganic sulfur in the form of pyrite or gypsum have been injected into or added to the original coal bed. By the time coalification has reached the stages of bituminous coal or anthracite, the bed has usually developed a system of vertical or nearly vertical fractures, these being very thin cracks which are often filled with coatings of pyrite, calcite, kaolinite or other minerals deposited therein from circulating groundwater. Although not usually a principal source of ash in coal, these minerals nevertheless may affect the combustion or other utilization of a particular coal-bed product. The amount of sulfur from the pyrite or gypsum, of course, materially affects the problems of air pollution and the solution thereof.

After cessation of the particular period of time during which the geological and climatic conditions were particularly favorable for accumulations of plant material which eventually became coalified, there were commonly one or more geological periods during which the entire series of sedimentary strata containing these coalifying beds were subjected to structural deformation caused by differential stresses during periods of uplift, compression, or tension which affected large portions of the then-continental masses. The results of such deformative stresses may have been relatively local or broadly regional in effect. Depending on their proximity to locations of maximum stress during such periods of diastrophism, the coal-bearing strata may have been but gently warped into broad undulations over areas of varying size, compressed into upward and downward folds of varying extent and degree, or sharply deformed into steeply slanted inclinations.

The more pronounced stresses inducing folding and sharp deformation likewise frequently resulted in specific dislocations or "faults" through the coal-bearing strata, where beds on one side of the fault were actually moved upward or downward from the horizons of the same beds on the other side. Such dislocations may range from only a foot or two to hundreds of feet in vertical distance and may range up to thousands of feet in linear extent. Where more than one fault was thus induced, they may occur in parallel series or may exist in an intersecting pattern.

Very generally, the coal and lignite fields within the Great Plains Province are relatively undisturbed. The coal-bearing areas in the Western and Eastern Interior Regions are prevailingly gentle in structure, but include a relatively few areas of locally sharp disturbances. The coal-bearing strata in the Appalachian Province are relatively flat along their northwestern margin, but increase in intensity of relatively mild but significant folding toward the southeast at right angles to the regional northeast-southwest trend of the component coal fields. The coal beds in the various basins comprising the Rocky Mountain Province range from comparatively gentle slopes of but a few degrees over areas of broad extent to areas of similar extent with prevailing dips of up to 20° or 30°, along with a few areas of limited extent where the coal beds are highly deformed.

Regardless of degree of coalification, geologic structure or proposed methods of mining (whether underground or surface), the composition, hardness, and regularity of the strata both overlying and underlying a coal bed are highly important in assessment of minability. The nature of the overlying material determines the comparative strength or weakness of the "roof" which must be controlled in underground mining and of the overburden which must be removed in surface mining. It may present relatively easy or relatively difficult mining conditions and is often one of the most determinative criteria in mine planning. Although perhaps not as significant in effect, the character and thickness of the underlying stratum, usually but not always a "seat" rock representing the original soil supporting plant growth, may affect mining conditions if it should become unduly soft or sticky when wet or excessively plastic under differential pressure during or after the mining operations.

Where exploration for a potential new coal mine is conducted in relative proximity to areas of past or present mining, the coal-bed characteristics revealed in such mining provide a useful background for guiding the new exploration program, although conditions can and do change within relatively short distances. Accordingly, even in such areas and most certainly in new areas, exploration programs must be designed to determine the presence or absence of any one or more of the irregularities described above and, if present, their extent and degree of influence on economic minability and commercial quality of the coal.

EXPLORATORY METHODS

Exploration encompasses all the activities of data gathering, compilation, and interpretation that are necessary to evaluate the economic and production viability of a mineral deposit.

Although in coal mining the percentage of salable product that is recovered from operations which may also include beneficiation is usually much higher than in the mining of most ores, coal remains a relatively low-bulk-value commodity. The realization (price) that a product ton of coal can command at the mine is governed by market conditions, including its suitability to a specific utilization and often the distance between the mine and the point of intermediate or ultimate utilization.

The coal prospector's art crosses several disciplines, since economic viability almost always includes the consideration of reserve availability, facility of recovery by mining, coal chemistry, and transportation.

CORE DRILLING

Historically, the diamond-core drill has been the most extensively used tool in coal exploration. Cores of coal, properly recovered, enable accurate seam descriptions and measurements and also provide material for chemical analysis.

Core drilling is done with a double-tube core barrel using conventional or "standard tools" or the newer "wire-line" method. In order to recover the core by means of standard tools, the entire drill string must be removed from the hole and the core barrel partially disassembled. The wire-line system is designed so that the inner tube of the core barrel can be recovered through the drill rods by a retriever attached to a small-diameter cable, without the drill rods being removed from the hole.

The time performance of the two systems is about the same to around 200 feet from the surface, at which point the advantage shifts to the wire-line method and increases with depth.

Coal cores recovered using standard tools are normally 2⅛ inches in diameter (NX size). Wire-line cores are usually 1⅞ inches (NQ) or 2½ inches (HQ). In general, the physical nature of coal is such that coring with tools much smaller than 2 inches in diameter results in poor core recovery.

Cores of rock strata are also required for reserve definition. For underground mining they are necessary to evaluate the mining conditions that can be expected from the roof and floor material. In these evaluations, judgment, experience, and a knowledge of nearby operations are more reliable than physical testing procedures. Comprehensive core logging and inspection by experienced personnel are required.

In a surface mining prospect, drillability tests are conducted on rock cores to determine factors that will affect drilling, blasting, and excavating practices, such as drill capacity required, unit performance, and bit life. Chemical testing is done on overburden cores for reclamation and revegetation studies.

In coal exploration, cores are normally pushed from the core barrel into core troughs or boxes in order to permit examination and measurement of the material in as nearly an in-situ condition as possible. Vertical fractures in coal, swelling clays, and subfreezing weather often prohibit the careful and orderly removal of the cores from the core barrel. These problems have been eliminated by the development of the split-tube or triple-tube core barrel, in which the innermost tube is split lengthwise so that one-half of the tube can be removed to permit the measurement and description of soft or badly broken strata in a relatively undisturbed state.

SAMPLING AND ANALYSES

Since the market value of coal is governed in part by its suitability to a specific utilization, exploration methodology must define the quality of a given prospect by chemical analyses of representative samples of the deposit.

The size of a sample must be sufficient to provide the necessary amount of material for the testing procedure. The sample should be representative of that portion of the seam which will be mined in case of splits or of selective mining. The number of samples must be sufficient to closely define chemical variability throughout the project area.

In order to maximize the value of the analyses, a coal sample should be properly described, carefully handled, preserved against loss of moisture and promptly delivered to the laboratory. The most widely accepted authority on sampling methods is the *U.S. Geol. Surv. Bull.* 1111-B (Schopf, 1960).

The results of coal analyses are used to determine the market on which a particular mined product can be sold, the degree of beneficiation that is required or desirable, and the specifications for utilization equipment.

For example, in analyzing a steam coal, the test results showing percentage of ash and sulfur content and calorific value will establish this coal's relative competitive position. However, washability tests are required to determine whether or to what extent beneficiation will reduce ash and sulfur and enhance calorific values. Further analytical determination required for the design of firing equipment may include mineral analysis of ash, fusibility of ash, grindability, ultimate analysis, and fouling and slagging indices.

The uses and limitations of most coal analyses are presented in the *Ill. State Geol. Surv. Rep. Invest.* 220 (Rees, 1966).

Rotary drilling, including churn or cable-tool methods, by itself is a most unreliable method of subsurface data gathering. The interpretation or log of the results of a rotary hole is reliant on the identification of cuttings and variations in drilling time. However, rotary drilling aided by geophysical hole-logging methods is a basic exploration tool, particularly in the Western states.

Although rotary drilling cannot provide the detail to be had from cores, it is a faster and a cheaper method of obtaining some types of data. It can be done using compressed air as the circulating medium (within some depth limits), thereby eliminating water sources and water haulage.

A recent improvement in rotary methods is the development of the reverse circulation-type drill rig. It employs double-tube rods and the circulating fluid (usually air, but water can be used) is pumped down the annulus between the inner and outer pipe wall and returned through the inner pipe. Samples are collected from a cyclone installed in the return circuit.

The advantages of this drilling method include the prevention of loss of circulating fluid, less contamination in the cuttings, and better overall performance in caving conditions.

Samples of cuttings from rotary holes are examined and described only for lithologic hole logging. Cuttings of coal seams are of no value in analytical programs.

LARGE SAMPLES

Although, by far, most coal testing is done on material recovered by diamond drilling, coal cores are not suitable for many analytical procedures because of the size limitation of the core diameter or the lack of a sufficient volume of the sample. Such procedures include washability tests to determine parameters for the design of preparation plants, coking tests, and even test firings or other "process" testing. The sample material requirements in each case may vary from several pounds to many tons.

In a few cases the volume requirement may be met by combining a number of core samples. But since a coal core of approximately 2 inches in diameter cannot be crushed to a size consist that would approximate raw run-of-mine coal (that may include 6-inch lumps), recovery of large samples is often included in project requirements.

Bulk sampling is time-consuming and expensive, and so various procedures that would simulate mine-run coal from small-diameter cores have been researched (Leonard and others, 1968). Such simulations have been largely experimental in nature and are not widely used in the industry.

Optimum bulk samples are obtained from fresh coal faces reached by driving adits or short slopes from the outcrop or locations having shallow depths of overburden, or by excavating shallow pits. In rarer instances, shafts have been sunk to some depth to acquire large bulk samples.

Amounts of material collected from these faces range from channel samples (which are full-seam vertical columns, square or rectangular in horizontal cross section and weighing several pounds) to bulk samples of many tons.

Face samples provide the opportunity for good examination and description of seam, immediate roof, and floor strata, and permit close control of volume, sample size consist, and contamination.

When practicable, face samples from adits or shallow pits produce the lowest cost per unit volume of all bulk sampling methods.

Unfortunately, in many instances the coal seam is too deep to permit direct access to coal faces from the surface. Consequently, large samples must be and are recovered by various drilling techniques.

In the past, the calyx core drill enjoyed limited success in this area of use. This drilling method, primarily employed for shaft sinking, could obtain cores up to 72 inches in diameter using steel shot as a cutting medium. Because of costs and depth limitations this method in recent years has been all but forgotten.

Recently, bulk samples for washability tests have been obtained by drilling large vertical holes with 30- and 36-inch auger bits. The size consist of these samples often includes an abnormally high percentage of fines, and dilution from overlying strata is a problem. Depth capability is about 200 feet or less, depending on the hardness of the strata overlying the seam.

In most cases where direct access to coal seams is impossible, large-diameter cores are recovered by conventional core-drilling methods. The most popular core sizes are the 6-inch from a 7¾-inch hole and the ZWF size, which recovers a 6½-inch core from an 8-inch hole. These cores provide between 15 and 20 lb of coal per foot of seam thickness.

Engineers from Amax Coal Company have developed a method of obtaining bulk samples at depth from small-diameter drill holes (Parks and others, 1973). This method consists of drilling a conventional hole and coring of the coal bed. The hole is then reamed, and casing is set from surface to the top of the coal. The cavity in the coal bed is enlarged by a combination of blasting and hydraulic jetting. Coal cuttings are brought to the surface with the returns of the high-pressure hydraulic system and recovered by means of a cyclone. Casing prevents dilution from up-hole material.

In order to obtain a truly representative sample using this approach, the resultant cavity must be a symmetrical cylinder. In some coals, the distribution and thickness of partings and the vertical variability of hardness, strength, and friability would tend to diminish the effectiveness of this technique.

Difficulty in the control of sample size consist may also tend to limit the utility of this method. However, since several hundred pounds of sample can be recovered by this system in situations where other methods would be economically prohibitive, this technique ranks as a significant development in bulk sampling which will experience increased use.

REMOTE GEOPHYSICAL LOGGING

Geologging or electric logging, long a basic tool of the oil and gas industry, has only recently become commonplace in coal-exploration technology.

Basically, the system involves hoisting a sensor up the length of a drill hole while electric pulses are transmitted through the hoist cable to a console mounted in a truck on the surface, which records the variations in properties of the rock strata as a function of depth of hole. The print-out is presented as curves on a graph with depth as the ordinate.

The electrical curves normally run in coal exploration are resistivity and spontaneous potential. Radiometric or nuclear curves include gamma ray, neutron, and the density or gamma-gamma log.

Although no one suite of geologs is ideal for every area of the country, the three most popular are gamma ray, density, and resistivity. Coal has a low natural radiation, low bulk density, and high apparent resistivity. Some logging units are designed to produce all three curves on a three-pen recorder in one traverse of the hole.

The impurities in coal beds, present either as distinct partings or disseminated throughout, are composed of clay minerals having a high density and a high natural radioactivity relative to coal. Consequently, the gamma ray and density curves, invaluable for bed correlation and thickness determination, can also be used as semiquantitative indices of coal quality. Figure 1 shows these two curves on the log of a hole drilled in a Western state.

Figure 2 shows a typical plot of coal-ash content versus natural gamma radiation.

The capabilities of new geologging methods have changed the makeup of drilling programs. Since good subsurface data on coal-bearing rocks can now be obtained from rotary-drilled holes, the usually more expensive core drilling can be minimized.

However, core holes are also routinely logged, because they can provide guidelines for the interpretation of the geologs of rotary holes in the field, present unbiased interpretation of the subsurface, and provide checks on depth measurements and core recovery, and because folded geologs are much easier to store than coreboxes.

EXPLORATION PROGRAMS

In the United States, coal-bearing rocks are well mapped, and principal deposits are well known. The main purpose of coal exploration is to define an area or prospect in an attempt to fit it to a specific economic model having parameters, which usually include an overall reserve life suitable to a specified annual rate of production, a physical configuration of reserves amenable to the desired production rate, coal quality necessary to a specific utilization, and often a profit floor or cost ceiling.

In order to minimize cost and risk in evaluating undeveloped coal lands, investigations are phased or staged. Progressive evaluations are made between phases. The more detailed, more time-consuming, and more expensive elements and methods are deferred until "first phase" investigation provides sufficient indication and reasonable assurance that the expenditure of additional funds and effort is warranted. Items most likely to be deferred (and these may be divided into more than one phase also) are large-scale topographic mapping, detailed (more closely spaced) drilling including additional core drilling, bulk sampling, additional and more esoteric analytical determinations, and large-scale testing.

DATA GATHERING

The initial step is always the assembly of base-line data to determine coal occurrence, relative coal quality (if information exists), and any history of previous exploration or nearby mining operations. Evaluation of these data will not only dictate the operations

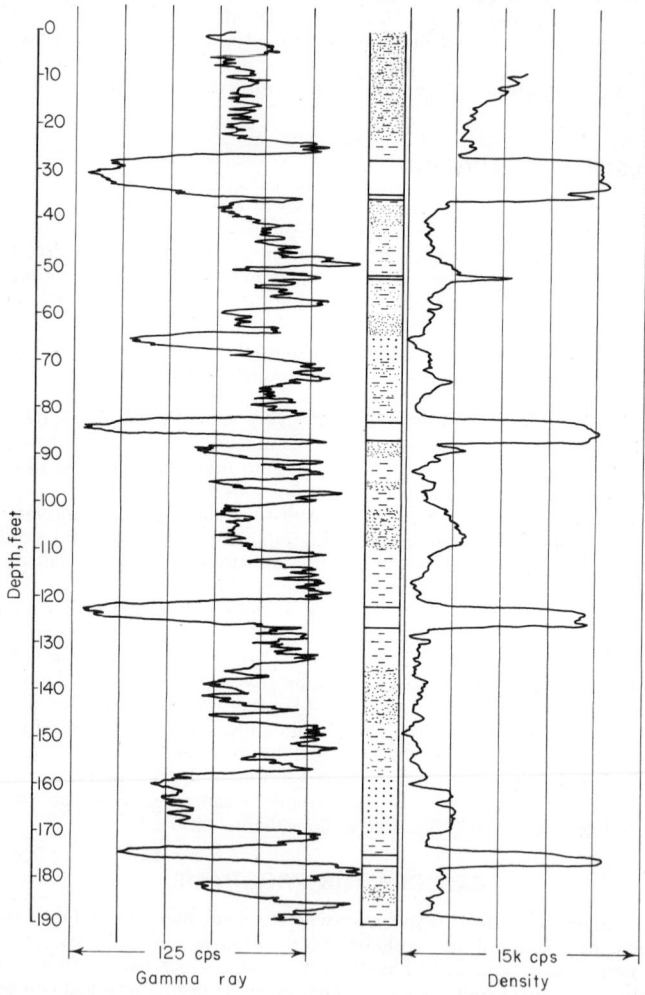

Fig. 1. Gamma ray and density curves on log of a hole drilled in a western state.

necessary to fulfill project requirements, but will in general show how the subsequent phases should be structured to conform to these requirements in an orderly and economical fashion.

Additional data on the physical and chemical conditions of the coal will always be required. The principal method of obtaining these subsurface data beyond the outcrop area is by a drilling program. Base-line data will determine to what extent core drilling, rotary drilling, or a combination thereof will be required to maximize the results of the initial field phase of the investigation.

Hole locations in any drilling program are controlled to some extent by topographic conditions, access, and surface ownership. In folded rocks, locations are chosen to determine the axes and limbs of structure, and these locations may require angle-hole drilling.

Most United States coal beds are reasonably planar, and proposed hole locations are laid out on a regular grid.

Square grids are often used in the Midwestern and Western states to take advantage of general land surveys, roads and property boundaries.

As shown in Fig. 3, an offset pattern of the square grid can effect a significant economy

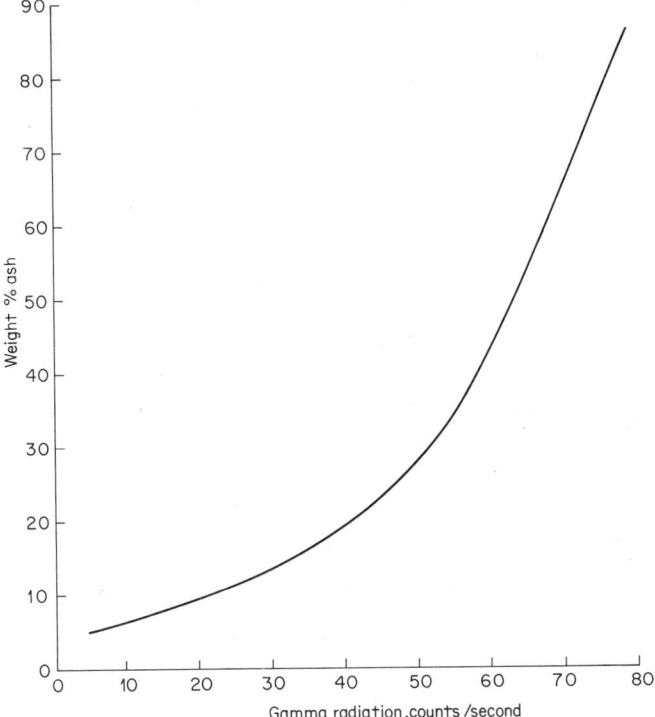

Fig. 2. Typical plot of coal ash content versus natural gamma radiation.

of drilling. In the square pattern (example A), the holes are spaced on d centers. In the offset pattern (example B), the center distance $d\sqrt{2}$ is an increase of only 40 percent, but the required number of holes is reduced by half.

The density of data points required to prove up a coal deposit is a function of the variability of physical and chemical properties of the coal bed and the subjacent and superjacent strata to the extent they tend to affect the facility of mining and product quality. The absolute spacing of these points varies from one deposit to another within a range of a few hundred feet to one mile.

In a multiphased program, the initial phase target may require only very widely spaced data points to determine order-of-magnitude estimates of reserve potential, approximate indices of coal quality, and the variability of critical parameters which will influence the design of subsequent phases or stages of the investigation.

Intermediate phases are only initiated when the results of initial work are inconclusive in terms of one or more of the key project elements.

The results of final-phase exploration programs form the bases for irrevocable decision making, which is accompanied by a commitment to capital spending. The total data grid, at this point, should raise the confidence level of the combined information so that the results of any additional, intermediate points of measurement or observation would show a predictable gradation or transition of features common to the grid points surrounding the new location.

At this juncture, the compilation and interpretation of data in hand should permit the determination of mining and economic feasibility.

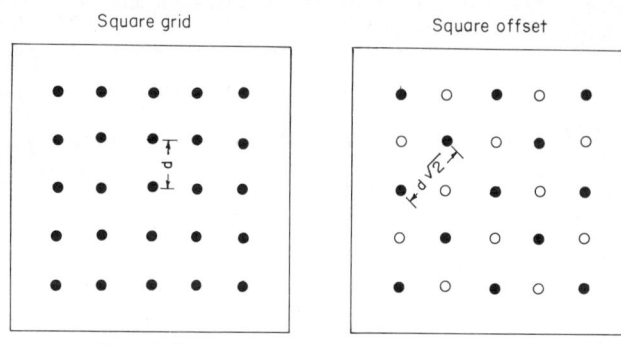

Fig. 3. Examples of use of square grids: (A) square pattern with holes placed on d centers; (B) offset pattern, where center distance is $d\sqrt{2}$.

DATA COMPILATION

Reserves of coal are usually compiled from graphic representations, such as maps or plots which depict bed thickness, structure, and quality. In a surface mining investigation, it is necessary to plot isopachs of overburden to be used in planning mining methods and procedures, equipment determinations and specifications, and to assess economic viability.

Since the values on which these various maps are based are derived from a grid net of data points which have a spatial interrelationship, the operations necessary to interpretation are well suited to data-processing methods and the use of mechanical plotters. Drill holes, outcrops, mine workings, or other locations of data points are assigned coordinate values. A third value is the variable to be plotted using the specific value indicated at each of the data points as control.

Volumes of overburden can be computed by digitizing the ground surface from large-scale aerial photogrammetric mapping and subtracting the surface representing the top of the minable coal bed.

The question of whether data-processing techniques and mechanical plotting are more or less accurate than manual methods is overshadowed by the speed of computer operations which, properly programmed, can expeditiously handle the ever-increasing volumes of data necessary to evaluate a coal prospect. These methods will increase in usage and in complexity of operation proportionately with the tonnage requirements of new mines, advances in mining technology, and increase in energy source demand.

POSTEXPLORATION DECISION MAKING

When reserves are properly defined and depicted by the geologists, the project area is subjected to a mining feasibility study. In this investigation, the prospect is literally mined out on paper by periods for the project or reserve life. The study also generates optimum rates of production, operating costs, capital requirements, and financial and economic analyses.

Although these factors are the determinants on which a decision to proceed with mining development is based, the reliability of the feasibility study is heavily dependent on the completeness of the reserve analysis presented as a result of exploration.

The fulcrum of the economic analysis of a project is unit productivity in terms of both manpower and machine capability. We have seen in recent years the effects of health, safety, and environmental legislation on productivity as well as on patterns of coal utilization. The magnitude of these effects was unforeseen.

In order to minimize the unpredictable downside effects on productivity by unforeseen occurrences, the importance of careful analysis of the geologic conditions affecting minability of a coal deposit is increased.

In the past, exploration has been a "poor relation" of the coal industry. Companies customarily utilized exploration methods only on a short-term or as-needed basis, while much of the work of exploratory nature was carried out by state and Federal surveys with, however, very general targets.

The recent resurgency in coal-mining geology and exploration is spurred by two factors. First, is the recognition that, with the increasing energy demand, coal (although our most abundant fuel source) does not exist in limitless supply. Second, because of the highly competitive nature of the coal-mining industry, reserves having the better mining conditions (and, therefore, the lowest costs of production) are constantly being depleted, leaving less desirable resources for a subsequent generation of economic parameters. Consequently, exploration will have to adopt more complex and intensified programs, and new techniques and methods, to define more accurately the factors that will affect the recoverability of future coal.

REFERENCES

Leonard, J. W., O. B. Bucklen and C. F. Cockrell: "Projecting Coal Drill Core Data for Use in Plant Design," part I, Coal Convention, American Mining Congress, 1968.

Parks, O. E., A. P. Widklund, and J. S. Osborne: "A Technique for Obtaining Bulk Samples from a Small Diameter Borehole," Annual Meeting, Society of Mining Engineers, 1973.

Rees, O. W.: Chemistry, Uses, and Limitations of Coal Analyses, *Ill. Geol. Surv. Rep. Invest.* 220, Urbana, Ill., 1966.

Schopf, J. M.: Field Description and Sampling of Coal Beds, *U.S. Geol. Surv. Bull.* 1111-B, 1960, U.S. Government Printing Office, Washington, D.C.

Coal Testing

BY KENNETH I. SAVAGE*

Coal may be defined geologically as a combustible, carbonaceous rock composed princi-pally of carbon, hydrogen, nitrogen, oxygen, sulfur, and mineral matter. Coal has eluded official chemical formulation because of the complex variety of chemical bond structures, and because of variations in the amounts of identifiable species from one coal deposit to the next. In addition to chemical heterogeneity, coals contain varying amounts of extra-neous mineral matter. Because these variations change both chemical and physical properties, current coal characterization analyses are very important to producers and consumers of coal.

Coal heterogeneity, both chemical and physical, reflects its history of formation. United States coal deposits originated about 300 million years ago in the swamp forests of the Carboniferous Age. At that time, the earth's mild climate encouraged vast growths of primitive trees and vegetation. Approximately three to seven feet of compacted plant material was required to form one foot of bituminous coal (Ref. 1). To form coal, chemical and physical transformations under heat and pressure were necessary. The progressive alteration from peat through a series of coals ranging from lignite to anthracite was a time-consuming process. Coal classification is discussed later.

Utilization of coal includes firing industrial and utility boilers, conversion to coke for chemical and metallurgical purposes, gasification, and liquefaction. Because of the heter-ogeneous composition of coal, not all coals are suitable for these purposes. Even within one of these general classifications, there are subdivisions. For example, coals suitable for grate-type boilers may not be suitable for cyclone firing. For these reasons, each coal must be subjected to a number of chemical and physical tests to determine its suitability for the desired market.

COAL CHARACTERIZATION TESTS

Standardized coal tests have been developed by the American Society for Testing and Materials (ASTM), the International Organization for Standardization (ISO), and the standardization bodies of other nations. ASTM standardization of tests for coal and coke started formally in 1904, and this work continues under Committee D.05.

Development of characterization tests required the work of many people over a long period. In the United States, test methods emerged from individual work supported by private industry, the U.S. Bureau of Mines (USBM), and several state agencies. These tests are accepted as standards on a consensus basis by ASTM Committee D. 05 which is composed of members from three groups: (1) coal producers (or owners), (2) consumers, and (3) those actively working with coal (but not owned by a producer or a consumer).

The applicabilities of proposed standards are carefully evaluated by testing coals of different rank. Comparisons are made within and between ASTM members' laboratories. Once established, these standards are constantly reviewed and improved to meet chang-ing market requirements. Also, new and improved analytical equipment and techniques are reviewed to assure that the best available methods are utilized.

Coal characterization tests commonly utilized to determine the suitability of each coal for a particular application are summarized in the following paragraphs. Selection of characterization tests is discussed in subsequent sections. Typical test results may be found in the USBM publications (Refs. 2, 6).

Proximate Analysis, ASTM D 3172, includes the determination of total moisture, volatile matter, and ash, and the calculation of fixed carbon for coals and cokes (Ref. 7). The historic term "proximate" should not be confused with the word "approximate," since all proximate analysis tests are performed according to rigid specifications and

---

*Assistant Technical Director, Commercial Testing & Engineering Co., Chicago, Illinois.

tolerances. Proximate analysis results may be used (1) to establish the rank of coals, (2) to show the ratio of combustible to incombustible constituents, (3) to provide the basis for buying and selling, and (4) to evaluate for beneficiation, or for other purposes.

Forms of moisture in coal have been studied by a number of investigators (Refs. 8, 9). Definitions of some moisture forms are described by Brown in ASTM D 3302 (Ref. 7). These forms might be (1) free or adherent moisture (essentially surface water), (2) physically bound or inherent moisture (that moisture held by vapor pressure and other physical processes), (3) chemically bound water (water of hydration or "combined" water).

ASTM Committee D.05 has defined *total moisture* as a loss in weight in an air atmosphere under rigidly controlled conditions of temperature, time, and airflow. At least in principle, total moisture represents a measurement of all the water not chemically combined. Traditionally, thermal treatment has provided the most commonly used basis for attempting to separate the nonchemically bound water from coal. The absolute separation of adsorbed moisture without loss of a portion of chemically bound water is difficult, especially with coals of lower rank.

Total moisture is determined by a two-step procedure (ASTM D 3302: Ref. 7). This procedure involves air drying for removal of surface moisture from the gross sample, division and reduction of the gross sample, and determination of residual moisture in the prepared sample (ASTM D 3173: Ref. 7). An algebraic calculation is used to obtain the total moisture value.

The as-received moisture is the total moisture at a given time. This term is commonly used in the trade to indicate the moisture present when the coal is delivered to a point of transfer.

Organic reflux moisture tests have been used to determine the residual moisture in the prepared sample (ISO R 348: Ref. 10). Studies of organic reflux techniques have been made in the United States (USBM RI 4969: Ref. 11). These methods may be used to determine moisture, and may be particularly applicable for our western low rank coals.

Equilibrium moisture is the moisture-holding capacity of coal at 30°C in an atmosphere of 97 percent relative humidity (ASTM D 1412: Ref. 7). The equilibrium moisture of a sample of coal is considered to be equal to the bed moisture for classification of coal by rank (ASTM D 388: Ref. 7). Also, equilibrium moisture results may be used for estimating the surface, or extraneous moisture of wet coal, especially where there appears to be excessive moisture.

Volatile matter is defined as the gaseous products, exclusive of moisture vapor, driven off during standardized test conditions (ASTM D 3175:Ref. 7). The combustible gases are carbon monoxide, hydrogen, methane, and other organic hydrocarbons. Those generally classified as noncombustible are carbon dioxide, ammonia, hydrogen sulfide, and some chlorides. Volatile matter is an empirically determined characteristic and is not a natural component of coal. Using the standard ASTM procedure, it is often difficult to obtain repeatable or reproducible results with many low-rank coals. To overcome these problems, low-rank coals have been blended with certain low-volatile bituminous coals. ASTM subcommittee D.05.21 has formed a task group to further investigate these western coals. Volatile-matter test results are used to establish the rank of coals, to indicate coke yield on carbonization processes, or to establish burning characteristics.

Ash is the noncombustible mineral matter left behind when coal is burned under rigidly controlled conditions of temperature, time, and atmosphere (ASTM D 3174: Ref. 7). The resulting ash obtained by this method differs in composition from the inorganic constituents present in the original coal. Burning causes the expulsion of water from the clays and calcium sulfate, and of carbon dioxide from carbonates, and the conversion of iron pyrites into ferric oxide. Each of these reactions involves a loss of weight from that of the original material. Formulas for correcting ash values to the original mineral-matter basis are presented in ASTM D 388 (Ref. 7).

Other ways in which ash values are used are the following: (1) to calculate other coal characterization values to an ash-free basis, (2) to evaluate the efficiency of coal cleaning or beneficiation processes, and (3) to estimate the amount of residue after coal is burned commercially.

Fixed carbon is the solid residue, other than the ash, resulting from the volatile-matter test. The value is calculated by subtracting moisture, volatile matter, and ash from 100% (ASTM D 3172: Ref. 7).

Another historic term, *ultimate analysis*, refers to the individual elements which are combined in the complex coal molecular structure. As defined in ASTM D 3176 (Ref. 7), these elements are total carbon, total hydrogen, total sulfur, total nitrogen, and oxygen. Ash is included in ultimate analysis as an estimate of the original mineral matter in coal so that it will be possible to calculate the oxygen content.

Total carbon is determined by catalytic burning of the sample in oxygen to form carbon dioxide, which can be readily measured (ASTM D 3178: Ref. 7). Total carbon includes both organic and carbonate carbon. *Total organic carbon* can be calculated by subtracting the carbonate carbon as determined by ASTM D 1756 from the total carbon (Ref. 7). For a given sample the total carbon content is always greater than the fixed carbon (see Proximate Analysis, discussed earlier).

Total hydrogen also is determined by the catalytic burning of a sample in oxygen to form water. This water is absorbed in a desiccant and weighed directly (ASTM D 3178: Ref. 7). Hydrogen results, as determined by ASTM D 3178, include the hydrogen present in both the sample moisture and the water of hydration. The hydrogen from the sample moisture can be removed stoichiometrically (ASTM D 3176: Ref. 7). Results can be calculated to other moisture bases according to formulas given in ASTM D 3176 (Ref. 7).

Total sulfur is another part of the ultimate analysis. Sulfur is generally present in three forms, and the sum of these is reported as the total sulfur. Total sulfur can be determined by the following three chemical methods (ASTM D 3177): (1) the Eschka method, (2) the bomb-washing method, and (3) the high-temperature combustion method (Ref. 7). In the Eschka method, the sample is ignited in a mixture of magesium oxide and sodium carbonate. The sulfur, now in a soluble form, is leached with water and precipitated from the resulting solution as barium sulfate ($BaSO_4$). The precipitate is filtered, ignited, and weighed. In the bomb washing method, sulfur is precipitated as $BaSO_4$ from the oxygen-bomb calorimeter washings. In the high-temperature combustion method, the sample is burned in a tube furnace, and sulfur oxides are collected and determined by an acid-base titration.

Total nitrogen is determined by chemical digestion (Kjeldahl-Gunning) methods (ASTM D 3179: Ref. 7). Total nitrogen is catalytically converted to ammonia. The ammonia is distilled, absorbed by an acid, and measured in an acid-base titration. A semimicro method similar to that described in ISO R 333 has proved to give equally accurate results (Ref. 12).

Ash is analyzed according to ASTM D 3174 (Ref. 7).

The amount of oxygen is calculated by subtracting total carbon, total hydrogen, total sulfur, total nitrogen, and ash from l00% (ASTM D 3176: Ref. 7). A method for the direct determination of oxygen is given in USBM RI 6753 (Ref. 13). However, the method is lengthy, and for most applications the increased precision is not considered to be in proportion to the effort intended.

Chlorine is commonly included as part of the ultimate analysis. The amount of chlorine may be determined by either the bomb combustion or the Eschka method (ASTM D 2361: Ref. 7). Other important chemical and physical tests performed to characterize coal are: (1) heating value (Btu content), (2) sulfur forms, (3) ash fusibility temperatures, (4) ash analysis, (5) trace-element analysis, (6) free swelling index, and (7) Hardgrove grindability.

The *heating value* (calorific value, or Btu per pound) is one of the most important determinations necessary in buyer-seller relationships. Several methods have been proposed, but the most common method is the adiabatic bomb calorimeter (ASTM D 2015: Ref. 7). The Btu value is determined by actually burning a coal sample in an oxygen bomb and measuring the temperature rise. The temperature rise is converted to Btu per pound by algebraic comparison with the heating value of benzoic acid. Corrections are made for the heat contributed by the ignition wire, and the heats of formation of nitric acid and sulfuric acid. See also *Bomb Calorimetry for Testing Solid and Liquid Fuels* in this section of the Handbook.

The *gross heating* value is the number obtained from the actual test (ASTM D 2015: Ref. 7). All the combustion-product water vapors have been condensed. The *net heating value*, a lower value, is calculated from the gross value by assuming that all water in combustion products remains in vapor form (ASTM D 407: Ref. 7).

Three *sulfur forms* recognized by ASTM D 2492 are: (1) sulfate sulfur which may be in the forms of calcium or iron sulfate, (2) pyritic sulfur, which is sulfur combined with iron in the form of minerals (pyrite and marcasite), and (3) organic sulfur, which is bonded to the carbon structure. The analysis used to designate each form is performed according to ASTM D 2492 (Ref. 7). Sulfate sulfur is extracted from the coal with dilute hydrochloric acid, precipitated as barium sulfate, and then ignited and weighed. After sulfate removal, the pyritic sulfur is extracted with nitric acid, and the extracted iron is measured by redox reaction. Organic sulfur is calculated by subtracting the sulfate and pyritic sulfurs from the total sulfur.

Ash fusibility can be defined broadly as the melting temperature of the ash. Small triangular pyramids (cones) prepared from coal or coke ash pass through certain defined stages of fusing, and flow when heated at a specified rate. A controlled reducing or oxidizing atmosphere is also required (ASTM D 1857: Ref. 7). The test method is empirical, and strict observance of the requirements and conditions is necessary to obtain reproducible fusion temperatures. Fusion temperatures in an oxidizing atmosphere are generally higher than those obtained in a reducing atmosphere. The melting or flux conditions at reported temperatures are: (1) initial deformation (cone tip starts to deform), (2) softening (cone height equals one-half the cone width), (3) hemispherical (cone height equals the cone width), and (4) fluid (melted cone spreads out into a flat layer). Examples are shown in Fig. 1.

The ash solid-liquid-phase diagram might be used to establish fusibility properties. Research has advanced slowly because of the high number of ash components and the

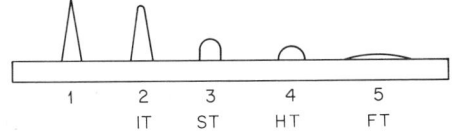

Fig. 1. Small triangular pyramids (cones) prepared from coal or coke ash pass through certain defined stages of fusing, and flow when heated at a specified rate. (*American Society for Testing and Materials.*)

effects of oxidizing and reducing atmospheres. Ash viscosity can be predicted from the ash analysis (Ref. 14).

Ash Analysis is the term used to designate analysis of the major elements commonly found in coal and coke ash (ASTM D 2795: Ref. 7). The elements, expressed as oxides, are SiO_2, Al_2O_3, Fe_2O_3, TiO_2, CaO, MgO, Na_2O, K_2O, P_2O_5, and SO_3. Phosphorus, a trace element, has been included historically because of its importance in the subsequent steelmaking processes. The sum of these oxides is generally more than 99 percent.

Interest in *trace-element analysis* has been increased by environmental concerns. Until recently the necessary sophisticated analytical instrumentation had not been developed. Currently, the U.S. National Bureau of Standards is developing coal-trace-elements standard samples. ASTM Committee D.05 is currently developing standard techniques. The main types of analytical equipment being considered are atomic absorption, spark source mass spectrophotometry, neutron activation, and X-ray fluorescence. Data on trace elements have been reported by the USBM and the Illinois Geological Survey (Refs. 15–18).

The *Free Swelling Index* (FSI) is determined by quickly heating a coal in a nonrestraining crucible (ASTM D 720: Ref. 7). The FSI values range from 0 to 9 as shown in Fig. 2. Noncaking and nonswelling coals are designated zero on the scale. The higher the FSI number, the greater the swelling properties of the coal.

The results may be used as an indication of the caking characteristics of the coal when burned as a fuel. ASTM D 720 does not recommend this method to determine the expansion of coals in coke ovens. For certain coals, a reduction of the FSI values over a period of coal storage indicates relative oxidation and possible deterioration of the coal.

The Hardgrove Grindability method is used to determine the relative ease of pulverization of coals in comparison with a series of standard coals (ASTM D 409: Ref. 7). A weighed amount of No. 16 by No. 30-mesh coal is ground for a set time in a ball race mill. The index is calculated from the amount of minus No. 200-mesh fines produced. A standard instrument is shown in Fig. 3. Manufacturers of commercial mills have determined the correlations for their various mill designs by comparing commercial mill performances with the Hardgrove grindability index. Mill capacity can be predicted from the laboratory-determined index.

Results may be used to approximate wear rates of crushing and grinding equipment.

For western low rank coals (subbituminous and lignite), the Hardgrove grindability index changes with the moisture content (Ref. 18). As lignite is ground, the moisture

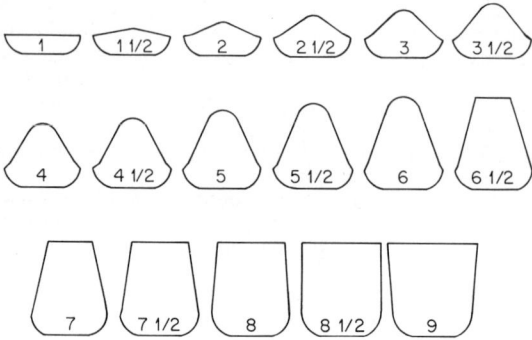

Fig. 2. Formats of various coals subjected to the free swelling index test in an electric furnace. FSI numbers are indicated. The higher the FSI number, the greater the swelling properties of the coal. *(American Society for Testing and Materials.)*

content may be reduced from 39 to 25 percent, and the Hardgrove grindability may be reduced 10 points. Therefore, a series of tests at different moisture levels is recommended.

The coal characteristics discussed so far are general for most coal uses. Additional tests required for metallurgical coals are the following: (1) coal petrography, (2) Gieseler plasticity, (3) Audibert Arni dilatometer, (4) sole-heated oven, and (5) movable-wall coke-oven test.

Coal Petrography Techniques are used to describe coals quantitatively in terms of optically distinct entities that constitute coal (Refs. 19–21). Extensive use has been made of reflectance measurements (ASTM D 2798) and reflectograms along with the physical microscopic components (ASTM D 2799) to characterize coking coals (Ref. 7).

The Gieseler Plasticity is the coal's tendency to fuse or soften and reach a fluid state

Fig. 3. Instrument for making the Hardgrove grindability test to determine the relative ease of pulverization of coals.

when heated (ASTM D 2639: Ref. 7). In operation, a constant torque is applied to a vertical stirrer embedded in the coal sample. Fluidity is measured by the rate of stirrer rotation as the coal is heated under a set temperature program, and the coal passes through its coking stage. An apparatus is shown in Fig. 4.

The *Audibert Arni Dilatometer* measures the contraction and expansion (dilation) of coal during carbonization by recording the rise and fall of a metal piston resting on a compressed sample of coal (ISO R 349: Ref. 22).

A *Sole Heated-Oven Determination* measures the expansion or contraction of coal under simulated coke-oven heating (ASTM D 2014: Ref. 7). The test sample consists of a

Fig. 4. Apparatus for making the Gieseler plasticity test on coal to determine the tendency of coal to fuse or soften and reach a fluid state when heated.

single coal or coal blends prepared as for use in a commercial coke oven. This test requiring 30 to 50 pounds of coal is one of the largest laboratory tests discussed. Its purpose is to determine the expansion or contraction properties of the coal. These data are important in predicting whether actual production coke can be removed without damage to the commercial oven.

Pilot-Plant Movable-Wall Coke-Oven Tests require 500 to 1000 pounds of coal for operation (Ref. 23). This test is performed for two purposes: (1) determination of wall pressures developed during carbonization, and (2) production of a sufficient quantity of coke to permit physical coke tests to be performed.

COAL CLASSIFICATION

Characterization results can be used to group or classify coals. ASTM has established classifications or rankings for coal according to the degree of metamorphism (progressive alteration in the coalification process) as shown in Table 1. The amount of volatile matter, fixed carbon, bed moisture, and oxygen all indicate rank, but no one item completely defines it. In the ASTM classification, the basic criteria are ash, fixed carbon, and the heating value. For higher-rank coals, ash and fixed carbon are sufficient criteria. However, these values alone are not suitable for designating the position of low-rank coals. For a low-rank coal, ash and heating value are the principal criteria.

Properties of a given coal can be generalized by referring to the classification system shown in Table 1. For example, anthracite is very low in bed (equilibrium) moisture and

TABLE 1. Classification of Coals by Rank for Analysis*

Class and group	Fixed carbon limits, percent (dry, mineral-matter-free basis)		Volatile matter limits, percent (dry, mineral-matter-free basis)		Calorific value limits, Btu/lb (moist,‡ mineral-matter-free basis)		Agglomerating character
	Equal or greater than	Less than	Greater than	Equal or less than	Equal or greater than	Less than	
I—Anthracitic							
1. Meta-anthracite	98			2			Nonagglomerating
2. Anthracite	92	98	2	8			
3. Semianthracite‡	86	92	8	14			
II—Bituminous coal							
1. Low volatile	78	86	14	22			Commonly agglomerating¶
2. Medium volatile	69	76	22	31			
3. High volatile A		68	31		14,000§		
4. High volatile B					13,000§	14,000	
5. High volatile C					{ 11,500 / 10,500	{ 13,000 / 11,500	Agglomerating
III—Subbituminous coal							
1. Subbituminous A					10,500	11,500	Nonagglomerating
2. Subbituminous B					9,500	10,500	
3. Subbituminous C					8,300	9,500	
IV—Lignite							
1. Lignite A					6,300	8,300	
2. Lignite B						6,300	

*This classification by ASTM D 388 does not include a few coals, principally nonbanded varieties, which have unusual physical and chemical properties and which come within the limits of fixed carbon or calorific value of the high-volatile bituminous and subbituminous ranks. All these coals either contain less than 48% dry mineral-matter-free fixed carbon, or have more than 15,000 moist mineral-matter-free Btu/lb.

†Moist refers to coal containing its natural inherent moisture, but not including visible water on the surface of the coal.

‡If agglomerating, classify in low-volatile group of the bituminous class.

§Coals having 69% or more fixed carbon on the dry mineral-matter-free basis shall be classified according to fixed carbon, regardless of calorific value.

¶It is recognized that there may be nonagglomerating varieties in these groups of the bituminous class, and there are notable exceptions in high volatile C bituminous group.

high in carbon. Lignite generally is high in moisture and low in heating value. Thus, the classification system is useful in forming general opinions about a coal before specific characterization properties are determined through tests.

EXPLORATION ANALYSES

Coal-exploration programs are being expanded and accelerated to satisfy demands resulting from long-term utility contracts (Ref. 24). Environmental requirements have led to the search for low-sulfur coals to meet air-pollution standards. Worldwide energy demands necessitated an increase in the export of coals, and expedited conversion of coal to pipeline gas. Specific questions being asked are: How much coal is available, and what is the coal quality? These and other basic mining engineering questions are answered in coal exploration programs.

In an exploration program, the subterranean deposit is probed by a series of drill holes. Holes ranging from 2 to 6 inches in diameter are drilled, and a solid core sample is obtained. The drilling process produces initially topsoil and then normally traverses several layers of stone formation, and finally a core of the coal is obtained. The driller retains a record of change of each formation with drill depth. Changes of coal seam(s), roof, coal depth, coal thickness, partings (rock and clay) in the seam and coal quality are noted. From elevation determinations contour maps can be drawn, showing rises and falls of the coal seam. Coal characterization data can be statistically treated to indicate gradation of a given coal property as shown in USBM RI 7827 (Ref. 25). An example of the usefulness would be the ability to locate high-sulfur areas before they are mined.

The coal core must be carefully handled and evaluated, for herein lies the key to the coal utilization and preparation. A full range of coal characterization tests may be used to evaluate a new property, so that alternative markets can be pursued. Additional cores are drilled to confirm the earlier core data or to establish variation trends.

The coal-core test should start with a macroscopic analysis to identify each change in strata properly with respect to type of material, location in the seam, thickness, approximate specific gravity, and quality. The macroscopic analysis provides a detailed evaluation of changes in seam structure, benches, and major partings (reported as a percentage of the total seam). The driller's record and the macroscopic analysis are used to establish minable portions of the seam and to calculate the scope of the required drilling program.

The next phase of the core test is to crush the full length of the core, or each of its benches, as determined from the macroscopic analysis. Washability tests are then performed (described later). Results of the washability test are used by engineers to evaluate the deposit for various markets.

Whenever possible, actual raw-coal studies are correlated with initial core-test data to finalize plant design and choice of market. One or more large raw-coal samples should be collected in a manner that will correspond to actual mining practice. The number of such samples varies according to the heterogeneity of the seam as determined by the core-test data. Washability tests should be made for each sample, using a full range of sizes, specific gravities, and analyses. All the tests described for the core evaluation should be repeated for the raw-coal studies.

WASHABILITY CHARACTERISTICS

Although the minable reserves of coal in the United States are tremendous, the reserves of the better coals are being exhausted, and future tonnages will come from seams with higher ash, higher sulfur, and other undesirable properties. One of the direct results of this trend has been a steady increase in the tonnage of mechanically cleaned coal (Ref. 26). These mechanically cleaned or beneficiated coals can qualify for more economically attractive markets.

Washability tests are performed to determine the quantity and qualities of "impurity free" or "cleaned" coal obtainable by various beneficiation techniques (Ref. 11). These tests quantitatively describe a coal deposit or a source of coal. Data serve as a basis for two types of judgment: determination (1) of preparation method and equipment to be used, and (2) of operating performance of equipment selected.

The principles of coal beneficiation are separation of impurities (1) by sizing, and (2) by specific-gravity differences. These are complicated by the fact that coal does not consist of pieces of pure coal mixed with pieces of stone (or clay), but rather each coal piece contains gradations or mixture of both. Screen sizes corresponding with size ranges are normally used in commercially available equipment. The combined sizing and specific-gravity separation tests to determine the washability characteristics of a coal, if properly planned and carried out, should yield valuable information, as follows: (1) the expected analysis of the product as to ash, sulfur, or other characteristics, if the cleaning is done efficiently, (2) the yield of the cleaned product and the amount of refuse to be expected, (3) an indication as to the ease of the cleaning problem, (4) the range of sizes necessary to be cleaned to meet any given market specification, (5) the type and amounts of the impurities in any tested size, with an indication as to whether certain middling gravity material should be crushed to increase the yield or the quality of the cleaned product, and (6) type of cleaning process best adapted for the coal. To determine the washability or cleaning characteristics of a coal, the following steps are necessary: (1) collection of an accurate sample, (2) screen analysis of the sample, (3) float-and-sink test of individual sizes, (4) chemical analysis of the gravity fractions and raw sizes not treated, (5) calculations and report of results obtained, and (6) interpretation of washability results. See Fig. 5.

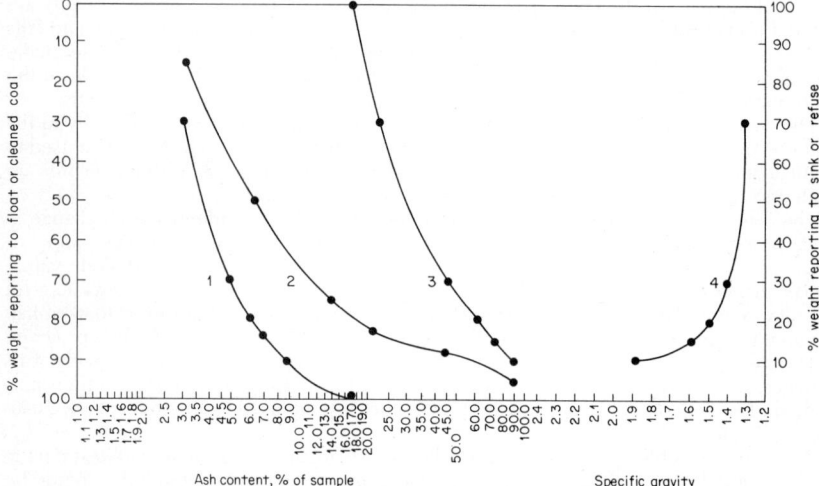

Fig. 5. Typical washability curves for coal. (1) Cumulative coal-ash; (2) coal characteristic; (3) cumulative refuse-ash; (4) yield-specific gravity. (*Commercial Testing & Engineering Co.*)

The specific-gravity test is also called the float-and-sink test. Two classes of liquids are usually employed in making float-and-sink tests: (1) an aqueous solution of salts such as calcium chloride and zinc chloride, and (2) organic liquids such as carbon tetrachloride, bromoform, ethylene dibromide, and naphtha.

After gravity-size fractions have been analyzed, washability tables are prepared. Typical washability curves are plotted in Fig. 6. The sulfur reduction test is a special-purpose test which utilizes washability and liberation techniques. Sulfur reduction tests are more widely known as mineral liberation tests where ash-forming minerals are liberated from the coal and separated in subsequent processes. In the sulfur reduction test, a specific undesired mineral, pyrite, is studied. Starting with a washability test, analysis reveals the magnitude of the various forms of sulfur and pattern of the occurrence of pyrite. If the pattern of occurrence shows, for example, that quantities of pyrite are present in larger sizes, these sizes (often only specified-gravity fractions of each size) are stage-crushed. The broken pieces are subjected to a washability test (again with sulfur analysis). This time the washability test is used to give a quantitative measure of the amount and degree of good coal recovered by liberating the pyrite.

STEAM-ELECTRIC COAL TESTING

The major interest in coal is the calorific (Btu) value. Today these values range from 6000 to 14,000 Btu per pound delivered (Refs. 2, 6, 27, 28). About 400 million tons of bituminous and lignite coals, nearly two-thirds of the annual production, were burned in steam-electric plants in 1973 (Ref. 29). These enormous amounts coupled with higher costs explain the importance of accurately determining the calorific value.

The ash content is important. As the ash content increases, the calorific value decreases. Also, ash is a waste product which must be collected. Disposal of the collected ash

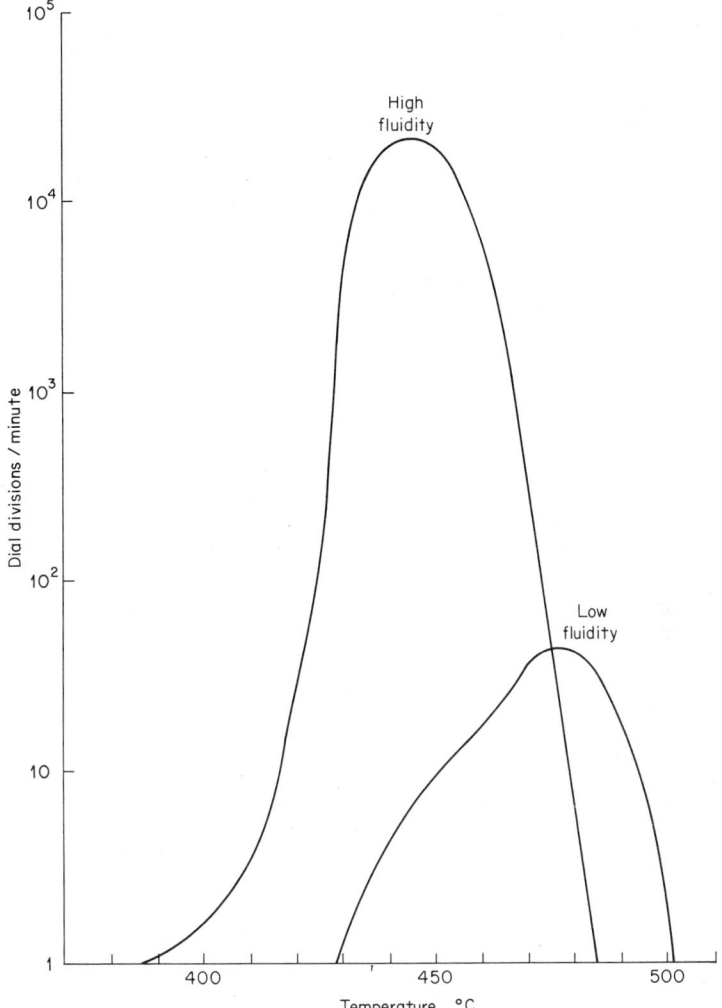

Fig. 6. Curves indicating Gieseler plasticity properties of coals.

presents another problem, and markets are rarely available for these products. Usage of fly ash for making bricks and lightweight concrete and as a Portland cement raw material has been proposed (Refs. 30–32). Research continues while much of the ash wastes are consumed as landfill (Refs. 33, 34).

Ash fusibility and ash analysis are important ash parameters. Ash fusibility is a determining factor in choosing the furnace design. Ash element analysis may be used to predict the viscosity of the slag so that coal may be matched to the operating characteristics of the equipment.

Analysis of common ash elements can also be used to predict corrosive and fouling properties of the coal (Refs. 14, 35). Trace-element studies around steam-electric plants have been conducted (Ref. 36).

Sulfur has been the environmental concern of recent years. It is present in the coal in three forms: pyritic, ferrous, and organic. One-half or more of the pyritic sulfur can be removed by conventional physical treatment methods. Additional pyritic sulfur can be removed by fine grinding and froth flotation techniques (Ref. 37). Coal liquefaction processes appear to be one way to separate the organic sulfur from the coal.

Moisture is an undesirable component of delivered coal. Not only does moisture add to the shipping weight, but also difficulties arise in handling and sampling coals with excessive moisture. In winter the coal freezes, and unloading is difficult.

Moisture originates from several sources. Coal or lignite has a "spongelike" moisture capacity often referred to as the "seam or bed moisture." The equilibrium moisture is an attempt to give a measure of the "natural bed moisture." Values of equilibrium moisture vary from 2 to 4 percent for certain bituminous coals to 27 to 44 percent for lignites. Water is added in the mining process to reduce coal dust at the coal face and at transfer points. Coal cleaning processes may use water, but because driers are employed, much of the added water can be removed. Additional moisture may be gained or lost by the type of weather experienced during transit.

The Hardgrove grindability index has been correlated with pulverizer capacity, fineness of grind, and mill wear (Ref. 18). Repairs and preventive maintenance can be scheduled on the basis of grindability characteristics.

Knowledge of the chlorine content in coal is important because it is converted into hydrogen chloride in the combustion gas. This very reactive gas produces corrosion in all parts of the combustion system (Ref. 38).

COKING COALS

Metallurgical coking coals must have unique chemical, petrographic, and carbonization characteristics. Chemically these coking coals must be low in ash, low in sulfur, low in phosphorus, and high in fixed carbon. A balance between "reactive" and "inert" petrographic constituents must be maintained.

Specific tests have been developed to ascertain carbonization properties. Plastic properties are evaluated in a Gieseler plastometer. A typical high-fluid high-volatile bituminous coal would have a high plastic curve as opposed to a low-fluid low rank coal (Fig. 6). A slight contraction during the coking process would be desirable in the sole-heated furnace test. A high FSI number is desirable for coking coals, but the FSI test was developed as a parameter in grate-furnace burning operations and is not recognized as a standard coking coal qualification.

The final test, before actual full-size coke operations, is often the movable wall oven test. It is performed to assure that a coal or coal blend will not physically damage the brick coke-oven walls. Also sufficient coke is obtained to perform physical tests. Some of these physical coke tests are the following (Ref. 7): (1) drop shatter test, (2) specific gravity, (3) porosity, (4) cubic foot weight, (5) sieve analysis, (6) tumbler test, (7) dustiness index, and (8) ash fusibility. Also all the other proximate, ultimate, and special coal tests can be performed.

After passing all the laboratory tests, the final coke evaluation is equated as the rate of production in the iron-ore blast furnace.

Few United States coals can be found to have all these desirable characteristics. Therefore, U.S. coke-oven practice is to blend several coals, often as high as 6 or 8 in order to make a desirable coke. Typical coking coal test results are indexed in *U.S. Bur. Mines Bulls.* 536 and 643 (Refs. 39, 40).

LIQUEFACTION AND GASIFICATION COAL TESTING

To date conventional testing methods have been used to characterize these coals. The important basic characteristics of coals appear to be the carbon and hydrogen contents

(Ref. 41). For example, ultimate analysis could be used to determine the amount of carbon and hydrogen necessary for the following methane production reaction:

$$C \text{ (coal)} + 2H_2 \rightarrow CH_4$$

Other ultimate analysis components, nitrogen, oxygen, chlorine and sulfur, will be needed. For the Lurgi gasification process, a coal should have a low FSI value. Proximate analysis tests would give ash and volatile matter values. Grindability characteristics can be predicted from the Hardgrove index. Ash fusibility temperatures and major ash elements could be correlated with gasifier temperatures, and possible problems averted. Trace-element material balances may be necessary for environmental concerns.

Future adaptations of the present tests and perhaps new tests will be added. One of these may be the modified Fisher assay test (Ref. 42). This laboratory test is used to determine the gaseous products of carbonization. The same principle is used in the Gray-King test, and this could be modified to quantify gaseous products (Ref. 43). Standardization of these and of a coke reactivity test is needed. Also, there is a need for a standard liquefication test.

PREPARATION-PLANT PERFORMANCE TESTING

When a preparation plant is built, it is common practice to write into the contract required plant-performance standards as part of the design specifications. Therefore, as a plant is completed, a performance test must be conducted on the entire plant, including individual unit operations. The feed and each product are sampled and subjected to a washability test. Calculations can then be made to establish the true operating gravity of separation and various measures of error of separation.

Once in operation, performance testing can be used to optimize separation conditions. As markets or coal characteristics change, additional testing is needed.

COAL SAMPLING

Data obtained from coal sampling systems are used to determine plant and component efficiency, allocate production costs, and establish price (Ref. 44). For these results to be meaningful, first the coal must be sampled correctly. If the sample does not represent the lot of coal under consideration, neither will the analysis. This potentially simple problem of sampling is first complicated by the heterogeneity of the coal. Mining methods, preparation, blending (different sizes or different seams), or segregation during handling (or shipping) add to the sampling difficulties (Ref. 45).

Sampling of coal is covered by ASTM D 2234 (Ref. 7). The introduction to the standard method reads:

> Data obtained from coal samples are used in establishing price, controlling mine and cleaning plant operations, allocating production costs, and determining plant or component efficiency. The task of obtaining a sample of reasonable weight to represent an entire lot presents a number of problems and emphasizes the necessity for using standard sampling procedures. Coal is one of the most difficult of materials to sample, varying in composition from noncombustible particles to those which can be burned completely, with all gradations in between. The task is further complicated by the use to be made of the analytical results, the sampling equipment available, the quantity to be represented by the sample and the degree of precision required. These standard methods give the overall requirements for the collection of coal samples. The wide varieties of coal handling facilities preclude the publication of detailed procedures for every sampling situation. The proper collection of the sample involves an understanding and consideration of the physical character of the coal, the number and weight of increments, and the overall precision required.

ASTM D 2234 covers procedures for the collection of a gross sample under various conditions of sampling (Ref. 7). The first basic criteria is the number of increments N_1 to be collected according to size and condition of the coal as presented in Table 2. This first value of the number of increments, N_1, may be increased for large lot sizes according to the following formula:

$$N_2 = N_1 \sqrt{\frac{\text{lot size, tons}}{1000}}$$

where N_2 is either equal to N_1 or larger (never smaller). N_2 represents the minimum number of increments to be collected. Minimum ASTM increment weights are also given in Table 2. Weights from mechanical samplers are much larger than those listed. If there is any doubt as to the condition of preparation of the coal (for example, partially cleaned coal), the minimum number of increments for raw uncleaned coal should be used.

**TABLE 2.   ASTM Sampling Requirements
(Number and weight of increments for general-purposes
sampling)**

Coal condition	Coal top size, in..		
	5/8	2	6
Mechanically Cleaned Coal			
N_1, minimum number of increments	15	15	15
Weight, minimum per increment, lb	2	6	15
Raw Uncleaned Coal			
N_1, minimum number of increments	35	35	35
Weight, minimum per increment, lb.	2	6	15

The sample collected by ASTM D 2234, called the gross sample, weighs from 30 pounds to 1 or 2 tons and usually has a top size of over 1 inch (Ref. 7). Procedures are included for reducing the amount of the gross sample. More and more frequently mechanical systems are used to collect samples. Procedures for preparing samples for analysis are given in ASTM D 2013 (Ref. 7).

MECHANICAL-SAMPLER BIAS TESTING

Mechanical sampling, often called automatic sampling, is being employed at many coal-transfer points. A mechanical device collects a primary sample by securing "cuts" or segments of the coal from the main conveyor belts—thereby systematically reducing a continuous flow of, for example, 700 tons/hour to a 3-ton sample. Secondary and tertiary sampling can be employed to mechanically reduce the primary sample to a convenient size for laboratory preparation.

Guidelines for the successful operation are given by ASTM D 2234, but experience is valuable also (Ref. 7). In any event, the new system must be tested initially and periodically to assure that the final sample represents the coal. This test is called the bias test.

Bias is defined by ASTM D 2234 as an error that is consistently positive or negative (Ref. 7). The error is the difference between the observed values and the best obtainable estimate of the true value. For example, Btu values may be consistently higher or lower than the heat value of the consignment being sampled.

Bias testing refers to the systematic collection of sets of samples, characterization of each sample (size, moisture, ash, sulfur, heating value, etc.), and a statistical evaluation. The reference sample is obtained by stopping the main conveyor and collecting specified amounts in a specified manner. Other samples in the set might include the primary cut, the secondary cut, the secondary reject, the tertiary cut, and the tertiary reject. All samples in one set are taken at the same time. It is recommended that a minimum of 30 sets of samples be collected.

The statistical significance of differences between analyses of corresponding samples can be determined, using analysis of variance techniques. Calculations required in making statistical comparisons of one sampling location with the others are not necessarily difficult, but they are extremely lengthy. For this reason, it is convenient to utilize a computer program for the calculations.

The conclusions of a series of these comparisons may be the following: (1) the magnitude of any bias that exists, (2) the practical and statistical significance of the bias, and (3) the components of the system responsible for creating the bias if a bias exists. Surveillance is required to assure that the mechanical sampling system remains unbiased.

QUALITY CONTROL

Some guidelines for calibration of equipment are given by ASTM. The laboratory chemist in charge can perform a series of routine inspections on balances, calorimeters, and other instruments. Time and temperature cycles should be checked. Inspections by manufacturers' representatives are also desirable.

Standard samples are available from the U.S. National Bureau of Standards for only three coal characteristics: total sulfur, ash, and mercury. Therefore participation in a sample exchange or cooperative program with other well-run laboratories is important. Additional advantages of cooperative participation are that samples are current with coal being marketed and are always fresh.

REFERENCES

1. Simon, J. A., and M. E. Hopkins: "Geology of Coal," chap. 2, Elements of Practical Coal Mining, American Institute of Mining, Metallurgical, and Petroleum Engineers, New York, 1973.
2. Aresco, S. J., and J. B. Janus: Analyses of Tipple and Delivered Samples of Coal, *U.S. Bur. Mines Rept. Invest.* 7219, Pittsburgh, Pa., 1968.
3. Staff: Analyses of Tipple and Delivered Samples of Coal, *U.S. Bur. Mines Rep. Invest.* 7346, Pittsburgh, Pa., 1969.
4. Staff: Analyses of Tipple and Delivered Samples of Coal, *U.S. Bur. Mines Rept. Invest.* 7490, Pittsburgh, Pa., 1970.
5. Staff: Analyses of Tipple and Delivered Samples of Coal, *U.S. Bur. Mines Rep. Invest.* 7588, Pittsburgh, Pa., 1971.
6. Janus, J. B., and B. S. Shirley; Analyses of Tipple and Delivered Samples of Coal, *U.S. Bur. Mines Rep. Invest.* 7712, Pittsburgh, Pa., 1972.
7. Consensus: "Gaseous Fuels, Coal, and Coke," part 26, American Society for Testing and Materials, Philadelphia, 1975.
8. Krumin, P. O.: The Determination of Forms of Moisture in Coal, *Ohio State Univ. Eng. Expt. Sta. Bull.* 195, Columbus, Ohio, 1963.
9. Rees, O. W.: Chemistry, Uses and Limitations of Coal Analyses, *RI 220, Illinois State Geol. Surv. Rep. Invest.*, Urbana, Ill., 1966.
10. Consensus: Determination of Moisture in the Analysis Sample of Coal by the Direct Volumetric Method, International Organization for Standardization, *ISO/R348*—1963 E, ISO Central Secretariat, Geneva, Switzerland.
11. Goodman, J. B., Manual Gomez, and V. F. Parry: Determination of Moisture in Low Rank Coals, *U.S. Bur. Mines Rep. Invest.* 4969, Pittsburgh, Pa., 1952.
12. Consensus: Determination of Nitrogen in Coal by the Semi-Micro Kjeldahl Method, International Organization for Standardization, *ISO/R333-E*, New York, 1963.
13. Abernethy, R. F., and F. H. Gibson: Direct Determination of Oxygen in Coal, *U.S. Bur. Mines Rep. Invest.* 6753, Pittsburgh, Pa., 1966.
14. Attig, R. C., and A. F. Duzy: Coal Ash Deposition Studies and Application to Boiler Design, *Amer. Power Conf. Proc.*, vol. 21, pp. 290–300, Illinois Institute of Technology, Chicago.
15. Kessler, T., A. G. Sharkey, Jr., and R. A. Friedel: Spark-Source Mass Spectrometer Investigation of Coal Particles and Coal Ash, *U.S. Bur. Mines TPR* 42, Pittsburgh, Pa., 1971.
16. Kessler, T., A. G. Sharkey, Jr., and R. A. Friedel: Analysis of Trace Elements in Coal by Spark-Source Mass Spectrometry, *U.S. Bur. Mines Rep. Invest.* 7714, Pittsburgh, Pa., 1973.
17. Ruch, R. R., H. J. Gluskoten, and N. F. Shimp: Occurrence and Distribution of Potentially Volatile Trace Elements in Coal, *Environ. Geol. Notes*, no. 61, Illinois State Geological Survey, Urbana, Ill., April 1973.
18. Savage, K. I.: Pulverizing Characteristics of Coal: The Hardgrove Grindability Index, "Keystone Coal Industrial Manual," McGraw-Hill, New York, 1974.
19. Schapiro, N., and R. J. Gray: The Use of Coal Petrography in Coke-Making, *J. Inst. Fuels*, vol. 37, pp. 234–242, 281, 1964.
20. Gray, R. J., and N. Schapiro: Petrographic Composition and Coking Characteristics of Sunnyside Coal from Utah, *Utah Geol. Mineral. Surv. Bull.* 80, pp. 55–79, University of Utah, Salt Lake City, Utah, 1966.
21. Gray, R. J., R. M. Patalsi, and N. Schapiro: Correlation of Coal Deposits from Central Utah, *Utah Geol. Mineral. Surv. Bull.* 80, pp. 81–86, Salt Lake City, Utah, 1966.
22. Consensus: Audibert-Arni Dilatometer Test for Coal, International Organization for Standardization, *ISO/R349-1963*, ISO Central Secretariat, Geneva, Switzerland.
23. Harris, H. E., W. L. Glowacki, and J. Mitchell: "Coking Pressure and Its Measurement—Current Theory and Practice," Blast Furnace Coke Oven and Raw Materials Conference, Chicago, April 1960, sponsored by American Institute of Mining, Metallurgical, and Petroleum Engineers, New York.

24. Wilson, I. H.: The Role of Testing in Coal Exploration, "Keystone Coal Industrial Manual," pp. 136–140, McGraw-Hill, New York, 1970.
25. Gomez, M.: Distribution of Sulfur and Ash in Part of the Pittsburgh Seam and Probable Mode of Deposition, *U.S. Bur. Mines Rep. Invest.* 7827, Pittsburgh, Pa., 1974.
26. Keller, G. E., and W. W. Anderson: Determining the Washability Characteristics of a Coal, *Com. Testing & Eng. Co. Publ.* 19, Chicago, 1938.
27. Hillier, L., and B. W. Rudgier: Lignite and the Cyclone Burner in a New 235 MW Generating Station, *Lignite Symp.*, Bismarck, N. Dak., 1973, Minnkota Power Cooperative, Grand Forks, N. Dak.
28. Vig, K. S.: Detailed Construction and Equipment Costs for the 235 MW Milton R. Young Generating Station, *Lignite Symp.*, Grand Forks, N. Dak., 1973.
29. Falkie, T. V.: Production of Coal—Bituminous and Lignite, *Miner. Ind. Surv. Rep.* 2950, U.S. Department of Interior, Washington, D.C., 1974.
30. Remirez, R.: Fly Ash Debut in Brick Manufacturing, *Chem. Eng.*, vol. 74, no. 5, 1967.
31. Staff: Lightweight Concrete in the West Virginia University Colosseum, National Ash Association, Washington, D.C.
32. Barton, W. R.: Fly Ash as a Portland Cement Raw Material, *Amer. Inst. Mining, Met., Petrol. Engrs.* Preprint 68F30, New York, 1968.
33. Minnick, L. J.: Reactions of Hydrated Lime with Pulverized Coal Fly Ash, *Miner. Process.*, February 1968, pp. 15–20; March 1968, pp. 12–19.
34. Boux, J. E.: Canadians Pioneer New Fly Ash Processing System, *Miner. Process.*, March 1969, pp. 16–19.
35. Borio, R. W., et al.: The Control of High Temperature Fire Side Corrosion in Utility Coal Fired Boilers, *Office Coal Res. R&D Rep.* 41, U.S. Department of Interior, Washington, D.C., 1967.
36. Billings, C. E., et al.: Mercury Balance on a Large Pulverized Coal-fired Furnace, *J. Air Pollut. Contr. Ass.*, vol. 23, no. 9, pp. 773–777, 1973.
37. Savage, K. I., and S. C. Sun: Flotation Recovery of Pyrite from Bituminous Coal Refuse, *Trans. Amer. Inst. Mining, Met., Petrol. Engrs.*, vol. 241, pp. 337–384, 1968.
38. Iapalucci, T. L., R. J. Demski, and D. Bienstock: Chlorine in Coal Combustion, *U.S. Bur. Mines Rep. Invest.* 7240, Pittsburgh, Pa., 1969.
39. Brown, R. L., et al.: Carbonizing Properties of American Coals, *U.S. Bur. Mines Bull.* 536, Pittsburgh, Pa., 1954.
40. Walters, J. G., C. Ortuglio, and J. Glaenzer: Yields and Analyses of Tars and Light Oils from Carbonization of U.S. Coals, *U.S. Bur. Mines Bull.* 643, Pittsburgh, Pa., 1967.
41. Rudolph, P. F. H.: "Coal Gasification—A Key Process for Coal Conversion," Conference on Synthetic Hydrocarbons, Lurgi Mineralotechnik, Frankfurt (Main), West Germany.
42. Hubbard, A. B.: Automated Modified Fisher Retorts for Assaying Oil Shale and Bituminous Materials, *U.S. Bur. Mines Rep. Invest.* 6676, Pittsburgh, Pa., 1965.
43. Consensus: Determination of the Gray King Coke Type of Coal, International Organization for Standardization, *ISO Draft* 552 (revised text), 1965, ISO Central Secretariat, Geneva, Switzerland.
44. Loeffler, F.: Bulk Sampling at the Navajo Mine, *Coal Age*, vol. 77, no. 8, p. 74, 1972.
45. Anderson, W. W.: "Evaluation of Coal Sampling Procedures," American Power Conference, Chicago, Mar. 31, 1959, sponsored by Illinois Institute of Technology, Chicago.

Bomb Calorimetry for Testing Solid and Liquid Fuels

BY M. R. STEFFENSON*

One of the most important characteristics of any combustible fuel is the quantity of energy or heat that it releases as it is burned. This value is referred to as either the *heat of combustion,* or the *calorific value* of the fuel and is usually expressed in *British thermal units (Btu)* per pound or ton, or in *calories per gram.* The heat of combustion of solid and liquid fuels is routinely determined in order to establish the price of the fuel, as well as to serve as a basis for calculating the overall efficiency of a power-generating facility or engine.

Calorimetry is the study of *heat* as contrasted with temperature. The oxygen bomb calorimeter, which is used to determine the heat of combustion of fuels, is only one of many types of calorimeters. Steam calorimeters, for example, are used to measure heat capacities, heat of reaction, or energy changes in biological processes. Instruments for differential thermal analysis are sometimes referred to as differential scanning calorimeters. The bomb calorimeter is a batch-type instrument which requires a discrete sample and, therefore, is used only for solid and liquid materials. Gaseous fuels (nondiscrete) are analyzed in flow-type calorimeters. Because of this basic difference in sample handling, instruments for gaseous fuels are described separately under *Gas Calorimetry* in the *Gas Technology* section of this handbook.

To determine the heat of combustion of a fuel, a representative sample is burned in a high-pressure oxygen atmosphere within a metal bomb or pressure vessel. The energy released by this combustion is adsorbed within the calorimeter and measured in terms of temperature change within the calorimeter. The heat of combustion of the sample is obtained by multiplying the temperature rise of the calorimeter by a previously determined energy equivalent or heat capacity for the instrument. Corrections are applied to adjust these values for any heat transfer occurring in the calorimeter as well as for any side reactions which are unique to the bomb combustion process.

The reliability of results obtained with bomb calorimetry depends upon a truly representative sample as well as a reliable calorimeter and proper operating techniques. A typical load of coal will include large lumps, fine powders, and particles varying in size between the two extremes. During loading and transit, the fines and smaller particles will work their way to the bottom of the shipment. A sample taken from the bottom would not be representative of the entire shipment inasmuch as it would be rich in fines and deficient in the larger particles. A sample from the top of the shipment obviously would be biased in favor of the larger particles. Similar problems can occur with liquid fuels. Because of the necessity that the sample used in the calorimeter be truly representative of the shipment, detailed procedures for sampling and sample reduction have been developed. See Ref. 1 and also article on *Coal Testing* in this section of the Handbook.

CHARACTERISTICS OF BOMB CALORIMETERS

Any oxygen-bomb calorimeter consists of four essential parts: (1) bomb or vessel in which the combustible charge is burned, (2) a bucket or container which holds the bomb as well as a measured quantity of water to absorb the heat released from the bomb and a stirring device to assure thermal equilibrium, (3) a jacket for protecting the bucket from transient thermal stresses, and (4) a calorimeter thermometer for measuring temperature changes within the bucket.

The bomb consists of a strong, thick-walled, metal vessel which can be opened for inserting the sample, for cleaning, and for recovering the products of combustion. Valves

*President, Parr Instrument Company, Moline, Illinois.

must be provided for filling the bomb with oxygen under pressure and for releasing residual gases after the combustion is complete. Electrodes to carry the ignition current to the fuse wire also are required. Since an internal pressure up to 1500 psig can be developed during combustion, most bombs are constructed to withstand pressures of at least 3000 psig.

In the moist, high-pressure oxygen environment within the bomb, some of the nitrogen present will be oxidized to form nitric acid. Similarly, any sulfur contained in the sample will be converted to sulfuric acid. Because of the formation of these hot and highly corrosive acids, the bomb must be made from materials that will not be attacked by these combustion products. Until Professor S. W. Parr developed a complex nickel-chromium alloy for use in oxygen bombs in 1912, linings of platinum and gold were the only means available for protecting the inside of the bomb. Although a few platinum-lined bombs are currently used for research applications, bombs for fuel testing are almost exclusively made of alloys similar to those developed by Professor Parr. Cross section of an oxygen combustion bomb is shown in Fig. 1.

The calorimeter bucket contains the bomb plus a sufficient quantity of water to immerse the bomb completely and to absorb the heat released from the combustion within the bomb. A stirrer is used in the bucket to rapidly bring the bucket and its contents to thermal equilibrium. Bucket systems must be carefully designed. The quantity of water must be sufficient to readily absorb the heat released, but not so large as to preclude acceptable sensitivity. Likewise, the stirrer and bucket geometry must enable rapid equilibration without introducing excessive heat in the form of mechanical energy. Buckets are commonly provided with a highly polished surface to minimize the absorption and emission of radiant heat.

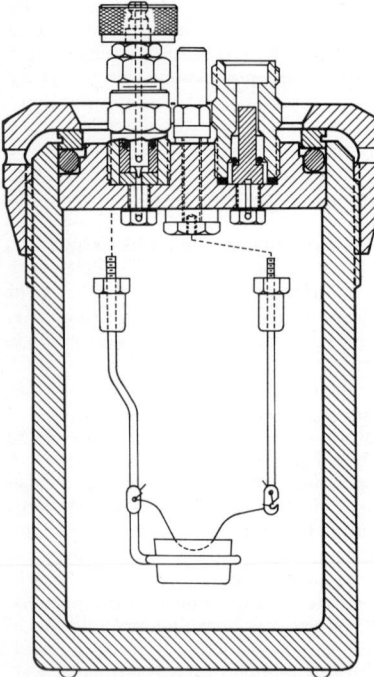

The jacket that contains the bucket with its bomb provides a thermal shield to control heat transfer between the calorimeter bucket and its surroundings. It is not necessary to prevent this transfer if a means of precisely deciding the amount of heat transferred during the determination can be established. Any effective jacket will, of course, minimize the effects of drafts, sources of radiant energy, and changes in room temperature during the test.

Two basic types of calorimeter jackets are commonly used on bomb calorimeters. The first is the isothermal system in which the jacket temperature remains constant while the bucket temperature rises. Isothermal jackets require that temperature readings be made to determine the net heat loss or gain from the bucket to its surroundings. Isothermal jackets are available in two models: the true isothermal jacket which is controlled

Fig. 1. Cross section of oxygen combustion bomb. *(Reproduced by permission of the source: Parr Instrument Company.)*

to a specific temperature, and the uncontrolled or plain jacket. The plain jacket assumes that the jacket temperature will remain sufficiently constant without controls. Because of its inherent simplicity, the plain jacket is popular for low-cost calorimeters. See Figs. 2 and 3.

The adiabatic system is the second type of jacket commonly used in bomb calorimeters. In the adiabatic system the jacket temperature is controlled during the determination to keep it equal at all times to that of the bucket. If temperature differences between the bucket and the thermal jacket can be eliminated, there will be no heat transferred between these components and the calculations and corrections required for the isothermal systems can be eliminated. Adiabatic jackets are widely used for fuel testing since they combine opportunities for high precision with speed and convenience. See Figs. 4 and 5.

The calorimetric thermometer measures temperature changes within the calorimeter bucket. It must be able to provide excellent resolution and repeatability. Interestingly, high single-point accuracy is not a requirement since temperature changes and not absolute temperatures are important in calorimetry. Mercury-in-glass thermometers, platinum-resistance bulbs, quartz oscillators, and thermistor systems have all been successfully used as calorimetric thermometers.

OPERATING PRINCIPLES

Before a material with an unknown heat of combustion can be analyzed in a bomb calorimeter, the energy equivalent or heat capacity of the calorimeter must first be determined. This value is, of course, dependent on the heat capacities of the materials within the calorimeter: notably the metal of the bomb and bucket, and the water in the bucket. Energy equivalents are determined empirically by burning a sample with a precisely known heat of combustion in the calorimeter under a carefully controlled and reproducible set of operating conditions. Benzoic acid is used almost exclusively as a reference material for fuel calorimetry because it is completely combustible and nonhygroscopic and is readily available in a very pure form.

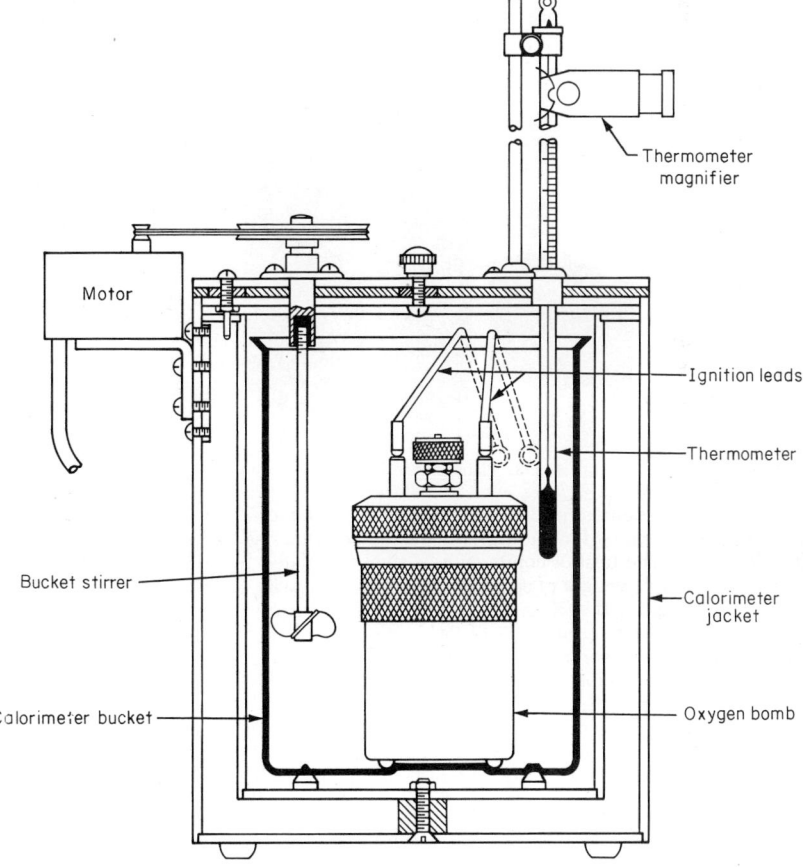

Fig. 2. Sectional view of plain oxygen bomb calorimeter. (*Reproduced by permission of the source: Parr Instrument Company.*)

The amount of heat introduced by the reference sample is determined by multiplying the heat of combustion of the standard material by the weight of the standard sample. If this value is divided by the net temperature rise produced in the calorimeter, the resultant is the energy equivalent under the specified operating conditions. For example, if 1.651 grams of benzoic acid with a heat of combustion of 6318 calories per gram were burned, a total of 7361 calories would be released. If this produced a temperature rise of 3.047°C, the energy equivalent for these conditions would be 2416 calories per degree Celsius.

Once the energy equivalent of the calorimeter has been determined, the calorimeter can be used for actual fuel testing. A sample of known weight is burned in the calorimeter

Fig. 3. Plain oxygen bomb calorimeter.

and the resulting temperature rise measured and recorded. The total energy released by the sample is determined by multiplying the temperature rise by the energy equivalent of the calorimeter. The heat of combustion is then calculated by dividing this total energy value by the sample weight to convert to a unit weight basis. Continuing the example, if a fuel sample weighing 0.9936 gram produced a temperature rise of 3.234°C in the calorimeter with an energy equivalent of 2416 calories per degree Celsius, a total of 7813 calories would have been released for a heat of combustion of 7863 calories per gram. A factor of 1.8 Btu per pound equals 1.0 calorie per gram can be applied to convert this value to 14,153 Btu per pound.

It is important to note that the energy equivalent for any calorimeter is dependent on a set of operating conditions, and these conditions must be reproduced when the fuel sample is tested if the energy equivalent is to remain valid. It is readily apparent that a difference of one gram of water in the calorimeter bucket will change the energy equivalent by one calorie per degree Celsius. Less obvious but equally important are the changes resulting from different bombs or buckets with unequal masses, different operating temperatures, different thermometers, or even the biases imposed by different operators.

Precision in bomb calorimetry is therefore dependent on a repeatable set of operating conditions applicable to both the standardization and the determination. Those factors which can not be held constant require corrections to compensate for their effects. The amount of fuse wire consumed in igniting the sample is one of the variables that cannot be controlled. A determination of the energy contributed by the burning fuse wire is therefore made and applied to each determination to compensate for this variance. Under the extreme conditions of combustion which occur within the bomb, molecular nitrogen is oxidized to form nitric acid, and sulfur is converted to sulfuric acid. Corrections are applied to compensate for these conditions which are peculiar to bomb combustions. Any heat transferred into the calorimeter or lost from the calorimeter to its surroundings must be accounted for if valid results are to be obtained.

In a bomb combustion the water produced by the oxidation of hydrogen condenses and liberates its latent heat of vaporization. The total heat produced is known as the gross heat of combustion at constant volume. In actual fuel-burning processes the water escapes as a vapor, and the total heat produced is known as the net heat of combustion at constant pressure. The net heat of combustion is the value of interest. It may be obtained from the

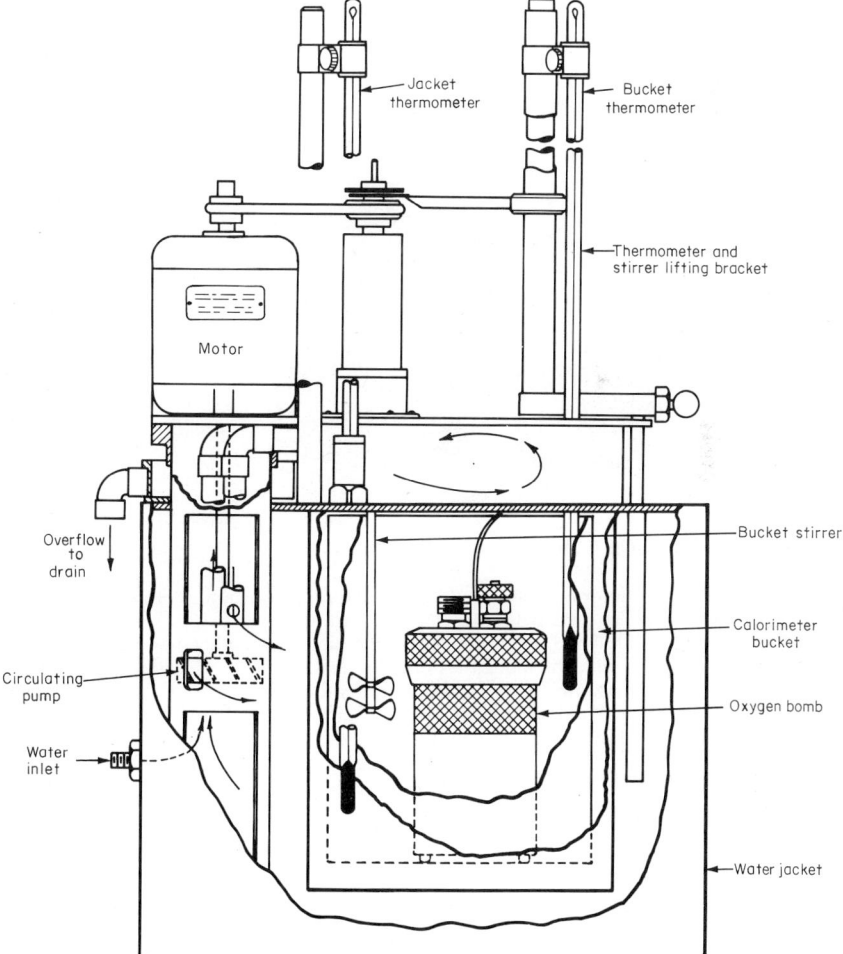

Fig. 4. Sectional view of adiabatic oxygen bomb calorimeter. *(Reproduced by permission of the source: Parr Instrument Company.)*

gross heat of combustion and the percent hydrogen in the sample by the following equation:

$$\text{Net heat of combustion} = \text{gross heat of combustion} - 91.23 \times (\text{wt. \% hydrogen})$$
$$\text{(Btu/lb)} \qquad\qquad\qquad \text{(Btu/lb)}$$

Because samples are completely oxidized during combustion in an oxygen bomb and because the combustion products are quantitatively retained within the oxygen bomb, procedures have been developed for determining sulfur, halogens, and other elements in

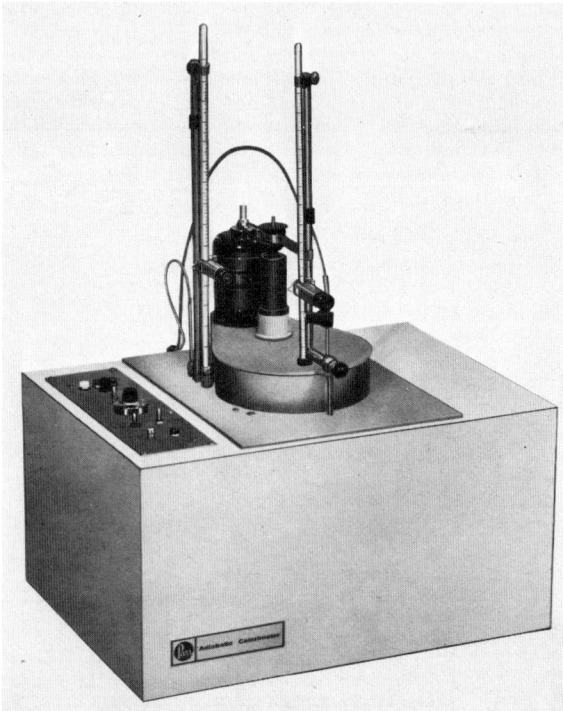

Fig. 5. Adiabatic oxygen bomb calorimeter.

conjunction with the determinination of the heat of combustion. Perhaps the most common of these auxiliary tests is the one for sulfur in fuels.

During the bomb combustion, sulfur present in the sample is oxidized to sulfur trioxide, which in turn combines with the moisture in the bomb to condense as sulfuric acid. The inner surfaces of the bomb are then washed with distilled water, and the washings are collected. After appropriate neutralization and preparation, the sulfate present is precipitated by the addition of barium chloride and determined by gravimetric procedures. The bomb-washing technique can also be used to determine chlorine, bromine, iodine, fluorine, arsenic, mercury, phosphorus, boron, lead, selenium, and other metals in combustible materials.

CALORIMETER SELECTION

The principal factors affecting the selection of a calorimeter for a specific application deal not with the type of fuel involved but rather with the precision required and the work load anticipated. The same oxygen bomb is used for testing solid and liquid fuels. Volatile liquids such as jet fuels and aviation gasolines do require special techniques for holding the sample, but these are actually concerned more with preventing vaporization losses while the sample is being weighed and handled than with differences in the combustion technique.

Commercial bomb calorimeters are available with either plain or adiabatic jackets. Most fuel testing is performed in adiabatic calorimeters. As mentioned earlier, the adiabatic system attempts to eliminate all heat transfer between the bucket and its surrounding by eliminating the temperature differential necessary for any heat transfer. Although no system can actually achieve this theoretical condition, modern adiabatic jackets and in particular those equipped with automatic control systems, can approach this goal and produce results with outstanding precision. In addition, since the adiabatic feature eliminates the need for heat-leak calculations and corrections, adiabatic systems also require less operator time per test. Because they produce excellent results in a minimum of time, the adiabatic system is the overwhelming choice for most fuel-testing laboratories.

Plain-jacket calorimeters are simpler and less expensive than adiabatic instruments. They are widely used for student instruction and other applications where limited calorimetric testing is performed. In these applications the lower initial cost of the plain-jacket calorimeter will compensate for the additional time required to perform each test. Adequate precision for most test methods can be achieved if the operator will apply reasonable care to obtain and apply the required heat-leak corrections.

Experienced operators will average approximately one test per hour with a plain-jacket calorimeter. Manually controlled adiabatic calorimeters will produce about two tests per hour. With automatic control of the adiabatic jacket, three tests per hour are commonly achieved. Approximately four tests per hour can be obtained by adding a recording thermometer and programmer to an automatic adiabatic calorimeter.

SAFETY CONSIDERATIONS

Oxygen-bomb combustions need not be considered unusually hazardous as they are carried out routinely in thousands of laboratories. Nevertheless, since they do involve nearly explosive reactions at high pressures, constant vigilance is required to assure safe operation of the calorimeter. Obviously the oxygen bomb must be maintained in good repair at all times. Any bomb that is leaking oxygen from a valve, an electrode, or another seal should never be fired since the combination of the flame front and the high-pressure oxygen may ignite the bomb itself in the area of the leak.

Care must be taken to ensure that the amount of sample utilized and the initial oxygen pressure do not exceed the limits for the bomb being used. A bomb should also be subjected to hydrostatic and proof tests on a regular basis to confirm that it has not been weakened by corrosion or wear and that it is suitable for continued service.

STANDARD METHODS

The American Society for Testing and Materials has developed a series of standard test methods for both solid and liquid fuels in oxygen-bomb calorimeters. These are available from the society (Philadelphia, Pennsylvania).

ASTM Designation D 3176 *(Ultimate Analysis of Coal and Coke)* and D 3180 *(Calculating Coal and Coke Analysis from "As Determined to Different Bases")* should be required reading for anyone concerned with solid fuels.

Most commercial analyses of solid fuels are conducted in accordance with ASTM Designation D 2015. This method deals with the calorific value of solid fuels as determined by the adiabatic bomb calorimeter. A unique feature of this method is an extensive discussion of the standardization and restandardization of the calorimeter. The basic method for determining the heat of combustion of liquid hydrocarbon fuels is ASTM Designation D 240. Either isotherm or adiabatic calorimeters may be used in D 240. Nonvolatile fuels are usually tested in accordance with this method.

ASTM Designation D 2382 is a high-precision method for the heat of combustion of hydrocarbon fuels by the bomb calorimeter. Although suitable for both volatile and nonvolatile fuels, this method finds its principal applications in the testing of aviation gasolines and jet fuels. The method calls for either an adiabatic or isothermal jacket, which is controlled by an automatic temperature-control system. An uncontrolled or plain jacket would not be acceptable for this method. As a high-precision method this procedure calls for platinum-resistance thermometers and precision balances with capabilities beyond those specified in the other three methods. Interesting techniques for handling volatile fuels were developed for this method and they will be of interest to anyone testing volatile samples, even though the extreme precision of D 2382 may not be required.

The various ASTM Test Methods also contain general information which will be of great value to anyone considering fuel testing in a bomb calorimeter. A discussion of units and definitions is always included in each method, as are the procedures for converting from gross to net heat of combustion. Also contained are cross-references to the various additional standards which have been published by other standards organizations. The discussions of precision, repeatability, and reproducibility will be invaluable to the analyst who wishes to evaluate the quality of his own work.

REFERENCES

1. Staff: Laboratory Sampling and Analysis of Coal and Coke, *ASTM Method* D 271–70, American Society for Testing and Materials, Philadelphia (periodically revised).
2. Staff: Method of Test for Heat of Combustion of Liquid Hydrocarbon Fuels by Bomb Calorimeter, *ASTM Method* D 240–64, American Society for Testing and Materials, Philadelphia (periodically revised).
3. Jessup, R. S.: Precise Measurement of Heat of Combustion with a Bomb Calorimeter, *U.S. NBS Monogr.* 7, Superintendent of Documents, U.S. Government Printing Office, Washington, D.C., 1970.
4. Rossini, Frederick D.: "Experimental Thermochemistry—Measurement of Heat Reaction," Interscience-Wiley, New York, 1956.

Roof Control for Underground Mining

BY ROBERT FLETCHER* AND JOHN BAKOSS†

Roof control is the technology of mine-roof support. Planning a roof-support system requires the careful integration of observed geological criteria with inferred mechanical principles. Thereby, a basis for selecting the method and equipment for support and support installation is established. In addition, practical experience and matters of local or general historical record should be considered.

Major factors of roof control are discussed in the following order here: Principles of roof mechanics in competent rock, types of roof failure, pillar support, timbering, rigid and yieldable steel arches, masonry support, history and techniques for roof bolting, elements of applied roof bolting, rock-drillability determination, roof-drilling equipment, and technological trends.

PRINCIPLES OF ROOF MECHANICS IN COMPETENT ROCK

The major rock types and structures associated with fossil fuel mining are defined in Table 1. Design criteria for underground openings may be established through idealization of roof configuration, structural behavior, and loading conditions. Of the several models of roof behavior that have been proposed, probably the most significant are: (1) the fixed-end beam concept, (2) the linear (voussoir) arch concept, and (3) the pressure-arch theory.

As a fixed-end beam, the roof depends on its internal resistance to bending to support its

TABLE 1. Major Rock Types and Structures Associated with Fossil Fuel Mining

ROCK TYPES

Sandstone: A cemented or otherwise lithified detrital sedimentary rock composed predominantly of sand-sized ($\frac{1}{16}$–2 millimeter) quartz grains.

Siltstone: A detrital sedimentary rock composed predominantly of silt-sized ($\frac{1}{256}$–$\frac{1}{16}$ millimeter) particles.

Shale: A laminated, detrital, sedimentary rock composed predominantly of clay minerals and clay-sized ($< \frac{1}{256}$ millimeter) particles.

Limestone: A sedimentary rock containing at least 80% of the carbonates of magnesium and/or calcium.

Clay: An unconsolidated sediment composed of clay-sized ($< \frac{1}{256}$ millimeter) particles. Or crystalline fragments of minerals that are essentially hydrous aluminum silicates or hydrous magnesium silicates.

Slate: A fine-grained metamorphic rock possessing a well-developed fissility (cleavage along closely spaced parallel planes).

STRUCTURES

Stratification: A structure produced by the deposition of sediments in beds or layers.

Fold: A bend or flexure in rock strata.

Joint: A break in a rock mass along which no relative movement of rock on opposite sides of the break other than separation has occurred.

Fault: A break in a rock mass along which there has been differential movement.

Lens: A tabular body of rock or soil (i.e., clay, coal, or sand) of a limited extent.

Facies Change: A lateral variation in the lithology of a stratigraphic unit.

Concretion: A nodular or irregular rock mass developed by the localized deposition of material from solution. Harder than enclosing rock.

Competent Rock: Rock which behaves as an elastic solid and in which it is possible to design self-supporting openings.

Pillar: A natural formation shaped like a pillar.

Pillar Mining: A method of mining coal and other bedded mineral deposits in which a network of passageways forms pillars of coal that are then extracted.

*President, †Staff Engineer, J. H. Fletcher & Co., Huntington, West Virginia.

own weight. General stress distribution for this behavioral model is shown in Fig. 1. The opening width which may be spanned by a fixed-end beam is directly proportional to its unit strength and thickness and inversely proportional to the density of the rock that it supports. Analytical methods derived from the fixed-end beam concept are described in Ref. 3.

The linear-arch concept may be used to describe the behavior of an opening spanned

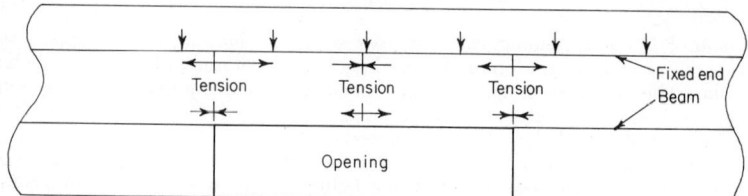

Fig. 1. General stress distribution for the fixed end beam. (After Woodruff—see references.)

by moderately jointed rock. It takes into account the redistribution of vertical pressures into horizontal and diagonal compressive forces which characterize arching action. The general stress distribution through the linear arch is illustrated in Fig. 2. Like the fixed-end beam, the span length of the linear arch is directly proportional to the arch thickness T and inversely proportional to the density of the rock that it supports. Unlike the beam, the linear arch depends on its unit strength in compression to resist vertical loading. Analytical methods for the linear arch are described in Ref. 3.

The pressure-arch theory may be used to investigate roof stability in bedded, nonfractured rock. The theory states that beds of the immediate roof deflect slightly to relieve

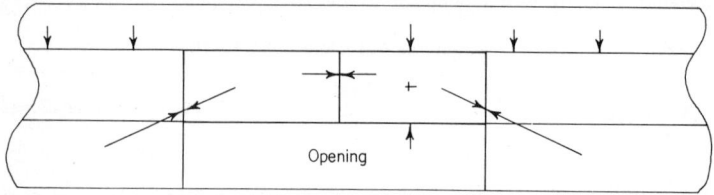

Fig. 2. General stress distribution for the linear arch. (After Woodruff—see references.)

themselves of vertical loading from the main roof. The main roof then stabilizes itself through arching action. (The natural tendency for openings to stabilize themselves through arching action manifests itself in the domed shape of many fall areas.) Observations indicate that the depth of overburden is the most significant factor in determining the width of the pressure arch. Further, the height of the pressure arch above an opening is approximately equal to twice the opening width. This is an important consideration inasmuch as some portion of the unloaded immediate roof must be analyzed as a beam, supported as such, or removed. Additional dimensional relationships attendant to the pressure arch are defined in Ref. 3. General stress distribution suggested by the pressure-arch theory is shown in Fig. 3.

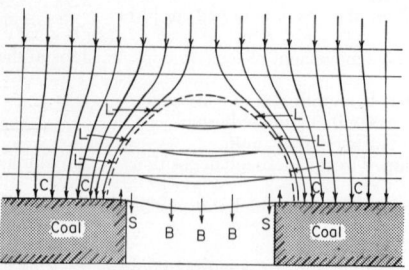

Fig. 3. General stress distribution for the pressure arch. B = bending stress; S = shearing stress; L = lateral compressive stress; C = vertical compressive stress. (After Stefanko—see references.)

ROOF SUPPORT

Roof support serves the twofold purpose of keeping entries and working places open for production and protecting men and equipment from debilitating roof failure. Planning a roof-support system normally involves an attempt to restore, insofar as possible, the stress equilibrium which the excavation of opening upsets. However, it is often expedient merely to delay the release of latent energy by using temporary supports.

TYPES OF ROOF FAILURE

The following general types of roof failure warrant consideration in planning roof-support systems:

1. Falls. Falls consist of the breaking away of rock from the mine roof. Their frequency and magnitude are highly variable and, to some extent, unpredictable. When unexpected falls occur in mines which are properly engineered and operated, they are usually attributable to some unforeseen structural defect in the roof itself. Falls may involve no more than a continual sloughing of small slabs of crumbly draw slate. On the other hand, it is possible for massive blocks of well-indurated rock to be released. Massive falls may involve thousands of tons of material and extend into the main roof.

2. Squeezing. Squeezing is the elastic yielding of the mine roof which results from insufficient pillar strength. The magnitude of squeeze may range from a fraction of an inch to several feet. Excessive squeezing may result in general mine-height reductions that are sufficient to render several sections unworkable.

3. Bursts (Bumps). Sudden explosionlike failures of coal or rock are called *bursts.* They are associated with the high compressive stresses which may occur in pillars or arched roof configurations.

4. Intentional Caving. During pillar recovery and in conjunction with long-wall mining, it is imperative that the roof in the mined-out area be induced to fall. This relieves pressure on supports in the working area which otherwise would be overloaded and would fail.

Pillars. For large mined areas having multiple openings, virtually the entire weight of overburden is supported by pillars.* Generally, where tectonic loading is negligible, pillar stress is essentially due to gravity, and large horizontal stresses are not present. However, accurate prediction of stress distribution is not possible unless field measurements are made. Field studies have indicated that concentration of stress may be expected in pillars adjacent to or otherwise near mine rib walls. Furthermore, it has been demonstrated that such concentrated stresses may be of sufficient magnitude to affect pillar strength. Under these circumstances, it may be economically, if not also structurally, expedient to take into account the strength of pillars in a failed condition as it affects general pillar stability.

Where prevailing horizontal field stresses are low, pillar orientation is not a significant structural factor. Orientation should be established on the basis of geological or operational considerations.

Correct procedures diminish the possibility of stress concentration in the pillars being worked and in those adjacent to the mined-out area. It is important that the immediate roof of the mined-out area be induced to fall. Improper extraction procedure can result in falls, bursts, and squeezing in the working areas.

Minimum requirements for pillar design and recovery are set forth in the Federal Register (Ref. 4).

Timbering. Timber has long been an important factor in the construction of permanent and temporary roof-support systems. It is relatively easy to handle and shape and is durable in the mine environment. Treated timber will provide about ten years of service, whereas untreated materials may be expected to last about three years. Timber is further suited to roof-control systems because the popping and cracking sound it emits as it begins to fail serves to warn of excessive ground pressures.

Used as posts or cribs, timbers will withstand compression loads. As crossbars, timbers may provide the additional strength necessary to span an opening. Elaborate roof-support structures have been built of timber, but these are not common to modern mining. Significant specific roof-control applications include:

1. Providing additional support at important intersections and haulage-ways.

---

*The brief description here is a condensation of Rock Mechanics Applications to the Design of Oil Shale Pillars, by Jose F. Agapito, *Mining Eng.*, pp. 20–25, May 1974.

2. Providing temporary support in newly excavated working places. (Also used are mechanical and hydraulic safety posts.)

3. Providing temporary support and helping to induce the desired breakline during pillar recovery.

4. Shoring up fall areas.

5. Providing supplemental support where pillar dimensions have been reduced by sloughing, rolls, or mining activity.

6. As headers or half headers, used with roof bolts to hold loose top.

Proper storage practice is essential to the preservation of the strength and longevity of timber. Important factors to consider regarding storage are: (a) Maintain well-drained timber yards free from vegetation and decaying wood; (b) stack timber on skids at least one foot above ground; and (c) stack timber so as to allow good air circulation through the stack.

Rigid and Yieldable Steel Arches. *   Steel-arch structures are generally applicable where (1) ground pressures are large, (2) the top is extensively fractured, or (3) ground movement (such as squeezing or heaving) is anticipated.

Rigid steel arches include continuous rib configurations and rib and post configurations. The former are suitable for controlling vertical and lateral pressures, while the latter are most applicable where vertical loads only are prevalent.

Common configurations of the yieldable arch include: (1) the three-segment arch with leg segments toed in for control of vertical pressures, (2) the three-segment arch with leg segments toed out for control of vertical and lateral pressure, and (3) the symmetrical ring (three or more segments) for control of pressures from all directions.

Some advantages of the yieldable system are: (a) Load capacities increase with the application of load, (b) joints yield at loads less than the yield load of the steel segments, (c) arches may be recovered and reshaped, and (d) they are almost maintenance-free.

Yieldable arches are generally used where ground pressures and movement would destroy rigid systems. The installation of yieldable structures requires more labor than that for similar rigid support.

Masonry Support.   Masonry piers and abutments are normally used to provide permanent relatively strong support at critical locations. Grouting will seal and strengthen permanent openings. Grouting is particularly applicable to stabilizing loose or permeable rock.

Lining with gunnite, concrete, or polymer resins also serves to seal and strengthen openings. For best results, lining should be accomplished as soon as possible after an opening is excavated. Lining material may be applied over reinforcing wire mesh and/or over bolted roof. Reinforced-concrete lining is used to provide permanent heavy-duty support.

HISTORY AND TECHNIQUES OF ROOF BOLTING

For the first half of the twentieth century, most mine roof was held in place by timber posts and crossbars. The theory was to wedge between the floor and the roof, carrying the load on the supports.

In early mining, when hand tools predominated, the loading was minimized by narrow working places and arched entries. With the advent of mechanized mining in the 1930s, rooms were widened and percent recovery of coal increased. Timber crews of four to six men were required for each production section.

In the 1940s, timbering machines mechanized timber installation, reducing most crews to two men, by mechanically lifting the bars and sawing the posts to length.

By 1950 a new theory was promulgated and roof bolting became established. This concept was tested in southern Illinois mines under the direction of C. C. Conway, and then further expanded and promoted by the U.S. Bureau of Mines.

Immediate results of the spread of roof bolting were seen in a dramatic decrease in roof-fall accidents. Production advantages were realized in faster and safer primary haulage due to the elimination of posts, greater recovery through wider working places, the

---

*Portions of the brief description here are a condensation of Roof Control, *Coal Age,* pp. 215–217, July 1967.

mining of areas that would not have been supportable by posting, and improved ventilation.

Early roof bolts consisted of steel bars 1 inch in diameter, split longitudinally 6 inches to receive a steel wedge. The lower end of the bolt was threaded. See Fig. 4. Installation was effected by drilling a 1¼-inch hole to a predetermined depth, less than the length of the bolt. The wedge was loosely inserted in the split of the bolt, and the assembly was pushed into the hole with the wedge contacting the top. Through impact, the split was driven over the wedge with either side cutting into the wall of the hole, thus forming a funnel-shaped cavity and holding the bolt in place. A plate washer was then slipped over the threaded end, extending from the hole—and a nut run over the threads, tensioning the bolt between the anchor point and the washer.

ELEMENTS OF APPLIED ROOF BOLTING

Both materials and tools for roof bolting rapidly improved. One design of the expansion shell which superseded the wedge bolt is shown in Fig. 5. This design necessitated a 1⅜-inch-diameter hole, but by only requiring that the hole be deep enough to accept the bolt,

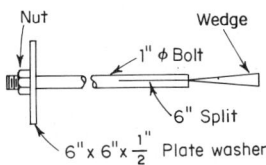

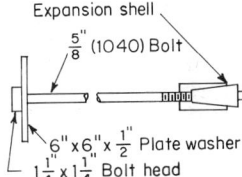

Fig. 4. The wedge bolt achieves anchorage as the split end of the bolt is driven over a wedge and thereby is forced against the bolt-hole wall.

Fig. 5. A modern roof bolt with expansion shell.

this eliminated the need for regulating depth. Mild steel bolts ¾ inch long were first used. Now standard are ⅝-inch-diameter, 1040 steel bolts, with rolled threads and 1¼-inch square head. The chemical, mechanical, and dimensional requirements for bolts and accessories are detailed in ASTM publication, "Standard Specifications for Roof and Rock Bolting and Accessories," obtainable from the American Society for Testing and Materials, Philadelphia, Pennsylvania.

The mechanical principles of roof support by vertically oriented bolts vary somewhat, depending on the type of roof structure being supported. For the fixed-end beam, bolts serve to increase the thickness factor T. The thickness of the beam will be approximately equal to the bolt length being used. In addition, bolts serve to resist shearing stress induced by flexure of the beam. This resistance to shearing is especially significant along bedding planes where differential movement of rock strata may occur.

When used to stabilize linear or pressure arches, bolts serve to resist shear and tension, which are associated with compressive forces in the arch. Shear stress normally occurs at slightly less than 45° with respect to the zone of compression in the arch. Tensile stress may be expected at about 90°. The response of a unit section of material to compressive loading is shown in Fig. 6, and exemplifies the distribution of shear and tensile stress in an element of an arch.

The compressive loading of mine rib results in stress distribution similar to that in an arched roof. Therefore, it is also possible to stabilize the rib with bolts.

Though the essential function of bolting is to maintain the structural integrity of mine roof, bolts also serve to suspend loose roof material, such as draw

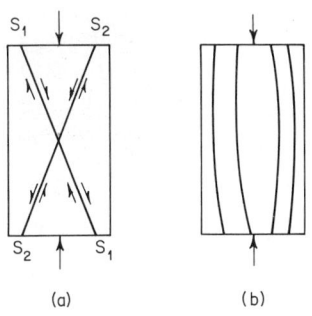

Fig. 6. Brittle material in compression, showing two types of failure: (a) S_1 and S_2 shear fractures; (b) T tension fractures. (After Hills— see references.)

slate, from more secure strata. In addition, holes drilled for bolt installation serve to relieve gas and hydrostatic pressure in the roof.

For application, bolts must be selected in terms of length, type of anchorage, and strength. The bolts must be sufficiently long to ensure that an adequately thick self-supporting beam is formed across a given opening. Where mechanically anchored, the anchorage must be in a secure stratum. If chemically grouted, length is primarily a function of the required beam thickness.

Panek (Ref.2) has proposed equations for the design of bolting systems. The procedure is applicable where competent roof is to be supported by conventional, vertically oriented roof bolts.

Because they may be installed more rapidly and at lower initial cost, bolts anchored by expansion shells are currently the most popular style. However, the chemically grouted (resin) bolt is beginning to find widespread acceptance in the mining industry. Some advantages of the resin bolt are: (1) Because anchorage is along the entire length, it is not necessary that the extreme upper end of the bolt be placed in hard rock. Therefore, it is often possible to use shorter bolts. (2) The shorter bolt holes result in reduced bit usage and drilling time. (3) Chemical grout seals the hole against water and air. (Under some conditions, however, the sealing effect of resin bolts means that additional holes must be drilled for the relief of gas and hydrostatic pressure in the roof.). (4) Pull tests are higher for chemically grouted bolts. (5) Chemical anchorage does not fracture or otherwise damage surrounding rock. (6) Bolt tension develops in response to loading by roof flexure and will not "bleed off." Some disadvantages of the resin bolt include: (1) Resin may drip from bolt holes and build up on equipment. (2) Installation takes longer. (3) The installation cycle cannot be interrupted. (4) Thrust exerted by the roof bolting machine is the only means of preloading the bolts. (5) Bolts cannot be recovered.

In strata where adequate anchorage cannot be found, truss bolting (Fig. 7) has proved effective, the anchor point being over the rib. Upward thrusts are produced at points A

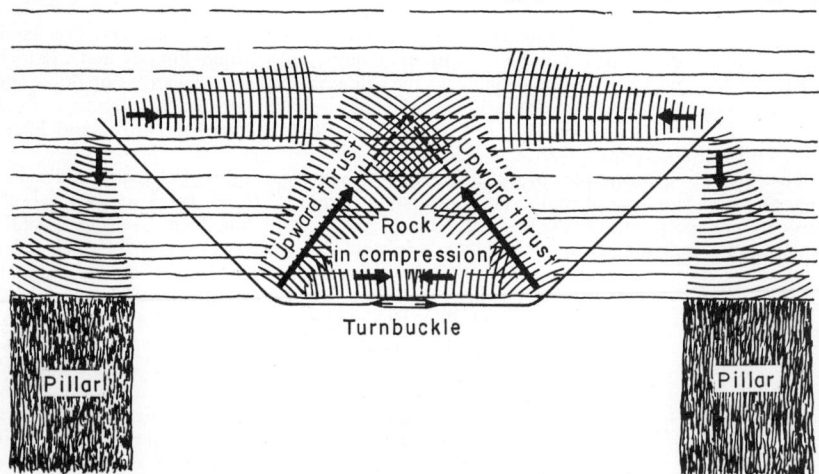

Fig. 7. The roof truss and associated stress distribution. Tension in the steel rods produces compression in all shaded areas.

and B when the turnbuckle is tightened. Downward reactions are transmitted to the rib by the bolts. Compressive loading induced by the truss system serves to counteract tensile stress in the lower portion of the roof "beam."

Roof bolts are also used in conjunction with mats and crossbars. In most cases the main objective is to build a beam of the roof strata. The U.S. Bureau of Mines standards for roof support by bolting are specified in the Federal Register (Ref.4).

ROCK DRILLABILITY DETERMINATION

In planning a mining operation, it is important to assess the economics of roof support. Although core samples from drill holes indicate the context of roof immediately above a seam, it has been difficult until recently to determine the drillability of this roof. However, an advanced technique has been developed for drillability determination. Correlation of microscopic structural details of selected rock specimens to actual drilling experience yields the following information:

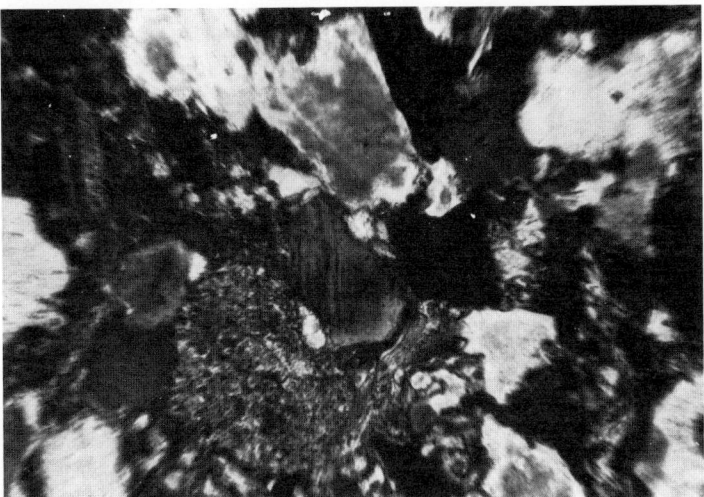

Fig. 8. Photomicrograph of a "soft" sandstone. (Magnification 25×.)

1. Observations indicate that poorly indurated sandstones (those with silt and clay matrix and lacking measurable amounts of crystalline cement) may be drilled with straight rotary units. Good penetration rates and bit life may be expected. See Fig. 8.

2. Dense sandstones (well indurated through mineral cementation) require rotary-

Fig. 9. Photomicrograph of a "hard" sandstone. Note that the individual particles are locked tightly together like pieces of a jigsaw puzzle. (Magnification 25×.)

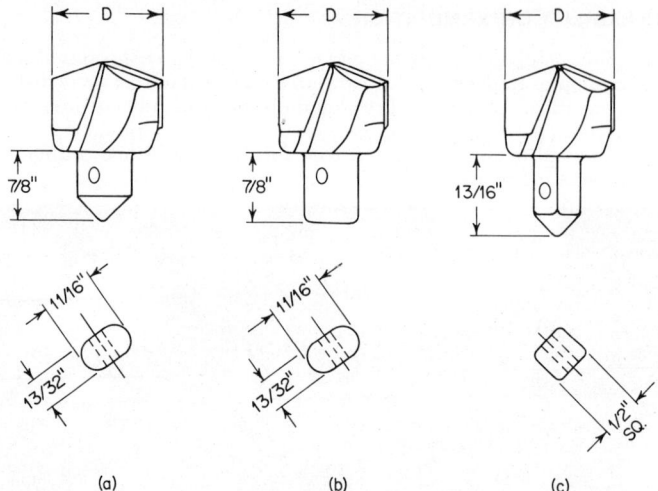

(a)   (b)   (c)

Fig. 10. Typical rotary roof bits: (*a*) FSP 1⅜-in. bit; oval shank design for extra strength and improved dust collection. (*b*) FSP 1½- and 1⅝-in. bits; oval shank design for extra strength and improved dust collection. (*c*) FVP 1⅜-in. bit with ½-in. square shank. (Kennametal Inc.)

impact drilling equipment. Reduced penetration rates and bit life may be anticipated. See Fig. 9.

This technique is especially applicable before property development when the only access to the proposed mine roof is through core drilling. Microscopic petrographic studies of roof cores may be used as a guide to drilling equipment selection and bit usage estimation. Findings may be further substantiated by mounting several feet of the roof core in concrete and test-drilling it with appropriate equipment.

Sedimentary rocks other than sandstone normally are not difficult to drill with conventional rotary drilling equipment.

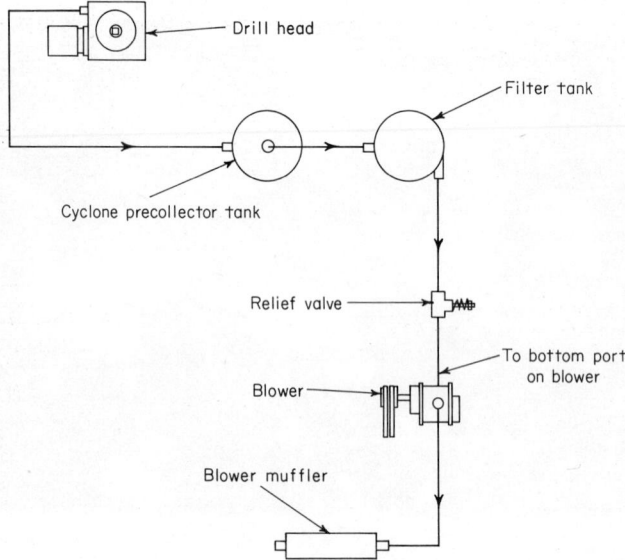

Fig. 11. Typical dust-collection pneumatic circuit.

ROCK DRILLING EQUIPMENT

It was first believed that rotary drilling was effective only in shale and soft rock. Sandstone and hard rocks were drilled by air-driven stopers. Gradually, carbides and bit configurations have been developed which enable rotary drills to cut the hardest rock successfully. See Fig. 10. Regulation of the rpm, control of the rate of penetration and thrust, and application of impact have resulted in rapid and economical drilling of hard rock.

Greater thrusts, a major factor in the drilling of harder rock, were attainable when hollow drill steels replaced scroll augers. The L/R (length-radius) ratio for hollow steels exceeds that for scroll augers for a given hole diameter.

The use of hollow steels was initiated by the development of internal dust collection. Early drill rods carried a scroll for transporting the cuttings from the hole. However, collection of the cuttings at the base of the hole was difficult. Using a hollow drill steel, a dust-collection system drawing the cuttings by vacuum into the steel just below the bit met safety and health regulations in a more satisfactory manner. This became known as internal collection, and presently predominates. A typical dust-collecting system is shown in Fig. 11.

Air-driven stopers predominated in the early stages because they were already in use for rock drilling. With the advent of rotary drilling and development of the expansion shell bolt, stopers were superseded in all but hard-rock roof conditions. Presently, even the hardest or roof rock can be rotary-drilled. With noise level becoming a serious factor, few stopers are used for roof bolting.

Most rotary drills are equipped with hydraulically driven drill heads. See Fig. 12. The hydraulically driven drill head generates up to 300 foot-pounds of torque. Two methods of feeding the drill head have evolved. The *mast feed* (Fig. 13) has the advantage of constant thrust, maintains a straight-line feed with properly placed drill guide, and can be tilted for angle bolting. The *boom feed* (Fig. 14) is most adaptable to varying seam heights.

Moving the entire machine to which it is attached is the simplest method of positioning the drill head. Position refinement is a combination of swinging, sliding, and extending motions (Fig. 15).

Although roof bolting reduced accidents from roof fall, there still exists an element of danger to the man working under unbolted roof, as during the setting of safety posts between the face and the last row of bolts.

Fig. 12. Rotary drill equipped with hydraulically driven drill head.

Of more recent development, a temporary roof-support system (TRS), as built into a dual-head roof drill (Fig. 16), has reduced this hazard. This system consists of a crossbar (or bars) placed by remote control to support the unbolted roof while it is being drilled. A hydraulic roof jack carried on an extending boom wedges the crossbar in place. The jack has a blocked loading of up to 30 tons. In conjunction with a safety post of 6-ton capacity mounted behind the drill head, the operator is working in a protected position.

In lieu of TRS, a canopy (Fig. 17), supported by the safety post, can shelter the operator from falling roof. The canopy must be capable of withstanding a load of 15 pounds per square inch, or a maximum of 18,000 pounds (U.S. Bureau of Mines standard).

TECHNOLOGICAL TRENDS

Improved technology is needed for rock stress-and-strain determination. A more complete and accurate knowledge of pressures affecting the stability of openings would facilitate

Fig. 13. Mast-feed drill unit that provides constant-thrust straight-line feed and can be tilted for angle bolting.

Fig. 14. Typical boom-feed single-head roof-bolting machine. Positioning of the drill head is accomplished by moving the entire machine.

Fig. 15. On more sophisticated equipment, drill-head position refinement is a combination of swinging, sliding, and extending motions.

the design of safer, more efficient support systems. Strain measurement techniques are currently being used to determine in-situ rock pressures and are contributing to the development of more precise quantitative design methods. Strain indicators also are being employed to monitor roof behavior and warn of impending roof failure. Some systems which have been used to indicate roof stress and/or strain and to locate hazardous top are outlined in Table 2.

New developments in the area of roof-control equipment technology include:

1. *Pumpable Bolts*. Polyester resin, catalyst, and fiberglass roving are pumped into roof bolt-holes and allowed to harden. Advantages of the pumpable bolt include high anchorage strength and sealing of the bolt-hole. Operational advantages include the elimination of need for bolt bending in low coal.

2. *Mechanical Bolt Bender-Straightener-Inserter Device*—being developed for use in low seam heights.

3. *Flexible Drilling Device*. When developed, this will enable long bolt-holes to be drilled continuously in low seam heights.

4. *Automation and Remote Control of Roof Drilling and Bolting Equipment*. Considerable research is proceeding in this field.

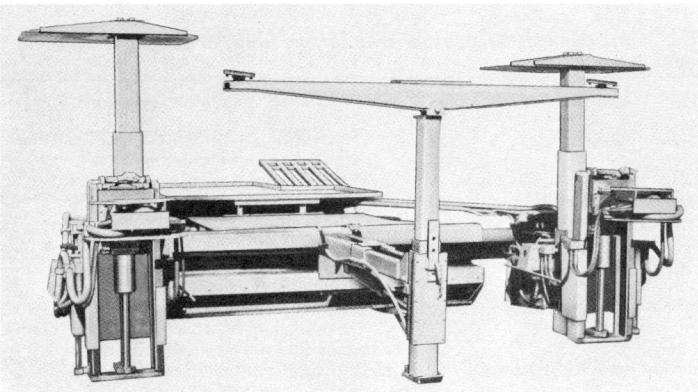

Fig. 16. Temporary roof-support system (TRS) protects personnel from exposure to the hazard of unsupported roof.

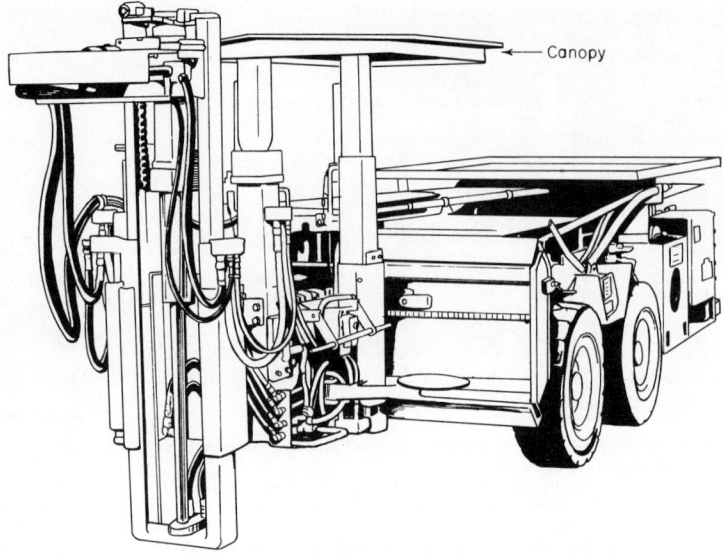

Fig. 17. Canopy for sheltering operators from falling roof. (J. H. Fletcher & Co.)

REFERENCES

1. Krynine, Dimitri P., and William R. Judd: "Principles of Engineering Geology and Geotechnics," McGraw-Hill, New York, 1957.
2. Panek, L. A.: Design for Bolting Stratified Roof, *Trans. Soc. Mining Eng.*, pp. 113–119, June 1964.
3. Woodruff, Seth D.: "Methods of Working Coal and Metal Mines," vol. 1, Pergamon, New York, 1966.
4. Federal Register, vol. 35, part II, U.S. Department of the Interior, Bureau of Mines, Washington, D.C., Aug. 14, 1970.
5. Given, I. E., (ed.): "SMF Mining Engineering Handbook," Society of Mining Engineers, New York, 1973.
6. Ground Control Aspects of Coal Mine Design, *U.S. Bur. Mines Inform. Circ.*, Washington, D.C., 1974.

Additional References and Bibliography
Wright, Fred D., and Jon E. Kelly: Investigations of Stresses near Cracks in a Bedded Mine Roof, *Proc. W. Va. Coal Mining Inst.*, pp. 1–9, 1969.

TABLE 2.   Some Systems Used to Indicate Roof Stress and/or Strain and to Locate Hazardous Top

System	Principle of operation
Infrared scanning	Senses temperature differential between unbroken and broken rock.
Strain gaging, including electrical, mechanical, and photoelastic methods	Provides direct or indirect measurement of: Strain at exposed rock surfaces. Borehole deformation. Convergence of opening perimeters. Pressures on material to be mined. Strain of supporting structures.
Indirect stress measurement	Seismic and sonic transmission and electrical resistivity characteristics of rock may be used as stress indicators.
Audible and subaudible vibration sensing	Vibrations originating with the rock indicate ground movement.
Load indicator installed between roof bolt head and plate	Monitors tension in roof bolts. Increased tension indicates roof flexure.

Beerbower, W. B.: Full-Length Resin Bonded Rock Bolts—A New Approach to Roof Support, *Proc. Ky. Mining Inst.*, pp. 61–68, 1971–1972.

Mitchell, D. W., and J. W. McCormick: Better Roof Bolting through Research, *Proc. Ky. Mining Inst.*, pp. 53–60, 1970–1971.

Kegel, W. G.: Roof Truss Installations, *Proc. Ky. Mining Inst.*, pp. 59–66, 1969–1970.

Schroder, John L.: Pillar Extraction with Continuous Miners in the High Splint Seam, *Proc. Ky. Mining Inst.*, pp. 71–85, 1966–1967.

Aughenbaugh, Nolan B.: Preliminary Findings in a Roof Shale Study in Coal Mines in the Illinois Basin, *Proc. Ill. Mining Inst.*, pp. 57–64, 1972.

Cassidy, Samuel M. (ed.): "Elements of Practical Coal Mining," Society of Mining Engineers, New York, 1973.

Merrill, R. H.: Design of Underground Mine Openings, Oil Shale Mine, Rifle, Colorado, *U.S. Bur. Mines Rep. Invest.* 5089, December 1954.

Evans, W. H.: The Strength of Undermined Strata, *Trans. Inst. Mining Met. Eng.*, pp. 475–500, 1940–1941.

Stefanko, R.: "Ground Control," lecture notes, Pennsylvania State University, 1973.

American Mining Congress Coal Division on Roof Action: Standard Roof Bolt Anchorage Testing Procedure, *Mining Cong.*, pp. 59–60, December 1959.

Barry, A. J., and J. A. McCormick: Evaluating Anchorage Testing Methods for Expansion Type Mine Roof Bolts, *U.S. Bur. Mines Rep. Invest.* 5649, 1960.

Panek, L. A.: Anchorage Characteristics of Roof Bolts, *Mining Cong. J.*, pp. 62–64, November 1957.

Anon.: Roof Control, *Coal Age*, pp. 214–219, July 1967.

Anon.: Pillaring Planbook, *Coal Age*, pp. 206–207, July 1967.

Anon.: Roof Control and Methane Monitors, *Coal Mining Process.*, pp. 50–51, August 1972.

Habberstand, J., C. Waide, and R. Simpson: The Pumpable Rock Bolt: A New Roof Control Concept, *Eng. Mining J.*, pp. 76–79, August 1973.

Chironis, Nicholas P. (ed.): New Infrared Scanner Helps Spot Hazardous Conditions in Mines, *Coal Age*, pp. 78–82, February 1972.

Anon.: Mining Follows the Quadrant Plan, *Coal Mining Process.*, pp. 40–43, June 1972.

Distler, William F., and Douglas E. Julin: Underground Mining—1972, *Mining Congr. J.*, pp. 26–31, February 1973.

Agapito, Jose F.: Rock Mechanics Applications to the Design of Oil Shale Pillars, *Mining Eng.*, pp. 20–25, May 1974.

Hills, E. Sherbon: "Elements of Structural Geology," Wiley, New York, 1963.

Ground Control Aspects of Coal Mine Design, *U.S. Bur. Mines Inform. Circ.*, 1974.

Stateham, Raymond M.: Field Studies on an Unsupported Roof, York Canyon Coal Mine, Raton, New Mexico, *U.S. Bur. Mines Rep. Invest.* 7886, 1974.

Coal Mining—Safety and Health

Although steady technological progress has taken place for several years toward making coal mining a less hazardous occupation, impetus was given to safety and health programs in the United States by passage of the Federal Coal Mine Health and Safety Act of 1969. The Act embraces not only promulgation and enforcement of regulations, but technical support, worker education and training, and a broad-based research program as well. The Act has been described as the most extensive and comprehensive industrial health and safety program ever to be undertaken to promote the welfare of a single class of industrial worker.

Generally, the allocation of Federal funds to these programs has reflected the much more severe problems of underground mining as contrasted with those of surface operations. The objectives of government-funded research since 1972 essentially have been representative of the major problem areas and have included:

1. Ground Control, for developing technology to prevent accidental falls of roof, rib, and face, and coal bumps. Study areas include (a) artificial support, (b) hazard detection, and (c) design of mine openings. Horizontal roof-strain indicators for detection of unstable roof conditions have been tested. Other research programs have included a microseismic fracture warning system, polymeric roof bolts, and chemical impregnation techniques. See article in this section of Handbook on *Roof Control for Underground Mining.*

2. Fire and Explosion Prevention. Study areas have included (a) ignition, (b) flame propagation, (c) fire detection and alarm, (d) suppression and extinguishment, and (e) methanometry. Devices and techniques tested have included explosion-proof bulkheads, coal-dust and rock-dust analyzers, ignition suppression devices for face equipment, and remote sealing techniques.

3. Industrial-Type Hazards, for identifying hazard sources in electrical, mechanical, illumination, and non-emergency communication fields. Developments have included (a) advanced remote surveillance and communications systems, (b) portable-area illumination systems, (c) trolley-phone wireless systems, and (d) protective canopies for use on underground low-coal machines.

4. Methane Control, for developing safe methods for mining methane-laden coal beds. Study areas have included (a) predictions of concentrations and flow, (b) control in advance of mining, and (c) control during mining. Techniques tested have included water infusion to reduce methane in the face area, degasification through vertical boreholes, the plugging of oil and gas wells which penetrate coal beds, and the complete degasification of operational mines.

5. Postdisaster Survival and Rescue, for developing emergency life support, communications, and rescue technology to improve the chances for a miner to survive a mine disaster. The development of technology for mine reopening operations also has been studied. Specific examples of improved techniques include (a) use of survey probes at the Blacksville and Sunshine mine disasters in 1972, (b) use of a wireless communication system for the rescue team at the Sunshine mine disaster in 1972, (c) the experimental employment of seismic methods for the location of trapped miners at the Nemacolin mine fire (1971) and the Blacksville mine fire (1972), and (d) use of electromagnetic techniques for location of trapped miners.

6. Respirable Dust, for providing improvements for protecting miners from exposure to respirable coal-mine dust. Study areas have included (a) dust formation, (b) dust control, and (c) dust measurement. Tests have included water infusion of coal beds for control of respirable dust, water-based high-expansion foaming systems in conjunction with continuous miners to reduce dust at the face, foam systems for dust suppression on conveyors and transfer points, and prototype dust meters. See subsequent section in this Handbook on *Monitoring of Hydrocarbon Emissions.*

7. Noise, with the assessment of permissible noise levels for communication and warning signals and the development of technology for noise abatement and control. Developments have included an audio dosimeter to replace conventional sound-level meters, discriminating earmuffs, and a noise-control muffler system to reduce pneumatic drill noise.

8. Industrial Hygiene, for developing instrumentation to detect and monitor toxic gases in underground mines; and to determine the requirements for safe operation of Diesel-powered equipment.

Other programs have included the development of advanced coal-mining systems and subsystems which are inherently safe and protect the health of miners. Specific areas investigated have included materials handling, hydraulic mining, remote control of mining equipment, and systems simulation studies.

ANALYSIS OF FATAL ACCIDENT REPORTS

In a federally funded study,* all accidents (except fires and explosions) reported to the U.S. Bureau of Mines for the years 1966 through 1971 were analyzed to determine which specific mining work functions had the highest number of deaths and injuries associated with them. Three sets of graphs summarize the findings. Figures 1 and 2 show the total number of fatal accidents for the six-year period by job classification of the principal victim. The analysis is further broken down to indicate in Figs. 3 and 4 roof-fall fatalities, and in Figs. 5 and 6 equipment-related fatal accidents. Figures 1, 3, and 5 rank the absolute (total) number of fatalities by job classification. The last year (1971) of the study is superimposed on each graph for comparison. Figures 2, 4, and 6 depict the same

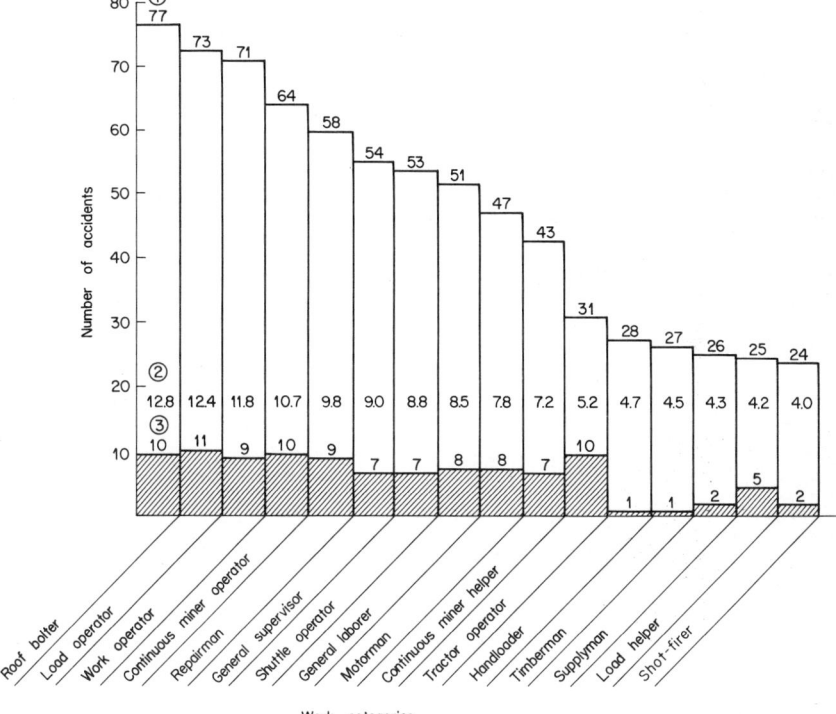

Fig. 1. Workers killed in underground mines (U.S.) during 1966–1971, classified by type of job. Of 850 total fatalities, 753 were analyzed in sample: (1) Total fatal accidents for 6-year period. (2) Average annual fatal accidents. (3) 1971 fatal accidents. *(Source: U.S. Bureau of Mines.)*

*Conducted by Theodore Barry and Associates. More details can be found in article by Kirk Prather and Kathy Bishop in *Coal Age*, vol. 78, no. 8, pp. 84–89, July 1973.

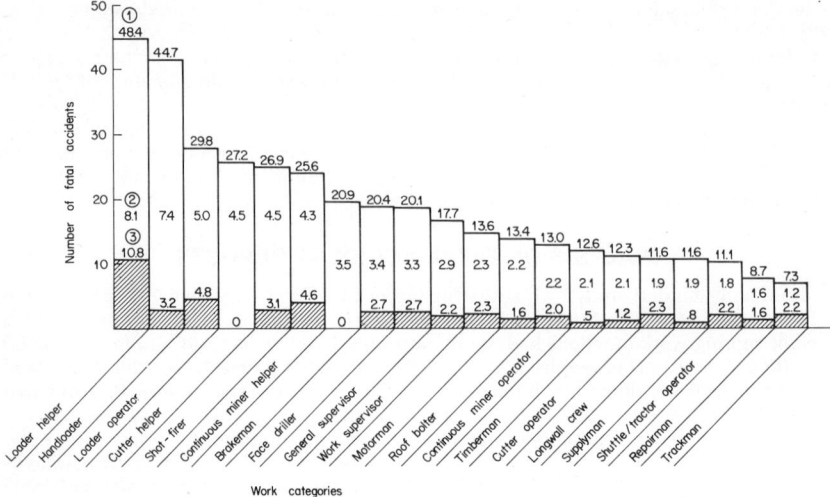

Fig. 2. Fatalities per thousand workers in underground mines (U.S.) during 1966–1971, classified by type of job: (1) Fatality rate for 6 years of fatalities. (2) Average annual fatality rate. (3) 1971 fatality rate. *(Source: U.S. Bureau of Mines.)*

information, normalized by the number of workers in each of the job categories.

In terms of absolute numbers, the roof bolter emerged as the number one fatal accident victim for the study period. During the period 77 roof bolters died, mostly in roof falls and related accidents. Loading-machine operators suffered 73 fatalities, followed by working supervisors (foremen) and continuous miner operators. These four classifications together accounted for 30 percent of all fatalities. A relatively different view of the danger associated with various job classifications emerges in Fig. 2, where the absolute number

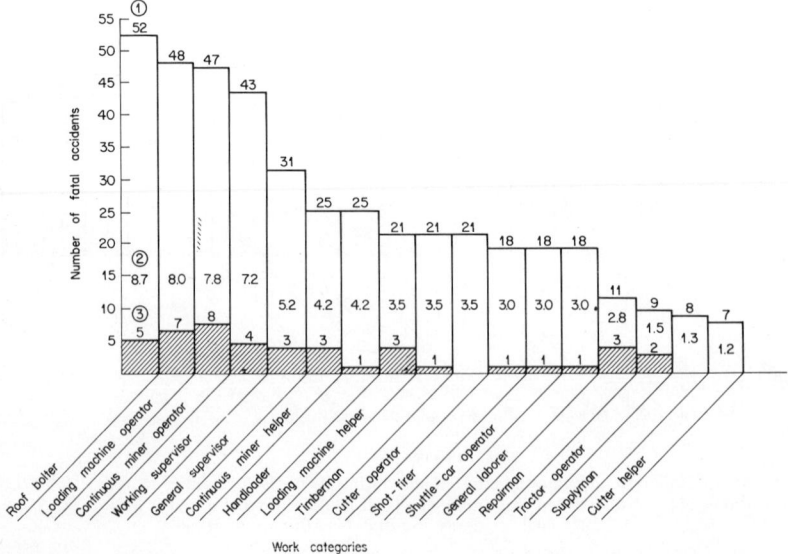

Fig. 3. Workers killed because of roof falls in underground mines (U.S.) during 1966–1971, classified by type of job: (1) Total fatal accidents for 6-year period. (2) Average annual fatal accidents. (3) 1971 fatal accidents. *(Source: U.S. Bureau of Mines.)*

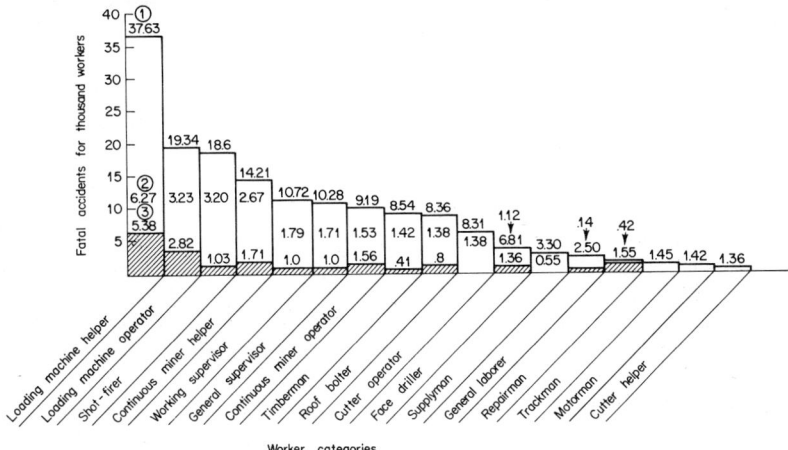

Fig. 4. Fatalities per thousand workers due to roof falls in underground mines (U.S.) during 1966–1971, classified by type of job: (1) Fatality rate for total fatalities in 6-year period. (2) Average annual fatality rate. (3) 1971 fatality rate. (*Source: U.S. Bureau of Mines.*)

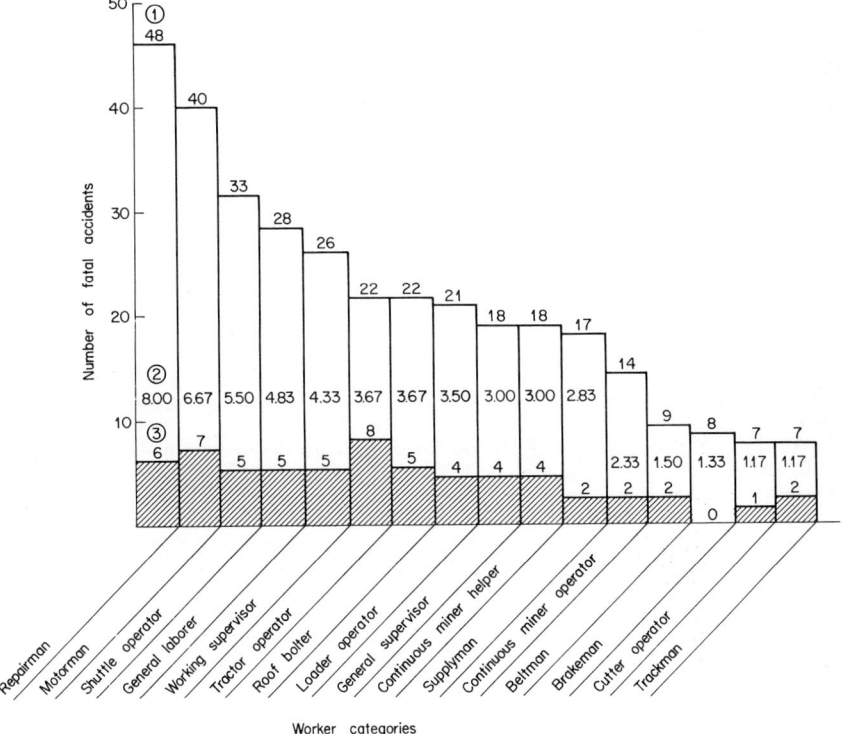

Fig. 5. Workers killed because of equipment-related accidents in underground mines (U.S.) during 1966–1971, classified by type of job: (1) Total for 6-year period. (2) Average annual fatal accidents. (*Source: U.S. Bureau of Mines.*)

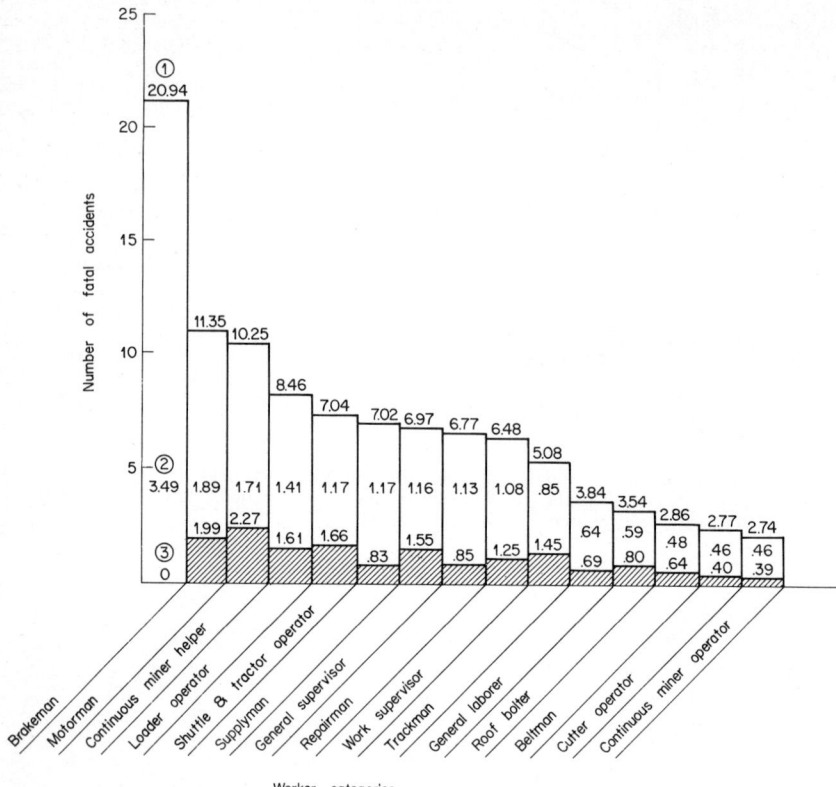

Fig. 6. Fatalities per thousand workers due to equipment-related accidents in underground mines (U.S.) during 1966–1971, classified by type of job: (1) Total for 6-year period. (2) Average annual fatality rate. *(Source: U.S. Bureau of Mines.)*

of fatal accidents is compared with the number of workers in each category. The rate per 1000 workers gives a better indication of likelihood of a particular job category being involved in a fatal accident. From this comparison, the classifications with the highest rate of fatal accidents were the loader helper and the handloader. Although both these classifications had only one-third of the absolute number of fatalities of the roof bolter, the classifications appear more dangerous. Notably, in this comparison, the roof bolter ranks twelfth in terms of fatality frequency per miner.

As indicated by Figs. 3 and 4, the job categories most frequently involved in fatal roof falls are all face workers. About one-half of the fatal roof-fall accidents occurred under unsupported roof. It is noteworthy, however, that three of the four highest classifications are all workers who technically should have less exposure to unsupported roof than roof bolters. Roof bolters have occasion to be under unsupported roof if they set temporary support. Nevertheless, the actual operations of supervision, machine loading, and continuous mining theoretically should take place under supported roof at all times. This indicates the requirement for continuing studies in minute detail of accident causes.

Fatalities due to equipment-related accidents are shown in Fig. 5. Although roof-fall fatal accidents have been declining since 1966, equipment-related accidents have been increasing. In 1971 the total number of equipment-related fatal accidents exceeded the number of roof-fall fatal accidents. Although the repairman and motorman job categories contributed the greatest absolute number of fatalities, the brakeman appeared to have the highest probability of being involved in a fatal equipment-related accident. The fatality rate of the brakeman was almost twice that of the next category, motorman. However, no brakeman accidents in 1971 represented an encouraging figure. The 1971 data indicate what investigators consider a general trend of increasing equipment-related accidents. The jobs of continuous miner helper, loading-machine operator, tractor and shuttle-car operator, general and working supervisor, roof bolter, and trackman all showed higher 1971 accident rates than the estimated rates for the six-year period of the study.

The analysis of fatalities alone can present an incomplete and potentially misleading picture of where emphasis should be placed in a safety program. A review of both fatal and nonfatal accidents is essential.

The nonfatal accident data for 1971 are given in Fig. 7. The data are normalized by the number of workers in each job classification. Roof bolters have the highest number of fatalities and a relatively low fatality frequency per 1000 workers, but they have the highest nonfatal injury frequency per 1000 workers. In 1971 the ratio of nonfatal to fatal accidents was 62.5 to 1. The handloader has one of the highest fatality frequencies, but the lowest injury frequency. The ratio for the handloader was 1.6 to 1. In summary, although the roof bolter may have the highest probability of being involved in some kind of accident resulting in death or a disabling injury, his likelihood of being killed in an accident is much less than that of the handloader.

Inexperience as an Accident Factor

Studies have indicated that a coal miner with less than one year of underground experience is ten times more likely to have an injury accident than a miner with five years of experience. Further, a miner on a new type of job is five times more likely to have a fatal accident during the first six months than during the second six months on the job. The importance of experience and thorough training is illustrated dramatically by mine accident statistics.

Most workers in the underground coal mines traditionally enter mining at the unskilled or laborer level and subsequently transfer to more skilled job classifications. Although some states have coal-mine apprenticeship requirements, they frequently do not provide for job skill development as compared with apprenticeship programs followed in several other industries and trades. Frequently, a miner will practice new skills on a casual basis before actual job assignment. He may perform a new task by substituting for an absent worker, or obtain a new assignment on the basis of seniority bidding. Thus, much training occurs on the job while under the pressure for getting out production. In some cases the miner may receive initial instruction and close supervision during the first few hours or

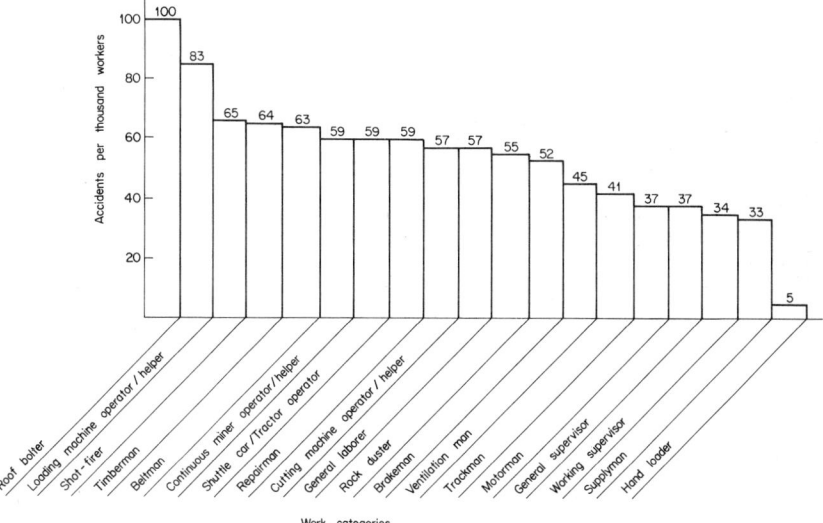

Fig. 7. Nonfatal accidents in underground mines (U.S.) during 1971 per thousand workers, classified by type of job. *(Source: U.S. Bureau of Mines.)*

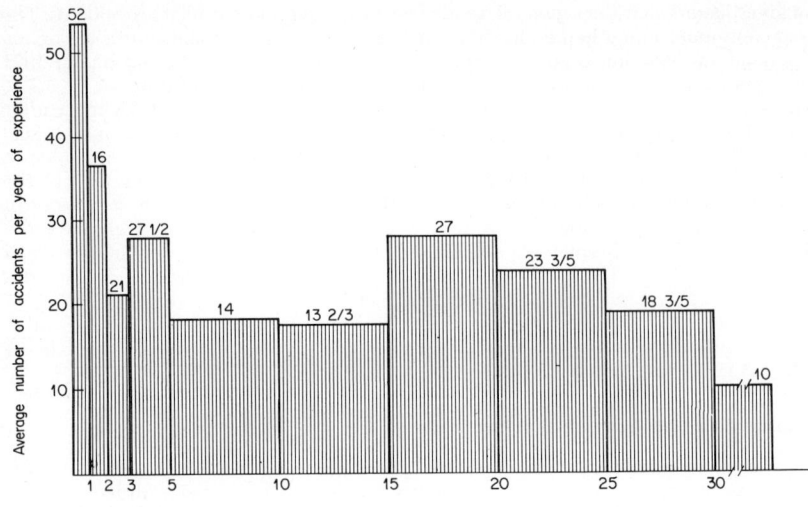

Fig. 8. Experience profile of 800 fatal accident victims in underground mines (U.S.) during 1966–1971. (*Source: U.S. Bureau of Mines.*)

days on an assignment; in other cases he may be left to cope with unusual problems or conditions prematurely.

In connection with the six-year study previously mentioned, Fig. 8 graphically shows the experience profile of eight hundred fatal-accident victims. The chart reveals that workers are killed with greater frequency in their first two years of underground exposure. In this study the higher frequencies could have resulted from a larger worker population in the higher groups. Therefore, the data shown in Fig. 9 were adjusted to relate the underground experience of the victims to the relative underground experience of the working population. With relatively minor changes, the data are similar. The importance of underground experience in connection with nonfatal accidents is illustrated in Fig. 10. Job mobility rather than job inexperience has been pointed out as a contributing factor to accidents, but detailed studies to date indicate that the job-inexperience factor cannot be explained by high job mobility.

Another variation of the low job-inexperience factor in accidents is evident from Fig. 11, which tabulates the percentage of substitute operators in equipment-related fatal accidents that had operator facilities. The percentages of 51 and 47 for shuttle-car and tractor

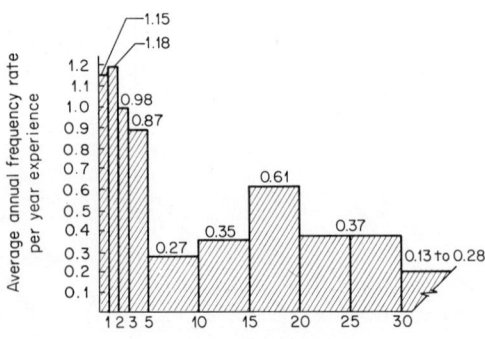

Fig. 9. Fatal accident frequency rate per thousand workers per year in underground mines (U.S.) for period 1966–1971. (*Source: U.S. Bureau of Mines.*)

operators is strikingly high. Doubtless, many substitute equipment operators are qualified, but these percentages indicate that a serious safety qualification problem exists for substitute equipment operators.

Upon inspection of the foregoing data, a mine manager may be inclined to advocate elimination of job posting or seniority bidding on jobs in order to cut down job mobility in underground mines. The number of accidents might be decreased by increasing the average job tenure. However, this does not solve the fundamental problem, that inevita-

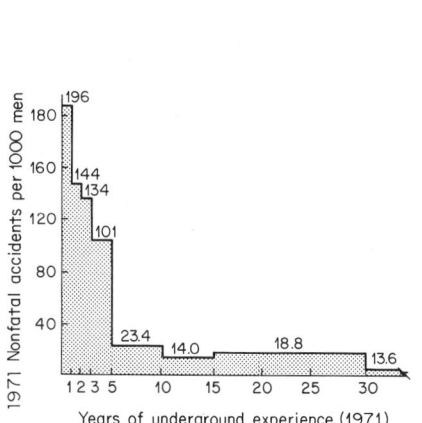

Fig. 10. Frequency rate distribution of nonfatal accidents per thousand in underground mines (U.S.) in 1971 by victim's total underground experience. Note: 795 accidents adjusted for 7979 accidents in 1971; 13,665 workers adjusted for 70, 000 workers. (*Source: U.S. Bureau of Mines.*)

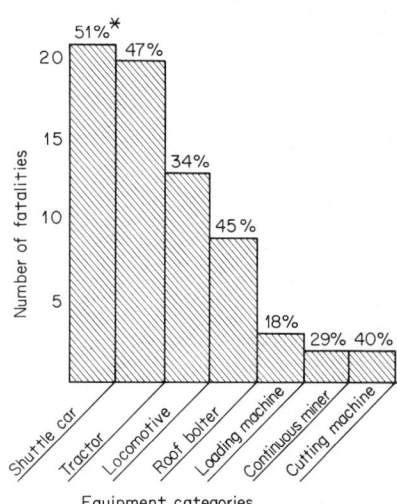

Fig. 11. Equipment-related fatal accidents in underground mines (U.S.) during 1966–1971. Note: 70 substitute operators out of 175 operator fatalities.

*Percentage of operator fatalities in which the operator was not the regular operator. (*Source: U.S. Bureau of Mines.*)

bly new men must be assigned to new jobs as men retire or otherwise leave the mine. And the promotion to higher skills is the inevitable goal of the better-than-average worker.

Some British and German coal-mining safety officials have observed that three programs were necessary to reduce the job-inexperience factor in underground coal-mining accidents—job training, job certification, and close supervision programs. There has been some reluctance on the part of United States mine operators to accept such extensive work training and certification, apparently out of fear that the productivity of U.S. mines might be lowered to the productivity figures of European mines. A comparison of the industries, however, indicates that these fears may not be warranted. Thinner seams, greater depth of coal requiring extensive ground support, and the exclusive use of longwall mining methods, not always fully mechanized, make European versus American productivity comparisons difficult. Conversely, there is evidence that the European training and certification practices have increased worker output and aided in the reduction of accidents.

British law requires that all underground personnel be properly trained. Although the National Coal Board hires school leavers as early as age 15, no person is permitted to work in a coal-getting operation until age 18. A youngster gets a minimum of 100 days of basic training, both classroom and practical, before being allowed to work underground. The law specifies that the individual must continue training under close personal supervision for at least 20 days. Under this program, one skilled worker or training officer supervises the training of one trainee. Additionally, at least 40 days of close personal supervision is provided for underground training in specific coal-getting operations.

An official of the National Coal Board ranks four elements in order of their permanent effectiveness in preventing mining accidents: (1) design, including not only machine design, but designs of entire mining systems which can eliminate a particular hazard; (2) protective devices for use where a hazard cannot be designed out of a system; (3) training, including refresher courses; and (4) exhortation to motivate the workers—mobile television campaigns, posters, and special warnings against special hazards. Motivation toward safety is short-lived and must be waged on a continuing basis.

The death rate in British mines in 1947 stood at 0.34 per 100,000 manshifts. (Multiply by 1.25 to convert the British 100,000 manshift figures to million man-hours.) Under

TABLE 1. Major Coal Mine Disasters in the United States*

Year	Location	Fatalities	Year	Location	Fatalities
1869	Plymouth, Pa.	110	1924	Castle Gate, Utah	171
1884	Pocahontas, Va.	112	1924	Benwood, W. Va.	119
1891	Mammoth, Pa.	109	1928	Mather, Pa.	195
1892	Krebs, Okla.	100	1930	Millfield, Ohio	82
1900	Scofield, Utah	200	1932	Moweaqua, Ill.	54
1902	Coal Creek, Tenn.	184	1940	Bartley, W. Va.	91
1902	Johnstown, Pa.	112	1940	Portage, Pa.	63
1903	Hanna, Wyo.	169	1943	Red Lodge, Mont.	74
1904	Cheswick, Pa.	179	1947	Centralia, Ill.	111
1905	Virginia City, Ala.	112	1951	West Frankfort, Ill.	119
1907	Monongah, W. Va.	361	1957	Bishop, Va.	37
1907	Jacobs Creek, Pa.	239	1958	Bishop, Va.	22
1908	Marianna, Pa.	154	1961	Terre Haute, Ind.	22
1909	Cherry, Ill.	259	1962	Carmichaels, Pa.	37
1911	Littleton, Ala.	128	1963	Dola, W. Va.	22
1913	Dawson, N. Mex.	263	1965	Redstone, Colo.	9
1914	Eccles, W. Va.	181	1966	Mt. Hope, W. Va.	7
1915	Layland, W. Va.	112	1968	Greenville, Ky.	9
1917	Hastings, Colo.	121	1968	Farmington, W. Va.	78
1922	Spangler, Pa.	77	1970	Hyden, Ky.	38
1923	Dawson, N. Mex.	120	1976	Oven Fork, Ky.	26

*SOURCE: U.S. Bureau of Mines.

government surveillance, the rate dropped to 0.25 fatalities in 1955 and has been below 0.20 since 1966 (equivalent to 0.25 million man-hours). Total fatalities in British mines dropped below 100 for the first time in 1970 when 91 miners were killed, 82 of them in underground accidents, bringing the death rate of 0.14 per 100,000 manshifts. In 1971, there were 72 mining deaths, including 66 underground, for a rate of 0.11. This is about one-fifth as high as the rate in the United States.*

Major coal-mine disasters in the United States are cited in Table 1. The worst recorded coal-mine disaster occurred in the Honkeiko Colliery in Manchuria on April 26, 1942, when about 1550 miners were killed.

COAL WORKERS' PNEUMOCONIOSIS

The Occupational Safety and Health Act of 1970 created in the United States the National Institute for Occupational Safety and Health. Among the objectives of this institute are (1) identification of the mechanisms involved in the development of coal workers' pneumoconiosis, and (2) determination of more effective methods for early detection of the disease. The Appalachian Laboratory for Occupational Respiratory Diseases, located in Morgantown, West Virginia, is conducting major studies of this disease, also known as "black lung." Best known of these activities are those dealing with autopsies and medical examinations of coal miners. Less generally well known are programs of coal-mine health research involving several fields of medicine and the physical sciences. Under the Occupational Safety and Health Act of 1970, miners are afforded an opportunity to have

*Further information on British mine-safety practices will be found in Coal Age, vol. 78, no. 8, pp. 106–109, July 1973.

TABLE 2.   Incidence of Coalworkers' Pneumoconiosis

State	Number examined	Percent of examinations indicating disease	
		Simple pneumoconiosis	Complicated pneumoconiosis†
Pa.‡	984	26.5	5.7
Md.	39	19.0	0
Ark.	593	18.5	3.7
N. Mex.	30	16.6	0
Pa.§	12,769	14.1	1.6
W. Va.	20,004	11.8	1.3
Tenn.	266	11.7	0
Ind.	141	11.4	0
Ala.	830	10.6	0.7
Ill.	3,517	10.5	0.7
Va.	4,466	8.6	1.0
Ky.	10,185	8.6	0.5
Colo.	722	6.9	0.8
Wyo.	33	6.0	0
Ohio	2,029	5.4	0.5
Okla.	20	5.0	0
Utah	593	4.0	0.4
Iowa	43	2.3	0
Mont.	14	0	0
Wash.	19	0	0

*SOURCE: National Institute for Occupational Safety and Health (U.S.).
†Complicated pneumoconiosis involves the development of fibrotic tissue in the lungs.
‡Anthracite.
§Bituminous.

chest X-rays and such other tests deemed necessary to research and protection. X-rays and tests may be made at intervals. New miners are required to be given a first examination no later than six months after commencing work, and a second examination three years later. If there is evidence of the development of coal workers' pneumoconiosis, another examination must be scheduled within two years. The examinations are made at no cost to the miner. The aforementioned Act also provides for autopsies of miners and former miners, made mainly to assist widows or next of kin in determining eligibility for black lung benefits, and to further medical research. In all cases to be autopsied, approval of the miner's widow or surviving next of kin is required. In 1972, 313 autopsies were submitted by pathologists. A Summary of the program as of September 1972 is given in Table 2.

Extensive research is being conducted in the design and application of dust monitors, respirators, and masks. The institute also is conducting studies of the psychological aspects of coal mining which influence coal miner health.

Contour Mining*

Appalachian contour mining has been seriously criticized in recent years, but as of the mid-1970s it continues to provide about one-fifth of all coal mined in the United States.

The contour-mining method, which is practiced in seven states stretching from Ohio and Pennsylvania into Alabama, has changed in recent years as miners attempt to maintain production while complying with more stringent reclamation laws.

CONVENTIONAL METHOD

Conventional contour mining used primarily large crawler tractors and rubber-tired loaders for overburden removal. See Fig. 1. The miner dozed material over the hill or stacked it along the outcrop and blended it into the outslope. The resulting spoil bank was sometimes unstable, and erosion and landslides sometimes occurred. In this method, many

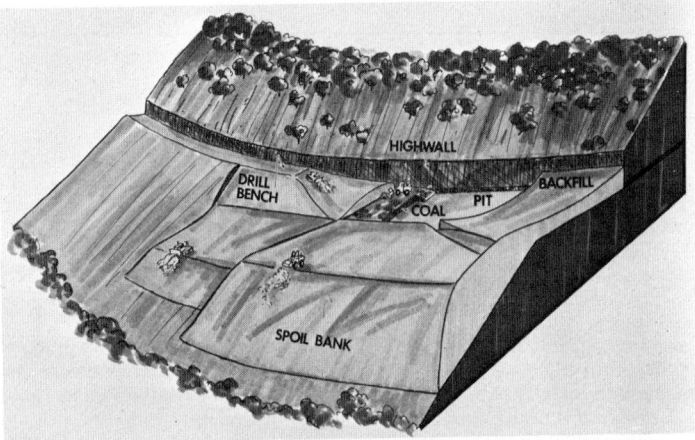

Fig. 1. Conventional method of contour surface mining. This was the predominant form of mining throughout Appalachia until the early 1970s when more stringent reclamation laws ushered in new integrated mining techniques. *(Caterpillar Tractor Co.)*

more acres were affected other than those disturbed by the original cut and overburden placement. Since this method moved overburden the shortest possible distance, however, its cost for overburden removal was low.

The reclamation phase generally included backfilling some material into the mined-out area to cover acidic material, the exposed coal bed, and any holes left from augering operations. The miner reshaped the area to control drainage and erosion and then seeded and planted the affected area.

NEW MINING METHODS

New methods in Appalachia correct many problems of the conventional method by consolidating extraction, backfilling, and reclamation operations into a continuous, integrated mining system. The miner should attempt to proceed simultaneously with all these operations in order to avoid second-handling of the overburden, which is very costly.

*Mining Section, Caterpillar Tractor Co., Peoria, Illinois.

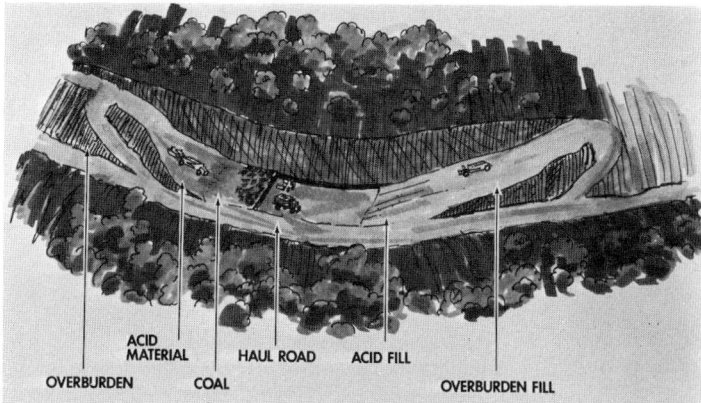

Fig. 2. Scraper haulback method. All spoil except that from the initial cut is moved along the bench instead of being placed on the outslope. *(Caterpillar Tractor Co.)*

Earthmoving requirements for the newer methods include burying all acidic and toxic materials, separating and replacing topsoil, and restructuring the land approximately to its original contour.

Some of the most promising of the new contour mining methods which are designed to meet all environmental regulations include (1) the *haulback,* (2) the *valley fill,* and (3) the *mountaintop-leveling* methods. Although they are significantly different in many aspects, all methods move material laterally along the bench instead of dumping it over the outslope; and all are designed to reduce or eliminate second-handling of overburden and waste material.

HAULBACK METHOD

The major operating principle of the haulback method is that all spoil except that from the initial cut is moved laterally along the bench instead of being placed on the outslope.

There are two basic options with the haulback method: (1) using *scrapers,* as shown in Fig. 2, or (2) using *trucks,* as shown in Fig. 3. No matter which method is used, nonacidic rock or clay overburden is placed over the acidic material so that trees, legumes, and other

Fig. 3. Truck haulback method. Used primarily in rocky overburden, which is uneconomical for loading by scrapers. With either truck or scraper haulback operations, nonacidic rock or clay overburden is placed over acidic material to facilitate revegetation. *(Caterpillar Tractor Co.)*

vegetation will survive when the miner completes the reclamation phase, and to prevent acidic contamination of runoff and drainage.

Overburden characteristics and the terrain must both be favorable to permit a scraper haulback operation. The scrapers load overburden from the advancing face of the pit and dump it at the backfill area. Ideally, scrapers should load and dump downhill for faster cycle times and less wear on the machines. When they reach acidic material overlying the coal, the scrapers load it out and spread it near the bottom of the pit so that it can then be covered with nonacidic material. As the overburden dump reaches the original level of the high wall, crawlers dress the slope and restore the area to its natural contour.

The truck haulback method utilizes crawler tractors, wheel loaders, and off-highway trucks. Crawlers doze overburden off a top bench down to the loader, which dumps it into the off-highway trucks. As with scrapers, acidic material is dumped at the base of the backfill area so that it can be covered with nonacidic material. Another crawler also works the overburden area, dressing and contouring the reclaimed area.

Some miners combine scrapers and off-highway trucks in the same haulback operation. When this is done, scrapers remove the easy-to-load surface material while trucks come in to haul away the underlying rock.

In mines where no augering is to be done and where the haul distance is kept to a minimum, wheel loaders can work a load-and-carry operation either by themselves or in conjunction with scrapers or trucks.

A variation of the haulback method, called the "block cut," is being practiced by some miners in the Appalachian area. Though basically the same in principle, it works best with cuts of only about 200 feet in length. After the initial block cut is made, the miner spoils each successive block into the previously mined block. Miners can successfully use wheel loaders and crawler tractors with this method since the blocks are kept fairly small.

VALLEY FILL

Also referred to as the "head of the hollow" method, the valley-fill method is relatively new to the Appalachian region. With it, the miner generally uses an equipment spread similar to that in a truck haulback operation. A crawler dozes overburden to the wheel loader which loads it into off-highway trucks, as shown in Fig. 4. The trucks haul the overburden along the bench to a selected hollow or valley and dump it over the side, building a waste dump at bench height. Scrapers can also be used to haul the overburden in a valley-fill operation. When such is the case, the scraper operator ejects overburden at bench height, and tractors doze it over the edge.

Fig. 4. Valley-fill method. The miner generally hauls overburden in trucks and constructs fills over the side at bench height as shown. When scrapers are used, the operator ejects overburden at bench height and tractors doze it over the edge. *(Caterpillar Tractor Co.)*

Valley fill is frequently ·combined with the haulback method, since there is normally extra material from the initial cut as well as continuous excess from material swell which cannot be stacked directly on the bench. Special precautions and construction to control erosion and drainage are necessary.

MOUNTAINTOP LEVELING

The mountaintop leveling method is quite similar to area stripping as practiced in the Midwest. As shown in Fig. 5, a part of the mountaintop is leveled to fill in an adjacent valley. The remainder is replaced on the mountaintop as mining progresses. This method can be employed only where topography, coal-seam position, and other physical and economical factors permit. It uses similar techniques and equipment to those of the haulback and valley-fill methods. Larger trucks and loaders, however, are adaptable.

Once the coal is loaded out, a level to gently rolling terrain results. The reclaimed flatland, which can be more valuable than in its original contour, can then be used for grazing as well as for institutional, residential, and industrial developments.

AUGERING

Augering is basically a horizontal drilling process in which auger sections are added to a drill head. Production per man-hour is very high, and the mining cost is generally very low. However, augering often only recovers about 30 percent of all the coal because some is left between, above, and below the holes.

Because augering requires stable high walls, level seams, and stable coal to support the, overburden, it accounts for only about 5 percent of all coal produced by surface mining.

Fig. 5. Mountaintop leveling. A part of the mountaintop is moved to fill up an adjacent valley and build a near-level fill in the mined area. The reclaimed flat or gently rolling land that results from this type of operation then can be used for a variety of purposes. Thus, the land value can increase. (*Caterpillar Tractor Co.*)

COAL LOADING

Wheel loaders of up to about 7 cubic yards capacity are the primary tool for coal loading in Appalachian contour mines. Besides being highly mobile, wheel loaders have a further advantage in that they do not severely disturb the coal surface when traveling.

Coal haulage is a significant expense in contour mining since the coal frequently must be hauled for distances ranging from 2 up to 25 miles each way. Also, since nearly all coal hauling requires some on-highway travel, tandem-axle on-highway dump trucks are the most frequent vehicles used in Appalachia.

OVERBURDEN HAULING UNITS

Miners who convert their operations to some of these newer methods are making greater use of wheel loaders in load-and-carry operations and are also using two machines, seldom seen previously in contour mining—wheel tractor-scrapers and off-highway trucks. Whether the miner chooses loaders, scrapers, or trucks as the prime overburden hauling unit depends on a multitude of factors.

LOAD AND CARRY

Introduction of the four- to six-yard loaders in the early 1960s brought about the first widespread use of these units in contour mining. In the mid-1970s, with the exception of crawler tractors, they are probably the most common machine in Appalachian surface mining.

Wheel loaders working a load-and-carry stripping operation with short cycle times have the obvious advantage of being the most economical of the three types of hauling units since one machine does it all. See Fig. 6. The loader is highly mobile, capable of negotiating most grades; and it can handle a variety of material from topsoil to shot rock.

Fig. 6. Wheel loaders working a load-and-carry stripping operation with short cycle times have the advantage of being the most economical of all hauling units. (*Caterpillar Tractor Co.*)

Loaders do, however, have some serious drawbacks in many contour mining situations. For example, they are limited to the ton-mile-per-hour rating of their tires, and they have a limited top speed.

The chief limiting factor for loaders is that the one-way haul distance must be kept relatively short. This distance generally should be 700 feet or less. Over that distance, scrapers or trucks become more economical and practical for hauling overburden.

WHEEL-TRACTOR SCRAPERS

Overburden composition is the key factor in determining whether the stripping fleet consists of scrapers or of loaders with off-highway trucks. Generally, scrapers are the most economical in dirt; and trucks, in rock.

Scrapers can be divided into four basic configurations: single-engine, tandem-powered, elevating, and push-pull. The most productive and economical configuration depends on three main factors: production requirements, haul conditions, and overburden characteristics.

PRODUCTION REQUIREMENTS

Overburden stripping requirements will usually determine the optimum scraper capacity. Generally, larger units offer a lower cost per yard, but their physical size limits them to mines where there is enough room for loading and maneuvering. Smaller scrapers, though

Fig. 7. Scrapers become more economical as cycle times extend, provided that the overburden can be economically handled by such units. *(Caterpillar Tractor Co.)*

they may have a higher theoretical cost per yard, would appear to be the choice where space is a factor and production requirements are not high.

HAUL CONDITIONS

The length, grade, and rolling resistance of the haul road greatly affect the choice of scraper configuration. Each of these greatly influences the cycle time and cost per yard of moving overburden.

The major impact of haul conditions is on single-engine scrapers. See Fig. 7. Depending on the particular model and the underfoot conditions, a single-engine unit can generally negotiate a 25 percent grade when empty or up to about 15 percent when loaded. By contrast, twin-engine scrapers are the most economical choice for steep grades like those in much of the contour mining region and for the changing haul-road conditions typical of contour mining.

Economic work ranges of scraper configurations in differing types of overburden are shown in Fig. 8. Conventional scrapers can handle material up to and including ripped rock, and when equipped with a special application bowl for severe loading conditions, they can load material such as well-shot rock. In poor underfoot conditions, though, a conventional scraper will have less traction, and this increases its boost time and total cycle time.

An elevating scraper is naturally limited by the capability of its elevator. It can work in overburden up to gravel on the scale, but beyond that it becomes marginal because of high maintenance costs on the elevating mechanism and on tires. Costs are generally not competitive when haul distances exceed 2000 feet.

Tandem-powered four-wheel drive scrapers can economically load about the same range of overburden as conventional machines. Twin-engine machines can usually work more days in a year than single-engine units, however, because of their ability to work in wet weather and poor underfoot conditions which are harder on two-wheel-drive machines. Extra performance is necessary to compensate for higher servicing and operating costs.

A push-pull scraper consists of a pair of tandem-powered machines working as a unit. The second machine push-loads the first, which when loaded, pulls the second scraper through its loading cycle. When both are full, they separate. When operating in the right conditions, the push-pull configuration has some notable advantages. It eliminates prob-

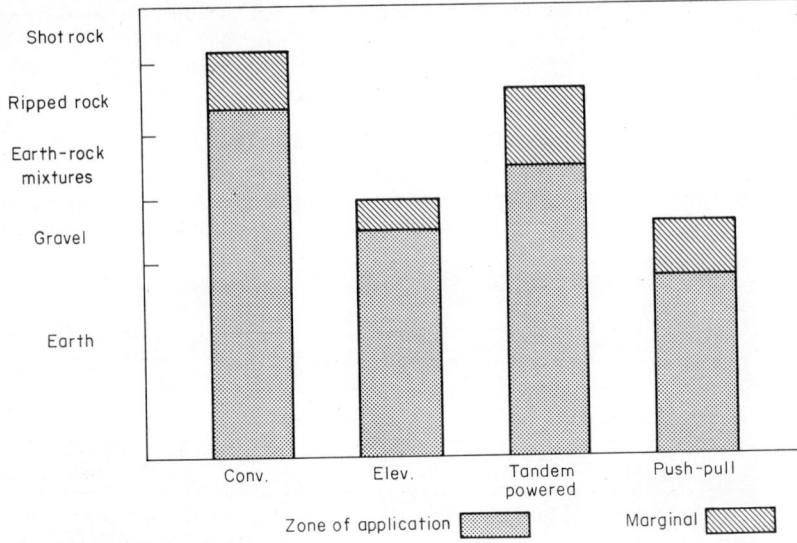

Fig. 8. Relative economic work ranges and marginal ranges for four scraper configurations based on the overburden characteristics.

lems of bunching, pusher-scraper mismatch, and pusher breakdown. But it requires a cut area twice as long as conventional scrapers, and tire wear can be costly in tough underfoot conditions.

Push-pulls also have the most limited economic overburden loading range. This is somewhat misleading, however, since these units can work individually with a pusher just like a standard tandem-engine scraper.

The effect of mine conditions on tires is always a factor in scraper selection. Internal heat, water, and sharp rocks are the chief enemies of tires. Tire life is usually lower on tandem-powered scrapers because of a tendency for more slippage. This is due to their lower weight-to-horsepower ratio and all-wheel drive, which sometimes lead operators to spin the tires more. This wear can be reduced, however, by having good loading conditions and a well-maintained haul road.

OFF-HIGHWAY TRUCKS

Trucks are the most suitable hauling units when the haul distance is too long for wheel loaders, and scrapers cannot economically load the overburden. See Fig. 9.

Fig. 9. The more rock present in the overburden and the longer the cycle time, the more economical trucks become as the optimum hauling unit. *(Caterpillar Tractor Co.)*

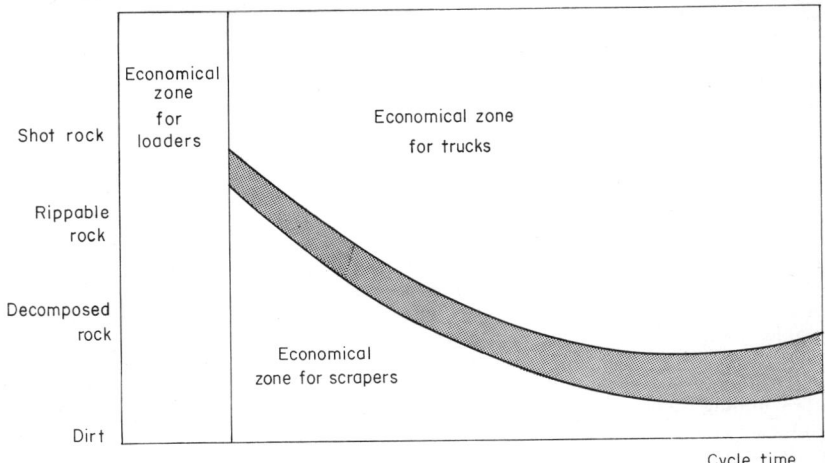

Fig. 10. Relative economical zones for trucks, loaders, and scrapers. Loaders prevail where short cycles are possible. As cycle time grows longer, scrapers and trucks are more economical. The quantity of rock in the overburden determines which of the latter two vehicles is the more economical.

Off-highway trucks range in capacity from about 15 tons up to giant 200-ton units. Among the key factors in selecting trucks for these newer contour-mining methods, the most obvious are space limitations and loader matching. Overburden loading, hauling, and dumping take place simultaneously in many of these methods, and the small operating area rules out the use of large trucks.

The haul road is another factor. Since haul roads for overburden have a relatively short life, miners cannot spend much money and time to construct permanent, quality roads. So the width and bearing capacity of the haul road will limit the gross vehicle weight and physical size of haul trucks.

Miners must also consider matching truck size with loaders for the lowest cycle times. It is imperative that a proper match of trucks with high output loaders exist for optimum productivity. Most contour miners are already using loaders in the 6- to 10-cubic-yard class; so 35- to 50-ton trucks are the most practical match for these loaders.

Another consideration in truck selection is to guard against congestion. Too many small trucks are inefficient because they must slow down whenever congestion develops anywhere on the mining site.

Another factor is a truck's braking capacity. Many Appalachian contour mines have steep grades to negotiate; so trucks with adequate braking and retarding systems can reduce downtime under such operating conditions.

HAULING UNIT ECONOMICS

The relative economic zones for each of the three types of hauling unit are summarized in Fig. 10. Loaders are the obvious choice where short cycle times are possible. As cycle times extend, scrapers become more economical so long as the overburden can be economically handled by such units. The more rock present in the overburden and the longer the cycle time, the more economical trucks become as the optimum hauling units.

Seeding and Planting to Achieve Land-Management Objectives

BY WILLIAM T. PLASS*

Extraction of coal by surface mining has disturbed about 2 million acres of land in 26 states. About half of this disturbance has occurred since 1965. Ninety percent is on land owned by mining companies, farmers, and other private interests. Most of the mining has been done under state laws and regulations.

Acreage disturbed by surface mining will increase at an accelerating rate because of greater dependence on coal to meet expanding energy demands. Land disturbed by surface mining is generally an undeveloped land resource. Evidence from many sources indicates that all but a small percentage of it is capable of producing some tangible or intangible benefit to society. This resource could contribute substantially to the regional economy if it were intensively developed and managed.

Mined sites, appropriately reshaped, can be developed for many purposes. In the midwestern and Appalachian coalfields, they have been used for successful and profitable production of agricultural and horticultural crops. Recreational and wildlife developments are popular. Industrial and residential sites have been established on areas disturbed by surface mining. Sites planted with trees yield various forest products. However, many of these are demonstration areas. A few progressive mining companies have identified opportunities and have developed areas that provide social and economic benefits.

Many landowners and mine operators are apathetic toward intensive land-management practices. Coal production and profits dominate their planning and development decisions. Many managers are unfamiliar with land-management techniques, and do not employ specialists to develop their properties. This may be attributed to the comparatively low profits from many land uses, and the relatively long time before tangible and intangible benefits accrue. Nevertheless, management should be encouraged to seek out opportunities for profitable land use and to develop them to their fullest potential.

Land-management options may be increased if consideration is given to potential spoil characteristics and landforms before mining begins. Basic to all uses are the physical and chemical characteristics of the spoil. Proper placement of spoil and overburden can minimize environmental problems and maximize land-management options. This cannot be overemphasized, because much of the expense of reclamation results from improper placement of overburden.

The expanding science of overburden analysis will provide guidelines for developing practical methods for movement and placement of overburden. These guidelines will enable managers to identify rock strata that provide the best medium for plant growth, material that carries a low risk of off-site pollution, or rock suitable for stabilizing fills and drainages. The objective will be to balance the increased costs of mining, reshaping, and vegetation against tangible or intangible benefits from land restoration, pollution abatement, and potential land use.

Many land-management objectives require the establishment of vegetation after mining and reshaping. The initial vegetative cover begins to restore the productive capacity of the disturbed land. It may also protect the site against erosion, the initial objective of most vegetation treatments.

Development of mined areas for maximum tangible and intangible benefits depends on the experience and training of those responsible for it. Selection of an appropriate land-management objective requires adequate site evaluation. Once an objective has been selected, planning for compatible vegetation proceeds. This includes: (1) soil preparation,

*Principal Plant Ecologist, Northeastern Forest Experiment Station, Forest Service, U.S. Department of Agriculture, Princeton, West Virginia.

(2) selection of species, (3) efficient seeding and planting, and (4) cultural treatments to maintain the vegetative cover or maximize yield.

SITE EVALUATION

The land-management options for any given area depend on the on-site and off-site evaluations. On-site factors are chemical and physical characteristics of the spoil, physical dimensions of the spoil area, configurations created by mining and reshaping, and local climate. Off-site factors are those features that affect the use of the area. These may include natural topographic features, geographical location, associated natural vegetation, or political, economic, and social considerations.

EVALUATION OF THE SPOIL

It is relatively simple to relate the physical dimensions and configuration of the disturbed area to land-use options. Evaluations of the chemical and physical properties of the spoil are based on field or laboratory analyses and an understanding of the relationship between spoil characteristics and revegetation or pollution control. Chemical factors include toxic ion concentrations and availability of nutrients. The important physical features are coarse fragments (particles larger than 2 millimeters), texture (particles 2 millimeters or less in size), and color.

Spoil evaluations are often complicated by the changes that occur in weathering. Predictions of the direction of change and the degree of change at various time intervals would improve the accuracy of these evaluations. They are often further complicated by the heterogeneity of the spoil material.

An understanding of the basic characteristics of the spoil will contribute to valid interpretations of site data. The high wall reflects the variation in the spoil. Each rock stratum has distinct chemical and physical characteristics. Many of these properties will change when fragments of the rock are exposed to moisture, light, air, and temperature variations. Release of iron, sulfur, manganese, and aluminum compounds may result in off-site pollution and failure of vegetation. The release of essential nutrients such as phosphorus, potassium, calcium, and magnesium may improve the growth of vegetation, reduce acidity problems, and increase the land-management options.

Physical Characteristics.   Physically, the spoil is a heterogeneous mass of rock fragments derived from the rock strata above the coal. Very simply, sedimentary rocks are composed of small mineral particles cemented together. The size of the rock fragments after mining and reshaping depends on the size of the basic particles and the strength of the cementing material. Fine-grained shales strongly bonded together may break into hard, platy fragments that resist weathering. Coarse-grained sandstones weakly cemented together may disintegrate rapidly into sand. Once rock fragments are exposed to the weathering processes, they will break down at rates that depend on the physical and chemical properties of the original material.

Spoil may be a dynamic material, more reactive than natural soils that have been exposed to the processes of soil genesis for centuries. The chemical and physical properties of a spoil will change rapidly for several years. The rate of change will gradually slow as characteristics of natural soils develop. Our knowledge of this process does not permit an estimate of the time it takes for fresh spoil to turn into a material resembling natural soil. The geographical location of the disturbed area, the position on the spoil surface, the chemical and physical characteristics of the spoil, and the density and composition of the vegetative cover all may affect the rate of change.

Chemical Characteristics.   Evaluations of the chemical properties of a spoil could be perplexing if an attempt were made to include all components that affect plant growth. Methods in use today rely on indicators for judging the interaction of many factors at one time. Soil reaction or hydrogen-ion concentration, expressed as pH, is the most widely employed indicator.

Acidity alone may or may not affect the establishment or growth of plants. Changes in acidity determine the concentration of toxic ions and nutrients in the soil solution. At pH levels below 4.0, two chemical changes may occur: toxic ions such as manganese, aluminum, and iron become more available to plants, and some essential nutrients become less so. When the pH of a spoil is between 5.5 and 7.0, the concentration of toxic ions in the soil solution decreases and more essential nutrients become available. Predicting plant response to the acidity of the spoil requires an understanding of the effects of pH on the components of the soil solution and the tolerance of the plants for toxic ion concentrations.

pH Categories and Other Chemical Criteria.   Systems for spoil classification have been developed, using pH categories that relate to predicted plant response. One system widely employed for the acidic spoils of the eastern United States is shown in Table 1. No system has been proposed for the alkaline or sodic spoils in the western coal fields. The principles would be the same, but the toxic ions or cations may differ.

Other chemical analyses have been considered for identifying characteristics that result in specific nutrient deficiencies or toxicities. Phosphorus is often deficient (0 to 7 ppm) in

TABLE 1.   A Soil Classification System for Acidic Spoils

		ACIDITY	
Class number	Description	pH value	Extent on area sampled
1	Toxic	Less than 4.0	More than 75%
2	Marginal	Less than 4.0	50 to 75%
3	Acid*	4.0 to 6.9	50 to 75%
4	Calcareous	7.0 or more	More than 50%
5	Mixed	(Too varied to be classified as any above)	—

	TEXTURE
Group	Description of texture
A	Chiefly sand, sandstones, or sandy shales
B	Chiefly loamy materials and silty shales
C	Chiefly clay and clay shales
	Combine acidity and textual classes to describe spoil type.

*Acid spoils may be subdivided into two classes: pH 4.1 to 5.4; and pH 5.5 to 6.9.

spoil material. Several laboratory analyses provide estimates of plant-available phosphorus. Total soluble salts in the soil solution may be important for the very strongly acid or alkaline spoils. This does not necessarily mean high concentrations of sodium salts, but salts of all anions and cations in the soil solution. Some consideration has been given to laboratory analyses that afford concentrations of exchangeable aluminum, manganese, and hydrogen. These analyses would identify specific toxicities, and may permit the selection of tolerant plant materials. Recently research has been initiated to determine whether the heavy metals (copper, zinc, nickel, mercury, cobalt) occur in concentrations toxic to plants.

Determination of Physical Characteristics.   Techniques for determining the physical characteristics of a site are not well established. There is no agreement on what characteristics are important or on what standards should be used in site evaluation. Research has shown that coarse fragments, texture, and color are important to plant establishment and growth. Texture refers to the soil-classification system in which particles 2 millimeters or less in size are classified according to the percentages of sand, silt, or clay-size material.

Coarse fragments are all components of the spoil over 2 millimeters in size—even massive rock fragments. The soil-moisture regime in spoil material is determined by texture and by the percentage of coarse fragments. A coarse-textured or stony spoil may not retain adequate moisture for plant growth.

Texture and clay mineralogy influence the degree of compaction that results from heavy equipment passing over the surface during mining and reshaping. There is evidence that compaction may limit plant growth by reducing water infiltration and nutrient release. Texture may also determine the rate of release of toxic ions or nutrients from soil particles. Smaller particles expose a larger surface area per unit volume to the forces of weathering than coarse fragments. This results in a more rapid release of chemicals.

Surface color is important because it influences heat exchange at the spoil surface. Dark materials absorb solar energy; so the spoil may attain temperatures lethal to plants. The degree of risk depends on the season of the year, the slope of the exposed surface, and the exposure of the site. The highest risk may occur during the summer months on steep slopes of black or dark gray material facing south or west.

IMPORTANCE OF SAMPLING AND MAPPING

The intensity of sampling and mapping needed to evaluate land-use capabilities is determined by the land-management options being considered. When the objective is simply to establish a vegetative cover, all that is necessary is to estimate the range in pH and find the areas of extremely low or high pH on the site. Often past experience or observation of similar mined areas that have been successfully revegetated will be sufficient. The objective of this type of evaluation is to select treatments and plant materials that will establish vegetation and provide acceptable site protection. Little consideration is given to maintaining the vegetative cover or developing cover for a specific use.

At the other extreme are sites that have a potential for intensive land management. These may be mapped to show the location and size of areas with distinctive spoil characteristics. Each area will be sampled, and laboratory analyses will be made to determine pH, availability of nutrients, and concentration of toxic ions. Treatments will be developed for each area to achieve the land-use objective.

PREPARATION OF THE SITE

Treatments to prepare the site for seeding or planting are determined after evaluation of site data and land-management options. The intensity of treatment may not relate to the intended land use. Sites that have restrictive toxic ion concentrations or physical properties unfavorable for plant growth may require intensive treatments even to establish a vegetative cover.

Grass and legumes often require more surface preparation than trees and shrubs. Uncompacted fill slopes and freshly reshaped surfaces may make acceptable seedbeds. Hard crusts form on many spoils: fine clay-size particles are consolidated by the drying action of wind and sun. These crusts may be broken by rainfall, frost, or mechanical scarification. In many instances, ground cover will be denser if it is seeded into fresh spoil or if spoil surfaces have been broken by natural or mechanical scarification.

Treatment to create special surface configurations may be necessary on toxic spoils, on steep slopes, or in geographical locations where precipitation is low or unfavorably distributed through the year. Furrows or depressions made by conventional or specialized machinery, trap precipitation, and moderate wind velocities reduce evapotranspiration rates, and moderate spoil temperature extremes. The orientation of the furrows with respect to direction of slope, exposure to the sun, or prevailing winds may determine the effectiveness of the treatment.

Amendments to modify acidity should be applied weeks or months before seeding in order to neutralize the acidity before the seedlings germinate. Scarification will mix the ameliorating material into the top 6 to 8 inches of spoil and create a neutralized layer for root development. On extremely rocky spoils where scarification is not practical, frost action and infiltrating water may carry the neutralizing materials below the spoil surface. The rate of application of neutralizing material depends on the type of vegetation to be seeded, the present and predicted acidity of the surface spoil, and the neutralizing capacity of the material used.

Agricultural limestone is preferred for most treatments, but it can be used only in areas accessible to application equipment. No practical system has been developed for applying large amounts of agricultural limestone to steeply sloping land.

Small quantities of lime can be applied to steep slopes with a hydroseeder. Finely ground limestone or hydrated lime may be mixed with water to form a slurry. Repeated applications may be necessary to achieve high rates per acre. The cost of repeated treatments could make this procedure impractical.

Alkaline power-plant fly ash and bottom ash are neutralizing materials but they are usually not as effective as limestone. Application rates may be much higher than those for agricultural limestone, depending on the acidity of the spoil and the neutralizing ability of the fly ash. Fly ash contains various quantities of essential plant nutrients: phosphorus, zinc, molybdenum, and boron have been identified in laboratory and field investigations.

Large quantities of fly ash may modify the surface texture of the spoil, which could increase infiltration and subsurface moisture. Symptoms of boron toxicity have been observed in plants growing on sites treated with large amounts of fly ash. It is believed that weathering and leaching will reduce boron concentrations rapidly.

Rock phosphate provides a slowly available source of phosphorus, and helps to neutralize acidity. On acidic spoils, rock phosphate at high application rates provides a large potential source of phosphorus. Spoil acidity may react with the rock phosphate to slowly release plant-available phosphorus.

FERTILIZATION

Fertilization is generally recommended for grass and legume crops. Nitrogen and phosphorus have been used to accelerate the growth of trees on spoils. Greenhouse and field tests show that most spoils are deficient in nitrogen. Phosphorus occurs in various concentrations, but deficiencies often limit plant growth. Potassium is usually adequate for all but the most intensive management. There is little information about the concentrations of other nutrients. If deficiencies occur, they are not easily identified, or they may not be important for the crops grown. More emphasis will be placed on broad-spectrum fertilization when the selected land use requires intensive scientific management.

For most land-management objectives today, the formulation of the fertilizer makes little difference when equivalent rates are applied. More important are the costs of storage, transportation, and application—all related to the analysis of the fertilizer. High-analysis fertilizers, ammonium nitrate, triple superphosphate, and ammonium phosphate are in demand for these reasons. Using low-analysis fertilizers on sites that require high rates of application may increase the total soluble-salt concentration to levels toxic to some plants.

Rates of fertilization vary with the seeded crop, the inherent fertility of the spoil, and the land-management objective. Nitrogen and phosphorus at 50 and 22 pounds per acre, respectively, are sufficient to establish grass on many spoils. Higher rates of phosphate are important for legumes. More consideration is being given to retreatment or treatment at regular intervals. Multiple treatments are attractive because they reduce the chance of unacceptable cover and increase the probability of achieving the intended land use within the shortest time.

Research results show that placing selected fertilizers in or near the planting hole increases the growth of black locust. This leguminous species responded to additions of nitrogen and phosphorus placed near seedlings on extremely acid spoil. In other trials, direct-seeded black locust made more rapid growth after surface applications of nitrogen and phosphorus.

Five tons of lime per acre per foot of soil increased the growth of loblolly, shortleaf, Virginia, and a hybrid *pitch × loblolly pine* planted in extremely acid spoil. Pitch pine did not respond to the liming treatments. Ten tons of lime per acre-foot reduced the growth of the pines.

Although research has shown that fertilizer and lime increase the growth of selected tree species, there is no evidence at this time that these treatments may be practical. Fertilizer applied to aid grass and legume growth could benefit interplanted or direct-seeded trees, but the advantages of fertilization would have to exceed the inhibiting effects of grass and legume competition.

Municipal waste products have been considered for surface-mine reclamation. Waste applications could improve soil texture, add essential plant nutrients, and provide mulch for seed and seedlings. Shredded or composted waste could be applied to the surface, and mechanical methods could be used to incorporate it into the spoil. Sewage sludge can be applied in a water slurry or in dried form. Mixtures of shredded or composted waste and sewage sludge offer another treatment possibility. The use of these materials has been restricted to relatively small demonstration areas for public health reasons.

SEEDING AND PLANTING

Selecting vegetation appropriate for site conditions and land-management objectives challenges the experience and imagination of the planner. Minimum treatments that

TABLE 2. Grass Species Used for Surface-Mine Revegetation in the West*

Scientific name	Common name	Scientific name	Common name
NATIVE GRASSES		INTRODUCED GRASSES	
Agropyron spicatum	Bearded bluebunch wheatgrass	*Agropyron cristatum*	Crested wheatgrass
		Agropyron dasystachyum	Thickspike wheatgrass
Agropyron latiglume	Parish wheatgrass	*Agropyron elongatum*	Tall wheatgrass
Agropyron trachycaulum	Slender wheatgrass	*Agropyron intermedium*	Intermediate wheatgrass
Alopecurus alpinus	Alpine foxtail	*Agropyron repens*	Quackgrass
Alopecurus pratensia	Meadow foxtail	*Agropyron riparium*	Streambank wheatgrass
Andropogon gerardi	Big bluestem	*Agropyron trichophorum*	Stiffhair wheatgrass
Andropogon scoparius	Little bluestem	*Alopecurus pratensis*	Meadow foxtail
Aristida longiseta	Red threeawn	*Bromus inermis*	Smooth brome
Bouteloua gracilis	Blue grama	*Dactylis glomerata*	Orchardgrass
Bouteloua curtipendula	Sideoats grama	*Elymus canadensis*	Canada wildrye
Bromus marginatus	Mountain brome	*Elymus junceus*	Russian wildrye
Bromus tectorum	Cheatgrass brome	*Phleum pratense*	Timothy
Calamagrostis purpurascens	Purple pinegrass	*Poa pratensis*	Kentucky bluegrass
Calamagrostis rubescens	Pinegrass	*Secale cereale*	Rye
Danthonia intermedia	Timber danthania	*Triticum aesticum*	Wheat
Deschampsia caespitosa	Tufted hairgrass		
Distichlis stricta	Inland saltgrass		
Elymus cinerus	Basin wildrye		
Festuca idahoensis	Idaho fescue		
Festuca ovina	Sheep fescue		
Festuca scabrella	Rough fescue		
Hesperochloa kingli	Spike fescue		
Koeleria cristata	Prairie junegrass		
Oryzopsis hymenoides	Indian ricegrass		
Phleum alpinum	Alpine timothy		
Poa alpina	Alpine bluegrass		
Poa nervosa	Wheeler bluegrass		
Poa secunda	Sandberg bluegrass		
Sitanion hystrix	Bottlebrush squirreltail		
Sporobolus crytandrus	Sand dropseed		
Stipa columbiana	Subalpine needlegrass		
Stipa comata	Neele and thread		
Stipa viridula	Green needlegrass		
Trisetum spicatum	Spike trisetum		

*List provided by USDA Forest Service, Intermountain Forest and Range Experiment Station, Logan, Utah.

provide only a protective cover require little planning, but comprehensive plans to achieve maximum benefits from a site require considerable expertise in surface-mine revegetation.

Unfortunately, much of the available information about revegetation is not fully utilized. Limited practical experience is passed around by word of mouth, but there is little effort to augment or improve it. This may change when the potential of surface-mined land is more widely demonstrated and adequate incentives stimulate greater interest in vegetation planning.

The list of plant species considered for surface-mine revegetation is long (see Tables 2 through 7). Many species that require intensive site preparation or management have not been widely used.

For each of the major coal-producing regions there is a group of preferred plant species. These may be arranged under four major categories: grasses, forbs, trees, and shrubs. The grasses and forbs may be further classified as temporary, semipermanent, or permanent, depending on the life expectancy of the plant. Temporary species give prompt and effective site protection by reason of their quick growth, fibrous roots, and ability to endure unfavorable site conditions. Semipermanent species are perennials that will be replaced by permanent vegetation. Under favorable conditions, permanent species will persist for many years.

GRASSES (See Tables 2 and 3.)

The grasses are a varied group of plant species well suited to surface-mine restoration. Experience has shown that many grasses are adapted to wide ranges of climate, spoil texture, nutrient regimes, and toxic ion concentrations. Germination is usually rapid, and growth often produces a crop, or at least site protection, during the first growing season.

Annual grasses often grown as agricultural crops may be used to provide quick site

TABLE 3. Grass Species Used for Surface-Mine Revegetation in the East*

Scientific name	Common name
Agrostis alba	Red top
Andropogon gerard	Big bluestem
Andropogon scoparius	Little bluestem
Archenatherum elatius	Tall oatgrass
Avena sativa	Oats
Bromus arvensis	Field brome
Bromus inermis	Smooth brome
Cynodon dactylon	Bermudagrass
Dactylis glomerata	Orchardgrass
Echinochloa crusgali	Japanese millet
Eragrostis curvula	Weeping lovegrass
Festuca arundinacea	Tall fescue
Festuca rubra	Red fescue
Lolium multiflora	Italian ryegrass
Lolium perenne	Perennial ryegrass
Panicum clandestium	Deer tongue
Panicum virgatum	Switchgrass
Paspalum notatum	Bahiagrass
Pennisetum glaucum	Pearl millet
Phalaris arundinacea	Reed canarygrass
Phyleum pratense	Timothy
Poa pratensis	Kentucky bluegrass
Secale cereale	Rye
Setaria italica	Foxtail millet
Sorghastrum nutans	Indiangrass
Sorghum vulgare (var)	Sorghum
Sorghum vulgare (var)	Sudangrass
Sorghum vulgare (var)	Sudangrass sorghum hybrid
Triticum aestivum	Wheat

*List provided by USDA Forest Service, Northeastern Forest Experiment Station, Berea, Ky.

protection. These temporary quick-cover crops may also serve as nurse crops for slower developing perennials. Some species are better adapted to summer seeding, whereas others should be seeded in the fall.

In selecting an annual grass for a site, consideration should be given to the seeding season, the chemical and physical properties of the spoil, and the compatibility of all the perennials seeded with the cover crop. Provisions for re-treatment should be made when seeding annuals. If annuals alone are used, perennials should be sown after the annuals mature. Perennials sown with the annuals may fail because the annuals compete aggressively for light, nutrients, and moisture. Any re-treatment should use as much seed and fertilizer as would be applied to bare spoil.

Perennial grasses are necessary components of most herbaceous mixtures. Many species could be employed, but the tendency is to use a limited number that are adapted to a wide range of site conditions. Grasses developed for range and forage crops are preferred. Native grasses grow well on surface-mined land, but are seldom used because of their slow early growth.

Considerations in the selection of perennial grasses include climatic factors, site characteristics, date of seeding, expected maintenance, and intended use of the grass cover. Care must be exercised to select species and seeding rates that are compatible with the other species in a mixture. Many grasses are aggressive and may eliminate less aggressive, more important components, such as leguminous forbs, if improperly managed.

FORBS. (See Tables 4 and 5)

Leguminous forbs are considered by many to be essential components of ground-cover mixtures. The fixation of atmospheric nitrogen by the legumes benefits associated plants. This assists in maintaining a vigorous ground cover and may reduce the need for re-treatment. The leguminous forbs generally are less tolerant of toxic ion concentrations and require more phosphorus than the grasses.

Annual legumes offer opportunities for quickly developing temporary cover that contributes to soil enrichment by nitrogen fixation and nutrient recycling. The germination of these species is seldom delayed by dormancy, hard or impermeable seed coats, or low germinative capacity. Germination is usually rapid, and the plants attain full growth in a few months. Under favorable conditions, some annual legumes reseed and perpetuate the species.

TABLE 4.   Species of Forbs Used for Surface-Mine Revegetation in the East*

Scientific name	Common name
Cornonilla varia	Crownvetch
Fagopyrum sagittatum	Buckwheat
Glycine max	Soybean
Helianthus spp.	Sunflower
Lathyrus sylvestais	Flat pea
Lespedeza stipulacea	Korean lespedeza
Lespedeza striata	Common lespedeza
Lespedeza striata (var)	Kobe lespedeza
Lespedeza cuneata	Sericea lespedeza
Lotus corniculatus	Birdsfoot trefoil
Lotus tennuis	Narrowleaf trefoil
Medicago sativa	Alfalfa
Melilotus alba	White sweet clover
Melilotus officinalis	Yellow sweet clover
Trifolium hybridium	Alsike clover
Trifolium repens (var)	Ladino clover
Trifolium repens (var)	White clover
Trifolium pratense (var)	Red clover
Vicia sativa	Common vetch
Vicia villosa	Hairy vetch
Vigna sinensis	Cowpea

*List provided by USDA Forest Service, Northeastern Forest Experiment Station, Berea, Ky.

TABLE 5. Forb Species Used for Surface-Mine Revegetation in the West*

Scientific name	Common name	Scientific name	Common name
NATIVE FORBS		**INTRODUCED FORBS**	
Archillea millefolium	Common yarrow	*Coronilla varia*	Crownvetch coronilla
Arnica cordifolia	Heartleaf arnica	*Helianthus annuus*	Common sunflower
Artemisia ludoviciana	Louisiana sagebrush	*Medicago sativa*	Alfalfa
Aster spp.	Aster	*Melilotus alba*	White sweetclover
Astragalus spp.	Milkvetch	*Melilotus officinalis*	Yellow sweetclover
Balsamorhiza sagittata	Arrowleaf balsamroot	*Taraxacum officinale*	Common dandelion
Castilleja spp.	Paintedcup		
Gaura coccinea	Scarlet gaura		
Geranium spp.	Geranium		
Geum rosii (Seversia montana)	Golden sieversia		
Grindelia squarrosa	Curlycup gumweed		
Kochia scoparia	Belvedere summercypress		
Lupinus alpinus	Lupine		
Lupinus spp.	Lupine		
Mertensia spp.	Bluebells		
Oxytropsis spp.	Crazyweed		
Penstemon spp.	Penstemon		
Potentilla spp.	Cinquefoil		
Salsola kali	Common Russian thistle		
Senecio spp.	Groundsel		
Solidago missouriensis	Missouri goldenrod		
Trifolium spp.	Clover		

*List provided by USDA Forest Service, Intermountain Forest and Range Experiment Station, Logan, Utah.

The perennial leguminous forbs used in mixtures with grasses require more intensive treatment to establish and maintain acceptable plant vigor. A few species have hard seed coats and dormancies that delay germination unless the seed is properly treated before sowing. Legumes require more phosphorus than grasses. If the land-management objective is hay or forage crops, lime and fertilizer must be applied on regular schedules to maintain acceptable yields and prevent deterioration of the established cover.

Other factors that affect the establishment and growth of leguminous forbs are the species composition and density of companion crops. Grasses may compete vigorously for moisture and nutrients. Legumes are generally less competitive and may not survive in dense grass covers. Manipulation of species composition and reduction in seeding rates of companion grass species could favor growth of legumes.

Forbs not classified as legumes are being evaluated in the Western states. The emphasis is on species that are components of natural vegetative covers and are common invaders of disturbed areas. Initial evaluations indicate that some species may be useful for surface-mine revegetation. Establishment and maintenance procedures need to be developed for the more promising forbs.

Inoculation with specific strains of rhizobium bacteria stimulates nodulation on leguminous forbs. Commercial inoculants are available for the important legume species. The viability of the inoculant may be reduced in hydroseeding, if extremely acid water is used or if the slurry of seed and fertilizer is allowed to stand in the hydroseeder for more than an hour. Rhizobium bacteria may not survive or produce effective nodules in acidic spoils with pH below 5.0.

TREES (See Tables 6 and 7.)

Trees are generally established by planting seedling nursery stock. Experience has shown that many species are adapted to surface-mine spoil. Species native to the geographical area are usually recommended. Species selection is based on the anticipated yield of merchantable forest products and on-site characteristics. Yield predictions may not be realistic, as they are based on the yields of stands growing on natural soils.

Spoil characteristics that cause mortality or limit growth are known for many species. There is little information about spoil characteristics or practical treatments that stimulate growth. Grass and legume ground covers reduce tree growth, but this is often unavoidable because state laws and regulations require a herbaceous cover. Manipulation of the species composition and reduction of ground-cover density may minimize the adverse effects.

SHRUBS (See Tables 6 and 7)

The planting of shrub species has not been emphasized in surface-mine revegetation. Shrubs have little tangible value; benefits accrue from site protection, wildlife food and cover, and aesthetics. A few species are widely used because they grow well on a range of sites and produce abundant wildlife food.

GENETICS

Genetic research may contribute to higher germination or survival and greater growth. Research has shown differences between varieties of grasses and legumes in toxic ion tolerance and growth on unfavorable sites. Some hybrid poplar clones grow more rapidly than others on the same site. A comparison of Virginia pine from selected seed sources indicates that seed source contributes to the rate of growth on extremely acid spoil. As the intensity of treatment increases, these and other factors will be taken into consideration in revegetation planning.

METHODS OF SEEDING AND PLANTING

Many species of plants may be seeded by conventional methods. Mechanical methods are preferred because of the size, configuration, and stoniness of many surface-mined sites. Dry seeding methods include hand dispersal, various types of mechanical spreaders operated on the spoil surface, and aerial seeding from helicopters or fixed-wing aircraft.

TABLE 6.   Tree and Shrub Species Used for Surface-Mine
Revegetation in the West*

Scientific name	Common name
NATIVE TREES AND SHRUBS	
Abies lasiocarpa	Alpine fir
Acer glabrum	Rocky mountain maple
Acer grandidentatum	Bigtooth maple
Alnus rubra	Red alder
Alnus tenuifolia	Thinleaf alder
Amelanchier alnifolia	Saskatoon serviceberry
Arctostaphylos uva-ursi	Bearberry
Artemisia tridentata	Big sagebrush
Artemisia cana	Silver sagebrush
Artemisia tripartita	Threetip sagebrush
Atriplex confertifolia	Shadescale saltbush
Betula occidentalis	Water birch
Berberis repens	Creeping barberry
Cassiops mertensiana	Mertens cassiope
Ceanothus velutinus	Snowbush ceanothus
Ceanothus prostratus	Squawcarpet ceanothus
Cercocarpus ledifolius	Curlleaf mountain mahogany
Chrysothamnus nauseosus	Rubber rabbittbrush
Cornus stolonifera	Red osier dogwood
Crataegus douglasii	Douglas hawthorn
Gutierrezia sarothrae	Broom shakeweed
Juniperus scopulorum	Rocky mountain juniper
Juniperus horizontalis	Creeping juniper
Pinus albicaulis	Whitebark pine
Pinus contonta	Shore pine
Pinus flexilis	Limber pine
Pinus jeffreyi	Jeffry pine
Pinus ponderosa	Ponderosa pine
Pachistima myrsinitis	Myrtle pachistima
Philadelphus microphyllus	Littleleaf mockorange
Physocarpus monogymus	Mountain ninebark
Populus tremuloides	Aspen
Potentilla fruticosa	Bush cinquefoil
Prunus emarginata	Bitter cherry
Prunus virginiana	Common chokecherry
Pseudotsuga menziesii	Douglas-fir
Pursha tridentata	Antelope bitterbrush
Quercus gambilii	Gambel oak
Rhus trilobata	Skunkbush sumac
Ribes cereum	Wax current
Rosa arkansasana	Arkansas rose
Salix spp.	Willow
Sambucus racemosa	European red elder
Sarcobatus vermiculatus	Black greasewood
Spirea canescens	Hoary spirea
Symphoricarpos occidentalis	Western snowberry
Vaccinium spp.	Blueberry
INTRODUCED TREES AND SHRUBS	
Betula papyrifera	Paper birch
Elaeagnus angustifolia	Russian olive
Fraxinus pennsylvanica	Green ash
Populus angustifolia	Narrowleaf poplar
Populus balsamifera	Southern poplar
Populus deltoides	Eastern poplar
Populus spp.	Poplar
Prunus americana	American plum
Robinia pseudoacacia	Black locust
Ulmus parvifolia	Chinese elm

*List provided by USDA Forest Service, Intermountain Forest and
Range Experiment Station. Logan, Utah.

TABLE 7.   Trees and Shrubs Used for Surface-Mine Revegetation in the East*

Scientific name	Common name
Acer rubrum	Red maple
Acer Saccharinum	Silver maple
Acer saccharum	Sugar maple
Alnus glutinosa	European alder
Amorpha fruticosa	False indigo
Betula nigra	River birch
Castenca mollissima	Chinese chestnut
Cornus amomum	Silky dogwood
Crataegus spp.	Hawthorn
Elaeagnus umbellata	Autumn olive
Fraxinus americana	White ash
Fraxinus pennsylvanica	Green ash
Juglans nigra	Black walnut
Larix decidua	European larch
Larix leptolepsis	Japanese larch
Lespedeza bicolor	Shrub lespedeza
Lespedeza thunbergii	Thunberg lespedeza
Liquidamber styraciflua	Sweetgum
Liriodendron tulipifera	Yellow-poplar
Lonicera maackii	Amur honeysuckle
Lonicera tartarica	Tartarian honeysuckle
Picea abies	Norway spruce
Pinus banksiana	Jack pine
Pinus echinata	Shortleaf pine
Pinus nigra	Austrian pine
Pinus palustris	Longleaf pine
Pinus resinosa	Red pine'
Pinus rigida	Pitch pine
Pinus strobus	Eastern white pine
Pinus sylvestris	Scotch pine
Pinus taeda	Loblolly pine
Pinus virginiana	Virginia pine
Plantanus occidentalis	American sycamore
Populus deltoides	Eastern cottonwood
Populus spp.	Hybrid poplar
Quercus alba	White oak
Quercus rubra	Northern red oak
Robinia fentilis	Bristly locust
Robinia hispida	Roseacacia
Robinia pseudoacacia	Black locust

*List provided by USDA Forest Service, Northeastern Forest Experiment Station, Berea, Ky.

Hydraulic seeding or hydroseeding is popular in many coal regions. Rough, stony, or steep areas may be treated if the hydroseeder is equipped with auxiliary hose lines.

Fertilizers may be mixed with the seed to expedite treatment and reduce the cost of dry seeding. Seed, fertilizer, lime, mulch, and soil stabilizers may be applied individually or in combination with a hydroseeder.

Mulches and soil stabilizers may aid the establishment of vegetation and reduce erosion. Mulches such as straw, hay, or bark are applied dry through mechanical spreaders. Wood fiber and chemical soil stabilizers may be applied with a hydroseeder.

In regions with abundant rainfall, mulches and soil stabilizers are applied to reduce erosion and prevent loss of seed and fertilizer by surface runoff. Mulching is recommended in regions of low rainfall or in areas where prolonged dry periods occur during the growing season. The mulch conserves moisture and protects emerging seedlings from microclimatic extremes. On dry sites, mulches may reduce damage to plants by wind-borne soil particles.

Irrigation is being evaluated in the western coal fields as a supplemental treatment for establishing vegetation. It is doubtful that it will be widely used to maintain an established cover.

Trees are often planted by hand, using one of several planting tools. On selected spoils, machine planting is possible with specially designed planters. Most planting stock is small, one to two years old, and bare-rooted. Spacing between trees varies by species, site characteristics, and product objectives. Plantings may be mixtures of several species or pure plantings of only one. The arrangement may be random or designed to protect the seedlings from environmental extremes.

There is increasing interest in the establishment of trees by direct seeding. Success has been achieved with several pine species in Alabama. Black locust is used in West Virginia and Kentucky. Research is evaluating other species of hardwoods and pine to expand the direct-seeding options. Direct seeding will cost less than planting, and require less hand labor.

LAND-MANAGEMENT OPTIONS

It would be idealistic to believe that all disturbed land should be developed and managed for tangible economic returns. The belief that all surface-mine spoil is infertile and must remain unproductive for many years is equally unrealistic. Actually, most surface-mined areas can be managed to provide some tangible or intangible benefit to society. The land-management options for many areas may be limited by technical, economic, or social factors. Progressive surface-mine reclamation goes beyond the establishment of a vegetative cover; every effort should be made to maximize the tangible and intangible benefits.

Land management begins before the coal is mined and continues until an appropriate objective is achieved. The spoil may not resemble the soil that was destroyed by mining, but under scientific management its productive capacity may equal or exceed that of the natural soil. The challenge to reclamation specialists is to use existing knowledge to recover our coal resources profitably by surface mining in such a manner as to maintain or enhance the productive capacity and social acceptability of the land.

Minimal treatment will satisfy state regulations and laws. Little or no planning may be involved; limited technical knowledge is required, and costs depend on the nature of the spoil. Society benefits by reduced erosion and aesthetic improvement of the site. Benefits may also accrue from improvement of wildlife habitat and production of forage or forest products. Seldom do these minimal treatments produce results comparable to those that could be realized by careful planning and somewhat higher treatment costs.

WILDLIFE HABITAT

Wildlife food and habitat development are attractive land-management options adapted to areas of all sizes. Benefits depend on the skill with which accepted wildlife-management techniques are used to support present and future wildlife populations.

Evaluations of existing habitat and food sources on natural soils surrounding the disturbed area are necessary to determine what treatments are appropriate. The extent of the inventory depends on the feeding preferences and foraging range of the wildlife species involved. In most cases treatments applied to surface-mined land will not supply all the food and cover required by one species of wildlife. The purpose will be to provide those portions of the habitat requirements that are lacking or deficient in the area. Because treatments begin with barren spoil, there are opportunities to provide preferred foods and special cover.

FOREST PRODUCTS

The development of an area to produce forest products goes beyond the decision to plant trees. Consideration must be given to future markets. Areas must be of sufficient size to form efficient management and marketing units. A management unit may be located entirely on surface-mined land, or it may involve both an area of mining disturbance and adjacent natural soils.

Once the decision has been made to develop an area for the production of forest products, treatments must be selected to achieve this objective in the most efficient manner. Reshaping must allow access to all parts of the area for establishment, management, and harvesting activities. Species selection depends on the chemical and physical characteristics of the spoil and predicted species preferences of future markets. Spacing between the planted trees may determine tree quality and rate of growth. Nitrogen-fixing trees or shrubs interplanted with the potential crop trees increase growth. Species mixtures can also be utilized to provide merchantable products at different stand ages. Rapidly growing species may be harvested first, while slower growing species mature later.

Yield data are lacking for many species planted on surface-mined land, but there is evidence that plantations on spoils with no predictable chemical or physical constraints can support yields equal to or higher than those on natural soils.

PASTURE AND FORAGE CROPS

Pasture and forage crops are realistic land-management objectives for many disturbed areas. Spoils with a wide range of physical and chemical characteristics may be considered for pasture. Forage crops are limited to the more productive sites that can be traversed by conventional farm machinery.

Many spoils considered for this use will have a high percentage of coarse fragments that would limit mechanical site preparation and maintenance treatments. The desired vegetative cover may be established in the first seeding, or it may be introduced after temporary covers achieve site protection and modification of potentially unfavorable climatic and spoil conditions. Maintenance treatments will increase forage yield. Future productivity depends on regulated use and continued maintenance.

In the midwestern and eastern coalfields, rainfall is generally adequate and well distributed through the growing season. In these regions, it is often possible to use reclaimed mined areas as supplemental pastures for established agricultural enterprises. Spoils in the low-rainfall areas of the West may also provide productive grazing land, although establishing an adequate cover will be more difficult. Grazing must be regulated to control seasons of use, herd size, and use by wild game.

HORTICULTURE

Horticultural crops offer financial opportunities in some regions. Many of these crops will produce high yields per acre; so small as well as large acreages can be utilized. Spoil characteristics are less restrictive, since most crops are produced from planted trees, shrubs, or vines. It may also be possible to select crops that are adapted to specific spoil conditions. Planned reshaping will provide access for establishment, maintenance, and harvesting. Beginning with barren spoil, the species composition and density of the ground cover may be controlled.

AGRICULTURE

Agricultural crops may be considered on selected spoils with physical properties that permit soil tillage. Sites should have dimensions compatible with the efficient operation of farm machinery as well as a minimum of potential toxicity problems. Crop selection should be based on spoil characteristics and regional agricultural practices. It is possible on some sites with a high percentage of coarse fragments to seed a small grain crop on barren spoil with a minimum of tillage or scarification. After this crop is harvested, a pasture or forage crop can be seeded as the permanent cover crop.

REFERENCES

"Research and Applied Technology Symposium on Mined-Land Reclamation," Bituminous Coal Research Inc., Monroeville, Pa., 1973. (355 pages.) May, Morton, R. Lang, L. Lujan, P. Jacoby, and W. Thompson: Reclamation of Strip Mine Spoil Banks in Wyoming, *Wyoming Univ., Agr. Exp. Sta. Res. J.,* vol. 51, 1971. (32 pages.)

"Rehabilitation Potential of Western Coal Lands," National Academy of Science, National Research Council, Washington, D.C., 1973. (144 pages.)

Plass, William T., and J. D. Burton: "Pulpwood Production Potential on Strip-mined Land in the South," *Soil Water Conserv.,* vol. 22, pp. 235–238, 1967.

"A Guide for Revegetating Bituminous Strip-Mine Spoils in Pennsylvania," Research Committee on Coal Mine Spoil Revegetation, Harrisburg, Pa., 1965. (46 pages.)

Riley, Charles V.: Revegetation and Management of Critical Sites for Wildlife, *North Am. Wildlife Conf. Trans.*, vol. 28, pp. 269–283, 1963.

Sindelar, B. W., R. L. Hodder, and M. E. Majerus: Surface-mined Land Reclamation Research in Montana, *Montana Agr. Exp. Sta. Res. Rep.* 40, 1973. (122 pages.)

Restoring Surface-mined Lands, *USDA Misc. Publ.* 1082, Washington, D.C., 1968. (18 pages.)

U.S. Department of Agriculture: "Reclamation of Surface-mined Lands in Montana," Soil Conservation Service, Washington, D.C., 1971. (22 pages.)

Coal Preparation Plants

BY ERNEST A. DRAEGER*

The primary function of the coal preparation plant is the delivery of a finished product which has been custom-prepared in terms of quantity and quality. Regardless of whether the coal is metallurgical quality or a power-plant fuel, the coal will be scrutinized by the consumer in terms of the raw-coal size, analysis, washability data, mining techniques, and the efficiency of the various types of coal-washing equipment.

Economic pressures have continually increased the need for improving coal preparation techniques. In general, these developing methodologies have been aimed at removing noncombustible impurities and minimizing product variation. In 1942, 24 percent of the total coal production was cleaned mechanically, as compared with 65 percent in 1965. The amount of refuse discarded during preparation for the same period increased from 13 to 21 percent. Cost factors associated with both fuel consumption and coal production are contributing to this change. Power plants and other consumers are currently using furnace and transport systems that demand very uniform feedstocks. Size, heat content, and handling characteristics are extremely important. Continued rising mine labor rates and more expensive safety features have accelerated the use of mechanization and increased the use of continuous mining equipment. These factors have resulted in raw coal with an enhancement of fines and impurities, and thus have required newer and better cleaning processes in order to maintain an economically competitive position for coal.

Coal from different areas generally possesses inherently different properties which mitigates the use of a unique combination of the various types of standardized preparation equipment to produce the product required. The design and construction of the preparation plant require prior knowledge of the physical and chemical properties of the coal which characterize its response to crushing, screening, blending, and cleaning operations. Basic units, engineered to perform these operations, are assembled to handle the coal in whatever sequence is necessary to furnish a product of predetermined specifications.

TYPICAL PREPARATION PLANTS

A typical plant utilizing a dense-media bath and a closed-water circuit is illustrated in Fig. 1. At the Virginia Pocahontas No. 3 Plant, the coal is received from the mine at a rate of 750 tons per hour. It is broken in a rotary breaker and stored in a 5000-ton raw-coal storage silo. From storage, the coal is withdrawn at a rate of 625 tons per hour, screened upon entering the plant to separate 4- by ¼-inch pieces for the dense-media circuit from the ¼- by 28-mesh portion for washing on tables. Cross-flow screens, located ahead of the tables, deslime the ¼-inch by 0 raw coal, and the 28-mesh by 0 underproduct is sluiced to flotation cells for recovery of the fine coal. The 28-mesh by 0 washed coal from flotation is dewatered in clean coal disk filters and combined with the ¼-inch by 28-mesh tabled washed coal, which was dewatered in centrifuges. Further dewatering is required, and therefore the ¼-inch by 0 washed coal is dried.† The 4- by ¼-inch washed coal product from the dense-media circuit is dewatered on vibrating screens and combined with the ¼-inch by 0 dried product from the *Flowdryer* before it is loaded in railroad cars. The 28-mesh by 0 refuse is directed to a refuse thickener, where the solids are thickened and pumped to refuse filters for additional removal of water. The clarified water from the thickener is recirculated through the plant as spray water. The 28-mesh by 0 refuse from the refuse filter is combined with the coarse refuse and conveyed to a truck bin. The equipment is arranged in the most appropriate circuit to ensure the maximum recovery of marketable coal from this particular raw coal.

---

*Project Engineer, McNally Pittsburg Manufacturing Corporation, Pittsburg, Kansas.
†McNally Flowdryer used in this installation.

Optimum conditions for other coals may require variations in both the equipment and circuits used. By comparison, a typical Baum jig-type plant is shown in Fig. 2.

At the Amax Coal Company's Leahy Mine, run-of-mine coal is delivered to a 600-ton-capacity truck hopper, from which it is fed by two 60-inch reciprocating feeders to a 60-inch belt that delivers the coal at a rate of 1800 tons per hour to a rotary breaker station. A bar screen removes some of the natural 5-inch by 0 pieces. The oversize fraction is fed to a

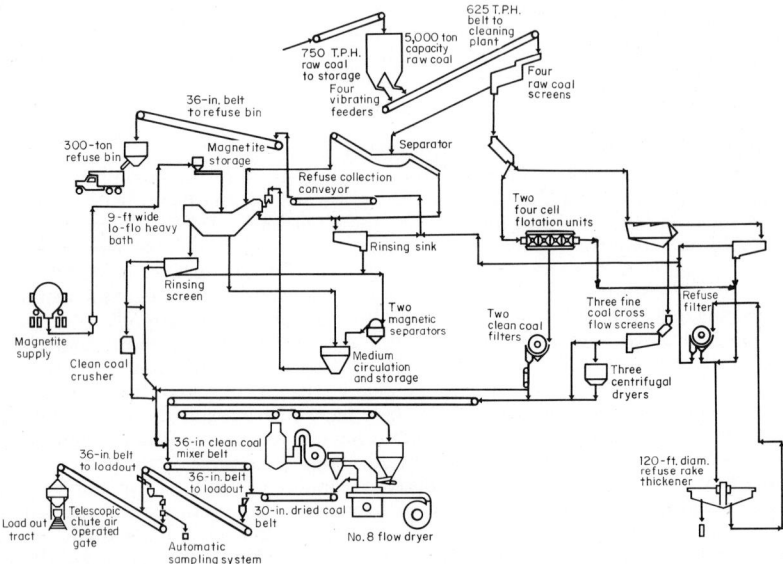

Fig. 1. Coal preparation plant of Island Creek Coal Company's Virginia Pocahontas No. 3, Buchanan County, Virginia. (T.P.H. = tons per hour)

rotary breaker (10 feet in diameter; 24 feet in length) which is equipped with screen plates with 5½-inch-diameter holes. The rock is removed and chuted to a 100-ton-capacity bin equipped with a motorized gate. The rock is ultimately disposed of in a strip pit area.

The minus 5-inch fraction from the bar screen and from the rotary breaker are combined on a 48-inch belt and discharged to an 8000-ton raw-coal storage silo. The silo has a diameter of 70 feet, walls that are 11¼-inch thick, and a height of 117 feet. Six 36-inch reciprocating feeders feed the 5-inch by 0 raw coal from the silo to a 48-inch belt that conveys the coal to the washery at a rate of 1500 tons per hour. A belt scale provides a record of the raw-coal input.

The 5-inch by 0 raw coal is processed in two 96-inch-wide Mogul wash boxes which produce clean coal, middlings, and refuse.

The refuse is dewatered on a single deck screen (6 by 16 feet) with ³⁄₁₆-inch openings in the screening surface. The refuse is then discharged via a 30-inch belt to a 400-ton-capacity refuse bin, from which it is removed to the strip pit area.

The middlings are dewatered on a single-deck screen (5 by 10 feet) using ½-inch openings in the screening surface. The 5- by ½-inch screen fraction is crushed to minus 1-inch size, then recycled through the wash boxes. The ½-inch by 0 screen fraction is pumped to a waste pond where the solids settle out and the clarified water overflows to the plant fresh-water supply pond.

The clean coal is discharged from the wash boxes over fixed screens with ³⁄₁₆-inch openings. The coal then passes to four double-deck washed-coal screens. Two screens are 8 by 16 feet, and two are 6 by 16 feet, with 1¼-inch screen cloth on the top deck and ⅜-inch screen cloth on the bottom deck.

The 5- by 1¼-inch screen fraction is reduced in two 36- by 60-inch Gearmatic crushers to 1½-inch by 0 size. The fraction then is discharged to the 48-inch belt which delivers it to a 15,000-ton-capacity washed-coal silo (70 feet in diameter, 192 feet in height).

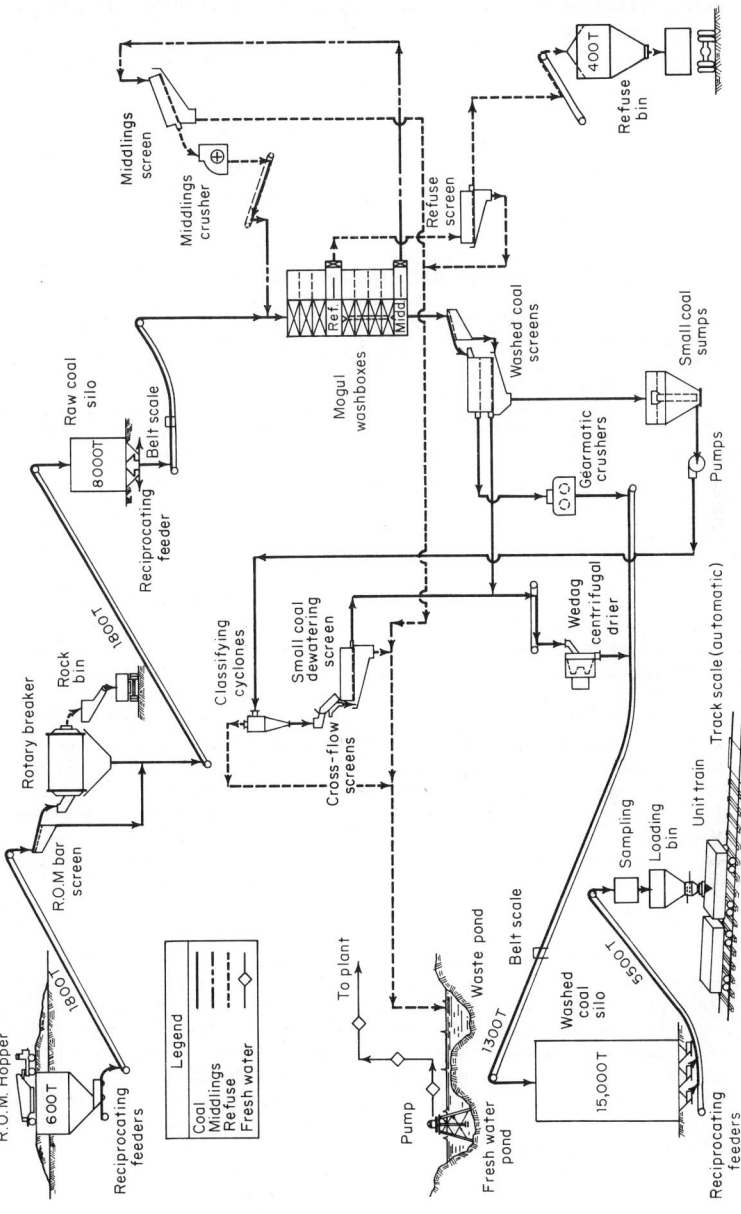

Fig. 2. Coal preparation plant of Amax Coal Company's Leahy Mine. (R.O.M = run-of-mine; T = ton)

The ⅜-inch by 0 fraction from the washed-coal screens is sent to two small coal sumps, from which it is pumped to eight 24-inch classifying cyclones. The cyclone underflow is discharged to four cross-flow screens, after which it passes over four single-deck dewatering screens equipped with ¾-millimeter openings. Three of the screens are 6 by 16 feet; and one is 8 by 16 feet.

The ⅜-inch by ¾-millimeter overproduct from the dewatering screens is discharged to the distributing conveyors that feed the centrifuges. The product from these centrifuges is discharged to the washed-coal silo. A bleed of solids from the cyclone overflow, the underflow of the cross-flow screens and the underflow of the small-coal dewatering screens, is pumped to the waste pond.

EQUIPMENT AND METHODOLOGY

The various kinds of coal-handling and cleaning equipment currently in use are categorized in Table 1. At one time, conveyors, screens, crushers, and picking tables comprised the entire machinery for handling and cleaning coal. More sophisticated equipment has been required in response to the demand for a cleaner, better-quality product.

Crushers, rotary breakers, Baum jigs, heavy-media systems, cyclones, and thermal driers are relatively complex units developed and operated using principles derived from detailed engineering studies and experiments. Although not in great detail, the description here is intended to illustrate the scope of equipment and methodology commonly found in a modern coal preparation plant.

SAMPLING

Reliable data as to the physical and chemical characteristics of coal are required by both coal producers and coal users. These data must be obtained by analyses made on samples which are representative of the coal. (See also earlier description in this Handbook on Coal Testing.) Automatic mechanical samplers are designed and built to yield data consistent with ASTM (American Society for Testing and Materials) sampling principles. To obtain completely representative samples, increments must be withdrawn from the full cross section of a flowing stream of coal. The mechanical sample cutter should accept with ease the full range of coal sizes to be found in the stream. Experience has shown that the cutter width should be 2.5 to 3 times the diameter of the largest possible piece. Also, design and velocity of movement of the cutter through the stream should create a minimal disturbance within the stream.

For a given accuracy, the number of increments or the quantity of material obtained by a single cut of the automatic sampler are increased with an increase in the top size of the coal and the expected ash content of the coal to be sampled.

Three different types of samplers are: (1) plate, (2) belt, and (3) rotary drum.

The automatic plate-type sampler is shown in Fig. 3 and consists of a chute with an integral plate cutter traveling at a predetermined rate of speed. The stainless steel cutter is supported by a plate which moves on rollers to cut through and beyond the coal stream with each cut. The sampler is provided with an adjustable timer for automatic sampling at intervals from one to 30 minutes. Constant intervals between cuts are maintained to obtain equal representation of the entire length of the coal stream.

The rotary-drum-type sampler may operate intermittently or continuously and is often used in conjunction with other types of samplers as a final sampler for reduction of the gross sample. See Fig. 4.

BREAKING AND CRUSHING

In an earlier era when there was extensive use of coal for home heating and hand-fed furnaces, the coal-plant operator attempted to recover as much lump as possible. In modern practice, high-quality market requirements dictate crushing to liberate the extraneous particles entrapped within the coal. Crushing allows the washing equipment to remove the extraneous material and to produce a high-quality coal. Selection of a crusher is very important because the crusher should produce a uniform top size with a minimum amount of extreme fines.

In general, pressure applied to coal will create a shear tendency along cleavage planes.

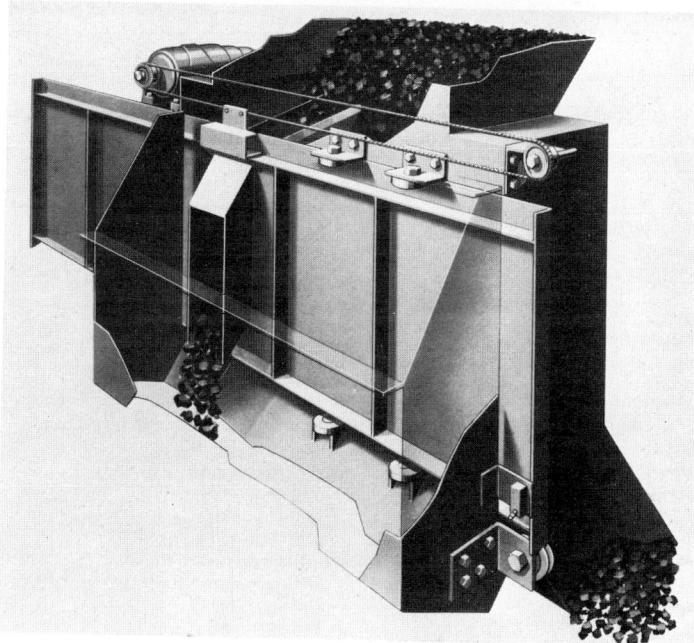

Fig. 3. Plate sampler. *(McNally Pittsburg Manufacturing Corp.)*

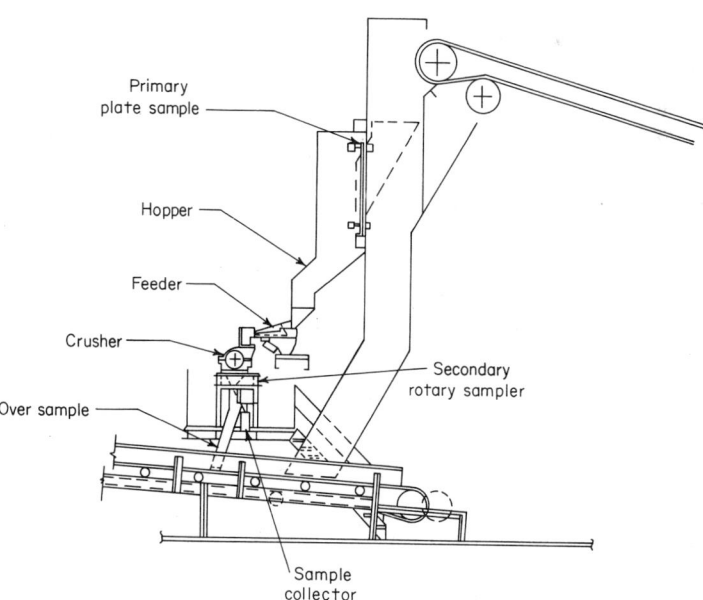

Fig. 4. Typical installation diagram of a plate and a rotary sampler for coal 2 inches by 0 and smaller. The plate-type sampler is used for the 2-inch by 0 and smaller coal; the rotary sampler is used for the sample after it is crushed to 4-mesh by 0.

TABLE 1.   Principal Categories of Coal Preparation Equipment

RAW COAL HANDLING

Bins and hoppers	Picking tables
Conveyors	Feeders
Screens	Storage and reclamation systems
Crushers	Weigh scales
Tramp-iron magnets	

COAL CLEANING

Jigs	Concentrating tables
Heavy-media systems	Hydroseparators
Cyclones	Froth-flotation equipment
Air cleaners	

CLEAN COAL HANDLING

Screens	Bins and hoppers
Crushers	Weigh scales
Spray systems	Car hauls
Conveyors	Chemical freezeproofing
Thermal driers	Oil-treating systems
Centrifugal driers	Storage and reclamation systems
Filters	

WATER HANDLING AND CLARIFICATION

Thickeners	Filters
Cyclones	Solids pumps
Flocculation units	Flotation units

REFUSE DISPOSAL

Bins and hoppers	Refuse pumps
Conveyors	Trucks
Larry systems	Stackers

OTHER PLANT EQUIPMENT AND SUPPLIES USED

Electric motors	Steel
Power transmission equipment	Wire and cable
Automatic control equipment	Conveyors and belting
Power-circuit equipment	Lubricating systems
Pumps	Plant heaters
Piping and valves	Laboratory equipment
Fans and blowers	Plastic pipe and hose
Plant lighting	Maintenance tools and equipment
Dust collection equipment	Monorail hoists
Scales	Plant elevators
Screen cloth	
Reagents and chemicals	

If these planes represent intrusion of mineral deposits into joint cracks or fissures, shearing under stress will yield a mixture of coal and mineral particles. The differences in specific gravities and other physical parameters provide the basis for a separation of the two.

Contact between the coal lumps or between the coal lumps and the moving parts of the crusher is required for successful size reduction. As the breaking action proceeds, the increased number of particles require more contacts per unit mass. The capacity of a crusher of fixed dimensions is therefore less for small sizes than for larger sizes, since it is necessary for the smaller particles to remain in the machine for longer periods of time in order to sustain the required number of contacts.

Run-of-mine breakers, generally located near the entrance of the preparation plant, receive the run-of-mine coal and initiate the size reduction. They should be capable of receiving the largest lumps that will come from the mine and reduce the coal to a top size of 1½ to 8 inches. The natural undersize coal is often screened out ahead of the breaker and added back to the breaker product.

After the coal is washed, the product is usually crushed to a top size of 2-inch or 1½-inch by 0 in a washed-coal crusher. This eliminates the need of crushing all the refuse to the smaller sizes. A thorough investigation is required before the design of a plant to determine whether additional recovery of washed coal could be obtained by crushing the run of mine to the smaller sizes ahead of the preparation plant.

TYPES OF CRUSHERS

For any particular job, one specific type of crusher will probably perform better than any other. The feed material, hardness, location, seam, volume, and finished-product data are the determining criteria in making the proper selection.

The following units are typical of the many crushers and breakers available:

The *rotary breaker* was invented by Hezekiah Bradford and is often referred to as the Bradford breaker. A modern version is illustrated in Fig. 5. It is simply a drum or cylinder of 6 to 12 feet in diameter, and length about 1.5 to 2.5 times the diameter. Lifters mounted on the inside of the drum, in conjunction with the rotary motion, lift and drop the coal for the gravity impact used in the breaking process. The perforated plate of the drum allows the passage of the broken coal particles, and a discharge chute at the end of the cylinder collects the large rock and other nonbreakable material. Since the drum rings of the rotary breaker ride on trunions and there is no center shaft or shaft bearings, it is relatively trouble-free and inexpensive to maintain.

The rotary principle has an additional advantage of low fines production. Undersize coal is immediately screened out at the feed end. This prevents the small particles from being ground down by larger lump and rock.

Both the single-roll and the double-roll crushers shown in Figs. 6 and 7 are commonly used in coal crushing.

The roll on the single-roll crusher has a series of long teeth spaced at intervals, with various short teeth covering the entire crushing surface. The long teeth act as feeders as the coal is squeezed between the revolving roll and the breaker plate, and also penetrate the lumps of coal, splitting them into smaller pieces. The smaller teeth complete the reduction at the trailing edge of the breaker plate. Size adjustments can be made while the crusher is running by changing the clearance between the fixed plate and roll.

The double-roll crusher relies mainly on the impact of specially designed teeth to split the coal, thereby reducing the particle size without producing excessive fines.

Fig. 5. Rotary breaker designed to handle run-of-mine coal. *(McNally-Pittsburg Manufacturing Corp.)*

Fig. 6. Sectional view of single-roll crusher. *(The Jeffrey Manufacturing Co.)*

Modern double-roll crushers, such as the McNally Gearmatic, are equipped with special power trains that will allow for up to an 11-inch adjustment of the rolls while the breaker is in operation. As the movable roll position is changed, the pinion yokes, which can be seen in Fig. 8, change their angle, always holding the pinions in mesh with the gears. The same action occurs when an excessively large rock thrusts back the movable roll. The shock is taken up by heavy-duty relief springs. An important feature of the Gearmatic breaker is that it can reduce lumps without producing excessive fines.

SCREENING

The grouping of coal particles within a range of particle sizes is one of the most important beneficiation operations performed within the coal preparation plant. Sizing is generally accomplished by passing coal over screens.

Separation by differential settling in air or water currents as well as the use of centrifugal forces have been adopted for the separation of very fine sizes, and these techniques will be discussed later.

The primary function of screening is to pass undersize particles through the apertures and restrain and reject the oversize particles. The efficiency of a screening operation is normally stated as the amount of undersize material which reports with the overproduct, divided by the total amount of undersize material in the feed, times 100. To accomplish this, the particles must be brought to the openings at a velocity and direction to least

Fig. 7. Cross section of double-roll crusher. *(The Jeffrey Manufacturing Co.)*

Fig. 8. Double-roll crusher with self-adjusting pinion yokes. *(McNally Pittsburg Manufacturing Corp.)*

hinder their passage through the openings. Collisions of the particles with one another and with the edges and walls of the holes reduce the probability of escape.

Screen surfaces may consist of parallel bars called grizzly screens, punched steel plates with a variety of opening shapes, and woven wire cloth with square or rectangular openings.

They may be stationary or moving; if moving they may shake, vibrate, or revolve. Shaking or vibrating the screen works in conjunction with the horizontal component of gravity to bring the particles to an escape opening. It is also responsible for the stratification process in which the larger-size particles rise to the top of a bed of material being shaken or vibrated, while the smaller-size particles sift through the voids and find their way to the bottom of the bed. This is illustrated in Fig. 9.

Vibrating screens find application in all phases of coal processing and are available in a variety of sizes and types. They have a wide range of uses; from scalping of raw-coal feed to dewatering of extremely fine sizes.

Two very commonly used units are shown in Figs. 10 and 11.

CLEANING

Since most impurities have specific gravities greater than coal, the density of a coal particle is a direct measure of its purity; and the differences in this physical parameter are the basis for the mechanical separation of coal from the noncoal refuse.

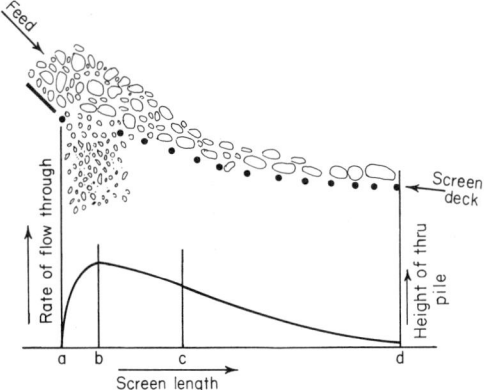

Fig. 9. Screening action in the longitudinal direction.

Both gravity and centrifugal-force devices are employed to effect a separation, and these may use air or liquids as washing media. Although not all coals can be beneficiated by air washing, the air washing of coals that are easy to clean can be readily proved advantageous.

However, actual plant operations have proved that there are a very few instances where air washing will work because of the moisture required on the coal to meet today's mining regulations. The few cases where the raw coal is dry enough for air washing require a complete dust collection system to meet today's air pollution standards. Therefore, the number of plants utilizing air washing is extremely small. The recent trend in mining the western coal fields may revive the interest in air washing because of the scarcity of water.

Fig. 10. Horizontal vibrating screen. *(Allis-Chalmers Manufacturing Co.)*

Water and water solutions, and/or suspensions with relatively low costs and a wide range of specific gravities, provide the basis for the majority of washing techniques currently in use.

In order to determine the proper method and equipment to be used in the cleaning operation, tests must be made to obtain data on size and specific-gravity distribution, moisture, ash, sulfur, and Btu contents.

A washability study of a coal is made by testing the coal sample at preselected, carefully controlled specific gravities. This is referred to as "float-and-sink" analysis. Heavy organic liquids are commonly used to obtain the desired specific gravities.

There are various methods of presenting the data, such as the washability curve shown in Fig. 12. The most important washability curves are: (A) yield-specific gravity, (B) culmulative float coal-ash, (C) plus and minus 0.10 near gravity material accumulation.

MATERIAL DISTRIBUTION

The yield–specific-gravity curve and the cumulative float-coal–ash curve are most important since they determine the theoretical cleaning possibilities. The yield–specific-gravity curve reveals the quantity of coal that can theoretically be obtained by cleaning at a certain specific gravity. The cumulative float-coal–ash curve shows the amount of ash contained in a particular quantity of floated coal. The plus-and-minus 0.10 near-gravity material distribution curve indicates the ease or difficulty of cleaning.

On the basis of the washability curves, one of several, or a combination of several, different techniques may be selected to effect an efficient coal-refuse separation. The available processes can be roughly classified as follows: (1) hydraulic separation, (2) dense medium separation, and (3) centrifugal (cyclone) separation. The basic principles utilized in (1) and (2) are also present in (3) and used in conjunction with the centrifugal force generated in the cyclone.

Hydraulic separation depends on a process referred to as jigging, which creates a particle stratification from an alternate expansion and compaction of a bed of particles by a pulsating fluid flow. As originally developed, a basket filled with material was moved up and down in a tank filled with water. The more modern Baum-jig process utilizes an air impulse concept in which the water is moved by air pressure from an adjacent sealed chamber. Typical of the many variations on this principle is the McNally Norton standard

Fig. 11. Inclined vibrating screen. *(Allis-Chalmers Manufacturing Co.)*

washer shown in Fig. 13. Raw coal is cleaned in two stages. A primary separation at the feed end of the washer removes the heavier refuse material. The secondary compartment divides the coal into a bottom layer of middlings, on which rides a second layer of quality coal. At the discharge end, the two layers are split, the good coal passing into a delivery sluice. Middling materials are discharged separately for rejection or reprocessing; or they may be delivered as a second-grade coal.

The washing jig is the most popular and least expensive coal washer commercially available but may not effect as accurate a separation as desired.

More accurate separations are made in dense-medium vessels. Coal is slurried in a medium with a specific gravity close to that at which the separation is to be made. The lighter coal tends to float and the refuse to sink. The two fractions can then be mechanically separated.

Theoretically, any size particle can be treated by dense-medium processes; practically the sizes range from about 0.5 millimeter to about 6 inches. Larger sizes are also occasionally washed.

Organic liquids, salt solutions, aerated solids, and water suspensions have found use as commercial media. However, water suspensions meet most of the practical requirements for a satisfactory inexpensive medium.

The bulk of coal mechanically cleaned by the dense-medium process is separated in suspensions of magnetite in water.

The McNally Tromp three-product vessel, illustrated in Fig. 14, consists of a shallow tank filled with high- and low-gravity medium consisting of a suspension of finely ground

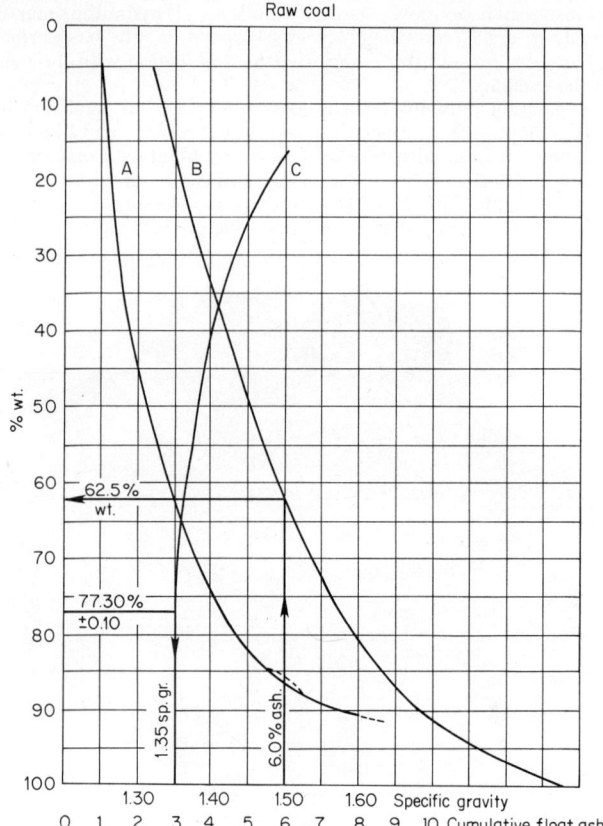

Fig. 12. Washability curve. *(Source: McNally Pittsburg Manufacturing Corp.)*

magnetite in water. A low-gravity medium is introduced through four horizontal headers and distributed in horizontal layers across the feed end of the bath by baffle plates. A high-gravity medium is introduced in the lower portion of the vessel by a fifth header and flows in a horizontal layer escaping through an adjustable underflow gate.

The material to be separated is distributed horizontally across the full width of the vessel in a uniform layer. On entering the low-gravity medium the coal floats and is removed by a scraper conveyor while the middlings and refuse sink to the high-gravity section, where the final separation of middling and refuse is made. This final separation is accomplished by a single scraper conveyor which carries the middlings and float material on the top flight, and the refuse and sink material in the bottom flight.

The laminar principle provides substantially horizontal currents from the feed end to the discharge end of the vessel in the low-gravity section, which compensates for any tendency of unstable medium to settle out. An air lift in the high-gravity section accomplishes the same for the high-gravity medium. There is a minimum of turbulence in the bath since the coal, medium, and conveyors move in mass in a substantially horizontal direction except for the refuse fractions which settle vertically.

The medium-density and level-control circuit is completely automatic.

The Barvoys vessel was originally designed to employ suspensions of barytes and clay which approached a true liquid. As now fabricated and marketed, it is designed to use a standard magnetite dense medium. The trough-type washer utilizes lifters to remove the clean coal out of the bath, thereby reducing the quantity of medium to be recirculated.

Another variation of the dense-medium technique is the *Wemco* separatory cone with

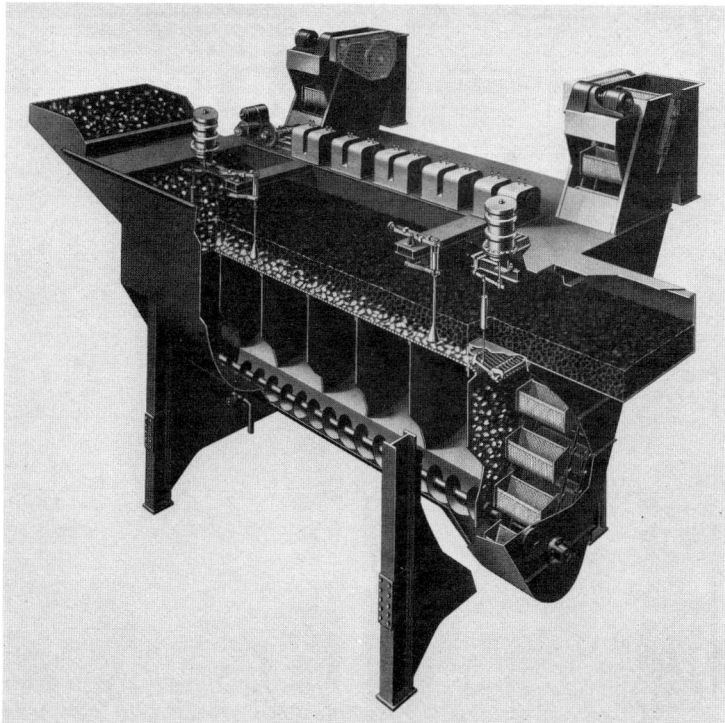

Fig. 13. McNally Norton standard washer.

an internal airlift illustrated in Fig. 15. The dense medium suspension is maintained by a mechanical stirring action. The raw coal enters the top of the cone near the center. The coal spirals out from the center to the periphery of the coal and overflows a weir with a quantity of the medium. The refuse sinks to the bottom of the cone from whence it is air-lifted to the sink launder. Medium enters the cone through a medium return pipe and is discharged at various levels to control the density of the gradient of the settling medium.

The use of centrifugal force as an aid in coal-refuse separation is a relatively recent addition to the coal cleaning processes. As originally developed, the device employed a

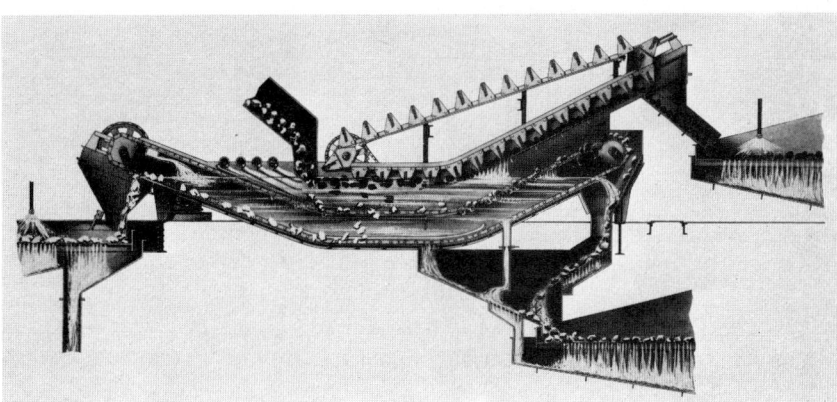

Fig. 14. McNally Tromp three-product dense-media vessel.

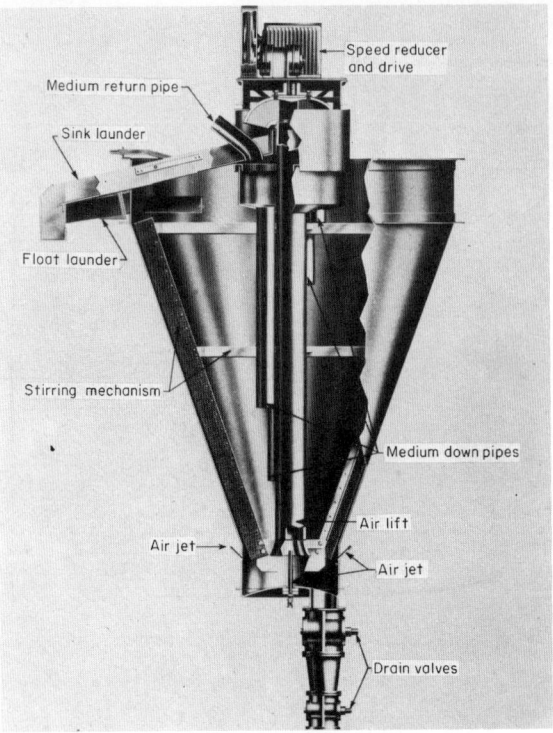

Fig. 15. Separator cone. *(Wemco Division, Envirotech Corp.)*

dense working medium, and was known as a "dense-medium cyclone." More recently, developed units for easy cleaning problems do not employ an artificial gravity suspension and are referred to as "hydrocyclones."

In a typical dense-medium cyclone illustrated in Fig. 16, the slurry of coal and medium is admitted at a tangent near the top of the cylindrical section, forming a strong vertical

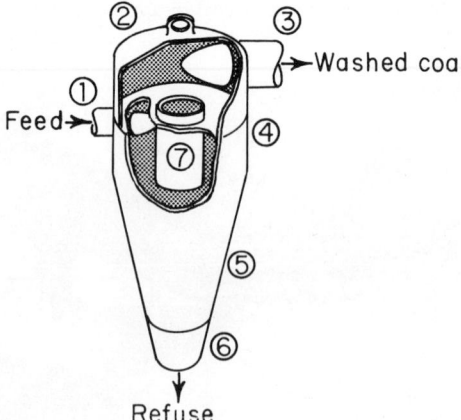

Fig. 16. Dense-medium cyclone: (1) Feed inlet. (2) Overflow chamber. (3) Washed-coal outlet. (4) Cylindrical section. (5) Conical section. (6) Replaceable underflow orifice. (7) Vortex finder.

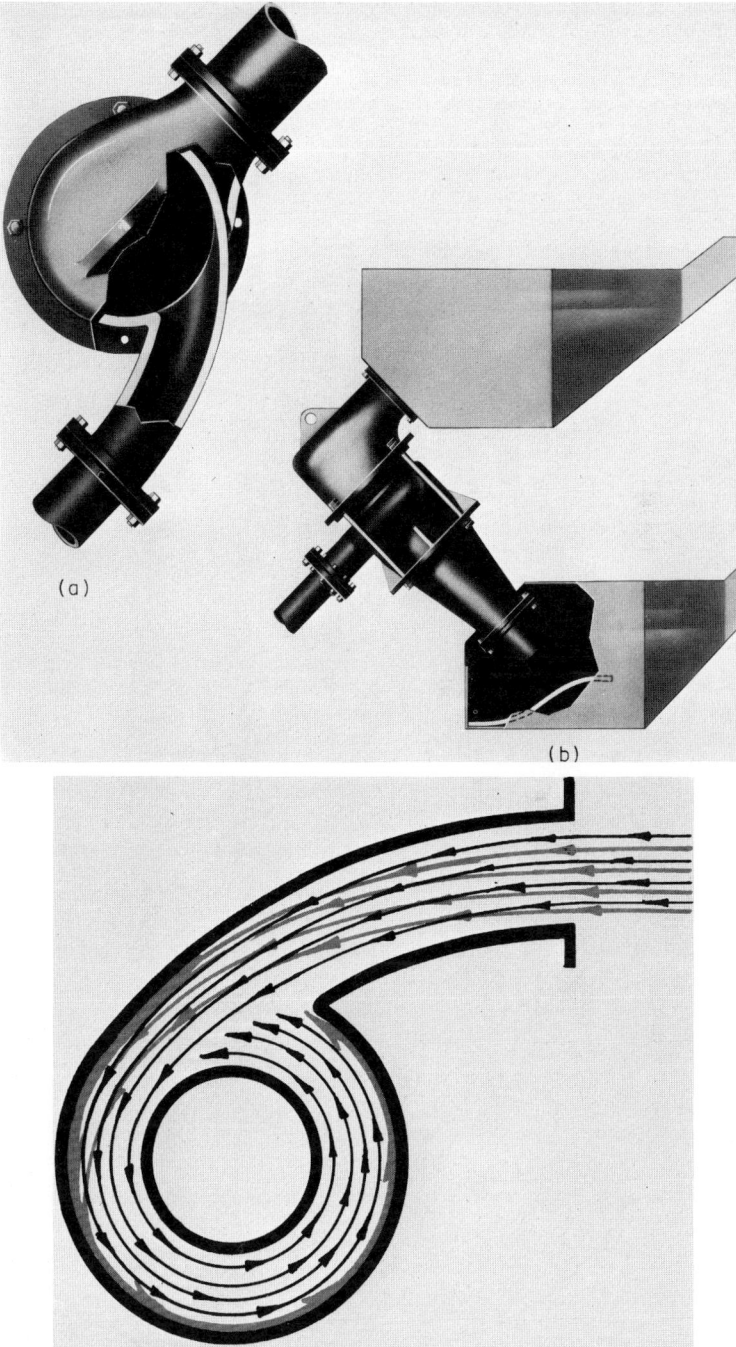

Fig. 17. Heavy-media cycloid: (*a*) Top view. (*b*) Side view. (*c*) Circulation pattern. (*McNally Pittsburg Manufacturing Corp.*)

flow. Under gravimetric forces the refuse with a higher specific gravity moves along the wall of the cone and is discharged at the apex. The coal particles of lesser specific gravity move toward the longitudinal axis of the cyclone and finally through the vortex finder and overflow chamber to the outside as clean coal.

The centrifugal force, the general flow pattern of the medium, and a progressive increase in specific gravity of the medium as it descends toward the apex all interact to provide an efficient separation.

An example of a dense-medium cyclone is shown in Fig. 17. These units are capable of

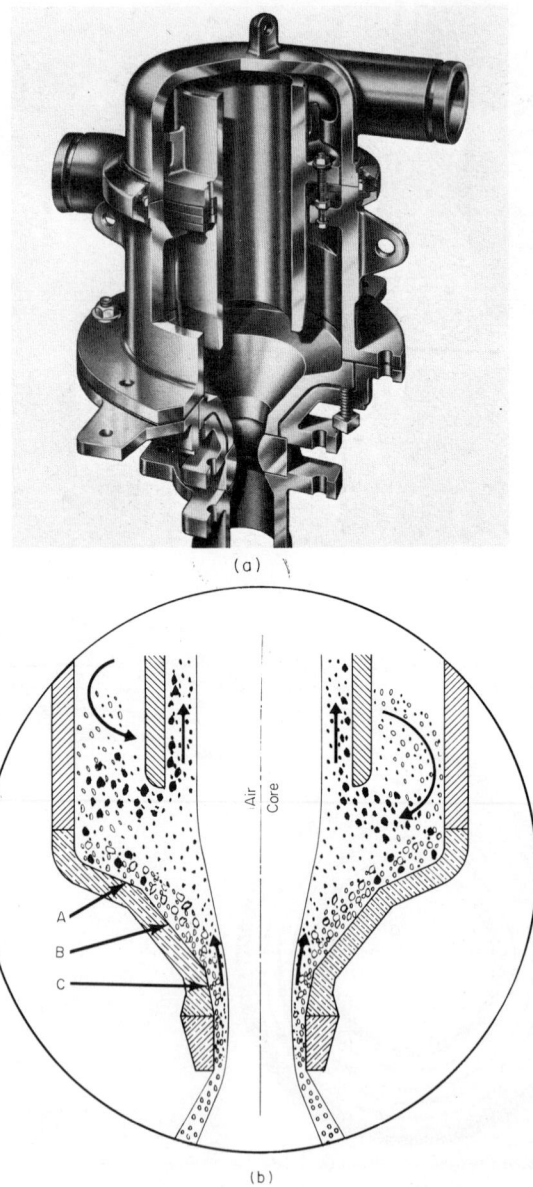

(a)

(b)

Fig. 18. McNally Visman Tricone: (a) Cross section. (b) Diagrammatic view of separation sequence.

washing coals in a size range of 1¾-inch top size to 28-mesh bottom size at specific gravities ranging from approximately 1.35 to 1.80, using magnetite and water as the dense medium. The dense-medium cyclone has proved to be a very efficient means of washing coal, regardless of the amount of near-gravity material in the feed.

Original research on cyclones led to the development of the hydrocyclone which is capable of performing a specific-gravity separation employing only water and centrifugal force. See Fig. 18.

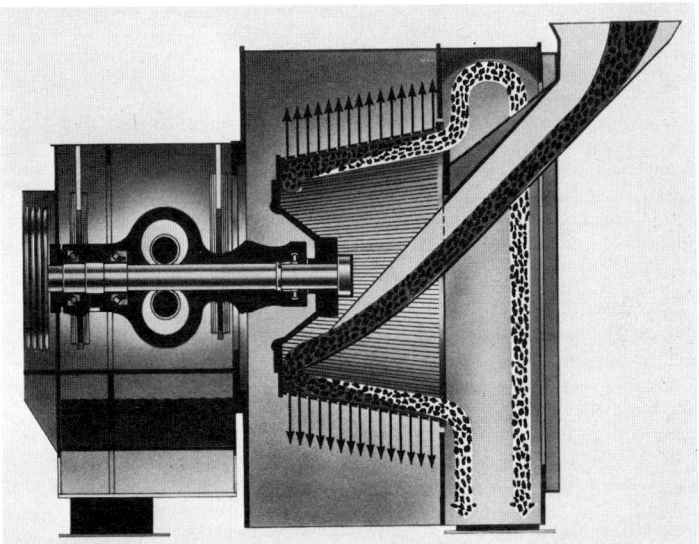

Fig. 19. *Wedag* vibrating-action centrifuge. *(McNally Pittsburg Manufacturing Corp.)*

The design feature of the hydrocyclone which permits the use of water only is the wide angle or angles in the cyclone bottom. This allows the formation of a hindered settling bed as the dense particles move down the side wall under the forces of gravity. Less dense particles are unable to penetrate this heavy bed and move back into the main hydraulic current and are discharged out the top of the unit through the vortex finder.

Hydrocyclones have been utilized to wash coals with a top-size range of 1¼ inches to 28 mesh. The units receiving the coarse top size are generally used as scalping devices to relieve the load on other equipment. The hydrocyclones washing 28-mesh by 0 coal are usually employed because of the presence of pyrite or oxidized coal which has proved difficult to wash by other means.

Tables are another method used to wash coal. The sizes usually washed in table installations are ½ inch by 0, ⅜ inch by 0, ¼ inch by 0, and finer fractions, or deslimed feeds of these top-size ranges. The feed to a table deck is normally in the ratio of two parts water to one part feed solids. The reciprocating action of a table stratifies the high-gravity coal particles on the bottom, and the low-gravity coal particles rise to the upper level of the bed. As the low-gravity particles rise, they are moved across riffles which separate the high- and low-gravity material by the water flowing to the low side of the table deck. The refuse is trapped in the riffle troughs, and the motion of the deck moves the refuse to discharge off the end.

Flotation is a process for separating fine (28-mesh by 0 or smaller) coal particles from refuse particles by using finely disseminated air bubbles passed through a coal slurry. The fine coal particles adhere to the air bubbles and rise to the top where they are removed as a concentrate, while the heavy refuse particles sink and are removed by the flow of water through the flotation cell. Flotation cells generally have some type of agitation, and a frothing reagent such as methylisobutylcarbinol (MIBC) is added to the feed.

DEWATERING AND DRYING

Water remaining on marketable coal is a contaminant as serious as the undesirable ash. It may cause problems in handling and shipping, increase the freight cost, and reduce the heating value of the fuel.

Coal dewatering is required, not only before the clean product is shipped, but it may also be required at various times during its movement through the preparation plant. For example, the raw coal being delivered to a heavy-media circuit should be sized, prewetted, and then dewatered. This preliminary treatment is a necessary aid in controlling the required specific gravity of the media.

There are two general processes that can be used to eliminate undesirable water. It may be removed mechanically or thermally; and practically speaking, both methods may be used successively on the same product.

Since the water is a surface contaminant, the difficulty of dewatering coal will increase with increases in the surface area of the material. The finer the coal, the greater the surface area available for the adherence of water.

There are many mechanical dewatering devices, most of which were designed to work most efficiently with coal particles within a certain size range. Overlapping applications of the various units are common but experience has matched the following coal and equipment:

Sizes larger than 1½ inch are dewatered sufficiently, using only vibrating screens.

Intermediate sizes down to ½ millimeter are usually dewatered on vibrating screens, with further dewatering required to produce an acceptable product. Vibrating-screen-type centrifuges are used to further dewater this size coal.

Disk filters and screen bowl centrifuges are used for removing some of the water from the minus 28-mesh solids.

Centrifugal driers, as mentioned above, are widely used in the coal industry and have proved to be reliable efficient machines. In general, centrifuges can be classified into two types: the vibrating-screen type and the high-speed positive-discharge type. The vibrating-screen type shown in Fig. 19 is becoming increasingly popular for the separation of coal solids and liquids in large quantities.

In this vibrating-screen-type centrifuge, material to be dewatered passes down inlet chute (1). Feed material is distributed onto inner surface of screen basket (2). The rotating

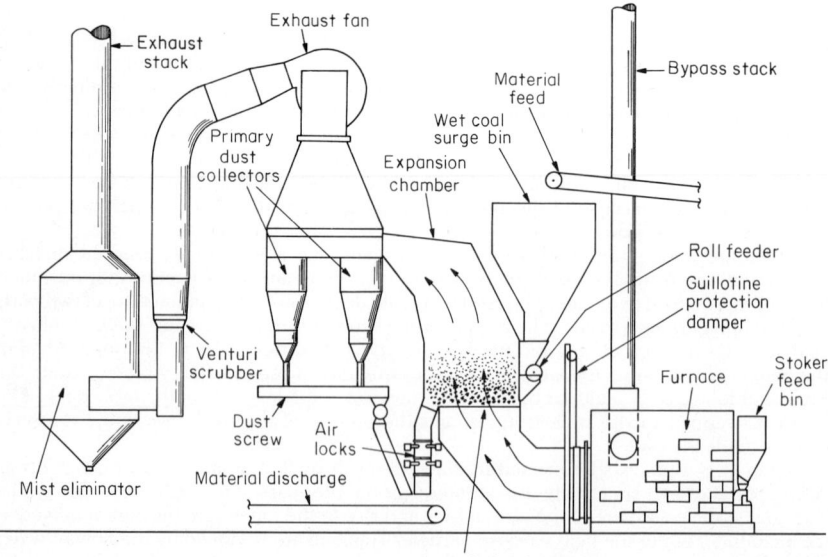

Fig. 20. "Fluid Flow" coal drier. *(FMC Corp.)*

Fig. 21. Large fluidized-bed drier. *(Flowdryer—McNally Pittsburg Manufacturing Corp.)*

screen basket is kept in axial vibratory motion by a vibratory unit. Then axial vibrations move the material toward the larger diameter of the basket. In addition, vibrating actions keep the basket openings clear. The feed material remains loose, which improves dewatering even with high throughput of solids.

The dewatered material which is thrown out at the large-diameter end of the screen basket falls downward, while the water, which is centrifuged out, is ejected at the side.

The high-speed positive-discharge-type centrifuges are used to dewater washed coals which have a top size of ⅜ inch. These machines will handle wet feeds with up to 40 percent surface moisture.

For the removal of very fine material (28 mesh and smaller) from a high percentage suspension, a filter process may be used. This is accomplished by placing the filter with a cloth or screen surface in the suspension and reducing the pressure to create a partial vacuum on the side away from the suspension. Both solid and liquid are drawn to the surface of the filter, with the liquid portion passing through, leaving a cake of the solid on the filter face. The filter is withdrawn from the suspension at intervals and the cake of coal mechanically removed.

Two basic types of filters are being used: the disk type and the drum type. The disks range in diameter up to 12.5 feet with multiple disks utilized when necessary. Drum-type filters of varying diameters and lengths are employed.

Sedimentation processes usually provide the initial step in a dewatering process where filters are required. A vessel called a thickener received a slurry and is arranged so that the settled solids or sediment may be collected and removed. Chemical flocculants are sometimes employed to assist the material settling process. The clarified decanted liquid may be used for recirculation to the preparation plant.

Dewatering of coal by many of the techniques cited is often followed by thermal drying for a final reduction in moisture content.

Coal driers can be grouped into six basic types. These are: (1) fluidized bed, (2) suspension on flash, (3) multi-Louvre, (4) vertical tray and cascade, (5) continuous carrier, and (6) drum type. All the driers employ convection as the major principle of heat transfer. Hot gases generated in a combustion chamber are brought into intimate contact with the wet coal.

For a number of years there has been an expanding use of coal-drying techniques and a specific expansion in the application of fluidized-bed coal drying.

The fluidized-bed driers use a rising current of hot gases to lift the solid coal particles and stir them into violent action. The mixture of coal and heated gas acts very much like a fluid. Particles of coal, up to 1¼-inch size, are suspended and bathed in the heated gas, which dries the coal and moves it to the discharge of the drier grate. The temperature of the heated gas is precisely controlled to obtain the amount of drying required.

The commercial application of this technique is illustrated in Figs. 20 and 21.

REFERENCES

1. Leonard, J. W., et al. (eds.): "Coal Preparation," 3d ed., American Institute of Mining Engineers, New York, 1968.
2. Anon.: Simplified Preparation of Utility Fuel, *Coal Mining Process.*, May 1972.
3. Eads, B. F.: Planning and Developing Monterey No. 1 Mine, *Coal Age*, February 1971.
4. Anon.: Storage Systems and Equipment, *Coal Age*, December 1959.
5. Packard, C. E., and P. T. Luckie: "Coal Preparation Ahead of Gasification," 1972 Coal Show, American Mining Congress, Pittsburgh, Pa., May 8–11, 1972.

Coal Handling and Storage Systems

BY PAUL SMARDO*

After World War II, there was a sharp increase in the use of natural gas and oil and a corresponding decline in coal consumption. The coal industry's response to remain competitive accelerated the development of techniques and transportation systems designed to lower costs and improve service. Pipelines, low-cost barge hauling, and large-volume rail units are now common practice. Both mine and user facilities have been created to accommodate the large-volume rapid movement of coal from producer to consumer.

FACILITIES FOR UNIT TRAINS (UNITRAINS)

The use of *unit trains,* composed of a hundred or more coal cars handled as an entity, has materially reduced freight rates. This system, however, calls for sufficient coal to load a unit train in the short time available, usually less than four hours, to secure the maximum freight cost reduction. Since few, if any, mines or preparation plants process coal at this rapid rate, the trend has been for the inclusion of large storage capacity in new facilities and the addition of this capability to existing plants.

These storage systems may take the form of open storage piles or enclosed bins and silos. Open storage piles may be conical-shaped, wedge-shaped, or kidney-shaped, depending on the stockpiling methods used. Bins and silos may be round or square, steel or poured concrete or precast concrete staves. Individual storage systems are usually of unique design to meet the specific needs of the producer. Factors may include, in addition to the desire for trainload surge and emergency storage, protection of low-moisture-content coal, protection from freezing weather, and suitable dust control.

The most common form of open storage is the conical-shaped stockpile shown in Fig. 1.

The "dead storage" is often replaced with a doughnut-shaped earth fill. This reduces the problem of spontaneous combustion in storage and also retains more of the coal in live

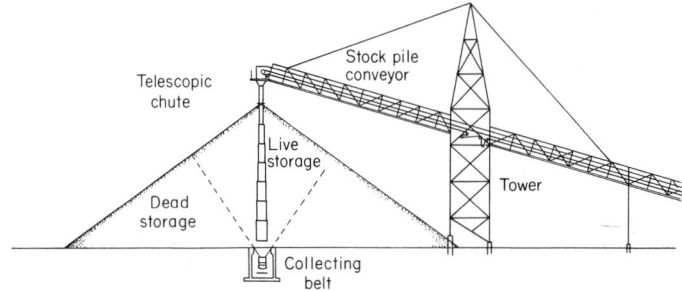

Fig. 1. Elevation of conical stockpile.

storage. A disadvantage with this arrangement is the limited withdrawal rate from the single recovery opening. By extending the reclaim system the full length of the diameter of an all-coal pile, as shown in Fig. 2, and adding several more openings, the withdrawal rate can be increased severalfold.

For the conical pile, the coal is generally delivered from a fixed, cantilevered stacker/conveyor normally equipped with a telescopic chute to restrict the dust at this point. A storage pile (Fig. 1) that is 70 feet high with a repose angle of 35° and a 200-foot diameter

*Assistant Manager, Coal Handling Systems, McNally Pittsburg Manufacturing Corp., Pittsburg, Kansas.

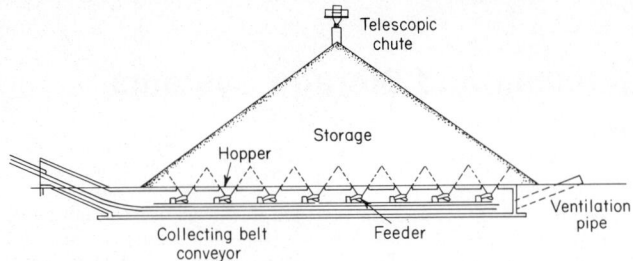

Fig. 2. Section through conical stockpile. Multiple recovery openings located below pile.

will contain approximately 18,000 tons. The coal used as dead storage is periodically reclaimed by moving it over a recovery opening with a bulldozer or other coal mover.

If very large storage capacity, ranging from 40,000 to 100,000 tons or more of total storage, is required, the long wedge-shaped pile shown in Fig. 3 is utilized. These wedge-shaped piles are normally fed from a traveling stacker that operates with a belt conveyor that runs parallel to the pile. This conveyor is often referred to as a trunk-line belt and is normally placed at an elevated position. Once the pile is built to its maximum height at one end, the tripper moves automatically a few feet at a time. This results in the coal being discharged at the top and leading edge of the wedge-shaped pile, with the coal cascading down.

An added feature of this type of pile is the ability to stow different types of coal in different sections of the pile, thus permitting loading each individually or both simultaneously as a blended coal.

Various feeder arrangements are used to withdraw the material from storage piles or bins. In the long tunnel, as indicated below the conical- and wedge-shaped piles shown in Figs. 2 and 3, electric or mechanical vibrating feeders lend themselves to in-line installation above the collecting belt. These feeders are usually spaced about 20 to 25 feet apart. This close spacing results in more active storage directly above the tunnel. Sloping chutes between the hopper outlets and the belt conveyor are often utilized to feed the material from a storage pile onto the collection conveyor. These chutes are either equipped with an overcut gate or hinged with the feed being cut off if it is raised with a hoist to a horizontal position.

The collecting-belt conveyor below the chutes is capable of a rapid delivery rate to the designated loading point. A variable withdrawal rate at this point is necessary to compensate for a nonuniform train speed, and additional facilities are also needed to accommodate the different types of cars (for example, gondola and hopper cars), as well as changing the coal flow from car to car as the train moves under the loading point. A large surge bin is

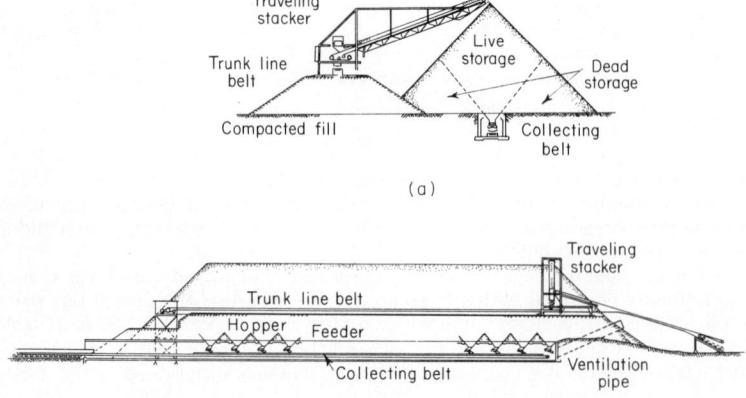

Fig. 3. Wedge-shaped stockpile: (*a*) section; (*b*) elevation.

ordinarily located at the loading point. A surge bin over two load points on adjacent tracks is shown in Fig. 4.

In this particular case, the capacity of the surge bin above each load point is approximately 1.5 times the railroad car capacity. If 100-ton cars are being loaded, the surge-bin capacity directly above the loading point would be approximately 150 to 175 tons.

The bins can be equipped with level detectors and indicating lights so that the operator

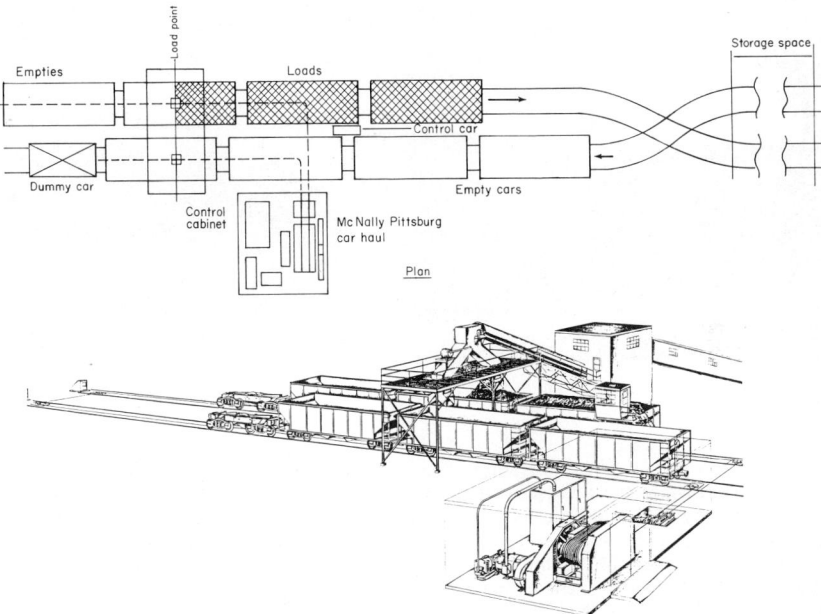

Fig. 4. Flood loading from surge bin utilizing car hauls for car movement.

of the locomotive or mechanical moving device can adjust the speed of the train to maintain approximately the same level in the bin. The chute attached to the hopper below the surge bin is designed to permit filling of the car at a uniform height without the danger of spilling the coal over the sides. The chute is equipped with a hoist to permit raising while not in use, and for adjusting to different car heights. However, the majority of the unit trains are equipped with cars of uniform height. A motorized gate is placed between the surge bin and the chute, and is closed during the time needed to replace a loaded car with an empty car. During the time the gate is closed, the level naturally rises in the surge bin. After the empty railroad car is in position under the chute, the gate is opened and the coal fills the end of the car immediately. This type of loading is referred to as "flood loading," as the movement of the car below the chute controls the rate of flow into the car.

A shallow track hopper located below the tracks at a loading point may be used to return the spillage, via a conveyor belt, back into the surge bin.

The surge-bin arrangement is only one of several methods used for railroad loading. A common alternative is the use of a traveling tripper, shown in Fig. 5. This arrangement consists of an elevated belt, with a belt tripper, running parallel to the loading tracks. The tripper is able to move back and forth over a distance of several hundred feet. To prevent spillage, a gate in the upper portion of the bifurcated spiral chute enables the operator to bypass the interval between cars. Two loading procedures are possible. The cars can be moved at near-constant speed, and the tripper moved slightly forward or backward to compensate for an uneven coal flow. The second procedure is to advance the train intermittently and allow the tripper to move and fill the cars.

An interesting variation of the "flood loading" technique has been incorporated in the Kaiser Steel Corporation's York Canyon mine in New Mexico. As shown in Fig. 6, a train tunnel 22 feet wide and 440 feet long runs underneath a conical-like storage pile. The

Fig. 5. Belt tripper loading system.

train consisting of 84 empty coal cars of 100-ton capacity per car is moved up a slight grade, and the cars are loaded from a single opening in two hours or less. A concrete structure approximately 24 feet square is located at the center of the tunnel. A tunnel arch 22 feet wide and 26 feet 8 inches high was utilized on both sides of the concrete structure. Earth fill on each side and over the tunnel provides a base for the coal storage pile.

   The coal is withdrawn from the center of the pile through a 21-foot-diameter steel cone having a 4½-foot opening at the bottom. Flow from the pile is regulated by a hydraulically operated undercut gate that serves a hinged loading chute which is raised and lowered by an electric hoist. Train loading is achieved by means of controls in a cab located within the concrete structure. The operator in the cab is in constant radio contact with the train crew

Fig. 6. Unit train loading at the York Canyon Mine (New Mexico).

and regulates the flow of coal into the cars as the train is pulled through the tunnel at a speed of about ¾ mile per hour.

At the point of delivery, facilities for the transfer of coal from the carrier to the user generally consist of a series of shallow track hoppers, located below the delivery tracks, into which the coal is spilled either with the use of a rotary dumper or through sets of gates in the bottom or sides of the coal cars. From the track hopper, the material is handled via feeders and belt conveyors into the user's plant area. See also *Railroad Transportation of Coal* in this Handbook section.

BARGE LOADING SYSTEM

There is a great variation in barge-loading plants. Each facility must be designed to meet a set of particular loading problems. Pertinent factors include tonnage, geographical considerations, size of the stream, and water-level fluctuations.

The kinds of systems currently being used are illustrated in Figs. 7, 8, and 9. The dock facility of the Badgett-Terminal Corporation at Grand Rivers, Kentucky, is an example of a stationary chute plant. This installation transfers coal from railroad hoppers to barges. From the single hopper below the tracks, the coal travels through a fixed chute into the barge. This installation can empty a 50-ton car in less than two minutes and can barge-load at a rate of 1800 tons per hour.

The elevating-boom type of load is the most commonly used configuration in the United States. It is a very flexible system, adapting easily to variations in water level, and frequently loads at rates in excess of 1500 tons per hour. In one installation of this type, reciprocating feeders discharge the coal from the hopper onto a conveyor belt that carries

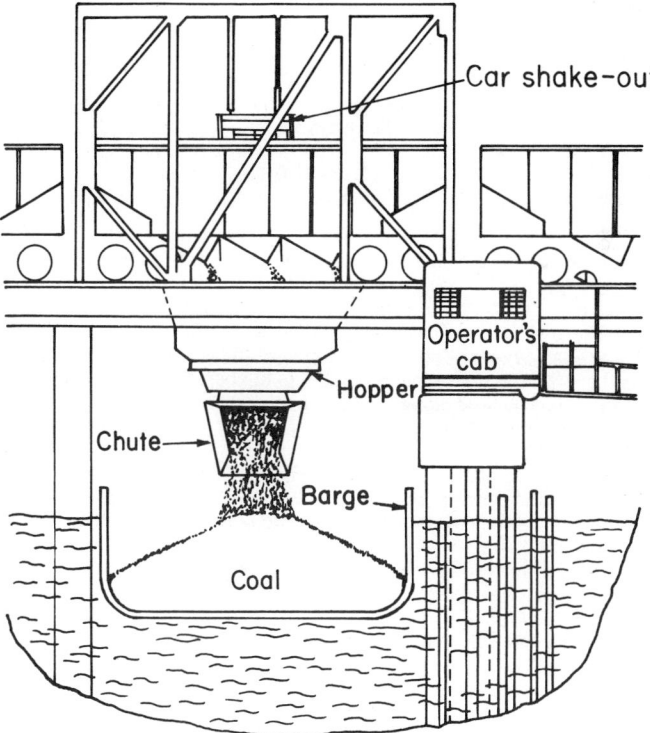

Fig. 7. Stationary chute barge-loading system.

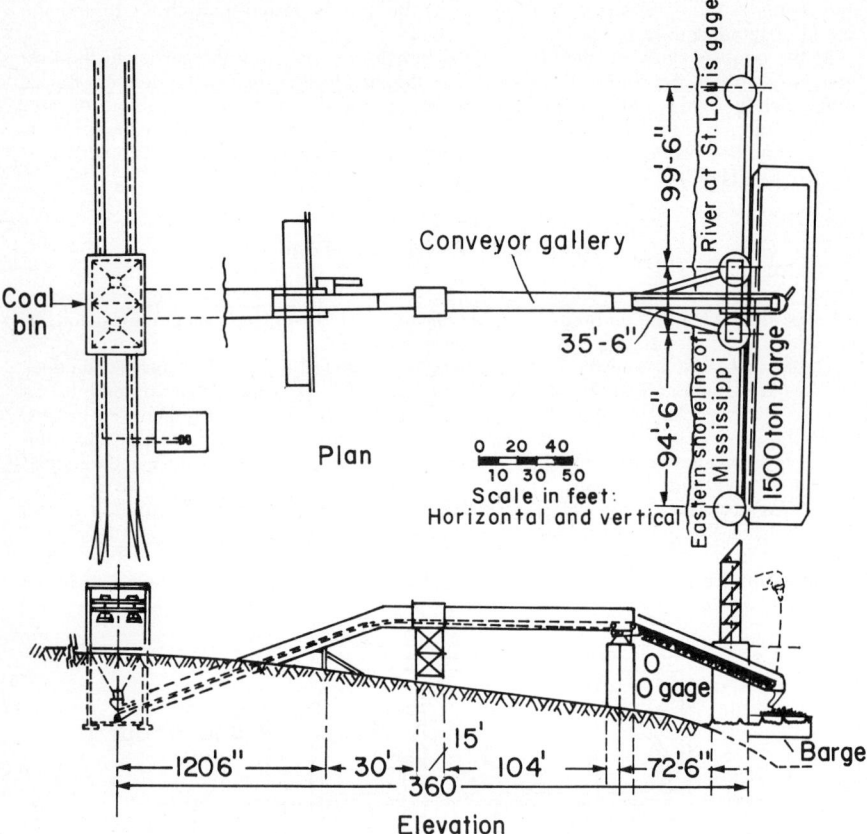

Fig. 8. Elevating-boom-type barge-loading system. Dimensions shown are for actual installation.

the coal 360 feet to the barges. Barges are moved through the loading point by a small switching boat.

The traveling tripper system, shown in Fig. 9, is being used by the Gibraltor Coal Company in western Kentucky. The system is fed from a 5000-ton conical-shaped stockpile with the use of a conveyor belt which discharges onto a tripper belt. The belt-tripper with an integral control travels on tracks parallel to the belt conveyor, both being installed atop a row of ten pillars located along the center line of a canal. The tripper feeds coal to a reversible traverse shuttle belt which discharges through telescoping chutes to barges on either side of the installation.

The barges are held in position for loading by a mooring system which is capable of moving vertically up and down as the barges take on weight. A weight-measuring device on the tripper belt tallies the coal being loaded. The belt tripper can load eight barges at a time, four on each side.

Early barge unloading was largely accomplished with the use of whirler-type cranes with clamshell buckets. The trend is now toward the use of clamshell buckets mounted on stationary, straightline-type towers, or on traveling tower barge unloaders. The twin clamshell buckets, shown in Fig. 10, can unload up to 3600 tons per hour, emptying a river barge in three passes. A barge-shuttle system moves barges under the loader. This type of installation is also known as a continuous-ladder unloader.

More recently, there has been growing interest in the continuous-bucket-type unloader, as shown in Fig. 11. These installations appear to be more efficient and less expensive than the normal clamshell bucket systems.

Fig. 9. Traveling tripper-type barge-loading system. *(McNally Pittsburg Manufacturing Corp.)*

Fig. 10. Twin clamshell buckets can unload up to 3600 tons of coal per hour. *(Dravo Corp.)*

HANDLING SYSTEMS—COAL USER FACILITIES

The great variety of coal handling and storage systems is a consequence of providing each user with the most efficient design for specific purposes. A complex system might include a coal preparation plant that would incorporate storage facilities to accommodate for the many differences between production and use. A simple system might provide transport of the fuel from public carrier to furnace silos and may or may not involve additional

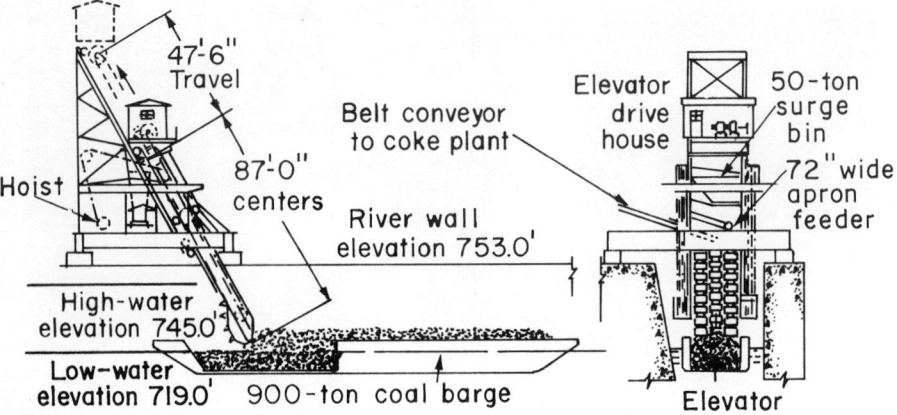

Fig. 11. Continuous bucket barge-unloading system.

storage. It should be pointed out that the descriptions in the two preceding articles of this Handbook also are concerned with complementary functions.

Martin Lake Steam Electric Station (Fig. 12) is typical of large handling systems. Coal (6 by 0) is received at the rate of 4800 tons per hour, stockpiled at 2600 tons per hour, and reclaimed at 1600 tons per hour. The coal is crushed to ¾-inch and delivered to silos at each of four unit boilers at the rate of 550 tons per hour.

At the Burlington Generating Station (Fig. 13), 1½- by 0-inch washed coal is received from central Illinois mines. Two strings of incoming cars move on parallel tracks beneath a car shaker and over a track hopper. A car haul can handle 21 loaded cars on one track and move out 21 empties on the adjacent track.

The 1½- by 0-inch coal is withdrawn from the track hopper by four vibrating feeders at a rate of 970 tons per hour and delivered to a transfer building. A portion of the stream can go to the crusher house and on to the silos—and the remainder to storage, or the entire flow can be delivered to the live storage pile.

Coal is delivered to the storage pile containing 3735 tons live storage by a traveling tripper centered directly above the pile. Coal is withdrawn from the storage pile at 420 tons per hour by six reclaim feeders beneath the pile. This flow, too, can be directed via the transfer house to the crusher to reduce any frozen lumps. The coal then passes over a belt-conveyor scale and is delivered to the tripper belt located above the silos.

Automatic sampling equipment, located in the transfer building, permits sampling of all coal delivered to the station.

A mine-mouth facility is illustrated in the aerial view of Fig. 14. The facility consists of two fully integrated coal-handling systems. The outlying section receives coal by truck from a nearby strip mine at a rate of 600 tons per hour. The run-of-mine coal is fed to a rotary breaker, and is sampled, weighed, and moved on to the generating station storage pile. The second system moves the coal from storage, crushes, weighs, samples, and discharges it into the generating station silos.

All types of conveyors (belt, apron, compartment, scraper, spiral, etc.), as well as reciprocating feeders, bearings, pulleys, sheaves, and sprockets, are vital components in coal-handling systems. Auxiliary, but often necessary equipment includes mobile coal movers, such as front-end and portable belt loaders, weighing and control devices.

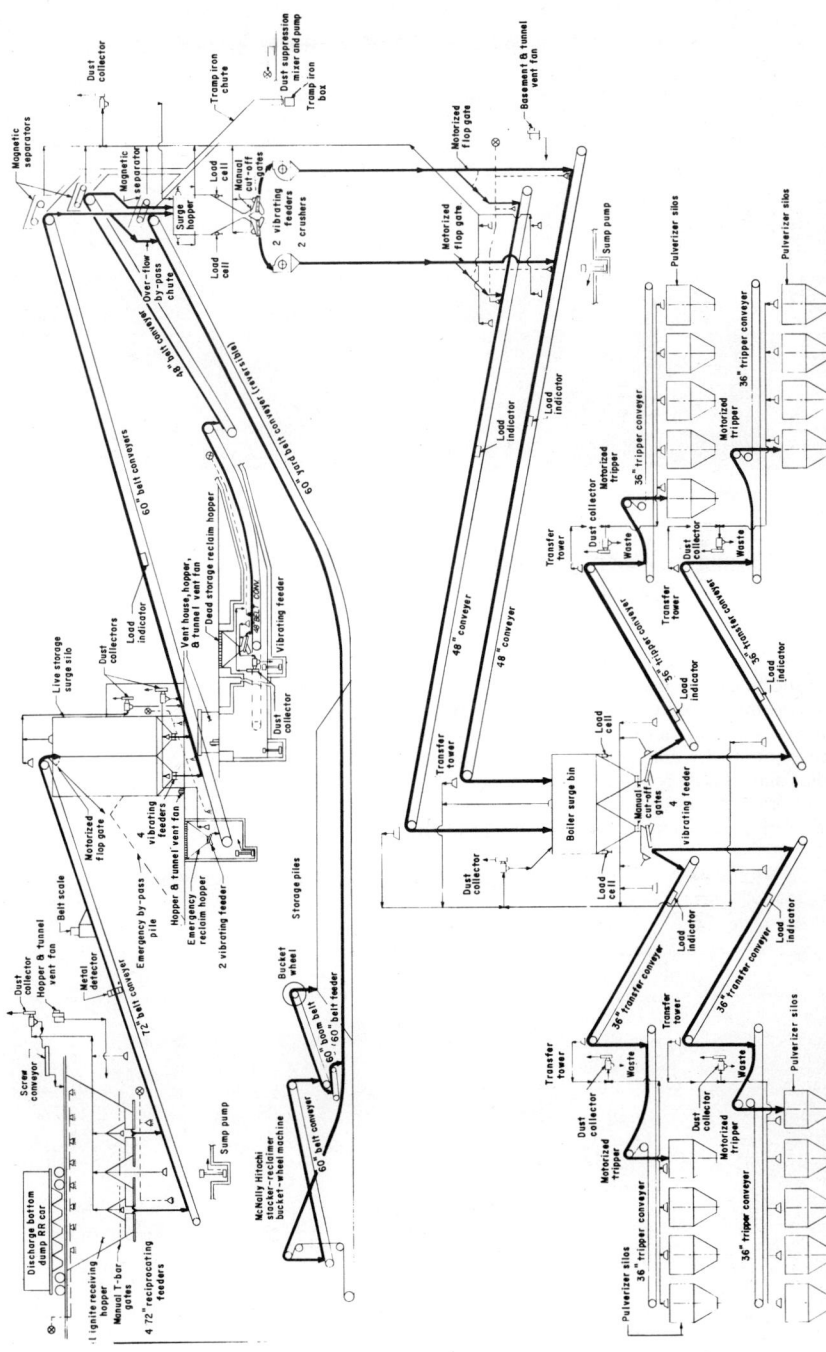

Fig. 12. Schematic diagram of Martin Lake Steam Electric Station. (*Dallas Power & Light Co.; Texas Electric Service Co.; Texas Power & Light Co.*)

Fig. 13. Aerial view of Burlington Generation Station, Iowa Southern Utilities.

The proper use and disposition of storage facilities require information concerning space availability, rate of deposition and use, and coal characteristics. Stacker/reclaimer equipment assures flexible operation and materials handling economies.

COAL TRANSPORT WITHIN USERS' FACILITIES

The short-range transportation considered here is generally effected with mechanical systems engineered to handle a wide range of particle sizes. Problems associated with the movement of coal include the degradation of size with handling, the flow characteristics of both wet and dry materials, and the suppression of dust. The flow of coal from bins and chutes and its adherence to conveyor and other handling equipment are problems chiefly arising from the nature of the material, but are usually amenable to solution on the basis of engineering experience. Both physical and chemical properties of the coal to be handled

Fig. 14. Aerial view of The Empire District Electric Company's power plant located near Pittsburg, Kansas. Long coal conveyors are clearly evident.

TABLE 1. Representative Actual Plant Situations

Coal characteristic (Ultimate analysis, percent by weight)	Plant 1 Expected range	Plant 1 Typical	Plant 2 Expected range	Plant 2 Typical	Plant 3 Combined range	Plant 4 Combined range	Plant 5 Expected range	Plant 5 Typical
Moisture	26.4–36.5	31.80	26.0–31.0	28.0	4.3– 7.8	13.1	7.46–12.00	10.00
Carbon	34.6–43.2	39.03	57.0–75.0	66.0	67.5–70.6	59.6	52.51–64.20	56.20
Hydrogen	2.6– 3.4	2.98	3.6– 6.5	4.74	5.0– 5.7	4.7	3.61– 4.28	3.80
Nitrogen	0.4– 0.8	0.54	0.5– 0.94	0.67	0.9– 1.3	1.2	0.73– 1.26	2.30
Chlorine	0.0– 0.05	0.01	0.0– 0.1	0.04	0.02– 0.22		0.00– 0.15	0.03
Sulfur	0.4– 1.0	0.62	0.25– 1.0	0.65	0.50– 0.56	1.2	2.70– 6.50	4.10
Ash	8.1–23.0	14.30	5.8–11.0	8.00	5.9– 8.5	7.7	9.10–24.00	19.72
Oxygen	3.0–12.2	10.72	15.0–20.0	19.90	9.0–11.5	12.5	3.86– 7.59	4.85
Grindability Index, Hardgrove	46.8–77.3	53.6	43–60	50	44–49	48.56	60.4–79.7 (As fired)	79.7
Heating value—as received, Btu/pound	5744 minimum	6711 typical	8150 minimum	8400 typical	12,200–13,000	—	9000–11,650	10,117
Density, pounds/cubic foot	45	—	45	—	50	50	45	—
Delivered size	6 × 0	—	2 × 0	—	1⅝ × 0	1½ × 0	18 × 0	—
Angle of Repose	—	—	—	—	—	35°	35° to 37½°	—

affect the selection of the most appropriate conveying equipment. As shown by Table 1, these properties vary considerably from one plant situation to the next.

High-speed transfer belt conveyors as illustrated in Fig. 15 have proved to be exceptionally effective coal-handling systems. They are long wearing and designed to reduce maintenance to a minimum. Belt-conveyor idlers are spaced to provide the maximum support and reduce the friction drag to a minimum.

Scraper conveyors of the type shown in Fig. 16 are designed to care for wide diversifica-

Fig. 15. High-speed transfer-belt conveyors are long-wearing with low horsepower consumption. *(McNally Pittsburg Manufacturing Corp.)*

tion in materials handling, and for single- or multiple-loading positions. They can be obtained in a wide range of sizes for converging materials on the horizontal or inclines ranging up to 45°, and are constructed with two runs, providing for material movement in opposite directions.

Other conveyors include the *compartment conveyor*, designed to deliver separate sizes of material to different loading points simultaneously; *spiral or screw conveyors* which are frequently used as feeder conveyors for the coal-plant boiler or for ash removal; and *heavy-duty bucket elevators*, designed to handle both wet and dry coal.

Hoppers and feeders are integral components of the transport system. The rate of delivery of hoppers and feeders must be closely matched with the performance of the conveyor. The total system must be sufficiently flexible to provide for a wide range of flow rates.

The reciprocating feeder shown in Fig. 17 is an example of one of the simplest and most accurate feeders for hopper service. These feeders are built of heavy steel skirt plates which attach to a bin or hopper, beneath which a roller-supported plate operates. The reciprocating motion of the feeder is transferred to the bottom plate through an eccentric. As the reciprocating plate moves forward, it carries forth the material with it. On the return stroke, the bottom plate slides beneath the material that accumulated. The material is fed from the hopper on the forward stroke and discharged from the feeder on the return stroke. The tonnage of feed can be controlled by speed changes or length of stroke. Speed variations can be controlled by an electric sensor. Heavy-duty drives are required, the type of drive and size of motor dependent on the tonnage of coal to be handled.

For efficient operation of the transport system, centralized electric control is mandatory. A panelboard equipped with push buttons and warning lights can provide minute-by-minute monitorship and control of all the moving equipment. Interlocking circuits assure the operator that the equipment will start in the correct succession and prevent the

Fig. 16. Scraper conveyers are constructed with two runs to provide for material movement in opposite directions.

Fig. 17. Reciprocating feeder. *(McNally Pittsburg Manufacturing Corp.)*

overloading of the system at any exchange point, such as the feeder and conveyor belt.

A means must be provided for overriding the automatic and interlocking circuits with manual control of the individual components. Accidents and situations have arisen which clearly indicated the need for this particular feature.

Cost-control purpose devices for monitoring and controlling the weights of materials and the level in hoppers and bins are very important. These may be classified in different ways; for example, static versus in-transit or gravimetric versus nongravimetric.

Generally the trend is toward weighing materials on the move, with the weighing equipment in line. If the device is dependent upon a gravimetric pull on the material being weighed, a measure of the counterbalancing force can be combined with speed measurements into a direct measurement of rate of flow, and a device for totaling the weight as a function of time.

Nongravimetric devices are not dependent upon the force exerted on the material, but instead might be measured with a nuclear-radiation device, wherein the greater the quantity of material placed in the path of a radiation beam, the greater the amount of absorption of radiation by the material. Properly calibrated, the readings from these devices may be combined with velocity measurements, and the weight totaled as a function of time.

COAL STORAGE WITHIN USERS' FACILITIES

The primary purposes of coal storage are well understood, namely: (1) to accommodate for the usual surge delivery of fuel to the user's facilities, and (2) to ensure a noninterruptible and flexible supply. Not the least among the many problems associated with storage, however, are oxidation and spontaneous combustion. The presence of moisture and pyrites in coal tends to trigger and accelerate a reaction with the oxygen in the air. For coal that is to remain in storage for several weeks or months, considerable care should be taken to exclude atmospheric oxygen from the stockpile. An approved method for coal stockpiling is shown in Fig. 18.

The coal is stored on a dry, well-drained location that has been previously cleared of all foreign matter, such as wood, rags, waste oil, or any other materials having a low-ignition

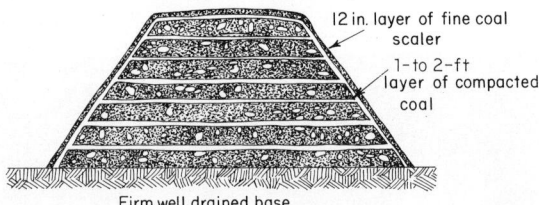

Fig. 18. An approved method for coal stockpiling. Diagram is not to scale.

temperature. The coal is laid down in layers of 1 to 2 feet thick, with each layer being compacted before the next one is added. The sloping sides and domed top act to shed moisture, and the 12-inch layer of fine coal added as a sealer prevents the penetration of air into the interior of the pile.

In general, there is little danger in the short-term storage of coals that have been screened or sized. A pile of uniform lumps is honeycombed with air passages, through which sufficient volumes of air can freely circulate and carry off any heat generated by the oxidation process.

Storage piles can be of various shapes: conical, kidney-shaped, or long and wedge-shaped; and they can be left entirely exposed, or enclosed in steel or concrete. The shape of the storage piles generally reflects the kinds of stacking and reclaiming equipment being used, and the shape and size of the available storage area.

Radial stackers are often employed at power-generating plants, because of their ability to build large, high coal-storage piles. Mounted on a tower and operating at ground level, they can be maneuvered at a wide angle to produce a kidney-shaped pile. See Fig. 19. In use with a kidney-shape storage pile, vibrating feeders can be placed under each half of the pile for an easily controlled withdrawal rate.

An advancement in stacker design and principle is embodied in the machine* shown in Fig. 20. Designed for large-volume work, a machine of this type is capable of stacking or reclaiming up to 3000 tons per hour of coal.

It will stack and reclaim on both sides of its supporting track. When stacking out, the boom is set at a predetermined angle to the track, and the reversing boom belt is traveling

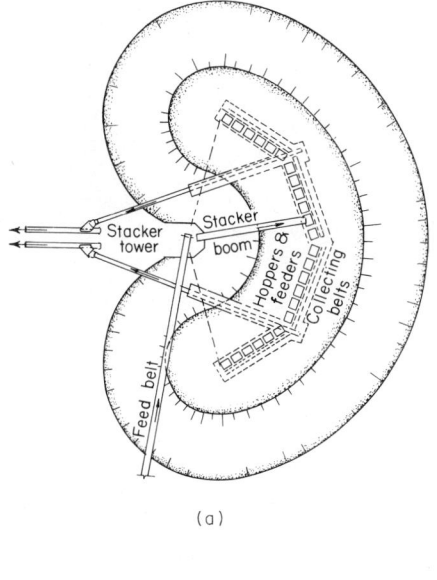

(a)

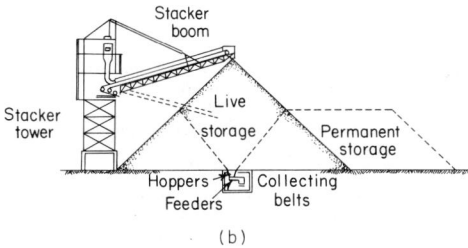

(b)

Fig. 19. Kidney-shaped coal storage pile: (a) Plan view showing feeder arrangement. (b) Section through pile. (*McNally Pittsburg Manufacturing Corp.*)

in the stackout direction; the machine then receives coal from the yard belt. When the pile reaches its predetermined height, the machine automatically advances, continuing to build the pile.

When reclaiming, the boom belt is reversed to the reclaiming direction, the bucket wheel is activated, and the boom begins its slewing motion through the stockpile as the bucket wheel begins reclaiming coal to predetermined steps, depending on the diameter of the bucket wheel and the tons per hour of material to be reclaimed. The bucket wheel dumps the material through a chute onto the boom belt, which transfers it to the belt feeder through the machine's surge hopper. The belt feeder discharges at a variable rate to the yard belt and on to its destination.

The machine can be operated manually or automatically 24 hours a day if needed. Separate modes are available to accomplish any normal operation: (1) Reclaiming only—reclaims from stockpile as needed to the designed rate. (2) Stacking only—stacks out material to stockpile at designed rate. (3) Excess demand—coal arriving at the stacker/

*McNally Stacker/Reclaimer.

Fig. 20. Rotary bucket stacker/reclaimer, capable of stacking up to 3000 tons per hour and reclaiming up to 2000 tons per hour of coal. Boom is 120 feet long. *(McNally Pittsburg Manufacturing Corp.)*

reclaimer is sent direct to the plant, and the bucket wheel will reclaim the difference to keep up with plant demands. (4) Excess feed—when more coal is arriving than is being consumed by the plant, the excess is stacked out automatically. All four modes may be accomplished manually or automatically.

Once the operator selects the mode to be used, he can rely on fully automatic operation through a numerical readout control system fed by a punched tape. A TV camera to monitor the stacking and reclaiming may also be included.

REFERENCES

Packard, C. E.: Stockpile Designs for Unit Train Loading, *Mining Eng.*, vol. 16, p. 8, August 1967.
Bucklen, O. B., and P. G. Merble: "Coal Preparation," 3d ed., chap. 15, Coal Storage and Loading, American Institute of Mining Engineers, New York, 1968.
Savanson, Gerald R.: Automated Railroad Car Loading, *Mining Congr.*, vol. 50, p. 1, January 1967.
Heers, Q. G.: Coal Handling at York Canyon Mine in New Mexico, *Coal Prep.* (British), March–April issue, 1968.
Anon.: The Changing Scene in Mine-to-Market Haulage, *Coal Age*, vol. 67, p. 12, December 1962.
Anon.: Slicta: Low Cost Barge Loading Savings: Coal to River Transport, *Coal Age*, vol. 61, p. 12, December 1956.
Anon.: Storage Systems and Equipment, *Coal Age*, December 1959.
Landers, W. S., and D. J. Donaven: in H. H. Lawry (ed.), "Chemistry of Coal Utilization," chap. 7, Storage, Handling, and Transportation, Wiley, New York, 1961.
Stahl, R. W., and C. J. Dulzell: Recommended Safety Precautions for Active Coal Stockpiling and Reclaiming Operation, *U. S. Bur. Mines Inform. Circ.* 8256, 1965.

Railroad Transportation of Coal

BY JOHN D. SWARD*

In 1971 approximately 513 million net tons of coal were shipped to various destinations in the United States. The railroads transported approximately 271 million net tons (53 percent of the total). An additional 153 million tons (30 percent) moved to destinations, using a coordinated rail-water transportation system. In total, the railroads transported approximately 83 percent of all coal in the United States. See Tables 1 and 2 (Refs. 1–3).

When the freight commodities transported by the railroads are ranked according to their gross freight revenues, carloadings, and tonnage originated, coal is the single most important commodity transported by the railroads (Ref.4). When the revenue and tonnage

TABLE 1. Movement of Bituminous Coal from Mine*
Millions of net tons

Year	Loaded at mine for shipment by rail	Loaded at mine for shipment by water	Shipment by motor vehicle	Used at mine	Total production
1955	356	47	52	10	465
1956	390	51	50	10	501
1957	380	51	50	11	492
1958	306	44	51	10	411
1959	301	46	53	13	413
1960	304	47	53	12	416
1961	294	46	51	12	403
1962	307	48	55	13	423
1963	334	51	61	13	459
1964	349	59	66	13	487
1965	372	60	68	12	512
1966	387	62	67	18	534
1967	405	67	62	20	554
1968	396	67	62	20	545
1969	398	71	66	26	561
1970	409	81	74	39	603
1971	382	59	60	51	552
1972	394	70	66	66	596

*Source of Information: National Coal Association, Washington, D.C.

derived from the transportation of coal are compared with the total tonnage and revenue obtained from the transportation of all commodities, bituminous coal accounts for approximately 23 percent of the industry's tonnage and 11 percent of its revenue.

Revenues from coal transportation of the largest coal-carrying railroads and the percent of their total revenue are given in Table 3 (Ref.3).

TYPES OF RAIL SERVICE FOR COAL TRANSPORTATION

Most American railroads offer four distinct types of road service for the transportation of coal. (1) single carload, (2) multiple carload, (3) trainload volume, and (4) unit train (unitrain). Each type of freight service has its own operating characteristics, which result in a distinct level of operating cost and freight rates.

The development of these four basic types of railroad service to transport bulk commodities has been an evolutionary process over the last 100 years. The first rail transportation of coal was in single-car shipments in conventional trains. As individual coal producers

*The Atchison, Topeka and Santa Fe Railway Co., Chicago, Illinois

TABLE 2.   Modes of Transporting Bituminous Coal to Destinations in the United States, Canada, and Mexico—1971*

Millions of net tons

Mode of transportation	Electric utilities	Coke and gas plants	Retail dealers	All other	Total
All-rail	176.8	46.3	5.7	42.1†	270.9†
River-rail	73.0	21.3	0.5	5.8	100.6
Great Lakes	20.2	12.9	2.2	8.9	44.2
Tidewater	3.5	4.7	—	0.1	8.3
Truck	43.6	1.9	2.8	16.1	64.4
Tramway, conveyor, and private railroad	24.6	(<0.1)	—	(<0.1)	24.6
Total	341.7	87.1	11.2	73.0†	513.0

*Source of Information: National Coal Association, Washington, D.C.
†Includes overseas exports from United States.

increased their output, multiple carload shipments in conventional trains were initiated. As production increased to where 75 to 200 carloads of coal were originated from one location on one day, the railroads initiated trainload volume movements for greater efficiency and economy. When potential competition from slurry pipelines became a serious threat in the early 1960s, unit train service was developed to allow the railroads to retain their existing coal traffic. Considering present railroad technology, railroad operating efficiency has been maximized in unit-train operations (Ref.5).

The terms single carload, multiple carload, trainload volume, and unit train have been incorrectly applied in certain railroad and traffic circles, and, consequently, they have lost much of their original meaning. As a result, confusion exists concerning various types of railroad service. Therefore, the basic types of railroad service are described in some detail here.

SINGLE CARLOAD

This service is the tendering of one carload of freight to a carrier for a movement to a given destination. A carload is a quantity of freight that can be loaded in or on a freight car, and is entitled to a particular rate because the weight of the lading is equal to, or exceeds that which is specified in the carrier's tariffs (Ref.6).

Single-carload movements are physically moved in the following procedure. The empty cars are accumulated in a railroad terminal, usually a division point, near the prospective loading point. After the cosignor has ordered railcars, the empty cars are switched into a

TABLE 3.   Bituminous Coal Hauled by Class I Railroads—1973

		Net tons (millions)		
Carrier	Revenue $ millions	Originated	Other	Total
Norfolk and Western	295.9	67.7	9.0	76.7
Penn Central	214.5	39.5	39.6	79.1
Chesapeake and Ohio	187.3	49.6	10.6	60.2
Baltimore and Ohio	110.7	26.7	13.8	40.5
Louisville and Nashville	110.3	48.7	2.4	51.1
Burlington Northern	90.0	24.7	2.0	26.7
Southern	70.1	23.9	10.8	34.7
Illinois Central Gulf	57.0	27.1	1.8	28.9
Union Pacific	36.3	7.5	3.0	10.5
Chicago and Northwestern	27.1	4.1	8.3	12.4
Seaboard Coast Line	24.9	. . .	15.8	15.8
Reading	23.0	2.2	9.0	11.2
Clinchfield	19.5	7.2	5.9	13.1
Denver and Rio Grande Western	17.1	7.9	2.1	10.0
Bessemer and Lake Erie	15.1	3.2	4.9	8.1

way freight train and delivered to the consignor for loading. After the cars have been loaded, a way freight train picks up the loaded railcars and delivers them to a railroad terminal or division point. Upon arrival at the railroad terminal, the loaded railcars are switched into *through freight trains.* The through freight trains move the loaded railcars from division point to division point until they arrive at a terminal point nearest the final destination. Upon arrival at this point, the railcars are switched out of the through freight train and moved into a way freight train for final delivery to the consignee. After the cars have been unloaded, they are released back into a car pool for subsequent use (Ref.7).

The carload movement of freight is an irregular movement of a commodity on an irregular schedule. Numerous complex switching and terminal operations are required to collect and distribute railroad cars. The cars are usually supplied from a car pool. The transportation movement is governed by tariff provisions requiring limited control over the loading and unloading of the railroad equipment, and the shipper has no commitment to an annual minimum tonnage. Considering all factors, single-carload movements are the highest cost method for the railroad to move freight (Ref.8).

MULTIPLE CARLOADS

Multiple-car service is the tendering of several cars at the same origin for movement to the same destination at the same time. The physical handling of the multiple-carload shipment is identical with the handling of single-carload shipments, except that the switching at the railroad terminals can be reduced. It is generally assumed that multiple-carload shipments generate economies in railroad operations which are reflected in incentive freight rates for multiple-carload movements. In terms of general rate levels, multiple-carload rates are usually lower than single-carload rates, but are higher than trainload volume or unit-train freight rates (Ref.9).

TRAINLOAD VOLUME

This service is the tendering of a sufficient number of carloads of freight on one day from one origin to one destination to permit the carrier to handle the movement in a special train. The number of carloads required to form a trainload will vary from one railroad to the next, depending on local operating conditions.

From an operational standpoint, a trainload volume movement will be handled in the following procedure. The required number of empty railcars will be accumulated at a railroad terminal near the origin. When the shipper is ready to load, a special train will be dispatched to the origin for loading. After the cars are loaded, the train will move over the road directly to the final destination. After the train has been delivered to the consignee, the locomotives and way car are returned to the nearest railroad terminal and released for other service. In like manner, when the railcars are unloaded, they will be returned to the nearest railroad terminal and released for other assignments.

The trainload volume movements are an irregular movement on an irregular schedule. The basic operation requires simplified switching and terminal operations, resulting in economies in rail operation. Trainload volume trains use railcars assigned to a car pool. The trains are governed by tariff provisions requiring a limited control over the loading and unloading of the railroad equipment and occasionally requiring a minimum annual volume.

Compared to single-carload or multiple-carload movements, the trainload volume movement has eliminated many of the costly switching and terminal operations. Unlike unit trains, however, the trainload volume movement makes only a limited attempt to control equipment utilization. At best, the trainload volume movements are a compromise between single- and multiple-carload movements on the one hand and unit trains on the other. In terms of freight rates, trainload volume movements usually have lower costs, but they do not have the low operating cost and attractive freight rates associated with unit trains.

There is a popular misconception that lower cost and freight rates are automatically associated with trainload volume movements. Depending on the railroad operating costs, there are conditions when trainload volume rates may be higher than single- or multiple-carload movements.

UNIT TRAINS

A unit-train movement is an integral movement of usually one commodity moving from a single origin to a single destination on a regularly scheduled train, avoiding all terminals and switching operations. Unit trains utilize specialized loading and unloading facilities and specialized railroad equipment assigned to dedicated service. The unit-train movement is governed by tariff provisions requiring both controlled loading and unloading of the railroad equipment and a minimum annual tonnage. See Fig. 1.

The operation of a unit train is quite simple. The railcars, locomotives, and way car are assigned permanently to the unit train. The unit train loads the coal at origin, proceeds

Fig. 1. Detroit Edison Company's unit train being loaded with a storage silo equipped with loading bin facility at Consolidation Coal Company's Blacksville No. 1 mine on Penn Central's Waynesburg Southern Railroad. *(Penn Central.)*

directly to the destination, unloads the lading, and returns directly to the origin for the next loading. The only time the unit train stops at a terminal is to service the locomotives and cars.

Unit trains develop intensive equipment utilization by eliminating costly and wasteful switching and terminal costs. The unit-train concept combines specialized railroad rolling stock with improved loading and unloading facilities and streamlined railroad operations. All elements of the unit-train operation (loading, haulage, and unloading) must be in

TABLE 4. Capital Cost of Owning a 100-Ton Hopper Car Costing $22,000

Estimated transit time, days	Equipment ownership cost/day, $	Type of rail service associated with	Car ownership cost/shipment, $	
			Cost/carload	Cost/net ton
1	10.90	Short-haul unitrain	10.90	0.109
4	10.90	Long-haul unitrain	43.60	0.436
7	10.90	Short-haul trainload volume	76.30	0.763
14	10.90	Long-haul trainload volume	152.60	1.526
21	10.90	Single or multiple carload	228.90	2.289
28	10.90	Single or multiple carload	305.20	3.052

balance and be properly coordinated to eliminate inefficiencies and to guarantee heavy loading of equipment. Above all, a unit train requires an obligation by shippers to change marketing practices, particularly in regard to terminal operations. Inventory must be carefully controlled. In terms of cost, unit trains are the most efficient method for the railroads to move freight (Ref.10).

Whenever unit-train service deviates from the basic shuttle pattern, its economics deteriorate rapidly. For example, when a railroad permits service outside the shuttle pattern to assemble or distribute cars, or whenever the carrier permits significant changes in the loading schedule, the economics essentially becomes that of a trainload volume movement. See Table 4.

RAILROAD CARS FOR COAL SERVICE

The railroads use four basic types of cars to transport coal: (1) Open-top hopper cars; (2) gondola cars; (3) gondola cars with rotary couplers; and (4) rapid-discharge bottom-unloading hopper cars. Because each type of car has its own advantages, there are many alternatives in selecting an unloading system. Each loading and unloading system must be evaluated independently, based on local conditions. A total physical distribution-cost approach should be used to intelligently determine the ultimate low-cost system.

OPEN-TOP HOPPER CAR

This is a self-cleaning car having fixed sides and ends, an open top, and two or more hopper doors hinged crosswise of the car, allowing the product to be discharged between the rails. See Fig. 2 (Ref. 11). Design of the car was adopted by the Association of American Railroads in the 1930s. At that time, the car was a 50-ton hopper. The shallow slope sheet and the small horizontal door opening made it very difficult for the product to flow from the car. Car shakers were usually employed to assist in the unloading, causing abuse of the equipment, increasing maintenance costs and shortening expected car life (Ref.12).

During the interim, many minor adjustments have been made, with the loading capacity increased from 50 to 80, 100, and 125 net tons. The number of pockets has been increased from 2 to 3, 4, or 5 pockets. The angle of repose has been adjusted to improve the flow of coal. Side sheets are embossed to reduce vibration in shaker unloading. As protection against corrosion, all plates that come in contact with the coal are of copper-bearing steel. Adjustments have been made in the hopper gates for easier unloading.

Fig. 2. The most common railcar used to transport coal, open-top hoppers are shown being loaded at a mine. *(Pullman-Standard.)*

However, beyond these minor adjustments, the open-top hopper car in basic design has not changed in over 40 years. The car is unloaded by opening the gates at the bottom of each hopper. The rate of unloading can be adjusted by selecting the number of gates that are opened at any one time.

GONDOLA CAR

This is the most versatile open-top car. It may have fixed or drop ends, steel or wooden floors, crosswise or lengthwise drop doors in the bottom, or it may have a solid bottom. The car comes in varied sizes in terms of length, height, and carrying capacity.

The crosswise drop doors will discharge the product between the rails, whereas the lengthwise drop doors will dump outside the rails. The gondolar car with solid bottom is unloaded with either a clamshell operation or a rotary dumping facility. None of the gondolar car designs are self-cleaning.

The most common type of gondola car used to transport coal is an open-top car, having fixed sides and ends, a solid bottom, and extended sides to increase the car's carrying capacity to 100 or 125 net tons. See Fig. 3 (Ref.13). Rotary-dumping facilities are generally used to discharge the product. The rotary-dump facilities require that each car be uncoupled from the train when placed in the rotary dumper, the car turned 180°, the car uprighted, and then moved out of the unloading facility. The time required to unload a 100-car train of gondola cars in a rotary-dump facility will range from 6 to 12 hours. See Fig. 4.

ROTARY-DUMP GONDOLA CAR

This is an open-top car having fixed sides and ends and a solid bottom. There are extended sides to increase capacity to 100 or 125 net tons (Ref.13). The car is unloaded in a rotary dumper. It is turned 135°, the coal unloaded, the car uprighted, and then moved on. The primary advantage of the rotary car is that a train can be unloaded without the cars being uncoupled. The time required to unload a 100-car unit train with this equipment is approximately 4 to 6 hours.

RAPID-DISCHARGE CAR

This is basically an open-top hopper car with improved unloading gates. It is engineered to achieve a maximum opening area for discharge of the coal. When the doors are opened, the support is removed from a large area under the load, and the material can discharge with little interference from the car structure. The doors are designed with opening and closing mechanisms that can be actuated automatically, permitting unloading in motion of the car. See Fig. 5.

Rapid-discharge cars are of two basic designs. Both Pullman Standard and Ortner have designed motion unloading cars which discharge the coal between the rails. Pullman Standard also has designed a rapid-discharge railcar which unloads the coal outside the rails (Ref.13).

Unloading in motion provides the ultimate utilization of railroad equipment. It is possible for 100 rapid-discharge coal cars of 100-ton capacity to be unladed in 15 to 20 minutes with a train moving at speeds from 4 to 8 miles per hour. This also enables optimum utilization of locomotive power inasmuch as the locomotive need not stop during the unloading process.

COST OF COAL CARS

At mid-1974 prices, the estimated cost of coal cars will range from $19,000 to over $30,000, depending on the type of car and the number purchased. See Table 5. The estimated cost for 100-ton cars in 1,000-car sets would be $21,800 for an open-top sawtooth hopper car, $21,250 for a gondola car, $21,800 for a gondola car with rotary couplers, $25,975 for an air-operated rapid-discharge car, and $23,670 for a mechanically operated rapid-discharge car. The foregoing costs are strictly estimates. However, the general cost relationships among types of cars are currently valid.

Fig. 3. The most universal coal car, a 100-ton solid-bottom gondola car. Several cars are shown here waiting in a railroad terminal for loading. *(Pullman-Standard.)*

Fig. 4. A rotary dumper for rapid emptying of 100 tons of coal from the Santa Fe's York Canyon unit train at the Fontana, California Kaiser Steel plant. This unit train transports 8400 net tons of coal over 1100 miles every four days. *(Santa Fe Railway.)*

Fig. 5. Pullman Standard's side-gate open-top hopper car, designed for unit train service. A 100-car train of motion unloading cars can unload 10,000 tons of cargo in less than 15 minutes while train is in motion.

SIZES OF COAL CARS

Cars designed to transport coal can be built in almost any capacity from 70 up to 125 tons. Most coal cars are currently being built for 100-ton capacity. Some railroads have purchased 125-ton cars for unit-train service. Their general acceptance has been restricted, however, because of uncertainties associated with maintenance-of-way costs for large-capacity cars. This is a matter of an economic tradeoff. Most likely, 100- and 125-ton cars do increase the maintenance-of-way cost. However, the larger cars decrease car maintenance costs, operating, and capital investment costs.

MATERIALS OF CONSTRUCTION FOR COAL CARS

Most commercial car builders will construct coal cars from either steel or aluminum. Some experts within the railway industry openly advocate the use of aluminum cars. Use of aluminum for car-body construction has been recognized by certain unit-train planners

TABLE 5.   Estimated Purchase Price of Railroad Cars for Hauling Coal (1974 Prices)

Type of car and capacity	Cost per car, $			
	100-car set	200-car set	500-car set	1000-car set
Open-top sawtooth hopper car:				
70-ton	19,985	19,490	19,270	19,160
100-ton	22,625	22,130	21,910	21,800
125-ton	24,825	24,330	24,110	24,000
Gondolar car:				
70-ton	19,545	19,050	18,830	18,720
100-ton	22,185	21,750	21,360	21,250
125-ton	24,385	23,945	23,560	23,450
Gondolar car—rotary coupler:				
70-ton	20,810	20,370	20,040	19,930
100-ton	22,735	22,295	21,910	21,800
125-ton	24,935	24,495	24,110	24,000
Rapid-discharge—air-operated:				
70-ton	26,175	24,825	24,175	23,775
100-ton	28,375	27,025	26,375	25,975
125-ton	30,575	29,225	28,575	28,175
Rapid-discharge—mechanically operated*				
70-ton	22,300	21,530	21,300	21,190
100-ton	24,500	24,000	23,780	23,670
125-ton	26,700	26,200	25,980	25,870

*Require track activity.

where equipment utilization is high. Transportation cost savings can be expected from the high net-lading-to-tare-weight ratio and the resulting lower operating cost for aluminum equipment. For example, if a railroad has a load limit of 263,000 pounds per car, any reduction in the weight of the railcar can be substituted by an increased loading of the car (Ref.14). See Table 6.

Since the mid-1960s, car builders have been using alloy steels where possible to obtain high strengths at minimum weight. *Cor-ten* and copper-bearing steel are used to ensure a

TABLE 6.   Weight Limitations—Load Limit

Journal size, inches	Total load limit		Remarks
	4-axle cars	6-axle cars	
4¼ × 8	103,000	154,000	Note 1
5 × 9	142,000	213,000	Note 1
5½ × 10	177,000	265,000	Note 1
6 × 11	220,000	330,000	Note 1
6½ × 12	263,000	394,000	Note 1
7 × 12	315,000	472,000	Note 2

NOTE 1: Cars can be operated in unrestricted interchange.
NOTE 2: Cars can only be operated under controlled conditions, agreed to by participating railroads.
The railroad industry has established rules governing the interchange of rail cars between the various railroads. Rule 91 of the *Field Manual of the AAR Interchange Rules* establishes weight limitation for railroad cars. The term *load limit* is the combined light weight of the car and the load limit which gives the maximum axle loading for a particular size car. The heaviest 4-axle railcar permitted in unrestricted interchange service has a journal size 6½ × 12 inches and a load limit of 263,000 pounds. This car usually weighs 63,000 pounds and can carry 200,000 pounds of lading. Although individual railroads may permit a heavier load limit, the 263,000 pounds represents the most common 100-ton coal car used on most railroads.

minimum of rusting. Many of the advantages originally anticipated for aluminum cars have been diluted by improvements in steel. As of the mid-1970s, very few railroads have purchased aluminum cars (Ref.15).

LOADING SYSTEMS

The usual list of components of a train-loading system includes stockpiles, storage silos, reclaiming equipment, scales, conveyors, surge bins, loading chutes, loading tunnels, loading structures, and energy inputs, both electrical and human labor. The manner in which these elements is combined will depend on conditions associated with each loading situation, with no universal solution available. The elements needed to load 100 net tons daily will be substantially different from those for 30,000 net tons daily (Ref.16).

MINIMUM CAPITAL-INTENSIVE LABOR

The minimum capital expenditure for loading coal cars on property already owned will probably be a lot of number 2 scoop shovels and the labor to man them. Assuming that one man can load three tons of coal per hour, the loading cost will be about $5.85 per hour times 34 hours, or $198.90 per carload. Thus, the loading cost will be $1.99 per ton of coal. Obviously, with such loading costs, little persuasion is required to accept large capital expenditures as a means of decreasing the unit loading cost.

FRONT-END LOADERS

The simplest form of substitution of capital for labor involves front-end loaders to move the coal from a storage pile to the railcar. Assuming that one front-end loader holds 10 cubic yards (7 tons of coal) and that the loading cycle is one minute, it will take approximately 15 minutes to load a 100-ton car. The time required to load a 100-car, 10,000-net-ton unit train is reduced to a matter of the number of men and machines. If only one man and one front-end loader are used, the time required will be 25 hours (15 minutes × 100 cars divided by 60 minutes equeals 25 hours). If five men and five front-end loaders are used, the loading time will be five hours. The loading cost for a 100-car train loading one 10,000-net-ton unit train per week, using five front-end loaders, five machine operators, and one foreman will be approximately $975, or about ten cents per ton.

Front-end loaders can be purchased in various sizes to meet specific loading conditions. The larger front-end loaders can handle payloads on the order of 30 tons and can load fifteen 100-ton railcars per hour per machine (Ref.17).

FLOOD LOADING

When large volumes are to be loaded within a short time period, some form of flood-loading facility is needed, permitting the coal to free-fall into the cars. The four basic types of flood-loading facilities are: (1) ground storage with loading tunnel, (2) ground storage with loading bin, (3) silo storage, and (4) silo storage with loading bin. These systems are described in further detail under *Coal Handling and Storage Systems* in this section of the Handbook.

UNLOADING SYSTEMS

Unloading of coal cars can be accomplished by several methods, depending on the type of car and the capital investment in specialized facilities. If volume is small and capital investment is minimized, several labor-intensive unloading systems can be used. If open-top hopper cars are employed, coal can be discharged into a pit, or a portable conveyor belt can be used to move the coal from underneath the hopper gates to storage. When large volumes must be unloaded in a short time, the consignee must make a capital investment in unloading facilities.

If the consignee must accept shipments in all kinds of gondola and/or hopper cars, a rotary-dump operation will be required. Also, the operation will be restricted to an unloading system in which each car must be uncoupled from the train before dumping because all cars will not be equipped with rotary couplers. Also, the rotary dumper must be large enough to accommodate railcars of varying sizes.

For a high-capacity unloading operation, the following unloading procedure would be used. The uncoupled cars are brought into the dumper. The incoming car pushes out the empty car, permitting the car to roll downgrade to a storage track. The rotary dumper turns the car 135°, permitting coal to free-fall into a pit. When unloaded, the car is returned to its upright position.

The majority of high-capacity unloading facilities are designed for unit-train service. With unit trains, all railcars are of the same design. The cars and unloading system can be designed to complement each other for maximum efficiency. There is a choice basically between two unloading systems: (1) conventional hopper cars or quick-opening bottom-drop cars, plus the required dump pit, tressels, and conveying equipment, or (2) fixed-bottom gondola cars with rotary couplers and a dump pit and associated conveyor equipment.

When using bottom-drop cars, a storage area capable of handling the entire capacity of the unit train must be built. This storage area can be a pit with approximately 10,000 to 15,000 net tons capacity; or it can be a tressel with comparable storage capacity.

A unit train with conventional hopper cars would be unloaded with the cars being spotted over the storage area. The gates on the cars would be opened by laborers stationed alongside the railcars. After the first cars over the unloading area are discharged, the train would move forward until the next loaded car is over the unloading area. This spotting, unloading, and spotting sequence is repeated until the entire train is unloaded.

A unit train with quick-opening bottom-drop cars would be unloaded by having the gates on the railcars opened by either a mechanical tripping mechanism or an electrical device as the cars roll over the pit or tressel. After the cars are unloaded, the gates on the cars would be closed by a similar mechanical or electrical mechanism.

When a rotary-coupler gondola car is used, the first car is stopped inside the dumper, the car is rotated 180°, permitting the coal to free-fall into the pit, the car is uprighted, and the unit train is moved forward until the second car is in the proper unloading position. The process is repeated until the entire train is unloaded. A rotary dumper can be built large enough to accommodate two cars, resulting in faster unloading time.

The main element in cost comparisons of systems is the number of cars assigned to the system. The number of cars required depends upon annual tonnage, unit-train schedule, and length of the haul. A single-car rotary dumper and train positioner installed in a concrete pit with all electric and conveying equipment to unload from 4000 to 6000 tons of coal per hour will cost approximately $2.4 million (1974). A comparable motion-unloading facility will cost approximately $1.1 million. The difference in original cost between the two facilities is approximately $1.3 million. The difference in cost between a 100-ton flat-bottom gondola rotary-dump car and a 100-ton motion-unloading car will range from $2400 up to $5000 per car. The break-even point ranges from 260 to 541 cars, depending on the purchase price of railcars: $1.3 mils per $5000 = 260 cars; $1.3 mils per $2400 = 541 cars. Thus the choice between rotary-dump and motion unloading systems is an economic tradeoff.

Operating personnel requirements for the two unloading systems are nearly equal. One person can run either operation. Maintenance costs usually will be higher for 260 to 541 rapid-discharge coal cars than for one rotary dumper and train positioner. Reliability is very high in either case, but when a rotary dumper is out of service, the whole operation ceases. Some operations can justify a second dumper side by side. In contrast, no appreciable percentage of the motion-unloading cars will likely be out of service at any given time.

Another factor favoring motion-unloading cars is the small turnaround time required at destination. Proceeding across the pit area at 4 to 5 miles per hour, a 100-car, 10,000-ton unit train can be unloaded in 15 minutes. Considering startup time, the total unloading time will be approximately one hour. The same 100-car unit train, in an efficient two-car rotary-dump facility, will require 4 to 5 hours to unload the train. If a single-car rotary-dump facility were used, the unloading time will range from 8 to 12 hours. Depending on the local conditions on the terminating railroad, the use of motion-unloading cars could permit the railroad to save unnecessary deadheading or crew cost, which may exceed $1 million throughout the life of a 15-year unit-train operation.

TERMINALS

In unit-train operations, terminals are facilities used to load and unload unit trains. Terminals are traditionally the responsibility of the railroad. With unit trains, terminal planning is most important because the time spent in loading and unloading railcars has a direct influence on the railroad's operating costs. In most unit-train operations, two terminals are required (origin for loading; destination for unloading). The loading terminal should consist of a main track, a flood-loading facility with a surge pile of at least 1.5 times the cargo capacity of the unit train, and a turnaround loop track capable of accommodating a complete unit train between the flood-loading facility and the main track. See Fig. 6.

The unloading facility should also consist of a main track, an unloading complex, a storage area of at least 1.5 times the cargo capacity of the unit train, and a turnaround loop track capable of accommodating the complete unit train between the unloading facility and the main track. See Fig. 7.

When space limitations at the loading or unloading facilities do not permit the construc-

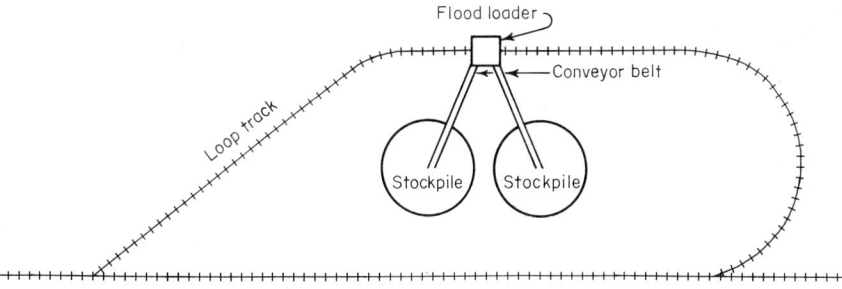

Fig. 6. Terminal facility at origin.

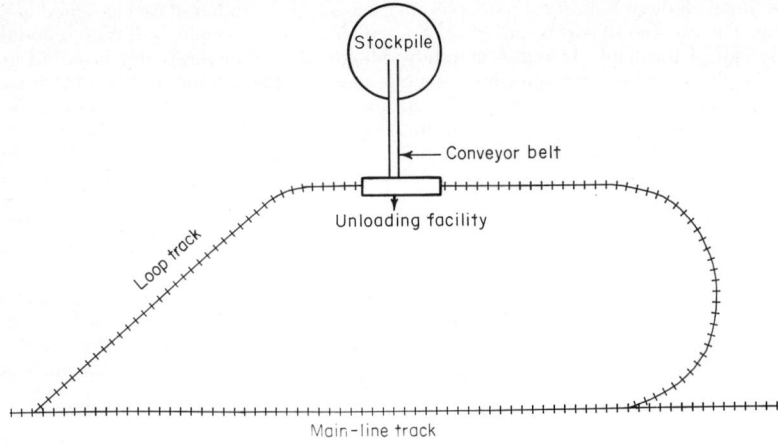

Fig. 7. Terminal facility at destination.

tion of a track loop, reasonably efficient loading or unloading can be developed by using either a stub-dead-end track, or a "Y" track. See Figs. 8 and 9. Utilization of these alternatives requires additional switching by the train crew, which could increase the cost of unit-train operations.

GEOGRAPHICAL ASPECTS OF COAL TRANSPORTATION

The characteristics of coal transportation vary significantly in different geographical areas. Two major areas have been considered for analytical purposes: (1) an Eastern Region; and (2) a Western Region. Both regions have differences which require specific analysis. See Fig. 9 and Addendum A.

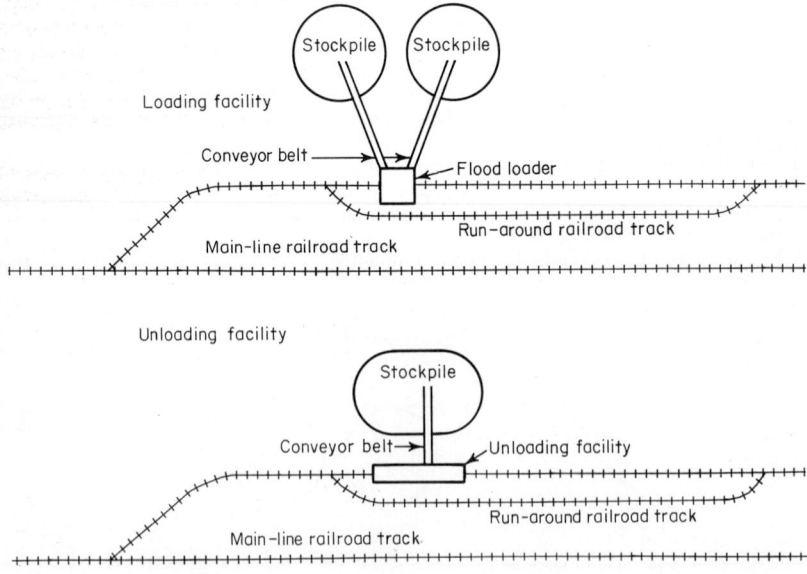

Fig. 8. Stub-end loading and unloading facilities.

EASTERN REGION

In this region, coal mines are located relatively close to consuming markets. In most situations they are less than 300 miles from the consumer. Long-haul transportation of coal is unusual (Ref.18). Eastern coal has two primary markets. Traditionally, the market has been the electric utility industry. Also, a substantial portion of eastern coal is used for metallurgical coke. Coking coal is a premium product capable of receiving a premium freight rate.

Major coal deposits in the East are predominantly located located either in the Appalachian Mountain or hill country, or under fertile farmland. Both situations create problems in developing large-scale mining operations to meet requirements of large power plants. Mines located in mountain areas are usually underground mines with high production cost. Also, they are usually small production units, seldom exceeding 1 million net tons of coal per year. Mines located under farmland are usually surface mines. Expansion of mining operations into established agricultural areas has proved difficult because of environmental considerations, particularly to the extent of a coal-mining operation being sufficiently large to meet the entire needs of a given power plant, which often will exceed 4 million tons per year.

The Eastern Region is laced with navigable rivers and waterways. Mines usually are located a relatively short distance from water. Coal transportation is often a coordinated effort among various modes of transportation. When coal is moved by way of an all-rail movement, the potential diversion of the traffic to water carriers has forced the general coal-freight rates to a low level (Ref.19).

Another major problem in the Eastern Region is an overexpansion of railroad facilities. Studies have indicated that perhaps 25 percent of the eastern capacity could be removed without significantly reducing rail service to the coal industry. A major indication of this is that when new mines are opened in the area, the additional investment in new road and structures required by the railroads to serve the new mine is relatively minor. In most

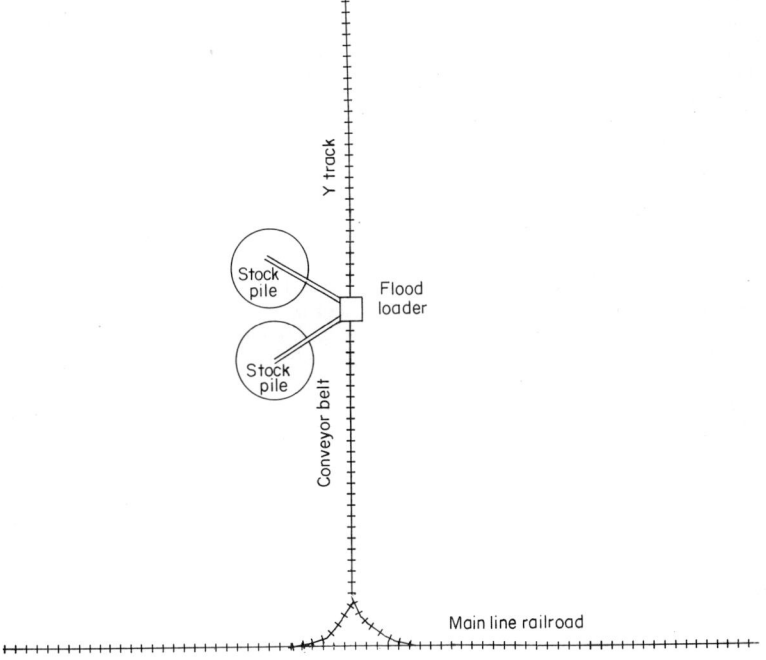

Fig. 9. "Y" track loading facility.

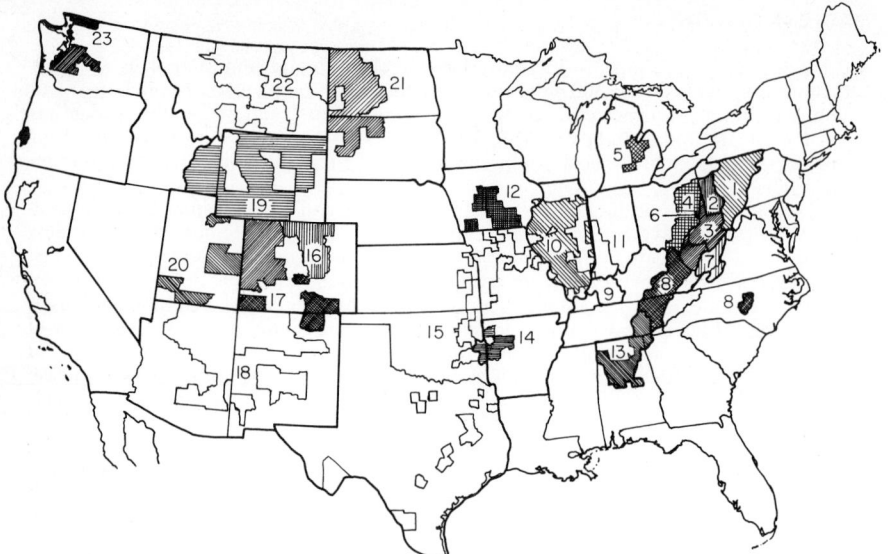

Fig. 10. Production Districts as set forth in the Bituminous Coal Act of 1937. (See also Addendum A.)

cases, cost of new construction will not exceed $5 million. The additional investment is largely associated with improving an existing right-of-way.

In summary, transportation of coal in the Eastern Region has the following characteristics: (1) Coal often is moved in single-carload or multiple-carload lots; (2) coal movements are usually short hauls; (3) coordinated transportation between rail and water modes is highly developed; (4) rail rates are usually depressed because of water competition; (5) capital investment in new road and structures is of minor importance in opening a new coal mine; and (6) unit trains, although pioneered in the Eastern Region, particularly on the southeastern railroads, have not become significantly important.

WESTERN REGION

Significant differences between the East and the West include: (1) Western coals are fundamentally different, and (2) the coal is remotely located, both from rail head and consumption areas.

Although quite variable in quality, western coals commonly have a sulfur content of less than 0.4 percent, a fact that has caused utilities in the East, Midwest, and South to search for supplies of western coal in the interest of meeting environmental standards.

Western subbituminous deposits are very large and readily accessible to surface mining in most cases. For example, the Gillette Field in northeastern Wyoming extends about 100 miles in a north-south direction, with portions that can be surface-mined often being more than a mile wide. The field will produce 100,000 to more than 175,000 tons of coal per acre. Several firms have reserves of a billion tons or more. One new mine will have a capacity of 12 million tons per year, and will possibly reach 15 million tons per year. Thus, the size of western mines will often be double or triple that of the largest mines in the East. Mining in the West is generally an application of modern mass-production technology to a very attractive mining situation.

However, the mines may be located from 1000 to 1500 miles from the consuming market. To market western coals, the railroads have had to keep transportation costs to a minimum. Thus, the western railroads have applied the unit-train concept to its maximum efficiency.

In the new coal areas of the West, there is a minimum of branch rail lines. Often a potential mine may be located 10 to 200 miles from the closest rail head. For western coal to be developed, it is usually necessary to build a new rail line through rugged topography, costing up to $400,000 per track-mile. In some cases, capital expenditures ranging from $15 million up to $120 million are required to build a new rail line into the coal mine. Frequently the capital cost associated with building a new rail line is the largest cost element in a proposed freight rate.

In summary, transportation of coal in the Western Region has the following characteristics: (1) Shipments are usually in unit trains. Single-carload, multiple-carload, and trainload volume movements are unusual. (2) Movements are of long distance, often between 1000 and 1500 miles. (3) The movement is usually all-rail. Coordinated rail-water movements are the exception. (4) Capital expenditures in new road and structures are of major importance in opening new coal mines.

REFERENCES

1. Bagge, Carl E.: Railroads—the Logistical Arm of the Coal Industry, *Ry. Manage. Rev.*, vol. 73, p. 3, 1973.
2. "Bituminous Coal Facts," National Coal Association, Washington D.C. Published every two years.
3. "Coal Traffic Annual," National Coal Association, Washington, D.C., 1972.
4. "Interstate Commerce Commission; Freight Commodity Statistics of Class I Railroads in the United States, Calendar Year 1970, U.S. Government Printing Office, Washington, D.C.
5. Glover, T. O., M. F. Hinkle, and H. L. Riley: "Unit Train Transportation of Coal," U.S. Government Printing Office, Washington, D.C., 1970.
6. Taft, C. A.: "Management of Traffic and Physical Distribution," 3d ed., Irwin, 1964.
7. Fair, M. L., and E. W. Williams, Jr.: "Economics of Transportation," Harper, New York, 1959.
8. Sward, John D.: "Unitrain Operating Methods and Cost," University of Michigan, Ann Arbor, Mich., 1972.
9. Wallace, G. R.: Marketing the Unit Train, "Unit Train Operations," Railway Systems and Management Association, Chicago, 1967.
10. Sward, John D.: "Changing Patterns in the World Market for Iron Ore," Society of Naval Architects and Marine Engineers, New York, 1974.
11. *Offic. Ry. Equip. Regis.*, vol. 87, no. 1, Railway Equipment and Publication Co., New York, July 1972.
12. Cooke, T. S., Jr.: New Technological Developments Affecting the Transportation of Coal, *Ry. Manage. Rev.*, vol. 73, 1973.
13. "Car Builder's Cyclopedia—1961," Simmons-Boardman, New York, 1961.
14. Morris, R. N.: The State of the Unit Train Art, "Unit Train Operations," Railway Systems and Management Association, Chicago, 1967.
15. Hanson, V. D.: Packaging the Unit Train Concept. "Unit Train Operations," Railway Systems and Management Association, Chicago, 1967.
16. Carlton, K.: Modern Railway Loading and Unloading of Coal, *Ry. Manage. Rev.*, vol. 73, p. 13, 1973.
17. Kneiling, J. G.: "Integral Train Systems," Kalmbach, Milwaukee, 1969.
18. "Interstate Commerce Commission, State to State Distribution Traffic and Revenue," U.S. Government Printing Office, Washington, D.C., 1968.
19. Wooters, R. B.: Regional Differences in the Coal Transportation Market—Eastern, *Ry. Manage. Rev.*, vol. 73, p. 3, 1973.

ADDENDUM A.   Schedule of Districts as Defined in the Bituminous Coal Act of 1937

DISTRICT 1

Eastern Pennsylvania. The following counties in Pennsylvania: Bedford, Blair, Bradford, Cambria, Cameron, Centre, Clarion, Clearfield, Clinton, Elk, Forest, Fulton, Huntingdon, Jefferson, Lycoming, McKean, Mifflin, Potter, Somerset, Tioga. Armstrong County, including mines served by the P. & S.R.R. on the west bank of the Allegheny River, and north of the Conemaugh division of the Pennsylvania Railroad.

Fayette County, all mines on and east of the line of Indian Creek Valley branch of the Baltimore and Ohio Railroad. Indiana County, north of but excluding the Saltsburg branch of the Pennsylvania Railroad between Edri and Blairsville, both exclusive.

Westmoreland County, including all mines served by the Pennsylvania Railroad, Torrance, and east.

All coal-producing counties in the state of Maryland. The following counties in West Virginia; Grant, Mineral, and Tucker.

DISTRICT 2

Western Pennsylvania. The following counties in Pennsylvania: Allegheny, Beaver, Butler, Greene, Lawrence, Mercer, Venango, Washington. Armstrong County, west of the Allegheny River and exclusive of mines served by the P. & S.R.R. Indiana County, including all mines served on the Saltsburg branch of the Pennsylvania Railroad north of Conemaugh River.

Fayette County, except all mines on and east of the line of Indian Creek Valley branch of the Baltimore and Ohio Railroad. Westmoreland County, including all mines except those served by the Pennsylvania Railroad from Torrance, east.

DISTRICT 3
Northern West Virginia. The following counties in West Virginia: Barbour, Braxton, Calhoun, Doddridge, Gilmer, Harrison, Jackson, Lewis, Marion, Monongalia, Pleasants, Preston, Randolph, Ritchie, Roane, Taylor, Tyler, Upshur, Webster, Wetzel, Wirt, Wood. That part of Nicholas County including mines served by the Baltimore and Ohio Railroad and North.

DISTRICT 4
Ohio. All coal-producing counties in Ohio.

DISTRICT 5
Michigan. All coal-producing counties in Michigan.

DISTRICT 6
Panhandle. The following counties in West Virginia: Brooke, Hancock, Marshall, and Ohio.

DISTRICT 7
Southern Numbered 1. The following counties in West Virginia: Greenbrier, Mercer, Monroe, Pocahontas, Summers. Fayette County, east of Gauley River and including the Gauley River branch of the Chesapeake and Ohio Railway and mines served by the Norfolk and Western Railway. McDowell County, that portion served by the Dry Fork branch of the Norfolk and Western Railway and east thereof.

Raleigh County, excluding all mines on the Coal River branch of the Chesapeake and Ohio Railway. Wyoming County, that portion served by the Gilbert branch of the Norfolk and Western Railway lying east of the mouth of Skin Fork of Guyandot River and that portion served by the main line and the Glen Rogers branch of the Norfolk and Western Railway.

The following counties in Virginia: Montgomery, Pulaski, Wythe, Giles, Craig. Tazewell County, that portion served by the Dry Fork branch to Cedar Bluff and from Bluestone Junction to Boissevain branch of the Norfolk and Western Railway and Richlands–Jewell Ridge branch of the Norfolk and Western Railway.

Buchanan County, that portion served by the Richlands–Jewell Ridge branch of the Norfolk and Western Railway and that portion of said county on the headwaters of Dismal Creek, east of Lynn Camp Creek (a tributary of Dismal Creek).

DISTRICT 8
Southern Numbered 2. The following counties in West Virginia: Boone, Clay, Kanawha, Lincoln, Logan, Mason, Mingo, Putnam, Wayne, Cabell. Fayette County, west of, but not including mines of the Gauley River branch of the Chesapeake and Ohio Railway. McDowell County, that portion not served by and lying west of the Dry Fork branch of the Norfolk and Western Railway.

Raleigh County, all mines on the Coal River branch of the Chesapeake and Ohio Railway and north thereof. Nicholas County, that part south of and not served by the Baltimore and Ohio Railroad. Wyoming County, that portion served by the Gilbert Branch of the Norfolk and Western Railway lying west of the mouth of Skin Fork of Guyandot River.

The following counties in Virginia: Dickinson, Lee, Russell, Scott, Wise. All of Buchanan County, except that portion on the headwaters of Dismal Creek, east of Lynn Camp Creek (tributary of Dismal Creek) and that portion served by the Richlands–Jewell Ridge branch of the Norfolk and Western Railway.

Tazewell County, except portions served by the Dry Fork branch of the Norfolk and Western Railway and branch from Bluestone Junction to Boissevain of the Norfolk and Western Railway and Richlands–Jewell Ridge branch of the Norfolk and Western Railway.

The following counties in Kentucky: Bell, Boyd, Breathitt, Carter, Clay, Elliott, Floyd, Greenup, Harlan, Jackson, Johnson, Knott, Knox, Laurel, Lawrence, Lee, Leslie, Letcher, McCreary Magoffin, Martin, Morgan, Owsley, Perry, Pike, Rockcastle, Wayne, Whitley.

The following counties in Tennessee: Anderson, Campbell, Claiborne, Cumberland, Fentress, Morgan, Overton, Roane, Scott.

The following counties in North Carolina: Lee, Chatham, Moore.

DISTRICT 9
West Kentucky. The following counties in Kentucky: Butler, Christian, Crittenden, Daviess, Hancock, Henderson, Hopkins, Logan, McLean, Muhlenberg, Ohio, Simpson, Todd, Union, Warren, Webster.

DISTRICT 10

Illinois. All coal-producing counties in Illinois.

DISTRICT 11

Indiana. All coal-producing counties in Indiana.

DISTRICT 12

Iowa. All coal-producing counties in Iowa.

DISTRICT 13

Southeastern. All coal-producing counties in Alabama. The following counties in Georgia: Dade, Walker. The following counties in Tennessee: Marion, Grundy, Hamilton, Bledsoe, Sequatchie, White, Van Buren, Warren, McMinn, Rhea.

DISTRICT 14

Arkansas-Oklahoma. The following counties in Arkansas: All counties in the state. The following counties in Oklahoma: Haskell, LeFlora, Sequoyah.

DISTRICT 15

Southwestern. All coal-producing counties in Kansas. All coal-producing counties in Texas. All coal-producing counties in Missouri. The following counties in Oklahoma: Coal, Craig, Latimer, Muskogee, Okmulgee, Pittsburg, Rogers, Tulsa, Wagoner.

DISTRICT 16

Northern Colorado. The following counties in Colorado: Adams, Arapahoe, Boulder, Douglas, Elbert, El Paso, Jackson, Jefferson, Larimer, Weld.

DISTRICT 17

Southern Colorado. The following counties in Colorado: All counties not included in northern Colorado district. The following counties in New Mexico: All coal producing counties in the state of New Mexico, except those included in the New Mexico District.

DISTRICT 18

New Mexico. The following counties in New Mexico: Grant, Lincoln, McKinley, Rio Arriba, Sandoval, San Juan, San Miguel, Santa Fe, Socorro. The following counties in Arizona: Pinal, Navajo, Graham, Apache, Coconino. All coal-producing counties in California.

DISTRICT 19

Wyoming. All coal-producing counties in Wyoming. The following counties in Idaho: Fremont, Jefferson, Madison, Teton, Bonneville, Bingham, Bannock, Caribou, Oneida, Franklin, Bear Lake.

DISTRICT 20

Utah. All coal-producing counties in Utah.

DISTRICT 21

North Dakota–South Dakota. All coal-producing counties in North Dakota. All coal-producing counties in South Dakota.

DISTRICT 22

Montana. All coal-producing counties in Montana.

DISTRICT 23

Washington. All coal-producing counties in Washington. All coal-producing counties in Oregon. The state of Alaska.

Barge Movement of Coal

BY WILLIAM L. PRICE*

The extent of the Inland Waterways System of the United States and its closeness to United States coal deposits are evident from an examination of the map of Fig. 1. The basic waterways available for easy access to coal reserves consist of the Mississippi/Ohio River system with its tributary streams. Large tows of river barges of coal from mine to power plant (Fig. 2) are an important part of normal river traffic. Since the Ohio River system tends to bisect the eastern bituminous coal reserves of the United States, it forms a

Fig. 1. Inland waterways in Eastern United States, with approximate locations of coal reserves. Lake ports coal transfer, rail to ship: (1) Conneaut, Ohio. (2) Ashtabula, Ohio. (3) Sandusky, Ohio. (4) Toledo, Ohio. (5) Chicago, Illinois.

natural supply artery. The river system offers a low-cost form of transportation and fortunately provides an easy source of natural cooling water for the fossil-fuel-fired power plant. With this incentive, a large number of coal-burning power plants have progressively appeared along this inland river system.

The demand for low-sulfur coal in the 1970s instigated a new rail movement of western

*Products Development Manager, Engineering Works Division, Dravo Corporation, Pittsburgh, Pennsylvania.

Fig. 2. Large inland waterways tow of coal barges. Twenty-two 1400-ton-capacity barges propelled by a 6400-horsepower towboat.

coals toward the closest possible available points on the river system for transfer to river barges. Supplementing the Mississippi River system running through the heartland of the United States is the Gulf of Mexico Intercoastal Waterway stretching from Brownsville, Texas, to the northern part of Florida. This auxiliary navigable access route is also available for future coal transportation. At present, the principal coal movement toward the Gulf area consists of movement of coal from the eastern coal reserves via the Ohio and Mississippi Rivers to the New Orleans area where it is normally transferred into special deepwater marine equipment. A high tonnage of coal is moved from New Orleans to the Tampa, Florida, area by means of 20,000-deadweight-ton and larger barges. (An important economic factor which helps make this long-distance movement of coal possible is a backhaul of phosphate rock suitable for fertilizer from Tampa to New Orleans for upriver

Fig. 3. Coal transfer terminal—from river barge to Gulf barge. A 3000-ton/hour continuous river barge unloader and 1400 ton/hour phosphate rock clamshell barge unloader.

shipment.) Figure 3 shows a transfer terminal in the New Orleans area: coal from river barges to Gulf barges and phosphate rock moving in the reverse direction.

The third major waterway presently utilized for coal transportation is the Great Lakes/ St. Lawrence Waterway opening up very large areas. Movement of coal along the Great Lakes traditionally was dependent on its position relative to the bituminous coal region. One traditional pattern for coal movement consists of a rail movement northward from the West Virginia/Ohio area to Lake Erie ports. There it is transferred to 26,000-deadweight lake boats by which it is moved to its ultimate power-plant destinations along the water's edge. Typical ports that handle great tonnages of coal in this manner are Conneaut, Sandusky, Ashtabula, and Toledo, Ohio. In addition, Illinois coal moves to Lake Michigan by rail and through the Chicago River link. A later development of the 1970s tied in with the desirability of low-sulfur coal is the planned movement of western coal to the lake head at Duluth and Superior. At these points, coal will be transferred from rail unit trains into lake boats and moved to power-plant locations along the Lakes.

The remaining waterway coal movement is that along the East Coast from Norfolk to Atlantic coastal power plants.

ECONOMIC FACTORS

On the inland river system, coal is generally moved in 1400-net-ton-capacity barges in fleets of 15 to 20 barges at a time. This efficient 25,000-ton tow, with its low manpower requirements, provides a low-cost coal movement, ranging from 3 to 10 mills per ton-mile (early 1975 figures). Comparable normal costs for rail movement are 3 cents/ton-mile; truck transportation can rise to 5 cents per ton-mile. This cost advantage, along with the very large tonnages required, makes barge movement attractive, especially to riverbank power plants.

From earliest United States history, power-plant locations have been geared to river transportation. There are dozens of power plants located on the Mississippi River system with a steady increase in this number taking place.

For movement of United States coal to power plants along the Canadian shores of the Great Lakes, the advantages of marine movement are obvious.

INLAND WATERWAYS SYSTEM

Since the normal governing depth of the channels on the inland river system is 9 feet, the coal barges used are normally designed for a working draft of about 8½ feet. The original barge sizes were designed around lock sizes at the navigational dams. Early barges that operated in the Allegheny and Monongahela River system and had locks 56 feet wide by 360 feet long, were normally 175 feet long, 26 feet wide, and 11 feet deep. The Ohio River navigation system of locks and dams was completed by 1930 with locks of 600 by 110 feet width. Starting in about 1950, the standard jumbo barge became one with a dimension of

TABLE 1.   Movement of Bituminous Coal by Inland
Waterways in the United States—1972

Inland waterways	Millions of tons
Mississippi River System	
Ohio River	66
Monongahela River	30
Mississippi	22
Green River	16
Tennessee River	12
Illinois River	7
	153
Gulf Intercoastal Waterway	5
Total marine coal movements	158
Total coal mined	595

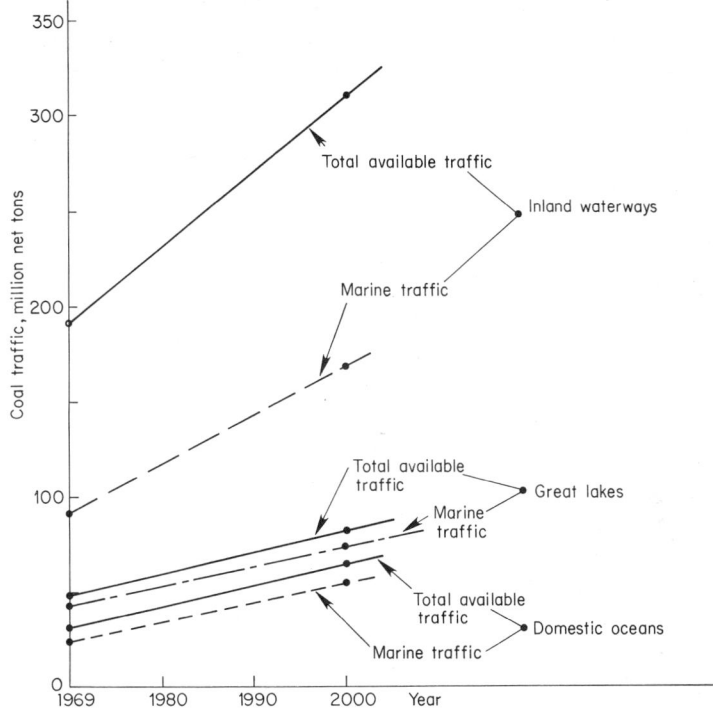

Fig. 4. Estimated growth in coal traffic. Marine coal traffic by location versus total coal traffic available for competitive market area—inland waterways, Great Lakes, and domestic oceans. *(Source: A. T. Kearney, "Domestic Waterborne Shipping Market Analysis—1974.")*

195 feet long and 35 feet wide. This allowed three lengths times three widths, or nine barge equivalents in a single locking. Under a major navigation modernization program to be completed before the end of the 1970s, the Ohio River system will be converted to locks 110 feet (width) by 1200 feet (length). This will permit more efficient continued use of this same size barge. The present 53 low dams will be replaced by 19 fixed type with pools of significantly greater length. A portion of the barge fleets is being constructed with units 200-feet long.

Diesel-powered, twin-screw towboats used on the river transportation system tend to be in the 2000 to 3000-horsepower class for general use, with larger towboats running up to 6000 horsepower—and a few boats for the lower, open Mississippi River system up to 10,000 horsepower (triple screw). For power-plant use, the greater tonnage of the coal is handled on the Ohio River and tributary system with the lower of the mentioned horsepowers in use.

The amount of coal handled on the inland waterways in the United States for the year 1972 is broken down in Table 1. River movement transported some 153 million tons out of a total of 600 million tons produced. Some concept of the projected growth in coal traffic can be visualized from Fig. 4 which predicts the growth until the year 2000 by various waterways as compared to the total projected traffic.

COASTAL/GULF OF MEXICO/GREAT LAKES MOVEMENT OF COAL

Of the nine navigation locks on the 1113-mile Gulf Intercoastal Waterway, two are 56 feet wide, one is 84 feet wide, and the remainder are 75 feet wide. River-type open barges can be moved in this protected waterway. For exposed waters, the barge traffic of coal along this waterway is handled by coastal barges with protected cargo, which allows for operation in rough water. For full deepwater movement of coal in the Gulf of Mexico

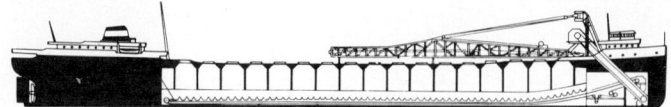

Fig. 5. Typical self-unloading 26,000-ton lake boat used for coal traffic on the Great Lakes. *(Hewitt-Robins unloader system.)*

proper, a considerable traffic exists between New Orleans and the Tampa area. Coal is transferred near New Orleans from the typical river coal barge into a special 20,000-deadweight-ton and even larger-capacity Gulf barge (approximately 420 feet long by 80 feet wide and 35 feet deep). These barges are normally towed singly across the Gulf open water to the Tampa area by a 6000-horsepower tug. At the Tampa destination, they are unloaded at the power plant, and a return cargo of phosphate rock is normally carried back to New Orleans, thus assisting in lowering the cost of the coal movement.

Coal movement on the Great Lakes is normally accomplished in a self-unloading-type lake boat about 700 feet long by 70 feet wide. See Fig. 5. Coal is loaded into this vessel through its numerous transverse hatches. The vessel is so constructed that the coal cargo can be discharged by gravity to long conveyor belts extending the length of the hull. These longitudinal conveyor belts, in turn, discharge the coal to a large, long, swinging unloading boom, which allows the coal to be discharged at the customer's dock into a receiving hopper from which it travels by conveyor belt to the power plant.

LOADING/UNLOADING BARGES ON INLAND RIVER WATERWAYS

River barge loading is normally accomplished by pulling the single empty barge along a dock beneath a projecting conveyor belt which is fed from the source of coal, from either railroad cars or a conveyor belt that extends from the coal mine. A later development is to load the barges in continuous strings of five barges, which are then ready for making up into a two- or three-barges wide-system. A high-speed barge loading system is shown in Fig. 6.

At the conclusion of the river trip, the barges are placed at the power-plant dock. They are then normally moved one at a time under unloading equipment. The traditional unloader on the inland river system has been the clamshell bucket-type unloader.

Fig. 6. River barge-loading arrangement: 1400-ton-capacity barges are loaded with coal via conveyor belt to three strings of five barges.

Fig. 7. Typical clamshell bucket coal-barge unloader, capacity 1400 tons/hour.

Capacities of this equipment vary from 400 tons per hour up to 1500 tons per hour. See Fig. 7. To assist in the complete cleanup of the open barges, a rubber-tired front-end loader is usually dropped into the barge and moves the remaining coal to a position in which the clamshell bucket may quickly complete the unloading.

A development of the 1960s for speeding up the unloading of coal barges at power

Fig. 8. Typical continuous barge unloader supplying power plant at free-digging rate of 3600 tons/hour. Twin 9-ft-wide bucket ladders empty open-type river barges in three passes.

Fig. 9. Typical open-hopper barge for transporting coal on inland waterways. 195 × 35 × 12-ft semi-integrated type.

plants is the continuous barge unloader of the type shown in Fig. 8. Here, a set of continuous bucket ladders are lowered into the barge, and the barge is pulled along, feeding coal to the buckets. Exceptionally high rates are possible with this system. Typical unloading rates will be 3600 tons per hour—and up to 5000 tons per hour (free digging rate). The average unloading rate for these barges will be 50 to 60 percent of this rating. The continuous unloader has the advantage of doing less damage to the barge bottom, and the initial cost is not significantly greater than that of a clamshell unloader. Basically, the clamshell unloader has been replaced for this type of service by the continuous unloader.

BARGE DESIGN

The normal 1500-ton-capacity coal barge used on the inland waterways is similar to that shown in Fig. 9, measuring 195 by 35 by 12 feet. A modification to this barge consists of having lift-off-type barge covers which can be stacked on both ends while the barge is used in the movement of coal. If another cargo is transported, such as grain, the covers are then set in place to protect the cargo. A barge used in this trade normally is built as a semi-integrated barge, or having one raked end with the other end normally a square end. Some double square-end barges are used along with occasional 200-foot-long barges. A typical twin-screw towboat is shown in Fig. 10.

Fig. 10. Typical inland-rivers towboat.

REFERENCES

Karney, A. T.: Domestic Waterborne Shipping Market Analysis—Inland Waterways Trade Area, *Report MA-RD-900-73007-A*, Maritime Administration, Washington, D.C., February 1974.

Kyle, B.: "Domestic Waterborne Commerce—Its Role and Future," Society of Naval Architects and Marine Engineers, Chicago Meeting, May 22, 1974.

Courtsal, D. P.: "The Marine Business in the Central United States," Society of Naval Architects and Marine Engineers, Annual Meeting, Nov. 11, 1971.

"Inland Waterborne Commerce Statistics—1971–1972," American Waterways Operators, Inc., Washington, D.C., 1973.

Waterborne Commerce of the United States, Supplement to part 5, "Domestic Inland Traffic (Annual)," U.S. Army Corps of Engineers, Washington, D.C.

Transport Statistics in the United States, part 5, "Carriers by Water (Annual)," U.S. Interstate Commerce Commission, Bureau of Transport Economics, Washington, D.C.

Using Waterways to Ship Coal, *Coal Age*, p. 122, July 1974.

Coal Transportation by Slurry Pipeline

BY R. B. JACQUES* AND J. G. MONTFORT†

In July 1970, Black Mesa Pipeline, Inc., began operation of an 18-inch-diameter slurry pipeline that transports 5 million tons per year of bituminous coal to a 1580-megawatt power generating station. Significantly, it marked initiation of the largest and longest coal slurry pipeline ever constructed, and for the first time that coal supplied in a slurry form comprised the total fuel supply for a large steam-generating plant.

The line, as shown in Fig. 1, originates at an altitude of 6500 feet in the northeastern part of Arizona in the heart of the Black Mesa coal fields, transports pulverized coal in a water carrier fluid 273 miles across some of the most remote and rugged country of the Western United States, and then terminates at the Mohave Generating Station at an elevation of 700 feet in the southern tip of Nevada. Operating with a reliability factor of 99 percent, the pipeline has demonstrated the feasibility of commercial long-distance transport of solids by pipeline.

The technology that served as the foundation for the Black Mesa line dates to the nineteenth century, when in 1891 Dr. Andrews of New York was issued the first patent in the United States covering the pumping of coal and water. In 1914 the first commercial transport of coal in water was conducted in England, when a short 8-inch pipeline was used to carry coal from river barges to a power plant. Thereafter, several proposals were submitted for the long-distance transport of coal from mine to market in the Eastern United States, but failed to materialize for several reasons, not the least of which were technical problems.

EARLY OHIO INSTALLATION

Intensive research into slurry transport was continued, and by 1957 technology and engineering had advanced to the point where the first long-distance transportation of coal in water was feasible. The result was the construction and operation of the Consolidation Coal Pipeline. This line, 10 inches in diameter and 108 miles long, transported 1.25 million tons of coal per year from Cadiz, Ohio, to an electric generating station 20 miles east of Cleveland on the shores of Lake Erie. The pipeline was powered by three pump stations, spaced 30 to 35 miles apart, in which discharge pressures reached 1000 psi. Coal with a graded size consist,‡ 8 mesh × 0, and a concentration of 50 percent solids was transported.

Although the line operated successfully, transporting 7 million tons of coal, Consolidation did experience some unexpected operating problems that were resolved by the usual intensive study and appropriate system modification approach. In addition, much investigation was conducted with variables, such as size consist and slurry concentration and the resultant effect on slurry stability.

After seven years of operation, the line was shut down in 1963, when the unit-train concept resulted in much lower freight rates on significantly higher tonnages. Improved rail rates resulted from the competitive effect of this pipeline. The Consolidation line was an economic and technical success and a pioneer effort in the field of long-distance slurry transportation. As indicated by Table 1, the Consolidation line forged a path that many would follow.

*System Engineer; †Vice-President and General Manager, Black Mesa Pipeline, Inc., Flagstaff, Arizona.
‡Consist means the size makeup of the solid phase of the coal slurry. The term 8 mesh × 0 indicates coal with a graded size makeup in the range of 8 mesh and zero (dust).

PRINCIPAL HYDRAULIC VARIABLES

The feasibility of slurry transportation depends on the resolution of a number of variables, the most important of which, from a hydraulic standpoint, are (1) size consist, (2) velocity, and (3) concentration.

The selection of a proper size consist (gradation) is important in order that homogeneous flow can be achieved at prudent operating velocities. For coal slurry, such a consist is on the order of 8 mesh × 0, i.e., 0.1 inch to dust. Homogeneous flow (solids evenly distributed across the pipe diameter) is important if excessive wear in the bottom of pipe is to be avoided and stable operation achieved.

Of equal importance and directly related to size consist is the proper selection of velocities for transport. The general relationship between velocity and pressure drop has been well documented and will not be detailed here. Suffice it to say that velocity determined by pipe diameter should be chosen so that it is not excessive and causes neither abrasion of pipe wall nor inordinately high pressure drops. Conversely, velocity should not be so low as to cause heterogeneous flow, with resultant excessive wear in the pipe bottom or bed formation which will cause unstable operation. Generally, practical operating velocities will be in the range of 4 to 7 feet per second.

Finally, the important parameter, slurry concentration, must be considered. The relationship between concentration and viscosity for any given slurry can be determined in laboratory bench testing. Although it varies for different slurries, all systems generally demonstrate a point in inflection where a small increase in concentration causes a large viscosity increase. Hence, it is important to maintain a concentration range below the inflection point in order to provide good operations without excessive velocities. Operation above the inflection point will result in necessarily higher operating velocities and much larger pressure drops, with only small gains in capacity. For coal, the practical concentration range is 45 to 55 percent solids.

THE BLACK MESA SYSTEM

In 1966, the Black Mesa Pipeline Inc., was organized to construct, own, and operate a pipeline to transport coal as previously described. Both pipeline and rail competed for the transport contract, but the pipeline was

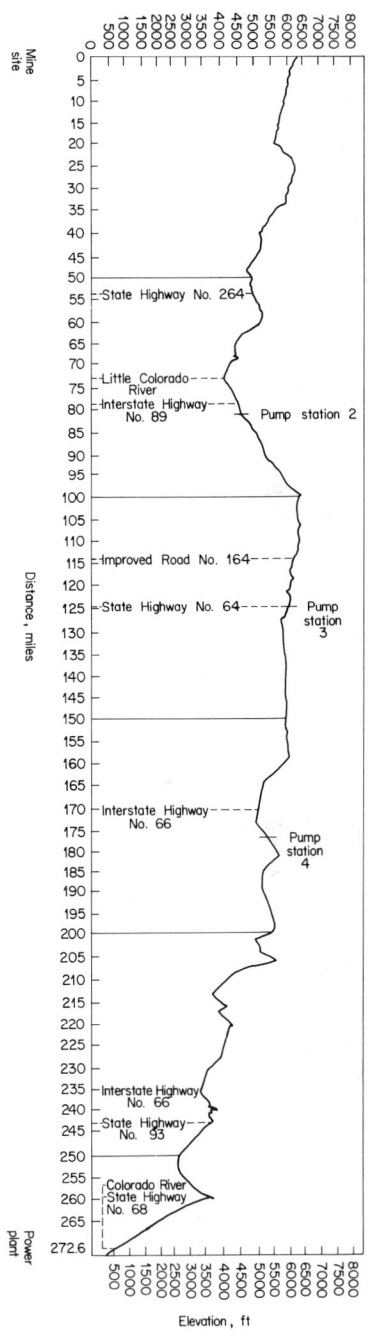

Fig. 1. Profile of the Black Mesa Pipeline.

TABLE 1. Principal Coal Slurry Pipelines

Operator and location	Pipeline length and (I.D.)*	Characteristics
Consolidation Coal Company. Ohio coalfield to Cleveland, Ohio, on Lake Erie (United States)	108 miles (10-inch)	Coal particles transported: up to 14 mesh (1.2 mm). Solids by weight: 50–60%. Specific gravity (slurry): 1.15 (solids): 1.4. Velocity in line: 4.5–5.25 feet/second. Pumping capacity: 1.3×10^6 tons/year. Types of pumps: duplex double-acting, three pumps/station. Stations: located at 30-mile intervals. Operated from 1956 to 1963. Closed because of advent of lowered rail-freight rates.
Black Mesa Pipeline, Inc. (subsidiary of Southern Pacific Pipelines, Inc.) and Peabody Coal Company. Arizona to Nevada-California border (United States)	273 miles (18-inch, and 12-inch)	Coal particles transported: up to 14 mesh (1.2 mm). Solids by weight: 45–55%. Specific gravity (slurry): 1.4–1.5. Velocity in line: 6 feet/second. Pumping capacity: 4.8×10^6 tons/year, increasing to 6×10^6 tons/year. Will carry 117×10^6 tons over 35-year period. Line has been in operation since July 1970. Total of thirteen 4500 gallons/minute duplex pumps. Four pumping stations.
Consolidation Coal Company and Texas Eastern Transmission Company. From West Virginia to New York City and Baltimore areas (United States)	350 miles (20-inch)	Proposed. Pumping capacity planned for 10×10^6 tons/year.
Novovolynskaya Mine (U.S.S.R.)	38 miles (12-inch)	Coal particles transported: $<\frac{1}{32}$-inch. Solids by weight: 50%. Velocity in line: 4.75 feet/second. Pumping capacity: 220 tons/hour. Constructed in 1957 and still operating.
Houillieres due Bassin du Lorraine (Carling, France)	51 miles (15-inch)	Coal particles transported: 16-mesh (1 mm). Solids by weight: 25–30%. Velocity in line: 7–10 feet/second. Pumping capacity: 250 tons/hour (1.5×10^6 tons/year). Line has been operating since 1952.
Cascade Pipeline Limited (East Kootenay, British Columbia to Vancouver, B.C., Canada)	490 miles (—)	Proposed.
Polish Central Mining Industry (Poland)	126 miles (10-inch)	Operating.

*Internal diameter (bore).

selected because of the lower cost per ton-mile of supplied fuel. In addition, a more-than-adequate supply of good water pumped from deep wells was found to be available.

The Black Mesa system consists of a coal preparation plant, pump stations, and 273 miles of telescoped pipe, of which 260 miles are 18 inches in diameter, with the remaining 13 miles 12 inches in diameter. The pipe-wall thickness varies from 0.219 to 0.469 inch. The line can transport 660 tons per hour at 48 percent solids and 4200 gallons per minute. Transit time is 3 days, and line fill requires 45,000 tons of coal. The entire system is automatically controlled, assisted by one operator, located on the Black Mesa.

Black Mesa receives coal from Peabody Coal Company, which conducts the mining by conventional strip-mining techniques. Before the delivery of coal at Black Mesa's preparation plant, the coal is sampled and weighed by an independent testing firm retained by the three parties involved in the project. An overview of the facilities located on the Black

Mesa is given in Fig. 2. The preparation plant is shown in Fig. 3. This plant consists of three identical open-circuit process lines of 300-tons-per-hour capacity each, comprised of a 2-hour storage bin, cage Paktor (mechanical impacting), rod mill, sump, and centrifugal pump.

Coal is delivered to one of the three storage bins in a 2-inch × 0 state. It is then conveyed from bin to cagemill where it is reduced in a dry crushing process to $\frac{1}{4}$ inch × 0. This product is then combined with the needed amount of water to achieve the desired grinding consistency for the rod mills, shown in Fig. 4. The mill product, typically 8 mesh × 0, is monitored for density and is then pumped to one of four 2-hour storage tanks.

Fig. 2. Aerial view of coal-preparation and other facilities located on the Black Mesa at the originating terminal of the Black Mesa Pipeline.

The rod-mill product is sampled on a bihourly basis and analyzed for size consist and concentration. These variables are carefully monitored and controlled to assure that slurry can be pipelined. In addition, the slurry passes through a test loop that monitors differential pressure, percent solids, and flow rate. For a given set of conditions, a certain differential pressure is generated. If the actual pressure deviates from the expected pressure, prompt action can be taken. With the foregoing information at hand, supervisory personnel can prevent nonspecification material from getting into the main line.

Four pump stations (Fig. 5) utilize large positive-displacement pumps with electric motor drive and hydraulic coupling for speed control. They supply the required energy to transport the slurry. Three of the stations involve three installed units, and the fourth station has four installed units. The interior of a typical pump station is shown in Fig. 6. At each location, a unit acts as a standby at all times. The three-unit stations operate at approximately 1000 psi discharge pressure, and the four-unit station operates at 1500 psi. The driver horsepower is 1500 to 1750, depending on location. All stations are operated on suction control to provide smooth, even response throughout the line to flow-rate changes.

Large water reservoirs are located at each station, with capacity sufficient to flush the downstream line section. In addition, dump ponds are located at each station. These are of sufficient capacity to take two emergency full dumps from the upstream line section.

The pipeline was designed to allow both operation over a range of conditions and shutdown of any or all sections if the need arose. The throughput of the line can be varied by changing the concentration of solids in the slurry and by altering the flow rate. Hence, some range of delivery rates is possible. Should emergency conditions occur, several alternatives are available. If either Station 3 or Station 4 is lost, the flow rate can be reduced in order to bypass that station for a limited period of time. If either of the other stations is lost, a maximum amount of slurry is kept moving by pumping out of the

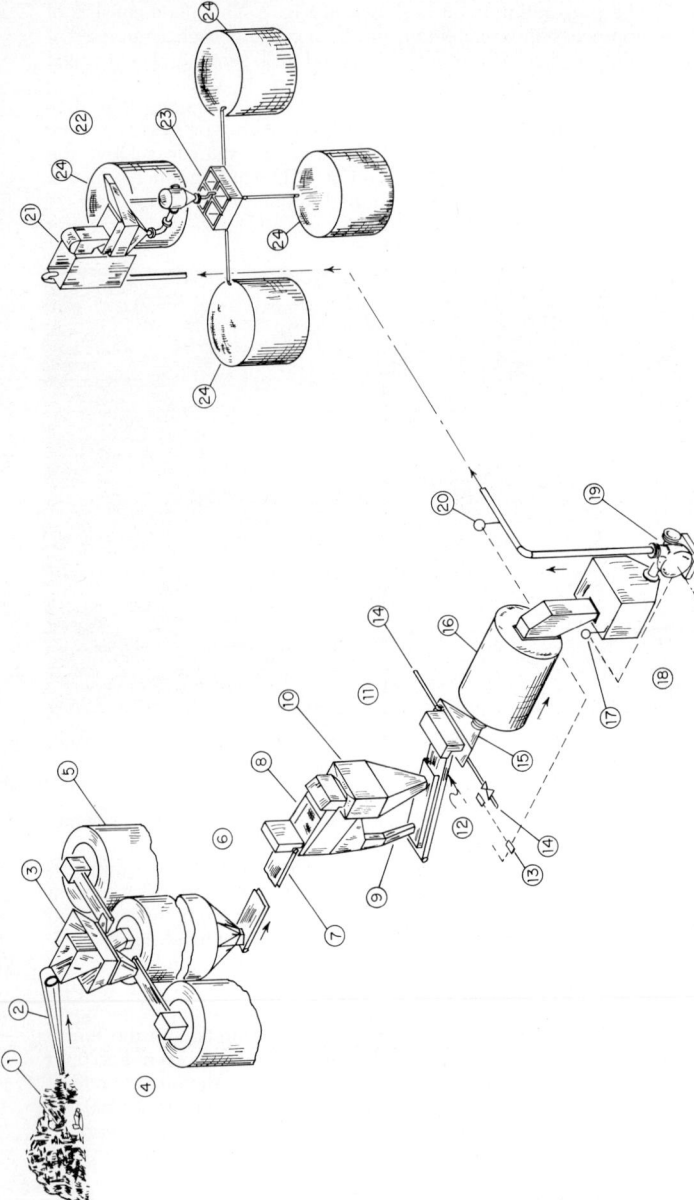

Fig. 3. Schematic diagram of one of three process lines in the slurry preparation plant for the Black Mesa Pipeline. (1) Coal stockpile from Peabody Mine (2-inch by 0 state); (2) 42-inch raw coal conveyor; (3) distribution chute; (4) 42-inch distribution conveyor (10 hp); (5) raw-coal bins (580 tons each—provide a 2-hour supply for one process line); (6) 54-inch screen belt feeder (10 hp); (7) feed rate of screen belt feeder is manually adjustable and controls rate to process line; (8) scalping screen (6 feet by 10 inches, double-deck, 15 hp); (9) screen undersize chute (fines ⅜-inch and smaller bypass); (10) 48- by 54-inch impact crusher (290 tons/minute, 350 hp); (11) 36-inch mill conveyor (10 hp); (12) weight scale; (13) density automatically controlled by data input from weight scale and density meter to weight-density ratio relay which, in turn, regulates amount of dilution water; (14) dilution water; (15) mill feed chute; (16) 13- by 18-foot rod mill (1500 hp); (17) level control determines speed of discharge pump; (18) rod mill discharge pump (125 hp); (19) rod mill discharge pump (125 hp); (20) density meter; (21) automatic sampler; (22) 4- by 3.5-feet safety screen (3 kilowatts); (23) swivel-joint distribution box; (24) 49- by 45-feet slurry storage tanks with 120 agitators (125 hp total). Three of the tanks are normally used. The fourth tank is for fines. Each tank provides a 2-hour supply.

downstream station reservoir, and the remaining line sections are shut down. Sections are shut down in a packed condition to prevent subcritical flows that can plug the line. Data are continuously monitored on the shutdown section. Operational techniques have been well developed for restarting the operations.

The pipeline delivers slurry at the power plant into any one of four large (7.5 million gallon) continuously agitated storage tanks. These tanks, many times larger than any agitated tanks previously built, have performed successfully. The slurry is withdrawn

Fig. 4. Rod mills for grinding coal to proper consistency for preparation of coal slurry.

from storage tanks by pump and fed to centrifuges, where 75 percent of the water is removed, clarified, and used in the power-plant cooling and steam-generating process. The cake-dry coal is then fed to bowl mills, where it is completely dried, pulverized to minus 200 mesh, and fired pneumatically to the combustion chamber. In addition to the active storage facilities, two large, dewatered *force majeure** stockpiles of coal (6 days' capacity each) are kept on hand at the plant for emergency supply.

OUTLOOK FOR COAL SLURRY PIPELINES

Prospects for further coal movements by slurry pipelines are good. The United States has very large coal reserves, mostly in the West. The amounts are sufficient for several hundred years and can serve for the decades needed to develop other long-term solutions to the energy problem. Western coals are low in sulfur and are less costly to mine than most other U.S. coals. Movement to major midwestern markets (800 to 1500 miles distant) by slurry pipeline is economically attractive. An indication of the transportation rates which are possible is shown in Fig. 7. A 1000-mile, 38-inch-diameter line is now planned to carry 25 million tons of coal per year.

A slurry pipeline will be attractive (1) when there is a large volume of coal (generally over 1 million tons per year) to move for a long term between stable locations, (2) when the mine or mines are in remote areas not served by existing rail lines, and (3) when the fine grind necessary for slurry transport does not conflict with end usage. High-volume slurry pipelines may be required eventually to avoid overloading rail lines.

The advantages of slurry pipelines are several. They are reliable because they are not

*Legal protection against acts of God, such as natural disasters, and strikes and other factors over which the party has no control.

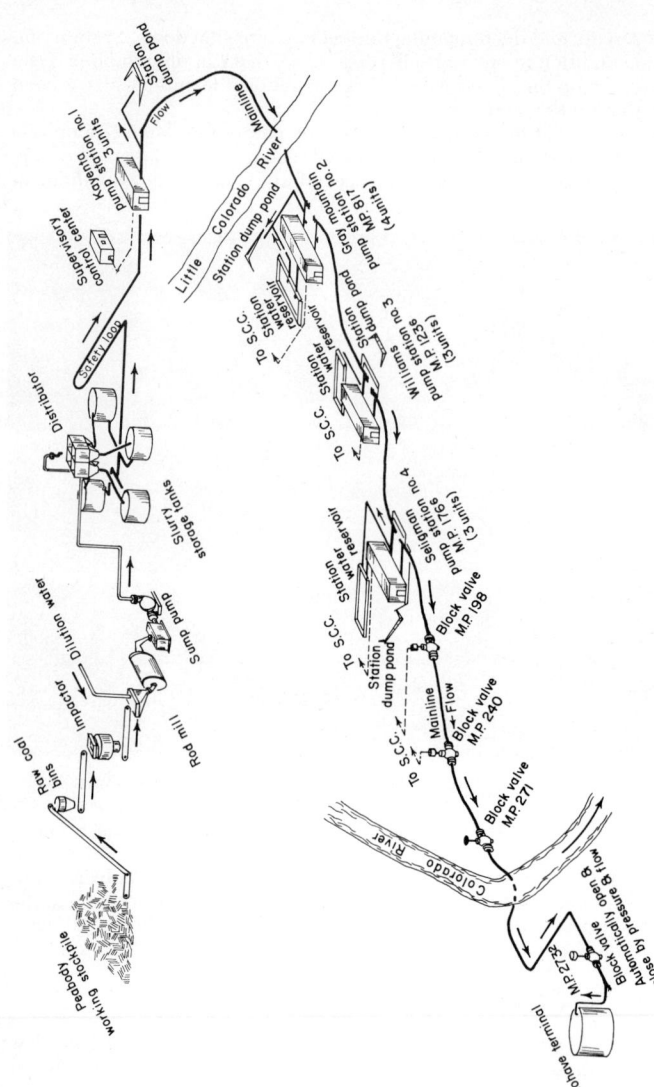

Fig. 5. Schematic diagram of Black Mesa Pipeline system. Total system covers a distance of 273.2 miles. The pipeline underpasses two rivers, the Little Colorado River and the Colorado River. The pipeline runs from the coal preparation plant on the Black Mesa at an altitude of 6500 feet in the northeastern part of Arizona to the Mohave Generating Station at an elevation of 700 feet in the southern tip of Nevada. The line transports 5 million tons of bituminous coal annually. The pipeline is buried below ground for its entire length. M.P. = mileage point from originating terminal of pipeline; S.C.C. = supervisory control center.

Fig. 6. Interior of a typical pump station on the Black Mesa Pipeline.

affected by severe weather or ground conditions. Because of automation and their inherent laborsaving characteristic, they are less subject to the effects of labor disputes. The Black Mesa Pipeline has demonstrated an availability of over 99 percent.

After installation, a coal pipeline is less severely affected by escalation of direct costs. This is so because this mode of transportation is capital-intensive. About 70 percent of the tariff is capital-related; about 15 percent is power-related; with only 15 percent attributable to labor and maintenance supplies. The effects of escalating costs on a slurry pipeline and on a railroad are shown in Fig. 8. In neither case illustrated is the inflationary effect on capital return reflected.

Coal slurry pipelines are environmentally superior to other forms of transportation because they are quiet and unobtrusive and because energy consumed by a coal pipeline on a ton-mile basis is significantly lower than by any other transportation mode, a factor that will become increasingly important.

The two prime disadvantages facing a coal slurry pipeline are: (1) An adequate and assured water supply is required. In the water-short areas of the Western United States, this is a major consideration, both physically and legally. (2) Dewatering slurry for consumption at a power plant, or for transshipment by barge is required. This is an important step and one that is developing toward better solutions. Centrifuging is the primary method used to date. Although reduction of coal particles to the very small size needed for movement by pipelines serves the requirements of a generating station for a finely ground coal, the fine size makes the problem of dewatering more difficult.

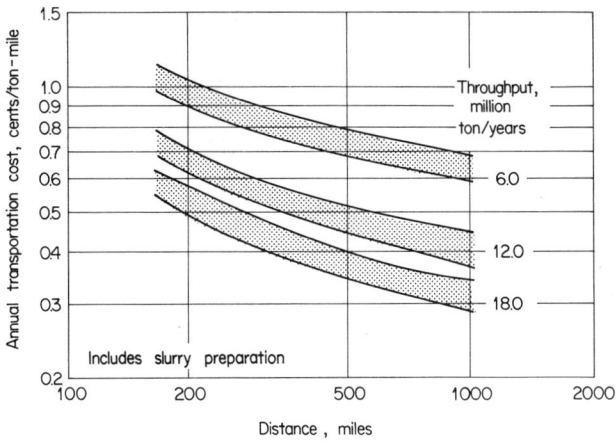

Fig. 7. Coal-slurry pipeline transportation costs.

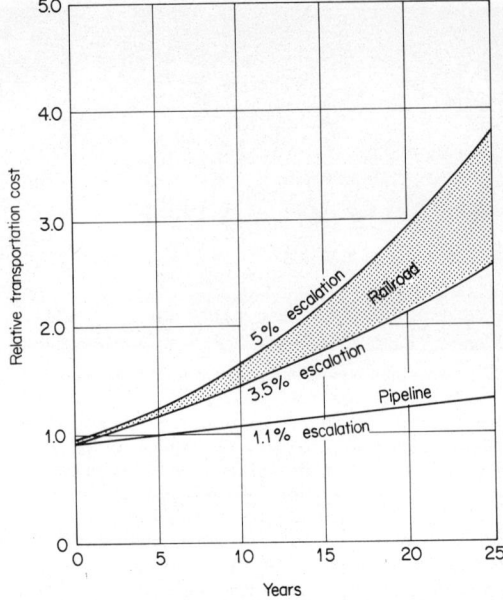

Fig. 8. Effects of inflation on coal-movement costs.

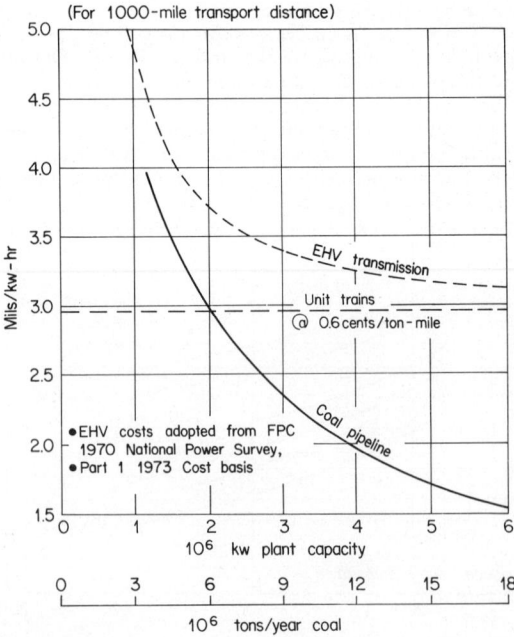

Fig. 9. Comparative costs for coal-energy transmission. EHV = extra-high voltage.

Transportation of coal from western sources can be accomplished by unit train, by pipeline, or, if generating stations are placed near the mines, by transmission of electricity via extra-high-voltage lines. If already in place, rail lines offer the most flexible and expeditious arrangements. A slurry line will offer the advantages of high reliability, low environmental impact, and, at high volumes, attractive economics.

Extra-high-voltage transmission lines have high environmental impact and can be used to serve at mine-mouth power plants only if large amounts of cooling water are available at the mine sites. The cooling-water requirement for a power plant is approximately eight times that for a slurry pipeline. Comparative economics of coal energy transmission by these modes are given in Fig. 9.

REFERENCES

Love, Frank H.: The Black Mesa Story, *Pipeline Eng.*, November, 1969.
Staff: Hydraulic Transport of Minerals, *Mining Mag.*, April, 1972.
Wasp, E. J., and T. L. Thompson: Slurry Pipelines, Energy Movers of the Future, *Oil & Gas J.*, Dec. 24, 1973.

The Lurgi Process for Coal Gasification

BY PAUL F. H. RUDOLPH*

Over forty years ago, the concept of complete gasification of coal with oxygen under pressure was put into practice by the pioneering work of Drawe, Danulat, and Hubmann, which ultimately led to the current Lurgi process for coal gasification. A very specific technology had to be developed for which at the time of the pioneers there were hardly any experiences or examples available from the other fields of chemical engineering.

When at that time the new field of town gas production by coal gasification was initiated in Germany, the processes available for coal upgrading were pyrolysis, the low-pressure air-blown gas producer, and the oxygen-blown atmospheric-pressure Winkler gasifier. The task was to produce substitute town gas from low-grade coal by complete gasification in a high-duty gasifier. This task was solved by using oxygen under pressure, a principle that remains valid today. As this gas had to have a minimum methane content of 15 percent, and the avoidance of by-product char was of utmost importance, gasification in countercurrent was the only solution possible. Although the air-blown gas producer, representing the countercurrent principle, was taken as a model, many parallel developments in various engineering disciplines had to be made to create the current Lurgi gasifier.

REACTOR DESIGN DEVELOPMENT

The technique of the application of pressure and the use of oxygen in an apparatus with arduous working conditions was developed stepwise to the present standard. All gasifier elements, such as coal and ash-lock chambers, coal distributor, revolving grate, and reactor and scrubbing cooler, had to be fully developed and then were refined and innovated parallel to the evolution of the other chemical and mechanical engineering aspects of the process. The principal phases in the development of the gasifier are shown in Fig. 1.

PROCESS DEVELOPMENT

Coal is a raw material of unique and variable nature, and its properties have a major influence on the design of the gasifier, its operating method, and the downstream units for gas conditioning and gas purification. Knowledge of operations in the gasifier was gained from experience with many different coals under varying operating conditions. Altogether some sixty grades of coal from all over the world, including coke, anthracite, semianthracite, and subbituminous coals including caking coals, lignite, and peat, have been processed successfully.

Lurgi has performed gasification using a wide variety of gasification agents, such as mixtures of steam with oxygen, air, and carbon dioxide, and with dry-ash removal as well as with liquid-slag removal.

DEVELOPMENT OF OVERALL PROCESS SCHEME

Although gasification is the key process, it must be stressed that it is only one step in the overall scheme comprising the processes for gas conditioning, gas purification, and by-product treatment. To a large extent these latter processes are governed by the properties of the coal and by the gasification process. The principal processes developed, which now are part of the present Lurgi process (overall), are shown in Fig. 2. In addition, plants for the supply of oxygen, power, steam, and water, and for the treatment and neutralization of waste products must be added. A *Clean Fuel Gas Process* also will be described later. It is

*Lurgi Mineralöltechnik GmbH, Fuel Technology Division, Frankfurt (Main), West Germany.

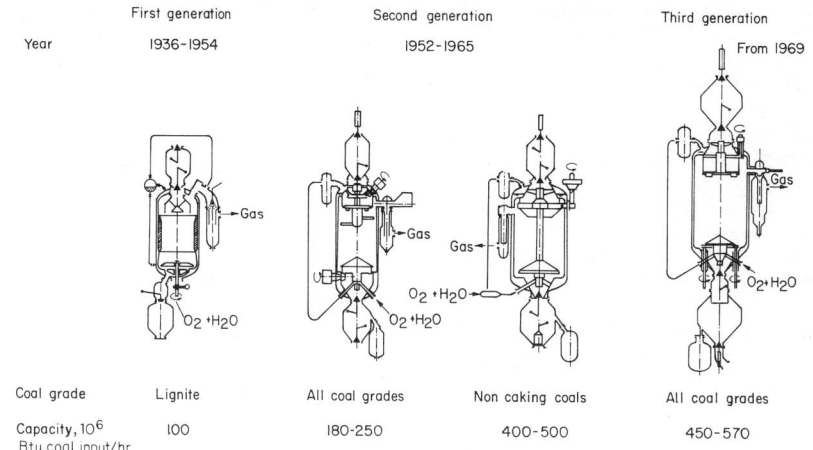

Fig. 1. Stages of gasifier development.

interesting to note that the gasification section amounts to only 15 percent of the total investment cost.

LURGI PRESSURE GASIFICATION

The Lurgi gasifier is a reactor for the performance of countercurrent gasification of coal in a moving bed. See Fig. 3. The principal mechanically operated devices include: (1) the automated coal-lock chamber for feeding coal from a coal bin to the pressurized reactor; (2) the coal distributor through which the coal is introduced into the reactor shaft in such a manner that uniform distribution of the coal across the shaft area is achieved (When processing caking coals, blades are mounted to the distributor which rotate within the fuel

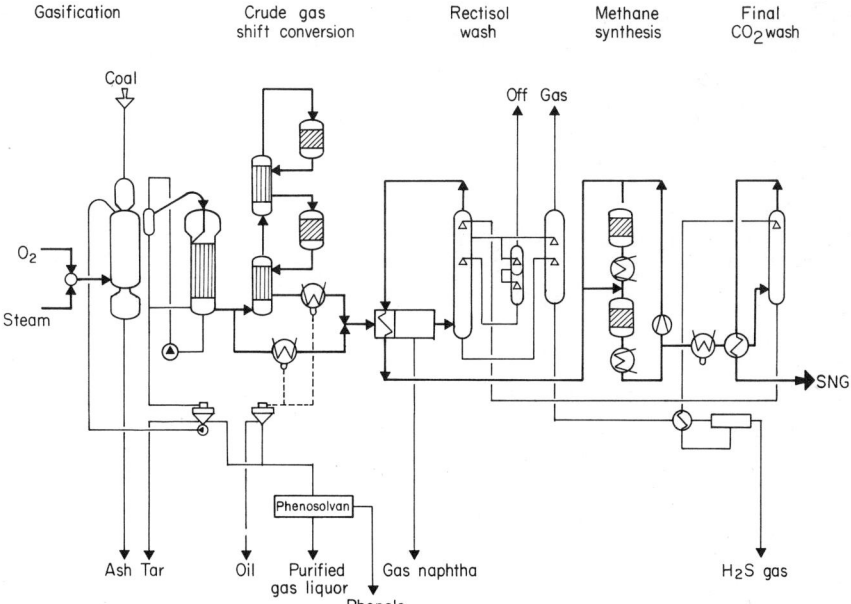

Fig. 2. Lurgi process for producing substitute natural gas (SNG) from coal.

bed); (3) the revolving grate introduces the gasification agent and extracts the ash; (4) the ash-lock chamber for discharging the ash from the pressure reactor to an ash bin from where the ash may be transported either hydraulically or mechanically; and (5) the scrubber in which the hot crude gas is quenched and washed before it is passed to the waste heat boiler.

The performance of the gasification process is shown in Fig. 4. It is a countercurrent operation in which the operating conditions are optimized for the various reactions.

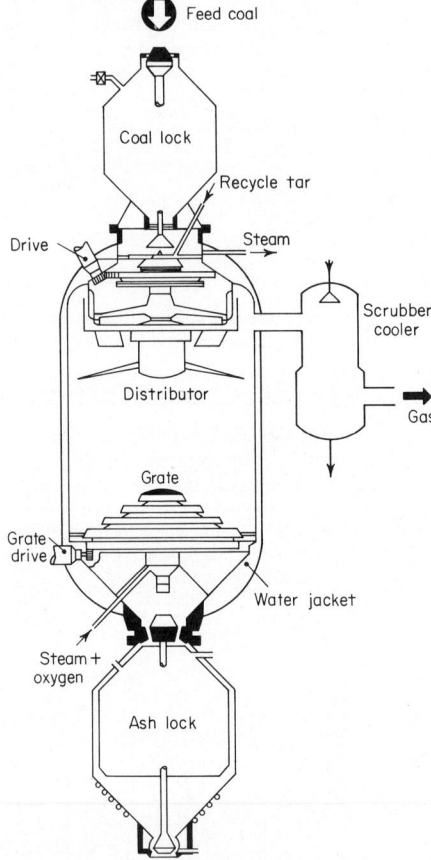

Fig. 3. Lurgi pressure gasifier.

The first step is preheating and drying of the coal. As the coal gravitates downward and is heated up, devolatilization commences, and from a temperature of 1150 to 1400°F (621 to 760°C) onward devolatilization is accomplished by gasification of the resulting char. The interaction between devolatilization and gasification is a determining factor for the kinetics of the complete gasification process.

The minimum residence time of a coal grain for good performance of the reactions at the desired temperature level of 1400 to 1600°F (760 to 871°C) is about one hour.

The diagram (Fig. 4) has been developed by the evaluation of various gasification tests in the Holten (Germany) pilot plant between 1950 and 1954. The gasifier was operated so that the coal feed was stopped while the gasification agent continued to be passed to the gasifier. This method yielded very accurate measurements.

A review of the gas analysis versus the height of the reactor reveals a very interesting phenomenon. That is, the equilibrium temperature of the heterogenous reaction, $C + 2H_2$ = CH_4, is approximately the same as the actual gas temperature in the respective area of the reactor, whereas the equilibrium temperature of the homogenous reaction, $CO + 3H_2$ = $CH_4 + H_2O$, is much lower than the actual gas temperature, which would not seem possible. What is the reason for this phenomenon? Is it correct to deduce that the methane has been formed according to the equation $C + 2H_2 = CH_4$? Attempts were made to gain more knowledge regarding the procedure of coal decomposition and the procedure of the resulting reactions on the surface of the coal grain by laboratory tests, simulating the various steps of the gasification process. Many additional interesting phenomena were observed, which generated even more questions. It is obvious that this area of coal science requires considerably more investigation. At this point, one can only deduce that

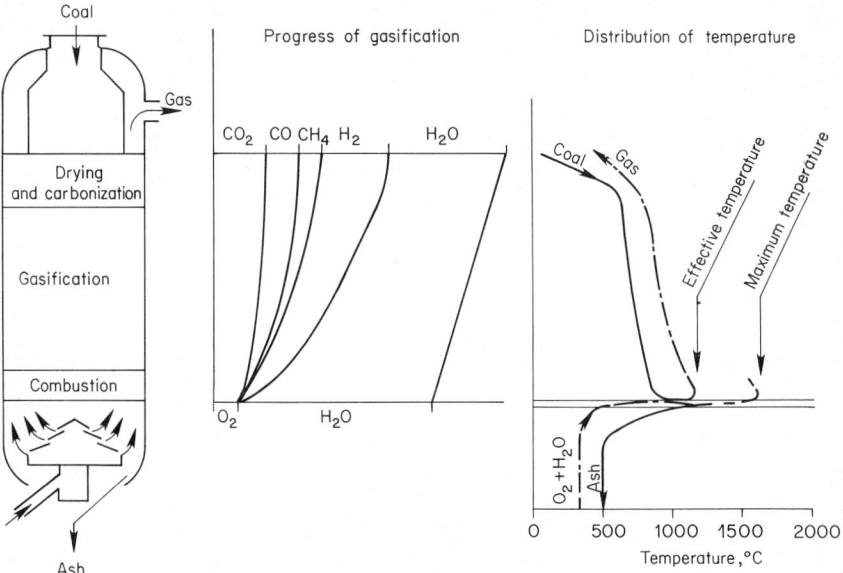

Fig. 4. Performance of gasification in a pressure gasifier.

the bulk formation of methane is not the result of hydrogenating the carbon, but rather a result of interactions between devolatilization and gasification, a factor that is utilized to advantage in countercurrent operation.

The crude gas leaves the gasifier at temperatures between 700 and 1100°F (371 and 593°C), depending on the type of coal used. The crude gas contains carbonization products, such as tar, oil, naphtha, phenols, and ammonia, and traces of coal and ash dust. This crude gas is passed through a scrubber where it is washed by circulating gas liquor and then cooled to a temperature at which the gas is saturated with steam. As higher-boiling tar fractions are condensed, the wash water contains tar to which the coal and ash dust is bonded.

The steam-saturated gas is passed to a waste-heat boiler in which waste heat at a temperature of 320 to 360°F (160 to 182°C) is recovered. The gas liquor condensed in this boiler is pumped to the scrubber, and surplus gas liquor is routed to a tar-gas-liquor separator. The mixture of tar and dust is returned to the gasifier for cracking and gasification.

A typical material balance of the process is given in Fig. 5. It is interesting to note that approximately 86 percent of the coal fed to the gasifier is gasified and the remainder, 14 percent, which is mostly carbon, is burned in the combustion zone with oxygen. Thus, the latent heat of this carbon is converted to sensible heat, which is transferred to the gasification agent flowing upstream countercurrently to the coal. This technique makes it possible to perform complete gasification of the coal. Only a negligible amount of unburned carbon remains in the ash. The sensible heat of the ash is utilized within the process.

INFLUENCE OF COAL PROPERTIES ON THE GASIFICATION PROCESS

Devolatilization is a process that proceeds parallel to gasification, and in this way the performance of the gasification process is influenced by interaction with the devolatilization step. This is the principal reason for the varied phenomena of coal gasification, some

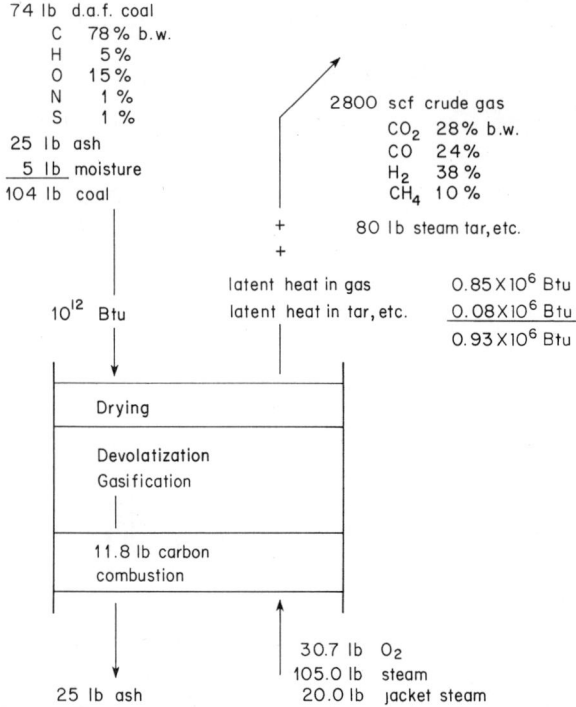

Fig. 5. Material balance for gasification: d.a.f. = dry ash-free. b.w. = by weight. scf = standard cubic foot.

of which factors include: (1) gas composition versus coal grade, (2) supply of reaction heat and phenomena in the combustion zone, (3) H_2/CO ratio, (4) reactivity of coal, and (5) influence of pressure.

GAS COMPOSITION VERSUS COAL GRADE

The different coals react in a varied manner. Figure 6 shows the combustible components of the gas versus coal grade, expressed by the volatile matter of the dry and ash-free coal. It should be noted that classification of coals by using volatiles as a characteristic feature is a rough approximation only. The plotted curves are average values of the results of more than 120 long-term gasification tests with 59 different coals. It has been found that the yield of methane, expressed in standard cubic feet of CH_4 per pound of dry and ash-free coal is a fairly constant figure for each type of coal. For subbituminous coals, the methane yield is approximately 3.9 standard cubic feet CH_4 per pound of dry and ash-free coal at an operating pressure of 350 psig.

SUPPLY OF REACTION HEAT

The method of supplying the heat for the performance of the reactions is an important item which has a decisive influence on the process and its economics. Gasification is a highly endothermic process, and, therefore, the heat requirements must be covered by the addition of heat which basically may be accomplished in four different ways:
1. Direct heat supply by partial combustion with oxygen
2. Direct heat supply by overlapping an adequate chemical exothermic reaction, such as $CaO + CO_2 = CaCO_3$, or $C + 2H_2 = CH_4$
3. Indirect heat supply by using a method similar to tubular reformers, or similar to coke-oven processes
4. Indirect heat supply by electrothermal heat or nuclear heat

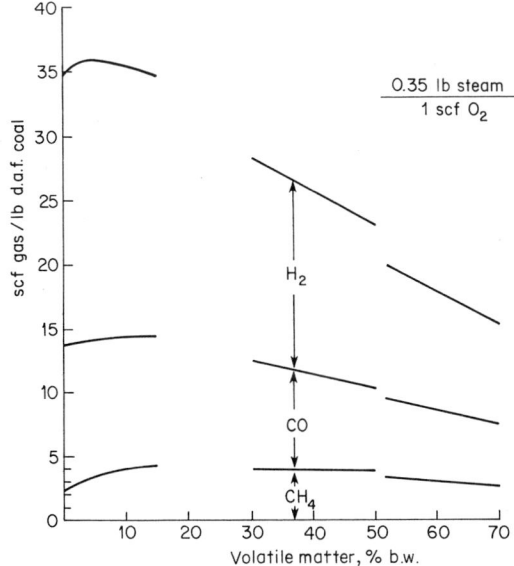

Fig. 6. Gas production versus coal grade: scf = standard cubic foot. d.a.f. = dry ash-free. b.w. = by weight.

5. Direct heat supply by heat carriers which are heated separately (recycled ash, solid particles, molten salts, and other materials which are liquid at the required reaction temperature)
All methods either have been applied or are available as conceptual processes.
The partial combustion with oxygen is represented by the Lurgi pressure gasification. By the partial combustion of the nongasified carbon with a steam-oxygen mixture, one obtains a steam-CO_2 mixture at high temperature which now serves as the gasification agent.

MAJOR PROCESS VARIABLES IN THE COMBUSTION ZONE

The steam-oxygen ratio in the gasification medium governs the temperature in the combustion zone. This ratio has to be so adjusted that (1) the ash does not fuse during combustion, and (2) the temperature is sufficiently high to ensure complete gasification of the coal.
Figure 7 shows the maximum combustion temperature t_{max} versus the steam-oxygen ratio. The t_{max} is the maximum attainable combustion temperature which would establish if the residual carbon entering the combustion zone were only converted to CO_2. However, t_{max} is, in fact, never reached and hence is a fictitious characteristic.

Inasmuch as the gasification reactions start in the combustion zone, the gas temperature is reduced considerably. Moreover, the combustion zone is relatively narrow. The height of the zone is about 5 to 10 times the diameter of the coal grains. This means that the residence time of coal and the ash in the combustion zone is short. It depends upon the ash content, the gasifier capacity, the grain size of the feed coal, and the reactivity of the char. Residence time generally is between 3 and 10 minutes. The short residence time of

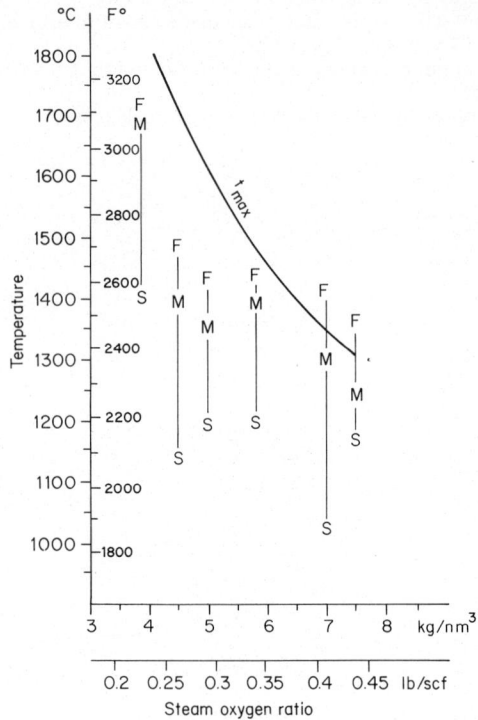

Fig. 7. Temperature in combustion zone: S = softening point. M = melting point. F = flow point.

the ash in the high-temperature zone is not sufficient to heat the ash to the gas temperature.

For the examination of the ash-fusion properties in the laboratory, the softening, melting, and flow points are usually determined by use of a heating microscope. Heating occurs in a CO_2-CO atmosphere. Figure 7 shows the results from the testing of some ashes versus the steam-oxygen ratio at which the gasifiers were operated. It is evident that the effective combustion zone temperature (temperature which the ash assumes) is far below the maximum temperature t_{max}. In all cases the ash in the gasifier was only slightly sintered, containing only a few slag lumps. It is further evident that a relationship between the ash-fusion properties, as determined in the laboratory, and the temperature to be adjusted in the combustion zone cannot be readily established. However, sufficient operational experience is available to judge from the coal and ash properties the influence the steam-oxygen ratio has on the behavior of the ash properties.

H₂/CO RATIO

The steam-oxygen ratio also has an influence on the gas composition. The amount of (CO + H₂) is dependent on the coal grade (Fig. 8), but is largely unaffected by the variation of the gasification agent composition. However, the H₂/CO ratio is mainly influenced by (1) the properties of the coal, and (2) the steam-oxygen ratio of the gasification agent, which is illustrated by Fig. 8, where the operating results of a selected number of various coals are plotted.

An interesting observation can be made if the curves of the various coals are extrapolated. They converge to an H_2/CO ratio of about 0.3 to 0.4 at a steam-oxygen ratio of 0.05 pound per standard cubic foot. This is the steam-oxygen ratio of the so-called hot operation attained in a slagging gasifier, and indeed this corresponds with the results of a slagging gasifier operation, both with coke and with subbituminous coal. Hot operation at

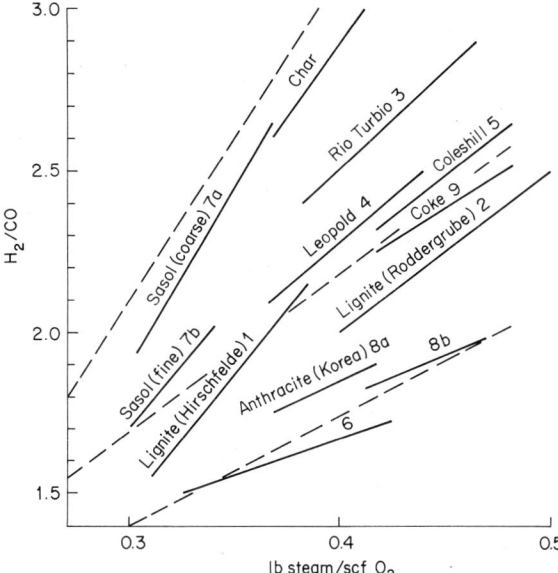

Fig. 8. H_2/CO ratio versus steam/oxygen ratio: (1) Lignite A. (2) Lignite B. (3) High-volatile coal. (4) Subbituminous coal A. (5) Subbituminous coal B. (6) Subbituminous coal C. (7) Subbituminous coal D. (8) Anthracite. (9) Coke.

high temperatures yields more uniform operation results and eliminates the influence of the various coal properties.

The area covered by the diagram (Fig. 8) represents the operation at lower temperatures, which should be just high enough so that the reactions can be performed at a reasonable rate. This results in a high methane yield and in good process performance. However, at low-temperature operation, the properties of the various coals appear in a manifold shape, differing from one coal to another.

This can be seen in the diagram in which, for example, coals with the same ultimate analysis have differing H_2/CO ratios. These are, for example, coals No. 6 and No. 4. To date, there is no scientifically justified rule which makes it possible to predict the H_2/CO function. This is simply a characteristic feature of each individual coal. For a process to produce high-Btu gas, the H_2/CO ratio must be adjusted by subsequent shift conversion, and, therefore, the H_2/CO ratio of the crude gas is not critically important. The diagram (Fig. 8) is presented simply to demonstrate one example of the influence of coal-property variations on the gasification process.

REACTIVITY

Another typical characteristic of a coal which defines gasification behavior is the reactivity. The higher the reactivity, the lower are the temperatures at which reactions still occur. Lower reaction temperatures favor the formation of methane, and, again, the exothermic methane reaction favors the other gasification reactions.

The lower the final reaction temperature, at which the reactions terminate, the more sensible heat of the gasification agent is available for conversion into latent heat. In other words, more coal is gasified per cubic foot of gasification agent.

The final reaction temperatures at which the reaction rate approaches zero are:

	Final reaction temperature	
Coal type	°F	°C
Lignite	1200	649
Subbituminous	1350	732
Semianthracite	1450	788
Coke	1550	843

Of course, these temperatures are not identical with the equilibrium temperature of the gas leaving the gasification zone because the approach to the gas equilibrium cannot be complete. The differential between the actual temperature of the gas and the coal and the differential between these temperatures and the various equilibrium temperatures constitute another measurement for the reactivity.

For a process calculation on a given coal, it is helpful to have a real scale for the reactivity. This is especially necessary to predetermine the consumption of oxygen. A well-known laboratory method is to determine the reactivity of coal by measuring CO_2 decomposition. The method can be modified so that CO_2 is passed through a fixed bed of char produced by carbonizing the respective coal under pressure in a crucible. This simulated gasification is done under pressure at a temperature of 1470°F (799°C) in a small reactor. The degree of CO_2 decomposition is a value representing the reactivity R. In Fig. 9 the composition of oxygen is plotted against the value R. The graph shows that there is a very reliable relationship between the two values. For Navajo coal and many other western (U.S.) coals, the value R is about 10, which represents a very high reactivity for these types of coals.

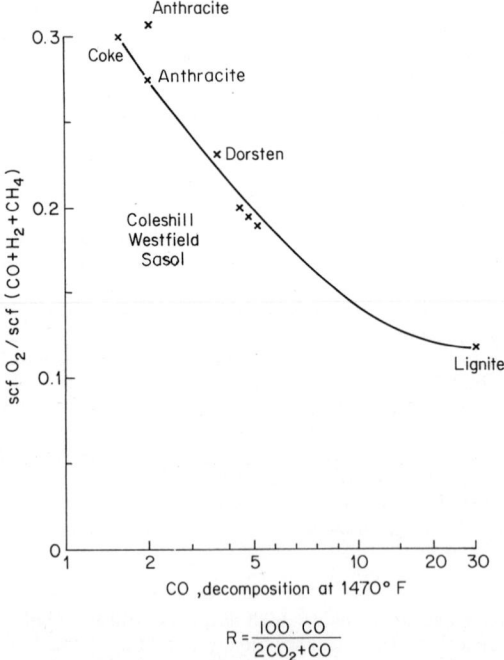

$$R = \frac{100 \cdot CO}{2CO_2 + CO}$$

Fig. 9. Consumption of oxygen versus reactivity of coal.

INFLUENCE OF PRESSURE

Although gasification should be performed under pressure, the pressure level that is most economic must be studied. There are several influencing factors. There is the influence of the operating pressure on the methane yield. Figure 10 illustrates the formation of methane in a steam-oxygen-operated pressure gasifier as a function of pressure. The figures up to 400 psi are based upon actual plant experience and include all losses which

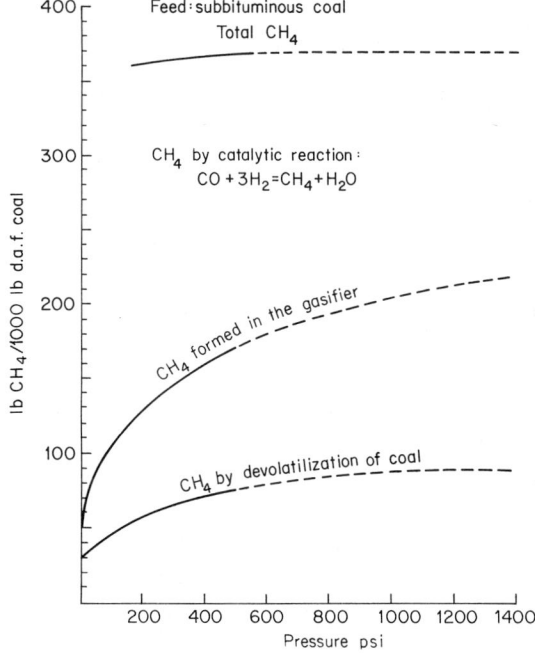

Fig. 10. Formation of methane as a function of pressure.

must be considered as, for example, hydrogen in H_2S and NH_3, lock gas of the coal-lock chambers, and so on. Furthermore, the tar in this case had not been processed to gas, but was disposed of as valuable by-product. On another coal, the absolute figures will differ, but the general tendency will be the same.

Beyond 400-psi pressures, the process had been calculated by extrapolation of equilibrium data from 400-psi operation. The graph illustrates clearly that beyond 400-psi operation, the incremental increase diminishes and that the total of methane from gasification and synthesis is nearly constant. Whether it would be more economical to generate more methane at higher pressure in the gasification reactor or to shift more methane formation to the synthesis because of a lower gasification pressure has been investigated. Many factors influence the economics. In terms of present technology, a pressure range of 350 to 450 psi appears to be most economical.

THE OVERALL LURGI PROCESS

GASIFICATION

Quickly summarizing this part of the process, the product of gasification is a mixture of various components, mainly CO_2, CO, CH_4, H_2, and H_2O, as well as products of coal carbonization, such as tar, oil, naphtha, phenols, and so on, plus sulfur compounds and other impurities. The first step of the process, gasification, produces only a crude gas, and, therefore, gasification must be followed by gas-treatment processes.

CRUDE GAS SHIFT CONVERSION

The gas composition leaving the gasifier is determined by the operating conditions in the gasifier which are set in such a way that the performance of gasification is optimized. The resulting H_2/CO ratio of the gas is not suitable for subsequent methane synthesis. Therefore, a shift-conversion reaction ($CO + H_2O = CO_2 + H_2$) is necessary to adjust the H_2/CO ratio.

The major feature of the crude-gas shift is the use of the undecomposed steam in the crude gas downstream of the waste-heat boiler in the gasification section for the performance of the shift reaction. However, the crude gas is contaminated with sulfur compounds and carbonization products.

By using a *Comox* catalyst and after solving the design problems of the heat exchanger, the crude-gas shift can fulfill this task. In addition, the catalyst has the advantage of having

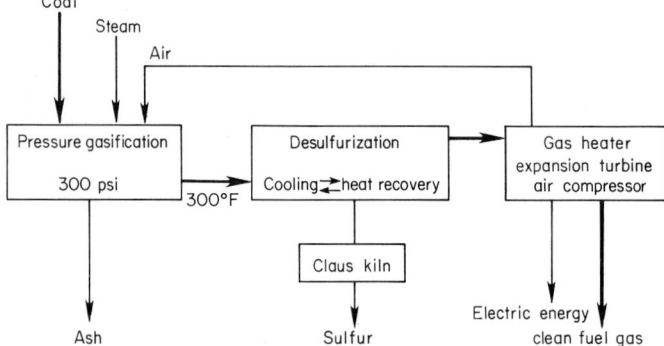

Fig. 11. Block diagram of clean fuel gas concept.

hydrogenating properties. Thus, the carbonization products are hydrogenated, and especially the gas naphtha is desulfurized. The amount of undecomposed steam and thus the output of gas liquor are reduced. The process has been commercially proved in several installations.

GAS PURIFICATION

Purification of gas from coal gasification to a degree of purity required for synthesis processes is difficult because of the large quantities of impurities and the variety of these impurities. The *Rectisol* process was developed for these conditions and is a physical gas absorption process which uses organic solvents, preferentially methanol, at low temperatures between 30 and − 80°F (− 1 and − 62°C). The process involves contacting the crude gas countercurrently with the solvent in a trayed absorption column and regenerating the spent solvent by flashing and subsequent stripping or reboiling in a desorption column, from which it is recycled to the top of the absorber.

For the substitute natural gas (SNG) plant described here, the *Rectisol* unit has three sections: (A) the prewash for the removal of gas naphtha, unsaturated hydrocarbons, and other high-boiling gas impurities; (B) the section for CO_2 removal and for the removal of H_2S and COS (The gas leaving this section is the "syngas" having the purity required for the methane synthesis); and (C) downstream of the methane synthesis section is a section where the last portion of the CO_2 is removed and the gas is also dehydrated.

In section A, the spent solvent is initially treated by liquid-liquid extraction with water to remove the dissolved hydrocarbons and then by two-stage distillation to eliminate the remaining hydrocarbons and to separate the water from the solvent. The hydrocarbons consisting mainly of light naphtha are recovered for use in the plant, or alternatively, used as a valuable low-sulfur, high-aromatic by-product.

Section *B* has a hot regeneration to boil out the sour gas components completely. In this solvent circuit between absorption and final regeneration a rectifying column (so-called reabsorber) can be installed to concentrate the H_2S.

Section *C* has no separate regeneration since the reboiled fresh solvent from section *B* is first used for CO_2 absorption in section *C* and then partially loaded with this CO_2 further used for H_2S and COS absorption in section *B*.

All impurities contained in the gas can be effectively and completely removed by the *Rectisol* process, which has been used in over 15 large commercial installations. The installations have demonstrated that a purity of less than 0.1 ppm sulfur can be continuously maintained at stream factors of more than 95 percent.

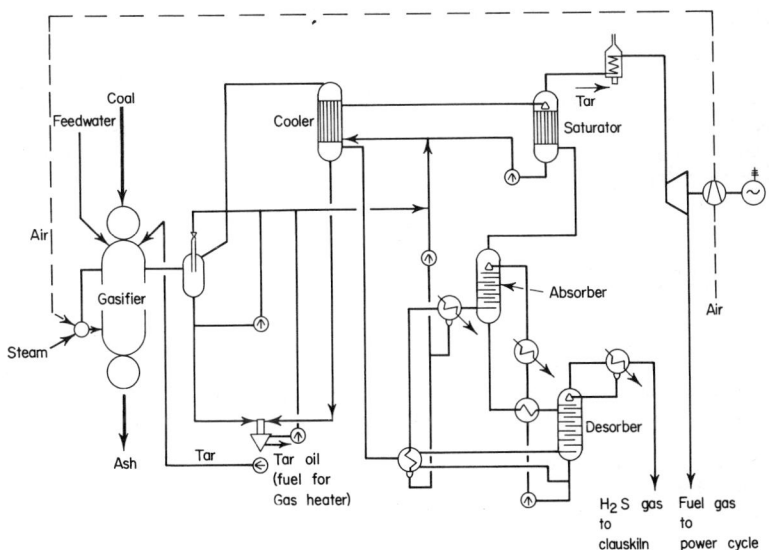

Fig. 12. Flow sheet of clean fuel gas concept.

METHANE SYNTHESIS

The synthesis gas from the Rectisol plant consists of CO_2, CO, and H_2 and is converted by catalytic reaction to CH_4 in a multiple-quench-type reactor system comprised of several fixed-bed reactors in series and with waste heat recovery. If the ratio $4CO_2 + 3CO/H_2$ is larger than one, the surplus portion of CO_2 will remain in the methanated gas and may be removed in a downstream CO_2 scrubbing process which preferably is combined with the upstream *Rectisol* plant.

Meanwhile, the viability of methanation of this synthesis gas and a lifetime of the catalyst of much more than one year have been proved in various demonstration plants.

GAS LIQUOR TREATMENT

In the Lurgi overall process, this phase of the procedure is termed the *Phenosolvan* process. Gas leaving a coal gasifier contains a moderately large amount of steam which is condensed in the subsequent processing steps. The resulting gas liquor has been in contact with coal carbonization products and consequently contains components that are soluble in water, mainly phenols, ammonia, and fatty acids.

The Phenosolvan process removes the phenols and the ammonia. The resulting purified effluent contains less than 20 ppm phenols and less than 60 ppm free NH_3. This effluent can be processed in a biological water-treatment plant, so that a final effluent will leave the plant which either meets the regulations for the disposal of waste liquor or can be used as raw water.

EFFICIENCY AND ECONOMICS

The process scheme just described must be supplemented by an oxygen plant, and the process units must be supported by a steam and power plant and by other utility plants. A thermal efficiency of 68 to 70 percent can be obtained for a fully self-supporting plant. This depends, however, very much on the properties of the coal, on the overall design of the plant, and on how far an approach to a highly sophisticated steam-power integration can be justified by the economics. The means of processing the by-products and the requirements of environmental protection are influential and vary from case to case.

Further attention to the material balance and economics of the Lurgi process is given in this section of the Handbook under article on *Coal Gasification at SASOL.*

CLEAN FUEL GAS PROCESS

The basic concept of what has been referred to as a *clean fuel-gas process* combines the gasification process just described with additional elements, as depicted in Fig. 11. The gas leaving the gasifier at a temperature of about 800°F (427°C) is cleaned in an attached scrubber to a degree of purity that makes it usable for gas turbines. The gas is saturated with steam at a temperature of 320°F (160°C). The ash is extracted in granular form which minimizes disposal problems. The thermal efficiency of the gasification process is high. More than 90 percent of the latent heat of the coal is recovered. During gasification and by quenching the resulting gas, an increase of volume under constant pressure occurs. This effect is utilized in the subsequent gas-turbine processes. See also Fig. 12.

After the gas is cooled in a heat-recovery system, H_2S can be removed by applying commercially available absorption processes. This is an appropriate approach to desulfurization since the sulfur compounds occur in a pressurized gas as H_2S and not as SO_2. The removed H_2S gas is converted in a Claus kiln to elemental sulfur. The energy required for performing the H_2S removal is supplied by the sensible heat of the gas. The result is a clean fuel gas at approximately 250 psi which may be used for furnaces, in existing steam boilers, and in advanced power cycles.

See also Handbook section on *Power Technology Trends.*

The Hygas Coal Gasification Process

BY J. HUEBLER,* F. C. SCHORA, JR.,† AND B. S. LEE‡

The Institute of Gas Technology (IGT) Hygas§ process for the hydrogasification of coal is tailored to maximize the direct production of methane in the hydrogasification reactor by the use of high temperature (1200 to 1700°F; 649 to 927°C) and high pressure (1000 psig). The process can use as feed, caking bituminous coals as well as noncaking lignite and subbituminous coals. Crushed coal is fed in an oil slurry to the gasifier where, in a hydrogen-rich atmosphere, two-thirds of the methane for the final product gas is produced. The gas produced in the hydrogasifier is purified and raised to pipeline quality by catalytic methanation. The hydrogasified char is used for hydrogen production. Three processes are being investigated for the production of hydrogen from coal char: (1) the steam-oxygen, (2) steam-iron, and (3) electrothermal systems.

The Hygas process is based on gasification studies that started at the Institute of Gas Technology in 1946. By the mid-1950s, two processes for converting coal to synthetic pipeline gas were being developed: (1) gasification of powdered coal in suspension with oxygen and steam to produce a mixture of carbon monoxide and hydrogen (synthesis gas), which was then methanated; and (2) direct hydrogenation of the coal at elevated temperature and pressure in a fluidized bed. The present Hygas process incorporates the principles developed in both these concepts.

Until 1964 this work was sponsored by the American Gas Association (AGA). Subsequently, it has been cosponsored by AGA and the Office of Coal Research of the U.S. Department of the Interior.¶ Under this joint sponsorship, construction of the Hygas pilot plant was completed in the spring of 1971. The $10.5 million plant is designed to deliver 1.5 million cubic feet of pipeline-quality gas from 75 tons per day of coal. In the early spring of 1973, large-scale conversion of coal to pipeline-quality gas was demonstrated for the first time. In the spring of 1974, IGT completed a run during which the plant was in operation for over 27 days. During this run, the methane produced by direct conversion of coal in the hydrogasifier exceeded the predicted 66.7 percent of the methane produced in the overall process; reaction rates and thermal effects were as expected. Plant modifications were then made to study the integration into the base of the pilot-plant hydrogasifier reactor of a steam-oxygen gasifier for the production of hydrogen-rich gas from hydrogasified coal char. A separate system was also constructed to investigate the steam-iron hydrogen system for producing a hydrogen-rich gas from char. The electrothermal gasification of char was successfully demonstrated in batch tests; that part of the pilot plant was then put on standby.

CHEMISTRY OF THE PROCESS

Methane can be formed directly from coal by

$$Coal + H_2 \rightarrow CH_4 \qquad (1)$$

or indirectly by

$$Coal + H_2O \rightarrow CO + H_2$$

$$CO + 3H_2 \rightarrow CH_4 + H_2O \qquad (3)$$

Since the overall process thermal efficiency is improved by maximizing direct methane formation, the conditions of the IGT Hygas process are tailored toward this end. High

*Senior Vice-President, †Vice-President, Process Research; ‡Assistant Vice-President, Process Research, Institute of Gas Technology, Chicago, Illinois.
§HYGAS is registered with U.S. Patent Office.
¶In January 1975, OCR sponsorship was assumed by ERDA (Energy Research and Development Administration).

temperature is needed to obtain reasonable reaction rates; high pressure is used to increase equilibrium methane yield. The most active fraction of coal is hydrogasified to form methane, while the less active fraction is used to generate hydrogen for hydrogasification. The raw gas from the hydrogasifier contains carbon monoxide and hydrogen. These are converted to methane through reaction 3 in the catalytic cleanup methanation unit to boost the heating value of the gas and to reduce the carbon monoxide content to the pipeline standard of less than 0.1 percent. Of the total methane formed, 65 to 70 percent is formed by the direct route in the hydrogasifier. This is a key feature that contributes significantly to the overall thermal efficiency of over 70 percent for the process. If the process were operated at a lower pressure, more methane would be made by the indirect than by the direct route, and losses in efficiency would result.

The second key feature of the process is the use of hydrogen and steam in hydrogasification. The coal-hydrogen reaction (1) is strongly exothermic, while the coal-steam reaction (2) is strongly endothermic. By using a mixture of hydrogen and steam instead of hydrogen alone, the heat released by reaction 1 is absorbed in situ by reaction (2), resulting in (a) built-in temperature control and (b) internal hydrogen generation.

PRINCIPAL PROCESS OPERATIONS

COAL PREPARATION

In the Hygas pilot plant, the coal-grinding mill dries and crushes the coal. Because the process can operate with all grades of coal, from lignite through bituminous, the mill must crush coals that differ greatly in grindability and moisture content. Thus, the mill operation is adjusted carefully to achieve the desired size distribution and degree of drying.

COAL PRETREATMENT

Although the Hygas system has a means for testing operation with raw, caking bituminous coals, a pretreatment step is currently provided. Pretreatment is performed in a fluidized bed where about 10 percent of the dried and sized coal is burned with air at atmospheric pressure and at 650 to 750°F (343 to 399°C). The temperature is controlled by limiting the oxygen and by removing the heat through steam production in tubes buried in the bed. Pretreatment volatilizes the lowest boiling constituents of the coal and oxidizes the outer skin of the particles. Lignite and subbituminous coal do not require pretreatment.

SLURRY FEED SYSTEM

The coal from the grinding and drying mill or from the pretreatment section must be pressurized to the pressure of the hydrogasification reactor and introduced into the first stage of the reactor. Because no one has had experience either with the use of lock hoppers at pressures of 1000 psig or over or with the finer consists of coal, IGT decided to pump coal slurry into the reactor. Calculations indicated that if water were used as the slurry medium, a large amount of heat would be required to evaporate the water at the top of the reactor. However, the light oil produced in the process requires only about one-sixth of the heat required to vaporize the water and was, therefore, an acceptable liquid. This system has been well proved in pilot-plant operation.

The dry, prepared coal passes on weigh belts into slurry tanks containing the light oil. The slurry is kept in constant circulation at low pressure through a pipe loop past the suction side of a conventional oil-field reciprocating mud pump and back to the slurry tanks to prevent solids settling. The mud pump pressurizes the slurry to the gasifier reactor pressure of 1000 psig. The coarse coal particles and low-viscosity oil, both of which make it easy for the coal to settle, cause this slurry to be more difficult to handle than others reported in the literature. In the pilot plant, IGT has successfully pumped slurry with concentrations of 45 percent by weight of solids. A heavy-duty centrifugal pump keeps the slurry in constant motion. Ni-Hard casting on all the wetted parts of the centrifugal pump successfully resists erosion by the coal particles.

HYDROGASIFIER REACTOR

In the pilot plant, the hydrogasifier reactor vessel, with four internally connected contact stages, is 135 feet high and has a 5.5-foot inside diameter. The slurry is sprayed into the top stage, a 2.5-foot-diameter 10-foot-high fluidized drying bed. The sensible heat in the gaseous reaction products from the later stages vaporizes the oil. At this point, the dry coal is heated to about 600°F (316°C).

The coal flows by gravity from the drying bed into a 3-inch-diameter, vertical-lift-line reactor in which the hot gases (1700°F; 927°C) from the reaction section below provide the lifting force, the heat to raise the solids temperature to 1200°F (649°C), and hydrogen that reacts with about 20 percent of the coal to produce methane. This is the first stage of hydrogasification. At the top of the lift line, the gas and coal disengage. The gas moves up to vaporize the oil in the slurry-drying stage. The partially reacted coal can be split into two streams. Part of it can be transferred to the base of the lift line to be mixed with the incoming fresh coal. By this means, it is expected to feed raw caking coal directly to the reactor without pretreatment. Eliminating pretreatment can reduce gas cost significantly.

The remainder of the partially reacted coal flows by gravity to the second-stage hydrogasifier. Here the solids are heated in a fluidized bed to about 1700°F and further gasified by the steam and hydrogen-rich gas rising from the steam-oxygen gasification stage below. (Or alternatively the hydrogen-rich gas may flow from the steam-iron reactor or the electrothermal generator.) The second-stage bed is 2.5 feet in diameter, is lined with refractory, and is 15 feet deep.

In this second stage of the reactor, the exothermic hydrogen reaction produces methane, and the endothermic steam reaction produces carbon monoxide and hydrogen. If the temperature rises, the steam-char reaction speeds up and prevents the temperature from rising any further. If the temperature drops, the steam-char reaction slows down and thus provides an automatic temperature control. About 25 percent more of the coal is converted in this reaction stage, making the total about 45 percent in the two stages.

From this reaction stage, the char goes to the hydrogen-producer stage, where, depending on the process being used, the char undergoes different degrees of additional gasification. The steam-oxygen gasifier being used as the hydrogen producer in the pilot plant is directly below the second-stage hydrogasifier. The steam and high-purity oxygen introduced into the gasifier convert char into hydrogen and carbon oxides at 1850°F (1010°C) in a fluidized bed 2 feet in diameter and 12 feet deep. Ash is discharged from this stage without being slagged.

SOLIDS DISCHARGE

The ash is discharged into a tank where water is added to make a slurry, which.is then depressurized. The ash is recovered by filtering and the water reused.

GAS PURIFICATION

The composition of the gas leaving the slurry drier at the top of the hydrogasification reactor depends on the type of hydrogen producer, as shown in Table 1. In addition to

TABLE 1. Gas Composition Leaving the Hydrogasification Reactor (Oil Free)

Component	Steam-oxygen	Steam-iron	Electrothermal
CO	18.0	7.4	21.3
CO_2	18.5	7.1	14.4
H_2	22.8	22.5	24.2
H_2O	24.4	32.9	17.1
CH_4	14.1	26.2	19.9
C_2H_6	0.5	1.0	0.8
H_2S	0.9	1.5	1.3
Other	0.8	1.4	1.0
Total	100.0	100.0	100.0

these major components, the gas contains the slurry oil, coal dust, and trace constituents such as ammonia and hydrogen cyanide.

In the pilot plant, this mixture at 600°F (316°C) passes to a baffle tower in which it is quenched and washed with water. This removes the dust and water-soluble trace components and condenses the water and light-oil vapors. The gas then flows to a conventional, packed-tower acid-gas removal system in which the carbon dioxide and the hydrogen sulfide are absorbed in a diglycolamine-water solution. When this solution is regenerated, the carbon dioxide and hydrogen sulfide are released and flow to a Claus plant for sulfur recovery. The amine purification system used in the pilot plant is not intended as a commercial design because it does not provide for separate collection of the various constituents. It was selected because it can handle the wide range of acid-gas concentrations arising from the various coals to be tested.

METHANATION

As it enters the methanation section, the purified gas would typically have the composition shown in Table 2. The methanation step has two purposes: one is to raise the heating

TABLE 2.   Gas Composition Entering the Methanator

Component	Steam-oxygen	Steam-iron	Electrothermal
CO	18.0	12.8	16.8
CO_2	2.0	2.0	2.0
H_2	54.0	38.5	50.5
CH_4	25.0	45.0	29.5
C_2H_6	1.0	1.7	1.2
Total	100.00	100.0	100.0

value of the gas to near that of methane; the other is to reduce the carbon monoxide concentration to the requisite 0.1 percent or less. This is accomplished by carrying out reaction (3) given earlier.

To obtain nearly complete elimination of carbon monoxide and low residual hydrogen (pipeline-quality gas) in the methanation section, the ratio of hydrogen to carbon monoxide is adjusted, usually by a water-gas shift reaction, to slightly above 3. High pressure and low temperature favor completion of the methanation reaction. Very reactive, high-nickel-content catalysts are generally preferred to make the reaction proceed rapidly at the low temperatures employed. The temperature of the catalyst must be above 450°F (232°C) to avoid formation of nickel carbonyl, which causes depletion of the nickel content of the catalyst, and below about 950°F (510°C) to avoid carbon deposition and catalyst sintering. The reaction is very exothermic. To avoid excessive temperature rises, the Hygas process uses a cold-gas recycle system.

HYDROGEN FOR THE HYGAS PROCESS

The Hygas process is very flexible in that it can successfully use any source of hydrogen. Three methods are being investigated for producing hydrogen for the process. The steam-oxygen version employs an oxygen system, whereas the electrothermal and steam-iron versions are air-based. The effect on material balance of these three systems in the overall process is indicated in Fig. 1.

STEAM-OXYGEN SYSTEM

The steam-oxygen gasifier (Fig. 2) produces a hydrogen-rich gas by reacting the hot char from the hydrogasifier with steam and oxygen in a high-pressure fluidized bed. An operating temperature of 1800 to 1900°F (982 to 1038°C) is maintained by controlling the quantities of steam and oxygen.

Multiple-feed gas cones inject premixed steam and oxygen at the bottom of the bed. Nonslagging conditions are maintained by a high fluidizing velocity at the reactor bottom and by the gas inlet design, which avoids stagnant areas and promotes rapid initial contact of oxygen, steam, and char. Ash is discharged through the space between the cones.

The hydrogen content of the gas from this gasifier is the lowest of the three systems. The mechanical simplicity of this system is partially offset by the cost of separating oxygen from air and removing additional carbon dioxide and by the need for a larger hydrogasifier reactor.

STEAM-IRON SYSTEM

The steam-iron process for making hydrogen is very old. In the past, the process was operated at atmospheric pressure using two beds of iron solids. One bed was reduced from iron oxide to iron by a suitable gas while the second bed was being oxidized by steam from iron to iron oxide. During the iron oxidation, the steam was converted to hydrogen. When the beds were fully reacted, valves switched the gases from one bed to the other, and the reverse operation was carried out. This process has been largely abandoned in favor of other hydrogen processes, especially steam reforming of natural gas.

The new IGT process replaces the cyclic operation with continuously flowing, fluidized-bed reactors. The process operates at Hygas pressure so that the hydrogen-steam mixture produced can be passed directly into the hydrogasification reactor. The steam-iron version of the Hygas process, shown in Fig. 3, is quite complex compared with the other versions. However, it is potentially superior in efficiency and in the cost of the gas produced.

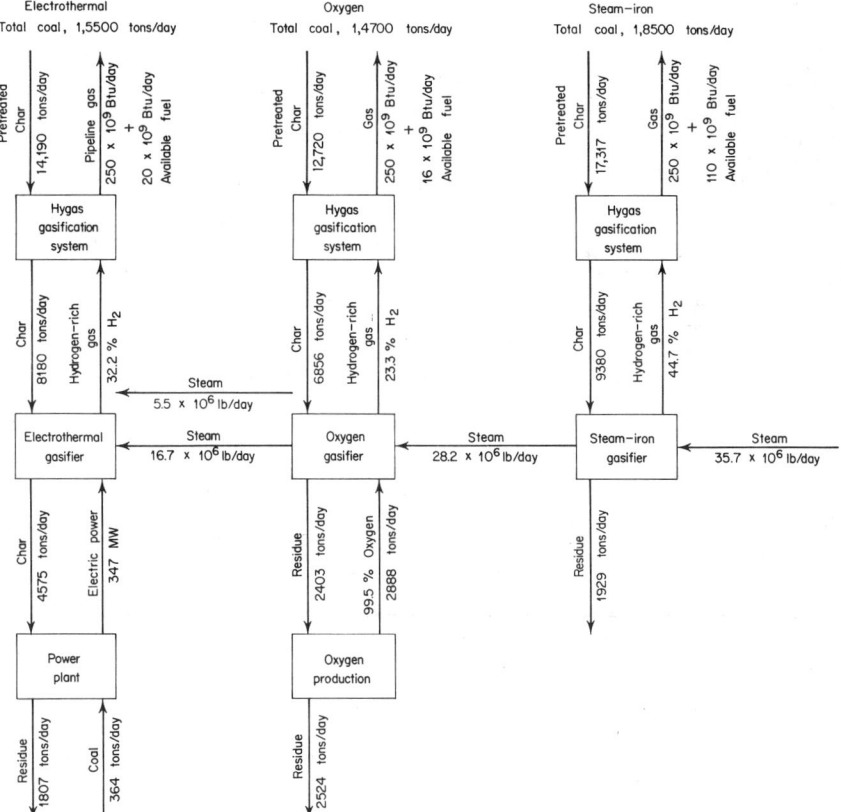

Fig. 1. Effect on the IGT Hygas process of using three different systems for producing hydrogen-rich gas. Plant capacity is 250×10^9 Btu/day of pipeline-quality gas.

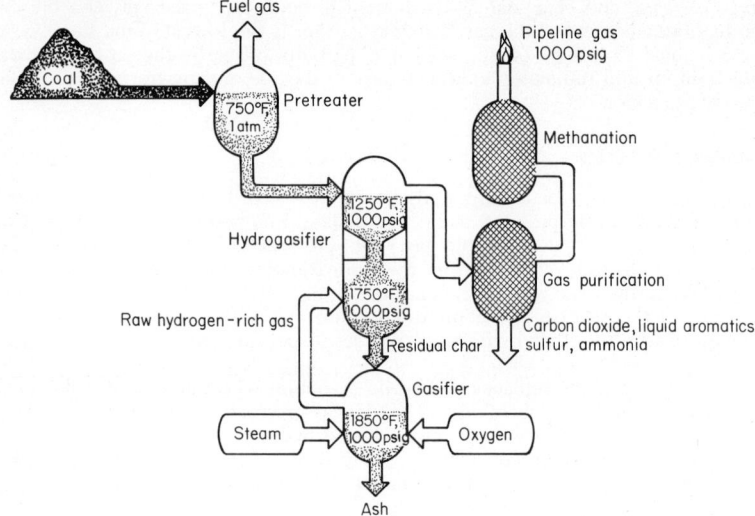

Fig. 2. The Hygas version using a steam-oxygen gasifier to produce hydrogen-rich gas.

Spent char from the hydrogasifier would be fed directly to the producer in which the char reacts with air and steam to generate a gas capable of reducing iron oxide to iron. Operation of the producer at temperatures near 2000°F (1093°C) should yield good reducing gas having CO/CO_2 and H_2/H_2O ratios exceeding 4:1. The hot reducing gases are fed directly to the reducer where they contact a recirculating stream of iron oxide and reduce it to iron. The iron is contacted with steam in the oxidizer and reoxidized to iron oxide, producing a hydrogen-steam mixture.

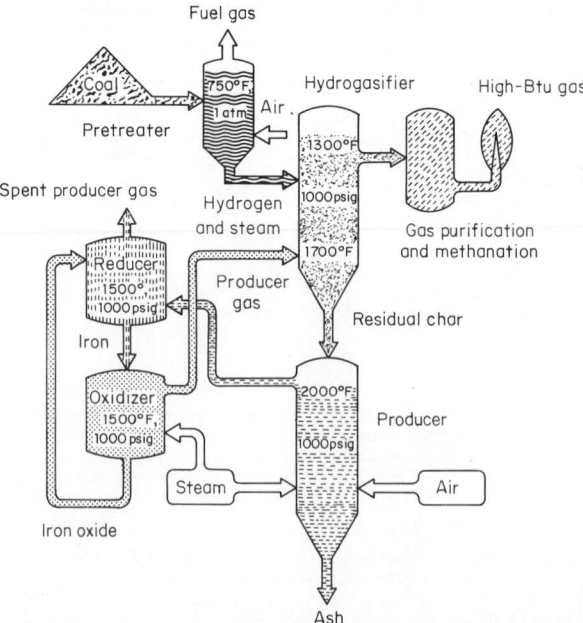

Fig. 3. The Hygas version using a steam-iron system to produce hydrogen-rich gas.

The reduction-oxidation reactor will be designed with four fluidized-bed stages, two each in the reducer and the oxidizer. The double beds ensure the highest steam and reducing-gas conversions for the operating conditions used. Locating the reducer on top of the oxidizer allows the spent reducing gases to convey the iron solids to the top of the reactor. The new steam-iron process is especially suited to the Hygas process because the product gas from the oxidizer can be fed directly at temperature and pressure into the second-stage hydrogasification reactor. Because the hydrogasification reactor operates well with a 60:40 steam/hydrogen ratio, it is not necessary to achieve high steam conversion in the steam-iron process. The very low carbon oxide content of the gas from the steam-iron reactor results in a very low carbon oxide content in the raw gas from the hydrogasifier. This, combined with the very high hydrogen content, permits a small hydrogasifier reactor size and much reduced scrubbing requirements.

ELECTROTHERMAL HYDROGEN

In this method for producing hydrogen-rich gas, as shown in Fig. 4, excess char from the Hygas hydrogasification reactor reacts with steam. Heat is furnished to the highly endo-thermic steam-carbon reaction by direct-current heating of a fluidized bed of char. Excess

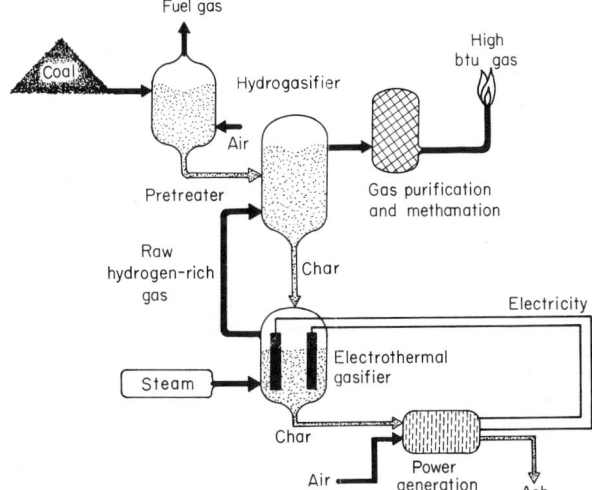

Fig. 4. The version of the IGT Hygas process using an electrothermal gasifier to produce hydrogen-rich gas.

char from this electrothermal reactor is used to produce the electric power required. In the years since development of this method was started, the cost of electric power has risen sharply. Therefore, this method is now the least economical of the three.

In the pilot plant, a 2-megawatt electrothermal gasifier was built and operated success-fully. Its diameter inside the refractory lining is 30 inches, and the reactor height is 70 feet. The electrode configuration is concentric. The gasifier has automatic and manual modes for controlling power input.

PROCESS ECONOMICS

The price of substitute pipeline-quality gas is a function of the coal cost, investment, other operating costs, and the financial method used to determine this price. The two major factors are coal cost and the costs associated with investment. Other operating costs, in addition to coal, are relatively minor items. The financial method can be a major factor if widely different assumptions are made. This is not discussed here. We have used a method now widely employed by workers in this field.

The costs presented here are derived from a large assemblage prepared by the Supply-

Technical Advisory Task Force—Synthetic Gas-Coal of the National Gas Survey. IGT contributed costs for the Hygas process to this effort. Costs are presented for the Hygas process with hydrogen from steam-oxygen, steam-iron, and electrothermal sources. Based on mid-1971 prices, costs were developed for both eastern bituminous and western subbituminous coals. The latter are noncaking and generally less expensive than the former.

PLANT CAPACITY AND COAL REQUIREMENTS

The standard capacity for gas-from-coal plants generally used is the equivalent of 250 million cubic feet per day of 1000 Btu per standard cubic foot (scf)* pipeline gas, or 250 billion Btu per stream-day. Calculations are based on a 90 percent annual stream factor, so that the annual production of pipeline gas is the equivalent of 82 billion cubic feet of 1000-Btu gas. Thus, the annual output of 12 such plants is needed to make 1 trillion cubic feet per year of pipeline-quality gas.

The accounting procedure used to determine the price of gas given here is based on a 20-year period. Thus a minimum 20-year supply of recoverable coal must be established for each plant. The mining and the cost of the coal present formidable problems. We have used 30 cents per million Btu for bituminous coal and 15 cents per million Btu for subbituminous coal. These represent costs per ton of $7.45 for deep-mined Pittsburgh Seam coal and $2.65 for a strip-mined Montana subbituminous coal. The cost of mining has been increased recently by inflation, high demand, mine safety provisions, and ecological requirements. The cost of coal can be expected to continue upward to perhaps twice the levels used here.

The logistics of bringing together the coal, water, and plant operations at a plant site are problems that can only be solved when the site for a particular plant is determined. Although gas can be moved more cheaply than coal, the difference is not necessarily enough to offset the other factors.

INVESTMENT COSTS

Table 3 summarizes plant investments for Hygas plants using the three hydrogen sources for eastern and western coals. These costs are extracted from the Synthetic Gas-Coal Task Force's assemblage of data referred to above. For comparison purposes, an adjustment in the total capital requirements was made to bring all investment totals to a common basis of 250 billion Btu per stream-day. Costs for pollution abatement include sulfur recovery from process gases and boiler flue gases and for cleanup and reuse of all plant waste-water streams plus makeup. These costs are not based on specific designs, but on rather broad generalizations. Total investments shown here represent unit costs of $965 to $1240 for 1000 cubic feet per day of 1000-Btu gas.

We assume that the gas plant will be located very near the coal-mining site, so that transportation costs are only a minor part of the cost of the coal at the gasification plant. Because coal is treated as an operating cost, the capital investment for coal mining is not included here. A representative minimum number might be $7.50 per annual ton. Applying this factor to the coal rates for these plants indicates initial investments of $37 to $53 million that must be raised in addition to that for the gas plant itself.

ANNUAL REVENUE AND GAS PRICE

Table 4 summarizes the components of gas price. A utility-type accounting procedure is used to determine 20-year average returns on the rate base plus Federal income tax, which are added to the annual operating costs to give the annual revenue from product gas. The debt/equity ratio is 75:25. Interest on the debt is 9 percent, and return on equity is 15 percent, giving a gross return on the rate base of 10.5 percent. Depreciation is 20-year straight line, and Federal income tax is 48 percent.

For the examples shown here, coal represents 40 to 45 percent of the gas price for the eastern bituminous coals and 25 to 30 percent of the gas price for the western subbituminous coals. For the processes shown here, gas prices with western coals range from 25 to

*scf = standard cubic foot of gas measured at 60°F (15.6°C) and 30 inches of mercury.

TABLE 3. Plant Investment and Capital Requirements for Hygas Process with Lock Hopper Feed for Synthetic Gas-Coal Task Force Review (in $ millions)

Time base of costs, mid-1971

Coal type	Pittsburgh No. 8 Ireland Mine			Montana subbituminous		
Hydrogen Generation Method	Steam-oxygen	Steam-iron	Electro-thermal	Steam-oxygen	Steam-iron	Electro-thermal
Plant size 10^9 Btu stream-day pipeline gas	248.4	253.9	254.6	250.6	250.7	254.2
INVESTMENT BREAKDOWN						
Process on-sites investment:						
Coal storage and preparation	8.7	11.3	9.5	14.8	18.4	15.5
Pretreatment	4.6	5.6	5.2	—	—	—
Feed system	8.8	12.0	9.9	5.2	6.7	5.5
Gasification and CO shift	25.9	56.6	29.7	26.3	61.5	27.6
Gas purification	26.9	17.0	22.0	28.2	16.2	24.0
Methanation	9.5	6.3	8.6	8.8	4.7	7.7
Compression	—	—	—	—	—	—
Total process on site	84.4	108.8	84.9	83.3	107.5	80.3
Capacity ratio	1.006	0.985	0.982	0.998	0.997	0.984
Auxiliary on-sites investment:						
Oxygen manufacture	18.5	—	18.5	19.0	—	8.4
Sulfur recovery	15.2	19.7	10.0	7.4	10.2	10.0
Water pollution control	10.0	10.0	10.0	10.0	10.0	10.0
Steam and power plant investment	13.4	6.2	66.9	18.6	12.8	73.9
General utilities investment	9.7	9.0	12.7	10.0	8.6	13.2
General off-sites investment	10.0	10.0	10.0	10.0	10.0	10.0
Subtotal excluding contingencies	161.2	163.7	203.0	158.3	159.1	195.8
Project contingency	24.2	24.5	30.5	23.8	23.9	29.4
Development contingency	11.3	11.5	14.2	11.1	11.1	13.7
Total plant investment	196.7	199.7	247.7	193.2	194.1	238.9
CAPITAL REQUIREMENT BREAKDOWN						
Total plant investment	196.7	199.7	247.7	193.2	194.1	238.9
Interest during construction	33.2	33.7	41.8	32.6	32.7	40.3
Start-up costs	11.3	13.9	12.9	7.5	8.4	8.5
Working capital	11.8	14.4	13.2	8.6	9.3	9.3
Total capital requirements (T.C.R.)	253.0	261.7	315.6	241.9	244.5	297.0
Adjusted T.C.R. for 250×10^9 Btu/stream-day	254.5	257.8	309.9	241.4	243.8	291.9

TABLE 4. Gas Cost Breakdown for Hygas Process with Lock Hoppers for Synthetic Gas-Coal Task Force Review
Time base of costs, mid-1971

Coal type	Pittsburgh No. 8 at 30¢/10^6 Btu			Montana subbituminous at 15¢/10^6 Btu		
Hydrogen generation method	Steam-oxygen	Steam-iron	Electro-thermal	Steam-oxygen	Steam-iron	Electro-thermal
	GAS COST BREAKDOWN: ¢/10^6 BTU					
Coal feed	44.8	59.7	49.0	22.63	27.38	24.10
Other raw materials	0.1	0.1	0.1	0.06	0.07	0.06
Catalysts and chemicals	1.5	1.2	1.3	1.06	1.05	0.93
Purchased utilities						
Electric power	—	—	—	—	—	—
Raw water	0.8	0.6	1.0	0.69	0.55	1.00
Labor						
Process operating labor	2.4	2.5	2.4	2.38	2.47	2.39
Maintenance labor	3.6	3.6	4.4	3.52	3.54	4.29
Supervision	0.9	0.9	1.1	0.88	0.90	1.00
Administration and general overhead	4.2	4.2	4.8	4.07	4.15	4.61
Operating and maintenance supplies	4.4	4.3	5.2	4.23	4.28	5.01
Local taxes and insurance	6.5	6.5	8.0	6.34	6.36	7.72
Total gross operating cost	69.2	83.6	77.3	45.86	50.75	51.11
By-product credits						
Sulfur*	1.9	2.3	2.2	0.32	0.51	0.34
Ammonia at $25/ton	0.7	1.0	0.8	0.75	0.96	0.78
Light oil (B-T-X) at 15¢/gal	2.6	3.8	3.1	2.72	3.57	2.84
Heavy oil (tars) at 30¢/10^6 Btu	1.1	1.4	1.2	2.66	3.50	2.78
Char at 90% of coal value	—	5.3	2.8	—	—	—
Total net operating cost	62.9	69.8	67.2	39.41	41.75	44.37
Capital charges (20-year average)						
Depreciation	14.8	14.8	18.1	14.17	14.28	17.23
Return on rate base	17.0	17.4	20.6	15.98	16.18	19.26
Federal income tax	5.6	5.7	6.8	5.27	5.33	6.35
20-year average gas cost	100.3	107.7	112.7	74.83	77.54	87.21

*Sulfur credit includes elemental sulfur at $10/long ton plus $4/ton for SO_2 recovered in the Wellman-Load process above that which can be converted to elemental sulfur in the Claus plant.

33 cents per million Btu less than those with eastern coal. However, most of this difference is due to the price of the coal rather than to the properties of the coal, although there is a small process advantage for the western noncaking coal.

WATER REQUIREMENTS

Pipeline-gas plants obtain hydrogen by the decomposition of water in gasifiers. Other water is needed for cooling, gas scrubbing, and other process uses. The amount of net water makeup for a given size of plant can vary widely, depending on the amount of air cooling, the process itself, and the amount of water reuse. The cost of the water is a very small part of the cost of gas from coal; the real problem is its availability at the plant site. If there is not enough water at the gas-plant site, pipelining it may become necessary because the costs for water transportation are much less than those for coal transportation. Typical makeup water requirements for the Hygas process with substantial use of air cooling for process streams, water-cooled power plants, and maximum reuse of water would be about 7 million gallons per stream-day, or an annual usage, at 90 percent stream factor, of 7060 acre-feet per year.

Acknowledgements. The work reported in this paper was supported by the Office of Coal Research of the U.S. Department of the Interior and the American Gas Association. The authors gratefully acknowledge their guidance throughout the program and their kind permission to publish the paper.

REFERENCES

1. Huebler, J., and C. L. Tsaros: "Coal Gasification Developments," National Meeting of American Association of Petroleum Geologists, Anaheim, Calif., May 15, 1973.
2. Matthews, C. W., and F. C. Schora, Jr: "Analysis of a HYGAS Coal Gasification Plant Design," Conference Preprint *70a*, 65th Annual Meeting, American Institute of Chemical Engineers, New York, Nov. 26–30, 1972.
3. Pyrcioch, E. J., et al.: Production of Pipeline Gas by Hydrogasification of Coal, *IGT Res. Bull.,* no. 39, Chicago, 1972.
4. Schora, F. C., Jr.: Technical and Historical Background, in "Symposium Papers: Clean Fuels from Coal," pp. 43–48, Institute of Gas Technology, Chicago, 1973.
5. Schora, F. C., Jr.: "Clean Energy From Coal," 9th Intersociety Energy Conversion Engineering Conference, San Francisco, Aug. 26–30, 1974.
6. Schora, F. C., Jr.: B. S. Lee, and J. Huebler: "The HYGAS Process," Conference Preprint *IGU/B* 3-73, 12th World Gas Conference and Exhibition, Nice, France, June 5–9, 1973.
7. Schora, F. C., Jr., D. J. Tobin, and R. L. Mount: "Progress in Coal Gasification," Institute of Electrical and Electronic Engineers Power Engineering Society Summer Meeting and Energy Resources Conference, Anaheim, Calif., July 14–16, 1974.

BI-GAS Process for Production of High-Btu Pipeline Gas from Coal

BY ROBERT J. GRACE*

The principal objective of the Bituminous Coal Research, Inc., gas generator research and development program is the development of the promising BI-GAS process for the production of high-Btu pipeline gas from coal. The initial phase of the program was a worldwide survey of available and proposed coal gasification processes. This resulted in a two-volume summary (Ref. 1), published in 1965, of these processes and in the conceptual design of a two-stage, entrained superpressure, oxygen-blown system for producing high-Btu gas from coal, now referred to as the BI-GAS process.

Development work in this area has proceeded from batch-type experiments in rocking autoclaves, through continuous-flow experiments in a 5 pound-per-hour externally heated reactor (CFR, continuous-flow reactor), to operation of a 100-pound-per-hour internally fired, Stage 2, process and equipment development unit (PEDU). Using North Dakota lignite, Wyoming subbituminous C coal, and Pennsylvania high-volatile A bituminous coal, the research has confirmed the basic assumption that high yields of methane could be obtained from coal at elevated temperatures and pressures. The essential conclusion of the experiments (Ref. 2) was that for residence times longer than a fraction of a second, methane yields depended on coal rank and processing conditions, such as temperature and hydrogen partial pressure.

With the completion of the Stage-2 PEDU test program, the correlations obtained were valuable for planning and designing a larger-scale, fully integrated 5-ton-per-hour gasification plant (Refs. 3, 4, 5) at Homer City, Pennsylvania. Figure 1 is a block flow diagram of this facility.† The final objective of the pilot plant was to provide sufficient design data for construction of a commercial plant.

TWO-STAGE GASIFIER

The heart of the BI-GAS process is the two-stage gasifier shown in Fig. 2, which uses pulverized coal (70 percent minus 200 mesh) in entrained flow. Fresh coal and steam are introduced into the upper section (Stage 2) of the gasifier at pressures in the range of 70 to 100 atmospheres. In Stage 2, the coal comes in contact with a rising stream of hot synthesis gas produced in the lower section (Stage 1) and is partially converted into methane and more synthesis gas. The residual char entrained in the raw product gas is swept upward and out of the gasifier. The char is separated from the product gas stream and recycled to Stage 1 of the gasifier.

In Stage 1, the char is completely gasified under slagging conditions with oxygen and steam, producing both the synthesis gas and the heat required in Stage 2 for the partial gasification of the fresh coal.

The raw gas from Stage 2, typical analysis of which is shown in Table 1, leaves the char cyclone and passes through a water-wash column where the gas is further cooled and the dust is removed. The wash column is operated at conditions to maintain only the moisture as needed for the downstream shift conversion.

GAS PROCESSING

The remainder of the process steps with the exception of methanation are existing commercial operations. The cleaned raw gas is partially shifted to provide the proper $H_2/$

*Process Engineering Manager, Bituminous Coal Research, Inc., Monroeville, Pennsylvania.

†The design of a pilot plant was completed under contract with the Koppers Company, Inc., in December 1971. The responsibility for constructing the pilot plant was awarded to Stearns-Roger, Inc., of Denver, Colorado, on July 11, 1972.

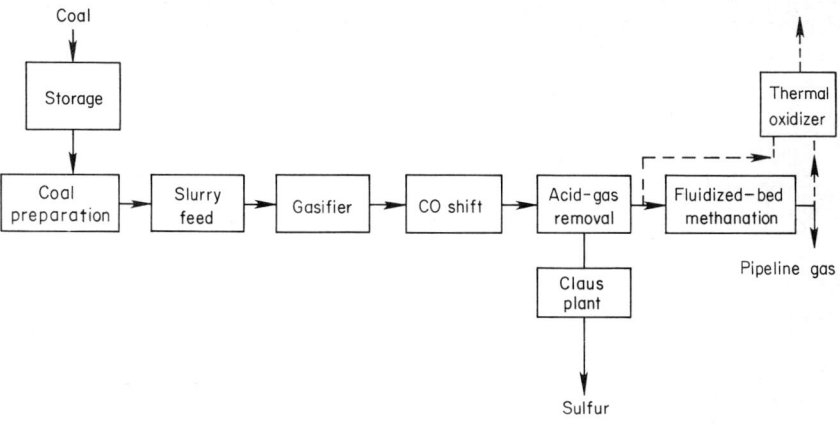

Fig. 1. Bi-Gas 5-ton/hr pilot-plant flow diagram.

CO ratio. Hydrogen sulfide and carbon dioxide are removed selectively from the entire gas stream in a Selexol unit. The H_2S is then converted to a sulfur product in a Claus unit. However, after the H_2S has been removed, provisions are made to bypass part or all of the raw gas containing CO_2 direct to methanation. A fluidized-bed catalytic methanation system using embedded water-cooled tubes is being installed in the pilot plant. Concurrently, laboratory and PEDU studies are being continued by BCR and will be discussed later. The methanated gas in the pilot plant will receive a final treatment for CO_2 removal in the SELEXOL unit to produce the high-Btu pipeline gas. Table 1 also shows the analysis of the gases after various stages of treatment and of the final pipeline gas.

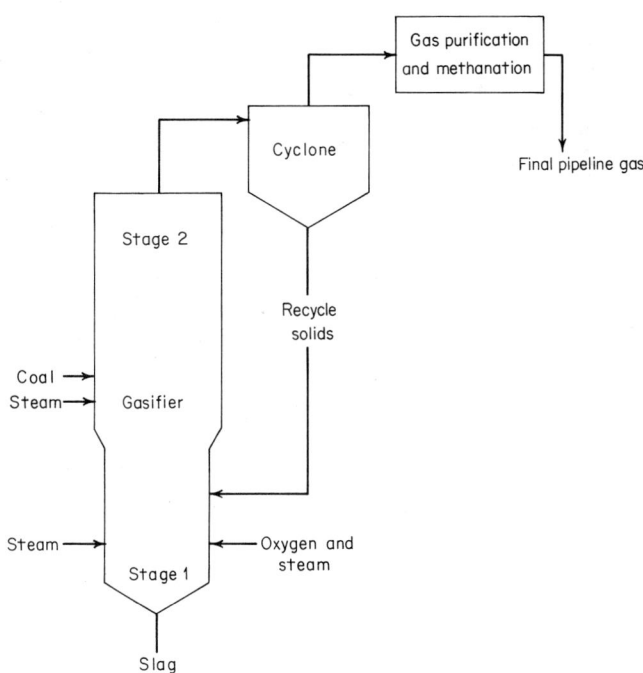

Fig. 2. Simplified flow diagram for two-stage superpressure gasifier.

TABLE 1. Gas Stream Composition for the BI-GAS Process
Volume percent

Component gas	1 Gasifier product	2 CO-shift feed	3 Acid gas removal plant feed	4 Acid gas removal plant product	5 Final pipeline gas
CO_2	16.2	12.6	31.0	0.2	0.5
CO	22.1	17.2	13.2	19.3	0.1
H_2	24.2	19.0	40.7	59.5	4.6
CH_4	11.8	9.2	13.7	20.0	92.7
N_2	0.5	0.4	0.6	0.9	2.1
H_2S	0.6	0.5	0.7	0.0	0.0
H_2O	24.6	41.1	0.1	0.1	0.0
Total	100.0	100.0	100.0	100.0	100.0

APPRAISAL OF THE PROCESS

The BI-GAS process offers several advantages in the production of synthetic natural gas.

1. A high yield of methane is obtained directly from coal, and subsequent processing of the product gas is minimized.

2. All types of coal can be gasified without prior treatment, since the process utilizes an entrained instead of a fixed or fluidized-bed system.

3. The reaction conditions in the upper stage of the gasifier are such that no tar and oils are formed in the gasification process.

4. All the coal charged to the process is consumed with the principal by-products, consisting of slag for disposal and sulfur for sale.

5. The two-stage gasifier, being an integral unit, is relatively simple in design and amenable to scale-up to larger sizes.

6. Gas can be generated at high pressures adequate for transport in existing gas distribution systems without further compression.

7. When the gasifier is operated on air, instead of oxygen, at moderate system pressures, a gas is produced which may be readily desulfurized and cleaned to yield a pollution-free fuel gas with a heating value of about 175 Btu per standard cubic foot.

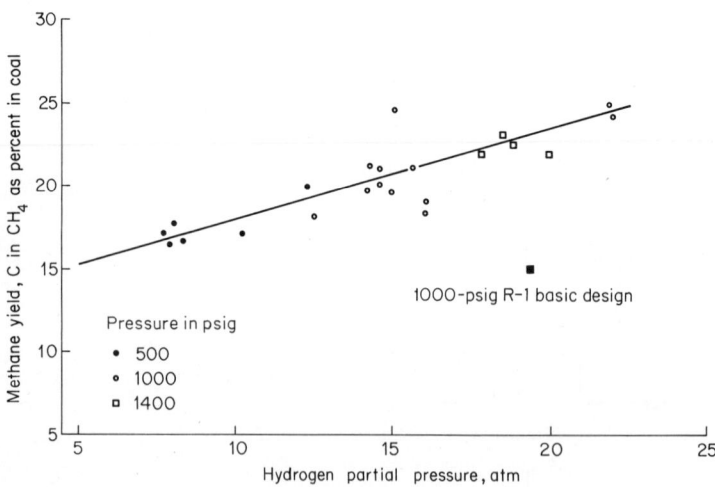

Fig. 3. Effect of hydrogen partial pressure on methane yield from tests in 100-lb/hr PEDU using Pittsburgh Seam coal.

DESIGN AND TESTING PROGRAM

The results obtained in the 100-pound-per-hour Stage 2 PEDU, particularly for Pittsburgh Seam coal (Fig. 3) showed that operating temperature and pressure as well as hydrogen partial pressure influence the methane yield obtained in Stage 2.

The physical effects, inherent in the design of the Stage 2 PEDU, also influenced the methane yields from coal (Ref. 6). This was particularly true with regard to the mixing patterns established between the hot synthesis gas from Stage 1 and the fresh coal stream in Stage 2. There was evidence that the momentum of the coal feed stream injected into the PEDU set up secondary recirculating flows which caused some of the methane initially formed in the direct methanation reaction to be introduced into the higher-temperature regions of the PEDU, where it was destroyed by the reaction with steam. In order to further investigate means for improving the flow patterns in Stage 2 and also to establish design criteria for the as-yet-untested slagging, Stage 1 section of the gasifier, a cold-flow model program was developed and carried out at BCR (Bituminous Coal Research, Inc).

This program consisted of a series of tests made in a Plexiglas model of the 5-ton-per-hour gasifier. A diagram of the cold-flow model is shown in Fig. 4. The model was full scale with respect to diameter and Stage 1 length.

Many facets of the gasifier design were tested on the cold model, but major emphasis was placed on determining the effects of nozzle position and orientation on gasifier operation, particularly with regard to slag deposition in Stage 1 and Stage 2 mixing patterns.

Air from a large blower was used to simulate the process gases, and the various nozzle diameters were adjusted to preserve momenta similarity.

Atomized glycerin was sprayed through the three Stage 1 char nozzles to simulate slag

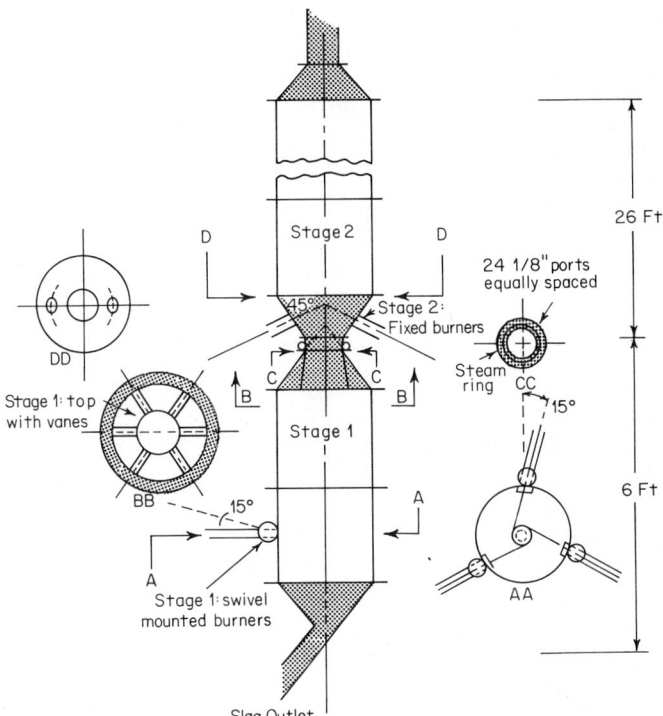

Fig. 4. Diagram of model for testing flow under cold conditions.

droplets. Percentage carry-over and uniformity of deposition were two criteria used to evaluate the Stage 1 nozzle locations. Various types of solids, injected through the two Stage 2 coal-feed nozzles, were used to simulate coal and to make visible the resultant flow patterns.

All nozzle and burner orientations and locations, as well as throat diameters and volumes for both Stages 1 and 2, were designated by BCR as a result of information developed before and during the cold-model program. During the model tests, it was also demonstrated that slag carry-over into Stage 2, should it occur, can be minimized in two ways: by installing vanes in the top of Stage 1 or by admitting some Stage 2 steam at the throat which divides the two stages. Provision for steam jets has been made.

As a result of the cold-model program, several design decisions were reached and incorporated in the final design of the 5-ton-per-hour gasifier.

The design of the 5-tph gasifier (Fig. 5) is as follows:

The outside diameter of the gasifier is 5.5 feet, and that inside the refractories is 2 feet. The heights of Stages 1 and 2 are 6 and 26 feet, respectively. The two stages are separated by a throat 9 inches in diameter. The slag outlet port connecting Stage 1 with the slag quenching zone is 6 inches in diameter. In Stage 1, char, oxygen, and steam are injected through each of the three horizontal burners, aimed slightly off-center. The resultant vortex flow thus created is sufficient to cause the heavier slag particles to deposit on the walls of Stage 1 and then to flow downward and into the slag quenching zone. In Stage 2, fresh coal transported with recycle gas and steam is injected through two opposed nozzles which are inclined at 30° to the horizontal. The momentum of the coal stream is kept low to minimize the creation of strong recirculating flow patterns, and the coal nozzles are located near the throat region in order to maximize the contact between coal and the hot Stage 1 gas.

The gasifier is designed for coal-feed rates up to 5 tons per hour and a maximum operating pressure of 1500 psi. Speed of the coal particle entering Stage 2 will range from 20 to 30 feet per second. The resultant char particles will slow down to about 2 feet per second in their passage through and exit from Stage 2. Typical residence times will be 2 seconds in Stage 1, and 4 to 6 seconds in Stage 2. The gas exit temperature from Stage 2 will be about 1700°F (927°C).

Concurrent with the work on the oxygen-blown gasifier, laboratory work continues to be conducted in support of the total program. Fluidized-bed catalytic methanation of the product gas to raise the calorific value to that of pipeline gas is being studied. Three process variations are under investigation: (1) methanation following shift conversion and acid gas removal, (2) hydrogen sulfide removal before and carbon dioxide removal after methanation, and (3) methanation of the synthesis gas containing all acid gas components.

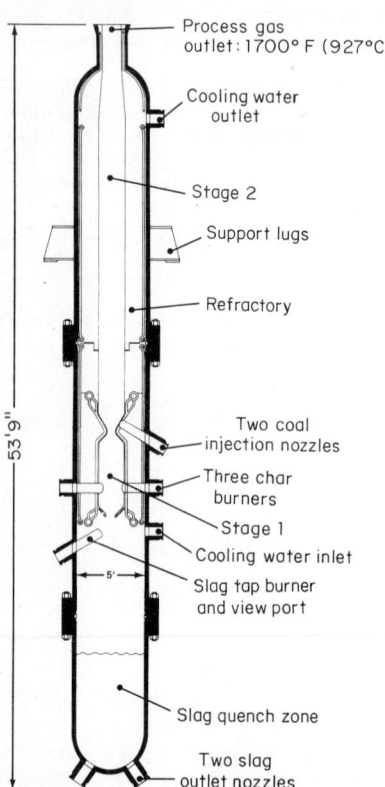

Process gas
outlet: 1700° F (927°C)

Cooling water
outlet

Stage 2

Support lugs

Refractory

Two coal
injection nozzles

Three char
burners

Stage 1

Cooling water inlet

Slag tap burner
and view port

Slag quench zone

Two slag
outlet nozzles

53' 9"

5'

Fig. 5. Bi-Gas reactor. Design data: Operating pressure, gas side, 500–1500 psig. Process gas outlet temperature, 1700°F. Shell design pressure, 1650 psig. Shell design temperature, 650°F. Shell dimensions: 60 in. internal diameter, 3⁹⁄₁₆ in. thick. Material of construction, 2¼% chromium, 1% molybdenum steel. Total gasifier weight, 280,000 lb.

The design and construction of a process-and-equipment development unit (PEDU) for studying fluidized-bed catalytic methanation were completed in May 1973 (Ref. 7). Process research is expected to continue through fiscal year 1977.

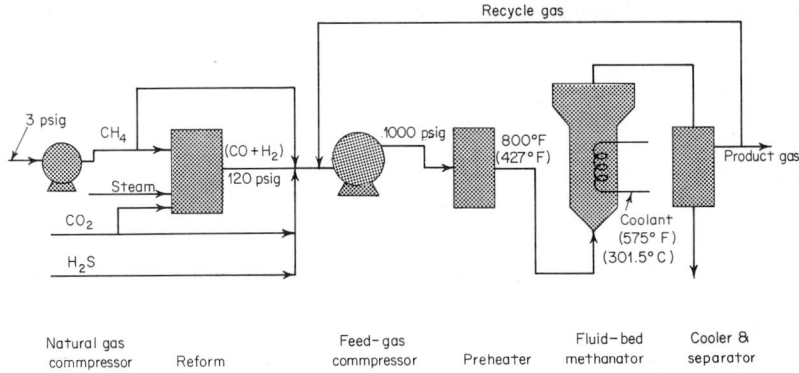

Fig. 6. Process flow diagram of methanation PEDU.

In-house research is divided into two separate phases: (1) laboratory-scale studies to investigate methanation catalysts under conditions imposed by the BI-GAS process, and (2) use of the PEDU to test selected catalysts under continuous or semicontinuous operation, and to develop optimum operating conditions for the pilot plant.

The PEDU has sufficient built-in flexibility to permit operation on a wide range of feed-gas compositions and operating parameters. At the nominal 6000-cubic-foot-per-hour capacity, the unit can operate at reactor temperatures to 1050°F (566°C), pressures to 1550 psig, and with a range of space velocities.

A process flow diagram of the methanation PEDU is shown in Fig. 6. Figure 7 is a diagram of the methanator. In the PEDU, natural gas will be reformed with steam or a mixture of carbon dioxide and steam to yield a synthesis gas having a hydrogen/carbon monoxide ratio ranging from about 1 to 10. Methane, carbon dioxide, and hydrogen sulfide may then be blended with the reformer effluent to yield the desired feed mix. The gas is compressed and heated in a lead-bath preheater to 800° (427°C). The methanator, the design of which is shown in Fig. 7, is 6 inches in diameter with a maximum possible bed height of 10 feet. The reactor holds about 2 cubic feet of catalyst; at 6000-standard-cubic-feet-per-hour input, the design inlet space velocity is thus 3000 cubic feet per hour. A desirable catalyst should convert at least 80 percent of the carbon monoxide to hydrocarbons at a 1000-cubic-feet-per-hour space velocity. The heat of reaction is removed by means of a coolant circulated through ¾-inch vertical tubes submerged in the fluidized catalyst bed. The methanated gas is cooled in a hairpin condenser, and the product gas and water are separated for analysis. The entire system is designed such that a PDP8 computer will sample data on line and perform all necessary heat and material-balance calculations for characterizing the process.

The BCR system contains the following features which make it unique in the field of methanation development:

1. It makes use of a high-pressure fluidized-bed reactor.

2. Minimum gas recycle will be used. The process catalyst must be able to operate under a concentration of carbon monoxide as high as 25 percent.

Recycle rate cannot be estimated at this point since the heat-transfer characteristics have not been studied. The PEDU program will provide data in this area.

3. The use of non-nickel catalyst types permits high-temperature operation and the production of other light hydrocarbons.

Acknowledgements. This paper is based on work conducted at Bituminous Coal Research, Inc., with support from the Office of Coal Research, U.S. Department of the Interior, under Contract No. 14-32-0001-1207, and the American Gas Association.

REFERENCES CITED

1. "Gas Generator Research and Development—Survey and Evaluation," Bituminous Coal Research, Inc., Report to Office of Coal Research, Washington, D.C., 650 pages, U.S. Government Printing Office, Washington, D.C., 1965. (Out of print; copies on file at Office of Coal Research repository libraries.)

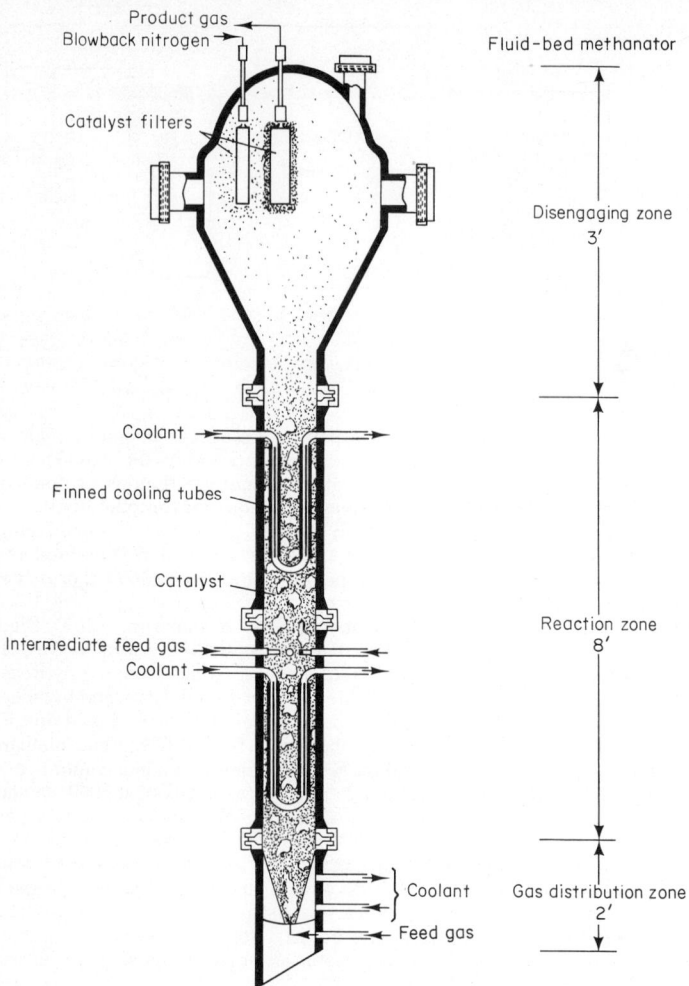

Product gas

Blowback nitrogen

Catalyst filters

Coolant

Finned cooling tubes

Catalyst

Intermediate feed gas

Coolant

Coolant

Feed gas

Fluid-bed methanator

Disengaging zone
3'

Reaction zone
8'

Gas distribution zone
2'

Fig. 7. Fluidized-bed methanator.

2. Glenn, R. A.: "Status of the BCR Two-Stage Super-Pressure Process," 3d Synthetic Pipeline Gas Symposium, Chicago, Illinois, 1970, American Gas Association, Arlington, Va.

3. Grace, R. J., and R. L. Zahradnik: "BI-GAS Program Enters Pilot Plant Stage," American Gas Association and Office of Coal Research, 4th Synthetic Pipeline Gas Symposium, Chicago, 1972, American Gas Association, Arlington, Va.

4. Grace, R. J., Brant, V. L., and V. D. Kliewer: "Design of BI-GAS Pilot Plant," American Gas Association and Office of Coal Research, 5th Synthetic Pipeline Gas Symposium, Chicago, 1973, American Gas Association, Arlington, Va.

5. Probert, P. B.: "Design and Fabrication of BI-GAS Reactor," American Gas Association and Office of Coal Research, 5th Synthetic Pipeline Gas Symposium, Chicago, 1973, American Gas Association, Arlington, Va.

6. Zahradnik, R. L., and R. J. Grace: "Chemistry and Physics of Entrained Coal Gasification," Coal Gasification Symposium, Division of Fuel Chemistry, American Chemical Society, Dallas, Tex., 1973, American Chemical Society, Washington, D.C.

7. Graboski, M. S., and E. K. Diehl: "Design and Operation of the BCR Fluidized Methanation PEDU," American Gas Association and Office of Coal Research, 5th Synthetic Pipeline Gas Symposium, Chicago, 1973, American Gas Association, Arlington, Va.

The CO_2 Acceptor Coal Gasification Process

BY CARL E. FINK,* JOHN D. SUDBURY,† AND GEORGE P. CURRAN‡

Production of pipeline gas from lignite or subbituminous coal is the objective of the CO_2 Acceptor Process.§ The process (late-1974) is in the pilot plant phase with a 40-ton-per-day plant located in Rapid City, South Dakota. Laboratory and continuous bench-scale work was done in Library, Pennsylvania, during 1964–1968 to obtain data for the pilot-plant design. The early continuous unit operated at pressures up to 20 atmospheres, and temperatures up to 1950°F (1066°C). Feasibility and economic studies also were performed during the early development phase. The results of these studies are reported in Refs. 1 through 5.

PROCESS FUNDAMENTALS

The principal steps involved in the process are (1) coal preparation, (2) coal preheating, (3) coal pressurization, (4) gasification, (5) gas purification, (6) methanation, and (7) dehydration and compression.

In the coal preparation section, the coal is ground and dried to produce a feedstock sized at approximately 8 × 100 Tyler mesh and having substantially zero moisture. Lignite and subbituminous coals are noncaking, and therefore when used require no pretreatment.

The feedstock is pressurized to system pressure by means of lock hoppers. These were chosen because their operability is commercially established, and they present a low investment and operating cost. Reliable operation of lock hoppers at process conditions has been demonstrated in the pilot plant.

The gasification section for the process is shown in Fig. 1. There are two reactors: (1) the gasifier, and (2) the regenerator.

The carbonaceous feedstock, be it lignite or subbituminous coal, is fed from the lock hoppers into the gasifier where it is devolatilized and then gasified at temperatures in the range of 1500 to 1550°F (816 to 843°C).

The feed introduction is at the bottom of the gasifier fluidized bed to allow adequate vapor retention time for cracking the coal volatile matter. The volatile hydrocarbons, which contain about 35 percent of the coal carbon, are converted to methane, carbon monoxide, carbon dioxide, and hydrogen. Devolatilization is essentially thermoneutral— the heat required is only that required to heat the coal to the gasifier temperature.

The devolatilized char is fluidized with steam, which reacts with the char carbon to form carbon monoxide, hydrogen, and some methane. The pertinent reactions are:

		Heat of reaction Btu/mole at 298.16°K
(1)	$C + H_2O(g) \longrightarrow CO + H_2$	56,500 endothermic
(2)	$CO + H_2O(g) \longrightarrow CO_2 + H_2$	17,700 exothermic
(3)	$CaO + CO_2 \longrightarrow CaCO_3$	76,200 exothermic
(4)	$CaCO_3 \longrightarrow CaO + CO_2$	76,200 endothermic
(5)	$2C + H_2 + H_2O(g) \longrightarrow CH_4 + CO$	24,300 endothermic

Carbon dioxide is formed by water-gas shift reactions [Eq. (2)]. The heat for the steam-carbon reaction is supplied by the reaction of CO_2 with lime-bearing material which is called an acceptor. An acceptor can be either limestone or dolomite. The product gas, containing the products of gasification and devolatilization, leaves the top of the reactor.

*Project Manager of Pilot Plant, Conoco Coal Development Company, Rapid City, South Dakota.
†General Manager, Research Division, Conoco Coal Development Company, Library, Pennsylvania.
‡Manager of Gasification Research, Conoco Coal Development Company, Library, Pennsylvania.
§Developed under contract with the office of Coal Research and the American Gas Association.

The residual char which contains about 33 percent of the coal carbon is transferred to the regenerator where it is burned with air to supply the heat required to reverse the acceptor reaction. The char combustion maintains the regenerator, which consists of a fluidized-bed acceptor, at a temperature of about 1850°F (1010°C). The residual ash, containing about 5 percent carbon, is elutriated from the acceptor bed and carried out of the regenerator with the flue gas. This ash contains 80 percent of the sulfur fed with the coal. An external cyclone separates the residual ash from the gas.

Hot acceptor in the calcium oxide form flows by gravity from the regenerator to the gasifier, where it is introduced above the gasifier char bed. The acceptor then showers through the char bed, supplying both chemical and sensible heat, and collects in a reduced cross-section boot at the bottom of the gasifier. The fluidizing velocity in the boot is such that finer-sized lower-density char is stripped from the acceptor. Char-free acceptor is fed by gravity to a pickup point from which it is carried pneumatically back to the regenerator.

In order to maintain acceptor activity, which declines with the number of calcination-recarbonation cycles, continuous makeup is added. The makeup requirement is on the

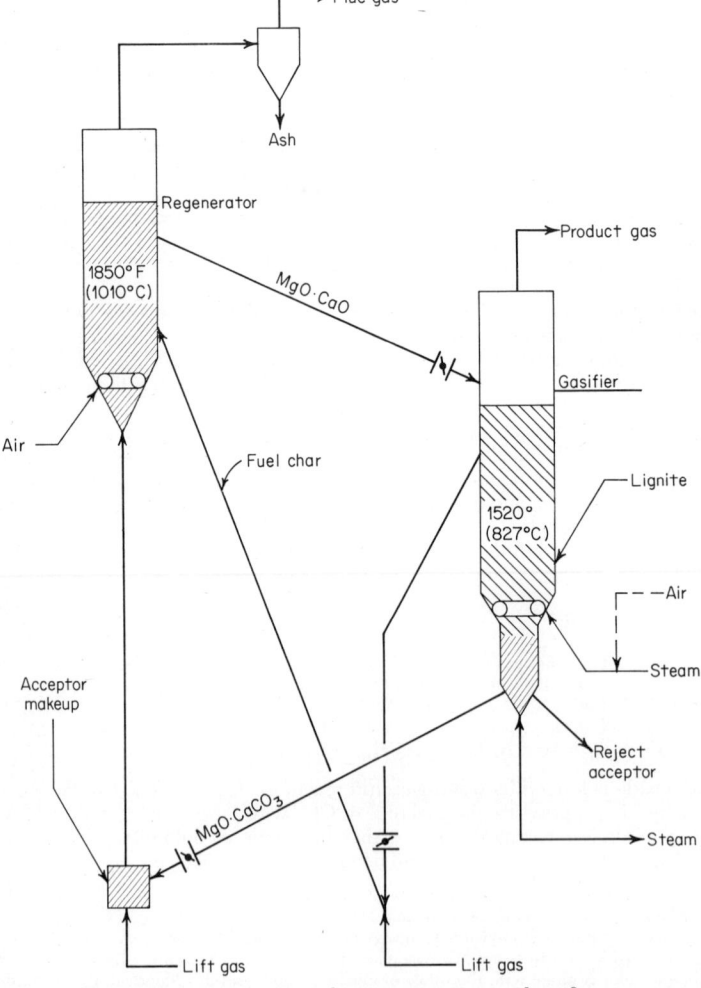

Fig. 1. CO_2 acceptor process showing regenerator and gasifier.

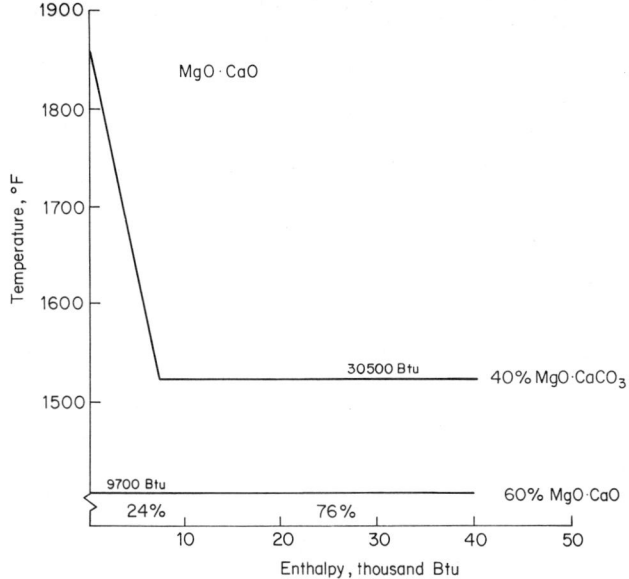

Fig. 2. Heat supply of acceptor at 40% activity. Basis: 1 lb-mol of dolomite acceptor.

order of 2 percent of the acceptor circulation rate. Since the acceptor attrition losses are essentially zero, purposeful withdrawal is necessary. A slip stream of acceptor is withdrawn from the boot to maintain a constant acceptor inventory.

Major Advantages of the Acceptor.   The use of the acceptor offers the following advantages:

1. The acceptor is heated and calcined in a separate reactor, where air can supply the oxygen for combustion without contaminating the product gas. This eliminates the need for an oxygen plant.

2. The required circulation rate is lower than in other heat-carrier processes inasmuch as most of the heat is supplied chemically. The temperature-enthalpy diagram of Fig. 2 shows the distribution of sensible heat and heat of reaction as supplied by one pound mole of dolomite acceptor with 40 percent activity and 1850°F/1520°F (1010°C/827°C) temperature differential between the gasifier and regenerator. Only 24 percent of the heat supplied is sensible heat; the remainder is heat of reaction.

3. The acceptor reacts with both hydrogen sulfide and carbon dioxide, the principal impurities in the gasifier product gas, thus reducing the gas cleanup requirements.

4. The raw gasifier product gas is hydrogen-rich, compared with all other gasification processes. No water-gas shifting is required before methanation. The raw gas contains sufficient hydrogen to methanate all the carbon monoxide and part of the carbon dioxide present.

Gas Cleanup and Methanation.   These steps are similar to those in other gasification processes. Shift conversion and carbon dioxide removal, however. may not be required. Because of the CO$_2$ acceptor reaction, the gases leaving the gasifier contain a high hydrogen-to-carbon oxides ratio. Existing information incorporated into a mathematical model of the process shows that the hydrogen content is sufficient to convert both CO and CO$_2$ to methane.

Feedstock and Acceptor Characteristics.   Both lignite and subbituminous coal can be used and have the following advantages: (1) low sulfur content, (2) high reactivity, (3) noncaking. Abundant reserves of readily extractable lignite and subbituminous coals exist in North Dakota, Montana, and Wyoming.

To avoid deposit formation in the gasifier, the feedstock must not have a high sodium or pyritic iron content in the ash.

The acceptor can be either a limestone or a dolomite. The CO$_2$ capacity per pound of stone, after repeated cycling, is the same for both materials. The main process criterion for stone selection is its physical strength, as measured by the quantity of fines produced during a calcination-recarbonation cycle.

REFERENCES

Consolidation Coal Company, Research & Development Report 16 to the Office of Coal Research, U.S. Department of Interior, under Contract 14-01-0001-415. Interim reports are listed as Refs. 1 through 7.

1. Interim Report 1: "Pipeline Gas from Lignite Gasification—A Feasibility Study," U.S. Department of Commerce, National Technical Information Service PB-166817 (feasibility study), PB-166818 (appendix), February 1965.
2. Interim Report 3: "Phase II—Bench-Scale Research on CGS Process, Book 1, "Studies on Mechanics of Fluo-Solids Systems," U.S. Government Printing Office, Washington, D.C., Catalog No. 163.10:15, January 1970.
3. Interim Report 3, "Phase II—Bench-Scale Research on CSG Process," Book 2, "Laboratory Physico-Chemical Studies," U.S. Government Printing Office, Washington, D.C., Catalog No. 163.10:16, January, 1970.
4. Interim Report 3: "Phase II—Bench-Scale Research on CSG Process," Book 3, "Operation of the Bench-Scale Continuous Gasification Unit," U.S. Government Printing Office, Washington, D.C., Catalog No. 163.10:16, January, 1970.
5. Interim Report 4: "Pipeline Gas from Lignite Gasification—Current Commercial Economics," U.S. Government Printing Office, Washington, D.C., Catalog No. 163.10:16.
6. Interim Report 5: "Phase III and Phase IVa—A Design and Construction of the CSG Pilot Plant, Rapid City, S. Dak.," issued by Office of Coal Research, Washington, D.C., Nov. 30, 1973.
7. Interim Report: "Consol Synthetic Gas Pilot Plant Operations," January 1972–June 1973, submitted to Office of Coal Research, American Gas Association, August 1973.
8. Wyllie, P. J., and U. F. Tuttle: *J. Petrology*, vol. 1, no. 1, 1960.
9. Curran, G. P., J. T. Clancey, B. Pasek, M. Pell, G. D. Rutledge, and Everett Gorin: "Production of Clean Fuel Gas from Bituminous Coals," EPA-650/2-73-049, prepared for Office of Research and Development, U.S. Environmental Protection Agency, Washington, D.C., December 1973.
10. Fink, Carl E., John D. Sudbury, and George P. Curran: "CO_2 Acceptor Gasification Process," 77th National Meeting, A.I.Ch.E., June 3–7, 1974, Pittsburgh, American Institute of Chemical Engineers, New York.

Production of High-Btu Gas by the Hydrane Process

BY JAMES A. GRAY,* HERMAN F. FELDMANN,† AND PAUL M. YAVORSKY‡

The Hydrane (HYDRANE) process is a promising coal gasification process for producing high-Btu gas (900 to 1000 Btu per standard cubic foot) from any raw coal of bituminous or lower rank and is based on the two-stage hydrogasifier concept developed at the U.S. Bureau of Mines (Ref. 1). Hydrogasification is the reaction of coal or char directly with hydrogen to produce methane. The hydrogen-coal reaction consisting of pyrolysis and hydrogasification becomes significant at about 400°C and at higher temperatures is initially very rapid. As conversion increases, the char residue becomes less reactive because of a competing polymerization reaction forming a more graphitic ring structure (Ref. 2).

There are two major differences between the hydrogasification approach and the synthesis gas approach used by the steam/oxygen-based processes. First, no preliminary partial oxidation or carbonization pretreatment of the coal is necessary to make it nonagglomerating. Prevention of agglomeration is of great importance because most of the eastern and midwestern American coals soften, swell, and become sticky at temperatures above 400°C, especially in the presence of hydrogen, and are therefore difficult to process. A unique feature of the Hydrane process is the first stage of the hydrogasifier called the free-fall dilute-phase reactor (FDP). Unpretreated pulverized coal is fed into the dilute-phase reactor in such a way that the particles free-fall in a dilute cloud through a hydrogen-rich atmosphere. During free-fall, the coal particles are rapidly heated to 750°C at 70 atmospheres pressure, causing 20 to 30 percent carbon conversion by pyrolysis and hydrogasification. This level of conversion makes the particles nonagglomerating and produces a very porous char which is highly reactive. The second difference is that 95 percent of the product methane is produced directly from coal in the two-stage hydrogasifier, thus requiring only a cleanup catalytic methanation step to convert the less than 4 percent carbon monoxide remaining in the raw-product gas. In contrast, other processes make as much as 50 percent of the product methane by catalytic methanation of carbon monoxide.

PROCESS ADVANTAGES

In 1956 Channabasappa and Linden (Ref. 3) pointed out that producing methane directly from coal by hydrogasification is more thermally efficient and hence less expensive than steam/oxygen gasification followed by catalytic methanation. More recently, Henry and Louks (Ref. 4) and Wen and coworkers (Refs. 5, 6) compared various processes, both commercial and those under development, and arrived at the same conclusion: i.e., production of methane directly in the gasifier is more thermally efficient and economical than catalytic methanation. In Wen's study, the Hydrane process had the highest thermal efficiency (78 percent) and the lowest relative gas price of all of the processes considered.

The high thermal efficiency¶ of the Hydrane process results from several process features. First, coal pretreatment is eliminated since pulverized caking coal is fed to the

*Supervisory Chemical Engineer, Coal Gasification—Hydrane Process, Pittsburgh Energy Research Center, Energy Research and Development Administration, Pittsburgh, Pennsylvania.

†Manager, Coal Research Programs, Battelle Columbus Laboratories, Columbus, Ohio.

‡Research Supervisor of the Exploratory Engineering Section, Pittsburgh Energy Research Center, Energy Research and Development Administration, Pittsburgh, Pennsylvania.

¶Thermal efficiency $= \dfrac{\text{Btu value of pipeline gas product}}{\text{Btu value of total coal entering plant}} \times 100$

(Total coal includes that needed to make the plant self-sustaining.)

dilute-phase reactor and decaked during free-fall by devolatilization. Thus the hydrogen-rich volatile matter is converted to methane, utilizing about an additional 10 to 15 percent of the coal. In most other processes, the 36 to 40 percent volatile matter must be reduced to, at most, 26 percent so that the gasifier does not plug from agglomerated char. Even if the pretreatment gases are recovered, a large amount of carbon and hydrogen will have been converted to carbon dioxide and water. Second, less process hydrogen is needed to produce methane when hydrogen reacts directly with carbon in coal rather than with

TABLE 1.   Comparison of Oxygen and Coal Requirements for Producing 250 Billion Btu per Day of Pipeline Gas*

Process description	Fraction of methane produced by hydrogasification†	Total plant coal, tons/hour	Oxygen, tons/hour
Direct hydrogasification of raw coal.	0.944	512	148
Steam-oxygen gasification of pretreated coal in two-stage nonslagging fluid-bed gasifier. 65% carbon conversion in gasification system.	0.678	622	217
Steam-oxygen gasification of char in slagging-bed gasifier, followed by entrained hydrogasification of raw coal. 100% carbon conversion in gasifier.	0.467	585	233

*Source: Ref. 5
†The remaining methane is produced by methanation.

carbon in carbon monoxide during catalytic methanation. The steam/oxygen-based processes, on the other hand, oxidize carbon in the coal to carbon monoxide in the gasifier, and then require hydrogen to convert it back to a reduced state as methane during catalytic methanation.

In hydrogasification no oxidation occurs, and hydrogen is directly added to the already partially reduced carbon in the coal, resulting in a lower hydrogen requirement. The decreased need for hydrogen appears in lower steam and coal requirements. Third, sensible heat in the residual char which is removed from the hydrogasifier is utilized in the endothermic carbon-steam reaction used to produce synthesis gas for making process hydrogen. This is accomplished by transferring the hot residual char from the hydrogasifier directly into the steam/oxygen gasifier. In processes that rely heavily on very exothermic catalytic methanation reaction to produce methane, a large amount of low-temperature heat (480°C catalyst temperature limit) is generated in the methanation reactor which is not utilizable in the gasifier. Consequently, the gasification heat is supplied by char combustion using oxygen, a relatively more expensive heat source. The foregoing advantages of hydrogasification thus minimize the coal and oxygen consumption for some commercial-size units, as illustrated in Table 1.

CONCEPTUAL FLOW SHEET

Reference is made to Fig. 1 which is a conceptual flow sheet of the Hydrane process. Hydrogen (16 standard cubic feet per pound of dry coal) is injected into the bottom stage of the hydrogasifier to fluidize the char bed. The intermediate gas from the fluid bed (975°C, 70 atmospheres) flows up through the annular space around the top stage to maintain the dilute-phase reactor wall temperature above 725°C, preventing particles from sticking to the reactor wall. The gas is then fed concurrently with pulverized raw coal downward through the free-fall dilute-phase reactor (top stage). The product gas, having a methane content of about 70 percent, is drawn off near the bottom of the dilute-phase reactor, and the char drops into the fluid-bed reactor (bottom stage) via a seal leg (Ref. 7). Residual char from the fluid-bed section is drawn off through a standard pipe and is fed directly into the hydrogen plant synthesis gas generator.

The methane and carbon monoxide produced in the dilute-phase reactor are typically 4.07 and 0.69 standard cubic feet per pound dry coal, respectively. Approximately 31 percent of the carbon in the dilute-phase char is converted in the fluid-bed reactor,

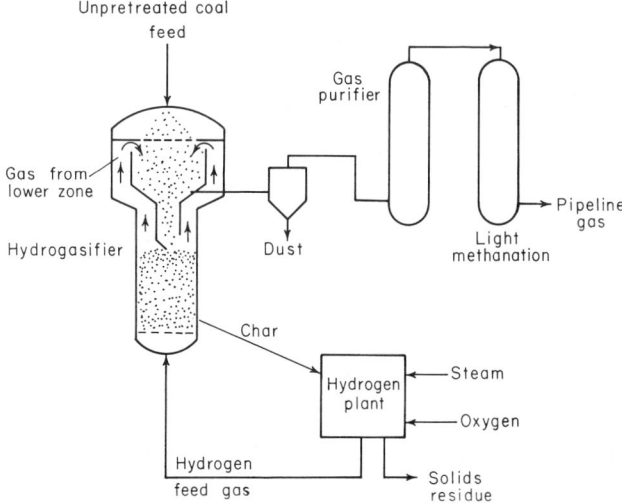

Fig. 1. Conceptual flow sheet of Hydrane process.

yielding an overall carbon conversion of about 46 percent and fluid-bed methane of 6.12 standard cubic feet per pound dry coal. The hydrogasifier total methane yield is 10.10 standard cubic feet per pound dry coal. An overall carbon conversion of 46 percent is required for balanced plant operation; so there is enough carbon left in the char to produce the process hydrogen in the hydrogen plant.

A more detailed view of the Hydrane process is shown in Fig. 2. The gas purification step downstream of the hydrogasifier is very simple because the raw-product gas usually contains about 1 percent carbon dioxide, thus permitting the recovered hydrogen sulfide to be fed directly to a Claus plant without the need to reduce the carbon dioxide concentration. The purification system in the hydrogen plant is also simple since the recovered acid gas is almost all carbon dioxide and less than 1.5 percent hydrogen sulfide. The sulfur in the char fed to the gasifier is below the limit (Environmental Protection Agency) for solid fuel (1.2 pounds sulfur dioxide per million Btu) in most cases because over 75 percent desulfurization occurs in the hydrogasifier (Ref. 8). The residual char from the hydrogen plant (0.137 pound per pound dry coal) may be used as fuel for steam and power generation since it contains enough carbon to be burnable. The purification, shift conversion, and methanation steps will utilize commercial technology, such as that available for ammonia plants.

PROJECT STATUS

The free-fall dilute-phase reactor has been operated as a separate laboratory-size unit at pressures of 35 to 205 atmospheres and at temperatures up to 900°C. Carbon conversion was more than adequate, and the kinetics have been well defined (Refs. 9, 10). The second-stage reactor has also been operated as a separate unit on a bench scale with either a moving bed (Ref. 1) or a fluidized bed (Ref. 11). Carbon conversion was more than needed for balanced plant operation. The major emphasis during the mid-1972 to mid-1974 period has been the operation of an integrated dilute-phase fluid-bed two-stage hydrogasifier. Design data for production of hydrogen-rich synthesis gas from Hydrane char is not available. Thus, bench-scale kinetic studies to define the reactivity of Hydrane char are ongoing and will be used in designing a pilot-scale gasifier to be incorporated in a Hydrane pilot plant.

LABORATORY DILUTE-PHASE REACTOR

The reactor and procedure used for most of the dilute-phase experiments have been discussed in Ref. 10. To summarize, the heated pipe reactor had an I.D. of 3.26 inches and an effective length of 5 feet, and was enclosed in a 10-inch pressure vessel. Pulverized coal was fed into the reactor from a coal hopper by means of a rotary vane feeder and a

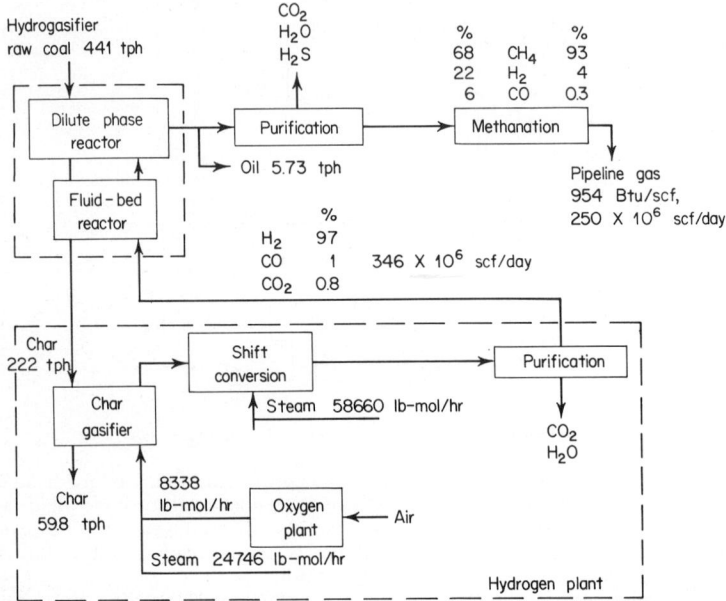

Fig. 2. Hydrane process, indicating quantities and compositions of materials: tph = tons per hour; 10^6 scf = million standard cubic feet.

nozzle at rates ranging from 5 to 18 pounds per hour. Char was collected in an air-cooled receiver positioned below the reactor. Both proximate and ultimate analyses were run on char samples. Samples of the product gas were taken periodically for complete chromatographic analysis. There were no thermocouples directly inside the reactor because of the possibility of char sticking on them. Instead, thermocouples were mounted on the outside wall of the reactor pipe, and these temperature readings are those that are reported.

Typical analyses of coals used in this work are shown in Table 2. The controllable parameters that determine the heating value of the raw-product gas are (1) percent methane in feed gas, (2) feed gas/coal ratio, (3) reactor pressure, and (4) reactor length (residence time). A large number of runs have been made and have been reported elsewhere along with the kinetic model (Refs. 9, 10). Some representative data from these tests are shown for various coals in Table 3. The product gas-heating value after methanation, which is shown in Table 3, was calculated on the assumption that the carbon monoxide concentration in the raw-product gas amounted to 4 percent. The carbon monoxide concentration in the product gas samples from the dilute-phase-reactor experiments was lower than 4 percent because there was no carbon monoxide contribution from a second-stage reactor, nor from the feed gas which would contain residual carbon monoxide from the hydrogen generation plant. The feed gas rate was an equimolar hydrogen/methane mixture flowing at approximately 150 standard cubic feet per hour, and the coal feed rate was 12 pounds per hour (207 pounds per hour × square feet). Coal feed rates beyond 18 pounds per hour led to severe particle-particle agglomeration.

Dilute-phase-reactor data for carbon conversion to methane has been correlated, using the rate equation

$$\frac{dz}{dt} = kP_{H_2}(1 - Z)$$

where z = fraction of carbon converted to methane
P_{H_2} = hydrogen partial pressure at the outlet of the reactor
k = rate constant
t = residence time

TABLE 2. Typical Analyses of Coals Used in Study Described
Percent

	Pittsburgh Seam hvAb coal	Illinois No. 6 hvCb coal	North Dakota lignite
PROXIMATE ANALYSIS			
Moisture	1.2	1.4	7.8
Volatile matter	36.4	36.8	39.7
Fixed carbon	56.7	55.9	46.9
Ash	5.7	5.9	5.6
ULTIMATE ANALYSIS (DRY BASIS)			
Carbon	79.09	75.45	64.64
Hydrogen	5.22	5.12	4.48
Nitrogen	1.60	1.72	0.76
Sulfur	1.10	1.32	0.76
Oxygen (by difference)	7.22	10.41	23.29
Ash	5.77	5.98	6.07
	100.00	100.00	100.00

hvAb coal from U.S. Bureau of Mines experimental mine, Bruceton, Pa.
hvCb coal from Orient No. 3 mine, Freeman Coal Co., Waltonville, Ill.
Lignite from Baukol-Noonan mine, Burke Co., N. Dak.
hvAb (high volatile A bituminous coal)
hvCb (high volatile C bituminous coal)

The resulting dependence of k on temperature for raw coal up to about 25 percent carbon conversion is given by the equations

$$\ln k = -15 \times 10^3 \left(\frac{1}{T}\right) + 21.22 \qquad \text{for } T \leqslant 853°K$$

$$\ln k = -3.2 \times 10^3 \left(\frac{1}{T}\right) + 7.43 \qquad \text{for } T > 853°K$$

Most char yields fell in the range 0.6- to 0.7-pound char as received per pound dry coal fed. Char-yield data correlated well with carbon conversion for the Illinois and Pittsburgh coals as given by the equation

Char yield = 0.892 − 0.00765 (% carbon conversion)

Similar results were obtained for coal hydrogen, nitrogen, and sulfur conversions, ranging from 18 to 62 percent:

% hydrogen conversion = 0.887 (% carbon conversion) + 40.3
% nitrogen conversion = 1.83 (% carbon conversion) − 21.8
% sulfur conversion = 0.738 (% carbon conversion) + 26.9

For carbon conversions above 20 percent, the oxygen conversion was usually greater than 90 percent and was therefore assumed to be complete for reactor design purposes.

Removal of sulfur in lignite was not as good as with the bituminous coals used because a large amount of the sulfur in lignite is inorganic and consequently is more difficult to remove (e.g., formation of stable sodium sulfides).

Organic liquid yields ranged from less than 0.01 to 0.06 pound per pound dry coal; the

TABLE 3. Selected Experiments Establishing the Feasibility of Producing Pipeline Quality Gas*

Test No., IHR-	156	166	160	157	164	184
Pressure, psig	1000	1200	1500	2000	1200	1000
Coal	hvAb	hvAb	hvAb	hvAb	hvCb	Lignite
Feed gas/coal ratio, scf/pound	12.3	12.3	13.1	12.4	14.0	12.3
Feed gas composition, % (volume)						
Hydrogen	49.0	49.2	53.8	49.9	50.1	50.1
Methane	49.4	48.7	43.4	48.4	47.7	47.8
Nitrogen	1.6	2.1	2.6	1.7	2.2	1.8
Carbon monoxide	0.0	0.0	0.0	0.0	0.0	0.2
Carbon dioxide	0.0	0.0	0.2	0.0	0.0	0.0
Carbon conversion, % (weight)	25.0	25.6	24.2	30.0	27.8	32.1
Product gas (water-free)/coal ratio, scf/ pound	13.4	13.3	13.5	13.9	16.9	10.6
Product gas composition, water-free, %						
Hydrogen	22.4	22.7	19.7	18.1	21.9	27.9
Methane	71.4	72.2	75.2	79.0	72.8	57.5
Ethane	0.5	0.1	0.3	0.1	trace	0.1
Carbon monoxide	3.2	2.3	1.4	0.5	2.4	6.3
Carbon dioxide	0.7	0.5	0.8	0.4	0.7	5.9
Nitrogen	1.4	2.3	2.4	1.7	1.9	2.1
Hydrogen sulfide	0.4	0.2	0.1	0.2	0.3	0.1
Methane/hydrogen ratio in product	3.19	3.18	3.82	4.36	3.32	2.06
Heating value, as-received, Btu/scf	817	815	835	863	818	695
Heating value with 4% CO methanation, Btu/scf	918	914	928	948	914	902
Percent methane equivalent ($CH_4 + C_2H_6$) made directly	94.7	94.8	95.0	95.2	94.8	90.1

*Using free-fall dilute-phase reactor.
hvAb = high-volatile A bituminous coal.
hvCb = high-volatile C bituminous coal.
scf = standard cubic feet.

yield depended on the gas-phase residence time and hydrogen partial pressure. Mass spectrometric data show the liquids to be highly aromatic, containing four to seven condensed rings. The organic liquid was free of sulfur, had a carbon/hydrogen content at least as high as the parent coal, and therefore would be suitable as a fuel.

Water recoveries for the Pittsburgh coal ranged from 0.01 to 0.08 pound per pound dry coal and for the Illinois coal from 0.05 to 0.09 pound per pound dry coal. The main contaminants in the water were phenols, cresols, xylenols, naphthalene, anthracene, indole, and traces of many other organic species.

Hydrogen consumption, which depends on reaction conditions, ranged from 0.6 to 1.0 standard cubic foot of hydrogen per standard cubic foot methane. At 900°C reactor wall temperature, the stoichiometric coefficient was typically 0.8 in contrast to 1.0 at a wall temperature of 750°C. At the lower reactor wall temperature more hydrogen was required because less hydrogen-rich material was generated by devolatilization.

Dilute-phase char particle characteristics varied according to heatup rate, the amount of hydrogen in the feed gas, and the reactor pressure. For example, the particles in a pulverized feed of Pittsburgh hvAB (high volatile A bituminous) coal explode like popcorn on exposure to rapid heating, thus yielding char particles having diameters five to six times the original particle diameter. The bulk density of the char can be six to seven times lower than the original bulk density of the feed coal. The particles are highly porous and often have a hooklike shape, therefore requiring higher than normal velocities to fluidize. The generalized correlation for predicting the minimum fluidization velocity of the char was found to be

$$(N_{Re})_{mf} = 0.0135(N_{Ga})^{0.73}$$

where N_{Re} = particle Reynolds number at minimum fluidization
     N_{Ga} = Galileo number*

*Galileo number is defined as $N_{Ga} = d_p^3 p_f(p_s - p_f)g/\mu^2$, where d_p is mean particle diameter, p_f is fluid density, ρ_s is solid density, g is gravitational constant, and μ is fluid viscosity.

The relationship of char bulk density to reactor pressure is linear over the range 35 to 232 atmospheres (50 × 100 mesh coal feed) and is

$$\rho = 7.00 \times 10^{-4}P + 0.0239 \text{ g/cm}^3$$

when the feed gas is pure hydrogen, and is

$$\rho = 8.15 \times 10^{-4}P + 0.0359 \text{ g/cm}^3$$

when the feed gas is an equimolar mix of hydrogen/methane. These relationships apply when reactor wall temperature is in the range 800 to 900°C.

FLUID-BED REACTOR

In order to verify that adequate carbon conversion could be obtained in the second stage of the hydrogasifier, char produced in the laboratory dilute-phase reactor was further hydrogasified in a batch fluid-bed reactor having a ⅝₆-inch inside diameter and a length of 32 inches. An 8-gram charge was reacted under a pressure of 69 atmospheres in a 1.3-standard-cubic-feet-per-hour stream of hydrogen. At 900°C, the carbon conversion was over 52 percent within 2 minutes and reached 54 percent after 10 minutes. At 800°C, the carbon conversion was 17 percent after 2 minutes, 30 percent after 12 minutes, and 35 percent after 20 minutes. These results indicate that the 31 percent carbon conversion needed in the second-stage fluid-bed reactor could be achieved. Lewis and coworkers (Ref. 1) in the early 1960s used a ½-inch I.D. moving-bed reactor to hydrogasify dilute-phase char and obtained carbon conversion beyond 31 percent and a product gas containing over 60 percent methane.

The kinetics and design of a fluid-bed reactor are summarized in Ref. 7, where work performed by organizations other than the Bureau of Mines is included. The same basic rate equation developed for the hydrogasification of raw coal in the dilute-phase reactor has been applied as well to the hydrogasification of chars, such as occur in the fluid-bed section of the hydrogasifier. For coal char above 25 percent carbon conversion, the dependence of k on temperature is given by

$$\ln k = -10.45 \times 10^3 \ (1/T) + 7.08 \text{ for } 973°K \leq T \leq 1173°K$$

LABORATORY SIMULATION OF AN INTEGRATED TWO-STAGE REACTOR

Process development emphasis in the past few years has been on the operation of a two-stage hydrogasifier, the latest of which is shown in Fig. 3. The dilute-phase reactor was almost identical with that used in single-stage experiments except slight nozzle modifications. For the integrated experiments, the dilute-phase reactor (top stage) was operated at 69 atmospheres and 800 to 900°C with coal feed rates ranging from 5 to 12 pounds per hour. The feed gas to the dilute-phase reactor usually consisted of a 46 percent methane–54 percent hydrogen mixture and was fed concurrent with the pulverized coal. The mixture simulates the second-stage product gas that would be obtained in a larger unit.

After falling through the dilute-phase reactor, the char dropped into the second-stage reactor which was operated as either a moving bed or a fluidized bed at 69 atmospheres and temperatures from 700 to 900°C. The effective reactor length could be varied up to 10 feet. The product gas from both stages was drawn off from a disengaging zone between the reactors, and each stage had a gas sample line for determining the respective methane yields. The second-stage reactor was 3.26 inches in diameter and had a removable sleeve for operation as a 2.0-inch I.D. reactor. This laboratory scheme was different from that of the conceptual hydrogasifier simply because it was impossible to operate a fluid bed with an interstage char transfer leg on this scale of equipment.

At the reactor operating temperature and pressures, the char particle size of normal operation (1.0 to 1.4 mm diameter), a very long fluid-bed reactor (20 to 30 feet) having a diameter of 1 inch or less would be required to give a high enough velocity to produce fluidization. Not only would the gas-solid mixture slug due to a large length-to-diameter ratio, but also the char would bridge across and plug the reactor.

Because of the equipment scale, early attempts at operating the second stage as a fluid bed were plagued with basic design problems. Pressure upsets caused by plugging of small gas offtake lines from entrained solids and condensation of volatile material (e.g., pyrene, phenanthrene) in critical places were additional problems (Ref. 8). Despite these setbacks, steady state was attained for short periods in several runs. Total carbon conversion was over 50 percent, well over the needed 45 percent for both stages. Because an excessive amount of hydrogen was needed to fluidize the bed in the second-stage reactor, the methane content of the fluid-bed gas product was lower than desired. Two key results

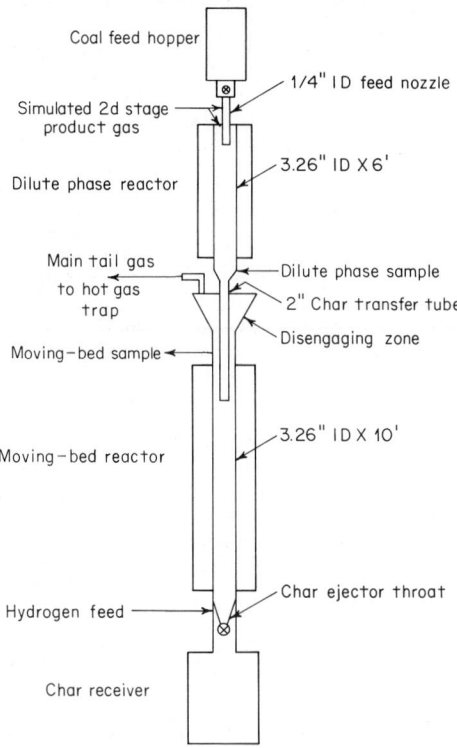

Fig. 3. Integrated two-stage hydrogasifier.

were obtained in addition to the needed carbon conversion: (1) Hydrogen consumption per unit of methane produced was quite low at 1.38 for the overall hydrogasifier, and (2) desulfurization of Illinois coal was high, ranging from 72 to 90 percent for the overall hydrogasifier.

FURTHER DEVELOPMENTS

Experiments with the laboratory two-stage hydrogasifier are aimed toward developing a smooth-running unit capable of prolonged steady-state operation. These experiments are needed to generate basic data on the second stage of the hydrogasifier using freshly produced dilute-phase reactor char, to obtain accurate liquid-product yield data and to prove operability. Previous tests were too short in duration to measure liquid yields accurately. The basic data for the second-stage reactor will consist of an optimum operating temperature, the limiting operating temperature, and the maximum methane concentration that can be produced.

Three secondary objectives are to determine more fully the desulfurization capability of the two-stage hydrogasifier, evaluate countercurrent operation of the dilute-phase reactor, and determine the lowest practical operating pressure. The last-mentioned objective is important because of the lack of a proved method for feeding dry pulverized coal into a reactor system at a pressure of 70 atmospheres.

Although a large amount of kinetic data has been obtained for the laboratory dilute-phase reactor, scale-up to a very large diameter reactor is risky because an accurate heat balance cannot be determined. In the 3.25-inch laboratory unit, radiant heat from the reactor walls dominates the heat balance and is responsible for the rapid heat-up of the coal (Refs. 8, 12). In a vessel whose diameter is 2 feet or more, the radiative heat transfer has only a minor influence. Consequently, the rapid heat-up and final temperature of the coal particles will depend on heat recuperated from the second-stage product gas and heat generated during devolatilization and hydrogasification. The heat generated from reaction in the dilute-phase reactor is unknown and may be so small that heat from the second stage alone may not be sufficient to operate the dilute-phase reactor. Thus, the next-scale hydrogasifier will have to be at least a 2-foot-diameter dilute-phase reactor in order to demonstrate its operability without radiative heat. In order to partially solve this problem, a bench-scale study of devolatilization and hydrogasification of raw coals is under way at temperatures and heat-up rates that may be typical for large-diameter dilute-phase reactors.

REFERENCES

1. Lewis, P. S., S. Friedman, and R. W. Hiteshue: Fuel Gasification, *Advan. Chem. Ser.*, vol. 69, pp. 50–63, American Chemical Society, Washington, D.C., 1967.
2. Lowry, H. H. (ed.): "Chemistry of Coal Utilization," Supplementary Volume, pp. 939–943; 994–1007, Wiley, New York, 1963.
3. Channabasappa, K. C., and H. R. Linden: *Ind. Eng. Chem.*, vol. 48, pp. 900–905, 1956.
4. Henry, J. P., Jr., and B. M. Louks: Preprints of Fuel Division Meeting, ACS, Chicago, American Chemical Society, Washington, D.C., September 1970.
5. Wen, C. Y., et al.: *Office Coal Res. R&D Rep.* 66, *Interim Reps.* 1–3, U.S. Department of Interior, Washington, D.C.
6. Wen, C. Y., C. T. Li, S. H. Tscheng, and W. S. O'Brien: Paper presented at 65th Annual A.I.Ch.E. Meeting, New York, Nov. 26–30, 1972, American Institute of Chemical Engineers, New York.
7. Feldmann, H. F., C. Y. Wen, W. H. Simons, and P. M. Yavorsky: Paper presented at 71st National A.I.Ch.E. Meeting, Dallas, Tex., Feb. 20–23, 1972, American Institute of Chemical Engineers, New York.
8. Feldmann, H. G., and P. M. Yavorsky: Paper presented at 5th AGA/OCR Synthetic Pipeline Gas Symposium, Chicago, Oct. 29–31, 1973, American Gas Association, Arlington, Va.
9. Feldmann, H. F., J. A. Mima, and P. M. Yavorsky: Preprints of Fuel Division Meeting, ACS, Dallas, Tex., April 1973, American Chemical Society, Washington, D.C. ·
10. Feldmann, H. F., W. H. Simons, J. A. Mima, and R. W. Hiteshue: Preprints of Fuel Division Meeting, ACS, Chicago, September 1970, American Chemical Society, Washington, D.C.
11. Pyrcioch, E. J., and H. R. Linden: *Ind. Eng. Chem.*, vol. 52, pp. 590–594, 1960.
12. Feldmann, H. F., W. H. Simons, C. Y. Wen, and P. M. Yavorsky: Paper presented at 4th International Congress of Chemical Engineering, Chemical Equipment Design and Automation, Marianske, Czechoslovakia, Sept. 11–15, 1972.

The Synthane Coal Gasification Process

BY JOHN F. S. FRITH*

The Synthane process was developed to produce high-Btu gas from coal as a substitute for natural gas. The product will be a clean gas with a calorific value of about 1000 Btu per standard cubic foot. The gas is produced at high pressure and, therefore, is available for direct injection into a pipeline system. The key steps of the Synthane process were developed by the U.S. Bureau of Mines. These are: (1) pretreatment of pulverized coal to render it suitable for fluidized-bed gasification, (2) the gasification operation itself, and (3) processes to methanate the gas produced.

These processing steps have been combined with other, more conventional process operations, in a continuous plant located at Bruceton, Pennsylvania (near Pittsburgh). This plant will be operated by the (ERDA) Energy Research and Development Administration, Bureau of Mines, to demonstrate the process. The plant is capable of processing all grades of bituminous coal or lignite, containing up to 5 percent sulfur at a design throughout of three tons of coal per hour.

In general, the processing operations and mechanical designs established in the prototype plant are adaptable to future scale-up.

The plant consists of the following sequence of process unit operations:

1. Coal handling and preparation to provide material suitable to feed the reaction steps

2. Coal pressurization and conveying to introduce the coal into the high-pressure system

3. Coal treatment to reduce the agglomerating tendency of certain coals

4. Coal gasification to produce raw gas plus residual char

5. Char removal to extract char from the high-temperature/high-pressure environment

6. Gas purification to remove solid fines, char, and steam from the raw gas

7. Shift conversion to adjust the CO/H_2 ratio in the clean gas

8. Acid gas removal to eliminate CO_2 and H_2S from the converted gas

9. Removal of trace sulfur compounds from the clean gas

10. Methanation to increase the calorific value of the gas by conversion of CO and H_2 to CH_4

11. Sulfur removal from the acid gas regenerator effluent to make it acceptable for discharge to the atmosphere

These operations and their novel features are described in the following paragraphs. See flow sheet of Fig. 1; also Tables 1 through 5 which give material balance data.

COAL HANDLING

At Bruceton, coal will be delivered by truck. It will be minus ¾ inch and, therefore, will require no precrushing and can be loaded directly into storage hoppers. Coal handling is entirely closed to minimize dust emissions. The minus ¾-inch coal must be reduced to a size range suitable for fluidization. Minus 20 mesh, the size used in the ERDA (U.S. Bureau of Mines) pilot plant, was selected as the design size for the demonstration plant; however, other sizes may be tried. It is also necessary to dry the coal so that it is free flowing and can be handled in subsequent operations. To meet these requirements, a pneumatic drier integral with the crushing mill is used. This arrangement has many advantages for this process where it is necessary to prepare material with a minimum of fines. The drying duct, carrying coal from the mill, extends to the top of the structure serving the drier, the coal feeding facilities, and the gasification equipment. It, therefore, provides a means of lifting the coal to an elevated storage bin, without the need to supply special conveying facilities.

*Assistant Manager, Chemical Engineering Department, The Lummus Company, a subsidiary of Combustion Engineering, Inc., Bloomfield, New Jersey.

TABLE 1.   Material Balance—Pretreatment and Gasification
Quantities shown are pounds/hour

In		Out	
COAL		RAW GAS	
H_2O	150	H_2	170
C	4427	CO	1550
H	312	CO_2	5650
O	480	CH_4	1350
N	91	C_2H_6	80
S	96	N_2	20
Ash	444	H_2S	90
		NH_3	90
Total	6000	H_2O	5900
		Tar	200
Pressurizing CO_2	1280	Total	15,100
Transport steam	1600		
Pretreatment			
oxygen	450	Char	1830
Gasification steam	5900		
Gasification			
oxygen	1050		
Steam purges	650		
Total	16,930	Total	16,930

A 20-mesh screen is used for classification. This provides efficient size separation and avoids introducing any oversized material into the fluidized gasifier, where it could cause problems. The oversized coal is recycled to the mill, and, in this way, a closed-circuit crushing operation is provided, which is the best way of obtaining a close size range without producing a large quantity of fines. Flexibility needed for experimental purposes in the prototype plant operations is provided in this system, since alternative size ranges, from minus 10 to minus 30 mesh, can be prepared by the simple expedient of changing the screen cloth.

The coal drying and crushing facilities will deliver the product into a large storage bin. This will be the source of feed for the pretreating and gasifier system.

COAL FEEDING SYSTEM

A lock-hopper system is used to feed coal under pressure to the pretreater.

In this system small batches of coal are transferred from the pulverized-coal storage bin, at essentially atmospheric pressure, to a pressurized feed hopper, which is at a high

TABLE 2.   Material Balance—Raw Gas Scrubbing
Quantities shown are pounds/hour

In			Out	
Component	Raw gas	Waste water	Tar	Clean gas to shift
H_2	170			170
CO	1550			1550
CO_2	5650	50		5600
CH_4	1350			1350
C_2H_6	80			80
N_2	20			20
H_2S	90			90
NH_3	90	40		50
H_2O	5900	5200		700
TAR	200		200	
Total	15,100	5290	200	9610

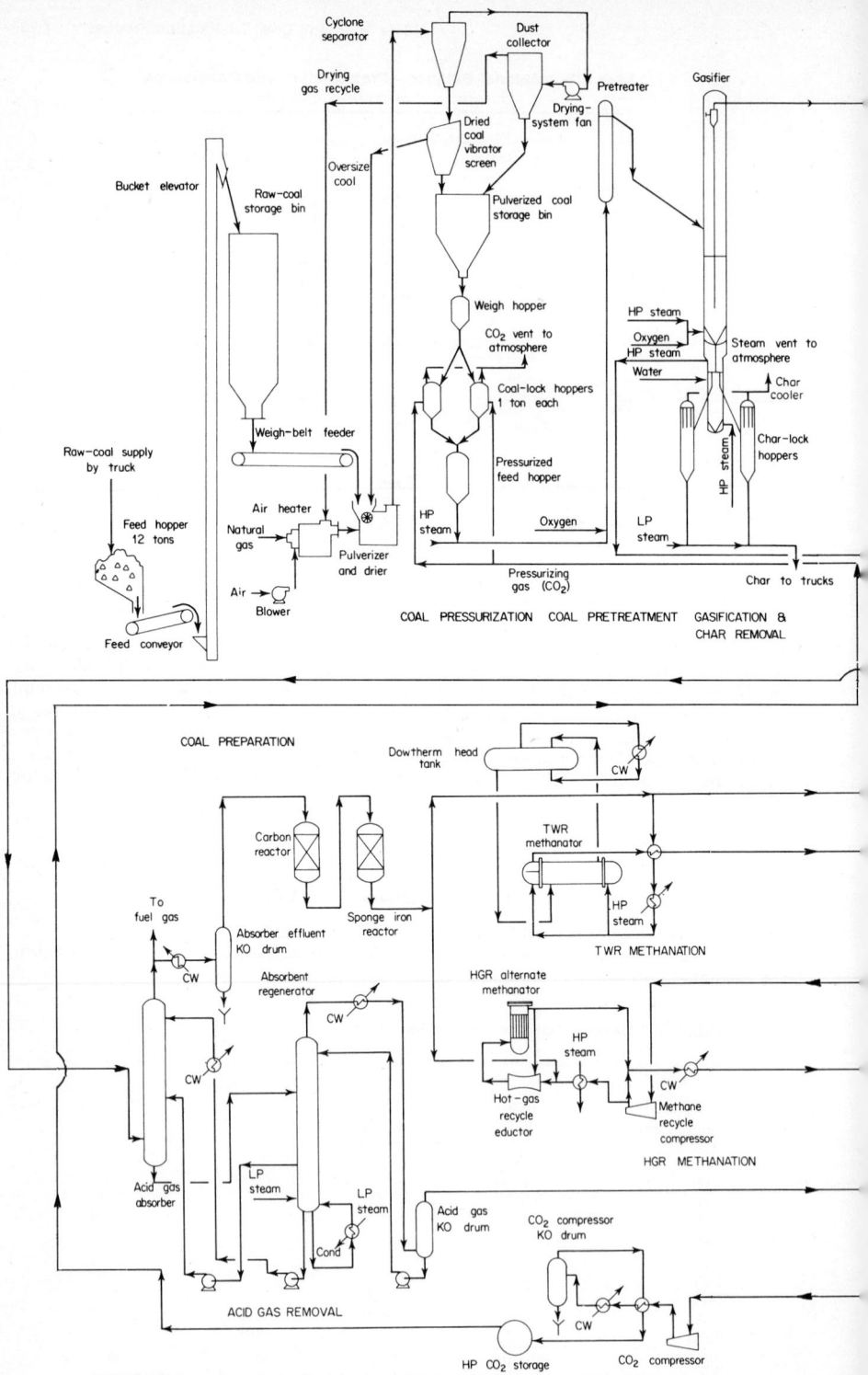

Fig. 1. Synthane coal gasification process: HP = high pressure; LP = low pressure; CW = cooling water;

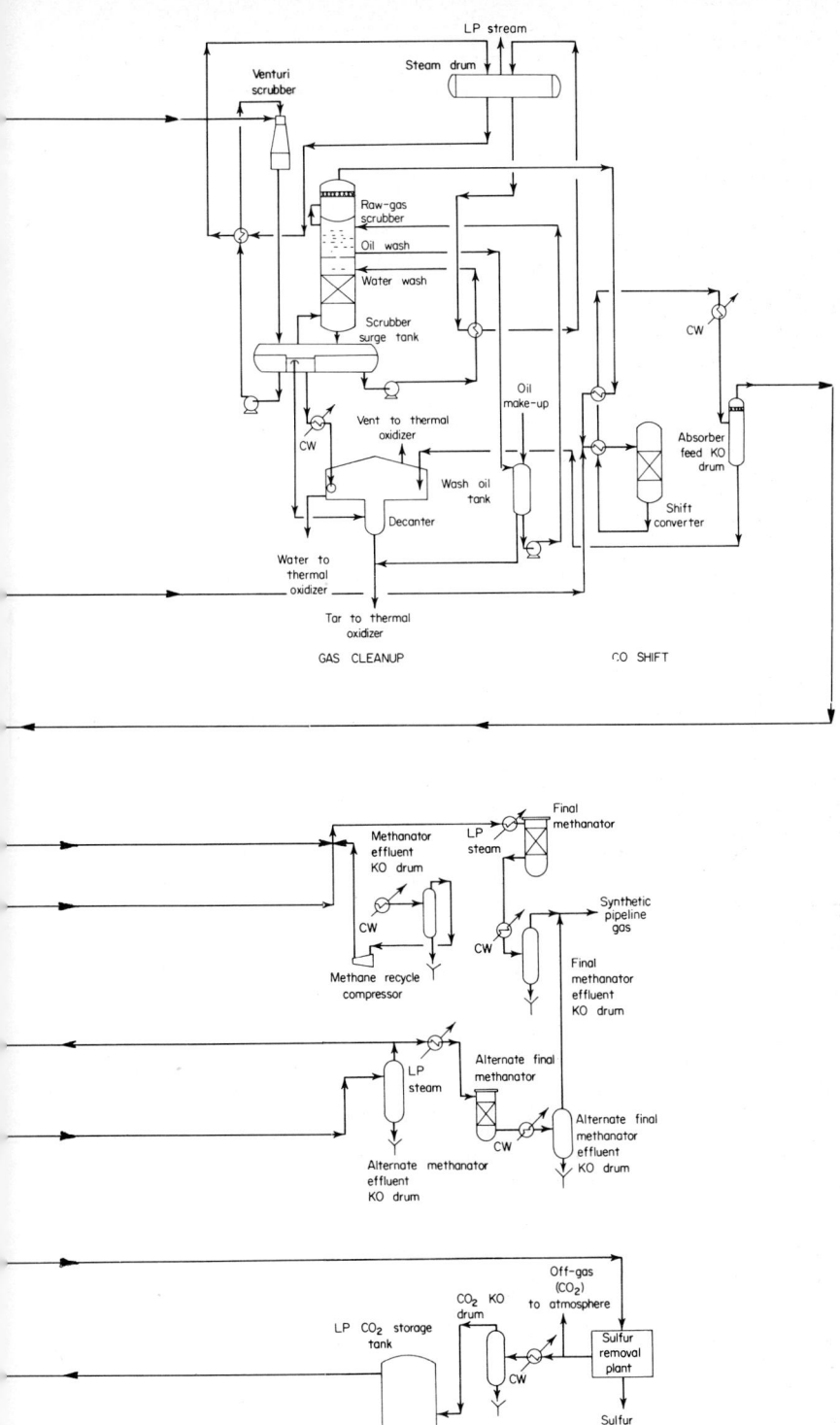

KO = knockout; Cond. = condensate; TWR = tube-wall reactor; HGR = hot-gas recycle.

enough pressure to feed continuously into the transport line to the pretreater. A weigh hopper is used to measure the batches, and each batch falls into a lock hopper. This vessel is then isolated by pneumatically operated valves, and pressurizing gas is introduced to raise its pressure to that of the pressurized feed hopper. Valves below the lock hopper are then opened to allow the batch to fall into the pressurized hopper. The lock hopper, now empty, is again isolated, and the pressurizing gas is vented to permit the introduction of the next batch, at atmospheric pressure, and then the cycle is repeated.

The main drawback of such a system is that it involves a very high power consumption, since the pressurizing gas must be vented and its energy is lost. To reduce this, dual lock

TABLE 3.   Material Balance—Shift Conversion

Quantities shown are pounds/hour

| | In | | Out | |
Component	Clean gas to shift	Steam	Converted gas	Condensate
H_2	170		215	
CO	1550		970	
CO_2	5600		6400	110
CH_4	1350		1350	
C_2H_6	80		80	
N_2	20		20	
H_2S	90		90	
NH_3	50			50
H_2O	700	2200	15	2510
Total	9610	2200	9140	2670

hoppers are provided so that the venting of one can partially pressurize the second. Ultimately, on a commercial plant with several parellel streams, this principle can be carried further with several stages for pressure equalization, resulting in a much higher thermodynamic efficiency. The dual system is installed in the prototype plant in order to prove its reliability.

Although the scheme is relatively simple, the pressure must be raised to 1000 psi, which is much higher than most lock-hopper systems presently operated commercially. This high-pressure operation introduced several problems which had to be solved in the specification and design of the coal-feeding system.

A suitable medium for transport of the coal had to be selected. The choice was limited to an inert gas which could be removed easily from the product gas. This eliminated manufactured inert gas, which is predominantly nitrogen and cannot be separated easily from the product. The choice, therefore, was between carbon dioxide and steam. The use of recycled carbon dioxide would have increased greatly the operating capacity of the carbon dioxide removal facilities. It would also have required a large compressor, since the carbon dioxide is exhausted from the acid-gas regenerator at only a few pounds above atmospheric pressure. Steam was therefore chosen as the transport medium. This is provided from the high-pressure boiler, needed to supply steam for gasification.

There is, however, a potential problem in using steam because of the very high operating pressure. At 1000 psig, steam will condense at about 550°F (288°C). For this reason, the transport system must be maintained at a high temperature. The transport line itself can easily be kept hot by suitable tracing, but the steam would be cooled by the coal. Although the steam is superheated, quite rapid heat interchange with the coal is expected, and the resultant equilibrium conditions could result in some condensation of steam, which is clearly unacceptable.

Experience gained by the Bureau of Mines in their work on entrained pretreatment showed that when oxygen is injected into the transport lines there is an instantaneous reaction which liberates heat and raises the temperature of the coal. Therefore, a small controlled injection of oxygen into the transport steam is used to heat the coal up to a temperature of about 600°F (316°C), and in this way any condensation is avoided.

Similar considerations were necessary in the selection of a suitable gas for pressurizing the lock hoppers. Clearly, steam could not be used for this purpose because it would condense. Inert gas generated from burning natural gas was considered. However, on analysis it was found to be unacceptable because a small quantity of the pressurizing gas leaves the pressurized feed hopper with the coal. The nitrogen in this gas would eventually find its way into the product gas. Although the quantity is very small, it would have an adverse effect on the calorific value of the product gas, reducing it below the required 900 Btu per cubic foot. Therefore, carbon dioxide had to be used. The quantity that must be recirculated through the process does not put a large additional load on the acid-gas removal system. There is one complication, however. Carbon dioxide will liquefy at ambient temperature and 1000 psi. It is, therefore, necessary to ensure that it is hot, and to maintain the lock-hopper system warm. Calculations showed that by eliminat-

TABLE 4.   Material Balance—Acid Gas Removal and Sulfur Recovery
Quantities shown are pounds/hour

| | In | | Out | | | |
| | Converted | | Sulfur plant | | | Purified |
Component	gas	Steam	off gas	Condensate	Sulfur	gas
H_2	215		2			213
CO	970		3			967
CO_2	6400		6300			100
CH_4	1350		8			1342
C_2H_6	80					80
N_2	20					20
H_2S	90					0
S					85	
H_2O	15	2000	210	1800		10
Total	9140	2000	6523	1800	85	2732

ing the aftercooler on the CO_2 compressor the gas will be available at $300°F$ ($149°C$), which is sufficient to prevent condensation.

PRETREATING

In order to broaden the application of the Synthane process to a wide range of coals, it is necessary to pretreat some types of coal to destroy their natural agglomerating tendencies. If this were not done, particles in close proximity in the fluidized gasifier would stick together and form large lumps that would settle out, and possibly be the source of hot spots or blockages in the bed.

Pretreatment provides a mild oxidation of the coal-particle surface. The Bureau of Mines has used successfully three types of pretreatment. These are the free-fall process, the fluidized-bed process, and the entrained process. For the demonstration plant, the fluidized-bed process has been selected. It is more easily controlled than free fall, and its stage of development is more advanced than that of the entrained process. The pretreatment can almost be considered part of the gasifying system, since the pretreater itself is closed coupled to the gasifier. Coal and fluidizing gas, which consists of oxygen diluted with high-pressure steam, enter the bottom of the pretreater. The pretreated coal leaves by an overflow line which goes directly into the gasifier. This line also caries the gas into the top of the gasifier. Thus, no cyclones or other separation devices are required in the pretreater.

The mode of operation of the pretreater is extremely simple. Premixing of the fluidizing gas and coal eliminates the need for a special distributor at the bottom. The overflow into the gasifier eliminates the need for level control or any gas-solids separation. The pretreater operates at the same pressure as the gasifier, and therefore no heating, cooling, or intermediate transport facilities are required.

GASIFICATION

The gasifier is a fluidized-bed reactor in which coal is introduced at the top, and a mixture of steam and oxygen in the bottom. Product gas leaves overhead, and unconverted coal, or char, is withdrawn from the bottom.

The gasifier operates at high pressure which favors the chemical equilibrium toward the formation of methane. The pilot-plant work was limited to an operating pressure of 40 atmospheres; however, the prototype unit has been designed for somewhat higher pressure operation, 60 atmospheres or 1000 psig, which offers the possibility of improving the yield of methane. It is also capable of operating at lower pressures similar to those that the pilot plant uses.

The depth of the fluidized bed in any fluid reactor is important since it is a measure of the residence time for the particles. Bureau of Mines data show that methane yield increases on a linear basis as bed volume is increased. Because of height limitations, the pilot plant was not tall enough to enable the Bureau of Mines to carry their operations to a point where this trend reached its maximum. The prototype therefore has been designed, with more flexibility, to achieve longer residence times, with a deeper bed.

TABLE 5.   Material Balance—Methanation
Quantities shown are pounds/hour

In		Out	
Component	Purified gas	Condensate	High Btu product gas
H_2	213		5
CO	967		4
CO_2	100		100
CH_4	1342		1897
C_2H_6	80		80
N_2	20		20
H_2O	10	620	6
Total	2732	620	2112

The pilot plant exhibits a characteristic which is rather unusual for fluidized beds without internals. That is, a temperature gradient exists from bottom to top. This is attributed partly to the fluidization regime existing under the high-pressure operation and also to the relatively high length-to-diameter ratio for the bed. Both of these minimize the amount of top-to-bottom mixing. Whatever the reason, it is advantageous because the lower temperature at the top of the bed again favors the equilibrium toward more methane formation. An attempt has been made to achieve a similar gradient in the scaled-up reactor. The pilot-plant fluid bed appears to operate in the regime of particulate fluidization where a minimum of bubbles are formed. The demonstration plant gasifier has been designed to provide the same type of fluidization. However, this may not be sufficient alone to restrict the top-to-bottom mixing which tends to produce a uniform bed temperature. The vessel is, therefore, provided with some internals to help reduce mixing and achieve the temperature gradient.

The pilot-plant reactor has operated with a free-fall zone above the fluid bed. Coal is introduced at the top of the vessel and falls, by gravity, countercurrent to the gases leaving the bed. It is believed that the devolatilization occurring in this region contributes significantly to the methane yield. A similar zone is provided in the prototype, by placing the coal inlet some ten feet above the highest fluid-bed level.

The fluidizing gas must be evenly distributed at the bottom of the gasifier in order to minimize the possibility of developing localized hot spots. The distributor used is conical, with a multitude of inlet nozzles. The premixed steam and oxygen enters a plenum chamber around the cone. Orifices provided at each nozzle create a pressure drop, and through this, equalize the gas flow to all inlets. The gas enters horizontally, at a fairly high velocity, which gives good mixing with the solids in the fluid bed. The conical base has been chosen to ensure that solids cannot settle in any dead spots. There are no projections that would provide ledges. Even if agglomerates form, they will flow downward to the outlet.

Entrainment must be minimized to avoid problems in the gas cleanup equipment. Work by both the Institute of Gas Technology and the British Gas Council has indicated that entrainment from a high-pressure fluidized bed would be very small. Calculations, based on extrapolation of correlations published by Wen and Hashinger and Zenz and Weil, tend to confirm this. However, this is a relatively unknown area, and so provision is made for installation of an internal cyclone. This introduces design problems since at high pressure the gas is very dense, and therefore pressure drop through the cyclone is high.

This pressure drop must be balanced by a dip leg to seal it and return the fines to the bed. The fines in turn have an extremely low bulk density, and, therefore, to balance the high-cyclone pressure drop, an extremely long leg is needed, which must be aerated to maintain free-flowing conditions. Facilities have been incorporated into the demonstration plant to study the effect of operation both with and without a cyclone.

CHAR REMOVAL

Although high conversion of coal to gas in the gasifier is expected, it is a single-stage fluid bed, and therefore the gasification can never be 100 percent because of the mixing in the bed. For this reason, some char must be removed from the bed, in order to maintain sufficient concentration of carbon in the reactor. With this necessity, it was decided, as a basic concept behind the process, that for a commercial plant sufficient char should be removed and burned to raise all the steam needed to provide the power requirements for the entire process. It is estimated that approximately one-third of the carbon entering as coal will be needed to supply the total energy requirements. The demonstration plant has therefore been designed to permit the removal of up to 30 percent of the carbon as char. This is not a limit on the conversion efficiency of the process, since the Bureau of Mines has successfully gasified more than 90 percent of the coal, and in the demonstration plant, with its greater residence time, even higher conversions are expected.

It has been decided that the best means of burning the char will be investigated in other facilities. For this reason the demonstration plant discharges the excess char as a product. Although this removes one step from the integrated process, it does not circumvent one of the more difficult engineering problems. The hot char must still be removed from the high-pressure reactor. It must be cooled and placed in a storage hopper at atmospheric pressure. If this can be done successfully, there will certainly be no trouble in feeding any kind of boiler or other processing facility in a commercial plant.

In order to release the pressure on the char, the gas associated with it must be throttled down to atmospheric pressure. This requires valves or restriction orifices. The very high temperature of the char in the gasifier is an extreme complication for the selection of equipment to handle the gas. For this reason it was decided that the first step must be to cool the char. The method chosen involves a further fluidized bed. The char from the gasifier flows continuously into the fluid-bed cooler located immediately below. No valves are required for this operation, flow being controlled by pressure differential between the gasifier and the char cooler. The cooler is fluidized by steam, and water is injected either directly into the bed, or as an atomized spray with the fluidizing steam. The quantity of water is carefully controlled, and in this manner it is possible to achieve a really significant degree of cooling, down to a temperature close to the dew point of steam at 1000 psi. This cooling from 1800 to 600°F (982 to 316°C) puts the design of subsequent equipment into a conventional day-to-day experience level.

A bonus for this system is that all the heat removed from the char, which is not insignificant, is converted to high-pressure steam.

Steam is required for carbon monoxide shift, at a pressure close to the char cooler pressure. Approximately two-thirds of the steam from the cooler can be used for this purpose. If the char removal is reduced to less than 30 percent, this steam production could well balance the shift reactor requirements, and thereby the entire system would be extremely efficient.

Continuous withdrawal of char and associated steam from the cooler was considered. To accomplish this, the vapor and solid stream would have to be throttled, through a valve, to dissipate the pressure. However, calculations indicated that the size of the opening for this valve, or an orifice, would be extremely small if continuous operation were employed. Intermittent operation would therefore be necessary for using a reasonable size orifice, and this would require frequent shutting and opening of valves located in the high-velocity solid-gas stream. This would probably result in the need for a high degree of maintenance; and the efficiency of shutoff, after a period of operation, would likely be impaired. It was decided, therefore, that though the system might be feasible for a large plant, the potential problems and high maintenance would make it unreliable.

The system installed has been selected to avoid the problems of the continuous letdown system by imposing the following design criteria:

1. The main throttling of gas to reduce pressure is performed on a clean gas, after separation from solids.

2. The valves to be used for shutoff are located where they can be blown free of solids before being shut.

3. Valves handling solid gas mixtures operate under very low pressure drop conditions.

These provisions led to a dual lock-hopper system for the removal of char. The system consists of two lock hoppers, located directly below the char cooler, into which char can flow by gravity as an overflow from the fluidized bed in the cooler. After one hopper is filled, the valves to the other are opened. A slide valve on the line to the first hopper will be shut, and the main isolating valve will be blown free of solids by steam before being shut. Removal of gas, to reduce the pressure, is via a filter in the top of the vessel. The steam released from the char will therefore be clean and filtered before passing through the throttling valve. After pressure is released, the char will be transferred to a storage hopper, using low-pressure steam for pneumatic transfer. The lock hopper will be repressured with steam again before the next switch.

GAS PURIFICATION

The raw gas, made in the gasifier, contains steam, some heavy hydrocarbons, hydrogen, hydrogen sulfide, and a relatively large quantity of carbon monoxide and carbon dioxide, in addition to the desired methane. It must, therefore, be passed through a series of cleanup and conversion operations, in order to produce a clean, high-Btu gas. The first step in the gas-treating train is removal of any fine solid particles and any tars or oils. A venturi scrubber is used for the first stage. This serves a dual purpose. It will trap any fines, and it will also quench the very hot gas leaving the gasifier. With the venturi virtually close-coupled to the gasifier, it was possible to eliminate the use of refractory-lined pipes and special valves and fittings. The gas is then scrubbed in two further stages: first with water and then with oil.

Water from the venturi scrubber and the water scrubber accumulates in a high-pressure vessel, equipped with baffles to provide for separation of oil and water. Although a generous time is provided for separation in this high-pressure vessel, a second separator at atmospheric pressure is also provided to permit the complete oil/water separation, with the addition, if necessary, of emulsion breakers. The oil/water separation at high pressure is provided because it is undesirable to reduce the pressure of the mixture before separation because of the possibility of emulsion formation in letdown valves.

SHIFT CONVERSION

After removal of tar and solids, the gas passes through a carbon monoxide shift in which the H_2/CO ratio is adjusted to 3:1 to be suitable for subsequent methanation. Shift conversion is accomplished by means of an iron-chrome-type shift catalyst in a packed-bed reactor.

ACID-GAS REMOVAL

The acid gases, carbon dioxide and hydrogen sulfide, must be removed, and this is done in a hot potassium carbonate (HPC) plant. The off gas from the HPC solution regenerator of this plant contains most of the sulfur which came in originally with the feed coal. In order to meet environmental requirements, this sulfur must be removed from the off gas. It cannot be vented directly to the atmosphere. A Stretford plant is provided to recover the sulfur.

TRACE SULFUR REMOVAL

The catalyst used in the methanation operation has a very low tolerance for sulfur compounds. A cleanup of the gases leaving the acid-gas-removal operation is required to remove traces of sulfur. A series of fixed-bed reactors containing activated carbon and sponge iron (in that order) are provided.

METHANATION

This is a difficult operation because the concentration of CO and hydrogen to be converted to methane is high. The reaction is highly exothermic, and therefore large amounts of heat must be removed. With high CO concentration, the possibility of carbon formation on the catalyst is high unless the reaction temperature is controlled within the rather narrow range of 700 to 790°F (371 to 421°C).

The Bureau of Mines has developed two novel methanation reactors. These are the TWR, or tube-wall reactor, and the HGR, or hot-gas recycle. Both reactors use Raney nickel as a catalyst coating, on tubes in one and on plates in the other. The advantage of this type of catalyst is its low cost, because it is very active, and relatively small quantities are required. On the other hand, because of the small quantities, it cannot tolerate poisons, and the gas must be extremely low in sulfur content.

The Bruceton prototype plant has two methanators, a TWR and an HGR, so that the two systems can be studied on a larger scale and their merits compared. The tube-wall reactor was developed for this process where the high concentration of CO (15 percent) results in a large quantity of heat to be removed. This is achieved by having the catalyst applied to heat-exchanger tubes so that the heat can be removed as it is generated and the reaction temperature remains essentially constant. The Bureau of Mines has developed a technique for spraying Raney nickel on the inside of tubes. This means that conventional heat-exchanger construction can be used. Spent catalyst can be replaced simply by removing the exchanger heads to obtain access to the coated surface.

The purified gas is preheated by heat exchange with high-pressure steam before entering the methanator. Heat is removed from the methanator by a thermosyphon Dowtherm circuit. The Dowtherm vapor is condensed by generating steam.

Water vapor produced by the reaction must be kept low, as otherwise it will suppress the forward reaction to methane. In order to accomplish this, some of the product gas is cooled, to condense water, and recycled to the methanator. The recycle ratio is in the range of 0.5 to 1.0. As a means of ensuring complete CO conversion, and to prolong the run lengths of the main reactors, the product gas is passed through a small vessel packed with conventional methanation catalyst before being cooled for delivery as product gas.

The alternative methanation scheme uses a vessel filled with plates which are coated with Raney nickel. No heat is removed in the reactor itself, and therefore the temperature control must be external. The temperature rise in the reactor must be limited, and this is accomplished by dilution of the feed gas to reduce the CO and hydrogen concentration. A large volume of the product gas is recycled to achieve this dilution. A portion of the recycle must be cooled to condense water and thereby remove the water vapor produced in the reaction, but most of the recycle is not cooled. The total recycle ratio is 16:1, of which 13:1 is hot recycle and 3:1 is cold recycle. In this way, the gas entering the reactor contains only 1 percent CO instead of 15 percent.

The major engineering problem in this system is the selection of a compressor to use for the hot gas which is at 765°F (407°C). For the prototype plant, an eductor is utilized for this purpose, employing the cold recycle as the motive fluid. This eliminates the use of machinery for handling the hot gas. On a larger plant it may be possible to use a centrifugal compressor at these temperatures. The fresh gas is at a sufficiently high pressure to mix with the cold recycle before the eductor. The temperature of the gas entering the reactor is controlled by preheating the cold recycle as necessary. Again, a final cleanup methanator using conventional catalyst is provided for the product gas.

Because of inherent size limitations in the fabrication of shell and tube equipment, it should be possible to build much larger capacity plate reactors than TWR, and a smaller number will be required on a commercial plant. Catalyst replacement will be easier. On the other hand, the hot-gas recycle is not as efficient thermodynamically as the tube-wall reactor.

In-Situ Coal Gasification

BY D. R. STEPHENS*

In-situ gasification of coal offers four potential advantages: (1) The product may be cheaper owing to lower capital investment, (2) environmental damage may be less, (3) the hazards to miners are avoided, and (4) it may be possible to utilize coal resources found at depths too great to be economically attractive for conventional deep-mining operations.

In-situ coal gasification was first suggested by Siemens in 1868, and the first patent was granted to Betts in 1909. The first field experimentation was conducted in England before World War I, and in the U.S.S.R. in 1933, where semicommercial plants were in operation. After World War II, nearly every western country experimented with in-situ coal

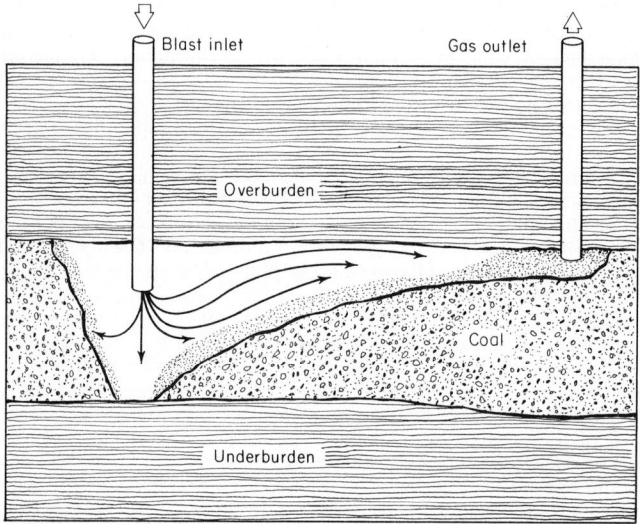

Fig. 1. Basic aspects of the percolation technique for in-situ coal gasification.

gasification. However, the leading nation in this work has been the U.S.S.R., which developed most of the technology available today.

A representative technique developed by the U.S.S.R. in the 1930s, known as the percolation method (Refs. 1, 2), is shown in Fig. 1. Communication is established between two boreholes by the natural permeability of the coal, by electrolinking, or by fluid or air fracturing the coal between the boreholes. In almost all cases, reverse combustion was used to increase the permeability between boreholes. The U.S.S.R. conducted a number of semicommercial operations between 1935 and 1965, mainly using hydraulic or air fracturing and the percolation technique, producing as much as 45 million standard cubic feet of gas. Generally air was injected, but sometimes oxygen or oxygen-enriched air was used. Steam also was injected in a few instances.

These installations made a combustible gas of the order of 100 Btu per standard cubic foot (about 10 percent of the heating value of natural gas). This low-Btu gas was primarily

*Project Leader, In-Situ Coal Gasification, Lawrence Livermore Laboratory, University of California, Livermore, California. (This work was performed under the auspices of the United States Atomic Energy Commission)

used for electric power generation. In the early experiment, the product-gas flow rate fluctuated, and the heating value generally decreased with time. The interpretation of the decreasing heating value is that the reaction zone invariably rose to the top of the bed. When the top of the bed was gasified, opening up a wide channel, short circuiting of the oxygen, and combustible gas products would occur, thus producing a low-heating-value product. In addition, approximately 30 percent of the gases were lost underground (Ref. 3).

More recent Russian work showed that, if the reverse combustion link was established near the *bottom* of the coal bed, override could be minimized and higher heating gases could be produced with higher resource efficiencies (Ref. 4).

Perhaps because of the much more encouraging recent work in the U.S.S.R., Texas Utilities Services, Inc., recently signed a contract with V.O. Licensingtorg (U.S.S.R.) to apply Russian underground coal gasification technology in deep lignite coal deposits in East Texas. The contract was arranged by the Resource Sciences Corporation, which will provide engineering services.

The development project is intended to verify the applicability and economic feasi-

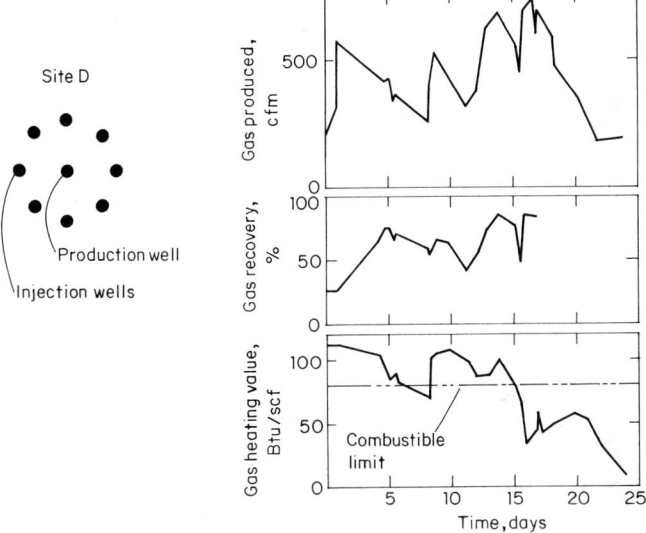

Fig. 2. Arrangement of injection and production wells and graphs of results of U.S. Bureau of Mines, Site D, Gorgas, Alabama, in-situ coal gasification experiments conducted in 1955.

bility of Soviet processes for producing low-Btu gas from the Texas deposits. The gas would be used for generating electricity. Drilling began in March 1975 as the first step in construction of a 5 mW pilot plant, which is scheduled for operation in 1976. If successful, commercial operations will follow.

Early American experiments, by the U.S. Bureau of Mines during the 1950s in Gorgas, Alabama, using U.S.S.R. techniques were not successful. In one experiment (Ref. 5), for example, several boreholes were hydrofractured in a 3-feet-thick coal deposit at 176-feet depth. The percolation technique (see Fig. 2) was used. Coal in the central borehole was ignited, and air injected in the peripheral boreholes surrounding the ignited well, which served as the production well. A gas of heating value of 100 Btu per standard cubic foot or greater was produced for the first 15 days of operation, and the recovery of product gas averaged 70 percent, just as in the U.S.S.R. experience. After 15 days, however, short circuiting of the air and products occurred. By the end of 24 days the gas consisted essentially of combustion products and nitrogen. Analysis showed that the primary reaction was carbonization of coal, *not* gasification. Less than 4 percent of the heating value of the coal within the circle of boreholes was recovered in the produced gas.

In 1969 experiments were conducted by the Gulf Research and Development Company in Kentucky and were reported by Raimondi, Terwilliger, and Wilson (Ref. 6). The percolation technique was used, and high-pressure air was injected to fracture the coal. A gas of heating value of 150 to 270 Btu per standard cubic foot was generated for 47 days. The gas was not produced, but was allowed to spread out into the formation. The experiment was terminated when noxious gases made their way to an adjacent strip-mining operation. After the zone was quenched, the overburden was stripped away from the site which showed the coke outlines. See Fig. 3.

The tendency to gasify the top of the coal bed during lateral gasification is shown quite dramatically. The coal-coke interfaces become closer and closer to the top of the bed with distance from the injection well, as can be seen from the side view in Fig. 3. The problem

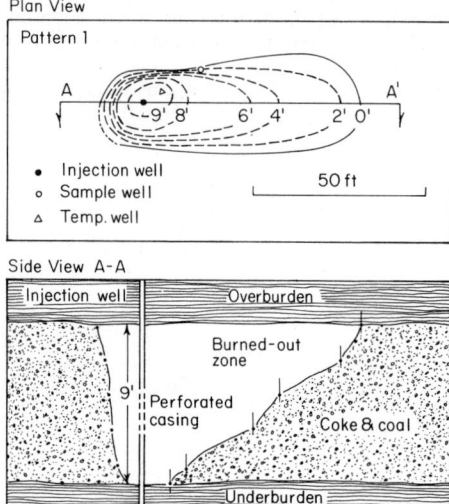

Fig. 3. Plan and section view of the Gulf Research and Development Company in-situ coal gasification experiments at Madisonville, Kentucky, in 1969. Air fracture and percolation techniques were used.

of a lateral reaction moving to the top of a zone is also a common occurrence for in-situ combustion experiments in tertiary recovery. Usually, the lower portion of the oil reservoir is left unaffected by the in-situ combustion sweep. It should also be noted that the pattern is oblong, not circular, thus complicating resource recovery. Most of the coal was reported to be carbonized, not gasified, and thus similar to the U.S. Bureau of Mine experiments at Gorgas, Alabama. Resource recovery cannot be evaluated for this experiment because the product gases were not recovered. However, it is significant that the operation produced a combustible gas for a total of 47 days and was voluntarily stopped.

In 1973 the U.S. Bureau of Mines initiated an underground gasification experiment near Hanna, Wyoming, in a 30-foot coal seam at 400 feet depth with results reported by Schrider and Pasini (Ref. 7). The field plan is shown in Fig. 4. The percolation technique was used with well number 3 being hydrofractured, causing the resultant major fracture pattern also shown. The coal was ignited at well number 3. From April to May air was injected in well number 3 and gas produced at wells numbers 5 through 8, more or less along the major hydrofracture direction. Owing to leaks in the well casing, the initial performance was poor, with 90 percent loss of products. At the end of May field improvements were made, and air injection was changed over to well number 5, with primary production at well number 3. In late September injection was changed over to wells numbers 9 and 15.

Combustion was sustained until March 22, 1974, when the experiment was shut down to allow for data interpretation and because objectives had been met.

From mid-September 1973 through February 1974 relatively stable gas production rates and heating values were maintained. Gas production averaged 2 million scf/day at an average heating value of 126 Btu/scf (Ref. 8).

Material balance calculations indicate approximately 20 tons/day of moisture-free coal were affected during this 165-day period. Energy recovery efficiency ranged from 30 to 50 percent during this period. There was no measurable gas leakage from the underground system. The Hanna number 1 seam is the major near-surface aquifer. It is theorized that the seam water acted as a gas seal so long as injection pressures were maintained below hydrostatic pressure. From 40 to 50 barrels of water entered the reaction zone each day with 15 to 20 barrels being consumed in the water gas shift reaction prior

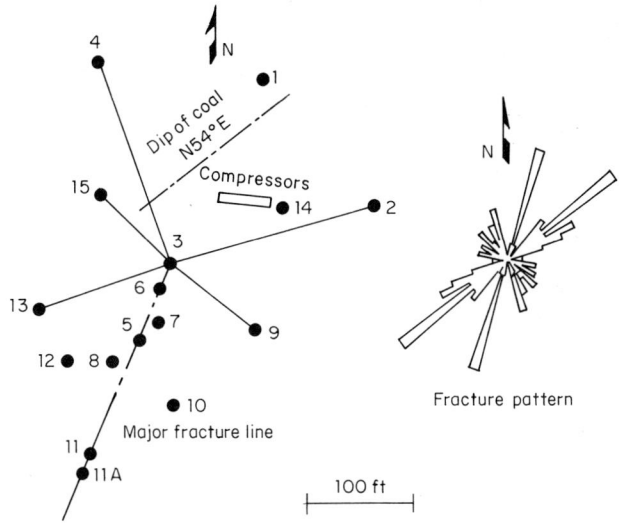

Fig. 4. Field plan of U.S. Bureau of Mines underground gasification experiment near Hanna, Wyoming. Fracture pattern is shown at right.

to product gases exiting the underground system and the rest appearing in the product gases as water vapor.

The coal utilization efficiency is yet to be determined. Coring and geophysical assessment of the well pattern are planned in the near future. This information should indicate the relative contributions of gasification and carbonization to the process, the size and shape of the affected area, the areal sweep efficiency, and the coal utilization efficiency.

The current experiment at Hanna can be described as follows. Refer to Fig. 5. During the initial phase, air was injected into well number 1 to determine the preferred flow direction (Fig. 5a). It was expected from laboratory studies that linkage would be established more quickly between wells numbers 1 and 2 (Fig. 5b). However, communication was first established by reverse combustion (backwards burning) between wells numbers 1 and 3 (Fig. 5c). The coal was ignited in well number 1 while maintaining air injection in the same hole. After a five-day ignition period the reverse combustion process was established by injecting air in well number 3 and producing the gas from well number 1. Linkage occurred ten days after the initiation of reverse combustion and was accompanied by dramatic changes in operating conditions. Injection pressure dropped from 250 psig to 141 psig and, within one day, to 30 psig. Air injection increased from 100 scfm to 500 scfm and then to 1200 scfm. Both the ignition and air injection wells were cased 20 feet into the seam in an attempt to create the link at or below the midpoint of the seam to minimize override. The thermocouple data from the linkage phase of the operation indicate that the link was effected near midstream.

Since linking was established, gasification via forwards burning proceeded very smoothly (Fig. 5d); gas of about 160 Btu/scf was produced. Another set of wells, 7 and 8, in a pattern parallel to 1-3 and 5-6 will be drilled (Fig. 5e), then after linkage is established between wells 5-6 and 7-8 the experimenters will try to propagate a line sweep by injecting air into 7-8 (Fig. 5f).

The "long wall generator" concept, shown in Fig. 6, is to be tested by the ERDA Morgantown Energy Research Center in West Virginia. The concept involves drilling parallel, relatively long directional holes in a horizontal plane in the coal, perpendicular to the direction of maximum natural permeability. The coal would then be ignited along the length of the borehole and gasification would proceed as indicated. Field work and laboratory work are well underway (Ref. 9).

Consideration of prior in-situ processing work led to the development of a new concept for in-situ coal gasification, designed to minimize the problems encountered in previous approaches. This concept was developed by Gary Higgins (Ref. 10) and one version is shown in Fig. 7.

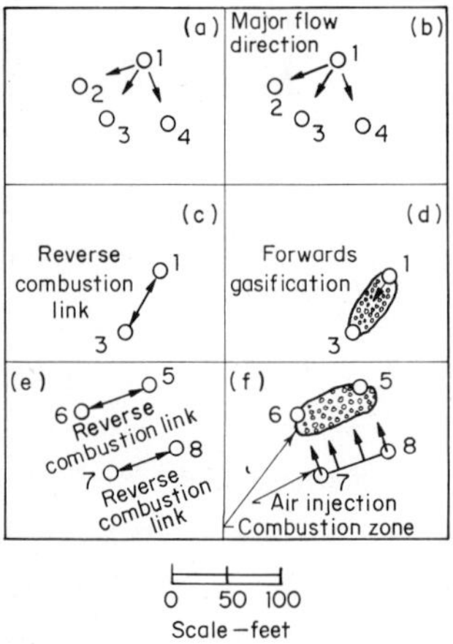

Fig. 5. LERC—sequential diagram of the underground gasification experiments at Hanna, Wyoming.

Our approach to in-situ coal gasification involves creating a selectively enhanced permeable reaction zone within a thick (10 m or more) coal bed using an array of chemical explosives. Our objective is to develop a commercial process for gasifying deep (150 to 1000 m), thick, western coals to produce pipe-line-quality synthetic natural gas. In 1974 we initiated a major program aimed at exploiting the new concept: laboratory experiments, computational studies, and an extensive field effort are in progress.

To create the permeable reaction zone, chemical explosives are emplaced and detonated in an array of drilled holes, fracturing the coal and providing selectively enhanced permeability. A permeable, fractured coal bed within a relatively impermeable medium should permit intimate mixing of the coal and reactants (oxygen and steam), allowing heat transfer and reactant access to the coal. The low-permeability surroundings should minimize leakage of reactants and products from the fractured zone. In essence, the LLL concept is an underground packed-bed reactor.

Access to the in-situ reactor will be through collection wells drilled to the bottom of the fractured coal and through reentered explosive holes. We plan to inject oxygen rather than air. After a suitable combustion zone has been established, the injected oxygen will be replaced by an appropriate mixture of oxygen and steam. Under suitable conditions, the gases produced underground should consist primarily of methane (the principal constituent of natural gas), carbon monoxide, carbon dioxide, and hydrogen. These underground products will be cleaned and upgraded in a surface facility to provide pipeline-quality gas.

The surface facility for pipeline gas production will include oxygen, water, and gas-treatment plants. Water would be obtained from a suitable aquifer, stored in a pond, and pumped underground as needed. An in-situ gasifier may not require the high-quality water necessary for surface gasifiers, and water requirements in general are expected to be less than for other fuel recovery processes.

Cost estimates (Refs. 11 and 12) suggest a selling price of pipeline gas produced in this way from one-half to two-thirds of the projected cost of gas produced by Lurgi coal gasification or by modern U.S. surface processes now under development.

Resources potentially available for this process are not accurately known. However, in the Powder River Basin in Wyoming and Montana over 100 billion tons of coal have been identified which are potentially suitable for the packed-bed process: the other thick deposits in the Rocky Mountain States would probably double this estimate.

Our initial work has been very encouraging. (Refs. 13 through 15). We conducted a successful deep exploratory drilling program delineating potential coal sites in southeast Montana and both northeast and southwest Wyoming. A deep site in the Powder River

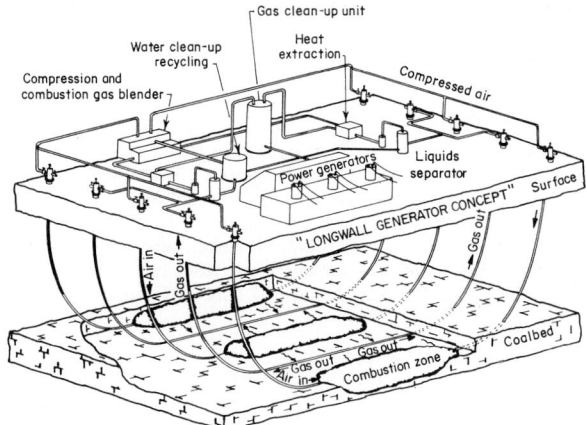

Fig. 6. MERC's longwall generator concept.

Basin of Wyoming was chosen for further characterization and a shallow site at Hoe Creek was also selected. Characterization of the shallow site proceeded during the fiscal year for future experiments in fiscal 1976.

We conducted a number of coal fracture experiments including a field coal outcrop experiment in Kemmerer, southwest Wyoming. Permeabilities and fracturing measured from this shot were in good agreement from preshot predictions with stress strain one- and two-dimensional codes and will serve as process design information for our future field fracture experiments.

We gasified coal in 1-foot-long packed-bed reactors with steam and oxygen and obtained gas qualities almost equivalent to the Lurgi coal gasification process. We measured the progress of thermal waves in such reactors and compared those with our modeling calculations. We found that Wyoming subbituminous coals shrink rather than swell upon heating and do not appear to soften. Thus these coals do not appear to be a problem with in-situ process operations due to either swelling or softening. Modeling studies were initiated considering convective instabilities, coal plasticity, and resource recovery

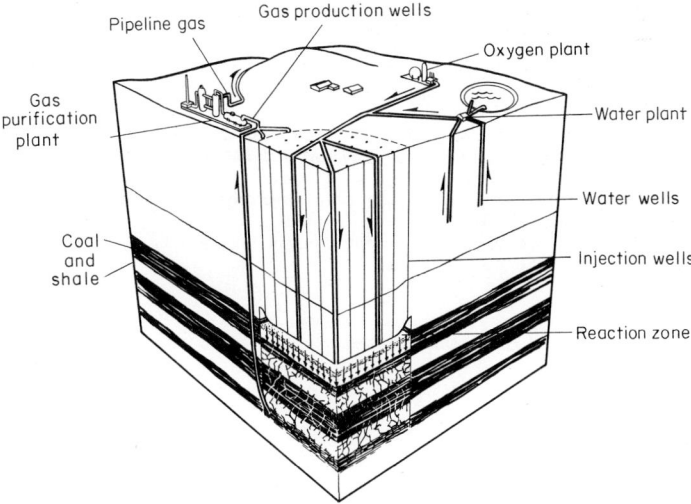

Fig. 7. Innovative in-situ coal gasification concept which is designed to minimize problems encountered in previous percolation approaches.

due to spatial distribution of collection and production pipes, and a one-dimensional reaction code including eight chemical reacting species was developed. These laboratory experimental and modeling efforts, combined with the modeling and experimental fracturing work conducted at Kemmerer, provided the design bases for our future gasification experiments.

Two feasibility studies were conducted during the fiscal year involving ERDA, LLL, and industry. The objectives of the studies were to evaluate the commercial potential of the LLL packed-bed concept, to examine coal resources available for the concept, and to make recommendations for the R&D programs to demonstrate the process. The feasibility study with Gulf Research and Development Company was completed with a positive review (Ref. 12). The second study, which is with Resource Sciences Company, is in the final stages. Four other firms are cooperating with RSC in the latter study. These are duPont, American Oil Company, Rocky Mountain Energy Company, and Pacific Gas and Electric Company.

REFERENCES

1. Elder, J. L.: Underground Gasification of Coal, in H. H. Lowry (Ed.), "Chemistry of Coal Utilization," pp. 1023–1041, Wiley, New York, 1963.
2. "A Current Appraisal of Underground Coal Gasification," *Rep.* C-73671, Arthur D. Little, Inc., Cambridge, Mass., December 1971.
3. Pitin, R. N.: Certain Problems Involved in the Reduction of Losses of Blast and Gas in Underground Gasification of Brown Coal from Moscow Area, pp. 109–118, in "Underground Processing of Fuels," translation TT-63-11063. Available from the National Science Foundation, Washington, D.C.
4. Skafa, P. V.: "Underground Gasification of Coal," Gosudarstvennoe Hauchno-Teknicheskoe Izdatel'stvo Literatury Po Garnomu Delu, Moscow (1960). Translated in *UCRL-Trans. 10880* (1975).
5. Capp, J. H., R. W. Lowe, and E. F. House: Underground Gasification of Coal: Operation of Multiple-Path System, *U.S. Bur. Mines Rep. Invest.* 5830, Pittsburgh, Pa.
6. Raimondi, P., P. L. Terwilliger, and L. A. Wilson, Jr.: "A Field Test of Underground Combustion and Gasification of Coal," Society of Petroleum Engineers, Dallas, Tex.
7. Schrider, L. A., and J. Pasini III: Preliminary Results Released for Wyoming In-Situ Gasification Test, *Coal Age*, vol. 78, no. 13, pp. 58–61, 1973.
8. Schrider, L. A., *et al.*: "The Outlook for Underground Coal Gasification," presented at Lignite Symposium, Grand Forks, N. Dakota, May 1975.
9. Overbey, W. K., C. A. Komar, and J. Pasini III: "Directional Properties of Coal: Proposed Concept for Directional Control of the Combustion Zone," presented at 1973 Annual Meeting of Geological Society of America, Dallas, Texas, November 1973.
10. Higgins, G. H.: A New Concept for In-Situ Coal Gasification, *UCRL Rept.* 51217, Rev. 1, Lawrence Livermore Laboratory, Livermore, Calif., 1972.
11. Stephens, D. R.: Economic Estimates of the Lawrence Livermore Concept of In-Situ Coal Gasification, *UCRL Rept* 51578, Lawrence Livermore Laboratory, Livermore, Calif., 1974.
12. Arscott, L., *et al.*: "In-Situ Coal Gasification in Perspective—A Joint LLL-GRDC Rept.," TID 26825 (1974).
13. Stephens, D. R., and A. Pasternak: "LLL In-Situ Coal Gasification Program, Quarterly Progress Rept.," July–September 1974. Lawrence Livermore Laboratory, Rept. URCL-50026-74 (Feb. 5, 1975).
14. Mead, W.: "LLL In-Situ Coal Gasification Program, Quarterly Progress Rept.," January–March 1975, URCL-50026-75-1 (May 26, 1975).
15. Stephens, D. R.: "LLL In-Situ Coal Gasification Program, Quarterly Progress Rept.," April–June 1975, UCRL-50026-75-2 (Sept. 22, 1975).

Carbonization/Hydrogenation Method for Producing Metallurgical Coke

BY K. A. SCHOWALTER* AND N. S. BOODMAN†

As a major producer of steel, the United States Steel Corporation currently uses approximately 25 million tons per year of coal to produce the coke for its blast furnaces. In addition, the company is one of the nation's leading holders of coal reserves. Accordingly, it has devoted considerable effort to research and development on processes for coke making and for more effective utilization of the coal. In looking at the problems that confront the steel industry, it is apparent that significant changes in present practices will be required to find solutions to such problems as atmospheric pollution, dwindling reserves of low-sulfur metallurgical coals, and decreasing availability of low-cost energy sources.

To find solutions to these problems, United States Steel's research laboratory has been

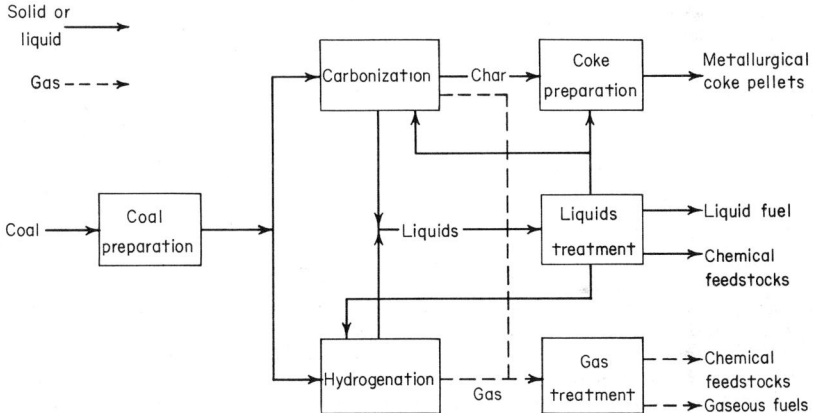

Fig. 1. Simplified flow diagram of the U.S. Steel Corporation's clean coke process.

evaluating existing technology and exploring newer concepts of coal processing. The result of these studies was a conceptualized process designed to (1) permit the manufacture of blast-furnace coke with a minimum of atmospheric pollution, (2) permit the use of high-sulfur high-ash coals that are normally not suitable for coke making, and (3) provide low-sulfur liquid and gaseous fuels as by-products.

The coal-conversion process‡ is shown in Fig. 1. The coal, after beneficiation and sizing in a coal preparation plant, is split into two fractions. Part of the coal is processed through a carbonization unit where it is devolatilized and partially desulfurized to produce the char that serves as the base material for production of the metallurgical coke. The second portion of the coal is slurried with a process-derived carrier oil and is hydrogenated to convert most of the coal to liquids.

Liquid products from both carbonization and hydrogenation are composited and pro-

*Chief Staff Engineer, Chemicals and Plastics, Research Laboratory, United States Steel Corporation, Monroeville, Pennsylvania.

†Section supervisor and manager of development work on Clean-Coke process, Research Laboratory, United States Steel Corporation, Monroeville, Pennsylvania.

‡Development work on the process is being conducted under a joint-funding contract between the Energy Research and Development Administration (ERDA) and the United States Steel Corporation.

cessed through a central liquids-treatment unit. In this unit, the liquids are processed into low-sulfur liquid fuels, chemical feedstocks, and three oil fractions that are recycled to other areas of the process.

One of the recycle fractions is used primarily as a carrier oil for the hydrogenation reaction. A second recycle oil is sent to the carbonizer where it is converted to pitch coke. The pitch coke and char mixture is blended with the third recycle oil that serves as a binder, and this mixture is formed into pellets in the coke preparation unit. These pellets

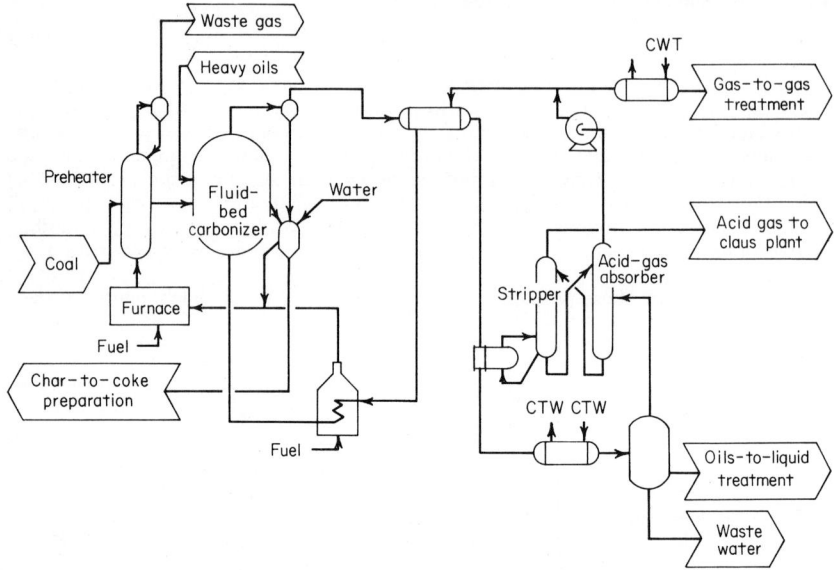

Fig. 2. Coal carbonization unit.

are subsequently baked to produce a formed metallurgical coke with strength properties equivalent to those of blast-furnace coke made by conventional coking.

The coke-preparation cycle, from char production to final coke, is carried out in a closed system with the off-vapors collected and returned to the process. Thus, there are no significant emissions of volatile matter during these operations, and atmospheric pollution is practically nonexistent. Gaseous products from all operations are processed through a common system to provide chemical feedstocks, low-sulfur gaseous fuels, and hydrogen for recycle to hydrogenation and liquids treatment.

COAL PREPARATION AND CARBONIZATION

In the coal preparation unit, the raw Illinois coal, which contains approximately 3 percent sulfur and 15 percent ash, is first processed into three fractions: (1) a *carbonization fraction* (42 percent of the coal) containing 2 percent sulfur and ~6 percent ash, (2) a *hydrogenation fraction* (42 percent) containing 3 percent sulfur and 14 percent ash, and (3) a *refuse or fuel fraction* (16 percent) containing 5 to 6 percent sulfur and approximately 43 percent ash.

The carbonizer feed coal, after pulverization and sizing to minus ⅛ inch by plus 100 mesh, is fed to the carbonization section, the details of which are shown in Fig. 2. The coal is first fed to a fluid-bed preheater where it is dried, preheated and mildly oxidized to about 400°F (204°C), utilizing stack gases from the main carbonizer heater which are boosted in temperature by passing them through an additional furnace. The off-gases from the preheater, after removal of particulate matter in a cyclone, go to a stack.

Preheated coal, along with heavy oils from the liquids-treatment unit, then enters the fluid-bed carbonizer, where it is heated by the fluidizing gas and carbonized at temperatures from 1200 to 1400°F (649 to 760°C) at about 120 to 150 pounds per square inch pressure. Because of the hydrogen content (about 33 percent) and the low sulfur content of the fluidizing gas, the coal is simultaneously desulfurized. After separation of particulates, the carbonizer off-gas is cooled in three steps, and desulfurized to provide the gas for recycle to the carbonizer and the surplus gas which goes to the gas-treatment plant. The condensed tars and moisture are separated from the system, and the tars are sent to the liquids-treatment plant.

COAL HYDROGENATION

Details of the coal-hydrogenation portion of the plant are given in Fig. 3. The crushed coal from the coal-preparation unit is combined with slurry oil from the liquids-treatment unit to provide the slurry that is fed to hydrogenation. This slurry is pumped, together with a stream of recycled slurry, into the hydrogenator with hydrogen at a pressure of between 3000 and 4000 pounds per square inch. Additional slurry oil, introduced through a second slurry oil pump, along with the slurry recycle, aids in controlling the exothermic reaction at a temperature of about 850 to 900°F (454 to 482°C).

The hydrogenator product is transferred to the vapor-stripping vessel where the hydrogenate is contacted countercurrently with a stream of hot process gases. The hot gases serve to vaporize volatile matter and thus separate the liquids from the unconverted coal and ash. Unconverted coal and ash then pass through a lock system into a quench vessel,

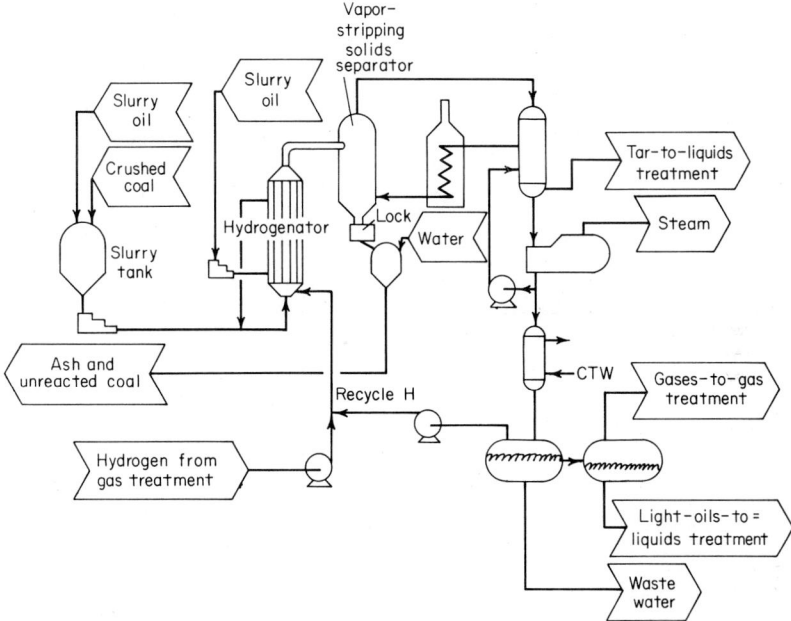

Fig. 3. Coal hydrogenation unit.

where the material is quenched with water and withdrawn from the system. The overhead gases and vapors from the vapor stripper are cooled, first by exchange with recycle gases, and second in a waste-heat boiler to produce process steam.

From this point, a portion of the gas is recycled to the vapor stripper, and the surplus gas is further cooled to cause liquids and moisture to condense. A portion of the gas is withdrawn at this point to provide recycle hydrogen which, together with makeup hydrogen from the gas-treating unit, serves as the hydrogen supply to the hydrogenator. The remaining gases go to gas treatment for processing. A light-oil fraction is separated and goes to liquids treatment for recovery of aromatics.

The flow sheet of the liquids-treatment unit is shown in Fig. 4. Heavy liquids from coal

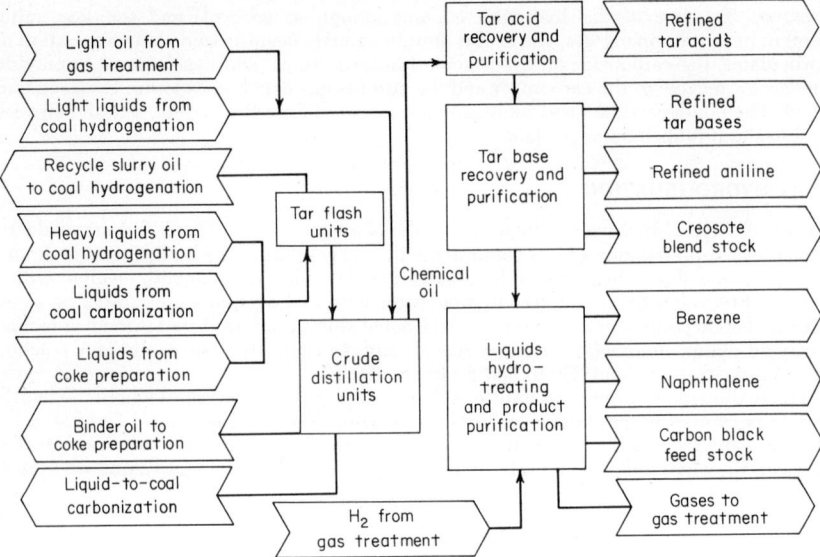

Fig. 4. Liquids treatment unit.

hydrogenation and liquids from coal carbonization and from coke preparation comprise the feed to the tar-flash units. In these units, the lower-boiling components are flashed overhead, and the remaining liquid is utilized as recycle oil in the coal-hydrogenation unit. The low boilers from the tar-flash units are combined in the crude distillation units with light oils from the gas-treatment unit and the coal-hydrogenation unit. This combined feedstock is distilled to supply chemical oil, the heavy oil for coal carbonization, and the binder oil for coke preparation.

The chemical oil is further processed for tar-acid recovery to provide phenol, cresols, and xylenols; and, for tar-base recovery, to provide the pyridine bases and aniline. The

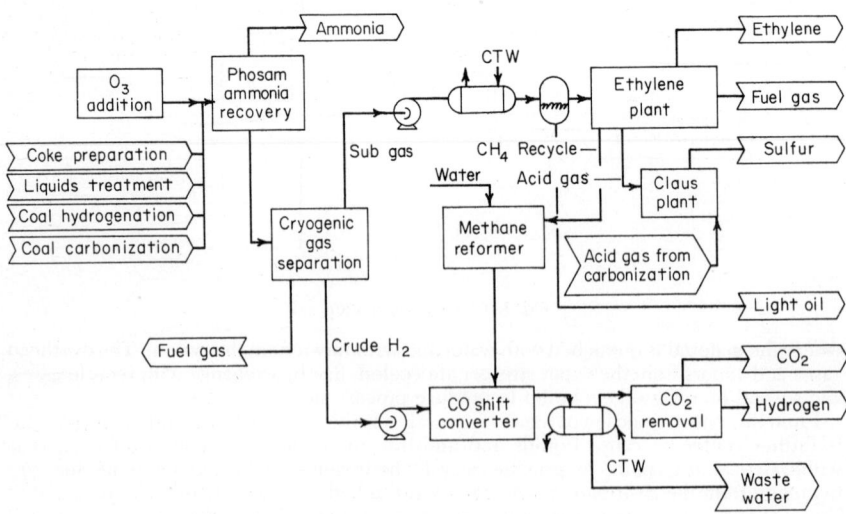

Fig. 5. Gas treatment facility.

remaining neutral oil would be suitable for blending into creosote for wood preservation or for use as a low-sulfur liquid fuel. The remaining chemical oil is then sent to a hydrotreating facility where the material is dealkylated to provide benzene and naphthalene, a heavy fraction which would be suitable for use as a carbon-black feedstock or as a low-sulfur liquid-fuel stream, and a gas stream which is recycled to the gas-treating unit. The hydrotreating plant uses hydrogen from the gas-treating facility to accomplish the hydrodealkylation. Details of the gas-treatment facility are given in Fig. 5.

COKE PREPARATION

The equipment involved in coke preparation is shown in Fig. 6. Char from coal carbonization is combined with solids recovered from the cyclones of the curing oven and the

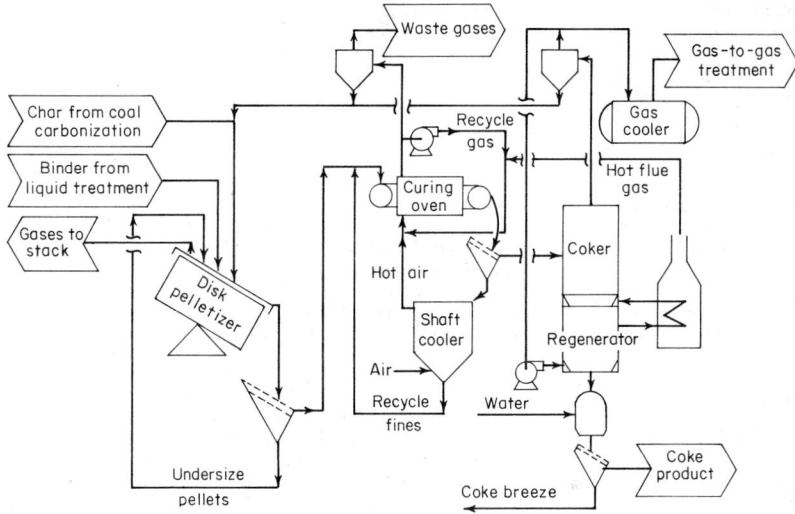

Fig. 6. Coke preparation facilities.

coker, and the mixture is fed to a disk pelletizer, along with binder from the liquids-treatment facility.

The mixture is formed into pellets, and as the pellets reach the proper size, they overflow from the pelletizer onto a screen, which is provided to separate any undersize pellets for recycle to the disk. Pellets of the proper size are then combined with recycle fines and distributed on the belt conveyor of a continuous curing oven. The purpose of the recycle fines is to support and separate the pellets so as to prevent slumping or agglomeration during curing.

In the curing oven, the pellets are heated to about 500°F (260°C) in the presence of air to promote oxidation and hardening of the binder and to provide "green strength" for the pellets. Upon discharge from the curing oven, the pellets are separated from the recycle fines and transferred to the coker. In the coker, they form a moving bed of pellets and are gradually heated from 500°F to the calcining temperature of 1800°F (982°C). The bottom section of the coker serves as a regenerator to recover heat from the pellets and, at the same time, to heat the recycle gases which serve to coke the pellets. Off-gases from the coker, after separation of particulates in a cyclone, are recycled to the regenerator section. The surplus gas is withdrawn through a gas cooler and then sent to the gas-treatment unit. The coke pellets from the regenerator are given a final quench with water, and then screened to separate any undersized product known as coke breeze. The product is equivalent in properties and chemical composition to coke made by the oven process.

Additional coke processes include the Atlantic Seacoke process which involved combining coal and petroleum residuum in a coking unit to extend the production of petroleum coke and to increase the yield of liquids, which could subsequently be upgraded to synthetic crudes. Literature on this process* includes:

1. Project Seacoke, vol. I with appendices A, B, and C, *Office Coal Res. R&D Rep.* 29 (refer to titled report and GPO Catalog no. 163.10:29/vol. 1), January 1970.
2. Project Seacoke, vol. II with appendix D, *Office Coal Res. R&D Rep.* 29 (refer to titled report and GPO Catalog no. 163.10:29/vol. 2), January 1970.

    Both available from the Superintendent of Documents, U.S. Government Printing Office, Washington, D.C. ($1.25 and $4.00, respectively).
3. Catalytic Hydrotreating of Coal-derived Liquids, Project Seacoke, Phase II, June 1967, *Office Coal Res. R&D Rep.* 29, *Interim Rep.* 1 (refer to titled report and PB-174926), available from the National Technical Information Service, U.S. Department of Commerce, Storage and dissemination Section, Springfield, Va. (6.00) ($6.00).

    The Consolidation Coal/Itoh formcoke process comprises charring of coking-grade coal, admixing the char with additional coking coal, nodulizing, curing, and finally calcining the mixture. The plasticized, sticky coal is considered here to be the principal source of binding force, although tar recovered from the charring process can be added as a supplemental binder if needed. A somewhat similar process has been developed by Bergbau-Forschung of Essen, Germany—which utilizes a briquetting roll press to agglomerate the formcoke. The briquettes thus formed are firm and strong and may or may not be calcined.

    Still another process is described in:
4. Szikla-Rozinek Coal Gasifier, *Office Coal Res. R&D Rep.* 47 (1), *Interim Rep.* 1, available from the Office of Coal Research, Publications Office, Department of the Interior, Washington, D.C. ($3.00).

---

*Editor's Note: This is a condensed description of article in *Chem. Eng. Progr.*, vol. 70, no. 6, pp. 76–82. This process is described in even more detail in an article on "Aromatic Chemicals from Coal" by K. A. Schowalter and E. F. Petras, in the *Chemical Engineering Progress, Technical Manual—Coal Processing Technology*, vol. 2, 109–113 (1975).

As pointed out in the aforementioned reference, economic and material balance information gained to date has been based upon batch-type bench-scale studies and thus is considered as preliminary. Revised material balance and product mix will be reported when data become available from the 500-pound/day process-development units being brought on stream as of 1976.

The yield of products and the economic information are based on a commercial-sized unit processing about 17,000 tons/day of washed and size Illinois No. 6 coal. The material-balance information indicates that the commercial unit on an annual basis would utilize 5.8 million tons of washed coal to produce 2.22 million tons of coke pellets, 2.42 billion pounds/year of chemicals, and 0.67 million ton of hydrogenation residue (fuel). The estimated total investment, including utilities, off-sites, engineering, and contingencies, was $740 million (1975 dollars).

Although this would seem to be a large investment, it should be emphasized that the plant is actually a complex of many plants. The liquids-treating facility, for example, is a combination refinery and tar plant, sized to process about 40,000 barrels/day. The cokemaking unit is equivalent in production to three or four modern coke batteries with tall ovens. The gas-treating plant includes units for removal of ammonia, H_2S, and CO_2, a Claus plant, a cryogenic separation plant, and an ethylene-propylene unit. Because of methods by which investment was developed, the estimate is considered to be conservative and on the high side.

The chemical products list includes substantial quantities of ethylene, propylene, phenols, cresols, benzene, and naphthalene, with a total annual value of over $380 million.

The overall economic attractiveness of the process depends on the cost of the coal and the values placed on the coke and chemicals. Using a coal cost of $25/ton, 1974 published prices for the chemicals, and a value of $120/ton for the coke (some imported coke in 1974 sold as high as $146/ton F.O.B. New York), the project would return the capital investment in less than 4 years. With an $80 value for a ton of coke, the investment would be returned in less than 5 years.

Char-Oil-Energy Development (Project COED)

BY JOHN F. JONES*

The objective of project COED is to expand the use of coal for clean energy by converting it to a synthetic crude oil and other clean fuels. The COED pilot plant shown in Fig. 1 is the result of a 12-year program between the FMC Corporation and the U.S. Office of Coal Research (OCR) of the Department of the Interior. The process is a major step forward toward the utilization of coal and is one of several cooperative programs pursued by OCR.

The COED project proceeded through several years of bench-scale research studies to the installation of a 100-pound per hour process development unit.† In this small unit, a careful evaluation was made of 12 different types of coal.‡ The results were so encouraging that piloting of the process was justified. This led to the design and construction of the

Fig. 1. COED pilot plant located at Princeton, New Jersey.

pilot plant with a capacity of 36 tons of coal per day. The pilot plant has operated successfully at design capacity and several scheduled 30-day runs have been made routinely. Approximately 20,000 tons of bituminous, subbituminous, and lignite coals have been processed.

THE COED PROCESS

The basic COED process upgrades coal to three useful products by heating in multistage, fluidized beds, as shown in Fig. 2. In each bed, pulverized coal behaves as a fluid because of gases passed upward through the bed.

In these stages, coal is heated rapidly to successively higher temperatures until essentially all the volatile matter (oil and gas content) of the coal is evolved. The optimum temperature and the number of stages vary with the coal. Stagewise heating is used to prevent agglomeration of the coal and plugging of the reactor system. The stage tempera-

*Technical Director of Coal Research, FMC Corporation, Princeton, New Jersey.
†Installed at the FMC Chemical Research and Development Center, Princeton, New Jersey.
‡As a result of this work, FMC has been granted U.S. Patents 3,375,175 and 3,453,202.

tures are selected just short of the maximum temperature to which the coal can be heated without agglomerating. Typical temperatures are: 600 to 650°F (316 to 343°C), first stage; 800 to 850°F (427 to 454°C), second stage; 1000°F (538°C), third stage; and 1600°F (871°C), fourth stage.

Heat for the pyrolysis is primarily provided by burning a portion of the char with oxygen in the last stage. Hot gases from the last stage then flow countercurrently to serve as the fluidizing gas and the heat supply for the third and second stages. Hot char can be recycled from one stage to another to supplement the heat from the gases. The first-stage fluidizing medium is supplied by burning a portion of the product char or gas.

The volatile products released from the coal in the fluidized-bed reactors pass to a

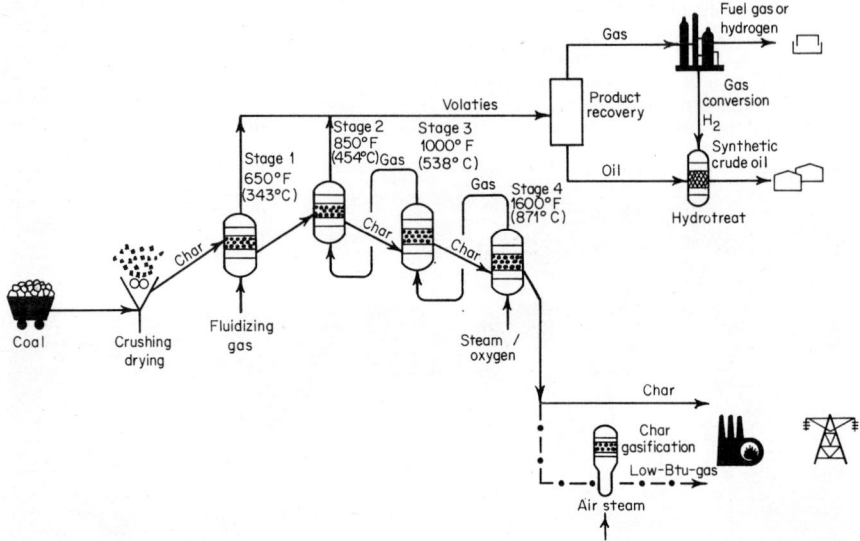

Fig. 2. Schematic flow diagram of the COED process. *(Reproduced by permission of source: FMC Corp., and Office of Coal Research.)*

product recovery system for condensing the oil and cooling the gases. In this system, these hot gases are directly quenched with liquor (water generated in the process) in a venturi-separator system. Water is decanted from the oil before the oil passes to a dehydrator for final drying. A rotary pressure precoat filter for removing the fine dust particles from the dried oil is also part of the recovery system. After filtering, the solids-free oil is pumped up to pressure and mixed with hydrogen for hydrotreating in a fixed-bed catalytic reactor operating at 750°F (~400°C) and 2500 psig. Hydrotreating removes sulfur, nitrogen, and oxygen from the coal oil and produces a 25°API synthetic crude oil, one of the final products from the COED process.

The gas is cooled and treated to remove carbon dioxide and hydrogen sulfide. It then can be sold as fuel gas or converted by application of additional technology to pipeline gas or hydrogen. The residual char can be used as power-plant fuel or gasified with the application of further technology to produce a fuel gas. The selection of a final gas product and end use of the char depend on the coal used in the process and the market for the products at the plant site.

MAJOR RESULTS

Synthetic crude-oil yields from the COED process vary from 1 to 1.5 barrels per ton of bituminous coal, depending on the feed coal type. The off-gas yield is 8000 to 10,000 standard cubic feet per ton of coal, with a heating value of 500 Btu per standard cubic foot. The char yield is about 0.5 ton per ton of coal. Conversion of this char with further technology to a clean fuel yields about 56,000 standard cubic feet per ton of coal of 220 Btu per cubic foot of gas for use as fuel to a power plant.

TABLE 1.   Some Average Operating Conditions of Pyrolysis Pilot Plant

Vessel or stage	1	2	3	4
Temperature, °F:				
Vessel—process	550	810	1050	1500
Fluidizing gas	760	1030	1160	830
Feed coal (char)	151	530	750	970
Recycle char	—	1100	1190	—
Vessel pressure, psig	7.8	6.1	8.4	9.5
Flows:				
Coal, tons/hr	1–1.5			
Fluidizing steam, lb/hr	—	—	—	500
Gas inputs, scfh (@ 60°F, 14.7 psia):				
Oxygen	—	—	—	3350
Recycle gas	90,500	8000	11,000	3500
Miscellaneous				
Bed density, pcf	43.5	15.7	12.9	17.1
Bed level, ft	1.5	7.6	9.0	9.4
Fluidizing velocity, ft/sec	1.4	1.0	1.1	1.1, 4.5

Significant advances achieved in operation of the COED pilot plant are:

1. Coal Pyrolysis.   Approximately 20,000 tons of coal were processed in the pilot plant. High-volatile bituminous coals from Colorado, Utah, Illinois, and Kentucky; a subbituminous coal from Wyoming; and a lignite from North Dakota were tested. Reliable operation of the four fluidized beds with the transfer of solids and gases between them was demonstrated over a variety of processing conditions.

2. Filtration.   Operating conditions were found for filtering the solids from the heavy, viscous coal oil of pyrolysis from the various coals in a rotary-pressure precoat filter. The coal oil containing solids concentrations of 5 to 10 percent (wt.) was filtered to produce a product with a solids content of less than 0.1 percent (wt.) at filtration rates up to 10 gallons per hour per square foot of filter surface. Major process variables that must be controlled to achieve best operating conditions for filtering the oils from different coals were defined. This processing step is ready (mid-1974) for scale-up to production equipment. See Fig. 3.

3. Hydrotreating.   Synthetic crude oil was produced from the filtered oil derived from the various coals processed in the pyrolysis section of the plant. Fixed-bed hydrotreating with commercially available catalyst was used as the processing step. A 30-barrel-per-day pilot plant was operated successfully and can be scaled up to a commercial plant similar to hydrotreaters used in the petroleum industry. See Fig. 4.

TECHNOLOGY

COAL PYROLYSIS

The unique character of the COED process lies in its ability to continuously process agglomerating coals without previously preoxidizing the coal. Preoxidation of the coal reduces the yield of useful products, mainly oil, upon subsequent pyrolysis. The continuous processing is accomplished by using multiple reactors which consist of fluidized beds in series.

Some typical operating conditions for the pyrolysis of a high-volatile bituminous coal are given in Table 1. Further, the coal conversion section of the COED process is a low-pressure process with vessel pressures varying between 6 to 10 psig. For this reason, the mechanical equipment for the process development and proven in-service in the pilot plant can be scaled up to commercially sized plants.

OIL FILTRATION

The vapors driven off the coal during pyrolysis are quenched directly with water to condense the oil. Occluded with this oil during condensation are tiny char particles. These char particles escape collection by the internal and external cyclones associated

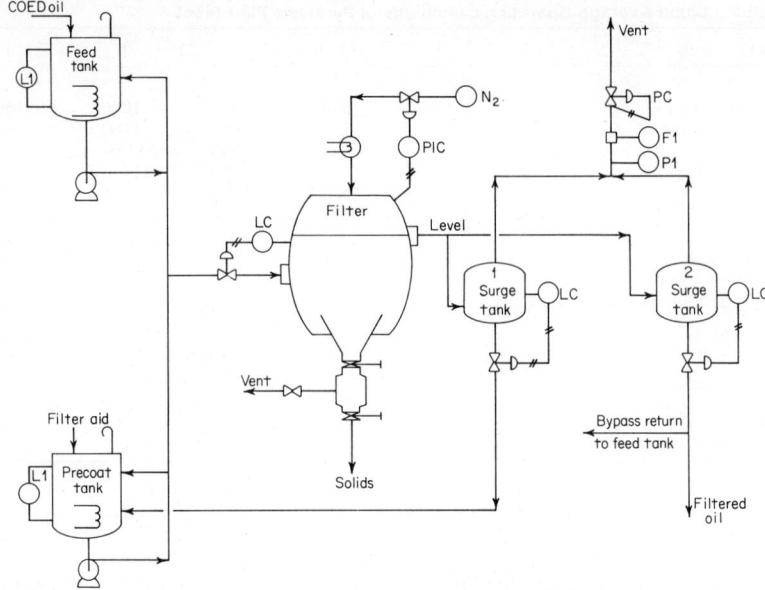

Fig. 3. Oil-filter system of COED process.

with the fluidized-bed reactors. The size of the occluded char particles ranges from 0.2 to 2.0 micrometers. These char particles must be removed from the raw pyrolysis oil before passage through a hydrogenation processing step used to upgrade the raw oil to synthetic crude oil.

A rotary-pressure precoat filter is employed to remove the char particles. Filtration operating conditions are 350°F (177°C) and 40 psig. The 350°F temperature is necessary to

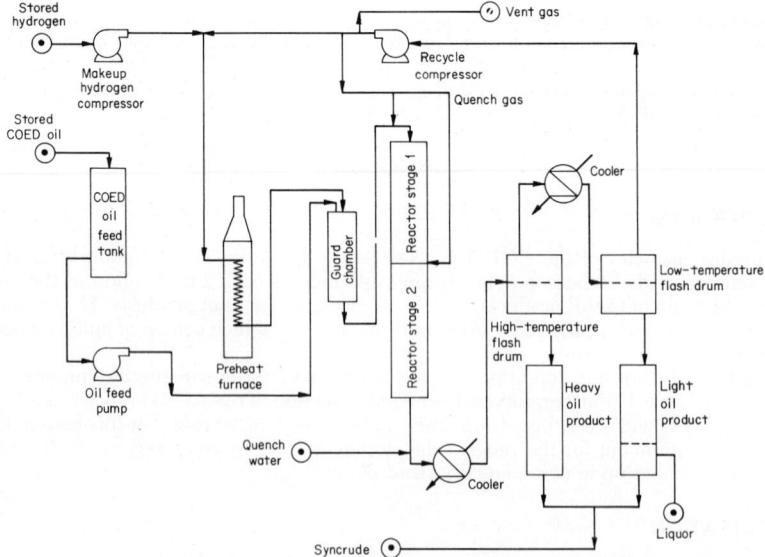

Fig. 4. Hydrotreating section of COED process.

TABLE 2.　Some Average Operating Conditions for Hydrotreating Pilot Plant

Catalyst	American Cyanamid AERO HDS–3A			
Average operating conditions	Oil–H_2 preheater	Hydrotreating reactor Section		
Temperature, °F:				
Vessel inlet	100	725	735	750
Vessel middle	195	745	765	780
Vessel outlet	585	745	750	796
Vessel inlet pressure, psig	2300	1775	1725	1725
Differential pressure, psig	525	48.5	1.0	0.3
Feed oil, lb/hr	200			
Hydrogen consumption, scf/bbl		3000		
Overall space velocity, lb oil/lb cat./ hr		0.50		

keep the oil liquid and pumpable. The filtration conditions suitable for scaling up to a commercial-size operation have been demonstrated within the existing mechanical technology.

HYDROTREATING

The hydrotreating operation removes the unwanted elements of nitrogen, sulfur, and oxygen from the coal oil. Also, the gravity, pour point, and viscosity are all upgraded to properties comparable to a petroleum crude oil. Some average operating conditions for the hydrotreating plant are given in Table 2. These conditions result in a product oil with the properties shown in Table 3.

PRODUCT USAGE

The products from a COED plant are the synthetic crude oil, a medium-Btu gas, and char. Additional technology can be applied to each one of these streams to arrive at the ultimate end use.

Synthetic Crude Oil.　The synthetic crude oil produced in the COED pilot plant can be used as a feed to a normal petroleum refinery. Experimental work* to define those refinery processes where the oil from coal could best be used to produce a typical refinery product mix has been performed. Alternatively, the oil produced from coal can be distilled to a naphtha fraction distilling up to 390°F (~200°C) to end point. Combustion tests† have been performed on this latter fraction and showed that the oil met all the specifications of a No. 4 oil and could be burned cleanly in industrial boilers. Further, the U.S. Navy used the synthetic crude oil from coal as a marine fuel in a destroyer.

Gas Product.　The gas as produced has an approximate heating value of 500 Btu per

TABLE 3.　Full-Range COED Synthetic Crude from an Illinois No. 6 Seam Coal

Test	Result
Flash point, cc, °F	46
Pour point, °F	<-36
Water and sediment, vol. %	Trace
Ash, wt. %	<0.005
ASTM Distillation, °F:	
1% bp (Initial boiling point)	190
10%	273
50%	518
90%	684
95%	720
Viscosity, centistokes at 100°F	3.4
API gravity	27.2

*By Atlantic-Richfield Company.
†By the American Oil Company.

standard cubic foot after the removal of carbon dioxide and hydrogen sulfide. Part of this gas would be converted into hydrogen for use in hydrotreating the oil. The remainder of the gas could be converted to hydrogen, sold as fuel gas, or converted to high-Btu gas. The decision would be based on process economics and a market for the gas products at the plant site.

Char.   In the original COED concept, char was to be used as a fuel for a power plant. Its use for this purpose was demonstrated by burning char in a large power-plant boiler. This was a successful test. However, char contains the same percent sulfur as the coal from which it is derived. Thus, the high-sulfur chars have the same limitations as high-sulfur coals as fuel for power plants. The most promising method of utilizing high-sulfur char is to convert it to a fuel gas and scrub the fuel gas free of hydrogen sulfide before burning it in a power plant. The conversion of char to fuel gas can be done with added technology. Work is planned to move this additional technology ahead.

COMMERCIAL OUTLOOK

Complete operating and yield data, as well as performance of equipment obtained while piloting several types of coal, confirmed the design scale-up factors used from the smaller experimental units. The COED process can reopen a potential for utilizing currently restricted high-sulfur coals for the production of clean energy products. The economies of the process are dependent on the price of coal and the value that can be obtained from clean fuels.

REFERENCES

Schoemann, F. H., L. Ford, and J. F. Jones: "Pressurized, Rotary-Drum Precoat Filtration of COED Oil from Coal," American Institute of Chemical Engineers Symposium, September 1973.

Jones, J. F.: "Project COED (Char-Oil-Energy Development)," Institute of Gas Technology, "Clean Fuels from Coal Symposium," Chicago, Sept. 12, 1973.

Greene, M. I., L. J. Scotti, and J. F. Jones: "Low Sulfur Synthetic Crude Oil from Coal," American Chemical Society, Division of Fuel Chemistry Meeting, Los Angeles, Calif., April 1974.

Terzian, H. D., J. A. Hamshar, N. J. Brunsvold, and J. F. Jones: "Processing Coal to Produce Synthetic Crude Oil and a Clean Fuel Gas," American Institute of Chemical Engineers, Southern California Section Meeting, Los Angeles, Calif., Ap. 16, 1974.

COGAS Process for Conversion of Coal to Gas and Oil Products*

BY RALPH BLOOM, JR.† AND HOWARD L. MALAKOFF‡

The COGAS Process (Ref. 1) for the conversion of coal to gas and oil products takes advantage of the high efficiency of multistage coal pyrolysis and the steam reactivity of the char from this pyrolysis. As a result, with raw oil and gas produced in a low-pressure system, the process can be economically competitive with more complex, high-pressure coal-gasification processes.

MAJOR ELEMENTS OF COGAS PROCESS

Two versions of the process are shown in block flow diagrams: (1) for production of pipeline quality or substitute natural gas (SNG) and oil, given in Fig. 1; and (2) for production of moderate Btu gas and oil, given in Fig. 2. The oil product may be a medium-heavy fuel oil, or a synthetic crude oil, depending on the conditions of hydrogenation of the raw oil from the coal pyrolysis.

PYROLYSIS

A variation of the low-pressure, multistage fluidized-bed pyrolysis of coal, described in the article on Char-Oil-Energy Development (Project COED) in this Handbook section, is employed in the COGAS process. The heat production, high-temperature stage of COED, on which a portion of the char is burned with oxygen, can be eliminated and heat for pyrolysis supplied by recycling hot char and synthesis gas from the gasifier. The oil produced from pyrolysis would be similar to that described for Project COED. The hydrogen for hydrotreating is supplied by reforming a portion of the gas product as shown by the flow diagrams. Typical analyses of the pyrolysis gas are given in Table 1. This gas is stripped of light hydrocarbons and then processed along with the synthesis gas from the char gasification. The light hydrocarbons can be a marketable product, or blended back to increase the quality of the product gas. The hot char from the 1000°F (538°C) pyrolysis stage is fed to the gasifier.

GASIFICATION-COMBUSTION

The hot burden process is designed to eliminate the need for bulk oxygen and to provide for air combustion of a portion of the char without introduction of the nitrogen from the air into the synthesis gas. The gasification-combustion step is illustrated in Fig. 3. A U.S.

*FMC Corporation developed a concept for char gasification using air combustion of a portion of the char, which is a variation of the ICI hot burden process (Ref.2). To develop the process, a joint venture company, the COGAS Development Company, was formed in 1972 with the following partners: Consolidated Natural Gas Service Company, FMC Corporation, Panhandle Eastern Pipe Line Company, Republic Steel Corporation, Rocky Mountain Energy Company, Tennessee Gas Transmission Company.

Process development includes pilot studies of two approaches to gasification—combustion, process engineering, and preliminary design of a commercial plant; and economic analyses. As of this writing, two pilot plants have been built and are in the early stages of operation. One at the FMC Chemical Research Center, Princeton, New Jersey, is designed for 2.5 tons per day char feed; the other at British Coal Utilization Research Associates laboratory in Leatherhead, England, is designed for 50 tons per day char feed. Presently, both are operated on char from Illinois No. 6 coal, but since the process can employ a wide range of coal feed, operation on other chars will be evaluated.

†Manager, Economics & Planning, COGAS Development Company, Princeton, New Jersey.
‡General Manager, COGAS Development Company, Princeton, New Jersey.

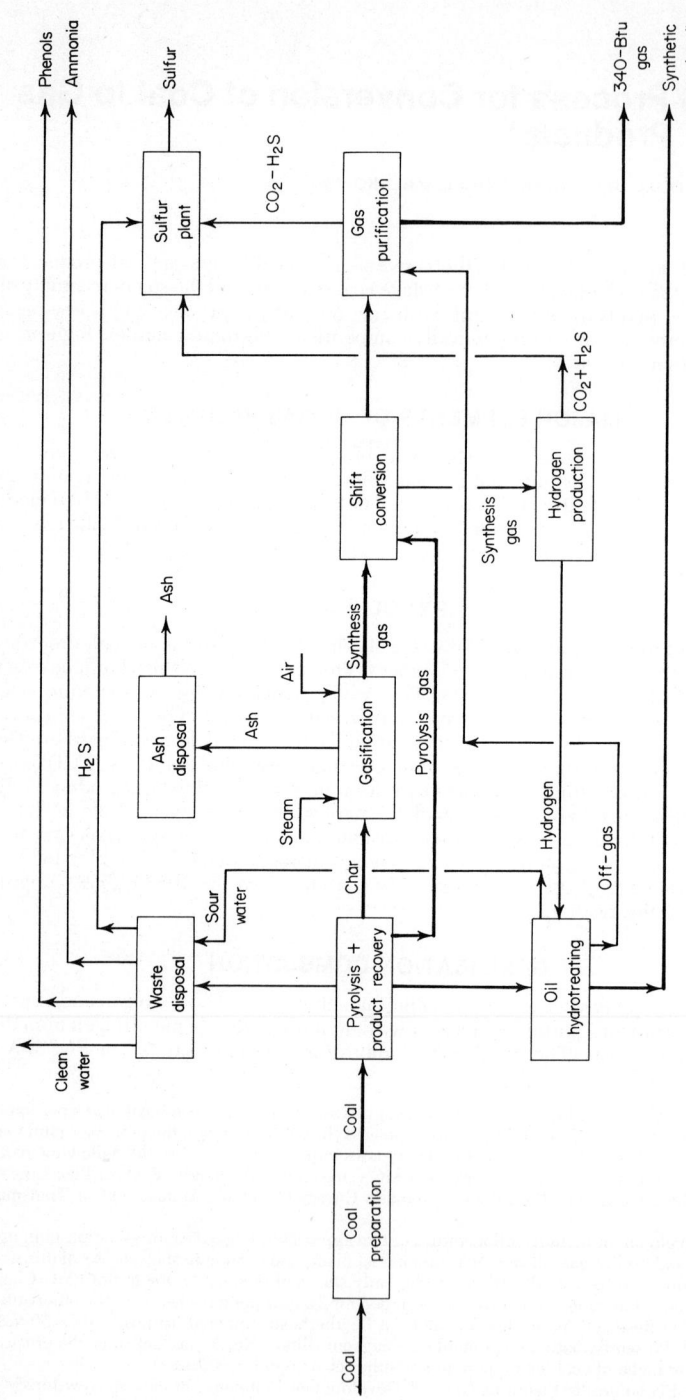

Fig. 1. COGAS process block diagram: medium-Btu gas.

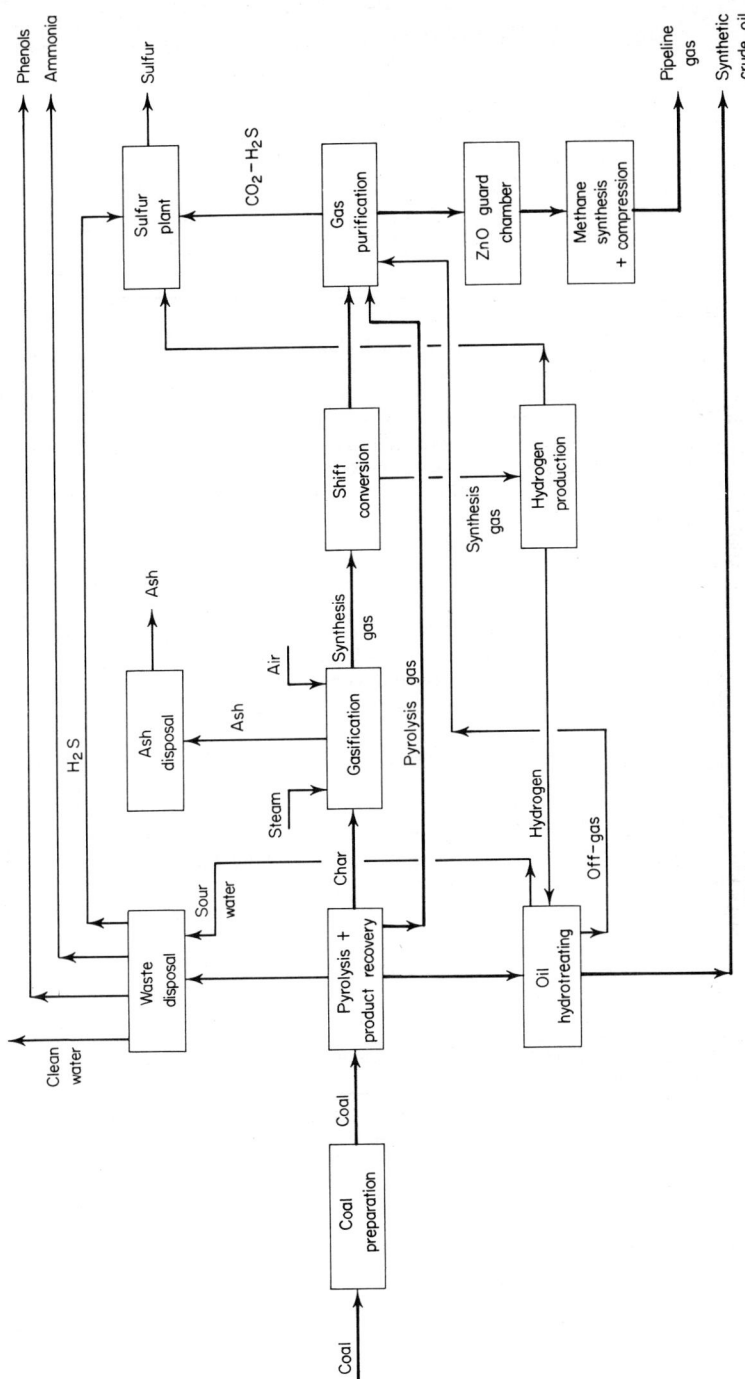

Fig. 2. COGAS process block diagram: pipeline gas.

TABLE 1. Pyrolysis Gas Analyses
Mole percent

Component	Illinois No. 6 bituminous	Wyoming subbituminous
CO	7.4	17.2
H_2	26.2	16.6
CH_4	34.0	34.9
CO_2	9.3	20.1
C_2–C_4	8.4	7.6
H_2S	10.8	1.2
N_2	4.0	2.4

patent* was recently issued covering the approach shown. Use of this method of supplying gasification heat, combined with fluidized-bed operation of the gasifier, is expected to result in a high degree of utilization of the carbon, while eliminating the problem of fines buildup in the gasifier bed which has plagued similar approaches in the past.

Low-pressure gasification with no superficial hydrogen partial pressure sacrifices methane production in the gasification step for operating advantages. Equilibrium concentrations of CH_4 for the steam-carbon reaction over a range of pressure up to 70 atmospheres absolute at two temperatures are given in Fig. 4. At the COGAS pressure of 100 psia or less, methane in the equilibrium composition is nil.

Confining the gasifier size to practical limits at low pressures is dependent on the reactivity of the char. Bench-scale tests in a 4-inch inside-diameter fluidized-bed batch reactor have been performed on chars from a number of coals. The relative reactivities of several chars at 0 psig total pressure are given in Table 2. These data show the higher reactivity of char produced by multistage fluidized-bed pyrolysis and the superior reactivity of char from lignite. The char reactivity studies have shown that for a given char, the rate of carbon reaction with steam at low pressures is correlated by the following equation:

$$R = 1334e^{-18700/T}(P_{H_2O})^{0.63}(u)^{0.73}$$

where R = specific gasification rate, pounds of carbon gasified per hour per pound of carbon in bed (average values over the 30 to 50% carbon burnoff range)

P_{H_2O} = inlet partial pressure of steam, atmospheres absolute

u = inlet superficial gas velocity, feet/second

T = temperature, °Rankine

Synthesis gas is the product of the gasification-combustion step. A typical analysis is given in Table 3. This gas, combined with the stripped pyrolysis gas, is purified and

TABLE 2. Relative Reactivities of Several Chars at 1600°F (871°C) and 0 psig

Type of char	Gasification rate, pounds of carbon per hour per pound of carbon in bed
From pyrolysis:	
Western bituminous	0.17
Illinois bituminous	0.10
Western subbituminous	0.34
Lignite	0.6
From other processing:	
Lignite	0.19
Eastern bituminous	0.15
Western bituminous	>0.10

*Patent No. 3,850,839.

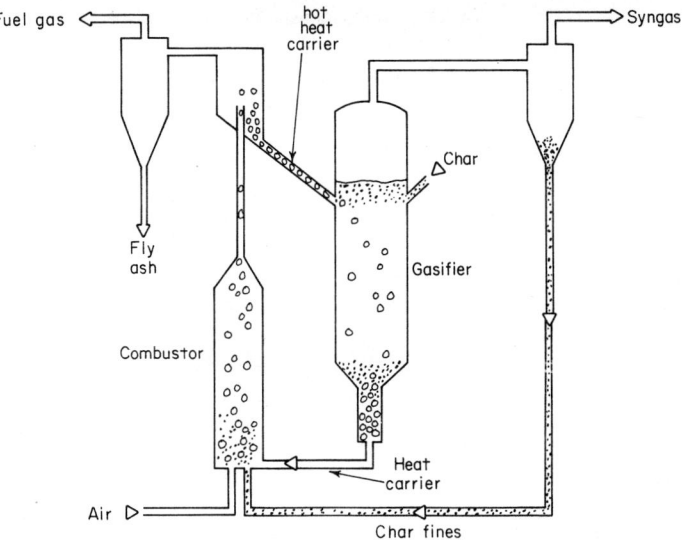

Fig. 3. Gasification-combustion step of COGAS process.

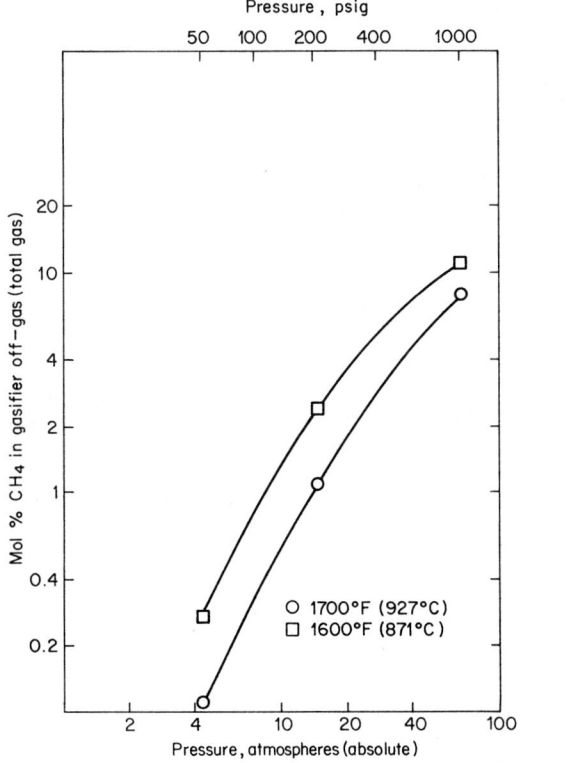

Fig. 4. Steam-carbon equilibrium. Methane content of gasifier off-gas; estimated effect of gasifier pressure (total gas basis). Basis (lb-atom/hr): C − 29, 852, H 91, 567, O 47, 604.

TABLE 3. Dry-Synthesis Gas Analysis

Component	Mole percent
CO	32
H_2	57
CO_2	9.4
CH_4	0.6
N_2	0.3
H_2S	0.7

processed to the product gas desired. A portion of the synthesis gas is recycled to the pyrolysis step and thus evolves mixed with pyrolysis gas.

HEAT AND POWER RECOVERY

It is essential for process efficiency to utilize all the heat available in the gases to the maximum extent. Such uses are to: (1) provide heat to pyrolysis, (2) raise steam in waste-heat boilers on both the synthesis gas stream and the combustor flue-gas stream, (3) preheat the combustion air in an economizer, (4) expand the hot flue gas through power recovery turbines which can drive the air compressors and other rotating power units, and (5) dry the coal. In the COGAS process design, all the heat recovery approaches mentioned are employed.

GAS PROCESSING

The objective of the COGAS process development program is to utilize gas-processing steps that have been commercially proved, or at least demonstrated on a large scale. Commercial processes are available for sulfur purification, CO-to-H_2 shift, high-purity hydrogen production, and bulk CO_2 removal. Since these processes all are more efficient at elevated pressure, the first step after gas production must be compression. At this stage, however, the compression ratio must be minimized because of the large gas volume. The gas-processing steps are undergoing intense engineering study in order to obtain an optimum arrangement.

For moderate-Btu gas production (Fig. 1) gas is raised to a minimum pressure and cleaned for reduction of particulates and sulfur compounds to a level acceptable for pollution control. As a suitable fuel for combustion processes, this gas can contain the CO_2 and may or may not include the light hydrocarbons. A typical moderate-Btu gas analysis is shown in Table 4.

Pipeline-quality gas requires methane synthesis and a final pressure of 1000 psig to be compatible with natural gas in pipelines. The COGAS process objective is to produce a gas of 950 Btu per standard cubic foot which meets the natural-gas quality criteria. To do this, the processing steps shown in Fig. 2 include compression to about 500 psig and gas removal plus shift to provide an H_2/CO mole ratio of 3, methane synthesis, and final dehydration and compression to 1000 psig.

The methane synthesis step may employ any of several process approaches for high CO feed that are being demonstrated or developed by others. COGAS has participated in a program which demonstrated a pilot-scale, long-term operation of a catalyst and process approach that was very successful. Presently the most active catalysts for methane

TABLE 4. Dry Moderate-Btu Gas Analysis

Component	Mole percent
CO	31.2
H_2	57.9
CH_4	4.0
CO_2	6.6
N_2	0.3
S	0.002

TABLE 5. Dry Pipeline-Quality Gas Analysis

Component	Mole percent
CH_4	93
CO	0.7
H_2	0.4
Inerts	3.3
C_2-C_4	2.7

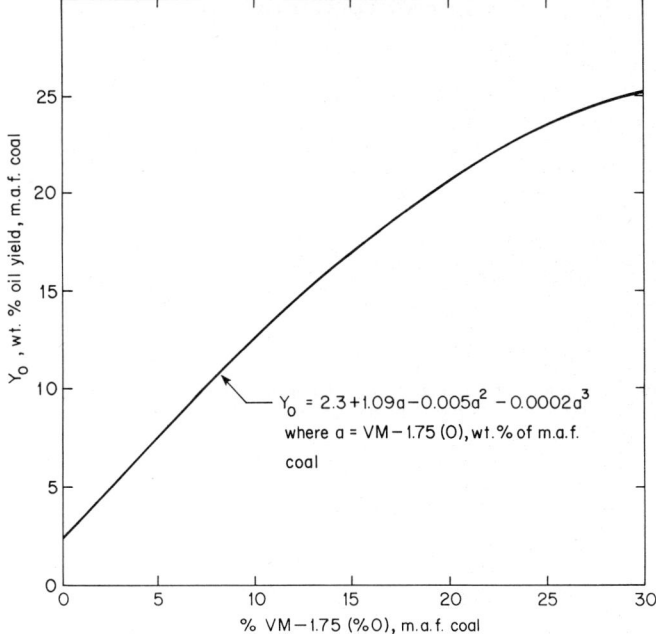

Fig. 5. Oil yield correlation for COGAS coal pyrolysis.

synthesis are all based on nickel. Since nickel is very sensitive to sulfur poisoning, it is necessary that the purification step reduce the sulfur content of the gas to less than 0.1 ppm. To ensure this, a zinc oxide (ZnO) guard chamber is shown immediately ahead of the methane synthesis step.

A typical pipeline-quality gas analysis is given in Table 5.

MATERIAL BALANCES

Using a preliminary process model, material and energy balances for the COGAS process have been prepared for a variety of coal feeds. The first step in the program is prediction of pyrolysis yields, using correlations based on the ultimate analysis of the coal. The correlation curve for oil yield is shown in Fig. 5.

Estimated coal requirements and product yields are given in Table 6 for COGAS plants producing 250×10^9 Btu per day of pipeline-quality gas and plants producing 140×10^9 Btu per day of moderate-Btu gas for a 1000-megawatt power plant.

TABLE 6.   Calculated Material-Balance Data

Type of coal	Tons/day	Synthetic crude, barrels/ day	C_2–C_4, million pounds/ day
250×10^9 BTU/DAY, PIPELINE GAS			
Illinois No. 6	27,500	28,400	400
Eastern bituminous	27,400	32,700	1,200
Western subbituminous *A*	29,000	15,500	450
Western subbituminous *B*	35,500	13,700	450
MODERATE BTU GAS, 140×10^9 BTU/DAY			
Illinois No. 6	11,800	12,900	
Western subbituminous *B*	16,000	6,200	

COMMERCIAL OUTLOOK

From a technology standpoint, the timetable for proceeding to a commercial scale is dependent on the progress of the pilot-plant studies, followed by construction and operation of a large pilot or demonstration scale plant. Because the latter plant will be large, and because process plant equipment delivery projections are very long, design and construction time estimates are quite long. Thus, it is difficult to predict at this time the completion of a commercial-scale plant before the early 1980s.

REFERENCES

1. Dierdorff, L. H., Jr., and R. Bloom, Jr.: "The COGAS Project—One Method of Coal to Gas Conversion," Society of Automotive Engineers meeting, August 1973.
2. Rayner, J. W. R.: "Gasification by the Moving-Burden Technique," American Institute of Mining Engineers Symposium on Gasification and Liquefaction of Coal, February 1952.
3. Bloom, Ralph, Jr., and R. Tracy Edinger: "Status of the COGAS Process," 6th Synthetic Pipeline Gas Symposium, pp. 51–70, October 28–30, 1974.
4. Sacks, M. E., and R. Tracy Edinger: "Development of the COGAS Process," American Institute of Chemical Engineers 79th National Meeting, paper No. 26a, March 1975.

Synthoil Process for Converting Coal to Nonpolluting Fuel Oil

BY SAYEED AKHTAR* AND PAUL M. YAVORSKY†

High-sulfur coals may be converted economically to clean-burning liquid fuels by the Synthoil process. A slurry of coal in recycle oil is reacted with hydrogen in the presence of a catalyst in a turbulent-flow, packed-bed reactor at 450°C and 2000 to 4000 psi. Under these conditions, coal is converted to liquid hydrocarbons, and sulfur is eliminated as gaseous hydrogen sulfide (H_2S). The liquids and unreacted solids are separated from the gases and centrifuged to obtain a low-sulfur, low-ash liquid fuel suitable for power generation. Although designed primarily for the conversion of coal to a utility fuel, this Bureau of Mines process is easily adaptable to the production of synthetic crude, distillate fuels, or naphtha if the economics justify.

The key to the long-term operability of a packed-bed reactor with coal slurry is the turbulent flow of hydrogen. Turbulence prevents plugging of the reactor and facilitates mass and heat transfer.

ROLE OF COAL IN PROJECT INDEPENDENCE

The primary fuel for 52 percent of the electricity generated in the United States in 1972 was coal (Ref. 1). About 351 million tons of coal were consumed in power generation. The coal reserves of the United States are ample for all present and projected future needs, but the distribution of the reserves is such that low-sulfur coals are located in the western parts of the country, whereas high-sulfur coals are located in the industrialized and heavily populated eastern parts. By and large, the eastern coals contain too much sulfur to meet air quality standards, and since it is uneconomical to transport western coals to the industrial and urban centers of the east, large quantities of power in the eastern states are generated from imported low-sulfur, petroleum-derived fuel oils. The importation puts considerable strain on the United States balance of payments, and the situation will worsen if the current trend of increased imports continues. In addition, heavy dependence on imported fuel for energy will have undesirable political ramifications. Accordingly, the President of the United States initiated *Project Independence* with the objective of making the United States self-sufficient in energy by 1980. Coal has a major role in the project; by converting the local eastern coals to nonpolluting fuels, the need for importing low-sulfur fuel oils can be eliminated.

ADVANTAGES OF LIQUEFACTION

Although coal can be converted to nonpulluting fuels by liquefaction or gasification, the liquefaction route has higher thermal efficiency, lower process water requirement, and less severe environmental impact. Furthermore, liquid fuels have far higher energy density than gaseous fuels and, therefore, are cheaper to store and transport. Liquefaction plants will also provide the base for developing the technology for converting coal to gasoline, an important consideration since the known petroleum reserves of the country are certain to run out far sooner than the coal reserves.

THE SYNTHOIL PROCESS

A flow diagram of the Synthoil process is shown in Fig. 1. Hydrogen and a slurry of powdered coal in a portion of the product oil are introduced concurrently in a reactor

*Project Leader, Synthoil Exploratory Engineering; † Research Supervisor of the Exploratory Engineering Section, Pittsburgh Energy Research Center, Energy Research and Development Administration, Pittsburgh, Pennsylvania.

packed with pellets of Co-Mo/SiO₂-Al₂O₃ catalyst. The product stream is led to a gas disengager where the liquids and unreacted solids are separated from the gases. The liquid stream is passed through a centrifuge to remove the unreacted solids, consisting of mineral matter and refractory coal substance. The centrifuged liquid product is a nonpolluting fuel oil.

The solids from the centrifuge go to a pyrolyzer, which yields an additional quantity of nonpolluting fuel oil and a residue that consists mostly of mineral matter with some carbonaceous material. This residue is fed to a gasifier to prepare makeup hydrogen for

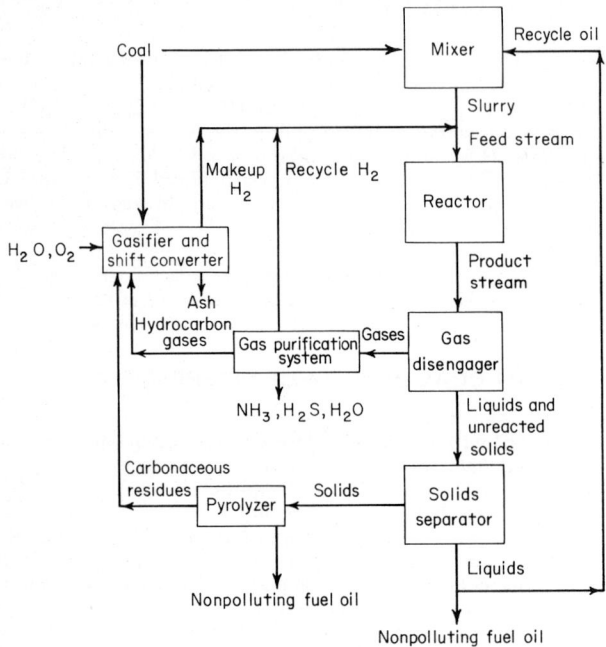

Fig. 1. Main elements of Synthoil process.

the process. Some coal may be added to the gasifier to prepare sufficient hydrogen for the process requirement.

The gases from the gas disengager are led through a purification train to remove ammonia (NH₃), hydrogen sulfide, water, methane (CH₄), ethane (C₂H₆), and other gaseous compounds formed during the liquefaction of coal; and the purified hydrogen is recycled to the reactor. The gas purification is conducted at the plant pressure to minimize the cost of recycling the purified hydrogen. Some of the impurities separated from the recycle gas will be useful by-products: NH₃ and H₂S may be converted to ammonium sulfate for the fertilizer market, and CH₄ and C₂H₆ may be sold to natural-gas distributors or steam-reformed to hydrogen for in-plant use. The ash from the gasifier may be disposed of as mine fill.

PROCESS DEVELOPMENT

The first phase of research was conducted with a $\frac{5}{16}$-inch internal-diameter reactor in a 5-pound (slurry) per hour plant (Refs. 2–4). Coals containing 3.0 to 4.6 percent sulfur were converted to oils containing 0.2 to 0.3 percent sulfur. No sensible attrition of the catalyst or loss of its desulfurization activity was observed in continuous runs extending up to 6 weeks.

A scaled-up reactor of 1.1 inches internal diameter in a ½-ton (slurry) per day plant is now in operation. In continuous experiments with a Kentucky strip coal containing 5.5 percent sulfur and 16.5 percent ash, the plant was operated at 4000 psi and 2000 psi for

TABLE 1.   Processing of Kentucky Coal by Synthoil Process
At 4000 and 2000 pounds per square inch

Sulfur in feed coal 5% (weight)	Product oil from	
Ash in feed coal 17.5% (weight)	4000 psi	2000 psi
Sulfur, % (weight)	0.2	0.5–0.7
Ash, % (weight)	0.1–0.2	1.3–2.9
Viscosity, SSF* at 77°F (25°C)	26–440	—
SSF at 180°F (82°C)	—	13.5–97.6
Specific gravity, 60°F/60°F (15.6°C/15.6°C)	1.020–1.082	1.060–1.148
Calorific value, Btu/lb	17,400	16,640
Yield, barrels/ton of coal (as received)	3.27	2.99
Consumption of hydrogen, standard cubic feet/barrel	4375	3450

*Saybolt Furol scale.

500 hours before voluntary shutdown. The results from these experiments are given in Table 1. The product oil from the 4000-psi operation was a premium-grade fuel containing 0.2 percent sulfur and 0.1 to 0.2 percent ash. It was fluid at room temperature and had a calorific value of 17,400 Btu per pound. The yield was 3.27 barrels per ton of coal (as received), and the consumption of hydrogen was 4375 standard cubic feet per barrel.

The product oil from the 2000-psi operation contained 0.5 to 0.7 percent sulfur and 1.3 to 2.9 percent ash. The yield was 2.99 barrels per ton of coal (as received), and the consumption of hydrogen was 3450 standard cubic feet per barrel. It will be an acceptable fuel in many communities with less stringent environmental protection regulations where combustion of the coal with 5.5 percent sulfur will not be legal. In view of the lower operating pressure and the lower consumption of hydrogen, the oil at 2000 psi will be cheaper to produce than the oil at 4000 psi.

The construction of a pilot plant which will process 8 tons per day of coal is in advanced stages of planning. In this plant, reactors of 4 to 6 inches internal diameter will be operated.

REFERENCES

1. U.S. Department of Commerce: "Statistical Abstracts of the United States," table 839, p. 512, Washington, D.C., 1973.
2. Yavorsky, Paul M., Sayeed Akhtar, and Sam Friedman: "Process Development: Fixed-Bed Catalysis of Coal to Fuel Oil," 65th Annual A.I.Ch.E. Meeting, Nov. 26–30, 1972, American Institute of Chemical Engineers, New York.
3. Akhtar, Sayeed, Sam Friedman, and Paul M. Yavorsky: "Low-Sulfur Liquid Fuels from Coal." Symposium on Quality of Synthetic Fuels, ACS Meeting, Apr. 9–14, 1972, Boston, Mass., American Chemical Society, Washington, D.C.
4. Akhtar, Sayeed, Sam Friedman, and Paul M. Yavorsky: "Process for Hydrodesulfurization of Coal in Turbulent-Flow, Fixed-Bed Reactor," 71st National Meeting of A.I.Ch.E., Dallas, Tex., Feb. 20–23, 1972, American Institute of Chemical Engineers, New York.

The Toscoal Process for Pyrolysis of Coal

BY FRANKLIN B. CARLSON*

The Toscoal process is an offspring of the Tosco II oil-shale pyrolysis process developed by The Oil Shale Corporation. The Toscoal process objective is to upgrade low-heating-value coal, especially low-sulfur western coal, through pyrolysis.

Pyrolysis, carbonization, or retorting are terms used in a general way to describe the same basic operation of heating coal to effect some degree of decomposition and separation of the coal into various products. Pyrolysis can be conducted over a wide range of processing conditions and can produce products with a wide range of characteristics. The physical and chemical changes occuring during pyrolysis are complex but do not involve large exchanges of heat, such as the steam-carbon reaction common to gasification. Pyrolysis is a relatively simple technique, in terms of energy requirements, control, and

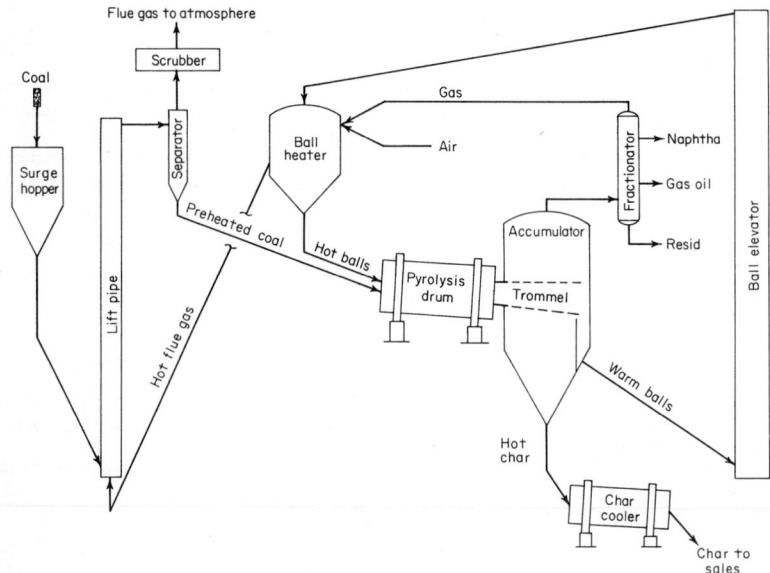

Fig. 1. Major elements of the Toscoal process.

processing complexity, for separating coal into several components of potentially improved value.

The products of pyrolysis are char, tar, and gas, listed in order of decreasing volume. An aqueous fraction containing some organic materials, such as carboxylic acids, is also produced; but, generally, this fraction does not have a positive economic value. With some coals, a considerable concentration or increase in heating value of the char, tar, and gas can be accomplished through pyrolysis by removing the inherent moisture of the coal. Some subbituminous coals contain up to 30 percent inherent moisture. The increase in product fuel value over the parent, high-moisture coal, especially the char, has the

*Manager, Pilot Plant and Engineering, The Oil Shale Corporation, Rocky Flats Research Center, Golden, Colorado.

TABLE 1. Toscoal Pilot Plant Test Results

Feedstock: Wyoming subbituminous coal			
Proximate analysis, wt. %		Ultimate analysis, wt. %	
Moisture	30.0	Carbon	46.4
Ash	5.3	Hydrogen	2.8
Volatile matter	30.7	Oxygen	14.7
Fixed carbon	34.0	Nitrogen	0.7
Gross heating value, Btu/pound	8140	Sulfur	0.3
		Chlorine	0.0
		Moisture	30.0
		Ash	5.3

RETORT PRODUCT YIELD (POUNDS/TON OF "AS-MINED" COAL)			
Retort Temperature	800°F	900°F	970°F
	(427°C)	(482°C)	(521°C)
Char	1049	1012	969
Gas (C₃ and lighter)	119	157	126
Standard cubic feet/ton	1250	1777	1625
Oil (C₄ and heavier)	114	143	186
Gallons/ton	13.2	17.4	21.7
Water*	702	702	702
Unaccounted	16	14	17

*Yield of retort water estimated from Fischer assay of feed coal. Total water yield in pilot plant includes both retort water and steam condensate from retort seals.

TABLE 2. Toscoal Char Properties

	Retort temperature		
	800°F	900°F	970°F
Char properties	(427°C)	(483°C)	(521°C)
Proximate wt. %			
Moisture	0.0	0.0	0.0
Ash	12.4	10.0	9.8
Volatile matter	25.3	19.7	15.9
Fixed carbon	62.3	70.3	74.3
Total	100.0	100.0	100.0
Ultimate wt. %			
Carbon	68.8	74.7	77.5
Hydrogen	3.4	3.0	2.9
Oxygen	13.3	11.8	8.3
Nitrogen	1.0	1.2	1.3
Sulfur	0.5	0.2	0.3
Chlorine	0.0	0.0	0.0
Moisture	0.0	0.0	0.0
Ash	12.4	10.0	9.8
Total	99.4	100.9	100.1
Other data			
Equilibrium moisture, wt. %	10.0	10.8	9.9
Hardgrove grindability	78.2	49.1	45.6
Heating values:			
Gross, Btu/pound	11,826	12,560	12,963
Net, Btu/pound	11,516	12,280	12,693
Bulk density (¼ inch × 0)			
packed, pounds/cubic foot	51.2	48.8	47.8

potential for substantially reducing the transportation cost for fuel produced for distant consumption. The tar produced by pyrolysis can be further processed to produce fuel oils. Low-temperature tar contains many chemical species and may be valuable as a source of chemicals in the future. At present, separating the chemicals (usually occurring in low concentrations) from the tar is not economically attractive.

The Toscoal process scheme utilized in a 25 ton-per-day pilot plant is shown in Fig. 1. In the process, coal is heated and pyrolyzed by contact with heated ceramic balls in a rotating pyrolysis drum or retort. The ceramic balls are separated from the pyrolyzed coal (char) in a trommel screen at the drum exit. The ceramic balls are conveyed to the ball heater and reheated for another cycle through the retort. The hot char from the retort is cooled in a rotary-tube cooler.

Feed coal is dried and preheated in the lift pipe by entrainment in a stream of hot flue gas from the ball heater. The preheating step reduces the thermal load on the retort and increases retort throughput for a given circulation of ceramic balls. The preheat step also provides an efficient way to recover waste heat from the ball-heater flue gas. Supplemental fuel is required in the preheater to dry coals with high moisture content. The optimum coal preheat temperature is generally the highest temperature attainable without excessive hydrocarbon vapor emission from the preheat scrubber.

Coal feed size of ¾ inch × 0 has been used in the pilot plant, although coal sized to pass ½ inch is preferred. Upper feed size was limited in the pilot plant by the lifting capacity of the entrained preheater. Coal fines do not adversely affect retort operation, but crushing with minimum fines production is desirable to avoid heavy dust-collector loading. The coal decrepitates and is attrited in both the preheater and retort so that the average particle size of the char is less than ¼ inch.

TABLE 3. Toscoal Oil Properties

| | Retort temperature | | |
| | 800°F | 900°F | 970°F |
Oil properties	(427°C)	(482°C)	(521°C)
Ultimate wt. %			
Carbon	81.4	80.7	80.9
Hydrogen	9.3	9.1	8.7
Oxygen	8.3	9.4	9.3
Nitrogen	0.5	0.7	0.7
Sulfur	0.4	0.2	0.2
Chlorine	0.0	0.0	0.0
Ash	0.0	0.2	0.1
Total	99.9	100.3	99.9
Heating values			
Gross Btu/pound	16,590	16,217	15,964
Net, Btu/pound	15,740	15,372	15,160
API gravity			
Primary oil	7.9	4.5	1.9
Primary oil Calculated, with C_4 and heavier components of gas added	13.2	12.1	6.2
Pour point	90°F	100°F	95°F
	(32.2°C)	(27.8°C)	(35.0°C)
Conradson carbon, wt. %	7.6	9.9	11.4
Distillation, vol. %			
2.5	413°F(211.8°C)	420°F(216°C)	390°F(199°C)
10	490 (254)	475 (246)	405 (207)
20	575 (301.5)	550 (288)	455 (235)
30	645 (340.5)	625 (329.5)	545 (285)
40	710 (371)	700 (371)	640 (388)
50	765 (407)	775 (413)	725 (385)
Viscosity (Saybolt Universal Seconds)			
180°F (82°C)	122	123	128
210°F (99°C)	63	66	69

TABLE 4.   Toscoal Gas Analysis

	Retort temperature		
	800° F	900°F	970°F
Component, mole %	(427°C)	(483°C)	(521°C)
H_2	0.8	1.0	7.8
CO	18.0	17.3	18.4
CO_2	51.1	42.3	36.4
H_2S	1.7	1.3	0.3
C_1	16.9	22.0	24.9
C_2	3.6	4.7	4.4
C_2-	1.9	1.9	2.4
C_3	1.3	2.2	1.2
C_3-	1.6	3.7	1.6
iC_4	0.1	0.1	0.0
C_4	0.3	2.0	1.1
C_5	1.0	1.8	0.7
C_6	0.7	0.6	0.4
C_7	0.5	0.1	0.3
C_8+	0.2	0.0	0.1
Total	99.7	101.0	100.0
Average molecular weight	35.9	35.0	30.6
Weight percent carbon	40.5	45.9	44.7
Heating values, calculated:			
Gross, Btu/standard cubic foot	534	717	630
Net, Btu/standard cubic foot	494	663	580
Heating values, calculated with CO_2 and H_2S removed:			
Gross, Btu/standard cubic foot	1,113	1,234	995
Net, Btu/standard cubic foot	1,029	1,138	920

Tar vapor is cooled and condensed in the fractionator in which top temperature is maintained so that all pyrolysis water and steam condensate are collected in the overhead accumulator. The aqueous phase from the accumulator is treated for chemical recovery and/or for disposal. Gas from the accumulator can be treated for sulfur removal and sold or used as fuel for the ball heater.

The Toscoal process has considerable flexibility in pyrolysis temperature and in the distribution and characteristics of the products produced. Pyrolysis temperature of 800 to 1000°F (427 to 538°C) is the range of general interest, however. At retorting temperatures above 1000°F, the char (from subbituminous coal) contains less than 16 percent volatile matter and would require modifications in most combustion systems to be used satisfactorily. Retort throughput declines and operating costs increase as retort temperature is increased. Tar yield increases with retort temperature, but the rate of increase declines considerably at temperatures above 1000°F. Gas yield also increases with temperature, and the rate increase above about 1000°F corresponds with probable cracking of the tar.

At temperatures below 800°F, tar and gas yields diminish sharply, and the principal product is essentially dried coal, containing substantially all the initial volatile matter.

The Toscoal process has been applied only to noncaking coals. Caking coals would require some oxidative pretreatment before processing in the Toscoal process.

Results of tests conducted on a Wyoming subbituminous coal in the Toscoal 25-ton-per-day pilot plant are shown in Tables 1 through 4. The chars produced from subbituminous coals in the temperature range of 800 to 1000°F are relatively reactive, and require care in storage and transportation to avoid spontaneous ignition.

Since air or flue gas is excluded from the Toscoal retort, the gas produced is not diluted and has a high heating value. The exclusion of air also simplifies tar condensation.

The direct solid-to-solid heat transfer employed in the Toscoal retort makes large-scale, continuous operation feasible. The Tosco II shale process, which is basically the same as the Toscoal process, has been operated at a 1000-ton-per-day level. A 66,000-ton-per-day plant to be constructed near Grand Valley, Colorado, for the recovery of oil from oil shale is in the final design stages.

The Consol Synthetic-Fuel Process (CSF Process)

BY JOHN A. PHINNEY*

The Consol synthetic-fuel process (CSF process) consists of the partial conversion of coal into an extract and a by-product solid residue, followed by hydrogenation of the extract to yield a synthetic crude. Further conversion of the crude to marketable distillates and the manufacture of the hydrogen needed for the process are presumed to use commercially proved methods and have not been part of the CSF development. Most of the development work has been applied to caking, high-sulfur eastern (United States) coals.

The development of the CSF process, under the sponsorship of the Office of Coal Research (OCR), U.S. Department of the Interior, was initiated on August 30, 1963. Major periods of activity have included: (1) prepilot plant research (September 1963 through July 1968); (2) design and construction of a pilot plant at Cresap, West Virginia (June 1964 through May 1967); (3) dedication of the pilot plant on May 27, 1967; (4) operation of the pilot plant (November 1966 through February 1970); and (5) contract termination on August 30, 1972.

Based on the work to date, the CSF process has been judged technically feasible by various independent appraisals. It is recognized that modifications and mechanical revisions are needed at the Cresap pilot plant to achieve better onstream performance. As of mid-1974, the Cresap pilot plant was the only existing large-scale installation owned by the OCR which has the facilities for optimizing the production of low-sulfur utility fuels from highly caking eastern coals. Commercial interest in the project continues.

SUMMARY OF PROCESS

COAL EXTRACTION

Continuous solvent extraction of highly caking Pittsburgh Seam coal was carried out successfully in bench-scale (0.12 ton per day) and pilot-scale (20 tons per day) runs. Kinetic data, yield structures, and process operability were completely consistent at the two scales of operation, providing a sound basis for the design of a semicommercial or commercial plant. The coal-extraction technology is ready to be utilized for the conversion of those caking eastern coals, which are high in pyritic sulfur and relatively low ($\sim$1.0 percent) in organic sulfur, into a utility fuel which is acceptable under the Environmental Protection Agency emission standards for new stationary sources.

EXTRACT HYDROGENATION

Continuous hydrogenation of extract to a 0.2 percent sulfur distillate over ebulliated contact catalysts was carried out successfully in bench-scale (0.10 ton per day) runs and pilot-scale (3 tons per day) H-oil runs. Because of many mechanical problems, comparable data were not obtained during the limited operating period with the 13 tons per day extract hydrogenation pilot plant at Cresap. Further pilot-scale studies of extract hydrogenation are needed to define the optimum conditions for producing acceptably low-sulfur products from coals of high organic sulfur content.

ZINC CHLORIDE HYDROCRACKING

A part of the OCR-sponsored bench-scale program was devoted to the study of molten zinc chloride as a hydrocracking catalyst for coal and coal extract, with the objective of

*Consultant, Conoco Coal Development Company, Library, Pennsylvania.

developing improvements over the CSF technology incorporated in the Cresap facility. This development effort was suspended in mid-1967 for lack of funding.

The $ZnCl_2$ concept is one of the few truly innovative departures from prior German art which has evolved from American work in the coal liquefaction field. The concept offers the possibility of significant reductions in the cost of manufacturing distillate fuels, particularly from western (United States) coals.

ECONOMIC POTENTIAL

The economics of gasoline manufacture through use of the CSF process were appraised periodically. On the basis of eastern deep mine coal prices and at refinery capacities of 40,000 to 60,000 barrels per day, the synthetic product was only marginally competitive at the prices prevailing during the period for comparable petroleum products. The process was judged to be competitive, however, assuming a 250,000-barrel-per-day installation and the use of lower-cost western coal. In the meantime, crude prices have increased substantially, obviously affecting earlier economic appraisals, and now put the process in a more favorable light economically.

The 1972 appraisal of the process contemplated the production of a 0.2 percent sulfur distillate for utility station use. It was shown that the product from a 50,000-barrel-per-day refinery could not compete with the delivered cost of 0.3 percent sulfur imported oil (at 1972 prices)—if high-sulfur eastern deep-mine coal were used as feedstock. However, the projected costs were found to be competitive with the published costs for other methods of converting the same coal products of equal sulfur content, when compared on the same bases of coal price, capital costs, and design conservatism.

It is apparent that the conditions demonstrated at the Cresap pilot plant and chosen as the basis for the 1972 appraisal are too severe for the optimum production of a low-sulfur utility boiler fuel. Significant cost reductions could result from the use of milder conditions.

THE CRESAP PILOT PLANT

The basic flow diagram of the CSF process is shown in Fig. 1. The operations enclosed within dashed lines were not included in the pilot plant. The Cresap pilot plant was designed to permit a study of the variables influencing the amount and quality of extract recoverable from coal as well as the variables influencing the quality and yields in extract hydrogenation.

Nonconventional items of equipment installed at the pilot plant included:
1. A 12-stage, stirred extractor vessel designed for 800°F (427°C) and 500 psig.
2. A rotary-pressure precoat filter designed for 650°F (343°C) and 150 psig.
3. A continuous centrifuge designed for the same conditions as the filter.
4. Hydrogenation recycle pumps designed for a suction condition of 5000 psig and 850°F (454°C).
5. Resilient high-pressure piping closures designed to remain leakproof through temperature cycles from ambient to 850°F (454°C).

The Cresap pilot plant, after revisions had been made in 1970, is shown in Fig. 2. The initial precoat filters in Section 300 were replaced by a two-stage hydroclone system. In addition, heated tankage was provided for extract-solvent storage, permitting the extract production system and the extract hydrogenation system to operate independently.

Major specific objectives of the pilot plant included:
1. Operating the five process sections involved in extract production as an integrated unit, using initially a startup extraction solvent and subsequently a natural donor solvent from hydrogenation.
2. Upon completion of the first objective, verifying or refuting the extraction kinetic and yield correlations which were developed in bench-scale work.
3. Simultaneous with objective 1 and 2, operating the process sections involved in extract hydrogenation as an integrated unit using initially startup heavy oils and subsequently coal extract.
4. Upon completion of objective 3, verifying or refuting the hydrogenation kinetics and yield correlations which were developed in bench-scale work.
5. Upon completion of the first four objectives, integrating the extract production and

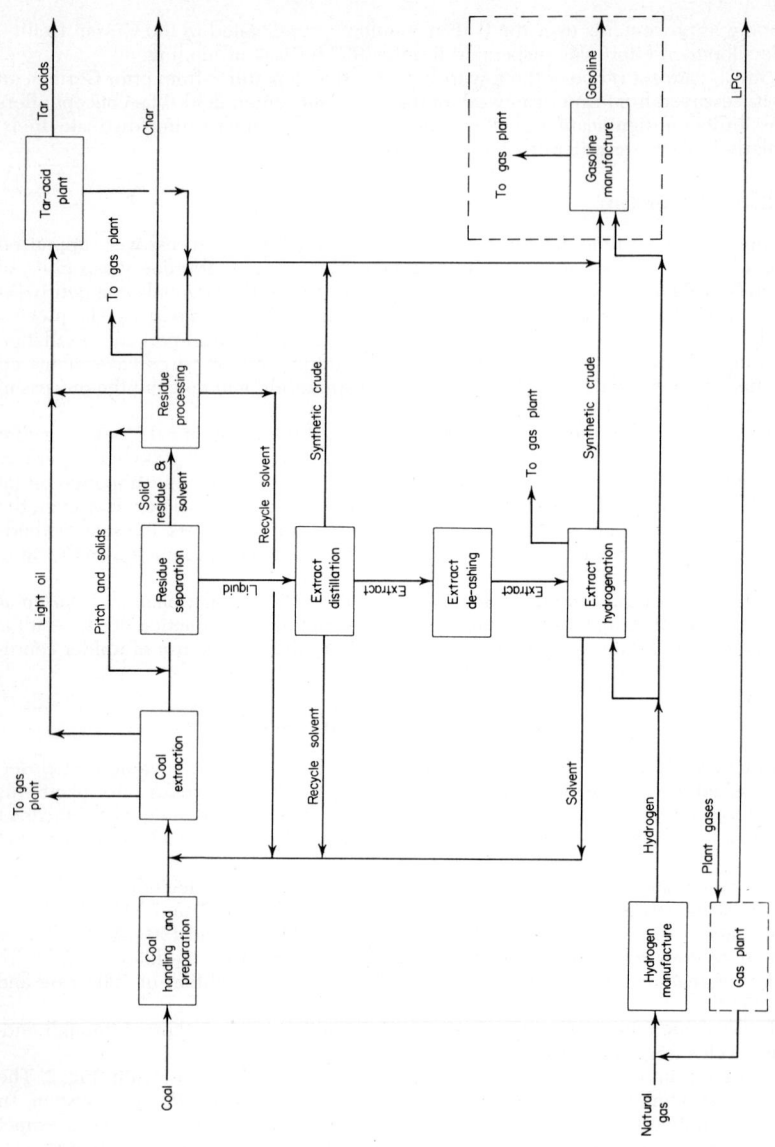

Fig. 1. Schematic flow diagram of the Consol synthetic fuel process. Operations enclosed within dashed lines were not included in the Cresap pilot plant.

hydrogenation sections and operating the plant for extended periods to evaluate various methods of maintaining hydrocracking catalyst activity.

6. Producing representative quantities of synthetic crude distillates for pilot-scale demonstration (at some other facility) of gasoline manufacture by commercial technology.

The pilot plant was operated over a series of 50 runs for an aggregate period of approximately 2100 hours. As shown in Table 1, Cresap pilot-plant results were fully consistent with bench-scale results using the same coal and natural solvents derived from extract hydrogenation. This demonstrated a successful 150-fold capacity scale-up of the

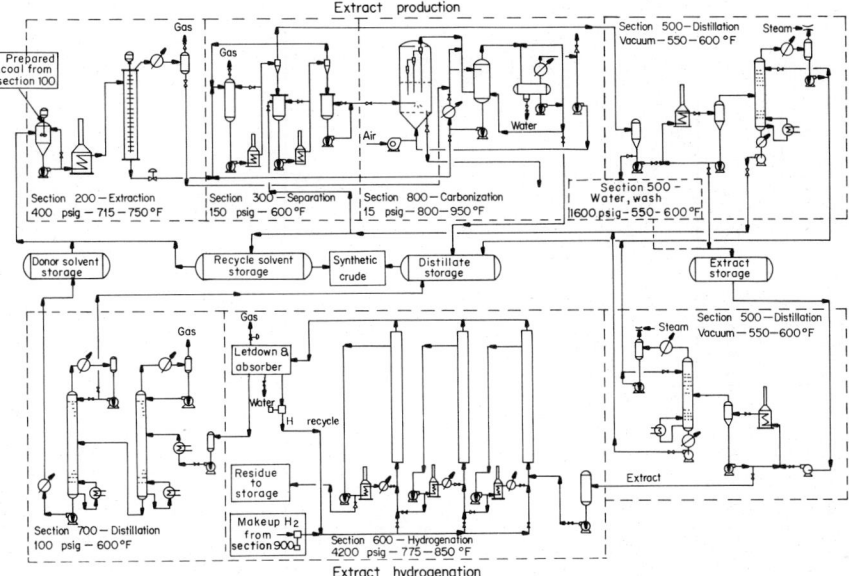

Fig. 2. Flow diagram of the Cresap pilot plant after modifications were completed in 1970.

extraction step. A commercial reactor would require a factor of the same magnitude to scale up from Cresap data. The operability of the extraction system was excellent in runs of up to 12 days' uninterrupted duration.

Hydroclones (liquid cyclones) were installed as replacement of rotary filters shortly after commencement of pilot-plant operations. These units, used to process extraction effluent, were essentially trouble-free, and some 2000 hours of performance on extraction effluent was completed in 46 runs for uninterrupted periods as long as 12 days. Solids removal via hydroclones was not as complete as by filtration; the ash content of filtered extract averaged about 0.35 percent, whereas the ash content of hydrocloned extract averaged about 0.90 percent. However, based on chemical and size analyses of the solids, and on prior bench-scale hydrogenation studies, it was concluded that the incremental ash present in hydrocloned extract would have little or no influence on the life of catalyst in the extract hydrogenation system.

The reaction kinetics and catalyst activity decline observed in the Cresap pilot plant were in substantial agreement with the prior bench-scale results. The pilot plant represented a 150-fold scale-up over bench scale. A valid basis exists for further extrapolation of Cresap results to a larger scale. The decline in hydrogenation catalyst activity is clearly related to the extent of exposure to the "soluble," that is, nonfilterable mineral matter contained in the extract.

The yield structure in the Cresap operation was, in general, poorer than that observed in prior bench-scale tests, in that gas yields were higher; hetero atom rejection was less. This result appeared to be related to adverse processing conditions in the Cresap tests instead of an inherent consequence of larger-scale operations. Sample of Cresap extract were processed to test the applicability of the H-oil process to extract conversion. The H-

TABLE 1. Comparison of Pilot and Bench-Scale Operations
Continuous extraction with natural solvent

Scale	Pilot Plant			Bench-Scale		Bench-Scale	
Run	55-VII	55-VIII	55-X	CE2-116A	CE2-117A	197-13	197-16
Coal feed rate, lb/hr	1261	1288	1700	8.1	8.0	6.7	12.5
Solvent/MAF coal, lb/lb:							
Donor	0.59[c]	0.43[a]	0.40[c]	0.43[c]	0.52[c]	0.41[e]	0.41[e]
Recycle	3.05[b]	3.20[b]	2.05[b]	1.31[d]	1.22[d]	1.25[f]	1.25[f]
Average temperature, °F	738	732	731	746	747	730	746
Final temperature, °F	731	725	724	747	750	737	750
Residual time, minutes	31.2	31.0	32.7	31.7	32.1	37.0	20.0
Coal Conversion, % MAF coal	79.2	78.4	77.6	83.2	80.2	80.0	79.0
Yields, % MAF coal							
Gas + water[g]	3.2	3.1	3.1	3.7	6.1	2.0	3.7
Extract	(76.0)	75.3	74.5	79.5	74.1	78.0	76.3
Residue	20.8	21.6	22.4	16.8	19.8	20.0	21.0
Solvent Polymer							
% MAF coal	—	1.7[h]	3.9[h]	3.3	12.4	4.1	4.9
Solvent	—	0.5[h]	1.6[h]	1.9	7.1	2.5	3.0
Operability	Good	Good	Good	Good	Good	Good	Good

Solvent Properties:
a. Solvent produced by pilot-scale hydrogenation of extract.
b. Solvent recycled from pilot-scale extraction (after depletion of start-up solvent).
c. Solvent produced by bench-scale hydrogenation of extract.
d. Solvent recycled from bench-scale extraction (after depletion of start-up solvent).
e. Solvent produced by pilot plant runs 55 and 56.
f. Solvent produced by pilot plant run 56.
g. Correlated yields.
h. Adjusted yields.
MAF = moisture and ash free.

oil yield data were consistent with prior bench-scale data in the CSF process. The Cresap data, the H-oil data, and the bench-scale data all indicate that a 90 percent (weight) yield of synthetic crude liquid averaging less than 0.2 percent sulfur can be produced by hydrogenation of the extract. The corresponding hydrogen consumption is 5 to 7 percent (weight) of the feed extract.

ZINC CHLORIDE PROCESS

In the early 1960s, exploratory work was initiated by Consolidation Coal Company to develop an improved catalyst for hydrocracking extract to replace the contact catalyst developed by the petroleum industry for oil hydrocracking. This work led to the conception and preliminary testing of molten zinc chloride for this purpose. Expanded bench-scale effort on this concept was conducted in the 1963–1967 period under OCR sponsorship.

Experimental data indicate that molten zinc chloride catalyzes the conversion of extract to distillate at much higher rates than contact catalysts. Most of the extract is converted in a single stage directly to gasoline of high quality and lead susceptibility at relatively mild conditions of temperatures and pressure. The saturated product fractions are characterized by a high degree of branched-chain structures.

The hydrocracking and isomerization activity of zinc chloride declines as a result of poisoning by nitrogen compounds in the feed extract. In addition, zinc chloride is depleted by reaction with sulfur and oxygen compounds in the feed. Thus, a continuous addition of regenerated catalyst to the reaction zone is required. Several alternative regeneration routes were studied. Combustion of the spent catalyst to reject SO_2 and NH_3 appeared to be the preferred alternative. Recovery of hydrogen chloride (HCl) from the combustion off-gases is an economic necessity to regenerate $ZnCl_2$ from the zinc oxide (ZnO) produced by combustion.

Corrosion studies revealed that 316 stainless steel and various high-nickel alloys show adequately low corrosion rates and are not subject to stress-corrosion cracking in the reaction environment at reactor temperatures. The critical points in the system are those where condensation of aqueous hydrochloric acid can occur. Plastic, ceramic, or Hastelloy linings are required. It was concluded that further bench-scale study of regeneration would be desirable before initiation of pilot-scale studies of zinc chloride hydrocracking.

In connection with the use of zinc chloride in coal processing, batch tests with eastern bituminous coals revealed the same superior reaction rates and yields as were observed with extract. The primary problem with coal processing is the chemical interaction between zinc chloride and the iron pyrites in the coal. Conceivably for Western coals (low pyrites content), partial conversion at very mild operating conditions (1000 psig) would be attractive.

REFERENCES

An extensive list of references, patents issued, and patents applied for, as well as more detailed appraisal of the CSF process, from bench-scale to pilot plant, is included in the Final Report, "Development of CSF Coal Liquefaction Process," prepared for the Office of Coal Research, U.S. Department of Interior, Washington, D.C., by the Consolidation Coal Company, Research Division, Library, Pa.—OCR R&D Rept. 39, vol. V, covering the period 8/30/63 to 8/30/72.

Coal Purification by Solvent Refining

BY W. G. JOHNSON* AND P. L. THIGPEN†

One approach to the problem of meeting air-quality standards when coal is used as a fuel is to remove contaminants from the coal before burning it. The process begins with coal washing and preparation at the mines. The first major extension to further purification of coal is the solvent refined coal (SRC) process, which has been under development since the early 1970s and for which a 50-ton-per-day demonstration plant is being built (1974) by the Rust Engineering Company at Fort Lewis, Washington (under contract to the Pittsburgh and Midway Coal Mining Company). This process produces a water-free, low-sulfur, low-ash coal that can be burned as a liquid or a solid. In its solid state, the coal is sufficiently brittle to grind readily into powder. The heating value of the refined coal is about 16,000 Btu per pound without regard to the original coal from which it is produced. Original work on the solvent refining process was done by Spencer Chemical Company and was carried forward by Pittsburgh and Midway Coal Mining Company under contract with the U.S. Office of Coal Research.

MAJOR PROCESS STEPS

Somewhat simplified, the process consists of the following steps: (1) coal preparation and pulverizing to a fine particle size (approximately 200 mesh); (2) blending of the fine coal with a coal-based hydrocarbon solvent that is produced internally in the process; (3) reaction at a high temperature (up to 900°F, 482°C) and high pressure (up to 2400 psig) of the coal-solvent blend with hydrogen or a hydrogen/carbon monoxide synthesis gas mixture; (4) mineral separation to remove the ash by hot, pressure filtration, followed by drying to recover the solvent; (5) solvent recovery, utilizing high-temperature vacuum flashing, and distillation for separation of various hydrocarbon oil fractions, which can be returned to the process or sold; (6) removal of a gas containing the evolved sulfur compounds; and (7) product solidification to a flake or prill form; or the low-ash, low-sulfur product can be used directly from the process as a hot liquid. Test work has indicated that coal of almost any quality except possibly anthracite is amenable to refining by the SRC process.

PROCESS FLOW

A schematic flow sheet of the SRC process is given in Fig. 1. Coal preparation for the process is minimal. In the demonstration plant, flash drying is done in the grinding process. Moisture in the coal, however, does not appear to hinder its solubility and is removed in subsequent process steps. No effort is made to reduce mineral matter before solvent treatment. The coal, ground to about 200 mesh, is mixed with an anthracene oil to a slurry consistency of 25 percent. The slurry recirculates through an eductor where the coal powder is introduced.

Slurry is injected into the high-pressure dissolver by means of duplex plunger pumps through a furnace, where it is heated to about 850°F (454°C) at a pressure of about 1100 psig. Hydrogen or hydrogen/carbon monoxide synthesis gas is added at the furnace inlet. Much less hydrogen is used than in most synthetic coal processing (less than 1 percent). It is theorized that hydrogen transfer from solvent to coal occurs, assisting and stabilizing dissolution. After a residence time of about 30 minutes in the dissolver, about 90 percent of the carbon in the coal is dissolved. Organic sulfur in the coal is converted to hydrogen sulfide during this period.

*Formerly Manager, Chemical Process Department, †Principal Process Engineer, Chemical Process Department, The Rust Engineering Company (a subsidiary of Wheelabrator-Frye Inc.), Birmingham, Alabama.

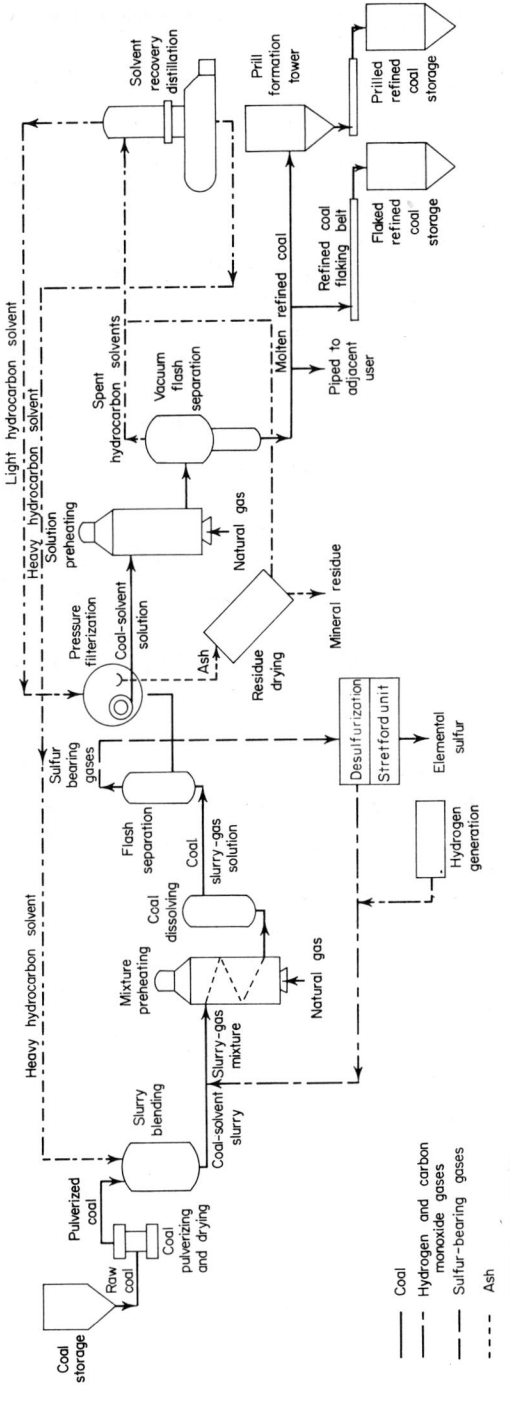

Fig. 1. Solvent refined coal process (*Pittsburgh and Midway Coal Mining Company*) for 50-ton-per-day pilot plant. (*The Rust Engineering Company.*)

The resulting mixture of solution, mineral residue, and dissolved gases is reduced in pressure through a series of pressure letdown valves, whereupon flashing of volatile gas occurs. Next, the coal solution, with the ash and pyritic sulfur as well as undissolved carbon, is filtered. In the demonstration plant, a rotary precoat, drum-type filter, operating at about 200 psig, pressurized with inert gas, is used to effect the solids removal at a temperature of 700°F (371°C). Effective removal of the suspended mineral matter without loss of solvent or coal is one of the critical operations of the project.

Solvent oil with the wet filter cake is removed in a rotary, indirect-fired drier in an inert-gas atmosphere and is returned to the solvent processing system. The carbon char that remains undissolved could find use in either the on-site hydrogen generating unit, or possibly as a fuel/raw material in cement manufacture.

After filtration, the coal solution is reheated and vacuum-flashed to remove the solvent oil. Since some oil is produced in the process, there is a net gain of solvent of about 15 percent. The solvent is refined by conventional distillation equipment to the degree necessary to provide recycle solvent oil. The melting point of the refined coal can be controlled by the degree of removal of solvent oil. The excess oil can be further refined to provide chemical feedstock.

After flashing of the solvent, the refined coal may then be stored molten at temperatures over 300°F (149°C). The coal also can be stored solid as a prill, or flake, or in large-bulk solid blocks, similar to the method of storing sulfur.

Solvent refined coal has been produced—with laboratory and pilot-plant tests showing a sulfur content as low as 0.5 percent and with indication that the sulfur content can be kept under 1 percent, a value that makes the product environmentally acceptable in most states. Ash content is reduced to about 0.1 percent from most coals.

Organic sulfur in the raw coal (but no pyritic sulfur) is converted mostly to hydrogen sulfide and disengages with the evolved gas containing recycle hydrogen and carbon dioxide, as well as some light hydrocarbons. Conventional proven gas-desulfurization processes, such as Stretford, Benefied, or Claus, can remove the sulfur compounds from the gas stream.

APPRAISAL

Production of an attractive fuel by solvent refining has a number of advantages, particularly for the interim period before gasification is established as an economical, proven operational process. The SRC process is attractive to the electric generating utility industry because existing steam generators can be retrofitted to use this material with a minimum change-over. In addition to purifying the coal, the energy value is concentrated to a consistent 16,000 Btu per pound, thus minimizing transportation charges. As contrasted with an on-site low-Btu gasifier, solvent-refined coal can be stockpiled to accommodate peaking requirements. No catalyst is required, and the quantities of hydrogen required are low. A 6-ton-per-day pilot plant has been operated for a short time by The Southern Company in Alabama. Process data for its design were based on the original Spencer Chemical and OCR (Office of Coal Research) sponsored work. Operation of the later 50-ton-per-day plant previously mentioned will further advance solvent refined coal-process technology.

Coal Gasification at Sasol (Republic of South Africa)

BY JAN C. HOOGENDOORN*

Editor's Note: This technical paper is noteworthy on several counts: (1) The planning that ultimately brought this facility into being commenced as early as the mid-1920s. (2) The production of synthetic motor fuels, pipeline gas, ammonia, and chemicals, where coal gasification is the key to successful production, is being conducted in a privately operated and managed plant which was completed in 1955, thus representing a major pioneering venture in the area of coal technology. (3) Extensive experience has been gained from a plant that has been operating for some twenty years, as contrasted with numerous coal-conversion programs and processes, some of which are still on the drawing board; others are in a pilot-plant phase or very early stage of full production. (4) The plant demonstrates the advantages of a close geographical tie between source of coal and coal conversion. (5) Cost figures and technical expertise gained over the years reveal interesting concepts in technological trade-offs, notably important because of attention being given to continuing improvements and expansions by this South African firm.

A new plant, Sasol II, got underway in 1975. The operations will be based upon the coal fields in the Eastern Transvaal Highveld. The new plant will produce more than ten times the volume of the first plant. Together, the two plants will produce approximately 40 percent of South Africa's petrol requirements in the late 1970s. In addition to fuels, Sasol II will produce ammonia, sulfur, ethylene, and propylene, with special emphasis on producing materials for the manufacture of fertilizers.

South Africa has an abundance of minerals, but in exploration activities to date, there are no signs of important oil deposits. The possibility of producing hydrocarbons from coal has attracted the attention of scientists and economists practically worldwide for many years. In 1927, a White Paper was already published discussing the processes then available for production of oil from coal. Developments in Germany were closely followed and especially the Fischer-Tropsch process, as its operating conditions did not appear to be very extreme and the process had already been demonstrated in a number of plants. A South African mining corporation, the Anglo Transvaal Consolidated Investment Co., better known as Anglo Vaal, acquired in 1935 the South African rights to the German Fischer-Tropsch process.

During the next few years Anglo Vaal devoted much attention to the development of a scheme for the production of oil from coal. Tenders were asked for, but because of the complications of the war, no orders were placed. However, during the war and in the postwar years, Anglo Vaal remained in close contact with developments. In 1943 negotiations were held in America which led to the procurement of the rights of the American variation of the Fischer-Tropsch process. In 1946 a new study was undertaken, and an application was made to the government to create a suitable framework within which a long-term industry could be established. During 1947 the liquid fuel and oil act was passed, and in 1950 an agreement was reached between the South African government and Anglo Vaal in which the Anglo Vaal rights were taken over by the government. The South African Coal, Oil and Gas Corporation Ltd. was formed and incorporated under the companies' act as an ordinary public company. The government appoints the majority of directors, including the chairman, and the remaining directors are appointed by the Industrial Development Corporation, which is a government-owned organization with the objective, as the name implies, to stimulate industrial development in the country. Sasol operates like a normal business concern, with an autonomous board of directors and is subject to South African company law and taxation like any other company.

A site for the plant was selected close to the banks of the Vaal River which is South Africa's major source of water supply, 50 miles south of Johannesburg and on top of a vast coal field. Sasol acquired approximately 8000 acres for the plant and its township which was to have the name of Sasolburg. The site was in the middle of an area where cattle grazing and corn production were the only activities and Sasol had to create its own infrastructure from scratch.

*Manager, Research and Development, South African Coal, Oil and Gas Corporation Limited, Sasolburg, Republic of South Africa.

SELECTION OF PROCESS

Although it was clear that the plant would be based on the synthesis of hydrocarbons from hydrogen and carbon monoxide as invented and developed by Fischer and Tropsch, it still had to be decided which processes to choose for the individual steps in this integrated complex. For gasification, the Lurgi pressure gasification with steam and oxygen was selected because this process had already been demonstrated in gasifiers of a smaller size. It had the advantage of being able to work on the rather low grade, high ash

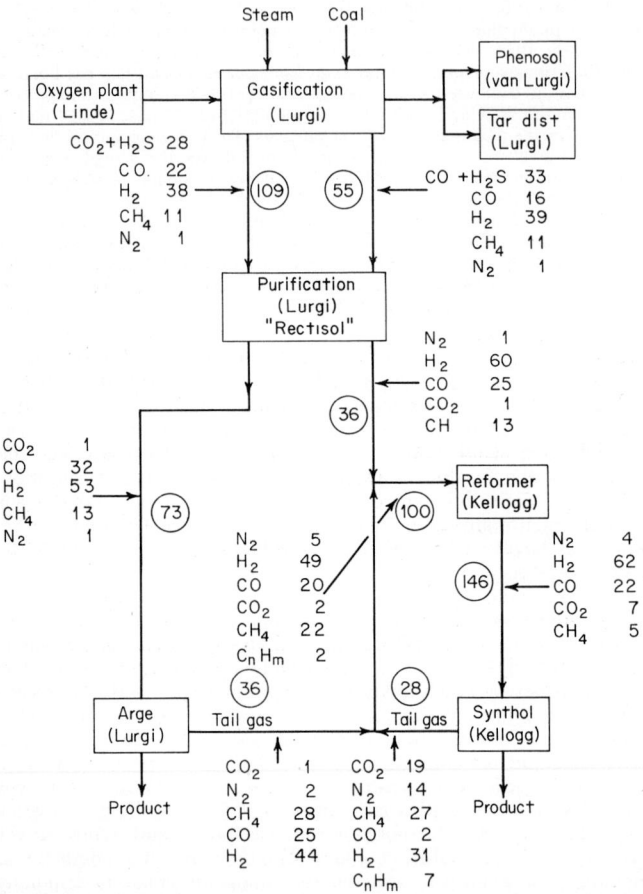

Fig. 1. Basic flowsheet of original plant at Sasol. Compositions are given in volume percent. Volumes (within circles) are 10^6 standard cubic feet per day. *(Reproduced by permission of the source: South African Coal, Oil and Gas Corporation Limited.)*

coal available to Sasol. The fact that it operated at a pressure of approximately 350 psi, which is also the desired operating pressure for the Fischer-Tropsch plant, was an additional advantage. This avoids cumbersome compression of large volumes of gas arising from low-pressure gasification.

The raw gas from such a gasification system, of course, contains, apart from the hydrogen and carbon monoxide, appreciable quantities of undesired components, such as unsaturated hydrocarbons, sulfur compounds, and so on. Moreover, the raw gas contains a large percentage of carbon dioxide which must be brought down to a lower level. A number of possibilities for purifying the gas existed, all involving at least two or three different process steps. However, in the late 1940s, Lurgi in cooperation with Linde of Germany, had developed a combined gas-purification process which used one solvent, methanol, at low temperatures as the single absorption agent for removing all undesired components from the gas. A full-size plant did not exist, but the pilot-plant work was convincing enough to justify its selection for the Sasol plant.

The gasification process produces, as a side stream, a gas liquor in which components, such as ammonia and phenols, are dissolved. Obviously, such a gas liquor cannot be disposed of before treatment, and for this treatment another "first" was chosen, the Phenosolvan process in which phenols are extracted from the water with a solvent, such as butyl acetate. The ammonia can then be recovered by stripping with steam and converted into, for instance, ammonium sulfate. An additional advantage of having the gasification, gas-purification, and gas-liquor treatment all from one supplier was that the responsibility for the performance of these plants, which are all to a certain extent interrelated, was concentrated with one company.

On the Fischer-Tropsch process itself, the choice was not easy. On the one hand, there was the German design as developed by the joint venture of Ruhrchemie and Lurgi of Germany, which went under the name of Arge. It used a fixed-bed reactor system which was developed by Lurgi in Germany and was known to work. The reaction took place in long tubes surrounded by a bath of boiling water for temperature control, and the only difference between the small demonstration reactors and the proposed reactors for the Sasol plant was the number of tubes in one shell. This was not expected to give scale-up problems. On the other hand, there was the American-developed moving-bed reactor type, using a fluidized catalyst on which only pilot-plant data were available, but which offered the opportunity of building reactor units with a much higher capacity. Although the basic chemistry for both reactor types is the same, the fixed-bed reactor produces, in general, straight-chain hydrocarbons with a high average molecular weight, and most of the production is in the range of Diesel oil and paraffin waxes. The fluid-bed process produces branched olefins of a low average molecular weight, and the production is mainly in the range of LPG and gasoline. In view of the uncertainties, the decision was taken to build two synthesis plants in parallel, using both systems.

The basic flow-sheet of the original plant appears in Fig. 1, in which the main process units are shown and the main gas flows are indicated with volumes in standard cubic foot (SCF) per day. This figure also shows the approximate gas compositions in volume percent. As the Lurgi gasification operates under pressure, a fair amount of methane is produced. With Sasol coal, this amounts to approximately 12 percent of the pure synthesis gas. Both synthesis plants also produce some methane as the lowest hydrocarbon in the range, and there will be a tendency for methane to build up in the cycle. To make full use of the gas, it was necessary to reform methane back to hydrogen and carbon monoxide for recycling into the system. Partial oxidation was selected for this process step, as in the early 1950s technology and tube materials for catalytic reforming in tubular furnaces under pressure of 300 psi were not yet available. Another complication is the presence of nitrogen and argon in the gas, most of which is introduced into the gas stream with the 98 percent oxygen. The nitrogen and argon act as inerts in the system and have to be removed as a purge to keep them within an acceptable level.

The original flow sheet made provision to send approximately two-thirds of the pure synthesis gas to the fixed-bed reactors, and to send the tail gas of that system, with the increased methane content, to the reforming plant where it was converted, together with the remaining one-third of the pure gas, into feed gas of the right composition for the fluid-bed plant. The tail gas of this plant is also recycled to the reforming units after removal of carbon dioxide. Also indicated in Fig. 1 are the companies that did the process design. Construction started toward the middle of 1952, and the first units were put into operation toward the end of 1954. By the end of 1955 all the main construction work was completed. The real cost with the full plant operating, including the capitalized cost of start-up and essential modifications and additions, but excluding cost of later improvements, was $142 million in 1956. The first estimated cost of the project in 1951, based on contractors' information and presented for approval to the government, was $75 million. A later estimate in 1952, using more complete information on off-sites, and so on, came to $101 million, including $9 million as a provision for escalation. All figures, of course, are preinflation dollars and are not valid for current evaluation, but they do show clearly the financial pitfalls of transforming a conceptual process into an operating unit.

SCOPE AND START-UP

More than 16 million man-hours and more than 58,000 cubic yards of concrete went into construction of the plant. Approximately 32,000 tons of equipment were imported, and an equivalent amount of structural and reinforcing steel and equipment was supplied by South African industries. A total of some 200 miles of pipe was laid, ranging from 5 ft in diameter to a few fractions of an inch. During the peak of construction activities, up to 3700 workers were engaged on the site. The first half-year of the start-up period was used for bringing on line the steam station, power generation, oxygen plant, gasification, gas purification, Phenosolvan, and the supporting services, such as tar distillation, biological treatment of effluent, and so on. In August 1955 the synthesis plants were brought on line. It became clear very soon that the decision to build two synthesis plants was indeed a wise one. The fixed-bed plant came on line with only minor problems and started to contribute immediately to the income of the company, but the fluid plant appeared to behave quite differently to what was predicted in the pilot-plant programs. After some modifications to the equipment, it became possible to make short production runs and to get a contribution to the income, although the production per reactor was far below design. Because of the process and mechanical problems, the availability of the two reactors was also low, and it was necessary to add, as soon as possible, a third reactor to reach the design production and to make full use of the available synthesis gas.

The catalyst for both synthesis plants is based on iron activated with certain promoters. Originally, the catalyst for the fixed-bed plant was bought from Germany, but in 1969 Sasol commissioned its own catalyst preparation plants, and the German production facilities were closed down. The catalyst preparation plant for the fluid-bed system was part of the original project and was based on the use of certain iron ores. It soon became obvious that the iron ore specified for the plant was not of a very constant composition, and there was not sufficient control on the level of impurities. Intensive research work in the Sasol laboratories, aimed at establishing a process with economic run lengths and control over the product composition, led to the development of a completely new catalyst, which was still based on iron, but could be produced with consistent qualities so that the performance of each run could be accurately predicted and which offered the possibility of modifying the product properties at will by changing the recipe. Other factors, such as operating temperature, pressure, and gas composition, also play an important part, and so by choosing the proper combination of process variables and catalyst composition it is possible to get a wide range of selectivity.

PROCESS COMPONENTS AND VARIATIONS

The possibilities for varying production are illustrated in Table 1. In the first column, the selectivity is shown for the conditions as normally used at Sasol for producing hydrocarbons mainly in the range of gasoline. However, the right-hand columns give indications of what has been obtained in a pilot plant where the objective was to produce a range of hydrocarbons that could be of importance to certain areas in the United States, where a high production of methane would be useful for SNG, and the light hydrocarbons would be useful as feedstocks for the petrochemical industry.

At the end of the 1950s, when the initial problems of the plant had largely been overcome, plans for further expansion took shape. Sasol is close to the industrial area of South Africa, which is concentrated in the Witwatersrand (center of gold mining activity), and inasmuch as no natural gas is available, it was decided to supply this area with a clean fuel gas of constant composition obtained from coal. A distribution company for operating the pipeline system was created as a subsidiary of Sasol, and it was decided to supply a gas with a heating value of 500 Btu per standard cubic foot, which is the approximate heating value of the tail gas of the Fischer-Tropsch synthesis plants.

By using this tail gas for industrial gas, the thermally inefficient reforming step could be partially eliminated. To keep the synthesis units supplied with sufficient feed gas, a further three gasifiers and another Rectisol stream were added. Also, the tail gas from the Synthol reactors has a high concentration of hydrogen, and, as previously stated, some of the gas must be purged to keep the nitrogen percentage at an acceptable level. This gas which is extremely pure is a good feedstock for ammonia-synthesis gas production, and in

TABLE 1. Synthol Selectivities:* Comparison of Pilot-Plant and Commercial Results

Component	Commercial run			Pilot plant		
	Start	End	Average	A	B	C
CH_4	7	13	10	30	50	70
C_2H_4	4	3	4			
				15	17	12
C_2H_6	3	9	6			
C_3H_6	10	13	12			
				15	11	6
C_3H_8	1	3	2			
C_4H_8	7	9	8			
				16	13	6
C_4H_{10}	1	2	1			
C_{5+}	6	9	8			
L.O.	40	30	35	22	8	6
D.O.	14	2	7			
N.A.C.	6	6	6	2	1	0.2
Acids	1	1	1	<0.05	<0.05	<0.01

*Selectivity is C atoms converted.
L.O. = Light oil: initial boiling point, 50°C; boiling range, 50–290°C.
D.O. = Decanted oil: initial boiling point, 160°C; boiling range, 160–490°C.
N.A.C. = Nonacid chemicals.
Acids = Water-soluble acids.

1964 an ammonia plant was commissioned which obtained its feed gas from a low-temperature separation unit in which part of the Synthol tail gas is split into its components. The methane fraction from this low-temperature unit is used as a blending gas stream for the gas distribution system to help control the heating value of the pipeline gas. In the same period it was decided to expand the Sasol activities further into petrochemicals by adding an ethylene plant which uses imported petroleum naphtha as feedstock, a styrene plant, and a butadiene plant. These plants are of the conventional type and not based on coal. An exception, however, must be made for the butadiene plant which gets its feed partly from the C_4 fractions from the naphtha cracker, but also from a dehydrogenation plant where normal butylenes from the Fischer-Tropsch synthesis are catalytically dehydrogenated. The butadiene is extracted from the crude C_4 streams with acetonitrile.

Further improvements in the equipment and improved knowledge of operating methods have continuously led to higher capacity in terms of instantaneous production per unit as well as increased availability, and the present gas production has reached a monthly average of 330 million standard cubic feet of raw gas per day from 13 installed gasifiers. A typical daily gas balance for the use of this raw gas in the synthesis sections for hydrocarbons and ammonia and for pipeline gas is shown in Fig. 2. The methane concentration in the feed to the reformers is now 38 percent compared with 22 percent in the original flow sheet (Fig. 1). The thermal efficiency of the overall process is, therefore, improved not only by decreasing the total amount of methane to be reformed, but also by feeding the balance of the methane to the reformers in a higher concentration.

Coal Supply.   Sasol is supplied with coal from its own coal mine, Sigma Colliery. The first coal was brought to the surface in 1953, and since then 57 million short tons have been mined. Current production amounts to 5 million tons per year. The colliery was originally designed for 39,000 tons per week, but mining operations have gradually been expanded and the present maximum production capacity is an average rate of 110,000 tons per week. Coal rights are owned over an area of some 16,000 acres. Mining operations are concentrated in two main areas, each of which is equivalent to a large colliery, producing in excess of 44,000 tons per week, respectively. Each area has its own network of service, incline and ventilation shafts, conveyors, underground crushers, and surface storage bunkers.

Sigma produces a low-grade coal with an inherent ash content which may exceed 35 percent, while the average calorific value is about 8300 Btu per pound. See Table 2. The in-situ coal reserves are of the order of 1 billion tons. The coal is present in three seams. Coal in the bottom seam is the oldest of the formation, the average minable height of the seam being 10 feet. The middle seam is separated from the bottom seam by a layer of conglomerate, the thickness of which varies considerably. The middle seam appears on two horizons, being split in two by a parting of mudstone. The average mining heights for these two portions of the middle seam are 8 and 10 feet. Most of the coal for the Sasol plant is produced from the middle seam. The top seam is separated from the middle seam by layers of shale, mudstone, and sandstone, varying in total thickness between 40 and 90 feet. The average mining height is 9 feet. The roof strata of the top seam consist of mudstone grading into carbonaceous shale, followed by shale overlain by 100 to 200 feet of white sandstone.

At present, the mechanized board-and-pillar mining method is standard practice. Roadways, on average 20 feet wide and 9 feet high, are mined parallel to each other on a horizontal plane in the coal seam. The roadways are spaced 65 feet apart, center to center, and as they advance, interconnecting splits are mined at 65-foot centers. Thus, pillars of

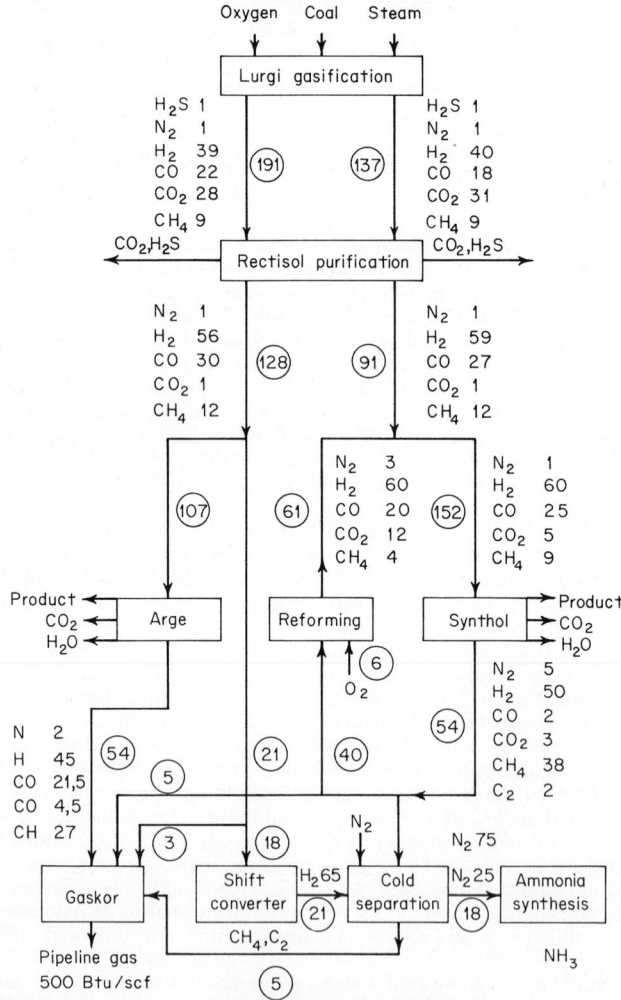

Fig. 2. Simplified gas flow at Sasol installation. Compositions are given in volume percent. Volumes (within circles) are 10^6 standard cubic feet per day. (*Reproduced by permission of the source: South African Coal, Oil and Gas Corporation Limited.*)

TABLE 2. Analysis of Sasol Coal

	% weight
Moisture as received	10.7
Volatiles (dry basis)	22.3
Ultimate analysis (dry basis):	
Ash	35.9
Sulfur	0.5
Nitrogen	1.2
Carbon	50.8
Hydrogen	2.8
Oxygen	8.8
Ash properties:	°C
Softening point	1340
Melting point	1430
Fluid point	1475
Heating value (gross)	8380 Btu/lb (dry)

46 feet square are left between the network of roads and splits and serve as one medium of support. The coal requirements of the factory are provided by eight producing sections, each mining on a double-shift basis.

The complete mining cycle comprises six activities: (1) Cut face with coal cutter, (2) load back duff from face, (3) drill shot holes, (4) charge up holes and blast, (5) load coal with loader into shuttle cars and transport to feeder, and (6) support exposed roof with roof bolts.

The output per section is determined by (1) geological conditions of coal seam, (2) pillar centers, (3) board dimensions, and (4) combination of equipment used. The production is further determined by effective available working time which is influenced by traveling time between sections and shafts. Because of geographical layout, traveling time can vary considerably between different areas. The average section productivity is 1120 tons per section per shaft.

Experiments have been conducted at Sigma regarding the feasibility of longwall mining practice (1965–1969), but because of high unit cost and unsuitability of available equipment at the time, the experiments were abandoned. However, as a result of subsequent advances in longwall equipment and its application, the possibility of reintroduction of longwall mining practice is being studied. The layout of the primary underground entries is suitable for longwall mining, thus ensuring sufficient flexibility to revert to the longwall method at this stage. A longwall face has since been started.

The colliery is served by eight shafts: three vertical shafts for ventilation, three vertical shafts for service, both men and material, and two inclined shafts for coal hauling, material, and equipment. Underground, the coal is transported on a 48-inch-wide trunk conveyor system in each area which delivers the coal to primary crushers situated at the bottom of the inclined shafts. Conveyors of 48-inch width transport the coal to the surface and discharge into surface storage bunkers of 12,000 tons total capacity. On the surface, conveyor belts feed the coal from the two storage bunkers to the secondary crushers and screens. The final product consists of two coal sizes: −0.4 inch and between 0.4 and 2 inches, which are transferred separately on twin 32-inch conveyor systems to the Sasol factory storage bunkers which have a capacity of 60,000 tons. The −0.4-inch product is used in the power station, while the +0.4-inch product is used in the gasification section. The distribution between the two fractions is roughly 50:50. The total staff of Sigma colliery, including workshop personnel and office staff, is 1600.

Synthesis Units.   A schematic view of a Synthol reactor is given in Fig. 3. See also Fig. 4. Fresh feed, combined with recycle gas, enters at a temperature of approximately 320°F (160°C) and pressure of approximately 300 psig. Catalyst at about 660°F (349°C) flows down a standpipe through the slide valves and is mixed with the gas at point C. The temperature of the gas and catalyst mixture rapidly comes to equilibrium by the intimate mixing of the gas and catalyst. The temperature of the mixture increases owing to the heat liberated as the Fischer-Tropsch and water-gas shift reaction is removed in the two coolers. In the settling hopper, the product gas separates from the catalyst, and the gas

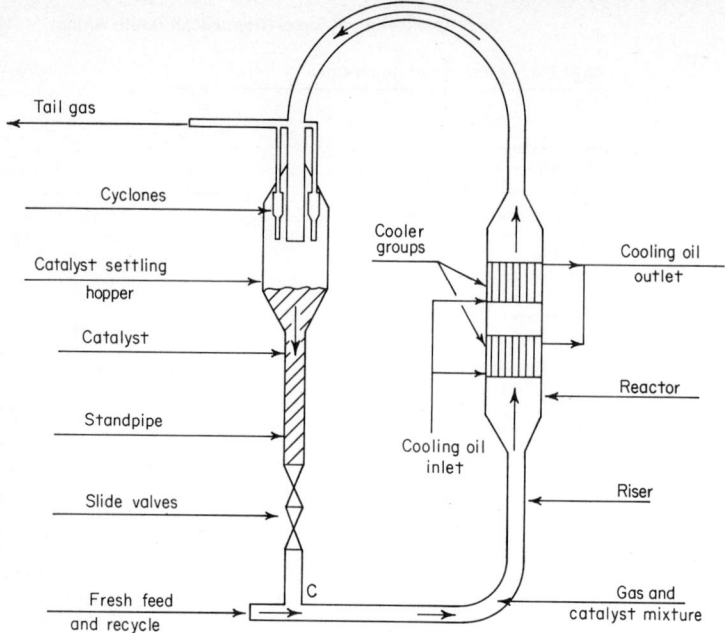

Tail gas

Cyclones

Catalyst settling
hopper

Catalyst

Standpipe

Slide valves

Fresh feed
and recycle

Cooler
groups

Cooling oil
outlet

Cooling oil
inlet

Reactor

Riser

Gas and
catalyst mixture

C

Fig. 3. *Synthol* reactor. *(Reproduced by permission of the source: South African Coal, Oil and Gas Corporation Limited.)*

Fig. 4. *Synthol* reactors at Sasol.

stream leaves the reactor system and enters the scrubbing tower, which is not shown in the diagram. The catalyst flows down to complete the cycle. The catalyst, which consists of iron particles with promoters, functions both as a catalyst and a heat-transfer medium. The tail gas with the hydrocarbon products is cooled, whereby most of the hydrocarbon products condense. Part of the tail gas is recycled to the reactor so that the combined feed to the reactor consists of a mixture of fresh feed and recycle gas. The lighter hydrocarbons do not condense completely and an absorber system is used for their recovery.

The fixed-bed reactor system is shown in Fig. 5. Each reactor consists of a shell of

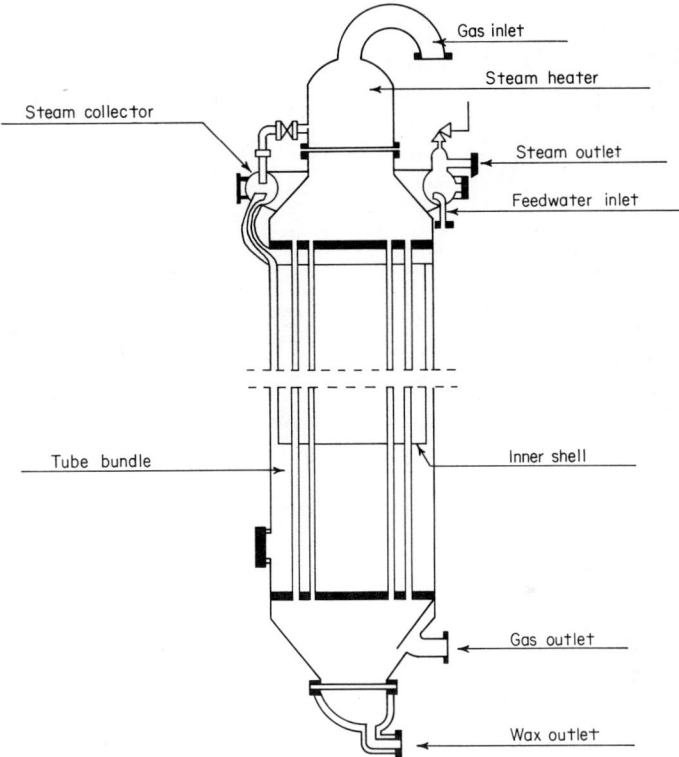

Fig. 5. *Arge reactor. (Reproduced by permission of the source: South African Coal, Oil and Gas Corporation Limited.)*

approximately 10 feet diameter and a height of approximately 42 feet. Inside the shell there are 2000 tubes with a diameter of 2 inches which contain the pelletized iron catalyst. The tubes are surrounded by boiler feedwater, and steam is raised in the reactor. By controlling the steam pressure, the temperature of the boiler feedwater is controlled, which in its turn controls the rate of heat transfer from the catalyst through the tube well into the water. In this way, a very fine control of the reaction temperature is possible, and overheating of the catalyst can be avoided. The operating temperature has a great influence on the conversion per pass, as well as on the product selectivity and is of extreme importance. The fresh gas produced by the Rectisol purification plant (see Fig. 6) is mixed with recycle gas and, after compression by the recycle compressor and preheating in a heat exchanger, enters the reactor at the top. At the bottom of the reactor, the gas separates from the heaviest hydrocarbons which are obtained as reactor condensate, and the hot gas changes heat with the incoming feed gas and is further cooled in condensers. The heat exchangers and the condensers produce hydrocarbon condensates which are, after pressure release and recovery of the dissolved gas, sent to the refinery. A part of the

Fig. 6. *Rectisol* plant (gas purification) in Sasol complex.

TABLE 3.   Comparison of Arge and Synthol Processes, including Selectivities*

Characteristic	Process	
	Arge	Synthol
Temperature, °F	Avg. 450	625
Pressure, psig	370	320
Conversion of $(CO + H_2)$ entering	65%	85%
H_2:CO ratio	1.7	2.8
Selectivities: CH_4	5.0	10.0
C_2H_4	0.2	4.0
C_2H_6	2.4	6.0
C_3H_6	2.0	12.0
C_3H_8	2.8	2.0
C_4H_8	3.0	8.0
C_4H_{10}	2.2	1.0
C_5 +	3.5	8.0
Petrol: C_5-C_{12}	19.0	31.0
Diesel: $C_{13}-C_{18}$	15.0	5.0
Heavy oil $C_{19}-C_{21}$	6.0	1.0
and wax $C_{22}-C_{30}$	17.0	3.0
$C_{31}\longrightarrow$	18.0	2.0
N.A.C.	3.5	6.0
Acids	0.4	1.0
	100.0	100.0

*Selectivity is C atoms converted. N.A.C. = Nonacid chemicals.

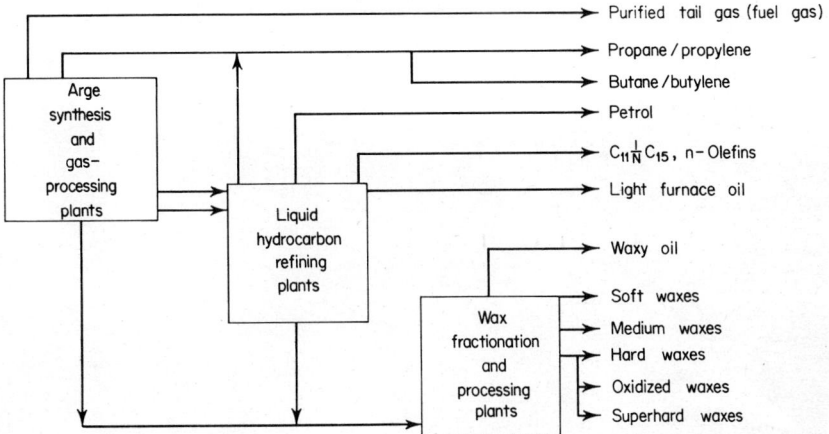

Fig. 7. Final products of the *Arge* process. *(Reproduced by permission of the source: South African Coal, Oil and Gas Corporation Limited.)*

tail gas leaving the condensers is used for recycle, whereas the net tail gas made is either used for supply to the gas distribution system, or during periods of low demand, returned to the methane reforming section.

Although the chemistry of the Fischer-Tropsch reaction is the same in the fixed-bed reactor and in the fluid-bed plant, and the catalyst in both plants consists of iron with promoters, the different reaction conditions, pressure, temperature, feed-gas composition, cause not only a difference in product selectivity, but also a difference in the properties of hydrocarbon products with the same boiling range.

A comparison of possible operating conditions for the two reactor types is shown in Table 3. The selectivity for the Synthol process is that normally obtained in the Sasol plant. The objective here is to produce mainly hydrocarbon liquids in the gasoline boiling

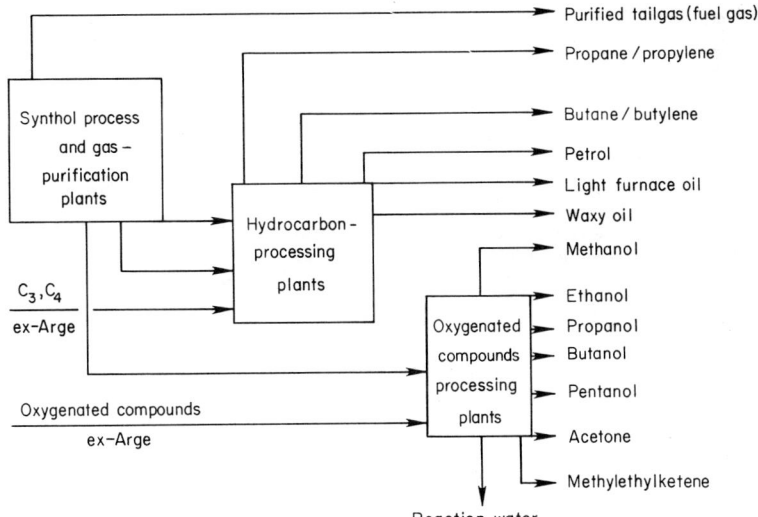

Fig. 8. Final products of the *Synthol* process. *(Reproduced by permission of the source: South African Coal, Oil and Gas Corporation Limited.)*

Fig. 9. General view of part of Sasol plant.

range. Propylene and butylene can also be converted to gasoline in the refinery. If so desired, the selectivity could be increased toward the formation of more methane and light hydrocarbon gases, and such a plant could be a source of petrochemical feedstocks as well as substitute natural gas (SNG). The fixed-bed plant at Sasol is operated for a maximum of heavy hydrocarbons, and in this operation relatively little gasoline is pro-

TABLE 4.   Comparison of Products Produced by Arge and Synthol Processes

Classes of compounds, %	Process			
	Arge		Synthol	
	C_5–C_{12}	C_{13}–C_{18}	C_5–C_{10}	C_{11}–C_{14}
Paraffin	53	65	13	15
Olefins	40	28	70	60
Aromatics	0	0	5	15
Alcohols	6	6	6	5
Carbonyls	1	1	6	5
n-Paraffin	95	93	55	60

duced which, because of the straight-chain nature of the product, has a low octane number. It produces, for the same reason, an excellent Diesel fuel. The straight-chain hydrocarbons are also an interesting feedstock for the production of chemicals, such as biodegradable detergents. The properties of typical synthesis products are compared in Table 4.

Schematic summaries of the commercial products made in the Sasol plant from the two synthesis systems are given in Figs. 7 and 8. A general view of the Sasol plant is given in Fig. 9.

Gas Technology

A. B. CROSSMAN, B.A., B.S.C.E. *Vice President, Brown & Root International Ltd., Houston, Texas. Registered Professional Engineer (Alaska, California, Louisiana, Texas) (Marine Pipelines)*

RICHARD K. DORAN, M.E., M.S. *Associate Director, Gas Requirements Agency, Denver Research Institute, University of Denver, Denver, Colorado. Member, American Economics Association, National Association of Business Economists, and American Institute of Mining, Metallurgical, and Petroleum Engineers. (Future Gas Consumption of the United States)*

HENRY E. DUCKHAM, JR., B.S.Ch.E. *Process Manager in Cryogenics, The M. W. Kellogg Company, a division of Pullman Incorporated, Houston, Texas. Member, American Institute of Chemical Engineers. Registered Professional Engineer (New York). (Chemical Feedstocks from Natural Gas)*

E. J. FISCH, B.S. *Staff Engineer, Process Engineering-Licensing Department, Shell Oil Company, Houston, Texas. (Absorption of Acidic Gases from Natural Gas and SNG)*

P. N. GLOVER, *Former Chairman, Editorial Committee of Potential Gas Committee, Mineral Resources Institute, Colorado School of Mines Foundation, Inc., Golden, Colorado. (Potential Supply of Natural Gas in the United States)*

E. S. HILL, Ph.D. *Consultant, Houston, Texas. Fellow, American Association for the Advancement of Science. Member, American Institute of Chemical Engineers. Registered Professional Engineer (California). (Absorption of Acidic Gases from Natural Gas and SNG)*

G. C. HUMPHREYS, Bach. Tech. Sci. *Head of Methanol Group, Process Engineering Department, Davy Power-*

gas Ltd, London, England. Member, Institution of Chemical Engineers. Registered Chartered Engineer in Great Britain. (Methanol Production and Fuel Methanol vs LNG)

J. C. JANKA, B.S., M.S.Ch.E., Supervisor, Hydrocarbon Processing Research, Institute of Gas Technology, Chicago, Illinois. Member, American Institute of Chemical Engineers. (Hydrocracking-Hydrogasification Process for Producing Pipeline Gas from Crude Oil)

J. KEVIN JONES, First Class Honors Degree (Chem. Eng.). Managing Director, Chem Share Process Systems Ltd., London, England. Member, Institute of Chemical Engineers, London. (Cryogenic Upgrading of Low-Btu Gases)

C. J. KUHRE, B.S., M.S.Ch.E. Process Engineer, Shell Oil Company, Houston, Texas. Member, American Chemical Society, American Institute of Chemical Engineers. Registered Professional Engineer (California, Texas). [Noncatalytic, Partial-Oxidation Gasification Process (Shell)]

JOSEPH F. McMAHON, B.S., M.S.Ch.E. Manager, Process Technology Department, Process Plants Division, Foster Wheeler Energy Corporation, Livingston, New Jersey. Member, American Chemical Society. Licensed Professional Engineer (New Jersey). (The Fluidized Bed Hydrogenation Process for Substitute Natural Gas Production)

SHOGO NOJIMA. Director, Japan Gasoline Co., Ltd., Tokyo, Japan. (Methane-Rich Gas Process for Substitute Natural Gas)

E. A. RASSINIER, B.S. Chairman, Potential Gas Committee, Mineral Resources Institute, Colorado School of Mines Foundation, Inc., Golden, Colorado. (Potential Supply of Natural Gas in the United States)

J. G. SEAY, B.S., M.S.Ch.E. Senior Advisor, Management Science, Institute of Gas Technology, Chicago, Illinois. Member, American Institute of Chemical Engineers, Society of Plastics Industries. (Hydrocracking-Hydrogasification Process for Producing Pipeline Gas from Crude Oil)

J. A. SYKES, JR., Ph.D. *Formerly with Shell Oil Company, Houston, Texas. Member, American Institute of Chemical Engineers, Registered Professional Engineer (California, Texas). [Noncatalytic, Partial-Oxidation Gasification Process (Shell)]*

P. B. TARMAN, B.S.,M.S.Ch.E. *Director, Chemical Processing Research, Institute of Gas Technology, Chicago, Illinois. Member, American Chemical Society, American Institute of Chemical Engineers. (Hydrocracking-Hydrogasification Process for Producing Pipeline Gas from Crude Oil)*

COLIN H. TAYLOR, B.Sc. *Chief Process Engineer (Gas), Woodall-Duckham Limited, Crawley, Sussex, England. Member, Institution of Chemical Engineers. Chartered Engineer in Great Britain. (The Catalytic Rich Gas (CRG) Process for Gasification of Light Hydrocarbons)*

R. W. VAN SCOY, Ph.D. *Process Engineer, Retired from Shell Oil Company, Houston, Texas. (Absorption of Acidic Gases from Natural Gas and SNG)*

C. W. WARNER, B.S.,M.S.E.E. *Development Consultant, Cutler-Hammer, Inc., Milwaukee, Wisconsin. Member, American Society for Testing and Materials, Institute of Electrical and Electronics Engineers. (Calorimetric Measurement of Gaseous Fuels) (Gas-mixing Control Systems)*

Natural Gas Production and Reserves in the United States

The workings of the American Petroleum Institute (API) in the collection, tabulation, and dissemination of statistics pertaining to petroleum production and reserves are described in the Handbook section on *Petroleum Technology,* under "Petroleum Production and Reserves in the United States." In a very similar fashion, the American Gas Association (AGA) collects and tabulates statistics on natural gas and natural gas liquids. Other agencies, including the U.S. Bureau of Mines and the U.S. Geological Survey, also compile natural gas statistics. In all summaries of statistics in this Handbook, it is stressed that, at best, estimates of reserves must be considered approximations, constantly subject to improved collection of data, new discoveries of resources, and factors that disturb the relationship between rate of discovery (new additions) and rate of use (production). Because of the great importance of understanding the guidelines and definitions which apply to the collection of estimates, a discussion of these factors appears later.

Although crude petroleum and natural gas are frequently associated, there are not exact parallels. Hence the structure of the Subcommittee Organization of the AGA for collection of gas statistics differs from that of the API for crude petroleum. The AGA Reserve Committee geographical areas are:

1. Appalachian Area.   Kentucky, Maryland, New York, Ohio, Pennsylvania, Tennessee, Virginia, and West Virginia.

2. Southeast Area.   Alabama, Florida, and Mississippi.

3. South Central Area.   Arkansas and North Louisiana.

4. South Louisiana Area.   South Louisiana including the Offshore Area.

5. Texas Gulf Coast Area.   Texas Districts 1, 2, 3, and 4, including the Offshore Area.

6. Northeast Texas Area.   Texas Districts 5, 6, 7B, and 9.

7. West Texas–South New Mexico Area.   Texas Districts 7C, 8, 8A, and South New Mexico.

8. Mid-Continent Area.   Illinois, Indiana, Iowa, Kansas, Oklahoma, Michigan, Minnesota, Missouri, Texas District 10, and Northeast New Mexico.

9. Rocky Mountain Area.   Colorado, Idaho, Montana, Nebraska, Northwest New Mexico, North Dakota, South Dakota, Utah, and Wyoming.

10. Pacific Coast Area.   Alaska, Arizona, Nevada, Oregon, Washington, and California (subdivided into San Joaquin Valley including the Sacramento Valley area, Coastal Region, and Los Angeles Basin) including Offshore Area.

The AGA Subcommittees are responsible for the determination of reserves and are composed of geologists and engineers from all segments of the oil and gas industry. They are experienced in reserves determination and/or producing capacity estimation and have intimate knowledge of the areas assigned to them. They also have available in company records detailed information of industry operations and developments.

NATURAL GAS RESERVES STATISTICS

The estimated total proved reserves of natural gas in the United States (as of December 31, 1974) and included in the API-AGA Report, vol. 29, released May 1975, are tabulated by major producing states in Table 1. This information is shown graphically in Fig. 1. Definitions of *nonassociated, associated-dissolved,* and *underground storage* are described later. Because gas is commonly transported from producing fields to available underground storage areas many hundreds of miles away from the source and hence closer to areas of consumption, the total reserves shown in Table 1 must be considered in light of the storage figure to indicate the native reserves of a given state.

It is interesting to note that Texas leads with slightly under one-third of all reserves. Louisiana holds well over one-fourth of the reserves; and Alaska well over one-tenth of the reserves. Taken together, these three states hold about three-fourths of all natural gas

TABLE 1.   Estimated Total Proved Reserves of Natural Gas in the United States*
Millions of Cubic Feet—14.73 psia at 60°F

State	Nonassociated	Associated-dissolved	Underground storage†	Total
Alabama	464,760	42,610	0	507,370
Alaska	5,436,788	26,429,824	0	31,866,612
Arkansas	1,934,788	155,058	23,558	2,113,404
California‡	2,273,602	2,588,435	332,555	5,194,592
Colorado	1,609,917	247,421	24,357	1,881,695
Florida	0	308,866	0	308,866
Illinois	1,103	20,741	377,570	399,414
Indiana	2,078	2,648	59,415	64,141
Kansas	11,435,236	170,992	98,503	11,704,731
Kentucky	683,888	42,717	117,397	844,002
Louisiana‡	52,929,627	10,949,026	173,792	64,052,445
Michigan	477,613	455,230	525,411	1,458,254
Mississippi	869,964	117,721	91,735	1,079,420
Montana	676,749	81,272	143,239	901,260
Nebraska	11,884	11,304	31,421	54,609
New Mexico	9,483,527	2,433,917	27,458	11,944,902
New York	62,191	25	103,330	165,546
North Dakota	6,059	426,623	0.	432,682
Ohio	786,412	174,250	347,548	1,308,210
Oklahoma	10,670,558	2,506,095	213,659	13,390,312
Pennsylvania	933,752	11,881	546,512	1,492,145
Texas‡	55,723,891	22,672,909	143,917	78,540,717
Utah	553,738	474,629	3,042	1,031,409
Virginia	44,707	0	0	44,707
West Virginia	1,869,226	51,917	344,438	2,265,581
Wyoming	3,239,711	624,754	52,922	3,917,387
Miscellaneous§	10,453	1,325	156,306	168,084
Total: United States	162,192,222	71,002,190	3,938,085	237,132,497

*Source: AGA–API Report, vol. 29, May 1975. Reserve conditions as of 12/31/74.

†Gas held in underground reservoirs (including native and net injected gas) for storage purposes.

‡Includes offshore reserves.

§Includes Arizona, Iowa, Maryland, Minnesota, Missouri, South Dakota, Tennessee, and Washington.

reserves in the United States. If the three other very large producing states, Oklahoma, Kansas, and New Mexico, are added to the first group of three states, the total of six states (or 22 percent of all states with some production), account for over 89 percent of the total gas reserves. Four other states, California, Wyoming, West Virginia, and Arkansas, account for nearly 6 percent of the remaining 11 percent of total reserves. The geographic distribution of major natural gas fields in the United States (Alaska not shown) is depicted by the map in a subsequent article in this Handbook section on "Gas Pipelines and Underground Storage."

In American Gas Association reporting, the natural gas reserves existing in the Gulf of Mexico are included with the figures for Louisiana and Texas. However, the Gulf reserves also are broken out separately. As of December 31, 1974, the total natural gas reserves estimated for the Gulf of Mexico fields were 35,347,841 millions of cubic feet (30,873,975 for non-associated; 4,473,866 for associated).

ULTIMATE RECOVERY OF NATURAL GAS RESERVES

The categorizing of ultimate recovery of gas reserves by reservoir lithology involves the compilation of such data by three types of reservoirs under the general headings, *sandstone, carbonate,* and *other.* The three types of reservoir lithology are defined as follows:

Sandstone Reservoir.   A reservoir consisting of a sedimentary rock composed predominantly of quartz grains or other noncarbonate mineral or rock detritus. Included in this reservoir type are unconsolidated sand, sandstone, siltstone, graywacke, arkose and

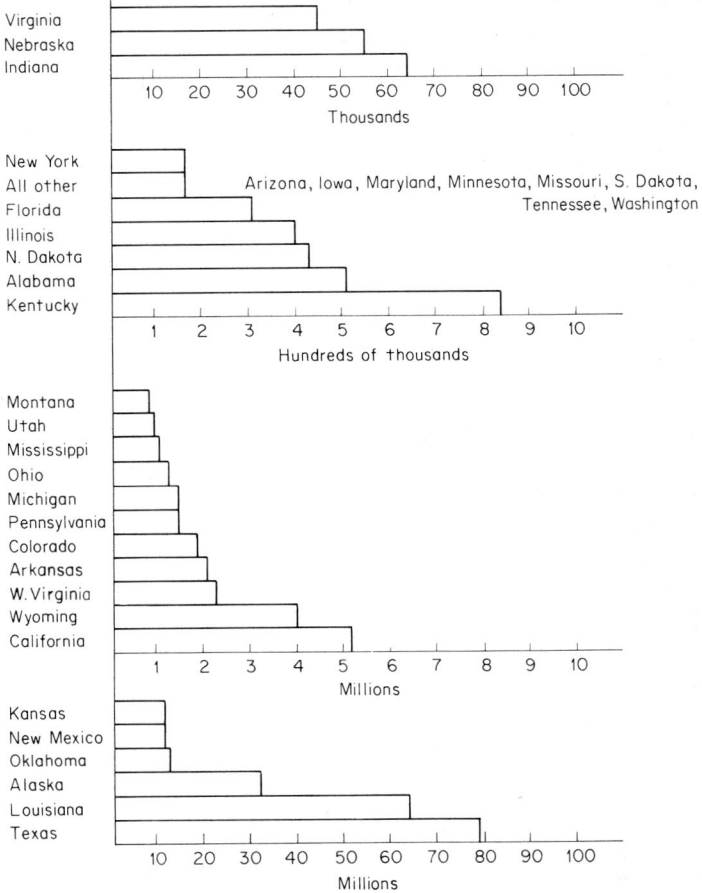

(All millions of cubic feet, 14.73 psia at 60°F)

Fig. 1. Estimated total proved reserves of natural gas in the United States (December 31, 1974).

granite wash, conglomerate and breccia. The cross section of a representative sandstone reservoir is shown in Fig. 2.

Carbonate Reservoir.   A reservoir composed of sedimentary rock made up predominantly of calcite (limestone) and/or dolomite.

Other Reservoirs.   These are reservoirs that are not accurately covered by the foregoing two definitions. Included in this type are igneous and metamorphic rocks and some sedimentary rocks, such as fractured shale.

The estimated ultimate recovery of natural gas in the United States in accordance with reservoir lithology is given in Table 2.

Estimated ultimate recovery is also reported by type of entrapment, of which there are two major types:

Structural Trap.   An entrapment in which migration of hydrocarbons in the reservoir rock has terminated primarily because of closure induced by structural deformation, such as folding or faulting. Within this category also should be included entrapments attributed to hydrodynamic forces.

Stratigraphic Trap.   An entrapment in which migration of hydrocarbons has terminated because of the pinchout of reservoir rock due to either truncation or nondeposition

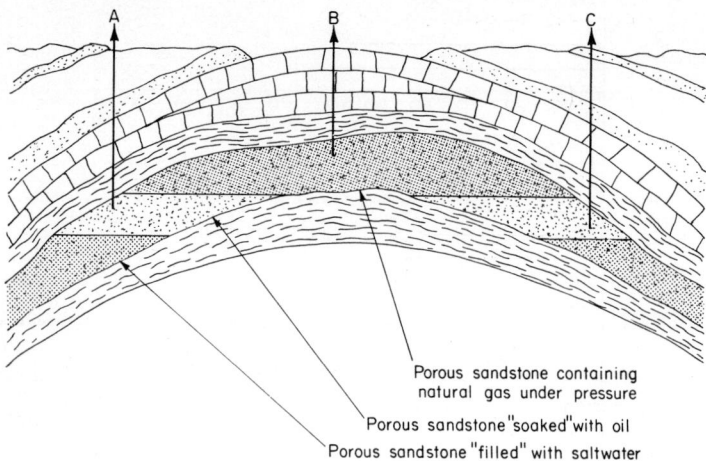

Porous sandstone containing natural gas under pressure
Porous sandstone "soaked" with oil
Porous sandstone "filled" with saltwater

Fig. 2. Cross section of a typical oil field, showing the distribution of natural gas, petroleum, and saltwater in the porous sandstone reservoir. Note that a well drilled at Point B, which penetrates the reservoir, would produce gas only, while holes drilled at Points A and C would yield not only oil, but oil mixed with gas. (*Battelle Memorial Institute.*)

TABLE 2.   Estimated Ultimate Recovery of Natural Gas in the United States by Reservoir Lithology*

Millions of Cubic Feet—14.73 psia at 60°F

State	Sandstone	Carbonate	Other	Total
Alabama	125,195	442,235	0	567,430
Alaska	32,776,795	0	0	32,776,795
Arkansas	3,612,437	1,256,368	156,417	5,025,222
California†	30,778,020	0	954,573	31,732,593
Colorado	4,250,083	109,329	3,884	4,363,296
Florida	0	401,010	0	401,010
Illinois	1,020,883	456,215	0	1,477,098
Indiana	103,938	70,977	0	174,915
Kansas	3,954,420	28,977,255	462,429	33,394,104
Kentucky	574,310	380,403	2,910,261	3,864,974
Louisiana†	175,863,134	4,497,352	0	180,360,486
Michigan	20,009	1,472,793	1,417	1,494,219
Mississippi	5,373,202	515,425	16,946	5,905,573
Montana	1,836,503	181,695	223,344	2,241,542
Nebraska	314,040	24	0	314,064
New Mexico	21,568,245	15,859,619	12,851	37,440,715
New York	497,872	17,130	73,014	588,016
North Dakota	3,688	451,144	703,941	1,158,773
Ohio	4,738,577	1,276,898	52,254	6,067,729
Oklahoma	41,044,652	14,614,800	53,036	55,712,488
Pennsylvania	9,783,820	500	0	9,784,320
Texas	156,002,919	112,283,430	445,730	268,732,079
Utah	1,467,437	643,332	5,175	2,115,944
Virginia	45,387	45,028	23,755	114,170
West Virginia	10,170,416	1,146,456	5,072,975	16,389,847
Wyoming	10,342,052	582,778	6,727	10,931,557
Miscellaneous‡	70,534	18,786	2,600	91,920
Total: United States	516,338,568	185,700,982	11,181,329	713,220,879

*Source: AGA–API Report, vol. 29, May 1975. Reserve conditions as of 12/31/74.
†Includes offshore reserves.
‡Includes Arizona, Iowa, Maryland, Minnesota, Missouri, South Dakota, Tennessee, and Washington.

or a facies change in the form of diminished permeability of reservoir rock. Also included in this category are entrapments, in which a pinchout of facies change provides part of the barrier to migration of hydrocarbons, and structural elements provide the remaining closure for the entrapment. In these instances it is recognized that the dominant cause of the accumulation is the lenticularity or truncation of the reservoir rock.

It is recognized that in some multireservoir fields, where more than one of the two types of entrapment occur, it may be impractical to segregate production and reserves by reservoirs, and difficult to identify the exact characteristics of each such reservoir. Then the reserve of the field should be assigned to the dominant type of entrapment.

The estimated ultimate recovery of natural gas in the United States in accordance with reservoir entrapment is given in Table 3.

Estimate of ultimate recovery is also reported by the geologic age of the reservoir, as shown in Table 4. It is recognized that problems may arise when the geologic age of a reservoir cannot be determined specifically, such as Permo-Pennsylvanian and Cambro-Ordovician, or when production from reservoirs of different geologic age are combined. Then the reserves are assigned to the youngest-age category.

In the full AGA–API report, historical actual production and reserves are reported by years. This information is depicted in summary form in Figs. 3 and 4.

NATURAL GAS LIQUIDS

Natural gas liquids occur in either the gaseous phase or in solution with crude oil in the reservoir. They are recovered at the surface as liquids by separation from produced natural gas by such processes as condensation and absorption in field separators, gasoline

TABLE 3.   Estimated Ultimate Recovery of Natural Gas in the United States by Type of Entrapment*

Millions of Cubic Feet—14.73 psia at 60°F

State	Structural	Stratigraphic	Total
Alabama	561,632	5,798	567,430
Alaska	32,764,470	12,325	32,776,795
Arkansas	3,701,837	1,323,385	5,025,222
California†	26,837,177	4,895,416	31,732,593
Colorado	2,758,463	1,604,833	4,363,296
Florida	398,869	2,141	401,010
Illinois	1,020,978	456,120	1,477,098
Indiana	43,643	131,272	174,915
Kansas	3,187,473	30,206,631	33,394,104
Kentucky	0	3,864,974	3,864,974
Louisiana†	169,535,937	10,824,549	180,360,486
Michigan	166,167	1,328,052	1,494,219
Mississippi	5,789,432	116,141	5,905,573
Montana	1,337,359	904,183	2,241,542
Nebraska	157,763	156,301	314,064
New Mexico	11,589,064	25,851,651	37,440,715
New York	169,690	418,326	588,016
North Dakota	590,730	568,043	1,158,773
Ohio	0	6,067,729	6,067,729
Oklahoma	16,357,059	39,355,429	55,712,488
Pennsylvania	2,449,335	7,334,985	9,784,320
Texas†	172,843,861	95,888,218	268,732,079
Utah	992,182	1,123,762	2,115,944
Virginia	552	113,618	114,170
West Virginia	2,257,504	14,132,343	16,389,847
Wyoming	8,228,857	2,702,700	10,931,557
Miscellaneous‡	67,206	24,714	91,920
Total: United States	463,807,240	249,413,639	713,220,879

*SOURCE: AGA–API Report, vol. 29, May 1975. Reserve conditions as of 12/31/1974.

†Includes offshore reserves.

‡Includes Arizona, Iowa, Maryland, Minnesota, Missouri, South Dakota, Tennessee, and Washington.

TABLE 4.   Estimate of Ultimate Recovery of Natural Gas in the United States by Geologic Age of Reservoir*
Millions of Cubic Feet—14.73 psia at 60°F

Geologic age	Nonassociated	Associated–dissolved	Total
Cenozoic	233,928,814	86,629,470	320,558,284
Quaternary	6,162,580	425,399	6,587,979
Recent	1,198	0	1,198
Pleistocene	6,161,382	425,399	6,586,781
Tertiary	227,766,234	86,204,071	313,970,305
Pliocene	14,077,652	11,354,326	25,431,978
Miocene	108,876,439	34,693,932	143,570,371
Oligocene	67,649,905	28,830,584	96,480,489
Eocene	35,455,016	11,289,662	46,744,678
Paleocene	1,707,222	35,567	1,742,789
Mesozoic	76,760,309	42,887,123	119,647,432
Cretaceous	67,843,823	14,413,455	82,257,278
Jurassic	8,756,530	2,454,756	11,211,286
Triassic	159,956	26,018,912	26,178,868
Paleozoic	181,519,655	91,230,102	272,749,757
Permian	76,961,987	37,298,074	114,260,061
Pennsylvanian	38,829,217	27,783,730	66,612,947
Mississippian	16,340,883	8,038,664	24,379,547
Devonian	26,878,783	4,320,721	31,199,504
Silurian	6,707,727	1,793,927	8,501,654
Ordovician	15,762,800	11,921,750	27,684,550
Cambrian	38,258	73,236	111,494
Pre-Cambrian	265,406	0	265,406
Total: All Ages	492,474,184	220,746,695	713,220,879

*SOURCE: AGA–API Report, vol. 29, May 1975. Reserve conditions as of 12/31/1974.

plants, and other surface facilities. Since natural gas liquids become available solely by separation from produced natural gas, their rate of availability depends directly on the rate of production of gas from crude oil or natural gas reservoirs.

The productive capacity of natural gas liquids as reported by the AGA Committee is defined as the amount of hydrocarbon liquids that would be produced coincident with the estimated productive capacity of natural gas based on unit recoveries at normal producing rates. Such estimated capacities are not limited by lack of capacity of processing plants or other surface facilities (in making estimates), and it is emphasized that adequate facilities would be required to effect the recovery of liquids from the natural gas produced at these rates.

Estimators should recognize that such facilities cannot be enlarged quickly. Therefore, the estimated natural gas liquid capacities, which relate to increased production of gas

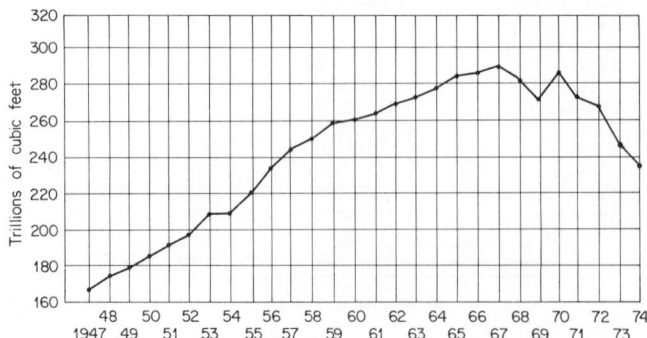

Fig. 3. Natural gas reserves (proved) in the United States (1947–1974).

from oil and gas wells operating at their productive capacities, are theoretical, and may not be realized in event of an emergency. In an emergency requiring capacity production of hydrocarbon liquids, both oil and natural gas liquids, the capacities of existing processing plants would limit the amount of natural gas liquid capacity realized. If increases in solution gas from oil wells operating at high rates are to be absorbed into existing markets with no venting, then gas production from nonassociated gas wells will have to be reduced by an equal amount, thus reducing the quantity of liquids obtainable from such gas well gas. Should there be an increase in gas demand equal to the increase in gas

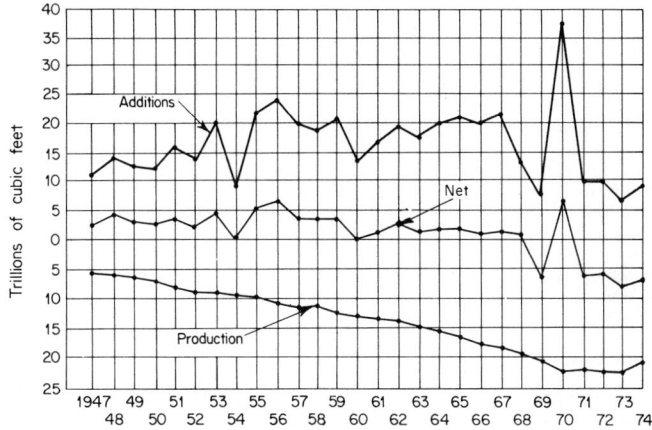

Fig. 4. Additions to reserves versus production of natural gas in the United States (1947–1974).

produced by oil wells, in such an emergency the capacities of existing processing plants would limit the amount of such liquid capacity that is realized. Although the estimates of productive capacity of gas liquids as determined under the definitions used by the AGA Committee are theoretical, such estimates are useful in determining the geographical location of potential availability.

Estimated total proved reserves of natural gas liquids in the United States are given in Table 5.

PRODUCTION OF NATURAL GAS AND NATURAL GAS LIQUIDS

Table 6 lists by states the production of natural gas and natural gas liquids in the United States (1974). Estimates of the *daily* productive capacity of natural gas and natural gas liquids, also by states, are given in Table 7.

DEFINITIONS OF TERMS

Natural gas is defined to be a mixture of hydrocarbon compounds and small quantities of various nonhydrocarbons existing in the gaseous phase or in solution with oil in natural underground reservoirs at reservoir conditions. The principal hydrocarbons usually contained in the mixture are methane, ethane, propane, butanes, and pentanes; and typical nonhydrocarbon gases which may be contained in reservoir natural gas are carbon dioxide, helium, hydrogen sulfide, and nitrogen.

The portions of the reservoir hydrocarbon gas recovered in liquid form in surface separators or plant facilities are reported as natural gas liquids. The statistics on natural gas reserves and production reported by the AGA Committee take into account the shrinkage of the reservoir gas volume resulting from the removal of the liquefiable portions of the hydrocarbon gases and the reduction of volume due to the exclusion of nonhydrocarbon gases where they occur in sufficient quantity to render the gas unmarketable.

Natural gas is found in underground rock formations which are usually sedimentary in origin. The natural reservoirs are composed of porous rock which provide space for accumulation of hydrocarbons. Economically recoverable quantities of hydrocarbons occur in the porous reservoir rock where depositional conditions or deformation of the strata have resulted in the formation of traps which terminate underground migration and

TABLE 5. Estimated Total Proved Reserves of Natural Gas Liquids in the United States*

Thousands of Barrels of 42 U.S. Gallons

State	Nonassociated	Associated–dissolved	Total natural gas liquids
Alabama	116,998	1,625	118,623
Alaska	0	845	845
Arkansas	3,124	1,624	4,748
California†	3,004	95,228	98,232
Colorado	10,867	9,822	20,689
Florida	0	0	0
Illinois	0	0	0
Indiana	0	0	0
Kansas	386,892	7,527	394,419
Kentucky	44,676	0	44,676
Louisiana†	1,607,399	274,982	1,882,381
Michigan	4,906	15,358	20,264
Mississippi	7,021	5,572	12,593
Montana	459	2,487	2,946
Nebraska	384	671	1,055
New Mexico	288,303	108,726	397,029
North Dakota	79	51,777	51,856
Ohio	0	0	0
Oklahoma	191,087	99,240	290,327
Pennsylvania	580	0	580
Texas†	1,320,969	1,476,019	2,796,988
Utah	645	51,709	52,354
West Virginia	81,755	0	81,755
Wyoming	39,980	38,109	78,089
Total: United States	4,109,128	2,241,321	6,350,449

*SOURCE: AGA–API Report, vol. 29, May 1975. Reserve conditions as of 12/31/1974.
†Includes offshore reserves. The remaining proved natural gas liquids reserves in the Gulf of Mexico are estimated to be 816,023,000 barrels; of which 727,405,000 barrels are nonassociated; and 88,618,000 barrels are associated–dissolved.

cause accumulations of hydrocarbon fluids and gases. Under reservoir conditions, natural gas and the liquefiable portions thereof occur either in a single gaseous phase in the reservoir or in solution with crude oil and are not distinguishable at that time as separate substances. Natural gas is classified by the AGA Committee in two categories based on the type of occurrence in the reservoir, as follows:

1. *Nonassociated gas* is defined as free natural gas not in contact with crude oil in the reservoir.

2. *Associated-dissolved gas* is the combined volume of natural gas which occurs in crude oil reservoirs either as free gas *(associated)* or as gas in solution with the crude oil *(dissolved)*.

Associated gas is free natural gas, commonly known as gas cap gas, which overlies and is in contact with crude oil in the reservoir. Dissolved gas is natural gas which is in solution with crude oil in the reservoir at original reservoir conditions.

Production of associated gas usually affects to some degree the ultimate recovery of crude oil by reduction of the gas cap volume and loss of reservoir pressure. Where the production of associated gas from gas wells does not have a significant effect on the crude oil recovery, such associated reserves may be classified as nonassociated gas by regulatory body ruling or company operating practice. In such instances this classification is also followed in the report of the AGA Committee.

If the ultimate recovery of the crude oil is significantly affected by the production of the associated gas from both oil and/or gas wells completed in the reservoir, such gas production is usually limited by regulatory body ruling or company operating practice to assure maximum recovery of crude oil.

In reservoirs containing associated gas and crude oil, the oil, dissolved gas, and

TABLE 6   Production of Natural Gas and Natural Gas Liquids in the United States (1974)*

State	Natural gas, millions of cubic feet (14.73 psia at 60°F)	Natural gas liquids, thousands of barrels of 42 U.S. gallons
Alabama	4,985	2,651
Alaska	18,626	98
Arkansas	12,815	876
California†	181,398	10,727
Colorado	26,089	3,485
Florida	39,481	0
Illinois	3,011	0
Indiana	269	0
Kansas	35,895	30,718
Kentucky	691	3,292
Louisiana†	1,006,336	233,482
Michigan	31,727	1,367
Mississippi	18,356	1,362
Montana	11,601	679
Nebraska	2,397	313
New Mexico	367,374	38,373
New York	45	0
North Dakota	31,309	1,829
Ohio	16,228	0
Oklahoma	482,167	41,778
Pennsylvania	169	79
Texas†	1,791,826	333,023
Utah	33,769	3,354
West Virginia	438	5,542
Wyoming	110,111	11,071
Miscellaneous‡	210	. . .
Total: United States	4,227,323	724,099

*SOURCE: AGA–API Report, vol. 29, May 1975.
†Includes offshore facilities.
‡Includes Arizona, Iowa, Maryland, Minnesota, Missouri, South Dakota, Tennessee, and Washington.

associated gas may be produced concurrently from the same well bore. This gas is measured and reported by the operator as one volume under the term *casinghead* or *oil well gas*. As oil is produced from many reservoirs, pressure is reduced below the saturation pressure of the crude oil, and the gas originally dissolved in the oil is released as free gas in the reservoir. At critical gas saturation of the reservoir, dissolved gas begins to flow and will either be produced with the crude oil, resulting in higher gas-oil ratios, or will migrate into primary gas caps or form secondary gas caps.

Production and productive capacity of associated and dissolved gas are generally more closely related to the production of crude oil than to the available market for gas. Since only rough estimates for the separation of the production of commingled associated and dissolved gas produced from oil wells can be made as a portion of the dissolved gas may change classification in the reservoir during depletion of the oil reserve, the AGA Committee has combined the reserves and productive capacity of associated and dissolved gas into a single category, known as associated-dissolved gas. Free gas contained in the reservoir after depletion of the oil reserve is reclassified to the nonassociated category. In such cases the gas reserves are determined on the basis of the remaining gas reservoir volume and conditions existing at the time of depletion of the oil reserve.

Volumes of gas reserves are determined by geological and engineering analyses of reservoir data, including structural interpretation, well tests, core analysis, pressure production data, and gas analysis, available at the time the estimate is made. Recoveries of gas are estimated on the basis of formation evaluation of the reservoir rock, the producing mechanism of the reservoir, and pressure production performance.

Because of certain physical and economic conditions, a part of the original gas volume in the reservoir is nonrecoverable. More specifically, these limitations relate to the

TABLE 7.   Estimate of the Daily Productive Capacity of Natural Gas and Natural Gas Liquids in the United States*

State	Natural gas, millions of cubic feet per day (14.73 psia at 60°F)	Natural gas liquids, thousands of barrels of 42 U.S. gallons per day
Alabama	27	2
Alaska	657	0
Arkansas	788	5
California†	2,016	50
Colorado	474	8
Florida	64	1
Illinois	12	0
Indiana	4	0
Kansas	6,932	228
Kentucky	200	6
Louisiana†	27,613	884
Michigan	264	7
Mississippi	324	5
Montana	208	2
Nebraska	16	2
New Mexico	3,763	168
New York	12	0
North Dakota	105	5
Ohio	290	0
Oklahoma	8,706	204
Pennsylvania	224	0
Texas	31,185	1,229
Utah	196	7
Virginia	17	0
West Virginia	559	15
Wyoming	1,330	39
Miscellaneous‡	12	0
Total: United States	85,998	2,867

*SOURCE: AGA–API Report, vol. 27, May 1973. Capacity is for heating season immediately following 12/31/72. These statistics not given in 1975 Report.

†Includes offshore productive capacity.

‡Includes Arizona, Iowa, Maryland, Minnesota, Missouri, South Dakota, Tennessee, and Washington.

inability to produce gas because of heterogeneity of the reservoir, gas trapped in the reservoir due to water influx, and the inability to produce gas economically because of depleted reservoir pressure. Since the energy that causes gas to flow from a reservoir is derived from the pressure at which the gas exists in the reservoir, all the gas originally contained in the reservoir cannot be produced in economic quantities within any reasonable time. The quantities that are recoverable during the economic life of a reservoir are quite high, several times more than in the case of crude oil. This higher recovery is to be expected because of the difference in viscosity of gas and crude oil and the fact that gas has less tendency to wet the reservoir rock and remain in situ.

In view of the small quantities of unrecoverable gas left in a reservoir during its economic productive life, the change in nature of the hydrocarbons as they are produced and the fact that gas recoveries cannot be increased by improved recovery techniques successfully applied to crude oil, no effort is made in the AGA report to provide any data on the reservoir space originally occuped by the mixture of gaseous hydrocarbons which eventually are recovered in part as natural gas and in part as natural gas liquids.

Proved Reserves. The AGA Committee's definition of proved reserves defines the current estimated quantity of natural gas and natural gas liquids which analysis of geologic and engineering data demonstrate with reasonable certainty to be recoverable in the future from known oil and gas reservoirs under existing economic and operating conditions. Reservoirs are considered proved that have demonstrated the ability to produce by either actual production or conclusive formation test.

The *area of a reservoir* considered proved is that portion delineated by drilling and defined by gas-oil, gas-water contacts or limited by structural deformation or lenticularity of the reservoir. In the absence of fluid contacts, the lowest known structural occurrence of hydrocarbons controls the proved limits of the reservoir. The proved area of a reservoir may also include the adjoining portions which are not delineated by drilling, but can be evaluated as economically productive on the basis of geological and engineering data available at the time when the estimate is made. Therefore, the reserves reported by the AGA Committee include total proved reserves which may be in either the drilled or the undrilled portions of the field or reservoir.

In general, the definitions of proved reserves contained in *API Tech. Rep.* 1, Definitions for Petroleum Statistics, are followed. See references at end of this Handbook section. It should be noted that in order to maintain a consistent continuing series, gas in underground storage is included in the total gas reserves in AGA reports. Anyone desiring a value for gas in storage may obtain such by subtraction.

Attention is called to the fact that natural gas is a mixture of hydrocarbon compounds and small quantities of various nonhydrocarbons. Generally the quantities of nonhydrocarbons are *de minimis* and do not affect the marketability of the gas. Then no reduction in volume for the theoretical removal of such nonhydrocarbons is made. In any reservoir where the quantity of nonhydrocarbons is sufficient to render the particular gas unmarketable, an appropriate reduction in the reservoir gas volume is made to cover the exclusion of such nonhydrocarbons.

RELATED STATISTICS

Other statistics in this Handbook which relate to natural gas and associated products include the following:
In the Handbook section on *Gas Technology:*
  "Natural Gas Reserves in Canada"
  "World Natural Gas Production and Reserves"
  "Potential Supply of Natural Gas in the United States"
  "Future Gas Consumption in the United States"
In the Handbook section on *Petroleum Technology:*
  "Petroleum Production and Reserves in the United States"
  "Petroleum Reserves in Canada"
  "World Petroleum Production and Reserves"

REFERENCES

American Gas Association: "Reserves of Crude Oil, Natural Gas Liquids, and Natural Gas in the United States and Canada and United States Productive Capacity," issued annually, AGA, Arlington, Va.
Am. Pet. Inst. Tech. Reps, **1, 2** (and subsequent), issued periodically, API, Washington.
NOTE: Also obtain lists of available publications from U.S. Bureau of Mines, Pittsburgh, Pa., and U.S. Geological Survey, Reston, Va.

Potential Supply of Natural Gas in the United States

By E. A. RASSINIER* AND P.N. GLOVER†

An appraisal of the long-range prospects of the natural gas industry in the United States—which is useful for financial, managerial, and government purposes—requires scientific, authoritative, and objective estimation of the potential supply of natural gas that may become available to the industry, in addition to proved recoverable reserves of natural gas currently available. The Committee on Natural Gas Reserves of the American Gas Association, an industry committee, is fulfilling part of this requirement by providing periodic estimates of proved recoverable reserves of natural gas immediately available to the industry. However, such estimates do not reflect potential supplies of natural gas that may be discovered as a result of future exploratory drilling.

For many years, estimates of future supply of natural gas have been developed by numerous individuals and organizations. Since these estimates of future supply of natural gas varied widely and often had entirely different objectives and meanings, they served to confuse rather than clarify the longrange outlook for natural gas supply. With the view to achieving a more unified industry-sponsored effort, many of the leaders of the natural gas industry expressed a strong desire for a continuous long-range study of the requirements and supply of natural gas in the United States. As a result, in 1960 the Future Requirements and Supply Committees were formed with the responsibility to develop estimated future requirements and supply. The Supply Group was organized into 12 area subcommittees that were composed of many qualified geologists and engineers, each expert in the occurrence of natural gas in a particular area. Each area subcommittee chairman was responsible for the estimate of potential supply of natural gas in a geologically significant area of the United States. Participation by producers, distributors, pipeliners, state organizations, academicians, and consultants assured the objectivity of the study.

In June 1966, the Gas Industry Committee, representing the American Gas Association, the American Petroleum Institute, and the Independent Natural Gas Association of America, suggested supporting a study of the future natural gas supply. In October 1966, an organizational structure was set up which provided for even broader and more diversified representation from other organizations interested in gas supply. Active supervision by the Mineral Resources Institute of the Colorado School of Mines Foundation, Inc., also was provided by the new structure. Formal announcement of the reorganized structure of the several committees was made in February 1967. The initial report of the Potential Gas Committee was published in 1967.

The Future Gas Requirements Committee was reorganized in a similar manner and has become the Future Requirements Committee, supervised by the Denver Research Institute of the University of Denver. The Future Requirements Committee and its directing agency have no connection whatever with the Potential Gas Committee and its supervising agency. The organization for the supply and requirements studies is shown in Fig. 1.

The highlights of the Potential Gas Committee Report, issued in November 1973, reflecting the potential supply of natural gas in the United States as of December 31, 1972, are summarized here. (As of early-1976, it is planned to issue report revisions at periodic intervals.)

*Chairman, Potential Gas Committee (PGC), supervised by the Colorado School of Mines Foundation, Inc., Golden, Colo.
†Former Chairman, Editorial Committee and former Vice-Chairman of PGC.

DEFINITIONS

Natural Gas.   As used in the PGC report, natural gas is any gas of natural origin that is composed primarily of hydrocarbon molecules producible from a borehole. Most natural gas contains some nonhydrocarbon components, such as carbon dioxide, nitrogen, hydrogen sulfide, and helium. However, in estimating potential supply, it is not feasible to separate small volumes of these components from the hydrocarbons. Areas of forma-

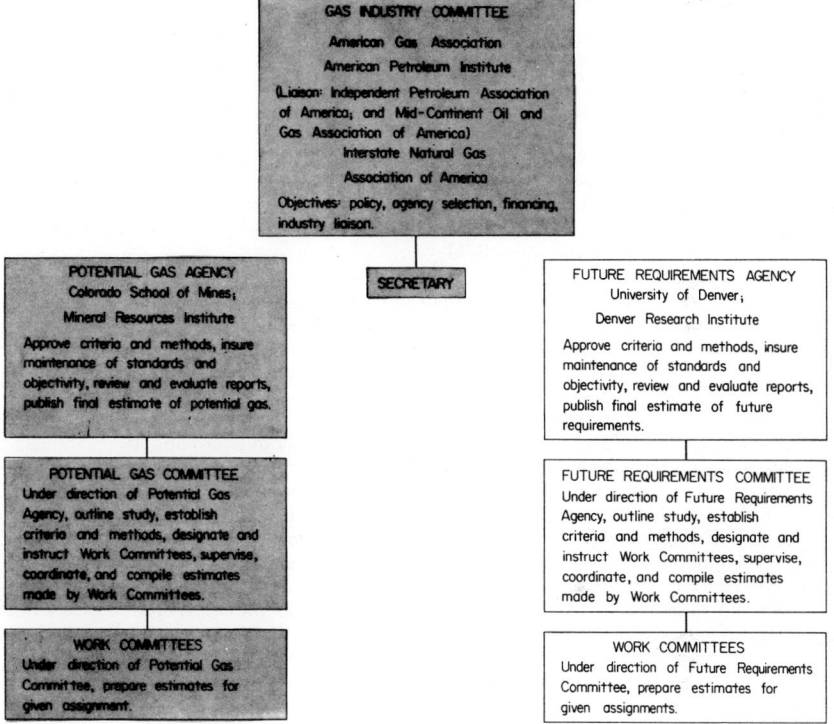

Fig. 1. Structure of organizations under cognizance of the Gas Industry Committee.

tions that are believed to contain large volumes of such nonhydrocarbon components are not counted as potential supply.

Potential Gas.   As used in the PGC report, the phrase "potential supply of natural gas" means: *That prospective quantity of natural gas yet to be discovered, or to be added in existing fields, which will be commercially recoverable in the future, assuming adequate but reasonable field prices and normal advancements in technology.* These estimates are made for specified geographical areas at given dates. No attempts are made to predict either the rates of discovery of gas or the level of future field prices. The phrase "adequate prices" does not mean current prices as of the date when the estimates are made.

Potential gas estimates cover quantities in addition to and not duplicative of proved reserves estimated by the American Gas Association.* Proved reserves have been discovered and are estimated from demonstrated producing capabilities. Most of the potential supplies have not yet been discovered and must therefore be estimated indirectly, as discussed in the later section on "Technique for Estimating Potential Supply of Natural Gas." An estimate of potential supply must take into consideration the criteria used by the

*AGA, "Proved Reserves Report," vol. 27, May 1973.

Committee on Natural Gas Reserves of the American Gas Association in preparing its annual estimates of proved recoverable reserves. †

The Committee on Natural Gas Reserves includes in proved reserves all gas estimated to be producible from tested formations under existing operating and economic conditions without regard to the size, use, or disposition of any production. Proved reserves in an undrilled area, however, must be so related to the developed or tested leases and to known field geology that its productive ability is assured.

CATEGORIES OF POTENTIAL GAS SUPPLY

Accuracy of the estimates of gas volumes included in the potential supply of a given area are dependent on geological conditions and the extent to which the area has been explored and developed. Using available geological data and the American Association of Petroleum Geologists classification of wells (Fig. 2), estimates are divided into three broad categories:

1. *Probable potential gas supply* (associated with existing fields):
   a. Supply from known accumulations obtained by:
      (1) Future extensions of *existing pools,* in known productive reservoirs.
      (2) Future *new* pool discoveries, within existing fields, in reserviors productive elsewhere within the same field.
   b. Supply from *new pool* discoveries obtained by:
      (1) Future shallower and/or deeper new pool discoveries, within existing fields, in formations productive elsewhere within the same geologic province or sub-province, under *similar* geologic conditions.
      (2) Future shallower and/or deeper new pool discoveries, within existing fields, in formations productive elsewhere within the same geologic province or sub-province, under *different* geological conditions.
2. *Possible potential gas supply* (associated with productive formations):
   a. Supply from *new field* discoveries obtained by:
      (1) Future new field discoveries, in formations *productive* elsewhere within the same geologic province or subprovince, under *similar* geological conditions.
      (2) Future new discoveries, in formations *productive* elsewhere within the same geologic province or subprovince, under *different* geological conditions.
3. *Speculative potential gas supply* (associated with nonproductive formations).
   a. Supply from *new pool* discoveries in formations not previously productive within a productive geologic province or subprovince.
   b. Supply from *new field* discoveries, obtained by:
      (1) Future new field discoveries in formations *not previously productive* within a productive geologic province or subprovince.
      (2) Future new field discoveries within a geologic province *not previously productive.*

A *geologic province* is defined in "Glossary of Geology and Related Sciences, by the American Geological Institute, as "A large area or region unified in some way considered as a whole." Hence, the Gulf Coast geosyncline and the Appalachian geosyncline often are referred to as *provinces.* Large provinces, such as those cited, are often divided into

---

†The Committee on Natural Gas Reserves defines proved recoverable reserves as follows: The current estimated quantity of natural gas and natural gas liquids which analysis of geologic and engineering data demonstrate with reasonable certainty to be recoverable in the future from known oil and gas reservoirs under existing economic and operating conditions. Reservoirs are considered proved that have demonstrated the ability to produce by either actual production or conclusive formation test.

The area of a reservoir considered proved is that portion delineated by drilling and defined by gas-oil, gas-water contacts or limited by structural deformation or lenticularity of the reservoir. In the absence of fluid contacts, the lowest known structural occurrence of hydrocarbons controls the proved limits of the reservoir. The proved area of a reservoir may also include the adjoining portions which are not delineated by drilling, but can be evaluated as economically productive on the basis of geological and engineering data available at the time when the estimate is made. Therefore, the reserves reported by the Committee include total proved reserves which may be in either the drilled or the undrilled portions of the field or reservoir. (Definition taken from American Gas Association, "Proved Reserves Report," vol. 27, p. 102, May 1973.)

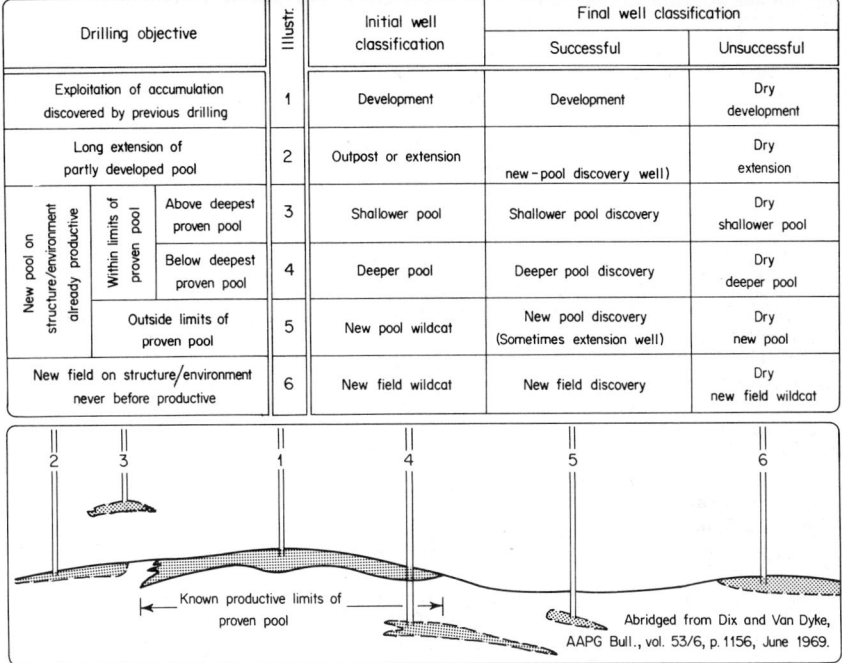

Drilling objective			Illustr.	Initial well classification	Final well classification	
					Successful	Unsuccessful
Exploitation of accumulation discovered by previous drilling			1	Development	Development	Dry development
Long extension of partly developed pool			2	Outpost or extension	new-pool discovery well)	Dry extension
New pool on structure/environment already productive	Within limits of proven pool	Above deepest proven pool	3	Shallower pool	Shallower pool discovery	Dry shallower pool
		Below deepest proven pool	4	Deeper pool	Deeper pool discovery	Dry deeper pool
	Outside limits of proven pool		5	New pool wildcat	New pool discovery (Sometimes extension well)	Dry new pool
New field on structure/environment never before productive			6	New field wildcat	New field discovery	Dry new field wildcat

Known productive limits of proven pool

Abridged from Dix and Van Dyke, AAPG Bull., vol. 53/6, p. 1156, June 1969.

Fig. 2. Classification of wells (American Association of Petroleum Geologists).

subprovinces in order to recognize geological homogeneity. Examples of subprovinces are the Mississippi embayment of the Gulf Coast geosyncline (province) and the Delaware basin within the Permian basin (province).

The term *basin* is avoided in PGC estimating and reporting, except in the foregoing illustration—because it has topographic and geomorphic meanings, as well as geologic. These meanings are often different and can lead to misinterpretations.

It is evident from the foregoing that, as drilling progresses, gas volumes estimated to be in a particular reservoir will move from one category to another, as shown schematically in Fig. 3. Since any projection of potential supply lacks the accuracy of the proved reserve figures, particularly in the speculative category, it must be recognized that these estimates of potential supply will also be subjected to upward and downward revisions when new geological and engineering data are provided by exploratory drilling. When gas is finally classified in the *Proved* category, it will no longer be included in the estimated potential supply.

TECHNIQUE FOR ESTIMATING POTENTIAL SUPPLY OF NATURAL GAS

Basic Approach. The basic technique for estimating potential gas supply is to compare the factors that control known occurrences of gas with factors present in prospective areas. Known occurrences are expressed as the volume of natural gas ultimately recoverable per unit volume of reservoir rock within an adequately explored portion of a geologic province. Such known relationship, with

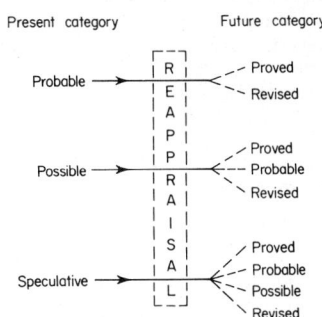

Fig. 3. Flow chart illustrating progression of potential gas supplies as a result of drilling.

appropriate adjustment for variations in geological and reservoir conditions, is then attributed to incompletely explored sedimentary rocks in the same or in a similar geologic province.

A general outline of the attribution technique for estimating potential supply of natural gas follows:

1. *For productive formations:*
    a. Within a productive province or subprovince estimate the volume of:
        (1) Productive gas-bearing rock.
        (2) Potential gas-bearing rock:
            (*a*) Associated with existing fields for estimating Probable Potential Supply.
            (*b*) Associated with productive formations for estimating Possible Potential Supply.
    b. Add cumulative production and proved reserves to obtain the total volume of ultimate recoverable gas for the adequately explored portion of the productive gas-bearing rocks.
    c. Divide this figure by the volume of adequately explored productive gas-bearing rocks to obtain the ultimate recoverable gas per unit volume of productive rocks.
    d. Adjust the unit volume figure for variations in geological and reservoir conditions in the Probable and Possible gas-bearing rocks.
    e. Using these adjusted figures:
        (1) Estimate *Probable Potential Supply* in extensions and new pools associated with existing fields.
        (2) Estimate *Possible Potential Supply* by applying these adjusted figures to the wildcat traps and structures assumed to be present in the inadequately tested portion of the province or subprovince.
2. *For nonproductive formations and nonproductive provinces:*
    a. Estimate the volume of untested sediments in nonproductive provinces and the volume of potential but nonproductive sediments in productive provinces.
        (1) Estimate the *Speculative Potential Supply* in these sediments by comparing them with similar sediments in other provinces or subprovinces where their productive characteristics are known.

Role of Judgment in Estimating Potential Supply.   In proceeding from the known to the unknown, the judgment of the estimator is the most significant factor in making calculation of potential supply, particularly in the categories of *Possible* and *Speculative* supply. The appropriate adjustments referred to previously under "Basic Approach" are direct reflections of the estimator's judgment. Only the estimator has the detailed knowledge and the experience necessary to select appropriate adjustments for the geological provinces for which he is responsible. In all respects, the estimates of potential supply by the Potential Gas Committee reflect an objective, scientific approach to the problem. An attempt is made to use all pertinent geological and engineering data.

Limiting Considerations in Making Estimates.   At the present, no potential supply of gas is considered at depths greater than 30,000 ft or in offshore areas, where water depths are in excess of 1500 ft. Economic, technological, and governmental policy considerations in estimating potential supply also impose restrictions, and to that extent they are considered limitations of the estimates. Some of these limitations are described in the following paragraphs.

Economic and Technological Aspects of Estimating Potential Supply.   Economic, technological, and governmental policy considerations that are taken into account in the Potential Gas Committee's estimate of potential gas supply are related to (1) past production and proved reserves of natural gas, and (2) all wells which would be drilled in the future under the assumed conditions that there will be adequate but reasonable prices for gas and oil and normal improvements in technology.

Fundamental economic considerations—which include, but are not limited to prices, costs, rates of take, recovery factors, abandonment pressures, and so on—are inherent in the definition of proved reserves (excerpted from prior footnote definition): " . . . recoverable in the future from known oil and gas reservoirs under existing economic and operating conditions." Some gas that is actually present in known reservoirs is not producible because of the relationship between costs and prices, and therefore is not included in recoverable reserves. Furthermore, there are other known gas accumulations of such small size that none of the gas can be produced in commercial quantities under " . . . existing economic and operating conditions." Such considerations result in proved reserves being something less than the total volumes of gas existing in known reservoirs. These same limitations are applicable to the estimates of potential supply. If a fundamental change in economics or technology occurs, estimates of potential supply will be changed accordingly.

The attribution technique previously given under "Basic Approach" describes relationships for known occurrences of ultimately recoverable volumes of gas to prospective gas accumulations, after appropriate adjustment for variations in geological and reservoir conditions are made. The explicit assumption of adequate but reasonable prices and normal improvements in technology in the definition of potential supply relates to improvements in exploration and production techniques. Adequate prices, normal technological improvements, and reasonable governmental policies are required to bring about the drilling necessary to prove the potential supply. These assumed conditions permit estimates of potential supply to be made by the Potential Gas Committee on the basis of relevant past history and experience concerning recovery factors, as well as the size and type of reservoirs that have been found, developed, and produced, without speculation concerning future levels of prices and costs.

GEOGRAPHIC DISTRIBUTION OF WORK COMMITTEES

Following the system long used by the petroleum industry, each Work Committee reporting to the Potential Gas Committee is assigned geologically significant areas (without regard to state lines) in which the Committee members have expert working knowledge. These assignments are shown in Fig. 4. These Committee members, with their fund of information (often confidential), are the source of knowledge and judgment for estimating the undiscovered gas supply in each area. Because geologic instead of geographic divisions are used, publication of estimates by individual states is not feasible. The publication areas of the Potential Gas Committee are essentially the same as the areas shown in Fig. 4, with the exception that Area J is divided into a northern (J_n) and a southern (J_s) area. The latter are indicated in Fig. 6.

AREA AND DEPTH LIMITS

The estimate presented in the Potential Gas Committee Report is limited to the contiguous United States, Alaska, and adjacent offshore areas. Hawaii, the island territories, and their adjacent offshore areas are not included.

The area considered in the current estimate is limited to offshore water depths of 1500 ft and to drilling depths of 30,000 ft. These limitations, together with the assumptions that adequate but reasonable prices and normal improvements in exploration and drilling technology will prevail, permit the application of estimating techniques without speculation concerning the future level of prices, costs, and trends in technology. Current techniques and normal improvements are assumed to prevail in the future; unusual improvements in exploration, drilling, development, and recovery techniques will be reflected in subsequent estimates as they are proved.

ESTIMATE OF POTENTIAL SUPPLY OF NATURAL GAS

The potential supply of natural gas in the United States, excluding Hawaii and island territories, as of December 31, 1972, is estimated to 1146 trillion ft³. See Table 1 and Fig. 5. The generally unfavorable areas are clearly indicated in Fig. 6. Recognizing the importance of accessibility (some areas are less accessible than others because of technological, governmental, or economic limitations), the potential reserves are summarized in Table 2 by depth increments.

Within the 48 states onshore area, 25 percent of the potential supply is below 15,000 ft. Within the 48 states offshore areas, 12 percent of the potential supply lies in water depths of 600 to 1500 ft. The Alaskan potential supply is listed separately because of the unique limitations on its accessibility for the United States market. Of the total potential supply,

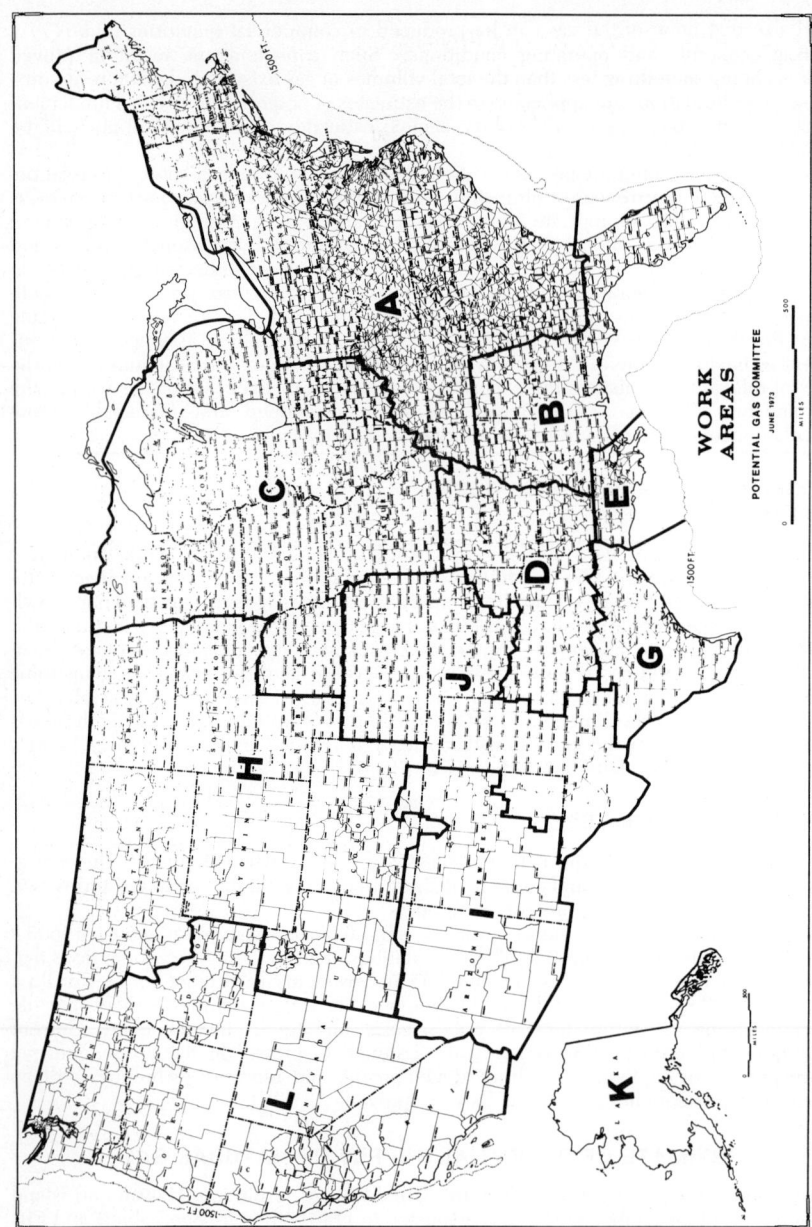

Fig. 4. Major areas covered by the Work Committees that report to the Potential Gas Committee.

WORK AREAS

POTENTIAL GAS COMMITTEE

JUNE 1972

TABLE 1.   Estimated Potential Supply of Natural Gas
Trillions of cubic feet at 14.73 psia and 60°F

	Probable	Possible	Speculative	Total
48 states	212	290	278	780
Alaska	54	94	218	366
Total United States	266	384	496	1146

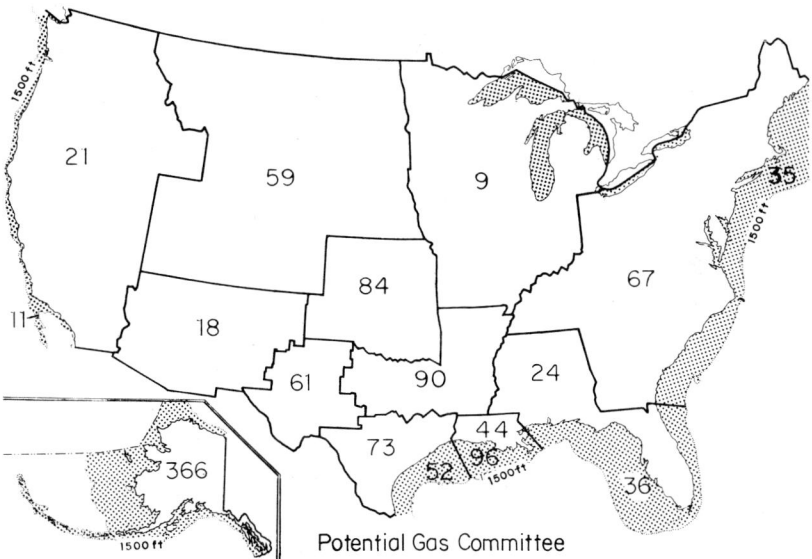

Potential Gas Committee

Fig. 5. Total potential gas supply as of December 31, 1972 (1146 trillion cubic feet). Large numbers indicate trillions of cubic feet in each area.

TABLE 2.   Summary of Estimated Potential Supply of Natural Gas in the United States by Depth Increments as of December 31, 1972
Trillions of cubic feet at 14.73 psia and 60°F

	Area Totals			
	Probable	Possible	Speculative	Total
Onshore (Drilling depth):				
0–15,000 ft	121	153	139	413
15,000–30,000 ft	33	45	59	137
Subtotal	154	198	198	550
Offshore (Water depth):				
0–600 ft	58	74	71	203
600–1500 ft	*	18	9	27
Subtotal	58	92	80	230
Total: 48 states	212	290	278	780
Alaska†	54	94	218	366
Total United States	266	384	496	1146

*Less than 1 trillion cubic feet.
†Not available by depth increments.

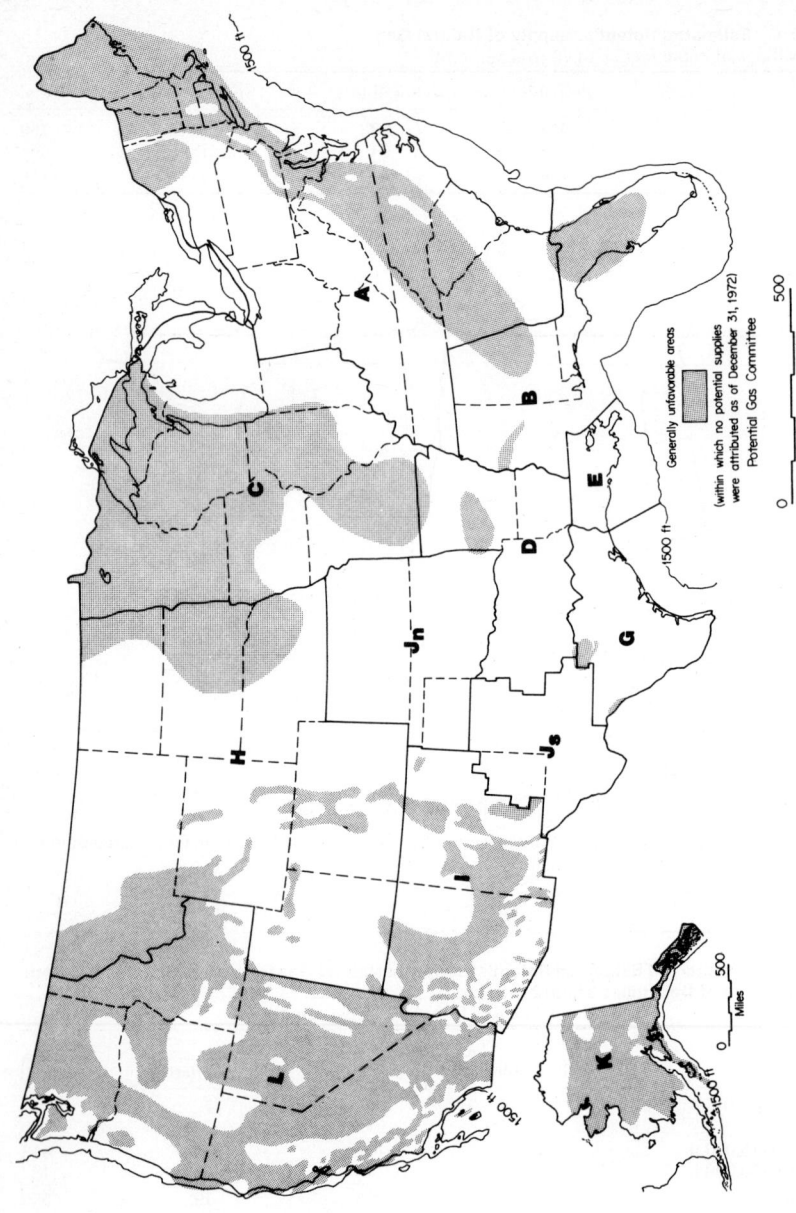

Fig. 6. Generally unfavorable areas for potential natural gas are shown shaded.

Generally unfavorable areas

(within which no potential supplies were attributed as of December 31, 1972) Potential Gas Committee

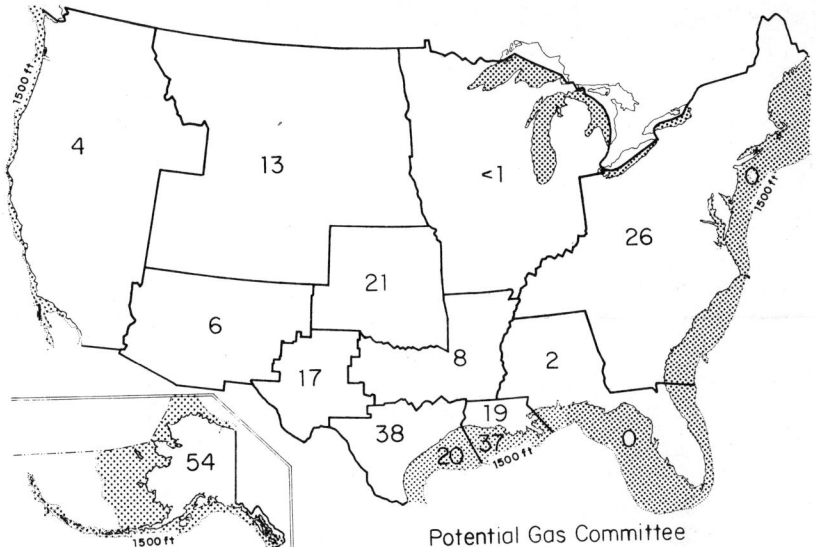

Fig. 7. Probable category of potential gas supply as of December 31, 1972 (266 trillion cubic feet). Large numbers indicate trillions of cubic feet in each area.

32 percent is in Alaska. With reference to the areas shown in Fig. 4, the Probable Category, the Possible Category, and the Speculative Category are shown in Figs. 7, 8, and 9, respectively.

To place these estimates in their proper perspective, Figs. 10 and 11 show a similar geographic distribution of Proved Reserves and Cumulative Production, as reported by

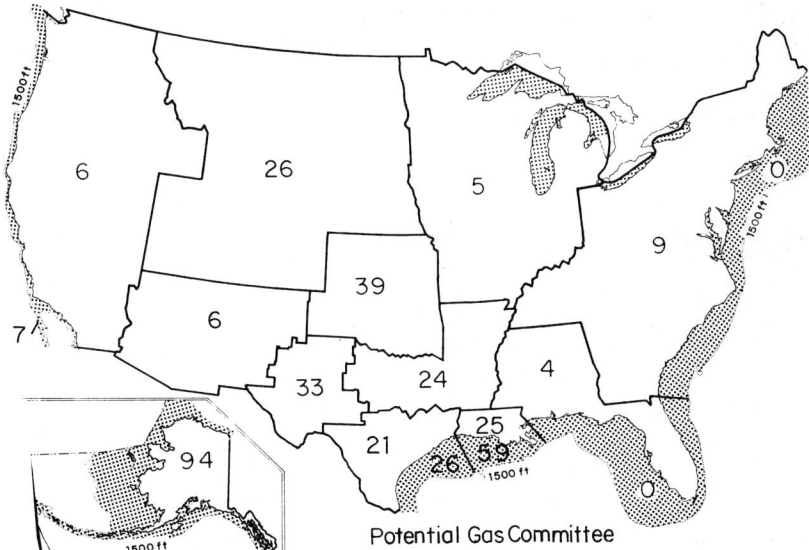

Fig. 8. Possible category of potential gas supply as of December 31, 1972 (384 trillion cubic feet). Large numbers indicate trillions of cubic feet in each area.

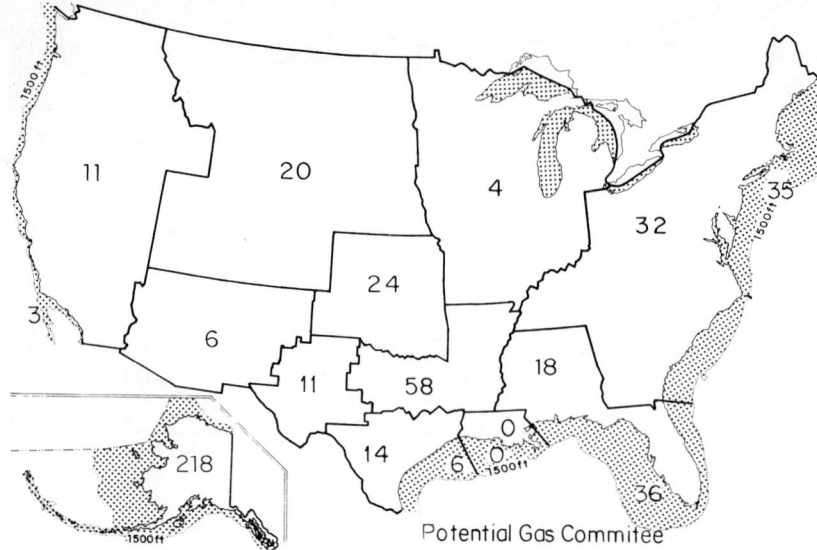

Fig. 9. Speculative category of potential gas supply as of December 31, 1972 (496 trillion cubic feet). Large numbers indicate trillions of cubic feet in each area.

the Committee on Natural Gas Reserves of the American Gas Association.* Breakdown of these two categories into onshore and offshore estimates is not available. Figures for cumulative production represent best estimates because historical data are incomplete. The estimated ultimately discoverable volume for each area is itemized in Table. 3.

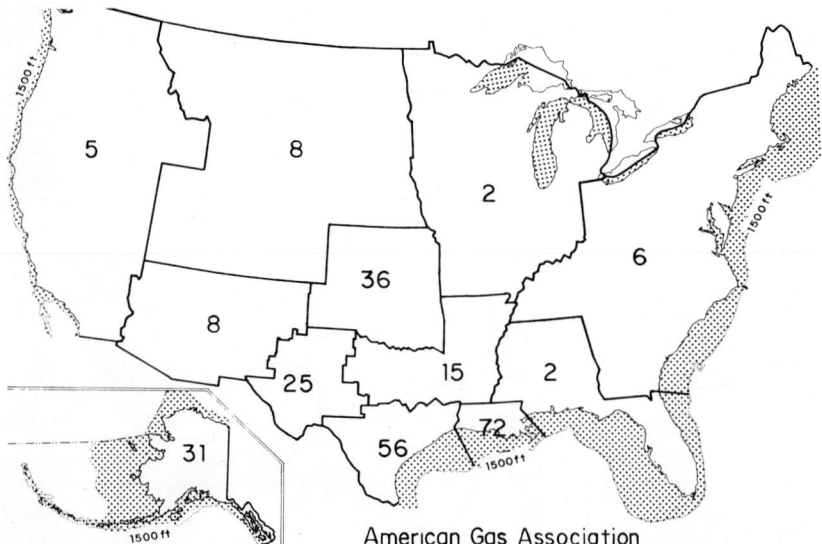

Fig. 10. Proved natural gas reserves as of December 31, 1972 (266 trillion cubic feet). Large numbers indicate trillions of cubic feet in each area.

*American Gas Association, "Proved Reserves Report," vol. 27, pp. 114–218, May 1973.

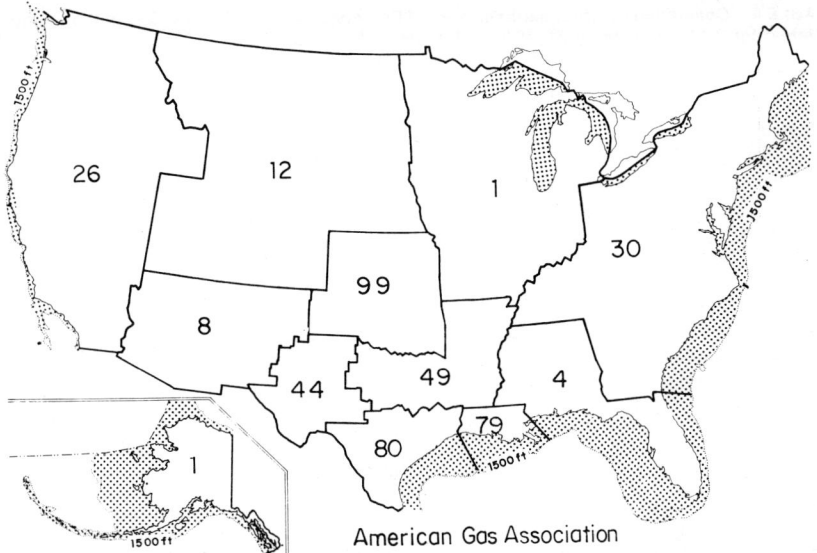

Fig. 11. Cumulative natural gas production as of December 31, 1972 (433 trillion cubic feet). Large numbers indicate trillions of cubic feet in each area.

TRENDS IN ESTIMATES

Tables 4 and 5 are included to present a comparison of the current estimate with estimates made previously.

A detailed analysis of the trends in each reporting area is contained in the "Potential Supply of Natural Gas in the United States," published by and obtainable from The Potential Gas Agency, Mineral Resources Institute, Colorado School of Mines Foundation, Inc., Golden, Colorado.

TABLE 3. Summary of Ultimately Discoverable Volumes of Natural Gas in the United States by Geographic Areas as of December 31, 1972.

Trillions of cubic feet at 14.73 psia and 60°F

Area	(a) Cumulative production*	(b) Proved reserves†	(c) Production plus reserves (a + b)	(d) Potential supply	(e) Ultimately discoverable volume (c + d)
A	30	6	36	102	138
B	4	2	6	60	66
C	1	2	3	9	12
D	49	15	64	90	154
E	79	72	151	140	291
G	80	56	136	125	261
H	12	8	20	59	79
I	8	8	16	18	34
J—North	99	36	135	84	219
J—South	44	25	69	61	130
L	26	5	31	32	63
48 states subtotal	432	235	667	780	1447
K—Alaska	1	31	32	366	398
Total United States	433	266	699	1146	1845

*Excluding stored gas.
†Including stored gas (AGA, "Proved Reserves Report," vol. 27, 1973).

TABLE 4.   Comparison of Biennial Estimates of Potential Supply of Natural Gas in the United States* for Period December 31, 1966, to December 31, 1972

Trillions of cubic feet at 14.73 psia and 60°F

	Probable	Possible	Speculative	Total
48 states				
As of 12/31/1966	300	210	180	690
1968	238	317	245	800
1970	218	326	307	851
1972	212	290	278	780
Total United States				
As of 12/31/1966	†	†	†	†
1968	260	335	632	1227
1970	257	387	534	1178
1972	266	384	496	1146

*These estimates do not include the State of Hawaii, the island territories of the United States, and their adjacent offshore areas.

†Alaska not estimated as of 12/31/66.

REFERENCES

Rept.: "Reserves of Crude Oil, Natural Gas Liquids, and Natural Gas in the United States and Canada, and the United States Productivity Capacity as of December 31, 1972," published jointly by the American Gas Association, Arlington, Va.; American Petroleum Institute, Washington; and Canadian Petroleum Association, Calgary, Alberta, vol. 27, 1973.

Averitt, Paul, and M. Devereux Carter: "Selected Sources of Information on United States and World Energy Resources—An Annotated Bibliography," *U.S. Geol. Surv. Circ. 641*, Reston, Va., 1970.

Beebe, B. W., and B. F. Curtis (eds.): "Natural Gases of North America," Memorandum 9, vol. 1, Gases in Cenozoic and Mesozoic Rocks; vol. 2, Gases in Paleozoic Rocks—plus general papers; American Association of Petroleum Geologists, Tulsa, Okla., 1968.

Rept.: "Potential Reserves of Oil, Natural Gas, and Associated Sulphur in Canada," Canadian Petroleum Association, Calgary, Alberta, 1969.

Crews, George C., James A. Barlow, Jr., and John D. Haun: "Natural Gas Resources—Green River Basin, Wyoming," WGA Symposium on Green River Basin, Wyoming Geological Association, Casper, Wyo., 1973.

Staff: "National Gas Reserves Study," Federal Power Commission, Washington, 1973.

TABLE 5.   Comparison of Biennial Estimates of Ultimately Discoverable Volumes of Natural Gas in the United States* for Period December 31, 1966 to December 31, 1972

Trillions of cubic feet at 14.73 psia and 60°F

	(a) Cumulative production†	(b) Proved reserves‡	(c) Production plus reserves (a + b)	(d) Potential supply	(e) Ultimately discoverable volume (c + d)
48 states					
As of 12/31/1966	308	286	594	690	1284
1968	346	282	628	800	1428
1970	388	260	648	851	1499
1972	432	235	667	780	1447
Total United States					
As of 12/31/1966	308	289	597	§	§
1968	346	287	633	1227	1860
1970	388	291	679	1178	1857
1972	433	266	699	1146	1845

*These estimates do not include the State of Hawaii, the island territories of the United States and their adjacent offshore areas.

†Excluding stored gas.

‡Including stored gas (AGA "Proved Reserves Report", vol. 27, 1973.)

§Alaska not estimated as of 12/31/66.

Staff: "National Gas Supply and Demand, 1971–1990," Staff Report No. 2, Federal Power Commission, Washington, 1972.

Rept.: "Future Natural Gas Requirements of the United States," Report of the Future Requirements Committee, Future Requirements Agency, Denver Research Institute, University of Denver, Denver, Colo., 1971.

Haun, John D., James A. Barlow, Jr., and Donald E. Hallinger: "Natural Gas Resources—Rocky Mountain Region," vol. 54, no. 9, pp. 1706–1718, American Association of Petroleum Geologists, Tulsa, Okla., 1970.

Theobold, P. K., S. P. Schweinfurth, and D. C. Duncan: "Energy Resources of the United States, *U.S. Geol. Surv. Circ.* 650, Reston, Va., 1972.

Winger, J. G., J. D. Emerson, G. D. Gunning, R. C. Sparling, and A. J. Zraly: "Outlook for Energy in the United States to 1985," Chase Manhattan Bank, New York, 1972.

Natural Gas Reserves in Canada

The Reserves Committee of the Canadian Petroleum Association prepares annual reserves information in an approach similar to that used by the American Petroleum Institute and described under "Petroleum Production and Reserves in the United States," included in the Handbook section on *Petroleum Technology*. Some of the more frequently used terms and definitions which apply to these data are given later in this description. No reserves are included for the Arctic Islands and Sable Island discoveries. Natural gas and natural gas liquid reserves for the Mackenzie Delta area are included.

Reserves of natural gas in Canada are summarized in Table 1. The historical record of remaining natural gas reserves in Canada is given in Fig. 1. The relationship of gross additions (proved and probable) with net production over the period, 1955–1974, is illustrated in Fig. 2. A tabulation of natural gas reserves—original in-place and ultimate in Canada—is presented in Table 2, again by provinces.

Reserves of natural gas liquids in Canada are summarized in Table 3. The historical record of remaining natural gas liquids in Canada appears in Fig. 3, The relationship of gross additions (proved and probable) with net production over the period, 1952–1974, is shown in Fig. 4.

DEFINITIONS OF TERMS

Proved Reserves. This defines the estimated quantity of natural gas and natural gas liquids which analysis of geological and engineering data demonstrate with reasonable certainty to be recoverable from known fields under existing economic and operating conditions.

Probable Reserves. These are figures which represent a realistic assessment of the reserves that will be recovered from known fields based on the estimated ultimate size and reservoir characteristics of such fields. Probable reserves include those reserves shown in the proved category.

TABLE 1. Remaining Marketable Natural Gas Reserves in Canada*
Millions of Cubic Feet at 14.65 psia and 60°F

Province	Non-associated	Associated	Dissolved	Underground storage	Total
	PROVED RESERVES				
Territories	4,815,535	....	....	....	4,815,535
British Columbia	6,946,299	266,437	92,112	....	7,304,848
Alberta	35,338,732	5,543,190	2,473,797	21,240	43,376,959
Saskatchewan	744,702	155,571	61,363	10,894	972,530
Ontario	88,104	....	....	149,038	237,142
Other Eastern Canada	977	95	....	....	1,072
Total: Canada	47,934,349	5,965,293	2,627,272	181,172	56,708,086
	PROBABLE RESERVES (including proved reserves)				
Territories	5,723,077	....	....	....	5,723,077
British Columbia	7,484,153	280,844	94,612	....	7,859,609
Alberta	40,874,240	5,630,043	2,852,016	21,240	49,377,539
Saskatchewan	1,019,119	156,008	70,218	10,894	1,256,239
Ontario	92,510	....	....	149,038	241,548
Other Eastern Canada	1,026	95	....	....	1,121
Total: Canada	55,194,125	6,066,990	3,016,846	181,172	64,459,133

*SOURCE: Canadian Petroleum Association Report, vol. 29, May 1975. Incorporated as part of API Report, vol. 29, May 1975. Reserves conditions as of 12/31/1974.

TABLE 2.   Natural Gas Reserves in Canada—Original In-Place and Ultimate*
Millions of Cubic Feet at 14.65 psia and 60°F

Province	Original raw gas in-place		Ultimate raw gas reserves		Ultimate marketable gas reserves	
	Proved	Probable†	Proved	Probable†	Proved	Probable†
Territories	6,553,374	7,714,148	5,445,280	6,431,800	4,882,998	5,790,540
British Columbia	15,856,423	16,717,224	12,754,257	13,435,080	10,633,276	11,188,037
Alberta	106,729,622	116,429,900	78,506,667	86,134,393	61,984,109	67,984,689
Saskatchewan	4,302,747	4,777,725	2,351,718	2,703,372	1,729,839	2,013,548
Manitoba	58,287	59,238	12,707	13,631	....	....
Ontario	946,231	950,990	902,423	906,962	876,140	880,546
Other Eastern Canada	34,592	34,645	33,072	33,123	30,600	30,649
Total: Canada	134,481,276	146,683,870	100,006,124	109,658,361	80,136,962	87,888,009

*SOURCE: Canadian Petroleum Association Report, vol. 29, May 1975. Incorporated as part of API Report, vol. 29, May 1975. Reserves conditions as of 12/31/1974.
†Includes proved reserves.

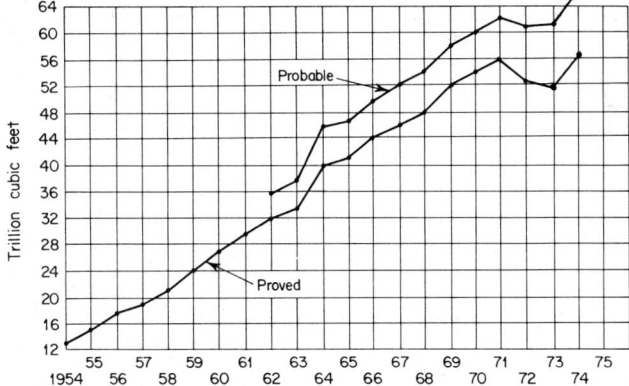

Fig. 1. Marketable (proved and probable) reserves of natural gas in Canada (1954–1974). Included in the figures for 1974, for the first time in CPA reporting, natural gas and natural gas liquids reserves are included for the Mackenzie Delta area, based upon data submitted by Canadian Arctic Gas Study Limited to the National Energy Board.

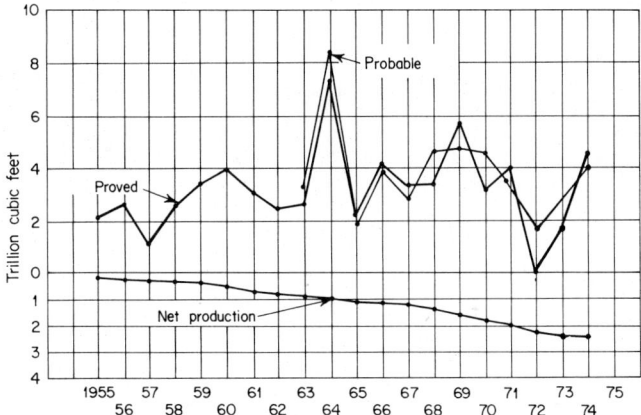

Fig. 2. Production versus gross additions (proved and probable) in Canada added each year during period (1955–1974).

TABLE 3. Natural Gas Liquids Reserves in Canada—Ultimate, and Remaining*

Thousands of Barrels

Province	Ultimate reserves		Remaining reserves	
	Proved	Probable†	Proved	Probable†
	32,000	38,000	32,000	38,000
British Columbia	61,402	62,620	37,508	38,726
Alberta	2,041,011	2,213,470	1,295,627	1,468,086
Saskatchewan	28,135	29,787	7,320	8,972
Total: Canada	2,162,548	2,343,877	1,372,455	1,553,784

*SOURCE: Canadian Petroleum Association Report, vol. 29, May 1975. Incorporated as part of API Report, vol. 29, May 1975. Reserves conditions as of 12/31/1974.
†Includes proved reserves.

Original In-Place Reserves.   These are defined as the total quantity of raw gas initially in place within the estimated area of a field from which production has been obtained, or for which reserves have been credited. This term represents the sum of (1) the anticipated ultimate raw gas production and (2) the raw gas estimated to be nonrecoverable under existing economic and operating conditions.

Ultimate Recoverable Reserves.   These reserves are defined as the total quantity of natural gas or natural gas liquids to be ultimately producible from a field, as determined

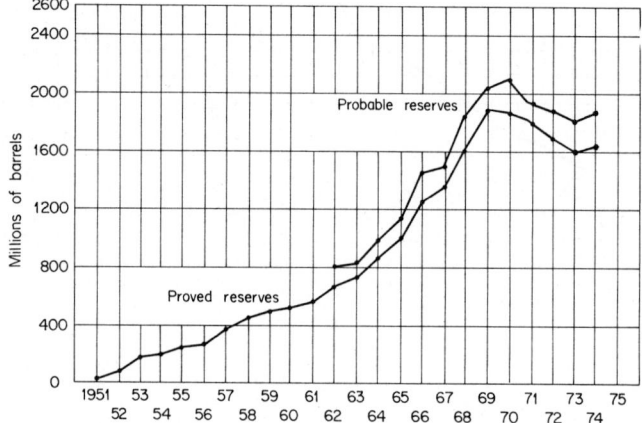

Fig. 3. Proved and probable reserves of natural gas liquids in Canada (1951–1974).

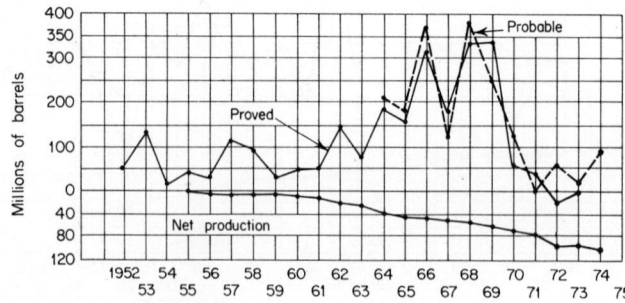

Fig. 4. Production versus gross additions (proved and probable) of natural gas liquids in Canada added each year during period (1952–1974).

by an analysis of the current geological and engineering data. This includes any quantities already produced up to the respective date of the estimate.

Remaining Reserves.   These are the quantities of natural gas or natural gas liquids as estimated under proved or probable reserves, after deduction of those quantities produced up to the respective date of the estimate.

Raw Gas.   This is natural gas, in its natural state, existing in or produced from a field.

Marketable Natural Gas.   This is defined as raw gas from which certain hydrocarbons and nonhydrocarbons have been removed or partially removed by processing. Marketable natural gas is often referred to as pipeline gas, residue gas, or sales gas.

Natural Gas Liquids.   These are defined as the hydrocarbon components—propane, butanes and pentanes-plus (also referred to as condensate), or a combination of them— that are subject to recovery from raw gas liquids by processing in field separators, scrubbers, gas processing and reprocessing plants, or cycling plants. The propane and butanes components are often referred to as liquefied petroleum gases or LPG.

RELATED STATISTICS

Other related statistics in this Handbook are listed at the end of the description, "Natural Gas Production and Reserves in the United States," in this Handbook section. References are also given in that description.

Annual reports of reserves of crude oil, natural gas liquids, and natural gas in Canada are issued by the Canadian Petroleum Association, Calgary, Alberta.

World Natural Gas Production and Reserves

Prior to the mid-1970s, the gathering of data on natural gas reserves and production on a worldwide basis was arduous and time-consuming. Stimulated by international concern over energy resources, a World Energy Conference organization was established and a concerted effort was made during the 1973–1975 period to survey the worldwide natural gas supply situation. The objective of this survey was to present comprehensive world-wide figures as of 1974. The first drafts of some reporting countries, including the United States and Canada, were completed as of early-1976. These drafts follow a very carefully prescribed format based upon thoughtfully worked out definitions. Other countries are following suit and a full first report probably will be available in 1977. Contact in the United States is the Executive Director, National Committee, World Energy Conference, Washington, D.C. 20006.

As of late-1976, the most comprehensive "official" report probably is Geological Survey Professional Paper 885, published in 1974 by the U.S. Geological Survey, Reston, Virginia 22092, authored by Sherwood E. Frezon, entitled "Summary of 1972 Oil and Gas Statistic for Onshore and Offshore Areas of 151 Countries," Stock No. 2401-02518, available from the U.S. Government Printing Office, Washington, D.C. 20402.

In the early 1940s, the American Gas Association and the Canadian Petroleum Association expanded their reporting of natural gas and natural gas liquids statistics for the United States and Canada, respectively. Today, these reliable figures are published in a joint report, updated annually. But, most other gas-producing nations over the years were much less formal about such reporting. With staggering volumes of natural gas vented to the atmosphere in some of the mideastern countries, for example, there only has been recent interest in conservation and hence a need for reliable statistics for long-range planning.

Annual natural gas production for the period 1960 through 1972 for 151 countries, essentially based on information in the Geological Survey Professional Paper 885, is given in Table 1. The probable recoverable reserves are given in Table 2. A brief chronology of the development of offshore oil and gas fields is given in the article on "World Petroleum Production and Reserves," contained in the section on *Petroleum Technology* in this Handbook.

The U.S. Geological Survey has placed the potential resources (ultimate recoverable resources) of natural gas into numbered categories. These are defined in Table 3, along with an evaluation of these resources for 151 countries investigated.

RELATED STATISTICS

Other statistics in this Handbook which relate to natural gas and petroleum include the following:
In the Handbook section on *Gas Technology:*
  "Natural Gas Production and Reserves in the United States"
  "Natural Gas Reserves in Canada"
  "Potential Supply of Natural Gas in the United States"
  "Future Gas Consumption in the United States"
In the Handbook section on *Petroleum Technology:*
  "Petroleum Production and Reserves in the United States"
  "Petroleum Reserves in Canada"
  "World Petroleum Production and Reserves"

REFERENCES

American Gas Association: "Reserves of Crude Oil, Natural Gas Liquids, and Natural Gas in the United States and Canada, and United States Productive Capacity," issued annually, AGA, Arlington, Va.
Am. Pet. Inst. Tech. Reps. **1, 2,** (and subsequent), issued periodically, API, Washington.

TABLE 1. Annual Natural Gas Production (1960–1972)
Billions of Cubic Meters

	1960	1961	1962	1963	1964	1965	1966	1967	1968	1969	1970	1971	1972
AFRICA													
Algeria	na	0.2	0.4	0.4	0.8	1.8	2.0	2.2	9.6	9.9	9.6	3.0	3.1
Angola	na	na	na	na	na	na	na	na	.1	0.4	0.8	1.1	1.1
Egypt (U.A.R.)	na	na	na	na	na	na	na	na	0.5	0.8	1.0	1.1	4.6
Gabon	0.0	0.0	0.0	0.0	0.0	0.0	0.0	0.0	0.0	0.0	0.1	0.0	0.9
Libyan Arab Republic	0.0	0.0	0.0	0.0	6.5	8.6	10.2	11.8	17.6	18.9	20.1	15.8	14.0
Nigeria	0.1	0.4	0.5	0.6	1.0	2.2	2.9	2.6	1.5	4.1	8.1	12.4	17.1
Total: Africa	0.1	0.6	0.9	1.0	8.3	12.6	15.1	16.6	29.3	34.1	39.7	33.4	42.5[a]
ASIA													
Afghanistan	0.0	0.0	0.0	0.0	0.0	0.0	0.0	0.3	1.6	2.2	2.7	2.7	2.6
Bahrain	0.1	0.2	0.2	0.2	0.4	0.7	0.8	1.2	0.9	0.9	0.7	0.5	1.8
Bangladesh	na	na	na	na	na	na	na	na	na	na	na	na	na
Burma	2.6	2.7	2.9	2.9	3.1	0.4	0.5	0.6	3.3	3.1	3.1	3.4	3.9
China, People's Republic of	na	na	na	na	na	na	na	na	na	na	11.4	4.2	na
China, Republic of (Taiwan)	0.0	0.0	0.0	0.1	0.2	0.3	0.4	0.5	0.7	0.9	0.9	1.1	0.1
India	0.1	0.2	0.2	0.2	0.4	0.7	0.8	1.2	1.3	1.4	1.4	1.4	1.6
Indonesia	2.6	2.9	2.9	3.1	3.1	3.7	4.5	4.5	3.3	3.1	3.1	3.4	3.9
Iran	7.5	8.4	9.2	10.3	11.8	14.4	17.9	20.1	22.7	25.3	31.0	37.0	42.4
Iraq	na	na	na	na	na	0.1	0.1	0.1	6.0	6.0	6.0	6.0	neg
Israel	na	0.6	0.9	1.4	1.3	1.3	1.3	1.3	1.4	1.5	1.7	2.4	0.1
Japan	0.4	0.9	1.1	1.1	1.2	1.3	1.7	1.9	2.0	2.1	2.1	2.4	2.5
Kuwait	9.9	10.1	11.4	12.1	13.4	13.7	14.5	14.5	15.4	16.3	17.4	18.6	18.2
Oman	0.0	0.0	0.0	0.0	0.0	0.0	0.0	0.5	1.9	2.5	2.6	2.2	1.5
Pakistan	0.7	0.9	1.1	1.3	1.6	1.8	2.1	2.3	2.4	2.8	3.6	3.4	3.5
Qatar	1.0	1.0	1.1	1.1	1.2	1.3	1.7	1.9	2.0	2.1	2.1	2.5	2.5
Saudi Arabia	7.7	8.6	9.6	10.4	11.0	12.8	15.1	16.3	17.7	18.7	22.1	27.7	32.3
Turkey	na	na	na	na	na	na	na	na	na	na	0.6	na	na
U.S.S.R.	50.0	64.0	79.0	96.0	109.0	129.0	145.0	159.0	171.0	183.0	198.0	212.0	224.0
United Arab Emirates	0.0	0.0	0.1	0.4	1.4	2.2	2.8	3.0	5.4	6.7	8.2	7.2	na
Total: Asia	79.9	96.5	115.5	136.2	154.4	178.0	202.2	221.6	253.8	273.9	314.3	333.1	337.6

TABLE 1. (continued)

	1960	1961	1962	1963	1964	1965	1966	1967	1968	1969	1970	1971	1972
OCEANIA													
Australia	0.0	0.0	0.0	0.0	0.0	0.0	0.0	0.0	0.0	0.3	1.5	2.2	2.8
New Zealand	0.0	0.0	0.0	0.0	0.0	0.0	0.0	0.0	0.0	0.0	0.1	0.3	0.4
Total: Oceania	0.0	0.0	0.0	0.0	0.0	0.0	0.0	0.0	0.0	0.3	1.6	2.5	3.2
EUROPE													
Austria	1.5	1.6	1.6	1.7	1.8	1.7	1.9	1.8	1.6	1.5	1.9	1.9	2.0
Bulgaria	na	na	na	na	na	0.1	0.1	0.1	0.5	0.5	0.5	0.4	0.4
Czechoslovakia	na	na	na	na	na	na	na	na	na	1.0	1.0	1.0	1.2
France	4.4	6.0	7.0	7.5	8.0	7.9	7.9	8.6	8.6	9.8	10.2	8.1	10.9
Germany, Federal Republic of (West)	0.5	0.5	0.6	0.9	1.5	2.2	2.8	3.7	5.8	8.1	12.0	13.8	17.7
Hungary	0.3	0.3	0.3	0.6	0.8	1.1	1.6	2.0	2.7	3.2	3.5	3.7	4.0
Italy	6.4	6.9	7.2	7.3	7.7	7.8	8.8	9.4	10.4	11.7	13.1	13.0	14.2
Netherlands	0.4	0.5	0.5	0.6	0.9	1.8	3.6	7.9	16.1	25.3	33.4	46.2	57.9
Poland	0.6	0.8	0.8	1.0	1.2	1.4	1.4	1.6	2.6	3.9	5.2	5.4	5.7
Romania	6.9	7.6	9.3	10.7	12.1	13.6	14.1	15.8	21.9	24.1	25.8	26.7	33.1
United Kingdom	0.0	0.0	0.0	0.0	0.0	0.0	0.0	0.5	2.0	5.1	11.2	18.7	26.1
Yugoslavia	0.1	0.1	0.1	0.2	0.3	0.3	0.4	0.5	0.6	0.7	1.0	1.2	1.2
Total: Europe	21.1	24.3	27.4	30.5	34.3	37.9	42.6	51.9	72.8	94.9	118.8	140.1	708.4[b]
NORTH AMERICA													
Canada	14.8	18.0	25.3	28.1	32.1	35.0	38.0	41.7	48.0	56.0	64.5	70.8	82.5
Mexico	10.2	10.8	11.1	12.0	13.7	14.0	15.0	16.2	16.3	17.2	18.8	18.2	18.7
United States	427.0	438.0	454.0	481.0	494.0	509.0	539.0	573.0	604.0	642.0	674.0	683.0	679.3
Total: North America	452.0	466.8	490.4	521.1	539.8	558.0	592.0	630.9	668.3	715.2	757.3	772.0	780.5
SOUTH AMERICA													
Argentina	3.5	4.9	6.2	6.0	6.6	6.2	6.0	6.5	7.1	7.0	7.7	8.1	7.6
Bolivia	0.0	0.0	0.0	0.0	0.0	0.0	0.0	0.0	0.9	0.8	0.8	2.3	3.4
Brazil	0.6	0.6	0.5	0.6	0.5	0.7	0.8	0.9	1.0	1.3	1.3	1.2	1.2

Chile	2.2	2.5	3.6	5.2	6.3	6.2	6.7	7.0	7.0	7.5	7.6	8.0	8.1
Colombia	2.3	2.2	2.2	2.4	2.4	2.7	2.8	2.8	2.7	2.9	3.0	2.4	3.3
Ecuador	na	na	na	na	na	na	na	0.2	0.2	0.2	0.3	0.1	0.2
Peru	1.3	1.3	1.4	1.4	1.7	1.8	1.8	1.9	2.1	2.1	2.1	1.9	1.9
Trinidad and Tobago	3.0	3.0	3.0	3.0	3.0	3.0	3.0	4.0	4.0	4.0	3.0	3.0	2.9
Venezuela	31.6	33.1	36.3	37.5	39.3	40.8	41.3	45.8	46.3	47.4	48.4	47.6	47.2
Total: South America	44.5	47.6	53.2	56.1	59.8	61.4	62.4	69.1	71.3	73.2	74.2	74.6	75.8
Total—151 countries	597.6	635.8	687.4	744.9	796.6	847.9	914.3	990.1	1095.5	1191.6	1305.9	1355.7	1948.0
							13,111.3 (cumulative)						

na = no information available.

neg = negligible.

[a]Includes production by Tunisia of 1.7 billion cubic meters.

[b]Includes production by the German Democratic Republic (not previously reported) of 5.4 billion cubic meters. Also, includes production of Norway of 528.6 billion cubic meters, reflecting burgeoning production from North Sea offshore gas and oil fields.

Notes:

Because the oil and gas fields of the U.S.S.R. are located on two continents and because of the difficulty of obtaining an accurate continental breakdown, figures traditionally have all been included under the Asian continent. However, there is a trend among statisticians to report U.S.S.R. figures in a separate category.

African countries with no reportable natural gas production: Botswana, Burundi, Cameroon, Central African Republic, Chad, Congo (na), Dahomey, Equatorial Guinea, Ethiopia, Gambia, Ghana, Guinea, Ivory Coast, Kenya, Lesotho, Liberia, Malagasy Republic, Malawi, Mauritania, Mauritius, Morocco (neg), Mozambique, Niger, (Portuguese) Guinea, Rwanda, Senegal, Sierra Leone, Somalia, Republic of South Africa, Southwest Africa, Sudan, Swaziland, Tanzania, Togo, Uganda, Upper Volta, Zaire, Zambia.

Asian countries with no reportable natural gas production: Bhutan, Brunei (na), Cyprus, Jordan, Khmer Republic, Korea, Laos, Lebanon, Malaysia, Maldives, Mongolia, Nepal, Philippines, Singapore, Sri Lanka, Syrian Arab Republic (na), Thailand, Vietnam, Yemen Arab Republic, Yemen People's Republic.

Oceanian countries with no reportable natural gas production: Fiji, Nauru, Tonga, Western Samoa. North American countries with no reportable natural gas production: Bahamas, Barbados (neg), Belize, Costa Rica, Cuba (na), Dominican Republic, El Salvador, Guatemala, Haiti, Honduras, Jamaica, Nicaragua, Panama.

South American countries with no reportable natural gas production: Guyana, Paraguay, Uruguay.

TABLE 2.   Proved Recoverable Natural Gas Reserves—1972
Billions of Cubic Meters

Continent and country	Onshore	Offshore	Total
	AFRICA		
Algeria	3,000	0	3,000
Angola	na	na	39.7
Egypt (U.A.R.)	na	na	213
Gabon	na	na	181
Libya	779	na	>779
Morocco	0.4	0	0.4
Nigeria	840	280	1,120
Tunisia	43	na	>43
Total: Africa	4,662.4	280	5,376.1[a]
	ASIA		
Afghanistan	142	*	142
Bahrain	23.8	0	23.8
Bangladesh	262	0	262
Burma	2.8	0	2.8
China, People's Republic of	na	na	133.3
China, Republic of (Taiwan)	14	0	14
India	42	0	42
Indonesia	na	na	156
Iran	na	na	57,000
Iraq	566.6	0	566.6
Israel	1.5	0	1.5
Japan	<11	na	11
Kuwait	na	na	1,190
Malaysia	na	na	284
Oman	na	na	53.7
Pakistan	550	0	550
Qatar	na	na	266.6
Saudi Arabia	na	na	1,541
Turkey	5	0	5
U.S.S.R.	<20,000	na	20,000
United Arab Emirates	na	na	33.2
Total: Asia	21,640.5[b]	na	82,298.3[b]
	OCEANIA		
Australia	na	na	1,068
New Zealand	na	na	>6,000
Total: Oceania	na	na	7,068
	EUROPE		
Albania	8.4	0	8.4
Austria	15	*	15
Bulgaria	28	0	28
Czechoslovakia	15	*	15
Denmark	0	14	14
France	187	0	187
Germany, Federal Repub. of (West)	351	0	351
Hungary	119	0	119
Italy	na	na	169.9
Netherlands	2,493	na	2,493
Norway	0	1,420	1,420
Poland	140	0	140
Romania	na	na	283.3

Billions of Cubic Meters

Continent and country	Onshore	Offshore	Total
Spain	na	na	15
United Kingdom	na	1,275	>1,275
Yugoslavia	48.2	0	48.2
Total: Europe	3,418.8^c	2,709	6,596^c
	NORTH AMERICA		
Canada	1,570	na	>1,570
Mexico	na	na	326
United States	d	d	7,537
Total: North America	1,570	na	9,433
	SOUTH AMERICA		
Argentina	190	na	190
Bolivia	135.9	*	135.9
Brazil	21	5	26
Chile	50	0	50
Colombia	70	0	70
Ecuador	170	0	170
Peru	na	na	67
Trinidad and Tobago	25.8	113	138.8
Venezuela	<1,020	na	1,020
Total: South America	1,682.7	118	1,867.7^a
Total: 151 countries	32,974.4	3,107	112,639.1

na = no information available.

^{a}Total column does not equal sum of onshore and offshore totals because of lack of information on split for some countries.

^{b}Includes reserves of the Syrian Arab Republic (19.8 billion cubic meters).

^{c}Includes reserves of the German Democratic Republic (14.2 billion cubic meters).

^{d}Full breakdown of United States onshore and offshore reserves under investigation.

*Landlocked.

Notes:

Because the oil and gas fields of the U.S.S.R. are located on two continents and because of the difficulty of obtaining an accurate continental breakdown, figures traditionally have been all included under the Asian continent. However, there is a trend among statisticians to report U.S.S.R. figures in a separate category.

African countries with no reportable proved recoverable reserves of natural gas: Botswana, Burundi, Cameroon, Central African Republic, Chad, Congo (negligible), Dahomey, Equatorial Guinea, Ethiopia, Gambia, Ghana (na), Guinea, Ivory Coast, Kenya, Lesotho, Liberia, Malagasy Republic, Malawi, Mali, Mauritania, Mauritius, Mozambique, Niger, (Portuguese) Guinea, Rwanda, Senegal, Sierra Leone, Somalia, Republic of South Africa, Southwest Africa, Sudan, Swaziland, Tanzania, Togo, Uganda, Upper Volta, Zaire, Zambia.

Asian countries with no reportable proved recoverable reserves of natural gas: Bhutan, Republic of China (Taiwan), Cyprus, Jordan, Khmer Republic, Korea, Laos, Maldives, Mongolia, Nepal, Philippines, Singapore, Sri Lanka, Vietnam, Yemen Arab Republic, Yemen People's Republic.

Oceanian countries with no reportable proved recoverable reserves of natural gas: Fiji, Nauru, Tonga, Western Samoa.

European countries with no reportable proved recoverable reserves of natural gas: Belgium, Finland, Greece, Iceland, Ireland (na), Lichtenstein, Luxembourg, Malta, Portugal, San Marino, Sweden, Switzerland.

North American countries with no reportable proved recoverable reserves of natural gas: Bahamas, Barbados (na), Belze, Costa Rica, Cuba (na), Dominican Republic, El Salvador, Guatemala, Haiti, Honduras, Jamaica, Nicaragua, Panama.

South American countries with no reportable proved recoverable reserves of natural gas: Guyana, Paraguay, Uruguay.

Some of the foregoing countries do possess potential resources, i.e., ultimate recoverable resources. See Table 3.

TABLE 3. Worldwide Ultimate Recoverable (Potential) Natural Gas Resources

Continent and country	Onshore	Offshore	Total
AFRICA			
Algeria	II	0	II
Angola	III	III	III
Botswana	III	*	III
Burundi	0	*	0
Cameroon	IV	IV	IV
Central African Republic	V	*	V
Chad	IV	*	IV
Congo	III	V	III
Dahomey	V	V	V
Egypt	II	III	II
Equatorial Guinea	V	V	V
Ethiopia	III	IV	III
Gabon	III	III	III
Gambia	V	V	V
Ghana	IV	IV	IV
Guinea	IV	IV	IV
Ivory Coast	V	IV	IV
Kenya	III	V	III
Lesotho	VI	*	VI
Liberia	0	V	V
Libyan Arab Republic	II	II	II
Malagasy Republic	IV	IV	IV
Malawi	0	*	0
Mali	IV	*	IV
Mauritania	IV	IV	IV
Mauritius	0	0	0
Morocco	IV	IV	IV
Mozambique	IV	IV	IV
Niger	IV	*	IV
Nigeria	III	III	III
(Portuguese) Guinea	V	V	V
Rwanda	0	*	0
Senegal	IV	IV	IV
Sierra Leone	V	V	V
Somalia	V	V	V
Republic of South Africa	III	IV	III
Southwest Africa	V	V	V
Sudan	V	V	V
Swaziland	VI	*	VI
Tanzania	IV	IV	IV
Togo	V	V	V
Tunisia	IV	IV	IV
Uganda	0	*	0
Upper Volta	0	*	0
Zaire	III	IV	III
Zambia	0	*	0
ASIA			
Afghanistan	III	*	III
Bahrain	III	IV	III
Bangladesh	III	III	III
Bhutan	0	*	0
Brunei	IV	V	IV
Burma	III	III	III
China (People's Republic of)	II	III	II
China (Republic of—Taiwan)	IV	IV	IV
Cyprus	V	IV	IV
India	III	III	III
Indonesia	II	III	II
Iran	II	II	II
Iraq	III	IV	III
Israel	IV	IV	IV

Continent and country	Onshore	Offshore	Total
Japan	III	IV	IV
Jordan	IV	*	IV
Khmer Republic	IV	IV	IV
Korea (Democratic People's Republic)	0	IV	IV
Korea (Republic of)	0	III	III
Kuwait	III	IV	III
Laos	IV	*	IV
Lebanon	V	IV	IV
Malaysia	III	III	III
Maldives	VI	VI	VI
Mongolia	III	*	III
Nepal	V	*	V
Oman	III	IV	III
Pakistan	III	IV	III
Philippines	III	III	III
Qatar	III	III	III
Saudi Arabia	II	II	II
Singapore	0	V	V
Sri Lanka	0	V	V
Syrian Arab Republic	IV	*	IV
Thailand	IV	IV	IV
Turkey	IV	IV	IV
United Arab Emirates	III	III	III
U.S.S.R.	II	II	II
Vietnam	V	V	V
Yemen Arab Republic	VI	VI	VI
Ymen People's Republic	V	IV	IV
OCEANIA			
Australia	II	II	II
Fiji	V	V	V
Nauru	IV	IV	IV
New Zealand	IV	III	III
Tonga	IV	IV	IV
Western Samoa	IV	IV	IV
EUROPE			
Albania	IV	IV	IV
Austria	IV	*	IV
Belgium	V	IV	IV
Bulgaria	IV	IV	IV
Czechoslovakia	IV	*	IV
Denmark	IV	IV	IV
Finland	0	V	V
France	III	IV	III
German Democratic Republic	II	IV	III
Germany (Federal Republic of)	III	IV	III
Greece	V	IV	IV
Hungary	IV	*	IV
Iceland	0	0	0
Ireland	IV	IV	III
Italy	IV	III	III
Lichtenstein	0	*	0
Luxembourg	0	*	0
Malta	VI	IV	IV
Netherlands	VI	IV	IV
Norway	0	III	III
Poland	IV	V	IV
Portugal	V	V	V
Romania	III	IV	III
San Marino	0	*	0
Spain	IV	V	IV
Sweden	VI	V	V
Switzerland	V	*	V
United Kingdom	V	III	III
Yugoslavia	IV	IV	IV

TABLE 3. (continued)

Continent and country	Onshore	Offshore	Total
NORTH AMERICA			
Bahamas	VI	VI	VI
Barbados	V	V	IV
Belize	V	V	V
Canada	V	V	V
Costa Rica	VI	IV	IV
Cuba	V	IV	V
Dominican Republic	V	V	V
El Salvador	0	V	V
Guatemala	IV	IV	IV
Haiti	V	V	V
Honduras	V	IV	V
Jamaica	V	IV	IV
Mexico	II	III	II
Nicaragua	V	IV	IV
Panama	V	IV	IV
United States	I	I	I
SOUTH AMERICA			
Argentina	III	III	III
Bolivia	II	*	II
Brazil	II	III	II
Chile	IV	III	III
Colombia	III	III	III
Ecuador	III	III	III
Guyana	IV	V	IV
Paraguay	IV	*	IV
Peru	III	IV	III
Trinidad and Tobago	IV	IV	IV
Uruguay	IV	IV	IV
Venezuela	IV	IV	IV

*Landlocked.

Categories of Potential Resources

Symbol	Trillions of cubic feet	
I	1,000	−10,000
II	100	− 1,000
III	10	− 100
IV	1	− 10
V	0.1	− 1
VI	0.01−	0.1

U.S. Bureau of Mines, Pittsburgh, Pa. (Obtain list of available publications.)

Am. Assoc. Pet. Geol. Bull., **45–56** (1965–1972).

Chase Manhattan Bank, New York, "Capital Investments of the World Petroleum Industry," 1971.

Hendricks, T. A.: "Resources of Oil, Gas, and Natural Gas Liquids in the United States and the World," *U.S. Geol. Surv. Circ.* 522, Reston, Va., 1965.

U.S. Bureau of Mines, Pittsburgh, Pa.: "Minerals Yearbook," published annually.

"Summary of 1972 Oil and Gas Statistics for Onshore and Offshore Areas of 151 Countries," Geol. Surv. Prof. Paper 885, 163 pp. Stock No. 2401-02518, U.S. Government Printing Office, Washington, D.C., 1974.

"World Energy Conference Survey of World Energy Resources (as of 1974)," National Committee, World Energy Conference, Washington, D.C. (in preparation)

"The Worldwide Search for Petroleum Offshore," Geol. Surv. Circular 694, U.S. Geological Survey, Reston, Virginia, 1974.

Future Gas Consumption of the United States

By Richard K. Doran*

The total gas consumption in the United States *including field use* is estimated to decline from 22.8 trillion cubic feet in 1974 to 20.6 trillion cubic feet in 1985, a decrease of 2.2 trillion cubic feet, or an annual decline rate of 0.9 percent over the eleven-year period. Field use in the summary given here includes only lease and plant use. Extraction loss due to shrinkage is not included as part of the gas consumption forecast. Total consumption *excluding field use* is estimated to decrease from 20.9 trillion cubic feet in 1974 to 18.4 trillion cubic feet in 1985, a decrease of 2.5 trillion cubic feet, or an annual decline rate of 1.2 percent for the entire period. These numbers are given in Table 1 and charted in Fig. 1. These estimates were originally made by the Gas Requirements Committee, formerly known as the Future Requirements Committee, an industry-sponsored forecasting activity.†

The estimated changes in gas consumption excluding field use during the eleven-year period by classification of use are also shown in Table 1. Annual growth (decline) rates for this period range from a high of 2.0 percent in the commercial sector to a decline of rate of 7.9 percent in the interruptible power generation sector. The largest volumetric increase is expected to occur in the residential sector.

Table 2 indicates the shift that is expected to take place during the next eleven years with respect to changing market shares. Significant increases will take place in the firm residential and commercial categories. The other and field use categories are also expected to gain, although not as significantly.

Residential consumption is estimated to reach 5.9 trillion cubic feet in 1985, an increase of 1.0 trillion cubic feet, or a 2.0 percent annual growth rate over the 1974 consumption of 4.9 trillion cubic feet. The majority of users in this category are individual residences, small multiple dwellings, and larger residential units with individually metered apartments. Some commercial usage occurring in these units may be included, where such usage is subordinate to residential purposes and is not metered separately. Both space heating and other uses (such as gas burned in ranges, water heaters, driers, and air conditioners) are included.

Commercial consumption was 2.2 trillion cubic feet in 1974 and is expected to rise to 2.7 trillion cubic feet in 1985, a 2.0 percent compound annual growth rate for the eleven-year period. The users in this category are customarily trade establishments, service enterprises, business offices, centrally metered larger apartment dwellings, hotels, warehouses, and similar establishments. In some cases, smaller multiple dwelling units, public institutions, or small industrial customers may be included in this category. All forms of use, process heat, space heat, appliances, and air conditioning may be included within the category.

Firm industrial consumption was 6.5 trillion cubic feet in 1974 and is forecast to

*Associate Director, Gas Requirements Agency (GRA), University of Denver Research Institute, Denver, Colorado.

†This article summarizes the more detailed material found in Vol. 6, *Future Gas Consumption of the United States,* prepared by the Gas Requirements Committee. Mr. Walter E. Caine, former Chairman of the Gas Requirements Committee, provided valuable material and assistance in preparing this article on future gas consumption.

The Gas Industry Committee, representing the American Gas Association, the American Petroleum Institute, and the Interstate Natural Gas Association of America, selected the Denver Research Institute in 1967 to form the Future Requirements Agency, renamed the Gas Requirements Agency in 1974, to "approve criteria and methods, insure maintenance of standards and objectivity, review and evaluate reports, publish the final estimates." In performing this assignment, the Agency appoints the Chairman, the Vice Chairman, and the Secretary of the Gas Requirements Committee, and then works closely with the officers, the Committee, and the regional work groups in establishing the objectives of each study and carrying the work to completion.

TABLE 1

GAS CONSUMPTION IN THE UNITED STATES BY CLASS OF SERVICE
ACTUALS 1973 AND 1974 AND ESTIMATED FOR 1975 THROUGH 1980 AND 1985

(Billions of Cubic Feet at 1000 Btu Per Cubic Foot at 14.73 psia)

	Firm			Electric Power Generation		Interruptible[1]		Other	Total Excluding Field Use	Field Use	TOTAL
	Residential	Commercial	Industrial	Firm	Interruptible	Commercial	Industrial				
ACTUALS											
1973	4,983	2,248	6,216	2,104	1,547		2,778	1,451	21,327	1,964	23,291
1974	4,887	2,215	6,445	2,187	1,255	295	2,258	1,365	20,907	1,859	22,766
ESTIMATED											
1975	5,155	2,319	5,577	2,145	883	347	2,022	1,432	19,880	1,762	21,642
1976	5,252	2,359	5,544	2,188	705	327	1,823	1,430	19,628	1,726	21,354
1977	5,326	2,400	5,384	2,136	589	316	1,691	1,430	19,272	1,699	20,971
1978	5,404	2,442	5,440	2,042	544	289	1,593	1,437	19,191	1,708	20,899
1979	5,488	2,482	5,360	1,858	498	296	1,515	1,422	18,919	1,767	20,686
1980	5,579	2,519	5,200	1,730	448	334	1,541	1,420	18,771	1,817	20,588
1985	5,933	2,691	4,829	1,084	507	357	1,549	1,457	18,407	2,196	20,603

[1] Does not include Interruptible Electric Power Generation.

TABLE 2

GAS CONSUMPTION IN THE UNITED STATES BY CLASS OF SERVICE AS A PERCENT OF TOTAL
ACTUALS 1973 AND 1974 AND ESTIMATED FOR 1975 THROUGH 1980 AND 1985

	Firm			Electric Power Generation		Interruptible[1]		Other	Total Excluding Field Use	Field Use	TOTAL
	Residential	Commercial	Industrial	Firm	Interruptible	Commercial	Industrial				
ACTUALS											
1973— % of Total	21.4	9.7	26.7	9.0	6.6	11.9		6.2	91.6	8.4	100
1974— % of Total	21.5	9.7	28.3	9.6	5.5	1.3	9.9	6.0	91.8	8.2	100
ESTIMATED											
1975— % of Total	23.8	10.7	25.8	9.9	4.1	1.6	9.3	6.6	91.9	8.1	100
1976— % of Total	24.6	11.1	26.0	10.3	3.3	1.5	8.5	6.7	91.9	8.1	100
1977— % of Total	25.4	11.4	25.7	10.2	2.8	1.5	8.1	6.8	91.9	8.1	100
1978— % of Total	25.9	11.7	26.0	9.8	2.6	1.4	7.6	6.9	91.8	8.2	100
1979— % of Total	26.5	12.0	25.9	9.0	2.4	1.4	7.3	6.9	91.5	8.5	100
1980— % of Total	27.1	12.2	25.3	8.4	2.2	1.6	7.5	6.9	91.2	8.8	100
1985— % of Total	28.8	13.1	23.4	5.3	2.5	1.7	7.5	7.1	89.3	10.7	100

[1] Does not include Interruptible Electric Power Generation.

decline to 4.8 trillion cubic feet in 1985. This represents an annual decline of 2.6 percent. The limited supply of gas is expected to have a substantial effect on industrial gas consumption during the forecast period. Firm industrial consumption is expected to reduce its share of total gas consumption from 28.3 percent in 1974 to 23.4 percent in 1985. This category includes all use of gas by manufacturing plants, process companies, etc., under schedules or contracts which anticipate no interruptions whatever, as well as under

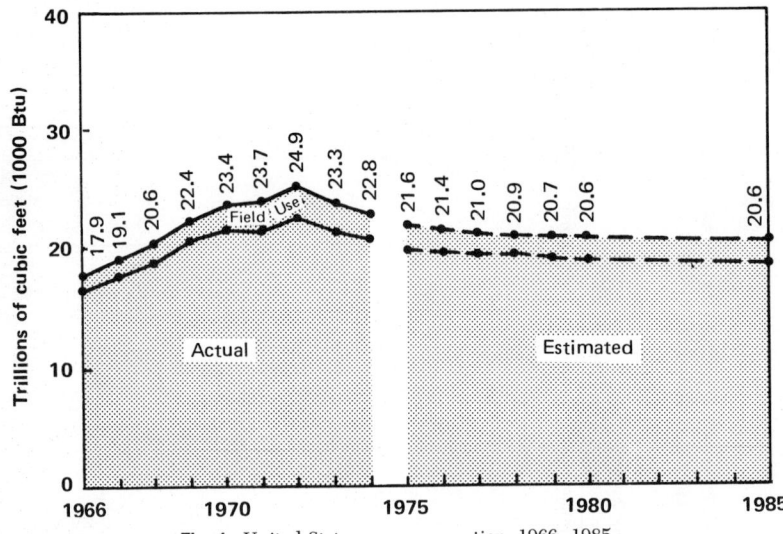

Fig. 1. United States gas consumption, 1966–1985.

those that provide for firm deliveries only during specified seasonal or other off-peak periods.

Firm electric power generation consumption is expected to decline from 2.2 trillion cubic feet in 1974 to 1.1 trillion cubic feet in 1985, which is a 6.2 percent annual decline rate. Share of market for this category of consumption is anticipated to decline from 9.6 percent in 1974 to 5.3 percent in 1985.

Interruptible consumption is gas delivered, most frequently to power generation or large industrial consumers, under arrangements that permit interruption on short notice. Traditionally, these interruptions have been due to adverse weather conditions; however, in recent years actual shortages in gas supply have been a contributory factor to these interruptions. The interruptible customer is curtailed on gas delivery so that service can be maintained to firm residential, commercial, and industrial customers. Notification of interruption may be written or oral and may vary, according to local or gas utility practice, from a few hours to a month in advance. Some large commercial users and public institutions may also fall into this category.

Interruptible power generation consumption is expected to be down 60 percent by 1985 compared to the 1.3 trillion cubic feet reported in 1974. The annual decline rate is 7.9 percent for the 1974–1985 period. The interruptible share of gas consumption will drop from 5.5 percent of the total in 1974 to 2.5 percent in 1985. The present natural gas shortage and company and regulatory policies are contributing factors to the expected decline in this consumption. Gas which normally would have been available for this use is being shifted to other, higher priority classes of service.

Total commercial and industrial interruptible consumption in 1974 was 2.6 trillion cubic feet and is estimated to decline to 1.9 trillion cubic feet in 1985. Because of the gas supply shortage, gas available for interruptible use has diminished, even though a much larger potential demand exists. It should be recognized, however, that availability of storage and economy in operations will continue to necessitate the sale of some volumes of interruptible gas, even though this type of service is generally regarded as having a lower priority.

Other use classification is defined as all gas not accounted for in the previous specific categories, including pipeline fuel, gas utility company use unaccounted for, transmission loss, stored input cushion gas, and any other sale or use not otherwise classified. This classification is estimated to remain at approximately the same level over the eleven-year period, reaching 1.5 trillion cubic feet in 1985.

Field use is gas consumed as fuel on lease for pumping, drilling, and other field

TABLE 3. United States Gas Consumption by Region, 1974 and 1985
Billions of cubic feet (Bcf) at 1000 Btu per cubic foot at 14.73 psia dry

No.	Region	1974 actual Bcf	Percent of total	1985 estimate Bcf	Percent of total	11-year compound growth (decline) rate, percent
1	New England	276	1.3	324	1.8	1.5
2	Appalachian	3,715	17.8	3,786	20.6	0.2
3	Southeast	1,514	7.2	971	5.3	(3.9)
4	Great Lakes	3,036	14.5	3,094	16.8	0.2
5	Northern Plains	970	4.6	783	4.3	(2.0)
6	Mid-Continent	1,520	7.3	1,350	7.3	(1.1)
7	Gulf Coast	6,683	32.0	4,823	26.2	(2.9)
8	Rocky Mountain	571	2.7	573	3.1	0.03
9	Pacific Southwest	2,224	10.7	2,180	11.8	(0.2)
10	Pacific Northwest	323	1.5	404	2.2	2.1
11	Pacific	77	0.4	117	0.6	3.8
	Total*	20,909	100.0%	18,405	100.0%	(1.2)

*Excluding field use.

activities; fuel used by field or mainline processing plants; and quantities vented and flared on producing properties. This category of gas consumption is shown in Tables 1 and 2 as well as in Table 7 as a separate item.

Extraction loss is defined as losses of volumes caused by field or mainline plant extraction of such products as propane, butane, natural gasoline, sulfur, or helium from the wet gas stream as it is received into the gas utility plant. These losses are shown historically in Table 8.

REGIONAL CONSUMPTION OF NATURAL GAS

Five of the eleven Gas Requirement Committee regions anticipate declines in gas consumption through 1985. Region 3 (Southeast) expects the largest annual percentage decline of 3.9 percent during the period. Region 7 (Gulf Coast) will experience the largest volumetric decline, going from 6.7 trillion cubic feet in 1974 to 4.8 trillion cubic feet in 1985. Region 11 (Pacific) is not significantly hampered by gas supply shortages. Gas consumption in the two states comprising this region (Alaska and Hawaii) will grow at 3.8 percent per year. Comparison of the regions is shown in Table 3.

Industrial consumption of natural gas by region and by Standard Industrial Classification (SIC) for the year 1974 is summarized in Table 5. The national totals for the SIC breakdown are shown in Table 4.

SUPPLEMENTAL GAS SOURCES

In their original format, GRC studies portrayed market requirements* for natural gas, aggregated for 20 to 25 years which would indicate the demands that could be anticipated to be made against the natural gas reservoirs within reach of the continental pipeline

*Requirements are an estimate of the future national need for gaseous hydrocarbons from all sources for all purposes, fuel and nonfuel, by the United States' economy under specifically stated assumptions as to the supply and price of gas.

TABLE 4. United States Gas Consumption by Regions and by Standard Industrial Classification (SIC), 1974*

Millions of cubic feet at 1000 Btu per cubic foot at 14.73 psia dry

Group	STANDARD INDUSTRIAL CLASSIFICATION — Description	REGION 1 Firm	REGION 1 Interruptible	REGION 1 Total	REGION 2 Firm	REGION 2 Interruptible	REGION 2 Total	REGION 3 Firm	REGION 3 Interruptible	REGION 3 Total
10-14	Mining	—	—	—	22,433	25	22,458	805	1,795	2,600
20	Food and Kindred Products	436	254	690	33,939	14,925	48,864	9,100	27,677	36,777
21	Tobacco Manufactures	—	—	—	1,104	435	1,539	205	898	1,103
22	Textile Mill Products	766	319	1,085	7,716	5,329	13,045	14,541	54,625	69,166
23	Apparel and Other Textile Products	431	728	1,159	2,027	854	2,881	3,647	24,501	28,148
24	Lumber and Wood Products	40	—	40	3,440	718	4,158	2,590	10,903	13,493
25	Furniture and Fixtures	24	—	24	2,712	220	2,932	1,465	950	2,415
26	Paper and Allied Products	503	859	1,362	24,873	9,372	34,245	17,021	61,257	78,278
27	Printing and Publishing	104	5	109	3,982	138	4,120	512	154	666
28	Chemicals and Allied Products	790	114	904	137,943	25,925	163,868	93,359	50,732	144,091
29	Petroleum and Coal Products	28	6	34	35,016	2,409	38,425	12,256	18,897	31,153
30	Rubber and Misc. Plastic Products	112	2	114	15,765	3,172	18,937	9,668	7,583	17,251
31	Leather and Leather Products	4	38	42	540	—	540	1,914	361	2,275
32	Stone, Clay, Glass & Concrete Products	1,177	65	1,242	150,902	24,687	175,589	26,883	81,005	107,888
331-332	Primary Iron and Steel	2	—	2	270,171	38,053	308,224	34,280	18,931	53,211
333-339	Primary Nonferrous Metals	2,943	—	2,943	86,596	8,041	94,637	17,994	6,688	24,682
34	Fabricated Metal Products	780	159	939	74,551	5,673	80,224	8,064	10,554	18,618
35	Machinery, except Electrical	2,185	303	2,488	19,190	4,521	23,711	1,716	596	2,312
36	Electric and Electronic Equipment	1,042	488	1,530	19,977	4,914	24,891	1,673	3,715	5,388
37	Transportation Equipment	427	873	1,300	36,500	1,823	38,323	2,159	2,507	4,666
38	Instruments and Related Products	28	58	86	3,371	546	3,917	106	576	682
39	Other Manufacturing	1,043	159	1,202	9,743	5,614	15,357	7,664	2,345	10,009
	Other SIC Classes Not Listed Above	—			33,348	7,513	40,861	5,601	3,309	8,910
	Not classified according to SIC and/or unreported	24,971	13,024	37,995	56,229	23,207	79,436	55,360	39,233	94,593
	TOTAL CONSUMPTION	37,836	17,454	55,290	1,052,068	189,114	1,241,182	328,583	429,792	758,375

STANDARD INDUSTRIAL CLASSIFICATION

Group	Description	REGION 4			REGION 5			REGION 6		
		Firm	Interruptible	Total	Firm	Interruptible	Total	Firm	Interruptible	Total
10-14	Mining	5,659	2,693	8,352	51	2,075	2,126	3,664	4,635	8,299
20	Food and Kindred Products	53,047	20,749	73,796	14,973	31,832	46,805	2,035	14,429	16,464
21	Tobacco Manufactures	25	15	40	—	—	—	—	—	—
22	Textile Mill Products	1,518	94	1,612	6	100	106	118	89	207
23	Apparel and Other Textile Products	513	116	629	253	366	619	2	165	167
24	Lumber and Wood Products	3,377	1,201	4,578	212	2,721	2,933	53	1,727	1,780
25	Furniture and Fixtures	4,594	783	5,377	150	343	493	41	3	44
26	Paper and Allied Products	26,655	40,537	67,192	14,229	5,260	19,489	313	10,424	10,737
27	Printing and Publishing	3,563	1,468	5,031	330	1,899	2,229	51	24	75
28	Chemicals and Allied Products	104,752	13,367	118,119	27,815	7,237	35,052	18,611	43,494	62,105
29	Petroleum and Coal Products	14,352	6,670	21,022	8,247	3,951	12,198	26,472	71,682	98,154
30	Rubber and Misc. Plastic Products	6,224	5,775	11,999	389	1,779	2,168	673	3,996	4,669
31	Leather and Leather Products	536	995	1,531	26	107	133	31	46	77
32	Stone, Clay, Glass & Concrete Products	41,082	18,265	59,347	2,979	16,162	19,141	14,445	35,603	50,048
331-332	Primary Iron and Steel	249,216	10,450	259,666	30,958	2,501	33,459	1,301	7,549	8,850
333-339	Primary Nonferrous Metals	64,640	5,019	69,659	5,080	1,579	6,659	460	2,283	2,743
34	Fabricated Metal Products	36,612	10,244	46,856	1,383	3,675	5,058	1,497	1,178	2,675
35	Machinery, except Electrical	36,638	11,382	48,020	2,138	5,026	7,164	111	83	194
36	Electric and Electronic Equipment	14,786	4,009	18,795	1,103	2,520	3,623	550	689	1,239
37	Transportation Equipment	91,062	29,383	120,445	203	1,178	1,381	2,149	4,840	6,989
38	Instruments and Related Products	2,282	56	2,338	71	556	627		56	56
39	Other Manufacturing	14,326	1,862	16,188	254	2,234	2,488	1,531	8,328	9,859
	Not classified according to SIC and/or unreported	29,558	7,951	37,509	3,121	5,091	8,212	4,909	2,386	7,295
	TOTAL CONSUMPTION	895,624	234,524	1,130,148	154,055	127,203	281,258	138,349	237,471	375,820

TABLE 4. (Continued)

STANDARD INDUSTRIAL CLASSIFICATION

Group	Description	REGION 7			REGION 8			REGION 9		
		Firm	Interruptible	Total	Firm	Interruptible	Total	Firm	Interruptible	Total
10-14	Mining	49,125	439	49,564	2,267	9,949	12,216	47,758	10,439	58,197
20	Food and Kindred Products	16,651	12,451	29,102	6,896	11,732	18,628	8,623	70,224	78,847
21	Tobacco Manufactures	–	–	–	–	–	–	–	–	–
22	Textile Mill Products	221	487	708	10	–	10	1,347	2,845	4,192
23	Apparel and Other Textile Products	53	925	978	30	18	48	582	173	755
24	Lumber and Wood Products	14,407	3,592	17,999	231	1,214	1,445	275	4,193	4,468
25	Furniture and Fixtures	25	552	577	416	–	416	556	229	785
26	Paper and Allied Products	55,440	1,978	57,418	273	5,229	5,502	1,350	21,035	22,385
27	Printing and Publishing	42	1,402	1,444	658	9	667	779	492	1,271
28	Chemicals and Allied Products	1,139,476	17,950	1,157,426	8,198	21,554	29,752	8,528	75,063	83,591
29	Petroleum and Coal Products	647,682	6,687	654,369	3,010	25,348	28,358	12,730	134,916	147,646
30	Rubber and Misc. Plastic Products	16,711	6,030	22,741	214	2,657	2,871	2,119	5,466	7,585
31	Leather and Leather Products	–	192	192	49	1,368	1,417	18	364	382
32	Stone, Clay, Glass & Concrete Products	30,126	50,129	80,255	1,757	19,040	20,797	15,163	87,898	103,061
331-332	Primary Iron and Steel	48,518	1,023	49,541	10,576	9,487	20,063	3,597	29,634	33,231
333-339	Primary Nonferrous Metals	205,426	4,540	209,966	447	26,110	26,557	7,703	6,337	14,040
34	Fabricated Metal Products	17,268	3,213	20,481	1,156	110	1,266	9,502	5,177	14,679
35	Machinery, except Electrical	595	1,804	2,399	431	530	961	1,193	1,929	3,122
36	Electric and Electronic Equipment	265	2,697	2,962	586	771	1,357	3,353	1,878	5,231
37	Transportation Equipment	136	4,438	4,574	240	720	960	3,063	5,539	8,602
38	Instruments and Related Products	17	433	450	152	30	182	599	58	657
39	Other Manufacturing	408	5,491	5,899	1,065	526	1,591	841	1,468	2,309
	Other SIC Classes Not Listed Above	637	10,028	10,665	1,414	10,085	11,499	294	22,479	22,773
	Not classified according to SIC and/or unreported	1,277,351	134,681	1,412,032	65	548	613	61,997	10,006	72,003
	TOTAL CONSUMPTION	3,520,580	271,162	3,791,742	40,141	147,035	187,176	191,970	497,842	689,812

STANDARD INDUSTRIAL CLASSIFICATION

		REGION 10		
Group	Description	Firm	Interruptible	Total
10-14	Mining	1,929	1,679	3,608
20	Food and Kindred Products	18,050	11,156	29,206
21	Tobacco Manufactures	—	—	—
22	Textile Mill Products	17	—	17
23	Apparel and Other Textile Products	7	118	125
24	Lumber and Wood Products	7,525	5,097	12,622
25	Furniture and Fixtures	—	—	—
26	Paper and Allied Products	16,154	36,118	52,272
27	Printing and Publishing	—	—	—
28	Chemicals and Allied Products	12,039	6,903	18,942
29	Petroleum and Coal Products	7,889	9,668	17,557
30	Rubber and Misc. Plastic Products	—	—	—
31	Leather and Leather Products	—	—	—
32	Stone, Clay, Glass & Concrete Products	2,249	11,885	14,134
331-332	Primary Iron and Steel	4,004	7,684	11,688
333-339	Primary Nonferrous Metals	3,599	6,663	10,262
34	Fabricated Metal Products	4,597	220	4,817
35	Machinery, except Electrical	136	28	164
36	Electric and Electronic Equipment	81	237	318
37	Transportation Equipment	558	4,834	5,392
38	Instruments and Related Products	73	—	73
39	Other Manufacturing	170	240	410
	Other SIC Classes Not Listed Above	831	1,543	2,374
	Not classified according to SIC and/or unreported	6,279	2,695	8,974
	TOTAL CONSUMPTION	86,187	106,768	192,955

NOTE: Of the volumes for which classification was reported and included in this table, chemicals and allied products were the major consumers of gas. The chemical industry accounted for almost 30% of the total classified industrial use. Petroleum and coal products were the second most important category in total and firm deliveries, with iron and steel in the third position. However, in the interruptible class of service, stone, clay, glass, and concrete products, food and kindred products, and paper and allied products ranked ahead of petroleum and coal and iron and steel. These patterns were generally reflected in most regions, with the exception of the major iron and steel centers, the Great Lakes and Appalachian regions, where ferrous products were the most important industrial use of gas.
*For identification of numbered regions, see Fig. 3. Region 11 not included here.

TABLE 5. **United States Gas Consumption by Standard Industrial Classification, 1972**
Millions of cubic feet at 1000 Btu per cubic foot at 14.73 psia

Description, and SIC group	Firm Volume	Firm % of volume reported	Interruptible Volume	Interruptible % of volume reported	Total Volume	Total % of volume reported
Mining 10–14	114,665	2.5	30,310	1.8	144,975	2.4
Manufacturing, nondurable						
Food and kindred products 20	152,611	3.4	192,657	11.4	345,268	5.6
Paper and kindred products 26	183,352	4.1	181,353	10.7	364,305	5.9
Chemicals and allied products 28	1,525,600	34.3	287,906	17.0	1,813,506	29.5
Petroleum and coal products 29	737,078	16.6	180,429	10.6	917,507	14.9
Other nondurable 21–23, 27, 30, 31	126,086	2.8	116,192	6.9	242,278	3.9
Manufacturing, durable						
Stone, clay, glass, and concrete products 32	278,163	6.3	273,105	16.1	551,268	9.0
Primary iron and steel 331–332	630,486	14.2	140,072	8.3	770,558	12.5
Primary nonferrous metals 333–339	243,114	5.5	62,846	3.7	305,960	5.0
Other durable 24, 25, 34–39	460,661	10.3	230,126	13.5	690,787	11.3
Total industrial reported	4,451,816	100.0	1,694,996	100.0	6,146,812	100.0
Reported	4,451,816	71.1	1,694,996	58.8	6,146,812	67.2
Not classified above or unreported	1,805,180	28.9	1,189,430	41.2	2,994,610	32.8
Total volume consumed in class of service	6,256,996	100.0	2,884,426	100.0	9,141,422	100.0
1974 Totals	6,445,393		2,258,365		8,703,758	

Note: It is interesting to observe considerable reduction of gas consumption subsequent to 1972. Figures for 1974 reflect a decrease of approximately 10% for the food industries; a decrease of approximately 4% for the paper industries; and a decrease of about 50% for the unclassified and unreported categories. Increases during the 2-year period included: mining up 15%; petroleum and coal products up 14%; stone, clay, glass, and concrete products up 15%.

system. Because the gas industry will be relying upon a more diverse array of gas sources over future years, the planned future gas deliveries of the gas industry no longer represent a requirement to be filled solely by natural gas. To determine the amount of natural gas consumption in the United States, it is now necessary to account for that portion of future gas consumption which may be satisfied by other gases supplementing the natural gas deliveries. For this purpose, the GRC, beginning in 1973, included a section in its questionnaire to the gas industry requesting all responding companies to report on the supplemental gas they put into their pipeline systems either for sale to final consumers or for resale by other companies. They were also asked to estimate how much supplemental gas they were including in their plans from 1973 through 1980 for the survey in 1973; and their plans from 1975 through 1985 in the latest survey.

Manufactured, synthetic, and substitute gases are not new to the United States energy economy. Although natural gas was utilized in the early 1800s as an illuminant, the use of gas from coal, coke ovens, and refineries was very important in the early history of the gas industry. After World War II, natural gas began to move to markets from distant gas fields, and the other forms of gas gradually decreased to less than 1 percent of peak-shaving uses instead of major sources of supply.

Supplemental gas consumption in 1974 was more than double that of 1973, increasing from 61.9 billion cubic feet to 152.4 billion cubic feet. Most of this was in the form of substitute gas produced from other liquid hydrocarbons. The importance of supplemental gas is expected to increase sharply between the mid-1970s and 1985.

Although propane air systems will continue to serve as peak-shaving sources for the rest of the 1970s, more than half of the supplemental fuel volumes in 1985 may be produced from coal, crude oil, and other liquid hydrocarbons. Imported LNG is expected to be second in importance as a supplemental source.

The role of supplemental gas in the United States during 1973 and 1974, by major regions, is shown in Table 6. The supplemental gas estimate for the period 1985, is also given in Table 6.

ROLE OF THE FUTURE REQUIREMENTS COMMITTEE

The background of the GRC and the role that it plays in forecasting future gas consumption are described and graphically depicted in each of its reports (published in 1964, 1967, 1969, 1971, 1973, and 1975), available through the Denver Research Institute, University of Denver. The current report results from the work of 140 individuals who planned and produced it through their membership on the GRC and its eleven regional work committees, who were made available by gas distribution companies, municipally owned gas distribution departments, gas pipeline companies, oil companies, associations, and agencies of the federal government.

Forecasts published by the GRC are basically a summation and analysis of the thinking of the gas industry itself. As the first step in its preparation of Volume 6, for example, the Committee mailed survey questionnaires to nearly 1500 companies and municipalities selling gas to ultimate consumers. Responses were received from approximately 1100 companies, accounting for approximately 93 percent of total United States usage. Estimates were made by the regional work committees for the nonreporting companies so that the total national consumption would be represented. Consumption data for 1974 nonrespondents were estimated by the work committees on the basis of data provided by pipeline companies serving those distributors, or reports filed with state utility commissions, or both. Forecasts of their future consumption were then prepared by the work committees on the basis of economic trends and other data for the appropriate areas. More specific details on forecasting methodology will be found in Monograph 1, *Procedure and Methodology Used in Forecasting United States Gas Requirements*, available from the Denver Research Institute, University of Denver. There is a work committee for each of the eleven geographical regions represented in Table 3.

The GRC program includes a number of special activities, such as:

1. *Governmental liaison:* maintenance of formal liaison with a number of governmental agencies with the objective of cooperating fully with governmental units having an interest in gas consumption statistics, such as the Federal Power Commission, Environ-

TABLE 6. United States Supplemental Gas: 1974, 1976, 1980, 1985
Millions of cubic feet at 1000 Btu per cubic foot at 14.73 psia dry

1974

Regions	1	2	3	4	5	6	7	8	9	10	11	Distribution Co. Total	Pipeline Co. Total	National Total
Source of Gas:														
Propane Air Utilized	5,081	3,705	616	11,391	584	—	—	—	33	41	—	21,451	1	21,452
LNG Imported	2,294	680	—	140	—	—	—	—	—	—	—	3,114	—	3,114
Substitute Gas Produced from:														
Coal	—	—	—	1,590	—	—	—	—	—	—	—	1,590	—	1,590
Crude Oil	—	6,402	—	—	—	—	—	—	—	—	—	6,402	—	6,402
Other Liquid Hydrocarbons	1,260	17,499	—	70,012	—	—	—	—	—	—	3,691	92,462	26,553	119,015
Other SNG	—	813	—	—	—	—	—	—	—	—	—	813	—	813
Total	8,635	29,099	616	83,133	584	—	—	—	33	41	3,691	125,832	26,554	152,386

1976

Regions	1	2	3	4	5	6	7	8	9	10	11	Distribution Co. Total	Pipeline Co. Total	National Total
Source of Gas:														
Propane Air Utilized	5,886	15,057	3,172	13,686	2,840	1,816	9	282	534	72	—	43,354	300	43,654
LNG Imported	5,006	1,385	—	—	—	—	—	—	—	—	—	6,391	70,000	76,391
Substitute Gas Produced from:														
Coal	—	—	—	1,600	—	—	—	—	—	—	—	1,600	—	1,600
Crude Oil	—	10,950	—	—	—	—	—	—	—	—	—	10,950	—	10,950
Other Liquid Hydrocarbons	3,503	37,398	—	196,020	—	—	—	—	—	—	3,517	240,438	83,697	324,135
Other SNG	—	18,136	—	—	—	—	—	—	—	—	—	18,136	—	18,136
Total	14,395	82,926	3,172	211,306	2,840	1,816	9	282	534	72	3,517	330,869	153,997	474,866

1980

Regions	1	2	3	4	5	6	7	8	9	10	11	Distribution Co. Total	Pipeline Co. Total	National Total
Source of Gas:														
Propane Air Utilized	7,209	9,193	6,307	7,444	3,644	1,952	9	285	534	134	—	36,711	72,300	109,011
LNG Imported	21,159	89,868	—	—	—	—	—	—	—	14,000	—	125,027	533,658	658,685
Substitute Gas Produced from:														
Coal	—	—	—	1,600	—	—	—	—	—	—	—	1,600	96,148	97,748
Crude Oil	—	4,000	—	—	—	—	—	—	—	—	—	4,000	—	4,000
Other Liquid Hydrocarbons	3,518	61,016	—	196,120	—	—	—	—	—	—	4,260	264,914	183,264	448,178
Other SNG	—	50,050	—	14,600	—	—	—	—	256,200	—	—	320,850	—	320,850
Total	31,886	214,127	6,307	219,764	3,644	1,952	9	285	256,734	14,134	4,260	753,102	885,370	1,638,472

1985

Regions	1	2	3	4	5	6	7	8	9	10	11	Distribution Co. Total	Pipeline Co. Total	National Total
Source of Gas:														
Propane Air Utilized	9,265	7,531	6,663	7,892	4,439	2,032	9	285	534	196	—	38,846	72,300	111,146
LNG Imported	31,360	131,902	—	—	—	13,038	—	—	—	14,000	—	177,262	1,225,680	1,402,942
Substitute Gas Produced from:														
Coal	—	—	—	1,600	263	—	—	—	182,500	12,000	—	209,401	323,565	532,966
Crude Oil	—	4,000	—	—	—	—	—	—	—	—	—	4,000	—	4,000
Other Liquid Hydrocarbons	3,648	61,949	—	196,020	—	—	—	—	—	—	5,180	266,797	182,941	449,738
Other SNG	3,000	40,767	—	29,500	—	—	—	—	584,000	—	—	657,267	—	657,267
Total	47,273	246,149	6,663	235,012	4,702	15,070	9	285	767,034	26,196	5,180	1,353,573	1,804,486	3,158,059

TABLE 7. Field Use of Natural Gas in the United States
Billions of cubic feet

| Year | As-reported basis | | | 1000 Btu/cubic foot Committee estimates |
	Lease	Plant use	Total	Total
1971*	1125	647	1772	1905
1972*	1116	658	1774	1907
1973	1173	654	1827	1964
1974	1085	644	1729	1859
Estimate				
1975	1027	612	1639	1762
1976	1009	597	1606	1726
1977	996	585	1581	1699
1978	1001	588	1589	1708
1979	1038	606	1644	1767
1980	1072	618	1690	1817
1985	1320	723	2043	2196

*Figures for 1971 and 1972 are as revised by U.S. Bureau of Mines.

TABLE 8. Extraction Loss Volumes*
Billions of cubic feet

Year	As reported by U.S. Bureau of Mines	1000 Btu/cubic foot Committee estimates
1965	753	1852
1966	739	1962
1967	785	2144
1968	812	2283
1969	867	2381
1970	906	2488
1971	883	2524
1972	908	2557
1973	917	2538
1974	887	2464

*Not included in field use.

mental Protection Agency, Department of the Interior, National Association of Regulatory Utility Commissioners, and the Federal Energy Administration.

2. *Workshops* conducted to improve company forecasting efforts.

3. *Special studies*, including (a) measurement of the temperature-sensitive portion of the gas market, i.e., gas consumption directly related to temperature changes, (b) data collection for improvement of the knowledge of industrial use of gas, (c) study of environmental effects on gas requirements, and (d) curtailment studies of the impact of gas shortages on demands for alternative fuels, particularly during the winter heating season.

Gas Pipelines and Underground Storage

For gas transportation planning purposes, normally included in the category of *gas*, are: (1) natural gas (gas phase); (2) liquefied natural gas (LNG); (3) substitute natural gas (SNG); and (4) liquefied petroleum gas (LPG). Gas sold as natural gas is primarily methane with small amounts of ethane. LNG is natural gas which has been liquefied at a temperature of approximately $-258°F$ ($-161°C$) for ease of storage and handling in certain modes of transportation. SNG is gas made synthetically from petroleum liquids, such as naphtha and methanol, or coal, and consists mostly of methane with small amounts of ethane and carbon dioxide. LPG is ethane, propane, butane, or a mixture of these gases. LPG is obtained primarily by extraction from natural gas, or as a by-product of the refining process. The mode selected for gas transportation depends mainly on (1) the distance over which the gas must be moved; (2) the geographical and geological characteristics of the terrain, considering both overland and overseas (underseas) across which the gas must be moved; (3) the complexity of the distribution system for which a gas transportation system is designed (few or many source points, few or many consumption terminals); (4) environmental factors directly associated with the gas transportation mode; (5) the physical characteristics of the gas to be transported, notably, the phase—whether gaseous or liquid; and (6) the construction and projected operating costs of the transportation system, based upon trading off the advantages and limitations over which some flexibility of selection may be present. Aside from economic factors, a system can be engineered to transport either the gaseous or the liquid phase, thus giving rise to considerable flexibility in certain situations.

The technology of marine pipelines and seagoing fuel carriers is described later and separately in this section of the Handbook.

Overland Pipelines. Detailed maps of gas pipelines in the United States and other parts of the world can be found in several references, particularly among the periodicals that serve the pipeline industry. Notable among these references is the international petroleum encyclopedia and atlas issued periodically by Petroleum Publishing Co., Tulsa, Oklahoma. Numerous trade associations serving the pipeline industry are also excellent sources on pipeline statistics.

Historically, Texas, Louisiana, Oklahoma, and New Mexico have been large producers of natural gas, as well as some significant fields in the West Virginia–Ohio–Pennsylvania area. New developments in Alaska are influencing and will continue to influence the gas transportation and distribution pattern. The general gas distribution and pipeline pattern for the lower 48 states which has been maintained for a number of years is evidenced by the map (Fig. 1). This map is not intended to delineate in detail the hundreds of specific pipelines, but rather to provide an overall view of how the large gas fields are connected with consuming areas.

In 1970 the National Petroleum Council (NPC), an officially established industry advisory board to the Secretary of the Interior, undertook a comprehensive study of the energy outlook for the United States (1971–1985). A number of reports were issued about three years later, including "U.S. Energy Outlook—Gas Transportation," (219 pages), available from the National Petroleum Council, Washington (Library of Congress Catalog Card Number 72-172997). As part of this study and to assist in making prognostications of the future gas balance (supply/demand) and capital requirements for meeting future demands, the NPC established a series of supply regions, termed NPC petroleum provinces, which are indicated in Fig. 2. On the demand side, a series of districts established by the Petroleum Administration for Defense (PAD) were used, as shown in Fig. 3.

In an effort to determine gas transmission costs, the major pipelines were assigned to corridors selected to fit the patterns of existing systems. These corridors are shown in Fig. 4. In the NPC analysis it was assumed that the pipelines within the corridors that were serving existing markets would continue to do so for some time to come. The assumption was made that market expansion would generally be accomplished by expanding the

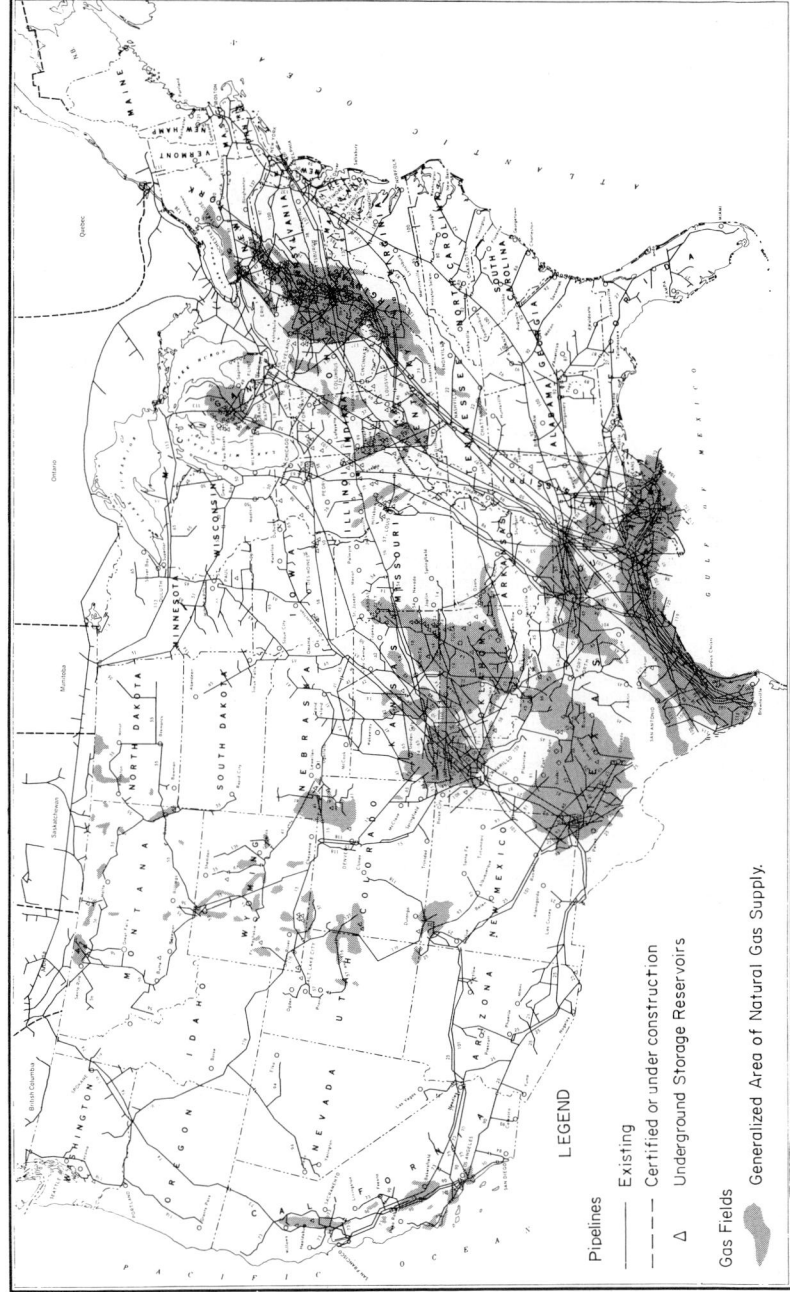

Fig. 1. Major natural gas producing areas and pipelines in the lower 48 states of the United States *(American Gas Association.)*

LEGEND

Pipelines

———— Existing

– – – – Certified or under construction

△ Underground Storage Reservoirs

Gas Fields

Generalized Area of Natural Gas Supply.

Fig. 2. National Petroleum Council petroleum provinces of the United States—Supply Regions: Regional boundaries: 1, Alaska and Hawaii, except North Slope; 2, Pacific Coast States; 2A, Pacific Ocean, except Alaska; 3, Western Rocky Mountains; 4, Eastern Rocky Mountains; 5, West Texas and Eastern New Mexico; 6, Western Gulf Basin; 6A, Gulf of Mexico; 7, Midcontinent; 8, Michigan Basin; 9, Eastern Interior; 10, Appalachians; 11, Atlantic Coast; 11A, Atlantic Ocean.

existing systems. In their analysis, NPC also designated pipeline companies as long line or feeder systems to minimize intercompany transactions. The company assignments to the corridors shown in Fig. 4 are given in Table 1.

Although the detailed cost analysis and forecast cannot be presented here, a brief review of the capital cost factors involved may be of interest. The inter PAD transmission

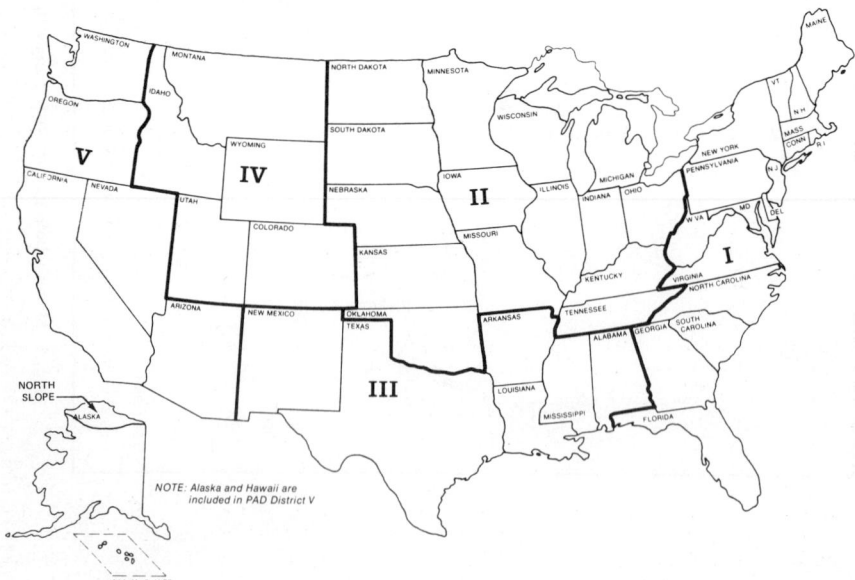

Fig. 3. Petroleum Administration for Defense (PAD) districts—Demand Regions.

costs were developed by averaging the incremental increases in gas plants in service for each of the corridors for the years 1966 through 1970. The average costs were escalated to put all dollars on a 1970 basis. The incremental volumes for each corridor were analyzed to eliminate as many intercompany transactions as possible. The intra-PAD transmission costs were based on a ratio of intra-PAD to inter-PAD mileages. The resulting incremental volumes to be moved between PAD districts were multiplied by the average costs on a 1970 basis obtained above to arrive at the cost of new transmission pipeline facilities.

The cost of incremental underground storage was handled in the same way.

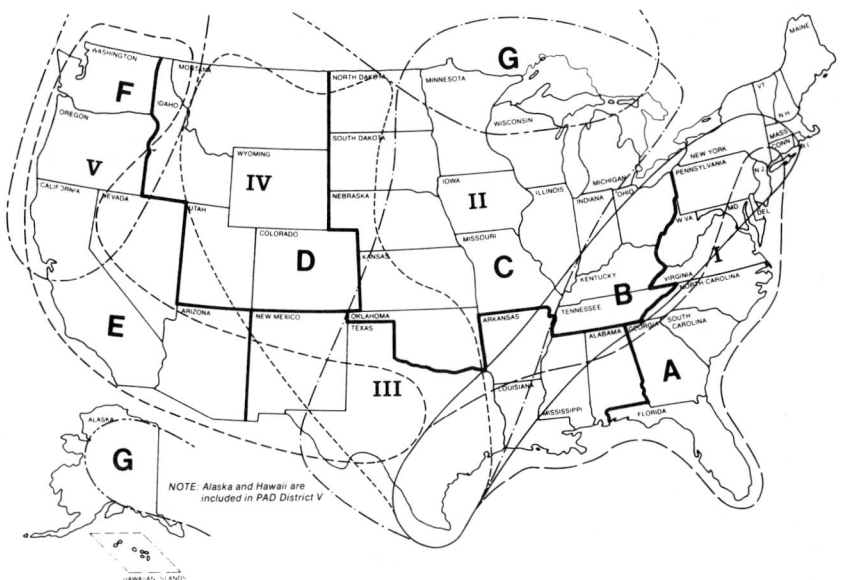

Fig. 4. Gas transmission corridors designated by National Petroleum Council. Numbers refer to PAD Districts (separated by heavy lines). Letters refer to gas transmission corridors (outlined by lighter solid or dashed lines).

Onshore attachment costs were developed by summing the incremental production and gathering costs for all corridors for 1966–1970. The offshore attachment costs were estimated, based on historical costs of existing and proposed offshore systems. LNG attachment costs were based on an equivalent 100-mile pipeline system required to connect the vaporization facilities to the existing gas transmission or distribution systems. The cost of the equivalent system was developed by estimating the cost of a 30-inch-diameter and a 36-inch-diameter line, each 100 miles long. The cost of each pipeline was divided by its capacity and then averaged to obtain a cost factor in dollars per million cubic feet.

SNG attachment costs were based on an equivalent 50-mile connection line from each plant to an existing system. The equivalent length was determined by scaling the distance between proposed plant sites and existing pipeline facilities. Because of the wide range of volumes from SNG plants, three pipeline systems—20-inch, 24-inch and 30-inch diameters—were used to develop the cost factor. Costs were adjusted to reflect different conditions in the Midwest and the East Coast.

Coal gas attachment costs were based on connecting the coal gas plants to existing transmission systems. The plant locations were estimated based on the locations of coal reserves by state. No specific plant locations were given. Two different line sizes—24-inch and 30-inch—were utilized to estimate the costs of transporting gas from the areas of coal reserves. Plants were assumed to be connected to the nearest major transmission system. It was estimated that coal gas would flow to the West Coast and the Midwest from New Mexico, to the West Coast and the Midwest from Montana and Wyoming, and to the

TABLE 1.   Pipeline Companies Assigned to NCP Corridors
Shown in Fig. 4

Company	Type of line LL = long line; FL = feeder line
CORRIDOR A	
Florida Gas Transmission Co.	LL
The Jupiter Corp.	FL
Sabine Pipe Line Co.	FL
South Texas Natural Gas Gathering Co.	FL
Southern Natural Gas Co.	LL
Transcontinental Gas Pipeline Corp.	LL
United Gas Pipeline Co.	LL
West Texas Gathering Co.	FL
CORRIDOR B	
Humble Gas Transmission Co.	LL
Kentucky Gas Transmission Corp.	FL
Manufacturers Light and Heat Co. (The)	FL
Ohio Fuel Gas Co. (The)	FL
Tenneco Inc. (Tennessee Gas Pipeline)	LL
Texas Eastern Transmission Corp.	LL
Trunkline Gas Co.	LL
United Fuel Gas Co.	LL
Algonquin Gas Transmission Co.	FL
Atlantic Seaboard Corp.	FL
Columbia Gulf Transmission Co.	LL
Consolidated Gas Supply Corp.	FL
CORRIDOR C	
Cities Service Gas Co.	LL
Great Lakes Gas Transmission Co.	LL
Michigan Gas Storage Co.	FL
Michigan Wisconsin Pipe Line Co.	LL
Midwestern Gas Transmission Co.	FL
Mississippi River Transmission Corp.	FL
Natural Gas Pipeline Company of America	LL
Northern Natural Gas Company	LL
Panhandle Eastern Pipe Line Co.	LL
CORRIDOR D	
Colorado Interstate Corp.	LL
El Paso Natural Gas Co.	LL
Transwestern Pipeline Co.	LL
CORRIDOR E	
Southwest Gas Corp.	FL
El Paso Natural Gas Co.	LL
Transwestern Pipeline Co.	LL
CORRIDOR F	
Pacific Gas Transmission Co.	LL
CORRIDOR G	
Not used due to the limited amount of information available.	

PAD DISTRICT-CORRIDOR RELATIONSHIP

PAD	Corridors
I	A and B
II	C
III	C, D, and E
IV	D and E
V	E and F

Midwest from North Dakota. The line capacities and horsepower required, assuming a 1.5 compression ratio and 100-mile station spacing, were calculated for each of the line sizes. The cost of the pipeline and compressor facilities was estimated, using Federal Power Commission data for 1970.

In terms of Alaskan production and Canadian Frontier production, all the Alaskan gas and a portion of the Canadian gas must be transported across Canada for delivery to the lower 48 states. Three major pipeline systems were assumed for the analysis, as shown in Fig. 5. A main trunk system which would be constructed from Alaska's North Slope to Emerson, Manitoba, with a bifurcation line that splits off at approximately Edmonton and runs toward Spokane, Washington. This system would handle all the Alaskan and Canadian gas produced in the Mackenzie Delta. The length of the system is 2400 miles with a

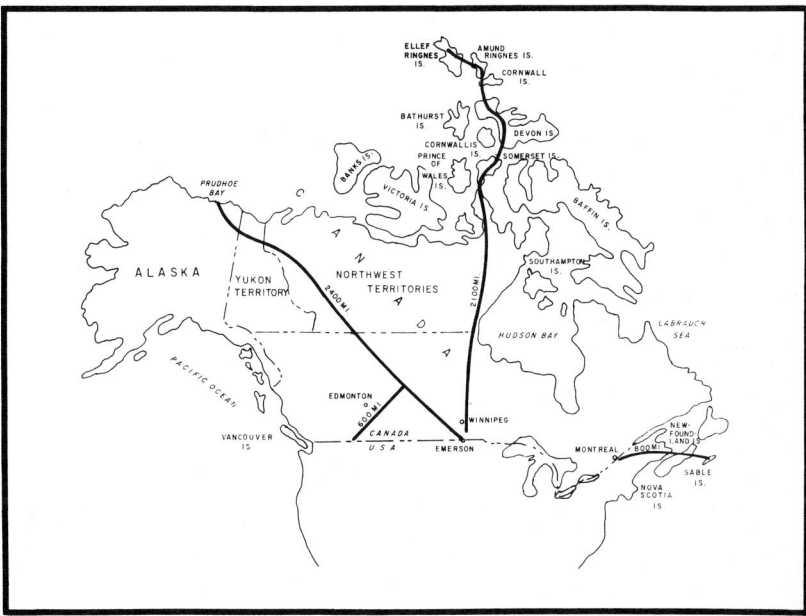

Fig. 5. Routings of three major pipeline systems used in the National Petroleum Council analysis and specifically for Alaskan/Canadian frontier projections.

600-mile bifurcation line. It is postulated that three complete 48-inch pipelines would have to be built between 1976 and 1985 for this system.

The second system is an Arctic Island system to transport gas from the Canadian Arctic Islands. For this study, a line was routed from Ellef Ringnes Island to Winnipeg, using an island stepping-stone route. The total line distance is 2100 miles. Because of the volumes involved, a 48-inch pipeline system was assumed. A route direct to Montreal was also considered. This would add about 500 miles to the estimated system length and about $600 million to the cost. The assumed routing interconnects with TransCanada Pipeline's existing system near Winnipeg.

The third system considered is a Canadian Atlantic offshore system. This was estimated as an 800-mile system from Sable Island to Montreal. Because of the volumes, a 48-inch pipeline was used in the estimates.

Pipeline Construction Projections.   The NPC study, undertaken for the U.S. Department of the Interior, considered four conditions of growth of the gas industry. These rates of growth are reflected in Table 2 in terms of trillions of cubic feet (TCF) of gas to be delivered over specific time periods (spans of years). The principal elements of capital requirements include: (1) cross-country natural gas pipelines; (2) natural gas pipelines from Alaska and the Canadian Arctic; (3) gas processing plants on pipelines from Alaska and Canada; (4) gathering lines to connect new wells to pipeline systems; (5) underground storage facilities; (6) pipelines to connect regasified LNG and SNG plants and nuclear stimulation projects to existing pipeline networks; (7) LNG facilities including liquefaction plants on soil outside the United States, LNG tankers, and domestic port facilities for receiving, storing, and regasification; (8) LNG pipelines; (9) ships and barges

TABLE 2 Total U.S. Natural and Synthetic Gas Requirements Versus Gas Supply*

CASE I

	1971		1975		1980		1985	
	TCF	BTU x 10^{15}	TCF	BTU x 10^{15}	TCF	BTU x 10^{15}	TCF	BTU x 10^{15}
Gas Supply								
Conventional Domestic	19.97	20.61	21.74	22.44	22.34	23.05	24.17	24.94
Alaska North Slope	0	0	0	0	1.30	1.34	3.00	3.10
Canadian Imports	0.90	0.93	1.00	1.03	1.60	1.65	2.70	2.79
Mexican Imports	0.05	0.05	0.05	0.05	0	0	0	0
Total Natural	**20.92**	**21.59**	**22.79**	**23.52**	**25.24**	**26.04**	**29.87**	**30.83**
LNG Imports †	0	0	0.24	0.26	2.28	2.51	4.11	4.52
Coal Gasification	0	0	0	0	0.56	0.52	2.48	2.29
Liquid Gasification	0	0	0.64	0.64	1.32	1.32	1.32	1.32
Total Syngas	**0**	**0**	**0.64**	**0.64**	**1.88**	**1.84**	**3.80**	**3.61**
Nuclear Stimulation	0	0	0.01	0.01	0.19	0.20	1.20	1.24
Grand Total—Gas Supply	**20.92**	**21.59**	**23.68**	**24.43**	**29.59**	**30.59**	**38.98**	**40.20**
Requirements ‡		20.27		25.56		30.89		36.99
(Shortage) or Surplus		1.32		(1.13)		(0.30)		3.21

	Case II 1971 TCF	1971 BTU x 10¹⁵	1975 TCF	1975 BTU x 10¹⁵	1980 TCF	1980 BTU x 10¹⁵	1985 TCF	1985 BTU x 10¹⁵
Gas Supply								
Conventional Domestic	19.97	20.61	21.55	22.24	20.99	21.66	21.16	21.84
Alaska North Slope	0	0	0	0	1.20	1.24	2.40	2.48
Canadian Imports	0.90	0.93	1.00	1.03	1.60	1.65	2.70	2.79
Mexican Imports	0.05	0.05	0.05	0.05	0	0	0	0
Total Natural	**20.92**	**21.59**	**22.60**	**23.32**	**23.79**	**24.55**	**26.26**	**27.11**
LNG Imports†	0	0	0.24	0.26	2.28	2.51	4.11	4.52
Coal Gasification	0	0	0	0	0.36	0.33	1.31	1.21
Liquid Gasification	0	0	0.64	0.64	1.32	1.32	1.32	1.32
Total Syngas	**0**	**0**	**0.64**	**0.64**	**1.68**	**1.65**	**2.63**	**2.53**
Nuclear Stimulation	0	0	0	0	0.09	0.09	0.73	0.75
Grand Total—Gas Supply	**20.92**	**21.59**	**23.48**	**24.22**	**27.84**	**28.80**	**33.73**	**34.91**
Requirements‡		20.27		25.56		30.89		36.99
(Shortage) or Surplus		1.32		(1.34)		(2.09)		(2.08)

Case III

	1971 TCF	1971 BTU x 10¹⁵	1975 TCF	1975 BTU x 10¹⁵	1980 TCF	1980 BTU x 10¹⁵	1985 TCF	1985 BTU x 10¹⁵
Gas Supply								
Conventional Domestic	19.97	20.61	20.17	20.82	17.60	18.16	16.11	16.63
Alaska North Slope	0	0	0	0	1.00	1.03	2.00	2.06

TABLE 2. (Continued)

	1971 TCF	1971 BTU x 10^{15}	1975 TCF	1975 BTU x 10^{15}	1980 TCF	1980 BTU x 10^{15}	1985 TCF	1985 BTU x 10^{15}
Canadian Imports	0.90	0.93	1.00	1.03	1.60	1.65	2.70	2.79
Mexican Imports	0.05	0.05	0.05	0.05	0	0	0	0
Total Natural	**20.92**	**21.59**	**21.22**	**21.90**	**20.20**	**20.84**	**20.81**	**21.48**
LNG Imports†	0	0	0.24	0.26	2.28	2.51	4.11	4.52
Coal Gasification	0	0	0	0	0.36	0.33	1.31	1.21
Liquid Gasification	0	0	0.64	0.64	1.32	1.32	1.32	1.32
Total Syngas	**0**	**0**	**0.64**	**0.64**	**1.68**	**1.65**	**2.63**	**2.53**
Nuclear Stimulation	0	0	0	0	0.09	0.09	0.73	0.75
Grand Total—Gas Supply	**20.92**	**21.59**	**22.10**	**22.80**	**24.25**	**25.09**	**28.28**	**29.28**
Requirements‡		20.27		25.56		30.89		36.99
(Shortage) or Surplus		1.32		(2.76)		(5.80)		(7.71)
Gas Supply								
Conventional Domestic	19.97	20.61	19.86	20.50	15.81	16.32	12.13	12.52
Alaska North Slope	0	0	0	0	0	0	1.20	1.24

CASE IV

Canadian Imports	0.90	0.93	1.00	1.03	1.60	1.65	2.70	2.79
Mexican Imports	0.05	0.05	0.05	0.05	0	0	0	0
Total Natural	20.92	21.59	20.91	21.58	17.41	17.97	16.03	16.55
LNG Imports†	0	0	0.24	0.26	2.28	2.51	4.11	4.52
Coal Gasification	0	0	0	0	0.18	0.17	0.54	0.50
Liquid Gasification	0	0	0.64	0.64	1.32	1.32	1.32	1.32
Total Syngas	0	0	0.64	0.64	1.50	1.49	1.86	1.82
Nuclear Stimulation	0	0	0	0	0	0	0	0
Grand Total—Gas Supply	20.92	21.59	21.79	22.48	21.19	21.97	22.00	22.89
Requirements‡		20.27		25.56		30.89		36.99
(Shortage) or Surplus		1.32		(3.08)		(8.92)		(14.10)

* Conversion factors:	
All Natural Gas	1,032 BTU/cu.ft.
LNG Imports	1,100 BTU/cu.ft.
Coal Syngas	925 BTU/cu.ft.
Liquid Syngas	1,000 BTU/cu.ft.

† These figures include gas from South Alaska.

‡ From Gas Demand Task Group.

These figures do not include gas consumed in production and distribution as this report is primarily concerned with logistics. Consequently, these figures will not coincide in all respects with those in Chapter Four of *U.S. Energy Outlook*.

for the import of non-U.S. supplies of LPG as well as for local transportation; and (10) railroad tankcars and trucks for local transportation of both LPG and LNG.

A breakdown of the total capital requirements (Table 3) for the various sources of supply and modes of transportation is shown in Table 4. The bases on which these capital requirements were derived included:

1. The location of new natural gas discoveries in the lower 48 states will result in the construction of new gathering and feeder line facilities, even though total supplies from this source may remain constant or may decrease.

2. Unit costs of pipeline facilities generally will increase because of (a) more difficult terrain, (b) deeper water offshore, (c) new and greater environmental restrictions, and (d)

TABLE 3.   Summary Projection of Capital Requirements for Gas Transmission Facilities*

Period	\$ millions			
	Case I	Case II	Case III	Case IV
1971–1975	6,146.5	5,894.9	4,035.3	3,101.7
1976–1980	16,007.1	13,406.9	10,702.1	5,075.1
1981–1985	22,023.5	16,214.7	13,775.7	10,267.6

*Figures are given in 1970 constant dollars. SOURCE: *U.S. Energy Outlook—Gas Transportation,* National Petroleum Council, 1973.

pipeline safety and other government regulations. (Reference to the Gas Pipeline Safety Act of 1968 and subsequent hearings and legislation is suggested.)

3. The total costs of pipeline capacity needed to transport gas from Alaska's North Slope to the lower 48 states.

4. Costs of pipeline capacity from Canadian Arctic areas to the United States border are included to transport the projected increases in Canadian imports.

5. Processing costs for stripping plants at or near the United States/Canadian border are included on the assumption that the pipelines from Arctic areas will be designed to carry as much of such liquids as temperature conditions will permit.

6. LNG costs including all necessary facilities from the inlet side of the liquefaction plant to the outlet side of the regasification plant.

7. Costs of pipelines from projected coal SNG plants to nearest major pipeline network are included.

In Case II, as one example, transportation to United States and Canadian markets of the gas volumes projected to be available from Alaska and from Canadian frontier areas will necessitate the construction of the equivalent of some 10,000 miles of 48-inch pipeline by 1984. Approximately 75 percent of this capacity will be needed for projected United States markets. At least 10 million tons of steel pipe and fittings will be required in sizes for which there are no presently existing manufacturing facilities in the United States or Canada. Capital requirements for this transportation are estimated at approximately \$15 billion, 80 to 85 percent of which will be invested in Canada.

PIPELINE REGULATIONS AND CONSTRUCTION

Natural gas pipeline companies must meet legal, engineering, safety, and economic requirements before they are permitted to build and operate pipelines. Such requirements, set up and enforced by the Government, serve both the companies and their customers. It would be foolhardy to build several pipelines getting supplies from a producing area that might have only enough gas underground to supply a single line. Also, it would be most uneconomical to build several lines to supply a city that might be served satisfactorily by one or two lines. The United States Congress has given to the Federal Power Commission (FPC) the right to regulate the interstate transportation of natural gas. Before a company can build a pipeline, or enlarge an existing one, it must obtain from the FPC what is called a *Certificate of Public Convenience and Necessity.* This permission means that the company has proved to the FPC that the proposed pipeline company has contracts with producers of natural gas for a sufficient supply to last a reasonable length of time, usually 20 years. The company must also prove that it has a

TABLE 4. Required Capital Expenditures for Gas Supply*
$ millions—1970 dollars

| | Gas Pipelines | | | | | LNG | | | LPG | | | | |
Period	1 Storage & Transmission Lower 48	2 Transmission Alaska	3 Transmission Canada	4 Attachments- New Production Coal Gas, LNG & Syngas	5 Extraction Plants	6 Plants	7 Ships	8 Terminals & Storage	9 Pipelines	10 Ships & Barges	11 Railroad Cars	12 Trucks	13 Total
						Case I							
1971-1975	4,888.4	0	0	1,258.1	0	131.0	150.0	49.0	195.0	50.0	0	92.3	6,813.8
1976-1980	6,027.8	5,576.0	1,711.0	2,527.9	164.4	2,035.0	2,179.0	701.0	123.0	77.0	44.7	144.9	21,311.7
1981-1985	8,854.8	6,919.0	3,569.0	3,425.9	254.8	1,833.0	2,570.0	672.0	123.0	73.0	55.9	180.9	28,531.3
Total	19,771.0	12,495.0	5,280.0	7,211.9	419.2	3,999.0	4,899.0	1,422.0	441.0	200.0	100.6	418.1	56,656.8
% of Total	34.9	22.1	9.3	12.7	0.7	7.1	8.6	2.5	0.8	0.4	0.2	0.7	100.0
						Case II							
1971-1975	4,676.0	0	0	1,218.9	0	131.0	150.0	49.0	180.0	50.0	0	92.3	6,547.2
1976-1980	4,552.0	5,049.0	1,743.0	1,906.7	156.2	2,035.0	2,179.0	701.0	108.0	77.0	38.8	138.7	18,684.4
1981-1985	5,768.7	4,548.0	3,499.0	2,185.3	213.7	1,833.0	2,570.0	672.0	104.0	73.0	45.9	168.3	21,680.9
Total	14,996.7	9,597.0	5,242.0	5,310.9	369.9	3,999.0	4,899.0	1,422.0	392.0	200.0	84.7	399.3	46,912.5
% of Total	32.0	20.5	11.2	11.3	0.8	8.5	10.4	3.0	0.8	0.4	0.2	0.9	100.0
						Case III							
1971-1975	3,153.4	0	0	881.9	0	131.0	150.0	49.0	170.0	50.0	0	87.6	4,672.9
1976-1980	2,977.7	4,506.0	1,743.0	1,335.7	139.7	2,035.0	2,179.0	701.0	67.0	77.0	22.0	127.7	15,910.8
1981-1985	4,510.0	3,896.0	3,499.0	1,681.6	189.1	1,833.0	2,570.0	672.0	69.0	73.0	35.4	151.0	19,179.1
Total	10,641.1	8,402.0	5,242.0	3,899.2	328.8	3,999.0	4,899.0	1,422.0	306.0	200.0	57.4	366.3	39,762.8
% of Total	26.8	21.1	13.2	9.8	0.8	10.1	12.3	3.6	0.8	0.5	0.1	0.9	100.0
						Case IV							
1971-1975	2,298.1	0	0	803.6	0	131.0	150.0	49.0	170.0	50.0	0	85.1	3,736.8
1976-1980	1,858.4	0	2,283.0	884.4	49.3	2,035.0	2,179.0	701.0	37.0	77.0	5.4	119.1	10,228.6
1981-1985	1,968.8	4,370.0	3,135.0	588.3	205.5	1,833.0	2,570.0	672.0	46.0	73.0	26.5	134.6	15,622.7
Total	6,125.3	4,370.0	5,418.0	2,276.3	254.8	3,999.0	4,899.0	1,422.0	253.0	200.0	31.9	338.8	29,588.1
% of Total	20.7	14.8	18.3	7.7	0.9	13.5	16.5	4.8	0.9	0.7	0.1	1.1	100.0

2-71

market for the gas at the other end of the line and that there is a sufficient demand for more energy by present and future consumers. The company must show that it has the engineering skill and financial resources to build a line that will operate safely and efficiently.

The Federal Power Commission has the authority to determine the prices charged for the gas when the producing companies sell it to the pipeline company, and when the pipeline company sells it to its customers. These price controls are considered to be in the public interest.

Additional protection for the general public is provided by the Gas Pipeline Safety Act of 1968. This law gives to the Secretary of the Department of Transportation the right to establish federal safety standards for the transportation of natural gas by pipeline. Committees of the Senate and the House of Representatives held a long series of public hearings before they recommended passage of the Pipeline Safety Act. The natural gas industry was commended at that time for its excellent safety record.

After the pipeline company has met the requirements of the Federal Power Commission and has been given a certificate as previously mentioned, the company's next problem is to find the best route for the pipeline from the producing wells to the communities where the natural gas will supply needed energy. An aerial survey is the first step. By observing and photographing from the air, engineers can learn quickly the best route that will reach the communities with the least disruption to farms and villages along the way. The route should avoid lakes and ponds wherever possible to avert obviously greater construction difficulties. It should cross as few highways and railroads as possible. Obviously, an ideal route would cross only firm, dry, flat land. Particularly, as new reserves are found or exploited, notably those in Arctic regions, entirely new methods of construction are under development.

The aerial survey is followed by careful study on the ground. Experts travel by automobile and on foot to inspect every mile of the proposed route. Detailed records are maintained so that the construction crews will know in advance the kind of ground, both on the surface and in the trench, that must hold the pipeline. The on-the-ground survey results in minor changes in route, avoiding difficulties not noted from aerial photographs.

In the mid-1970s, additional legal and social impediments may delay the start of construction of an interstate pipeline. In the absence of clearly spelled-out state and community codes and guidelines, environmental interest groups, for example, may petition the various courts for delaying a project, pending the preparation of sometimes seemingly endless environmental impact studies, some of which may be sound and objective, others of which may incorporate strong pro or con biases. Thus, the legal sorting-out process can cause delays of months and years.

Next, the right-of-way must be purchased. This does not mean that the company actually buys a strip of land many hundreds or thousands of miles long. Rather, it means that the company pays the landowners a fee for the use of a corridor, usually about 100 feet wide. This width generally is needed to accommodate construction and pipe-laying and welding equipment. An easement grant is obtained from the owner of each piece of land along the route, which allows the company to use the corridor for the single purpose of constructing and maintaining a pipeline. The fee paid to the landowner is based on the length of the strip of land used. The easement grant also includes the right to install valves or other equipment needed to control the flow of gas and operate the line. When requested, the company also agrees to supply the landowner with gas taken from the line crossing his property. A farm tap of smaller pipe is welded to the side of the pipeline, a regulator is provided to reduce the pressure of the gas to safe limits, and a meter measures the gas used by the landowner, who pays for the fuel. The pipeline company also guarantees that it will cause the least possible damage to crops, buildings, fences, or streams on the right-of-way, and will pay the landowner for any such harm to his property. The construction crew fills in the trench after the pipe is laid, covers it with the same earth that was removed, and may even plant the strip with grass or a crop so that the land is restored completely to its original condition. At intervals along the right-of-way, the pipeline company may purchase, outright, tracts of land where it can erect compressor stations, regulator and valve housings, or other buildings needed to operate the line.

In clearing the right-of-way before the pipe-laying crews arrive, the company also exercises care to avoid destruction of valuable shrubs and trees. Rocks are removed to a location agreeable to the owner, and brush is safely burned to clean up the strip of land.

If a landowner refuses to allow the pipeline to cross his property, or cannot agree on a price for the easement, the pipeline company may ask the state to exercise the right of eminent domain. (The right of eminent domain is based on the principle that one man should not obstruct a project that will work for the good of all.)

Much of the installed gas pipeline ranges from 14 to 30 inches in diameter, the most common ranging from 20 to 36 inches, but there is a strong trend toward larger-diameter lines, from 42-inch upward. Line pipe is made from high-strength plates, ⅜ to 1 inch in thickness. Sections of pipe are usually 40 feet long, minimum, ranging up to 60 or 80 feet. Lengths of pipe arrive at the scene most often by truck and are strung out by special pipe carriers along the right-of-way so that the construction crews will find them near the place where they are to be installed. Helicopter delivery of pipe is sometimes used where it is impossible for trucks to do the job. The total weight of steel going into a long-distance pipeline is impressive. For example, a pipe with a wall thickness of ½ inch and a diameter of 30 inches will weigh more than 400 tons per mile.

In building very long pipelines, the pipeline company usually employs several construction contractors. The total length of line is divided into a number of sections with separate equipment and crews. Usually, each crew works on not more than 100 miles. By partitioning the construction task, the entire operation can be speeded up, particularly important in areas where freezing temperatures or rain and mud may interfere with the work.

The numerous machines needed to dig the trench, weld the sections of pipe, apply protective coating to prevent corrosion, lower the pipe into the trench, and cover the trench with earth are known collectively as a *mainline spread.* The trench is usually 3 or more feet in depth, sufficiently deep to prevent damage by plowing and earthmoving equipment. Depending on the size of the pipe, the trench will range from 2 to 4 feet or more in width.

Teams of welders join the pipe sections into a continuous tube. The most modern welding techniques involve automatic welding machines. X-ray equipment is used to inspect welds. When several sections of pipe have been welded together, the continuous tube is lowered gently into the trench by *side-boom tractors.* These machines have cranes or derricks slanted over to one side so that they can pick up the pipe and lower it several feet away from the tractor itself. Pipe purchased from steel mills may come with a coating and wrapping already applied. The thick coating may be of coal tar or asphaltic material, which is then covered with heavy paper or fiber glass. This protective coat-and-wrap is needed to prevent rusting. If bare pipe is used, there are special machines that coat and wrap right on the job just before the pipe is lowered into the trench.

A special piece of equipment, known as the *holiday detector,* is a hoop of metal which is placed around the pipe after it is coated and wrapped. A small electric current flows through the hoop. If there is a "holiday," i.e., a spot where there is no coating, the detector alerts the operator. This is brought to the attention of a special crew which coats and wraps bare spots in the pipe.

Since pipelines do not follow an absolutely straight line, bending machines are used to curve the pipe in the vertical, horizontal, or both directions. When a pipeline must cross a river, the contractor will dig or dredge a deep trench in the river bed. The pipe is then surrounded by heavy weights and encased in concrete so that it will not be carried away by the current. If there is a suitable bridge across the river, the pipeline may be hung from the underside of the steel girders of the bridge. In some cases, a special bridge is constructed to carry the pipe across the stream. In crossing a highway or railroad, the pipeline must be put through a tunnel under the structure. A giant auger will be used to bore under the road to accommodate a section of somewhat larger-diameter pipe, forming the tunnel through which the main pipeline passes.

Gas pressures in long-distance pipelines may range from 500 to 5000 pounds per square inch, with 1000 psi being quite common. Pressure is boosted to make up for frictional losses by use of compressor stations located every 50 to 100 miles along the pipeline. In terms of lineal velocity, natural gas may travel at a rate of about 15 miles per hour; thus about three days are required to move a molecule of gas over a distance of 1000 miles.

All along the pipeline are valves and regulators that may be opened or shut to control the internal pressure, or to cut off the flow entirely if an unexpected break in the line is caused by a flood, an earthquake, or other disaster. The valves and regulators can be operated by microwave radio long before any crew could reach them. Stations for reducing the pressure, located near points of consumption, frequently are called *city gates*. These stations measure the amount of gas leaving the main pipeline at this point as well as reducing the pressure.

For comparison only, in a typical five-year period (1966–1970), the Federal Power Commission issued 1366 certificates authorizing the construction of 27,541 miles of pipeline at an estimated cost of $5.5 billion. The average cost per mile during that period was $200 thousand. Where pipelines must enter densely habitated areas so that various underground structures must be avoided (subways, water and sewer pipes, etc.), the cost may be as high as $1 million or more per mile.

Pipeline construction, excluding utility distribution and water lines, in non-Communist areas of the world during 1975–1976 is given by the following figures:

	Miles	
	1975	1976
United States	7511	7569
Middle East	4128	1253
Latin America	3213	1113
Canada	2808	676
Africa	2426	1488
Europe	1891	2989
Far East	1548	1178

The new lines in 1976 (estimate) will cost approximately $6.76 billion. U.S. construction will account for $3.32 billion, only 15 percent less than the combined total for the rest of the non-Communist world. Nearly one-third of this construction value will go into Alaska. That project, in 1975, spent about $2.125 billion laying 368 miles of pipe, together with associated facilities. As of early-1976, the project is approximately 36 percent complete, with movement of Prudhoe Bay crude expected some time during the first half of 1977.

UNDERGROUND STORAGE

The largest additional supply of natural gas for peak demands comes from underground storage reservoirs located, for example, close to the northern cities, as compared with the producing wells which may be located in the southwestern area of the country. Some of the storage pools are operated by pipeline companies, but most of the gas in underground storage is owned by the local gas companies that serve metropolitan areas.

The underground reservoirs are filled with gas from the pipelines during the summer months, when all the fuel that the lines can deliver is not consumed. This method allows the producing wells and the pipelines to operate at fairly steady rates at all times of the year. Also, it is established that a gas field will produce more gas over a longer period if the gas is withdrawn at a steady rate. See Fig. 6.

Of the lower 48 states, the 15 states with the largest underground storage reservoirs are listed in Table 5. Four of these states, Pennsylvania, Michigan, Illinois, and Ohio, have half the total underground capacity presently used in the United States. As of the first of 1975, there were 341 underground gas storage pools in the United States, with a total capacity of 3.9 trillion cubic feet. In a typical year, about one-fourth of this volume was used during cold waves to furnish the additional gas needed to supply approximately 36 million homes and apartments.

The most common type of underground reservoir now storing gas is a previously producing gas or oil field. The supplies remaining in these pools are too small, and at too low a pressure, to justify continued production. But, the reservoir rock can hold gas pumped down through the same wells that once took gas out of the ground. Of the 325 storage pools being used, 279 once produced gas or oil. The Commonwealth of Pennsylvania is fortunate in having 65 such pools close to the large industries and centers of population. There are 33 such pools in West Virginia; 29 in Michigan; 21 in Ohio; 16 in Kansas; 15 each in Indiana, New York, and Kentucky. The remaining pools are located in 13 other states. The gas to be stored is pumped into the old wells by compressors similar to those used to move gas in pipelines. The gas is stored under about the same pressure

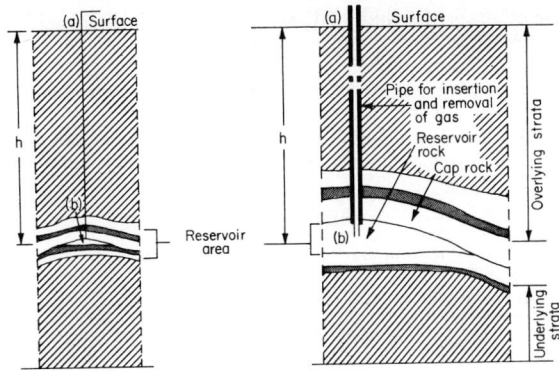

Fig. 6. *(a)* Typical anticlinal reservoir. *(b)* Enlarged view of reservoir area. *(American Gas Association.)*

that originally existed in the field. In developing a gas storage reservoir, a company obtains a lease from the landowners in much the same manner that gas producers do.

The gas industry has been developing underground storage reservoirs for more than 50 years. The first known experiment in storing gas underground was conducted in 1915 in Welland County, Ontario, Canada, by the National Fuel Gas Company. The success of this effort prompted the Iroquois Gas Corporation, a subsidiary of National Fuel, to develop, in 1916, the Zoar field south of Buffalo, New York. It was the first storage operation in the United States and is the oldest continuously used reservoir.

In the last half-century, over 80 companies have invested more than $2 billion in underground storage facilities.

Another kind of underground storage reservoir is called an aquifer. An aquifer is an underground rock structure holding large quantities of water. The underground rock is porous and permeable. The pore spaces are filled with water; an impermeable rock covers the porous rock. Wells are drilled into such formations, and gas is forced into the pores under pressure. As the gas pressure increases, the gas pushes the water farther down into the porous rock, making room for the gas.

There are 44 aquifers in the United States: Illinois, 19; Indiana, 11; Iowa, 6; and

TABLE 5. Underground Storage of Natural Gas*

State	Billions of cubic feet
Pennsylvania	547
Michigan	525
Illinois	378
Ohio	348
West Virginia	344
California	333
Oklahoma	214
Louisiana	174
Texas	144
Montana	143
Kentucky	117
New York	103
Kansas	99
Mississippi	92
Indiana	59
Other reservoir capacity in the United States	318
Total reservoir capacity in the United States	3,938

*Source: American Gas Association, 1975.

Kentucky, 4. The others are located in Minnesota, Missouri, Utah, and Washington. Three unusual storage reservoirs have been developed: an abandoned coal mine in Colorado; and salt domes in Michigan and Mississippi.

Other means of storing gas, notably liquefaction, are described in articles in the Handbook dealing with LNG (Liquefied Natural gas).

Acknowledgements.   Appreciation is extended to the American Gas Association and to the National Petroleum Council for much of the basic information provided here.

REFERENCES

Pertinent publications available from the American Gas Association, Arlington, Va., include:
L00007, "Steady Flow in Gas Pipelines," 1967.
L00024, "Pipeline Research Summary," 1971.
L00070, "Flow of Natural Gas through High-Pressure Transmission Lines," 1935.
L00240, "Research on the Properties of Line Pipe; Summary Report," 1973.
L00320, "Measurements of Secondary Stresses in Pipelines; Report 2," 1973.
L20010, "Transient Flow of Gas," 1964.
L20020, "Steady State Flow Computation Manual for Natural Gas Transmission Lines," 1964.
L20040, "Natural Gas Transmission System Optimization," 1973.
L20269, "Design Manual for Gas-Liquid Two-Phase Flow in Pipelines," 1973.
L30000, "Symposium on Line Pipe Research," 1973.
L30005, "Analysis of Causes and Determination of Possible Means for Prevention of External Damage to Pipelines," 1973.
L40000, "Materials of Construction for Use in an LNG Pipeline," 1968.
H20000, "Gas Engineers Handbook," 1965.
L00400, "New Concepts in Underground Storage of Natural Gas," 1966.
XU0272, "Underground Storage of Gas in the U.S.: Statistics," 1971.
X51465, "Bibliography on Underground Storage," 1973.
"U.S. Energy Outlook—Gas Transportation," National Petroleum Council, Washington, 1973.
Katz, D., D. Cornell, R. Kobayashi, F. Poettmann, C. Weinaug, J. Vary, and J. Elenbaas: "Handbook of Natural Gas Engineering," McGraw-Hill, New York, 1959.
Pipeline technology is regularly covered by such publications as: *Alaska Industry*, Anchorage; *Canadian Petroleum*, Don Mills, Ontario; *Gas*, Houston; *Oil & Gas Journal*, Tulsa Okla., *Pipeline Industry*, Houston; and *Pipeline & Gas Journal*, Dallas, Tex.

Marine Pipelines

By A. B. CROSSMAN*

The term *marine pipeline* is applied here to those pipelines constructed in a marine environment for the transportation of a fluid, such as gas or oil. Generally speaking, these lines could be under water in a river, marsh, or ocean, but the predominant industry effort in recent years, and consequently the object of this discussion, is the construction of pipelines in the open ocean at increasingly deeper levels.

The trend toward deepwater pipelining and construction in harsher environments follows naturally the expansion of the search for offshore oil and gas. This search began in earnest after World War II and is expanding at an ever-increasing rate. Worldwide energy needs have caused oil companies to move into areas that only a few years ago would have been too expensive to develop on a practical basis. Lines are now being laid in water depths of over 400 feet and cover distances as long as 200 miles from field to shore. These longer lines are major trunk-lines bringing the oil or gas to land terminals. Other lines are necessary out at the field to connect platforms to each other, or possibly to connect platforms to sea berths.

The sizing of the pipeline, the design of the pump and compression equipment needed to move the products, the design of the automation systems, and many of the corrosion control procedures are the same, regardless of whether the pipeline location is on land or at sea. This discussion will concentrate on those factors that are unique to marine lines.

DESIGN FACTORS

The two major areas of design difference between land and marine pipelines are (1) the stresses incurred in getting the pipeline to the sea bottom, and (2) the necessity of keeping the line stable and in place while it is exposed to forces induced by current and wave.

The stability problem is theoretically simple but is complicated somewhat by the uncertainty of precise values for some of the coefficients used in the calculations. Basically, it is a matter of providing enough weight in the pipe and pipe coating system to provide a net downward force when balanced against the buoyancy and the lift force caused by the seawater moving by the pipe. This net downward force, in conjunction with the coefficient of friction for the particular pipe-soil combination under examination, can then mobilize a horizontal resisting force. This should be somewhat larger than the drag force exerted on the pipe by the water motion in order to give the desired safety factor.

Different safety factors or horizontal water velocities may be utilized, depending on the operating conditions that will be encountered during the life of the pipeline. For example, many pipelines will be buried beneath the sea bottom at some time interval ranging from a few weeks to a year or two after their construction. The exposure of such a line to maximum horizontal water velocities caused by storm current and waves is obviously much less than that of a line that will remain on the surface of the sea bottom. It is also obviously necessary to consider whether the line will contail water, oil, gas, or some other substance at the time when the design loads may occur.

It is important to consider carefully the aforementioned points in the design of the weight coating, since the ability of a contractor to construct the lines safely relates very closely to the negative buoyancy of the pipe and coating. This leads to the second area of design difference from a land pipeline: getting the line from the work platform on the surface of the sea to the sea bottom. Again, the concept of what is required is simple; the accomplishment of the task is much more complex.

The most common method of marine pipeline construction utilizes a floating vessel on which the pipe is assembled in a horizontal position. As additional joints or sections of pipe are added to the already completed segment, the barge is moved forward, actually

---

*Vice President, Brown & Root International Ltd., Houston.

moving out from under the completed pipeline. This is sometimes called the "stovepipe" method, named after the manner in which the pipe sections are added one after another. This pipeline extends off the stern of the vessel and spans down to the sea bottom. It is supported part of the way down by a construction aid called a "pontoon" or "stinger." This is described in more detail later, but is basically a slender structure pinned to the vessel on one end and with built-in buoyancy that can be controlled so that it floats at the proper angle to the water surface to provide support to the pipeline.

In shallow water, the pipeline is then allowed to span from the end of the pontoon to the sea bottom as a simple beam. As water depths increase, it becomes necessary to add tension to the pipe on the barge. This, of course, changes the analytical problem from one of a simple beam to one of a beam under tension. This analysis must take into account the weight of the pipe, the wall thickness and type of steel in the pipe, the tension on the pipe, the support of the pontoon, the geometrical configuration of the tension on the pipe, the geometrical configuration of the pipe-pontoon-barge system, and the pipe end condition at the sea bottom.

It should be obvious from this oversimplified discussion why the negative buoyancy of the pipe is so important in this phase of the work. The heavier the pipe, the higher the simple beam stress, the longer or more complex the pontoon, and the higher the tension that is required for the same stress level. Excessive conservatism in the weight coat design can result in a line that cannot be practically constructed by present methods because of inability to physically handle in the field the theoretically required pontoon or to apply the theoretically required amount of tension.

CONSTRUCTION EQUIPMENT

To understand properly the limits of what can or cannot be constructed and what some of the difficulties are in this field, it is important to be familiar with the equipment used. In the more severe environmental areas the size of line and the water depth in which it can be laid are usually controlled by the capabilities of this equipment.

On occasion, a line is assembled on shore and simply winched from shore or pulled to the final location. Auxiliary buoyancy is added so that either the line can be floated to final position or the friction between the line and the sea bottom can be reduced to enable it to be pulled with a minimum of force. It can be seen that this method offers advantages in minimum equipment requirements but disadvantages in the requirements for very sheltered waters, relatively short line distances, and an onshore assembly area.

Another approach that is currently gaining some usage is the "reel" method. In this, pipe is fabricated at an onshore base and then spooled onto a large-diameter reel. This reel is placed on a vessel which then goes to the jobsite. The pipe, up to several miles in length per reel, can then be spooled quickly off the barge and onto the sea bottom. This quickness, resulting from the elimination of offshore welding and associated tasks, minimizes the exposure of the project to the elements, a major factor affecting costs. Currently, pipe in diameters of up to 12 inches can be constructed in this manner with proposed equipment to handle up to 24-inch pipe. This system has primarily been used for relatively short lines such as those between two offshore platforms.

Most modern lay barges capable of laying the longer, larger lines that now make up the majority of the projects are barge-type vessels laying pipe by the "stovepipe" method. These vessels are typically in the size range of $400 \times 100 \times 30$ feet, and are manned by a work force of approximately 200 people. They are completely self-contained construction facilities.

Figure 1 shows a barge with the pipe-lay ramp on the starboard side. Supply boats or pipe haul vessels tie up on the port side, and one or more cranes on the lay barge deck lift the pipe aboard, placing each joint in a storage rack. From the storage rack, the pipe is lifted by crane to a conveyor, which in turn feeds the pipe to a lineup station where each joint is aligned with the pipeline and the initial weld pass is made, followed by a "hot pass" at the first weld station. Simultaneously, other stations on the ramp are applying additional weld passes. Nondestructive testing is conducted on finished welds, and field joints are installed. As each section of pipe is added on, the barge moves ahead one length of pipe, and the pipeline moves down the ramp and stinger to the seabed. Components of this system are discussed in more detail in the following paragraphs.

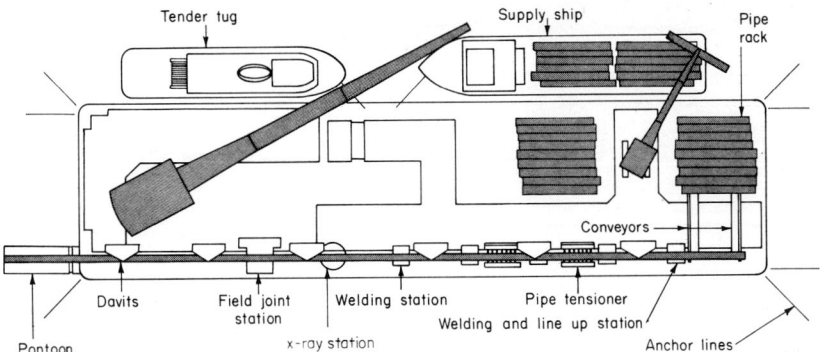

Fig. 1. Lay barge-pipe assembly operation (top view).

PIPE-LAY VESSELS

There are three basic configurations of pipe-lay vessels in common use today; barge-type hull, ship-shape hull, and semi-submersible. The barge-type hull (Fig. 2) is by far the most common because of its economy and simplicity of construction, its ability to provide the space and stability for heavy lifts and deck cargo including pipe, and its shallow draft, permitting work close inshore. The primary disadvantage of this type of hull is its relative sensitivity to sea conditions. In particular, roll and heave motions will shut down pipe-lay operations in 6-foot to 14-foot waves, depending on wave direction and period.

There are relatively few ship-shape hulls (Fig 3) in use as pipe-lay vessels, although their popularity is increasing. The original pipe layers of this type were conversions, but some have now been specifically designed for the purpose of laying pipe and performing derrick work. Self-propulsion is a common feature of this type of vessel, and this permits reduction in the number or size of support tugs. Because of draft and freeboard, mooring forces are high, and the draft limits the minimum working water depth. Steadiness and stability of ship-shape hulls are difficult to compare with those of a conventional barge because vessel dimensions have as big an influence as hull shape, if not more.

The column-stabilized semisubmersible hull (Fig. 4) experiences the least motions due to waves. In areas where even summertime conditions are persistently rough, this type of hull would experience far less weather downtime than other hulls. Depending on other factors, such as daily cost and progress rate while working, it may prove to be the best solution for consistently poor environments. Drawbacks to this type of equipment are its high initial cost and the fact that progress in poor conditions may be limited by the ability of support vessels to work rather than by the characteristics of the lay barge.

PONTOON AND RAMP TYPES

The state of the art in joining sections of pipe requires that multiple-weld passes be used, followed by nondestructive examination of the joint and the addition of protective coating

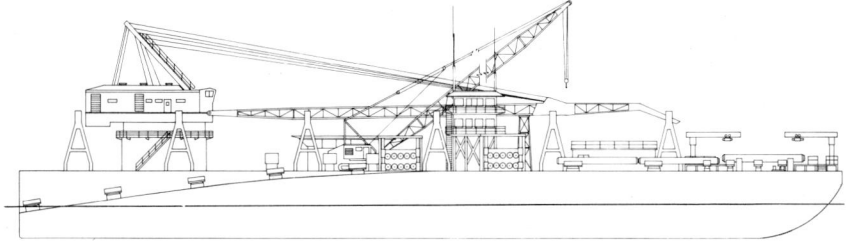

Fig. 2. Pipe-lay vessel with barge-type hull.

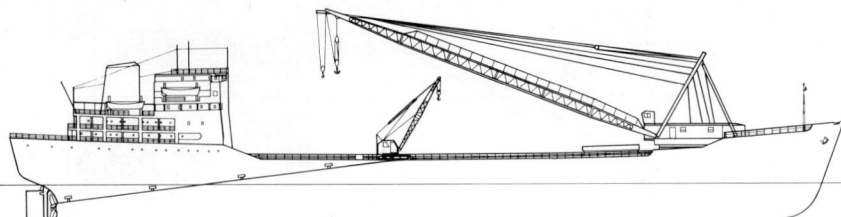

Fig. 3. Pipe-lay vessel with ship-shape hull.

and weight coating over the finished joint. Efficiency requires that several stations be used for these operations so that several joints can be worked on simultaneously. These stations are normally 40 feet apart, the length of one section of pipe as normally produced by the pipe mills.

All the above takes place on the barge "ramp." From the number of operations that must be performed, it is obvious that the length of ramp is an important consideration in lay barge design. The modern lay barge ramp will have six to eight or more stations spaced at 40-foot intervals, the length of a "joint" of pipe. Therefore, the ramp is normally horizontal, slightly inclined, or built to a large radius curve, with the first station being approximately horizontal at the barge main deck level.

The types of pontoons employed are quite varied and are functions of the barge design, the pipe to be laid, the environmental conditions, and numerous other factors. Most common until recent years has been the straight, rigid pontoon built of pipe members (Fig. 5). This pontoon is best suited for large-diameter pipe in water depths up to about 150 feet. Compartmented tanks in the pontoon may be ballasted to provide the proper support for the pipeline and to control the stress in the pontoon. The long rigid pontoons are difficult to handle and sensitive to rough environments. In order to shorten the length of pontoons and extend water depth capabilities, pipe tensioning is now commonly used.

The length of the pontoon can also often be shortened by changing from a straight to a curved configuration (Fig. 6). The rigid curved pontoon has the advantages of being relatively simple to operate and control, as changing the ballast in only one tank can cause the pontoon to pivot about its upper end. Also, since the entire pontoon can be built to the minimum allowable bending radius for the pipeline being laid, often it is the shortest

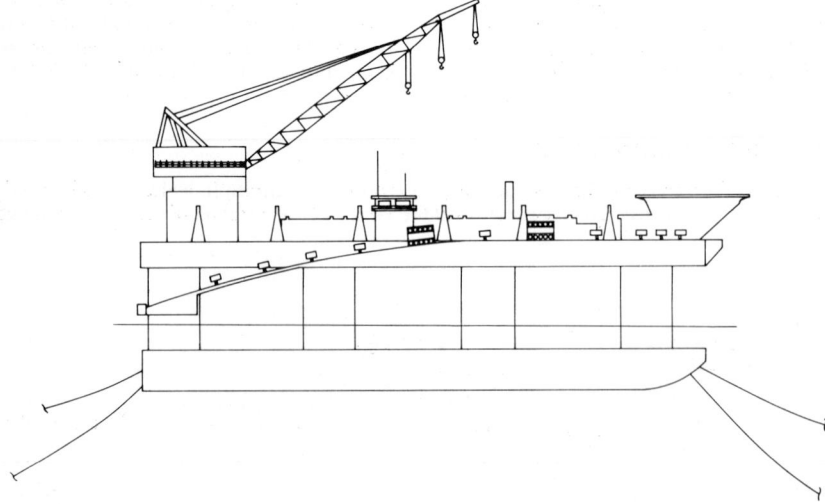

Fig. 4. Pipe-lay vessel with semisubmersible hull.

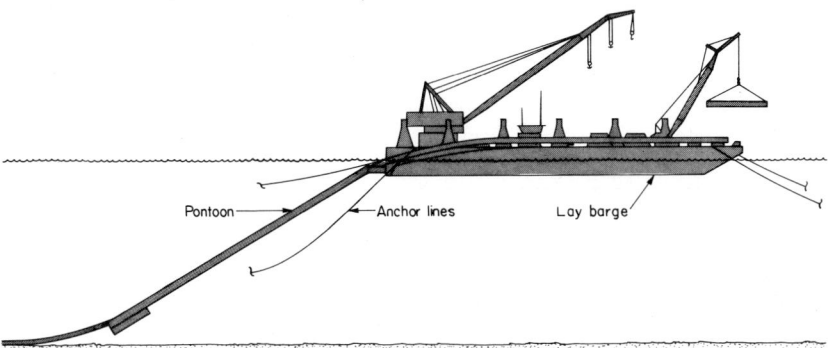

Fig. 5. Lay barge with straight pontoon.

possible pontoon. However, because the radius may need to change with factors such as water depth, it is not always the most practical type of pontoon.

To overcome this disadvantage, and to reduce stresses in the pontoon, the articulated pontoon was developed (Fig. 7). This type of pontoon, which has a number of variations, is made up of individual sections which are hinged together. Because of the hinges, changes in pontoon radius are easily accomplished, and sections can be added or removed for different job requirements.

TENSIONERS

The ability to lay pipe in deep water today is possible primarily because of the development of pipe tensioning equipment. As is true for most pipeline laying equipment, there are a number of different types and designs of tensioners.

All tensioners presently in use apply the tension to the pipe through pneumatic tires, solid elastomeric wheels, or tracks with resilient pads. In its simplest form the tensioner may be merely a braking device to hold the pipe while tension is applied by means of the barge anchor system. When the barge moves ahead, the braking force is overcome by

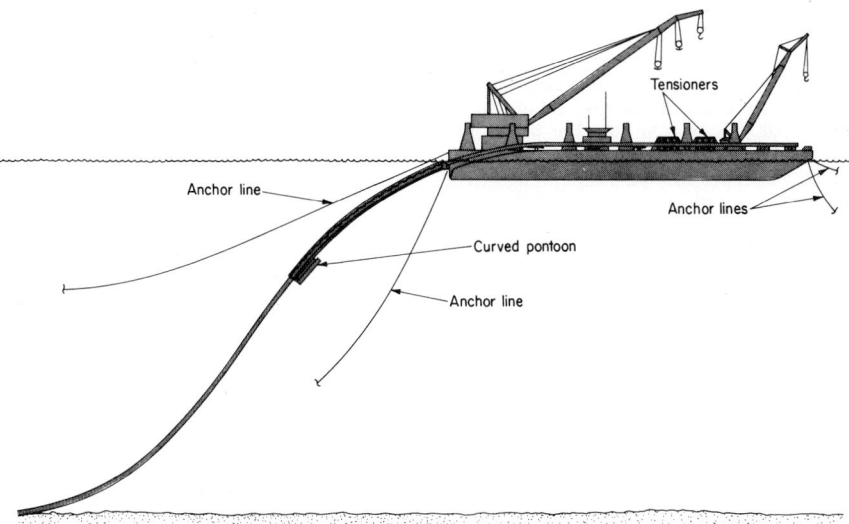

Fig. 6. Lay barge with curved pontoon.

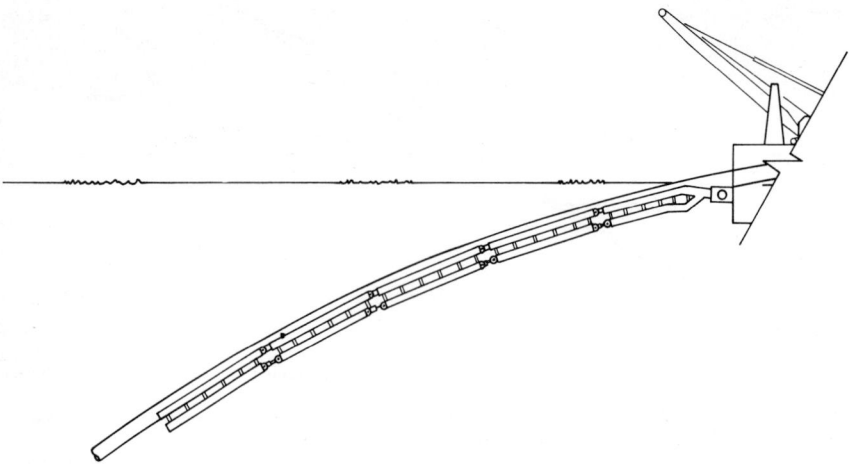

Fig. 7. Articulated pontoon.

higher anchor line tension, and the pipe "slips" through the tensioner. Most modern tensioners use a feedback control system to attempt to minimize tension variations due to barge motions and barge move-up.

ANCHORING AND POSITIONING

The functions of the lay barge anchoring system are to permit moving the barge ahead within specified tolerances as each joint of pipe is stalked on, and to hold the barge in a given position after moving ahead, within specified tolerances. A lay barge anchor system differs from drilling vessels, construction barges, and other types of vessels in its requirement to perform the first of these functions in an efficient manner. This precludes the use of anchor chains, clump anchors, piling or drilled-in anchors, or other systems that may work well in holding a vessel on station but do not permit efficient continued move-up of the barge.

The number of anchors that a lay barge has may vary from six to fourteen (Fig. 8). In general, it is necessary to maintain a minimum of eight when working in environments that are severe. Even in this case, it may be necessary to use a tug to help hold barge position while any one anchor is being moved ahead. Anchor sizes commonly used range up to 40,000 pounds and anchor wire up to 3½ inches in diameter.

The length of anchor wire is primarily dependent on the maximum expected water depth. Speaking very generally, longer wire permits working in deeper water. Although this sounds simple, the investment that must be made to provide long anchor wires can be considerable. The anchor winch drum size becomes larger, the prime mover for the winch increases in size, the space provided on the barge for the winches become larger, possibly the hull itself must be increased in size to accommodate the larger equipment, and the anchors become more difficult to move.

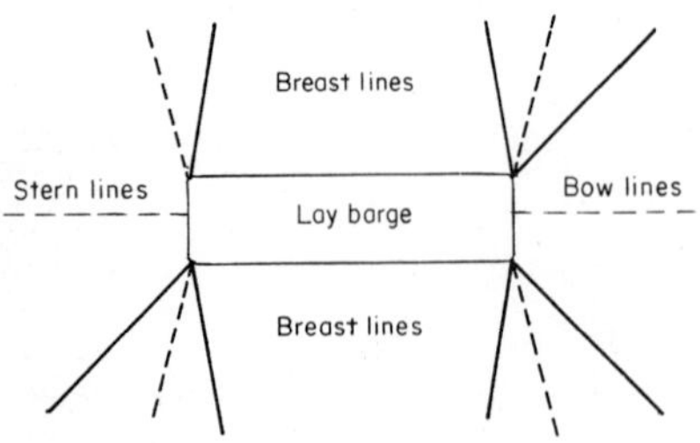

At some water depth, which is difficult to predict at present, it will probably become necessary to go to a dynamic positioning system. In this system, the barge is positioned with thrusters, and no anchors are used. Of course, it would be possible to have a combination system with varying numbers of anchor lines and thrusters.

The control systems used with the

Fig. 8. Typical lay-barge anchor spread.

thrusters will be extremely important. It is well to recognize the difference in control required for a pipe-lay barge compared to a dynamic positioned drilling vessel. A drilling vessel remains on one location, and the control system merely has to sense and respond to errors in location. A lay barge must travel along a given route, and even if exactly at the desired location, it is possible to buckle the pipe. A control system for a pipe-lay barge must not only sense the position, but also account for other parameters such as pipe tension, lateral loading on the pipe, and barge heading.

CONSTRUCTION PROCEDURES

Pipe Coating and Transportation.   Once the decision to build a marine pipeline is made, the first order of business is to purchase the pipe. The wall thickness and grade of this pipe will often depend on the laying stresses, and so this factor must be evaluated immediately. The owner will usually order this material ahead of time because of the long lead time required.

The pipe will be delivered to a pipe coating yard. Here the protective coating and weight coating will be applied. The protective coating is similar to that employed on a land line. Conventional coal-tar primer, enamel, and wrappers are frequently used. Sacrificial anode "bracelets" are placed around the pipe at frequent intervals for cathodic protection.

The most common method in use today for weight coating is the "impact method." Here the pipe is rotated while a stream of very dry mix concrete is discharged at high velocity against the pipe. This adheres to the pipe and the wire reinforcing wrapped around it. The density of this concrete can be varied from 135 up to 210 pounds/cubic foot. Once the desired concrete thickness has been reached, the pipe is carefully transported to a curing yard where the coating is allowed to gain strength and not moved again for a minimum of five days.

On occasion, special cross-sectional shapes are required in order to change the drag and lift characteristics of the pipe. Here, as may be the case if special reinforcing steel is required in the concrete coating, or if the contractor does not have the special equipment required for the impact method, forms must be made and the concrete poured in a more or less conventional manner. The biggest cost factor associated with these special coatings is not the additional coating itself but the possibility of extra coating weight and unusual shape that can adversely affect the laying of the line itself.

Generally, all the pipe will be coated and stored before the laying operation begins. This pipe must then be transported to the lay barge. The usual method is by large deck cargo barge. The most commonly used size is approximately 140 × 40 feet, and can carry approximately 140 joints of 30-inch pipe to the lay barge. This might supply the lay barge for one day. In very rough waters, however, cargo barges become difficult to handle and pipe-haul boats are utilized instead. These are large vessels up to 180 feet long that can carry 35 or 40 joints of 30-inch pipe. This may be only 5 hours supply for the lay barge. Since the lay barge itself can only store approximately 90 joints of this size pipe, logistics is an important factor. The last statement demonstrates, particularly in rough waters, the need for a large fleet or spread of equipment to support a lay barge. In remote North Sea waters this can amount to two anchor handling tugs, one standby tug, one supply vessel, and ten pipe haul boats.

Once the pipe haul boat or barge is tied up alongside the lay barge, the crane on the lay barge transfers the pipe aboard, usually one joint at a time. The crane may be a very large fixed pedestal crane also used for platform construction, a gantry type that traverses the length of the barge, or a conventional crawler.

On most lay barges, the pipe goes from the storage rack to a conveyor, which carries the pipe to a location adjacent and parallel to the line-up station. In the lineup operation, the new section of the pipe is supported on two or more support shoes, which are adjustable vertically and laterally. In addition, powered rollers move the new section axially until it abuts the end of the pipeline. In conjunction with an internal pipe alignment device, the lineup of the pipe for welding is thus accomplished.

Welding, X-ray Inspection, and Coating.   The number of welding stations on the pipe ramp is important because of the influence on pipe lay rate. Automatic welding equipment can completely weld out ¾-inch wall pipe in five stations. Most manual welding

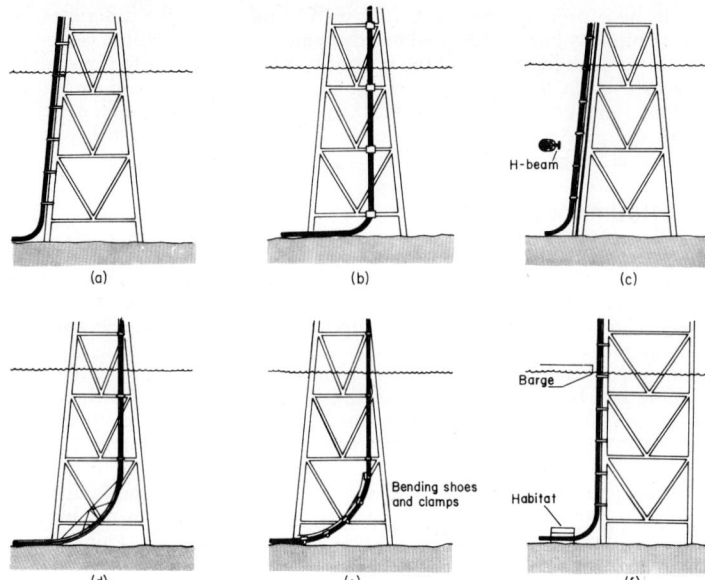

Fig. 9. Riser types. (*a*) Conventional clamp on platform leg. (*b*) Conventional clamp on platform braces. (**c**) Monorail. H-beam allows clamps to be installed on riser above water. (*d*) J-tube riser. (*e*) Bend-up riser. (*f*) Underwater habitat. Habitat allows dry weld beneath surface.

processes require either additional stations or the adding of more than one weld pass at a station.

In addition to the welding electrode leads, each welding station is usually provided with controls and monitoring equipment for the welding machines, which may be located elsewhere, such as below decks. If a gas metal arc welding system is used, provisions must be made for the supply, control, and monitoring of the shielding gas, such as CO_2.

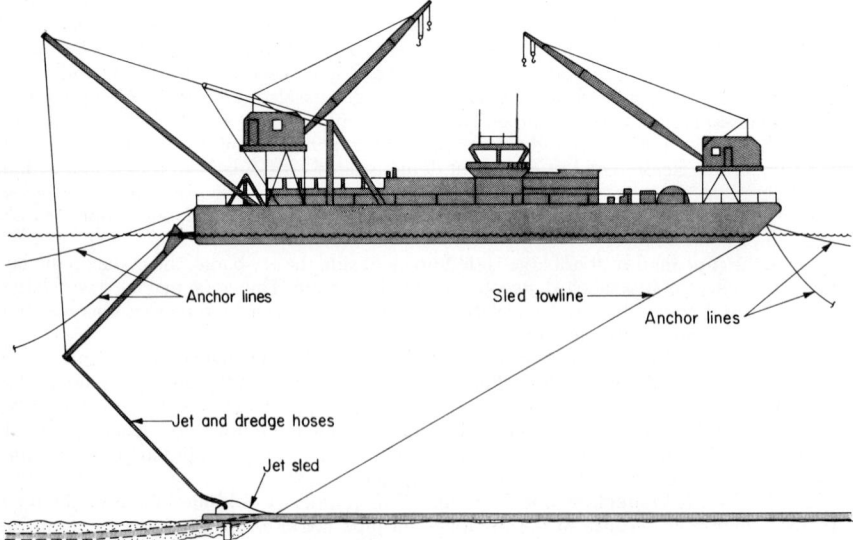

Fig. 10. Jet-suction dredge for pipe trenching operation.

Because of the high cost of repairs to a submarine pipeline after it is laid, every effort is made to ensure that the welds made on a lay barge are sound. It is common practice to radiograph each field girth weld for its entire length. Radiography equipment is commonly either X ray or gamma ray and may be either external or internal to the pipe.

After all welding and inspection is completed, it is necessary to apply a corrosion protection coating to the area of the pipe that was left uncoated for welding. The most common "field jointing" method is to use a sheet metal form banded to the pipe, and to fill this form with a hot asphaltic compound. On large-diameter pipe, a considerable volume of material is required to fill the joint. This field joint material is one of the biggest supply items that requires continual supply boat service to a lay barge.

Navigation. As the line is laid, great care is taken to determine the exact location of the lay barge. Repeatability and absolute accuracy in navigation are important, not only from the standpoint of original location but also for locating the line in the future. This need for exact location becomes particularly critical when the line must be restarted after being laid down, because of poor weather.

For many years, control of a pipe lay barge position was maintained by following a line of buoys strung along the right-of-way by a survey boat equipped with electronic navigation gear. Within recent years more sophisticated systems of navigation have been utilized by some contractors. Such a system might consist of shore or platform mounted responder units and barge mounted computer, plotters, and telewriters. A preprinted chart of the route can be printed on the plotter, and the location of the barge can be checked against

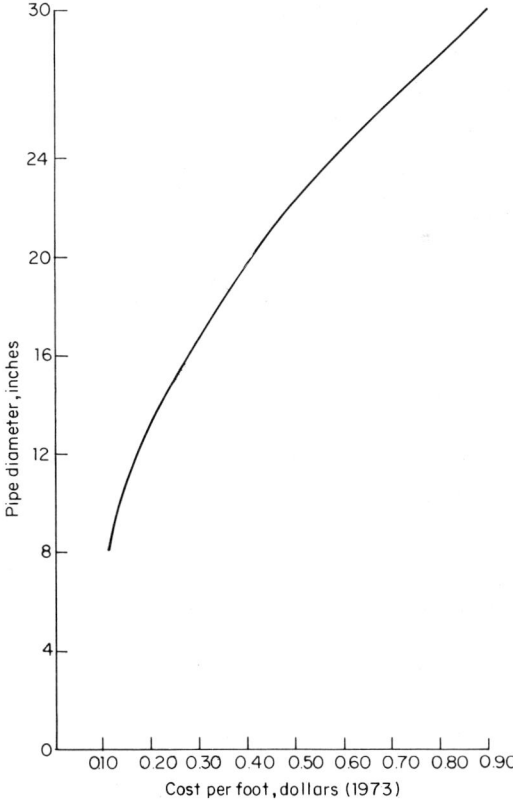

Fig. 11. Costs of unloading and receiving bare pipe and loading out weight-coated pipe: (1) Specific gravity of coated pipe = 1.25 (2) Pipe wall thickness = 0.375 in. for 16-in. pipe diameter and less. (3) Pipe wall thickness = 0.500 in. for over 16-in. pipe diameter.

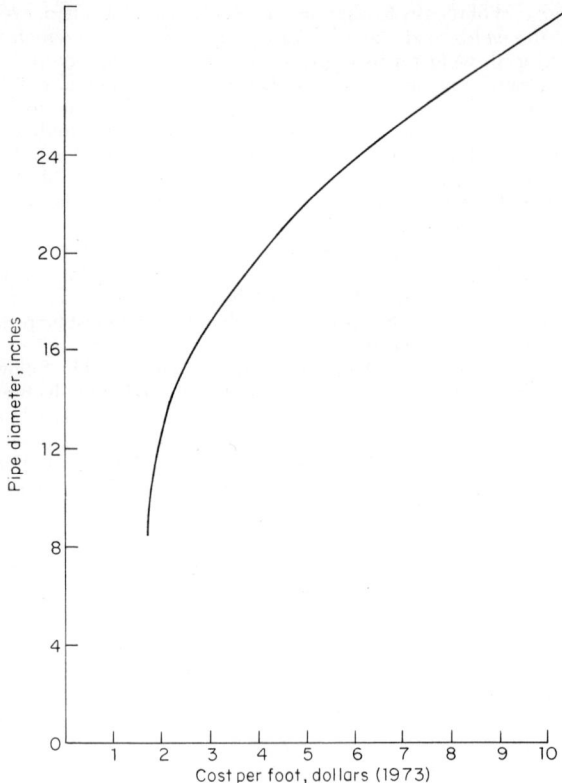

Fig. 12. Cost of combined protective and weight coating: (1) Specific gravity of coated pipe = 1.25. (2) Pipe wall thickness = 0.375 in. for 16-in. pipe diameter and less. (3) Pipe wall thickness = 0.500 for over 16-in. pipe diameter.

this at any time by the operator by interrogating the computer. Position accuracies of up to ±2 meters are possible at a range of 60 miles, and acceptable repeatability has been achieved up to 200 miles.

RELATED CONSTRUCTION OPERATIONS

Risers. Pipeline problems are not limited to the actual laying operation. Installation of the vertical connection (riser) from the seabed pipeline to the platform production deck is a problem which increases in complexity as water depths increase (Fig. 9). In relatively shallow water the most common method of riser installation is to lift the end of the pipeline to the surface, weld the riser onto the line, and then lower the line back to the bottom, allowing the riser to be secured to a structure leg or braces by diver-installed clamps. In deeper water, the riser may be installed on the structure before being connected to the pipeline. Then the riser may either be clamped to the structure as previously described, or may use a monorail-type connection whereby guide clamps on the riser are slipped over an H-beam rail on the structure as the riser is lowered. The riser is then connected to the submarine pipeline by underwater "dry habitat"-type welding techniques, by a flanged connection, or by patented mechanical devices which allow for relative misalignment between riser and pipelines.

Other innovations in riser installation for small-diameter lines include the "J-tube" method, in which risers are literally pulled from the sea floor to the platform deck through curved pipe guides which are built into the platform, and the "bend-up" method in which the riser is laid past the platform and then bent up to the deck around preinstalled bending shoes.

Diving.   As work moves into deeper waters, the importance of divers and diving increases. Divers must be used for repair work if any is needed on the line. Therefore, they must be able to function and perform meaningful work at whatever depth the pipeline is to be constructed.

In water over 150 feet it is necessary to use saturation diving if much work is to be performed. In this type of diving, the diver is actually "saturated": i.e., his physical system is subjected in a diving chamber to the pressure of the water at the sea bottom where the work will be performed. Special mixtures of gases are used at these pressures. The divers actually live in a pressurized chamber on the deck of the barge and are lowered to the sea bottom in a diving bell. They can then perform any necessary tasks at the desired depth.

At the greater depths at which present lines are being laid, this diving apparatus and team of divers and support personnel become a major operation in themselves. For example, for recent projects in the northern North Sea it was necessary to have a pressure chamber on the barge deck in which nine men could live for weeks subjected to pressure, and a diving bell designed for these depths that can carry the men down to the bottom. Such a bell also may include complex gas storage, and mixing, and cleansing facilities, as well as backup crews and support personnel. As many as 50 such diving company personnel may be on a lay barge under such conditions.

Trenching.   Submarine pipelines are often lowered into trenches beneath the sea floor

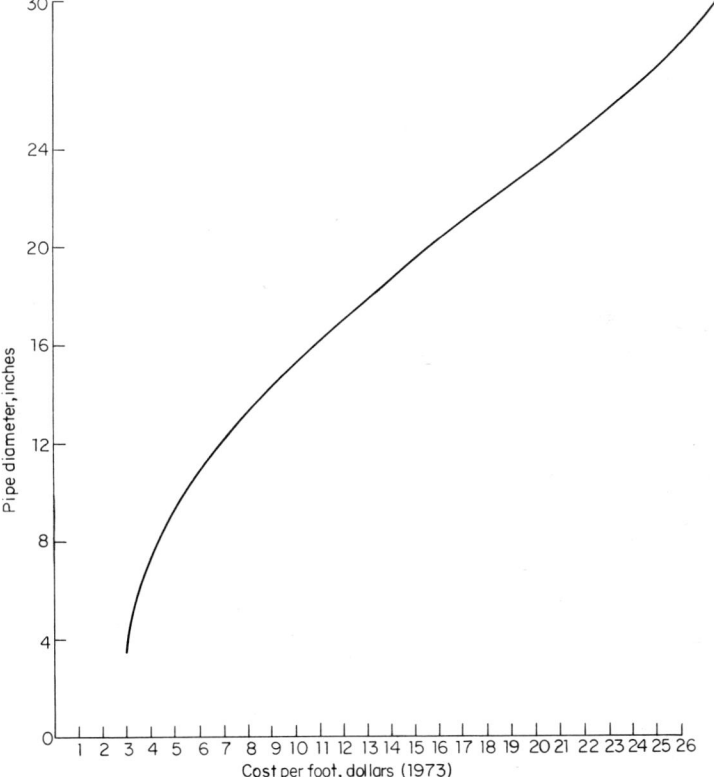

Fig. 13. Approximate cost of laying pipe in Gulf of Mexico: (1) Cost based on project at least 10 miles in length. (2) Material costs not included. (3) Estimates based on moderate water depths; no saturation diving costs included. (4) Estimates based on 1973 costs.

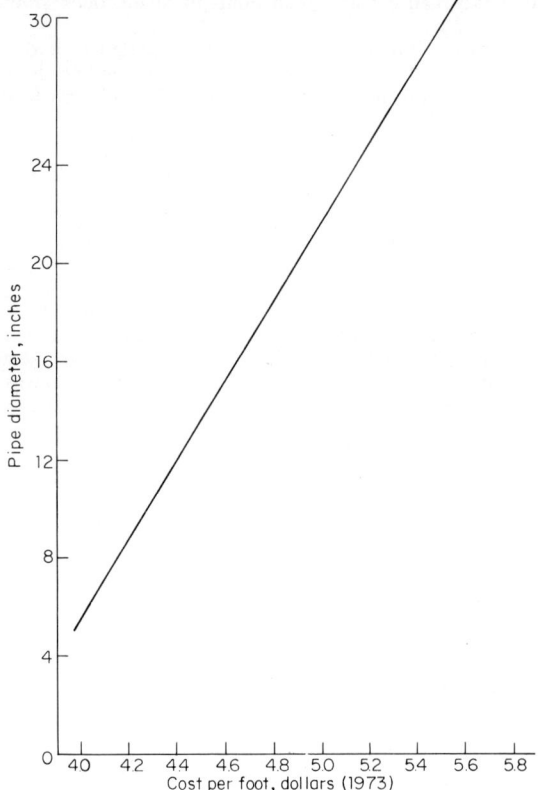

Fig. 14. Approximate cost of burying pipe in Gulf of Mexico. Three-foot depth of cover assumed over pipe.

to provide protection against extreme environmental conditions and against damage from dragging anchors and fishing operations. Such trenches usually provide approximately three feet clearance between the sea floor and the top of the line; however, in areas of increased risk such as shore approaches, shipping lanes, and around platforms, ten feet of clearance is common.

The trenching operation is normally accomplished by use of a jet-suction dredge specifically designed for this purpose (Fig. 10). The dredge, or "bury barge," is equipped with high-capacity pumps which furnish high-pressure water to a jet sled straddling the pipe on the sea floor. As the sled is pulled along the pipe by the barge anchor system, the jet nozzles cut the sea floor from beneath the pipe. The jet slurry is then removed by a high-volume dredge pump, or by air lift, and discharged to the side of the line, allowing the line to settle into its buried position. As the sled is pulled along the pipe, pressure-sensing cells mounted in the sled roller guides indicate the degree of pressure being exerted on the line, thus indicating the extent of settlement and guiding the operator in controlling the rate of movement of the sled along the line. Diver inspection is normally used to verify proper pipe cover and other features of operation.

Shore Approach.   Burying of pipelines near the shoreline often poses special problems. The usual method of bringing a line to shore is to anchor the lay barge as close in as its draft will allow, and then pull the line ashore as it is welded up on the barge. Since the bury barge is also limited by its draft, it is necessary to precut a trench for that portion of the line which cannot be jetted in. This trench may be cut by draglines working from land or from small work barges. If soil conditions or current and/or wave action are such that a trench is difficult to keep open for a long enough period to install the line, sheet piles may be driven before excavation. After the pipeline is pulled ashore, the sheet piles may be pulled or cut off below the mud line and the trench backfilled. If the shore approach contains coral or rock, controlled intensity explosives may be used in cutting the trench.

APPROXIMATE COST DATA

The cost of marine pipelines varies considerably from one part of the world to another and even from one project to another in the same area. This is due to the variables involved in the construction, such as pipe material and wall thickness, welding requirements, weight-coating requirements, environmental effects, such as the magnitude and direction of wind and well, the water depth, the sea bottom conditions, the time of year, the distance of the jobsite from the normal base of operations of the barges, and the type of equipment being utilized. Nevertheless, the cost data presented in Figs. 11 through 14 provide some concept of the magnitude of the costs and aid in preparing budget estimates. It should be noted that the data are for an average case in the Gulf of Mexico, and the data should be used carefully for unusual situations in the Gulf, or for remote or other world locations.

The charts show the primary cost areas of marine pipeline construction. These are the costs for receiving, coating, loading out, laying, and burying the pipeline. While materials, with the exception of those for welding or field joints, are not included, the costs would provide for normal contingencies. Items that cannot be generalized are risers, underwater valves, shore approaches, crossing other lines, unusual weather, and other special factors. These are not considered in the data given.

Calorimetric Measurement of Gaseous Fuels

By C. W. WARNER*

To accurately define the energy content of gaseous fuels, the calorific value (heating value) per unit volume must be measured. The instrument used for this purpose is called a *gas calorimeter.*

The need for such measurement may include: (1) *sale and purchase* to establish quality and range in value, particularly if billing is on a *therm* basis (therms = (volume in cubic feet)(Btu per cubic foot)/100,000); (2) *record or compliance* with government regulations; (3) *process monitoring or control:* e.g., natural gas stripping plants, synthetic or substitute natural gas production (SNG), liquid natural gas operations (LNG), and standby or peak-shaving plants; (4) *industrial applications* where process requirements depend on calorific value; and (5) *manufactured gas operations:* e.g. blast furnace gas, producer gas, coke oven gas. See separate description in this Handbook section on "Gas Mixing Control Systems."

Terms Used in Gas Calorimetry. For complete definition of terms in English units, see the ASTM Standards listed in the references at the end of this subsection. English and metric units are the same for several of the common terms used, including total calorific value, net calorific value, observed calorific value, theoretical air, excess air, combustion air, products of combustion; and flue gases. However, there are some exceptions, such as:

English units	*Metric units*
British thermal unit, Btu	Kilocalories
Standard temperature, °F	Standard temperature, °C
Standard pressure, inches of Hg	Standard pressure, millimeters of Hg
Standard cubic foot	Cubic meter

Typical Gas Calorific Values (Total Calorific Value) in Btu. Blast furnace gas (70 to 95); producer gas (125 to 160); carbureted water gas (475 to 575); coke oven gas (490 to 560); natural gas (975 to 1180); LNG (1000 to 1170); SNG (970 to 985); refinery gas (1200 to 2400); propane (2400 to 2600); and butane (3100 to 3400). Values are per standard cubic foot.

GAS CALORIMETERS

There are three basic classifications of gas calorimeters:

1. *Total Calorific Value Types.* An instrument of this type makes a direct determination of total calorific value by means which conform quite closely to the definition.

2. *Net Calorific Value Types.* Net calorific value is less than the total calorific value by an amount equal to the latent heat of vaporization of the water formed during combustion. An instrument of this type uses means that give results more nearly related to the net value. Thus, these types are affected by gas composition and must be calibrated for the gas to be tested.

3. *Inferential Types.* Included in this category are comparison types and those instruments that depend upon some other characteristic of the gas, such as flame appearance, maximum flame temperature, specific gravity, or gas analysis as indicative of calorific value.

Summary of Gas Calorimeter Characteristics and Features. Information on the most generally recognized instruments in each of the foregoing three categories is summarized in Table 1. There follows a brief description of the use and/or principle of operation of the most commonly used of these instruments.

*Development Consultant, Cutler-Hammer, Inc., Milwaukee.

WATER FLOW CALORIMETER

Historically, the manual type was the first instrument to be widely accepted by the gas industry. However, in the United States and in some other countries, the manual type has been almost completely replaced by the continuous recording type.* Details of the manual type are contained in the ASTM Standard D900-55(70) listed in the references.

RECORDING CALORIMETER (CUTLER-HAMMER TYPE)

In principle, this instrument is of the flow type in which air is used as the heat-absorbing medium. Schematic detail of the instrument is shown in Fig. 1. The calorific value of the gas is determined by imparting all the heat of combustion of a metered quantity of gas to a metered quantity of air. The temperature rise of this heat-absorbing air is sensed by a pair of nickel wire resistance thermometers, forming two legs of a self-balancing Wheatstone bridge-type strip-chart recorder. The scale and chart are both calibrated in Btu per cubic foot (or kilocalories per cubic meter), and the readings require no corrections or computations (to be performed on the results of the measurement) on the basis of any standard volume conditions, including the metric units listed previously. Complete information is given in ASTM Standard D1826-64(70) listed in the references. Details of the recorder portion of the Cutler-Hammer type calorimeter are given in Fig. 2.

THERMETER

This instrument is a modification of the principle of the calorimeter of the Cutler-Hammer type. The heat-absorbing air is not separated from the products of combustion, and temperature of this mixture is sufficiently high so that none of the water formed in combustion is condensed. Thus, the indication is more nearly a function of net calorific value although calibration is usually in terms of total calorific value. The recorder is a round chart (24 hours) potentiometer type with input from a form of iron-constantan thermocouple. Scale and chart have a suppressed zero.

SIGMA CALORIMETER

This instrument, also of the net calorific value type, consists of the gas flow regulator which maintains constant flow of gas to burner in the recorder, essentially independent of changes in specific gravity and atmospheric pressure. The ambient temperature should be maintained relatively constant for accurate results. The recorder senses the calorific value by measuring the differential expansion of two concentric steel tubes responding to the temperature of the mixture of flue products and excess air.

REINEKE CALORIMETER

This is another example of the net calorific value type in which the heat released by the instrument burner raises the temperature of fan-supplied room air, intimately mixed with the hot flue gases. The rate of air flow (required to keep a constant mixture temperature) is used as the indication of the sample calorific value.

UNION CALORIMETER

In this form of net calorific value type instrument, the hot products of combustion transfer their heat either to ambient air supplied through natural draft or by fan cooling. The temperature rise of the mixture is sensed by a thermocouple arrangement, and the output is converted into the calorific value reading.

*In the United States, by the Cutler-Hammer type; in Great Britain, a recording type (Fairweather) is the generally accepted standard instrument.

TABLE 1. Gas Calorimeter Characteristics and Features

Type or name of gas calorimeter	Total calorific value type	Net calorific value type*	Inferential type, total or net	Detailed calculations and corrections required	Operation				Application		Ranges, Btu full-scale values
					Manual	Spot sample	Continuous sample	Continuous recording of calorific value	Laboratory only	General use	
1. Water-flow manual	×	—	—	×	×	×	—	—	×	—	475 to 3300
2. Water-flow Fairweather	×	—	—	—	—	—	×	×	—	×	Usual fuel gases
3. Cutler-Hammer type, air as heat absorbing medium	×	—	—	—	—	—	×	×	—	×	120 to 3600
4. Thomsen type	×	—	—	×	—	×	—	—	×	No	Usual fuel gases
5. Bomb type, combustion at constant volume	×	—	—	×	×	×	—	—	×	No	Usual fuel gases
6. Thermeter, air as heat absorbing medium	—	×	—	—	—	—	×	×	—	×	150 to 3600
7. Sigma, heated air, expansion of a material	—	×	—	—	—	—	×	×	—	×	120 to 3300
8. Reineke, heated air, controls temperature of products of combustion	—	×	—	—	—	—	×	×	—	×	Usual fuel gases
9. Union, heated air, thermoelectric or thermophysical	—	×	—	—	—	—	×	×	—	×	Usual fuel gases
10. Calorimixer, Cutler-Hammer type controller	—	×	—	—	—	—	×	×	—	×	300 to 1500 mixtures
11. Parr, comparison type	—	—	×	—	×	×	—	—	×	No	Usual fuel gases
12. Caloroptic visual appearance of flame	—	—	×	—	×	—	×	—	—	×	500 to 2500
13. Gas analysis (Orsat) volume, chemical	—	—	×	×	×	×	—	—	×	No	Usual fuel gases
14. Gas analysis chromatograph	—	—	×	×	×	×	—	—	×	No	Usual fuel gases
15. Gas analysis mass spectrometer	—	—	×	×	×	×	—	—	×	No	Usual fuel gases
16. Specific gravity	—	—	×	×	—	—	×	—	—	×	Natural gas
17. Precision type, maximum flame temperature	—	—	×	—	—	—	×	×	—	×	Natural gas

*Usually calibrated to give results in total calorific value.
1. Requires skilled operator, laboratory environment, very steady room, and water temperature [ASTM D900-55(70)].
2. Requires controlled temperature of room, steady water temperature, combustion air, humidity and temperature control.
3. Greatest accuracy with room temperature steady in range from 72 to 77°F [ASTM D1826-64(70)].
4. Skilled operator, laboratory environment.
5. Skilled observer, laboratory conditions closely controlled.
6. Best results under uniform room temperature conditions can be standardized using hydrogen.
7. Requires uniform room temperature conditions. Standardizing gas must be same as test gas composition.

Calibration				Compensation provided in instrument					Features						
Requires tests of individual parts	By overall tests	By comparison with more accurate calorimeter	Use of standard gas sample or hydrogen	Ambient temperature	Gas composition	Specific gravity	Barometric pressure	Must be standardized for gas to be tested	Retransmitting for remote reading, computer input, etc.	Expanded range of scale and chart	Wobbe index version, $Btu/(sp\ gr)^{1/2}$	Gas mixing control use	Accuracy guarantee ± percent full scale	Attainable accuracy depends on frequent calibration, percent	Time lag of complete instrument start, 66⅔% of calorific value change typical for 1200 range
×	—	—	—	No	—	—	No	—	—	—	—	No	—	±0.3	15 min per determination
×	—	—	×	Lim.	×	×	Lim.	—	×	—	—	Lim.	—	±0.5	6 min, 12 min
—	×	—	×	60 to 85°F	×	×	×	—	×	×	—	Lim.	±0.5	±0.3	3 min, 7.8 min
×	—	—	×	No	—	—	No	—	—	—	—	No	—	Good	Long calculation time
×	—	—	—	No	×	×	×	—	—	—	—	No	—	Very good	Long operation calculation time
—	×	×	×	60 to 90°F	—	—	×	×	×	50% to 100%	—	Yes	±2.0	±1.0	1.7 min, 4.4 min
—	—	×	×	—	—	—	×	×	×	—	×	Yes	±2.0	±1.0	3.5 min, 6.0 min
—	—	×	×	50 to 90°F	—	—	×	×	×	×	×	Yes	—	±1.0	8 sec
—	—	×	×	—	—	×	—	×	×	—	×	Lim. and Yes	—	±2.0	15 sec
—	—	×	×	60 to 90°F	—	×	×	×	—	−20 0% +20	—	Yes	—	—	24 sec, 1.4min
×	—	—	×	—	—	—	—	—	—	—	—	No	—	Good	Long operation time
—	—	×	—	—	—	—	—	—	—	—	—	No	—	±5.0	Fast
×	—	×	—	—	—	—	—	—	—	—	—	No	—	Fair	Long analysis calculation time
×	—	—	—	—	—	—	—	—	—	—	—	No	—	Very good	Long analysis calculation time
×	—	—	—	—	—	—	—	—	—	—	—	No	—	Very good	Long analysis calculation time
—	—	—	—	—	—	—	—	—	—	—	—	No	—	Good	Sp. gr. response fast
—	—	×	×	55 to 109°F	—	—	—	—	×	—	—	Yes	—	±1.0	Fast

8. Time lag of burner only. Requires uniform room temperature conditions. Standardize against more accurate instruments.
9. Same as 8.
10. Controller only.
11. Laboratory instrument.
13. Laboratory method only. Analysis of natural gas [ASTM D1136-53(70)].
14. Laboratory method only. Analysis of natural gas (ASTM D1945-64).
15. Laboratory method only. Analysis. [ASTM D1137-53(70)].
16. Accuracy on natural gas determined by accuracy of specific gravity determination.
17. Limited to natural gases. Presence of H_2 will materially affect accuracy.

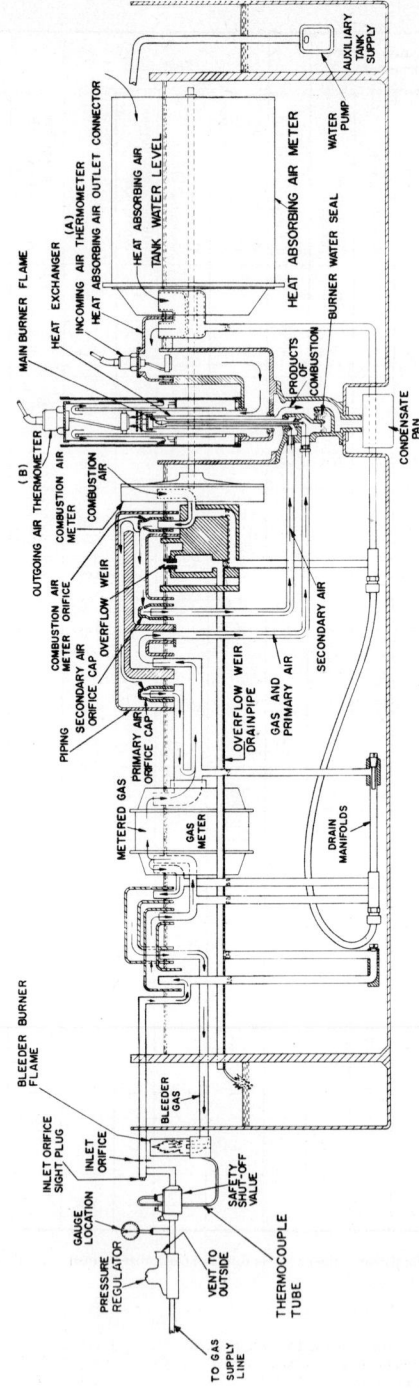

Fig. 1. Principal elements of Cutler-Hammer calorimeter. Recorder portion is shown in Fig. 2.

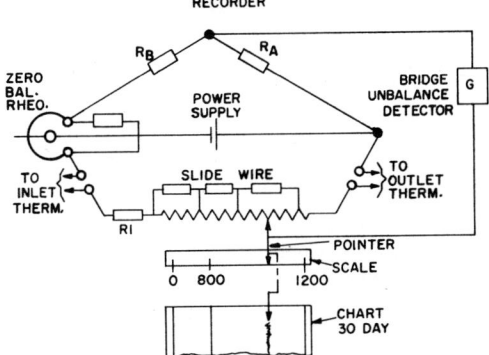

Fig. 2. Recorder portion of Cutler-Hammer calorimeter shown schematically.

EFFECTS OF GAS COMPOSITION

All the net-heating-value types of instruments are affected by gas composition and must be calibrated for the gas on which they will be used. The Sigma, Reineke, and Union instruments can be provided to read the Wobbe Index (calorific value divided by square root of specific gravity).

INFERENTIAL TYPES

The gas chromatograph is probably the most commonly used of the inferential-type calorimeters. The gas chromatograph identifies and quantifies the constituents of the gas sample. Calculation of the calorific value is made, based on the percentage of each combustible constituent multiplied by the calorific value of each constituent when pure. The end results are affected by (1) the accuracy of the chemical analysis performed, and (2) reliability of the calorific values of the pure constituents.

SAMPLING

To ensure that the gas sample received by the calorimeter is truly representative of actual line conditions, the following factors should be considered:
1. Tap must be on an active line.
2. Sample line time lag must be reduced to a minimum.
   a. Locate calorimeter close to sample point.
   b. Use small size sample line, operating at minimum pressure.
   c. Provide additional purge burners.
   d. Recirculate sample line to low pressure point, or use a sample line pump.
3. Avoid any condition that can result in condensation of any constituent.

REFERENCES

ASTM Standard D900-55(70), "Calorific Value of Gaseous Fuels by the Water Flow Calorimeter," American Society for Testing and Materials, Philadelphia.

ASTM Standard D1826-64(70), "Calorific Value of Gases in Natural Gas Range by Continuous Recording Calorimeter."

Melrose, D. C.: "Comparison of Calculated and Measured Heating Value of Natural Gas," presented at AGA 1972 Distribution Conference 72-D-2, American Gas Association, Arlington, Va., 1972.

Warner, C. W.: "The Cutler-Hammer Calorimeter Accuracy and Maintenance Requirements," AGA 1972 Distribution Conference 72-D-22, American Gas Association, Arlington, Va., 1972.

Armstrong, G. T., et al.: "Standard Combustion Data for the Fuel Gas industry," AGA 1972 Distribution Conference 72-D-76, American Gas Association, Arlington, Va., 1972.

Gas-Mixing Control Systems

By C. W. WARNER*

Gas mixing can become a necessity for one or more of the following reasons: (1) Permitting the use of gas available under special conditions; (2) blending the various sources of natural gas which can have different calorific values, (3) use of liquefied natural gas which can have varying calorific values, (4) eliminating the need for costly storage or permitting more advantageous use of existing storage, (5) peak-shaving or standby installations, and (6) stabilization of a gas supply to meet a specialized requirement of certain industrial uses.

Since the proper solution of a gas-mixing problem involves the measurement of certain variables, the system must be based on final control of one or more variables, of which the most significant are: (1) volume (rate of use or production), (2) density or specific gravity, (3) combustion characteristics, and (4) calorific value.

Analysis of Gas-Mixing Problem. The ease with which gas-mixing systems can be controlled depends to a great extent on local circumstances, including: (1) send-out requirements: (*a*) variations in rates of gas demand, and (*b*) permissible variations in calorific value or other gas characteristics; (2) available plant equipment; (3) final mixture calorific value; and (4) the calorific value of the constituent gases.

Portion of Each Constituent in the Desired Final Mixture. For mixtures of two gases, the general expression in terms of calorific value is given by

$$H_R(X) + (1 - X)H_L = H_M$$

where H_R = calorific value of rich gas
$\quad X$ = decimal part volume of constituent H_R
$\quad H_L$ = calorific value of lean gas
$\quad H_M$ = calorific value of desired mixture
$\quad 100X$ = percent by volume

If the mixture is to be gas and air, the formula can be simplified to

$$H_R(X) = H_M$$

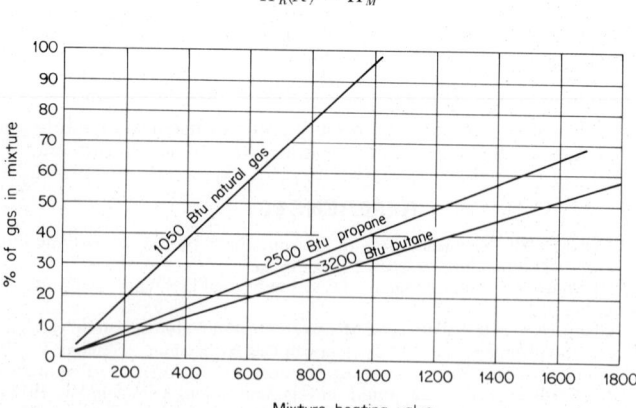

Fig. 1. Mixture heating value versus percentage of gas in mixture for various gas-air mixtures.

*Development Consultant, Cutler-Hammer, Inc., Milwaukee, Wis.

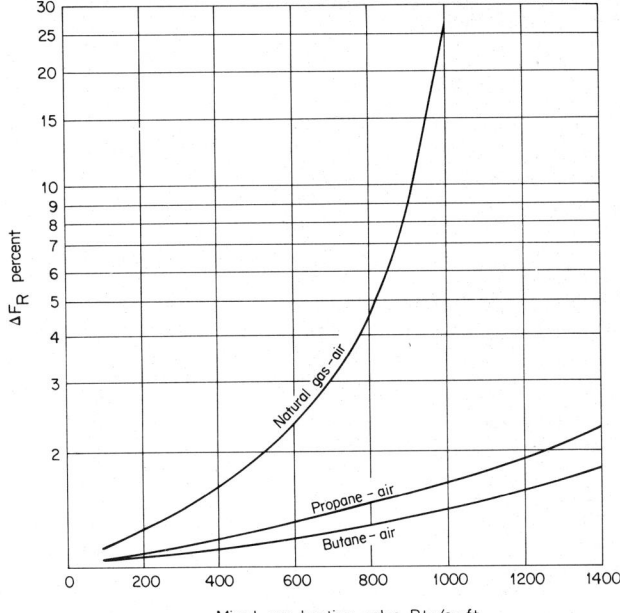

Fig. 2. Curves of ΔF_R for gas-air mixtures. (1) Natural, 1050 Btu. (2) Propane, 2500 Btu. (3) Butane, 3200 Btu.

In most instances, if more than two gases are involved, the system can be subdivided into a series of two-gas mixtures.

EVALUATION OF ROLE OF PROPORTIONING EQUIPMENT

Required is an expression of the change in flow of either constituent in a two-gas mixture to produce a fixed change in mixture calorific value. For the sake of uniformity, and on the

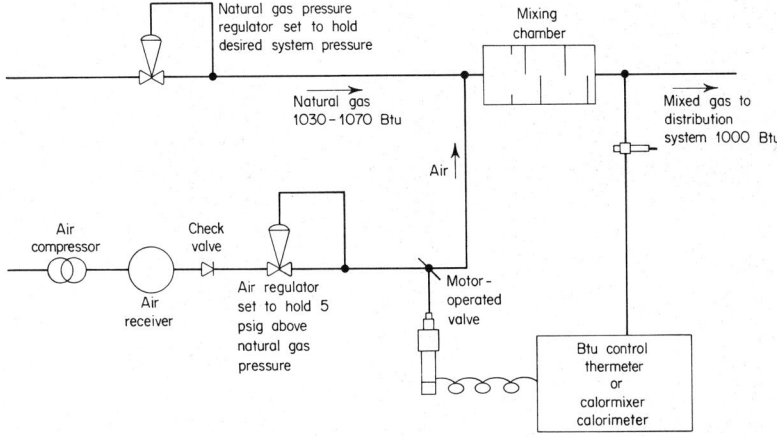

Fig. 3. Simple stabilization control. This type of plant is suitable when the base gas flow is steady or changes very slowly—and when the heating values of constituents change very slowly.

basis that a 1.0 percent change in mixture calorific value is a good criterion of performance, the following equations apply:

$$\Delta F_R = \frac{H_M}{X(H_R - 1.01H_M)}$$

where ΔF_R = percentage change in flow of rich gas H_R to produce a 1 percent change in mixture calorific value H_M, and

$$\Delta F_L = \frac{H_M}{(1 - X)(H_L - 1.01H_M)}$$

where ΔF_L = percentage change in flow of lean gas H_L to produce a 1 percent change in mixture calorific value.

Effect of Variations in Calorific Value.   It is also very important to determine the effects of changes to the calorific value of the mixture which will result from changes in

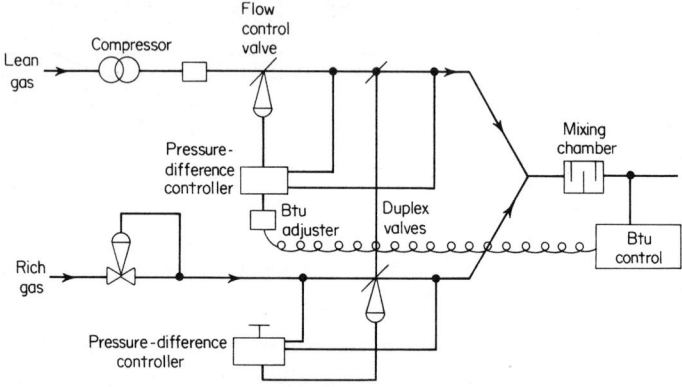

Fig. 4. Mixing system using automatically set duplex valves and two pressure-difference controllers. This type of plant is particularly suited for installations in which system pressure is held in accordance with the setting of the pressure regulator in the rich gas line, and is useful in standby, peak-shaving, or complete replacement plants. The system is suitable for flow ranges up to 30 to 1 and for a wide range in calorific values of the constituent gases.

the calorific value of the constituents. On the basis of the two-gas mixture, the following equations apply:

$$\Delta H_R = \frac{H_M}{X(H_R)}$$

where ΔH_R is the percentage change in calorific value of H_R which will produce 1 percent change in the calorific value of the mixture. Similarly,

$$\Delta H_L = \frac{H_M}{(1 - X)H_L}$$

Typical curves from gas-mixing formulas are shown in Figs. 1 and 2.

CONTROL SYSTEM

In general, a gas-mixing control system consists of the following elements:

1. A system of volumetric proportioning to establish and maintain an approximate proportion of the constituents in the final mixture for the complete range of sendout requirements.

2. Proper calorimetric equipment to indicate and record the mixture calorific value and to permit automatic control of the value to some preselected value.

3. A Btu-adjusting mechanism which couples the calorimetric proportioning equipment to provide the means for automatically maintaining the mixture calorific value at the desired control point.

Two examples of gas-mixing control systems are illustrated in Figs. 3 and 4.

Liquefied Natural Gas (LNG)

Liquefaction of natural gas provides the needed concentration of this energy medium for practical transportation, notably overseas, and for readily available, convenient storage, particularly in areas of high and seasonal demands that are located considerable distances from natural sources. Pipeline transmission of natural gas, in either the gaseous or the liquid phase, is not practical for a number of consuming nations.

Emphasis on Safe Handling. The technology of liquefied natural gas (LNG) and, in fact, the cryogenic handling and processing of numerous gases has reached a degree of proficiency and safety not visualized in the aftermath of the tragic LNG accident in Cleveland, Ohio, on October 20, 1944, which claimed 135 lives and involved several hundred injuries. At that time, a cylindrical tank with a capacity of 150,000 cubic feet of LNG failed (147,000 cubic feet at time of accident), and released the liquid, part of which overflowed an enclosed area and entered surrounding storm sewers of the adjacent neighborhood where there were numerous ignition sources. Flames were reliably reported to have reached a height of some 2800 feet. Subsequent fires damaged an area in excess of one-quarter mile from the tank site. At the same facility were three additional spherical tanks, each with a capacity of 83,000 cubic feet of LNG. About 20 minutes after collapse of the cylindrical tank, heat from the conflagration caused one of the spherical tanks to fail.

This accident was analyzed thoroughly and is regarded by numerous authorities as having set back LNG technology for a minimum of 20 years. But during the 1940s and 1950s there was not nearly the incentive in any event to pursue LNG technology, as later developed in the 1960s and 1970s. Much was learned from the Cleveland accident and other LNG accidents elsewhere, admittedly at a high price, but nonetheless resulting in a keener awareness of the hazards of LNG and how to deal with them, now many times more important in view of the huge quantities shipped and stored.

It is not out of order here to summarize the findings of the U.S. Bureau of Mines Report R.I. 3867, dated February 1946, on the results of investigating the accident at the Cleveland liquefaction, storage, and regasification (LS&R) plant. Even with its thoroughness, the report left some questions unanswered; doubtless with better instrumentation and methodology available 30 years later, the investigation, if conducted today, would yield more definitive information. The essence of the report findings was: (1) poor selection of site location, being immediately adjacent to a residential area; (2) poor selection of tank material, 3½ percent nickel steel, which reportedly had the lowest impact strength of all materials tested for the job; (3) inadequate dikes; and (4) nearness to numerous and not easily extinquished sources of ignition. The investigators found no evidence to indicate an operating or personnel failure in the shutdown of the liquefaction plant, which was in progress at the time when the tank failed. There was no evidence of a gas-air explosion preceding failure of the tank. There was no evidence of sabotage.

In a much later report (1967), M. G. Zabetakis of the U.S. Bureau of Mines, after analyzing a number of accidents involving not only LNG, but also liquid hydrogen and liquid oxygen, emphasized the following factors as being of primary importance as causes of system failures in cryogenic plants:

1. Mechanical failure of the containment vessel, piping, or auxiliary components (brittle failure, hydrogen embrittlement, freeze-up, etc.).

2. Reaction of the fluid with a contaminant (e.g., oxygen plus a hydrocarbon).

3. Reaction of the fluid with the confining vessel or auxiliary equipment (e.g., fluorine plus a freshly exposed metallic surface).

4. Failure of a safety device to operate properly.

5. Operator error.

Zebetakis stresses that whenever any of the foregoing conditions occur, the plant must be designed so that a series of chain reactions is not initiated. Needless to say, the operating personnel must be familiar with the emergency procedures to be utilized under such conditions.

The original Bureau of Mines report also carried a number of details pertaining to the safety of LNG plants, most of which remarks are applicable today. However, new materials of construction and construction configurations tend to obsolete a few of these observations. All of them, however, are worthy of note by designers and operators of LNG facilities:

1. Plants in which large quantities of flammable gases are liquefied and stored should be isolated, and activities not directly related to the operation of the liquefaction and storage plant should be prohibited within the plant area. The distance between the boundary of such a plant and the nearest inhabited building should be greater than half a mile.

2. Storage containers for liquefied gases should be isolated from other parts of the plant and should be provided with dikes large enough to confine the entire contents of the tank in the event of a failure. The construction of dikes and the distances between tanks should conform with established fire underwriting codes.

Because of the pronounced volatility of liquefied petroleum gases, dikes are not normally effective; hence their general requirement may not be justified as in the case of gasoline or similar flammable liquids. When, however, in the opinion of the inspection department having jurisdiction, owing to the slope of the ground or other local conditions, aboveground containers are liable, in case of rupture or overflow, to endanger adjacent property, then each container shall be surrounded by a dike of such capacity as may be considered necessary to meet the needs of the situation under consideration.

3. All pipelines carrying cold liquid or cold gas should be furnished with suitable bolted flanged joints. Expansion loops capable of taking care of pipe movement due to changes in temperature should be provided.

4. Storage tanks for liquefied gases should be provided with independent inlet and outlet lines for the liquefied gas. The inlet line should discharge at a point near the maximum liquid level in the tank.

5. Extreme precaution should be taken to prevent spilled liquefied gas from entering storm sewers or other underground conduits.

6. The appearance of frost spots on the outer shell of storage containers for LNG should be regarded with suspicion, and the tank should be drained and thoroughly inspected, unless the reason for the appearance of the frost spots can be definitely established.

7. Efforts should be made to prevent wide variations of temperature in the inner shell of storage containers when they are filled for the first time with liquefied gas.

8. Although closely coupled equipment is desirable in low-temperature refrigeration processes, nevertheless some dispersion is indicated when such hazardous materials as highly flammable gases in the liquid state are handled.

9. Design of storage containers for liquefied natural gas should make provision for remote closing of the foot valve from a point at ground level.

10. Positive and foolproof liquid-level indicators should be supplied for each storage container. These indicators should be equipped with automatic high- and low-level alarms.

11. When possible, the tops of the vent gas lines of all storage tanks for liquefied gas connected to the same manifold should be at the same level, so that accidental leakage of liquid from one tank to another would never result in the overflowing of liquid into the vent gas system. When this is not possible, extreme precautions should be taken to prevent leakage of liquid from one storage container to another in which the level of the liquid is lower. Containers for storage of liquefied gases should not hold liquid in excess of their rated capacity. The liquid level in any tank connected to a common liquid manifold should always be at least one foot below the top of the lowest overflow in the group of tanks.

12. All outer portions of storage containers for liquefied natural gas should be readily accessible to inspection and should be in the open to permit proper ventilation.

13. The gas purged through the annular space of storage containers for LNG should be metered both into and out of these containers.

14. All sources of electrical ignition should be eliminated in and around gas-liquefaction plants. This hazard should be safeguarded to the extent considered necessary in explosive plants. Precautionary measures in such plants include protection against lightning; elimination of static charge on machinery, equipment, and persons; and the use of explosion-proof electric equipment and wiring throughout hazardous areas.

15. Means should be provided for rapid egress of personnel from the plant area in case of emergency. Escape drills should be held frequently, and damage-control drills should be conducted after types of damage that might possibly be repaired with safety have been established.

LNG Spillage.   In addition to the possibility of fire and/or an explosion of natural gas mixed with air, there is the possibility of a low-temperature, nonchemical reaction of LNG with water. Tests indicate that this nonchemical reaction is due to the superheating of a thin layer of LNG on the water surface. When this layer reaches its limit of superheat, it apparently undergoes spontaneous homogeneous nucleation. This rapid vaporization with its associated shock wave has come to be known as *vapor explosion.* Although the vapor explosions result in shock waves, they are orders of magnitude less destructive than what one usually associates with a chemical explosion. It is believed that the greatest danger from an LNG vapor explosion is the subsequent dispersion of the gas which might, if ignited, cause substantial damage. This type of explosion was first reported by the U.S. Bureau of Mines in 1970. At that time, it was reported that several moderately severe explosions had occurred during tests in which LNG had been spilled on water. Subsequent research has been conducted by Shell Pipe Line Corp., Exxon, Massachusetts Institute of Technology, and others. Explosions have not been reported where the methane content of LNG exceeds 40 percent methane. This would explain, of course, why few accidental industrial spills have resulted in vapor explosions (LNG usually contains 90 percent or more of methane). It has been found, however, that LNG with more than 40 percent methane will explode when spilled on water which has a thin layer of hexane on the surface. It has been found that LNG does not explode when spilled on normal nonane and other hydrocarbons with high freezing points.

An excellent summary of the known circumstances of LNG vapor explosions (as of 1973) is available from Massachusetts Institute of Technology.* This document also incorporates an excellent bibliography.

LIQUEFACTION PROCESSES

In the early 1950s, there was a proposal to Union Stock Yard and Transit Company to liquefy Gulf Coast natural gas and barge it to Chicago with the concept of freezing meat as part of the LNG regasification process. The program was not completed, but valuable experience was obtained, notably in the design and insulation of gas storage tanks. Continental Oil Company and Union Stock Yard joined in a program in 1955, forming the Constock Liquid Methane Corporation. This in turn formed the Behamian corporation, British Methane, Ltd., with the British Gas Council. As a result, a liquefaction unit and storage facility were constructed at Lake Charles, Louisiana. A 5100-ton dry-cargo ship *(Methane Pioneer)* was converted to carry 2100 tons of LNG. The ship delivered its first of seven LNG cargoes at Canvey Island, in the Thames Estuary, in early 1959. At that point, the LNG was stored, regasified, reformed, mixed with manufactured gas, and distributed

TABLE 1.   Representative LNG Plants of the Early 1970s

| Location | Number of modules | Million standard cubic feet/day | | Refrigeration cycle used |
		Nominal capacity per module	Nominal plant capacity	
Arzew, Algeria	3*	....	150	Standard cascade
Kenai, Alaska	3*	....	140	Standard cascade
Marsa el Brega, Libya	4	86.3	345	MRC (mixed refrigerant cascade), single-pressure
Skikda, Algeria	3	150	450	MRC, two-pressure
Brunei, Borneo	5	140–150	>700	Propane–MCR

*Not entirely modular.

*Doctorate Proposal of W. Purteous, Massachusetts Institute of Technology, Cambridge, Mass., 1973.

by the London Gas Board. Voyages of the *Methane Pioneer,* costing about $11 million, created worldwide interest in the potential of overseas shipping of LNG.

There followed a decision by the British to import LNG from North Africa and of Gaz de France to import LNG from North Africa as well. In making its decision, the British Gas Council at the time compared the cost of LNG (before reforming) of $0.87 per million cubic feet, to oil gas (range of $1.10 to $1.50), to water gas (range of $1.10 to $1.75), to coal carbonization (range of $1.40 to $1.75).

There is much similarity among cryogenic gas processes. See also article on "Chemical Feedstocks from Natural Gas" later in this section of the Handbook.

Among the first of the latest vintage of LNG plants were the Brunei plant for exporting LNG to Japan; and the first of a series of Sonatrach's Skikda, Algeria, complex for shipping LNG to France and elsewhere. These plants are large and far from fabrication facilities and thus could not be constructed as single units because of limitations imposed on components by manufacturing and transport facilities. The principal components are the heat exchangers, the refrigeration compressors and the drivers for the latter. Large LNG projects require the subdivision into and the parallel assembly of several units, which are often referred to as liquefaction modules or strings. According to Kniel,* the characteristics of these base load plants are given in Table 1.

LNG plants utilize one of three basic processes, as illustrated by Figs. 1, 2, and 3. The plants at Arzew and Kenai use a standard cascade with three refrigerants (methane, ethylene, and propane), all circulating in closed cycles. As shown by Fig. 1, this cycle requires three separate compressors for the aforementioned refrigerants. In the interest of economy, each subcycle must be divided into several stages. Thus, compressors designed for side loads are required.

Methane and propane are available from the feed gas. Ethylene as a working fluid in a

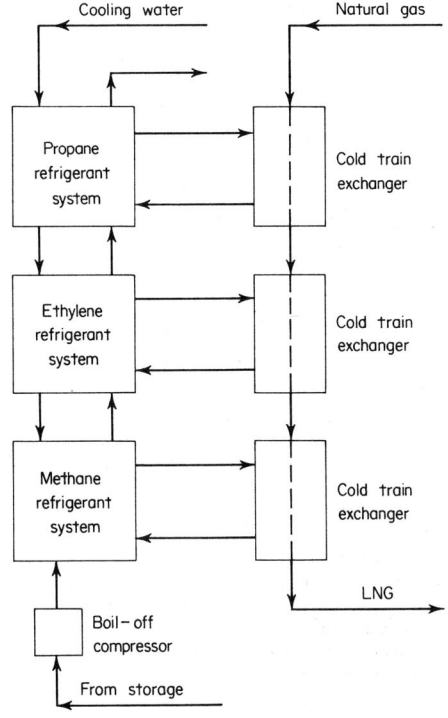

Fig. 1. Standard cascade cycle used in LNG plants.

*Energy Systems for LNG Plants, by L. Kniel, *Chem. Eng. Prog.* **69** (10): 77–84, October 1973.

plant where none exists has not proved to be a handicap. Experience has shown that by good housekeeping refrigerant losses can be held to a point where replenishment is required only infrequently. Ethane also may be used in place of ethylene, at a subatmospheric suction pressure. It is generally acknowledged that the cascade process ranks highest in thermal efficiency.

In the early 1960s, the mixed refrigerant process was developed with the aim of

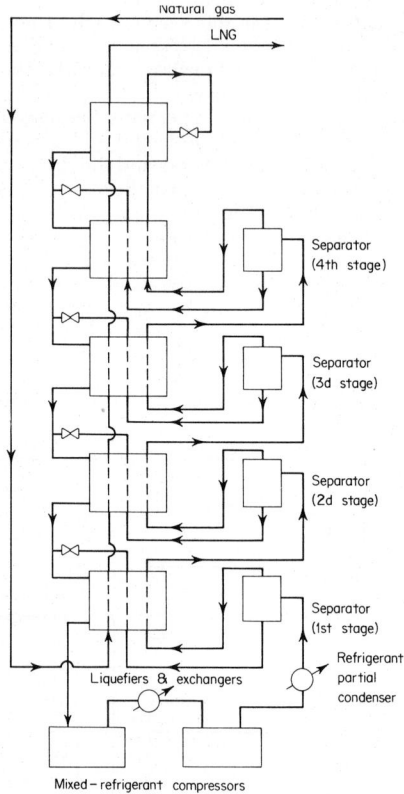

Fig. 2. Single-pressure MRC cycle used in LNG plants. *(U.S. Patent No. 3,593,535.)*

improving on the cascade. It was reasoned that the use of a single multicomponent working fluid would reduce the number of refrigeration compressors and thus reduce plant costs. In principle, only one compressor-drive set would be needed. However, the cost advantage of this design had to be weighed against the potential disadvantages of large, yet unproved, compressors and drivers, and the complexities of control which could put obstacles in the way of realizing a surprisingly simple concept. The single-pressure MRC (mixed refrigerant cascade) is in operation in some plants, as shown in Fig. 2. This process makes use of a hydrocarbon-plus-nitrogen mixture of rather wide boiling range (N_2 through C_5) as the refrigerant. All components can be recovered from the natural gas in separate equipment. By successive partial condensation of the refrigerant mixture, first against water, then against itself; separating liquefied fractions of consecutively lower boiling points and recombining these fractions at proper points on the cooling side of the liquefaction train; an evaporation curve (heat content versus temperature) can be obtained which approaches closely a particular LNG cooling curve and thus minimizes transfer irreversibilities.

The sum of the irreversibilities over the cooling curve or the irreversible work due to the finite temperature differences may be obtained by an integration

$$W = T_0 \int_{T_2}^{T_1} \frac{dq \times \Delta T}{T^2}$$

where T may be taken as the arithmetic mean of the temperatures of the hot and cold media at any point of the cooling curve, and T_0 represents the absolute temperature of the surroundings.* The irreversible work over the refrigerant condenser may be calculated in the same manner.

The two-pressure process is a variant of this process. In this technique the liquefaction load is divided vertically (in reference to a vertical temperature scale) over two liquefiers; one to stepwise condense the refrigerant mixture, and the second to condense and

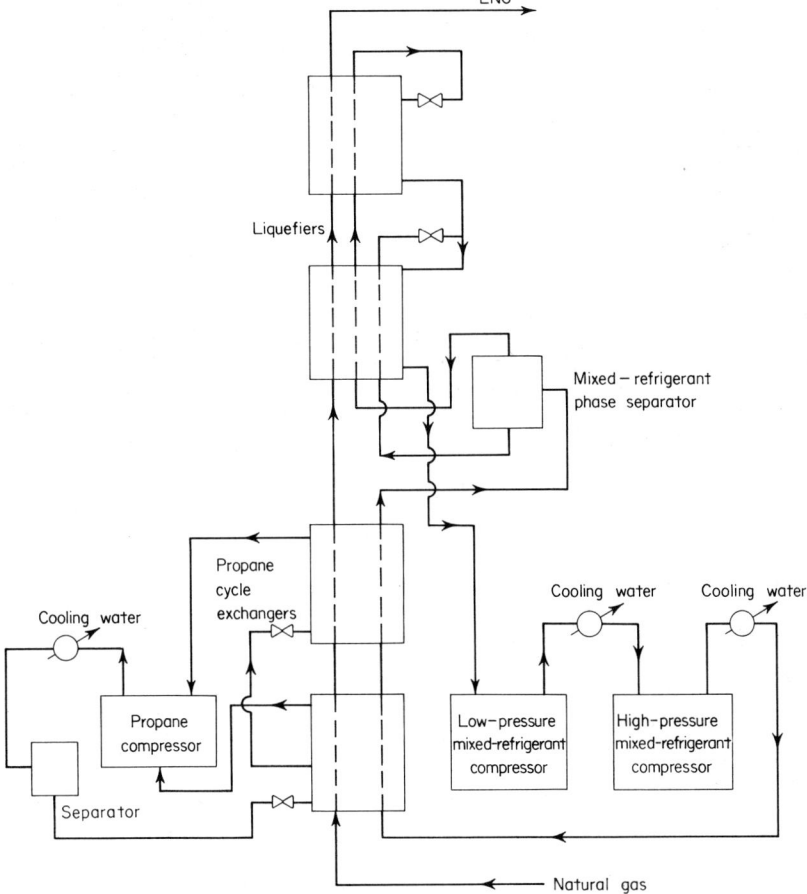

Fig. 3. Propane-MRC cycle used in LNG plants.

*The highlights of this analysis were extracted from aforementioned reference (L. Kniel) with permission.

subcool the natural gas. This has the effect of reducing liquefier dimensions so that larger module sizes can be used. The two-pressure process also shows a slightly better thermo-dynamic efficiency than the single-pressure process. Both the advantages and disadvantages of this approach are pointed out by Kniel in the aforementioned reference.

A third process makes use of a propane and mixture refrigerant cycle combination. See Fig. 3. In this propane-MR process, the cooling load is divided horizontally at about $-30°F$ into an upper portion absorbed by propane and a lower portion absorbed by the mixed refrigerant. As in the two-pressure MRC cycle, this division results in a reduction of the liquefier size.

The propane-MR design cannot be classified as a mixture refrigerant process in the sense that the term has come to be used, but it represents a dual refrigerant cascade in which the lower-boiling fluid is a mixture refrigerant. The cascade combination with propane makes it possible to reduce the boiling range of the mixture refrigerant substantially, which improves the thermodynamic efficiency over that of the straight MRC process. The propane cycle cools the natural gas to $-30°F$ and desuperheats (except for superheat passed directly to cooling water) and partially condenses the mixture refrigerant. The latter takes care of cooling needs from -30 to $-262°F$ without any additional partial condensation stages (as provided in the straight MRC process), which greatly improves operational suitability and control.

Thermodynamic Efficiencies. The cycle efficiency is

$$E = \frac{(AL)_{rev}}{(AL)_{act}}$$

The subscripts designate reversible and actual work. The reversible or Carnot work for a single component gas may be calculated from

$$(AL)_{rev} = T_0(S_1 - S_2) - (h_1 - h_2)$$

The subscripts 1 and 2 refer to terminal conditions, and T_0 is the absolute temperature of the surroundings. The reversible work may also be obtained from a Mollier h-s diagram by a construction as shown by Fig. 4. The reversible work of liquefaction versus inlet pressure for methane at $t_0 = 90°F$ is indicated in Fig. 5. For hydrocarbon mixtures that are apt to deviate from the behavior of methane, the calculation becomes more involved. [See M. G. Zellner et al., *IEC Fund.*, **9**, 549 (1970).]

The reversible work depends also on the gas density. This dependence is shown in Fig. 6. When taken together, Figs. 5 and 6 permit a reasonably accurate estimate of the reversible work for gas mixtures.

The entropy increase furnishes an indication of the work lost in the process as the reversible work produces no change of entropy. Thus:

$$(AL)_{act} = (AL)_{rev} + (AL)_1$$

and

$$(AL)_1 = T_0 \times \Delta S$$

where ΔS refers to an envelope that includes the environment.

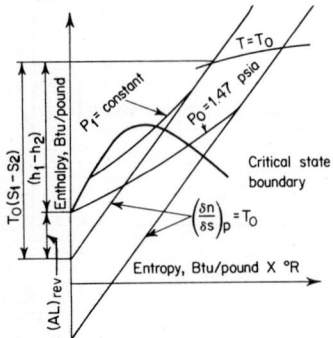

Fig. 4. Graphical construction of reversible work $(AL)_{rev}$.

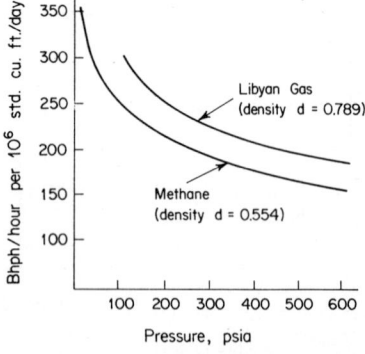

Fig. 5. Reversible work of liquefaction, $t_0 = 90°F$.

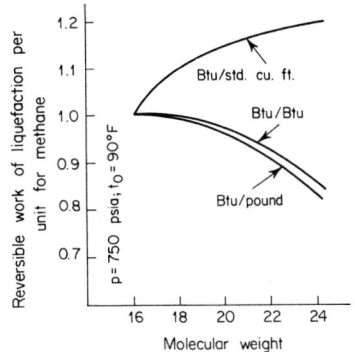

Fig. 6. Reversible work of liquefaction versus density.

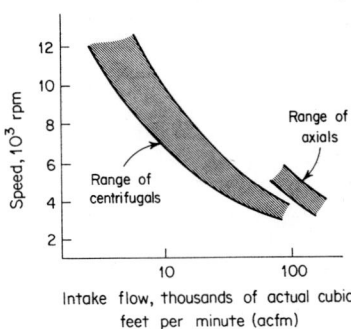

Fig. 7. Compressor capacities.

TABLE 2. Carnot Cycle Efficiencies

| Parameter | Type of process | | |
	Standard cascade	MRC, single-pressure	Propane–MRC
Inert gas:			
N_2 content, vol %	1.68	1.20	0.20
C_4+content, vol %	0.00	4.70	1.40
Molecular weight	17.42	22.73	19.13
Inlet pressure, psia	625	620	735
Carnot cycle efficiency, %	40.6	29.0	34.1

MRC = mixed refrigerant cascade.

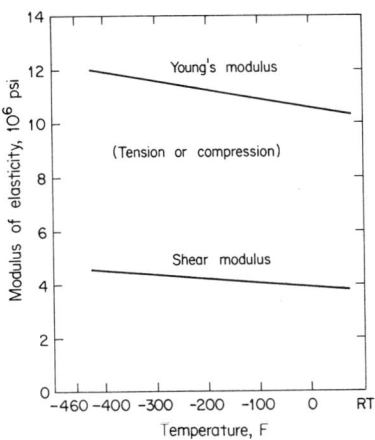

Fig. 8. Moduli of elasticity of aluminum alloy 5083 at low temperatures. (*Aluminum Company of America*)

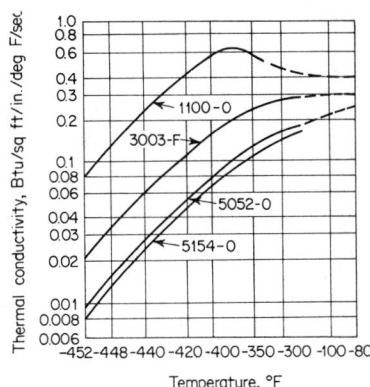

Fig. 9. Thermal conductivities of some aluminum alloys. (*Data from W. J. Hall et al., Thermal Conductivities of Common Commercial Aluminum Alloys, Nat. Bur. Stand. Proc. 1957 Cryog. Eng. Conf.*)

TABLE 3. Refrigerant Flows for 250 Million Standard Cubic Feet/Day Module

Number	Type of refrigeration cycle	Fluid	Type of compressor	Suction volume, acfm	Power rating, bhp	Suction pressure, psia
....	Standard cascade	Propane	Centrifugal	94,000	42,000	31.6
		Ethylene	Centrifugal	67,000	47,600	26.6
		Methane	Centrifugal	13,500	18,400	16.0
3	MRC, single-pressure	MR, l.p.	Axial	288,000	89,700	28.0
		MR, h.p.	Axial	......	86,300	...
5	Propane–MRC	Propane	Centrifugal	103,000	49,000	16.0
		MR, l.p.	Centrifugal or axial	110,000	43,600	37.7
		MR, h.p.	Centrifugal or axial	....	45,000	...

TABLE 4.   Nominal Chemical Composition of Aluminum Alloys Frequently Used in Cryogenic Equipment

Alloy designation	Percent					
	Silicon	Copper	Manganese	Magnesium	Chromium	Zinc
2014	0.8	4.4	0.8	0.50		
2021*		6.3	0.3			
2219†		6.3	0.30			
3003		0.12	1.2			
5052				2.5	0.25	
5083			0.6	4.45	0.15	
5086			0.45	4.0	0.15	
5154				3.5	0.25	
5454			0.8	2.7	0.12	
5456			0.8	5.1	0.12	
6061	0.6	0.27		1.0	0.20	
6063	0.40			0.7		
6351	1.00		0.6	0.6		
7005‡			0.45	1.4	0.13	4.6

*Titanium 0.06; vanadium 0.10; zirconium 0.18; cadmium 0.15; tin 0.05
†Titanium 0.06; vanadium 0.10; zirconium 0.18
‡Titanium 0.04; zirconium 0.13
NOTE: Aluminum and normal impurities constitute remainder.

The reversible work plus the losses shown by the foregoing equation in caloric units must match the actual work input to the compressors.

Carnot cycle efficiencies for the three basic processes are summarized in Table 2.

Refrigerant flows for a 250 million standard cubic feet per day LNG module are given in Table 3. The range of available compressor capacities is summarized in Fig. 7.

MATERIALS OF CONSTRUCTION

Aluminum alloys have been widely used in cryogenic services, ranging from tanks and vessel mobile Dewars to trailers and marine tankers, welded heat exchanger shells, and rolled and welded pipe. The nominal chemical composition of some of the most popular aluminum alloys for these services are given in Table 4. Design stress for aluminum alloys, welded construction, is given in Table 5. The results of tests of representative lots of aluminum alloys for cryogenic service are given in Table 6. Additional physical properties of some aluminum alloys for cryogenic applications are charted in Figs. 8 through 11.

Nine percent nickel steel, available in the form of plates, forgings, pipe, tubing, bars, sheets, and rolled sections, when properly heat treated, provide a useful combination of high strength and resistance to brittle fracture at low temperature. The chemical composi-

TABLE 5.   Design Stress for Aluminum Alloys, Welded Construction*

Alloy designation	ASME design stress, 100°F†
3003	3,350
5052	6,250
5154	7,350
5086	8,700
5083	10,000
5456	10,500
6061-T6	6,000
6063-T6	4,250

*SOURCE: ASME Boiler and Pressure Vessel Code, section VIII, division 1.
†In some cases these vary with thickness.

TABLE 6. Results of Tests of Representative Lots of Aluminum Alloys for Cryogenic Service

Alloy and temper	Temperature F	Tensile strength psi	Yield strength psi	Elongation in 2 in. or 4 D %	Reduction of area %	Notch tensile strength psi	NTS/TS	NTS/YS	Weldability (inert gas arc)	Applicability
2014-T651	RT	69,000	63,500	10.2	24	82,800	1.20	1.30	weldable with special techniques	aerospace—Saturn V propellant tanks (LH2 and LOX); high strength
	—320	84,000	76,100	12.0	22	99,600	1.19	1.31		
	—423	95,600	80,200	15.0	23	103,000	1.08	1.28		
	—452	95,600	81,800	12.5	19	102,200	1.07	1.25		
2021-T8151	RT	74,400	66,600	8	8	84,600	1.14	1.27	readily weldable	aerospace—high strength; more weldable than 2014
	—112	79,400	69,500	9.5	17	95,800	1.21	1.38		
	—320	90,200	78,900	11.0	19	106,100	1.18	1.34		
	—423	101,600	86,100	12.5	20	116,400	1.15	1.35		
2219-T851	RT	67,600	53,800	11.0	27	79,400	1.12	1.48	readily weldable	aerospace—tougher than 2014 and more weldable
	—112	71,400	57,600	11.5	28	84,300	1.18	1.46		
	—320	82,500	63,800	13.8	30	94,500	1.15	1.48		
	—423	95,600	68,800	16.0	28	103,500	1.08	1.50		
	—452	96,100	69,000	14.0	26	102,100	1.06	1.48		
3003-H14	RT	22,900	21,100	16.8	68	—	—	—	excellent	pipe and tube brazed heat exchangers
	—18	24,000	21,900	15.0	58	—	—	—		
	—112	25,300	22,300	18.5	59	—	—	—		
	—320	36,600	25,900	32.5	56	—	—	—		
	—452	58,100	30,100	32.0	49	65,100	1.12	2.16		
5052-0	RT	28,000	13,000	30		not available			good	See 5083-0
	—18	28,000	13,000	32						
	—112	29,000	13,000	35						
	—320	44,000	16,000	46						
5083-0	RT	46,800	20,400	19.5	26	54,000	1.16	2.65	excellent	Cryogenic tanks & vessel mobile Dewars; trailers and marine tankers; welded heat exchanger shells; rolled & welded pipe
	—320	63,000	23,000	34.0	34	61,000	0.97	2.65		
	—452	80,800	25,800	32.0	33	62,300	0.77	2.42		
5083-H321	RT	48,600	34,100	15.0	23	61,100	1.26	1.80	excellent	same as 5083-0
	—320	66,100	39,700	31.5	33	70,400	1.06	1.77		
	—423	90,000	41,800	30.0	24	72,800	0.81	1.74		
	—452	85,800	40,500	29.0	33	73,700	0.86	1.82		

Alloy	Temp								Weldability	Applications
5086-0	RT	38,000	17,000	30						
	−18	38,000	17,000	32						
	−112	39,000	17,000	35		not available			good	same as 5083-0
	−320	55,000	19,000	46						
5454-0	RT	35,800	16,700	24.5	48	48,100	1.34	2.88		
	−18	37,200	16,600	26.5	57	—	—	—		
	−112	38,200	16,800	29.5	57	—	1.11	3.04	good	same as 5083-0
	−320	54,300	19,400	39.5	49	60,200	0.89	2.72		
	−452	73,900	24,100	34.3	35	65,600				
5454-H32	RT	40,900	28,900	15.7	32	56,200	1.37	1.94		
	−18	42,200	28,800	19.5	44	—	—	—		
	−112	43,700	29,200	23.0	48	—	1.13	2.00	good	same as 5083-0
	−320	61,100	34,500	32.0	40	69,200	0.94	1.97		
	−452	82,300	39,400	28.6	31	77,700				
5456-0	RT	49,000	23,200	21.8	31	50,900	1.04	2.19		
	−18	49,000	23,200	24.0	35	—	—	—		
	−112	49,100	23,600	26.5	43	—	0.89	2.29	good	same as 5083-0
	−320	66,000	26,100	34.5	35	59,600	0.72	2.02		
	−452	84,400	29,500	30.7	24	60,900				
6061-T6	RT	44,900	42,200	16.5	50	69,200	1.54	1.64		pipe, pipe fittings, tube
	−320	58,300	48,900	23.0	48	83,400	1.43	1.71	good	extrusions
	−452	70,100	55,000	25.5	42	89,900	1.28	1.63		
6063-T6	RT	35,000	31,000	18						
	−18	36,000	32,000	19		not available			good	pipe, tube extrusions
	−112	38,000	33,000	20						
	−320	47,000	36,000	24						
6351-T5	RT	45,000	41,000	11						
	−18	48,000	43,000	11		not available			good	pipe, tube extrusions
	−112	51,000	44,000	11						
	−320	59,000	47,000	17						
7005-T5351	RT	62,000	55,000	15.0	43	86,200	1.39	1.59		best combination of strength
	−112	67,800	58,600	14.0	30	92,100	1.36	1.57		and toughness in 7xxx series
	−320	83,900	67,500	17.0	27	99,100	1.18	1.47	good	
	−423	102,500	73,400	18.5	28	107,500	1.05	1.46		
	−452	97,600	75,600	17.0	22	106,900	1.09	1.41		

*General Information—refer to applicable design and construction codes for equipment design data and procedures.

SOURCE: Aluminum Company of America.

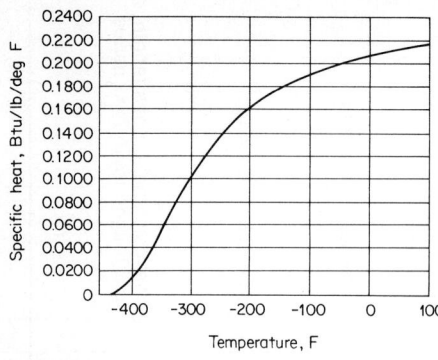

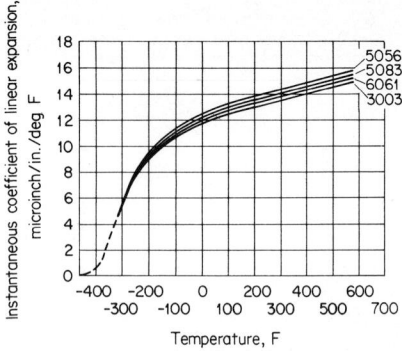

Fig. 10. Specific heat of aluminum and most of its alloys.

Fig. 11. Coefficients of expansion of some aluminum alloys. *(Aluminum Company of America.)*

TABLE 7. Chemical Compositions (ASTM Standards) of 9% Nickel Steel for Low Temperature Service

Chemical element	Plates	Other configurations
Carbon	0.13 max.	0.13 max.
Manganese	0.90 max.	0.90 max.
Phosphorus	0.035 max.	0.040 max.
Sulfur	0.040 max.	0.040 max.
Silicon	0.15–0.30	0.15–0.30
Nickel	8.50–9.50	8.50–9.50

TABLE 8. Compact Tension Fracture Toughness Evaluation of Quenched and Tempered 9% Nickel Steel at −260 and −320°F

Temperature, °F	K_q, ksi $\sqrt{\text{in.}}$	$K1st$ pop, ksi $\sqrt{\text{in.}}$	Km, ksi $\sqrt{\text{in.}}$
−320	129.0	136.0	—
−320	133.0	—	197
−260	98.8	—	194
−260	139.0	—	185

SOURCE: The International Nickel Company, Inc.

TABLE 9. Fracture Toughness Test Results of Quenched and Tempered 9% Nickel Steel in 3-Inch Plate Thicknesses at 75, −100, and −321°F

Temperature, °F	0.2% yield strength, ksi	K_Q,* ksi $\sqrt{\text{in.}}$
75	92.3	95.0
		101.5
		100.5
		101.4
−100	96.6	99.8
		98.5
		96.3
−321	124.5	114.1
		100.8
		106.5

SOURCE: The International Nickel Company, Inc.
*Conditional plane stain fracture toughness.

TABLE 10. Effect of Low Temperature Cycling and Tensile Stress on Impact Properties of Double Normalized and Tempered 9% Nickel Steel

Applied tensile stress, psi	No. of cooling cycles, 68 to −320°F	Charpy impact (V-notch)			
		At 68°F		At −320°F	
		Energy absorbed, ft-lb	Fibrous fracture, %	Energy absorbed, ft-lb	Fibrous fracture, %
0	0	111	100	44, 48	63, 67
0	20	110	100	48, 52	67, 68
22, 250*	20	109	100	46, 52	67, 68

SOURCE: The International Nickel Company, Inc.
*89% of allowable design stress.

TABLE 11. Effect of Temperature on Hardness of Quenched and Tempered 9% Nickel Steel

Temperature, °F	Hardness	
	Diamond Pyramid	Rockwell C*
75	255	23
0	265	25
−100	285	28
−200	325	33
−300	367	37
−320	375	38

SOURCE: The International Nickel Company, Inc.
*Converted from Diamond Pyramid Hardness measurements.

tions of these materials, according to ASTM standards, are summarized in Table 7. Selected physical properties of 9 percent nickel steel are summarized in Tables 8 through 14.

REFERENCES

Kniel, L.: Energy Systems for LNG Plants, *Chem. Eng. Prog.* **69** (10): 77–84, October 1973.
French Patent 2,074,594, to Technip.
U.S. Patent 3,593,535, to Air Products Co.
Schlatter, R. G., et al.: LNG III, Section II, paper no. 3, Washington, September 1972.

TABLE 12. Thermal Expansion of Double Normalized and Tempered 9% Nickel Steel

Temperature interval, °F	Mean coefficient of linear expansion per °F	Temperature interval, °F	Mean temperature, °F	Instantaneous coefficient of linear expansion per °F
−320 to 70	4.9×10^{-6}	−320 to −200	−260	3.4×10^{-6}
−200 to 70	5.5	−200 to −100	−150	5.0
−100 to 70	5.9	−100 to 0	−50	5.7
−100 to 200	6.2	0 to 100	50	6.2
70 to 200	6.5	100 to 200	150	6.5
70 to 300	6.6	200 to 300	250	6.8
70 to 400	6.7	300 to 400	350	7.0
70 to 500	6.8	400 to 500	450	7.3
70 to 600	6.9	500 to 600	550	7.5
70 to 700	7.1	600 to 700	650	7.7
70 to 800	7.2	700 to 800	750	7.9
70 to 900	7.3	800 to 900	850	8.1
70 to 1000	7.4	900 to 1000	950	8.2

SOURCE: The International Nickel Company, Inc.

TABLE 13.   Thermal Conductivity of Double Normalized and Tempered 9% Nickel Steel

Mean temperature		Thermal conductivity	
°F	°C	(Btu)(in.)/(hr)(sq ft)(°F)	(cal)(cm)/(sec)(sq cm)(°C)
−320	−196	91	0.031
−243	−153	134	0.046
−189	−123	149	0.051
−99	−73	172	0.060
−9	−23	190	0.066
81	27	203	0.071
261	127	226	0.078
441	227	236	0.082
621	327	238	0.082
801	427	232	0.080
981	527	224	0.077

SOURCE: The International Nickel Company, Inc.

Zellner, M. G., et al.: *IEC Fund.*, **9**, 549 (1970).
Kinard, G. E., et al.: *Chem. Eng. Prog.*, **69**, 56 (1973).
Gaumer, L. S., et al.: LNG III, Section II, paper no. 15, Washington, September 1972.
Barrington, E. A., A.I.Ch.E. 74th National Meeting, paper no. 86b, New Orleans, La., March 1973.
O'Neill, P. S., et al.: LNG III, Section II, paper no. 4, Washington, September 1972.
Barber, H. W., and S. J. Orloski: *Pet. and Petrochem. Int.*, **12**, 8, 46–48 (1972).
Hallock, D. C., et al.: A.I.Ch.E. 71st National Meeting, paper no. 10a, Dallas, Tex., February 1972.
Bourguet, J. M., et al.: LNG II, Section 2, paper no. 2, Paris, France, October 1970.
Vrancken, P. L. L., *Pet. and Petrochem. Int.*, **12** (8): 26–31 (1972).
Material Specifications, Part A—Ferrous, ASME Boiler and Pressure Vessel Code, Section II, 1971.
Rules for Construction of Pressure Vessels, ASME Boiler and Pressure Vessel Code, Section VIII, Division 1, 1971.
Case 1499 (Special Ruling), Improved Stress Value for Welded 8% and 9% Nickel Steel, Section VIII, Division 1, Interpretations of ASME Boiler and Pressure Vessel Code, 1971.
Low-Pressure Storage Tanks for Liquefied Hydrocarbon Gases, Appendix Q, API Standard 620, 5th ed., 1973.
Studies of 9% Nickel Steel for Moss-Type Liquefied Natural Gas Carriers, Cat. No. EXE 338, Nippon Steel Corp., Tokyo, Japan, March 1973.
Vishnevsky, C., and E. A. Steigerwald: Plane Strain Fracture Toughness of Some Cryogenic Materials at Room and Subzero Temperatures, ASTM STP, 496, 1971.
Tenge, P., and O. Solli: 9% Nickel Steel in Large Spherical Tanks for Moss-Rosenberg 87600 M³ *LNG Carrier. A Fracture Mechanical Approach to Testing and Design*, International Institute of Welding, Annual Assembly, 1971.
Tenge, P., and O. Solli: Fracture Mechanics in the Design of Large Spherical Tanks for Ship Transport of LNG, *Norw. Marit. Res.*, **1**, 2, (1973).
Tharby, R. H., D. J. Heath, and J. W. Flannery: The Welding of 9% Nickel Steel—A Review of the Current Practices, for presentation at the Welding Institute's International Conference, Welding Low Temperature Containment Plant, London, England, Nov. 20–22, 1973.
Operation Cryogenics, The International Nickel Company, Inc., Chicago Bridge and Iron Co., and United States Steel Corp., 1961.
Kaufman, J. G., and M. Holt: Fracture Characteristics of Aluminum Alloys, *ALCOA Res. Lab. Tech. Pap.* 18 (1965).

TABLE 14.   Specific Heat of Double Normalized and Tempered 9% Nickel Steel

Temperature range, °F	Mean specific heat, Btu/(lb)(°F)
−320 to 80	0.0878
80 to 700	0.119
80 to 1000	0.131
700 to 1000	0.156

SOURCE: The International Nickel Company, Inc.

Martin, H. L., P. C. Miller, A. G. Imgram, and J. E. Campbell: Effects of Low Temperatures on the Mechanical Properties of Structural Metals, National Aeronautics and Space Administration, 1968.

Kaufman, J. G., M. Holt, and E. T. Wanderer: Aluminum Alloys for Cryogenic Temperatures, Aluminum Company of America, 1967.

Kaufman, J. G., K. O. Bogardus, and E. T. Wanderer: Tensile Properties and Notch Touchness of Aluminum Alloys at −452°F in Liquid Helium, *Adv. Cryogenic Eng.*, **13**, 1968.

Van Horn, K. R. (ed.): Aluminum, 3 vols. American Society for Metals, Metals Park, Ohio, 1967.

Elliott, M. A., C. W. Seibel, F. W. Brown, R. T. Artz and L. B. Berger: Report on the Investigation of the Fire at the Liquefaction, Storage, and Regasification Plant of the East Ohio Gas Co., Cleveland, Oct. 20, 1944, *U.S. Bur. Mines Rep.* R.I.3867, February 1946.

Zabetakis, M. G.: Safety with Cryogenic Fluids, Plenum Press, New York, 1967.

Cryogenic Upgrading of Low-Btu Gases

By J. KEVIN JONES*

Worldwide, there are substantial reserves of natural gas in which the reservoir formation hydrocarbons are contaminated with nonburning components. The presence of components, such as helium, nitrogen, or carbon dioxide, reduces the heating value of the gas mixture and can result in the gas being unsuitable for existing transmission and distribution systems. Such contaminated mixtures are termed low-Btu gases if their heating values fall below the minimum standards, regulations, or contract heating-value requirements.

Cryogenic processing can be used to upgrade some of these low-Btu gases so as to produce an acceptable high-Btu product. Cryogenic upgrading is a physical process in which subambient temperatures are employed to bring about a separation between the hydrocarbons and nonhydrocarbons in the mixture. The reduction of temperature occurring during cryogenic processing produces a two-phase (gas-liquid) mixture. The relative volatilities between the components in the mixture result in selective mass transfer between the two phases. One phase becomes enriched with hydrocarbons and then has a heating value higher than the original gas. The second phase becomes denuded of hydrocarbons and, of course, has a heating value below that of the original gas mixture. Frequently, the mass transfer operation requires several theoretical stages in order to achieve the desired product heating value and high hydrocarbon recoveries. Although cryogenic upgrading can be applied to gas mixtures containing carbon dioxide or hydrogen sulfide, it has so far been applied commercially only to those hydrocarbon mixtures contaminated with nitrogen and helium.

Pretreatment of Low-Btu Gas. A main consideration in the design and operation of cryogenic upgrading plants is to identify and remove any component from the gas which could adversely affect the operation of the cold sections of the plant. Such components are carbon dioxide, water vapor, and heavy hydrocarbons which have high solidification temperatures and low solubilities. In general, if these components are allowed to remain in the gas and reach the cryogenic unit, they will form solids during the cooling process. These solids will be deposited on the heat-exchanger surfaces. This will lead to falloff in performance and, possibly, blockages and plant shutdown.

Particular attention should be paid to identifying any high-freezing-point components in the low-Btu gas. There exists a range of absorption and adsorption processes for pretreating the low-Btu gas to remove these undesirable components. Allowable concentrations of impurities in the feed to the cryogenic unit, typically, would be: carbon dioxide 5 vpm (parts per million by volume); water, 1 vpm; pentanes, 30 vpm; and heavier hydrocarbons, 1 vpm.

Typical Process. The simplified flowsheet of a large cryogenic upgrading plant is shown in Fig 1. The plant, consisting of two identical trains, is capable of processing 260 million standard cubic feet per day of low-Btu gas (580 Btu/standard cubic foot) and upgrading this into 143 million standard cubic feet per day of high-Btu gas (980 Btu/standard cubic foot). The plant simultaneously recovers helium. Plant stream parameters are given in Table 1.

The low-Btu gas is available at 800 psig and is mainly a nitrogen-methane mixture. In addition to a small quantity of helium, the gas also contains small quantities of carbon dioxide, water vapor, and heavy hydrocarbons. The carbon dioxide is removed by washing with monoethanolamine (MEA), the water is taken out on molecular sieve, and the heavy hydrocarbons by adsorption on activated carbon. The gas is then cooled in aluminum plate-fin exchangers against the returning high-Btu product gas and vent gas. The gas is then expanded to 380 psig, and a vapor-liquid mixture passes into the high-pressure (HP) fractionator. The purpose of this fractionator is to bring about an initial separation of

*Managing Director, ChemShare Process Systems Ltd., London, England.

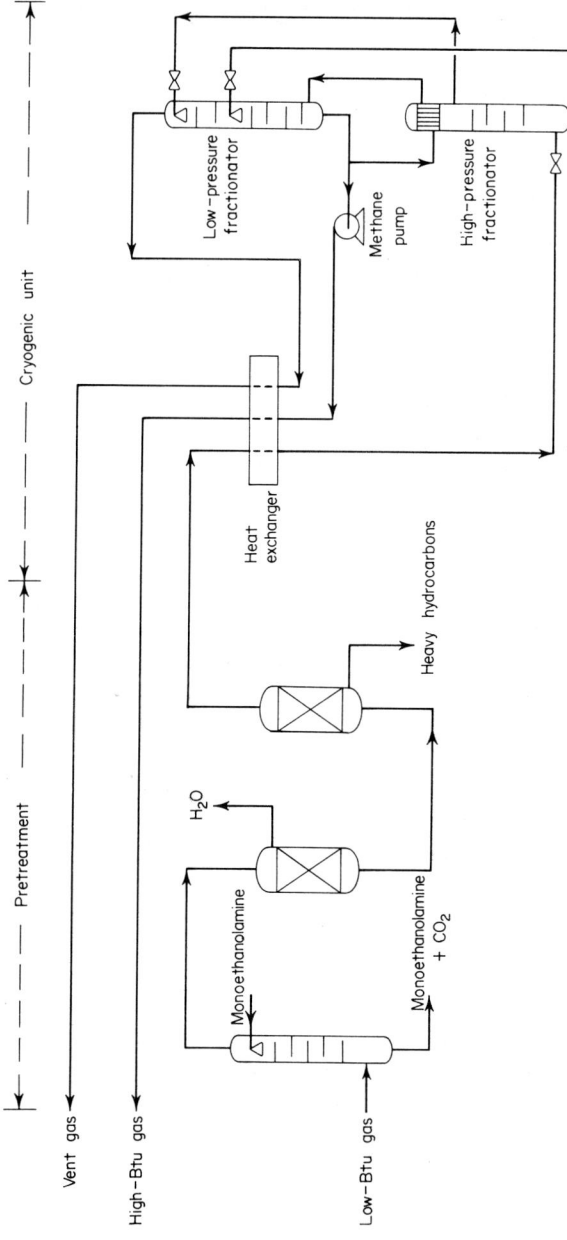

Fig. 1 Plant for nitrogen removal from natural gas using cryogenic upgrading. Plant, located in Poland, built by Petrocarbon Developments Limited, Manchester, England.

TABLE 1.   Cryogenic Upgrading of Low-Btu Gases—Plant Stream Parameters

| | Composition, mole % | | | |
Component	Low-Btu gas	High-Btu gas	Vent gas	Helium
Helium	0.40	—	0.09	100.00
Nitrogen	42.75	4.00	98.95	—
Methane	56.02	95.09	0.96	—
Ethane (plus)	0.53	0.91	—	—
Carbon dioxide	0.30	—	—	—
Flow, millions of standard cubic feet per day	246	143	100	0.43
Heating value, Btu per standard cubic foot	580	980	—	—

the nitrogen-methane and to produce a liquid nitrogen reflux for the low-pressure (LP) fractionator.

A nitrogen-enriched vapor flows up the HP fractionator, while methane is returned to the sump of this column by a nitrogen reflux stream produced in the tubes of the overhead condenser. The refrigeration required to produce this nitrogen reflux is provided by evaporating some of the liquid methane from the LP column in the shell of the overhead condenser.

Two liquid streams are taken from the HP fractionator, and these become the feed and reflux for the LP fractionator. The LP feed is an enriched methane stream taken from the base of the HP fractionator. The LP reflux is a high-purity nitrogen liquid taken off the HP fractionator just below the condenser. The upgrading is completed in the LP fractionator. The feed stream is stripped to produce a high-Btu liquid containing 4 percent nitrogen and having a heating value of 980 Btu/standard cubic foot. The liquid is pumped from the column sump, evaporated, and superheated against the incoming low-Btu gas. The gas from the top of the LP column is mainly nitrogen and is also heated to ambient temperature against the incoming low-Btu gas. By this arrangement of two distillation columns, the separation of nitrogen and methane can be achieved using only the pressure energy available in the low-Btu gas.

A list of cryogenic upgrading plants which have been built to remove nitrogen from natural gas is given in Table 2. Some of these plants also recover helium. Additional smaller-size units (up to 10 million standard cubic feet of low-Btu gas per day) are in the design or installation stage.

Engineering Summary.   As mentioned earlier, the pretreatment of the low-Btu gas needs to be conducted to a much higher degree of impurity removal than in conventional gas processing plants.

Data Prediction.   The successful design and operation of the cryogenic unit itself is dependent upon the accurate prediction of thermodynamic data (phase equilibria and enthalpy) and also transport properties (viscosity, conductivity, and density). The thermodynamic data for these multicomponent systems, operating at high pressure and low temperature, are normally produced from equations of state. The equations of state are usually computerized and cross-correlated against experimental data.

TABLE 2.   Cryogenic Upgrading Plants for Nitrogen Removal

Low-Btu gas rate, millions of standard cubic feet per day	Nitrogen in gas, mole %	High-Btu flow rate, millions of standard cubic feet per day	High-Btu heat value, Btu/standard cubic foot	Helium recovery	Year built
850	9–14	760	950	Yes	1963
168	15–30	146	940	Yes	1968
120	14	50	980	Yes	1968
12	60	4	950	No	1971
5	14	2	800	No	1971
260	43	143	980	Yes	1974

Fractionation Column Hydraulics.   When high-pressure fractionation columns are used on cryogenic upgrading plants, they operate in a different regime as compared with those conventionally used. The low temperatures and high pressures involved result in small density differences between the vapor and liquid phases (of the order of 20 pounds/cubic foot) and in low surface tension values (0.5 dyne/centimeter). The column hydraulic design needs to take these factors into account.

Materials of Construction.   Dependent on the temperature levels involved, different materials can be used. Below $-55°F$ (about $-43°C$), the most common materials are aluminum, stainless steel, and 9 percent nickel steel.

Insulation.   All the equipment operating below ambient temperature is normally insulated to conserve refrigeration. On smaller plants, the cold equipment is usually installed inside a cold box. The cold box, which is a steel enclosure, is then filled with a powdered insulant such as expanded perlite, and purged with nitrogen. On larger plants, the major items, such as columns, will be free-standing and insulated with polyurethane and then clad with aluminum sheeting.

FEATURES OF PROCESS

The cryogenic upgrading of low-Btu gas to produce a high-heating-value product results in: (1) the gas being more readily utilized; (2) the elimination of impurities which improves (a) the gas-burning characteristics and (b) the transport properties (high-Btu gas will have a lower density and viscosity than low-Btu gas), resulting in less horsepower and compression equipment for transmission; (3) increased thermal delivery of pipelines previously transporting low-Btu gases, or smaller pipe diameter for a required thermal delivery; (4) allowance for the extraction of heavier hydrocarbons and natural gas liquids; and (5) higher hydrocarbon partial pressures when the gas is used in chemical reactions.

Cryogenic upgrading plants can be designed (1) to accept a wide variation of gas composition and flow rate; (2) to produce a controlled, high-heating-value product (1000 Btu/standard cubic foot) with a high hydrocarbon recovery (up to 99 percent); (3) to recover natural gas liquids as an integral part of the process; (4) to produce liquefied natural gas (LNG) for peak shaving or for distribution to satellite stations and beyond-the-mains service; and (5) to produce liquid cryogens, such as liquid nitrogen.

REFERENCES

Mullins, P. V., and R. W. Wilson: Prospective Methods and Estimated Costs for Removing Excess Nitrogen from Natural Gas, Project NGD-6, American Gas Association, Arlington, Va., 1952.

Jones, J. K.: The Cryogenic Upgrading of Low-BTU Gases Containing Air, Carbon Dioxide and Nitrogen, Gas Producers Association Annual Conference, Denver, Colo., 1974.

Jones, J. K.: The Upgrading of Low-BTU Natural Gas, *Oil Gas Jr.*, July 2, 1973.

Methanol Production and Fuel Methanol vs. LNG

By G. C. HUMPHREYS*

Growing domestic shortages of natural gas and fossil fuels and environmental controls in the highly industrialized nations of the world led in recent years to a renewal of interest in the use of methanol (methyl alcohol, wood alcohol) as a clean-burning industrial fuel, as a gasoline additive (or substitute), for the production of substitute natural gas (SNG), and as a gas turbine fuel. Other possible applications include single-cell protein synthesis, fuel cells, and sewage treatment. See also descriptions of methanol in the *Chemical Fuels Technology* section of this Handbook. In particular, there is much current interest in the economic and technical comparisons of fuel methanol with liquefied natural gas (LNG) (Ref. 1).

Historically, methanol was used in the United States and Western Europe for many years for domestic heating, lighting, and cooking. Eventually, kerosine supplanted methanol in these domestic applications. Then, methanol was produced by the destructive distillation of wood in forest areas and was transported to city users, the transportation of methanol being less costly than the transportation of timber. Perhaps to a degree history may repeat itself when one considers the analogous situation of transporting LNG over the oceans against the cheaper transportation costs for methanol.

METHANOL PRODUCTION

Methanol today can be produced from a large range of feedstocks by a variety of processes. Natural gas, liquefied petroleum gas (LPG), naphthas, residual oils, asphalt, oil shale, and coal are in the forefront as feedstocks for producing methanol, with wood and waste products from farms and municipalities possible additional feedstock sources.

In order to synthesize methanol the main feedstocks are converted to a mixture of hydrogen and carbon oxides (synthesis gas) by steam reforming, partial oxidation, or gasification.

The hydrogen and carbon oxides are then converted to methanol over a catalyst:

$$2H_2 + CO \rightleftharpoons CH_3OH$$
$$3H_2 + CO_2 \rightleftharpoons CH_3OH + H_2O$$

Synthesis Gas Production. The major part of methanol produced today comes from plants where synthesis gas is produced by steam reforming of hydrocarbons (natural gas, LPG, or light naphthas) over a nickel catalyst at elevated temperatures (up to 900°C, 1650°F).

The synthesis gas is produced by the following mechanisms:

$$C_xH_y + xH_2O \rightleftharpoons \left(x + \frac{y}{2}\right)H_2 + xCO$$
$$CO + H_2O \rightleftharpoons CO_2 + H_2$$

The proportions of H_2, CO, and CO_2 in the synthesis gas are dependent on the approach to equilibrium achieved from the reforming furnace, the reforming temperature and pressure, and the steam-to-hydrocarbon ratio utilized. Some residual methane will also be present in the synthesis gas.

With the present tendency toward shortages of the feedstocks that can be handled by steam reforming, it is likely that synthesis gas will be provided by partial oxidation of heavier oils or by gasification of coal. Again the synthesis gas will consist of H_2, CO, and CO_2 with some residual methane.

Methanol Synthesis Processes. Large-scale synthesis of methanol began by utilizing

*Engineering Staff, Davy Powergas Ltd., London.

catalysts based on zinc/chromium mixtures. The activity of such catalysts requires high synthesis pressure (up to 5000 psi) to be used in order to achieve economic conversion of synthesis gas to methanol. Operating temperatures in the range 305 to 400°C, 580 to 750°F are necessary.

The effect of pressure on conversion of synthesis gas to methanol over the zinc/chromium catalyst is illustrated by the following data 1): At 5000 psi a conversion of approximately 5.5 percent is achieved, at 3000 psi approximately 2.2 percent, and at 750 psi not enough methanol is synthesized for condensation.

Licensed processes are available which utilize this high-pressure route to methanol using zinc/chromium catalysts or zinc/chromium with added copper (Ref. 2).

A flow diagram of a typical high-pressure process is shown in Fig. 1.

In 1966, a different type of methanol process* was announced. The process centers around a copper catalyst. Historically, the favorable properties that copper possesses for catalytically promoting synthesis of methanol from hydrogen and carbon oxides have been well known. However, copper is very susceptible to poisoning by sulfur, chlorides, and other materials that may be present. Early tests on catalysts showed unacceptably short catalyst lives because of the impure sources of synthesis gas then available.

The breakthrough in terms of being able to utilize copper-based catalysts occurred when steam/hydrocarbon reformers were developed for which very low levels of poison in feedstock were essential for adequate operation of the reforming catalyst. Synthesis gas from the reformers thus contained minute poison levels and, as such, proved ideal for the operation of a commercial copper catalyst suitable for methanol synthesis.

The activity of the ICI copper catalyst is such that synthesis can be conducted at low pressure (down to 750 psi) and temperature (~245°C, 475°F) and still obtain satisfactory yields of methanol (Ref. 3). Other firms in addition to ICI have now developed low-pressure catalysts. A typical flow diagram of a low-pressure methanol plant is shown in Fig. 2.

The similarity between the flow diagrams for high-pressure and low-pressure processes is apparent. However, centrifugal synthesis gas compression is feasible for plant capacities down to 150 tons per day with the low-pressure process, whereas the minimum plant sizes utilizing centrifugal compression in the high-pressure processes lie in the range 600 to 800 tons per day. An important feature of the low-pressure processes, especially relevant to producing methanol for fuel, is that carbon dioxide need not be imported (if unavailable near the fuel methanol plant), or recovered from reformer flue gases for

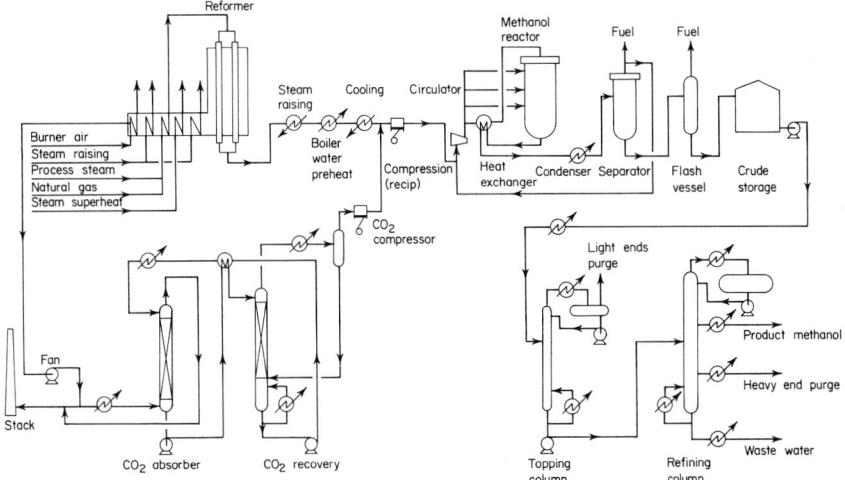

Fig. 1. High-pressure process for producing methanol from natural gas.

*Imperial Chemical Industries Ltd. (ICI).

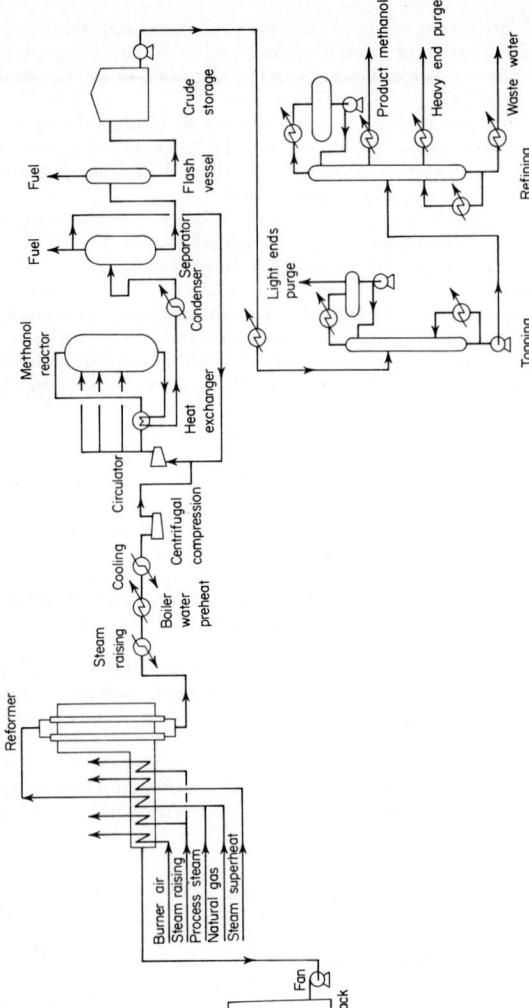

Fig. 2. ICI low-pressure process for producing methanol from natural gas. (*Davy Powergas Ltd.*)

reinjection to the synthesis section. This "low carbon" concept (Ref. 4) enables methanol to be produced from nonstoichiometric synthesis gas.

For correct overall stoichiometry of the methanol reaction, the hydrocarbon feedstock to the reformer should have an empirical formula close to CH_2:

$$CH_2 + H_2O \rightarrow CO + 2H_2 \rightarrow CH_3OH$$

Naphtha feedstock, with an approximate empirical formula of C_7H_{15}, is an almost ideal feedstock from an H_2/C balance viewpoint. With natural gas, however, the H_2/C ratio is too high for production of a perfect synthesis gas, and conventionally a gas stream containing excess carbon (in most cases CO_2) was added to the system.

Methanol producers in the past have therefore gone to some length to locate their production facilities adjacent to ammonia plants for a supply of CO_2. Not all such producers were in the position to have their own captive sources of CO_2, and when this was so, reformer flue gases were scrubbed and CO_2 was recycled. This system involves a high premium in initial investment and utility requirements in the form of steam and cooling water. Additional problems arise in that it is necessary to ensure that the CO_2 produced satisfies all the conditions of purity laid down for reformer feedstock or synthesis gas and in the continuing degradation of wash chemicals due to the oxygen content of the flue gases.

Despite these factors, the economics of the high-pressure synthesis route normally dictate that adoption of this or a similar technique is necessary.

An alternative to the addition of CO_2 from external sources is to pass the reformed gas directly into the synthesis loop and purge out the resultant excess hydrogen from the synthesis section itself.

The loss of hydrogen as a process reactant means that the reformer size is greater than it would be if CO_2 were added into the makeup gas stream, and also that the compression duty on the synthesis gas machine increases. The additional hydrogen purge gas flow from the loop is burned in the reforming furnace as fuel, and the overall process thermal efficiency is maintained at a high level.

Methanol for Fuel.   The concept of utilizing associated or natural gas for production of methanol which could be transported more economically than liquefied natural gas from areas of surplus to areas of shortage, was being examined in 1968 (Ref. 5), but at that time the biggest single stream plant designed had a capacity of 1000 short tons per day (STD). A fuel plant that might need to produce, say, 25,000 STD of methanol was assessed as 25 streams of 1000 STD. Methanol fuel delivered, for example, to the United States from the Middle East could not compete with local natural gas supplies which were then available at approximately 25 cents/million Btu.

More recently, with the growing shortage of natural gas supplies in the United States and orders being placed for substitute natural gas plants [which would produce SNG at an approximate cost of $1.13/million Btu in 1972 (Ref. 6)], the possibility of economic large-scale fuel methanol production was investigated anew. Results of feasibility studies for long-distance transportation of fuel methanol, as an alternative to LNG, looked encouraging.

In order to minimize the capital investment in the battery limits,* fuel methanol plants, which is greater than for LNG plants with an equivalent energy capacity, there is an incentive to scale up single-stream plant capacity to a maximum. One engineering firm† is offering a 5000 STD single-stream fuel methanol production unit and multiples of those 5000 STD modules. The flow diagram (Fig. 2) illustrates a fuel methanol plant with the exception that single-column distillation is sufficient to produce the product grade required. Other methanol plant contractors offer fuel methanol plant designs based on both low- and high-pressure technology.

Fuel Methanol Plant.   With reference to Fig. 2, the following description applies. The first unit is a steam reformer using natural gas as a feedstock to produce synthesis gas. The heat in the flue gas and in the reformed gas is utilized in raising steam for driving the

*Process plant plot limits.
†Davy Powergas Ltd., London (licensee offering ICI low-pressure process).

compressor and circulator. The system of heat recovery shown consists primarily of steam-raising and boiler feedwater preheating. After final cooling, the reformed gas is compressed and delivered to the synthesis loop.

The compressor is driven by a turbine utilizing steam at 900 to 1500 psi (depending on the plant steam balance) with steam passout at reforming pressure for process use and at low pressure for use in the distillation plant.

The synthesis loop comprises a circulator, a catalytic reactor, a feed/effluent exchanger (interchanger), a condenser, and a separator. Compared with a high-pressure synthesis loop, the most significant change is that the low operating temperature and pressure enable a hot wall converter vessel to be used. Temperature control is effected by injecting cold gas at appropriate levels into the single catalyst bed, using specially developed distributors which provide excellent gas mixing.

The gases leaving the converter are cooled in an interchanger, bringing the reactor inlet gases up to reaction temperature. The crude methanol is condensed and cooled in the condenser and separated from unreacted gas in the crude methanol separator.

The uncondensed gases, after the addition of makeup gas, are recycled to the converter by a rotary circulator. To maintain the inert gas level in the synthesis loop at a predetermined optimum level, a purge stream is removed before the point of makeup gas addition.

The distillation of the crude methanol to produce fuel methanol consists simply of the removal of light impurities, dissolved gases, and water, in a single column.

A typical product grade fuel methanol from this type of plant has the following specifications:

Methanol + higher alcohols	99.5 wt%
Water	0.5 wt%
Higher heating value (liquid)	9600 Btu/pound (min)
Specific gravity	0.793

The quantities of heavier alcohols in the methanol fuel produced by the low-pressure processes are very low, owing to the high selectivity of the copper-based catalysts. The proposition that methanol with a higher proportion of higher alcohols would improve the calorific value in terms of Btus available per unit weight, has been put forward (Ref. 7). However, as of 1974, the process has not been proved commercially.

Fuel Methanol vs. LNG.   The transportation of natural gas and crude oil associated gas from areas of abundant supply but low utilization to locations with deficient supplies is becoming an increasingly important factor in the bulk shipment of energy around the world. The drawback in transporting liquefied natural gas is the need for extremely expensive cryogenic tankers for shipment: The cost of these is reflected in the high cost of the energy when delivered over long-haul routes such as from the Middle East to the United States or Japan.

Fuel methanol, considered as an additional means of transportation of natural gas from supply areas to consumers, has particular advantages in reduced delivery costs over these long distances.

With a subatmospheric vapor pressure, fuel methanol is a liquid at ambient temperatures, and as a result can be carried in standard oceangoing tankers. The initial investment in new tankers is significantly less than it would be for the equivalent cryogenic LNG tankers, giving substantial savings in transportation costs.

The advantages of fuel methanol over the transport of LNG include:

1. Lower cost and less hazardous transportation. (Methanol is totally miscible with water, and spillage from tankers would be much less of a problem than with oil spillage. On the other hand, LNG spillage is a problem of major concern in transportation, with the potential hazards of nondispersion and explosion.)

2. No cryogenic storage at the supply and receiving terminals required for handling methanol.

3. The versatility of methanol:

   a. As a liquid fuel, the combustion characteristics of methanol are superior (even as compared with natural gas), including undetectable SO_x levels, lower NO_x levels than natural gas, and no smoke emission.

  b. Fuel methanol can be regasified into substitute natural gas of a quality suitable for pipeline transmission when desired.

  c. Potentially fuel methanol can be either substituted for gasoline motor fuel, or used as a nonpolluting component of the gasoline pool.

Fuel methanol has a Research Octane Number of 105 plus. See also description of alcohol and alcohol blends as transportation fuels under *Chemical Fuels Technology* in this Handbook.

Methanol can be produced from natural gases containing high carbon dioxide concentrations. The extra carbon in the CO_2 is beneficial to methanol production in that it is synthesized to methanol. In LNG production, CO_2 cannot be tolerated and must be removed before the liquefaction plant. Similarly, nitrogen in natural gas need not be removed and can be tolerated when producing methanol. Two ICI low-pressure plants in Europe are fed with natural gas containing over 14 percent nitrogen. In producing LNG, nitrogen must be removed.

The utilization of natural gas fields, so far uneconomic by reason of high CO_2 and N_2 concentrations, is made possible by production of fuel methanol. Purge gas from the methanol synthesis loop is available as a source of low-cost hydrogen for ammonia production, for hydrodesulfurization of crude oil, and for direct reduction of metal ores.

ECONOMIC TRADEOFFS

In considering the choice of process route—LNG or fuel methanol— the distance between the point of gas supply and the consumer is a major influence. The longer the supply route, the greater the number of expensive cryogenic tankers required for LNG and the more the total scheme investment (onshore plant and ships) turns in favor of fuel methanol.

However, fuel methanol plants are more expensive to build and less thermally efficient. Also the way a project is financed (the debt:equity ratio), the interest on the debt, and the return required on the equity portion of the investment, all affect the final price of fuel methanol and LNG in the consumer pipeline.

There are therefore three major influences on the choice between fuel methanol and LNG:

  1. Shipping distance
  2. Equity participation
  3. Wellhead gas costs

Longer shipping distances, higher equity participation and lower wellhead gas costs influence the project choice in favor of fuel methanol.

The following example illustrates a typical calculation comparing LNG and fuel methanol delivered costs.

The basis used for the comparison is that both the fuel methanol and the LNG plants will deliver 165×10^{12} Btu (HHV—high heating value) per annum to the user pipeline in the consumer country.

This delivered fuel corresponds to the output from a 25,000-STD fuel methanol plant (equivalent to 480×10^9 Btu/day) where the product fuel is burned directly as a liquid. The equivalent LNG plants will need a greater battery limits output to allow for evaporation, revaporization and return voyage heel losses.

For a one-way shipping distance of 6800 nautical miles an LNG plant will have a battery limits output of 520×10^9 Btu/day, and for 11,800 nautical miles distance, 545×10^9 Btu/day.

The following assessment can only be regarded as typical for a hypothetical scheme. The capital cost data cannot be absolute, as all comparisons will depend on the specific site location, gas analyses, infrastructure requirements, etc. The results of this particular assessment do, however, indicate why fuel methanol is being very carefully compared with LNG for shipment of energy. The capital cost data applied at the end of 1973. In the example, a financing of 30 percent equity, 70 percent debt is taken with 30 percent return required on equity and debt interest of 8.5 percent.

Shipping. Consider two shipping routes, Middle East Gulf to the United States (approximately 11,800 nautical miles one way) and Middle East Gulf to Japan (approximately

	LNG	Fuel methanol
Capital costs for battery limits plant, offsites, storage, loading facilities, receiving terminal revaporization (for LNG only) plus working capital: Million U.S. dollars	358	470

	Cents (U.S.) per million Btu product LNG or fuel methanol	
	LNG	Fuel methanol
Operating labor, supervision, maintenance, and overhead	8.30	13.45
Insurance, depreciation, interest on loan, and return on equity	46.05	60.50
Total capital costs	54.35	73.95
Operating costs: Natural gas for battery limits plant plus offsites at 30¢/million Btu	35.30	57.00
Catalysts and chemicals	—	4.00
	35.30	61.00
Total operating capital costs for installations (excluding shipping)	**89.65**	**134.95**

6800 nautical miles one way): For fuel methanol, six 200,000 DWT (deadweight tonnage) conventional tankers would be required for the Middle East Gulf to Japan route, and ten 200,000 DWT tankers for the Middle East Gulf to United States route.

For LNG, six (120,000 cubic meter) cryogenic tankers would be needed for the Gulf-to-Japan run, and ten (125,000 cubic meter) cryogenic tankers for the Gulf-to-United States run.

	Nautical miles		Nautical miles	
	6,800	11,800	6,800	11,800
Capital invested, million U.S. dollars	600	1050	240	400
Operating costs: Fuel at $70/ton, port charges, crew maintenance, and insurance, ¢/million Btu	26.70	45.55	24.25	40.70
Capital charges, ¢/million Btu	72.30	126.50	28.90	48.20
Transportation losses, loading and revaporization, ¢/million Btu	7.20	10.55	—	—
Total operating and capital costs, ¢/million Btu	106.20	182.60	53.15	88.90

The grand total of operating and capital costs for product energy delivered to the user pipeline in this particular example, expressed in cents per million Btu (HHV), is thus:

LNG, nautical miles		Fuel methanol, nautical miles	
6,800	11,800	6,800	11,800
89.65	89.65	134.95	134.95
106.20	182.60	53.15	88.90
195.85	272.25	188.10	223.85

Similar calculations for different natural gas supply charges and equity/debt ratios can be plotted graphically as in Figs. 3 and 4.

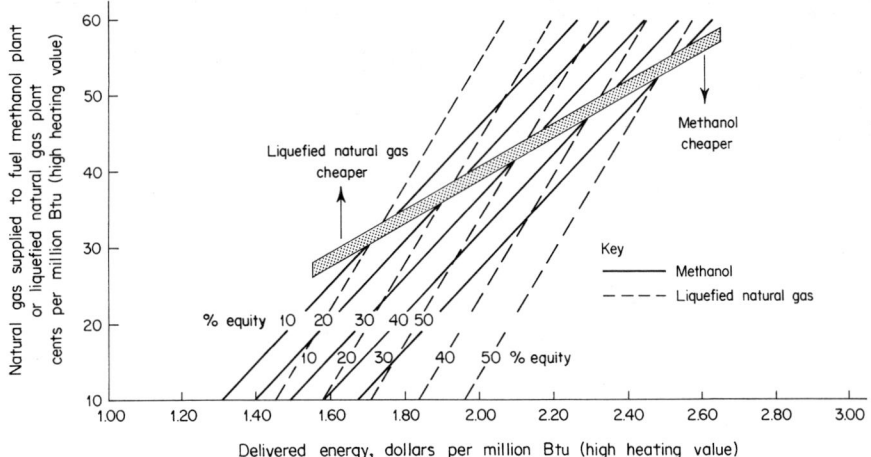

Fig. 3. Comparative delivered costs of liquid fuel methanol and liquefied natural gas in consumer pipeline for differing equity participation in project financing. Situation: Middle East Gulf to Japan, 6800 nautical miles.

Outlook.   There may be near-term applications of methanol fuel for utilization in modified carburation systems, particularly for vehicular fleets. The most likely application for methanol fuel on a relatively large scale, however, is for gas turbines employed in electricity "peaking" duties. Trials using methanol fuel have shown the clean-burning characteristics which meet with environmental control requirements. Power outputs and thermal efficiency are also improved by a small amount. As of end-1973, the worldwide installed methanol plant capacity was approximately 25,000 STD. Single fuel methanol

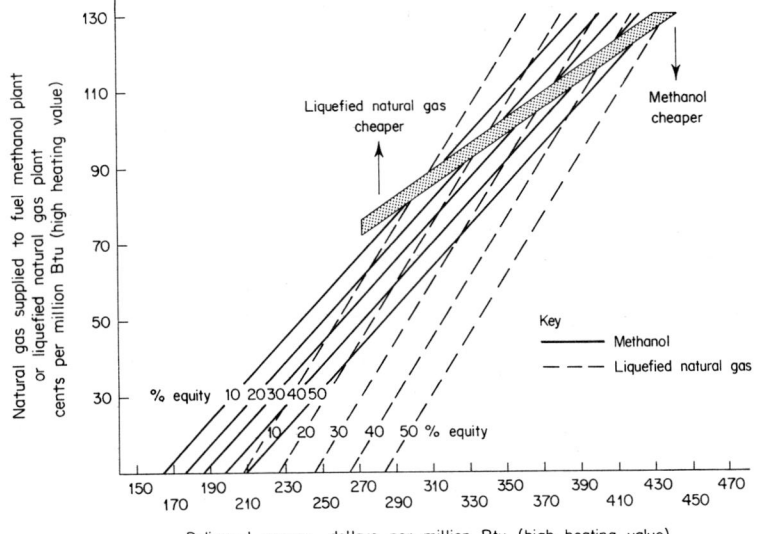

Fig. 4. Comparative delivered costs of liquid fuel methanol and liquefied natural gas in consumer pipeline for differing equity participation in project financing. Situation: Middle East Gulf to the United States, 11,800 nautical miles.

schemes are being assessed in the Middle East equal to and even greater than this capacity.

REFERENCES

1. Royal and Nimmo: Why LP Methanol Costs Less, *Hydrocarbon Process.*, March 1969.
2. Morita, Takahashi, and Koseki: Methanol Production by JGCC Process, *Chem. Econ. Eng. Rev.,* **3,** 9, September 1971.
3. Rogerson, P. L.: "ICI's Low-Pressure Methanol Plant," 64th National Meeting, American Institute of Chemical Engineers, New Orleans, La., Mar. 16–20, 1969.
4. Kenard and Nimmo: "Present Methanol Manufacturing Costs and Economics Using the ICI Process," 64th National Meeting, American Institute of Chemical Engineers, New Orleans, La., Mar. 16–20, 1969.
5. Humphreys: Methanol—The Fuel of the Future? *Nitrogen,* **83,** May–June 1973.
6. Finneran: SNG—Where Will It Come From, and How Much Will It Cost? *Oil Gas J.,* July 17, 1972.
7. Staff: Methanol—Alcohol Fuel Gets New Orleans Tryout, *Oil Gas J.,* Oct. 9, 1972.

Chemical Feedstocks from Natural Gas*

In addition to its use as a fuel because of its clean-burning characteristics, natural gas is a major raw material for numerous chemical processes, notably as a feedstock for the production of ammonia, hydrogen, and methanol, which, in turn, are extremely high-tonnage raw materials for thousands of synthetic materials.

For chemical processing, natural gas normally is processed to recover components heavier than methane. The constituents of natural gas vary with geographic location. No one composition can be considered typical. The analysis of a gas sample taken from the Panhandle natural gas field in Texas is given in Table 1.

Depending upon the constituents and percentage recovery required, two general techniques are available: (1) an absorption-oil process for low ethane recoveries, and (2) a cryogenic process for high ethane recoveries.

The basic *absorption process* is illustrated in Fig. 1. Rich natural gas is fed into the bottom of an absorption tower, where the gas is contacted countercurrently by a lean

TABLE 1. Analysis of a Sample of Natural Gas*

Component	Mole %
Methane	76.2
Ethane	6.4
Propane	3.8
n-Butane	1.3
Isobutane	0.8
n-Pentane	0.3
Isopentane	0.3
Cyclopentane	0.1
Hexane + hydrocarbons	0.3
Nitrogen	9.8
Oxygen	Trace
Argon	Trace
Hydrogen	0.0
Hydrogen sulfide	0.0
Carbon dioxide	0.2
Helium	0.45

*Panhandle natural gas field (Texas).

presaturated absorption oil. The lean tail gas exits from the top of the absorber and is directed back to the pipeline. The lean-oil circulation rate and temperature are determined by the product spectrum and the required percentage recovery. High recovery percentages require large lean-oil circulation rates. If ethane recovery is desired, the lean oil must be refrigerated. Hence the feed gas must be dehydrated to prevent plant freeze-ups.

The absorption oil removes the desired hydrocarbons from the incoming gas stream and then is directed into a demethanizing tower, where absorbed methane is removed from the absorption oil by fractionation and exits from the top of the tower. The methane is further refrigerated, combined with regenerated lean oil, and sent to a presaturator. The vapor phase containing mostly methane is then combined with the tail gas from the absorber overhead. The presaturated liquid is used for reflux in the demethanizer and as lean oil to the absorber.

The bottoms of the demethanizer are directed into the lean-oil still, where the desired

*Prepared originally by Henry E. Duckham, Jr., The M. W. Kellogg Company, a division of Pullman Incorporated, Houston—for the "Chemical and Process Technology Encyclopedia," D. M. Considine, Editor-in-Chief, McGraw-Hill, New York, 1974.

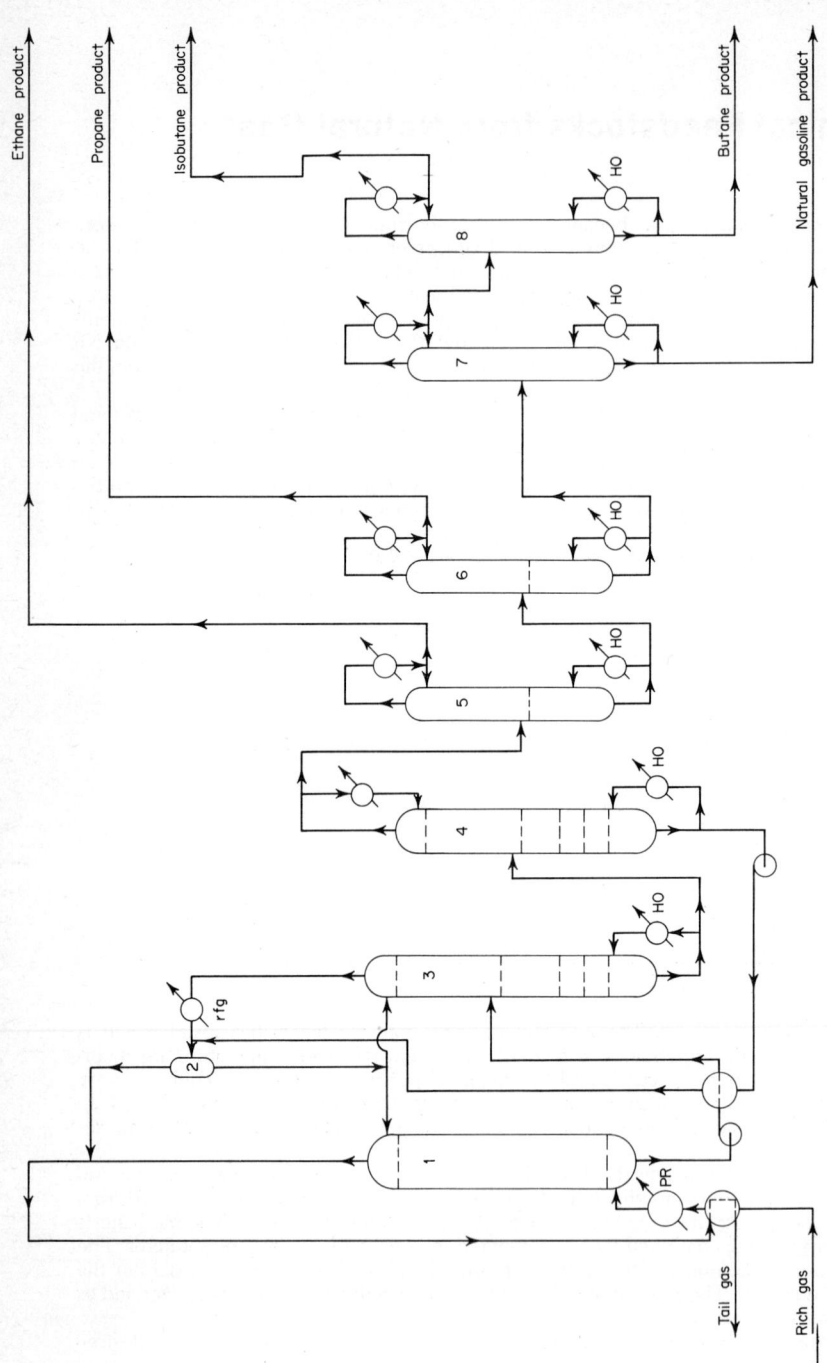

Fig. 1. Representative absorption-oil process for recovery of ethane and heavier hydrocarbons from natural gas: (1) Absorber. (2) Presaturator. (3) Demethanizer. (4) Lean-oil still. (5) Deethanizer. (6) Depropanizer. (7) Debutanizer. (8) C₄ splitter. rfg = refrigerant, PR = propane refrigerant, HO = hot oil. *(The M. W. Kellogg Company, a division of Pullman Incorporated.)*

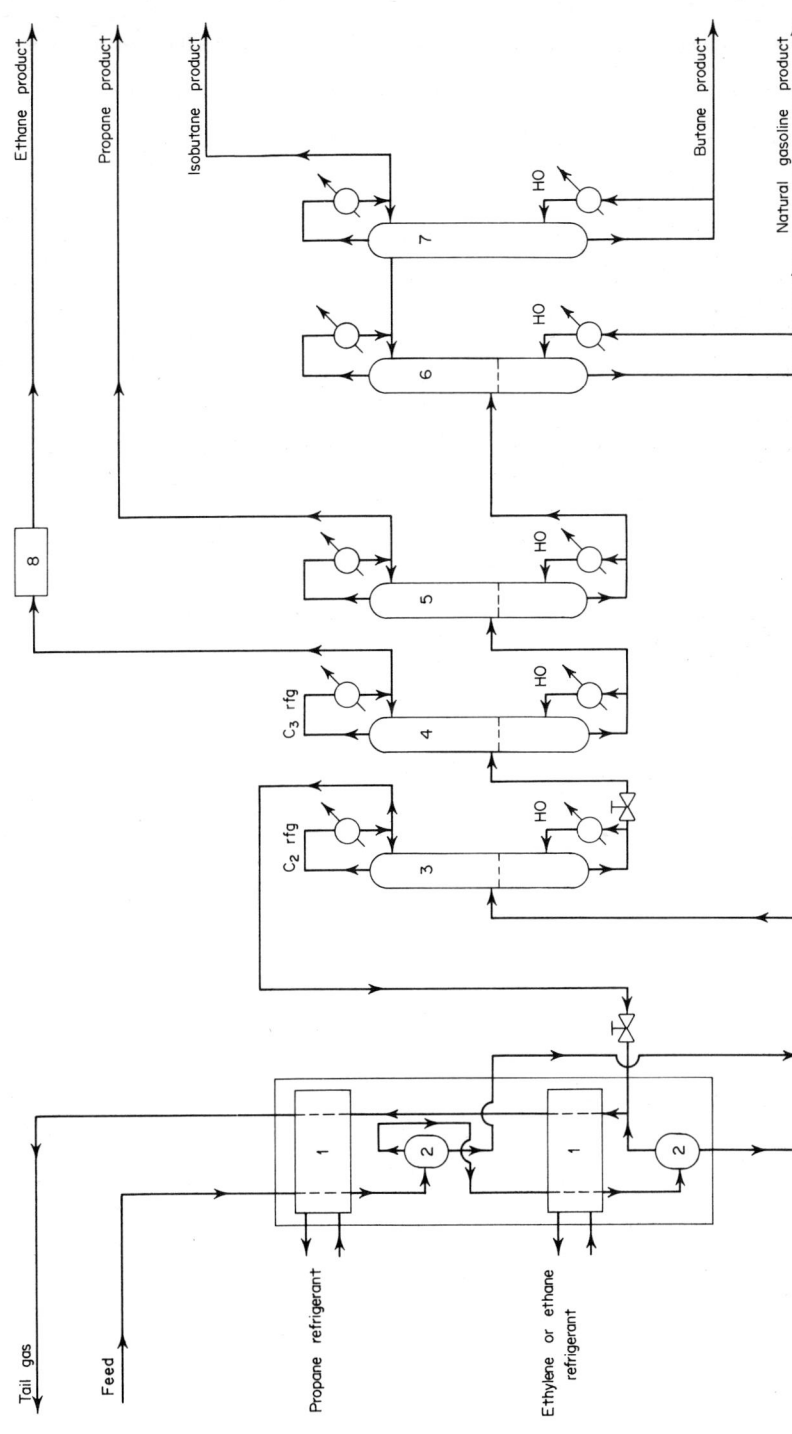

Fig. 2. Cryogenic recovery of hydrocarbons from natural gas: (1) Exchangers. (2) Knockout pots. (3) Demethanizer. (4) Deethanizer. (5) Depropanizer. (6) Debutanizer. (7) C_4 splitter. (8) Carbon dioxide removal. rfg = refrigerant, HO = hot oil. (*The M. W. Kellogg Company, a division of Pullman Incorporated.*)

hydrocarbon components are separated from the lean oil. The overhead vapors containing ethane and heavies are directed into a fractionating train, where the desired product spectrum is obtained. The bottoms liquid then is recycled through a number of heat exchangers back to the presaturator.

Reboiling heat can be supplied by hot oil or steam. Tower reflux can be generated by using refrigeration, water, or air coolers. Gas turbines can be used to drive the necessary rotating equipment.

The basic *cryogenic process* is shown in Fig. 2. A compressed and dried natural-gas stream is directed into a cryogenic processing unit, where its temperature is progressively lowered, with the subsequent effect of condensing the desired hydrocarbons. Various temperature-level liquid dropouts are situated in the cold box to optimize the hydrocarbon liquid recovery. One such level is maintained by propane at $-37°C$ ($-35°F$). Ethane or ethylene refrigerant can be used to stabilize the lowest-temperature-level dropout.

The condensed hydrocarbons are separated from the vapor-liquid mixture in the knockout drums provided. The vapor effluent from the knockout drums (the Btu-controlled tail gas) is heat-exchanged against the incoming feed and exits from the cold box at essentially ambient temperature.

The condensed hydrocarbon phases are directed to the high-pressure demethanizer, where the methane is fractionated from the product hydrocarbon stream. The cold high-pressure methane vapor is isenthalpically expanded and directed into the heat exchangers, where it is combined with the cold tail gas. The bottoms of the demethanizer contain the recovered hydrocarbons and subsequently are fractionated into the desired hydrocarbon products, as shown by the flow diagram.

If the feed contains CO_2 and it is considered a contaminant, the CO_2 is removed from the ethane product stream. The refrigeration compressors may be driven by a gas turbine. Waste heat is recovered by a circulating hot-oil system, which in turn supplies reboiler heat to the fractionating train. Fractionating-tower condenser duties are handled by refrigeration or air coolers, depending upon the towers involved

Absorption of Acidic Gases from Natural Gas and SNG

BY E. J. FISCH,* E. S. HILL,† AND R. W. VAN SCOY*

Absorption processes frequently are used (1) to sweeten sour natural gas containing relatively large concentrations of sulfur compounds or mixtures of sulfur compounds and carbon dioxide; and (2) to remove hydrogen sulfide, carbonyl sulfide, and carbon dioxide from crude synthesis gas made by partial oxidation of high-sulfur fuel oil. These processes also are used to purify crude hydrogen generated by a number of processes in mixture with carbon dioxide, and to sweeten some sour gases in oil refinery operations.

One process of this type, the Sulfinol‡ process, uses a conventional solvent absorption and regeneration cycle, as shown in Fig. 1. The sour gas components are removed from the feed gas by countercurrent contact with a lean solvent stream under pressure. The absorbed impurities are then removed from the rich solvent by stripping with steam in a heated regenerator column. The hot lean solvent is then cooled for reuse in the absorber. Part of the cooling may be by heat exchange with the rich solvent for partial recovery of heat energy. In some applications where there is a large amount of hydrogen sulfide in the feed, the overall Sulfinol/Claus plant heat balance permits the omission of the heat exchanger. The gas flashed from the rich solvent after partial depressuring is shown in Fig. 1 as *fuel gas*. In some cases, it is desirable to treat this gas stream with Sulfinol solvent to control the acid gas content of the plant fuel supply.

The Sulfinol solvent reclaimer shown is a small ancillary facility for recovering solvent components from higher boiling products of alkanolamine degradation or from other high-boiling or solid impurities. A Sulfinol reclaimer is similar to a conventional MEA (mono-ethanolamine solvent) reclaimer, but is much smaller than that in an MEA plant of comparable gas treating capacity. Usually the Sulfinol reclaimer need not be started up until several months after the treating plant is started up.

Solvent. The Sulfinol solvent consists of a mixture of water, an alkanolamine (usually diisopropanolamine is used), and sulfolane (tetrahydrothiophene dioxide). The alkanolamine is an alkaline agent which absorbs acid gases by chemical combination, just as in other treating processes using alkanolamines. The sulfolane constituent, however, imparts the added dimension of significant physical absorbent properties to the solvent and is, in part, responsible for the high solution loadings and low regeneration heat requirements which are characteristic of the Sulfinol system for many cases. The Sulfinol process is believed to be the only commercially used gas-treating process which utilizes a solvent comprised of three liquid components. This factor permits tailoring of the solvent composition to meet a wide range of operating conditions, feed gas composition, and treated gas specifications, either at the time of new plant design or upon conversion from another solvent to Sulfinol.

Range of Applications. The treating requirements in applications of the Sulfinol process vary widely—from removing a few percent of carbon dioxide from dry methane or ethane gases to removing over 50 percent of hydrogen sulfide and carbon dioxide plus carbonyl sulfide and mercaptans. In other instances, natural gas containing 28 percent carbon dioxide and a trace of hydrogen sulfide, and natural gas containing 12 percent hydrogen sulfide plus carbon dioxide and over 600 ppm carbonyl sulfide and mercaptans, have been economically treated.

The process can be designed to achieve a wide range of treated gas purity depending on the requirements of a given application. Typically, pipeline gas specifications will require

*Staff Engineer, Process Engineering-Refining Department, Shell Oil Company, Houston.
†Consultant. Retired from Shell Oil Company, Houston.
‡Trade name, Shell Oil Company.

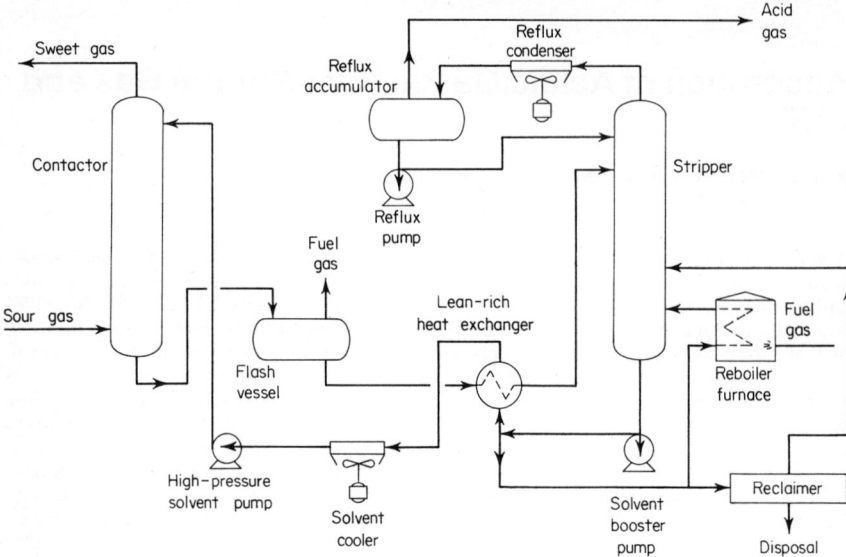

Fig. 1. Sulfinol process (trade name, Shell Oil Company) for treatment of gases to remove acidic components.

that the hydrogen sulfide content not exceed 0.25 grain hydrogen sulfide per 100 standard cubic feet (4 parts per million, ppm). In addition, mercaptan sulfur and total sulfur maxima as low as 0.2 grain/100 standard cubic feet and 1 grain/100 standard cubic feet, respectively, may be specified for treating feed gas which contains mercaptans, carbonyl sulfide, and other sulfur compounds. The Sulfinol process can be designed to achieve these low levels of impurities. If required, the level of hydrogen sulfide can be reduced to less than 1 ppm in the treated gas in most instances. Similarly, a carbon dioxide content of less than 500 ppm is regularly achieved in synthesis gas plants which operate at moderate pressure. If required, carbon dioxide removal to less than 100 ppm can generally be achieved. Table 1 illustrates an example of the cleanup obtained in one plant processing three streams derived from a natural gas with high levels of organic sulfur compounds.

TABLE 1. Removal of Organic Sulfur Compounds Attained in a Test Run on a Commercial *Sulfinol Unit

Contactor	High pressure	Low pressure	Fuel gas
Feed gas:			
Carbon dioxide, mole %	4.16	4.29	7.88
Hydrogen sulfide, mole %	7.90	21.13	30.74
C_1–C_5 mercaptans, ppm	261 (488)†	427 (2850)	830
Carbon disulfide, ppm	10	10	10
Thiophene, ppm	10	10	10
Unidentified sulfur compounds, ppm	283	563	651
Carbonyl sulfide, ppm	N.A.‡ (125)	50 (84)	14
Treated gas:			
Carbon dioxide, mole %	—	—	—
Hydrogen sulfide, ppm	1	1	100
C_1–C_5 mercaptans, ppm	1, each	1, each	299+
Carbon disulfide, ppm	1	1	10
Thiophene, ppm	N.A.	N.A.	10
Unidentified sulfur compounds, ppm	3	8	237+
Carbonyl sulfide, ppm	1	2	N.A.

*Trade Name, Shell Oil Company.
†N.A. Not analyzed.
‡( ) Design.

Originally, it appeared that the Sulfinol process might be applicable primarily where the hydrogen sulfide/carbon dioxide ratio was 1:1 or greater and where the acid gas partial pressure in the feed was about 110 psi or greater. As more experience was gained from pilot and commercial plant operations,* numerous applications were found where the process could be used at hydrogen sulfide/carbon dioxide ratios far below 1:1, and at acid gas partial pressures substantially lower than 110 psi.

Process Principles.   The Sulfinol process uses a solvent that combines the physical solvent-absorption capacity of sulfolane for acid gases with the chemical absorption capacity of diisopropanolamine. This combination of absorption capabilities offers advantages for both absorption and regeneration. The diisopropanolamine combines with sour components in an acid-base reaction, essentially nonsensitive to pressure, and sulfolane adds physical ability which is proportional to pressure. The net result is a solvent having good absorption for sour components at low to medium partial pressures, and very high absorption of these components at high partial pressures. A substantial portion of the absorbed gases is released from the solution on pressure reduction, requiring less heat for stripping.

The process is nonfoaming in that it can be operated with high contactor loadings on the solution gas from crude oil without foaming problems, even when some crude oil enters the contactor. A few pipeline corrosion inhibitors have caused foaming in Sulfinol contactors, and a satisfactory antifoam agent has been used until the type of corrosion inhibitor could be changed to one of the many that have no effect on Sulfinol solvent.

The process is noncorrosive to carbon steel in applications where some hydrogen sulfide is present. If no hydrogen sulfide is present, a limited amount of stainless steel is required. A corrosion inhibitor is not used in the Sulfinol process. The results of measurements on corrosion in Sulfinol systems were published by MacNab and Treseder (Ref. 1).

The heat requirements for regenerating the solvent are low, with the steam requirement per pound of acid gas removed somewhat less than half the quantity used for monoethanolamine solutions employed for acid-gas treating.

Auxiliary facilities used in the process are a small side-stream filter to remove solids from the solvent and a small reclaimer. The stable degradation products have boiling points higher than sulfolane and diisopropanolamine so that the solvent components can be recovered in a reclaimer.

Losses of solvent components are distributed between vapor, chemical, and mechanical. Sulfolane does not degrade, and typical losses of sulfolane in operating plants have been substantially less than 1 pound per million standard cubic feet of sour gas treated, some plants adding little or no sulfolane during the first year of operation. Carbon dioxide reacts to form diisopropanolamine-oxazolidone in quantities varying from less than 5 to 20 lb/million standard cubic feet of carbon dioxide; therefore losses vary considerably between plants.

Some typical applications of the Sulfinol process for natural-gas treating are shown in Table 2 by Goar (Ref. 2) for four different gases. These illustrations show the wide range of acid-gas solubility in the Sulfinol solvent and the low utility requirements. The reboiler steam rates depend on the amount of heat recovered in a lean-rich solvent heat exchanger if one is used. Generally, no lean-rich exchangers are needed for gases B and C because the steam generated in the sulfur plant is adequate for regenerating the Sulfinol solvent without any process heat recovery.

The removal of carbonyl sulfide and acid gases from synthesis gas generated by partial oxidation of heavy fuel oil is reported by Klein (Ref. 3). The sour synthesis gas contained 0.3 to 1.2 percent hydrogen sulfide and 100 to 500 ppm carbonyl sulfide in addition to carbon dioxide. The removal of the carbonyl sulfide and sour-acid-gas components from typical crude synthesis gases is shown in Table 3.

Other processes are available for removing acid gases from natural and synthetic gases. Some of these depend solely on chemical adsorption in a solvent and some solely on physical adsorption in a solvent. The Sulfinol solvent is unique in utilizing both types of adsorption. Physical adsorption processes using molecular sieves could also be used. Differences in capital cost, operating cost, solvent degradation, corrosion rates, and other factors exist among the processes available. Therefore, prudent selection of a process requires careful evaluation of the likely alternatives.

*Over 100 units in commercial operation throughout the world.

TABLE 2   Typical Circulation and Reboiler Steam Rates of Sulfinol Units for Natural Gas Applications

Item	Gas A	Gas B	Gas C	Gas D
CONDITIONS FOR ALL GASES				
Contactor pressure, 1000 psia				
Feed-gas temperature, 110°F (43°C)				
Feed-gas flow (dry), 100 million standard cubic feet/day				
Feed-gas composition, mole %:				
Hydrogen sulfide	0.65	20.1	51.5	0.10
Carbon dioxide	8.73	2.0	3.5	18.00
Nitrogen	2.37	1.4	8.6	0.70
Carbon, C_1	87.90	71.5	25.8	80.94
C_2	0.35	2.0	5.8	0.17
C_3	—	1.7	3.2	0.05
C_4	—	1.1	1.6	0.04
C_5		0.2		
	100.00	100.00	100.00	100.00
Carbonyl sulfide, grains/100 standard cubic feet	3.0	7.3	8.4	
Mercaptans, grains/100 standard cubic feet	2.1	1.5	3.1	
Hydrogen sulfide/carbon dioxide ratio	0.0744	10.05	14.71	0.0056
Sweet-gas hydrogen sulfide content, grains/100 standard cubic feet	<0.25	<0.25	<0.25	<0.25
Sweet-gas carbon dioxide content, mole %	<1.0	<1.0	<1.0	2.0*
Sweet-gas total sulfur, grains/100 standard cubic feet	<1.0	<1.0	<1.0	<1.0
Solution circulation rate, gallons/minute at 110°F (43°C)	1,483	1,748	2,366	2,167
Solution net pick up, standard cubic feet acid gas/gallon	4.39	8.78	16.14	5.14
Reboiler steam rate, pounds/hour	69,740	111,440	156,010	99,600
Reboiler steam rate, pounds/gallon solution	0.78	1.06	1.10	0.77

*The Sulfinol system for gas D was designed to leave 2.0 mole % carbon dioxide in the sweet gas, while reducing hydrogen sulfide content to 0.25 grain/100 standard cubic feet.
Reproduced by permission of source: *Oil Gas J.*, June 30, 1969 (B. G. Goar).

TABLE 3.   Typical Operating Data of Sulfinol Units for Synthesis Gas Applications

Factor	Case 1	Case 2
Gas throughput, million standard cubic feet per day	13.71	4.48
Hydrogen sulfide content, vol %	0.36	0.87
Carbon dioxide content, vol %	5.20	5.20
Carbonyl sulfide content, ppm	125	155
Absorption pressure, psia	588	485
Absorption temperature	104°F (40°C)	120°F (49°C)
Number of trays	30	45
Purified gas:		
Hydrogen sulfide, ppm	0.5	2
Carbonyl sulfide, ppm	0.3	5
Carbon dioxide content, ppm	25	—
Steam consumption at 51.5 psia, pounds/pound of acid gas removed	1.9	1.5
Electricity consumption, kilowatt-hours/pound of acid gas removed	0.015	0.011

REFERENCES

1. MacNab, A. J., and R. S. Treseder: Materials Experience in the Sulfinol Gas-Treating Process, *Mater. Protect. Performance*, **10**, 21–26, January 1971.
2. Goar, B. G.: *Oil Gas J.*, pp. 117–120, June 30, 1969.
3. Klein, J. P.: Developments in Sulfinol and ADIP Process Increase Uses, *Oil Gas Int.*, **10**(9): 109–112, September 1970.

The Catalytic Rich Gas (CRG) Process for Gasification of Light Hydrocarbons

BY COLIN H. TAYLOR*

The catalytic rich gas (CRG) process was developed by the British Gas Council (now British Gas Corporation) for the production of a gas of relatively high calorific value from light hydrocarbons. The reactions are those employed in conventional steam reforming of natural gas (e.g., for production of hydrogen), but they take place at a lower temperature and steam-feedstock ratio. Under these conditions, the overall reaction is slightly exothermic, and can, therefore, be made self-sustaining in an adiabatic reactor.

BRIEF HISTORY

The CRG process was developed from the work of a team at the Gas Council's Midlands Research Station (MRS) led by Dr. F. J. Dent. In the late 1950s, it became apparent that, due to the postwar increase in refining capacity in Europe, naphtha was becoming available as a potential feedstock for gas making and that its use would be more economic than coal carbonization, which was then the major source of fuel gas in the United Kingdom.

The first semicommercial plant, producing 4 MMSCFD of rich gas, was commissioned in 1964. In the next five years, nearly 40 units were installed in the United Kingdom for production of rich gas and town gas (470 to 500 Btu per standard cubic foot) (Ref. 1). Plants have also been installed in Japan, Italy, Brazil, and the United States.

Operating experience of plants incorporating the CRG process has been very satisfactory. Apart from relatively minor mechanical troubles at startup, the problem areas were chiefly in naphtha vaporization and CO_2 removal. The reaction stages themselves have performed as predicted by the laboratory and pilot plant experience, and the expected catalyst life has in most cases been achieved. An availability of better than 99 percent of scheduled onstream time has been reported for many plants.

The process has also proved to be very flexible, to the extent that some plants are operated to match gas demand, the output varying from 40 to 100 percent and back again over a 24-hour period. Indeed, one plant was specially designed to operate over the range 10 to 100 percent. Most plants incorporate hold hot circuits to facilitate rapid gas making upon increase in demand. Also, the basic philosophy has been to design plants which are easy to operate, and above all reliable, instead of to strive after the ultimate in efficiency. Even so, the thermal efficiency at nominal throughput is normally 90 percent or greater.

As a result of operating experience, catalyst loadings have been progressively increased. Finally, the first plants in the United States have demonstrated operation at pressures higher than were necessary in the United Kingdom.

CHEMISTRY AND THERMODYNAMICS

The overall reactions which occur in the steam reforming of naphtha may be represented as follows:

$$4C_6H_{14} + 10H_2O \rightarrow 19CH_4 + 5CO_2 \qquad \text{Exothermic} \qquad (1)$$
$$CH_4 + H_2O \rightleftharpoons CO + 3H_2 \qquad \text{Endothermic} \qquad (2)$$
$$CO_2 + H_2 \rightleftharpoons CO + H_2O \qquad \text{Slightly endothermic} \qquad (3)$$

At all practical temperatures, reaction (1) proceeds almost to completion; no significant quantities of higher hydrocarbons exist at the outlet of the CRG reactor.

Reactions (2) and (3) are reversible; the concentrations of the five components CH_4,

*Chief Process Engineer (Gas), Woodall-Duckham Limited, Crawley, Sussex, England.

H_2O, CO_2, CO, and H_2 which result are governed by thermodynamic equilibrium. Raising the reaction temperature shifts the equilibrium for both reactions to the right. Thus at low temperatures the exothermic reaction (1) predominates, while at high temperatures the overall reaction is endothermic. At approximately 500 to 550°C the reaction is thermally neutral. Figure 1 shows this effect and contrasts the reforming of methane, which is

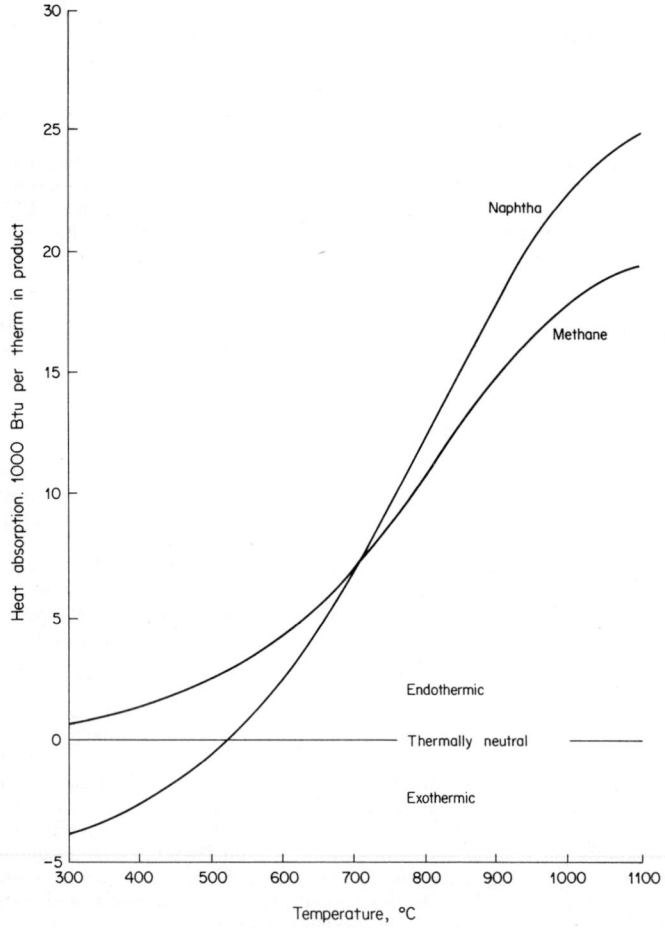

Fig. 1. Effect of temperature on overall heat of reaction.

endothermic at all temperatures. A more detailed analysis and a discussion of the reasons for the choice of optimum operating conditions for the reaction are given by Dent (Ref. 2).

A typical temperature profile through the catalyst bed is shown in Fig. 2. The reaction is completed within a finite zone which moves through the bed as the catalyst becomes deactivated.

The small "endothermic dip" at the top of the reaction zone should be noted. When the reaction starts, the methane formed by Eq. (1) is immediately converted into CO and H_2, and the overall reaction is endothermic. As the concentrations of CO and H_2 increase and that of H_2O decreases, reaction (2) is displaced to the left so that an increasing proportion of the methane formed does not react, and the exothermic reaction (1) predominates.

FEEDSTOCK SPECIFICATIONS

The CRG process is suitable for methane, LPG, and naphtha feedstocks including raffinates. The range of recommended naphthas is as follows:

Final boiling point	240°C max.
Olefins	1% vol max.
Aromatics	10% vol max.
Chlorine	1 ppm max.
Lead	1 ppm max.

Naphthas boiling up to 185°C can be reformed at pressures up to 600 psig. Naphthas with final boiling point up to 240°C may be reformed at lower pressures.

Higher olefin contents may be accepted provided that sufficient hydrogen is available in the recycle gas to saturate the feed in the desulfurization section.

Higher aromatic contents may be accepted but the catalyst life will be reduced.

If the chlorine content is greater than 1 ppm, special precautions must be taken to remove it in the desulfurization section.

The nickel-based CRG catalyst is deactivated by sulfur. However, a British Gas Corporation desulfurization process, which reduces the total sulfur content to less than 0.2 ppm, is always used in conjunction with the CRG catalyst.

USES OF RICH GAS

A typical rich gas leaving the CRG reactor has the following composition:

CO_2	23.0 mole % (dry)
CO	0.7
H_2	12.8
CH_4	63.5
	100.0
Calorific value,	
Btu/standard cubic foot	675

Higher hydrocarbons are present in negligible quantities.

The calorific value of this gas is too high for direct use as town gas (470 to 500 Btu/

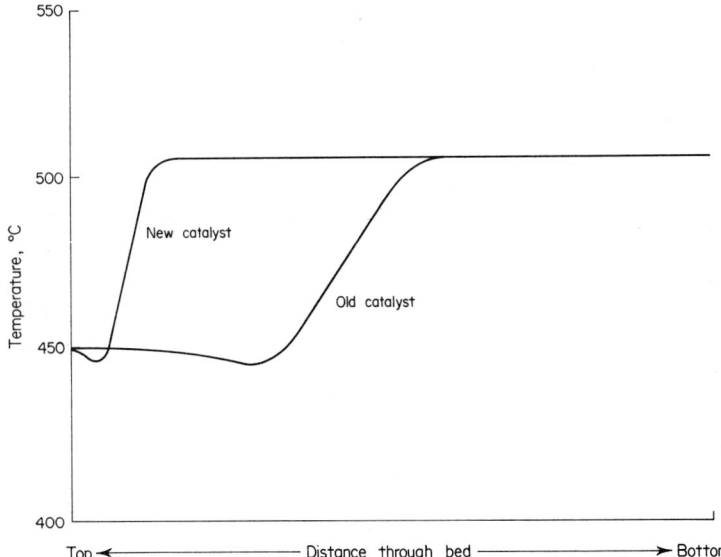

Fig. 2. Temperature profile through CRG reactor.

standard cubic foot in the United Kingdom) and too low for SNG. However, by removing the CO_2 the calorific value is increased to about 870 Btu/standard cubic foot, which is useful for enriching lean gas (e.g., from an ICI naphtha reformer) to town gas quality. Long (Ref. 3) has suggested that by enriching this gas with LPG a satisfactory SNG may be obtained.

Alternatively, the calorific value may be changed by bringing the components to a new equilibrium at a different temperature. In the Series A process, part of the rich gas is further reformed at high temperature and remixed with the remaining rich gas. After water gas shift and partial CO_2 removal, a 500 Btu/standard cubic foot product is obtained, which is fully interchangeable with the town gas distributed in the United Kingdom.

If the subsequent stage is at a lower temperature, carbon oxides and hydrogen recombine to methane, increasing the calorific value. After CO_2 removal, very little enrichment is required to achieve a product fully interchangeable with natural gas.

REMOVAL OF SULFUR AND CHLORINE COMPOUNDS FROM FEEDSTOCKS

CRG catalyst is deactivated by low concentrations of sulfur and chlorine compounds.

To achieve removal of sulfur to very low concentrations (less than 0.2 ppm), British Gas developed their own process, which is always used in association with CRG catalyst (Ref. 4).

Organic sulfur compounds are hydrogenated to H_2S over nickel-molybdenum catalyst at about 380°C. The hydrogen is usually generated by reforming rich gas from the CRG reactor with added steam in a tubular reformer. Alternatively, gas from the reactor may be used directly, whereas in series A plants it is normal to use CO_2-free town gas. The H_2S is then absorbed on zinc oxide or, in the case of many United Kingdom town gas plants, Luxmasse (hydrated ferric oxide).

Because of the large stream sizes involved, many of the SNG plants in the United States incorporate a bulk sulfur removal stage using a hydrofining process. The H_2S produced is commonly recovered as elemental sulfur by a Stretford plant, in which the hydrofiner off gas is washed with an aqueous alkaline solution, which is regenerated by oxidation with air (Ref. 5).

Chlorine compounds, if present in concentration higher than 1 ppm, not only deactivate the CRG catalyst but also interfere with the absorption of H_2S. Therefore they are removed by hydrogenation to HCl, which is absorbed on a proprietary absorbent (Ref. 6).

THE CRG REACTOR

Reforming in the CRG process occurs adiabatically at 450 to 550°C at pressures up to about 600 psig. The reactor is a vertical cylindrical pressure vessel containing a bed of the special high-nickel catalyst, which is supported on a grid or on inert ceramic balls. The gas flow is downward through the bed, and distributors are provided at inlet and outlet. A layer of ceramic balls on top of the bed prevents disturbance of the catalyst by the entering gas.

Because of the temperature and hydrogen concentration, CRG reactors in the United Kingdom are of stainless steel or $2\frac{1}{4}$% Cr 1% Mo alloy. However, in some of the large units in the United States, refractory-lined carbon steel vessels were chosen for reasons of cost.

Catalyst bed temperatures are monitored by thermocouples arranged at close intervals in a vertical thermowell which extends downward through the bed. In large reactors two or three widely spaced thermowells are installed to detect any tilt in the plane of the reaction zone, indicating poor flow distribution or nonuniform deterioration of catalyst.

Normal practice is to install two reactors in parallel, of which one is working at any time. The catalyst charge in each vessel is designed for 3 to 6 months operation at full load. This system avoids unnecessary exposure of catalyst to high temperatures, minimizes the catalyst loss in the event of damage by maloperation, and provides instant standby if such damage occurs.

By plotting the temperature profile through the catalyst at regular intervals, the end of the life of a charge can be anticipated and the standby charge can be heated up ready for changeover "on the run."

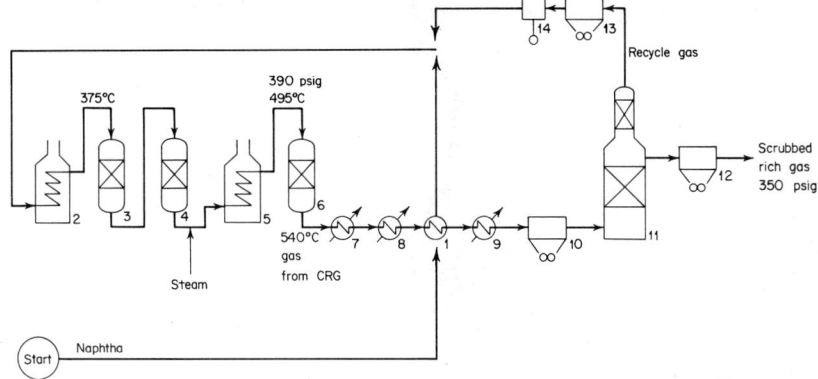

Fig. 3. Rich gas plant: (1) Naphtha preheater. (2) Naphtha vaporizer. (3) Sulfur hydrogenator. (4) Sulfur absorber. (5) Reactants superheater. (6) CRG reactor. (7) Waste heat boiler. (8) Feedwater heater. (9) Carbonate reboiler. (10) Cooler. (11) CO_2 absorber. (12, 13) Coolers. (14) Compressor.

RICH GAS PRODUCTION

The flow diagram for a rich gas plant producing gas with a calorific value of 710 Btu/ standard cubic foot is shown in Fig. 3. The product is used to enrich lean gas from an ICI naphtha reformer which has a calorific value of about 320 Btu/standard cubic foot to the town gas standard of 500 Btu/standard cubic foot.

Typical gas analyses are given in Table 1.

TOWN GAS PRODUCTION BY THE SERIES A PROCESS

Part of the gas from the CRG reactor is reformed with additional steam, and the resulting lean gas is reblended with the remaining rich gas. The mixed gas is then subjected to water gas shift and partial CO_2 removal, yielding a product with a calorific value of 470 to 500 Btu/standard cubic foot. The flowsheet is shown in Fig. 4, and typical gas analyses appear in Table 2.

By varying the proportion of gas which flows to the tubular reformer and the degree of CO_2 removal, the characteristics of the product gas can be made interchangeable with any of the different standards employed by the United Kingdom Area Boards and Japanese and European gas companies.

LNG DERICHMENT

A variation of this process was employed at the La Spezia, Italy, plant installed by Esso Research and Engineering for SNAM (Ref. 7). The calorific value of imported Libyan LNG (1395 Btu/standard cubic foot) is too high for direct use as pipeline gas. The LNG is

TABLE 1.   Gas Analyses in Rich Gas Plant

	Recycle gas	Gas from CRGR*	Scrubbed rich gas
CO_2, mole %	0.9	20.3	13.5
CO	1.8	1.4	1.5
H_2	23.3	19.1	20.7
CH_4	74.0	59.2	64.3
	100.0	100.0	100.0
CV† (Btu/SCF)	816	654	710

*Catalytic rich gas reactor.
†CV = calorific value.

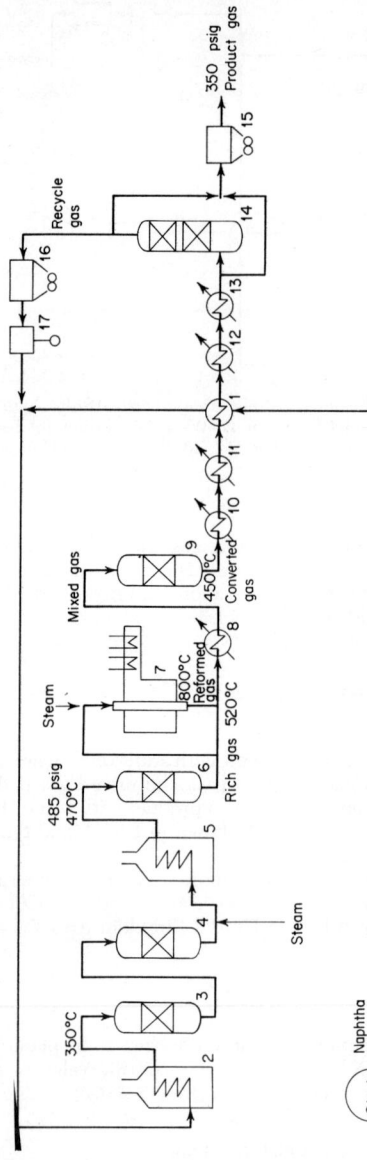

Fig. 4. Typical town gas plant—series A process: (1) Naphtha preheater. (2) Naphtha vaporizer. (3) Sulfur hydrogenator. (4) Sulfur absorber. (5) Reactants superheater. (6) CRG reactor. (7) Tubular reformer. (8) Waste heat boiler. (9) CO converter. (10) Waste heat boiler. (11) Feedwater heater. (12) Carbonate reboiler. (13) Makeup water heater. (14) CO_2 absorber. (15, 16) Coolers. (17) Compressor.

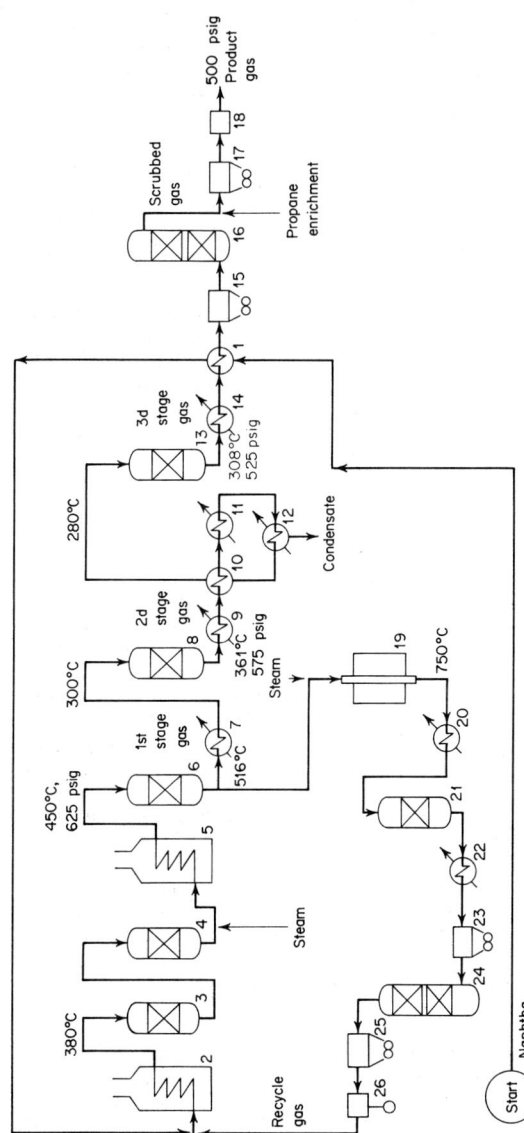

Fig. 5. SNG plant—double methanation route: (1) Naphtha preheater. (2) Naphtha vaporizer. (3) Sulfur hydrogenator. (4) Sulfur absorber. (5) Reactants superheater. (6) CRG reactor (7) Waste heat boiler. (8) Primary methanator. (9) Waste heat boiler. (10) Methanator exchanger. (11) Feedwater heater. (12) Carbonate reboiler. (13) Secondary methanator. (14) Waste heat boiler. (15) Cooler. (16) Main CO_2 absorber. (17) Cooler. (18) Drying plant. (19) Tubular reformer. (20) Waste heat boiler. (21) CO converter. (22) Waste heat boiler. (23) Cooler. (24) Recycle CO_2 absorber. (25) Cooler. (26) Compressor.

TABLE 2.   Gas Analyses in Series A Plant

	Recycle gas	Rich gas	Reformed gas	Mixed gas	Converted gas	Product gas
CO_2, mole %	1.0	21.6	13.4	16.1	21.3	13.5
CO	3.4	0.9	13.7	9.6	2.7	3.0
H_2	60.9	15.3	59.0	44.9	48.4	53.2
CH_4	34.7	62.2	13.9	29.4	27.6	30.3
	100.0	100.0	100.0	100.0	100.0	100.0
CV* (Btu/SCF)	550	670	370	466	438	480

*CV = calorific value.

therefore fractionated and the heavy ends (C_2H_6–C_6H_{14}) are fed to a CRG unit. Part of the rich gas is reformed to provide hydrogen to increase the flame speed of the product. After CO_2 has been removed from the mixed gas, it is reblended with the light fraction of the LNG. The product gas, with a calorific value of 1024 Btu/standard cubic foot is fully interchangeable with the Po Valley natural gas distributed in the area.

SNG PRODUCTION BY THE METHANATION ROUTE

If the rich gas from the CRG reactor is passed over another bed of high-nickel catalyst at a lower temperature, the equilibrium of the five components is reestablished. Carbon oxides react with hydrogen to form methane, and the calorific value of the gas is increased. It should be noted that this methanation step differs from that encountered in ammonia synthesis gas production; because of the high steam content the temperature rise is reduced, and there is no possibility of temperature "runaway" as the exit temperature can never rise above the temperature corresponding to equilibrium at the inlet composition, i.e., the CRG exit temperature.

In order to minimize cold enrichment and to achieve a very low carbon monoxide content in the product, a second methanation stage is frequently employed. To achieve sufficient "driving force" to make the reaction proceed, the water vapor content is reduced by cooling the gas, rejecting condensate, and reheating to the required reaction temperature.

Figure 7 shows the effect of the second methanation stage on product calorific value. While consumption of LPG is minimized, the capital cost is increased and the overall thermal efficiency is slightly reduced (Ref. 8).

A typical flowsheet is shown in Fig. 5, and gas analyses appear in Table 3.

The life of the catalyst in the methanation stages is expected to be substantially in excess of one year, and so no standby reactors are normally installed.

SNG PRODUCTION BY THE HYDROGASIFICATION ROUTE

If part of the purified naphtha vapor from desulfurization is allowed to bypass the CRG reactor, it can be fully gasified by reaction with the hydrogen and steam in the rich gas. This reaction occurs at lower temperatures than the CRG reaction, and is known as hydrogasification. It has the advantage that the total steam requirement for the process is

TABLE 3.   Gas Analyses—SNG Production by Double Methanation

	Recycle gas	1st-stage gas	2d-stage gas	3d-stage gas	Scrubbed gas	Product gas
CO_2, mole %	0.5	21.7	22.0	21.9	0.5	0.50
CO	1.5	0.8	0.1	0.1	0.1	0.04
H_2	86.5	12.8	3.7	0.4	0.5	0.53
CH_4	11.5	64.7	74.2	77.6	98.9	97.98
C_3H_8	—	—	—	—	—	0.95
	100.0	100.0	100.0	100.0	100.0	100.0
CV* (Btu/SCF)	395	687	750	773	986	1000

*CV = calorific value.

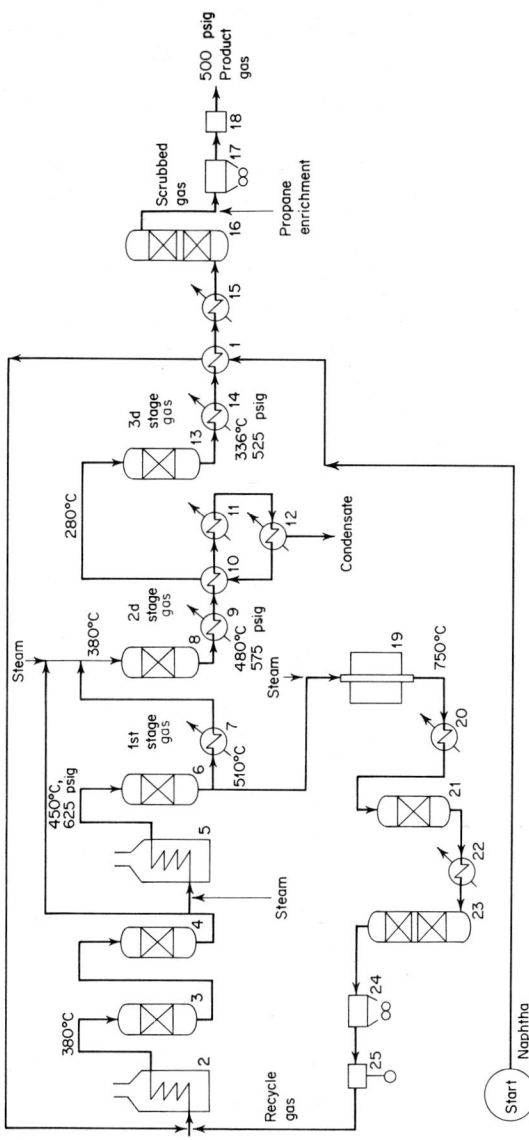

Fig. 6. SNG plant—hydrogasification route: (1) Naphtha preheater. (2) Naphtha vaporizer. (3) Sulfur hydrogenator. (4) Sulfur absorber. (5) Reactants superheater. (6) CRG reactor. (7) Waste heat boiler. (8) Hydrogasifier. (9) Waste heat boiler. (10) Methanator exchanger. (11) Feedwater heater. (12) Carbonate reboiler. (13) Methanator. (14) Waste heat boiler. (15) Feedwater heater. (16) Main CO_2 absorber. (17) Cooler. (18) Drying plant. (19) Tubular reformer. (20) Waste heat boiler. (21) CO converter. (22) Feedwater heater. (23) Recycle CO_2 absorber. (24) Cooler. (25) Compressor.

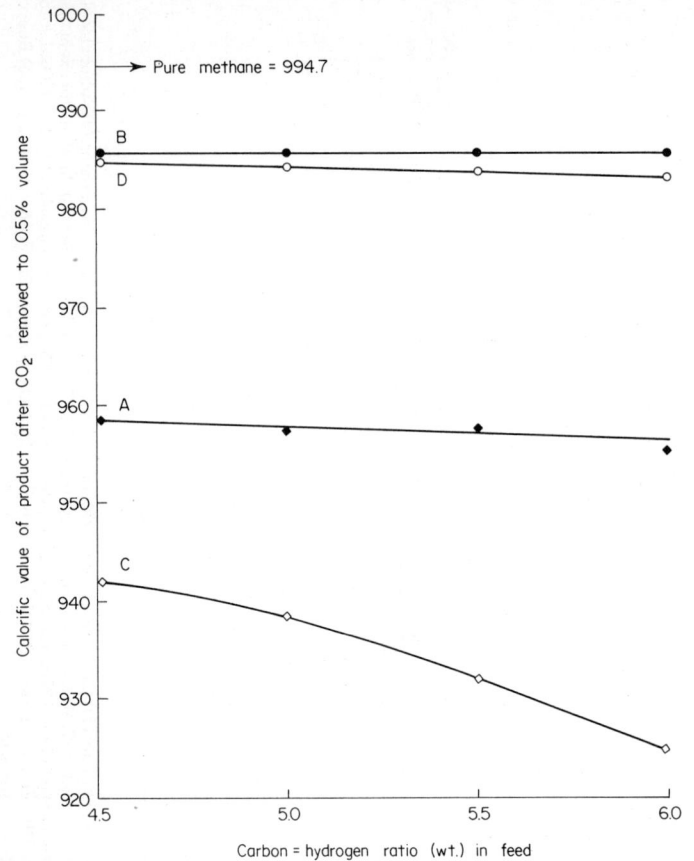

Fig. 7. Calorific value of scrubbed gas: *A*, CRG reactor + single methanation. *B*, CRG reactor + double methanation. *C*, CRG reactor + hydrogasifier. *D*, CRG reactor + hydrogasifier + methanation. CRG reactor pressure, 600 psig.

reduced, although with heavier feedstocks it may be necessary to add a little steam to the hydrogasifier in order to ensure that carbon is not formed by the Boudouard reaction (Ref. 9). Since less makeup steam has to be generated from fired boilers, the overall efficiency is improved by 1 to 2 percent. The capital cost is slightly lower than that of the methanation route.

The calorific value of the product from hydrogasification is lower than that from single methanation (Fig. 7), particularly with high carbon/hydrogen feedstocks because of the

TABLE 4.   Gas Analyses—SNG Production by Hydrogasification

	Recycle gas	1st-stage gas	2d-stage gas	3d-stage gas	Scrubbed gas	Product gas
CO_2, mole %	0.5	21.8	21.7	21.9	0.5	0.50
CO	1.6	0.7	0.6	0.1	0.1	0.07
H_2	86.5	13.3	6.2	0.6	0.8	0.79
CH_4	11.4	64.2	71.5	77.4	98.6	97.56
C_3H_8	—	—	—	—	—	1.08
	100.0	100.0	100.0	100.0	100.0	100.0
CV* (Btu/SCF)	395	683	733	772	984	1000

*CV = calorific value.

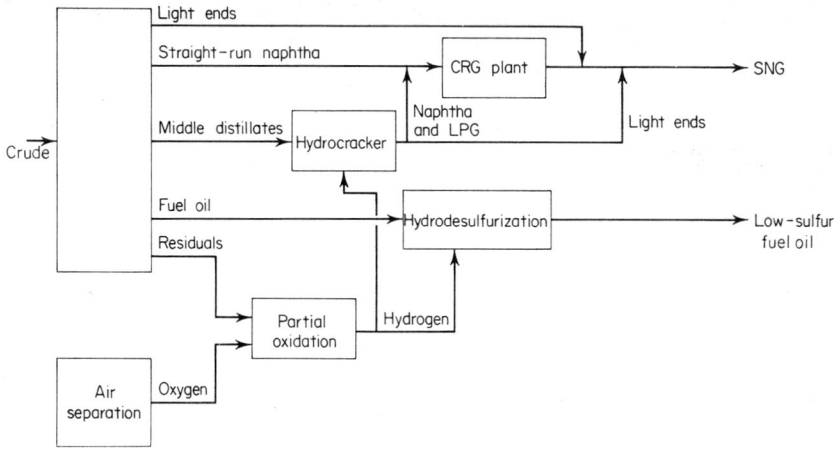

Fig. 8. SNG and low-sulfur fuel oil from crude.

additional steam required. However, by adding a final methanator the calorific value can be increased to that obtained from double methanation, again with increased capital cost and reduced efficiency. This process (flowsheet, Fig. 6) is used in the first operational SNG plant in the United States at Harrison, N.J. (Ref. 10). Typical gas analyses are given in Table 4.

Because of the lower temperature, the catalyst in the hydrogasifier, which is the same as that in the CRG reactor, is slowly deactivated by polymer formation. The activity may be recovered in situ by heating in hydrogen. Two hydrogasifiers are therefore provided in parallel so that regeneration can be performed without interrupting production.

THE ENERGY REFINERY

The CRG process is one of a range of processes developed by the British Gas Corporation for production of fuel gases. The range and application of these processes and their impact have been described by Hebden and Percival (Ref. 11), illustrating the effect on capital cost of increasing carbon/hydrogen ratio of the feedstock.

An alternative method of handling crude oil is the "energy refinery" in which crude is split into a number of fractions which can be treated by proved processes to yield two products, SNG and low sulfur fuel oil. One such scheme is shown in Fig. 8. Others are described by Hazelton and Tennyson (Ref. 12).

The advantages of using the CRG process as the final stage in the production of SNG are high efficiency and low capital cost, the predictable quality of the product gas, and the absence of by-products.

NOMENCLATURE

"Calorific value" in this section refers to the higher (gross) heating value in Btu per cubic foot measured at 60°F (15.6°C), 30 inches Hg, saturated with water vapor.

REFERENCES

1. Cairns, J., and I. J. Hartill: British Gas Experience of High Pressure Gas-Making Processes, AGA/OCR/IGU Fifth Synthetic Pipeline Gas Symposium, Chicago, October 1973.
2. Dent, F. J.: Principles of the New Gas-Making Processes, *Gas J.*, **320**, 301–305, 335–341 (1964).
3. Long, G.: Why Methanate SNG, *Hydrocarbon Process.*, p. 91, August 1972.
4. Lacey, J. A., and S. K. Mukherjee: Pilot-Plant Investigations of the Desulphurization of Light Distillates, *J. Inst. Gas Eng.*, **7**, 473 (1967).
5. NG/LNG/SNG Handbook, *Hydrocarbon Process.*, p. 109, April 1973.
6. Bechtel Associates Professional Corp.: Chrloride Removal from Liquid Feedstocks, *Pipeline Gas J.*, **200**, 32, May 1973.

7. LNG Receiving Facility at La Spezia Nears Start-up, *Oil Gas Int.*, **9**, 85, November 1969.
8. Hart, F. E., N. C. Baker, and I. Williams: CRG Route to SNG, *Hydrocarbon Process.*, p. 94, April 1972.
9. Davies, H. S., and J. A. Lacey: CRG Process is Ideal Route to SNG, part II, *Pet. Petrochem. Int.*, **13**, 50, April 1973.
10. DiRienzo, J. B.: The Harrison SNG Plant, AGA/OCR/IGU Fifth Synthetic Pipeline Gas Symposium, Chicago, October 1973.
11. Hebden, D., and G. Percival: New Horizons for Pressure Gasification, *J. Inst. Gas Eng.*, **12**, 229 (1972).
12. Hazelton, J. P., and R. N. Tennyson: SNG Refinery Configurations, *Chem. Eng. Prog.*, **69**, 97, July 1973.

Methane-Rich Gas Process for Substitute Natural Gas

BY SHOGO NOJIMA*

The methane-rich gas (MRG) process, which produces methane gas from feedstock hydrocarbons, such as naphtha, liquefied petroleum gases (LPG), and refinery gas, was developed by Japan Gasoline Co., Ltd., in collaboration with its affiliate, Nikki Chemical Co., Ltd., for application to town-gas facilities. The MRG process had its origin in the high-temperature hydrocarbon steam reforming technology. First efforts culminated in a successful installation in 1956.

In the tendency to employ heavier hydrocarbons as feedstock, carbon formation in the low-temperature range posed a problem. The Japan Gasoline Co. (JGC) took this up as a main research subject and continued the study of low-temperature-range reaction, concentrating especially on the difference of product gas properties and carbon formation according to reaction conditions and catalyst specifications. As a result, a new catalyst was developed which converts butane or naphtha to a gas consisting mainly of methane, hydrogen, and carbon dioxide, with a negligible amount of carbon monoxide. This was the first stage of development of the present MRG process.

In 1964, while developing practical applications for the town-gas industry, a pilot plant with a daily capacity of 15,000 cubic meters† was built. Continuous test runs were conducted over a long term, in cooperation with Osaka Gas Co., Ltd., thus starting commercial production of equipment for the process.

Based on the results of the aforementioned test runs, a commercial-size town-gas plant (200,000 cubic meters daily capacity) was constructed at the Hokkoh plant of Osaka Gas Co., Ltd. This was followed by two plants each of 500,000 cubic meters daily capacity in 1967 and 1969. The three plants are currently enjoying successful production performance.

Additional plants followed not only for town-gas uses, but also for petrochemical needs. Late in 1971, an MRG plant incorporating a wet methanation system went into operation at the facility of Keiyo Gas Co., Ltd., near Tokyo, with a capacity of 105,000 cubic meters per day. In late 1972, a complete MRG-based SNG plant consisting of gasification, methanation, and CO_2 removal sections was completed for the same firm, with a capacity of 200,000 cubic meters per day. In early 1974, Boston Gas Company (U.S.) started up a 1,070,000 cubic meters per day SNG plant which employs a two-stage MRG gasification system. All these plants are performing well.

Chemistry of the MRG Process. The basic reactions of the MRG process consist of three stages: (1) hydrodesulfurization of sulfur compounds in the hydrocarbon feedstock, (2) low-temperature steam reforming (gasification) of desulfurized hydrocarbons, and (3) methanation reaction between hydrogen and carbon dioxide in methane gas available by gasification.

Hydrodesulfurization. Sulfur compounds contained in hydrocarbon feedstock vary, depending on the types of crudes and their boiling points. Naphtha, for example, contains mainly mercaptans, disulfides, and thiophenes. Such sulfur compounds deteriorate the activity of the low-temperature steam-reforming catalyst (MRG catalyst). They should be removed to some degree before the hydrocarbon feedstock enters the MRG system. This is one of the important steps of the MRG process.

---

*Director, Japan Gasoline Co., Ltd., Tokyo.
†Cubic meters described in this article are Nm³ (a gas volume at 0°C (32°F) and 1 atmosphere (14.7 psi). 1 Nm³ = 37.3 SCF (standard cubic feet).

Major reactions of the hydrodesulfurization step are:

$$RSH + H_2 \rightarrow RH + H_2S$$
$$R\text{-}S\text{-}R' + 2H_2 \rightarrow RH + R'H + H_2S$$
$$R\text{-}S\text{-}S\text{-}R' + 3H_2 \rightarrow RH + R'H + 2H_2S$$

$\underset{S}{\boxed{}}\!\!R + 3H_2 \rightarrow RC_4H_9 + H_2S$ (alkyl group R not fixed to a specific carbon)

Inasmuch as these are all exothermic reactions, low ambient temperatures are favorable from the standpoint of equilibrium theory, but in consideration of reaction rates, general processes are operated in a range of 350 to 400°C (662 to 752°F) with the aid of highly active catalyst (like Co-Mo or Ni-Mo), involving the side reactions:

$$CO_2 + 4H_2 \rightleftharpoons CH_4 + 2H_2O$$
$$CO + 3H_2 \rightleftharpoons CH_4 + H_2O$$

The foregoing reactions are highly exothermic and significantly raise reaction temperatures. The MRG process, however, does not involve such adverse side reactions with use of a special, selective hydrodesulfurizing catalyst.* The MRG process uses part of product gas for hydrodesulfurization, and even if it contains only 20 to 25 percent hydrogen and as high as 20 to 23 percent carbon oxides, only the proper hydrodesulfurization reactions take place. The MRG process features a recycle use of product gas for hydrodesulfurization purposes without any special treatment.

To eliminate hydrogen sulfide formed in hydrodesulfurization reactions, two solutions are available: (1) fixation of H_2S by contact with an adsorbent (zinc oxide) via the reactions: $ZnO + H_2S \rightarrow ZnS + H_2O$; and (2) physical removal by stripping. The selection of the particular solution depends on the feedstock sulfur content.

The JGC hydrodesulfurization system is most economically applicable to feedstocks containing less than about 200 ppm sulfur. The removal of H_2S by stripping after hydrodesulfurization with an external hydrogen supply may be applied to naphtha stocks contaminated by trace metals as well as those high in sulfur.

Low-Temperature Steam Reforming Reaction.   Gasification by low-temperature steam-reforming reactions, the heart of the MRG process, is conducted between liquid hydrocarbons and steam over catalyst to form methane, hydrogen, and carbon oxides. The procedures are not clearly known, but the essence is as follows:

$C_nH_m + H_2O \rightarrow CO + H_2$	(gasification)	(1)†
$CO + H_2O \rightleftharpoons CO_2 + H_2$	(shift reaction)	(2)
$CO + 3H_2 \rightleftharpoons CH_4 + H_2O$	(methanation)	(3)
$C_nH_m + H_2 \rightleftharpoons CH_4$	(hydrocracking)	(4)

The gasification reaction (1) is endothermic, while the others, particularly the methanation reaction (3), are exothermic. When the four types of reactions are balanced by the adjustment of temperatures, steam/hydrocarbon ratio, and pressures, the gasification reaction can be in balance thermodynamically and no outside heat sources are needed. Typical temperature profiles of the catalyst bed are shown in Fig. 1.

Alteration of one of the reaction conditions, such as decreasing the H_2O/C ratio or lowering the inlet temperature, makes the exothermic reaction become dominant, and the total reaction process displays an exothermic tendency. If the conditions are reversed, the tendency turns to endothermic. In many cases fluid temperature sharply drops just after it enters the reactor. A remarkable rise in temperature follows and, subsequently, the temperature becomes constant. This may be explained as follows. The first temperature decrease results because of the endothermic reforming reaction (1) which is dominant in the reactor. As steam reforming of hydrocarbons progresses, (a) the partial pressure of hydrogen increases; (b) hydrocracking reaction (4) goes on; and (c) heat generation via methanation reaction (3) proceeds, resulting in a marked rise in temperature. At the end of the reactions, equilibrium is reached, keeping the temperature constant thereafter.

Figure 2 shows the process of the gasification of feed hydrocarbons observed in a pilot

---

*Jointly developed by JGC and Nikki Chemical.
†C_nH_m indicates a feed hydrocarbon and is used for convenience to express the feed hydrocarbon composed of various hydrocarbons derived from petroleum.

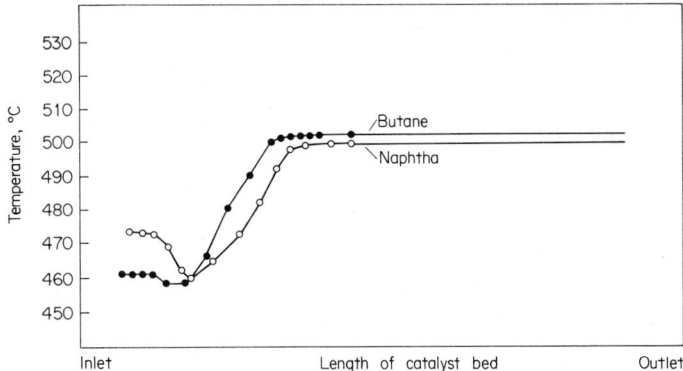

Fig. 1. Temperature profile of catalyst zone of MRG reactor. (1 kg/(cm²)(G) = 14.22 psig.) Butane: 20 kg/(cm²)(G), 1.5 H₂O/C mol/atom. Naphtha: 20 kg/(cm²)(G), 1.7 H₂O/C mol/atom.

plant. The practical gas composition deviates from the calculated equilibrium composition—higher in methane content and catalyst bed temperature, 20 to 40°C (68 to 104°F). When feed hydrocarbons are completely gasified, the figure is consistent with the calculated composition and temperature. This signifies that during the gasification process the effect of hydrocracking is strong. This function permits the gasification reaction to proceed, keeping the catalyst bed temperature at a range of 350 to 550°C (662 to 1022°F) by the control of heat absorption by steam reforming.

The function of hydrocracking effects a reduction in gasification catalyst required by increasing the hydrogen-to-feed hydrocarbons ratio. Thus, the initial gasification of feedstock plus subsequent gasification of the remaining feedstock, together with hydrogen available from the initial gasification, sets up optimal conditions for hydrocracking.

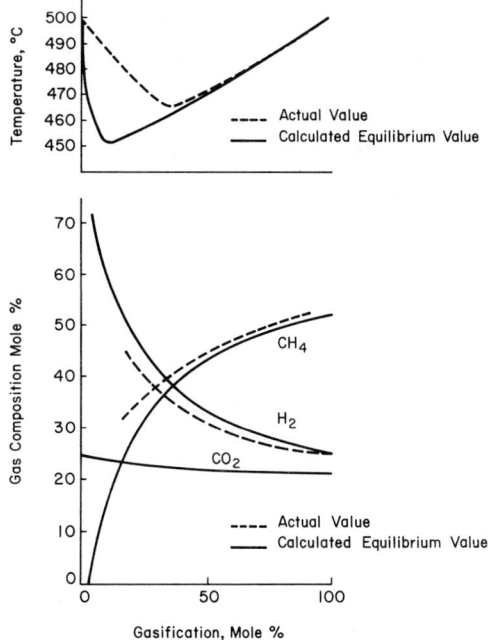

Fig. 2. Relationship of hydrocarbon gasification to temperature and gas composition.

This is called a two-stage gasification process and is used when high-methane-content gas is desired, as in SNG production. Steam/feedstock ratio in terms of weight is 1.8 to 2.5 when a single gasifier is used. The two-stage gasification process reduces the ratio to 1.2 to 1.4. Direct supply of hydrogen to the initial gasifier from outside sources produces the same effects. This is particularly effective in the gasification of hydrocarbons having high boiling points. As catalyst performance deteriorates, the width of the depressed portion of the curve of Fig. 1 expands toward the reactor outlet. But, because reaction is completed at the top of the ascending curve, the composition of the product gas remains at equilibrium so long as the flat portion is observed on the outlet side. This state remains unchanged so long as the feedstock and reaction condition are kept constant, with no unprocessed feedstock detected.

If reaction conditions (temperature, pressure, and steam ratio) are given, product gas composition can be estimated by equilibrium calculations. Figure 3 shows the equilibrium composition of product gas using naphtha (with a mean molecular formula $C_{5.9}H_{13.4}$) as feedstock. The diagram indicates obvious effects of reaction temperature, reaction pressure, and steam ratio on the composition and calorific value of the product gas.

Methanation Reaction.   In order to increase the calorific value of product gas to the values similar to those of natural gas methanation reactions are required. Hydrogen in product gas is reacted with CO_2 and CO to form methane, with only a small portion unconverted. Methanation reactions are:

$$CO + 3H_2 \;\rightleftharpoons\; CH_4 + H_2O \;-\; 49.3 \text{ kcal/g-mole at } 25°C$$
$$CO_2 + 4H_2 \;\rightleftharpoons\; CH_4 + 2H_2O \;-\; 9.8 \text{ kcal/g-mole at } 25°C$$

The relationships of reaction temperatures, reaction pressures, and product gas methanation are given in Fig. 4. After methanation, the gas goes to a scrubber to remove CO_2 for further purification.

Commercially, both of Keiyo Gas Co's No. 1 and No. 2 MRG plants perform gasification with a single reactor and methanation by a wet methanation process with a single reactor. Figures 5a and b show axial temperature profiles of the methanators. The process conditions are given in Table 1, along with the actual product gas composition and the equilibrium composition calculated on the basis of API Project 44 data.*

Table 1 apparently shows that methanation reactions reach chemical equilibrium. Further evidence of equilibrium is the axial temperature profile of Figs. 5a and b. The constant temperature after the characteristic rises indicates that no further reactions are occurring in the methanators.

The second-stage methanator can be used when higher gas calorific value is desired.

Figure 6 and Table 2 show the axial temperature profile and practical product gas composition obtained from JGC's adiabatic pilot plant, as compared with the predicted temperature profile and the calculated gas composition, respectively. These data are nearly consistent.

PROCESS FLOW

The MRG process consists of three basic sections: (1) desulfurization, (2) gasification, and (3) gasification plus CO_2 removal and heat recovery. A schematic flowsheet is given in Fig. 7.

Desulfurization.   Feedstock (e.g., naphtha) is preheated and vaporized by heat exchange with hot effluent gas from the methanation reactor. Mixed with part of product gas from the MRG reactor as a source for hydrodesulfurization, the vaporized feedstock naphtha is heated to 350 to 400°C (662 to 752°F). The heated naphtha is then fed into the hydrodesulfurization reactor where sulfur compounds contained in the naphtha are converted into hydrogen sulfide. In this process, a special, selective catalyst is used. The desulfurizer effluent is passed over a zinc oxide bed which adsorbs H_2S. This desulfurization scheme is especially suitable for "grass roots" and "peak-shaving" facilities. Virtually, feedstocks of any sulfur content may be processed by replacing zinc oxide according to feed sulfur concentrations. The desulfurized effluent is ready for gasification.

Gasification.   Desulfurized/refined naphtha, after being mixed with a preset quantity of superheated steam, is heated to reaction temperature of 480 to 550°C (896 to 1022°F)

*Data on physical and chemical properties of hydrocarbons specified by the American Petroleum Institute. Formally called American Petroleum Institute Research Project 44.

Feed...Straight-run Naphtha H/C...2.27 H₂O/C...1.7

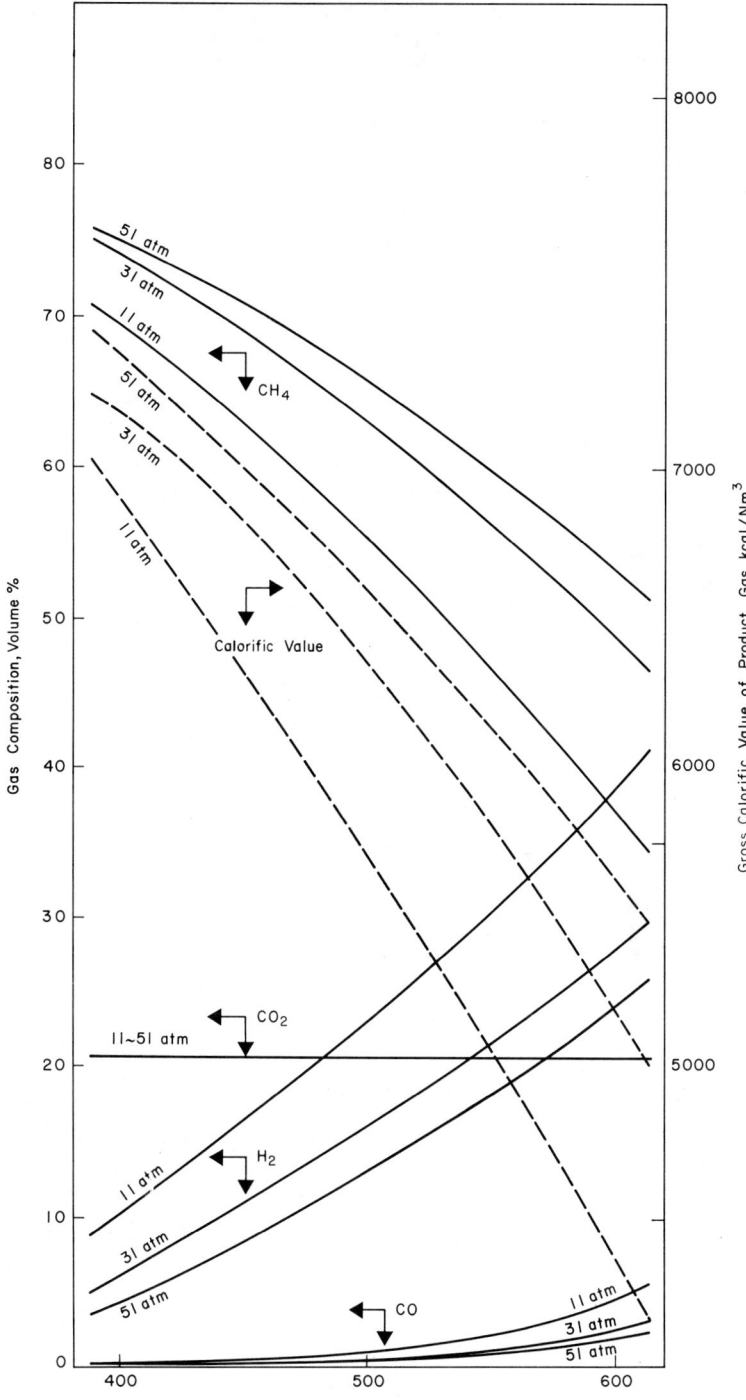

Fig. 3. Equilibrium composition and calorific value of MRG reactor effluent. Nm³ = cubic meters at 0°C (32°F) and 14.7 psi; 1 Nm³ = 37.3 standard cubic feet.

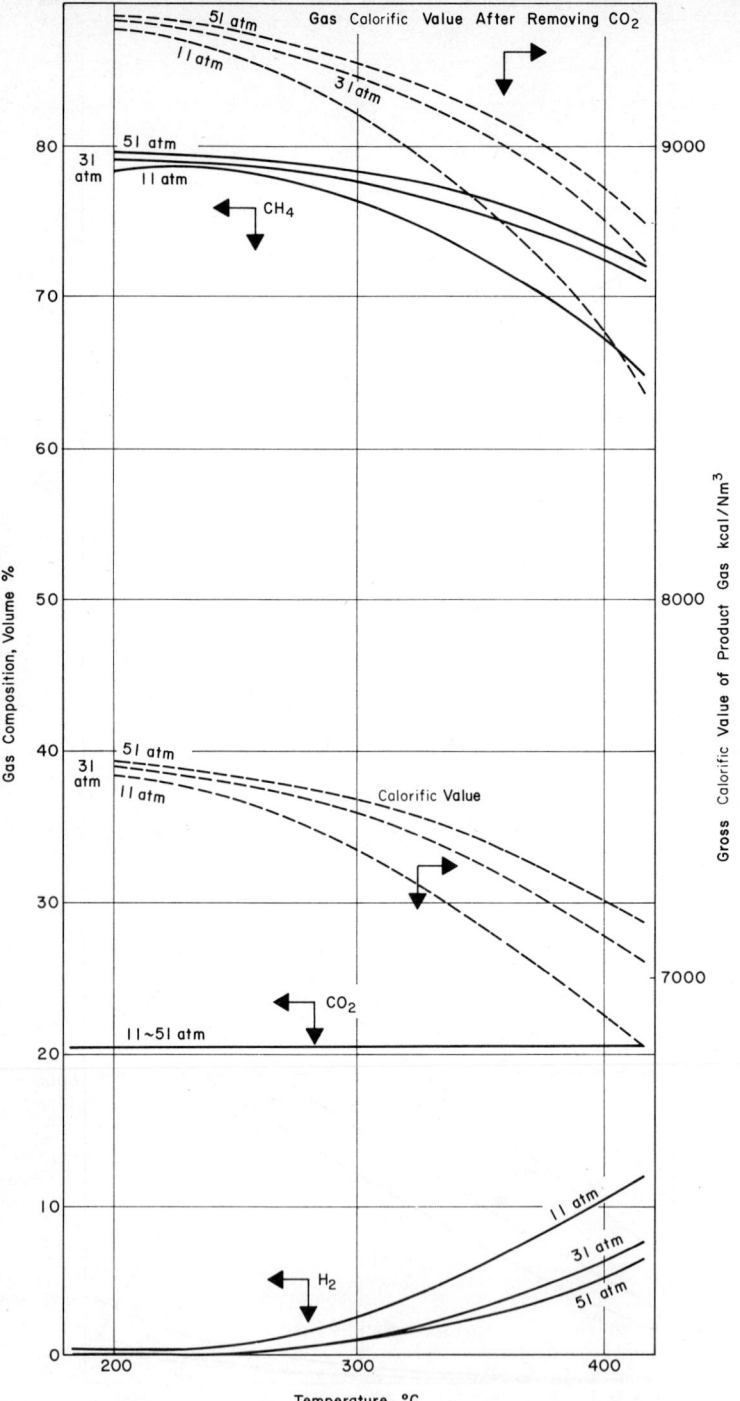

Fig. 4. Equilibrium composition and calorific value of methanator effluent. (See note on Fig. 3.)

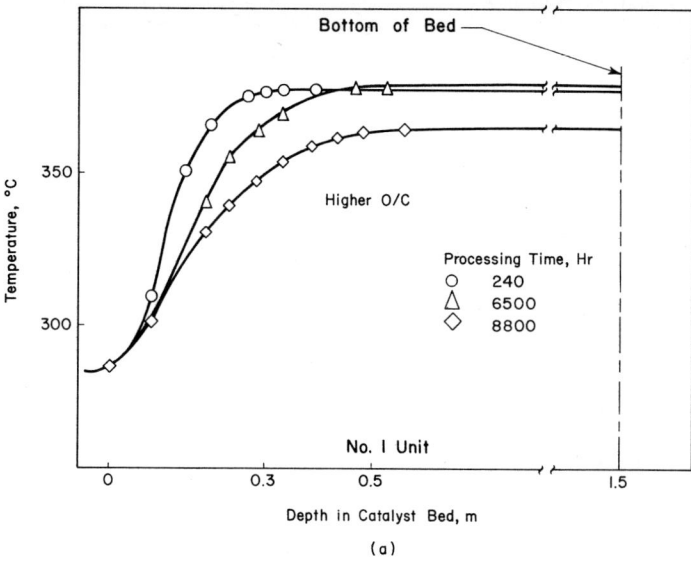

(a)

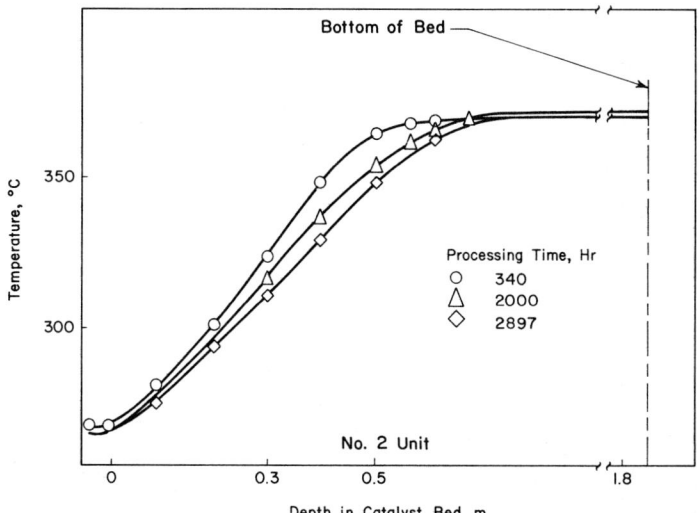

Fig. 5. Axial temperature profile in methanator: (*a*) In no. 1 unit. (*b*) In no. 2 unit. (*Keiyo Gas.*)

and then goes to the MRG reactor. The number of reactor stages depends on feedstocks, plant size, and other economic factors. However, generally, the two-stage reactor design (shown in the flowsheet) effects reduction in the total steam/naphtha ratio. This reduces utility requirements and increases thermal efficiency. The low steam/naphtha ratio leads to increased methane yields.

In the No. 1 reactor, about 60 percent of the entire feed naphtha is charged and mixed with steam (approximately 1.8 to 2.5 times the weight of the naphtha). The No. 1 reactor effluent gas is mixed with the remaining 40 percent naphtha and goes to the No. 2 reactor. In the No. 2 reactor, no additional steam is required. The steam entrained in the No. 1 reactor effluent also serves for gasification in the No. 2 reactor, thus reducing feed steam volume by 40 percent as compared with the single-reactor system.

TABLE 1. Process Conditions and Product Composition of Methanation Section (Keiyo Gas)

	No. 1 unit Design	No. 1 unit Practice	Equilibrium	No. 2 unit Design	No. 2 unit Practice	Equilibrium	After CO_2 removal	Final product
Process condition:								
Inlet temperature, °C	311	286		266	259			
Outlet temperature, °C	400	364		371	377			
Outlet pressure, kg/(cm²)(G)	11.8	9.7		11.5	11.3			
Oxygen/carbon, atom/atom	1.65	2.40		1.70	1.76			
Inlet stream composition, mole %:								
Methane	29.95	22.42		29.23	28.37			
Hydrogen	12.13	10.80		12.20	12.34			
Carbon monoxide	0.64	0.29		0.63	0.62			
Carbon dioxide	11.14	9.04		10.98	10.89			
Steam	46.14	57.45		46.96	47.78			
Product gas composition mole %—dry basis:								
Methane	68.96	69.37	69.30	70.87	71.08	71.00	89.07	87.19
Hydrogen	9.97	9.59	9.61	8.16	7.96	7.90	9.97	9.76
Carbon monoxide	0.18	0.050	0.068	0.09	0.068	0.094	0.09	0.09
Carbon dioxide	20.89	20.99	21.02	20.88	20.88	21.01	0.87	0.85
Butane	—	—	—	—	—	—	—	2.11

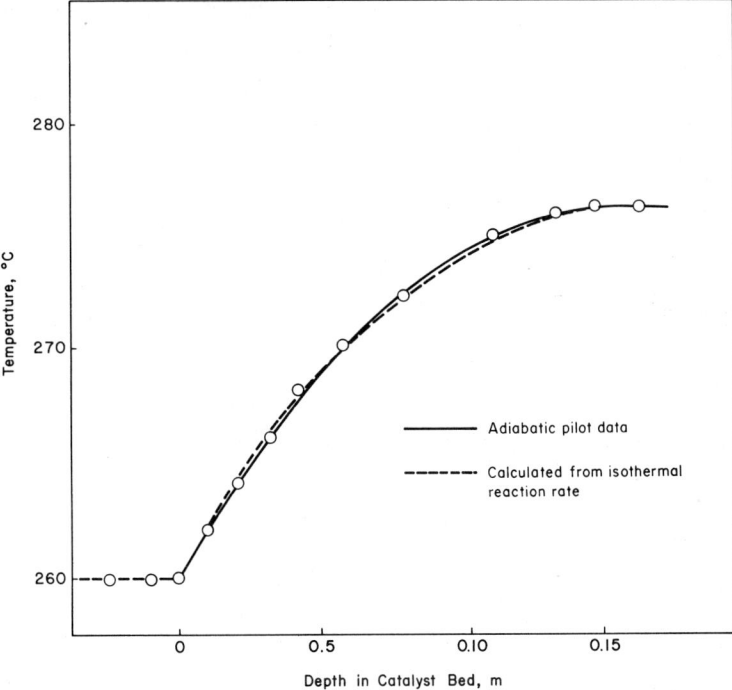

Fig. 6. Temperature profile in second-stage methanator of Japan Gasoline Co., Ltd. pilot plant.

In both reactors, stoichiometric balance and thermodynamic equilibrium are reached, requiring no additional heat to maintain reaction conditions. Pressure and temperature are adjusted to favor methane production. Lower temperature, higher pressure, and lower steam/hydrocarbon ratio accelerate methane production while suppressing hydrogen and carbon monoxide formation.

Methanation.   The MRG reactor effluent is cooled by generating steam and enters the methanator where it is methanated by reactions among hydrogen, CO_2, and carbon monoxide. Lower temperature favors the progress of methanation reactions.

TABLE 2.   Experimental Results on Second-Stage Methanator

		Equilibrium
Process condition		
Inlet temperature, °C	260	
Outlet temperature, °C	278	
Outlet pressure, kg/(cm²)(G)	15.8	
Inlet stream composition, mole %		
Methane	49.60	
Hydrogen	2.67	
Carbon monoxide	0.06	
Carbon dioxide	14.74	
Steam	32.93	
Product gas composition, mole %—dry basis		
Methane	77.02	77.12
Hydrogen	0.94	0.92
Carbon monoxide	0.01	0.01
Carbon dioxide	22.03	21.95

To lower the level of residual hydrogen in the product gas, one or more methanators may be used. Hydrogen content of the product gas is as low as 3 percent where a single methanator is used. Intermediate cooling plus a second methanator allow further low-hydrogen content.

Carbon Dioxide Removal.   The methanator effluent contains about 20 percent surplus CO_2, which is absorbed by monoethanolamine or hot potassium carbonate solution to

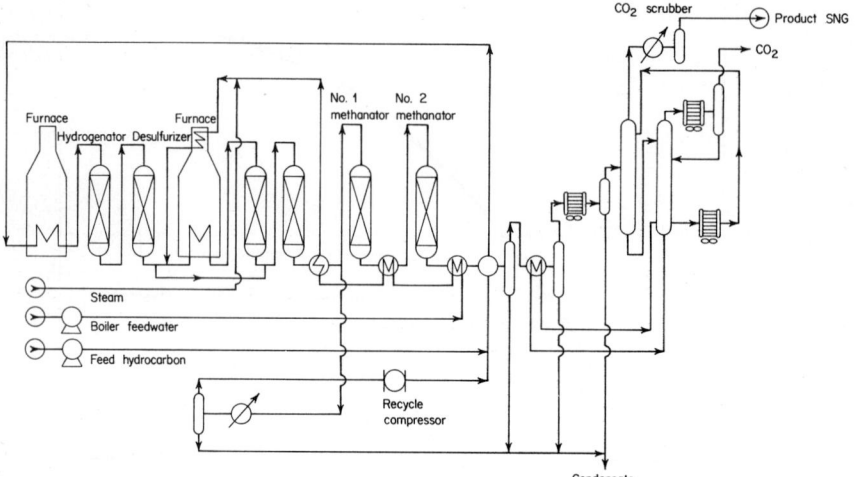

Fig. 7. Main steps in MRG process for production of substitute natural gas.

produce high-purity methane gas. By cooling before entering the CO_2 stripping section, steam is condensed in the reboiler, where it is used to heat the fat absorber solution. It is possible to reduce CO_2 content to as low as 0.1 percent, but levels from 1 to 2 percent are considered most economical.

Purified product gas is dried, as necessary. For this purpose, glycol solution (for scrubbing), refrigeration, or a combined hygroscopic agent of activated alumina and silica gel may be used. Inasmuch as the discharged condensates from the process contain no aldehyde, organic acid, polymer, or dissolved catalyst, they are recycled as boiler feedwater.

Control Factors.   Because the MRG process is mainly based on steam reforming, the success of the process hinges on the reliable availability of steam. Steam should be controlled at a constant level somewhat above the projected requirements to assure continuous, effective reforming. Adequate steam must be available at all times to assure effective control of the steam/naphtha ratio.

Changes in pressures in the process lines threaten the maintenance of desired gas composition, and thus proper control systems should be provided to avoid pressure fluctuations.

If there is power or equipment failure, the plant should be safely shut down by means of a carefully engineered instrumentation system. It is necessary to follow a sequence of shutdown operations so that the catalyst will not be affected.

Care must be exercised to avoid a reactor inlet temperature that may be too low or too high. Desulfurizer inlet temperature should not exceed 390°C (734°F) because naphtha heated beyond a certain point will cause coking. The gasification reactor inlet temperature should be held within a range of 350 to 550°C (662 to 1022°F). The methanator inlet temperature should be above the fluid dew point. Operation of the CO_2 removal section involves no particular difficulties. It is necessary to maintain the proper concentration of chemicals to assure maximum absorption efficiency and prevention of corrosion to process lines. The solutions should be kept as clean as possible.

FEEDSTOCK SPECIFICATIONS AND THE MRG
PROCESS IN THE REFINERY

It is preferable for an SNG process to be capable of using as feedstocks as many kinds of hydrocarbons as possible and as heavy fractions as possible. Initially, available feedstock to the MRG process was straight-run naphtha up to 185°C end point. Today, the availability is as wide as to cover kerosine fractions up to 315°C end point. This has been attached by improvements in gasification catalyst. Aside from the foregoing, the MRG process can handle, as feedstock, many other fractions compatible with the MRG catalyst characteristics: olefin-rich hydrocarbons like butane-butene fractions available as by-product from ethylene and fluid-catalytic cracking plants; return naphtha after aromatics extraction, concentrated hydrocarbons separated from natural gas, and crude benzene produced as by-product from coke ovens.

SUMMARY OF TECHNICAL FEATURES

For the proper feedstocks and economic situations, the MRG process offers the following characteristics: (1) a wide selection of feedstock; (2) a broad selection of calorific value of product gas (product gas is available in a range of calorific values from 5500 to 9400 kcal/cubic meter); (3) high-pressure operation [Many conventional town-gas plants which operate at near atmospheric pressure require an additional compressor to convey product gas through pipelines. The MRG process does not require any additional compressor, but does permit operation at high pressures—up to 80 kilograms/square centimeter gage (approximately 1140 pounds/square inch gage), enabling high-pressure, long-distance transportation of product gas]; (4) noncomplex equipment (The use of a drum-type reactor makes the reactor design quite simple, resulting in a compact design of the overall system. In terms of product gas calorific value, the MRG reactor requires only one-sixth to one-eighth of the area of a general coke oven and about one-third that of a conventional high-temperature steam-reforming plant); (5) sulfur-resistant catalyst (Up to several parts per million, sulfur compound content of feedstock hydrocarbons has no effect on the highly active nickel-based catalyst over a temporary time span); and (6) high thermal efficiency. Operation at low temperatures and high pressures permits thermal efficiency as high as 92 to 96 percent, depending on the desulfurized feedstock used.

REFERENCES

Ishiguro, T.: Make Town Gas from Naphtha, *Hydrocarbon Process*, **47** (2), February 1968.
Nojima, S.: *Chem. Econ. Eng. Rev.*, **4** (2), February 1972, Chemical Economy Research Institute, Tokyo, Japan.
Okagami, Uemoto, and Morikawa: *ACS Preprint* **18** (2): 410–413, March 1973.
Ward, D. J. and S. Nojima: SNG by the MRG Process, *Gas*, May 1973.
Ward, D. J.: SNG Future Tied to Heavy Feeds, *Hydrocarbon Process.*, **53** (4), April 1974.
Thornton, Ward, and Erickson: MRG Process for SNG, *Hydrocarbon Process.*, **51** (8): 81, August 1972.
Methanation Experience at Keiyo Gas Co., 5th Synthetic Pipeline Gas Symposium, Oct. 29–31, 1973, Chicago.
Richardson, J. T.: SNG Catalyst Technology, Hydrocarbon Process., **52** (12), December 1973.

The Fluidized Bed Hydrogenation Process for Substitute Natural Gas Production

By JOSEPH F. McMAHON*

The fluidized bed hydrogenation (FBH) process involves the noncatalytic reaction of hydrocarbons and hydrogen to produce a methane- and ethane-rich gas which is further processed to substitute natural gas (SNG). The FBH process is particularly suited for the gasification of whole crude oil or crude oil fractions boiling above naphtha.

The British Gas Corporation started work on crude oil gasification in the mid-1950s, based on earlier studies of coal gasification. A recirculating fluidized bed reaction was developed using pilot plants with gas capacities up to 1 million standard cubic feet per day. A semicommercial FBH plant with a rich gas capacity of 5 million standard cubic feet per day was built and is being operated by the Osaka Gas Company. This plant, operating at pressures of 750 to 825 psig, has gasified whole crude oil and crude oil distillates to produce gases with heating values of 780 to 900 Btu per standard cubic foot. Development of the FBH process is continuing at the Osaka semicommercial plant and at the laboratories of the British Gas Corporation.†

Reaction of hydrocarbons with a hydrogen-containing gas produces methane according to the reaction

$$C_nH_{2n+b} + \frac{2n - b}{2} \ H_2 \rightarrow nCH_4$$

At elevated pressure and in the temperature range of 1300 to 1470°F (704 to 799°C), a rapid thermal reaction occurs between hydrogen and the paraffinic and naphthenic compounds in the oil feedstock. Ethane is also produced since it is an intermediate in the conversion of higher hydrocarbons to methane and reacts relatively slowly with hydrogen. Small quantities of olefinic compounds also appear in the product gas, particularly when the reaction is conducted at low pressures, or at low hydrogen/oil ratios.

Aromatic compounds in the oil feedstock are extensively dealkylated under normal reaction conditions, but the basic ring structures survive. Reaction at temperatures above about 1470°F (799°C) leads to aromatic ring decomposition.

A simplified flow diagram of an FBH process plant is shown in Fig. 1. The unit consists of a reaction and quench section, gas purification, secondary hydrogenation, and gas drier. Oil and hydrogen gas are preheated and then reacted in a fluidized bed of coke particles, producing a rich gas which contains a high concentration of methane and ethane. The reaction is carried out at temperatures of about 1400°F (760°C) and pressures up to 1000 psig or higher.

The extent of gasification of the hydrocarbon feed varies with operating conditions and the properties of the feedstock. Under normal conditions, 50 to 90 percent of the carbon contained in the feed is converted to gaseous hydrocarbons, depending on the paraffinicity of the hydrocarbons used. Feedstock that is not gasified is converted primarily to an aromatic liquid which contains benzene, toluene, xylenes, naphthalene, and higher-boiling aromatics. A small amount of coke, generally less than 5 wt% of the feed, is withdrawn from the FBH reactor through a cooler and lock hopper system.

Product gas from the FBH reactor is quenched with a circulating stream of light aromatics to effect a separation of normally liquid products from the rich gas.

A portion of the sulfur contained in the oil feed is converted to hydrogen sulfide in the

---

*Manager, Process Technology Department, Process Plants Division, Foster Wheeler Energy Corporation, Livingston, N.J.

†Foster Wheeler Energy Corporation is licensed directly by the British Gas Corporation to design and construct FBH process plants.

FBH reactor. Hydrogen sulfide is removed from the rich gas in a conventional absorption-stripping system and sent to a sulfur recovery unit for conversion to elemental sulfur. After desulfurization, in addition to methane and ethane, the rich gas contains 10 to 40 mole percent hydrogen, depending on the feedstock and operating conditions used. The higher heating value of the rich gas ranges from about 750 to above 900 Btu per standard cubic

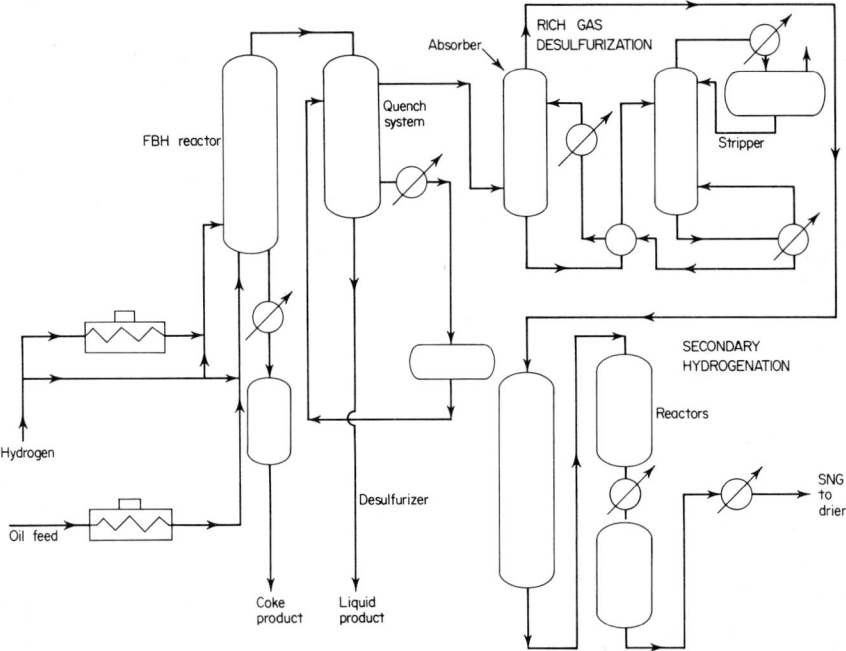

Fig. 1. The fluidized bed hydrogenation (FRH) process *(British Gas Corporation.)*

foot on a dry basis. In order to produce substitute natural gas, the hydrogen concentration must be reduced to bring the gas Wobbe number* and flame speed factor† within acceptable limits.

The final conversion step can be performed in a secondary hydrogenation section. The gas, after purification to remove traces of sulfur compounds, is contacted with a catalyst in a fixed-bed reactor. Ethane contained in the rich gas reacts with hydrogen to produce methane. If there is an excess of hydrogen in relation to ethane in the gas, carbon dioxide or benzene recovered from the FBH reactor effluent would be added upstream of the secondary hydrogenation step. These materials would then react with excess hydrogen to produce gas containing less than 5 mole percent hydrogen.

The secondary hydrogenation reaction is exothermic although the heat of reaction is considerably less than that of the carbon monoxide–hydrogen methanation reaction. Limited steam addition is used to moderate the reaction temperatures and avoid carbon deposition.

An alternative to secondary hydrogenation is the physical separation of hydrogen from the rich gas. This method separates unreacted hydrogen for recycle to the FBH reactor and leaves a product gas containing substantial amounts of both methane and ethane.

Product gas is dried conventionally to meet pipeline water vapor specifications. Since the FBH reactor can operate at a pressure of 1000 psig, product gas is available at a

---

*Ratio of the higher heating value of a gas divided by the square root of the specific gravity of the gas relative to air.

†Ratio, expressed as a percentage, of the flame speed of a gas-air mixture to the flame speed of a hydrogen-air mixture.

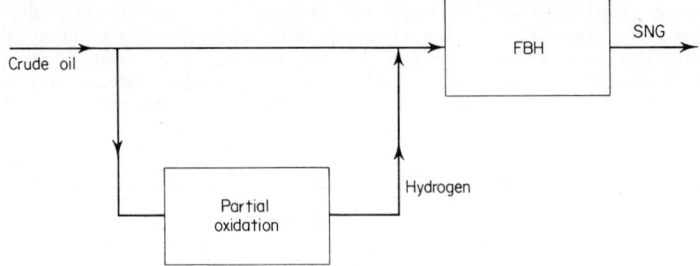

Fig. 2. Production of SNG from whole crude oil by FBH process.

pressure of 850 to 900 psig so that final gas compression for pipeline transmission is not required.

A wide range of oil feedstocks can be gasified in the FBH process to produce SNG. Whole crude oil, distillate oils, and residual oil fractions can be gasified. Desulfurization of the feedstock before gasification is not required since the FBH reaction is carried out noncatalytically. High-boiling compounds which may be present in the FBH reactor feed are adsorbed on coke particles in the fluidized bed during reaction. As a result, deposition of coke on reactor internals does not occur.

In addition, hydrogen required for the FBH reaction need not be of high purity. It can be supplied admixed with other compounds, such as methane or carbon monoxide, allowing the effective use of hydrogen-containing gases from a variety of sources. Two examples which suggest a wide range of application include:

1. Substitute natural gas may be produced from whole crude oil or atmospheric residues. This type of operation, shown in Fig. 2, would apply to situations where SNG is the only product desired, in contrast with fuels refining plants where both gas and liquid oil products must be produced.

2. Very heavy residual oils are difficult to hydrogasify because of the high concentration of polynuclear aromatic compounds in these materials. A combination processing sequence for these feedstocks, shown in Fig. 3, involves initially a solvent deasphalting of the residue. The deasphalted oil, having a higher paraffinicity than the residue, is then charged to the FBH process. Hydrogen required is produced by partial oxidation of deasphalter bottoms. In this way, high sulfur vacuum residue is converted to SNG.

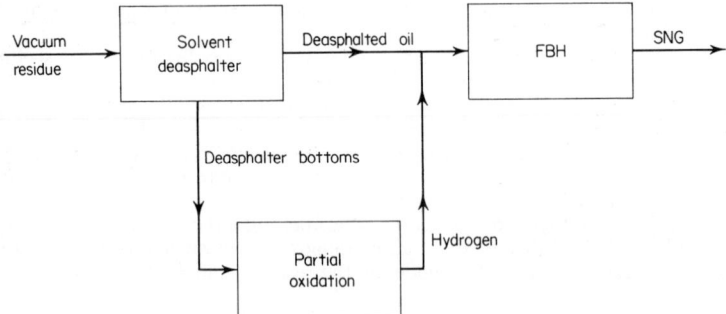

Fig. 3. Production of SNG from crude-oil vacuum residue by FBH process.

REFERENCES

McMahon, J. F.: Fluidized Bed Hydrogenation Process for SNG, *Chem. Eng. Prog.,* **67** (12): 51, December 1972.

Thompson, B. H., and C. T. Brooks: The Production of Methane by the Thermal Hydrogenation of Hydrocarbon Oils, paper presented at the 165th ACS National Meeting, Dallas, Tex., Apr. 8–13, 1973, American Chemical Society, Washington.

Ohoka, I., and H. L. Conway: Progress in the Gasification of Heavy Hydrocarbon Oils in a Recirculating Fluidized Bed Hydrogenator, paper presented at the 5th Synthetic Pipeline Gas Symposium, Chicago, Oct. 29–31, 1973, sponsored by the American Gas Association, Arlington, Va.

Dent, F. J., R. F. Edge, D. Hebden, F. C. Wood, and T. A. Yarwood: Experiments on the Hydrogenation of Oils to Gaseous Hydrocarbons, *Trans. Inst. Gas Eng.*, **106**, 694 et seq., London, 1966.

Thompson, B. H., Majumdar, B. B., and H. L. Conway: The Hydrogenation of Oils to Gaseous Hydrocarbons, *J. Inst. Gas Eng. (London*, **6**, 415 et seq., 1966.

Davies, H. S., J. A. Lacey, and B. H. Thompson: Processes for the Manufacture of Natural Gas Substitutes, *J. Inst. Gas Eng. (London)*, **9**, 375 et seq., 1969.

Hydrocracking-Hydrogasification Process for Producing Pipeline Gas from Crude Oil

BY J. C. JANKA,* J. G. SEAY,† AND P. B. TARMAN‡

In a continuous, two-step process developed by the Institute of Gas Technology (IGT), crude oil can be hydrocracked to approximately Diesel oil weight and then made to react noncatalytically with hydrogen at elevated pressures (500 to 1500 psig) and temperatures (1100 to 1400°F, 593 to 760°C) to produce methane-rich gas containing about 30 vol % hydrogen and 10 vol % ethane. This gas can be desulfurized and methanated to yield pipeline gas. By adjusting the conditions of hydrocracking, enough heavy fuel oil can be produced to provide feedstock for hydrogen production by partial oxidation, or a low-sulfur fuel oil product can be made if desired.

Interest has centered on plants to produce substitute natural gas (SNG) from light distillate feedstocks, such as naphtha. However, when naphtha is in short supply, a process capable of converting the more plentiful crude and residual oils to pipeline gas is needed. A number of processes were developed over 20 years ago to provide supplemental gas in winter periods when demand is high. Because the supplemental gas was required for only 20 or 30 days during the year, only cyclic thermal cracking at atmospheric pressure was used, in order to avoid the high capital cost of more complex continuous processes. However, with the need for base-load gas, a continuous process would be feasible.

The major difficulty associated with the production of pipeline gas from crude and residual oils is carbon deposition during the gasification step, which leads to reactor vessel plugging. One approach to this problem has been to conduct pressure gasification with hydrogen in a fluidized bed of coke, allowing carbon to deposit on the coke particles. To avoid accumulation of these deposits, a small amount of coke is continuously withdrawn from the bed. This technique, developed by the British Gas Council (see Ref. 2), gasifies the oil in a single reaction step. One approach to the problem of carbon deposition taken by IGT (see Ref. 3) is to eliminate the carbon-forming materials in the heavy oil before hydrogasification by catalytic hydrocracking.

In developing the IGT concept, experiments were conducted on both the hydrocracking and the hydrogasification operations for a variety of feedstocks, ranging from kerosine to Bunker-C fuel oil. Distillate feeds required no hydrocracking.

RESULTS OF HYDROCRACKING EXPERIMENTS

In the late 1950s, IGT conducted tests to determine the operating conditions required to produce an oil suitable for hydrogasification. Operating conditions were varied as follows: (1) pressure 500 to 2500 psig; (2) average bed temperature 735 to 895°F (391 to 479°C); (3) feed oil space velocity 0.21 to 0.87 volume oil/(volume catalyst bed)(hr); and (4) hydrogen/oil feed ratio 240 to 320 standard cubic feet/gallon (referred to 60°F and 30 inches of mercury pressure).

Researchers found that the depth of hydrocracking required by the process design described here could be achieved at 825°F (441°C) with a space velocity of 0.42 vol/(vol)(hr) and 260 standard cubic feet/gallon hydrogen feed at 550 psig. The experimental results generally parallel contemporary commercial experience.

Because the main purpose of the hydrocracking tests was to determine the severity of hydrocracking required to produce an acceptable feedstock for hydrogasification, it was particularly important to determine the effect of hydrogen treatment on the Conradson Carbon Residue (CCR)-forming materials in the oils tested. The CCR test measures the

*Supervisor, Hydrocarbon Processing Research, †Senior Advisor, Management Sciences, ‡Director, Chemical Processing Research, Institute of Gas Technology, Chicago.

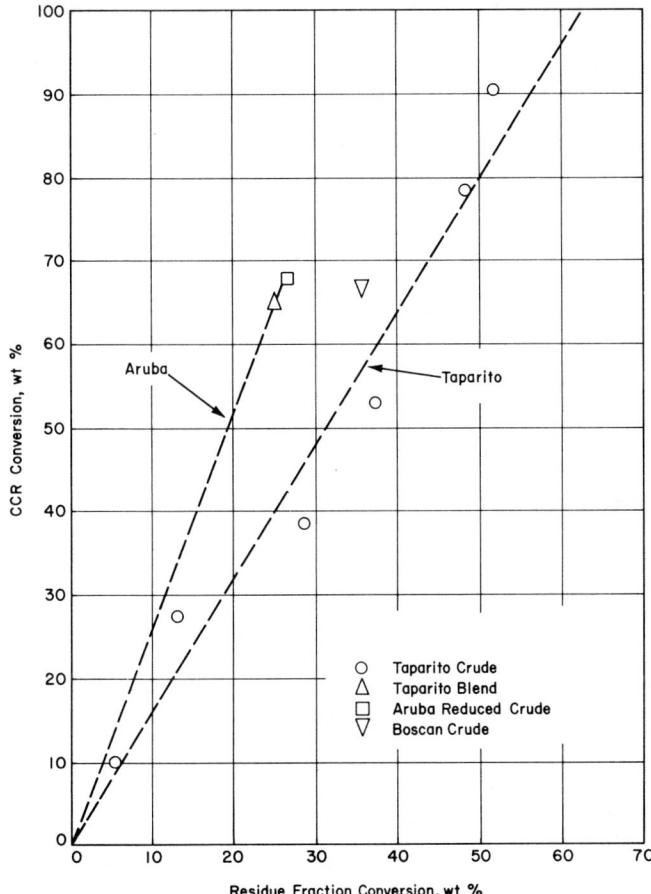

Fig. 1. Relation of Conradson carbon conversion (CCR) to residue fraction conversion.

amount of carbon deposition expected in thermal-cracking operations and can, therefore, be used to indicate the relative tendency of oils to deposit carbon in hydrogasification. Conversion of CCR-forming materials ranged from 1.5 to 2.5× the conversion of the high-boiling residue fraction (+680°F, 360°C) of the crude oils tested (Fig. 1). This indicates that it is not necessary to completely convert all the high-boiling residue fraction in order to produce an oil suitable for hydrogasification. Hydrocracking conditions can be relatively mild, especially if, as in the design given here, the heavy product oil, which contains most of the remaining CCR-forming materials, is used for hydrogen production.

RESULTS OF HYDROGASIFICATION EXPERIMENTS

Comprehensive tests were conducted to determine the reaction conditions required to eliminate carbon deposition and maximize gas product yields. Feedstocks used included kerosine, Diesel oil, and hydrocracked product oils. Experimental conditions were varied as follows: (1) temperatures from 1100 to 1400°F (593 to 760°C), (2) pressure from 500 to 1500 psig, (3) 50 to 100 percent of the stoichiometric hydrogen requirements for methane formation, and (4) 50-second residence time based on product gas flow for paraffinic oils to 750 seconds for the heavier oils.

The most important variables affecting product distribution are the stoichiometric hydrogen/oil ratio and the gasification temperature. Figure 2 summarizes the effects of stoichiometry with Diesel oil at constant operating temperature (1400°F, 760°C). The amount of carbon deposited from Diesel oil was not influenced by system pressure in the range studied, but was significantly affected by the hydrogen/oil feed ratio, dropping from 12 percent of the carbon in the feedstock at 50 percent stoichiometric ratio to 0 percent before a 100 percent stoichiometric ratio was attained.

The relative quantities of gaseous and liquid products at a given stoichiometric ratio depend primarily on temperature. An increase in operating temperature from 1100 to

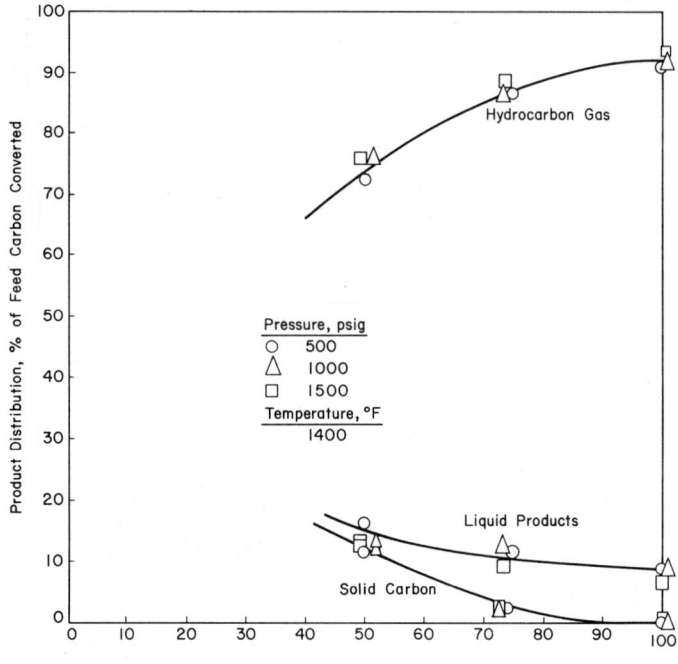

Fig. 2. Increased hydrogen-oil ratio eliminates carbon deposition.

1400°F (593 to 760°C) at a 100 percent stoichiometric ratio (Fig. 3) decreased liquid product yield from 25 to 8 wt% of the feed oil.

At a given stoichiometry and temperature, the primary gasification reactions occur rapidly, being over 90 percent completed in the first 100 to 200 seconds (Fig. 4). Additional reaction time adds only slightly to further gasification, but allows the gas-phase conversion to ethane to methane. A nominal residence time of 140 seconds, based on gas outlet conditions, was used for process design.

PROCESS DESIGN

For process design purposes, a pressure of 550 psig was chosen, delivering pipeline gas at 500 psig. A higher pressure could have been used, but because oil can be shipped more cheaply than SNG, the gasification plant is intended to be located near the consumer instead of the oil field. Therefore, the minimum pressure consistent with process considerations was used.

The hydrocracker operates at temperatures up to 850°F (454°C) to maximize residue conversion. Higher temperatures may lead to carbon laydown in the catalyst. The hydrogasifier operates at 1400°F (760°C) to maximize gas production and at a 100 percent stoichiometric hydrogen/oil ratio because lower ratios would lead to carbon deposition and higher ratios would result in a low-heating-value gas. A residence time of 140 seconds was chosen to ensure a high methane yield in a reasonably sized reactor. Longer times

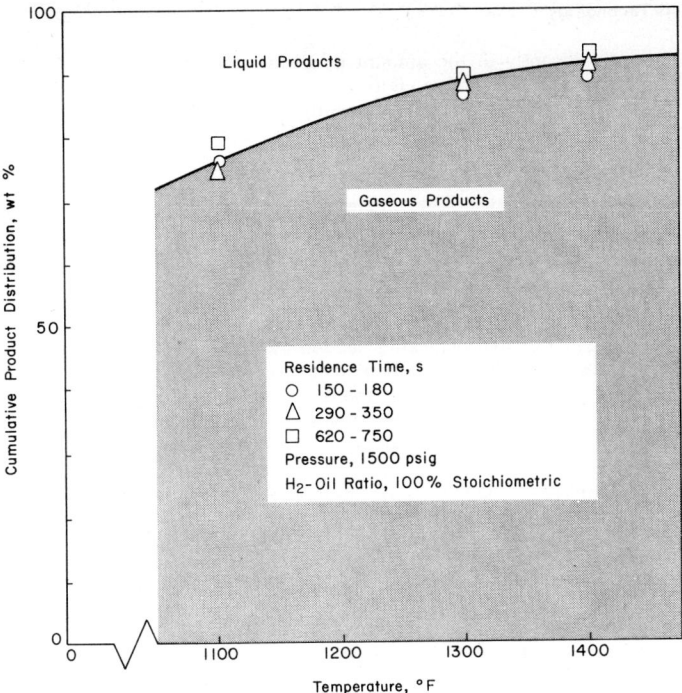

Fig. 3. Effect of temperature on liquid product formation.

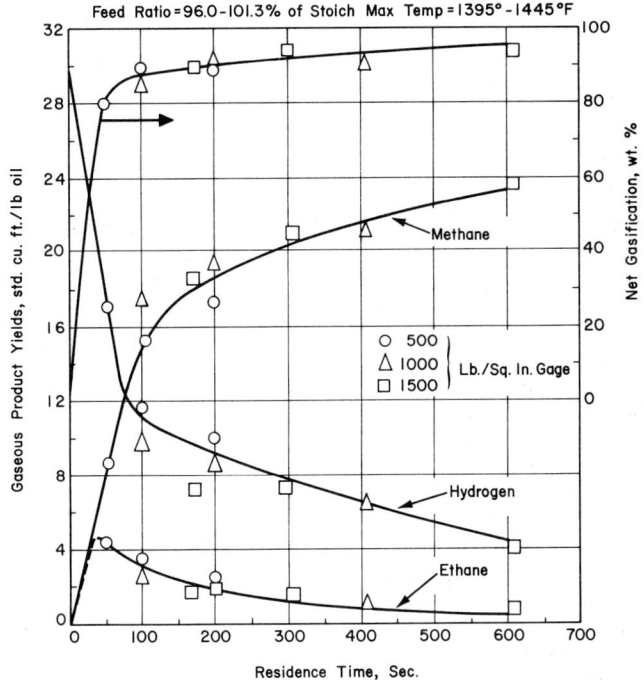

Fig. 4. Effect of residence time on gaseous product yields from Diesel oil.

would not add significantly to the amount of gasification and would only marginally increase the relative methane/ethane yield.

A simplified flowsheet based on this concept (hydrocracking, separation, and hydrogasification) is shown in Fig. 5, where 250 billion Btu/day of pipeline-quality gas is produced from 59,350 barrels/day of Taparito crude. The overall fuel efficiency of the process is 67 percent, allowing for all utility requirements, including oxygen production.

The design of hydrogen plants based on partial oxidation of residual oils (see Ref. 4) and the design of hydrocracking operations (see Ref. 1) are well-established arts. The only unusual component of the process is, therefore, the hydrogasification reactor itself.

Offsites not shown on the flowsheet include an oxygen plant, sulfur-recovery facilities (Claus plant), power and steam generation equipment, and water treatment facilities. Following is a description of plant operation.

Hydrocracking.   In the hydrocracking section, 59,350 barrels/day of process oil is hydrocracked at a space velocity of about 0.42 vol/(vol)(hr) over a cobalt-molybdate catalyst. This low space velocity gives about 85 percent desulfurization at the low operating pressure used. Chemical hydrogen consumption is about 1260 standard cubic feet/barrel; approximately 10,920 standard cubic feet of hydrogen per barrel of oil passes through the bed.

Total product oil from hydrocracking is distilled at atmospheric pressure in a conventional column to yield three product fractions. The heaviest fraction is used for hydrogen production by partial oxidation; the middle fraction, containing 0.3 percent sulfur, is used for boiler fuel; and the lightest fraction, from which carbon-forming constituents have been removed, is hydrogasified.

Hydrogasification.   The light oil fraction from fractionation (C/H ratio of 6.63, API gravity of 36.5°) is pumped up to 550 psig and made to react with hydrogen in the hydrogasification reactor. This reactor is similar in construction to a fire-tube boiler with sufficient heat-exchange surface to remove the exothermic heat of reaction. For design purposes, the overall progress of the gasification reactions can be indicated by the methane yield, which has been expressed as a function of residence time as in Fig. 4, and residence time and temperature as in Fig. 6, for a 100 percent stoichiometric feed ratio. Figures 4 and 6, together with data on gas stream composition, can be used to design a hydrogasification reactor.

In a 250 billion Btu/day pipeline gas plant, the total heat load imposed on the hydrogasification reactor is about 421 million Btu/hour, or 1110 Btu/pound of oil feed. The remainder of the heat load is removed as sensible heat in the product gas. Since gas backmixing is not used in this design, the reactants must be preheated to the autoignition temperature, which is estimated to be about 1050°F (566°C). The hydrogasification reactors employed in this design generate about 360,000 pounds/hour of 600-psig saturated steam.

Gas Cleanup and Methanation.   At the hydrogasification conditions used, about 90 percent of the 680°F (360°C) end-point feed oil is gasified, yielding a raw gas containing about 52 percent methane and 10 percent ethane, with the remainder principally hydrogen. About 60 percent of the liquid products is benzene. Total liquid products are removed first by separation in a knockout drum and then by straw oil scrubbing. Benzene, the last traces of which are removed from the gas by activated carbon, is recovered and sold. Very heavy oil is used for plant fuel. Since the excess hydrogen in the gas is to be methanated with carbon dioxide, a small amount of carbon dioxide from the hydrogen plant is added to the gas before removal of hydrogen sulfide. Most of the hydrogen in the gas at this point will react with ethane during methanation.

Because carbon dioxide is to be utilized for methanation, hydrogen sulfide must be selectively removed from the gas before methanation. A solvent having a fairly large difference in carbon dioxide and hydrogen sulfide solubility should be employed. A single-stage recycle-quench methanation system is used, with a recycle/feed ratio of 1.36:1. After methanation, the pipeline gas is dried to −40°F (−40°C) dewpoint, producing a final heating value of 950 Btu/standard cubic foot.

Hydrogen Production.   Hydrogen for hydrocracking and hydrogasification is produced from the heaviest fraction of the hydrocracked oil (C/H ratio of 7.63, API gravity of 15.3°C, and CCR of 9.3 percent) by partial oxidation. A sulfided shift catalyst is employed for water-gas shift. Two-stage hot potassium carbonate scrubbing is used to remove hydrogen

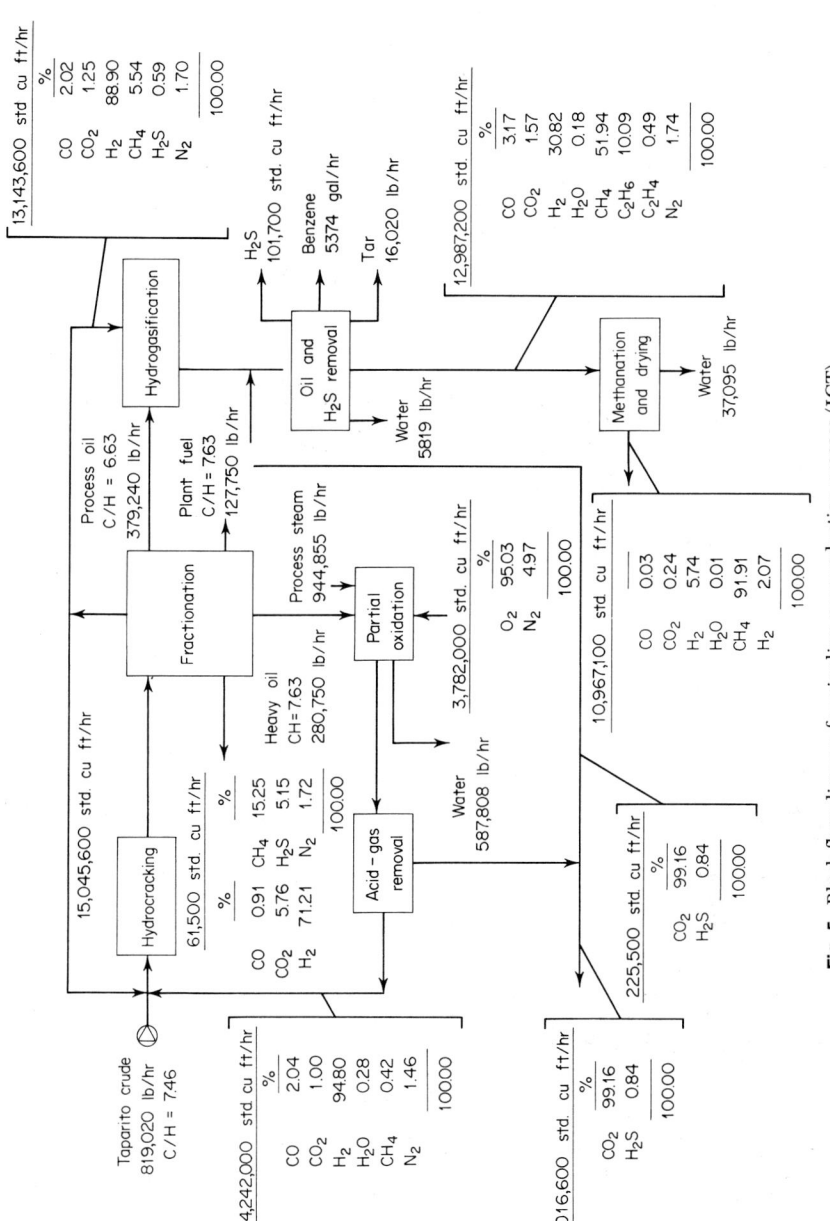

Fig. 5. Block flow diagram for pipeline gas production process (IGT).

2-169

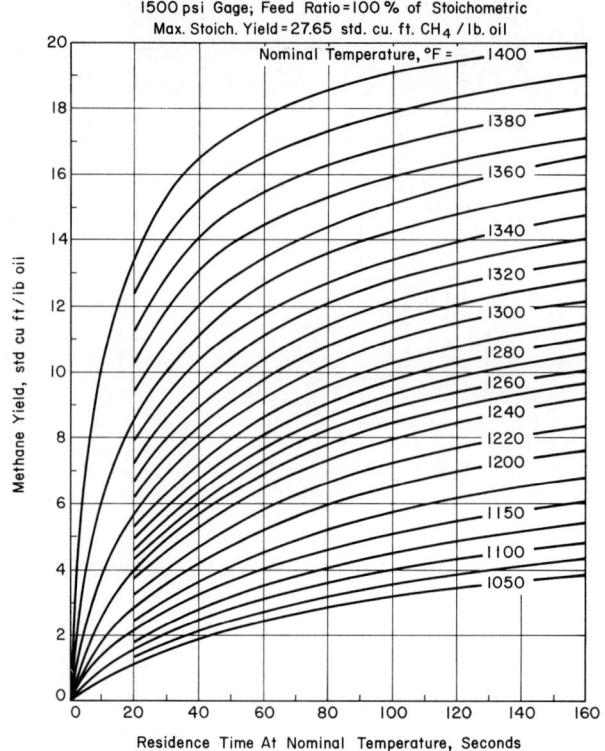

1500 psi Gage; Feed Ratio = 100 % of Stoichometric
Max. Stoich. Yield = 27.65 std. cu. ft. CH_4 / lb. oil

Fig. 6. Design chart for hydrogasification of 0 to 360°F ($-17.8 + 182$°C) fraction of upgraded Taparito crude.

sulfide and carbon dioxide from the shifted product gas, yielding a 95 percent pure hydrogen stream for hydrocracking and hydrogasification. A hydrogen sulfide-free carbon dioxide stream and a hydrogen sulfide-rich stream suitable for feed to a Claus unit are produced.

Sulfur Recovery. Approximately 48,400 long tons/year of elemental sulfur are recovered for disposal from waste gas streams in the plant. The waste gas streams from the acid-gas removal unit upstream of the methanator and from the first stage of the acid-gas removal unit in the hydrogen plant are sent to a Claus plant.

REFERENCES

1. Barnet, W. I., et al.: Extend Hydrocracking to Heavy Stocks, *Pet. Refiner*, **40**, 131–136, April 1961.
2. Dent, F. J., et al.: *Brit. Gas Counc. Res. Commun.* **GC 37**, London, 1956.
3. Huebler, J., J. C. Janka, J. G. Seay, and P. B. Tarman: Pipeline Gas from Crude Oil—Combined Hydrocracking-Hydrogasification, Preprint No. 66b presented at the 65th Annual Meeting of the American Institute of Chemical Engineers, New York, Nov. 26–30, 1972.
4. Singer, S. C., and L. W. terHaar: Reducing Gases by Partial Oxidation of Hydrocarbons, *Chem. Eng. Prog.* **57**, 68–74, July 1961.

Noncatalytic, Partial-Oxidation Gasification Process

BY C. J. KUHRE* AND J. A. SYKES, JR.†

The SGP‡ process is designed for the partial combustion or oxidation of hydrocarbons and is particularly suitable for the partial combustion of heavy, sulfur-containing residual fuels and heavy crude oils to produce a mixture of hydrogen, carbon monoxide, and small amounts of other gases.

The process is conducted by injecting oil and air (or oxygen) through a specially designed burner assembly into a closed combustion vessel, where partial oxidation occurs at about 2400°F (1316°C). The term *partial oxidation* describes the net effect of a number of component reactions that occur in a flame supplied with less than stoichiometric oxygen. This net effect can be approximated by

$$C_nH_m + \left(\frac{n}{2}\right)O_2 \; \rightleftharpoons \; nCO + \left(\frac{m}{2}\right)H_2 \tag{1}$$

and is a combination of several reactions that occur within the reactor.

Heating-up and Cracking Phase. In the fuel injection region of the reactor, hydrocarbons leaving the atomizer at about preheat temperature are intimately mixed with the oxidant in the form of air or oxygen. The atomized hydrocarbon is heated and vaporized by back radiation from the flame and the reactor walls. Some cracking of the hydrocarbons to carbon, methane, and hydrocarbon radicals may occur during this brief phase.

Reaction Phase. When the fuel and oxidant reach the ignition temperature, part of the hydrocarbons reacts with oxygen according to the highly exothermic reaction,

$$C_nH_m + \left(n + \frac{m}{4}\right)O_2 \; \rightleftharpoons \; nCO_2 + \left(\frac{m}{2}\right)H_2O \tag{2}$$

As the equilibrium is far to the right, practically all available oxygen is consumed in this phase. The remaining hydrocarbons which have not been oxidized react with steam and the combustion products from reaction (2) according to the endothermic reactions,

$$C_nH_m + nCO_2 \; \rightleftharpoons \; 2nCO + \left(\frac{m}{2}\right)H_2 \tag{3}$$

$$C_nH_m + nH_2O \; \rightleftharpoons \; nCO + \left(\frac{m}{2} + n\right)H_2 \tag{4}$$

To prevent excessive local temperatures, it is essential that all reactants of Eqs. (2), (3), and (4) be intimately mixed so that the endothermic reactions tend to locally balance the exothermic reactions. In this way, the complex of reactions is brought to a thermal equilibrium resulting in a measured temperature of about 2350 to 2550°F (1288 to 1399°C) in a very short residence time.

Soaking Phase. This phase takes place in the rest of the reactor where the gas is at a high temperature. The final gas composition is determined by secondary reactions of methane, carbon, and the water-gas shift equilibrium.

Methane produced by cracking decreases according to

$$CH_4 + H_2O \; \rightleftharpoons \; CO + 3H_2 \tag{5}$$

$$CH_4 + CO_2 \; \rightleftharpoons \; 2CO + 2H_2 \tag{6}$$

*Process Engineer, Shell Oil Company, Houston.
†Formerly with Shell Oil Company, Houston.
‡Shell Gasification Process.

As the reaction rate is relatively low, the methane content is higher than would be expected from equilbrium.

During the soaking phase, a portion of the carbon also disappears, according to the reactions,

$$C + CO_2 \rightleftharpoons 2CO \tag{7}$$

$$C + H_2O \rightleftharpoons CO + H_2 \tag{8}$$

Some carbon, however, is always present in the product gas from the reactor in a quantity equivalent to about 1 to 3 percent wt of the oil feed.

The composition of the fuel gas is determined by the water-gas shift equilibrium,

$$CO + H_2O \rightleftharpoons CO_2 + H_2 \tag{9}$$

which becomes fixed as the gas leaves the reactor at 2200 to 2400°F (1204 to 1316°C) and then cools rapidly in the waste heat boiler.

SGP PROCESS CONFIGURATION

With reference to Fig. 1, the hydrocarbon charge (e.g., a heavy residuum) and the oxidant are first preheated and fed to the reactor where the reactions just described take place. See Table 1. The hot reactor-effluent gas containing some of the ash from the feed and 1 to 3 percent by weight of the hydrocarbon feed as soot is passed to a specially designed waste heat boiler which produces high-pressure steam. High heat transfer rates are achieved, with the result that the temperature of the gas leaving the waste heat boiler can closely approach that of the steam produced in the boiler with economically sized equipment. The design and construction of the waste heat boiler are such that the surface remains clean for an indefinite period without requiring any external cleaning devices. A prototype unit* has been in operation for nearly 20 years without requiring cleaning on the gas side. The waste heat boiler can be designed for steam pressures up to about 1500 psig. Commercial units are operating above 1100 psig in oxygen gasification service.

The waste heat boiler of the SGP process is an important element in a clean fuel process—since the desulfurized, particulate-free gas product contains about 70 percent of the hydrocarbon feedstock's LHV (lower heating value) (cold gas efficiency equals 70 percent), while the high-pressure steam recovers an additional 20 to 25 percent of the fuel LHV in a useful form. Thus, the overall thermal efficiency of the fuel conversion is 90 to 95 percent (usually 93 to 94 percent). The high thermal efficiencies are obtained by inclusion of an economizer, such as a boiler feedwater preheater, on the crude gas from the waste heat boiler.

The crude gas leaving the waste heat boiler and its economizer at temperatures around 325°F (163°C) is next passed to the carbon removal system, consisting of (1) a bulk removal of the carbon by means of a special method of contacting the gas with water, and (2) a final water wash. The product gas is virtually free of carbon (less than 5 ppm), ash, and any other particulates, and may be safely used in high velocity, high temperature service, such as gas turbines.

The carbon produced during the gasification is recovered as a soot-in-water slurry (carbon content of 1 to 2 wt%). Since it is usually not possible or desirable to dispose of the carbon slurry directly, a special process has been developed for removing the carbon from the slurry, resulting in carbon-free water for reuse. Depending on the metals content of the feedstock and the economics and maintenance policy of the process operator, up to 100 percent of the soot can be recycled to extinction with the fresh feed.

Depending on the desired LHV in the product gas, either oxygen or air (enriched or unenriched) may be used as the oxidant. Nitrogen present in the air acts as a moderator for temperature control in the reactor and does not enter into the reactions—as might be expected, considering the strongly reducing atmosphere of the reactions. When either oxygen or air enriched with oxygen is used as the oxidant, a quantity of steam must be injected into the reactor for temperature moderation. Air oxidation alone requires no steam. The latter method produces a low heating-value gas (~ 120 Btu/SCF) because of the presence of nitrogen. Oxygen feed produces a medium heating-value gas (~ 300 Btu/

*Prototype unit at a Shell-affiliated plant.

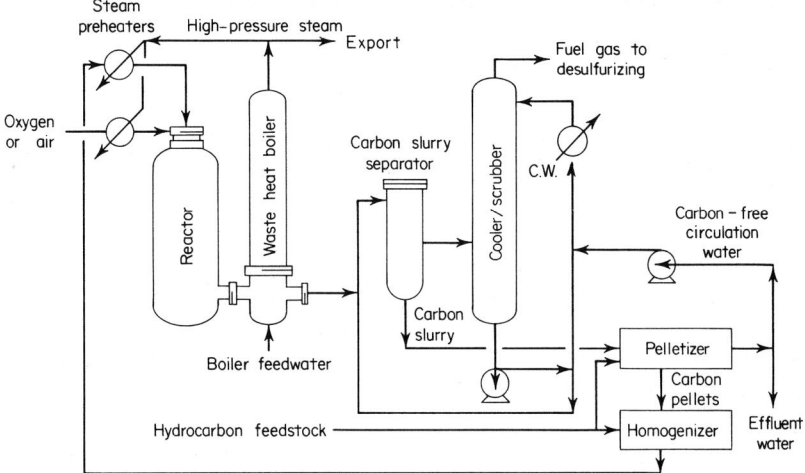

Fig. 1. SPG for fuel gas manufacture. *(Shell Oil Company.)*

SCF). Typical product gas compositions for air and oxygen gasification are shown in Table 2 for the residual feedstock shown in Table 1.

PRODUCTS OF THE SGP PROCESS

The three net products are (1) high-pressure steam, (2) clean waste water, and (3) carbon-free fuel gas. While the high-pressure steam is saturated, pressures over 1100 psig have been commerically demonstrated, and increases to substantially higher pressures in commercial practice are anticipated. Under any condition, using superheating, this steam is easily converted into an attractive feed for steam turbines. With appropriate design, oxygen-based partial oxidation units can be made almost entirely energy self-sufficient (Ref. 1). As a result of either the exothermic reactions during air oxidation, or the moderating steam used in the reactor during oxygen gasification, from 20 to 35 lb of waste water are generated for each 100 lb of heavy hydrocarbon gasified.

TABLE 1. Properties of Typical Residual Feed to SGP

Gravity, °API, 60°F(15.6°C)	12.0	
Specific gravity, 60/60	0.986	
Composition, wt %:		
Carbon	86.00	
Hydrogen	10.73	
Sulfur	2.65	
Nitrogen	0.30	
Oxygen	0.30	
Ash	0.02	
Total	100.00	
Viscosity, centistokes:		
470°F (243°C)	4	
202°F (99.4°C)	55	
100°F (37.8°C)	800	
	ppm in feed	**% of ash**
Ash analysis:		
Nickel	30	15.0
Vanadium	100	50.0
Sodium	1	0.5
Iron	4	2.0
Others	65	32.5
Total	200	100.0

TABLE 2.   Typical Product Gas Composition from SGP

| Component in gas | Vol %, dry basis* | |
	Oxygen oxidation	Air oxidation
Hydrogen	48.0	12.0
Carbon monoxide	51.0	21.0
Methane	0.6	0.6
Nitrogen	0.2	66.0
Argon	0.2	0.4
Sulfur	5 ppm	5 ppm
Total	100.0	100.0

*After desulfurization by the Shell Sulfinol or ADIP process.

In addition to the carbon, hydrogen, and oxygen contained in the feedstock, other contaminants, such as metals, nitrogen, and sulfur, are often found. Metals in the feedstock are liberated primarily as an easily removed residue in the reactor bottom when there is total soot recycle. A small amount of the very heavy metals ultimately appears in the waste water. The levels are low, and they can be removed by existing water-treatment methods. If extensive treatment proves necessary, it may be advisable to consider use of the water as boiler feedwater to the waste heat boiler.

Nitrogen in the feedstock is almost entirely converted by the SGP to elemental N_2, whereas in conventional combustion a significant amount of the nitrogen in the feedstock is converted to NO_x. A small part of the fuel's nitrogen content will appear as ammonia and hydrogen cyanide, which are scrubbed quantitatively into the carbon slurry. Where these materials cannot be handled by the regular process waste-water treatment system, low-cost processes are available for their destruction.

FUEL GAS DESULFURIZATION

Sulfur in the feedstock is converted primarily to hydrogen sulfide and carbonyl sulfide in an H_2S/COS ratio of approximately 24:1. These may be removed by acidic gas absorption processes. See also "Absorption of Acidic Gases from Natural Gas and SNG" in this Handbook section.

POWER GENERATION USING GASIFICATION PROCESS

Gas from the SGP process, after post treatment for desulfurization (sometimes involving absorption processes, Claus units, and Claus off-gas treating) is suitable for several applications. The product fuel is easily burned in CO-boiler type furnaces with low NO_x emissions, acceptable sulfur emissions, and low particulate, carbon monoxide, and hydrocarbon emissions. In refineries and chemical plants, the fuel can be utilized directly for process heating or steam generation. Several studies show that the fuel can be used, in conjunction with the steam generated, for the production of electric power. See "Gasification as Basis for Combined-Cycle Power Generation" in the *Power Technology Trends* section of this Handbook.

REFERENCES

1. Van den Berg, G. J., E. Reinmuth, and E. Kapp: Hydrogen from Heavy Residues, *Chem. Proc. Eng.*, October 1971.
2. Goar, B. G.: Sulfinol Process Has Several Key Advantages, *Oil Gas J.*, June 30, 1969.
3. Dravid, A. N., C. J. Kuhre, and J. A. Sykes, Jr.: Power Generation Using the Shell Gasification Process, technical paper, *3d Int. Conf. Fluidized Bed Combustion*, October 1972.
4. Van den Berg, G. J.: The Economics of Power Generation via SGP and Town Gas Manufacture via SGP, technical paper, Japan Petroleum Institute, February 1972.
5. Kuhre, C. J., and C. L. Reed: "Shell Noncatalytic Gasification Process Leads to SNG," *Oil and Gas Journal*, 110–118, January 12, 1976.

Petroleum Technology

MARK T. ATWOOD. *Manager of Laboratories, The Oil Shale Corporation, Golden, Colorado. (Oil Shale Retorting)*

P. BARRY BALDWIN, B.A., M.A. *Assistant to Vice-President, Marketing, Marine Transport Lines, Inc., a subsidiary of General American Transportation Corporation, New York. (Seagoing Vessels for Transporting Liquid and Gaseous Fuels)*

T. C. BUECHLEY, B.S., M.S.Ch.E. *Development Engineer, Halliburton Services, Duncan, Oklahoma. Member, American Institute of Chemical Engineers, American Institute of Mining, Metallurgical and Petroleum Engineers. (Drilling and Production)*

MAURICE CARRIGY, M.Sc. *Currently, Vice-Chairman, Alberta Oil Sands Technology and Research Authority, Edmonton, Alberta, Canada. (Fuel from Tar Sands)*

HON. BILL DICKIE, Q.C. *Minister of Mines and Minerals, Alberta, Edmonton, Alberta, Canada. Retired from Government, March 1975. (Fuel from Tar Sands)*

S. FRIEDMAN, B. Ch. E. *Assistant Research Supervisor and Supervisory Chemical Engineer, Pittsburgh Energy Research Center, U.S. Bureau of Mines, Department of the Interior, Pittsburgh, Pennsylvania. Member, American Institute of Chemical Engineers. (Wood-to-Oil Process)*

KENNETH E. HASTINGS, Ph.D. *Manager, Synthetic Gas Projects, Cities Service Research and Development Company, Cranbury, New Jersey. Member, American Chemical Society, American Institute of Chemical Engineers. (Hydrodesulfurization of Petroleum Crude Oil Fractions and Petroleum Products)*

H. J. KLOOSTER, B.S.Ch.E. *Manager, Process Engineering, Fluor Engineers and Constructors, Inc., Los Angeles, California. Member, Air Pollution Control Association, American Institute of Chemical Engineers, Water Pollution Control Federation. (The Total Energy Concept in Refinery and Process Plant Planning)*

JOHN A. KNOX, B.S., M.S. *Chemical Service Coordinator, Environmental Control Department, Halliburton Services, Duncan, Oklahoma. Member, American Chemical Society, American Institute of Mining, Metallurgical and Petroleum Engineers, American Petroleum Institute, American Society for Testing and Materials, National Association of Corrosion Engineers. (Acidizing)*

DAVID L. LORD, B.S.Ch.E., M.S.Pet.Eng. *Group Leader, Halliburton Services, Duncan, Oklahoma. Member, American Institute of Chemical Engineers, American Institute of Mining, Metallurgical and Petroleum Engineers. Registered Professional Engineer (Oklahoma). (Drilling and Production)*

J. L. LUMMUS, B.S. *Special Research Group Supervisor, Research Department, Amoco Production Company, Tulsa, Oklahoma. Member, American Institute of Mining, Metallurgical and Petroleum Engineers, American Petroleum Institute. (Drilling Optimization)*

H. C. McLAUGHLIN, B.S., M.S. *Senior Research Chemist, Halliburton Services, Duncan, Oklahoma. Member, American Chemical Society, American Institute of Mining, Metallurgical and Petroleum Engineers, National Society of Professional Engineers, Oklahoma Society of Professional Engineers. Registered Professional Engineer (Oklahoma). (Drilling Techniques—Water Problems) (Well Completion—Water Control)*

A. S. MEHTA, B.S., M.S. *Formerly Staff Engineer, Chemical Process Department, The Rust Engineering Company, a subsidiary of Wheelabrator-Frye Inc., Birmingham, Alabama. Member, American Institute of Chemical Engineers. (Wood-to-Oil Process)*

F. POSTHUMA. *Executive Director (retired), Rotterdam-Europoort. Port Consultant, Rotterdam, The Netherlands. (Superports)*

ROGER F. RENSVOLD, M.A. *Development Chemist, Halliburton Services, Duncan, Oklahoma. (Work-Overs—Water Control)*

EDWIN J. STAHL, JR., B.S.Ch.E. *Section Supervisor, Fracturing Section, Chemical Research and Develop-*

ment Department, Halliburton Services, Duncan, Oklahoma. Member, American Institute of Mining, Metallurgical and Petroleum Engineers, Oklahoma Society of Professional Engineers, National Society of Professional Engineers. Registered Professional Engineer (Oklahoma). (Hydraulic Fracturing) (Fracturing in Work-Overs)

DONALD J. SZOSTAK, B.S. (Naval Architecture and Marine Engineering). Vice-President, Engineering, Marine Transport Lines, Inc., a subsidiary of General American Transportation Corporation. Member, American Petroleum Institute (Tanker Committee), National Cargo Bureau (New York), Naval Architects and Marine Engineers, Royal Institution of Naval Architects (London). Registered Professional Engineer (Pennsylvania). (Seagoing Vessels for Transporting Liquid and Gaseous Fuels)

P. L. THIGPEN, B.S.Ch.E. Principal Process Engineer, Chemical Process Department, The Rust Engineering Company, a subsidiary of Wheelabrator-Frye Inc., Birmingham, Alabama. (Wood-to-Oil Process)

DAVID P. THORNTON, JR., B.S.Ch.E. Manager of Technical Publications, Universal Oil Products Company, Des Plaines, Illinois. Member, Association of Petroleum Writers, Society for Technical Communication. (Petroleum)

ROGER P. VAN DRIESEN, B.S.ChE. Vice-President, Cities Service Research and Development Company, Cranbury, New Jersey. Member, American Institute of Chemical Engineers, American Petroleum Institute. (Hydrodesulfurization of Petroleum Crude Oil Fractions and Petroleum Products)

BILL M. YOUNG, B.S., M.S.Ch.E. Group Leader, Water and Sand Control Section, Chemical Research and Development, Halliburton Services, Duncan, Oklahoma. Member, American Chemical Society, American Institute of Mining, Metallurgical and Petroleum Engineers (Sand Control—Well Completions) (Sand Control Work-Overs)

MARIO ZAMORA, P.E. Manager, Staff Engineering, IMCO Services, a division of Halliburton Company, Houston, Texas. (Geopressured Drilling) (Well Control)

Petroleum*

The definitions of petroleum, also variously referred to as *crude oil* or *rock oil,* are numerous, but may be summarized as follows: an *oily, flammable liquid of widely varying viscosity, possessing a unique, characteristic heavy odor, varying in color from yellow to dark reddish-brown or black, but usually exhibiting a distinct greenish fluorescence,* believed to have been formed from marine and semimarine near-shore sediments by geological action (precise mechanism unknown) over millions of years. Crude oil is found throughout the world, from deserts to arctic regions to the continental shelves. Source configurations range from surface oozes, to subsurface seeps and tar sands and pits, to rock strata which may be buried from a few hundred feet to several miles deep. Petroleum is produced in commercial quantities on virtually every land mass of significant size on the earth with the present exception of Antarctica and Greenland and their adjacent offshore areas.

BASIC CLASSES

Petroleums are nonuniform, highly complex mixtures of paraffinic, naphthenic, and aromatic hydrocarbons. Small amounts of sulfur and even smaller amounts of nitrogen and oxygen compounds are present. The terms *paraffinic, aromatic,* and *asphaltic-base* (or *naphthenic*) are applied to designate the most prevalent classes of chemical constituents found in specific crude oils from various localities. Similarly, the terms *sour* and *sweet* are used. Sour crudes contain sulfur and have an unpleasant, sometimes sickening, odor of garlic or rotten eggs. The odorous sulfur usually exists in the form of mercaptans or hydrogen sulfide. Sweet crudes contain very little sulfur and have comparatively pleasant odors.

The oils derived from *oil shales* are not true petroleum, although they are petroleum-like products after specialized chemical processing. Shales are sedimentary rocks which have a relatively high content of a bituminous substance named *kerogen* and 30 to 60 percent organic matter and fixed carbon. Kerogen, although not a definite chemical compound, yields an oily substance when heated (retorted) in the absence of air. Extraction of oil shale with ordinary solvents produces no oil, and its solubility in solvents is low. This evidence supports the conclusion that the "oil" is the result of a chemical change, the thermal cracking or fragmenting of the molecules that make up kerogen, i.e., pyrolysis. See *Oil Shale Retorting* later in this Handbook section.

Natural gas is not formally defined as a component of crude petroleum, although natural gas commonly exists in the same geological formations, often directly in contact with crude petroleum. Natural gas is described in the Handbook section on *Gas Technology.*

COMPOSITION OF CRUDE OILS

Petroleum oils vary considerably in composition, even when closely associated geographically. In some areas of the United States, for example, crude oils near the surface may have quite a different chemical composition from those in deeper strata. Depth alone, however, does not correlate significantly with composition.

Analyses of typical crude oils found in representative areas of the United States are given in Table 1. It may be generalized that crudes found in the eastern and midwestern portions of the United States are predominantly sweet and paraffinic; those found along the Gulf Coast usually are naphthenic; those occurring in the inland southwest are sour

*Prepared originally by David P. Thornton, Jr., Universal Oil Products Company, Des Plaines, Ill., for the "Chemical and Process Technology Encyclopedia," D. M. Considine, Editor-in-Chief, McGraw-Hill, New York, 1974.

and naphthenic; and those found along the West Coast are asphaltic. The waxy, sweet paraffinic oils found in Pennsylvania first became prominent because of the high quality of lubricating oils and greases that could be made from them. The severe stresses imposed by the bearings and close-fitting reciprocating surfaces of machinery led to the development of refining processes and the discovery of additive materials whereby many other crude oils also can be transformed into excellent lubricants. Even Pennsylvania oils require special refining and additives to meet modern quality specifications.

Analyses of some crude petroleums found outside the United States are given in Table 2, which illustrates the variety of crudes existent but is not intended to give a typical or representative picture of worldwide petroleum source compositions.

SIGNIFICANCE OF COMPOSITION

Although greatly abridged in analytical detail, the tabular information provides a basis for understanding the importance of composition differences, particularly for the types of processing operations employed in converting various crudes to useful end products. Interpretation of some of the key variations would include the following:

API Gravity.   This parameter (API stands for American Petroleum Institute), expressed in degrees, is mathematically related to specific gravity and usually is determined by a hydrometer. The specific gravity of water (arbitrarily defined as unity) is 10.00 when expressed as degrees API. As used in Tables 1 and 2, API gravity usually, although not infallibly, indicates the gasoline and kerosine contents of the crude. As an example, the Mississippi, Texas, New Mexico, and Louisiana crudes have API gravities between approximately 35 and 40; as do the Arabian, Iranian, and Colombian crudes. In checking the distillation figures, it will be noted that the gasoline content (that fraction boiling below about 400°F (204°C) of these crudes ranges from about 25 to over 35 percent by volume. The kerosine portions of such "light" crudes also are usually high. In contrast, Wyoming sour crude with an API gravity of 17.9 will be noted to contain but 6 percent gasoline and about 40 percent asphalt. California crude has an even greater content of residuum and almost no gasoline.

Sulfur Content.   The amount of sulfur in crude is important in terms of handling the crude within the refinery and the undesirable effects of sulfur in finished products. High-sulfur crudes require special materials of construction for refinery equipment because of their corrosiveness. Certain refinery processes require desulfurization of sour charge stocks before use as a feedstock, not only because of their corrosiveness, but also because of the effect of sulfur-bearing compounds on expensive catalysts. From the standpoint of the consumer, sulfurous gasoline has an unforgettably offensive odor unless specially sweetened, and may corrode the fuel system and engine parts as well as polluting the atmosphere after it has been burned. See also *Hydrodesulfurization of Petroleum Crude Oil Fractions and Petroleum Products* in this Handbook section.

Other factors indicated in the data of Tables 1 and 2 include:

Distillation Range, which indicates what fractions and how much of each fraction is present.

Pour Point, defined as the lowest temperature at which the material will pour and a function of the composition of the oil in terms of waxiness and bitumen content.

Sediment and Water, which are measures of dirt and other foreign matter, as well as water.

Salt Content, which is not confined to sodium chloride, but usually is interpreted in terms of NaCl. Salt is undesirable because of its tendency to obstruct fluid flow and to accumulate as an undesirable consitutent of residual oils and asphalts, as well as a tendency of certain salt compounds to decompose when heated and cause corrosion of refining equipment.

Metals Content.   The heavy metals, such as vanadium, nickel, and iron, tend to accumulate in the heavier gas oil and residuum fractions where the metals may interfere with refining operations, particularly by poisoning catalysts. The heavy metals also contribute to the formation of deposits on heated surfaces in furnaces and boiler fireboxes, leading to permanent failure of equipment, interference with heat-transfer efficiency, and increased maintenance.

TABLE 1. Analyses of United States Crude Oils

	McComb, Mississippi		Southwest Texas		East Texas	Wyoming (Sour)	New Mexico	N. Kenai Penin., Alaska	San Ardo, California	Ospelousas, Lousiana		Velma, Oklahoma
Gravity, °API	40.7		36.5		39.1	17.9	37.5	25.9	13.3*	38.2		29.1
Distillation, °F: Type	D 86¶	UOP 76§	D 86	UOP 76	D 86	D 86	D 86	D 86	UOP	D 86	UOP 76	D 86
IBP (initial boiling point)	152	152	158	158	125	306	118	160	IBP 180 0.3†	127	127	145
5% over	192	192	208	208	191	408	162	196	180– 380 1.6†	206	206	224
10	224	224	244	244	211	476	224	220	380– 550 16.1†	256	256	482
30	344	344	461	461	355	633	354	514	550– 650 10.6†	438	438	659
50	504	506	700	770	539	675	530	660	650– 900 17.3†	545	575	712
70	655	703	760+	1075	702	710	626	690	900–1000 6.0†	672	720	748
90	760+	977	...	...	760	732‡	728	710‡	1000+ 48.1†	760+	905	748+
EP (end point)	...	1062	...	...	760+	732+	734+	710+		...	978	
% recovered	98.5	95.5	98.5	90.5	97.0	89.0	93.0	80.0	....	98.5	96.0	95.5
% bottoms	...	4.5	...	9.5	...	11.0	5.6	...	....	...	4.0	
% coke, wt.	1.6	...	1.8	...	2.8	...	...	...	....	0.7		
% recovered At 400°F	37.0	...	32.0	...	34.5	4.5	35.0	...	1.9*	26.5		22.5
525°F	53.0	...	48.0	...	54.0	15.0	55.0	...	17.9*	46.0		33.5
572°F	60.0	...	54.0	...	...	20.0	...	...	....	54.0		39.0

Property									
Total sulfur, wt %	0.07	0.45	0.2	3.33	1.0	1.036	1.93*	0.08	1.13
Reid vapor pressure, psi	4.6	3.4	...	0.2	8.8	4.2	...	8.4	4.0
Pour point, °F	60	30	55	−5	25	...	...	40	<−30
Bottom sediment and water, vol %	0.10	0.10	0.1	0.3	0.3	2.5	...	0.15	0.2
Conradson carbon residue, wt %	0.79	1.74	...	...	...	...	...	0.52	...
Salt (as NaCl) lb/1,000 bbl	4	<0.5	31	0.6	14.0	76.0	...	5.0	78.0
Gasoline, vol%	35.5	32.0	29.0	6.3	37.8	14.4	1.9	26.1	22.3
Kerosine, vol %	18.1	12.1	10.1	9.1	...	18.0	16.1	18.9	17.3
Diesel fuel, vol %	14.6	38.0	13.8	14.0	41.2	18.4	10.6	22.9	8.5
Gas oil, vol %	28.1	12.6	...	30.7	20.8	22.3	23.3	27.9	31.9
Asphalt (bottoms), vol %	3.7	5.3	(47.1)	39.9	...	25.7	48.1	4.1	20.0
Metals (in gas oils), ppm:									
Nickel	0.06	...	...	...	...	...	0.15	...	...
Vanadium	0.08	...	...	...	...	...	<0.1	...	...

From Douglas M. Considine (ed.), "Chemical and Process Technology Encyclopedia," McGraw-Hill Book Company, New York, 1974.

*Calculated.
† % over in indicated boiling range.
‡ Cracked at 80% over.
¶ Designates data from Method of Test D 86, Committee D-2 on Petroleum Products and Lubricants, ASTM.
§ Designates data from UOP Laboratory Test Methods for Petroleum and Its Products, No. 76, Universal Oil Products Company.
Copyright © 1973 Universal Oil Products Company.

TABLE 2. Analysis of World Crude Oils

	Arabian	Minas, Central Sumatra, Topped		Putomayo, Colombia		Gulf Nigeria		Zulia, Venezuela		Iran		Kuwait	
Gravity, °API	30.0	35.3		35.0		34.7		25.2		36.6		31.5	
Distillation, °F: Type	D 86	D 86*	UOP 76†	°F	Vol %	°F	Vol %	Hempel‡	Vol %	Hempel‡	Vol %	Hempel‡	Vol %
IBP (initial boiling point)	77	173	594	IBP 400	34.1	IBP 140	6.3	IBP 122	0.1	IBP 122	1.5	IBP 122	2.8
5%	160	216	662	400–500	9.3	140–170	1.8	122–167	1.2	122–167	2.8	122–167	2.5
10	231	246	699	500–650	20.3	170–310	16.8	167–212	2.3	167–212	4.5	167–212	3.2
20	287	295	750	650–750	9.0	310–520	26.5	212–257	3.7	212–257	6.3	212–257	4.2
30	...	341	792	750–900	11.4	520–680	19.3	257–302	3.9	257–302	7.1	257–302	4.3
50	330	427	890	900+	17.2	680+	30.9	302–347	4.1	302–347	5.6	302–347	4.5
70	435	497	1042	...	...	...	...	347–392	3.6	347–392	5.4	347–392	4.0
90	452	575	Cracked	...	...	...	...	392–437	3.4	392–437	5.4	392–437	4.0
95	526	619	...	...	...	...	...	437–482	4.7	437–482	5.5	437–482	4.2
EP (end point)	526	639	1042+	...	...	...	...	482–527	6.0	482–527	7.4	482–527	5.5
								527–583	2.2	527–583	2.6	527–583	2.6
								583–633	5.1	583–633	6.6	583–633	7.0
								633–687	4.9	633–687	5.6	633–687	4.4
								687–738	5.2	687–738	5.1	687–738	4.8
								738–790	7.3	738–790	5.9	738–790	5.3

% recovered	...	99.0	72.5	101.3	101.6	57.7	77.3	63.3
% residue	...	1.0	27.5	...	...	42.1	22.5	34.8
Total sulfur, wt %	3.05		0.2	0.49	0.16	1.69	1.12	2.62
Reid vapor pressure, psig	3.8							
Pour point, °F	−33		-0	45	20	<5	5	<5
Gasoline, vol %	29.1		11¶	34.1	24.9	18.9	32.2	25.5
Kerosine, vol %	16.0		16¶	9.3	26.5	14.1	18.3	13.7
Gas oils, vol %	12.5		14¶	40.7	19.3			
Residuum, vol %	42.4		59¶	17.2	30.9			
Metals in gas oils, ppm:								
Vanadium	0		...	25§	0.7§			
Nickel	0		...	11	5.1§			
Iron	3		...					
Salt, lb/1,000 bbl	12		...	Trace	5			

From Douglas M. Considine (ed.), "Chemical and Process Technology Encyclopedia," McGraw-Hill Book Company, New York, 1974.
*IBP to 650°F.
†On 650°F+ bottoms.
‡Bureau of Mines Hempel, vol % at stated cut points.
§Estimated.
¶In crude oil.

TABLE 3. Derivation and Use of Major Petroleum Products

Product	Refining Processes Employed	Typical Uses
Light gases	Distillation of crude petroleum	Chemical manufacturing, gasoline manufacturing, fuels (LPG)
Gasoline	Distillation of crude petroleum Cracking of heavy fractions (thermal; catalytic; hydrocracking; coking of pitch, residues, etc.) Reforming (catalytic and thermal) Polymerization of light olefins Isomerization of C_5 and C_6 paraffins Alkylation of olefins with isoparaffins	Automotive and aircraft fuels, solvent, chemical manufacturing, illuminant, cooking fuel
Kerosine	Distillation of crude petroleum Cracking of heavier fractions (thermal, catalytic, etc.)	Jet-aircraft fuel, illuminant, cooking and space-heating fuel, solvent
Gas oils	Distillation of crude petroleum (atmospheric and vacuum) Cracking of heavier fractions (thermal and catalytic, etc.) Hydrocracking of residual oils Vacuum distillation	Domestic and light industrial fuels, diesel fuels, chemical manufacturing, gasoline manufacture, solvents, road oils
Lubricating oils	Distillation of crude petroleum (atmospheric and vacuum) Hydrocracking special residual oils Solvent refining of residual oils	Various lubricating oils, pharmaceutical white oils, sources of waxes and petrolatums, petroleum jelly, asphalt, lubricating greases
Residua	Vacuum distillation of petroleum Distillation of synthetic petroleum made by cracking	Manufacturing of fuel oils and gasoline, chemical manufacturing, source of coke, asphalt
Coke	Thermal cracking of residuums and pitches	Fuel, metallurgy, industrial electrodes

Copyright © 1973 Universal Oil Products Company.
From Douglas M. Considine (ed.), "Chemical and Process Technology Encyclopedia," McGraw-Hill Book Company, New York. 1974.

TABLE 4. Chronology of Petroleum Industry Progress

Problem	Solution
1. Better gasoline quality and greater quantity needed	1. Thermal cracking processes (about 1910)
2. Need to improve odor and stability of gasoline and kerosine	2. Refining with chemical solutions and synthesis and use of oxidation inhibitors; started in late 1920s
3. Better gasoline quality and greater quantity needed	3a. Discovery of tetraethyl lead (1921)
	b. Polymerization of light olefins to make "poly" gasoline by catalysis (mid-1930s)
	c. Catalytic cracking invented and improved (late 1930s)
4. Combat grade aviation gasoline testing above 100 octane needed for World War II	4. Alkylation of light olefins with light isoparaffins by catalysis; discovered in 1932; commercialized in early 1940s
5. More aromatic hydrocarbons needed, especially toluene for TNT; benzene, toluene, and other high-octane aromatics needed for combat grade aviation gasoline and for chemical synthesis	5a. Catalytic reforming to make toluene from petroleum naphthas (early 1940s), using non-noble metal catalyst
	b. Extractive distillation of toluene from reformate with phenol and other materials (early 1940s)
	c. Extraction with SO_2, suggested in 1907 to purify kerosine, applied to secure aromatics from reformate (early 1940s)
	d. Alkylation of propylene with benzene using solid H_3PO_4 catalyst to make cumene (early to mid-1940s)
6. Butadiene needed for synthetic rubber in wartime	6. Thermal and catalytic processing applied to petroleum distillates, "quicky" butadiene program (early 1940s)
7. More isobutane for alkylation in wartime aviation-gasoline program	7. Isomerization of n-butane (early 1940s)
8. Improve quality of straight-run gasoline	8. Catalytic reforming using noble-metal catalyst (1949)
9. Remove catalyst poisons and sulfur compounds from gasoline and naphtha	9. Catalytic hydrotreating (early 1950s)
10. Increase supplies of pure aromatic hydrocarbons and aromatic concentrates	10. Liquid-liquid solvent extraction processes using aqueous glycols and improved contacting means (1952)

TABLE 4. Chronology of Petroleum Industry Progress *(Continued)*

11. Purify kerosines and light and heavy distillates	11. Modified catalytic hydrotreating (mid-1950s)
12. Improve quality of light hydrocarbons used in gasoline	12. New catalytic isomerization processes using noble-metal catalysts, converting C_4, C_5, and C_6 n-paraffins to isoparaffins (mid-1950s)
13. Increase production of light fuels and gasoline; reduce production of heavy fuels	13. Development of catalytic hydrocracking processes having great flexibility (1959–1960)
14. Ethylbenzene for styrene manufacture	14. Catalytic alkylation process developed, uniting benzene directly with dilute ethylene in refinery gases (1958)
15. Separation of normal paraffins from mixtures with isoparaffins	15. Molecular sieves used as solid adsorbants (1959, but not commercialized until late 1960s)
16. Increase benzene supply and decrease toluene	16. Hydrodealkylation of toluene; produce naphthalene from alkyl naphthalenes (early 1960s)
17. Synthesize cyclohexane for nylon	17. Catalytic hydrogenation of benzene (early 1960s)
18. Improve quality of heavy fuel oils	18. Hydrodesulfurization of heavy fuels, also by hydrocracking (mid-1960s)
19. Biodegradable synthetic detergents	19. Development of processes of dehydrogenate n-paraffins to n-olefins and alkylate benzene with them (mid-1960s)
20. Increase production of p-xylene	20. Isomerization of C_8 aromatics to p-xylene (late 1950s)
21. Improve supplies of individual pure xylene isomers	21. Adsorptive separation of p-xylene in high yield and purity, making possible separation of other isomers by precise fractionation (early 1970s)

From Douglas M. Considine (ed.), "Chemical and Process Technology Encyclopedia," McGraw-Hill Book Company, New York, 1974.

DIVERSITY OF PROCESSING ROUTES

Although there is a wide variation in crudes, processing operations including pretreating units are available for handling a reasonable variety of charge stocks within a given refinery. A given petroleum refinery may process a number of crudes with varying characteristics over a period of time.

The initial step in refining a crude petroleum or a gas condensate liquid is to separate it into fractions by first distilling at or near atmospheric pressure. That portion boiling above approximately 650°F (343°C) and called *reduced crude* is subjected to vacuum flashing or vacuum distillation if several fractions are desired—as in the manufacture of lubricating oils.

Light gases, straight-run gasoline, kerosine, and distillate fuel oils are the primary products of atmospheric distillation, as shown in Table 3, but generally are unsuited for direct use without further processing. Subsequent processing steps are dictated in part by the amount and quality of the respective fractions, equipment available in the refinery, and the particular markets that the individual refiner supplies.

The products from vacuum distillation can be used directly if only heavy fuel oils and pitch are to be made; otherwise, the distillables are taken as column overhead and side cuts for further processing. The residuum is cut back with ligher oils to form a salable heavy fuel oil (as for ship's bunkers). Alternatively, it may be thermally cracked or hydrocracked.

Refiners may vary their processing schemes according to the season of the year or other marketing and demand factors. Traditionally, about midsummer of each year, most refiners in the United States start reducing their gasoline production by adjusting cracking operations to produce more fuel oils in anticipation of winter needs. Some proprietary hydrocracking processes can have special capability designed into them to produce all gasoline under certain conditions; and to maximize fuel oils under other conditions.

Since requirements for petrochemicals and for gasoline, for example, often compete for the same hydrocarbon molecule, it is common for certain process units in some refineries to be run *blocked operation*. For a specified period, the unit is adjusted to produce the maximum of aromatic hydrocarbons—the product being stored and withdrawn as needed. The unit then is readjusted to make the desired quality of gasoline-blending component and run for the required time, when the unit is returned to making aromatics. Segregated storage and the unit's charge capacity are coordinated so that the alternating service of the same refinery units has no undesired effect on overall refinery operation.

A chronology of petroleum industry progress over the years is given in Table 4.

Further information on petroleum processing is given later in this Handbook section. Consult contents page.

REFERENCES

Staff: "Vapor and Gravity Control in Crude Oil Production," Petroleum Extension Service, Division of Extension, University of Texas, Austin, Tex., 1973.

Staff: "Methods of Test for Water and Sediment in Crude Oils," *API Stand.* 2542 (ASTM D 96–98) (ANSI Z11.8-1969), American Petroleum Institute, Washington, 1968–1969.

Staff: "Density, Specific Gravity, or API Gravity of Crude Petroleum and Liquid Petroleum Products," *API Stand.* 2547 (ASTM D 1298-67) (ANSI Z11.84-1968), American Petroleum Institute, Washington, 1967–1968.

Staff: "Method of Test for Water and Sediment in Crude Oils and Fuel Oils by Centrifuge," *API Stand.* 2548 (ASTM D 1796-68) (ANSI Z11.115-1969), American Petroleum Institute, Washington, 1968–1969.

Staff: "Test for Sediment in Crude and Fuel Oils by Extraction," *API Stand.* 2561 (ASTM D 473-69) (ANSI Z11.58-1970), American Petroleum Institute, Washington, 1969–1970.

Petroleum Fuels Characteristics

Since petroleum fuels essentially are mixtures of various hydrocarbons, plus various additives and minor impurities, their performance characteristics are not evaluated on the basis of chemical analysis alone, but rather a large number of physical tests and properties, such as octane numbers, Reid vapor pressure, distillation temperature, kinematic viscosity, specific gravity, pour point, flash point, color, and corrosiveness, are required to fully describe what might be termed effective performance of the fuels. Further, there are specific chemical data of importance, notably the content of sulfur, lead, phosphorus, water, sediment, and gum-forming materials.

Because petroleum fuels are refined from a rather broad range of feedstocks, as noted in the description on *Petroleum* given earlier in this Handbook section, and are refined and distributed by a number of suppliers, large and small, and because efforts are made by suppliers to adapt their products to the specific needs of both season and local geography and terrain, a national average of the properties of a given fuel category can be obtained only by conducting periodically rather massive samplings of fuel products in which all consuming areas and user segments are represented in a proportionate manner.

This sampling technique has been used for many years by the Bureau of Mines, U.S. Department of the Interior, and the American Petroleum Institute as a joint, cooperative effort. Such annual reports are available from the Bureau of Mines. Other private surveys are made from time to time.

Four major categories of petroleum fuels are described briefly here: (1) motor gasolines, (2) Diesel fuel oils, (3) aviation turbine fuels, and (4) burner fuel oils. A few of the principal physical properties and tests are also defined.

MOTOR GASOLINES

Historical data going back to 1946 to illustrate the trend of key physical properties of so-called regular-price gasoline and premium-price gasoline are presented in Figs. 1 and 2, respectively. Formal cooperative reporting by the Bureau of Mines and API (American Petroleum Institute) in the gasoline category dates back to the December 1964 report covering the summer season of 1964. Since that time, a summer season report has been issued each year, usually in the December or January following the season. Similarly, a winter season report has been issued each year, usually in the midsummer following a particular winter season. See list of references at end of this description.

In recent years, close to 5000 samples of the products of about 50 refiners are collected from service stations. The products of both large and small suppliers are collected, and the laboratories of the various refiners, motor manufacturers, and chemical companies furnish data to the Bureau of Mines for analysis and compilation. Generally, the testing procedures recommended by the American Society for Testing and Materials (ASTM) are followed. See list of references.

For a summary of values of the motor gasoline survey see Table 1.

In the early days of automotive transportation, gasoline was a relatively simple mixture of petroleum fractions derived from straight-run and thermally cracked stocks. In contrast, the modern fuel is a complex mixture of blends derived from catalytic cracking, alkylation, catalytic reforming, polymerizations, isomerizations, and hydrocracking, plus small amounts of additives designed to further improve the overall efficiency and reliability of the internal combustion engine.

From 1925 to 1950 there was a gradual but steady increase in compression ratio. During the 1950s this increase was more rapid, with a final leveling off in the late 1960s. This increase in compression ratio was aimed at improving overall engine performance and efficiency. Its effect on fuel composition was direct, since increased compression ratio requires an increase in fuel octane number to prevent knocking. The octane increase was achieved partly by the addition of lead alkyls, but mostly via the changes in the petroleum refining processes. Without these changes the amount of gasoline that could be obtained from a barrel of crude would have been greatly reduced.

During the 1950s and 1960s, other changes also were made in automotive gasoline to help achieve the objectives of greater efficiency and reliability. Additives were developed that helped to minimize such problems as carburetor icing and fouling, valve and engine

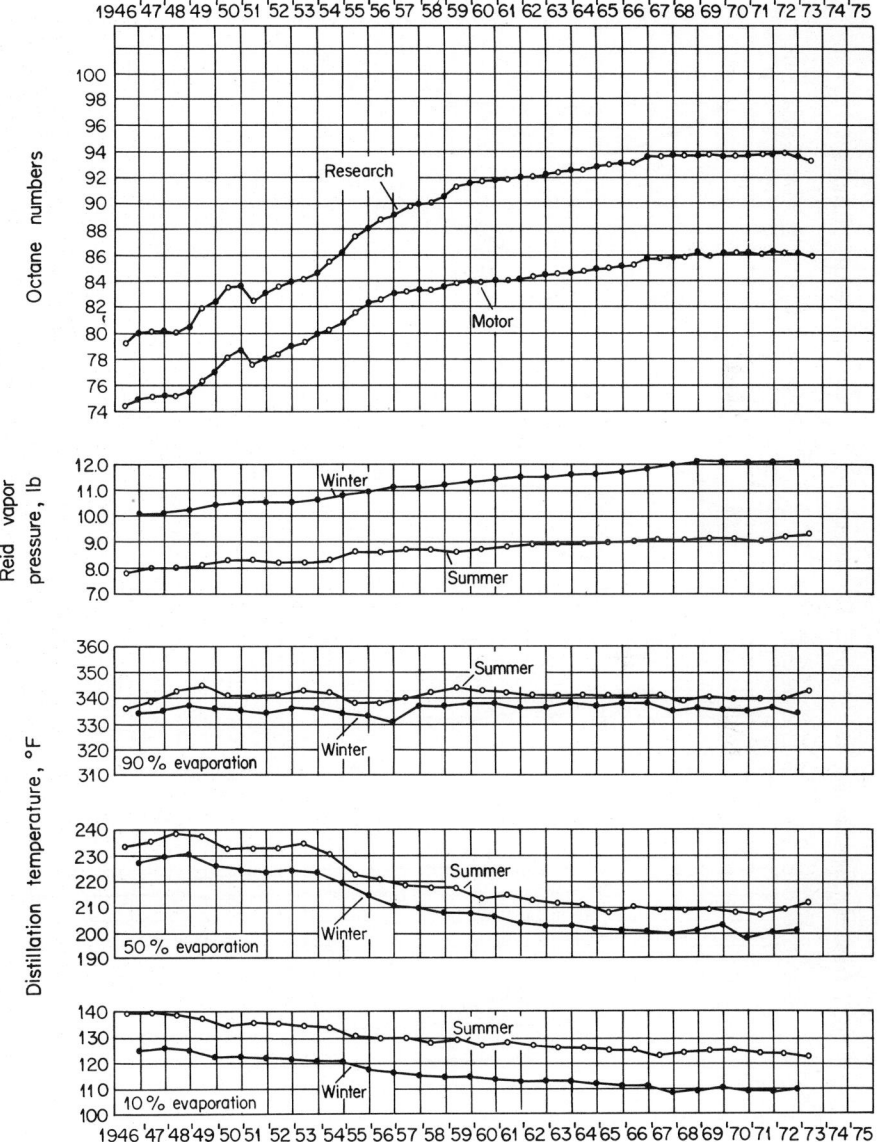

Fig. 1. Trends of certain characteristics of regular-price gasoline.

deposits, spark plug fouling, fuel system corrosion, and poor fuel distribution. All these factors have contributed to modern highly complex gasoline formulations. Commencing in the early 1960s, ecological problems also entered into the gasoline picture. Small amounts of carbon monoxide and nitrogen oxides are formed and emitted to the atmosphere, along with some unburned hydrocarbons when gasoline is burned in an internal combustion engine. The predominant products of combustion, of course, are carbon dioxide and water. When stress on atmospheric pollution coupled with increased demand and lower fuel efficiencies arrived at about the same time, conventional new oil sources were being discovered at a lessened rate.

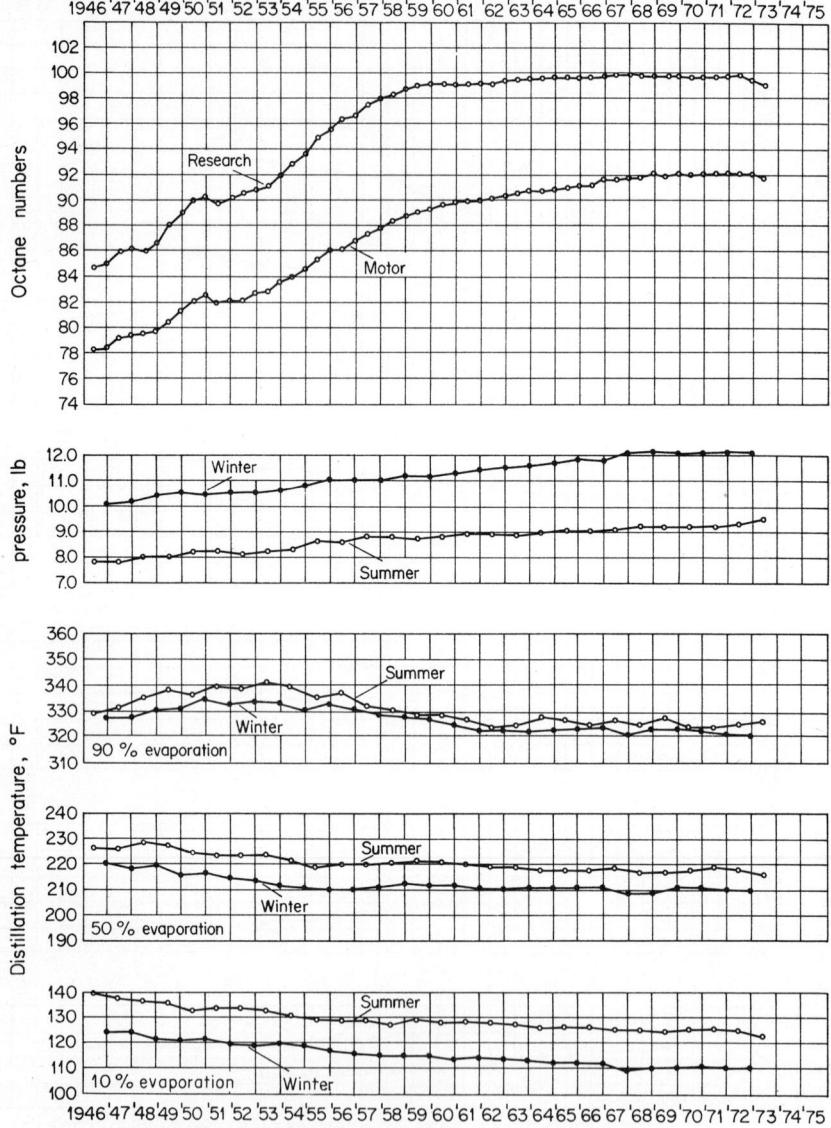

Fig. 2. Trends of certain characteristics of premium-price gasolines.

The mid-to-late 1970s is a period that requires innovation in fuel technology. As newer automobiles with new fuel requirements are phased in, fuels must still be produced to satisfy the older cars. In addition to the requirement for current grades of gasoline, one or more new grades will be needed.

Gasoline Antiknock Quality. With the long-term trend toward more efficient utilization of fuels in spark ignition, compression ratios reached a peak in the 1965–1970 period. In 1971 and 1972, a sharp decrease in compression ratio occurred in passenger car engines in the United States. This decrease was made to permit the engines to operate on unleaded gasolines of an octane quality, which could be produced by petroleum refineries in reasonably large volumes. This quality was calculated to be 91 Research Octane Number (RON) and 83 Motor Octane Number (MON). Octane number is described later.

Lead Alkyl Levels and Other Catalyst Poisons. In addition to the apparent need to furnish unleaded gasolines for new cars having lead-sensitive catalytic reactors, there is

TABLE 1.   Summary of Value, Motor Gasoline Survey (Summer 1972)

Test	ASTM method	Regular-price gasoline Average	Premium-price gasoline Average
Gravity, °API	D 287	60.3	61.7
Corrosion, No.	D 130	1	1
Sulfur content, wt %	D 1266	0.040	0.026
Gum, mg/100 ml	D 381	1	1
Phosphorus, g/gal	D 3231	0.004	0.003
Lead, g/gal	D 526	2.01	2.42
Octane number, Research	D 2699	93.5	99.3
Octane number, Motor	D 2700	86.1	91.9
Research + motor octane Nos./2		89.8	95.6
Reid vapor pressure, lb	D 323	9.3	9.5
Vapor-liquid ratio of 20, °F	D 439	136	137
Distillation temp., °F, IBP	D 86	91	90
5% evaporated		108	107
10% Do.		121	121
20% Do.		142	146
30% Do.		163	171
50% Do.		211	215
70% Do.		265	255
90% Do.		342	325
95% Do.		378	361
End point		417	405
Residue, vol %		1.0	1.0
Loss, vol %		1.6	1.7

an objective to reduce lead particulate emissions from automotive engines. Two additional elements which are believed to poison catalysts are phosphorus and sulfur. Phosphorus additives in the past have been beneficial to engine performance. Sulfur levels in gasoline already are low and are declining slowly.

Gasoline Mileage. The progressively lower limits set on hydrocarbon, carbon monoxide, and nitrogen oxide emissions from automobile exhausts have a significant effect on gasoline mileage relative to older cars (1970) for a number of reasons: (1) Reduced compression ratios lessen gasoline mileage and power output. (2) Increased engine size or increased gear ratios (to regain power lost by reduced compression ratios) lower gasoline mileage. (3) Use of exhaust gas recirculation to reduce nitrogen oxide emissions lowers gasoline mileage. (4) Operation at nonoptimum spark advance settings to reduce emissions lowers gasoline mileage. The combined effects of these changes are indicated in Table 2.

Refinery Pool Octane Levels. The trends of high compression ratios through 1970 and lower compression ratios from 1971 onward have created a car population in the United States with widely different and constantly changing octane-number requirements. To satisfy this variety of varied octane requirement levels and lead antiknock additive limits requires the marketing of several additional grades of gasoline. Several independent studies have indicated that the overall effect of the new marketing requirements is to raise the unleaded pool octane-number requirements of petroleum refineries in the United States. It is projected that the average refinery pool clear RON will have to increase from a value of 88–89 which prevailed in 1971 to approximately 92–94 between 1977 and 1980 to accommodate projected lead removal schedules. These RON levels are influenced by several factors, including: (1) Removal of lead reduces some of the freedom now used in blending the various refinery streams. Thus, an overall increase in average RON will be required to maintain adequate quality control at any given RON level. (2) The average RON requirement of used automobiles is generally somewhat higher than the automobile manufacturer's target for new cars—because of the influence of deposit accumulation, average timing maladjustments, and overall engine condition. (3) Automobile manufacturers, once having achieved desired emission controls, will most likely begin to design engines having higher octane requirements in order to regain efficiencies lost because of the emission controls. (4) There is a general tendency for a certain

TABLE 2.   Effect of Emission Standards and Engine Modifications on the Average Gasoline Mileage of Passenger Cars Operated in the United States

Model years	Percent reduction in gasoline mileage	Investigator
1968–1973	7.7	1
1968–1972	5.0	2
1972–1973	10.0	2
1968–1973	15.0	2
1970–1973	17.0*	3
1965–1975	26.0*	4
1974–1975	6.3	4

*Includes effect of car weight and other design changes, in addition to the effect of emission standards and engine modifications.
1. Environmental Protection Agency, EPA, Office of Air and Water Programs, Mobile Source Pollution Control Program, "Fuel Economy and Emission Control," 1972.
2. Chrysler Corporation, by G. J. Huebner, Jr., and D. J. Gasser, General Factors Affecting Vehicle Fuel Consumption, presented at Society of Automotive Engineers, National Automobile Engineering Meeting, Detroit, Mich., 1973.
3. E. I. duPont De Nemours & Co., Inc., Petroleum Chemicals Division, "1973 Technical Conference," p. PE-11.
4. Ford Motor Company, remarks of Harold C. MacDonald, vice-president, Product Development Group, at 1973 Society of Automotive Engineers, National Automobile Engineering Meeting, Detroit, Mich., 1973.

percentage of motorists to purchase a higher octane level than recommended because of their desire for a "safety factor."

Additives.   These are ingredients that are added at concentrations ranging from a few to several hundred parts per million (ppm), and many have become essential ingredients of modern gasolines. Gasoline additives now must be registered with the air pollution authorities. Manufacturers and users of gasoline additives must provide the authorities with information on the function, chemical composition, recommended dosage, and emission products of the additives that are registered.

Gasoline additives are categorized according to type and function in Table 3. The number, type, and quantity of each additive will vary among individual marketers of gasoline. In some instances several functions can be combined in one chemical compound to provide "multifunctional" additives.

The increasingly stringent control of automotive emissions has stimulated the development and use of "extended-range" detergents, which are designed to promote peak engine performance by maintaining engine cleanliness.

Volatility.   Optimizing volatility for peak engine performance has become increasingly difficult to achieve because of seemingly incompatible performance goals. Minimizing automotive emissions while maintaining satisfactory driveability, octane quality, and volume of gasoline production requires fuels having divergent volatility characteristics. It has been suggested that refiners decrease front-end volatility during much of the year to reduce the loss of vapors during handling and to avoid overloading of the vapor recovery systems on emission-system equipped cars. This can have an adverse effect on starting and warm-up and, in some cars, can cause a definite reduction in the amount of high-octane, clean-burning butane that can be blended into automotive fuels. In the mid- and upper-boiling ranges, on the other hand, it has been suggested that refiners increase volatility to reduce final boiling points. These changes make it easier to design cars to give reduced emissions during the warm-up period and have a beneficial effect on the overall driveability of the cars. From the refinery viewpoint, however, these changes reduce the amount of high-octane components that can be blended into gasolines and cause a marked reduction in the volume of gasoline that can be produced by current refineries. Additionally, these volatility changes involve large expenditures for refinery processing equipment and increased refinery operating costs.

DIESEL FUELS

In a cooperative effort similar to the gasoline survey, the Bureau of Mines has published the results of Diesel fuel surveys since 1950. In a typical recent survey, some 250 samples of Diesel fuel, manufactured by over 30 refiners in over 100 refineries, large and small and

TABLE 3.   Automotive Gasoline Additives and Functions

Additive	Function
Antiknock compounds	Increase octain number
Scavengers	Remove combustion products of antiknock compounds
Combustion chamber deposit modifiers	Suppress surface ignition and spark plug fouling
Antioxidants	Provide storage stability
Metal deactivators	Supplement storage stability
Antirust agents	Prevent rusting in gasoline-handling systems
Anti-icing agents	Suppress carburetor icing and fuel system freezing
Detergents	Control carburetor and induction system cleanliness
Upper cylinder lubricants	Lubricate upper cylinder areas and control intake system deposits
Dyes	Indicate presence of antiknock compounds and identify makes and grades of gasoline

widely distributed throughout the United States, were reported. Trends dating back to 1960 for four categories of Diesel fuels are shown in Figs. 3 to 6. The categories are:

   Type C-B, Diesel fuel oils for city bus and similar operations
   Type T-T, Fuels for Diesel engines in trucks, tractors, and similar service
   Type R-R, Fuel for railroad Diesel engines
   Type S-M, Heavy-distillate and residual fuels for large stationary and marine Diesel engines

For those readers interested in historical data, it should be noted that the four aforementioned groups replace the Class 1, 2, 3, and 4 categories used in survey reports before 1960.

Diesel fuel surveys are published annually, usually near the end of each year. The annual summaries are broken down into four geographical regions of the United States, as noted in Tables 4 to 7. The reporting regions are delineated by the map of Fig. 7.

Although some evolution in Diesel fuel composition has occurred since the early 1960s, the net effect has not resulted in markedly improved performance levels. Major performance factors of Diesel fuel are characterized by sulfur content, cetane number, viscosity, volatility, and pour point.

Recent changes in Diesel fuel properties have resulted from competition for available distillate blending stocks for use in other fuels or as process materials. For example, the rapid increase in commercial jet fuel consumption has significantly reduced availability of the more volatile straight-run components. Thus, the fraction of cracked stocks in Diesel fuel continues to increase. Hydrogenation of cracked stocks reduces sulfur content, improves stability, and upgrades rather poor performers to satisfactory Diesel components. Most properties have leveled off, but cetane number is slowly declining. This trend is expected to continue as the demand for energy progresses. Cetane number is defined later.

Most Diesel engines in truck service can operate satisfactorily on truck and tractor fuels currently available. However, a wide range of fuels is possible under this classification, and variation within the classification can have a pronounced effect on a given engine's performance.

Emphasis on improving air quality is bringing about engine design changes to reduce exhaust emissions. Previously preeminent factors, such as increased power output and improved fuel economy, are yielding in importance to reduction of smoke, nitrogen oxides, carbon monoxide, and hydrocarbon emissions. Engine modifications designed to reduce

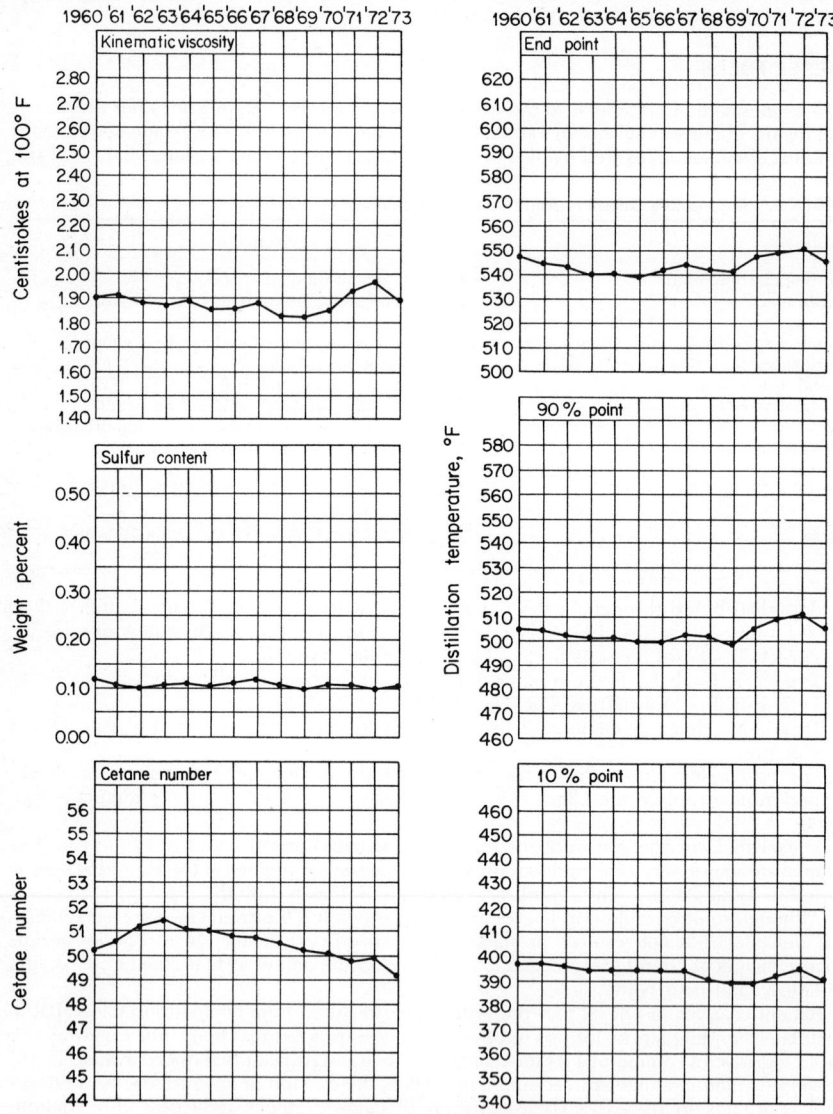

Fig. 3. Trends of some properties of Type C-B Diesel fuel oils.

exhaust emissions probably will result in a narrowing of the range of fuel specifications in regard to cetane number and volatility.

The quality of railroad fuels has not changed significantly over the years. Some railroads operate on special economy-grade fuels that have much broader volatility and lower cetane numbers and always contain large percentages of cracked stock. The large Diesel engines in railroad service are less sensitive to fuel properties than their smaller truck or tractor counterparts and can operate satisfactorily on fuels with less exacting specifications.

Of the total Diesel fuel oil sales in the United States, about 75 percent is for use in transportation. Trucks and buses consume about 45 percent, railroads about 25 percent; and marine uses amount to about 5 percent. Industrial plants, utilities, and the military

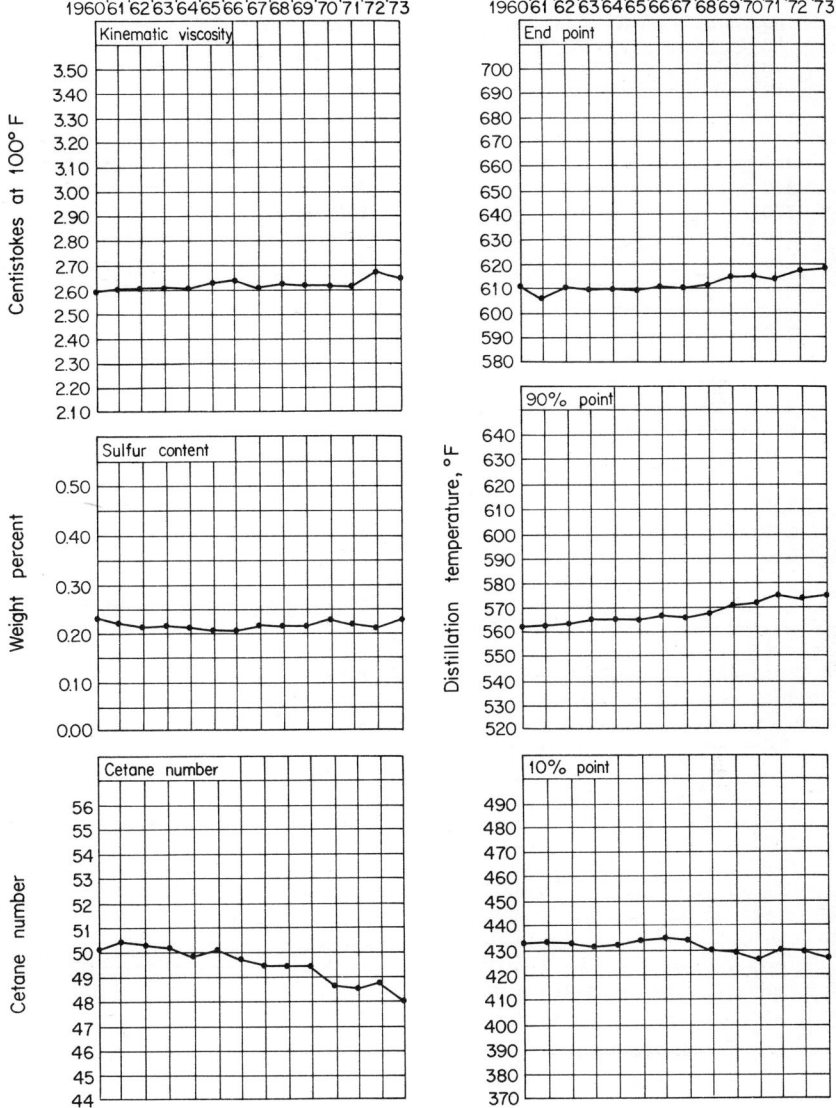

Fig. 4. Trends of some properties of Type T-T Diesel fuel oils.

consume the remainder. In metropolitan areas, electric power companies are installing gas turbines that burn Diesel fuel for power generation. Gas turbines can be installed quickly at relatively low initial costs and are reasonably free of undesirable emissions. The fraction of Diesel fuel used by utilities is increasing.

As in gasolines, the use of additives has become much more common in Diesel fuels. Cetane improvers, largely alkyl nitrates, provide ignition-quality improvement. Ignition quality influences ease of starting and smoothness of operation. A variety of additives are employed to improve storage stability and permit the use of otherwise unstable stocks. Polymeric and other types of additives have been utilized as detergents and dispersants. The detergents have the ability to maintain fuel-injection nozzle cleanliness and will markedly increase fuel-filter life; many Diesel fuels also contain rust inhibitors.

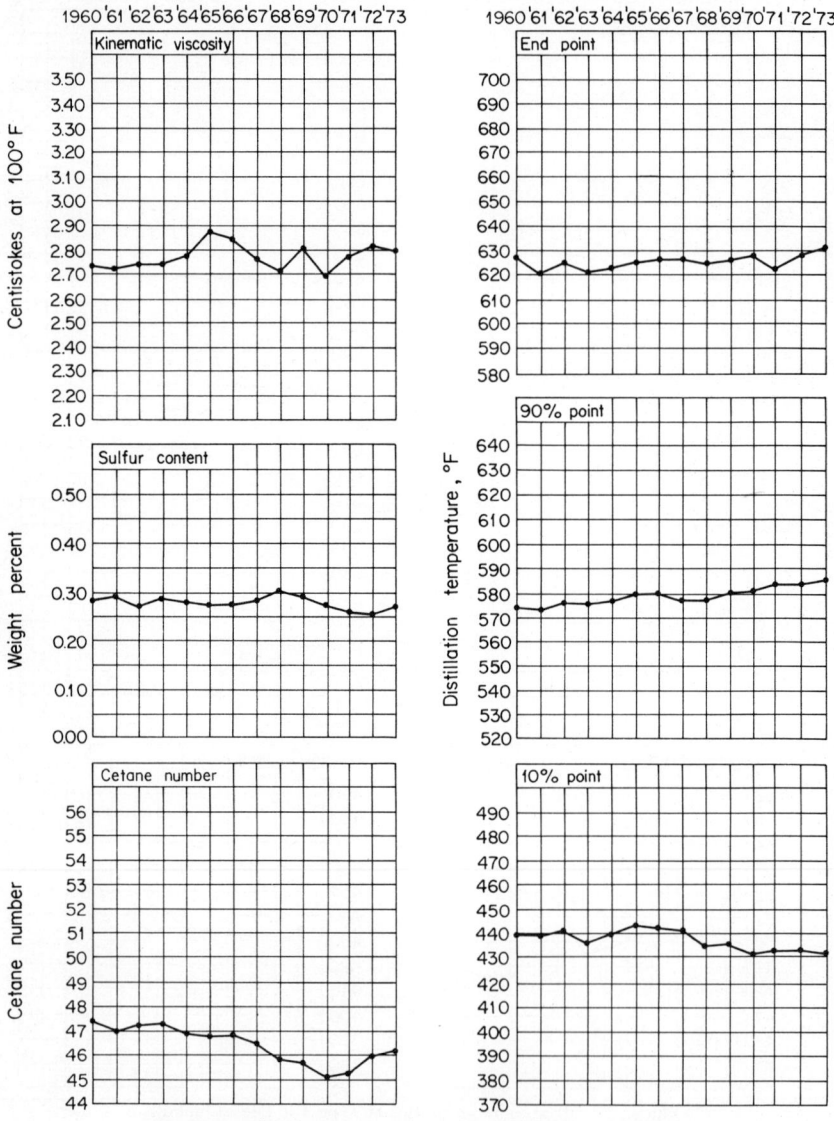

Fig. 5. Trends of some properties of Type R-R Diesel fuel oils.

The high demand for Diesel and jet fuels has made it difficult to obtain appropriate low-temperature flow characteristics by base stock selection. This has led to increased use of pour-point depressant additives. Several of these materials, both polymeric and nonpolymeric, have the ability to reduce pour points, resulting in substantial improvement in flow through distribution and truck piping systems. However, their effect on cloud point (the first appearance of wax crystals which can cause filter plugging under cold conditions) is small. Fuel system modifications, such as mounting filters in warm locations or installing filter heaters, have been used to provide for the use of pour-point depressants.

Smoke emission laws have led truck operators to investigate antismoking additives. The most functional of these are barium-based organic compounds which are effective at

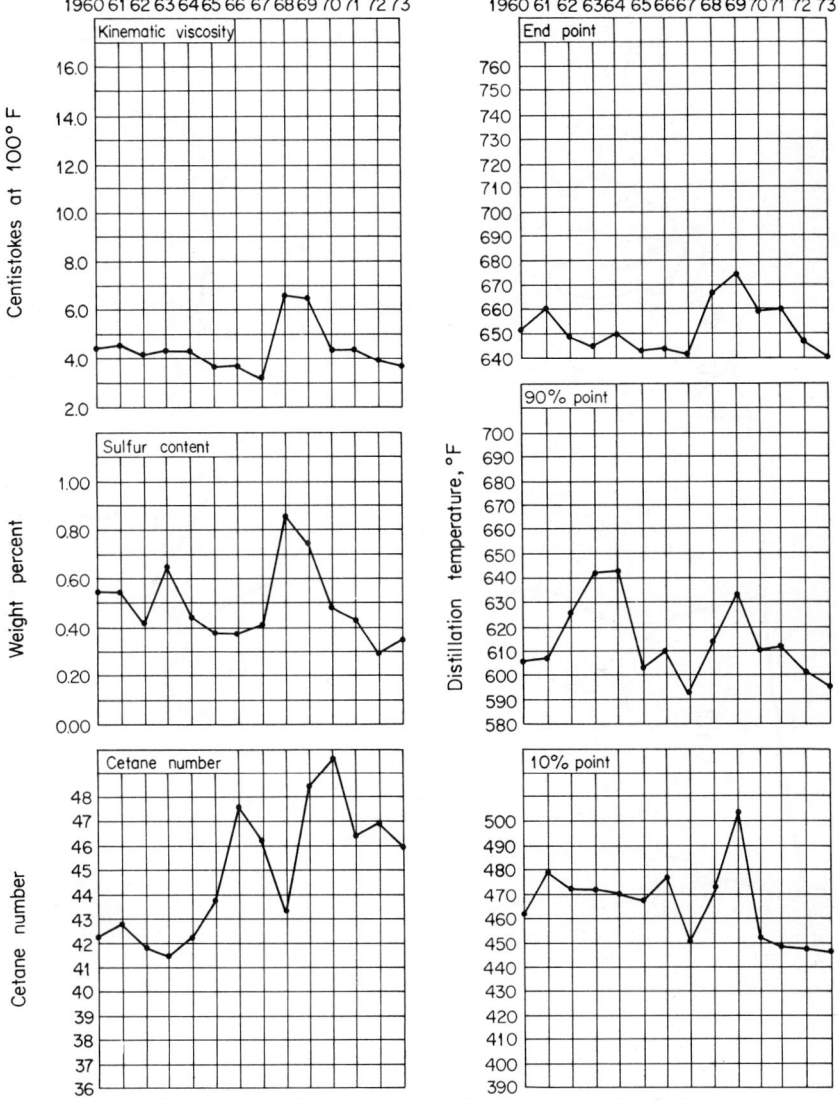

Fig. 6. Trends of some properties of Type S-M Diesel fuel oils.

concentrations of about 1000 ppm. Effective concentrations of the barium additive are costly, and the ash contributed by the barium salts has caused a problem in some engines. Thus, antismoke additives generally are not used in Diesel fuels by manufacturers. Two other types of additives are emerging from the development stage, namely, biocides to prevent bacterial attack on the fuel and masking agents to improve the odor of exhaust fumes in city bus service.

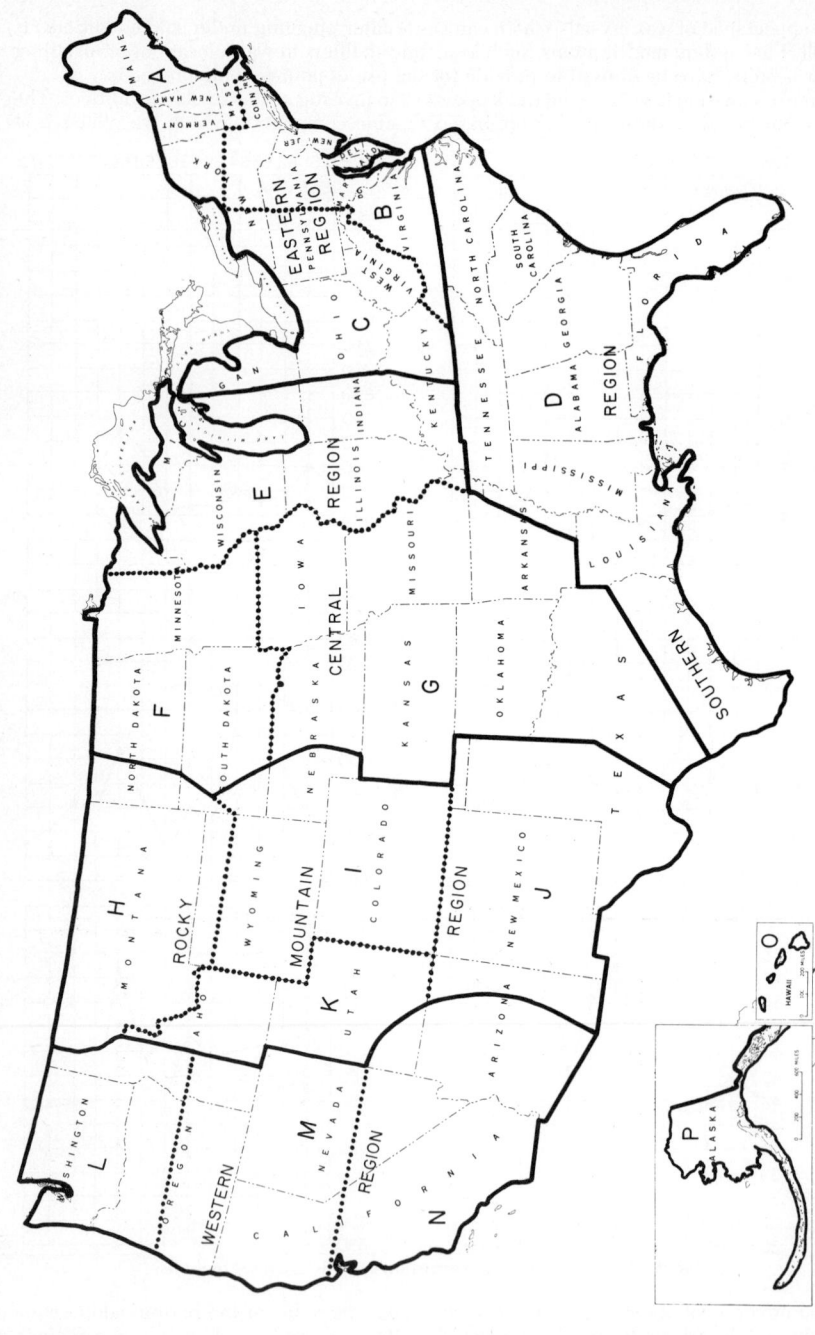

Fig. 7. Geographical areas upon which national surveys of Diesel fuel oils and burner fuel oils are based.

TABLE 4. Characteristics of Type C-B Diesel Fuel

Geographic distribution of Diesel fuels		Eastern region A,B,C D,E,F,G 38			Southern region D A,B,C,E,F,G,I,J 15			Central region E,F,G A,B,C,D,H,I,J,K,L 34			Rocky Mountain region H,I,J,K D,E,F,G,L,M,N,O,P 13			Western region L,M,N,O,P F,H,I,J,K 14		
Test	ASTM	Minimum	Average	Maximum	Minimum	Average	Maximum	Minimum	Average	Maximum	Minimum	Average	Maximum	Minimum	Average	Maximum
Gravity, °API	D 287	32.3	41.4	47.9	35.4	41.0	42.7	32.3	40.9	44.7	35.0	40.9	43.8	33.4	42.7	47.6
Flash point, °F	D 93	122	…	194	126	…	176	122	…	194	122	…	176	122	…	178
Color:																
ASTM	D 1500	L0.5	…	L1.0	L0.5	…	…	L0.5	…	1.0	L0.5	…	L1.0	L0.5	…	L1.0
Saybolt chromometer	D 156	+30	…	+11	+30	…	+21	+30	…	+17	+30	…	+20	+30	…	+18
Viscosity at 100° F:																
Kinematic, cs	D 445	1.50	1.90	3.42	1.50	1.85	2.92	1.45	1.93	3.05	1.50	1.93	3.05	1.44	1.77	3.01
Saybolt Universal, sec	D 88	…	32.3	37.3	…	32.1	35.7	…	32.4	36.1	…	32.4	36.1	…	…	36.0
Cloud point, °F	D 2500	<-66	…	20	-60	…	6	-56	…	10	-62	…	10	-62	…	14
Pour point, °F	D 97	<-65	…	0	<-65	…	-5	-60	…	-5	-65	…	-5	-65	…	5
Sulfur content, wt %	D 129	0.011	0.100	0.34	0.01	0.050	0.222	0.015	0.130	0.464	0.001	0.112	0.464	0.001	0.089	0.38
Aniline point, °F	D 611	130.5	149.1	178.0	138.0	146.9	160.6	130.5	146.8	160.6	138.0	148.3	160.6	139.9	147.9	161.9
Carbon residue on 10%, wt %	D 524	0.00	0.076	0.19	0.005	0.073	0.19	0.03	0.081	0.19	0.00	0.058	0.090	0.01	0.055	0.120
Ash, wt %	D 482	0.000	0.001	0.007	0.000	<0.001	0.002	0.000	0.001	0.007	0.000	0.001	0.002	0.000	0.001	0.005
Cetane number	D 613	43.0	49.8	65.3	43.0	47.1	54.6	42.3	49.1	54.6	39.0	48.6	54.6	39.0	50.8	65.3
Distillation temp., °F, volume recovered:	D 86															
IBP		318	352	412	324	347	387	324	353	412	320	352	398	320	350	408
10%		350	394	460	364	388	442	350	393	460	360	393	442	360	387	454
50%		407	445	530	411	437	514	408	446	514	404	449	514	404	438	525
90%		444	506	604	465	497	590	465	510	590	472	516	590	456	498	574
End point		490	545	645	499	533	637	498	549	637	496	556	634	496	537	608

SOURCE: Mineral Industry Surveys, *U.S. Bur. Mines Petrol. Prod. Surv.* 82.

TABLE 5. Characteristics of Type T-T Diesel Fuel

Geographic distribution of Diesel fuels Districts within region Additional districts Number of fuels Test	ASTM	Eastern region A,B,C D,E,F,G 57			Southern region D A,B,C,E,F,G,I,J 30			Central region E,F,G A,B,C,D,H,I,J,K,L 75			Rocky Mountain region H,I,J,K D,E,F,G,L,M,N,O 38			Western region L,M,N,O,P F,H,I,J,K 32		
		Minimum	Average	Maximum	Minimum	Average	Maximum	Minimum	Average	Maximum	Minimum	Average	Maximum	Minimum	Average	Maximum
Gravity, °API	D 287	31.5	36.1	44.7	31.5	35.8	42.5	30.0	36.6	44.7	30.0	36.6	44.3	30.0	36.7	44.7
Flash point, °F	D 93	122	...	194	128	...	190	122	...	226	136	...	206	110	...	216
Color, ASTM	D 1500	L0.5	...	1.5	L0.5	...	1.5	L0.5	...	2.0	L0.5	...	12.0	L0.5	...	2.5
Viscosity at 100°F:																
Kinematic, cs	D 445	1.50	2.56	3.54	1.50	2.73	3.77	1.50	2.61	3.77	1.66	2.67	3.90	1.27	2.79	4.25
Saybolt Universal, sec	D 88	...	34.5	37.7	...	35.1	38.4	...	34.7	38.4	...	34.9	38.9	...	35.3	40.0
Cloud point, °F	D 2500	-42	...	24	-26	...	12	-50	...	16	-40	...	30	<-70	...	38
Pour point, °F	D 97	-60	...	15	-60	...	5	-50	...	10	-50	...	25	<-70	...	30
Sulfur content, wt %	D 129	0.02	0.192	0.50	0.02	0.235	1.1	0.002	0.236	0.76	0.02	0.251	0.648	0.003	0.230	0.49
Aniline point, °F	D 611	122.5	142.7	178.0	126.0	145.1	160.6	122.5	145.7	172.0	130.0	148.3	160.6	127.5	146.5	160.0
Carbon residue on 10%, wt %	D 524	0.00	0.109	0.33	0.01	0.102	0.237	0.015	0.110	0.33	0.00	0.091	0.237	0.04	0.097	0.15
Ash, wt %	D 482	0.000	0.002	0.06	0.000	<0.001	0.004	0.000	0.001	0.009	0.000	0.001	0.006	0.000	0.002	0.006
Cetane number	D 613	39.0	46.9	63.1	40.1	47.0	56.0	39.0	48.6	56.0	40.6	48.7	54.8	41.0	48.7	63.7
Distillation temp., °F volume recovered:	D 86															
IBP		309	366	412	324	368	412	326	377	463	341	378	436	319	375	450
10%		364	423	466	364	425	466	366	425	483	376	429	489	346	430	489
50%		408	493	533	417	497	534	408	493	535	428	493	544	389	498	545
90%		465	572	628	467	579	648	465	571	648	484	574	656	455	580	626
End point		507	618	698	507	624	692	500	613	692	509	619	700	489	624	670

SOURCE: Mineral Industry Surveys, *U.S. Bur. Mines Petrol. Prod. Surv.* 82.

TABLE 6. Characteristics of Type R-R Diesel Fuel

Geographic distribution of Diesel fuels		Eastern region A,B,C D,E,F,G 29			Southern region D A,B,C,E,F,G,I,J 16			Central region E,F,G A,B,C,D,H,I,J,K 40			Rocky Mountain region H,I,J,K D,E,F,G,L,M,N,O,P 22			Western region L,M,N,O,P H,I,J,K 14		
Test	ASTM	Minimum	Average	Maximum	Minimum	Average	Maximum	Minimum	Average	Maximum	Minimum	Average	Maximum	Minimum	Average	Maximum
Gravity, °API	D 287	30.0	35.7	43.3	30.0	34.6	39.0	28.5	35.0	41.7	28.3	33.5	41.7	28.3	34.0	37.3
Flash point, °F	D 93	144	…	194	158	…	186	130	…	198	148	…	206	156	…	216
Color, ASTM	D 1500	L0.5	…	L2.0	L0.5	…	L1.5	L0.5	…	2.5	L0.5	…	L2.0	L0.5	…	4.0
Viscosity at 100°F:																
Kinematic, cs	D 445	2.00	2.72	4.30	2.47	2.86	3.20	1.80	2.63	3.41	1.81	2.91	3.85	2.49	3.15	3.85
Saybolt Universal, sec	D 88	32.6	35.0	40.2	34.2	35.5	36.6	…	34.7	37.3	32.0	35.7	38.7	34.3	36.4	38.7
Cloud point, °F	D 2500	-20	…	30	-18	…	10	-36	…	10	-36	…	30	-6	…	30
Pour point, °F	D 97	-40	…	25	-40	…	0	-40	…	0	-40	…	25	-20	…	25
Sulfur content, wt %	D 129	0.07	0.178	0.444	0.10	0.223	0.461	0.09	0.281	0.70	0.02	0.361	0.74	0.019	0.359	0.74
Aniline point, °F	D 611	124.3	146.7	186.0	120.0	141.7	157.0	111.0	139.7	162.0	111.0	140.0	157.0	120.0	144.4	155.0
Carbon residue on 10%, wt %	D 524	0.06	0.113	0.33	0.014	0.112	0.237	0.00	0.120	0.33	0.010	0.110	0.237	0.04	0.110	0.20
Ash, wt %	D 482	0.000	0.001	0.0085	0.000	0.001	0.0085	0.000	0.001	0.008	0.000	0.001	0.005	0.000	0.001	0.002
Cetane number	D 613	36.4	47.8	70.5	36.1	45.5	53.6	35.8	46.2	54.2	35.8	44.8	53.6	37.0	45.5	53.8
Distillation temp,, °F, volume recovered:	D 86															
IBP		334	378	494	354	373	402	337	377	416	355	384	420	332	392	450
10%		392	432	524	410	427	474	384	426	460	409	440	489	409	448	489
50%		464	501	562	480	501	536	442	497	525	442	505	543	477	520	543
90%		536	580	610	550	587	610	490	580	618	490	594	664	575	604	664
End point		592	627	664	608	634	650	524	627	662	524	635	707	602	646	707

SOURCE: Mineral Industry Surveys, *U.S. Bur. Mines Petrol. Prod. Surv.* 82.

TABLE 7. Characteristics of Type S-M Diesel Fuel

Test	ASTM	Eastern region A,B,C / D / 7			Southern region D / A,B,G / 8			Central region E,F,G / D / 4			Rocky Mountain region H,I,J,K / L,M,N,O,P / 3			Western region L,M,N,O,P / H,I,J,K / 7		
Geographic distribution of Diesel fuels / Districts within region / Additional districts / Number of fuels		Minimum	Average	Maximum	Minimum	Average	Maximum	Minimum	Average	Maximum	Minimum	Average	Maximum	Minimum	Average	Maximum
Gravity, °API	D 287	30.1	35.1	42.3	24.5	32.6	35.8	32.3	34.5	37.7	33.5	34.3	35.8	28.4	33.5	36.3
Flash point, °F	D 93	132	...	174	164	...	286	144	...	180	156	...	206	156	...	266
Color, ASTM	D 1500	0.5	...	L2.0	L1.0	...	2.0	L1.0	...	L2.0	L0.5	...	L1.0	L0.5	...	L1.5
Viscosity at 100°F:																
Kinematic, cs	D 445	1.60	2.56	3.54	2.70	5.31	21.1	2.71	2.99	3.77	2.80	3.30	3.85	2.80	3.75	6.40
Saybolt Universal, sec	D 88	...	34.5	37.7	34.8	35.7	102.5	35.0	35.9	38.4	35.1	36.9	38.7	35.1	38.4	46.9
Cloud point, °F	D 2500	-44	...	2	-4	...	8	2	...	8	16	...	26	-4	...	26
Pour point, °F	D 97	-50	...	-10	-10	...	65	-10	...	-5	-5	...	25	-20	...	25
Sulfur content, wt %	D 129	0.06	0.156	0.29	0.12	0.56	2.35	0.18	0.40	0.76	0.02	0.23	0.48	0.019	0.32	0.62
Aniline point, °F	D 611	135.0	140.5	146.0	135.0	142.2	151.0	134.0	143.7	153.0	148.5	150.3	151.5	148.0	150.6	155.0
Carbon residue:	D 524															
on 10%, wt %		0.070	0.123	0.18	0.11	0.13	0.18	0.11	0.15	0.18	0.11	0.12	0.12	0.04	0.10	0.12
on 100%, wt %					0.11	...	0.11							0.09	0.13	0.16
Ash, wt %	D 482	0.00	...	0.00	0.00	...	0.00	0.000	...	0.000	0.000	...	0.000	0.000	0.001	0.002
Cetane number	D 613	36.4	44.7	51.7	40.0	44.4	49.8	43.6	47.6	54.2	47.0	48.7	50.0	41.0	47.5	53.8
Distillation temp., °F, volume recovered:	D 86															
IBP		334	358	400	340	387	535	341	366	384	368	394	416	368	420	522
10%		386	428	453	425	453	595	420	426	434	420	458	489	420	469	537
50%		439	499	533	511	539	680	506	513	524	497	526	543	497	534	576
90%		489	569	604	591	607	648	587	604	648	600	605	613	575	607	637
End point		546	617	644	638	652	692	628	651	692	636	646	661	602	647	674

SOURCE: Mineral Industry Surveys, *U.S. Bur. Mines Petrol. Prod. Surv.* 82.

AVIATION FUELS

National aviation fuel survey reports have been issued by the Bureau of Mines since March 1951, the first report covering 1950. Again, there is a cooperative agreement between the Bureau of Mines and the API in terms of gathering and compiling information. For a typical survey, well over 100 samples will be taken and analyzed, representing the products of form 15 to 20 manufacturers. As will be noted in Table 8, data are collected for several categories of aviation fuels, including both military and commercial aviation fuels.

TABLE 8.   Characteristics of Aviation Turbine Fuels

	Military turbine fuels		Commercial jet fuels	
	JP-4	JP-5	A	B
Number of samples tested	35	12	64	5
Gravity, °API	54.6	41.6	43.0	54.2
Distillation temperature, °F:				
10% evaporated	215	388	372	243
50% evaporated	285	419	415	301
90% evaporated	394	460	474	405
Evaporated at 400°F, %	87.5	25.0	35.7	85.9
Reid vapor pressure, lb	2.5	......	0.2	2.6
Freezing point, °F	−84	−59	−50	−80
Viscosity, kinematic, −30°F,				
centistokes	3.01	10.1	9.38	
Aniline point, °F	130.7	139.7	144.8	136.0
Aniline-gravity constant, No.	7,136	5,714	6,241	7,371
Water tolerance, milliliter	0.4	......	0.3	
Sulfur, total, wt %	0.032	0.037	0.048	0.022
Sulfur, mercaptan, wt %	0.0005	0.0009	0.0004	0.0005
Naphthalenes, wt %	0.74	1.21	1.79	
Aromatic content, vol %	10.7	15.7	16.1	9.5
Olefin content, vol %	0.9	0.6	1.1	1.6
Smoke point, mm	28.0	21.7	23.2	30.0
Smoke volatility index	64.9	32.2	37.5	66.1
Gum, mg/100 ml				
Existent, at 450°F	0.7	1.3	0.8	0.5
Potential, at 212°F	1.1	2.2	1.6	0.9
Heat of combustion, net, Btu/lb	18,725	18,515	18,589	18,744
Luminometer number	63	......	50	
Thermal stability, pressure drop,				
inches mercury	0.06	0.14	0.23	0.03

SOURCE: Mineral Industry Surveys, *U.S. Bur. Mines Petrol. Prod. Surv.* No. 79.

Aviation Gasoline.   Quality control of aviation gasoline is even more critical than that of motor gasoline for obvious reasons of safety. Antiknock control is especially critical because, unlike the motorist, the pilot is unable to hear an engine knock over the high ambient noise level. Volatility, freezing point, heat of combustion, and oxidation stability are all very important to the AVGAS user. Quality control techniques and close control of processing have made aviation gasoline a reliable premium refinery product.

Aviation gasolines contain up to 4.6 milliliters of tetraethyl lead per gallon. Ethylene dibromide is added to scavenge the lead and has been found more effective under high-load aircraft conditions than the chloride/bromine scavenger mixtures used in motor gasolines. Other alkyl leads, tetramethyl lead or methylethyl lead compounds, are not used in aviation gasolines.

Some hydrocarbon constituents of AVGAS tend to oxidize during storage at ambient temperatures. The products of oxidation, fuel-soluble and fuel-insoluble gums, interfere with metering of the fuel to the engine and must be controlled. Certain amine and phenolic chemical compounds have been found effective in this service.

Jet Fuels.   Commercial kerosine was used as a fuel in early developmental work on jet aircraft in the United States. The choice of kerosine over gasoline was based on its low volatility, to avoid occurrence of vapor lock under certain flight conditions, and its availability as a commercial product of uniform characteristics. JP-1, the first military jet fuel, was highly refined kerosine having a very low freezing point ($-76°F$, $-60°C$). Kerosine from selected crudes high in naphthenes was the only fuel having this low freezing point. As the demand for the fuel increased, the Military Petroleum Advisory Board recommended development of a military jet fuel having greater availability in wartime than either JP-1 or AVGAS. The second candidate jet fuel was JP-2, but it did not have the desired availability. JP-3 fuel was another possibility which included the total boiling range of kerosine and gasoline. A cooperative program of testing by the Coordinating Research Council demonstrated that the high vapor pressure of JP-3 (Reid vapor pressure of 5 to 7 pounds) resulted in vaporization of the fuel during climb to altitude. In addition, some fuels foamed excessively during vaporization, so that very large losses of liquid could occur along with the vented vapors.

To overcome the disadvantages of JP-3, the Reid vapor pressure was reduced to 2 to 3 pounds, and JP-4 was developed in 1951. This fuel is a blend of 25 to 35 percent kerosine and 65 to 75 percent gasoline components, and has proved satisfactory for military requirements. An important Navy turbine fuel, developed for carrier operation during the Korean War, was a mixture of a special kerosine and AVGAS. The latter was stored in tanks in the central zone of carriers to minimize the possiblity of hazardous fuel leaks in event of battle damage. But, space was limited for such storage. Thus JP-5 fuel was developed for aircraft carriers. This was a special $140°F$ $(60°C)$ flash-point kerosine. Because of its low volatility, it could be stored safely in outer tanks of carriers. When mixed with AVGAS, JP-5 produced a fuel similar to JP-4. Later, the Navy eliminated the use of the AVGAS mixture and used JP-5 alone.

Commercial airline jet fuels in the United States fall within the general framework of ASTM Jet A, A-1, and B fuels. Jet A and A-1 are of the kerosine type. Jet B corresponds to the military JP-4 fuel. Volume demands for Jet A and A-1 are large, but small for Jet B.

Jet fuel fulfills a dual purpose in the aircraft. It provides the energy and serves as a coolant for lubricating oil and other aircraft components. Exposure of the fuel to high temperatures may cause the formation of oxidation materials (gums) which reduce the efficiency of heat exchangers and clog filters and valves in aircraft fuel-handling systems. Thermal stability is the resistance to formation of gums at high temperature. The JP-4, JP-5, and equivalent commercial fuels have satisfactory thermal stability for aircraft operating at speeds up to about Mach 2.0. Future jet aircraft operating at higher speeds, e.g., Mach 3, may expose the fuel to greater thermal stresses and, therefore, may require a more stable fuel. The development of Mach 3–4 turbojets, Mach 6+ ramjets, and rockets using hydrocarbon fuels will pose additional problems on fuel stability characteristics.

Additives provide an attractive method of improving jet fuel quality. Common applications of additives include the following:

Antioxidants.   Some jet fuels may oxidize during storage at ambient temperature to form fuel-soluble and fuel-insoluble gums. These oxidation products may cause clogging of filters in the aircraft fuel-distribution systems or coking of engine burner nozzles. The same antioxidants approved for AVGAS are used in jet fuels.

Copper Deactivator.   Traces of dissolved copper in jet fuel accelerate oxidation as just described. A copper chelating agent, *N, N*-disalicylidene-1, 20-propanediamine, has been approved for addition to jet fuel to deactivate the prooxidant effects of copper.

Corrosion Inhibitors.   Fuel-soluble corrosion inhibitors have limited specific approval for jet fuels. The inhibitors are added to protect pipelines against corrosion by occluded water and indirectly to reduce contamination of the fuel with rust. Several commercial inhibitors of this type are available.

Anti-Icing Additive.   The major hazard of free water in jet fuels is plugging of fuel lines and filters by ice during flight. Fuel heaters are used in civilian aircraft to prevent ice formation. The military adopted the approach of adding 0.1 to 0.15 percent of an anti-icing inhibitor comprising a mixture of 99.6 percent (wt) ethylene glycol monomethyl ether and 0.4 percent (wt) glycerol. The additive has the extra feature of having biocidal properties. At above freezing temperatures, liquid water can cause erratic operation of electric fuel gages and provides a suitable environment for growth of bacteria and fungi. More recently, the anti-icing additive has been changed to eliminate the glycerol.

Antistatic Additive.   High-speed flow (600 to 800 gallons per minute) through fill lines, filters, and valves generates electrostatic charges in jet fuels. Electrostatic potential differences between the fuel and fuel tank walls may cause sparking in the vapor spaces of fuel tanks. Intensive investigations have shown that additives which decrease resistivity of the fuel can reduce electrostatic hazards.

Contaminants.   In handling and delivery of jet fuels, there are several places where traces of contaminants can enter the fuel. Particulate matter and free water can be removed by bulk settling or filtering. Removal of solids down to 5 micrometers is generally desired to avoid plugging of filters on fuel control valves and burner nozzles. Surface-active agents (surfactants) in jet fuels are highly undesirable because they promote the formation of water-in-fuel emulsions and reduce the efficiency of filter/coalescers. Surfactants produced by refining operations are removed by neutralization, water washing, clay treating, filtration, or settling. Polar additives that are added to jet fuel are chosen on the basis of having minimum surfactant properties.

Future Fuels.   By proper selection of structural materials and aircraft design, it may be feasible to use conventional hydrocarbon fuels up to flight speeds approaching Mach 4. Flight speeds higher than Mach 4 will require more heat-sink capacity than normally provided by fuels in the liquid state. Vaporization of the fuel can increase the overall heat-sink capacity by about 25 percent. A promising alternative is to allow endothermic fuels to undergo mild thermal cracking which can absorb several times the heat picked up in the liquid state alone. Cryogenic fuels (liquefied hydrogen, methane, and propane) also offer an attractive way of cooling aircraft, although these fuels have the disadvantage of a low volumetric energy content.

Low-altitude, high-speed, ramjet-powered aircraft require a fuel with a high volumetric heat content. The U.S. Air Force has been pursuing the development of slurry fuels in which metals are burned to take advantage of the high heat of formation of metal oxides. Research efforts are continuing to develop fuels having the greatest possible metal content and to overcome problems, such as poor pumpability, abrasiveness, and low combustion efficiency. Another promising area includes the high-density fuels (aromatics and condensed polycyclic hydrocarbons. A desirable material would have a heating value of about 150,000 Btu per gallon or higher, with a freezing point of $-50°F$ ($-45.6°C$) or lower.

BURNER FUEL OILS

The first annual survey of burner fuel oils was made and reported by the Bureau of Mines and the API in 1955. The work was initiated at the request of the National Oil Fuel Institute, Inc., to provide information on the characteristics of burner-fuel oil marketed in the United States. The analytical data are used by manufacturers of heating appliances, consumers, and various governmental agencies. In a typical survey, over 300 samples of burner-fuel oil are represented, covering the products of about 30 petroleum refiners from over 100 domestic refineries located throughout the United States. This survey involves six grades of fuels. Data collecting and reporting are done on a regional basis, as with Diesel fuel oils, with the same regional breakdown as indicated by the map of Fig. 7.

Home Heating Oil.   Domestic heating oil must be a clean product and it should not form sediment in storage nor leave a measurable quantity of ash or other deposit upon burning. Since it may be stored at low temperatures, it should be fluid at storage conditions encountered outdoors during winter months. The chemical composition of the product must be controlled to assist in reducing smoke emission. Sulfur content, at one time not considered a major problem, is now important.

Grade 2 fuel oil is the designation given to the heating oil most commonly used for domestic and small commercial space heating. This product is a distillate product, normally fractionated to a boiling range of 350 to 650°F (177 to 343°C). Since about 1950, significant improvements have been made in both the quality of home heating oils and in the manufacturing techniques used in producing them. Originally, grade 2 heating oil was composed of selected refinery straight-run stocks, blended to meet product standards. The resultant product had a good stability and was very satisfactory in performance. As the refining industry was required to make larger amounts of motor gasoline at higher octane levels, cracking processes were developed to convert virgin gas oils to lighter boiling products. This necessitated the use of increasing amounts of both catalytically and thermally cracked gas oils in finished heating oil blends. Heating oil blending became more complex in order to maintain a satisfactory quality level without excessive treating expense.

The industry worked in several directions to correct the quality problems associated with the extensive use of cracked distillates. New treating processes, burner alterations, and the development of additives all progressed together. Additives to improve stability also were developed. Caustic washing processes were developed as an improvement over acid washing. In the early 1950s, reforming of straight-run gasoline became widespread. This process made available large volumes of hydrogen, which previously had been costly to produce. With hydrogen available, catalytic hydrogen treating was used to further improve fuel oil quality. The primary objective in hydrogen treating is enhancement of quality through a reduction in sulfur and removal of small, but objectionable amounts of nitrogen compounds. The treatment also reduces carbon residue, improves burning characteristics, and reduces sludging tendencies. Sulfur reduction is generally 70 to 80 percent complete, but can be as high as 90 to 95 percent if needed. Carbon residue on 10 percent distillation bottoms is reduced to less than 0.10 percent.

Residual Fuels.   Grades 4, 5, and 6 are the designations given to the fuels most commonly employed for commercial, industrial, marine, and other purposes involving larger installations than those used for domestic and small commercial establishments. Typically, these fuels provide steam and heat for industry and large buildings, generate electricity in competition with gas and coal, and power ships. Most users of residual fuels have converted their equipment to handle higher-viscosity grade 6, which is less costly because it requires less of the distillate "cutter" stocks which can be converted more readily into gasoline. In the shipping industry, heavy bunker fuels are known as Bunker C and generally correspond to grade 6 fuel oil.

The largest single user of residual fuels is the electric power generating industry which consumes about 40 percent of the available residual fuel. Because of air pollution measures which tend to favor residual fuels over coal, this segment of the market demand has been the most rapidly growing segment in recent years. Growth has been to the extent of creating a degree of disequilibrium in the supply/demand picture. Future growing demand for low-sulfur fuel oils can be met, in theory, but each strategy has its own particular drawbacks. Adding desulfurization capacity to the existing array of refining equipment involves a substantial time requirement, and desulfurization of fuel oils, especially the heavier grades, is an expensive process. The alternatives of selecting low-sulfur crudes or of importing large amounts of low-sulfur residual fuels are no longer practical because of limited supplies.

Residual fuels are, by their nature, high-boiling and contain stocks which are difficult to burn quickly under "cold" conditions. Accordingly, such fuels are generally burned in equipment which permits relatively steady operation in an environment where firebox temperatures can be high.

Since residual fuels compete directly with other fuels in many areas, the price of the fuel must be competitive. Accordingly, it has not been economically practical to improve the quality of residual fuels to levels that are possible in theory.

The steady increase in the use of catalytic cracking since the mid-1960s has had the effect of decreasing the percentage yield of residual fuels as well as changing their makeup. As more high-boiling materials were charged to catalytic cracking, the remaining oil which was sold as residual fuel became heavier. Common industry practice was to blend these heavy stocks with a distillate to lower their viscosity. Continued work in this field led to the use of mild thermal cracking of the vacuum still bottoms which yielded a small additional amount of distillate product and reduced the viscosity of the remaining bottoms. Such bottoms required less distillate cutter stock to produce a salable residual fuel oil. This modest advance was developed before World War II and was known as visbreaking.

Most advances in residual fuel oil technology since World War II have led to improvements in its use rather than in oil quality. The oil industry and boiler manufacturers have increased their efforts in the areas of desulfurizing fuel oil and flue gas and reducing fuel oil metals content. A large number of additives have been developed for reducing residual fuel oil sludge, tube deposit formation and corrosion, and for increasing combustion efficiency.

Trends in the key characteristics of burner fuel oils, dating back in most instances to 1955, are shown in Figs. 8 through 13.

BASIC PHYSICAL CHARACTERISTICS OF
PETROLEUM FUELS—DEFINITIONS

Octane Number.   The octane rating of a motor fuel is defined in terms of its knocking characteristics relative to those of blends of isooctane (2, 2, 4-trimethylpentane) and n-heptane. Arbitrarily, an octane number of zero has been assigned to n-heptane, and a rating of 100 to isooctane. The octane number of an unknown fuel is numerically equal to the volume percent of isooctane in a blend with n-heptane, which has the same knocking

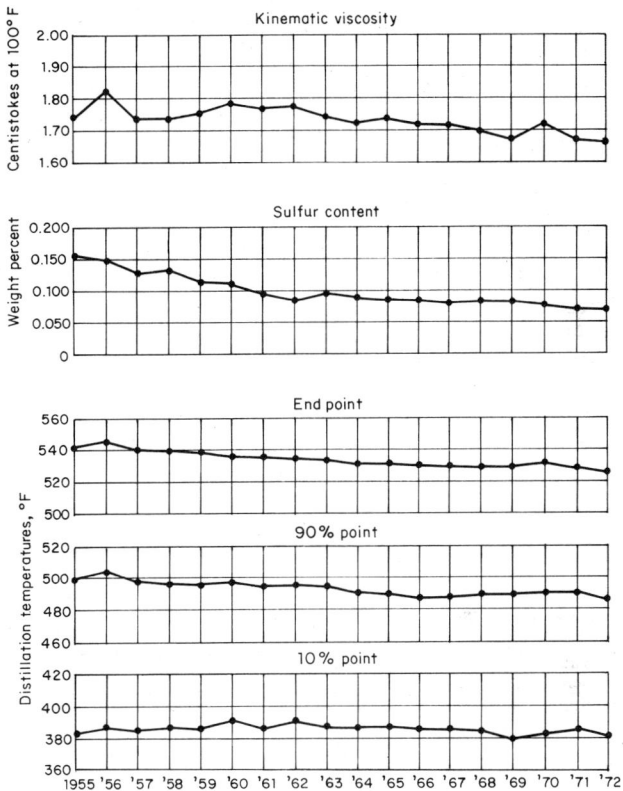

Fig. 8. Trends of some properties of Grade 1 burner fuel oils.

tendency as the unknown fuel when both the unknown and the reference blend are run in a standard single-cylinder engine operated at specified conditions. Motor method octane numbers are measured at more severe engine conditions and are numerically lower than those determined by the milder Research method. The difference between the two numbers is termed *sensitivity*.

Equilibrium Volatility of a Gasoline.   The volatility of a gasoline is determined by the Reid vapor pressure and the ASTM distillation data. The Reid vapor pressure is the vapor pressure of a gasoline at 100°F (37.8°C) under specified conditions. The distillation curve of a fuel indicates the temperatures at which the various amounts of a given sample are distilled under specified test conditions. However, gasoline will be completely evaporated in the presence of air at a temperature much lower than the end point of the distillation curve.

According to O. C. Bridgeman (*Nat. Bur. Stand. Res. Paper* 694), the volatility of a gasoline is the temperature at which a given air-vapor mixture is formed under equilibrium conditions at a pressure of one atmosphere, when a given percentage is evaporated. According to this definition, one gasoline is more volatile than another for any given percentage evaporated if it forms the given air-vapor mixture at a lower temperature. Distillation temperature curves, for a given test sample, plot amount of sample distilled over (percentage of sample) at the time a given temperature has been reached.

Specific Gravity.   The specific gravity of a petroleum fuel is the ratio of the weight of a given volume of the product at 60°F to the weight of an equal volume of distilled water at

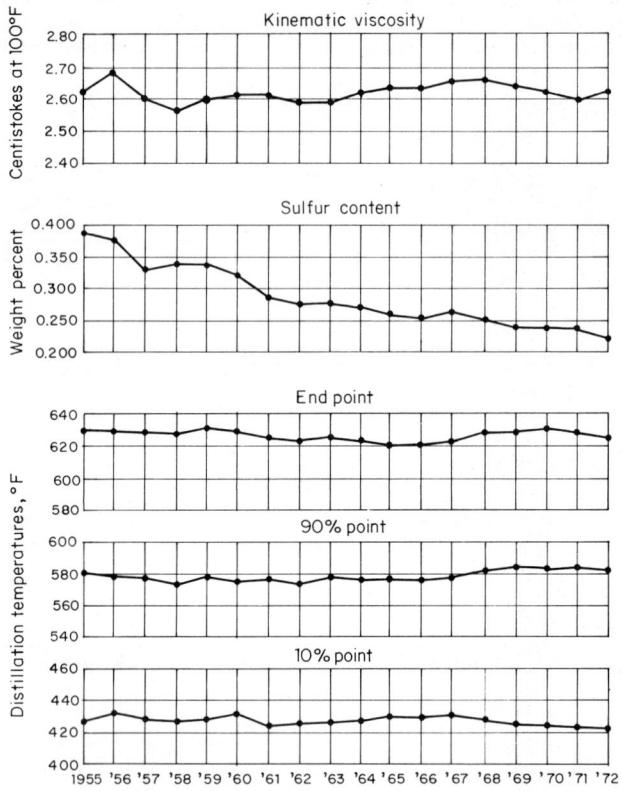

Fig. 9. Trends of some properties of Grade 2 burner fuel oils.

the same temperature, both weights corrected for air buoyancy. The relation between the API gravity scale and specific gravity is given by

$$°API = \frac{141.5}{sp\ gr\ 60/60°F} - 131.5$$

Pour Point.   This property is defined as the lowest temperature at which the fuel will pour and is a function of the composition of the fuel. Normally, the pour point of a fuel should be at least 10 to 15°F below the anticipated minimum operating temperature.

Cloud Point.   The aniline cloud point is a measure of the paraffinicity of a fuel oil, a high value indicating a straight-run paraffinic oil and a low value indicating an aromatic, a naphthenic, or a highly cracked oil.

Flash Point.   The lowest temperature at which a flash appears on the fuel surface when a test flame is applied under specified test conditions is an approximate indication of the tendency of the fuel to vaporize.

Fire Point.   The lowest temperature at which a fuel ignites and burns for at least 5 seconds under specified test conditions.

Smoke Point.   The smoking tendency of a fuel is indicated by the smoke point, which is the maximum height of a specified type of flame in a given wick lamp that results in no visible smoke.

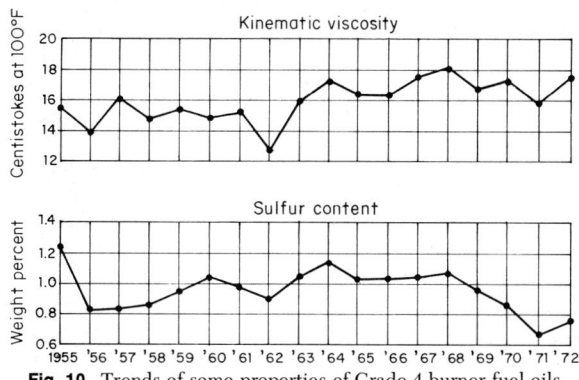

Fig. 10. Trends of some properties of Grade 4 burner fuel oils.

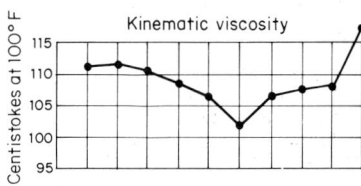

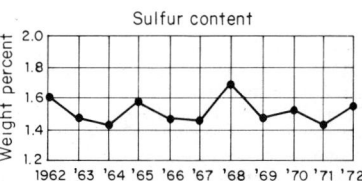

Fig. 11. Trends of some properties of Grade 5 (light) burner fuel oils.

Fig. 12. Trends of some properties of Grade 6 (heavy) burner fuel oils.

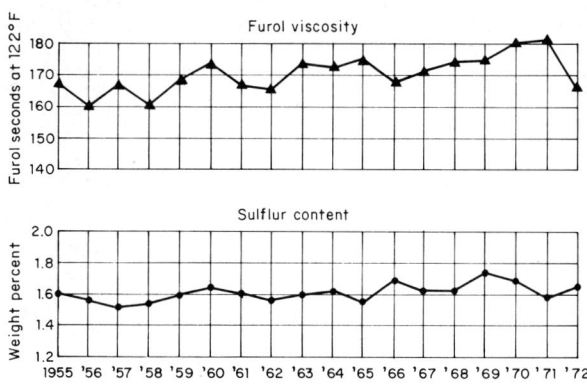

Fig. 13. Trends of some properties of Grade 6 burner fuel oils.

Viscosity. This is generally expressed in terms of the time required for a given quantity of fuel to flow through a capillary tube under specified conditions.

Kinematic viscosity v is the dynamic viscosity divided by the density, or $v = \mu/p$. The unit in the cgs system is the *stoke*. The customary unit is the *centistoke*. The value of the kinematic viscosity in stokes can be obtained from the following approximation equations where t is the efflux time in seconds.

$$\text{Saybolt Universal, when } 32 < t < 100 = 0.00226t - 1.95/t$$
$$\text{Saybolt Universal, when } t > 100 \quad = 0.00220t - (1.35/t$$
$$\text{Saybolt Furol, when } 25 < t < 40 \quad = 0.0224t - 1.84/t$$
$$\text{Saybolt Furol, when } t > 40 \qquad = 0.0216t - 0.60/t$$

An approximate kinematic conversion table is given in Table 9.

Cetane Number. The cetane number (C.N.) of a fuel is the percentage by volume of normal cetane in a mixture of cetane and α-methylnaphthalene which matches the unknown fuel in ignition quality when compared with a standard Diesel engine under specified conditions. The C.N. scale ranges from 0 to 100 C.N. for fuels equivalent in ignition quality to α-methylnaphthylene and cetane, respectively. For routine testing, secondary reference fuels having cetane values of about 25 and 74 are blended in any desired proportion.

TABLE 9.   Kinematic Viscosity Approximation Conversion Table

Centistokes	2	6	10	20	30	40	50	60	70
Saybolt at 100°F	32.60	45.50	58.80	97.50	140.9	185.7	231.4	277.4	323.4
Saybolt at 210°F	32.83	45.82	59.21	98.18	141.9	187.0	233.0	279.3	325.7
Redwood at 70°F	30.20	40.50	51.70	85.40	123.7	163.2	203.3	243.5	283.9
Redwood at 200°F	31.20	41.50	52.55	86.90	126.0	166.7	208.3	250.0	291.7

REFERENCES

"Motor-Gasoline Survey Reports," (Mineral Industry Surveys), issued semiannually by Bureau of Mines, U.S. Department of Interior, Bartlesville Energy Research Center, Bartlesville, Okla., published since 1964.

"Diesel Fuel Oils Survey Reports," *ibid.*, published annually since 1950.

"Aviation Turbine Fuels Survey Reports," *ibid.*, published annually since 1951.

"Burner Fuel Oils Survey Reports," *ibid.*, published annually since 1955.

"Annual Book of ASTM Standards, par' 17, Petroleum Products—Fuels; Solvents; Burner Fuel Oils; Lubricating Oils; Cutting Oils; Lubricating Greases; Hydraulic Fluids," revised annually, American Society for Testing and Materials, Philadelphia.

"Factors Affecting U.S. Petroleum Refining—Impact of New Technology," National Petroleum Council, Washington, 1973.

Acknowledgment. The cooperation of the National Petroleum Council and the American Petroleum Institute in providing much of the basic information for this description is gratefully acknowledged. The availability of data from the U.S. Bureau of Mines also contributed to the scope of this description.

See also *Petroleum Refining and Processing* later in this Handbook section.

Petroleum Production and Reserves in the United States

In compiling and presenting data bearing upon the reserves of great natural resources, such as petroleum and natural gas, it cannot be overstressed that we are dealing with vast geographical areas and with geological formations and phenomena that cannot be inventoried by a physical count as one may prepare a detailed accounting of the merchandise in a store. By the time that data of this nature find their way into impressive reports of a permanent-appearing nature and, particularly, if garnished with the "infallibility" of a computer printout, one may temporarily lose sight of the fact that reserves data are indeed estimates. Consequently, when they are used for serious planning, it must be kept constantly in mind that reserves data are approximations, be they presented in ever so much detail. The data are and should be subjected to constant reappraisal and revised upward or downward as better judgments can be made from improved inputs.

Because of the geographical scope involved in establishing petroleum reserves for a large country, made up of several petroleum producing areas, hundreds of people must be called upon to assess the reserves situation in highly localized areas with which they are professionally and intimately familiar. Experience with similar geophysical situations must be extrapolated with the assistance of geologists, other specialized scientists, oil field engineers, and practical operating oil field people.

The large numbers of people who furnish the inputs must operate within a tight framework of guidelines and definitions. Such terms as proven, probable, speculative, potential, etc., are each subject to a wide degree of personal interpretation. Thus, rules and, in particular, definitions must be drawn up and followed meticulously. Great care must be exercised in the appointment of the persons who will be furnishing the inputs and the persons who administer and finally present the statistical information.

Equal responsibilities lie with those persons who use reserves data. If they are seeking other than gross approximations, they should take sufficient time to understand the modus operandi of the data collection agency and, above all, fully comprehend the definitions and guidelines used in the gathering and establishment of the data. The reserves data user, if in doubt concerning statistics, should develop sufficient personal familiarity with the statistical background so that intelligent questions can be posed to the sources of statistics for further clarification. It must be admitted that planning which deals in billions of barrels, trillions of cubic feet, etc., requires more than a trifling mental acclimation to its scope.

When one contemplates the worldwide resources situation, there are additional complexities, such as conversions not only between metric and English units, but different conditions of measurement (notably affecting gases), and conversion factors involving marketing practices within different countries. For example, to convert a metric ton to 42-gallon barrels (U.S.) from Albanian data requires a factor of 6.672; 6.929 for Egypt; 7.352 for Japan, 7.428 for Canada, 8.043 for New Zealand, etc.

PETROLEUM RESERVES IN THE UNITED STATES

The petroleum industry and the general public have long been alert to the need for information pertaining to reserves of crude oil. In 1915, Ralph Arnold prepared an estimate of crude oil reserves. Similar estimates were prepared by the U.S. Geological Survey in 1916 and 1919. For the year 1922, estimates of crude oil reserves were prepared jointly by the U.S. Geological Survey and the American Association of Petroleum Geologists. Estimates of crude reserves were prepared by the American Petroleum Institute in 1925, and estimates for 1927 and 1933 were made by the Federal Oil Conservation Board.

The petroleum reserves of the United States have been systematically reviewed and reported on an annual basis by industry technical groups since 1936. Initially, the estimations were made by the Committee on Petroleum Reserves of the American

Petroleum Institute, and were limited to reserves of crude oil, including lease condensate. In 1966, the API Committee's name was changed to the "Committee on Reserves and Productive Capacity." In addition to continuing its work with respect to proved reserves, the Committee's responsibilities were expanded to include the development of considerable supplementary data. The Committee operates through a series of District Subcom-

TABLE 1. Estimated Reserves of Crude Oil in the United States

Thousands of barrels of 42 U.S. gallons

State	Proved reserves as of 12/31/1974
Alabama	68,718
Alaska	10,094,099
Arkansas	106,336
California†	3,557,036
Colorado	289,333
Florida	302,709
Illinois	159,789
Indiana	24,351
Kansas	395,107
Kentucky	36,572
Louisiana†	4,226,514
Michigan	82,299
Mississippi	261,408
Montana	207,389
Nebraska	26,779
New Mexico	624,968
New York	10,898
North Dakota	172,794
Ohio	123,871
Oklahoma	1,232,377
Pennsylvania	50,414
Texas†	11,001,506
Utah	250,648
West Virginia	32,210
Wyoming	903,360
Miscellaneous*	8,471
Total—United States	34,249,956

SOURCE: *API Rep.*, vol. 29, May 1975.
*Includes Arizona, Missouri, Nevada, South Dakota, Tennessee, and Virginia.
†Includes offshore reserves.
NOTE: This is a portion of Table I of the API Report referred to later under discussion of Definitions of Terms.

mittees which bear the primary responsibility for the determination of estimates of reserves and productive capacity. In discharging these responsibilities within the Subcommittees, a special effort is made to assign the more significant sized fields to individual members who are knowledgeable thereof, either directly or through their affiliations. Additionally, the Subcommittees are expected to make multiple assignments of selected fields to their members where it will beneficially contribute to the quality of reserve and productive capacity estimates and promote the exchange of expert views important thereto.

The geographic districts of the United States assigned to members of the Committee on Reserves and Productive Capacity and their respective subcommittee organizations are indicated later in Fig. 4.

PETROLEUM RESERVES STATISTICS

The estimated proved reserves of crude oil in the United States (as of December 31, 1974) and included in the API Report, vol. 29, released in May 1975, are tabulated by major producing states in Table 1. This information is shown graphically in Fig. 1. It is interesting to note that Texas holds approximately one-third of the estimated reserves, while Alaska holds over one-fourth of the reserves, together accounting for over 60 percent of the petroleum reserves. If the reserves of Louisiana, California, Oklahoma, and Wyoming are considered, these four states, plus the aforementioned two states, account

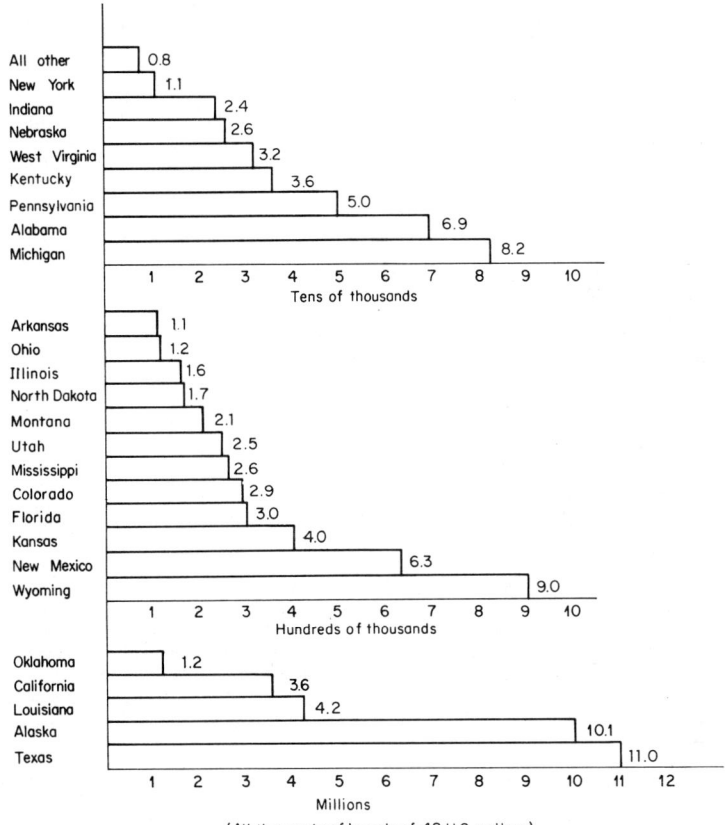

Fig. 1. Estimated reserves of crude oil in the United States (December 31, 1974). Total proved reserves: 34,249,956 thousand barrels (42 U.S. gallons per barrel). States with no reportable crude oil reserves: Connecticut, Delaware, Georgia, Hawaii, Idaho, Iowa, Maine, Maryland, Massachusetts, Minnesota, New Hampshire, New Jersey, North Carolina, Oregon, Rhode Island, South Carolina, Vermont, Washington, Wisconsin.

for about 90 percent of the reserves. In other words, while petroleum reserves are held in a large number of states, only about 19 percent of the states with such reserves account for 90 percent of the total and, in this figure, are included about ten states with very low reserves.

A breakdown of the reserves in four states, in accordance with the structure of the API District Subcommittees, is delineated in Table 2.

The effect of new data and refinements in petroleum reserves estimating, principally governed by exploratory activity and the discovery of new sources, is dramatically indicated by Fig. 2. The relationship of production and new reserves added over the time span 1945 through 1974 is shown in Fig. 3, and further indicated by Table 3.

The geographic distribution of the API District Subcommittees is mapped in Fig. 4, for the entire United States, and for principal regions in Figs. 5 through 8.

TABLE 2.   Breakdown of Estimated Reserves of Crude Oil by Major Holding States in the United States
Thousands of barrels of 42 U.S. gallons

State and reporting region		Proved reserves as of 12/31/1974
California		3,557,036*
Coastal Region	627,221	(See map, Fig. 2)
Los Angeles Basin	1,113,260	
San Joaquin Basin	1,816,555	
Louisiana		4,226,514*
North	273,948	(See map, Fig. 3)
South	3,952,566	
New Mexico		624,968
Northwest	27,207	(See map, Fig. 4)
Southeast	597,761	
Texas		11,001,506*
District 1	133,389	(See map, Fig. 5)
District 2	622,886	
District 3	1,352,103	
District 4	237,192	
District 5	110,892	
District 6	1,913,658	
District 7-B	250,387	
District 70C	199,587	
District 8	3,064,516	
District 8-A	2,578,203	
District 9	369,822	
District 10	168,871	

SOURCE: *API Rep.* vol. 29, May 1975.

*Includes offshore reserves. Total Gulf of Mexico reserves (included with Texas and Louisiana figures is 2,212,008 thousand barrels).

GEOLOGICAL INFORMATION ON OIL OCCURRENCE

Current estimates of original oil-in-place and the ultimate recovery of crude oil from known reservoirs (including those considered to be depleted) are differentiated by age, lithology, and type of entrapment. See Tables 4 through 7.

To the extent possible, the occurrence of oil is classified on the basis of a review of individual reservoirs. Where single reservoirs overlap various categories or cumulative production from several reservoirs cannot be separately identified, the district subcommittees exercise judgment in (1) allocating estimates of production, original oil-in-place, and ultimate recovery to the various categories; or (2) assigning all estimates to the category that dominates the ultimate recovery from the field. New information on reservoirs regarding age, lithology, and type of entrapment may cause some adjustments in

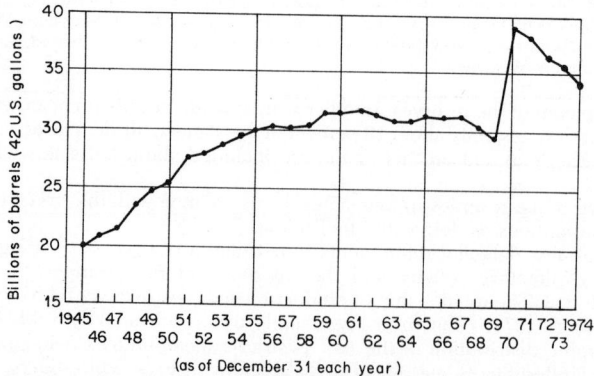

Fig. 2. Proved reserves of crude oil in the United States (1945–1974).

TABLE 3. Annual Estimates of Proved Crude Oil Reserves in the United States (1946–1974)

Thousands of barrels of 42 U.S. gallons

Year	Proved reserves at beginning of year	Net change from previous year ( ) denotes downward change
1946	19,941,846	931,714
1947	20,873,560	614,125
1948	21,487,685	1,792,759
1949	23,280,444	1,369,045
1950	24,649,489	618,909
1951	25,268,398	2,199,633
1952	27,468,031	492,523
1953	27,960,554	984,274
1954	28,944,828	615,918
1955	29,560,746	451,424
1956	30,012,170	422,479
1957	30,434,649	(134,244)
1958	30,300,405	235,512
1959	30,535,917	1,183,430
1960	31,719,347	(106,136)
1961	31,613,211	145,294
1962	31,758,505	(369,282)
1963	31,389,223	(419,233)
1964	30,969,990	20,520
1965	30,990,510	361,881
1966	31,352,391	99,736
1967	31,452,127	(75,457)
1968	31,376,670	(669,553)
1969	30,707,117	(1,075,255)
1970	29,631,862	9,369,473
1971	39,001,335	(938,378)
1972	38,062,957	(1,723,549)
1973	36,339,408	(1,039,569)
1974	35,299,839	(1,049,883)

Source: *API Rep.*, vol. 29, May 1975.
NOTE. This is a portion of Table II of the API Report referred to later under discussion of Definition of Terms.

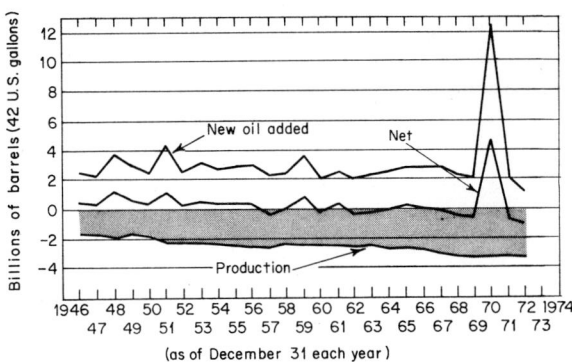

Fig. 3. Production versus new oil added during each year in the United States (1945–1974). Oil added represents total of discoveries, revisions, and extensions. Net amounts are shown by middle curve.

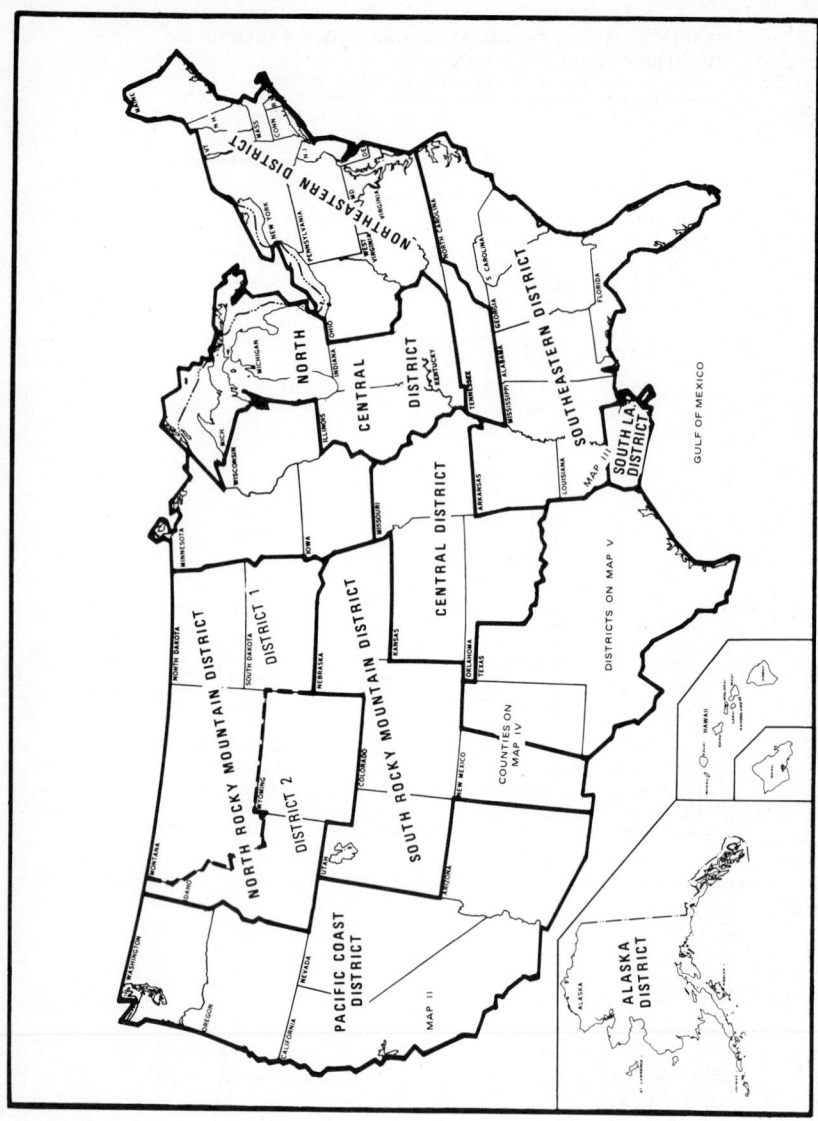

Fig. 4. Geographic districts of the United States covered by API District Subcommittees. For subdivisions of California, Louisiana, New Mexico, and Texas, see Figs. 5 to 8.

previously published data; however, major annual revisions usually reflect revisions in estimates of original oil-in-place and ultimate recovery.

Oil by Age of Reservoir Rock.   The distribution of crude oil by geologic age of the reservoir rock is reported in terms of estimated ultimate recovery (cumulative production from both producing and depleted fields plus remaining proved reserves) and the estimated original oil-in-place which is related to such recovery. The published compilation is on a total basis for the United States and according to geologic age units as follows:

Era	System	Series
Cenozoic	Quaternary	Recent
		Pleistocene
	Tertiary	Pliocene
		Miocene
		Oligocene
		Eocene
		Paleocene
Mesozoic	Cretaceous	
	Jurassic	
	Triassic	
Paleozoic	Permian	
	Pennsylvanian	
	Mississippian	
	Devonian	
	Silurian	
	Ordovician	
	Cambrian	
Pre-Cambrian	—	

It should be recognized that the age period of a limited number of reservoirs has not been firmly established, particularly where dual classifications such as Permo-Pennsylvania, Cambro-Ordovician, etc., have been traditionally applied. In such instances the younger age is assigned unless reasonable evidence is available to the contrary. Additional problems are presented for multireservoir fields when production is commingled or not identifiable from the individual reservoirs of different ages. When a reasonable basis for assignment to different ages is not available, the age of the reservoir which dominates as to ultimate recovery is used.

Oil by Reservoir Lithology.   The occurrence of crude oil according to lithology of the reservoir rock is reported in terms of estimated ultimate recovery (cumulative production plus remaining proved reserves) and the estimated original oil-in-place of both depleted and currently proved reservoirs.

The three lithologic categories used in reporting oil occurrence are defined as follows:

1. *Sandstone.*   A sedimentary rock composed predominantly of quartz grains or other noncarbonate mineral or rock detritus. Included in this reservoir type are unconsolidated sand, sandstone, siltstone, graywacke, arkose, granite wash, conglomerate, and sedimentary breccia.

2. *Carbonate.*   A sedimentary rock composed predominantly of calcite (limestone) and/or dolomite. Clastic carbonates are included in this category.

3. *Other.*   All reservoirs not fitting the definitions of the sandstone and carbonate categories. Included in this reservoir type are igneous and metamorphic rocks and some sedimentary rocks (i.e., fractured shale and chert).

In multireservoir fields having more than one type of reservoir lithology, reservoirs are individually handled to the extent possible. Where cumulative production from different types of reservoirs cannot be estimated with reasonable accuracy, the ultimate recovery and original oil-in-place are credited to the lithology which is believed will supply the greatest recovery for the field.

Oil by Entrapment.   Crude oil occurrence according to two broad categories of entrapment types is reported in terms of estimated ultimate recovery (cumulative production plus remaining proved reserves) and original oil-in-place for the depleted and currently proved reservoirs.

The two types of entrapment are defined as follows:

1. *Structural.*   An entrapment in which fluid migration of hydrocarbons in the reservoir rock has terminated primarily because of closure induced by structural deformation

and/or hydrodynamic forces. Most Gulf Coast piercement salt dome reservoirs are included in this category.

2. *Stratigraphic.*   An entrapment in which fluid migration of hydrocarbons has terminated because of (*a*) the pinch-out of a reservoir rock due either to truncation or nondeposition, or (*b*) facies change in the form of diminished permeability of the reservoir rock. Also included in this category are entrapments in which a pinch-out or facies change provides part of the barrier to migration of hydrocarbons, and structural elements provide the remaining closure for entrapment.

In multireservoir fields where both of the two types of entrapment occur, estimates of ultimate recovery and original oil-in-place for each individual reservoir are assigned to the appropriate category, if it is reasonable to do so. Where production from different types of

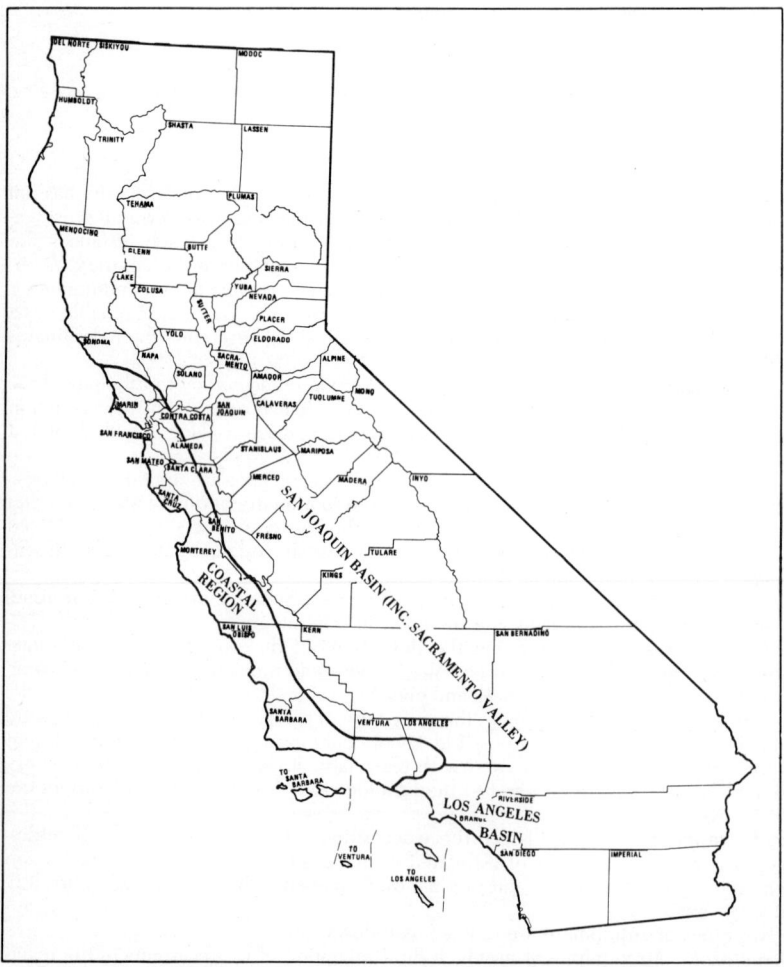

Fig. 5. Subdivisions of California used in reporting reserves data.

Fig. 6. Subdivisions of Louisiana used in reporting reserves data.

entrapments cannot be separately identified, ultimate recovery and original oil-in-place for a combination of reservoirs are assigned to the category expected to provide the larger ultimate recovery.

PETROLEUM PRODUCTION IN THE UNITED STATES

Estimates of the daily productive capacity of crude oil in the United States are presented in Table 8. These data are plotted in Fig. 9. As would be expected, there is a reasonable correlation between reserves and productive capacity, Alaska representing the major divergence reflects the current high rate of activity in readying the Alaskan resources. In order of descending positions:

	Petroleum reserves	*Daily productive capacity*
No. 1	Texas	Texas
No. 2	Alaska	Louisiana
No. 3	Louisiana	California
No. 4	California	Oklahoma
No. 5	Oklahoma	Wyoming
No. 6	Wyoming	New Mexico
No. 7	New Mexico	Alaska
No. 8	Kansas	Kansas
No. 9	Florida	Mississippi
No. 10	Colorado	Florida
No. 11	Mississippi	Utah
No. 12	Utah	Colorado
No. 13	Montana	Montana
No. 14	North Dakota	Illinois

On April 15, 1975, the American Petroleum Institute issued a list of the 100 largest fields in the United States showing proved reserves of crude oil as of December 31, 1974, and productive capacity that might be achieved within 90 days from December 31, 1974. See Table 9.

Texas and Louisiana have 62 of the nation's 100 largest fields. All told, the 100 largest fields contain 24.3 billion barrels of crude oil, representing 71.1 percent of the nation's

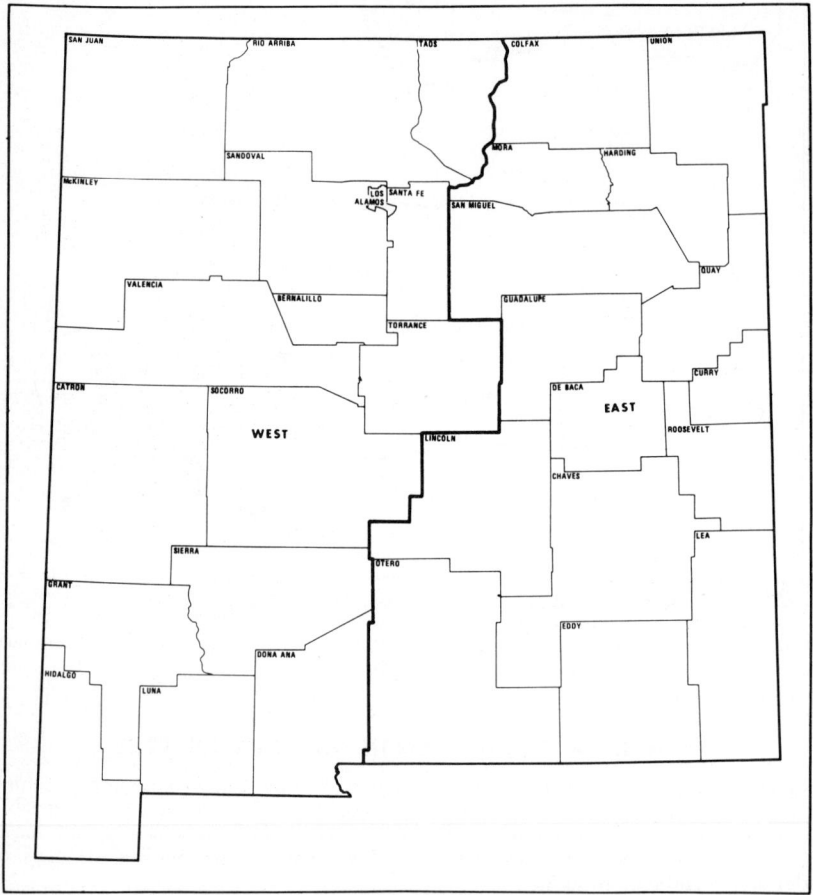

Fig. 7. Subdivisions of New Mexico used in reporting reserves data.

total proved reserves of 34.25 billion barrels as of December 31, 1974. The figures include both onshore and offshore fields. Thirteen states are represented in the list. The largest field is the Prudhoe Bay Field on Alaska's North Slope, which has an estimated 9.6 billion barrels of oil. The smallest field on the list is the Lake Barre Field in southern Louisiana's Terrebonne Parish. Of the 100 leading fields, 35 are in Texas, 27 in Louisiana, and 13 in California. Alaska, Oklahoma, and Wyoming have five each, New Mexico has four, and Utah has two. One field each is listed for Colorado, Montana, and North Dakota. The Jay Field, which is partly in Florida and partly in Alabama, is the only large field listed in those two states. Except for the Prudoe Bay Field in Alaska, only two fields, both in Texas, have reserves exceeding one billion barrels.

The guidelines and definitions affecting the assembly of information on productive capacity of crude oil are given in the following section.

DEFINITIONS OF TERMS

In vol. 29 (May 1975) on "Reserves of Crude Oil Natural Gas Liquids and Natural Gas in the United States and Canada and United States Productive Capacity as of December 31, 1974," some of the terms used are defined as follows:

Crude Oil.   Crude oil is technically defined as a mixture of hydrocarbons that exists in the liquid phase in natural underground reservoirs and remains liquid at atmospheric pressure after passing through surface separating facilities. For statistical purposes, volumes reported as crude oil include:

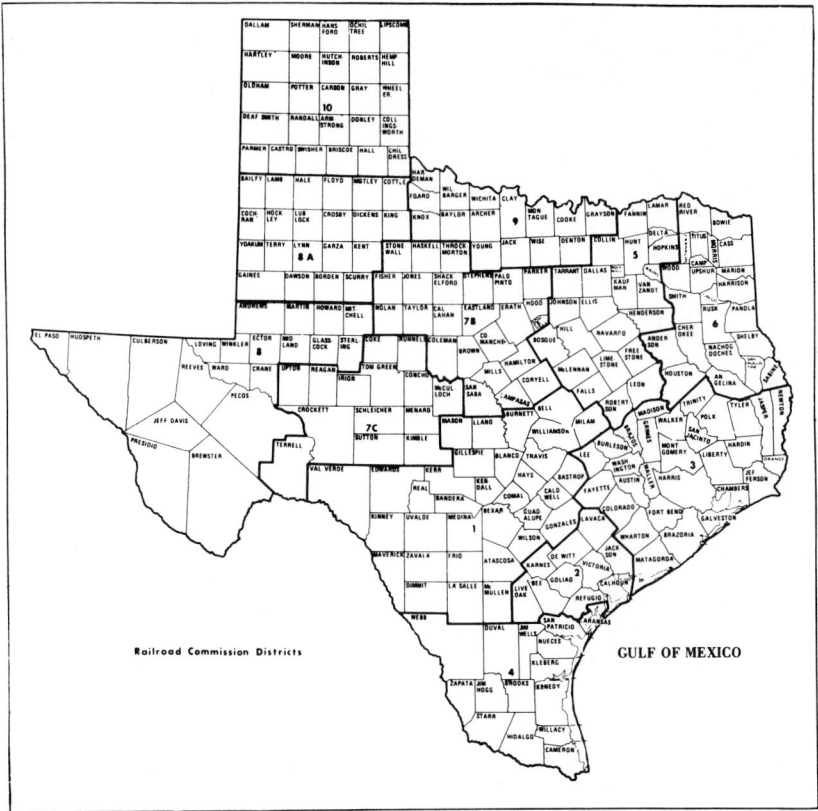

Fig. 8. Subdivisions of Texas used in reporting reserves data.

1. Liquids technically defined as crude oil;
2. Small amounts of hydrocarbons that exist in the gaseous phase in natural underground reservoirs but are liquid at atmospheric pressure after being recovered from oil well (casinghead) gas in lease separators*; and
3. Small amounts of nonhydrocarbons produced with the oil.

*From a technical standpoint, these liquids are termed "condensate"; however, they are commingled with the crude stream and it is not practical to measure and report their volume separately. All other liquids recovered from natural gas (including lease condensate) are included in natural gas liquid volumes reported by the American Gas Association. (See American Petroleum Institute Technical Report No. 1, "Standard Definitions for Petroleum Statistics," First Edition, July 1, 1969, pp. 16 and 17.)

Statistical data pertaining to crude oil production, reserves, and productive capacity are reported as liquid equivalents at the surface (excluding basic sediment and water) measured in terms of stock tank barrels of 42 U.S. gallons at atmospheric pressure, and corrected to 60°F.

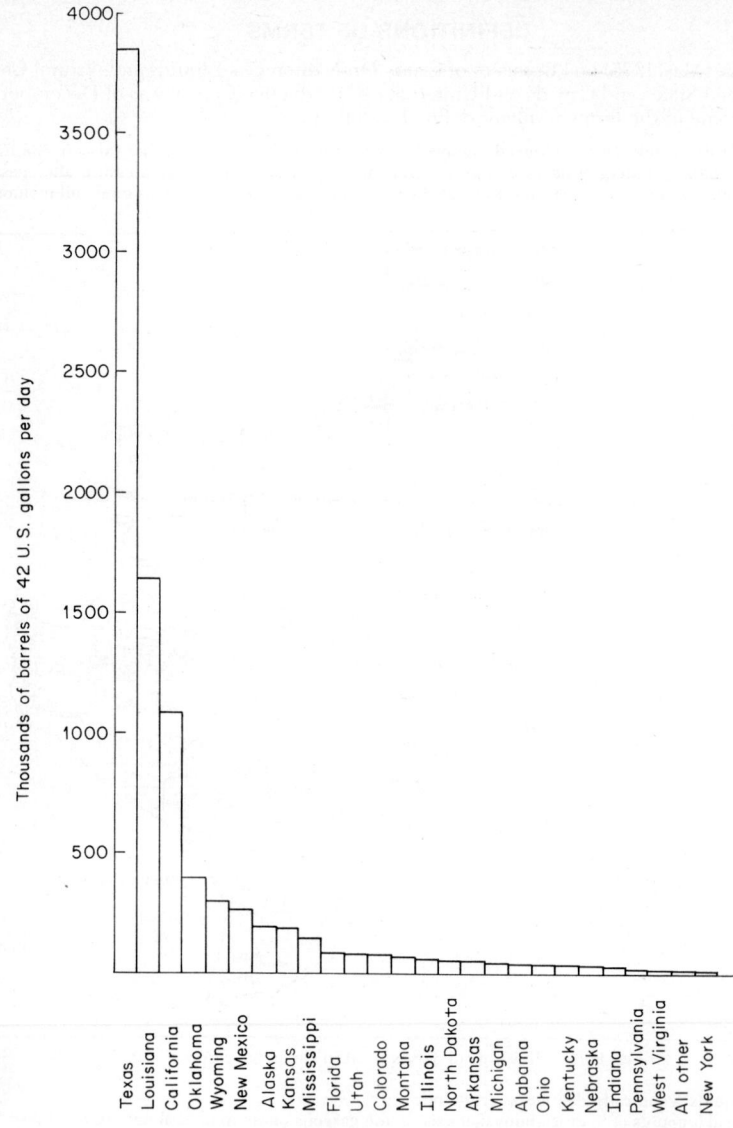

Fig. 9. Estimate of daily productive capacity of crude oil, by states, in the United States (1st quarter 1975).

TABLE 4.   Original Oil-in-Place, Cumulative Production, Ultimate Recovery, and Proved Reserves of Crude Oil in the United States

Thousands of barrels of 42 U.S. gallons

State	Estimated original oil-in-place	Cumulative production	Estimated ultimate recovery of crude oil from proved reserves	Proved reserves	Ultimate recovery as % of original oil-in-place
Alabama	742,212	154,654	223,372	68,718	30.1
Alaska	25,531,540	610,355	10,704,454	10,094,099	41.9
Arkansas	4,246,688	1,372,675	1,479,011	106,336	34.8
California*	83,236,631	16,753,123	20,310,159	3,557,036	24.4
Colorado	4,370,176	1,079,310	1,368,643	289,333	31.3
Florida	968,762	110,333	413,042	302,709	42.6
Illinois	8,884,754	2,998,721	3,158,510	159,789	35.5
Indiana	1,564,814	445,219	469,570	24,351	30.0
Kansas	15,357,836	4,559,315	4,954,422	395,107	32.3
Kentucky	2,030,461	619,131	655,703	36,572	32.3
Louisiana*	39,118,529	13,846,505	18,073,019	4,226,514	46.2
Michigan	2,291,479	646,187	728,486	82,299	31.8
Mississippi	4,713,782	1,464,366	1,725,774	261,408	36.6
Montana	4,862,930	859,721	1,067,110	207,389	21.9
Nebraska	1,325,151	347,255	374,034	26,779	28.2
New Mexico	14,921,333	3,055,703	3,680,671	624,968	24.7
New York	1,121,739	225,690	236,588	10,898	21.1
North Dakota	2,456,479	434,523	607,317	172,794	24.7
Ohio	6,857,347	803,264	927,135	123,871	13.5
Oklahoma	37,650,845	11,012,298	12,244,675	1,232,377	32.5
Pennsylvania	6,678,167	1,283,311	1,333,725	50,414	20.0
Texas*	150,101,448	38,509,288	49,510,794	11,001,506	33.0
Utah	3,487,490	511,909	762,557	250.648	21.9
West Virginia	2,606,516	510,083	542,293	32,210	20.8
Wyoming	15,426,191	3,894,838	4,798,198	903,360	31.1
Miscellaneous†	169,241	27,156	35,627	8,471	21.1
Total—U.S.	440,722,541	106,134,933	140,384,889	34,249,956	31.9

SOURCE: *API Rep.*, vol. 29, May 1975.
*Includes offshore reserves. Gulf of Mexico included with Texas and Louisiana.
†Includes Arizona, Missouri, Nevada, South Dakota, Tennessee, Virginia, and Washington.
NOTE: This is a portion of Table IV of the API Report referred to later under discussion of Definition of Terms.

Production.   Crude oil production is the volume of liquids statistically defined as crude oil, which is produced from oil reservoirs during given periods of time. The amount of such production for a given year is generally established by measurement of volumes delivered from lease storage tanks (i.e., the point of custody transfer) to pipelines, trucks, or other media for transport to refineries or terminals with adjustments for (1) net differences between opening and closing lease inventories, and (2) basic sediment and water (BS&W).

For purposes of the annual reserves reports, the subcommittees need production data for individual fields and for the specific geographic areas for which they are responsible. Since "official" sources such as state agencies and the U.S. Bureau of Mines do not provide the required detail, the subcommittees must analyze all available data (including company records, commercial services, state records, and Bureau of Mines reports) and make such adjustments as may be necessary to develop production series which satisfy their particular requirements. Because of differences in definitions and differences in data collection procedures used by various sources, and because of the variety of adjustments which must be made, production data used in annual reserves reports should not be expected to agree precisely with that published by sources such as state agencies and the U.S. Bureau of Mines.†

†It should be noted that the difference between the final production data used by the Committee on Reserves and Productive Capacity and "official" sources are small. For example, the total U.S. production for 1973 in the report for 1974 is 99.1% of the total U.S. production reported by the U.S. Bureau of Mines.

TABLE 5.   Estimated Original Oil-in-Place and Ultimate Recovery of Crude Oil in the United States by Reservoir Lithology

Thousands of barrels of 42 U.S. gallons

State	Estimated original oil-in-place			Estimated ultimate recovery		
	Sandstone	Carbonate	Other	Sandstone	Carbonate	Other
Alabama	548,017	194,195	....	168,073	55,299	....
Alaska	25,530,000	....	1,540	10,704,300	....	154
Arkansas	3,147,451	1,099,237	....	1,073,651	405,360	....
California*	79,252,001	....	3,984,630	19,544,378	....	765,781
Colorado	4,118,777	113,745	137,654	1,321,634	24,575	22,434
Florida	....	968,762	....	....	413,042	....
Illinois	6,423,819	2,460,935	....	2,367,770	790,740	....
Indiana	725,226	839,588	....	288,267	181,303	....
Kansas	4,275,576	11,076,514	5,746	1,499,871	3,453,324	1,227
Kentucky	1,126,875	903,586	....	376,708	278,995	....
Louisiana*	34,717,832	4,400,697	....	17,433,515	639,504	....
Michigan	8,082	2,283,397	....	2,338	726,148	....
Mississippi	4,285,325	373,047	55,410	1,597,926	100,896	26,952
Montana	2,357,649	2,505,281	....	570,335	496,775	....
Nebraska	1,226,700	98,451	....	343,339	30,695	....
New Mexico	3,442,044	11,323,169	156,120	693,105	2,964,865	22,701
New York	1,121,739	....	....	236,588	....	
North Dakota	230,210	2,226,269	....	57,074	550,243	....
Ohio	3,012,297	3,840,450	4,600	451,023	475,652	460
Oklahoma	33,208,855	4,441,310	680	11,197,312	1,047,180	183
Pennsylvania	6,678,167	....	....	1,333,725	....	....
Texas*	79,812,644	69,560,025	728,779	28,885,375	20,478,503	146,916
Utah	1,975,735	1,472,190	39,565	333,050	424,332	5,175
West Virginia	2,593,396	11,490	1,630	540,623	1,480	190
Wyoming	11,296,555	3,937,953	191,683	3,746,312	1,037,918	13,968
Miscellaneous†	11,620	102,621	55,000	2,190	16,937	16,500
Total—U.S.	311,126,592	124,232,912	5,363,037	104,768,482	34,593,766	1,022,641

SOURCE: *API Rep.*, vol. 29, May 1975.

*Includes offshore reserves. Gulf of Mexico included with Texas and Louisiana.

†Includes Arizona, Missouri, Nevada, South Dakota, Tennessee, Virginia, and Washington.

In addition to the problems outlined above, it should be noted that when reserve estimates are prepared for a current year, the subcommittees only have access to actual production data for the first nine or ten months of that year. Consequently, production totals reported for the current year in an annual report are preliminary *estimates* prepared by the subcommittees on the basis of incomplete information. However, by the time each annual report is prepared, the subcommittees have access to the "actual" total production for the preceding year and the figures shown in Table III* ("Historical Record of Production and Proved Reserves; Also the Ultimate Recovery and Original Oil-in-Place by Year of Discovery") are revised accordingly.

The difference between the preliminary estimate for any given year and the actual production for that year, as subsequently determined by the subcommittees, is treated as a "revision" in Table I† ("Estimated Reserves of Crude Oil in the United States") of the following year's report. It should be noted that the original annual estimates of production used in Table II‡ ("Annual Estimates of Proved Crude Oil Reserves in the United States 1946 through 1974") are not revised since this would disturb the internal balance and consistency of the reserve estimates.

Cumulative Production.   The sum of the estimated crude oil production for the current year and the actual production for each of the prior years is the cumulative production reported by the Committee.

Proved Reserves of Crude Oil.   Proved reserves of crude oil as of December 31 of any given year are the estimated quantities of all liquids statistically defined as crude oil, which geological and

*Refers to Table III in actual API Report. Figure 3 in this Handbook description is an abridgment of the data in that table.

†Refers to Table I in actual API Report. Table 1 in this Handbook description is an abridgment of that table.

‡Refers to Table II in actual API Report. Table 3 in this Handbook description is an abridgment of that table.

TABLE 6.   Estimated Original Oil-in-Place and Ultimate Recovery of Crude Oil in the United States by Geologic Age of Reservoir

Thousands of barrels of 42 U.S. gallons

Age	Estimated original oil-in-place	Estimated ultimate recovery
Cenozoic (total)	150,910,342	51,717,361
Quaternary (total)	6,713,922	1,491,978
Pleistocene	6,713,922	1,491,978
Tertiary (total)	144,196,420	50,225,383
Pliocene	33,818,113	9,875,373
Miocene	70,041,984	23,650,817
Oligocene	23,713,975	10,483,704
Eocene	16,422,804	6,157,748
Paleocene	199,544	57,741
Mesozoic (total)	69,348,288	27,977,965
Cretaceous	42,561,846	16,812,366
Jurassic	4,101,809	1,473,458
Triassic	22,684,633	9,692,141
Paleozoic	220,463,911	60,689,563
Permian	78,042,028	18,991,989
Pennsylvanian	67,924,913	20,559,718
Mississippian	20,904,671	6,016,279
Devonian	16,196,948	4,093,841
Silurian	5,083,028	1,159,923
Ordovician	31,660,813	9,696,380
Cambrian	651,510	171,433
Total	440,722,541	140,384,889

SOURCE: *API Rep.*, vol. 29, May 1975.

TABLE 7.   Estimated Original Oil-in-Place and Ultimate Recovery of Crude Oil in the United States by Type of Entrapment

Thousands of barrels of 42 U.S. gallons

State	Estimated original oil-in-place Structural	Stratigraphic	Estimated ultimate recovery Structural	Stratigraphic
Alabama	727,088	15,124	220,401	2,971
Alaska	25,530,000	1,540	10,704,300	154
Arkansas	3,934,295	312,393	1,355,480	123,531
California*	70,453,634	12,782,997	17,189,856	3,120,303
Colorado	2,946,812	1,423,364	993,888	374,755
Florida	899,312	69,450	391,128	21,914
Illinois	5,154,734	3,730,020	2,056,719	1,101,791
Indiana	151,145	1,413,669	33,041	436,529
Kansas	6,232,080	9,125,756	2,034,357	2,920,065
Kentucky	306,589	1,723,872	103,336	552,367
Louisiana*	34,108,366	5,010,163	17,068,171	1,004,848
Michigan	1,362,369	929,110	486,626	241,860
Mississippi	4,064,135	649,647	1,496,396	229,378
Montana	2,935,608	1,927,322	634,245	432,865
Nebraska	84,431	1,240,720	24,192	349,842
New Mexico	8,084,098	6,837,235	1,901,954	1,778,717
New York	....	1,121,739	....	236,588
North Dakota	1,695,972	760,507	431,221	176,096
Ohio	587,570	6,269,777	88,206	838,929
Oklahoma	16,586,635	21,064,210	5,748,143	6,496,532
Pennsylvania	....	6,678,167	....	1,333,725
Texas*	75,050,543	75,050,905	26,989,973	22,520,821
Utah	337,867	3,149,623	98,402	664,155
West Virginia	....	2,606,516	....	542,293
Wyoming	11,091,202	4,334,989	3,694,457	1,103,741
Miscellaneous†	41,060	128,181	8,110	27,517
Total—U.S.	272,365,545	168,356,996	93,752,602	46,632,287

SOURCE: *API Rep.*, vol. 29, May 1975.
*Includes offshore reserves. Gulf of Mexico included with Texas and Louisiana.
†Includes Arizona, Missouri, Nevada, South Dakota, Tennessee, Virginia, and Washington.

engineering data demonstrate with reasonable certainty to be recoverable in future years from known reservoirs under existing economic and operating conditions.

Reservoirs are considered proved if economic producibility is supported by either actual production or conclusive formation tests. The area of an oil reservoir considered proved includes: (1) that portion delineated by drilling and defined by gas-oil or oil-water contacts, if any; and (2) the immediately adjoining portions not yet drilled but which can be reasonably judged as economically productive on the basis of available geological and engineering data. In the absence of information on fluid contacts, the lowest known structural occurrence of hydrocarbons controls the lower proved limit of the reservoir.

TABLE 8.   Estimate of the Daily Productive Capacity of Crude Oil in the United States*

Daily average in thousands of barrels of 42 U.S. gallons

State	Estimated production
Alabama	30
Alaska	198
Arkansas	42
California†	1,055
Colorado	100
Florida	113
Illinois	75
Indiana	12
Kansas	164
Kentucky	20
Louisiana†	1,670
Michigan	60
Mississippi	131
Montana	95
Nebraska	17
New Mexico	261
New York	2
North Dakota	55
Ohio	25
Oklahoma	438
Pennsylvania	9
Texas†	3,813
Utah	102
West Virginia	7
Wyoming	377
Miscellaneous‡	6
Total—U.S.	8,877

SOURCE: *API Rep.*, vol. 29, May 1975.
*Attainable after 90 days immediately following 12/31/1974.
†Includes offshore reserves. Gulf of Mexico included with Texas and Louisiana.
‡Includes Arizona, Missouri, Nevada, South Dakota, Tennessee, and Virginia.

Reserves of crude oil which can be produced economically through application of improved recovery techniques (such as fluid injection) are included in the "proved" classification if successful testing by a pilot project, or the operation of an installed program in the reservoir, provide support for the engineering analysis on which the project or program was based.

Estimates of proved crude oil reserves do not include the following: (1) oil that may become available from known reservoirs but is reported separately as "indicated additional reserves", (2) natural gas liquids (including lease condensate); (3) oil the recovery of which is subject to reasonable doubt because of uncertainty as to geology, reservoir characteristics, or economic factors; (4) oil that may occur in untested prospects; and (5) oil that may be recovered from oil shales, coal, gilsonite, and other such sources.

At the close of each year, estimates of proved reserves are made for each new pool and new field discovery. In addition, previous estimates of proved reserves for fields and/or reservoirs discovered prior to the current year are reviewed and adjusted for (1) changes in proved areas, (2) revisions in recovery estimates based on better defined performance of the reservoirs or other geologic and engineering factors, and (3) the effect of the current year's production.

Proved Developed Reserves.   Proved developed reserves as of December 31 of any given year are proved reserves estimated to be recoverable through existing wells. Reserves in proved reservoirs penetrated by wells but currently not being produced are classified as "developed" if it is anticipated that such reserves will be recovered through existing wells requiring no more than workover operations.

Proved Underdeveloped Reserves.   Proved underdeveloped reserves as of December 31 of any given year are defined as economically recoverable reserves estimated to exist in proved reservoirs which will be recovered from wells to be drilled in the future. Reserves in undrilled areas are included in proved reserve estimates if they are considered proved by geologic analysis of the current well information.

Indicated Additional Reserves.   With the present state of industry technology, certain quantities of crude oil (other than those defined and reported as proved reserves) may be economically recoverable from the following potential sources:

Known productive reservoirs in existing fields expected to respond to improved recovery techniques such as fluid injection where (a) an improved recovery technique has been installed but its effect cannot yet be fully evaluated; or (b) an improved technique has not been installed but knowledge of reservoir characteristics and the results of a known technique installed in a similar situation are available for use in the estimating procedure.

Crude oil potentially available from these sources is reported as "indicated additional reserves." The economic recoverability of these reserves is not considered to be established with sufficient conclusiveness to allow them to be included in proved reserves; however, if and when improved recovery techniques are successfully applied to known reservoirs, the corresponding indicated additional reserves will be reclassified and added to the inventory of "proved" reserves.

Indicated additional reserves do not include reserves associated with acreage that may be added to the area of a proved reservoir as the result of future drilling.

Discoveries.   Discoveries reported as of December 31 for any given year are proved reserves credited to new fields and new pools in old fields as the result of successful exploratory drilling and associated development drilling during the current year.

For definitions of field, pool, development drilling, exploratory drilling, etc., see aforementioned Technical Report No. 1, pages 18, 26, and 27.

The reliability of estimates of the proved productive area of new discoveries or partially developed reservoirs varies in relation to the amount of geological information available at the time the estimate is prepared. Important factors such as the areal extent of the structure, the average thickness of the producing reservoir, the oil column within the reservoir, and the continuity and characteristics of the reservoir formation cannot be determined accurately unless sufficient subsurface information is available.

The ultimate size of newly discovered reservoirs, whether in new fields or old fields, is seldom determined in the year of discovery. Therefore, first-year estimates of proved reserves in new reservoirs are often only a small part of the total that will be ultimately assigned to the new reservoirs. It follows that reserves credited to discoveries in any given year are usually less than total extensions and revisions for the same year which represent adjustments of reserves in reservoirs discovered in all prior years.*

*Subcommittees are not necessarily aware of all new reserve estimates at the time reserve estimates are prepared. This is especially true if a discovery is made late in the year for which a report is being prepared. In such cases, new proved reserves are reported in Table I (title given earlier) as discoveries in new fields or new pools in old fields for the year in which the discovery becomes known. In Table III (title given earlier) these reserves are assigned to the year in which the field was actually discovered. The classification of discoveries as "new field" discoveries and "new pool" discoveries is based on the final classification of wells drilled. Well classifications are defined in previously mentioned Technical Report No. 1, Part II.

Current-year estimates of discoveries in new fields and new pools in old fields are reported in Table I (title given earlier) of the annual report. These estimates are also included in Table III (title given earlier) of the annual report where they are credited to the year in which the field was initially discovered under the headings of "Ultimate Recovery" and "Original Oil-in-Place." (This latter term is defined shortly) These two columns in Table III provide a historical record of all discoveries classified according to the year in which each field was first discovered. Discoveries in new fields are credited to the years in which the new fields were found. New pool discoveries in old fields are credited to the years in which the old fields were found. Exceptions are made when new

TABLE 9. The One Hundred Largest Oil Fields in the United States
(Figures are barrels of 42 gallons)

Name of field	Location	Remaining proved crude oil reserves (December 31, 1974)	Daily crude oil productive capacity—90 days from December 31, 1974
Prudhoe Bay	North Slope Borough, Alaska	9,598,511,000	1,500
Yates	Pecos County, Texas (District 8)	1,398,000,000	250,000
East Texas	Cherokee, Gregg, Rusk, Smith, and Upton Counties, Texas (District 6)	1,287,165,000	400,000
Elk Hills	Kern County, California (San Joaquin Region)	705,754,000	160,000
Wilmington	Los Angeles County, California (Los Angeles Region)	705,400,000	180,000
Wasson	Gaines County, Texas (District 8A)	636,000,000	247,500
Kelly Snyder	Scurry County, Texas (District 8A)	477,000,000	210,000
Slaughter	Cochran County, Texas (District 8A)	349,000,000	130,000
Tom O'Connor	Refugio County, Texas (District 2)	314,669,000	122,000
Hawkins	Wood County, Texas (District 6)	304,762,000	133,050
Midway-Sunset	Kern and San Luis Obispo Counties, California (San Joaquin Region)	303,794,000	98,000
Sho-Vel-Tum	Carter and Stephens Counties, Oklahoma	297,651,000	95,000
Jay	Escambia and Santa Rosa Counties, Florida and Alabama	267,936,000	93,606
Hastings, West	Brazoria County, Texas (District 3)	253,212,000	75,000
Kern River	Kern County, California (San Joaquin Region)	250,528,000	73,000
McArthur River	Kenai Peninsula Borough, Alaska	249,660,000	113,000
Webster	Harris County, Texas (District 3)	225,960,000	68,924
Caillou Island	Terrebonne Parish, Louisiana (South)	205,086,000	43,175
Rangely	Rio Blanco County, Colorado	198,710,000	55,000
Conroe	Montgomery County, Texas (District 3)	196,774,000	60,827
Cowden, North	Ector County, Texas (District 8)	165,000,000	42,100
Seminole	Gaines County, Texas (District 8A)	162,000,000	60,000
Delta, West, Block 58	Plaquemines Parish, Louisiana (South) (offshore)	161,337,000	28,000
Fullerton	Andrews County, Texas (District 8)	153,000,000	17,800
Bay Marchand, Block 2	Lafourche Parish, Louisiana (South) (offshore)	126,053,000	84,000
Panhandle	Hutchinson County, Texas (District 10)	124,860,000	31,000
Levelland	Cochran County, Texas (District 8A)	122,000,000	35,000

Field	Location		
Huntington Beach	Orange County, California (Los Angeles Region)	121,530,000	51,000
Grand Isle, Block 43	Plaquemines Parish, Louisiana (South) (offshore)	119,776,000	65,000
Eugene Island, Block 330	Iberia Parish, Louisiana (South) (offshore)	116,348,000	73,300
Delta, West, Block 30	Plaquemines Parish, Louisiana (South) (offshore)	110,125,000	46,700
Cote Blanche Bay, West	St. Mary Parish, Louisiana (South)	103,138,000	21,500
Delhi	Franklin, Madison, and Richmond Parishes, Louisiana (North)	102,359,000	17,968
San Ardo	Monterey County, California (Coastal Region)	101,079,000	37,000
Middle Ground Shoal	Kenai Peninsula Borough, Alaska	97,836,000	24,000
South Pass, Block 27	Plaquemines Parish, Louisiana (South) (offshore)	97,161,000	31,000
West Ranch	Jackson County, Texas (District 2)	96,266,000	39,838
Garden Island Bay	Plaquemines Parish, Louisiana (South)	94,968,000	21,500
Hondo	Santa Barbara County, California (Coastal Region)	94,000,000	0 (undeveloped)
Thompson	Fort Bend County, Texas (District 3)	93,717,000	37,836
Ventura	Ventura County, California (Coastal Region)	91,835,000	31,000
Goldsmith	Ector County, Texas (District 8)	91,000,000	40,500
Empire	Eddy County, New Mexico	90,697,000	48,019
Van	Van Zandt County, Texas (District 5)	89,997,000	55,901
Fairway	Anderson and Henderson Counties, Texas (Districts 5 and 6)	89,922,000	31,020
South Pass, Block 24	Plaquemines Parish, Louisiana (South) (offshore)	85,621,000	40,000
Vacuum	Lea County, New Mexico	83,459,000	38,453
Salt Creek Light Oil Unit	Natrona County, Wyoming	82,487,000	32,400
Belridge South	Kern County, California (San Joaquin Region)	80,718,000	23,000
Howard Glasscock	Howard County, Texas (District 8)	80,000,000	18,100
Midland Farms	Andrews County, Texas (District 8)	79,000,000	14,500
Ship Shoal, Block 207	Terrebonne Parish, Louisiana (South) (offshore)	78,266,000	14,000
South Pass, Block 61	Plaquemines Parish, Louisiana (South) (offshore)	77,157,000	20,000
Cogdell	Kent County, Texas (District 8A)	76,000,000	33,100
Main Pass, Block 41	Plaquemines Parish, Louisiana (South) (offshore)	74,847,000	22,000
Dos Cuadras	Santa Barbara County, California (Coastal Region)	72,977,000	40,000
Diamond M	Scurry County, Texas (District 8A)	72,000,000	18,400

TABLE 9 The One Hundred Largest Oil Fields in the United States (Continued)
(Figures are barrels of 42 gallons)

Name of field	Location	Remaining proved crude oil reserves (December 31, 1974)	Daily crude oil productive capacity—90 days from December 31, 1974
Coalinga	Fresno County, California (San Joaquin Region)	71,934,000	17,000
Lafitte	Jefferson Parish, Louisiana (South)	71,828,000	15,050
Elk Basin	Park County, Wyoming	68,950,000	20,800
Swanson River	Kenai Peninsula Borough, Alaska	68,878,000	26,000
Main Pass, Block 306	Plaquemines Parish, Louisiana (South) (offshore)	67,289,000	11,170
Sooner Trend	Kingfisher County, Oklahoma	64,145,000	27,000
Andector	Ector County, Texas (District 8)	63,000,000	17,200
Eunice	Lea County, New Mexico	61,966,000	11,000
Golden Trend	Garvin County, Oklahoma	61,780,000	19,000
Lake Pasture	Refugio County, Texas (District 2)	61,766,000	15,000
Neches	Anderson and Cherokee Counties, Texas (District 6)	61,702,000	16,728
Oregon Basin	Park County, Wyoming	61,268,000	33,100
Lake Washington	Plaquemines Parish, Louisiana (South)	61,170,000	18,000
Altamont	Duchesne County, Utah	58,814,000	32,000
Tule Elk	Kern County, California (San Joaquin Region)	58,232,000	30,000
Healdton	Carter County, Oklahoma	58,119,000	20,000
Foster	Ector County, Texas (District 8)	58,000,000	17,600
Anahuac	Chambers County, Texas (District 3)	56,390,000	24,339
Bay St. Elaine	Terrebonne Parish, Louisiana (South)	55,172,000	11,140
Oyster Bayou	Chambers County, Texas (District 3)	54,848,000	16,046
Bluebell	Duchesne and Uintah Counties, Utah	53,882,000	32,000
Granite Point	Kenai Peninsula Borough, Alaska	53,864,000	12,200
Timbalier Bay, Block 21	Lafourche Parish, Louisiana (South) (offshore)	53,641,000	32,200
Talco	Franklin and Titus Counties, Texas (District 6)	53,117,000	10,360
Salt Creek	Kent County, Texas (District 8A)	53,000,000	36,500
Venice	Plaquemines Parish, Louisiana (South)	51,952,000	12,165
Bell Creek	Powder River County, Montana	51,094,000	25,500
Cote Blance Island	St. Mary Parish, Louisiana (South)	51,002,000	14,300
Delta, West, Block 73	Plaquemines Parish, Louisiana (South)	50,901,000	18,750
Grand Isle, Block 16	Jefferson Parish, Louisiana (South) (offshore)	50,530,000	31,600

Field	Location		
Hobbs	Lea County, New Mexico	50,321,000	12,600
McKittrick	Kern County, California (San Joaquin Region)	49,427,000	16,000
McElroy	Crane County, Texas (District 8)	49,000,000	30,500
Ship Shoal, Block 298	Terrebonne Parish, Louisiana (South) (offshore)	48,574,000	25,700
Bay de Chene	Jefferson and Lafourche Parishes, Louisiana (South)	47,067,000	14,392
Sprayberry Trend	Reagan County, Texas (District 7C)	45,635,000	14,438
Postle	Texas County, Oklahoma	44,021,000	16,000
Prentice	Yoakum County, Texas (District 8A)	44,000,000	16,200
Main Pass, Block 69	Plaquemines Parish, Louisiana (South) (offshore)	43,871,000	19,000
Lost Soldier	Sweetwater County, Wyoming	43,299,000	9,100
Beaver Lodge	Williams County, North Dakota	43,136,000	8,500
Teapot Dome	Natrona County, Wyoming	42,515,000	2,500
Lake Barre	Terrebonne Parish, Louisiana (South)	42,254,000	10,000
Total United States		34,249,956,000	8,877,000
Total one hundred largest fields		24,345,462,000	4,712,995
Percent of total United States		71.1%	53.1%

pools have special exploratory significance or discovery results from the application of new exploration concepts as compared to those applied in the discovery of the old fields. In such situations, new discoveries are credited to the years in which the new pools were found. Special decisions as to the year to be credited for a particular discovery involve geologic and exploratory judgments made by subcommittee members.

Estimates of ultimate recovery and original oil-in-place for fields discovered in the most recent years are often subject to substantial revision in future years based on information provided by additional drilling, production performance, and the successful installation of improved recovery techniques. For this reason, caution should be exercised in the interpretation of the most recent data of this kind.

The generalized procedures used in classifying "new field" and "new pool" discoveries are described later.

Extensions. The ultimate size of newly discovered fields, or newly discovered pools in old fields, is normally determined by drilling in years subsequent to discovery. Wells drilled in subsequent years usually add to the proved area of previously discovered reserviors, thereby serving to increase estimates of proved reserves. The reserves credited to a reservoir because of enlargement of its proved area are classified as "extensions."

Revisions Both development drilling and production history add to the basic geological and engineering knowledge of a petroleum reservoir and provide the basis for more accurate estimates of proved reserves in years following discovery. Changes in earlier estimates, either upward or downward, resulting from new information (except for an increase in proved acreage) are classified as "revisions." Revisions for a given year also include (1) increases in proved reserves associated with the successful installation of improved recovery techniques; and (2) an amount which corrects the effect on proved reserves of the difference between estimated production for the previous year and actual production for that year.

The drilling of additional wells in a reservoir better defines the productive area and provides additional basic geological and engineering data pertaining to the reservoir. Estimates of porosity, interstitial water, pay thickness, and other reservoir factors may be revised on the basis of information provided by additional drilling.

Analysis of the producing history of a reservoir, including production of oil, gas, and water and pressure performance, results in more accurate knowledge of the producing mechanism, recovery efficiency, and the performance of the reservoir. This new and improved information provides the basis for more accurate estimates of ultimate recoveries and remaining reserves, and results in revisions to previous estimates, either upward or downward.

Changes in reserve estimates brought about by the successful application of fluid injection or other improved recovery techniques are classified and reported as "revisions" to proved reserves. Changes in reserves resulting from a reduction in the estimate of the proved area are reported as "revisions," but changes due to an increase in the proved area are classified and reported as "extensions."

Proved Acreage. Proved acreage is the area which has been credited with proved reserves. Acreage is credited with proved reserves if the presence of a productive formation has been verified by drilling and testing. Undrilled acreage adjacent to drilled acreage and certain other undrilled acreage are also credited with proved reserves if geological and engineering information demonstrate with reasonable certainty that the underlying formations are continuous and productive.

Improved Recovery Techniques. Improved recovery techniques include all methods of supplementing natural reservoir forces and energy, or otherwise increasing ultimate recovery from a reservoir. Such techniques include: (1) pressure maintenance; (2) cycling; and (3) secondary recovery in its original sense (i.e., fluid injection applied relatively late in the productive history of a reservoir for the purpose of stimulating production after recovery by primary methods of flowing or artificial lift have approached an economic limit). Improved recovery techniques also include thermal methods and the use of miscible displacement fluids.

Reserves resulting from the application of any of the methods listed above are reported as "revisions" to proved reserves for the year in which successful testing by a pilot project, or the operation of an installed program in the reservoir, provides support for the engineering analysis on which the project or program was based.

Additional reserves related to the application of improved recovery techniques are reported as "revisions" to proved reserves, or an "indicated additional reserves," in accordance with the definitions of these terms.

Since it is not possible to separate primary production from production resulting from improved recovery techniques, no attempt is made to classify production and remaining reserves according to the method of recovery.

Original Oil-in-Place.   The estimated number of stock tank barrels of crude oil in known reservoirs prior to any production is defined as "original oil-in-place." Known reservoirs include (1) those that are currently productive; (2) those to which proved reserves have been credited but from which there has been no production; and (3) those that have been depleted.

Original oil-in-place is not to be confused with recoverable oil-in-place. Original oil-in-place is a gross quantity independent of recovery efficiency or economics of operation; recoverable oil-in-place is a net quantity which is dependent upon recovery efficiency and economics of operation.

The estimation of original oil-in-place is based on calculations using the volumetric method or the material balance method when sufficient factual data are available concerning reservoir rock, fluid properties, reservoir limits, and production performance. Where such data are not available, the estimation of original oil-in-place may be based on information and performance characteristics of reservoirs believed to be comparable.

Oil-in-place estimates are limited to the reservoir area and volumes associated with proved reserves and past production.

Ultimate Recovery.   Ultimate recovery represents the estimated quantity of crude oil which has been produced from a reservoir and is expected to be produced in the future if there are no substantial changes in current economic and operating conditions.

Ultimate recovery also may be expressed as the percentage of original oil-in-place which is expected to be eventually produced. This percentage will vary from one reservoir to another in accordance with the reservoir fluid, rock characteristics, and the producing mechanism or drive which is present.

Estimates of ultimate recovery from a given reservoir may be revised in subsequent years if there is (1) a successful application of an improved recovery technique, (2) an increase or decrease in the extent of the reservoir, or (3) information which indicates that recovery mechanisms are performing more or less efficiently than previously estimated.

Productive Capacity of Crude Oil.   Estimates of productive capacities of crude oil developed by the American Petroleum Institute Committee on Reserves and Productive Capacity represent the maximum daily rates of production which can be attained under specified conditions on March 31 of any given year.

The definition of productive capacity used by the Committee is as follows: "The ninety-day crude oil productive capacity is the maximum daily crude production rate, at the point of custody transfer, that could be achieved in ninety days (following December 31 of any given year) with existing wells, well equipment, and surface facilities—plus work and changes that can be reasonably accomplished within the time period using present service capabilities and personnel and with productivity declining as it would under capacity operation."

Estimates of the productive capacity for particular fields or reservoirs are based on proved acreage wells, well equipment, and surface production facilities as of the previous December 31, with adjustments for (1) increases in productive capacity which would result from alterations and improvements in existing facilities and programs for development drilling and improved recovery techniques, which could be completed within the ninety-day period with existing capabilities and personnel; and (2) the natural decreases in productive capacity resulting from capacity operations during the ninety-day period. It should be noted, however, that there is no adjustment for additions to reserves and increased productive capacity that might result from exploratory drilling during the ninety-day period. Furthermore, estimates do not include quantities of crude oil in lease storage on March 31 which could be drawn upon at the time of capacity operation.

Estimates prepared by the Committee are based on the following assumptions:

1. There will be no restrictions on production resulting from a lack of markets for crude oil.
2. There will be no change in crude oil prices or the unit cost of materials, equipment, and labor within the ninety-day period allowed for the buildup of capacity.
3. There will be no statutory restrictions on production, but gas and water production will be controlled according to prudent and accepted engineering practices, where appropriate to prevent the significant reduction of crude oil recovery. The only other production restrictions applicable would be those which prohibit the pollution of water and those which prohibit air pollution with gas or the creation of fire hazards from gas by operations up to the point of transferring the gas to market or to gas processing facilities.
4. There will be no restrictions on production resulting from the inadequacy of storage or transportation facilities beyond the point of custody transfer.
5. Intrafield equity considerations will be satisfactorily resolved so that production for given fields can be maximized.

For the guidance of subcommittee members, individual elements of the working definition of productive capacity may be interpreted as follows:

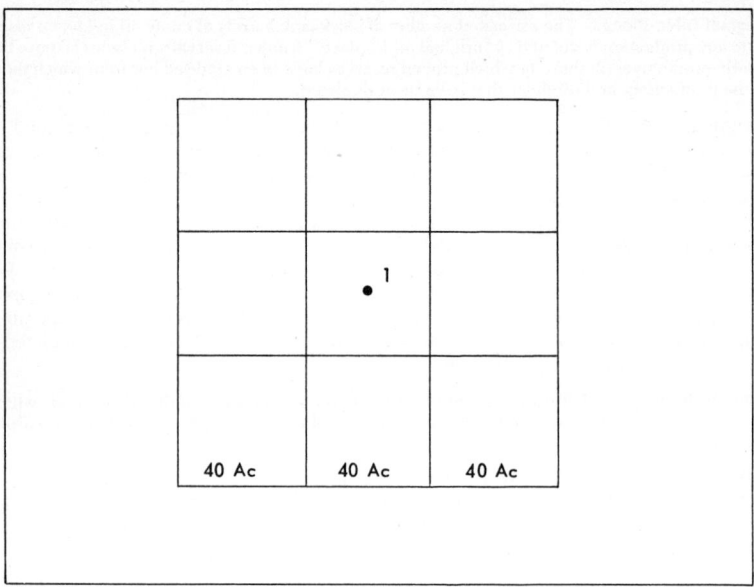

Fig. 10. Example of a new field discovery. Ac = acres.

1. Maximum daily crude production rate.   The figures reported in most cases reflect production that could be obtained from the proved and, in some instances, the indicated additional reserves as defined and reported by the API Committee on Reserves and Productive Capacity. The definition excludes reserves and producing capacity that might be developed by exploratory drilling during the 90-day period. The concept implies nothing about the determined rate being sustainable over any specified period of time—it is simply a point on a continuous curve. The production decline rates applicable in each specific field are considered in determining the capacity at the end of the 90-day period. Data are reported as of December 31 and reflect the capacity that would be developed by March 31, based on field conditions and information as of December 31.

2. Point of Custody Transfer.   The point in the production system at which capacity is estimated is that point at which the oil is transferred from the producing function to the transportation function. Generally the point of transfer would be where the oil is put into another system (pipeline, truck, barge, etc.) for movement to refineries or terminals. The selection of this point to measure capacity implies that not only may reservoir characteristics and down-hole equipment place limitations on capacity, but wellhead equipment separators, flow lines, lease tanks, intrafield barges, and other oil handling facilities may also create limitations.

3. That could be achieved in 90 days with existing wells, well equipment, and surface facilities—plus work and changes that can be reasonably accomplished within the time period, using present service capabilities and personnel.   The wells considered include those already producing, those shut in that could and would be put on stream, and development wells on proved acreage which would be completed and put on stream within 90 days. Various changes in, or additions to, down-hole and wellhead equipment and lease facilities could be made: e.g., pumps, tubing, flow lines, separators. Also, certain steps could conceivably be taken to stimulate or improve production from the reservoir; these include various types of formation treatments and work-overs. The limitation imposed by the definition of such changes or additions is that they must be capable of being completed in 90 days with present services, personnel, material, and equipment capabilities. For example, although it may be desirable to work over most of the wells in a given field, work-over equipment and personnel, reasonably expected to be available, may be capable of handling only a few wells; or it may be desirable to install larger pumping units or flow lines in a field, but the equipment cannot be obtained and installed in this short a time. It should be assumed that all fields would be undergoing such servicing and improvement and that no one field could expect a larger share of services, equipment, or personnel than it would receive in normal times.

4. With productivity declining as it would under capacity operation: It was mentioned under item 1, that consideration should be given to the declining production capability of a reservoir over time—in this case, 90 days. The specific rate of decline selected will vary from field to field and will be determined by the particular set of circumstances in each field. Both reservoir characteristics and ability of well and surface equipment to handle maximum production will influence the choice of a decline rate; however, in all instances, it is assumed that production over the time period will be limited only by facilities and equipment or the reservoir itself. For example, two fields with similar reservoir characteristics may have drastically different producing rates if one has a restriction imposed by separator capacity and the other does not; as a result the two fields may have drastically different decline rates.

5. It is assumed that there would be no change in crude oil prices or costs of material, equipment and/or labor.   This is a simplifying assumption to avoid predicting move-

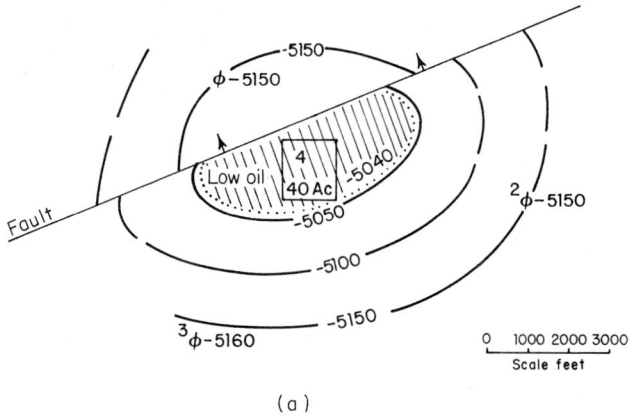

(a)

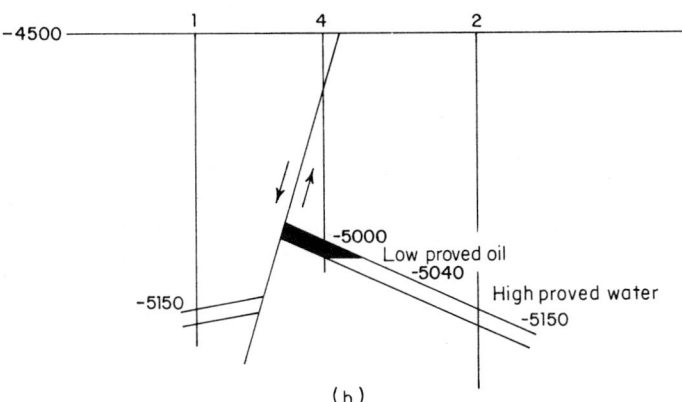

(b)

Fig. 11.   (a) New field discovery in area of additional well. (b) Cross section of structure shown in (a).
Ac = acres.

ments of crude oil prices and operating costs. The intent is to emphasize the capacity that would be developed because of incentives accruing from additional production and not from incentives resulting from increased crude oil prices. It is conceivable that *total* operating expenses might be increased, but the increase would result from the use of more labor, services, equipment, etc., and not from a rise in the prices of these items.

6. *No statutory restrictions on producing rates (but adhering to judicious operating practices):* All market demand restrictions are removed and the maximum daily producing rates established for a field by regulatory agencies can be increased if in the judgment of the subcommittee these rates may be exceeded without causing a significant reduction in ultimate crude oil recovery. It would be assumed that intrafield equity considerations can be satisfactorily resolved so that a *field's* production can be maximized. Gas-oil and water-oil ratio limitations cannot be ignored if they would violate prudent and accepted engineering practices and result in a significant reduction in ultimate crude oil recovery.

7. *No restrictions on storage or transportation beyond the point of custody transfer and no marketing constraints:* It is assumed that transportation facilities, storage, refineries, terminals, and markets are adequate, and so situated as to accommodate all the oil made available at the point of custody transfer if the current market prices for crude oil persist.

Estimates of productive capacity are, to the extent practicable, handled on a field-by-field basis, consideration being given whenever possible to basic production units within a field. Priority attention is given to those fields believed to have significantly large excess productive capacities.

GENERALIZED PROCEDURES IN CLASSIFYING A NEW FIELD

Figures 10 through 14 serve to illustrate (1) the generalized procedures used in classifying "new field" and "new pool" discoveries; (2) the allocation of acreage between the "drilled" and "undrilled" proved categories; (3) the range in estimates that may be expected under varying conditions and the effect of limited data on the estimates; and (4) the fact that original estimates of discoveries are not restricted to blocked-out developed areas if geological information warrants an interpretation of a larger area with proven reserves.

One type of new field discovery situation is shown in Fig. 10. No additional well control is in the immediate vicinity and few data are available on type and extent of structure. Two situations are examined:

Case I: Discovery in a reservoir known to have a wide areal extent (blanket sand or porosity): (*a*) Discovery has reasonably thick producing section, 30 feet, all

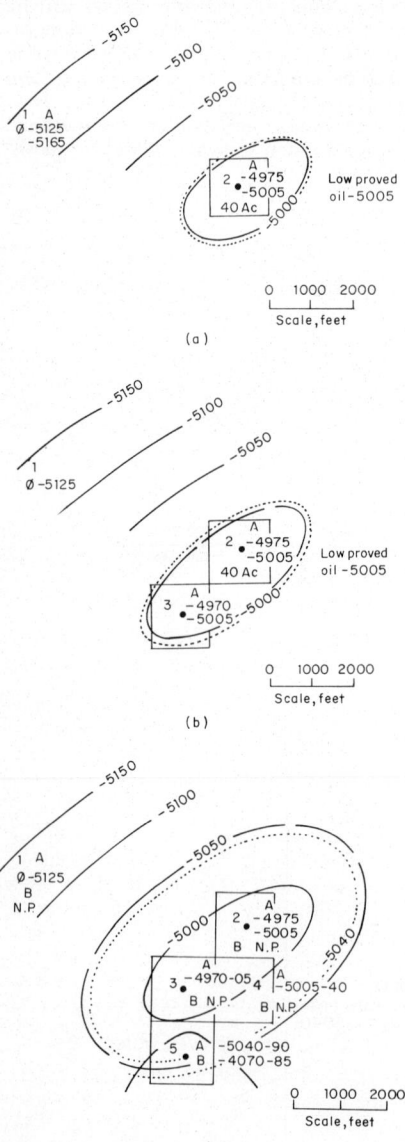

Fig. 12. Progression of a new field situation in which there is a sequence of drilling. Ac = acres.

saturated. Regional rate of dip 150 to 200 feet per mile. *Prove 60 to 100 acres on initial estimate.* (*b*) Thick producing section, as in example (*a*), but dips are steep. *Prove 20 to 40 acres on initial estimate.* (*c*) Thin reservoir, 6 to 10 feet of saturation, normal rate of dip. *Prove 20 to 40 acres on initial estimate.*

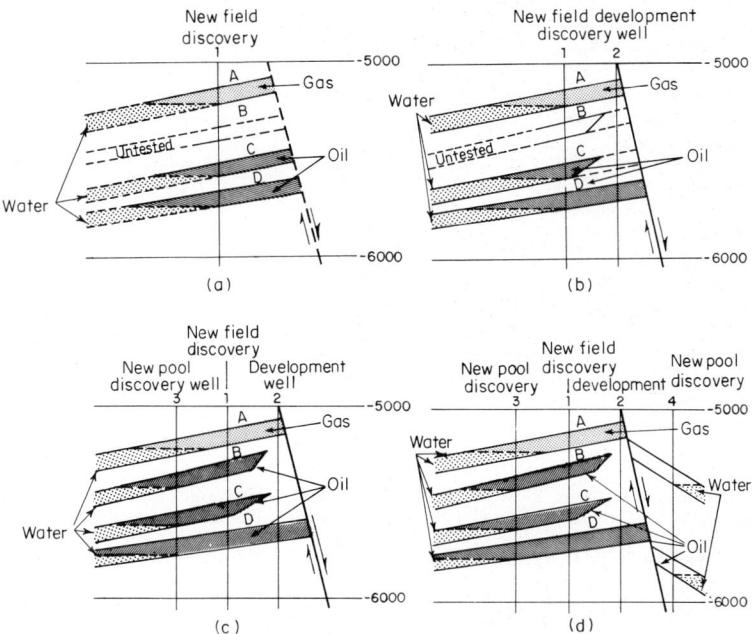

Fig. 13. Cross section of field discovery. Wells drilled in numerical sequence. (*a*) New field discovery well no. 1 tested gas in reservoir A and oil in C and D. Reservoir B not tested. Well completed as dual oil well in reservoirs C and D. (*b*) Development well no. 2 completed as an oil well in reservoir D. Reservoirs B and C not present. Productive area of reservoir C initially considered proved between well no. 1 and fault, revised downward on basis of new information. (*c*) Development well no. 3 extends the proved area and defines the limits of reservoir A. Tests oil in reservoir B. Completed as dual well in C and D. Classified as new pool discovery in reservoir B. Reserves in reservoirs A and B are classified as proved, but undeveloped if they are not to be produced by wells 1, 2, or 3. (*d*) New pool discovery well no. 4 drilled at location across fault found in wells 1 and 2 is an exploratory well which tests gas in reservoir A and oil in reservoir D. This well is classified as new pool discovery by proving production in separate structural segment.

Case II: Discovery in a reservoir known to have lenticular or variable porosity and permeability. *Prove 10 to 20 acres on initial estimate.*

In the situation shown in Fig. 11*a*, *Discovery Well No. 4* proved production in reservoir from subsea 5000 to 5040. Proved area limited by structural position of low oil occurrence. Proved drilled area 40 acres, proved undrilled area 170 acres. *Total proved area on initial estimate 210 acres.* The cross section of the structure is shown in Fig. 11*b*. Wells drilled in numerical sequence. Reservoir volume between fault and low proved oil occurrence considered proved. Reservoir volume between low proved oil and highest proved water is prospective.

Progression of a new field situation in which there is a sequence of drilling is shown in Fig. 12*a* to *c*. In (*a*) is shown reservoir A with 30 feet of oil-saturated reservoir. One oil completion, *Well No. 2*. Rate of dip on structure, 225 feet per mile indicated by data on *Well No. 1*, dry hole. Proved drilled area 40 acres. *Total proved area* 90 acres on initial estimate, including 50 acres proved undrilled.

Figure 12*b* shows the completion of extension *Well No. 3* to discovery shown in Fig. 12*a*. Proved area of reservoir A increased to 140 acres. Oil column is 35 feet, from subsea

4970 to 5005 as reservoir was found to be 5 feet thicker than in Well No. 2. Developed of proved drilled area is 80 acres. *Total proved area,* 140 acres.

Figure 12c shows the completion of Well No. 4, which confirms the structural interpretation of Fig. 12a, and extends oil column in reservoir A to subsea 5040. *Proved drilled area* 120 acres, *total proved area* 425 acres. *Well No.* 5 tested water in reservoir A below subsea 5040 and proved the limits of the reservoir on the south side of the structure; also, discovered oil accumulation in new reservoir B from subsea 4070 to 4085. New pool discovery assigned 15 to 20 acres on initial estimate. Reservoir B not present in *Wells 1, 2, 3, and 4.*

A cross section of field discovery is shown in Fig. 13a through d. The situations prevailing in each example are described in captions. Note that wells are drilled in numerical sequence.

Related Statistics.   Other statistics in this Handbook which relate to petroleum and associated products include:

In the Handbook section on *Gas Technology,* see:
Natural Gas Production and Reserves in the United States
Natural Gas Reserves in Canada
World Natural Gas Production and Reserves
Potential Supply of Natural Gas in the United States
Future Gas Consumption in the United States
In the Handbook section on *Petroleum Technology,* see:
Petroleum Reserves in Canada
World Petroleum Production and Reserves

REFERENCES

American Petroleum Institute: "Reserves of Crude Oil, Natural Gas Liquids, and Natural Gas in the United States and Canada and United States Productive Capacity," issued annually, API, Washington.

American Petroleum Institute: Technical Reports, Nos. 1, 2 (and subsequent), issued periodically, API, Washington.

NOTE Also obtain lists of available publications from U.S. Bureau of Mines, Pittsburgh, and U.S. Geological Survey, Reston, Va.

Petroleum Reserves in Canada

The Reserves Committee of the Canadian Petroleum Association prepares annual reserves information in an approach similar to that used by the American Petroleum Institute and described under "Petroleum Production and Reserves in the United States," included earlier in this Handbook section. Some of the more frequently used terms and definitions which apply to these data are given later in this section. No reserves are included for the Mackenzie Delta, the Artic Islands, or East Coast offshore discoveries.

Reserves of crude oil in Canada are given in Table 1. The historical record of remaining crude oil reserves in Canada is given in Fig. 1. The relationship of gross additions (proved and probable) with net production over the period, 1950 through 1974, is illustrated in Fig. 2. Crude oil reserves in Canada attributable to pressure maintenance are presented in Table 2. A tabulation of oridinal crude oil-in-place in Canada by geological age is given in Table 3. Ultimately recoverable crude oil reserves in Canada are given in Table 4.

DEFINITIONS OF TERMS

Proved Reserves.   This term defines the estimated quantity of crude oil which analysis of geological and engineering data demonstrates with reasonable certainty to be recoverable from known oil fields under existing economic and operating conditions.

Probable Reserves.   Probable reserves are a realistic assessment of the reserves that will be recovered from known oil fields based on the estimated ultimate size and reservoir characteristics of such fields. Probable reserves *include* those reserves shown in the proved category.

Original In-Place Reserves.   These reserves are defined as the total quantity of crude oil initially in place within the estimated area of an oil field from which production has been obtained or for which reserves have been credited. This term represents the sum of (1) the anticipated ultimate crude oil production and (2) the crude oil estimated to be nonrecoverable under existing economic and operating conditions.

Ultimate Recoverable Reserves.   These are defined as the total quantity of crude oil to be ultimately producible from an oil field as determined by an analysis of current geological and engineering data. This includes any quantities already produced up to the respective date of the estimate.

Remaining Reserves.   These are those quantities of crude oil as estimated under proved or probable reserves after deducting those quantities produced up to the respective date of the estimate.

Non-Conventional Reserves.   (A term included by the Canadian Petroleum Association for the first time in its 1975 Report.) At the present time only those reserves attributable to the operational Athabasca oil sands project are included in the CPA estimates of non-conventional reserves. Normally those quantities of synthetic crude oil and sulfur would be included which the Committee estimates could be recovered by established production projects within an economic radius of the project from deposits having equivalent or better characteristics than the deposit under development.

In this Handbook, statistics on oil sands are given in article on "Fuel from Tar Sands," which appears later in this section.

REFERENCE

Canadian Petroleum Association: "Report of the Reserves Committee of the Canadian Petroleum Association," issued annually. Available as part of American Petroleum Institute Report on "Reserves of Crude Oil, Natural Gas Liquids, and Natural Gas in the United States and Canada and United States Productive Capacity," API, Washington, or separately from Canadian Petroleum Association, Calgary, Alberta.
For other statistical information in this handbook on petroleum, natural gas, and associated resources, see listing at end of prior Handbook section on Petroleum Production and Reserves in the United States.

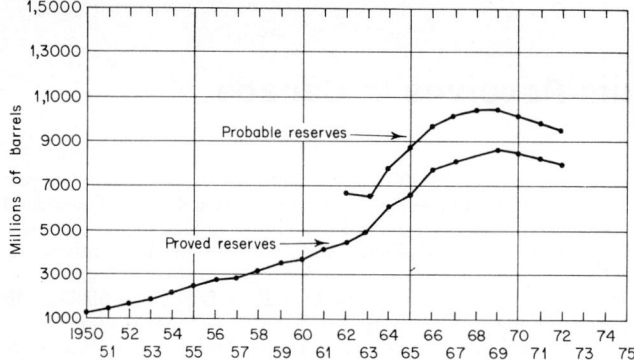

Fig. 1. Proved and probable reserves of crude oil in Canada (1951–1974). Mackenzie Delta included.

TABLE 1.   Crude Oil Reserves in Canada, Original Oil-in-Place, Ultimate, and Remaining
Thousand barrels

Province	Original-in-place		Ultimate reserves		Remaining reserves	
	Proved	Probable*	Proved	Probable*	Proved	Probable*
Territories	500,000	500,000	60,000	90,000	40,795	70,795
British Columbia	1,261,261	1,271,883	410,752	452,891	160,637	202,776
Alberta	33,700,260	34,266,214	11,447,581	12,403,141	6,351,437	7,306,997
Saskatchewan	9,742,117	10,031,874	1,938,439	2,100,920	566,265	728,746
Manitoba	668,970	679,430	146,716	156,356	40,782	50,422
Ontario	192,390	199,845	62,668	65,096	11,244	13,672
Other Eastern Canada	18,000	18,000	850	2,324	69	1,543
Total—Canada	46,082,998	46,967,246	14,067,006	15,270,728	7,171,229	8,374,951

SOURCE: *Can. Petrol. Ass. Rep.*, vol. 29, May 1975. Incorporated as part of *API Rep.*, vol. 29, May 1975. Reserves conditions as of 12/31/1974.
*Includes proved reserves.
NOTE: This table does not include the recovery of bitumen from the athabasca Oil Sands, which are described later in this Handbook section.

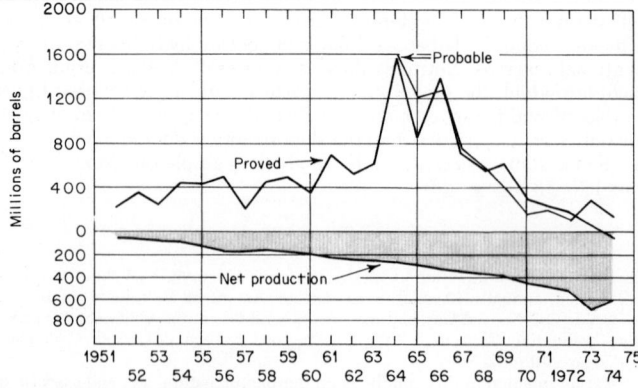

Fig. 2. Production versus gross additions (proved and probable) added during each year in Canada (1950–1974). Mackenzie Delta included.

TABLE 2. Crude Oil Reserves in Canada Attributable to Pressure Maintenance*

Thousand barrels

Province	Proved ultimate reserves			Probable† ultimate reserves		
	From natural depletion	From pressure maintenance	Total	From natural depletion	From pressure maintenance	Total
Territories	60,000	....	60,000	60,000	30,000	90,000
British Columbia	220,315	190,437	410,752	223,765	229,126	452,891
Alberta	7,550,647	3,896,934	11,447,581	7,742,215	4,660,926	12,403,141
Saskatchewan	1,173,334	765,105	1,938,439	1,196,628	904,292	2,100,920
Manitoba	84,641	62,075	146,716	86,531	69,825	156,356
Ontario	56,011	6,657	62,668	56,939	8,157	65,096
Other Eastern Canada	850	....	850	850	1,474	2,324
Total—Canada	9,145,798	4,921,208	14,067,006	9,366,928	5,903,800	15,270,728

SOURCE: *Can. Petrol. Ass. Rep.*, vol. 29, May 1975. Incorporated as part of *API Rep.*, vol. 29, May 1975. Reserves conditions as of 12/31/1974.
*Figures include all enhanced recovery operations.
†Includes proved reserves.

TABLE 3. Original Crude Oil-in-Place in Canada by Geological Age
Thousand barrels

Geological age	Territories	British Columbia	Alberta	Saskatchewan	Manitoba	Ontario	Other Eastern Canada	Total Canada
			PROVED RESERVES					
Cenozoic	......	......	......	......	......	......	......	......
Quaternary								
Mesozoic								
Upper Cretaceous	......	......	9,816,366	......	......	......	......	9,816,366
Lower Cretaceous	......	38,343	4,694,075	2,779,833	......	......	......	7,512,251
Jurassic	......	......	13,500	2,176,486	......	......	......	2,189,986
Triassic	......	1,139,809	208,978	......	......	......	......	1,348,787
Paleozoic								
Permian	......	19,344	500	......	......	......	......	19,844
Pennsylvanian	......	......	......	......	......	......	......	......
Mississippian	......	63,765	2,613,401	4,777,398	668,970	......	18,000	8,141,534
Devonian	500,000	......	16,353,440	7,300	......	127,896	......	16,988,636
Silurian	......	......	......	......	......	36,855	......	36,855
Ordovician	......	......	......	1,100	......	3,110	......	4,210
Cambrian	......	......	......	......	......	24,529	......	24,529
Total—All ages	500,000	1,261,261	33,700,260	9,742,117	668,970	192,390	18,000	46,082,998
		PROBABLE RESERVES (Includes proved oil-in-place)						
Cenozoic	......	......	......	......	......	......	......	......
Quaternary								
Mesozoic								
Upper Cretaceous	......	......	9,935,470	......	......	......	......	9,935,470
Lower Cretaceous	......	38,343	5,021,595	3,011,104	......	......	......	8,071,042
Jurassic	......	......	13,500	2,220,605	......	......	......	2,234,105
Triassic	......	1,150,431	222,728	......	......	......	......	1,373,159
Paleozoic								
Permian	......	19,344	500	......	......	......	......	19,844
Pennsylvanian	......	......	......	......	......	......	......	......
Mississippian	......	63,765	2,665,201	4,791,765	679,430	......	18,000	8,218,161
Devonian	500,000	......	16,407,220	7,300	......	130,113	......	17,044,633
Silurian	......	......	......	......	......	39,069	......	39,069
Ordovician	......	......	......	1,100	......	3,122	......	4,222
Cambrian	......	......	......	......	......	27,541	......	27,541
Total—All ages	500,000	1,271,883	34,266,214	10,031,874	679,430	199,845	18,000	46,967,246

SOURCE: *Can. Petrol. Ass. Rep.*, vol. 29, May 1975. Incorporated as part of *API Rep.*, vol. 29, May 1975. Reserves conditions as of 12/31/1974.

TABLE 4. **Ultimate Recoverable Crude Oil Reserves in Canada by Geological Age**
Thousand barrels

Geological age	Territories	British Columbia	Alberta	Saskatchewan	Manitoba	Ontario	Other Eastern Canada	Total Canada
			PROVED RESERVES					
Cenozoic	......	......	....	....	......	......	......	....
Quaternary	......	......	....	....	......	......	......	....
Mesozoic								
Upper Cretaceous	......	......	2,085,853	....	......	......	......	2,085,853
Lower Cretaceous	......	10,182	925,313	233,878	......	......	......	1,169,373
Jurassic	......	......	3,068	523,231	......	......	......	526,299
Triassic	......	387,063	44,396	....	......	......	......	431,459
Paleozoic								
Permian	......	2,029	50	....	......	......	......	2,079
Pennsylvanian	......	......	....	....	......	......	......	....
Mississippian	......	11,478	431,495	1,177,674	146,716	......	850	1,768,213
Devonian	60,000	......	7,957,406	3,400	......	41,660	......	8,062,466
Silurian	......	......	....	....	......	12,005	......	12,005
Ordovician	......	......	....	256	......	1,013	......	1,269
Cambrian	......	......	....	....	......	7,990	......	7,990
Total—All ages	60,000	410,752	11,447,581	1,938,439	146,716	62,668	850	14,067,006
			PROBABLE RESERVES (Includes proved reserves)					
Cenozoic	......	......	....	....	......	......	......	....
Quaternary	......	......	....	....	......	......	......	....
Mesozoic								
Upper Cretaceous	......	......	2,125,998	....	......	......	......	2,125,998
Lower Cretaceous	......	10,182	992,135	273,427	......	......	......	1,275,744
Jurassic	......	......	3,068	612,120	......	......	......	615,188
Triassic	......	420,913	47,691	....	......	......	......	468,604
Paleozoic								
Permian	......	2,029	50	....	......	......	......	2,079
Pennsylvanian	......	......	....	....	......	......	......	....
Mississippian	......	19,767	473,302	1,211,717	156,356	......	2,324	1,863,466
Devonian	90,000	......	8,760,897	3,400	......	42,382	......	8,896,679
Silurian	......	......	....	....	......	12,726	......	12,726
Ordovician	......	......	....	256	......	1,017	......	1,273
Cambrian	......	......	....	....	......	8,971	......	8,971
Total—All ages	90,000	452,891	12,403,141	2,100,920	156,356	65,096	2,324	15,270,728

SOURCE: *Can. Petrol. Ass. Rep.*, vol. 29, May 1975. Incorporated as part of *API Rep.*, vol. 29, May 1975. Reserves conditions as of 12/31/1974.

World Petroleum Production and Reserves

For many years, the gathering of data on crude oil reserves and production statistics on a worldwide basis has been very difficult. Although reasonably reliable figures could be obtained on an individual basis from most major oil-producing countries, this was an arduous and time-consuming task, frequently resulting in data 4 to 6 years old prior to final publication. A few countries, including the United States and Canada, have developed highly refined data gathering and reporting systems as the result of concerted actions on the part of state and national government departments and of leading trade associations. The highly refined reporting systems of the American Petroleum Institute and the Canadian Petroleum Association are described in prior articles in this Handbook section. The U.S. Bureau of Mines and the U.S. Geological Survey have been very active in turning out energy resources information for a number of years. Reports of United States and Canada findings have been made available worldwide. Some other countries have developed reports of their own crude oil statistics, but have not always made these widely available. And still other producing countries have had inadequate reporting systems even for their own planning.

Until the mid-1970s, probably the most comprehensive and reliable summary of world-wide crude oil statistics was the annual reports of the U.S. Geological Survey pertaining to mineral statistics. The first concerted international effort for reliable statistical reporting

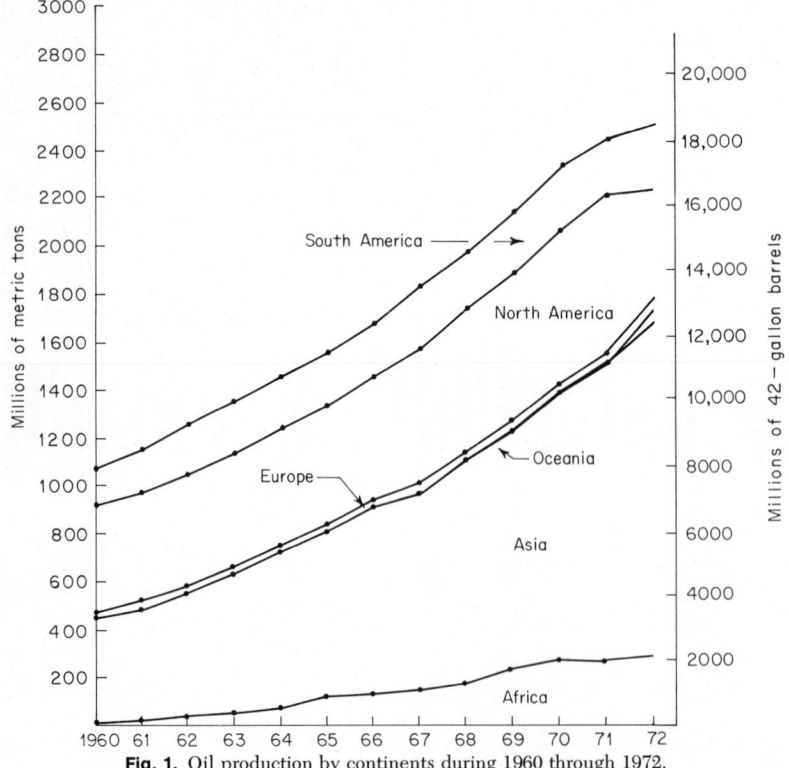

Fig. 1. Oil production by continents during 1960 through 1972.

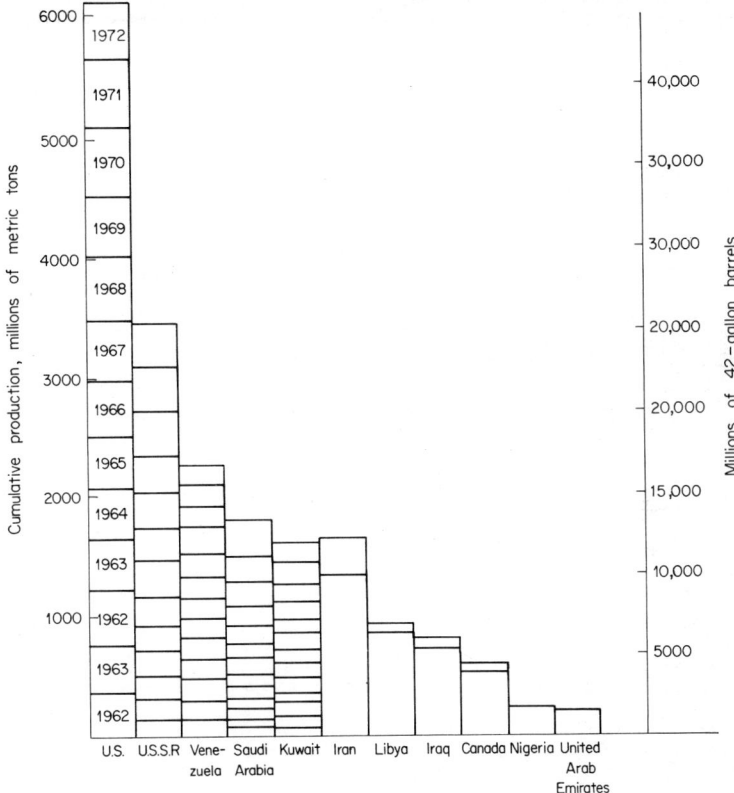

Fig. 2. Cumulative oil production by major producing countries during 1960 through 1972.

and analysis was commenced by the World Energy Conference Survey of World Energy Resources. The target of this survey is to present comprehensive worldwide figures as of 1974. The first drafts of some reporting countries, including the United States and Canada, were completed as of early-1976. These drafts follow a very carefully prescribed format based upon thoughtfully worked out definitions. Other countries are following suit and a full first report probably will be available in 1977. Contact in the United States is the Executive Director, National Committee, World Energy Conference, Washington, D.C. 20006.

As of late-1976, the most comprehensive "official" report probably is Geological Survey Professional Paper 885, published in 1974 by the U.S. Geological Survey, Reston, Virginia 22092, authored by Sherwood E. Frezon, entitled "Summary of 1972 Oil and Gas Statistics for Onshore and Offshore Areas of 151 Countries," Stock No. 2401-02518, available from the U.S. Government Printing Office, Washington, D.C. 20402.

Annual oil production for the period 1960 through 1972 for 151 countries is given in Table 1. These data are also summarized by continent in Fig. 1 and plotted by leading producing nations in Fig. 2. This information reflects the degree to which a country participates in crude oil production and the accuracy of the data (as of 1972) permits comparisons between countries with an acceptable degree of assurance. The resource and reserve estimates of Table 2 are useful in determining the future ability of a country to participate in the economy, but they must be used with caution. Reserve estimates are transitory numbers subject to upward change with new discoveries and downward change resulting from production. Resource estimates are less accurate because in many areas they are based on a minimum of information and some data are highly conjectural. The transitory and interpretive nature of reserve and resource estimates, combined with the

TABLE 1.   Annual Oil Production (1960–1972)
Millions of metric tons

	1960	1961	1962	1963	1964	1965	1966	1967	1968	1969	1970	1971	1972
AFRICA[a]													
Algeria	8.8	15.8	20.5	23.9	26.5	26.7	34.0	38.6	42.9	44.8	48.1	36.3	52.0
Angola	0.1	0.1	0.5	0.8	0.9	0.7	0.6	0.5	0.7	2.4	4.9	5.7	7.1
Egypt (U.A.R.)	3.5	3.8	4.7	5.6	6.3	6.6	6.4	5.7	9.0	12.9	17.2	15.5	11.4
Gabon	0.8	0.8	0.8	0.9	1.0	1.2	1.4	3.4	4.6	5.0	5.5	5.8	6.3
Libyan Arab Republic	0.0	0.9	8.8	22.2	41.4	58.5	72.3	83.6	125.0	149.1	158.8	132.3	107.8
Nigeria	0.9	2.3	3.3	3.7	5.9	13.4	20.6	15.7	7.0	26.6	53.4	75.4	89.8
Tunisia	0.0	0.0	0.0	0.0	0.0	0.0	0.6	2.2	3.2	3.6	4.4	4.2	4.1
All other countries[b]	0.2	0.2	0.2	0.2	0.2	0.2	0.2	0.2	0.2	0.1	0.0	0.0	0.3
Total—Africa	14.3	23.9	38.8	57.3	82.2	107.3	136.1	149.9	192.6	244.5	292.3	275.2	278.8
ASIA[c]													
Bahrain	2.2	2.2	2.2	2.2	2.5	2.8	3.1	3.5	3.8	3.8	3.8	3.7	3.5
Burma	0.5	0.6	0.6	0.6	0.6	0.5	0.6	0.6	0.8	0.9	0.9	0.9	1.0
China, People's Republic of	4.5	4.8	5.3	6.0	7.5	9.0	10.4	11.0	13.0	16.0	20.0	25.0	26.0
India	0.5	0.5	1.1	1.7	2.3	3.0	4.6	5.8	6.0	6.9	7.1	7.4	7.7
Indonesia	20.8	21.1	22.8	22.5	23.3	24.3	23.0	25.3	30.0	36.9	42.4	44.3	54.1
Iran	52.3	58.6	65.4	73.0	83.9	93.4	104.6	128.6	141.0	167.1	189.6	224.2	249.4
Iraq	47.6	49.2	49.3	56.9	62.1	64.9	67.9	60.3	73.8	74.6	76.6	84.0	71.2
Israel	0.0	0.2	0.2	0.2	0.2	0.2	0.2	1.2	2.0	2.5	4.4	6.0	6.0
Kuwait	85.3	87.2	98.3	105.0	115.8	118.2	125.0	125.8	132.8	140.7	150.1	160.7	166.5
Oman	0.0	0.0	0.0	0.0	0.0	0.0	0.0	3.1	11.9	16.2	16.5	14.2	14.1
Qatar	8.2	8.4	8.8	9.1	10.1	10.9	13.7	15.4	16.1	16.8	17.2	20.4	22.9
Saudi Arabia	65.6	73.7	81.7	88.8	94.6	109.7	129.5	139.7	151.8	160.1	188.9	237.3	319.0
Turkey	0.4	0.4	0.6	0.7	0.9	1.5	1.9	2.7	3.1	3.6	3.5	3.4	3.4
U.S.S.R.	147.9	166.1	186.2	206.1	223.6	242.9	265.1	288.1	309.2	328.0	352.5	369.5	394.0
United Arab Emirates	0.0	0.0	0.8	2.3	8.9	13.6	17.4	18.6	24.1	29.4	37.7	51.2	57.7
All other countries[d]	0.9	1.2	1.2	1.3	1.2	1.3	1.2	1.2	1.5	1.7	2.3	4.6	21.2
Total—Asia	436.7	474.2	524.5	576.4	637.5	696.2	768.2	830.9	920.9	1005.2	1113.5	1256.8	1417.7
OCEANIA[e]													
Australia	0.0	0.0	0.0	0.0	0.2	0.3	0.4	1.0	1.8	2.1	8.5	14.8	18.4
New Zealand	0.0	0.0	0.0	0.0	0.0	0.0	0.0	0.0	0.0	0.0	0.1	0.1	0.1
Total—Oceania	0.0	0.0	0.0	0.0	0.2	0.3	0.4	1.0	1.8	2.1	8.6	14.9	18.5

EUROPE[f]													
Albania	0.8	0.8	0.8	0.8	0.8	0.9	0.9	1.0	1.1	1.4	1.5	1.5	2.2
Austria	2.4	2.4	2.4	2.6	2.7	2.9	2.6	2.7	2.7	2.8	2.8	2.5	2.5
France	1.9	2.1	2.3	2.5	2.8	3.0	2.9	2.8	2.7	2.5	2.3	1.9	2.3
Germany, Fed. Repub. of (West)	5.5	6.2	6.8	7.4	7.7	7.9	7.9	7.9	8.0	7.9	7.5	7.4	7.1
Hungary	1.2	1.4	1.6	1.8	1.8	1.8	1.7	1.7	1.8	1.8	1.9	2.0	2.0
Italy	2.0	2.0	1.8	1.8	2.7	2.2	1.8	1.6	1.5	1.5	1.4	1.3	1.2
Netherlands	1.9	2.0	2.2	2.2	2.3	2.4	2.4	2.3	2.1	2.0	1.9	1.7	1.6
Romania	11.5	11.6	11.8	12.2	12.4	12.6	12.8	13.2	13.3	13.2	13.4	13.5	14.2
Yugoslavia	0.9	1.3	1.5	1.6	1.8	2.1	2.2	2.4	2.5	2.7	2.9	3.0	3.2
All other countries[g]	0.6	0.7	0.7	0.7	0.8	0.8	1.1	1.4	1.5	1.2	1.2	1.4	2.6
Total—Europe	28.7	30.5	31.9	33.6	35.8	36.6	36.3	37.0	37.2	37.0	36.8	36.2	38.9
NORTH AMERICA[h]													
Canada	25.5	29.7	32.9	34.7	37.0	39.9	43.1	47.3	51.1	55.3	62.1	66.3	75.8
Mexico	13.9	15.0	15.7	16.2	16.3	16.6	17.1	18.7	19.9	20.6	25.0	25.0	26.0
United States	391.9	402.2	411.0	425.1	431.5	443.5	471.3	502.9	521.5	533.7	555.9	552.8	466.9
Total—North America	431.3	446.9	459.6	476.0	484.8	500.0	531.5	568.9	592.5	609.6	643.0	644.1	568.7
SOUTH AMERICA[i]													
Argentina	9.1	12.1	14.0	13.9	14.3	14.1	15.0	16.4	17.9	18.6	20.3	21.7	22.7
Bolivia	.4	.4	.4	.4	.4	.4	.7	1.8	1.9	1.8	1.1	1.7	1.9
Brazil	4.0	4.8	4.6	4.9	4.6	4.7	5.8	7.5	8.0	8.6	8.2	8.7	8.4
Chile	.9	1.2	1.5	1.7	1.8	1.6	1.6	1.6	1.8	1.7	1.6	1.7	1.6
Colombia	7.9	7.5	7.4	8.5	8.9	10.3	10.1	9.8	9.0	10.9	11.3	11.1	10.2
Ecuador	.4	.4	.3	.3	.4	.4	.3	.3	.2	.2	.2	.2	3.8
Peru	2.6	2.6	2.8	2.9	3.1	3.1	3.1	3.4	3.6	3.5	3.5	3.0	3.1
Trinidad and Tobago	6.0	7.0	7.0	7.0	7.0	7.0	8.0	9.0	10.0	8.0	7.0	7.0	7.3
Venezuela	148.7	152.1	166.7	169.3	177.2	180.9	175.6	184.6	188.4	187.3	193.2	185.0	168.3
Total—South America	180.0	188.1	204.7	208.9	217.7	222.5	220.2	234.4	240.8	240.6	246.4	240.1	227.3
Total—151 countries	1091.0	1163.6	1259.5	1352.2	1458.2	1562.9	1692.7	1822.1	1985.8	2139.0	2340.6	2467.3	2549.9
						22,884.7 (cumulative)							

TABLE 1.   Annual Oil Production (1960–1972)   *(Continued)*

[a]African countries with no reportable oil production: Botswana, Cameroon, Central African Republic, Chad, Equatorial Guinea, Gambia, Ghana, Guinea, Ivory Coast, Kenya, Liberia, Malagasy Republic, Malawi, Mali, Mauritania, Mauritius, Mozambique, Niger, Rwanda, Senegal, Sierra Leone, Somalia, Republic of South Africa, Sudan, Swaziland, Tanzania, Togo, Uganda, Upper Volta, Zaire, Zambia.

[b]African countries reported under All other countries: Dahomey (0.3 million metric ton), Morocco (negligible).

[c]Asian countries with no reportable oil production: Afghanistan, Bangladesh, Bhutan, Jordan, Khmer Republic, Laos, Lebanon, Maldives, Mongolia, Nepal, Philippines, Singapore, Sri Lanka (Ceylon), Thailand, Vietnam, Yemen.

[d]Asian countries reported under All other countries: Brunei (9.2 metric tons), Japan (0.7), Malaysia (4.4), Pakistan (0.5), Syrian Arab Republic (6.3), Republic of China (Taiwan) (0.1). Total of these countries, 21.2 metric tons.

Because the oil fields of the U.S.S.R. are located on two continents and because of the difficulty of obtaining an accurate continental breakdown, figures traditionally have all been included under the Asian continent. However, there is a trend among statisticians to report U.S.S.R. figures in a separate category.

[e]Oceanian countries with no reportable oil production: Fiji, Nauru, Tonga, Western Samoa.

[f]European countries with no reportable oil production: Belgium, Denmark, Finland, Greece, Iceland, Ireland, Lichtenstein, Luxembourg, Malta, Portugal, San Marino, Sweden, Switzerland.

[g]European countries reported under All other countries: Bulgaria (0.2 metric ton), Czechoslovakia (0.2), German Democratic Republic (0.2), Norway (1.6, but with increased activity since 1972, Poland (0.3), Spain (0.1), United Kingdom (negligible, but with increased activity since 1972). Total of these countries, 2.6 metric tons.

[h]North American countries with no reportable oil production: Bahamas, Beize, Costa Rica, Dominican Republic, El Salvador, Guatemala, Haiti, Honduras, Jamaica. There is a very small amount of production in Barbados; also an estimate of 0.2 metric ton in Cuba.

[i]South American countries with no reportable oil production: Guyana, Paraguay, Uruguay.

TABLE 2.   Proved Recoverable Reserves of Crude Oil—1972
(millions of metric tons)

Continent and country	Onshore	Offshore	Total
AFRICA[a]			
Algeria	1,532.5	0.0	1,532.5
Angola	na	na	166.7
Congo (Brazzaville)	<571.0	>98.0	699.0
Egypt (U.A.R.)	na	na	550.7
Gabon	38.0	90.0	128.0
Ghana	0.0	0.1	0.1
Libya	4,000.0	na	4,000.0
Morocco	0.2	0.0	0.2
Mozambique	0.0	na	0.0
Nigeria	1,243.0	459.0	1,702.0
Tunisia	na	na	130.0
Total—Africa	<7,384.7	>647.1	8,909.2[b]
ASIA[c]			
Afghanistan	12.0	0.0	12.0
Bahrain	70.4	0.0	12.0
Bangladesh	na	330.0	330.0
Burma	5.4	0.0	5.4
China, People's Republic of	< 2,700.0	na	2,700.0
China, Republic of (Taiwan)	2.5	2.5	2.5
India	112.0	0.0	112.0
Indonesia	1,410.0	55.0	1,465.0
Iran	8,141.0	678.0	8,819.0
Iraq	4,400.0	0.0	4,400.0
Israel	1.2	0.0	1.2
Japan	1.0	2.0	3.0
Kuwait	na	na	10,750.0
Malaysia	na	na	205.0
Oman	na	na	684.9
Pakistan	5.0	0.0	5.0
Qatar	na	na	909.0

Continent and country	Onshore	Offshore	Total
Saudi Arabia	11,500.0	7,879.0	19,373.0
Thailand	1,044.0	0.0	1,044.0
Turkey	77.5	0.0	77.5
U.S.S.R.	10,273.0	na	10,273.0
United Arab Emirates	na	na	2,695.0
Total—Asia	<39,755.0	8,946.5	63,878.5[b]
OCEANIA[d]			
Australia	na	na	268.0
New Zealand	na	na	31.3
Total—Oceania	na	na	299.3
EUROPE[e]			
Albania	13.6	0.0	13.6
Austria	26.7	0.0	26.7
Bulgaria	38.0	0.0	38.0
Czechoslovakia	2.0	0.0	2.0
Denmark	0.0	34.0	34.0
France	13.0	0.0	13.0
Germany, Federal Republic of (West)	76.0	0.0	76.0
Hungary	27.6	0.0	27.6
Italy	na	na	32.0
Netherlands	39.0	0.0	39.0
Norway	0.0	960.0	960.0
Poland	8.0	0.0	8.0
Romania	na	na	41.9
Spain	41.7	na	41.7
United Kingdom	na	694.0	694.0
Yugoslavia	48.0	0.0	48.0
Total—Europe	333.6	1,688.0	2,095.5[b]
NORTH AMERICA[f]			
Barbados	0.1	0.0	0.1
Canada	1,378.0	na	1,378.0
Mexico	na	> 272.4	758.8
United States	4,107.0	1,035.0	5,143.0
Total—North America	5,485.1	>1,307.4	7,279.9[b]
SOUTH AMERICA[g]			
Argentina	> 357.7	na	357.7
Bolivia	24.0	0.0	24.0
Brazil	106.3	2.7	109.0
Chile	12.9	0.0	12.9
Colombia	227.0	0.0	227.0
Ecuador	758.0	0.0	758.0
Peru	na	na	66.6
Trinidad and Tobago	39.4	186.2	225.6
Venezuela	< 1,971.0	na	1,971.0
Total—South America	<3,496.3	188.9	3,751.8[b]
Total—121 countries	<56,454.7	>12,777.9	86,242.6[b]

na = data not available.

[a]African countries with no reportable proved recoverable reserves include: Botswana, Burundi, Cameroon, Central African Republic, Chud, Dahomey, Equitorial Guinea, Ethiopia, Gambia, Ghana, Guinea, Ivory Coast, Kenya, Lesotho, Liberia, Malagasy Republic, Malawi, Mali, Mauritania, Niger, (Portuguese) Guinea, Rwanda, Senegal, Sierra Leone, Somali, Republic of South Africa (na), Southwest Africa, Sudan, Swaziland, Tanzania, Togo, Uganda, Upper Volta, Zaire, Zambia.

[b]Total column does not equal sum of onshore and offshore totals because of lack of information on split for some countries.

[c]Asian countries with no reportable proved recoverable reserves include: Bangladesh (na), Bhutan,

TABLE 2.   Proved Recoverable Reserves of Crude Oil—1972   (Continued)

Cypress, Jordan, Khmer Republic, Korea, Laos, Lebanon, Maldives, Mongolia, Nepal, Philippines, Singapore, Sri Lanka, Thailand (0.1 metric ton possible), Vietnam, Yemen Arab Republic, Yemen People's Republic.

Because the oil fields of the U.S.S.R. are located on two continents and because of the difficulty of obtaining an accurate continental breakdown, figures traditionally have all been included under the Asian continent. However, there is a trend among statisticians to report U.S.S.R. figures in a separate category.

[d]Oceanian countries with no reportable proved recoverable reserves include: Fiji, Nauru, Tonga, Western Samoa.

[e]European countries with no reportable proved recoverable reserves include: Belgium, Finland, German Democratic Republic (1.5 metric tons possible), Greece, Iceland, Ireland, Lichtenstein, Luxembourg, Malta, Portugal, San Marino, Sweden, Switzerland.

[f]North American countries with no reportable proved recoverable reserves include: Bahamas, Belze, Costa Rica, Cuba (1.3 metric tons possible), Dominican Republic, El Salvador, Guatemala, Haiti, Honduras, Jamaica, Nicaragua, Panama.

[g]South American countries with no reportable proved recoverable reserves include: Guyana, Paraguay, Uruguay.

complexity of political and economic factors, reduce the certainty of any predictions made using reserve and resource data. Complexity of data compilation is highlighted by the frequently necessary requirement of converting metric tons into barrels. The number of barrels in a metric ton of oil is dependent upon the specific gravity of the oil. A conversion table, prepared by the U.S. Bureau of Mines in 1974, is given in Table 3.

Table 1 illustrates the steep rise in petroleum production throughout the world over the period 1960 through 1972. Although one must be wary of an analysis of this type, taking into consideration full reliability of the data, it is interesting to compare the worldwide cumulative oil production figure of 22,885 million metric tons for the aforementioned period with the current (1972) proved recoverable reserves figure of 86,242 million metric tons. Obviously, going back to the start of 1960, the interim production through 1972

TABLE 3.   Conversion of Metric Tons to Barrels

Abu Dhabi	7.612	Japan	7.352
Albania	6.672	Kuwait	7.261
Algeria	7.661	Libyan Arab Republic	7.615
Angola	7.206	Malaysia	7.709
Argentina	7.172	Mexico	7.104
Australia	7.775	Mongolia	7.300
Austria	6.974	Morocco	7.602
Bahrain	7.335	Netherlands	6.816
Bolivia	8.086	Neutral Zone	6.825
Brazil	7.315	New Guinea	7.468
Brunei	7.334	New Zealand	8.043
Bulgaria	7.300	Nigeria	7.410
Burma	7.464	Norway	7.444
Canada	7.428	Oman	7.390
China		Pakistan	7.308
People's Republic of China	7.300	Peru	7.517
Republic of China (Taiwan)	7.419	Poland	7.419
Colombia	7.054	Qatar	7.704
Cuba	6.652	Romania	7.453
Czechoslovakia	6.782	Saudi Arabia	7.338
Denmark	7.650	Senegal	7.535
Ecuador	7.580	Spain	7.287
Egypt	6.929	Syrian Arab Republic	6.940
France	7.287	Trinidad and Tobago	6.989
Gabon Republic	7.245	Tunisia	7.709
Germany (Federal Republic of)	7.223	Turkey	7.161
Hungary	7.630	United Arab Emirates	7.549
India	7.441	United Kingdom	7.279
Indonesia	7.348	United States	7.418
Iran	7.370	U.S.S.R.	7.350
Iraq	7.453	Venezuela	7.005
Israel	7.253	Yugoslavia	7.407
Italy	6.813	Zaire	7.478

would have represented reserves in 1960. Thus, one could estimate that the reserves at the start of 1960 were about 109,000 million metric tons. Further, this could indicate that during the 13-year period, about 21 percent of the 1960 reserves were consumed. At an annual production rate of 2,500 to 3,000 million metric tons per year, one could estimate that current reserves of conventional crude oil will last for another 29 to 34 years (from 1973), or until about the year 2002 or 2007. It is likely, of course, that a substantial percentage of probable and speculative reserves will fall into the proved category, that entirely new fields may be discovered, and that improved recovery techniques will increase the overall figures, thus extending the practical demise of crude oil as an energy source. But, demand also may increase at an even more accelerating rate.

With regard to Table 2, it is interesting to note that of the proved recoverable reserves, only two countries (Saudi Arabia and Kuwait) account for over one-third of the reserves. Both the U.S.S.R. and Iran each account for over 10 percent of the reserves, while the United States holds only about 6 percent and Canada less than 2 percent of the reserves. It is also interesting to note the appearance on the following list of countries quite new to the oil-producing industry, as examples, Norway and the United Kingdom. The latter countries, of course, reflect the discovery and accelerating exploitation of nearby offshore sources.

Country	Percent of total proved recoverable reserves
Saudi Arabia	22.5
Kuwait	12.5
U.S.S.R.	11.9
Iran	10.2
United States	6.0
Iraq	5.1
Libyan Arab Republic	4.6
United Arab Emirates	3.1
China (People's Republic of)	3.1
Venezuela	2.3
Nigeria	2.0
Algeria	< 1.8
Indonesia	1.7
Canada	1.6
Syrian Arab Republic	1.2
Norway	> 1.1
Qatar	< 1.1
Mexico	< 0.9
Ecuador	< 0.9
United Kingdom	0.8
Oman	< 0.8
All others	~5.0

It is also interesting to note a trend in production, as evidenced by the "doubling time" (time required to double production using 1960 as the base production period). The doubling time is approximately 2 years for Africa, 7 years for Asia, indeterminate for Europe, 2 years for Oceania, and an estimated 20 years for North and South America. The relatively short doubling period and large production for Africa and Asia suggest that the producing centers of oil are rapidly shifting to Asia and Africa.

The footnotes to Tables 1 and 2 provide an interesting summary of the "have" and "have not" countries in terms of crude oil. A large number of the developing nations in Africa have no known sovereign crude oil resources. As these countries progress economically, on their own and also as the result of assistance from the more developed nations of the world, their demand for crude oil hydrocarbons will increase. Notably, the Republic of South Africa took note of the energy situation many years ago and constructed the first commercially practical coal gasification plant in 1955, reflecting planning commenced in the mid-1920s. See "Coal Gasification at Sasol" in the *Coal Technology* section of this Handbook. Even taking into consideration the development of offshore oil fields, it is evident that much of Europe will continue to place high demands on crude oil from other continents. Equally evident is the dependence of Japan on outside sources of crude oil and hydrocarbon products.

Again, with caution, if one compares the production figures of Table 1 for the United States over the period 1960 through 1972 with the proved recoverable reserves figures for the United States of Table 2, it is evident that the production of 6,110 million metric tons of crude oil just about equals the current reserves estimate of 6,116 million metric tons. With a decelerating production rate and an increasing consumption rate in the United States, the present well-established dependence of the United States on imported crude oil is dramatized once again.

TABLE 4. Production and Exploratory Wells Completed (1971)
Millions of metric tons

Continent and country	Production wells	Exploratory wells	Total wells
AFRICA[a]			
Algeria	84	25	109
Angola	26	29	55
Cameroon	0	2	2
Congo (Brazzaville)	14	2	16
Egypt (U.A.R.)	21	22	43
Gabon	33	11	44
Ghana	0	2	2
Kenya	0	2	2
Liberia	0	4	4
Libya	46	41	87
Malagasy Republic	0	8	8
Morocco	1	8	9
Mozambique	0	6	6
Nigeria	172	55	227
Senegal	0	5	5
Republic of South Africa	0	16	16
Togo	0	1	1
Tunisia	3	13	16
Zaire	0	4	4
Total—Africa	400	256	656
ASIA[b,c]			
Afghanistan	na[c]	na[c]	6
Bahrain	7	0	7
Republic of China (Taiwan)	2	10	12
India	37	22	59
Indonesia	152	135	287
Iran	55	10	65
Iraq	17	0	17
Israel	1	6	7
Japan	9	4	13
Kuwait	14	2	16
Malaysis	na[c]	21	21
Oman	44	11	55
Pakistan	na[c]	4	4
Philippines	0	17	17
Qatar	0	5	5
Saudi Arabia	56	7	63
Thailand	na	2	2
Turkey	28	24	52
United Arab Emirates	27	17	44
Total—Asia	449	297	752[d]
OCEANIA[e]			
Australia	31	65	96
New Zealand	0	9	9
Total—Oceania	31	74	105
EUROPE[f,g]			
Austria	28	25	53
Denmark	2	3	5
France	5	7	12
Germany, Fed. Repub. of (West)	35	27	62
Greece	0	1	1
Ireland	0	7	7

Continent and country	Production wells	Exploratory wells	Total wells
Italy	53	39	92
Malta	0	1	1
Netherlands	55	29	84
Norway	1	13	14
Spain	3	11	14
Sweden	0	4	4
United Kingdom	42	32	74
Yugoslavia	11	6	17
Total—Europe	235	205	440
NORTH AMERICA[h]			
Bahamas	0	1	1
Barbados	0	2	2
Canada	1,544	1553	3,097
Honduras	0	1	1
Jamaica	0	1	1
Mexico	387	119	506
Nicaragua	0	3	3
Panama	0	1	1
United States	18,929	6922	25,851
Total—North America	20,860	8,603	29,463
SOUTH AMERICA[i]			
Argentina	459	145	604
Bolivia	14	4	18
Brazil	88	87	175
Chile	61	20	81
Colombia	68	17	85
Ecuador	16	15	31
Guyana	0	1	1
Peru	100	26	126
Trinidad and Tobago	185	35	220
Venezuela	462	44	506
Total—South America	1,453	394	1,847
Total—121 countries	23,428	9,829	33,263[j]

[a]African countries not exhibiting any activity during 1971: Botswana, Ethiopia, Guinea, Ivory Coast, Malawi, Mali, Mauritania, Mauritius, Rwanda, Sierra Leone, Somalia, Sudan, Tanzania, Uganda, Upper Volta, Zambia.

[b]Asian countries not exhibiting any activity during 1971: Cyprus, Republic of South Korea, Lebanon, Nepal, Singapore, Sri Lanka (Ceylon), Republic of South Viet-Nam, Yemen (Sanaa).

[c]Asian countries for which information is not available: Bangladesh, Burma, People's Republic of China, U.S.S.R.) na = not available.

[d]Total column does not equal sum of production and exploratory wells because of lack of information on split for some countries.

[e]Fiji did not report any activity during 1971.

[f]European countries not exhibiting any activity during 1971: Belguim, Finland, Iceland, Portugal, Switzerland.

[g]European countries for which information is not available: Albania, Bulgaria, Czechoslovakia, Hungary, Poland, Romania.

[h]North American countries not exhibiting any activity during 1971: Costa Rica, Dominican Republic, El Salvador, Guatemala, Haiti, Martinique and Guadeloupe.

[i]South American countries not exhibiting any activity during 1971: Uruguay.

[j]Total column does not equal sum of production and exploratory wells because of lack of information on split for some countries.

New production and exploratory activities (1971) are demonstrated by Table 4. Such activities were greater in the United States by a wide margin (over eight times) than they were in the next country (Canada), the number of production and exploratory wells drilled in the United States and its offshore waters being 25,851. It is interesting to note the large number of countries not engaged in any of these activities during the reporting period. It is to be expected that increased drilling activity will be reported when the findings of the World Energy Conference Survey are published. See also Table 5.

TABLE 5.   Producing Wells Worldwide as of December 31, 1972

Country	Onshore	Offshore	Total
Afghanistan	na	na	na
Albania	na	na	na
Algeria	574	0	574
Angola	57	101	158
Argentina	5,660	0	5,660
Australia	na	na	474
Austria	1,461	*	1,461
Bahamas	0	0	0
Bahrain	221	0	221
Bangladesh	6	0	6
Barbados	5	0	5
Belgium	0	0	0
Belize	0	0	0
Bhutan	0	0	0
Bolivia	120	0	120
Botswana	0	0	0
Brazil	748	495	1,243
Brunei	233	198	431
Bulgaria	na	na	na
Burma	712	0	712
Burundi	0	0	0
Cameroon	0	0	0
Canada	17,101	0	17,101
Central African Republic	0	0	0
Chad	na	na	na
Chile	325	0	325
China (People's Republic of)	na	na	na
China (Republic of—Taiwan)	51	0	51
Colombia	2,026	0	2,026
Congo	8	12	20
Costa Rica	0	0	0
Cuba	na	na	na
Cyprus	0	0	0
Czechoslovakia	na	na	na
Dahomey	0	0	0
Denmark	0	5	5
Dominican Republic	0	0	0
Ecuador	773	0	773
Egypt	<166	>54	220
El Salvador	0	0	0
Equatorial Guinea	0	0	0
Ethiopia	0	0	0
Fiji	0	0	0
Finland	0	0	0
France	340	0	340
Gabon	140	31	171
Gambia	0	0	0
German Democratic Republic	na	0	na
Germany (Federal Republic of)	3,318	0	3,318
Ghana	0	0	0
Greece	0	0	0
Guatemala	0	0	0
Guinea	0	0	0
Guyana	0	0	0
Haiti	0	0	0
Honduras	0	0	0
Hungary	na	na	na
Iceland	0	0	0
India	1,190	0	1,190
Indonesia	2,404	23	2,427
Iran	224	89	313
Iraq	na	na	na
Ireland	0	0	0
Israel	27	10	37

Country	Onshore	Offshore	Total
Italy	na	na	na
Ivory Coast	0	0	0
Jamaica	0	0	0
Japan	2,601	133	2,734
Jordan	0	0	0
Kenya	0	0	0
Khmer Republic	0	0	0
Korea (Democratic People's Republic)	0	0	0
Korea, Republic of	0	0	0
Kuwait	na	na	na
Laos	na	*	na
Lebanon	0	0	0
Lesotho	0	0	0
Liberia	0	0	0
Libyan Arab Republic	1,010	0	1,010
Lichtenstein	0	*	0
Luxembourg	0	*	0
Malagasy Republic	0	0	0
Malawi	0	*	0
Malaysia	88	45	133
Maldives	0	0	0
Mali	0	*	0
Malta	0	0	0
Mauritania	0	0	0
Mauritius	0	0	0
Mexico	3,640	50	3,690
Mongolia	1,461	*	1,461
Morocco	37	0	37
Mozambique	0	0	0
Nauru	0	0	0
Nepal	0	*	0
Netherlands	na	0	na
New Zealand	4	0	4
Nicaragua	0	0	0
Niger	0	*	0
Nigeria	425	185	610
Norway	0	4	4
Oman	109	0	109
Pakistan	23	0	23
Panama	0	0	0
Paraguay	0	*	0
Peru	2,139	237	2,376
Philippines	0	0	0
Poland	0	*	0
Portugal	0	0	0
(Portuguese) Guinea	0	0	0
Qatar	na	na	na
Romania	na	na	na
Rwanda	0	*	0
San Marino	0	*	0
Saudi Arabia	395	203	598
Senegal	0	0	0
Sierra Leone	0	0	0
Singapore	0	0	0
Somalia	0	0	0
South Africa, Republic of	0	0	0
Southwest Africa (Namiba)	0	0	0
Spain	na	na	na
Sri Lanka	0	0	0
Sudan	0	0	0
Swaziland	0	*	0
Sweden	0	0	0
Switzerland	0	*	0
Syrian Arab Republic	120	*	120

TABLE 5. Producing Wells Worldwide as of December 31, 1972 *(Continued)*

Country	Onshore	Offshore	Total
Tanzania	0	0	0
Thailand	25	0	25
Togo	0	0	0
Tonga	0	0	0
Trinidad and Tobago	na	na	2,932
Tunisia	49	0	49
Turkey	292	0	292
Uganda	0	*	0
U.S.S.R.	60,000†	na	60,000†
United Arab Emirates	na	na	na
United Kingdom	59	140	199
United States	495,500†	8,000†	503,500†
Upper Volta	0	0	0
Uruguay	0	0	0
Venezuela	11,299	0	11,299
Vietnam	0	0	0
Western Samoa	0	0	0
Yemen Arab Republic	0	0	0
Yemen (People's Republic of)	0	0	0
Yugoslavia	na	na	na
Zaire	0	0	0
Zambia	0	0	0

na = information not available.
*Landlocked.
†Estimate.

In the aforementioned U.S. Geological Survey Professional Paper 885, the potential resources (ultimate recoverable resources) are placed into numbered categories. These are defined in Table 6, together with an evaluation of these resources for 151 countries investigated.

By early-1973, some 780 oil and gas fields had been discovered on the submerged continental margins of the world. Of these, 493 were cataloged as oil fields. It is interesting to note that petroleum was produced offshore as early as 1896 by way of seaward extensions of the Summerland field, California. Drilling was accomplished from wooden piers built outward from shore. In 1923, the Venezuelan Gulf Oil Company acquired concessions extending 1 kilometer off the shores of Lake Maracaibo. In 1926, the initial discovery was made in the Lagunillas field in the shallow waters of Lake Maracaibo. Lagunillas proved to be a giant field, and its discovery indicated that additional oil and gas deposits lay farther out in deeper water. Methods developed for offshore operation at Lagunillas became world famous and helped to pave the way toward future offshore production. Discovery of oil on Bahrain Island in 1932 brought the realization that substantial amounts of petroleum might lie beneath the waters of the Persian Gulf. The occurrence of large and productive salt domes along the coast of the Gulf of Mexico in Texas and Louisiana suggested that the oil-bearing province extended offshore. In 1938, oil was discovered off Louisiana in 8 meters of water a mile (1.6 kilometers) from shore. By 1946, nine wells were drilled in the Gulf of Mexico, five off Louisiana, and four off Texas. All were in shallow water close to shore and were drilled from rigid platforms.

Developments in offshore petroleum exploration essentially trace back to the Gulf of Mexico and in particular to work done there during the summer and fall of 1947. The first subsea well completed from a mobile platform was in the Gulf of Mexico. Drilling thereafter was no longer limited to the water-depth restrictions of the rigid platform that had to be built in place. Using two World War II surplus Navy barges and a surplus Navy LCT converted to form a drilling platform, Kerr-McGee Oil Industries, Inc., discovered oil in Ship Shoal Block 32 on October 4, 1947, 12 miles (19.2 kilometers) from the Louisiana shore. The ability to drill successfully from a mobile platform was demonstrated. The era of large-scale exploration and production from the offshore was launched.

During a period after 1947, controversies arose between the maritime states and the federal government over jurisdiction of the seabed of the continental shelf. The Outer

Continental Shelf Lands Act was signed by President Eisenhower on August 7, 1953. This action provided for the immediate leasing of Federal offshore areas by the Department of the Interior and the validation of leases previously issued by the states. In decisions in 1954, the U.S. Supreme Court settled the jurisdictional boundary disputes by giving proprietorship of the inner 3 miles (5.8 kilometers) of the shelf to the maritime states. In separate decisions, Texas was given jurisdiction 3 leagues seaward (10.3 miles; 16.5 kilometers) and Florida was likewise granted a 3-league jurisdiction over that part of its offshore in the Gulf of Mexico. Federal jurisdiction was established for that part of the shelf seaward from the state boundaries and the first federal lease sale was held in 1954.

As of the mid-1970s, offshore exploration involved some 80 different countries. Several factors have influenced the rates of discovery and development of petroleum offshore: (1) technical capability, including exploration techniques and engineering methods, (2) onshore oil production adjacent to the offshore area, (3) local physical environmental conditions which include sea-floor topography, (4) political and legal circumstances, including boundaries of proprietorship, regulations and attitudes for granting licenses, concessions, and leases, attitudes toward environmental effects, and attitudes about the petroleum company ownership, and (5) capital available for investment.

The search for offshore petroleum began in areas having established production along the adjacent shoreline, where geologic conditions governing the petroleum occurrences were known, and those regions now have the largest number of offshore fields. Some authorities estimate that nearly one-third of all producing basins involve some offshore production, and 60 percent represent extensions of established onshore production. The offshore giants may be categorized on the basis of available data as follows:

1. 22 percent are directed offshore extensions of fields that were first discovered onshore.

2. 56 percent have geologic conditions that relate closely to the producing areas onshore.

3. 22 percent are in geologic settings either not related closely to a petroleum-producing area onshore, or not predictable on the basis of onshore geology.

Of the 70 major finds in the offshore fields as of the mid-1970s, some 78 percent are in geologic settings that are related closely to known onshore conditions. All of the offshore extensions were discovered before 1965, whereas those fields not related closely geologically to the onshore conditions have been found since 1964.

In the Persian Gulf area, reserves of petroleum on land are so large that there has not been great urgency to proceed hastily in developing the offshore resources. In Nigeria, where a large petroleum potential was not identified until the 1960s, exploration and development of producing capacity have proceeded rapidly both on land and offshore. Drilling activity has been high. The impetus for rapid demand has come from the market demand for Nigeria's low-sulfur oil and a desire on the part of the Nigerian government to expand production rapidly. Generally, fields discovered offshore have been brought to production in periods of 2 to 4 years. As would be expected, the discovery rate on the shelf off Nigeria has fallen rapidly in the past several years—from 85 percent in 1967 to 14 percent in 1971. Exploitation of the Nigerian Continental Shelf has been rapid because physical constraints have been minimal, production to date coming from fields in relatively shallow water, and because of the large demand for low-sulfur oil.

Recognition of the North Sea as a new source of petroleum came with the onshore discoveries of gas at Groningen (Netherlands) in the late 1950s. Although geophysical exploration of the offshore followed shortly thereafter, drilling was delayed until 1964 when jurisdictions of the various countries bordering the North Sea were established and boundaries of proprietorship drawn. The time expended in development of the North Sea as a petroleum province has been influenced mainly by two factors: (1) geological interpretations that guided initial selection of sites for detailed investigation and drilling, and (2) the rigorous physical conditions at sea. In the early stages of exploration, effort was directed to the southern periphery of the basin nearest areas of known geologic conditions. In the late 1960s, however, exploration was directed to the central part of the basin where the really large oil discoveries have been made. In dealing with the North Sea, rigs and platforms must be built to withstand possible wave heights of 100 feet (30 meters). Consequently, platforms must be built that can withstand an overturning force exerted by waves and currents that is 10 times greater than that exerted on platforms in the Gulf of Mexico. Exploration and development drilling under rigorous North Sea conditions thus

TABLE 6.   Worldwide Ultimate Recoverable (Potential) Crude Oil Resources

Continent and country	Onshore	Offshore	Total
AFRICA			
Algeria	II	0	II
Angola	IV	IV	IV
Botswana	VI	*	VI
Burundi	0	*	0
Cameroon	V	V	IV
Central African Republic	V	*	V
Chad	IV	*	V
Congo	IV	V	IV
Dahomey	V	V	V
Egypt	IV	III	III
Equatorial Guinea	V	V	V
Ethiopia	V	V	V
Gabon	IV	IV	III
Gambia	V	V	V
Ghana	V	IV	IV
Guinea	V	V	V
Ivory Coast	V	IV	IV
Kenya	IV	V	IV
Lesotho	VI	*	VI
Liberia	0	V	V
Libyan Arab Republic	II	II	II
Malagasy Republic	IV	IV	IV
Malawi	0	*	0
Mali	IV	*	IV
Mauritania	IV	IV	IV
Mauritius	0	0	0
Morocco	IV	IV	IV
Mozambique	IV	*	IV
Niger	IV	*	IV
Nigeria	III	III	III
(Portuguese) Guinea	V	V	V
Rwandi	0	*	0
Senegal	IV	IV	IV
Sierra Leone	V	V	V
Somalia	IV	V	IV
South Africa (Republic of)	VI	V	V
Southwest Africa	V	V	V
Sudan	V	V	V
Swaziland	VI	*	VI
Tanzania	IV	V	IV
Togo	V	V	V
Tunisia	IV	IV	IV
Uganda	0	*	0
Upper Volta	0	*	0
Zaire	V	V	V
Zambia	0	0	0
ASIA			
Afghanistan	IV	*	IV
Bahrain	IV	IV	IV
Bangladesh	IV	IV	IV
Bhutan	0	*	0
Brunei	V	IV	IV
Burma	III	IV	III
China (People's Republic of)	III	IV	III
China (Republic of—Taiwan)	V	IV	IV
Cyprus	V	IV	IV
India	III	III	III
Indonesia	III	III	III
Iran	II	III	II
Iraq	III	IV	III
Israel	IV	IV	IV

Continent and country	Onshore	Offshore	Total
Japan	IV	IV	IV
Jordan	IV	*	IV
Khmer Republic	0	IV	IV
Korea (Democratic People's Republic)	0	IV	IV
Korea (Republic of)	0	III	III
Kuwait	II	III	II
Laos	IV	*	IV
Lebanon	V	IV	IV
Malaysia	III	III	III
Maldives	VI	VI	VI
Mongolia	IV	*	IV
Nepal	V	*	V
Oman	III	IV	III
Pakistan	III	IV	III
Philippines	IV	IV	IV
Qatar	III	III	III
Saudi Arabia	II	II	II
Singapore	0	V	V
Sri Lanka	0	V	V
Syrian Arab Republic	IV	*	IV
Thailand	IV	IV	IV
Turkey	IV	IV	IV
United Arab Emirates	III	III	III
U.S.S.R.	II	II	II
Vietnam	V	V	V
Yemen Arab Republic	VI	IV	VI
Yemen People's Republic	V	IV	IV
OCEANIA			
Australia	III	III	III
Fiji	V	V	V
Nauru	IV	IV	IV
New Zealand	V	V	V
Tonga	IV	IV	IV
Western Samoa	IV	IV	IV
EUROPE			
Albania	III	IV	III
Austria	IV	*	IV
Belgium	V	IV	IV
Bulgaria	IV	IV	IV
Czechoslovakia	IV	*	IV
Denmark	IV	IV	IV
Finland	0	V	V
France	IV	IV	IV
German Democratic Republic	IV	V	IV
Germany (Federal Republic of)	IV	IV	IV
Greece	V	IV	IV
Hungary	I	*	IV
Iceland	0	0	0
Ireland	V	V	V
Italy	IV	IV	IV
Lichtenstein	0	*	0
Luxembourg	0	*	0
Malta	IV	IV	IV
Netherlands	IV	IV	III
Norway	0	III	III
Poland	IV	V	IV
Portugal	IV	V	V
Romania	IV	IV	IV
San Marino	0	*	0
Spain	IV	IV	IV
Sweden	VI	V	V

TABLE 6.   Worldwide Ultimate Recoverable (Potential) Crude Oil Resources   (Continued)

Continent and country	Onshore	Offshore	Total
Switzerland	V	*	V
United Kingdom	V	III	III
Yugoslavia	IV	V	IV
NORTH AMERICA			
Bahamas	VI	VI	VI
Barbados	V	V	IV
Belze	V	V	V
Canada	III	III	III
Costa Rica	VI	V	V
Cuba	V	V	V
Dominican Republic	V	V	V
El Salvador	0	V	V
Guatemala	IV	V	IV
Haiti	V	V	V
Honduras	V	IV	IV
Jamaica	V	IV	IV
Mexico	III	II	II
Nicaragua	V	IV	IV
Panama	V	IV	IV
United States	II	II	II
SOUTH AMERICA			
Argentina	III	III	III
Bolivia	III	*	III
Brazil	IV	IV	IV
Chile	IV	IV	IV
Colombia	III	IV	III
Ecuador	III	IV	III
Guyana	IV	V	IV
Paraguay	IV	*	IV
Peru	III	IV	III
Trinidad and Tobago	IV	IV	IV
Uruguay	V	V	IV
Venezuela	III	III	II

*Landlocked.

Categories of potential resources

Symbol	Billions of barrels	
I	1,000	–10,000
II	100	– 1,000
III	10	– 100
IV	1	– 10
V	0.1 –	1
VI	0.01–	0.1

were time-consuming and costly. About 10 years were required to yield a first-producing well.

The United States Continental Shelf in the Gulf of Mexico has had the greatest concentration of exploration and drilling and includes by far the greatest number of producing fields of any offshore area in the world. Consensus from a variety of sources indicates an average time of between 2 and 5 years to bring a field to significant production after initial discovery. Other offshore petroleum sources, in various stages of study or development, include the Atlantic continental margin, southern California, Alaskan waters, including the Gulf of Alaska, the Bering Sea, the Chukchi Sea, and the Beaufort Sea shelf off northern Alaska, and Prudhoe Bay.

Excellent background information on offshore petroleum technology can be found in "The Worldwide Search for Petroleum Offshore," Geological Survey Circular 694, by Henry L. Berryhill, Jr., some of the highlights of which have been summarized here.

In summary, crude oil reserves and production figures can be very valuable to long-range planning provided that economic and political factors are woven into the analysis.

The latter factors, of course, are among the most difficult to forecast. Consequently, a trend has developed among energy resources statisticians to present several forecasts rather than a single forecast. Reasonable, but nevertheless hypothetical, cases (assumptions) are made pertaining to economic and political factors, usually two of these cases of an extreme nature. In the one extreme case, for example, are all economic and political factors that would motivate greater demand (lack of conservation, etc.) and lesser production (using the underside of reserves figures, lack of incentive for exploration and drilling, etc.). In the other extreme case, assumptions would be made of factors that would minimize demand and accentuate production. In a sense, this enables the statistician to "tool up" for making current forecasts that reflect present and near-future economic and political forecasts.

Related Statistics.   Other statistics in this Handbook which relate to petroleum and associated products include:

In the Handbook section on *Gas Technology,* see:
Natural Gas Production and Reserves in the United States
Natural Gas Reserves in Canada
Potential Supply of Natural Gas in the United States
World Natural Gas Production and Reserves
Future Gas Consumption in the United States
In the Handbook section on *Petroleum Technology,* see:
Petroleum Production and Reserves in the United States
Petroleum Reserves in Canada

REFERENCES

"Summary of 1972 Oil and Gas Statistics for Onshore and Offshore Areas of 151 Countries," *Geol. Surv. Prof.* Paper 885, 163 pp., Stock No. 2401-02518, U.S. Government Printing Office, Washington, 1974.
American Association of Petroleum Geologists: Foreign Developments Issues, *Amer. Assoc. Petrol. Geol. Bull.,* vols., 49–56, 1965–1972.
American Petroleum Institute: World Crude Oil Production by Countries, 1857–1967, in "Petroleum Facts and Figures," pp. 548–557, API, Washington, 1971.
Chase Manhattan Bank: "Capital Investments of the World Petroleum Industry, 1971," New York, 1973.
Hendricks, T. A.: "Resources of Oil, Gas, and Natural Gas Liquids in the United States and the World," *U.S. Geol. Surv. Circ.* 522, 1965.
International Monetary Fund, Exchange Rates, *Intern. Finan. Statist.,* vol. 25, no. 9, p. 27, 1972.
McCaslin, John: Worldwide Offshore Oil Output Nears 9 Million b/d, *Oil Gas J.,* vol. 70, no. 18, pp. 196–198, 1972.
Worldwide Oil at a Glance, *Oil Gas J.,* vol. 69, no. 52, pp. 72–73, 1971.
Worldwide Production, *Oil Gas J.,* vol. 69, no. 52, pp. 86, 91–106, 1971.
Statistics—Worldwide Crude Production: Worldwide Gas Production, *Oil Gas J.,* vol. 70, no. 9, p. 103, 1972.
"World Energy Conference Survey of World Energy Resources (as of 1974)," National Committee, World Energy Conference, Washington (in preparation).

Petroleum Requirements

In 1970, petroleum products accounted for over 40 percent of the total energy utilized in the United States. Since World War II petroleum products demand has grown at an average annual rate of 4.7 percent (through 1970). In 1946 the total oil demand in the United States was less than 5 million barrels per day. By 1970 the demand had grown to 14.7 million barrels per day. It is the intent here simply to review briefly a few of the highlights which have precipitated increased demands. Extensive statistical and economic analyses are available from the sources listed in the references at the end of this description.

The demand by major categories of petroleum products is summarized in Table 1 and charted in Fig. 1. Some of the principal factors that have contributed to demand include: (1) increased population (from 142 million people in 1946 to about 205 million people in 1970, a 44 percent increase, equivalent to 1.5 percent per annum); (2) industrial expansion (from a gross national product in 1946 of $313 billion to $722 billion in 1970, a growth rate of 3.7 percent, calculated on the basis of constant 1958 dollars; (3) increased per capita income from $1500 in 1950 to $3935 in 1970 (actual dollars), which, coupled with revised credit policies, has encouraged people to improve their living standards (more automobiles, more vacations, more appliances—nearly all factors contributing to an increased demand for petroleum products). The United States per capita use of oil products is charted in Fig. 2. The foregoing factors also are summarized in Tables 2 and 3. Projecting future petroleum requirements for the United States is complex and at times tenuous. Some of the fundamental factors that have been used in making projections are summarized in Table 4.

Some sources project a 2.7 percent per year growth rate for petroleum products from 1976 to 1980, following the greater growth rate of close to 6 percent during the 1971–1975 period. These sources further project a growth rate of about 3 percent for the 1981–1985 period. Total demand is projected at about 22.6 million barrels per day required in 1980 and 26.2 million barrels per day required in 1985. As in prior decades, gasoline is expected to be the principal petroleum product, with consumption reaching about 8.5 million barrels per day by 1980, representing a growth rate of about 3.4 percent. Factors

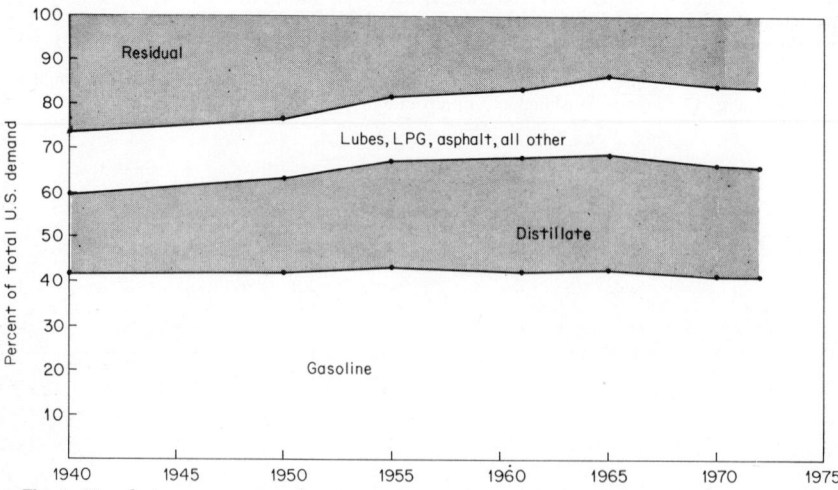

Fig. 1. Trends in consumption of major categories of petroleum products in the United States.

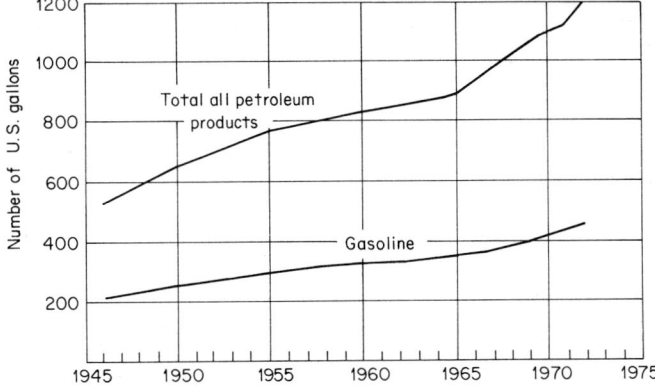

Fig. 2. Per capita use of oil products in the United States.

such as increased earnings, longer vacations, earlier retirements, and greater longevity are expected to fully offset the lower projections for population growth, thus with no relief in terms of a lesser rate for petroleum products consumption. Further, the anticipated growth in domestic demand takes into consideration the lower efficiencies of motive power brought about by modifications resulting from air-pollution regulations.

The growth rate for commercial turbine fuel is projected, conservatively, at 5.2 percent per year because of increased air travel in larger and more comfortable aircraft. See Fig. 3. The projections include the assumption that the travel industry will continue to aggressively promote both domestic and international travel. Projections of distillate fuel consumption are even more difficult to make because these interlock with the supplies and costs of natural gas. If no crisis in the natural gas situation occurs, distillate fuel consumption is not expected to grow appreciably.

Diesel fuel consumption will continue to expand, with the growth rate of highway Diesel consumption approaching 10 percent per year. If there are slowdowns in the completion of nuclear or fossil-fueled electric generation facilities or dislocations in the conventional (coal or gas) fuel supply to existing power generators, this would result in throwing a greater load on turbine-fuel-generated power for peak shaving, thus increasing demand for furnace oil/Diesel fuel type fuels.

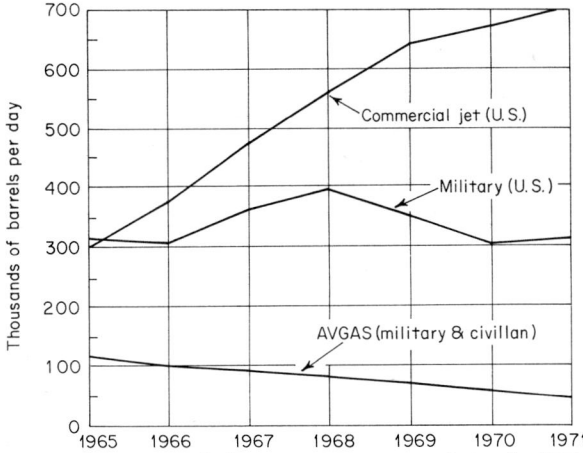

Fig. 3. Important trends in aviation fuels consumption were set during the 1965–1971 period.

TABLE 1. United States Domestic Petroleum Products Demand by Uses
Thousands of barrels per day

Product	1946	1950	1955	1960	1965	1970	1971	1972	Average annual increase, %	
									1946/1970	1960/1970
Gasoline										
Automotive	1,920	2,534	3,413	3,845	4,586	5,785	6,015	6,377	4.7	4.7
Aviation	35	108	132	161	120	55	49	45	1.9	(14.6)
Special naphthas*	60	82	110	124	147	243	238	246	4.6	3.9
Total gasoline	2,015	2,724	3,655	4,130	4,853	6,083	6,302	6,668	4.7	4.4
Jet Fuels										
Naph-type	....	....	154	280	271	249	259	242	....	(1.3)
Kero-type	....	....	....	91	333	718	741	803	....	16.6
Total jet fuels	....	....	154	371	604	967	1,010	1,045	....	9.9
Intermediates										
Kerosine (ex Jet)	244	323	320	271	267	263	249	235	0.3	(0.3)
Distillate										
Heating	406	646	978	1,195	1,304	1,440	1,435	} na†	5.4	2.0
Diesel	148	287	464	563	340†	534	586		....	2.0
Other distillate	111	149	150	114	482	538	640		....	2.3
Total distillate	665	1,082	1,592	1,872	2,126	2,512	2,661	2,913	5.7	3.4
Total intermediates	909	1,405	1,912	2,143	2,393	2,775	2,910	3,148	4.8	3.0

Residual Oils									
Heating	136	199	236	342	428	511	500	nat	5.7	3.6
Other residual	1,179	1,318	1,290	1,186	1,179	1,724	1,796	nat	1.6	7.9
Total residual oils	1,315	1,517	1,526	1,528	1,607	2,235	2,296	2,529	2.2	6.8
Lubricants	96	106	117	117	129	136	136	144	1.5	1.1
Asphalt and road oil	135	179	254	302	368	446	457	468	5.1	5.4
LPG									
Chemical use	20	41	97	196	343	556	575	nat	14.9	10.1
Residential & commercial use	49	132	183	275	312	494	500	nat	10.1	9.6
Other LPG	40	61	124	150	186	174	176	nat	6.3	(1.3)
Total LPG	109	234	404	621	841	1,224	1,251	1,420	10.6	7.7
All Other	333	342	438	449	508	843	863	957	4.3	12.5
Total Domestic Demand—All Petroleum Products	4,912	6,507	8,460	9,661	11,303	14,709	15,225	16,379	4.7	5.4
Net Imports	(42)	545	881	1,604	2,282	3,160	3,702	4,518		
Net Imports as Percent of Total Domestic Demand	....	4.4	10.4	16.6	20.2	21.5	24.3	27.6		

SOURCE: *U.S. Bur. Mines, Mineral Ind. Surv.*: Monthly and Annual Petroleum Statements; Annual Sales of Fuel Oil & Kerosine; Annual Sales of Liquefied Petroleum Gases.

*1965 and 1970 include naphtha 400° for petrochemical use. Earlier years estimated at 3% of total gasoline demand.

†Diesel reporting was changed in 1968 to show only on and off highway use, excluding railroad, vessel, etc. Figures for 1965 and 1970 shown on that basis.

‡na = not available.

TABLE 2. Factors Affecting Demand for Petroleum Products in the United States

Factor	1946	1950	1955	1960	1965	1970	1971	1972
Total population (thousands)	141,936	152,271	165,931	180,864	194,572	204,835	207,049	208,837
Per capita income (1973 dollars)	1,249	1,496	1,876	2,215	2,746	3,935	4,160	4,481
(constant 1958 dollars)	......	1,810	2,027	2,157	2,507	3,043	2,940*	3,070*
Gross national product								
(billion 1973 dollars)	211	285	398	504	676	976	1,050	1,152
(billion constant 1958 dollars)	313	355	438	488	618	722	742	790
Federal Reserve Board Index								
(1957–1959 = 100)	60	75	97	109	147	176	176	188
(1967 = 100)	......	45	59	66	89	107	107	114
Motor vehicle registrations								
(millions)	34.2	49.2	62.6	73.7	90.1	108.6	112.9	117.6†
(passenger cars)	28.2	40.3	52.0	61.5	75.1	89.8	92.8	96.4†
(trucks and buses)	6.2	8.9	10.6	21.2	15.0	18.8	20.1	21.2†

*Calculated, using GNP price deflator.
†Approximation.

TABLE 3. Gasoline Demand for Passenger Cars in the United States

Factors affecting demand	1946	1955	1965	1970	1975*
Annual passenger car gasoline					
Barrels per person	4.9	7.4	8.6	10.3	12.6
Barrels per household	19.2	27.4	29.3	33.6	39.6
Barrels per passenger car	16.7	16.2	15.3	15.8	17.4
Passenger cars (millions)	28.2	52.0	75.1	89.8	106.3
Persons per car	5.1	3.2	2.6	2.3	2.0
Cars per household	0.8	1.1	1.3	1.4	1.7
Gasoline demand (millions of barrels per day per billion $ GNP)†	9.6	9.2	7.2	6.2	5.2
Population (millions)	141.9	165.9	194.6	204.8	214.8

*Estimated.
†GNP = gross national product.

TABLE 4. Economic Assumptions Sometimes Used for Projecting the Demand for Petroleum Products

Factor	Level		Per annum % gain, 1980/1970
	1970	1980	
Population (millions)	204.8	226.7*	1.1
Households (millions)	62.9	74.7	1.7
Gross national product (billions of constant 1958 dollars)	722	1,090	4.2
Federal Reserve Board Index of industrial production (1957–1959 = 100)	176	273	4.5

*Census Bureau, Series D projection, U.S. Department of Commerce.

Because of sulfur pollution regulations, tighter natural gas availability, and a gain of more than 7 percent per annum for electricity, residual fuel oil requirements have risen sharply over several years, with an increase of 71 percent predicted by 1980. This is translated into a residual fuel demand of 4 million barrels per day.

The demand for liquefied petroleum gas (LPG) is expected to rise 5 percent per annum. Petrochemical feedstock requirements are expected to show a gain of 10 percent per year.

REFERENCES

Anon.: "Factors Affecting U.S. Petroleum Refining—Impact of New Technology," National Petroleum Council, Washington, September 1973. NOTE: Numerous other publications of an economic, statistical, and technical nature are available from the National Petroleum Council, not all listed here because such materials are constantly being updated.

Bureau of Mines, U.S. Department of the Interior, Washington: Numerous publications, including the constantly updated series of *Mineral Industry Surveys*. Latest publication list is available upon request.

Bureau of the Census, U.S. Department of Commerce, Washington: Summaries available on an annual basis for the sale of numerous petroleum products.

American Petroleum Institute, Division of Statistics, Washington: Several annual, monthly, and weekly reports are available, including: "Annual Statistical Review," an annual report containing operating statistics for the United States petroleum industry by years, and months for the latest two years for which data are available. Usually issued in April of each year. "Joint Association Survey (sections I and II)," published jointly with the Independent Petroleum Association of America, and the Mid-Continent Oil & Gas Association. Section I includes the estimated cost of drilling oil wells, gas wells, and dry holes by depth range for major areas in the United States. Section II includes estimated expenditures for oil and gas exploration, development, and production in the United States.

"Liquefied Petroleum Gas Report," a monthly publication which includes inventories of liquefied petroleum and liquefied refinery gases located at plants and refineries, and in underground storage, by areas, and by individual products.

"Monthly Report on Drilling Activity in the United States," which shows by states the number of oil wells, gas wells, dry holes, stratigraphic and core tests, and service wells, reported as completed to the American Petroleum Institute; and a table showing exploratory wells (oil, gas, and dry) by states. Also includes a summary of oil wells, gas wells, and dry holes in the development and exploratory categories.

"Quarterly Review of Drilling Statistics," includes quarterly and annual tabulations of United States drilling statistics compiled by the American Petroleum Institute. Wells and footage drilled are reported by type of well for each state and major areas of states. Special tabulations are also shown for multiple completion wells.

"Reserves of Crude Oil, Natural Gas Liquids, and Natural Gas in the United States and Canada and the United States Productive Capacity," published annually and jointly with the American Gas Association and the Canadian Petroleum Association.

"Weekly Statistical Bulletin," containing United States data relating to refinery activity and principal inventories, crude oil and product imports, crude oil production, and gasoline consumption by states.

Technical Report No. 1: Standard Definitions for Petroleum Statistics, 1st ed. (no date given).

Technical Report No. 2: Organization and Definitions for the Estimation of Reserves and Productive Capacity, 1st ed. June 1970.

Drilling and Production

There are general theories for explaining the origin of petroleum and natural gas. The most generally accepted explanation is the *organic theory*, which can be quickly summarized. Over millions of years, rivers flowed to the seas, carrying large volumes of mud and sand to be spread out by currents and tides over the sea bottoms near the gradually changing shorelines. New deposits were distributed, layer upon layer, over the floors of the seas. Because of the increasing weight of these accumulations, the sea floors slowly sank, building up a thick series of mud and sand layers. High pressure and chemical forces ultimately converted these layers into sedimentary rocks of the type that often contain petroleum—the sandstones, shales, limestones, and dolomites. The organic theory further stipulates most importantly that tiny marine life organisms were buried with the silt. In an airless environment and under high pressures and elevated temperatures, these carbon- and hydrogen-containing miniscule life-forms were converted over an extremely long time span into hydrocarbons. This theory, of course, requires acceptance of the concept of drastically altered shorelines because obviously oil deposits are found in many parts of the world hundreds of miles from the present coastlines.

Geologists find it particularly difficult to trace the history of a given hydrocarbon deposit because the oil and gas may have been moved as the result of numerous seismic events, again occurring over an extremely long time span. A past requisite for commercially exploitable hydrocarbon deposits has been prior movement and concentration of large quantities in various forms of traps. In contrast with oil shales and tar sands, natural gas and petroleum flow relatively easily in permeable underground structures and, consequently, tend to concentrate, greatly assisting the economic exploitation of these materials.

The movement of petroleum from the place of its origin to the traps where accumulations are found is believed to have occurred in an upward direction. This movement took place as the result of the tendency for oil and gas to rise through the ancient seawater with which the pore spaces of the sedimentary formations were filled when originally laid down. An underground porous formation or series of rocks which occur in some shape favorable to the trapping of oil and gas must also be covered or adjoined by a layer of rock that provides a covering or seal for the trap. A seal of this type, frequently called a *cap rock*, stops further upward movement of petroleum through the pore spaces.

As oil and gas gathered in the upper part of a trap, because of differences in weight of gas, oil, and salt water, these fluids also separated vertically, much in the same manner as though these materials were all present in a bottle.* Thus gas, if any is present, is found in the highest part of the trap, followed by oil, and oil with gas below the gas, and finally salt water below the oil. Experience has indicated that the salt water seldom was completely displaced by oil or gas from the pore spaces, even within the trap. Even in the midst of oil and gas accumulation, pore spaces within the trap may contain from 10 to 50 percent or more of salt water. It appears that the remaining water (termed *connate water*) fills the smaller pores and also exists as a coating or film, covering the rock surfaces of the larger pore spaces, and thus the oil and/or gas are apparently contained in water-jacketed pore spaces. The geological structures called traps are petroleum reservoirs; i.e., they are the oil and gas fields that are explored and produced. All oil fields contain some gas, but the quantity may range widely.

Primary requirements for the formation of an oil or a gas field thus may be summarized as: (1) a source of carbon and hydrogen, (2) conditions of pressure and temperature and chemical environment such that the carbon and hydrogen recombined to form the various hydrocarbons that make up petroleum, (3) a porous rock or series of such rocks within which the petroleum was able to migrate upward through the seawater with which the

*Some of the fundamentals described here were derived with permission from "Primer of Oil and Gas Production," American Petroleum Institute, Washington, 1973.

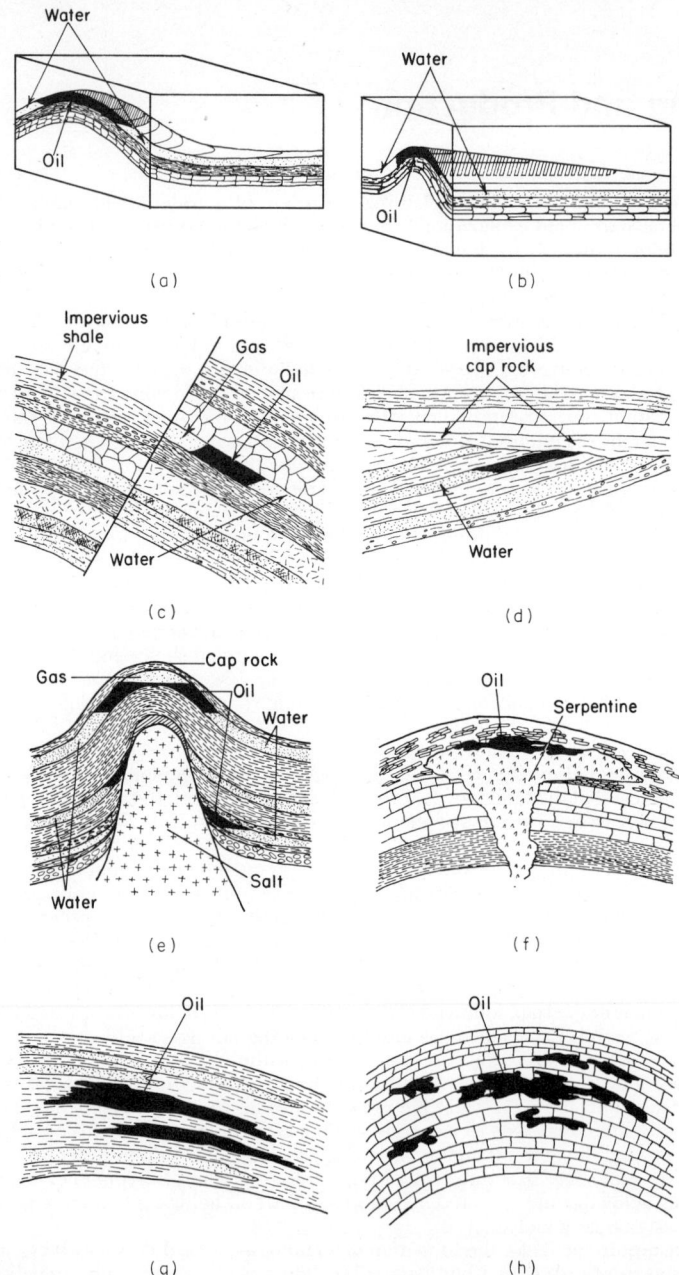

Fig. 1. Geological types of oil and gas reservoirs. (Source of sketches: Exxon.) (*a*) Accumulation of oil in a dome-shaped structure. The dome is circular in outline. (*b*) Anticlinal type of folded structure. This differs from a dome in being long and narrow. Reservoirs formed by folding of the rock layers or strata usually have the shapes indicated in (*a*) and (*b*). These traps were filled by upward migration or movement of oil or gas (or both) through the porous strata or beds to the location of the trap. Further movement was arrested by a combination of the form of the structure and the seal or cap

source material and rocks were originally deposited, and (4) a local structure or trap, having a cap rock seal, that forms a reservoir where petroleum has gathered.

Geology of Reservoirs.   There is a broad range of shapes, sizes, and types of traps, of which six principal types are illustrated and described in Fig. 1.

Production Processes.   It is convenient to classify oil and gas reservoirs in terms of the type of natural energy and forces available to produce the oil and gas. At the time oil was forming and accumulating in reservoirs, pressure and energy in the gas and salt water associated with the oil were also being stored, which would later be available to assist in producing the oil and gas from the underground reservoir to the surface. Oil cannot move and lift itself from reservoirs through wells to the surface. It is largely the energy in the gas or the salt water (or both) occurring under high pressures with the oil that furnishes the force to drive or displace the oil through and from the pores of the reservoir into the wells.

In nearly all cases, oil in an underground reservoir has dissolved in it varying quantities of gas that emerges and expands as the pressure in the reservoir is reduced. As the gas escapes from the oil and expands, it drives oil through the reservoir toward the wells and assists in lifting it to the surface. Reservoirs in which the oil is produced by dissolved gas escaping and expanding from within the oil are called *dissolved-gas-drive reservoirs.* See Fig. 2.

Often more gas exists with the oil in a reservoir than the oil can hold dissolved in it, under the existing conditions of pressure and temperature in the reservoir. This extra gas, being lighter than the oil, occurs in the form of a cap of gas over the oil. This condition is illustrated by Fig. 1c, Fig. 1e, and Fig. 3. Such a gas cap is an important additional source of energy because, as production of oil and gas proceeds and as the reservoir pressure is lowered, the gas cap expands to help fill the pore spaces formerly occupied by the oil and gas produced. Where conditions are favorable, some of the gas coming out of the oil is conserved by moving upward into the gas cap to further enlarge the gas cap. As compared with the dissolved-gas drive, the *gas-cap drive* is more effective, yielding indicated oil recoveries ranging from 25 to 50 percent.

The gas-drive processes just described are typically found with the discontinuous, limited, or essentially closed reservoirs of the types shown in Fig. 1f, Fig. 1g, and Fig. 3.

Where the formation containing an oil reservoir is quite uniformly porous and continuous over a large area, as compared with the size of the oil reservoir per se, very large quantities of salt water exist in surrounding parts of the same formation, often directly in contact with the oil and gas reservoir. This condition is demonstrated by Fig. 1a through e. These large quantities of salt water occur under pressure and provide a large additional store of energy to assist in producing oil and gas. A situation like this is termed a *water-drive reservoir* and is shown in Fig. 4.

The energy supplied by the salt water comes from expansion of the water as pressure in

---

rock provided by the formation covering the structure. Examples of domal structures are the Conroe Oil Field in Montgomery County, Texas, and the Old Ocean Gas Field in Brazoria County, Texas. An example of a reservoir formed by anticlinal structure is the Ventura Oil Field in California. (c) A fault trap. Reservoirs are formed by breaking or shearing and offsetting of strata (faulting). The escape of oil from such a trap is prevented by nonporous rocks that have moved into a position opposite the porous petroleum-bearing formation. The oil is confined in traps like this because of the tilt of the rock layers and faulting. Examples of fields of this type exist along the Mexia fault zone of East-Central Texas. (d) Unconformity. Upward movement of oil has been halted by the impermeable cap rock laid down across the cutoff (possibly by water or wind erosion) surfaces of the lower beds. An example of this type of reservoir is the great East Texas field. (e) Salt dome. These often deform overlying rocks to form traps like this. A nonporous salt mass has formed dome-shaped trap in overlying and surrounding porous rocks. An example of a salt-dome field is the Sugarland Oil Field in Fort Bend County, Texas. (f) Serpentine plug. These sometimes form reservoirs similar to this one. A porous serpentine plug has formed a reservoir within itself by intruding into nonporous surrounding formations. An example of a serpentine plug field is the Hilbig Field in Bastrop County, Texas. (g) Lens-type trap formed in sand. An example is the Burbank Field in Osage County, Oklahoma. (h) Lens-type trap formed in limestone. Examples of limestone reservoirs of this type are found in the limestone fields of west Texas. In the lens-type trap, the reservoir is sealed in its upper regions by abrupt changes in the amount of connected pore space within a formation. This may be caused in sandstones by irregular depositing of sand and shale at the time the formation was laid down. In these cases oil is confined within porous parts of the rock by the nonporous parts of the rock surrounding it. In limestone formations they often have areas of high porosity with a tendency to form traps.

Fig. 2. Dissolved-gas drive. *(Texas Mid-Continent Oil and Gas Association.)*

Fig. 3. Gas-cap drive. *(Texas Mid-Continent Oil and Gas Association.)*

the petroleum reservoir is reduced by production of oil and gas. Water will compress, or expand, to the extent of about one part in 2500 per 100 psi change in pressure. Although this effect is slight with reference to small quantities, the phenomenon becomes of importance when changes in reservoir pressure affect large volumes of salt water that are often contained in the same porous formation adjoining or surrounding a petroleum reservoir.

The expanding water moves into the regions of lowered pressure in the oil- and gas-saturated portions of the reservoir caused by production of oil and gas, and retards the decline in pressure. In this way, the expansive energy in the oil and gas is conserved. As shown by Fig. 4, the expanding water also moves and displaces oil and gas in an upward direction out of lower parts of the reservoir. By this natural process, the pore spaces vacated by oil and gas produced are filled with water, and oil and gas are progressively moved toward the wells.

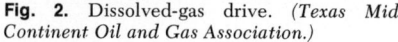

The water drive is generally the most efficient oil-production process. Oil fields in which water drive is effective are capable of yielding recoveries ranging up to 50 percent of the oil originally in place, if (1) the physical nature of the reservoir rock and of the oil are conducive to the process, (2) care is exercised in completing and producing the wells, and (3) depending on the rate of oil and gas production from the field or reservoir as a whole. These factors also affect the oil-recovery efficiency in gas-cap-drive reservoirs. However, rate of production seems to exert only minor effect on oil recoveries obtainable from dissolved-gas-drive-type fields except where conditions are favorable for gas caps to form. In many instances, reservoirs may possess the potential for either water drive or gas drive. Then the kind of operation and total rate of production will determine which type of drive

Fig. 4. Water drive. *(Texas Mid-Continent Oil and Gas Association.)*

TABLE 1.   Derrick Sizes and General Dimensions

1 Derrick size no.	2 Height A		3 Nominal base square B		4 Opening D		5 V window opening C		6 Drawworks window opening C		7 Gin pole clearance E	
	ft	in.	ft	in.	ft	in.	ft	in.	ft	in.	ft	in.
10	80	0	20	0	5	6	23	8	7	6	8	0
11	87	0	20	0	5	6	23	8	7	6	8	0
12	94	0	24	0	5	6	23	8	7	6	8	0
16	122	0	24	0	5	6	23	8	7	6	8	0
18	136	0	26	0	5	6	23	8	7	6	12	0
18A	136	0	30	0	5	6	23	8	7	6	12	0
19	140	0	30	0	7	6	26	6	7	6	17	0
20	147	0	30	0	6	6	26	6	7	6	17	0
25	189	0	37	6	7	6	26	6	7	6	17	0

Tolerances: $A = \pm 6$ in.; $B = \pm 5$ in.; $C = \pm 3$ ft, 6 in.; $D = \pm 2$ in.; $E = \pm 6$ in.

will be most effective, and accordingly will affect the oil recovery. Artificial lifting is described later.

DRILLING AND WELL SERVICING STRUCTURES

The American Petroleum Institute Committee on Standardization of Drilling and Servicing Equipment has established specifications for suitable steel structures for drilling and well servicing operations. These specifications are given in detail in *API Spec.* 4E, 2d edition, March 1974, issued by the API Division of Production, Dallas, Texas.

The dimensions of a standard derrick are given in Fig. 5 and Table 1. Specific loads on drilling and servicing structures include: (1) *dead load*—the weight of the steel and of all equipment and material fastened thereto and supported therefrom; (2) *wind loads*—resulting from wind from any direction acting on all exposed elements of the structure; (3) *dynamic load*—the force produced as a result of the pitch, roll, and heave of a floating vessel in floating rigs; (4) *earthquake load*—the product of a numerical constant C and the total deadweight of the point under consideration (when connected to another structure, such as a platform, $C = 0.05$; when on ground, $C = 0.025$); (5) *static hook load*—the weight that is applied at the hook for the designated location of the deadline

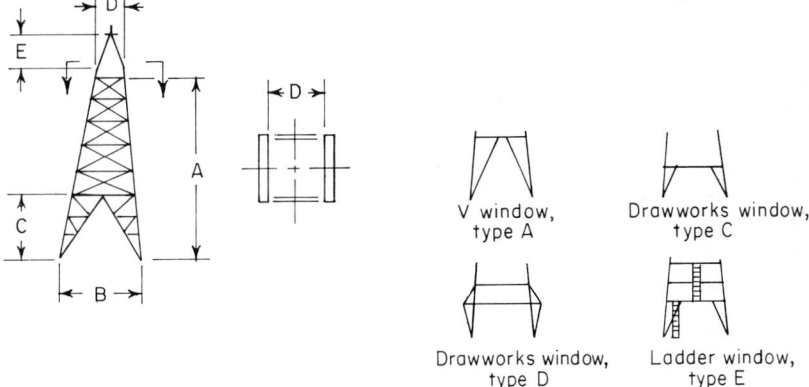

V window,
type A

Drawworks window,
type C

Drawworks window,
type D

Ladder window,
type E

Fig. 5. Standard derrick with principal dimensions and derrick window configurations. $A = $ vertical distance from top of base plate to bottom of crown block support beam. $B = $ distance between heel and heel of adjacent legs at top of base plate. $C = $ window opening measured in the clear and parallel to center line of derrick side from top of baseplate. $D = $ smallest clear dimension at top of derrick that would restrict passage of crown block. $E = $ clearance between horizontal header of gin pole and top of crown support beam. See also Table 1. (*API Spec.* 4E, March 1974.)

anchor and in the absence of any pipe setback, sucker rod, or wind loadings; the rated static hook load includes the weight of the traveling block and hook; (6) the *setback load*—the deadweight of the maximum amount of drill pipe with tool joints and drill collars set back on the substructure; (7) the *rotary table load*—the maximum weight supported by the rotary table support beams; (8) *erection load*—that stress produced in the mast and its supporting structure during the raising and lowering operation; (9) *guy line loads*—imposed from attached guy lines; (10) *floor loads*—imposed by equipment, material, and personnel on floor. The dynamic loading conditions for floating rigs are shown in Fig. 6.

$$FP = \left(\frac{WL_1}{32.2} \times \frac{4\Pi^2}{T_p^2} \times \frac{\Pi\Phi}{180}\right) + W \sin \Phi$$

$$FA = \left(\frac{W}{32.2} \times \frac{4\Pi^2}{T_r^2} \times \frac{\Pi\Theta L}{180}\right) + W \sin \Theta$$

$$FH = \frac{W2\Pi^2 H}{T_h^2 g}$$

where: F = force, pounds
P = pressure, pounds per square foot
A = projected area, square feet (of all the exposed surfaces in either the upright or heeled condition)
W = dead weight of the point under consideration, pounds
L_1 = distance from pitch axis to the center of gravity of the point under consideration, feet
L = distance from roll axis to the center of gravity of the point under consideration, feet
H = heave (total displacement), feet
T_p = period of pitch, seconds
T_r = period of roll, seconds
T_h = period of heave, seconds
Φ = angle of pitch, degrees
Θ = angle of roll, degrees
g = acceleration due to gravity, 32.2 feet per second per second

Platforms.   The American Petroleum Institute (in *Recommended Practice* 2A, 5th edition, January 1974) defines three types of offshore platforms:

(1) Fixed Platform: a platform extending above and supported by the seabed by means of piling, spread footings, or other means with the intended purpose of remaining stationary over an extended period.

(2) Manned Platform: a platform that is actually and continuously occupied by persons accommodated and living thereon. See Fig. 7.

(3) Unmanned Platform: a platform upon which persons may be employed at any one time, but upon which no living accommodations or quarters are provided.

In designing a platform, the following factors must be taken into consideration:

Winds.   Forces are exerted by wind on that portion of the structure above the water as well as on the equipment on the platform. The wind velocity that creates the force may be classified as either: (1) gusts which are of brief duration and intermittent in action with varying periods of occurrence, or (2) average wind speeds over specified intervals of time. The average wind velocity is used where storms are classified by the intensity of the wind velocity. Average wind velocities are computed for continuous periods equal to or greater than one minute. A concise definition of the time interval associated with the duration of gusts and average wind speeds should be given. All wind data should be corrected to a standard elevation, such as 10 meters.

For extreme conditions, projected extreme wind speeds from specified directions should be developed and presented graphically versus their expected recurrence intervals. Also data should be available giving: (1) the measurement site, date of occurrence, magnitude of measured gusts and other wind speeds, and wind directions for the recorded

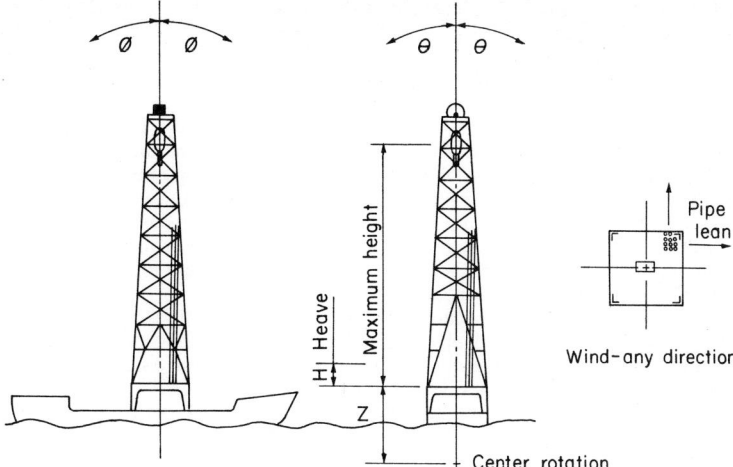

Fig. 6. Dynamic loading conditions for floating rigs. (*API Spec.* 4E, March 1974.)

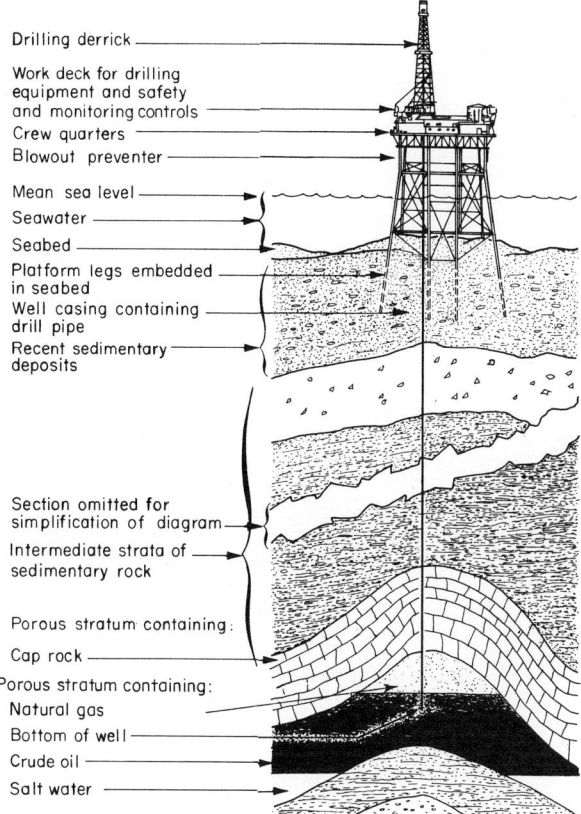

Fig. 7. Simplified schematic diagram of an undersea drilling platform. *(American Gas Association.)*

wind data used during the development of the projected values; and (2) the projected number of occasions during the specified life of the structure when average wind speeds from specified directions should exceed a specific lower bound (e.g., 50 miles per hour).

Waves. Wind-driven waves are a major source of environmental forces on offshore platforms. Such waves are irregular in shape, can vary in height and length, and may approach a platform from one or more directions simultaneously. For these reasons the intensity and distribution of the forces applied by waves are difficult to determine. Because of the complex nature of the technical factors that must be considered in developing wave-dependent criteria for the design of platforms, experienced specialists in meteorology, oceanography, and hydrodynamics should be consulted. In those areas where prior knowledge of oceanographic conditions is insufficient, the development of wave-dependent design parameters should include the following steps at a minimum:

1. Development of all necessary meteorological data
2. Theoretical projection of surface wind fields
3. Theoretical prediction of deepwater general sea states along storm tracks
4. Definition of maximum possible sea states consistent with geographical limitations
5. Delineation of bathymetric effects on deepwater sea states
6. Introduction of probabilistic techniques to predict sea state occurrences at the platform site against various time bases
7. Development of design wave parameters through physical and economic risk evaluation

For operating conditions (for both seas and swells), the following specifics should be developed: (1) For each month and/or season, the probability of occurrence and average duration (persistence data) of various sea states (e.g., waves higher than 8 feet) from specified directions in terms of general sea state parameters (e.g., the significant wave height/average height of the highest one-third of the waves in the train, and the average wave period during a certain duration of time); (2) the wind velocities, tides, and currents occurring simultaneously with the wave trains mentioned in (1); and (3) the percentage of waves having heights and directions within specified ranges (e.g., 10 to 12 feet high waves from SSE $\pm$ 11.25°) during each month and/or season.

For extreme conditions, definitions of the extreme sea states should provide an insight into the number, height, and crest elevations of all waves above a certain height which might approach the platform site from any direction during the life of the structure. Projected extreme wave heights from specified directions should be developed and presented graphically versus their expected average recurrence intervals. Other data should include:

1. Probable range and distribution of wave periods associated with extreme wave heights
2. Projected distribution of other wave heights in the wave train producing an extreme wave height(s)
3. Maximum crest elevation of the extreme sea states
4. The tides, currents, and winds that potentially occur simultaneously with the wave trains producing the extreme waves
5. Nature, date, and place of the event that produced the historical sea states, as for example a specific hurricane

Tides. These are obviously an important consideration in platform design. Tides may be classified as: (1) lunar or astronomical tide, (2) wind tide, and (3) pressure differential tide. The sum of these three tides is called the storm tide. In the design of a fixed offshore platform, the storm tide elevation is the datum upon which storm waves are superimposed. The variations in elevations of the daily lunar tides, however, determine the elevations of the boat landings, barge fenders, and the splash zone treatment of the steel members of the structure.

Currents. These are important to platform designs as they affect: (1) location and orientation of boat landings and barge bumpers, and (2) forces on the platform. Boat landings and barge bumpers should be located, where possible, to allow the boat to engage the platform as it moves against the current.

Data on currents associated with specific sea states should include an appreciation of the magnitude and direction of the water motion at specified elevations below the sea surface. Directional data on currents which exist in the absence of waves should be provided for each month and/or season. Current velocity must be considered in computing the hydrodynamic forces on the structure.

In the absence of field data, current variation with depth for shallow channels can be calculated by

$$V_x = V_s \left(\frac{X}{d}\right)^{1/7}$$

where V_x = current velocity, feet per second, at distance X feet above mud line
V_s = current velocity, feet per second, at water surface
X = distance, feet, above mud line
d = distance, feet, from water surface to mud line.

Where current is acting alone (i.e., no waves) the force is given by

$$F_L = 0.5C_L pV^2A$$
$$F_D = 0.5C_D pV^2A$$

where F_L = lift force per unit length
F_D = drag force per unit length
C_L = lift coefficient
C_D = drag coefficient
p = mass density
V = current velocity, feet per second
A = projected area per foot of length

C_D and C_L should be based on best information available for the current velocity and type of structural member involved. For long cylindrical members with large length-diameter ratios, the lift force may be critical and should be checked in the design. All symmetrical members exposed to the current should be investigated for the possibility of flutter due to von Kármán vortex shedding.

Ice. Where petroleum developments are being carried out in very cold regions, ice, in the form of fields of floating, grinding packs, moving back and forth on the tidal currents which ebb and flood, produce formidable design problems. Total ice forces vary, depending on such factors as severity of the winters, location of platform, size and configuration of platform, and unit ice strength. Unit ice strength depends on temperature, speed of load applications, composition of the ice, and other factors. To reduce forces and the possibility of local damage, a platform configuration should be selected which limits the number of elements in the ice zone. Appropriate steels for use in low temperatures should be specified.

Ice is assumed to exert a force on a structure equal to

$$F = CfcA$$

where F = ice force, pounds
C = coefficient
fc = compressive strength of ice, pounds per square inch
A = area struck by ice, square inches

Ice strengths fc varies with temperature, salt content, load rate, etc. However, values in the range between 200 and 500 psi may be expected. The coefficient C depends on shape, speed of application, etc., and will range between 0.3 and 0.7.

Earthquakes. In areas that are seismically active, seismic forces must be considered in platform design. Seismic investigation to determine forces for a given site must rely on existing seismic information that is available for the general area under consideration.

Marine Fouling. In tropic and semitropic oceans, fouling of all structures built in the sea is an ever-present problem. Marine growth is confined more or less to the zone of light infiltration, and limited air entrapment. This growth may be strong enough to create added wave resistance.

Soil Conditions. On-site soil investigations that outline the various soil strata and strength parameters should be made if sufficient knowledge has not been gained in a particular locality from prior soil investigations. Soundings should also be gathered during the on-site studies. These data should be combined with an understanding of the geology of the region to develop the required foundation design parameters. These studies should extend throughout the depth of soils to be affected by installation of the foundation elements. The bearing capacity of mat and spread footing foundations and the lateral capacity of pile foundations are largely determined by the strength of the soils close to the sea floor. Consequently, particular attention should be paid to developing complete information on these soils.

The on-site soil investigation should be a core boring that tests the soil in as nearly an undisturbed state as feasible. The foundation investigation for pile-supported structures should yield at least the soil test data necessary to prepare load-deflection $(p - y)$ curves of lateral resistance, and axial capacity of piles in tension and compression.

Scour.   This is removal of sea floor soils caused by currents and waves. Such erosion can be a natural geologic process, or can be caused by structural elements near the sea floor. The potential for scour is particularly great in locations with sand and silt close to the sea floor. Scour can result in removing vertical and lateral support for foundations, causing undesirable settlements of mat foundations and overstressing foundation elements. Where scour is a possibility, prevention should receive special consideration.

Submarine Slides.   These are lateral movements of large portions of the sea floor caused by waves, earthquakes, and gravity. Such movements of the soils can cause significant lateral forces against foundations. Submarine slides can occur in many types of soils. The weak, unconsolidated soils surrounding actively growing deltas are susceptible to such movements. Detailed information, such as the contour of the bottom, rate of deposition, and gas content are required for the evaluation of suspect areas.

Structural Classification of Platforms.   The API categorizes fixed offshore platforms as follows:

1. *Template-Type Platform* consisting of:
    a. The jacket or welded tubular space frame which is designed to serve as a template for pile driving, and as lateral bracing for the piles
    b. Piling that permanently anchors the platform to the ocean floor, and carries both lateral and vertical loads
    c. A superstructure consisting of the necessary trusses and deck space for supporting operational and other loads.

2. *Tower Platform:* A platform that has relatively few large-diameter, e.g., 15-foot legs. The tower may be floated to location and placed in position by selective flooding. Tower platforms may or may not be supported by piling. Where piling is used, it is driven through each leg of the tower into the soil foundation to support the platform. The piling may also serve as well conductors. If the tower's support is furnished by spread footings instead of by piling, the well conductors may be installed either in or outside the legs.

3. *Caisson Platform:* A platform whose foundation support consists of one large member. Such structures range from free-standing caissons which support only one well (e.g., 30 inches O.D.) to large structures several feet in diameter.

Marine pipeline connections are described in the article on Marine Pipelines in the Handbook section on *Gas Technology*. Power systems for platforms are discussed in the article on Gas and Expansion Turbines in the Handbook section on *Power Technology Trends*.

DRILLING GEOMETRY

Deviation surveying was introduced into oil-well drilling technology in 1929. Before that time, it was generally assumed that a hole properly started as a vertical hole would remain essentially vertical. In many instances, that was not a realistic assumption because lots of "vertical" holes were found to be quite crooked. Crooked holes not only caused operation problems, but also resulted in false indications of depth. Since the early 1930s, drilling contracts usually have contained a specification that the maximum deviation shall not exceed so many degrees—usually 3 to 5 degrees. The problem of drilling a straight hole usually is simpler with uniform materials, such as limestone, and more difficult when laminar formations of sandstone and shale are encountered. Often of even greater concern than a crooked hole is an irregular "jagged" hole that does not have a graceful bending contour in the vertical. The presence of abrupt changes in angle interfere with the casing program and ultimately with production. Deviation can contribute to casing wear, key seating, stuck pipe, and costly "fishing" jobs. See Fig. 8. Of course, in directional drilling, described later, deviation is desired. The cause and effect are the same. In the one case, undesirable; in the other case, carefully planned for and desirable.

The mechanics involved in causing a bit to "drill into the hill" so to speak are not fully understood. From a practical standpoint, it is known that the bit will build up angle when heavy weight is loaded on the bit; the reverse being the case with a lightly loaded bit. Low weights on a bit, of course, result in slow drilling.

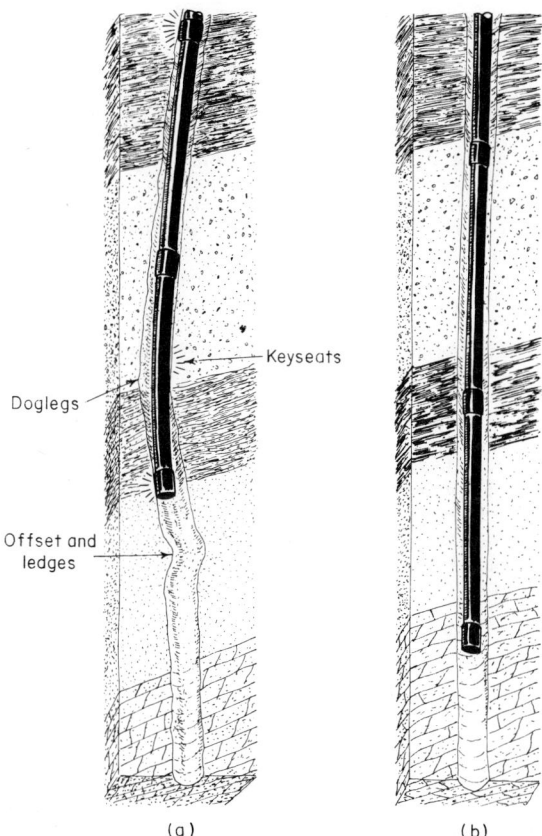

Keyseats

Doglegs

Offset and ledges

(a)   (b)

Fig. 8. Example of crooked hole (a) drilled without a square collar, and a relatively straight hole (b) drilled with a square collar. (Drilco.)

Deviation appears to result from flexibility of the drill string (drill collars) and the forces acting on the string that cause it to bend. Normally, one does not consider drill collars to be very flexible. However, if one would take a 90-foot string of 10-inch-diameter collars lying horizontally and supported at both ends, the change in slope of the collars from one end to the other would amount to a dogleg (bend) of about 7 degrees per 100 feet of length. The relative stiffness of steel drill collars varies with the collar diameter. Using a figure of unity for the relative stiffness of a 6-inch O.D. × 2-inch I.D. steel collar, the relative stiffness of a 12-inch collar is 16; of a 10-inch collar, 8; of an 8-inch collar, 3; and of a 4-inch collar, 0.2. (These figures from W. B. Bradley in SPE preprint, October 1974).

Apparently, the length of the drill string contributing to deviation is determined by the hole geometry, the weight on the bit, and the physical characteristics of the drill string. Normally, only the bottom 150 feet of the drill string is involved. However, this can be 60 feet or less where higher weights on the bit and greater hole angles are concerned. With reference to Fig. 9, Bradley gives the following formulas:

$$\frac{1}{p} = \frac{M}{EI} \qquad V = \frac{dM}{dS}$$

where
- p = radius of curvature of the borehole and the drill string ($p = \infty$)
- M = bending moment
- V = internal shear force
- E = modulus of elasticity
- I = area moment of inertia
- S = distance along drill string

As pointed out by Bradley, for $p = \infty$, these equations yield $M = 0$ and $V = 0$. For curved hole with $p = C$, M is constant, and $V = 0$. Therefore, for purposes of analysis, the portion of the drill string above the tangency point lying along the low side of the borehole can be removed and replaced by the axial force transmitted by the drill string at that point. Increases in weight on bit and/or hole angle move the tangency point closer to the bit. Bradley also points out that as the drill string becomes more complex with addition of stabilizers, square drill collars, etc., or with changes in hole geometry (e.g., washouts), the determination of the active length of the drill string involved in the process becomes more difficult.

Bit-rock interaction also contributes importantly to crooked holes. H. M. Rollins in a 1963 paper introduced the concept of the pendulum effect which he defined as the tendency of the unsupported length of drill collar to swing over against the wall of the hole. It is the only force tending to straighten up a crooked hole. See Fig. 10.

In his 1974 paper, Bradley concludes: (1) Only that portion of the drill string below the tangency point is effective in controlling deviation. (2) Maximizing drill collar stiffness and drill collar weight can lead to a substantial increase in penetration rate while holding a constant hole angle. Therefore, in most cases, the largest diameter (stiffest and heaviest) collar that can be safely tripped in and out of the hole should be used. (3) Effective placement of multiple stabilizers can improve penetration rates by a factor of four over poorly stabilized strings. (4) Downhole motor case flexibility can reduce deviation performance by as much as 30 percent. (5) Additional research is needed to understand the

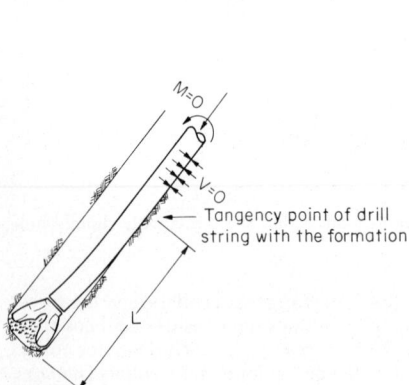

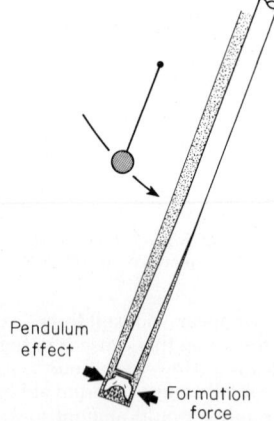

Fig. 9. Simple example of a uniform unstabilized drill string lying on the low side of an inclined straight hole. Active length of the drill string is the portion of the string, L, below point of tangency. Above the tangency point, the drill string is supported by the borehole wall, resulting in zero internal shear and bending moment. (*Bradley.*)

Fig. 10. Relationship between pendulum effect and formation force. When pendulum force equals formation reaction, which tends to force the bit to increase angle, an equilibrium exists. Under these conditions, hole will continue straight, but will be inclined. If pendulum force is greater, hole angle will decrease. If formation reaction is greater, hole angle will increase. (*Rollins.*)

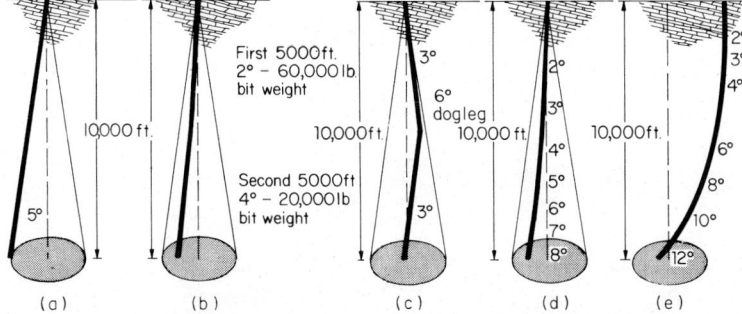

Fig. 11. If a 10,000-foot well is specified for a 5-degree maximum deviation, the well bore must be kept within a 1744-foot circle directly below the floor of the rig (*a*). (*b*) Shows the situation where the first 5000 feet were drilled with only a 2-degree deviation. The remainder of the hole was drilled with lighter loading so as to remain within the 5-degree maximum deviation. (*c*) Indicates severe dogleg that can result even if total deviation remains within specifications. (*d*) Indicates a situation where deviation is permitted to increase gradually. (*e*) Indicates movement of rig from normal surface location in anticipation of severe deviations. (*Rollins.*)

deviation process (bit-rock interaction) in drilling anisotropic rock. Emphasis needs to be directed at developing techniques to maintain hole angle without sacrificing penetration rate.

The effects of angles of deviation and various forms of compensation are illustrated schematically in Fig. 11.

In a 1970 paper of a highly analytical nature, Rollins concluded: There are two basic considerations for controlling hole deviation: (1) Where the hole angle has not approached the maximum permissible angle, a stiff bottom assembly will permit the use of heavy weight with an assurance that the rate of angle change will be small. The hole may still build angle, however. This stiff bottom assembly may consist of reamers, short collars, and stabilizer, or these tools in conjunction with a square drill collar; and (2) where the maximum hole angle is approached, it is necessary to remove the stiff bottom assembly, and return to the pendulum technique using a bit, drill collars, and a stabilizer placed at the ideal position. It will be necessary to reduce weight to find the weight that will produce an equilibrium condition just below the maximum permissible angle. If a section of the hole is drilled without a stiff assembly, it will be necessary to rotate and ream through that section when the stiff assembly is run again later.

CONTROLLED DIRECTIONAL DRILLING

Controlled directional drilling is the planned and intentional drilling of a well bore along a course to a target located at a given distance from the true vertical line running below the rig. Some of the reasons for directional drilling are depicted in Fig. 12. Other reasons for directional drilling not illustrated include (1) To sidetrack or drill around an obstacle (fish) in the hole: After a cement plug is set above the fish, the well bore is deviated around it and drilling continued. (2) Where a straight hole has not been successfully drilled and it is necessary to use directional tools to turn the hole back toward the target area. (3) In connection with drilling through some fault planes, it may be more economical to deflect the well bore into the pay zone from either side of the fault.

Regardless of what tool may be used to deflect the hole, the first step in drilling a controlled directional well is to drill the hole as straight as possible to the depth where the initial deflection is to be made. A survey is then run at that point to determine where the bottom of the hole is in regard to its surface location, its present inclination, and direction. If it is found that the vertical hole has a drift angle, and because deflection tools have deflection angles built into them, the problem of turning or facing a tool in the direction it should drill to reach the target presents some complications. A mathematical calculation called vector analysis is used to calculate the direction in which the tool should face.

The three most commonly patterned directional holes are illustrated in Fig. 13.

Before a directional well is started, a horizontal and vertical plan of the course of the well bore must be drawn up. This planned course depends on several factors, including rig capacity, hole size, mud program, types of formation, and casing program. The target is designated at a certain depth and direction from the rig.

This guides the directional engineer in keeping the hole on course during drilling operations. By running a special surveying instrument in the hole, the engineer can plot

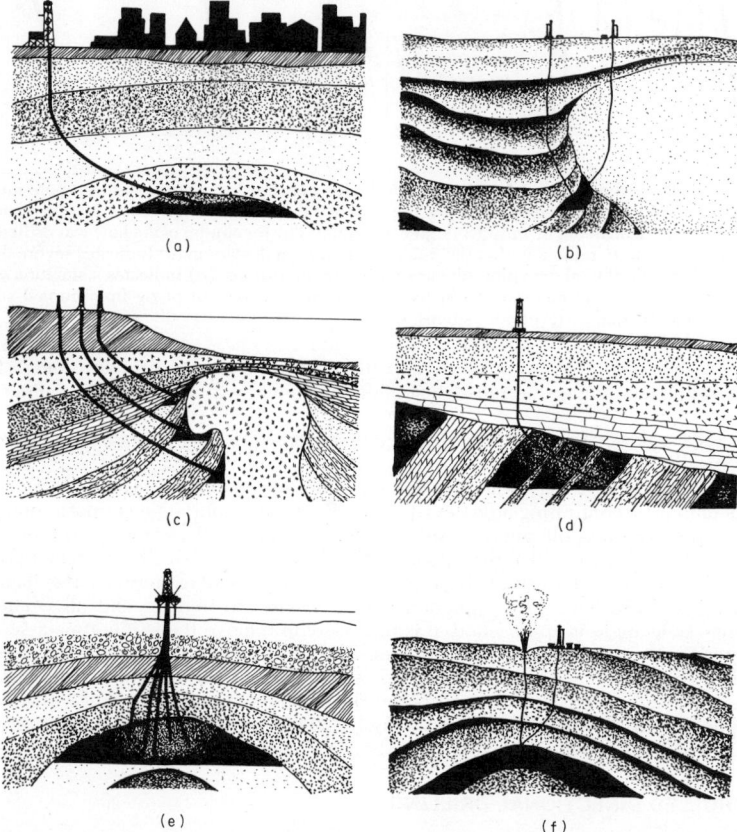

(a)

(b)

(c)

(d)

(e)

(f)

Fig. 12. Some applications for controlled directional drilling. (*a*) Reaching formations which lie below inaccessible locations, such as towns, rivers, and lakes. (*b*) Formations sometimes are found below the overhanging cap of a salt dome. A well may be drilled around this cap, or through the salt and deflected into the productive formation. (*c*) Formations below harbors or the ocean floors sometimes can be reached from rigs located on the shore. (*d*) Directional drilling into the intersection of several oil sands from a single wellbore. Obviously, a straight hole would be less effective in this type of situation. (*e*) Offshore drilling is usually most economic when several directional wells can be drilled from a single platform. As many as 20 or more wells can be drilled from a small area. (*f*) Drilling of a relief well to intersect a wild, cratered well near the source of pressure. Mud and water can be pumped in to kill the blowout. This technique, first used in 1934, helped to establish the importance of directional drilling. (*Petroleum Extension Service, The University of Texas at Austin.*)

the actual course of the hole and compare it with the planned course. Three factors are required for each point in the survey: (1) compass direction, (2) hole angle, and (3) depth. The rig is operated by the driller and crew. Usually, from start to final depth, a nonmagnetic drill collar of Monel or stainless steel is carried at the lower end of the string. Magnetic survey instruments are seated in this special collar when readings are made. Since most instruments contain magnetic compasses, an ordinary drill collar cannot be used. An ordinary collar becomes magnetized and causes the magnetic compasses to behave erratically. The surveying instrument may be run in on wire line, or dropped through the drill string to seat in the nonmagnetic drill collar.

Basically, a magnetic surveying instrument consists of a magnetic compass, floating in a liquid, and photographic equipment. The latter is used for taking pictures of the compass. The instrument may make a single picture or multiple pictures. Gyroscopic units are also means for obtaining directional surveys. Since the gyroscopic compass is not affected by magnetism, this instrument is valuable for taking surveys and orienting deflection tools inside casing and in places where the earth's magnetic forces are irregular.

To achieve hole deflection, one of several deviation tools may be used. One method is built around the positive-displacement down-hole hydraulic motor, and drills a full gage

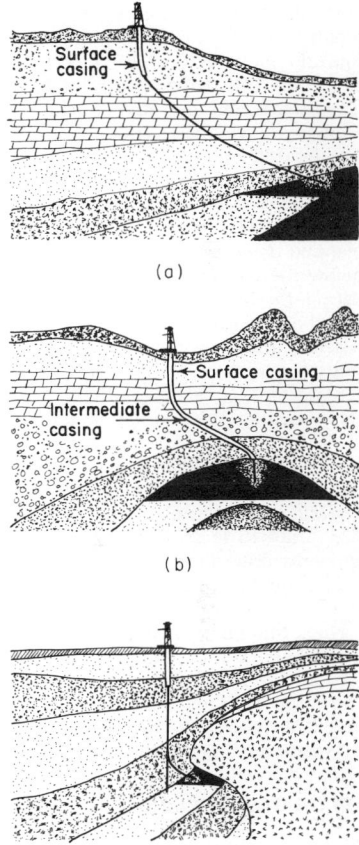

(a)

(b)

Fig. 13. Principal types of directional drilling patterns. (*a*) The most widely used directional drilling pattern is one in which the initial deflection is obtained at relatively shallow depth. Then, surface casing is set and cemented through the deviated section of hole. From that point, the angle is maintained as a straight line to the target zone. (*b*) This pattern is also initially deflected at shallow depth. Surface casing is set and cemented. Drilling continues on a straight line to a point where the hole is gradually returned to vertical. After intermediate casing is set, drilling is continued to final depth. This type hole is used when undesirable formations must be penetrated and isolated with an intermediate casing string. (*c*) In this pattern, deflection commences at a greater depth. Drift angle is maintained on a straight line to the target. This type hole may be used for exploratory drilling from a dry hole. Normally, the deflected part of this hole is not protected by casing during drilling operations. (*Petroleum Extension Service, The University of Texas at Austin.*)

(c)

hole. The down-hole hydraulic motor is operated by either mud or air pressure supplied by the rig's circulating system. When operating, the drill pipe, motor housing, and stator remain stationary; and only the rotor, connecting rod, drive shaft, and bit sub rotate. This motor, by itself, will not deviate the hole off vertical, but several things can be done to make it do so. The manufacturer can supply a hydraulic motor with the housing bent one or two degrees off vertical at the point where the rotor and drive shaft are connected with a U joint. The face of the tool, in this case, is the direction in which the case is bent. Another method used to cause the hydraulic motor to drill off vertical is to make up a one-to-three-degree bent sub on top of the motor. Again, the face of the tool is in the direction in which the sub is bent. When the bent housing or bent sub is used, the assembly must be run into the hole very carefully because the bit will tend to dig into the wall of the hole because of the bend in the bottom-hole assembly. The rather complex operating and controlling details of this system are very well summarized in "Controlled Directional Drilling," published by the Petroleum Extension Service, The University of Texas at Austin, 1973.

Another assembly that can be used to drill a full-gage, deviated hole is the down-hole turbine. Because of the physical characteristics of the turbine, it is best used where light weight and high rotating speeds are desired. The down-hole turbine can be operated only by a liquid circulating medium. While operating, the drill string and turbine housing, to which stators are attached internally, are held stationary. Drilling mud, pumped down the drill string, is directed by the stators onto the turbine blades, which turn a drive shaft and bit sub below. A portion of the drilling mud is used to lubricate the turbine bearings. Like the down-hole hydraulic motor, the turbine must have some device to deflect the bit off vertical. The bent sub or hydraulic deflecting tool is used. Both the down-hole hydraulic motor and the down-hole turbine have one common characteristic that other deflecting tools do not have. Both produce a reverse torque. This reverse torque is great enough to cause the drill string to turn away from the direction in which the hole is to be drilled. Experience and the advent of new orienting devices have reduced this problem of allowing for reverse torque to a routine procedure.

A jet deflection bit will also deviate a well bore and maintain a full-gage hole. The general principle is to employ a full-gage jet bit with one large nozzle—$\frac{5}{8}$ or $\frac{3}{4}$ inch in diameter. The other jets are blocked off or equipped with smaller nozzles. Using a drill collar and stabilizer, the large nozzle is oriented in the direction necessary to obtain the planned deviation. A high rate of mud circulation erodes the well bore on the side and below the large nozzle. The stabilizer keeps the bit centered in the hole. Continually jetting, the bit is lowered slowly without rotation. It follows the path of least resistance into the deviated hole. Jetting continues until a directional survey indicates that the bit has set a proper course. Then rotation is started to drill ahead with a full-gage hole. In soft formation, this method saves time and equipment costs.

Spud bits may occasionally be used for drilling directional full-gage holes in relatively soft formation. When oriented in the proper direction, this bit employs an eccentric action to drill the curved section. Deflection can be checked and the tool reoriented without removal from the hole. The whipstock is an old reliable deflection tool, but has one major disadvantage over newer methods—it does not allow a full-gage hole to be drilled. Therefore, it requires more trip time because the small rat hole must be reamed to full size. Only 15 to 20 feet of hole can be drilled at one time. A bottom-jet whipstock is generally used if the hole is to be deflected with a whipstock. With this approach, drilling mud or air can be circulated through the tool itself. Fluid is circulated through a hollow shear pin and out the bottom. The advantage of this whipstock is that bridges can be washed through and cuttings can be pumped from the bottom of the hole. Therefore, the whipstock has a firm, clean bottom on which to seat. After the bottom has been cleaned satisfactorily, a ball is dropped down the drilling string to seat in a slotted sleeve assembly above the bit. This blocks the flow of drilling fluid through the whipstock shear pin and diverts it through the bit. Then weight is applied to set the whipstock and shear the pin. Bit rotation is begun, and the bit is moved down the whipstock face.

AIR OR GAS DRILLING*

Air or gas drilling possess great advantages over conventional drilling in penetration rate, lack of lost circulation problems, and mud expenses. Air or gas drilling would soon replace a high percentage of conventional drilling if it were not for the major problem of drilling into formations that produce water. Large volumes of water flood out the operation. Small amounts of water tend to ball up the cuttings and make it difficult to withdraw a worn bit from the hole.

Water problems are often avoided by air-drilling only formations where water production is unlikely. Sometimes water is handled by using mist or foam drilling with surfactants to wash out the cuttings. Frequently water cannot be tolerated. Conditions such as too much water to mist or foam-drill, disposing of produced water, or penetrating water swelling formations that stick the drill pipe in the hole must be corrected by shutting off the water.

*Prepared by H. C. McLaughlin, Haliburton Services, Duncan, Okla.

Slurry-type sealants such as portland cement may shut off water if the water source is from cracks and fissures in impermeable formations, but slurries cannot penetrate the capillary system of the rock. Therefore, chemical sealant solutions must be used. Sealant solutions commonly employed to shut off water are acrylamide with crosslinker, and sodium silicate. These solutions have waterlike viscosity and are injected easily into formation capillaries. They produce firm, semisolid gels capable of withstanding 700 to over 10,000 psi water pressure, depending on formation permeability and barrier thickness.

A technique for water shutoff is: Clean the formation face by water circulation without the bit and collars on the drill string, circulate sealant solution to the formation, raise the drill string above the sealant solution, close the blowout preventers and gradually raise pressure until sealant solution enters the formation, and stage the sealant solution in until most of it is in place, or gel time has expired. Allow one hour after gel time, release pressure, replace bit and drill collars, and use surfactant to foam out the water and excess sealant. Caution should be used so as not to break or fracture the formation by excess treating pressure. Breaking the formation down has a history of reducing treatment effectiveness.

GEOPRESSURED DRILLING*

Abnormally high subsurface pressures are common in most producing areas of the world. (Refs. 1, 2 under Geopressured Drilling in reference section.) Their anticipation, detection, and control are important to the safe and economic execution of drilling projects in these areas.

Regardless of their origin (Refs. 3, 4), abnormal pressures (or geopressures) are characterized by a pressure gradient greater than 0.465 psi per foot in salt water basins, or 0.429 psi per foot in fresh or brackish water basins. These accepted standards define normal pressures for their respective areas. The maximum gradient for geopressured formations is limited by the overburden gradient. This gradient, derived from the average bulk density for rock, is generally assumed to be 1.0 psi per foot.

Many techniques (Refs. 5–8) for identifying and measuring geopressures are available in the industry today. Basically, they can be divided chronologically in the life of a well: (1) before drilling, (2) during drilling, and (3) after drilling. Table 2 outlines most of the present methods. The critical nature of the parameters during drilling has precipitated the development and refinement of sophisticated wellsite surveillance units and related equipment.

The importance of carrying sufficient mud weight to overbalance formation pressures is self-evident. But it is equally important that precautions be given to excessive mud weights that would induce or extend fractures in the weaker formations. The minimum fracture gradient (Refs. 6, 9) dictates the maximum mud weight that can be safely carried without casing off the weak zones. This knowledge, coupled with the projected formation pressures, is used to select optimum casing, drilling, and mud programs.

The most effective method to plan and drill a geopressured well is to construct a pressure profile (Ref. 10). The profile, shown schematically in Fig. 14, compares the relationships among the expected formation pressure gradient, fracture gradient, and mud weight at depth. For convenience, the gradients are shown in equivalent pounds per gallon.

The dashed lines on Fig. 14 show the procedure for determining casing setting depths. Also note that in the transition zone just below 8000 feet, the mud weight is less than the formation pressure gradient. This prevents fracturing at the surface casing shoe and avoids masking identification of the zone. For best use, the pressure profile should be reevaluated as the well is drilled.

WELL CONTROL*

The drilling of wells will always carry a potential blowout hazard. But the industry is pursuing a concentrated effort to greatly minimize or eliminate the possibilities. Industry programs stress the importance of proper well planning and training of rig crews, espe-

*Prepared by Mario Zamora, IMCO Services, Houston, Tex.

TABLE 2.   Detection and Evaluation of Geopressures—Checklist

Before Drilling
  Mud histories, bit records, and drilling reports from offset wells
  Geologic correlation to similar areas and formations
  Evaluation of offset wire-line logs:
    Induction (conductivity)
    Electrical (resistivity)
    Acoustical (interval transit time)
    Gamma-gamma (density)
    Neutron-gamma (porosity)
  Geophysical aspects:
    Seismic data (interval transit time)
    Gravity data (bulk density)
During Drilling
  Kick
  Increase in background gas and connection gas
  Abnormal trip fill-up behavior
  Penetration rate/SP correlation
  Periodic log runs
  Paleontology
  Change in size and shape of shale cuttings
  Increase in fill on bottom
  Increase in drag and torque
  Increase in shale penetration rate
  Decrease in d-exponent trend
  Decrease in shale bulk density trend
  Increase in flow-line temperature
  Increase in chloride content in mud filtrate
After Drilling
  Drill stem tests
  Shut-in pressure tests
  Down-hole pressure bombs
  Wire-line log evaluation

cially in the fundamental concepts and maintenance and use of blowout prevention equipment. (Refs. 1, 2 under Well Control in reference section.)

Well control is divided into three categories:

1. Primary Control: The proper use of hydrostatic head to overbalance the formation and prevent unwanted formation fluids from entering the well bore. More simply, primary control is the prevention of kicks. Any event or chain of events that create a negative differential pressure between the hydrostatic head of the drilling mud and the formation can lead to a kick. The most common cause is a failure to keep the hole properly full on trips. Other major causes are insufficient mud weight, swabbing, and lost circulation due to surge pressures or excessive mud weight.

2. Secondary Control. The use of equipment, techniques, and hydrostatic head to maintain control of the well after primary control is lost. A kick that is not properly circulated out can deteriorate to a blowout, or an uncontrolled situation. The key at the secondary control level is recognition of kick warning signs and rapid action by the rig crew to shut the well in. After the necessary calculations are made, the kick is circulated out, using a constant bottom hole pressure slightly greater than the formation pressure. This is done by maintaining a constant, but reduced pump rate and varying the casing pressure with an adjustable choke. If the casing pressure is excessive, the formation may fracture and create an underground blowout. Well control tests and sophisticated simulators (Refs. 3, 4) have proved effective in understanding secondary control techniques.

3. Tertiary Control: The proper use of equipment, techniques, and hydrostatic head to regain control once a blowout has occured. This often involves drilling of relief wells (Ref. 5).

DRILLING OPTIMIZATION*

There is no such thing as a "true" optimum drilling program. Invariably compromises must be made because of limitations beyond our control that result in something less than optimum. Perhaps it can be explained this way—for years it has been found that rate of

*Prepared by J. L. Lummus, Research Department, Amoco Production Company, Tulsa, Okla.

penetration could be increased by drilling with water, by rotating the bit faster, and by increasing flow velocity through jets in the bit. Lack of sufficient mechanical and hydraulic horsepower, however, often prevents the proper balancing of variables to obtain maximum drilling efficiency. Also, there has always been a limiting value over which an increase in revolutions per minute, weight on bit, and pump rate does little or no good. New technology has raised these limits, but they are still there. The limits set for drilling variables are all influenced by the resulting bit life, the first major factor. Some are set by the second major factor—stability of the well bore.

Although water is the fastest drilling liquid, in many areas some colloidal solids in the fluid are necessary to provide hole stability. In other areas, weight on bit is limited by the deviation characteristics of the formation. Rotary speed and pump pressure are generally limited by equipment capability and resulting maintenance costs. Therefore, the lowest-cost drilling will result when limits are imposed that maximize not only drilling rate, but also equipment life and well-bore stability. In some cases, if well-bore stability and equipment life are maximized, a decreased penetration rate will have to be accepted. In other words, a balanced program must be developed—one in which the drilling variables being considered are at their most effective level.

From a practical viewpoint, the general idea of optimized drilling can be expressed by the series of curves in Fig. 15. A semiwildcat well is drilled at the lowest bid of $12.00 per foot. It takes 100 days and costs $180,000. From the experience gained on this well, the

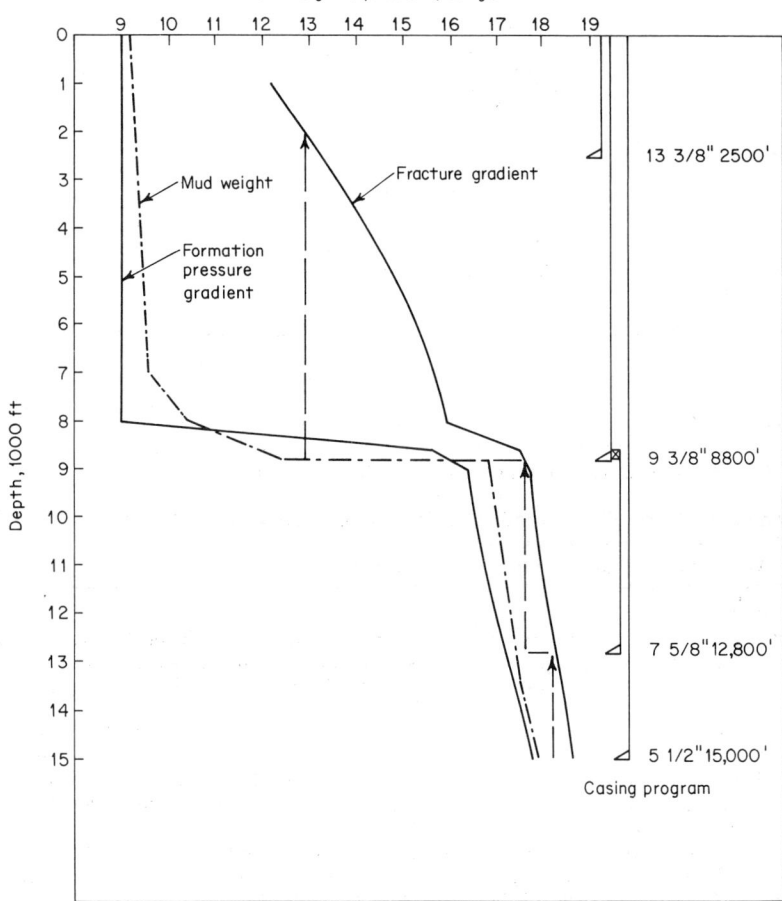

Fig. 14. Pressure profile of a typical geopressured well.

next well is negotiated at $10.80 per foot. It takes 90 days and costs $162,000. This process continues, with each succeeding well costing a little less, because of experience previously gained, until after many wells the average drilling time has been reduced to 50 days, and the cost has been reduced to $90,000, at a standard field price of $6.00 per foot.

The philosophy of optimized drilling is to use the record of the first well as a basis for calculations and to apply optimum techniques to the second and third wells, thus arriving at a negotiated field price of $6.00 per foot much sooner. If drilling costs could be reduced by utilizing optimum techniques, an operator could drill more wells per year in a given

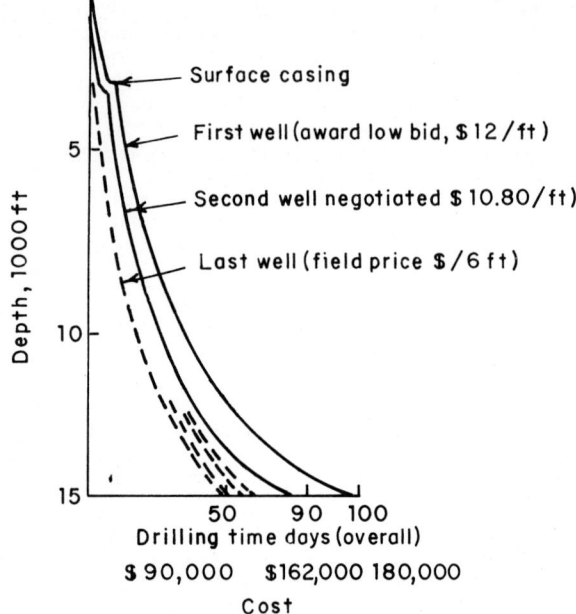

Fig. 15. Demonstration of optimum drilling philosophy.

area than otherwise might be done, or the operator might be able to drill wells that would otherwise be uneconomical from a cost-to-production standpoint.

The reduction of costs through the use of optimum techniques does not stem from faster penetration rates alone. An optimum drilling program anticipates possible well problems and provides methods for handling these problems as they arise, thus reducing overall rig days.

General Concept of Optimization as Applied to Rotary Drilling.   Optimized drilling can be divided into three distinct phases:

1. Assembling of data and preparation of program
2. Implementation
3. Postappraisal

Each of these phases is equally important, but, because of space limitations, this summary will emphasize preparation and mention some aids in implementation, but gloss over postappraisal with the warning: After each operation, ask, "What did we do right? What did we do wrong? Why? Is there a better way?"

Importance of approaching the optimization of drilling variables in a logical sequence cannot be overemphasized. First and probably the most important variable is the drilling fluid. This can be divided into two main parts: (1) selection of a fluid in terms of its relative ability to drill a particular formation(s); and (2) obtaining of effective hole cleaning and well-bore stabilization.

The second variable is hydraulics, and this should consider the type of mud to be used for providing adequate cleaning of the hole and maintaining hole stability. Also, the amount of horsepower that will be available at the bit should be balanced against the availability and effectiveness of the fluid for cleaning the hole and in stabilizing the well bore.

Selection of the bit should be the third aspect of optimization considered. The proper bit(s) for drilling certain sections of the hole should be programmed in terms of the drilling mud that will be used and the available hydraulics.

Once the foregoing three conditions are met, then optimum weight and revolutions per minute can be provided for the particular bit selected, and the proper interaction of the mud and hydraulics can be obtained.

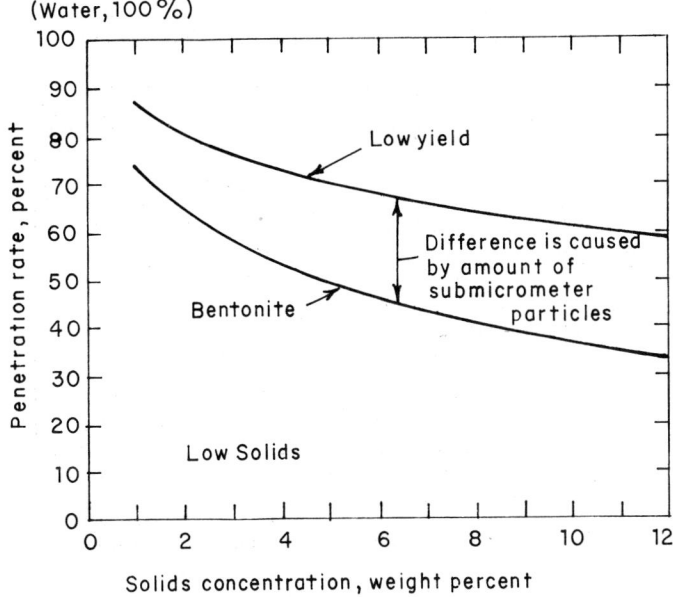

Fig. 16. Clay solids versus penetration rate.

Effect of Mud.   Controlled drilling rate tests in various rocks have confirmed that air or gas is a faster drilling fluid. Most mud systems are water-based, and drilling tests show that most commonly used additives have a detrimental effect on the drilling rate of water. Effect of mud solids on drilling rate is a well-known phenomenon, and, as shown in Fig. 16, different clays have different effects on the penetration rate for a given solids concentration. *It was found that colloidal-size particles which are less than 1 μm in size have 12 times more effect on drilling rate than particles coarser than 1 μm.*

A mathematical model that accurately predicts the effect of mud on penetration rate must consider the type and amount of solids present, particularly the particle-size distribution of the colloidal solids present. The total clay solids content can be easily calculated from graphs or equations based on measurements of the mud weight, the salt content, and the oil content, and if it is a weighted mud system, the total solids content as determined by the mud report. The Methylene Blue Test (MBT) is a field test for the quantitative determination of the base exchange capacity of clay materials in mud systems. Particle-size distributions were obtained on bentonite slurries by means of a supercentrifuge, and the results were related to MBT values. It was found that approximately 13 percent of the particles in aged bentonite slurries were less than 1 μm. When these same bentonite slurries were dispersed with lignosulfonate, the amount of submicron particles increased to 80 percent. This explains why dispersed muds generally drill much more slowly for a given solids content than nondispersed water-bentonite systems.

However, for a given solids content, the penetration rate increased when a beneficiating-type polymer was used, even though the polymer increased the viscosity of the drilling fluid. This was a difficult problem to resolve since it is known that viscosity has a detrimental effect on penetration rate. Particle-size distribution tests indicated that addition of a polymer to bentonite slurries decreased the amount of submicron particles from 13 to 6 percent. Drilling tests conducted with these slurries showed that the effect of increasing the size of suspended clay particles influences penetration rate to a much greater extent that the viscosity increase associated with polymer addition.

Knowledge of this mechanism, coupled with definitive data from the MBT test and particle-size distribution studies, provides the technology to build a mathematical model for predicting the effect of mud on penetration rate.

Effect of Hydraulics.   Sufficient flow rate is needed to clean the bit, and that jet velocity is needed to free the drilled chips held by differential pressure to the bottom of the hole. This jet velocity has a greater effect on penetration rate than the flow rate alone. However, for a given hole size, there is a limited range of flow rates that can be used for

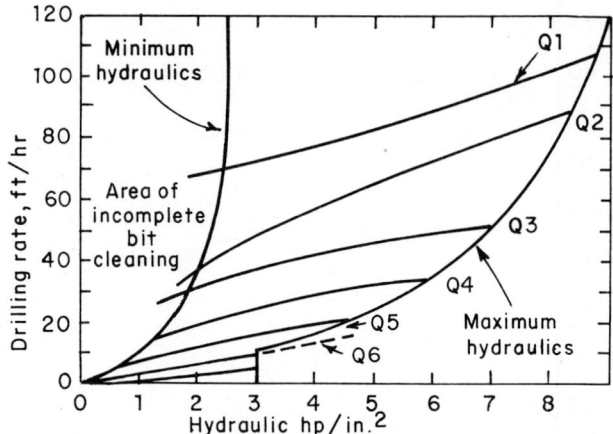

Fig. 17. Hydraulic drillability curves. *Note:* Place point where control bit run is located. Using ship curve, fit Q line below point. Move ship curve up to point. Effect of a change in hydraulics can then be traced.

hole cleaning. Therefore jet velocity for a given hole size is somewhat dependent on the hydraulic horsepower (HHP) at the bottom of the hole. A relationship of drilling rate versus hydraulic HHP per square inch at the bit can be developed for various formations from very soft to very hard. Such a relationship has developed from analysis of field data and is shown in Fig. 17.

This chart can be employed to estimate the relative effect on drilling rate by a change in hydraulic conditions at the bottom of the hole. The chart is utilized as follows: Place a point where the control-bit conditions are located, i.e., the intersection of the drilling rate and the bottom-hole hydraulic horsepower per square inch (HHP per square inch) for the control-bit run. In this analysis, control-bit run constitutes data from the last bit used in the well being optimized. By means of a ship's curve, set on the Q line (formation drillability) below the point and move curve up to the point, keeping it parallel with the line below it. The effect of a change in hydraulics can then be traced. Note that the point of complete bit cleaning and formation cleaning varies with the hardness of the formation and is shown by the right-hand envelope of the curves. The area of incomplete bit cleaning is shown at the left-hand envelope of the curves.

However, the effect of drilling rate versus hydraulic horsepower is not sufficient for optimization, since the basic concept is to optimize cost, and this does not necessarily mean increasing the drilling rate. It is apparent that several mechanisms affect drilling at the bottom of a hole: (1) insufficient hydraulics which causes incomplete bit cleaning, (2) the effect that mud solids have on penetration caused by the bit–tooth–rock interaction during chip formation, and (3) differential pressure tending to hold drill chips down, and the removal of these with varying amounts of jet velocity impinging on the bottom of the hole.

Optimizing both the mud and the hydraulics involves all three of these interactions.

Mud-Hydraulic Interactions. Complicated effects of these mud and hydraulic variables have been programmed into an interactive mathematical model available in a time-sharing computer program called MUDHY.

The advantage of the MUDHY drilling assistance program and other programs which are described below is that the user can analyze each bit run as drilling progresses, immediately implement improvements, and, hopefully, see the results of applying improved technology.

Bit Selection. Bit selection is probably the most difficult phase of drilling optimization. It requires more expertise and availability of background information to achieve efficiency than, for example, specifying running conditions for the bit. Current and future drilling trends indicate that as drilling locations and rigs become more expensive, and deeper holes are drilled, the relative cost of the drill bit in comparison to other costs will probably decrease. This statement refers only to the cost and is not intended to indicate that bit selection will become less important. Selecting the proper bit and a well engineered mud system and intelligent application of hydraulic parameters provide the foundation for building a successful optimized drilling program.

The development of practical nondispersed, low-solids polymer muds and the proper application of the rig hydraulic horsepower considerably improved the penetration rate of the original insert bits and spurred the development of the new type of chisel and projectile tooth insert bits.

Just two decades ago, 1000 feet of medium-hard, abrasive formation required 20 to 30 milled-tooth steel cutter bits. The tungsten carbide button bits were then introduced and the same 1000 feet could be drilled with only 10 bits, but usually at a higher cost per foot because of the very slow drilling rates of these bits. The advent of sealed bearings decreased the number of bits required to five. In 1968, a single sealed bearing bit with "tooth-shaped" tungsten carbide inserts could drill the entire 1000-foot formation unit. Today, a single journal bearing insert bit can drill 2500 to 4500 feet through these same formations.

It should be noted that today's "super bits" cost about 10 times what yesterday's milled-tooth steel cutter bits did, but in many cases they deliver well over 10 times the performance. The development of these high-performance bits is indeed timely, because the demand on the oil industry to find, drill, and produce oil and gas is at an all-time high, and there does not appear to be a slowing down of technology in this area.

Amoco has conducted considerable research to evaluate the response of popular bit sizes to weight and rotary speed, and to determine the effect of the mud environment, hydraulics, and the interactions of these variables on the cutting action of the bit. This provides one phase of the information needed to mathematically define the characteristics of a bit. The other aspect of modeling bit performance is assembling information on bearing life based on a statistical analysis of a wide variety of bit runs in different areas. This determines how long a bit should drill in a particular formation.

The accumulation of a large number of data from bit runs in the field, which now total over 160,000 bits, provides the basis for statistically determining which bit should drill a formation most efficiently. Individual runs can be assessed, taking into account such parameters as area, bit size, bit series, depth, footage, or hours. These data can provide invaluable aid in planning a new well.

Besides being of critical importance, bit selection is somewhat controversial. Some people who select bits tend to have preferences for one manufacturer over another, while others favor high rates of penetration over fewer trips or vice versa. Some optimum drilling programs are designed to maximize bit life, whereas the contrasting philosophy is to sacrifice bit life to gain penetration rate. Amoco's approach is that some drilling situations call for maximizing bit life whereas in others lower footage costs can be obtained by wearing out the bit faster.

Table 3 is a brief outline of the procedure used to develop a mathematical drilling model. Laboratory data provide information on the effect of mud and hydraulics on drilling rate. For estimating the life expectancy of a particular bit extensive field data are collected.

The drilling engineer employs offset well data as input to provide the necessary scaling factors for the mathematical model utilized by the computer to arrive at optimum conditions. While the data are being processed by the computer, the engineer makes an analysis of logs from offset wells and adjusts for the major formation tops expected in the proposed well. The output data from the computer are then used to select bits, estimate the depths at which they should be placed, and specify optimum running conditions for these bits.

The selection of the bit cannot be accomplished without good data. Good logs and bit records are a must, along with information on the mud and hydraulics that were used in

TABLE 3. Development of Op Drilling® Programs

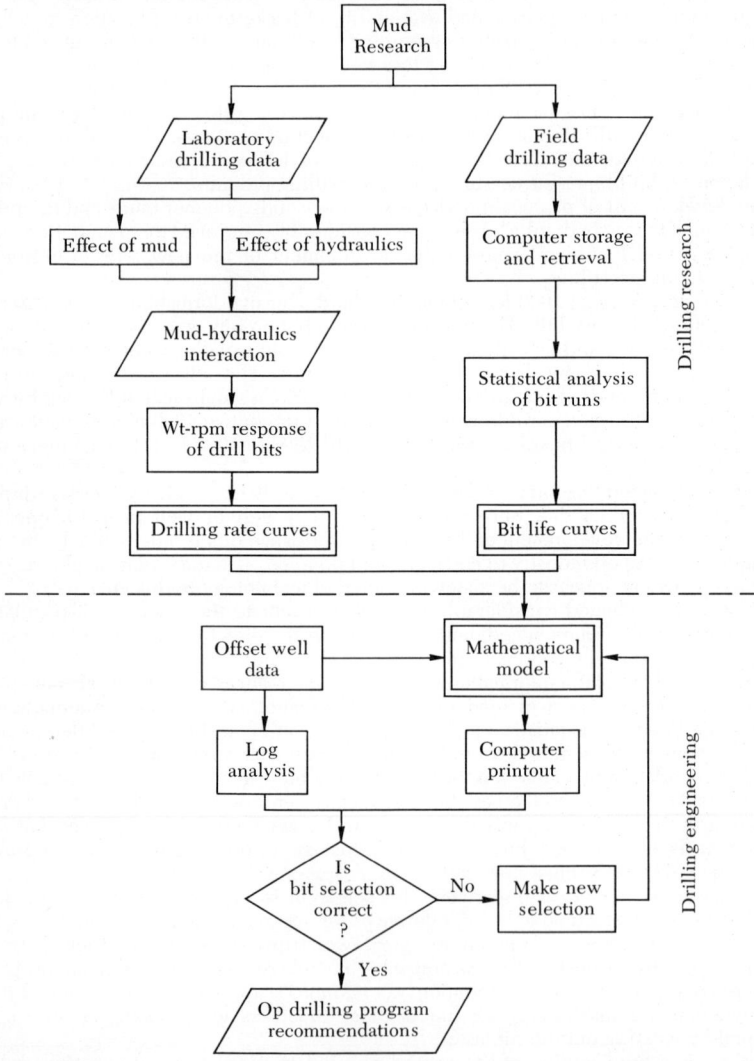

the wells selected as a control. It is often obvious at this optimization phase that some of the offset well bits were not the best possible selection for certain formation units. This requires the engineer to make a new selection and rerun portions of the computer program. The end result is the *Op Drilling® Program* recommendations which are posted on a copy of the offset well log, upon which are also posted the adjusted depths of major formation tops.

The drilling research indicated on the top part of Table 3 provides continuous support of the mathematical models used by the drilling engineer, who may be miles away from the computer in some field location or area office.

Drilling Assistance Programs.   These programs were developed to provide the drilling engineer, tool pusher, or drilling foreman with the necessary tools for making logical day-by-day decisions. They are based on large data banks which contain drilling and geological information, but have the capability of generating recommendations based on a wide range of considerations.

Table 4 lists examples of drilling assistance programs that are available on a time-share network for analyzing daily drilling operations. These programs are divided into the following categories:

1. Drilling operations
2. Equipment evaluation
3. Economics
4. Deviation control
5. Well control/safety

Certain drilling assistance programs are used to maintain a basic drilling plan, once the parameters have been established and a detailed plan has been developed on how to approach the drilling on a specific well.

The basic purpose of the drilling assistance programs is to provide the operator-contractor with the capability of analyzing conditions as drilling progresses so that adjustments can be made in the basic program to take into account unexpected formation changes or handle problems that inevitably arise in drilling.

Figure 18 shows how the system works: contact can be made directly to the time-shared computer and new information on drilling parameters can be obtained for implementation at the rig.

The drilling assistance programs make decisions such as best bit weight, bit type, jet nozzle size, pump pressure.

These programs are designed to take into account changing conditions. For example, if the mud type, hole size, pump pressure, bit type, etc., are changed, what difference can be expected in drilling cost per foot? If the mud type is changed, the bit may drill more effectively. However, if the hole size is increased, the larger bit will cost more, heavier collars will slow round trips, hydraulics will have to be recalculated. The bit must move more earth, but may have longer teeth to cut farther each turn, and larger bearings which may last longer. All these factors must be considered in determining the overall changes to be expected in drilling costs per foot of hole.

Equipment evaluation answers such questions as: How efficient is a centrifuge? Should equipment rental be continued, should it be fixed, or should it be abandoned? Which of several rigs with different costs and capabilities has the best chance of drilling the cheapest hole?

Well control and safety programs provide information for preventing troubles before and during drilling operations, and instruct how to minimize the hazard of a kick after it occurs. For example, one program will calculate the amount of casing required below a

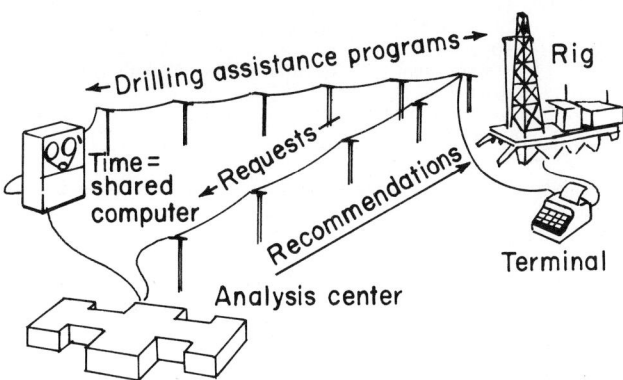

Fig. 18. Drilling assistance program projects technology to the rig.

TABLE 4.   Drilling Assistance Programs

DRILLING OPERATIONS	
HYDRAUL	Optimized rig hydraulics for drilling
HYD-ANAL	Computes conditions for the analysis of the drilling hydraulics for a particular bit run
MUDHY	Effects of mud and hydraulics on penetrating rate and drilling cost
BUTTON	Compares mill-tooth bits with an insert bit
DIABIT	Compares roller rock bits with a diamond bit
DRILLOFF	Drill-off test to determine actual optimum weight-on-bit
COSTCAL	Calculates the economic life of a bit or when the bit should be replaced
WTRPM	Best bit weight and rotary speed to drill a particular section
BITOP	Bit optimization program for sealed bearings and journal bearing tungsten carbide insert bits
EQUIPMENT EVALUATION	
CNTRFUGE	API bulletin for evaluation of centrifugal separators
CYCLONE	API bulletin for evaluation of hydrocyclones
RIGSEL	Compares the capabilities and probable costs to drill a well with three different rigs
SHAKER	API bulletin for evaluation of vibrating screens
DRILPIPE	Designs drill pipe per Youngstown catalog and API RP-7G
ECONOMICS	
COST. SUM	Identifies and prices bits. Also computes cost per foot, trip times and net days for a single bit or an interval
MUDADD	Calculates costs to treat and maintain a mud system with specific products
MUDCOST	Calculates the amount and the cost to treat a mud system with dry mud materials. Based on materials usage in pounds per barrel and pounds per foot of hole drilled
WELLCOST	Calculates the cost of drilling a gas, oil, or dry well for a given year. Based on industry data from joint association survey and petroleum engineering deep well report
DEVIATION CONTROL	
ASSEMBLY	Bottom hole assembly rating—rigidity, stickability, fishability
DEV. PLAN	Calculates and plots relative location of up to 23 wells to a reference well over any desired depth interval
TRAJECT	Solutions and file creation for wellbore trajectory having either "S" shape or build and hold configurations
TVD. INCR	Calculates wellbore location at specified true vertical depth increments (up to 200) from standard survey file
VERTSECT	Deviation survey calculations and plotting. Either Tangential, Balanced Tangential, Angle Averaging, or Radius of Curvature methods available
WELL CONTROL—SAFETY	
D. EXP	Calculates standard D-exponent and modified D-exponent which makes provisions for changes in bit type, bit wear, mud weight, and over pressure
KICK	Gives IADC consistent instructions for preventing a kick from becoming a blowout
SHOEFRAC	Calculates fracture conditions and casing-depth requirements for offshore operations
SURGE	Pressure surge and pipe lowering speeds. Both hydraulic and impact dynamic forces
CASING	Designs casing from either United States or Canadian stocks
MUDMIX	Calculates the amount of dry barite to increase the mud weight

jack-up rig to safely circulate 12 pounds of mud when drilling in 230 feet of water. Another question which can be resolved is how fast the drill pipe can be lowered in this well with a packed bottom-hole assembly without causing lost circulation due to fracturing.

To keep computing costs at a minimum, drilling assistance programs were designed so that they do not actually call data files but work from mathematical expressions developed from 15 years of research and analysis of massive data files. Amoco's in-house computers are used to access and analyze data files in a background batch process. The technical assistance programs are backed up by these large computer files containing location and formation top data on over a million wells, 160,000 complete bit runs, detailed cost information on approximately 5000 wells, and daily drilling reports on 800 completed wells.

Many of the drilling assistance programs calculate answers no different from those that can be done by hand or by using many other types of computer programs. A few of the programs are unique, and hopefully they all share a common difference from most available computer programs in that they are ready for the field man to use now and give a maximum amount of accurate information with a minimum of effort. These programs are all written in the very popular question-and-answer format. A person can sit down at the computer terminal and, with a minimum of instruction, can input information that he has assembled concerning the drilling operations and can ask questions pertaining to his particular problem, whether it be on muds, bits, or what drilling rate should be expected for a given set of conditions. This provides him with guidelines that he can use in conducting his daily overview of the drilling program.

WELL COMPLETION

An oil well may be described as a pipeline reaching from the top of the ground to the oil-producing formation. It is through this pipeline that oil is brought to the surface.* This pipeline is a series of joints of a special kind of pipe (casing) screwed together to form a continuous tube or string for the oil and gas to flow through. A simplified diagram is shown in Fig. 19. It is necessary to protect the hole from underground water and from loose earth falling from the surface. Also, the fresh water zones must be protected from the drilling and produced fluids. To provide this protection, usually two or more strings of casings are cemented in the hole. The first and larger of the casings is called the surface string. This casing will extend from the surface to a depth great enough to keep surface waters and loose earth from entering the well. The length of the surface string will vary from 200 to 1500 feet, depending on local conditions. See Table 5.

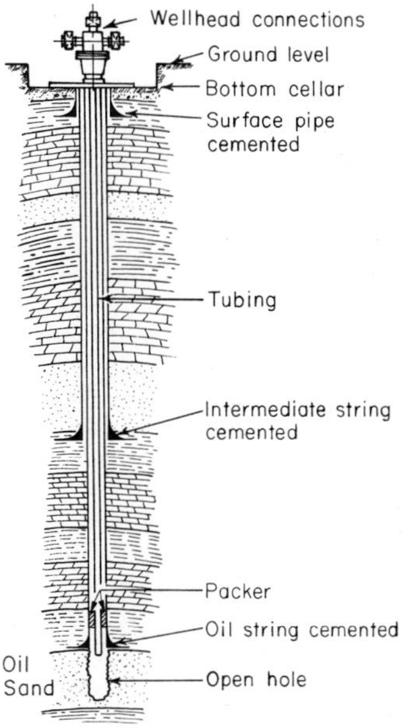

A second protective string may be used. This casing is called the intermediate string or "salt string" because it generally is run to a depth sufficient to case off salt and anhydrite formations. Such a string may be set at a depth of 5000 feet or more. The final string of casing, called the oil string, will usually extend from the surface through the surface and intermediate pipe, to the top of and sometimes through the producing zone at total depths of 20,000 feet or more. Because both strings of casing are subjected to large pressures and forces, it is necessary that the casing string be carefully designed and properly run into the well.

The methods of preparing an oil well to produce are many and are governed by the kind of oil reservoir. If the well is bottomed in hard formations, the oil-producing zone may be left entirely open, with no perforated casing or liner used to protect the hole. In loose, soft sands, it may be necessary to cement the oil string at the top of the producing zone

Fig. 19. Simplified diagram of an oil well.

and use a slotted or even a gravel-packed liner set through the oil sand. This liner is a string of casing that does not reach to the surface and is usually suspended, or hung, from the bottom of the oil string. Four types of completion are shown in Fig. 20.

The purpose of the liner is to keep sand and solids out of the well, yet allow the passage of oil and gas into the well. If there are several oil-producing zones at different depths, the

---

*A world record drilling depth was achieved in mid-1973 with completion of the 30,050-foot E. R. Baden well in Oklahoma of the Lone Star Producing Co. The Bertha Rogers near Burns Flat, Okla., is somewhat in excess of 29,000 feet in depth. Costs for the latter well exceed $5 million (1974 figures).

TABLE 5. Common Sizes of Casing and Line Pipe

		CASING*		
Outside diameter, in.	Weight, lb/ft	Resistance to collapse, psi	Internal yield pressure, psi	Joint strength tension (1000 lb)
4½	9.50	3310	4380	112
5½	17	4500	5320	272
7	23	3290	4360	344
7⅝	26.40	3010	4140	378
8⅝	36	2740	3930	437
9⅝	40	2770	3950	521
10¾	40.50	1730	3130	450
13⅜	54.50	1140	2730	545

		LINE PIPE		
Nominal size, in.	Outside diameter, in.	Inside diameter, in.	Weight, lb/ft	Average length, ft/joint
1	1.315	1.049	1.70	22
2	2.375	2.067	3.75	22
2½	2.875	2.469	5.90	22
3	3.50	3.068	7.70	22
4	4.50	4.026	11.00	22
6	6.625	6.065	19.45	22

*Sizes given above are Grade K-55, seamless steel, threaded and coupled, API joint. Other grades of casing are available. Common lengths are: range 2: 25 to 34 ft; range 3: 34 ft and longer.

oil string of casing may be run the entire depth of the well, and then be gun-perforated or jet-perforated opposite the horizon to be produced. Gun perforating is done by shooting metal bullets through the casing so that holes are left in the casing at the desired depth. Jet perforating uses specially shaped charges instead of bullets to perforate the casing. A common type of completion not shown in Fig. 20 consists of setting the oil string of casing through the producing formation, cementing it in place, and then perforating the casing and cement into the producing formation. The method is similar to that shown in Fig. 20b. Another type is multiple completion, a process by which it is possible to produce from different pay zones through the same well bore. This affords a means of obtaining the maximum amount of oil with the minimum use of casing. See Fig. 21.

Because the casing and liner must remain in a well for a long time and their repair or replacement would be costly, another string of pipe is placed in the well through which the oil is produced. This is called tubing and is used as the flow string. During the later life of the well, the same tubing may be used to support a pump or for other means of artificial lift. Tubing sizes generally range from 1¼ to 4½ inches in diameter. The tubing is suspended from the well head (surface) and usually reaches to within a few feet of the bottom of the well. Tubing is employed as the flow string because casing is usually too large to permit the well to flow efficiently, or, in some cases, to maintain continuous flow. Tubing is also required in a pumping well to support the pump.

Tubing packers are sometimes used in the tubing string to seal off the space between the tubing and the oil string of casing. This is done particularly in wells where there are high reservoir pressures. By sealing off this space, the casing is not exposed to high pressure, and the chances of a casing failure are reduced. Tubing anchors and packers also support part of the weight of the tubing in the casing and prevent the tubing string from moving up and down.

Sometimes it is both practical and economical to drill a small-diameter hole and use conventional tubing as casing in completing a well. This is called a tubingless completion since no retrievable inner string of tubing is used to conduct fluids to the surface. The casing is cemented from bottom to top and perforated opposite the pay zone. The equipment used is essentially the same as a conventional well, including a float collar, guide shoe with back pressure valve, and landing nipple. Tubingless completions with pipe as small as 2⅞ inches O.D. provide for well control, well stimulation, sand control, work-over, and an artificial lift system.

WATER CONTROL*

The primary means of water control in the well completion stage is properly cemented casing. This will help in stopping water flow between the formation strata via the annulus between casing and hole. Within the formation, flow is normally limited by laminations between strata. Unfortunately, the flow within strata known as coning is common. Coning is a response by a fluid above or below the completed interval to the pressure drop caused by oil production. The fluid travels to the completed interval in a path shaped like a cone. Where oil is produced by water drive, wells drilled near the oil-water contact line in the field are likely to start coning soon after production commences. Other wells located farther from the oil-water contact will develop coning later as the contact advances through the field.

Coning can be delayed by placing a barrier between the water and the oil- or gas-producing interval. Cement barriers are not successful because cement tends to enter the formation in a sheet vertically oriented instead of horizontally, and so chemical sealing solutions are preferred. The sealing solution can be any fluid of low viscosity that will set up to an immobile sealant within the formation. Two sealants in use are sodium silicate and acrylamide plus crosslinking chemical. These two produce firm, semisolid gels within a capillary system, withstanding considerable differential pressure, 700 to over 10,000 psi, depending on the thickness of barrier and permeability of rock.

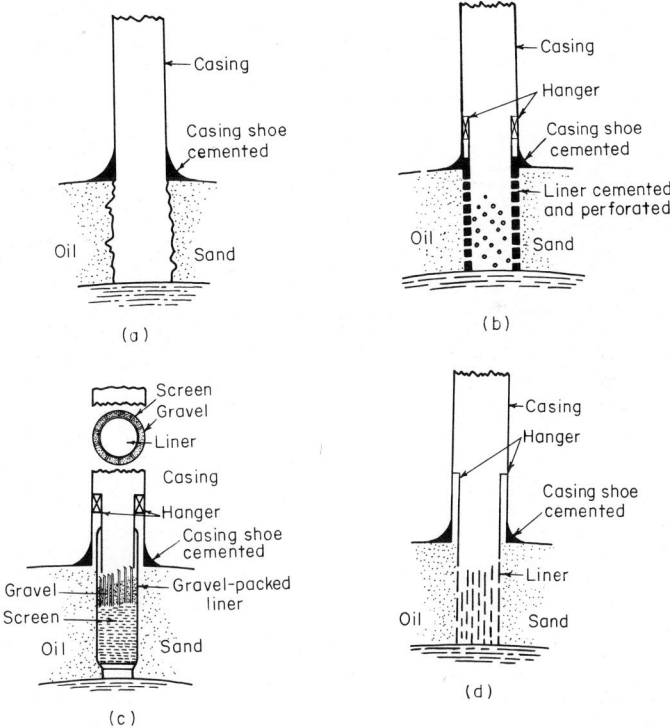

Fig. 20. Some types of well completions. (*a*) A "barefoot," in which the hole is left open to the formation. This can be done when oil-producing zone is not loose. (*b*) Gun-perforated liner prepared by shooting bullets or jets through a section of cemented casing so that oil will flow into the wellbore. (*c*) Gravel-packed liner, used where producing sands are fine-grained and loose. The gravel in space around liner acts as a filter to keep sand out of well. (*d*) Vertically slotted liner used where producing sand is loose and of fine-to-coarse grains. The grains of sand form a bridge across the slot, through which oil can enter the well. (*Source of sketches: American Petroleum Institute.*)

*Prepared by H. C. McLaughlin, Halliburton Services, Duncan, Okla.

In planning water control measures for the well, drilling is continued past the oil zone into the water zone. After the casing is cemented, perforations are placed in the top of the oil zone and in the top of the water zone. It is advisable to leave space between the two sets of perforations to set a packer. The packer is set between the two sets of perforations. For thin strata or high vertical permeability, it is best to simultaneously inject into both sets of perforations. The lower stream is the sealing solution which is pumped through the tubing and into the water-bearing formation. The upper stream is an inert fluid, preferably oil, which is pumped through the annulus between tubing and casing to help in preventing upward migration of the sealing solution into the oil strata. With adjustment for differences in hydrostatic column weights between annulus and tubing, the two streams are pumped at equal pressure, but not necessarily equal volumes.

The treating pressure must be maintained below fracturing pressure because fracturing has been found to be detrimental to complete sealing of the formation. Also, fracturing will waste the sealing solution by depositing it in and along the crack.

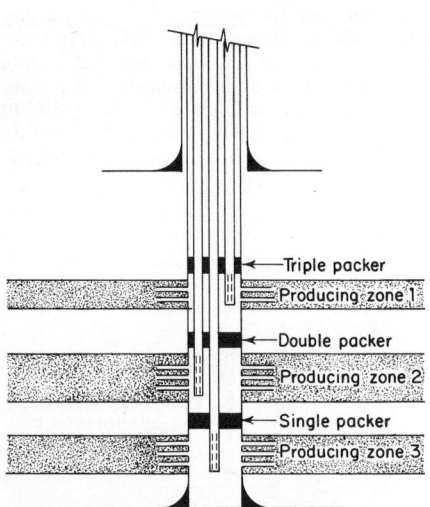

Fig. 21. Subsurface details of a triple (multiple) type of well completion: highly schematic. *(American Petroleum Institute.)*

Triple packer
Producing zone 1
Double packer
Producing zone 2
Single packer
Producing zone 3

SAND CONTROL*

Many newly completed oil- and gas-bearing formations throughout the world are incompetent. Formation sand migrates into the well bore during initial withdrawal efforts, in some instances causing production to cease completely or be reduced to an uneconomical level. Therefore, it is often considered essential to include sand control measures as a part of the original well completion procedure. In other situations, new completions are treated for sand control even though immediate sand problems do not exist. The previous production history of some areas indicates that sand may eventually become a problem sometime during the life of the well.

The effective application of sand control measures depends on properly preparing the well for treatment. The well should be conditioned to help ensure that each perforation and/or all the sanding formation accepts treating fluids. For instance, perforating fluids, usually filtered and containing no solids, are used to minimize permeability damage to the formation during perforating. All formation solids and plugging debris should be removed from across the perforated intervals. Special surge and wash tools are often used to clean perforation of solids. Properly cemented casing tends to prevent channeling of treating fluids to unwanted zones. Producing intervals should be selectively perforated, avoiding shale strata if possible. Narrow isolation of zone with special tools is also considered important to improve diversion of treating fluids into the sanding zone.

Often special preflush solutions are run ahead of sand control applications. These include clay-treating chemicals to shrink or stabilize clays; surfactants to break emulsions and remove water blocks; solvents for cleaning formation solids of oil and water; and acids which react with clays and calcareous components to improve formation permeability resulting in easier fluid injection.

A variety of sand control techniques are used by the industry on new completions. These may be classified into two general types.

Mechanical Technique.   In one method sand control is achieved by using only an in-hole screening device designed to retain the formation solids while permitting sand-free

*Prepared by Bill M. Young, Water and Sand Control Section, Halliburton Services, Duncan, Okla.

production. In another technique a specially sized pack sand (designed to bridge or retain the formation solids) is left placed around an in-hole screening device which spans the producing zone. The pack sand is transported in oil- or water-based carrier fluids in a variety of concentrations, ranging from less than 1 to 15 pounds of solids per gallon of carrier fluid. Carrier fluids of different viscosities, sometimes gelled, are available for use, depending on the process involved.

The gravel pack completion permits the formation fluids to be routed through the pack sand and screen, and to the surface while the pack and formation sands are retained in place. Open-hole and perforated-casing-type conpletions are suitable for application of these pack procedures.

A less commonly used gravel pack procedure, applicable only in perforated completions, requires no screening device in the borehole. With this method, a small percentage, usually less than 25 percent, of a larger sand is incorporated with the recommended-size pack sand. The larger solids bridge on the formation side of the perforations as fluid flow into the well bore is resumed. This bridge retains the pack sand in place during production. With this procedure, perforation sizing must be carefully controlled. Fracturing pressures may be used in placing sand behind perforated casing, particularly on new completions.

Chemical Technique.   Organic resins, e.g., furan, epoxy, and phenol formaldehyde types, are used to consolidate the unstable formation solids. The liquid resins, sometimes containing clay-treating chemicals, silane resin-to-sand coupling agents, surfactants, and diverting agents, are injected into these formations in a variety of procedures. The formation particles are left thinly coated with resin which is caused to harden. This results in the bonding of the solids together at the grain-to-grain contact points. The desired objective is to leave the resin-sand consolidation sufficiently porous to permit acceptable production of hydrocarbons therethrough. Sand consolidating materials applied in this manner do not require a screening device in the borehole. This can be an advantage if multiple and/or small hole completions are involved. In-hole devices may hamper future well work-over operations.

ACIDIZING*

One of the major contributions to increased production of both existing wells and new wells was made by Herman Frasch, inventor of acidizing (Ref. 1 under "Acidizing" in reference section.) On October 10, 1895, the Oil City, Pennsylvania "Derrick" gave a glowing report of what is thought to be the first acid treatment of an oil well. Only a few wells were acidized at that time, and not until 1928 did the Gypsy Oil Company, a division of Gulf, use inhibited hydrochloric acid for scale removal in oil wells. (Refs. 2, 3) In December 1931, W. A. Thomas of Pure Oil Company, independent of prior art, decided that hydrochloric acid could be used to stimulate oil wells producing from limestone formation. Grebe and Poffenberger of The Dow Chemical Company had previously stimulated injection and brine wells with acid and worked with Pure Oil on their oil wells. In February 1932, Pure Oil Company successfully treated its first well. In June 1932, The Oil Makers Company was organized in Michigan. Between 1932 and 1935, a number of oil well acidizing companies were formed, several of which are still in business. From that time to the present, acidizing has been a major stimulation tool for calcareous formations.

Sandstone Acidizing.   The first wells treated by Frasch were primarily limestone and, although a great deal of production was from sand, it was not until 1933 that a mixture of hydrochloric acid and hydrofluoric acid, especially designed to dissolve silicious minerals, was developed. Two applications were submitted to the U.S. Patent Office in March 1933 from different inventors (Refs. 4, 5), but Halliburton conceived of the technique early in 1933 and actually conducted the first job on May 3, 1933 (Ref. 6). The technique did not gain widespread use until the early 1940s and is only now beginning to be

*Prepared by John A. Knox, Chemical Services, Environmental Control Department, Halliburton Services, Duncan, Okla.

understood (Refs. 7 to 9). Hydrofluoric acid is used primarily for removal of mud damage or damage from other causes. Naturally occurring silicates or clay minerals are often contributory to damage. Nonaqueous mud removal solutions may be required when oil base muds are encountered.

Inhibitors.   An inhibitor containing arsenic was used as early as 1928. Until 1971, no inhibitor comparable to arsenic for 15 percent hydrochloric acid at elevated temperatures had been developed (Ref. 10).

Early claims were made that a scientific study of the properties of limestone resulted in the choice of 15 percent hydrochloric acid as the preferred acid strength for stimulation. Actually, two other factors, neither of which have anything to do with limestone, probably resulted in the choice. First, arsenic inhibitor is not effective in hydrochloric acid if the concentration is much higher than 15 percent, and, since arsenic was the prime inhibitor until about 15 years ago, acid concentrations above 15 percent were used only on rare occasions. The second reason was probably cost and expediency. Most commercial muriatic acid is about 30 percent HCl by weight. By diluting it one-half with water, a larger volume of a less expensive acid solution was available which would hopefully penetrate further into the formation and provide better results. The value of higher-strength acid is now recognized and the kinetics of the reaction show that deeper penetration can be achieved with stronger acid. However, the use of stronger acid had to await the development of better organic inhibitors.

Organic inhibitors have been vastly improved since their conception. It was not until the late 1950s that a well hotter than $200°F$ could be safely treated with hydrochloric acid unless arsenic was used as the inhibitor (Ref. 11). Today, 30 percent HCl can be inhibited with organic systems up to about $325°F$, and 15 percent HCl can be inhibited up to $400°F$. By using cooling preflushes of water and other fluids, wells much hotter than $400°F$ can be considered as candidates for acidizing. Careful job planning is important any time acid is used in a well, but even more care must be taken when treating hot wells to preclude the possibility of trapping acid in the pipe for a period longer than it can be safely inhibited.

The presence of hydrogen sulfide, H_2S, in many wells is a factor that must be considered when planning acid treatments. Hydrogen sulfide cracking of high-strength steels is intensified as the pH of the fluid contacting it is lowered. An interesting development in acid corrosion inhibitors is one that will prevent cracking of these high-strength tubular goods when contacted with HCl saturated with H_2S (Ref. 13).

Surfactants.   One of the first additives for acid, developed in the mid-1930s (Ref. 6), was a surfactant to lower the surface and interfacial tension to allow easier penetration into the tighter formations. This type of material is still in use today, particularly in gas wells and water injection wells. Surfactants are employed today as foaming agents, suspending agents for fine solids released by the acid, antisludging agents for asphaltic oils, retarders, aids for mud removal, and emulsifiers. Emulsifiers were recognized as a problem during the early 1930s (Ref. 6), but it was not until 1938 that nonemulsifiers were incorporated into the acid-treating solutions. Today they are an integral part of any acidizing treatment in oil wells. Mutual solvents have been used with acid, primarily in damage removal treatments (Ref. 9). Acid solutions containing alcohol are also widely used in cleaning up gas wells.

Diverting Agents.   These were first developed in 1936 (Refs. 14 to 16). They were utilized to allow more uniform distribution of the acid throughout the producing interval. Their use has gone through three or four revivals, in the early 1950s and 1960s, and now again in the 1970s, but each time utilizing new technology and providing a better product. The use of solids having wide particle-size distribution in conjunction with viscous gels is now enjoying success (Ref. 17).

In the middle 1950s, perforation ball sealers (Ref. 16) took the place of other diverting materials, but they have not proved to be the panacea once thought.

A few years ago, limited-entry-type perforating (Ref. 19), combined with adequate horsepower and fluid properties, promised to do away with any type of diverting material, including perforation ball sealers, but proved to have some shortcoming as it relates to stimulation. At the present time, a modified-limited entry perforating technique, which utilizes clusters of perforations situated in the most highly permeable sections and separated by fairly long intervals of casing, promises to provide control of stimulation fluids for better overall formation coverage. During stimulation, balls or diverting agents are used in conjunction with acid or fracturing fluid to divert from zone-to-zone.

Modified limited-entry perforating applies primarily to fracturing stimulation, but also has application in high-permeability unconsolidated sands where the fracpac technique is utilized.

Fracturing. The 1940s provided no developments of consequence until 1948 when fracturing formations with sand-laden fluids was developed (Refs. 20, 21). This proved not only to be a boon to low-permeability sandstone formations, but to acidizing limestone and dolomite formations as well. Until this time acid was primarily used for damage removal.

The concept of fractures being present in calcareous formations was accepted in the early 1940s, but it was assumed that they existed naturally and that acid only cleaned them out and enlarged them. No other explanation could be given for the very large production increases observed with some acid treatments. Based on Muskat's work (Ref. 32), the diameter of a well bore would have to be increased to 20 feet to obtain a twofold production increase—but much larger increases were common. During this time better pumping equipment was developed, allowing treatment at higher pump rates and pressures, but until it was accepted that fractures could be created, the real potential of acid was not utilized.

Retarded Acids. The desirability of retarded acid systems to provide deeper penetration had been recognized earlier, but until about 1951 no effective process had been found.

Both gelled acid and emulsified acid became available about that time. Gelled systems trap the acid in the cellular structure of the gel and interfere with the mobility of the hydrogen ions. Emulsified acid consists of droplets of acid wrapped in oil which is the external phase, and the spending time of the acid is a function of the stability of the emulsion. Some acid external emulsions are being used for stimulation purposes at the present time.

Gels show friction reduction and are easily pumped, but do not show appreciable retardation above about $150°F$ in static tests. Although stable emulsions showing retardation at $300°F$ are available, they have high friction pressures and are not widely used since high-rate treatments are the vogue.

No new retarding systems were developed until about 1960 at which time a study of mixtures of organic acids and hydrochloric acid was presented (Ref. 23). Hydrochloric acid mixed with acetic acid is now one of the preferred treating solutions. In addition to the longer reaction time of acetic-hydrochloric acid mixtures, unusual etching properties are also observed. Special dynamic etching equipment is now available (Ref. 24) to compare the etching properties of various types of acid. With many formations, the advantages of the hydrochloric-acetic acid mixture are obvious. Organic acids alone are used occasionally to stimulate high-temperature wells, but their etching properties are not outstanding.

Two other developments in 1960 were the use of high concentrations of calcium chloride in HCl to provide retardation (Ref. 25) and the addition of carbon dioxide (Ref. 26), which also retards the acid slightly. Both were replaced by concentrated HCl when adequate inhibitors were developed. However, carbon dioxide can probably still aid in providing more effective stimulation even with concentrated acid systems. Nitrogen also has application in gas well stimulation.

Differential Etching. Chemically retarded acid, which provides not only retardation but also differential etching was reported in 1964 (Refs. 27 to 29). This technique relies on the adsorption of an anionic surfactant on the calcareous rock surface which renders the rock oil wettable. Oil can then adsorb on the rock and prevent the hydrogen ions from reaching it. Until the film degrades, little or no reaction can take place. When the film degrades, it does so in an uneven manner, resulting in uneven etching and high fracture flow capacity. Systems are available today which effectively retard not only 15 percent HCl, but also acetic acid-hydrochloric acid mixtures and concentrated acid up to about $250°F$. These are primarily applicable in oil-producing zones.

Reactions. By 1940 the reactions of HCl on limestone and dolomite were thought to be well defined (Ref. 30). Nougaro and Labbe (Ref. 31) penetrated the phenomenon further in 1955 when they described how far acid could penetrate into the pores of limestone. Rowan (Ref. 32) further defined acid penetration by observing that most limestone and dolomite porosity is vugular rather than intergranular, and that different principles apply in this instance which can result in deeper penetration, and consequently greater production increases. Lasater (Ref. 33) and Dunlop (Ref. 34) developed the kinetics of the reaction of HCl and limestone further in the early 1960s, while Barron et al. (Ref. 35) described the effect of flow on acid reactivity in fractures.

A number of years passed with little new work being published until Schechter, Gidley, Williams, and Guin gave a series of papers on matrix acidizing, worm-holing, and characterization of the reactions between HCl and limestone (Refs. 36 to 38). Reactions of dolomite were characterized by Lund et al. (Ref. 39). The American Institute of Chemical Engineers sponsored a symposium at the 74th National Meeting (New Orleans, Louisiana, March 11 to 15, 1973) on "Surface Reactions in Porous Media," (Refs. 40 to 44). This renewed interest in matrix acidizing and reaction kinetics has been accompanied by new developments in fracture acidizing.

Reactivity in Fractures.   There is still much disagreement on just what happens while acid is rapidly moving down fractures. Nierode and Williams presented a kinetic model for the reaction of HCl with limestone and utilized this to predict acid reaction during a fracturing operation (Ref. 45). They carried this work further into a design of "Fracture Acidizing" treatments (Ref. 46). This latest and most comprehensive study on acid reaction in fractures includes factors, such fracture geometry, reaction rates, fluid mixing, fluid leak-off, temperature, acid concentrations, and others. The work uses the premise of an infinite surface reaction. The investigators suggest that longer conductive fractures and hence higher productivity can be obtained by increasing injection rate, viscosity of acid, fracture width, and acid concentration. Also suggested is the use of large preflushes to cool the fracture face and viscous preflushes to provide viscous fingering.

A supporting paper showing field results of treatments based on the design parameters previously described was presented at the CIM meeting in 1972 (Ref. 47). At the SPE symposium on oil field chemistry in Denver (May 24 to 25, 1973), Roberts and Guin described a mathematical model which was formulated for predicting the penetration distance of live acid in a fracture for situations in which the overall reaction rate is influenced by the surface kinetics (Ref. 48). It is based on an effective diffusion coefficient and finite surface kinetics. This has been further developed in a paper which uses data taken from rotating disks to predict acid penetration distance (Ref. 49). Computerized designs of this type, but utilizing different assumptions, have been available for some time. These designs include well-bore cooling, fracture face cooling, fracture geometry, viscous fingering, and acid penetration.

Fluid Loss Additives.   The importance of acid fluid loss additives to effective stimulation cannot be overemphasized. Although they have been available and recommended for many years, fluid loss additives are not widely used. The same principles apply to fracturing with acid that apply to fracturing with any other fluid. If fluid leak-off is not controlled, shorter and thinner fractures will result, consequently accompanied by lower production. In addition, a great deal of expensive acid will be more or less wasted near the well bore when it could be utilized to provide long, highly conductive fractures. If it were necessary to place sand along with the acid, the importance of the fluid loss additive would be quickly realized because a sandout would occur without it. The problem with acid is actually more difficult because the acid is eating out the pore space which the fluid loss additive is trying to cover. Usually a larger amount of fluid loss additive is required to provide control in acid than in other fluids.

Viscous Preflushes.   Several highly viscous fracturing fluids have been developed in the last few years. Their viscosity is much higher than anything used as a sand-carrying fluid before—and they are said to have perfect particle transport. This allows placement of proppant throughout the fracture. These highly viscous fluids have application in fracture acidizing as well allowing the development of much wider fractures. Also, when a thinner fluid, such as acid, follows the thicker fluid into a fracture, a phenomenon described as viscous fingering occurs. The acid may contact only 30 to 60 percent of the exposed fracture, resulting in deeper penetration with a smaller amount of acid.

According to Tinsley et al. (Ref. 50), it is not necessary for a conductive fracture to penetrate the whole producing horizon vertically to obtain significant production increases. Contacting this lesser amount of the formation can conserve acid while resulting in favorable stimulation. An additional advantage can result—it is no longer necessary to depend upon differential etching of a fracture face for conductivity since 40 to 70 percent of the fracture face may be untouched by acid. This provides a pillar and post effect to prop the fractures open.

HYDRAULIC FRACTURING*

Hydraulic fracturing is the technique of applying sufficient pressure on the formation to cause it to fail or fracture. Pressure is transmitted from the surface pumping equipment through the tubular goods to the bottom of the well by the fracturing fluid. Once the fracture is initiated, this fluid is pumped at sufficiently high rates to open and extend the fracture as well as carry propping agents into the fracture. Propping agents, such as sand, UCAR® (Union Carbide), or other granular materials are used to provide a conductive path for the fluids in the reservoir to reach the well bore.

Formation stimulation by hydraulic fracturing is the utilization of this technique to improve the initial production rate from the well, together with the possible increased total recovery from the reservoir.

Hydraulic fracturing has done much to alert the oil and gas industry to the need for better well completions. Many of the first wells fractured were older wells completed with lightweight casing which was often inadequately cemented. Even some newer wells had low-strength casing set with limited amounts of cement. Thus, the maximum pressure permitted on a casing string, which seldom exceeded 3000 psi, usually was governed by the test pressure of the pipe, or by limitations of the surface pack-off equipment. Because of these completion methods, treating pressures encountered with the fluids were sufficiently high to warrant the use of packers with the treatments conducted through tubing.

Because of the low injection rates that could be achieved through tubing, many operators began preparing wells for fracturing at the time of completion by setting better grades of casing, using improved cementing practices and fracturing with fluids requiring lower surface-treating pressures. However, as wells were drilled deeper, it was necessary to develop higher-strength tubing to permit stimulation at pressures of 15,000 to 20,000 psi to achieve sufficient injection rates through the smaller tubular goods.

A large percentage of the wells completed today will probably require some form of stimulation at completion or at some later date. With this in mind, consideration should be given to the hole size, mud program, casing and tubing schedule (strength and size) logging techniques, cementing programs, perforating practices, and how these factors may affect future treatments. The production interval of current interest should not be the only one to receive attention. Other zones not exposed may be future candidates for stimulation and production.

If a well is to produce additional fluid or gas as a result of fracturing, one of the first factors to determine is that the reservoir has additional recoverable reserves. It must also have sufficient reservoir pressure to allow flow into the fracture. A drill stem test may be used on initial completion to determine the ability of a well to respond to stimulation. It can also be employed in older wells to determine the amount of driving force or pressure remaining in the reservoir. Information from a drill stem test may also be used to determine the presence of a permeability restriction, such as well-bore damage, around or near the well bore.

A well penetrating a low-permeability pay zone should also be considered for stimulation since some of these wells will not produce into the well bore until fractured because of large-pressure drawdown at the desired production rate.

Hydraulic fracturing may also be of benefit in wells that are completed in "dirty" zones of medium-to-high permeability. These zones may contain large amounts of water-sensitive or water-swelling clays, together with water-migratable clays; thus the term "dirty" formation. This type of formation may have been blocked or damaged during drilling because of the loss of drilling mud to the permeability or the drilling fluid contacting the clays. If the depth and type of damage is not severe, a simple "clean-up" treatment may produce the desired results. However, if the damage is severe, it may be necessary to create a fracture through the damaged zone to allow the well to produce.

Some formations contain existing fracture networks, and production is through these small hairline fractures. Then a fracturing treatment may be used to connect these natural fissures in sufficient quantity to increase production.

However, fracturing is not necessarily limited to producing wells. This technique has been successfully employed in fracturing injection or disposal wells. The injection profile of a well in a waterflood unit may be improved or altered by fracturing.

*Prepared by E. J. Stahl, Section Supervisor, Halliburton Services, Duncan, Okla.

In selecting a formation for stimulation, the properties and characteristics of the formation and the fluid in the reservoir should be considered. Formations classified as medium to hard in strength generally show the best response to stimulation. This is probably due to the ability of the propping agent to keep the fracture open and maintain some fracture flow capacity or conductivity path for fluid in the reservoir to reach the well bore. Formations that are classified as soft or unconsolidated generally produce the least response. In this type of formation it may not be possible to prop the created fracture because of the embedment of the propping agent. Also very hard formation encountered in deep wells may cause the proppant to crush and thus decrease the fracture flow capacity.

Hydraulic fracturing in highly vugular or naturally fractured limestones may require special techniques, depending on the constituents within the formation. If acid tends to dissolve the limestone evenly along the face of the fracture, then a stimulation treatment using water and a propping agent may produce better results. However, if there are insolubles in the formation, the acid will etch the fracture face unevenly, creating a pillar and post effect which should result in high-fracture flow capacity after treatment.

If the formation contains clays, then care should be taken in the selection of the fracturing fluid. Some clays are water-sensitive and have a tendency to swell or slough when contacted with certain aqueous base fluids. When this occurs, fines may be released which can migrate to plug formation permeability or fracture flow capacity. Therefore, any incompatibility between the fracturing fluid and formation should be evaluated.

The success of the fracturing treatment then depends on several items: (1) well cleanup and preparation before fracturing, (2) type and properties of the base fracturing fluid, (3) additives selected to condition the fluid, and (4) volume of fluid and effective fracture flow capacity created in the formation.

Perforations may become plugged with paraffin or other material that should not be pumped into the formation during the treatment. If this should happen, it could cause high treating pressures on a fracturing treatment, and a chemical treatment to clean the perforations may be all that is required before fracturing. If the well is damaged near the well bore because of the loss of drilling mud or contamination of formation clays through invasion of a foreign water, then attempts should be made to remove this potentially damaging material before fracturing.

In all these instances, it is important to determine the type of material to be removed or treated so that the correct type and amount of fluid can be used, together with the proper treating procedures. Fluids normally used in the fracturing processes will consist of oil, water, or acid-base fluids; and in some instances, nitrogen or carbon dioxide may be added during the treatment. Tests should be conducted to determine that the fluid selected for the fracturing treatment is compatible with the formation to be treated. Some of the characteristics of the formation which should be considered are solubility, salt content, and clay minerals. In addition, the fluids in the reservoir should be evaluated to determine whether the formations of emulsions, paraffin plugging, or other solids could be precipitated when contacted by the fracturing fluid.

The fracturing fluid should possess properties of low friction loss in the tubing and fluid loss control within the fracture. Once formation breakdown is achieved, the purpose of the fracturing fluid is to create new fracture area as well as to generate fracture flow capacity by transporting propping agents into the fracture or through acid etching of a soluble formation. The extent of the fracture area created is controlled by the efficiency of the fluid, which is a function of the fluid leak-off during the treatment. Fluid loss additives are used for this purpose, and specific additives depend on the characteristics of the formation. Other factors that should be considered in selection of the fracturing fluid are temperature versus viscosity properties, chemicals needed to break a thickened fluid, or the degree to which a fluid might damage the formation.

In summary, the result of a fracturing treatment should be the development of a fracture with sufficient propped width, length, and height to provide adequate fracture flow capacity in order to effectively increase and sustain the production from the reservoir.

WORK-OVERS

A work-over may be defined as the process or technique of cleaning out or otherwise working on a well in order to increase or restore production.

Water Control.* Excessive water production with oil occurs when: (1) the reservoir is near depletion, (2) highly permeable zones connect water-bearing strata with the borehole, (3) natural or induced fissures connect the borehole with underlying water-bearing formations, and (4) a combination of any of the foregoing conditions.

In almost all cases, the water produced with oil is brine containing mostly chlorides, sulfates, and carbonates of alkaline earth metals. Elimination or reduction of this water production is desirable to conserve formation pressure for oil production, reduce lifting costs, avoid produced water-disposal problems, minimize water blockage of the oil-producing intervals, and reduce equipment corrosion and scale formation.

Water shutoff chemicals are applied as solutions that invade and plug formation matrix permeability by precipitation, gelation, or polymerization to form solids. Such solutions are often made into slurries by the addition of finely divided inert solids which render them useful for treatment of formations containing both fissure and matrix permeability. When both types of permeability are present in the same formation, the solids portion of the slurry plugs the fissures, while the filtrate from the slurry invades and plugs the matrix.

Proper placement in the formation is controlled by utilizing down-hole tools and/or regulating pumping rates. Down-hole pack-off devices may be used to isolate the water-producing interval from the oil zone. When such tools cannot be used, a balanced pumping technique may be employed, wherein communication between the oil zone and the borehole is preserved by pumping oil-compatible radioactively tagged fluids down the annulus between the casing and the tubing, while simultaneously pumping the water shutoff material through the tubing and into the water-producing interval. Pump rates for the two fluids are controlled from the surface, and monitored by a gamma-ray detector suspended by cable in the tubing, to maintain the interface between the two fluids at the proper level.

General rules for water control treatments are: (1) never exceed formation fracturing pressure, and (2) do not overdisplace the treating fluid. Factors that govern the choice of treatment procedures and material quantities are: (1) formation temperature, (2) interval length, (3) injection rate, (4) interwell distance, and (5) type of permeability.

Sand Control Work-overs.† Unconsolidation of many oil- and gas-bearing formations can occur after various periods of sand-free production. Some reasons for the delayed breakdown of these formations are: (1) Water encroaches in with oil production, causing disintegration of natural cementing materials within the sand grain network. (2) Depletion of reservoir fluids permits more overburden stresses to be placed on the natural consolidation. (3) Excessive production rates result in the occurrence of damaging pressure differentials across the stabilized formation near the well-bore vicinity. (4) Exposure of formation to certain well work-over fluids can destroy the natural consolidating materials.

Remedial work-over sand control measures often involve sand packing in a formation to replace the produced sand and to impart some restressing of the zone near the well bore. Commonly used sand control methods for work-over wells that have produced significant quantities of sand are classified as mechanical or chemical-mechanical. Mechanical methods were described earlier under Sand Control in connection with well completion. The chemical-mechanical procedures are described in the following paragraphs. The well preparation procedures and the use of preflush solutions as previously described apply also with these methods.

Resin-coated sand is positioned across the producing zone by means of a viscous oil carrier fluid. A sand concentration of 10 to 15 pounds per gallon of oil is normally used. Once the resin, which contains an internal curing agent, has hardened, the placed sand becomes consolidated. A hole is drilled through the consolidated sand pack in the well bore to permit production of formation fluids into the borehole and to the surface. In effect, a consolidated sand screen is formed within the borehole.

Another method employs low concentrations of resin-coated pack sand (½ to 2 pounds per gallon of carrier fluid) to pack a formation behind the perforations. Additional resin-carrier dispersion may be placed through the pack sand to cause consolidation of a portion of the formation particles. An overflush catalyst solution is injected through the resin-coated sand to cause resin hardening. No borehole screening device is generally needed.

---

*Prepared by Roger F. Rensvold, Development Chemist, Halliburton Services, Duncan, Okla.
†Prepared by Bill M. Young, Water and Sand Control Section, Halliburton Services, Duncan, Okla.

Still another method involves the gravel packing of a formation with untreated sand. The placed sand pack is then consolidated by treatment with resinous materials, as described previously under Sand Control in connection with well completion.

Sand control involving a long interval usually favors the use of pack methods because of generally more favorable treating economics as well as the likelihood of more complete treatment of all zones.

Sand consolidation methods wherein resinous materials are injected into a formation are sometimes considered for a work-over well if the interval is less than 10 to 15 feet in length, and if the formation has produced only a small quantity of sand.

Fracturing in Work-overs.*   The use of fracturing techniques in connection with well completion has been previously described. In the following paragraphs, the use of hydraulic fracturing treatments on wells that have been previously fracture-stimulated will be described. These *refrac* treatments are used as a work-over procedure on a large number of wells each year.

Refrac treatments may be required if a well is not producing or accepting fluid at the anticipated rate. Production rates may have declined after the previous treatment to the point where they are below the economic limit, or the well may have never produced at the anticipated rate.

In 1949, the hydraulic fracturing process was introduced to the petroleum industry on a commercial basis. Since 1949, more than a half-million hydraulic fracturing treatments have been conducted, and about 35 percent of these have been refract treatments on wells previously fractured. This process is employed to increase the production or injection rate of liquid or gas from or into a subsurface formation. Hydraulic pressure is used to cause failure or fracturing of the formation, at which time a fracture or crack is created in the formation. The effective radius of the well bore is increased because of the fracture, thus making possible larger production or injection rates. To keep the fracture open after the treatment, a propping agent, such as sand, is normally employed.

There are several reasons why a well, even after a stimulation treatment, may not be producing or accepting fluid at the expected rate. These include:

1. Because of a limited drainage area, the well may have been depleted before or after the previous treatment.

2. The formation does not contain the anticipated producible reserves.

3. The previous fracturing treatment may have been too small to create a fracture of sufficient length to drain the formation effectively at the expected rate.

4. Or the treatment may have generated a fracture of sufficient length to drain the producing horizons effectively, but of insufficient propped height.

5. If an extensive perforated interval was selected, some of the perforated intervals may not have been processed by previous treatments, thus limiting production.

6. Because of salt or paraffin deposition during production, or incompatibility of the fracturing fluid and the formation, the permeability of the formation at the fracture faces may have been reduced owing to this damage, thus limiting production.

7. The fracture flow capacity of the proppant system may not be sufficiently high to allow the formation to produce at the expected rate. This lack of sufficient fracture flow capacity could be due to a small size proppant having been used in a fracture treatment on a high permeability formation. This type of formation would have required a high fracture flow capacity proppant system to sustain adequate production rates.

Also, the proppant may have generated adequate fracture flow capacity initially, but has decreased with time because of plugging of the proppant system caused by scale deposition, proppant crushing or embedment due to the closure pressure, and the resultant release of fines from the proppant or the formation.

When contemplating refrac treatment for a specific well, one of the first considerations should be a review of reservoir energy, recoverable reserves, and economics. It is possible that sufficient reservoir energy and recoverable reserves exist and that the first treatment was not an optimum treatment. Before a refrac treatment, it should be determined whether sufficient oil is available for recovery by reviewing original estimates of oil-in-place, amount of oil produced, expected recovery factor, and the bottom-hole pressure. Also, methods of estimating whether the entire pay section was originally fractured should be discussed.

*Prepared by E. J. Stahl, Section Supervisor, Halliburton Services, Duncan, Okla.

If it is determined that the entire net pay interval was not fractured on the initial treatment, then creation of new fractures may be necessary. These would not come under the classification of a refrac or work-over treatment in the strictest sense. However, the use of diverting agents or mechanical isolation, since these zones were not stimulated originally, may be employed to bridge off or divert the fracturing fluid from the old fracture system, allowing a new fracture to be created.

Assuming that the created fracture from the original treatment was vertical and extended throughout the net pay thickness and that the reservoir energy and economically producible reserves are present, it becomes necessary to evaluate the propped fracture length, flow capacity, and distribution of the proppant used in the original treatment. By knowing the type of proppant used in the previous treatment, the fracture closure pressure, the concentration of the proppant in the fracture, and the type of formation fractured, it is possible through laboratory tests or from published information to estimate the flow capacity of the propped fracture system. The effects of embedment, proppant crushing, and fines in the proppant system should be considered. In addition, the propped fracture height and length obtained on the previous treatment can also be ascertained by means of published proppant transport equations.

With information on flow capacity and propped fracture height, a relative capacity obtained by the treatment may be calculated. Therefore, it should be possible to decide whether sufficient relative capacity resulted from the initial treatment to adequately stimulate the well.

If it is determined from these calculations that the relative capacity is low and a large portion of the net pay zone was propped, then this proppant system should be removed or redistributed and a higher flow capacity proppant system should be placed in the fracture.

In those instances where it was decided that the old proppant system had sufficient relative capacity, but was of insufficient length or height, it would be a matter of proppant transport studies to place the proppant to greater depths and to greater heights in the formation.

If it has been established that production from the well may be helped by another treatment, then the following procedure may be of assistance in evaluating the design of the refrac treatment:

1. Determine the propped fracture length and height resulting from the previous treatment, using published equations for fracture width, fracture area, and proppant transport. By knowing the propped fracture height and created fracture height, the position of the proppant system in relation to net pay intervals may be evaluated. If it is found that very little of the proppant covered the net pay intervals, then perhaps all that will be required in the retreatment will be to extend the propped bed height to cover these zones. If the calculated propped fracture length penetrated only a small percentage of the drainage radius of the well, then the retreatment will require extension of the propped length.

2. If it is found from the calculations that most of the net pay interval was vertically covered with proppant, then the fracture flow capacity and relative capacity of the fracture system should be examined.

3. Based on the theoretical production increase curves, if the relative capacity is initially low or minimal, a reduction of flow capacity with time would be detrimental. Thus, a refrac treatment may be required to place the proper propping agent in the fracture to establish again a high fracture flow capacity and a favorable relative capacity to sustain production. However, if the relative capacity was initially high, then a decrease in flow capacity may still leave sufficient relative capacity for sustained production increase.

FLUID INJECTION

The practice of injecting or returning gas and water underground has become an important part of oil-producing operations. The injection of gas into an oil pool increases recovery of oil, and sometimes results in a saving of natural gas. Water is injected into underground formations for two reasons: (1) to dispose of salt water that is produced with oil, and (2) to increase the amount of oil recovered by injecting water into a producing formation.

Injection of gas requires the use of compressor plants to raise the pressure of the gas so that it will go into the wells, which take the gas to the producing formation. Water can sometimes be injected into an underground formation by gravity, but often a pump is necessary in order to inject water at the rate desired. Pumping is usually required where water is injected into a producing formation for a waterflood or pressure-maintenance program.

Primary pressure maintenance is the application of fluid injection early in the producing life of a reservoir, when there has been little loss of natural reservoir energy. Fluids of

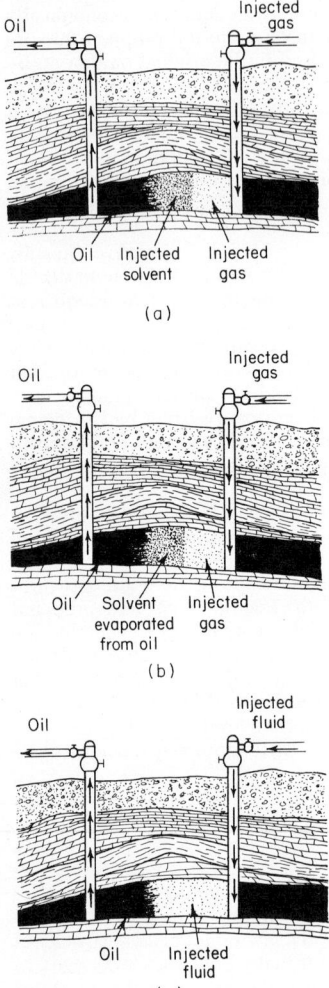

(a)

(b)

(c)

Fig. 22. Principal types of fluid injection systems. (*a*) Miscible flooding—also called solvent-flood or miscible-slug method—involves the injection of a slug of solvent, capable of being mixed with the reservoir oil. This is followed by the injection of a displacing fluid which mixes with the solvent. The injected displacing fluid moves the solvent through the reservoir. The solvent, in turn, removes the oil from the portion of the reservoir through which it passes. Solvents used include propane and butane, and mixtures of these with or without methane. The displacing fluid is usually natural gas under sufficient pressure so that both the solvent mixture and the displacing fluid become liquid. When water is used as the displacing fluid, some alcohols which have a high degree of solubility with both oil and water have been used. (*b*) High-pressure gas drive. This method involves the injection of natural gas at high pressures (above 3000 psi). When the injected gas acts on the reservoir fluids, the gas becomes changed so as to become mixable with the reservoir oil, forming a solvent zone. This solvent zone, which is formed using elements already in the reservoir oil, results in almost complete displacement of the reservoir oil in that part of the formation swept by the injected gas. (*c*) Enriched-gas drive. This method involves injection of gas enriched with propane and butane into the reservoir. As this gas contacts the reservoir oil, some of its elements condense into a solvent, swelling the oil, reducing its viscosity, and changing its flow properties. The swollen oil becomes mixed with the injected gas and moves toward the producing well to increase recovery. *Note:* The injection of salt water into the producing formation is commonly used to increase the ultimate recovery of oil—and it is generally agreed that better results are obtained from this method than may be obtained from the injection of gas. The water which is injected is obtained either from wells that produce water along with the oil, or from water-supply wells drilled for this purpose. Water injected into formations must meet certain requirements—whether injection is for disposal, for pressure maintenance, or for water flood. Mainly, the injected water must be clear, stable, similar to water in the formation, not severely corrosive, and free of materials that may plug a formation. Consequently, injection water often will require pretreatment, including deaerating, softening, filtering, chemical treating, stabilizing, and careful testing.

controlled composition are put into selected wells at or near discovery pressure to achieve maximum recovery efficiency. Some of the more common methods of fluid injection are miscible flooding, high-pressure gas drive, and enriched-gas drive—as illustrated and described briefly in Fig. 22.

Waterflooding. Waterflood projects are used principally in fields found before full benefits of pressure maintenance could be realized. Some or all of the existing wells may be utilized, plus carefully located new wells. Water is injected into certain wells to flush oil from the reservoir rock. In operation, water is pumped through these injection wells into the oil reservoir, spreads out from the injection wells and moves toward the oil wells, driving reservoir oil ahead of it. This system is continued until the fluid taken from the producing wells becomes uneconomically high in water content.

Thermal Processes.   One recorded technique proposed is to burn some of the oil in a reservoir (in situ) as an energy source (increased pressure) to increase total amount of oil recovered. The process would actually use some of the oil that could not be produced by any known method and hence would not be wasteful, as it might appear. Other thermal methods use either heated water or steam to provide both lowered viscosity and reservoir drive. Steam injections may be used as a displacement mechanism or to stimulate production from individual wells.

All improved recovery techniques include methods for supplementing natural reservoir forces and energy, or otherwise increasing ultimate recovery from a reservoir. Such techniques cover secondary recovery in its original sense, i.e., fluid injection applied relatively late in the productive history of a reservoir for the purpose of stimulating production after recovery by primary methods of flowing or artificial lift has approached an economic limit.

ARTIFICIAL LIFT

When pressures in an oil reservoir have fallen to the point where a well will not produce by natural energy, some method of artificial lift must be used. Oil-well pumps are of three general types: (1) pumps at the bottom of the hole run by a string of rods, (2) pumps at the bottom of the hole run by high-pressure liquids, and (3) bottom-hole centrifugal pumps. Another method involves the use of high-pressure gas to lift the oil from the reservoir.

OTHER PRODUCTION OPERATIONS

Space limitations do not permit a full description of several other oil production operations, such as well logging, well testing, separation and treatment of oil and gas, field storage systems, gaging and switching, lease and oil rights administration, government regulations and controls.

REFERENCES

Drilling

1. Bradley, William B.: Factors Affecting the Control of Borehole Angle in Straight and Directional Wells, SPE paper 5070 presented at Society of Petroleum Engineers of the American Institute of Mining, Metallurgical, and Petroleum Engineers, 49th Annual Fall Meeting, Houston, Tex., Oct. 6–9, 1974.
2. Rollins, H. M.: "How to Drill a Better Hole," part I (January), part II (February), part III (March), part IV (April), *World Oil,* 1970.
3. Rollins, H. M.: "Straight Hole Drilling," part I (March), part II (April), *World Oil,* 1963.
4. "Drilling and Well Servicing Structures," API specification 4E, 2d. ed., American Petroleum Institute, Division of Production, Dallas, Tex., March 1974.
5. "Planning, Designing, and Constructing Fixed Offshore Platforms," API Recommended Practice RP 2A, 5th ed., American Petroleum Institute, Division of Production, Dallas, Tex., January 1974.
6. Jackson, R. A.: Insert Bits Can Cut Drilling Costs, *Oil Gas J.,* Sept. 17, 1973.
7. Knowlton, Jack: Console Gives Driller Linear, Digital Data, *Oil Gas J.,* June 11, 1973.
8. Kennedy, J. L.: Oil Firms Join for High-Pressure Drilling Tests, *Oil Gas J.,* July 23, 1973.

Geopressured Drilling

1. Fertl, Walter H.: Worldwide Occurrence of Abnormal Formation Pressures, part I, paper no. SPE 3844, presented at the Abnormal Subsurface Pressure Symposium of SPE–AIME, Baton Rouge, La., May 15–16, 1972.
2. Rehm, Bill: Worldwide Occurrence of Abnormal Pressures, part II, presented at same symposium as Ref.1.
3. Smith, N. E., and T. H. George: The Origins of Abnormal Fluid Pressures, "Abnormal Subsurface Pressure," a study group report, Houston Geological Society, Houston, Tex., 1971.
4. Thomeer, J. H. M. A., and J. A. Botteman: Increasing Occurrence of Abnormally High Reservoir Pressures in Boreholds, and Drilling Problems Resulting Therefrom, *Amer. Ass. Petrol. Geol. Bull.,* 45, 10, October 1961.
5. Hottman, C. E., and R. K. Johnson: Estimation of Formation Pressures from Log-derived Shale Properties, *J. Petrol. Technol.* June 1965, pp. 717–722.

6. Pennebaker, E. S.: An Engineering Interpretation of Seismic Data, *Drilling*, SPE Reprint Series No. 6a, pp. 21–32, 1973.
7. Matthews, W. R.: Drilling Variables Show Transition Zone, *Oil Gas J.*, Nov. 3, 1969, pp. 68–76.
8. Zamora, Mario: Slide-Rule Correlation Aids '*d'* Exponent Use, *Oil Gas J.*, Dec. 18, 1972.
9. Eaton, B. A.: Fracture Gradient Prediction and Its Application in Oilfield Operations, *Drilling*, SPE Reprint Series No. 6a, pp. 7–14, 1973.
10. Eaton, B. A.: How to Drill Offshore with Maximum Control, *World Oil*, October 1970, pp. 73–77.

Well Control

1. "Blowout Prevention," Lessons in Rotary Drilling, unit III, lesson 3, Petroleum Extension Service, The University of Texas at Austin, July 1974.
2. Bell, F. S.: High-Pressure Drilling and Blowout Prevention, *Oil Gas J.*, Oct. 14, 1957.
3. Drilling Well Simulator Teaches Blowout Control, *World Oil*, March 1972, p. 53.
4. Zamora, M., and G. Gau: Computerized Simulator Aids Well Control Training, *Oil Gas J.*, Aug. 19, 1974, pp. 61–62.
5. Bruist, E. H.: A New Approach in Relief Well Drilling, *Drilling*, SPE Reprint Series No. 6a, pp. 97–106, 1973.
6. Hamby, T. W., and J. R. Smith: Contingency Planning for Drilling and Producing High-Pressure Sour Gas Wells, *Drilling*, SPE Reprint Series No. 6a, pp. 107–116, 1973.

Acidizing

1. Frasch, Herman: Increasing the Flow of Oil Wells, U.S. Patent 556,669 issued Mar. 17, 1896.
2. Fitzgerald, P. E.: A Review of the Chemical Treatment of Wells, *J. Petrol. Technol.*, September 1953, p. 198.
3. "History of Petroleum Engineering," 598 pp., American Petroleum Institute, Division of Production, Dallas, Tex., 1961.
4. Wilson, J. R.: Well Treatment, U.S. Patent 1,990,969, issued Feb. 12, 1935.
5. Heath, S. B., and W. Fry: Well Treating Method, U.S. Patent 2,011,579, issued Dec. 17, 1935.
6. Roberts, H: "Creative Chemistry," a history of Halliburton Research Laboratory (1930–1958), Halliburton Services, Duncan, Okla.
7. Smith, C. F., and A. R. Hendrickson: Hydrofluoric Acid Stimulation of Sandstone Reservoirs, *J. Petrol. Technol.*, February 1965, pp. 215–222.
8. Gatewood, J. R., et al: Predicting Results of Sandstone Acidizing, *J. Petrol. Technol.*, June 1970, pp. 693–700.
9. Gidley, J. L.: Stimulation of Sandstone Formations with the Acid Mutual Solvent Method, SPE paper 3007, *J. Petrol. Technol.*, May 1971, p. 551.
10. Keeney, B. R.: Acid Corrosion Inhibition Using Metal Halide-Organo Inhibitor Systems, paper No. 106 presented at National Association of Corrosion Engineers, St. Louis, March 1972.
11. Funkhouser, J. G.: Acid Corrosion Inhibition with Secondary Acetylenic Alcohols, presented at South Central Regional meeting of National Association of Corrosion Engineers, Oct. 25–27, 1960.
12. Billings, W. E., and David Morris: Effect of Acid Volume and Inhibitor Quantity on Corrosion of Steel Oil Field Tubing in Hydrochloric Acid, "Corrosion," 208 pp., 1960.
13. Keeney, R. B., Lasater, R. M., and J. A. Knox: "New Organic Inhibitor Retards Sulfide Corrosion Cracking in Acid," *Mater. Prot.*, April 1968, p. 23.
14. Clason, C. E.: Acid Treatment of Wells, U.S. Patent 2,122,452, issued July 5, 1938.
15. Menaul, Paul L. Method and Means of Acidizing Wells, U.S. Patent 2,122,483, issued July 5, 1938.
16. Hower, F. W.: Process for Increasing Permeability of Underground Formations, U.S. Patent 2,803,306, issued Aug. 20, 1957.
17. Dill, W. R.: Effect of Bridging Agents and Carrier Fluids on Diverting Efficiency, SPE paper 2331, November 1968.
18. Derrick, J. V., et al: Method of Temporarily Closing Perforations in the Casing, U.S. Patent 2,754,910, issued July 17, 1956.
19. Lagrone, K. W., and J. W. Rasmussen: Better Completion by Controlled Fracture Placement Limited-Entry Technique, presented at Southwest District Division of Production, American Petroleum Institute, Mar. 21, 1962.
20. Clark, J. B.: A Hydraulic Process for Increasing the Production of Wells, *Trans. AIME*, vol. 186, pp. 1–8, 1949.
21. Fast, C. R. and G. C. Howard: Hydraulic Fracturing (a History), SPE–AIME, "Henry L. Doherty Series Monograph," vol. 2, 1948.
22. Muskat, M.: "Physical Principles of Oil Production," chap. 6, McGraw-Hill, New York, 1949.
23. Dill, W. R.: Reaction Times of Hydrochloric-Acetic Acid Solutions on Limestone, presented at 16th Southwest Regional Meeting of American Chemical Society, Oklahoma City, Okla., Dec. 1–3, 1960.
24. Broaddus, G. C., J. A. Knox, and S. E. Fredrickson: Dynamic Etching Tests and Their Use in Planning Acid Treatments, SPE paper 2362, presented at Stillwatter, Okla., Oct. 25, 1968.
25. Dunlop, Pegg M., and J. S. Hegwer: An Improved Acid for Calcium Sulfate Bearing Formations, *J. Petrol. Techn.*, January 1960, p. 67.

26. Dill, W. R.: Acidizing Wells, U.S. Patent 3,076,762, issued Feb. 5, 1963.
27. Knox, J. A., R. W. Pollock, and W. H. Beecroft: The Chemical Retardation of Acid and How It Can Be Utilized, *J. Can. Petrol. Technol.*, vol. 4(I), pp. 5–12, January–March 1965.
28. Knox, J. A., R. M., Lasater, and W. R. Dill: A New Concept in Acidizing Chemical Retardation, SPE paper 975, presented Oct. 11–14, 1964.
29. Knox, J. A.: Well Acidizing Method, U.S. Patent 3,319,714, issued May 16, 1967; Canadian Patent 800,386, issued Dec. 3, 1968.
30. Stone, J. B., and D. G. Hefley: Basic Principles in Acid Treating Limes and Dolomites, *Oil Weekly*, Nov. 11, 1940, pp. 32–44.
31. Nougaro, J., and C. Labbe: Étude de Lois de l'Acidification dans le Cas d'un Calcaire Vacuolaire, *Rev. inst. Franc, pétrole*, vol. 10(I), p. 354 (January–June 1955).
32. Rowan, G.: Theory of Acid Treatment of Limestone Formations, presented to Institute of Petroleum, London, Sept. 2, 1959.
33. Lasater, R. M.: "Kinetic Studies of the HCl–CaCO$_3$ Reaction," presented at the 18th Southwest Regional Meeting of American Chemical Society, Dallas, Tex., Dec. 8, 1962.
34. Dunlop, P.: Kinetics of Acid Reaction on Limestone, presented at 17th Southwest Regional Meeting of American Chemical Society, New Orleans, La., December 1961.
35. Barron, A. N., A. R., Hendrickson, and D. R. Wieland: Effect of Flow on Acid Reactivity in a Carbonate Fracture, *J. Petrol. Technol.*, April 1962, p. 409.
36. Williams, B. B., J. L. Gidley, J. A. Guin, and R. S. Schechter: The Change in Pore Size Distribution from Surface Reactions in Porous Media," *A.I.Ch.E. J.*, vol. 15, p. 339, 1969.
37. Williams, B. B., J. L. Gidley, J. A. Guin, and R. S. Schechter: Characterization of Liquid Solid Reactions, *Ind. Eng. Chem. Fund.*, vol. 9, p. 589, November 1970.
38. Guin, J. A., and R. S. Schechter: Matrix Acidization with Highly Reactive Acids, *SPE J.*, December 1971, p. 390.
39. Lund, Kasper, and H. S. Fogler: Dissolution Kinetics of Selected Minerals in Mud Acid, paper 52c, presented at 74th National Meeting, American Institute of Chemical Engineers, New Orleans, La., Mar. 11–15, 1973.
40. Williams, B. B.: Fundamentals of Matrix Acidization, paper 52a (same meeting as Ref. 39).
41. Nierode, D. E.: Acid Reaction Kinetics of Carbonates, paper 52b (same meeting as Ref. 39).
42. Guin, J. A.: Permeability Changes in Dissolving Porous Media, paper 52d (same meeting as Ref. 39).
43. McCune, C. C.: An Experimental Technique for Obtaining Permeability-Porosity Relationships in Acidization, paper 52e (same meeting as Ref. 39).
44. Fass, S. M.: The Consequences of Different Temperatures on Pore Structure Development in Carbon, paper 52f (same meeting as Ref. 39).
45. Nierode, D. E., and B. B. Williams: Characteristics of Acid Reaction in Limestone Formations, SPE paper 3101, published in *Soc. Petrol. Eng. J.*, December 1971, p. 406.
46. Williams, B. B., and D. E. Nierode: Design of Acid Fracturing Treatments, *J. Petrol. Technol.*, July 1972, pp. 849–859.
47. Nierode, D. E.: Prediction of Stimulation from Acid Fracturing Treatments, paper No. 7228 presented at CIM meeting, Calgary, Canada, May 14–19, 1972.
48. Roberts, L. D., and J. A. Guin: The Effect of Surface Kinetics in Fracture Acidizing, *Soc. Petrol. Eng. J.*, SPE paper 4349, Oilfield Chemistry Symposium, Denver, Colo., May 24–25, 1973.
49. Roberts, L. D., and J. A. Guin: A New Method for Predicting Penetration Distance, SPE paper 5155 presented at Houston, Tex., Oct. 6–10, 1974.
50. Tinsley, J. M., et al: Vertical Fracture Height—Its Effect on Steady State Production Increase, SPE paper 1900 presented at Houston, Tex., October 1967.

Drilling Optimization

1. "Impact of New Technology on the U.S. Petroleum Industry," National Petroleum Council, Washington (covers 1946–1965).
2. Lummus, J. L., and L. J. Field: Non-Dispersed Mud: A New Drilling Concept, *Petrol. Eng.*, March 1968.
3. Field, L. J.: Low Solids Non-Dispersed Mud Usage in Western Canada, paper presented at Spring Meeting of Rocky Mountain District CIM, Division of Production, American Petroleum Institute, Calgary, Canada, May 1968.
4. Galle, E. M., and H. B. Woods: Best Constant Weight and Rotary Speed for Rotary Rock Bits, paper presented at Spring Meeting of Pacific Coast District, Division of Production, American Petroleum Institute, Los Angeles, Calif., May 21, 1963.
5. Estes, J. C.: Selecting the Proper Rotary Rock Bit, *J. Petrol. Technol.*, November 1971.
6. Lummus, J. L.: Analysis of Mud-Hydraulics Interactions, *Petrol. Eng.*, February 1974.
7. Lummus, J. L.: Weight-Rotary Speed Considerations, *Petrol. Eng.*, May 1974, p. 84.
8. Lummus, J. L.: Bit Selection, *Petrol. Eng.*, March 1974, p. 82.
9. Rader, D. W., and A. T. Bourgoyne: Factors Affecting Bubble Rise Velocity of Gas Kick, SPE paper 4647, Las Vegas, Nev., Sept., 30–Oct. 3, 1973.
10. Matthews, W. R., and J. Kelly: How to Predict Formation Pressure and Fracture Gradient from Electric and Sonic Logs, *Oil Gas J.*, Feb. 20, 1967.

11. Christman, S. A.: Offshore Fracture Gradients, *J. Petrol. Technol.*, August 1973.
12. Clark, E. H., Jr.: Bottom-Hole Pressure Surges While Running Pipe, *Petrol. Eng.*, vol. B-68, January 1955.
13. Burkhardt, J. A.: Well Bore Pressure Surges Produced by Pipe Movement, *J. Petrol. Technol.*, June 1961, p. 595.
14. Moore, P. L.: Pressure Surges and Their Effect on Hole Conditions, *Oil Gas J.*, Dec. 13, 1965, p. 90.
15. Schuh, F. G.: Computer Makes Surge-Pressure Calculations Useful, *Oil Gas J.*, Aug. 3, 1964, p. 96.
16. Melrose, J. C., et al. A Practical Utilization of the Theory of Bingham Plastic Flow in Stationary Pipes and Annuli, *Petrol. Trans. AIME,* vol. 213, p. 316, 1958.
17. Fontenot, J. E., and R. K. Clark: An Improved Method for Calculating Swab/Surge and Circulating Pressures in a Drilling Well, SPE paper 4521, Las Vegas, Nev., Sept. 30–Oct. 3, 1973.
18. O'Brien, T. B., and W. C. Goins: "The Mechanics of Blowouts and How to Control Them," American Petroleum Institute, Division of Production, Southern District, March 1960.

Acknowledgments.   The editors extend their particular appreciation to David L. Lord, Group Leader, and Mr. T. C. Buechley, Development Engineer, Halliburton Services, Duncan, Oklahoma, for their efforts and excellent cooperation in coordinating the presentation of much of the information in this article.

Oil Shale Retorting*

BY MARK T. ATWOOD†

The term *oil shale* refers to a carbonaceous rock that can produce oil when heated to pyrolysis temperatures (800–1000°F, 427–538°C), but not when extracted with organic solvents at room temperature. The oil precursor in the rock is a high-molecular-weight organic polymer called kerogen. Kerogen from the upper zones of Colorado and Utah oil shales has the following average composition (Ref. 1):

Element	% (weight)
Carbon	80.5
Hydrogen	10.3
Nitrogen	2.4
Sulfur	1.0
Oxygen	5.8
	100.0

The host rock consists mainly of dolomite, calcite, quartz, and clays (Ref. 2).

The oil shale area of most significance in the United States is the Green River formation of Colorado, Utah, and Wyoming. These reserves are described in terms of their potential to produce oil by a laboratory retorting procedure called Fischer assay (Ref. 3). A recent resource estimate was described by Henry O. Ash, based on a Federal Interagency Task Force Report (Ref. 4), and is given in Table 1.

A small percentage of this oil shale may be mined by surface techniques. Most of it, however, will be recovered by underground mining in large room-and-pillar mines.

RETORTING OIL SHALE

The major emphasis in the development of commercial oil shale retorting technology has been on underground mining of oil shale and above-ground retorting. In a sense then, the oil shale operation would resemble conventional minerals mining with above-ground beneficiation, such as is carried out in the production of metals. The recent pilot-scale work of Occidental Petroleum Corp. has, however, involved a potentially very interesting combination of underground mining and underground retorting.

Retorting requires that the shale be heated to around 900°F (482°C) to cause decomposition of the kerogen and evolution of the oil either in liquid form or as a vapor. Retorting has been accomplished, through large-scale precommercial development operations, in both vertical and horizontal retorts.

In the vertical retorting mode, processing can be divided into schemes in which the oil condenses and falls to the bottom of the retort and those in which it rises as a vapor and mist to be collected in a condenser.

TABLE 1.   Oil Shale Resources in Beds at Least Ten Feet Thick
Billions of Barrels

	25 gal/ton shale		30 gal/ton shale	
	In place	Recoverable	In place	Recoverable
Colorado	607	202	355	118
Utah	64	21	50	17
Wyoming	60	20	13	4
	731	143	418	139

*This article is confined to technology and practice in the United States.
†Manager of Laboratories, The Oil Shale Corporation, Golden, Colorado.

VERTICAL MODE (THE OIL FALLS)

N-T-U Retorting.   When the U.S. Bureau of Mines resumed oil shale investigations in 1944, two 40-ton-capacity batch N-T-U (Nevada-Texas-Utah) retorts were built and put into service in 1947. Upon completion of the test program in 1951, these two retorts had been used for 920 runs with a total consumption of 37,500 tons of raw shale.

A description of the process is shown in Fig. 1, taken from a report of investigations by J. R. Ruark, et al. (Ref. 5).

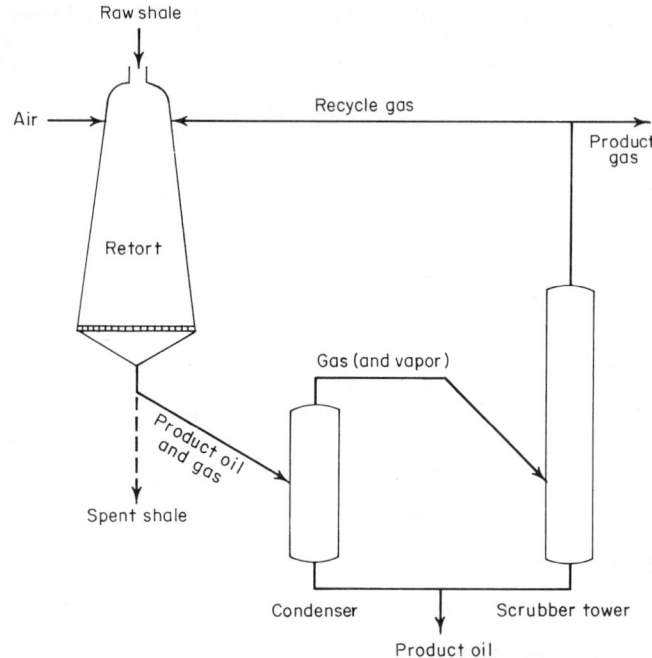

Fig. 1. N-T-U retorting plant flow diagram.

Forty tons of shale were placed in the retort by a feed mechanism operating from the top. The ignition was started at the top of the oil shale column. The proper flow of air and recycle product gas was adjusted by use of gas and air blowers; the downflowing hot gases, preceding the combustion zone, heated the oil shale and converted most of the kerogen to oil and gas. An organic residue was left behind on the processed shale, and it served as a fuel for heating the downflowing gases. Oil in both liquid and vapor forms exited from the bottom of the retort. As the retorting zone progressed to the bottom of the retort, the remaining shale bed became hotter, and more and more of the oil left the bottom of the retort as a vapor. Most of the oil was collected in the contact condenser, and the gas was scrubbed and conducted back to the top of the retort to provide fuel for the process. Oil yields averaged about 80 percent of that predicted by Fischer assay.

After completion of the runs, the spent ash was discharged by gravity from the bottom of the retort.

The shale feed utilized varied in size from ½ to 3½ inches. The presence of fines, that is, finer than ½ inch, would tend to restrict gas flow through the retort.

Product oil properties are given in Table 2 (Ref. 5). Comparison with the properties of other shale oils shows similarities in gravity, pour point, and elemental analyses. The naphtha yield is lower than that produced by the TOSCO II process.

A major operating difficulty with the N-T-U retort was buildup of coke deposits in the oil-recovery system. More importantly, however, it soon became obvious that there was no place for batch processing of oil shale in commercial operations.

Union Oil Process.   An ingenious and technically successful, continuous version of the N-T-U retort was developed by Union Oil Company. Union Oil acquired an oil shale property north of Grand Valley, Colorado, in the 1920s and has been engaged in oil shale research since that time. Development efforts were accelerated in 1943, and a small two-ton-per-day unit was designed. Later, a larger 30-to-50-ton-per-day pilot plant was constructed and operated in California. After successful results, a large retort was constructed and operated on the Colorado property.

The process employs countercurrent continuous flow of shale and air. The ash and klinkers are removed from the top, and the oil is drained from the bottom after being cooled by heat exchange with incoming, cold oil shale (Ref. 6). The process is described in Fig. 2 (Ref. 7).

Oil shale is fed at the bottom of a retort by an oscillating rock pump which alternates

TABLE 2.   Properties of Typical N-T-U Crude Shale Oil

Gravity, °API	20.3
Pour point, °F	90
Viscosity: SUS at 130°F	130
SUS at 210°F	48.4
Conradson carbon residue, wt %	4.5
Ash	0.02
Elemental analysis, wt %:	
Carbon	84.6
Hydrogen	11.4
Nitrogen	2.1
Oxygen (calculated)	1.1
Sulfur	0.8
	100.0
Distillation (Hempel), °F	
IBP	350
5 vol %	465
10	520
20	600
30	670
40	738
50	800

between loading and discharging operations. The bottom part of the retort is built of 70-degree cones, the outer part of which consists of an assembly of narrow slots through which oil and gas are discharged, but which inhibit, for the most part, bottom discharge of raw shale. The product liquids are collected in an annular section surrounding the slotted area. Any solids that do get through the slots drop down through the liquid oil seal and are reintroduced through the oscillating rock pump feeder. The preferred raw shale feed size is ⅛ to 2 inches, although larger sizes can be tolerated.

The retorting section is cooled by fins which are inside a cooling assembly through which cooling air flows constantly. This cooling assembly maintains a maximum metal temperature of 750°F (399°C) at the burning zone. At the top of the retort, it is important to avoid formation of large klinkers which would impede air distribution. A plow mechanism was used in the earlier pilot plants which consisted of spiral metal plows containing a liquid sodium coolant in their internals. Later, this spiral plow was replaced by a horizontally acting set of rams which removed the spent shale effectively.

In operation, the top zone effects heat exchange between incoming air and the hot spent ash. In the lower portion of the top zone, combustion of carbon residue on the spent shale produces flue gas at a temperature of 2,000°F (1,093°C), which moves down into the lower retort zone, causing pyrolysis and release of oil and gas. In the condensation zone, the incoming cold shale is preheated while it condenses the oil vapors. A major advantage of this process, in water-short western Colorado, is the use of the incoming raw shale for condensation instead of conventional utilization of cooling water.

Recent announcements have shown that the Union retorting technology is developing in the direction of using hot recycle retort gas to cause retorting instead of an internal combustion zone. R. F. During and G. E. Irish patented a scheme (Ref. 8) in which combustion retorts are combined with recycle gas retorts. The combustion retorts produce low Btu gas for indirect heating of rich recycle retort gas, which is used in separate units for hot gas retorting. A later patent (Ref. 9) describes retorting with exclusive use of hot recycle gas.

Finally, in June 1974, Union Oil announced the design of a commercial-scale plant employing the Steam Gas Recirculation (SGR) process. In this scheme the basic retort, with the rock pump, is unchanged, but the carbon on the spent shale is gasified with steam, along with either air or oxygen (Refs. 10, 11). The hot gas produced provides sensible heat to the retort to cause shale retorting in the absence of a combustion zone. It is also used as fuel gas to heat recycle product gas to provide further sensible heat to the retort. A 15 percent increase in shale oil yield over earlier methods is claimed.

In the SGR process, coking of the heaviest portion of raw shale oil will not be part of the prerefining step, and whole oil will be hydrogenated (Ref. 11).

The properties of the oil produced are given in Table 3. The gas produced consists mainly of nitrogen and carbon dioxide, but has sufficient combustible components to have a gross heating value of around 100 Btu per standard cubic foot. This low Btu gas can be utilized to provide process heat to the retort.

In a late-1974 recent announcement, Union Oil Company stated that they would begin operation of a 1500-ton-per-day retort on their property north of Grand Valley in the near future. The retort which was utilized in operations in 1956 and 1957 operated at feed rates of 1200 tons of shale per day for continuous periods up to six weeks. A commercial plant is considered by Union to consist of clusters of 3000-ton-per-day retorts.

Occidental Petroleum Process. This firm has been conducting large-scale tests on the southern rim of the Piceance Basin. The test site is located at the head of Logan Wash, north of DeBeque, Colorado (Ref. 12). The processing used could be viewed as the operation of a series of large-scale N-T-U retorts underground. The "retort" is constructed by driving horizontal drifts into the formation and removing about 15 to 20 percent of the total amount of oil shale which will be ultimately processed. The oil shale removed may be stored or processed by other means. From the base of the drifts, hole are drilled up and down into the area to be fractured. Explosives are inserted and subsequently detonated with "tiny time delays." The retort is then filled with rock in the form of a confined chimney. The void volume of the chimney will be the same as that of the rock removed, and the broken rock will support the walls and ceiling, but will permit gas flow.

As in the operation of the N-T-U retort, the broken shale rock is ignited from the top, and oil flows cocurrently with gas to the bottom of the broken rock chimney. Cooling and condensation occur in the lower zones. Liquid oil is pumped from a sump at the bottom of the broken rock chimney. Excess carbon remaining from pyrolysis provides fuel for the flame front. About 30 percent of the low Btu product gas is recycled to the top of the formation and utilized as fuel for the retorting operation. The other 70 percent of the gas must be consumed in some manner such as in the production of electric power. Broken rubble heights in early tests were 100 to 150 feet. In August of 1974, the firm was constructing a room 300 feet high and more than 100 feet square (Ref. 13).

The advantages of this process, compared with aboveground retorting, are that it would involve savings on handling of raw shale, would not involve surface disposal of spent shale, and, presumably, would function with lower capital investment and operating costs. The pilot operation in Colorado has produced substantial quantities of oil, 25 to 30 barrels per day, and has been considered a success by Occidental Petroleum Corporation.

A possible disadvantage of the process is the problem of disposal of the 15 to 20 percent of the shale which must be mined and removed from the formation before retorting.

The oil produced by this operation is described as very similar to that produced by the U.S. Bureau of Mines aboveground retorting operation conducted at Laramie, Wyoming. These runs were conducted at Laramie in large-scale batch retorts, again simulating the N-T-U concept, to study retorting of a broken shale chimney which may ultimately result from utilization of nuclear explosives. An analysis of the Bureau of Mines shale oil properties is given in Table 4 (Ref. 14).

VERTICAL MODE (THE OIL RISES)

Gas Combustion Process (U.S. Bureau of Mines). The U.S. Bureau of Mines developed the Gas Combustion retort process at the Anvil Point experimental station near Rifle, Colorado, and concluded a 12-year experimental program in 1956. In 1964, the facility was reactivated under lease to the Colorado School of Mines Research Foundation in order to

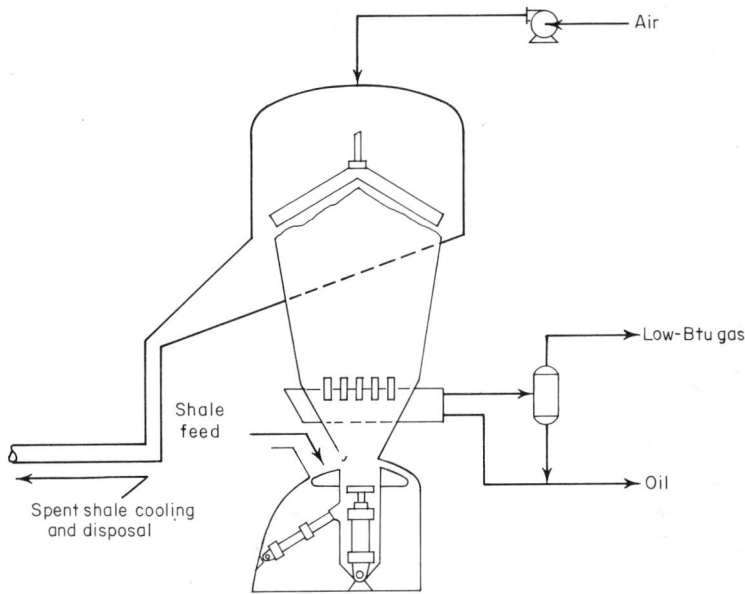

Fig. 2. Union Oil retorting process.

TABLE 3.   Union Oil Company Retort Crude Shale Oil Properties

Property	Retort A*	SGR
Gravity, °API	19.7	21.5
Sulfur, wt %	1.0	0.7
Nitrogen, wt %	1.8	1.8
Conradson carbon, wt %	5.6	1.8
Pour point, °F	85	70

*Retort A is illustrated in Fig. 2.

TABLE 4.   U.S. Bureau of Mines In Situ Oil Properties

Property	U.S.B.M. 150-ton retort	U.S.B.M. 10-ton retort
Oil yield, vol. % of Fischer assay	60.4	71.1
Specific gravity	0.909	0.910
API gravity	24.2	24.0
Pour point, °F	60	70
Viscosity, SUS at 100°F	98	77
Nitrogen, wt %	1.59	1.68
Sulfur, wt %	0.94	0.91
Naphtha, wt % of crude	6	6
Light distillate, wt % of crude	29	27
Heavy distillate, wt % of crude	37	36
Residuum, wt % of crude	28	31

proceed with a research program which was supported by six oil companies. These were Mobil Oil Corporation, acting as project manager; Humble Oil and Refining Company; Pan American Petroleum Corporation; Sinclair Research, Inc.; Continental Oil Company; and Phillips Petroleum Company.

Test programs were conducted on a six-ton-per-day retort, followed by further development work on a 25-ton-per-day retort. Finally, extensive development work was conducted on a 150-ton-per-day retort. This experimental work is described in two extensive reports by the Bureau of Mines (Refs. 15, 16).

The process is illustrated in its simplest form in Fig. 3 of Report of Investigations No. 7303 (Ref. 15). Raw shale is fed continuously into the retort and proceeds by continuous gravity flow. A combustion zone is maintained around the air-gas distributor, providing the necessary heat for retorting. Product oil rises as a mist, countercurrently to the downward flowing raw shale, and while preheating the incoming raw shale, avoids, for the most part, adsorption on the raw shale particles. The shale oil mist carried by product gas proceeds through a cyclone, a knockout pot, and finally to an electrostatic precipitator.

As the preheated raw shale moves down through the retort, it is heated to the retorting temperature range of 850–1,000°F (427–538°C) by the hot gases rising from the combustion zone. In the combustion zone, part of the required heat is provided by burning the carbonaceous residue on the spent shale. The retorted and partially burned spent shale continues to fall through the retort, providing sensible heat to the incoming air and recycled gas. The spent shale is finally discharged through a feeder. The product gas passes through the oil recovery system and is partially recycled to the bottom of the retort. The main portion is removed as low-Btu product gas.

In operation of this process, it is necessary that the raw shale and processed shale move smoothly in plug flow from the top to the bottom of the retort and, further, that the upward-moving gases flow smoothly without channeling. To facilitate uniform flow of gases, it is necessary that no raw shale fines be present, and, for this reason, raw shale pieces smaller than ¼ inch in size were rejected before retorting.

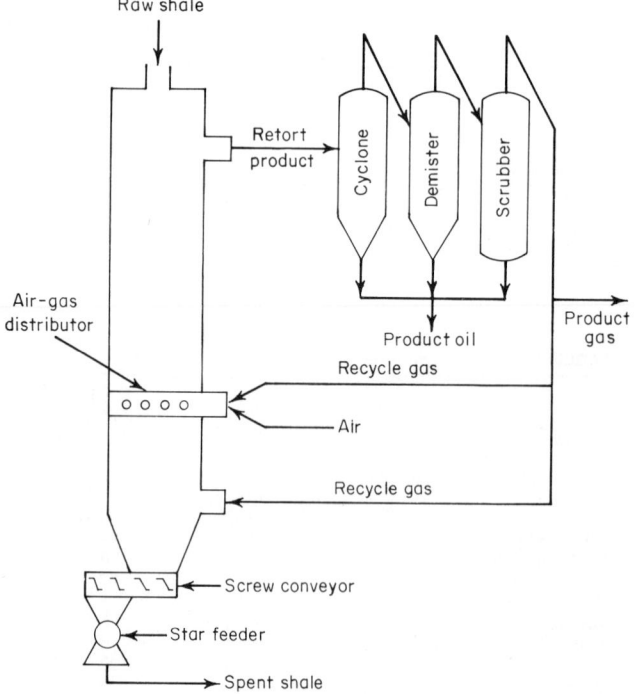

Fig. 3. Flow diagram of Gas Combustion Retort No. 1.

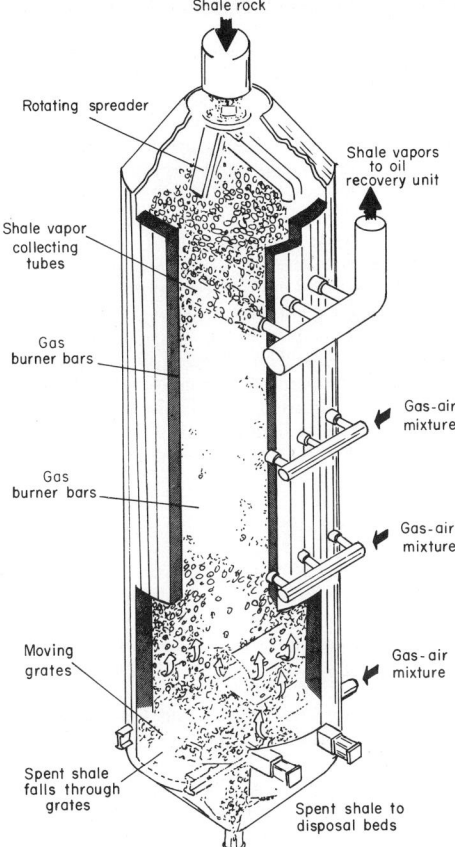

Shale rock

Rotating spreader

Shale vapors
to oil
recovery unit

Shale vapor
collecting
tubes

Gas
burner bars

Gas
burner bars

Gas-air
mixture

Gas-air
mixture

Moving
grates

Gas-air
mixture

Spent shale
falls through
grates

Spent shale to
disposal beds

Fig. 4. Paraho retort. *(Paraho Oil Shale Demonstration, Inc., Grand Junction, Colorado.)*

In a series of demonstration runs on the 150-ton-per-day retort, the average oil recovery was about 86 weight percent of that expected from Fischer assay of the raw shale feed. The product oil had the properties given in Table 5.

Serious problems with bridging and formation of clinkers in the retort handicapped pilot plant operations and discouraged further commercialization plans.

Paraho. Paraho Oil Shale Demonstration, Inc., was formed in 1973 to develop commercial oil shale applications of the DEI (Development Engineering, Inc.) vertical kiln. Three commercial installations had been successfully operated for the pyrolysis of limestone. A long-sought goal has been the application of the kiln to the retorting of oil shale. Under the leadership of the Standard Oil Company (Ohio), more than 15 participants were enlisted to support a $7.5 million development program. The Anvil Points Experimental and Demonstration Facilities near Rifle, Colorado, were leased for five years from the U.S. Bureau of Mines as a site for the work.

Figure 4 depicts a relatively simple cylindrical, vertical shaft vessel. Shale rock, sized from ¼ to 3 inches, is fed at the top and distributed evenly by the rotating spreader. The shale mass then moves downwardly through the retort in plug flow and is discharged uniformly by the unique moving grates located at the bottom of the kiln. Mixtures of low-Btu product gas and air enter the retort at the positions indicated. The gas entering at the bottom position is preheated by the downward moving spent shale (Ref. 18).

The amount of gas-air mixture is restricted to that required for maintaining the desired temperature while burning off some of the carbon from the spent shale. Excessive temperatures would result in decomposition of some of the mineral carbonates in the spent shale and thus wasted energy.

Rising hot gases cause retorting of the shale above the gas burner bars and, higher up, cause preheating of the incoming raw shale. Hydrocarbon products, as vapor and mist, are removed to an electrostatic precipitator for separation of liquid and gaseous products. Good heat exchange in the upper part of the kiln between hot product vapors and cold, raw shale utilizes waste heat for shale preheating and limits the need for cooling water. All materials will enter and leave the retort at a relatively low temperature of less than 200°F (93°C).

Product Oil properties are given in Table 6 (Ref. 19).

Two retorts were being tested (November 1974) at Anvil Points; one 4¼ feet in diameter and 60 feet high, and the other 10½ feet in diameter and 75 feet high. Cold flow tests have been followed by retorting operations. Combustion retorting will be tried first, and then retorting with externally heated recycle gas (Refs. 20, 21). The entire program is to be concluded in early 1976.

The White River Oil Shale group [The Standard Oil Company (Ohio), Phillips Petroleum Company, and Sun Oil Company] has announced that the Paraho process will be used in future commercial operations on the Utah prototype Federal test lease tracts which this group leased in March and April of 1974. Total investment is expected to be about $1.4 billion for production of 100,000 barrels per day of shale oil (The Denver Post, August 25, 1974).

HORIZONTAL MODE

TOSCO II Process. Of all processes being considered for commercial oil shale processing in the United States, the TOSCO II is the most advanced from the viewpoint of large-scale field testing and engineering design. Pilot plants, approximating one-ton-per-hour feed capacity, have been operated by The Oil Shale Corporation and its partners since the late 1950s. The first pilot plant was built and operated by the Denver Research Insitute. Semiworks operation at the 1000-ton-per-day level began in 1965 at a facility north of Grand Valley, Colorado; and intermittent test work was completed in 1972. Commercial design has been nearly completed.

A schematic diagram of the TOSCO II retort is shown in Fig. 5. Raw shale is fed through a surge hopper and preheated by dilute-phase fluid bed techniques. The preheated feed is then transported to a pyrolysis drum where it is contacted with heated ceramic pellets. The solid, processed shale leaves the pyrolysis drum and passes through the trommel screen, is cooled, and goes to storage. The cooled ceramic pellets pass over the trommel screen and are returned to the pellet heater by elevator. Pyrolysis vapors are condensed and fractionated. Uncondensed hydrocarbon gases can be utilized as an in-plant fuel or can be processed and sold.

This process offers a number of advantages. Since sensible heat is put into the process circuit by combustion external to the retort, the product gas is undiluted with nitrogen and combustion-derived carbon dioxide; also more naphtha is present in the product oil compared to combustion retort product slates. Combustion retorting tends to burn up naphtha components. Further, the process functions with a high throughput at steady operating conditions and accepts the total raw shale size consist produced by crushing to minus ½ inch. Since the TOSCO II process is not dependent on the flow of gas as a heat carrier inside the retort, it works well with the usual 15 percent of fines, i.e., less than ¼ inch, usually present in this raw shale feed. The TOSCO II process is energy-efficient since the hot flue gas, from the ball heater, is utilized to preheat the raw shale.

Typical TOSCO II retort gas and oil analyses are shown in Tables 7 and 8 (Ref. 22).

The material balance results from operation of the TOSCO II semiworks plant in July 1967 have been previously published (Ref. 23). About 4700 tons were retorted, and good material and elemental balance data were obtained. These data are shown in Table 9.

Information in Table 9 shows that retorting 33 gallons per ton of oil shale gives a processed shale yield of 82.4 percent and a combined oil and gas yield of 16.1 percent. Furthermore, about 25 percent of the organic carbon in the raw shale is left in the processed shale. One-half or more of the sulfur and nitrogen in the raw shale remains in the processed shale.

TABLE 5. Properties of Crude Oil from Oil Shale Retorting in 150-Ton-per-Day Gas Combustion Retort

Property	Retort
API gravity	19.7
Nitrogen, wt %	2.18
Conradson carbon, wt %	4.5
Ash, wt %	0.06
Sulfur, wt %	0.74
Pour point, °F	80
Viscosity, SUS at 100°F	256

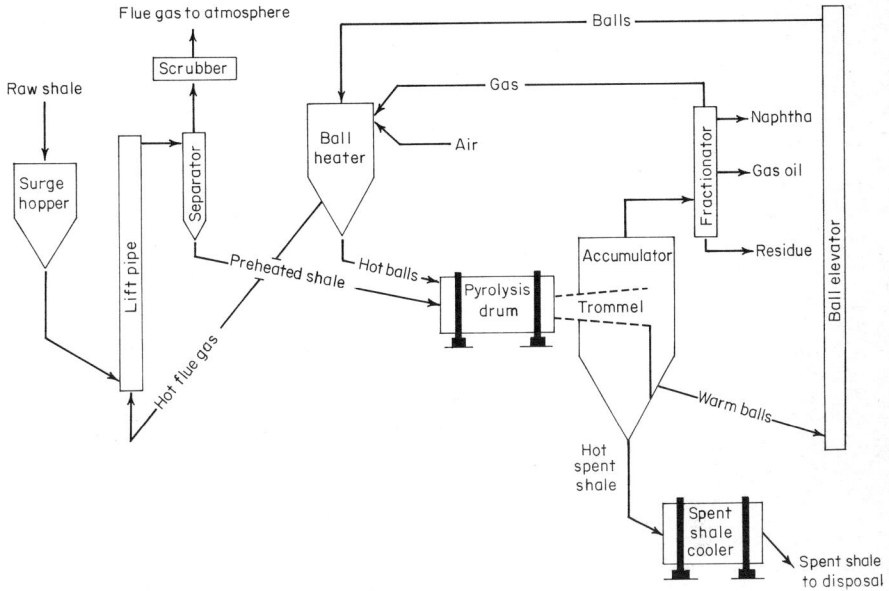

Fig. 5. TOSCO II process.

TABLE 6. Paraho Oil Shale Demonstration, Inc., Projected Product Oil Properties

Property	Crude shale oil	Syncrude
Gravity, °API	25	46.2
Pour point, °F	80	50
Sulfur, wt %	0.7	0.005
Nitrogen, wt %	1.8	0.035
Fractions, vol %:		
C_4	nil	9.0
IBP to 750°F		
naphtha	6.0	27.5
350–500°F distillate	19.0	41.0
550–850°F distillate	50	22.5
850 + residue	25.0	nil

TABLE 7. TOSCO II Process Typical Gas Analyses

33 gallons per ton raw shale

Component	Weight Percent	Mole Percent
H_2	1.53	22.44
CO	3.37	3.56
CH_4	8.25	15.23
C_2H_6	10.53	10.31
C_2H_4	5.07	5.36
C_3H_8	6.00	4.03
C_3H_6	5.25	3.71
i-C_4H_{10}	0.49	0.26
n-C_4H_{10}	2.54	1.30
Butenes	5.11	2.70
C_5's	6.18	2.55
C_6's	4.35	1.54
C_7's	2.81	0.83
C_8's	1.35	0.36
C_8+	0.36	0.09
CO_2	31.84	21.43
H_2S	4.97	4.30
	100.00	100.00

REFERENCES

1. Smith, J. W.: Ultimate Composition of Organic Material and Green River Oil Shale, *U.S. Bur. Mines Rep. Invest.* 5725, 1961.
2. Smith, J. W.: Theoretical Relationship between Density and Oil Yield for Oil Shales, *U.S. Bur. Mines Rep. Invest.* 7248, April 1969.
3. Goodfellow, L., and M. T. Atwood: Fischer Assay of Oil Shale Procedures of The Oil Shale Corporation, 7th Oil Shale Symposium, Apr. 18–19, 1974.
4. Ash, H. O.: Federal Oil Shale Leasing and Administration, in D. K. Murray (ed.), "Rocky Mountain Association of Geologists—1974 Guidebook," pp. 185–191.
5. Ruark, J. R., K. L. Berry, and B. Guthrie: Description and Operation of the N-T-U Retort on Colorado Oil Shale, *U.S. Bur. Mines Rep. Invest.* 5279, November 1956.
6. Reed, H. C., and C. Berg: "Engineering Features of the Union Oil Shale Retort," paper no. 52-PET-2, American Society of Mechanical Engineers Conference, Kansas City, Mo., Sept. 22–24, 1952.
7. Private communication from Arnold Kelly, Union Oil Company. See also U.S. Patents 2,501,153; 2,640,014; and 2,640,019.
8. During, R. F., and G. E. Irish: U.S. Patent 3,058,904, Oct. 16, 1962.
9. Irish, G. E.: U.S. Patent 3,228,869, Jan. 11, 1966.
10. *Oil Gas J.*, June 17, 1974, pp. 26–27.
11. Address by Fred L. Hartley: Oil Shale: Another Source of Oil for the United States, Oil Daily's Third Annual Synthetic Energy Forum, New York, June 10, 1974, available from Union Oil Company.
12. Chew, R. T. III: Geology, Hydrology, and Extraction Operations at the Occidental Petroleum Corporation Oil Shale Pilot Plant near DeBeque, Colorado, in D. K. Murray (ed.), "Rocky Mountain Association of Geologists—1974 Guidebook," pp. 135–140.
13. National Science Foundation: "Conference on In-Situ Oil Shale Methods," San Diego, Calif., September 1974.

TABLE 8. Typical Shale Oil from the TOSCO II Process

Gravity, °API	22.0
Pour point, °F	80.0
Sulfur, wt %	0.8
Nitrogen, wt %	1.8
Carbon, wt %	84.7
Hydrogen, wt %	11.3
Carbon to hydrogen, weight ratio	7.5

Distillation	Volume percent
IBP to 400°F	18.0
400 to 600°F	24.0
600 to 900°F	34.0
900°F and heavier	24.0

TABLE 9. Product and Elemental Balances

Basis: 2000 pounds of raw shale (33 gallons per ton)

Material	Weight of components, pounds	Distribution of components, weight %	Weight of organic carbon, pounds	Distribution of carbon, %	Weight of sulfur, pounds	Distribution of sulfur, %	Weight of nitrogen, pounds	Distribution of nitrogen, %	Weight of hydrogen, pounds	Distribution of hydrogen, %
Feed										
Raw Shale	2000		330.6		15		9.2		43.0	
Products										
Processed shale	1648	82.4	81.4	24.6	10.2	68.0	4.6	50.0	4.4	10.2
Oil	252	12.6	213.4	64.6	2.1	14.0	4.6	50.0	28.4	66.0
Gas	71	3.5	34.7	10.5	3.1	20.7	nil		7.0	16.3
Water	23	1.2	nil		0.1	0.7	0.3	3.3	2.6	6.0
Total	1994	99.7	329.5	99.7	15.5	103.4	9.5	103.3	42.4	98.5
Recovery %	99.7		99.7		103.4		103.3		98.6	

14. Harak, A. E., et al.: Preliminary Design and Operation of a 150 Ton Oil Shale Retort, *Quart. Colo. Sch. Mines,* vol. 65, no. 4, p. 52, October 1970.
15. Ruark, J. R., H. W. Sohns, and H. C. Carpenter: Gas Combustion Retorting of Oil Shale under Anvil Points Lease Agreement: Stage I, *U.S. Bur. Mines Rep. Invest.* 7303, 1969.
16. Ruark, J. R., H. W. Sohns, and H. C. Carpenter: Gas Combustion Retorting of Oil Shale under Anvil Points Lease Agreement: Stage II, *U.S. Bur. Mines Rep. Invest.* 7540, 1971.
17. Annual Report, The Standard Oil Company, Ohio.
18. Pforzheimer, H. L.: *Chem. Eng. Prog.,* vol. 70, no. 9, pp. 62–65, September 1974.
19. Paraho Oil Shale Demonstration, Inc., "Project Brochure," p. 64, 1973.
20. Barnett, F. M.: "Paraho—New Prospects for Oil Shale," American Institute of Chemical Engineers, Los Angeles, Apr. 16, 1974.
21. Jones, J. B., and A. A. Reeves: U.S. Patent 3,841,992, Oct. 15, 1974.
22. Atwood, M. T.: Production of Shale Oil, *Chemtech,* October 1973, pp. 617–621.
23. Lenhart, A. F.: The Oil Shale Corporation, TOSCO Process Shale Oil Yields, 97th AIME Annual Meeting, Feb. 27, 1968.

Fuel from Tar Sands

BY THE HON. BILL DICKIE,* Q. C., AND MAURICE CARRIGY, M.SC.†

Tar Sands is an expression commonly used in the petroleum industry to describe sandstone reservoirs impregnated with a very heavy viscous crude oil which cannot be produced through a well by conventional production techniques. Two other terms, *bituminous sands* and *oil sands*, are rapidly gaining favor and may ultimately replace the older term as the true nature of these deposits is more widely recognized. The heavy viscous petroleum substances impregnating the "tar sands" are called asphaltic oils. Other names used to describe these oils are *maltha, brea,* and *chapapote.* Asphaltic petroleums are most commonly confused with, but are not related to, asphaltities (gilsonite, glance pitch, and grahamite); the *asphaltic pyrobitumens* (elaterite, wurtzilite, albertite, and imponsonite); the *native mineral wax* (ozokerite); and the *pyrogenous distillates* of bituminous substances (tar and pitch).

OCCURRENCE

The most extensive deposits of "tar sands" in the world (Ref. 1) are located in Canada in the northeastern part of the province of Alberta. The largest and best known of these is the Athabasca deposit which outcrops in the valley of the lower Athabasca River for a distance of 100 miles (160 km). In addition to these surface "showings," heavy oil impregnates the lower Cretaceous McMurray Formation in the subsurface under an area of more than 10,000 square miles (26,000 km²) mainly to the west and south of the outcrops. These spectacular petroleum beds have been known since 1788 and have been extensively explored by core drilling. More recently (Ref. 2) other reservoirs of similar oil have been discovered in the subsurface of northern Alberta in lower Cretaceous strata down to depths of 2500 feet (760 m). These are known as the Peace River, Wabasca, and Cold Lake Deposits. The total amount of oil in all these deposits exceeds 890 billion (8.9×10^{11}) barrels of oil. See Fig. 1.

In 1962, reports were received of "tar sands" in the Bjorne Formation of Triassic age on Melville Island in the Canadian Arctic Archipelago near the southern margin of the Sverdrup sedimentary basin (Ref. 3). Subsequent investigation of these deposits by officers of the Geological Survey of Canada (Ref. 4) revealed them to be a seepage derived from the oxidation and polymerization of a 19 to 31° API gravity oil, with total in-place reserves of only 30 million barrels.

In the United States, many small "tar sand" deposits are known (Ref. 5), and a high proportion of these were quarried and sold under the name "asphalt rock," as paving material for road construction in the early part of this century. At the present time, most of the workings are abandoned. There has been a revival of interest in these tar sand deposits as sources of fuels due to oil export policies of various oil-rich nations (Ref. 6). The most extensive areas of tar sands in the United States are located in Utah at Asphalt Ridge near Vernal on the northern rim of the Uinta sedimentary basin and in the Book Cliffs near Sunnyside on the southern margin of the same sedimentary basin. See Fig. 2. Next in importance are the tar sand deposits in the Tertiary sedimentary basins of California. Other minor deposits are found in Kentucky, Oklahoma, Kansas, Montana, Missouri, New Mexico, Wyoming, Alabama, Arkansas, Louisiana, and Texas.

In South America, virtually unexplored tar sand deposits said to contain 692 billion barrels (Ref. 7) of heavy oil are present in the "Orinoco tar belt," 25 miles (40 km) wide

*Minister of Mines and Minerals, Alberta, Canada; retired from Government, March 1975; †currently, Vice-Chairman, Alberta Oil Sands Technology and Research Authority, Edmonton, Alberta, Canada.

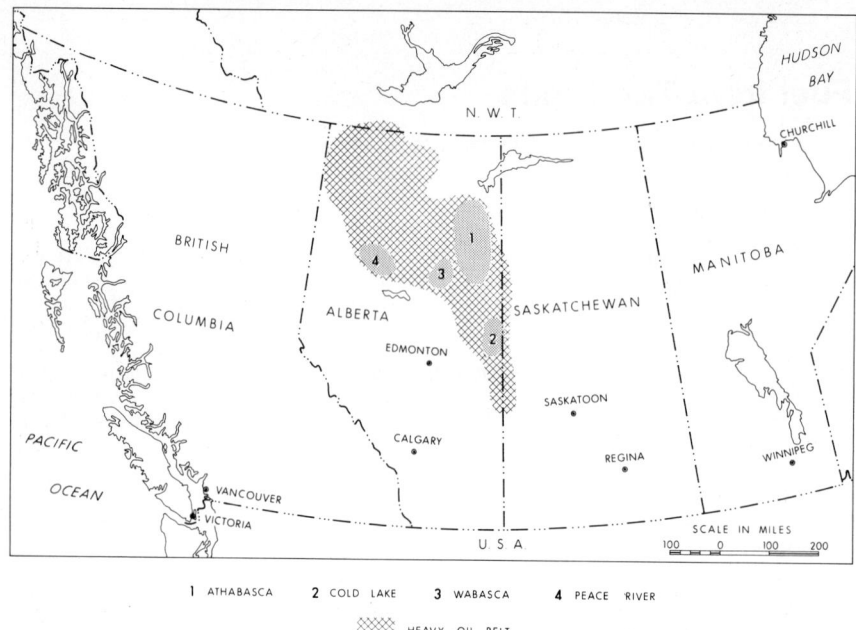

1 ATHABASCA   2 COLD LAKE   3 WABASCA   4 PEACE RIVER

░░░ HEAVY OIL BELT

Fig. 1. Map of western Canada showing location of major tar (oil) sand deposits.

south of the Oficina producing trend in the eastern Venezuelan sedimentary basin. See Fig. 2. In addition to these tar sands, there are numerous large seepages of asphaltic oil from pitch lakes in the island of Trinidad and on the mainland at Quanoco and in the Orinoco delta northeast of Pedernales. Another large reservoir of heavy oil in Venezuela is present in the northeastern part of the Maracaibo sedimentary basin. Other deposits of heavy oil in South America are reported to be present in the sedimentary basins of Brazil, Argentina, Colombia, Ecuador, and Peru.

A very large subsurface deposit of tar sands said to rival the Athabasca deposit in size has been reported on the Olenek anticline in northeastern Siberia in beds of Permian age.

Outcroppings of tar sand or asphaltic rock have been extensively quarried in Europe and Asia for many centuries (Ref. 8). However, only two deposits have been exploited successfully as a source of crude oil. One is located at Pechelbronn in Alsace and the other at Wietze in Hanover. At Pechelbronn, Oligocene sands deposited in the Rhine graben were drained by an underground mining method. Until 1952, these deposits had yielded 21 million barrels of oil. At Wietze, oil sands varying in age from Rhaetic to Senonian had been exploited by drilling in the conventional manner before mining was started. Until 1951, approximately 2 to 5 million metric tons of oil had been produced; one mine having produced over 700,000 tons of this total from the Wealden beds over a period of 50 years.

In the early part of this century, large quantities of a semiliquid asphalt were produced from the tar sands located at Tataros in beds of Pliocene age in western Romania.

In the Middle East, an extensive tar belt containing large reservoirs of heavy sulfurous oils is believed to exist in the Mesopotamian-Persian Gulf sedimentary basin parallel to the Euphrates River. Heavy oils have been found in Miocene and Upper Cretaceous limestone reservoirs at Quaiyarah and in seepages at Hit on the Euphrates and at many other localities. Bitumen from these deposits was extensively used in antiquity for building, waterproofing, medicine, and magic.

Lesser known tar sands are present in Nigeria, on the west side of the island of Madagascar (Malagasy Republic), on the island of Leyte in the Philippines, in Japan, on Boeton Island (Indonesia), and on the island of Sakhalin, and many other localities throughout the world (Ref. 1).

NATURE OF THE DEPOSITS

"Tar sands" or petroleum reservoirs impregnated with heavy asphaltic oils are commonly found at or near the surface of the ground in stratigraphic traps in the up-dip margins of large sedimentary basins. The reservoir sand is often very porous and permeable, and even in the oldest rocks it commonly lacks a mineral cement. The associated formation waters are low in chlorides and high in carbonates. The oil varies in consistency, but is mostly soft and viscous with rheological properties intermediate in nature between a flowing crude oil and semisolid asphalt and is probably best described as maltha (Ref. 9). See Fig. 3.

Although the oil is heavier than pure water (Fig. 4), it always overlies the water at the edge of the reservoir. "Tar" seals are reported to be present at the edgewater of many oil fields.

The heavy asphaltic oils of tar sands commonly contain a higher proportion of sulfur, nitrogen, oxygen, and trace metals, such as vanadium, nickel, iron, and uranium, than

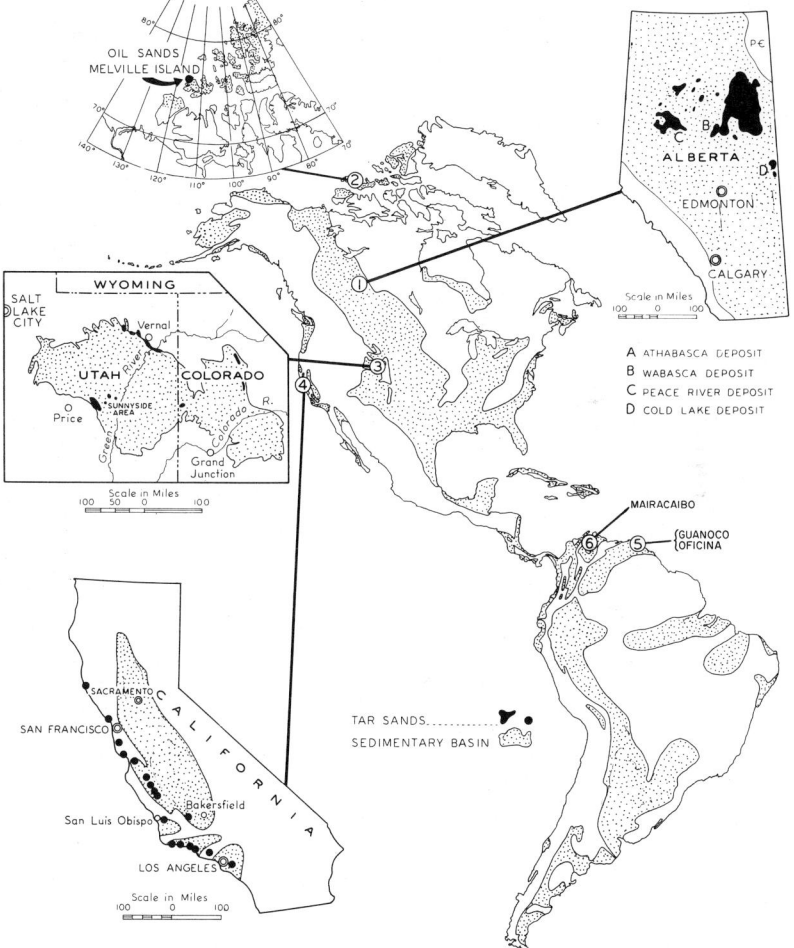

Fig. 2. Western Hemisphere showing locations of the major tar sand deposits in North and South America.

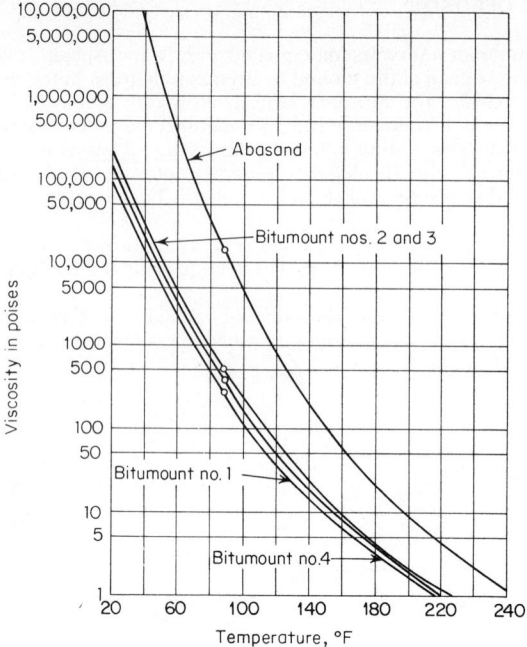

Fig. 3. Plots of viscosity versus temperature for Athabasca bitumen from various localities (Refs. 9 and 14).

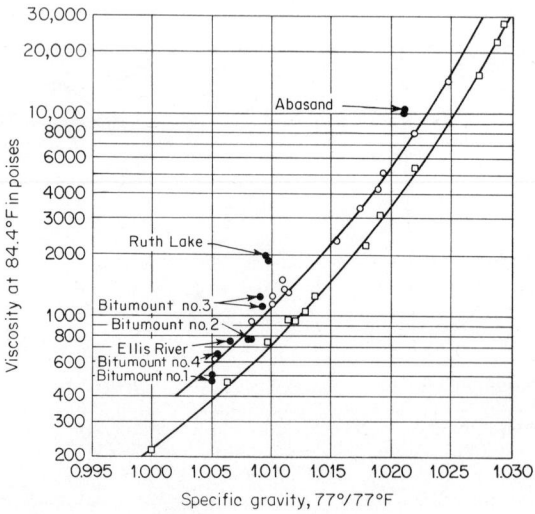

Fig. 4. Plots of viscosity at 84.4°F (29.1°C) versus specific gravity for Athabasca bitumen from various localities (Refs. 9 and 14). Upper curve: derived from bitumen extracted by hot-water method and thickened by heating. Lower curve: same bitumen after solution in benzene and recovery by evaporation.

lighter oils and are also richer in prophyrin pigments and their metallic complexes (Ref. 10). See Table 1.

Because the tar sands of the world contain the largest accumulations of liquid hydrocarbon in the earth's crust, their origin and relationship to the crude oils of conventional oil fields are of considerable interest to geologists and geochemists. It has been established that a continuous series of petroleum hydrocarbons exists with physical and chemical properties intermediate between the heaviest asphalts and the lightest crude oils. Thus the oil of the tar sands probably has the same origin as the crude oil produced from wells.

Several hypotheses have been advanced to explain the origin of the heavy oils. One theory postulates that they are not far removed from the hypothetical protopetroleum, the light crude oils being derived from them by physical and chemical changes induced by deep burial and tectonic forces. Another hypothesis postulates that they are the most

TABLE 1.   Properties of Athabasca Bitumen (Ref. 10)

Gravity at 60°F (15.6°C)	6.0°API
UOP characterization factor	11.18
Pour point	+50°F (10°C)
Specific heat	0.35 cal/(g)(°C)
Calorific value	17,900 Btu/pound
Viscosity at 60°F (15.6°C)	3000–300,000 poise
Carbon/hydrogen ratio	8:1
Components, percent:	
Asphaltenes	20.0
Resins	25.0
Oils	55.0
Ultimate analysis, percent:	
Carbon	83.6
Hydrogen	10.3
Sulfur	5.5
Nitrogen	0.4
Oxygen	0.2
Heavy metals, ppm:	
Nickel	100
Vanadium	250
Copper	5

mobile hydrocarbons that have been polymerized and immobilized (inspissated) by low-temperature chemical changes, aided by mineral catalysts or by biochemical activities, as they migrated laterally toward the margins of the sedimentary basins. Whatever the origin of the heavy asphaltic oil in these deposits, once formed it is very resistant to attack by mineral acids and alkalies and to normal weathering processes. It readily forms emulsions with water and migrates freely, filling newly formed reservoirs, or forming asphalt lakes.

EXPLOITATION

Although tar sands or tar belts are more widespread than is generally recognized, only those exposed at the surface are of economic value at the present time. In the past, most of the applied research was devoted to the development of a method of separating the oil from sands obtainable by mining or quarrying. Hot or cold water seems to have been the most widely used separating agent, and it has been employed with varying success on deposits in France (Pechelbronn), Germany (Wietze), Romania (Tataros), Nigeria, the United States (Oklahoma, Utah, California), and Canada (Athabasca). One ton of tar sand will yield 0.5 to 1.0 barrel of oil, depending on the degree of impregnation and efficiency of the separation process. A commercial hot-water separation plant has been in operation in the Athabasca deposit of northern Alberta since 1967 and is currently producing over 50,000 barrels of synthetic crude oil per day. A second plant, designed to produce 125,000 barrels per day, is scheduled to begin operations in 1978, and applications have been received for the construction of three other hot-water plants to produce in excess of 300,000 barrels of synthetic oil in the early 1980s. Originally, the bitumen from tar sands was extracted and used for road building. It was not until the late 1930s, when hydrogenation proved feasible, that oil companies became interested in these deposits as a substitute for crude oil.

The extraction of bitumen from the Athabasca deposit is a two-stage process (Ref. 11). The primary extraction step is a continuous process (Fig. 5), in which the bituminous sand passes through a conditioning drum where makeup water, steam and caustic soda are added to adjust the solids content to about 70 percent, the temperature to 180–190°F (82–88°C), and the pH to 8.0–8.5. The pulp is passed over a screen to remove oversize material and then into a primary separation cell where the oil floats to the top in the form of froth and the sand falls to the bottom and is removed while part of the middlings are recycled and part drawn off to be treated by air flotation in scavenger cells. The bitumen froths from separation cell, and scavenger cells are combined, and diluted with naphtha; and the fine mineral matter and water are removed in a two-stage centrifuge operation (Fig. 6), after which the naphtha is recovered and recycled. The recovered bitumen is then subjected to an upgrading step to produce a marketable crude oil substitute.

UPGRADING OF BITUMEN

The major deficiencies of the bitumen extracted from the Alberta Oil Sands are the low ratio of hydrogen to carbon and the high sulfur content. See Table 1. Hydroprocessing and desulfurization are necessary to produce a satisfactory synthetic crude oil. To be acceptable, the overall upgrading scheme must give a high yield of synthetic oil, high thermal efficiency, low sulfur emissions, and low rejection of solid waste. Four major primary upgrading schemes are currently available: (1) delayed coking, (2) fluid coking, (3) residual hydrocracking, and (4) solvent deasphalting.

Delayed Coking.   This process is used in the existing plant. Bitumen is charged directly to the coker units and heated. However, the quantity of coke produced is in excess of the plant fuel requirement. A moderately high-gravity synthetic crude oil can be produced by this process without hydrocracking. An assay of the synthetic oil produced from the Athabasca deposit by this process is given in Table 2.

Fluid Coking.   This process is performed in a fluidized bed of coke particles with a portion of the coke burned to supply the heat of reaction. This process is more flexible than delayed coking and gives higher yields, including a larger yield of olefins. The process is to be used in the second plant in the Athabasca area. See Fig. 7.

Hydrocracking.   The direct addition of hydrogen has been proposed as a method of correcting the low hydrogen content of bitumen. This can be accomplished catalytically as in the H-Oil process or by a noncatalytic thermal, high-pressure process. The presence of mineral matter (clay) in the bitumen from the Athabasca deposit is believed to favor the noncatalytic process in this deposit. The major advantage of hydrocracking is its higher liquid yield.

Solvent Deasphalting.   This process has been chosen for the third plant in the Athabasca area. Asphaltenes and metals in the bitumen are precipitated by treating the vacuum distillation residue with a paraffinic solvent. The deasphalted oil is hydrocracked to reduce its viscosity and to remove sulfur. The yield distribution of synthetic crude oil from this process depends on the severity of the hydrocracking step. The residual asphaltic material is gasified, and the gas is desulfurized and used as a low-Btu fuel for steam generation.

FUTURE OF SYNTHETIC OIL PRODUCTION FROM
THE ALBERTA OIL SANDS

In 1973, the Alberta Energy Resource Conservation Board added to the recoverable oil reserves of the province 38×10^9 barrels of bitumen in the Athabasca deposit from which it is estimated that 26.5×10^9 barrels of synthetic crude oil will ultimately be produced. This synthetic oil will come only from surface mining and hot-water washing, and the rate at which this synthetic crude oil will arrive on the market depends on a number of factors other than the availability of a satisfactory extraction technology and massive reserves (Ref. 12).

In broad terms, these factors are:
1. The price of competitive oil
2. The availability of capital

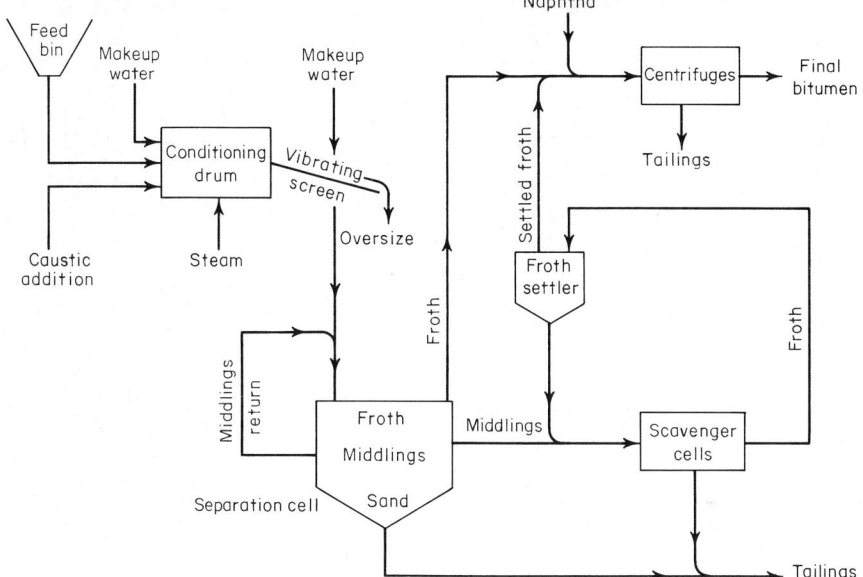

Fig. 5. Primary extraction process for Clark hot-water method (Ref. 10).

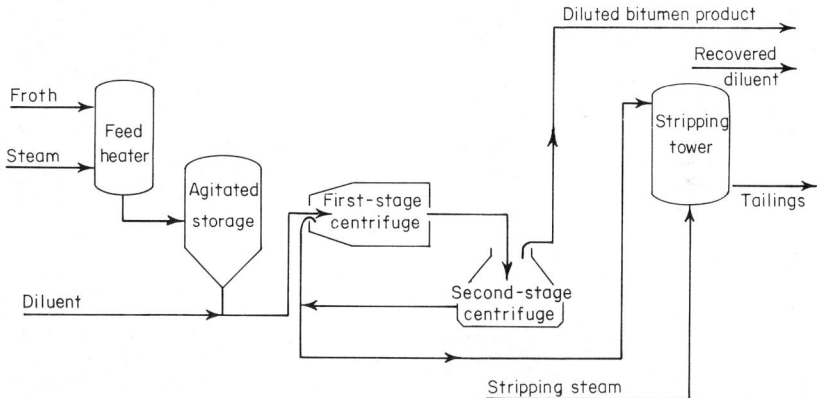

Fig. 6. Schematic flow diagram of final extraction process for Clark hot-water method (Ref. 10).

TABLE 2.   Assay of Typical Blend of Synthetic Crude Oil Derived from Athabasca Bitumen

Gravity, °API	36.0
Sulfur, net %	0.15
Distillation characteristics:	
Temperature	Cumulative yield (vol. %)
0–200°F (93°C)	10.5
380°F (193°C)	29.4
650°F (343°C)	71.4
1050°F (566°C)	100.0

3. The availability of technical personnel

4. The satisfactory solution to a large number of environmental problems, such as reduction of sulfur dioxide emissions from the burning of high-sulfur fuels in power and steam generators, the reclamation of the vast areas of sterile sand remaining after the oil has been extracted, and the efficient use of recycled water to reduce the volume of water in tailings ponds

5. Guarantees of assured market for the oil for a period of 30 years or more from the commencement of production

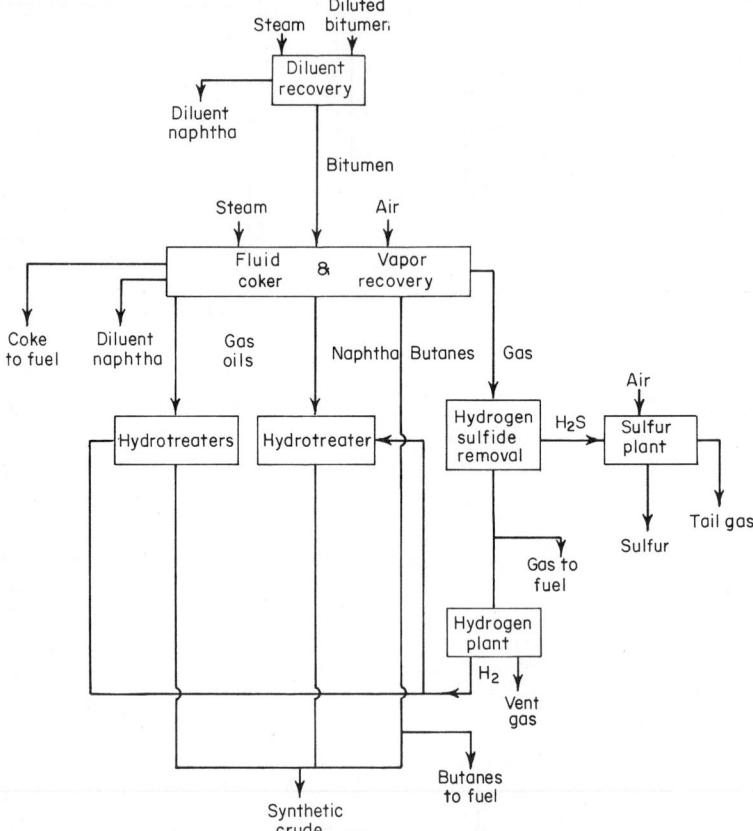

Fig. 7. Block diagram showing flow in an upgrading scheme for Athabasca bitumen based on fluid coking.

Many forecasts of the rates of synthetic crude oil production from the Alberta oil sands over the next 15 years have been made. Committed projects will bring the level to approximately 175,000 barrels per day by 1982. Two or three additional plants planned for completion during the balance of the decade could bring that figure to about 500,000 barrels per day. Higher levels of production may prove to be difficult to obtain due to the long lead time required to bring new projects to commercial fruition.

Although the magnitude of the in-place reserves in the Alberta Oil Sands is impressive, the quality of the oil is very poor. It commonly contains a high proportion of sulfur, nitrogen, oxygen, and trace metals and other heavy metals, as previously mentioned, which make refining difficult. Another feature preventing rapid development is that, apart from that portion of the Athabasca deposit accessible to open-pit mining and hot-water washing, the greater part of the oil is locked in the reservoirs at depth and is not economically available with any current technology. To produce oil from the subsurface deposits, many methods of in-situ extraction have been proposed. All of these involve production from a well after some form of stimulation (Ref. 13). Among the many novel methods tried or proposed are: injection of gases, steam, and solvents; underground combustion; and nuclear explosions. However, it is estimated by people most familiar with these developments that it will take at least 10 years of experimental work in the field before any of these methods will result in commercial production.

A number of major experimental in-situ projects have been undertaken in the Alberta oil sands, and several are still operating. The Alberta government has established the Alberta Oil Sands Technology and Research Authority to foster the development of new in-situ processes, and its is expected that the level of field experimental activity will increase markedly during the next few years. The bitumen produced by these methods will have to be upgraded to a synthetic oil in a manner comparable to that used from the bitumen produced by the open-pit mining hot-water method.

REFERENCES

1. Phizackerley, P. H., and L. D. Scott: Major Tar Sand Deposits of the World, *Proc.* 7th *World Petrol. Congr.*, vol. 3, pp. 551–571, 1968.
2. Pow, J. R., G. H. Fairbanks, and W. J. Zamora: Descriptions and Reserve Estimates of the Oil Sands of Alberta *Alberta Res. Counc. Inform. Ser.* 45, pp. 1–14, Edmonton, Alberta, 1963.
3. Sproule, J. C., and G. V. Llloyd: A Note on the Comparison of McMurray and Melville Island Oil Sands of Alberta, *Alberta Res. Counc. Inform. Ser.* 45, pp. 1–14, Edmonton, Alberta, 1963.
4. Trettin, H. P., and L. V. Hills: *Lower Triassic Tar Sands of Northwestern Melville Island, Arctic Archipelago, Geol. Surv. Can. Paper* 66-34, 122 pp., Ottawa, Canada, 1966.
5. Surface and Shallow Oil-Impregnated Rocks and Shallow Oil Fields in the United States, *U.S. Bur. Mines Monogr.* 12, 375 pp., Pittsburgh, Pa., 1965.
6. Staff: BuMines Plans Utah Tar-Sands Pilot, *Oil Gas J.*, vol. 72, no. 15, pp. 34–35, Apr. 15, 1974.
7. Galavis, Jose, and H. M. Velarde: Geological Study and Preliminary Evaluation of Potential Reserves of Heavy Oil of the Orinoco Tar Belt Eastern Venezuelan Basin, *Proc.* 7th *World Petrol. Congr.*, vol. 3, pp. 229–234, 1968.
8. Abraham, H.: "Asphalts and Allied Substances," vol. 1, 370 pp., Van Nostrand Reinhold, New York, 1960.
9. Carrigy, M. A.: The Physical and Chemical Nature of a Typical Tar Sand: Bulk Properties and Behaviour, *Proc.* 7th *World Petrol. Congr.*, vol. 3, pp. 573–581, 1968.
10. Carrigy, M. A., and J. W. Kramers: "Guide to the Athabasca Oil Sands Area," Alberta Research Council, 213 pp., Edmonton, Canada, 1973.
11. Innes, E. D., and J. V. Fear: Canada's First Commercial Tar Sands Development, *Proc.* 7th *World Petrol. Congr.*, vol. 3, pp. 633–650, 1968.
12. Carrigy, M. A.: Alberta Oil Sands: Fuel of the Future? *Geosci. Can.*, vol. 1, pp. 41–44, 1974.
13. Doscher, T. M.: Technical Problems in In-Situ Recovery Methods for Recovery of Bitumen from Tar Sands, *Proc.* 7th *World Petrol. Congr.*, vol. 3, pp. 229–234, 1968.

Wood-to-Oil Process

BY S. FRIEDMAN,* A. S. MEHTA,† AND P. L. THIGPEN‡

In a continuing search for alternative supplementary energy sources, scientists at the Pittsburgh Energy Research Center of the Bureau of Mines, U.S. Department of the Interior, have successfully converted wood wastes, bovine manure, and municipal organic refuse into oil in the laboratory. Actually, any cellulosic material can be converted to oil by this process. This is an important technological development in view of the energy shortage as well as the waste-disposal problems. It must be noted that as of late-1974 only bench-scale work had been carried out.

With the technical feasibility demonstrated in the laboratory, a preliminary economic feasibility study was made by an engineering firm for the Bureau of Mines. The results of the study were encouraging, and, as a result, a pilot plant is being constructed at the Albany Metallurgical Research Center of the Bureau of Mines in Albany, Oregon. The Rust Engineering Company was awarded a contract to design this pilot plant, which will further define the process and reveal any operational problems that may occur upon scale-up so that remedial technology can be developed for the design of large plants. Wood chips were selected as feedstock for this pilot plant.

Economics. The total of various solid organic wastes generated yearly in the United States is about 3 billion tons (Ref. 1). The population is rising, and so is the amount of solid wastes per person. Two billion tons of waste per year, containing approximately 50 percent organic matter, could yield some two billion barrels of oil annually. This is approximately 50 percent of the 1970 United States demand for oil (Ref. 2). In essence, this is a means of utilizing solar energy which is, of course, the basis of cellulose production. Looking into the future, it is possible to cultivate fast-growing crops on a large scale to serve as feedstock for such an oil-producing plant. No economics are available at the present time for such a large-scale, integrated agricultural and oil production facility. The oil produced by this process has a low sulfur content and is suitable for use by power plants or for conversion to gasoline and Diesel fuels.¶

According to the preliminary economic feasibility study, a 3000-ton-per-day plant will cost approximately $51.4 million and will have a breakeven operating cost of $7.24 per barrel. It might be pointed out that either a city having a population of 300,000 or a cattle feed lot having 200,000 cattle can feed a plant that would just break even on total costs, including a 20-year amortization of investment (Ref. 1). The foregoing are 1974 dollars.

Process. Basically, the process involves reacting carbon monoxide with the wood wastes in the presence of a sodium carbonate solution as a catalyst at up to a temperature of 700°F (371°C) and up to 4000 psig. Carbon monoxide reacts with sodium carbonate in the presence of water to form sodium formate which, in turn, reacts with cellulose in the wood wastes to form oil and regenerate sodium carbonate. The following reactions are believed to be typical:

$$Na_2CO_3 + H_2 + 2CO \rightarrow 2HCOONa + CO_2$$

$$2C_6H_{10}O_5 + 2HCOONa \rightarrow 2C_6H_{10}O_4 + H_2O + CO_2 + Na_2CO_3$$

---

*Assistant Research Supervisor and Supervisory Chemical Engineer, Pittsburgh Energy Research Center of the U.S. Bureau of Mines.

†Formerly Staff Engineer, Chemical Process Group, The Rust Engineering Company (a subsidiary of Wheelabrator-Frye Inc.), Birmingham, Alabama.

‡Principal Process Engineer, Chemical Process Group, The Rust Engineering Company (a subsidiary of Wheelabrator-Frye Inc.), Birmingham, Alabama.

¶See also articles on *Biochemical Sources of Fuels* and *Photosynthesis* in the *Chemical Fuels Technology* and the *Solar Energy Technology* sections of this Handbook, respectively.

The major problem in this process is anticipated to be the introduction of wood wastes into the reactor against a pressure of 4000 psig. An overall process flow diagram is given in Fig. 1.

Drying and Grinding.   Wood wastes have a high moisture content (up to 50 percent) and practically no ash. For this pilot plant, it was believed necessary to dry the wood wastes to approximately 4 percent moisture content because (1) excessive moisture going into the reactor with the wood effectively reduces the reactor capacity, and (2) dry wood

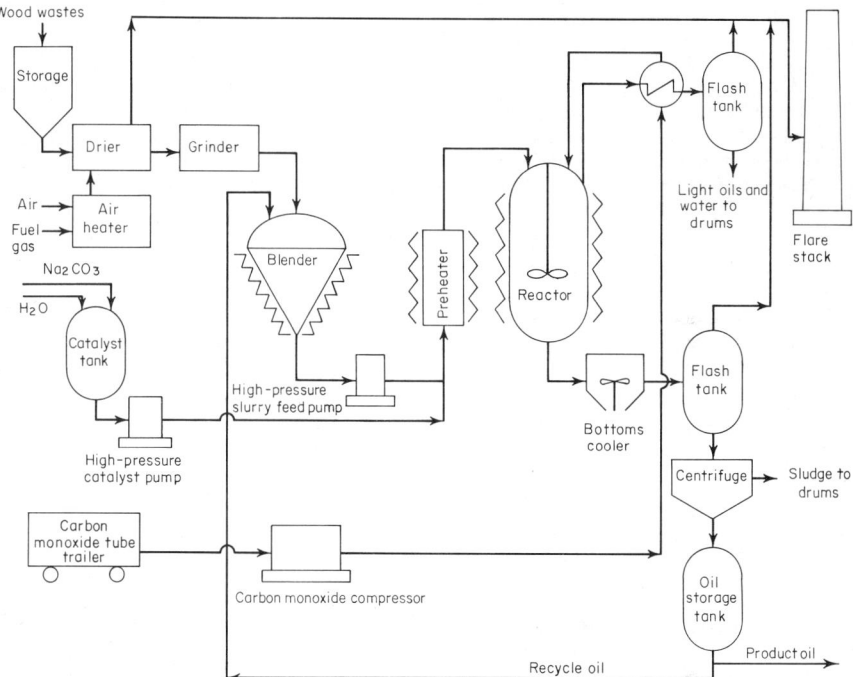

Fig. 1. General flow diagram of wood wastes-to-oil process.

grinds much more easily than wood with a high moisture content. Grinding dry wood to a fine powder is necessary to facilitate pumping a wood-oil slurry. It is anticipated that the process improvement in this pilot plant will allow deletion of the drying step in a commercial plant and also that hardware will be developed to eliminate fine grinding of the wood.

Feed Systems.   Because of the difficulties experienced in feeding wood-oil slurry during bench-scale operations, the Bureau of Mines specified that three different feed systems be designed and incorporated for this pilot plant in order to select the best system for commercial use: (1) wood-oil slurry feed, (2) solids feed, and (3) pretreated wood-oil slurry feed. These three feed systems and other process areas are described in the following paragraphs.

Wood-Oil Slurry Feed System.   This feed system is shown as part of the overall process illustrated in Fig. 1. The dried and ground wood is continuously mixed with oil to form a slurry with up to 30 percent solids content. For initial start-up, anthracene oil is used; after the plant is in operation, oil produced in the process is recycled for slurry preparation. This slurry is extremely thick and poses a problem in pumping at 4000 psig. A plunger pump with special knife-edged spring-loaded cone valves will be used to develop the required pressure. It is expected that a pump will be developed that will have either mechanically or electrically operated suction and discharge valves to provide positive opening and closing actions and thus minimize plugging problems. A pump of this type already has been built by the Bureau of Mines. Further interest may be generated when this process nears commercialization.

The catalyst used in this process is a 20 percent sodium carbonate solution. A high-pressure plunger pump will be used to develop 4000 psig pressure. Such a pump is not expected to pose any particular problems because the catalyst solution is clear. Although the amount of catalyst is not great, it is anticipated that the aqueous stream recovered from the reaction mixture will be partially recycled in a commercial plant for economy and to minimize waste effluent.

Carbon monoxide will be compressed in a diaphragm compressor to a pressure of 4000 psig and introduced into the reactor through a dip pipe. For this pilot plant, carbon monoxide will be purchased in tube trailers. But in a commercial plant, part of the wood wastes will be used to produce synthesis gas containing hydrogen and carbon monoxide that will be used in the reactor. Synthesis gas is actually better suited for the reaction because its hydrogen content removes sulfur and nitrogen from the wood wastes, forming hydrogen sulfide and ammonia, respectively.

Solids Feed System.   The solids feed system is shown in Fig. 2. The direct solids feed system has a major advantage of achieving the greatest throughput for a given size reactor. It also simplifies hardware problems of pumping a slurry. However, the solids feed system is quite complex and expensive.

Two methods were considered for feeding solids directly into the reactor: (1) pneumatic, and (2) mechanical feeding. Because of partial pressure considerations in the reactor, carbon monoxide must be used to pneumatically convey and feed the dry solids. Preliminary results of tests conducted in a vendor's laboratory indicated that excessive quantities of carbon monoxide would be required. Because of the gas consumption and the fact that the reactor off-gases will not be recycled or utilized in any manner in this pilot plant, pneumatic feeding was not selected. However, for a commercial plant, where the

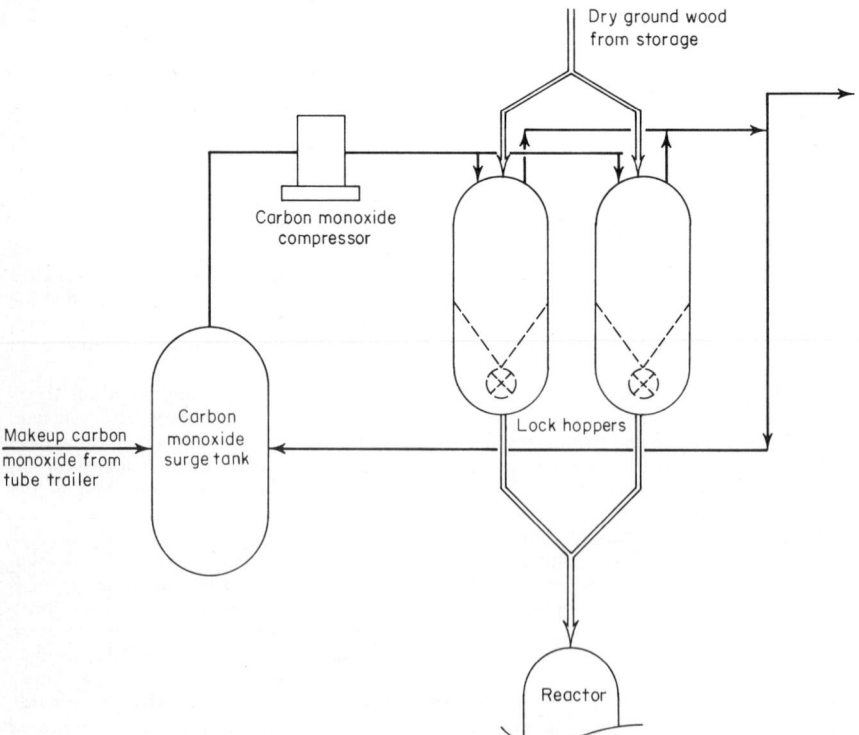

Fig. 2. Solids feed system for wood-to-oil process.

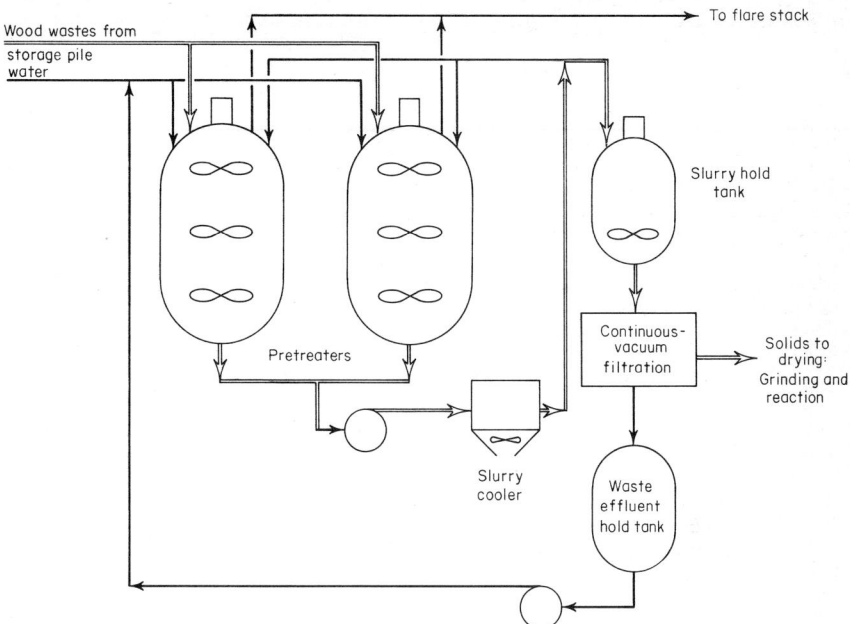

Fig. 3. Pretreatment of wood wastes in wood-to-oil process.

reactor off-gases could be used for some purpose, as mentioned later, pneumatic feeding would be a possibility.

For this pilot plant, a mechanical feed system incorporating a rotary feeder was designed. The feed system basically consists of two pressure-balanced lock hoppers which will be used alternately. To be able to use conventional rotary feeders, the lock hoppers were designed so that the feeders are installed inside and, therefore, are not exposed to high differential pressure. However, a rotary feeder capable of withstanding a 4000-psig differential is preferred for better access, improved maintenance, and a simpler lock-hopper design. Vendor interest for building such a feeder has been evidenced, but details are not immediately available.

The lock-hopper system will function in the following manner. A lock hopper is filled with wood wastes at atmospheric pressure, pressurized to 4000 psig using carbon monoxide, and the solids are fed to the reactor with the help of the rotary feeder. When the lock hopper is empty, the pressure is reduced, first by venting to a carbon monoxide surge tank, and then to a flare stack. Once the lock hopper is vented to atmospheric pressure, it is ready for the next cycle. The purpose of the carbon monoxide surge tank is to collect most of the carbon monoxide from the lock hoppers, so that it can be recompressed and reused in the next cycle.

Pretreated Wood-Oil Slurry System. The overall pretreatment process is shown in Fig. 3. This is another alternative to feeding wood wastes into the reactor. It is similar to the wood-oil slurry feed system except that it involves pretreatment of wood wastes with water at up to 500°F (260°C) and up to a pressure of 700 psig. Under these conditions, wood is converted into material similar to charcoal. Two pretreaters are provided for this pilot plant for alternate use. A pretreater will be loaded with wood wastes and water and heated up to 500°F. This will build up pressure to 700 psig because of the vapor pressure of water. After maintaining the process conditions for the required length of time, the pretreater and its contents, which comprise a 15 percent slurry of pretreated wood in water, are cooled to 150°F (66°C) through an external pump-around loop, including an air-cooled heat exchanger, and then transferred to a holding tank. The pretreated wood is filtered out, dried, ground, mixed with either anthracene or recycle oil, and fed to the reactor as previously described. The advantage of pretreating the wood wastes is that it can be slurried in oil up to 50 percent.

Reaction.   The reaction is carried out continuously in a stirred pressure vessel at 4000 psig and 700°F (371°C). The process conditions will be optimized in this pilot plant to achieve the lowest operating pressure and temperature. The pressure in the reactor will be controlled by the pressure-reducing valve in the reactor off-gas line. This is an angle valve with Kennametal K-703 trim and seat. This material was selected on the basis of success previously experienced by the Bureau of Mines.

The level in the reactor is maintained by regulating the flow of liquid from the reactor. The level detection in the reactor is a major consideration because of severe process conditions and a 4½-inch-thick stainless steel wall for the 24-inch inside-diameter vessel. A nuclear level measurement system was investigated. Because of the geometry of the reactor, the required radiation source level would be high. This could pose a radiation hazard, and, therefore, the method was not selected. Two methods of level detection are designed for the pilot plant: (1) Multiple thermocouples are provided inside the reactor to detect the temperature difference between the gas and liquid phases. This method will function only if there is a measurable temperature difference between the two phases. (2) The second method utilizes a differential-pressure transmitter isolated from the process by a chemical seal. These two methods are provided only to determine their performance.

For control purposes, an overflow loop system, as shown in Fig. 4, is provided downstream of the bottoms cooler. An enlarged spool piece in the downstream side of the overflow loop seal will maintain a liquid pool. A differential-pressure transmitter will detect the liquid level in this enlarged pool piece. This transmitter is believed to be more reliable because the liquid is cooled down to 400°F (204°C). The control valve will be similar to the valve used in the reactor off-gas line previously described.

Reactor Off-Gases.   After exchanging heat with the incoming carbon monoxide, and after the pressure reduction, the reactor off-gases (consisting primarily of carbon monoxide, carbon dioxide, and water vapor) are cooled and flashed to recover light oils and water. The noncondensible gases are burned in a flare stack. In a commercial plant, it is anticipated that the noncondensible gases will be purified to remove ammonia and hydrogen sulfide and then burned along with a portion of the synthesis gas in a Dowtherm

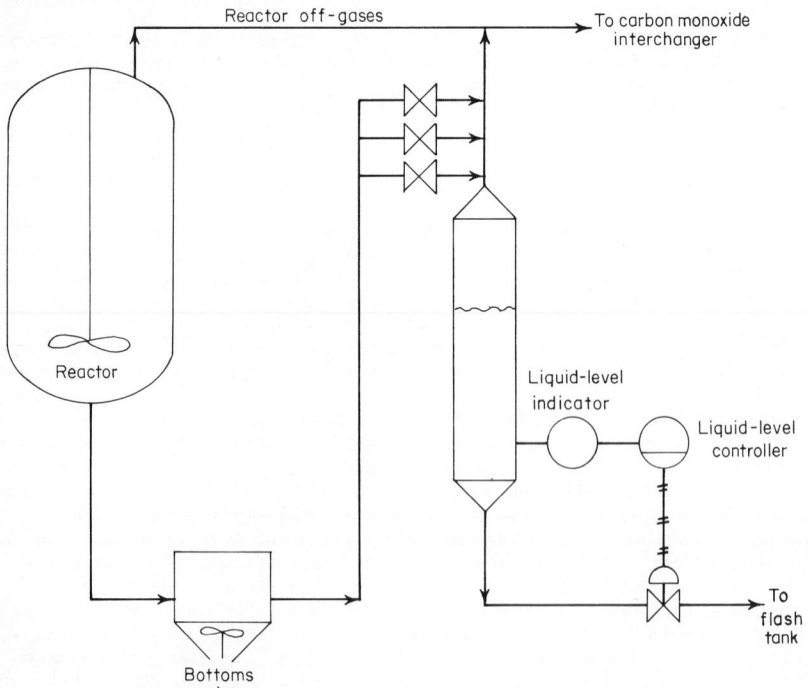

Fig. 4. Reactor level control scheme for wood-to-oil process.

system to provide approximately 90 percent of the heat requirements of the plant. The waste gases will then be vented to the atmosphere.

Oil Recovery.   The liquid phase from the reactor is also flashed after it is cooled. Light oils and water are removed from the flashed gases, and the noncondensable gases are disposed of in the same way as the reactor off-gases. The liquid consists mainly of the oil, water, unconverted wood, and sodium salts of organic acids. The liquid phase is centrifuged to remove suspended solids and water. The oil, thus recovered, is recycled back to the processes, as needed, and the remainder is stored as product.

The oil produced has the following characteristics:

Specific gravity	1.1
Viscosity, cp (140°F, 60°C)	515
Heating value, Btu per pound	15,000
Composition, % Carbon	76.62
Hydrogen	7.05
Nitrogen	0.13
Sulfur	0.14
Oxygen	20.05

Environmental Considerations.   A flare stack is furnished for this pilot plant to handle unreacted carbon monoxide, hydrogen, and miscellaneous gases. For a commercial plant, a gas-purification system is expected to remove noxious components from the gas stream before discharging to the atmosphere. There will be some liquid and solid wastes. No provisions are made in this pilot plant for their disposal. The quantity and nature of such wastes are not fully known at the present time. It is anticipated that the wastes from this pilot plant will be fully analyzed, and means for their disposal will be developed at a later date.

REFERENCES

1. Knapp, E. C.: Agriculture Poses Wastes Problems, *Environ. Sci. Technol.*, vol. 4, pp. 1098–1100, December 1970.
2. Converting Organic Wastes to Oil, *U.S. Bur. Mines Rep. Invest.* RI-7560, Pittsburgh, Pa., 1971.
3. Friedman, Sam, Henry Ginsberg, Irving Wender, and Paul Yavorsky: Continuous Processing of Urban Refuse to Oil Using Carbon Monoxide, paper presented at Third Mineral Waste Utilization Symposium, IIT Research Institute, Chicago, Mar. 14–16, 1972.

Evolution of Petroleum Pipelines in the United States

Pipeline technology for land and marine pipelines is covered in the Handbook section on *Gas Technology.* Following are key statistics as they relate historically to pipelines for petroleum and related products.

Period (1900–1919). Pipeline mileage in 1900 was 6800 miles; by 1919 the mileage had increased to 24,400 miles. There were two principal technological trends during this period: (1) More widespread use of larger-diameter pipe. Initial pipelines were 3 to 6 inches in diameter—steadily moving up to 8, 10, and 12 inches in diameter. (2) More efficient Diesel pumps were installed at stations along the line, replacing steam-driven models. In 1906 interstate pipelines came under the jurisdiction of the U.S. Interstate Commerce Commission and has remained there ever since.

Period (1920s). During this decade, an additional 19,872 miles of pipeline were added to the network. Principal advancements included (1) introduction of welded instead of screwed-into couplings, (2) high carbon steel began to replace lap-welded pipe, and (3) electrically powered pumps replaced Diesel equipment.

Period (1929–1941). During this period approximately 20,000 more miles of interstate trunk line were constructed. These pipelines were used chiefly to transport crude oil to the owners' refineries located strategically throughout the country.

The transportation of refined petroleum products by pipeline during the 1930s was accompanied by a decline in railroad shipments. A cost comparison made in 1937 explains this trend: Cost per ton mile: by truck 4.873 cents, by rail 1.640 cents, by tanker or barge 0.063 cents, by product pipeline 0.527 cents.

The emergence of product pipelines in the early 1930s marked a new development in transportation and commenced to neutralize the advantages of refineries located at market centers over those located closer to the crude producing centers, such as Texas and Oklahoma. However, despite the impressive expansion of product pipelines, these lines handled only a small proportion of the refined products. For example, in 1941, pipeline movements constituted only 5 percent of the amount then transported by rail tankcar.

Period (World War II Years). On June 10, 1942, the "Big Inch" pipeline was authorized from Longview, Texas, to Norris City, Illinois, with an extension to Linden, New Jersey. These were only two of 33 lines completed by September 1945, adding approximately 14,000 miles to the network during the war years. By 1945 pipeline deliveries accounted for 40 percent of the petroleum deliveries to the eastern United States whereas rail shipments had dropped to 27 percent.

Period (1962–1972). Using 1962 as a base, during this decade pipeline growth was as follows: Natural gas pipelines 52 percent, refined products 138 percent, and crude oil pipelines, 134 percent. In 1971 the combined mileage of interstate petroleum products pipelines moved ahead of crude oil lines for the first time. This growth is shown in greater detail in Table 1.

Proportionately, the big gains have been in jet fuel, up 701 percent; and natural gas

TABLE 1.   Comparison of Petroleum Pipelines: 1962–1972
Volume is given in 1000 barrels

Product transported	1962 Volume	1962 Percent	1972 Volume	1972 Percent	10-Year Change Percent
Gasoline	775,296	62.3	1,638,756	55.2	+111
Distillates—fuel oil	264,444	21.3	659,409	22.2	+149
Kerosine	75,248	6.0	46,132	1.6	−39
Jet fuel	28,261	2.3	226,371	7.6	+701
Natural gas liquids	100,807	8.1	397,330	13.4	+294

liquids, up 294 percent. The former gain, of course, reflects the almost complete replacement of piston by jet-powered aircraft in commercial transportation.

Two large liquefied petroleum gas (LPG) pipelines commenced operation during the 1960s: *Mapco* from the southwest to Minneapolis, and *Dixie* from the southwest to the Georgia area. Both operations have since expanded, and Texas Eastern systems have expanded. These firms account for a large portion of the increase in gas liquids transportation.

In the 1970s, much attention has been concentrated on the construction of pipelines from offshore facilities, with entirely new techniques in connection with marine pipelines, and on the construction of long lines to bring crude from the Arctic regions.

In the mid-1970s, pipeline statistics in the United States were approximately as follows: crude oil pipelines, 63,533 miles (trunk), 42,240 miles (gathering); refined products, 66,291 miles—or a total of 173,064 miles. Deliveries (annual) of crude by pipelines in the United States amounted to 5,509,663,000 barrels; of refined products, 3,336,872,000 barrels; or a total of 8,846,535,000 barrels.

Railroad Transportation of Liquid and Gaseous Fuels*

In 1974, the American Petroleum Institute had registered members who operated approximately 40,000 tankcars in their petroleum railcar fleet. In addition, the Manufacturing Chemists Association members and the Compressed Gas Association members, along with the independent firms, had numerous additional compressed gas tankcars and other types of railcars which normally are used in petroleum service. Although tankcar operations do not account for a high percentage of the transportation needs required for the distribution of liquid and gaseous fuels in the United States, tonnage figures are nevertheless significant (26 million tons in 1971). The energy crisis of 1973 arrested a downward trend in the rail shipment of liquid and gaseous fuels.

Factors that favor the increased use of rail tankcars in the United States for energy products distribution include: (1) the world situation pertaining to crude oil, including prices, uncertain policies of oil-rich exporting nations, and the dependency of the United States upon petroleum imports, with possible unscheduled disruptions of normal delivery and distribution channels—rail transportation offering greater flexibility in such situations; (2) ecological factors, including environmental concerns and delays in connection with the construction of pipelines, superports, seaport storage tankage, and greater use of inland waterways; and (3) dislocations caused by the overall energy deficiency, involving such factors as much higher over-the-road long-distance fuel hauls by motor trucks, the probable increased handling of oil imports at locations in the East as contrasted with the heavy past activity on the West and Gulf Coasts, the construction of new refineries in locations where rail transportation may be required for stop-gap operations, and a switch on the part of numerous users from natural gas to oil, again where stop-gap tankcar operations may be required for a number of years and, in some instances, on a long-term basis. See Table 1.

EVOLUTION OF THE TANKCAR IN THE UNITED STATES

The history of the tankcar began with the drilling of the country's first successful oil well in Titusville, Pennsylvania, on August 27, 1859. The railroads immediately built spurs into the producing regions and soon were hauling oil in small wooden barrels on ordinary flatcars.

Charles P. Hatch of the Empire Transportation Company of Philadelphia should be credited with the invention of the first railroad tankcar, when he mounted three iron-banded wooden tubs with a total capacity of 3500 gallons on a flatcar in early 1865. Others, however, credit Amos Densmore, an oil well owner, as the inventor. Densmore attached two 1700-gallon wooden tubs with iron hoops over the trucks of a railroad flatcar and shipped crude oil from Millers Farm, Pennsylvania, to New York City on September 1, 1865.

In 1869 the Empire Transportation Company, in order to overcome the leakage problem of wooden tanks, designed a railroad tankcar with a heavy, iron tank set horizontally on its own wooden underframe. These tanks were riveted, had domes and headblocks, and introduced for the first time a double-beamed wooden rectangular frame for mounting the tank instead of mounting it on a flatcar. These Empire cars served for many years as the standard method of shipping crude oil and refined oil products, such as kerosine, which was then quickly replacing whale oil as illuminating lamp fuel.

The railroads until this time had made it a practice to supply their customers with all the rolling stock needed to move freight. However, because of the early obsolescence of the wooden tanks and the rapid change from iron to steel equipment, the railroads sought

*Staff, Shippers Car Line Division, ACF Industries, Incorporated, St. Charles, Missouri.

TABLE 1. Transportation of Petroleum Products by Mode of Transport
Millions of tons

Mode	1939		1947		1955		1960		1965		1970		1971	
Pipeline	148	39.2%	238	38.4%	413	42.8%	468	43.1%	588	44.4%	790	46.9%	806	46.9%
Water Carrier	148	39.2%	209	33.7%	284	29.5%	318	29.2%	324	24.5%	403	23.9%	417	24.2%
Truck	22	5.8%	106	17.1%	223	23.2%	270	24.8%	385	29.1%	466	27.6%	471	27.4%
Railroad	60	15.8%	67	10.8%	43	4.5%	32	2.9%	26	2.0%	27	1.6%	26	1.5%
Totals	378	100. %	620	100. %	963	100. %	1088	100. %	1323	100. %	1686	100. %	1720	100. %

SOURCE: Petroleum Facts and Figures—API, 1971.

ways to avoid the obligation to invest in new tankcar equipment. They argued that it was impractical and economically unsound for each railroad to maintain a fleet of tankcars to satisfy the needs of all shippers along their routes, particularly when a large portion of the cars might be idle during slack periods.

In 1888 the Interstate Commerce Commission agreed with the railroads, and thus the securing of tankcar equipment became a shipper's responsibility and has remained so ever since. As a result, private tankcar companies were born, and shippers or builders henceforth invest their capital in the acquisition and maintenance of tankcars for their own use, or else have this service performed under leasing arrangements.

During the late 1800s and early 1900s, an ever-increasing number of tankcars were injected into rail service, as tankcar design remained relatively stable. The standard sizes were 6000, 8000, and 10,000 gallons; but a few 13,000-gallon cars began to make their appearance for specialized services.

During and after World War I, there was a very large increase in technology in all fields. It was during 1917 and 1918 that the concept of tankcar design to fit the needs of certain specific products was formally implemented by the specifications for tankcars. Welding was first authorized in 1920.

In 1928 the first tankcar for liquid propane was built for Phillips Petroleum Company. It consisted of 30 large ICC-27 cylinders mounted vertically on a flatcar and had a capacity of 8000 gallons. In 1938 The Association of American Railroads (AAR) Committee on Tank Cars authorized the construction of 523 fusion-welded tankcars for experimental services. As a result, in 1941, a whole new series of tankcar specifications were established.

In the late 1950s, based on technological advancements derived from World War II, tankcar designs improved, with great emphasis on increased capacity. By 1960 the majority of tankcars were in the 10,000-gallon range (as compared with the early 3500-gallon cars). During the 1960s, after AAR Committee approval of the larger-size experimental cars, shippers started to modify their tankcar fleets in the direction of jumbo-size cars. This conversion was further motivated by the publishing of incentive rail freight rates applicable to jumbo-tankcar shipments.

In 1970, as a restraining factor, the U.S. Department of Transportation (DOT) limited the size of new tankcars to 34,500 gallons and to 263,000 pounds over-the-rail weight, which includes the weight of the car and lading. This rule applied specifically to materials subject to the *Hazardous Materials Regulations** of the Department of Transportation, but it also has had a restricting influence on the construction of superjumbo-size tankcars for nonregulated products. In 1974 the conventional jumbo-size tankcar moving flammable liquids and corrosives was generally in the 20,000-gallon range.

GASEOUS FUEL MOVEMENTS

The growth of the liquefied compressed gas industry has depended heavily on the development and improvement of transportation facilities. The principal methods of transport include:

1. Cylinders. The portable metal cylinders of limited capacity are transported to users for replacement of exhausted supplies. The empties are returned to the manufacturer for refilling. These containers may be shipped on the railroads, by truck carriers, or by freight or express, subject to regulations approved by the Interstate Commerce Commission.

2. Tank Trucks. Trucks are used for both local and intermediate hauling from manufacturing points and for distance hauling from manufacturing and terminal points to individual domestic, commercial, and industrial consumers who maintain storage tanks on their premises. Where these trucks are required to cross state lines, their equipment is subject to ICC regulations. Local and state regulations also apply.

3. Marine Shipments. Since the end of World War II, marine shipments of liquefied petroleum gas (LPG), both in bulk and in portable containers, have increased rapidly. Marine shipments and vessel design are subject to the regulations of the American Bureau of Shipping and the U.S. Coast Guard and, in the case of portable containers, the ICC. See article on *Seagoing Vessels for Transporting Liquid and Gaseous Fuels* in the Handbook section on *Petroleum Technology.*

*The transportation of flammable liquids, liquefied gases, and other hazardous materials by rail is regulated by the Hazardous Control Act of 1970, which is administered by the Hazardous Material Regulations Board (HMRB) of DOT, Washington.

4. Pipelines.   For moving large quantities of liquid and gaseous fuels, the products pipelines often serve as the most practical and economical method of transportation. See articles on *Gas Pipelines and Underground Storage,* and *Marine Pipelines* in the Handbook section on *Gas Technology.*

5. Tankcars.   Special tankcars must be used to ship bulk quantities of liquefied petroleum gas and other compressed gas fuels by rail. These tanks are of sturdier construction and differ in other respects from general-service tankcars. Tankcars offer a less expensive method than cylinders for long-distance shipping. In the last several years, the efficiency of tankcar shipments has been improved by the design and construction of larger cars and through the use of uninsulated cars on which the deadweight is reduced. Sizes up to 33,600 gallons without insulation are now in use.

INTERCONNECTED TANK TRAIN

An interesting European innovation now being experimentally introduced in United States markets is the interconnected tank train.* The basic idea calls for a system of interconnects, mainly flexible hoses and special valves, which permit a million-gallon 40-car tank train to be filled or emptied from one connection. In operation, one person connects the first car of each 40-car unit to the fluid supply and turns on a pump. Loading is automatically leveled from the first car to the last to the prescribed outage.† The interconnecting hoses have a 10-inch diameter and are reinforced with steel mesh. This system appears to hold promise for improving distribution of bulk-liquid commodities such as petroleum and petrochemicals.

FUELS SHIPPED BY RAIL AND PRINCIPAL EQUIPMENT USED

The principal gaseous and liquid fuels shipped by rail and the types of railcars generally recommended for use in the United States are summarized in Table 2. The use of rail transportation for energy products in Europe and the United Kingdom is described later.

COMPRESSED GAS TANKCAR (CLASS DOT-112A340W)

A car of this type is illustrated in Fig. 1. The car is shown diagramatically in Fig. 2. General specifications of the car are given in Table 3.

The ACF‡ 33,600-gallon tankcar in this class is equipped with a single safety relief

Fig. 1. A standard tank car for the transportation of compressed gases. Capacity 33,500 gallons. *(Shippers Car Line Division, ACF Industries, Incorporated.)*

*General American Transportation Corp.
†Outage is the unfilled, open space in a tankcar containing a liquid and usually is expressed as a percentage of the total useful volume of the tank.
‡Shippers Car Line Division, ACF Industries, Incorporated.

TABLE 2.   Major Fuels Shipped by Rail and Equipment Used

BUTANE

Colorless gas with characteristic natural gas odor
Weight/gallon, 4.86 pounds
Melting point, −217°F (−138°C)
Boiling point, −31°F (+0.6°C)
Vapor pressure, 51 psig at 105°F (40.5°C)
DOT Hazardous Classification, Flammable Gas
DOT Section, 173.306; 173.304; 173.314; 173.315
Freight rate minimum, 138,000 pounds
Recommended car to be used:
   Capacity, 33,500 gallons; 100 tons, DOT 112A340W
   Filling density, April–October .... 54.7% (153,206 pounds)
                           November–March .... 56.13% (157,068 pounds)
Tankcars of Class DOT 105A100W and Class DOT 111A100W4 are also authorized.
Necessary appurtenances:
   Interior loading and unloading lines must be equipped with excess flow valves.

PROPANE

Colorless gas with characteristic natural gas odor
Weight/gallon, 4.23 pounds (specific gravity, 0.509)
Melting point, −310°F (−190°C)
Boiling point, −44°F (−44.5°C)
Flash point, −156°F (−104°C)
Vapor pressure, 187 psig at 105°F (40.5°C)
DOT Hazard Classification, Flammable Gas
DOT Section, 173.306; 173.304; 173.314; 173.315
Car type, DOT 112A340W
Freight rate minimum, 127,000 pounds
Recommended car to be used:
   Capacity, 33,500 gallons; 100 tons
   Filling density, April–October .... 45.9% (128,441 pounds)
                           November–March .... 47.88% (133,982 pounds)
Tankcars of Class DOT 105A300W are also authorized

LIQUEFIED PETROLEUM GAS (LPG)

Liquefied gas usually consisting of pure propane and butane, or a butane containing both normal
   and isobutanes.
Physical properties vary with composition, but generally are in the ranges bracketed by butane and
   propane as previously given.
DOT Hazard Classification, Flammable Gas
DOT Section, 173.306; 173.304; 173.314; 173.315
Freight rate minimum, 127,000 pounds
Recommended tankcar to be used:
   Capacity, 33,500 gallons; 100 tons, DOT 112A340W
   Filling density and maximum payload. Varies, but exceeds 127,000 pounds
Necessary appurtenances:
   Interior loading and unloading lines must be equipped with excess flow valves.

KEROSINE

A distilled hydrocarbon from petroleum or shale oil
Weight/gallon, 6.75 pounds
Flash point, 160–165°F (71–74°C)
Vapor pressure, negligible psig at 100°F (38°C)
DOT Hazardous Classification, Combustible Liquid
DOT Section, 173.115; 173.118; 173.119
Freight rate minimum, 132,000 pounds
Recommended tankcar to be used:
   Capacity, 26,800 gallons; 100 tons, DOT 111A100W1 (noncoiled, noninsulated)

FUEL OIL

Any liquid or liquefiable petroleum product used as a fuel. The various types of fuel oil are
   described in this *Handbook* under Petroleum Fuels Characteristics.
DOT hazard Classification, None
Feight rate minimums, 132,000; 150,000; and 160,000 pounds

Suggested tankcars to be used		*Payload, lb*
NI/C-73-20-SBC	20,433 gal × 7.4 (lb/gal)	151,204
NI/C-73-23-SBC	22,860	169,186
NI/C-73-26-SBC	26,060	192,844

FUEL OIL			
EC/Ins. 73-21-SBEC	20,580		152,292
EC/Ins. 77-23-SBEC	23,030		170,422

AVIATION GASOLINE AND JET FUELS

Highly volatile fuel. Several grades, including:

Weight/gallon, 6.46–6.99 pounds	6.49–6.99 pounds	6.25–6.68 pounds
Flash point, 105–150°F	105–150°F	(Relatively wide
(40.5–66°C)		boiling range)

DOT Hazard Classification, Combustible Liquid
DOT Section, 173.115 through 173.119
Freight rate minimum, 132,000 pounds
Suggested cars to be used:
  Capacity, 26,800 gallons; 100 tons (noninsulated; noncoiled)
          30,000 gallons; 100-ton trucks (noninsulated; noncoiled)

GASOLINE

Weight/gallon, 6.00 pounds
Flash point, 50°F (10°C)
Vapor pressure, 15 psig maximum at 100°F (37.8°C)
DOT Hazard Classification, Flammable Liquid
DOT Section, 173.115; 173.118; 173.119
Freight rate minimum, 132,000 pounds
Recommended tankcar to be used:
  DOT 111A100W1 (noncoiled, noninsulated)
  Capacity, 26,800 gallons; 100 tons

DOT: Department of Transportation (United States).

TABLE 3.   General Specifications—Pressure Car for Compressed Gas Service

Example of car with tank shell capacity of 33,500 gallons is used here. Other capacities are available.

Tank shell capacity	33,500 gallons
Light weight (unloaded)	97,500 pounds
Truck capacity	100 tons
Rail load limit	263,000 pounds
Maximum payload	165,500 pounds
Length over strikers	63 feet, 1 inch
Truck centers	52 feet, 2¾ inches
Length over outside wheel centers (car capable of being weighed on a 60-foot track weigh scale)	58 feet, ¾ inch
Extreme height	15 feet, 1 inch
Maximum width (at center of car)	10 feet
Maximum width (at end platform)	10 feet, 6½ inches
Maximum curve (uncoupled)	159 feet (36 degrees)
Maximum curve (coupled)	260 feet (22 degrees)
Tank Data	
Outside diameter	120 inches
Plate thickness	0.7317 inch
Tank plate material	AAR TC-128, Grade B
Interior cleaning	Light grit blast, sweep clean
Body	
Design	Underframeless
Running boards	Longitudinal, along top center line of tank
Ladders	At each end of car
Safety platform	Two-level with safety railing at manway housing
Air brakes	AB, truck-mounted
Hand brake	Vertical handwheel
Trucks	
Type	Ride control, 100 tons
Journal bearings	Roller bearings
Wheels	Steel, 36-inch diameter, one-wear

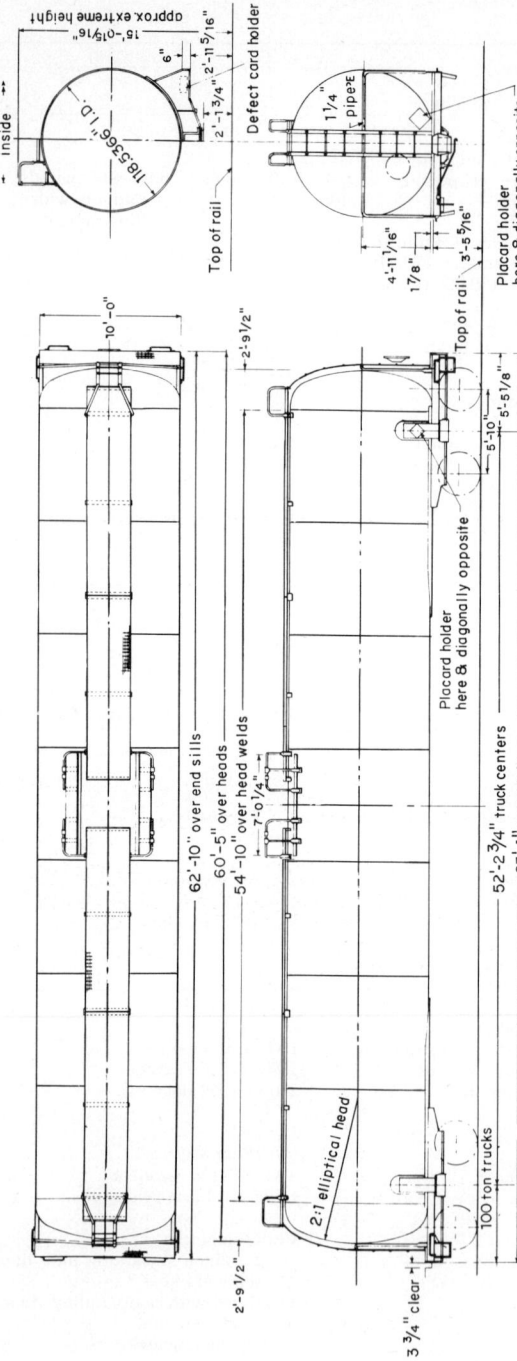

Fig. 2. Top and side views of compressed-gas tank car. Capacity 33,500 gallons. *(Shippers Car Line Division, ACF Industries, Incorporated.)*

valve that meets all DOT and AAR requirements, both for materials of construction and discharge capacity. DOT regulations specify that the discharge capacity of the valve shall be sufficient to prevent the buildup of pressure in the tank in excess of 82½ percent of the test pressure (340 psig). The car is widely used for the transportation of propane, butane, LPG, and other compressed gases. The safety relief valve must be sized to meet the discharge capacity required for each liquefied gas the tank is authorized to carry. Appendix A of the AAR Specifications for Tank Cars establishes the formula for calculating the required discharge capacity, and also sets forth the requirements for materials of construction, testing, and other factors.

For propane service, the required and actual discharge in cubic feet per minute (air) of the safety valve are: Required flow rate 27,765 cfm, actual flow rate 36,640 cfm, excess 9095 cfm. The DOT Hazardous Materials Regulations specify that the basic start-to-discharge setting or the safety relief valve shall be 75 percent of the test pressure, i.e., 225 psig. However, the DOT has granted special provisions that authorize this class of car to have the start-to-discharge setting of the safety relief valve set at 280.5 psig. Most cars in service probably have safety relief valves set to discharge at 280.5 psig.

Manway and Fittings Arrangement.   Cars containing compressed gases do not have expansion domes or bottom outlets. All connections for loading, unloading, sampling, gaging, and temperature reading are attached to the manway cover under a protective steel housing at the top center of the car. See Fig. 3. Following are descriptions of the principal hardware in this system.

1. Liquid and Vapor Valves.   Two liquid valves and one vapor valve are required.

When loading or unloading, the product line connection is made to one or both of the 2- or 3-inch liquid angle valves. Either of the two types of valves shown in Fig. 4 may be used. The valves are bolted to the manway cover on the longitudinal center line of the car. Directly below the valves, connected to the underside of the manway cover and running to the bottom of tank, are the liquid (or siphon) lines which are used for top loading and unloading.

The vapor (or gas) valve is located on the transverse center line of the manway cover and is identical with the two liquid valves. During loading, a return vapor line may be

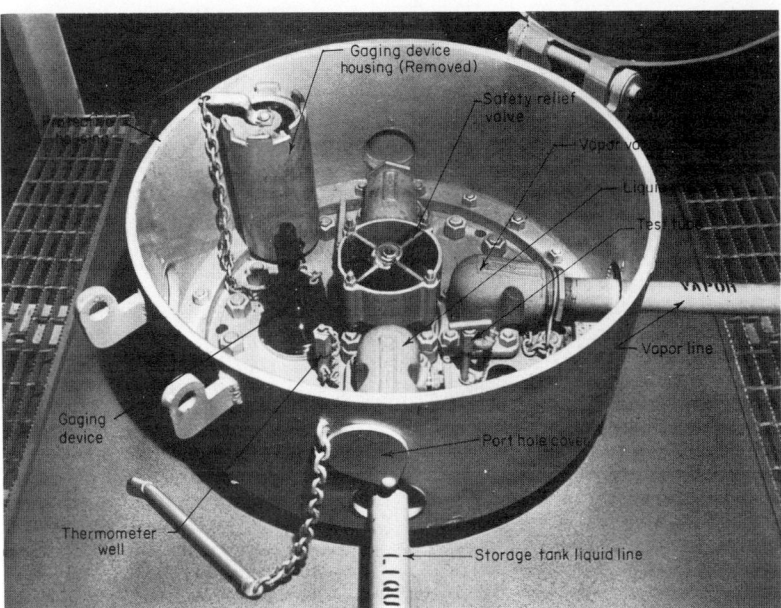

Fig. 3. Manway and fittings arrangements for compressed-gas and LPG tank cars. *(Shippers Car Line Division, ACF Industries, Incorporated.)*

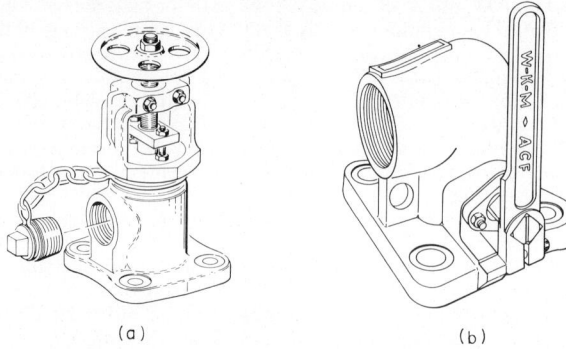

Fig. 4. Valves used for both liquid and vapor service on compressed-gas and LPG tank cars: (*a*) ACF 1312—2-inch plug-type angle valve. (*b*) WKM ball-type angle valve.

attached to the vapor valve to assist the loading rate by returning vapors from within the tank to the storage tank or other sources.

2. Gaging Device.   The slip tube gaging device, as shown in Fig. 5, is located on the transverse center line (or at a right angle to the length of the car) opposite the vapor valve and is protected by a steel housing which should be removed only when loading and unloading the car. This device is used to determine the amount of product in the car by measuring the unfilled space (outage) above the liquid in the tank. Use of the gaging device is illustrated in Fig. 6.

3. Safety Relief Valve.   This valve is located in the center of the manway and fittings arrangement. The valve relieves when the pressure within the tank exceeds a preset level (225 or 280.5 psig) to prevent pressure buildup in the tank. As soon as the pressure in the tank drops below the start-to-discharge setting, the valve will close. Maintenance of safety relief valves or adjustment of start-to-discharge settings should be performed only by the valve manufacturer, or by properly trained and experienced personnel.

4. Optional Fittings.   A test tube fitting can be used to determine whether any liquid product remains in the car. It will emit liquid if liquid product is in the tank, or gas if all the liquid product has been unloaded. The fitting is installed on the manway cover plate between a liquid valve and the vapor valve. It is usually closed by a ¼-inch needle angle

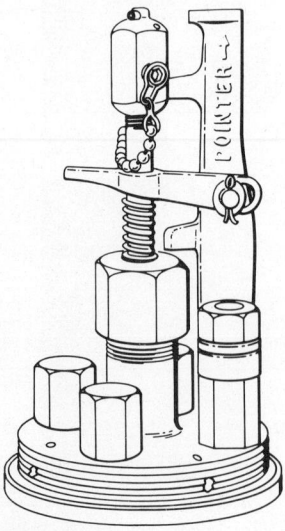

valve fastened to the top of the manway cover plate. Below the plate is a tube which extends into the sump in the bottom of the tank. By removing the plug and opening the valve handle, a sample of the product which remains in the tank will be emitted.

Optional also is a thermometer well which is used to determine the temperature of the product. During loading this is necessary so that proper calculations can be made to determine the correct outage as required by DOT Hazardous Materials Regulations. The well extends from the manway cover plate to within 18 inches of the bottom of the tank and is filled with a permanent-type antifreeze. The armored thermometer which is used to take readings is furnished by the user.

TANKCARS FOR LIQUIDS AND GENERAL SERVICE

These cars are available in both insulated and non-insulated designs. They transport such products as gasoline, lubricating oil, fuel oil, and other flammable liquids and fuels. A DOT-111A100W-1 noninsulated tankcar is illustrated in Fig. 7, with the car shown diagrammatically in Fig. 8. General specifications for the car are given in Table 4. The general specifications for an insulated car of this type are given in Table 5.

Fig. 5. Slip-tube gaging device.

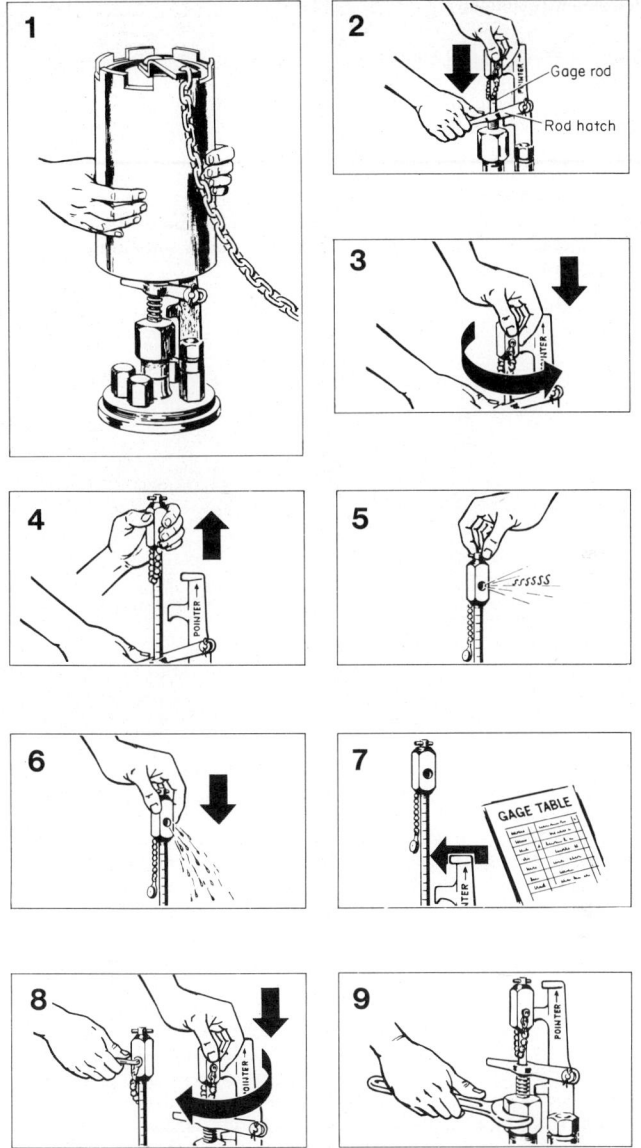

Fig. 6. On pressure-type tank cars transporting LPG, measurement of the tank's contents is usually achieved by use of the car's gaging device. The following steps illustrate the use of this device: (1) Unscrew and remove the gaging device housing. (2) Holding the top of the rod down firmly with one hand, depress the rod latch downward to permit the rod to move up and down. (Never have your head over any of the manway fittings!) (3) Push down on the gage rod, and turn counterclockwise to release valve plug from beneath the arm on the pointer. If rod does not move freely, loosen packing gland nut slightly. (4) Raise the gage rod up as far as it will go. (5) Remove valve plug, and open control valve slowly. LPG vapor will hiss out of control valve, indicating that bottom of gage rod is in vapor space. (6) Slowly push down on gage rod until liquid squirts out of nozzle. Immediately close control valve. This indicates that bottom of gage rod has contacted the liquid-vapor interface. (7) Read the outage marking on gage rod opposite pointer. This figure, used in conjunction with gage table for the individual car, will enable calculation of volume of LPG in car. (8) Replace valve plug, and push gage rod down as far as it will go. Turn clockwise to engage catch beneath pointer arm. Lift arm up against catch. (9) Make sure that packing gland is tight. If there is any evidence of leakage, gasketing around packing gland should be replaced when car is empty before the next loading.

Loading and Unloading of Tankcars. Most nonpressure tankcars are equipped with a siphon arrangement which can be used for both loading and unloading the cars. The system consists of a connection nozzle and fitting on the top of the tankcar and the siphon (or eduction) pipe itself, which extends to the bottom of the tank. A typical siphon arrangement is shown in Fig. 9. The siphon arrangement can be used for loading a tankcar by connecting a loading line from storage to the siphon pipe fitting on the dome. The product then can be pumped into the car. The siphon arrangement makes it possible to unload tankcars from the top—either by pressure (which forces the liquid up through the siphon pipe to storage) or by using a pump (which draws the lading through the siphon pipe and then pumps it to storage).

Fig. 7. A noninsulated general-service slope-bottom tankcar for transporting various petroleum products, such as gasoline, fuel oil, and other flammable fuels and liquids. (*Shippers Car Line Division, ACF Industries, Incorporated.*)

For pressure unloading, the tankcar is equipped with an air inlet pipe through which air or gas is introduced to pressurize the tank. Although the air pressure will not be influenced by the viscosity of the lading and the desired rate of flow, it should not exceed 90 percent of the start-to-discharge setting on the car's safety relief valve. For pump unloading, the pump is placed in the line between the storage tank and the siphon pipe connection on the tankcar. During pump unloading, pressure within the tank must be adjusted to prevent collapse of the tank. If the car is equipped with a vacuum relief valve, the adjustment will be made automatically. If it is not so equipped, the air inlet or the dome cover should be opened to permit air to enter the tank to balance the pressure. Siphon loading is advantageous when protection of the lading is important. Also, since it is a closed system, it provides for safe handling of toxic or similarly hazardous ladings. Because all necessary connections are made on the top of the tankcar, top unloading through a siphon is frequently less difficult than bottom unloading. The siphon arrangement makes the tankcar more versatile because it makes the car compatible with more loading and unloading situations

Bottom Unloading is common and is accomplished by a bottom outlet valve which is located on the bottom center line of the tank and is opened or closed by a valve rod which extends through the tank to an operating handle (valve wrench) at the external top of the car. Extending below the outlet valve is an outlet chamber which channels the product from the valve to the unloading lines. This unit (see Fig. 10) may be either a casting or a built-up welded chamber and is a permanent part of the car's outlet arrangement. It is designed with a breakage groove so that if it is impacted in an accident it will be severed from the tank without the valve unsetting causing loss of the tank's contents. The ordinary closure for the outlet chamber is a 4-inch outlet valve cap which is chained to the car to prevent its possible loss in transit. Screwed into the main outlet cap is a 2-inch outlet plug used for checking purposes to assure that the valve is seated and not leaking.

Three principal types of outlet valves are used: (1) the internal positive self-locking

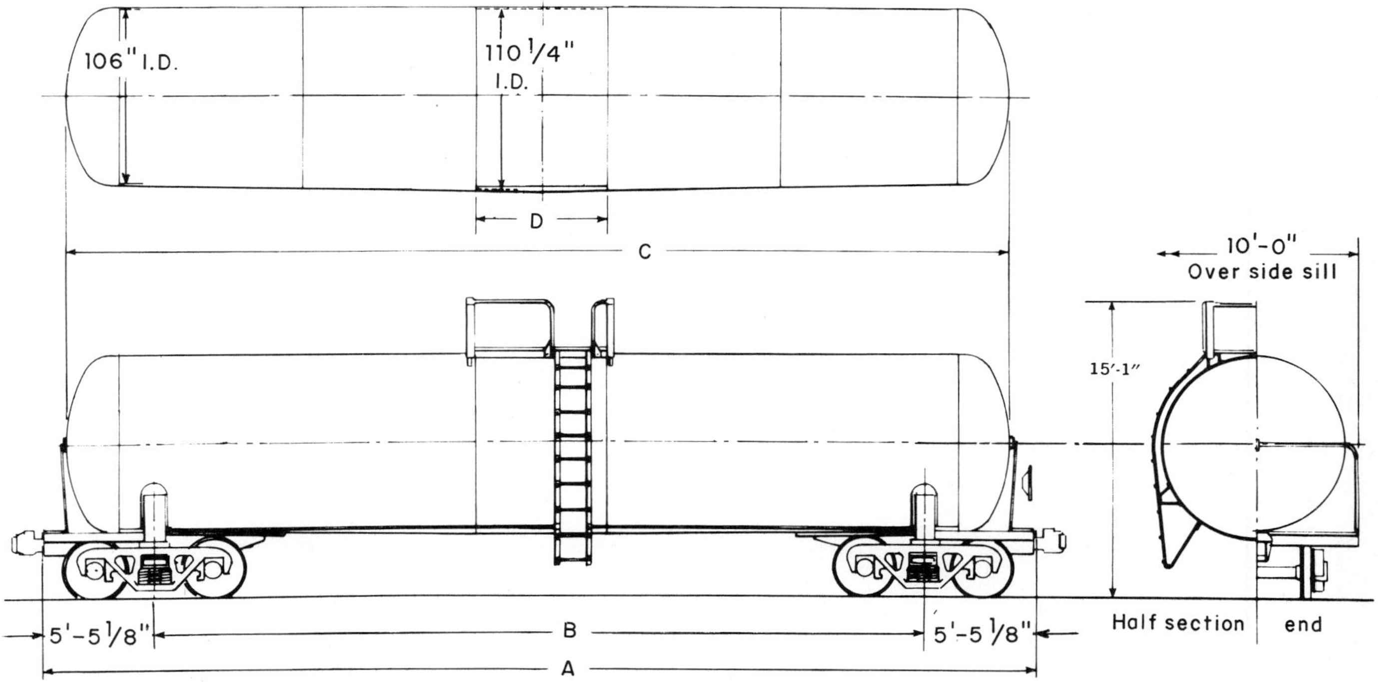

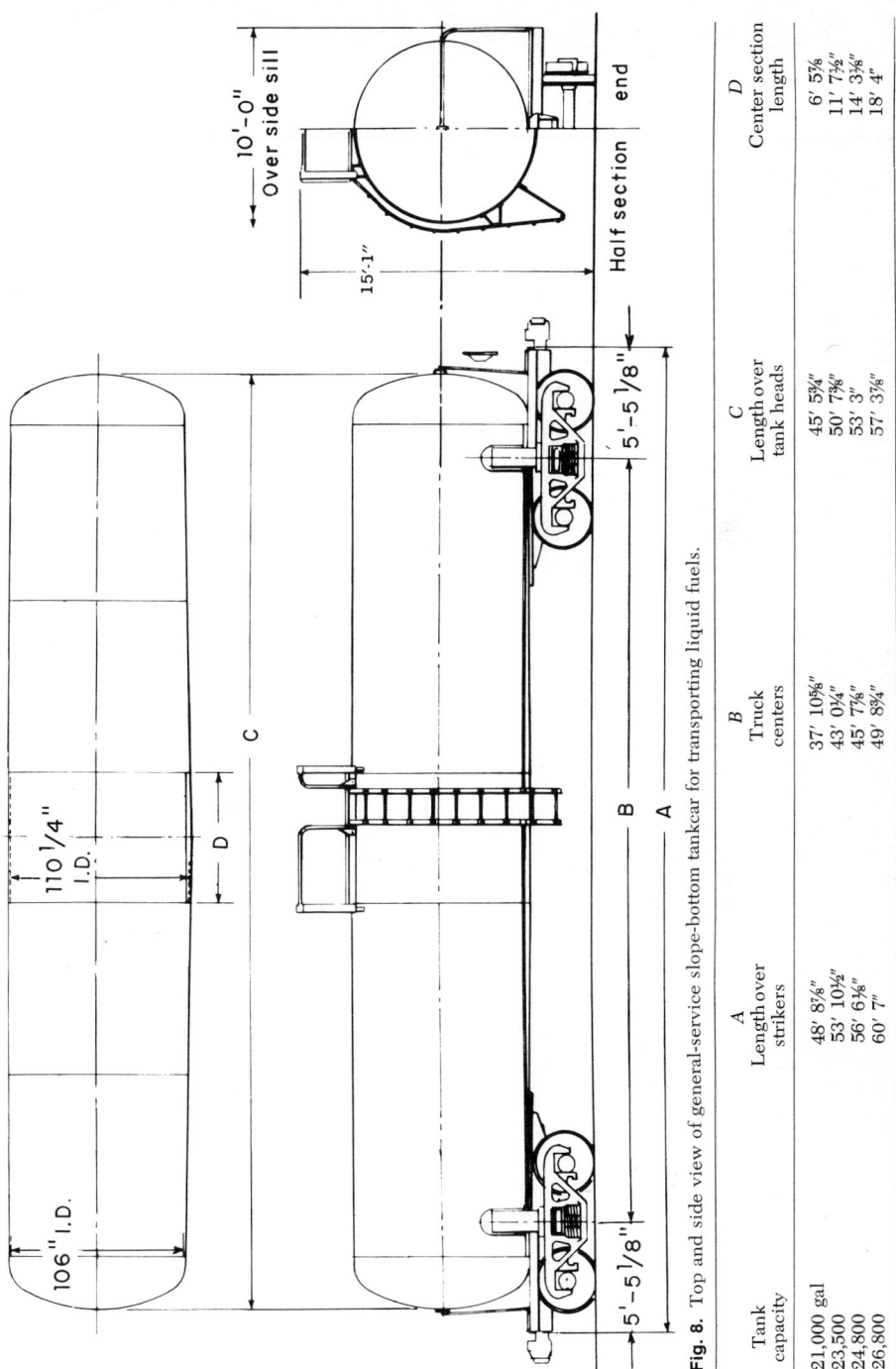

Fig. 8. Top and side view of general-service slope-bottom tankcar for transporting liquid fuels.

Tank capacity	*A* Length over strikers	*B* Truck centers	*C* Length over tank heads	*D* Center section length
21,000 gal	48′ 8⅞″	37′ 10⅝″	45′ 5¾″	6′ 5⅞
23,500	53′ 10½″	43′ 0¼″	50′ 7⅜″	11′ 7½″
24,800	56′ 6⅛″	45′ 7⅞″	53′ 3″	14′ 3⅛″
26,800	60′ 7″	49′ 8¾″	57′ 3⅞″	18′ 4″

TABLE 4. General Specifications—Slope-Bottom Tankcars (Noninsulated), Class DOT 111A100W1 for Transporting such Products as Gasoline, Lubricating Oil, Fuel Oil, Flammable Liquids

Tank shell capacity (shell full), gallons	20,850	21,000	23,500	23,330	26,800	26,600
Lading capacity (less 2% outage), gallons	20,433	20,850	23,030	22,860	26,264	26,060
Railroad limit, pounds			263,000			
Light weight, pounds	62,600	59,600	61,400	66,000	66,900	69,700
Maximum payload, pounds	200,400	203,400	201,600	197,000	196,100	193,300
Length over strikers (feet, inches)	48, 8⅞	48, 8⅞	53, 10½	53, 10½	60, 7	60, 7
Truck centers (feet, inches)	37, 10⅝	37, 10⅝	43, ¼	43, ¼	49, 8¾	49, 8¾
Extreme height			15 feet, 1 inch			
Maximum width over end platform			10 feet			
Maximum curve (uncoupled), feet	155	115	131	131	160	160
Maximum curve (coupled), feet	198	198	221	221	249	249
Maximum product density, pounds/gallon	9.8	9.8	8.75	8.61	7.46	7.41
Safety platform			Single level			
Heater coils	16-line internal	None	None	16-line internal	None	16-line internal

Except as noted, the following specifications cover cars of all capacities listed above:

Tank Data

Inside diameter	106–110¼ inches
Slope	¼ inch per foot
Plate thickness	⁷⁄₁₆ inch
Tank plate material	Steel (ASTM A-515, Grade 70). TC-128, Grade B for intermediate shell sheets

Trucks

Type	Ride control, 100-tons
Journal bearings	Roller bearings
Wheels	Steel, 36-inch diameter, one-wear

Body

Design	Stub sill
Running boards	At end platforms only
Air brakes	AB, truck-mounted
Hand brake	Vertical handwheel type

Tank Fittings

Manway cover	Cast steel (hinged and bolted)
Manway nozzle	20-inch inside diameter
Siphon pipe	Schedule 80 carbon steel (2-inch or 3-inch)
Siphon valve	Ball valve (2-inch or 3-inch)
Air inlet	Ball valve (1-inch)
Safety valve	75 psi
Safety vent	Optional (75 psi)
Vacuum relief valve	Optional
Gaging device	Visual marker gage
Bottom outlet valve (with universal flange)	ACF type 1561–6-inch (optional) WKM 4-inch ball valve (optional) Other ball valves (optional)
Outlet valve chamber	To suit outlet valve

TABLE 5. General Specifications—Slope-Bottom Tankcars (Insulated)—DOT 111A100W-1

Tank shell capacity (shell full), gallons	21,000	21,000	20,850	23,500	23,500	23,330
Lading capacity (less 2% outage), gallons	20,580	20,580	20,433	23,030	23,030	22,860
Railroad limit, pounds	263,000					
Light weight, pounds	68,300	71,200	72,100	72,800	76,100	76,100
Maximum payload, pounds	194,800	191,800	190,700	190,200	186,900	186,900
Length over strikers (feet, inches)	48, 8⅞	48, 8⅞	48, 8⅞	53, 10½	53, 10½	53, 10½
Truck centers (feet, inches)	37, 10⅝	37, 10⅝	37, 10⅝	43, ¼	43, ¼	43, ¼
Extreme height	15 feet, 1 inch					
Maximum curve (uncoupled), feet	115	115	115	131	131	131
Maximum curve (coupled), feet	198	198	198	221	221	221
Maximum product density, pounds/gallon	9.46	9.31	9.39	8.25	8.11	8.16
Safety platform	None		Two-level	None		
Heater coils*	None	16-line external	16-line internal	None	16-line external	16-line internal
Insulation	4-inch Fiberglas					
Insulation jacket	Steel					

*Exterior heater coil systems are recommended over interior heater coil systems because: (a) The heat-up efficiency of the exterior coil system is equal to an equivalent interior heater coil system; (b) in the event of a coil leak, the product is not contaminated; (c) an exterior coil system is preferred if the tank is to be lined.
Except as noted, the following specifications cover cars of all capacities listed above:

Tank Data

Inside diameter	106–110¼ inches
Slope	¼ inch per foot
Plate thickness	7/16 inch
Tank plate material	Steel (ASTM A-515 Grade 70). TC-128 Grade B for intermediate sheet sheets

NOTE: Trucks, body, and tank fittings specifications are same as those shown for noninsulated cars in Table 4.

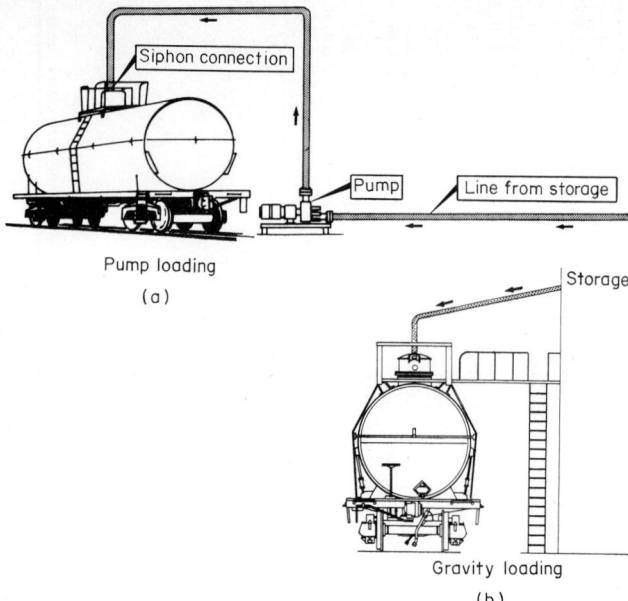

Fig. 9. Loading arrangements for nonpressure tankcars: (*a*) Siphon loading. (*b*) Gravity loading, which is convenient, but not always practical.

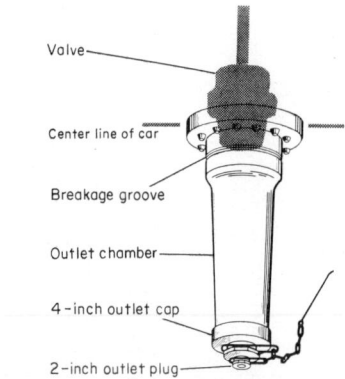

Fig. 10. Outlet chamber that channels product from valve to unloading lines.

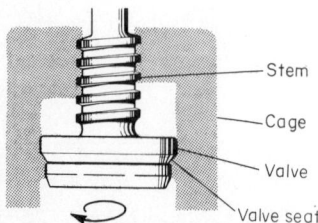

Fig. 11. Positive self-locking-type bottom outlet valve. This valve has a turning action when opening or closing.

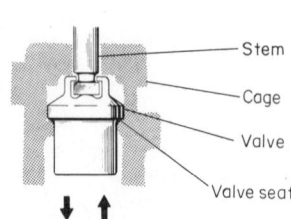

Fig. 12. Positive clapper-type bottom outlet valve. This valve has a rising and falling action when opening or closing.

valve, (2) the internal positive clapper valve, and (3) the bill valve. A positive self-locking outlet valve is shown in Fig. 11. The valve has a one-piece stem and valve construction and is designed so that the valve seat is wiped clean each time the valve is opened or closed. This type of valve normally is used in cars transporting free-flowing liquids, such as gasoline or alcohol. A positive clapper outlet valve is shown in Fig. 12. The valve in this design is opened or closed by the action of the stem, but does not rotate with the stem. Positive clapper valves are designed for the transport of viscous ladings which tend to clog other types of valves and hinder normal operation.

Both types of valves are approved by the AAR Committee on Tank Cars and are available in 4- and 6-inch sizes, with either a threaded or a bolted cage and with a metal-to-metal or a gasketed seat. Gasketed seats are preferred when an especially tight closure at the outer seal is required. The gasket material is dictated by the lading being transported.

Before loading any tankcar, the outlet valve must be examined by the shipper in accordance with the requirements of DOT Regulation 173.31. After it is certain that the outlet valve is closed tightly, the outlet cap or the outlet cap plug must be removed from the outlet chamber so that the valve can be checked during loading. The cap or plug must be left off during the entire loading process and must be replaced by the shipper after the car is loaded. If the valve permits more than a minor dripping of product, loading must be stopped, and the car is not to be offered for transportation until proper repairs are made. The bill of lading certifies that this inspection has been made, and, in the event the car is shipped, the outlet valve is presumed to be operable, based on the certification contained on the bill of lading.

Tankcar Gaging. Three terms are commonly used in connection with gaging of nonpressure tankcars, namely, shell capacity, shell innage, and shell outage. These terms are illustrated and defined in Fig. 13. Also, with most tankcars in this category, there are three general dome or manway arrangements, as illustrated and described in Fig. 14. The short-pole method is another commonly used way to measure a tank shell's outage. It also is possible to determine a tank shell's innage, using either

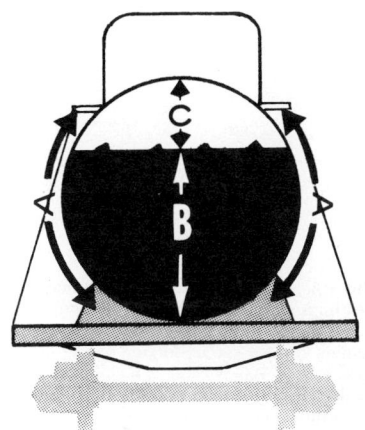

Fig. 13. Terms commonly used in tankcar gaging practice: *A. Shell capacity*—the amount of liquid in a car when the shell is completely filled. *B. Shell innage*—the amount of liquid in a car, measured from the bottom of the tank to the top of the liquid. *C. Shell outage*—the unfilled portion of the tank, measured from the top of the liquid to the underside of the top of the tank shell.

the innage or outage tables for that car. The short-pole method may be used when the liquid level does not exceed a distance of 12 inches below the gaging point. The method is illustrated and described in Fig. 15. The long-pole method is another method of measuring tank shell's innage. The method is illustrated and described in Fig. 16.

Some general-service cars are equipped with a telltale pipe, as shown in Fig. 17. The pipe is set at the same level as the gage marker. The device ejects liquid when the load level of the car is reached. However, for the telltale pipe to perform properly, the dome or manway cover must be closed securely to permit sufficient pressure buildup in the tank to raise the liquid.

Safety Procedures. All personnel involved in the loading and unloading of tankcars should receive thorough and continued training. Just a few examples of safety procedures which might be included in a program for handling flammable and explosive ladings include: (1) The tank should be grounded electrically. (2) All possible sources of ignition within the vicinity of the tank should be eliminated before and while working on the car. (3) Nonsparking tools should be used to prevent sparks when opening covers or valves. (4) Open lights must be prohibited. (5) Remaining vapors in a car should be eliminated by using a mechanical exhaust approved for use in explosive atmospheres or by completely filling the car with water.

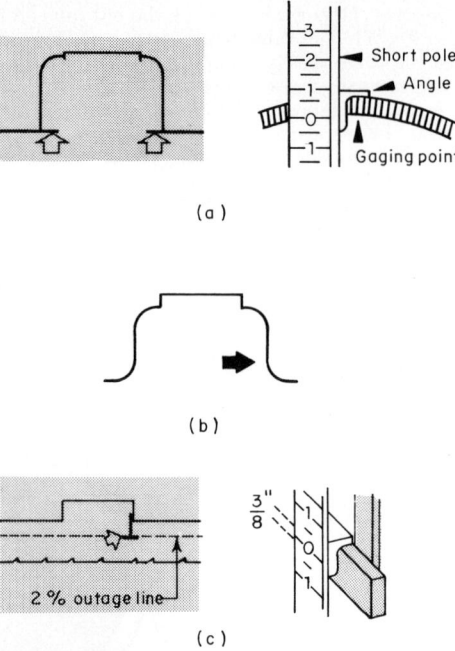

(a)

(b)

(c)

Fig. 14. Three general dome arrangements require an indication of the point on each where readings are made in the *short-pole* method of measurement: (*a*) Dome shelf (DOT 103 and 103-W cars). The short-pole angle is rested on the edge of the tank shell which protrudes into the dome. The gaging point is the underside of the top of the tank shell. (*b*) Marker bar in cars with fueled dome (some DOT 103-W cars). A metal bar is welded to the dome to indicate the shell full line. When the tank shell is less than full, it is recommended that the *long-pole* method of measurement be used to determine shell innage. (*c*) Gage marker in domeless car (DOT III-A-100-W series). Expansion is provided for within the tank. Such cars are equipped with a gage marker to indicate 2 percent outage line. In short-poling such cars, the pole angle is rested on the marker gage. However, ⅜ inch must be subtracted from the reading in order to determine the correct level line.

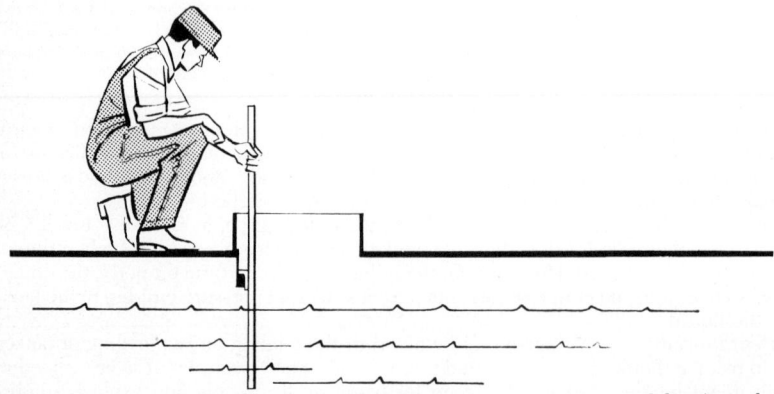

Fig. 15. Short-pole method. Pole is lowered straight down into tank, short end first (0 mark is 12 inches from short end), resting pole angle on either dome shelf or gage marker, depending upon type of car. See Fig. 14. When the pole is withdrawn, the point where liquid cuts the pole is noted. This measurement (in quarter-inches) is subtracted from measurement at the end of the gage table designated as Shell Full or Top. Resulting measurement and corresponding gallon reading on the innage table provide the car's shell innage.

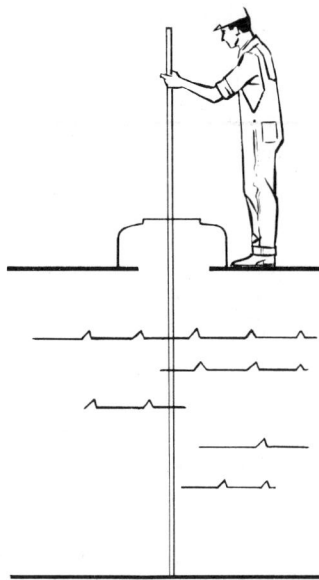

Fig. 16. Long-pole method. Pole is lowered straight down into tank, making certain it touches bottom of tank. Pole is withdrawn, and point where liquid cuts pole is noted. The corresponding reading on innage table for that car is the car's shell innage.

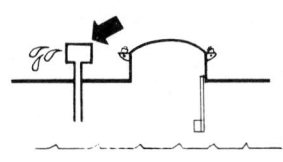

Fig. 17. Telltale pipe.

TANKCAR PROCUREMENT

Tankcars are most commonly acquired either by purchase or by leasing on full-service basis.

Purchasing. Acquisition with company funds may be advisable when these conditions exist:

1. If an administrative and technical staff already exists, that staff can be better utilized managing more cars.

2. If repair facilities are owned by purchaser, this provides for competent periodic repairs and maintenance of the cars.

3. The car is judged to be of such a capacity and design that a variety of commodities can be handled, reducing the risk of obsolescence and/or contamination.

4. Adequate funds are available and return on investment meets company requirements.

Leasing. Some of the major benefits of full-service leasing are:

1. Freedom from maintenance of the technical and administrative staff required to manage the cars including repair arrangements.

2. All record keeping, including tax, regulatory and mileage compensation requirements, is performed by the lessor.

3. Lessee avoids the risk of technical and economic obsolescence of the equipment. The lease term can be tailored to match the expected product life cycle.

4. Creates a hedge by fixing equipment costs during the lease period, assuming persistent inflation and rising costs.

5. Conservation of capital so that it can be employed in the lessee's primary line of business.

Purchasing is like net leasing (as opposed to full-service leasing) in that they are almost identical from an operating point of view. The primary distinction between the two is financial and relates only to the cost of the capital needed to acquire the cars.

The lease versus purchase decision can be quite complex and must always be tailored to the unique conditions of the firm. A number of computer programs are commercially available to assist in the quantitative or cost analysis side of the decision.

Since the railroads do not supply tankcars, they, in turn, give the shipper an allowance (called mileage compensation) for providing the car, depending on the loaded miles each car travels over their lines. The allowance is based on the depreciated value of the car. On cars less than 30 years of age, the following allowances are paid on loaded miles traveled. The depreciated value of the car is determined by applying a 3 percent annual depreciation rate from the date built.

Depreciatated value, dollars	Allowance per mile, cents
0,000– 4,999	14.90
5,000– 9,999	16.63
10,000–14,999	17.63
15,000–19,999	17.87
20,000 and over	19.44

For example, the most popular of the current cars would fall into the $20,000 and over category, and, consequently, the loaded mileage allowance would be 19.44 cents. On a 600-mile trip, the allowance would amount to $116.64. The compensation is forwarded to the owner of the tankcar by the railroads involved and, in the case of a leased car, to the lessor. The lessee, in turn, receives compensation by the adjustment of this amount on the next rental invoice from the lessor. Major lessors of tankcars in the United States include: Shippers Car Line, ACF Industries, Incorporated, St. Charles, Missouri; Tiger Leasing Group, Chicago, Illinois; General American Car Company, Chicago, Illinois; and Trans Union Corporation, Chicago, Illinois.

EUROPEAN RAIL SITUATION

Europe's longest pipeline is the 772-kilometer South European pipeline which operates from the Marseilles region in southern France to Karlsruhe in southwest Germany and carries up to 34 million tons of crude oil annually. Piped oil costs at average flow levels are way below rail and waterway rates.

British Railway's sales methods are typical of the approach to pipeline competition now being adopted all over Europe. It basically is forceful marketing which highlights the advantages of high-speed regular block train service linked to low tariff rates. Many oil companies, after taking a careful look at the economics involved, are using rail tanker services.

SNCF, like most other European railways, is fighting for the short-distance less competitive petroleum traffic. Their weapons are similar to those of the British Railways.

Deutsche Bundesbohn and Italian State Railways are pursuing similar policies.

HIGHLIGHTS ON RAIL TRAFFIC IN GREAT BRITAIN

Much of Britain's rail freight does not pay its way, nor is it very likely to do so in the future.

Approximately 90 percent of all Britain's freight movements are less than 80 kilometers, and most of them neither start nor finish near a railway. The remaining 10 percent represents the maximum that is economical for British Railways. All 16 of Britain's major oil companies have signed up to use company train facilities. Approximately 90 percent of all oil shipped in Britain moves via rail.

Daily block trains are taking fuel oil loads from the Stanlow refinery in Cheshire to Cahemore Terminal in Birmingham. Others take products from refineries on the Thames to terminals at Thame and Royston which were specially designed for service by rail.

One of the busiest block train operations, 30 trains a week, carries refined oil from Heyshan to Shell-Max and BP's supply terminal at Haydock or direct to the sidings of principal customers. In just over five years, rail oil traffic out of Heyshan has grown from 94,000 to 800,000 tons, with an eventual traffic of a million tons a year forecast.

There are now more than 10,000 railtank wagons in British railway service, nearly all of them owned by the oil companies. Increasing use of rail facilities has resulted in spectacular advances in tanker design and capacity. The average BR rail tanker once carried approximately 16,000 liters, but now there are giant tankers of 100 tons gross laden weight in service, capable of carrying 100,000 liters.

The British Railways have been luckier than most other European systems. Early development of cheap block train services made a number of pipeline proposals uneconomical. The systems in the rest of Western Europe have been up against fast-growing pipeline competition since the early 1960s.

LPG TANK WAGONS WITH NO UNDERFRAME

Unsupported tank barrels which carry all longitudinal and vertical forces have been introduced on British tracks by Procor (UK) Limited, which has placed the first of a batch of four-axle wagons of 20,360 gallons capacity in service with Esso Petroleum Limited. These tank wagons were built by R. X. Pickering of Glasgow with Swedish barrels and have Schlieren M25D primary suspension bogies conforming to British Railway requirements for 25-axle loads.

The wagons are transporting propane or butane from refineries at Fawley and Milford Haven to various distribution points and customers throughout Britain. Tare weight is 35.4 tons, and a full load of butane weights 47.3 tons. Since these wagons must also be capable of transporting anhydrous ammonia, 25-ton axle load bogies were specified.

REFERENCES

References on statistics, standards, and regulations governing the railroad transportation of liquid and gaseous fuels are periodically updated, commonly on an annual basis. Thus reference should be made to the "Annual Statistical Review," published by American Petroleum Institute, Washington; the "Annual Book of ASTM Standards," part 17, published by the American Society for Testing and Materials, Philadelphia; "Statistics of Railroads Class I," Association of American Railroads, Washington; periodic coverage by Association of Oil Pipelines, the *International Railway Journal* (see, in particular, November 1973 issue); the U.S. Department of the Interior, Washington, which published "Energy through the Year 2000" in December 1972; and the U.S. Department of Commerce, which publishes U.S. Industrial Outlook annually.

Seagoing Vessels for Transporting Liquid and Gaseous Fuels

BY DONALD J. SZOSTAK* AND P. BARRY BALDWIN†

The process involved in the evaluation of a shipping project presents the same set of complex problems and difficulties that must be confronted in evaluating a new refinery, power, petrochemical plant, or other capital-intensive energy-related facility. A shipping project is further complicated, however, in that a ship moves from one location to another, frequently across international boundaries, and the labor force is obliged to move and live with the plant. Finally, the survival of both the vessel and the labor force is dependent on the ability of the structure to withstand the most violent of physical conditions.

The evaluation of a shipping project involves both technical and commercial consideration. Various of the most basic topics and fundamental definitions presented here are intended as a starting point for planning or evaluating an ocean transportation project.

TECHNICAL AND BUSINESS CONSIDERATIONS IN OPERATING AND LEASING VESSELS

Size and Number of Vessels. Each shipping project involves various time, location, and product characteristics which in the proper combination begin to describe the ultimate size of the vessel in terms of deadweight tons and the number of vessels required.

Factors that must be considered include: (1) The amount of product required in terms of the pertinent unit of measure and/or time frame involved (e.g., Btu per day), (2) the voyage distance and desired speed of the vessel, (3) the physical characteristics of the loading and discharging facilities, (4) the water depth at loading and discharge ports, and (5) the limits imposed on ship size by virtue of cost or technology.

The value of the product is also considered in terms of a decision concerning the speed of the ship and the possible loss of cargo resulting from an extended voyage time, or the number of potential trips that the ship can make. Thus, a liquefied natural gas (LNG) carrier tends to be considerably faster than a crude carrier.

Regulatory Bodies. In addition to the factors to be evaluated in determining the physical characteristics of the ship, based on the demands of the trade, certain regulatory and class requirements must be dealt with.

The classification societies are private organizations, each with standards for the construction of a vessel as well as for its maintenance and seaworthiness.

It is not compulsory by law that a shipowner have a vessel built to a given class. However, the advantages in terms of securing satisfactory insurance rates make it a practical reality. Some of the better-known classifications include American Bureau of Shipping, Lloyds Register of Shipping, and Det Norske Veritas.

In planning for the physical requirements of a given type of vessel, the regulations of the maritime authorities of a particular country must be determined as well as of the international regulatory groups, such as the United Nations–sponsored Inter-Governmental Maritime Consultative Organization (IMCO).

Recommendations and requirements of all such groups, as well as the advantages and disadvantages involved, should be ascertained. Guidance in such matters can be obtained from reputable naval architects, shipyards, or established shipping companies, and law firms specializing in admiralty law.

Registry. All ships that engage in foreign trade are registered in a particular country in order to secure the privileges of the ships of the nation to which they belong. Registry of a

*Vice-president, Engineering, †Assistant to Vice-president, Marketing, Marine Transport Lines, Inc., a subsidiary of General American Transportation Corporation, New York.

vessel in any country automatically conveys various implications with respect to such matters as crew nationality, ship standards, operating areas, and tax regulations.

The decision concerning country of registry carries economic and political considerations which must be weighed into the total shipping project. Some countries require that the crew go with the flag, and in the United States, for example, a United States flagship generally must have an American crew. Furthermore, if a ship is to trade between United States ports, the vessel must be under the United States flag. Other national requirements vary. Although the course of action selected can have a very large economic impact on the project, the degree of the impact may be reduced as the total capitalized cost of the vessel increases.

For example, in a highly competitive trade, such as crude oil, the nationality of the vessel and the crew employed can have a much greater impact on the cost of the project than in an LNG movement where the total capitalized cost of the vessel is so great that even the most expensive crew costs are not a significant percentage of the total project cost.

Alternatives for Obtaining Vessels.   Having determined the general physical characteristics of the vessel required and having decided which flag and crew to employ, a specific vessel selection must be made. Two basic options are available if the movement is to be continuous and not of a transient nature: (1) Construct a new vessel tailored to the specific movement, or (2) obtain an existing ship. If it is desirable that the full useful life of the vessel be available, and if there is sufficient lead time, then new construction offers the greatest number of options and the widest latitude in negotiations.

If, on the other hand, time is critical and the project must move rapidly, at the market price, then the purchase of existing tonnage may be desirable, provided that the decision is compatible with the desired movement.

The decision concerning method of acquisition involves needed competence in relationship to shipping activities. Dependent upon the degree of familiarity with shipping, the services of a marine consultant, sales and purchase broker, or shipping companies may be desirable. The shipping department of a major commercial bank also can be of service.

In all these areas there are well-established and reputable firms offering varying services, some overlapping. In each instance, however, a clear statement of available services, experience of the firm, and cost of services should be obtained before entering into a contract. The services to be evaluated, especially when considering a new construction, are: (1) vessel design concept, (2) design plan approval, (3) shipyard selection and negotiations, (4) construction supervision, (5) financing strategy, (6) insurance, and (7) manning as well as general operating experience and capability of a potential shipping company.

Financial Considerations.   Since the ownership and operation of ships constitute a discipline with its own set of characteristics, it is generally not advisable to seek to own a modern energy transporter. Thus, a potential shipper generally enters into a time charter agreement ranging in length from a few months to the life of the ship. This arrangement is, in effect, an operating/lease deal where, among other factors, a large initial capital outlay is not required in favor of periodic lease-type payments.

It is possible to obtain a straight bareboat or financial lease structure and contract separately for the operation of the vessel. The cost areas involved in a time charter are the payment necessary to service the loan outstanding on the vessel, the owner's earnings, crew wages, maintenance, vessel stores and provisions, insurance and management fee. Cost escalation clauses are standard in the shipping industry with long-term charters.

The methods and techniques available to finance ships are always changing, but the most basic method is to obtain sufficient funds to cover the total capitalized cost of the vessel, using the ship or the charter, or both, as guarantees. Various national shipbuilding subsidies and mortgage guarantee plans are available and, if obtained, can impact favorably on a shipping program. However, the use of subsidies often requires the elimination of certain options (for example, to obtain attractive United States construction subsidies, the vessel must be United States flag, thus requiring a United States crew).

A United States levered lease, or British lease, can be used as well as various national financial plans, such as that provided by the Canadian Export Development Corporation.

Since nearly all shipping projects tend to be unique in some respects, various combinations of ownership/lease arrangements and financing strategies should be evaluated in order to determine that arrangement which is most appropriate for the product movement under consideration.

CLASSIFICATIONS AND DESCRIPTIONS OF VESSELS

Crude Petroleum Carriers—Tankers.   Crude oil is transported as liquid bulk cargo aboard oceangoing tankers. Since the amounts of cargo to be transported far exceeds the capacity of a single vessel, these ships are usually built to maximum size consistent with limiting port facilities. An approximation of the expected amount of traffic between major exporting and importing areas of the world during the latter years of the 1970s to be handled by oceangoing vessels is given in Table 1.

Early tankers were small and were arranged with many tanks. During World War II numerous vessels of a type known as T-2 were built in relatively large numbers. This vessel had a length of 504 feet, a depth of 39 feet, 3 inches, and was approximately 69 feet wide. The deadweight was 16,500 long tons (107,250 barrels)* on a draft of 30 feet, 2 inches. Installed main engine horsepower was 7240. Later ships as well as fleet sizes for individual owners, and even nations, were referenced in terms of T-2 equivalents.

As individual ships became larger, new colloquial terminology developed, probably first among brokers, intended succinctly and with pride to describe their size. The terms *supertanker, VLCC* (very large crude carrier), and *ULCC* (ultralarge crude carrier) came into common usage but have reference only to approximate size and no real technical validity.

The relative sizes of these three categories, comparing cargo tank sections, are illustrated in Fig. 1. Note the relative size of the man in the lower right-hand corner of the diagram. The trend of vessel dimensions of the "world's largest tanker" (for a given year) is plotted against year of delivery in Fig. 2.

Trends in draft, depth, beam, length, and installed main propulsion horsepower are illustrated in Fig. 3. The arrangement of a 230,000 DWT (deadweight tonnage) tanker is shown in Fig. 4. Basic design considerations for the cargo spaces normally include integral tanks of regular length, usually with one center-line tank and wing tanks port and starboard. Tank size is dictated by practical considerations of structural requirements, cargo lot size, free-surface effects, and safety of ship and environment in the event of vessel collision. All tank boundaries are oiltight with access from the main deck through raised oiltight hatches. Venting is arranged through preset pressure vacuum valves. Interior tank design must take into consideration ease of cargo flow to discharge piping, ease of cleaning, and corrosion control. Tankers carry their own pumps for cargo discharge.

Ecological interest has resulted in tank arrangements that incorporate segregated

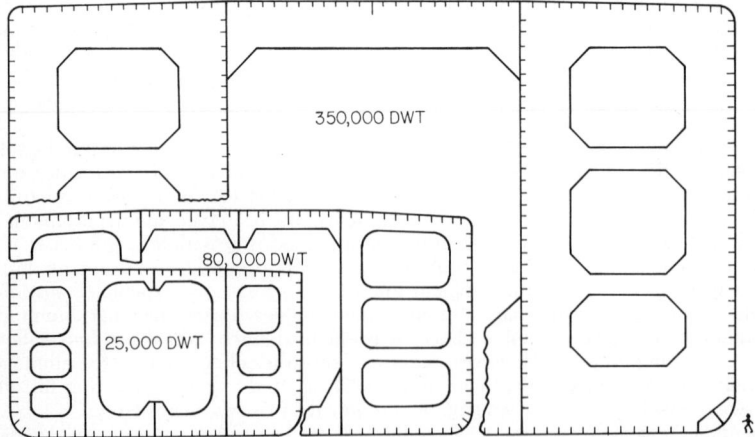

Fig. 1. Relative cargo tank sizes. DWT = deadweight tonnage. For size comparison, note figure of person in lower right-hand corner of diagram.

*Crude capacity of any given ship can be approximated by multiplying the ship's deadweight by 6.5 barrels per ton.

TABLE 1.   Approximation of Expected Seaborne Petroleum Traffic between Major Exporters and Importers in Latter Years of the 1970s

Thousands of barrels per day

Importers	Exporters North and West Africa, Eastern Mediterranean, Persian Gulf	Caribbean	Far East	U.S.S.R.	Canada	Europe
Europe	17,000	500		1000		200
North America	3,000	5,000	300	100	400	1300
Japan	6,000		1500			
South America	3,700	20		160		40
Far East	1,900			40		20
Africa	900	20		60		1000
Australasia	400		40			20
Other areas	600					
Total	33,500	5540	1840	1360	400	2580

.... 45,220 ....

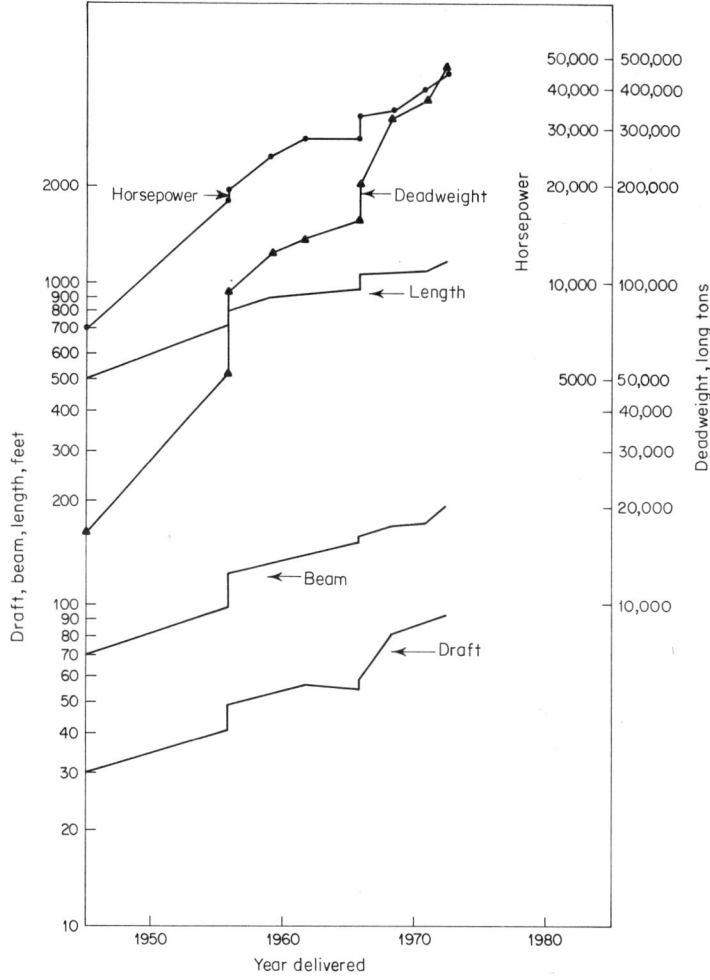

Fig. 2. Trends of vessel dimensions.

ballast and slop tanks. Safety developments in large tanker design include inert gas systems for cargo tank inerting. Gas·may be obtained from boiler flue gas, or from a separate inert gas generator.

More recently, there have been various energy-related projects under review that involve the transportation of fuel-grade methanol and naphtha. Standard tankers are contemplated for general use in this service. Perhaps the most significant factors to

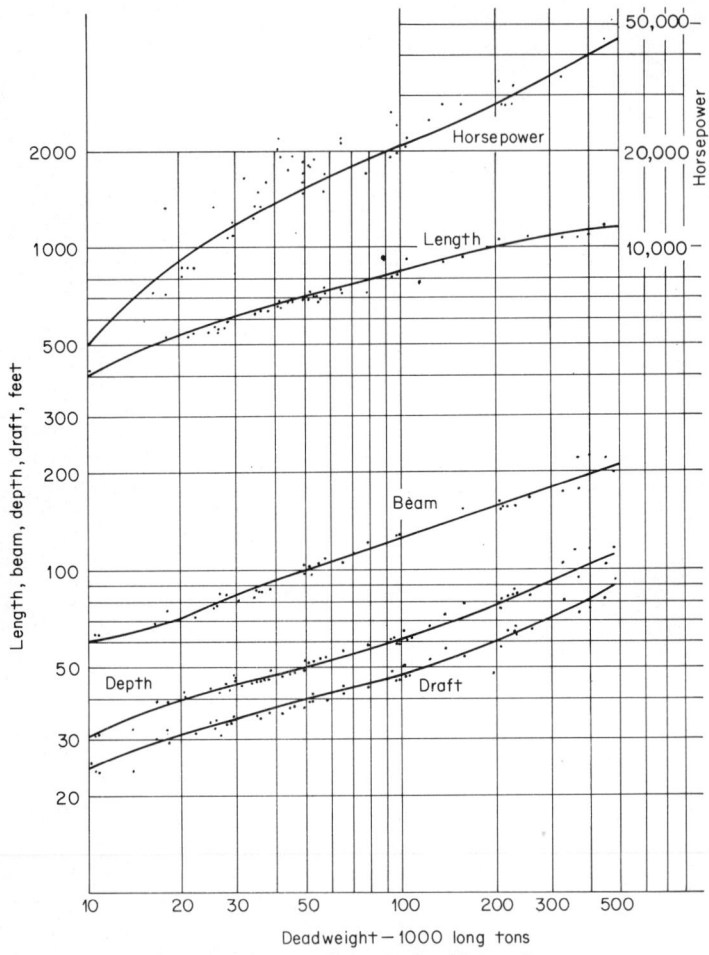

Fig. 3. Dimensions versus deadweight—tankers.

consider in this latter trade when selecting tonnage is concern with product contamination and with the fact that less product can be carried in any given tanker because of the lower specific gravity of these products.

Propulsion power selection for tankers has been geographically influenced: steam turbines in the United States; Diesel propulsion elsewhere. The trend toward ULCCs, with large associated power requirements, has necessitated steam installations worldwide in order to maintain single screw capability. For smaller vessels, there has been renewed interest in Diesel installations in the United States, mostly because of good fuel economy. However, such benefits are offset somewhat by more expensive fuel and greater maintenance requirements.

Combination Carriers. A combination carrier is a tanker designed to carry oil or alternatively solid cargoes in bulk. The characteristics and utilization of this type of vessel

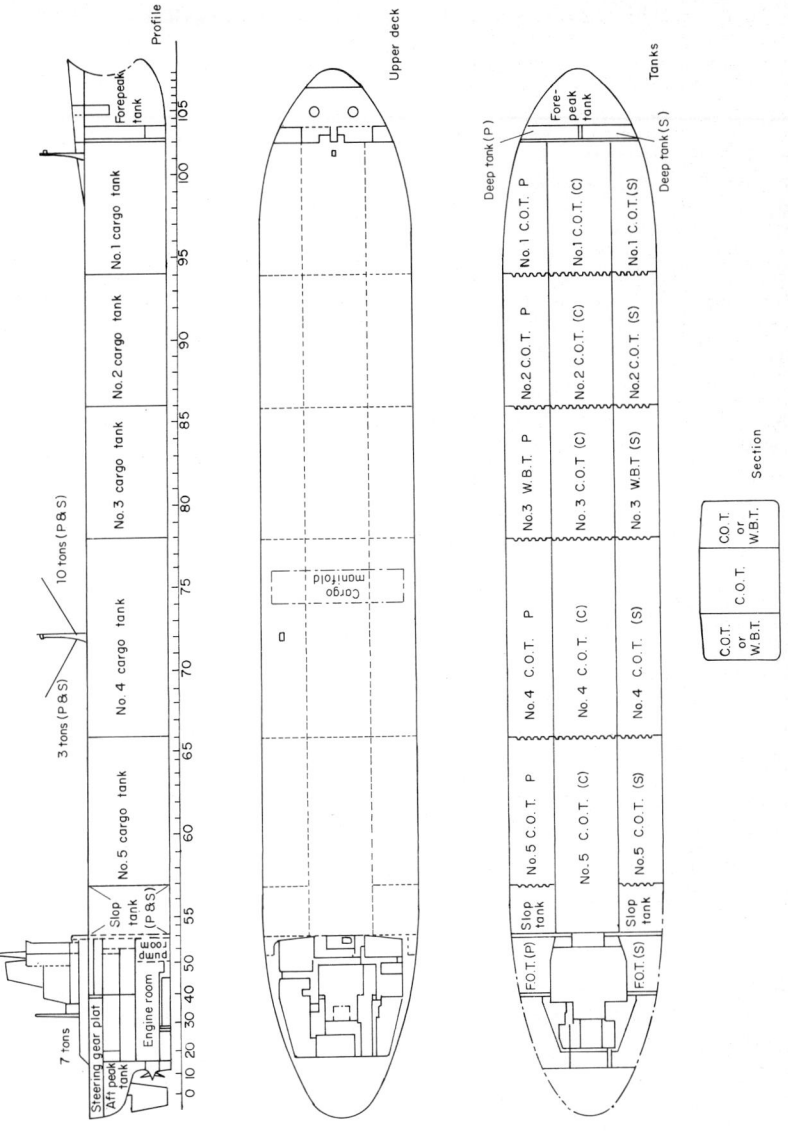

Fig. 4. Arrangement of a 230,000 DWT tanker. P = port, C = center, S = starboard, C.O.T. = cargo oil tank, W.B.T. = water ballast tank.

are well outlined in Ref. 1 and also described in the article on Superports which follows in this Handbook.

Product Carriers.   The bulk transportation of products presents essentially the same problems as crude oil transport. Most products do not have as much protective influence on steel as is provided by the heavier fuels or crudes; thus the design of products carriers demands consideration of anticorrosive measures, mostly in the form of special tank coatings of epoxy, inorganic zinc, or other materials with a long life. Surface preparation and coating application are expensive considerations in the capitalized cost of these vessels.

The size and basic dimensions of product carriers are similar to those of crude carriers.

Gas Carriers.   Gaseous fuels are transported in oceangoing vessels either under pressure at ambient temperatures (butane and propane), or at atmospheric pressure, with the cargo refrigerated and carried at its boiling-point temperature (liquefied natural gas (LNG) and liquefied petroleum gas (LPG)). Designs incorporating pressure and refrigeration are also feasible, but not commonly considered, particularly where large capacities are contemplated.

The trend of principal dimensions as a function of cubic capacity for LNG and LPG vessels is shown in Fig. 5. The greater density of LPG accounts for the deeper drafts indicated.

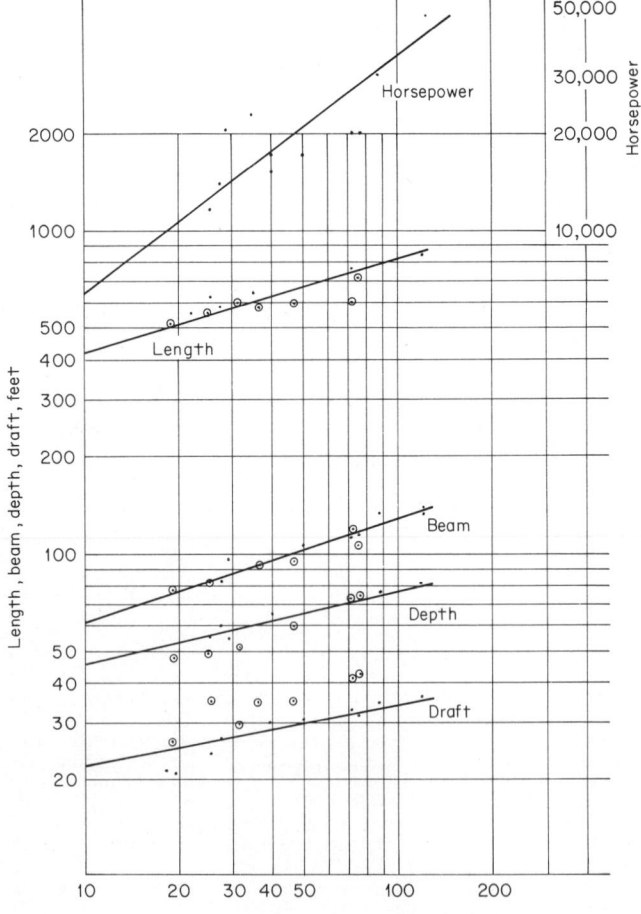

Fig. 5. Trends of principal dimensions versus cubic capacity for LNG and LPG carriers. Dots are plots for LNG carriers. Dots within circles are plots for LPG carriers.

Cargo tanks in gas carriers are normally independent tanks. When carried at ambient temperature, cylindrical tanks are normally used. The current trend is toward low-temperature transportation, incorporating well-insulated independent gravity tanks. The tanks may be free standing or the membrane type. Many patented designs exist, utilizing special materials, including stainless steel, nickel steels, and aluminum. Similarly, many insulation methods exist, utilizing balsawood, fiberglass, polyurethane foam, polyvinyl chloride, perlite, and others. See Ref. 2 and article on Cryogenic Materials for Liquefied Gases, in the section on *Gas Technology* in this Handbook.

In low-temperature designs, the boil-off of cargo may be reliquefied by refrigeration. This is normally accomplished for LPG and has been proposed for LNG, but reliquefaction equipment for LNG is large and costly and generally has not been adopted. Instead, boil-off is burned as fuel for vessel propulsion. The arrangement of an LPG-ammonia tanker of 20,000 cubic meters is shown in Fig. 6. A special-purpose tanker used to carry ammonia-LPG and methanol is shown in Fig. 7.

Barges.   Barge designs exist to accomplish transportation of any of the aforementioned liquid or gaseous fuels. Safety considerations of conditions encountered in the open sea limit the practical application of barges for oceangoing service to relative near-shore operations, sometimes referred to as coastwise service.

Future Developments.   Integrated tug-barges are a form of barge wherein the design incorporates close marrying of the tugboat to the stern of the barge. This type of transportation has become quite important in United States coastwise service. Self-propelled barges represent a concept that envisions simplified machinery with maximum automation. Early units incorporated gas-turbine electric main propulsion machinery and contemplate minimal use of personnel for operation.

The integrated tug-barge is the simplest form of what is called a *compoundable ship*. With this configuration, barges are fastened together, by elastic coupling methods. The scheme, though in a conceptual stage for use in the open ocean, shows promise.

Long espoused by the nation's leading submarine designer, the submarine tanker concept received renewed interest after discovery of Alaskan oil for possible application in under-ice crude oil transportation.

DEFINITIONS

Ballast: Any solid or liquid weight placed in a ship to increase the draft, to change the trim, or to regulate the stability.

Ballast, Clean: Ballast water in a tank which if discharged would not cause pollution.

Ballast, Segregated: Ballast water that is introduced into a tank that is completely separated from the cargo and fuel systems and which is permanently allocated to the carriage of ballast.

Bareboat Charter: Basically a financial arrangement whereby the vessel is hired with no services rendered by the owner.

Beam, Molded: The maximum breadth of the hull measured between the inboard surfaces of the side shell plating.

Bulk Cargo: Cargo made up of commodities such as oil, coal, ore, grain, etc., and not shipped in bags or containers.

Charter Party: An agreement whereby a shipowner agrees to place his vessel at the disposal of another party and the terms under which the ship will function.

Combination Carrier: A tanker designed to carry oil or alternatively solid cargoes in bulk.

Deadweight: Difference between displacement of a ship at summer load waterline and the light weight of the ship. Can serve as a guide as to the carrying capacity of the vessel.

Depth, Molded: The vertical distance from the top of the keel to the top of the main deck at side, measured at midlength of the ship.

Draft: The depth of the ship below the waterline measured vertically to the lowest part of the hull, propellers, or other reference point. When measured to the lowest projecting portion of the vessel, it is called the extreme draft, when measured at the bow, it is called the forward draft, and when measured at the stern, the after draft, the average of the draft, forward, and the draft, aft, is the mean draft, and the mean draft when in full load condition is the load draft.

Length, Overall: The extreme length of a ship measured from the foremost point of the stern to the aftermost point of the stern.

Length between Perpendiculars: The length of a ship between the forward and after perpendiculars. The forward perpendicular is a vertical line at the intersection of the fore side of the stem and the summer load waterline. The after perpendicular is a vertical line at the intersection of the summer loadline and the after side of the rudder post or sternpost, or the center line of the rudderstock if there is no rudder post or sternpost.

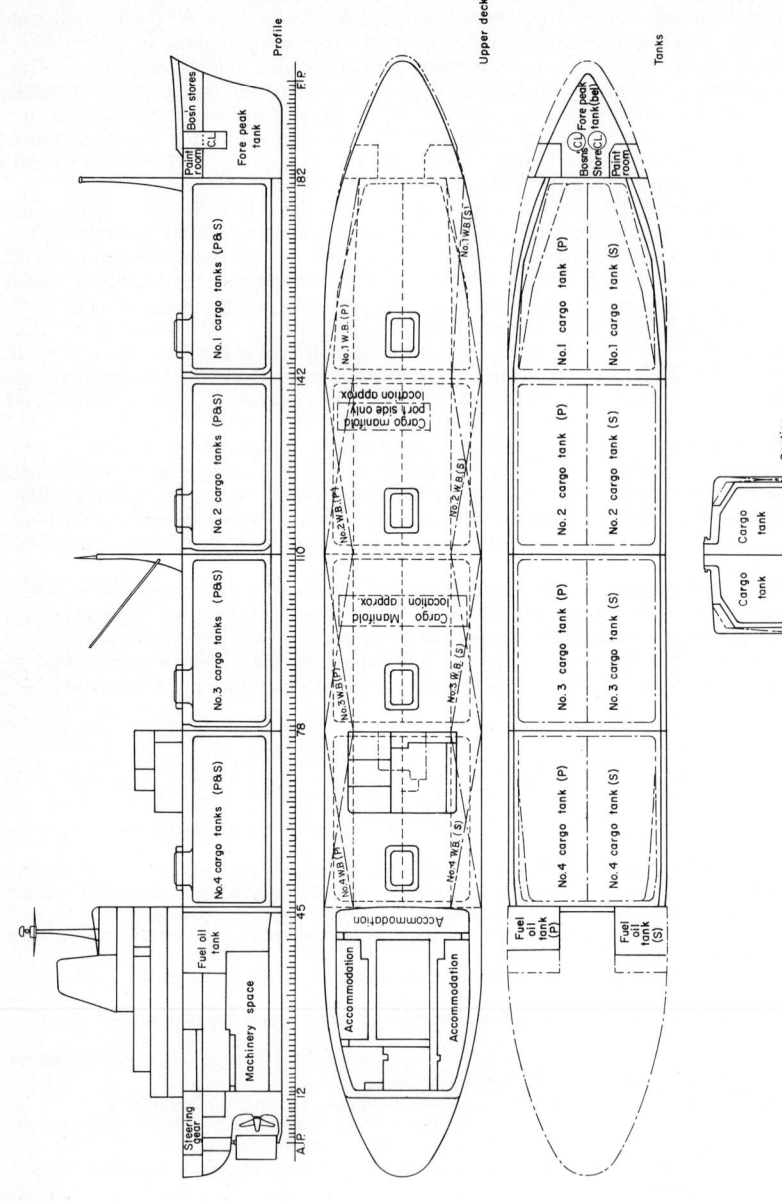

Fig. 6. Arrangement of an LPG-ammonia tanker of 20,000 cubic meters capacity.

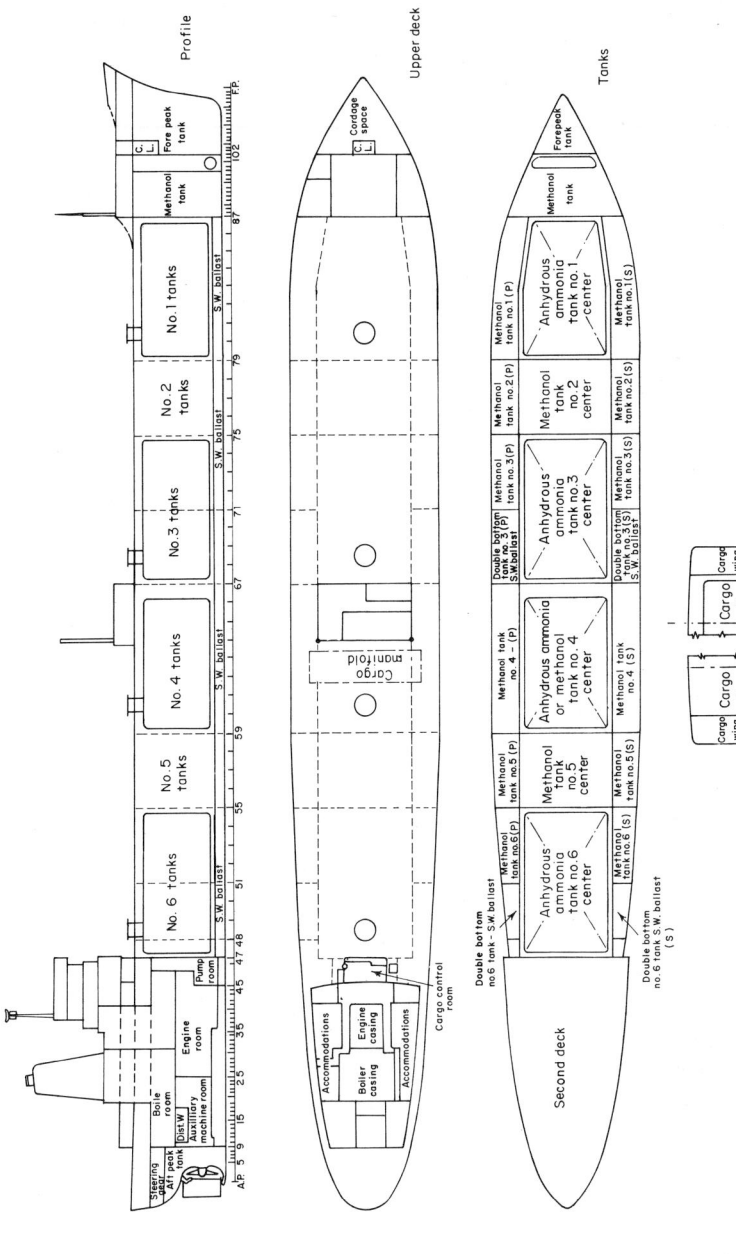

Fig. 7. Special-purpose tanker for carrying ammonia-LPG and methanol (26,000 deadweight tonnage). S.W. = salt water.

Lightweight: Displacement of a ship without cargo, lube oil, ballast and fresh water in tanks, stores and crew and their effects.

O-B-O: A combination carrier designed to carry ore, bulk, or oil cargoes.

O-O: A combination carrier designed to carry ore-oil.

PROBO: Gotaverken registered trademark used to designate a combination carrier designed to carry Product-Oil-Bulk-Ore.

Product Tanker: Refers to carriers of refined products and special-purpose fuels such as gasoline, fuel oil, jet fuel, methanol, and benzene.

Registry: A duty imposed on ship owners in order to secure the priviledges of ships of the nation to which they belong.

Supertanker: Over 30,000 deadweight long tons (DWT). This size was developed after World War II and the term is usually limited to vessels over 30,000 DWT up to about 150,000 DWT.

Tank, Ballast: Watertight compartment to hold water ballast.

Tank, Center: Any tank inboard of a longitudinal bulkhead.

Tanker: A vessel that is specifically constructed or converted to carry liquid bulk cargo in tanks.

Tank, Gravity: A tank having a design pressure not greater than 10 psig at the top. Gravity tanks may be independent or integral.

Tank, Independent: A cargo containment envelope which is not a continuous part of the hull structure. An independent tank is built and installed so as to eliminate whenever possible (or in any event, to minimize) its stressing as a result of stressing or motion of the adjacent hull structure. An independent tank is not essential to the completeness of the ship's hull.

Tank, Integral: A cargo containment envelope which forms part of the ship's hull and may be stressed in the same manner and by the same loads which stress the contiguous hull structure. An integral tank is essential to the structural completeness of the ship's hull.

Tank, Slop: A tank specifically designated for the collection of cargo drainings, washings, and other oil mixtures.

Tank, Wing: A tank that is located adjacent to side shell plating.

VLCC: Very large crude carrier—over 150,000 DWT. This term developed in the mid 1960s as tankers became "very large" and is commonly limited to vessels over 150,000 DWT to approximately 300,000 DWT.

ULCC: Ultralarge crude carrier—over 300,000 DWT. This term is an obvious extension of VLCC meant to emphasize the extraordinary size of the largest tankers built in the later 1960s and in the 1970s.

REFERENCES

1. Dorman, W. J.: "Combination Bulk Carriers," Society of Naval Architects and Marine Engineers, New York, 1966.
2. Thomas, W., and A. Schwendtner: LNG Carriers, the Current State of the Art, *Trans. Soc. Nav. Architects and Mar. Eng.*, vol. 79, New York, 1971.
3. Central Committee on Transportation by Water: *Annu. Tanker Conf. Proc.*, American Petroleum Institute, Washington.
4. Benford, H.: "On the Rational Selection of Ship Size," Society of Naval Architects and Marine Engineers, New York, 1967.
5. Benford, H.: "Engineering Economy in Tanker Design," Society of Naval Architects and Marine Engineers, New York, 1957.
6. Bes, B. M.: "Tanker Shipping," Drukkerij V/H C. De Boer Jr., Hilversum, Netherlands, 1972.
7. "Liquid Gas Carrier Register," H. Clarkson & Company Limited.
8. "The Tanker Register," H. Clarkson & Company Limited.
9. D'Arcangelo, A. M.: "Ship Design and Construction," Society of Naval Architects and Marine Engineers, New York, 1969.
10. deKerchove, R.: "International Maritime Dictionary," Van Nostrand Reinhold, New York, 1961.
11. Gannett, Ernest: "Tanker Performance and Cost: Measurement, Analysis and Management," Cornell Maritime Press, New York, 1969.
12. "Ship Types of the Future," *IMAS* 1969 *Proc.*, sec. 2, Institute of Marine Engineers, London, England.
13. "International Petroleum Encyclopedia," The Petroleum Publishing Co., Tulsa, Okla., 1972.
14. King, G. A. B.: "Tanker Practice," 6th ed., W. S. Heinman, New York, 1971.
15. Lom, Walter L.: "Liquefied Natural Gas," Applied Science Publishers, Ltd., Essex, England, 1974.

Superports

BY F. POSTHUMA*

A superport, within the current framework of development, may be defined as a port that can accommodate carriers of 200,000 DWT (deadweight tonnage) and over. The minimum draft of such ships, when fully loaded, is about 18.3 meters (60 feet) insofar as tankers are concerned. These tankers are known as VLCCs (very large crude carriers).

With the exception of 26 vessels, the 190 tankers in service and the 231 tankers on order as of January 1, 1974, all between 100,000 and 200,000 DWT, have a draft of less than 18.3 m (60 ft). The draft of the other 26 tankers, also of 100,000 and 200,000 DWT, lies between 18.3 and 18.9 m (60 and 62 ft). In Table 1, tankers and carriers with a maximum draft of 18.3 m (60 ft) and over are divided according to their different drafts and their scheduled years of delivery.

From this table it appears that of the total of 856 tankers in service, under construction, or on order, and with a draft of at least 18.3 m (60 ft), only 47 vessels have a maximum draft in excess of about 22.9 m (75 ft). It is reasonable to conclude that the 47 tankers with a maximum draft of over 22.9 m (75 ft) are designed for special trade routes. One also can conclude that a superport that can accommodate tankers with a maximum draft of 22.9 m (75 ft) should be in a good competitive position for the next several years. The table indicates that the fastest growing group of ships is comprised of tankers with a draft between 22.0 and 22.9 m (72 to 75 ft).

Possibly the next best alternative in considering superport design would be a superport that would accommodate VLCCs with a draft up to 21.4 m (70 ft), because out of the 856 VLCCs indicated in Table 1, no fewer than 586 of these vessels have a draft between 18.3 m (60 ft) and 21.4 m (70 ft). Moreover, such a superport could also accommodate the 421 tankers with deadweights between 100,000 and 200,000 tons, as well as many of the 375 tankers which already are in service.

In the planning of each superport, the maximum allowable draft must be economically justified. For superports that serve large consuming centers, a planning study should be made in close cooperation with the future users of the superport, notably the oil companies that own, operate, or lease vessels and that have an intimate knowledge of their planned movements of oil and petroleum products over a period of years. These kinds of studies have progressed in connection with already existing superports and in particular have emphasized discussions between the oil companies and the port authorities involved. A large part of these discussions has dealt with the possible reduction of the maximum draft of the VLCCs, often the major problem with which superport designers must cope. Past discussions have resulted in studies pertaining to the feasibility of the restricted-draft (RD) tanker. As an outcome of these studies, one firm commenced to order RD tankers, and other firms followed suit.

With reference to Table 2, this is evident when the maximum drafts of the VLCCs are related to their deadweight tonnages. Of the 841 tankers with a draft of 18.3 m (60 ft) and over, 470 are VLCCs with a deadweight tonnage of 200,000 tons and more. With further reference to Table 2, several other interesting observations can be made:

1. Of the 58 tankers between 300,000 and 350,000 DWT, three vessels have a draft of no less than about 24.9 m (81.6 ft). These 320,000-DWT ships are all owned by one firm which also owns some very deep terminals in Bantry Bay, Ireland, and Okinawa, and thus were built for these special trade routes. Compared with these three tankers, many of the 53 VLCCs between 300,000 and 350,000 DWT will be reduced-draft tankers and will be designed specifically with the RD concept in mind.

2. Of the 45 VLCCs between 400,000 and 450,000 DWT, 12 vessels have a draft between 24.4 and 25.0 m (80 and 82 ft), and all are owned by ship operating firms (Onassis, 2; Marine Transport Lines, 4; Andreades Group, 2; National Bulk Carriers, 4).

*Executive Director (retired), Rotterdam-Europoort, Port Consultant, Rotterdam.

TABLE 1. Tankers and Carriers with a Draft of 18.3 M (60 Ft) and Over—Worldwide

In service, under construction, or on order as of January 1974

Range of draft		Total ships		In service		Under construction or on order		Scheduled for delivery in										
Meters	Feet	T	C	T	C	T	C	1974 T	1974 C	1975 T	1975 C	1976 T	1976 C	1977 T	1977 C	1978 T	1978 C	1979 T
18.3–18.9	60–62	24	7	14	4	10	3	2	2	3	1	4		1				
18.9–19.5	62–64	122	6	113	6	9		5		2		2						
19.5–20.1	64–66	213	20	125	18	88	2	43	2	20		14		11				
20.1–20.7	66–68	149		57		92		31		28		23		10				
20.7–21.3	68–70	78	2	19	1	59	1	23	1	19		8		9				
21.3–22.0	70–72	49	5	15	2	34	3	13	2	12	1	7		2				
22.0–22.9	72–75	174		19		155		21		39		47		40		8		
22.9–24.4	75–80	13				13				2		4		4		3		
24.4–27.5	80–90	28		11		17		2		2		3		7		2		1
27.5–30.5	90–100	6		2		4				1		2		1				
Tankers		856		375		481		140		128		114		85		13		1
Carriers			40		31		9		7		2							
Total ships		896		406		490		147		130		114		85		13		1

NOTE: T = tanker, C = carrier.

TABLE 2. **Maximum Drafts of VLCCs Related to Their Deadweight Tonnages**
Under construction or on order as of January 1974

Deadweight, thousands of tons, DWT	Total number of ships	Draft categories							
		Number	Draft	Number	Draft	Number	Draft	Number	Draft
200–240	82	70	19.8–20.4 m (65–67 ft)	9	20.7 m (68 ft)	3	21.4 m (70.1 ft)		
240–260	87	71	19.8–20.7 m (65–67.9 ft)	16	20.7–21.0 m (68–69 ft)				
260–280	125	82	20.4–21.0 m (67–69 ft)	43	21.4–22.1 m (70–72.4 ft)				
280–300	14	14	22.0–22.3 m (72–73 ft)						
300–350	58	53	22.0–22.6 m (72–74 ft)	2	23.2 m (76 ft)	3	24.9 m (81.6 ft)		
350–400	47	43	22.0–22.6 m (72–74 ft)	4	22.3 m (76.5 ft)				
400–450	45	28	22.0–22.9 m (72–75 ft)	5	23.6 m (77.5 ft)	12	24.4–25.0 m (80–82 ft)		
450–500	9	4	22.3 m (73 ft)	2	24.1 m (78.9 ft)	2	25.0 m (82 ft)	1	28.2 m (92.6 ft)
500–550	3	3	28.5 m (93.6 ft)						

Not one is owned by an oil company. Although the exact trade routes over which these vessels will be used cannot be forecast accurately at this time, it may be reasonable to assume that they are scheduled for use between the Persian Gulf and Japan, perhaps a few calling at Le Havre or Marseilles.

3. The 28 ships with a draft between about 22.0 and 22.9 m (72 and 75 ft) can, in principle, be considered as RD tankers.

4. The one tanker with a draft of about 28.2 m (92.6 ft) is jointly owned by a Japanese ship operator (Tokyo Tankers) and a shipyard (I.H.I.), with construction in the I.H.I. shipyard. It is interesting to note that this vessel with 483,000 DWT has almost 6.1 m (20 ft) greater draft than four other tankers of about the same DWT.

5. Two of the three tankers with a draft of about 28.5 m (93.6 ft) are owned by Société Maritime Shell (540,000 DWT); the other is owned by Compagnie Nationale de Navigation S.A. (550,000 DWT); all were built in the Chantiers de l'Atlantique (Le Havre).

Possibly the most important conclusion that can be drawn from Table 2 is that any superport that can accommodate VLCCs with a draft of about 22.9 m (75 ft) will be able to receive the 53 ships between 300,000 and 350,000 DWT, the 43 ships between 350,000 and 400,000 DWT, the 28 ships between 400,000 and 450,000 DWT, and even the 4 ships of 470,000 DWT. The latter four vessels are built by Howaldtswerke-Deutsche Werft and were designed as RD tankers.

Special RD Tanker Study.   A study of restricted-draft tankers was undertaken under the supervision of the Rotterdam-Europoort Port Authority, together with the Department of Waterways of the Dutch Government, the Shipping Laboratory of Wageningen, and Verolme Shipyards. The effect of the size of ships on freight costs is depicted graphically in Fig. 1, a chart first released by Exxon in September 1972. The graph indicates that savings are especially important in connection with VLCCs up to about 300,000 DWT. These tankers have a draft of approximately 22.0 m (72 ft). It will be noted, however, that savings on still larger conventional tankers are so small that they become insignificant when compared with the enormous expenditures required by ports to accommodate the vessels. For example, a ship of 1 million DWT would have a draft ranging from about 36.6 to 45.8 m (120 to 150 ft). Even a ship of 500,000 DWT would have a draft ranging from about 27.5 to 30.5 m (90 to 100 ft). Moreover, such mammoth ships are under all kinds of restrictions while at sea and in channels. The most important restrictions may stem from the recent I.M.C.O.* regulations. Reopening and deepening of the Suez Canal perhaps to accommodate vessels of 250,000 DWT also may have a large influence on the size of tankers and the world tanker fleet—with the possibility that an overcapacity at some future date may be the result.

An interesting study in comparative freight costs per ton has been made. A tanker was designed of 425,000 DWT and with a "restricted draft" of 22.0 m (72 ft), but with a width greater than conventional. This RD tanker was compared with VLCCs of various sizes and drafts in two common services—with the results indicated in Table 3.

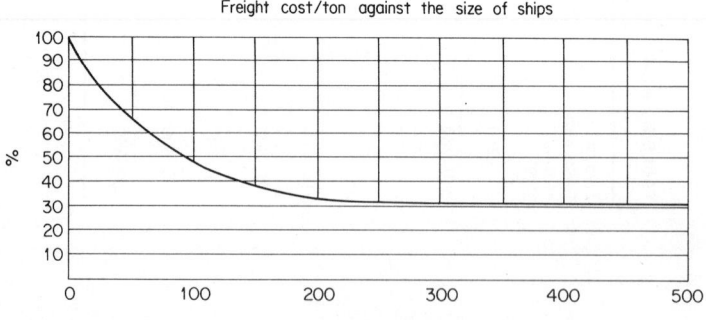

Freight cost/ton against the size of ships

Fig. 1. Tanker cost versus size. (*De Essobron*, September 1972.)

*Intergovernmental Marine Consultative Organization of the United Nations, especially dealing with safety at sea.

TABLE 3.   Comparison of Freight Costs per Ton—RD Tanker versus VLCC

RD Costs = 100

Type of vessel	Service between	
	Persian Gulf and Rotterdam-Europoort	Libya and Rotterdam-Europoort
RD tanker, 425,000 DWT (22.0 m) (72 ft) draft	100	100
VLCC, 250,000 DWT (19.8 m) (65 ft) draft	110	107
VLCC, 310,000 DWT (22.3 m) (65 ft) draft	106	105
VLCC, 425,000 DWT (26.8 m) (88 ft) draft	101	101

Similar RD tanker studies have been made and are being continued by oil companies and shipyards. In January 1973, six RD tankers were ordered from the A. G. Weser yard by Hapag Lloyd, Niarchos Tankers, and others. Two were ordered from the Itachi yards in Japan by Exxon.

The numbers of vessels and the drafts of the 360 VLCCs with a draft of about 18.3 m (60 ft) and over and *in actual service* as of January 1974 are given in Table 4.

RELATED SHIP AND PORT DESIGN CRITERIA

It is imperative that the draft of the VLCCs must be of the utmost importance to the designer of a superport in an effort to select the size of tankers that can be accommodated. As previously mentioned, of course, the draft also is extremely important to the operators, owners, lessors, and so on of such vessels. Several observations concerning draft have been given, and the topic is described in even greater detail later in this article under Design Criteria for Superports.

TABLE 4.   Very Large Crude Carriers in Service—Draft of 18.3 M (60 Ft) and Over

As of January 1974

Deadweight, thousands of tons, DWT	Total number of ships	Draft categories					
		Number	Draft	Number	Draft	Number	Draft
200–240	221	152	18.6–19.8 m (61–65 ft)	69	19.8–20.8 m (65.1–68.1 ft)		
240–260	76	67	19.8–20.7 m (64.8–68 ft)	8	20.7–21 m (68–69 ft)	1	21.7 m (71.1 ft)
260–280	26	21	20.1–21.3 m (66–70 ft)	5	21.6–22.1 m (71–72.5 ft)		
280–300	24	24	21.3–22 m (70–72.1 ft)				
300–350	10	10	24.6–24.8 m (80.8–81.5 ft)				
350–400	1	1	27.1 m (88.9 ft)				
400–450 450–500	2	2	28.2 m (92.6 ft)				

NOTES:

As observed from Table 1, the number of tankers in service as of January 1, 1974, with a draft between 21.9 and 22.9 m (72 and 75 ft) is relatively small. As no fewer than 155 tankers in this class are under construction or on order, it may be concluded that between 1974 and 1978 a new class of VLCCs will be sailing. This is caused partly by the introduction of the RD tanker. Many large RD tankers, even up to 470,000 DWT belong to this class.

Of the ten VLCCs with a draft between 24.6 and 24.8 m (80.8 and 81.5 ft), six belong to the Universe class of 327,000 DWT with a draft of 24.8 m (81.5 ft) and were especially designed for the (Gulf) terminal in Bantry Bay, Ireland, from where the crude is transshipped in smaller tankers. The other four vessels are named the *Argeaga* and the *Butron* (both 318,000 DWT) and the *Venoil* and *Venpet* (both 326,000 DWT). The one tanker with a draft of 27.1 m (88.9 ft) is the *Nisseki Maru* (367,000 DWT), and the two vessels with a draft of 28.2 m (92.6 ft) are the *Globtik London* (483,000 DWT) and the *Globtik Tokyo* (476,000 DWT).

Some attention also must be given to the length and width of the VLCCs. Length is important in terms of the distance required between berths, the diameter of turning basin, the maneuverability, the radius of bends, and so on. The width is important in the dimensioning of channels, in determining the distance between berths running parallel to each other, and so on. Often model tests are made to find the best design for a superport. Model tests, moreover, are almost always needed for the design in general in order to study the influence of currents, waves, and winds, and to study the amount of maintenance dredging that may be required. Models also provide much other important detailed information.

Following are a few highlights pertaining to length and width of VLCCs:

1. *In the range of 200,000 to 300,000 DWT*, the length overall (loa) varies from about 320 to 348 m (1,050 to 1,140 ft). Ships of 285,000 DWT have an loa of about the maximum figure. The width of ships in this category varies from about 47 to 55 m (155 to 180 ft).

2. *In the range of 300,000 to 350,000 DWT*, the length varies from about 323 to 367 m (1060 to 1200 ft); and the width varies from about 53 to 61 m (175 to 200 ft).

3. *In the range of 350,000 to 400,000 DWT*, "normal" tankers with a draft up to about 27 m (88.9 ft) have an loa between 329 and 372 m (1080 to 1220 ft); and a width between 55 and 63 m (210 and 230 ft).

RD tankers in this DWT category have an loa about the same as "normal" tankers, but a greater width, ranging between 64 and 70 m (210 and 230 ft); all with a draft up to about 22.6 m (74.2 ft).

4. *RD tankers between 400,000 and 420,000 DWT with a maximum draft up to 22.8 m (74.8 ft)* have an loa up to 378 m (1240 ft) and a width between 68.6 m and 70.2 m (225 to 230 ft).

5. *The largest RD tankers under construction* are four 470,000 DWT (Germany) with an loa of about 390 m (1280 ft); and a width of almost 72 m (235 ft). The maximum draft is about 22.3 m (73 ft).

6. The dimensions of two tankers (Shell) of 540,000 DWT are about: length, 412 m (1,350 ft); width, 63 m (206 ft); draft, 28.5 m (93.6 ft). Two tankers (Globtik) of 477,000 DWT are about: length, 381 m (1250 ft); width, 62 m (203 ft); draft, 28.2 m (92.6 ft). Two tankers (N.B.C.) of 446,500 DWT are about: length, 360 m (1180 ft); width, 68 m (223 ft); draft, 25 m (82 ft).

Dry-Bulk Carriers.   Most of the dry-bulk carriers have a draft of less than 18.3 m (60 ft). The maximum draft of these ships does not exceed 22 m (72 ft). These large carriers with a draft in these ranges do not transport dry cargo exclusively, but they also carry oil. It is unlikely that the draft of these ships will increase much beyond the 22-m (72-ft) figure in the future. Just so long as ore and coal are not pumped ashore as slurry, single-point moorings at sea cannot be used, and thus installations ashore on piers or on quays will remain necessary for exporting as well as for importing dry bulk cargo.

As of January 1974, carriers of this type with a draft of 18.3 m (60 ft) and over which were under construction or on order worldwide are described in Table 5.

Thirty-one ships that can carry dry-bulk cargo as well as oil, have a draft of over 18.3 m (60 ft), and were *in actual service* as of January 1974 can be categorized in terms of their drafts and deadweight tonnages as follows:

5 carriers (165,000–170,000 DWT)	Draft:	18.3 to 18.9 m (60 to 62 ft)
5 carriers (about 215,000 DWT)		18.9 to 19.2 m (62 to 63 ft)
15 carriers (215,000–245,000 DWT)		20.4 to 20.7 m (67 to 68 ft)
3 carriers (260,000–265,000 DWT)		20.6 m (67.7 ft)
3 carriers (265,000–285,000 DWT)		20.8 to 21.7 m (68.11 to 71.2 ft)

From the foregoing data, it would appear that an ore port which can accommodate ships with a draft of 20.7 m (68 ft) is in a very good competitive position, particularly because it is not to be expected that many ore carriers with a draft exceeding this figure will be built. Also, oil/ore carriers cost approximately 10 percent or more to construct than a conventional ore-carrying ship or tanker. Until recently, the possibilities of the oil/ore/coal carriers have not been fully exploited, but the existing fleet of these carriers may well be of a greater advantage to the owners in the future because, with the expected surplus of VLCCs, they can be switched over to the dry-bulk trade.

DESIGN CRITERIA FOR SUPERPORTS

Major design factors for planning a superport include: (1) Depth of access lanes, (2) width of access channels, (3) maneuvering area, (4) maneuverability of very large crude carriers, (5) stopping distance, (6) navigational aids, and (7) entrance channel considerations. Descriptions of these factors are limited here to highlights.*

Depth of Access Lanes. The *admissible draft* is based on the draft of the highest-class vessels to be received by the port. Corrections must take into account (1) water salinity and (2) loading conditions.

The *gross underkeel clearance* may be defined as the reserved margin under the keel of the vessel, considering (1) the sea level during the vessel's navigation within the port, (2) the maximum draft of the ship at rest in quiescent water, and (3) the nominal channel bed level.

The *net underkeel clearance* may be defined as the *minimal margin* remaining under the keel of the vessel when it is moving at planned passage speed under the action of the most severe planned tolerable wind and wave conditions.

TABLE 5. Dry Bulk Carriers with Draft of 60 Feet and Over
As of January 1974

Number of ships	Shipyard	Shipowner	Deadweight tonnage	Draft, feet per inch	Type
1	Gotaverken	Soc. Cameli	223,000	67/9	Ore/oil
1	I.H.I.	Petrobras	265,300	68/11	Ore/oil
2	Swan Hunter	Thornhope Shipping Co.	167,000	60/8	O.B.O.
1	Swan Hunter	Bibby Line	167,000	60/8	O.B.O.
1	Uljanik	Sig. Bergesen d.y. & Co.	225,100	66/10	Ore/oil
2	Uljanik	Gränges A/B	265,000	71/6	Ore/oil
1	Uljanik	Valeco tankers	265,000	71/6	Ore/oil

The *nominal channel bed level* is established by allowing a net underkeel clearance at least equal to 0.50 meter (1.64 ft) if the bottom is *sandy;* and 1 meter (3.28 ft) if the bottom is *rocky.*

Within the same port, the nominal channel-bed level may differ from one water area to another, caused by such factors as (1) the sea level variations which may occur during a particular maneuver, (2) the natural sea level variations from one water area to another, and (3) the variations of the draft of the vessel as it may proceed from one water area to another because of the vessel's admissible speed, the swell present, and the water salinity.

Squat. To determine the squat of a vessel due to speed, the highest speed allowed must be taken into account, noting that the speed in a channel should be increased with the speed of the cross current. Various estimates of squat are given in the aforementioned Bulletin No. 16.

German Report: Ships of 200,000 DWT at speed of 15 knots allow 1.50 meters (4.9 ft); at speed of 10 knots, 0.75 meter (2.5 ft).

French Report: Ships of 500,000 DWT at speed of 10 knots, allow 1.40 meters (4.6 ft).

Swedish Report: Ships of 200,000 DWT at speed of 5 knots, allow 0.4 meter (1.3 ft).

Guliev Study: Ships of 200,000 DWT at speed of 5 knots, allow 0.25 meter (0.8 ft); at speed of 10 knots, 0.5 to 1.0 meter (1.7 to 3.3 ft).

Other observations pertaining to squat include: (1) the squat is more important at the bow of the ship than at the stern; (2) the bulbous bow seems to diminish this phenomenon; (3) the squat lessens when the ratio of the wet section of the channel to the main frame of the ship increases, or when the keel clearance increases.

It would appear that for the usual navigation speeds in a channel (from 5 to 10 knots), the squat for a tanker of 200,000 DWT would range between 0.3 and 1.3 meters (1 ft to 4.3 ft).

*Based mainly on Bulletin No. 16 of the Permanent International Association of Navigation Congresses.

Swell. The vertical movements of a vessel occurring under the effect of the swell must be calculated, taking into account the amplitude, period, and direction of the swell; the dimensions and speed of the vessel; and water depth.

In general, the movements of ships due to the swell depend on:

1. *Amplitude of the swell:* For a swell of constant period, movements vary roughly in proportion to the amplitude.

2. *Period of the swell:* Movements are slight where the period is about 6 seconds or less, greater for periods of about 10 seconds, and can reach several meters for swells with periods of 15 seconds or greater. Inasmuch as the natural period of tankers in the 200,000 to 500,000 deadweight-tonnage range, for both roll and pitch motions, varies from about 12 to 15 seconds, large motions can occur for swell periods of 10 to 15 seconds, because of resonance, even though the amplitude of the swell may be small.

3. *Ratio of the draft to the depth:* As the keel clearance grows less, movements due to the swell grow less.

4. *Length of the ship:* Sensibility to the swell increases with the ratio: *length of the wave/length of the ship.*

5. *Direction of the swell,* as compared with the direction of ship movement: It should be pointed out that an oblique swell on the aft sector is frequently met by ships approaching a port.

Portion of Channel in Open Sea. According to a Dutch study, fully loaded 200,000-DWT tankers in the open-sea portion of a channel leading to a port, the following increases in draft may occur:

Increase in Draft in Channels in Open Sea with Varying Weather

Cause	Fair weather	Extremely bad weather (when ships stay at sea)
Pitching	Up to 0.3 m (1 ft)	Up to 1.2 m (4 ft)
Rolling	Up to 1.5 m (5 ft)	Up to 3.7 m (12 ft)
Squat (at 10 knots)	Up to 0.9 m (3 ft)	Up to 0.9 m (3 ft)
Squat (at 14 knots)	Up to 1.5 m (5 ft)	

From the foregoing table, it will be seen that a large ship may be admissible during fair weather, whereas it may not be admissible during a period of bad weather.

When the vessel arrives between the breakwaters or in the basins, the channel dredged level can be less because the vertical ship movements are smaller and, if the bottom is sandy, the net underkeel clearance can be reduced to one foot.

A list of g degrees and a width of the ship of w will increase the draft by $\frac{1}{2}w \times \sin g$.

Depth-of-Channel Calculations. As an example, the calculation of the gross underkeel clearance in the reception of ships of 200,000 to 250,000 DWT with a draft of 20 meters (65.6 ft) would include:

Z_1 = level of reference of water plane during maneuvering. (In a sea with tides, this level takes into account the security margin. Where the level can sustain contingent variations, such as winds and flood, the reference Z level is characterized by the appearance probability of the contingent variable.)

Z_2 = nominal dredging level

t = admissible draft, which will be assumed here as 20.5 meters (67.3 ft).

Underkeel clearance as previously defined corresponds to

$$\text{Underkeel clearance} = Z_1 - Z_2 - t$$

The calculation is performed by considering successively two hypotheses: (1) the first concerning light-limit conditions corresponding to a favorable site (for example, a swell not exceeding 2 meters (6.6 ft) of amplitude and 7 seconds of period; negligible current; and short channel, allowing navigation at a speed equal to or less than 5 knots); and (2) the second concerning more difficult conditions (for example, a swell of an amplitude of 4 meters (13.1 ft) and of a period of 10 seconds; a transverse current or long channel, requiring a navigation speed ranging between 8 and 10 knots).

In the first instance the keel clearance would be as follows:

	Range	Mean value	Range	Mean value
Motions due to the swell	0.40–0.80 m	0.60 m	1.3–2.6 ft	2.0 ft
Squat	0.30–0.50 m	0.40 m	1.0–1.6 ft	1.3 ft
Net underkeel clearance	0.30–0.50 m	0.55 m	1.0–2.6 ft	1.8 ft
Gross underkeel clearance	1.00–2.10 m	1.55 m	3.3–6.8 ft	5.1 ft
Percentage of draft	5–10.5	7.5	5–10.5	7.5

In the second the keel clearance would be:

	Range		Range	
Motions due to the swell	0.80–1.50 m		2.6–4.9 ft	
Squat	0.80–1.50 m		2.6–4.9 ft	
Net underkeel clearance	0.50–1.00 m		1.6–3.3 ft	
Gross underkeel clerance	2.10–4.00 m		6.8–13.1 ft	
Percentage of draft	10.5–20		10.5–20	

Thus, as mentioned in numerous studies, the gross keel clearance values range between 10 and 20 percent of the draft. It should be noted, however, that the preceding calculations are given only by virtue of this particular example and do not form limits in any way.

For larger VLCCs and reduced-draft (RD) tankers, new tests, verification trials in situ, and new calculations must be made.

To assess the required *channel dredged level,* the water reference level, the ship's size, the vertical ship movement, the difference between nominal channel bed and channel dredged level, and a net underkeel clearance of 0.5 meter (sandy bottom) and 1 meter (rocky bottom) must be taken into account. The *nominal channel bed level* is, by definition, the level in comparison to the datum, above which no obstacles for navigation remain. The channel dredged level is determined from the nominal channel bed level by considering: (1) the height of deposits in the channel between two maintenance dredgings, (2) the tolerance applying to the dredgings, and (3) the accuracy of the soundings. Particular attention must be given to minimal sea level.

In operational use, consideration must be given on each occasion to the appropriateness of the design criteria, especially in regard to: (1) actual water levels, (2) actual ship's draft, including effects of hogging, sagging, list, and trim, (3) anticipated vertical movement of the ship under the action of currents, wind, and wave forces prevailing at the actual time of passage, and (4) actual bed levels derived from the latest surveys as amended for sedimentation occurring after the survey and for sounding precision.

Width of Access Channels.   The nominal width of a channel is the minimum width of the channel over which prevails the nominal depth. In particular, the dredging must take into account the uncertainty of the position of the dredgers during their work.

Factors that must be considered in establishing the nominal width of an access channel with one circulation lane include:

1. *The accuracy with which the difference* between the vessel's position and the channel axis can be determined.

2. *The greatest difference* that may occur from the moment a difference is noted to the moment when correction of the resulting trajectory takes effect.

3. *The half-width of the vessel and the area that the vessel sweeps* on account of drifting due to currents. The effects of crosswinds also should be taken into account in the same manner.

4. *The allowance of a margin* no less than half the width of the vessel.

The fact that the accuracy of navigational aids decreases with an increasing distance from shore also should be taken into consideration.

Recommendations have been made to use a nominal channel width of no less than five times the width of the vessel, even when there is no crosscurrent. When there are crosscurrents, the ship must regulate its speed such that, allowing for the crosscurrent, tan *d* (drift angle) must be less than 0.25. With 0.30 as a maximum instantaneous value of the drift angle tangent, one discovers that the crosscurrent necessitates an extra width of about 0.30 *L* (length of ship), or thus approximately 2*B* (breadth of ship).

Generally speaking, then, the width of the channel should be between five and seven times the width of the ship, depending on the intensity of the crosscurrents.

It should also be taken into account that the banks and slopes of channels and canals create sheering for ships. The stern can indeed tend to move aside from the bank because the section of water by the propeller, between the ship and the bank, can create an opposite movement of the aft part of the ship.

In channels subjected to the effects of long swells or tail currents, the width should be increased because of sheering due to swell or current action.

In visual navigation in a channel that is marked by buoys, the difference between the actual positions of the buoys that may be subject to a swinging motion and the theoretical positions of the buoys must be taken into account.

Curved Channels.   A straight-line layout of a channel is, of course, preferable, but when curves are unavoidable, the width of the channel must be greater and must take into account:

1. *The necessary extra width* for the path of the ship while in the curve. This may be calculated by: *Extra width* $= L^2 / 8R$, where L = vessel length and R = radius of curve.

2. *A supplemental margin for maneuvering difficulties.* Factors to take into consideration include: (*a*) The vessel's path may be unknown. (*b*) The vessel does not respond immediately and consequently the pilot must anticipate the maneuver. This margin is much more important when: (*c*) The radius is short. (*d*) The center angle is open. (*e*) The current and wind intensities are high.

The ends of zones having different widths will be jointed together by straight lines, giving a total width variation of at most 40 m (131 ft) for a stretch of 100 m (328 ft) long.

Characteristics of Some Specific Channels. ,Following are values for channels open to ships of 200,000 to 250,000 deadweight tonnage:

Gothenburg: Width, 210 m (689 ft) straight line. Crosscurrent 0.5 knot. 250–300 m (820–984 ft) curve. One-way traffic (channel reserved for VLCCs during passage).

Le Havre: Width, 300 m (984 ft). Crosscurrent 1.5 to 2 knots. Priority for VLCCs; other ships give way.

Marseille-FOS: Width, 250 m (820 ft). No current.

Milford-Haven: Width, 375 m (1230 ft) outer; 180 m (591 ft) inner. Priority for VLCCs; other ships give way.

Point Tupper: 5 times the width of 300,000 DWT ship.

Rotterdam: Width, 1200 m (3937 ft) at the extremity of the first section; for second section, variable between 600 m (1969 ft) outer; and 400 m (1312 ft) inner. Cross current of 2 to 3 knots. Two-way traffic, including VLCCs in the 1200- and 600-m sections.

Wilhelmshaven: Width, 300 m (984 ft); widening in curves with response to the center angle will be provided (future). Cross component of current is 1 knot. Priority for laden VLCCs; other ships give way.

With reference to existing channels, it appears that the minimum width of a channel to accommodate one-way navigation for ships of 200,000 DWT and without a crosscurrent lies between 180 and 250 meters (591 and 820 ft). In other cases, the width is about 300 meters (984 ft).

Maneuvering Area.   The maneuvering area comprises (1) the necessary area to allow vessels to reduce their speed and (2) the necessary swinging area. Generally, these water areas are protected. The turning of a VLCC at the Port of Rotterdam is illustrated in Fig. 2.

In determining the depth of the maneuvering area, the previously established criteria for channels should apply—except: (1) The motion due to the swell may be neglected if the water is protected. (2) The squat due to speed is negligible.

The water area necessary to reduce the vessel's speed includes (*a*) a *speed-decreasing area* in which the vessel's speed decreases to 3 knots and the ropes are reeved, and (*b*) a *stopping area* in which the speed is reduced to zero, the vessel standing in under the action of the tugs. These factors are not to be confused with Stopping Distance, which is described later.

The *length of the necessary water area* equals the *stopping distance of the vessels,* owing to their speed in the access channel, *increased by a margin.* The speed in the approach channel will be determined mainly by the maximum value of the drift angle.

The *width of the water area* must take into account the drift during the stopping maneuver, due in particular to the reverse of propulsion with single-screw vessels.

When a vessel enters a calm water area at the end of a channel under the effect of cross swell, the width must be enlarged, taking into consideration the sheering to which the vessel is subjected. In this condition, the bow is no longer subjected to the action of perpendicular stresses, but the stern is still so affected.

Stopping maneuvers preferably will be carried out where they are sheltered from currents; or, if possible, with the vessel facing the current. If this is not feasible, the water area necessary for braking must be sheltered from important crosscurrents. It is recommended that the cross component and the longitudinal component coming from the aft should not exceed 0.15 meter per second (0.5 ft per second), each.

Normal conditions for maneuvering may be described as follows:

Fig. 2. Turning a VLCC (very large crude carrier) at the Port of Rotterdam.

1. The vessel is assisted by tugs adequate in number and power.
2. The current is less than 0.10 meter per second (0.3 ft per second), and the wind in the case of turning (the vessel being on light waterline) is less than 10 meters per second ($\sim$ 3 ft per second).

For all practical purposes, if the foregoing conditions are met, the area required for maneuvering is a turning circle with a diameter equal to about twice the length of the vessel.

Should the foregoing conditions not be fulfilled (or because of an incident or accident cannot be fulfilled), the turning axis should be determined by a circle, the diameter of which is at least three times the length of the vessel, permitting the vessel to turn by her own means.

MANEUVERABILITY OF VERY LARGE CRUDE CARRIERS

In the maneuvering of VLCCs, changes of course must be sufficiently anticipated with respect to rudder setting to take into account the response time of the vessel to the change in rudder angle. The degree of accuracy that can be obtained in anticipating the response time depends mostly on the familiarity of the navigator with the response characteristic of his ship. However, a margin of error should always be allowed.

The response time of a 280,000-DWT tanker to the rudder is about 1.5 minutes. More than twice this response time is required to come from a straight course into a circle by using constant rudder angle. Thus, when entering or leaving a curve, the pilot must anticipate his rudder setting for a considerable time.

A margin of error in timing the maneuvering of about 45 seconds should be assumed for all conditions where expert pilotage is available. Where such expertise is not available, the margin should be increased to at least 60 seconds.

These matters must be considered in determining the width of a channel. The extra width needed at the entrance and exit of a curve to accommodate the foregoing condition depends on the speed with which the vessel is expected to pass through the curve. An accurate plotting of the advance of the vessel is needed in order to determine the width requirement in the curve because of misalignment of the vessel with the curve as caused by overswing during changes of rudder setting.

STOPPING DISTANCE

Varying opinions have been expressed by P.I.A.N.C. and others concerning the recommendations relative to the sizes of the water area necessary to stop a ship. These variations arise from important differences in the moves conducted in various ports.

As an example, in the port of Gothenburg, tugs intervene far out in the channel. Therefore, it is not necessary to provide for a particular water area for the reeving operation. Also the speed of the convoy is reduced. The intervention of tugs, with taut towlines, is hardly possible when the speed of the ship exceeds 3 knots. The distance necessary for its stoppage is, therefore, not important.

In contrast, however, the intervention of tugs is generally not possible in water areas which are not sheltered and where the swell amplitude exceeds 2 meters (~6.6 ft). Thus, this type of move is subject to the type of operational limits which apply when the channel is situated in an open sea.

In other instances, the tugs intervene only in a sheltered water area, or immediately outside the port. Thus it is advisable to make an allowance for stoppage so that the passing of the towlines can be effected. Since the time necessary for this operation can be as much as 15 minutes, the minimum length of the water area needed thus corresponds to the distance traversed during this time by the ship. The speed of the ship at the origin of the move is that of navigation in the channel. When the towlines are passed, the speed must be at least 3 knots because, at a lower speed, the ship would be subject to significant sheers if not being held by the tugs.

Sometimes available water areas for stoppage of the ship are very long, the ship maintaining her headway. Sometimes, on the other hand, the stoppage maneuver must be carried out half astern, or even fullspeed astern. In the first case, the ship maintains herself easily on her axis. In the second case, important castings can occur. Thus, this is a maneuver that can be considered only in an emergency.

With the foregoing wide variations in conditions, only general recommendations can be given here.

The dimensions of the water area necessary for the stoppage of a ship depend on the distance required to slow the ship and, consequently, the initial speed of the ship. Obviously, the more this speed can be reduced, the less important and costly the provision of the water area necessary for the stoppage. However, it cannot be overstressed that an underestimate of this water area can bring about catastrophic consequences.

There exists a minimum value of speed below which a ship must not, making allowance for the wind and current, sail in full safety in the approach channel. The tangent of the drift angle should not exceed 0.25. Thus, a cross current of 2 knots would make it necessary that the speed of the ship be at least 8 knots. In the choice of navigation speed, consideration must also be taken of the fact that the speed of the engines must be increased to augment the maneuverability of the ship when, in passing from a water area controlled by the action of current or swell to a calm water area, it is subjected to different transverse effects fore and aft.

Numerous observations have been made pertaining to the stopping distance of ships with reference to their speed. For an initial speed of 5 knots, the approximate stopping distance for a 200,000-DWT ship can be expected to be about as follows:

With full astern .... 850 m (2,789 ft)
With half astern .... 1,250 m (4,101 ft)
With slow astern .... 2,050 m (6,726 ft)

Other preliminary data include:

Deadweight tonnage	Stopping distance	
	Engines completely stopped and not put astern	Crash stop. Engines full astern, initial speed, 15 mi/hr
100,000	5.0 mi (~8 km)	2.0 mi (~3.2 km)
150,000	5–6.5 mi (~8–10.5 km)	2.4 mi (~3.9 km)
200,000	5–7.0 mi (~8–11.3 km)	2.5 mi (~4.0 km)
250,000	5–7.0 mi (~8–11.3 km)	2.6 mi (~4.2 km)
325,000	About 6.0 mi (~9.7 km)	2.0 mi (~3.2 km)
400,000	6–7.0 mi (~9.7–11.3 km)	About 2.5 mi (~4.0 km)
483,600 (Globtik Tokyo)	About 7.5 mi (~12.1 km)	About 2.4 mi (~3.9 km)

NAVIGATIONAL AIDS

Modern techniques make it easier to control the navigation of large tankers within channels. See Fig. 3. These means include:

1. *Port radar,* furnishing knowledge at every moment of the traffic in the access channel and its approaches so as to establish circulation rules for vessels during the whole duration of a VLCC's passage in the access channel; and, in certain cases, during their maneuver. Regulation of the latter may be achieved by using VHF radiotelephony communication.

2. *Navigational devices,* which allow masters and pilots to know clearly the position, direction, and trajectory of the vessel in the channel (channel aids) and in the maneuvering areas (turning and berthing aids), whatever the visibility and light conditions may be (daytime, and particularly nighttime).

Channel aids include (1) far-reaching leading lights, (2) radiolocation systems, such as DECCA, and (3) harbor radars equipped with a very sharp range and bearing antenna, located across the channel and with radar equipment developed for effective display. Turning and berthing aids include (1) leading lights, (2) harbor radar, (3) Doppler radar for

Fig. 3. Radio and radar post for controlling the waterways at the Port of Rotterdam.

screening the vessel's berthing speed, and (4) ultrasonic devices for screening the distance between the ship and wharf.

3. *Meteorological and hydrological data systems* for providing data on visibility, wind, water level, current, swells, and other important navigational factors. These data can be transmitted to vessels by means of VHF.

4. *Channel markings* should include a sufficient number of buoys, allowing masters and pilots to observe at least two pairs of buoys under normal operating circumstances.

Shipowners' Equipment.   All large tankers should be equipped with instruments for measuring: (1) the longitudinal as well as the transverse components of the ship's speed when traveling at low speed in a channel and during the berthing maneuver; (2) the drift, that is, the angle between the speed vector and the axis of the ship, and slight variations of this angle with the purpose of establishing the yaw rate; (3) the distances between stern and bow of the ship and the berths; and (4) the true draft at bow, stern, and midship starboard and portside.

All the data resulting from these measurements should be grouped in a visual display located near the navigation control panel.

TRADEOFFS IN SUPERPORT DESIGN

Most entrance channels to superports are limited in some way. Such limitations may include depth of water, channel width, or frequent adverse weather conditions. Port designers will attempt to compensate one or more of these disadvantages by an added provision for another. In particular, the unavailability of, or underprovision for traffic control, or of accurate navigational aids or of hydrographic information frequently necessitates the choice of design data which ultimately results in overdesigning the dimensions of the entrance channel. In a practical sense, the most important factors to be noted are the maximum size of the ships that will use the channel and the regularity of use. One should ask the question, for example: Should it be necessary to bring ships in with a draft of 20 meters (65.6 ft) with a regularity of 99 percent, and at wind velocities of up to 20 meters (65.6 ft) per second? Emphasis on size and when coupled with regularity most always leads to greatly increased construction costs. These costs obviously must be compared with the commercial and operating economic advantages to be gained.

BUILDING OF A SUPERPORT

The development of shipping and industry after World War II led to the need for improving the accessibility of the Rotterdam port area from the sea. Adaptation of the port within the existing contours and entrance sufficed until about 1965, at which time it became apparent that greater measures must be taken, with one objective being that of accommodating VLCCs at Rotterdam. A multidisciplinary approach to the problem was undertaken* via formation of a work group, including contractors. Conclusions pertaining to objectives included: (1) the new harbor area must be accessible to the largest ships that can navigate the North Sea, and (2) the harbor must be accessible from the sea via a separate entrance.

Initial studies of the plan, particularly in terms of the hydraulic engineering aspects, showed that a solution needed to be sought in a fusion of the two entrances. This was culminated in a final plan, with the configuration depicted in Fig. 4. The design is such that ships with a draft of 19.8 meters (65 ft) can be accommodated. Simply by deepening the channel, the draft can be increased to about 22.9 meters (75 ft) and even up to about 27.5 meters (90 ft) should the demands by newly designed ships be presented.

The harbor entrance comprises a north pier, an extension of about 3 km (~ 1.9 mi) to the existing north jetty, and a south pier about 13 km (~ 8 mi) long, consisting of a 4.5-km (~ 2.8-mi) sand dam, a 6-km (~ 3.7-mi) curved stone jetty, and a 2- to 3-km (1.2- to 1.9-mi)

*This description of a major port expansion and modernization program involved most, if not all, of the types of problems that are encountered in altering and developing a superport for the handling of VLCCs. The description is primarily based on a report by the chief engineer in charge of construction of the new entrance for VLCCs (J. van Dixhoorn).

sandy beach along the southern shore of the outer harbor (parallel to the extended north pier).

The sand dam which begins at the former Westplaat and runs in a northwesterly direction into the sea consists of a sand body built to a height of 7 m (~ 23 ft) above the mean sea level (M.S.L.), a width at the crest of 200 m (~ 656 ft), with the slopes above and below the waterline contoured to the prevailing natural situation. The sand body, completed in the autumn of 1969, has been seeded and planted.

The stone dams, on both the north and south sides of the harbor entrance, consist of various layers of stony materials, covered with concrete blocks of 43 tons each. The stone

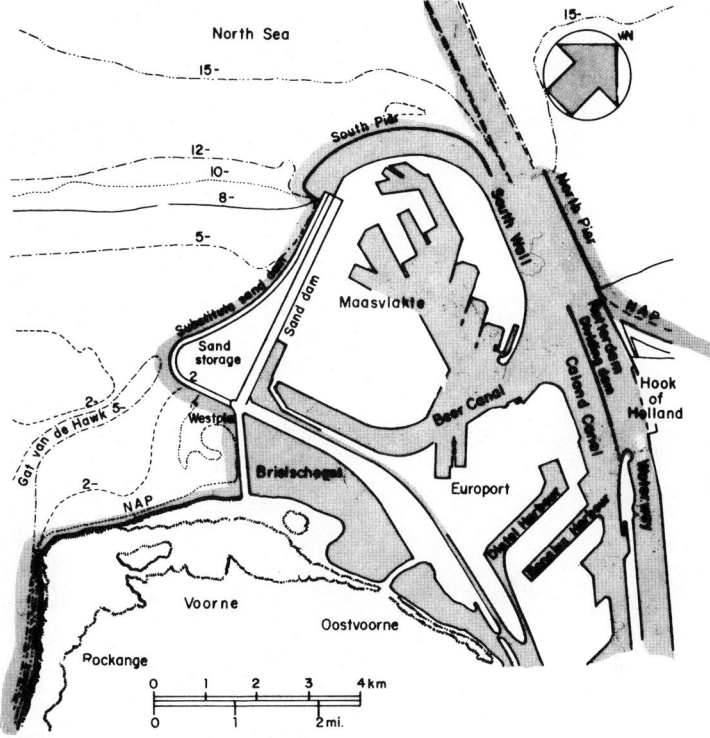

The harbor entrance

Fig. 4. Plan view of Europoort and associated facilities at the Port of Rotterdam.

dam has a crest of about 2 m (6.6 ft) plus M.S.L. and slopes that vary from 1:1.5 to 1:6, depending on the height at which they occur in the cross section.

The sandy beach which forms the southern boundary of the new outer harbor was originally designed as a sand body rising up to a height of about 6 m (19.7 ft) plus M.S.L. with natural slopes above and under water. This was considered the logical termination of the harbor site and a means for achieving a wave-deadening, gradual-slope beach. Later investigations showed that in order to achieve a nautically acceptable current situation in the outer harbor, the sandy beach had to have a greater guiding effect. Therefore, the beach had to be made steeper. Consequently, it had to be constructed partly of stone and gravel. With these alterations made, the once-termed sandy beach is now called the South Wall.

The plan shows a definitive entrance of 350 m (~ 1148 ft) width with a radius of 1800 m (5905 ft) to the Beer Canal (expandable to 500 m (1641 ft) width and 2500 m (8203 ft) radius), an extremely narrow dividing dam, and a small opening for service vessels.

A far-reaching investigation was made into the depth and width which very deep-drawing ships require for safe passage, partly in relation to the available navigational aids. Nautical and hydraulic research using models, and in practice and perception investigations (including the help of a simulator) provided new insights in this respect whereby it appeared that the availability of adequate navigational aids and a sound source of metohydro information could have a reducing effect on the sizes of the channel and approach. See Fig. 5.

Fig. 5. Hydraulic laboratory model of the first phase of the Europoort scheme.

During construction, in view of the large amount of work traffic (some 40,000 movements), a traffic post (using radar and radio) was established for guiding the work movement. This post was situated on the dam between the Caland Canal and the Rotterdam Waterway to guide work activity as well as normal shipping, the latter comprising some 70,000 movements per year.

Project operations in the sea are difficult because position finding is affected by distance and frequently by restricted visibility. A DECCA high-fix position-finding system with three transmitters, available at the outset of the project, proved to be inadequate in terms of range and accuracy. A radio position-finding system in the southern part of the North Sea could be used for work far out at sea, but for more accurate position finding during construction of the dams, an additional Cubic Auto Tape position-finding system, operating with transmitters on board the work ships in cooperation with responders on shore, was found necessary. For testing radio position-finding systems and for indicating axis lines, lasers installed on the north bank of the Rotterdam Waterway found incidental use.

The weather, water level, and wave information essential for the work was supplied by the Royal Netherlands Meteorological Institute (KNMI), directly from observation posts at sea, including measuring posts and measuring buoys located in the region.

All information is received in a monitor room of the Harbor Entrance Department in the Hook of Holland where a close watch is maintained over all position-finding systems, shown previously in Fig. 3.

For control and supervision, use is made of such modern techniques as industrial television on a goniometric basis (Artemis), wherever possible, whereby supervision can be made from a distance on several points simultaneously. Aircraft are used for making measurements. By this means, the South Wall was "water-leveled" in 20 minutes, and the data were processed by computer. Stone dams were controlled "photographically" by means of helicopters equipped with stereo cameras. The depth sounding vessels are automated. Much of the processing of this latter information is done by computer. During a large construction project such as this, the sequence and tempo of the work must be such that regular shipping is not impeded. To assure the continuity of normal operations, the various phases of the project are studied on an operational model of a scale 1:640, located in the Delft Hydraulics Laboratory, in advance of their execution. Each phase is also subsequently checked by use of the model. Although this tool is expensive, costing 2.5 million guilders ($\cdot \sim$ \$1 million) per year, it has proved to be essential in the realization of a safe, optimal, yet hydraulically and nautically sound project.

The complex of projects has been designed to allow for future developments. For example, construction of the harbor dams is such that they will, in principle, function well with a depth of 30 m (98 ft), should this occur naturally, or should it prove necessary for shipping. In such cases, only supplementary measures will be necessary to the toe of the dams.

The design provides for accommodating tankers of 500,000 DWT in their entrance to Europoort—and perhaps tankers even larger. This design planning takes into account the runout length, turning circles, and maneuvering characteristics of newer, larger ships. The definitive entrance, dividing dam, and South Wall are so designed that finally an entrance of 500 m (1641 ft) width is possible—width now under construction is 350 m (1148 ft); so a bend to the Beer Canal of 2500 m (8203 ft)—now 1800 m (5906 ft) can be realized. Sufficient runout length is available for the very large tankers in the Beer Canal and the Caland Canal. By the use of continuing improvements in navigational aids and communications, the manageable number of ships using the harbor entrance can be increased.

As a result of the program just described, not only is the harbor entrance (Hook of Holland), in principle, suitable for the largest ship that can navigate the North Sea, but also a further extension of the harbor area behind the harbor entrance is possible, and a further increase of traffic can be absorbed.

To what extent the harbor entrance will be used by ships larger than about 22.9 m (75 ft) draft will depend on the transportation economics of the future. Obviously, this factor is also linked with the navigability and accessibility of the North Sea to very large ships. This question is being studied all the way to the English Channel, including the work of specialists in the navigational, nautical-hydraulic, marine-topographical, geological, and dredging fields. The 30 to 40 m (98 to 131 ft) deep southerly North Sea is accessible via the Channel along the presently recommended (I.M.C.O.) route for ships with maximum 22-m (72-ft) draft, a route that possibly can be deepened to a maximum for accommodating a 27.5-m (90-ft) draft. The distance from the harbor entrance (Hook of Holland) to the 30-m-deep (98-ft) North Sea is about 40 km ($\sim$ 25 mi), a gap that can be bridged by a technically sound artificial channel, suitable for drafts of 22 m (72 ft) and later 27.5 m (90 ft).

In addition, attention also is being given to traffic density and traffic behavior at sea. Among the tools used are modern aerial observation techniques, such as SLAR, infrared methods, and so on, particularly in cooperation with Great Britain.

In planning the total number of ships to be accommodated versus the entrances to be used for the Port of Rotterdam, three alternative methods have been carefully studied and projected as follows for the conditions that may exist in 1990.

Method	New entrance	Old entrance	(Annual) Total no. of ships
I	12,000	31,500	43,500
II	20,000	25,000	45,000
III	12,500	25,000	37,500

From the foregoing table, the following conclusions can be drawn: (1) The old entrance will, in the most unfavorable conditions, hardly experience any extra burden in the future. (2) In the most unfavorable instance, the new entrance will have to handle 20,000 ships, which would point to a significant overcapacity of the entrance by 1990.

GENERAL CHARACTERISTICS OF SUPERPORTS

The basis of design of any port is to provide the ship a safe berth where a quick turnaround can be performed. Some of the factors that apply to VLCCs in particular are related almost entirely to local circumstances and thus requiring intensive specific analysis. Some questions to be answered include: Is protection by breakwaters necessary? Are jetties necessary? Should an island port be preferred? Is one, or are two, three, or four point moorings necessary? But, even though each situation is highly specific, it is also advantageous to study ports built under partly similar circumstances.

Petroleum Refining and Processing

Although no two petroleum refineries are exactly alike, a typical integrated refinery that uses representative refining processes is shown schematically in Fig. 1. For simplicity, the diagram is confined to operations that are concerned with the production of fuels. Manufacturing operations involving lubricating oils, waxes, solvents, road oils, asphalt, petrochemicals, and other nonfuel products are omitted. The diagram is based on United States refinery practice. Outside the United States, the gasoline-creating processes, such as catalytic cracking, alkylation, and catalytic hydrocracking are less common.

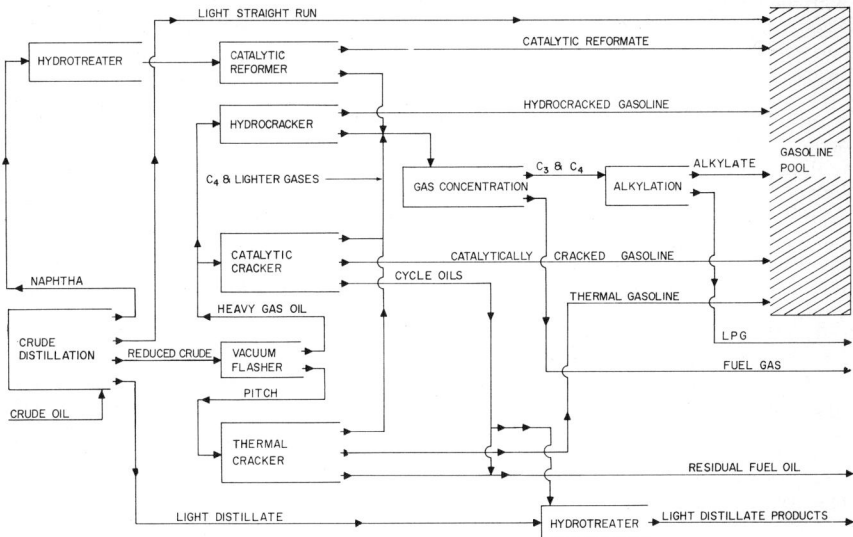

Fig. 1. Major paths of materials flow in an integrated refinery for producing fuels.

The processing units basic to the manufacture of fuel products in the refining industry include: (1) crude distillation, (2) catalytic reforming, (3) catalytic cracking, (4) catalytic hydrocracking, (5) alkylation, (6) thermal cracking, (7) hydrotreating, and (8) gas concentration. Petroleum refineries will also use many auxiliary processes, such as treating units to purify both liquid and gas streams, waste-management and pollution-control systems, cooling-water systems, units to recover hydrogen sulfide from gas streams and convert it to elemental sulfur or sulfuric acid, electric-power stations, steam-producing facilities, and provisions for storage of crude oil and products.

Key statistics on refining capacities are given later in this description.

CRUDE OIL DISTILLATION

To minimize corrosion of refining equipment, a crude-oil distillation unit generally is preceded by a *desalter,* which reduces the inorganic salt content of raw crudes. Salt concentrations vary widely (from nearly zero to several hundred pounds, expressed as NaCl per 1000 barrels). The crude unit functions simply to separate the crude oil physically, by fractional distillation, into components of such boiling range that they can be processed appropriately in subsequent equipment to make specified products. Major petroleum products are tabulated in description of Petroleum which appears earlier in this Handbook section on *Petroleum Technology.*

Although the boiling ranges of these components (or fractions) vary between refineries, a typical crude unit will resolve the crude (as indicated by Fig. 1) into the following fractions:

A. By distillation at atmospheric pressure:
   1. A light straight-run fraction, consisting primarily of C_5 and C_6 hydrocarbons but also containing any C_4 and lighter gaseous hydrocarbons dissolved in the crude
   2. A naphtha fraction having a nominal boiling range of 200–400°F (93–204°C)
   3. A light distillate with boiling range of 400–650°F (204–343°C)
B. By vacuum flashing:
   1. Heavy gas oil having a boiling range of 650–1050°F (343–566°C)
   2. A nondistillable residual pitch

In the atmospheric-pressure distillation section of the unit, the crude oil is heated to a temperature at which it is partially vaporized and then introduced near, but at some distance above, the bottom of a distillation column. This cylindrical vessel is equipped with numerous trays through which hydrocarbon vapors can pass in an upward direction. Each tray contains a layer of liquid through which the vapors can bubble, and the liquid can flow continuously by gravity in a downward direction from one tray to the next one below. As the vapors pass upward through the succession of trays, they become lighter (lower in molecular weight and more volatile), and the liquid flowing downward becomes progressively heavier (higher in molecular weight and less volatile). This countercurrent action results in fractional distillation, or separation of hydrocarbons based on their boiling points. A liquid can be withdrawn from any preselected tray as a net product: the lighter liquids, e.g., naphtha, from trays near the top of the column, and the heavier liquids, e.g., Diesel oil, from the trays near the bottom. The boiling range of the net product liquid depends on the tray from which it is taken. Vapors containing the C_6 and lighter hydrocarbons are withdrawn from the top of the column as a net product, but a liquid stream boiling higher than about 650°F (343°C) is removed from the bottom of the crude distillation column.

This bottom liquid stream, called the *atmospheric residue,* is further heated and introduced into a vacuum column operated at an absolute pressure close to 50 mm Hg maintained by the use of steam ejectors. In this vacuum column, a flash separation is made to produce the heavy gas oil and the nondistillable pitch products previously described. Although the vacuum column contains certain internal hardware to minimize the entrainment of pitch in the rising vapors and to aid in heat transfer between vapor and liquid, it is more nearly a chamber in which vapor and liquid are separated by a single-stage flash than a fractional-distillation column.

The crude oil and atmospheric residue are brought to their desired temperatures in tubular heaters. Oil is pumped through the inside of the tubes contained in a refractory combustion chamber fired with oil or fuel gas in such manner that heat is transferred through the tube wall in part by convection from hot combustion gases and in part by radiation from the incandescent refractory surfaces.

Light Straight-Run Gasoline.   This fraction generally contains all hydrocarbons lighter than C_7 in the crude, and consists primarily of the native C_5 and C_6 families. This light fraction is stabilized (not shown in Fig. 1) to remove the C_4 and lighter hydrocarbons which are routed to a central gas-concentration unit for further resolution. The stabilized C_5/C_6 blend usually contains odorous mercaptans, which are treated for odor improvement before delivery to the refinery gasoline pool.

Of the components in modern gasoline pools, the light straight-run fraction has the lowest octane number* (antiknock rating). Its unleaded octane number, in a typical case, will be just under 70, while the unleaded octane number for the entire refinery pool (on a United States average basis) will be about 89. The light straight-run fraction has a good octane-number response to the additions of lead alkyls. Isomerization is also used to improve its octane rating.

NAPHTHA

Catalytic Reforming.   The chemical composition of the naphtha fraction, and therefore its octane number, varies with the crude source, but in an average case it will be in the

---

*In this description, Research Method octane numbers are used.

range of 40 to 50 octane. To become a suitable component for blending into finished gasoline pools, its octane number must be raised by changing its chemical composition. Nearly all refineries of the world accomplish this change by catalytic reforming.

Practically all naphtha feedstocks to catalytic-reforming units are hydrotreated first to prolong the processing life of the reforming catalyst. An important by-product of catalytic reforming is hydrogen, which is used in hydrotreating and whatever hydrocracking may be practiced in the refinery. In some cases, supplementary hydrogen is produced by the steam reforming of natural gas or light naphtha cuts.

HEAVY GAS OIL AND PITCH

Catalytic Cracking. The primary function of catalytic cracking is to convert into gasoline those fractions having boiling ranges higher than that of gasoline. An important secondary function is to create light olefins, such as propylene and butylenes, to be used as feedstocks for motor-fuel alkylation and petrochemical production. Isobutane, a necessary reactant for the alkylation process, also is an important product of catalytic cracking.

Although the principal feedstock is the gas oil separated from the crude by vacuum distillation, this feed often is supplemented with portions of light distillates and with distillate fractions resulting from thermal coking operations. These options are not shown in Fig. 1.

For practical reasons, the conversion of distillate feedstocks to lighter materials is not carried to completion. The remaining, uncracked distillates (cycle oils) are used as components for domestic heating fuels (generally after hydrotreating) and to blend with residual fractions to reduce their viscosity to make acceptable heavy fuel oil, as shown in Fig. 1. In some refineries, cycle oils are hydrocracked to complete their conversion to gasoline.

Unleaded octane numbers of catalytically cracked gasolines fall in the range of 89 to 93. After treatment for odor control, they are blended directly into the refinery gasoline pool.

Hydrocracking. In a sense, hydrocracking is complementary and supplementary to catalytic cracking, in that hydrocracking processes over a catalyst in a hydrogen environment heavy distillates and, in some instances, cycle oils which are impractical to convert completely in catalytic cracking units. The main products are gasoline or jet fuels and other light distillates.

An important secondary product is isobutane.

Generally, the C_5/C_6 fraction is blended into the gasoline pool. In some instances, the heavier portion of the gasoline is also blended into the gasoline pool; in other cases, this portion may be reformed first to improve its octane number. The flow diagram of Fig. 1 shows only heavy gas oil as a feedstock, and the entire liquid product as gasoline is routed directly to the refinery gasoline pool, even though the aforementioned options are widely performed in various combinations.

Thermal Cracking. The pitch, as produced by most vacuum-flashing units, is too viscous to be marketed as a heavy fuel oil without further treatment. In some refineries, the pitch is processed further in a thermal cracking unit *(visbreaking)* under relatively mild conditions to reduce its viscosity. In many instances, the thermal cracking does not reduce the viscosity sufficiently, and, as shown in Fig. 1, additional viscosity reduction is obtained by blending in a required amount of catalytic cycle oil to produce marketable residual fuel oil.

In certain situations, it is more economical to process the pitch in a thermal coking unit, from which the main products are gasoline, distillates, and coke. The gasoline from a coking unit is handled as previously described. The coke is useful, after calcination, for electrode manufacture where it meets certain purity specifications, but it is employed principally as a metallurgical coke or as fuel. Distillates from thermal coking operations may be utilized as feedstock for catalytic cracking, or the lighter distillates may be routed to the refinery distillate product pool after hydrotreatment.

A few refiners obtain additional feedstock for catalytic cracking or hydrocracking operations by the solvent extraction of the vacuum pitch, usually with propane as the solvent. The extract is relatively free of organometallic compounds and highly condensed aromatic structured hydrocarbons. Thus, the extract is suitable for handling by catalytic units. Extracted pitch is subsequently processed in thermal units or converted to asphalts.

The small amount of thermal gasoline which is made as a by-product is routed, after treatment, to the gasoline pool or to catalytic reforming through a hydrotreating unit because its octane number is relatively low.

Hydrotreating.   As a processing tool, hydrotreating has numerous applications in a refinery, where its principal function is to purify, cleanse, and improve the quality of the feedstock. The process employs hydrogen and a catalyst. The use of hydrotreating for pretreating naphthas before catalytic reforming has already been mentioned.

Figure 1 shows all the crude light distillate and the net catalytic cycle oil being hydrotreated in a single block before being routed to the refinery light distillate pool. Sometimes the light distillate in the crude may be sufficiently low in sulfur content to bypass hydrotreating; in other instances only a portion of the stream is hydrotreated to remove native sulfur compounds. Some refiners hydrotreat portions of their catalytic cracking feeds, particularly if they originate from thermal operations or if they are inordinately high in sulfur content.

Desulfurization is also an objective in the production of low-sulfur residual fuel oils. Reduced crudes which are especially high in sulfur (of the order of 4 percent or more) can be brought to sulfur levels of the order of 1 percent by vacuum flashing, hydrodesulfurizing the overhead vacuum-distilled gas oil, and blending the gas oil with a very low sulfur content with the untreated pitch to obtain a reconstituted low-sulfur fuel oil.

Gas Concentration.   The gas-concentration system, as shown in Fig. 1, collects gaseous product streams from various processing units and physically separates the components to provide, in the usual case, a C_3/C_4 stream as a feedstock for alkylation and a C_2 and lighter stream that is almost always used in its entirety to supply process heat requirements within the refinery.

Hydrogen sulfide is removed from gas streams in which it occurs by selective absorption in liquid solutions (usually organic amines). The H_2S released from the rich solution is converted by further processing into elemental sulfur or H_2SO_4.

Alkylation.   In motor-fuel refineries, the alkylation units produce a high-quality paraffinic gasoline by the chemical combination of isobutane with propylene and/or butylenes. A small amount of pentenes also are alkylated. The alkylation is accomplished with the catalytic aid of hydrofluoric or sulfuric acid to produce a gasoline having octane numbers in the range of 93 to 95.

Propane and n-butane associated with the olefins in the feedstocks are withdrawn from alkylation units as by-products. A portion of the n-butane is routed to the gasoline pool to adjust the vapor pressure of the gasoline to a level that permits prompt and easy starting of engines. The remainder of the n-butane and the propane are available for LPG (liquefied petroleum gas), a clean fuel which is easily distributed even to remote points as bottled gas for heating purposes.

CATALYTIC REFORMING

The purpose of reforming is to rearrange or reform the molecular structure of hydrocarbons, particulary with the objectives of upgrading naphthas having poor antiknock characteristics to premium-quality motor fuels or producing aromatics, notably benzene, toluene, and C_8 aromatics, from selected naphtha fractions.

The Catalytic Reforming Process.   The essential components of a particular reforming process* are shown in Fig. 2 and include (1) reactors that contain the catalyst in fixed beds, (2) heaters to bring the naphtha and recycle gas to reaction temperature and to supply heats of reaction, (3) product cooling system and a gas-liquid separator, (4) hydrogen-gas recycle system, and (5) stabilizer to separate light hydrocarbons dissolved in the receiver liquid.

Practically all naphtha feedstocks to catalytic reforming units are hydrotreated to remove nonhydrocarbons which would adversely affect the stability of noble-metal reforming catalysts from the standpoint of their activity and selectivity. Materials removed include sulfur, nitrogen, oxygen, and organic compounds of arsenic and palladium; all are catalyst poisons.

The catalyst is contained as a fixed bed in three or more separate adiabatic reactor vessels with feed and hydrogen-recycle-gas preheating before the first and reheating

---

*Platforming, Universal Oil Products Co. The original of the family of processes using noble-metal catalysts.

between the subsequent reactors. Because of the rather large endothermic heats of the dehydrogenation reactions, there is a substantial drop in temperature of the flowing and reacting stream, particularly in the first reactor, where the rapid naphthene dehydrogenation occurs. Therefore, the effluents from the first and second reactors are reheated to bring them to the proper inlet temperature for the subsequent reactor. Often, the charge heater and the interheaters are contained in the same furnace housing.

Effluent from the last reactor in the train is cooled and led to a receiver in which the product mixture is separated into a liquid and a gas stream. Most of the separated gas stream (largely hydrogen) is compressed and recycled to the reactors to provide the protective hydrogen partial pressure in the reaction environment. A net hydrogen-rich product stream is withdrawn from the system by pressure control, as shown in Fig. 2.

The receiver liquid which contains dissolved light hydrocarbons is routed to a fractionator to produce a stabilized reformate suitable for blending into finished gasoline. This

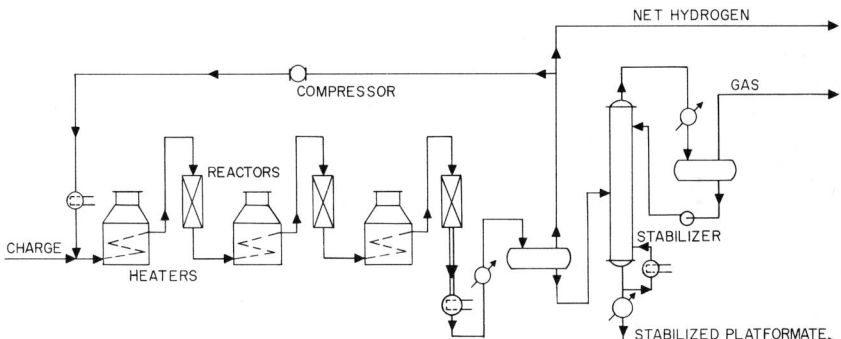

Fig. 2. Platforming catalytic reforming process. (Universal Oil Products Company.)

liquid is generally free of hydrocarbons lighter than C_5. The C_4 and lighter hydrocarbons, separated as an overhead product stream by the stabilizer fractionator, usually are routed to a gas-concentration system within the refinery.

CRACKING PROCESSES

In petroleum technology, cracking denotes reactions in which a hydrocarbon molecule is fractured or broken into two or more smaller fragments. Other terms used to describe these reactions are cleavage, decomposition, fragmentation, pyrolysis, rupture, and scission. Chemical cleavage may occur at a carbon-hydrogen bond; a bond between carbon or hydrogen and an inorganic atom, such as sulfur or nitrogen; or at a carbon-carbon bond. Since the main objective of cracking is reduction in size of hydrocarbon molecules, the principal reaction involves the breaking of carbon-carbon bonds. There are three principal types of cracking: thermal cracking, catalytic cracking, and hydrocracking.

Thermal Cracking.   Developed during the early part of the twentieth century, thermal cracking was the first of the major cracking processes. Largely replaced by fluid catalytic cracking for increasing gasoline production and improving its quality, thermal cracking is used principally in coking and viscosity breaking. Both processes are used to convert nondistillable residues into more valuable products.

Thermal Coking.   This process converts heavy residual stocks into gas, gasoline, distillates, and coke with the objective of maximizing the yield of distillates and minimizing the yields of gas, gasoline, and coke. The light distillates are used in domestic and industrial heating oils. The heavy distillates are appropriate feedstocks for catalytic cracking. Gasoline from thermal coking units can be blended into motor fuels after suitable treating, or further processed for octane number improvement by catalytic reforming after purification by hydrotreatment. Although only a by-product of the process, the petroleum coke is useful as a fuel in steam generation, and if it meets purity and other required specifications, it is one of the major raw materials for the manufacture of carbon electrodes in the aluminum and other electrometallurgical industries.

Two types of thermal coking processes are currently employed: (1) a continuous fluid coking technique and (2) a cyclic, semicontinuous process, variously named *delayed coking, decarbonizing,* or *low-pressure coking.* In the cyclic processes, the coke is alternately formed in, and removed from, a vessel or drum so that with two or more such manifolded drums in a unit, one drum can be filling while the others are being emptied.

A delayed coking unit consists of the three sections shown in Fig. 3 plus coke removal and handling equipment. Ordinarily, the residual feedstock is charged to the fractionator, where it encounters the hot vapors from the coke drum. Here any light components are first flashed from the feed before the feed joins with recycle and is charged from the bottom of the fractionator as a combined feed to the furnace. The furnace quickly heats the charge to temperatures in the vicinity of 480°C (896°F). The heated, combined feed then is introduced to the bottom of one of two or more insulated vessels where the residence

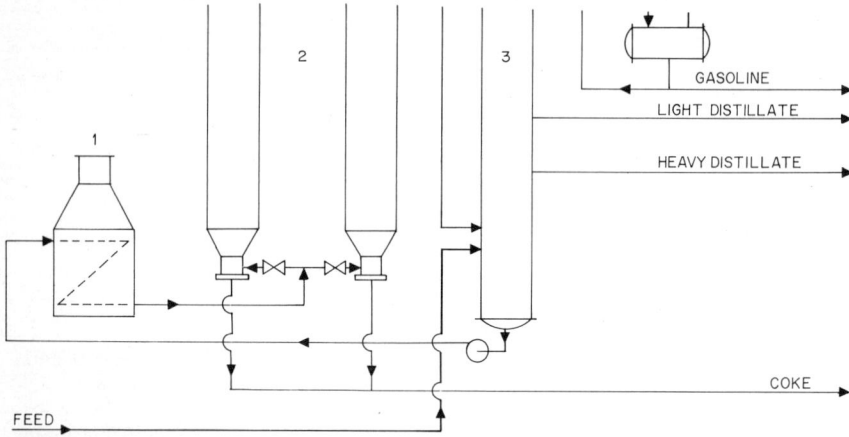

Fig. 3. Delayed coking unit: (1) Furnace, (2) Coke drums, (3) Fractionator. (*Universal Oil Products Company.*)

time is sufficient for the material to crack as the result of its contained heat and to form a solid coke residue. Simultaneously, lighter cracked products are evolved. Since the reaction in the drum is endothermic, the resulting temperature is in the region of 425°C (797°F). The coke remains in the drum, and the cracked products leave as vapors. The vapors are routed to the fractionation section, where they are separated into gas, gasoline, distillates of desired boiling ranges, and a heavy fraction, which is recycled to be cracked to extinction in the coking section of the unit.

When the coke accumulates to a predetermined level in one of the drums, the flow is switched to another empty, preheated drum so that operation of the furnace and fractionation sections is continuous. The filled drum just taken out of service is steamed to remove vapors, and water is admitted to cool the coke so that it can be removed safely. In modern designs, the coke is removed from the drum by hydraulic cutting, in which high-velocity water jets bore a hole downward from the top of the coke bed. After this operation, laterally directed water jets remove coke by a cutting action from the bottom of the bed. The coke falls with the water into cars or is sluiced by water to storage.

In order to minimize the yield of coke, drum pressures are relatively low—in the general range of 10 to 70 psig. There can be a wide variation in yields of products, depending on the nature of the feedstock and the boiling range of the liquid distillates produced.

The Fluid Coking^{SM} process is performed in equipment similar to that used in fluid catalytic cracking but operates by circulating small particles of coke as seed, which are formed by the process itself in one of two vessels to which the feed is charged. A small portion of the circulating coke stream is burned in the other vessel to raise its temperature enough to provide for the heat of reaction and feed preheat in the reactor vessel. The latter is maintained at temperatures of 480–565°C (896–1050°F). The net coke-product particles can be transported to storage by a stream of air.

Other Thermal-Cracking Operations.   Viscosity breaking is used to reduce the viscosity of heavy residues by a mild thermal cracking with a minimum yield of gasoline. Feedstock is passed through a tubular furnace with a minimum of recycle. The slightly cracked products are separated by distillation into gas, gasoline, a light distillate with a nominal end point of about 345°C (650°F), and a fuel-oil residue having a considerably lower viscosity than the feedstock. The visbroken residue is often further reduced in viscosity by blending with catalytic cycle oils to produce a marketable fuel oil which meets viscosity specifications. Although visbreaking units consist basically of a cracking section and a separation section, a number of variations in flow arrangements exist.

Fluid Catalytic Cracking.   Introduced during World War II, fluid catalytic cracking progressively displaced the earlier thermal-cracking processes to a very large extent. Early catalytic-cracking processes used a fixed-bed cyclic system. That arrangement has been displaced by fluid catalytic-cracking units.

Catalytic cracking is employed mainly to create gasoline, C_3/C_4 olefins, and isobutane primarily by the selective decomposition of heavy distillates. Because the cracking reactions are directed by specially prepared catalysts, the gasoline produced contains substantial proportions of high-octane-number hydrocarbon components, such as aromat-

TABLE 1.   Typical Proportions of C_1, C_2, C_3, and C_4 Hydrocarbon Fragments in Products of Thermal and Catalytic-Cracking Processes

Mole percent in C_1-C_4 fraction	C_1	C_2	C_3	C_4	Total
Thermal cracking	32	21	24	17	100
Catalytic cracking	12	11	29	48	100

Adapted from Douglas M. Considine (ed.), "Chemical and Process Technology Encyclopedia," McGraw-Hill Book Company, New York, 1974.

ics, branched paraffins, and olefins. Since the cracking reaction proceeds in accordance with the carbonium-ion mechanism, there are relatively minor amounts of fragments lighter than C_3 in the products. This result contrasts with the decomposition of hydrocarbons in thermal cracking by the free-radical mechanism, in which relatively large proportions of fragments lighter than C_3 are produced. This difference is illustrated by Table 1, in which the typical proportions of C_1, C_2, C_3, and C_4 hydrocarbon fragments contained in the products of the two processes are shown. Some hydrogen is produced by both processes in varying amounts.

Cycle Oil.   Catalytic cracking also produces a substance known as *cycle oil*. This is the distillate which boils above gasoline. A part of the cycle oil can be considered as a synthetic, cracked material that boils between the end point of the gasoline and the initial boiling point of the feedstock. The heavier portion of the cycle oil that falls within the boiling range of the feedstock represents the more refractory, uncracked components of the feed which are predominantly aromatic in nature. The cycle oils, withdrawn as net products from catalytic-cracking operations, are useful as components in heating oils, feedstocks to hydrocracking units, and for blending with heavy residuals to reduce viscosity; highly aromatic cycle oils are appropriate feeds for carbon-black manufacture.

Catalysts.   Materials used as catalysts in modern catalytic-cracking units generally are crystalline in nature and sometimes are referred to as *zeolitic catalysts*, since they are modified hydrated alumina silicates. They have proprietary compositions. These catalysts, introduced in the early 1960s, offer improved stability over the powders, pellets, extrudates, and beads, which were synthetic amorphous silica-alumina composites or specially treated natural clays.

The catalysts used by fluid units are spray-dried microspheres. The average particle size of the equilibrium catalyst inventory in a fluid unit is typically 60 μm in diameter with nominally 10 percent by weight of the particles being smaller than 40 μm and 10 percent being larger than 105 μm. In circulation within the catalyst system, the microspheres gradually are reduced in size by formation of fines which leave the unit in the regenerator combustion-gas stream. Fresh catalyst is added to make up for these losses. In many units, such replacement is close to the economic optimum for sustaining the catalyst activity and selectivity.

As shown by Fig. 4, a typical fluid catalytic-cracking unit comprises (1) reactor, (2) regenerator, (3) main fractionator, (4) air blower or compressor, (5) spent-catalyst stripper, (6) catalyst recovery equipment, including (*a*) cyclones internal in the reactor and regenerator, (*b*) slurry settler, (*c*) an optional electrostatic precipitator, and (7) gas-recovery unit.

The feedstock, which may be preheated by exchange or in some cases by a fired heater, along with recycle from the fractionation section meets a controlled stream of hot, regenerated catalyst. The resulting mixture of vaporized oil and catalyst ascends in the riser at a velocity such that the catalyst particles are suspended more or less discretely in a

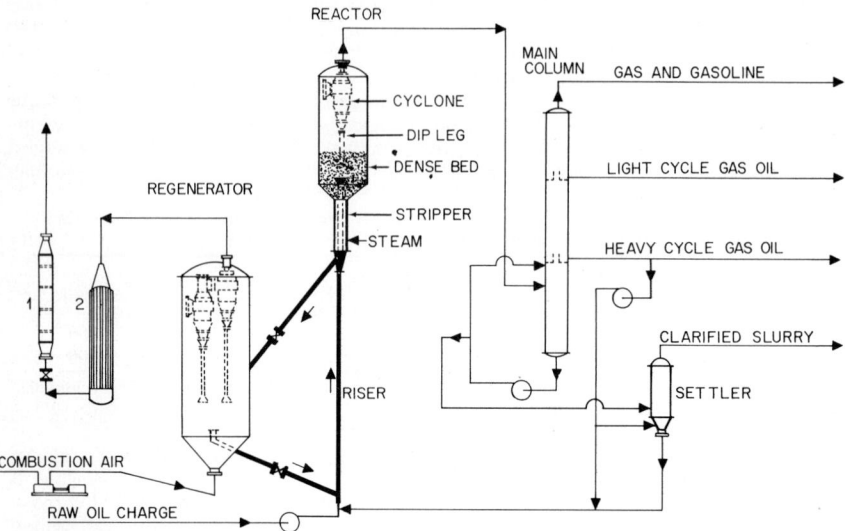

Fig. 4. Fluid catalytic-cracking process: (1) Pressure-reducing orifice chamber. (2) Flue-gas steam generator. The gas and gasoline from the main column go to the gas-concentration plant. (*Universal Oil Products Company.*)

dilute phase. A large portion of the cracking occurs in the riser. Upon reaching the reactor, the linear velocity of the hydrocarbon vapors is reduced to such an extent that most of the catalyst settles out to form a *dense-phase* bed, the amount or height of which can be controlled by the valve in the spent-catalyst line leading to the regenerator. The remainder of the desired conversion is accomplished as the hydrocarbon vapors pass through the dense bed.

Above the dense bed in the reactor is another dilute phase of suspended fine catalyst particles, most of which are separated from the cracked vapors by one or more stages of cyclones in the upper portion of the reactor vessel. The separated particles are returned to the dense phase through a dip leg. The cracked vapors, carrying a minor concentration of catalyst fines, pass to the main fractionator, where they are distilled into several products. The column overhead product is separated by flashing into gas and unstabilized gasoline streams. These overhead products are routed to a gas-recovery unit, in which can be produced a debutanized gasoline, a C_3/C_4 cut as feed for alkylation, and a fuel-gas stream containing C_2 and lighter components. The steam used for stripping the spent catalyst and cycle oil is removed as water from the overhead separator (not shown).

The total cycle oil can be resolved in the main fractionator into a light cycle oil, which usually boils between 205 and 345°C (401 and 653°F), for ultimate use as a heating oil or Diesel fuel. The light cycle oil generally is steam-stripped to remove light ends for flash-point control. The heavy cycle oil, boiling above 345°C (650°F) and removed as a side cut from the column, generally is recycled to the reactor, as shown in Fig. 4. When the conversion is 75 percent or more, the heavy material is recycled to extinction, although some may be withdrawn as a net product. In catalytic cracking, the percent conversion of the feedstock has been reckoned by tradition as 100 minus the volumetric percentage yield of the total net cycle-oil product (100 percent minus the sum of light cycle oil, heavy cycle oil, and clarified slurry oil).

The column bottoms contain whatever catalyst fines escaped recovery by the cyclones. Generally, a stream is recycled to a higher point in the column to wash catalyst fines from the incoming vapors and then in most cases through heat exchange to heat colder liquid streams or to generate steam. A portion of the column bottoms stream is routed to a settler, in which catalyst fines are separated for return to the reactor as a concentrated settler bottoms stream. The settler overhead is a catalyst-free clarified oil which is withdrawn as a net product, generally in an amount up to about 5 volume percent of the feedstock. This is a highly aromatic, refractory product that resists cracking. The clarified oil is useful as a cutter stock to reduce viscosities of heavy fuel oils.

The spent catalyst, which is the dense bed in the reactor, descends by gravity into a stripper where hydrocarbon vapors that are adsorbed, or within the interstices of the catalyst particles, are removed by a countercurrent flow of steam. The stripped catalyst descends through a control valve into the regenerator, in which the catalyst deposit is burned off. Air is introduced by a centrifugal compressor. The residual carbon on the regenerated catalyst is about 0.2 weight percent of the catalyst.

As in the reactor, the dilute phase above the dense bed contains suspended fine catalyst particles, most of which are removed by two stages of cyclones and returned to the dense phase through dip legs. The combustion products, traditionally termed *flue gas*, are virtually free of oxygen and contain nearly equal amounts of CO and CO_2, along with a certain amount of steam formed by the combustion of a low-hydrogen-content catalyst deposit, brought into the regenerator as occluded stripping steam and as humidity in the combustion air.

The flue gas that leaves the regenerator through a pressure-control valve can be routed through a variety of auxiliary equipment, such as a heat exchanger for steam generation (shown in Fig. 4). A CO boiler (not shown) is sometimes used, in which steam is generated by the combustion of the CO in the flue gas and other external fuel. Or the flue gas may be routed through an electrostatic precipitator for final removal of catalyst particulates where air pollution is a critical matter. In some instances, the cleaned flue gas may go to gas turbines for pressure letdown and the recovered energy used for driving electric generators.

The fluid catalytic-cracking unit is operated as a heat-balanced system. The chemical heat of combustion of the catalyst deposit supplies (1) the heat required to vaporize and crack the oil coming to the riser, (2) the heat required to raise the temperature of the air to the flue-gas temperature, and (3) the heat required to make up losses due to radiation. If the amount of catalyst deposit burned, the amounts and temperatures of flowing liquid and gas streams, and the temperature of the two dense catalyst beds are known, heat balances can be written for the reactor and regenerator and solved for the rate of catalyst circulation required to satisfy the heat balance. A unit processing 40,000 bbl per day of feedstock will circulate about 5 million lb per hr of catalyst between the reactor and regenerator in a typical instance.

Hydrocracking.   This process differs from catalytic cracking in using a different catalyst and an environment of hydrogen at total pressures of 800 to 2500 psig. Since the accumulation of carbonaceous deposit on the catalyst is extremely slow, process periods on-line range from several months to a year or more. In some cases, the need to burn off the catalyst deposit is not the chief cause for shutdown.

Generally, the hydrocracking processes can accommodate a wider range of feedstocks than catalytic cracking. Feedstocks include not only heavy distillates but also solvent extracts from residuals that contain several parts per million organometallic complexes. In some designs, residual oils can be processed economically if the metals content is not too high.

Products.   Hydrocracked products differ from those made by catalytic cracking in that they are not olefinic. The gasoline products are not as high in octane number. Although the C_5/C_6 fraction can be blended directly into the gasoline pool, the C_7 + naphtha generally is used as a feed to catalytic reforming, since it is high in naphthene content. The distillate products that are heavier than gasoline are not so aromatic as those from catalytic cracking. Thus the distillates are appropriate components for jet fuels and other uses where low aromatic content is a requirement. All hydrocracked products are low in sulfur. The light hydrocarbons have a predominance of branched isomers. The C_4 fraction is a valuable alkylation feed because of its high isobutane content. Hydrocracking is particularly applicable for the production of specialty products, e.g., low-pour-point low-sulfur-content Diesel fuels, jet fuels, LPG, high-viscosity-index lube bases, and a wide variety of low-sulfur-content fuels.

A variety of proprietary catalysts are used and are supplied by the process licensors or through formulations furnished by the licensors.

Fixed-Bed Units.   The fixed-bed process is one of two basically different techniques of hydrocracking, the other being the ebullating bed. Fixed-bed processes can be classified by the number of reactors used and the configuration of flow with respect to the reactor sequence and fractionation of reactor products. In the simplest arrangement, one reactor is followed by fractionation of the reactor effluent into net products and a recycle stream to be cracked further. A process of this type is shown in Fig. 5. In another design, two reactors are used in series, each reactor employing a different catalyst, with the entire effluent from the first reactor being processed in the second reactor. Another two-reactor system, with different catalysts in each reactor, employs fractionation of the effluent from the first reactor to remove certain products before the remainder is processed in the second reactor.

Figure 5 illustrates a single-reactor unit* designed to hydrocrack a heavy vacuum gas oil into gasoline and light distillate products. In this particular arrangement, the combined feed—comprising fresh feedstock, recycle and makeup hydrogen, and recycled uncoverted distillate—is preheated by exchange with reactor effluent before it is brought to the desired reactor inlet temperature in the heater.

The hydrocarbons, which are cracked in a downflow, fixed-bed reactor, are cooled and brought to a drum, in which the recycle hydrogen is separated. The equilibrium liquid phase from the separator contains (in solution) some hydrogen as well as hydrogen sulfide and ammonia (formed from the sulfur and nitrogen in the feed) and net hydrocracked light hydrocarbons. These dissolved, light materials are removed in a debutanizer.

The stabilized liquid then is led to a fractionator for splitting into a C_5/C_6 fraction, which is suitable for blending into finished gasoline pools, and into a naphtha, which in most refining situations is used as a feedstock to catalytic reforming to improve its octane number. In some instances, the naphtha may be suitable as a blending component in finished gasoline pools.

A middle distillate is shown in Fig. 5 as a net product. This material is suitable for use as a blending component for jet fuel, kerosine, Diesel fuel, and a variety of heating fuels. The end point of the middle distillate can be adjusted to suit the specific purpose for which it is produced. The remainder of the reactor effluent is withdrawn as a higher-boiling column bottoms stream, which is recycled to the reaction zone for further hydrocracking.

In some designs, the relative positions of the debutanizer and the splitter (as shown in Fig. 5) are reversed, so that only the light gasoline overhead from the first column is stabilized in the second column. Some designs in which relatively large amounts of light hydrocarbons are formed employ two stages of flashing of the reactor effluent before further separation by fractional distillation. The secondary flash helps reduce the vapor load in the first fractionator and leads to a more economical overall design. In some instances, the heat of reaction is high enough to make it practical to employ quenching at intermediate points in the reactor and thus keep the temperature profile of the reactor more nearly isothermal.

When the sulfur content of the feedstock is especially high, hydrogen sulfide (H_2S) removal from the recycle hydrogen stream may be economical or H_2S removal from a slipstream of the recycle may be used, especially if H_2S recovery is routinely performed in the refinery.

In a unit of the type described, the relative proportions of gasoline and light distillate may be varied, either (1) by changing the operating conditions that influence the conversion per pass or (2) by choice of catalyst. Some catalysts selectively convert heavy distillates into light distillates without further converting the primary product into gasoline; other catalysts convert any distillate selectively into gasoline; and still others are capable of hydrocracking naphthas (particulary) into LPG fragments, that is, C_3 and C_4.

Ebullating-Bed Reactor.   In this design, the upward linear velocity of hydrogen and

*UOP Isomax process.

hydrocarbons is sufficiently high to expand the bed of catalyst particles and thus induce a condition of continuous, random motion. The upward linear velocity is not so high, however, that catalyst particles will be carried out of the reactor with the hydrocracked products.

Two catalyst sizes and configurations can be used. When a powdered catalyst is employed, the recycle hydrogen and hydrocarbon feed is sufficient to ebullate the catalyst bed. In another design, where extruded catalyst particles are used, an internal ebullating pump is utilized to recirculate clear liquid from above the top of the expanded bed downward through a draft tube and upward through the bed. Provisions are made in the ebullating-bed design to withdraw spent catalyst and replace it with fresh catalyst at such

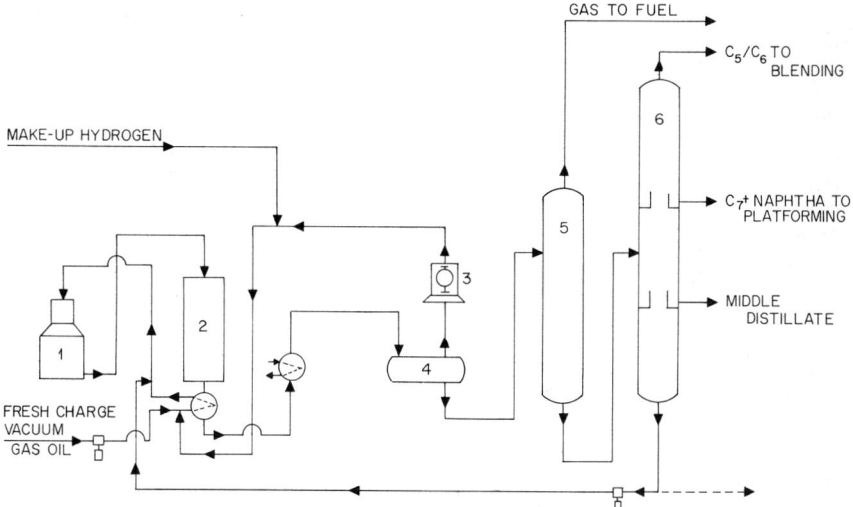

Fig. 5. Hydrocracking process: (1) Heater, (2) Reactor, (3) Compressor, (4) Separator, (5) Debutanizer, (6) Splitter column. (*UOP Isomax, Universal Oil Products Company.*)

rates that the catalyst inventory in the reactor can be maintained at a constant performance level—or at a new level if desired.

Since the ebullated catalyst beds are in a state of agitation, isothermal conditions are realized. Back mixing of hydrocracked products does occur. To compensate for diminished hydrocracking selectivity due to back-mixing effects, more than one reactor is used in series to obtain the benefit of staging.

Hydrotreating.   Generally understood to include a variety of applications in which the quality of liquid hydrocarbon streams is improved by subjecting them to mild or severe conditions of hydrogen pressure in the presence of a catalyst, hydrotreating may be regarded as a rather specialized kind of *hydrogenation*. The primary purpose of hydrotreating is to selectively convert to a desirable material or eliminate from the system one or more unwanted materials in the feedstock. The use of hydrotreating is extensive, being involved in over 30 percent of the crude refined in the United States. Although the catalysts and technique were known earlier, the availability of a ready supply of by-product hydrogen from catalytic reforming accelerated the use of hydrotreating in the early 1950s.

Applications of hydrotreating are numerous, and the feedstocks processed range from light fractions of gasoline to heavy residual stocks, as evidenced by the objectives of hydrotreating which include (1) pretreatment of naphtha feeds for catalytic reforming units; (2) desulfurization of distillate fuels; (3) improvement of burning quality of jet fuels, kerosines, and Diesel fuels; (4) improvement of color, odor, and storage stability of various fuels and petroleum products; (5) pretreatment of catalytic cracking feeds and cycle oils by removal of metals, sulfur, and nitrogen, and reduction of polycyclic aromatics; (6) upgrading of lubricating-oil quality; (7) purification of light aromatic by-products from pyrolysis operations; and (8) reduction in sulfur content of residual fuel oils.

Reactions. Some of the reactions commonly employed in hydrotreating processes include:

1. Removal of sulfur from its organic combinations in various types of sulfur compounds by hydrodesulfurization to form H_2S.

2. Removal of nitrogen as ammonia (NH_3) from its organic combinations.

3. Hydrogenation of diolefins and olefins to paraffins or naphthenes.

4. Hydrogenation of monoaromatics to naphthenes to improve the burning quality of certain fuels.

5. Hydrogenation of polycyclic aromatics so that only one aromatic ring remains in the molecule; or, if desired, all the aromatic rings can be saturated.

6. Removal of oxygen from its organic combinations as H_2O.

7. Decomposition and removal of organometals, e.g., arsenic compounds in naphthas, by retention of these metals on the catalyst. Vanadium and nickel can be removed from gas oils intended as feedstocks for catalytic cracking.

Hydrogen sulfide, ammonia, and water are removed from the hydrotreated liquid product by stripping in the stabilization section of the unit.

As shown in Fig. 6, which is a generalized representation of the majority of hydrotreating processes, the essential components are (1) heaters and heat-exchange equipment; (2) the fixed-bed reactor section, which contains the catalyst and operates at pressures ranging widely from 100 to 3000 psig, depending on the requirement of the treatment; (3) a gas-liquid separation section; (4) the hydrogen recycle system; and (5) a liquid product stripper or stabilizer.

Liquid feed is preheated by exchange with the reactor effluent and brought to a controlled reactor-inlet temperature in a fired heater. A recycle hydrogen stream joins the feedstock. The amount of the hydrogen recycle stream is in excess of that required for the chemical reactions in order to suppress the accumulation of deactivating carbonaceous deposits on the catalyst. Some of the cold recycle hydrogen may be brought into the reactor at intermediate points in the reaction zone to serve as a heat-absorption medium, thereby making the temperature profile through the reactor more nearly isothermal than it would be without the cold-gas quench.

To provide hydrogen consumed by the reactions and by solution losses, a stream of fresh makeup hydrogen is brought into the system either before or after the recycle gas compressor and in sufficient quantity to maintain the desired pressure in the unit. In many hydrotreaters, particularly in units used to pretreat naphthas for catalytic reforming, the output of hydrogen from the reformer is sufficient and can be routed directly through the hydrotreater without need for a recycle compressor. In these cases, the once-through hydrogen separated from the cooled reactor effluent liquid is released from the hydrotreater by pressure control and routed to other units where it can be used or to the refinery fuel-gas system.

The cooled reactor effluent stream is brought to a separator vessel, where the recycle or net hydrogen is removed. The liquid is routed to a stripper or stabilizer, which functions to remove H_2, H_2S, NH_3, H_2O, and light hydrocarbons dissolved in the separator liquid. When relatively large amounts of gases are dissolved in the separator operated at plant pressure, there is an advantage in routing the liquid from the primary separator to another separator maintained at a lower pressure. In this latter unit, a portion of the dissolved gases is flashed from the liquid and removed at this point of lower pressure, thus relieving the load on the stabilizer. The stabilized hydrotreated liquid, free of dissolved, unwanted contaminants, is routed to subsequent processing or to product fuel blending.

Alkylation. For the production of a motor-gasoline blending component, alkylation means the chemical combination of *isobutane* with any one or a combination of propylene, butylenes, and amylenes to form a mixture of highly branched paraffin that has a high antiknock rating and good stability. Alkylation reactions are conducted at or slightly below normal cooling-water temperatures and at pressures sufficiently high to keep the feed and reaction mixture in a liquid phase. Either sulfuric acid or hydrofluoric acid is used as a catalyst.

Most all alkylation practiced by the petroleum refining industry employs blends of butylenes and propylenes as the olefin portion of the feedstock. Only a minor amount of amylene is alkylated. The resulting alkylate consists of a mixture of isoparaffins, ranging from pentanes to decanes and higher, regardless of which olefin is used as a reactant. The simple addition of isobutane to an olefin does not explain the formation of this wide range

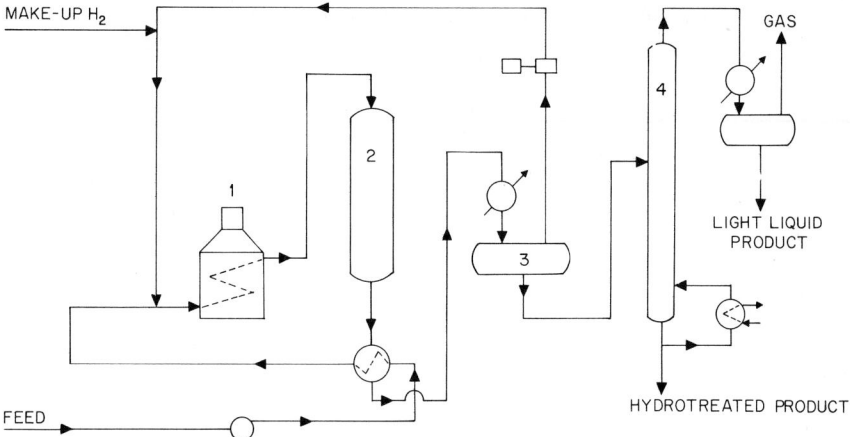

Fig. 6. A representative hydrotreating unit: (1) Heater, (2) Reactor, (3) Separator, (4) Stripper.

of compounds found analytically in alkylates. The overall reaction mechanism is exceedingly complex.

Properties of Alkylates. Motor fuel alkylates can be typified as having unleaded octane ratings* in the low and middle 90s, a low sensitivity (Motor and Research Methods octane numbers nearly equal), and an excellent octane number response to the addition of lead alkyls. These properties characterize alkylates as being a high-quality blending stock for the refiner's gasoline pools and especially for premium-grade gasolines. Even though motor fuel alkylates are a mixture of paraffinic isomers having a wide range of molecular weights, they do fall within the boiling range of commercial gasolines. The composition by carbon number of a particular depentanized alkylate, made from a mixed propylene-butylene feed in a commercial unit using hydrofluoric acid as a catalyst, is shown in Table 2 to give an indication of the range in molecular weights. Although the alkylate is typically predominant in C_8 paraffins, the C_9 and heavier portion ranged up to dodecanes, which were present in the small amount of 0.1 percent.

Also of interest is the isomeric composition of the C_8 paraffins in this alkylate, as given in Table 3. Well over one-half of the C_8 fraction in the sample was found to be 2,2,4-*trimethylpentane*, which by definition has a numerical value of 100 on the octane number scale. The highly branched trimethylpentanes comprised nearly 90 percent of the C_8 fraction, and no *n*-octane was found.

Octane numbers and other properties of a typical hydrofluoric acid (HF) alkylate are given in Table 4. On an unleaded basis, the Motor octane number is only slightly less than the Research octane number. But with 3 ml of tetraethyllead per gallon, the Motor octane number exceeds the Research octane number. This relationship is typical of alkylates, which, by this characteristic, are noted to have a low octane number sensitivity, as distinct from gasolines that contain olefins and/or aromatics whose Motor octane numbers can be

TABLE 2. Composition and Carbon Number of a Particular Depentanized Alkylate*

Carbon no.	Composition, vol %
C_6	2.7
C_7	26.8
C_8	63.8
C_9 (and heavier)	6.7

Adapted from Douglas M. Considine (ed.), "Chemical and Process Technology Encyclopedia," McGraw-Hill Book Company, New York, 1974.

*Made from a mixed propylene-butylene feed in a commercial unit using hydrofluoric acid as a catalyst.

*Octane number is defined under Petroleum Fuels Characteristics in this Handbook section.

TABLE 3.   Composition of C_8 Fraction in an Alkylate*

C_8 isomer	Vol %
2,2,4-Trimethylpentane	58.6
2,3,4-Trimethylpentane	19.3
2,3,3-Trimethylpentane	9.3
2,2,3-Trimethylpentane	0.8
2,4-Dimethylhexane	3.9
2,2- and 2,5-Dimethylhexane	2.3
3,4-Dimethylhexane	0.5
3,3-Dimethylhexane 4-Methylheptane 2-Methylheptane	0.8
3-Methylheptane 2,3-Dimethylhexane	4.5

Adapted from Douglas M. Considine (ed.), "Chemical and Process Technology Encyclopedia," McGraw-Hill Book Company, New York, 1974.
*Same sample as described in Table 2.

TABLE 4.   Properties of a Typical Hydrofluoric Acid Alkylate

Gravity, °API	71.5	
ASTM distillation	°F	°C
Initial boiling point	116	(46.7)
10% point	163	(72.6)
30% point	210	(98.9)
50% point	220	(104.4)
70% point	226	(107.8)
90% point	246	(118.9)
End point	380	(193.3)
Octane ratings:		
Research Method, unleaded	95.0	
Research Method, with 3 ml tetraethyllead/gal	106.3	
Motor Method, unleaded	92.8	
Motor Method, with 3 ml tetraethyllead/gal	107.4	

Adapted from Douglas M. Considine (ed.), "Chemical and Process Technology Encyclopedia," McGraw-Hill Book Company, New York, 1974.

TABLE 5.   Composition of Typical C_3/C_4 Feedstock Produced by Catalytic Cracking*

Component	Vol %
Propane	12.7
Propylene	23.6
Isobutane	25.0
n-Butane	6.9
Isobutylene	8.8
1-Butylene	6.9
2-Butylene	16.1

Adapted from Douglas M. Considine (ed.), "Chemical and Process Technology Encyclopedia," McGraw-Hill Book Company, New York, 1974.
*Composition of feedstock will vary to some extent, depending upon stock being cracked, the catalyst, and operating conditions.

several numbers lower than their Research octane ratings. Octane quality of alkylates can vary over a small range of up to four or five numbers, depending on the relative amounts of the several olefins usually present in the feedstock as well as on the catalyst condition and the plant design and operating conditions used to produce the alkylate.

Alkylate-Production Capacity.   Since its introduction during the early 1940s to produce a high-quality aviation-gasoline blending component, alkylation capacity has increased steadily. About 71 percent of the alkylation plants use sulfuric acid as a catalyst; the remainder use hydrofluoric acid.

Feedstocks for Alkylation.   Catalytic cracking units are the major source of olefinic feeds for alkylation in the refining industry. A typical C_3/C_4 feedstock produced by catalytic cracking will have the composition indicated in Table 5. Alkylates produced from butylenes will have octane numbers about three units higher than those made from propylene. From this standpoint, 2-butylene is superior to 1-butylene. C_3/C_4 fractions from catalytic cracking are considered to be excellent alkylation feedstocks because the butylenes amount to nearly 60 percent of the total olefin content. The isobutane in the C_3/C_4 fraction from cracking, however, is insufficient in amount to alkylate the olefins. Additional quantities must be supplied from catalytic reforming, hydrocracking, or natural-gas sources. Under normal HF alkylation conditions, about 1.15 and 1.33 volumes of isobutane are required for reaction with 1 volume of C_4 and C_3 olefins, respectively. The resulting yield of alkylate will be 1.75 and 1.77 volumes per volume of the respective olefins. These factors for the isobutane requirement and alkylate yield vary, depending on operating conditions employed, and perhaps reflect a more meaningful practical description of the overall reaction than statements of chemical mechanisms, which are complex in accounting for the broad spectrum of hydrocarbons found in alkylates. For instance, to account for the appearance of slightly greater amounts of propane and n-butane in the total alkylation-unit products than are brought in with the feed, obviously the principle of hydrogen transfer is involved as a part of the overall reaction mechanism.

HF Alkylation Process.   One of several possible configurations of an HF alkylation process is illustrated in Fig. 7. The principal components of this unit are (1) the reactor, in which intimate contact is brought about between the mixture of C_3 and C_4 olefins, the isobutane, and the HF catalyst and in which the heat of reaction is removed; (2) the acid system, which includes a settler to separate the acid and hydrocarbon phases, provision to recycle acid to the reactor, an acid regenerator, and a stripper to remove acid from the net propane product; and (3) the hydrocarbon fractionation system, which provides a stream rich in isobutane for recycle to the reactor and which yields n-butane and propane as separate products.

Before introduction to the alkylation unit, the feedstocks are treated, primarily to remove sulfur compounds and water. The olefin-containing stream, the HF acid recycle, and a recycle stream that contains a controlled excess of isobutane over that necessary to react with the olefins are fed in various proprietary arrangements to the reactor vessel. In the reactor, the hydrocarbons and acid are intimately contacted to form an emulsion, and herein the reactions occur to produce the alkylate. A substantial heat of reaction is removed in one of several ways. In HF alkylation, cooling water in a heat-exchange-tube bundle within the reactor is commonly used, and thereby controls the emulsion temperature.

The emulsion effluent from the reactor passes to a settler in which the hydrocarbon phase rises and is routed to the fractionation system. The acid phase is recycled to the reactor to maintain the desired ratio of acid to hydrocarbons. A small portion of the recycled acid is diverted to a regenerator in which relatively pure HF is distilled from a minor amount of heavy organic compounds and water. In many units, the acid-regeneration system is operated on an intermittent schedule.

Hydrocarbons leaving the settler are fed to a fractionator that separates an overhead stream rich in isobutane. This stream is recycled to the reactor system to supply isobutane considerably in excess of that required to react with the feed olefins. If the olefin-bearing feed system is deficient in isobutane, an outside saturate stream is brought to the alkylation unit, generally as a blend of isobutane and normal butane. This stream is fed to the isostripper column as shown in Fig. 7. Normal butane that comes to the alkylation unit with the olefin feed or in the saturate stream is rejected from the system by the isostripper column as a side cut near the bottom. In some cases, this n-butane is processed in an isomerization unit from which the effluent mixture of n- and isobutane is brought back to

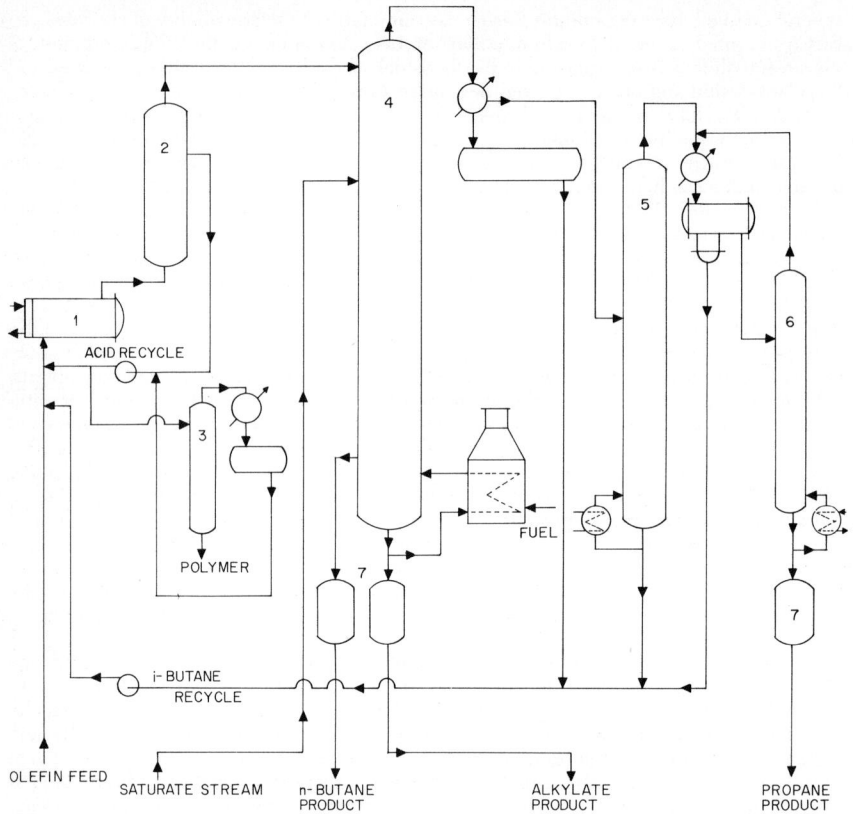

Fig. 7. Hydrofluoric acid alkylation unit: (1) Reactor, (2) Settler, (3) Regenerator, (4) Isostripper, (5) Depropanizer, (6) Hydrofluoric and stripper, (7) Potassium hydroxide treaters. *(Universal Oil Products Company.)*

the isostripper as the saturate stream. The alkylate is produced as a bottom stream by the isostripper column.

When the olefin feed contains propylene and propane, a portion of the isostripper overhead is routed to a depropanizer which separates propane as an overhead stream. The bottoms stream contains butanes and joins the isobutane recycle stream as shown in Fig. 7. Because the propane stream will contain a small amount of dissolved HF, an HF stripper is required to recover the acid so that it may be returned to the reactor section with the isobutane recycle. Propane product is removed from the unit as a bottoms stream from the HF stripper.

HF alkylation processes differ from one another primarily in the reactor-section design: in the manner in which acid and hydrocarbons are brought together; in the mechanics of producing an emulsion; in the design, configuration, internals, and arrangement of the reactor and settler; and in the technique for removing exothermic heats of reaction. Some units separate isobutane recycle of relatively high purity in a center-fed, refluxed fractionator, while other units employ a top-fed, unrefluxed isostripper to provide a recycle of somewhat lower isobutane purity, but also at a lower capital and operating cost.

Sulfuric Acid Alkylation.   The predominant difference between HF and H_2SO_4 alkylation processes, of course, is the type of acid catalyst used. Other differences include the manner of producing the emulsion, with provision for an extended interfacial surface for the reaction; the way in which the reactor section is designed to accept the hydrocarbon feed and acid; and the geometrical relationship between the reactor and the settler. Important differences result from the technique of removing the heat of reaction and of controlling the reactor temperature by means of evaporative refrigeration. In earlier designs, external refrigerant and a tubular exchanger were used. More modern, proprietary designs involve a refrigerated cascade reactor in one instance, and in another, a portion of the reactor effluent is vaporized by pressure reduction to provide cooling for the reactor.

Mixed Olefinic Feeds. In the alkylation of mixed olefinic feeds, both H_2SO_4 and HF processes generally employ fractional distillation to separate propane and butane as separate product streams. Unlike H_2SO_4, HF is slightly soluble in hydrocarbons. Thus, in the HF process, provision is made, as shown by Fig. 7, to strip and recover HF from the hydrocarbon product for return to the reactor section.

Aromatic Alkylation. Although not intended for motor fuels, the refining industry makes a number of petrochemical intermediates by the use of aromatic alkylation with olefins. These applications include (1) alkylation of benzenes with ethylene to produce ethylbenzene, by means of special catalysts; (2) benzene alkylation with propylene, using a catalyst to make cumene; and (3) detergent alkylate production by reacting benzene with tetramer (a selected cut from a wide-boiling-range polymer), or with *n*-olefins of appropriate molecular weights for use in biodegradable-detergent manufacture.

Isomerization. Isomerization can be defined as the rearrangement of the structural configuration of a molecule without changing its molecular weight. Although structural changes of this type occur in other processes, e.g., catalytic reforming and cracking, this description considers only processes in which isomerization is the principal reaction.

In petroleum refining, isomerization processes are used to change the structural configuration of C_4 paraffins (*n*-butane) into isobutane in order to supplement other sources to provide enough isobutane for alkylation with olefins in the production of motor fuel. C_5 and C_6 paraffins are isomerized to the more highly branched structures to improve their antiknock ratings. Isomerization also is applied to a much lesser extent to C_8 aromatic hydrocarbons.

Any significant increase in the production of unleaded motor fuel will accelerate the need for C_5/C_6 isomerization. Isomerization first was used extensively in the early 1940s because of the demand for isobutane in the manufacture of alkylate needed for aviation gasoline. Among the components in motor gasoline, the one having the lowest unleaded octane number is the light, straight-run gasoline, which is composed generally of the C_5 and C_6 paraffins native in the crude. The octane numbers of these native paraffin fractions can be improved markedly by isomerization.

Table 6 lists the octane numbers and normal boiling points of the several paraffin isomers in the C_5 and C_6 fractions arranged in order of their boiling points. A typical, light, straight-run gasoline would contain (in addition to the paraffins listed) minor amounts of cyclopentane, methylcyclopentane, cyclohexane, and benzene and would have a clear Research Method octane rating of 68 to 70. The C_5 paraffin fraction would contain about 60 percent *n*-pentane. The *n*-hexane content of the C_6 paraffin fraction would be about 48 percent, with the high-octane dimethylbutanes representing only 4 percent of the straight-run C_6 paraffin fraction. Thus, the very low octane *n*-paraffins are predominant in these two light, straight-run fractions.

An ideal situation would exist if a catalyst could be found to convert the paraffins

TABLE 6.   Characteristics of C_5 and C_6 Paraffin Isomers

Paraffin isomer	Bp °F	Bp °C	Approximate Research-Method octane number, clear
Isopentane	82.0	27.8	93
n-Pentane	98.0	36.7	62
2,2-Dimethylbutane	121.5	49.7	93
2,3-Dimethylbutane	136.4	58.0	104
2-Methylpentane	140.5	60.3	73
3-Methylpentane	145.9	63.3	74
n-Hexane	155.7	68.7	30

Adapted from Douglas M. Considine (ed.), "Chemical and Process Technology Encyclopedia," McGraw-Hill Book Company, New York, 1974.

completely to the highest-octane-number structures, i.e., to convert, say, to isopentane and 2,3-dimethylbutane. Unfortunately, thermodynamic equilibrium limits the extent of conversion possible, and this extent is highly dependent on the temperature at which the conversion occurs. The equilibrium is more favorable to the appearance of a greater proportion of the more highly branched isomers at low temperatures. Thus, it is of great interest to devise a catalyst that has sufficient activity to provide a high rate of isomerization at a low temperature. In a similar way, low reaction temperatures favor higher proportions of isobutane in the C_4 system at equilibrium. The influence of temperature on these equilibria in the vapor phase is illustrated by Table 7, which reveals that equilibria at low temperatures favor the appearance of the high-octane-number structures of the C_4, C_5, and C_6 paraffins.

With respect to the C_5 and C_6 fractions, the refiner is interested in octane numbers instead of isomer distribution. For the C_5 fraction, the Research Method clear octane

TABLE 7.   Equilibrium Vapor-Phase Compositions at Two Temperatures for C_4, C_5, and C_6 Paraffin Hydrocarbons

Component	Approximate composition, wt %	
	200°F (93°C)	400°F (204°C)
Butanes:		
Isobutane	75	56
n-Butane	25	44
Pentanes:		
Isopentane	86	70
n-Pentane	14	30
Hexanes:		
Methylpentanes	43	48
2,2,-Dimethylbutane	43	29
2,3-Dimethylbutane	9	9
n-Hexane	5	14

Adapted from Douglas M. Considine (ed.), "Chemical and Process Technology Encyclopedia," McGraw-Hill Book Company, New York, 1974.

numbers corresponding to equilibrum compositions at 200°F (93°C) and 400°F (204°C) are 89 and 84, respectively. These octane numbers for the C_6 equilibrum mixtures are 82 and 76 at 200 and 400°F respectively. Thus, an improvement of five to six octane numbers can be realized at equilibrium by dropping the temperature by 200°F in this region.

Catalysts.   The early catalysts used during the 1940s for butane isomerization were of the expendable Friedel-Crafts type, e.g., aluminum chloride on a support or dissolved in molten antimony trichloride. These systems were promoted with HCl and operated in the general range 200–300°F (93–149°C), and the reaction environment was rather corrosive. More recently developed catalysts may be classified as *hydroisomerization* types in that they perform in the presence of a hydrogen atmosphere in order to minimize formation of carbonaceous deposits, which tend to deactivate the catalyst. The newer catalysts are usually a supported noble metal employed in fixed-bed reactors. The catalysts have a service life of several years.

Isomerization Process.   One isomerization process* is shown in Fig. 8. This unit is arranged to process a C_5/C_6 mixture with fractionation facilities to provide for the recycling of both n-pentane and n-hexane. A desulfurized C_5/C_6 blend first is fractionated to remove the native isopentane as a net product. The deisopentanizer bottoms are desiccant-dried before being joined by n-hexane recycle and brought to reaction temperature by heat exchange and suitable preheating. Before entering the reactor, the combined feed stream is joined by hydrogen recycle gas, which functions to suppress catalyst-deposit formation.

The fixed-bed reactor effluent is cooled and passed to a high-pressure separator. Gas from the separator, along with a small quantity of dried makeup hydrogen, is recycled to the reactor. The separator liquid is stabilized as a next step to remove any C_4 and lighter hydrocarbons that may be introduced with the makeup hydrogen, plus a very minor

*UOP Penex unit.

amount of light hydrocarbons formed by hydrocracking in the reactor. Hydrogen dissolved in the separator liquid is also removed by the stabilizer.

The next fractionator in series receives the stabilized liquid, from which it separates an equilibirum isopentane-*n*-pentane mixture that is routed back to the deisopentanizer for separating the isopentane as a net product. Thus, the *n*-pentane content of the feed is converted entirely to isopentane in the flow arrangement shown.

As the final step in the fractionation sequence, the hexane fraction is separated into a dimethylbutanes concentrate as a net overhead product and a *n*-hexane-rich bottoms stream to be recycled for the further isomerization of the *n*-hexane and methylpentanes. With economically practical fractionation, the methylpentanes split between the overhead and bottoms of the deisohexanizer column. For the C_5 fraction, the boiling points of

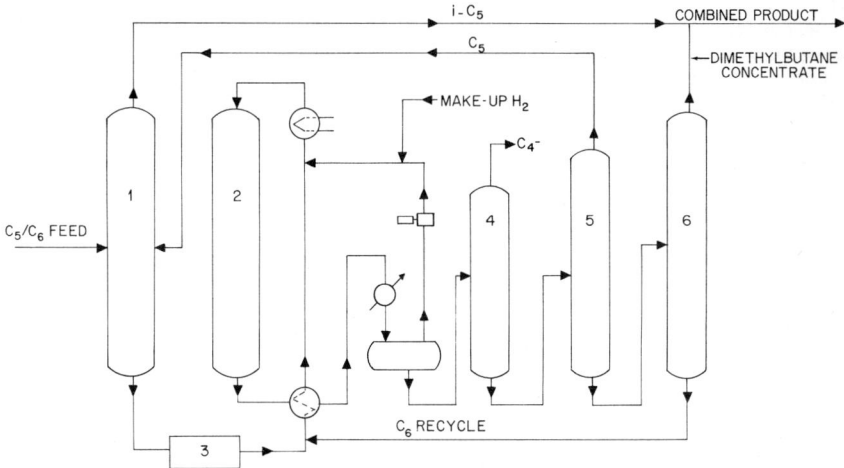

Fig. 8. C_5/C_6 isomerization unit: (1) Deisopentanizer, (2) Reactor, (3) Drier, (4) Stabilizer, (5) Depentanizer, (6) Deisohexanizer. (*UOP Penex Process, Universal Oil Products Company.*)

the two isomers are far enough apart to make a relatively clean split economically feasible. For the C_6 fraction, the greater number of isomers and the bunching of some of their boiling points preclude precision separations in columns having a reasonable number of plates.

Organic chloride promoter is added continuously and is converted to HCl in the reactor, but since the catalyst functions with only parts per million of the promoter, it is not necessary to provide separate facilities to recover and recycle the HCl. The HCl leaves the system by way of the stabilizer overhead product, which is treated with caustic before use as a refinery fuel. Because the system is dry and the concentration of HCl is low, the environment is noncorrosive. Carbon steel construction is permissible.

TRENDS IN PETROLEUM PROCESSING

Technological trends that commenced as early as the 1940s are continuing. These include catalytic cracking, catalytic reforming, alkylation, light hydrocarbon (C_4) isomerization, and distillate hydrotreating. The use of these processes has succeeded in producing larger volumes of higher-quality products in the gasoline, kerosine, and furnace-oil boiling ranges. More recent developments include hydrocracking and heavier hydrocarbon (C_5/C_6) isomerization, as well as process and catalyst improvements to the catalytic cracking and catalytic reforming processes. Processes for stack gas scrubbing and extension of the hydrotreating process to residue desulfurization have undergone extensive development, but some of these methods still require a demonstration of commercial suitability on a large scale.

The marked trend to larger individual processing units continues. Recently built units have capacities of up to five times the size of the largest units of the 1940s. To some extent, the economies resulting from larger and improved processes have offset continuing increases in crude oil, labor, and transportation costs.

The requirement for increasing the yield of light products over the last several years has led to a continually evolving process technology. The demand, as a percent of the market, for gasoline and distillates in the United States grew steadily from 1940 through the mid-1960s, with a steady decline in demand for residual fuel. However, in the mid-1960s the market for residual began to increase, mainly as a result of the environmental factors limiting coal in industrial and utility uses and a growing shortage of natural gas. This trend, coupled with increased demand for liquefied petroleum gas (LPG), petrochemical feed, and other products, led to a reduction in the gasoline and distillate share of the total market. During the entire period, of course, the total demand for petroleum products increased. See Fig. 9. See also article on *Petroleum Requirements* given earlier in this Handbook section.

Size of Facilities.   In a continuing effort to reduce unit capital and operating costs, the petroleum refining industry continues to build operating units and complete grass roots refineries of increasing capacity, utilizing larger single train equipment, based on the principle that it costs less to build and operate one unit double the size of two smaller units. As a result, refinery throughput in the United States has risen by a factor of 240 percent during the 1945–1971 period, during which period the number of operating refineries dropped by 33 percent, increasing the size of the average refinery from 12,400 to 41,500 barrels per day. See Fig. 10. Whereas the size of the average refinery has increased markedly in recent years, indicative of concentrating operations in fewer refineries, the size of individual units and throughput of grass roots refineries appears to be stabilizing, with most of the recent larger grass roots facilities being built in the 140,000- to 160,000-barrel-per-day range. It should be noted that the technology and physical construction

Fig. 9. Trends in the markets for major petroleum products.

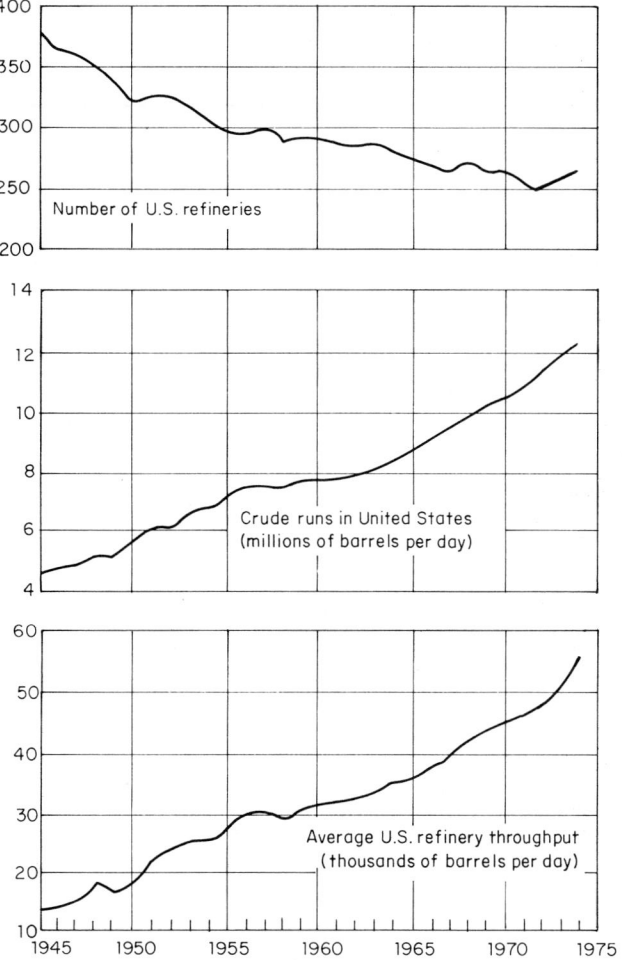

Fig. 10. Trends in the number and size of petroleum refineries.

capabilities exist to allow building single-train 500,000-barrel-per-day refineries at a substantial capital savings over multiple-train refineries. However, the economics of downtime associated with maintenance shutdowns and crude and product distribution logistics, plus operating problems associated with running at reduced throughput, remain to be evaluated. As energy demands rise, it is most likely that single-train refineries will increase in size to the 200,000- or 250,000-barrel-per-day range, but the economics of the 500,000-barrel-per-day single-train refinery remain uncertain.

The trend to larger units in terms of crude and catalytic cracking units is clearly shown in Fig. 11.

Specific Process Trends. In contemplating future new refineries and refinery expansions, an observation of past trends can be helpful. Some of the major petroleum processes, previously described in this article from a technological standpoint, are further described from a more generalized standpoint in the following paragraphs, and their growth and/or decline curves are given in Fig. 12.

Distillation. There is an overall trend toward larger, more highly automated crude distillation units of increasing complexity. Distillation capacity has more than doubled since World War II. The capacity of the average atmospheric distillation unit built in 1945

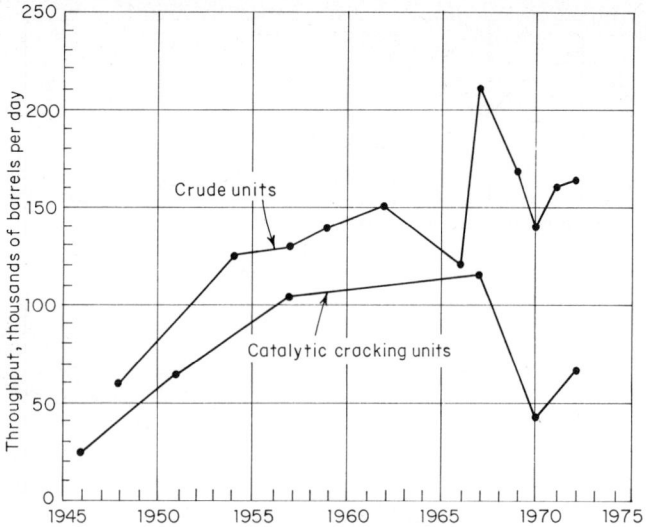

Fig. 11. Trends in the size of crude units and catalytic-cracking units.

was about 20,000 barrels per day. Several units are now in operation with capacities in excess of 150,000 barrels per day. During the last 30 years, both process and mechanical design improvements, notably tray designs, have resulted in better separation efficiency and increased operating flexibility.

With high-boiling stocks, significant decomposition by thermal cracking begins at about 800°F (427°C). For this reason, heavy crude-oil fractions must be distilled at reduced pressure. The economic incentive for converting more of the residual crude fraction to gasoline, heating oil, and jet fuel has promoted increasing use of vacuum distillation. Since 1945, vacuum distillation has been widely used to augment the yield of catalytic cracking feedstocks. Total vacuum distillation capacity in 1974 corresponded to 37 percent of atmospheric distillation capacity. Again, the trend has been toward larger-capacity vacuum distillation towers. Units of over 50,000 barrels per day capacity today compare with units of less than 10,000 barrels per day in 1945. Design improvements include better fractionation, improved control of recycle streams by way of better reflux operation, and more effective equipment for minimizing entrainment of heavy asphalt and metal contaminants. The control of entrainment in vacuum operations has become increasingly important because of the heavier feedstocks being processed and because of the importance of minimizing contaminants in heavy-distillate catalytic cracking feedstocks, a small fraction of which can poison catalysts and reduce gasoline yields in downstream cracking units. In late 1974, atmospheric distillation capacity in the United States was about 15 million barrels per day; vacuum distillation capacity, about 5.5 million barrels per day.

Thermal Cracking. Before World War II thermal cracking of gas oil was a major source of gasoline. This process began to decline in use as catalytic cracking was commercialized. In 1945 the thermal cracking process represented 36 percent of the crude throughput, whereas in the mid-1970s the percentage had dropped below 1 percent.

Visbreaking. This is a mild thermal cracking process, but is usually considered separately, since it is confined to lowering the viscosities and pour points of crude oil residua and thus to decrease the ultimate yield of heavy marketable fuel oil. In 1955, about 7.5 percent of the crude processed was subjected to visbreaking; by the early 1970s, the percentage had dropped below 2 percent. For the near term, it appears that visbreaking will continue to decline. However, if the economic climate should be such that it would encourage domestic refining to supply an increasing demand for low-sulfur heavy fuels, then visbreaking could assume a revival of importance.

Coking. This is a well-established technique for decreasing residual fuel yield. Delayed coking, which is still the predominant process, was commercialized in 1930. Continuous operation is achieved by using two or more drums in rotation, the cycle of each being on the order of one to three days. The continuous fluid coking process was announced in 1953, and the first commercial unit was brought onstream by the end of 1954. In this latter process, the heavy feed is cracked by contacting it with a fluidized bed of coke particles at a relatively high temperature and essentially at atmospheric pressure. The latest development in coking is a process that combines fluid coking with coke gasification to convert about 98 percent of normal crude residuum into liquid and gaseous products. This process has its greatest advantage in the processing of high-sulfur, high-metal-content crude oils. The liquid products from this process are low in metals and can be more readily hydrotreated to produce high-quality fuel oils or feedstocks. Approximately 95 percent of the total sulfur in the coker feed can be recovered as elemental sulfur through conventional sulfur recovery processes. The process is most likely to be used in future processing schemes where high-sulfur crude is the primary feedstock and high-cost fuel is an incentive to convert low-quality coke to a usable fuel.

Coking yields vary widely with the nature of the feed and are also influenced by the amount of recycling and other conditions. The maximum yield of catalytic cracker feed is

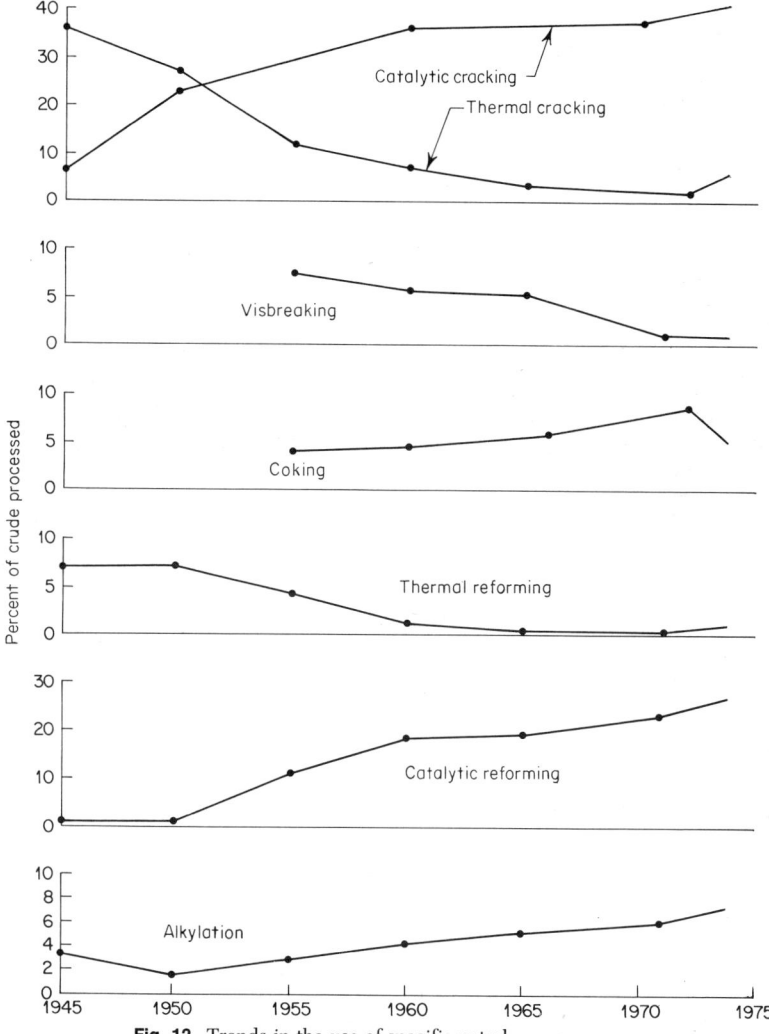

Fig. 12. Trends in the use of specific petroleum processes.

obtained by operating with little or no recycling and at low pressure, but the gas oil thus produced has a higher end point. The combination of vacuum distillation and coking of vacuum residue produces materials of higher quality as compared with the direct coking of an atmospheric residuum. This procedure results in increased yields of better-quality catalytic cracker feed and also minimizes the yield of coker naphtha, which is of poor quality.

Approximately 25 percent of all petroleum coke produced is utilized as fuel. The other major uses are for the manufacture of carbon electrodes employed in the metallurgical industries. In the last 17 years, coking capacity had more than doubled because of the lack of demand for residual fuel oil, a condition fostered by the abundant supply of inexpensive natural gas. Coke production had also increased faster than coker capacity because of the increased use of vacuum distillation for increasing gasoline production and the resulting production of heavier coker feedstocks. These trends are likely to be reversed because of the expected increase in demand for heavy fuel oils brought about by less abundant and higher-cost natural gas, also possibly coupled with restrictions on the use of high-sulfur coal.

Catalytic Cracking.   This process has continued to be the most important conversion process within the refining industry in terms of capacity and gasoline production. During the 1945–1972 period, catalytic cracking capacity in the United States increased steadily to a level of about 4.6 million barrels per stream-day, based on fresh feed. However, in refineries where a major product objective is heavy fuels, catalytic cracking, like other heavy-oil conversions processes, will decline in relative importance. The fluid catalytic cracking process also has some unique atmospheric pollution problems, i.e., catalyst and carbon monoxide discharges from the regenerator. Some refineries have burned the carbon monoxide in an external boiler or combustion chamber and installed tertiary catalyst collection systems external to the regenerator, thus eliminating the need for external carbon monoxide combustion equipment. This requires changes in regenerator metallurgy as well as additional heat-recovery equipment.

Sulfur dioxide emissions from catalytic cracking flue gases can be reduced by hydrodesulfurization of the cracking unit feed. Catalytic cracking feed hydrotreating has been practiced to improve the yield of high demand products instead of to control sulfur emissions. This secondary advantage will make hydrotreating even more important.

A major objective in the modern United States refinery has been and continues to be conversion of the heavy end of the barrel into gasoline and other distillate products. Therefore, the trend in catalytic cracking has been to feed even heavier, less valuable gas oils, and to operate at higher temperatures to achieve increased conversion and improvement of gasoline octanes.

Of the many important technological and mechanical advances made in catalytic cracking since the end of World War II, the development of zeolitic cracking catalysts is outstanding. The greater activity and selectivity of these catalysts give higher yields than were obtainable with natural clay or synthetic silica-alumina catalysts. Only about 10 percent of the units in operation still employ the older catalysts.

Hydrocracking.   This is one of the oldest (developed in Germany in 1927) and yet one of the most versatile of refinery processes. In terms of contributing to refinery flexibility, it may be noted from Table 8 that hydrocracking will handle a variety of charge stocks to produce gasoline, jet fuels, and distillates. The heavy emphasis on hydrocracking capacity in recent years has been to maintain a balance between the production of gasoline and middle distillate fuels. Gasoline demand was increasing at a rate exceeding that of middle distillate, and hydrocracking provided a valuable means of converting the distillates to gasoline. Additionally, some refinery expansion projects utilized hydrocracking as a supplement to fluid catalytic cracking by including various gas oil fractions in the hydrocracker feed stream, with resultant conversion to gasoline and/or distillate or jet fuel. This strategy increased fluid catalytic cracking capacity as well, since the poorest-quality feedstocks for fluid cracking were actually the best stocks for hydrocracking. Removing the poor-quality stocks from fluid cracking feed, therefore, increased the processing capacity for high-quality stocks. Hydrocracking units may be designed to provide variable product yields ranging from 50 to 60 percent conversion to gasoline, with the remaining product being high-quality distillate or jet fuel—up to 100 percent conversion to gasoline and light hydrocarbons. The design allows the refiner to provide a wider range in the final product slate, but does increase manufacturing cost for the overall operation.

The most promising area for future development in the hydrocracking process appears to be residuum feedstocks. Conversion of high-sulfur, high metal-content, heavy-boiling fractions to a wide range of low-sulfur, high-quality fuel-oil products will be a persistent goal for refinery processing. Unfortunately, metals contaminants of catalyst are presently an obstacle to the utilization of hydrocracking as an economical solution to this problem.

Thermal Reforming.   At the end of World War II, octane improvement of virgin heavy napththa depended primarily on thermal reforming, a severe pyrolytic process introduced in the early 1930s. Octane improvement was limited to about an 80 Research octane number, and yields were relatively low. The product gasoline from this process is inadequate for modern gasoline production. Thermal reforming has been rendered obsolete by catalytic reforming.

Catalytic Reforming.   This process has undergone many improvements since it was first commercialized in 1940. Moving-bed techniques were introduced, and significant improvements have been made in catalysts, including the introduction in 1969 of bimetallic catalysts. Continuous catalyst regeneration units more recently have been introduced

TABLE 8.   Some Applications of Hydrocracking

Charge stock	Products
Naphtha	Propane and butane (LPG)
Kerosine	Gasoline
Straight-run Diesel	Gasoline and/or jet fuel
Atmospheric gas oil	Gasoline, jet fuel and/or distillates
Natural gas condensates	Gasoline
Vacuum gas oil	Gasoline, jet fuel and/or distillates
Propane deasphalted gas oil	Gasoline, jet fuel and/or distillates
Catalytically cracked heavy-cycle oil	Gasoline and/or distillates
Catalytically cracked light-cycle oil	Gasoline
Coker distillate	Gasoline
Coker heavy gas oil	Gasoline and/or distillates
Residuum	Gasoline, jet fuel and/or distillates

in place of cyclic regenerative processes. The hydrogen by-product of this process is becoming increasingly attractive. Fortunately, the trend toward higher-severity operation, combined with the use of improved catalyst, makes more hydrogen available. With the prospect of lower lead levels, reforming capacity should grow at least as fast as crude capacity with perhaps an upsurge occurring if there is a need to reform naphtha from catalytic cracking.

Polymerization.   Catalytic polymerization reached a peak capacity about 1960, but its use has decreased steadily and rapidly since that time. The process represented one means of converting unsaturated C_3 and C_4 hydrocarbons into a liquid of good research octane quality in the gasoline boiling range. The poor motor octane of the polymer product and the improved quality and yields achieved from alkylation of the C_3/C_4 fraction have combined to render catalytic polymerization an obsolete process.

Alkylation.   Alkylate is a premium-quality gasoline component. It is low in sulfur and gum contents and has excellent stability, high energy content, and high octane number. Although trimethylpentanes predominate in butane alkylate, and dimethylpentanes in propene alkylate, the product contains many different isoparaffins ranging from C_5 to C_{10} and higher, regardless of the olefin feed. Newer designed units, improved control of operations, and various additives have made it possible to reduce acid comsumption significantly in the sulfuric acid units. Improvements have also been realized in both the sulfuric acid and hydrofluoric acid alkylation processes—in alkylate yield and quality.

Alkylation capacity will continue to be installed in light-fuels refineries where catalytic cracking is the major heavy-oil conversion process. In the mid-1970s, the combined capacity for both sulfuric acid and hydrofluoric acid units is about 869 thousand barrels per stream-day. In the last several years, there has been a swing toward the hydrofluoric acid process, possibly accounted for by (1) lower investment and operating costs, (2) more flexibility in olefins processed, and (3) lower acid consumption.

Continued emphasis will be placed on propylene and butylene alkylation because of favorable economic incentives in both leaded and unleaded gasolines.

Isomerization.   Many of the isomerization units were shut down when the military need for aviation gasoline declined after World War II. Interest in isomerization was revived in the middle 1950s, both as a means of augmenting the supply of isobutane for alkylation and as a way of improving the octane number of light ends in straight-run naphtha. Refinements were made in aluminum chloride processes to improve operations, extend catalyst life, and decrease cost. Several new processes were developed in which the paraffinic feed and recycle hydrogen were passed through fixed beds of solid catalyst at pressures of several hundred pounds per square inch, in a manner similar to that in catalytic reforming. The new processes avoid corrosion and catalyst handling and disposal problems inherent in the aluminum chloride catalyst processes. Operating temperatures have been lowered.

Capacity for C_5 and C_5/C_6 isomererization is less than 1 percent of crude capacity. Inasmuch as light straight-run gasoline has a very high lead susceptibility, isomerization has not been economically attractive as compared with tetraethyllead. With the reduced lead levels projected, the process may gain wider application. If lead-free gasoline represents a sizable portion of the market by 1980, isomerization capacity should increase. An important advantage of isomerates is their very low sensitivity. With lower engine compression ratios, Motor octane number becomes a more important factor in road performance than Research octane number. This, together with any restriction in the aromatic content of gasoline, would stimulate activity in isomerization of pentanes and hexanes.

Hydrotreating.   In the mid 1950s, under the general name of hydrotreating, a number of hydrogenation processes became popular with refiners. The principal ‚processes involved were hydrodesulfurization, saturation of diolefins, and hydrodealkylation. In 1957 installed capacity of hydrotreaters represented 8 percent of refining capacity in the United States. Current applications for hydrotreating cover a wide range, including residual desulfurization, heavy gas oil desulfurization, middle distillate desulfurization, residual visbreaking utilizing hydrogen, catalytic cracking fresh feed and recycle feed pretreatment (sometimes referred to as hydrorefining), pretreatment of catalytic reformer feeds, naphtha desulfurizing, naphtha treating for olefin or aromatic saturation, light distillate treating, and lube oil treatment. In 1974 combined hydrotreating capacity reached nearly 6.9 million barrels per stream-day, nearly 36 percent of crude capacity. This has probably been the most rapidly growing refining process and is likely to maintain that status, particularly in view of the restrictions on sulfur content of fuel oils coupled with rising demand. See following article on Hydrodesulfurization.

Hydrogen Production.   Even in the late 1950s hydrogen was a surplus by-product from catalytic reforming. Hydrotreating consumed some of the hydrogen, but in 1961 about 50 percent of the hydrogen produced from reforming was burned as fuel. Currently, less than 25 percent of the by-product hydrogen is burned as fuel. Because of numerous installations of hydrocracking, hydrodealkylation (such as toluene to benzene), ammonia manufacture and other petrochemical uses, hydrogen production has become a common refinery process. The major sources of hydrogen are catalytic reforming, steam reforming, and partial oxidation. The refinery normally uses all possible hydrogen from catalytic reforming before installing any process to manufacture hydrogen as a primary product.

Steam reforming is the principal means for producing hydrogen in the refinery. Hydrocarbons from natural gas, naphtha, etc., are mixed with steam in excess of the stoichiometric quantities. The mixture then is reformed over a nickel catalyst packed inside tubes in a furnace. Plant sizes vary from 0.1 to over 100 million cubic feet per day. Pressures range from 50 to 400 psi and can be extended to 500 psi. Final product purity ranges from 90 to 98 percent hydrogen, with methane as the major impurity. Present and future requirements for low-sulfur fuel oils will require increased distillate hydrotreating and either residuum or reduced crude desulfurization. Therefore, increase hydrogen requirements will result. In some cases, even refineries without hydrocracking will require hydrogen plants to provide sufficient hydrogen for operating the extensive hydrotreating and hydrodesulfurization facilities necessary for producing low-sulfur fuels.

Acknowledgment.   The editors extend their appreciation to the National Petroleum Council, the American Petroleum Institute, and others for the basic information provided for presentation here. Descriptions of specific processes given early in this description were adapted from various entries by Melvin J. Sterba, formerly Assistant to the Vice-president, Universal Oil Products Company, Des Plaines, Illinois, which appear in the "Chemical and Process Technology Encyclopedia," D. M. Considine (ed.), McGraw-Hill, New York, 1974.

Hydrodesulfurization of Petroleum Crude Oil Fractions and Petroleum Products

BY KENNETH H. HASTINGS* AND ROGER P. VAN DRIESEN†

Sulfur removal as practiced in petroleum refineries can take the form of chemical removal (acid treating or sweetening), concentration in refinery products (high sulfur coke) or hydrodesulfurization. The purpose of this description is to review how hydrodesulfurization is used to remove sulfur compounds from petroleum crude oils or fractions, and to comment on how molecular structure and organometals content affect sulfur removal. Sulfur removal is practiced in petroleum refineries for four reasons: Sulfur compounds (1) lower the quality of some refinery products, (2) pollute the atmosphere as sulfur oxides when they are burned, (3) corrode refinery process equipment, and (4) poison the catalysts used in some refinery processes.

Hydrodesulfurization is a catalytic process whereby a hydrocarbon feedstock and hydrogen are passed through a catalyst bed at elevated temperatures and pressures. Some of the sulfur atoms attached to hydrocarbon molecules react with hydrogen on the surface of the catalyst to form hydrogen sulfide. Thermodynamic equilibrium calculations show that these reactions could be driven to almost 100 percent completion, but economic considerations normally limit the commercial application of hydrodesulfurization.

DEFINITION OF TERMS

The reaction of molecular hydrogen with a hydrocarbon, such as petroleum is called hydrogenation (Ref. 1). As shown in Table 1, the field of hydrogenation is further subdivided into more specific types of reactions. *Saturation* is the addition of hydrogen to aromatic compounds or compounds with double or triple bonds without bond cleavage. When hydrogen reacts with bond cleavage, the reaction is called *hydrogenolysis*. Hydrogenolysis reactions with bond cleavage of carbon-sulfur or sulfur-sulfur bonds are called *hydrodesulfurization*. Hydrogenolysis reactions with cleavage of carbon-carbon bonds are referred to as *hydrocracking*. Cleavage of carbon-carbon, carbon-sulfur, and sulfur-sulfur bonds at high temperatures in the absence of hydrogen is also found in petroleum refineries. This type of bond cleavage is called *cracking*. Two examples of cracking are catalytic cracking and visbreaking (noncatalytic).

Hydrogenolysis reactions are applied in petroleum processes for the purpose of hydro-desulfurization, hydrocracking, or a combination of both. The difference between these

TABLE 1. Mechanisms of Hydrogenation

Saturation

| Benzene | Hydrogen | Cyclohexane |

$$\text{Benzene} + 3H_2 \longrightarrow \text{Cyclohexane}$$

Hydrodesulfurization

$$R-SH + H_2 \longrightarrow RH + H_2S$$
$$R-S-S-R' + 3H_2 \longrightarrow RH + R'H + 2H_2S$$

Hydrocracking

$$R-R' + H_2 \longrightarrow RH + R'H$$

*Manager, Synthetic Gas Projects, †Vice-president, Cities Service Research and Development Company, Cranbury, New Jersey.

processes is found in the choice of catalyst and process conditions selected to achieve the desired result. Hydrocracking will be covered here only as it relates to hydrodesulfurization processes (i.e., vacuum gas oil or residuum hydrodesulfurization).

PROPERTIES OF PETROLEUM CRUDE OILS

Petroleum crude oils are composed of hundreds of different organosulfur compounds dispersed with thousands of different hydrocarbon compounds. The major elements in petroleum crude oil are carbon and hydrogen. Khafji crude oil (Ref. 2) has been chosen as an example, and an analysis is shown in Table 2. Khafji crude oil contains 82.2 weight percent carbon and 11.5 weight percent hydrogen. The remaining 6.3 weight percent of Khafji crude oil is composed of oxygen, sulfur, nitrogen, and trace metals, such as aluminum, barium, calcium, chromium, cobalt, copper, iron, lead, magnesium, manganese, molybdenum, nickel, silicon, strontium, tin, titanium, vanadium, and zinc. The major metal contaminants are nickel and vanadium. Table 2 shows how properties vary from fraction to fraction in a crude oil. Low-boiling fractions such as naphthas are low in density, viscosity, and sulfur. Naphthas contain no measurable metals content to poison desulfurization catalysts. This is why naphthas with low concentrations of easy-to-remove sulfur compounds require only mild severity process conditions for hydrodesulfurization. High-boiling-point fractions such as vacuum residuums are high in density, viscosity, and sulfur content. They often contain appreciable quantities of vanadium and nickel. The vanadium and nickel deposit on the catalyst, lowering the catalyst activity during hydrodesulfurization. This is why severe process conditions are required in order to maintain the catalyst at a high enough activity to be useful in residuum hydrodesulfurization.

Crude oils also contain water and salt. The salt in crude oil is reduced to an acceptable level by mixing the crude with water and separating the crude/water mixture in electrostatic settling tanks. The excess water is then removed from the crude in an atmospheric distillation tower.

Examples of Petroleum Sulfur Compounds.   Sulfur compounds (Refs. 3, 4) have different structures and molecular weights. They react at various rates of reaction to produce H_2S and different hydrocarbon products. During the process of hydrodesulfurization, nonsulfur-containing molecules also undergo reaction. Oxygenated hydrocarbons react with hydrogen at a faster rate than organosulfur compounds, and nitrogen-containing compounds react at a slower rate. Table 3 contains a partial list of some of the types of sulfur compounds that exist in crude oils. Besides this list, crude oils also contain elemental sulfur, dissolved hydrogen sulfide, benzonaphthothiophenes, and many other classes of sulfur compounds. Petroleum crude oils contain hundreds of sulfur compounds of which only 200 plus compounds have been identified. The easier-to-remove (more reactive) sulfur compounds are found in the distillate fractions. The less-easy-to-remove (less reactive) cyclic sulfur compounds are found in the residue portion of crude oils. The list of sulfur compounds in Table 3 is shown in order of decreasing reactivity in hydrodesulfurization.

TYPICAL HYDRODESULFURIZATION PROCESS CONFIGURATION

All hydrodesulfurization processes react hydrogen with a hydrocarbon feedstock to produce H_2S and a desulfurized hydrocarbon product. The basic flow sheet for a typical hydrodesulfurization process is shown in Fig. 1. Liquid petroleum feedstock is first pumped up to a pressure which is slightly higher than that of the reactor section. The pressurized feedstock is then mixed with hot recycle gas and preheated to reactor inlet conditions. The hot feedstock and recycle gas are passed over catalyst in the reactor section at temperatures 550–850°F (288–454°C) and pressures 150–3000 psig. The reactor effluent is then cooled by heat exchange, and desulfurized liquid hydrocarbon product and recycle gas are separated at a pressure slightly below that of the reactor section. The recycle gas is then scrubbed and/or purged of the H_2S and light hydrocarbon gases, mixed with fresh hydrogen makeup, compressed to a pressure above that of the reactor section, and mixed with liquid hydrocarbon feedstock.

The recycle gas scheme is used in the hydrodesulfurization process to minimize physical losses of expensive hydrogen. Hydrodesulfurization reactions require a high

TABLE 2. Properties of Khafji Crude Oil

Boiling range, °F	Volume %	Gravity,* °API	Kinematic viscosity, centistokes		Carbon residue, Ramsbottom, wt %	Sulfur, wt %	Vanadium, WPPM	Nickel, WPPM
			At 100°F	At 210°F				
Straight-run light								
Naphtha IBP–185	7.3	92.3	.....	.....	...	0.03	.....	.....
Naphtha 185–360	14.1	59.8	.....	.....	...	0.04	.....	.....
Kerosine 360–500	10.9	45.3	1.5	.....	...	0.32	.....	.....
Light gas oil 500–700	17.0	33.3	4.6	1.6	0.08	1.65	0.1	0.2
Vacuum gas oil 700–1000	23.7	21.0	87.0	8.0	0.16	2.60	0.2	0.2
Vacuum residuum 1000–EP	27.0	3.5	......	4,940	21.7	5.71	128	47
Whole crude	100.0	29.0	19.2	13.2	6.3	2.90	41	15
Atmospheric residuum 700–EP	50.7	11.2	10,360	147.4	12.6	4.74	71	26

*Specific gravity = 141.5/(°API + 131.5).

TABLE 3. Examples of Sulfur-containing Hydrocarbons in Petroleum Crude Oil

Name	Structure	Typical reaction
Thiols (mercaptans)	$R–SH$	$R–SH + H_2 \longrightarrow RH + H_2S$
Disulfides	$R–S–S–R'$	$R–S–S–R' + 3H_2 \longrightarrow RH + R'H + 2H_2S$
Sulfides	$R–S–R'$	$R–S–R' + 2H_2 \longrightarrow RH + R'H + H_2S$
Thiophenes		$+ 4H_2 \longrightarrow n\text{-}C_4H_{10} + H_2S$
Benzothiophenes		$+ 3H_2 \longrightarrow CH_3CH_2\text{–} $ $ + H_2S$
Dibenzothiophenes		$+ 2H_2 \longrightarrow$ $ + H_2S$

hydrogen partial pressure in the gas phase to maintain high desulfurization reaction rates and to suppress carbon laydown. The high hydrogen partial pressure is maintained by supplying hydrogen to the reactors at several times the chemical hydrogen consumption rate. The majority of the unreacted hydrogen is cooled to remove hydrocarbons, recovered in the separator and recycled for further utilization. Hydrogen is physically lost in the process by solubility in the desulfurized liquid hydrocarbon product, and from losses during the scrubbing or purging out of H_2S and light hydrocarbon gases from the recycle gas.

In order to maintain a high hydrogen partial pressure in the reactor, the H_2S and light hydrocarbon gases that are produced in the reaction section must be scrubbed or purged from the recycle gas. A lean oil absorption system is one choice for scrubbing light hydrocarbons and H_2S out of the recycle gas. A monoethanolamine (MEA) or equivalent

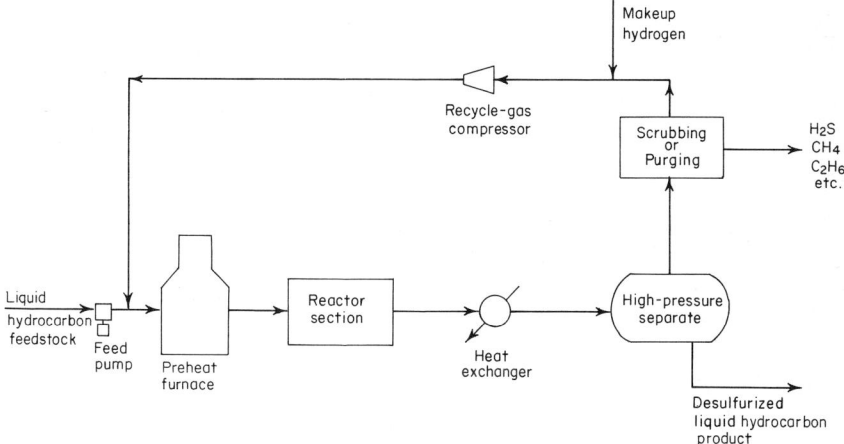

Fig. 1. Typical hydrodesulfurization process.

scrubbing system may also be employed for scrubbing H_2S out of the recycle gas. If neither of these systems is utilized, then purging of part of the recycle gas may be used. For distillate hydrotreating where very little methane is made, a gaseous purge is not necessary. The solubility of methane and inerts in the liquid product is sufficient to purge the system.

Makeup hydrogen for hydrodesulfurization would typically come from a hydrogen plant or as a by-product from catalytic reforming. In catalytic reforming, paraffinic naphthas are converted to aromatics by dehydrocyclization, and naphthenic naphthas are converted to aromatics by dehydrogenation. These reactions produce a by-product hydrogen with an 85 to 95 volume percent purity. A hydrogen plant would typically use steam/hydrocarbon reforming to make hydrogen with 95+ volume percent purity. The hydrocarbon feedstocks normally employed in steam reforming are methane (natural gas), propane/butane (LPG), or naphthas.

DOWNFLOW FIXED-BED REACTOR DESIGN

The most common reactor design used in hydrodesulfurization is the fixed-bed reactor design. The feed enters at the top of the reactor and the product leaves at the bottom of the reactor (Fig. 2). The catalyst remains in a stationary position with hydrogen and petroleum feedstock passing in a downflow direction through the bed of catalyst. The hydrodesulfurization reaction is exothermic, and the temperature rises from the inlet to the outlet of each catalyst bed. With a high hydrogen consumption and subsequent large temperature rise, the reaction mixture is quenched with cold recycle gas at intermediate points in the reactor system. This is done by dividing the catalyst charge into a series of catalyst beds. The effluent from each catalyst bed is quenched with cold recycle gas to the inlet temperature of the next catalyst bed. This sequence continues from catalyst bed to catalyst bed until the oil passes through the reactor or in some cases a series of reactors.

At a fixed set of process conditions, the catalyst charge would lose activity and the sulfur content of the product would increase with time. The sulfur content of the product is maintained at a constant level by slowly increasing the bed inlet temperature which slowly increases the average catalyst bed temperature. The bed inlet temperature is controlled to maintain constant catalyst activity. The bed inlet temperature at the beginning

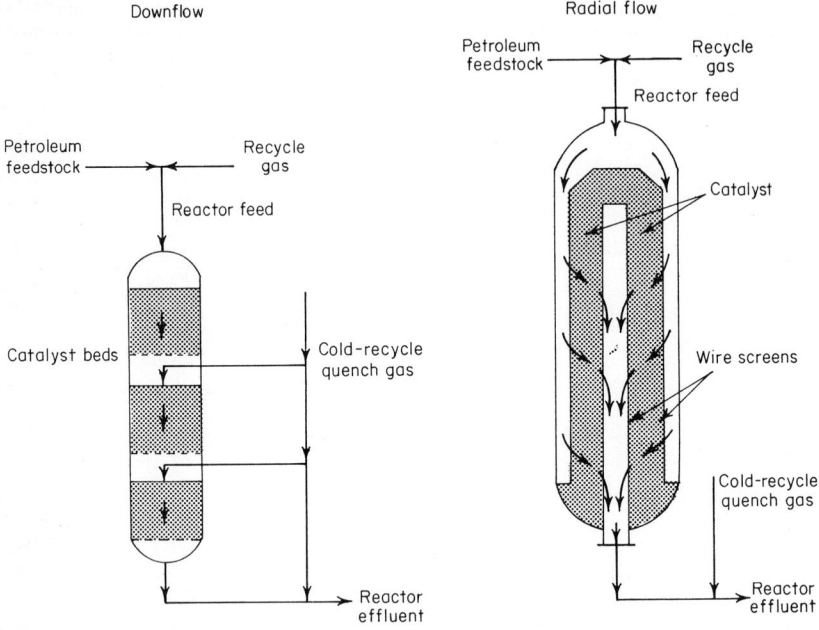

Fig. 2. Fixed bed hydrodesulfurization reactor design.

of the run is referred to as the *start-of-run temperature* and the bed inlet temperature at the end of the run is called the *end-of-run temperature.*

Fixed-bed reactors are mathematically modeled as plug-flow reactors with very little backmixing in the catalyst beds. The first catalyst bed is poisoned with vanadium and nickel at the inlet to the bed. As the catalyst is poisoned in the front of the bed, the active catalyst zone moves down the bed. At end of run the catalyst in each bed is either dumped or regenerated in situ in the reactor. For in-situ regeneration one method is to burn the carbon off the catalyst with low concentrations of oxygen in nitrogen. This is accomplished by injecting small quantities of air into large quantities of inert recycle gas (mostly nitrogen).

An in-situ method is to use steam in place of inert gas. The purpose of the high inert content of regeneration gas is to dissipate the heat of combustion and to maintain the catalyst particles below temperatures where the catalyst would sinter and lose catalyst surface area. After in-situ catalyst regeneration the reactors are opened and inspected, and the high-metal-content catalyst layer at the inlet to the first catalyst bed may be discarded and replaced with fresh catalyst. The catalyst loses activity after a series of regenerations. Then it is necessary to replace the complete charge catalyst. With very high metal content feedstocks such as residuums it is necessary to replace the catalyst instead of regenerating it. This is due to the fact that the metal contaminates cannot be removed by economical means during rapid regeneration. Moreover, the heavy metals have been reported to interfere with the combustion of carbon and sulfur so as to form sulfates that have a permanent poisoning effect.

Fixed-bed hydrodesulfurization units are generally used for distillate hydrodesulfurization. They are also employed for residuum hydrodesulfurization, but require special precautions in processing. The residuums must undergo two-stage electrostatic desalting so that salt deposits do not plug up the inlet to the first catalyst bed. The residuums must be low in vanadium and nickel content so as not to plug up the beds with metal deposits.

During the operation of a fixed-bed reactor, contaminants entering with fresh feed are filtered out and fill the voids between catalyst particles in the bed. The buildup of contaminants in the bed can result in channeling of reactants through the bed, which reduces hydrodesulfurization efficiency. As the flow pattern becomes distorted or restricted, the pressure drop through the catalyst bed increases. If the pressure drop becomes high enough, physical damage to the reactor internals can result. When high-pressure drops are observed through any portion of the reactor, the unit is shut down, and the catalyst bed is skimmed and refilled.

RADIAL-FLOW FIXED-BED REACTOR DESIGN

The radial-flow reactor design is a special type of fixed-bed reactor. As shown in Fig. 2, fresh feed enters the top of the reactor and flows downward around the outside of the catalyst bed. The feed then flows in a radial direction through the catalyst bed to a vertical standpipe where it is collected and transported in a downward axial direction through the bottom of the reactor. This type of reactor is normally used in series with other radial-flow reactors. Its main advantage is very low pressure drop. A radial-flow fixed-bed reactor has a larger catalyst cross-sectional area and a shorter bed depth than an equivalent downflow reactor. The shorter bed depth leads to a smaller pressure drop across the catalyst bed, and the larger cross-sectional area of the bed is more resistant to pressure drop increases caused by contaminants plugging the voids between catalyst particles. The disadvantages of a radial-flow design are:

1. More chance of hot spots in the catalyst bed
2. More difficult to skim catalyst from the bed to remove contaminants on turnaround
3. More expensive reactor design per unit volume of catalyst bed

The application of radial-flow reactors is generally limited to gas-phase hydrodesulfurization of low-boiling feedstocks such as naphthas and kerosines.

UPFLOW EXPANDED-BED REACTOR DESIGN

Expanded-bed reactors are commercially used for very heavy feedstocks and for dirty feedstocks having extraneous fine solids material. They operate in such a way that the catalyst is in an expanded state so that the extraneous solids pass through the catalyst bed without plugging. They operate in an isothermal fashion which conveniently handles the exothermic heat of reaction associated with hydrogen consumption. The heat of reaction is dissipated in raising the temperature of fresh feed to that of the reactor contents. Since the catalyst is in an expanded state of motion, it is possible to treat it as a fluid and to withdraw and add catalyst during operation. Commercial units do not have to be shut down for catalyst replacement. This makes the ebullated reactor system applicable to the high metal feedstocks which rapidly poison catalyst.

Expanded beds of catalyst (Fig. 3) are referred to as particulate fluidized. This means that the petroleum feedstock and hydrogen·flow upward through an expanded bed of catalyst with each catalyst particle in independent motion. The catalyst migrates throughout the entire catalyst bed. Expanded-bed reactors are mathematically modeled as back-mix reactors with the entire catalyst bed at one uniform temperature. Spent catalyst may be withdrawn with the expanded-bed reactor in operation and replaced with fresh catalyst on a daily basis. Daily catalyst addition and withdrawal eliminates the need for costly shutdowns to change out catalyst and also results in a constant-equilibrium catalyst activity and product quality. The catalyst is withdrawn daily with a vanadium, nickel, salt and carbon content which is representative on a macro scale of what is found throughout the entire reactor. On the micro scale individual catalyst particles have ages from that of fresh catalyst to as old as the start of run in the unit, but the catalyst particles of each age group are so well dispersed in the reactor that the reactor contents appear uniform on a macro scale.

Referring again to the expanded-bed reactor in Fig. 3, petroleum feedstock and recycle gas enter the bottom of the reactor, pass up through the expanded catalyst bed and leave from the top of the reactor. Commercial expanded-bed reactors normally operate with $\frac{1}{32}$-inch extrudate catalysts. With extrudate catalysts of this size the upward liquid velocity based on fresh feedstock is not sufficient to keep the catalyst particles in an expanded state. Therefore, for each part of the fresh feed several parts of product oil are taken from the top of the reactor, recycled internally through a large vertical pipe to the

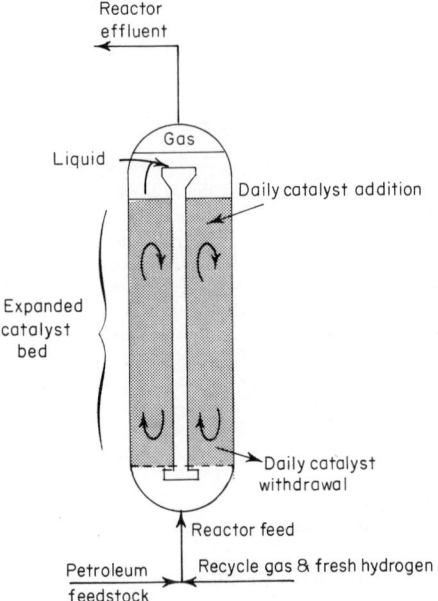

Fig. 3. Expanded-bed hydrodesulfurization reactor design.

bottom of the reactor, and pumped back up through the expanded catalyst bed. The amount of catalyst bed expansion is controlled by the recycle of product oil back up through the catalyst bed.

The expansion and turbulence of gas passing through the expanded catalyst bed are sufficient to cause almost complete random motion in the bed (particulate fluidized). This effect produces the isothermal operation. However, it also causes almost complete back-mixing. Consequently, in order to effect near complete sulfur removal, over 75 percent, it is necessary to operate with two or more reactors in series. This ability to operate at a single temperature throughout the reactor or reactors and to operate at a selected optimum temperature instead of at an increasing temperature from the start to end of run results in more effective use of the reactor and catalyst contents. When all these factors are put together for a given catalyst, i.e., isothermal, fixed temperature throughout run, backmixing, daily catalyst addition, and constant product quality, the reactor size required for an expanded bed is often smaller than that required for a fixed bed to make the same product goals. This is generally true when the feeds have high initial boiling points or the hydrogen consumption is very high or any combination of the two.

Expanded-Bed Reactor Design with Finely Divided Catalyst.   It was previously mentioned that commercial expanded-bed units normally use $\frac{1}{32}$-inch extrudate catalyst. It was further pointed out that this catalyst could not be completely expanded by the velocity of fresh feedstock. With extrudate, liquid product oil was recycled to create a high enough liquid velocity to expand the catalyst bed. By using finely divided catalyst in the 50- to 200-micrometer size it is possible to operate an expanded bed without recycle of liquid product oil. The velocity of fresh feedstock is sufficient to expand the bed of fine catalyst. A commercial operation was employed with fine catalyst several years ago (Ref. 5).

In addition to eliminating the cost of a liquid recycle pump the fine catalyst system has several other advantages over extrudate catalyst. The average diffusional path in a catalyst particle is shorter. The relative number of pores on the external surface of the fine catalyst is larger. Therefore, the fine catalyst is less subject to metal contaminants plugging up external pores. Since the fine catalyst has a shorter diffusional length and less chance of pore blockage, the sulfur removal is greater for fine catalyst for a given set of process conditions, i.e., pressure, temperature, reactor volume, and catalyst usage. Offsetting these demonstrated advantages are restrictions on reactor size, configuration, and turn-down of the fresh feed rate. With the fine catalyst the catalyst bed expansion is controlled by the fresh feed rate and not by recycle of product oil. If inexpensive catalyst can be developed of intermediate size range between the fine catalyst and the $\frac{1}{32}$-inch extrudate, then this type of processing might find wider usage.

HYDRODESULFURIZATION CATALYSTS

All hydrodesulfurization catalysts consist of metals supported on a porous alumina base. Almost all the catalyst surface area is found in the pores of the alumina (200 to 300 square meters per gram). The metals are dispersed in a thin layer over the entire alumina surface within the pores. This type of catalyst displays a huge catalytic surface for a small weight of catalyst, but it suffers a slight disadvantage due to mass-transfer diffusional limitations within the catalyst pores.

Cobalt and molybdenum are the two most popular metals used for desulfurization catalysts. These are normally used together in a desulfurization catalyst. This combination of metals is the least sensitive to poisons and is the most universally applied catalyst for hydrodesulfurization of everything from residuums to naphthas.

Another metals combination is nickel and molybdenum on alumina. This catalyst is more active for hydrogenation and consumes more hydrogen per mole of sulfur removed than cobalt molybdate catalysts. It is useful for hydrodesulfurization of catalytic cracking feedstocks where maximum hydrogen consumption is desirable. It is also more selective for removing nitrogen in feedstocks and has special application where nitrogen removal is important (i.e., certain naphthas).

Nickel and tungsten on alumina is the most active catalyst of the above group for hydrogenation. It is also the most easily poisoned catalyst and displays the greatest hydrocracking activity among the above group of catalysts. This is why this type of catalyst is generally used where high conversion of middle distillate cuts is desired.

Reaction rates with hydrodesulfurization catalysts are limited by the diffusion of reactants into and products out of the catalyst pores. Reaction rates may be increased by making the catalyst particles smaller (i.e., smaller-diameter extrudates or finely divided catalyst) which shortens the average diffusional path length within the catalyst particle.

With fixed-bed reactors a balance must be reached between reaction rate and pressure drop across the catalyst bed. As catalyst particle size is decreased, the desulfurization reaction rate increases, but so does the pressure drop across the catalyst bed. Expanded-bed reactors do not have this limitation and small $\frac{1}{32}$-inch extrudate catalysts or finely divided catalysts may be used without increasing pressure drop.

HYDRODESULFURIZATION OF CRUDE OIL FRACTIONS

Fractionation (distillation) and hydrodesulfurization (hydrotreating) of Khafji crude oil are shown in Fig. 4. This process flow sheet is not meant to depict a particular petroleum refinery, but is a listing of some intermediate refinery products which can be hydro-treated. The relative flow rates of Khafji crude oil fractions are based on the crude oil analysis shown in Table 2.

The atmospheric distillation tower separates Khafji crude oil into C_4 minus gas, distillates, and atmospheric residuum. The vacuum distillation tower separates atmospheric residuum into vacuum gas oil (another distillate) and vacuum residuum. All these intermediate refinery streams can be hydrodesulfurized and converted into refinery products. The process conditions for hydrotreating these streams are shown in Table 4.

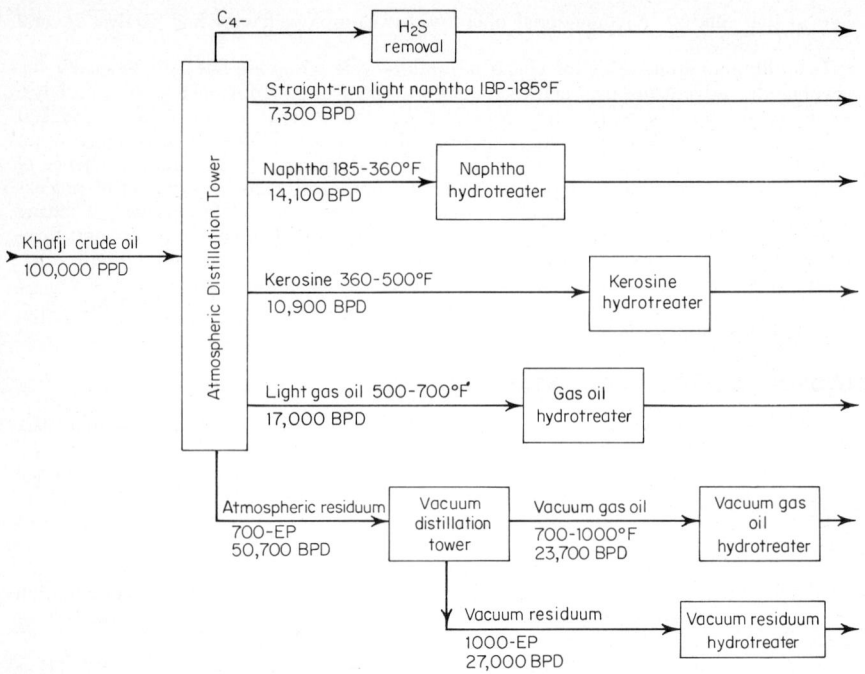

Fig. 4. Hydrodesulfurization of Khafji crude oil fractions.

NAPHTHA

Hydrotreated naphthas are normally used as feedstocks for catalytic reforming. Catalytic reforming is a process that converts paraffinic and naphthenic naphthas into highly aromatic reformates which are used as gasoline blending stocks or sources of benzene, toluene, and xylenes. Hydrotreating of catalytic reforming naphtha feedstocks is required because sulfur compounds poison the precious-metal catalysts used in reforming. Platinum catalysts can tolerate sulfur levels of 20 ppm and above, but the new platinum-rhenium reforming catalysts are poisoned by sulfur levels as low as 1 to 5 ppm. Hydrotreating of naphthas is also useful for improving the octane boost in gasoline with addition of tetraethyllead, reduction in gasoline gum content, and improvement in the stability of gasoline.

Naphthas are commercially hydrotreated (Ref. 6) in the gas phase with fixed beds of catalyst. With all reactants in the gaseous phase only minimal diffusion problems are encountered within the catalyst pores. Desulfurizations of 99+ percent are carried out commercially at low cost. Process conditions are adjusted to give high extents of desulfurization with very low conversions of naphtha to C_4 minus gases (< 0.5 weight percent).

TABLE 4. Process Conditions for Hydrodesulfurization

	Naphtha	Kerosine	Gas oil	Vacuum gas oil	Residuums
Temperature °F	550–750	600–800	650–800	650–850	700–850
°C	288–399	316–427	343–427	343–454	371–454
Hydrogen partial pressure, psig	100–450	150–500	150–700	450–800	750–2250
Space velocity, V/hr/V	10–2	6–1	6–1	5–1	2–0.2
Hydrogen rate, standard cubic feet per barrel	250–1500	500–1500	1000–2000	1000–4000	1500–10,000
Ultimate catalyst life, barrels per pound	1200–500	600–300	400–200	350–50	50–2

KEROSINE

Kerosine is mainly used today as a jet fuel. As with naphtha, it theoretically would be more efficient to treat kerosine in the vapor phase. The resistance to diffusion through the liquid phase would thus be avoided. However, it is difficult to achieve complete varporization of the kerosine at mild process conditions (low temperatures). At mild process conditions, a fine mist of liquid droplets and solid sediment would form. The fine mist could settle out in the preheaters, causing fouling problems. For this reason, kerosine is hydrotreated at process conditions which leave part of the kerosine in the gas phase and part of the kerosine in the liquid phase. Sufficient liquid phase is maintained in the catalyst bed to keep the liquid in continuous phase and to carry the largest kerosine molecules through the bed. Insufficient liquid phase would create stagnant pockets where the largest kerosine molecules could breed carbon deposition precursors. Even with the liquid phase present, kerosine is desulfurized commercially to remove 85 to 99 percent of the sulfur with very little conversion to lower-boiling stocks (0.5 Wt.%).

Kerosine is desulferized to improve thermal stability, to reduce acidic components, to improve color, and to reduce sulfur oxide pollution of the environment. Hydrodesulfurization of kerosine at higher pressures than that normally employed will saturate aromatics and improve smoke point.

Heating oils and Diesel fuels are two examples of petroleum products that boil in the range from kerosine to gas oil. Higher percentages of these stocks are found in the liquid phase during hydrodesulfurization than kerosine. Heating oils are desulfurized to reduce odor (mercaptans), to reduce sulfur oxide pollution of the atmosphere, to improve the burning characteristics of the fuel, to reduce corrosion in home heating equipment and to improve color.

Diesel fuels have higher end points than heating oils and are desulfurized to reduce sulfur oxide pollution and to improve stability. Hydrodesulfurization of Diesel fuels at higher than normal pressures will saturate aromatics and improve cetane number.

ATMOSPHERIC AND VACUUM GAS OILS

Atmospheric and vacuum gas oils are not produced as refinery products, but are hydro-treated (Ref. 7) and converted in other refinery processes to lower boiling stocks. Hydro-desulfurization of a gas oil reduces its carbon residue, sulfur, and nitrogen contents. Gas oils are commercially desulfurized to remove 80 to 95 weight percent of the sulfur content with less than 0.5 weight percent conversion of atmospheric gas oils and less than 10 weight percent conversion of vacuum gas oils to lower boiling stocks. The hydrotreated gas oils are then converted to lower boiling stocks in processes such as catalytic cracking and hydrocracking. Hydrodesulfurization of feedstocks to these processes improves catalyst life and increases volumetric product yield.

INDIRECT HYDRODESULFURIZATION OF RESIDUUMS

Atmospheric and vacuum residuums are indirectly desulfurized by separating the resids into high- and low-boiling-point fractions. The low-boiling-point fractions are hydrodesul-furized and mixed back with undesulfurized high-boiling fractions. Indirect desulfuriza-tion processes are limited to a maximum of 45 to 50 percent sulfur removal.

Where low extents of desulfurization are desired, indirect hydrodesulfurization is a possible route. This is due to the fact that low-boiling fractions such as vacuum gas oils or deasphalter gas oils are low in metals and are easy to desulfurize.

DIRECT HYDRODESULFURIZATION OF RESIDUUMS

Direct hydrodesulfurization of residuums is employed to reduce sulfur, nitrogen, and metal contents of residual fuel oils. Sulfur contents are kept low so as not to pollute the atmosphere with sulfur oxides during combustion of the residual oils. Metal contents are kept low to minimize corrosion in boilers and furnaces during combustion.

Processes (Ref. 8) selected for the direct hydrodesulfurization of atmospheric or vacuum residuums depend on the properties of the resid and the end use of the products. If the resid contains high concentrations of vanadium and nickel, then high catalyst consumption can be anticipated. If a fixed-bed process is to be used with a resid containing high salt and suspended solids content, then desalting and filtering must be done before processing.

Hydrodesulfurization of residuums is normally accompanied with a high degree of hydrocracking (Ref. 9). The process conditions required to achieve up to 90 percent desulfurization will also produce considerable conversion of vacuum residuums to lighter boiling stocks. The extent of hydrocracking (conversion) relative to hydrodesulfurization is controlled by the temperature of the reaction. At low temperatures (less than 750°F, 399°C) there is a minimum of hydrocracking, and considerable desulfurization can be accomplished at low space velocities. As temperatures are increased (800–850°F, 427–454°C) the rates of both hydrocracking and hydrodesulfurization increase. However, the hydrocracking rate increases more rapidly than the hydrodesulfurization rate. Higher temperatures also produce a higher rate of carbon deposition, which lessens the rate of desulfurization by covering catalytic sites on the catalyst with carbon. Hydrocracking reactions which occur in a hydrodesulfurization process are not strongly influenced by carbon deposition because they occur mainly as a result of noncatalytic thermal energy. In processes where hydrodesulfurization is not the goal, such as distillate hydrocracking, specific catalysts with strong acid sites in the support are used to catalytically improve hydrocracking. The activity of distillate hydrocracking catalysts would be reduced by carbon deposition. This is not true of hydrodesulfurization catalysts which employ low-acidity alumina supports.

DEMETALLIZATION OF RESIDUUMS

Petroleum residuums contain many impurities other than sulfur and nitrogen. The most troublesome are the organo compounds of nickel and vandium. Depending on the crude source, the metals content of residuum can be anywhere from a few parts per million to over one thousand parts per million. In hydrodesulfurization the metals will also react with hydrogen. Generally it is easier to crack a bond between carbon and a heavy metal than a carbon-sulfur bond. Consequently, the rate of metals conversion from organometals to inorganic sulfides is faster than the rate of organosulfur conversion to hydrogen sulfide. The free metals or metal sulfides are solids, and deposit on the catalyst surface. The metals cover the catalyst surface, filling the pore volume or blocking the pores, which "poisons" the catalyst at least in the sense that active sites are made inaccessible to the oil and hydrogen.

The stocks that have relatively high metal contents, 100 ppm and higher, pose a problem for commercial operation. The actual catalyst consumption is high because the metals poison the catalyst, and the catalyst must be replaced often. Also, the cost of frequent shutdowns to change out catalyst is high. The cost of change-out can be eliminated by use of daily on-stream catalyst addition and withdrawal with an expanded-bed system. The actual catalyst consumption is approximately as high with daily replacement in an expanded bed as for complete change-out in a fixed bed. The usual desulfurization catalysts are relatively expensive for these consumption rates. There are proprietary catalysts (Ref. 8) which are very inexpensive and can be used in a first reactor to remove a large percentage of the metals. Subsequent reactors downstream of the first reactor would use normal hydrodesulfurization catalysts. Since the catalyst materials are somewhat proprietary, it is difficult to identify them here. However, the catalysts contain little or no metal promoters, i.e., nickel, cobalt, molybdendum. Vanadium plus nickel removals of the order of 90 percent have been shown with these materials.

COMMERCIAL APPLICATION OF CATALYTIC HYDROTREATING

Catalytic hydrotreating is used extensively in refineries in the United States and throughout the free world. Table 5 has been compiled to show an estimate of this usage. This table is not necessarily 100 percent complete, but it does include those units which the authors could readily identify. Note that over 6 MM BPSD (million barrels per stream day) are hydrotreated in the United States and Canada. This throughput is equivalent to more than one-third of the crude oil refined in the United States today. Residual hydrodesulfurization has not been widely used in the United States. However, in other parts of the world, notably the Middle East and Japan, there are residual hydrodesulfurization or hydrocracking units with a combined capacity of over 400,000 BPSD. Also, in the Caribbean area there is considerable application of the indirect method of hydrodesulfurization where gas oil is desulfurized and blended back with undesulfurized vacuum bottoms to make a medium-sulfur-content fuel oil. The heavy-gas-oil desulfurization capacities in either the Caribbean or Japan exceed those of the United States.

TABLE 5.   Commercial Application of Catalytic Hydrogenation

CATALYTIC HYDROTREATING IN THE UNITED STATES AND CANADA		
	Total capacity, barrels per stream-day (BPSD)	Number of installations
Naphtha desulfurizing	668,130	41
Naphtha olefin or aromatics saturation	77,070	12
Distillate treating	2,054,250	106
Lube oil "polishing"	139,000	12
Catalytic reformer feed pretreating	2,788,300	152
Catalytic cracking and cycle stock feed pretreating	367,200	19
Heavy gas oil desulfurizing	160,000	5
Residual desulfurizing	6,000	1
Other	230,850	25
Total	6,490,800	373
CATALYTIC HYDROTREATING IN THE FREE WORLD		
	Total capacity, BPSD	Number of installations
Hydrotreating	10,586,320	481
Hydrodesulfurization	8,036,160	328
Total	18,622,480	809
RESID HYDRODESULFURIZATION IN THE FREE WORLD		
	BPSD	
Far East	401,000	
Mid East	63,800	
U.S.A. and Mexico	24,500	
Total	489,300	

FUTURE RESEARCH AND DEVELOPMENT EFFORT

Hydrodesulfurization processes are among the most widely used refinery processes. They use catalysts of various sizes and shapes which for the most part have the common characteristics of being physically rugged, relatively nonsusceptable to poisoning, and easily regenerable. The chemical species found in crude oils mount up to a staggering number, many of which have not been precisely identified. Therefore, hydrodesulfurization of crude oil fractions and petroleum products is far from an explicit science. Consequently, there is still room for improvement.

There are two major areas where research and development should be productive in improving the art of hydrodesulfurization. The first is in improving the commercial process technology of high-temperature high-pressure design. The improvements will be a result of the most economical and metallurgically sound vessel and piping designs as well as the best conceptual ideas in process flow planning. These improvements will be made by commercial unit designers, operating personnel and maintenance groups.

The second major area for improvement is in catalyst development. In the immediate future, the ability to tailor-make catalysts for specific feedstocks or applications will be important. At the present time, the catalysts used for residual hydrodesulfurization are quite different from the catalysts used for light distillate hydrotreating. In the near future, the chemical compositions of residuum processing catalysts may not change, but the pore sizes and pore size distributions will be modified to improve the mass transfer of heavy oil molecules within the catalyst pores. This development will generate a series of catalysts particularly suited for residuums from different crude sources at various process conditions.

In the longer term, future catalysts will be developed which will improve hydrodesulfurization activity by either better accepting or totally rejecting metals that cause catalyst deactivation. Also, if a catalyst could be developed that would perform at a lower carbon deposition rate, it should be possible to maintain higher catalyst activity. As always in the case of catalyst development, one would hope to find a catalyst of higher intrinsic activity and more selectivity for the specific reaction required, in this case, hydrodesulfurization.

At this point in time there is no inexpensive process for regenerating the activity of hydrotreating catalysts which have been heavily poisoned with metals. If such a regeneration technique could be developed, then catalyst costs for residual hydrotreating could be substantially reduced.

REFERENCES

1. Schuman, S. C.: "The Desulfurization of Petroleum Stocks, *Am. Chem. Soc. Div. Petrol. Chem.*, vol. 13, no. 4, pp. D19–D29, 1968.
2. Cities Service Assay on Khafji Crude Oil.
3. Schuit, G. C. A., and B. C. Gates: Chemistry and Engineering of Catalytic Hydrodesulfurization, *A.I.Ch.E. Journal*, vol. 19, no. 3, pp. 417–438, 1973.
4. Schuman, S. C., and H. Shalit: "Hydrodesulfurization," *Cata. Rev.*, vol. 4, no. 2, pp. 245–318, 1970.
5. McFatter, W., E. Meaux, W. R. Mounce, and R. P. Van Driesen: Advanced H-Oil Techniques, 34th *Amer. Petrol. Inst. Midyear Meet. Div. Refining*, vol. 49, pp. 489–503, 1969.
6. Roeder, R. A.: Hydrotreating Catalyst Capabilities—Desulfurization of Naphthas and Furnace Oils, 39th *Amer. Petrol. Inst. Midyear Meet. Div. Refining*, vol. 53, pp. 146–155, 1973.
7. Metzger, K. J., R. D. Christman, and J. F. Marmo: "Gulfining—A Flexible Distillate Desulfurization Process," 36th *Amer. Petrol. Inst. Midyear Meet. Div. Refining*, vol. 51, pp. 72–88, 1971.
8. Hastings, K. E., L. C. James, and W. R. Mounce: "Demetallization Cuts Desulfurization Costs," *Oil and Gas Journal*, vol. 73, no. 26, pp. 122–130, 1975.
9. Galbreath, R. B. and R. P. Van Driesen: Hydrocracking of Residual Petroleum Stocks, Manufacturing *Proc.* 8th *World Petrol. Congr.*, vol. 4, pp. 129–137, 1971.

Petrochemical Complex

The production of petrochemicals from petroleum and natural gas, as was a few decades ago the preparation of important organic chemicals from coal tar, exemplifies the dual role played by fossil-type natural raw materials. Natural gas and crude oil must be regarded in the planned use of these resources not only as excellent energy sources, but also as chemical raw materials. Over the years, the use of natural gas and crude oil essentially as fuels, hence the term fossil fuels, has eclipsed their uses as chemical raw materials in terms of quantities, but not necessarily in terms of importance. At the same time that the percentage of these materials for raw-material purposes is increasing, the outlook for future supplies is becoming less optimistic. Particularly in the case of natural gas, which can be used as a chemical raw material with very minimal preprocessing, the intrinsic worth of natural gas as a raw material is emerging—as contrasted with its excellent qualities as a fuel. Because of the appearance of substitute natural gas (SNG) and the inevitable improvements both in quality and availability of SNG (largely from coal as a raw material) which will occur during the next few years, it is likely that the marketplace will bid up the price of natural gas, thus favoring a greater percentage of its use in the manufacture of chemicals and synthetics. Although economic forces are difficult to forecast, it will be fortunate if the system operates in a manner to conserve not only natural gas, but certain petroleum fractions as well, so that these materials can be available as chemical raw materials for scores of years to come.

Because of the important nonfuel uses of materials that have preponderantly been considered as fuels in past years, this brief summary of petrochemicals is included in this energy-oriented Handbook.

In a generic sense, *petrochemicals* are those chemicals which are derived in whole or in part from petroleum or natural gas constituents. Certain petrochemicals are separated in their final product form directly from petroleum without undergoing changes in chemical composition. Other petrochemicals may experience several intermediate steps in their synthesis before becoming final useful products. This fact has led to the term *petrochemical intermediate* to designate chemical compositions at stages between one or more raw materials and the final products of commerce.

The term *petrochemical complex* applies to an assortment of aggregated processing units which are engaged in the conversion of several raw materials derived from petroleum into a variety of petrochemical intermediates or final products. The complex usually is associated with, is near, or in some fashion receives its raw materials largely from a petroleum refinery which produces a diversity of fuels as its primary products. Most of the raw materials used by a petrochemical complex arise as by-products or coproducts from the petroleum industry. Propylene from catalytic cracking or aromatics from catalytic reforming are examples. Or the complex may generate some of its own raw materials, such as ethylene and hydrogen, from natural gas components. On a worldwide basis, 80 percent of the feedstocks for petrochemicals originates from petroleum refinery gases and liquids, with natural gas as the second most important raw material.

Classes of Petrochemicals. The principal end-use products of the petrochemical industry include: (1) synthetic elastomers; (2) plastics and resins; (3) synthetic fibers and films; (4) detergents; (5) solvents, paint vehicles, and plasticizers; (6) agricultural chemicals, such as fertilizers, pesticides, and herbicides; (7) automotive chemicals, including antifreeze agents and lead alkyls; and (8) pharmaceuticals.

The original, primary sources of raw materials for the organic chemical industry were coal, coal tar, and other substances of animal or vegetable origin, occurring in nature or agriculturally grown. The impetus toward petroleum and natural gas sources began during World War II, when the demand for aromatics rose more rapidly than the ability to supply them as by-products from the coal-coking industry, which economically was tied to steelmaking. It is significant that the production of former natural sources of materials, including coal mining and natural rubber plantations, were essentially labor-intensive enterprises, whereas the petroleum refinery, involving continuous and automated operations achieved through high capital investment, was not so sensitive to labor costs. Thus, as labor costs increased in other areas, an economic incentive triggered the transition from conventional raw materials to petroleum sources.

The period following World War II saw intensive development of the petrochemical industry. Petroleum replaced coal as a main source of aromatic chemicals, and the petrochemical industry took over a large part of the production of alcohols formerly secured only from fermentation of carbohydrates. World War II also introduced the era of synthetic polymer substitutes for inorganics, such as metals and glass, and for many other natural substances, including leather, wood, rubber, fibers, glues, waxes, and gums. The large-scale production of these new substances necessitated supplies of raw materials *far in excess of those available from refinery off-gases.* The additional requirements for olefins (primarily ethylene) began to be supplied by the steam pyrolysis of light hydrocarbons recovered from natural gas. A parallel phenomenon was the rapid rise in demand for ammonia and its derivatives for use in fertilizers on a worldwide scale. The steam reforming of methane to produce synthesis gas for ammonia manufacture was practiced *wherever natural gas was available.*

During the decades that followed World War II, numerous factors motivated a continuing, rapid expansion of petrochemical activity. Expanded knowledge of surface phenomena led to the discovery and development of a variety of synthetic detergents, adhesives, and water repellants. A better understanding of macromolecules and polymerization led to entirely new classes of materials, including plastics, fibers, foams, rubbers, and coating materials. Accelerated scientific research, coupled with a growing dependence on petroleum as a new source of raw materials, had a dramatic effect on the economy of certain industries, including natural-rubber production in the Far East, the effect of synthetic detergents on conventional soap production, the effect of synthetic nitrates on the former Chilean monopoly, and the effects of synthetic fibers on natural materials, such as Australian wool, worldwide cotton production, and Japanese silk.

Of large significance was the synthesis of aromatics by catalytic reforming and the development of processes for the low-cost extraction from the reformates of building-block aromatics—benzene, toluene, and the C_8 aromatic family. To adjust the imbalance between benzene and toluene, processes were developed for the dealkylation of toluene (and C_8 aromatics) to produce benzene. Thus, dependence on coal as a source of raw materials for the synthetic chemical industry no longer prevailed.

Interlocking Production Complex.   To illustrate the possible arrangement of a number of processing units that might constitute a petrochemical complex, the flow sheet of Fig. 1 is included. This diagram shows a particular combination of processes that receives certain raw materials from a petroleum refinery and converts them mainly into petrochemical intermediates. The flow sheet is not complicated by the inclusion of additional processing units for converting the intermediate building blocks into finished products. Many petrochemical complexes which are associated with refineries carry their processing sequence to about the stage of intermediate manufacture as shown in Fig. 1.

Processing arrangements within petrochemical complexes can be quite similar on a worldwide basis, since the variety of intermediate or finished products is rather common wherever petrochemical manufacture occurs on a large scale. Certain raw-material sources can differ from one nation to another, however, as exemplified by the fact that in the United States catalytic cracking units provide almost all the propylene and butylene requirements, but because catalytic cracking is employed to a much lesser extent outside the United States, petrochemical manufacturers must provide these olefinic raw materials, along with ethylene, largely by the pyrolysis of naphthas. These pyrolysis units also provide an abundant amount of their light-aromatic raw-material needs. See article on Methane-Rich Gas Process for Substitute Natural Gas in *Gas Technology* section of this Handbook.

The present trend is for petrochemical complexes to extend their manufacture farther toward the finished product. Particularly among the larger organizations, the petrochemical complex produces and markets a great number of finished products. There is no typical petrochemical complex inasmuch as each differs from the others both in the variety of raw materials that it digests and in the number and kind of products (intermediate or finished) that it markets.

The hypothetical complex of Fig. 1 receives the following materials from a petroleum refinery:

1. Ethane, propane, or a light naphtha for olefin manufacture
2. A C_6–C_8 cut from a catalytic reformer as a source of aromatics
3. A hydrotreated kerosine as a source of n-paraffins for detergent manufacture
4. Propylene from a catalytic cracking unit
5. Hydrogen as a by-product from catalytic reforming and other manufactured sources.

See also Petroleum Refining and Processing in this section of the Handbook.

In addition, an input of natural gas is shown and designated for the manufacture of ammonia and its derivatives. The flow sheet is devised so that the primary products from the first step in the conversion of raw materials just listed interact with one another to a

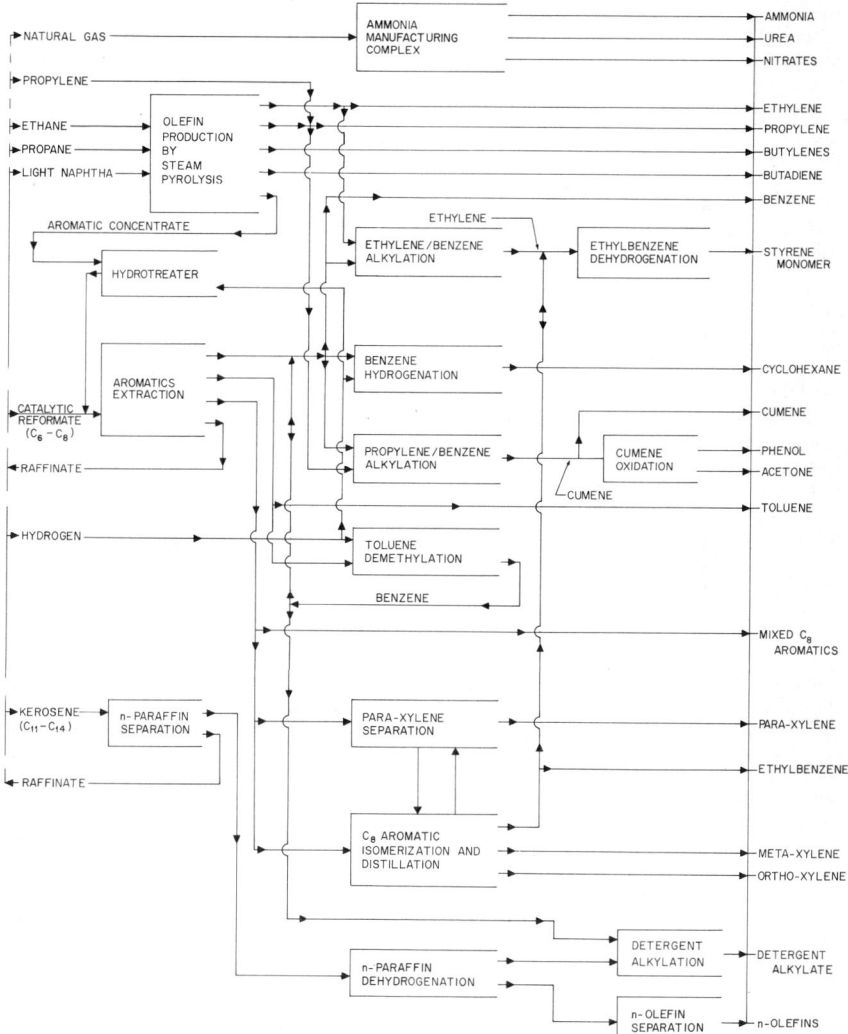

Fig. 1. Interlocking processes and flow of materials in a petrochemical complex. (*Universal Oil Products Company.*)

great extent, a practice quite common to those complexes which carry their processing sequence at least one step beyond the first. At the same time, these complexes (as shown on the flow sheet) market their primary products to other firms for further subsequent syntheses.

A steam-pyrolysis unit is shown for the conversion of ethane, propane, or a light naphtha primarily into ethylene, but with propylene, butylene, butadiene, and an aromatic concentrate shown as products. With ethane as the feedstock, the primary product is ethylene. Propane can be made to yield both ethylene and propylene in varying ratios. Light naphtha yields a spectrum of light olefins and a coproduct which is rich in light aromatics.

The flow sheet shows the ethylene as a feed for ethylbenzene manufacture within the complex and as a net product for sale and subsequent conversion to such important petrochemicals as polyethylene, ethanol, acetic acid, ethylene oxide, and vinyl chloride. In recent years about 65 percent of the ethylene has been produced by cracking ethane or propane, 26 percent obtained from refinery off-gases; and 9 percent by cracking liquid feeds. Also, about 34 percent of the ethylene is converted into polyethylene, 19 percent into ethylene oxide, 11 percent into ethylene dichloride, and 8 percent into styrene.

Propylene produced by steam pyrolysis is shown supplanted by propylene from the refinery catalytic cracking unit and (on the flow sheet) is to be used within the complex for the manufacture of cumene as well as being a net product for sale. Of the propylene used in the United States for petrochemicals, 88 percent originates from refineries; the remaining 12 percent is produced by steam pyrolysis. Further conversion of propylene (not shown on the flow sheet) results in such important products as polypropylene, acrylonitrile, propylene oxide, glycerin, isopropanol, and acetone.

The C_4 olefins are shown in Fig. 1 as net products. Butadiene condensed with styrene finds its way largely into SBR rubber, while isobutylene is converted into butyl rubber. Normal butane is used for the manufacture of alcohols, plastics, maleic anhydride resins, and methyl ethyl ketone.

The aromatic concentrate, shown as a product from the pyrolysis unit when light naphtha is the feedstock, contains mostly benzene, toluene, and C_8 aromatics, but associated with it are mono- and diolefins. It is common practice to hydrotreat this stream (shown on the flow sheet) to saturate the olefins and to purify it of sulfur and nitrogen compounds before it is routed to an extraction unit for recovery of the contained aromatics in pure form.

An aromatics extraction unit* is shown yielding pure benzene, toluene, and C_8 aromatics as extracts from a C_6–C_8 catalytic reformate cut and the hydrotreated pyrolysis unit coproduct. The benzene product is used, as shown within the flow-sheet complex, for ethylbenzene manufacture, for hydrogenation to produce cyclohexane, for cumene manufacture, for the production of detergent alkylate, and as a net product. These are all quite common secondary processing steps for benzene conversion to the next intermediate in many petrochemical complexes. Benzene derivatives, of course, are many in number and include rubber products, polystyrene plastics, fibers, detergents, surface coatings, aspirin, week killers, insecticides, dyes, sulfa drugs, solvents, films, and resins.

The extracted toluene is shown on the flow sheet as routed to demethylation for benzene manufacture within the complex and as a net product for sale. Among the derivatives of toluene in subsequent processing are explosives, solvents, plastics, dyestuffs, pharmaceuticals, surface-coating vehicles, and food perservatives.

The extracted C_8 aromatic family would contain the distribution of isomers in a typical catalytic reformate as shown by Table 1, arranged in the order of increasing normal boiling points.

The mixture is useful as a solvent and as a high-octane-number blending component for motor fuels. For chemical syntheses, however, it is necessary to separate the isomers in relatively pure form. From their boiling points, it is noted that it is possible to separate the ethylbenzene or the o-xylene by precision fractional distillation. On a commercial scale, however, these columns are huge, and this technique is not practical in separating the p- and m-xylenes.

Until 1971 p-xylene was separable from the mixture (from a practical standpoint) only

---

*Typified by a Sulfolane (proprietary name) or Udex^{SM} unit (both processes licensed by UOP Process Division, Universal Oil Products Company).

by fractional crystallization on the basis of its high melting point relative to its isomers. The first unit of a novel technique† for the continuous adsorptive separation of p-xylene in pure form and with high recoveries from its isomers was put into commercial operation in 1971. Isomerization techniques‡ also are available to bring any C_8 aromatic mixture to an equilibrium composition. The combination of separation and isomerization operations, appropriately integrated, can convert a C_8 aromatic mixture into predominantly any one or any desired distribution of isomers. The flow sheet of Fig. 1 shows all four isomers as net products in pure form, available for further synthesis. Figure 1 also shows the separated ethylbenzene being routed to a dehydrogenation unit for styrene production or as a net product from ethylbenzene alkylation. The direction of flow will be determined by varying market demands.

The usual next step in the conversion of o-, m-, and p-xylenes is to phthalic anhydride and isophthalic and terephthalic acids, respectively. These materials, in turn, can be further converted into alkyd resins and polyesters, useful in surface coatings and fibers as end products.

Figure 1 also shows a sequence of processing steps to produce a biodegradable detergent alkylate and n-olefins, starting with a hydrotreated kerosine fraction containing,

TABLE 1.   Isomers of C_8 Aromatic Family

Isomer	Percent	Normal boiling point		Melting point	
		°F	°C	°F	°C
Ethylbenzene	18	277.1	136.1	−138.9	−94.9
p-Xylene	18	281.0	138.3	+55.9	+13.3
m-Xylene	44	282.4	139.1	−54.2	−47.9
o-Xylene	20	291.9	144.4	−13.3	−25.2

Adapted from Douglas M. Considine (ed.), "Chemical and Process Technology Encyclopedia," McGraw-Hill Book Company, New York, 1974.

typically, C_{11}–C_{14} hydrocarbons. The primary processing step is the separation of the n-paraffins in pure form from the kerosine mixture by molecular-sieve adsorption.* The raffinate from the separation process normally is returned to the refinery, where it finds its way into fuel blends.

The n-paraffins are catalytically dehydrogenated as a next step.¶ Because of chemical equilibrium, the product is a mixture of n-olefins and unconverted n-paraffins. A biodegradable detergent can be prepared by alkylating the olefins with benzene (as shown) and the n-paraffins returned to the dehydrogenation step for further conversion.

Similarly, if n-olefins are desired as a net product, they can be separated in pure form by a continuous-adsorptive technique** and the n-paraffin stream returned to the dehydrogenation unit for further conversion.

In the hypothetical complex of Fig. 1, benzene is shown as:

1. Alkylated with ethylene to produce ethylbenzene, which subsequently is dehydrogenated to a styrene monomer as a net product. A number of petrochemical complexes include further processing to produce a styrene polymer for sale as a net product, and in some instances molded articles are manufactured as finished products.

2. Hydrogenated to cyclohexane, shown as a net product on the flow sheet but converted in subsequent steps, mainly into nylon.

3. Alkylated with propylene to form cumene, which is oxidized within the complex to phenol and acetone.§

About 90 percent of the benzene produced in the United States is converted into styrene, cyclohexane, phenol, and detergent alkylate. Thus, the complex flow sheet in Fig. 1 represents a processing arrangement that accounts for most of the important outlets for benzene.

†Parex^{SM} process, licensed by UOP Process Division, Universal Oil Products Company.
‡Isomar^{SM} process, licensed by UOP Process Division, Universal Oil Products Company.
*Molex^{SM}.
¶Pacol^{SM}.
**Olex^{SM}.
§Unisir^{SM}; all processes (UOP Process Division, Universal Oil Products Company) are typical.

Many petrochemical enterprises are involved in the manufacture of ammonia. Natural gas is shown on the flow sheet as a raw-material feed to a stream reforming system that provides the synthesis gas for the production of ammonia. Although not detailed, the ammonia complex is shown to produce urea and nitrates as derivatives, so that a diversified package of agricultural fertilizer components is a typical net product for an enterprise of this nature.

Interdependence of Plants and Material Resources.   It is evident from Fig. 1 that the several processing units are integrated in such fashion that the relative amounts of products can be varied over wide ranges as desired or as demanded by economic forces—acting in the marketplace, in the area of raw-material supply, or within the complex itself. An ability to shift product distribution is characteristic of petrochemical complexes largely because of the diversity of the processing units and their interlocking nature.

Petrochemical entities of the kind described have the choice of expanding their operations in at least three different directions:

1. Processing units can be added to increase the number of intermediate products for sale: e.g., units might be added to manufacture chlorobenzenes or nitrobenzenes; the butylenes or butadiene might be appropriately polymerized.

2. Intermediates already being made can be further processed in the direction of the finished final product: e.g., the complex depicted on the flow sheet might produce polystyrene from the styrene monomer, convert phenol to phenolic resins, or produce acrylonitrile from ammonia and propylene. Because of wider profit margins which generally result from considerably higher unit prices for the finished product than for the intermediate, there often is an economic incentive in this direction.

3. The raw-material situation may be improved. This usually is accomplished by acquiring a small refinery or by joint ventures between a refiner and a petrochemical manufacturer. Joint ventures seem to be especially attractive because each partner has a special contribution to make to the combined operation in terms of technology, markets, and financial matters.

Still another popular route for expansion is for the petroleum refiner to diversify his operations in the direction of undertaking petrochemical manufacture. Most frequently this is done by performing petrochemical manufacture within the framework of the operation and business structure of the refinery itself. A petrochemical division or subsidiary may be formed as expansion proceeds. This is an easy entry for the refiner because he has captive raw materials and expertise in catalysis and automated continuous-flow processing. Further, the refiner is accustomed to the tradition of high capital investment in equipment per employee, a characteristic of the petrochemical industry. In addition, petroleum refiners are research- and development-oriented with a background of accelerated discoveries of new and less costly processing methods—desirable attributes for the petrochemical industry.

Some important petrochemicals are listed below. Even this rather long list is not fully representative of the many hundreds of compounds that could be classified as petrochemicals.

Acetic acid	Cyclooctadiene (COD)
Acetone	Cymene
Acrylonitrile	Diacetone alcohol (DAA)
Adipic acid	Dimethylformamide (DMF)
Ammonia	Dimethylmetadioxane (DMD)
Amylenes	Dimethylterephthalate (DMA)
Benzene	Epichlorhydrin
Benzoic acid	Ethane
Bisphenol A	Ethanolamines
Butadiene	Ethyl alcohol
Butylene	Ethylbenzene
Butyraldehydes	Ethylene
Caprolactam	Ethylene glycol
Cumene	Ethylene oxide
Cumene hydroperoxide (CHP)	Formaldehyde
Cyclododecatriene (CDT)	Fumaric acid
Cyclohexane	Glycerin

Glycol ethers
Hexamethylenetetramine
Hydrazine
Hydrogen
Isobutylene
Isophthalic acid
Isoprene
Isopropyl alcohol
Maleic anhydride
Mesitylene
Methane
Methyl alcohol
Methyl chloride
Methyl ethyl ketone (MEK)
Methylisobutylcarbinol (MIBC)
Methyl isobutyl ketone (MIBK)
Monochlorostyrene
Naphthalene
Natural gas

Perchloroethylene
Petroleum
Phenol
Phthalic anhydride
Potassium terephthalate (TPAK)
Propane
Propylene
Propylene oxide
Styrene
Synthesis gas
Terephthalic acid (TPA)
Toluene
Tolyene diisocyanate (TDI)
Urea
Vinyl chloride
Vinyl cyclohexane (VCH)
Xylenes
Xylene diamine

Acknowledgment.   The fundamentals of this description are based upon an entry on this topic by Melvin J. Sterba, formerly Assistant to the Vice-president, Universal Oil Products Company, Des Plaines, Illinois, which appeared in the "Chemical and Process Technology Encyclopedia," McGraw-Hill, 1974.

The Total Energy Concept in Refinery and Process Plant Planning

BY H. J. KLOOSTER*

The basic design objective for all modern integrated process plants is reliable production with low initial and operating costs and compliance with environmental regulations. Achievement of these objectives requires dilegent attention in applying basic engineering principles to compensate for wide variations in environmental needs, in energy costs and availability, and in plant complexity and capacity. In the economy of present times and the foreseeable future, one of the most critical needs is to minimize energy consumed. This can be accomplished most effectively by applying a *total energy concept* whenever a new plant is designed. Implementation of this concept first requires the selection of the available fuel or fuels to be employed to provide the energy needed for plant operation.

Approximately 28 percent of all the energy consumed in the United States (1975) was for use by the industrial sector. The number and size of integrated plants under design and planned for the future is constantly increasing. The conversion of feed material to products meeting modern environmental requirements and the increased demand for more highly refined products require that much additional hydrogen processing be used (particularly in oil refineries) with a consequent increase in overall plant energy needs. A survey (Ref. 2) of overall refinery energy use indicates that energy consumption decreased from 800,000 Btu per barrel of crude in the early 1930s to about 550,000 Btu per barrel in the early 1950s. Then, in 1964, energy consumption increased to 724,000 Btu per barrel, primarily because of more complex processing. This consumption is expected to exceed 810,000 Btu per barrel within the near future.

The foregoing figures become more significant when quoted as a percentage of the crude oil feed required to furnish the necessary energy. For instance, the low point of energy use in the early 1950s was 9.2 percent of crude oil feed, and the predicted new, higher level is 13.5 percent. In a recent large, integrated refinery design, the total energy concept was used to (1) minimize energy consumption, and (2) meet air, water, and noise environmental constraints. The consequent design resulted in a refinery which consumed only 9 percent of the crude oil for energy requirements. This 4.5 percent savings over the predicted average of 13.5 percent will conserve over 3 million barrels of crude oil per year in a typical 200,000 barrel-per-day refinery (operating 92 to 95 percent of the time).

The total energy concept begins at the inception of plant design with an examination of various available energy-saving ideas and process systems which can be used to conserve energy. This is closely followed by an evaluation of both the traditional and new methods for energy conservation which are available to best accomplish the design objective.

GRASS ROOTS OR MODERNIZATION

The first step in the decision-making process is to determine whether to expand and/or modernize existing facilities, or to build a new integrated plant. Some advantages and disadvantages for either decision are summarized in Table 1. As this table implies, if space is available for the expansion/modernization of existing facilities and the existing plant is readily adaptable, the modernization approach is economically and environmentally more attractive. This is evidenced by the numerous process plants which have been revamped and updated in recent years. The problems of siting caused by objections of the local populace is a very strong factor as well.

The comparisons given in Table 1 are most applicable to refineries, chemical plants, petrochemical systems, and like facilities. Energy producers, such as the gas and power industries, have different perspectives because they are regulated pricewise by governmental agencies. These industries must apply a formula to new plant construction which

*Manager, Process Engineering, Fluor Engineers and Constructors, Inc., Los Angeles, California.

allows an established fixed return on invested capital. Consequently, the major cost factor in these industries is not capital cost, but the cost of feedstocks and fuels.

ESTABLISHMENT OF ECONOMIC PARAMETERS

The next step in applying the total energy concept is to establish the economic parameters that are to become the bases for subsequent evaluations. These factors include: (1) Assigned or calculated utility values, considering expected future energy costs; (2) bases for cost estimates, including such matters as the time frame of the project, the desired starting date of construction, the project location, labor rates in the construction area,

TABLE 1. Comparison of Grass Roots versus Modernization

Advantages	Disadvantages
MODERNIZATION	
1. Lower capital per unit of fuel	1. Higher operating costs
2. Better return on investment	2. Less flexibility for future
3. Fewer siting problems	3. Not easily adaptable to modern environmental rules
4. Less stringent environmental controls	4. Limited utilities availability
5. Optimum use of existing operating personnel within union constraints	5. Limited plot space
6. Only incremental increase in feed is required	
7. Only incremental increase in products is needed	
GRASS ROOTS PLANT	
1. Higher operating efficiency and lower operating costs	1. Higher capital costs
2. Maximum flexibility	2. More siting problems
3. Environmentally advanced	3. Requires new operating personnel
4. Maximum production per unit feedstock	4. Unemployment resulting from possible shutdown of existing plants
5. More flexibility of plot space	
6. Can be located close to desired markets	

methods of accounting, expected plant technology life, among others; (3) the time value of money; and (4) the projected rate of inflation for energy.

Some estimated utility costs for 1968, 1972, and 1974 are given in Table 2, and the predicted rates of escalation in fuel costs through 1990 are given in Fig. 1.

INTEGRATED SYSTEMS ANALYSIS

After the economic parameters are established, the next step is to select the most effective flow scheme from among available alternate process combinations, each of which can yield the desired overall design objectives. During this selection, consideration also must be given to plot layouts which conserve energy. A typical application of this concept is grouping heaters in a single area to best recover the waste heat and to minimize fuel consumption.

Several other factors have an important bearing on the conceptual design of total energy systems, including: (1) Must the plant be self-sufficient in utilities? or are economical purchased power and fuels available? (2) What is the dependability of outside purchased power and fuel? (3) Will local regulations permit burning fuel containing sulfur?

Also, capacity, complexity, and process-unit makeup of the plant are factors and include: (1) What utilities will be available from energy recovery within the units? (2) How much fuel gas, if any, will be generated? Can it be used for gas turbine or other fuel uses, or must it be conserved for manufacture, or combusted in critical process heaters?

Factors in connection with operation to consider include: (1) Can orderly startup and shutdown procedures be worked out for the proposed integrated recovery and supply system? (2) Is sufficient flexibility provided for day-to-day operations?

TABLE 2. Estimated Utility Costs*

Commodity	1972	1974	1978
Water, cents/1000 gallons			
Cooling water	3.0	3.5	4.0
Treated boiler feedwater	40.0	57.0	63.0
City water	21.0	23.0	32.0
Power, cents/kilowatt-hour			
Coal based	1.2	1.5	1.7
Oil based	1.2	2.0	3.4
Fuel			
Fuel gas, dollars/million Btu (low heating value, LHV)	0.78	1.6	2.90
Fuel oil, dollars/barrel (No. 6 fuel oil, HHV)	5.40	9.00	14.00
Steam, dollars/1000 pounds			
High pressure, superheated	1.40	2.30	3.60
Medium pressure, saturated	1.25	1.95	3.10
Low pressure, standard	1.10	1.60	2.50
Condensate, dollars/1000 pounds			
High pressure	0.45	0.70	0.83
Medium pressure	0.30	0.50	0.60
Low pressure	0.25	0.35	0.45

*Average values based on data from numerous sources.

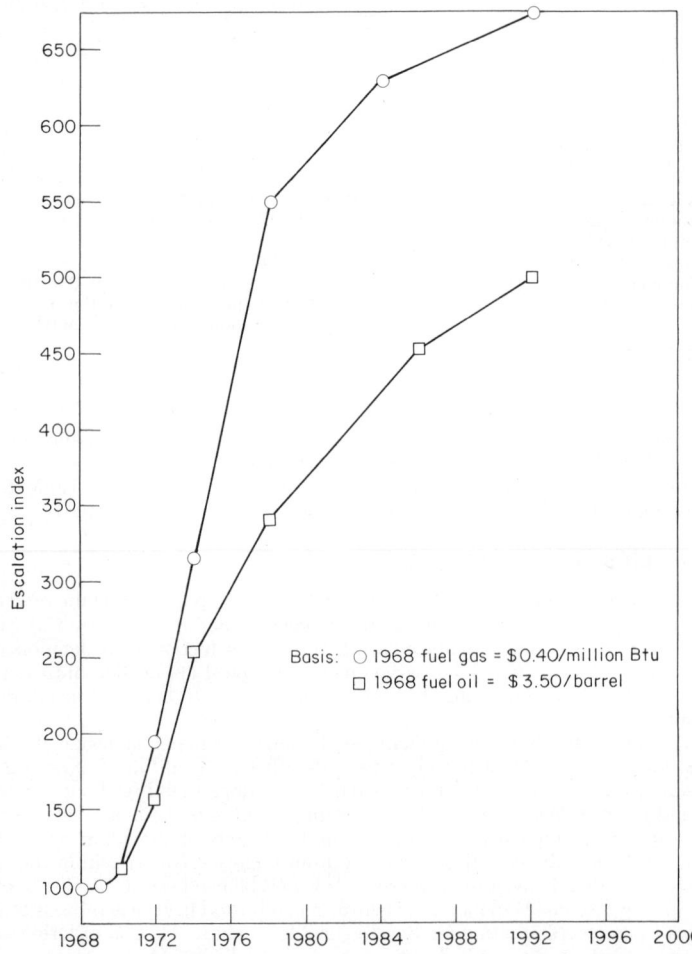

Fig. 1. Estimated escalation of fuel costs (averaged curves based on literature survey).

In terms of the degree of energy integration: (1) Between what units should cross heat exchange be allowed? (2) Are these arrangements justified from a heat-recovery standpoint, or are nearly equivalent designs available that are simpler to operate? (3) Are operationally simple, energy-conserving processes available, although these may be new to the plant design personnel?

Obviously, there are no simple process answers to any of the foregoing questions. Each requires individual, careful attention and technical expertise to develop alternate cases from which a final optimum solution can be selected.

The overall problem cannot be solved immediately because the questions just mentioned must first be answered individually and then collectively. Development of an energy balance must proceed simultaneously with other phases of the design throughout the planning and design periods. Thus, a development plan should be outlined before work is started on a new project. In the past, industry has depended too much on "battery limits" energy balances supplied by the individual unit process designer. A total energy balance was obtained by simply summing up the parts. As industry passes further into the era of integrated plant systems, design of the energy system requires a stepwise plan. At job inception, an individual should be appointed who fully understands and is responsible for the design of all plant energy systems.

Practical Design Problem.   To illustrate the foregoing observations, let us consider a practical design problem, namely, the role of a *fluid catalytic cracking unit* in supplying energy to the remainder of an integrated refinery. A catalytic cracking regenerator which has an expander recovering energy from the regenerator flue gas is shown in Fig. 2. The recovered energy drives the air compressors which supply regeneration air to the catalytic cracking unit. The expander also provides added power through a motor-generator set. In addition, exhaust gas from the expander supplies oxygen for combustion in the carbon monoxide boiler. Up to 90 percent of the driver and process steam requirements for the remainder of the refinery might be supplied by the high-pressure steam from the carbon monoxide boiler. And, finally, heat recovery from the cat-cracker product-fractionator reflux and product streams may supply additional steam and reboil heat for an integrated light-ends recovery unit.

Another possible application is shown in Fig. 3, which defines a partial-oxidation unit in which treated, hot, high-pressure gases are let down through an expander turbine to a gas turbine which drives the air compressors supplying the partial-oxidation unit. The gas turbine also provides hot exhaust gases for generating steam in a boiler.

Such arrangements can influence the design of entire process plants and supporting systems.

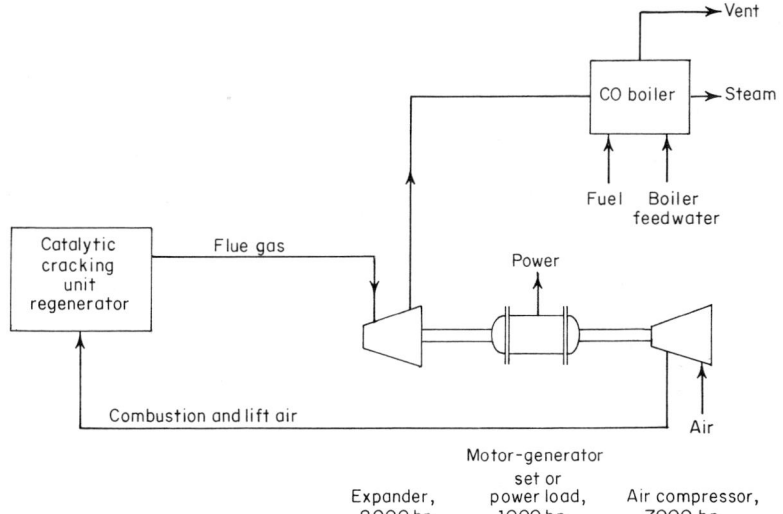

Fig. 2. Fluid catalytic cracking unit energy system: CO = carbon monoxide.

The total energy approach to optimization of energy consumption accomplishes the following:

1. Provides early indication of the significance of a problem.

2. Provides for logical development to a planned solution reasonably approaching the economical optimum.

3. Provides for the detailed checks necessary to ensure reliable operation.

Some of the steps in a logical energy-use optimization plan are given in the checklist of Table 3 and example of Table 4.

TABLE 3.   Checklist of Steps in a Logical Energy-Use Optimization Plan

1. Start planning early. The initial needs are a list of the process units to be installed with their approximate capacities and fairly reliable information on external factors.

2. Determine which units will operate together, or whether several blocks, each consisting of an integrated group of units, will be planned. Most complex plants follow the second pattern.

3. Gather data on process heating loads, compressor loads, total pump and fan horsepower requirements, and possible energy-recovery sources for the various process units. If process data are not yet available, data from similar units will serve as a preliminary source. Do not forget off-site loads which today are substantially larger than in the past because ecological regulations must be met.

4. Use judgment in selecting preliminary information. Try to select data from similar situations, but do not let former experiences lead to premature conclusions. It is well to keep a file of pertinent information for units frequently encountered. Table 4 is a typical example.

5. When data are summarized, start preparation of preliminary steam, fuel, electric power, and cooling water balances. Look for combinations of larger loads that will improve energy utilization and initial cost, but are within the bounds of reasonable operating judgment. For example, in a recent refinery design, the hydrogen-recycle compression services for three high-pressure hydrotreaters (kerosine, light-Diesel, and heavy-Diesel) were combined in a common circulation system so that a single turbine-driven centrifugal compressor could be used.

6. Make preliminary assumptions concerning the types of compressors to be used. Compare the cost of energy consumed for different assumed driver systems that are consistent with many other concomitant factors such as fuel and electric power availability. Select systems that merit further study, and reject alternatives that show no advantage.

7. As soon as more reliabile process information becomes available, revise the preliminary data. Review the process flow sheets for consistency with improving energy system patterns. The reasons for starting early studies now become apparent—to suggest and guide process alternatives so that integrated systems will be optimized before plans become too crystallized for change.

8. Review the process flow sheets to be sure that all significant energy users have been considered, and that promising combinations have not been overlooked in preliminary studies. Recheck assumed steam pressures to determine whether or not they fit process needs.
Review the process heater conditions for efficiency maximization. If process inlet temperatures are high and loads are large, supplemental heat recovery may be economical.

9. Review the process flow sheets critically, considering the use of cross heat exchange among units. There are so many possibilities for cross exchange among individual units that it is difficult to generalize on the approach to the problem. Unnecessary cross exchange may complicate operations. Nearly equivalent arrangements from a heat-recovery stand-point may reduce such complications. One example comes from the design of a four-unit integrated addition to an already integrated refinery.
The new units were: crude and vacuum, hydrogen generation, hydrocracking, and catalytic reforming. After alternatives were investigated, it was found unnecessary to cross exchange at any point among the four units. The only heat integration practiced involved heat transfer between hot oil streams in different units and collection of steam produced in the convection recovery coils in heater stacks.

10. Carry out economic studies of systems selected for investigation, securing preliminary cost estimates as required. Vendors of major equipment can often make very helpful suggestions if they are given a chance to understand the objectives and the freedom to answer inquiries in a broad sense. Approximate quotations are usually sufficient at this stage.

11. Make final selections based on the latest information from all sources. Bring all fuel, steam, electric power, and cooling water balances up to date. Review startup, shutdown, and operating problems presented by the selected combinations. Discuss these problems with those who will be responsible for operations. Check to be sure that all process, off-site and external information is reliable. Decisions made at this stage must be firm to avoid costly changes and delays later in the project.

12. Prepare and release final specifications for equipment procurement. Proceed with detailed mechanical design.

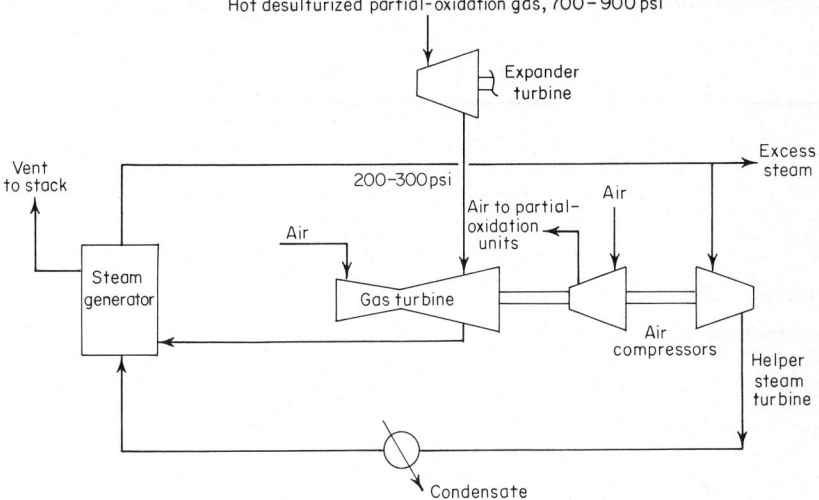

Fig. 3. Partial-oxidation unit energy system.

TABLE 4.   Hydrogen Unit Summary

Feed and product gas, vol %

	Feed	Product
Hydrogen	44.6	97.0
Methane	30.7	3.0
Ethane	15.9	
Propane plus	8.8	
	100.0	100.0

Conditions
    Feed gas at 350 psia, 250°F. Product at 235 psig, 250°F (121°C)
    Deaerated boiler feedwater in at 700 psig, 240°F (116°C)
Overall water balance (per million standard cubic feet of product hydrogen)

	Pounds		Pounds
Feedwater	63,000	Reaction consumes	23,000
		Export 640 psig steam	24,000
		Blowdown and losses	16,000
			63,000

Steam balance (per million standard cubic feet of product hydrogen)

Generated	Pounds	Used	Pounds
Reformer heater	39,000	In process	89,000
Waste heat boilers	74,000	Export (640 psig)	24,000
	113,000		113,000

Heat and power (above feedwater at 240°F) (per million standard cubic feet of product hydrogen)

	Million Btu
Fuel fired (lower heating value)	175.8
Used for process	125.5
With 640 psig export steam	23.8
Heater losses	26.5
	175.8

	Kilowatts/hour
Power required (pumps and fans)	400
(compressors)	None
Cooling water	None

ENERGY UTILIZATION AND CONSERVATION

Concomitant with developing the various design configurations and energy alternatives is the need for economic and ecologic evaluation of the alternate types of energy conservation methods which are to be included in the final design for an optimized plant. These alternate energy utilization and conservation methods may be categorized as (1) thermal energy conservation, (2) process energy conservation, (3) electric energy conservation, (4) fuel and products conservation, and (5) miscellaneous other energy-conserving systems. Present energy efficiency checklists for use in designing a process plant are given in Appendixes A, B, and C. Some of the points checklisted are further described in the text which follows.

THERMAL ENERGY CONSERVATION

Increased Fired Heater Efficiency.   The largest energy consumers in a process plant, of course, are the heaters and boilers. As such, these items of equipment are primary targets for energy conservation. Methods most used in industry to increase the efficiency of fuel in fired heaters and boilers are outlined briefly in the following paragraphs.

1. Steam Generation from Waste Heat.   The most utilized method for waste heat recovery in furnaces and boilers is the use of economizer coils in convection sections to produce steam. As much as 35 to 50 percent of the heat generated in the radiant section of a heater can be recovered as waste heat in this manner. The actual quantity of heat conserved depends on such variables as (1) temperature of the feed to the radiant section, (2) heat flux specified, (3) amount of excess air of combustion, and (4) type of fuel burned.

A typical refinery application is shown in Fig. 4, which outlines the conditions for a

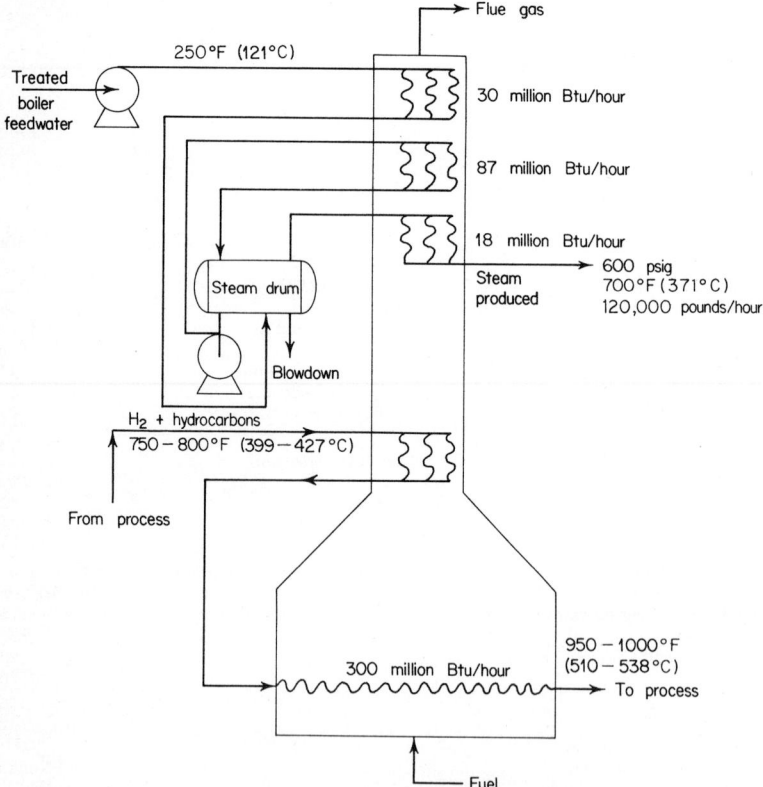

Fig. 4. Heat recovery in a catalytic reformer heater.

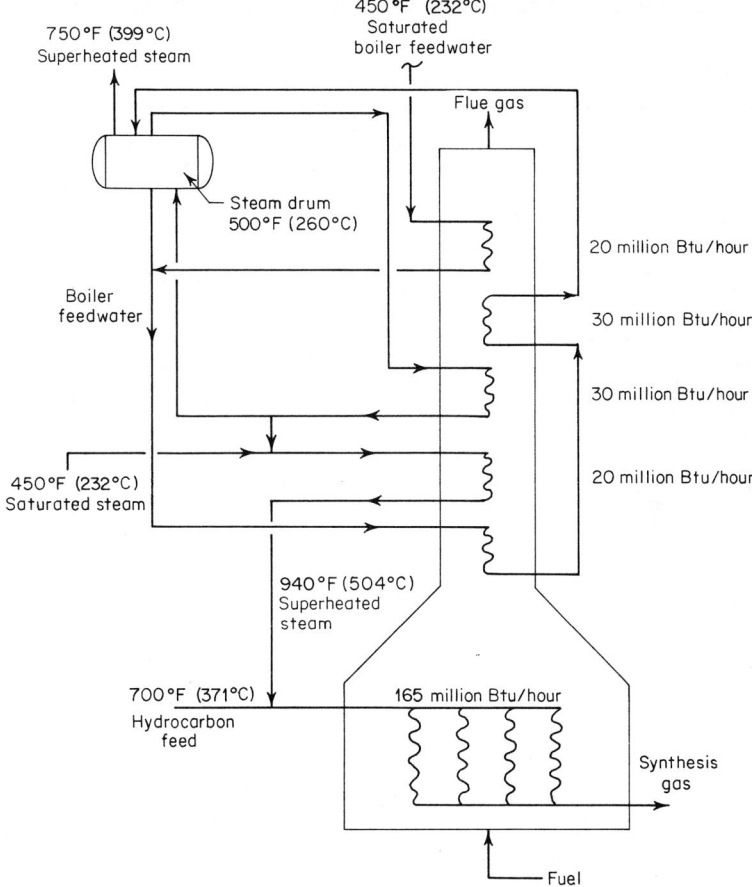

450°F (232°C)
Saturated
boiler feedwater

750°F (399°C)
Superheated steam

Flue gas

Steam drum
500°F (260°C)

20 million Btu/hour

Boiler
feedwater

30 million Btu/hour

30 million Btu/hour

450°F (232°C)
Saturated steam

20 million Btu/hour

940°F (504°C)
Superheated
steam

700°F (371°C)
Hydrocarbon
feed

165 million Btu/hour

Synthesis
gas

Fuel

Fig. 5. Heat recovery in a steam-methane reformer heater.

catalytic reformer containing one preheat and three reheat coils in the radiant section; and heat conservation (recovery) coils in the convection section. In this case, the use of recovery coils increased the heater efficiency from about 58 to 85 percent, while producing 120,000 pounds per hour of high-pressure superheated steam.

Another common application of this principle is shown in Fig. 5, which illustrates an integrated steam-methane reformer heater in a hydrogen plant in which sufficient heat is recovered in convection section coils to produce not only all the high-pressure superheated steam required for the steam-hydrocarbon reaction, but also an excess steam production for use in the rest of the process plant. In a similar manner, power boilers and steam boilers utilize this basic recovery method to maximize boiler efficiency.

In the United States today many types of industrial heaters are operating inefficiently and exhausting large quantities of recoverable waste heat to the atmosphere because initially it was not economical to recover this heat (when energy costs were lower). With the advent of high fuel costs, however, heat recovery by steam generation may be economical in a majority of the relatively low-efficiency heat applications.

2. Preheating of Combustion Air. About 1 percent of fuel consumption can be saved for every 40°F (~ 5.5°C) increase in air temperature to the furnace or boiler, based upon 20 percent excess air. Thus, the preheating of combustion air is a major potential source of energy conservation. A good example of this can be noted from Table 5 which outlines the operation of a heater designed for use either with or without an air preheater system. This

TABLE 5.   Comparison of Furnace Efficiency with and without an Air Preheating System

Operating factor	Without air preheat system	With air preheat system
Absorbed duty, million Btu/hour	83	83
Ambient air temperature	80°F, 26.7°C	80°F, 26.7°C
Air to furnace (temperature)	80°F, 26.7°C	660°F, 349°C
Gas to air heater (temperature)		815°F, 435°C
Gas leaving air heater (temperature)		319°F, ~ 160°C
Gas to stack (temperature)	815°F, 435°C	291°F, ~ 143°C
Excess air, percent	30	15
Furnace efficiency, percent	75.7	90.2
	(lower heating value)	(lower heating value)
Million Btu fired	109.6	92.0

SOURCE: Air Preheater Co.

system is shown schematically in Fig. 6 and includes a Ljungstrom air preheater, induced- and forced-draft fans, ducts, dampers, controls, and insulation. As shown by Table 5, furnace efficiency increased 14.5 percent with the use of air preheat in this example.

This energy conservation can be translated into operating cost savings and/or energy savings by using the following calculation:

Assume

1974 cost basis
Heater fired with No. 6 heavy fuel oil (10°API)
Then
From Table 2, power cost $0.017/kilowatt-hour
From Table 2, No. 6 fuel oil cost ······························· $9.00/barrel
From Table 6 (or Fig. 7), No. 6 fuel oil heating value............ 152,000 Btu/gallon
From Fig. 8, No. 6 fuel oil .... $1.43/million Btu

Example

$$\text{Total savings} = \frac{17.6 \text{ million Btu}}{\text{hour}} \times \frac{8000 \text{ hours}}{\text{year}} \times \frac{\$1.43}{\text{million Btu}} = \$200,000/\text{year}$$

Operating Costs
Forced-draft fan .... 70.8 bhp
Induced-draft fan .... 36.8 bhp
Air heater .... 1.5 bhp
Total .... 109.1 bhp

Total Operating Costs =
$$109.1 \times \frac{0.746 \text{ kilowatts}}{\text{bhp/hour}} \times \frac{8000 \text{ hours}}{\text{year}} \times \frac{\$0.017}{\text{kilowatt}} = \$10,000/\text{year}$$

Net Savings .... = $190,000/year

bhp = Brake horsepower

TABLE 6.   Typical Btu Values of Fuels (High Heating Value)

Coal	*Btu/pound*
Anthracite	13,900
Bituminous	14,000
Subbituminous	12,600
Lignite	11,000
Heavy fuel oil	*Btu/gallon*
No. 4 heavy fuel oil	144,000
No. 5 heavy fuel oil	150,000
No. 6 heavy fuel oil, 2.7% sulfur	152,000
No. 6 heavy fuel oil, 0.3% sulfur	143,000
Gas	*Btu/cubic foot*
Natural	1,000
Liquefied gases	*Btu/gallon*
Butane	103,300
Propane	91,600

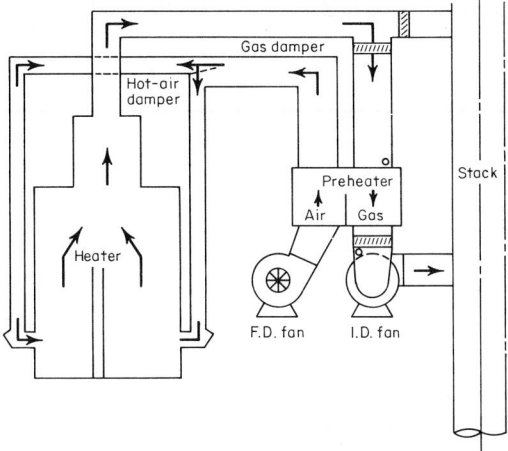

Fig. 6. Air preheater application: F.D. = forced-draft. I.D. = induced-draft.

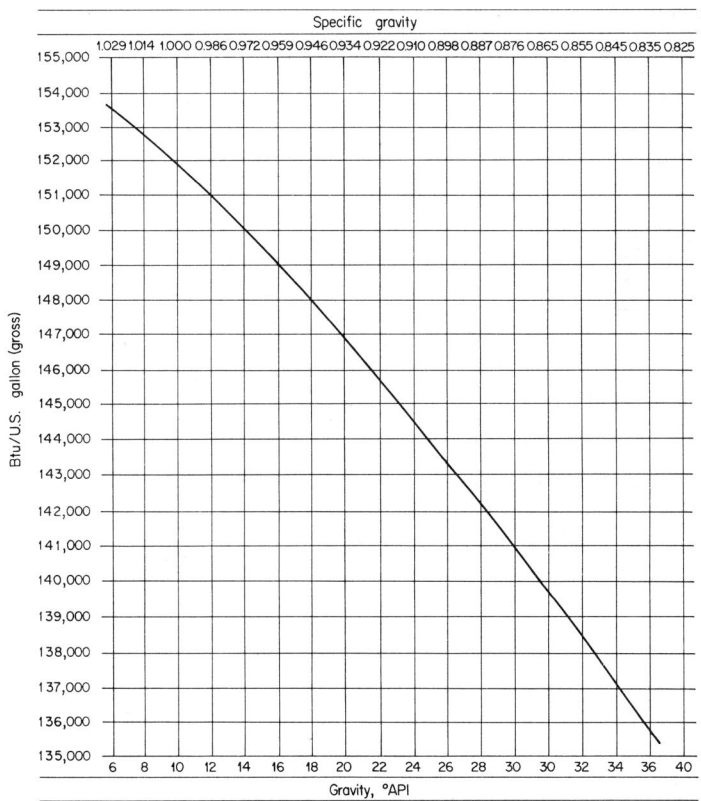

Fig. 7. Heat of combustion of fuels (approximate Btu-gravity relation). Graph does not allow for variance due to sulfur content.

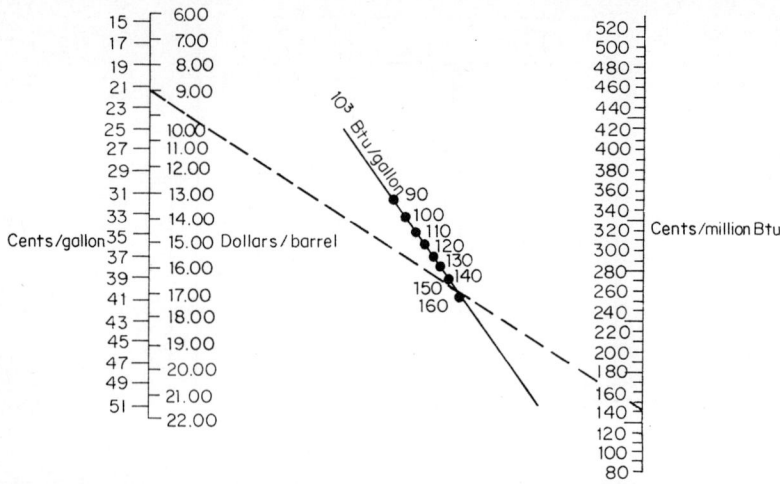

Fig. 8. Nomograph for converting fuel cost from cents per gallon and dollars per barrel to cents per million Btu. To determine the cost of a fuel in cents per million Btu, locate the fuel cost in cents per gallon or dollars per barrel on the left scale. Draw a straight line from that point through the heating value of the fuel on the center scale. Extend the line until it intersects the right scale. The point where the line intersects the right scale is the cost of the fuel in cents per million Btu. In the example illustrated: Line has been drawn for no. 6 heavy fuel oil (normal sulfur) at $9.00 per barrel. The line crosses the center scale at 152,000 Btu per gallon, the higher heating value of no. 6 heavy fuel oil, and intersects the right scale at 143 cents per million Btu.

The net savings converted into equivalent fuel oil results in the conservation of about 21,000 barrels per year. The installed 1974 cost for the system outlined in Table 5 would be approximately $160,000, which would be paid out in less than one year with the net savings in fuel.

Another air preheat system proposed for heat conservation in integrated process plants is the use of hot exhaust air from large air coolers as the combustion air for heaters. See Fig. 9. Many air coolers exhaust air at elevated temperatures. At 1 percent fuel savings per 40°F ($\sim$5.5°C) of increased air temperature, exhaust air at 280°F (138°C) would conserve 5 percent of the fuel required in a heater, with inlet air at 80°F (26.7°C), for just the cost of ductwork and a forced-draft fan from the air cooler to the heater burners.

A third application of air preheat is shown in Fig. 10, which illustrates the use of an air

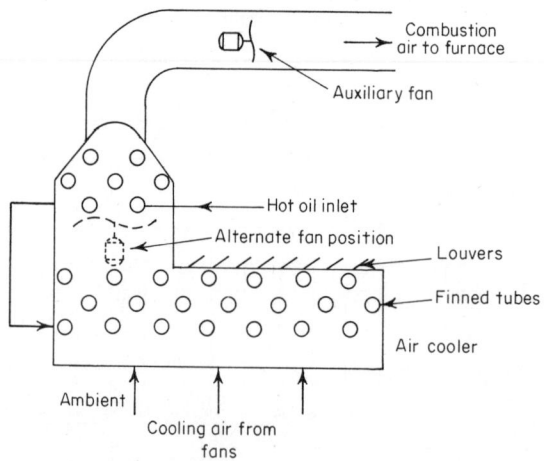

Fig. 9. Combustion air preheat from air coolers.

preheater in a steam-methane reformer flue gas stream after it has given up part of its heat content for steam generation in the convection section.

As the price of other fuels (i.e., gas, distillate fuel) becomes more expensive, power plants and process plants will be required to fire increasing amounts of low-sulfur/heavy fuel oils or coal in their furnaces. When firing heavy oils some added benefits derived

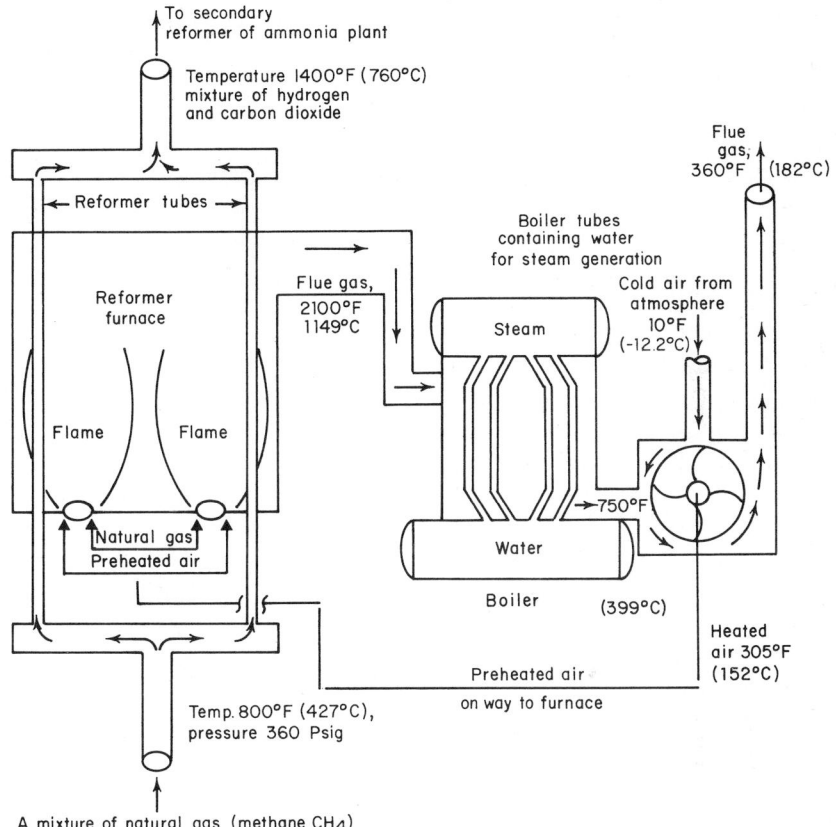

Fig. 10. Heat conservation with an air preheater. (*Air Preheater Co.*)

from using air preheat include (1) less coking of the burner tips, (2) more complete combustion which results in cleaner furnace-side tubes and a concomitant reduction in soot-blowing requirements, (3) cleaner furnace-side tubes which produce more evenly distributed heat, resulting in less coking on the tubes and consequently increasing the potential for a better-quality product from the heater, and (4) possible reduction in sulfur oxides formation because of the potential reduction of excess air required.

A number of detrimental effects are also encountered, however, such as (1) higher potential for nitrogen oxides formation because of higher flame temperatures, (2) increased complexity of equipment with associated increased replacement and maintenance costs, and (3) potential for upsetting the process system if the preheat system fails.

RECOVERY OF ADDITIONAL WASTE HEAT

1. Waste Heat Recovery from Process Streams. This is another large source of energy conservation. Typical of this approach is the utilization of hot fractionator tower circulating reflux streams for such services as: (*a*) direct heat exchange against another process stream, (*b*) reboiler heat for other fractionating systems, and (*c*) a source of heat for generating steam. Examples of all three of these applications are shown in Fig. 11, which compares a fractionation system using air cooling for heat removal and the same fractionation system employing heat-recovery techniques to provide maximum energy conservation for the same heat removal.

2. Closer Heat Exchanger Approach Temperatures.   This is another area that lends itself to conservation and optimization techniques because of the large increases in

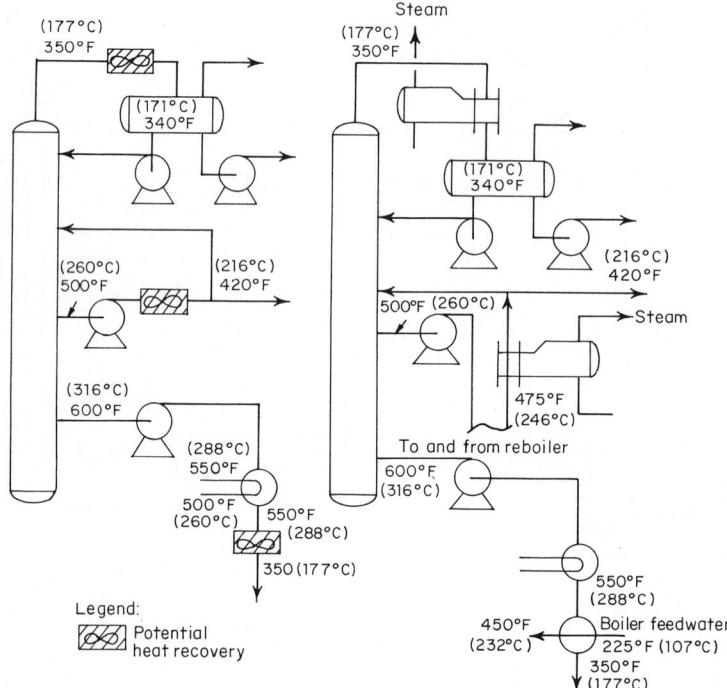

Fig. 11. Maximizing steam generation from process heat.

energy costs. This approach requires the application of basic engineering skills to calculate an economic optimum. To illustrate the methods used in this procedure, let us compare the heat recoveries in an existing system which was installed in 1965 based on an economic optimization of heat-exchanger approach temperatures. Reference is made to the flow diagram of Fig. 12. Assume that the fuel oil fired in the heater is a 10° API, No. 6 fuel oil, which determines the incremental exchanger surface that can be economically added, based on a 1974 fuel oil cost of $9 per barrel. See Table 2. Using Figures 7 and 8, this converts into $1.43 per million Btu in 1974. A simple economic calculation is performed, based on:

1. A two-year payout for new exchanger surface
2. 8000 hours per year operation
3. $45 per square foot installed cost for all added exchanger surface up to four shell panes

It can readily be determined that about 508 square feet of added exchanger surface can be installed for every million Btu per hour saved in the heater—as follows:

$$\frac{\dfrac{1\ \text{million Btu}}{\text{hour}} \times \dfrac{8000\ \text{hours}}{\text{year}} \times \dfrac{\$1.43}{\text{million Btu}} \times 2\ \text{years}}{\$45/\text{square feet}} = 508\ \text{square feet}$$

Thus, up to the point where more than four exchanger shells would cause overcomplexity of piping and potential space limitations, the system under comparison can economi-

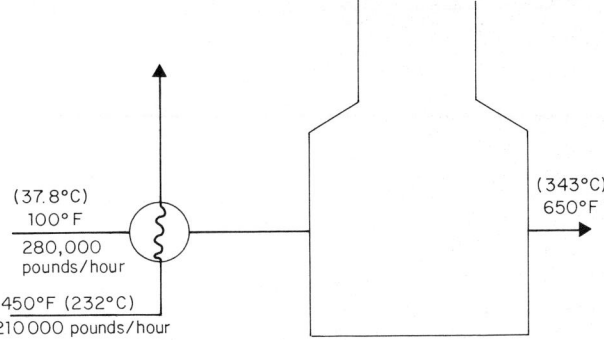

Fig. 12. Heat-exchanger optimization.

cally add exchanger surface at the rate of 508 square feet for every million Btu per hour saved in heater operation. Some typical heat-transfer coefficients are given in Table 7. A log mean temperature difference chart to assist the designer in an evaluation similar to that just described is given in Fig. 13.

3. Optimum Ratio of Air to Oil.   Probably the simplest and most effective means of improving overall operating efficiency of burners, whether they be fired boilers, process equipment, or kilns, is to maintain the optimum ratio of air to oil during combustion. This, in simplest terms, means burning the fuel with the absolute minimum of excess air consistent with complete combustion of the fuel. If, for example, a 50-million-Btu-per-hour furnace is designed to operate at 30 percent excess air, but instead operates at 50 percent excess air, the operation wastes about 3 percent of the fuel burned in the furnace. In the course of a year, this represents a loss of over 80,000 gallons of oil. Approximate fuel losses as a function of excess air are shown in Table 8.

4. Viscosity Changes.   Another factor affecting combustion efficiency in oil-fired furnaces is the need to adjust the fuel feed rate to compensate for changes in oil viscosity. The viscosity of residual fuel oils can vary considerably. The primary cause of these viscosity fluctuations is related to the manufacture of the low-sulfur fuels, which are required by air-quality regulations. Low-sulfur residual fuel can be produced in many ways, and, depending upon the method, the viscosity of the fuel can vary considerably. Variations in oil viscosity also occur frequently when there is a fuel shortage. In past years, fuels were made at a particular refinery from a consistent type of crude oil. However, when crude supplies are limited, it is often necessary to use oils from many sources. Consequently, oil viscosities may vary from delivery to delivery.

Burner nozzles are designed for a specific viscosity and any variation from that viscosity will result in poor performance of the burner. Correct oil viscosity is reached by preheating the oil, and therefore it is a basic prerequisite to know the design viscosity of a burner and the correct level of preheat needed to maintain that viscosity. If the preheat is too high, the viscosity will be too low, and, in most cases, the feed rate will be too low. The flame will be noisy and unstable, and its cone can collapse. If the preheat is too low, the

TABLE 7.   Typical Heat-Transfer Coefficients
   U = Btu/(hour)(square foot)(°F)

Equipment	Butane	Light naphtha	Heavy naphtha	Light gas oil	Heavy gas oil
Reboilers					
Steam-heated	155	140	95	70	60
Light gas oil-heated	90	60	45	70	40
Condensers					
Water-cooled	90	85	75	50	45
Light gas oil-cooled	40	40	35	30	30
Exchangers					
Heavy gas oil in tubes	75	55	50	50	45

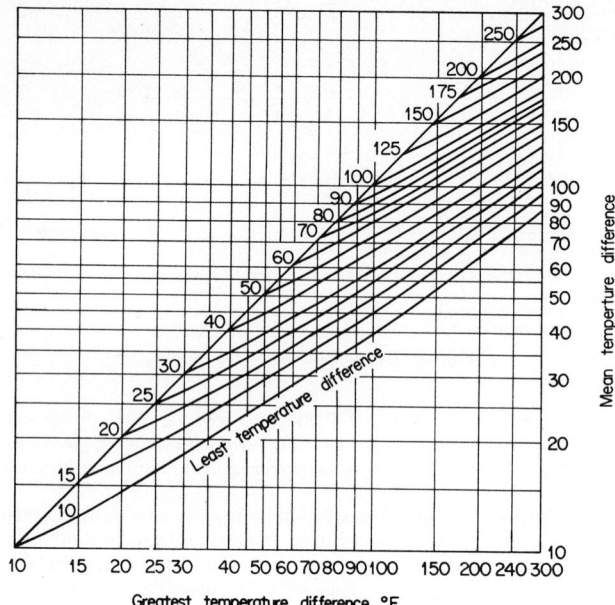

Fig. 13. Log mean temperature difference chart for use in making heat-exchanger evaluations.

TABLE 8.   Fuel Loss as a Function of Excess Air

Excess air, %	Characteristic loss, %	Avoidable loss, %
10	18.79	
20	20.25	1.46
30	21.70	2.91
40	23.16	4.27
50	24.62	5.83
60	26.08	7.29

TABLE 9.   Examples of Fuel Conservation by Use of Combustion Analysis

Operating factor	Boiler no. 7	Boiler no. 8	Boiler no. 9
Fuel basis	No. 6 fuel oil	No. 6 fuel oil	Gas
Exit-gas oxygen, %	9.10	8.50	9.50
Exit-gas temperature	290°F, 143°C	440°F, 227°C	842°F, 450°C
Furnace efficiency, %	85.88	31.31	64.66
Less radiation, %	0.76	1.13	4.75
Unit efficiency, %	85.12	80.18	59.91
Input, million Btu/hour	84.00	35.67	6.12
Output, million Btu/hour	71.50	28.60	3.67
Goal exit-gas oxygen	3.00	3.00	2.00
Goal exit-gas temperature	290°F, 143°C	396°F, 202°C	714°F, 379°C
Goal furnace efficiency, %	88.16	85.64	76.09
Less radiation, %	0.76	1.13	4.75
Expected unit efficiency, %	87.40	84.51	71.33
Goal input, million Btu/hour	81.81	33.84	5.14
Input savings, million Btu/hour	2.19	1.83	0.98
Annual 1974 savings	$25,027.20	$20,914.29	$10,192.00

viscosity will be too high and the feed rate will be too high, in most cases. The droplet size will be too large, causing incomplete combustion and soot.

The correct nozzle viscosity can be maintained with any oil, regardless of its viscosity at ambient temperature, by establishing procedures for making frequent tests of the viscosity of the fuel in storage, and consulting a viscosity-temperature chart to determine the correct level of preheat. Procedures should be set up to make frequent tests of fuel oil viscosity to ensure the proper preheat temperature.

5. Combustion Analyzers.   The use of combustion analyzers to measure and report the efficiency of furnaces is becoming increasingly necessary as fuel costs escalate—so that heat losses may be found and corrective action taken. Control devices can be installed to automatically monitor the oxygen concentration in the intake to the burners and thereby provide maximum combustion of the fuel. This assumes, of course, that the characteristics of the fuel remain essentially constant.

A computer calculation system has been developed by Dupont (Ref. 3) which produces a report showing the annual savings obtainable by operating a furnace at its optimum specified condition, as shown in Table 9. In boiler No. 7, it was determined that the exit-gas oxygen was 9.1 percent and should have been 3.0 percent. This reduced the efficiency from 87.4 to 85.1 percent and resulted in an annual loss of about $25,000 per year. The example shows two additional furnaces which list potential savings, if operated properly, of about $21,000 and $10,000 per year, respectively. The total fuel saved is the equivalent of 263,000 barrels per year of No. 6 fuel oil.

There are industrial plants where annual savings in excess of $300,000 per year have been attainable through a program to control excess air for their furnaces.

Gas Turbines with Heat Recovery.   Whenever a suitable fuel is available, gas turbines should always be considered as drivers for rotating equipment which have large horse-power consumption to provide for optimization of energy use at an economic cost. Although gas turbines have run for long periods on residual fuel oil, the use of heavy fuel oil in this service has resulted in high maintenance and low reliability. Clean, distilled fuels are satisfactory, but the price may be too high compared with that of other available fuels. Therefore, most gas-turbine installations in the process industries burn natural gas or a fuel gas produced in plant operations.

Gas turbines, when properly designed and installed for heavy-duty industrial service, are among the most reliable of driver systems. The practical economics of gas-turbine applications in most refineries depends on the utilization of the heat, or better the heat and oxygen, contained in the exhaust gases. A typical flow scheme is shown in Fig. 14. The quantity of heat transferred in the boiler or process heater operating on the exhaust from a 9000-bhp gas turbine, for example, varies from 53 million Btu per hour with no

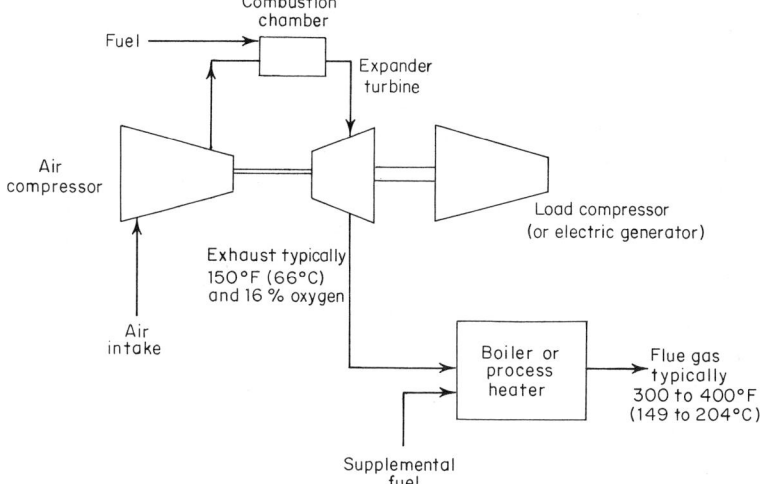

Fig. 14. Typical gas-turbine installation in process plant.

supplemental firing to a maximum of 375 million Btu per hour with full supplemental firing. An integrated plant design concept allows maximum use of the potential for highly efficient mechanical and heat energy generation in combined gas-turbine and exhaust heat-recovery systems. A trend toward such systems is apparent in recent large refinery installations in the United States.

A gas turbine requires a starting driver of relatively small size. However, it is often advantageous to use a steam-turbine driver capable of delivering up to 30 percent of the total driver requirement, to provide maximum flexibility and efficient operation of the gas-turbine cycle.

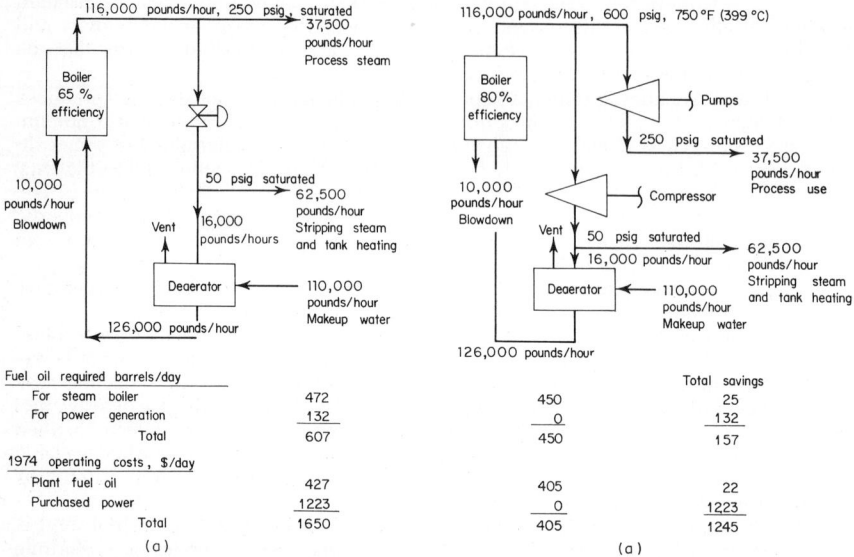

Fuel oil required barrels/day			Total savings
For steam boiler	472	450	25
For power generation	132	0	132
Total	607	450	157
1974 operating costs, $/day			
Plant fuel oil	427	405	22
Purchased power	1223	0	1223
Total	1650	405	1245
(a)		(a)	

Fig. 15. Steam generation efficiency: (a) With 300-kilowatt purchased electricity, (b) With no purchased electricity. Fuel oil required to generate 3000 kilowatts of power = 132 barrels per stream day.

A system related to gas-turbine technology has been developed* which uses an expander driver for energy recovery from catalytic cracking unit flue gases. See Fig. 2. In most recent installations, the expander driver normally furnishes all the power required for the axial-flow air blower, plus some additional power converted to electricity.

Steam Generation Efficiency.   The generation of steam at higher pressures can be another source of energy conservation. A simple example is given. Assume that a plant requires 3000 kilowatts for the mechanical drivers on pumps and compressors; 37,500 pounds per hour of 250-psig saturated steam for process use; and 62,500 pounds per hour of 50-psig saturated steam for stripping steam, tank heating, and deaeration of boiler feedwater. The simplest approach would be to purchase the power needed from a local utility company, generate 100,000 pounds per hour of 250-psig steam on plot, and let down 62,500 pounds per hour of steam through a desuperheating station to 50 psig to provide the low-pressure steam requirements. If the same plant requirements are taken, but instead generate the steam at a higher pressure and utilize it to provide both the motive force for the mechanical drivers and for the steam required for process use, a subtantial savings in both overall fuel oil consumption and plant operating costs can be realized. As shown in Fig. 15, the aforementioned application of steam pressure optimization will result in a savings of about $415,000 per year in operating costs, and an overall conservation of fuel oil of 2,260,000 gallons per year. This approach to conservation is necessary if industry is to maximize energy conservation.

Another application of the use of higher-pressure steam levels to conserve energy in a power plant is shown in Fig. 16. The more recent large power plant designs utilize the

*Shell Oil Company.

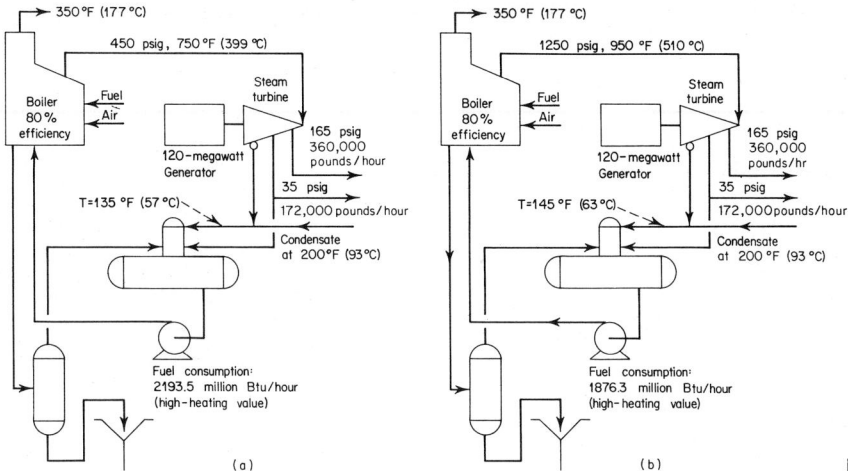

Fig. 16. Power-plant conservation: (*a*) Lower-pressure steam system. (*b*) Higher-pressure steam system.

availability of industrial *gas turbine-power generator-steam turbine* sets in combined-cycle systems to conserve fuel.

Reduction of Heat Losses from Equipment. Motivated by higher-energy costs and need for energy conservation, a more effective approach should be taken in the applications involving the use of insulation on storage tanks and transmission lines that handle hot materials, such as asphalt, fuel oil, waxes, steam, and the like. Large amount of heat energy may be lost if inefficient or improper insulation is installed. The estimated heat loss from various sizes of pipe, valves, and flanges for a range of operating temperatures is indicated in Fig. 17. Similar charts which depict the estimated heat losses for various thicknesses of insulating materials are available from insulation suppliers. Use of these charts enables an economic evaluation for determining the optimum insulation type and thickness. In addition, the traditional concept of not insulating flanges and valves must be carefully reevaluated to include better attention to energy costs and conservation.

A typical evaluation is shown by Table 10, which describes the method used to select the economical insulation thickness for a two-tank asphalt storage system, while maintaining a constant oil temperature using fuel-oil-fired bayonet heaters. Using 1976 economics, this tabulation shows that four-inch insulation on the tank side and two-inch insulation of the roof is the economical selection, based on a simple two-year payout. The economic insulation thickness with a 40-cent per million Btu fuel cost (1970) would have been two inches on both side and roof.

TABLE 10.   Cost Evaluation of Tank Insulation

	Side	2-inch	3-inch	4-inch	5-inch
Insulation	Roof	2-inch	2-inch	2-inch	2-inch
Heat loss from side, million Btu/hour					
At −20°F(−28.9°C), 90 hours/year		1.063	0.732	0.559	0.427
At 50°F(10°C), 4000 hours/year		0.928	0.651	0.490	0.375
At 85°F(29.4°C), 4670 hours/year		0.852	0.589	0.450	0.344
Heat loss from roof, million Btu/year					
At −20°F(−28.9°C), 90 hours/year		0.591	0.580	0.573	0.566
At 50°F(10°C), 4000 hours/year		0.500	0.495	0.490	0.485
At 85°F(29.4°C), 4670 hours/year		0.448	0.446	0.444	0.442
Total still air heat loss, million Btu		11,932	9,496	8,197	7,200
Heat loss at 10 miles/hour wind		12,767	9,876	8,525	7,637
Installed insulation cost		$125,000	$130,500	$136,000	$142,500
Two year fuel cost at $2/million Btu		73,000	57,000	49,000	44,000
Total two-year cost		$198,000	$187,500	$185,500	$186,500

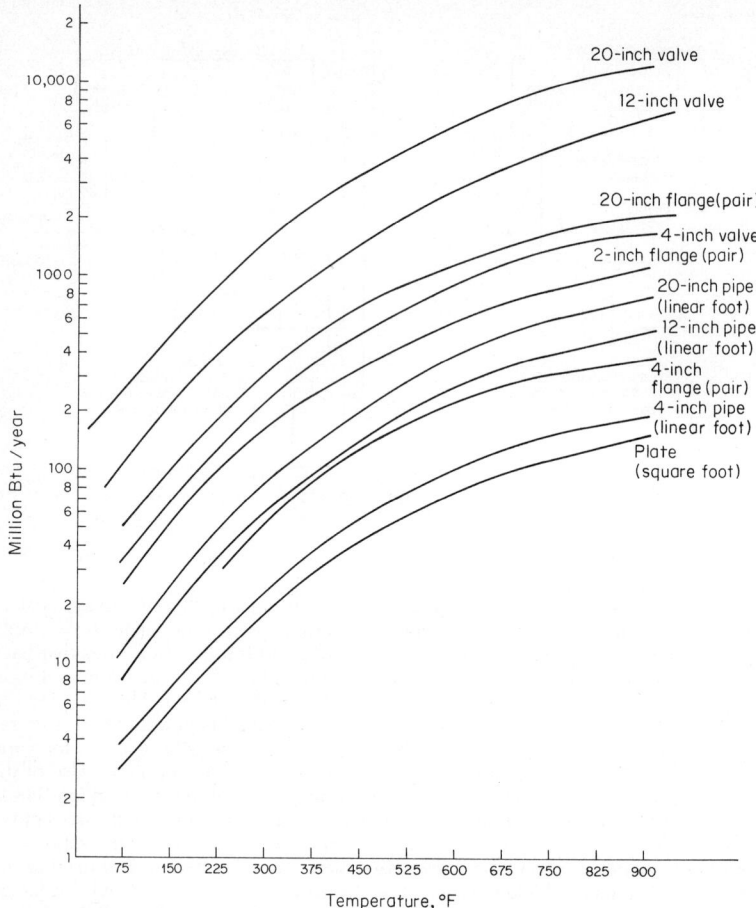

Fig. 17. Heat loss from uninsulated pipe, flanges, and valves. *(Exxon Lubtext D250.)*

PROCESS ENERGY CONSERVATION

Recovery from Waste Heat.   The use of integrated process systems for heat recovery is a major area in which a high quality of engineering skill is required to optimize energy use. Some typical applications of total systems have been shown before in this text (Figs. 2 and 3). In addition to total plant systems such as those just cited, numerous other individual heat-recovery applications exist. Some of these are described in the following paragraphs.

1. Cold Process Streams.   Passing a cold process stream through heater convection coils is a means of heat recovery. A typical application is shown in Fig. 18. Naphtha produced from a new crude unit required preheating to satisfy the requirements for feed to a naphtha desulfurizer unit. Pumping naphtha through crude heater convection coils not only met the process condition, but also substantially improved the heater efficiency.

2. Preheating Process Gases.   Process gases which are to be burned in incinerators or thermal oxidizers may be preheated as a means of heat recovery. The heat recovered reduces the fuel needed for combustion in a ratio of almost one Btu for each Btu saved. In many cases where the gas being heated is part of the fuel supply, sufficient heat is conserved to reduce the fuel makeup to essentially zero. Applications of such concepts are shown in Figs. 19 and 20.

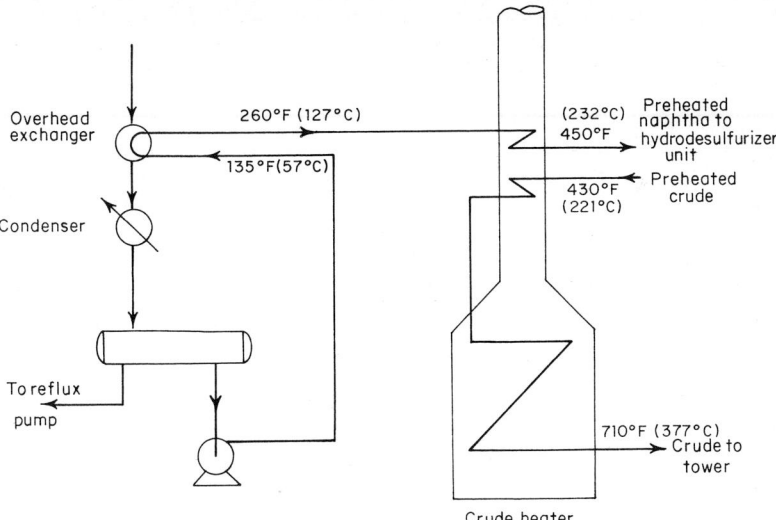

Fig. 18. Process heat recovery in a heater.

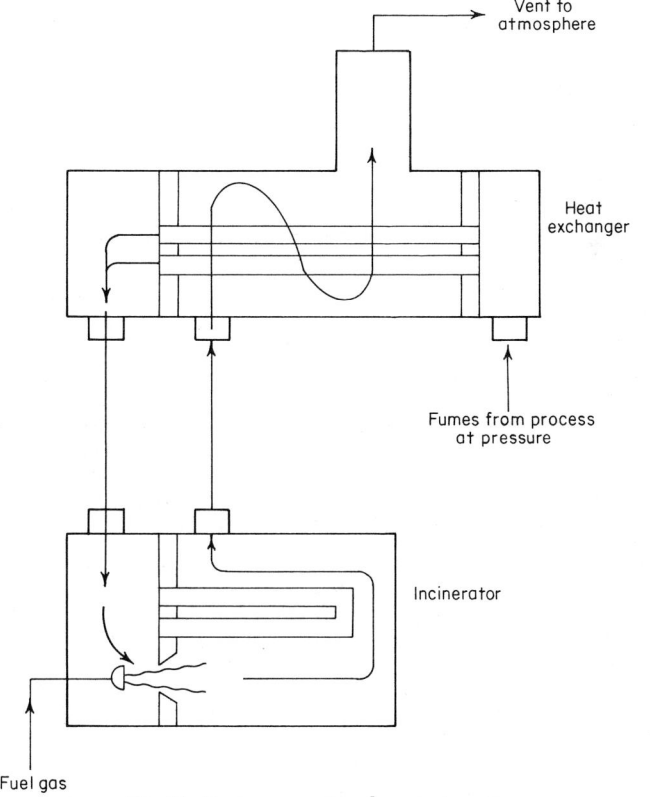

Fig. 19. Heat recovery in a fume incinerator.

3. Innovative Heater Design Concepts.   Heaters can be utilized to perform numerous process services involving innovative design concepts, as outlined in Fig. 21. The sketch depicts a three-tower fractionation system in which the heater is used to directly reboil one of the towers, indirectly reboil a second column, and utilizes the convection section waste heat to reboil a third tower.

Recovery through Rotating Equipment.   Rotating equipment usually represents the second largest energy consumer in a process plant. The selection of rotating equipment drivers includes the evaluation of such factors as the recovery of energy from the pressure reduction of liquid or vapor process streams. This energy may be recovered by the use of hydraulic turbines, expander turbines, and/or gas turbines, individually or collectively, which, in turn, can drive pumps or compressors.

1. Hydraulic Turbines.   A typical illustration of this application is given in Fig. 22, which portrays the actual installation in a 20,000 barrels per stream day, two-stage

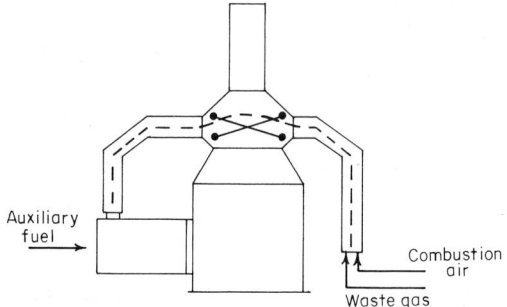

Fig. 20. Preheat system for thermal oxidizer.

hydrocracker. Liquid streams from the high-pressure separators in both the first and second stages are combined after passing through liquid-level control valves. A single hydraulic turbine recovers energy from the combined liquid stream. The hydraulic turbine, coupled with a supplementary motor, drives the second-stage feed pump. The first-stage feed pump, not shown in Fig. 22, is motor-driven. A common spare feed pump is also motor-driven.

2. Gas Expanders.   Recovery of energy from gas at elevated pressures provides another opportunity for conservation in process plant design. Two examples of this application are shown in Figs. 23 and 24. A turboexpander cycle (Fig. 23) in which "free" pressure drop available from a high-pressure gas that is intended for use at low pressure is utilized to recover refrigeration energy and develop added power for other uses. The use of a spent-air stream (Fig. 24) can provide supplemental motive power through a gas turbine to help drive the fresh air compressor.

3. Combined Systems.   A collective application of both a hydraulic turbine and a simple gas expander in a combined cycle to maximize the energy recovery available in the liquid and gas streams of a gas-treating plant is shown in Fig. 25. The application of these energy-saving techniques has resulted in the conservation of more than 500 kilowatts per hour in a relatively small process unit.

Recovery by Application of Process Design Techniques.   The economic balance in large fractionating towers has shifted from the use of a few trays and high reflux ratios to the use of more trays and lower reflux ratios because of the rapidly escalating cost for fuel. Therefore, designers must evaluate each system to determine the maximum energy conservation when it is integrated into an overall energy balance for the entire plant. The results of an economic comparison of reflux ratio versus annual cost for a typical debutanizer are shown in Fig. 26.

Another process technique for conserving fuel comprises using the heat provided by vapor compression for reboiling towers having low enough bottom temperatures to utilize the low-level heat. A typical application of this approach is a demethanizer in an ethylene plant in which the heat of compression generated by the overhead product compressor is used to reboil the tower in a closed cycle.

In another example, in connection with a recent gas-treating plant design, it was found

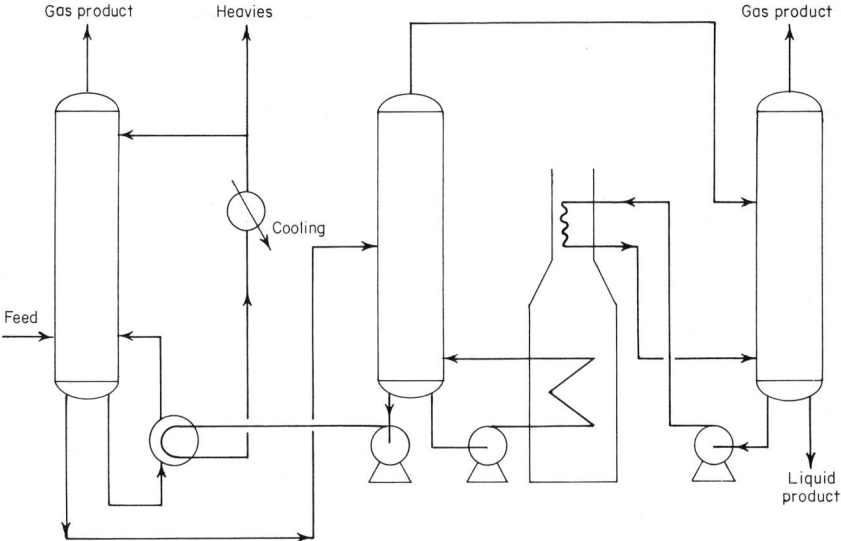

Fig. 21. Maximizing heat recovery in a process scheme.

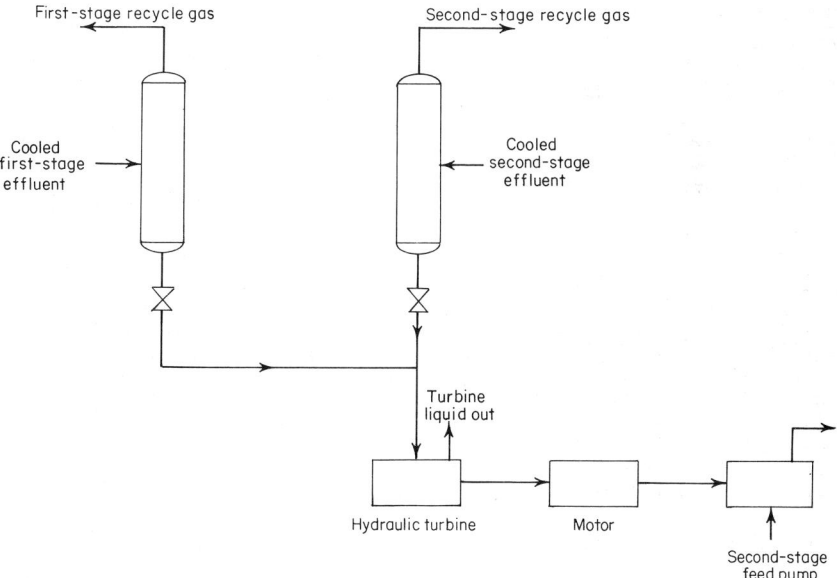

Fig. 22. Hydraulic turbine system.

Rating	Hydraulic turbine	Motor	Second-stage feed pump
Normal, bhp	800	500	1300
Installed, bhp		700	

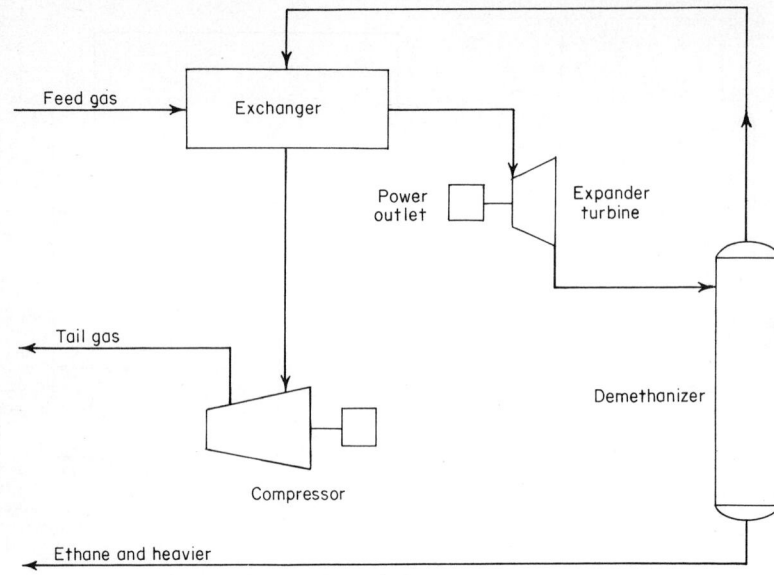

Fig. 23. Turboexpander cycle for cryogenic gas processing.

that a two-tower stripper configuration provided substantial savings in reboiler steam over a single-tower system because both a high- and a low-level residual hydrogen sulfide content could be tolerated in the treated-gas product from each of two hydrogen sulfide absorbers. This reduced the stripping steam required to produce the high-level hydrogen sulfide product gas with a resultant energy savings. The two systems are compared in Fig. 27.

Reduction of Air-System Losses. The motors and turbines used to drive air compressors in utility and instrument air systems are potential sources of energy losses, particu-

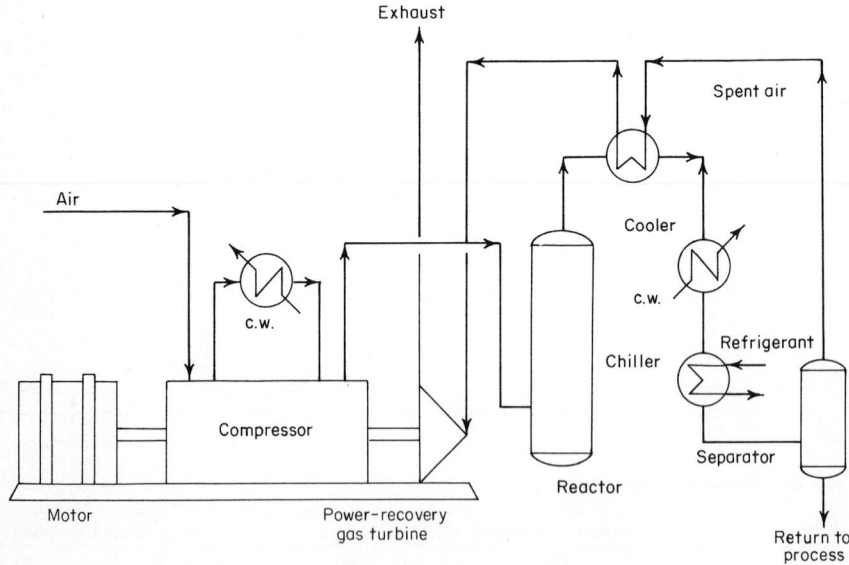

Fig. 24. Spent-air power-recovery system: C.W. = cooling water.

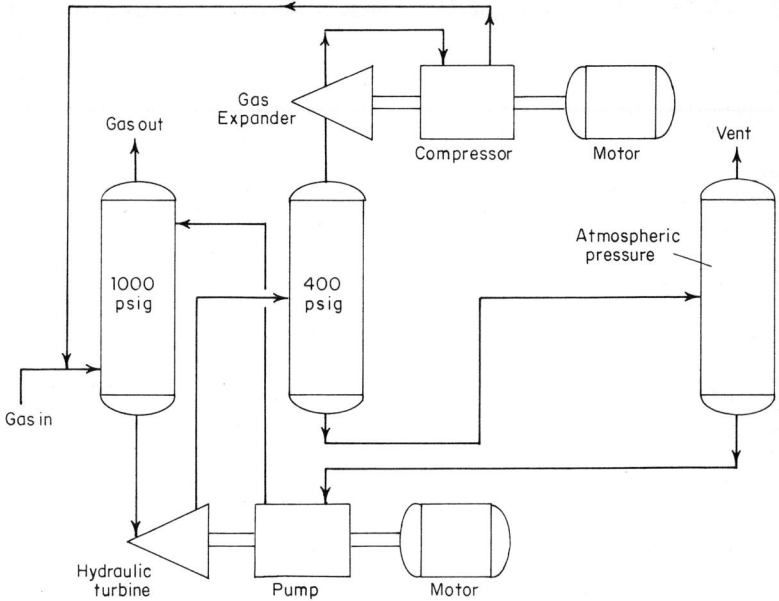

Fig. 25. Combined hydraulic-turbine–gas expander system.

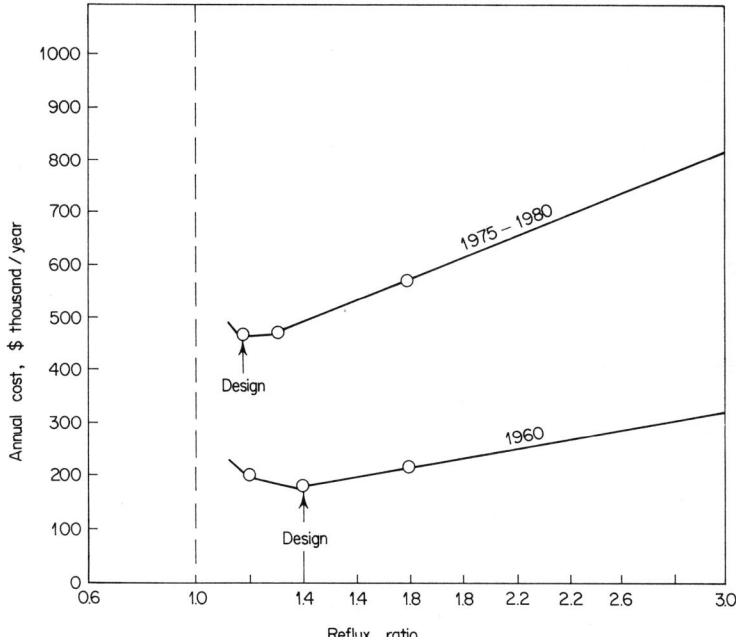

Fig. 26. Economic comparison of reflux ratio versus annual cost for a typical debutanizer.

larly where the air systems are used in poorly designed, constructed, and/or poorly maintained plants.

1. Air Leakage.   A major offender in the waste of energy in a compressed air system is air leakage. This leakage can result from leaky shutoff valves, leaky valves in air tools, poor-quality air hoses and clamps, leaking valve-stem packing, excessive instrument vents, and many other like factors. The "air leak" type of energy loss can usually be determined only after the plant is in operation. Then, by operating an air compressor of known capacity while production is shut down, the percentage of time the compressor operates under load, as compared to the time of one complete load-unload cycle, indicates the percentage of the compressor capacity that is being used to supply air leakage. After

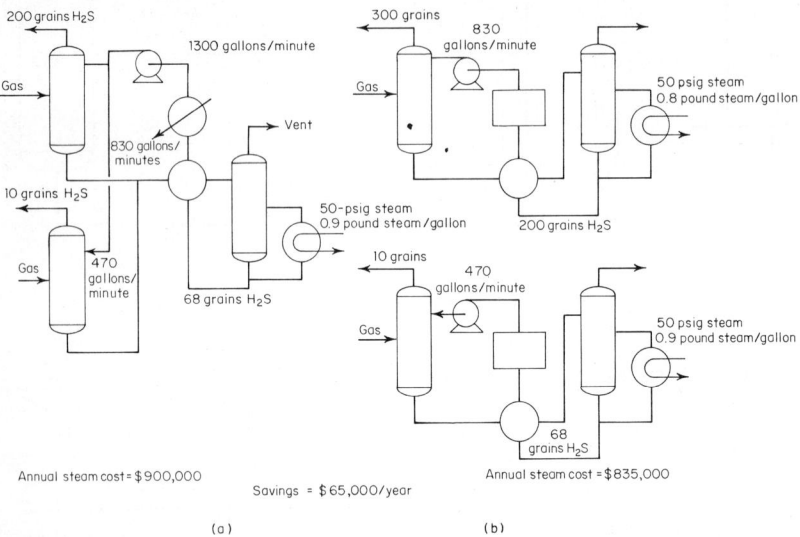

Annual steam cost = $900,000   Annual steam cost = $835,000

Savings = $65,000/year

(a)   (b)

Fig. 27. Comparison of gas-treating systems: (*a*) Traditional system. (*b*) Energy-saving system. H_2S = hydrogen sulfide.

this test is run, a comparison between this percentage and the average total demand for air will indicate the percent of plant air leak. If this is more than 10 percent of plant demand, leaks should be located and repaired.

2. Overcompression.   The overcompressing of air to accommodate the pressure drop in poorly designed air piping systems is another waste of energy which can be easily prevented. The energy waste from excessive line pressure drop can result in such factors as high energy cost for wasted horsepower and/or low air pressure at utility stations which can cause pneumatic tools to operate very inefficiently. The drastic reduction in tool output caused by low air supply pressure is indicated in Fig. 28 (Ref. 4).

CONSERVATION OF ELECTRIC ENERGY

Optimization of Plant Power Factor.   Excessive voltage drop from the incoming power source at the transformers to the point of application can be a waste of energy. Excessive voltage drop ($E = IR$) can be caused by undersized wiring, high resistance, or excessive current.

An element that is often overlooked in the operation of a plant electric power distribution system is the power factor, which is defined as

$$\text{Power factor, \%} = \frac{\text{useful power consumed}}{\text{total power transmitted}} = \frac{\text{Kilowatts}}{\text{Kilovolt-amperes}}$$

As this ratio indicates, if the power factor of an industrial plant electric distribution system is held at 100 percent at all times, then all the power transmitted is used. Practically, few industrial plants operate near a 100 percent power factor.

Plants whose production processes require a very high percentage of electrical resistance load can operate with a very satisfactory power factor. Induction motors are usually the greatest consumers of plant electric energy, and result in a high consumption of inductive power. This results in plant power factors much lower than 100 percent. It is not unusual to find overall power factors as low as 60 to 70 percent.

Installation of synchronous motor drivers, instead of induction motors, is one way to prevent this powerloss, especially when the horsepower required is 200 hp or above. Synchronous motors provide leading reactive kilovolt-amperes when operating with full-field excitation that offsets the lagging kilovolt-amperes inherent with poor or low power factor.

Installation of static capacitors (which provide reactive leading kilovolt-amperes) to directly offset the lagging kilovolt-amperes with inductive electric equipment is another conservation approach for upgrading a low power factor. Static capacitors have the advantage of availability in a multiple of standard kilovolt-ampere ratings, have no moving parts, and can be installed at desirable locations in the plant distribution system. They also can be wired directly in connection with a major potential power loss source, such as a large horsepower, lightly loaded induction motor.

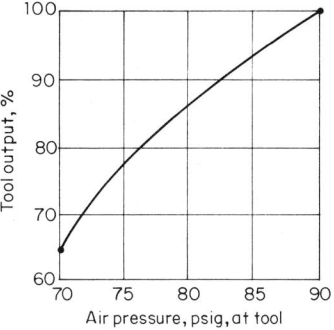

Fig. 28. Effect of inadequate air pressure on operation of a pneumatic tool.

OPTIMIZATION OF PLANT-LIGHTING SYSTEM

Light abatement can be included in the design philosophy to minimize energy consumption by specifying pendant, hanging reflective and flood-lighting fixtures for general use and defining these as the color-corrected, mercury-vapor type. This practice results in the need for fewer fixtures as compared with incandescent lighting for the same amount of illumination. The lighting fixtures can be equipped with reflectors to minimize uplighting (light loss diffused behind the fixture).

Further, photoelectric cells may be used to automatically turn lights on at dusk and off at dawn; and local switches may be installed with instructions to turn lights off unless required for safety. Light abatement is also achieved by optimizing the mounting height and focus angle of all fixtures.

FUEL AND PRODUCTS CONSERVATION

Proper Tank Design.   The design of storage tanks for feed, rundown, and product storage requires careful consideration of the vapor pressures of the material being stored, the storage temperature and pressure, toxicity, and other factors. With the advent of environmental constraints for air and water emissions and the constantly increasing value of the stored materials, it becomes mandatory to utilize either the floating roof and weather-master tanks for all materials with a vapor pressure at storage temperature of 1.12 to 11.5 psia; or to provide vapor recovery systems if other types of tanks are used. A listing of expected losses from all types of tanks in refinery systems is contained in "API Guide for Tank Venting," periodically updated by the American Petroleum Institute, Washington.

Vapor-Recovery Systems.   In past years, valuable vapors and liquids have been allowed to vent to the atmosphere or drain to sewers because recovery costs were too high. The high cost of feed, products, and fuels now makes recovery economical. Gas and liquids are lost in many areas of a process plant. Three major areas are (1) storage tank vents; (2) relief systems losses; and (3) tank truck/tankcar-loading station losses.

1. Gas Recovery from Storage Vessels.   Light fuel gas or inert gas may be used to fill the vapor space above the stored liquid in a storage tank or drum. These gases may be contained in a closed system designed for distributing blanket gas and for containment, treatment, and reuse of the vent gases. The vent gases can then be collected into a variable-volume gas holder, compressed, and sent to a fuel gas system. A typical arrangement of this type is shown in Fig. 29.

2. *Gas Recovery from Relief Effluent Systems.* Liquids and vapors are lost in large amounts in most relief and flare systems whenever an upset or emergency condition occurs in a process plant. In the past, these systems were kept as simple as possible because they were provided as "safety" systems at minimum cost. As a result, many relief effluent streams were vented to the atmosphere, discharged to sewers, or simply dumped on the ground with a consequent loss of valuable, energy-rich materials. These materials may be recovered in various ways by systems that contain, retain, and recirculate a

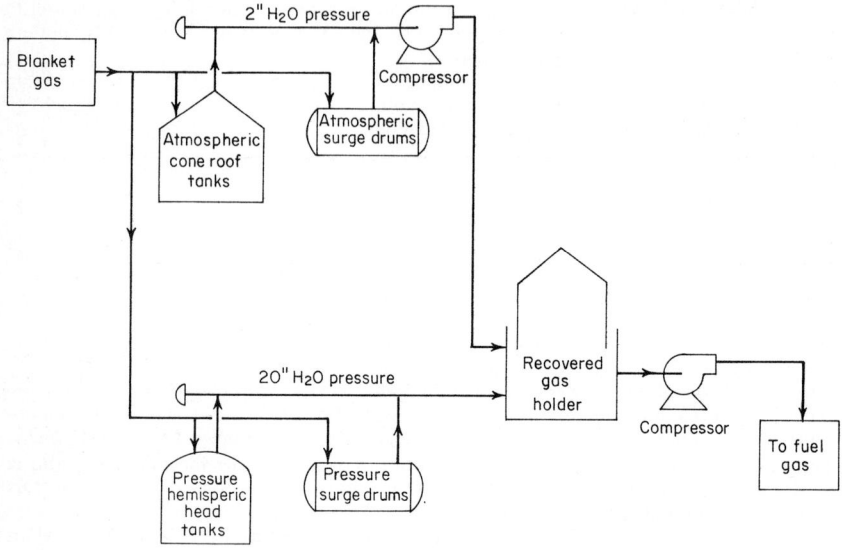

Fig. 29. Gas recovery from storage vessels.

maximum quantity of these effluent streams. Some of the design parameters that can be considered in the design stage to minimize the quantities of gases and liquids vented into the flare system include:

1. Maximum practical air cooling to provide partial cooling of the relieved stream during power outages or cooling water failure.

2. Cascading relieved materials to lower-pressure and lower-temperature levels within the unit.

3. Locating relief valves on the cold side of equipment.

4. Increasing mechanical design pressures (relief-valve set pressures).

5. Multiple, parallel-operating, reflux systems with alternate steam- and electric-power-driven pumps.

6. Automatic pressure maintenance of the fuel gas system by an LPG (liquefied petroleum gas) or equivalent fuel vaporizer, which allows heaters and other equipment to be removed from service without inadvertent venting of excess fuel gas to the flare.

Once an upset condition occurs and gases and liquids are vented into the flare header system, numerous other design parameters can be used to reduce or practically eliminate continuous losses from the flare, including (1) quenching of the flared gases (and liquids), using oil, water, or other compatible media to remove heat and thereby recover the condensable portion of the effluent stream for reuse in the plant as fuel oil or incremental feed/product, (2) liquid recovery by installation of a separate header for liquid relief streams, and (3) containment and recovery of the noncondensed gases in gas spheres or gas holders with subsequent recycle of these gases to the fuel gas system.

A system that embodies all the foregoing criteria is shown in Fig. 30. The resultant economies which may be realized when this system is used in an integrated optimized design for a 250,000 barrel-per-calendar-day refinery are summarized in Table 11.

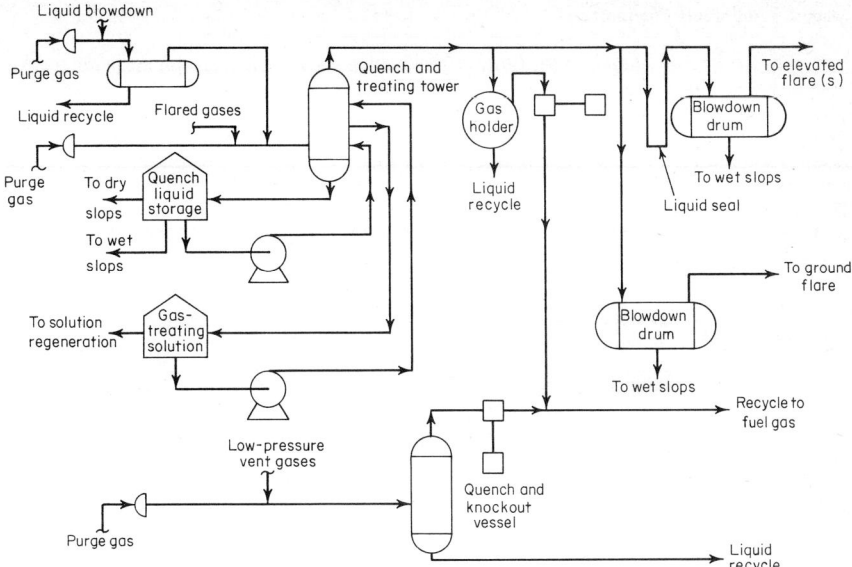

Fig. 30. Modern flare system configuration for maximum liquid and gas recovery.

TABLE 11. Return on Investment and Payout Period for Modern Flare System Configuration for Maximum Liquid and Gas Recovery

Cost, profitability, and investment elements	Case A: With no revenue for increased efficiency*	Case B: With revenue for increased efficiency†
Differential depreciable investment (D.I.), $ million	8.20	8.20
Differential operating costs (O.C.), $ million/year		
Utilities	−(0.10)	−(0.10)
Maintenance @ 2.5% of D.I.	+0.20	+0.20
Operating Labor	+0.04	+0.04
Plant general @ 40% of total labor	+0.07	+0.07
Total	0.21	0.21
Differential fixed costs, (F.C.), $ million/year		
Instruments @ 5.0% of D.I.	+0.41	+0.41
Taxes & insurance @ 2.5% of D.I.	+0.20	+0.20
Accelerated depreciation‡	+1.37	+1.37
Total	1.98	1.98
Profitability, $ million/year		
Gross differential revenue	+3.00	+4.10
Deduct differential O.C.	−0.21	−0.21
Deduct differential F.C.	−1.98	−1.98
Gross profit	+0.81	+1.91
Deduct Taxes @ 52%	−0.42	−0.99
Net profit	+0.39	+1.37
Add depreciation as income	+1.37	+1.37
Total annual cash accumulation	+1.76	+2.29
Return on investment, %	4.8	11.2
Payout Period, Years	4.6	3.6

*Recovery of flare purge gases, relief valve leak gas and small intermittent relief streams
†Takes into account the added reliability built into the refinery for a "maximum refinery" design with a resultant increase in plant operating efficiency
‡Based on legislated 6-year payout for environmental systems

3. *Gas Recovery at Loading Stations.*   The loading of volatile material into tank trucks and tankcars, much like filling an automobile gasoline tank at a service station, typically results in emission losses to the atmosphere. These emission losses can be recovered in systems of the type shown in Fig. 31, which illustrates a gasoline-loading operation. This recovery is conducted in a vapor-tight system which prevents the escape of vapors to the atmosphere by condensing the gasoline vapors and returning the condensed liquid to product storage tanks.

Although this article has concentrated on energy planning and conservation in the

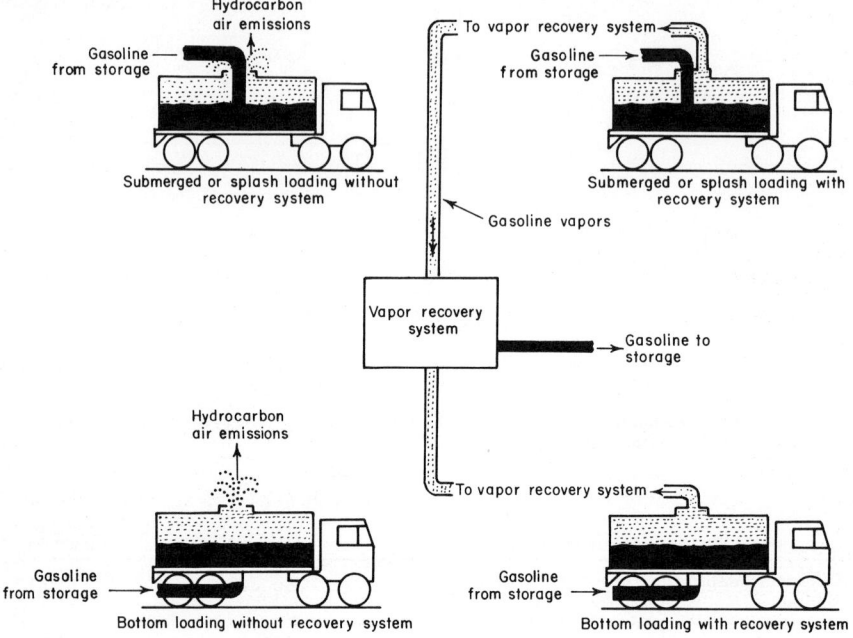

Fig. 31. Gasoline-loading operations.

petroleum processing industry, it is of interest to point out some specific actions (albeit some of these may be rather obvious actions) taken in other fields.* These include:

GENERAL INDUSTRY

Sprayed 1.5 inches of urethane foam insulation on 55,000-barrel tanks storing heavy fuel oil at 150°F (66°C) in winter. The foam is covered with polyvinyl chloride to protect against weathering.

Planned computer setup which will briefly shut down equipment when electric load reaches a preset maximum.

Staggered startup times on equipment to minimize use of utilities at peak demand periods.

Utilized exhaust steam to supplement propane in heating paper driers. This cut propane use from 6500 to 4200 gallons per day.

Changed to set process in grinding phosphate rock for phosphoric acid production, saving 8 to 9 gallons of oil per ton of phosphorus. This reduced ball mill size 30 to 40 percent and decreased horsepower 30 to 40 percent, saving about 35 kilowatt hours per ton—reducing operating costs by $3 to $4.25 per ton and significantly reducing air pollution.

Rescheduled operations to ease demand for electric power in peak periods and began using carbon monoxide and methane waste from a manufacturing process as fuel.

Installed ultraviolet lights for paint curing on 12 production lines, saving 365 trillion Btu and $300,000 per year.

Replaced a 50-year-old periodic kiln with a large insulated kiln. Energy savings, 54 percent.

Used high-pressure gas stream for mechanical power. Fuel gas stream was being throttled across a valve. Installation of an expander turbine recovered 1200 horsepower, worth $96,000.

*Courtesy of source: *Oilways*, vol. 3, 1974. Exxon Company, U.S.A., Houston, Texas

Changed process operating using a series of driers to dry foam and latex carpet backing. Investigation revealed that the process was really one of curing. Drying required large volumes of hot air, whereas curing required only the maintenance of a lower temperature. Previously 500 percent more air was heated than necessary.

Installed hydraulic returns on all outside doors to assure closing.

CHEMICAL INDUSTRY

Stopped leaks in steam traps, saving energy equivalent to heat 5000 homes.

Substituted liquid heat-transfer medium for steam in tracing systems. Cut steam requirements by 50 percent in one plant.

Used latent energy in olefinic feedstocks.

Substituted steam turbines for electric motor drives wherever turbine exhaust could be used for process heat.

Shut down extra pumps, fans, and compressors so remaining units operate at higher efficiency.

Began operating distillation equipment at near minimum steam consumption.

Recovered heat from effluent air in hot processes.

Began maintaining steam jets in vacuum systems. Jet wear increases steam consumption.

Matched electric motors to job requirements so that motors operate at full loads.

Made changes to avoid subcooling of process streams that have to be reheated.

Modified power plants, adding gas turbines, waste-heat recovery boilers and back-pressure steam turbines. Combination increases capacity by 100,000 kilowatts with lower fuel consumption.

Installed coalescer to reduce oil content to 1 ppm in condensate. Savings 10 million Btu per hour.

Began using gas residues from operations as fuel supplements.

Installed flowmeter on steam stripping column previously set for 20,000 pounds of steam per hour. Meter determined that 14,000 pounds was proper quantity.

Invested $600,000 in furnace to burn scrap tires.

Used new dimensionally-stable anode to cut electric cell power consumption in chlorine plants.

METALS INDUSTRY

Moved hot quench oil tank indoors. The heat it gives off contributes to space heating needs.

Began loading annealing furnaces to full capacity, eliminating frequent heats for partial loads.

Changed to lower viscosity cutting oil and used a chip wringer to reduce cutting oil carry-off.

Started transferring metal in molten state between processes to avoid remelting; delivered aluminum in molten state to die casters.

Began producing calcined coke by using volatile by-products of coke-making as fuel. Energy savings totals more than 80 percent.

Adopted new process for smelting aluminum. Requires 30 percent less electricity.

Installed solvent recovery system at cost of $25,000 to recover solvent used in washing foil. Saves 20,000 to 25,000 gallons of solvent per year.

Began using blast-furnace gas and coke-oven gas for heating rolling mills.

Insulated water-cooled pipe support for slab-reheat furnaces. Saved two trillion cubic feet of gas in three years.

Provided portable enclosures to reduce heat loss when moving ingots that have to be reheated.

Substitute coal tar for fuel oil in some furnaces.

PULP AND PAPER INDUSTRY

Improved insulation of digesters and steam lines.

Replaced compressed air with low-pressure blowers where feasible and required justification for all uses of compressed air.

Redesigned refiner plate.

Reduced refiner horsepower by 5 percent, saving $350,000 per year.

Installed process steam turbines. Exhaust steam is used for drying.

Began removing more moisture in press section to reduce dryer requirements. Removing only 1 percent moisture reduces steam demand by 4 percent.

REFERENCES

1. Staff: Local Trash Cuts Downtown Fuel Bills, *DMGH, Environ. Techno.*, vol. 6, no. 9, p. 780, American Chemical Society, Washington, 1972.
2. Huneke, J. M.: Conservation, *Hydrocarbon Process*, vol. 45, no. 7, pp. 119–124, 1966.
3. Moore, G. F.: How to Conserve Energy in the Design of Process Plants, *Actual Specifying Eng.*, September 1973.
4. Staff: Saving Energy in Industrial Plants, *Compressed Air*, June 1974.

APPENDIX A.   Energy Efficiency Checklist for Thermal Energy Conversion

Increase Fired-Heater Efficiency:
  Preheat combustion air.
  Generate steam.
  Design for less excess air.
  Use combustion analyzers.
Recover Additional Waste Heat:
  Generate steam.
  Use closer heat-exchange approach temperature.
  Drive absorption refrigeration system.
  Integrate larger heat train groupings.
Consider Applications of Gas Turbines with Heat Recovery:
  For large gas compressor drives.
  For high horsepower pump drives.
Increase Steam Generation Efficiency:
  Consider use of reheat cycles.
  Generate steam at higher-pressure level.
Reduce Heat Losses from Equipment:
  Increase vessel insulation thickness.
  Increase piping insulation thickness.
  Police quality of applied insulation.
  Consider insulating feed storage tanks.
  Consider insulating flanges and valves.
  Optimize steam tracing design.
Employ Vapor Recompression:
  Potential low-level heat source.
Optimize Fractionation System Designs:
  Use less reflux; more trays.
Design Buildings to Conserve Energy:
  Reduce heating/cooling loads:
    Optimum door and window arrangement
    Insulation.
    Sun shielding.
    Efficient lighting system.
  Use outside air for cooling when possible:
    Select most effective hot water plumbing system.
    Consider process streams to supply building heating/cooling needs.

APPENDIX B.   Energy Efficiency Checklist for Process and Electric Energy Conservation

Recover Potential Energy from Process:
  With hydraulic turbines.
  With gas expansion turbines.
Use More Efficient Rotating Machinery:
  Pumps
  Compressors.
  Steam turbines.
  Gas turbines.
  Electric motors.
Optimize Power Utilization for Process Cooling:
  Use air coolers where temperature permits.
  Consider two-speed motors for fan drives.
  Minimize cooling water circulation:
    Use maximum allowable outlet cooling water temperature.
    Consider reuse of cooling water.
Minimize Electrical Distribution Losses:
  Consider optimization of system power factor.
  Optimize plant lighting system.
Increase Steam Utilization Efficiency:
  Use more efficient steam drives.
  Minimize amount of letdown steam.
Reduce Losses with Properly Designed Instrumentation Systems:
  Police pressure-drop basis used for sizing:
    Control valves.
    Orifice meters.
  Consider use of an annular averaging element in place of orifices.
Reduce Instrument Air Consumption.

APPENDIX C.   Energy Efficiency Checklist for Fuel and Products Energy Conservation

Fuel/Products Conservation:
    Reduce product losses with proper instrumentation:
        Maintain product quality with process analyzers.
        Use feedforward and adaptive control systems.
        Use relief valve isolators and protectors.
    Use floating-roof tanks more extensively:
        Recover storage tank breathing losses.
        Recover relief valve discharges.
        Recover loading-station vent losses.
    Maximize slop oil recovery in treating facilities
Environmental Systems:
    Optimize pollution abatement process design:
        Sour water strippers.
        Tail gas treating units.
        Waste water treating facilities.
    Monitor overall energy conservation during preparation of environmental impact reports.

Chemical Fuels Technology

O. J. ADLHART, Ph.D. *Research and Development Department, Engelhard Minerals & Chemicals Corporation, Menlo Park, New Jersey. Member, Electrochemical Society. (Fuel Cells)*

RICHARD A. CHAPMAN, B.S., M.S.C.E. *Senior Civil Engineer. Fuels Technology Branch, Energy Systems Environmental Control Division, Industrial Environmental Research Center, U.S. Environmental Protection Agency, Cincinnati, Ohio. (Solid Waste Systems for Energy Generation)*

J. L. S. DALEY, B.S. *Deceased, formerly Development Associate, Battery Engineering Department, Union Carbide Corporation, Cleveland, Ohio. (Batteries)*

JOHN D. KEENAN, Ph.D. *Assistant Professor of Civil and Urban Engineering, University of Pennsylvania, Philadelphia, Pennsylvania. (Biochemical Sources of Fuels)*

ALAN PASTERNAK. *Formerly with Lawrence Livermore Laboratory, University of California, Livermore, California. (Methyl Alcohol—Potential Fuel for Transportation)*

ROBERT M. REED, Ph.D. *Consulting Chemical Engineer, Louisville, Kentucky. Fellow, American Institute of Chemists. Member, American Chemical Society, American Institute of Chemical Engineers. Registered Professional Engineer (California, Georgia, Indiana, Kentucky, Maine, Wyoming). (Hydrogen).*

DONALD M. ROSS. *Consulting Engineer, Lancaster, California. Formerly Deputy Director, U.S. Air Force Rocket Propulsion Laboratory, Edwards, California. (Propellants)*

S. SENDEROFF, Ph.D. *Senior Research Associate, Battery Products Division, Parma Technical Center, Union Carbide Corporation, Cleveland, Ohio. Member, American Chemical Society, Electrochemical Society. (Batteries)*

Biochemical Sources of Fuels

BY JOHN D. KEENAN*

Fuels considered here are those elaborated by or otherwise rendered available by living systems. The discussion is limited to a short geological time scale, thus eliminating the fossil fuels from direct consideration. Fossil fuels are described in considerable detail in earlier portions of this Handbook.

Energy and carbon are obtained by organisms either directly or indirectly via the photosynthetic conversion of solar energy. These organisms have evolved metabolic machineries for the photochemical reduction of carbon dioxide to organic matter and/or for the subsequent utilization of the organics for biosynthesis and controlled energy liberation. These metabolic routes can be exploited to provide fuels from biochemical sources.

The majority of the bioengineering strategies for biochemically derived fuels involve options for the disposition of organic matter produced via photosynthate. The bulk of the presently exploited photosynthate is directed toward the production of wood, food, and feed. During processing and consumption, waste organic materials are generated which can be used for energy production via combustion, pyrolysis or biochemical conversions to ethanol, hydrogen, methane, and isopropanol. A second option is to engineer the photosynthetic apparatus to provide hydrogen. The third strategy is the cultivation of crops as energy sources, i.e., the farming of an energy crop which can be used as an energy source via the foregoing processes.

The discussion here follows this general outline—photosynthesis to organic matter, photosynthesis to hydrogen, and biochemical conversion of organic matter to fuels. Several other topics are discussed briefly, including the microbiological extraction of shale oil, conversion economics, and the application of biochemical engineering principles to fuel generation. Cross references at the end of this discussion are given to other areas of this Handbook where some of these topics are covered in even greater detail.

PHOTOSYNTHESIS

The photosynthetic apparatus and the mechanisms by which it operates have been intensively investigated over the past 30 to 40 years. The current understanding (Ref. 1) is that of three series of interconnected oxidation-reduction reactions: The first involves the evolution of oxygen from water. The second is the transfer of H atoms to a primary hydrogen acceptor. The third is the reduction of CO_2 to carbohydrates by the primary hydrogen acceptor. The light energy required for photosynthesis is used to drive the H atoms against the potential gradient.

The photochemical stage of photosynthesis consists of two separate steps, I and II. The products of light reaction II are an intermediate oxidant and a strong oxidant which is capable of oxidizing water to oxygen. An intermediate oxidant and a strong reductant that can reduce carbon dioxide are produced in light reaction I. The two light reactions involve two pigment systems, photosystems I and II, interconnected by enzymatic reactions coupled with photophosphorylation yielding adenosine triphosphate (ATP). ATP is one of several high energy (7 to 8 kcal liberated upon hydrolysis per mole) compounds used in biological systems for chemical energy storage.

PLANT CULTIVATION

Agricultural productivities can be used to estimate photosynthetic energy conversion efficiencies. Odum (Ref. 2) summarizes net primary production rates for a number of cultivated crops. The productivities range from 344 to 6700 grams dry weight per square

---

*Assistant Professor of Civil and Urban Engineering, University of Pennsylvania, Philadelphia, Pa.

meter per year. Assuming the plant material to be 90 percent organic matter at 4.68 kcal per gram organic matter (Ref. 3), and assuming incident solar radiation to be 500 cal per square centimeter per day, the photosynthetic efficiencies range from 0.07 to 1.6 percent. Energy crop production could be based on terrestrial, fresh-water, or marine systems. The major problems involved are the large land areas required and the cost and energy expense of harvesting (Ref. 4).

PHOTOSYNTHETIC HYDROGEN

Several schemes have been proposed to utilize the photosynthetic apparatus, at the cellular and molecular level, for hydrogen production. Gest (Ref. 5) has recommended the use of photosynthetic purple bacteria for the production of hydrogen. These bacteria, such as *Rhodopseudomanus* species, evolve hydrogen when grown in the presence of an organic carbon source. The production of hydrogen is dependent on the presence of light and an electron donor, the proper nutritional history especially with respect to nitrogen, and conditions of metabolic imbalance between energy conversion and biosynthesis.

Other proposed biosynthetic hydrogen methods involve the production of the highly reduced hydrogen carrier coupled with an hydrogenase enzyme system for removal of the H atoms as hydrogen gas. Krampitz (Ref. 5) has developed laboratory-scale reactors coupling spinach chloroplast with the hydrogenase from the bacterium *Escherichia coli*. Alternately, the hydrogen may be produced via photoreduction of carbon dioxide to formate followed by enzymatic conversion to hydrogen and carbon dioxide.

Among the problems involved in the application of hydrogen production via photosynthesis are maintenance of reducing conditions, land area requirements, and system management. The appropriate redox conditions can be maintained within the chloroplast environment. The oxygen sensitivity of hydrogenases may be overcome by mechanical, chemical, or enzymatic removal of oxygen, mutagenesis to yield less sensitive hydrogenases, use of hydrogenases from aerobic organisms, and use of genetic manipulation to regulate hydrogenase activity.

The areal efficiency of hydrogen generation has been estimated (Ref. 5). An optimistic production estimate is 9 moles H_2 per day per square meter, corresponding to a gross energy yield of 1.92×10^5 kcal per year per square meter, or 1.97×10^{12} Btu per year per square mile. The calculations assume that 25 percent of these undergo primary conversion, and that four quanta are required per hydrogen molecule. The energy conversion efficiency of this system would be on the order of 10 percent at a daily incident solar radiation of 500 cal per square centimeter.

BIOCHEMICAL OXIDATIONS

Respiration refers to those biochemical processes in which organisms oxidize organic matter and extract the stored chemical energy needed for growth and reproduction. Respiration patterns may be subdivided into two major groups, based on the nature of the ultimate electron acceptor.

Although alternative pathways exist for the oxidation of various organic substrates, it is convenient to consider only the degradation of glucose. (The metabolic routes provide the means for metabolism of pentoses and for interconversions between sugars and other metabolites.) The breakdown of glucose is via the Embden-Meyerof-Parnas glycolytic pathway which yields 2 moles each of pyruvate, ATP, and reduced nicotinamide adenine dinucleotide (NAD) per mole of glucose.

Under aerobic conditions, the pyruvate is oxidized to CO_2 and H_2O via the tricarboxylic acid or Krebs cycle and the electron transport system. The net yield for glycolysis followed by complete oxidation is 38 moles ATP per mole glucose, although there is evidence that the yield for bacteria is 16 moles ATP per mole glucose (Ref. 6). Thus, 673 kcal are liberated per mole glucose, much of which is stored as ATP.

Under anaerobic conditions, various pathways exist for pyruvate metabolism which serve to reoxidize the reduced hydrogen carriers formed during glycolysis. The ultimate acceptor builds up as a waste product in the culture medium. The end products of the pathways are: (1) CO_2, ATP, and acetate; (2) CO_2 and ethanol; (3) H_2 and CO_2; (4) CO_2 and 2,3-butylene glycol; (5) CO_2, H_2, acetone, ATP, and butanol; (6) succinate; and (7) lactate. The pathway that occurs depends on the microorganism cultivated and the culture conditions.

In terms of energy liberation, the anaerobic fermentations are inherently inefficient. The end products of these metabolic activities are reduced and possess high heats of combustion. Several examples are shown in Table 1. It is the value of these products for various purposes including fuels which makes the anaerobic oxidation of organic substrates attractive.

TABLE 1. Energy Liberation during Microbial Oxidation of Glucose
Heats of combustion for theoretical oxidation of glucose by various routes are shown as kcal per mole of glucose fermented

Products	Heat of Combustion
$H_2O + CO_2$	0
$2\ CO_2 + 2\ C_2H_5OH$	654
$3\ CH_4 + 3\ CO_2$	634
Mixed acid (Escherichia)	633
2 lactic acid	652
Lactic acid	654

ETHANOL FERMENTATION

Ethyl alcohol is produced biologically by the well-known yeast fermentation (Refs. 7 to 9). Alcohol-tolerant strains of *Saccharomyces cerevisiae* are usually used. *S. cerevisiae* converts hexose sugars to ethanol and carbon dioxide, theoretically yielding 51 and 49 percent by weight, respectively. *S. anamensis* and *Schizosaccharomyces pombe* are also used. *Candida pseudotropicalis* is utilized for the ethanol fermentation from lactose, and *C. utilis* from pentoses.

Ethanol can be fermented from any carbohydrate, although starchy or cellulosic materials require a pretreatment step for hydrolysis. The usable raw materials can be categorized as saccharin (sugarcane, sugar beets, molasses, and fruit juices), starchy (cereals and potatoes), or cellulosic (wood and waste sulfite liquor). Pretreatment methods are discussed elsewhere (Refs. 7 to 9).

TABLE 2. Heats of Combustion and Costs of Various Fuels
Cost Basis: See references cited; otherwise mid-1970s

Fuel	kcal/gram*	Btu/pound	$/million Btu	Cost references
Ethanol	327.6	12,790		
Synthetic	....	....	6.54–10.70	10–13
Fermentative			17.82–23.80	13
Hydrogen	68.4	61,500	0.89– 1.02	14–15
Methane	210.8	23,600		
Natural gas—wellhead	....	....	0.20– 0.25	
Consumers	....	....	0.75– 1.00	
Anaerobic digestion	....	....	0.40– 2.00	
Substitute natural gas			0.52– 1.50	16–19
Methanol	170.9	9,990		
Natural	....	....	14.68	10
Synthetic	....	....	3.86	10
Isopropanol	474.8	14,210		
Synthetic	....	....	5.18	10

*Refs. 10–11.

The environmental conditions of the alcoholic fermentation vary somewhat, depending primarily on the strain of yeast. Acidic conditions are used to inhibit bacterial contaminants. The initial pH is in the range of 4.0 to 5.5. Suitable temperatures are on the order of 20 to 30°C. Industrial alcoholic fermentations are normally operated on a batch basis, the process being completed within 50 hours. Yields are in excess of 90 percent of theoretical, based on fermentable sugars. The concentration of alcohol in the culture medium depends on the alcohol tolerance of the yeast. Typically, this is on the order of 10 to 20 percent which is increased by distillation and other techniques.

The economics of the ethanol fermentation depend on the cost associated with the carbohydrate feed material and the market for nonalcoholic by-products. These by-products consist of grain residues, recovered carbon dioxide, and the residual cells. Recovered grain and cells are normally sold as feed materials. In recent years, chemosynthesis has largely displaced fermentation for the industrial production of ethyl alcohol.

Synthetic ethanol is manufactured from ethylene by absorption in concentrated sulfuric acid followed by hydrolysis of the ethyl sulfates to ethyl alcohol, or by the direct catalytic hydration of ethylene. As of the mid-1970s, 80 percent of the ethanol synthesized in the United States is via the catalytic process (Ref. 10). The synthetic processes yield 0.25 gallon ethanol per pound of ethylene and 0.58 gallon per gallon of ethyl sulfate (Ref. 10). Mid-1970s prices for industrial ethyl alcohol are summarized in Table 2. Goldstein (Ref. 13) has estimated that for corn at $1.80 per bushel (1974 support price was $1.30 per bushel), fermentation is competitive when ethylene exceeds $0.18 per pound, approximately triple the 1974 price (Ref. 10).

BUTANOL-ISOPROPANOL FERMENTATION

The butanol-isopropanol fermentation (Refs. 7 to 9) is mediated by the anaerobic bacterium *Clostridium butylicum*. A wide variety of carbohydrate feeds may be used. Saccharin feeds yield 30 to 33 percent mixed solvents, based on the original sugars. At 33 to 37°C, the fermentation is complete within 30 to 40 hours. Product ratios vary with the strain and with culture conditions, but are normally in the range 53 to 65 percent *n*-butanol, 19 to 44 percent isopropanol, 1 to 24 percent acetone, and 0 to 3 percent ethanol. This fermentation has been supplanted by petrochemical synthetic processes.

METHANE FERMENTATION

Methane and carbon dioxide are the primary gaseous end products of the anaerobic digestion process which has been widely used for many years in the stabilization of organic sewage solids. The quality of the digester off-gases is dependent upon feed composition. Mixed feeds normally yield approximately 65 percent methane and 35 percent carbon dioxide. Approximately equal volumes arise from carbohydrates, and the methane yield increases with proteins and lipids. In addition, the product gases contain small volumes of hydrogen sulfide and nitrogen.

The generation of methane occurs as the last step of a series of biochemical reactions. The reactions are divided into three groups, each mediated by heterogeneous assemblages of microorganisms, primarily bacteria. A complex feed, consisting of high-molecular-weight bipolymers, such as carbohydrates, fats, and proteins, undergoes exocellular enzymatic hydrolysis as the first step. The hydrolytic end products are the respective monomers (or other low-molecular-weight residues), such as sugars, fatty acids, and amino acids. These low-molecular-weight residues are taken up by the bacterial cell before further metabolic digestion.

The second step is acid production in which the products of hydrolysis are metabolized to various volatile organic fatty acids. The predominant fatty acids are acetic and propionic acids. Other low-molecular-weight acids, such as formic, butyric, and valeric acid, have been observed. Additional end products of the acid production step include lower alcohols and aldehydes, ammonia, hydrogen sulfide, hydrogen, and carbon dioxide.

The products of the acid generation step are metabolized by the methane-producing bacteria to yield carbon dioxide and methane, and, in addition, methane arises from metabolic reactions involving hydrogen and carbon dioxide. The current understanding of the biochemistry and kinetics of the methane fermentation is summarized in Refs. 20 and 21.

Anaerobic digestion of organic solids wastes has been investigated as an alternative methane source. Various cost estimates have been made which indicate production costs, including gas purification and compression, in the range of $0.40 to $2.00 per million Btu. The major cost items, and sources of variability in the estimates, are the digester capital costs, waste sludge disposal cost, and the credit or debit associated with the collection and preparation of the solid waste feed material. Multiple staging and separate optimization of anaerobic digestion may provide reduced capital costs through lower detention times and reduced operation and maintenance costs by improved process stability.

HYDROGEN FERMENTATION

Hydrogen gas is a product of the mixed acid fermentation of *Escherichia coli*, the butylene glycol fermentation of *Aerobacter*, and the butyric acid fermentations of *Clostridium spp.* (Ref. 6). A possible fruitful research approach would be to seek methods of improving the yield of hydrogen.

CELLULOSE DEGRADATION

A significant portion of the organic matter suitable for fermentation to fuels is cellulosic. Cellulosic materials tend to resist biochemical degradation. A system has been described which utilizes fungal cellulase for the hydrolysis of cellulose (Refs. 22 and 23). Wilke (Ref. 24) described a system for the net production of 429 tons of glucose (5.88% w/w) per day from 885 tons of waste paper at a cost of $18.56 per ton of glucose, excluding credits or debits associated with the disposal of the paper.

BIOCHEMICAL FUEL CELLS

Young et al. (Ref. 25) have discussed the possibilities of utilizing biological processes as an integral part of fuel cells. They define three basic types of biochemical fuel cells: (1) *depolarization cells* in which the biological system removes an electrochemical product, such as oxygen; (2) *product cells* in which an electrochemically active reactant, such as hydrogen, is biologically produced; and (3) *redox cells* (oxidation-reduction) in which electrochemical products are converted to reactants (ferricyanide/ferrocyanide system) by the biological system. Young et al. concluded that application of biochemical fuel cells will most probably involve immobilized enzymes as a method of increasing efficiency and decreasing costs.

EXTRACTION OF SHALE OIL

Several biological approaches have been suggested as adjuncts in the extraction of shale oil and subsequent processing of kerogen (Ref. 5). One procedure would utilize sulfur-oxidizing bacteria. A metabolic end product of certain sulfur oxidizers is sulfuric acid which would help to break down the inorganic matrix and thus improve the shale porosity.

An alternative strategy is the use of bacteria to partially degrade the organic kerogen phase to a more usable low-molecular-weight form. Various microorganisms capable of oxidizing hydrocarbons have been described. The general metabolic pathway is ring cleavage, if required; oxidation to fatty acid; and then degradation via the beta-oxidation route (Ref. 26). Biological methods have also been proposed for the secondary recovery of petroleum (Refs. 27 and 28). The applicability of biological processes to the development of shale oil deposits is speculative, and considerable work must be undertaken to approach the biological and engineering problems involved.

CONVERSION ECONOMICS

In some cases the cost of biochemical processes is fairly well understood, whereas in others very little is known. In Table 2 some mid-1970s costs are summarized for biochemical and other sources. Widespread application of these processes will be a function of competition which can occur at any of three levels. At the first level is competition for raw materials. Strong pressure will exist for utilization of photosynthate for food and feed (Ref. 29). Waste materials also face competition for alternative uses. For example, waste newsprint is $40 per ton (mid-1970s, Philadelphia area). Demand may force decisions to direct fermentation toward food and feed production instead of fuel generation. The third level of competition is alternative uses of the end product, such as synthetic feedstock and solvents.

Other factors impinging upon cost decisions are markets for by-products (feed residue, cells, carbon dioxide), land area requirements, and product extraction procedures. Cell yields are on the order of 0.119 gram per kcal taken up from the medium (Ref. 30). The biochemical fuels could relieve shortages in the petrochemical industry and would provide environmental advantages in that they generally are clean burning and renewable, and do not alter the planetary heat balance.

The biologically derived products will complement the existing energy structure. Methane gas is easily transportable in the well-developed natural gas distribution system. Ethyl and isopropyl alcohols have been utilized as gasoline additives for internal combustion engines (Ref. 31). Suggestions for widespread utilization of hydrogen fuel have been made (Ref. 32).

There are a number of other fermentations in which various metabolites accumulate which can be used to alleviate petrochemical shortages. These include acetone and *n*-butanol from *Clostridium acetobutylicum;* fats from molds; glycerol; 2-3 butylene glycol; and various acids, such as formic, lactic, citric, fumaric, and glutamic acid (Refs. 7 to 9).

It is apparent that the production of fuels by biochemical means is feasible and desirable. Process economics and efficiencies require improvement which, in turn, necessitates a concerted and coordinated research effort on the part of the biologist and the engineer. Enzyme and genetic engineering hold the key to improved process efficiencies. The immobilization of enzymes onto solid matrices has promise for increasing efficiencies by eliminating competing reactions and bringing reactants close together. Greater yields may result from the selection of better strains, alteration of genetic makeup, and taking advantage of existing enzyme regulatory mechanisms, such as enzyme inhibition, repression, and depression.

Editor's Notes.   Numerous other descriptions in this Handbook touch upon several of the topics described here. In particular, further reference is suggested to Chemical Feedstocks from Natural Gas, LNG versus Methanol Route, Wood-to-Oil Process, Petrochemical Complex, Oil Shale Processing and Fuels, Alcohol and Alcohol-Blend Fuels, Hydrogen, Batteries, Fuel Cells, Photosynthesis, Solar Physics. Consult Contents pages in front of book and alphabetical index at end of book.

REFERENCES

1. Rabinowitch, E., and Govindjee: "Photosynthesis," Wiley, New York, 1969.
2. Odum, E. P.: "Fundamentals of Ecology," Saunders, Philadelphia, 1959.
3. Prochazka, G. J., W. J. Payne, and W. R. Mayberry: Calorific Contents of Microorganisms, *Biotech. and Bioeng.*, **15**, 1007–1010 (1973).
4. Oswald, W. J., and C. G. Golueke: Biological Transformation of Solar Energy, *Advan. Appl. Microbiol.*, **2**, 223–262 (1960).
5. Hollaender, A. K., J. Monty, R. M. Pearlstein, F. Schmidt-Bleek, W. T. Snyder, and E. Volkin: "An Inquiry into Biological Energy Conversion," University of Tennessee, Knoxville, Tenn., 1972.
6. Aiba, S., A. E. Humphrey, and N. F. Millis: "Biochemical Engineering," 2d ed., Academic, New York, 1973.
7. Prescott, S. C., and C. G. Dunn: "Industrial Microbiology," McGraw-Hill, New York, 1959.
8. Rhodes, A., and D. L. Fletcher: "Principles of Industrial Microbiology," Pergamon, New York, 1966.
9. Peppler, H. J.: Ethyl Alcohol, Lactic Acid, Acetone, Butyl Alcohol, and Other Microbial Products, in H. J. Peppler (ed.), "Microbial Technology," Van Nostrand Reinhold, New York, 1967.
10. "Chemical Economics Handbook," Stanford Research Institute, Menlo Park, Calif., 1974.
11. "Handbook of Chemistry and Physics," 53d ed., Chemical Rubber Co., Cleveland, 1972.
12. "Handbook of Tables for Applied Engineering Science," Chemical Rubber Co., Cleveland, 1970.
13. Goldstein, S.: "The Economics and Problems of Ethanol from Starches," presented at Symposium on the Progress of Research in Enzyme Technology and Its Application, University of Pennsylvania, Philadelphia, May 3, 1974.
14. Van den Berg, G. J., E. Reinmuth, and E. Kapp: Hydrogen from Heavy Residue, *Chem. & Process Eng.*, **52**, 10, 49–55 (1971).
15. Anon.: Hydrogen Plants—Where They Are and Where They Are Going, *Oil Gas Jl.*, **69**, 176–182, June 21, 1971.
16. Anon.: Coal Gasification Circulates Molten Salt, *Oil Gas Jl.*, **69**, 103, Sept. 27, 1971.
17. Anon.: Synthetic Gas Processes Pushed in the U.S., *Oil Gas Jl.*, **69**, 86–87, Jan. 25, 1971.
18. Anon.: Coal Gas at 66¢/MM BTU Is Estimated by Kellogg Company, *Oil Gas Jl.*, **70**, 46, Dec. 4, 1972.
19. Finneran, J. A.: SNG—Where Will It Come from and How Much Will It Cost, *Oil Gas Jl.*, **70**, 83–88, July 17, 1972.

20. Gould, R. F. (ed.): Anaerobic Biological Treatment Processes, in *Advan. Chem. Ser.*, vol. 105, American Chemical Society, Washington, 1971.
21. Stadtman, T.: "Methane Fermentation," in C. E. Clifton, S. Raffel, and M. P. Starr (eds.) *Ann. Rev. Microbiol.*, **21**, 121–142, Annual Reviews, Inc., Palo Alto, Calif., 1967.
22. Mandels, M., and J. Weber: The Production of Cellulases, in R. F. Gould (ed.) Cellulases and Their Applications, in *Advan. Chem. Ser.*, **95**, 391–414, American Chemical Society, Washington, 1969.
23. Reese, E. T., M. Mandels, and A. H. Weiss: Cellulose as Novel Energy Source, in T. K. Ghose, A. Fiechter, and N. Blakebrough (eds.) *Advan. Biochem. Eng.*, **2**, 181–200, Springer-Verlag, New York, 1972.
24. Wilke, C.: "Conversion of Cellulosic Materials to Sugar and Alcohol," presented at Symposium on Progress of Research in Enzyme Technology and Its Application, University of Pennsylvania, Philadelphia, May 3, 1974.
25. Young, T. G., L. Hadjipetrou, and M. D. Lilly: Theoretical Aspects of Biochemical Fuel Cells, *Biotechnol. and Bioeng.*, **8**, 581–93 (1966).
26. Abbott, B. J., and W. E. Gledhill: The Extracellular Accumulation of Metabolic Products by Hydrocarbon-degrading Microorganisms, *Advan. Appl. Microbiol.*, **14**, 249–388, 1971.
27. Beerstecher, E., Jr.: "Petroleum Microbiology," Elsevier, New York, 1954.
28. Davis, J. B.: "Petroleum Microbiology," Elsevier, New York, 1967.
29. Bhattacharjee, J. K.: Microorganisms as Potential Sources of Food, *Advan. Appl. Microbiol.*, **13**, 139–161 (1970).
30. Payne, W. J.: Energy Yields and Growth of Heterotrophs, in C. E. Clifton, S. Raffel, and M. P. Starr (eds.) *Ann. Rev. Microbiol.*, **24**, 17–52, Annual Reviews, Inc., Palo Alto, Calif., 1970.
31. Pleeth, S. J. W.: "Alcohol—A Fuel for Internal Combustion Engines," Chapman & Hall, Ltd., London, 1949.
32. Sailor, V. L.: "Hydrogen Economy—Based on Hydrogen as Energy Source," Brookhaven National Laboratory, Upton, N.Y., 1973.

Solid Waste Systems for Energy Generation

BY RICHARD A. CHAPMAN*

The United States combustible solid waste production for 1974 is estimated to be about 566×10^9 kilograms (624×10^6 tons) having an energy value of about 2.37×10^{18} Calories (9.42×10^{15} Btu). If all the solid waste were collected and used as a fuel, it could supply about 12 percent of the total energy requirements of the United States or 109 percent of the energy from coal used for the generation of electricity in 1973.

TABLE 1. Summary of Combustible Solid Waste Discarded in the United States in 1974

Source of waste	Quantity discarded			Energy value	
	Percent of total weight	10^9 kilograms	10^6 tons	10^{15} Cal	10^{15} Btu
Refuse and sewage, urban	16.3	92	101	413	1.64
Manufacturing and processing	5.8	33	36	144	0.57
Agricultural	77.9	441	487	1817	7.21
Total	100.0	566	624	2374	9.42

The quantity and energy value of dry combustible solids discarded in the United States in 1974 are summarized in Table 1. The quantity estimates include only the *combustible portion* of the solid waste (inert materials and water not included). Unfortunately, not all the solid waste is collected and disposed of in central locations and, therefore, is unavailable for use as a fuel. This is particularly true of agricultural wastes. Estimates of the

TABLE 2. Summary of Solid Waste Availability

Source of waste	Number of plants	Waste per year		Energy value per year	
		10^9 kilograms	10^6 tons	10^{15} Cal	10^{12} Btu
SOLID WASTE PLANT SIZE: 90,700 KILOGRAMS (100 TONS) of COMBUSTIBLES PER DAY					
Urban	1598	52.9	58.3	23.8	943
Manufacturing	579	19.1	21.1	80	319
Agricultural	1026	34.0	37.5	134	532
Total	3203	106.0	116.9	237.8	1794
SOLID WASTE PLANT SIZE: 453,000 KILOGRAMS (500 TONS) of COMBUSTIBLES PER DAY					
Urban	252	41.7	46.0	187	744
Manufacturing	9	1.6	1.8	7	27
Agricultural	5	0.9	1.0	4	15
Total	266	44.2	48.8	198	786

quantity of solid waste available in concentrations sufficient to fuel a solid waste energy-producing system are given in Table 2. These estimates do not include waste that is currently being recycled or used as an energy source.

Nature of Wastes.   Urban waste includes household, sewage, commercial, institutional, manufacturing plant, and demolition waste. The availability of this waste is directly

---

*Senior Civil Engineer, Fuels Technology Branch, Energy Systems Environmental Control Division, Industrial Environmental Research Center, U.S. Environmental Protection Agency, Cincinnati, Ohio.

related to the population living in urban areas of adequate size to support a given size system.

Manufacturing and processing wastes include all residuals generated from material inputs that leave the plant as product output. Office and packaging wastes associated with this sector are included in the urban waste sector. The majority of these wastes are from pulp and paper manufacturing, primary and secondary wood manufacturing, and the construction industry. The availability of these wastes is related primarily to manufacturing plant size and its proximity to an urban area.

Agricultural wastes include animal manures, crop wastes, and forest and logging residues. The only available wastes in this sector are those animal wastes generated on large feedlots and dairies, and a portion of those crop wastes (bagasse and fruit tree prunings) not readily recycled back to the soil.

From Table 2 it can be concluded that if all the readily available waste were used as an energy source, an additional 452×10^{15} Cal (1794×10^{12} Btu) of fuel would have been available in 1974. This is equivalent to approximately 2 percent of the total United States energy requirement for 1973.

SOLID WASTE PREPARATION

The energy recovery system selected dictates the extent that solid waste must be prepared. Some systems require nothing more than the removal of massive noncombustibles, such as kitchen appliances from the refuse, whereas other processes require extensive shredding, air classification, reshredding, and drying. In conjunction with the fuel preparation, it is usually worthwhile to reclaim the metals and glass in the waste for recycling. The following unit operations may be required for the preparation of solid waste for energy recovery.

Shredding.   This processing step is employed in most of the developing solid waste energy conversion systems to reduce the incoming solid waste to a relatively constant particle size. One-stage shredding is often used to reduce the waste to a nominal size as small as 2.5 centimeters (1 inch). When finer-sized fuel is required, a second shredding step is usually employed after air classification has removed many of the noncombustibles.

Air Classification.   This unit operation is often used to remove the larger noncombustibles from the shredded waste. Smaller noncombustibles such as metal foils and fine glass are often not removed by air classification. Both vertical and horizontal air classifiers depend on the heavy noncombustibles settling out by gravity in a moving airstream while the lighter combustibles are pneumatically transferred through the air classifier. Often the denser combustibles such as rubber and leather are removed with the heavy fraction while some of the fine glass and metal foils are carried with the combustibles.

Drying.   Some energy recovery systems employ this operation to remove excess fuel moisture. Almost without exception drying is required when animal manures or sewage sludge are used as a fuel. Generally waste heat from the process is utilized in the drying operation.

Storage.   Solid waste is stored in live bottom bins of various designs. Bins successfully used for the storage of wood chips are usually employed. The size of storage bins depends primarily on the solid waste delivery and shredding schedule.

Material Recovery.   Systems are available for removing steel, aluminum, other nonferrous metals, and glass from the heavy-product air-classifier stream. Material recovered from air-classified refuse has a higher value than that recovered from the waste after the energy conversion system, usually in the form of slag or frit.

PROCESSES THAT PRODUCE GAS AND OIL

Garrett Pyrolysis Process.   This process employs low-temperature flash pyrolysis to produce char and a highly viscous, highly oxygenated fuel oil having a heating value of about 4800 Cal per gram (10,500 Btu per pound). The system, illustrated in Fig. 1, employs two stages of shredding, air classification, and drying to produce a minus-24-mesh fuel for the pyrolysis reactor. The heat required for pyrolysis is derived from the combustion of the pyrolysis off-gas and from a portion of the char produced; it is transferred by means of a heat exchanger of proprietary design. From the pyrolysis unit the gases are exhausted through a cyclone to remove the char and then scrubbed to remove the oil, water, and other solids and liquids.

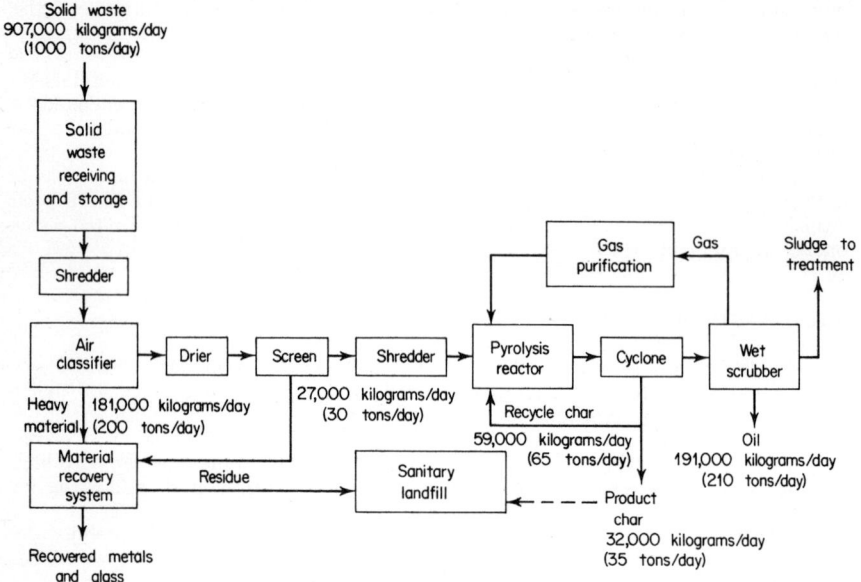

Fig. 1. Garrett pyrolysis process.

The properties of the pyrolytic oil are shown in Table 3. The effect of the 0.3 percent chlorine, ash, and the organic acids present in the oil on corrosion have not been determined.

Residues from a 907,000-kilogram-per-day (1000-ton-per-day) plant include any unrecycled solid waste inerts, 61,700 kilograms (68 tons) of solid residue in slurry form from the scrubber, and about 102,200 liters (27,000 gallons) of process water high in biological oxygen demand (BOD) and methyl chloride that must be treated.

TABLE 3.   Typical Properties of Garrett Pyrolytic Oil

Carbon, wt %	57.5
Oxygen, wt %	33.4
Hydrogen, wt %	7.6
Nitrogen, wt %	0.9
Chlorine, wt %	0.3
Sulfur, wt %	0.1–0.3
Ash, wt %	0.2–0.4
Cal/gram	5830
Btu/pound	10,500
Sp. gr.	1.30
Pour point, °C (°F)	32 (90)
Flash point, °C (°F)	56 (133)
Viscosity SSU @ 88°C (180°F)	3150
Pumping temperature, °C (°F)	71 (160)
Atomizing temperature, °C (°F)	116 (240)

A 3628-kilogram-per-day (4-ton-per-day) pilot plant has been developed and tested in La Verne, California. An 18,140-kilogram-per-day (200-ton-per-day) plant is being constructed in San Diego, California, under the sponsorship of an Environmental Protection Agency grant and will be used to verify predicted product costs and product yields.

Projected system economics (see editorial note at end of this description) are summarized in Table 4 for a 907,000-kilogram-per-day (1000-ton-per-day) system—about 454,000 kilograms (500 tons) per day of dry combustibles—operating 24 hours a day, 300 days a year.

TABLE 4.   Projected Economics for a 907,000-Kilogram-per-Day (1000-Ton-per-Day) Garrett Pyrolysis System

Capital cost (1974, CE plant cost index = 154) = $18,000,000
Annual oil production = 333×10^{12} cal (1.323×10^{12} Btu)

Cost	Municipal ownership	Private ownership
Annual capital cost	20 years at 6%: $1,569,000	20 years at 15%: $2,876,000
Annual operating cost	2,300,000	2,300,000
Total annual cost	$3,869,000	$5,176,000
OIL PRODUCTION COSTS PER MILLION BTU		
No dumping fee, no material recovery	$2.92	$3.91
$5 per ton dumping fee and material income*	1.79	2.78
$10 per ton dumping fee and material income*	0.66	1.65

*Net income for recovered steel, aluminum, glass, and other nonferrous metals may be up to $5 per ton of incoming waste, depending on local markets and waste composition.

Union Carbide Purox System.   This system employs a high-temperature partial-oxidation process to produce a medium-Btu gas. Oxygen is used to partially combust the solid waste, which provides the heat required for pyrolysis. By using oxygen, high process temperatures and a relatively small quantity of exhaust gas requiring cleaning are achieved.

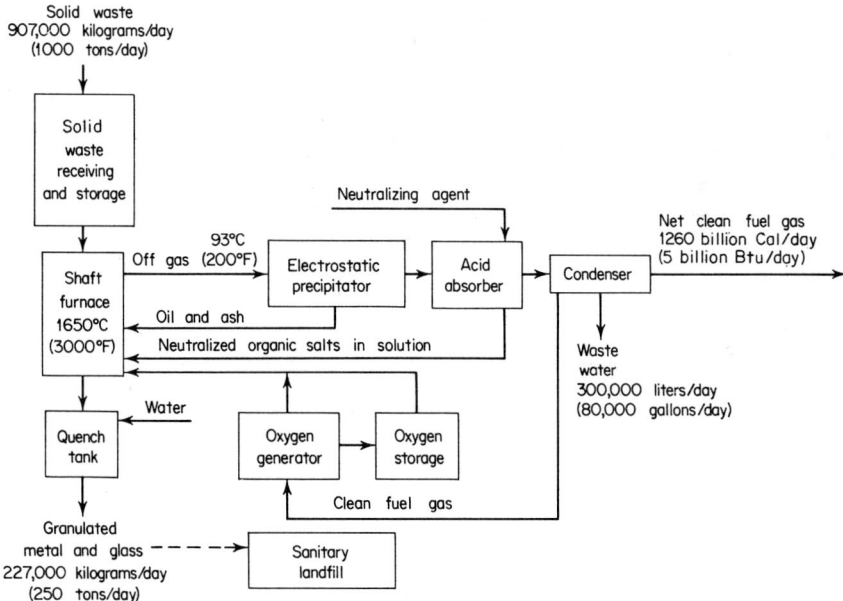

Fig. 2. Union Carbide Purox system.

As illustrated in Fig. 2, municipal refuse is charged at the top of a shaft furnace and is pyrolyzed as it passes downward through the furnace. Oxygen enters the furnace through tuyeres near the furnace bottom and passes upward through a 1425 to 1650°C (2600 to 3000°F) combustion zone. The products of combustion then pass through a pyrolysis zone and exit at about 93°C (200°F). The off-gas then passes through an electrostatic precipitator to remove the fly ash and oil formed during pyrolysis, both of which are recycled to the furnace combustion zone. The gas then passes through an acid absorber and a condenser. The clean fuel gas has a heating value of about 2.7 cal per cubic centimeter (300 Btu per cubic foot) and a flame temperature equivalent to that of natural gas. As the solid waste passes downward through the furnace, it contacts the exiting pyrolysis products and traps a portion of the oil and fly ash while itself losing moisture. After passing through the pyrolysis and combustion zones, the remaining solid waste is removed as a slag from the furnace bottom.

The Purox system has a net thermal efficiency of about 65 percent in converting solid waste to fuel gas. Process losses include energy losses in the conversion process and energy required for the operation of the on-site cryogenic gas separation unit which produces 95 percent pure oxygen for use in the system.

Table 5 describes the composition of the clean fuel gas.

The clean fuel gas is low in sulfur (about 15 ppm) and essentially free of nitrogen oxides. It can be burned in existing utility plants or by other large fuel consumers as an auxiliary fuel without contributing to corrosion of the boiler or undesirable stack emissions.

Process residues from a 907,000-kilogram-per-day (1000-ton-per-day) system include granulated metal and glass from the quench tank and about 300,000 liters (80,000 gallons) per day of condensed water containing organics that must be treated or discharged to a sanitary sewer.

A 4500-kilogram-per-day (5-ton-per-day) pilot plant has been in operation since 1971 in Tarrytown, New York; and a 181,000-kilogram-per-day (200-ton-per-day) plant began operation in 1974 in Charleston, West Virginia, to verify plant performance and cost estimates.

Estimated system economics are summarized in Table 6 for a 907,000-kilogram-per-day (1000-ton-per-day) system operating 24 hours a day, 300 days a year.

TABLE 5.   Purox System Fuel Gas Composition

Component	Volume%
CO	50
H_2	30
CO_2	15
CH_4	3
C_2H_x	1
N_2	1
	100

TABLE 6.   Projected Economics for a 907,000-Kilogram-per-Day (1000-Ton-per-Day) Union Carbide Purox System

Capital cost (1974, CE plant cost index = 154) = $18,000,000
Annual gas production = 378×10^{12} Cal (1.5×10^{12} Btu)

Cost	Municipal ownership	Private ownership
Annual capital cost	20 years at 6%: $1,569,000	20 years at 15%: $2,876,000
Annual operating cost	1,700,000	1,700,000
Total annual cost	$3,269,000	$4,576,000
GAS PRODUCTION COSTS PER MILLION BTU		
No dumping fee, no material recovery	$2.18	$3.05
$5 per ton dumping fee and material income*	1.18	2.05
$10 per ton dumping fee and material income*	0.18	1.05

*Net income for recovered metal and glass granules may be up to $1.00 per ton of incoming waste.

Biological Methane Production.   This process involves the anaerobic digestion of a solid waste and water or sewage sludge slurry at 60°C (140°F) for five days to produce a methane-rich gas. As illustrated in Fig. 3, solid waste is prepared by shredding and air classification before being blended with water or sewage sludge to a 10 to 20 percent solids concentration. The slurry is heated and placed in a mixed digester at 60°C (140°F) for 5 days' detention. The digester gas is drawn off and separated into CO_2 and methane. The spent slurry from the digester is pumped through a heat exchanger to partially heat the incoming slurry before filtration. The filtrate is returned to the blender, and the sludge is landfilled. Heat addition to the refuse slurry is required to maintain the necessary digester temperature. This process is very well suited for use on sewage sludge, animal manures, and other high-moisture-content solid wastes.

It is estimated that the process will reduce the volume of the volatile solids by 75 percent while producing about 85 cubic meters (3000 cubic feet) of methane per 907 kilograms (1 ton) of incoming solid waste.

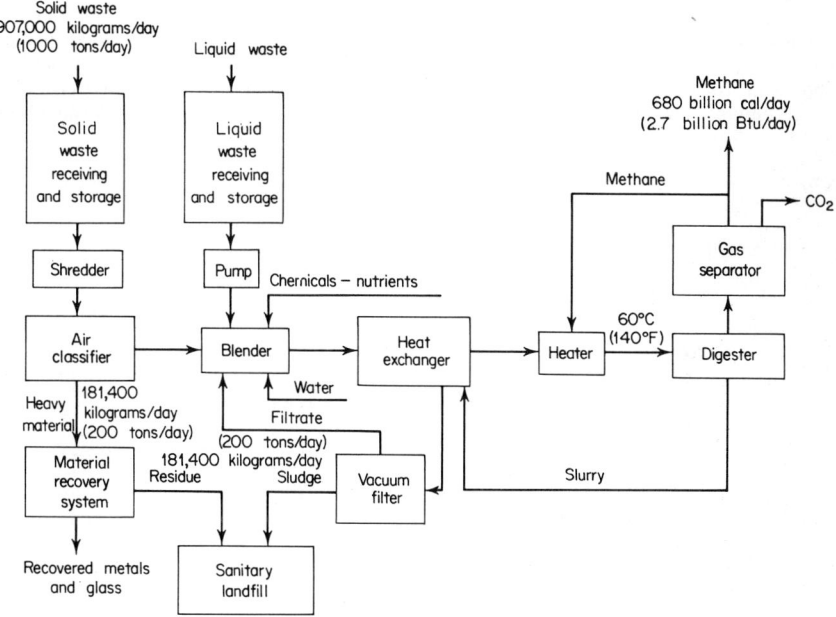

Fig. 3. Biological methane production.

The major residue from a 907,000-kilogram-per-day (1000-ton-per-day) plant will be about 80,000 kilograms (200 tons) of solids in a 25 to 50 percent sludge that will require landfilling or incineration.

It is estimated that about 10 percent of the methane will be consumed in heating the digester feed.

TABLE 7. Projected Economics for a 907,000-Kilogram-per-Day (1000-Ton-per-Day) Biological Methane Process

Capital cost (1974, CE plant cost index = 154) = $10,000,000
Annual methane production = 204×10^{12} Cal (8.1×10^{11} Btu)

Cost	Municipal ownership	Private ownership
Annual capital cost	20 years at 6%: $ 872,000	20 years at 15%: $1,598,000
Annual operating cost	4,000,000	4,000,000
Total annual cost	$4,872,000	$5,598,000
METHANE PRODUCTION COSTS PER MILLION BTU		
No dumping fee, no material recovery	$6.01	$6.91
$5 per ton dumping fee and material income*	4.16	5.06
$10 per ton dumping fee and material income*	2.31	3.21

*Net income for recovered steel, aluminum, glass and other nonferrous metals plus sewage sludge disposal credit may be up to $7 per ton of incoming waste, depending on local markets and waste composition.

This process has been studied at the laboratory scale. Pilot scale units are being considered.

Estimates of system economics in Table 7 for a 907,000-kilogram-per-day (1000-ton-per-day) plant operating 24 hours a day, 300 days a year, are based primarily on sewage sludge digestion costs and include the disposal of the thickened refuse slurry.

PROCESSES THAT PRODUCE STEAM

Monsanto Landgard Process. This system uses a rotary kiln pyrolyzer followed by an afterburner and boiler to produce steam from shredded solid waste. A system schematic is shown in Fig. 4. The pyrolysis process in the kiln is operated countercurrently. Solid waste enters at one end and pyrolyzed residue is discharged at the other. External fuel

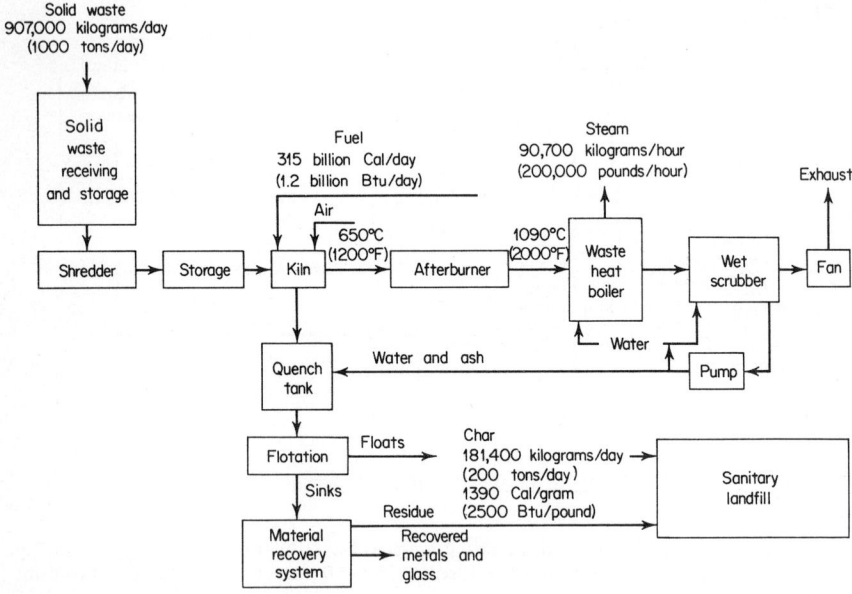

Fig. 4. Monsanto Landgard system.

and air are introduced at the residue discharge area and combustion products and pyrolysis gases leave the kiln at the feed opening. This configuration causes the solid waste to be exposed to progressively higher temperatures as it passes through the kiln. The kiln off-gases pass through a refractory-lined afterburner into which air is introduced

TABLE 8.   Projected Economics for a 907,000-Kilogram-per-Day (1000-Ton-per-Day) Monsanto Landgard System

Capital cost (1974, CE plant cost index = 154) = $19,000,000
Annual steam production = 6.8×10^8 kilograms (1.5×10^9 pounds)

Cost	Municipal ownership	Private ownership
Annual capital cost	20 years at 6%: $1,656,000	20 years at 15%: $3,035,000
Annual operating cost	2,600,000	2,600,000
Total annual cost	$4,256,000	$5,635,000
STEAM PRODUCTION COSTS PER THOUSAND POUNDS		
No dumping fee, no material recovery	$2.84	$3.76
$5 per ton dumping fee and material income*	1.84	2.76
$10 per ton dumping fee and material income*	0.84	1.76

*Net income for recovered char, steel, and aggregate may be up to $2 to $3 per ton of incoming waste, depending on local markets and waste composition.

to allow complete combustion prior to passing through the waste heat boiler. A wet scrubber is used for air pollution control while an induced draft fan is used to draw the gases through the system. One ton of solid waste together with about 315 million calories (1.25 million Btu) of auxiliary fuel and 55 kilowatt-hours of electricity will produce about 2160 kg of 2274 kPa (4800 lb of 330 psig) steam and 181 kilograms (200 pounds) of char in this process. Process residues from a 907,000-kilogram-per-day (1000-ton-per-day) plant include any unrecycled solid waste inerts, unsold char, and wet scrubber sludge. A 32,000-kilogram-per-day (35 ton-per-day) pilot plant was operated in St. Louis County, Missouri, for two years; and a 907,000-kilogram-per-day (1000-ton-per-day) system is scheduled for completion in late 1974 in Baltimore, Maryland. The steam from the Baltimore plant will be sold to Baltimore Gas and Electric Company for $0.81 per 454 kilograms (1000 pounds).

Estimated system economics are summarized in Table 8 for a 907,000-kilogram-per-day (1000-ton-per-day) system operating 24 hours a day, 300 days a year. The operating costs include auxiliary fuel at $1.00 per million Btu.

American Thermogen Process.   This process is similar to the Union Carbide Purox system in that unshredded solid waste is pyrolyzed in a shaft furnace at high temperature.

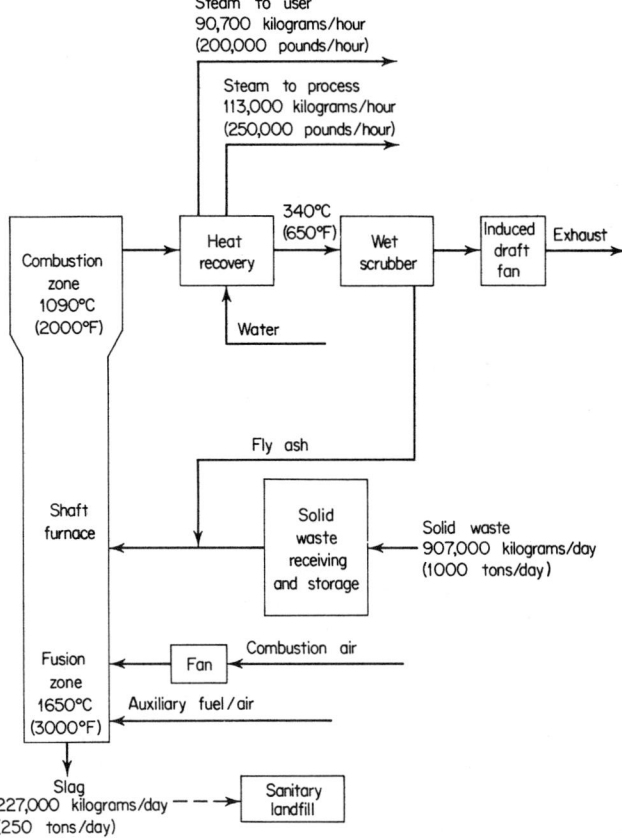

Fig. 5. American Thermogen high-temperature incinerator.

This system, however, uses air instead of oxygen for the combustion/pyrolysis step and burns the off-gas to produce steam. The process, illustrated in Fig. 5, includes a shaft furnace into which solid waste is fed at the top. As the waste moves downward through the furnace, it is dried, heated, pyrolyzed, and finally combusted in the 1650°C (3000°F) fusion zone. The molten residue is drawn off as slag and quenched. Auxiliary fuel and air are introduced at the bottom of the furnace, and the pyrolysis gas is combusted in the top portion of the furnace. A heat recovery boiler is used for the generation of steam, and a wet scrubber is used for air pollution control. The fly ash from the scrubbers is injected into the furnace and leaves the process in the slag. The Torrax system is similar to this one except that the auxiliary fuel is used to preheat the pyrolysis/combustion air in a silicon carbide heat exchanger and the pyrolysis gases are combusted in a separate vessel before passage through a steam boiler.

TABLE 9. Projected Economics for a 907,000-Kilogram-per-Day (1000-Ton-per-Day) American Thermogen System

Capital cost (1974, CE plant cost index = 154) = $18,000,000
Annual steam production = 6.8×10^8 kilograms (1.5×10^9 pounds)

Cost	Municipal ownership	Private ownership
Annual capital cost	20 years at 6%: $1,569,000	20 years at 15%: $2,876,000
Annual operating cost	2,300,000	2,300,000
Total annual cost	$3,869,000	$5,176,000
STEAM PRODUCTION COSTS PER THOUSAND POUNDS		
No dumping fee, no material recovery	$2.58	$3.45
$5 per ton dumping fee and material income*	1.58	2.45
$10 per ton dumping fee and material income*	0.58	1.45

*Net income for recovered slag may be up to $1.00 per ton of incoming waste.

A 907,000-kilogram-per-day (1000-ton-per-day) American Thermogen system produces about 204,000 kilograms (450,000 pounds) per hour of steam. About 90,700 kilograms (200,000 pounds) per hour are available for sale; the remainder is used in the process. Slag is the only process residue.

A small pilot plant has been operated for a number of years.

Projected system economics are summarized in Table 9 for a 907,000-kilogram-per-day (1000-ton-per-day) system operating 24 hours a day, 300 days a year.

Waterwall Incinerators. Waterwall incinerators generate steam by burning unprepared solid waste on a grate and passing the hot products of combustion through a boiler.

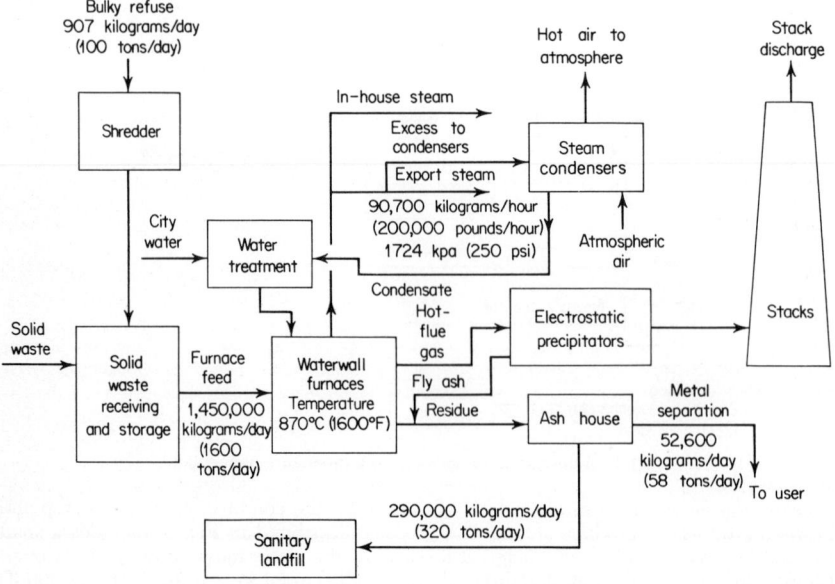

Fig. 6. Chicago Northwest incinerator.

Numerous waterwall incinerators have been built in Europe and the United States. Figure 6 illustrates the Chicago Northwest incinerator whose design is based on the Martin Incinerator System widely used in Europe, and includes four 363,000-kilogram-per-day (400-ton-per-day) waterwall furnaces having a combined capacity of 1,452,000 kilograms (1600 tons) per day. The unprepared refuse is taken from the storage pits and charged directly into the incinerator feed hopper. From there, the refuse drops onto a feed chute and then is fed automatically onto the stoker by means of a hydraulic feed ram. Temperatures in the 870°C (1600°F) range effectively burn the solid waste. Before the flue gas enters the boiler, secondary air is added to produce temperatures in the 1090°C (2000°F) range. The boiler is constructed of membrane waterwalled tubes with extruded fins. After passing through the boiler, the gases travel through an economizer section and then into an electrostatic precipitator for particle removal. Each of the four 400-ton-per-day furnaces produces 50,000 kilograms (110,000 pounds) per hour of 1724 kPa (250-psig) steam. A typical 907,000-kilogram-per-day (1000-ton-per-day) waterwall incinerator produces about 136,000 kilograms (300,000 pounds) per hour of steam, one-half of which is usually available for sale.

Process residues include ash and unrecycled solid waste inerts that must be landfilled.

System economics for a typical 907,000-kilogram-per-day (1000-ton-per-day) waterwall incinerator operating 24 hours a day, 300 days a year, are given in Table 10.

TABLE 10.   Economics for a Typical 907,000-Kilogram-per-Day (1000-Ton-per-Day) Waterwall Incinerator

Capital cost (1974, CE plant cost index = 154) = $22,000,000
Annual steam production = 4.9×10^8 kilograms (1.1×10^9 pounds)

Cost	Municipal ownership	Private ownership
Annual capital cost	20 years at 6%: $1,918,000	20 years at 15%: $3,515,000
Annual operating cost	2,000,000	2,000,000
Total annual cost	$3,918,000	$5,515,000
STEAM PRODUCTION COSTS PER THOUSAND POUNDS		
No dumping fee, no material recovery	$3.56	$5.01
$5 per ton dumping fee, and material recovery*	2.20	3.65
$10 per ton dumping fee and material recovery*	0.84	2.29

*Steel recovery income may be up to $1.00 per ton of incoming waste, depending on local markets and waste composition.

PROCESSES THAT GENERATE ELECTRICITY

Supplemental Utility Boiler Fuel.   In this process solid waste is prepared and fired as a supplemental fuel in a suspension-fired utility boiler. Solid waste provides 10 to 20 percent of the boiler fuel requirement, while powdered coal or other fuel provides the remainder. By weight, solid waste provides up to 40 percent of the fuel input and by volume up to 80 percent of the boiler input.

As illustrated in Fig. 7, solid waste is received, shredded, air-classified, and stored before being fired in the boiler. Two-stage shredding with air classification as an intermediate step may be employed. In the event that the boiler is not located near the solid waste generation center, transfer stations are often used to minimize transportation costs. In that case, the shredding and air classification are generally accomplished at the transfer station, and additional storage facilities and truck transport are required.

At the boiler site, if independent of the solid waste processing site, receiving and storage facilities are necessary. A feeding system, usually pneumatic including feeder valves and a blower, is also essential for transporting the solid waste into the boiler. Bottom-ash-handling facilities are required, and reserve pollution control equipment capacity may be needed to handle increased fly ash loadings.

The composition and properties of typical municipal solid waste that has been shredded and air-classified for use as a supplemental fuel are given in Table 11.

Typically, 20 percent of the input solid waste is removed by the preparation system, leaving about 80 percent for use as a fuel.

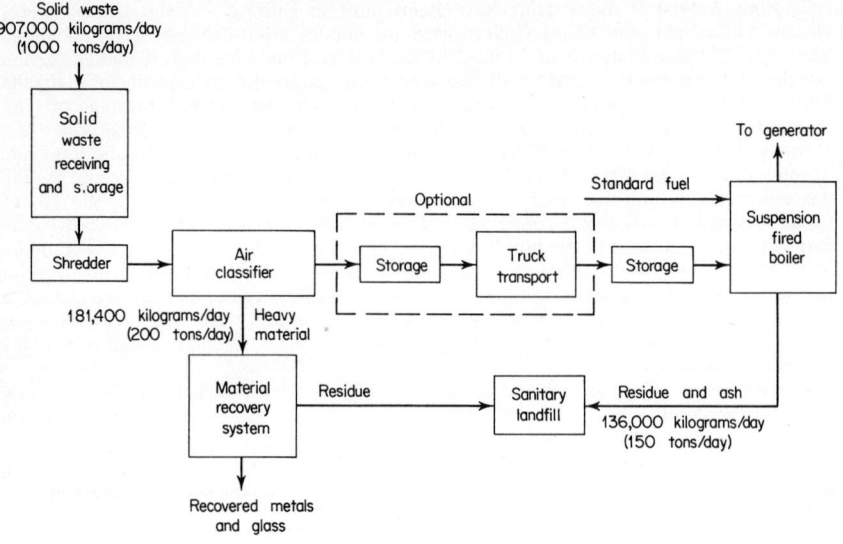

Fig. 7. Supplemental utility boiler fuel.

TABLE 11.   Composition and Properties of Typical Shredded and Air-classified Solid Waste

REFUSE COMPOSITION		
	Weight percent	
Component	Wet basis	Dry basis
Glass	5	5
Metal	2	2
Paper	45	34
Plastics	4	4
Leather, rubber, textiles	4	3
Wood	4	3
Food wastes	17	7
Yard wastes	18	11
Miscellaneous	1	1
Moisture	—	30
Total	100	100

ULTIMATE ANALYSIS	
Component	Weight percent
Carbon (total)	29.5
Carbon (fixed)	7.3
Hydrogen	3.9
Oxygen	23.4
Nitrogen	0.7
Sulfur	0.1
Chlorine	0.2
Ash	12.0
Moisture	30.2

Heating value = 2775 cal per gram (5000 Btu/pound)
Bulk density = 96–100 kilograms per cubic meter (6–10 pounds per cubic foot).

Residues from this process include unrecycled air-classified as heavy product, bottom ash, and fly ash. It is estimated that about 9 percent of the supplementary fired waste becomes bottom ash and about 5 percent becomes fly ash. The bottom ash may require special handling as it contains some organic material.

This system has been demonstrated under the sponsorship of the U.S. Environmental Protection Agency and the Union Electric Company in St. Louis, Missouri, for a number of years. The 125-megawatt Meramec Unit No. 1 is being supplied with 10 to 20 percent of its fuel requirements by 272,000 to 544,000 kilograms (300 to 600 tons) per day of shredded and air-classified solid waste provided by the city of St. Louis. Tests are being conducted to determine the effect of solid waste firing on the performance, emissions, and maintenance requirements of the Meramec Plant.

A number of cities and utilities; including Chicago; Bridgeport and Berlin, Connecticut; and Union Electric Company of St. Louis; have announced plans to burn up to 6,350,000 kilograms (7000 tons) per day of solid waste as supplemental fuel in suspension-fired utility boilers. Also, a number of companies have announced plans to construct systems to prepare solid waste for sale to local utilities for use as a supplemental fuel.

Projected economics for a 907,000-kilogram-per-day (1000-ton-per-day) supplemental fuel system operating 24 hours a day, 300 days a year, are presented in Table 12.

TABLE 12.   Projected Costs for a 907,000 Kilogram-per-Day (1000-Ton-per-Day) Supplemental Fuel System

SOLID WASTE PREPARATION FACILITIES		
Capital cost (1974, CE plant cost index = 154) = $7,000,000		
Annual fuel production = 6×10^{14} cal (2.4×10^{12} Btu)		
Cost	Municipal ownership	Private ownership
Annual capital cost	20 years at 6%: $ 610,000	20 years at 15%: $1,118,000
Annual operating cost	1,500,000	1,500,000
Subtotal annual cost	$2,110,000	$2,618,000
FIRING FACILITIES		
Capital cost (1974, CE plant cost index = 154) = $4,000,000		
Cost	Municipal ownership	Private ownership
Annual capital cost	20 years at 15% $ 639,000	$ 639,000
Annual operating cost	500,000	500,000
Subtotal annual cost	$1,139,000	$1,139,000
Total annual cost	3,249,000	3,757,000
SUPPLEMENTAL FUEL COSTS PER MILLION BTU		
No dumping fee, no material recovery	$1.35	$1.57
$5-per-ton dumping fee and material income*	0.73	0.94
$10 per ton dumping fee and material income*	0.10	0.32
Cost per mile of transit if required	0.01	0.02

*Net income for recovered steel, aluminum, glass, and other nonferrous metals may be up to $5 per ton of incoming waste, depending on local markets and waste composition.

CPU-400 System. This system uses a gas turbine, together with a fluidized-bed combustor and three stages of particle collection, to convert shredded solid waste directly into electricity. As illustrated in Fig. 8, shredded and air-classified waste, together with liquid waste, are fired in a fluidized-bed combustor to produce hot products of combustion for expansion through the gas turbine. The turbine compressor provides pressurized air for combustion and for transporting the shredded waste from beneath the rotary air-lock feeder valves into the combustor. Three stages of cyclones are employed to remove fly ash from the hot gas stream before expansion through the gas turbine. A waste heat boiler is utilized to produce steam for the generation of additional electricity. A completely automatic process-control computer system is used to run the process. The primary control loop regulates the fuel feed rate in response to turbine temperatures.

Liquid waste firing is not required, but is available as a desirable option because the addition of water to the system allows more solid waste to be consumed and greater amounts of electricity to be produced at a lower than normal inlet temperature in the same-sized capital equipment. System efficiency is reduced somewhat by the addition of water, but with a negative-value fuel that is not undesirable. The energy used to evaporate the water is partially recovered by the expansion of the resulting steam through the gas turbine.

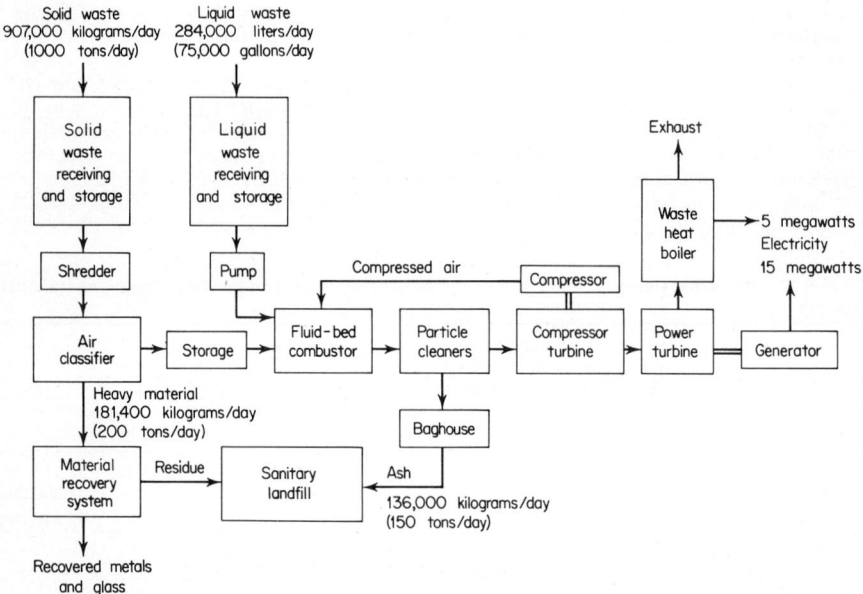

Fig. 8. CPU-400 system.

This system is very flexible regarding the type of fuel that can be burned (up to 60 percent H_2O) and the mix of steam and electricity produced. Residues from a 907,000-kilogram-per-day (1000-ton-per-day) system include unrecycled solid waste inerts and about 136,000 kilograms per day (150 tons per day) of dry sterile ash.

The CPU-400 has been under development by Combustion Power Company, Inc., since 1967 on a contract basis with the U.S. Environmental Protection Agency. A 90,700-kilogram-per-day (100-ton-per-day) pilot plant using a 4-atmosphere Ruston Hornsby TA 1500 turbine to generate 1000 kilowatts of electricity has been developed and tested. Deposits of aluminum oxide in the gas-turbine nozzles have temporarily stopped pilot-plant testing with solid waste. A moving-bed granular filter that will collect the deposit causing aluminum is being developed to replace the third-stage cyclones. The system is scheduled to resume solid waste testing in early 1975.

Currently the pilot plant is being used to burn high-sulfur coal in a program sponsored by the Office of Coal Research. Dolomite is successfully being employed as a fluidized-bed additive to suppress sulfur oxide emissions. The pilot plant has also been utilized successfully to burn hog fuel in a 100-hour test sponsored by a major forest products industry.

Preliminary prototype designs for solid waste burning systems include numerous power modules that incorporate a combustor, particle-removal stages and a 10-atmosphere turbine. Each power module will be capable of consuming 19,400 kilograms (200 tons) per day of solid waste and 56,775 liters (15,000 gallons) per day of liquid waste while generating 4000 kilowatts of electricity. Multiple-power modules served by a common solid-waste processing facility and automatic control system will provide redundancy and disposal capability in multiples of 91,400 kilograms (200 tons) per day.

Projected system economics for a 907,000-kilogram-per-day (1000-ton-per-day) system operating 24 hours a day, 300 days a year, are shown in Table 13.

Editor's Note.   Projected system economics included in the foregoing description are based on estimates compiled from the available literature and adjusted to mid-1974 costs. Capital cost and product yield estimates are, in many cases, projected from small pilot-scale systems. It is assumed that no major system changes are required to obtain the predicted product yields. Annual operating cost estimates assume that no unexpected maintenance requirements will be encountered during continuous operation of full-size systems. Mention of commercial products and systems does not constitute endorsement by the U.S. Environmental Protection Agency.

TABLE 13.   Projected Economics for a 907,000-Kilogram-per-Day (1000-Ton-per-Day) CPU-400 System

Capital cost (1974, CE plant cost index = 154) = $18,000,000
Annual electricity production = 1.5×10^8 kilowatt-hours

Cost	Municipal ownership	Private ownership
Annual capital cost	20 years at 6%: $1,569,000	20 years at 15%: $2,876,000
Annual operating cost	2,100,000	2,100,000
Total annual cost	$3,669,000	$4,976,000
ELECTRICITY PRODUCTION COSTS MILLS PER KILOWATT-HOUR		
No dumping fee, no material recovery	24	33
$5 per ton dumping fee and material income*	14	23
$10 per ton dumping fee and material income*	4	13

*Net income for recovered steel, aluminum, glass, and other nonferrous metal plus sewage sludge disposal credit may be up to $7 per ton of incoming waste, depending on local markets and waste compositions.

REFERENCES

1. Kos, P., P. M. Meier, and J. M. Joyce: "Economic Analysis of the Processing and Disposal of Refuse Sludges," Curran Associates, Inc., Contract 68-03-0183 Final Report, Office of Research and Monitoring, U.S. Environmental Protection Agency, Washington, March 1974.
2. Lowe, R. A.: "Energy Recovery from Waste, Solid Waste as Supplementary Fuel in Power Plant Boilers," U.S. Environmental Protection Agency, Publication no. SW-36d.ii, Washington, 1973.
3. Office of Solid Waste Management Programs: Second Report to Congress, "Resource Recovery and Source Reduction," U.S. Environmental Protection Agency, Publication no. SW-122, Washington, 1974.
4. "Problems and Opportunities in Management of Combustible Solid Wastes," International Research and Technology Corporation, Contract 68-03-0060, Final Report, Office of Research and Development, U.S. Environmental Protection Agency, Washington, October 1972.
5. *Proc. 1970 Nat. Incinerator Conf.*, American Society of Mechanical Engineers, New York, United Engineering Center, May 1970.
6. *Proc. 1972 Nat. Incinerator Conf.*, American Society of Mechanical Engineers, New York, United Engineering Center, May 1972.
7. *Proc. 1974 Nat. Incinerator Conf.*, American Society of Mechanical Engineers, New York, United Engineering Center, May 1974.
8. Resource Recovery, "Catalogue of Process," Midwest Research Institute, Project no. 3634-D, Final Report, Council on Environmental Quality, Washington, February 1973.
9. Resource Recovery, "The State of Technology," Midwest Research Institute, Final Report to The Council on Environmental Quality, Washington, February 1973.
10. Schulz, H. W.: "Pollution-free System for the Economic Utilization of Municipal Solid Waste for The City of New York—Phase I, A Critical Assessment of the Advanced Technology; Thermal Oxidation Processes," Columbia University, New York, National Science Foundation, Apr. 15, 1973.

Hydrogen

BY ROBERT M. REED*

Essentially with the advent of the petrochemical industry and the appearance of advanced petroleum processing methods in the early 1950s, hydrogen, as a chemical raw material, commenced its steady growth in industrial importance and high-tonnage production. Also, for many years, hydrogen has served as a specialized fuel for certain applications, such as oxyhydrogen cutting and welding torches. Even during the last century the advantages, disadvantages, and broad implications of hydrogen as a fuel received the attention of some scientists. The rediscovery of energy imbalances and the increasing long-range concern over fuel supplies accelerated both scientific and economic interest in hydrogen in the early 1970s. Some readily apparent advantages of hydrogen as both a direct and an indirect fuel were extrapolated into the terms of a future so-called *hydrogen economy*. As the result of vast research and development, coupled with severe engineering and economic changes in the total energy concepts of a plant, nation, or the world, a hydrogen economy may emerge.

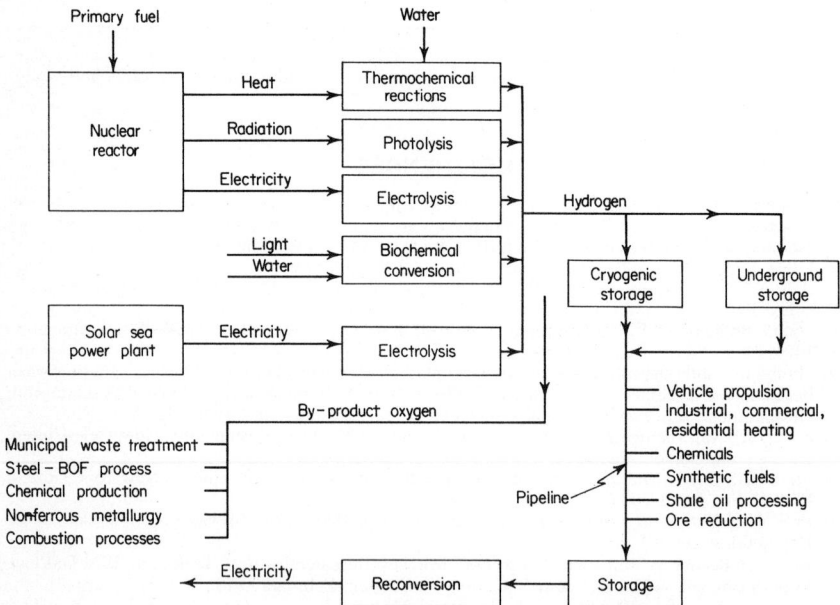

Fig. 1. Some elements of a hydrogen fuel economy.

Like other energy proposals that are basically sound scientifically, essentially three factors will determine the pace of hydrogen energy technology: (1) the manner in which, step by step, hydrogen-oriented systems and subsystems will compete economically and environmentally with other energy-source, conservation, and utilization proposals; (2) the pace of technological advancements in related fields, such as nuclear engineering, upon which hydrogen systems may depend; and (3) the pace of unilateral efforts on behalf of hydrogen-oriented systems, including the refinement of current planning-purpose data

*Consulting Chemical Engineer, Louisville, Kentucky.

and opinions into actual operating information relating to hydrogen generation, transportation, conversion and/or end utilization, and safety. Without the funding of a series of "crash programs," unilateral developments probably will be relatively slow. Most likely, the information bank for hydrogen systems will stem from an increasing awareness of the energy characteristics of hydrogen and the progressive use of hydrogen subsystems in situations where they are eminently superior.

The present concept of a hydrogen fuel economy includes a primary energy source, such as a nuclear fission or fusion reactor, a geothermal source, or a solar-powered source, with hydrogen being produced as the portable energy carrier. See Fig. 1. Thermal energy from nuclear sources would be employed to generate electricity, which would then be utilized to electrolyze water for the production of hydrogen and oxygen. The hydrogen would be distributed by pipeline to distant points of use, with storage provided by underground gas storage, or by liquefaction and refrigerated storage. Several processes have been studied for splitting water into hydrogen and oxygen thermally instead of by electrolysis, in an effort to achieve much improved conversion efficiency. These proposals remain to be proved, particularly on a large, practical scale.

BASIC CHARACTERISTICS OF HYDROGEN

Hydrogen, atomic symbol H, atomic weight 1.00797, atomic number 1, is the simplest and lightest element and is considered to have been the primordial substance from which all other elements in the universe developed. Stars were formed by gravitational forces from a rotating mass of hydrogen; the resultant high temperatures led to the fusion reaction converting hydrogen to helium, releasing thermal energy, as in the sun, and leading to the formation of the rest of the elements found on earth. Jupiter, Saturn, and Uranus still have substantial amounts of free hydrogen, whereas free hydrogen has long since escaped from the earth's lower atmosphere. However, at an altitude of 1000 miles above the earth, hydrogen atoms are more abundant than either nitrogen or oxygen.

Hydrogen was identified by Cavendish in 1766 and named by Lavoisier in 1783. In 1931, Urey discovered a second isotope, deuterium, with mass 2; a third isotope, tritium, mass 3, was prepared synthetically in 1934 by Rutherford, Oliphant, and Harteck. Libby then detected the presence of tritium in water. Hydrogen, mass 1 (protium), and deuterium, mass 2, are stable isotopes, while tritium, mass 3, is radioactive, with a half-life of 12.26 years.

Although the abundant hydrogen isotope protium is the simplest known atom, it forms two diatomic molecules, namely, *orthohydrogen*, in which the two atomic nuclei spin in the same direction, and *parahydrogen*, in which the nuclei spin in opposite directions. While the equilibrium composition of hydrogen gas is 75 percent ortho at ambient temperature, it changes to 99.8 percent para in the liquid state. The transition from ortho- to parahydrogen is exothermic (168 cal per gram), so that the heat released is more than enough to revaporize liquid hydrogen (heat of vaporization 107 cal per gram). Recognition of the existence of the ortho-para transition and the development of catalysts to equilibrate the liquid during liquefaction have made possible the large-scale production, use, and storage of liquid hydrogen.

Hydrogen molecules dissociate to atoms endothermally at high temperatures (heat of dissociation about 103 cal per gram mole), in an electric arc, or by irradiation. This property is used to effect atomic-hydrogen arc welding, in which hydrogen gas is dissociated by an a-c electric arc between two tungsten electrodes, the hydrogen atoms recombining at the metal surface to provide the heat required for welding.

The major properties of hydrogen are given in Table 1.

Actual and potential uses for hydrogen can be predicted by inspection of its properties. Its low density, 7 percent that of air, plus its high thermal conductivity, 6.7 times that of air, have led to its use as a coolant in large rotating electric equipment. The low density reduces windage friction losses to less than 10 percent of those with air, while its high thermal conductivity and heat capacity permit more efficient heat transfer, the result being an overall increase in generator efficiency of as much as 1 percent. The high heats of reaction of hydrogen with oxygen or fluorine, plus the low molecular weights of the product gases, have made hydrogen a prime fuel for rocket propulsion, since rocket thrust increases directly with the temperature and inversely with the molecular weight of the exhaust gases. Liquid hydrogen and oxygen were used in the second- and third-stage

Saturn engines in the Apollo moon flights. The low atomic weight of hydrogen has made it the preferred propellant for nuclear rockets, in which nuclear emission provides heat for exhausting hydrogen gas at high temperatures.

Some studies have indicated that the cost of transporting and distributing hydrogen by pipeline may be less than the cost of transporting and distributing electric power. Presumably existing natural gas pipelines and distribution systems can be adapted to the use of hydrogen. Although hydrogen has a net heating value of only 275 Btu per cubic foot, compared with 913 Btu per cubic foot for methane, the lower density and viscosity of

TABLE 1.   Properties of Hydrogen

Atomic weight (1961 basis C^{12})	1.00797
Atomic number	1
Melting point, K	13.96
Heat of fusion at 14.0 K, cal/g	14.0
Boiling point at 1 atm, K	20.39
Heat of vaporization at 20.4 K, cal/g	107
Density, g/cm³:	
Solid at 4.2 K	0.089
Liquid at 20.4 K	0.071
Critical temperature, K	33.3
Critical pressure, atm abs.	12.8
Critical volume, cm³/mole	65.0
Critical density, g/cm³	0.031
Heat of transition, ortho to para at 20.4 K, cal/g	168
Specific heat: At constant pressure C_p, cal/g	
Liquid at 17.2 K	1.93
Solid at 13.4 K	0.63
0–200°C	3.44
Specific heat: At constant volume C_v (0–200°C), cal/g	2.46
Specific heat: Ratio C_p/C_v (0–200°C)	1.40
Gas density at 0°C and 1 atm, g/l	0.0899
Gas specific gravity (air = 1.0)	0.0695
Gas thermal conductivity at 25°C (cal)(cm)/(s)(cm²)(°C)	0.00044
Gas viscosity at 25°C and 1 atm, cP	0.0089
Coefficient of thermal expansion per °C	0.00356
Heat of combustion at 25°C, kcal/g mole	
Gross	68.3174
Net	57.7976
Energy release upon combustion: cal/g	29,000
cal/cm³	2,050
joule/g	1.21×10^5
Flame temperature, K	2,483
Autoignition temperature, K	858
Heat of formation of HF at 25°C, kcal/g mole ΔH	−64.2
Flammability limit, %:	
In oxygen	4–94
In air	4–74

hydrogen make it possible for a pipeline to deliver about the same amount of thermal energy as with methane, at a somewhat greater compression cost. The thermal energy in hydrogen can be utilized more efficiently in home heating than natural gas, because hydrogen can be burned in nonvented heaters, with no loss of heat, since its only primary combustion product is water. By using flameless catalytic heaters, nitrogen oxide formation can be fully eliminated.

One advantage of hydrogen as a source of thermal energy, compared with electricity, is that it can be stored for later use, whereas electricity must be used as it is generated. Hydrogen, like natural gas, may be stored and transported as a refrigerated liquid, or stored as a gas under pressure in underground storage. Hydrogen may also be stored as a metallic hydride, such as magnesium hydride, or the hydride of a rare-earth metal plus nickel, such as $LaNi_5$. The absorption of hydrogen by the metal is exothermic, which requires cooling, and heat is required to release the hydrogen—factors which lower the efficiency of this means of storage. See Appendix A to this section.

OCCURRENCE

Hydrogen is the ninth in abundance by weight of the elements in the earth's crust (third in number of atoms), most of it being found in water, which contains 11.2 percent hydrogen. It is present in acids, bases, hydrocarbons, all living organisms, and most organic compounds. Natural hydrogen contains about 0.015 mole percent of deuterium and infinitesimal traces of tritium.

CHEMICAL BEHAVIOR

Hydrogen is a colorless, odorless, tasteless gas, unreactive at ordinary temperatures in the absence of catalysts. It will react with oxygen at room temperature in the presence of a platinum catalyst, or explosively with a flammable mixture in the presence of a flame or spark, forming water. Hydrogen also reacts with sulfur, selenium, and tellurium.

Hydrogen reacts with all the halogens, forming acids. The reaction is violent and exothermic with fluorine, mild and endothermic with iodine. Hydrogen reacts directly at elevated temperatures with most alkali and alkaline-earth metals to form hydrides, decomposable by water. Many metal oxides, halides, and sulfides react with hydrogen at elevated temperatures to form the free metals and thus it plays a reducing agent role.

In the presence of platinum, palladium, or nickel catalysts, hydrogen reacts with various organic compounds to saturate double or triple bonds and to convert aldehydes, ketones, and esters to alcohols. Hydrogenation is generally considered as the addition of hydrogen with other materials. Vegetable and fish oils are hardened or solidified by catalytic hydrogenation. Partial hydrogenation clarifies some oils and makes them odorless. Fatty acids, such as oleic acid, are converted into stearic acid by hydrogenation. Coconut oil, peanut oil, and cottonseed oils can be made to appear, taste, and smell like lard; or they can be made to resemble tallow. Hydrogenation reactions account for a major portion of the industrial use of hydrogen as a reactive chemical raw material.

ENERGY-RELATED USES OF HYDROGEN

The probable functions of hydrogen in future energy technology may be put into two major categories: (1) *direct functions* in which hydrogen serves as a fuel, that is, as the source of heat, power, and light without prior conversion to some other energy form; and (2) *indirect functions* in which hydrogen is an important component of the total energy system, but before the final end use of that energy, the hydrogen is involved in some conversion, possibly chemically used in the creation of a synthetic fuel, such as substitute natural gas, or possibly converted into electric energy which becomes the final end energy used. One of the major indirect or secondary roles proposed for hydrogen is that of an energy transporter, wherein in one scheme, other forms of energy would be consumed to generate hydrogen which then would be pipelined and stored at distant points available for another conversion step—for example, converted into electric energy as needed.

HYDROGEN AS AN ENERGY SOURCE FOR MOTIVE POWER

Aside from their relatively low costs over past years, the hydrocarbon fuels, notably gasoline and kerosine, have offered convenience in handling and transportability for use in connection with powered vehicles.

Hydrogen, were it available at a reasonable cost, permits these advantages, plus the additional attraction of decreasing air pollution. As mentioned earlier, where cost was not a major objective, liquid hydrogen along with liquid oxygen has been a favorite propellant for rockets and space vehicles.

Numerous tests have checked the performance of hydrogen as a fuel for an internal combustion engine. Efficiencies as much as 50 percent greater than obtainable with gasoline have been reported. A test motorcar obtained 19 miles per pound of hydrogen. However, since liquid hydrogen weighs only 0.58 pound per gallon, the mileage figure was 11 miles per gallon of liquid hydrogen. The use of liquid hydrogen as a motor fuel presents several major problems despite its basic attraction.

Because of its low density, the net storage volume required would be at least as much as for gasoline. The storage tank must be maintained at a temperature of $-423°F$ ($-253°C$), which is the boiling point of hydrogen at atmospheric pressure. This would require insulation that would increase the overall size of the storage container. Vaporization losses from the storage tank, amounting to perhaps 2 percent or more per day, must be vented so that no ignition of the vented hydrogen gas can occur, and no accumulation of explosive hydrogen-air mixtures can be possible. The lower explosive limit of hydrogen in air is 4 percent; so adequate ventilation must be provided. Fortunately, hydrogen gas, being the lightest gas, with a specific gravity of 0.07 referred to air, will rise and diffuse rapidly, and thus can be easily dispersed. Service stations for dispensing liquid hydrogen will require more expensive storage and pumping facilities than are required for gasoline.

Battery-powered electric vehicles have been reported as operating with considerably higher efficiencies than either hydrogen- or gasoline-fueled internal-combustion engine types, and so could provide transportation with smaller total energy requirements than is presently possible with hydrogen-fueled internal-combustion engines.

In addition to much basic research required to develop reasonably economical sources of hydrogen, motive power use of hydrogen will require, among other factors, (1) much attention to the alteration and adaptation of current internal-combustion engine designs, (2) the possible development of new engine concepts, and (3) concerted study of hydrogen storage (transportable), dispensing systems, all with careful attention to a new order of safety problems.

As will be evident from a number of terse summaries of research efforts past and present in this area, as given in Appendix A following this article, currently there is no large, concerted effort directed toward the use of hydrogen as a fuel for motive power, but rather the efforts are scattered and, although of high caliber in most instances, they may be described as modest in terms of funding and manpower. The thrust of most of the R&D on the part of engine and vehicle manufacturers has had to be directed to finding relatively short-term solutions to emission problems associated with conventional, readily obtainable fuels.

As may be evident from Appendix A, it is highly likely that the first major use of liquid hydrogen as an energy source for motive power will be jet aircraft, largely because of the excellent weight advantage and the less serious nature of the boil-off loss and distribution problems, as compared with other forms of transportation. City buses and long-haul motor trucks may follow as candidates, in which refueling may be effected through replacement of entire storage tanks (dewars). Because the private motorcar presents the most crucial logistics problems, including the small-capacity fuel system, concern with safety, boil-off loss of fuel even when vehicle is not in use, and the education and acceptance involving millions of users, it probably will follow rather than lead the use of hydrogen in other modes of transportation.

HYDROGEN AS A HEATING FUEL

As just described, the attraction of hydrogen as the basic, if not exclusive fuel for jet aircraft, trucks, buses, and ultimately private motorcars is strong. On the other hand, the routine use of hydrogen as a heating fuel for industry and commercial-residential installations entails greater complications and would appear to be much more dependent upon the overall economic and technical aspects of a so-called hydrogen fuel economy. A hydrogen fuel economy for transportation presents fewer impacts on the established fuel distribution system. All these matters, of course, are interlocked, but the major point for consideration is that an effective hydrogen fuel economy for transportation most likely can be effected without necessarily involving the wide use of hydrogen for industrial-commercial-residential applications.

Strictly from the standpoint of technological qualifications, assuming availability, hydrogen can be an excellent fuel for almost any heating application. Hydrogen can be used in the home for cooking and heating (and even lighting) and likewise in commerce and industry. Compared with natural gas, hydrogen burns with a faster, hotter flame. Hydrogen-air mixtures are flammable over wider limits of mixtures. Hydrogen burns without producing noxious exhaust products, allowing the use of unvented appliances except where water vapor and the resulting increased humidity may be objectionable. In the winter heating season, the additional humidity can, in fact, be highly desirable. In excessively humid locations in summer, the additional water vapor coming from cooking appliances, for example, could be objectionable.

But, generally because of the absence of hazards from carbon monoxide and other fumes, large savings could be achieved from the elimination or at least simplification of flues. Some experts suggest that not only construction costs could be lowered as the result of clean burning, but that an increase of some 30 percent in the efficiency of a gas-fired home heating system could be achieved. The concept of peripherally placed unflued devices, particularly through the use of catalytic "flameless" heaters, could ultimately lead to a serious revision of the widely accepted central heating concept. By maintaining the temperature of a catalytic bed as low as 212°F (100°C), the production of nitrogen oxides would be virtually eliminated.

Because hydrogen burns with a hotter flame, some design features of heating apparatus would require change. The energy content per unit mass of liquid hydrogen is about 2.75 times greater than that of hydrocarbon fuels. On the other hand, there are only 325 Btu per standard cubic foot of hydrogen compared with 1000 Btu per standard cubic foot of natural gas, thus dictating further design changes. The ignition energy of hydrogen is about 0.02 millijoule, which is less than 7 percent that of natural gas, a major factor in making low-temperature catalytic burners possible.

Despite the numerous advantages of hydrogen as a direct heating fuel, particularly in the home, the application of hydrogen should be viewed in terms of the total energy concept of an exclusively hydrogen-supplied (all-hydrogen home) installation. Where the direct use of hydrogen for heating is large, the economy will be most favorable. If a substantial amount of the hydrogen must be converted into electric energy, as by a fuel cell, then economic justification becomes more difficult.

Lighting in the all-hydrogen home may be accomplished by condoluminescence, a cold process. A phosphor is spread on the inside of a tube similar to the conventional fluorescent lamp. Upon coming in contact with the phosphor, small amounts of hydrogen combine with the oxygen in the air to excite bright luminescence in the phosphor.

Conversion of burners and other design aspects of heating systems and appliances to pure hydrogen or to a hydrogen-enriched natural or substitute natural gas supply, although costly and inconvenient, is certainly not in the economically insurmountable category. Similar alterations over the years were made in the United States when communities switched from manufactured gas (about 50 percent hydrogen) to natural gas. Such switchovers are even more recent in European communities.

As more hydrogen becomes available for transportation use and as more hydrogen is pipelined regionally or transcontinentally, depending largely on the demand placed upon the supply of hydrogen for industrial uses, it may be that the hydrogen content of community gas supplies will be progressively enriched with hydrogen (in a periodic, stepwise manner because of switchover problems) and thus contribute in a gradual manner to less pollution and to the conservation of natural gas.

HYDROGEN AS AN ENERGY TRANSPORT

With the possible exception of the use of hydrogen as a source of motive power in the transportation field, the *non*chemical interest in hydrogen in the total energy picture is directed to the use of hydrogen as a means or mode of storing and transporting energy. It is in this area that hydrogen directly confronts the past ever-increasing trend toward a fully electric energy economy. Undeniably, hydrogen energy has a major starting advantage over electric energy; namely, hydrogen is a storable energy form. Investigations also are showing that hydrogen in pipelines may cost less to transport than electricity flowing over long power lines. Thus, hydrogen may play an important future role simply as a mode of storage and transport, even though source and terminal energy conversions (such as electricity → hydrogen → electricity) may be required.

Electric power plants are most efficient when operated at constant output at full-rated load. Because of wide fluctuations in consumer load (daily and seasonally), generating rates require constant adjustment. Communication systems and some emergency systems employ batteries for interim storage of electric energy, but these applications are minuscule when compared with the total electric generating and distribution system. The principal means of large-scale storage is the use of pumped storage, i.e., in essence a reversible hydroelectric station in which electric energy is temporarily converted to a hydraulic head by pumping water to an elevated reservoir. Unfortunately, the topography must be suitable for such an installation, and thus this approach is limited to comparatively few power-generating sites.

The high-voltage cables required to transmit electricity from generating stations to load centers are costly in terms of equipment, installation, maintenance, and land costs. There is growing resistance among environmentalists to the extension of overhead cable systems. The cost of going to underground cables for transmitting bulk current ranges from 9 to 20 times that of overhead configurations. The effective use of cryogenic superconducting cables may lower underground costs considerably, but much research remains to be completed before this is possible. See also Trends in Electric Power Transmission in Handbook section on *Power Technology Trends*.

Because of the tremendous volumes of fuel that can be moved in pipelines, the construction, maintenance, and operating costs of a buried pipeline are much less in terms of a percentage of the total product moved. Pipeline operations have been profitable even at the relatively low price ranges for liquid and gaseous fuels prevailing before the early 1970s. Pipeline technology, of course, is well established—with hundreds of thousands of miles of trunklines installed and operating in the United States as of 1970. These lines annually transport 23 trillion cubic feet of gas. Typical pipelines range from 600 to 1000 miles in length and are up to 48 inches in diameter. Line pressures may range from 600 to 800 psi, but go up to 1000 psi. A representative 36-inch pipeline will carry a gaseous fuel with the equivalent of 37,500 million Btu per hour. The electric energy equivalent would be 11,000 megawatts. By comparison, this is ten times the energy-carrying capacity of a single-circuit 500-kilovolt overhead transmission line.

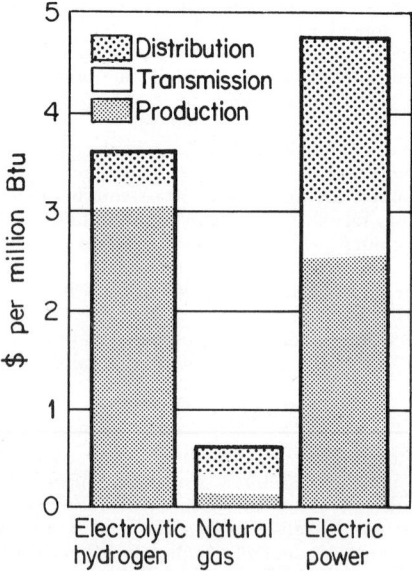

Fig. 2. Major segments of cost for electrolytic hydrogen, natural gas, and electric power. *(Institute of Gas Technology.)*

The figures for pipeline transportation of pure hydrogen are not quite so attractive at present, but nevertheless the comparison with electric transmission costs remains highly significant.

One study* shows that the pipeline transmission costs for hydrogen will range from 30 to 50 percent more than for natural gas. In another study,† assuming that the average

*B. J. Berkowitz, General Electric-TEMPO.
†G. Beghi, Ispra (Italy) Research Center.

temperature in hydrogen pipelines will be about 50°F (10°C), with a pressure of 925 psig, and a 91 percent utilization factor, it is shown that the cost of transporting hydrogen will be about 3 cents per million Btu per 100 miles, an increase of 30 to 50 percent over transporting methane. Thus, the two studies are essentially in agreement on this point. Since pumping stations for hydrogen could be established at greater space intervals, some of this cost differential could be made up.* But, to convert an existing natural gas line to hydrogen service would require a rise of compressor capacity by a factor of 3.8 and compressor horsepower by 5.5.

Projections of hydrogen transportation costs, as compared with other energy, are given in Figs. 2 and 3.

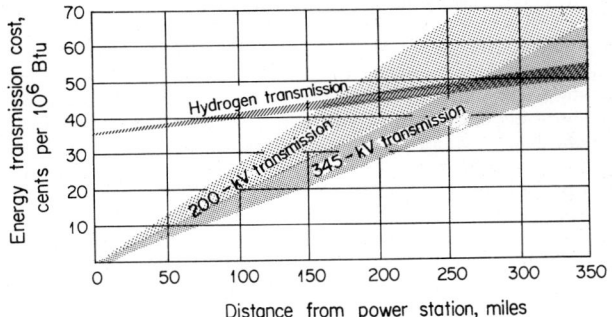

Fig. 3. Costs of hydrogen pipeline versus electric power transmission as a function of distance from power station. (*Institute of Gas Technology.*)

Obviously, hydrogen transmission costs represent but one part of a total system. Should the costs of generating hydrogen in the first place, and the subsequent conversion of hydrogen into electricity at the terminal end of the system, remain excessively high, then the savings in energy transportation costs, of course, become academic.

SOURCES OF HYDROGEN

As will be pointed out later under chemical uses of hydrogen, with few exceptions, the past major source of hydrogen has been natural gas. In strictly terms of chemical needs, where economic factors are favorable, natural gas has served this need well. However, in terms of total energy conservation, where hydrogen is looked to as a means of conserving fossil-fuel sources, a much less costly and more abundant hydrogen-containing raw material must be sought. Obviously, this raw material is water. Particularly in areas of the world where hydrocarbons have not been readily available, reasonably large water electrolysis installations have been made, notably in locations with low electricity costs.

In addition to electrolysis, the principal means under consideration for deriving hydrogen from water is thermochemical splitting. The waste heat and high temperature available from certain types of nuclear reactors would effect a series of chemical reactions, still in the research phase as of the mid-1970s, to free hydrogen and oxygen from water. Additional proposals have included the use of ultraviolet radiation from the plasma of a fusion reactor for the direct photolysis of water vapor †, and the use of some forms of algae, under the stimulation of light, to convert hydrogen ions to hydrogen gas by a complex chain of biochemical reactions. ‡

---

*An interesting set of cost comparisons is given by W. E. Winsche, K. C. Hoffman, and F. J. Salzano in *Science*, **180**, 4093, 1325–1332 (1973).

†Proposal of B. L. Eastlund, Atomic Energy Commission. It is claimed that injection of elements, such as aluminum, into a hydrogen plasma will produce photons of desired wavelength for photolysis, which would occur in water vapor surrounding the plasma chamber. This scheme must await development of an operating fusion reactor.

‡Proposal of L. O. Krampitz, Case Western Reserve University, Cleveland, under a National Science Foundation grant.

Electrolysis   Because of years of operating experience, electrolysis is possibly an order of magnitude ahead of other proposals from a technological standpoint. Although simple in concept, electrolysis is expensive—hence the research efforts to find other ways of splitting water. Nevertheless, this side of one or more breakthroughs in the other areas, most likely electrolysis operations will have to continue to serve as the basis for costs in extending the use of hydrogen in the near term.

As of 1974, industrial electrolyzers ranged in size from 500 standard cubic feet of hydrogen production per day, consuming 3 kilowatts of electricity, to more than 40 million standard cubic feet of hydrogen per day, consuming 240,000 kilowatts. Most common installations are from 10,000 to 500,000 standard cubic feet of hydrogen per day. Two factors generally characterize an electrolyzer installation: (1) access to comparatively low-cost electricity, as found in some areas served by hydroelectric installations; and (2) need for the oxygen which accompanies the production of the hydrogen. Industrial electrolyzers usually operate at efficiencies of about 60 to 70 percent. Some high-pressure prototype models have reached 85 percent. It has been pointed out* that, in theory, electrolyzers can approach a maximum electrical efficiency of nearly 120 percent as the result of the ideal unit absorbing ambient heat and also converting this energy into hydrogen. A reasonable, practical target for an improved electrolyzer appears to be around 100 percent. Thus, the production of electrolytic hydrogen would be limited only by the efficiency of electric current generation, namely, between 35 and 45 percent. An estimate has been made that the overall conversion efficiency of electricity-to-hydrogen-to-electricity will thus approximate 38 percent.† The theoretical power required to produce hydrogen from water is 79 kilowatthours per 1000 cubic feet of hydrogen gas. One of the largest electrolyzers operating commercially is that of Cominco, Limited (British Columbia). This is a 90-megawatt installation that produces approximately 36 tons of hydrogen per day for use in ammonia synthesis. Other large plants are located in Norway and Egypt.

Two principal types of electrolyzers are in commercial use: (1) Tank cells are used with monopolar electrodes. Porous diaphragms separate the alternate cathodes and anodes to prevent gas mixing. The anodes and cathodes are connected in parallel to keep the required voltage at approximately 2 volts and to permit high current densities. The arrangement requires a large floor area. (2) Bipolar electrodes, connected in series and suitably insulated, are used. The electrodes are cathodic on one side; anodic on the other side. This arrangement, requiring less floor space, is more complex and requires high voltages.

High pressure can increase efficiency, and this concept has been under development for many years. A commercial electrolyzer‡ is available which operates at a pressure of 30 atmospheres and 194°F (90°C), requiring 300 amperes of electric current at 217 volts.

In the mid-1960s, bipolar cells of porous nickel electrodes were developed, which operate at current densities of 800 and 1600 amperes per square foot. However, present commercial practice limits current to about 200 amperes per square foot. As of mid-1972, the cost of a unit of this type, operating at 800 amperes per square foot and producing 4400 pounds of hydrogen per hour, was estimated at about $5.5 million. However, by increasing the current density to 1600 amperes per square foot, the estimated cost would be about $4.5 million.

In the mid-1960s, electric-high-temperature, vapor-phase electrolysis§ was developed. In this process, the electrolyte is solid, porous zirconia which contains dopants. Operating temperature ranges from 932 to 1472°F (500 to 800°C). A modification of the process is under development which will produce only hydrogen by consuming the by-product oxygen.

Among electrolyzer design improvements that may occur are better electrodes which may result as a spin-off from fuel-cell work. There are indications that electrode improvement could cut the costs of electrolytic hydrogen by about 20 to 25 percent. Electrolysis looms high in considering the utilization of ocean thermal gradients as an energy source. See also article in Handbook section on *Solar Technology*.

---

*D. P. Gregory, Institute of Gas Technology, Chicago.
†E. C. Tanner, Princeton University, and R. Huse, Public Service Electric & Gas Company.
‡Lurgi, GmbH.
§General Electric Co.

Thermochemical Splitting. The major target here, of course, is finding one or more series of chemical reactions that will result in the satisfactory separation of hydrogen (and oxygen) from water. Considerable work has been going forth at the Nuclear Research Center, Julich, West Germany, where much attention has been given to sulfur- and chlorine-based thermochemical cycles. Other researchers* have been probing various combinations of at least 56 chemical elements, including over 700 different compounds, that may show promise in various schemes for a closed-water-splitting cycle. It is understood that approximately twenty promising schemes emerged from this research, mainly centered in chlorine compounds. The most frequent flaw encountered among prospective reactions is the large amount of free energy required to force one or possibly two of the series of reactions; and the appearance of reactions that produce stable compounds incapable of regeneration.

Chemically based processing sequences also have been under intensive investigation by other organizations.† These reactions which simply would rely upon a nuclear reactor as a heat source would not have to await the emergence of a practical, operating fusion reactor. Ispra, for example, has worked out some chemical-reaction sequences that operate at temperatures no higher than about $1346°F$ $(730°C)$. One of these consists of four reactions:

$$CaBr_2 + 2H_2O \rightarrow Ca(OH)_2 + 2HBr$$
$$Hg + 2HBr \rightarrow HgBr_2 + H_2$$
$$HgBr_2 + Ca(OH)_2 \rightarrow CaBr_2 + HgO + H_2O$$
$$HgO \rightarrow Hg + \frac{1}{2}O_2$$

A drawback to this sequence is its use of highly corrosive hydrogen bromide. The scheme also requires a large inventory of mercury. A more-recent Ispra-developed sequence lessens these problems. It is based on compounds of chlorine and iron.

Of major concern to all investigators working on thermochemical splitting schemes is the availability of appropriate materials of construction. Heat exchangers between the nuclear side and the chemical side must withstand both corrosion and radioactive contamination. The conventional nickel-chromium alloys are capable up to about $1050°K$; exotic, but available alloys up to about $1400°K$. Above these temperatures, ceramics and new alloys may have to be used. Considerable materials research along these lines is going forth at the Los Alamos Scientific Laboratory.

CHEMICAL USES OF HYDROGEN

Even before its serious consideration in the fuel economy, the demand for hydrogen has been growing at a rate of about 15 percent annually since World War II. About 3 trillion standard cubic feet of hydrogen (8 million tons) were produced in the United States in 1970. Not including energy applications, the chemical requirements for hydrogen are expected to increase by at least 7 percent per annum through the year 2000. Among demands include petroleum refining, plastics, elastomers, increased desulfurization of fuel oils, increased usage in iron ore reduction, aerospace applications, and hydrogen/air fuel cells. Approximately 42 percent of the hydrogen produced is consumed in ammonia production; about 38 percent in petroleum refining. Other large consumers are metallurgical and food processing.

In terms of presently nonconventional fuels that will require increasing quantities of hydrogen as new processes develop, it is estimated that: (1) synthetic crude oil from coal will require 6500 standard cubic feet of hydrogen per barrel of oil; (2) 1300 SCF per barrel of oil from shale; (3) 1500 SCF per every 1000 SCF of synthetic pipeline gas produced from the gasification of coal. Petroleum refining use of hydrogen is expected to increase to 610 SCF per barrel of crude refined. Direct iron ore reduction use of hydrogen is expected to increase to 20,000 SCF per ton of iron. If there were no other source, all of these and current needs as well could be met by using approximately 10 percent of the natural gas production. ‡

*General Atomic Co., San Diego, Calif.
†Institute of Gas Technology, Chicago; General Electric Co., San Jose, Calif.; European Atomic Energy Community, Ispra, Italy.
‡U.S. Bureau of Mines.

HYDROCARBON SOURCES OF HYDROGEN

The catalytic reaction of hydrocarbons with steam (steam reforming) is one of the most widely used processes for producing hydrogen commercially. Hydrocarbons, from methane to petroleum naphtha, are vaporized, mixed with steam, and passed over a nickel catalyst at temperatures of 1200–1800°F (649–982°C), producing carbon oxides and hydrogen. A second widely used process is the noncatalytic partial oxidation of hydrocarbons, exemplified by two processes,*·† which can utilize any liquid hydrocarbon, including crude oil and residual fuel oil, and hence can be used in locations where light hydrocarbons are unavailable or expensive. Preheated hydrocarbon, steam, and oxygen are fed through a burner to a refractory-lined combustion chamber, in which they react at temperatures of 2300–2700°F (1260–1482°C) to form carbon oxides and hydrogen. The two processes are similar chemically but differ in mechanical features. Most plants of the one type‡ use a direct quench for cooling reaction products, while plants of the other type§ use indirect cooling in waste-heat boilers.

Steam reforming and partial oxidation have both been operated at pressures of 600 psig or higher, which permits more effective heat recovery and reduces compression costs. The partial-oxidation process has one important advantage, namely, the ability to utilize any available pumpable hydrocarbon, including those with high sulfur contents, but this is offset by the requirement for an air-liquefaction plant to supply the oxygen used in the process. Another advantage of partial oxidation is that no catalyst is needed. However, carbon removal from the product gas is required, an operation superfluous with steam reforming.

Other hydrogen manufacturing processes include catalytic partial oxidation of hydrocarbons, the steam-water-gas process, the steam-iron process, the electrolysis of water already discussed, ammonia dissociation, the steam-methanol process, and the thermal dissociation of hydrocarbons. Before 1940, the steam-water-gas process dominated hydrogen manufacture, but it has since been supplanted by the steam reforming of hydrocarbons. The steam-water-gas process uses coke as raw material, reacting it with steam to produce water gas, a mixture of carbon monoxide and hydrogen. This is then reacted with steam over an iron oxide catalyst to convert the carbon monoxide to carbon dioxide and hydrogen, which then is purified to produce hydrogen. The high cost of coke, the large steam requirement, and cyclic operation at atmospheric pressure have rendered this process obsolete.

The steam-iron process produces hydrogen by reacting steam at high temperature with reduced iron oxide to give hydrogen and then re-reducing the iron oxide with a reducing gas, such as water gas or producer gas, in a cyclic operation at atmospheric pressure. This process also is obsolete, although attempts have been made to modernize it with fluid-bed techniques.

Catalytic partial oxidation of hydrocarbons, in which hydrocarbons up to and including petroleum naphtha are reacted with steam and oxygen over a nickel catalyst, is in commercial use mainly to produce synthesis gas for ammonia or methanol. Like steam reforming, this process can be operated under pressures up to 30 atmospheres or more. The necessary equipment is simpler than for steam reforming, since the catalyst chamber is not externally heated, but an oxygen supply is required.

The steam-methanol and ammonia-dissociation processes have been used only in relatively small installations for hydrogen production, where the cost of the equipment compared with steam reforming outweighs the cost of ammonia or methanol versus the cost of naphtha or other hydrocarbons.

A simplified flow sheet of hydrogen production by the steam reforming of hydrocarbons is shown in Fig. 4. The feed gas is desulfurized by activated carbon, mixed with steam, reacted over a nickel catalyst in an externally heated reformer at 1600°F (871°C), quenched to 450°F (232°C) with condensate, passed over a carbon monoxide conversion catalyst, cooled by heat exchange with gas en route to the methanator and monoethanolamine solution, further cooled to 120°F (49°C), passed through the carbon dioxide

*Texaco process.
†Shell process.
‡Texaco process.
§Shell process.

absorber, in which carbon dioxide is absorbed in monoethanolamine solution,* then through the heat exchanger and into the methanator at 500°F (260°C), where carbon monoxide and carbon dioxide are hydrogenated to methane. The product hydrogen then is cooled.

Large quantities of hydrogen are recovered as a by-product in certain oil-refining operations, principally catalytic dehydrogenation processes to convert aliphatic and naphthenic hydrocarbons to aromatics. Formerly, much of this by-product hydrogen was burned as fuel or utilized for ammonia synthesis, but present refining processes have such large hydrogen requirements for hydrodesulfurization and other hydrogen-consuming processes that all by-product hydrogen is used and, in fact, often supplemented by hydrogen produced by steam reforming or partial oxidation.

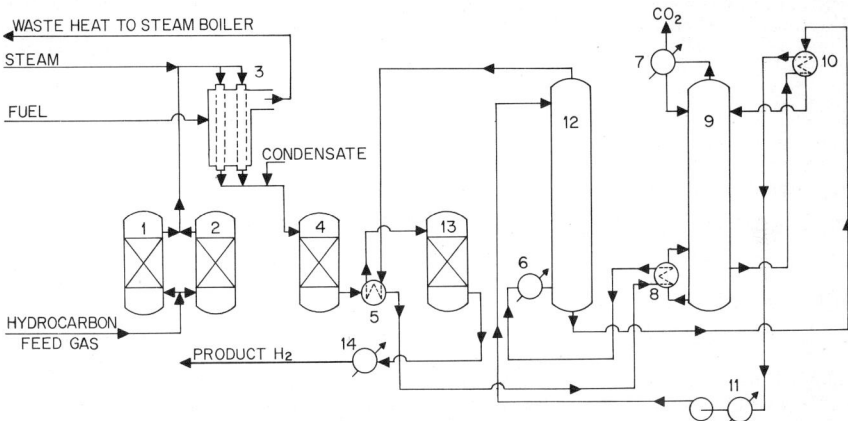

Fig. 4. Production of hydrogen by steam reforming of hydrocarbons: (1) and (2) carbon drums; (3) reformer; (4) carbon monoxide converter; (5) to (8), (10), (11), and (14) exchangers; (9) amine regenerator; (12) carbon dioxide absorber; (13) methanator. (*C & I/Girdler Corporation.*)

Purification.   The crude hydrogen produced by steam reforming or partial oxidation contains impurities, such as carbon dioxide, carbon monoxide, methane, carbon, hydrogen sulfide, and water vapor. Carbon monoxide is converted to carbon dioxide and hydrogen by reaction with steam (water-gas shift reaction), followed by carbon dioxide removal by various absorption processes, with trace amounts of carbon oxides being hydrogenated to methane (methanation). Methane in small amounts (1 percent or less) is unobjectionable in certain uses of hydrogen, such as ammonia or methanol synthesis, but must be removed in other cases, e.g., where liquid hydrogen is desired. Methane may be removed by cryogenic means, by adsorption, or by re-reforming with steam. Carbon is removed by water-washing or filtering. Hydrogen sulfide is removed by absorption or adsorption processes. Water vapor may be removed by absorption, adsorption, or refrigeration. Synthetic zeolites (molecular sieves) have been used in adsorption processes for removing carbon oxides and water vapor from hydrogen.

Large quantities of extremely pure hydrogen (99.9 percent) are produced by subjecting impure hydrogen to diffusion through palladium films. At a temperature of about 600°F (316°C), molecular hydrogen will dissociate to atoms on palladium, the atoms then diffusing through the palladium and recombining on the opposite surface to yield pure hydrogen.

REFERENCES

Chopey, Nicholas P.: Hydrogen: Tomorrow's Fuel?, *Chem. Eng.*, pp. 24–26, Dec. 25, 1972.
Gregory, Derek P.: The Hydrogen Economy, *Sci. Amer.*, **228**, 1, 13–21, January 1973.
Staff: Hydrogen: Synthetic Fuel of the Future, *Science,* pp. 849–852, Nov. 24, 1972.

*Girbitol process.

Staff: Hydrogen: Likely Fuel of the Future, *Chem. Eng. News*, pp. 14–17, July 3, 1972.

Staff: Hydrogen Fuel Use Calls for New Source, *Chem. Eng. News*, pp. 16–18, July 3, 1972.

Staff: Hydrogen Fuel Economy: Wide-ranging Changes, *Chem. Eng. News*, pp. 27–30, July 10, 1972.

Staff: Hydrogen Generation May Get a Lift, *Ind. Res.*, pp. 31–33, May 1974.

Schoeppel, R. J.: Prospects for Hydrogen-fueled Vehicles, *Chem. Tech.*, pp. 476–480, August 1972.

Jones, L. W.: Liquid Hydrogen as a Fuel for the Future, *Science*, pp. 367–370, Oct. 22, 1971.

Bockris, J. O. M.: A Hydrogen Economy, *Science*, p. 1323, June 1, 1972.

Winsche, W. E., K. C. Hoffman, and F. J. Salzano: Hydrogen: Its Future Role in the Nation's Energy Economy, *Science*, pp. 1325–1332, June 29, 1972.

Adt, Robert R., Jr.: Hydrogen Utilization, in *Proc. Solar Sea Power Plant Conf. and Workshop*, National Science Foundation, Washington, June 27–28, 1973, published by Carnegie-Mellon University, Pittsburgh, Pa.

Staff: Hydrogen Economy Concept Gains Credence, *Chem. Eng. News*, pp. 15–16, Apr. 1, 1974.

Staff: Cesare Marchetti: Hydrogen Economy Proponent, *Chem. Eng. News*, p. 34, Apr. 22, 1974.

NOTE: See also references listed with Appendix A.

Appendix A
Hydrogen as Energy Source for Motive Power

AUTOMOTIVE ENGINES*

In recent years there have been a number of automobiles converted to run on hydrogen. The impetuses behind this are the air pollution problem and the realization that eventually we will have to find some substitute for our reliance on petroleum products as fuel for our ground vehicles. The research effort required to develop an automobile which uses hydrogen as fuel is in two basic areas. The first is concerned with the engine development and the second with the storage problem. We at the University of Miami have been concerned with the engine development. Others, such as the Brookhaven National Laboratory and the Phillips Company, have been concerned with the development of chemical fuel storage systems. This is in contrast to the usual method of storing hydrogen as a compressed gas or in a liquid state. They have developed hydrides from which the hydrogen is liberated as required by the engine. This is advantageous from a safety point of view as the amount of hydrogen present at a given time is minimal.

Both hydrogen-air and hydrogen-oxygen breathing piston engines have been tested. The concept of using hydrogen as a fuel for piston-type engines is not a new one. In the early nineteenth century hydrogen was combusted with oxygen, and the vacuum created on cooling was used to move the piston. At the Seventh Intersociety Energy Conversion Engineering Conference (IECEC) in San Diego, Professor Weil gave a history of the use of hydrogen for internal-combustion engines. He described how in 1927 the German Zeppelin Company used hydrogen that was normally vented to compensate for the increase in lift as a result of fuel consumption to fuel their internal combustion engines. The Great Britain Royal Airship Works did a similar thing in 1935. Contrary to this use where hydrogen was used as an addition to liquid and gaseous fuels, Rudolf Erren worked with hydrogen-air and hydrogen-oxygen Diesel type engines in the 1920s and 1930s.

Other early experimenters using hydrogen fuel-air mixtures indicated that there was a great tendency for the engine to knock and preignite unless low compression ratios and lean mixtures were used. Around 1955, however, tests by Captain King in Canada showed that a hydrogen fueled Cooperative Fuel Research (CFR) knock rating engine could be run at stoichiometric fuel-air ratios with compression ratios of up to 12 to 1 without combustion knock or preignition provided the combustion chamber was free of fluffy carbon deposits, a low operating-temperature sparkplug was used, and the exhaust valve temperature was kept low.

In 1971, the Perris, California Smogless Automobile Association fueled an engine with a hydrogen rich mixture of hydrogen and oxygen. In addition to claiming to overcome the problems of preignition and detonation, they also eliminated nitric oxides from their exhaust which consisted of condensed water. The excess hydrogen was separated from this water and recycled. The advantage of this scheme

---

*Excerpts of remarks by R. R. Adt, Jr., Associate Professor of Mechanical Engineering, School of Engineering and Environmental Design, University of Miami, Coral Gables, Florida, given at Workshop sponsored by the National Science Foundation, Washington, June 27–28, 1973. Reproduced with permission of publisher, Carnegie-Mellon University, Pittsburgh, Pennsylvania, publisher of proceedings.

is the elimination of nitric oxides, but the disadvantages are the need to carry an oxygen supply and to install a condenser.

In another study at about the same time, Murray and Schoeppel of the Oklahoma State University tested an airbreathing hydrogen-fueled single cylinder internal combustion engine. It had a 6.5 to 1 compression ratio and the hydrogen was injected at high pressures directly into the cylinder and ignited by a sparkplug. A second cam shaft was used to actuate the hydrogen injection valve and their

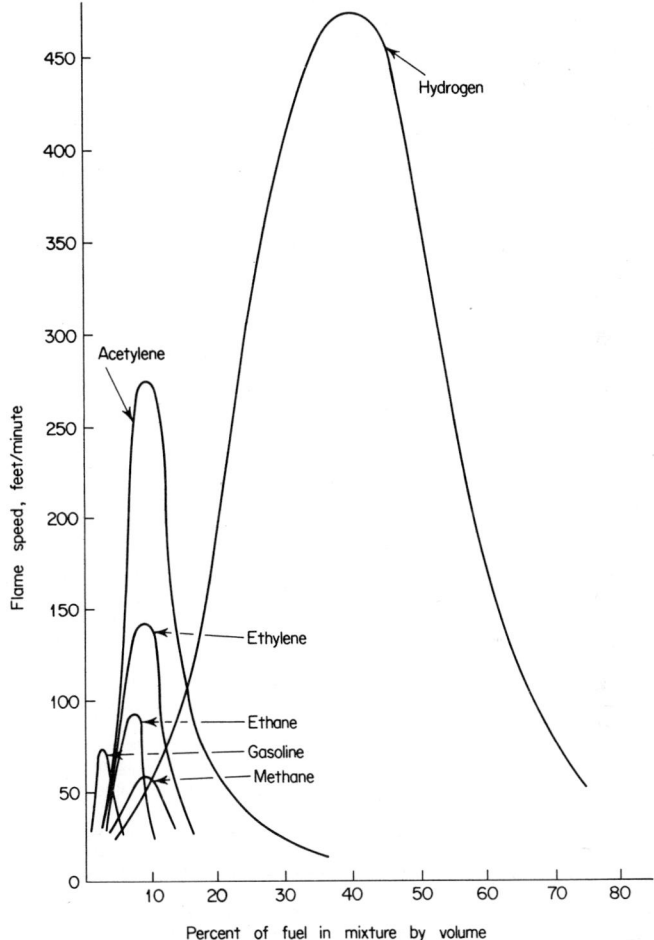

Fig. A-1. Flame speed of various fuels.

hydrogen pressures varied from approximately 200 psi at low power output to 1000 psi at high output. The engine was capable of operating over a wide range of conditions with very low nitric oxide emissions. Their brake specific fuel consumption (bsfc) varied from 0.36 to 0.4 lb H_2/bhp HR at wide open throttle for engine speeds between 2000 and 4000 rpm. They felt this fuel consumption was high and could be improved.

Two entries in the 1972 Urban Vehicle Design Competition, UCLA and Brigham Young University, were fueled by hydrogen. These were "carbureted" spark ignition engines in that hydrogen was mixed with air in the intake manifold quite similar to the way gasoline is mixed with air in a carbureted spark ignition. The unique feature about the UCLA car was that it cooled and recycled to the intake manifold the exhaust gases from 2 of 8 cylinders (25 percent exhaust gas recycling). This reduced their nitric oxide levels below that of the 1976 standards. Fuel consumption was reported as 10 miles per pound of hydrogen. The unique feature of the Brigham Young engine was the use of massive water injection to lower nitric oxide emissions. This entry was the only one in the competition which passed the 1976 standards. The reason the UCLA entry did not pass the test was because of excessive oil leakage past the intake valve stems which resulted in unacceptable levels of hydrocarbon emissions.

The University of Miami has converted two spark ignition engines to run on a hydrogen-air mixture. The design used incorporates a unique method of fueling the engine which is termed the Hydrogen Induction Technique. This design modification was made assuming the hydrogen fuel to be available in a gaseous state. Modifications were made to take advantage of the characteristics of hydrogen fuel as compared to gasoline. For example, hydrogen-air mixtures burn with high flame speed over a very wide range of fuel-air ratios as compared to gasoline-air mixtures (see Fig. A-1). Because of this, the brake horsepower of the hydrogen engine can be varied by varying the fuel-air ratio rather than by varying the density of the charge with the usual carburetor butterfly valve. This is similar to the way the load is controlled in a Diesel engine. There is a difference, however, in that the fuel is not injected at high pressures. Instead the fuel is inducted as the piston descends during the intake stroke. The system is thus called the Hydrogen Induction Technique (HIT).

The HIT is illustrated schematically in Fig. A-2. The engine goes through the normal spark ignition Otto cycle. On the intake stroke, unthrottled air is inducted into the cylinder. At the same time, when the intake valve opens, the cylinder end of the hydrogen flow tube is opened and hydrogen is inducted into the cylinder. The amount of hydrogen inducted is controlled by the throttle valve. For a fixed setting of the throttle valve a fixed amount of hydrogen is inducted during each intake stroke. For a greater opening of the throttle valve, the amount of hydrogen inducted is increased. This results in a richer fuel-air ratio and thus higher power output.

The HIT method of load control results in a substantial increase in efficiency at part load operation because pumping losses are minimized. This effect is illustrated in Fig. A-3 where the ratio of pumping horsepower to brake horsepower is plotted as a function of load. Curve A is for a butterfly-valve load controlled, normally carbureted gasoline fueled CFR engine. At part load, the pumping losses are observed to be a substantial portion of the brake horsepower. In fact under some light load conditions the pumping power may equal the brake horsepower in passenger cars. Curve B shows the pumping loss curve for the engine using the HIT. The pumping losses are seen to be much less than that of the gasoline fueled engine over the load range. As expected, at wide open throttle and 100 percent load curves A and B coincide since the pumping work is due to the pressure drop in the intake manifold with the butterfly valve fully opened. There may be some pressure drop in the normally carbureted engine due to blockage which is present even for a fully opened valve but this is neglected. Curve C is for the HIT engine but advantage was also taken of the fact that the hydrogen fuel, being in a vapor state, poses none of the atomization problems which occur in liquid fueled engines. Thus, large intake valves and smooth wall intake porting can be employed which result in an increase in volumetric efficiency. This explains why for curve C hydrogen engine pumping losses are less than that of the gasoline engine even at full load operating conditions. There are other aspects of interest associated with the HIT. Since quenching will not be a source of pollution in the hydrogen fueled engine, cylinders of large bore to stroke ratio can be used. This will allow the use of larger or possibly multiple intake and exhaust valves thereby resulting in a larger volumetric efficiency. Multiple exhaust valves will also run cooler and help alleviate the problem of preignition reported by King without having to use sodium filled valves. In addition, large bore to stroke ratios also result in lower mechanical frictional losses. Another advantage of the HIT is that the engine can be designed to operate at the most prevalent part load operating condition at fuel-air ratios of 55 to 70 percent lean which coincides with the range of fuel-air ratios where efficiency is highest. Additionally, the flow of hydrogen can be shut off completely on deceleration. This also results in greater fuel economy. Under such operation balking is not a problem because throttle response is very good. Good throttle response is attributed to the changing of the fuel-air ratio at the cylinder rather than at the upstream end of the intake manifold where the fuel and air are mixed in a typical gaseous fueled engine conversion. In the latter case, the response to a change in upstream mixture ratio is like that of an exponential dilution flask where the downstream mixture of ratio approaches that of the upstream ratio exponentially.

The HIT is also advantageous from a safety point of view as the hydrogen and air are separate and do not form a combustible mixture except in the cylinder. While the Murray and Schoeppel engine have this feature, they require a pressurized source of hydrogen. This presents problems when the pressure of gaseous stored hydrogen drops with consumption and may be a significant problem if low pressure methods of hydrogen storage, such as hydrides, are used.

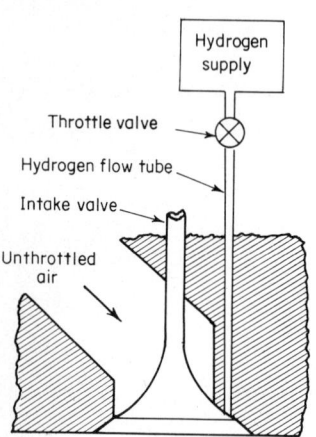

Fig. A-2. The Hydrogen Induction Technique (HIT).

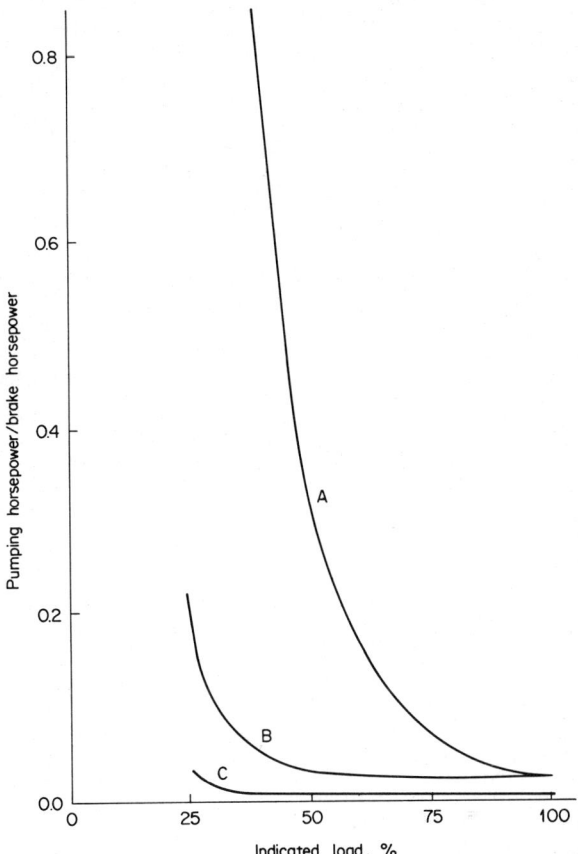

Fig. A-3. Ratio of pumping to brake horsepower versus indicated load.

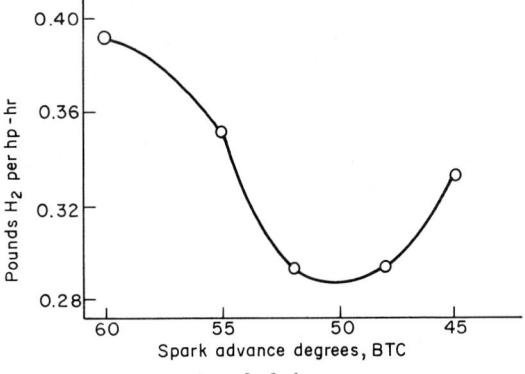

Fig. A-4. Specific fuel consumption.

Test Engines. A 1971 Model 1600 Toyota Corolla station wagon with a four speed manual transmission served as the test vehicle for our first HIT conversion. The four cylinder, 97.6 cubic inch displacement, 9 to 1 compression ratio engine, was modified to accommodate the HIT. The anticipated increase in efficiency over a throttled engine was realized during the testing of the Toyota. While cruising on a level road at 40 miles per hour, the fuel consumption, based on lower heating values, was

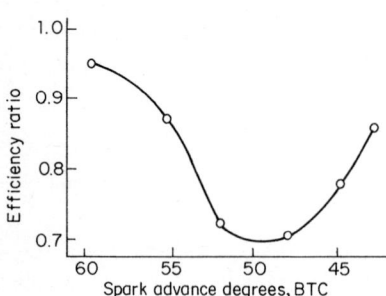

Fig. A-5. Efficiency ratio, gasoline to hydrogen.

equivalent to 42 miles per gallon of gasoline. This represents an approximately 50 percent increase in brake thermal efficiency over the throttled gasoline fueled engine. In terms of miles per pound of hydrogen, the HIT Toyota performance was approximately 19 miles per pound at 40 miles per hour cruise. The second engine to be converted to the HIT was a 4 cylinder, 196 cubic inch Pontiac engine. Figures A-4 and A-5 show the data obtained. Again, high efficiencies have been attained.

MOTORCAR HYDROGEN STORAGE OPTIONS*

The estimated weights and volumes expressed in the following table are relative to the same energy content of gasoline. Relative weight includes that of containers. The data indicate that magnesium hydride would be at a 4.6 weight disadvantage and thus would require four times the tankage in comparison with the use of gasoline in a conventional automobile. This would be equivalent to 450 pounds of added vehicle weight and 60 more gallons ($2 \times 2 \times 2$ feet storage) over that required for a conventional vehicle with a 20-gallon gasoline tank. Burst upon collision for liquid storage can be overcome by using containers capable of withstanding 30 Gs, which are presently available.

Hydrogen Storage Options

Hydrogen storage system	Relative system weight[a]	Relative contained volume[a]
Gas at 2000 psi	15.0	24.5
Solid as metal hydride[b]	4.6	4.0
Liquid at 37°R	2.4	3.8

[a]Relative to gasoline, as unity for same energy content.
[b]Magnesium hydride, 40% porosity.

VEHICULAR STORAGE OF HYDROGEN

Two fundamentals are apparent in connection with the use of hydrogen in vehicles: (1) Even when highly pressurized, hydrogen in the gas phase requires entirely too much volume to provide a vehicle with an adequate range, and (2) the use of hydrogen in the liquid phase necessitates a rather expensive, possibly potentially dangerous cryogenic storage vessel. However, cryogenics technology has advanced considerably in recent years. Liquid hydrogen can be stored in superinsulated vacuum dewars with loss rates of only 2 percent per day for 150-liter containers. If the jacket were cooled to liquid-nitrogen temperatures, losses could be reduced to 1 percent per day. As the storage vessels become larger, they have correspondingly smaller losses because the ratio of surface area (heat loss) to volume decreases. Thus fractional loss is approximately proportional to volume. Stationary storage dewars of modern design and 5000-liter capacity have a loss rate of 0.85 percent per day. Currently, liquid hydrogen is shipped by truck in semitrailer dewar tanks with a capacity of 8300 pounds per dewar. The tanks frequently are closed during shipment, and thus the boil-off gas builds up static pressure. The rates of heat loss correspond to a boil-off of about 0.5 percent per day. Liquid hydrogen also is shipped in railroad tankcars with a capacity of 17,000 pounds each.

Metal hydrides offer an alternative approach to hydrogen storage. It has been established for a number of years that hydrogen, because of its small molecular size and high diffusivity, will penetrate the lattice structure of solid metals and alloys, and in so doing will bind at various sites within the crystal. In some metals, such as titanium, the penetration is such that the concentration of hydrogen per unit volume is actually greater than in liquid hydrogen. Hydrides may be formed simply by exposing the metal to hydrogen under pressure. Although hydride formation is exothermic, the phenomenon is

*Data presented at the 1972 Intersociety Energy Conversion Engineering Conference, San Diego, California, by F. A. Martin, Linde Division, Union Carbide Corporation.

reversible through the application of heat. In a vehicle, waste heat from combustion could be used to free the hydrogen from the hydride. Hydrides remain stable at ambient temperature and pressure, but begin to dissociate to hydrogen gas and metal at about 500°F (260°C).

Research is going forth on hydrides, particularly at the Brookhaven National Laboratories. Magnesium hydride was among the earliest to be studied. Among others that show promise are the monohydrides and dihydrides of vanadium and niobium (columbium); hydrides based on a rare-earth metal together with iron, cobalt, nickel, or copper; and hydrides formed with an intermetallic compound containing iron and titanium. Mg_2Cu, Mg_2Ni, and Mg will combine with hydrogen, binding it as Mg_2NiH_4, for example. Part of the research at Brookhaven is comprised of a hydride reservoir plus a power source (Wankel engine) that operates on the hydrogen. The systems proposed thus far involve high-cost metals which would translate into expensive storage tanks, particularly for private vehicles.

STOICHIOMETRY AND TEMPERATURES

A stoichiometric mixture of hydrogen and air consists of 2 parts of hydrogen to 1 part of oxygen (or 5 parts of air). Thus, a representative gasoline-air mixture is 1 part of heptane vapor to 11 parts of oxygen (55 parts of air). Heptane contains more atoms per unit volume at a given temperature and pressure than hydrogen. Thus, the combustion chamber for burning hydrogen must be larger than that for a gasoline-burning engine for the same output of power and compression ratio. Hydrogen-oxygen flame temperatures are about 2483°K and thus comparable with the flame temperatures of gasolines. The peak temperature of the Otto cycle is about 3100°K.

AIRCRAFT USE OF HYDROGEN

The largest single use of liquid hydrogen as a fuel has been in rocket propulsion in the space program, where the combustion of liquid hydrogen and liquid oxygen has been employed in various rocket stages. Liquid hydrogen is also the propellant selected for nuclear powered rockets. In this case, the hydrogen is not burned, but is heated to a high temperature (3500°F, 1927°C) and exhausted at high velocity.

Hydrogen has the advantage of being 2.5 times lighter than jet fuel. However, hydrogen occupies three times the volume of jet fuel. It has been estimated that the use of liquid hydrogen can increase the range of a Mach 3 supersonic aircraft by 30 percent over that of a jet-fueled aircraft. The use of liquid hydrogen also gives the designer a good sink to cool the engine and structural parts of hypersonic aircraft. As early as 1957, liquid hydrogen was successfully used to fuel a B-57 aircraft at Mach 0.7 by the National Aeronautics and Space Agency's Lewis Research Center. As an interim measure, pollution at takeoff and landing could be alleviated by using relatively small wing-tip liquid hydrogen tanks. It has also been suggested that one-man jet helicopters could be practical where liquid hydrogen would provide a 300 percent reduction in weight of fuel per unit of energy as compared with gasoline.

Use of batteries and fuel cells in vehicles is discussed elsewhere. Consult subject index.

REFERENCES

Winsche, W. E., K. C. Hoffman, and F. J. Salzano: Hydrogen: Its Future Role in the Nation's Energy Economy, *Science,* **180,** 4093, 1331 (1973).

Daunt, J. G., and W. S. Goree: Miniature Cryogenic Refrigerators, Stevens Institute of Technology and Stanford Research Institute preprint, Stanford Research Institute, Menlo Park, Calif. (1970).

Erren, R. A., and W. H. Campbell: Hydrogen: A Commercial Fuel for Internal Combustion Engines and Other Purposes, *J. Inst. Fuel,* **6,** 277–290 (1933).

Lichty, L. C.: "Combustion Engine Processes," McGraw-Hill, New York, 1967.

Murray, R. G., and R. J. Schoeppel: A Progress Report on the Development of OSU's Hydrogen-burning Engine, *Proceedings Frontiers of Power Technology Conference,* Oklahoma State University, Stillwater, Okla., 1969.

U. S. Patent 3,572,297, Hydrogen Fueled Internal Combustion Engine (Mar. 23, 1971).

Murray, R. G., and R. J. Schoeppel: Emission and Performance Characteristics of an Air-breathing Hydrogen-fueled Internal Combustion Engine, *Proceedings 1971 Intersociety Energy Conversion Engineering Conf.,* Boston, Aug. 3–6, 1971, p. 38, SAE Paper 719009.

Schoeppel, R. J., and R. G. Murray: The Development of Hydrogen-burning Engines, *Proceedings Frontiers of Power Technology Conference,* Oklahoma State University, Stillwater, Okla., October 1970.

The Search for a Low-Emission Vehicle, Staff Report, Prepared for the Committee on Commerce, U. S. Senate, Government Printing Office, Washington, 1969.

Witcofski, R. D.: Hydrogen Fueled Hypersonic Transports, presented at American Chemical Society Symposium on Non-Fossil Chemical Fuels, Boston, Aug. 3–6, 1971, American Chemical Society, Washington.

Becker, J. V., and F. S. Kirkham: Hypersonic Transports, NASA (ST-Set 292, paper 25), Dec. 4, 1971, National Aeronautics and Space Administration, Washington.

Methyl Alcohol–A Potential Fuel for Transportation*

BY ALAN PASTERNAK†

Transportation is unique among the energy-consuming sectors of the economy in that it is largely dependent on one source of fuel, *crude oil*. Other energy consuming sectors, particularly those that use electricity, have greater flexibility. But cars, trucks, and planes move on products refined from oil.

Interest in alcohols as automotive fuels is not new. At various times, both ethyl alcohol (ethanol) and methyl alcohol (methanol) have been used, either mixed with gasoline or as pure fuels. The current interest is primarily a result of petroleum prices rising to the point where alternative fuels may be economically competitive with gasoline from crude oil, and the fact that methanol can be produced from coal with known technology.

There are several possible process routes from coal to liquid fuel: (1) coal liquefaction by thermal cracking (pyrolysis) or by its reaction with high-pressure hydrogen to produce liquid products, which can be further refined to synthetic gasoline; (2) gasification of coal to a synthesis gas of carbon monoxide and hydrogen, which can be subsequently converted to liquid hydrocarbons by high-pressure catalytic processing (Fischer-Tropsch process); and (3) coal gasification and subsequent conversion of synthesis gas to methanol by high-pressure catalytic processing.

As described in the Handbook section on *Coal Technology*, a number of coal liquefaction processes are under development; and some are in the pilot-plant stage. Fischer-Tropsch synthesis of synthetic gasoline from coal is a commercial process (as of the mid 1970s) in South Africa. Production of methanol from coal was a commercial process in the United States for a brief period of time in the mid-1950s (Ref. 1). Thus, there are potentially several routes by which coal can be converted to liquid transportation fuels. Two of these processes, coal liquefaction and Fischer-Tropsch synthesis, will yield products that are chemically similar to gasoline. Methyl alcohol is different from gasoline in many significant ways. This article discusses those differences and the potential of methanol as an automotive fuel.

Two routes have been proposed to introduce methanol into the transportation system. First, methanol can be mixed with gasoline and the resulting blend used as the fuel for existing cars with no engine modifications. There is disagreement about the upper limit of methanol blending possible before changes in carburetion are required, but the limit is probably between 10 and 15 percent. Second, essentially pure methanol ("neat" methanol) can be used to fuel cars which have appropriate adjustments. Each route has its advantages and disadvantages. In the remainder of this description, both routes will be discussed. Data on performance, fuel economy, and emissions will be cited. Problem areas will be indicated where further development is required before methanol can be put into the general market as a fuel for automobiles.

PROPERTIES OF METHANOL—RELATION TO ITS POTENTIAL AS AUTOMOTIVE FUEL

Physico-Chemical Properties. Table 1 lists some important properties of methanol, gasoline, isooctane, and benzene (Ref. 2). Isooctane is a branched-chain hydrocarbon frequently blended in gasoline to boost the octane level, particularly in unleaded gasoline.

The most important difference between gasoline and methanol is the stoichiometric air/fuel ratio requirement 6.45 for methanol and 14.2 to 14.8 for gasoline. In practical terms,

---

*Work performed under the auspices of the U.S. Energy Research and Development Administration.
†Formerly with Lawrence Livermore Laboratory, University of California, Livermore, California.

this means that before a car can run on pure methanol, the carburetor must be rejetted to permit a greater volume of liquid per unit volume of air than with gasoline. This is a direct result of the fact that, as shown in Table 1, one gallon of methanol contains one-half as much energy as one gallon of gasoline. Another important difference between gasoline and methanol is the much higher latent heat of vaporization for methanol. For gasoline, the heat of vaporization is less than 1 percent of the heat of combustion, but for methanol it is about 6 percent. Therefore a second modification is required.

TABLE 1.   Properties of Some Liquid Fuels

Property	Methanol	Isooctane	Benzene	Gasoline
Formula	CH_3OH	C_8H_{18}	C_6H_6	Mixture of C_6 to C_{14} hydro-carbons
Lower heating value:				
Btu/lb	8,570	19,070	17,260	18,900 avg
Btu/gallon	56,560	110,000	126,700	115,400 avg
Stoichiometric mass air/fuel ratio	6.45	15.1	13.2	14.2–14.8
Energy of standard stoichiometric mixture at 68°F Btu/ft³	87.3	93.1	95.1	~95
Latent heat of vaporization:				
Btu/lb at 68°F	506	135	188	~150
Btu/gal at 68°F	3,340	779	1,380	~900
Octane number				
Research ON	106	100	101	91–100
Motor ON	92	100	105	82–92

SOURCE: Ref. 2.

Octane Ratings and Octane Boosting.   Table 2 presents information on the octane rating of unleaded gasoline, methanol, and unleaded gasoline plus 10 percent methanol. The table is taken from a paper by Ingamells and Lindquist of Chevron Research Co. (Ref. 3).

Methanol's high octane value is impressive and provides an incentive to design an optimized engine to run on the pure alcohol. Because of methanol's high octane rating,

TABLE 2.   Octane Rating and Octane Boosting

OCTANE RATING		
	Fuel	
Testing method	Gasoline*	Methanol†
Research octane rating	91–100	106–115
Motor octane rating	82–92	88–92
OCTANE BOOSTING WITH METHANOL		
	Fuel	
Testing method	No methanol†	10% methanol
Research octane rating	91.1	95.5
Motor octane rating	82.5	84.5
Road octane rating	89.2	92.0

SOURCE: Ref. 3.
*Unleaded gasolines.
†Numbers depend on the investigator.

such an engine could have a higher compression ratio than existing engines, resulting in higher efficiency.

Methanol's value as an octane booster in blends will depend on the composition of the base gasoline. For late-model cars (mid-1970s), motor octane numbers are more meaningful than research octane numbers. Thus, the improvement in octane due to methanol additions will not be as great as indicated by the research octane number (Ref. 4).

VEHICLE OPERATION ON METHANOL-GASOLINE BLENDS

Fuel Economy and Emissions.   There are conflicting reports on fuel economy and driveability obtained when cars are operated on blends of methanol and gasoline. Since a gallon of methanol has only one-half the energy of a gallon of gasoline, a 90 percent gasoline–10 percent methanol blend has only 95 percent of the energy of an equal volume of gasoline. Therefore, fuel economy in miles per gallon obtained on the blend might be expected to be only 95 percent of that obtained operating on gasoline. But there are some data which show better fuel economy, not only on an energy basis, but on a miles per gallon basis as well (Ref. 5). One explanation for improved efficiency with methanol gasoline blends has been offered by Wigg and Lunt of Exxon Research and Engineering Co. (Ref. 6). Since the stoichiometric air/fuel ratio for methanol (6.5) is lower than that for gasoline (14.5), addition of methanol to gasoline, without a change in carburetion, has the same effect as operating on gasoline with slightly leaner carburetion. For pre-1972 cars, whose carburetors were adjusted to run slightly rich, addition of methanol to the fuel has the effect of raising the air/fuel ratio closer to the stoichiometric value, and this improves efficiency.

The reason for rich carburetion is to improve driveability, and the argument is made that "leaning-out" the engine with additions of methanol will cause some loss of performance. Since 1972, automobile manufacturers have adjusted cars to run leaner than the pre-1972 cars. This has been done in an effort to reduce emissions of carbon monoxide and unburned hydrocarbons. Thus, operation of late-model cars on blends of methanol and gasoline may cause excessive "leaning" and result in more driveability problems than when older model cars are operated on the blends.

Reed and Lerner (Ref. 5) of M.I.T. have reported tests in which unmodified cars from the 1966–1972 model years were operated on blends containing 5 to 30 percent methanol. It was found that fuel economy increased by 5 to 13 percent and CO emissions were reduced by 14 to 72 percent compared to operation on gasoline. Data for a 1969 Toyota showed that optimum fuel economy and CO reductions were obtained with a 15 percent methanol-gasoline blend.

Perfley (Ref. 7) of the University of Santa Clara has reported improvements in thermal efficiency and reduction of pollutants as methanol additions to gasoline are increased to a maximum of 20 percent.

Colucci of General Motors Research Laboratories has reported somewhat different results operating cars on blends of 10 percent methanol in unleaded gasoline (Ref. 4). The cars ranged from a 1966 model through a 1975 prototype equipped with a catalytic converter. Compared with operation on unleaded gasoline, emissions of CO were lower, but emissions of NO_x and unburned hydrocarbons were about the same. Fuel economy was reduced proportionally with the lower thermal energy content of the blend. Operating performance suffered with the blend. Although older model cars show less impairment of performance and can make better use of methanol's octane boosting qualities than can late model cars, Colucci points out that since it will be some time before methanol can be introduced as an automotive fuel, vehicle tests with methanol fuels should concentrate on late model and future cars and engines.

Wigg and Lunt (Ref. 6) have studied the operation of cars on methanol-gasoline blends and have reported data in the areas of fuel economy, exhaust emissions, and driveability. They conclude that the benefits in fuel economy and emissions reduction (CO and hydrocarbons) which are observed with a 15 percent blend can be explained by methanol's leaning effect on carburetion. (It should be noted that the data of Wigg and Lunt show that the use of a catalyst markedly reduces emissions of CO and hydrocarbons with or without methanol additions. The catalyst tested had no pronounced effect on NO_x emissions nor did the choice of fuel: gasoline or the 15 percent methanol blend.) Wigg and Lunt also report a 25 to 50 percent increase in formaldehyde emissions with the use of methanol-gasoline blends. One of the most interesting findings of this study is the increased potential for vapor lock with the blend.

Vapor-Lock Problems.   Several investigators (Refs. 3 and 6) have reported that solutions of methanol and gasoline exhibit positive deviations from ideality. That is, the vapor pressure of the blend is greater than the vapor pressure of either gasoline or methanol alone. This property can lead to "vapor lock," which is the obstruction of the fuel line by formation of vapor at the intake of the fuel pump. This is the point in the fuel supply system where the pressure is the lowest. Wigg (Refs. 6 and 14) has reported that in order to compensate for the high vapor pressure of the blends and to reduce vapor pressures to the same value as gasoline alone, it will be necessary to remove the more volatile components, butane and pentane, from gasoline stocks. But this might result, according to Wigg, in the loss of as much or more energy from the gasoline pool as will be provided by the methanol additions. However, butane is an important petrochemical feedstock and has other uses as well. As a practical matter, omitting butane from gasoline stocks will create problems for some refiners and alleviate problems for others.

Phase Separation.   Probably the most critical problem in the use of methanol-gasoline blends is phase separation due to contamination by very small amounts of water. Methanol and water are completely miscible, but gasoline can dissolve only about 200 ppm of water. When a small amount of water (less than 1 percent) is added to a blend of methanol and gasoline, phase separation occurs. The denser phase consists of methanol, plus the aromatic components of gasoline such as benzene and toluene, plus water. The lighter phase consists of most of the gasoline hydrocarbons. A car that is carbureted for gasoline will be unable to operate on the methanol-rich fuel from the dense phase at the bottom of the fuel tank. Ingamells and Lindquist (Ref. 3) report data on water tolerance for blends as a function of temperature and percentage of methanol in the blend. Wigg and Lunt (Refs. 6 and 14) present data which show that higher alcohols (propanol, butanol, and pentanol) have only limited ability to stabilize methanol-gasoline blends.

Possible solutions to the phase separation problem are to maintain dry fuel distribution systems or to blend the gasoline with anhydrous methanol at the fuel pump.

VEHICLE OPERATION ON NEAT METHANOL

Several investigators have reported data obtained operating cars and engines on pure methanol (Refs. 3 and 7 to 9). Most observers report reductions in emissions of CO, unburned hydrocarbons, and NO_x. Increases in aldehyde emissions are frequently reported. There are some data which show improved efficiency operating on methanol as compared to gasoline. Problem areas are corrosion of fuel tanks and poor compatibility of methanol and some plastic parts.

Fuel Economy and Emissions.   Adelmann et al. (Ref. 8) modified a 1970 American Motors Gremlin to run on pure methanol. The modifications included a rejetted carburetor to lower the air/fuel ratio and an exhaust-heated intake manifold. A catalytic muffler was used to reduce emissions of CO and unburned hydrocarbons. Thus outfitted, the car, operating on methanol, met or surpassed the 1975–1976 federal emissions standards. Adelmann, et al. claim that there were four reasons for reduced emissions with the modified Gremlin operating on methanol.

1. Methanol has a lower "lean misfire limit" than gasoline. This means that with methanol it is possible to operate further on the lean side (excess air) of the stoichiometric air/fuel requirement than with gasoline. This tends to reduce emissions of CO, NO_x, and unburned hydrocarbons.

2. Methanol has a higher flame speed than gasoline, which permits spark retardation and lower NO_x emissions.

3. Combustion temperatures are about 100°C lower with methanol. This lowers the rate of formation of NO, reducing No_x emissions.

4. The catalytic muffler reduced emissions of CO and unburned hydrocarbons.

Jackson and Tillman (Ref. 10) of the Continental Oil Co. have reported data obtained operating a 1973 Ford on both gasoline and pure methanol. With methanol, they found lean operation lowered emissions significantly compared with gasoline. Of particular interest in the Continental Oil studies was the finding that in road tests with methanol as the fuel, higher efficiencies were obtained than with gasoline. When gasoline operation was adjusted to reduce NO_x emissions to the same level as experienced with methanol, there were large losses in fuel economy.

Pefley has reported a total of 45,000 miles of vehicle operation (6 vehicle-years) on pure methanol with reduced emissions (Ref. 7).

In general, operation on pure methanol results in lower emissions of CO, NO_r, and unburned hydrocarbons than are experienced with gasoline. It is expected that methanol itself will be less reactive in the photochemistry of smog production and in the production of eye irritants than are unburned hydrocarbons (Ref. 8). But the photochemical reactivity of aldehydes may be significant (Ref. 8). Further research in this area is needed.

Corrosion and Compatibility. Corrosion is probably the most serious problem experienced in the operation of modified vehicles on pure methanol. Ingamells and Lindquist (Ref. 3) report severe corrosion of the terneplate linings* of gasoline tanks by pure methanol (and also by some methanol-gasoline blends). In addition, methanol softens and expands certain plastics, causes swelling of the Viton tips of float valves in carburetors and softens fiber gaskets in fuel pumps.

Other researchers have also reported corrosion and compatibility problems (Refs. 7 and 10).

Cold Starting. Methanol has a lower vapor pressure than gasoline, and this may lead to difficulty in starting an engine in cold weather. Jackson and Tillman (Ref. 10) claim that addition of 3 percent butane to methanol has permitted starting at temperatures as low as 10°F.

PROBLEM AREAS

The studies reported to date have brought to light the problem areas cited in Table 3. Some of these areas may require further research and development. For example, the corrosion problem may be solved by development of an appropriate corrosion inhibitor which could be added to the fuel.† Other compatibility problems may require the use of substitute materials. The most difficult problem may be water-induced phase separation. This is essentially a fuel-handling and distribution problem which requires maintenance of dry conditions.

TABLE 3.   Problem Areas

Methanol-gasoline blends	Neat methanol
Driveability	Corrosion and compatibility
Corrosion and compatibility	Cold start
Vapor lock	
Phase separation	

The problems of vapor lock with blends and poor cold starting with pure methanol seem amenable to a systems treatment. The blend approach and the pure fuel approach for introducing methanol into the transportation system are not mutually exclusive. Blends of methanol and gasoline can be introduced at the same time that selected captive fleets (e.g., government or commercial vehicles) are adapted to run on pure methanol. As indicated earlier, removing volatile butane from gasoline stocks destined for blending with methanol can solve the vapor-lock problem. At the same time, addition of butane to neat methanol can solve the cold-start problem. An overall transfer of butane from one fuel to the other may therefore solve both problems with no loss of fuel energy.

In addition to the above areas of research, a thorough analysis is needed of methanol toxicology, and the effects of emissions from methanol-fueled automobiles on air quality (Refs. 11 and 12). These studies should be on a comparative basis so that methanol gasoline, and hydrocarbons refined from shale oil and coal liquids can all be evaluated.

Providing liquid fuels for transportation from domestic sources may be the most difficult part of achieving energy independence for the United States (Ref. 13). Methanol from coal is one way to provide clean liquid fuels from a large domestic resource. Certain problem areas have been identified in adapting methanol to an automotive fuel. But national interests provide strong incentive to carry out the research and development necessary to solve the problems.

---

*Terneplate is the name given to soft steel sheets which have been coated with a thin layer of a 90% lead/10% tin alloy. The alloy acts as a solid lubricant during fabrication.

†Atlantic Richfield Corporation has developed a special nonphosphate inhibitor which is used in blends of 5% tertiary butyl alcohol in gasoline.

SUMMARY

Methanol is attractive as an alternative transportation fuel because it is a liquid and is therefore easy to handle and because it can be made from coal with known technology.

A number of reports describe operating experience for cars running on methanol-gasoline blends and also neat methanol. At a limit of 10 to 15 percent methanol in gasoline, no modifications are necessary to the carburetion of existing automobiles although late model cars may suffer some loss of performance. For operation on pure methanol, several modifications are necessary: carburetors must be rejetted to lower the air/fuel ratio, and exhaust heat must be recycled to compensate for methanol's lower energy content.

Operation of vehicles on pure methanol has some potential advantages:

1. Methanol has a high octane rating which may permit increases in engine compression ratios and higher efficiencies.

2. There are data which show that existing engines can operate at higher efficiencies on methanol than on gasoline.

3. Methanol's high heat of vaporization makes possible the recycle of exhaust heat and should lead to higher thermodynamic efficiency.

4. There are data which show that emissions of CO, NO_x, and unburned hydrocarbons are lower when cars are operated on methanol than on gasoline.

REFERENCES

1. Remarks made by R. Busche on the experience of DuPont, at a U.S. Bureau of Mines meeting on "Methanol from Coal," Feb. 13, 1974, Washington.
2. White, J. R., C. N. Rowe, and C. N. Koehl: "Physico-Chemical Properties of Methanol Related to Fuel Use," paper delivered at the 1974 Engineering Foundation Conference on "Methanol as an Alternate Fuel," Henniker, N.H., July, 1974.
3. Ingamells, T. C., and R. H. Lindquist: "Methanol as a Motor Fuel," paper delivered at the 1974 Engineering Foundation Conference on "Methanol as an Alternate Fuel," Henniker, N.H., July 1974.
4. Colucci, J. M.: "Methanol Gasoline Blends—Automotive Manufacturer's Viewpoint," paper presented at 1974 Engineering Foundation Conference on "Methanol as an Alternate Fuel," Henniker, N.H., July 1974.
5. Reed, T. B., and R. M. Lerner: "Methanol: A Versatile Fuel for Immediate Use," *Science,* **182,** 1299 (1973).
6. Wigg, E. E. and R. S. Lunt: "Methanol as a Gasoline Extender—Fuel Economy, Emissions, and High Temperature Driveability," paper 741008, Society of Automotive Engineers, Toronto, Canada, Oct. 21–25, 1974.
7. Pefley, R. K.: "Methanol-Gasoline Blends—University Viewpoint," paper delivered at 1974 Engineering Foundation Conference on "Methanol as an Alternate Fuel," Henniker, N.H., July 1974.
8. Adelman, H. G., D. G. Andrews, and R. S. Devoto: "Exhaust Emissions from a Methanol-fueled Automobile," paper 720693, Society of Automotive Engineers, San Francisco, Aug. 21–24, 1972.
9. Ebersole, G. D., and F. S. Manning: "Engine Performance and Exhaust Emissions: Methanol versus Iso Octane," paper 720692, Society of Automotive Engineers, San Francisco, Aug. 21–24, 1974.
10. Jackson, R. G., and R. M. Tillman: "Automotive Uses of Methanol Fuel," paper presented at 1974 Engineering Foundation Conference on "Methanol as an Alternate Fuel," Henniker, N.H., July 1974.
11. Berger, B. J.: "Environmental Aspects of Methanol as Vehicular Fuel: Health and Environmental Effects," UCR L-76076, Sept. 25, 1974, paper presented at 1974 Engineering Foundation Conference on "Methanol as an Alternate Fuel," Henniker, N.H., July 1974.
12. Pollack, R. I.: "Environmental Aspects of Methanol as Vehicular Fuel: Air Quality Effects," UCRL-76064, Sept. 24, 1974, paper presented at 1974 Engineering Foundation Conference on "Methanol as an Alternate Fuel," Henniker, N.H., July 1974.
13. Anderson C. J., et al.: "An Assessment of U.S. Energy Options for Project Independence," UCRL-51638, Sept. 1, 1974.
14. Wigg, E. E.: Methanol as a Gasoline Extender: A Critique, *Science,* **186,** 4166, 785 (1974).

Batteries

BY S. SENDEROFF*

A battery consists of two or more series- or parallel-connected galvanic cells. A primary galvanic cell converts chemical energy directly into electric energy and consists of two electrodes of dissimilar material isolated from each other electronically, in a common ionically conductive electrolyte. The electrolyte may be solid or liquid, but usually is an aqueous salt solution. In recent years, considerable attention has been given to nonaqueous electrolytes. If the cell is of a secondary, or rechargeable, variety, input electric energy can be converted to chemical energy and thus stored. Generally, it is impractical and, in some cases, *potentially dangerous* to attempt to recharge a cell which is intended for primary use.

For any cell system to be commercially attractive, the electrode-electrolyte combination must be such that it will deliver a reasonable quantity of electric energy at useful voltage and current. The voltage exhibited by a particular system is the algebraic sum of the observed individual electrode emf values, compared with a standard in the chosen electrolyte. As will be observed from Table 1, the range of nominal voltages available per cell is not wide, all major cells being in the 1-2 volt per cell span. The quantity of energy available per unit cell volume is almost solely a function of the electrochemical equivalents and densities of the active electrode materials selected. As the table shows, there is a much wider range among the major cells in this regard.

Aside from the effects of cell size current is limited throughout discharge by cell-element configuration, activities of the electrode materials, reaction-product solubilities, and both ionic and electronic conductivities of all components.

The two electrodes of the galvanic cell are termed negative and positive here to avoid confusion which can result from use of the terms anode and cathode, the latter often being used loosely in the industry. The negative electrode of the primary (or charged secondary) cell is metallic and is oxidized (increased in valence) during discharge, giving up electrons to the external circuit. The positive electrode in an aqueous system initially is an oxygen donor, usually a metal oxide, which is reduced as it receives electrons from the external circuit. Charge transfer from one electrode to the other within the cell is via the ions of the electrolyte salt. In certain cells, the electrolyte serves only as a charge carrier. In other systems, the electrolyte enters further into reactions and actually changes in composition during use. The overall equations of reaction for major primary and rechargeable cells are given in Table 1.

Major points of difference between primary and secondary cells include: (1) features of electrode design (including choice of active materials used) which assist in maintaining secondary-electrode integrity through extended cycling of charging and discharging; (2) use of auxiliary structural materials which are especially oxidation-resistant and durable in secondary cells; and (3) in the most refined, hermetically sealed secondary cells, inclusion of materials in the system to prevent excessive gas-pressure buildup during any phase of use. Features designed to enhance rechargeability generally add less by proportion to consumer cost than they add to overall usefulness.

Cells designed for complete portability range in utility from the commonplace inexpensive Leclanche primary cell to the hermetically sealed nickel-cadmium cell which will serve very efficiently through hundreds of full-capacity cycles.

Cells are marketed which deliver as little as a few milliampere-hours at 1.2 volts (0.05 ounce) to as much as several thousand ampere-hours at 2 volts (upward of a ton). Most cells larger than about 25 ampere-hours capacity require correct orientation during use, and the secondaries usually must be serviced (water addition) occasionally. More recently, some lead-acid batteries for automotive use have appeared which do not require servic-

*Technology Department, Union Carbide Corporation, Cleveland. Based on original article by J. L. S. Daley (deceased), also of Union Carbide Corporation.

ing. Probably the largest battery in common use is that for submarine propulsion, a battery weighing several hundred tons which will deliver more than 5 million watthours.

In recent years, there has been a concerted effort to develop batteries with higher energy densities to meet new applications requiring lighter and/or smaller batteries. Outstanding as one such application is highway vehicle propulsion, although there are many other applications, including portable power tools, toys, photographic and communications equipment, and other uses where higher-energy-density batteries are required. One direction of present development effort is the modification and redesign of the highest-energy-density system shown in Table 1, namely, zinc-air, so that it may be used at high rates and be recharged. It is believed that if its power density can be increased to approach that of the lead-acid battery while maintaining its excellent energy density, it may prove useful for vehicle propulsion.

Except for some study of the iron-air system in aqueous potassium hydroxide and the zinc-chlorine hydrate system in aqueous zinc chloride, most other high-energy-density systems now under study involve nonaqueous electrolytes.

Nonaqueous Electrolytes.   The main advantage to be derived from nonaqueous electrolytes is the possibility of utilizing as negative electrodes more reactive metals, such as sodium or lithium, which cannot be used with water. In this way, an increase of cell voltage from the maximum of 2 volts (see Table 1) to a value of greater than 3 volts is possible.

The nonaqueous electrolytes under consideration may be divided into three classes: (1) molten salts, (2) organic or inorganic solvents rendered conductive by dissolved salts, and (3) solid electrolytes.

In molten salt electrolytes the systems which have received most attention are (a) lithium-chlorine in molten lithium chloride as electrolyte, operated at 680°C at a voltage of 3.5 volts and an expected (but not yet attained) energy density of about 150 watthours per pound; and (b) lithium-sulfur in a molten mixture of alkali halides, operated at 2.3 volts at 347°C and an expected energy density (also not yet attained) of about 125 watthours per pound. Both systems are rechargeable. Neither these nor other molten salt systems have yet emerged from their developmental stages, but both systems have received consideration as power sources for vehicles. (See brief description of Electric Automobile that follows this article.)

Systems involving nonaqueous solvents which have been described in the literature are very diverse and have as the only common element the almost exclusive use of lithium as the negative electrode. For positive electrodes a wide variety of materials has been used. Among them are carbon monofluoride and transition metal halides, oxides, and sulfides. The electrolytes studied include a very large number of organic solvents in which lithium salts, such as the tetrachloroaluminate, hexafluorophosphate, hexafluoroarsenate, or perchlorate (among others) are dissolved. Among the more popular organic solvents under study are propylene carbonate, tetrahydrofuran, v-butyrolactone dioxolane, and many others, alone and in mixtures. Except for limited use in a few special military applications, all these systems are in various precommercial stages of development. In general, their energy densities are high, but their power densities are low because of high internal resistance. Although occasionally some reference may be seen pertaining to their possible use in vehicle propulsion, the most likely applications of these systems appear to be as power sources for electrical, electronic, and other portable equipment. Also, almost all systems of this type are primaries, at least at their present stage of development. A typical system of this type, a lithium-carbon monofluoride cell of 5-ampere-hour capacity, operates at about 2.4 volts and has an energy density of about 100 watthours per pound and 9 watthours per cubic inch.

Systems using inorganic solvents, such as liquid sulfur dioxide, thionyl chloride, sulfuryl chloride, or phosphorus oxychloride, containing dissolved salts such as lithium tetrachloroaluminate, have a unique feature. With a lithium negative electrode, the electrolyte solution is also the active positive material and is reduced on discharge. The external electric circuit is completed with an inert current collector inserted into the electrolyte as the positive pole of the cell. This is sometimes referred to as a "soluble cathode" cell. The active cathode material is in direct contact with the lithium, and it is believed that a film formed on contact between the materials acts as an effective separator. A lithium-liquid sulfur dioxide battery has been offered for sale commercially in the form of single cells similar in size and shape to the usual flashlight batteries. A D-size cell with

TABLE 1.   Commercially Available Batteries and Characteristics

Electrochemical System	Negative Electrode	Positive Electrode	Electrolyte	Type	Overall Equations of Reaction
Zinc-manganese dioxide (usually called Leclanché or carbon-zinc)	Zinc	Manganese dioxide	Aqueous solution of ammonium chloride, zinc chloride	Primary	$2MnO_2 + 2NH_4Cl + Zn \rightarrow$ $ZnCl_2 2NH_3 + H_2O + Mn_2O_3$
Zinc–zinc, chloride–manganese dioxide	Zinc	Manganese dioxide	Aqueous solution of zinc chloride	Primary	$8MnO_2 + 4Zn + ZnCl_2 + 9H_2O \rightarrow$ $8MnOOH + ZnCl_2 \cdot 4ZnO \cdot 5H_2O$
Zinc-alkaline-manganese dioxide	Zinc	Manganese dioxide	Aqueous solution of potassium hydroxide	Primary	$2Zn + 2KOH + 3MnO_2 \rightleftharpoons$ $2ZnO + 2KOH + Mn_3O_4$
				Rechargeable	$Zn + KOH + 2MnO_2 \rightleftharpoons$ $ZnO + Mn_2O_3 + KOH$
Zinc-mercuric oxide	Zinc	Mercuric oxide	Aqueous solution of potassium hydroxide	Primary	$Zn + HgO + KOH \rightarrow ZnO + Hg + KOH$
Zinc-silver oxide	Zinc	Monovalent silver oxide	Aqueous solution of potassium hydroxide or sodium hydroxide	Primary	$Zn + Ag_2O + KOH \rightarrow ZnO + 2Ag + KOH$
		Divalent silver oxide	Aqueous solution of potassium hydroxide	Rechargeable	$Zn + Ag + KOH \rightleftharpoons ZnO + Ag + KOH$
Lead-lead dioxide (usually called lead-acid	Lead	Lead dioxide	Aqueous solution of sulfuric acid	Rechargeable	$2Pb + 2PbO_2 + 2H_2SO_4 + H_2O \rightleftharpoons$ $PbSO_4 + 2PbO + 3H_2O$
Nickel-cadmium	Cadmium	Nickelic hydroxide	Aqueous solution of potassium hydroxide	Rechargeable	$Cd + 2NiOOH + KOH + 2H_2O \rightleftharpoons$ $Cd(OH)_2 + 2Ni(OH)_2 + KOH$
Nickel-iron	Iron	Nickelic hydroxide	Aqueous solution of potassium hydroxide	Rechargeable	$Fe + 2NiOOH + KOH + 2H_2O \rightleftharpoons$ $Fe(OH)_2 + 2Ni(OH)_2 + KOH$
Magnesium-manganese dioxide	Magnesium	Manganese dioxide	Aqueous solution of magnesium perchlorate	Primary	$2Mg + 2MnO_2 + 4H_2O + Mg(ClO_4) \rightarrow$ $Mn_2O_3 H_2O + 2Mg(OH)_2 + Mg(ClO_4) + H_2$
Silver-cadmium	Cadmium	Divalent silver oxide	Aqueous solution of potassium hydroxide	Rechargeable	$Cd + AgO + KOH + H_2O \rightleftharpoons$ $Cd(OH)_2 + Ag + KOH$
Zinc-air (oxygen)	Zinc	Oxygen	Aqueous solution of potassium hydroxide	Primary	$2Zn + O_2 + 4KOH + 2H_2O \rightarrow 2K_2Zn(OH)_4$

SOURCE: Union Carbide Corporation.
Adapted from Douglas M. Considine (ed.), "Chemical and Process Technology Encyclopedia,"
McGraw-Hill, New York, 1974.
NOTE: Wh = watthours
        Ah = ampere-hours

a capacity of 8 ampere-hours is said to operate at about 2.8 volts and have an energy density of about 120 watthours per pound and 7 watthours per cubic inch.

Another class, high-energy-density solid-electrolyte batteries, is based on the use of β-alumina as a solid electrolyte with a molten sodium negative electrode and a positive consisting of sodium polysulfide held in a porous inert collector. This is usually called the sodium-sulfur battery. It operates at about 300°C, is rechargeable, and has been studied for application in vehicle propulsion or power station load leveling. Although still in the developmental stage, this battery is expected to operate at about 2 volts and may attain an energy density of 100 to 150 watthours per pound.

Another development involving the β-alumina solid electrolyte is its suggested use at room temperature with a sodium amalgam negative and bromine positive as a heart-

Nominal-Voltage-per Cell	Typical Commercial Service Capacities	Input if Rechargeable	Energy Density (Commercial)		Features	Limitations
			Wh/lb	Wh/in.³		
1.5	Several hundred mAh to 30 Ah		5-40	1-3	Low cost;variety of shapes and sizes; excellent shelf life	Efficiency decreases at high current drains; poor low-temperature performance
1.5	Several hundred mAh to several Ah		15–30	1–3	Good service at high current drain, leak resistant, good low temperature performance	Relatively expensive for low drains
1.5	Several hundred mAh to 23 Ah		20-40	2-3	High efficiency under moderate continuous drain conditions; good low-temperature performance; low impedance; long shelf life	Expensive for low drains
		Approximately 100% of energy withdrawn	10	1.0-1.2		Rechargeable-limited taper current charging
1.35	16 mAh-14 Ah		10-50	4-8	High service capacity/volume ratio; flat voltage discharge characteristic; good high-temperature performance; good storage life	Poor low-temperature performance on some types
1.5	38-190 mAh		30-60	4-8	Moderately flat voltage discharge characteristic; good storage life	
1.8/1.5 (two-step)	Vented, 5 Ah to several thousand Ah	Minimum of 110% of energy withdrawn	40-70	2-8	High-energy density	Expensive; short storage life after activation; short cycle life; two-step discharge curve
2	Vented, 1-10,000 Ah	Minimum of 110% of energy withdrawn	Sealed, 10-15	0.8-1.1	Spill-resistant	Limited low-temperature performance; vented cells require serving
			Vented, 7-12	0.5-2	Low cost	
1.25	Sealed, 20 mAh-100 Ah	Sealed, minimum of 140% of energy withdrawn	Sealed, 12-17	Sealed, 1-1.5	Excellent cycle life; flat voltage discharge characteristic; good high- and low-temperature performance; high resistance to shock and vibration; can be stored indefinitely in any charge state	High initial cost; only fair charge retention
	Vented-few Ah to over 500 Ah	Vented, minimum of 125-150% of energy withdrawn	Vented, 12-20	Vented, 1-1.5		
1.2	Vented, several hundred Ah	Minimum of 125% of energy withdrawn	8-14	0.9	Excellent cycle life; rugged	Poor charge retention; usually restricted to low-rate use; heavy gassing on overcharge
1.8	Sealed, up to 20 Ah		15-50	1-3	Excellent high-temperature storage life; good in moderate-drain continuous uses	Poor low-temperature performance; delay after circuit is closed before battery operates; not good for intermittent use
1.4	Sealed, up to 300 Ah	Minimum of 110% of energy withdrawn	22-34	1.8-2.5	Good energy/weight ratio; good charge retention; long wet-stand life	Expensive; poor low-temperature performance; two-step discharge curve
1.25	Vented, ½-2,000 Ah		80-100	3.2	Flat voltage discharge characteristic	Restricted to low-rate use

pacer battery. Operation at 2.8 volts with energy densities of 200 watthours per pound and 20 watthours per cubic inch at low drain rate is claimed.

Solid electrolyte batteries based on compounds, such as silver rubidium iodide, silver potassium iodide cyanide, and a few other double salts of silver iodide as electrolyte, and using silver metal as negative and polyiodides as positive, have been described. These systems do not have high-energy densities, but they have long shelf life, are leak-free (since they contain no liquids), and are usable over a wide temperature range. These batteries have been developed and offered for sale, but their low operating voltage (~ 0.6 volt per cell) has limited their application.

See also description of fuel cells later in this Handbook section; and descriptions of solar cells and arrays given in Handbook section on *Solar Energy Technology*.

REFERENCES

Heise, G. H., and N. C. Cahoon (eds.): "The Primary Battery," vol. 1, Wiley, New York, 1971.

Vinal, G. W.: "Storage Batteries," 4th ed., Wiley, New York, 1955.

Falk, S. O., and A. J. Salkind: "Alkaline Storage Batteries," Wiley, New York, 1969.

Fleischer, A., and J. J. Lander (eds.): "Zinc-Silver Oxide Batteries," Wiley, New York, 1971.

Kordesch, K. V. (ed.): "Batteries," vol. I "Manganese Dioxide," Dekker, New York, 1974.

Jasinski, R.: "High Energy Batteries," Plenum, New York, 1967.

Compton, W. D.: Energy Conversion and Storage Technology—The Sodium-Sulfur Battery, in Energy, Environment, Productivity, *Proc. 1st Symp. Res. Appl. to Natl. Needs*, National Science Foundation. U.S. Government Printing Office, Stock no. 3800-00177, Washington, 1973.

Electric Automobile

Records show that 1575 electric automobiles were built in the year 1900, compared with 939 cars equipped with gasoline engines. Electric automobiles were particularly popular with the women of that time because of the difficulty in starting piston engines with a crank. Improvements in gasoline-powered engines and the invention of the self-starter eclipsed the future of the electric automobile for many years. By 1930, the electric automobile was largely regarded as a museum piece.

The principal reason behind the lack of competitiveness of the electric automobile was the limitation of batteries. A 20-gallon tank of gasoline can provide about 2.4 million Btus; lead-acid storage batteries, weighing about the same as that of a tank of gasoline, can provide only some 7700 Btus, or about 2¼ kilowatthours. The ratio, then, of energy in a tank of gasoline to the energy in the same weight of conventional storage batteries is more than 300 to 1. However, an automobile engine can convert only about one-fifth of the energy in the gasoline into driving power, whereas the electric motor can produce motive effort from nearly all the electricity delivered by a battery. Even with this advantage, however, electric drive suffers badly in performance compared with gasoline.

The problem is to provide an automobile with sufficient electricity to give it the same range and performance as a car using gasoline. A number of exotic battery systems are under development, including zinc-air rechargeable cells, hot sodium-sulfur, lithium-sulfur, and lithium-chlorine secondary batteries, and other high-performance electrochemical devices. As yet, none of these devices has proved practical for wide public use. Rechargeable batteries, such as nickel-cadmium, iron-nickel, and silver-zinc, are far too costly. Only the conventional lead-acid battery, similar in construction and chemistry to the starting-lighting-ignition (SLI) battery in present automobiles, has proved practical. Essentially all the electric vehicles on the road today—some 50,000 vehicles in England alone—are powered with lead-acid batteries.

Battery developments are described in more detail under Batteries, the article immediately preceding this one.

SIMPLIFIED COST COMPARISONS

If it is first assumed that lead-acid batteries are used to power passenger cars, experience indicates that an electric automobile in city traffic at speeds up to 35 miles per hour uses about ¼ kilowatthour per 10 miles traveled. Assume further that the car weighs 3,000 pounds, half of that weight allowed for the battery. The very best lead-acid batteries currently available can deliver 15 watthours per pound if discharged over 20 hours. Thus, the battery can deliver, at the maximum, 22.5 kilowatthours. However, it will be discharged over a 1- to 2-hour period, and thus its capacity will be about halved. Thus, the useful available output at most will be 11 kilowatthours from a 1500-pound battery pack. Inasmuch as the car is assumed to weigh 1½ tons, thereby using 0.375 kilowatthour per mile, it will have a maximum range of 30 miles at 35 miles per hour with a fully charged battery. If, instead of 35 miles per hour, the operator elects to travel at a rate of 55 miles per hour, the range would be cut in half, or approximately 12 to 15 miles per charge.

Some experts believe* that there will be no essential difference between the cost of present gasoline-power systems and an electric motor, including its controls and simplified drive train. The battery, however, will be an additional cost. Based on prices† paid by the United States government for lead-acid batteries, the least expensive model is a 12-volt, 200 ampere-hour-capacity, SLI type, costing $48. Since each such battery can deliver about 2.4 kilowatthours at a 20-hour discharge rate, ten batteries would be required for a car, for a total battery cost of $480. Because these batteries can provide about 500

*Not a universal opinion.
†Late 1974 costs.

discharge cycles, with 30 miles each cycle, the battery cost alone would be about 3 cents per mile. Power costs at 3 cents per kilowatthour to the battery charger and an overall efficiency of 50 percent from power in to power out would add up to about 2¼ cents per mile. Thus, total operating costs, including battery replacement, would approximate 5 to 6 cents per mile, or about double the cost for gasoline vehicles of comparable performance with gasoline estimated at 50 cents per gallon. As all other energy comparisons indicate, of course, when the cost of one fuel rises, the cost of a substitute fuel may or may not be more attractive. The total system cost must be taken into consideration, including in this case, for example, the costs of electric energy for battery recharging, such costs also increasing.*

In another specific cost estimation, a 3200-pound vehicle, powered in one mode by an internal-combustion engine and in another mode by a dc electric motor, was used as the basis for comparison.† This weight class was selected arbitrarily, but it reasonably represents a compact class vehicle. The projected operating characteristics are shown in the two columns on the left of Table 1 for an electric vehicle with lead-acid batteries and on the right for a standard 2.3-liter internal-combustion engine with automatic transmission.

TABLE 1.   City Vehicle Performance
(3200 pounds gross weight)

Performance factor	Electric drive (Lead-acid battery)		Internal-combustion engine
Payload, pounds	300		800
Motor	30 hp	17hp	2.3 liters*
Range per charge, miles:			
At 20 miles per hour	48	58	273
City driving	28	34	233
Acceleration distance in			
10 seconds, feet	365	230	373

SOURCE: Ref. 1.
*With automatic transmission.

An electric vehicle with the characteristics described in the far-left column has reasonably high performance, as measured by the maximum distance traveled in 10 seconds, starting from rest. As this suggests, it is technically possible to build an electric vehicle that is essentially equivalent to the internal-combustion vehicle by this single criterion, but the range and payload of the electric vehicle are very limited. The middle column of Table 1 indicates the limited tradeoffs that can be effected between performance and range for the same vehicle—still powered with lead-acid batteries. Not only are the range and payload of this vehicle limited, but also it would have unacceptable performance. The column on the right gives the data for this vehicle weight when powered by an internal-

TABLE 2.   City Vehicle Operating Costs
(3200 pounds gross weight)

Cost factor	Electric drive (Lead-acid battery)		Internal-combustion engine
Motor	30hp	17hp	2.3 liters*
Fuel cost, cents per mile†	1.1	0.9	1.4
Operating cost	4.2	3.8	2.1
Total vehicle cost, cents per mile	8.3	7.8	5.6

SOURCE: Ref. 1.
*With automatic transmission.
†Taxes not included.

*The foregoing cost estimates were abstracted from "Energy Perspectives," Battelle Memorial Institute, Columbus, Ohio, April 1974.
†"Energy Conversion and Storage Technology—The Sodium-Sulfur Battery," by W. Dale Compton, Ford Motor Company, presented at 1st Symposium on Research Applied to National Needs, sponsored by National Science Foundation, Nov. 18–20, 1973, Washington.

combustion engine. Obviously, neither of the electric vehicles has the desirable operating characteristics of the internal-combustion engine.

Projected operating costs for these same vehicles are given in Table 2. For purposes of this comparison, the gasoline tax has been removed from the figures on the right. The limitation in the performance of the electric vehicle, although somewhat affected by motor design and controller design, is determined almost entirely by the battery system. This poor performance can be traced directly to the problem of how much battery weight can be accommodated in the electric vehicle. Thus, it relates directly to the low-energy density of the lead-acid battery. The total operating cost of the system also is affected strongly by the relatively limited number of charges and discharges that the lead-acid batteries can sustain, because this determines the life-span of the battery. Thus, one is led to believe that only through the development of battery systems superior to those of lead-acid can a major adjustment in electric propulsion be achieved.

Electric Car Battery Systems.   Some researchers* have concentrated on the sodium-sulfur battery as a substitute for the lead-acid battery in terms of vehicle propulsion use. (See separate article on Batteries in this Handbook section.) The characteristics of the inert electrolyte used in the sodium-sulfur battery make possible important trade-

TABLE 3.   Energy Densities of Some Battery Systems
(Comparison with gasoline)

System	Theoretical	Actual
Lead-acid	49	10
Silver-zinc	230	50
Sodium-sulfur	350	100*
Zinc-oxygen	495	60*
Lithium-sulfur	660	100*
Lithium–copper fluoride	646	100
Lithium-chlorine	990	250
Gasoline	6000	1200

SOURCE: Ref. 1.
*Projected.

offs between energy and power. See Table 3. In the sodium-sulfur battery, the total stored energy depends only on the total weight of the sodium and sulfur, whereas the power density is related directly to the total surface area of the ceramic electrolyte. For a vehicle, the stored energy is related directly to the achievable range; whereas the power density is related to the achievable acceleration of the vehicle. Since one of the most costly components of the sodium-sulfur battery is the ceramic electrolyte, it will be advantageous economically to tailor the battery to the amount of ceramic that is used. Where weight is critical, as in a vehicle, a high energy and power density are required.

The objectives of the development of the sodium-sulfur battery have included: (1) an

TABLE 4.   Sodium-Sulfur Battery Development Objectives

Characteristics	Electric vehicles	Load leveling
Energy density, watthours per pound	100	25
Power density, watts per pound	100	25
Durability, years	5*	25
Cost, dollars per kilowatthour	20	5–15

SOURCE: Ref. 1.
*1000 cycles.

energy density of 100 watthours per pound; (2) a power density of 100 watts per pound; (3) a durability of 5 years (about 1000 charge/discharge cycles); and (4) a cost of $2 to $3 per pound. See Table 4.

When used for load leveling in a generating and transmission system, it is believed that the energy density and power density can be somewhat lower, inasmuch as weight is not

*Development of sodium-sulfur battery began at Ford Motor Company in 1963.

such a problem. However, the durability should be about 25 years. The desired cost for this application has been projected to be between $5 and $15 per kilowatthour of stored energy—somewhat lower than the projected cost of batteries that are to be used for automotive propulsion.

TABLE 5.   City Vehicle Performance
(3200 pounds gross weight)

Performance factor	Electric drive (Sodium-sulfur battery)	Internal-combustion engine
Payload, pounds	300	800
Motor	30 hp	2.3 liters*
Range per charge, miles:		
At 20 miles per hour	483	273
City driving	280	233
Acceleration distance in 10′ seconds, feet	365	373

SOURCE: Ref. 1.
*With automatic transmission.

Based upon the availability of a satisfactory sodium-sulfur battery (not yet achieved), the performance of a vehicle with (1) an internal-combustion engine and (2) sodium-sulfur battery propulsion is outlined in Table 5. An estimate of the operating cost is given in Table 6.

TABLE 6.   City Vehicle Operating Costs
(3200 pounds gross weight)

Cost factor	Electric drive (Sodium-sulfur battery)	Internal-combustion engine
Motor	30 hp	2.3 liters*
Fuel cost, cents per mile	1.0†	1.4†
Operating cost, cents per mile	1.7	2.1
Total vehicle cost, cents per mile	8.2	5.6

SOURCE: Ref. 1.
*With automatic transmission.
†Taxes not included.

It will be noted that for a comparable performance, as determined by the acceleration, a range can be achieved that is at least as large as that of an internal-combustion engine with a normal-sized tank of gasoline. Although the operating cost per mile appears to compare favorably with that of the internal-combustion engine, the total vehicle operating cost per mile in city driving would be higher because it is expected that the cost of the electric motor controller and the sodium-sulfur battery system would be higher than the cost of an internal-combustion engine. Because the cost figures of operating an internal-combustion engine are taken from Department of Transportation data showing averages for the past 10-year period, they do not reflect increases in the cost of gasoline and, therefore, must be considered relatively low.

Continuing development work on the sodium-sulfur battery is being undertaken by Ford Motor Company in cooperation with the University of Utah and Rensselaer Polytechnic Institute.

REFERENCES

1. Compton, W. Dale: "Energy Conversion and Storage Technology—The Sodium-Sulfur Battery," presented at 1st Symposium on Research Applied to National Needs, Nov. 18–20, 1973, Washington; published in Energy, Environment, Productivity, Stock no. 3800–00177, U.S. Government Printing Office, Washington, 1973.
2. Anon.: The Electric Automobile, in *Energy Perspectives*, no. 9, Battelle Memorial Institute, Columbus, Ohio, April 1974.

Fuel Cells

BY O. ADLHART*

The fuel cell is basically a galvanic energy conversion device and as such circumvents the limitations of the Carnot cycle. Conversion efficiency, therefore, can be high as compared with other conversion methods. The concept of the fuel cell has been known for well over a century, but only in recent years has the technology been brought into the realm of practicality. It was in an exotic mission, the manned space exploration, that fuel cells were first deployed. The space program has materially aided the development, but advances in material technology of the last decade were major factors as well.

Modern fuel cell work dates back to the late 1930s when F. T. Bacon (Ref. 1) developed a cell operative on high-purity hydrogen and oxygen. Bacon's development was the basis for the Apollo power plant and represented the starting point for a sophisticated technology based on the hydrogen-oxygen cell with alkaline electrolyte. The adaptation of this technology to the seemingly endless possibilities in terrestrial applications proved to be difficult. Not only did cost prevent its use, but operation on hydrogen of lesser purity and on air rather than oxygen was not possible with available cells. This capability, however, is a prerequisite for broad terrestrial application.

There are a number of considerations which favor fuel cells in addition to high conversion efficiency. Being primarily a low-temperature conversion device, the fuel cell is largely without polluting emissions. Noteworthy also is the versatility of the fuel cell in respect to size and power level which is partly reflected by an adaptability to a modular design. Also, very important, the fuel cell requires few moving parts, and, therefore, promises to be a quiet, reliable, and comparatively maintenance-free source of electric power.

PRINCIPLES OF FUEL CELL OPERATION

As an energy conversion device, the fuel cell has unique design and operating features. It is distinguished from a conventional battery, generally speaking, by the fact that the electrodes are invariable and catalytically active. Current is generated by reaction on the electrode surfaces which are in contact with a suitable electrolyte. As a rule, fuel and oxidant are not an integral part of the cell, but are supplied as required by the current load, and reaction products are continuously removed.

The nature of the fuel and oxidant is a critical aspect of cell operation. Hydrogen combustion with air or oxygen is by far the most important reaction for power generation in a fuel cell, although fuels such as hydrazine or oxidants such as hydrogen peroxide are considered for special uses.

Single Cell. Under load, the voltage of one individual fuel cell element is less than one volt. Therefore, the assembly of many cells, connected in series as a cell stack, is necessary. Each individual cell contains the elements needed for sustained operation. This implies means for feeding reactants to the electrode surface, and removal of reaction products and waste heat from the cell. The technology for satisfying these often conflicting requirements is discussed later. The principles involved in assuring sustained cell operation vary from one system to the next.

A useful example, demonstrating the components and their functions, is the hydrogen-air cell with acid electrolyte. Considerable development effort is in progress to achieve a practical cell of this type.

A hydrogen-air cell is shown schematically in Fig. 1. The cell consists of a pair of porous catalyzed electrodes separated by an acid electrolyte. The electrode reactions are comprised of the oxidation of hydrogen on the anode (the negative electrode) to hydrated

*Research and Development Department, Engelhard Industries Division, Engelhard Minerals & Chemicals Corporation, Menlo Park, New Jersey.

protons with the release of electrons; and on the cathode the reaction of oxygen with protons to form water vapor with the consumption of electrons. Electrons flow from the anode through the external load to the cathode, and the circuit is closed by ionic current transport through the electrolyte. In an acid cell, the current is carried by protons.

Reactants in this cell need not be pure. Hydrogen may be extracted from fuel mixtures and oxygen from air. Since product moisture is formed in an acid cell on the cathode, the air depleted in oxygen can be conveniently used for water removal. For this purpose, the cell is operated at a sufficiently high temperature to vaporize the water formed.

The electrode obviously has a central function in cell operation. In its catalyzed layer, it provides a large number of sites where gases and electrolyte can react. By virtue of the porous configuration, fast reactant transport and removal of inerts and product moisture are possible.

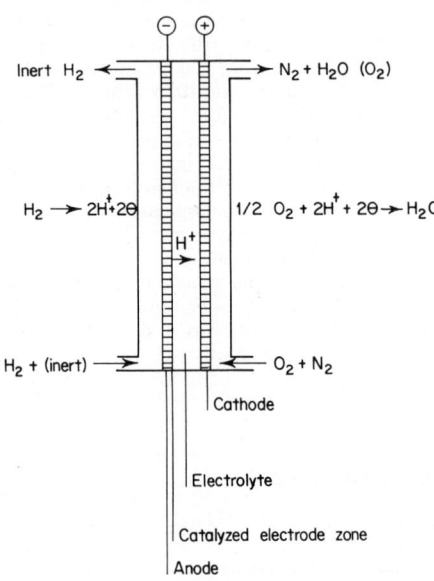

Fig. 1. Principles of operation of the hydrogen-air cell with acid electrolyte. Product water is removed by the flowing air. Reactants pass through the cell essentially at atmospheric pressure.

The electrode also must provide a path for current to flow to the terminals, and it often serves to contain the electrolyte. The latter, of course, not only provides ionic conduction, but also assures separation of the reactants.

Cell Voltage.   The cell voltage and the free energy of the underlying reaction are defined by

$$U = \frac{\Delta F}{nF}$$

where U = theoretical cell voltage
n = number of electrons transferred in the reaction
F = Faraday constant

Since $\Delta F = \Delta H - T \Delta S$, it follows that, depending on the value of ΔS, the electric energy to be derived from the cell can be larger or smaller than the energy ΔH obtained by direct combustion of the fuel.

ΔH = reaction enthalpy for the current generating reaction
ΔS = entropy change
T = absolute temperature

The quotient $\Delta F/\Delta H$ provides a measure for the conversion efficiency and has been defined as the thermodynamic conversion efficiency. It determines the theoretical limits to which the heat of reaction can be recovered as electric energy. For the hydrogen-oxygen couple, this quotient is 0.84 (at 25°C, 1 atmosphere) if the reaction is carried to liquid water. The corresponding theoretical cell voltage is 1.23 volts. Cell voltage is 1.18 volts if water vapor is the reaction product.

The thermodynamically possible conversion efficiency is only partly realized in a practical fuel cell. This is related to an "intrinsic" inefficiency of the conversion process which results in a reduction in cell voltage. This reduction is to be distinguished from losses associated with the operation of the device, as, for example, any parasitic power consumption of auxiliary components.

Two basic losses are encountered in the galvanic cell—the *ohmic loss* and the *electrode polarization,* that is, the deviation of the actual from the thermodynamic electrode potential. The ohmic loss occurs primarily in the electrolyte. Electrical resistance in the electrodes and conductors leading to the cell terminals is of some significance also, since fuel cells are a low-voltage device and high currents have to be conducted. The polariza-

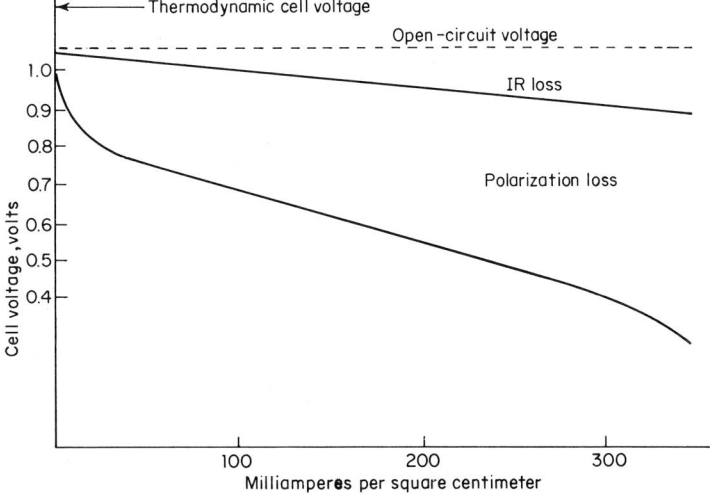

Fig. 2. Current-voltage characteristic of hydrogen-air cell with phosphoric acid electrolyte. Operating temperature is 125°C.

tion itself is the result of the irreversibility of the electrode process, the activation polarization, and the voltage loss which develops from concentration or activity gradients of the reactants. This leads to a unique current-voltage characteristic shown in Fig. 2. When current is not drawn from the cell, the open-circuit voltage is established as the differential of equilibrium electrode potentials. This, however, is lower than that given by the thermodynamic equilibrium. As previously stated, in the hydrogen-oxygen cell, this voltage is well below the thermodynamic value due to irreversible effects. As current is drawn from the cell, the voltage declines significantly. At low current densities, the decline is steep and this is the region where activation polarization is predominant. It is followed by a flatter current-voltage gradient where ohmic losses and concentration polarization, caused by depletion of reactants in the electrode, materially affect cell voltage.

Cell Technology.   The technology of the fuel cell battery as a building block for a device or power plant has been perfected to a considerable degree in recent years. Of the many options that present themselves for cell development in terms of electrolyte choice, operating temperatures, fuel, electrode configuration and activation, a select group has been emphasized. This includes cells with aqueous electrolyte, fused salt electrolyte, and those cells operative at very high temperatures in which ionic conduction is provided by oxygen ion mobility in the solid state. See Table 1.

At present, cells operative below 175°C using aqueous or quasi-aqueous electrolyte are the most advanced: specifically, cells with concentrated phosphoric acid electrolyte. A comparable state of development has been reached with cells using aqueous alkaline electrolyte (Refs. 2 to 5).

For these cell types, viable solutions have been found. This refers to the electrodes per se and their activation. The containment of the electrolyte within single cells, or the construction of complete batteries by stacking of individual cells, and, important for operation, the means for maintaining thermal and material balance—all have been thoroughly investigated.

TABLE 1.   Major Electrochemical Systems for Fuel Cells

Electrolyte	Current transport	Operating temperature, °C	Electrode catalyst	Reactants		State of development
				Fuel	Oxidant	
Aqueous potassium hydroxide (KOH)	OH^-	20–90	Nickel, silver, platinum metals	Hydrogen, hydrazine	Oxygen, scrubbed air, H_2O_2	Multikilowatt systems developed by several manufacturers.
Aqueous sulfuric acid (H_2SO_4)	H^+	20–80	Tungsten carbide, platinum metals, carbon	Impure hydrogen*	Air	Long life demonstrated in laboratory cells.
Concentrated phosphoric acid (H_3PO_4)	H^+	70–175	Platinum metals	Impure hydrogen*	Air	Multikilowatt system developed and larger systems in development. Long life has been demonstrated.
Fused alkali carbonate	CO_3^{--}	600–800	Nickel, silver	Impure hydrogen*	Air	Several months' life for small cells demonstrated.
Stabilized zirconium oxide	O^{--}	700–1000	Base metal oxides	Impure hydrogen*	Air	10,000 hours' life demonstrated in single cells. Multikilowatt systems in design.

*Refers to hydrogen-containing mixtures, such as steam reformate.

Historically, the aqueous electrolyte cells were not the first to be perfected. On the contrary, the first fuel cell for a practical application was built under the National Aeronautics and Space Administration's *Gemini* Program by the General Electric Company and relied on an unconventional electrolyte, namely, a solid polymer electrolyte membrane. This so-called ion-exchange membrane consists of a lacelike organic structure with anionic groups firmly bonded, and hydrogen ions being loosely held in the polymer chain. This provides sufficient mobility required for ionic transport. A main advantage of the ion-exchange membrane is the elimination of the need for electrolyte containment because of its well-defined boundaries. Separate electrodes are not required since a layer of platinum black bonded to the surface serves this function. Presently, the merits of cells with ion-exchange membranes are not visualized primarily for use as fuel cells, but rather for water electrolysis.

HIGH-TEMPERATURE FUEL CELLS

Cells operative at comparatively high temperatures (600 to 800°C for systems with fused alkali carbonate electrolyte, 1000°C for cells with solid ceramic electrolyte) are still in an early stage of development. The main incentive for development of these cell systems is the expectation that favorable electrode kinetics will permit the use of inexpensive catalyst materials. Also, it has been assumed that fossil fuels can be reacted directly, a goal which thus far has not been realized (Refs. 6 to 9).

In cells with solid ceramic electrolyte, Brown-Boveri & Cie. has an active development program. Zirconia stablized in its cubic crystal form is used. Studies have shown that the cell functions as an oxygen concentration cell, and overvoltage is almost exclusively resistive. In this development, conical tubes 1.6 centimeters high with a wall thickness of 0.5 millimeter and a diameter of 1 centimeter are used. The electrode materials, nickel on the anode and doped oxides on the cathode, are deposited on the inside and outside surfaces of the ceramic tubes. At 1000°C, 0.3 watt per square centimeter is obtained at 0.5 volt in single cells. As an oxygen concentration cell, the solid electrolyte cell has already found an important application for the determination of oxygen in gases. A device based upon this principle is expected to be widely used for the control of the air-to-fuel ratio of automotive engines.

In cells with fused carbonate electrolyte, an active program is in progress at the Institute of Gas Technology. Although little information has been published about this specific effort, earlier work in the Netherlands and by Texas Instruments Incorporated in the United States relied on molten lithium-sodium carbonate melts contained in porous alumina or magnesia as electrolyte. Nickel and silver are the catalyst materials mentioned.

In general, the problems in fused carbonate and solid electrolyte cells relate to material stability. Matching of coefficients of thermal expansion, chemical interaction between electrode and electrolyte, corrosion and sealing, and separation of reactants are more difficult to achieve at elevated operating temperatures.

CELLS WITH AQUEOUS ELECTROLYTE

Development of cells with aqueous electrolyte has been in favor in the past and is also receiving present emphasis. In this area, a well advanced technology is at hand. See Table 2. The preference for aqueous systems is partly due to the high specific conductivity of the electrolyte and the higher cell performance at ambient or near-ambient temperatures. Material stability is another important factor.

The classic example, the hydrogen-oxygen fuel cell with aqueous potassium hydroxide as electrolyte, is still preferred for operation with pure hydrogen and oxygen, or with liquid reactants, such as hydrazine and hydrogen peroxide. However, since the electrolyte is sensitive to carbonation by the carbon dioxide contained in air, acid cells have a distinct advantage for operation on air and impure hydrogen, which may contain carbon dioxide as derived from carbonaceous fuels.

In the development of practical cell systems based on aqueous electrolyte, several approaches have been taken which may be differentiated by the mode of electrolyte confinement. A number of manufacturers have elected to base their cell development on a free-flowing electrolyte contained by the electrodes or porous membranes adjacent to the electrode, as shown schematically in Fig. 3a.

Another approach is to render the electrode hydrophobic. This provides the advantage that cells can operate at atmospheric pressure (Ref. 10). In the development by Union

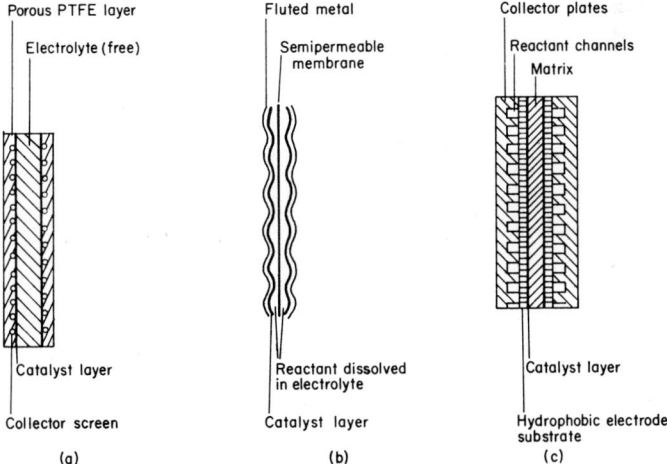

Fig. 3. Representative cell types: (a) Free-electrolyte cell with screen reinforced electrodes with polytetrafluoroethylene (PTFE) liquid barrier. (b) Alstholm cell with nonporous electrodes and semipermeable membranes for reactant separation. (c) Matrix-type cell—electrodes and matrix are sandwiched between grooved collector plates.

TABLE 2. Advanced Fuel Cell Systems

Source	Electrolyte	Reactant	Electrode catalysts	Heat and water removal	Cell construction and observations
Alstholm (France)	Potassium hydroxide (KOH)	N_2H_4–H_2O_2	Silver/cobalt	Circulating electrolyte	Nonporous electrodes densely packed. Surface coating of catalyst.
Engelhard Minerals & Chemicals Corp.	Concentrated phosphoric acid (H_3PO_4)	Impure hydrogen–air	Platinum metals	Air or circulating coolant	Hydrophobic electrodes, matrix-type construction, bipolar stacking. Operates at atmospheric pressure.
Siemens AG (Germany)	Aqueous potassium hydroxide (KOH)	Hydrogen–oxygen (air)	Raney nickel/silver	Circulating electrolyte	Catalyst layer supported on asbestos sheet. Operating pressure is one atmosphere.
Varta AG (Germany)	Aqueous potassium hydroxide (KOH)	Hydrogen–oxygen (air)	Raney nickel/silver	Circulating electrolyte	Electrodes of carbonized nickel. Substrates with embedded Raney catalyst. Operating pressure is one atmosphere.
Pratt & Whitney Division, United Technology Corp.	Potassium hydroxide (KOH)	Hydrogen–oxygen	Platinum metals, silver	Evaporative cooling	Matrix-type construction.
Pratt & Whitney Division, United Technology Corp.	Concentrated phosphoric acid (H_3PO_4)	Impure hydrogen–air	Platinum metals	Air or air with circulating cooling	Details not published

Carbide Corporation, layers of precious-metal-activated carbon of graded hydrophobicity are pressed onto collecting screens. This establishes a liquid/gas interface in the electrode and provides containment for the electrolyte. For the same purpose, the American Cyanamid Co. has developed a hydrophobic electrode structure with a polytetrafluoroethylene (PTFE) layer bonded to the back side of a catalyzed structure as a liquid barrier. In either case, porosity and thickness of the electrodes must be carefully controlled to prevent electrolyte leakage.

(a)

(b)

Fig. 4. (*a*) Battery module with electrolyte regenerator, electrolyte container, and pump. Aqueous caustic is used as electrolyte. With hydrogen and oxygen, 2.3 kilowatts are generated at 24 volts. (*b*) A fully instrumented system. (Siemens AG.)

Free-flowing electrolytes also are being used by Siemens AG and Varta AG (Germany), but the electrodes are hydrophilic.

In order to contain the electrolyte and establish an interface between gas and liquid, both of these designs are operated at elevated gas pressure (Ref. 11). For operation with hydrogen and oxygen, Siemens AG and Varta AG have well-developed cells which make use of relatively low cost, yet active Raney metal catalysts. See Fig. 4. Raney nickel is used for hydrogen oxidation, and Raney silver on the oxygen electrode. Raney-type catalysts are obtained in high-surface area form by alloying the active components with aluminum, which is subsequently dissolved. Siemens AG uses an inexpensive approach in the electrode fabrication by applying the catalyst onto a microporous asbestos sheet on the gas side. These "supported electrodes" are backed up by a screen and encapsulated in a plastic frame. The performance with hydrogen-oxygen is high. For example, at 80°C, 0.7 volt is obtained at 340 milliamperes per square centimeter.

(a)

(b)

Fig. 5. (a) Phosphoric acid matrix-type hydrogen-air cell. The unit is air-cooled and produces 750 watts at 24 volts. The only auxiliary component is an integral air blower. (b) Unit with fiber glass housing removed. (*Engelhard Minerals & Chemicals Corp.*)

A unique approach deviating from the concept of the porous high-surface-area electrode has been taken by Alstholm (France). The system has low specific current density,
but dense component stacking. Nonporous electrodes of finely fluted metal sheets are
catalyzed by a silver or cobalt catalyst which is applied by an adhesive resin. The
electrodes are separated by semipermeable membranes; and the reactants, dissolved in
the electrolyte, are circulated through respective half-cells. Complete cells are only 0.5
millimeter thick, and power densities are accordingly high. This type of cell is shown
schematically in Fig. 3b.

The matrix type cell is compact and has a low fabrication cost. This type of cell was first
used by Allis-Chalmers for an alkaline hydrogen-oxygen cell. The electrolyte is retained

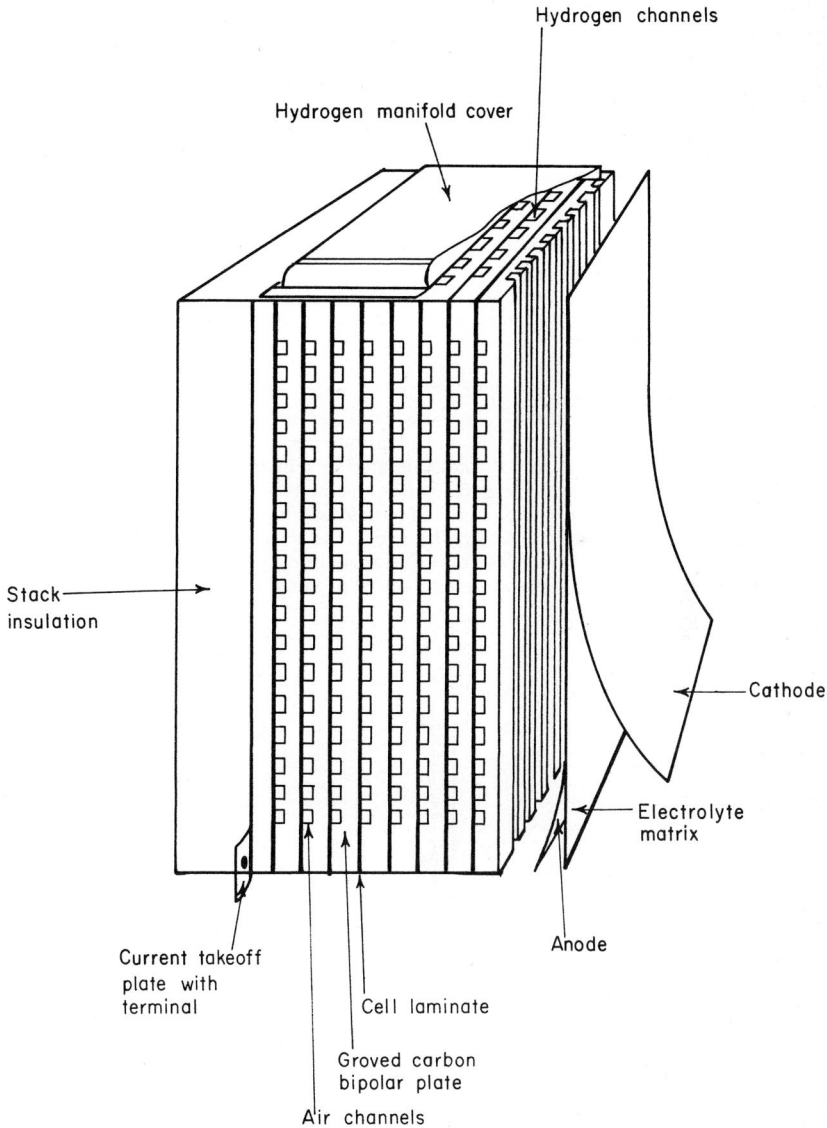

Fig. 6. Stack section of phosphoric acid matrix cell. Operation and design of cell are exceedingly
simple. Product water is removed with air.

in a microporous matrix, such as asbestos, by capillary forces. Hydrophobic electrodes are preferably used in this cell construction. Since electrolyte containment is not a main function, the electrodes can be made thinner and more porous than free electrolyte cells. The design principle of this cell is shown schematically in Fig. 3c.

Earlier American Cyanamid Co. developed an electrode for the matrix cell comprising an intimate mixture of catalyst and PTFE bonded to a fine screen for current collection and structural support without a distinct Teflon barrier (Ref. 12). This fonfiguration proved very useful for alkaline matrix-type cells. It was replaced, however, in cells with acid electrolyte by a hydrophobic, paperlike structure of carbon fiber with the catalyst bonded to one surface (Ref. 13). This electrode configuration is used successfully in the matrix-type phosphoric acid cell, a cell that excels in operational simplicity. See Fig. 5.

With the development of this particular cell, a concern often voiced regarding component cost and stability in acid electrolyte fuel cells was dispelled. The "all carbon" construction has become the accepted concept. All current-carrying components, with exception of the electrode catalyst itself, are carbon-based. The electrodes just mentioned consist of thin fibrous carbon structures activated by a layer of precious-metal catalyst bonded to the surface which faces the electrolyte. The thickness of a single electrode is only 0.5 millimeter, and that of the laminate of both electrodes with the matrix, which represents the entire current-generating unit, is 1-1.5 millimeters. These elements are sandwiched between nonporous carbon plates with a suitable pattern for current collection and grooves for reactant distribution (Fig. 3c). Single cell assemblies are stacked in a series-connected bipolar mode with respective terminals at the end plates. The stack configuration is shown schematically in Fig. 6. The directions of flow channels for air and hydrogen fuel are perpendicular to each other. The reactants are manifoled on the stack sides. Air-cooled and liquid-cooled phosphoric acid cells are well developed and under evaluation at various installations.

MASS AND ENERGY BALANCE

A prerequisite for sustained cell operation is a continuous supply of reactants and removal of reaction products, as well as of the heat generated by conversion losses. This aspect of fuel cell operation has a major bearing on cell design and may take preference over purely electrochemical considerations. Although common to any fuel cell operation, the foregoing considerations become of greater significance in the aqueous electrolyte cell. This is so partly because a closer control of operating temperature is required and partly because the water generated in the reaction is a component of the electrolyte.

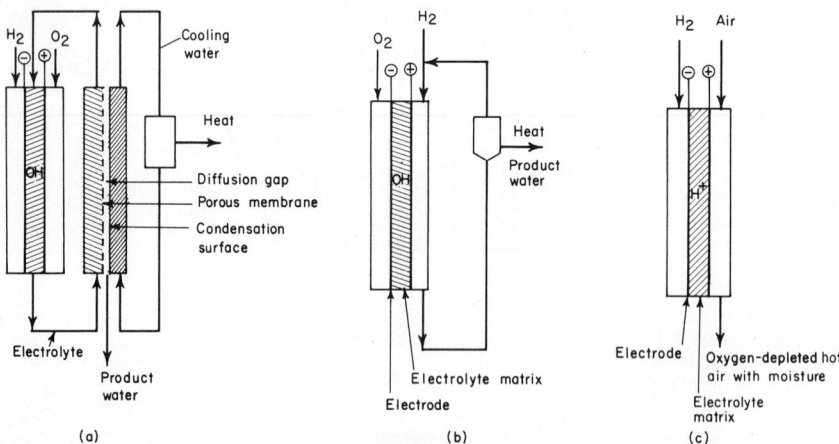

Fig. 7. Methods for waste heat and product water removal: *(a)* Free-electrolyte cell with electrolyte recirculation; water removal is based on vapor condensation principle. *(b)* Alkaline matrix cell with hydrogen recirculation for water and heat removal. *(c)* Air-cooled phosphoric acid matrix cell. In large cells, recirculating coolants are used for heat removal.

Specific methods had to be developed for the aqueous electrolyte cell to avoid imbalances in electrolyte volume, either by excessive water removal or by product water retention. These methods are a major aspect of fuel cell technology. See Fig. 7. Some of the representative approaches for maintaining the mass and energy balance are described in the next few paragraphs.

Siemens AG and Varta AG remove heat and product water by recirculation of the electrolyte. The hot electrolyte diluted with product water is circulated through a separation cell where water is transferred as vapor through a porous membrane to a cooled condensing surface. See Fig. 7a (Ref. 14). This diffusion-condensation principle is applicable only for free-electrolyte cells.

In matrix type cells the electrolyte is fixed and, furthermore, the quantity of electrolyte is smaller than in free-electrolyte cells. This reduces the allowable margin of variation which will cause drying out of the matrix or flooding of the electrode structure. One method for maintaining the water balance in matrix cells with alkaline electrolyte is hydrogen recirculation. See Fig. 7b. Hydrogen is circulated through the stack saturated with product water. The latter is condensed to a prescribed humidity level before the hydrogen reenters the stack. The recirculating hydrogen stream or evaporative cooling can be used for heat removal.

Reactant recirculation is not required in the matrix-type phosphoric acid cell. For this reason, this has been termed an "open cycle" system. See Fig 7c. This cell type can be more easily controlled because excessive water removal is not possible. The electrolyte concentration is high, and the water vapor pressure is correspondingly low. A large excess of air is used to remove the product water as it is generated. This air stream can also be utilized to remove waste heat (Ref. 15). In larger cells, however, air cooling is not practical, and liquid coolants are recirculated through cooling plates located in the stack. In very small fuel cells with phosphoric acid electrolyte, convective cooling can be employed (Ref. 16).

FUEL SOURCES FOR FUEL CELLS

The utility of the fuel cell as an electric power source depends critically on the fuel needed for its operation. Fuel cost, availability, energy density, toxicity, handling, shipping, and storage requirements are factors to be considered. Fossil fuels would be favored for several reasons and, in particular, natural gas, light petroleum fractions, or liquefied petroleum gas (LPG).

An overriding consideration in fuel selection, considering the present state of fuel cell technology, is conversion of the raw fuel into a hydrogen-containing stream for feed to the fuel cell.

For very large fuel cell power plants as may be considered for utility applications, numerous techniques are available or under development to convert liquid and solid fossil fuels to provide a range of synthetic gases and/or clean liquids of varying quality, such as desulfurized fuel oil or light naphtha. See appropriate articles under *Coal Technology, Gas Technology,* and *Petroleum Technology* in this Handbook. Gaseous and liquid hydrocarbons are preferably converted by catalytic steam-reforming to yield intermediate gases which contain as major constituents, hydrogen, carbon dioxide and carbon monoxide. They can be utilized most easily in the fuel cell after conversion of carbon monoxide to carbon dioxide and production of additional hydrogen by the catalytic water gas shift reaction with steam.

In general, the processing required for fossil fuels involves a degree of complexity which limits its application to large installations. In recognition of this fact, the use of synthetic fuels has been explored in recent years. Although not necessarily usable in the fuel cell directly, such fuels can be more readily processed into a hydrogen-containing stream than most fossil fuels. Synthetic fuels currently considered for fuel cells are summarized in Table 3.

Of extraordinary interest for cell operation, of course, is hydrogen per se. One consideration is the fact that the trend toward a nuclear-based energy system increases the potential availability of hydrogen fuel, either for use in an energy storage system, or as part of a more extensive hydrogen economy. For many uses, the storage density of hydrogen will be a deciding factor. See also Hydrogen in this Handbook section.

Low-cost synthetic fuels are methanol and, to a lesser extent, ammonia. The intriguing aspect of these fuels is the fact that they can be readily processed into a hydrogen-containing stream. Anhydrous ammonia, widely available and of assured purity, can be catalytically decomposed at 750°C into hydrogen and nitrogen. Methanol can be converted by steam reforming under milder conditions than those required for hydrocarbons. See Methanol Production and Fuel Methanol vs. LNG in the *Gas Technology* section of this Handbook.

TABLE 3.   Synthetic Fuels for Fuel Cells

Fuel	Storage form	Density, grams/ cubic centimeter	Interstate Commerce Commission classification	Gross heating value		Conversion process
				Kilowatt-hours/ kilogram	Kilowatt-hours/ liter	
Hydrogen See Note 1	Compressed gas	—	Hazardous flammable gas (Red Label)	0.298	0.71	None
Reversible hydride (iron titanium hydride) See Note 2	Solid	5.47	—	0.643	3.52	Hydrogen release 1.08 $FeTiH_{0.1} + H_2 \rightleftharpoons 1.08$ $FeTiH_{1.95}$
Nonreversible solid hydride (calcium hydride, CaH_2) See Note 3	Solid	1.70	Flammable solid (Yellow Label)	3.75	6.38	Reaction with water $CaH_2 +$ $2H_2O \rightarrow Ca(OH)_2 + 2H_2$
Methanol	Liquid	0.79	Hazardous flammable liquid (Red Label)	6.30	4.99	Steam reforming at 200°C $CH_3OH + H_2O \rightarrow CO_2 +$ $3H_2$
Anhydrous ammonia	Compressed liquid	0.60	Hazardous material— poisonous liquid	6.25	3.77	Thermal cracking at 750°C $2NH_3 \rightleftharpoons N_2 + 3H_2$
Hydrazine hydrate	Liquid	1.03		3.46	3.56	None

NOTES:
1. 200 atmospheres gage pressure; includes weight and volume of 1A-size cylinder.
2. Particular hydride cited as example.
3. Based on hydrogen generated by reaction with water.

OUTLOOK

The fuel cell was selected by NASA on the basis of the energy density offered by the hydrogen-oxygen cell with cryogenically stored reactants. An added feature was the fact that potable water was generated. The space program provided an opportunity to demonstrate the suitability of the fuel cell as an energy conversion device under technically difficult conditions. Its successful deployment is even more impressive, considering the inherent complexity of the fuel cell system selected at that time. A considerable advance in technology has subsequently occurred. A major reduction in system weight is one evidence of this. For example, NASA is currently considering a fuel cell for the space shuttle with a design goal of 4.5 to 13.5 kilograms per kilowatt. This compares with 74 kilograms per kilowatt for the Apollo power plant.

The Target Program (Advance Research for Gas Energy Conversion) launched in 1967 by a group of electric utility firms and the Pratt & Whitney Division of United Technology Corp. is worthy of note. The purpose of this program is to explore the applicability of the fuel cell as a prime power source for domestic use. The fuel cell ultimately is seen in a total energy-environmental conditioning role in which it will provide not only electricity but also heat. Estimates indicate a 20 to 25 percent higher energy utilization over conventional energy supplies, a figure that does not take into consideration the additional use of the fuel cell waste heat for hot-water and home heating. See Fig. 8. Over 60 units of approximately 13-kilowatt rating have been installed in homes and commercial buildings through the United States, Canada, and Japan. Natural gas or LPG are primarily considered as fuel sources.

Orders of magnitude higher in power rating are fuel cell power plants under design by Pratt & Whitney for dispersed power generation (Ref. 19). In this concept, fuel cells are intended to supplement conventional power plants within the existing network. This not only provides reduced transmission losses, but offers advantages in operating flexibility. The plan encompasses the use of fuel cells operative on fossil fuels including liquid hydrocarbons at various locations for base load, standby generation, backup or peak power. The competitiveness of the fuel cell in this use versus other means of power

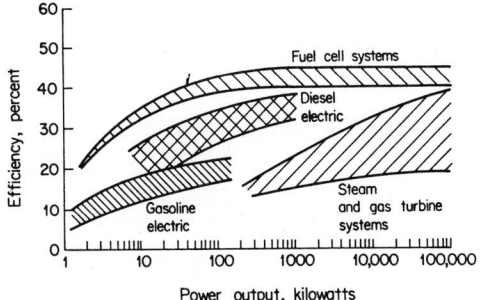

Fig. 8. The thermal efficiency of fossil fuel-operated fuel-cell power plants compares favorably with conventional means of energy conversion. The efficiency is reduced in small units mainly because of losses in the fuel processing. *(Pratt & Whitney Division, United Technology Corp.)*

generation remains to be demonstrated. The fuel cell offers attractions for urban locations where noise and pollution may be major concerns. See Fig. 9.

Additionally, even though less dramatic, there is a wide spectrum of other possible uses for the fuel cell. For example, the primary and secondary battery or the motor generator could be replaced by fuel cells. In these applications, synthetic fuels are most appealing since they can be processed into a hydrogen-containing stream comparatively easily, even

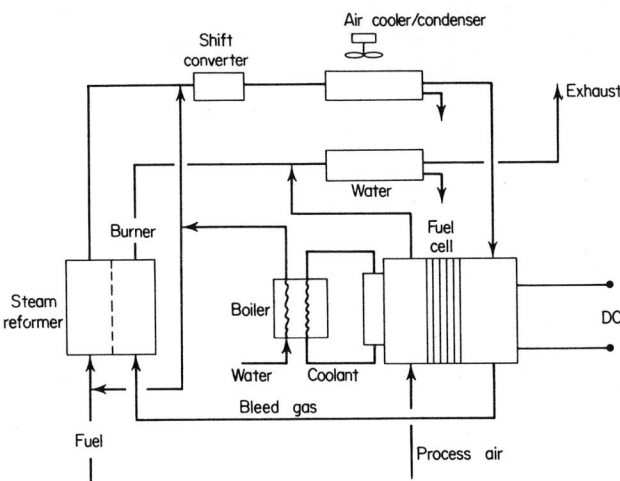

Fig. 9. Fuel-cell power plant with fuel conditioner for liquid fossil fuels. Phosphoric acid cell type with liquid cooling loop is used. *(Pratt & Whitney Division, United Technology Corp.)*

in smaller power plants without an excessive efficiency penalty. Thus, thermal efficiencies of fuel cells operative on methanol have been estimated to reach over 35 percent in power plants of only a few kilowatts rating (Ref. 20).

With anhydrous ammonia as fuel, an efficiency of over 30 percent has been demonstrated at the 300-watt level, and correspondingly higher values are expected in larger power plants with this fuel (Ref. 21). See Fig. 10. Compared with batteries, fuel cells operated with these fuels offer considerably higher energy densities, with the advantage that capacity is not reduced at low ambient temperatures. Appealing features are also the ability for instant refueling and favorable storage characteristics. Compared with motor generators, the reduction in emissions, silent operation, and reduced maintenance could become deciding factors in favor of fuel cells. Life, reliability, and fuel efficiency stand out as advantages.

Several tests have been initiated for the latter reasons to deploy fuel cells in accessible locations for navigational or communication devices (Ref. 22). A broader application is likely to be standby power or electrotraction, where the fuel cell provides a higher energy density than storage batteries and is less costly in longer missions.

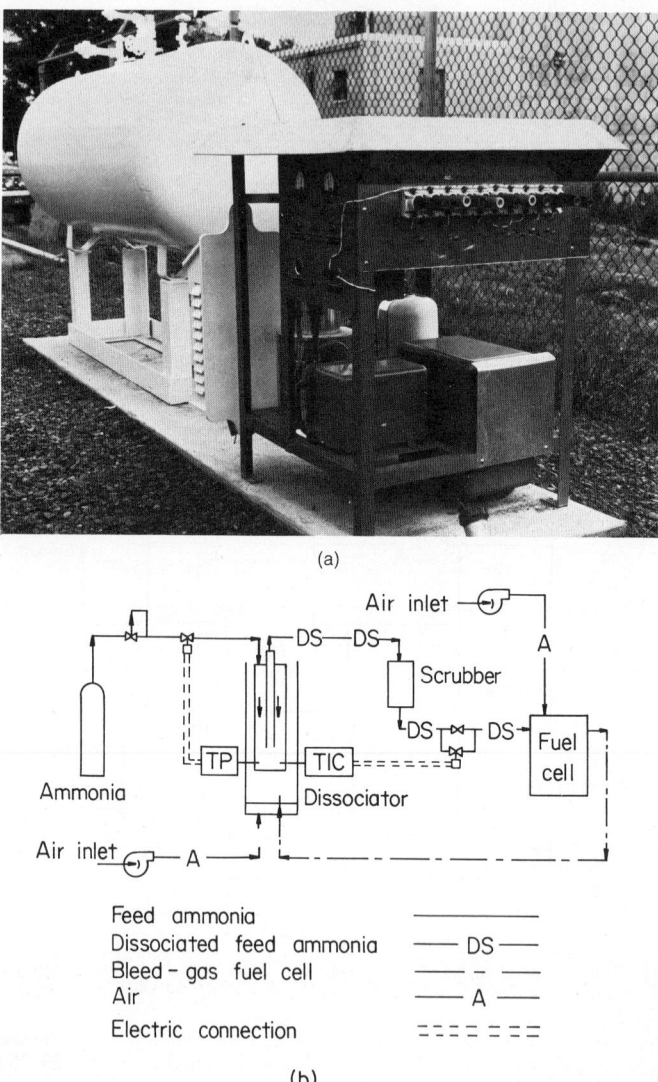

(a)

(b)

Fig. 10. (a) 300-watt ammonia-air fuel-cell system with fuel supply for three months of continuous operation. (b) Flow schematic diagram for ammonia-air fuel-cell system. Fuel-cell bleed gas is used to heat the ammonia dissociator. (*Engelhard Minerals & Chemicals Corp.*)

REFERENCES

1. Bacon, F. T.: High Pressure Hydrogen-Oxygen Fuel Cells, in G. J. Young (ed.), "Fuel Cells," Vol. 1, Van Nostrand Reinhold, New York, 1960.
2. Heath, C. E., B. Vurger, C. Hespel, and P. Faurel: "High Power Fuel Cells for Undersea Applications," 9th Intersociety Energy Engineering Conference, San Francisco, 1974.
3. van Dohren, H. H., and K. J. Euler: "Brennstoffelemente," Varta AG, Frankfurt, West Germany, 1971.
4. Adlhart, O. J.: "The Acid Cell, A Long Life Power Source for the Low to Medium Power Range," 7th Intersociety Energy Engineering Conference, San Diego, Calif., 1972.
5. Gutbier, H., D. Hasenauer, and K. Strasser: "A 2-KW Hydrogen-Oxygen Fuel Cell Battery," Third International Conference of Space Technology.
6. Archer, D. H., L. Elikaw, and R. L. Zahradnik: The Performance of Solid-Electrolyte Cells and Batteries on $CO-H_2$ Mixtures; A 100-Watt Solid Electrolyte Power Supply, in B. S. Baker (ed.), "Hydrocarbon Fuel Cell Technology," Academic, New York, 1965.
7. Sverdrup, E. F., C. J. Warde, and A. D. Glasser: A Fuel Cell Power System for Central Station Power Generation Using Coal as a Fuel, in "From Electrocatalysis to Fuel Cells," University of Washington Press, Seattle, 1972.
8. Tannenberger, H., and H. Siegert: Fuel Cell Systems—II, *Advan. Chem. Ser.*, **90**, 281, American Chemical Society, 1969.
9. Broers, G. H. J., and M. Scheuke: Long-Run Experiments on High-Temperature Molton Carbonate Fuel Cells, in B. S. Beler (ed.), "Hydrocarbon Fuel Cell Technology," Academic, New York, 1965.
10. Kordesch, K. VL: Low-Temperature-Low-Pressure Fuel Cell with Carbon Electrodes, in "Handbook of Fuel Cell Technology," Prentice-Hall, Englewood Cliffs, N.J., 1968.
11. Justi, F., M. Pilkuhn, W. Scheibe, and A. Winsel: Hochbelastbare Wasserstoff—Diffusionselektroden für Betrieb bei Ungebungstemperatur und Niederdruck, *Abhandl. Mainzer Akad.*, no. 8, Franz Steiner Verlag, Wiesbaden, Germany, 1959.
12. Haldeman, R. G., W. P. Colman, S. H. Langer, and W. A. Barber: Thin Fuel Cell Electrodes, in Fuel Cell System, *Advan. Chem. Ser.*, **47**, 106–115, American Chemical Society, 1965.
13. Kordesch, K. F., and R. F. Scarr: "Thin Carbon Electrodes for Acid Fuel Cells," 7th Intersociety Energy Engineering Conference, San Diego, Calif., 1972.
14. Gutbier, H.: Abtrennung des Reaktionswassers ans Brennstoff-Zellen durch Diffusion und Kondensation, *Chem.-Ing. Tech.*, **40**, 1209–1214 (1968).
15. Adlhart, O. J.: The Air-cooled Matrix Type Phosphoric Acid Cell, in "From Electrocatalysis to Fuel Cells," University of Washington Press, Seattle, 1972.
16. Adlhart, O. J.: "Mass and Energy Balance in Low Wattage Hydrogen-Air Fuel Cell Systems," Engelhard Industries Division, Engelhard Minerals & Chemicals Corporation, East Newark, N.J., 1975.
17. Strickland, G., J. J. Reilly, and R. H. Waswall, Jr.: "An Engineering Scale Energy Storage Reservoir of Iron Titanium Hydride," Theme Conference, Miami Beach, Fla., 1974.
18. King, J. M., Jr., R. J. Isler, and J. M. Burger: "Energy Storage for Utilities via Hydrogen Systems," 9th Intersociety Energy Engineering Conference, San Francisco, 1974.
19. Lueckel, W. J., L. G. Eklund, and S. H. Law: "Fuel Cells for Dispersed Power Generation," IEEE Power Engineering Society Meeting, Pratt & Whitney, East Hartford, Conn.; Northeast Utilities, Hartford, Conn., 1972.
20. Kurpit, S. S., and E. A. Gillis: "1.5-KW Indirect Methanol-Air Fuel Cell Power Plant," Engineering Foundation Conference on Methanol Fuel, New England College, Henniker, N.H., July 1974.
21. Collins, M. F., R. Michalek, and W. Brink: "Design Parameters of a 300-Watt Ammonia-Air Fuel Cell System," 7th Intersociety Energy Engineering Conference, San Diego, Calif., 1972.
22. Cnobloch, H., H. Nischik, and F. V. Sturm: Wasserstoff/Sauerstoff Brennstoffzellen für Wartungsarmen Betrieb, *Chem.-Tech.* **41**, 4, 146–154 (1969).

ADDITIONAL REFERENCES

Bockris, J. O. M., and T. Srinivsan: "Fuel Cells: Their Electrochemistry," McGraw-Hill, 1969.
Breiter, M. W.: "Electrochemical Processes in Fuel Cells," Springer-Verlag, Berlin, 1969.
Berger, C.: "Handbook of Fuel Cell Technology," Prentice-Hall, Englewood Cliffs, N.J., 1968.
Liebhafsky, H. A., and E. J. Cairns: "Fuel Cells and Fuel Batteries: A Guide to Their Research and Development," Wiley, New York, 1968.
Williams, K. R.: "Introduction to Fuel Cells," Elsevier, Amsterdam, 1966.

Propellants

BY DONALD M. ROSS*

This article describes chemical rocket propellants, outlines the general characteristics needed in propellants, lists important propellants and their applications, gives properties of production and experimental propellants, and identifies new scientific approaches to greatly increase the energy level of propellants. Included is a brief explanation of liquid, solid, and hybrid propellant rockets and other uses of rocket propellants.

CLASSIFICATION AND FUNCTION

Chemical rocket propellants can be classified in several ways, including liquid or solid, monopropellant, bipropellant, or tripropellant, cryogenic or storable, hypergolic or nonhypergolic, double-base, composite, or composite double-base. Propellant classification frequently influences the classification of rocket engines: for example, monopropellant rocket, liquid rocket, solid rocket, and hybrid rocket (generally using a liquid oxidizer and a solid fuel).

Propellants for chemical rockets serve two primary functions as contrasted to one for nuclear, solar, electrical, or laser-heated rockets. In a chemical propellant rocket, the propellant is both the energy source and the ejected mass or "working fluid."

Compared to the seemingly limitless number of chemical compounds that exist or can be formed, the number of chemical propellants in common usage are relatively few. This is because of criteria including costs, source availability, toxicity, resistance to shock, and other requirements imposed by the vehicle application and the propulsion system design. Another practical reason is that extensive overlap of physical, chemical, and economic properties is displayed by many of the theoretically possible propellants. During the 1960s, the universities, industry, and government in the United States pursued extensive research programs of synthesizing new chemical compounds viewed as candidate propellants which would increase the performance capability of chemical rockets. Although dozens of compounds were synthesized, few results reached the production line. This does not mean, however, that a scientific breakthrough in increasing the molecular energy of a propellant is impossible to achieve.

FUNDAMENTAL RELATIONSHIPS

The characteristics desired of a rocket propellant are many in number and can be divided into economic, safety, materials compatibility, engine-cycle needs, and vehicle requirements.

In general terms, the engine-cycle needs ideally are:

1. A propellant or propellant combination that has a high heat of reaction per unit weight (also called heat of combustion). Most vehicles add a requirement for high heat of reaction per unit volume of propellant to minimize vehicle size.

2. Reaction products that are all gaseous, have a very low molecular weight, and a very high temperature of dissociation.

In addition to specific impulse, the vehicle requirements usually influence propellant selection in terms of storability, density, toxicity, and other hazards, and other application-sensitive factors, sometimes even exhaust plume properties, including radar cross section and radiation emissions.

Other factors being essentially equal, the higher the heat of reaction of a propellant or

*Consulting Engineer, Lancaster, California. Credit is given Curtis C. Selph, Propellant Research Engineer, U.S. Air Force Rocket Propulsion Laboratory, Edwards, California, for confirming all the tabular performance data.

propellant combination, the more attractive the propellant. Sharp exceptions to this rule occur in some missiles because of volume limitations, the need for smokeless exhaust, or similar restraints. The heat released by a propellant is the difference in heat between the constituents and the end products of combustion:

$$H_r^\circ = \left[\sum n_k (\Delta H_f^\circ)_k - \sum n_j (\Delta H_f^\circ)_j\right]$$
$$\quad\quad\; k, \text{ products} \quad\quad j, \text{ reactants}$$

where ΔH_r° is the heat generated; ΔH_f° is the standard heat of formation of the constituent at reference temperature, 298 K; and n is the number of moles of each j reactant or k product. Large heat release is afforded by reaction products having large negative values, while the reactants should have positive, or at least small negative values, if possible. The heat of reaction is often noted in energy per weight units, such as kilocalories per gram.

Specific impulse, I_{sp}, the universally accepted measure of rocket engine performance, can also be used to indicate the performance of propellants. The most commonly stated expansion ratio is $1000 \rightarrow 14.7$ giving "sea-level specific impulse at 1000 psi chamber pressure." Sometimes the expansion ratio is $1000 \rightarrow 0.2$ to indicate specific impulse for high-altitude or space flight.

By definition, specific impulse is

$$I_{sp} = \frac{F}{\dot{w}}$$

with the I_{sp} units being seconds, the short designation for units of thrust (force) per units of propellant mass flow per second.

For an ideal rocket with the nozzle exhaust pressure being the ambient pressure, the thrust, according to Newton's second law of motion, is

$$F = \frac{\dot{w}c_e}{g}$$

where w is propellant flow rate in pounds per second, c_e is exhaust velocity in feet per second, and g is the gravitational constant in feet per second per second. In this equation pound force is equated to pound mass.

From the foregoing equations, specific impulse is

$$I_{sp} = \frac{c_e}{g}$$

By using the conservation of energy law, the relationship of specific impulse to the propellant heat of combustion and the molecular weight of the exhaust products can be shown to be

$$I_{sp} = \sqrt{2\frac{J}{g}(h_c - h_e)} = \sqrt{\frac{2J\,\Delta H}{M}}$$

where J is the mechanical equivalent of heat (778 foot-pounds per Btu), h_c is the specific enthalpy of the combustion chamber gases in Btu per pound, h_e is the specific enthalpy of the exhaust gases (nozzle exit) in Btu per pound, ΔH is the heat released per mole of propellants, and $\overline{M}$ is the average molecular weight of the exhaust gases in pounds per mole. In either form, the equation is rigorous.

Note that the specific enthalpy difference, $h_c - h_e$, is not equivalent to the heat of reaction discussed earlier, although it is affected by it. The connection between the two is sometimes pictured with the aid of an efficiency term η:

$$I_{sp} = \sqrt{2\frac{J}{g}(h_c - h_e)} = \sqrt{2\frac{J}{g}\frac{H_r^\circ}{M}}\,\eta$$

This shows that the enthalpy convertible to rocket specific impulse is limited to the enthalpy available chemically. The efficiency term η is a function of the amount of expansion accomplished in the nozzle, and certain properties of the combustion gases.

Although any real expansion introduces losses and irreversible processes, a reasonable approximation may be obtained with great simplification by assuming the expansion process to be isentropic. When this is done, the specific impulse function may be cast in the form

$$I_{sp} = \sqrt{\frac{2k}{k-1} \frac{RT_c}{\overline{M}} \left[1 - \left(\frac{p_e}{p_c}\right)^{k-1/k} \right]}$$

where R is the universal gas constant (1544 foot-pounds per mole per °F); g is the gravitational constant (32.2 feet/sec²); T_c is chamber temperature, °R; p_e is exhaust pressure, psi; p_c is chamber pressure, psi; k is the ratio of specific heat at constant pressure c_p to specific heat at constant volume c_v; and $\overline{M}$ is the mean molecular weight in pounds per mole. The foregoing equation is an approximation of the isentropic process since k and $\overline{M}$ are held constant by the equation throughout the expansion process. It is useful in making manual calculations of approximations with values for T_c, $\overline{M}$, and k being approximations themselves. Calculation of a correct T_c value is generally performed by large digital computers by iterative techniques, and calculation of the exhaust conditions then avails itself of the same rigor, instead of by means of the foregoing equation.

The relationship

$$I_{sp} \approx \frac{\Delta H_r^\circ}{\overline{M}}$$

is useful in screening propellant candidates. In some applications, particularly volume-limited air-launched missiles, propellant density is very important. Sophisticated analyses are employed in selecting propellant formulations, generally solid propellants.

Table 1 relates the heat of formation to the atomic number of chemical element fuels (reacted with oxygen) from Groups I, II, and III of the periodic chart. Above atomic number 10, aluminum and magnesium are the only elements worth noting as rocket fuels, since the others are either low in energy or high in atomic weight, or both. The same trend exists for fluorides, chlorides, and nitrides. Below atomic number 10, hydrogen, lithium, beryllium, boron, and carbon are the most promising fuels.

TABLE 1.   A Comparison of Energy Levels* and Atomic Numbers for Various Fuels in the Periodic Table (Reacted with oxygen)

Oxide	Atomic number	Heat of formation, cal/gram
Group IA		
Li_2O	3	−4790
Na_2O	11	−1610
K_2O	19	− 920
Rb_2O	37	− 420
Cs_2O	55	− 270
Group IIA		
BeO	4	−5830
MgO	12	−3570
CaO	20	−2710
SrO	38	−1370
BaO	56	−860
Group IIIA		
B_2O_3	5	−4370
Al_2O_3	13	−3930
Ga_2O_3	31	−1390
In_2O_3	49	− 800

*Since only elements are employed in this table, the heats of formation of the oxides listed are equivalent to the heats of reactions involved.

In practice, only about 10 percent of the elements of the periodic chart are adaptable to chemical rocket propellants. Propellants have made little use of elements other than hydrogen, carbon, nitrogen, oxygen, chlorine, fluorine, aluminum, boron, and beryllium.

Other Uses of Propellants.   In addition to being utilized by rocket engines for primary propulsion in missiles, space launch vehicles, spacecraft, and underwater vehicles, rocket propellants have many other applications, including gas generators for powering turbo-pumps and aircraft starters, for inflating crash bags (automotive) and evacuation chutes, and pressurizing propellant tanks.

Propellants, both solid and liquid, are used in many secondary propulsion applications, including crew capsule ejection, attitude control and station keeping of satellites, braking of reentry vehicles, and extravehicular space operations. Also, propellants are essential in

Fig. 1. Titan III-C space-launch vehicle and its payload. A major application of propellants with liquid N_2O_4/N_2H_4–UDMH (50–50) propellants in the main core engines (2 stages) and solid composite propellant (PBAN) in the 120-inch-diameter strap-on motors. *(Courtesy United Technology Center, Division of United Technology Corp.)*

rocket engine igniters, signal and illumination flares, and fuel-cell-type electric genera-tors. New developments, such as the high-powered gas dynamic laser (Ref.1), continue to broaden the field of applications.

Figure 1 shows the launching of an unmanned communication satellite into synchro-nous earth orbit, a major application of solid and liquid propellants. Figures 2 to 4 illustrate the basic elements of liquid, solid, and hybrid rockets.

LIQUID PROPELLANTS

Liquid propellants fall into two broad classes: (1) earth-storable (monopropellants and bipropellants); and (2) cryogenic, depending upon whether they can be kept in the vehicle tankage for months and years, or must be used in a few hours or days.

Some of the important properties and characteristics of liquid propellants, both storable and cryogenic, are listed in Table 2. Tables 3 and 4 list the theoretical performance of storable and cryogenic bipropellants combusting ideally at 1000 psia chamber pressure and expanding to sea-level pressure without loss, assuming shifting chemical equilibrium of the combustion products during expansion in the engine exhaust nozzle. See Ref. 2 for performance calculation method.

Table 5 lists a few properties of the more common monopropellants for purposes of comparison. Water (not listed) as a source of hydrogen and oxygen via electrolysis has merit as a propellant in long-life satellites (5 years or more), equipped with solar electric cells.

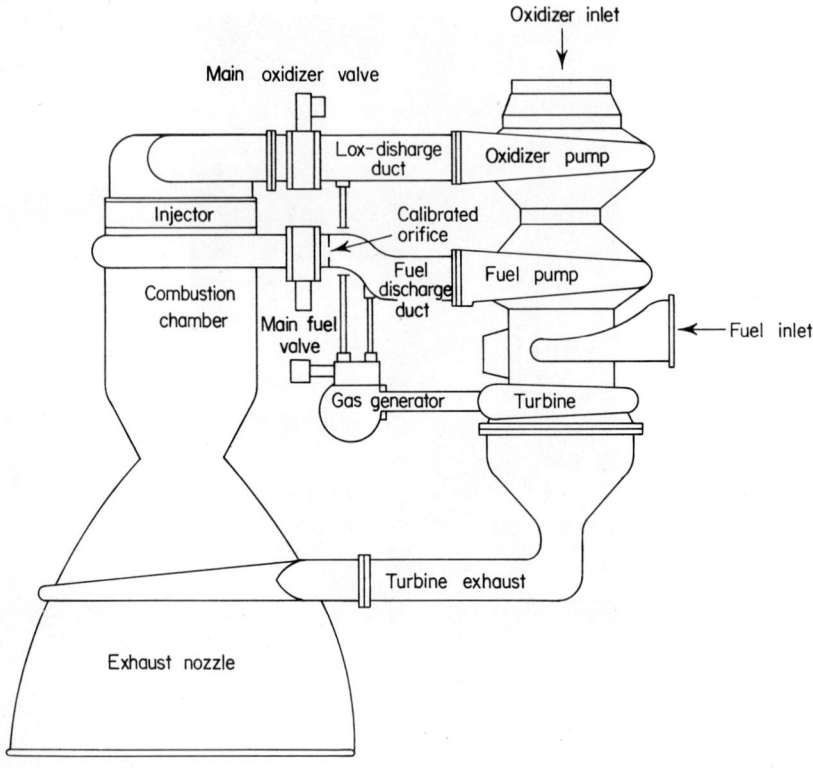

Fig. 2. Typical liquid-propellant rocket with turbopump feed system.

SOLID PROPELLANTS

Processed (including curing) propellants fall into three general types: (1) double-base; (2) composite; and (3) composite double-base.

A double-base propellant forms a homogeneous cured propellant, usually a nitrocellu-lose-type of gunpowder dissolved in nitroglycerin, plus minor percentages of additives. Both the major ingredients are explosives, and both contribute to the functions of fuel, oxidizer, and binder.

A composite propellant forms a heterogeneous propellant grain with the oxidizer crystals and a powdered fuel (usually aluminum), held together in a matrix of synthetic rubber (or appropriate plastic) binder, such as polybutadiene. Normally, composite pro-pellants are less hazardous to manufacture and handle than double-base propellants.

Composite double-base propellants are a combination of the double-base and compos-ite propellants, usually comprising a crystalline oxidizer (ammonium perchlorate) and powdered aluminum fuel, held together in a matrix of nitrocellulose-nitroglycerin. The hazards of processing and handling this type of propellant are similar to those of double-base propellants.

Representative formulations for the three basic types of propellants are given in Table

6. In actual practice, each manufacturer of a propellant has a proprietary precise formulation and processing procedure. The exact percentages of ingredients, even for a given propellant, such as PBAN, will not only vary from one manufacturer to another, but also often varies from one motor application to the next.

Table 7 lists several common solid propellants and their characteristics.

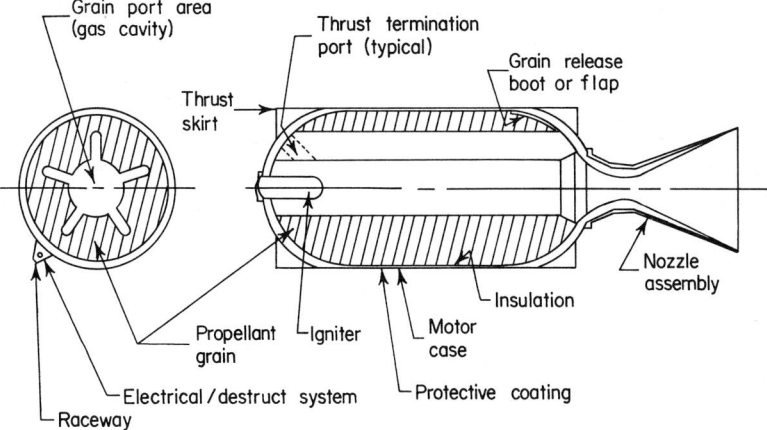

Fig. 3. Typical solid-propellant motor with case-bonded grain.

Classification of Ingredients.   Ingredients are generally classified according to their function: e.g., fuel, oxidizer, binder, curing agent, burn-rate catalyst.

Ingredients used in small amounts are called additives and generally have other functions than as a fuel, an oxidizer, or a binder.

For example, an additive can reduce the viscosity of the propellant during mixing and casting (pouring) of the propellant, increase the burning rate of the propellant, or improve the storage stability. Often an ingredient serves or affects more than one function, the

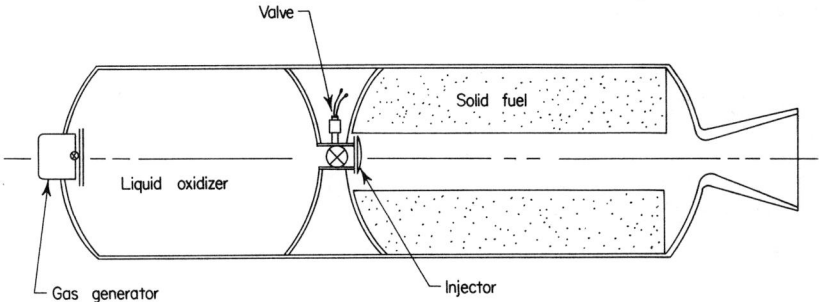

Fig. 4. Typical hybrid rocket arrangement.

most diffused situation relating to composite double-base ingredients where the binder is the nitrocellulose-nitroglycerin complex, with each of these two ingredients having their own fuel and oxidizer chemical elements. The binder contributes also as a fuel and, in some propellant formulations, such as asphalt-base nonmetalized propellants, the binder is the fuel.

Perchlorates.   Ammonium perchlorate, NH_4ClO_4, is the most widely used crystalline oxidizer in solid propellants. Because of its characteristics, including compatibility with the other propellant materials, specific impulse performance, quality uniformity and availability, ammonium perchlorate dominates the solid oxidizer field. Table 8 compares the properties of ammonium perchlorate with those of other crystalline oxidizers.

TABLE 2. Properties and Characteristics of Selected Liquid Propellants

Propellant	Formula	Approximate molecular weight	Density g/cm³ @ 72°F	Freezing point, °F	Boiling point, °F	Heat of combustion, kcal/g	Vapor pressure, (psia @ 68°F)
Ammonia (liquid)	NH_3	17.0	0.608	-107.9	-28.16	4.4	128
	Auto ignites at 1000 to 1500°F, 14.7 psia. Has narrow combustion limits: 17 to 28.5% (volume) in O_2 @ 72°F, 14.7 psia. Insensitive to impact. Stores very well in sealed containers. Decomposition starts above 600°F in stainless steel. Vapors are toxic, and 0.6 to 1% (volume) is lethal. Liquid mixes well with hydrazine.						
Chlorine trifluoride	ClF_3	925	1.85 (53°F)	-117.4	53.16	NA (oxidizer)	17.2 (60°F)
	Reacts very much like fluoride but is storable in sealed containers. Compatible with steel, nickel, stainless steel, copper. Vapors are extremely toxic.						
Ethyl alcohol	C_2H_5OH	461	0.79 (60°F)	-174	173	6.4	0.85
	Nonhypergolic with oxygen and air. Nontoxic and noncorrosive. Stores well in sealed containers. Addition of water improves cooling capacity.						
Fluorine (liquid)	F_2	38	1.505 (-306.5°F)	-360.4	-306.9	6.4 (oxidizer)	23 (-300°F)
	Highly reactive oxidizer, reacts violently with organic compounds. Makes all bipropellant combinations hypergolic and combusts very efficiently. Mixes well with O_2. Vapors are extremely toxic. Cryogenic storable in insulated tanks with boil-off.						
Hydrazine	N_2H_4	32	1.01	34.5	235.4	4.6	0.2
	Monopropellant or bipropellant fuel. Hypergolic with most oxidizers. Highly flammable in air (4.6 to 99+% by volume). Combusts very efficiently. Subject to catalytic decomposition. Decomposes above 320°F. Explodes above 450°F. Insensitive to impact or friction. Stores very well in sealed containers. Vapors are very toxic; liquid mildly attacks the skin. Compatible with stainless steel, glass, cadmium, some aluminum alloys, and several plastics.						
Hydrogen (liquid)	H_2	2	0.071 (-423°F)	-434.5	-423	28.5	1.2 (-435°F)
	Nonhypergolic with oxygen. Highly flammable in air (4 to 96% by volume). Combusts rapidly and efficiently (98 to 100%). Storable in insulated containers with controlled boil-off. Liquefaction of air in hydrogen is an explosion hazard. Compatible with most materials. Vapors are nontoxic, although asphyxiation is possible in confined space. Liquid freezes the skin.						
Hydrogen peroxide (100%)	H_2O_2	34	1.442 (77°F)	31.4	302.4	NA (monopropellant)	0.25 (130°F)
	Can be a monopropellant or a bipropellant-oxidizer. Hypergolic with fuels. Combusts efficiently. Decomposes slowly during storage. Passivation of metal surfaces in tanks, lines, valves, etc., reduces catalytic action and decomposition rate. Explodes above 350°F. Liquid burns the skin.						

Propellant	Formula	Molecular weight	Specific gravity	Freezing point, °F	Boiling point, °F		Vapor pressure	Remarks
Methane (liquid)	CH_4	16	0.415 (−263.2°F)	−299.2	−258.7	13.2	10.14 (263.6°F)	Flammable in air (5.3 to 14.0% by volume). Stores well in sealed containers. Compatible with most materials. Vapor is nontoxic although asphyxiation can occur. Liquid will burn the skin.
Monomethylhydrazine	CH_3NHNH_2		0.870	−62	190	NA (oxidizer)	0.957	Generally similar to hydrazine but not a good monopropellant because of carbon. Superior to hydrazine as a rocket fuel in terms of freezing point, stability, storability, and heat transfer, but gives lower specific impulse.
Nitrogen tetroxide	N_2O_4	92	1.44	11.08	70.1	NA (oxidizer)	13.9	Hypergolic with most fuels. Ignites spontaneously with paper, wood, oil. Stores very well in sealed containers. Liquid burns the skin, and vapors are highly toxic. Vapors are very corrosive in moist air.
Nitromethane	CH_3NO_2	61	1.15	−20	214	2.77	0.1 (102°F)	Can be a monopropellant and a bipropellant fuel. None hypergolic with oxygen. Storable in plastic-lined steel drums. Subject to catalytic decomposition. Confined liquid detonates at 500°F. Vapors are quite toxic.
Oxygen (liquid)	O_2	32	1.14	−361.8	−297.6	NA (oxidizer)	0.75 (−300°F)	Nonhypergolic at atmospheric pressure. Nontoxic and noncorrosive. Liquid freezes the skin. Storable in insulated tanks with boil-off. Compatible with clean materials. Detonation of mixtures of liquid oxygen and organic materials can occur upon impact.
Ozone (liquid)	O_3	48	1.573 (−297.2°F)	−315.0	−169.4	NA (oxidizer)	0.002 (−297.4°F)	Hypergolic with all fuels. Liquid is extremely shock-sensitive. Mixes well with O_2 up to 25% by weight of ozone. Difficult to store because of sensitivity to contaminants. Vapor is extremely toxic. Liquid will burn the skin.
Pentaborane	B_5H_9	63.2	0.627 (68°F)	−59.8	137.1	16.1	6.3	Hypergolic with most oxidizers including humid air. Combusts inefficiently (70 to 90%). Storable in sealed containers; slight solidification. Vapors are extremely toxic.
Petroleum fuel (RP-1)	Mil. Spec. F25576B	165 to 195	0.80 to 0.82 (60°F)	−47 to −64	342 to 507	5.7	0.3 (160°F)	Nonhypergolic with oxygen. Flammable in air. Autoignition 470°F. Stores very well (similar to kerosine). Vapors are midly toxic; liquid is a slight irritant to the skin.
Red fuming nitric acid (RFNA)	85% HNO_3, 15% N_2O_4	60	1.57	−56.3		NA (oxidizer)	2.0	Hypergolic with several fuels. Gives highest performance of the nitric acid group. When inhibited with hydrogen fluoride (0.4 to 0.6%), stores well in certain stainless steel and aluminum alloy containers. Highly toxic and corrosive. Liquid severely burns skin.
Unsymmetrical dimethylhydrazine	$(CH_3)_2 NNH_2$	60	0.79	−61.6	145.4	7.8	1.9 (60°F)	Characteristics of flammability, storability, toxicity, insensitivity to impact, and compatibility with metals are very similar to those of hydrazine. Not a monopropellant. Swells most plastics and synthetic rubbers.

The oxidizing potential of the perchlorates is generally high, which makes this material suited to high specific-impulse propellants. Both ammonium and potassium perchlorate are only slightly soluble in water, a favorable trait for propellant use. Nitronium perchlorate is objectionably hygroscopic, is relatively incompatible with available binders, and detonates easily.

TABLE 3.   Storable Liquid Bipropellant Combinations

Oxidizer	Fuel	Oxidizer-fuel ratio (weight) for maximum specific impulse	Bulk specific gravity (maximum specific impulse)	Theoretical specific impulse,* seconds
Nitrogen tetroxide	50/50 Hydrazine/ UDMH†	2.0	1.19	289
	Hydrazine	1.3	1.21	292
	UDMH	2.6	1.17	286
RFNA‡ (15% NO₂)	UDMH	3.4	1.28	266
	RP-1	5.6	1.37	257
MDFNA§	UDMH	2.9	1.29	278
MON-15¶	UDMH	2.6	1.15	288
Hydrogen peroxide	Hydrazine	2.0	1.24	287
	UDMH	4.2	1.22	284
Chlorine trifluoride	Hydrazine	2.8	1.64	295
	UDMH	3.0	1.39	280
Chlorine pentafluoride	Hydrazine	2.7	1.47	313
	UDMH	2.9	1.34	297
Hydrazine	Pentaborane	1.3	0.80	328

*1000 → 14.7 psia, shifting equilibrium (chemical composition of exhaust gases change during nozzle flow).
†Unsymmetrical dimethylhydrazine.
‡Red fuming nitric acid.
§Maximum-density red fuming nitric acid (a mixture of RFNA and N_2O_4).
¶N_2O_4 with 15% NO.

TABLE 4.   Cryogenic Liquid Bipropellant Combinations
At least one propellant is cryogenic.

Oxidizer	Fuel	Pounds oxidizer/pound fuel		Bulk specific gravity (maximum specific impulse)	Theoretical specific impulse,* seconds
		Stoichiometric	Maximum specific impulse		
Liquid oxygen	RP-1†	3.41	2.6	1.02	300
	Hydrazine	3.0	0.9	1.07	313
	UDMH‡		1.7	0.98	310
	Ammonia	2.37	1.4	0.89	294
	Ethyl alcohol	2.09	1.8	0.99	290
	Methane		3.3	0.82	311
	Liquid hydrogen	7.95	4.2	0.29	390
Liquid fluorine	Liquid hydrogen	19.0	8.0	0.46	412
	Hydrazine	2.71	2.3	1.31	365
	RP-1	4.07	2.6	1.21	322

*1000 → 14.7 psia, shifting equilibrium.
†A kerosine-type fuel.
‡Unsymmetrical dimethylhydrazine.

All the perchlorate oxidizers produce hydrogen chloride, HCl, and other chlorine compounds in their reaction with fuels. Their exhaust gases are toxic and corrosive to the extent that care is required in firing rockets, particularly the very large rockets, to safeguard operating personnel or communities in the path of the exhaust clouds.

Ammonium perchlorate is available in the form of small white crystals, and close control of the size range and the percentage of the several sizes present in a given quantity or batch is required. This is necessary because particle size influences propellant processing and the physical and ballistic properties of the finished propellant.

Nitrates. The inorganic nitrates are relatively low-performance oxidizers compared with perchlorates. However, ammonium nitrate is used in some applications for economy and because of its smokeless and relatively nontoxic exhaust. Ammonium nitrate is mainly suited to low-burning-rate, low-performance applications, such as gas generators for turbine pumps.

TABLE 5. Liquid Monopropellants

Propellant	Specific gravity at 68°F	Theoretical specific impulse,* seconds	Exhaust (average molecular weight)
Hydrogen peroxide	1.39	165	22.68
Hydrazine	1.01	199	12.77
Nitromethane	1.12	245	20.34
Ethylene oxide (CH_2CH_2O)	0.89	199	20.50

*1000 → 14.7 psia, shifting equilibrium.

HMX and RDX. One or two crystalline high explosives, such as HMX (cyclotetramethylene tetranitramine) and RDX (cyclotrimethylene trinitramine), are sometimes included in propellant formulation to achieve a specific performance characteristic. Depending on the objectives of the type of propellant, the percent of the explosive can range from 5 to 50 percent. Usage is quite common in double-base smokeless (nonmetal) propellants, and the percentage can be high (50 percent), depending on the specific impulse, burning rate, and other internal ballistic properties desired. Lower percentages are sometimes employed in composite and composite double-base propellants, particularly in highly metallized formulations to ensure efficient combustion, increase the burning rate, and gain a few seconds of specific impulse.

TABLE 6. Representative Propellant Formulations*

Double base (JPN propellant)	Wt %	Composite (PBAN propellant)	Wt %	Composite double base (CMDB propellant)	Wt %
Nitrocellulose	51.5	Ammonium perchlorate	70.0	Ammonium perchlorate	20.4
Nitroglycerin	43.0	Aluminum powder	16.0	Aluminum powder	21.1
Diethylphthalate	3.2	Polybutadiene–acrylic acid–acrylonitrile	11.78	Nitrocellulose	21.9
Ethyl centralite	1.0	Epoxy curative	2.22	Nitroglycerin	29.0
Potassium sulfate	1.2			Triacetin	5.1
Carbon black	<1%			Stabilizers	2.5
Candelilla wax	<1%				

*Furnished by U.S. Air Force Rocket Propulsion Laboratory, Edwards, Calif.

Powdered Aluminum. The one prominent solid fuel is powdered aluminum, and it is employed in a wide variety of composite and composite double-base propellant formulations, usually being 14 to 22 percent of the propellant by weight. Table 9 compares a few properties of aluminum with those of other solid fuels which have received extensive attention.

Boron. Even though it appears as one of the high-energy fuels and is lighter than aluminum, boron has not proved to be a practical fuel because it is so difficult to burn with high efficiency in combustion chambers of reasonable length.

Beryllium and Aluminum Hydride. Beryllium burns much more easily than boron and improves the specific impulse of a solid-propellant motor, usually by about 15 seconds;

TABLE 7. Characteristics of Selected Operational Propellants

See footnote for acronyms and symbols.

Propellant type	I_{sp} range seconds	Flame temperature, °F	Density, pounds/ cubic inch	Metal content, %	Burning rate, inch/ second	Hazards class (Military)	Stress/strain psi −60°F	Stress/strain % +150°F	Processing method
DB	255	5,340	0.057	0	0.45	7	4,600/2	490/60	Extruded
DB/AP/Al*	258	6,990	0.069	25	0.78	2	2,750/5	120/50	Extruded
DB/AP-HMX/Al*	272	6,630	0.067	20	0.55	7	2,375/3	50/33	Solvent cast
PVC/AP†	239	4,810	0.065	0	0.45	2	369/150	38/220	Solvent cast
PVC/AP/Al†	253	6,120	0.069	20	0.45	2	359/150	38/220	Cast or extruded
PU/AP/Al†	263	6,000	0.065	23	0.27	2	1,170/6	75/33	Cast
PBAN/AP/Al‡	265	5,600	0.063	19	0.55	2	520/16 (at −10°F)	71/28	Cast
CTPB/AP/AL‡	265	5,540	0.063	19	0.45	2	325/26	88/75	Cast
HTPB/AP/A‡	264	5,540	0.063	19	0.40	2	910/50	90/33	Cast
PBAA/AP/Al‡	265	5,660	0.063	20	0.32	2	500/13	41/31	Cast

NOTES:
*AP/Al optimized with 40% DB as binder.
†AP/Al optimized with 20% binder.
‡AP/Al optimized with 15% binder.

Acronyms and Symbols

Al	Aluminum.	HTPB	Hydroxy-terminated polybutadiene.
AP	Ammonium perchlorate.	PBAA	Polybutadiene-acrylic acid polymer.
CTPB	Carboxy-terminated polybutadiene.	PBAN	Polybutadiene-acrylic acid-acrylonitrile terpolymer.
DB	Double-base.	PU	Polyurethane.
HMX	Cyclo tetramethylene tetranitramine.	PVC	Polyvinyl chloride.

but as a powder or dust it is highly toxic to animals and human beings when ingested. The technology with composite propellants using powdered beryllium fuel is sufficiently advanced for vehicle application, space travel being the most likely application.

Theoretically, both aluminum hydride, AlH_3, and beryllium hydride, BeH_2, are attractive fuels because of their high heat release and gas volume contribution. Both are difficult to manufacture, and both deteriorate chemically during storage owing to the loss of hydrogen. Because of these difficulties, coupled with relatively modest specific impulse gains (10 to 15 seconds for AlH_3, 25 to 30 seconds for BeH_2), these compounds remain experimental.

Binding Ingredients.   The binding ingredient, usually a polymer, has primary effect on motor reliability, mechanical properties, propellant processing complexity, and cost. Some polymers undergo complex chemical reactions, crosslinking, and branch chaining during curing of the propellant. Others cure or plasticize without a molecular reaction.

Polymers can be grouped according to their effect on propellant processing as (1) plastisol binders, (2) oxygen-rich double-base propellant binders, (3) prepolymers or monomers for cast composite propellant binders, and (4) polymers from rubber gum stocks.

TABLE 8.   A Comparison of Crystalline Oxidizers

Oxidizer	Chemical formula	Molecular weight	Density, pounds/ cubic inch	Oxygen content, % wt	Remarks
Ammonium perchlorate	NH_4ClO_4	117.5	0.0704	34.0	Low cost, readily available
Potassium perchlorate	$KClO_4$	138.6	0.0910	46.2	Low burning rate, medium performance
Sodium perchlorate	$NaClO_4$	122.4	0.0729	52.3	Hygroscopic, high performance
Lithium perchlorate	$LiClO_4$	106.4	0.0877	60.2	Hygroscopic, high performance
Ammonium nitrate	NH_4NO_3	80.0	0.0625	20.0	Low cost, smokeless, medium performance
Potassium nitrate	KNO_3	101.1	0.0762	39.6	Low cost, low performance
Sodium nitrate	$NaNO_3$	89.0	0.0815	47.0	Low cost, low performance
Lithium nitrate	$LiNO_3$	68.9	0.0860	58.0	Hygroscopic, low cost
Nitronium perchlorate	NO_2ClO_4	145.5	0.0794	66.0	Very reactive, hygroscopic, high performance, unstable

With plastisol binders (Group I), the propellants are solidified by solvation of the polymerized resin particles in a nonvolatile liquid. With this procedure, all polymerization reactions are completed before propellant processing begins. Polyvinyl chloride, PVC, is a typical plastisol binder; the polymer suspended in a liquid that serves as the plasticizer.

The most common (Group II) double-base propellant binder is nitrocellulose, NC, a relatively low-cost commercial material. Even though its molecular structure is complex, NC has consistent chemical, mechanical, and ballistic properties.

Two polymers typifying Group III binders are: (a) polybutadiene-acrylic acid-acrylonitrile, PBAN, and (b) carboxy-terminated polybutadiene, CTPB. Both are suited to a wide range of storage and service temperatures. Cross-linking agents and other chemical additives are reacted during mixing and curing (heating after casting) to gain desirable physical and ballistic properties in the manufactured propellant. The industrial term for these complex composite propellant binders is *binder system.*

Neoprene, styrene-butadiene, and butyl, examples of Group IV, require special heavy-duty processing equipment to cope with their high viscosity. Although this group of binders is suited to molding, extrusion, or casting, applications are limited.

BURNING RATE

The burning surface of a propellant grain recedes in a direction essentially perpendicular to the surface. Burning rate is primarily a function of the propellant composition itself. Common ways of increasing the burning rate of a propellant are: (1) Add a burning rate catalyst, or increase percentage of catalyst; (2) decrease the oxidizer particle size; (3)

TABLE 9. A Comparison of Solid Fuels

Fuel	Chemical symbol or formula	Melting point, °F	Molecular weight	Density, grams/cubic centimeter	Representative theoretical I_{sp}, seconds ($1000 \rightarrow 14.7$ psia)	Remarks
None Aluminum	Al	1218	26.98	2.70	249 (PU/AP) 263 (PU/AP/Al)*	Included as reference base. Low cost. Burns well. Mildly toxic if inhaled in daily work.
Beryllium	Be	2330	9.01	1.8	284 (PU/AP/Be)*	Expensive. Highly toxic if inhaled even in low concentrations. Burns moderately well.
Boron	B	4180	10.81	2.3 (crystalline) 1.42	252 (PU/AP/B)*	Difficult to burn.
Aluminum hydride	AlH₃	Decomposes	30.00	1.42	282 (PU/AP/AlH₃),*	High purity (80%) is difficult to make. Short storage life.
Beryllium hydride	BeH₂	Decomposes	11.03	0.64–0.66 (compacted)	313 (PU/AP/BeH₂)*	High-density BeH₂ (0.35) requires compaction. Guard against toxic poisoning same as with Be.

*Fuel/oxidizer optimized with binder being 15%.

increase oxidizer percentage; (4) increase the binder heat of combustion; and (5) imbed wires or metal staples in the propellant (longitudinally placed wires or "paper staples" mixed in with the propellant). Composite propellants can be influenced by all the foregoing five factors. Items 2 and 3 are not applicable to double-base propellants. When burning in an actual rocket, the burning rate is also influenced by (1) combustion chamber pressure, (2) initial temperature of the propellant, (3) combustion gas temperature, (4) gas flow velocity, and (5) motor motion (acceleration and spin). The burning rate of propellant in a motor is a function of many parameters, and at any instant governs the rate of hot gas generated and flowing from the motor (stable combustion). The latter can be seen from

$$\dot{w} = A_b r \zeta_b$$

where w is the weight flow rate, A_b is the burning area of the propellant grain, r is the burning rate, and ζ_b is the propellant density.

The weight of propellant burned can be determined as

$$W = \int \dot{w}\, dt = p_b \int A_b r\, dt$$

where W is the total weight of effective propellant, and w is the propellant flow rate. A_b and r vary with time and pressure.

Analytical modeling and the supportive research have yet to adequately predict the burning rate of a new propellant in a new motor. Elemental laws and equations on burning rate usually deal with the influence of the important parameters individually, the most common being (1) burning rate-pressure relationship, and (2) burning rate-temperature relationship.

A propellant grain is the body of propellant with its material and shape inside a rocket motor, or with some designs it exists as a component ready for installation within the motor. Both the material (propellant) and geometry or shape of the grain are very important. These factors govern most of the performance characteristics of the motor. Some rocket motors employ a multiple propellant grain having sections of the grain made of propellant formulations with different burning rates to achieve a desired thrust profile. Others use two or more separate grains, each with its own igniter to give an on-off-on type of thrust capability, which can be programmed during flight.

CHARACTERISTICS OF A PROPELLANT GRAIN

The characteristics of a propellant grain are customarily categorized as follows:

1. Performance Properties: Usually confined to theoretical specific impulse, propellant density (pounds per cubic inch), and specific impulse efficiency (ratio of theoretical to delivered specific impulse). Propellants burn to varying degrees of completeness, depending on the fuel, oxidizer, their ratios, and the environment within the motor. Propellants with nonmetal fuels usually burn with an efficiency of 97 or 98 percent as contrasted to 90 to 94 percent for propellants with aluminum powder as the fuel.

2. Internal Ballistic Properties: Those parameters that govern the burning rate and mass discharge rate of the motor. These include the sensitivity of burning rate to propellant ingredients, chamber pressure, propellant grain temperature, gas velocity, oscillatory combustion, and acceleration forces.

3. Mechanical Properties: Commonly expressed in terms of stress/strain values over a range of propellant temperatures, usually at those temperatures, such as -65, 76, and $160°$F, with -65 and $+160°$F often being the lower and upper extremes expected during motor exposure. Propellant must be strong enough and have elongation capability sufficient to meet the high stress concentrations present during shrinkage of the grain at low temperature, and also under the dynamic conditions of ignition and motor operation. These properties, together with the grain design, determine the upper grain temperature limit, beyond which the grain strength is too low to withstand imposed loads, and the lower temperature limit, below which the grain would crack, thus exposing undesirable additional burning area and causing unexpected increases in chamber pressure.

4. Hazard Properties: Pertain to (a) individual ingredients of the propellant, (b) propellant undergoing mixing and other manufacturing processing, and (c) completed solid rocket motors. Combustion (deflagration), explosion, or detonation can occur, depending on the materials and their environment—with initiation of energy release from the discharge of static electricity, friction, hot embers or particles, impact, etc. Ingredients and motors subject to shipment are assigned an explosive hazard classification by the U.S. government, based upon standardized tests. The toxicity, both dermatological and inhalation effects, of some propellant ingredients, such as epoxide and amine crosslinking agents used with polybutadiene prepolymer binder systems, necessitates specific safeguards and training in handling and storing the ingredients, and in processing the propellants. For example, the dust toxicity problem with powdered beryllium fuel requires the complete isolation of personnel from dust particles during handling and processing operations, during motor firings and in postfiring operations around the motor and test facility.

HYBRID ROCKET PROPELLANTS

Hybrid rockets use various combinations of solid and liquid propellants, usually a solid fuel and a liquid oxidizer. Sometimes a third propellant, liquid hydrogen, is added, not for energy release, but as a low-molecular-weight working fluid. The main advantages of a hybrid rocket are: (1) use of liquid and solid propellant combinations offering the highest performance attainable with chemical rockets, (2) simplicity of a solid grain (usually fuel), (3) a liquid for nozzle cooling and thrust modulation (compared to a solid rocket), (4) restart capabilities, and (5) good storability traits, including safe storage. Table 10 lists several hybrid propellant combinations, some of which are storable and some of which are cryogenic. Their theoretical performance is also included.

TABLE 10.   Selected Hybrid Propellant Combinations

Propellant combinations	Theoretical specific impulse*	Density specific impulse†
N_2H_4, N_2O_4, 22% Al (powder)	303	397
N_2H_4, N_2O_4, 14% Be (powder)	328	398
H_2, O_2, 41% Al (powder)	396	107
H_2, O_2, 26% Be (powder)	458	106
Li, 71% F_2	382	377

*1000 → 14.7 psia, shifting equilibrium.
†Bulk density times specific impulse.

METASTABLE PROPELLANT INGREDIENTS

The chemical bond energy present in propellant molecules is the energy source employed by chemical rocket engines to date. This source affords energy densities of approximately 3 kcal per gram in the liquid H_2/O_2 combination and up to approximately 5.7 kcal per gram with the Li/F_2 combination.

Theoretically, supplemental energy can be added to molecules or molecular fragments which, upon recombination or relaxation to their normal energy state, release significant amounts of energy. For example, 52 kcal per gram is theoretically released when two hydrogen atoms (free radicals) recombine to form H_2. Even higher energy densities, as much as 100 kcal per gram, are theoretically available from lightweight molecules, such as helium, that are in an excited state.

Metastable, in the sense of propellant ingredients, means that the "energized" molecule, atom, or molecular fragment tends to return promptly to its normal state. Some molecular species distinctly assume a metastable state upon excitation with the lifetime at room temperature being 10^{-3} to 10^{-2} second, as compared to less than 10^{-6} for nonmetastable excited species. Atoms subjected to excitation move into a more energetic state of translational motion of vibration, or into a high-energy electron orbital state; diatomic molecules can do likewise. Molecules containing more than two atoms can experience higher translational rotational motion, as well as a higher electron orbital state.

Most of the research conducted to date on metastable propellants has been with gaseous atoms and molecules. Obviously, energized ingredients in a condensed phase, solid or liquid, would be needed for most rockets. The primary objectives to be reached, if metastable ingredients are to benefit rocket propulsion, are (1) An efficient process for energizing the ingredients, and (2) a means of storing the ingredients for days at a time without appreciable energy loss. Actual use of metastable ingredients in a rocket is envisioned in the company of liquid hydrogen, or other low-molecular-weight working fluid.

Limited research has been conducted on two approaches to generating and storing (stabilizing) metastable propellant ingredients: (1) free radicals, specifically atomic hydrogen, and (2) *helides* which are excited states of helium. In the late 1950s, the National Bureau of Standards produced low concentrations of free radicals and stored them in inert matrices at very low temperatures (Ref. 3). More recently (Ref. 4), an approach has been taken to generate hydrogen atoms, immediately condensed at liquid-helium temperature, in the presence of a high-density (70 to 100 kilogauss) magnetic field for the purpose of stabilizing the hydrogen atoms. Theoretically, the high-strength magnetic field is capable of aligning the spin of the electron of the hydrogen atom so as to prevent recombination into the hydrogen molecule.

Triplet helium has a theoretical energy level of 114 kcal per gram above the ground state. Assuming release of this energy and subsequent expansion through a rocket nozzle gives a specific impulse of 2800 seconds. Techniques for generating activated helium and other noble gases are well known, but concentrating and storing these metastable species are another matter since they revert to their ground state by collision processes. Experimental approaches to activating helium and trapping the helium molecules in a hydrocarbon wax are reported in Ref. 5.

The creation and use of metallic hydrogen, hydrogen derived from normal hydrogen subjected to about 2-megabar pressure, should release about 52 kcal per gram upon transitioning from the metallic to the normal solid form. The idea dates back to 1935, but interest has been renewed in recent years, in that some scientists believe that metallic hydrogen exists in some large planets.

ANTIMATTER ROCKETS

Sufficient atomic particle research has been accomplished to warrant discussion of possible methods of applying energy available from particle mass annihilation to rocket propulsion. Complete conversion of matter to energy would allow exhaust velocities near that of light to be obtained from a propulsion device. Anti-matter, by definition, is matter made up of antiparticles, such as antineutrons, negatrons (antiprotons), and positrons (anti-electrons). An annihilation property is known to exist between particles with one particle termed the *antiparticle* of the other.

Rocket design concepts envisioned for utilizing the reaction between atomic particles and antiparticles (matter and antimatter) are based upon the following postulations:

1. Annihilation products can be accelerated by means of electric and magnetic forces (consider the annihilation reaction of a neutrino with an antineutrino, yielding a proton and an electron).

2. Annihilation products can be used indirectly to heat a working fluid for thermal expansion through a nozzle (consider the annihilation reaction of hydrogen and antihydrogen, leaving high-energy gamma rays).

3. Antimatter possesses negative gravitational mass although its inertial mass may be positive. This could give rise to antigravity propulsion.

4. Annihilation products of ordinary quanta give rise to the possibility of a photon expelling beam for the direct generation of thrust.

Before any form of antimatter rocket can exist, a lightweight method must be developed for producing antiparticles at a flow rate of grams per second in contrast to the few dozen of antiparticles produced in research laboratory generators. Also, a practical storage or containment method must arise, since antiparticles explode violently upon contact with normal matter. Reference 6 gives a performance estimate of a specific impulse of 3.06×10^7 seconds for a rocket-propelled vehicle with a thrust/weight ratio of 10^{-7}.

REFERENCES

1. Gerry, E.: "Gas Dynamic Lasers," *IEEE Spectrum*, pp. 51–58, November 1970.
2. Sutton, George P.: "Rocket Propulsion Elements," chap. 6, Wiley, New York, 1960.

3. Bass, A. M., and H. P. Broida (eds.): "Formation and Trapping of Free Radicals," pp. 69, 292, 327, Academic, New York, 1960.
4. Cohen, William: New Horizons in Chemical Propulsion, *Astronaut and Aeron.* **2**, 12, 46–51, December 1973.
5. Quinn, L. P., et al: "High Energy Storage Investigations," *Rep. AFRPL TR-71-36*, U.S. Air Force Rocket Propulsion Laboratory, Edwards, Calif., October 1971.
6. Corliss, W. R.: "Propulsion Systems for Space Flight," McGraw-Hill, New York, 1960.

Section **5**

Nuclear Energy Technology

WAYNE T. BARRETT, Ph.D. *President, Foote Mineral Company, Exton, Pennslyvania, Member, American Chemical Society, American Mining Congress, Ferroalloys Association, Manufacturing Chemists Association. (Lithium for Thermonuclear Fusion Reactors)*

J. FILE, Ph.D. *Plasma Physics Laboratory, Princeton University, Princeton, New Jersey. Member, Cornell Society of Engineers. Registered Professional Engineer (New York) (Superconducing Magnets as Applied to Fusion Reactors)*

A. S. GIBSON, B.S.M.E. *Manager, Product Engineering, Fast Breeder Reactor Department, Energy Systems and Technology Division, General Electric Company, Sunnyvale, California. Registered Professional Engineer. (The liquid Metal Fast Breeder Reactor)*

ALFRED L. HABOUSH, B.S. *Formerly a Senior Project Manager, General Atomic Company, San Diego, California. Member, American Nuclear Society, American Society of Mechanical Engineers, California Society of Professional Engineers, San Diego Engineering Council. Registered Professional Engineer. (Nuclear Power Plant—Concept to Startup)*

IHOR A. KUNASZ, Ph.D. *Chief Geologist, Foote Mineral Company, Exton, Pennsylvania. American Institute of Mining, Metallurgical and Petroleum Engineers, the Geochemical Society, Geological Society of America. (Lithium for Thermonuclear Fusion Reactors)*

LAWRENCE M. LIDSKY, Ph.D. *Associate Professor of Nuclear Engineering, Massachusetts Institute of Technology, Cambridge, Massachusetts. (Fusion Power)*

RALPH W. MOIR, Sc.D. *Research Physicist, Lawrence Livermore Laboratory, University of California, Livermore, California. Member, American Nuclear Society,*

American Physical Society. (Direct Energy Conversion in Fusion Reactors)

C. O. PEINADO, B.S.,M.S. Director, Program Development, General Atomic Company, San Diego, California. Registered Professional Engineer (California). (High-Temperature Gas-cooled Reactors)

ROBERT H. SIMON. Director, Gas-cooled Fast Breeder Reactor Program, General Atomic Company, San Diego, California. (Gas-cooled Fast Breeder Reactor)

D. STEINER, Ph.D. Member of the Thermonuclear Division and Director of the Fusion Reactor Technology Program, Oak Ridge National Laboratory, Oak Ridge, Tennessee. (Tritium Breeding Requirements in Fusion Reactor Blankets)

L. W. VERMEULEN. Metallurgical Engineer, Rio Algom Mines Limited, Toronto, Canada. (Uranium Resources and Processing)

Nuclear Fission Reactors

The energy of a nuclear fission reaction can be computed from the change in mass between reactants and products according to Einstein's law,

$$\Delta E = \Delta mc^2$$

where E is the energy in ergs, m is mass in grams, and c is the velocity of light in centimeters per second.

For example, the mass difference in this equation is Δm = 0.2058 amu (atomic mass units). Therefore, ΔE = 931 MeV/amu × 0.2058 amu = 191.6 MeV. The average amount of energy released in the various fission reactions is about 200 MeV. This energy is distributed in the fission process as:

	MeV
Kinetic energy of fission fragments	165
Radioactive-decay energy	23
Kinetic energy of neutrons	5
Prompt gamma-ray energy	7

The energy of a chemical reaction, approximately 3 to 4 eV, is dramatically lower than that of a nuclear reaction. Hence, the fission of ^{235}U yields 2.5 million times as much energy as the combustion of the same weight of carbon.

The importance of fission in energy production lies in two facts: (1) An exceedingly large amount of energy is released in the fission reaction. (2) The production of excess neutrons permits a chain reaction. These two circumstances make it possible to design nuclear reactors in which self-sustaining reactions occur with the continuous release of energy.

Nuclear fission is not the only energy-releasing nuclear reaction. The *fusion* of light nuclides, like hydrogen, into heavier elements is also an energy-producing process. Nuclear fusion reactions, still in the research phase, are described later in this Handbook section.

The heat generated in nuclear power plants is transferred to a working fluid, and from this point on the nuclear power plant and the conventional fossil-fueled power plant are essentially similar.

Fission Reaction. In nuclear fission, the nucleus of a heavy atom is split into two or more fragments. The reaction is initiated by the absorption of a neutron. A typical reaction is

$$^{235}_{92}U + {}^{1}_{0}n \rightarrow {}^{137}_{56}Ba + {}^{97}_{36}Kr + 2{}^{1}_{0}n + \Delta E$$

In this reaction, a ^{235}U atom absorbs one neutron, becomes unstable, and subsequently fissions into two fission fragments plus two neutrons. This is just one of the many ways in which ^{235}U might fission. The number of neutrons produced in a fission reaction is usually two or three. The excess neutrons produced by the fission reaction provide the means of self-sustaining the chain reaction. Nuclides including ^{233}U, ^{235}U, and ^{239}Pu, which are fissionable by neutrons of all energies, are termed *fissile* nuclides.

Natural uranium consists of 99.3 percent ^{238}U and 0.7 percent ^{235}U. Under uncontrolled conditions, emitted neutrons from the fissioned ^{235}U nucleus emerge at very high speed. One property of ^{238}U, which makes up the bulk of natural uranium, is that it absorbs these fast neutrons to such an extent that those produced by one fission reaction are absorbed before finding a ^{235}U nucleus in which to produce another fission.

Thermal Reactors. The ^{235}U nucleus can be fissioned by neutrons of low speed and kinetic energy. Also, the absorption properties of ^{238}U are greatly reduced when bombarded by slower neutrons. Thus, by bombarding natural uranium with slow neutrons (the speed with which molecules move when in thermal equilibrium),^{235}U can undergo fission, and, consequently, high-purity ^{235}U is not required to sustain a chain reaction.

Reactors in which most of the fission reactions are induced by *thermal neutrons* are known as *thermal reactors*. Means must be used, of course, to slow down the neutrons.

Moderators.   The most important slowing-down mechanism is elastic scattering on elements of low mass number. Materials like light and heavy water, beryllium oxide, and graphite are used to slow down, or *thermalize*, the neutrons to an energy of about 0.025 eV. As neutrons collide with the nuclei of these atoms, their kinetic energy and speed are gradually reduced until thermal equilibrium is achieved with the reactor structure. The fewer such collisions before deceleration is complete, the less chance of ^{238}U atoms absorbing neutrons.

Critical Mass.   Thermal neutrons, which move like atoms in a low-pressure gas, diffuse throughout the reactor. They may be absorbed by a nucleus of the reactor structure, in which case they merely make that nucleus radioactive. Or they may strike a fissionable atom of ^{235}U, causing fission and, in turn, releasing more neutrons to maintain the reaction. Should the number of neutrons absorbed by the moderator and ^{238}U be greater than about 1.5 excess neutrons emitted from each fission, the chain reaction will not be maintained. Therefore, the reactor core must be designed so that the mass of fuel will be just sufficient to ensure one neutron from each fission causing fission in another atom. A mass and configuration of fissionable material in which this occurs is termed the *critical mass*—or a reactor in which this condition is achieved is said to have "gone critical."

To measure a chain reaction, a multiplication factor k is used to indicate the ratio of neutrons in one generation to those in the preceding generation. Thus, in a constant chain reaction where the total number of neutrons neither increases nor decreases, the heat output is constant, and $k = 1$. Should k rise above unity, the rate of fission, and hence rate of heat productivity, steadily rises. This is so even if k is held constant at its new value. Here lies one *major difference between nuclear reactors and conventional steam generators*. In the latter, heat output is proportional to firing rate. If firing rate is increased, steam output is increased, but it remains constant at its new level. In a nuclear reactor, an increase in k results in continuously rising heat output. Only by returning the rate of neutron production to its original ratio can heat output be maintained at its new level.

Reactivity Control.   Absorption of excess neutrons, above those needed to maintain a constant reactivity level, provides close control over the degree of reactivity. This is accomplished by inserting materials having a high neutron-capture rate into the core. Control rods of special alloy metals are moved into and out of the cores as required. To start the reactor from shutdown, control rods are partially withdrawn until k becomes greater than one. Neutron flux and heat output grow until the desired level is reached. At this point, control-rod movement is quickly reversed to keep k at unity. The reactor is shut down by inserting the rods to their full extent. In this position, the rods absorb more than 1.5 excess neutrons per fission, and the chain reaction quickly stops. Heat production continues for a time, but is usually dissipated by an auxiliary cooling system. See Figs. 1 and 2.

Nuclear Fuels.   There are two broad categories of nuclear fuels: (1) the fissile nuclides previously mentioned, and (2) the fissionable nuclides, ^{232}Th and ^{238}U. Thermal reactors use fissile nuclides as fuel, while *fast reactors* are designed to burn fissionable materials. In fast reactors, only a small portion of the ^{232}Th and ^{238}U are fissioned directly. A larger portion of these materials is converted into ^{238}U and ^{239}Pu, respectively, through neutron absorption. Thus, this type of reactor not only consumes fuel, but also produces (breeds) new fuel material. Hence, the term *breeder reactor* is used for reactors designed to take advantage of this phenomenon. Breeding is possible in thermal reactors also, but to a lesser extent. The fuel material in a fast reactor must contain a significant amount (about 10 percent) of one of the fissile materials. The remainder of the fuel must have a high mass number in order to avoid slowing down the neutrons. The natural reserves of fissionable materials are more than 100 times greater than the reserves of fissile materials. Consequently, from the viewpoint of utilization of available energy resources, fast reactors are of great importance. Breeder reactors are described later in this Handbook section.

Types of Reactors.   Most of the power reactors currently in service or under construction are thermal reactors. The many types of thermal reactors depend largely on the selection of the coolant and moderator. A majority of plants in the United States use water as coolant and moderator material. Great Britain played a pioneering role in the development of gas-cooled (CO_2) and graphite-moderated reactors. Canada specialized in reactors

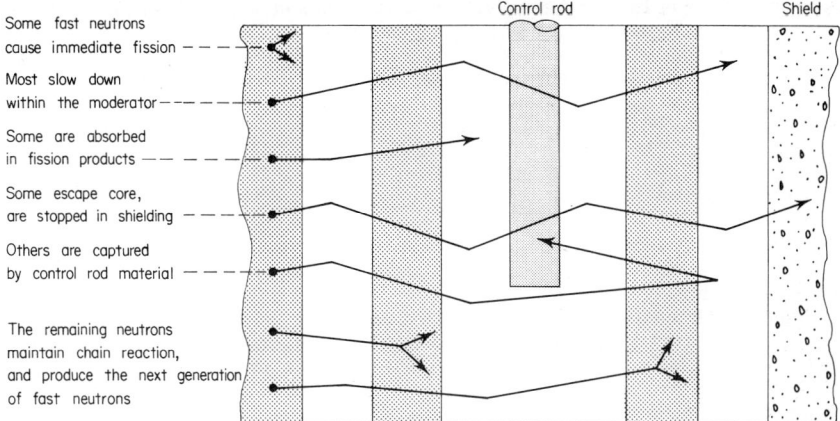

Some fast neutrons
cause immediate fission — — —

Most slow down
within the moderator — — — —

Some are absorbed
in fission products — — — —

Some escape core,
are stopped in shielding — — — —

Others are captured
by control rod material — — — —

The remaining neutrons
maintain chain reaction,
and produce the next generation
of fast neutrons

Control rod

Shield

Fig. 1. Neutron dispersal. Not all neutrons ejected from a ^{235}U nucleus cause further fission. Many are absorbed by nonfissionable nuclei in the fuel, or by core structural material, coolant, and moderator. Some neutrons are absorbed by fission products. In time, the number of neutrons captured becomes so great because of accumulation of these products that fission chain can no longer be sustained. Fuel is then said to be poisoned and must be reprocessed. The amount of practical heat released from a given fuel charge is below the theoretical figure, which is based upon complete fissioning of the uranium.

cooled and moderated by *heavy water*. In nuclear power terminology, ordinary water, in contrast to heavy water, is termed *light water*.

Light-water reactors are of two principal designs: (1) *pressurized water reactors* (PWR) and (2) *boiling-water reactors* (BWR). In a PWR, heat generated in the nuclear core is removed by water (reactor coolant) circulating at high pressure through the primary circuit. The water in the primary circuit both cools and moderates the reactor. Heat is transferred from the primary to the secondary system in a heat exchanger, or boiler, thereby generating steam in the secondary system. The BWR differs from the PWR primarily in that boiling takes place in the reactor itself. Comparable steam temperatures are possible at pressures of about 1000 psi, as contrasted with 2,000 psi for pressurized reactors.

Gas-cooled reactors must cope with the relatively poor heat-transfer characteristics of gases. The power required for gas circulation also is significant. The ideal gas-cooled power reactor operates at high temperatures to obtain reasonable thermodynamic efficiences. High gas-temperature rise must be obtained within the reactor core by reducing

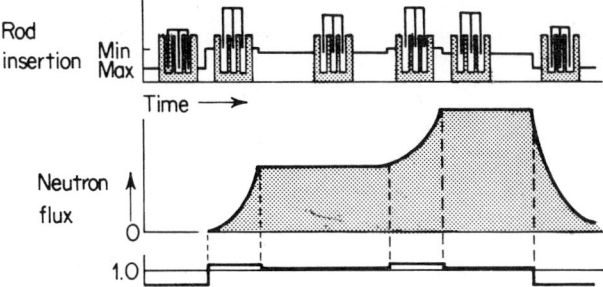

Rod
insertion   Min
             Max

Time ⟶

Neutron
flux   0

          1.0

Fig. 2. Regulation of neutron density. At reactor startup, rods of neutron-absorbing material are fully inserted in the core. To initiate chain reaction, these control rods are slowly withdrawn. Position indicators show degree of withdrawal. At some point, the percentage of neutrons absorbed in rods falls sufficiently for the chain reaction to be self-sustaining. The relationship between control-rod position and various power levels is shown schematically.

gas flow rate, but high fuel temperatures lead to difficulties with fuel materials and cladding. Advantages of this type of reactor include the use of relatively familiar fuels and materials. Steam pressures and temperatures are low, but gas coolant permits reasonably high reactor temperatures without pressurization. Carbon dioxide is particularly attractive because of its low cost, safety, and ease of handling.

Desire to achieve higher natural uranium fuel burnup than is possible with a graphite moderated reactor led to the use of heavy water as a moderating medium. Heavy water (D_2O), in which the normal hydrogen atom of water is replaced by the heavy hydrogen isotope, deuterium, has the desirable handling and thermal characteristics of light water, with the added advantage of low neutron absorption. The first nuclear reactor to operate outside the United States was a heavy water reactor at Canada's Chalk River experimental plant. Another Canadian D_2O cooled and moderated unit was the world's first reactor capable of refueling at full power.

Fuel and Material Handling Costs.   Experience with fossil-fueled power plants has shown a direct relationship between the total amount of power generated and the total amount of raw fuel required. Not so with nuclear reactors, particularly breeder reactors. Over 2 million tons of coal are consumed by a 1000-megawatt coal-fired power plant. A state-of-the-art thermal reactor, using uranium dioxide as fuel, of the same power output would require only about 35 tons of nuclear fuel. In the fossil-fueled plant, coal receipts are continuous. Further, a coal-fired plant requires a large reserve pile of coal in case of work stoppage at the mines or of transportation stoppages and emergencies. The nuclear plant requires only one delivery of fuel per year. The coal-fired plant generates thousands of tons of ashes to be hauled away, whereas nuclear wastes can be removed as spent fuel once per year. A nuclear plant requires no combustion air and, consequently, can be located underground if desired. By contrast, the coal-combustion plant of the stated capacity will generate around 10 million tons of carbon dioxide per year, accompanied by several hundred thousand tons of sulfur dioxide and nitrogen oxides, plus ash particles. The relatively low volume of nuclear fuel required permits a nuclear facility to obtain its fuel sources economically from great distances, whereas transportation costs loom high where coal must come from long distances. The foregoing comparisons with coal apply in large measure to the other fossil fuels, natural or synthetic gas, and liquid petroleum products.

It is interesting to note, too, that nuclear power essentially is independent of fuel costs, whereas fuel is a significant cost of production in fossil-fueled plants. Granted that uranium costs will rise as the demand grows, the fuel-cost factor of a nuclear plant will remain significantly less than for a fossil-fueled plant, assuming also that fossil fuels will rise in price. See Tables 1 and 2.

The principal types of nuclear fission reactors are described in further detail in subsequent portions of this Handbook section.

Probably because of the early association of nuclear fission and nuclear fusion processes with military weapons which dramatically demonstrated the tremendous explosive and destructive power of instantaneous energy release, and because of the earlier bad effects of radioactive fallout from fission products released during a nuclear explosion, the general public has developed fears of nuclear technology, which have carried over to the carefully controlled nuclear reactions which proceed within a commercial nuclear power plant. Consequently, even minor accidents that occur in a nuclear facility, although not related to the nuclear portion of the plant, often draw unduly intensive coverage by the public media.

Nuclear power reactors installed, on order, or planned in the United States as of September 30, 1974, are listed in Table 3. Site locations are shown in the map of Fig. 3. A list of reactors in actual operation in the United States as of February 1974, showing reactor types, is given in Table 4. An alphabetical list of nuclear plant names, showing power and reactor type for reactors ordered, announced, or planned, appears in Table 5.

A list of reactors operating, ordered, announced, or planned for countries other than the United States is given in Table 6. A listing and ranking of the various nations with nuclear power plants in operation as of September 30, 1974, appears in Table 7. A similar listing of nuclear power plants on order as of the same date is presented in Table 8. Further nuclear power plants in the planning stage by various nations as of the same date appear in Table 9. In Table 10, all categories—power plants in operation, on order, announced, or planned for the various nations are listed. It is interesting to note that approximately half this total

TABLE 1.   Projected Costs for a 940,000-kW Station—Nuclear versus Coal-fired Facility

Item	Nuclear plant	Coal-fired plant
Unit investment cost of plant, $/kW	255	202
Capital charge, rate/yr	0.13	0.13
KWh generated/yr/kW capacity	5,256	5,256
Heat rate, million Btu/kWh	0.0104	0.009
Cost of heat from fuel, $/million Btu	0.18	0.45
Cost of electricity, mills/kWh:		
Investment in plant	6.31	5.00
Operation and maintenance	0.38	0.30
Fuel	1.87	4.05
Total	8.56	9.35

It should be noted that the analysis does not make allowance for projected cost of sulfur removal from fuel or of sulfur dioxide from stack gas. Also, there is some unbalance in the figures because a coal-fired facility, in actual operation, would generate somewhat less electricity per kilowatt capacity than the nuclear plant.

capacity is attributed to the United States, with a significant percentage (13.4 percent to Japan), followed by The Federal Republic of Germany (West) with 6.8 percent. The top ten nations (United States, Japan, West Germany, United Kingdom, Canada, the U.S.S.R., France, Sweden, Spain, and Switzerland) account for nearly 90 percent of the total nuclear power scheduled or in operation.

Of the nuclear power plants on order for use in the United States, it is interesting to note that:

1. Pressure water reactors (PWR) account for approximately 65 percent of planned capacity—with nearly 100 installations.

2. Boiling water reactors (BWR) account for approximately 30 percent of planned capacity—with 40-plus installations.

3. High-temperature gas reactors (HTGR) account for nearly all the remaining 5 percent of planned capacity—with seven installations scheduled.

4. One liquid metal fast breeder reactor (LMFBR) is scheduled as of September 30, 1974.

It should be emphasized, of course, that as progress is made in the development of other reactor configurations and as improvements are made in the "conventional" approaches, some of the less firm situations may be altered.

Of a total of about 75 future installations announced and in the planning stage (less firm than orders), about 34 percent are pointing toward the PWR; less than 7 percent toward the BWR; two are HTGRs; and a preference for about 55 percent of the installations has not yet been formally stated.

A comparison of thermal reactor characteristics is presented in Table 11. Subsequent articles in this Handbook section describe specific reactor types.

TABLE 2.   Effect of Increased Costs of Uranium on Operation of Water Reactors versus Fast Breeders

Assumed cost of uranium, $/lb	Electricity generated, million kW-years		Cost increase of electricity, mills/kWh	
	Water reactor	Fast breeder	Water reactor	Fast breeder
10	5,500	720,000	0.1	0.0
15	8,480	1,120,000	0.4	0.0
30	13,100	1,720,000	1.3	0.0
50	58,300	7,700,000	2.5	0.0
100	146,000	19,200,000	5.5	0.0

NOTE: In the original presentation of this information (*Technology Review*, Alumni Association of the M.I.T., October–November 1971, uranium at a price of $8.00/lb was the starting base—with 0.0 cost increase of electricity in the column for water reactor.

TABLE 3. Nuclear Power Reactors Installed, on Order, or Planned in the United States

As of September 30, 1974

Site	Plant name	Capacity, net kilowatts	Utility	Commercial operation
Alabama				
Decatur	Browns Ferry Nuclear Power Plant: Unit 1	1,065,000	Tennessee Valley Authority	1974
Decatur	Browns Ferry Nuclear Power Plant: Unit 2	1,065,000	Tennessee Valley Authority	1974
Decatur	Browns Ferry Nuclear Power Plant: Unit 3	1,065,000	Tennessee Valley Authority	1975
Dothan	Joseph M. Farley Nuclear Plant: Unit 1	829,000	Alabama Power Co.	1976
Dothan	Joseph M. Farley Nuclear Plant: Unit 2	829,000	Alabama Power Co.	1977
Clanton	Alan R. Barton Nuclear Plant: Unit 1	1,159,000	Alabama Power Co.	1984
Clanton	Alan R. Barton Nuclear Plant: Unit 2	1,159,000	Alabama Power Co.	1985
Clanton	Alan R. Barton Nuclear Plant: Unit 3	1,159,000	Alabama Power Co.	1986
Clanton	Alan R. Barton Nuclear Plant: Unit 4	1,159,000	Alabama Power Co.	1987
Scottsboro	Bellefonte Nuclear Plant: Unit 1	1,213,000	Tennessee Valley Authority	1979
Scottsboro	Bellefonte Nuclear Plant: Unit 2	1,213,000	Tennessee Valley Authority	1980
Arizona				
Wintersburg	Palo Verde Nuclear Generating Station: Unit 1	1,270,000	Arizona Public Service	1981
Wintersburg	Palo Verde Nuclear Generating Station: Unit 2	1,270,000	Arizona Public Service	1982
Wintersburg	Palo Verde Nuclear Generating Station: Unit 3	1,270,000	Arizona Public Service	1984
Arkansas				
Russellville	Arkansas Nuclear One: Unit 1	850,000	Arkansas Power & Light Co.	1974
Russellville	Arkansas Nuclear One: Unit 2	912,000	Arkansas Power & Light Co.	1976
California				
Eureka	Humboldt Bay Power Plant: Unit 3	65,000	Pacific Gas and Electric Co.	1963
San Clemente	San Onofre Nuclear Generating Station: Unit 1	430,000	So. Calif. Ed. & San Diego Gas & El. Co.	1968
San Clemente	San Onofre Nuclear Generating Station: Unit 2	1,140,000	So. Calif. Ed. & San Diego Gas & El. Co.	1979
San Clemente	San Onofre Nuclear Generating Station: Unit 3	1,140,000	So. Calif. Ed. & San Diego Gas & El. Co.	1980
Diablo Canyon	Diablo Canyon Nuclear Power Plant: Unit 1	1,084,000	Pacific Gas and Electric Co.	1975
Diablo Canyon	Diablo Canyon Nuclear Power Plant: Unit 2	1,106,000	Pacific Gas and Electric Co.	1976
Clay Station	Rancho Seco Nuclear Generating Station	913,000	Sacramento Municipal Utility District	1975
*	—	1,128,000	Pacific Gas & Electric Co.	1981
*	—	1,128,000	Pacific Gas & Electric Co.	1982
Colorado				
Platteville	Ft. St. Vrain Nuclear Generating Station	330,000	Public Service Co. of Colorado	1976
Connecticut				
Haddam Neck	Haddam Neck Plant	575,000	Conn. Yankee Atomic Power Co.	1968
Waterford	Millstone Nuclear Power Station: Unit 1	652,100	Northeast Nuclear Energy Co.	1971

Location	Plant	Company	Capacity	Year
Waterford	Millstone Nuclear Power Station: Unit 2	Northeast Nuclear Energy Co.	828,000	1974
Waterford	Millstone Nuclear Power Station: Unit 3	Northeast Nuclear Energy Co.	1,156,000	1979
Delaware				
Summit	Summit Power Station: Unit 1	Delmarva Power & Light Co.	770,000	1981
Summit	Summit Power Station: Unit 2	Delmarva Power & Light Co.	770,000	1984
Florida				
Florida City	Turkey Point Station: Unit 3	Florida Power & Light Co.	693,000	1972
Florida City	Turkey Point Station: Unit 4	Florida Power & Light Co.	693,000	1973
Red Level	Crystal River Plant: Unit 3	Florida Power Corp.	825,000	1974
Ft. Pierce	St. Lucie Plant: Unit 1	Florida Power & Light Co.	810,000	1975
Ft. Pierce	St Lucie Plant: Unit 2	Florida Power & Light Co.	810,000	1979
*	—	Florida Power Corp.	1,300,000	—
	—	Florida Power Corp.	1,300,000	—
Georgia				
Baxley	Edwin I. Hatch Nuclear Plant: Unit 1	Georgia Power Co.	786,000	1974
Baxley	Edwin I. Hatch Nuclear Plant: Unit 2	Georgia Power Co.	795,000	1978
Waynesboro	Alvin W. Vogtle, Jr. Plant: Unit 1	Georgia Power Co.	1,113,000	—
Waynesboro	Alvin W. Vogtle, Jr. Plant: Unit 2	Georgia Power Co.	1,113,000	—
Illinois				
Morris	Dresden Nuclear Power Station: Unit 1	Commonwealth Edison Co.	200,000	1960
Morris	Dresden Nuclear Power Station: Unit 2	Commonwealth Edison Co.	809,000	1970
Morris	Dresden Nuclear Power Station: Unit 3	Commonwealth Edison Co.	809,000	1971
Zion	Zion Nuclear Plant: Unit 1	Commonwealth Edison Co.	1,050,000	1973
Zion	Zion Nuclear Plant: Unit 2	Commonwealth Edison Co.	1,050,000	1974
Cordova	Quad-Cities Station: Unit 1	Comm. Ed. Co.-Ia.-Ill. Gas & Elec. Co.	800,000	1972
Cordova	Quad-Cities Station: Unit 2	Comm. Ed. Co.-Ia.-Ill. Gas & Elec. Co.	800,000	1972
Seneca	LaSalle County Nuclear Station: Unit 1	Comm. Ed. Co.-Ia.	1,078,000	1978
Seneca	LaSalle County Nuclear Station: Unit 2	Comm. Ed. Co.-Ia.	1,078,000	1979
Byron	Byron Station: Unit 1	Comm. Edison Co.	1,120,000	1980
Byron	Byron Station: Unit 2	Comm. Edison Co.	1,120,000	1981
Braidwood	Braidwood: Unit 1	Comm. Edison Co.	1,120,000	1980
Braidwood	Braidwood: Unit 2	Comm. Edison Co.	1,120,000	1981
Clinton	Clinton Nuclear Power Plant: Unit 1	Illinois Power Co.	955,000	1980
Clinton	Clinton Nuclear Power Plant: Unit 2	Illinois Power Co.	955,000	1982
Indiana				
Westchester	Bailly Generating Station	Northern Indiana Public Service Co.	660,000	1979
Madison	Marble Hill Nuclear Power Station: Unit 1	Public Service Indiana	1,150,000	1983
Madison	Marble Hill Nuclear Power Station: Unit 2	Public Service Indiana	1,150,000	1984
Iowa				
Palo	Duane Arnold Energy Center: Unit 1	Iowa Electric Light and Power Co.	569,000	1974
*	—	Iowa Electric Light and Power Co.	1,000,000	1983

Site	Plant name	Capacity, net kilowatts	Utility	Commercial operation
Kansas				
Burlington	Wolf Creek Generation Station: Unit 1	1,150,000	Kansas Gas & Electric–Kansas City P & L	1982
Louisiana				
Taft	Waterford Generating Station: Unit 3	1,113,000	Louisiana Power & Light Co.	1980
St. Francisville	River Bend Station: Unit 1	934,000	Gulf States Utilities Co.	1980
St. Francisville	River Bend Station: Unit 2	934,000	Gulf States Utilities Co.	1981
Alliance	St. Rosalie Generating Station: Unit 1	1,160,000	Louisiana Power & Light Co.	1982
Alliance	St. Rosalie Generating Station: Unit 2	1,160,000	Louisiana Power & Light Co.	1984
Maine				
Wiscasset	Maine Yankee Atomic Power Plant	790,000	Maine Yankee Atomic Power Co.	1972
Maryland				
Lusby	Calvert Cliffs Nuclear Power Plant: Unit 1	845,000	Baltimore Gas and Electric Co.	1974
Lusby	Calvert Cliffs Nuclear Power Plant: Unit 2	845,000	Baltimore Gas and Electric Co.	1975
Douglas Point	Douglas Point Project Nuclear Gen. Station: Unit 1	1,178,000	Potomac Electric Power Co.	1982
Douglas Point	Douglas Point Project Nuclear Gen. Station: Unit 2	1,178,000	Potomac Electric Power Co.	1984
Massachusetts				
Rowe	Yankee Nuclear Power Station	175,000	Yankee Atomic Electric Co.	1961
Plymouth	Pilgrim Station: Unit 1	664,000	Boston Edison Co.	1972
Plymouth	Pilgrim Station: Unit 2	1,180,000	Boston Edison Co.	1980
Montague	—	1,150,000	Northeast Utilities	1981
Montague	—	1,150,000	Northeast Utilities	1983
Michigan				
Big Rock Point	Big Rock Point Nuclear Plant	75,000	Consumers Power Co.	1965
South Haven	Palisades Nuclear Power Station	700,000	Consumers Power Co.	1971
Lagoona Beach	Enrico Fermi Atomic Power Plant: Unit 2	1,093,000	Detroit Edison Co.	1976
Lagoona Beach	Enrico Fermi Atomic Power Plant: Unit 3	1,171,000	Detroit Edison Co.	1981
Bridgman	Donald C. Cook Plant: Unit 1	1,060,000	Indiana & Michigan Electric Co.	1974
Bridgman	Donald C. Cook Plant: Unit 2	1,060,000	Indiana & Michigan Electric Co.	1976
Midland	Midland Nuclear Power Plant: Unit 1	492,000	Consumers Power Co.	1980
Midland	Midland Nuclear Power Plant: Unit 2	818,000	Consumers Power Co.	1979
St. Clair County	Greenwood: Unit 2	1,200,000	Detroit Edison Co.	1982
St. Clair County	Greenwood: Unit 3	1,200,000	Detroit Edison Co.	1983

Location	Plant	Utility	Capacity	Year
Minnesota				
Monticello	Monticello Nuclear Generating Plant	Northern States Power Co.	545,000	1971
Red Wing	Prairie Island Nuclear Generating Plant: Unit 1	Northern States Power Co.	530,000	1973
Red Wing	Prairie Island Nuclear Generating Plant: Unit 2	Northern States Power Co.	530,000	1974
Missouri				
Fulton	Callaway Plant: Unit 1	Union Electric Co.	1,150,000	1981
Fulton	Callaway Plant: Unit 2	Union Electric Co.	1,150,000	1983
Mississippi				
Port Gibson	Grand Gulf Nuclear Station: Unit 1	Mississippi Power & Light Co.	1,250,000	1979
Port Gibson	Grand Gulf Nuclear Station: Unit 2	Mississippi Power & Light Co.	1,250,000	1981
Nebraska				
Fort Calhoun	Ft. Calhoun Station: Unit 1	Omaha Public Power District	457,400	1973
Fort Calhoun	Ft. Calhoun Station: Unit 2	Omaha Public Power District	1,150,000	1983
Brownville	Cooper Nuclear Station	Nebraska Public Power District and Iowa Power and Light Co.	778,000	1974
New Hampshire				
Seabrook	Seabrook Nuclear Station: Unit 1	Public Service of N.H.	1,200,000	1979
Seabrook	Seabrook Nuclear Staton: Unit 2	Public Service of N.H.	1,200,000	1981
New Jersey				
Toms River	Oyster Creek Nuclear Plant: Unit 1	Jersey Central Power & Light Co.	640,000	1969
Forked River	Forked River Generating Station: Unit 1	Jersey Central Power & Light Co.	1,070,000	1980
Salem	Salem Nuclear Generating Station: Unit 1	Public Service Electric and Gas, N.J.	1,090,000	1976
Salem	Salem Nuclear Generating Station: Unit 2	Public Service Electric and Gas, N.J.	1,115,000	1979
Salem	Hope Creek Generating Station: Unit 1	Public Service Electric and Gas, N.J.	1,067,000	1981
	Hope Creek Generating Station: Unit 2	Public Service Electric and Gas, N.J.	1,150,000	1983
Little Egg Inlet	Atlantic Generating Station: Unit 1	Public Service Electric and Gas, N.J.	1,150,000	1985
Little Egg Inlet	Atlantic Generating Station: Unit 2	Public Service Electric and Gas, N.J.	1,150,000	1987
*	—	Public Service Electric and Gas, N.J.		1990
*		Public Service Electric and Gas, N.J.		1992
New York				
Indian Point	Indian Point Station: Unit 1	Consolidated Edison Co.	265,000	1962
Indian Point	Indian Point Station: Unit 2	Consolidated Edison Co.	873,000	1973
Indian Point	Indian Point Station: Unit 3	Consolidated Edison Co.	965,000	1974
Scriba	Nine Mile Point Nuclear Station: Unit 1	Niagara Mohawk Power Co.	625,000	1969
Scriba	Nine Mile Point Nuclear Station: Unit 2	Niagara Mohawk Power Co.	1,080,000	1978
Ontario	R.E.Ginna Nuclear Power Plant: Unit 1	Rochester Gas & Electric Co.	490,000	1970
Brookhaven	Shoreham Nuclear Power Station	Long Island Lighting Co.	819,000	1977
Scriba	James A. Fitzpatrick Nuclear Power Plant	Power Authority of State of N.Y.	821,000	1973
*	Green County Nuclear Power Plant	Power Authority of State of N.Y.	1,160,000	1982
Jamesport	Jamesport 1	Long Island Lighting Co.	1,150,000	1981
Jamesport	Jamesport 2	Long Island Lighting Co.	1,150,000	1983
Oswego	Sterling Nuclear: Unit 1	Rochester Gas & Electric Co.	1,150,000	1982

TABLE 3. Nuclear Power Reactors Installed, on Order, or Planned in the United States (Continued)
As of September 30, 1974

Site	Plant name	Capacity, net kilowatts	Utility	Commercial operation
North Carolina				
Southport	Brunswick Steam Electric Plant: Unit 1	821,000	Carolina Power and Light Co.	1975
Southport	Brunswick Steam Electric Plant: Unit 2	821,000	Carolina Power and Light Co.	1974
Cowans Ford Dam	Wm. B. McGuire Nuclear Station: Unit 1	1,180,000	Duke Power Co.	1977
Cowans Ford Dam	Wm. B. McGuire Nuclear Station: Unit 2	1,180,000	Duke Power Co.	1977
Bonsal	Shearon Harris Plant: Unit 1	900,000	Carolina Power & Light Co.	1981
Bonsal	Shearon Harris Plant: Unit 2	900,000	Carolina Power & Light Co.	1982
Bonsal	Shearon Harris Plant: Unit 3	900,000	Carolina Power & Light Co.	1983
Bonsal	Shearon Harris Plant: Unit 4	900,000	Carolina Power & Light Co.	1984
Davie County	Perkins Nuclear Station: Unit 1	1,280,000	Duke Power Co.	1981
Davie County	Perkins Nuclear Station: Unit 2	1,280,000	Duke Power Co.	1982
Davie County	Perkins Nuclear Station: Unit 3	1,280,000	Duke Power Co.	1984
*	—	1,150,000	Carolina Power & Light Co.	1980
*	—	1,150,000	Carolina Power & Light Co.	1980
*	—	1,150,000	Carolina Power & Light Co.	1980
Ohio				
Oak Harbor	Davis-Besse Nuclear Power Station: Unit 1	906,000	Toledo Edison-Cleveland El. Illum. Co.	1976
Oak Harbor	Davis-Besse Nuclear Power Station: Unit 2	906,000	Toledo Edison-Cleveland El. Illum. Co.	1982
Oak Harbor	Davis-Besse Nuclear Power Station: Unit 3	906,000	Toledo Edison-Cleveland El. Illum. Co.	1984
Perry	Perry Nuclear Power Plant: Unit 1	1,205,000	Cleveland Electric Illuminating Co.	1979
Perry	Perry Nuclear Power Plant: Unit 2	1,205,000	Cleveland Electric Illuminating Co.	1980
Moscow	Wm. H. Zimmer Nuclear Power Station: Unit 1	810,000	Cincinnati Gas & Electric Co.	1979
Moscow	Wm. H. Zimmer Nuclear Power Station: Unit 2	1,170,000	Cincinnati Gas & Electric Co.	1979
Oklahoma				
Inola	Black Fox Nuclear Station: Unit 1	950,000	Public Service of Oklahoma	1982
Inola	Black Fox Nuclear Station: Unit 2	950,000	Public Service of Oklahoma	1984
Oregon				
Prescott	Trojan Nuclear Plant: Unit 1	1,130,000	Portland General Electric Co.	1975
*		1,260,000	Portland General Electric Co.	1981
*		1,260,000	Portland General Electric Co.	1983
Pennsylvania				
Peach Bottom	Peach Bottom Atomic Power Station: Unit 1	40,000	Philadelphia Electric Co.	1967†

Location	Station	Capacity	Company	Year
Peach Bottom	Peach Bottom Atomic Power Station: Unit 2	1,065,000	Philadelphia Electric Co.	1974
Peach Bottom	Peach Bottom Atomic Power Station: Unit 3	1,065,000	Philadelphia Electric Co.	1974
Pottstown	Limerick Generating Station: Unit 1	1,065,000	Philadelphia Electric Co.	1981
Pottstown	Limerick Generating Station: Unit 2	1,065,000	Philadelphia Electric Co:	1982
Shippingport	Shippingport Atomic Power Station: Unit 1	90,000	Duquesne Light Co.	1957
Shippingport	Beaver Valley Power Station: Unit 1	852,000	Duquesne Light Co.-Ohio Edison Co.	1975
Shippingport	Beaver Valley Power Station: Unit 2	852,000	Duquesne Light Co.-Ohio Edison Co.	1981
Goldsboro	Three Mile Island Nuclear Station: Unit 1	819,000	Metropolitan Edison Co.	1974
Goldsboro	Three Mile Island Nuclear Station: Unit 2	905,000	Jersey Central Power & Light Co.	1978
Berwick	Susquehanna Steam Electric Station: Unit 1	1,050,000	Pennsylvania Power and Light	1979
Berwick	Susquehanna Steam Electric Station: Unit 2	1,050,000	Pennsylvania Power and Light	1981
Fuller	Fulton Generating Station: Unit 1	1,140,000	Philadelphia Electric Co.	1984
Fuller	Fulton Generating Station: Unit 2	1,140,000	Philadelphia Electric Co.	1986
South Carolina				
Hartsville	H. B. Robinson S.E. Plant: Unit 2	700,000	Carolina Power & Light Co.	1971
Seneca	Oconee Nuclear Station: Unit 1	886,000	Duke Power Co.	1973
Seneca	Oconee Nuclear Station: Unit 2	886,000	Duke Power Co.	1973
Seneca	Oconee Nuclear Station: Unit 3	886,000	Duke Power Co.	1974
Broad River	Virgil C. Summer Nuclear Station: Unit 1	900,000	South Carolina Electric & Gas Co.	1978
Lake Wylie	Catawba Nuclear Station: Unit 1	1,153,000	Duke Power Co.	1979
Lake Wylie	Catawba Nuclear Station: Unit 2	1,153,000	Duke Power Co.	1980
Cherokee County	Cherokee Nuclear Station: Unit 1	1,280,000	Duke Power Co.	1982
Cherokee County	Cherokee Nuclear Station: Unit 2	1,280,000	Duke Power Co.	1983
Cherokee County	Cherokee Nuclear Station: Unit 3	1,280,000	Duke Power Co.	1984
Tennessee				
Daisy	Sequoyah Nuclear Power Plant: Unit 1	1,140,000	Tennessee Valley Authority	1976
Daisy	Sequoyah Nuclear Power Plant: Unit 2	1,140,000	Tennessee Valley Authority	1977
Spring City	Watts Bar Nuclear Plant: Unit 1	1,169,000	Tennessee Valley Authority	1978
Spring City	Watts Bar Nuclear Plant: Unit 2	1,169,000	Tennessee Valley Authority	1979
Oak Ridge	Clinch River Breeder Reactor Plant	350,000	U.S. Government	1980
Hartsville	—	1,205,000	Tennessee Valley Authority	1980
Hartsville	—	1,205,000	Tennessee Valley Authority	1981
Hartsville	—	1,205,000	Tennessee Valley Authority	1981
Hartsville	—	1,205,000	Tennessee Valley Authority	1982
Texas				
Glen Rose	Commanche Peak Steam Electric Station: Unit 1	1,150,000	Texas Utilities Services Inc.	1980

TABLE 3. Nuclear Power Reactors Installed, on Order, or Planned in the United States (Continued)
As of September 30, 1974

Site	Plant name	Capacity, net kilowatts	Utility	Commercial operation
Glen Rose	Commanche Peak Steam Electric Station: Unit 2	1,150,000	Texas Utilities Services Inc.	1982
Jasper	Blue Hills: Unit 1	918,000	Gulf States Utilities	1982
Jasper	Blue Hills: Unit 2	918,000	Gulf States Utilities	1982
Wallis	Allens Creek: Unit 1	1,150,000	Houston Lighting & Power Co.	1980
Wallis	Allens Creek: Unit 2	1,150,000	Houston Lighting & Power Co.	1982
Matagorda County	South Texas Project	1,250,000	Central Power & Lt.-Houston Lt. & Power	1980
Matagorda County	South Texas Project	1,250,000	Central Power & Lt.-Houston Lt. & Power	1982
Vermont				
Vernon	Vermont Yankee Generating Station	513,900	Vermont Yankee Nuclear Power Corp.	1972
Virginia				
Gravel Neck	Surry Power Station: Unit 1	788,000	Virginia Electric & Power Company	1972
Gravel Neck	Surry Power Station: Unit 2	788,000	Virginia Electric & Power Company	1973
Mineral	North Anna Power Station: Unit 1	898,000	Virginia Electric & Power Company	1975
Mineral	North Anna Power Station: Unit 2	898,000	Virginia Electric & Power Company	1976
Mineral	North Anna Power Station: Unit 3	907,000	Virginia Electric & Power Company	1977
Mineral	North Anna Power Station: Unit 4	907,000	Virginia Electric & Power Company	1978
Gravel Neck	Surry Power Station: Unit 3	859,000	Virginia Electric & Power Company	1980
Gravel Neck	Surry Power Station: Unit 4	859,000	Virginia Electric & Power Company	1981
Washington				
Richland	N-Reactor/WPPSS Steam	850,000	Atomic Energy Commission	1966
Richland	WPPSS No. 1	1,206,000	Washington Public Power Supply System	1980
Richland	WPPSS No. 2	1,103,000	Washington Public Power Supply System	1977
Satsop	WPPSS No. 3	1,242,000	Washington Public Power Supply System	1981
Richland	WPPSS No. 4	1,250,000	Washington Public Power Supply System	1982
Satsop	WPPSS No. 5	1,240,000	Washington Public Power Supply System	1983
Sedro Woolley	Skagit Nuclear Project: Unit 1	1,277,000	Puget Sound Power & Light	1982
Sedro Woolley	Skagit Nuclear Project: Unit 2	1,277,000	Puget Sound Power & Light	1983
Wisconsin				
Genoa	Genoa Nuclear Generating Station	50,000	Dairyland Power Cooperative	1971
Two Creeks	Point Beach Nuclear Plant: Unit 1	497,000	Wisconsin Michigan Power Co.	1970
Two Creeks	Point Beach Nuclear Plant: Unit 2	497,000	Wisconsin Michigan Power Co.	1972
Carlton	Kewaunee Nuclear Power Plant: Unit 1	541,000	Wisconsin Michigan Power Co.	1974

Ft. Atkinson	Koshkonong Nuclear Plant: Unit 1	900,000	Wisconsin Electric Power Co.	1981
Ft Atkinson	Koshkonong Nuclear Plant: Unit 2	900,000	Wisconsin Electric Power Co.	1982
Durand	Tyrone Energy Park: Unit 1	1,150,000	Northern States Power Co.	1985
Puerto Rico				
Arecibo	Norco Nuclear Power Station	583,000	Puerto Rico Water Resources Authority	1981
*	—	1,300,000	Tennessee Valley Authority	1982
*	—	1,300,000	Tennessee Valley Authority	1983
*	—	1,300,000	Tennessee Valley Authority	1983
*	—	1,300,000	Tennessee Valley Authority	1984
*	—	1,200,000	New England Electric System	1982
*	—	1,200,000	New England Electric System	1982

*Site not selected.
†Shut down in 1974 after seven years of successful operation.

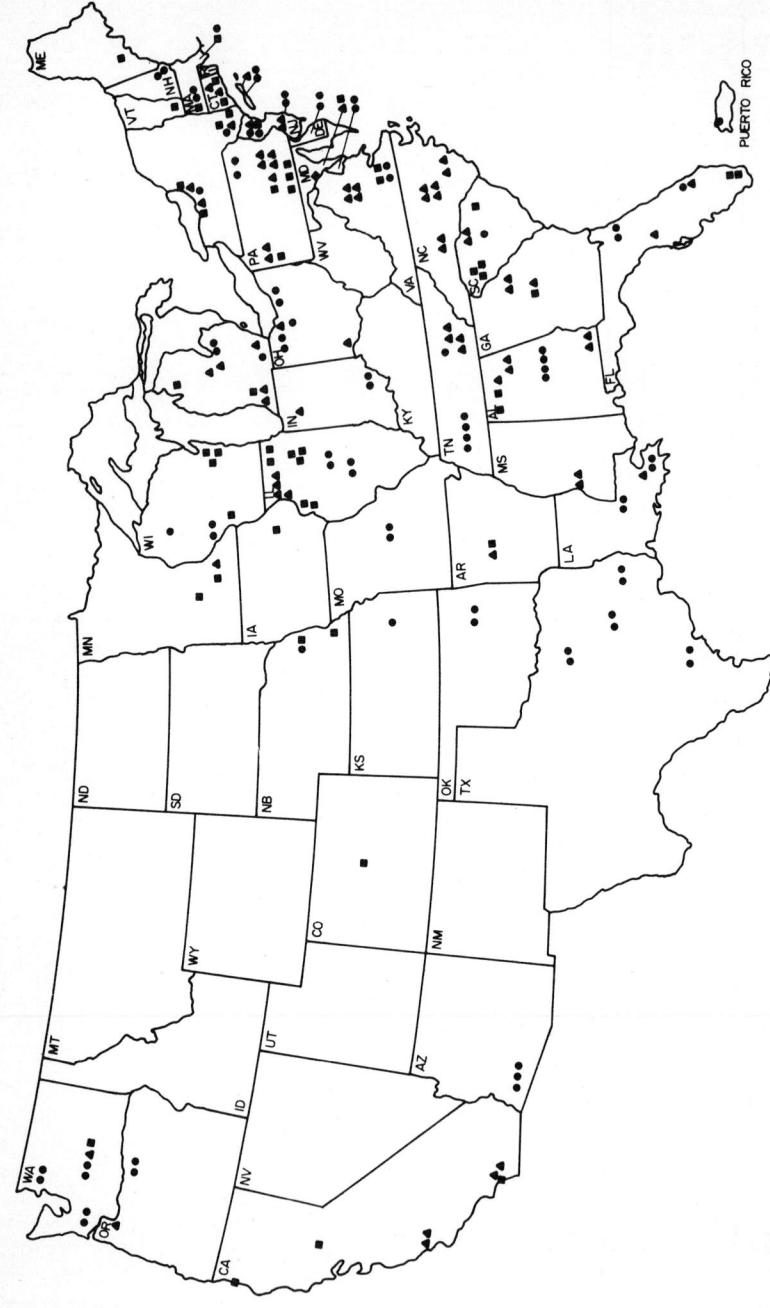

Fig. 3. Nuclear power reactors in the United States as of September 30, 1974. Operable: 33,618,000 kilowatts; under construction: 68,522,000 kilowatts; planning stage; 134,372,000 kilowatts; total: 236,512,000 kilowatts. Because of space limitations, symbols do not reflect precise locations. ■ Operable; ▲ under construction; ● planning stage. (*U.S. Energy Research and Development Administration.*)

TABLE 4.   Central Station Nuclear Power Reactors in Operation in the United States
February 1974

Plant name	Utility	Reactor type	Net power, MW(e)	Date of commercial operation Month	Date of commercial operation Year
Shippingport	Duquesne Light	PWR	90	12	1957
Dresden #1	Commonwealth Edison	BWR	200	8	1960
Yankee	Yankee Atomic Electric Company	PWR	175	2	1961
Indian Point #	Consolidated Edison	PWR	265	10	1962
Humboldt Bay #3	Pacific Gas & Electric	BWR	69	8	1963
Big Rock Point	Consumers Power	BWR	70	11	1965
N-Reactor	WPPSS	GR	850	7	1966
Peach Bottom #1	Philadelphia Electric	HTGR	40	6	1967
San Onofre #1	Southern California Edison	PWR	430	1	1968
Haddam Neck	Connecticut Yankee	PWR	575	1	1968
LaCrosse	Dairyland Cooperative	BWR	53	8	1969
Nine Mile Point #1	Niagara Mohawk	BWR	625	12	1969
Oyster Creek	Jersey Central Power & Light	BWR	650	12	1969
Ginna	Rochester Gas & Electric	PWR	470	7	1970
Dresden #2	Commonwealth Edison	BWR	809	8	1970
Point Beach #1	Wisconsin Electric Power	PWR	497	12	1970
Millstone #1	Millstone Point Company	BWR	652	3	1971
Robinson #2	Carolina Power & Light	PWR	730	3	1971
Monticello	Northern States Power	BWR	545	7	1971
Dresden #3	Commonwealth Edison	BWR	809	11	1971
Vermont Yankee	Vermont Yankee	BWR	514	11	1972
Palisades #1	Consumers Power	PWR	821	12	1972
Maine Yankee	Maine Yankee Atomic Electric	PWR	860	12	1972
Pilgrim #1	Boston Edison	BWR	664	12	1972
Point Beach #2	Wisconsin Electric Power	PWR	497	12	1972
Quad-Cities #1	Com. Ed./Ia.-Ill. G&E	BWR	800	12	1972
Quad-Cities #2	Com. Ed./Ia.-Ill. G&E	BWR	800	12	1972
Surry #1	Virginia Electric Power Co.	PWR	819	12	1972
Turkey Point #3	Florida Power & Light	PRW	725	12	1972
Surry #2	Virginia Electric Power Co.	PWR	819	5	1973
Turkey Point #4	Florida Power & Light	PWR	725	7	1973
Oconee #1	Duke Power	PWR	886	10	1973
Indian Point #2	Consolidated Edison	PWR	873	11	1973
Browns Ferry #1	TVA	BWR	1117	12	1973
Ft. Calhoun #1	Omaha Public	PWR	457	12	1973
Oconee #2	Duke Power	PWR	886	12	1973
Peach Bottom #2	Philadelphia Electric	BWR	1065	12	1973
Prairie Island #1	Northern States Power	PWR	550	12	1973
Zion #1	Commonwealth Edison	PWR	1080	12	1973
Arnold #1	Iowa Electric Light & Power	BWR	569	1	1974
Cooper	Nebraska Pub. Power District	BWR	778	1	1974
Calvert Cliffs #1	Baltimore Gas & Electric	PWR	875	2	1974

SOURCE: U.S. Energy Research and Development Administration.

NUCLEAR POWER OUTLOOK FOR THE UNITED STATES

The forecast of energy consumption and electric generating capacity in the United States to the year 2000 is given in Table 12. Four situations are reflected in this table, cases A through D.

Case A.   The lowest of the forecasts in terms of nuclear capacity. The assumption is made that delays in bringing nuclear plants on line will continue to plague the industry. Sources of delay include late equipment deliveries, construction delays, strikes, poor labor productivity, excessive and unrealistic environmental concerns, and regulatory problems and red tape. It is not assumed that any particular source of delay is predominant or that any particular source is corrected, but rather that some of these sources of

TABLE 5. Central Station Nuclear Power Reactors Ordered, Announced, or Planned in the United States
February 1974

Plant name	Utility	Reactor type	Net power, MW(e)	Status	Scheduled for commercial operation Month	Year
Allens Creek #1	Houston Lighting & Power	BWR	1150	Ord.		1980
Allens Creek #2	Houston Lighting & Power	BWR	1150	Ord.		1982
Arkansas Nuclear #1	Arkansas Power & Light	PWR	850	Ord.	3	1974
Arkansas Nuclear #2	Arkansas Power & Light	PWR	926	Ord.	10	1976
Atlantic #1	Public Service Elec. & Gas	PWR	1200	Ord.	5	1980
Atlantic #2	Public Service Elec. & Gas	PWR	1200	Ord.	1	1981
Atlantic #3	Public Service Elec. & Gas	PWR	1200	Ord.	1	1982
Atlantic #4	Public Service Elec. & Gas	PWR	1200	Ord.	1	1984
Bailly	No. Indiana Public Service	BWR	660	Ord.	3	1978
Beaver Valley #1	Duquesne Light	PWR	885	Ord.	5	1975
Beaver Valley #2	Duquesne Light	PWR	885	Ord.	6	1979
Bellefonte #1	TVA	PWR	1279	Ord.	9	1979
Bellefonte #2	TVA	PWR	1279	Ord.	6	1980
Blue Hills #1	Gulf States Utilities	PWR	918	Ord.	10	1981
Blue Hills #2	Gulf States Utilities	PWR	918	Ann.		1980
Boardman #1	Portland General Electric	PWR	1260	Ord.		1980
Boardman #2	Portland General Electric		1260	Plan.		1982
Braidwood #1	Commonwealth Edison	PWR	1120	Ord.	10	1980
Braidwood #2	Commonwealth Edison	PWR	1120	Ord.	3	1982
Browns Ferry #2	TVA	BWR	1117	Ord.	7	1974
Browns Ferry #3	TVA	BWR	1117	Ord.	12	1974
Brunswick #1	Carolina Power & Light	BWR	855	Ord.	12	1975
Brunswick #2	Carolina Power & Light	BWR	855	Ord.	12	1974
Byron #1	Commonwealth Edison	PWR	1120	Ord.	5	1979
Byron #2	Commonwealth Edison	PWR	1120	Ord.	3	1980
Callaway #1	Union Electric	PWR	1100	Ann.		1981
Callaway #2	Union Electric	PWR	1100	Ann.		1983
Calvert Cliffs #2	Baltimore Gas & Electric	PWR	875	Ord.	2	1975
Catawba #1	Duke Power	PWR	1180	Ord.	3	1979
Catawba #2	Duke Power	PWR	1180	Ord.	3	1980
Central Alabama #1	Alabama Power	BWR	1100	Ord.		1981
Central Alabama #2	Alabama Power	BWR	1100	Ord.		1982
Charlestown #1	Narragansett Electric		950	Ann.		1980
Charlestown #2	Narragansett Electric		950	Ann.		1982
Cherokee #1	Duke Power	PWR	1200	Ord.		1984
Cherokee #2	Duke Power	PWR	1200	Ord.		1985
Cherokee #3	Duke Power	PWR	1200	Ord.		1986
Clinch River	U.S. Government	LMFBR	400	Ord.		1980
Clinton #1	Illinois Power	BWR	950	Ord.		1980
Clinton #2	Illinois Power	BWR	950	Ord.		1982
Comanche Peak #1	Texas Power & Light	PWR	1160	Ord.	1	1980
Comanche Peak #2	Texas Power & Light	PWR	1150	Ord.	1	1982
Commonwealth Edison #1	Commonwealth Edison		1100	Plan.	3	1982
Commonwealth Edison #2	Commonwealth Edison		1100	Plan.	10	1982
Cook #1	Indiana & Michigan Elec.	PWR	1096	Ord.	10	1974
Cook #2	Indiana & Michigan Elec.	PWR	1096	Ord.	1	1976
Crystal River #3	Florida Power	PWR	825	Ord.	12	1974
Davis-Besse #1	Toledo Edison	PWR	906	Ord.	5	1975
Davis-Besse #2	Toledo Edison		906	Ann.	6	1981
Davis-Besse #3	Toledo Edison		906	Ann.	1	1983
Diablo Canyon #1	Pacific Gas & Electric	PWR	1131	Ord.	3	1975
Diablo Canyon #2	Pacific Gas & Electric	PWR	1156	Ord.	3	1976
Douglas Point #1	Potomac Electric Power	BWR	1237	Ord.	3	1980
Douglas Point #2	Potomac Electric Power	BWR	1237	Ord.		1981
Erie #1	Ohio Edison		1200	Ann.	1	1982
Erie #2	Ohio Edison		1200	Ann.	1	1983
Farley #1	Alabama Power	PWR	866	Ord.	12	1975
Farley #2	Alabama Power	PWR	866	Ord.	1	1977
Fermi #2	Detroit Edison	BWR	1171	Ord.	8	1976
Fermi #3	Detroit Edison	BWR	1125	Ord.	8	1981
FitzPatrick	Power Auth. St. of NY	BWR	853	Ord.	6	1974
Forked River	Jersey Central Po. & L.	PWR	1120	Ord.	11	1978
Ft. Calhoun #2	Omaha Public Power Dist.		900	Plan.		1981
Ft. St. Vrain	Colorado Public Service	HTGR	330	Ord.	4	1976
Fulton #1	Philadelphia Electric	HTGR	1140	Ord.	5	1981
Fulton #2	Philadelphia Electric	HTGR	1140	Ord.	5	1983
Grand Gulf #1	Mississippi Power & Light	BWR	1290	Ord.	6	1979

Plant name	Utility	Reactor type	Net power, MW(e)	Status	Scheduled for commercial operation Month	Year
Grand Gulf #2	Mississippi Power & Light	BWR	1290	Ord.	6	1981
Greenwood #2	Detroit Edison	PWR	1240	Ord.	8	1980
Greenwood #3	Detroit Edison	PWR	1240	Ord.		1981
Harris #1	Carolina Power & Light	PWR	915	Ord.	3	1978
Harris #2	Carolina Power & Light	PWR	915	Ord.		1979
Harris #3	Carolina Power & Light	PWR	915	Ord.		1980
Harris #4	Carolina Power & Light	PWR	915	Ord.		1981
Hartsville #1	TVA	BWR	1290	Ord.	12	1980
Hartsville #2	TVA	BWR	1290	Ord.	12	1981
Hartsville #3	TVA	BWR	1290	Ord.	6	1981
Hartsville #4	TVA	BWR	1290	Ord.	6	1982
Hatch #1	Georgia Power	BWR	822	Ord.	4	1974
Hatch #2	Georgia Power	BWR	825	Ord.	4	1978
Hope Creek #1	Public Service Elec. & Gas	BWR	1067	Ord.	5	1981
Hope Creek #2	Public Service Elec. & Gas	BWR	1067	Ord.	5	1982
Houston #1	Houston Light & Power		1150	Plan.	2	1981
Indian Point #3	Consolidated Edison	PWR	965	Ord.	12	1974
Inola	Public Service of Oklahoma	PWR	1100	Ord.		1982
Islote	Puerto Rico WRA	PWR	600	Ord.		1979
Jacksonville #1	Jacksonville Elec. Auth.	PWR	1110	Ord.	4	1982
Jacksonville #2	Jacksonville Elec. Auth.	PWR	1110	Ord.	4	1984
Jamesport #1	Long Island Lighting	PWR	1150	Ord.	6	1981
Jamesport #2	Long Island Lighting	PWR	1150	Ord.	6	1984
Kewaunee	Wisconsin	PWR	541	Ord.	6	1974
LaSalle County #1	Commonwealth Edison	BWR	1078	Ord.	10	1978
LaSalle County #2	Commonwealth Edison	BWR	1078	Ord.	10	1979
Limerick #1	Philadelphia Electric	BWR	1100	Ord.	4	1979
Limerick #2	Philadelphia Electric	BWR	1100	Ord.	6	1980
Lower Lehigh #1	Pennsylvania Power & Light		1100	Ann.		1983
Lower Lehigh #2	Pennsylvania Power & Light		1100	Ann.		1985
McGuire #1	Duke Power	PWR	1180	Ord.	3	1976
McGuire #2	Duke Power	PWR	1180	Ord.	3	1977
Mendocino #1	Pacific Gas & Electric	BWR	1128	Ord.	7	1983
Mendocino #2	Pacific Gas & Electric	BWR	1128	Ord.	7	1984
Metropolitan X-1	Metropolitan Edison		1150	Plan.	6	1983
Midland #1	Consumers Power	PWR	492	Ord.	3	1980
Midland #2	Consumers Power	PWR	818	Ord.	3	1979
Middle South #1	Middle South Utilities (La.)	PWR	1150	Plan.	6	1982
Middle South #2	Middle South Utilities (La.)	PWR	1150	Plan.	6	1984
Millstone #2	Millstone Point	PWR	828	Ord.	12	1974
Millstone #3	Northeast Utilities	PWR	1150	Ord.		1979
Montague #1	Northeast Utilities		1100	Ann.		1981
Montague #2	Northeast Utilities		1100	Ann.		1985
Nine Mile Point #2	Niagara Mohawk Power	BWR	1080	Ord.	11	1978
North Anna #1	Virginia Electric Power	PWR	934	Ord.	4	1975
North Anna #2	Virginia Electric Power	PWR	934	Ord.	4	1976
North Anna #3	Virginia Electric Power	PWR	938	Ord.	4	1977
North Anna #4	Virginia Electric Power	PWR	938	Ord.	4	1978
Northwest Illinois #1	Commonwealth Edison		1100	Plan.	3	1981
Northwest Illinois #2	Commonwealth Edison		1100	Plan.	10	1981
Oconee #3	Duke Power	PWR	886	Ord.	6	1974
Palo Verde #1	Arizona Public Service		1270	Ann.		1981
Palo Verde #2	Arizona Public Service		1270	Ann.		1983
Palo Verde #3	Arizona Public Service		1270	Ann.		1985
Peach Bottom #3	Philadelphia Electric	BWR	1065	Ord.	12	1973
Perkins #1	Duke Power	PWR	1200	Ord.		1981
Perkins #2	Duke Power	PWR	1200	Ord.		1982
Perkins #3	Duke Power	PWR	1200	Ord.		1983
Perry #1	Cleveland Elec. Illuminating	BWR	1205	Ord.		1979
Perry #2	Cleveland Elec. Illuminating	BWR	1205	Ord.		1980
Pilgrim #2	Boston Edison	PWR	1180	Ord.	8	1980
Pilgrim #3	Boston Edison	PWR	1180	Plan.	5	1982
Prairie Island #2	Northern States Power	PWR	555	Ord.	10	1974
Quanicassee #1	Consumers Power	PWR	1150	Ord.		1981
Quanicassee #2	Consumers Power	PWR	1150	Ord.		1982
Rancho Seco #1	Sacramento Mun. Util. Dist.	PWR	913	Ord.	6	1974
Rancho Seco #2	Sacramento Mun. Util. Dist.		1100	Plan.		1980

TABLE 5. Central Station Nuclear Power Reactors Ordered, Announced, or Planned in the United States (*Continued*)

February 1974

Plant name	Utility	Reactor type	Net power, MW(e)	Status	Scheduled for commercial operation Month	Year
River Bend #1	Gulf States Utilities	BWR	934	Ord.	2	1980
River Bend #2	Gulf States Utilities	BWR	934	Plan.	10	1982
Salem #1	Public Service Elec. & Gas	PWR	1131	Ord.	3	1975
Salem #2	Public Service Elec. & Gas	PWR	1156	Ord.	3	1976
San Joaquin #1	Los Angeles Dept. W&P		1100	Plan.	12	1981
San Joaquin #2	Los Angeles Dept. W&P		1100	Plan.	12	1982
San Joaquin #3	Los Angeles Dept. W&P		1100	Plan.	12	1983
San Joaquin #4	Los Angeles Dept. W&P		1100	Plan.	12	1984
San Onofre #2	Southern California Edison	PWR	1140	Ord.	9	1980
San Onofre #3	Southern California Edison	PWR	1140	Ord.	12	1981
Seabrook #1	Pub. Serv. of New Hampshire	PWR	1260	Ord.	11	1979
Seabrook #2	Pub. Serv. of New Hampshire	PWR	1260	Ord.		1981
Sequoyah #1	TVA	PWR	1177	Ord.	12	1975
Sequoyah #2	TVA	PWR	1177	Ord.	8	1976
Shoreham #1	Long Island Lighting	BWR	854	Ord.	7	1977
Skagit	Puget Sound Power & Light		1100	Ann.		1982
Somerset #1	N.Y. State Electric & Gas	BWR	1220	Ann.		1982
Somerset #2	N.Y. State Electric & Gas	BWR	1220	Ann.		1984
South Texas #1	Houston Lighting & Power	PWR	1250	Ord.		1980
South Texas #2	Houston Lighting & Power	PWR	1250	Ord.		1982
So. California Ed. HTGR #1	Southern California Edison	HTGR	1160	Plan.		1984
So. California Ed. HTGR #2	Southern California Edison	HTGR	1160	Plan.		1985
St. Lucie #1	Florida Power & Light	PWR	833	Ord.	6	1975
St. Lucie #2	Florida Power & Light	PWR	833	Ord.	12	1979
Sterling	Rochester Gas & Electric	PWR	1100	Ann.	9	1980
Summer #1	South Carolina Electric & Gas	PWR	915	Ord.	1	1978
Summer #2	South Carolina Electric & Gas		915	Ann.		1981
Summit #1	Delmarva Power & Light	HTGR	770	Ord.	5	1980
Summit #2	Delmarva Power & Light	HTGR	770	Ord.	6	1982
Surry #3	Virginia Electric Power	PWR	950	Ord.		1980
Surry #4	Virginia Electric Power	PWR	950	Ord.		1981
Susquehanna #1	Pennsylvania Power & Light	BWR	1120	Ord.	11	1979
Susquehanna #2	Pennsylvania Power & Light	BWR	1120	Ord.	5	1981
Three Mile Island #1	Metropolitan Edison	PWR	819	Ord.	8	1974
Three Mile Island #2	Metropolitan Edison	PWR	905	Ord.	9	1976
Trojan	Portland General Electric	PWR	1130	Ord.	7	1975
Tyrone #1	Northern States Power	PWR	1100	Ann.		1982
Tyrone #2	Northern States Power	PWR	1100	Ann.		1983
Vidal #1	Southern California Edison	HTGR	770	Ord.	6	1982
Vidal #2	Southern California Edison	HTGR	770	Ord.	6	1983
Vogtle #1	Georgia Power	PWR	1121	Ord.	4	1980
Vogtle #2	Georgia Power	PWR	1121	Ord.	4	1981
Vogtle #3	Georgia Power	PWR	1121	Ord.		1982
Vogtle #4	Georgia Power	PWR	1121	Ord.		1983
Waterford #3	Louisiana Power & Light	PWR	1165	Ord.	10	1977
Watts Bar #1	TVA	PWR	1219	Ord.	3	1978
Watts Bar #2	TVA	PWR	1219	Ord.	12	1978
Wisconsin #1	Wisconsin Electric Power	PWR	900	Ord.		1980
Wisconsin #2	Wisconsin Electric Power	PWR	900	Ord.		1982
Wisconsin #3	Wisconsin Electric Power	PWR	900	Ann.		1983
Wisconsin #4	Wisconsin Electric Power	PWR	900	Ann.		1985
Wisconsin #5	Wisconsin Electric Power	PWR	900	Ann.		1986
Wisconsin #6	Wisconsin Electric Power	PWR	900	Ann.		1988
Wolf Creek	Kansas Gas & Electric	PWR	1100	Ann.	4	1981
WPPSS #1	Wash. Pub. Pow. Sup. Sys.	PWR	1200	Ord.		1980
WPPSS #2	Wash. Pub. Pow. Sup. Sys.	BWR	1103	Ord.	9	1977
WPPSS #3	Wash. Pub. Pow. Sup. Sys.	PWR	1300	Ord.		1981
Zimmer #1	Cincinnati Gas & Electric	BWR	810	Ord.	8	1977
Zimmer #2	Cincinnati Gas & Electric	BWR	1170	Ord.		1982
Zion #2	Commonwealth Edison	PWR	1080	Ord.	5	1974

SOURCE: U.S. Energy Research and Development Administration.

TABLE 6. Central Station Nuclear Power Reactors Operating, Ordered, Announced or Planned for Countries Other than the United States

February 1974

Country	Plant name	Utility	Reactor type	Net power, MW(e)	Status	Scheduled for commercial operation	
						Month	Year
Argentina	Atucha #1	CNEA	HWR	318	Ord.	12	1973
	Atucha #2	CNEA		800	Plan.	12	1985
	Cordoba	EPEC	HWR	600	Ord.	12	1978
	Argentina #5			800	Plan.	12	1984
	Madrilena	SEGBA		600	Plan.	12	1983
	Bahia Blanca			600	Plan.	12	1980
Australia	Australia #1	Electric Comm. NSW		800	Plan.	12	1981
Austria	Tullnerfeld #1	Gemeinschaft Kernkraftwerke	BWR	700	Ord.	12	1976
	Austria #2		LWR	1200	Ann.	12	1980
Bangladesh	Roopur		PWR	200	Ann.	12	1979
Belgium	Doel #1	EBES	PWR	390	Oper.	12	1973
	Tihange	EDF/SEMO	PWR	870	Ord.		1975
	Doel #2	EBES	PWR	390	Ord.	5	1975
	Belgium #4			800	Plan.	12	1981
	Belgium #5			800	Plan.	12	1982
Brazil	Angra Dos Reis #1	FURNAS	PWR	626	Ord.	3	1977
	Angra Dos Reis #2	FURNAS		800	Ann.	12	1980
	Angra Dos Reis #3	FURNAS		1000	Plan.	12	1982
	Angra Dos Reis #4	FURNAS		600	Plan.	12	1984
Bulgaria	Koslokj #1		PWR	405	Ord.	7	1974
	Koslodj #2		PWR	405	Ord.	12	1975
	Koslodj #3		PWR	405	Ann.	12	1976
	Kodlodj #4		PWR	405	Ann.	12	1977
	Kodlodj #5		PWR	405	Ann.	12	1978
Canada	Douglas Point	Ontario Hydro	HWR	200	Oper.	1	1967
	Gentilly	Hydro Quebec	HWR	250	Oper.	12	1971
	Pickering #1	Ontario Hydro	HWR	508	Oper.	7	1971
	Pickering #2	Ontario Hydro	HWR	508	Oper.	12	1971
	Pickering #3	Ontario Hydro	HWR	508	Oper.	10	1972
	Pickering #4	Ontario Hydro	HWR	512	Oper.	6	1973
	Bruce #1	Ontario Hydro	HWR	750	Ord.	6	1977

TABLE 6. Central Station Nuclear Power Reactors Operating, Ordered, Announced or Planned for Countries Other than the United States (*Continued*)

February 1974

Country	Plant name	Utility	Reactor type	Net power, MW(e)	Status	Scheduled for commercial operation	
						Month	Year
Canada (continued)	Bruce #2	Ontario Hydro	HWR	750	Ord.	12	1975
	Bruce #3	Ontario Hydro	HWR	750	Ord.	12	1978
	Bruce #4	Ontario Hydro	HWR	750	Ord.	12	1979
	Bruce #5	Ontario Hydro	HWR	750	Plan.	12	1981
	Bruce #6	Ontario Hydro	HWR	750	Plan.	12	1982
	Bruce #7	Ontario Hydro	HWR	750	Plan.	12	1983
	Bruce #8	Ontario Hydro	HWR	750	Plan.	12	1983
	Gentilly #2	Hydro Quebec	HWR	500	Ord.	12	1979
	Gentilly #3	Hydro Quebec	HWR	1000	Ord.	12	1978
	Hydro 1000-1	Hydro Quebec	HWR	1000	Plan.	7	1978
	No Name #1	Hydro Quebec	HWR	500	Plan.	10	1978
	No Name #2	Hydro Quebec	HWR	500	Plan.	1	1979
	Hydro 1000-2	Hydro Quebec	HWR	1000	Plan.	10	1979
	Hydro 1000-3	Hydro Quebec	HWR	1000	Plan.	3	1980
	Bowmanville #1	Ontario Hydro	HWR	750	Plan.	12	1982
	Bowmanville #2	Ontario Hydro	HWR	750	Plan.	12	1983
	Bowmanville #3	Ontario Hydro	HWR	750	Plan.	12	1984
	Bowmanville #4	Ontario Hydro	HWR	750	Plan.	12	1985
Chile	Chile #1	ENDESA		400	Ann.	12	1979
	Chile #2	ENDESA		600	Ann.	12	1984
Czechoslovakia	Bohunice 1A		GCR	110	Oper.	12	1972
	Bohunice 2A		PWR	413	Ord.	12	1979
	Bohunice 2B		PWR	440	Ord.	12	1979
	Bohunice V1-1		PWR	761	Ord.	12	1977
	Bohunice V1-2		PWR	761	Ord.	12	1980
Finland	Loviisa #1		PWR	420	Ord.	6	1976
	Loviisa #2		PWR	420	Ord.	4	1978
	Finnish #3		BWR	660	Ann.	12	1978
	Vuosaari			600	Plan.	12	1982
	Meisaari			720	Plan.	12	1992
France	Chinon #2	EDF	GCR	200	Oper.	2	1965
	Chinon #3	EDF	GCR	480	Oper.	8	1966

Country	Plant	Utility	Type	MW	Status		Year
	St. Laurent #1	EDF	GCR	480	Oper.	6	1969
	St. Laurent #2	EDF	GCR	515	Oper.	8	1971
	SENA (Chooz)	EDF/EBES	PWR	266	Oper.	4	1967
	Bugey #1	EDF	GCR	540	Oper.	4	1972
	Phenix	CEA	FBR	250	Oper.	12	1973
	Bugey #2	EDF	PWR	930	Ord.	12	1976
	Bugey #3	EDF	PWR	930	Ord.	12	1977
	Fessenheim #1	EDF	PWR	890	Ord.	11	1975
	Fessenheim #2	EDF	PWR	890	Ord.	11	1976
	FDR-Demo #1	EDF/RWE	FBR	1000	Ord.	3	1981
	St. Laurent #3	EDF	BWR	995	Plan.	11	1979
	Dampierre	EDF	BWR	995	Plan.		1981
	Caux-1	EDF		750	Ord.	12	1982
	Caux-2	EDF		750	Plan.		1983
	Caux-3	EDF		750	Plan.		1984
	Caux-4	EDF		750	Plan.		1985
	Languedoc #1	EDF		1000	Ann.		1980
	Languedoc #2	EDF		1000	Plan.		1981
East Germany	Rheinsberg #1		HWR	70	Oper.	12	1966
	Nord 1 #1		PWR	365	Oper.	12	1973
	Nord 1 #2		PWR	365	Ord.	12	1975
	Nord 2 #1		PWR	365	Ord.		1977
	Nord 2 #2		PWR	365	Ord.		1978
	East Germany #1			1000	Plan.		1980
West Germany	Kahl VAK	RWE	BWR	15	Oper.	12	1962
	KRB Gundremmingen	KRB/RWE	BWR	237	Oper.	2	1967
	KWL Lingen	VEW	BWR	160	Oper.	4	1968
	KWO Obrigheim	KWO	PWR	328	Oper.	9	1968
	KKS Stade	NWK	PWR	630	Oper.	12	1972
	KKW Wuergassen	Pruessische EW	BWR	640	Oper.	3	1972
	KKN Niederachbach	GFN (Karlsruhe)	HWGCR	100	Oper.	3	1973
	KKI Isar	Isar Amperwerke	BWR	870	Ord.	4	1977
	Biblis #1—Unit A	RWE	PWR	1150	Ord.	12	1974
	Biblis #2—Unit B	RWE	PWR	1178	Ord.	6	1976
	KKB Brunsbuettel	NWK/HEW	BWR	770	Ord.	10	1975
	BASF #1	BASF	PWR	660	Ord.	2	1978
	BASF #2	BASF	PWR	660	Ord.	12	1979
	Unterweser	NWK/PEW	PWR	1230	Ord.	12	1979
	KKP Philipsburg #1	Nord Badenwerk/EVS	BWR	864	Ord.	1	1976
	KKP Philipsburg #2	Nord Badenwerk/EVS	BWR	864	Ord.	8	1978
	THTR	Consortium	HTGR	300	Ord.	12	1976

TABLE 6. Central Station Nuclear Power Reactors Operating, Ordered, Announced or Planned for Countries Other than the United States (Continued)

February 1974

Country	Plant name	Utility	Reactor type	Net power, MW(e)	Status	Scheduled for commercial operation Month	Year
West Germany (continued)	SNR Kalkar	PSB	FBR	300	Ord.	12	1978
	Leverkusen	Farbanfabriken		600	Ord.	12	1978
	GKN Neckarwestein	GKN	PWR	800	Ord.	1	1976
	Kaerlich (Koblenz)	RWE	PWR	1300	Ord.	12	1978
	FBR-Demo	EDF/RWE	FBR	1000	Ord.	12	1984
	Kruemmel (Geesthacht)	NWK/HEW	BWR	1260	Ord.	12	1977
	KKP Philipsburg #3	Nord Badenwerk EVS	BWR	864	Ord.	9	1979
	KKP Philipsburg #4	Nord Badenwerk EVS	BWR	864	Ord.	3	1980
	Biblis #3	RWE	PWR	1240	Plan.	6	1979
	KKB Brunsbuettel #2	NWK/HEW	BWR	770	Plan.		
	Breisach #1	EVS/Badenwerk	PWR	1240	Plan.	6	1979
	Breisach (KRB) #2	EVS/Badenwerk	PWR	1300	Plan.	12	1981
	Raum Grosswelzheim	RWE	LWR	1200	Ann.	12	1978
	Oberweser	Preussische EW	LWR	1300	Plan.	12	1980
	Grafenheinfeld	BWAG	LWR	1200	Plan.	12	1979
	PREAG	BWAG		860	Plan.	12	1978
	Lingen #2	VEW	HTGR	1100	Plan.	12	1978
	Oberrheim	KKW Sud/EVS	LWR	1300	Ord.	12	1979
	Emden	NWK	LWR	1300	Plan.		1981
	Buttel/St. Margarethen	NWK	LWR	1300	Plan.		1982
	Cuxhaven	NWK	LWR	1300	Plan.		1983
Greece	Greek #1	Public Power Corp.		600	Plan.	12	1982
	Greek #2	Public Power Corp.		600	Plan.	12	1983
	Greek #3	Public Power Corp.		750	Plan.	12	1984
Hong-Kong	Hong-Kong #1			500	Plan.	12	1979
	Hong-Kong #2			500	Plan.	12	1981
Hungary	Paks #1	MVMT	PWR	440	Plan.	6	1981
	Paks #2	MVMT	PWR	440	Plan.	6	1981
India	Tarapur #1	Ind. Dept. At. Energy	BWR	200	Oper.	4	1969
	Tarapur #2	Ind. Dept. At. Energy	BWR	200	Oper.	4	1969
	Rapp #1	Ind. Dept. At. Energy	HWR	200	Oper.	1	1973

Country	Unit	Owner/Operator	Type	MW	Status	Mo.	Yr.
	Rapp #2	Ind. Dept. At. Energy	HWR	200	Ord.	1	1975
	Kalpakkam #1	Ind. Dept. At. Energy	HWR	200	Ord.	3	1974
	Kalpakkam #2	Ind. Dept. At. Energy	HWR	200	Ord.	3	1977
	Narora #1	Ind. Dept. At. Energy	HWR	200	Ann.	1	1980
	Narora #2	Ind. Dept. At. Energy	HWR	200	Ann.	1	1981
	Indian #9	Ind. Dept. At. Energy		500	Plan.	1	1979
	Indian #10	Ind. Dept. At. Energy		500	Plan.	1	1980
	FBR Prototype		FBR	500	Plan.		1981
Iran	Iran #1			400	Plan.	6	1979
Ireland	Irish #1			500	Plan.	1	1980
	Irish #2			250	Plan.	12	1982
	Irish #3			250	Plan.	12	1980
Israel	Israel #1			400	Plan.	7	1980
Italy	SENN		BWR	150	Oper.	1	1964
	SELNI		PWR	247	Oper.	11	1964
	Latina		GCR	200	Oper.	1	1964
	Caorso		BWR	800	Ord.	4	1975
	Enel-5		PWR	1000	Ord.	6	1978
	Enel-6		LWR	800	Ord.	12	1978
	Enel-7		LWR	1000	Plan.	12	1980
	Sicily-1		LWR	700	Plan.	4	1982
Jamaica	Jamaica #1			200	Plan.	12	1980
	Jamaica #2			200	Plan.		1980
Japan	Tokai-Mura	JAPCO	GCR	157	Oper.	6	1967
	Tsuruga	JAPCO	BWR	340	Oper.	3	1970
	Fukushima #1	TEPCO	BWR	460	Oper.	3	1971
	Mihama #1	KEPCO	PWR	320	Oper.	11	1970
	Mihama #2	KEPCO	PWR	470	Oper.	7	1972
	Fukushima #2	TEPCO	BWR	784	Oper.	10	1973
	Shimane #1	CEPCO	BWR	439	Ord.	11	1973
	Mihama #3	KEPCO	PWR	781	Ord.	8	1976
	Fukushima #3	TEPCO	BWR	760	Ord.	12	1974
	Fukushima #4	TEPCO	BWR	760	Ord.	8	1976
	Fukushima #5	TEPCO	BWR	760	Ord.	12	1975
	Fukushima #6	TEPCO	BWR	1100	Ord.	3	1977
	Takahama #1	KEPCO	PWR	781	Ord.	7	1974
	Takahama #2	KEPCO	PWR	781	Ord.	5	1975
	Shimane #2	CEPCO	BWR	750	Ord.	5	1980
	Shimane #3	CEPCO	BWR	750	Ord.	5	1982
	Shimane #4	CEPCO	BWR	1000	Plan.	7	1983
	Hamaoka #1	Chubu EPCO	BWR	500	Ord.	11	1974

TABLE 6. Central Station Nuclear Power Reactors Operating, Ordered, Announced or Planned for Countries Other than the United States (Continued)

February 1974

Country	Plant name	Utility	Reactor type	Net power, MW(e)	Status	Scheduled for commercial operation	
						Month	Year
Japan (continued)	Hamaoka #2	Chubu EPCO	BWR	850	Ord.	6	1977
	Hamaoka #3	Chubu EPCO	BWR	750	Plan.		1980
	Genkai #1	Kyushu EPCO	PWR	529	Ord.	2	1975
	Genkai #2	Kyushu EPCO	PWR	529	Ord.	2	1978
	Genkai #3	Kyushu EPCO	LWR	826	Plan.	2	1980
	Genkai #4	Kyushu EPCO	LWR	826	Plan.	7	1982
	Genkai #5	Kyushu EPCO	LWR	826	Plan.	12	1983
	Onagawa #1	Tohoku EPCO	BWR	500	Ord.	3	1977
	Onagawa #2	Tohoku EPCO	BWR	750	Ord.	12	1979
	Onagawa #3	Tohoku EPCO	BWR	750	Ann.	12	1982
	Hamaoka #4	Chubu EPCO	BWR	1100	Plan.	1	1979
	Hamaoka #5	Chubu EPCO	HTG	1100	Plan.		1980
	Hamaoka #6	Chubu EPCO	BWR	1500	Ord.	12	1981
	Hamaoka #7	Chubu EPCO	BWR	1500	Ord.	12	1982
	Hamaoka #8	Chubu EPCO	BWR	1500	Ord.		1984
	Hokuriku #1 (Shiga)	Hokuriku EPCO	PWR	500	Ord.	12	1977
	Hokuriku #2	Hokuriku EPCO	PWR	800	Plan.	8	1979
	Hokuriku #3	Hokuriku EPCO	PWR	800	Plan.		1983
	Ohi #1	KEPCO	PWR	1122	Ord.	4	1977
	Ohi #2	KEPCO	PWR	1122	Ord.	8	1977
	Ikate #2	Shikoku EPCO	PWR	559	Ord.	4	1977
	Ikate #2	Shikoku EPCO	PWR	566	Ord.	7	1979
	N-1 (Tokyo #7)	TEPCO	PWR	1100	Ord.	1	1978
	N	Hokkaido EPCO	BWR	327	Ord.	11	1977
	N-3 (Tokyo #8)	TEPCO	BWR	1100	Ord.	3	1978
	ATR (Fugen)	Chubu EPCO	HTG	200	Ord.	3	1975
	N-2 (Tokyo #9)	TEPCO	BWR	1100	Plan.	1	1979
	N-4 (Tokyo #10)	TEPCO	BWR	1100	Plan.	7	1979
	N-5 (Tokyo #11)	TEPCO	BWR	1100	Plan.	10	1980
	N-6 (Tokyo #12)	TEPCO	BWR	1100	Plan.	7	1981
	N-8 (Tokyo #13)	TEPCO	BWR	1100	Plan.	1	1982
	N-9 (Tokyo #15)	TEPCO	BWR	1500	Plan.	6	1982

Country	Plant	Utility	Type	MWe	Status		Year
	N–10 (Tokyo #14)	TEPCO	BWR	1100	Plan.	8	1982
	N–11 (Tokyo #16)	TEPCO	BWR	1500	Plan.	7	1983
	N–12 (Tokyo #17)	TEPCO	BWR	1500	Plan.	8	1983
	N–13 (Tokyo #18)	TEPCO	BWR	1500	Plan.	7	1984
	N–14 (Tokyo #19)	TEPCO	BWR	1500	Plan.	8	1984
	Kansai-U	KEPCO	PWR	1200	Plan.	7	1977
	Kansai-V	KEPCO	PWR	1200	Plan.	1	1979
	X1 (Kansai #10)	KEPCO	PWR	1200	Plan.	7	1978
	X2 (Kansai #11)	KEPCO	PWR	1200	Plan.	1	1979
	X3 (Kansai #12)	KEPCO	PWR	1200	Plan.	7	1980
	X4 (Kansai #13)	KEPCO	PWR	1200	Plan.	1	1981
	X5 (Kansai #14)	KEPCO	PWR	1200	Plan.	7	1981
	X6 (Kansai #15)	KEPCO	PWR	1200	Plan.	1	1982
	X7 (Kansai #16)	KEPCO	PWR	1200	Plan.	7	1983
	X8 (Kansai #17)	KEPCO	PWR	1200	Plan.	1	1983
	Kansai #18	KEPCO	PWR	1500	Plan.	7	1984
	Kansai #19	KEPCO	PWR	1500	Plan.	1	1985
	Kansai #20	KEPCO	PWR	1500	Plan.	7	1985
	FBR Proto (Monju)		FBR	300	Plan.	6	1979
	Takai-Mura #2	JAPCO	BWR	1100	Ord.	8	1977
South Korea	Tonganae #1	KECO	PWR	564	Ord.	12	1975
	Tonganae #2	KECO	PWR	564	Plan.	12	1979
	Rok #3	KECO	LWR	600	Plan.	12	1979
Luxemburg	Lux #1		PWR	1100	Plan.		
Mexico	Laguna Verde #1	FURNAS	BWR	800	Ord.	12	1977
	Laguna Verde #2	FURNAS	LWR	800	Ord.	12	1978
	Sonora Project #1			1100	Plan.	12	1979
	Sonora Project #2			1100	Plan.	12	1980
Netherlands	Dodewaard		BWR	55	Oper.		1960
	Borselle	PZEM	PWR	450	Oper.	7	1973
	Dutch #2	GKN	LWR	600	Plan.	6	1978
	Dutch #3	GKN	LWR	1000	Plan.	6	1980
New Zealand	New Zealand #1			420	Plan.	12	1982
	New Zealand #2			420	Plan.	12	1984
Norway	Skiens Fjord	Norsk Hydro	LWR	600	Plan.		1979
	Oslo Fjord	NVE		800	Plan.		1981
Pakistan	Kanupp	Statakraftwerkene	HWR	125	Oper.	1	1969
	PAEC #1	PAEC		400	Plan.		
	PAEC #2	PAEC		500	Plan.		1979
Philippines	Philippine #1	Manila Electric		600	Ord.	6	1980

TABLE 6. Central Station Nuclear Power Reactors Operating, Ordered, Announced or Planned for Countries Other than the United States (Continued)

February 1974

Country	Plant name	Utility	Reactor type	Net power, MW(e)	Status	Scheduled for commercial operation Month	Year
	Philippine #2	Manila Electric		600	Plan.	6	1981
Poland	Poland #1			1000	Plan.	12	1980
Portugal	Portugal #1			500	Plan.	12	1979
Romania	Romanian #1		PWR	440	Ord.	12	1978
	Romanian #2		PWR	440	Plan.	12	1978
Singapore	Singapore #1			500	Plan.	12	1980
South Africa	Koeberg A	ESCOM	LWR	500	Plan.	9	1981
	Safr #2	ESCOM		600	Plan.	12	1982
	Safr #3	ESCOM		800	Plan.	12	1983
Spain	Zorita #1	UEM	PWR	153	Oper.	8	1969
	Sta. Maria De Garona	Nuclenor	BWR	460	Oper.	5	1971
	Vandellos	Hifrenze	GCR	487	Oper.	12	1972
	Zorita #2	UEM	LWR	500	Plan.	2	1980
	Lemoniz #1	IBUERCO	PWR	902	Ord.	4	1977
	Lemoniz #2	IBUERCO	PWR	902	Ord.	6	1978
	Almaraz #1	Hidroelec Espanola	PWR	902	Ord.	4	1977
	Almaraz #2	Hidroelec Espanola	PWR	902	Ord.	2	1978
	Cofrentes	Hidroelec Espanola		930	Ann.		1970
	Cataluna #1	FESCA	PWR	902	Ord.	9	1977
	Cataluna #2	FESCA	PWR	902	Plan.	12	1978
	Trillo #1	UEM		1200	Plan.	12	1981
	Trillo #2	UEM		1200	Plan.	12	1985
	Electra Del Viesgo #1	Electra Del Viesgo		900	Ord.	12	1980
Sweden	Oskarshamn #1	OKG	BWR	440	Oper.	8	1971
	Oskarshamn #2	OKG	BWR	580	Ord.	8	1974
	Oskarshamn #3	OKG		900	Plan.	6	1980
	Ringhals #1	Sw. State Power Bd.	BWR	760	Ord.	4	1974
	Ringhals #2	Sw. State Power Bd.	PWR	809	Ord.	11	1974
	Ringhals #3	Sw. State Power Bd.	PWR	900	Ord.	12	1977
	Ringhals #4	Sw. State Power Bd.	PWR	900	Ord.	7	1979
	Barseback #1	Syosvenska Kraft	BWR	580	Ord.	7	1975
	Barseback #2	Syosvenska Kraft	BWR	900	Ord.	7	1977

	Plant	Owner	Type	MW	Status	Mo.	Year
	Vastra Frolunda		LWR	750	Plan.	12	1980
	Hisingen		LWR	750	Plan.	12	1980
	Gauli #1	Krangede AB	LWR	500	Plan.	12	1980
	Gauli #2	Krangede AB	LWR	500	Plan.	12	1982
	Gauli #3	Krangede AB	LWR	500	Plan.	12	1984
	Gauli #4	Krangede AB	LWR	500	Ord.	7	1987
	Forsmark #1	Sw. State Power Bd.	BWR	900	Ann.	7	1978
	Forsmark #2	Sw. State Power Bd.	BWR	900	Plan.		1980
	Forsmark #3	Sw. State Power Bd.		800	Plan.		
	Forsmark #4	Sw. State Power Bd.		800	Plan.		
	Swede Hetr						
Switzerland	Beznau #1	NOK	PWR	350	Oper.	12	1969
	Beznau #2	NOK	PWR	350	Oper.	3	1972
	Muhleberg	BKW	BWR	306	Ord.	10	1972
	Leibstadt		BWR	875	Ord.	12	1977
	Kaiseraugst	ENK	BWR	850	Ord.	6	1979
	Gosgen	Consortium	PWR	920	Plan.	12	1978
	Graben #1	PC of Canton & Bern	LWR	880	Plan.	12	1980
	Graben #2	PC of Canton & Bern	LWR	880	Ann.		1981
	Ruethi #1	NOK		650	Plan.		
	Verbois			900	Ord.		1980
Taiwan	Chinshan #1	Taiwan Power	BWR	604	Ord.	12	1977
	Chinshan #2	Taiwan Power	BWR	604	Ord.	9	1978
	Chinshan #3	Taiwan Power	BWR	900	Ord.	9	1978
	Chinshan #4	Taiwan Power	BWR	800	Ord.	12	1980
	Northern 2 #1	Taiwan Power	BWR	900	Plan.	4	1978
	Northern 2 #2		BWR	900	Plan.		1979
Thailand	Phai Bay	EGAT	HWR	500		12	1980
Turkey	Turk #1		GCR	325			
United Kingdom	Calder Hall #1	UKAEA	GCR	50	Oper.	9	1956
	Calder Hall #2	UKAEA	GCR	50	Oper.	9	1956
	Calder Hall #3	UKAEA	GCR	50	Oper.	9	1956
	Calder Hall #4	UKAEA	GCR	50	Oper.	9	1956
	Chapel Cross #1	UKAEA	GCR	50	Oper.	11	1958
	Chapel Cross #2	UKAEA	GCR	50	Oper.	11	1958
	Chapel Cross #3	UKAEA	GCR	50	Oper.	11	1958
	Chapel Cross #4	UKAEA	GCR	50	Oper.	11	1958
	Berkeley #1	CEGB	GCR	138	Oper.	6	1962
	Berkeley #2	CEGB	GCR	138	Oper.	6	1962
	Bradwell #1	CEGB	GCR	150	Oper.	7	1962
	Bradwell #2	CEGB	GCR	150	Oper.	7	1962

TABLE 6. Central Station Nuclear Power Reactors Operating, Ordered, Announced or Planned for Countries Other than the United States (Continued)

February 1974

Country	Plant name	Utility	Reactor type	Net power, MW(e)	Status	Scheduled for commercial operation Month	Year
United Kingdom (continued)	Hunterston A #1	SSEB	GCR	160	Oper.	5	1964
	Hunterston A #2	SSEB	GCR	160	Oper.	6	1964
	Trawsfynyd #1	CEGB	GCR	250	Oper.	3	1965
	Trawsfynyd #2	CEGB	GCR	250	Oper.	3	1965
	Hinkley Point A #1	CEGB	GCR	250	Oper.	3	1965
	Hinkley Point A #2	CEGB	GCR	250	Oper.	3	1965
	Dungeness A #1	CEGB	GCR	275	Oper.	10	1965
	Dungeness A #2	CEGB	GCR	275	Oper.	10	1965
	Sizewell A #1	CEGB	GCR	290	Oper.	3	1966
	Sizewell A #2	CEGB	GCR	290	Oper.	3	1966
	Oldbury #1	CEGB	GCR	300	Oper.	12	1967
	Oldbury #2	CEGB	GCR	300	Oper.	4	1968
	Winfreth	UKAEA	HWR	94	Oper.	2	1968
	Wylfa #1	CEGB	GCR	590	Oper.	1	1971
	Wylfa #2	CEGB	GCR	590	Oper.	8	1971
	Windscale	UKAEA	AGR	35	Oper.	2	1963
	PFR	UKAEA	FBR	254	Oper.	10	1973
	Hinkley Point B #1	CEGB	AGR	625	Ord.	9	1973
	Hinkley Point B #2	CEGB	AGR	625	Ord.	12	1973
	Hunterston B #1	SSEB	AGR	625	Ord.	12	1973
	Hunterston B #2	SSEB	AGR	625	Ord.	12	1973
	Dungeness B #1	CEGB	AGR	625	Ord.	12	1974
	Dungeness B #2	CEGB	AGR	625	Ord.	12	1974
	Hartlepool #1	CEGB	AGR	625	Ord.	12	1974
	Hartlepool #2	CEGB	AGR	625	Ord.	12	1975
	Sizewell B #1	CEGB		660	Ann.		1977
	Sizewell B #2	CEGB		660	Ann.		1977
	Sizewell B #3	CEGB		660	Ann.		1977
	Sizewell B #4	CEGB		660	Ann.		1977
	Heysham #1	CEGB	AGR	625	Ord.	12	1975
	Heysham #2	CEGB	AGR	625	Ord.	12	1976
	Oldbury B	CEGB	HTG	650	Ann.		1976

Country	Plant	Utility	Type	MWe	Status		Year
	Portskewett #1	CEGB	FBR	625	Ann.	12	1979
	Portskewett #2	CEGB	FBR	625	Ann.	10	1979
	Stake Ness	NSHEB	LGR	1220	Plan.	12	1958
	CFR	CEGB	LGR	1300	Plan.	12	1964
U.S.S.R.	Troitsk		LGR	600	Oper.	12	1967
	Beloyarsk #1		PWR	94	Oper.	12	1964
	Beloyarsk #2		PWR	200	Oper.	12	1969
	Novo-Voronezh #1		PWR	210	Oper.	12	1972
	Novo-Voronezh #2		FBR	375	Oper.	12	1969
	Novo-Voronezh #3		FBR	440	Oper.	12	1973
	Bor-60		BWR	12	Oper.	7	1966
	BN-350		PWR	350	Oper.	1	1973
	VK-50		PWR	70	Oper.	12	1975
	Novo-Voronezh #4		LGR	410	Ord.	12	1973
	Novo-Voronezh #5		LGR	1000	Ord.	12	1974
	Lenin #1		PWR	1000	Ord.	12	1975
	Lenin #2		PWR	1000	Ord.	12	1975
	Kola #1		LWGR	440	Ord.	3	1977
	Kola #2		LGR	440	Ord.		
	Smolensk #1		LGR	1000	Ord.	12	1975
	Smolensk #2		GMR	1000	Ord.	7	1977
	Koursk #1		GMR	1000	Ord.	8	1978
	Koursk #2		LWGR	1000	Ann.		
	Tchernobylsk #1		PWR	1000	Ord.	5	1977
	Tchernobylsk #2		PWR	1000	Ord.	12	1975
	Bilibino		FBR	150	Ord.	12	1975
	Oktemberyan #1			410	Ord.	6	1974
	Oktemberyan #2			410	Plan.	12	1978
	BN-600			600	Plan.	12	1980
Yugoslavia	Videmkrsko	Electro Priureda	FBR	600			
	Prevlako		LWR	800			

SOURCE: U.S. Energy Research and Development Administration.

TABLE 7.   Nuclear Power Reactors Installed and Operating Worldwide
September 30, 1974

Country	Rank	Net power MW(e)	Percent of worldwide capacity	Number of installations
United States	1	25,784	52.5	42
United Kingdom	2	5,589	11.3	29
U.S.S.R.	3	2,761	5.6	10
France	4	2,731	5.5	7
Japan	5	2,531	5.1	6
Canada	6	2,486	5.1	6
Germany (West)	7	2,110	4.3	7
Spain	8	1,100	2.2	3
Switzerland	9	1,006	2.0	3
India	10	600		3
Italy	11	597		3
Netherlands	12	505		2
Germany (East)	13	435		2
Sweden	14	400		1
Belgium	15	390		1
Pakistan	16	125		1
Czechoslovakia	17	110		1
		49,260		127

SOURCE: U.S. Energy Research and Development Administration.

TABLE 8.   Nuclear Power Reactors on Order Worldwide
September 30, 1974

Country	Rank	Net power MW(e)	Percent of worldwide capacity	Number of installations
United States	1	152,599	60.5	147
Japan	2	23,014	9.1	28
Germany (West)	3	16,234	6.4	18
U.S.S.R.	4	11,450	4.5	15
Sweden	5	6,329	2.5	8
United Kingdom	6	6,250	2.5	10
Spain	7	5,410	2.1	6
Taiwan	8	4,708	1.9	6
France	9	4,640	1.8	5
Canada	10	4,500	1.8	6
Switzerland	11	2,645		3
Italy	12	2,600		3
Czechoslovakia	13	2,375		4
Mexico	14	1,600		2
Belgium	15	1,260		2
Germany (East)	16	1,095		3
Argentina	17	918		2
Finland	18	840		2
Bulgaria	19	810		2
Austria	20	700		1
Brazil	21	626		1
South Korea	22	564		1
Romania	23	440		1
		251,607		276

SOURCE: U.S. Energy Research and Development Administration.

TABLE 9.   Nuclear Power Reactors Announced or in Planning Stage Worldwide
September 30, 1974

Country	Rank	Net power Mw(e)	Percent of worldwide capacity	Number of installations
United States	1	50,989	27.4	47
Japan	2	39,978	21.5	35
Germany (West)	3	14,710	7.9	13
Canada	4	10,000	5.4	13
United Kingdom	5	7,060	3,8	9
France	6	6,990	3.8	8
Sweden	7	6,900	3.8	10
Spain	8	4,732	2.5	5
Switzerland	9	3,310	1.8	4
Argentina	10	2,800	1.5	4
Brazil	11	2,400		3
Mexico	12	2,200		2
Finland	13	1,980		3
Greece	14	1,950		3
South Africa	15	1,900		3
Italy	16	1,700		2
Belgium		1,600		2
Netherlands	17	1,600		2
India	18	1,500		3
Norway		1,400		2
U.S.S.R.	19	1,400		2
Yugoslavia		1,400		2
Bulgaria	20	1,215		3
Austria	21	1,200		1
Philippines		1,200		2
Luxemburg	22	1,100		1
Chile		1,000		2
Germany (East)		1,000		1
Hong-Kong	23	1,000		2
Ireland		1,000		3
Poland		1,000		1
Pakistan	24	900		2
Hungary	25	880		2
New Zealand	26	840		2
Australia	27	800		1
South Korea	28	600		1
Portugal		500		1
Singapore	29	500		1
Thailand		500		1
Romania	30	440		1
Iran	31	400		1
Israel		400		1
Jamaica		400		2
Turkey	32	325		1
Bangladesh	33	200		1
		185,899		211

SOURCE: U.S. Energy Research and Development Administration.

delay will remain. In essence, this case situation reflects a lack of dedication on the part of the citizens and government by stressing conservation instead of pioneering new technology as the means for overcoming continuing energy shortages. The conservation philosophy, of course, is tantamount to lower standards of living, demands severe alterations in life-style, and major shifts in employment, with consequent excessive unemployment during very long periods of readjustment.

As part of case A, the time required for a nuclear project is assumed to consist of about two years for planning and design, license application and environmental report prepara-

TABLE 10.   Consolidated Figures (Installed and Operating, Ordered, and in Planning Stage) for Nuclear Power Reactors Worldwide

September 30, 1974

Country	Rank	Capacity, MW(e)	Percent of capacity	Number of installations
United States	1	229,372	47.1	236
Japan	2	65,523	13.4	69
Germany (West)	3	33,054	6.8	38
United Kingdom	4	18,899	3.9	48
Canada	5	16,986	3.5	25
U.S.S.R.	6	15,611	3.2	27
France	7	14,361	3.0	20
Sweden	8	13,629	2.8	19
Spain	9	11,242	2.3	14
Switzerland	10	6,961	1.4	10
Italy	11	4,897		8
Taiwan	12	4,708		6
Mexico	13	3,800		4
Argentina	14	3,718		6
Belgium	15	3,250		5
Brazil	16	3,026		4
Finland	17	2,820		5
Germany (East)	18	2,530		6
Czechoslovakia	19	2,485		5
Netherlands	20	2,105		4
India	21	2,100		6
Bulgaria	22	2,025		5
Greece	23	1,950		3
Austria	24	1,900		2
South Africa		1,900		3
Norway	25	1,400		2
Yugoslavia		1,400		2
Philippines	26	1,200		2
South Korea	27	1,164		2
Luxemburg	28	1,100		1
Pakistan	29	1,025		3
Chile	30	1,000		2
Hong-Kong		1,000		2
Ireland		1,000		3
Poland		1,000		1
Hungary	31	880		2
Romania		880		2
New Zealand	32	840		2
Australia	33	800		1
Portugal	34	500		1
Singapore		500		1
Thailand		500		1
Iran	35	400		1
Israel		400		1
Jamaica		400		2
Turkey	36	325		1
Bangladesh	37	200		1
		486,766		614

SOURCE: U.S. Energy Research and Development Administration.

tion, two years for construction permit approval, and six years for construction and startup. As shown by Table 12, case A forecasts 85,000 MW of nuclear generating capacity to be on line at the end of 1980 out of a total electric generating system capacity of 655,000 MW. By comparison, present utility schedules indicate about 124,000 MW of nuclear capacity on line by 1980. Case A forecasts 231,000 MW of nuclear capacity for 1985.

Case B.   This case assumes that there will be some improvement over the experiences in construction and regulation that prevailed during the early 1970s. The total nuclear project time is indicated to average eight years, with about 15 months for planning and design, license application and environmental report preparation; 15 months for construction permit issuance; and about 5½ years for construction and startup. Nuclear capacity would then be 102,100 MW in 1980 out of a total system capacity of 700,000 MW. In 1985, the nuclear capacity is forecast to be 260,000 MW.

Case C. This case assumes additional improvements in construction performance and regulatory processes. New legislation and rules would permit construction to begin before completion of the construction permit and application safety review. The site environmental review would be completely separated from the safety review. This presupposes that standardized plant designs would be used in the license application. The total nuclear project time is indicated at about six years, with one year for design and planning, license application preparation and environmental review, and five years for construction and startup with concurrent operating license review and approval. Under these assumptions, the nuclear capacity in 1980 would be 112,400 MW out of a total system capacity of 770,000 MW. The 1985 nuclear capacity is forecast to be 275,000 MW.

Case D. This case assumes a general reduction in the growth rate of electricity use for the near term. The total electric generating capacity in 1980 is forecast to be 680,000 MW, compared to 700,000 MW for case *B*, with the same nuclear capacity of 102,100 MW and the same assumptions about nuclear project schedules as in case *B*. Any reduction in electricity production in the near-term would be effected by reducing use of oil- and gas-fired plants. The 1985 nuclear capacity is forecast to be 250,000 MW.

The Longer Term. Any forecasts beyond the early 1980s must be based upon assessments of possible changes in technologies and relative costs, structural changes in the economy, the attitudes of the American citizenry toward energy as related to living standards, and the worldwide political climate. With so many of these kinds of variables present in the mid-1970s, long-term projections, while of some value, tend to be academic. One estimate of the energy mix through the year 2000 is given in Table 13.

See also Electric Power Production and Requirements in Handbook section on *Power Technology Trends*.

ABRIDGED GLOSSARY OF NUCLEAR TERMS

Accelerator. A device that accelerates charged particles to high velocities so that they have high kinetic energy. It can be used for electrons, protons, deuterons, and ions. Also called particle accelerator. Some early references used the term "atom smasher."

Activation energy. The amount of energy that must be applied to initiate a reaction per molecule of a chemical reaction, or per atom of a nuclear reaction.

Alpha particle. The nucleus of a helium atom containing two protons and two neutrons. Alpha rays are streams of high-velocity alpha particles originating in radioactive atoms or in particle accelerators.

Atom. The smallest particle into which an element can be divided and still retain the chemical properties of that element. Each of the over-100 elements has a different arrangement of electrons and protons.

Atomic disintegration. Conversion of the nucleus of an atom of one element into that of some other element.

Atomic mass. The mass of a neutral atom of a nuclide, usually expressed in atomic mass units. Also called nuclidic mass. Formerly called isotopic mass.

Atomic mass unit (amu). One-sixteenth of the average mass of a natural mixture of oxygen atoms (chemical scale); one-sixteenth of the mass of oxygen 16 (physical scale). 1.000279 amu (physical) = 1.00000 amu (chemical). 1 amu (physical) = 1.65983×10^{-27} kg = 931.162 MeV of energy.

Atomic number Z. The number of protons in the nucleus of an atom. Atomic numbers of the known elements include all integers from 1 through 103.

Atomic weight. A relative number for which one specific element is used as a standard rather than an absolute number. Since the International Atomic Weight tabulation of 1961, carbon has been used as the reference element—with a relative mass of 12. Earlier tabulations were based upon oxygen with a relative mass of 16.0000. Atomic weight figures in most instances are average figures because most elements are composed of two or more stable isotopes, each with a different atomic mass.

BCD. Burst cartridge detection system used in gas-cooled reactors to sense gaseous fission products which may have escaped from exposed uranium.

Beta particles. An electron, positive or negative. Beta rays are high-velocity electrons, usually negative, originating in radioactive atoms or in particle accelerators.

Binding energy. The total amount of energy binding together the particles of a nuclide. The amount of energy released when a nuclide is formed from protium atoms and neutrons. The energy equivalent of the mass decrement of an atom.

Binding energy per nucleon. The total binding energy of a nuclide divided by its mass number.

TABLE 11. Comparison of Thermal Reactor Characteristics

Characteristic	Boiling water reactor	Pressurized water reactor	High-temperature gas-cooled reactor	Advanced gas reactor (British) Inner core	Advanced gas reactor (British) Outer core	Advanced thermal reactor (Japanese)	Heavy-water reactor (Canadian)
Thermal efficiency, %	34	33	39	42		31	30
Specific power, MW(th)/MTU[a]	28	38	82	13		13	22
Initial core, average:							
Irradiation level, MWD(th)/MTU	17,000	22,600	54,500	13,000		13,700	6900
Fresh fuel assay, wt% ^{235}U	2.03	2.26	93.15	1.46	1.75	1.2	0.71
Spent fuel assay, wt% ^{235}U	0.86	0.74	a	0.75	1.00	0.33	0.31
Fissile Pu recovered, kg/MTU[b]	4.8	5.8	a	2.5		4.3	1.7
Feed required, ST U$_3$O$_8$/MW(e)[c]							
0.3% tails	0.581	0.498	0.456	0.737		0.577	0.199
At 0.2% tails	0.494	0.422	0.367	0.640		0.516	0.199
Separative work required, SWU/MW(e)[c]							
0.3% tails	185	174	311	188		103	0
0.2% tails	239	222	366	249		142	0
Replacement loadings (Annual rate at steady state and 75% plant factor):							
Irradiation level, MWD(th)/MTU	27,500	32,600	95,000	20,000		20,000	9600
Fresh fuel assay, wt% ^{235}U	2.73	3.21	93.15	2.10	2.54	0.71	0.71
Spent fuel assay, wt% ^{235}U	0.84	0.90	a	0.59	0.78	0.15	0.2
Fissile Pu recovered, kg(MTU)[b]	5.9	7.0	a	4.0		e	2.3
Feed required, ST U$_3$O$_8$/MW(e)[c]							
At 0.3% tails	0.179	0.191	0.106	0.165		e	0.125
At 0.2% tails	0.144	0.154	0.085	0.133			0.125
Separative work required, SWU/MW(e)[c]							
At 0.3% tails	84	94	73	69			0
At 0.2% tails	105	117	85	89			0
Replacement loadings (annual rate with plutonium recycle, 75% plant factor):							
Fissile Pu recycled, kg/MW(e)	0.163	0.167				0.3	
Fissile Pu recovered, kg/MTU[b,a]	8.1	9.5				5.5	
Feed required, ST U$_3$O$_8$/MW(e)[c,a]							
At 0.3% tails	0.148	0.158				0.072	
At 0.2% tails	0.121	0.129				0.072	

Separative work required, SWU/

$MW(e)^c$		
At 0.3% tails	66	75
At 0.2% tails	82	93

	0
	0

[a]$MW(th)$ = thermal megawatts
$MW(e)$ = net electrical megawatts
$MWD(th)$ = thermal megawatt-days
MTU   = metric tons (thousands of kilograms of uranium)
$ST\ U_3O_8$ = short tons of U_3O_8 yellowcake from an ore processing mill
1 SWU = 1 kg of separative work

[b]After losses.

[c]For replacement loadings the required feed and separative work are net, in that they allow for the use of uranium recovered from spent fuel. Allowance is made for fabrication and reprocessing losses.

[d]All spent fuel and fissile production (primarily ^{233}U) are recycled on a self-generated basis. Only one cycle of ^{235}U is assumed.

[e]Includes natural uranium to be spiked with plutonium: 0.0087 ST U_3O_8/MW(e) for BWR and 0.0067 for PWR.

[f]Self-sustaining plutonium recycle is implicit in the design of this reactor.

[g]Plutonium available for recycle ratchets up each pass because not all the plutonium charged is burned. Therefore, more plutonium is recovered from mixed oxide fuel than from standard uranium fuel, and this increment increases with each cycle (5 to 6 years per cycle), requiring several passes to reach steady state. The data shown represent conditions for the 1980s when most reactors will be discharging fuel which has seen only one recycle pass.

[h]Average for all fuel discharged with full recycle of self-generated plutonium. For mixed oxide fuel (natural U spiked with self-generated plutonium) the spent fuel from BWRs contains 15.1 kg Pu per MTU and from PWRs, 18.7.

SOURCE: U.S. Energy Research and Development Administration.

5-41

TABLE 12. One Forecast of Energy Consumption and Electric Generating Capacity in the United States
Assumptions made retroactive to 1960 and extrapolated to 2000

	Case	1960	1970	1975	1980	1985	1990	1995	2000
Energy consumed, million Btu per capita	A	247.	329.	357.	378.	401.	429.	462.	499.
	B	247.	329.	372.	428.	485.	558.	635.	719.
	C	247.	329.	376.	434.	497.	569.	650.	737.
	D	247.	329.	364.	399.	438.	494.	563.	642.
Fraction for electricity generation	A	.18	.24	.29	.33	.37	.42	.46	.51
	B	.18	.24	.29	.31	.34	.40	.45	.50
	C	.18	.24	.29	.34	.38	.43	.49	.54
	D	.18	.24	.29	.32	.36	.41	.46	.50
Energy consumed for electricity generation, million Btu per capita	A	44.2	80.3	105.	125.	148.	180.	215.	253.
	B	44.2	80.3	107.	133.	166.	220.	283.	357.
	C	44.2	80.3	111.	147.	189.	246.	316.	399.
	D	44.2	80.3	107.	129.	156.	201.	257.	324.
Apparent capacity factor	A B C D	.49	.52	.50	.49	.50	.51	.51	.52
Heat rate, thousands Btu per kWh	A B C D	10.7	10.5	10.2	10.1	10.0	9.8	9.8	9.6
Total electric generating capacity per capita, kW per capita	A	.97	1.67	2.36	2.88	3.33	4.14	4.90	5.81
	B	.97	1.67	2.41	3.07	3.76	5.07	6.45	8.19
	C	.97	1.67	2.50	3.38	4.25	5.67	7.27	9.22
	D	.97	1.67	2.41	2.99	3.52	4.62	5.85	7.45
Total electric generating capacity, thousands of MW	A	168.	341.	510.	655.	800.	1040.	1280.	1575.
	B	168.	341.	520.	700.	903.	1275.	1685.	2220.
	C	168.	341.	540.	770.	1020.	1425.	1900.	2500.
	D	168.	341.	520.	680.	865.	1160.	1530.	2020.
Total nuclear generating capacity, thousands of MW	A	.02	5.8	43.3	85.0	230.9	410.	620.	850.
	B	.02	5.8	47.3	102.1	260.0	500.	820.	1200.
	C	.02	5.8	52.0	112.4	275.0	575.	960.	1400.
	D	.02	5.8	47.3	102.1	250.0	475.	760.	1090.

SOURCE: U.S. Energy Research and Development Administration.

TABLE 13. One Forecast of Energy Supply for the United States
1973–2000

		Quadrillion (10^{15}) Btu							
		Case A		Case B		Case C		Case D	
	1973	1985	2000	1985	2000	1985	2000	1985	2000
Electric									
Oil	3.4	6.2	4.1	6.4	8.3	7.2	8.3	6.1	6.5
Gas	3.9	3.1	1.0	3.3	2.1	3.7	2.1	3.0	1.6
Coal*	8.7	9.6	10.6	12.0	14.2	15.1	14.8	10.5	13.8
Nuclear	0.9	13.2	47.8	14.6	66.9	15.7	78.5	14.2	60.8
Other	2.9	3.6	5.3	3.6	5.3	3.6	5.2	3.6	5.2
Total	19.8	35.7	68.8	39.9	96.8	45.3	108.9	37.4	87.9
Nonelectric									
Oil	31.3	31.5	37.5	42.3	59.6	40.8	54.1	36.4	49.9
Gas	19.7	19.3	16.0	24.1	23.9	22.4	22.3	21.5	24.2
Coal*	4.8	9.7	13.0	10.3	14.7	10.9	14.3	9.5	12.3
Total	55.8	60.5	66.5	76.7	98.2	74.1	90.7	67.4	86.4
Summary									
Oil									
Domestic	22.2								
Imported	12.5								
Total	34.7	37.7	41.6	48.7	67.9	48.0	62.4	42.5	56.4
Gas									
Natural	23.6	22.4	17.0	27.4	26.0	26.1	24.4	24.5	25.8
Synthetic	–0–	(2.1)	(4.1)	(2.2)	(4.5)	(2.8)	(4.3)	(2.1)	(4.1)
Total	23.6	(24.5)	(21.1)	(29.6)	(30.5)	(28.9)	(28.7)	(26.6)	(29.9)
Coal*	13.5	19.3	23.6	22.3	28.9	26.0	29.1	20.0	26.1
Nuclear	0.9	13.2	47.8	14.6	66.9	15.7	78.5	14.2	60.8
Other	2.9	3.6	5.3	3.6	5.3	3.6	5.2	3.6	5.2
Grand total	75.6	96.2	135.3	116.6	195.0	119.4	199.6	104.8	174.3

*Includes coal used to provide synthetic gas.
SOURCE: U.S. Energy Research and Development Administration.

Biological shield.   The outer portion of shielding around a reactor designed to reduce the neutron and gamma fluxes to safe working levels.

Breeder reactor.   A converter reactor in which more new fuel is produced than the amount consumed.

Burnup figure of merit.   For a reactor, this may be expressed either as a percent of the fuel that is consumed before fuel elements must be replaced; or in terms of the megawatt-days of energy obtained per unit mass of fuel in a charge.

BWR.   Boiling water reactor.

Cermet.   A combination of the words ceramic and metal to identify mixtures of these materials for use at high temperatures. Used in fabricating ceramic fuels, such as uranium-plutonium cermets, which offer the dual advantages of high-temperature use and resistance to radiation damage.

Chain reaction.   In the neutron-fission chain reaction, a neutron plus a fissionable atom cause a fission, resulting in a number of neutrons which, in turn, cause other fissions. A self-sustaining reaction.

Chemical shim.   A chemical, usually boric acid, placed in the coolant system of a nuclear reactor to serve as a neutron absorber which compensates for fuel burnup during normal operation. A chemical shim can also compensate for temperature changes in the coolant, for buildup and decay of various elements, and for depletion of fissionable material.

Claddant.   A coating used, particularly on reactor fuel elements, to protect a substance from corrosion and erosion.

Containment structure.   A massive reinforced and prestressed concrete structure, usually of a right circular cylindrical shape or light-bulb shape depending upon type of reactor, in which the reactor is housed. Structures are designed to withstand high pressure and high temperature and, in an emergency, to contain the radioactivity released.

Control rod.   Any rod used to control the reactivity of a nuclear reactor. It may be a fuel rod or a part of the moderator; in a thermal reactor it is commonly a neutron absorber. The control rod changes the effective multiplication constant and hence the time derivative or reactivity.

Coolant.   A material, liquid or gaseous, circulated through a reactor in order to remove heat.

Conversion of nuclear fuels.   The process by which nonfissionable fertile materials (^{232}Th, ^{238}U) are converted into fissionable material (^{233}U, ^{239}Pu) in a reactor.

Conversion ratio.   The ratio of the number of atoms of fissionable fuel produced to the number of atoms of fissionable fuel consumed in a converter or breeder reactor.

Converter reactor.   A reactor in which new fissionable fuel is produced from nonfissionalbe fertile material.

Core, reactor.   The active portion of a nuclear reactor, containing the fissionable material.

Critical mass.   The smallest mass of fuel that will support a self-sustaining chain reaction for a particular geometry and combination of materials.

Critical size.   The smallest size for a particular geometry and combination of materials in which a self-sustaining chain reaction can be maintained.

CVCS.   Chemical and volume control system.

Decay probability.   The number of disintegrations per second per nucleus of a radioactive substance. In equation form $\lambda = -(1/N)(dN/dt) = 0.6931/T$, where T is the radioactive half-life.

Delayed neutron.   A neutron released from fission products after the initial fission.

Deuterium.   The stable isotope of hydrogen having mass number 2. The atomic abundance of deuterium in natural hydrogen (ocean water) is 0.0149 percent.

Deuteron.   The nucleus of deuterium (heavy hydrogen). A particle containing one proton and one neutron.

D_2O.   Heavy water.

Electromagnetic pump.   A pump in which electromagnetic fields are used directly to pump a liquid metal.

Electron.   A subatomic particle carrying one negative quantum of electric charge (1.6021×10^{-19} coulomb) and having a mass of 0.00054876 amu. Electrons make up the shell structure of atoms.

Electron volt.   A unit of energy equal to the work required to transport an electron through a potential difference of one volt.

Enriched uranium.   Uranium containing a higher portion of ^{235}U than the 0.7 atomic percent found in natural uranium.

Fast neutron.   A neutron with energy greater than 0.1 MeV.

Fast reactor.   A reactor in which the chain reaction is sustained by fast neutrons.

Fissile.   Fissionable.

Fission, nuclear.   The splitting of the atomic nucleus, usually into two or three large pieces, plus two or more neutrons, with the release of relatively large amount of energy.

Fuel conversion factor.   Often used to compare converter reactors, in which new fuel is produced from fertile material. A conversion factor of 1 indicates that one atom of new fuel is produced for every atom of fuel fissioned. Leakage and absorption reduce the conversion ratio in most thermal reactors. When the conversion factor is greater than 1, signifying that more fuel is produced than is consumed, the reactor is a breeder type, and this ratio is the breeding factor.

Fuel cycle.   The full series of events commencing with the mining and processing of uranium to put it into appropriate reactor fuel form, the makeup of new fuel as required, the reprocessing of spent fuel, and the storage of radioactive wastes arising from spent fuels.

Fusion, nuclear.   The combining of light into heavier nuclei, usually with relatively large energy release.

Gamma ray.   A photon emitted from a nuclear or annihilation reaction. Like x-rays, to which their properties are similar, gamma rays have sufficient energy to penetrate through considerable thicknesses of materials.

Gaseous diffusion.   A process used to separate ^{235}U from ^{238}U and thus provide an enriched ^{235}U fuel. The process, requiring extensive equipment and huge structures when done on a large scale, separates the materials on the basis of their differences in diffusivity.

GCFR.   Gas-cooled fast breeder reactor.

Half-life.   The time required for one-half the atoms of a radioactive nuclide to decay.

Heavy water.   Also known as deuterium oxide, D_2O, heavy water is water in which the hydrogen of the water molecule consists entirely of the heavy-hydrogen isotope having a mass number of 2. Density of heavy water is 1.1076 at 20°C. Heavy water is used as a moderator and coolant in heavy water nuclear reactors.

HTGR.   High-temperature gas-cooled reactor.

Intermediate neutron.   A neutron with energy between 100 eV and 0.1 MeV.

Intermediate reactor.   A reactor in which the chain reaction is sustained by intermediate neutrons.

Isotopes.   Atoms of an element differing in the number of neutrons.

LMFBR.   Liquid metal fast breeder reactor.

LWR.   Light water reactor.

Magnetic jack.   A type of linear motor. Flexible cables of control rod are moved in precise steps by switching coils in sequence, and then held to walls by friction. Used in reactor control systems.

Magnox.   A magnesium alloy used to clad fuel elements for reduction of neutron absorption in gas-cooled reactors.

Mass decrement.   The difference between the mass of an atom and the mass of the numbers of free neutrons and protium atoms contained therein.

MeV.   Million electron volts.

Moderator. The material employed in a nuclear reactor to reduce (moderate) the energy of neutrons.

Multiplication factor, effective k_{eff}. The ratio of the number of fissions in one generation to the number in the previous generation in a chain reaction.

Multiplication factor, excess k_{ex}. The difference between the effective multiplication factor and 1.

Neutron. A subatomic particle carrying no electric charge and having a mass of 1.008986 amu. Neutrons together with protons make up the nuclei of atoms.

Neutron cross section. A measure of the probability of a reaction of a neutron with a single atomic nucleus (microscopic cross section σ); or with one of the nuclei in a unit volume of material (macroscopic cross section Σ).

Neutron flux. The number of neutrons passing through a unit volume per unit of time.

Nucleon. A nuclear particle. A proton or neutron.

Nuclide. A kind of atom distinguished by the construction and energy of its nucleus.

PCS. Primary coolant system of a reactor.

Period, reactor. The length of time in which the reactor power level increases by the factor $e = 2.718$.

Plutonium. A transuranic radioactive element (symbol Pu) formed by the decay of certain isotopes of neptunium. Plutonium does not occur in nature. Plutonium as made in reactors is a mixture of ^{239}Pu, ^{240}Pu, ^{241}Pu, and ^{242}Pu. ^{239}Pu can be used as a fuel in fast breeder power reactors in which ^{238}U will be converted to plutonium to replace fuel used in the chain reaction.

Poison, fuel. Materials (fission products) formed during a nuclear reaction which reduce the reactivity of a nuclear reactor by absorbing neutrons unproductively. Two nuclides among the fission products have large capture cross sections for thermal neutrons and hence have a bearing on reactor design. These are ^{135}Xe and ^{149}Sm. As fission products accumulate in a reactor, they capture an increasing fraction of all fission neutrons. This neutron loss causes a reduction in reactivity which must be compensated for if the reaction rate is to be kept constant.

Power density. Power per unit volume.

Primary system, nuclear power plant. The nuclear portion of the plant, including the reactor, moderating, and cooling system. The generation of steam and electricity comprises the secondary system.

Prompt critical. That condition in which a fission assembly is critical because of prompt neutrons alone. The reaction rate in such an assembly increases very rapidly.

Protium. The isotope of hydrogen having the mass number 1. Light hydrogen.

Proton. A subatomic particle carrying one positive quantum of electric charge (1.6021×10^{-19} coulomb) and having a mass of 1.007596 amu. Protons together with neutrons make up the nuclei of atoms.

PWR. Pressurized water reactor.

Quantum. An indivisible quantity or unit of energy or electric charge.

Radioactivity. The property of certain nuclides characterized by the spontaneous emission of alpha or beta particles or capture of an orbital electron by the nucleus, with the release of energy and change of atomic number.

Radioisotopes. Isotopes that are distinguishable from other species of atoms with the same atomic number by radioactive transformation. Some radioisotopes occur in nature; others are produced in particle accelerators and nuclear reactors. Radioisotopes are present in spent nuclear fuels, reactor structures, and wastes.

Reactivity. The ratio of the excess to the effective multiplication constant of a nuclear-fission reactor.

Regenerative reactor. A converter or breeder reactor in which part or all of the fuel produced is used in the reactor.

Regulating rod. A fine control rod used to regulate the chain reaction in a reactor.

Relative biological effectiveness (RBE). The same number of roentgens (R) from various types of radiation produces different amounts of body damage. The roentgen equivalent man (rem) is used in stating allowable radiation exposure values. rem = R × RBE, where RBE has a value of 1 for x- and gamma radiation and beta rays, 5 for thermal or slow neutrons, 10 for fast neutrons, and 20 for alpha rays.

Roentgen. That quality of gamma or x-radiation which produces 2.083×10^{9} ion pairs (one electrostatic unit of charge) per cm^3 of free air at a temperature of 0°C and a pressure of 1 atm.

Safety rod. A control rod with large effect on the reactivity of a reactor and used to ensure the immediate halt of the chain reaction.

Secondary system. The steam- and electricity-generating (nonnuclear) portion of a nuclear power plant.

Self-sustaining reaction. See *Chain reaction.*

Shield. Material placed around a radiation source to attenuate the radiation.

Shim rod. A coarse control rod used in a reactor to compensate for fission-product poisoning, fuel burnup, and, in a research reactor, for experiments that have a gross effect on the reactivity.

Slow neutron. A neutron with energy less than 100 eV.

Thermal neutron. A neutron with energy and velocity determined by the temperature of the surroundings. At 21°C, the average energy is 0.025 eV, and the most probable velocity is 2200 m/s.

Thermal reactor. A reactor in which the chain reaction is sustained by thermal neutrons.

Thermonuclear reaction. A nuclear fusion reaction in which the required activation energy is supplied thermally.

Thorium. An element, symbol Th. ^{232}Th absorbs neutrons to yield fissionable ^{233}U.

Tritium. A radioactive isotope of hydrogen with mass number 3.

Uranium. A radioactive element, symbol U. Naturally occurring uranium is a mixture of 99 percent ^{238}U and 0.7 percent ^{235}U, with a trace of ^{234}U. The nucleus of ^{235}U is capable of absorbing a neutron and thereupon undergoing fission into two highly radioactive fragments, with release of large energy and additional neutrons. ^{235}U is the fuel used in thermal nuclear reactors.

Zircaloy. An alloy of 98 percent Zr, 1.5 Sn, 0.35 Fe-Cr-Ni, and 0.15 O which optimizes the requirements of a low neutron cross section, reliable corrosion resistance in very hot water, and the strength required for cladding uranium fuel used in light water reactors.

REFERENCES

See lists of references that accompany subsequent articles in this Handbook section.

Nuclear Power Plant—Concept to Start-Up

BY ALFRED L. HABOUSH*

Editor's Note: As an example of one of the most recent and significant nuclear power installations in the United States, this description of the Fort St. Vrain Nuclear Generating Station portrays, in part, the extensive planning, problems, both engineering and regulatory-administrative, solutions, and actions which make up the rather extensive time span between concept and startup in the nuclear power field in the United States. The overall background and theory of HTGRs will be found in the article on High-Temperature Gas-cooled Reactors given later in this Handbook section.

The Fort St. Vrain Nuclear Generating Station (Fig. 1) which General Atomic Company (GA) has built for the Public Service Company of Colorado (PSC) is at the time of this writing† in the final stages of startup. The station will have a net electric output of 330 MW(e), 330 million watts electric. The U.S. Atomic Energy Commission issued the license to load fuel in December 1973. Initial criticality was achieved in January 1974 and was followed by physics testing. The rise to power is presently in progress. Commercial operation is expected in 1976.

The prime contractor for Fort St. Vrain, responsible for the design, procurement of equipment and materials, and construction of the nuclear steam supply system and the turbine plant is GA. PSC is responsible for site preparation, including grading, roads, fencing, railroads, landscaping, yard piping and lighting, makeup water systems, including chemical treatment facilities and the settling ponds, main power transformer, and switchyard. GA has retained Sargent and Lundy as the architect-engineer and has contracted the construction to EBASCO Services.

Except for the addition of reheat, primary system operating conditions of the Fort St. Vrain HTGR will closely parallel those of the 40 MW(e) Peach Bottom prototype, which operated successfully since its startup in June 1967.‡ In the Fort St. Vrain HTGR, as in the Peach Bottom plant, high-temperature helium coolant will produce steam at 1000°F (538°C) to match modern turbine conditions. The plant will have a net overall efficiency of approximately 39 percent.

PLANT DESCRIPTION

The Fort St. Vrain system uses a uranium-thorium fuel cycle; graphite for the moderator, fuel cladding, core structure, and reflector; and helium for the primary coolant. The plant incorporates a number of significant design features new to power reactor systems. The most prominent are the prestressed concrete reactor vessel (PCRV), once-through modular steam generators with integral superheaters and reheaters, steam-driven axial flow helium circulators, and hexagonal graphite fuel elements incorporating improved carbon-coated fuel particles.

Figure 2 is a simplified flow diagram of the plant, which generates 842 MW(th) (842 million watts thermal) to achieve a net output of 330 MW(e). The flow diagram points out the major operating parameters. The helium coolant at a pressure of about 700 psia flows downward through the reactor core, where it is heated to 1430°F (777°C). The coolant flow can be trimmed by the use of orifice valves located at the top of the core that are integral with the control rod drive mechanisms. From the reactor core, the coolant flows through the steam generators. After passing through the steam generators, the helium is returned to the core at a temperature of about 760°F (404°C) by four steam-turbine-driven helium circulators. Two identical loops are used, each including a six-module steam generator

---

*Formerly a Senior Project Manager, General Atomic Company, San Diego, Calif.
†Fourth quarter, 1974.
‡Philadelphia Electric Co., Peach Bottom, Pa. Plant shutdown in 1974 after 7 years of successful operation.

and two helium circulators. Each loop contributes half the total output of the nuclear steam supply system, which produces steam at 2400 psig and 1000°F (538°C) with single reheat to 1000°F. The helium circulators are driven by the exhaust steam from the high-pressure turbine. This steam is then reheated and returned to the intermediate-pressure turbine. The circulators are also equipped with a Pelton waterwheel drive so that they may be driven by the boiler feed pumps for emergency conditions.

The PCRV is 31 feet in internal diameter with a 75-foot internal height. The upper and lower heads are nominally 15 feet thick, and the walls have a nominal thickness of 9 feet. Thus, the PCRV provides the dual function of containing the coolant at operating pressure and also providing radiological shielding. The exterior vertical surface of the vessel may

Fig. 1. Exterior view of the Fort St. Vrain Nuclear Generating Station, Platteville, Colorado. *(Public Service Company of Colorado.)*

be described as a hexagonal prism with vertical pilasters at each corner. It is 61 feet across pilasters, 49 feet across flats, and 106 feet high.

The concrete walls and heads of the PCRV are constructed around a carbon steel liner which is nominally ¾ inch thick. The liner is anchored to the concrete and provides a helium-tight membrane. A system of water-cooled tubes welded to the concrete side of the liner provides a heat removal system to control concrete temperature. In addition to the coolant tubes, a thermal barrier is provided on the inside surface of the liner to limit the flow of heat to the liner walls. The top head has refueling penetrations that also house the control rod drives. It also incorporates penetrations and wells to house the helium purification system and neutron detection chambers. The bottom head has penetrations for each of the steam generator modules and helium circulators, plus a large central opening for access to the main cavity. All the penetrations through the PCRV are provided with two independent closures. The PCRV inner cavity and the primary closures serve as primary containment for the reactor; the massive PCRV and the secondary closure act as secondary containment.

The vessel is prestressed to place the concrete structure in compression before service. The prestressing system used is known as a *linear tendon posttensioning system.* During construction, steel tubes are embedded in the vessel concrete for later insertion of the tendons, each of which consists of up to 170 ¼-inch-diameter "thermalized" wires. The wires are anchored by means of cold-deformed button heads to a washer assembly that transfers and distributes the loads over split shims and a steel bearing plate into the concrete. Three tendon arrangements are used: 90 longitudinal (vertical) tendons, 310 circumferential tendons, and 48 crosshead tendons.

Inside the PCRV is the core support floor, a water-cooled structure of steel and reinforced concrete 5 feet thick. It is supported from the bottom of the PCRV cavity by 12 water-cooled steel columns. Twelve ducts conduct reactor outlet helium through the core support floor to the 12 steam generator modules located below.

Development Program.   This included fullscale testing of a helium circulator, control rod drive, and fuel handling machine; and model testing of the prestressed concrete reactor vessel. Extensive testing was also performed on materials for reactor internal structures and the steam generators. An extensive irradiation testing program was accomplished on the fuel.

Operator Training.   Public Service Company recognized at the beginning of the project that operators with specialized training would be required to operate the Fort St. Vrain Station. In anticipation of this need, a nuclear training program was initiated in

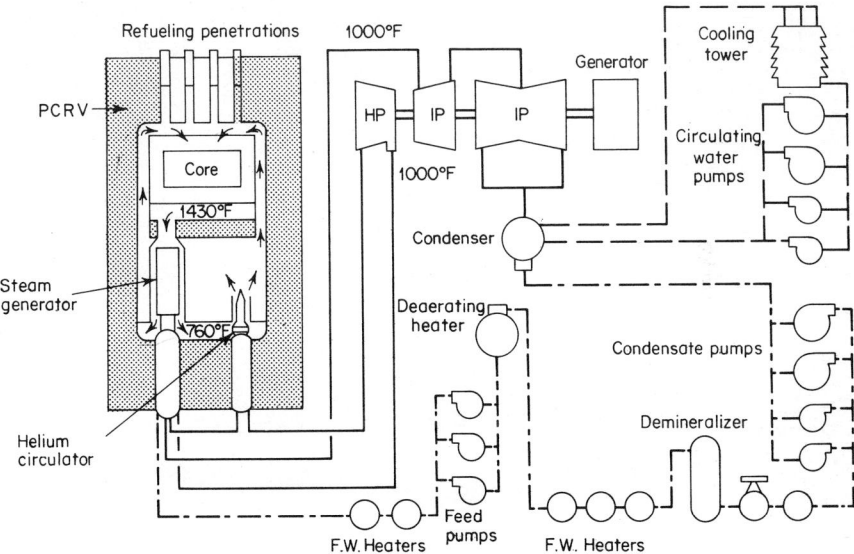

Fig. 2 Simplified flow diagram of Fort St. Vrain Nuclear Generating Station.

September 1966 to train company operating and engineering personnel. This program was conducted by Colorado State University and consisted of 22 full-day sessions conducted over a 19-week period. The class size varied from 12 to 16 persons, and a total of 75 Public Service employees completed this training.

LICENSING

As for any nuclear project, certain governmental approvals are required. The key regulatory dates related to the Fort St. Vrain project were:

1. October 1966	Application for construction permit filed with the USAEC.
2. April 1968	Colorado Public Utilities Commission issued Certificate of Public Convenience and Necessity.
3. July 1968	Public hearing before Atomic Safety and Licensing Board of Construction Permit application.
4. September 1968	Construction permit issued by USAEC.
5. November 1969	Application for operating license submitted to USAEC.
6. August 1972	Final Environmental Statement issued by USAEC.
7. December 1973	Operating license issued.
8. January 1974	Reactor achieved criticality.

CONSTRUCTION

Preliminary on-site construction work started in April 1968. This consisted of grading, fencing, railroad spur construction, temporary building erection, concrete batch plant erection, domestic water service, and construction power service. The excavation for the reactor plant building was also begun. Since the water table at the plant site was at a depth of 20 to 25 feet and since there was a considerable flow of groundwater, the freeze wall technique was used to prevent water from flowing into the excavation and to provide a vertical wall to support cranes and other heavy equipment required. The frozen wall was created by pumping refrigerated brine through 360 U-tubes sunk to the bedrock depth of approximately 55 feet. See Fig. 3.

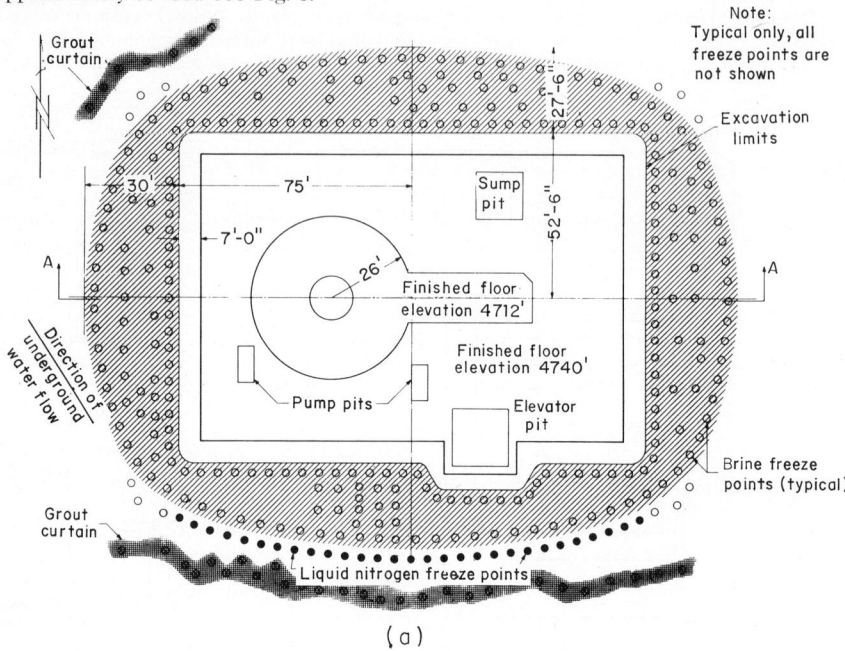

(a)

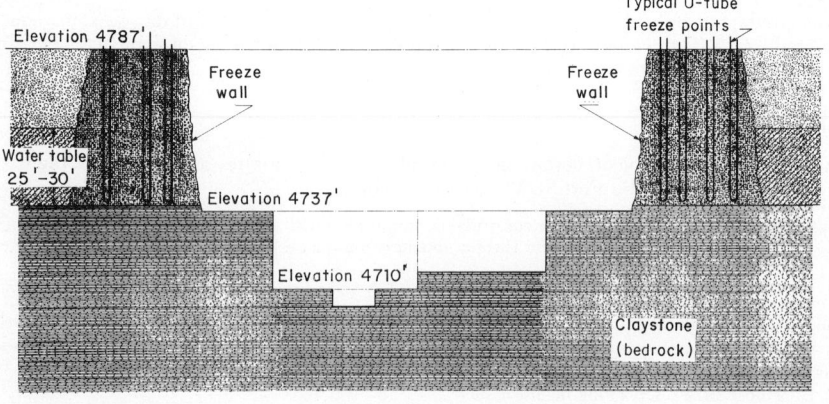

(b)

Fig. 3. Freeze-wall technique for providing cofferdam for the foundations during construction of the Fort St. Vrain Nuclear Generating Station: (a) Plot plan; (b) section through A–A. (General Atomic Company.)

The first permanent concrete was poured September 18, 1968. The reactor building basement walls were up to ground level by February 1969. In the center of the excavation, the reinforcing steel for the 50-foot-diameter support ring for the PCRV was being formed. The ring is 3½ feet thick and 32 feet high, and contains 575 cubic yards of concrete, which was poured in an 18-hour period. The unique shape of this ring and the high density of reinforcing steel required made this task extremely difficult and took more time than was scheduled.

Before construction of the PCRV, special partial mock-ups of the vessel were built at the site to ensure that the placement of the tendon tubes and reinforcing steel would not interfere with each other, or with the filling of open spaces with concrete. Concurrently with the foundation and support ring work, field fabrication of the PCRV liner and penetrations and their associated cooling water tubes were conducted adjacent to the reactor building location. In February 1969, the bottom head assembly, which weighed some 400 tons including a temporary erection jig, steel forms for the bottom head soffit, and some of the reinforcing steel, was moved on rails by a modified turbine generator stator lifting rig and was lowered into place on temporary support steel erected within the support ring. See Fig. 4. A preplaced aggregate concrete technique was used for the PCRV bottom head. The aggregate was positioned by hand, in layers, to avoid large voids and to achieve intimate contact with all embedments. Grout was pumped into the aggregate in a continuous operation through previously embedded pipes. The remainder of the concrete for the PCRV was conventionally mixed and placed. This concrete, designed for a strength of 6000 psi, was carefully cooled to minimize shrinkage and high-temperature-gradient cracks. After placement of rebar, tendon tubes and instrumentation wiring, and erection of the necessary forms, the first PCRV sidewall lift was poured on July 17, 1969. The fourteenth lift was completed January 7, 1970. In the meantime, the top head liner was completed and moved on to the PCRV.

Pouring of the concrete for the top head was completed February 13, 1970, ahead of schedule, despite the time lost during construction of the support ring. The total concrete placed in the PCRV was approximately 6500 cubic yards. Installation of the tendons in the completed portion of the PCRV was started in October 1969. The first tendon was stressed January 16, 1970, and stressing of all the tendons was completed by May 1, 1970.

During the next 12 months, the 12 steam generator modules, the vessel liner insulation, and the outer graphite reflector blocks were installed inside the PCRV. Meantime, in the conventional side of the plant, the main steam turbine, boiler feed pumps, and other conventional power plant equipment were set up. The four helium circulators, after completion of their 100-hour test runs using steam produced at the Valmont Steam Electric Station of PSC, were readied for installation in the PCRV. After some minor modifications to provide proper alignment in the penetrations, these circulators were placed in the vessel. In July 1971, the installation of instrumentation and testing equipment required for the PCRV pressure test was completed. The PCRV pressure test, using nitrogen, was started July 29, 1971, and the proof test pressure of 970 psi was achieved August 6, 1971. The test was concluded with an 84-hour leak test which was completed August 14, 1971, on schedule.

PLANT TESTING PROGRAM

After construction testing, preliminary operating tests were begun in August 1971. A preoperational test program was devised for Fort St. Vrain, comprising some 65 individual systems tests procedures. These tests were designed to demonstrate that plant systems would function as designed, to the extent practicable, before fuel loading. The preliminary operating tests were performed by PSC operators under the technical direction of GA startup engineers. The procedures call for filling systems and operating pumps and compressors in all the normal and abnormal plant operating conditions as can be simulated without having nuclear heat available. The operation of instruments, controls, alarms, and interlocks was checked by manipulating the process variable (level, flow, pressure, etc.) to determine whether the installed systems worked as designed.

The plant has a control room (Fig. 5) from which all normal and emergency operation may be accomplished, including startup and shutdown. The control room instruments are all electronic; pneumatic operation is limited to smaller control valves, small remotely actuated on/off valves, and some local control loops. The central control boards are duplex

Fig. 4. Bottom head assembly being lifted into position during construction of the Fort St. Vrain Nuclear Generating Station.

Fig. 5. Control room at the Fort St. Vrain Nuclear Generating Station.

walk-in type with a bench board in the front and vertical panels in the rear. No control console is provided, as the bench board contains all the controls.

The major instrument systems are: (1) the plant control system, (2) the plant protection system, (3) nuclear instrumentation, (4) radiation monitoring, and (5) analytical instrumentation. A separate digital logger is also provided to make periodic logs of vital plant data; to do simple power, flow, and efficiency calculations; and to provide pretrip and posttrip review capabilities.

During February and March 1972, the reliability of the standby Diesel generator sets was affirmed by the completion of more than the required 298 successful starts required to achieve the required level of reliability. Also, in March 1972, the PCRV was filled with helium in preparation for running the hot flow test. In April and May 1972, preparation for the hot flow test continued with the testing of the helium circulator auxiliary systems. Such testing included rotating all four circulators, using normal, backup, and emergency bearing water systems.

The basic objectives of the hot flow test are to evaluate the performance of the primary coolant system and the internals of the prestressed concrete reactor vessel. This test includes operation of the four helium circulators under various system conditions and combinations, measurement of steam generator characteristics on the helium side, and a determination of the PCRV thermal barrier insulation effectiveness. Furthermore, it provides a means of operating with all PCRV internal components at essentially full-load cold-side primary coolant temperature. After cooldown and depressurization, a detailed posttest inspection inside the vessel is made, and then commencement of fuel loading and subsequent nuclear operation can be undertaken.

The four helium circulators are driven by temporary special oversize Pelton water turbines during the hot flow test, so that the heat of compression will provide the required energy input to raise the primary coolant temperature to approximately 700°F (371°C). Before the helium temperature is raised, however, a series of lower-temperature tests are performed on the circulators and steam generators to determine flow characteristics using various combinations of operating circulators. These tests also serve to evaluate acoustic response of the system and to determine whether potential flow-induced vibration problems exist in various areas of the steam generator modules.

The hot flow test was begun June 28, 1972. Helium temperature and pressure during the hot flow test had reached approximately 450°F (232°C) and 400 psig, respectively, in July 1972, when temporary special oversize prenuclear Pelton wheels used to drive the helium circulators experienced severe cavitation damage, which in two of the four wheels caused mechanical failure. The water used to drive these Pelton turbines was delivered by the boiler feed pumps at 3000 psig pressure and temperatures of 210–220°F (99–104°C). The Pelton turbine cavity was maintained at approximately atmospheric pressure, and it was determined that flashing of the water produced the cavitation. It was found that the design of the Pelton water nozzles and the straightening vanes behind them was partially to blame for the cavitation. These were redesigned and a Pelton turbine drain cavity pressurization system was installed in the plant. This system is designed to maintain the drain cavity at varying pressures (15 to 60 psi) above the saturation pressure of the water to suppress cavitation. The helium circulators were refurbished, and the hot flow test was resumed in May 1973. It was completed in August 1973. All the results were satisfactory.

FUEL LOADING

Following completion of the hot flow test, all efforts were directed toward preparations for fuel loading. All the fuel had been delivered to the plant site. The reactor was initially loaded by hand in air. Fuel loading involved 1482 fuel elements and approximately 2000 reflector elements that were moved from storage, inspected, and placed in the reactor vessel. Fuel loading was started in December 1973 and was completed in January 1974. It was done on a 24-hour-per-day basis and took three weeks.

START-UP

After fuel loading, two series of start-up tests are run. The first series, which includes 10 tests, is used to determine various core parameters and compliance with technical

TABLE 1.   Principal Characteristics and Operating Parameters—Fort St. Vrain Nuclear Generating Station, Platteville, Colorado (Public Service Company of Colorado)

Power output:	
Thermal rating, KW/(th)	841,000
Gross electric, kW(e)	342,000
Net electric, kW(e)	330,000
Net plant efficiency, %	39.23
Helium circuit:	
Temperature, °F:	
Core inlet	760 (404°C)
Core outlet	1430 (777°C)
Pressure, psia:	
Circulator inlet	686
Circulator outlet	700
Flow rate, pounds/hour	3.4×10^6
Core design:	
Effective diameter, feet	19.6
Height, feet	15.6
Reflector thickness inches:	
Side	47
Top	39
Bottom	47
Number of fuel elements	1482
Number of control rod drives	37
Control material	B_4C
Initial ^{235}U loading, kilograms	1,020
Initial ^{232}Th loading, kilograms	19,200
Initial ^{238}U loading, K	75
Equilibrium-annual ^{235}U, kilograms*	219
Equilibrium-annual ^{232}Th, kilograms	2,460
Average moderator temperature, °F	1380 (749°C)
Heat flux, peak, Btu/hour/square foot	120,000
Heat flux, average, Btu/hour/square foot	45,000
Heat capacity, Btu/cubic foot/°F	35
Neutron flux (initial):	
Thermal (E <2.38 eV), nv†	5.5×10^{13}
Fast (>0.18 MeV), nv	3.3×10^{13}
Total, nv	1.8×10^{14}
Excess reactivity (Cold, clean), % ΔK	13
(Hot, clean), % ΔK	6
Control rod worth (74 rods, hot), % ΔK	20
Scram insertion time, seconds	180
Power density, kW(th)/liter	6.3
Specific power avg. (equilibrium), kW(th)/kg	1,100
Fuel life (full power), years	5.1
Average burnup (equilibrium), megawatt-days/metric ton (U + Th)	100,000
Conversion ratio (equilibrium)	0.62

*Before ^{233}U recycle.
†nv = neutron flux (neutrons × velocity).

Prestressed concrete reactor vessel (PCRV):	
Overall height, feet	106
Overall diameter, feet	61
Internal height, feet	75
Internal diameter, feet	31
Liner material	Carbon steel
Liner thickness (nominal), inch	¾
Concrete, cubic yards	6,480
Number of tendons	448
Load per tendon:	
Heads, kips	1,400
Circumferential, kips	1,250
Longitudinal, kips	1,400
Reference pressure, psig	845
Normal working pressure, psia	700
Penetrations	Double closures

Helium circulators:
Number	4
Horsepower (each)	5,500
Speed (full power)	9,550

Compressor:
Type	Single stage, axial
Pressure rise, psi	14.0
Flow rate (each), pounds/second	242

Turbine:
Type	Single stage, axial
Pressure drop, psi	195
Flow rate (each), pounds/second	151
Inlet pressure, psia	843
Inlet temperature, psia	648

Steam generators:
Type	Helical coil, once-through
Number	2
Modules per steam generator	6
Feedwater temperature, °F	403 (~208°C)
Feedwater pressure, psia	3,100
Feedwater flow rate (total), pounds/hour	2.31×10^6
Outlet steam pressure, psia	2,512
Outlet steam temperature, °F	1005 (540.5°C)
Reheat flow rate (total), pounds/hour	2.25×10^6
Reheat inlet pressure, psia	649
Reheat inlet temperature, °F	670 (354°C)
Reheat outlet pressure, psia	600
Reheat outlet temperature, °F	1002 (~540°C)

Materials:
Economizer	SA 213 T-2
Evaporator	SA 213 T-22
Superheater	SB 163 Gr 11
Reheater	SB 163 Gr 11
Height of tube bundle, feet	25
Diameter of tube bundle, feet	5.5

Containment:
Type	PCRV with double penetration closures
Reactor building space	Slight negative pressure with filtered discharge

Main turbine generator:
Type	Tandem compounds, two flow, 33.5-inch blade, 3600 rpm
Rating	403 MVA at 0.85 power factor

Main condenser:
Type	Two-pass, divided water box; 2.5 inches of mercury absolute

Feedwater heaters:
Type	Closed and deaerating

Stages:
Low pressure	3
Deaerating	1
High pressure	2

specifications. These tests are designed to determine rod worth, core excess reactivity, shutdown margin, temperature coefficients, and flux distribution. In this phase of testing, the fuel-handling machine was used to remove and reinsert one fuel region of the core in helium. The results of the physics tests satisfactorily confirmed the design.

The approach to power consists of a second series of startup tests. These tests are designed to measure control rod worths at power, temperature coefficients, and the performance of major plant systems. In addition, xenon buildup and decay will be measured. Xenon spatial stability will also be determined. At various levels of power, the integrated control system will be tested for response and will be tuned as required. Several transients will be imposed to check the response of both the control system and the plant. The approach to full power is scheduled to take eight weeks. It will be followed by a 72-hour full-power demonstration run.

SUMMARY OF PLANT CHARACTERISTICS

The principal characteristics and operating parameters of the Fort St. Vrain Nuclear Generating Station at Platteville, Colorado, are summarized in Table 1.

For background and theory of high-temperature gas-cooled reactors (HTGR) see article on this topic later in this section of Handbook.

Boiling Water Reactor

BY MEMBERS, TECHNICAL STAFF, NUCLEAR ENERGY DIVISIONS, GENERAL ELECTRIC
COMPANY, SAN JOSE, CALIFORNIA

Aside from its heat source, the boiling water reactor (BWR) generation cycle is substantially similar to that found in fossil-fueled power plants. One of the first commercial BWRs was the Vallecitos BWR, a 1000-psi reactor which powered a 5-megawatt electric generator and provided power to the Pacific Gas & Electric Company grid from 1957 through 1963. Power output capabilities now range from about 650 to about 1300 MW(e)* gross, and over 40 boiling water reactors are installed worldwide, with plans and commitments from utilities for more than 100 additional reactors of this type.

Some of the more recent design advancements include: compact jet pumps with increased coolant circulation capability; increased capacity from steam separators and driers; more fuel bundles in standard pressure vessels and improvements in reactor internals arrangement; and smaller-diameter fuel rods, longer in active fuel length and arranged in an array of 64 rods (8×8) per bundle with the same external outline as the previous 7×7 design. The reduced diameter rod and increased heat-transfer surface permit increasing the heat output per bundle, while at the same time reducing the peak heat fluxes and peak linear heat-generation rate (reduced kW per foot).

The direct-cycle boiling water reactor nuclear system (Fig. 1) is a steam generating system consisting of a nuclear core and an internal structure assembled within a pressure vessel, auxiliary systems to accommodate the operational and safeguard requirements of the nuclear reactor, and necessary controls and instrumentation. Water is circulated through the reactor core, producing saturated steam, which is separated from the recirculation water, dried in the top of the vessel, and directed to the steam turbine-generator. The turbine employs a conventional regenerative cycle with condenser deaeration and condensate demineralization. The direct-cycle system is employed because of its inherently simpler design, which ultimately should mean a more reliable system with higher availability.

The steam from a boiling water reactor is, of course, radioactive. The radioactivity is primarily ^{16}N, a very short-lived isotope (7 seconds half-life) so that the radioactivity of the steam system exists only during power generation. Extensive generating experience has fully demonstrated that shutdown maintenance on a BWR turbine, condensate, and feedwater components can be performed essentially as at a fossil-fuel plant. Carry-over of long-lived radioactive particles from the primary system to the turbine/feedwater system is virtually nonexistent.†

The reactor core, the source of nuclear heat, consists of fuel assemblies and control rods contained within the reactor vessel and cooled by the recirculating water system. A 1220-MW(e) BWR/6 core consists of 748 fuel assemblies and 177 control rods, forming a core array 16 feet in diameter and 14 feet high. The power level is maintained or adjusted by positioning control rods up and down within the core. The BWR core power level is further adjustable by changing the recirculation flow rate without changing control rod position, a feature that contributes to the superior load-following capability of the BWR.

The BWR is the only light water reactor system that employs bottom-entry control rods. From the very first BWRs, bottom-entry control rods have been used because reactivity and moderator density is highest in the lower part of the core. They provide optimum power shaping characteristics for the type of core in which moderator density is varied as a function of power level. Moreover, bottom-entry and bottom-mounted control rod drives

---

*MW(e)= megawatt electric; the electric output in megawatts at the generator terminals of a nuclear power plant.

†Over 289 billion kWh of operating experience with the GE BWR (through September 1975).

allow refueling without removal of rods and drives, and allow drive testing with an open vessel before initial fuel loading or at each refueling operation.

The hydraulic system, using water at pressures higher than reactor system pressure, provides rod insertion forces far greater than any gravity or mechanical system.

The boiling water reactor requires substantially lower primary coolant flow through the core than pressurized water reactors. The core flow of a BWR is the sum of the feedwater flow and the recirculation flow, which is typical of any boiler.

Unique to the BWR is the application of jet pumps inside the reactor vessel. The jet pumps derive their driving force from the external recirculation pumps and generate about two-thirds of the recirculation flow within the reactor vessel. See Fig. 2. The jet

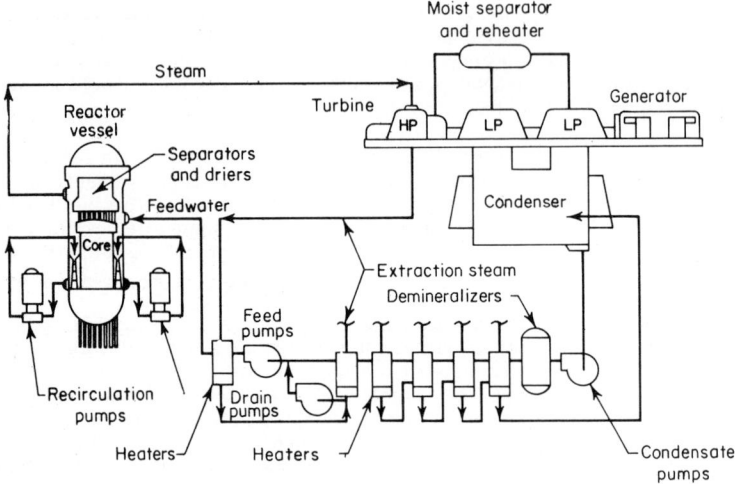

Fig. 1. Direct-cycle BWR nuclear system.

pumps also contribute to the inherent safety of the BWR design under loss-of-coolant emergency conditions because they continue to provide internal circulation with one or both external recirculation loops out of service. Moreover, the BWR can deliver about one-fourth power through this natural jet pump circulation mode, a vital capability in effecting a "black restart" of the plant without external power.

The BWR operates at constant pressure and maintains constant steam pressure similar to most fossil-fueled boilers. The BWR primary system operates at pressure about half that of a PWR primary system, while producing steam of equal pressure, enthalpy, and quality.

The integration of the turbine pressure regulator and control system with the reactor water recirculation flow control system permits automated changes in steam flow to accommodate varying load demands on the turbine. Power changes of up to 25 percent can be accomplished automatically by recirculation flow control alone, at rates of 15 percent per minute increasing and 60 percent per minute decreasing. This provides a load-following capability that can track rapid changes in power demand.

Several auxiliary systems are used for normal plant operation: (1) reactor water cleanup (RWCU) system, (2) shutdown cooling function of residual heat removal (RHR) system, (3) fuel and containment pools cooling and filtering system, (4) closed cooling water system for reactor service, and (5) radioactive waste treatment systems.

The following auxiliary systems are used as backup (standby) or emergency systems: (1) standby liquid control (SBLC) system; (2) reactor core isolation cooling (RCIC) system; (3) residual heat removal (RHR) system with (a) containment cooling function, and (b) low-pressure coolant injection (LPCI) function; (4) high-pressure core spray (HPCS) system; (5) low-pressure core spray (LPCS) system; and (6) automatic depressurization function.

NUCLEAR BOILER SYSTEM

The nuclear boiler system consists of the equipment and instrumentation necessary to produce, contain, and control the steam power required by the turbine generator. The principal components of such a nuclear boiler system are: (1) *reactor assembly*—reactor pressure vessel, jet pumps for reactor water recirculation, steam separators and driers, and

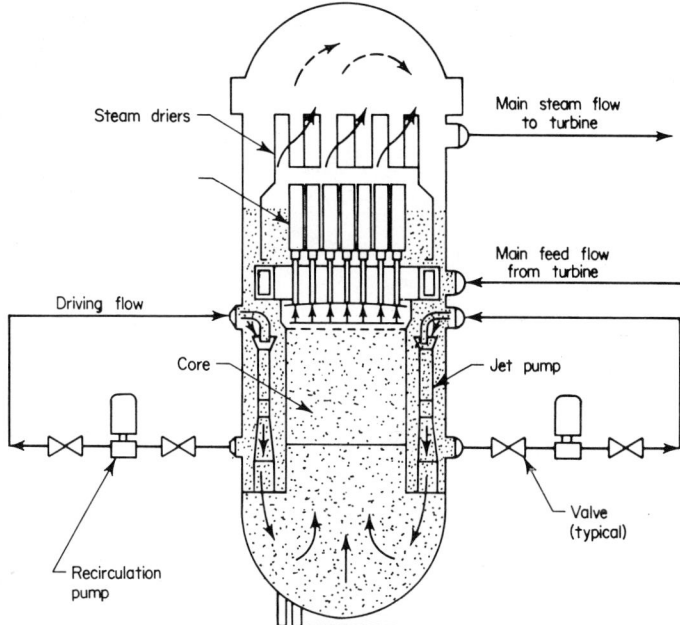

Fig. 2. Steam and recirculation water flow paths of the BWR.

core support structure; (2) *reactor water recirculation system*—pumps, valves, and piping used in providing and controlling core flow; (3) *main steam lines*—main steam safety and relief valves, piping, and pipe supports from reactor pressure vessel up to and including the isolation valves outside the primary containment barrier; (4) *control rod drive system*—control rods, control rod drive mechanisms and hydraulic system for insertion and withdrawal of the control rods; and (5) *nuclear fuel and in-core instrumentation*.

REACTOR ASSEMBLY

The reactor assembly (Fig. 3) consists of the reactor vessel, its internal components of the core, shroud, top guide assembly, core plate assembly, steam separator and drier assemblies, and jet pumps. Also included in the reactor assembly are the control rods, control rod drive housings, and control rod drives.

Each fuel assembly that makes up the core rests on an orificed fuel support mounted on top of the control rod guide tubes. Each guide tube, with its fuel support piece, bears the weight of four assemblies and is supported by a control rod drive penetration nozzle in the bottom head of the reactor vessel. The core plate provides lateral guidance at the top of each control rod guide tube. The top guide provides lateral support for the top of each fuel assembly.

Control rods occupy alternate spaces between fuel assemblies and may be withdrawn into the guide tubes below the core during plant operation. The rods are coupled to control rod drives mounted within housings which are welded to the bottom head of the reactor vessel. The bottom-entry drives do not interfere with refueling operations. A flanged joint is provided at the bottom of each housing for ease of removal and maintenance of the rod drive assembly.

Except for the *Zircaloy* in the reactor core, these reactor internals are stainless steel or other corrosion-resistant alloys. All major internal components of the reactor can be removed except the jet pump diffusers, the core shroud, the jet pump, and the high-pressure coolant injection inlet piping. The removal of the top guide assembly and the core plate assembly is a major task, and it is not expected that these components would need to be removed during the life of the plant. The removal of other components, such as fuel assemblies, in-core assemblies, control rods, fuel support pieces, and so on, is accomplished on a routine basis.

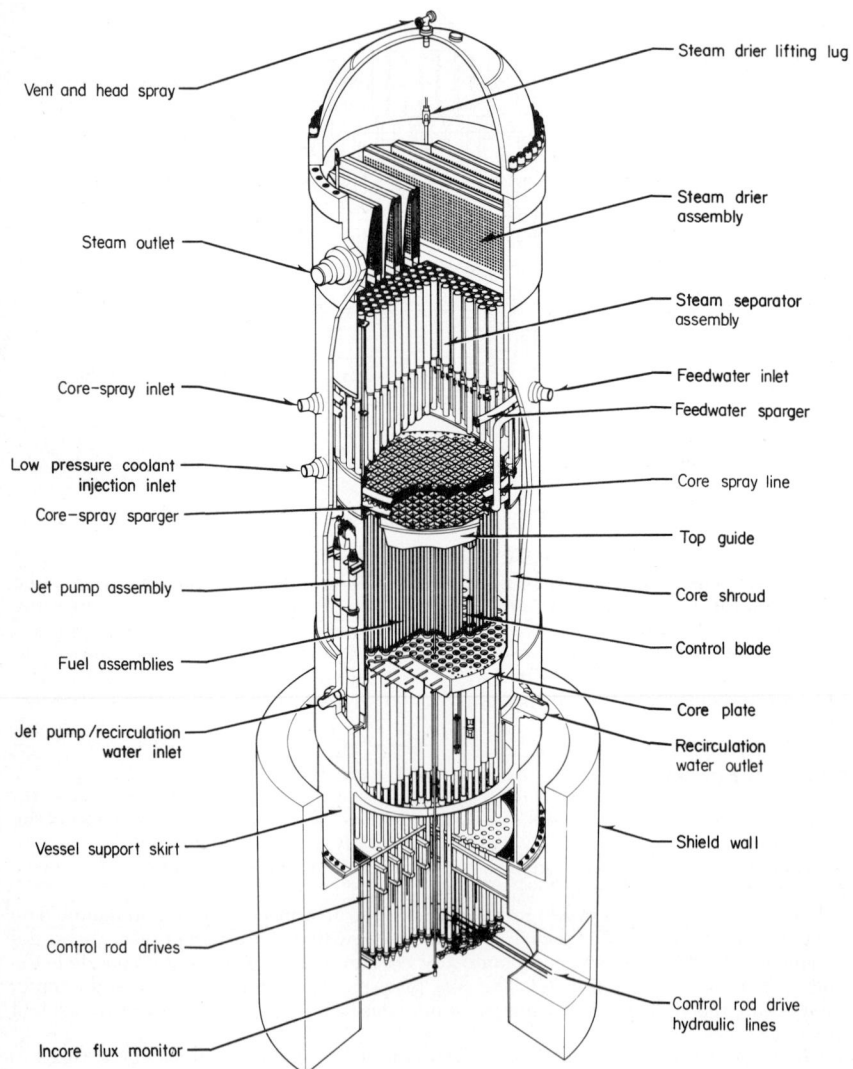

Fig. 3. Reactor assembly of the BWR.

REACTOR VESSEL

The reactor vessel is a pressure vessel with a single full-diameter removable head. The base material of the vessel is low alloy steel which is clad on the interior except for nozzles with stainless steel weld overlay to provide the necessary resistance to corrosion. Since the vessel head is exposed to saturated steam environment throughout its operating lifetime, stainless steel cladding is not used over its interior surfaces.

Fine-grained steels and advanced fabrication techniques are selected to maximize structural integrity of the vessel. BWR vessels have the lowest neutron exposure of any light water reactor. Furthermore, the annulus space which carries recirculating water and feedwater downward between the core shroud and the vessel, reduces radiation damage to the vessel wall material. Vessel material surveillance samples are located within the vessel to enable periodic monitoring of exposure and material properties. Provisions are made for irradiating both tensile and impact specimens so that the user can carry out a program of monitoring and evaluating radiation-induced changes in the vessel.

The vessel head closure seal consists of two concentric metal O rings. This seal system has been demonstrated to perform without detectable leakage at all operating conditions. These conditions include cold hydrostatic testing, heating and cooling, and power operation. To monitor seal integrity, a leak detection system is provided.

Vessel supports, internal supports, their attachments, and adjacent shell sections are designed to take maximum combined loads, including control rod drive reactions, earthquake loads, and jet reaction thrusts. The vessel is mounted on a supporting skirt which is bolted to a concrete and steel cylindrical vessel pedestal which is integral with the reactor building foundation. Steam outlet lines are welded to the vessel body, thereby eliminating the need to break flanged joints in the steam lines when removing the head for refueling. A seal between the vessel and the surrounding drywell permits flooding of the space (reactor well) above the vessel.

CORE SHROUD

The shroud is a cylindrical, stainless steel structure which surrounds the core and provides a barrier to separate the upward flow through the core from the downward flow in the annulus. A flange at the top of the shroud mates with a flange on the top guide, which, in turn, mates with a flange on the steam separator assembly to form the core discharge plenum. The jet pump discharge diffusers penetrate the peripheral shelf of the shroud support below the core elevation to introduce the coolant into the inlet plenum. The peripheral shelf of the shroud support is welded to the vessel wall to prevent the jet pump outlet flow from bypassing the core and to form a chamber around the core which can be reflooded in the event of a loss-of-coolant accident. The shroud support is designed to carry the weight of the shroud, the steam separators, the jet pump system, and the seismic and pressure loads, in both normal and fault conditions of operation.

Two ring spargers, one for low-pressure core spray and the other for high-pressure core spray are mounted inside the core shroud in the space between the top of the core and steam separator base. The core spray ring spargers are provided with spray nozzles for the injection of cooling water under emergency conditions. The core spray spargers and nozzles do not interfere with the installation or removal of fuel from the core. A nozzle for the emergency injection of neutron absorber (sodium pentaborate) solution is mounted below the core in the region of the recirculation inlet plenum.

STEAM SEPARATOR ASSEMBLY

The steam separator assembly consists of a domed base on top of which is welded an array of standpipes with a three-stage separator located at the top of each standpipe. The steam separator assembly rests on the top flange of the core shroud and forms the cover of the core discharge plenum region. The seal between the separator assembly and core shroud flanges is a metal-to-metal contact and does not require a gasket or other replacement sealing devices. The fixed axial flow type steam separators have no moving parts and are made of stainless steel.

In each separator, the steam-water mixture rising through the standpipe impinges on vanes, which give the mixture a spin to establish a vortex wherein the centrifugal forces separate the water from the steam in each of three stages. Steam leaves the separator at the top and passes into the wet steam plenum below the drier. The separated water exits from the lower end of each stage of the separator and enters the pool that surrounds the standpipes to join the downcomer annulus flow. An internal steam separator is shown schematically in Fig. 4.

STEAM DRIER ASSEMBLY

The steam drier assembly is mounted in the reactor vessel above the separator assembly and forms the top and sides of the wet steam plenum. Vertical guides on the inside of the vessel provide alignment for the drier assembly during installation. The drier assembly is supported by pads extending inward from the vessel wall and is held down in position

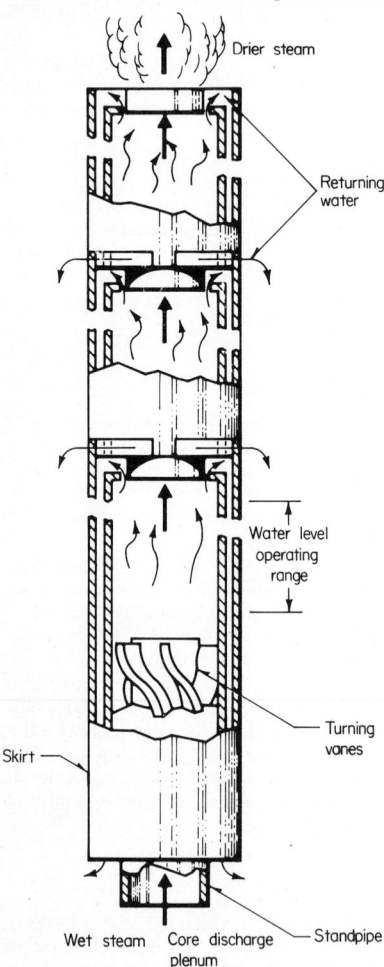

during operation by the vessel head. Steam from the separators flows upward and outward though the drying vanes. These vanes are attached to a top and bottom supporting member forming a rigid, integral unit. Moisture is removed and carried by a system of troughs and drains to the pool surrounding the separators and then into the recirculation downcomer annulus between the core shroud and the reactor vessel wall.

REACTOR WATER RECIRCULATION SYSTEM

The function of the reactor water recirculation system is to circulate the required coolant through the reactor core. The system consists of two loops external to the reactor vessel, each containing a pump with a directly coupled, water-cooled (air-water) motor, a flow control valve and two shut-off valves. High performance jet pumps located within the reactor vessels are used in the BWR recirculation system. These pumps provide a continuous internal circulation path for a major portion of the core coolant flow.

The jet pump recirculation system (see Fig. 5) provides forced circulation flow through BWR cores. The recirculation pumps take suction from the downward flow in the annulus between the core shroud and the vessel wall. Approximately one-third of the core flow is taken from the vessel through the two recirculation nozzles. There, it is pumped at a higher pressure, distributed through a manifold to which a number of riser pipes are connected, and returned to the vessel inlet nozzles. This flow is discharged from the jet pump nozzle into the initial stage of the jet pump throat where, because of a momentum exchange process, it induces

Fig. 4. Internal steam separator of the BWR.

surrounding water in the downcomer region to be drawn into the jet pump throat where these two flows mix and then diffuse in the diffuser, to be finally discharged into the lower core plenum. Operation of the jet pump principle is illustrated in Fig. 6.

MAIN STEAM LINES

Steam exits from the vessel several feet below the reactor vessel flange through four nozzles. Carbon steel steam lines are welded to the vessel nozzles, and run parallel to the vertical axis of the vessel, downward to the elevation, where they emerge from the containment. Two air-operated isolation valves are installed on each steam line, one inboard and one outboard of the primary containment penetration. The safety relief valves

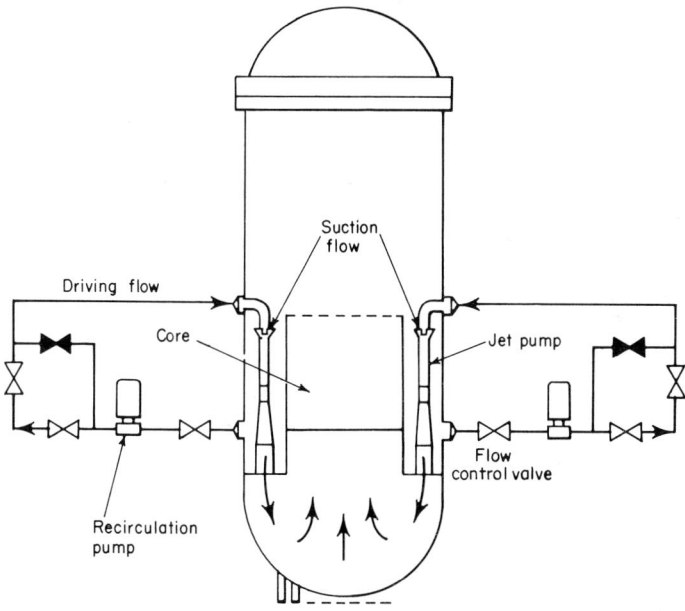

Fig. 5. Jet pump recirculation system of the BWR.

are flange-connected to the main steam line for ease of removal for test and maintenance. A flow-restricting nozzle is included in each steam line as an additional safeguard to limit the blowdown rate in the unlikely event of a break in a main steam line.

Safety/Relief Valves. The safety/relief valves are dual function valves discharging directly to the pressure suppression pool. The safety function includes protection against overpressure of the reactor primary system (in accordance with ASME Boiler and Pressure Vessel Codes). The relief function provides power-actuated valves opening to depressurize the reactor primary system. The valves are sized to accommodate the most severe of the following two pressurization transient cases determined by analysis: (1) turbine trip from turbine design power, failure of direct *scram** on turbine stop valve closure, failure of the steam bypass system, and reactor scrams from an indirect scram, or (2) closure of all main steam line isolation valves, failure of direct scram based on valve position switches, and reactor scrams from an indirect scram.

For the safety function, the valves open at spring set point pressure and close when inlet pressure falls to 96 percent of spring set point pressure. For the pressure relief function, the valves are power-actuated manually from the control room or power-actuated automatically upon high pressure (70 to 90 psi above the rated operating pressure). Each valve is supplied by separate power circuits.

Isolation Valves. The isolation valves for the main steam line are spring-loaded, pneumatic piston-operated globe valves designed to fail closed on loss of pneumatic pressure or loss of power to the pilot valves. The isolation valves close upon: (1) low water

---

*Scram signifies prompt shutdown.

in the reactor vessel, (2) high radiation from the steam line, (3) a break in the main steam line, or (4) low pressure at inlet to the turbine. The signal for closure comes from two independent channels; each channel has two independent tripping sensors for each measured variable. Once isolation is initiated, valves continue to close and cannot be opened except by manual means.

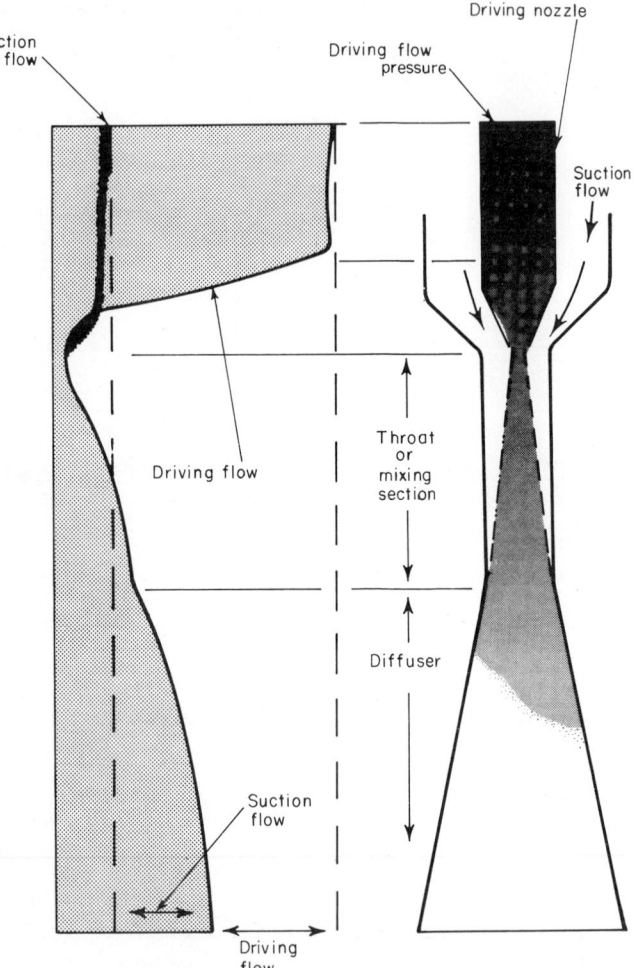

Fig. 6. Jet pump principle used in the BWR.

CONTROL ROD DRIVE SYSTEM

Positive core reactivity control is maintained by the use of movable control rods interspersed throughout the core. These control rods thus regulate the overall reactor power level and provide the principal means of quickly and safely shutting down the reactor. The rods are vertically moved by hydraulically actuated, locking piston type drive mechanisms. The drive mechanisms perform both a positioning and latching function, and a scram function, with the latter overriding any other signal. The drive mechanisms are bottom-entry, upward scramming drives which are mounted on a flanged housing on the reactor vessel bottom head. Here they cause no interference during refueling, but are readily accessible for inspection and servicing.

The control rod drive system consists of a number of locking piston control rod drive mechanisms, a hydraulic control unit for each drive mechanism, a hydraulic power supply for the entire system and instrumentation and controls. See Fig. 7. The locking-piston-type control rod drive mechanism is a double-acting hydraulic piston which uses condensate water as the operating fluid. Accumulators provide additional energy for scram. An index tube and piston, coupled to the control rod, are locked at fixed increments by a collet mechanism. The collet fingers engage notches in the index tube to prevent unintentional withdrawal of the control rod, but without restricting insertion. The drive mechanism can position the rods at intermediate increments over the entire core length.

For manual position of the control rods, the operator selects one or a group of rods (known as "gangs") to be positioned, and then by means of a control switch activates appropriate relays and valves to move (known as "jogging") the rod or gang one notch in the selected direction (only one rod or gang can be moved at a time until the movement cycle is completed). By operating a notch override switch, the rod or gang can be withdrawn several notches continuously. The use of gangs assists in the more rapid recovery from scrams, and more rapid ascensions to power from any type of restart condition; however, normal adjustments of power level may still be made with single rods.

Rod position is sensed by a series of sealed glass reed type switches. They are spaced one every 3 inches and contained within a tube inside the drive. The switches are actuated by a magnet located in the main piston. The status of all scram valves, accumulators, and drives "in" and "out" limit positions, is indicated by panel lights in the control room.

REACTOR CORE DESIGN

The design of the BWR core and fuel is based on the proper combination of many design variables and operating experience. Design parameters include moderator-to-fuel volume ratio, core power density, thermal-hydraulic characteristics, fuel exposure level, nuclear characteristics of the core and fuel, heat transfer, flow distribution, void content, cladding stress, heat flux, and operating pressure.

A number of important features of the BWR core design are summarized as follows:

1. The BWR core mechanical design is based on conservative application of stress limits, operating experience, and experimental test results. The moderate pressure level characteristics of a direct-cycle reactor (approximately 1000 psia) reduce cladding temperatures and stress levels.

2. The low coolant saturation temperature, high heat-transfer coefficients, and natural water chemistry of the BWR are significant, advantageous factors in minimizing Zircaloy temperature and associated temperature-dependent corrosion and hydride buildup. This results in improved cladding performance at long exposures. The relatively uniform fuel cladding temperatures throughout the BWR core minimize migration of the hydrides to cold cladding zones and reduce thermal stresses.

3. The basic thermal and mechanical criteria applied in the BWR design have been proved by irradiation of statistically significant quantities of fuel. The design heat fluxes and linear thermal outputs (approximate maximum of 13.4 kW per foot) are similar to values proved in fuel assembly irradiation.

4. The design power distribution used in sizing the core represents a worst expected state of operation. Provisions for nonoptimum operation ensure operational flexibility and reliability.

5. The reactor is designed so that the peak heat fluxes at rated conditions are significantly less than the critical heat flux limit (approximately 55 percent of the limiting values).

6. Because of the large negative moderator density (void) coefficient of reactivity, the BWR has a number of inherent advantages. These are the use of coolant flow as opposed to control rods for load following, the inherent self-flattening of the radial power distribution, the ease of control, the spatial xenon stability, and the ability to override xenon in order to follow load.

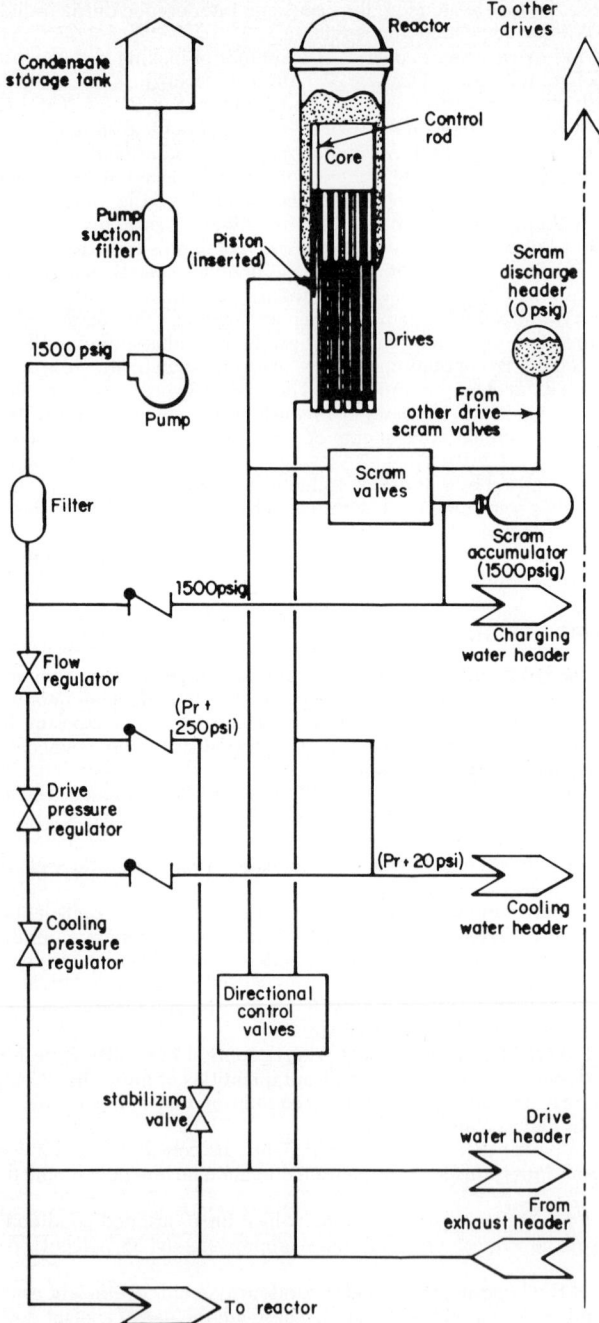

Fig. 7. Basic control-rod drive system used in the BWR. Pr = reactor pressure.

The inherent spatial xenon stability of the BWR is particularly important for large-size plants. For example, the Dresden 1 reactor has been operated for several years on full power during the day and on half power at night load schedule. This produces maximum xenon concentration gradients; yet no xenon instabilities have been observed.

Core Configuration.   The reactor core of the boiling water reactor is arranged as an upright cylinder containing a large number of fuel assemblies and located within the reactor vessel. The coolant flows upward through the core. Plan view of a typical core arrangement of a large BWR is shown in Fig. 8. The lattice configuration is shown in Fig.

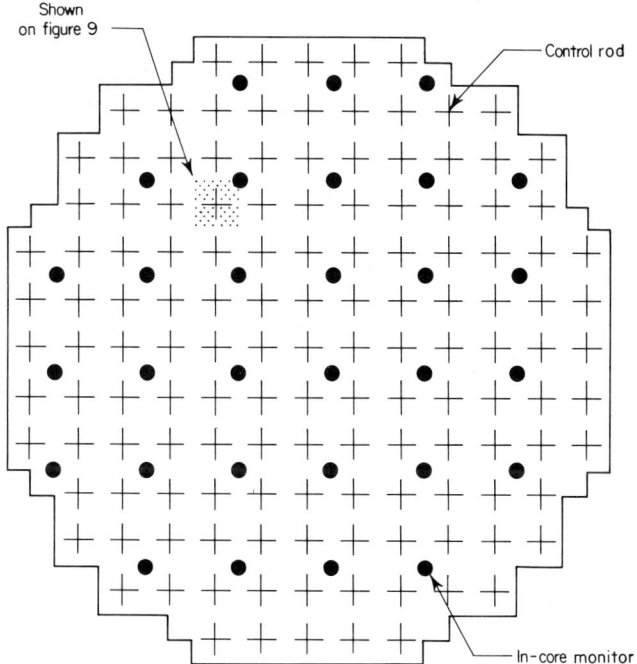

Fig. 8. Typical BWR core arrangement.

9. These figures illustrate that the BWR core comprises essentially only two components: fuel assemblies and control rods.

Fuel Rod.   A fuel rod consists of uranium dioxide (UO_2) pellets and a Zircaloy 2 cladding tube. The UO_2 pellets are manufactured by compacting and sintering UO_2 powder into cylindrical pellets and grinding to size. The immersion density of the pellets is approximately 95 percent of theoretical UO_2 density. A sample of BWR fuel pellets is shown in Fig. 10.

A fuel rod is made by stacking pellets into a Zircaloy 2 cladding tube which is evacuated, back-filled with helium to atmospheric pressure, and sealed by welding Zircaloy end plugs in each end of the tube. The pellets are stacked to an active height of 150 inches, with the top 10 inches of tube available as a fission gas plenum. A plenum spring is provided in the plenum space to exert a downward force on the pellets; this plenum spring keeps the pellets in place during the preirradiation handling of the fuel bundle.

Fuel Bundle.   Each fuel bundle contains 62 fuel rods which are spaced and supported in a square (8 × 8) array by a lower and upper tie plate. The lower tie plate has a nosepiece which fits into the fuel support piece and distributes coolant flow to the fuel rods. The upper tie plate has a handle for transferring the fuel bundle. Three types of rods are used in a fuel bundle: tie rods, water rods, and standard fuel rods. The third and sixth fuel rods

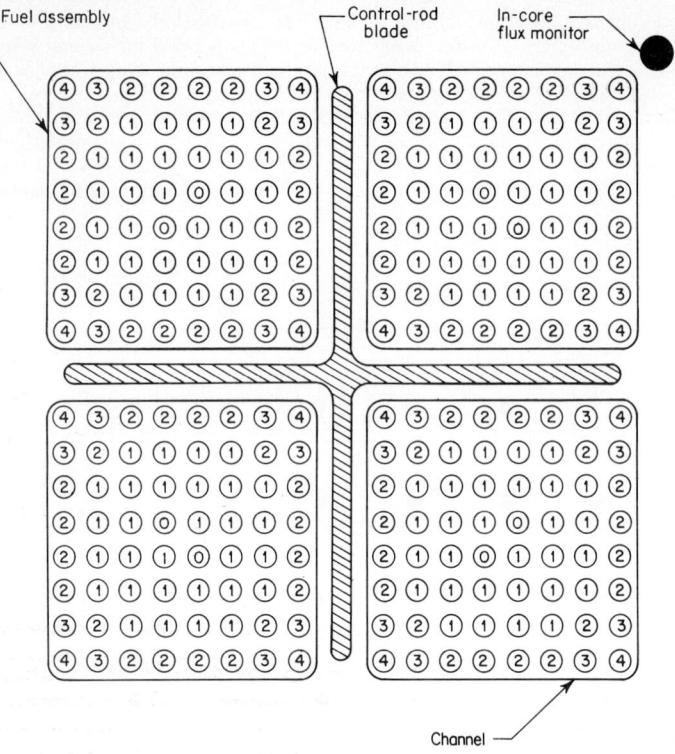

Fig. 9. Core lattice of the BWR.

along each outer edge of a bundle are tie rods. The eight tie rods in each bundle have threaded end plugs which screw into the lower tie plate casting. A stainless steel hexagonal nut and locking tab is installed on the upper end plug to hold the assembly together. The water rod not only serves as a spacer support rod, but also provides a source of moderator material near the center of the fuel bundle. This flattens the neutron flux across the bundle, and leads to lower local peaking factors and better utilization of uranium in the interior rods of the fuel assembly.

The initial core will contain fuel assemblies having a common average enrichment ranging from approximately 1.7% wt^{235} U to 2.1% wt ^{235} U, depending on initial cycle requirements. Each assembly will contain different enrichment rods. Selected rods in each assembly will, in addition, be blended with gadolinium burnable poison. The reload fuel will also contain several different enrichment rods with an average enrichment in the range of approximately 2.5 to 3.1 percent.

Fig. 10. Typical BWR fuel pellets.

Different ^{235}U enrichments are employed in fuel assemblies to reduce the local power peaking. Low enrichment uranium rods are utilized in the corner rods and in the rods nearer the water gaps; higher enrichment uranium is used in the central part of the fuel bundle. The fuel rods are designed with characteristic mechanical end fittings, one for each enrichment. End fittings are designed so that it is not mechanically possible to complete assembly of a fuel bundle with any high enrichment rods in position specified to receive a lower enrichment. A completed fuel bundle is shown in Fig. 11.

Fuel Channel. A fuel channel encloses the fuel bundle; the combination of a fuel bundle and a fuel channel is called a *fuel assembly*. See Fig. 12.

The channel is a square-shaped tube fabricated from Zircaloy-4; its outer dimensions

Fig. 11. Completed fuel bundle for the BWR.

are ~ 5.5 inches × ~5.5 inches by 167 inches long. The reusable channel makes a sliding seal fit on the lower tie plate surface. It is attached to the upper tie plate by the channel fastener assembly, consisting of a spring and a guard, and a cap screw secured by a lock washer. The fuel channels direct the core coolant flow through each fuel bundle and also serve to guide the control rods.

The use of the individual fuel channel greatly increases operating flexibility because the fuel bundle can be separately orificed, and thus the reload fuel design can be changed to meet the newest requirements and technology. The channels also permit fast in-core sampling of the bundles to locate possible fuel leakers.

NEUTRON SOURCES

Several antimony-beryllium start-up sources are located within the core. They are positioned vertically in the reactor by "fit up" in a slot (or pin) in the upper grid and a hole in the lower core support plate. The active portion of each source consists of a beryllium sleeve enclosing two antimony-gamma sources. The resulting neutron emission strength is sufficient to provide indication on the source range neutron detectors for all reactivity conditions equivalent to the condition of all rods inserted before initial operation.

The active source material is entirely enclosed in a stainless steel cladding with an outside diameter of approximately 0.7 inch. The source is cooled by natural circulation of the core leakage flow in the annulus between the beryllium sleeve and the antimony-gamma sources.

CORE DESIGN MARGINS

The reactor core is devised to operate at rated power with sufficient design margin to accommodate changes in reactor operations and reactor transients without damage to the

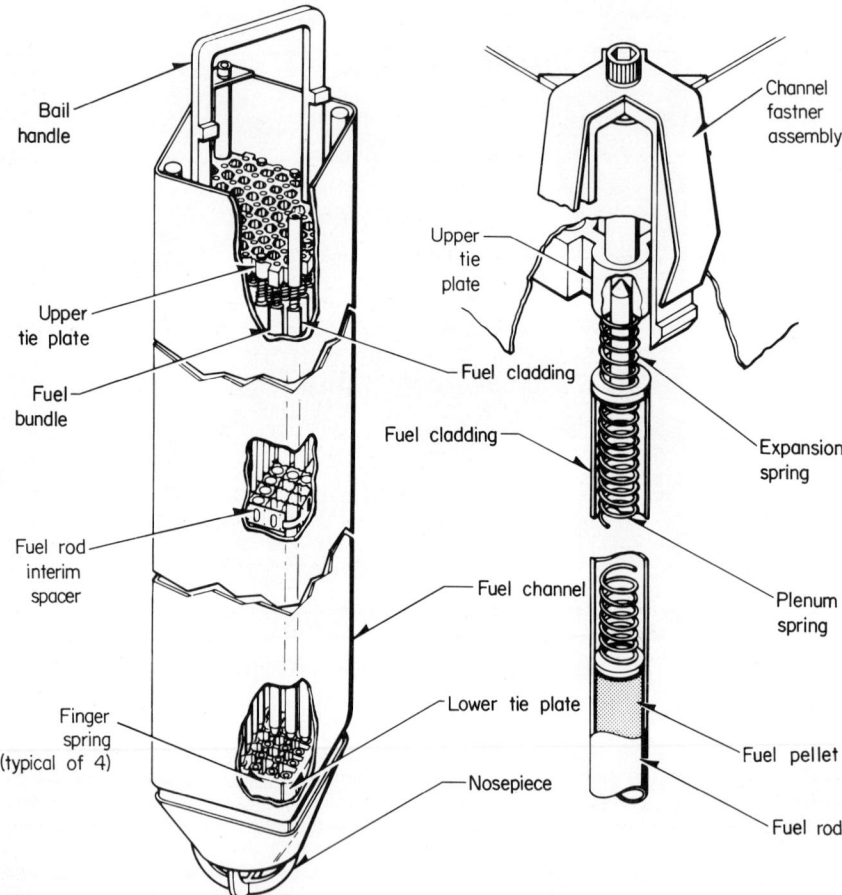

Fig. 12. Fuel assembly for the BWR.

core. In order to accomplish this objective, the core is designed, under the most limiting operating conditions and at 100 percent of rated power, to meet the following bases: (1) The maximum linear heat generation rate, in any part of the core, is always less than 13.4 kW per foot. (2) The minimum ratio between critical power and power in the bundle will not result in transition boiling during a transient.

Power Distribution.   The design power distribution is divided for convenience into several components: (1) relative assembly power, (2) local, and (3) axial. The relative assembly power peaking factor is the maximum fuel assembly average power divided by the reactor core average assembly power. The local power peaking factor is the maximum fuel rod average power in an assembly divided by the assembly average fuel rod power. The axial power peaking factor is the maximum heat flux of a fuel rod divided by the average power in that rod. Peaking factors vary throughout an operating cycle, even at steady-state full-power operation, since they are affected by withdrawal of control rods to compensate for fuel burnup. Typical target values for maximum power peaking are:

Relative assembly power	1
Axial peaking	1
Local peaking	1
Total maximum-to-average peaking	2.

It should be noted, however, that these target peaking values are not design or operational limits. These values may be exceeded so long as the basic design and operational limits (maximum linear heat generation rate and minimum critical power ratio) are not exceeded.

These design peaking factors have been selected on the basis of analysis of performance data from operation of large boiling water reactors.

Because of the presence of steam voids in the upper part of the core, there is a natural characteristic for a BWR to have the axial power peak in the lower part of the core. During the early part of an operating cycle, bottom-entry control rods permit a partial reduction of this axial peaking by locating a larger fraction of the control rods in the lower part of the core. At the end of an operating cycle, the higher accumulated exposure and greater depletion of the fuel in the lower part of the core reduce the axial peaking. The operating procedure is to locate control rods so that the reactor operates with approximately the same axial power shape throughout an operating cycle.

REACTIVITY CONTROL

The movable boron-carbide control rods are sufficient to provide reactivity control from the cold shutdown condition to the full-load condition. Supplementary reactivity control in the form of solid burnable poison is used only to provide reactivity compensation for fuel burnup or depletion effects. The movable control rod system is capable of bringing the reactor to the subcritical when the reactor is at ambient temperature (cold), zero power, zero xenon, and with the strongest control rod fully withdrawn from the core. In order to provide greater assurance that this condition can be met in the operating reactor, the core design is based on obtaining a reactivity less than 0.99, or a 1 percent margin on the "stuck rod" condition.

Supplementary solid burnable poisons are used to assist in providing reactivity compensation for fuel burnup. For all operating cycles, the supplementary control is provided by gadolinia‡ mixed into a portion of the UO_2 reload fuel rods. Only a few materials have nuclear cross sections that are suitable for burnable poisons. A satisfactory burnable poison must deplete completely in one operating cycle so that no poison residue exists to penalize initial ^{235}U enrichment requirements. It is also desirable that the positive reactivity from poison burnup match the almost linear decrease in fuel reactivity from fission product buildup and ^{235}U depletion. The concentration of gadolinia is selected so that the poison depletes in a one-year operating cycle. It is also possible to improve power distributions by spatial distribution of the burnable poison.

The control rods perform dual functions of power distribution shaping and reactivity control. Power distribution in the core is controlled during operation of the reactor by manipulation of selected patterns of rods. The rods, which enter from the bottom of the near-cylindrical reactor, are positioned in such a manner to counterbalance steam voids in the top of the core and effect significant power flattening. These groups of control elements, used for power flattening, experience a somewhat higher duty cycle and neutron exposure than the other rods in the control system.

‡(Gd_2O_3)

The reactivity control function requires that all rods be available for either reactor scram or reactivity regulation. Because of this, the control elements are mechanically designed to withstand the dynamic forces resulting from a scram.

The cruciform control rods (see Fig. 13) contain stainless steel tubes (19 tubes in each wing of the cruciform) filled with boron carbide power compacted to approximately 65 percent of theoretical density. The tubes are seal-welded with end plugs on each end. The individual tubes act as pressure vessels to contain the helium gas released by the boron-neutron capture reaction. The control rods have an active length of 144 inches of boron carbide, a span of 9.75 inches, and an overall length of 173.75 inches. They can be positioned at 6-inch steps and have a nominal withdrawal and insertion speed of 3 inches per second. They are cooled by the core leakage (by-pass) flow. In addition to satisfying initial control effectiveness requirements, it is expected that the control rods will have an average lifetime of approximately 15 full-power years.

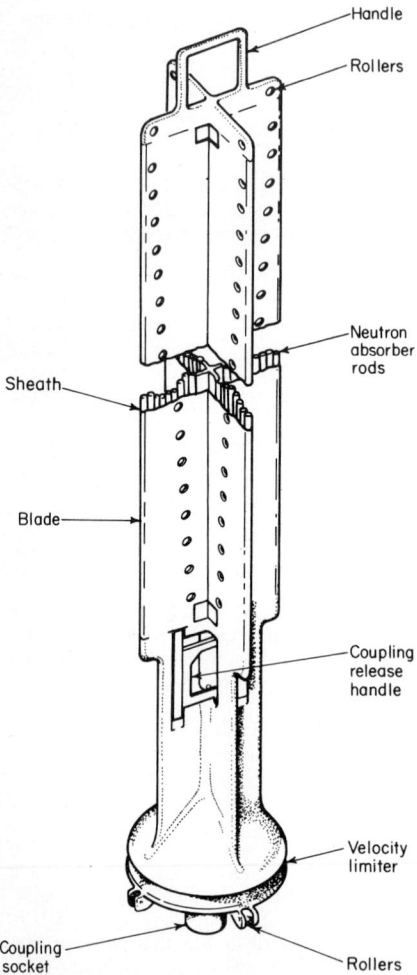

Fig. 13. Control rod used in the BWR.

REACTOR AUXILIARY SYSTEMS

Because the reactor is basically a water boiler, process systems are required which clean and control the chemistry of the water in the reactor vessel as well as protect the reactor core. Called reactor auxiliary systems, they may be divided into two general categories: (1) systems necessary for normal nuclear boiler operations, including start-up and shut-down; and (2) systems that accommodate or provide backup in case of an abnormal condition.

Reactor Water Cleanup System. The purpose of this system is to maintain high reactor water quality by removing fission products, corrosion products, and other soluble and insoluble impurities. In addition, it provides a means for water removal from the primary system during periods of increasing water volume. The cleanup system is sized to process the water volume of the reactor system approximately every 4½ hours.

Fuel and Containment Pools Cooling and Cleanup System. This system accommodates the beta and gamma radiation heating from the fission products that remain in the spent fuel, as well as dry-well heat transferred to the upper containment pool. Of equal importance is the cleanup function of the system. Because reactor refueling is accomplished through visual location and fuel bundles are removed by means of grapples, the water must be extremely clear. Thus, the filtration and demineralization functions of the system permit good visibility, even at great depths, during refueling.

Closed Cooling Water System for Reactor Service. This system consists of a separate, forced circulation loop. It uses water piped from the site service water source to provide a heat sink for selected nuclear system equipment. Its purpose is to provide a second barrier between the primary system containing radioactive products and the service water system that is the final heat sink and, therefore, eliminates the possibility of radioactive discharge into plant effluents that could result from heat exchanger leaks. Any possible radioactive leakage from the selected nuclear equipment would be to, and would be confined in, the closed-loop cooling water system which is monitored continuously for radioactivity.

Emergency Equipment Cooling System.   This system services certain equipment required for normal and emergency shutdown of the plant. It provides cooling water for the residual heat removal system pump motor and pump seal cooler and the high-pressure and low-pressure core spray systems pump motors and pump seal coolers.

Standby Liquid Control System.   This is a redundant control system capable of shutting the reactor down from rated power operation to the cold condition in the postulated situation that the control rods cannot be inserted. The operation of this system is manually initiated from the reactor control room.

Reactor Core Isolation Cooling System.   This system maintains sufficient water in the reactor pressure vessel to cool the core and then maintain the nuclear boiler in the standby condition in the event that the vessel becomes isolated from the turbine steam condenser and from feedwater makeup flow. The system also allows for complete plant shutdown under conditions of loss of the normal feedwater system by maintaining the necessary reactor water inventory until the reactor vessel is depressurized, allowing the operation of the shutdown cooling function of the residual heat removal system. The system delivers rated flow within 30 seconds after initiation.

Emergency Core Cooling System (ECCS).   The boiling water reactor ECCS utilizes several alternative systems to provide a high degree of redundancy, and ample core cooling is supplied by two separate mechanisms: flooding and spraying. It comprises the low-pressure coolant injection function of the residual heat removal system, and the high- and low-pressure core spray systems, all of which are assisted by automatic depressurization systems. The aggregate of this system is designed to protect the reactor core against fuel cladding damage (fragmentation) across the entire spectrum of line break accidents. Additionally, adequate safeguards are provided by any one system operating by itself. Any two systems operating provide more than adequate protection, and with all three systems operating there would be very little perturbation of the core and essentially no overheating of the core.

High-Pressure Core Spray System.   The purpose of this system is to depressurize the nuclear boiler system and to provide makeup water in the event of a loss of reactor coolant inventory. In addition, it prevents fuel cladding damage if the core becomes uncovered because of loss of coolant inventory by directing this makeup water down into the area of the fuel assemblies.

Low-Pressure Core Spray System.   This system also functions to prevent fuel cladding damage if the core is uncovered by the loss of coolant.

Automatic Depressurization System.   Blowdown, through selected safety/relief valves, in conjunction with the operation of the low-pressure coolant injection mode of the residual heat removal system and/or low-pressure core spray system, functions as an alternative to the operation of the high-pressure core spray system for protection against fuel cladding damage upon loss of coolant over a given range of steam or liquid line breaks. A delay of approximately 2 minutes after receipt of the relevant signals allows the operator time to bypass this system if the signals are erroneous or if the condition has corrected itself.

Residual Heat Removal System.   This system removes residual heat generated by the core under normal (including hot standby) and abnormal shutdown conditions.

REFERENCES

Kramer, A. W.: "Boiling Water Reactors," Addison-Wesley, Reading, Mass., 1958.

Argonne National Laboratory, Lemont, Ill.: "The Experimental Boiling Water Reactor," 1957.

Levy et al.: Large Boiling Water Reactors—Operations Confirm Designs, *Proc. Amer. Power Conf.*, **33**, 207, Chicago.

Ybarrondo et al.: The *Calculated* Loss of Coolant Accidents: a Review, *A.I.Ch.E. Monogr.* 7, pp. 9, 12–41, American Institute of Chemical Engineers, New York, 1972.

General Electric Co., San Jose, Calif.: "General Description of a Boiling Water Reactor," 1975.

"General Electric Standard Safety Analysis Report," U.S. Atomic Energy Commission, Document Room, Washington, 1973.

Pressurized Water Reactor

In a typical pressurized water reactor system, heat generated in the nuclear core is removed by water (reactor coolant) circulating at high pressure through the primary circuit. The water in the primary circuit cools and moderates the reactor. The heat is transferred from the primary to the secondary system in a heat exchanger, or boiler, thereby generating steam in the secondary system. The steam produced in the steam generator, a tube-and-shell-type heat exchanger, is at a lower pressure and temperature than the primary coolant. Therefore, the secondary portion of the cycle is similar to that of the moderate-pressure fossil-fueled plant.

In contrast, in boiling-water or direct-cycle systems, steam is generated in the core and is delivered directly to the steam turbine. See also Boiling Water Reactor in this Handbook section.

The similarities of basic pressurized water reactor design from one manufacturer to another are more striking than the differences. Therefore, the description of one particular configuration* can suffice to convey the general operating principles. The major components of a pressurized water reactor (PWR) are (1) the reactor vessel which contains the oxide fuel core, core internals, control element assemblies, and in-core instruments; (2) the electrically heated pressurizer; (3) the electric-motor-driven primary coolant pumps; and (4) the U-tube-type steam generators. Typically, the 3800-megawatt system† consists of two coolant loops, each containing a steam generator and two primary coolant pumps. See Fig. 1. Improvements within the primary loop system, as compared with earlier designs, include a redesigned rugged core internals structure, a fuel assembly with substantially lower linear heat rating, improved and simplified control element assemblies, a U-tube steam generator producing 1070 psia steam at better than 99.75 percent quality, an improved pressure vessel support system, an improved bottom-mounted incore instrumentation system, and improved instrumentation and control systems. The overall characteristics of the system are given in Table 1.

The primary coolant system layout can be fitted into a variety of containment types and concepts. A pre-stressed cylindrical containment is common. Figure 2 shows the arrangement in a spherical containment. This type of building lends itself to separation of safeguards equipment, steam lines, and emergency power supplies.

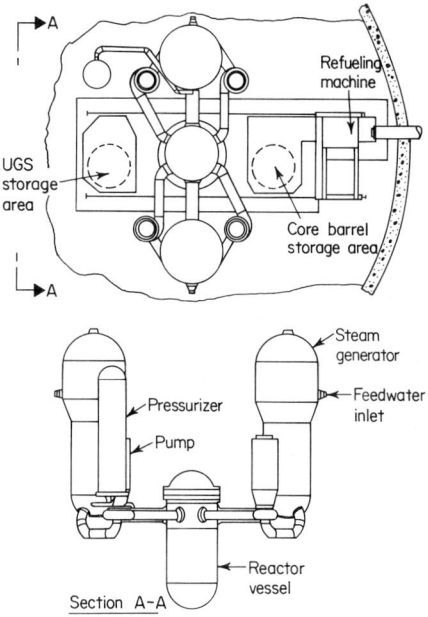

Fig. 1. Nuclear steam supply system. (*Combustion Engineering, Inc.*)

STEAM GENERATORS

The general design parameters of the two U-tube steam generators used are given in Table 2. The basic geometry is shown in Fig. 3. With the nuclear steam supply

---

*The assistance of Combustion Engineering, Inc., Windsor, Conn., in providing technical data and diagrams is much appreciated.
†Combustion Engineering System 80.

TABLE 1. General Characteristics of Pressurized Water Reactor*

Core power, megawatts	3800
Number of fuel assemblies	241
Active length of core, inches	150
Power density, kilowatts/liter	95.9
Fuel assembly dimensions, inches	7.98×7.98
Number of fuel rods per assembly	236
Fuel rod outside diameter, inch	0.382
Clad thickness to outside diameter ratio	6.5×10^{-3}
Maximum linear power density, kilowatts/foot	12.5
Average kilowatts/foot	5.2
H_2O/UO_2 volume ratio	2.02
Specific power, kilowatts/kilograms uranium	37.0
Number of control rod drives	89
Number of control elements fingers	708
Primary system pressure, psia	2250
Reactor coolant average temperature	594°F, 312.2°C
Coolant flow rate, pounds/hour	164×10^6
Maximum core heat flux, Btu/(hour)(square feet)	425,400
Minimum departure from nucleate boiling ratio (DNBR)	2.13
Secondary steam pressure, psia	1070

*Combustion Engineering System 80.

system operating at 3817 megawatts, two steam generators produce a total of 17.18×10^6 pounds of steam per hour at 1070 psia. The steam generators are constructed using carbon steel pressure-containing members and Inconel-600 tubes. The tube sheet is clad by weld

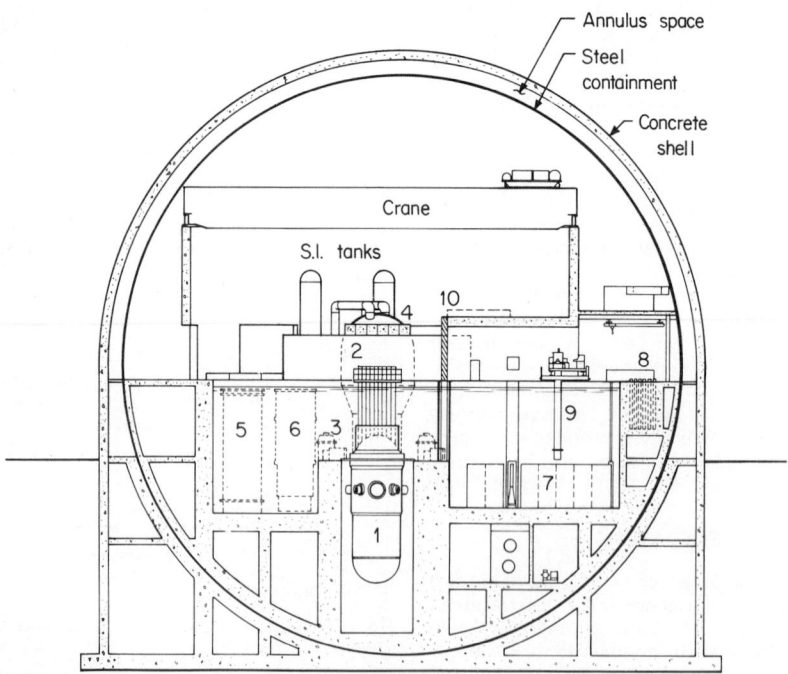

Fig. 2. Spherical containment for pressurized water reactor: (1) Reactor vessel. (2) Steam generator. (3) Reactor coolant pump. (4) Missile shield. (5) CSB storage. (6) UGS storage. (7) Spent fuel. (8) New fuel. (9) Refueling machine. (10) Removable shield wall.

deposit for maximum strength; tongue and groove construction of the divider plate places no stress on the tube-sheet cladding.

Fusion welding of the end of each tube to the tube-sheet primary cladding provides an effective seal for leakage control, and "expanding" (explosively expanding) the tubes in the full length of the tube sheet eliminates corrosion-prone crevices.

An economizer section on these units improves heat transfer by preheating the incoming feedwater using the low (primary side) temperature heat transfer area of the U-tubes. Multiple feed nozzles allow the economizer flow distribution to be optimized for each power level.

The steam generator is supported at the bottom by a sliding base bolted to an integrally attached conical skirt. The sliding base rests on low friction bearings which allow unrestrained thermal expansion of the reactor coolant system. Two keyways within the sliding base guide the movement of the steam generator during expansion and contraction of the reactor coolant system and, together with a stop and anchor bolts, prevent excessive movement of the bottom of the steam generator during seismic events and after a loss-of-cooling accident (LOCA). A system of keys and snubbers located on the steam drum guide the top of the steam generator during expansion and contraction of the reactor coolant system and provide restraint during seismic events and after a LOCA or a steam line break.

REACTOR COOLANT PUMPS

As indicated in Fig. 1, four reactor coolant pumps are used, two for each steam generator. The pumps are vertical, single-bottom-suction, horizontal-discharge, motor-driven centrifugal units. Typical coolant pump parameters are given in Table 3. The pump impeller is keyed and locked to its shaft. Pump shaft alignment is maintained by a water-lubricated bearing within the pump and by radial and thrust bearings. The pump and motor shafts are directly connected by a coupling.

The shaft seal assembly consists of three face-type mechanical seals in series, with controlled leakage bypass to provide the same pressure differential across each set. The seals are designed for 2500 psi differential, and reduce the leakage pressure from primary coolant pressure to volume control tank pressure. A fourth face-type, low-pressure vapor seal at the top is designed to withstand system operating pressure when the pumps are not operating. The leakage past the third pressure seal and the controlled leakage are piped to the volume control tank in the chemical and volume control system, and leakage past the

TABLE 2.   Steam Generator Parameters*

Number of units	2
Heat transfer rate, each, Btu/hour	6.512×10^9
Primary side:	
Coolant inlet temperature	621.2°F, 327.3°C
Coolant outlet temperature	564.5°F, 295.8°C
Coolant flow rate, each, pounds/hour	82×10^6
Coolant volume at 68°F (20°C), each, cubic	
feet	2,158
Tube size, outside diameter, inch	0.75
Tube thickness, nominal, inch	0.042
Secondary side:	
Steam pressure, psia	1070
Steam flow rate (at 0.25% moisture), each,	
pounds/hour	8.59×10^6
Feedwater temperature at full power	450°F, 232°C
Moisture carryover, % (weight), maximum	0.25
Primary inlet nozzle, No./inside diameter, inches	1/42
Primary outlet nozzle, No./inside diameter,	
inches	2/30
Steam nozzle, No./inside diameter, inches	2/28
Feedwater nozzles, No./size	1/14
No./size	2/12
Auxiliary feedwater nozzle, No./size	1/6

*Combustion Engineering System 80.

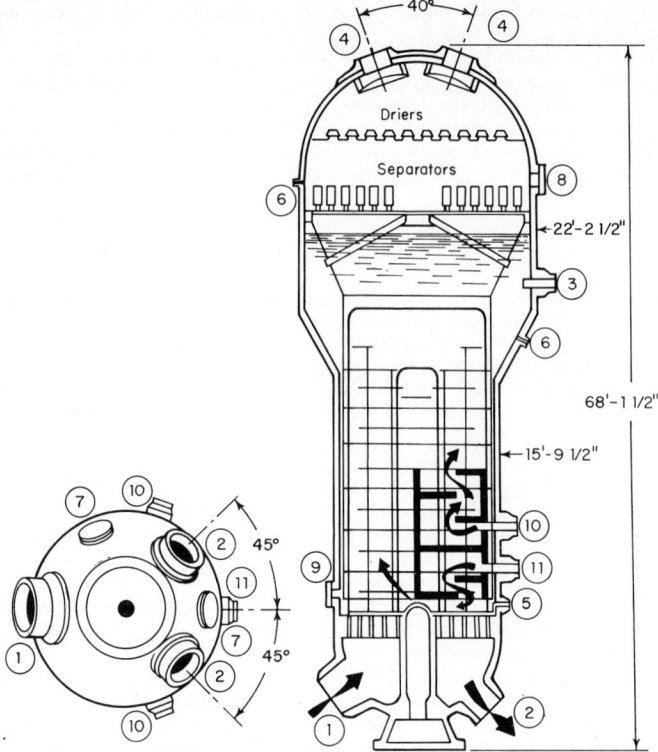

Fig. 3. Steam generator of the PWR: (1) Primary inlet. (2) Primary outlet. (3) Auxiliary feedwater. (4) Steam outlet. (5) Blowdown. (6) Liquid level. (7) Primary manway. (8) Secondary manway. (9) Handhole. (10) Upper feedwater. (11) Lower feedwater.

low-pressure vapor seal is collected and piped to the reactor drain tank. This maintains essentially zero leakage to the containment. In the event of seal malfunctions, instrumentation is provided to alert the operator.

The motors are designed to start and accelerate to speed under full load with a drop to 80 percent of normal rated voltage at the motor terminals. Each motor is provided with an antireverse rotation device. The device is designed to withstand the following conditions:

TABLE 3. Reactor Coolant Pump Parameters*

Number of units	4
Type	Vertical, single-stage, centrifugal
Design total dynamic head, feet	363
Design flow, gallons/minute	111,400
Design pressure, psia	2,500
Design temperature	650°F, 343°C
Normal operating pressure, psia	2,250
Normal operating temperature	565°F, 296°C
Shaft seals	Mechanical face seals
Pump speed, rpm	1,190
Motor synchronous speed, rpm	1,200
Motor type	Ac induction
Horsepower, hot	8,991
cold	12,230

*Combustion Engineering System 80. Parameters are related to four-pump, full-power operating conditions.

(1) motor starting torque if the motor is incorrectly wired for reverse rotation, (2) reactor coolant flow through the pump in the reverse direction of up to 60 percent rated capacity with the motor deenergized, and (3) reverse flow due to LOCA at the suction leg.

Each motor also has a flywheel to increase the rotating inertia of the pump and maintain flow in the event of loss of power to the motor during power operation. The design overspeed of the flywheel is 125 percent of rated speed. An overspeed test of each flywheel is performed before assembly.

Each reactor coolant pump is provided with four vertical support columns, four horizontal support columns, and one vertical snubber. The structural columns provide support for the pumps during normal operation, earthquake conditions, and any hypothetical loss-of-coolant accident in either the pump suction or discharge line.

PRESSURIZER

The pressurizer is a cylindrical pressure vessel, vertically mounted and bottom-supported. Energy to the water is supplied by replaceable, direct-immersion electric heaters which are inserted from the bottom head of the pressurizer. Nozzles are provided for spray, surge, relief, and instrumentation connections. A schematic diagram of the pressurizer is shown in Fig. 4.

The pressurizer maintains reactor coolant system operating pressure and, in conjunction with the chemical and volume control system, compensates for changes in reactor coolant volume during load changes, heat-up and cool-down.

During full-power operation, the pressurizer is about one-third full of saturated steam. Reactor coolant system pressure may be controlled automatically or manually by maintaining the temperature of the pressurizer fluid at the saturation temperature corresponding to the desired system pressure. A small continuous spray flow is maintained in the pressurizer to avoid stratification of pressurizer boron concentration and to maintain the temperature in the surge and spray lines. A number of the heaters are connected to proportional controllers, which adjust the heat input to account for steady-state losses and to maintain the desired steam pressure in the pressurizer. The remaining backup heaters are connected to on-off controllers. These heaters are normally deenergized, but are automatically turned on by a low pressurizer pressure signal. A low-low pressurizer water level signal deenergizes all heaters before they are uncovered, to prevent heater damage. A small continuous flow is maintained through the spray lines at all times to keep those lines and the surge line warm, thereby reducing thermal shock as the spray control valves open. Typical pressurizer parameters are given in Table 4.

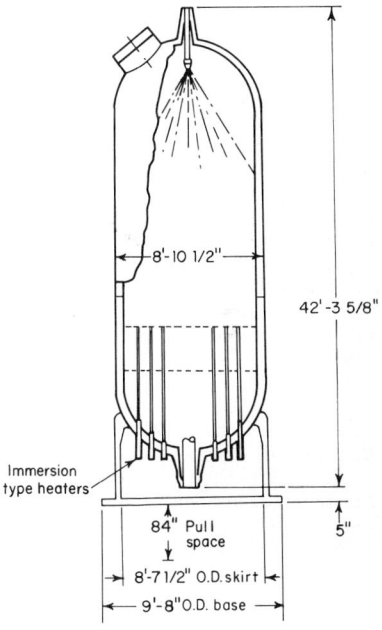

Fig. 4. Pressurizer used in the PWR.

REACTOR VESSEL

The reactor vessel is designed to contain the fuel bundles, the control element assemblies, and the internal structures necessary for support of the core. The reactor is a stainless clad, thick-walled, carbon steel pressure vessel comprised of a cylinder with two hemispherical heads. The lower head is integrally welded to the vessel shell and contains in-core instrumentation nozzles; the upper closure head, containing the control element drive mechanism nozzles, is attached to the vessel by means of a bolted flange, thus permitting the head to be removed to provide access to the reactor internals. The head flange is drilled to match the vessel flange stud bolt locations.

The vessel flange is a forged ring with a machined ledge on the inside surface to support the core support barrel. The flange is drilled and tapped to receive the closure studs and is machined to provide a mating surface for the reactor vessel closure seals. Sealing is accomplished by using two silver-plated, NiCrFe-alloy self-energizing O rings. The space between the two rings is monitored to detect any inner-ring coolant leakage.

The inlet and outlet nozzles are located radially on a common plane below the vessel flange. Extra thickness in the vessel course supplies the reinforcement required for the nozzles. Additional reinforcement is provided for the individual nozzle attachments.

Snubbers built into the lower portion of the reactor vessel shell limit the amplitude of any displacement of the core support barrel. Core stops are also built into the reactor vessel to limit the downward displacement of the core support barrel.

Cladding for the reactor vessel is a continuous integral surface of corrosion-resistant material, having $\frac{7}{32}$-inch nominal thickness, and a $\frac{1}{8}$-inch minimum thickness. The

TABLE 4. Pressurizer Parameters*

Design pressure, psia	2500
Design temperature	700°F, 371°C
Operating pressure, psia	2250
Operating temperature	652.7°F, 344.5°C
Internal volume, cubic feet	1800
Heaters:	
Type/rating, kilowatts	Immersion/50
Installed heater capacity,	
kilowatts	1800

*Combustion Engineering System 80.

detailed procedures used (i.e., type of electrodes, welding position, speed of welding, and nondestructive testing requirements) are in compliance with ASME Section III.

The reactor vessel is supported by four vertical columns located under the vessel inlet nozzles. These columns are designed to flex in the direction of horizontal thermal expansion and thus allow unrestrained heat-up and cool-down; the columns also act as holddown devices for the vessel. The supports are designed to accept normal loads and seismic and pipe rupture accident loads.

Reactor Arrangement. The reactor arrangement is shown in Fig. 5. In comparison with earlier designs, (1) the core support assembly has been redesigned to lower the reactor core and contain bottom-mounted in-core instrumentation nozzles, and (2) the upper guide structure has been redesigned as a barrel-calandria arrangement to accommodate a new concept in control element arrangement and shrouding.

Lowering the reactor core within the reactor vessel decreases the amount of emergency core cooling system (ECCS) water needed before the water level reaches the bottom of the fuel during post-LOCA refill, thereby improving core cooling.

The high coolant flow passes out of the vessel through the calandria structure of the upper guide structure which protects the individual fingers of the control element assembly. Coolant flow in the barrel region above the calandria is restricted to a level sufficient to ensure mixing and minimize blowdown forces. Simple, open-structure, perforated shrouds are used to guide the control element assembly in the upper region.

Upper Guide Structure. The barrel-calandria guide structure is a rugged (3-inch-thick barrel section) unit which can withstand the effect of, and protect all control element fingers from, the combined effects of seismic and blowdown loads resulting from a LOCA. The calandria structure fits over the control element guide tubes of the fuel assemblies, aligning all fuel assemblies, and laterally restraining the top ends of the fuel assemblies. With the upper guide structure in place, a continuous guide tube for each control finger is formed, extending from the top of the tube sheet to the bottom of the fuel assembly. Because of this feature, which isolates every control finger from the coolant crossflow, flexibility is obtained in the number of control fingers that can be attached to one control assembly: i.e., one control element assembly can serve more than one fuel assembly.

Fuel and Control Rods. Severe emergency core cooling system criteria require that the builders of water reactors increase the linear feet of fuel in the reactor core for the same power in order to reduce LOCA fuel temperatures. In the unit described here, an assembly with a 16 × 16 fuel rod array of smaller-diameter rods is used in the same assembly envelope as previously occupied by a 14 × 14 assembly in an earlier design. This results in a maximum linear heat rate decrease in the assembly of about 25 percent.

As shown in Fig. 6, the active core is made up of 241 fuel assemblies, all of which are mechanically identical. As indicated by Fig. 7, each fuel assembly contains 236 Zircaloy-

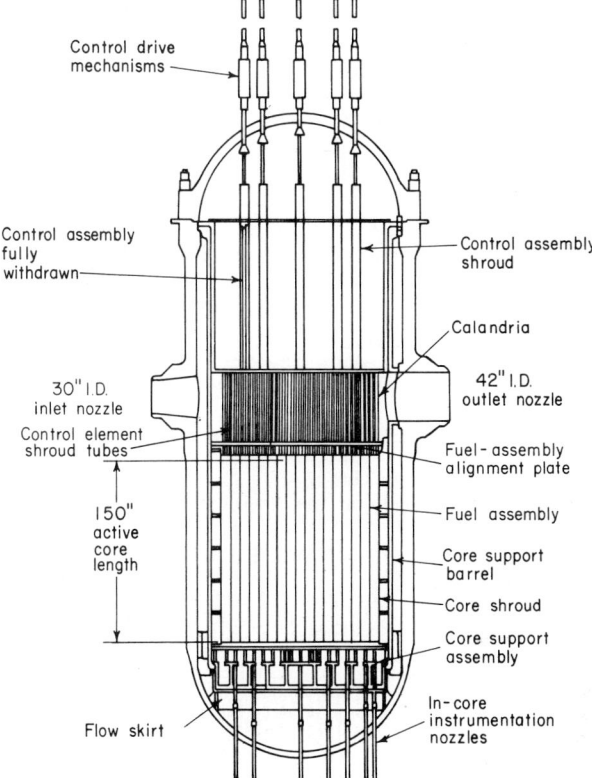

Fig. 5. Reactor arrangement used in the PWR. (*Combustion Engineering, Inc.*)

clad, UO_2 fuel rods retained in a structure consisting of Zircaloy spacer grids welded at about 15-inch intervals to five Zircaloy control element assembly guide tubes which, in turn, are mechanically fastened at each end to stainless steel end fittings. The overall length of the fuel assembly is about 177 inches, and the cross section is about 8 by 8 inches. Each fuel assembly weighs about 1450 pounds.

All the structural materials in the active core zone, including the control element assembly guide tubes and the spacer grids, are Zircaloy, which minimizes the enriched uranium requirements of the core and eliminates concern over the formation of a lower-melting-point eutectic during a LOCA as a result of using dissimilar metals.

Each rod is free to grow axially through the spacer grids, whose function is to restrain the fuel rods laterally by means of a spring support.

Fuel Rod. Typical design of a fuel rod is shown in Fig. 7. The fuel rods, consisting of uranium dioxide (UO_2) pellets of low enrichment canned in thin-walled Zircaloy-4 tubing, are designed to achieve average burnups of about 33,000 MWD/MTU (thermal megawatt-days per metric tons of uranium) and peak burnups of about 50,000 MWD/MTU. The design factors limiting burnup of the fuel are the effects on the clad of volumetric changes of the fuel pellet and fission gas release.

As indicated in Fig. 7, the fuel rod consists essentially of 0.325-inch-diameter, 0.390-

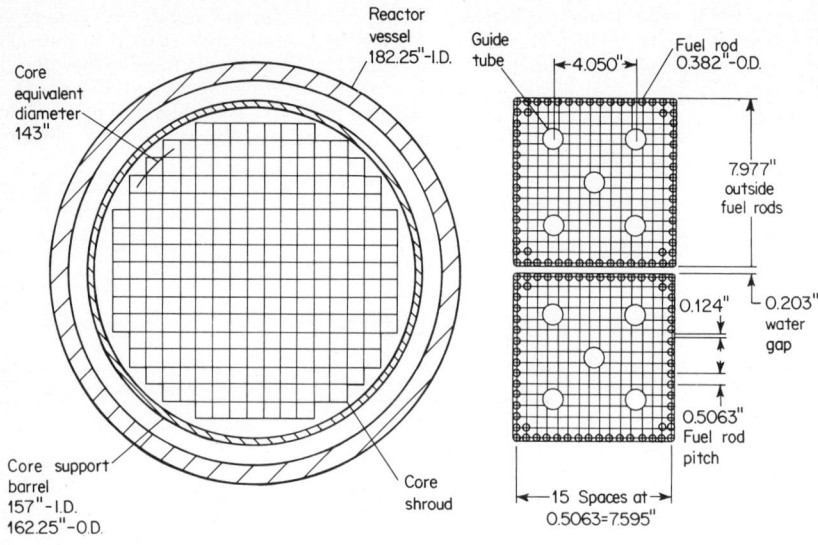

Fig. 6. Reactor core cross section comprising 241 fuel assemblies (System 80).

inch-long UO_2 pellets canned in a 0.382-inch-outside-diameter Zircaloy-4 tube. The high-density fuel pellets are dished at both ends to allow for axial differential thermal expansion and fuel volumetric growth with burnup.

Selection of Zircaloy-4 for the cladding was based on (1) low thermal neutron absorption cross section, (2) good corrosion resistance in the water coolant, (3) improved hydriding behavior compared to that of Zircaloy-2, and (4) good strength properties. In

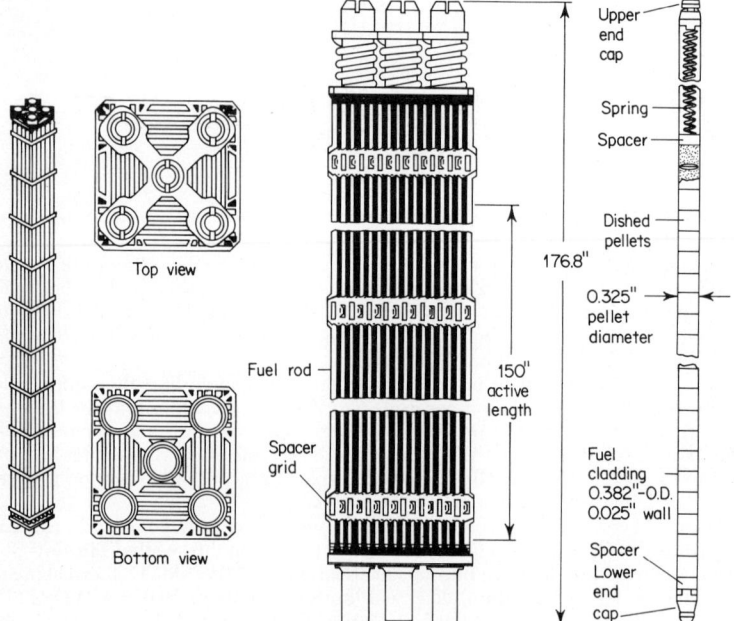

Fig. 7. Fuel assembly used in the PWR (System 80).

addition to accommodating differential thermal expansion effects, the cladding must also be designed to accept without failure radial fuel volumetric growth. The design criterion generally accepted as satisfactory is to limit the clad net unrecoverable circumferential strain to 1 percent. In applying this criterion, consideration must be given to the linear heat rate as a function of burnup.

Control Element Assemblies.   The control element assemblies consist of an assembly of 4, 8, or 12 fingers approximately 0.8 inch outside diameter arranged as shown in Fig. 8.

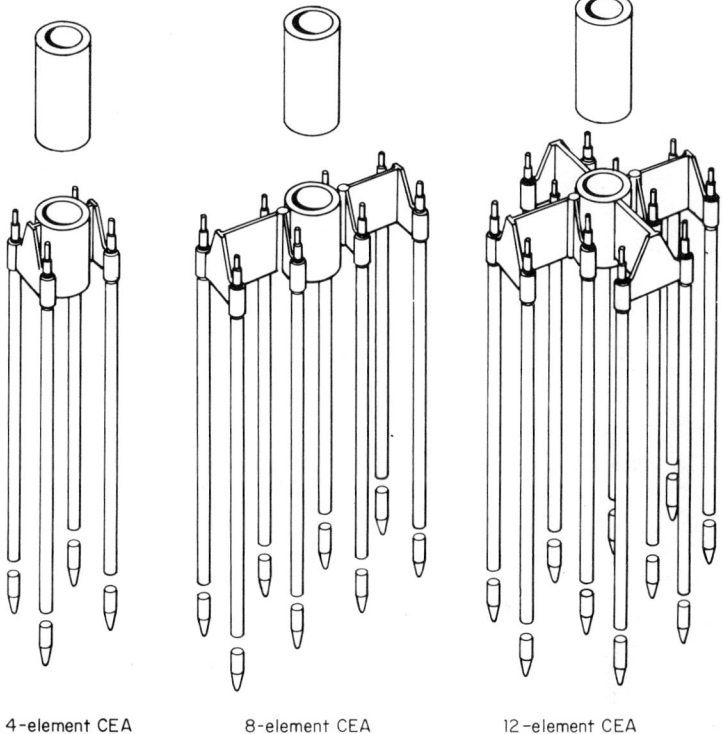

4-element CEA   8-element CEA   12-element CEA

Fig. 8. Four-, eight-, and twelve-element control element assemblies used in the PWR (System 80).

Control Rod Local Peaking.   The use of cruciform control rods, as in boiling water and in early pressurized water reactor designs, necessitates large water gaps between the fuel assemblies to ensure that the control rods will scram* satisfactorily. These gaps cause peaking of the power in fuel rods adjacent to the water channel compared to fuel rods some distance from the channel.

A five-hole fuel assembly design was evolved from consideration of the lower peaking effect of smaller (removal of one fuel rod) water holes versus the mechanical advantages of larger (four fuel rods removed) water holes. The larger water holes allowed the use of rugged, 0.9-inch outside-diameter by 0.035-inch-thick Zircaloy guide tubes for the fuel assembly structure. These water holes are distributed relatively uniformly in the reactor core when placed in the 16 × 16 fuel rod lattice. The particular arrangement of water holes was selected in consideration of the water gap between fuel assemblies; the effect of the central water hole in the fuel assembly is balanced by the water gap between fuel assemblies. The mechanical simplicity and ruggedness outweigh the advantage of obtaining a small decrease in local peaking by using very small fingers. The slightly higher peaking associated with the design can be compensated, to a large extent, by varying the

*Scram signifies prompt shutdown.

enrichment of the fuel in the rods adjacent to the control channel and/or by using water displacers in local hot spots.

The control element assembly, shown in Fig. 9, consists of 0.8-inch outside diameter Inconel tubes containing boron carbide pellets as the neutron absorbing material. A gas plenum is provided in order to limit the maximum stress due to generation of internal gas pressure.

The individual control fingers are attached mechanically and locked to the various spider assemblies. This allows for simplifications in manufacture, shipment, and collection of the control element assembly. Because all fingers are removable and replaceable, servicing and disposal problems are decreased. It is intended that the spider assembly and its extension shaft be reused whenever possible. Because of their long length, the fingers have adequate flexibility.

A buffer is incorporated in the extension shaft of the control element assemblies. This buffering arrangement, essentially a hollow tube in the extension shaft mating with a buffer piston on the upper guide structure base plate, puts the control fingers into tension during buffering which is preferable to placing them in compression as occurs when buffering takes place within the fuel assembly guide tubes.

Control Rod Drive Mechanism. All three types of the control element assembly (4-, 8-, and 12-finger designs) are accommodated by a magnetic jack. For part-length control element assemblies, which are designed not to scram, a special mechanically actuated, magnetically removed non-scram feature is added to the magnetic jack.

Control Assembly Patterns. Design of the upper guide structure permits flexibility in the number of control fingers that can be attached to one control assembly. The standard pattern of control assemblies is shown in Fig. 10. In the standard design, power changes at close to full power, shaping of the radial power distribution, and control of the axial power distribution are best handled by the low worth 4-finger control element assembly entering a single fuel assembly. Shutdown reactivity control in the peripheral region of the core is handled by the 8-finger control element assembly, and in the central region of the core by the 12-finger control element assembly. The need for the two types of shutdown control element assemblies is to obtain "stuck rod worths" in the high reactivity fuel on the periphery of the core, which are about equal to the control element assembly control worth in the lower reactivity central zone of the core.

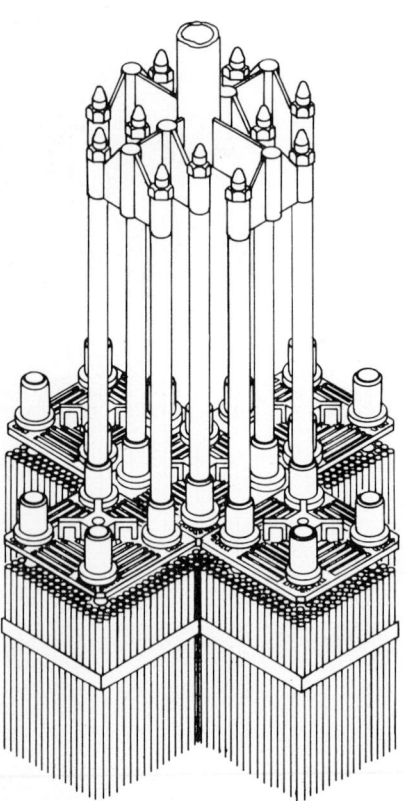

Fig. 9. Control element assembly and fuel used in the PWR. *(Combustion Engineering, Inc.)*

A feature of the design is the high shutdown worth that can be obtained, since control fingers can be placed in every fuel assembly, if desired, while a conservative nozzle head pattern is maintained. This is a distinct advantage in considering the recycling of plutonium. Although it is possible to minimize control worth requirements by positioning the control rods in non-plutonium-recycle fuel assemblies, such restrictions are not desirable from an operating and fuel management viewpoint. Thus, the unique patterns available provide sufficient shutdown worth to accommodate open-market plutonium recycle. In addition, sufficient low worth control element assemblies are provided to maintain suitable power distribution in the recycle mixed oxide cores without severly restricting fuel management.

INSTRUMENTATION AND CONTROL

The large size of the present water reactors and the nuclear effects that can occur, such as xenon redistribution, stuck rods, reactivity anomalies, and so on, require that great

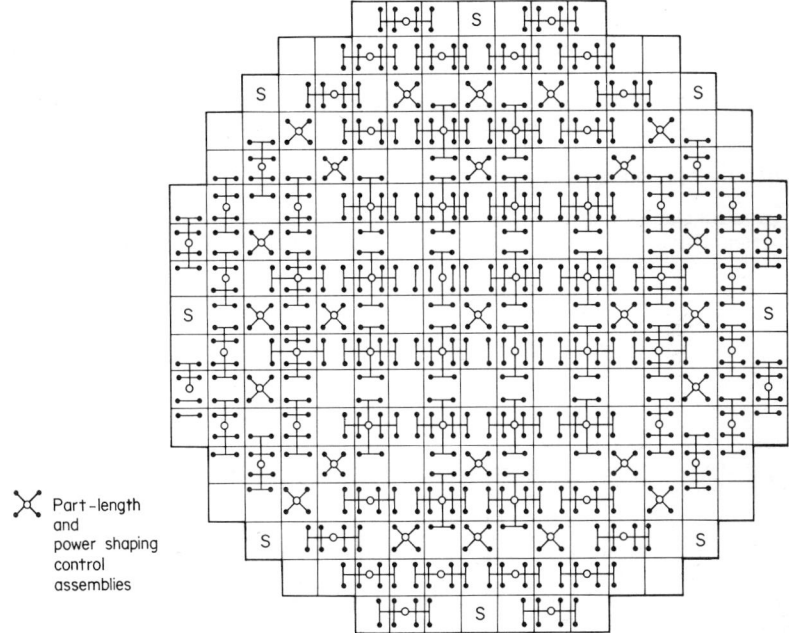

Fig. 10. Standard pattern of control assemblies in the System 80 core. The pattern provides more than sufficient control for self-generated plutonium recycle. For complete open-market plutonium recycle, four-element control assemblies are added in positions marked "S." (*Combustion Engineering, Inc.*)

emphasis be placed on instrumentation and control systems if high plant availability is to be maintained, while the necessary protection due to abnormal occurrences is provided.

In-Core Instrumentation. Plant availability of any process is greatly enhanced when on-line data are available for analysis. Because there are many reactivity effects that can produce changes in the reactor power distribution, more reliable operation can be obtained by on-line monitoring of the reactor. This is achieved with in-core instrumentation. The system described here has provision for up to 61 in-core instrument (ICI) assemblies which enter from the bottom of the reactor vessel. Radial distribution of the ICI is such that every type of fuel assembly, rodded and unrodded, is instrumented, assuming symmetrical core power distribution. Five sets of four symmetrically located ICI assemblies are included to monitor core power tilts. Also, every instrumented fuel assembly is either immediately adjacent to or diagonally adjacent to an instrumented fuel assembly to obtain good radial coverage of the core.

Each of the ICI assemblies contains five self-powered fixed detectors distributed axially along the length of the core, a thermocouple at the end of the assembly to monitor outlet temperature, and a dry-well instrument tube which can accommodate a movable detector; this allows for high measurement accuracy of the on-line fixed detector. Two movable detector systems are provided, each system being capable of inserting the movable detector into every ICI assembly.

The continuous monitoring and processing of the data from over 300 fixed detectors by the core monitoring computer provide the operator with information on core power distribution, maximum linear heat rate in the fuel, departure from nucleate boiling ratio (DNBR), and fuel exposure. These data can then be used to obtain improved maneuvering of core power by means of the relative low-worth 4-finger control element assemblies for power changes and power distribution control.

Out-of-Core Nuclear Instrumentation.   Eight separate channels are used, consisting of:

1. Two start-up channels using proportional counters and logarithmic amplifiers to cover a range of 1 to 10^5 cps ($\sim10^7$ percent full power).

2. Four safety channels using three vertically stacked fission chambers and a combination of logarithmic and linear amplifiers to cover a range of 10^{-8} to 200 percent full power.

3. Two control channels using fission chambers and linear amplifiers to cover the reactor regulating system range of 1 to 125 percent full power.

PLANT PROTECTIVE SYSTEM

The plant protective system consists of two independent systems: (1) the reactor protection system (RPS), and (2) the engineered safety features actuation system (ESFAS).

Details on the instrumentation system, refueling, and a direct injection cooling system can be found in Ref. 1.

REFERENCES

1. Bevilacqua, F., and J. F. Gibbons: "System 80: Combustion Engineering's Standard 3800-Mwt PWR," presented at American Power Conference, Chicago, Apr. 29, 30, May 1, 1974; Combustion Engineering, Inc., Windsor, Conn.
2. Dietrich, J. R.: "The PWR—Present and Future," presented at Technische Vereinigung der Grosskraftwerksbetreiber E.V., Vienna, June 27, 1973. (Copies available from Combustion Engineering, Inc.)
3. Holzer, R. and H. Knabb: Experience with Fuel Assemblies for Pressurized Water Reactors, *Kerntech.*, **14**, 519–525, 1972.
4. Abbott, W. E., W. Braun, W. P. Chernock, and A. Martin: "PWR Practice in the United States and Europe," presented at American Power Conference, Chicago, May 8–10, 1973.
5. Bevilacqua, F.: "Design Features of Large Pressurized Water Reactor Cores," presented at American Nuclear Society 17th Annual Meeting, Boston, June 13–17, 1971.
6. Meijer, C. H., E. M. Brown, and J. G. Brooks: "Use of Solid State Technology in Advanced PWR Protection Systems," presented at IEEE Nuclear Power Systems Symposium, San Franciso, Nov. 16, 1973.
7. Bevilacqua, F., and E. A. Yuile: "On-Line Computer Application to a PWR Nuclear Power Plant," presented at NUCLEX 69, Basle, Oct. 6–11, 1969.
8. Martin, A.: "Pressurized Water Cooled Reactors—Performance and Product Improvement," *Kerntech.*, **14**, 427–447, 1972.

High-Temperature Gas-Cooled Reactors

BY C. O. PEINADO*

The high-temperature gas-cooled reactor (HTGR) is an advanced thermal reactor that produces modern steam conditions. See Fig. 1. Helium is used as the coolant. Graphite, with its superior high-temperature properties, is employed as the moderator and structural material. The fuel is a mixture of enriched uranium and thorium in the form of particles clad with ceramic coatings.

The high-temperature conditions and high thermal efficiency (approximately 39 per-

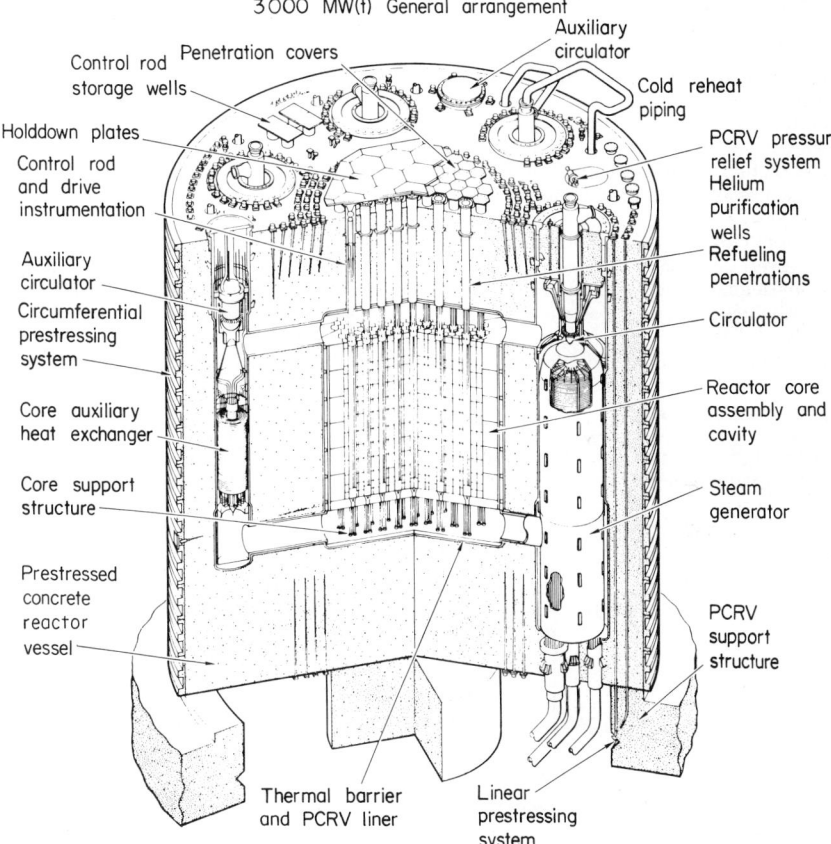

Fig. 1. Nuclear steam system of a 3000 MW (thermal) high-temperature gas-cooled reactor (HTGR) (*General Atomic Company.*)

cent) of the HTGR result in high performance through conservation of fuel, competitive cost, and the use of conventional turbine-generating equipment. The amount of cooling water required to carry away the waste heat is significantly less than in a light-water

*Director, Program Development, General Atomic Company, San Diego, Calif.

reactor (LWR). The use of thorium in the fuel cycle decreases fuel cost, improves the conservation of fuel, and adds the large deposits of thorium available to fuel reserves. The HTGR has significant environmental advantages including: (1) lower thermal discharge because of its high efficiency, (2) low release of radioactive waste because of the high-integrity fuel and the inert coolant, and (3) low consumption of raw materials because of high efficiency and use of thorium in the fuel cycle.

High operating temperatures at moderate pressures are achieved through the use of helium as the coolant. Helium has the fundamental advantage that it always remains in the gas phase, making complete loss of coolant no longer a problem. The attractive features of helium as a coolant include: (1) it is chemically inert, (2) it absorbs essentially no neutrons, (3) it makes no contribution to the reactivity of the system, and (4) it will continue to be readily available to meet the requirements of the HTGR.

Graphite is used as the moderator and core structural material because of (1) excellent mechanical strength at high temperatures, (2) very low neutron-capture cross section, (3) good thermal conductivity, and (4) high specific heat. Graphite has proved to be an excellent moderator, with a long history of use in thermal reactors. Its low neutron-capture cross section places it high among moderator candidates. No neutrons are lost within the core through absorption in metallic fuel cladding or structural supports. In addition to its nuclear characteristics, graphite is ideally suited to high-temperature operation, since, unlike most materials, it increases in strength at higher temperatures, reaching a maximum at about 4500°F (2482°C), well above the reactor operating range, and continues to maintain significant strength at much higher temperatures. The tensile strength of graphite is plotted against temperature in Fig. 2.

The use of the thorium/uranium fuel cycle in the HTGR provides improved core performance over the plutonium/uranium low-enrichment cycle used in LWRs. The principal reason for this is that fissle ^{233}U produced from neutrons captured in thorium during reactor operation is neutronically a better fuel than ^{239}Pu, produced from ^{238}U in the low-enrichment cycle. The excellent neutronic characteristics of the graphite-moderated thorium/uranium cycle lead directly to high conversion ratios and low fuel inventories. Reduced ^{235}U inventories and makeup requirements mean reduced sensitivity to changing ore costs and to possible future increases in uranium prices.

EARLY DEVELOPMENT

The HTGR has been under development in England since 1950. Work began in the United States on this concept in the late 1950s, and Germany and other European countries came into the program in the early 1960s. Three prototype installations have been operated successfully for at least six years: (1) the 20-MW (thermal) Dragon reactor, (2) the 15-MW (electric) German AVR, and (3) the 40-MW (electric) Peach Bottom 1

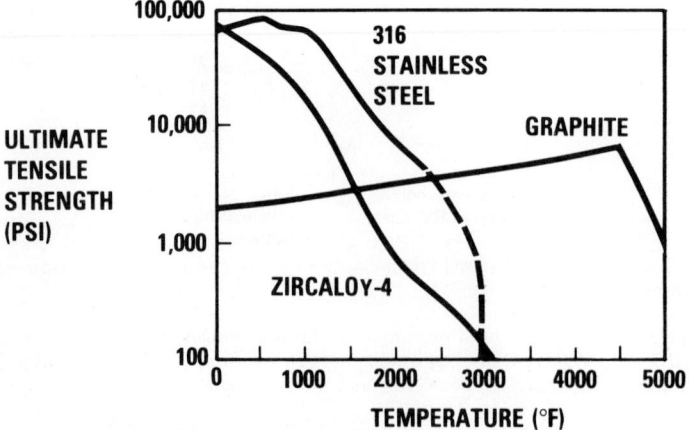

Fig. 2. Tensile strength of graphite versus temperature.

reactor. Two demonstration plants are under construction or nearing completion as of early 1975. The first is the 330-MW (electric) Fort St. Vrain station, on which construction, preoperational testing, core loading, and initial core criticality tests have been completed. Rise to power is scheduled during the first half of 1976. The second demonstration plant is the 300-MW (electric) Schmehausen reactor, scheduled for completion in Germany in 1979. In addition to these prototype HTGRs, much directly applicable experience has been gained from other gas-cooled reactors. The first of these, built at Calder Hall in the United Kingdom, achieved full-power operation in 1956. Gas-cooled reactors in operation or under construction worldwide are listed in Table 1.

Dragon Reactor.   This installation was started in June 1965 and reached its design output of 20 MW (thermal) in April 1966. The unit has been operated exclusively as an experimental facility. The operating record and availability have been excellent, and much basic information has been obtained on fuel materials, production, control methods, behavior of various types of fuel under irradiation, fission product migration, plateout in the primary circuit, coolant chemistry, water and chemical impurity monitoring, failed-fuel detection, transient codes, component design, and coolant flow. Of particular interest

TABLE 1.   Gas-Cooled Reactors in Operation or Under Construction

Name	Location	Power, MW (electric)	Power date	Coolant used
Calder Hall	U.K. (UKAEA)	4×45	1956	CO_2
Chapelcross	U.K. (UKAEA)	4×45	1959	
Berkeley	U.K. (CEGB)	2×138	1962	
Bradwell	U.K. (CEGB)	2×150	1962	
Hunterston A	U.K. (SSEB)	2×180	1964	
Hinkley Point A	U.S. (CEGB)	2×250	1965	
Trawsfyndd	U.S. (CEGB)	2×250	1965	
Dungeness A	U.K. (CEGB)	2×275	1965	
Sizewall	U.K. (CEGB)	2×290	1966	
Oldbury	U.K. (CEGB)	2×300	1967	
Wylfa	U.K. (CEGB)	2×590	1971	
Latina	Italy	1×210	1963	
Tokai-Mura	Japan	1×154	1965	
G2, G3, Marcoule	France	2×40	1958	
EDF1, Chinon	France	1×62	1963	
EDF2, Chinon	France	1×213	1965	
EDF3, Chinon	France	1×476	1966	
St. Laurent 1	France	1×487	1969	
St. Laurent 2	France	1×516	1971	
Bugey I	France	1×488	1972	↓
Windscale AGR	U.K. (UKAEA)	1×28	1963	
AVR Pebble Bed	Germany	1×15	1966	He
THTR	Germany	1×300	1978	
Dragon	U.K. (OECD)	1×20(th)	1966	
Peach Bottom	U.S.A. (PE)	1×40	1967	
Fort St. Vrain	U.S.A. (PSC)	1×330	1975	
Bohunice	Czechoslovakia	1×150	1972	CO_2
Brennilis EL4	France	1×73	1967	
Lucens	Switzerland	1×7	1967	
Vandellos	Spain	1×500	1972	
Niederaichbach	Germany	1×100	1973	
Hartlepool	U.K. (CEGB)	2×660		
Hinkley Point B	U.K. (CEGB)	2×625	1974	
Hunterston B	U.K. (SSEB)	2×625	1975	
Dungeness B	U.K. (CEGB)	2×600	1975	
Heysham	U.K. (CEGB)	2×625		↓

NOTE:
CEGB (Central Electricity Generating Board)
OECD (Organization for Economic Co-operation and Development)
PE Philadelphia Electric Co.
PSC Public Service Company of Colorado
SSEB (South of Scotland Electricity Board)
UKAEA (United Kingdom Atomic Energy Authority)

is the circulating activity, which has remained two or more orders of magnitude below the specified operational convenient limit. The reactor has a semihomogenous graphite-moderated core based on the uranium/thorium cycle and is capable of producing helium temperatures of 1375°F (746°C). It is similar in many respects to the Peach Bottom 1 reactor.

AVR.   This installation went critical in August 1966 and reached full-power operation of 15 MW (electric) in December 1967. This is a pebble-bed reactor using helium gas cooling and graphite moderation. The uranium/thorium dicarbide fuel in particles is disbursed in a graphite matrix contained in machined graphite spheres approximately 1½-inches in diameter. On-line refueling is a design feature of the reactor. The performance of the pebble-bed fuel has been excellent. It has achieved an average availability of 74 percent, outstanding for a prototype station.

Peach Bottom 1 Reactor.   This was the first HTGR to be constructed in the United States (built by General Atomic Company for the Philadelphia Electric Company). This installation achieved full-power operation in 1967 and was shut down in October 1974. It was a graphite-moderated reactor using the uranium/thorium cycle. The station produced 40 MW (electric) with steam delivered to the turbine at 1450 psig and 1000°F (538°C). Operation of the plant successfully demonstrated the fundamental features of the HTGR concept. Helium was used successfully and routinely as a power reactor coolant at temperatures exceeding .1400°F (760°C). Core physics and control predictions were verified, and the uranium/thorium fuel cycle obtained high conversion ratios and high burnup. The primary system activities were maintained at very low values, because of the use of the coated fuel particles and the all-graphite core structure. The nuclear steam supply system availability was excellent, varying from 80 to 95 percent.

FORT ST. VRAIN GENERATING STATION

The first commercial HTGR is the 330-MW (electric) Fort St. Vrain station, located near Platteville, Colorado, approximately 35 miles north of Denver (built by General Atomic Company for Public Service Company of Colorado). An external view of this plant is given in the article on Nuclear Power Plant—Concept to Startup appearing earlier in this Handbook section. The reactor arrangement is given in Fig. 3. The station, while retaining the fundamental features demonstrated in Peach Bottom 1, also incorporates certain components and design features of larger HTGRs. These include (1) prestressed concrete reactor vessel containing the entire primary coolant system, (2) series steam-driven axial-flow helium circulators, (3) once-through steam generators with integral reheaters, (4) top-mounted control rod drives, and (5) improved uranium/thorium coated fuel particles capable of fission-product retention and loaded in hexagonal graphite fuel elements.

LARGE HIGH-TEMPERATURE GAS-COOLED REACTORS

Large HTGRs are considered to approximate or exceed a capacity of 1000 MW (electric).

Reactor Primary Components.   All the major components of the primary system, including steam generators, are housed in a steel-lined prestressed concrete reactor vessel (PCRV). The PCRV also provides the necessary biological shielding. A cross section through the PCRV showing primary system component cavities is given in Fig. 4. This arrangement permits compact installation and eliminates major external primary coolant piping.

The PCRV has numerous advantages. It is field-erected, thus minimizing transportation problems and site limitations associated with shop-fabricated steel reactor vessels. Of greatest importance, however, are the safety features. A PCRV has within and around its concrete walls a large number of steel tendons that are tensioned before the vessel is put into service. Pretensioning creates compressive stresses in the concrete and liner greater than the tensile stresses generated by the internal system pressure under service conditions. Because of the multiplicity and redundancy of the tendons, the PCRV is a very safe structure.

The tendons and their anchorage are positioned near the outer surfaces of the concrete walls and hence are shielded from neutron damage and thermal transients. Also, they are readily accessible for surveillance over the lifetime of the plant and can be retensioned or

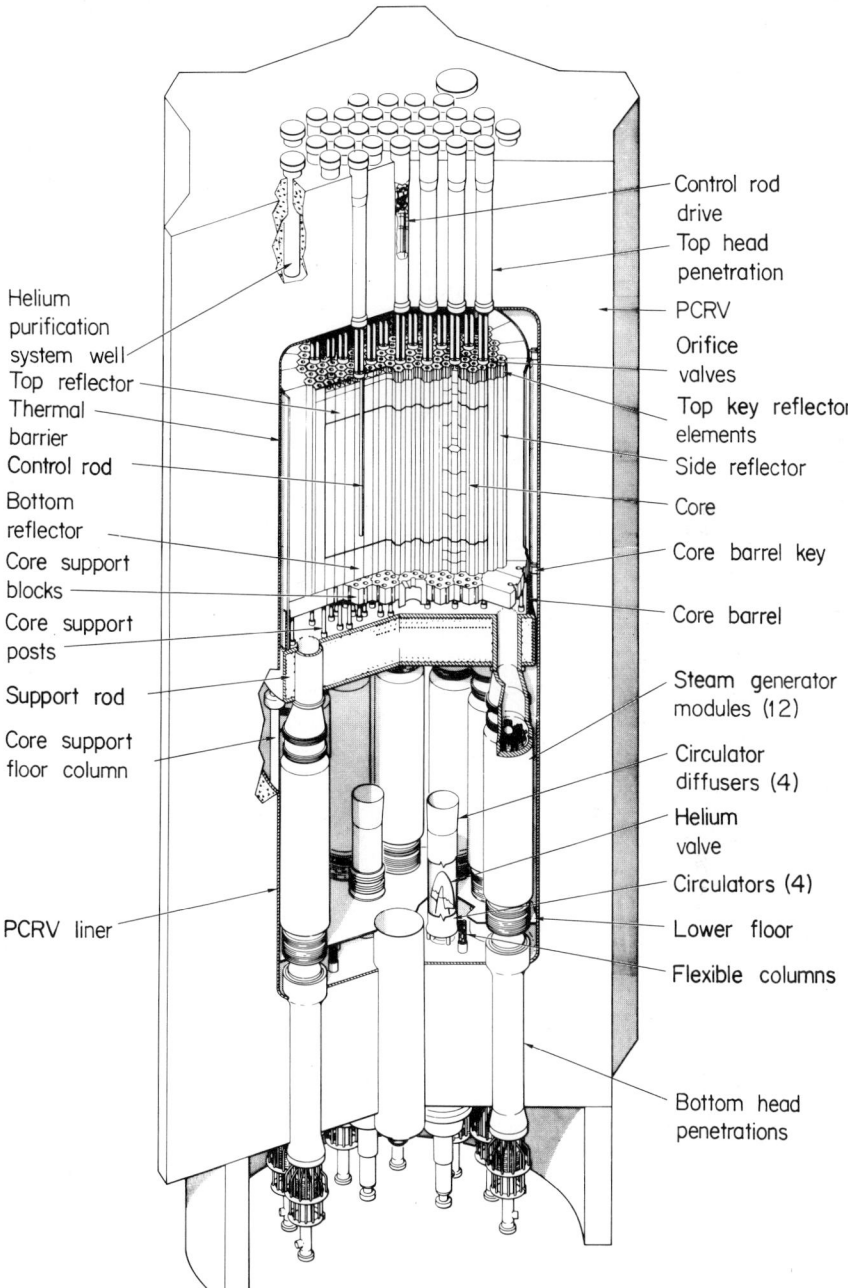

Fig. 3. Reactor arrangement at the Fort St. Vrain Nuclear Generating Station. (*General Atomic Company.*)

replaced. The PCRV concrete is cooled by cooling tubes welded to the exterior surfaces of the PCRV liner. The interior of the liner is covered by a thermal barrier consisting of a ceramic fiber blanket held in place by a steel cover plate (Fig. 5), which reduces the heat conducted to the liner cooling tubes and protects the concrete from the high helium temperatures.

The nuclear steam supply system contains several independent primary coolant loops—four for a 2000-MW (thermal) plant and six for a 3000-MW (thermal) plant—each

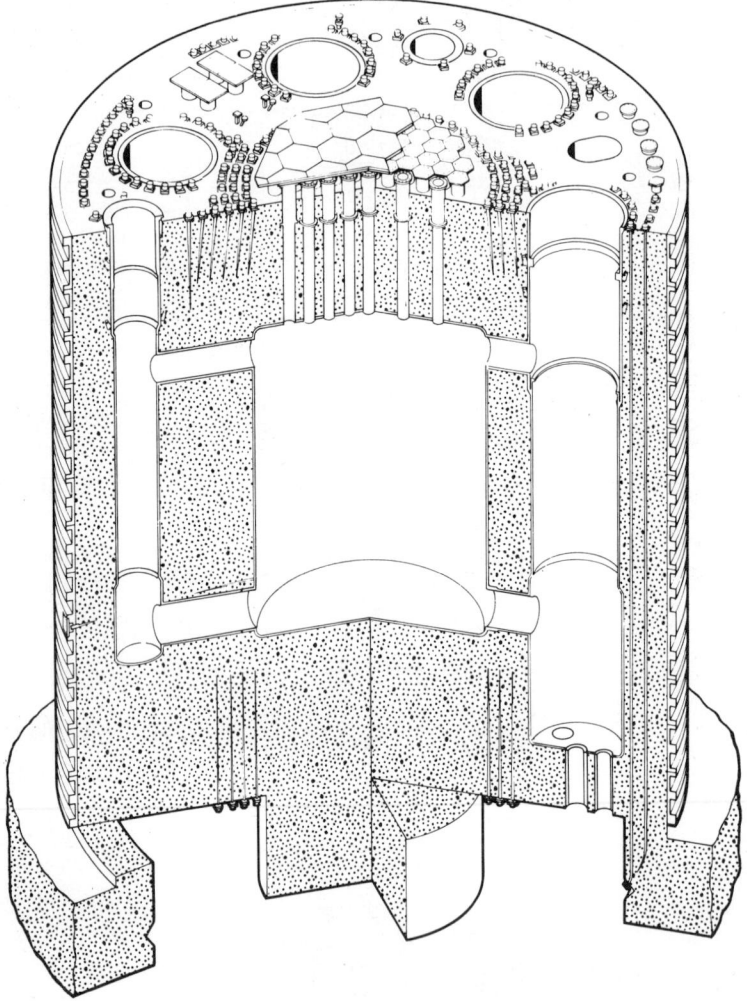

Fig. 4. Prestressed concrete reactor vessel.

containing a single helium circulator and a steam generator. Helium flows downward through the core and up through the steam generators and circulators. The steam generator and circulator pairs are located in individual cavities around the central or core cavity in the PCRV.

Circulators are single-stage, steam-turbine-driven, axial-flow machines. See Fig. 6. Circulator output is regulated by throttling flow to its steam-driven turbine and bypassing cold reheat steam around the turbine. In the primary loop, a helium shutoff valve is provided at the discharge of each circulator. By preventing any significant backflow of

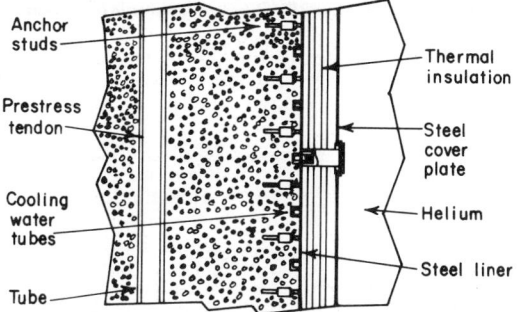

Fig. 5. Cross section of PCRV wall used with HTGR.

helium through the loop, these valves allow one or more loops to be shut down while the others are operating.

Steam generators are of the once-through type, each composed of two bundles, one for the main steam section, encompassing the economizer-evaporator and superheater, and the other for the reheater section. The feedwater flow is directed upward through the economizer portion of the main steam section, converted to steam in the evaporator, and raised to its final temperature in the superheater. Superheater steam is brought out through the bottom penetration of the PCRV and the main steam piping headers that run to the main turbine in the turbine building. After expansion in the high-pressure section of the main turbine, the steam is returned to the circulator turbine. The cold reheat from the circulator turbine is then directed to the reheater section of the steam generator. The reheater sections are mounted below the main steam section and exposed to the high-temperature helium within the core. A schematic flow diagram showing flows through the

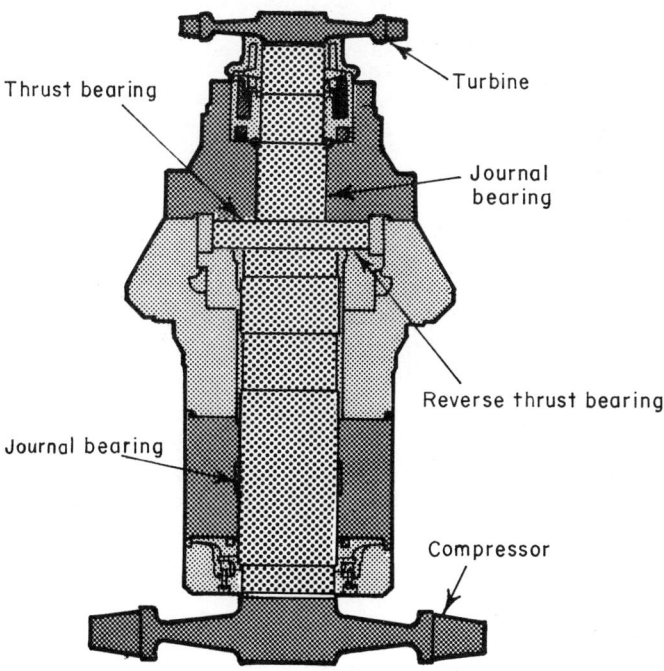

Fig. 6. Helium circulator.

steam generator is given in Fig. 7. The primary and secondary coolant flows are illustrated in Fig. 8.

Several auxiliary cooling loops are provided as the engineered safeguard to carry out residual heat removal functions, following a design-basis depressurization accident. Each auxiliary cooling loop contains an electric-motor-driven auxiliary circulator, a helium shutoff valve, and a water-cooled heat exchanger, all installed within the PCRV, as shown in Fig. 9.

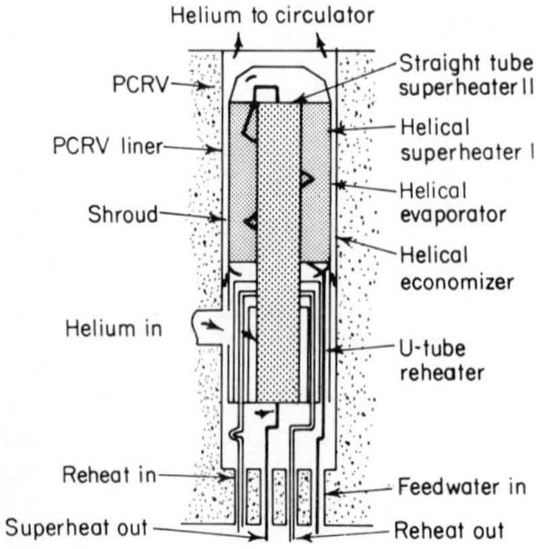

Reactor Core.   The HTGR fuel element is a graphite block, hexagonal in cross section and having a grid of longitudinal fuel holes and coolant channels. See Fig. 10. The fuel element blocks are stacked in columns of eight blocks each and grouped into fuel regions consisting of a central column surrounded by six columns. Each region rests on a large core support block, which in turn, rests on graphite posts standing on the liner of the central cavity. Hexagonal graphite reflector elements are located above, below, and around the active core. These elements are surrounded by permanent side-reflector blocks to give the entire assembly a circular configuration, as indicated by Fig. 11. The fuel holes contain a rod consisting of ceramic-coated fuel particles in a graphite matrix. The coatings, applied by pyrolitic techniques, are multilayered to ensure a high degree of fission-product confinement. A porous inner layer, or buffer zone, accommodates the expansion of the irradiated fuel and provides storage space for gaseous fission products. The outer layers act as a fission-product retention barrier and provides structural strength. In effect, the particle coating functions as a miniature spherical pressure vessel.

Fig. 7. Steam generator of HTGR. *(General Atomic Company.)*

To achieve a fuel management scheme with the lowest fuel cycle cost consistent with the current thermal and material performance limits, the following parameters are selected: (1) a fuel cycle incorporating uranium/thorium, (2) a fuel lifetime of four years, (3) an average power density of 8.4 W/cm^3, and (4) a refueling frequency of once a year. The HTGR core can be loaded to yield conversion ratios that are very high by comparison

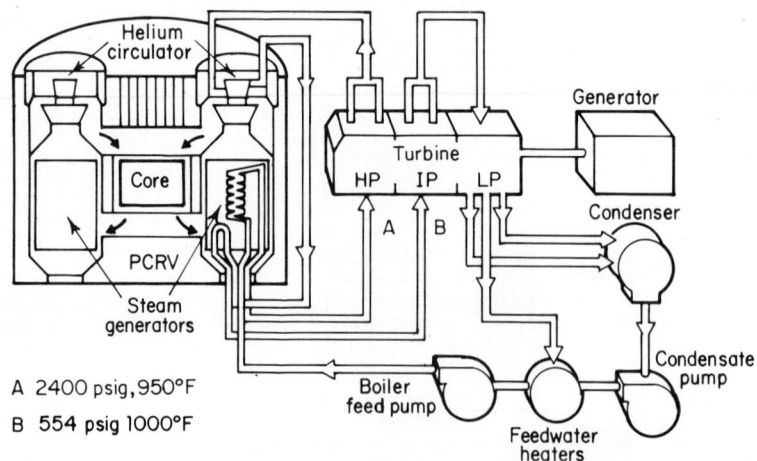

Fig. 8. Primary and secondary coolant flows of HTGR. HP = high-pressure; IP = intermediate pressure; LP = low-pressure. *(General Atomic Company.)*

A 2400 psig, 950°F

B 554 psig 1000°F

with those of the LWR, thus allowing the option of using fertile material in a way that utilizes resources efficiently. The tradeoffs between resource utilization and economics can be optimized, depending on the cost of working capital. For example, a public utility with a lower working capital rate can have a higher conversion ratio than a privately owned utility.

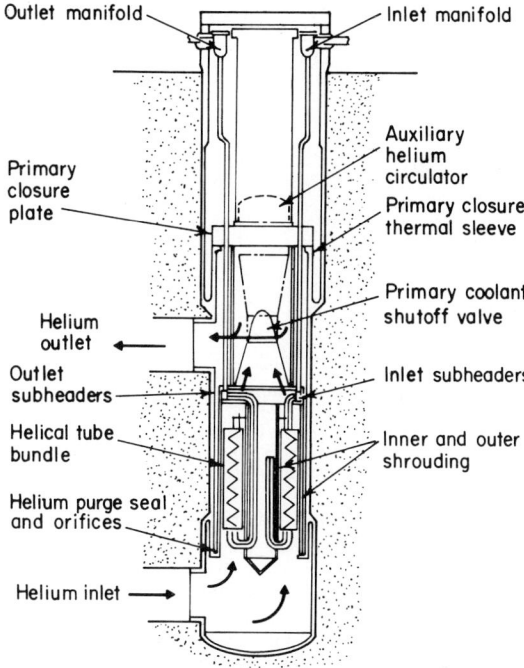

Fig. 9. Auxiliary cooling loop for HTGR.

The reactor is controlled by two control rods located in each refueling region. All control rod pairs have scram (quick shutdown) capability and are driven by gravity. A backup reserve shutdown system is included. This consists of boronated graphite pellets that can be introduced from hoppers located in each refueling penetration into the core via the cylindrical channels in the central fuel element of each refueling region.

Safeguard Systems.   The design of the HTGR incorporates many inherent safety features and a number of engineered safeguards. The inherent safety characteristics include negative power and temperature coefficients, assured by the thorium content of the fuel. In addition, the high heat capacity of the large mass of graphite ensures that any core temperature transient resulting from reactivity insertions or interruptions in cooling will be slow and readily controllable. This important safety feature eliminates the need for an emergency core cooling system and only a residual heat removal system is required for the long-term decay heat, and control of the HTGR is inherently easier than in reactors in which the coolant functions as the moderator. The uranium/thorium fuel contained in the ceramic-coated particles is not susceptible to sudden release of the stored-up fission products as a result of melting. Since the entire primary coolant system is contained within the PCRV, external piping, which might be subject to sudden rupture, is eliminated. Structural strength and integrity of the PCRV is enhanced by the redundant reinforcing steel and prestressing wire tendons. At the maximum credible pressure, the prestressing elements are not stressed above levels experienced during their initial tensioning. As a result, sudden loss of coolant due to prestress failure is not credible. In addition to these inherent safety features and engineering safeguards, the following engineered safeguards are included in the design:

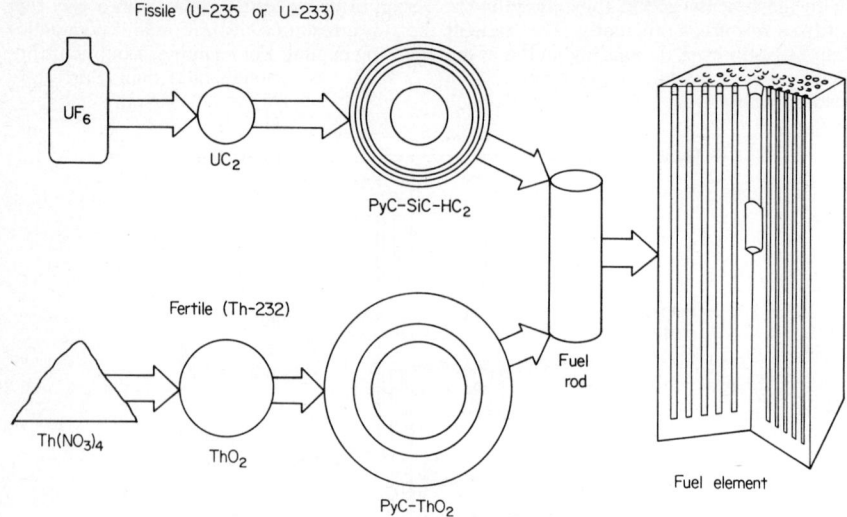

Fig. 10. Fuel element components for HTGR.

1. PCRV Penetration Flow Restrictor Devices.   These are located in the large PCRV penetration to limit the rate of pressurization in the event of a hypothetical failure of the closure of the penetration.

2. Reactor Reserve Shutdown System.   This system is a diverse redundant means of shutting down the reactor.

3. PCRV Safety Valves.   These valves are provided to prevent overpressurization of the PCRV resulting from hypothetical large steam and water leaks into the primary circuit.

4. Steam Water Dump System.   This system is provided to minimize the amount of water that could leak into the primary coolant as a result of steam generator tube rupture.

5. Core Auxiliary Cooling System.   This system is provided as redundant diverse means of providing core cooling in the event that use of the main loops for heat removal is assumed inoperative.

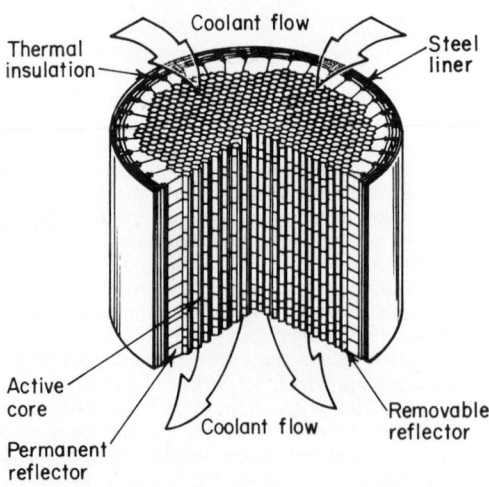

Fig. 11 Reactor core arrangement in HTGR. (*General Atomic Company.*)

Radioactive Waste Mangement.   During normal operation, an HTGR discharges only very small amounts of radioactive wastes to the environment. Two of the principal reasons for this are: (1) The helium coolant is essentially free of induced radioactivity. (2) Several barriers and sinks are effective in preventing release of significant quantities of radioactive fission products from the fuel to the environment. The most important barrier is the coating on each individual fuel particle. If the particle coating fails, the particle kernel is an effective diffusion barrier for all the fission products. The fuel rod matrix and the fuel

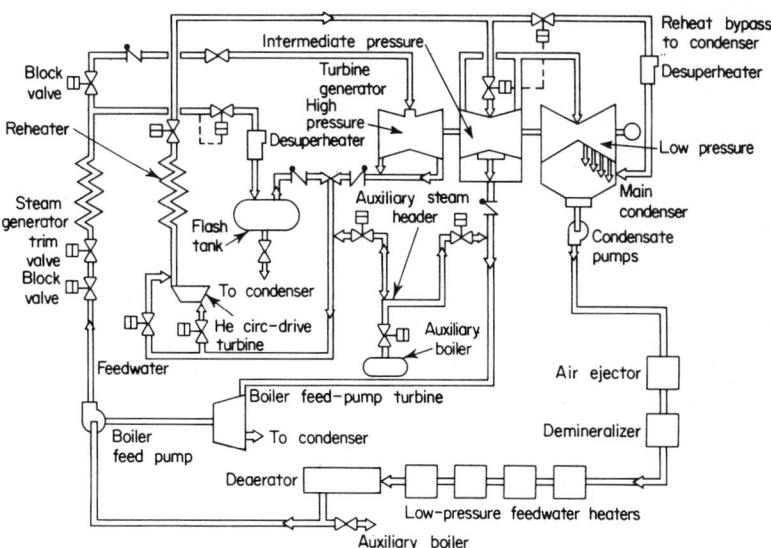

Fig. 12. Normal operation of secondary coolant system of HTGR. *(General Atomic Company.)*

element graphite are relatively porous to gas and, therefore, significantly delay only the short-lived noble gas nuclides that escape from the fuel particle. On the other hand, the matrix and graphite materials act as sinks for the fission product metals, since the graphite is an effective diffusion barrier for the metals. The next barrier is the PCRV steel liner, which forms the primary coolant boundary.

The circulating activity in the primary coolant is controlled by the helium purification system, which removes the noble gas nuclides that escape from the fuel, and also removes particulates and chemical contaminants. Tritium, produced principally from ^{3}He in the primary coolant, is removed in solid form through absorption in a titanium sponge. After several months' decay, contaminants are extracted from the system by periodic regeneration and sent to a recovery system, in which the noble gases of long half-life are separated from the nonradioactive gases, which are released. The noble gases, almost exclusively ^{85}Kr (krypton), can be recycled back to the helium purification system or released to the atmosphere. In either event, the gaseous effluent from the plant is only a few hundred curies per year.

Liquid wastes result chiefly from decontamination of primary system components prior to maintenance. Both quantity and activity are low, about 2500 gallons and 20 curies per year. Liquid wastes can be satisfactorily handled by packaging and offsite burial if required. A 1160-MW (electric) HTGR generates about 610 cubic feet, less than 20,000 curies per year, of solid wastes. These are handled by packaging and offsite burial.

Nuclear Steam Supply System Instrumentation.   This has five major instrumentation, protection, and control systems:

1. *Overall Plant Control.*   Maintains reactor power and main turbine inlet steam conditions at the required values.

2. *Nuclear Instrumentation and In-Core Monitoring.*   Provides assistance in plant

control, start-up information, reactor protection information, and other core data for efficient operation.

3. Plant Protection. Includes the reactor protection circuits and the instrumentation and control for the engineered safeguards and safety-related instrumentation.

4. Analytical Instrumentation. Measures chemical contaminants in the primary coolant system, helium purification system, and radioactive gas wastes system, and provides sample conditioning for the radiation monitors.

5. Radiation Monitoring. Monitors the activity and radiation levels for the nuclear steam supply system.

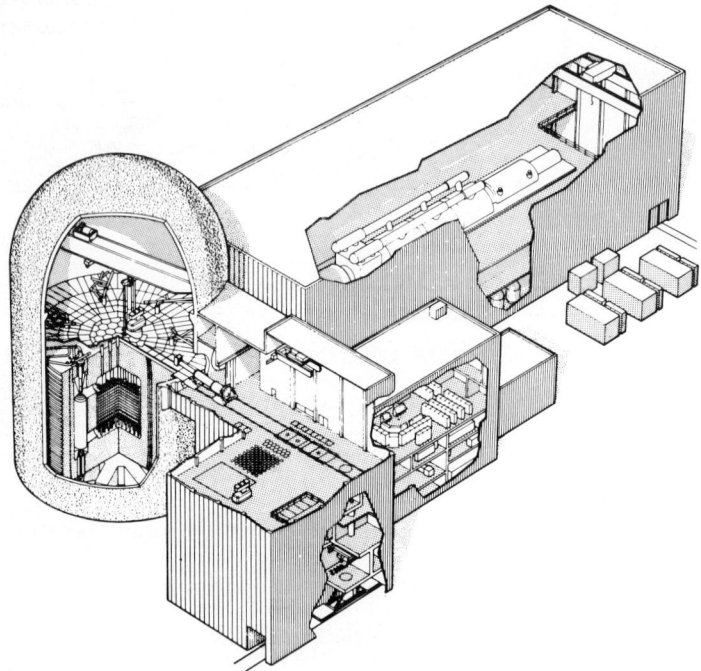

Fig. 13. Principal buildings and overall arrangement of HTGR power plant. *(General Atomic Company.)*

Plant Operation. Start-up and load changes of the HTGR plant are now described with the assistance of the flow schematic diagram given in Fig. 12. At plant start-up, feed pumps are placed in operation to circulate feedwater through the steam generators. Water leaving the steam generators flows through back-pressure control valves to flash tanks. Water is drained from the flash tanks through the feedwater heaters for heat recovery and then flows to the condenser hot well. During the initial phase of start-up, steam supplied by an auxiliary boiler drives the helium circulator turbines and serves other auxiliary *balance of plant* uses, such as the boiler feed pump turbines.

With feedwater flow and helium flow established, the reactor is started by withdrawal of the control rods. The rods are withdrawn symmetrically to achieve balanced flux distribution during approach to power. As reactor power is increased, heat is added to the high-pressure water circulating through the steam generator, and saturated steam is produced in the flash tanks. Flash-tank steam then supplements auxiliary boiler steam to meet increased demands from the nuclear steam supply system and balance of plant users. Steam leaving the circulator turbines flows through the reheater and hot reheat lines and is then dumped into the main condenser through the hot reheat steam bypass valves and desuperheaters.

Further increases in reactor power and helium temperature lead to steam production in the steam generators. When dry steam at suitable pressure and for turbine starting is available, the main turbine is placed into service after prior warm-up. Main and reheat steam temperatures are brought to their normal operating values as turbine load is increased to approximately 25 percent of rated capacity. At this point, automatic control is established, permitting a normal increase to full power.

During normal operation, an electric load increase results in opening of the turbine-generator throttle valves. Increased turbine steam flow decreases the steam pressure and acts through the control system to increase the feed pump speed. The resulting higher feedwater flow signal is applied to the circulator speed control. This action anticipates the main steam temperature decrease due to increased steam flow and increases the helium flow. A small automatic adjustment of the helium flow returns the main steam temperature to the desired value. The increased helium flow tends to reduce the core outlet helium temperature slightly and, together with the increased reheat steam flow, causes a decrease in the reheat steam temperature. The reheat steam temperature signal, plus the main turbine steam flow signal, minus the neutron flux signal, are summed by the neutron flux controller, which generates control rod motion to increase the reactor power. This feedback control sequence will return the reheat steam temperature to the desired value. In the case of load reduction, these operations are reversed. A normal plant shutdown is initiated from the minimum point, following procedures that are generally the reverse of plant start-up.

The nuclear steam supply system is capable of automatically making step changes of 10 percent of rated load at any power level between 25 and 100 percent. The system is capable of following a daily load cycle covering the range of 50 percent of full power, at 5 percent of rated power per minute. It will accommodate step reductions in load, such as loss of load of any magnitude up to and including a complete turbine trip. Power reduction is accomplished automatically by control rod runback from any power level to 25 percent and feedwater flow rate reduction to hot standby. Hot standby for an HTGR is defined as a steaming condition at 25 percent of rated steam flow, with steam going through the bypass line to the flash tank and the condenser rather than to the turbine.

Plant Facilities.   These are similar to those for conventional fossil-fuel power plants. The principal buildings, shown in Fig. 13, include:

1. *Reactor Containment Building.*   This provides a leakage barrier to prevent significant fission-product release to the atmosphere following any accident. This building contains auxiliary systems for removal of postincident radioactivity from the containment atmosphere and for ultimate reduction of containment pressure to atmospheric pressure.

2. *Reactor Service Building.*   This structure houses the fuel shortage and shipping facilities, equipment service facilities, and certain portions of the reactor auxiliary systems.

3. *Plant Auxiliary and Control Building.*   This building houses the nuclear steam supply system control and instrumentation and equipment as well as the balance of plant control and instrumentation equipment.

4. *Penetrations Building.*   This protects both safety and nonsafety related penetrations through the secondary containment.

5. *Diesel Generator Buildings.*   These structures house the emergency Diesel required to operate the core auxiliary cooling system and other essential loads during the design-basis accident.

6. *Turbine Building.*   This houses the superheated steam turbine generator and auxiliary equipment.

The parameters for two HTGR plants are summarized in Table 2.

REFERENCES

Birely, W. C.: Operating Experience of the Peach Bottom Atomic Power Station, *Nucl. Eng. Des.*, January 1974.

Habush, A. L., and R. F. Walker: Fort St. Vrain Nuclear Generating Station Construction and Testing Experience, *Nucl. Eng. Des.*, January 1974.

Landis, J. W., and J. F. Watson: Materials and Design Concepts of Gas-cooled Reactor Systems, *Nucl. Eng. Des.*, January 1974.

Dahlberg, R. C., K. Asmussen, D. Leed, L. Brooks, and R. K. Lane: HTGR Fuel and Fuel Cycle, *Nucl. Eng. Des.*, January 1974.

Wessman, G. L., and T. R. Moffette: HTGR Plant Safety Design Bases, *Nucl. Eng. Des.*, January 1974.

Boyer, V. S., J. P. Gibbons, T. A. Johnston, R. J. Hoe, D. K. Feldtmose, and W. C. Drotleff,: Fulton Station HTGR, *Nucl. Eng. Int.*, August 1974.

Chapman, B. G.: "Operating and Maintenance Experience with the Dragon Reactor Experiment," UKAEA paper given at ANS Topical Meeting (HTGR and GCFBR), Gatlinburg, Tenn., May 7–10, 1974.

Ivens, G.: "AVR-HTR-Operating Experience," paper given at ANS Topical Meeting (HTGR and GCFBR), Gatlinburg, Tenn., May 7–10, 1974.

TABLE 2.   Parameters for Two HTGR Plants

Parameter	Output capacity	
	770 MW(e) 2000 MW(th)	1160 MW(e) 3000 MW(th)
Capacity		
Nominal net electrical output, MW	770	1160
Overall station net efficiency, %	39	39
Net heat rate, Btu/kWh	8896	8843
Plant auxiliary power, MW	13.4	18.0
Reactor Core		
Reactor core output, MW(th)	2000	3000
Net NSS* output, MW(th)	1980	2980
Core dimensions, diameter/height, feet	23.1/20.8	27.7/20.8
NSS helium inventory, pounds	15,000	21,000
Number of fuel elements/columns	2744/343	3944/493
Primary coolant flow, 10^6 pounds/hour	7.480	11.230
Primary coolant inlet pressure, psig	710	710
Avg. coolant temp., reactor inlet, °F (°C)	605 (318)	605 (318)
Avg. coolant temp., reactor outlet, °F (°C)	1366 (741)	1366 (741)
Core pressure drop, psi	9.7	10
Total initial neutron flux, nv†	2.4×10^{14}	2.4×10^{14}
Maximum fast fluence (E > 0.18 MeV), nv(th)†	8×10^{21}	8×10^{21}
Avg. power density, kW(th)/liter	8.1	8.4
Avg. heat flux, Btu/hour-ft²	62,000	65,000
Max. heat flux, Btu/hour-ft²	177,000	185,000
Fuel life, full power years	3.2	3.2
Fraction of core replaced each year (80% capacity)	¼	¼
Avg. conversion ratio (equilibrium)	0.68	0.68
Fuel and Thermal Data		
Number of refueling regions, full/partial	37/18	61/24
Element (hexagonal across flats), inches	14.17	14.17
Element (length), inches	31.22	31.22
Fuel holes per element, standard/control	132/76	132/76
Fuel rod diameter, inches	0.618	0.618
Fuel hole diameter, inches	0.624	0.624
Coolant channel diameter, inches	0.826	0.826
Coolant channels per element, standard/control	72/43	72/43
Reflector thickness, inches: top/bottom/side (mean)	46.8/46.8/41.9	46.8/46.8/40.5
Replaceable reflector lifetime, years	8–12	8–12
Total quantity of uranium (^{235}U) thorium, kg (initial core)	1149/26,100	1617/37,487
Total weight reactor graphite, 10^6 pounds	0.63	0.91
Avg. Fuel burnup, MW/metric ton	95,000	95,000
Max. fuel center-line temp. (short term), °F (°C)	2470 (1354)	2570 (1410)
Avg. fuel temp., °F (°C)	1612 (878)	1634 (890)
Fuel melting point, °F (°C)	4440 (2449)	4440 (2449)
Avg. moderator temp., °F (°C)	1348 (731)	1362 (739)
Reactor Vessel		
Internal clearance dimensions (ID × IH), feet-inches	32–8 × 47–4	37 × 47–4
Max. external dimensions (OD × OH), feet-inches	96 × 91–2	100–6 × 91–2
Min. PCRV‡ wall thickness, feet-inches	15–6	15–6
Normal working pressure, psig	710	710

Design pressure, psig (setting of second relief valve)	765	765
Vertical prestress tendons, number	336	414
Wires per vertical tendon	169	169
Posttensioning force per vertical tendon (initial jacking), tons	700	700
Circumferential tendons or channels, number	24 channels	24 channels
Wire layers per channel	12–18	12–18
Volume of concrete in PCRV‡ and PCRV support, cubic yards	20,000	22,000
Weight reinforcing steel in PCRV and PCRV support, tons	1050	1200
Average concrete design compressive strength, psi	6500	6500
Liner thickness, core cavity/penetrations, inches	0.75/0.5	0.75/0.5
Liner temp. normal (avg.)/hot spot, °F	150/250	150/250
°C	66/121	66/121
Helium Circulators (Each)		
Dimensions (length × diam.), feet-inches	10 × 5–6	10 × 5–6
Number of circulators	4	6
Steam flow, including bypass, 10^6 pounds/hour	1.32	1.32
Speed, revolutions/minute	6750	6750
Circulating capacity, 10^6 pounds/hour	1.870	1.870
Static press. rise (helium), psi	20.7	20.7
Compressor inlet pressure, psig	690	690
Weight (each), pounds	21,000	21,000
Power, horsepower, for 20.7 psi	14,500	14,500
NSS* power to drive each circulator, MW(th)	10.8	10.8
Thermal power returned to NSS* from each circulator, MW(th)	10.5	10.5
Steam Generators (per Module)		
Steam generator modules, number	4	6
Dimensions (height by diameter), feet-inches	72–2½/12–8	72–2⅛/12–8
Module dry weight assembled, pounds	525,000	525,000
Number of tubes, main steam/reheat	328/228	328/228
Heat transfer, main steam/reheat Btu/hour	14.49 × 10^8/2.79 × 10^8	14.49 × 10^8/2.79 × 10^8
Heat losses from NSS heat transfer system, MW(th)	17	19
Bulk gas inlet temp., °F (°C)	1340 (727)	1340 (727)
Coolant mass flow, 10^6 pounds/hour	1.860	1.860
Superheater steam flow, 10^6 pounds/hour	1.34	1.34
Superheater outlet pressure, psia/°F (°C)	2515/955 (513)	2515/955 (513)
Reheater steam flow, 10^6 pounds/hour	1.33	1.33
Reheater inlet pressure, psia/°F (°C)	645/635 (335)	
Reheater outlet pressure, psia/°F (°C)	585/1000 (538)	
Feedwater pressure, psia/°F (°C)	3150/370 (188)	
Reactivity Control Systems		
Control rods (2 per drive), number	98	146
Active control rod length/diameter, inches/inches	250/3.25	250/3.25
HTGR Auxiliary Systems		
Core Auxiliary Cooling Systems		
Core auxiliary circulators, number	2	3
Helium flow rate (each), pounds/hour	137,000	101,000
Depressurized PCRV,‡ at 8.2 psig	385	385
Pressurized PCRV, at 685 psig	188,000	139,000
	570°F (299°C)	560°F (293°C)
Core auxiliary heat exchangers, number	2	3
Feedwater flow rate (each), pounds/hour	871,000	653,000
Total core auxiliary cooling system heat load, 10^6 Btu/hour:		
Normal operation	31.4	29.1
Emergency-pressurized PCRV‡	231	346
Emergency-depressurized PCRV	75	112
Refueling operations (24 hours after shutdown)	44	67

*NSS = nuclear steam supply system.

†nv = neutron flux (neutrons × velocity).

nv(th) = thermal neutron flux.

‡PCRV = prestressed concrete reactor vessel.

The Liquid Metal Fast Breeder Reactor

BY A. S. GIBSON*

The fast breeder reactor derives its name from its ability to breed, that is to create more fissionable material than it consumes; and from the fact that its neutrons travel faster than they do in a thermal reactor. The breeding process depends, in part, upon the neutrons maintaining a high speed, or high energy. If their speed or energy is allowed to degrade, as occurs in thermal reactors, the number of neutrons produced per absorption in uranium or plutonium decreases. Furthermore, at lower velocities, neutrons tend to be captured in various structural materials of the reactor, and this further reduces the breeding potential. It is important, therefore, in fast reactors to keep the velocity of the neutrons high. Water, which is used as a coolant in some thermal reactors, tends to slow the neutrons down and thus prevent efficient breeding. Therefore it is necessary to use a coolant which does not slow the neutrons or capture them as they travel through the coolant. One such coolant is sodium, and it is widely used in fast reactors. Thus, the term, liquid metal fast breeder reactor (LMFBR).

Characteristics of Sodium. Sodium is an alkali metal which melts at about 210°F (99°C). Its cross section for absorbing and thermalizing neutrons is very low (i.e., it does not capture or slow large amounts of them), and its ability to transfer heat is excellent. It has a very high boiling point (1640°F; 893°C) and a very low vapor pressure at most temperatures. These properties make it nearly ideal as a reactor coolant. Unlike water, it can be heated to very high temperatures without generating pressure, and its excellent ability to transfer heat makes it much less sensitive to short-term disturbances in the surfaces from which heat is being transferred. Because the coolant systems operate at low pressure, the liquid does not escape so rapidly, in the event of a leak from a pipe or another piece of equipment containing sodium, as with high-pressure systems.

The chemical reactivity of sodium is a safety asset in one respect—its ability to combine with or retain other elements. For example, during irradiation of fuel, many radioactive isotopes are formed which are known as fission products. Some of these isotopes are radioactive and unstable species of elements which decay gradually to stable isotopes of elements.

Typical fission products which are produced and have a higher radiological significance in terms of their potential hazard to human beings are ^{131}I, ^{137}Cs, and [95]Nb. In some types of fast reactors, these fission products are deliberately vented or discharged from fuel in the reactor into the sodium coolant. In other fast reactors, failure of the fuel outer cladding or covering can release the fission products to the sodium. Fortunately, because of its unique chemical properties, sodium tends to retain many of these fission products (Ref. 1), and so they are not readily released to the inert gases (such as helium and argon) which normally are used to blanket the sodium. Radioactive iodine, for example, combines with the sodium to form sodium iodide, and cesium is retained in solution. Niobium and certain other solid fission products tend also to be retained in the sodium. Even direct release from the fuel cladding of particles of fuel such as uranium or plutonium tends to be controlled because these particles settle and collect in relatively stagnant regions of the sodium system. This does not mean that sodium retains all fission products.

Nearly all the radioactive xenon and krypton gases bubble up through the sodium and are released into the inert cover gas. However, some of the more radiologically significant fission products are retained in the sodium (such as iodine). This property of sodium to retain these materials, although complicating maintenance, acts as a safety advantage since accidental spillage of sodium does not free large quantities of the fission products. Even if the sodium were to violate all its containers and burn in air, the burning would be at a controlled rate (2 to 14 pounds per hour per square foot of exposed surface, Ref. 2), and

*Manager, Product Engineering, Fast Breeder Reactor Department, Energy Systems & Technology Division, General Electric Company, Sunnyvale, Calif.

the fission products would not be released rapidly. This allows time for coping with the problem, such as containing the fire or smothering it. When exposed to air, sodium oxidizes rapidly if it is in the solid state; in the liquid state it burns. This burning is at a controlled rate and is easily extinguished by eliminating oxygen. When exposed to water, sodium reacts violently and forms hydrogen. The hydrogen in turn can combine with oxygen and increase the reaction energy.

Sodium has certain other features which are undesirable for a reactor coolant. Under irradiation it forms radioactive isotopes (Na 22 and Na 24) which emit gamma radiation. Fortunately, most of this radiation decays within a few days. However, this characteristic of sodium to become radioactive and to contain radioactive products from other sources makes it potentially hazardous. Under properly controlled conditions sodium is not dangerous, but care must be taken to assure that the sodium is properly controlled.

The practical effect of radioactivity in sodium is to limit it from direct human access while it is in a radioactive condition. The means for accomplishing this in a fast breeder reactor is to include two separate cooling circuits containing sodium and one containing water. The first circuit circulates sodium through the reactor core and becomes highly radioactive. This circuit is shielded from human access; any maintenance is accomplished by means of remote mechanisms. The second circuit picks up heat from the first, and in turn transfers its heat to the water circuit without becoming radioactive. Because sodium has such excellent heat-transfer characteristics, it is possible to use two circuits and still have an economically attractive system. Nevertheless, the extra sodium loop is a safety feature which is included at extra cost.

LMFBR FLOW CIRCUIT

Figure 1(a) schematically shows the flow circuit for an LMFBR where the two sodium circuits are included. The reactor is cooled by the primary sodium, which becomes radioactive as it picks up heat in passing through the core or fueled region. In this particular arrangement, the sodium is heated to 1050°F (560°C) and flows through pipes (schematically shown as a single line in the figure) to the intermediate heat exchangers. In the heat exchangers, the primary sodium transfers heat to the nonradioactive sodium.

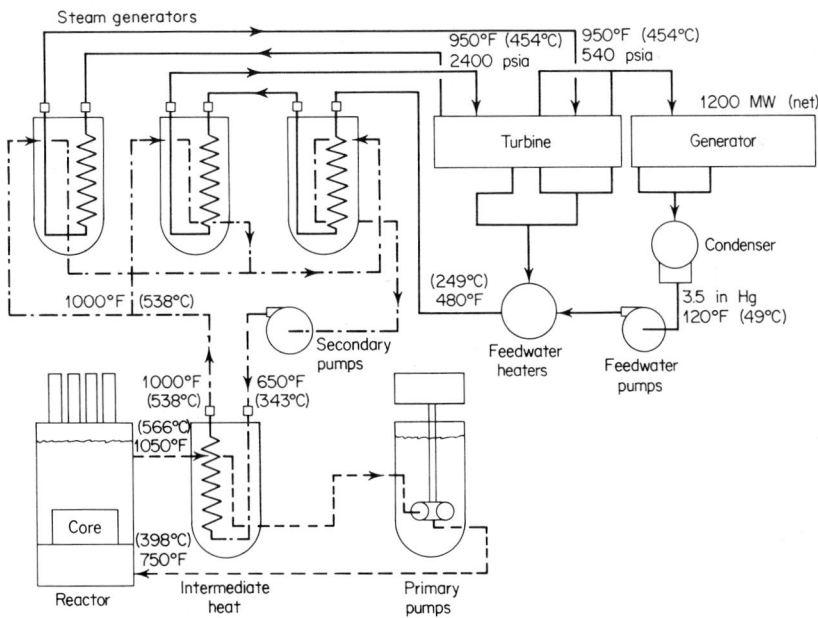

Fig. 1(a). Liquid metal fast breeder reactor flow circuit for high-temperature cycle.

After being cooled to 750°F (393°C) in the heat exchangers, the primary sodium is pumped back into the reactor where it again repeats the circuit. The nonradioactive secondary sodium is circulated from the intermediate heat exchangers through steam generators where the heat from the sodium is transferred to water, which becomes superheated steam for use in the turbine. The cooled secondary sodium is pumped back through the intermediate heat exchangers where the process is repeated. Steam from the steam generators is used to turn the rotor of the turbine generator and to generate electricity. In the arrangement shown, 1200 megawatts of electricity are generated at a net overall efficiency of 39 percent.

Figure 1(b) shows an alternative flow circuit for an LMFBR with lower temperatures than system shown in Fig. 1(a). Although this circuit results in a lower thermal efficiency

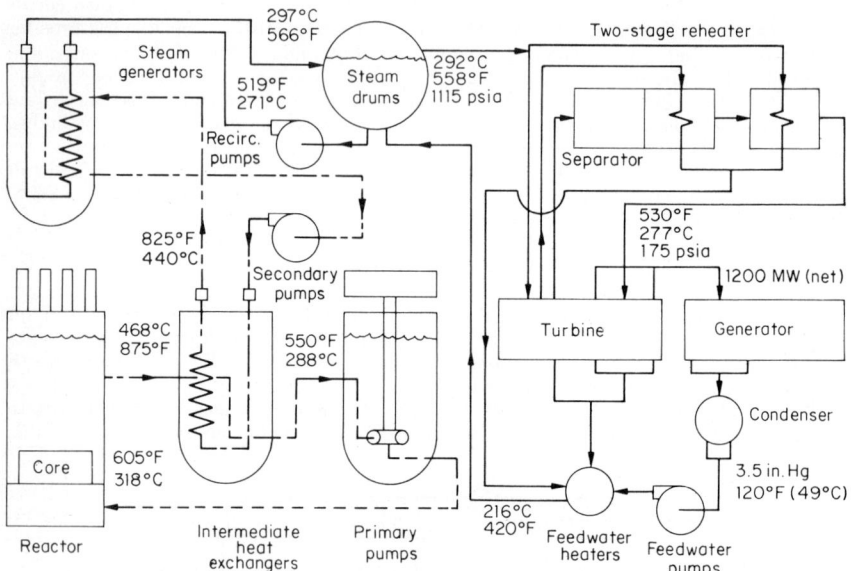

Fig. 1(b). Liquid metal fast breeder reactor flow circuit for low-temperature cycle.

than that of Fig. 1(a), the lower system temperatures allow the use of less advanced materials for the piping and components. Stainless steels and other high-temperature alloys are used extensively for the high-temperature circuit, whereas ferritic steels can be used in the low-temperature circuit. This reduces the need for scarce resources, such as chromium and nickel, which are used extensively in the fabrication of stainless steel. Also, the lower temperatures make design of components, such as pumps and heat exchangers, much simpler. Both types of circuits are being considered for future LMFBRs. It is not clear at this point which type of circuit will be economically most attractive. Fortunately, because of the efficient use of fuel in an LMFBR, either circuit will produce electricity with a minimum use of scarce fuel resources.

In the LMFBR, the sodium is normally safely contained within the piping and equipment. In order to protect against the possibility of leaks in equipment or piping, all equipment with radioactive sodium is contained within shielded vaults which exclude oxygen by substituting an inert gas like nitrogen for air so that any escaping sodium merely spills harmlessly into these vaults. Again, the exclusion of oxygen from the vaults is a safety feature which is included at extra cost.

In summary, the undesirable aspects of sodium are treated in two ways. First, equipment containing radioactive sodium is placed in gastight cells which exclude oxygen. Second, water is used only to transfer heat from nonradioactive sodium circuits, and these circuits are designed to withstand the effects of a sodium-water reaction.

NEUTRON LIFETIME

Two characteristics of fast reactors which are not shared by thermal reactors are a short neutron lifetime and the possibility of secondary criticality. The neutron lifetime is the length of time between the birth of a neutron during fission, and its capture either in uranium or other materials. In a thermal reactor, the neutrons are slowed by bouncing off hydrogen atoms (if the reactor is a pressurized water reactor or a boiling water reactor), and they "live" longer than in a fast reactor where there is no hydrogen or other "moderator" material to slow them down before they are captured. This short neutron lifetime was originally thought to be an undesirable feature. Subsequent research has demonstrated that a short lifetime is not a disadvantage providing that the instantaneous power coefficient is strongly negative. With a strong negative instantaneous power coefficient, the length of the neutron lifetime has little effect on the amplitude or duration of energy ramps emanating from the reactor (Ref. 3).

SECONDARY CRITICALITY

The issue of secondary criticality is somewhat more complex. In any reactor system it is necessary for a minimum quantity of fissionable materials to be present before a self-sustaining chain reaction can take place. The self-sustaining chain reaction occurs when the number of neutrons lost from the system or captured is exactly balanced by the number of neutrons being created in the fission process. A thermal reactor is designed so that this balance occurs only after the neutrons have been slowed down or thermalized. This thermalization requires a moderator such as water so that the neutrons can give up some of their energy against hydrogen atoms by bouncing off them. In a fast reactor, since the neutrons are not slowed down, criticality is achieved without a moderating material.

Normally in fast reactors the fuel is arranged in rods with space between the rods reserved for the sodium coolant. Since the sodium serves no function other than to remove the heat, the coolant can be removed and the self-sustaining chain reaction can still continue. Indeed, under certain circumstances the reaction can actually be accelerated. This particular characteristic of fast reactors to be able to be critical without the coolant present can result in "secondary criticality." If for example some of the fuel were to melt and fall to the bottom of the reactor while at the same time rearranging itself into a more dense configuration by filling up the passages normally occupied by coolant, then a critical mass could be present and this new configuration would in effect become an uncontrolled reactor. This particular characteristic has resulted in considerable study with the result that fast reactors are designed with great care so as to avoid possibilities that could lead to a rearrangement of the core into a more reactive configuration. This is accomplished by designing the coolant system so that the probability of the loss of a large fraction of the coolant capacity is very low and by selecting a geometric arrangement which makes the reassembly into a more reactive configuration very difficult. Additionally, instrumentation is included to detect the onset of abnormal circumstances which might lead to meltdown of large portions of the core.

Two fast reactors have actually experienced partial core meltdown, and in both of them the coolant systems and core geometry were such that secondary criticality did not occur. Nevertheless, although the probability of secondary criticality is very low, most fast reactor systems designed to date have included provisions for accommodating the energy released during an uncontrolled transient from secondary criticality.

One of the most important ways to avoid secondary criticality is to assure that coolant integrity is always maintained. To achieve such assurance, reactor systems engineers generally take great care in the design of the primary coolant circuit. For example, the reactor vessel, which contains the primary sodium that cools the fuel, is designed, fabricated, and monitored in service with great care to assure that no leaks occur. Additionally, an extra safety vessel surrounds the primary reactor vessel so that even if the first vessel should develop a leak, it would be contained within the second vessel in such a manner that the core could not be uncovered. In addition to these precautions to assure that coolant integrity is maintained, the maximum temperature of the sodium ($1050°F$, $566°C$) is very much below the boiling point of the sodium ($1640°F$, $893°C$). In a practical sense this means that a rather substantial margin exists before the sodium could be overheated to the point where it might boil from the core and thus allow the fuel to melt. It should be noted that other mechanisms besides boiling sodium exist which could melt fuel, and these are also considered in the design of the core. The principal means of assuring that the temperature of the coolant leaving the core does not exceed a safe margin is to include instruments that can sense abnormal conditions within the core. All these techniques are combined to make the possibility of a secondary criticality accident extremely remote.

POWER COEFFICIENT

It was previously observed that an instantaneous or prompt negative power coefficient was desirable. A power coefficient is simply a mathematical term which describes the response of the reactor core to certain input stimuli. For example, if the reactor power is increased by withdrawing control rods which regulate the nuclear chain reaction, this normally causes the fuel to increase in temperature, and to expand physically. As the core expands from the higher temperature, its height grows slightly, and its outside surface area becomes larger. This allows a greater number of neutrons to leak out of the core and to be lost from the system, thus tending to reduce the number of neutrons that are fissioning and liberating energy. This is turn causes the reactor power increase to be reduced compared with what it would have been if thermal expansion had not occurred.

This entire effect is described as a negative thermal expansion power coefficient. It is negative because the total power increase is reduced compared to that implied by the control rod movement, and it is instantaneous or prompt because it occurs with the same velocity that neutron movement occurs. If the coefficient were positive instead of negative, the opposite effect would occur; namely, as power increase was fed into the reactor by withdrawing control rods, this increase would be amplified beyond the movement implied by the control rods in a nonexpansive core. For reasons having to do with the stability and safety of the reactor it is now known that it is desirable to have the combination of all the reactor power coefficients negative. One early fast reactor which had a prompt positive power coefficient was EBR-I (the net coefficient was negative, but a prompt portion was positive). This was a major factor contributing to a core meltdown in EBR-I.

LONG FUEL LIFETIME

During early development of the fast reactor it became apparent that two particular reactor characteristics were desirable. One is a long fuel lifetime, and the other is the negative power coefficient just discussed. A long fuel lifetime, which results from leaving the fuel in the reactor for an extended period, can yield a low fuel cost. Most of the early fast reactor designs included uranium fuel in the form of metal. Under irradiation this metal is gradually damaged until it must be removed from the reactor. However, by changing the form of the uranium or plutonium metal to uranium or plutonium oxide, it was found that the lifetime of the fuel could be extended substantially (10,000 to 20,000 MW-days per ton for metal fuel compared with 100,000 MW-days per ton burnup for oxide fuel).

Metal fuel expands as its temperature increases in response to a power increase, and this results in a prompt negative power coefficient. Oxide or ceramic fuel also expands with increasing temperature, but because of the way in which the fuel is fabricated, this expansion is less predictable than that of the metal fuel. Fortunately, it was found that using ceramic fuel not only improved the fuel lifetime characteristics, but also introduced a prompt negative power coefficient which was as predictable as the expansion coefficient in metal fuel. This is known as the Doppler coefficient. Since the ceramic fuel is a high-temperature material, in order for the fuel to undergo damage it must reach very high temperatures. It is the change in temperature from the operating point to some higher temperature which produces the Doppler effect. The effect is caused by the heating up of the atoms of uranium in the fuel, thus causing them to move faster. Neutrons that are passing through the fuel tend to be captured by some of the ^{238}U atoms at what is known as a "resonant" energy. The increased velocity of the uranium atoms increases the number of these atoms which are at the resonant capture energy relative to the passing neutrons. These ^{238}U atoms therefore stop some of the neutrons which otherwise would have continued their travel until they were captured in the fission process, and this tends to reduce reactivity and power. Again a negative reactivity or negative power coefficient results.

DOPPLER EFFECT

Discovery of the Doppler effect in fast reactors was an extremely important development. By using ceramic fuel* it was found that the fuel lifetime could be extended, and at the same time a mechanism was found to exist which would instantaneously reduce the effect of inadvertently adding large amounts of reactivity which could cause the reactor to become "prompt critical" or uncontrolled. In a typical large sodium-cooled fast breeder reactor, for example, it is possible to imagine a very improbable event in which one of the control rods is instantaneously removed while the reactor is running at full power. If we further imagine that the many automatic safety devices fail to function so as to rapidly insert the other safety rods, then the fuel temperature will rise and hesitate at a value below the point at which damage to the fuel would occur (Ref. 4). This hesitation for one second or so allows corrective action to be taken. The major factor in this hypothetical accident which prevents the reactor from becoming uncontrolled is the Doppler effect. Similarly, if a hypothetical and extremely unlikely meltdown and secondary criticality accident is assumed to occur, a fast reactor with a typical Doppler coefficient ($-T\, dK/dT = 0.003$ or larger) would have an energy release for doing damage five times smaller than a reactor that had no Doppler effect (Ref. 3).

SODIUM VOID COEFFICIENT

One of the power coefficients in a sodium-cooled fast breeder reactor which may not be negative is the sodium void coefficient. If the sodium were to boil, it could be expelled from the coolant channels. Depending on the geometry of the fast reactor core and the manner in which the sodium is removed, this can result in a positive reactivity effect. This occurs because sodium tends to slow neutrons down and reduce the number of fast neutrons available for fissioning. Therefore, when sodium is removed from the core by boiling, not as many neutrons are slowed down, and more fast neutrons are available for the fission process. An offsetting effect is that removal of sodium also tends to allow more neutrons to leak from the core (because of less scattering), and this results in a decrease in the total number of neutrons. The net result of these two competing effects is dependent upon the geometric pattern of the sodium being removed from the core. Under the proper circumstances, the net effect can be to increase the number of neutrons available for fission, with a consequent reactivity increase and an increase in the power level of the reactor. As previously observed, the sodium operates very much below the boiling point in the reactor, and this reduces the likelihood of boiling. Furthermore, instruments are present to detect conditions that might cause boiling, and the reactor can be shut down if anomalies appear. Finally, in order for the boiling to introduce large amounts of positive reactivity, it must be generated in a uniform manner, starting in the center of the core, and progressing gradually outward as a sphere. Experiments and analyses have shown that, in fact, boiling cannot occur in this manner but would develop in a manner that is noncoherent and does not produce as much positive reactivity as would otherwise be the case (Ref. 4).

DESIGN FACTORS AND SAFETY CONSIDERATIONS

Recent work has shown that it is possible to design an LMFBR core so that an additional negative power coefficient can be obtained which is analogous to the thermal expansion of fuel in early metal-fueled reactors. This is accomplished by allowing the oxide fuel to transfer its heat to the surrounding metal cladding and core structure during a rapid temperature increase. As the structure heats up, it expands. If the supporting structure has been designed in a particular way, this expansion results in an outward bowing or "flowering" of the fuel assemblies, which in turn creates a strong negative coefficient due

---

*It should be noted that it is the dilute mixture of uranium and moderating atoms like oxygen in the ceramic which produces the Doppler effect.

to increased leakage of neutrons from the core. The French took advantage of this effect in their design of the Phenix reactor.

From the foregoing discussion of the characteristics of fast breeder reactors, it is clear that some of these characteristics have a salutary effect on the safety of the reactor, and others have a deleterious effect. The intelligent designer selects parameters and design features so as to amplify the desirable characteristics, and so as to deemphasize or to properly cope with the undesirable characteristics. By this means the designer can achieve a system which responds in a conservative manner with considerable design safety margin for both normal and abnormal operating conditions.

By designing to have a safety margin for abnormal operating conditions, such as might occur in an accident, the designer does not expect to be able to identify exactly how an accident will occur. Indeed, if the exact sequence of events of an accident could be foreseen, then the designer could incorporate features to prevent these events from occurring. Instead, the designer considers a sufficiently broad range of unlikely events that he is able to identify an envelope of safe and unsafe conditions. He can then choose design features and parameters so that the unsafe conditions have an extremely low probability of harming human beings.

MIT–AEC Safety Study. Just how low this probability is has been the subject of considerable speculation until recently. Two major attempts have been made to determine quantitatively what is the probability of harming human beings from accidents in nuclear power plants. By far the most ambitious effort to date was the reactor safety study performed under the direction of Professor Norman C. Rasmussen of the Massachusetts Institute of Technology (Ref. 5). The work was performed in 1973 and 1974 for the U.S. Atomic Energy Commission.

The objective of the Rasmussen study was to use methods developed by the Department of Defense and NASA to make a realistic estimate of the risks of damage and death from presently operating commercial nuclear plants. A secondary objective was to compare these risks with other nonnuclear risks to which society and its individuals are already exposed.

To determine the nuclear risks, a detailed analysis was made of the probability of occurrence and associated consequences of a very large number of different accident possibilities in presently designed light water reactors. The basic conclusion of the study was that the risk to the public from potential accidents in light water nuclear power plants is very small.

The results of the Rasmussen study are summarized in Figs. 2 and 3. Table 1 shows these risks in another way. In the table, the risk of fatality or injury for 15 million people living within 20 miles of United States reactor sites is shown, assuming that there are 100 reactors operating. As can be seen from these data, the risk from nuclear accidents is trivial in comparison with other risks such as those from earthquakes, automobile accidents, falls, and tornados.

The Rasmussen study was performed on specific light water reactors. Moreover, the investigators made the point that the computed risks were very much a function of the specific designs being evaluated and did not necessarily apply to other systems. They made no attempt to identify risks with other reactor systems such as LMFBRs. They did, however, make the point that the existing low level of risk for water reactors has been achieved principally by the efforts of industrial design, construction, and operation, and by the efforts of the AEC's regulatory process. Those same techniques which resulted in this very low level of risk for light water reactors are being applied to LMFBRs. Indeed, if anything, they are being made even more stringent.

UCLA Study. Another major effort to quantify nuclear risks was produced by Chauncey Starr and his colleagues at UCLA (Ref. 6). In this instance, Dr. Starr compared the public health risks of 1000-MW(e) nuclear versus oil-fired power plants in the Los Angeles basin. The general conclusions of the study were that, for both types of power plants operating within the range of regulatory limits, the public health risks were within the range of other activities of man, which have general acceptance by society. For the oil-fired plant these risks are low compared with other risks generally accepted by society, and for the nuclear power plant the risks are negligible (a factor of 60 better than the oil-fired plant).

In the Rasmussen study, fault tree analysis, which is commonly used to study potential accidents and their probabilities, was found to be inadequate for evaluating the detailed

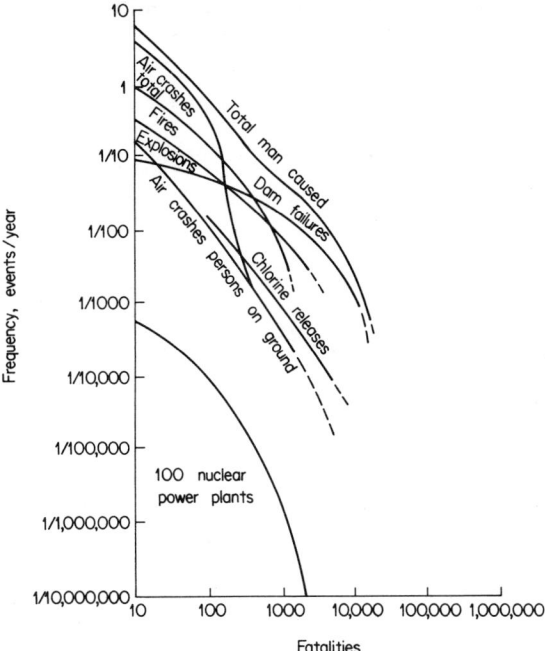

Fig. 2. Frequency of fatalities due to person-caused events.

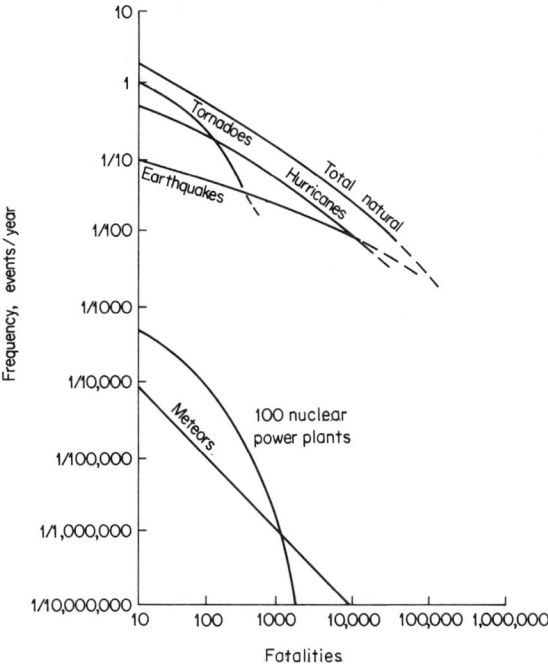

Fig. 3. Frequency of fatalities due to natural events.

events postulated to occur in different accident situations. Therefore, a new technique was developed for tracing the mechanistic possibilities of different events in an accident chain. This technique entitled "event tree" analysis, when coupled with the more traditional fault tree methods, allowed the interactions between events to be properly considered. The Starr study was less extensive and used fault tree methods. In one respect the Starr study was more comprehensive than the Rasmussen study; in addition to comparing the risks from water reactors and oil-fired power plants with other societal risks, the Starr study examined risks for a hypothetical 1000-MW(e) LMFBR (Ref. 7). In this instance, since the LMFBR was in an earlier state of design and development than water reactors, fault trees were screened for accuracy and completeness by application of the Delphi method. The Delphi method is a technique for obtaining improved estimates of probability from a group of experts.

The results of the Starr study on the LMFBR showed that a median curve for total

Table 1.   Annual Fatalities and Injuries Expected among the 15 million People Living within 20 Miles of U.S. Reactor Sites*

Accident type	Fatalities	Injuries
Automobile	4200	375,000
Falls	1500	75,000
Fire	560	22,000
Electrocution	90	
Lightning	8	
Reactors (100 plants)	0.3	6

*Data from Ref. 5.

cumulative mortality risk was at least as low as that of a water reactor. However, since the LMFBR has not yet reached the state of commercialization of water reactors, it is too early to establish its ultimate safety relative to these reactor systems. The present approach to development and licensing of the LMFBR, and studies such as those noted above give encouragement that it will have characteristics which result in very low risks to the public when compared to other risks to which society is exposed.

EARLY HISTORY AND LATE 1975 STATUS OF LIQUID METAL FAST BREEDER REACTORS

The nuclear industry has been working for more than 20 years on the development of the liquid metal fast breeder reactor. Table 2 shows some of the early fast reactors in the United States and overseas. It is clear that a substantial effort has been under way since the 1940s. As observed previously, most of the objectives of these early reactors have been accomplished—with one of the most important, the demonstration of the Doppler effect, being completed by the successful conclusion of the SEFOR (Southwest Experimental Fast Oxide Reactor) program in 1971.

Table 3 shows sodium-cooled fast breeder reactors now under construction or being planned for near future construction throughout the world. The dates correspond to published information on startup times. Undoubtedly some of the future dates will slip as the programs proceed. It is noted from Table 3 that the European countries and the U.S.S.R. have planned completion of demonstration plants on an earlier schedule than the United States. Also, the United States program plans include a large irradiation test facility (the Fast Flux Test Facility), while the French, British, and Russians do not. These differences of approach, plus a broader base development program in the United States, will result in the accumulation of a larger informational base in our country, but possibly at the expense of early application of the breeder reactor. By their more aggressive construction program, the Europeans and Russians are implicitly accepting the increased economic risk commensurate with the accelerated schedule, in return for the possibility of achieving technological breakthroughs leading to earlier beneficial use of the breeder system.

The French have been particularly successful with an aggressive approach to breeder development with operation of Rapsodie commencing in 1967, and most recently with their demonstration plant Phenix in operation. They started preliminary design of Phenix in 1966. From the start of site preparation in October 1968 until completion of construction and full-power operation in March 1974 was less than six years. This is less time than is normally used to construct a water reactor power plant. Figure 4 shows the Phenix plant in a middle stage of construction. Up to the latter part of 1975, Phenix operation has been

Table 2   Early Liquid-Metal-cooled Fast Reactors

Country	Reactor	Reactor power, thermal MW	Fuel type	Coolant	Initial operation
U.S.A.	Clementine	0.025	Plutonium metal	Mercury	1949
U.S.A.	EBR-1	1.2	Uranium oxide	Sodium–potassium	1945
U.S.A.	Submarine prototype SIG*		Uranium oxide	Sodium	1955
U.S.A.	Sea Wolf, S2G		Uranium oxide	Sodium	1956
U.S.A.	EBR-2	62.5	Uranium metal	Sodium	1965
U.S.A.	Enrico Fermi	430	Uranium metal	Sodium	1966
U.S.A.	SEFOR	20	Uranium–Pu oxide	Sodium	1969
U.S.S.R.	BR-1	0	Plutonium metal		1955
U.S.S.R.	BR-2	0.1	Plutonium metal	Mercury	1956
U.S.S.R.	BR-5	5	Plutonium oxide and carbide	Sodium	1959
U.K.	Dounreay	72	Uranium metal	Sodium–potassium	1963
France	Rapsodie	20	Uranium–Pu oxide	Sodium	1967

*Intermediate spectrum reactor.

very successful, with only minor start-up difficulties. As shown in Table 3, the French are now completing the detailed design of the 1200-MW(e) Super-Phenix and plan to start construction in 1976.

All the reactors shown in Tables 2 and 3 use liquid metal as the primary coolant, and most use uranium or plutonium oxide or some mixture of the oxides of plutonium and uranium as fuel. These materials have been selected for the coolant and fuel for most of the fast reactor development throughout the industrialized world because of the favorable neutronic and thermal properties of sodium coolant and oxide fuel, as explained earlier.

Reactor Plant Design Differences.   Despite the unanimity of choice on fuel and coolant, there are differences in design in the various reactor plants. The main differences are in the following areas: (1) primary coolant system arrangement, (2) refueling mechanism design and arrangement, (3) steam generator type and arrangement, (4) core support method, (5) structural material choices, and (6) safety features.

Perhaps the most noticeable difference is that of the primary system arrangement. This is schematically illustrated in Figs. 5 and 6. Figure 5 corresponds to a "loop", or "piped" arrangement, where the reactor, pumps, and intermediate heat exchangers are located separate from one another, and piping carries the sodium from one point to another. The "pool" or "tank" arrangement of Fig. 6 includes the reactor, intermediate heat exchangers, and pumps in one large pool of sodium which is contained in a separate tank. Each concept has certain advantages and disadvantages. The pool concept is somewhat easier to design for certain hypothetical accident situations, and the loop concept is easier to construct and to maintain. Primary interest in the United States is centered on the loop concept, and in Europe and Asia interest is divided: the French and British have selected the pool concept for their large plants, and the Germans and Japanese have selected the loop concept. The Russians are trying one of each.

Table 3. Sodium-cooled Fast Breeder Reactors
Operating, under Construction, or Planned

Name, country, purpose	Power, megawatts		Initial operation	Status (December 1975)
	Thermal	Electrical		
BN-350 U.S.S.R. Demonstration, electrical-desalination	1000	150	1973	Operating
PFR U.K. Demonstration, electrical	600	254	1974	Low power operation
Phenix France Demonstration, electrical	563	233	1973	Operating
BN-600 U.S.S.R. Early evolution, electrical	1500	600	?	Under construction
FFTF U.S.A. Irradiation test facility	400		1979	Under construction
Joyo Japan Irradiation test facility	100		1976	Functional tests
PEC Italy Irradiation test facility	140		1977	Under construction
SNR West Germany Demonstration, electrical	736	282	1981	Under construction
Clinch River Project U.S.A. Demonstration, electrical	975	350	1983	Detailed design
Monju Japan Demonstration, electrical	750	300	1981	Detailed design
Phenix-4 France Power	~ 1080	450	?	Detailed design
Super Phenix France Prototype commercial	~ 2900	1200	~ 1983	Detailed design
CFR-1 U.K. Prototype commercial	~ 3000	1300	~ 1983	Preliminary design
SNR-2 West Germany Prototype commercial	~ 4800	2000	?	Conceptual design

The refueling mechanisms can be roughly described as under-the-plug schemes and refueling cell schemes. Under-the-plug arrangements use rotating reactor plugs or covers in conjunction with mechanisms which go through these plugs and grapple fuel for movement to various positions in the core. The refueling cell arrangement permits complete removal of the reactor closure for movement of fuel. The entire upper region of the reactor is enclosed with a refueling cell which has thick walls for radiation shielding and is filled with an inert gas. There are numerous variations to these two basic schemes, and each has certain advantages and disadvantages.

Many steam generator designs have been used or are proposed for use in LMFBRs. In all these designs, a major emphasis is placed on providing extreme reliability to minimize

Fig. 4. Phenix power plant utilizing a liquid metal fast breeder reactor. Photo taken during construction phase.

the possibility of water leaking into the liquid metal region and having a reaction with sodium. Despite these precautions, systems are also included to limit the effects of possible sodium-water reactions. Most of the designs are arranged in either of two arrangements: a once-through design or a recirculating design. The once-through arrangement heats the water and produces steam in one pass through the steam generator; the recirculating units recirculate water through the boiler region to a steam drum, where steam is removed for further heating before being used in the turbine.

Core support methods also vary considerably, but most designs include some provision for irradiation-induced steel swelling. This is done by leaving some space between fuel assemblies in the region where the steel swelling occurs.

Structural materials choices are largely influenced by operating temperatures and pressures and by the fluids being contained. A reactor with a sodium outlet temperature in excess of 1000°F (538°C), for example, might use different materials than one that has an outlet temperature of 900°F (482°C). These temperature and pressure choices are determined by economic optimization of many competing effects.

The safety characteristics of the world's LMFBRs differ mainly by reason of different national philosophies. On one end of the spectrum, for example, the Russians take a very relaxed view concerning special safety features to be included in the reactor plant. They do not include the containment provisions or other features to accommodate very low probability accidents as do the Western nations. The United States tends to be quite restrictive in this regard, and its designs include additional conservatism and safety features (at extra cost).

From the foregoing discussion it is clear that development of the LMFBR has reached

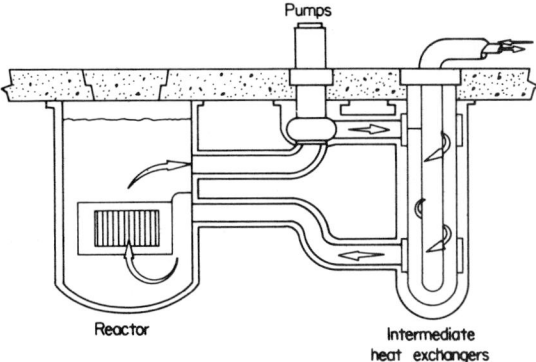

Fig. 5. Loop arrangement in liquid metal fast breeder reactor.

the point where large plants with numerous variations in design have been and are being designed and built. Figure 7 shows the conceptual design of a 1000-MW(e) LMFBR with a loop-type primary system and a refueling cell arrangement for fuel handling. It seems very likely that plants such as this will soon be built in many of the industrialized nations of the world.

The reasons for interest in the LMFBR have been explained previously. In areas of the world where fuel resources are less abundant than in the United States, these reasons are even more compelling. This explains the very aggressive programs that the British, French, Germans, and particularly the Japanese have. In the United States, the LMFBR can act as an insurance policy for assuring adequate future energy resources. Fortunately, its state of development is such that it should be ready when needed. If aggressive moves

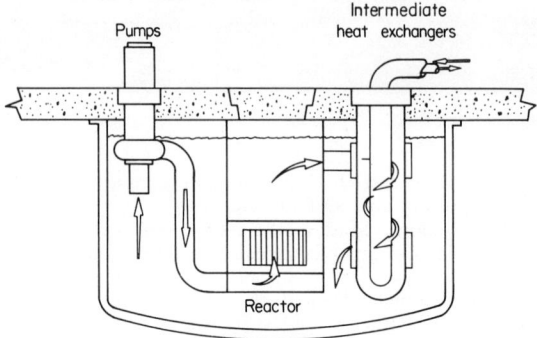

Fig. 6. Pool arrangement in liquid metal fast breeder reactor.

are made, a prototype commercial LMFBR could be in operation in the United States by 1987 or 1988.

FUEL CYCLE CONSIDERATIONS

Approximately 99.3 percent of uranium as it is found in nature is the isotope ^{238}U and 0.7 percent is ^{235}U. Uranium-235 is a fissile isotope; that is, if it is struck by a neutron it will split, or fission, yielding on the average approximately two neutrons and 200 MeV of energy. This amount of energy corresponds to approximately 78 million Btu for every gram of uranium which fissions (3.5×10^{10} Btu per pound). Most reactors that exist today are largely dependent on ^{235}U for their energy. However, some of the neutrons released in fission of ^{235}U also are absorbed in nonfissionable ^{238}U. As the ^{238}U absorbs a neutron, it is transformed into fissionable ^{239}Pu (plutonium). Thus, while the reactor is sustaining the fission process and thereby creating energy, it is also generating fresh fuel which can later be used to create more energy. Unfortunately, this is an inefficient process in present thermal reactors where the neutron velocity is established by the temperature, or thermal energy so only limited amounts of additional energy are made available by transformation of ^{238}U into ^{239}Pu.

The fast breeder reactor makes possible the recovery of most of the available energy in uranium. This occurs because during fission in the fast breeder nearly three neutrons are released for every neutron absorbed, as compared with only approximately two neutrons in a thermal reactor. On the average, between one and two neutrons are necessary for sustaining the fission process, and the extra neutron in a fast reactor can be absorbed in nonfissionable ^{238}U and thereby transformed into fissionable ^{239}Pu. Reactors that have a breeding ratio greater than one create more fuel than they need for their own purpose, and the extra plutonium can be used to fuel new breeder reactors. By this means, 80 percent or more of the available energy in uranium can be recovered and used in reactors.

In a typical fast breeder, most of the fuel is ^{238}U (90 to 93 percent). The remainder of the fuel is in the form of fissile isotopes which sustain the fission process. The majority of these fissile isotopes are in the form of ^{239}Pu and ^{241}Pu, although a small portion of ^{235}U can also be present. Normally, the fissile isotopes are located in a central "core" region, which is surrounded by the fertile isotopes in the "blanket" region. This is illustrated in Fig. 8.

When the fuel is initially loaded into the reactor, the core region will typically contain 10 to 15 percent fissile isotopes, the remainder being ^{238}U. Essentially all the blanket will be ^{238}U. As energy is extracted from the fissile isotopes, they become depleted (the initial plutonium is gradually used up). However, in a breeder reactor, new plutonium will be

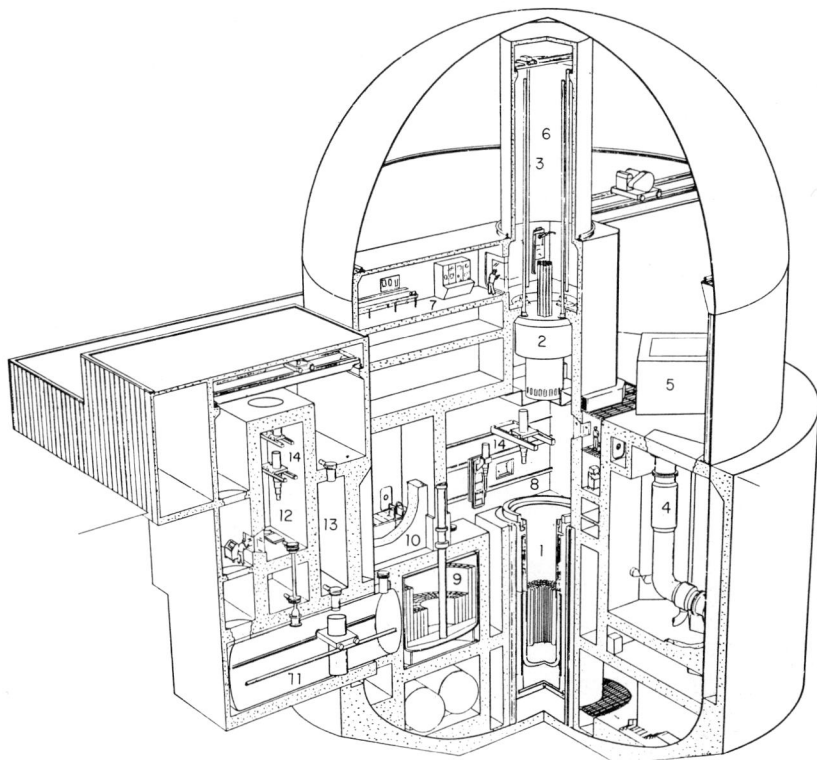

Fig. 7. A 1000-MWe LMFBR conceptual design: (1) Reactor vessel. (2) Reactor closure (in raised position). (3) Reactor closure lifting screws (four in number). (4) Primary sodium intermediate heat exchanger. (5) Elevator, stairs, and equipment access. (6) Reactor closure maintenance cell. (7) Drive and instrument maintenance room. (8) Refueling cell. (9) Carrousel fuel-storage tank. (10) Fuel-transfer cell. (11) Fuel-transfer lock. (12) Irradiated fuel-handling cell. (13) New fuel-handling cell. (14) Fuel-handling hoists. (15) Sodium drain tanks. *(General Electric Company.)*

formed in the core and blanket regions faster than it is consumed. Additionally, undesirable fission products are formed which must ultimately be removed. This process is schematically illustrated in Fig. 9. The "before" chart represents the new fuel condition, and the "after" chart corresponds to the situation when the fuel is removed for reprocessing. Typically, the fuel removed for reprocessing will contain from 1 to 3 percent new plutonium. It is in this manner that the fast breeder can recover 80 to 90 percent of the available energy in uranium resources.

With the assumption of a ten-year nuclear capacity doubling time noted in Fig. 10, it is possible to compute the cumulative use of uranium ore for different time periods and for different assumptions concerning the type of reactor systems being added to the electric network. The top curve of Fig. 10 corresponds to the situation where no LMFBRs are added; only the present type of water reactors or high-temperature gas-cooled reactors are used. The second curve shows the decreased ore usage if plutonium generated in the present type of reactors is recycled back into those reactors. The other curves show the decreased ore usage possible with breeders having different doubling times (i.e., different abilities to transform more fertile materials into fissile materials). Even with a relatively modest doubling time of 20 years, it is seen that the breeder can significantly reduce ore usage.

Decreased ore usage is ecologically important since only about two million tons of

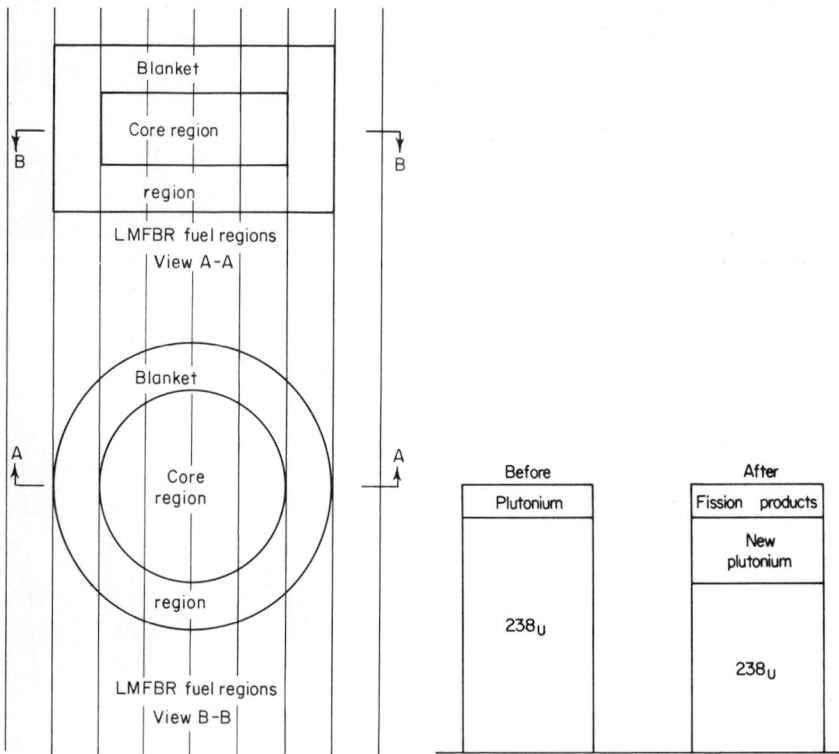

Fig. 8. LMFBR core and blanket arrangement.

Fig. 9. Basic operation of the breeder. This figure does not attempt to show geometrical disposition of fuel in the core and blanket regions.

uranium ore is of high-grade quality. This high-grade ore has an energy content in present reactors equivalent to 200,000 to 600,000 Btu for every pound of ore mined. For the lower-grade ores the energy content is about the same as coal, namely, 10,000 to 12,000 Btu per pound of material mined. Thus, after we have mined about two million tons of uranium, the mining disturbance to supply uranium for the present type of reactors will be roughly equivalent to that caused by mining coal. The LMFBR can significantly alter that picture.

High-grade ore used in fast reactors as a source of fertile material to be converted into fissile materials has an evergy content equivalent to about 217 million Btu per pound of material mined. This high energy content for fast breeders coupled with the reduced need for ore to fuel the present type of reactors materially reduces the mining disturbance. A recent study quantitatively determined this reduced mining effect by computing the annual fuel materials mined for electric power plants in the United States through the year 2020 for an assumed electrical growth rate of 6 percent both for the case where breeders are available and where they are not. (Ref. 8). The results of this computation are shown in Fig. 11. It is apparent that the breeder significantly reduces the mining requirements compared with the other alternatives studied.

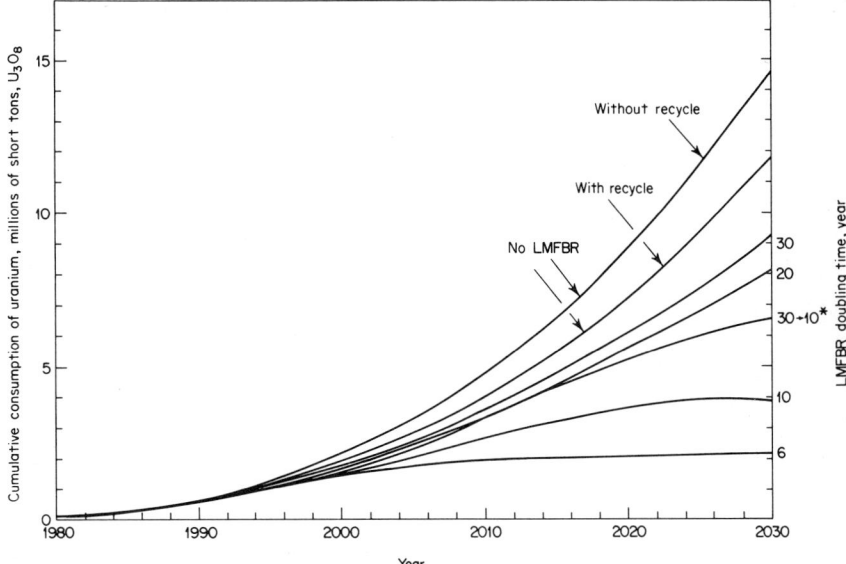

Fig. 10. Potential effect of LMFBR doubling time on uranium consumption through the year 2030. Basis: Nuclear capacity doubling time = 10 years. Assumptions: Introduction of the LMFBR in 1985. Plutonium-limited growth after 1987. *30-year LMFBR introduced in 1985; 10-year LMFBR introduced in 2005.

Most present reactors require some enrichment of the ^{235}U isotope used to fuel them. This enrichment process requires plants which, in turn, use large amounts of electric energy. Because the fast breeder converts the fertile isotope ^{238}U into the fissile isotope ^{239}Pu, no enrichment plant is necessary. The fast breeder serves as its own enrichment plant. The need for electricity for supplemental uses in the fuel cycle process is thus reduced.

ECONOMIC COMPARISONS WITH OTHER POWER SOURCES

In comparing the breeder economics with other power sources it should be borne in mind that, unlike a coal-fired power plant or a water reactor power plant, the breeder has not yet demonstrated its economic potential in a practical application. It *has* demonstrated the technical feasibility of designing and building safe reactors which have operated successfully in numerous installations in the United States and abroad. Economic demonstration must await the current generation of breeder reactors being designed and built throughout the world. The economic data for the breeder used in this article are based on design work performed by the General Electric Company.

The operating costs for nuclear plants are slightly higher than for fossil-fired plants because of the need for special personnel with nuclear experience, but this is a minor difference. The major differences between systems occur in the fuel and capital cost categories. It is possible to compare *relative* costs of one power system with another. However, it should be noted that such comparisons are only illustrative of cost trends since inflation effects, and material and fuel scarcities, are causing rapid changes in the costs of all power systems. Most of these changes are in the direction of increasing costs with the breeder being less sensitive to future cost increases because of its more efficient utilization of scarce fuel resources.

Figure 12a shows the fuel costs* for a typical coal-fired power plant located in the

*Representative of 1974 time frame.

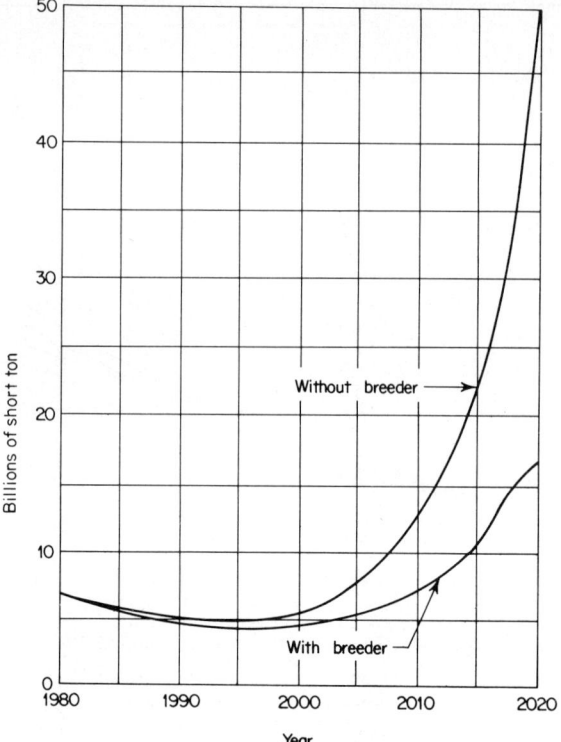

Fig. 11. Annual coal and uranium ore mined.

midwestern United States. The largest element of its costs is from the mining and transportation of coal. The sulfur stack gas removal portion of the cost is to pay for equipment and operation required to meet current laws and regulations on sulfur emission. The land reclamation costs have not been included because it is not yet known what the costs of reclaiming strip-mined areas will be. The mining and transportation portion of

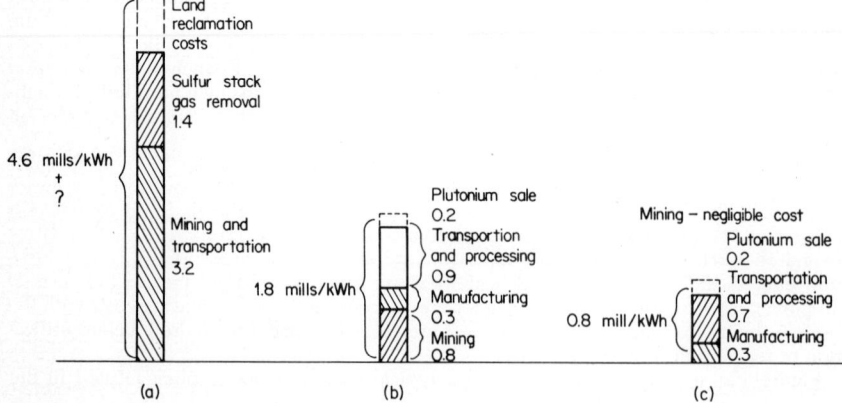

Fig. 12. Representative fuel costs in 1974 for: (a) Coal-fired power plant; (b) water reactor power plant; (c) breeder reactor power plant.

the costs are labor-intensive, and as such are very sensitive to future increases in the cost of labor.

Figure 12b shows the water reactor fuel cost. Its cost of about 1.8 mills per kWh is only 39 percent of the coal-fired plant fuel cost of 4.6 mills per kilowatt-hour. Because the water reactor fuel cost elements are less labor-intensive than the coal-fired fuel, it is less sensitive to future cost increases.

Figure 12c illustrates the anticipated fuel cost for a fast breeder reactor. Its cost of 0.8 mills per kilowatt-hour is only 44 percent of the water reactor fuel cost, or 17 percent of

Fig. 13. Sodium valves for Phenix liquid metal fast breeder reactor.

the coal-fired power plant fuel cost. It is even less labor-sensitive than the water reactor fuel and is very resistant to future cost increases. This is particularly true because of the enormous quantities of energy available to the breeder from our uranium fuel reserves.

The capital costs of power plants are dependent on the plant arrangement and equipment used. Because the LMFBR uses nuclear fuel and other radioactive systems, its plant arrangement and equipment more nearly resemble those of a water reactor than a coal-fired plant. It is therefore convenient to determine its cost relative to that of a light water reactor. By examining those parts of the design which are similar to the design of a light water reactor, and by reviewing the details of differences of other parts of the design, it is possible to determine the relative capital cost of the LMFBR. A number of such studies have been made (Refs. 9 to 11). These studies show that if certain technological advances are achieved, and if the LMFBR is sold in multiple units as present light water reactors (LWR) are, then its capital cost can be expected to be between $50 and $140 per kilowatt more expensive than that of a light water reactor. If the light water reactor has a cost in 1974 dollars of $400 per kilowatt, then the breeder would cost from $450 to $540 per kilowatt.

From the above discussion, it is clear that the fuel cost for the LMFBR is lower than that of the water reactor and the capital cost is higher. The fuel cost for the breeder is lower than that of the light water reactors because of the more efficient way that the breeder uses neutrons. The higher capital cost results from additional equipment needed in the breeder compared with that in light water reactors. A similar analogy can be made for coal-fired plants which have higher fuel costs than water reactors, but lower capital costs—again because of poorer fuel economics and decreased equipment.

If the LMFBR is to compete with water reactors or coal-fired plants, then its lower fuel costs must more than offset the higher capital costs. A recent study was completed in

Fig. 14. Intermediate heat exchanger for Phenix liquid metal fast breeder reactor.

which *all* the power plant costs for the LMFBR and for other projected power sources in the United States* were added up into the early part of the next century (2020) and compared with one another (Ref. 8). To determine the effect of different economic climates, alternative assumptions were made on capital costs, uranium ore and coal costs, discount rate, inflation rate, and breeder introduction date. For the reference case it was found that on a discounted present value basis† the LMFBR was cheaper than the next best alternative by 157 billion dollars. Even for a case where the breeder capital costs were assumed to be twice those of light water reactors, the LMFBR showed a discounted net benefit of 90 billion dollars.

These values, then, represent the savings in power costs that the United States can realize by introducing the LMFBR into its electric network if the LMFBR is developed to its full commercial potential. The incentive for such development appears to be in excess of 100 billion dollars.

---

*The costs included fuel, operating, capital, and money for all of the power plants assumed to be built, whether LWRs, coal plants, LMFBRs, or high-temperature gas-cooled reactors. Development costs were excluded.

†For a discount rate of 6.36% per year plus inflation. Inflation in the reference case was taken to be 6% per year.

STATE OF LMFBR TECHNOLOGY

In the prior description of the early history of LMFBR technology, it was noted that a number of small special-purpose reactors have been built, including some in the United States. Some of these early reactors were in service as early as 1949. Examination of this history shows that the nuclear community has been engaged in a long and careful development of the fast reactor. This has resulted in a stepwise progression from one objective to another. A partial listing of these objectives includes: (1) demonstration of breeding, (2) demonstration of the safety of liquid metal coolant systems, (3) determination of reactor and system characteristics, (4) demonstration of the Doppler effect and

TABLE 4. Advantages of Fast Breeder Reactors

Improved Utilization of Resources:
    Fossil fuel reserves equal about 35 Q (1 Q = 10^{18} Btu). 35 Q will last from 100 to 200 years with shortages sooner.
    Thermal reactors (boiling water reactors and pressurized water reactors make available additional 1 to 4 Q.
    Fast breeder reactors make available additional 600 to 1000 Q, or greater.
    Thus, energy independence is made more probable by breeder.
Low Ecological Disturbance:
    Negligible atmospheric pollution (from radioactivity or other sources).
    Reduced ore mining requirements.
    Reduced energy required for fuel preparation (e.g., enrichment).
Economic Incentives:
    Power costs are potentially lower than other power options.
    Breeder reduces capital requirements since diffusion plants are not necessary and mining requirements are reduced.
Technology Is Reasonably Well Understood:
    Under development since 1940s.
    Most industrialized nations have high-priority development programs.
    Closer to practical reality than other potentially very large energy sources (greater than 500 Q).

other power coefficients important to safety, and (5) determination of the effects of irradiation on fuel, coolant, and reactor structural materials.

These objectives have been largely satisfied by the early reactors and by supplemental development programs. Still, however, substantial development work is being conducted to better understand the effects of irradiation on fuel and structural materials, and to assure the safety characteristics of LMFBRs.

In recent years, the technology for LMFBRs had advanced beyond the feasibility issues which characterize the foregoing objectives. It has evolved to the point where large components have been fabricated and tested in prototypical environments. Successful operation of these components has made it possible to fabricate and put into service such items as pumps, valves, heat exchangers, reactor components, and other equipment in large demonstration plants throughout the world. Typical of this equipment are the large sodium valves shown in Fig. 13 and the intermediate heat exchangers shown being fabricated in Fig. 14. These particular components were fabricated for the French Phenix plant previously described.

The state of technology in liquid metal fast breeder reactors today has reached the point where the major remaining uncertainty involves how to reduce the known technology to economic practice. The possible advantages of accelerating progress in the design and construction of LMFBRs are summarized in Table 4. Environmental comparisons of three energy plants are given in Table 5.

REFERENCES

1. Silberberg, M., et al.: AEC Task 12-B, Sodium Chemistry, Fission Product Contamination, Quarterly Technical Progress Report, Fuels, Materials, and Coolant Chemistry Programs, *U.S. At. Energy Comm. Publ.* AI-AEC-12817, Washington, January–March 1969.
2. Koontz, R. L., et al.: AEC Task 6-D & E, Nuclear Safety, Gaseous Effluent, Sodium Fires, Quarterly Technical Progress Report, LMFBR Safety Programs, *U.S. At. Energy Comm. Publ.* AI-AEC-12816, Washington, January–March 1969.

TABLE 5.   Environmental Effects of Different Types of 1000-MW(e) Power Plants
Capacity factor for all plants = 0.8

Mining requirements	Coal-fired plant	Water reactor	Liquid metal fast breeder reactor
Coal			
Mass, tons/year*	3,000,000		
Volume, cubic feet/year	120,000,000		
Uranium Ore			
Mass, tons/year as U_3O_8		130	1 to 2
Mass, tons/year as 0.21% ore		52,000	~400
Volume, cubic feet/year of ore		1,390,000	~11,000
Gaseous and Liquid Wastes			
Carbon monoxide, tons per day	2	0	0
Carbon dioxide, tons per day	21,000	0	0
Sulfur oxides, tons per day			
Nonregulated release for 1% low-sulfur coal	140	0	0
Regulated release with sulfur removal	100	0	0
Nitrogen oxides, tons per day			
Nonregulated release	82	0	0
Regulated release with controlled firing	57	0	0
Particulates released to atmosphere, tons per day†	0.4	0	0
Radioactive gases or liquids equivalent dose, mrem/year:			
At power plant boundary	Minor amounts	~5‡	<1
At reprocessing plant boundary		~5	~5
Solid Wastes			
Collected fly ash or slag:			
Mass, tons/year	330,000	0	0
Volume, cubic feet/year	7,350,000	0	0
Radioactive wastes:			
Volume, cubic feet/year	0	~15,000	~15,000
Thermal Wastes §			
Discharge to cooling water, thermal megawatts	940	1684	1244
Discharge to atmosphere, thermal megawatts	320	16	16
Total waste heat, thermal megawatts	1260	1700	1260

*For net heat rate of 8760 Btu/kilowatt-hour and 10,200 Btu/pound of coal.
†For coal with total ash content of 11% and fly ash precipitator efficiency of 99.5%.
‡For present low release systems. Future systems can be designed to match the fast breeder.
§For net plant efficiencies of 32% for water reactor; 39% for coal plant; and 39% for LMFBR, and at 80% capacity factor.

The effluent wastes for the three power plants shown in the table have been categorized as liquid and gaseous, solid, and thermal. Reviewing first the liquid and gaseous wastes, it is clear that the coal-fired plant releases large quantities of material compared with both the water reactor and the LMFBR. From a public health point of view, the sulfur dioxide, nitrogen oxides, and radioactive gases constitute the greatest hazard. Effort has been and is being put forth to reduce these effluents from both the nuclear plants and the fossil-fired plants. The 1% sulfur fuel for the coal plant represents low sulfur fuel such as that found in the Western United States. Environmental Protection Agency regulations require that the sulfur dioxide and nitrogen oxides be reduced to the regulated values shown. Techniques are being developed to accomplish this through the use of limestone scrubbers, controlled firing, and other means. These means are somewhat developmental, and pose questions concerning sludge removal and operating reliability. These questions will be addressed in pilot installations which will be tested over the next few years. Even with these new cleanup systems, however, it should be noted that 100 tons per day (or 232 million cubic feet per day under STP conditions at 70°F and one atmosphere pressure) of sulfur dioxide gas are released, and 57 tons of nitrogen oxides.

Both the water reactor and the LMFBR are seen from the table to release some gases or liquids that are radioactive. However, these quantities are very small (5 mrem per year or less at the plant boundary). If an individual were to reside continuously at the nuclear plant fence boundary for one year, he would receive less radiation exposure than from two airplane flights across the continent.

The solid wastes shown in the table also reveal substantially greater volume for the coal-fired plant than for the nuclear plants. Again, as with the gaseous wastes, the solid wastes from the nuclear plants

3. Meyer, R. A., and B. Wolfe: Fast Reactor Meltdown Accidents Using Bethe-Tait Analysis, *Advan. Nucl. Sci. Technol.*, **4** (1968).
4. Murphy, P. M., D. B. Sherer, and A. S. Gibson: "The Effect of Using Different Safety Criteria and Selected Safety Studies for a 1000 MWe Sodium Cooled Fast Reactor," International Conference on Sodium Technology and Large Fast Reactor Design, Argonne National Laboratory, Chicago, Nov. 7–9, 1968.
5. WASH-1400, Reactor Safety Study—an Assessment of Accident Risks in U.S. Commercial Nuclear Power Plants, *U.S. At. Energy Comm. Publ.* WASH-1400, August 1974.
6. Starr, C., et al.: A Comparison of Public Health Risks: Nuclear vs. Oil Fired Power Plants, *Nuclear-News*, October 1972.
7. Starr, C., P. Godbout, et al.: Appendix III, "Probabilistic Safety Analysis of a Hypothetical 1000 MWe Liquid Metal Fast Breeder Reactor—Public Health Risks of Thermal Power Plants," UCLA-ENG-7242, University of California at Los Angeles, May 1972.
8. Stauffer, T. R., R. S. Palmer, and H. L. Wyckoff: "An Assessment of the Economic Incentive for the Fast Breeder Reactor," American Society of Mechanical Engineers, paper 75-WA/NE-5, Houston, Texas Meeting, November 30, 1975.
9. GEAP-5678: "Conceptual Plant Design, System Descriptions, and Costs for a 1000 MWe Sodium Cooled Fast Reactor," p. 28, U.S. Atomic Energy Commission, Clearing House for Federal Scientific and Technical Information, Springfield, Va., September 1968.
10. Palmer, R. S., and K. M. Horst: Cost Targets for Commercial Breeder Reactors, *Publ. Util. Fortn.*, July 18, 1974.
11. Horst, K. M., F. W. Sciacca, and A. S. Gibson: "Uranium Utilization and the Commercial LMFBR," ASME Winter Annual Meeting, American Society of Mechanical Engineers, New York, November 1974.
12. Perge, A. F., and R. W. Ramsey, Jr.: "Plans for the Management of Radioactive Waste from Reprocessing Fuel in Commercial Operations," International Symposium on Management of Radioactive Wastes from Fuel Reprocessing, Paris, November–December 1972.
13. De Buchananne, George D.: "Activities of the Geological Survey, United States Department of the Interior, in the Search for a Geologic Environment Suitable for the Management of High-Level Radioactive Waste," International Symposium on Management of Radioactive Wastes from Fuel Reprocessing, Paris, November–December 1972.
14. NASA TMX-2912: "Feasibility of Space Disposal of Radioactive Nuclear Wastes," National Aeronautics and Space Administration, Washington, May 1974.
15. Platt, A. M., and R. W. Ramsey: "Long-Term Waste Management Methods," International Symposium on Management of Radioactive Wastes from Fuel Reprocessing, Paris, November–December 1972.

ADDITIONAL READING

Theobald, P. K., S. P. Schweinfurth, and D. C. Duncan: Energy Resources of the United States, *Geol. Surv. Circ.* 650, Government Printing Office, Washington, 1972.
"U.S. Energy Outlook," a Report of the National Petroleum Council's Committee on U.S. Energy Outlook," National Petroleum Council, Washington, December 1972.
Hubbert, M. King: Energy Resources (chap. 8) in "Resources and Man," National Academy of Sciences, W. H. Freeman, San Francisco, 1969.
"Civilian Nuclear Power—a Report to the President—1962," and "The 1967 Supplement to the 1962 Report to the President," U.S. Atomic Energy Commission, Washington.
Ross, Philip N.: "Development of the Nuclear-Electric Energy Economy," Westinghouse Electric Corporation, East Pittsburgh, Pa., 1973.

are radioactive. In this instance, because the majority of the nuclear wastes are collected in the fabricating and reprocessing of the fuel, they are highly radioactive. For that reason they are treated in a special process to bind them in a solid ceramic material which is very impervious to intrusion. Fortunately their quantities are small, and means are being devised for their safe long-term storage. Nevertheless, because of the long radiation life of some of these wastes the nuclear systems present a problem of waste management that has no analogy with the coal-fired plant. Most of the radioactive materials generated during operation of the reactor decay shortly after shutdown (within days to weeks). Some persist for months, however; and a few tenacious species exist for hundreds of years. It is these latter long-lived species known as actinides which cause the biggest problem of nuclear waste management. A number of ways for handling this problem are under study, including concrete storage facilities (Ref. 12), underground storage in rock or salt mines (Ref. 13), shooting the wastes into outer space (Ref. 14), and transmutation in fast reactors (Ref. 15). Because of their small volumes, it is likely that the final solution for waste disposal will not have a significant economic impact on nuclear power.

The significantly reduced mining requirement shown in the table for the LMFBR relative to the LWR and coal-fired plants has important long-term implications with regard to environmental disturbance and achievement of energy independence. This is further indicated in Figs. 10 and 11.

Gas-cooled Fast Breeder Reactor

BY ROBERT H. SIMON*

The gas-cooled fast breeder reactor (GCFR) is a nuclear power plant that produces more nuclear fuel than it consumes; i.e., it breeds fuel. As there is no moderator in the core, the neutrons produced in fission retain their high energy (speed); hence the classification "fast" reactor. The heat generated in the reactor is transferred by the coolant, which is pressurized helium, to the steam generators which, in turn, supply superheated steam to drive the turbine generator for electric power production.

Initial studies on the possibility of using helium as a coolant for fast reactors were conducted in the early 1960s in Germany at the Karlsruhe Nuclear Research Center (Gesellschaft für Kernforschung) under the direction of Professor Karl Wirtz and at General Atomic Company in San Diego, California, by Dr. Peter Fortescue (Ref. 1). These studies indicated that at pressures of 1000 psi or higher helium could adequately remove the heat generated in a fast reactor core.

The advantages of helium as a reactor coolant have been discussed in the literature (Refs. 2 and 3). Most of the advantages stem from its inertness and the fact that it is a single-phase coolant:

1. Helium does not interact significantly with the neutrons generated in the fission process. Thus, a very hard spectrum is possible in the gas-cooled fast reactor, which leads to breeder reactor performance in terms of fuel doubling time and fuel conservation that is unmatched by any other reactor concept. An additional advantage is that the helium does not become radioactive, and thus the heat absorbed by the primary coolant may be transferred directly to the water cycle in a steam generator.

2. The chemical inertness of helium allows a wide choice of materials, including carbon steels, to be used for the primary system components.

3. Helium does not undergo any phase changes within the range of operating temperatures and pressures used in nuclear power reactors. Thus, hydrodynamic and thermodynamic instabilities that can be associated with changes in phase of a coolant cannot occur. This feature of gas coolants reduces operational risks because of the increased ability to predict thermodynamic conditions during normal and off-normal operation.

4. Pressurized helium has adequate heat-transfer properties, although not as good as those of a liquid-sodium coolant. Thus, wherever high heat-transfer coefficients are needed, increased coolant flow velocity and, if necessary, artificial roughening of the heating surface in the core yield the desired heat transfer. Where low heat-transfer coefficients or even insulation effects are needed, low-velocity flow or stagnant helium provide these properties, which is an obvious advantage in reducing thermal shock.

The GCFR is a direct descendant of the high-temperature gas-cooled reactor (HTGR) (see prior article in this Handbook section), which is a thermal converter system that uses helium as the primary system coolant (Ref. 4). The GCFR is thus based on the established primary coolant and components technology of the HTGR system. The use of prestressed concrete reactor vessels (PCRVs) with their multiple, inspectable, and replaceable tendons has eliminated the possibility of gross failure of the primary containment envelope. To date, the application of PCRVs has been limited to gas-cooled reactors because of the difficulty of providing an adequate thermal barrier between the concrete and hot liquid primary coolants.

The similarities of the GCFR and HTGR are evident in Fig. 1, which shows a conceptual design of the primary system of a 300-MW(e) GCFR demonstration plant. The fast reactor core, surrounded by breeding blankets and thermal shielding, is located in the center cavity of a multicavity PCRV. The helium coolant, at a pressure of 1305 psia, flows downward through the core; passes through once-through steam generators to top-mounted, series steam-turbine-driven helium circulators; and then flows back through the

*Director, Gas-Cooled Fast Breeder Reactor Program, General Atomic Company, San Diego, Calif.

core. The designs of the steam generators and helium circulators are based on the corresponding HTGR components. Furthermore, the auxiliary circulators, heat exchangers, helium purification and handling systems, instrumentation, and other components are similar to those used in the HTGR.

These similarities to the HTGR are important because the development work required for the GCFR is much less than that normally considered necessary for new reactor concepts.

In addition, the GCFR, as presently conceived, utilizes the fuel and physics technology being developed under the LMFBR program [See later article in this Handbook section; (Ref. 5)]. A compilation of several key fuel-rod parameters for the LMFBR and the corresponding values for a 300-MW(e) GCFR are shown in Fig. 2. For both types of fast reactors, mixed uranium and plutonium oxide pellets in stainless-steel cladding tubes are used as the fuel.

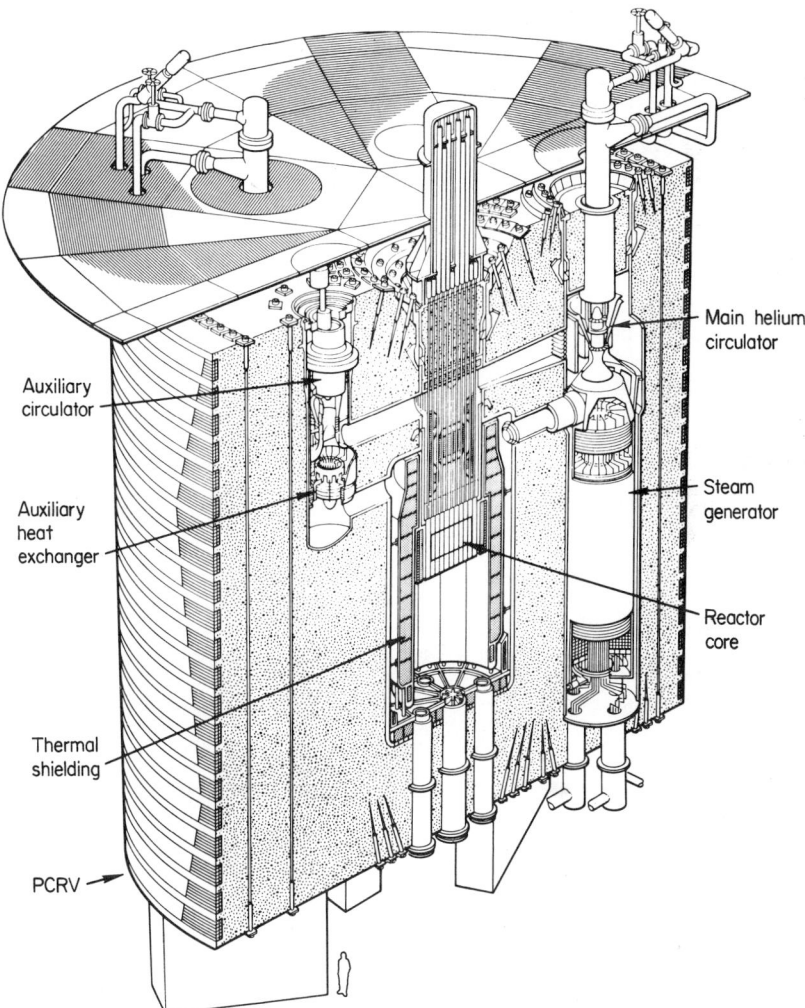

Auxiliary circulator

Auxiliary heat exchanger

Thermal shielding

PCRV

Main helium circulator

Steam generator

Reactor core

Fig. 1. Nuclear steam supply system of GCFR demonstration plant. For size comparison, note person in lower left-hand portion of diagram.

DESCRIPTION OF THE GCFR

Although a number of possible GCFR concepts have been proposed, the only one under active development in Europe and in the United States is based on the high-temperature gas-cooled reactor system in which the ceramic core and graphite reflector have been

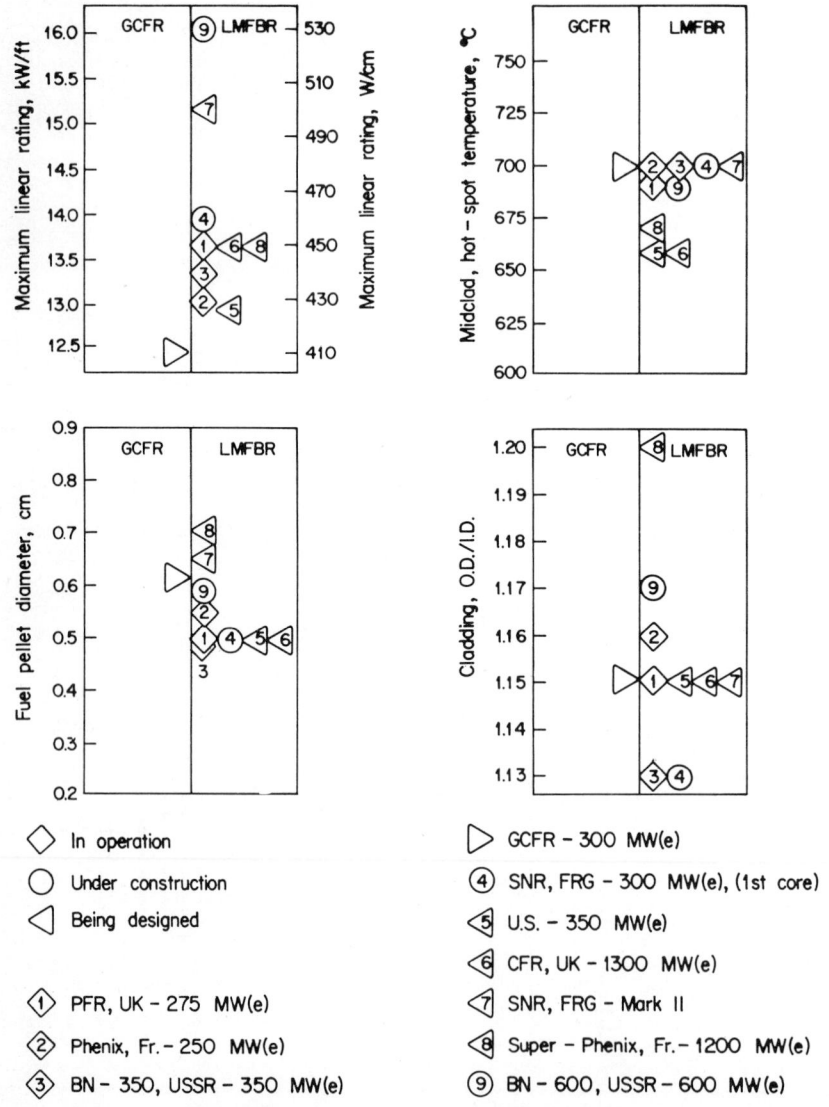

Fig. 2. Fuel characteristics of fast breeder reactor plants.

replaced by a fast reactor core and blankets. As in all large gas-cooled reactors, the entire primary system of the GCFR is enclosed within a PCRV. A central cavity in the PCRV contains the reactor core and shielding and cavities in the concrete sidewall around the reactor cavity contain the steam generators and helium circulators and the auxiliary core cooling systems (Fig. 1; Ref. 6). The fuel elements consist of bundles of stainless-steel-clad mixed-oxide fuel rods within a stainless-steel hexagonal wrapper. Although the same materials are used in the GCFR core as in the LMFBR core, two modifications have been provided to improve the heat-transfer performance of the helium coolant: three-fourths of the surface of the cladding in the core region is artificially roughened to reduce the coolant film temperature drop (Ref. 7), and a pressure-equalization system, PES (Refs. 5 and 8), to equalize the internal and external gas pressure on the cladding (see Fig. 3). Therefore, gas plenums are not required in the fuel rods and little, if any, mechanical stress on the fuel cladding is expected to occur during the entire fuel lifetime. A coolant pressure of about 89 atmospheres has been chosen for the GCFR design, as compared to about 50 atmospheres for the HTGR, because the much higher power density in the core of fast breeders requires enhanced cooling. The outlet gas temperature in the GCFR is lower than in HTGRs (550 to 600°C versus 750 to 800°C) because of the limitations of presently

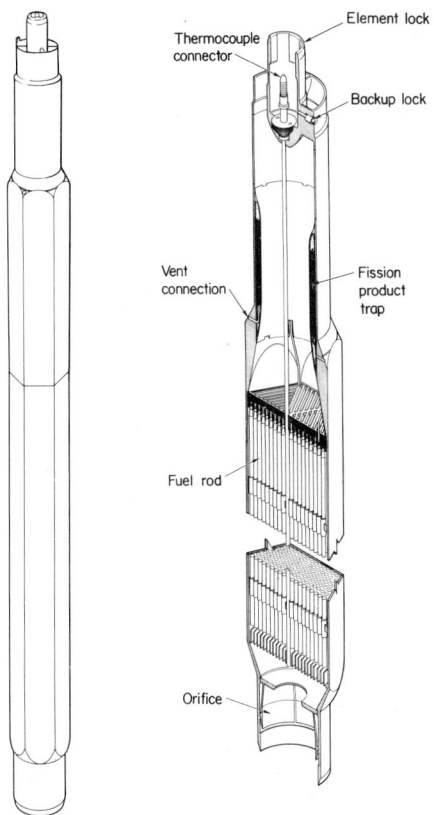

Fig. 3. Vented fuel element used in GCRF. (*General Atomic Company.*)

available metallic cladding for the fuel rods (700 to 750°C cladding mid-wall hot-spot temperature). Even so, net cycle efficiency of 36 to 38 percent is achieved.

The nuclear steam supply system of a 300-MW(e) GCFR demonstration plant has three main loops and three auxiliary loops (see Fig. 1). Pressurized helium at the reactor inlet temperature enters the core inlet plenum and flows downward through the core support grid. All the core support provisions and control-rod drives are located at the cooler end of the reactor, and thus gravity trip is permitted without the existence of any shear plane where the control rods enter the core. The hot helium from the core enters the reactor outlet plenum, flows upward on both sides of the thermal shield, and leaves the reactor cavity through horizontal hot coolant ducts to each of the three main-cooling-loop cavities.

The thermal shield over the floor of the reactor cavity is designed to be impact-resistant and sufficiently level to permit recovery of any dropped object. It is an advantage of refueling from below the core that a dropped object cannot cause damage by falling on the core.

The fuel elements contain 270 fuel rods 7.2 millimeters in root diameter. In addition to the fuel and upper and lower axial blankets, there is a 203-millimeter section of neutron shielding in the upper end of each fuel rod to assure a minimum residual ductility of the core support grid of 10 percent at the end of 30 years. In addition to venting the fission-product gases to the helium purification system, the venting system provides a means for locating and monitoring a leaking fuel or blanket element. The coolant outlet temperature from the core is measured by three thermocouples installed in a tube in the center of each fuel element.

Fixed, replaceable coolant orifices are installed in the outlet end of the fuel and blanket elements to equalize the maximum cladding temperature among the elements. These orifices are changed when the plant is shut down for the annual partial reloading of one-third of the core to compensate for the change in power sharing when new fuel is installed adjacent to partially spent fuel. The fuel and blanket elements are held firmly in place in the grid plate by locking mechanisms that are actuated from above the PCRV. In addition to the primary element locking device, a backup element lock and support is provided. No other core clamping devices are used. The core design is such that the resulting power coefficient of reactivity is negative. Analyses of the effects of seismic excitation on the plant structure have been made, and the resulting reactivity effects have been found to be small.

A control element is a fuel element with the center 37 fuel rods replaced by a guide tube and a 48-millimeter-diameter control rod containing boron carbide, which is vented directly to the helium coolant. There are 21 control rods, each having a reactivity worth of 85 cents. Each rod is positioned by a mechanism with a stepping-motor powering a ball-nut, lead-screw device. A magnet supports the control-rod extension to permit gravity trip capability. An inertial snubber converts the linear velocity to rotational velocity at the end of the rod travel. A completely independent set of 6 shutdown rods of a diverse design from the control-rod insertion mechanism is also capable of reactor shutdown. Each shutdown rod has a worth of $1.60. The shutdown rods are designed to have a larger clearance in the guide tube, their insertion mechanisms are made simpler by elimination of the trip capability, and a different electronic control system is used to operate the faster-acting constant-speed drive motors.

The three main-coolant-loop cavities and the three smaller core auxiliary-cooling-loop cavities are located in the wall of the PCRV. The overall height of the steam-generator–resuperheater assembly is within that feasible for its replacement within the reactor containment building.

All piping connections to the steam generators terminate at tube sheets at the bottom of the PCRV. The main coolant circulators are single-stage axial-flow compressors driven by single-stage impulse steam turbines. This machine is simple, compact, and can be removed through a central hole in the main-loop cavity penetration closure. A helium-coolant-flow isolation (check) valve is located between the main circulator outlet and the duct leading to the reactor inlet. The auxiliary core-cooling loops are used for long-term shutdown cooling and as backup for the main cooling loops or for emergency cooling. Electrically driven centrifugal circulators powered by separately driven high-frequency alternators are employed in the auxiliary loops. The auxiliary-loop heat exchangers are cooled by a pressurized water loop that transfers its heat to the air through a forced-air heat exchanger.

The PCRV is reinforced with steel rods and is prestressed by longitudinal tendons and circumferential wire wrapping, as in large HTGRs. The major penetrations are closed by concrete cavity closures designed to be always in compression and held in place by two structurally independent and redundant means. The PCRV liner is insulated on the helium side by a thermal barrier. Cooling of the liner and penetrations in the PCRV is provided by cooling tubes on the concrete side of the steel liner. The conservative design of this typical PCRV, with inspectable and replaceable redundant tension members, precludes a gross failure of the pressure vessel. The core is refueled through the bottom of the PCRV by a fuel-transfer machine, a fuel-lifting machine, and a fuel-transfer cask which conveys spent fuel from the PCRV to the fuel storage pool where it is stored under water.

A preliminary design of the balance of plant (BOP) for the 300-MW(e) gas-cooled fast breeder reactor (GCFR) demonstration plant (see Fig. 4) has been developed by the Bechtel Corporation for General Atomic Company.

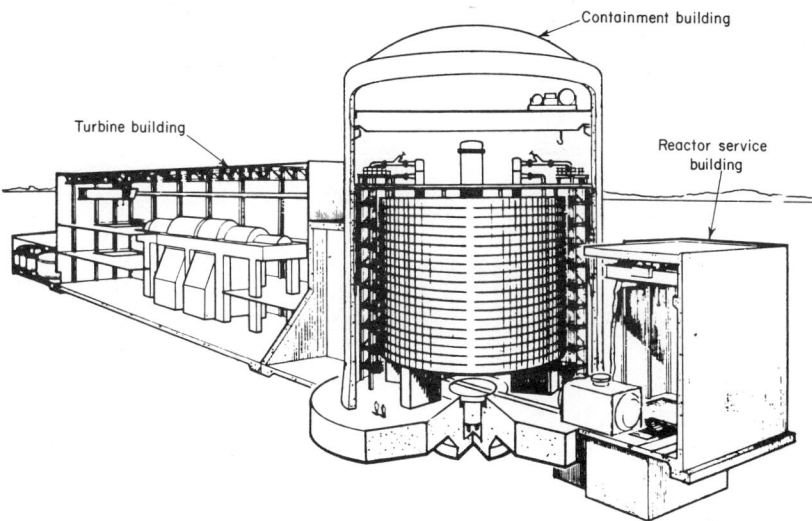

Fig. 4. A 300-MWe gas-cooled fast breeder reactor plant. (*General Atomic Company.*)

A hypothetical site was assumed for the BOP study. The necessary site data, such as soil conditions, seismic requirements, and meteorology, were based on this site. The resulting plant design is not site-sensitive, and thus the plant could be located at other sites with appropriate modifications.

The general plant arrangement shown in Fig. 5 has a peninsular turbine building orientation and incorporates a communication corridor to provide accessibility to all major equipment, and ensure a relatively clear space for routing piping and cable trays. All buildings are located at grade except for pits in the containment building and the reactor service building.

The general plant arrangement and the internal arrangement of some of the buildings are patterned after a generic BOP design being developed by Bechtel. The basic design philosophy and much of the specific arrangement is readily adaptable to GCFR. In those areas unique to the GCFR, much of the equipment for the nuclear steam supply system (NSSS) is in modular form, which minimizes space requirements and facilitates equipment equipment location.

The containment building is a conventionally reinforced concrete 116-foot (35-meter) inner-diameter cylinder with a flat sphere-torus dome and a flat circular base slab. The design pressure is 23 psig (1.6 atmospheres), and the net free volume is 1.21×10^6 ft³ (3.42×10^4 m³), which results in an equilibrium pressure of 1.83 atmospheres absolute following a depressurization. The 84-foot (25.7-meter)-diameter PCRV is supported above the base slab on six 23-foot (7.0-meter)-high

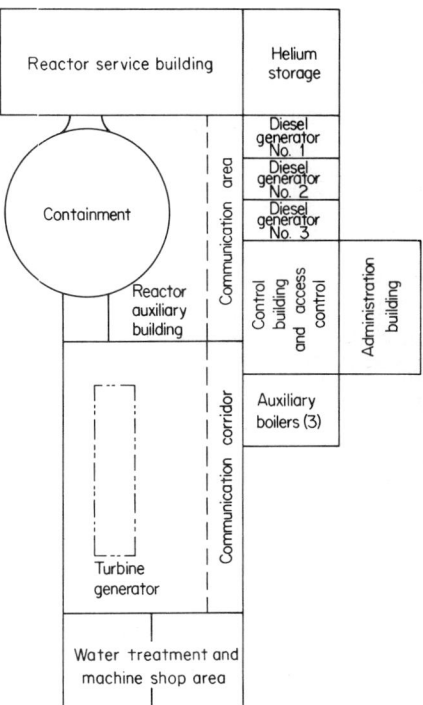

Fig. 5. General plant arrangement of GCFR demonstration plant. (*General Atomic Company.*)

supports. The 16-foot (4.9-meter) annulus between the PCRV and the containment building wall contains support equipment and piping, and the 23-foot (7.0-meter)-high area
beneath the PCRV accommodates the equipment for refueling the reactor. A pit is
located in the center of the base slab beneath the PCRV so that handling equipment
can be used to remove shield plugs from the PCRV and insert refueling machines. An

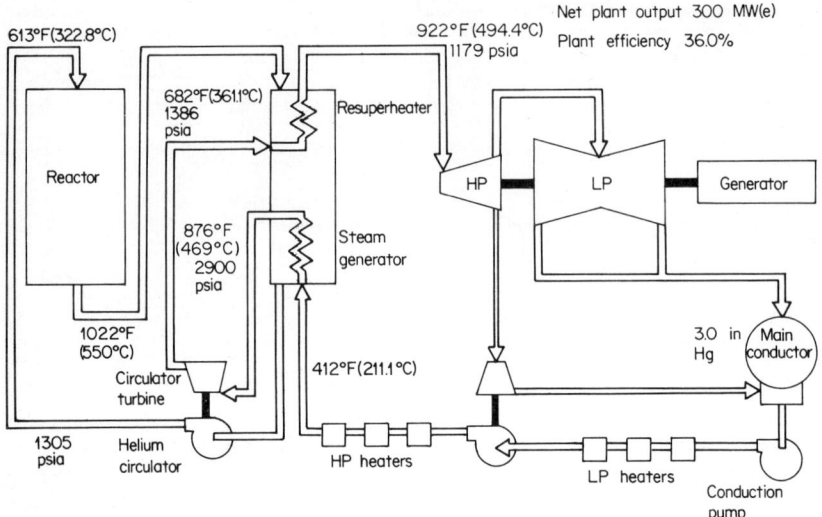

Fig. 6. Simplified flow diagram of the 300-MWeGCFR demonstration plant. (*General Atomic Company.*)

equipment hatch approximately 20 feet (6.1 meters) in diameter is located in the lower
containment wall for transfer of the fuel-handling machines and other equipment into the
adjacent reactor service building. The equipment hatch is also used to isolate the fuel
transfer cask from the fuel service building during transfer of spent fuel to the storage
pool.

The reactor service building contains the fuel service facilities, the radioactive-waste
processing equipment, and the main plant ventilation filters and blowers. The fuel service
facilities include a spent-fuel storage pool similar to that in a light-water reactor plant, a
new-fuel storage area, a railroad-car loading area, and a facility for servicing and checking
the functioning of the refueling equipment.

The reactor auxiliary building is located between the containment building and the
control building and contains the safety-related BOP and NSSS equipment that must be
protected and located close to the reactor. Areas are provided to accommodate the
electrical, instrument, and piping penetrations through the containment wall and to
maintain physical separation of multiple safety channels.

The turbine building houses the turbine generator and all associated power-generating
equipment and auxiliaries. The Diesel generator building contains three emergency
Diesel generators and associated auxiliary equipment in three separate compartments.
Other buildings are the control building, administration building, helium storage building, auxiliary boiler building, water-treatment building, and a machine shop.

The steam cycle of the plant is based on the use of a tandem compound, four-flow 25-
inch (635-millimeter) main steam turbine, using 2.62×10^6 pounds per hour (1.19×10^6
kilograms per hour) of steam at 1179 psia (80 atmospheres) and 922°F (495°C) at the
throttle (see Fig. 6). The maximum guaranteed turbine-generator rating at these conditions is 304.2 MW(e). Six feedwater heating stages are provided, two extractions from the
high-pressure cylinder, three from the low-pressure cylinders, and one at the crossover.
Two separate feedwater strings, each containing a steam-turbine-driven main feed pump,

are used. Each string normally operates at 50 percent of full load but is capable of operating at 75 percent if the other is shut down for any reason.

The steam produced in the superheater section of each of the three steam generators is used to drive the main helium circulator in the corresponding loop. Steam exhausted from the circulator turbine is reheated in the resuperheater section of the steam generator before it goes to the main steam turbine. Two bypass circuits are provided in the steam system to permit operation in a number of modes, including running the helium circulators on steam produced by decay heat following turbine trip and reactor shutdown. Auxiliary steam is provided by auxiliary boilers to operate the helium circulators during plant start-up and for long-term decay heat removal.

A number of other auxiliary and service systems were developed as part of the BOP preliminary design. For example, the auxiliary cooling-loop water system consists of three independent closed loops which are used to reject core decay heat to the atmosphere through air coolers if the primary coolant and turbine steam systems are out of service. The service-water system provides two loops of cooling water for normal and emergency cooling of the NSSS and BOP systems. The reactor-plant-cooling water system is used to cool the PCRV, other essential NSSS equipment, and the spent-fuel storage pool and is in turn cooled by the service-water system. The helium storage system and the nitrogen system provide important services to the NSSS.

The station electric system consists of the generation system, the essential auxiliary power system, and the nonessential power system. Three independent essential systems are provided; these correspond to the three auxiliary core-cooling loops and the normal two-out-of-three safety philosophy used in the plant. There are three emergency Diesel generators, one capable of supplying the entire load on each essential bus.

PERFORMANCE OF THE GCFR

One of the main advantages of helium as coolant for fast breeder reactors is its minimal interaction with the neutronics of the reactor. For instance, helium does not moderate and absorb neutrons, which leads to a hard neutron spectrum and thus a high breeding gain. With mixed plutonium-uranium oxide fuel, the breeding ratio for the GCFR is approximately 1.45 (Ref. 9). Since the specific inventory for large GCFRs is on the order of 4 kilograms fissile per MW(e), the doubling time of the GCFR system is expected to be approximately 10 years. Thus, with present-day mixed-oxide fuels, the GCFR could provide system doubling times compatible with the requirements of the energy-generating capacity.

Another attraction of the high breeding gain of GCFRs is the possibility of using excess neutrons to produce ^{233}U from ^{232}Th instead of ^{239}Pu from ^{238}U. For this purpose, ThO_2 instead of UO_2 would be used in the radial blanket of the GCFR. The change to ThO_2 would not appreciably affect the neutronics of the reactor (Ref. 10) and would lead to a significant production of ^{233}U, which is much more valuable than ^{239}Pu in thermal systems (perhaps by as much as a factor of $2\frac{1}{2}$). The GCFR core and axial blankets of (PuO_2-UO_2) and UO_2, respectively, are designed to produce sufficient plutonium for recycling in the core (see Fig. 7). The ^{233}U produced in the radial blanket of the GCFR would be fed into HTGRs operating on the $^{232}Th/^{233}U$ cycle; the excess ^{233}U from the GCFR would be employed as makeup fuel, along with the recycled ^{233}U bred in the HTGR (Ref. 11). With the present HTGR reference design, the use of recycled ^{235}U would lead to a conversion ratio of 0.78, which could be raised to about 0.84 with the present fuel-element concept and semiannual refueling. The HTGR conversion ratio could be nearer to 0.90 with on-line refueling, which has already been proposed in some European designs. Thus, as shown in Fig. 8, one GCFR could support about three HTGRs in a self-sufficient static economy. The plant mix of GCFR and HTGRs will vary according to the annual growth rate of the energy requirements. For instance, for an annual growth rate of 4.5 percent, the plant mix would be about one HTGR for each GCFR of the same power output. Even if the capital costs for the GCFR are slightly higher than for thermal reactors (HTGR or light-water reactor), the fuel-cycle cost is so much lower, especially with a thorium radial blanket, that the overall power cost is expected to be the lowest of all thermal or fast systems for a ^{235}U price of \$15 per gram or higher (Ref. 12).

Typical GCFR plant design parameters for a 300-MW(e) demonstration plant (three loops) and a 1200-MW(e) commercial plant (six loops) are shown in Table 1. If semiannual

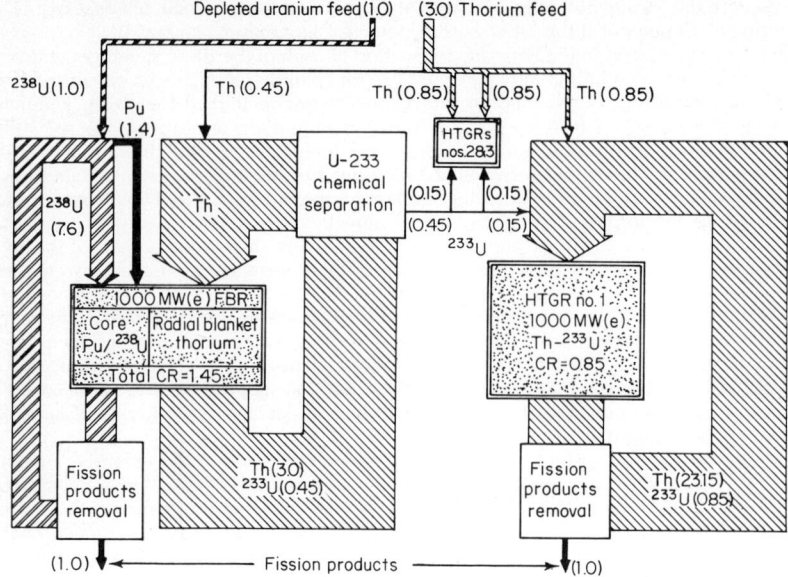

Fig. 7. Symbiosis between fast breeder reactors and HTGRs.

reload were used in the large GCFR plant, an even lower doubling time would be expected because of the lower plutonium inventory required.

SAFETY ASPECTS OF GCFR

As previously indicated, the basic component technology of GCFRs has been chosen to be very similar to HTGR component technology. The only disadvantages of helium are that it must be pressurized to attain the required heat removal (to about 89 atmospheres in the GCFR) and that natural convection of the coolant cannot be relied on for decay-heat removal when the reactor is depressurized. But these pressures, which are much lower

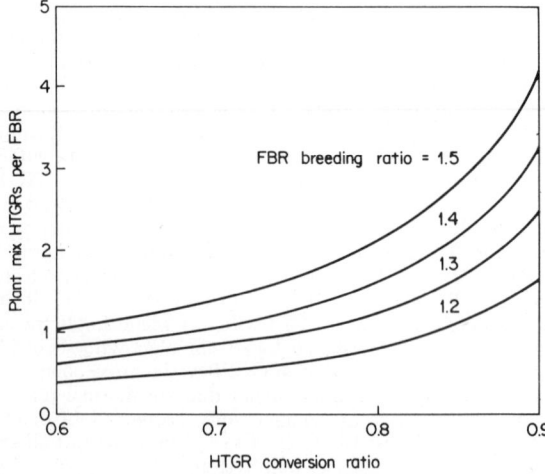

Fig. 8. HTGR and FBR combination in a self-sustaining system at equilibrium.

than in a pressurized-water reactor, can be safely accommodated by the PCRVs; the lack of natural convection means only that forced circulation must be maintained by a highly reliable system.

A comparison of the inherent and design safety features of gas-cooled reactors shows the similarities of the HTGR and GCFR systems and the significant differences due to the high power density metal-clad core of the GCFR (Ref. 13). The PCRV, the technology for which had been well developed in France and England before it was introduced in the United States, ensures a highly reliable reliable primary containment. Flow restrictors are designed to provide an acceptable maximum rate of helium leakage in the event of failure of a PCRV penetration closure; this flow leakage is limited to about 160 square centime-

Table 1.   Typical GCFR Plant Design Parameters

	Demonstration plant	Large plant
Plant output, MW(e)	300	1200
Number of loops	3	6
Loop capacity, MW(th)/loop	278	530
PCRV outside diameter, meters	25.6	31.8
PCRV overall height, meters	24.4	28.6
Reactor helium pressure, bars	89	89
Reactor outlet temperature, °C	550	595
Turbine throttle pressure, bars	81	91
Condenser pressure, bars	0.1	0.06
Cycle efficiency, %	36.0	37.8
Maximum midcladding temperature, °C	700	730
Reload interval	Annual	Annual
Radial blanket material	ThO_2	ThO_2
Breeding ratio	1.4	1.45
Doubling time, years	12	10

ters for the main penetration closure in the PCRV for the GCFR. Furthermore, a leak-tight secondary containment building provides a back pressure of 1.2 to 2 atmospheres after thermal equilibration.

Residual heat removal (RHR) by the forced-convection cooling required for most gas-cooled reactors is achieved in the HTGR and the GCFR by two separate systems: (1) an operational RHR system that utilizes the main cooling loops and normal steam-power conversion system components, and (2) an independent and diverse safety RHR system, called the core auxiliary cooling system. Design features and performance requirements for both systems are, in general, similar in the HTGR and the GCFR, some of the differences being those related to the thermal response time characteristics in the reactor coolant systems. Steam-driven helium circulators are utilized in the steam power-conversion system and for operational RHR. Electric-motor-driven circulators are used in the safety RHR system. The steam generators and condensate feedwater system serve as the heat dump for an operational RHR, whereas pressurized water loops that exhaust heat to the atmosphere or to water are employed in the safety RHR system.

The safety RHR system is provided as an independent means of cooling the reactor in the event that main-loop cooling is not available. The safety RHR is designed as a seismic category 1 system and as an "engineered safeguard" to cool the reactor core and remove decay heat following all anticipated transients and postulated accidents, including the safe shutdown earthquake, design-basis depressurization accident (DBDA), and failures leading to loss of main-loop cooling. The capability of this system is such that adequate cooling is provided to keep core temperatures and gas temperatures below prescribed limits so that a safe cooldown of the reactor and primary coolant system is ensured.

The tolerable temporary stoppage of core heat removal in the GCFR is on the order of tens of seconds, and so considerable analytical and design effort is being devoted to the prevention of a very rapid loss of the normal core cooling system. The main-loop helium circulator pumping power in the GCFR is provided by using a series flow of the circulator steam turbine and main steam turbine, where, in effect, the circulator turbine replaces high-pressure stages of the main turbine. This arrangement allows steam from each GCFR steam generator to pass directly to the circulator turbine in that loop, and thus for operational RHR the main cooling loops are independent of each other on the steam side. Adequate protection can thus be provided against the rapid loss of all main-loop cooling so as to ensure adequate time for start-up of the safety RHR system.

The ^{238}U in the GCFR fuel, through the Doppler coefficient, ensures negative prompt and overall temperature coefficients of reactivity throughout reactor life at all temperatures of interest. All anticipated reactor shutdown functions are performed in the GCFR by a control-rod system that is also used for shim, burnup, and power-distribution control. In the GCFR, the control rods move in a central guide channel in a fuel element; the rods, which are normally motor-driven, fall rapidly into the core under gravitational forces when they are released from their holding magnet by the plant protection system. Reactivity control requirements are small in the GCFR and can be provided by a number of small-worth control rods (there are 21 rods, each worth $0.85 in the 300-MW(e) demonstration plant).

An independent and diverse backup shutdown-rod system is provided that is able to shut the reactor down in adequate time in the event of any postulated total failure of the control-rod system to effect reactor shutdown. This backup shutdown system consists of 6 rods each worth $1.60. A motor-driven backup shutdown-rod system has been chosen after a comparative evaluation of possible designs.

Anticipated operational occurrences are accommodated relatively well, even in the event of failure of subsequent control or protective actions expected during such occurrences. For example, a turbine trip or a main loop isolation and shutdown requires only a power reduction. The minimum time available for reactor shutdown in any anticipated operation occurrence is on the order of minutes.

The overall reliability of reactor shutdown provided by the control-rod system and the diverse backup shutdown-rod system, particularly considering the time available for rod insertion, is sufficiently high that accident sequences involving failure of all protective actions have a probability well below that which would require their consideration in design. The reliability and diversity is more than sufficient to meet the reliability requirements for shutdown following anticipated transients without scram, as recently established for light-water reactors by the U.S. Atomic Energy Commission (Ref. 14).

Pressure equalization provides significant safety benefits for the GCFR in that it eliminates fuel-rod failure modes by cladding creep collapse early in life or by internal fission-gas pressure buildup late in life. Incorporation of charcoal traps within the fuel element ensures that only noble-gas fission products will be vented from the element. Furthermore, recent experimental data indicate that the blanket regions of the fuel rod are effective traps for volatile fission products. Fuel-rod venting does not, therefore, affect the containment of most fission products within the fuel-rod cladding. Those gaseous fission products vented from the element are swept directly into a helium purification system where they are trapped on low-temperature delay beds.

Because of the relatively high power density and low core heat capacity, the detailed behavior of the reactor during and immediately following helium depressurization is of particular interest in the GCFR (Ref. 15). Typical reactor transient results for a GCFR depressurization accident are shown in Fig. 9. The core temperature response is primarily determined by the coolant mass flow through the core. The small quantity of heat stored in the core is removed in about the first 10 seconds after reactor trip. The helium circulators are initially slowed after reactor trip as a normal control procedure and then the control system reaccelerates the circulators to maintain core mass flow as the helium pressure decreases. Fuel-cladding temperatures initially fall but then rise to a maximum value (below the damage level) near the time when depressurization is complete. The steam generators have a relatively large stored energy and serve both as good heat sinks and as sources of energy, in the form of steam, to drive the helium circulators during a depressurization accident.

Extensive studies of depressurization accidents have shown that the fuel and coolant temperatures can be maintained at acceptable levels by the use of either the operational RHR system or the safety RHR system. Any activity that should escape from the PCRV would be confined within the containment building, and operation of the containment cleanup system would reduce airborne activities to low levels. Doses to the public would be maintained well below those specified by the licensing authorities.

DEVELOPMENT AND TEST PROGRAMS

The components for the primary cooling loops in the GCFR are essentially the same scale as the corresponding components in the HTGR, and this justifies confidence in engineering a reliable large-scale primary coolant system. The fuel development program is based on materials, temperatures, and irradiation characteristics typical of LMFBR fuel technology, except for the differences associated with helium cooling. A rather large plant is required to provide a prototypical fuel irradiation environment for demonstrating economic viability, whereas a small plant would neither simulate fuel operating conditions realistically nor fully demonstrate the key system provisions for commercial plant safety.

The core development program includes successful vented-fuel-rod irradiations in the Oak Ridge Research Reactor (up to 53,000 MWD/Te) and fuel capsule irradiations to high

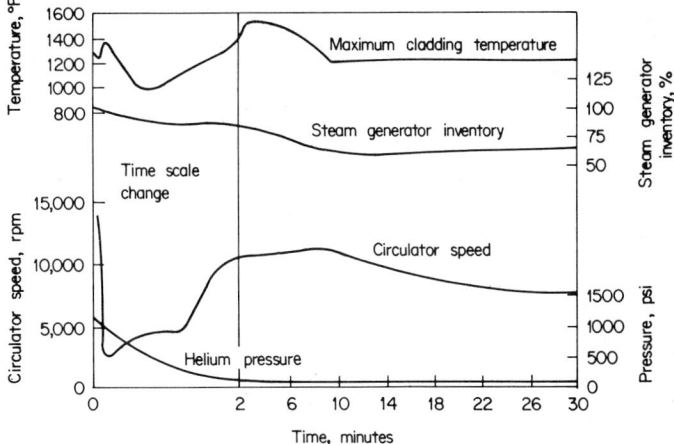

Fig. 9. Reactor transients in depressurization.

temperatures and significant fast fluences in EBR-II (over 110,000 MWD/Te). A vented-fuel-rod bundle irradiation program is being conducted by Kraftwerk Union (KWU) and Kernforschungsanlage, Jülich, Germany, in cooperation with General Atomic; the 12-rod bundle will be tested in 1976 in a high-pressure helium loop in the BR-2 Reactor at Mol, Belgium. A cooperative heat-transfer and fluid-flow development program, which was started in 1967 at the Eidgenossisches Institüt für Reaktorforschung (EIR) in Switzerland, has demonstrated conditions under which surface-roughened fuel rods will produce the heat-transfer enhancement on which the core design is based. In addition, out-of-pile and in-pile tests in the United States are planned for investigating transient fuel-rod-bundle behavior, fuel lifetime properties, and flow and vibration characteristics (Ref. 16). Further capsule experiments in the core of the EBR-II are in progress.

A design verification and support program is planned for carrying out the first-of-a-kind engineering for the primary coolant system components. The main helium circulator and its control system will be rigorously tested to ensure proper response under transient conditions, including the design-basis accident, which is depressurization of the primary coolant circuit. The circulator is operated at higher power and higher steam pressure than its counterpart in the HTGR, although its power is the same as or less than that for steam-driven gas-coolant circulators employed in other nuclear power stations.

The steam-generator operating conditions will be less demanding than for the HTGR, and design verification test programs will be mainly directed toward investigation of stability of steam-water flow at very low flow rates, helium-side flow distribution measurements, and tube support and vibration testing. Model tests of the PCRV and its closures and full-size tests of the fuel-handling system are part of the test program.

OUTLOOK

Because the GCFR promises to be a low-power-cost, high-gain breeder based on commercial HTGR technology, there is increased interest in the United States and abroad in the timely development of the GCFR as a parallel, although much less costly, effort to LMFBR development. A 1973 report on the GCFR by the Edison Electric Institute's Reactor Assessment Panel (Ref. 17) recommends increased support by industry and government toward a GCFR demonstration plant. A typical summary schedule for such a plant shows that a project commitment in 1976 could lead to plant operation in 1986.

REFERENCES

1. Fortescue, P., R. Shanstrom, L. Meyer, W. Simon, and P. Fischer: "Gas Cooling for Fast Reactors," Third United Nations International Conference on the Peaceful Uses of Atomic Energy, A/Conf.28/P/694, 1964.
2. Fortescue, P.: The Case for Gas Cooling, *Power Eng.*, **76** (7) 42, July 1972.
3. Fortescue, P., The Gulf General Atomic Gas-cooled Fast Breeder Reactor System, *Proc. Int. At. Energy Agency Study Group Meeting on Gas-cooled Fast Reactors, Minsk, Byelorussian SSR*, July 24–28, 1972.
4. Simon, R. H., and G. J. Schlueter: From High-Temperature Gas-cooled Reactors to Gas-cooled Fast Breeder Reactors, *Proc. 1972 Joint Power Generation Conf.*, 1972.
5. Lindgren, J. R., R. J. Campana, S. Langer, A. F. Weinberg, and R. H. Simon: Irradiation Testing in the Development of Gas-cooled Fast Breeder Reactor Fuel Elements, *J. Brit. Nucl. Energy Soc.*, **12** (4), pp. 395–408, 1973.
6. Simon, R. H., J. B. Dee, and W. I. Morita: Gas-cooled Fast Breeder Reactor Demonstration Plant, *Proc. Am. Nucl. Soc. Topical Meeting on Gas-cooled Reactors: HTGR and GCFBR*, 1974.
7. Markoczy, G.: "Core Heat Transfer and Fluid Flow in the GCFR," *Trans. Amer. Nucl. Soc.*, **14** (1), pp. 275–276, June 1971.
8. Campana, R. J.: Pressure Equalization System for Gas-cooled Fast Breeder Reactor Fuel Elements, *J. Nucl. Tech.*, **12**, p. 185, October 1971.
9. Oppenheim, C., and R. J. Cerbone: International Benchmark Calculations on a Typ cal 1000-MW(e) Gas-cooled Fast Breeder Reactor, *Proc. Amer. Nucl. Soc. Topical Meeting on   anced Reactors; Physics, Design, and Economics* (to be published).
10. Cerbone, R.: Thorium Blankets in the GCFR, *Trans. Amer. Nucl. Soc.*, **18**, 182, 1974.
11. Fortescue, P.: Fast Breeders and High-Temperature Gas-cooled Reactors: a Profitable Partnership, in A Reactor Strategy: FBRs and HTGRs, *Nucl. News*, **15** (4) 36, April 1972.
12. Brogli, R., and K. Schultz: Thorium Utilization in a FBR/HTGR Power System, *Proc. Amer. Power Conf.*, 1974.
13. Moffette, T. R., and J. A. Larrimore: Safety Evaluations for Gas-cooled Reactors, *Proc. Amer. Nucl. Soc. Topical Meeting on Gas Cooled Reactors: HTGR and GCFBR*, 1974.
14. Anticipated Transients without Scram for Water-cooled Power Reactors, Regulatory Staff, *U.S. At. Energy Comm. Publ.* WASH-1270, September 1973.
15. Milioti, S. J., and J. A. Larrimore: Status and Safety Aspects of the 300-MW(e) GCFR Demonstration Plant, *Proc. Amer. Power Conf.*, 1973.
16. Weinberg, A., et al.: Gas-cooled Fast Breeder Reactor Fuel Development, *Proc. Amer. Nucl. Soc. Topical Meeting on Gas-cooled Reactors: HTGR and GCEBR*, 1974.
17. Everett, J. L., et al.: "Report of the Edison Electric Institute Reactor Assessment Panel on the Gas-cooled Fast Reactor," prepared for the EEI Policy Committee on Atomic Power by the Reactor Assessment Panel, Edison Electric Institute, May 1973.

Fusion Power

BY LAWRENCE M. LIDSKY*

The search for economical controlled fusion power is a scientific hunt for the Lost Dutchman Mine. Only a few true believers are absolutely certain that the goal exists, but the search takes place over interesting terrain, and the rewards for success are overwhelming. In the case of fusion power, the potential long-term societal rewards are so enticing and the possibility of success is so high that a major, truly international research effort has developed over the last two decades. The United States has allocated over \$400 million for research in controlled thermonuclear reactors (C.T.R.) to date, and the U.S.S.R. more than twice that amount.

There seems little question that eventually either fusion energy or solar energy will be called upon to deliver the enormous quantities of "environmentally gentle" power that people will need. We should be able within five years to say whether significant amounts of energy can be obtained from controlled fusion within the next 20 to 30 years (before an irreversible commitment to an economy based on fission breeder reactors) or whether fusion power development will have to wait much longer until the technological and economic considerations are even more favorable than they are at this moment.

In the following pages, the potentialities of controlled fusion power will be enumerated, the physical conditions that we have designated as significant milestones will be described, experiments now or soon to be in operation will be described, the engineering problems of controlled fusion (which may be far more difficult of solution than the physics problems) will be discussed—and, finally, a possible scheme for relaxing both the physics and engineering constraints of an economical system will be described.

THE POTENTIAL OF FUSION POWER

The character of the atomic nucleus is such that the individual nuclear particles are most tightly bound in elements of intermediate atomic number. Thus when we seek energy, we focus our attention on the more loosely assembled elements, releasing energy by splitting (fissioning) the heavy isotopes and proposing to do so by joining (fusing) the lighter ones. There is less energy release per fusion reaction than there is per fission reaction, but the reactants are more plentiful and easier to handle.

In general, a particular fusion reaction is interesting if the power produced can be large enough to offset the power consumed in generating and maintaining the reacting medium, and if the relevant rates can be large enough so that economically interesting regimes are accessible to modern technology.

There are, in fact, over 30 such reactions possible. The most interesting of the fusion reactions as possible routes to fusion energy are those that involve the heavy hydrogen isotopes deuterium ($^2H_1$ or D) and tritium ($^3H_1$ or T). These tend to have the largest fusion reaction probability (cross section) at the lowest energies. Deuterium is an abundant, naturally occurring isotope in wide use now as D_2O in heavy-water-moderated reactors. Tritium is a radioactive isotope with a 12.3-year half-life (it emits an electron and decays to stable helium-3) that does not occur in nature.

The deuterium (D–D) reaction chain is

$$D + D \rightarrow {}^3He + n + 3.2 \text{ MeV}$$
$$D + D \rightarrow T + p + 4.0 \text{ MeV}$$
$$D + T \rightarrow {}^4He + n + 17.6 \text{ MeV}$$
$$\underline{D + {}^3He \rightarrow {}^4He + p + 18.3 \text{ MeV}}$$
$$6D \rightarrow 2{}^4He + 2p + 2n + 43.1 \text{ MeV}$$

*Associate Professor of Nuclear Engineering, Massachusetts Institute of Technology, Cambridge, Mass.

The first two equations represent the fact that the D–D reaction can follow either of two paths, producing tritium and one proton or helium-3 and one neutron, with equal probability. The products of the first two reactions form the fuel for the third and fourth reactions and are burned with additional deuterium. The net reaction consists of the conversion of six deuterium nuclei into two helium nuclei, two hydrogen nuclei, and two neutrons along with a net energy release of 43.1 MeV (one million electron volts). The reaction products—helium, hydrogen, and neutrons—are patently harmless (in contrast to some of the myriad fission products in a fission reactor), and the neutrons may in fact be used in a variety of ways. One very simple possibility is their absorption in sodium to produce an additional 25 MeV per cycle. Therefore, the D–D reaction produces at least 7 MeV per deuterium atom (deuteron), and with absorption in sodium more than 10 MeV per fuel atom.

The peak reaction rate coefficient of the D–D reaction is considerably less than that of the deuterium-tritium (D–T) reaction occurring within the D–D cycle; thus attention tends to focus on the latter. However, because tritium does not occur naturally, the reaction must be supplemented by one using lithium to reproduce the tritium fuel:

$$D + T \rightarrow {}^4He + n + 17.6 \text{ MeV}$$
$$n + {}^6Li \rightarrow {}^4He + T + 4.8 \text{ MeV}$$
$$D + {}^6Li \rightarrow 2{}^4He + 22.4 \text{ MeV}$$

This reaction is tritium-regenerating and produces only helium as a reaction product.

The D–T reactor is technologically more complicated than the D–D reactor because of the need to facilitate the second reaction (which takes place outside the plasma) and because very energetic neutrons must be slowed down to allow the reaction with lithium to take place. Nonetheless, the conditions needed to achieve net power output are much less demanding than for the D–D fuel reactor.

Deuterium occurs naturally in seawater in a ratio such that one out of every 6500 hydrogen atoms in H_2O is in fact the heavy isotope. Isotopic separation is relatively straightforward because the mass difference between H_2O and HDO exceeds 5 percent. At 7 MeV/D, the energy attainable by fusion of all of the deuterium in a cubic meter of seawater is 12×10^{12} joules. This value is easier to visualize when compared to the energy content of a barrel of crude oil (6×10^9 joules). A cubic meter of seawater, then, corresponds to 2×10^3 barrels of oil and a cubic kilometer to 2000×10^9 barrels. This last number is coincidentally nearly equal to recent estimates of the earth's total oil reserves. The oceanic volume is approximately 1.5×10^9 cubic meters, and so there is more than one billion times the energy content of the world's oil reserve available to us through deuterium fusion. There are many interesting ways to illustrate the magnitude of this number (for example, it could support a world population of 7×10^9 people at 20 kilowatts per capita—ten times the current United States rate—for 3×10^9 years), but such exercises are, in fact, just exercises. It is far simpler and just as accurate to say that *fusion of deuterium represents an essentially inexhaustible supply of energy.*

The D–T reaction will probably be exploited first, but its use will be limited by the availability of lithium. For such a light element, lithium turns out to be surprisingly rare. Recent estimates of lithium reserves place them at several times 10^7 metric tons, which limits the total energy release through the D–T–${}^6Li$ cycle to approximately 5×10^{23} joules. The total energy content of the world's fossil fuel is 2.6×10^{23} joules.

It is easy to dispense with detailed computation of energy utilization trends here also. All reliable estimates of energy use conclude that we have sufficient fossil fuels to last for 150 to 350 years. The lithium reserves will thus last for at least that long, and it is hard to believe that we will not have gained sufficient expertise in plasma physics during that time to enable us to tap the energy content of the D–D reaction.

FUSION HAZARDS

The end products of the fusion reaction—helium, hydrogen, and neutrons—can hardly be better chosen from the point of view of easing environmental pollution. It has even been proposed that the neutrons be used to clean the environment by transmuting certain particularly dangerous long-lived radioactive by-products of fission reactors to harmless stable nuclei. Two environmental hazards still remain to be considered, however. There is the tritium that occurs as an intermediate reactant in both the D–D and D–T–^{6}Li cycles, and there is the problem of radiological activation of the fission plant structure by neutron bombardment.

Tritium must be recycled in both fusion chains considered above and is by no means a waste product. Instead, economical design for a fusion plant demands that the inventory of tritium in the plant complex be kept as small as possible and be circulated through the plasma reaction zone as rapidly as possible. Several studies of the inventory problem have resulted in estimates of the total inventory for a 5000-thermal-megawatt plant ranging from 2 to 10 kilograms. Most of the inventory is tritium dissolved in the circulating coolant and the metallic structure of the reactor. The larger value above corresponds to a radioactive burden in tritium of 10^8 curies (Ci), which is 40 percent of the ^{131}I activity predicted for a similarly sized fast breeder reactor. However, the inventories cannot be compared on so simple a basis. The maximum permissible concentration of tritium in the environment is 2×10^{-13} Ci/cm^3, whereas the allowable concentration of iodine is 10^{-16} Ci/cm^3. The relative biological hazard attributable to escape of the volatile inventory in a fusion reactor is nearly four orders of magnitude smaller than in a comparably sized fission reactor.

Because a single reactor is thus relatively safe from the standpoint of catastrophic tritium release, the tritium problem resolves itself to the minimization of leakage from all the reactors in a hypothetical fusion-power-based economy. For example, if all the world's population were using fusion-produced electricity at the current United States consumption rate (2 kilowatts per capita), then an inventory leakage of 0.01 percent per year would contribute less than 0.1 percent increase to the naturally occurring radioactive burden. This level of containment is easily achieved.

The other radiological hazard, activation of the reactor structure itself, is less easily quantifiable. The afterheat and residual activity depend critically on the construction materials. In the worst case studied to date, that of a niobium structure, the decay power is 10 percent and the activity 100 percent of the power and activity of comparable pressurized-water-reactor plants. In the best case studied to date, a vanadium structure, the decay power is 1 percent and the activity only 0.1 percent that of a comparable pressurized-water reactor plant. But, as in the case of tritium activity, a simple comparison is misleading; the activity in a fusion reactor is confined to the few activation products of a single material, whereas that of a fission reactor resides in a spectrum of fission products. Thus it appears that the structural activity of a fusion plant will certainly be no greater than and probably much less than that of a comparable fission plant.

Rejected waste heat has more recently become appreciated as a significant environmental burden imposed by electric power generation. The ratio of energy rejected to the environment to energy transmitted as electric power is a sensitive function of the power plant efficiency. An increase in efficiency from 32 percent (typical of existing fission reactors) to 48 percent (typical of proposed fusion reactor designs) would reduce the waste heat rejected to the environment by a factor of two. This difference exists in large measure because the fusion reactor, with separate reaction and heat removal zones, allows higher operating temperature than the single-zone fission reactor. This return to the apparently primitive concept of "boiler and firebox" permits the design of liquid-metal-cooled systems with exit temperatures as high as 850 to 950°C.

Efficiencies much higher than 48 percent can be contemplated for some fuel cycles by direct conversion of the high-energy charged reaction products to electricity, but the D–T reaction which will be the bulwark of the supply in the near future does not lend itself to direct conversion.

SCIENTIFIC FEASIBILITY

Fusion reactions can take place only when the nuclei of the fuel atoms are brought into close enough conjunction. The nuclei are, of course, positively charged and so repel each other. This repulsion is equivalent to an energy barrier which can be penetrated with reasonable efficiency only if the reacting nuclei have kinetic energy comparable to the barrier height. The level of kinetic energy required depends on the particular reaction and the desired reaction rate, but in general, plasmas of interest to C.T.R. have average energy per particle in excess of 5 keV.

A collection of particles with average energy of 5 keV has an effective temperature of nearly 40 million degrees Kelvin. At these temperatures, the gas is completely dissociated into its constituent positively charged nuclei and free electrons. The electric charge density is such that the behavior of the collection of particles is completely dominated by electrostatic and electromagnetic phenomena. Such a charge-dominated collection of ionized matter is known as a plasma. Plasma at high temperatures cannot be confined by material walls but does respond to electromagnetic forces. The central problem in the search for controlled fusion energy has been the design of a magnetic field configuration that would allow containment of plasma in more or less stable equilibrium.

As in any gas of energetic particles, the ions and electrons in the plasma are continually undergoing collisions with each other. The collision rate is much higher than the fusion rate at reasonable energies, and so the plasma particles in any magnetic trap undergo many collisions and rearrange their energy in a statistical fashion. Accordingly, there will be present in a magnetically confined plasma a spectrum of particle energies reaching up to very high values with a finite probability of fusion reactions occurring even in relatively low-temperature low-density plasma. The occurrence of fusion reactions therefore furnishes no milestone in itself, and other more meaningful criteria must be sought. This is in contrast to fission reactors, in which the achievement of a multiplication coefficient greater than one is a clear dividing line between success and failure.

If the existence of fusion reactions in a contained plasma is not particularly noteworthy, the resultant neutrons do furnish a very useful means of describing the plasma energy distribution. These neutrons are surprisingly copious. For example, the Russian T-3 tokamak emits more than 10^{11} neutrons per second from its 500 eV plasma, whereas the hotter, much denser Scylla IV plasma at Los Alamos Research Laboratory actually produces 40 to 50 watts of D–D fusion energy. If fueled with a D–T mixture, this device would generate nearly 200 kilowatts of fusion power for a three-microsecond pulse period.

What plasma regimes must be reached to yield usable power? The so-called "Lawson Criterion," one attempt to define a milestone, is developed by considering the energy balance per unit volume of plasma. One assumes that the plasma is brought together at temperature T, held there in equilibrium for time τ, and then allowed to disperse. The energy of this expanding plasma is recovered. If the operation is considered to occur in cyclic fashion, the energy balance equation becomes

$$3nkT + W_r\tau = \eta\epsilon(W_n\tau + W_r\tau + 3nkT)$$

where n is the density, W_r is the radiation emission per unit volume, and W_n is the fusion energy generation rate per unit volume. If η is the efficiency, then ϵ is the fraction of the power recovered that is needed to achieve the steady state. If $\epsilon = 1$, then all the energy is required just to keep the reaction going. If ϵ is less than one, then some power is left over, which presumably can be considered as output. The left side of this equation represents the energy needed to assemble the plasma ions and electrons $[2 \times \frac{3}{2}(nkT)]$ and maintain the plasma in the face of radiation losses. The right side represents the energy recovered, assuming that all forms of energy are handled with the same efficiency. When the proper functional forms are inserted for the radiation and fusion rates, the equation can be solved in terms of the $n\tau$ product, as shown in the adjacent diagram. The lowest $n\tau$ system is not necessarily the lowest cost system, and, in fact, economic studies of toroidal reactors predict optimum operation for those devices at temperatures much closer to 10 keV than the 25 keV which corresponds to the $n\tau$ minimum. See Fig. 1.

Once the confinement geometry is determined, one can separate the terms in the $n\tau$ product by an appeal to engineering judgment. In steady-state devices, the requirement that the overall energy density in the reactor be in the range of several watts per cubic centimeter leads to $n = 5 \times 10^{14}/\text{cm}^3$ and $\tau = 0.5$ second. But the fast "pinch" device would operate near $n = 5 \times 10^{16}/\text{cm}^3$, $\tau = 10^{-2}$ second. Yet another class of reactors, "stabilized mirrors," would necessarily operate far to the right of the minimum point; such reactors, if they are economical at all, will be so only if they utilize direct conversion.

There is clearly not a single goal whose attainment guarantees success but rather a set of goals corresponding to different confinement schemes. It is the object of controlled fusion researchers to predict on experimental and theoretical grounds which, if any, path is likely to reach its individual goal.

Success from the standpoint of plasma conditions alone is not sufficient, because the question of whether the particular confinement scheme is compatible with economic and engineering requirements remains to be answered. However, it is useful to remember as a touchstone that reasonable goals for a steady-state reactor are: plasma density $n = 5 \times 10^{14}/cm^3$; pulse duration $\tau = 0.5$ sec; and plasma temperature $T_i = 8$ to 10 keV. The experiments now being conducted are not aimed at these values, but rather at producing plasma conditions that allow all relevant interactions.

Experiments in Plasma Confinement. The experiments now in operation or just being completed are probably the last exploratory plasma physics devices that will be constructed for some years. Experiments now entering the design phase will be much more ambitious; indeed, they will aim at proof of "scientific feasibility" by generating plasma conditions so similar to those of a prototype reactor that the results can be extrapolated with reasonable confidence.

The three concepts with the greatest potential are the tokamak, the stabilized mirror, and the theta pinch. These three devices are clearly differentiated from one another with respect to the dominant plasma processes, and they all scale differently with variations in plasma density, temperature, and, most importantly, in the size of the plasma.

Many ways have been proposed to tap controlled fusion energy in addition to the ones described here. Most discussed among these is laser-induced fusion, which requires no confinement whatever. In the simplest form of laser-induced fusion, a focused, very energetic laser beam is brought to bear on a small deuterium-tritium fuel pellet. If the laser pulse is energetic enough and the energy is delivered in a short enough time, the pellet can be heated to fusion temperatures. The fusion energy is released while the particle is in the process of rapid, uncontrolled expansion.

Computer studies indicate that the energy released in the "microexplosion" can be larger than the energy needed to generate the laser pulse. The key to attaining energy multiplication at physically realizable laser power levels lies in the compression of the fuel pellet to very high densities by the action of the incoming laser pulse. This compres-

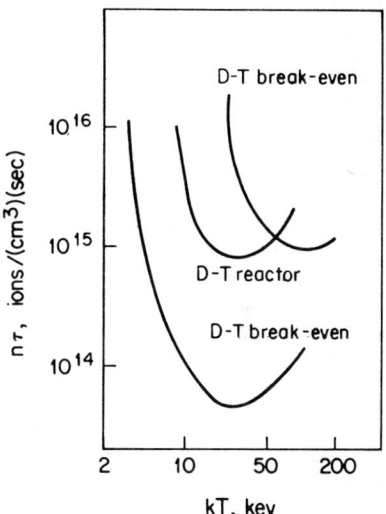

Fig. 1. The numerical values for the Lawson criterion, described in the text as the boundary at plasma regime, which might be expected to yield usable power, are plotted here for two values corresponding to a plasma system in which the thermal efficiency is 33 percent and either all the energy is returned to the system ($\eta\epsilon = 0.33$), or only 10 percent is utilized to maintain the plasma ($\eta\epsilon = 0.033$). From the plot, it is evident that the minimum values of $\eta\tau$ are 4.5×10^{13} second/cm^3 for power break-even; and 8.3×10^{14} second/cm^3 for a system of possible economic interest. (*Culham Lab. Rep. CLM-R85.*)

sion process is very complicated, and there are many doubts about the possibility of achieving stable uniform compression without imposing very difficult constraints upon the laser beam geometry. These doubts can only be resolved by experiments which will be performed as soon as appropriately powerful lasers can be developed.

The energy release is explosive, and so there is a limit to the maximum amount of energy that can be usefully generated in a single pulse. The interval between pulses (determined by the time needed to clear away the residue of the explosion and to recharge the lasers) is much longer than the duration of the power pulse. This inherently low duty cycle will raise difficult questions of power output versus capital cost, pulse system reliability, and system scaling. Laser-induced fusion is an interesting and potentially powerful concept, but it is still in the early stages of research, and the economic and engineering problems inherent in its scaling to the requirements of central station power plant use seem more severe than for other techniques of fusion energy release.

Tokamaks. The T-3 device in operation at the Kurchatov Institute of Technology since 1962 is the prototypical example of toroidal confinement. The magnetic field lines in such closed geometries are constrained to follow toroidal surfaces, and the plasma particles (to first approximation) spiral along the field lines. But simple toroidal fields cannot confine a plasma in equilibrium, and one stabilizing scheme or another must be employed. The tokamaks, of which T-3 is an example, supply equilibrium by means of a large circulating current induced around the torus. This current also serves to heat the plasma by resistive (I^2R) heating.

The other commonly explored toroidal devices supply the required equilibrium by means of externally imposed twisted multipole magnetic fields. This class is represented by the stellarator design developed at the Princeton Plasma Physics Laboratory. Although one theoretically expects many equivalences between tokamaks and stellarators, the tokamaks to date have yielded far better results.

The induced plasma current in the tokamak generates a magnetic field that loops the minor axis of the torus; the field lines form helices along the toroidal surface, and the plasma must cross the lines to escape. It does so through the cumulative action of many random displacements caused by interparticle collisions, in effect, diffusing across the field lines and out of the system. Thermal energy is transported by much the same process.

Particle orbits in toroidal fields are exceedingly complicated. Because of the spatially varying field and the acceleration experienced in moving along the curved field lines, the particles drift away from and return to the original magnetic field lines. These excursions are quite large, and the particle and energy diffusion are enhanced by large factors. The principal goal of the tokamak program has been to measure particle and energy confinement times under many conditions and to predict how these relationships may change as the size of the torus is increased. It is also apparent that the simple resistive heating of T-3 will not scale up to allow ignition of a toroidal reactor, and another major goal of the tokamak program is the development of an additional heating scheme. See Fig. 2.

There are several candidate heating schemes, including plasma compression, the induction of moderate turbulence, absorption of high power oscillations at various characteristic frequencies, and heating by neutral beam injection. This last scheme is presently the most promising method. Beams of neutral atoms can easily penetrate the magnetic field surrounding the plasma only to be ionized with very high efficiency upon entering the plasma itself. The resultant trapped ions are a significant energy source and also provide additional particles to offset plasma losses. Neutral beam injectors have been the subject of intense developmental effort with the result that 125-kilowatt beams are available now, and experiments leading to the "mass production" of modular 1-megawatt injectors are in progress. Neutral beam injection heating is the key to the TCT break-even experiment planned for operation in the early 1980s.

Stabilized Mirrors. If some way can be found to stop the charged particles of a plasma in their motion along the magnetic field lines, the magnetic field lines need not be closed within the plasma region, and many problems associated with toroidal systems can be avoided. The "magnetic mirror" is based on a fundamentally simple phenomenon: charged particles will reflect from a region of increasing magnetic field if their encounter with the field lines is close enough to perpendicular, and they will not be reflected if they are moving nearly parallel to the lines of force. "Mirror" devices take advantage of this by trapping the plasma in a bulge of the field, where the lines are spread. When the field increases, the lines reapproach each other, and a reflecting region is formed. See Fig. 3.

The 2X mirror experiment at the Lawrence Radiation Laboratory uses a carefully shaped set of pulsed magnetic field coils. The plasma is injected and then trapped by these coils when the magnetic field is weak; it is subsequently heated by compression as the field grows with time. The containment region has the typical stabilized mirror shape and a volume of about 16 liters; the resulting plasma has density $n = 5 \times 10^{13}$/cm^3, ion temperature $T_i = 8$ keV, and confinement time of 0.2 to 0.4 millisecond. This is a very interesting hot, dense plasma with confinement time only a factor of three to five below the theoretically expected value.

But the success of a laboratory-scale experiment does not guarantee success of a reactor concept based on that experiment. There are very strong theoretical predictions that the mirror reactor will suffer increasingly severe plasma instabilities as the device is made larger. One could ordinarily tolerate a moderate level of instability or turbulence in a plasma because the effect is simply to increase to some extent the rate of particle loss from the trap. Unfortunately, the mirror reactor has very little margin for error.

Particles escaping from a toroidal reactor must move across magnetic field lines; but in a mirror device it is required only that a collision send a particle in a direction nearly aligned with the magnetic field in order for it to escape. This can happen with reasonable probability in a single collision. The mirror device is thus inherently more leaky than the various closed-line systems, and this leakage manifests itself in particles streaming out the ends of the trap. If this escaping particle stream could be recaptured with high efficiency, the energy could be reinjected with minimal effect, and the development of electrostatic direct conversion was an effort to accomplish this purpose. Economic estimates show that

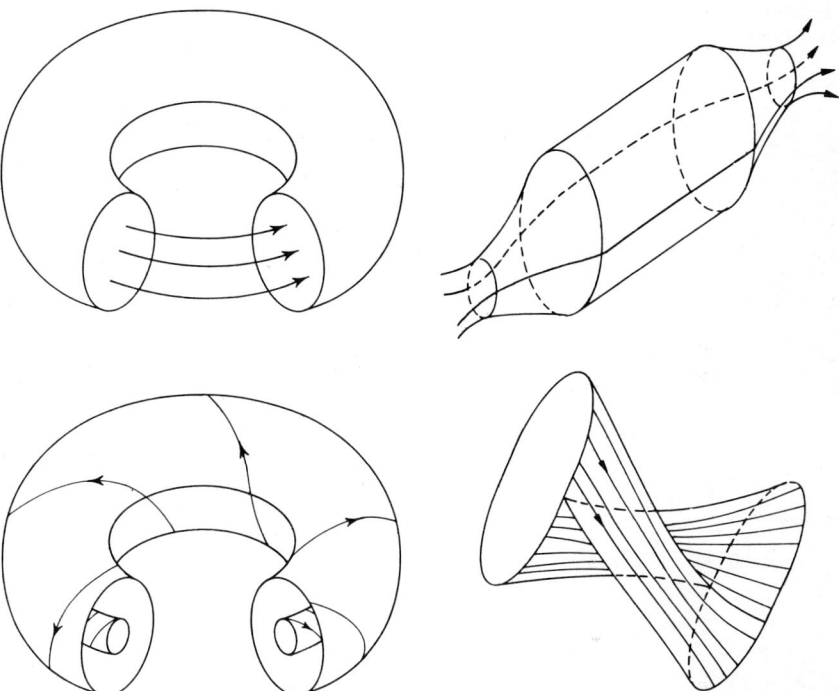

Fig. 2. The twisted field lines result in very complex particle orbits. The twist of the field lines is greatly exaggerated in the figure. In actuality, the magnetic field lines trace several orbits along the major circumference of the toroid before their paths close around the smaller circumference.

Fig. 3 The top diagram shows the basic concept of the mirror device. Plasma is kept within the central region because the particles are repelled by the two areas of relatively greater magnetic field at the ends. Where magnetic lines of force crowd together, a plasma-reflecting "mirror" is formed, and a pair of these mirrors contains the plasma. However, plasma researchers soon found that the shape of the magnetic field around the central region bulges outward, causing the plasma to be unstable. The solution now generally accepted is to make the magnetic field lines bulge inward toward the plasma, as in the lower diagram, which shows the central region of one type of "stabilized mirror"

this is sensible if the reactor performs up to its theoretical limit, but that even a small increase in the particle loss rate would doom the concept. The 2X experimental results, although not conclusive, appear to show that the mirror reactor does not perform quite up to its theoretical limit.

   Theta Pinch—Scylla IV.   The theta pinch experiments have been exploring yet another regime of plasma parameters—very high density. The object in these experiments is the extremely rapid compression of an existing low-density, low-temperature plasma. If the heating pulse is fast enough, a two-step process can follow: the plasma is heated first by the shock wave generated by the rapidly rising field and then heated further by adiabatic compression as the field continues to grow more slowly in time.

The technology needed to perform these experiments is impressive indeed. Capacitors storing nearly a megajoule of energy at 50 keV are discharged through hundreds of parallel paths (to reduce the inductance) into a massive single-turn compression coil to generate fields in excess of 100 kilogauss. In the Scylla IV experiment, for example, the current rises to 8.6 million amperes in only 3.7 microseconds. The resulting plasma has density $n = 5 \times 10^{16}/cm^3$ and ion temperature $T_i = 3.2$ keV. The plasma lifetime is very short, because the plasma particles simply stream out the ends of the device, though the plasma is in stable equilibrium occupying a thin cylinder in the center of the coil during its residence there. See Fig. 4.

Plasmas achieved in the theta pinch process are so hot and dense that the kinetic pressure is several hundreds of atmospheres, and the plasma energy density is very nearly

Fig. 4. View of experiment at Los Alamos Scientific Laboratory (University of California) to construct a toroidal theta pinch apparatus (Scylla torus) in which plasma may be confined and energized. The capacitor bank surrounds the Scylla torus, portions of which can be seen in this view.

equal to the magnetic field energy density. Plasma behavior under these conditions is not yet theoretically understood, and we are very dependent on experimental observations.

It is obvious that the way to avoid end losses in a theta-pinch device is to close the ends upon themselves—to generate a toroidal theta pinch. This immediately reintroduces the toroidal loss-of-equilibrium problem, but there are indications that this dilemma may be resolved through a proper combination of external fields which interact strongly with the high-density theta-pinch plasma.

In the United States, ERDA is funding a very large research program in toroidal theta-pinch apparatus (Scyllac) at the Los Alamos Laboratory. Because the experiment is so expensive, and because the theory is not very highly developed, an initial experiment is under way to provide "proof of principle." A 120° sector of the torus has been constructed and operated at reduced power. In this configuration, the stored energy is 3.5 megajoules and the peak current is 34.5 million amperes. Preliminary results (June 1971) indicate that the external fields can, in fact, supply at least some degree of toroidal equilibrium.

The toroidal theta pinch may well provide the first proof that the appropriate "Lawson Criterion" can in fact be achieved, but this method is possibly incapable of being scaled up to an economical power-producing reactor. The economic difficulty arises because of the great capital cost of the requisite energy storage and because the pinch tends to produce power in bursts, with relatively long periods of time between bursts for cooling, pumping out impurities, and recharging the capacitor banks.

EXPERIMENTAL SUMMARY

Table 1 summarizes the plasma conditions achieved in the experiments described above and their comparison with the "Lawson Criterion" values. Several points are worthy of particular note. First, it is obvious that, at least in terms of density and temperature, it is possible *now* to reach the lower fringes of the plasma regimes of thermonuclear interest. Certainly we are close enough to these regimes so that almost all the dangerous instabilities have manifested themselves. The Scylla IV containment time looks disconcertingly small, but this is inherent to the experiment and does not denote instability. In any event, the proper "Lawson Criterion" for such pinch-like devices calls for confinement time of only 2 to 10 milliseconds.

The various routes to proof of scientific feasibility are clear. The tokamak concept must

TABLE 1.   Plasmas at "the Lower Fringes of Thermonuclear Interest".

	Tokamak (Alcator)	Mirror (2X)	Theta pinch (Scylla IV)	"Lawson Criterion"
Plasma density, n/cm³	5×10^{14}	5×10^{13}	5×10^{16}	5×10^{14}
Maximum plasma temperature T_i, keV	10	8	3.2	10
Confinement time τ, msec.	25	0.4	0.01	200

NOTE: The fusion process is such that there is no sharp dividing line in a fusion reactor—as there is in a fission reactor—between "go" and "no-go." Instead, there is simply the question of achieving a high enough steady-state fusion reaction rate to provide a sufficient excess of energy for power generation. This condition is roughly expressed by the so-called "Lawson Criterion." As the table shows, it is now possible—at least in terms of density and temperature—to reach what the author calls plasmas at "the lower fringes of thermonuclear interest." The author suggests that five more years of research will reveal whether the goal is in fact attainable soon.

be tested at large radius to see whether the plasma loss scales inversely with the square of the plasma radius. If it does, then the development of an appropriate energy input scheme would lead to the goal. Several very large tokamaks have been proposed, and many heating methods are being tried on smaller scale experiments.

The TCT experiment, which is planned for operation at the Princeton Plasma Physics Laboratory in 1980, will utilize energetic neutral beam injection and possibly subsidiary heating by plasma compression. This is the first attempt to reach the conditions required for energy break-even and will actually utilize a deuterium-tritium plasma. If TCT works as planned, it will produce 10 megawatts of fusion power for periods approaching one-tenth of a second. The figure illustrates the scale of TCT itself; in operation it will of course be surrounded by extensive radiation shielding.

As for the theta pinch, it must be closed upon itself. If external coils can provide equilibrium, then nothing further will be needed. Experiments are now in progress at the Los Alamos Scientific Laboratory, but the results are still unclear.

Mirror traps will need to be shown to be less prone to instability and enhanced losses than is currently predicted by theory. Experience shows that the opposite is usually true—that plasmas are in fact more unstable than expected. However, a mirror reactor would have a much larger ratio of plasma pressure to magnetic field pressure than the experiments conducted to date, and this regime is not amenable to theoretical analysis. Proponents of mirror reactor systems argue that appropriate experiments should be conducted in this regime. Such experiments would be extremely expensive, and there can be no certainty that they will be funded in the near future.

For both the tokamak and the toroidal pinch, five years should be sufficient time either to demonstrate scientific feasibility or to find that unsuspected obstacles stand in our way. The distance remaining to the goal is relatively short; so the unexpected has a small domain in which to lurk; the chances for success are therefore relatively good.

ENGINEERING PROBLEMS OF C.T.R.

Even if and when we discover how the "Lawson Criterion" can be fulfilled, there will remain the task of engineering large-scale economic systems. And this is an issue of which the resolution is by no means certain.

We already know that in conventional power plants the greatest economic advantage can be obtained when materials are used near the limit of their capabilities. This will be true particularly of advanced power-producing schemes such as fast breeder reactors and controlled fusion reactors, because highly stressed mechanical elements allow the designer to relax requirements on the reactor core. In the fusion reactor, for example, higher heat loads at the interface between plasma and container would permit a lower ratio of container-to-plasma volume and thereby lower the unit capital cost of the plant.

There are important problem areas common to fission and fusion reactors. These include radiation damage, limiting heat fluxes and temperatures, and induced radioactive afterheat. The first of these will apparently be far more severe in fusion reactors; the other two will be comparatively less severe than in fission plants. (Other engineering problems of the fusion reactor which are specific to the particular design—tokamak, for example—will be ignored here because they are somewhat less fundamental and appear solvable by extensions of existing techniques.)

Many engineering features are surprisingly insensitive to details of the nature of the plasma. Consider for a concrete example, a steady-state, D–T-fueled toroidal reactor, and recall that 80 percent of the energy release is in 14.1-MeV neutrons. Elementary economic analyses of steady-state reactors show that the magnetic field must be generated by superconducting magnetic coils, because the power loss in normally conducting materials is prohibitively high. On this basis one arrives very quickly at a conceptual model of the main features of the cylindrical blanket that encircles the plasma column.

The shell immediately surrounding the plasma must be of a refractory material to withstand the enormous flux of neutrons and short-wavelength electromagnetic radiation. This first wall must be very well cooled because—although most of the energy passes through this wall—an appreciable quantity of energy (the amount depending on the thickness of the wall) will be deposited within it and must be removed. On the other hand, a well-chosen material will generate appreciable amounts of additional neutrons, and the final design thickness will be a compromise between neutron production and allowable thermal loads. The shell surrounding this one—the first wall coolant—will be devoted primarily to heat removal from the first wall, but proper material choice will allow neutron and tritium generation in this region. The third and subsequent shells will moderate the neutron flux, remove the neutron energy in the form of heat, generate tritium by the ^{6}Li absorption reaction, and shield the magnetic field coils from the neutron and gamma ray flux.

The bulk of the energy removal from the blanket occurs in the thermalization and tritium generation region, but it is not as highly stressed as the first wall because of the cylindrical geometry of the system. The radial size of the blanket will be determined by the need to use with high efficiency every neutron emitted from the plasma, because at least one triton must be generated for each neutron emitted. Almost all blanket designs call for a radial thickness of 1 to 1.5 meters. This requirement impinges, through economic considerations, on the plasma physics, because the plasma radial size must be comparable to the blanket thickness, or the power density of the system will be far too low.

Another strong constraint on the blanket thickness will be the necessity for using superconducing coils to generate the intense magnetic fields required. The coils operate near 4.2°K, but the energy deposited in them by neutrons and gamma radiation must be rejected at the temperature of the environment. When the thermodynamic Carnot factor and refrigeration efficiencies are considered, it becomes clear that only 0.01% of the total reactor power can be allowed to reach the magnetic field coils. Thus, it appears impossible to use a blanket much smaller in radial extent than 1.5 meters.

Several groups (most notably at the Oak Ridge National Laboratory, The United Kingdom's Culham Laboratory, and the University of Wisconsin) have published the results of careful studies of conceptually reasonable blankets. These have all been done in sufficient detail to point up the engineering problem areas.

A. Fraas and D. Steiner of the Oak Ridge National Laboratory have shown that the problem of material damage by radiation will be more severe for fusion reactors than for fission breeder reactors. The radiation damage effects, particularly swelling, are the limiting design considerations in breeder reactors. They may not be quite so important in fusion reactors because fission reactor cores are characterized by a large number of narrow cooling passages, while the blankets of fusion reactors can be designed with much larger coolant passages and mechanical tolerances. Although the swelling is certainly a major concern, there is a paucity of information regarding the effect of very high doses of 14-MeV neutrons, and no experimental facility is presently capable of measuring the effect. A key piece of engineering design data is simply unknown.

There is yet another problem inherent in the blanket design. The ideal material for use as heat transfer medium for the first wall is liquid lithium, because of its low vapor pressure at high operating temperatures and because lithium in this location takes full advantage of the high-energy neutrons for the ^{7}Li reaction. However, this means that an electrically conducting fluid must be moved through the steady magnetic field. One consequence of this is that considerable power must be available for pumping the coolant, and the cooling passages must be designed to minimize cross-field flow. A possibly more important consequence arises because the magnetic field suppresses the turbulence in the fluid and so reduces the heat-transfer coefficient. Experiments are now in progress to measure the magnitude of this effect, but the analysis is not complete. See Fig. 5.

To summarize: Many severe engineering problems, radiation damage chief among them, will limit the allowable heat flux at the blanket-plasma interface. The heat flux will

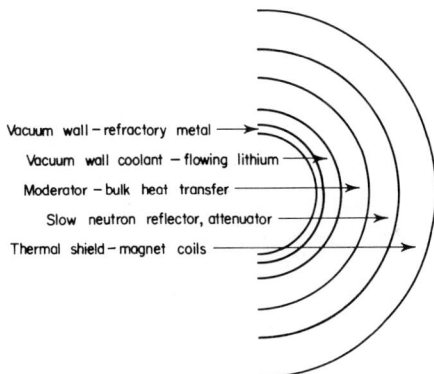

Vacuum wall – refractory metal
Vacuum wall coolant – flowing lithium
Moderator – bulk heat transfer
Slow neutron reflector, attenuator
Thermal shield – magnet coils

Fig. 5. Cross-sectional structure of a hypothetical fusion power generator. Around the plasma there must be: (*a*) Coils to create the magnetic field which contains the plasma; (*b*) coolant fluids to remove the heat generated; (*c*) radiation shielding. In the lithium version of this particular scheme, a nuclear reaction generates tritium and neutrons which are needed by the plasma in order to maintain the fusion reaction.

determine the capital cost per unit of power, and low heat flux might possibly raise the cost to economically uninteresting levels. It may well take longer to determine the allowable engineering parameters than it will take to prove the "scientific feasibility" of fusion power. It is in belated realization of this that the AEC is now starting large-scale funding of engineering research in problems of C.T.R.

FISSION-FUSION SYMBIOSIS

Those who attempt to assess the prospects of generating appreciable fusion power in the near future usually assume that the D–T–^{6}Li cycle will be utilized. Recall that this cycle demands that the tritium be regenerated by various reactions with lithium in the blanket. Engineering studies of such tritium regenerating blankets have been conducted in many laboratories, and all of them show that it is a relatively simple matter to breed more tritium than is needed. A typical calculation, for example, shows that 1.3 tritons are generated for each neutron incident upon the blanket. In other words, the cycle becomes $D + {}^6Li \rightarrow 2{}^4He + 22.4 \text{ MeV} + 0.3n$.

The neutron excess offers many intriguing possibilities. Of course, some of the excess neutrons will be needed to generate more tritium than is consumed in the reactor to allow the start-up of additional fusion plants. However, the reaction rate per unit of power is very high, and a neutron excess will clearly be available. This raises the intriguing question of how to exploit the resulting change from a "neutron-poor" to a "neutron-rich" economy.

One of the more provocative possibilities involves the absorption of the excess neutrons in either thorium-232 or uranium-238 to yield the fissionable isotopes uranium-233 or plutonium-239. Each of the fissionable nuclei that might be produced would represent at least 200 MeV when burned in a fission reactor and could result in a net yield of more than 1000 MeV in a reactor with reasonably high conversion efficiency. Thus the excess neutrons from a fusion reactor could in fact represent far more energy than is produced by fusion reactions in the core itself. Indeed, a new "generalized Lawson Criterion" can be defined in terms of this symbiotic fission-fusion scheme. Fusion reactors that would not be economically interesting by themselves become economically viable in the symbiotic system. From the point of view of plasma physics, the $n\tau$ curves are shifted to the lower values of the product.

In general, the technological characteristics of fast fission reactors and conceptually reasonable fusion reactors form a complementary set. Fission reactors produce low-cost power at high power density, but they can be made to breed only at the price of compromising either safety or economics. Fusion reactors may well have low power density and concomitant high costs; but because of the neutron excess, they will breed new fuel at a prodigious rate. A combined system in which fuel is bred primarily in the fusion reactor and power generated primarily in the fission reactor achieves properties attainable by neither alone—economical, safe power generation with almost arbitrarily short doubling times. Furthermore, the loosening of design constraints would allow other

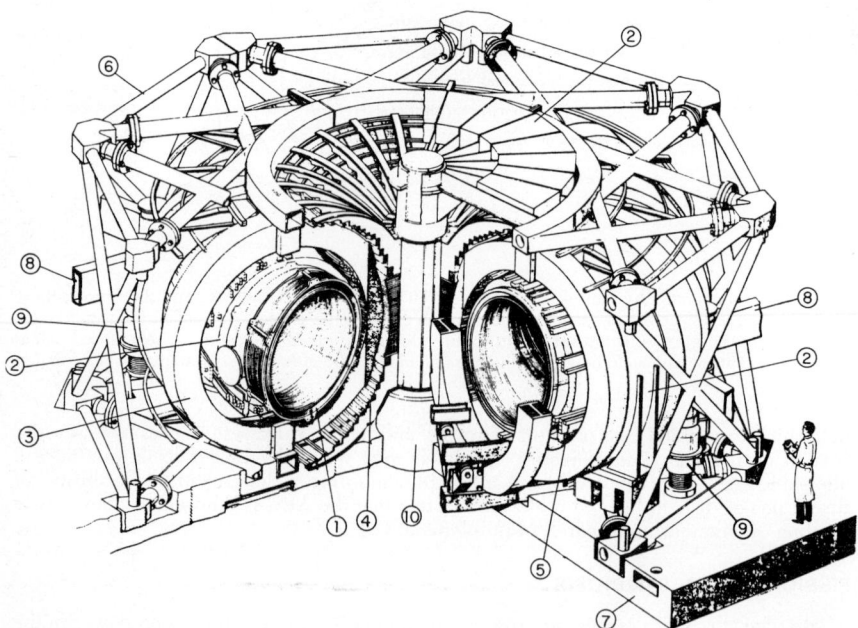

Fig. 6. Trimetric view of the TCT device. Scale is given by "standard 2-meter person" shown in lower right-hand corner of view. (1) Toroidal vacuum vessel. (2) Device shielding. (3) Toroidal field coils. (4) Ohmic heating field coil. (5) Equilibrium field coil. (6) Device torque frame. (7) Device substructure. (8) Neutral beam injection ducts. (9) Toroidal vessel vacuum pumps. (10) Central support column.

benefits. For example, to ease problems of thermal pollution, the fission reactor could be optimized for safe high-temperature operation, and any resulting penalty in neutron economy could be compensated for in the fusion portion of the system. In terms of such overall considerations, the distance between "scientific feasibility" and "economic feasibility" is materially reduced.

FINDING THE LOST DUTCHMAN MINE

The focus of the search for controlled fusion power has changed during the past several years. We have learned to confine plasmas at high temperature and high density for long periods of time. The densities and temperatures have nearly attained values of reactor interest, and energy confinement times have been stretched by many order of magnitude from the several microseconds typical of the early, unstable devices to several tens of milliseconds in tokamaks. See Fig. 6.

The question of whether we achieve fusion power in the near future is now cast into quantitative rather than qualitative terms. The full evaluation will depend at least as much on engineering limitations of the system as it will on the plasma physics, but there is some hope for relaxing these constraints in combined fission-fusion systems.

In terms of the original metaphor, we think we have found the Lost Dutchman Mine. The hoped-for proof of "scientific feasibility" will be needed to demonstrate its existence conclusively; and we are still unsure whether exploiting the mine will make economic sense. Although this is the subject of intense physics and engineering research—and far from solution—it seems highly unlikely that so glorious a resource as the vast oceans full of clean-burning deuterium will remain unavailable for use.

See also subsequent articles in this Handbook section.

SUGGESTED READING

Gough, William C., and B. J. Eastlund: Prospects of Fusion Power, *Sci. Ameri.*, pp. 50–64, February 1971.

Postma, Herman: Engineering and Environmental Aspects of Fusion Power Reactors, *Nucl. News*, pp. 57–62, April 1971.

Rose, David J.: Controlled Nuclear Fusion: Status and Outlook, *Sci.*, pp. 797–808, May 21, 1971.

Steiner, D.: The Technological Requirements for Power by Fusion, *Nuclear Science and Engineering*, bol. 58, pp. 107–165 (1975).

Direct Energy Conversion in Fusion Reactors*

BY RALPH W. MOIR†

Direct energy conversion‡ may play a major role in the development of a high-efficiency fusion reactor, here defined as a fusion reactor that converts fusion energy to electricity at an efficiency significantly greater than modern thermal-cycle efficiencies. Thermal efficiencies are now about 40 percent; 70 percent would be a high but achievable goal for a fusion reactor. Successful development of direct energy conversion could materially contribute not only to making fusion an abundant energy source but also to making it an environmentally outstanding energy source.

Four factors contribute to the efficiency of a fusion reactor:

1. Good plasma confinement in the sense of low recirculation power; i.e., $n\tau \gg n\tau$ Lawson, where n is the plasma density, and τ is the mean ion lifetime

2. A fuel cycle that primarily results in charged reaction products instead of in neutrons

3. An operating cycle and a containment device that minimize radiation by the plasma

4. A direct converter that can efficiently convert to electricity the energy released in the form of charged reaction products

Direct energy conversion thus offers both near- and long-term advantages. In near-term fusion reactors, it would improve the power balance by efficiently and cheaply recirculating power. For the long term, it would raise plant efficiency because fuel cycles that primarily result in charged fusion products can be used: i.e.,

$$
\begin{array}{llllll}
D + D & \to & T + & p & + & 4.04 \text{ MeV} \\
D + D & \to & {}^{3}\text{He} + & n & + & 3.27 \text{ MeV} \\
D + T & \to & {}^{4}\text{He} + & n & + & 17.58 \text{ MeV} \\
D + {}^{3}\text{He} & \to & {}^{4}\text{He} + & p & + & 18.34 \text{ MeV}
\end{array}
$$

$$
6D \longrightarrow 2{}^{4}\text{He} + \quad 2p \quad + 2n \quad + 43.23 \text{ MeV}
$$

Two classes of direct energy converters are being studied: electrostatic and magnetic. The electrostatic converter is essentially a linear accelerator run backwards. That is, fast ions from the fusion plasma enter the "exit" of the accelerator and are decelerated and finally collected. By this process, the kinetic energy of the ions is directly converted to electric potential energy. The magnetic direct energy converters are analogous to the internal-combustion engine. As the hot plasma expands against a moving magnetic-field front in a manner similar to that in which hot gases expand against a moving piston, part of the energy of the internal plasma is inductively converted to an electric current in the magnet (pickup) coil.

ELECTROSTATIC DIRECT ENERGY CONVERSION

As illustrated in Fig. 1, five processes (Ref. 2) are involved in the electrostatic direct conversion of the plasma energy that leaks out of a mirror fusion reactor:§

1. *Selective Leakage:* By means of magnetic and electrostatic fields, the ions and electron are made to leak selectively through limited regions of the plasma boundary.

2. *Expansion:* The plasma stream is guided and expanded in volume by a decreasing magnetic field that reduces the power density and converts rotational energy to directed energy.

3. *Electron Separation:* The electrons are separated from the plasma stream and

*Work performed under the auspices of the U.S. Energy Research and Development Administration.
†Lawrence Livermore Laboratory, University of California, Livermore, Calif.
‡For a complete discussion of this topic, see Ref. 1.
§It may be possible to convert the plasma energy directly, leaving a toroidal reactor via a diverter; however, a detailed technique has not yet been worked out.

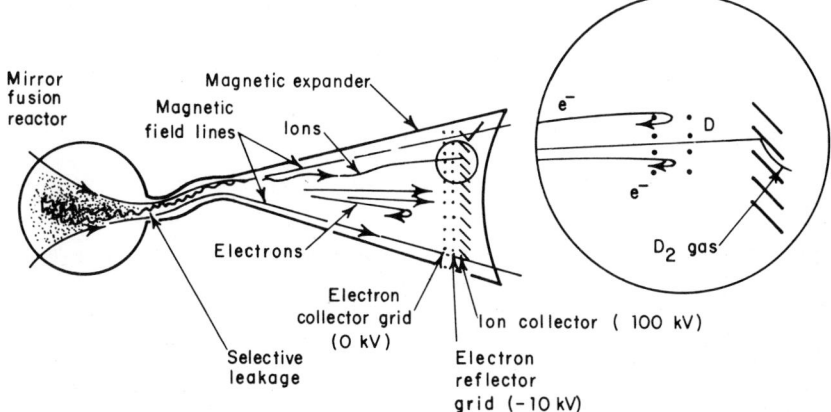

Fig. 1. One-stage direct energy converter with a conically-shaped magnetic expander.

collected on an electron collector grid, an electrode that forms the negative terminal of the power source of the direct energy converter.

4. *Deceleration:* The ions are decelerated by retarding electric fields; kinetic energy is thereby converted to potential energy.

5. *Collection:* The decelerated ions are collected on high-voltage electrodes that form the positive terminal of the power source of the direct energy converter.

Ions, as shown in Fig. 2, have a wide energy distribution (Ref. 3). Therefore, because it has only one collector electrode at only one potential, the direct energy converter shown in Fig. 1 is limited in efficiency to about 50 percent (Ref. 4). Higher efficiencies can be obtained by providing many collectors at different potentials so that the ions of different energies can be collected on electrodes where potentials (measured in volts) are near the initial energies of the ions (measured in electron volts). In the multistage (22-stage) collector (Ref. 2) shown in Fig. 3, the plasma stream is formed into a slab beam by a fan-shaped magnetic expander. A carefully controlled laboratory test of this multistage converter gave a measured efficiency of 86.5 ± 1.5 percent. This value compares well with a computer-simulation calculation of 88.5 ± 1.5 percent (Ref. 4).

These same principles are being used in several other applications. For example, NASA has developed a practical, multistage, direct energy converter to recover the energy of an

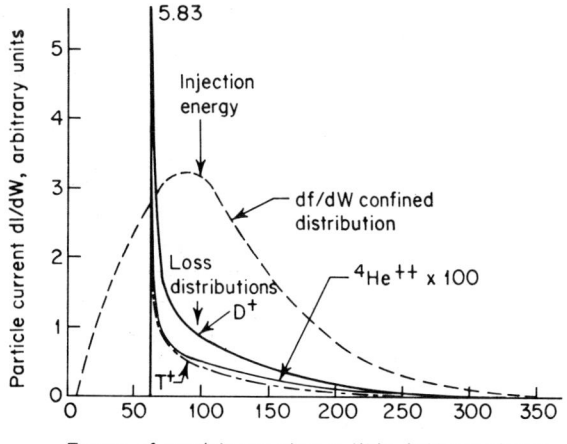

Fig. 2. Energy distribution of the ions leaking out of a mirror fusion reactor.

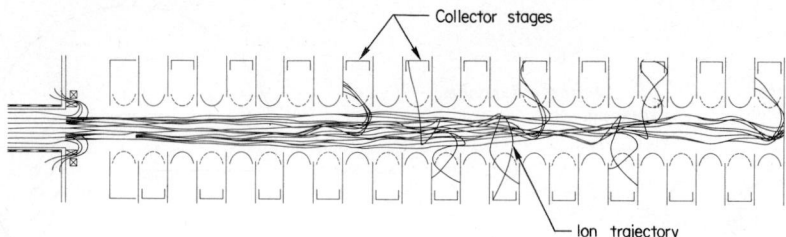

Fig. 3. Twenty-two stage direct converter, with ion trajectories inside the focusing and collecting system.

electron beam as it leaves a traveling microwave tube (Ref. 5). Also, the concept illustrated in Fig. 4 is being developed (Ref. 4) for mirror fusion reactors that are fueled and heated by the injection of energetic (100- to 200-keV) neutral deuterium and tritium beams. These neutral beams are formed by first accelerating either D^+ and T^+ or D^- and T^- ions to the desired energy, and then using a gas cell to convert a fraction of these ions to neutral atoms. To produce the neutral beam efficiently, it is highly desirable to directly convert the energy of those ions not converted to neutrals.

Electrostatic direct energy converters designed for fusion reactors will encounter various effects that must be considered in their design. For example:

1. The efficiency of the converter may be limited because of space charge in the collector regions, secondary electron-leakage currents arising from ion impact and the resulting x-rays, voltage holding and sparking damage, and charge exchange and ionization of background gas.

2. The lifetime and operation must allow for blistering and spalling of collector surfaces because of He^{++} bombardment, sputtering, tritium recovery, and cooling of the electrodes and possible recovery of this energy in a thermal bottoming cycle. To be practical, a direct converter must operate for about 10,000 hours between maintenance cycles.

MAGNETIC COMPRESSION-EXPANSION DIRECT ENERGY CONVERSION

As illustrated in Fig. 5, four steps are involved in the magnetic compression-expansion cycle for direct energy conversion:

1. *Compression:* A column of plasma is compressed by a magnetic field that acts like a piston.

2. *Burn:* The compression heats the plasma to the thermonuclear ignition temperature.

3. *Expansion and Energy Removal:* The thermonuclear burn (fusion reactions) increases the plasma pressure and pushes the magnetic field outward.

4. *Refueling:* After expansion, the old, partially burned fuel, D^+ and T^+, and ash $^4He^{++}$,

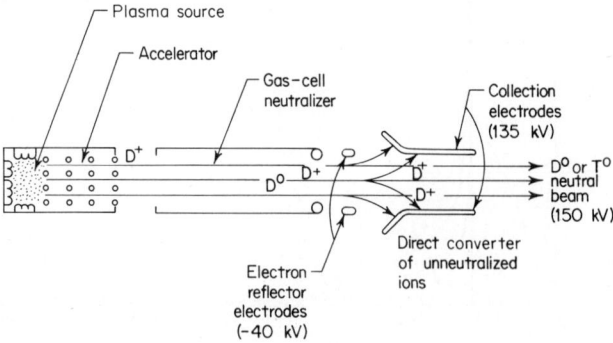

Fig. 4. Neutral beam injector system with "in-line" beam direct energy converter.

for example, are flushed out; new fuel in the form of gas is introduced and ionized; and the cycle is thus completed.

The magnetic compression-expansion concept is being developed by the Los Alamos Scientific Laboratory for the toroidal theta-pinch reactor (Ref. 6) and has also been suggested for an ATC-type tokamak reactor being developed at the Princeton Plasma Physics Laboratory (Ref. 7).

Although experimental verification of the compression-expansion cycle has not been

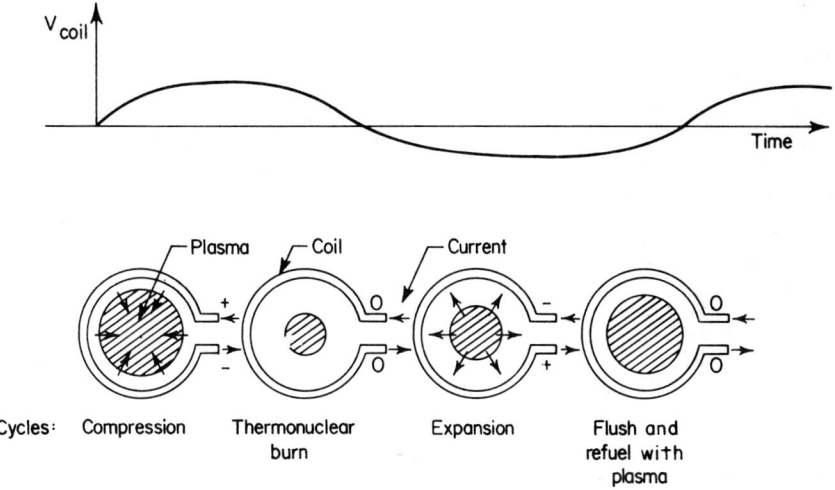

Fig. 5. Magnetic direct energy converter, with the compression, burn, expansion, and refueling parts of the cycle shown.

reported, successful compression heating in several different devices has demonstrated that the principle is sound.

Application of the magnetic compression-expansion cycle to reactors involves several considerations:

1. Because large expansion ratios are needed for high efficiencies, a relatively large vessel is required.

2. The rapid current changes place stringent requirements on superconducting magnets for the pickup coils.

3. Joule heat losses, switching losses, and storage of the large power pulses required are critical aspects.

4. To attain reasonable average-power levels, burn times must be maximized, and cooling and refueling times minimized.

5. As with the electrostatic converter, this direct energy converter must also operate about 10,000 hours between maintenance cycles.

REFERENCES

1. Miley, G. H.: "Fusion Energy Conversion," American Nuclear Society, Hinsdale, Ill. (in preparation).
2. Post, R. F.: "Mirror Systems: Fuel Cycles, Loss Reduction and Energy Recovery," in *Proc. Brit. Nucl. Energy Soc. Conf. Nucl. Fusion Reactors*, pp. 88–111, Culham Laboratory, Culham, England 1969, (UKAEA, 1970).
3. Futch, A. H., Jr., J. P. Holdren, Killeen, J., and A. A. Mirin: *Plasma Phys.*, **14**, 211 (1972).
4. Moir, R. W., W. Barr, and G. A. Carlson: Plasma Physics and Controlled Nuclear Fusion Research, *Proc. 5th International Conf. Plasma Phys. and Controlled Nucl. Fusion Res., Tokyo, Japan, 1974* (IAEA, Vienna), p. 583, vol. III.

5. Kosmahl, H. G., B. D. McNary, and O. Sauseny: High Efficiency, 200-Watt, 12-Gigahertz Traveling Wave Tube, Lewis Research Center, Cleveland, *NASA Rept.* TN D-7709.
6. Oliphant, T. A.: "Fuel Burn-up and Direct Conversion of Energy in a D-T Plasma," in *Proc. Brit. Nucl. Energy Soc. Conf. Nucl. Fusion Reactors,* pp. 306–321, Culham Laboratory, Culham, England, 1969 (UKAEA, 1970).
7. Miley, G. H.: Compression and Expansion Cycles for Toroidal Fusion Reactors, *Proc. 1st Topical Meeting on Technol. of Controlled Nuclear Fusion,* San Diego, Calif., 1974 (AEC, CONF-740402-Pl, Washing, 1974), vol. 1, pp. 448–463.

Tritium Breeding Requirements in Fusion Reactor Blankets*

BY D. STEINER†

The requirement that the blanket regenerate tritium implies that lithium in some form must be present in the blanket since the only neutron-induced reactions that offer any promise for tritium breeding are ^{6}Li$(n,\alpha)t$ and ^{7}Li$(n,\alpha n')t$. The ^{6}Li$(n,\alpha)t$ reaction cross section varies inversely with neutron velocity below ~ 0.3 MeV. In any practical blanket the ^{6}Li$(n,\alpha)t$ reaction must compete with parasitic absorption reactions in structural materials, e.g., (n,γ) reactions. The ^{7}Li$(n,\alpha n')t$ reaction has a threshold at ~ 2.8 MeV and, therefore, must compete with neutron down-scattering processes, i.e., elastic scattering, inelastic scattering, and $(n,2n)$ reactions. Neutron multiplication resulting from $(n,2n)$ reactions to some extent compensates for neutron losses due to parasitic absorption and leakage.

In addition to the breeding medium, the blanket will, in general, consist of structural material, coolant, and neutron moderator. In some designs beryllium is included for neutron multiplication via $(n,2n)$ reactions. The range of materials currently being considered for the various blanket functions is indicated in Table 1. Note that (1) the same material might perform several functions: e.g., liquid lithium could serve as the breeding, cooling, and moderating medium; and (2) both liquid and solid breeding materials are of interest.

The tritium-breeding performance of the blanket must be such that the tritium doubling time is consistent with electric energy growth requirements, about 7 to 10 years by present standards. An approximate expression for the tritium doubling time t_2 is given (Refs. 1 and 2) by

$$t_2 = \frac{1.9 \times 10^{-3}\,\text{EY}}{\text{SP(BR} - 1)} \tag{1}$$

where t_2 = doubling time in years

SP = specific power of the reactor in thermal megawatts per gram of tritium inventory (MW/g T)

BR = tritium breeding ratio, that is, the number of tritons produced in the blanket per fusion neutron, or equivalently, per D–T reaction

EY = energy yield in thermal megawatt-days per gram of tritium consumed (MWD/g T).

I shall assume that the energy yield is 8.32 MWD/g T consumed. This value is based on a total fusion energy release of 22.4 MeV per D–T fusion.

The value of the specific power of the reactor SP depends on the magnitude of the total tritium inventory in the plant, including the plasma, the plasma exhaust recovery system, the blanket and its associated tritium recovery system, the fueling system, the fuel storage system and the structural materials of these systems. The calculation of tritium inventories for each system depends on detailed design considerations (and assumptions) which are beyond the scope of this presentation. However, some general observations can be made regarding the anticipated tritium inventories. Consider a 1000-MW(e) plant.

1. The plasma will contain ~ 1 g of tritium.
2. At 1000 MW(e) and 30 percent overall plant efficiency, the reactor will consume

*Research sponsored by the U.S. Energy Research and Development Administration, under contract with Union Carbide Corporation.

†Oak Ridge National Laboratory, operated by Union Carbide Corporation, Nuclear Division, Oak Ridge, Tenn.

~ 0.4 kg T/day. Thus, the blanket tritium recovery system (including the blanket itself) must handle a tritium throughput of ~ 0.4 kg T/day. In current reactor design studies the estimated tritium inventory in the blanket tritium-recovery system is in the range of ~ 0.1 to 10 kg.

3. At a fractional burnup of 5 percent, the throughput of tritium in the plasma exhaust recovery system would be ~ 8 kg T/day. In current design studies the estimated tritium inventory in this system is in the range ~ 0.1 kg-4 kg.

4. If a one-third-day supply of tritium is assumed in the fueling system, the associated inventory is ~ 0.1 kg.

5. If a ten-day supply of tritium is assumed in the storage system, the associated inventory is ~ 4 kg.

6. In order to minimize the effects of tritium on the mechanical properties of structural

TABLE 1.   Summary of Materials Being Considered for Blanket Applications

Application	Materials
Breeding	Liquid lithium
	Molten salts (Li_2BeF_4, LiF)
	Ceramic compounds (Li_2O, Li_2C_2)
	Aluminum compounds (LiAl, $Li_2Al_2O_4$)
Structural	Refractory-based alloys (Nb, V, Mo)
	Iron-based alloys
	Nickel-based alloys
	Aluminum-based materials
	Silicon carbide
Coolant	Liquid lithium and potassium
	Molten salts
	Helium
Moderator	The breeding materials
	Graphite
Neutron multiplication	Beryllium

materials the concentration of tritium in these materials should be maintained at a few ppm. The associated inventory would be less than 0.1 kg.

Thus, the estimated total tritium inventory is in the range ~ 4 to 18 kg, and the associated specific power is ~ 0.8 to 0.2 MW/g T. For illustrative purposes I shall take the specific power as 0.5 MW/g T.

Using the above values for SP and EY, the tritium doubling time in years becomes

$$t_2 = \frac{3.2 \times 10^{-2}}{BR - 1} \qquad (2)$$

From this equation it is seen that a breeding ratio only slightly greater than unity will yield a doubling time of about 10 years. Doubling times of approximately 1 year might be desirable in the initial stages of fusion reactor deployment when the availability of tritium would be limited. Such doubling times would require a breeding ratio of ~ 1.1. It is important to note that the specific power anticipated in fusion reactors is several orders of magnitude greater than that calculated for fission breeder reactors. Thus, the doubling times of interest require significantly lower breeding ratios with fusion reactors than with fission reactors.

Current blanket designs yield calculated breeding ratios in the range ~ 1.1 to 1.5 (Refs. 3 and 4). Table 2 summarizes the characteristics and breeding performance of several proposed blankets. Note that designs 1 and 4 employ additions of beryllium and lithium in ^6Li. Considerations of nuclear data uncertainties (Refs. 8 and 9) and the idealized nature of the one-dimensional geometrical model used in these calculations (Ref. 10) indicate that the values of breeding ratio given in Table 2 might overestimate the actual situation by 10 to 20 percent. On the other hand, I emphasize that these blanket configurations have not been optimized for breeding. For example, the breeding performance of design 2 could be enhanced by the addition of beryllium. Therefore, I conclude

TABLE 2.   Characteristics and Breeding Performance of Several Blanket Designs

Design no.	1	2	3	4
Reference number for blanket design	4	5	6	7
Thickness of breeding region, cm	82	80	101	38
Breeding Material	LiAl	Li_2BeF_4	Li	Li*
(^{6}Li enrichment)	(90%)	(Natural)	(Natural)	(95%)
Volume % beryllium additions	16	None	None	6.6
Structural material	SAP†	PE16‡	Vanadium	Niobium
Volume % structural material	21	5	1.6	3.4
Coolant	Helium	Helium	Lithium	Lithium
Additional moderator	Graphite	None	Graphite	Graphite
Breeding ratio	1.49	1.07	1.56	1.12

*Both enriched and natural lithium are employed in this design.
†Sintered Aluminum Product consisting of 90% Al and 10% Al_2O_3.
‡NIMONIC alloy consisting of 43% nickel, 39% iron and 18% chromium.

that the required tritium-breeding performance can be satisfied by a variety of blanket configurations and materials choices. Optimization of the blanket design for commercial applications will require an improved nuclear data base.

REFERENCES

1. Steiner, D.: *Nucl. Appl. & Tech.*, **9**, 83 (1970).
2. Vogelsang, W. F.: *Nucl. Tech.*, **15**, 470 (1972).
3. Gough, W. C. (edited and compiled by): Fusion Reactors System Studies Review Meeting, *U. S. At. Energy Comm. Publ.* WASH-1278 (UC-20), 1973.
4. Powell, J. R. et al.: "Minimum Activity Blankets for Commercial and Experimental Power Reactors," BNL-18439, Brookhaven National Laboratory, Upton, Long Island, N.Y., 1973.
5. Greenspan, E., and W. G. Price, Jr.: Tritium Breeding Potential of the Princeton Reference Fusion Power Plant, *Princeton Univ. Plasma Phys. Lab. Rep.* MATT-1043, 1974.
6. Steiner, D.: *Nucl. Fusion*, **14**, 33 (1974).
7. Dudziak, D. J.: "Discrete Ordinates Neutronic Analysis of a Reference Theta-Pinch Reactor (RTPR)," LA-DC-72-1427, Los Alamos Scientific Laboratory, 1972, and Technology of Controlled Thermonuclear Fusion Experiments and the Engineering Aspects of Fusion Reactors, Symposium Proceedings, CONF-721111, Austin, Tex., 1972.
8. Bartine, D. E., et al.: *Nucl. Sci. & Eng.*, **53**, 304 (1974).
9. Steiner, D., and M. Tobias: *Nucl. Fusion,* **14**, 153 (1974).
10. Lee, J. D.: "Geometry and Heterogeneous Effects on the Neutronic Performance of a Yin Yang Mirror-Reactor Blanket," UCRL-75141, Lawrence Livermore Laboratory, Livermore, Calif., 1973.

Superconducting Magnets as Applied to Fusion Reactors

BY J. FILE*

The phenomenon of superconductivity was discovered by H. Kamerlingh Onnes (Ref. 1) in 1911. At that time, he was engaged in making routine experiments on the electrical resistance of metals cooled with the newly available liquid helium. At atmospheric pressure, liquid helium boils at 4.2 K. With a known current passing through a sample of the metal, a sensitive potentiometer was used to measure the potential drop across the specimen. Onnes (Refs. 2 and 3) found that the resistance of mercury dropped almost discontinuously (over several millidegrees) to an immeasurably low value at 4.26 K. For example (Ref. 4), a particular sample of the metal that had a resistance of 125 milliohms at a temperature slightly above 4.2 K dropped to an immeasurably low value of less than 3×10^{-6} ohm, the measuring limit of his equipment, when the temperature was decreased to slightly below 4 K. The resistance drop occurred within a temperature range of about 0.03 K (see Fig. 1). When the temperature was again raised to above 4 K, the resistance of 125 milliohms was fully restored.

In 1913, Onnes concluded (Ref. 4) that the metal had entered a new state, which he

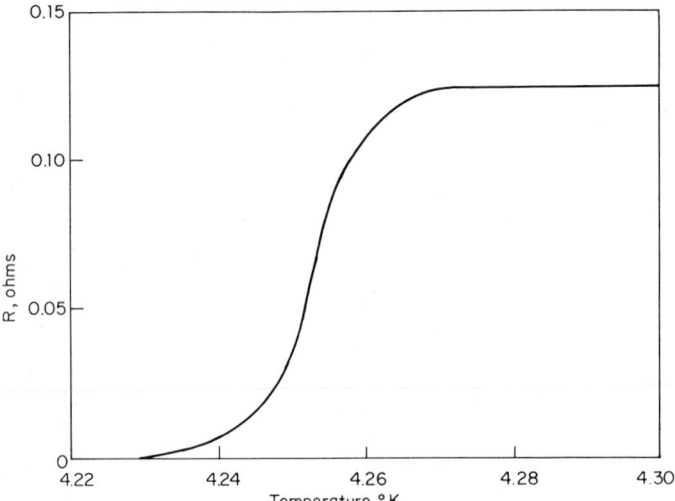

Fig. 1. Resistance of mercury versus temperature.

called the *superconducting state*. The temperature at which the transition takes place is usually called the *critical temperature* and is designated by T_c.

The property of superconductivity has been found to be quite common among metallic elements and their compounds and alloys. The critical temperatures of the various substances vary from a few millidegrees to about 23 K (Ref. 5), the critical temperature of the binary alloy, Nb_3Ge (niobium and germanium).

Onnes also discovered that a superconductor reverts to the normal state when it is placed in a sufficiently strong magnetic field. The metal returns to the superconducting

---

*Plasma Physics Laboratory, Princeton University, Princeton, N.J.

state when the field is removed. The minimum field required to destroy superconductivity is known as the *critical field* and is designated by H_c. The value of the critical field depends on the temperature. It is zero at the transition temperature T_c and rises to a maximum at 0 K. The relation $H_c(T)$ is a characteristic of any particular superconductor and is often represented with fair accuracy by the parabola

$$H_c(T) = H_c(0)\left[1 - \left(\frac{T}{T_c}\right)^2\right]$$

This equation has been verified experimentally for many superconductors, and some typical curves for three low-critical-field soft superconductors are shown in Fig. 2.

Having observed that the resistance of some metals displays an apparent discontinuous drop to zero at temperatures close to 0 K, it was correctly deduced that if supercurrents

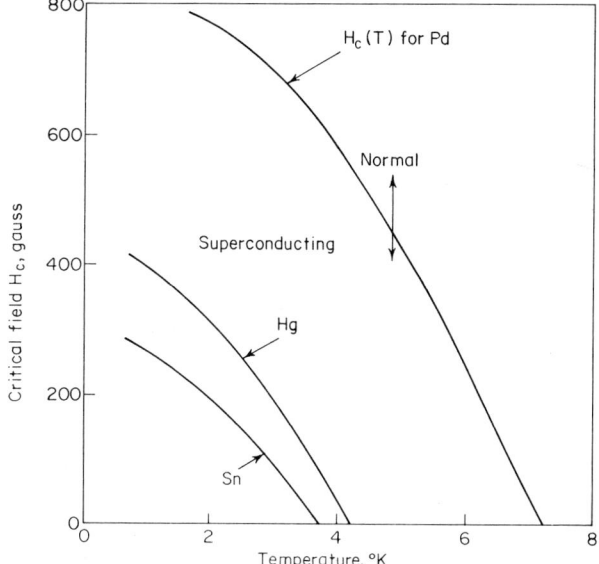

Fig. 2. Threshold curves for various superconductors.

were induced in a closed superconducting circuit, they would persist. If a ring of superconducting material is placed in a magnetic field perpendicular to the plane of the ring, and if the temperature of the ring is higher than transition T_c, and subsequently cooled below T_c, then the induced flux is trapped in the hole encircled by the ring until superconductivity is destroyed. The trapped flux is sometimes described as being "frozen in." If the external field is switched off before superconductivity is destroyed, the ring acts as a permanent magnet with enough induced current flowing to sustain all the trapped flux. On the other hand, if a superconducting ring at temperature below T_c and initially free of current is placed in a magnetic field, then the lines of force will not penetrate, and the hole will be field-free.

SOFT AND HARD SUPERCONDUCTORS

Superconductors are divided into two main classes. The first is soft (type I) superconductors; the second is hard (type II) superconductors. Type I superconductors include most of the superconducting metallic elements. They are usually pure materials, mechnically soft and malleable. They are generally characterized by a low critical field, H_c, of a few hundred gauss. Figure 2 indicates some typical examples. Type I superconductors are further distinguished by low critical temperatures, usually less than 4.2 K, and low critical

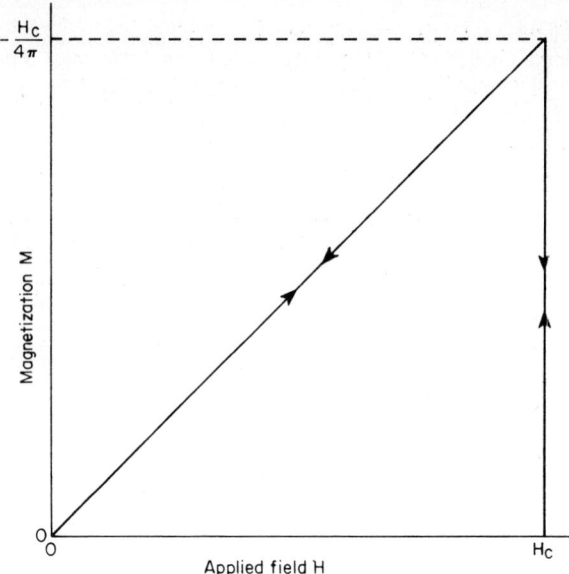

Fig. 3. Magnetization curve of an ideal Type-I superconducting cylinder with the axis parallel to the applied field H.

current densities. Their behavior is "ideal" in the sense that their electromagnetic behavior is predicted by London's phenomonological equations and that they display a pure Meissner effect (Ref. 6). The critical field behaves according to the parabolic law of the equation previously given. Type I superconductors react in essentially the same way as a perfectly conducting diamagnetic substance. Figure 3 shows the magnetization curve of a cylinder of Type I superconducting material whose axis is parallel to the magnetic field. The magnetic moment increases linearly up to the critical field H_c, at which point the magnetic moment drops discontinuously to zero.

Type II or hard superconductors differ from Type I in that they remain superconducting even after some magnetic field has penetrated the material. Below the lower critical field H_{c1} the applied magnetic field is excluded completely from the superconductor. This is

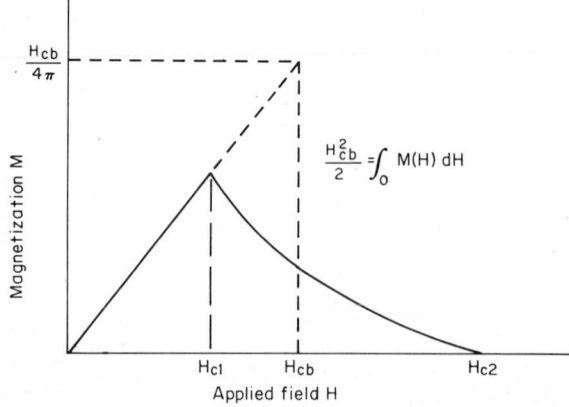

Fig. 4. Magnetization curve of an ideal Type-II superconducting cylinder with the axis parallel to the applied field H.

illustrated on the magnetization curve of Type II superconductors shown in Fig. 4. Between the limits of the lower and upper critical fields, H_{c1} and H_{c2}, the magnetic field penetration increases with applied field. Flux is not totally excluded. At the upper critical field H_{c2}, the flux penetration is complete and the superconductor reverts to its normal state. The thermodynamic critical field H_{cb} is determined by the area under the magnetization curve, as indicated in Fig. 4. The thermodynamic critical field H_{cb} indicates the theoretical reversible curve, i.e., what the critical field might be if this were a Type I superconductor. Experimentally, it has been found that this theoretical reversible curve is approached as strains and other imperfections in the material are eliminated. The lower and upper critical fields, H_{c1} and H_{c2}, are also functions of temperature and applied field, but generally they do not follow the parabolic law given by the previous equation. Type II superconductors generally have higher transition temperatures than Type I, up to 23 higher critical fields, up to hundreds of kilogauss; and high current densities, up to a million amperes per square centimeter.

Type II superconductors are usually hard, brittle materials. The magnitude of the critical current seems to increase with the following: hardness of material, degree of cold work, degree of lattice imperfection, and number of lattice defects.

MATERIALS

Because of the low transition temperatures, and low critical fields, soft superconductors are of little value in applications of high-field magnet technology. Therefore, attention will be focused on those materials which may be utilized in fusion reactors.

In 1961, Kunzler and his coworkers (Ref. 7) successfully demonstrated the use of the first high-critical-field Type II superconductor, an intermetallic compound of niobium

TABLE 1.   Commercially Available Superconducting Materials

Material	Critical temperatures T_c, K	Upper critical field, H_{c2}, at 4.2K, kG	Current density, A/cm²	Approximate costs/kAM	References
Nb, 25% Zr	11	70	5×10^4 at 4.2 K, 58 kG	$10.00	9
Nb, 33% Zr	10.7	80	5×10^4 at 4.2 K, 60 kG	$10.00	9
Nb$_3$Sn	18.2	245	5×10^5 at 4.2°K, 100 kG	$20.00	10, 11
Nb, 48% Ti	9.5	122	8×10^4 at 4.2°K, 75 kG	$ 4.00	12, 13
V$_3$Ga	14–16.8	208–210	2×10^6 at 4.2°K, 200 kG / 1×10^6 at 4.2°K, 40 kG / 4×10^5 at 4.2 K, 100 kG	$ 5.00	10, 14, 15

and tin, Nb$_3$Sn. Since that time, alloys of niobium and zirconium, niobium and titanium, and vanadium and gallium, as well as niobium and tin, have become available commercially. In addition, many other alloys and compounds are now being developed in the laboratories that may very well have commercial application in from 5 to 50 years. Table 1 lists those materials that are now available commercially, giving some of their superconducting properties and present (1975) approximate prices. Table 2 gives some of the most promising compounds and alloys that are currently being investigated and developed in various laboratories throughout the world and lists some of their superconducting properties. Some of the information in Tables 1 and 2 was obtained from Ref. 8.

THERMAL STABILIZATION

As has been described (see Fig. 4), flux penetrates inside a Type II superconductor. As the flux penetrates into the superconductor, it concentrates into bundles and becomes pinned to inhomogeneities, strains, dislocations, and other defects in the material. Because of the $J \times B$ Lorentz force, these bundles jump from barrier to barrier. This is known as *flux jumping* (Refs. 19 to 21), and if enough energy is released during a flux jump to bring the local temperature above the critical temperature T_c, a normal transition follows. The local

TABLE 2.   A Partial Listing of Developmental Superconducting Materials

Material	Critical temp., T_c, K	Upper critical field, H_{c2}, at 4.2 K, kG	Current density, A/cm^2	References
V_3Si	17	235	3×10^5 at 4.2 K, 40 kG	10, 15
PbBi (glass)	8.6	125	4×10^4 at 4.2 K, 50 kG	16
Nb_3Al	17.5–19	295	No information	5, 10
$Nb_3(Al_{0.8}Ge_{0.2})$	20.7	400	3×10^5 at 4.2 K, 100 kG	5, 10, 17
		410	1.5×10^4 at 16 K, 150 kG	5, 10, 17
$V_2Hf_{0.5}Zr_{0.5}$	10	230	10^5 at 4.2 K, 130 kG	18
Nb_3Ge	23	370	10^5 A 4.2 K, 210 kG	5
			10^5 at 10 K, 150 kG	5

transition usually propagates rapidly, thereby rendering the whole coil to its normal state. The effect of flux jumping can be minimized if one can quickly restore the local hot spot to a temperature below T_c before the normal transition spreads and becomes catastrophic. Superconductors have poor heat transport properties. In addition, the normal electrical resistivity of these materials is high, even at liquid helium temperatures. Therefore, if one places another material, usually called stabilizing material, of high thermal and electrical conductivity in intimate contact with the superconductor, the effect of flux jumping is minimized in the following ways:

1. As the temperature of a local area of the superconductor exceeds T_c owing to flux jumps, its electrical resistance becomes excessively high. The current in the superconductor then flows in the lower resistance parallel path made available by the stabilizing material.

2. At the same time, and because of the good conduction and thermal transport properties of the stabilizer, the heat produced in the superconductor is quickly conducted to the helium coolant.

3. If the I^2R ohmic heating in the stabilizer is minimal, and if the temperature of the superconductor attains a value less than T_c before the local normal transition spreads, then superconductivity is restored locally, and the current again flows in the superconductor.

There are three methods of stabilizing superconductors. These are known as cryogenic, dynamic, and adiabatic stabilization. They are defined by Hancox (Ref. 22) as:

1. *Cryogenic stabilization,* which depends on the provision of an alternative low resistance path for the current to flow. Sufficient stabilizing material must be provided so that the conductor is capable of carrying the current indefinitely without raising the temperature of the superconductor above T_c.

2. *Dynamic stabilization,* which is obtained by the addition of sufficient normal material to magnetically damp any flux jumps so that the energy released can be removed locally by conduction and possible instabilities can be prevented from growing.

3. *Adiabatic stabilization,* which requires the use of fine filaments of superconductor so that the energy associated with a flux jump is too small to drive the conductor normal.

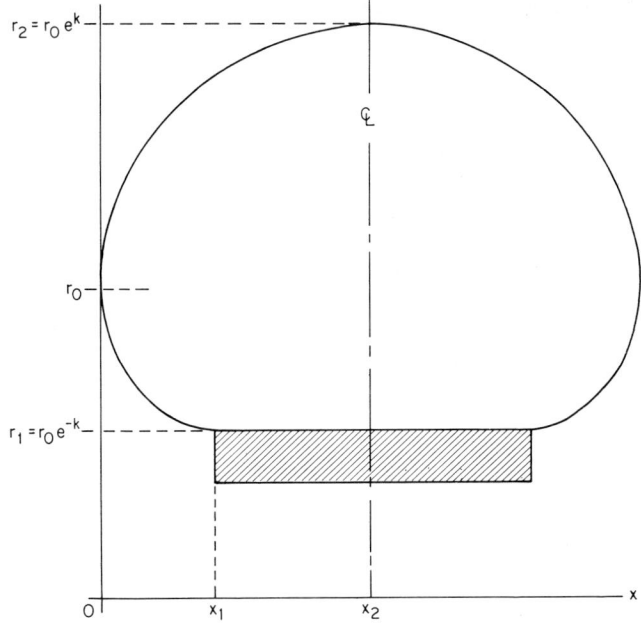

Fig. 5. Mathematical shape of a constant tension, current-carrying element in a $1/r$ field.

APPLICATION OF SUPERCONDUCTIVITY AND CRYOGENICS TO FUSION DEVICES

Superconducting Magnets.   The output power density of a fusion reactor increases as B^4, and the cost of the magnet increases at a somewhat lower rate (somewhere between B and B^2) (Refs. 23 and 24). Because of these relationships, it is generally agreed that fusion reactor designers intend to take advantage of the highest fields available, consistent with other mechanical constraints, such as first wall loading and stress limits of structural materials. Commercially obtainable superconductors are already capable of producing magnets with fields in excess of 160,000 gauss (Ref. 25) in relatively small bores. That field level taxes the known limits of structural design. In addition to high fields, fusion reactors will require larger working volumes (by several orders of magnitude) than any previously

TABLE 3.   Electrical and Mechanical Characteristics of the Princeton Reference Design Reactor Coil

Superconductor	Nb_3Sn
Width of conductor	4.5 cm
Thickness of conductor	0.48–1.52 cm
No. of pancakes per coil	7
Maximum field at conductor	160,000 gauss
Nominal plasma radius	1050 cm
Field at nominal plasma radius	60,000 gauss
Ampere turns per coil	6.57×10^6
Minimum current density (conductor area)	2170 amp/cm²
Current per conductor	10,000 amps
Energy stored in field	250×10^9 joules
Inductance of torus	5×10^3
Weight of cylinder	10^6 kG
Weight of toroidal coils	6.2×10^6 kG
Heat leak (structure, insulation and neutron heating)	2.8×10^2 kW

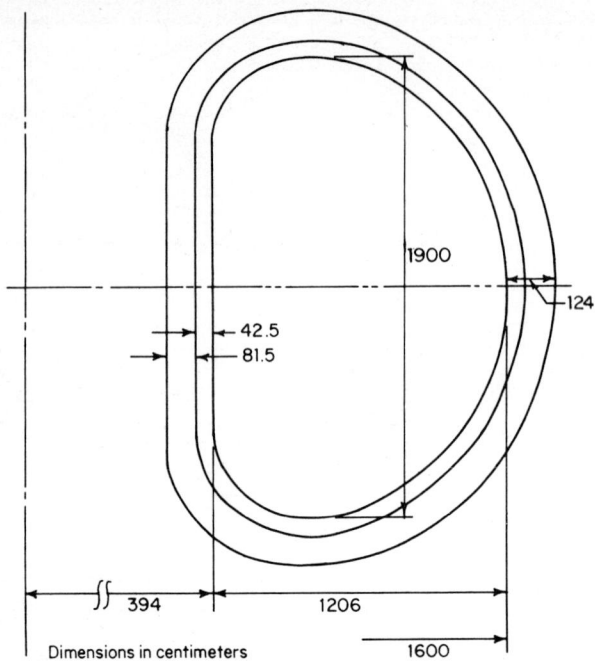

1900

124

42.5

81.5

394

1206

1600

Dimensions in centimeters

Fig. 6. Constant tension coil for proposed Princeton Fusion Reactor. Forty-eight coils in toroidal arrangement.

designed. The combination of high field and large working volume requires the design of superconducting magnets well beyond present known technology.

A novel design of a toroidal magnetic coil that partially accommodates the large forces generated by a high-field large-volume magnet was previously described (Refs. 26 and 27). In a toroidal magnet, the field strength within the useful volume varies inversely with the radius from the axis of symmetry, and in almost all instances the conductors generating such fields will be subject to bending moments in addition to the effective internal pressure. It was shown that a conductor tethered at either end and immersed in a toroidal field will be stable if it is in pure tension and, therefore, not subject to bending moments. Net forces are then taken on a cylindrical structural element to which the conductor is tangent. Except where the conductor lies flat against the support, it lies in a curve such that its radius of curvature, ρ, is proportional to the radius from the axis of the torus, r; $\rho = kr$. k is a constant depending on the geometry of the system. If one substitutes the expression for radius of curvature for ρ it is found that the equation for the equilibrium curve of a current-carrying conductor in a $1/r$ field is

$$ r \frac{d^2r}{d_x2} = \pm \frac{1}{k} \left[1 + \left(\frac{dr}{dx} \right)^2 \right]^{3/2} $$

The shape of the curve obtained from the solution of the foregoing equation is shown on Fig. 5. A coil of this shape, later modified (Ref. 27) for more practical type toroidal magnets, is commonly called a D-shape coil.

Figures 6, 7, and 8 show sections of the 48 toroidal coils required for the 2030-MW(e) reactor proposed by the Princeton Reactor Studies Group (Refs. 28 and 29). Table 3 gives some other mechanical and electrical characteristics of the coil.

Superconducting magnets used thus far in fusion research are very much smaller than the one shown in Fig. 8 and, in general, are used as some special part of a plasma experiment, such as the Princeton floating ring. Table 4 lists these magnets, giving some of their characteristics as compared with those of Fig. 6 and Table 3.

The 36-inch floating ring, no longer in operation, was the first levitated ring plasma experiment (the Princeton spherator) performed. This was replaced by the much larger

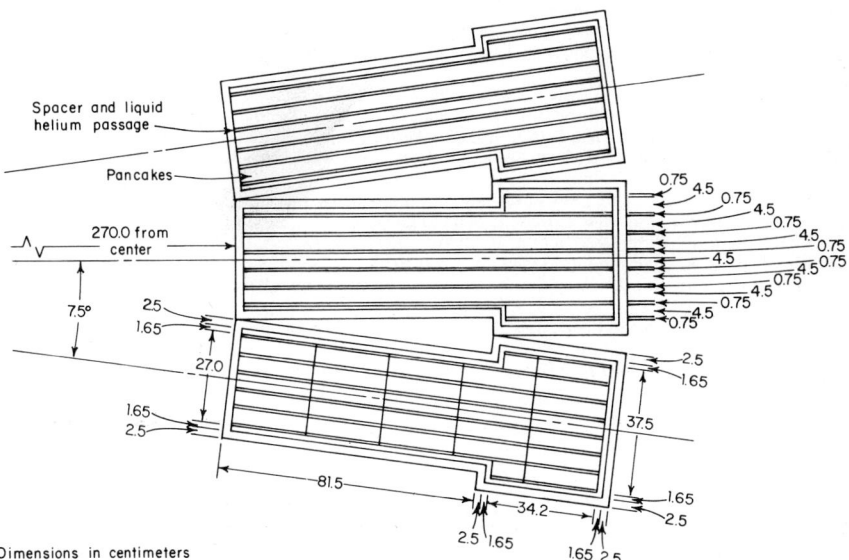

Fig. 7. Cross section of the constant tension coils.

60-inch-diameter ring in the Princeton FM-1 machine. This is one of the largest-bore Nb_3Sn magnets yet constructed. It is, however, of relatively low field and stored energy. The Lawrence Radiation Laboratory's floating ring is similar in characteristics to the foregoing two.

The Lawrence Radiation Laboratory's Baseball II, and the Oak Ridge National Laboratory IMP quadrupole are mirror type, open-ended devices and are characteristic of the size of devices thus far constructed for plasma experiments. When the five aforementioned devices are compared to a reactor-size magnet, it is seen that they are from one to three orders of magnitude smaller in bore and two to three orders of magnitude smaller in stored energy than is required for reactors.

Fusion Reactor Studies.   Of the many recent publications on the subject of fusion reactor studies, the two most comprehensive were contributed by the Plasma Physics Laboratory, Princeton University (Ref. 29) and the University of Wisconsin (Ref. 35). These two publications have highlighted the technologies that must be expanded or developed before such a device becomes operational. Figures 9 and 10 show the conceptual fusion power plants presented by Princeton University and the University of Wisconsin, respectively. Table 5 compares various parameters of both devices.

REFERENCES

1. Onnes, H. Kamerlingh: Comm Leiden No. 120b, *Kon. Ned. Akad. Wet.*, **13**, 1274 (1911).
2. Onnes, H. Kamerlingh: Comm Leiden No. 122b, *Kon. Ned. Akad. Wet.*, **13**, (1911).
3. Onnes, H. Kamerlingh: Comm Leiden No. 124c, *Kon. Ned. Akad. Wet.*, **14**, 818 (1911).
4. Onnes, H. Kamerlingh: Comm Leiden No. 133c, *Kon. Ned. Akad. Wet.*, **15**, 947 (1913).
5. Gavaler, J. R., M. A. Janocko, S. Foner, and E. J. McNiff: "Upper Critical Fields of Nb_3Ge *Thin Film Superconductors*," to be published.

6. Rickayzen, G.: "Theory of Superconductivity," chap. 1, Wiley, New York, 1965.
7. Junzler, J. E., E. Buehler, F. S. L. Hsu, and J. H. Wernick: *Phys. Rev. Letters* **6**, 89 (1961).
8. Lubell, M. S.: *Pro. Int. Working Sessions on Fusion Reactor Technol.*, Oak Ridge National Laboratory, Oak Ridge, Tenn., 1971.
9. Grigsby, D. L. (compiler): Niobium Zirconium, *EPIC Rep.* DS-152, November 1966.
10. Foner, S., E. J. McNiff, B. T. Matthias, T. H. Geballe, R. H. Willens, and E. Corenzwitt: *Phys. Rev. Letters*, **31A**, 349 (1970).
11. Grigsby, D. L. (compiler): Niobium Tin, *EPIC Reps.* DS-159, January 1968, and DS-160, July 1968.
12. Stekly, Z. J. J.: *J. Appl. Phys.*, **42**, 65 (1971).
13. Grigsby, D. L. (compiler): Niobium Titanium Data Table and Supplementary Bibliography, *EPIC Rep.* DS-148, October 1968.
14. Gregtak, T. J., and J. H. Wernich: *J. Phys. Chem. Sol.*, **25**, 535 (1964).
15. Otto, G., E. Saur, and H. Wizgall: *J. Low Temp. Phys.*, **1**, 19 (1969).
16. Watson, J. H. P.: *Appl. Phys. Letters*, **16**, 428 (1970).
17. Foner, S., E. J. McNiff, B. T. Matthias, and E. Corenzwitt: *Proc. 11th Int. Conf. Low Temp. Phys.*, **2**, 1025 (1968).
18. Inoue, K. Tachikawa, and Y. Iwasa: *Appl. Phys. Letters*, **18**, 235 (1971).
19. Anderson, P. W.: *Phys. Rev. Letters*, **9**, 309 (1962).
20. Kim, Y. B., C. F. Hempstead, and A. R. Strnad: *Phys. Rev. Letters*, **9**, 306 (1962).
21. File, J.: *J. Appl. Phys.* **39**, (5), 2335–2338, April 1968.
22. Hancox, R.: *Proc. Int. Working Session of Fusion Technol.*, Oak Ridge National Laboratory, Oak Ridge, Tenn., 1971.
23. Powell, J. R.: Design and Economics of Large DC Fusion Magnets, *Proc. 1972 Appl. Superconductivity Conf.*, pp. 346–353.

(References continued on p. 5-172)

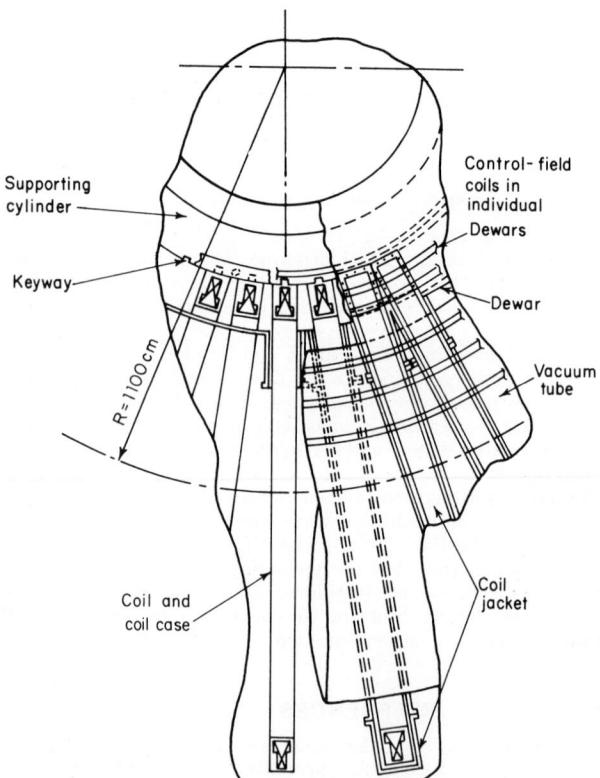

Fig. 8. Cross section of the Princeton Reactor showing the support cylinder and Dewar system.

TABLE 4. Superconducting Magnets Used Thus Far in Fusion Research (1975)

Magnet	Superconducting material	Geometric parameters, cm	Max. B, kG, at Superconductor	Current density, A/cm^2	Stored energy, kilojoules	Max. design current, kiloampere-turns	Remarks	References
Floating ring, 36" major diam.	Nb$_3$Sn	Major diam. 92, minor diam. 3.6	19.3	13,000	26	130	Used for plasma experiments in the Princeton spherator	30
Floating ring, 60" major diam.	Nb$_3$Sn	Major diam. 152, minor diam. 6.2	28	12,900	240	375	Used for plasma experiments in the Princeton FM-1 device	31
LRL floating ring	Nb$_3$Sn	Major diam. 80, minor diam. 9	20 36	12,200 21,900	75.5 242	334 600	Operating conditions and design parameters for ring used in plasma experiments in the LLL Levitron device	32
LRL Baseball II	NbTi	Average diam. of Winding—120	75	4,000	17,000	4800	Used for fusion research in a mirror device at LLL	33
ORNL IMP quadrupole	Quadrupole coils—Nb$_3$Sn Mirror coils—NbTi	Plasma dimensions 45 × 45	72 & 85 58.5	13,000 6,558	2,240 147	1500 950	Used for fusion research in a mirror device at ORNL	34
Proposed Princeton toroidal magnet	Nb$_3$Sn	See Fig. 7 for dimensions	160	2,100	250 × 10^6	3.15 × 10^6	Toroidal magnet made of 48 segments	28 29

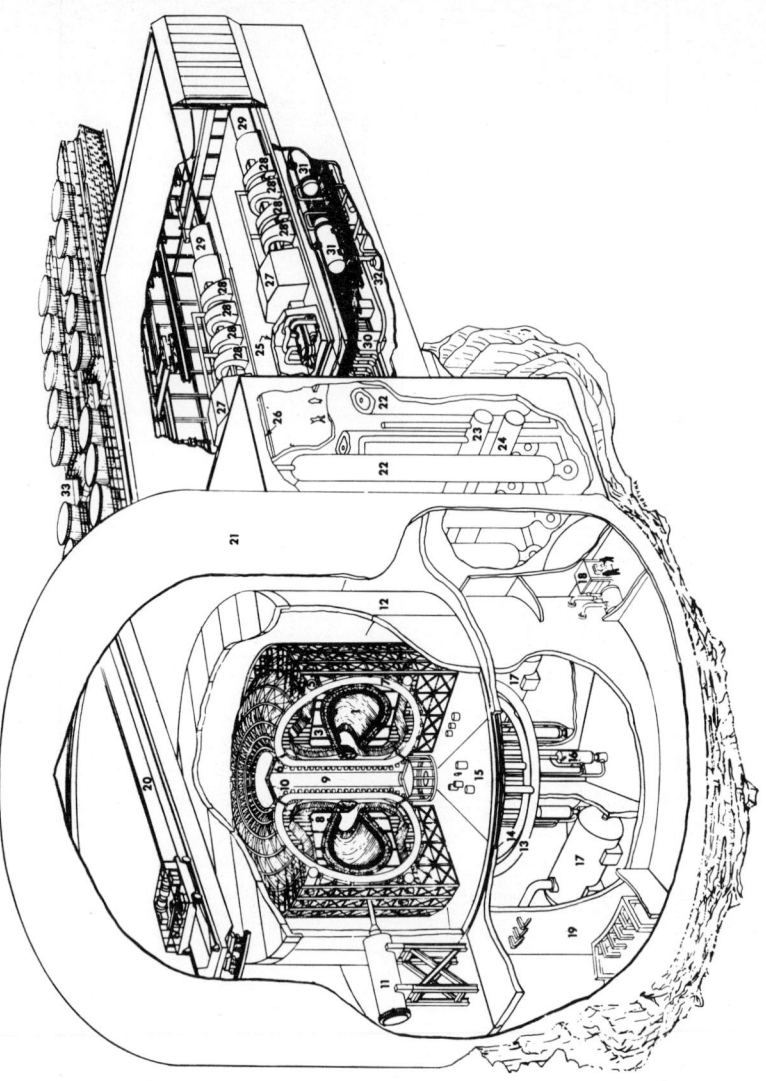

Fig. 9. Princeton Plasma Physics Laboratory reference design for a fusion reactor: (1) Vacuum chamber. (2) Blanket. (3) Primary shield. (4) Diverter. (5) Vacuum pumps and exhaust manifold. (6) Control field coils. (7) Toroidal field coils. (8) Poloidal field coils. (9) Liquid helium tank. (10) Toroidal coil support cylinder. (11) Fuel injector. (12) Biological shield. (13) Blanket helium outlet header. (14) Blanket helium inlet header. (15) Helium and flibe lines. (16) Flibe-TF disengagers. (17) Flibe storage tanks. (18) Cryogenic system. (19) Maintenance hot cell. (20) Polar crane. (21) Reactor containment. (22) Steam generators. (23) Hot helium supply header. (24) Cold helium return header. (25) Very high pressure turbine. (26) High pressure turbine. (27) Intermediate pressure turbines. (28) Low pressure turbines. (29) 3600-rpm generator. (30) Steam lines. (31) Feedwater heaters. (32) Feedwater treatment. (33) Cooling towers.

TABLE 5. Comparison of the Princeton Reference Design Power Plant with the University of Wisconsin UWMAK-I

	Princeton reference design	UWMAK-I
I. Plasma and magnetics:		
A. Plasma configuration:		
1. Tokamak	With single null poloidal divertor	With double null poloidal divertor
2. Nominal major radius	10.5 m	13 m
3. Nominal minor radius	3.25 m	5 m
4. Nominal volume	2190 m^3	6400 m^3
5. Confining field at center line	6 T	3.82 T
B. Plasma parameters:		
1. Mean hydrogen ion density	$5 \times 10^{19}/m^3$	$3.67 \times 10^{19}/m^3$
2. Mean ion temperature	30 keV	28 keV
3. Mean ion confinement time	3.83 sec	61 sec
4. Average β, maximum β	0.04, 0.13	0.052
5. Average β, θ	1.83	1.08
6. Plasma current	14.6 MA	25.7 MA
7. Ion ratios (D:T: argon)	(1:1:0.096)	(D+T):argon = 1:0.0365
8. Nominal safety factor q	2.067	1.75
C. Burn parameters:		
1. DT reaction rate	1.22×10^{21}sec	1.56×10^{21}/sec
2. Tritium burnup per pass	8.7%	30%
3. Deuterium feed rate	1.40×10^{22}/sec	0.52×10^{22}/sec
4. Tritium feed rate	1.40×10^{22}/sec	0.52×10^{22}/sec
5. Argon feed rate	0.134×10^{22}/sec	
6. Tritium consumption	0.527 kG/day	0.672 kG/day
7. Burn cycle length	100 min	35 min
8. Shutdown time per cycle	3 min	
D. Coil descriptions:		
1. Toroidal field coils	48	12
2. Material	Nb_3Sn, 10-kA/conductor	NbTi
3. Peak field	16 T	9 T
E. Cryogenic system:		
1. Coil heating	280 kW	11 kW
From pulsed fields	35 kW	0
From neutrons & gammas	95 kW	0.4 kW
Thermal conduction	150 kW	10.6 kW
2. Refrigeration power required	100 MW	4 MW
3. Helium inventory (STP)	4.55×10^5 m^3	250 m^3
4. Helium loss rate (STP)	8.4 m^3/day	Unknown
II. Energy and power systems:		
A. Reactor power:	5305 MW	5000 MW
1. Origin:		
Fusion alphas	688 MW	3.52MeV, 877 MW
Fusion neutrons	2752 MW	14.06MeV, 3503 MW
Neutron capture	1865 MW, 9.53 MeV	2.49MeV, 620 MW
2. Deposition:		
Radiation in divertor	134 MW	0 MW
Radiation on wall	554 MW	877 MW
Nuclear heating	4617 MW	4123 MW
B. Wall load:		
1. Nominal wall area	1563 m^2	2830 m^2
2. DT neutron production rate	1.22×10^{21}/sec	1.55×10^{31}/sec
3. DT neutron current through wall	$0.78 \times 10^{18}/m^2$sec	$0.55 \times 10^{18}/m^2$ sec
4. Total neutron flux at wall	$8.65 \times 10^{18}/m^2$sec	Unknown
5. Wall load due to DT neutrons	1.76 MW/m^2	1.25 MW/m^2
6. Wall load due to plasma radiation	0.35 MW/m^2	0.32 MW/m^2
C. Cooling:		
1. Gaseous helium	5.07	Lithium
2. Mean pressure	50 atm (5.07 MPa)*	27 atm (2.76 MPa)
3. Mean pressure drop	2.8 atm (0.29 MPa)	6.8 atm (0.69 MPa)
4. Outlet conditions	3.24 Mg/sec at 638°C	6 Mg/sec at 483°C
5. Inlet conditions	2.83 Mg/sec at 360°C	6 Mg/sec at 283°C
	0.41 Mg/sec at 66°C	6 Mg/sec at 283°C

TABLE 5.   Comparison of the Princeton Reference Design Power Plant with the University of Wisconsin UWMAK-I *(Continued)*

	Princeton reference design	UWMAK-I
6. Pumping power	241 MW	22 MW
7. Helium inventory (STP)	2×10^5 m^3	(Lithium cooled)
D. Steam System		
1. Energy inputs	5546 MW	5000 MW
From main steam generator	3364 MW	Unknown
From first reheater	780 MW	Unknown
From second reheater	474 MW	Unknown
From helium feedwater heaters	628 MW	Unknown
2. Heat exchanger area	2.1×10^2 m^2	2.8×10^4 m^2
3. Max. steam temperature	538°C	425°C
4. Max. steam pressure	3690 psi	600 psi
5. Steam flow rate	6.6 Gg/hr	7 Gg/hr
6. Energy to turboalternator shafts	2405 MW	1500 MW
To VHP turbine	498 MW	Unknown
To HP turbine	434 MW	Unknown
To IP turbines (2)	483 MW	
To LP turbines (4)	990 MW	Unknown
7. Energy to condenser	3141 MW	3500 MW
E. Electrical system:		
1. Gross electric generation	2405 MW	1500 MW
2. Station service	375 MW	Unknown
Helium circulator motors	251	22
Cryogenic refrigerators	100	Unknown
Cooling tower fans	15	Unknown
Miscellaneous	9	Unknown
3. Net electric output	2030 MW	1450 MW
4. Net efficiency	38.3%	30%
F. Cost evaluation:		
1. Capital cost	1215 M$	1455 M$
2. Capital cost per installed capacity	$598/kW(e)	$970/kW(e)
3. Cost of busbar energy	15.2 mills/kWh	21.4 mills/kWh
III. Materials and design:		
A. Fuel reactions:		
1. Tritium burnup	0.527 kg/day	0.672 kg/day
2. Deuterium burnup	0.352 kg/day	0.448 kg/day
3. Lithium burnup	1.104 kg/day	2.3 kg/day
4. Beryllium burnup	0.362 kg/day	
5. Helium production	10.2 m^3 (STP)/day	
B. Blanket materials:		
1. Flibe: eutectic mixture	53.1% BeF$_2$, 46.9% LiF, m.p.355°C, 1.89 Mg/m^3 at 800°C	Li
2. PE-16:austenitic nickel alloy	43% Ni, 39% Fe, 18% Cr	316 ss
C. Tritium handling:		
1. Tritons per fusion (breeding ratio)	1.04 (net)	1.49
2. Breeding rate	0.548 kg/day	1.001 kg/day
3. Inventory	2.57 kg	Unknown
(a) Main fuel loop (two hour cycle at feed rate of 4.21 g/min)	0.505 kg	12.2 kg
(b) Fuel reserve (8 hours of fuel feed, or 4 days of burnup)	2.020 kg	Unknown
(c) Dissolved in structure	0.030 kg	Unknown
(d) Dissolved in flibe	0.010 kg	In lithium 12.2 kg
(e) Entrained in helium	0.007 kg	
(f) Release rate	8 Ci/day	10 Ci/day
4. Doubling time (without compounding)	121 days	2–3 months
Energy amplification	67.8%	17.8%
D. Vacuum chamber:		
1. Nominal wall area	1563 m^2	2830 m^2

TABLE 5.   Comparison of the Princeton Reference Design Power Plant with the University of Wisconsin UWMAK-I *(Continued)*

	Princeton reference design	UWMAK-I
2. Actual wall area, main chamber	1635 m²	Unknown
3. Divertor wall area	1080 m²	Unknown
4. Mass of radiation shield wall, vacuum wall, and divertor	204 Mg	Unknown
E. Breeding blanket:		
1. Dimensions	74 cm thick, 1378 m³	
2. Average composition:		73.4 cm thick
PE 16	9.4% 129m³, 1.03 Gg	316 ss, 6.25 Gg
Flibe	69.8% 982 m³, 1.86 Gg	Li 1.42 Gg
Helium	20.8% 287 m³, 905 kg	
F. Primary shield:		
1. Dimensions:	80 cm thick, 1563 m³	77 cm thick
2. Average composition		
Heavy concrete	90%	316 ss, 2.03 Gg
Steel	5%	34 C, 1.62 Gg
Helium	5%	He, 666kg
3. Total mass	9.03 Gg	20.05 Gg
4. Dose rates		
While operating	3690 rem/hr	Unknown
While shut down	27 m rem/hr	Unknown
5. Biological shield	2 m additional concrete	Unknown

*MPa = mega pascal; 1 Pa = 10 dynes/cm².

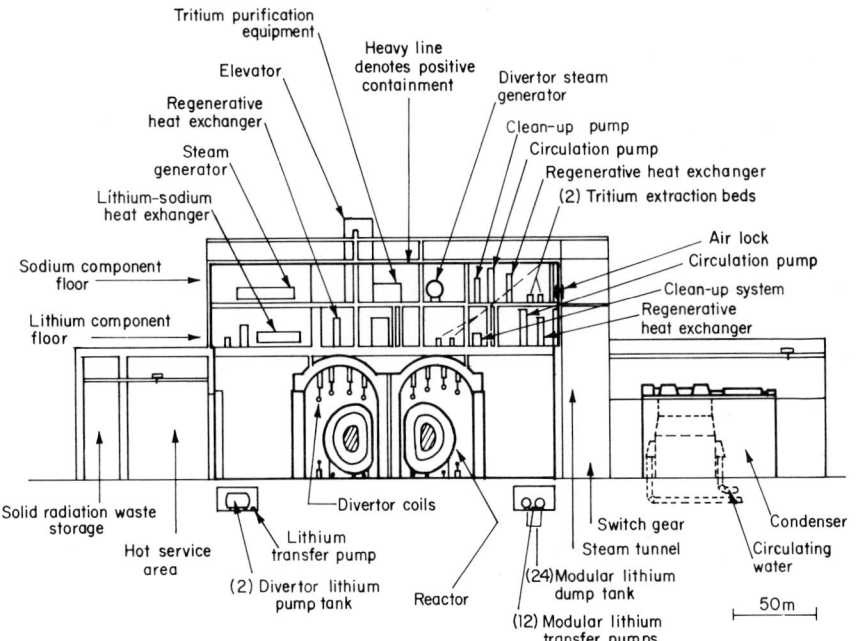

Fig. 10. University of Wisconsin reference design for a fusion reactor.

24. Lubell, M. S., W. F. Gauster, K. R. Efferson, A. P. Fraas, H. M. Long, J. N. Luton, C. E. Parker, D. Steiner, and W. C. T. Stoddart: *Plasma Phys. and Controlled Nucl. Fusion Res.*, **3**, 433–446, International Atomic Energy Agency, Vienna, 1971.

25. Fietz, W. A., P. S. Swartz, and E. F. Mains: Applied Superconductivity Conference, Oakbrook, Ill., September 1974.

26. File, J., R. G. Mills, and G. V. Sheffield: *IEEE Trans. Nucl. Sci.* **NS-18**, 277 (1971).

27. File, J., and G. V. Sheffield: *Proc. 4th Int. Conf. Magnet Technol.*, Brookhaven National Laboratory, pp. 240–243, 1972.

28. File, J.: Proposed Superconducting Coils for the Princeton Fusion Reactor, *Proc. 5th Symp. Eng. Probl. Fusion Res.*, pp. 90–95, IEEE, New York, 1973.

29. File, J.: Superconducting Magnets and Cryogenic Systems, in R. G. Mills (ed.), *Princeton Univ. Plasma Phys. Lab. Rep.* MATT-1050, pp. 323–361, 1974.

30. File, J., G. D. Martin, and K. E. Wakefield: *Princeton Plasma Phys. Lab. Rep.* MATT-794, 1970.

31. File, J., and P. Bonanos: *Princeton Plasma Phys. Lab. Rep.* MATT-898, 1972.

32. Taylor, C. E., T. J. Duffy, T. L. Rossow, D. R. Barnum, H. H. Sexton, and R. L. Leber: *IEE Trans. Nucl. Sci.*, **NS-18**, 69 (1971).

33. Henning, C. D., R. L. Nelson, M. O. Calderon, A. K. Chargin, and A. R. Harvey: In K. D. Timmerhaus (ed.), "Advances in Cryogenic Engineering," Plenum, New York, 1969.

34. Gauster, W. F., and D. L. Coffey: *J. Appl. Phys.*, **39**, 2647 (1968): also *Proc. 1968 Summer Study on Superconducting Devices and Accelerators*, Brookhaven National Laboratory, **3**, 929 (1969).

35. Efferson, K. R., D. L. Coffey, R. L. Brown, J. L. Dunlap, W. F. Gauster, J. N. Luton, and J. E. Simpkins, *IEEE Trans, Nucl. Sci.* **NS-18**, 265 (1971).

36. UWMAK-I, "A Wisconsin Toroidal Fusion Reactor Design," *Univ. Wisconsin Rep.* UWFDM (1973).

Uranium Resources and Processing

BY L. W. VERMEULEN*

The phenomenon of radioactivity, first recognized only as recently as 1896, when Becquerel noted that uranium compounds emit radiations resembling x-rays, themselves discovered only a short while earlier, soon received intensive study in many countries, and by 1904 Rutherford, in his book "Radioactivity" was able to write, "There is reason to believe that an enormous store of latent energy is resident in the atoms of radioactive elements." Only uranium, of the elements found in nature, was originally considered to be radioactive, but work, notably in France by the Curies, showed that natural uranium contains notable amounts of other substances more strongly radioactive than uranium which has been purified. The elements polonium and radium were subsequently discovered in natural uranium, followed by the other elements in the radioactive decay chain.

The two most commonly recognized radioactive elements are uranium and thorium; of these the utilization of uranium has received most attention.

Natural uranium contains three isotopes, of identical chemical properties, but of differing atomic weights, caused by slight differences in the composition of their atomic nuclei. In all natural uranium, 99.28 percent of the atoms consist of ^{238}U; 0.71 percent of ^{235}U; and less than 0.01 percent of ^{234}U. The radioactive decay of an atom of ^{235}U causes the emission of neutrons from its nucleus; these may then collide with another nucleus, of either ^{235}U or ^{238}U, causing the emission of further neutrons. In the case of neutron collision with a ^{235}U nucleus, the whole nucleus breaks down into the nuclei of atoms of elements of lower atomic weight, with the emission of more neutrons and of considerable amounts of heat. When a ^{238}U nucleus is struck by a neutron, the neutron may be captured, forming a new isotope of uranium, ^{239}U. This decays radioactively into a new element, neptunium ^{239}Np, and then into a further new element, plutonium ^{239}Pu. Like ^{235}U, ^{239}Pu is a fissile element, capable of breakdown into simpler elements with emission of neutrons; ^{238}U is a fertile element, capable of breeding other elements.

Depending on the spatial relation of the atoms, the breakdown of the fissile elements ^{235}U and ^{239}Pu can be made to take place in a self-sustaining chain reaction, when large amounts of heat will be continuously emitted. The rate of breakdown and therefore of heat emission can be moderated by surrounding the uranium with neutron-absorbing substances, and the process can then be used as a steady source of heat to convert water into steam and so produce power (the basis of nuclear power plants). If the reaction is not controlled, and if a sufficient mass is present to prevent excessive heat loss by radiation, an explosive situation in effect is produced. In the case of natural uranium (U), the concentration of ^{235}U present is too small to produce enough neutrons to give this result, but uranium enriched in ^{235}U must be handled with care.

^{235}U is the only fissile material occurring in nature, while ^{238}U and ^{232}Th (thorium), both of considerably greater abundance, are naturally occurring fertile materials. ^{232}Th, like ^{238}U, captures a neutron, and it then becomes ^{233}U, another fissile material. Its use in a chain reaction has not yet been developed to any extent, and natural uranium remains the prime source of both fissile and fertile materials.

URANIUM RESERVES AND OUTLOOK

Uranium has been known to be a distinct element since 1789, but apart from the small amount of its salts used in yellow pottery glazes, it remained more or less a laboratory curiosity until the 1920s, when the treatment of uranium ore for the recovery of its contained radium (for the treatment of cancer) began in Czechoslovakia and the Belgian Congo (now Zaire), followed by Canada in 1933. The separated uranium was mostly stockpiled or discarded.

*Metallurgical Engineer, Rio Algom Mines Limited, Toronto, Canada.

After the development and successful explosion of the atomic bomb toward the end of World War II, an urgent search for workable uranium deposits was set in motion all over the world. The only high-grade deposits known to the western world were those in the countries just named as radium sources, but in view of the limited demand previously, serious exploration for uranium had never been undertaken up to that time. However, the offer by the U.S. Atomic Energy Commission (AEC) of contracts for fixed quantities at stated prices stimulated exploration for this hitherto largely ignored material.

Uranium is actually widely distributed throughout the world, and most countries willing to pay the cost of production could find sufficient uranium in low-grade deposits within their borders to meet their own requirements, whether for weapons or for power generation. As uranium has been in free supply in recent years, however, most countries requiring uranium have been purchasing it from previously developed higher-grade deposits, notably those in the United States, Canada, and South Africa, where it is found in the gold ores of the Witwatersrand, as well as those more recently developed in France and French territories. No information is available about sources or supplies in countries behind the Iron Curtain; it has, however, been general policy in the Western world until recently not to sell uranium to countries under Communist domination, and it is therefore obvious that there must be significant uranium production in some parts of the Communist world, since the Peoples Republic of China as well as the U.S.S.R. have developed atomic weapons, and the U.S.S.R. at least has built nuclear power stations.

During 1974–1975, the outlook, after nearly 20 years of weak uranium demand, became confused. Assured supplies for power reactors already installed, under construction, or committed, appeared to be inadequate, which led to panic spot purchases at prices up to $40 per pound. Against this, construction of previously planned reactors was being postponed because of the economic situation and because of environmental restraints. As prospecting is again being actively pursued, because of the higher prices, supplies may well prove to be adequate after all.

WORLD RESOURCES OF URANIUM

The U.S. Energy Research and Development Administration has estimated reasonably assured uranium reserves in the United States (as of January 1, 1974) to be 277,000 short tons of U_3O_8 contained in 129×10^6 tons of ore with an average grade of about 0.21 percent U_3O_8 and recoverable at a cost of $8.00 per pound or less (1974 dollars). Potential reserves at $8.00 per pound U_3O_8 are estimated at 450,000 tons.

As of early 1975, estimates have been made of resources in countries lying outside the Iron Curtain, but including Yugoslavia, in short tons of U_3O_8, and these figures are given in Table 1.

The potential resources include about 150,000 tons contained in copper and phosphate deposits, from which uranium is not at present being recovered. With the exception of France, in countries outside the Iron Curtain, the production of uranium is mostly in the hands of private enterprise. There has been a buyers' market in uranium since the very early 1960s, and only previously existing low-cost private producers have therefore remained in production; some new mines have quite recently commenced operation, but considerable increases in production cannot be looked for until potential producers are assured of continuing prices high enough to encourage exploration and the construction of new facilities, as well as a sympathetic political climate. This statement is, of course, not applicable to countries where production is state-controlled; in such countries operations are apt to be continued, regardless of cost.

In the 1973 edition of "Uranium—Resources, Production and Demand," prepared by the Nuclear Energy Agency of the Organization for Economic Cooperation and Development (OECD) and the International Atomic Energy Agency (IAEA), current and estimated reserves of world uranium, both at current price ranges of less than $10 (1972 United States dollars) per pound of U_3O_8 and a projected price in the $10 to $15 range, are surveyed. See Table 2. In each price category, the reserves have been grouped into two categories: (1) reasonably assured resources (RAR), defined as "uranium occurring in known ore deposits of such grade, quantity, and configuration that it can, within the given price range, be economically recovered with currently proved mining and processing technology," and (2) estimated additional resources (EAR), defined as "uranium surmised

TABLE 1.   Estimate of Uranium Resources in Countries Lying Outside the Iron Curtain, but including Yugoslavia
Short Tons of U_3O_8

Cost of production per pound (1972 dollars)	Reasonably assured	Additional potential
Under $10	765,000	470,000
Under $15	1,375,000	775,000
Under $30	1,760,000	1,740,000

to occur in unexplored extensions of known deposits, or in undiscovered deposits in known uranium districts, which is expected to be economically exploitable in the given price range."

Because of the uncertainty regarding uranium supplies which was mentioned above, and which will be reflected in due course in the economics of production, Tables 1 and 2 have not been revised.

CONSUMPTION

Recent actions of the Middle East oil-producing countries have considerably stimulated interest in nuclear power production, but environmental concerns and technical problems will limit immediate construction of additional reactors in most Western countries. Demand for uranium as fuel before 1980 is therefore unlikely to exceed the requirements estimated before the energy crisis arose, even in Japan, which is probably more vulnerable to fossil fuel shortages than any other industrialized country. With these qualifications in mind, the following estimates are probably reasonably valid:

Annual world demand,[*] currently about 23,000 short tons of U_3O_8, will probably increase to nearly 80,000 tons by 1980, and to about 130,000 tons by 1985.

As a result, the supply of uranium in currently known or probable resources of $10 per pound price (1972 dollars) will be exhausted by 1980 and supplies of $15 per pound uranium by 1985.

In the United States alone, the number of reactors in operation, 42 in 1974, producing 25 GW(e), is projected, on the basis of orders already placed, to increase to 183, with capacity of 169 GW(e), in 1986. The urgency of exploration for additional resources therefore becomes apparent.

URANIUM PROCESSING AND PRODUCTS

Recovery from Ore.   Two basic processes are used for the extraction of uranium from its ores: namely, an acid leach (sulfuric acid is employed for reasons of cost); or an alkaline leach, utilizing a mixture of sodium carbonate and bicarbonate. From the alkaline leach solution, after separation of insoluble impurities, uranium is recovered as sodium diuranate by addition of caustic soda. From acid solutions the uranium is separated by solvent extraction or by adsorbing it on ion exchange resins. The uranium is then further concentrated from the loaded solvent or resin by stripping these with an acid or salt solution, and is recovered by being precipitated as ammonium diuranate by addition of ammonia. Both sodium and ammonium diuranates are yellow solids, which are filtered off and dried. Depending on the impurities extracted with the uranium, the dried product would contain 70 to 90 percent U_3O_8; additional processes before the diuranate precipitation stages to remove many of these impurities are customary, and generally result in a final mine product containing 85 to 88 percent U_3O_8.[†] See Fig. 1.

Refining.   For use in nuclear reactors, uranium, in whatever chemical combination it is employed, is required to be in a state of high purity. Most impurities permitted must lie within the parts per million range. A typical specification calls for maximum contents of manganese and molybdenum under 1.0 ppm; silver under 0.5 ppm; boron, cadmium, and

*Reference: "Uranium Resources, Production, and Demand," Organization for Economic Cooperation and Development, Publications Center, Washington.

†Additional details of processing can be found in entry on Uranium by L. W. Vermeulen in the "Chemical and Process Technology Encyclopedia," Douglas M. Considine, Editor-in-Chief, McGraw-Hill, New York, 1974.

TABLE 2. Estimated World Resources of Uranium
Thousands of short tons of U_3O_8*

Price range	$10/pound U_3O_8		$10–15/pound U_3O_8	
Country	Reasonably assured resources (reserves)	Estimated additional resources	Reasonably assured resources (reserves)	Estimated additional resources
Argentina	12	18	10	30
Australia	92	102	38.3	38
Brazil	—	3.3†	0.9	
Canada	241	247	158	284
Central African Republic	10.5	10.5		
Denmark (Greenland)	7.0	13		
Finland			1.7	
France	47.5	31.5	26	32.5
Gabon	26	6.5	—	6.5
India	—	—	3	1
Italy	1.6	—	—	
Japan	3.6	—	5.4	
Mexico	1.3	—	1.2	
Niger	52	26	13	13
Portugal (Europe)	9.3	7.7	1.3	13
(Angola)			—	17
South Africa	263	10.4	80.6	33.8
Spain	11	—	10	
Sweden			351	52
Turkey	2.8	—	0.6	
United States	337	700‡	183	300
Yugoslavia	7.8	13		
Zaire	2.3	2.2		
Total (rounded)	1130	1190	885	820

*From data as available January 1973.
†Plus 90,000 short tons by-product from phosphates.
‡Plus 90,000 short tons by-product from phosphate and copper production.

ytterbium under 0.2 ppm; and dysprosium under 0.1 ppm. Other rare earth metals are also undesirable.

Material of such high purity is produced by redissolving diuranate in nitric acid, adjusting the acidity, and extracting the uranium into an immiscible solvent, leaving the impurities in aqueous solution. The uranium may then be reextracted into aqueous

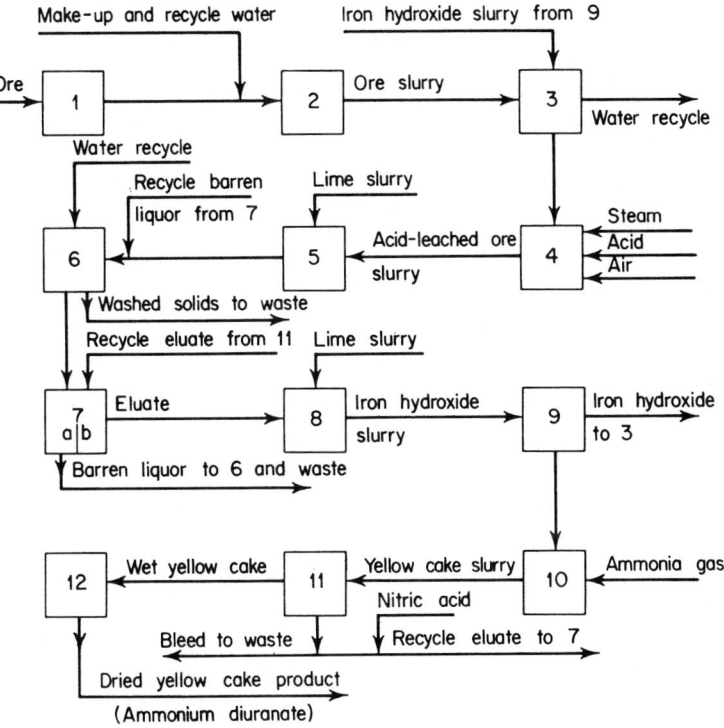

Fig. 1. Schematic representation of materials flow for uranium concentration by acid leach and ion exchange: (1) Dry crushing. (2) Grinding. (3) Dewatering. (4) Leaching. (5) Partial neutralization. (6) Solid-liquid separation and washing. (7) Ion exchange: *(a)* uranium absorption; *(b)* uranium elution. (8) Iron precipitation. (9) Solid-liquid separation. (10) Uranium precipitation. (11) Solid-liquid separation. (12) Drying. Nitric acid is added after step 11 to adjust pH of recycle eluate.

solution and recovered by evaporation as uranyl nitrate, or by again precipitating as diuranate. Both uranyl nitrate or ammonium diuranate can be converted into uranium oxide by calcining at comparatively low temperature.

Conversion to Desired Uranium Compound.   For use in power generation, the refined uranium compound is converted into metallic uranium or into a number of other compounds, such as the dioxide or carbide. If the uranium is to be enriched in its [235]U content, it is first necessary to convert it into the hexafluoride, which is a solid at ordinary temperatures but sublimes at 56°C under atmospheric pressure.

Uranium Dioxide.   If ammonium diuranate is calcined in air, it is converted into the oxide U_3O_8, which is the essential composition of the mineral uraninite (pitchblende) found in nature. By calcining in a current of hydrogen, however, at about 800°C, the dioxide UO_2 is formed. This is a fine black powder, and must be cooled under hydrogen or preferably carbon dioxide to prevent its reoxidation. If the reduction is not permitted to go to completion, so that the powder has a composition of approximately $UO_{2.09}$, the product is easier to handle; the completely reduced powder is pyrophoric and catches fire very readily.

UO_2 is used in power reactors in the form of pellets, formed by compressing the powder under high pressure and then sintering the "green" pellets at high temperature under hydrogen. The pellets are usually made slightly oversize, and the sintered pellets are ground to size in centerless grinders.

Uranium Tetrafluoride.   Uranium dioxide can be converted to uranium tetrafluoride, UF_4, by calcining the oxide in a stream of hydrogen fluoride containing some hydrogen. The product ("green salt") is a dark green powder. It can be produced from ammonium diuranate in the same manner as from the oxide.

Uranium Hexafluoride.   Uranium hexafluoride, UF_6, is produced from UF_4 by calcining the UF in a stream of mixed HF and fluorine gases, or with chlorine trifluoride. The hexafluoride is a yellow solid which is a vapor at temperatures above 56°C and is collected by cooling the vapor to a lower temperature. It reacts with water to form uranyl fluoride UO_2F_2 and HF, and must therefore not be exposed to water or water vapor.

Metallic Uranium.   The metal is made from the tetrafluoride by reacting it with metallic sodium, calcium, or magnesium, at high temperatures. The fluorides of the reacting metal are also formed and are separated mechanically from the uranium metal after the whole mass has solidified.

URANIUM ENRICHMENT

Most power reactors (and all weapons) require uranium enriched in ^{235}U above the natural content of 0.72 percent. The methods found practicable depend on the small differences between the densities of the vapors of the hexafluorides of ^{235}U and ^{238}U. The separation is done either by diffusion through porous diaphragms, or by centrifuging the gases. Many stages of separation are necessary because the density differences are very small, and because of high power requirements the process is very expensive. For power reactors, enrichments to ^{235}U content of 3 to 5 percent are adequate, but by using a sufficient number of stages a 100 percent content can be achieved.

After the desired degree of enrichment has been achieved, the enriched fraction is reacted with water to form UO_2F_2, precipitated as ammonium diuranate, and converted to the metal or to any required chemical compound by the same methods previously noted for natural uranium.

Care is required in the handling of more than minimal quantities of enriched uranium, to ensure that less than a critical mass is present in a single container. The actual critical mass in a given situation depends on the degree of enrichment, as well as the effect on a given mass of the shape of the containing vessel and on its environment.

As far as is known, until recently, only the United States, the United Kingdom, France, and the U.S.S.R. have had enrichment plants, although it may be assumed that the Peoples Republic of China and some other countries have had them also. In the next few years, some of the European Common Market countries and Japan will certainly be added to the known list, and possibly Canada, Australia, South Africa, and India. The techniques used in enrichment, although theoretically simple, require vast capital expenditures and consume enormous amounts of power. Enrichment plants to produce even ^{235}U contents still below 10 percent are beyond the financial and technical capabilities of most nations. This is likely to discourage other possible entrants into the field until some breakthrough leads to the development of a completely new process.

All existing enrichment plants have been either built or sponsored by the respective governments, but some at least of those projected will be constructed by private enterprise, although, because of military considerations, a measure of governmental control is certain to be exercised over them.

The U.S. Energy Research and Development Administration (successor to the Atomic Energy Commission) had received four proposals for new plants by its October 1, 1975 deadline. Three of these were for centrifuge plants and one for a diffusion plant.

Little has been reported further to the AEC's late 1971 meeting with representatives of 13 countries on possible foreign access to its gaseous-diffusion technology, although it now appears likely that the first new plant to be based on United States technology will be built in the United States. Japan has signed an agreement to study a possible United States–Japan enrichment joint venture, and also similar agreements with France and Australia, for the possible location of a diffusion plant in the Pacific area based on French technology. In Canada, a feasibility study of the possibility of building an enrichment plant there, using United States technology, has continued.

In Europe, industrial and state-owned companies of Belgium, Britain, Italy, the Netherlands, West Germany, and later Sweden, joined France for a two-year economic feasibility study (designated EURODIF) of a six to seven million separative work unit (SWU) per year diffusion plant in western Europe.

One of the two companies set up under the 1970 British–Dutch–West German Tripartite Agreement, has designed a 300,000 SWU per year centrifuge enrichment plant; production is scheduled for 1976. Meanwhile, the national plants of the three countries were in various stages of construction and commissioning. At year end, the "troika" set up a "study association" to make classified information on the gas centrifuge process available to agencies and enterprises interested in examining the economic, technological, and organizational possibilities for establishing additional centrifuge enrichment capacity.

Of considerable significance to the world enrichment situation, was the AEC's announcement March 7, 1972, that it planned to continue operating its enrichment plants, after June 30, 1973, with a tails-assay (0.275 to 0.30 percent ^{235}U) higher than the assay (0.20 percent ^{235}U) used as a basis for transactions with enrichment customers. The plants have been operating at a tails-assay of 0.30 percent since July 1, 1971. The higher tails-assay will require additional uranium feed, which will be supplied from the AEC surplus uranium stockpile. Contracting with enrichment customers, however, will continue to be on the basis of a 0.20 percent tails-assay. Under this arrangement, customers will provide less feed to produce their enriched uranium requirements at the higher operating tails-assay, but will pay for a greater number of SWUs than actually used in producing their enriched product; a portion of the cash payment will constitute payment to the AEC for such additional feed that it supplies from its stocks.

The effect of the arrangement is to decrease the requirement for separative work, allowing the AEC to further build up its inventory of enriched uranium, thus enabling the existing plants to meet the demand for enriched uranium until a later date, and deferring the time (by about a year) when new enrichment capacity will be needed. The industry had earlier been advised to plan for a tails-assay of 0.25 percent ^{235}U for transactions after fiscal year 1973.

PACKAGING AND SHIPPING OF URANIUM

Like most metals, uranium and its compounds are poisonous when ingested (the kidneys are affected), and the usual precautions against such intake should therefore be taken. The processing of uranium ores separates most of the radioactive breakdown products found in the ore, so that ammonium diuranate or other uranium compounds, when freshly prepared, contain only uranium of the radioactive decay series. The half-life of ^{238}U is 4.5 $\times$ 10^9 years; and of ^{235}U, 7.1 $\times$ 10^8 years, and so formation of breakdown products is slow, even over the number of years approaching the human lifespan. Natural uranium in any form except the hexafluoride, which is somewhat volatile and also reacts with water in any form, can be transported in unsophisticated packaging. Steel drums, with plastic liners to prevent contamination in the case of refined products, are most often employed. The hexafluoride is not often transported, but may be confined in steel bottles similar to those used for industrial gases.

As previously mentioned, in the case of uranium enriched in ^{235}U, it is essential that only subcritical masses should be transported. Special containers have been developed for such material, and many other conditions have been laid down to ensure safe handling.

URANIUM EXPLORATION

Although uranium has been known for many years, limited uses for the metal, all of them in the form of its compounds, precluded active exploration for it. Most production came as a by-product from other metal recovery operations until the development of the atomic bomb during World War II suddenly brought about an active interest in finding additional sources of the metal. Many deposits where the presence of uranium had been reported, as it were in passing, were reexamined, and active uranium prospecting commenced all over the world.

Earlier discoveries of uranium had been made visually, from hand specimens. Development of a portable Geiger counter permitted a new breed of prospectors to detect radioactivity with ease, and led to the discovery of new deposits of uranium minerals in many countries. Most of these were of low grade or limited extent, but some were the outcrops of workable deposits, which had to be further explored by diamond drilling or excavation.

After a period of low interest due to limited demand during the 1960s, active exploration resumed in 1972, using newer and more sophisticated prospecting methods developed in recent years, such as geophysical and geochemical techniques.

Geophysical Methods.   These methods detect ore bodies as such rather than individual elements, but prove useful in establishing the dimensions of an ore body. In the Elliot Lake area of Canada, uranium minerals are associated with disseminated sulfides; in these cases induced polarization methods can be useful. Mineralization associated with a fracture zone can be traced by electromagnetic or magnetic methods, and such zones may be traversed for a greater distance or to a greater depth than would be possible by radiometric or geochemical methods.

Geophysical methods specifically relating to uranium depend on the measurment of radioactivity, principally gramma radiation, but also alpha or beta radiation. Gamma radiation has a greater range than the others, up to 12 inches in solid rock, or hundreds of feet in air; and gamma radiations originating from different naturally occurring radioactive elements (potassium, uranium, and thorium) have different energies or wavelengths, which enable the originating element to be identified. In practice, neither uranium nor thorium are identified directly; gamma radiations from their respective decay products, ^{214}Bi (bismuth) and ^{208}Tl (thallium), have higher energies and travel further. It is usually assumed that the decay products are in a fixed relation to the parent element, and this is generally true of hard-rock uranium ores, such as those of Elliot Lake or South Africa, but does not necessarily hold for some of the sandstone ores of the western and southern United States, where leaching has taken place and the uranium and its decay products are not in equilibrium. For thorium ores, disequilibrium is rarely a problem.

In nature, potassium, thorium, and uranium occur together in most common rock types, and the relative amounts of each do not vary widely. Should there be any natural concentration of any of these elements, the value of gamma-ray spectrography depends as much on making relative as well as absolute measurements of uranium abundance.

The instruments generally employed for gamma-ray detection are the scintillometer, which counts all gamma radiation and still has uses in delineating areas of high and low radiation; and the gamma-ray spectrometer, which distinguishes the different types of gamma rays. These detectors can be used on the ground and also when airborne. The Geiger counter is now considered obsolete for exploration work.

Radon monitors, which are essentially alpha detectors, have also proved useful as supplementary tools in the search for uranium. Radon is found in ground air in the vicinity of uranium mineralization, and also finds its way into groundwater.

By the use of airborne detectors large areas can be explored for radiation emissions at relatively small cost by traversing such areas on quite widely separated lines. Promising areas can be examined at closer spacings, followed if necessary by ground level surveys.

Geochemical Methods.   Geochemical prospecting is based on the detection of trace amounts of elements removed from but related to the ore body being sought. Most commonly tested are soils, stream and lake sediments, surface water, soil gas, plant parts, and groundwater. In uranium, radioactive decay gives rise to a large number of daughter products; rapid analytical techniques are available for uranium, radium, and radon; and some research has been done with helium and the fission gas, krypton-85. Geochemical methods are not adversely affected by radiation from the thorium decay series or from potassium.

Radium is less mobile than uranium, and residual radium is often detected in deposits from which the uranium has been leached near the surface. Uranium is often more strongly enriched in organic material (plants) than radium, and is thereby a less reliable indicator.

Lake water samples can be collected by aircraft and analyzed for uranium or radon.

Radon determination equipment can be set up in field camps, and samples can be analyzed immediately, with rapid follow-up.

In soils radon gas is more penetrating than gamma radiation, and the radon counter therefore responds to mineralization at greater depths. Radon anomalies are found to coincide both in position and in orientation with the presence of mineralization to a greater extent than scintillometer anomalies.

Helium, an alpha particle which has picked up two electrons, is the lightest of the noble gases. It is not radioactive and is much less soluble in water than radon; it is therefore more mobile, and tends to migrate upward instead of horizontally. Since it is difficult to analyze, it has only recently been considered as a prospecting tool, but it is now receiving increasing attention for this purpose.

Lithium for Thermonuclear Fusion Reactors

BY WAYNE T. BARRETT* AND IHOR A. KUNASZ†

Lithium, element no. 3 (in increasing atomic number), is widely distributed in the earth's crust including the oceans. It is a member of the alkali metal group, having an atomic weight of 6.94. Natural lithium occurs as a mixture of two isotopes: (1) the more abundant ^{7}Li, and (2) ^{6}Li, which makes up 7.4 percent of the atomic weight of the natural substance. Although it occurs in some 145 minerals, only spodumene lepidolite, petalite amblygonite eucryptite, and brines are commercial sources of lithium. These sources will be described later.

There is little doubt that lithium will be used in thermonuclear fusion power plants. However, because of the wide diversity of design approaches and the early stage of many of these designs, it is impossible to predict with any accuracy how much lithium will be required. Initially, all fusion reactors probably will use deuterium plus tritium in the core.

$$^2D + {}^3T \rightarrow {}^4He + n + 17.6 \text{ MeV}$$

The tritium will be produced from lithium. The quantity of lithium necessary to produce the tritium is relatively small. For example; assuming 100 percent efficiency, the lithium consumption for a 1000-MW(e) fusion power plant would be approximately 200 kg per year.

Lithium will be used in the blanket surrounding the core. This probably will be liquid lithium, although it is possible it might be in the form of certain lithium compounds. The function of the lithium blanket is:

1. To bring about the breeder reaction

$$^6Li + n \rightarrow {}^4He + {}^3T + 4.8 \text{ MeV}$$

This, of course, is necessary to generate the tritium to be used in the core.

2. The blanket may also be used to bring about the reaction

$$^7Li + n \rightarrow {}^3T + {}^4He + n - 2.5 \text{ MeV}$$

This reaction produces additional tritium and supplementary neutrons so as to make up for inevitable neutron losses in the system. However, the quantity of lithium in the blanket could be reduced by the use of beryllium compounds to supply the supplementary neutrons.

3. The blanket is used for heat absorption to remove energy from the very high energy neutrons coming from the core.

Most designs involve the use of natural lithium metal in the blanket. However, it is possible that the lithium in the blanket might partially be enriched in ^{6}Li.

The quantity of lithium in the blanket might be as small as 50 metric tons for a 1000-MW(e) fusion power plant. This assumes the minimum use of lithium and the use of beryllium or some other light element for the production of supplementary neutrons. On the other hand, the quantity of lithium in the blanket might be 500 metric tons or even somewhat higher if it is assumed that ^{7}Li will be utilized for the production of tritium and supplementary neutrons. This quantity, again, is for a 1000-MW(e) fusion power plant.

It is possible that additional lithium might be employed as the cooling fluid to transfer heat out of the blanket to the steam-producing portion of the overall power plant.

Another uncertainty is how many fusion power plants wil be in operation at some future time. If the fusion power system is a technical and economical success, it will certainly be used for an appreciable fraction, but probably not all, of the total United States electric power requirements. Thus, within a few decades of the year 2010 we might expect to have as many as 500 one-thousand-MW(e) fusion power plants in operation. This corresponds to the total United States electric power production for the year 1970.

*President, †Chief Geologist, Foote Mineral Company, Exton, Pa.

Using this number as a reasonable guess, the quantity of lithium necessary would range from 25,000 to 250,000 tons. The lower quantity could be supplied readily by the present lithium industry from its existing reserves. The higher quantity would require the development of additional lithium sources.

LITHIUM RAW MATERIALS

Spodumene, a lithium aluminum silicate ($LiAlSi_2O_6$), is a monoclinic member of the pyroxene group. It has a very pronounced cleavage plane (110) which results in typically lath-shaped particles upon breaking. The color of spodumene is variable, being nearly white in low-iron variety, and dark green in iron-rich crystals. Sources of spodumene are described later.

Spodumene constitutes the most abundant commercial source of lithium. Theoretically, it may contain up to 8.03 percent Li_2O, but the actual lithia concentrations vary from 2.91 to 7.66 percent, probably as a result of sodium and potassium substitution for lithium. Spodumene concentrates typically contain 6.3 to 7.5 percent Li_2O.

Lepidolite is a phyllosilicate with the general formula

$$K_2(Li, Al)_{5-6} Si_{6-7}Al_{2-1}O_{20} (OH, F)_4$$

The lithia concentration in lepidolite varies between 3.3 percent and a possible theoretical maximum of 7.74 percent. In commercial deposits, the concentrations are more normally 3 to 4 percent Li_2O. In addition to lithium, lepidolites also carry substantial concentrations of rubidium (0.91 to 3.80 percent Rb_2O) and cesium (0.16 to 1.90 percent Cs_2O).

The major commercial occurrences of lepidolite are located in Rhodesia (Bikita), Southwest Africa (Karibib), Canada (Bernic Lake), and Brazil.

Petalite, $LiAlSi_4O_{10}$, is a monoclinic mineral with a framework silicate structure. Its color is grayish white and more rarely pinkish. It has two cleavage directions which form an angle of 38.5°. The basal cleavage is perfect.

The theoretical lithia content of petalite is 4.88 percent. In actual commercial deposits, the concentration varies from 3.0 to 4.7 percent Li_2O. Sizable deposits of petalite occur with lepidolite in Rhodesia, Southwest Africa, Brazil (Aracuai), Australia (Londonderry), the U.S.S.R. (eastern Transbaikalia), and Sweden (Utö).

Eucryptite is also a lithium aluminum silicate which is deficient in silica. It has a formula $LiAlSiO_4$ and may contain 11.88 percent Li_2O. The only large deposit of eucryptite is found in Rhodesia (Bikita) where its occurrence with quartz suggests a spodumene origin. The grade of eucryptite is 5.03 percent Li_2O.

Amblygonite, with the generalized formula $LiAl(PO_4)$ (F, OH), is the fluorine-rich end member of a phosphate series, while montebrasite represents the hydroxyl-rich end member. It occurs in white to gray masses. Basal cleavage planes are pearly; others are vitreous. Although amblygonite may contain as much as 10.2 percent Li_2O, commercial ores usually carry 7.5 to 9 percent Li_2O. Amblygonite has been mined in Canada, Brazil, Surinam, Rhodesia, Ruanda, Mozambique, Southwest Africa, and the Republic of South Africa.

Brines. Lithium is found in commercial quantities in certain brine deposits. The brines are present in desert areas and occur in playas and saline lakes where solutions have been concentrated by solar evaporation. In Searles Lake (California) where production of dilithium phosphate began in 1938, the lithium concentration is 0.015 percent Li_2O. In Clayton Valley (Nevada), lithium-bearing brines contain 0.065 percent Li_2O. Smaller concentrations of lithium (0.009 to 0.012 percent Li_2O) are found in the Great Salt Lake (Utah).

LITHIUM METAL

Lithium metal is the lightest of all metals, having a density of 0.54 gram per cubic centimeter. Lithium metal is readily produced from anhydrous lithium chloride by the fused salt electrolysis of the eutectic mixture of potassium chloride and lithium chloride. It is estimated that the presently installed capacity of the lithium industry to produce lithium metal is somewhat less than 500 tons per year.

The commercial lithium metal presently produced has a typical analysis:

Lithium	99.9%
Sodium	0.006%
Potassium	0.004%
Calcium	0.018%

This should be of sufficient purity to be used in thermonuclear fusion reactors.

Lithium chloride is produced by the reaction of hydrochloric acid with lithium hydroxide or with lithium carbonate. The aqueous lithium chloride solution must be evaporated for the preparation of dried anhydrous lithium chloride. The mid-1974 price for lithium metal was $9.50 per pound.

The ^6Li isotope has been separated from natural lithium at an Atomic Energy Commission (U.S.) plant at Oak Ridge, Tennessee. A published AEC estimate (1974) indicates that the cost of enriching ^6Li (93 percent enrichment) is $200 per kilogram.

LITHIUM CHEMICALS

Lithium chemicals of commercial significance are:

Lithium carbonate, which is produced from spodumene or from brines. Lithium carbonate is used by the glass, ceramic, and aluminum industries.

Lithium hydroxide. Lithium hydroxide monohydrate is generally produced from lithium carbonate. Its chief use is in the manufacture of high-quality lubricants.

Lithium bromide is produced from either lithium hydroxide or lithium carbonate. Its major use is in the form of an aqueous solution, as employed in absorption-type air conditioning equipment.

Lithium chloride is produced from either lithium hydroxide or lithium carbonate. Its major end use is in various fluxes and in dehumidification systems.

Lithium butyl is produced by the reaction of lithium metal with butyl chloride. Its major use is as a catalyst for the preparation of synthetic rubber.

The consumption of lithium chemicals in all forms has been growing at the annual rate of 10 to 15 percent for the past decade. The United States production of lithium chemicals in 1973 corresponds to about 38 million pounds, expressed as lithium carbonate equivalents. This corresponds to approximately 4000 tons of lithium.

NORTH AMERICAN RESERVES OF LITHIUM MINERALS

Presently Active Operations. There are four producing mines for lithium products in the United States. Two of these are operated by the Foote Mineral Company: one an open-pit spodumene mine at Kings Mountain, North Carolina; the other a brine deposit at Silver Peak, Nevada. Lithium Corporation of America, a wholly owned subsidiary of Gulf Resources & Chemical Corporation, operates an open-pit spodumene mine at Bessemer City, North Carolina. American Potash Company, a wholly owned subsidiary of Kerr-McGee Corporation, produces lithium from a brine deposit at Searles Lake, California. The reported reserves (measured and indicated) of lithium in these four mines are as follows:

Active	Reserves, tons of lithium	References
Kings Mountain, North Carolina	201,000	
Bessemer City, North Carolina	81,000	1
Searles Lake, California	42,000	2
Silver Peak, Nevada	45,000	
Total	369,000	

It should be noted that the figure given for Silver Peak reserves is very much less than that reported earlier in the literature. The present figure represents Foote's estimate of the amount of lithium economically recoverable as lithium carbonate by present-day technology. It represents only a small fraction of the total amount of lithium at Silver Peak, Nevada.

It should be further noted that reserves for Kings Mountain, Bessemer City, and Searles Lake represent lithium in the ground and not recoverable lithium.

Other Potential Sources.   Although there are many reported occurrences of spodumene and other lithium pegmatites in North America, the only ones of known significance are given below:

Location	Reserves, tons of lithium	References
Bernic Lake, Manitoba	68,000	3
Preissac-Lacorne, Quebec	84,000	4
Other Canadian pegmatites	124,000	4
Great Salt Lake, Utah brines	650,000	5
Imperial Valley, California Brines	930,000	6, 7
Total	1,856,000	

Sources Located outside North America.   The Bikita Mines in Rhodesia are reported to be an active source of petalite and a potential source of spodumene. Reserves are estimated at 81,000 tons of lithium contained in petalite, lepidolite, and spodumene (Ref. 8). There are also reported reserves in the Karibib district, Southwest Africa, of 16,000 tons contained in lepidolite (Ref. 9). Large reserves of spodumene are known to exist in the Manono region of Zaire. Recently, the occurrence of a lithium-bearing brine deposit has been reported in Chile.

It should be noted that none of these deposits are controlled by United States interests and, thus, may not be available to the United States as lithium sources for future thermonuclear fusion reactors.

REFERENCES

1. H. E. Uhland: Private communication, Lithium Corporation of America, 1972.
2. Norton, J. J., and D. M. Schegel: Lithium Resources of North America, *U.S. Geol. Surv. Bull.* 1027-G, pp. 325–350, 1955.
3. A. C. A. Howe; Private communication, Chemalloy Minerals, 1972.
4. Mulligan, R.: The Geology of Canadian Lithium Pegmatites, *Geol. Surv. Can. Econ. Rep.* 21, Ottawa, 1965.
5. Cummings, A. M.: Mineral Facts and Problems, *U.S. Bur. Mines Bull.* 650, Pittsburgh, Pa., 1970.
6. Koenig, J. B.: "Geological Setting of the Imperial Valley and Its Geothermal Resources," compendium of papers presented at Imperial Valley–Salton Sea Area Geothermal Hearing, Oct. 22–23, 1970, pp. E1–E5, Sacramento, Calif.
7. Hegelson, H. C.: Geologic and Thermodynamic Characteristics of the Salton Sea Geothermal System, *Amer. J. Sci.*, **266**, 129–166, March 1968.
8. Symons, R.: Operation at Bikita Minerals (Private), Ltd., Southern Rhodesia, *Bull. Inst. Mining Met.* no. 661, pp. 129–172, December 1961.
9. Anon.: "Klockner Buys SWA Lithium," *Metal Bull.*, no. 5288, p. 26, Apr. 5, 1968.

Solar Energy Technology

JAMES C. BURKE, B. S., M. S. *Arthur D. Little, Inc., Cambridge, Massachusetts. (Solar Climate Control)*

GORDON L. DUGGER, Ph.D. *Applied Physics Laboratory, The Johns Hopkins University, Laurel, Maryland. (Ocean Thermal Energy Conversion)*

JOSEPH W. GELZER, JR., B. S., M. S. *Manager of Product Development in the Software and Computer Systems Engineering organization, The Boeing Company, Seattle, Washington. (Solar Cells and Arrays)*

DANIEL L. GREGORY, B.S. *Specialist Engineer, Space-Based Power Study, Research and Engineering Division, Boeing Aerospace Company, Seattle, Washington. Member, Aerospace Industries Association of America. (Heat Engines for Solar Power) (Satellite Energy Systems)*

E. WENDELL HEWSON, Ph.D. *Professor and Chairman, Department of Atmospheric Sciences, Oregon State University, Corvallis, Oregon. (Electrical Energy from the Wind)*

A. F. HILDEBRANDT, Ph.D. *Director, Solar Energy Laboratory and Professor of Physics, University of Houston, Houston, Texas. (Solar Tower Energy Collector)*

H. OMAN, E.E. *Supervisor of Electric Power Research, Boeing Aerospace Company, Seattle, Washington. (Solar Cells and Arrays)*

LAWRENCE F. SMALL, Ph.D. *Professor, School of Oceanography, Oregon State University, Corvallis, Oregon. Member, American Association for the Advancement of Science, American Institute of Biological Sciences, American Society of Limnology and Oceanography, Ecological Society of America. (Photosynthesis)*

L. L. VANT-HULL, Ph.D. *Associate Professor of Physics and Solar Thermal Program Manager, Solar Energy*

Laboratory, University of Houston, Houston, Texas. (Solar Tower Energy Collector)

JESSE D. WALTON, JR., B.S. *Technical Manager, Solar Energy Programs, Applied Sciences Laboratory, Engineering Experiment Station, Georgia Institute of Technology, Atlanta, Georgia. (High-Temperature Solar Energy)*

GORDON R. WOODCOCK, B.S.,M.S. *Engineering Supervisor, Future Space Transportation Systems Analysis, Research and Engineering Division, Boeing Aerospace Company, Seattle, Washington. Member, Aerospace Industries Association of America. (Heat Engines for Solar Power) (Satellite Energy Systems)*

NOTE: *Tidal power is discussed in the* Hydropower *section of this Handbook.*

Heat Engines for Solar Power

BY GORDON R. WOODCOCK* AND DANIEL L. GREGORY†

A solar thermal power plant requires a conversion system to transform the collected and concentrated solar light energy into electricity. The power conversion system (engine) must include a means of converting light energy into heat energy (the absorber), of transporting heat between conversion system elements, of converting heat energy into electricity, of rejecting unusable heat to the environment (the cooler); and may include a storage system to deliver energy for use when sunlight energy is not available.

This discussion is directed primarily to solar heat engines as they may be applied to generation of utility electric power in the future. There is, of course, no technological barrier in the way of constructing a solar electric plant, unless one wishes it to produce power at an economically competitive cost. As of 1976, there are no operating solar electric utility power systems; and none are under construction. No convincing evidence has been developed that a solar plant could compete economically with conventional or nuclear plants, but recent studies of thermal engine solar plants sponsored by the National Science Foundation (Washington, D.C.), particularly those of the towertop concept (see "Solar Tower Energy Collector" covered later in this Handbook section), have begun to show promise of economical generation of solar electric power. Economic aspects of design and selection of heat engines for solar power plants are of principal importance.

ECONOMIC CRITERIA

Studies of the economics of solar power have provided general economic criteria and guidelines for plant design. These are highlighted in the several following paragraphs.

Plant Type.   Utility electric power plants are classed as (1) *base load,* (2) *intermediate load,* and (3) *peaking systems.* Base-load plants provide the majority of total energy demand and operate continously except during outages for planned maintenance or due to equipment failure. Intermediate-load plants operate during normal load demand periods, but not at low demand. Peaking plants operate only during periods of peak or near-peak demand. Economic criteria for each are designed to enable the power system to meet energy demand at the lowest possible cost. Thus, base-load plants aim for the combination of capital cost and efficiency that provides minimum busbar power cost, while peaking plants provide relatively little energy compared to their capacity and tend toward minimum capital cost.

Solar plants of any type are expected to have high capital cost. They will, therefore, serve primarily base-load or intermediate-load needs. Solar plants may be considered as either (1) *fuel displacement* or (2) *capacity displacement* systems. A capacity displacement solar plant provides generation capacity in lieu of conventional plants. It must be able to operate on demand and, therefore, must incorporate energy storage in order to provide power when the sun is not shining. The economic value of capacity displacement translates to an allowable solar plant capital cost of about $1000 to $1500 per kW of installed capacity (1975 dollars), including the storage system. A fuel displacement plant operates only when the sun shines and displaces (i.e., reduces consumption) of fossil fuels. It does not reduce requirements for conventional generating capacity. The value of fuel displacement, based on cost savings in fossil fuel consumption, translates to about $150 to $300 per kilowatt of solar electric output capacity, the upper end of the range giving some credit for peaking capacity displacement in areas where peak loads are correlated with sunshine.

Availability.   Widespread electric power outages caused by failure of a total generation and distribution network are rare enough to make national headlines when they occur.

---

*Engineering Supervisor, Future Space Transportation Systems Analysis, Research and Engineering Division, and †Specialist Engineer, Space-Based Power Study, Research and Engineering Division, Boeing Aerospace Company, Seattle, Wash.

Utilities aim for, and generally attain, a system availability of 0.9997 or better, equivalent to one or fewer major outages in ten years. Individual generation plants of the traditional type have availabilities on the order of 0.95. Unplanned shutdowns (forced outages) of individual plants do not shut down the system because a portion of available capacity is held in reserve. Some of this is in rotating reserve, i.e., plants operating at no load but ready to pick up load immediately in the event of a forced outage elsewhere in the system. High system availability is achieved through reserve capacity because forced outages of conventional (and nuclear) plants are random in nature. However, outage of solar plants due to loss of sunlight will be correlated. When the sun goes down, all the plants shut off. Therefore, solar plants intended to operate as a significant part of a utility system, displacing either base-load or intermediate-load capacity, must incorporate enough storage capacity so that the plant availability, with respect to loss of sunlight (e.g., due to a spell of bad weather), equals the system availability criterion. The storage system capacity requirement will depend upon regional weather statistics, but even in parts of the United States most suited for solar energy, several days capacity will undoubtedly be required.

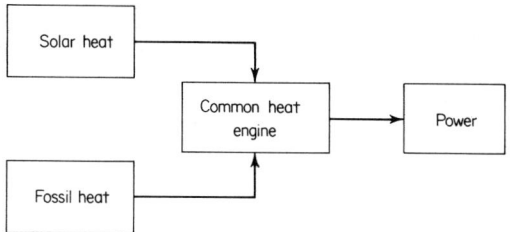

Fig. 1. Fuel replacement concept.

Hybrid Plants. Fuel displacement solar plants are to be economically justified by the value of fossil fuel saved, will not displace conventional capacity, and do not require storage. There are cost benefits in hybrid solar/fossil plants employing the common heat engine principle shown in Fig. 1, assuming the solar heat source meets the fuel-displacement economic criteria. (Ref. 1) The hybrid plant, incorporating solar heat for fuel displacement is then capable of capacity displacement. The heat engine must be compatible with both solar and fossil heat sources. If the relative economics of solar power continues to improve along trends of current studies and developments, the hybrid plant may be one of the first types to become economically practical.

Efficiency and Temperature. Heat engines are limited in efficiency to a maximum ideal of the Carnot efficiency determined by cycle heat input and rejection temperature,*

$$\eta_I = \frac{T_1 - T_2}{T_1} \qquad (1)$$

where T_1 and T_2 are input and rejection temperatures, respectively, in absolute units. Real heat engines approach the ideal to varying degrees, depending on cycle type and efficiency of the components.

High cycle efficiency is extremely important. Half or more of the cost of a solar heat source is expected to be in the cost of concentrator mirrors. The total reflecting area of mirrors required may be approximated as

$$A = \frac{P}{I \eta_c \eta_e} \qquad (2)$$

where P is plant output required; I is solar insolation in kilowatts per square meter, typically 0.8; and η_c and η_e are concentration-absorber and cycle efficiencies. In a hybrid plant, the heat engine must be competitive with those used in modern conventional plants, typically 40 percent efficiency or better, or else the solar portion of the plant must be economical enough (unlikely!) to overcome the fuel consumption excess due to inefficient use of the fossil fuel heat energy.

*Nomenclature—symbols and subscripts, defined in Table 1.

TABLE 1.   Nomenclature—Symbols and Subscripts

SYMBOLS

A	Area
C	Optical concentration ratio (geometric), cost: Eq. (12) only
Cp	Specific heat in kJ/(kg)(°K): 1 kcal = 4.186 kJ
D	Diameter in meters
F	View factor: see Eq. (4)
f	Fraction of absorber (cavity internal) wall area utilized for active heat exchanger
$f\hat{u}$	Utilization factor
G	Pressure drop factor
h	Heat-transfer coefficient in kW/(m²)(°K)
I	Solar radiation intensity in kW/m², measured in a plane normal to the incident sunlight, and direct only (i.e., excludes the sky brightness or diffuse component) unless otherwise stated.
I_o	Mean solar radiation intensity above the Earth's atmosphere (air mass zero), 1.353 kW/m²
k	Thermal conductivity in kW/(m)(°K)(sec)
Nu	Nusselt's number
P	Power, in kilowatts
Pr	Praundtl number
$\dot{q}$	Heat-transfer rate, kW ^{2}
R	Ratio of cavity interior wall area to aperture area
Re	Reynolds number
r	Radius in meters
S	Length in meters
T	Temperature, in absolute degrees unless otherwise stated
u	Velocity in m/sec
α	Absorbtivity, fraction of incident radiant energy absorbed
γ	Specific heat ratio (for a gas)
ϵ	Emissivity, fraction of radiant energy emitted as compared to ideal black body
ζ	$= (\gamma - 1)/\gamma$
η	Efficiency
μ	Fluid viscosity, kg/(m)(sec²)
ρ	Density in kg/m³
σ	Stefan-Boltzmann constant, 5.6697×10^{-11} kW/(m²)(°K⁴)
τ	Storage capacity factor

SUBSCRIPTS

a	Absorber
aw	Adiabatic wall
b	Compressor (blower)
c	Concentrator
d	Output (delivery)
e	Engine (cycle)
f	Flame (combustion)
g	Gas; also generator
H	Reflector (Heliostat or concentrator)
i	Enumerating index
I	Ideal input
o	Initial or unmodified condition
p	Power plant
r	Recuperator
T	Turbine
w	Wall
$\odot$	Sun
0, 1, 2, 3	—Designate points in a thermodynamic cycle at which the state is described.

Figure 2 shows the variation of ideal efficiency with cycle top temperature, assuming a rejection temperature of 320°K(116°F).

Also shown are typical real cycle efficiencies, temperature ranges for various solar collector/concentrator types, and representative fossil heat source efficiencies considering open cycle flue gas loss. The portion of total heat of combustion extractable from a fossil heat source is approximated by

$$\eta_g = \frac{T_f - T_g}{T_f - T_2} \tag{3}$$

where T_f is combustion temperature, T_g is flue gas temperature, and T_2 is ambient temperature. Flue gas loss may be reduced by a heat exchanger using residual flue gas heat content to heat air flowing into the combustor. The improvement is significant, and this technique is used in modern steam plants. Practical systems are limited in top temperature by one of several possible constraints: (1) heat source capability, (2) nature of cycle, (3) working fluid limits, (4) materials limits, and (5) efficiency considerations.

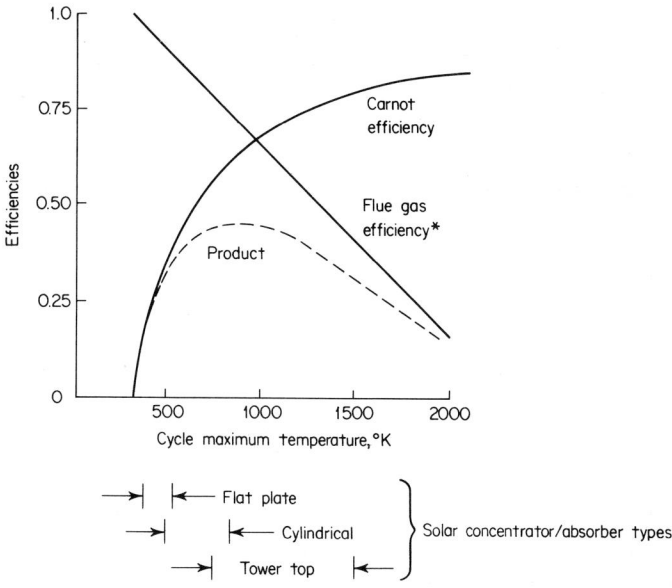

Fig. 2. Idealized efficiencies.

SOLAR ABSORBERS

Solar absorbers convert collected solar light energy to heat. Absorber geometry and characteristics are intimately related to collector geometry.

Flat Plate Absorbers.   The simplest (in principle) geometry is the flat plate absorber. Sunlight is not concentrated, and no collector is used. In a sense, any object illuminated by sunlight is a "flat plate" absorber, even if not flat. Flat plate absorbers designed for solar energy absorption generally employ selective coatings to improve efficiency. Typically, operating at temperatures of 400 to 450°K, their reradiation spectrum is shifted far into the infrared from the solar spectrum with its equivalent black body temperature on the order of 5800°K. It is, therefore, possible to apply coatings having a high average absorptivity for the solar spectrum, yet a low emissivity for reradiation.

Flat plate absorbers are well suited to solar heating and cooling of buildings. Since they do not employ optical concentration, they collect diffuse (sky brightness) as well as direct (solar image) energy. With a clear sky, they collect about 20 percent more energy per unit area than an optical concentrator. Flat plate absorbers can have useful building heating output under hazy or cloudy conditions when an optical concentrator is quite useless. Their limited working temperature, however, restricts ideal (Carnot) heat engine cycle efficiency to 25 percent or less, and actual system efficiencies to about 10 percent. The flat plate absorber is unattractive for utility electric power generation for three reasons: (1) Low cycle efficiency requires large total collection area: (2) The distributed nature of heat absorption requires an extensive (and expensive) heat-transport system: (3) Low cycle temperature is not compatible with the shared heat engine concept.

Absorbers for Optical Concentrators.   Cylindrical paraboloid optical concentrators for solar power were pioneered by Drs. Aden and Marjorie Meinel at the University of Arizona. (Ref. 2) Cylindrical paraboloids can achieve optical concentration ratios (ratio of light intensity on the absorber to unconcentrated sunlight intensity) exceeding 100. The concentrating reflector can be formed from a flat sheet, and this concentrator, if length along the axis is large compared to width, need be steered on only one axis (at some loss in efficiency compared to two-axis steering). The heat absorber can be a pipe of relatively simple form, since the sunlight is focused to a line image. With a simple (e.g., black pipe) absorber, the cylindrical concentrator can achieve temperatures on the order of 550°K. The Meinels utilized sophisticated absorbers consisting of selectively coated evacuated glass pipes surrounding the metal pipe carrying the heat-transport fluid, and in this way were able to achieve temperatures as high as 850°K, compatible in principle with modern high-pressure steam Rankine turbogenerators. Figure 3 illustrates a cylindrical concentra-

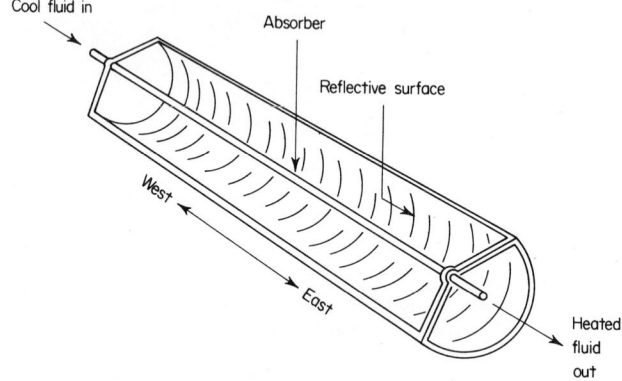

Fig. 3. Cylindrical concentrator with axial absorber.

tor/absorber concept of the Meinels. Cylindrical concentrators with simple absorbers show the same disadvantages, to a lesser degree, as flat plate absorbers. Sophisticated absorbers also exhibit practical problems that have not (as of 1976) been sufficiently resolved to show economic feasibility, principally because the sophisticated absorbers themselves are expensive, and the vacuums and selective coatings are difficult to maintain at the desired operating temperatures.

As with the flat plate absorber, an extensive heat-transport system is required to bring the heat energy to a central heat engine. Pressure drop considerations prohibit piping the heat engine working fluid itself through the absorber system. Pumping power requirements are impractical unless a liquid transport medium is used, and peak temperatures suitable to modern steam plants require use of costly liquid metal systems, e.g., sodium or potassium.

The tower-top optical transmission system pioneered by Vant-Hull, Hildebrandt, and Lipps at the University of Houston overcomes these difficulties to a large degree. Only the absorber alternatives are described here. See later article in this Handbook section for details. The simplest absorber is the direct absorber. Concentrated sunlight falls directly on the heat-transfer surface, in some design concepts protected by a selectively coated glass window and vacuum space. Figure 4 shows a typical direct absorber. An alternative is the cavity absorber; sunlight is concentrated on and enters the cavity aperture. The internal wall of the cavity is the heat-transfer surface. Figure 5 shows a typical tower-top cavity concept. The aperture is horizontally oriented at the bottom to minimize convective heat losses through air circulation.

Approximate Analysis for Absorbers.   The following are preliminary estimating relationships for sizing and heat-transfer conditions in direct or cavity absorbers. A more rigorous treatment is described under "Satellite Energy Systems," later in this Handbook section. Finite-element numerical thermal models should be used for design purposes.

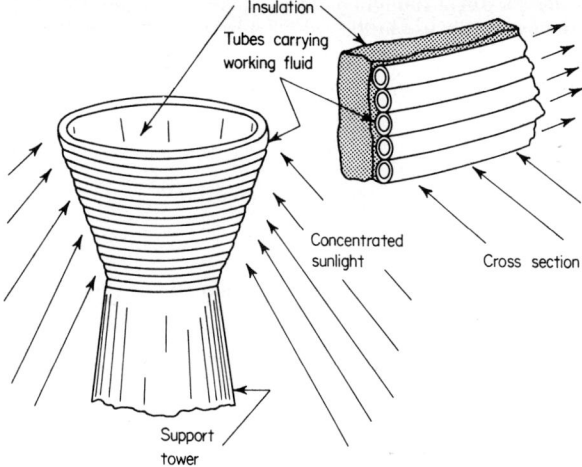

Fig. 4. Direct absorber.

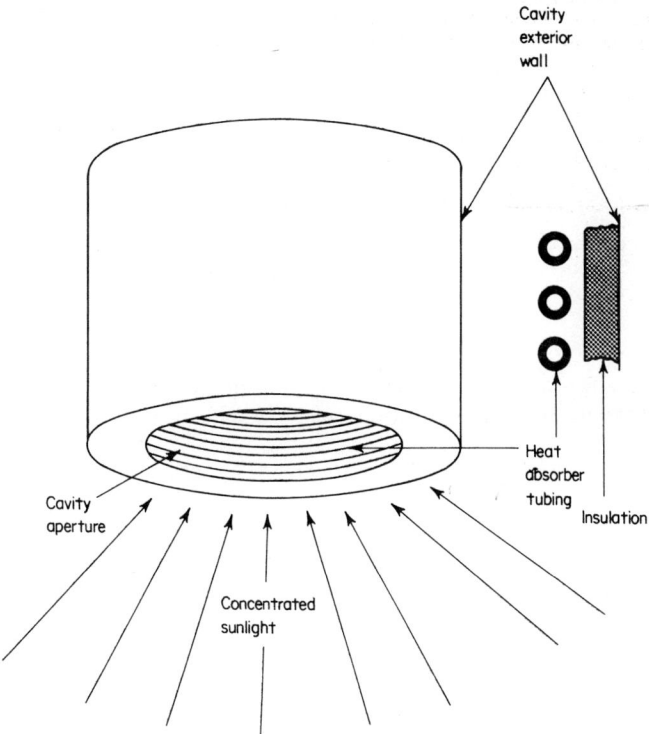

Fig. 5. Cavity absorber.

The *view factor* for direct sunlight overhead is approximately $F_\odot = 22 \times 10^{-6}$, where view factor is defined by its use in the radiation heat-transfer equation,

$$\dot{q} = \sigma F \alpha \epsilon T_\odot{}^4 \tag{4}$$

where σ is the Stefan-Boltzmann constant,
   F is view factor,
   α is absorptivity of the heat absorber,
   ϵ is emissivity of the emitter.
The emissivity of the sun is generally considered 1.0.

$$I_0 = 5.6697 \times 10^{-11} \times 22 \times 10^{-6} \times (5800)^4 = 1.4 \text{ kW/m}^2$$

(A more precise value is $I_0 = 1.395$ kW/m².) Equation (4) as written does not account for heat reradiated by the absorber.

The atmosphere (without clouds) absorbs and diffuses about 40 percent of the sun's energy; if this equation is used to calculate heat transfer to an absorber, atmosphere attenuation may be taken into account by including the factor I/I_0, typically 0.6.
The heliostat area required is

$$A_H = \frac{P}{I_0 \, (I/I_0) \, \eta_g \eta_e \eta_a \eta_H} \tag{5}$$

where I/I_0 = solar intensity ratio as described above
   η_g = generator efficiency
   η_e = heat engine cycle efficiency
   η_a = absorber efficiency
   η_H = collector/concentrator (heliostat) efficiency, including averaged aspect fac-
      tor, the total projected heliostat area normal to the sunlight divided by
      total heliostat area
The solar view factor required at the absorber is

$$F_a = \left(\frac{T_{aw}{}^4}{T_\odot{}^4}\right)\left(\frac{I}{I_0}\right) \eta_H{}^{-1} \tag{6}$$

where T_{aw} is the desired *adiabatic wall temperature* at the absorber (the equilibrium temperature that the absorber would reach if no heat were extracted from it), and $T_\odot$ is the equivalent solar black-body temperature, approximately 5800°K. It is assumed here that absorptivity and emissivity of the absorber are equal.

A geometric concentration ratio, (reflector area)/(absorber area), is then defined as

$$C = \frac{F_a}{F_\odot} = \frac{F_a}{22 \times 10^{-6}} \tag{7}$$

Heat-transfer calculations to determine heat exchanger parameters within the absorber are best done by numerical integration. Figure 6 illustrates a typical heat exchanger situation. Tube wall temperature must be found for each integration step by solving

$$\bar{h}(T_w - T) = \sigma \frac{r_2}{r_1}\left(\sum_i F_i \epsilon_i \alpha T_i{}^4 - \epsilon T_w{}^4\right) \tag{8}$$

where $\bar{h}$ is net heat-transfer coefficient.

$$\bar{h} = \frac{r_1 \log_e (r_2/r_1)}{k} + \frac{1}{h} \tag{9}$$

with k the tube material thermal conductivity, and h the fluid flow film coefficient, which may be computed from the pipe flow equation,

$$\text{Nu} = 0.023 \text{Re}^{0.8}\text{Pr}^{0.4} \tag{10}$$

where Nu = Nusselt's number, hD/k
   D = tube diameter
   k = fluid thermal conductivity
   Re = $\rho u D/\mu$

ρ = fluid density
u = fluid velocity
μ = fluid viscosity
$\mathrm{Pr} = C_p \mu / k$
C_p = fluid specific heat

With wall temperature known, heat transfer per unit area is

$$\dot{q} = \bar{h}(T_w - T) \tag{11}$$

The gas temperature rise along the tube length element δs is then given by

$$\delta T = \frac{4\dot{q}}{D\rho C_p u}\, \delta s \tag{12}$$

Typical numerical integration results are shown later in Fig. 16.

The minimum heat transfer rate is roughly

$$\dot{q} = I_0 \left(\frac{I}{I_0} \right) \frac{\eta_H \eta_a C}{Rf} \tag{13}$$

where R is ratio of cavity interior wall area to aperture area, and f is tube wall area/cavity interior wall area.

Figure 7 shows typical variation of $\bar{h}$ with gas velocity for helium at 3MN/m^2 (435 psia). A steam Rankine cycle would employ boiling water flow; any cycle type could utilize a liquid metal heat transport loop to deliver heat to the cycle working fluid or equipment.

The direct absorber with glass vacuum cover is relatively unattractive because of the difficulty of developing and maintaining the vacuum seal and selective coatings. Comparing the direct absorber with the cavity absorber:

1. The cavity will be somewhat more efficient because (a) its absorptivity will be higher, (b) its effective wall temperature will be lower for a given working fluid temperature, and (c) its convective losses will be less.

2. The cavity, because of its greater heat-transfer area, will tend to have a lower pressure drop.

3. The cavity, because it is larger, will tend to be heavier and more costly. Its improved performance is likely, however, to reduce *system* cost.

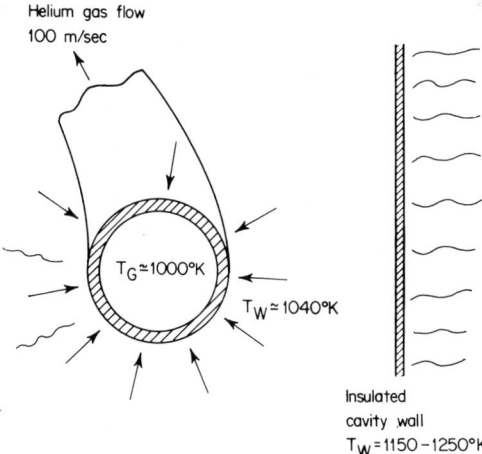

Fig. 6. Representative heat-transfer conditions in cavity. Radiation heat balance view factors (typical):

Sun seen through aperture	$F_1 = 0.0013$
Ambient temperature through aperture	$F_2 = 0.03$
Cavity wall	$F_3 = 0.8687$
Adjacent tube	$F_4 = 0.1$
	$\Sigma F = 1.000$

4. In situations where wall temperature is limited by materials considerations, the cavity permits somewhat higher working fluid temperatures.

Both cavity and direct absorbers are presently under study; the writers prefer the cavity approach.

HEAT ENGINE CYCLES

Heat engines for conversion of solar energy to electric power ideally should have all the following attributes: (1) low cost per kilowatt output capacity, (2) long life and reliable operation with minimal maintenance, (3) safe and environmentally acceptable operation, (4) characteristics compatible with cycle top temperatures up to 1000°K, and (5) efficiency approaching Carnot values.

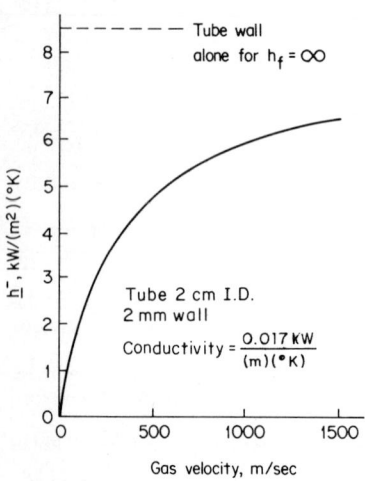

Fig. 7. Typical tube heat transfer.

Heat engines that are potential candidates for coupling to a solar heat source include thermoelectric, thermionics, thermochemical, magnetohydrodynamic, Rankine, Brayton (simple or recuperated), and cascaded cycles. Table 2 presents a summary evaluation of the cycle options against the foregoing criteria. A cycle receiving one or more poor or unacceptable ratings is not likely to enter widespread use. It is, of course, important to recognize that future technology developments could revise these evaluations significantly.

Thermoelectric devices employ the Seebeck effect wherein electric potentials occur at a junction of dissimilar electric conductors. A circuit made up of dissimilar conductors with a temperature difference between the two junctions produces electric power. Such a circuit designed to produce power is called a thermoelectric converter. Thermoelectric converters have been used in RTGs (radioisotope thermoelectric generators) for a number of space systems, heated by radioisotope fuel sources. Thermoelectric devices are relatively inefficient. Contemporary space-type systems would exhibit efficiencies of about 6 percent at a typical solar heat temperature of 1000°K; 10 percent devices are under development (Ref. 3). RTGs are attractive for applications in certain space systems because they are extremely reliable (no moving parts; highly redundant) and do not depend on the sun. But, barring significant (and unforeseen) new technology developments, thermoelectrics can be dismissed as practical utility heat engines on grounds of low efficiency and exotic materials leading to high cost.

See also article on "Thermoelectric Cooling" in Handbook section on *Power Technology Trends.*

Thermionics devices utilize the electric potential and current that results when electrons thermally evaporated from a hot cathode are collected by a cooler anode. Thermionics has been the subject of a great deal of research, but no practical power generation applications have yet emerged. Typical thermionic cells utilize thoriated cathodes and cesium vapor drift space between cathode and anode. The lower limit of practical operating temperatures (1200°C, 1473°K) is close to the upper limit of temperature capability of heliostat/tower-top solar heat sources. As of 1976, thermionics are less efficient (typically 15 percent at 1200°C) than mechanical heat engines. Hence thermionic devices are unlikely to be a part of utility electric power generation. The possibility of a high temperature cycle using a thermionic system cascaded with a more conventional system should not be entirely discounted.

See also article on "Magnetohydrodynamic Generators" in Handbook section on *Power Technology Trends.*

Thermochemical systems employing thermal dissociation of water into hydrogen and oxygen have been proposed by various authors. Thermal dissociation of water normally occurs at high temperatures, i.e., 4000 to 5000°K, beyond the reach of practical solar heat sources. A number of thermochemical cycles, however, have been proposed, employing a series of reactions, all occurring at or below practical (i.e., 800 to 1400°K) temperature limits. One such cycle currently under study at Euratom (Ref. 4), is as follows:

$$6FeCl_2 + 8H_2O \rightarrow 2Fe_3O_4 + 12HCl + 2H_2 \qquad (925°K)$$
$$2Fe_3O_4 + 3Cl_2 + 12HCl \rightarrow 6FeCl_3 + 6H_2O + O_2 \qquad (450°K)$$
$$6FeCl_3 \rightarrow 6FeCl_2 + 3Cl_2 \qquad (695°K)$$

Many other candidate reactions have been published. Insufficient research has been done to permit a selection. Practical thermal water cracking has not been demonstrated,

TABLE 2. Qualitative Ranking of Heat Engine Cycles

	Low cost potential	Long life, reliable operation	Safe and environmentally acceptable operation	Cycle temperature characteristics compatible with solar heat source	High efficiency
Thermoelectric	P	G	G	G	P
Thermionic	P	F	G	P	P
Thermochemical	?	?	?	G	?
MHD (liquid)	?	?	?	G	G
Rankine	G	G	G	G	G
Brayton	G	G	G	G	G
Cascaded cycles	G	G	G	G	G

P—Poor
F—Fair
G—Good
?—Unknown

although large-scale laboratory experiments are underway at Euratom. The theoretical efficiency of some of the water cracking cycles is attractive. The Carnot limit applies here as it does to all heat engines. An assessment of the concept cannot be made from the very limited data available.

The attractiveness of the thermochemical water cracker lies in its ability to produce hydrogen, the subject of some energy research and considerable speculation. Hydrogen has several potential uses, including: (1) fuel in liquid, gas, or hydride form, with very low pollution potential; (2) low-cost energy transmission medium (pumped as gas through pipelines); (3) low-cost energy storage medium; and (4) source of electricity produced by fuel cells. See also "Fuel Cells" in Handbook section on *Chemical Fuels Technology*.

This leads to concepts for solar-based integrated energy systems. See Fig. 8. Practicality and economic potential of thermochemical-based systems are at present unknown. See article on "Hydrogen" in Handbook section on *Chemical Fuels Technology*.

Magnetohydrodynamics (MHD).   Electricity is generated by the motion of a conducting fluid in a magnetic field in MHD devices. The conducting fluid may be either an ionized gas or a liquid metal. Gas MHD systems require very high temperatures in order to achieve sufficient ionization, and are generally incompatible with the temperature capabilities of practical solar heat sources. Research on liquid metal MHD systems has been under way at Argonne National Laboratory since the mid-1960s (Ref. 5). These systems utilize two-phase flow to accelerate the fluid. The fast-moving liquid metal/gas mixture then enters an MHD channel where its kinetic energy is converted to electricity. Figure 9 illustrates a typical cycle concept. Theoretical studies have indicated that

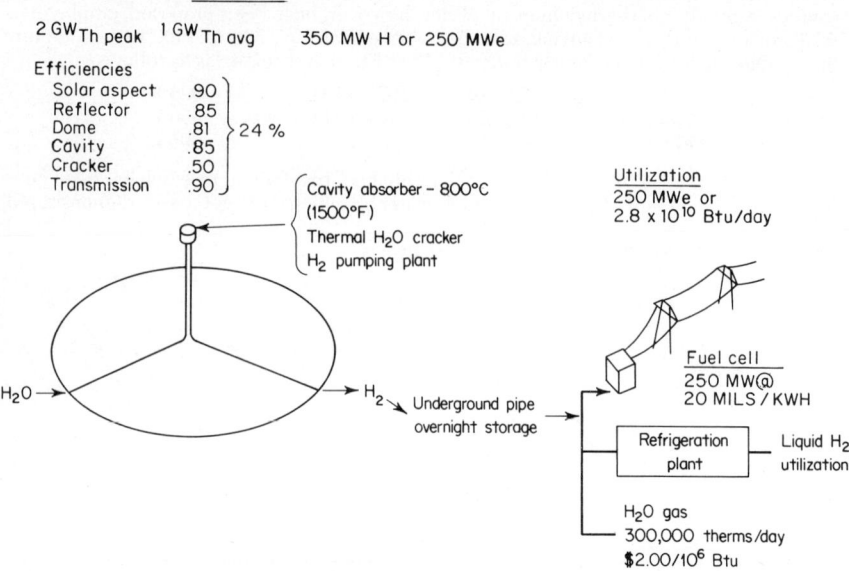

Fig. 8. Integrated hydrogen energy system concept. Advantages of system include: (1) It provides fuel and electricity from solar energy. (2) Thermal generation of hydrogen from water eliminates electrical generation at main plant. (3) Energy storage as hydrogen is cheaper per kilowatt hour than pumped water or flywheels. (4) Transmission of energy as hydrogen is cheaper than as electricity.

efficiencies of 40 percent or better should be attainable at 1500°F (816°C) cycle top temperature. Experimental work on key system components, such as the liquid/gas separator and the MHD channel, although not achieving forecast efficiencies, have tended to confirm predictions in that the experimental devices are not optimal designs and perform generally as expected. Closed cycles have not been operated.

It is premature to forecast the future of two-phase liquid metal MHD cycles as heat

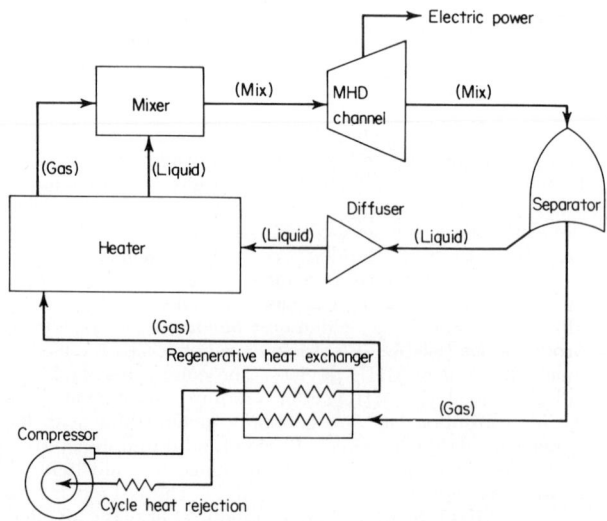

Fig. 9. Two-phase liquid-metal MHD cycle diagram.

engines for solar energy conversion. They have the potential for high efficiency and dependability, but their cost characteristics are hard to predict at present. It is quite unlikely that first-generation solar electric power plants will employ this cycle.

See also article on "Magnetohydrodynamic Generators" in Handbook section on *Power Technology Trends.*

Rankine Cycle.   The steam Rankine cycle employing steam turbines has been the mainstay of utility thermal electric power generation for many years. The cycle, as developed through the years, is sophisticated and efficient, and the equipment is dependable and available on the commercial market. A typical cycle (Fig. 10) employs superheat, reheat, and regeneration. Heat exchange between flue gas and inlet air adds several percentage points to boiler efficiency in fossil fueled plants. Modern steam Rankine

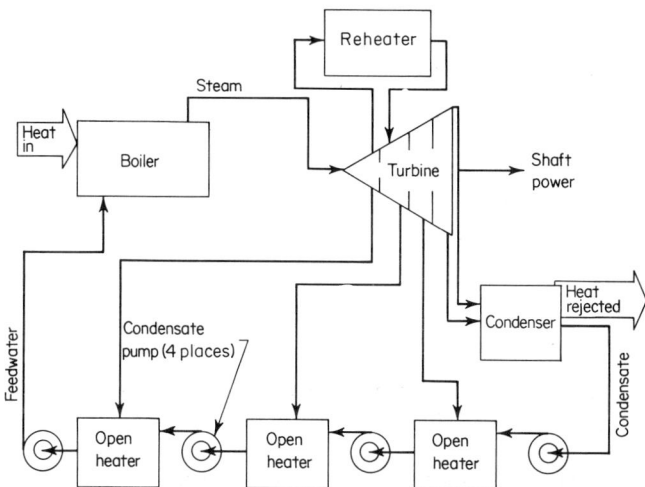

Fig. 10. Regenerative-reheat steam Rankine cycle shown schematically.

systems operate at a cycle top temperature of about 800°K with efficiencies of about 40 percent. All characteristics of this cycle are well suited for use in solar plants. Since this system is in widespread use and amply described in many texts, we will not elaborate here.

If the steam Rankine system could be improved on, it would be to ameliorate the following drawbacks:

1. The critical temperature for water is 650°K. Above this temperature, it behaves like a gas, and the cycle begins to have Brayton characteristics.

2. It is desirable to reject heat at as low a temperature as possible: 300 to 320°K can readily be attained by cooling towers. In this temperature range, the condensing pressure is on the order of 1 psia. Large equipment sizes are required to handle the resulting large volume flow.

3. Constant temperature heat rejection (condensing) requires either a source of water as a heat sink: i.e., a lake, river, etc., or a great flow of cooling air. Water is not likely to be conveniently available in the dry, sunny climates best suited to solar power.

4. Hot steam is corrosive.

Brayton Cycle.   Recently attention has been focused on the Brayton cycle as a potential practical alternative to the steam Rankine cycle for solar power and for high-temperature gas-cooled reactors. The Brayton cycle is most familiar in its open form as used in aircraft gas turbines. The open Brayton cycle cannot compete with steam Rankine in efficiency; in a power generation application, cycle efficiencies on the order of 20 percent would be expected. (Aircraft gas turbines driving alternators are sometimes used for peaking plants, where their low capital cost is more important than their relatively poor fuel economy.)

See also article on "Gas and Expansion Turbines" in Handbook section on *Power Technology Trends.*

The Brayton cycle can achieve high efficiency through recuperation, sometimes called regeneration. Figure 11 is a representative cycle diagram. The working fluid is an inert gas, typically helium. Inert gas mixtures such as helium–xenon have been studied and have potential advantages. Rarity of xenon and other heavy inert gases probably precludes their large-scale use.

The recuperated Brayton cycle approaches Carnot efficiency in the ideal limit. As compressor and turbine work are reduced, the average temperatures for heat addition and rejection approach the cycle limit temperatures. The limit is reached as compressor and turbine work (and cycle pressure ratio) approach zero, and fluid mass flow per unit power output approaches infinity.

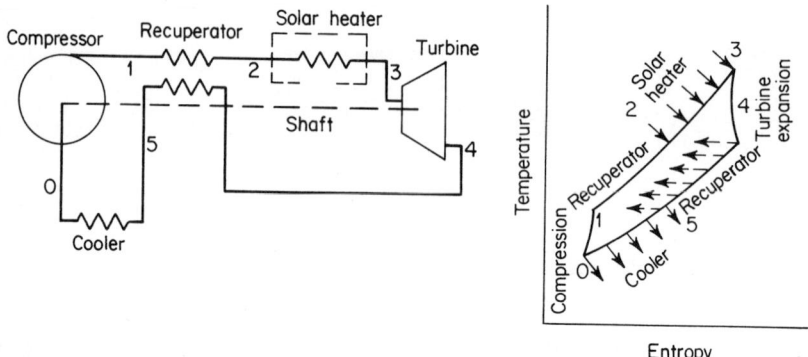

Fig. 11. Recuperated Brayton cycle. As compression work is reduced: (*a*) Ideal cycle efficiency approaches Carnot. (*b*) Heat into and rejected from cycle becomes small. (*c*) Heat transfer in recuperator becomes large. (*d*) Cycle pressure ratio approaches zero. (*e*) Cycle mass flow per unit shaft power approaches infinity.

One expects from this that practical recuperated Brayton cycles would operate at relatively low pressure ratios but be very sensitive to pressure drop. With the assumption of constant gas specific heat over the cycle temperature range, a very good assumption for helium, the cycle efficiency of a recuperated Brayton cycle may be expressed as

$$\eta_e = 1 - \left[\frac{(r_{pc}^\zeta - 1)/\eta_b + \Delta T_r/T_0}{(T_3/T_0)\,\eta_T\left[1 - (G/r_{pc})^\zeta\right] + \Delta T_r/T_0} \right] \tag{14}$$

where r_{pc} = compressor pressure ratio (>1)

η_b = compressor efficiency

ΔT_r = temperature difference across recuperator ($T_4 - T_2$ in cycle diagram)

T_0 = cycle lower limit temperature

T_3 = cycle top temperature

η_T = turbine efficiency

ζ = specific heat factor, $(\gamma - 1)/\gamma$, = 0.4 for $\gamma = 1.67$ as for helium

G = the product of the four pressure drop factors:

$$(P_1/P_2)(P_2/P_3)(P_4/P_5)(P_5/P_0)$$

This equation is plotted in Fig. 12 to show the sensitivity of the recuperated Brayton cycle to pressure drop and recuperator performance. The quasi-ideal line includes compressor and turbine efficiencies without pressure drop and with a perfect recuperator.

Experience in building small (1- to 10-kW) recuperated Brayton machines for space applications indicates that pressure drop of 1 percent per heat exchanger pass ($F = 1.04$) and recuperator ΔT's of 15° ($\Delta T_r/T_0 = 0.05$) are attainable. Cycle pressure ratios on the order of 2 are used. Pressure ratios somewhat greater than the maximum efficiency point are desirable, to reduce cycle mass flow. A parameter proportional to mass flow may be expressed as

$$Pm = \left\langle \eta_T\left[1 - \left(\frac{G}{r_{pc}}\right)^\zeta\right] - \left(\frac{r_{pc}^\zeta - 1}{\eta_b}\right)\frac{T_0}{T_3} \right\rangle^{-1} \tag{15}$$

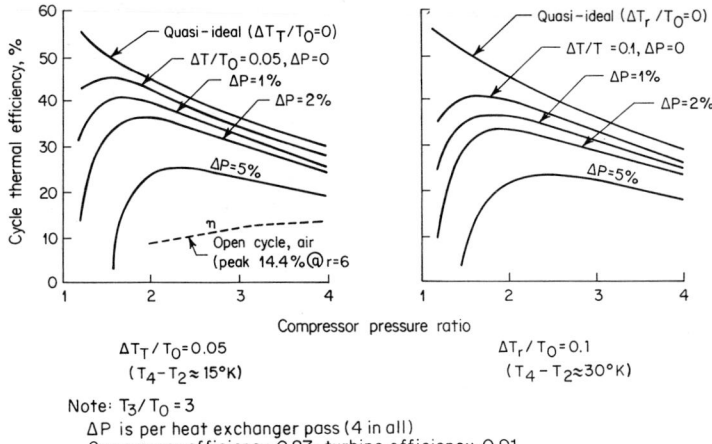

ΔT_T / T_0 = 0.05
(T_4 - T_2 ≈ 15°K)

ΔT_r / T_0 = 0.1
(T_4 - T_2 ≈ 30°K)

Note: $T_3/T_0 = 3$
ΔP is per heat exchanger pass (4 in all)
Compressor efficiency 0.87; turbine efficiency 0.91

Fig. 12. Recuperated Brayton performance characteristics. $T_3/T_0 = 3$; Compressor efficiency = 0.87; turbine efficiency = 0.91.

This parameter is plotted in Fig. 13 with cycle efficiency in an example case. Whereas peak cycle efficiency occurs at $r_{pc} = 1.5$, a reduction in mass flow of 40 percent can be achieved at a very small cost in efficiency by operating at $r_{pc} = 2$. Sensitivity of actual efficiency to pressure drop is also reduced. Mass flow in kilograms per second is obtained by multiplying shaft power in kilowatts by $Pm/(C_p T_3)$, where C_p is specific heat of the working gas in kilojoules per kilogram per degree Kelvin.

The example given employed a temperature ratio of 3.4, indicating a cycle top temperature of 1088°K(1500°F) (816°C) for a rejection temperature of 320°K. This is about the upper temperature limit for superalloy heat exchangers in cavity absorbers. The temperature limit for direct absorbers will be about 100°K less because of the higher heat flux and greater ΔT through the heat exchanger wall.

Cascaded Cycles. The operation of Rankine cycles over large temperature differences poses problems in selecting a working fluid with suitable properties, and with large pressure ratios. Cascaded cycles provide relief from these problems. Mercury/steam cycles were originally tried many years ago, but have not gained widespread use because

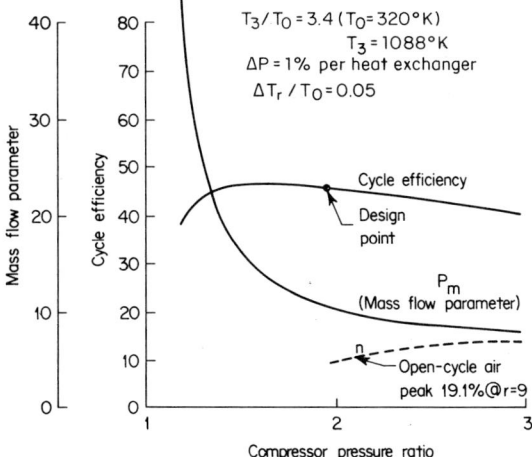

Fig. 13. Recuperated Brayton design-point selection.

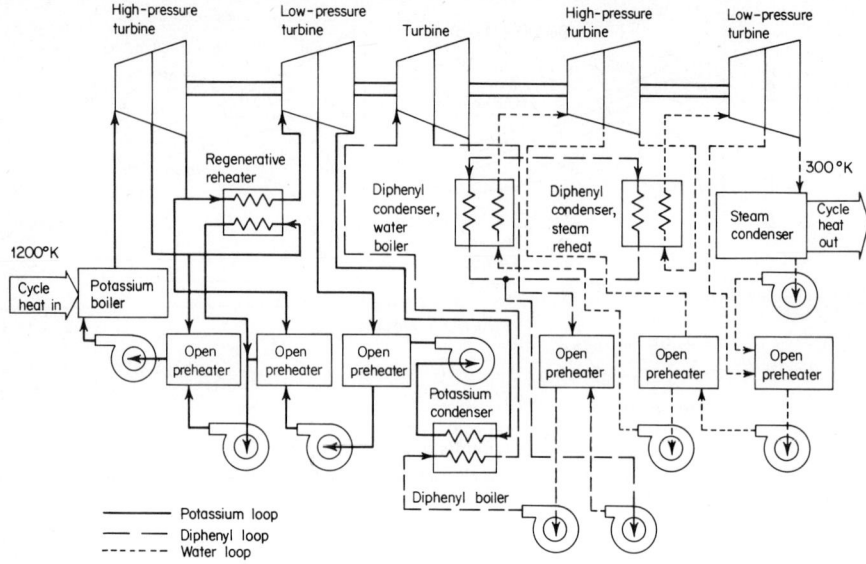

Fig. 14. Three-stage cascaded rankine cycle.

of the noxious properties of mercury. Recently, potassium/steam and potassium/organic/steam have been proposed, and potential efficiencies as high as 60 percent have been claimed.

In the cascaded cycle, the low temperature cycle operates on the rejected heat from the high temperature cycle. Figure 14 is a typical schematic diagram (Ref. 6).

EXAMPLE SYSTEM DESIGN POINT

The following example of a 250-MW plant module illustrates typical design point data. It does not represent an optimized design.

From Fig. 15, assume generator efficiency of 0.95 and Brayton cycle efficiency of 0.45, with $T_3 = 1088°$K and $T_o = 320°$K. From Fig. 16, assume that $T_{aw} = 2000°$K and $T_w = 1200°$K; then tube $s/d = 800$, and cavity efficiency ≈ 0.72. Assume that reflector efficiency $= 0.65$ and solar intensity factor $I/I_o = 0.6$. Then the heliostat area required is, by Eq. (5),

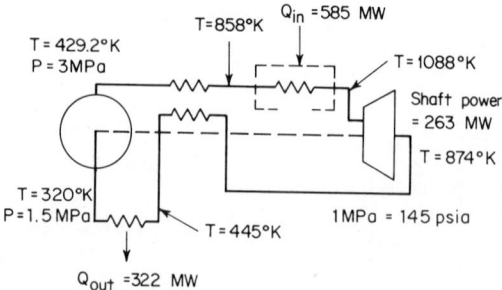

Fig. 15. Typical design data for 250 MWe plant module. Note the following: (a) *Solar intensity* = 0.8 kW/m² (normal). (b) Heliostat efficiency = 65% (including aspect factor). (c) Cavity efficiency = 72%. (d) Heliostat area = 1.5 × 10⁶m² = 18,000 10-m hexagonal heliostats. (e) Geometric concentration ratio = 1650. (f) Cavity aperture = 34 m diameter.

$$A = \frac{250,000}{1.353 \times 0.6 \times 0.95 \times 0.45 \times 0.72 \times 0.65} = 1.5 \times 10^6 \text{ m}^2$$

The solar view factor required is, by Eq. (6),

$$F_a = \left(\frac{2000°\text{K}}{5800°\text{K}}\right)^4 (0.6 \times 0.65)^{-1} = 0.0362$$

and geometric concentration ratio, by Eq. (7),

$$C = \frac{0.0362}{22 \times 10^{-6}} = 1650$$

Cavity aperture area, therefore, $= 1.5 \times 10^6/1650 = 909$ m², and diameter $= 34$ m. Assume cavity area ratio of 25; internal area $= 22,700$ m².

From Fig. 12, the mass flow parameter value is approximately 10, and for helium as a working fluid, per the discussion following Eq. (15),

$$\dot{m} = \frac{10 \times 250,000}{5.23 \times 1088 \times 0.95} = 462 \text{ kg/sec}$$

Assuming helium inlet velocity of 100 m/sec in 2-cm tubes, flow per tube is

$$\dot{m} = \frac{\pi D^2}{4} \frac{pMu}{RuT}$$

$$\dot{m} = 0.785 \times (0.02)^2 \times \frac{3 \times 10^6 \times 4 \times 100}{8300 \times 858} = 0.053 \text{ kg/sec} \qquad (16)$$

where M is molecular weight, Ru is universal gas constant, and other terms as defined before.

$$\text{Number of tubes} = \frac{462}{0.053} = 8700$$

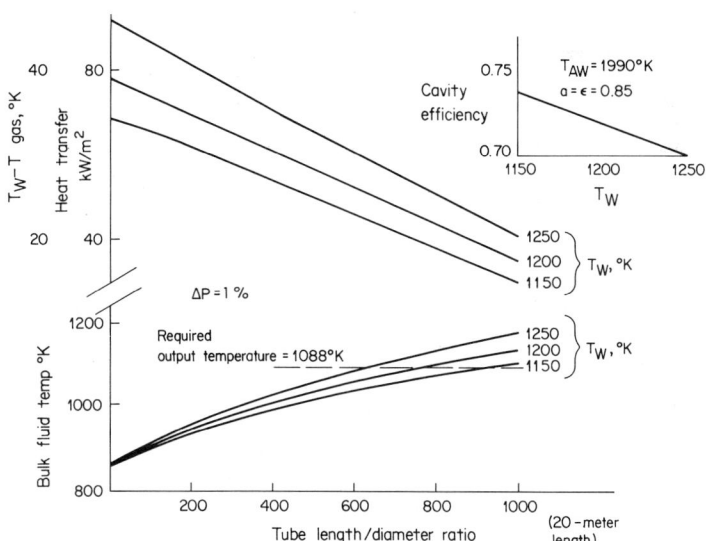

Fig. 16. Cavity absorber representative heat-transfer analysis. Numerical integration results: (a) $T_2 = 860°\text{K}$. (b) Inlet velocity $= 100$ m/sec. (c) Tube diameter $= 0.02$ m. (d) Wall $= 2 \times 10^{-3}$ m. (e) Helium gas. (f) $\eta_r \frac{I}{\gamma} = 0.45$

Fraction of cavity interior wall utilized:

$$f' = \frac{ND^2(s/d)}{A_c} = \frac{f}{\pi}$$

$$f' = \frac{8700 \times (0.02)^2 \times 800}{22{,}700} = 0.123 \qquad (17)$$

We can compare cavity heat transfer with direct absorber heat transfer by Eq. (13), with the result $\dot{q}_d = 400$ kW/m^2, $\dot{q}_c = 50$kW/m^2. Thus, in this rather typical case, the wall ΔT, $T_w - T_f$, will be several times higher for the direct absorber as compared to that of the cavity. Finally, recuperator heat flow is:

$$\dot{q}_r = C_p (T_2 - T_1) \, \mathring{m}$$
$$\dot{q}_r = 5.23 \times (858 - 429) \times 462 = 1035 \text{ MW}$$

Recuperator heat exchange area, assuming a heat-transfer coefficient of 1.2, is

$$A = \frac{1.035 \times 10^6}{1.2 \times 15°} = 57{,}500 \text{ m}^2$$

HEAT ENGINE AND ENERGY STORAGE INTEGRATION

To this point, heat engines have been considered in the context of fuel-replacement plants. Plants incorporating energy storage will now be briefly considered. Energy for storage may be extracted from the power system in the form of heat, mechanical energy, or electricity, depending on the storage form.

The storage problem is largely economic in nature: i.e., what it costs versus what it is worth. Capital cost versus worth depends on the characteristics of the storage system, primarily the following parameters:

1. The cost per kilowatt of the device to convert energy into the stored form.
2. The efficiency of this device; this influences solar engine plant capacity dedicated to storage charging.
3. The cost per kilowatt hour of storage capacity and the amount of capacity expressed as the duration τ during which the fully charged storage system can deliver energy at the plant average output.
4. The cost per kilowatt of the device to convert stored energy back to electricity.
5. The efficiency of this device.
6. The use function f_u described as the ratio of storage output to direct plant output as a function of storage capacity.

Given these parameters, the cost of storage in dollars per kilowatt of plant output rating may be approximated by the equations shown in Fig. 17. Rough estimates of the cost parameters for a variety of storage media are given in Table 3. Figure 17 shows the variation of f_u with τ as estimated from Aerospace Corporation data (Ref. 7).

Results of applying these estimating factors to the above equation are shown in Fig. 18. A rough estimate of the value of storage, based on increased fuel displacement plus capacity displacement, is also shown. The value estimate assumes that a solar plant with 100 hours storage will be sufficiently dependable to serve as a base-load plant. This is an estimate, and the correct requirement is not known. It will be necessary to analyze long-term weather statistics over tens of years to determine this factor with sufficient accuracy.

Three general observations may be drawn from Fig. 18:

1. Storage schemes do not appear to pay off for short-term (i.e., overnight) storage. All the storage alternatives increase in cost faster than the increase in value. This observation, however, is related to capability to displace base-load capacity. More detailed studies of short-term storage have found payoff in employing about 6 hours of storage to displace intermediate load capacity.
2. Long-term storage employing a hydrogen system may be promising. However, cost still appears to exceed value, even assuming a 20-mil/kWh competitive cost.
3. Hybrid systems, e.g., momentum plus hydrogen, are promising.

Considerable further study and analysis are needed to determine the best ways of

Cost of storage	=	cost of input system	+	cost of storage capacity	+	cost of output system	+	cost of increased plant capacity required to charge storage

$$C \;=\; \frac{C_i f_u}{\eta_{oh}(1-f_u)} \;+\; \frac{C_c \psi}{\eta_o} \;+\; C^o \;+\; \frac{C_e f_u}{n_i\, n_o (1-f_u)}$$

Value of storage	=	value of additional fuel displacement	+	value of plant capacity displacement

$$V_s \;=\; \frac{f_u(1-f_u^{\,*})}{1-f_u} \;+\; \frac{r}{r*}\,V_p$$

$f_u^{\,*} = 0.52$
$r^* = 100$ hr

f_u = fraction of plant output that comes from storage

τ = storage capacity in hours of output at plant rating

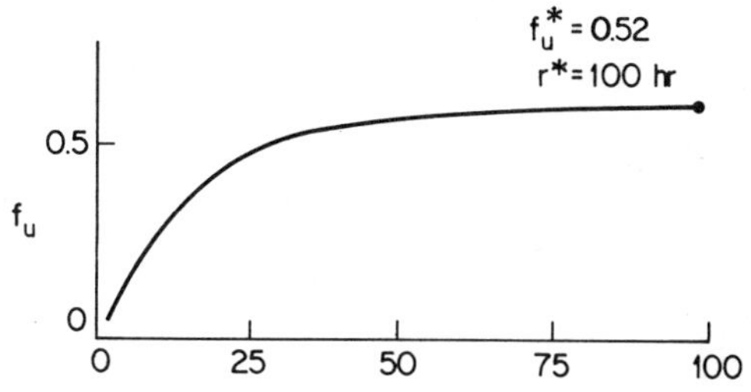

Fig. 17. Cost and value of storage.

integrating storage systems into solar plants. Figure 19 is a schematic diagram of a hypothetical solar thermal recuperated Brayton plant with momentum and liquid hydrogen energy storage.

OUTLOOK

The tower-top solar heat concentrator concept is sufficiently promising that we may expect engineering development of prototype solar utility electric power plants for fossil

TABLE 3. Storage Cost Factors Used for Illustration

	C_e (plant capacity) \$/kW	C_i (input) \$/kW	C_o (output) \$/kW	C_c (storage capacity) \$/kWh	η_o (output)	η_i (input)
Thermal, liquid metal	$100*	0	0	$20	0.45	1.0
Pumped water	$350	0†	$100	$10‡	0.8	0.8
Momentum§	$350	0†	$150	$10	0.95	0.95
Gaseous H₂¶	$350	$ 80	$150	0	0.7	0.75
Liquid H₂	$350	$230	$150	0.3	0.7	0.6

*Cost per kW(th) of solar heat source.
†Input and output, same device.
‡Assumes that reservoir must be constructed.
§Price based on steel fly wheels.
¶Assumes H₂ gas stored in underground wells (depleted gas wells). Fabricated gas H₂ storage is prohibitive in cost.

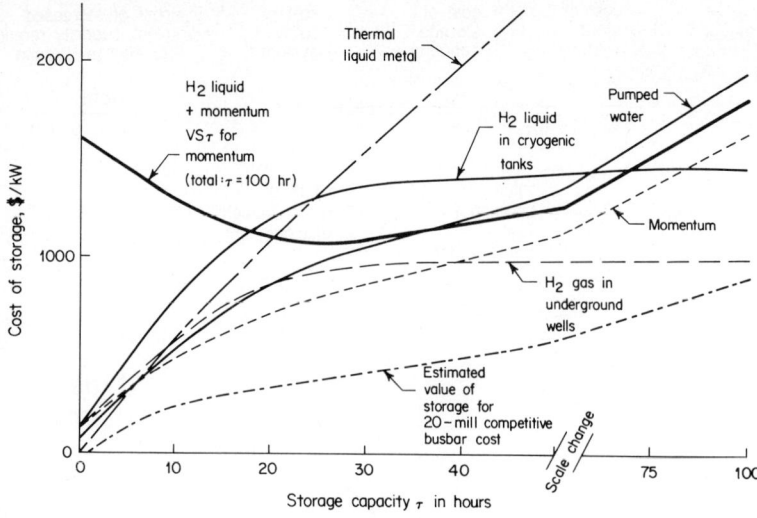

Fig. 18. Storage costs for ground-based solar plant. Example: 20 mills = $400/kW plant and $1.50/$10^6$ Btu. fuel.

fuel displacement in the near future. Open-cycle Brayton equipment may be used for early demonstrators, where its extreme simplicity (ambient air can be the working fluid) is important. Initial operational fuel-displacement plants may use steam Rankine or recuperated Brayton cycles. Later, as the technology matures, cascaded cycles, or liquid metal MHD may appear. The thermochemical water cracker, if it becomes practical, may appear at a still later date. It will, of course, integrate very well with hydrogen energy storage systems.

The future for ground-based solar plants is difficult to predict. Current estimates place the busbar power cost of intermediate-load solar plants in the vicinity of 50 mils/kWh. Eventually, this will become favorable with respect to the monotonically increasing cost

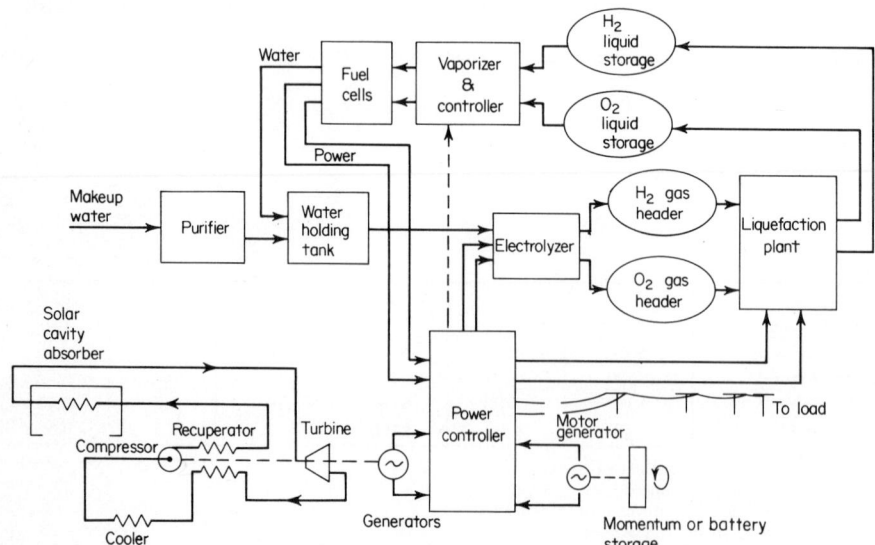

Fig. 19. Solar baseload plant concept with hybrid storage shown schematically.

of fossil fuel energy sources. Alternative nonfossil sources that may compete with solar plants include fission breeders, thermonuclear (fusion), and solar power from space.

REFERENCES

1. Woodcock, G. R., and D. L. Gregory: Economics Analyses of Solar Power, Ninth Intersociety Energy Conversion Engineering Conference (IECEC), San Francisco, Aug. 17, 1974.
2. Meinel, A. B., and M. P. Meinel: A Harvest of Solar Energy, *J. Opt. Sci.*
3. Lieberman, A. R., and W. E. Osmeyer: A 10% Efficient Economic RTG Design, Ninth IECEC, San Francisco, August 1974.
4. Savage, R. L., et al.: A Hydrogen Energy Carrier, NASA, ASEE, September 1973.
5. Cutting, J. C., and W. F. Amend: Status of the Two-Phase Liquid Metal MHD Program at the Argonne National Laboratory, Ninth IECEC, San Francisco, August 1974.
6. Rajakovics, G. E.: Energy Conversion Process with About 60% Efficiency for Central Power Station, Ninth IECEC, San Francisco, August 1974.
7. Anon.: Solar Thermal Conversion Mission Analysis, Aerospace Corporation report for the NSF, Contract NSF-C797, Jan. 15, 1974.

Solar Tower Energy Collector*

BY A. F. HILDEBRANDT† AND L. L. VANT-HULL‡

Solar energy can be usefully collected optically from a square mile area (2.6 square kilometers) or larger and concentrated onto a central receiver by an array of heliostats, i.e., independently steered flat mirrors. By judiciously spacing mirrors over 35 percent of the area, such a system in the Desert Southwest of the United States, for example, could collect 2800 megawatthours (thermal) per day in midwinter and almost twice this amount in midsummer. In order for the reflected radiation from this field to be efficiently intercepted, the central receiver would have to be placed atop a 330-meter tower. The authors propose that this heat be used when available to replace part of the fossil fuel burned in a conventional generating plant. An alternative is to produce a fuel, either in a closed-cycle transmission system or by the conversion of solid organic materials into oil.

The anticipated worldwide fuel shortages make desirable the development of energy alternatives (Ref. 1). Solar energy represents a large undeveloped resource that is both renewable and nonpolluting. There are no inherent technical problems preventing the use of solar energy, but because it is diffuse and intermittent, it has not generally been viewed to be competitive with other sources for uses other than home heating and cooling. In contrast, the authors find that the solar tower concept of concentrating solar energy for production of electricity, or possibly fuel, is economically viable. Although thermal storage of solar energy is possible, it is believed that the intermittent nature of solar energy can best be overcome by full integration into the electric and gas utility grids.

The direct solar energy irradiating a square kilometer of ground at noon on a clear day in the desert south of 35°N latitude produces over 400 MWth (megawatts thermal) of heat even in midwinter. We will now consider large-scale methods of optically concentrating this energy onto a central receiver. The system in Fig. 1 shows a large array of heliostats; the essentially flat mirrors are automatically steered to reflect or redirect the incident solar radiation to a central receiver atop a high tower, typically 330 meters high (Ref. 2). We assume that the terrain of the heliostat field is flat; however, a gentle south slope would be advantageous. After reflection from the mirrors, the redirected solar energy can be absorbed and converted to heat by a black body receiver placed in the focal region. The heat can be transported down the supporting tower via liquid metal and/or steam lines, and can be stored or used to operate a conventional turbine generating station. Alternative uses of the concentrated solar energy would be direct conversion to electricity via high power density solar cells placed in the focal region, or use of the heat to produce a fuel thermodynamically.

A typical heliostat system is shown in Fig. 2. Two-axis control can be obtained by either hydraulically or electrically operated servomechanisms that derive a signal from a simple position-sensing four-element phototube. The heliostats (4 to 10 m square) can be spaced over a field having a radius of two to three times the tower height so that the heliostats do not shade one another excessively and so that nearly all the reflected energy reaches the central receiver.

Solar collectors bearing an obvious resemblance to that described here have been studied by others (Refs. 3 and 4). Trombe has operated a solar furnace having 63 individually steered flat heliostats, each 6 × 7.5 meters (Ref. 5). He has achieved mirror aberrations and steering errors below 5 minutes of arc (1.4 mr). The energy from the 63 heliostats placed on a sloping terrain is directed onto a parabolic mirror formed on the

*Partially supported by NSF/RANN Grant GI-39456 to University of Houston and McDonnell Douglas Astronautics.

†Director, Solar Energy Laboratory and Professor of Physics; ‡Associate Professor of Physics and Solar Thermal Program Manager, Solar Energy Laboratory, University of Houston, Houston.

back of a building and brought to a focus. With this arrangement an equilibrium temperature, for 100 percent reradiation, of approximately 4100°K was attained—a truly remarkable feat inasmuch as the black body. temperature of the sun is 5762°K. Operating experience showed that the energy required to steer the heliostats was a fraction of a percent of the one megawatt of thermal energy collected.

An alternative approach involving linear concentrator arrays has been studied by Meinel and Meinel (Ref. 6) and more recently by Jordan, et al. (Ref. 7). The solar energy is concentrated onto a pipe inside an evacuated glass tube that is placed at the focus of a cylindrical mirror or lens. A "greenhouse effect" is produced within this pipe by coating it with a selective film that absorbs solar radiation, but does not reradiate readily in the infrared range characteristic of the operating temperatures. The heat is removed from the pipe and brought to a central point by a circulating fluid, either pumped flow or via a heat pipe, and is stored either as sensible or as latent heat in a reservoir of salts. Overall concentration ratios of about 100 are anticipated for such systems, which will result in

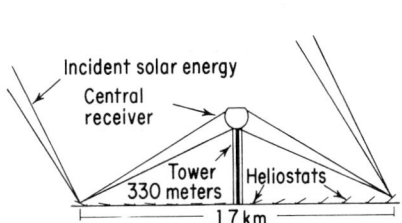

Fig. 1. Central receiver—solar tower energy collector.

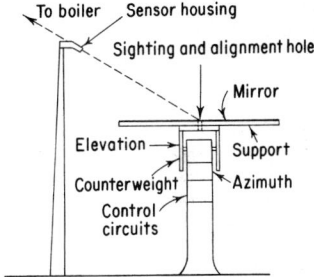

Fig. 2. Typical heliostat system.

operating temperatures of 500 to 600°K. Some difficulties expected for linear arrays are convective heat losses and fluid pumping losses, which would introduce a limit to plant size for efficient operation.

The solar tower collector outlined here avoids much of the heat transport problem expected with linear arrays by transporting the energy to a central point as light. In addition, the high concentration ratio obtained with the point-focus concept, makes possible operating temperatures of 1000°K or higher without excessive radiation losses and without utilizing selective coatings. In fact, the high power density may permit operation with the absorber exposed to the atmosphere, provided suitable long life absorbing surfaces can be found. It should be noted that only direct beam solar radiation is collected with any system utilizing a large geometric concentration factor; consequently operating efficiency is limited on hazy days.

The mirror surfaces can be made either of silvered float process glass or of metallized thin films with protective coatings. Each mirror would be mounted in a rigid support frame to form a heliostat. The angle subtended by the solar disk at the earth is about 0.5°, and we assume at present that the receiver presents twice this angle, or 1°, for the farthest mirror. This requires that optical deviations of the reflected light be limited to 0.1° from each of the principle sources; i.e., surface flatness, substrate deflections and pointing inaccuracy. The flatness tolerance gives an indication of the rigidity required in the mirror support frame to withstand effects due to gravity load and wind stress. The above tolerances are not as precise as those actually achieved by Trombe, but the temperatures we require are much lower than those he obtained.

In order to avoid deterioration of the surfaces due to sand and hail, a heliostat design is under study, in which the surface can be inverted or at least made vertical. Also, electrostatic methods are being investigated for keeping the surfaces clean. The frame structure is designed to withstand a 200 kilometers/hour wind; however, the heliostats need only have flexure properties adequate to avoid serious defocusing in winds below 65 kilometers/hour—appropriate for much of the time in the Desert Southwest.

The economics of solar energy collection using the tower concept depend very heavily on the ability to mass-produce high quality heliostats cheaply. Estimates show that 75 to

85 percent of the capital cost of heat production via the solar tower concept will lie in the cost of the installed heliostats.

The tower is a relatively straightforward structure which can be slip-cast of reinforced concrete and would require a year for construction. The important design parameters for the tower include wind forces, expected local seismic activity and weight of the receiver. The receiver and receiver-to-ground heat exchanger present some unique problems, e.g., absorbing the heat at a flux density which may be as large as 2 to 4 megawatts/m² because of the large concentration ratio we anticipate. The high heat flux can be handled with a liquid metal heat exchanger, or for steam the effective density can be lowered by multiple surfaces providing a larger effective area.

The question of scale or size has not been addressed directly and warrants comments. First, it requires the energy collected from an area comparable to a square mile (100 megawatts electric) to be of interest to power companies, and energy collection from hundreds of such plants will be required to impact the energy problem significantly. Second, if one replaces the 330-meter tower system with geometrically identical systems using 110-meter towers, nine such tower units would be required to provide the equivalent energy. Although the cost of the nine smaller collector units would be comparable to that of a single large tower unit, additional costs and losses would be incurred in connecting heat-transfer lines to a central generator in operating nine collectors. A further point results from the fact that heliostats smaller than about 20 m² are not economical because the cost per heliostat of the support, actuator, and steering systems is a substantial fixed component. Since the size of the heliostat contributes an effective optical aberration given approximately by its radius divided by the receiver distance, the use of large mirrors in a small-scale system causes large aberration which will lead to overly large receivers and less efficient operation. Curving the heliostat surface will produce a smaller focused spot; however, this has not yet been shown to be economically competitive with flat heliostats. Note that it is not economically sound to keep all dimensions fixed and simply reduce the height of the tower, as shading and blocking of the redirected power by neighboring heliostats soon become excessive. In addition, radiation losses from the receiver become comparable to the diminished solar energy abosrbed. There is in fact a minimum size tower for a 20-m² heliostat to be used for electric power generation which we find to be about 150 meters. The conclusion we reach is that bigger is better for the tower system, provided the tower costs either do not increase faster than the square of the tower height or are a small fraction of the capital costs of the collector. In the 100 to 500-m range these conditions are satisfied. Detailed calculations of the energy collection efficiency of the mirror field and the receiver are difficult, however, and require an extensive computer program. While not as economical as the large scale system, smaller scale systems using focusing heliostats (either curved surface or canted segments) are competitive with distributed collector systems down to the range of 1 MWe.

In this general concept, the authors assume that the heliostats will be permanently located and will be steered to direct the solar image toward a central tower. If γ is the (fixed) angle from a given mirror to the boiler, and ω is the (varying) angle from the mirror to the sun, then Snell's law requires the mirror to be lifted to the angle $\frac{1}{2}(\gamma + \omega)$. In general, this will raise one edge of the mirror above its center and change the area of ground shaded by the given mirror with respect to the sun and blocked from view from the boiler. Any part of any other mirror placed in such a shaded or blocked region will be of no use in producing redirected solar power. We thus define ρ, the average mirror utilization factor, i.e., the ratio of the energy reflected to the receiver by the mirror to the solar energy incident upon an equal area of level ground. If the mirrors were too close together, ρ could be less than one, either because of shading of the sun or blocking of the view of the tower. For an isolated mirror, ρ can be greater or less than one, depending on the angle of the sun above the horizon and the orientation of the mirror. ρ is the average for the entire array of mirrors.

In optimizing a mirror field, our procedure has been to first establish a criterion for the optimization, e.g., "minimize the summer noontime peak and maximize the energy collected over a winter day." For each hour of a typical day in the period of optimization (in this case, November 21 was selected, one month from the winter solstice), a few uniform density fields are computed to get preliminary data. On the basis of these runs, a preliminary taper, or spacing vs. distance, is chosen, which hopefully will give a *Useful Fraction of Mirror* greater than 0.9 everywhere in the field so long as the solar elevation exceeds about 15°. Normally the field should operate equally well in the morning and in the afternoon, so symmetric tapers are chosen, but this criterion can be relaxed if afternoon peaking requirements predominate. The output for this preliminary taper is studied critically, contours are plotted, and the taper is adjusted to give the desired result. Finally a set of four quarterly or twelve monthly runs is made, and curves of the redirected energy versus time are plotted as the primary output of this stage.

For a field optimized for winter operation, as previously discribed, a summary of the output information is shown in bar graph form in Fig. 3. Although a nominal tower height of 450 m is indicated, none of the results shown here are changed if the scale is changed. The height of the various bars appearing in the field represent the variation of the

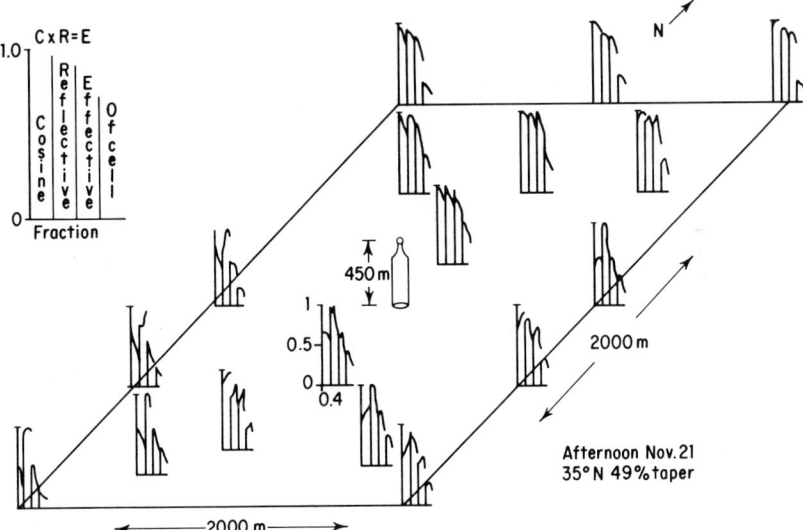

Fig. 3. Summary of a typical output from solar flux program, optimized for winter operation.

corresponding quantity with time for that region of the field. The quantities shown are plotted from noon to 4 P.M., but by reversing East and West the results are valid from 8 A.M. to noon as well. All the quantities plotted range from 0 to 1, and are C, cosine of angle of incidence; R, fraction of mirror usefully reflecting light; E, effective fraction of mirror; and finally P, the effective fraction of the cell. Thus $C \times R = E$ as indicated in the legend, and P is $\phi \times E$, where ϕ is the ground coverage factor for the mirror. Although relatively pictorial in nature, displaying the results in this form is very useful. Representative information from 20 pages of computer output is shown in this one display.

A collector field designed in this way could be extended indefinitely, but the density of mirrors at large distances from the tower must become very small in order to avoid excessive blocking. Simultaneously the image from a distant mirror eventually exceeds the receiver dimensions, and a new loss is introduced which ultimately becomes overwhelming and essentially defines the perimeter of the field, or the "trim." A first step in trimming the field is accomplished by choosing the location of the tower in response to the low values of the cosine of the angle of incidence on the southern boundary of the field. For the winter optimum we assume that the tower is located one-third of the way from the southern border of a field that is five tower heights on a side. The combination of all effects finally results in an essentially oval field, which has an area of 2.6 km² for a 330-meter tower height. Monthly curves of the power collected by this field are shown in Fig. 4. The performance of this collector field is determined by dividing the power redirected toward the receiver by the product of the mirror area and the insolation, and is, for example, 67 percent at noon on the winter solstice. At this point, effects of haze, mirror losses, and optical aberrations have not yet been included. Only shading and blocking,

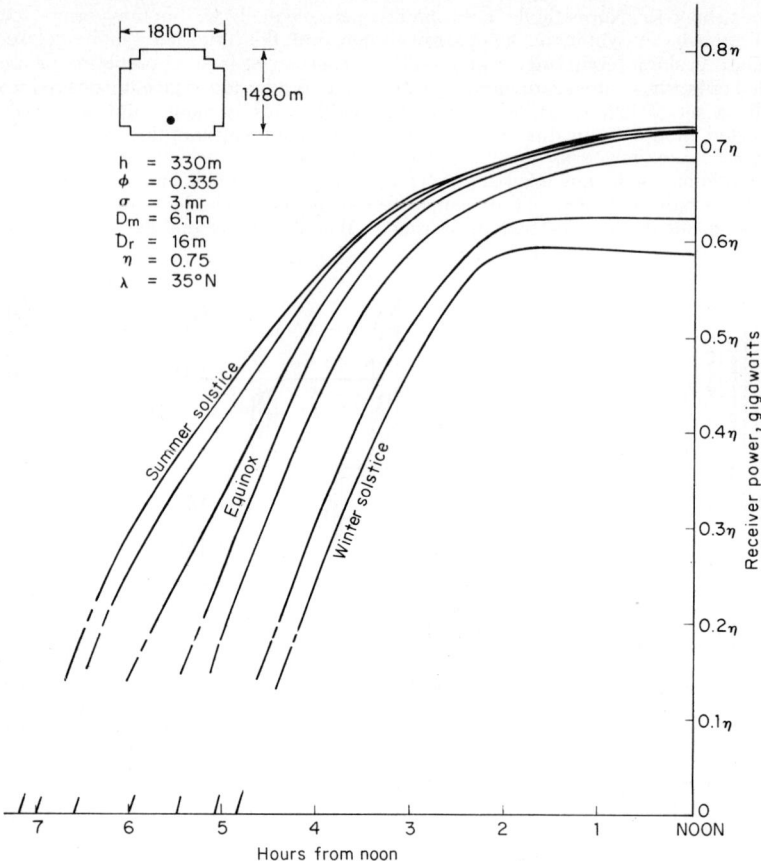

Fig. 4. Diurnal variation of power redirected to the receiver by the field for the 21st day of each month. A typical value of η is 0.75 as discussed in the text. This array could produce 150 to 200 MW of electricity during mid-day from a ground area of 2.5 km² and a reflective area of 0.82 km². The receiver radius is 8 meters and the optical errors are 3 mr. [Note: The direct solar beam intensity was derived from a model based on C. W. Allen's "Astrophysical Quantities," The Athlone Press, page 169, 1963, Clear Air Mode 1 (Ref. 8).]

foreshortening of the mirrors, and sunlight reaching the ground are included as losses here.

For the field described, winter operation is improved, but at the expense of peak summer performance; similarly, noontime performance is sacrificed for afternoon performance. Consequently, the seasonal and diurnal power curves are relatively flat, with the solar position having relatively little effect on the total energy collected. This mode of collection results in a minimization of the peak power, and thereby reduces the capital requirement for the receiver and the associated electric generator.

In order to evaluate the properties required of the receiver, let us consider the available power Q, including radiation and other losses from the receiver.

$$Q = \alpha \eta \phi \rho E \; A_l - \epsilon \sigma T \; A_r - Q_t \tag{1}$$

α is the receiver black-body absorption coefficient taken as 0.9. η taken as 0.75, is the fraction of the energy striking the mirrors which arrives at the receiver after a reasonable correction for haze, reflection and absorption by the mirrors, and net beam dispersion due to surface and aiming inaccuracies are accounted for. E_0 is the incident solar energy on flat ground in watts/m²; A_l is the area of land in the heliostat field. ϵ is the emissivity and is taken equal to α. σ is the Stefan-Boltzmann constant of 5.67×10^{-8} watt/(m²)(°K⁴). T is temperature in Kelvins. A_r is the radiating area of the receiver. The transport losses Q_t are due to pumping liquid and heat losses in carrying the energy collected to the central point of utilization. Ignoring the transport losses, which are only a few percent at most in this instance, the thermodynamic efficiency of a solar concentrator E_{sc} can be defined from Eq. (1) as

$$\frac{Q}{E_o \varphi A_l} = E_{\text{sc}} = \alpha \eta \rho \left[1 - \frac{\sigma T^4}{\rho E_o \eta} \cdot \frac{\varepsilon}{\alpha} \cdot \frac{A_r}{\varphi A_l} \right] \tag{2}$$

Now the overall thermodynamic efficiency of power production will be the product of Eq. (2) and the practical efficiency E_p of a particular engine. We take here the practical efficiency as $0.7E_c$, where E_c is the well-known Carnot efficiency. The overall efficiency is then a product of the practical efficiency and E_{sc}

$$E = E_p E_{\text{sc}} = 0.75 E_o E_{\text{sc}} \tag{3}$$

The efficiency given by Eq. (2) divided by $\alpha \eta \rho$, the efficiency of the Carnot cycle with an exhaust temperature of 50°C, and the practical efficiency E_p, are plotted in Fig. 5. Here ρE_0 is appropriate for noon in winter and is estimated from Fig. 4 to be 500 watts/m²; η is taken as 0.75. Note that the concentration ratio here is

$$C = \frac{\alpha \phi A_t}{\epsilon \ A_r} \tag{4}$$

where A_r is determined by the solar image size at the receiver. Thus, it is seen that for typical cyclindrical concentrators with $C = 100$ about 26 percent of the solar energy could be converted to electricity if the transport losses were truly negligible. For the point focus concentrator, the subject of this paper, a conversion efficiency in excess of 40 percent is expected for practical concentration ratios of order 1000 and larger. Operating temperatures of 550 to 650°C, manageable with present steam generating systems, are attainable without excessive radiation losses. In fact, radiation losses are sufficiently small that there is very little advantage in reducing the emissivity of the receiver; i.e., we use $\alpha = \epsilon$ and

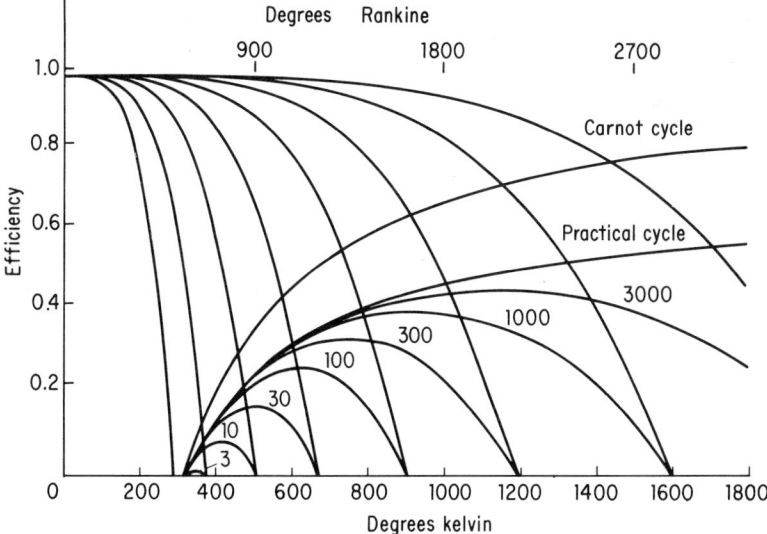

Fig. 5. Efficiency of the system is the product of solar collector efficiency E_{sc} and a practical cycle efficiency of seven-tenths of the Carnot cycle efficiency E_c. The Carnot cycle efficiency is taken as $(T - 323K)/T$. The collector concentration ratio appropriate for each concave product efficiency curve is noted at the vertex formed by the two respective curves. The concentrator and system efficiency are in units of $1/\alpha \eta \rho$. ρE_0 is taken conservatively as 500 watts/m² on flat ground in the winter at 35°N.

concentrate on achieving $\alpha \sim 1$. The power density at the receiver for concentration ratios of order 1000 to 2000 will be in the range of 2 to 4 MW/m² peak or 0.5 to 1 MW/m² average.

The first choice of heat transfer fluid is steam because it would appear not to require any new technology. However, because the flux density which must be absorbed and transferred to the fluid can be appreciably larger than in conventional steam plants, efficient operation may require developing new technology. Also, because of the large daily and seasonal heat flux variations the design of the receiver is not trivial and may ultimately be best accomplished by utilization of liquid metal such as sodium for heat transfer from the receiver surface to a steam line (Refs. 9 and 10). Liquid sodium technology has been developed for nuclear reactors, and operating temperatures of 550 to 650°C appear reasonable.

Sodium also presents a promising high temperature thermal storage medium. In general, due to the intermittent nature of solar energy, the thermal cycling of this high-temperature system will have to be considered carefully in any detailed design. The black body boiler surface should not deteriorate at high temperatures in the presence of air. If deterioration is a problem, and if convective losses are appreciable, an inverted cavity design can be used, as developed by Francia (Ref. 4), in order to avoid the use of vacuum jacketing but a present there seems to be no requirement for this.

SOLAR LIQUID-FUEL HYBRID

Solar energy may best be utilized at first by a steam generator operating in a solar-only mode with short-term storage of an hour or less to provide operational stability. Since the utilities require very high reliability, little new capacity credit would be given such a plant because of possible cloudiness. New plant credit could be given if there were possibilities of using liquid fuels, such as liquefied petroleum gas (LPG) or oil for backup. This could be accomplished by adding a simple low cost, but possibly inefficient, burner to add heat. Although liquid fuels may be in short supply, they are easily stored and afford an ideal way of giving a solar plant high reliability as an intermediate plant. Depending on actual operating conditions, it may be that very little liquid fuel is burned. Such a hybrid plant should be compared with a solar plant that has a gas turbine as a backup. An alternative approach might be to store solar energy as sensible heat, perhaps in underground cavities, or as proposed by Meinel and Meinel (Ref. 5), as latent heat.

Long-range development of solar energy, we believe, will be in the area of fuel production by thermodynamic means. Efficiencies of thermodynamic solar fuel cycles might be as large as 60 to 70 percent, and thus be twice as efficient as those of electrical-electrolysis cycles. The direct chemical conversion of water into hydrogen and oxygen in a single step requires temperatures that cannot be attained with the outlined collector, but excellent candidates for multiple-step processes are being considered (Ref. 11). Recently, the transmission of energy via a methane-water and hydrogen–carbon monoxide cycle (EVA-ADAM) has been discussed by Häfele (Ref. 12). Although this cycle is being studied in conjunction with nuclear reactors, it is well suited for solar energy if developed. Another candidate is the reversible reaction of methanol. The transmission of energy via a gas cycle, either closed (as in EVA-ADAM) or open (hydrogen), offers a possibility of integrating solar energy into gas utilities. The intermittent nature of solar energy can be overcome by storing gases in wells or by liquefaction of gases. It appears more desirable to store and transport solar energy in the chemical bond rather than in sensible heat. Thermal losses are negated, and there is a higher energy density in the chemical bond; furthermore, a thermodynamic fuel cycle is more efficient than the electrical cycle. A large fraction of our energy requirements are for process heat, such as could be supplied by a successful EVA-ADAM thermo-chemical cycle development at temperatures up to 500°C. The actual costs of a closed cycle gas transmission system have not been fully evaluated but are comparable to electric energy transmission costs.

Another attractive solar-fuel cycle is the conversion of organic waste to either oil or gas. There are large quantities of solid wastes that can be stored and converted when solar energy is available. Once a chemical such as oil is formed, it can also be stored economically because of its compactness. While gases, such as methane, hydrogen, and carbon monoxide may be produced, methyl alcohol is probably a more desirable fuel. Methyl alcohol is easily stored and can be used in the internal combustion engine with a minimum of modifications.

Finally, we would like to note that the environmental impact of solar energy is minimal. Removing solar energy from a region is no more severe than changing the landscape from green crops to desert sand or the inverse. For either electric power generation or a closed cycle gas transmission system, no air pollution would be encountered. Also for present energy production via the burning of fuel, about three gigawatts of new thermal heat is ultimately added to the biosphere for each gigawatt of electricity produced, whereas solar energy adds no new energy to the biosphere on a global basis. Sufficient raw materials are available for building thousands of solar tower systems, and studies to date show that only a very small fraction of the energy produced by a plant would be required to produce materials for a second plant. If solar energy is developed and proves economically competitive, it could have a significant impact on our energy production before the year 2000. Economic estimates now show that the heat produced with a solar concentrator is becoming competitive with rising oil costs. It appears reasonable to undertake building a small pilot plant of 0.2 km² and fully evaluate the technology and economics of integrating solar energy into the electric and gas utilities.

REFERENCES

1. Hubbart, M. King: Survey of World Energy Resources, presented at Plenary Session of the 75th Annual General Meeting of the Canadian Institute of Mines, Vancouver, Canada, Apr. 16, 1973. Published in *Mining and Metallurgical Bulletin*, July 1973.
2. Hildebrandt, A. F., G. M. Haas, W. R. Jenkins, and J. P. Colaco: Large Scale Concentration and Conversion of Solar Energy, *EOS Trans. Am. Geophys. Union*, 53(7) 684–692 (1972).
   Vant-Hull, L. L., and A. F. Hildebrandt: A Solar Thermal Power System Based upon Optical Transmission, *Proceedings of the Solar Thermal Conversion Workshop*, Arlington, Virginia, 1973; NSF-RA-N-74-125, November 1974.
   Hildebrandt, A. F., and L. L. Vant-Hull: A Tower Top Point Focus Solar Energy Collector, *Proc. Hydrogen Econ. Miami Energy Conf.* (THEME), University of Miami, Miami Beach, Fla., March 1974.
   Vant-Hull, L. L., and A. F. Hildebrandt: Solar Thermal Power System Based on Optical Transmission, *Solar Energy*, 18(1) (1976).
   Hildebrandt, A. F., and L. L. Vant-Hull: A Tower Top Focus Solar Energy Collector, *Mech. Eng.*, pp. 23–27, September 1974.
3. Baum, V. A., R. R. Aparase, and B. A. Garf: High Power Solar Installations, *Solar Energy*, 1(2) 6–13 (1957).
4. Francia, G.: Pilot Plants of Solar Steam Generating Stations, *Solar Energy*, 15(1) 57–61 (1973). Reference is also made to an unpublished report, dated June 25, 1973: "Studio Di Campo Specchi E Caldia per lo Sfruttamento in grande Dell' Energia Solare." Also reference is made to a private communication.
5. Trombe, F., and A. Le Phat Vinh: Thousand KW Solar Furnace Built by the National Center of Scientific Research in Odeillo (France), *Solar Energy*, 15(1) 57–61 (1973).
6. Meinel, A. B., and M. P. Meinel: Physics Looks at Solar Energy, *Physics Today*, 25(2) 44–50 (1972).
7. Jordan, J. C.: Private communication.
8. Allen, C. W.: "Astrophysical Quantities," p. 169, The Athlone Press, Washington, D.C., 1963.
9. Bauley, J. A., and B. V. Green: The Design and Development of Boilers for the Sodium-cooled Fast Reactor, *Proc. Symp. Progr. Sodium-cooled Fast Reactors*, pp. 527–548, International Atomic Energy Agency, Vienna, 1970.
10. Friedland, A. J.: Coolant Properties, Heat Transfer, and Fluid Flow of Liquid Metals, pp. 16–88 in J. Yevick (ed.), "Fast Breeder Reactor Technology Plant Design," M.I.T. Press, Cambridge, Mass., 1966.
11. De Beni, G., and C. Marchetti: Hydrogen, Key to the Energy Market, *Eurospectra*, IX(2) 46 (1970).
12. Häfele, W.: Energy Choices that Europe Faces: a European View of Energy, *Science*, 84(4134), 364 (Apr. 19, 1974).

Solar Absorption Coating and Heat-Pipe System

This proposed system uses a selective solar absorption coating capable of operating at high temperatures; and a heat pipe for absorbing and conveying the collected solar energy. See schematic diagram (Fig. 1). The main system components are: (1) a solar concentrator which focuses solar energy on the heat pipe, (2) the heat pipe with a selective optical coating, (3) a heat transfer loop connecting the heat pipe to a thermal

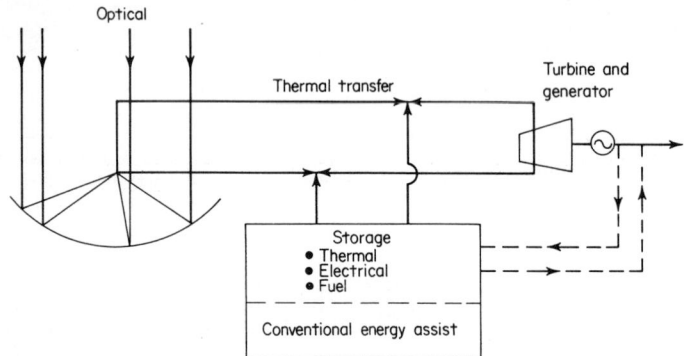

Fig. 1. Schematic diagram of a solar power plant. A parabolic mirror or an array of plane mirrors concentrates the solar radiation optically on the surface of a pipe or vessel. The radiation is absorbed there and converted to heat, which is picked up by a fluid and transported through pipes to a central location. It then is converted by a thermal power plant into electric energy. (Rann.)

storage unit, and (4) a thermal storage unit which receives thermal energy from the transfer loop and releases it to the power cycle working fluid.

Some of the specific tasks that have been undertaken in the development of this system* include: investigation of (1) solar reflector surface life, (2) solar absorber coating life, (3) gravity-aided long heat pipes, (4) heat transfer and pressure drop in a transfer loop, (5) heat transfer in a central storage facility, (6) heat transfer in a decentralized storage facility, (7) overall system design, and (8) module demonstration.

An underlying philosophy of the proponents of this system is the application of new materials and methodologies, such as selective solar absorber coatings, durable mirror coatings, low-temperature saturated steam turbines, and high-temperature gas heat transfer, which have come out of the aerospace and nuclear industries in very recent years— and hence were not available to the earlier pioneers of solar energy capture projects. It is claimed that these technological advantages can be coupled with the economics of automated mass production of several of the elements of the system for which thousands of like components would be required. Also, some investigators have recognized the likely attractiveness of a system of this type as a source of peak power, where the system can be used by an electric utility in conjunction with a nuclear or fossil-fuel-fired base-load generating facility. Studies have shown that a major cost of the solar thermal power plant is storage. Even as little as three hours' storage capability requires about 20 percent of the power plant capital investment, as indicated in Fig. 2. Storage of one or two days adds very considerably to the power plant costs. With only three hours' storage capacity, how can a solar power plant function over a daily cycle and cloudy weather? A solar plant

*Some of this work has been sponsored by National Science Foundation Grant GI-34871, initiated July 1, 1972.

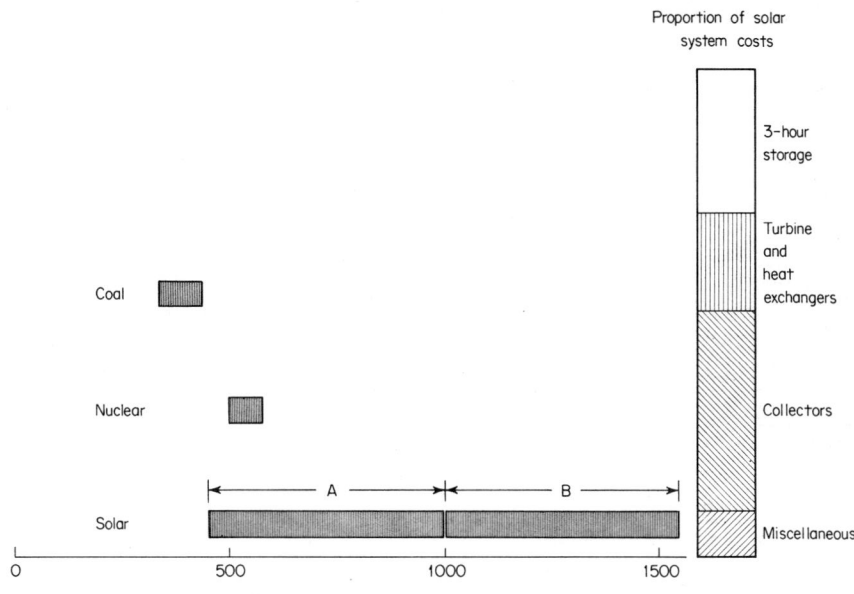

Fig. 2. A major cost of the solar thermal power plant is storage. As little as three hours of storage capacity requires about 20% of the power plant capital investment. *(A)* Few hours of storage. *(B)* 1 to 2 days of storage. (Rann.)

probably will not stand alone, but will be integrated in a power grid with a base-load facility. As such, it may be possible to use the solar power plant as a load-following plant. Figure 3 shows how load and insolation coincide in summer and winter. Thus, the solar plant can generate power during the peak load periods of the day and year. There is a correlation between the weather conditions and a load on a utility—air conditioning being a major draining factor. For example, a study in Dallas, Texas, showed that the utility load would drop 40 percent if the temperature dropped from 105 to 85°F. Peak load is considered to be more expensive for a utility to supply than base-load power. Utilities often lose money supplying this peak load because it costs more to generate the power than regulations permit them to charge for it. Peaking is often done with gas- or oil-fired turbines, the costs of which are relatively higher than energy costs in generating the electricity for base-load requirements.

In a RANN* study made by the Aerospace Corporation, it was found that solar thermal power has a higher probability of becoming cost-competitive with conventional power in the peaking and intermediate application than in a base-load operation, as shown in Fig. 4. The study further indicated that by 1990, a solar plant could be competitive if the collector field costs could be held to $15 to $25 per square meter.

Collector Module. The basic collector configuration proposed is a parabolic trough concentrator (overall length approximately 4 meters) mounted so as to track the sun. The focal line of the concentrator coincides

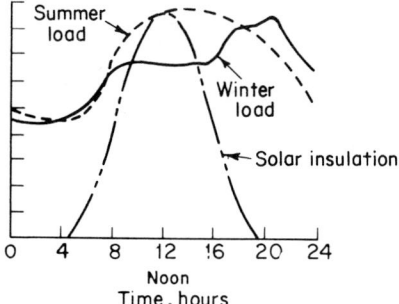

Fig. 3. Coincidence of load and solar insolation in summer and winter.

*Research Applied to National Needs.

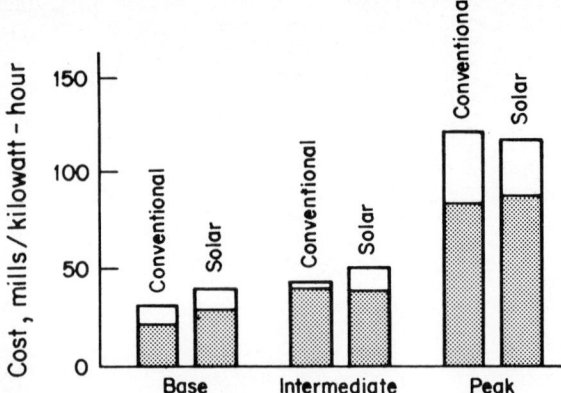

Fig. 4. Solar thermal power has higher probability of becoming cost-competitive with conventional power in peaking and intermediate applications than in base-load operation. (Rann.)

with the axis of rotation of the supports. The rotation is accomplished with a reversible motor controlled by a sun sensor. The schematic diagram of a concentrator is shown in Fig. 5.* A concentrator is an assembly of a substrate, two end plates, and a reflecting surface. The focal length is 19.4 centimeters, and the *f* number is 0.159. Each concentrator weighs about 34 kilograms. The substrate is the primary structural part of the collector and is a sandwich assembly of two epoxy skins and an aluminum core, as shown in Fig. 6. The skins are made from three plies of fiberglass cloth, wetted with room-temperature-curing epoxy resin and are each 0.127 centimeter thick. The core is 2.54-centimeter-thick aluminum flex core. The skins are bonded to the core with a room-setting epoxy adhesive. The reflecting surface is a highly specular deposition of aluminum and silicon oxide on an acrylic carrier. The finished surfaces are designed for integrated solar reflectances on the order of 85 to 87 percent.

The overall collector configuration (used for testing purposes) is shown in Fig. 7. A pair

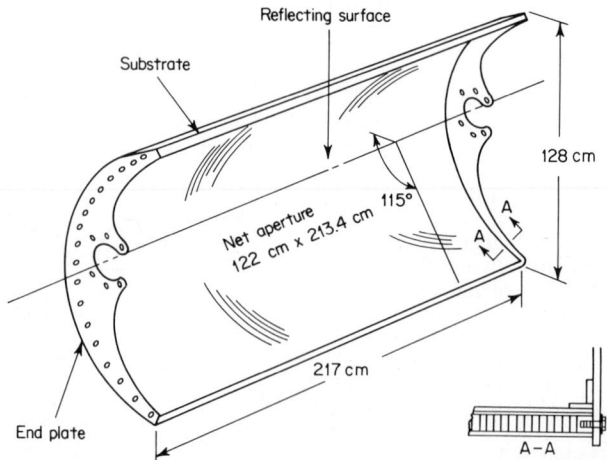

Fig. 5. Schematic diagram of concentrator for use in solar thermal system.

*Materials, dimensions, and specifications, of course, are subject to change as the result of continuing research and development.

of cylindrical parabolic trough concentrators are mounted on special trunnions having their center of rotation at the focal axis of the concentrators. The trunnions are supported by pedestals which are, in turn, mounted on a base made of channel iron.

A single-axis tracker (Fig. 8) has been designed, which will automatically keep the solar image centered on the heat pipe. The tracking loop consists of a sun angle sensor, control electronics, drive electronics, and a reversible motor drive. The sun angle sensor, (Fig. 9a) is a photopotentiometer mounted behind a long rectangular slit. The equivalent circuit of the photopotentiometer is shown in Fig. 9b. The band of light formed by the image of the slit falls on the photoconductor, forming an electrical path between the deposited resistor and the conducting bus. The position of the band of light is determined by the angle of incidence of the sun's rays. If a fixed voltage is applied across the resistor element, the voltage appearing on the conducting bus is proportional to the sun angle. At angles near normal incidence, the relationship between voltage and angle is nearly linear. This voltage is applied to the control electronics to servo the mirror position to the desired angle with respect to the sun. It is also fed to the instrumentation system to provide a record of the pointing offset. Acquisition angle, i.e., the angle within which automatic tracking can be maintained, is approximately plus or minus 15 degrees. A commercial dc servoamplifier provides the driving current for the reversible motor drive. To prevent the heat pipe temperature from rising above a dangerous level, a temperature limit switch can be used as a safety feature.

As shown by Fig. 7, the absorber (in the form of a heat pipe) is located at the focal line of the mirror. The fluid in the heat pipe transports the collected energy to a heat exchanger located at one end of the collector. The amount of energy transferred to the fluid in the absorber pipe depends upon the

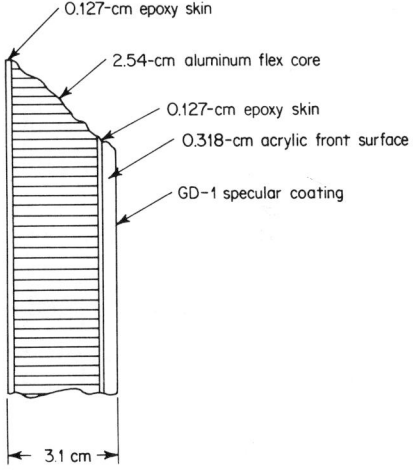

Fig. 6. Structure of concentrator.

optical properties of the absorber surface. During an early part of the investigation of this system, the input of energy to and output of energy from the absorber pipe were subjected to computer analysis. Input factors included: (1) latitude on earth; (2) time of day; (3) time of year; (4) orientation of tracking axis relative to local north and local vertical; (5) length and width of mirror; (6) focal length; (7) position of surrounding collectors (to evaluate shadowing); (8) the geometry of concentrator end obstructions, such as hubs and heat storage elements; (9) empirical data of mirror reflectivity (as a function of wavelength and incidence angle); (10) slope errors of mirror surface; (11) misalignment in pointing; (12) angular extent of sun; (13) optical properties of glass envelope with respect to focal axis of the mirror; (16) absorber (receiver) outside diameter; (17) empirical radiation properties of absorber surface (as a function of wavelength and incidence angle); (18) position of absorber with respect to focal axis of the mirror; (19) empirical solar flux as a function of time of year and day; and (20) number of rays to be traced by Monte Carlo ray trace methods.

Output factors included: (1) expected value of the energy rate absorbed at the specified time, (2) angular position of the sun relative to local north and vertical as a function of solar time, (3) optimal tracking angle of the concentrator about the specified tracking axis, (4) map of the collector surface showing the areas shadowed by surrounding collectors and end obstructions, (5) fraction of energy lost at each optical surface; and (6) fraction of energy not focused to the receiver.

The study has included investigations of the performance and life expectancy of various solar absorber coatings, including AMA* ($\sim$ 600Å Al_2O_3, $\sim$ 200Å MoO_x, $\sim$ 600Å Al_2O_3 on $\sim$ 6000Å Mo on SS). Composition-versus-depth profiles of AMA coatings were studied

*Honeywell Inc.

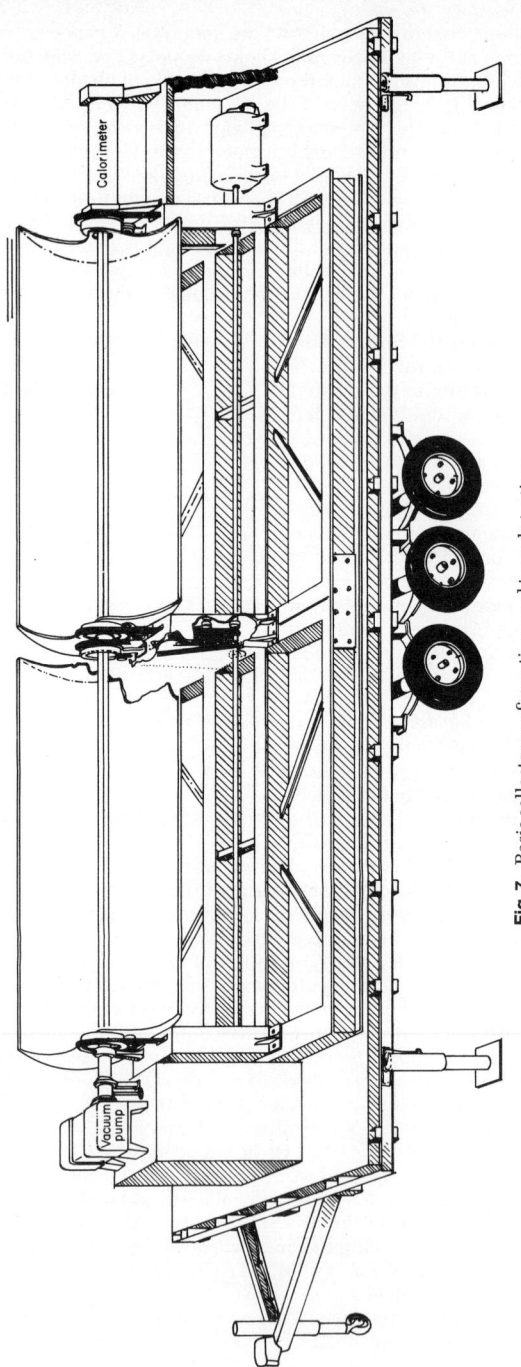

Fig. 7. Basic collector configuration used in early testing.

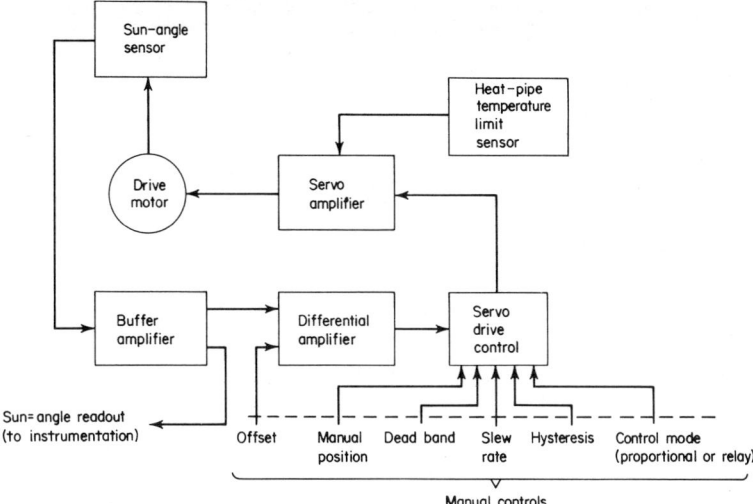

Fig. 8. Sun-tracking servo system.

after various temperature treatments to establish their diffusion behavior. Various film substrate combinations, as they appear in the solar absorber coating, were also investigated.

   Heat Pipe.   Studies in this area concentrated on the detailed design and fabrication of heat pipe for further testing. The compatibility of 347 stainless steel with water was studied; also with steel-clad copper heat pipes using water as the working fluid. The compatibility tests involving 347 stainless steel and water indicated that any gas generated within the heat pipe occurred primarily in the first few hundred hours of service. Such gas can be removed from the system, with little or no additional gas generated thereafter. Tests of steel-clad copper were discontinued because of material fatigue.

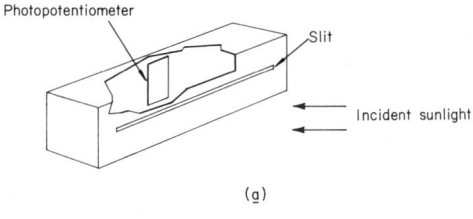

(a)

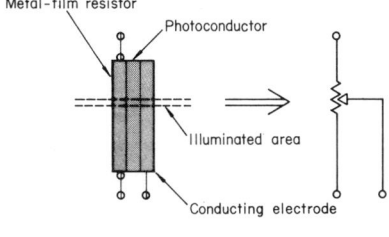

(b)

Fig. 9. Sun angle sensor: (a) Cutaway view. (b) Equivalent circuit of photopotentiometer.

Transfer Loop.   The nighttime cooling and the subsequent early morning warming of the pipelines and other fluid-carrying components of the transfer loop were analyzed. The transient cooling and warming studies were performed for three different configurations: (1) the pipelines and heat exchangers of a transfer loop in which the heat exchangers are arranged in parallel, (2) the pipelines and heat exchangers of a transfer loop in which the heat exchangers are arranged in series, and (3) the absorber tubes of a pressurized-water transfer loop.

In all three instances the flow circulation in the loop was assumed to be discontinued during the nighttime hours owing to heat losses to the surroundings, the metal walls of the pipelines and heat exchangers, the fluid contained therein, and the insulation cool down. In the morning, the return of insolation causes a reheating of the loop so that, eventually,

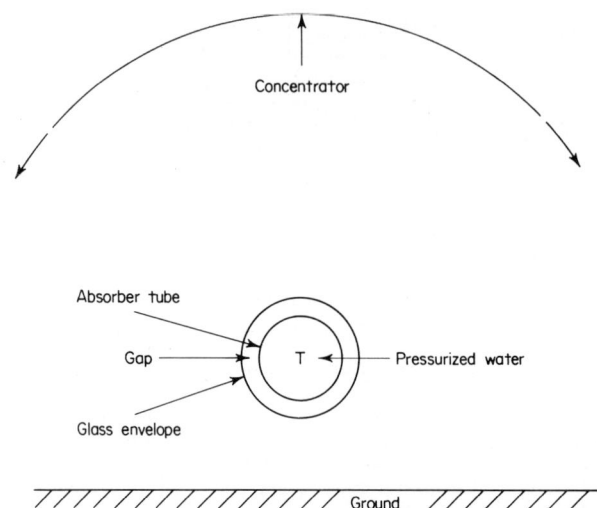

Fig. 10. Schematic diagram of computational model for cool-down of a forced-convection absorber tube.

the operating temperature is restored and flow circulation is reinitiated. In instances (1) and (2), where heat pipes are used as the solar absorber tubes, the nighttime heat losses from the collector modules are negligible, owing to the diode action of heat pipes. The computations were performed for a 0°C ambient temperature. The influence of a night-sky temperature of −40°C was also examined.

The schematic diagram of a computational model for cool down of a forced convection absorber tube is shown in Fig. 10, and that of a computational model for examining boundary condition at the outside of insulation is presented in Fig. 11. The transfer loop with heat exchangers arranged in series appears in Fig. 12.

There are two aspects of the cool-down and warm-up transient operations that are of particular relevance. The first is the extent of the loop temperature decrease during the nighttime hours and, in particular, how close the fluid temperature gets to its freezing point. The second is the number of sunshine hours in the early morning that are required to recoup the heat lost during the night.

For the loop with heat exchangers in parallel array and with a fluid temperature of 276°C in the hot pipelines and heat exchangers at the onset of nightfall, the calculations show that the water temperature in the loop remains above 200°C throughout the night, provided that insulation of good quality is used. The duration of the morning warm-up period is about 1.5 hours. In the series system, the water temperature drops to a somewhat lower level, but is still above 150°C. On the other hand, the morning warm-up time is somewhat shorter than for the parallel system.

The absorber tubes of a pressurized-water system reach lower temperatures during cool down than those just described, the reason being that these tubes are not insulated. To prevent freezing, it is necessary that a hard vacuum be maintained in the space between

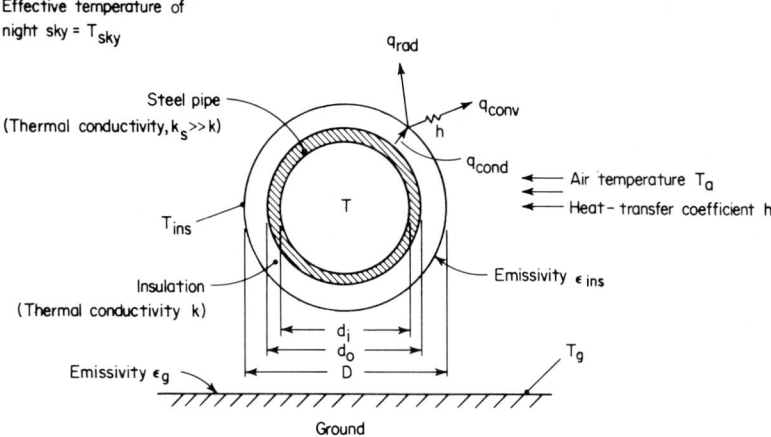

Fig. 11. Schematic diagram of computational model for examining boundary conditions at outside of insulation. Effective temperature of night sky = T_{sky}.

the absorber tube and its surrounding glass envelope. Morning warm-up times of about two hours are indicated by the calculations.

During daytime hours, there is a variation of time of the solar power absorbed by the collectors, the extent of the variation being strongly dependent on how the sun is tracked. For example, for a one-axis east-west tracker, the peak absorbed insolation is substantially greater than the daytime-mean value. Computations were performed for the parallel-array loop to examine how the heat loss and pumping energy results are affected by the timewise variation of the absorbed solar power. The daytime-integrated loss fraction was compared with the instantaneous loss fraction corresponding to the daytime mean insolation. It was found that the two loss fractions were in close accord, indicating that computations performed for daytime-mean conditions provide a useful assessment of the losses sustained during the entire day.

Steady-state operating characteristics of the three systems were examined. One of these systems is characterized by heat exchangers arranged in series along a feeder line. In each heat exchanger, vapor from a heat pipe condenses in an annular sleeve surrounding the

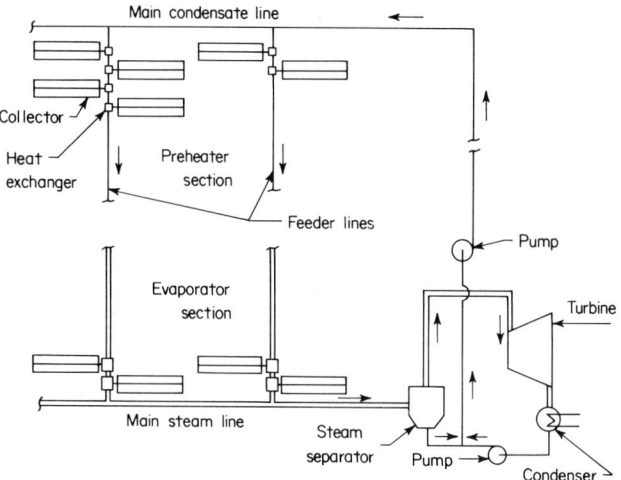

Fig. 12. Schematic diagram of a transfer loop with heat exchangers arranged in series.

feeder line. As the transfer loop fluid passes through the successive heat exchangers situated along a feeder line, it is initially heated to the saturation temperature and subsequently boils in a state of two-phase flow. The results of the computations indicated that the heat losses sustained by such a loop configuration are lower than those for a parallel-array system. However, the losses of both systems can be held to less than 5 percent of the absorbed solar power with the use of suitable insulation. Possible instabilities associated with boiling in two-phase flow have to be examined to ensure the viability of the series arrangement.

The use of pressurized water as the transfer loop fluid eliminates the vaporization process from the solar collection field. The hot pressurized water provided by the solar

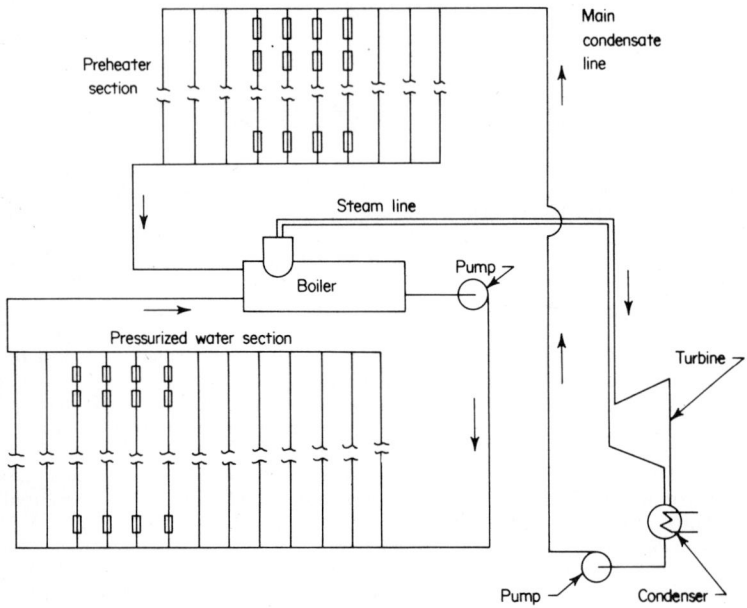

Fig. 13. Schematic diagram of a system having a pressurized-water transfer loop.

field can serve to heat and boil a second fluid, e.g., water, at a lower pressure for use in the power plant. See Fig. 13. The heat exchanger used to transfer energy from the pressurized water to the power-plant fluid has been analyzed. Calculations indicate that it is feasible to obtain power plant steam at 60 bars pressure from a pressurized water loop by means of a central boiler-heat exchanger. As shown by Fig. 13, there are two distinct flow loops. One is a pressurized-water loop which collects energy from the solar field. It operates at a maximum temperature of 300°C and a pressure of about 90 bars. The second is the water-steam loop of the power cycle. In this loop, feedwater coming from the main condensate lines is preheated in its own field of solar collectors to the saturated liquid state at a pressure of about 60 bars. It then enters the boiler where it is heated by tubes containing the pressurized water at 300°C. The saturated steam thus generated passes on to the turbine. Control of the flow leaving each feeder line allows regulation of the temperature produced by the preheater and pressurized-water sections. In such a system, the boiler itself, if of sufficient volume, could act as a steam accumulator for short-term storage of energy.

Boiler Concept.   The generation of steam is accomplished in a shell-and-tube heat exchanger having hot pressurized water flowing through the tubes. The preheated water of the power-cycle loop nearly fills the shell, completely covering the heating tubes. A conceptual design of such a boiler is given in Fig. 14. Boiling takes place on the tubes, and the steam generated leaves through a pipe at the top of the shell.

Considering the first law of thermodynamics, the required flow rate of the pressurized water can be determined by

$$\frac{\dot{m}_p}{\dot{m}_s} = \frac{h_{fg}}{c_p(T_{p1} - T_{p2})}$$

where c_p = 5.40kJ/(kg)(°C), specific heat of pressurized water
$\quad h_{fg}$ = latent heat of vaporization
$\quad \dot{m}_p$ = mass flow rate, steam
$\quad T_{p1}$ = temperature of entering pressurized water
$\quad T_{p2}$ = temperature of leaving pressurized water

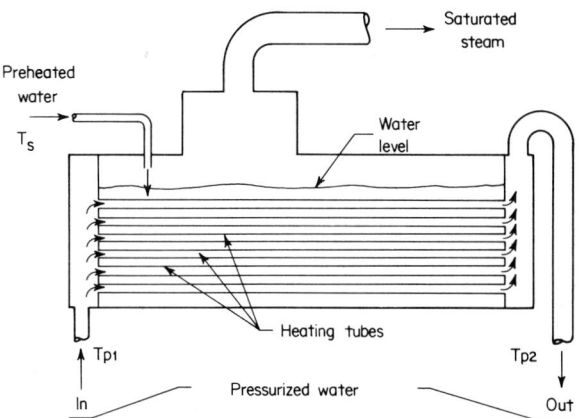

Fig. 14. Conceptual design of boiler heated by pressurized water.

The pressurized water must exit at a temperature slightly higher than that of the steam. This temperature difference $(T_{p2} - T_s)$ is a parameter that determines how effective the heat exchanger is and also influences the flow rate ratio $\dot{m}_p/\dot{m}_s$.

A plot of the mass flow ratio versus steam temperature with $(T_{p2} - T_s)$ as a parameter is shown in Fig. 15. Since it is desired to limit the fluid temperature in the pressurized loop to 300°C, T_{p1} has been held constant at this value. The figure shows that in order to produce saturated steam at a pressure of 60 bars (T_s = 275°C), the pressurized water flow rate is 15 times that of the steam if the outlet temperature difference is $T_{p2} - T_s = 5°C$. At higher outlet temperature differences, the mass flow ratios are even higher. The foregoing shows that the circulation system for the pressurized-water loop will necessarily have to pass very large flow rates, so that pumps and piping must be of larger size than for the parallel and series steam generation systems mentioned earlier.

HEAT STORAGE IN LIQUID WATER BY DIRECT CONTACT

Calculations have been performed to determine whether the heat stored in the loop piping, insulation, and fluid would be sufficient to enable the system to produce steam and continue to operate during short periods in which there is no receipt of energy from the sun. Such calculations indicated that system operation could be sustained by the stored heat for only a few minutes. Consideration was given to numerous possible storage media. Hundreds of eutectic mixtures which melt in the 300 and 500°C range were identified. Ten of them were elected as having desirable properties for a storage system utilizing a melting-solidifying process in the 300°C temperature range. Of these, sodium chloride–sodium nitrate and sodium chloride–sodium nitrate–sodium sulfate cost only about $0.14 per pound or $0.30 per kilogram in 100-pound quantities (mid-1974 figures). Such materials would have to be used in elements with thicknesses of a few centimeters to avoid excessive temperature variations during the operation of the heat-storage system.

One of the attractive candidate concepts for storage of thermal energy is to change the temperature of a single-phase medium, such as liquid water. During the time when energy flows into the storage medium, the temperature of the medium obviously increases. In solar energy collection fields where vapor is produced, the vapor can be introduced directly into the liquid-storage medium. The direct contact between the condensing vapor bubbles provides intimate contact with the liquid, and, in effect, there is a resultant heat-exchange mechanism which is much simpler and more efficient than that provided by conventional heat exchangers. The condensation of the vapor bubbles ensures that all the heat intended for storage is actually stored.

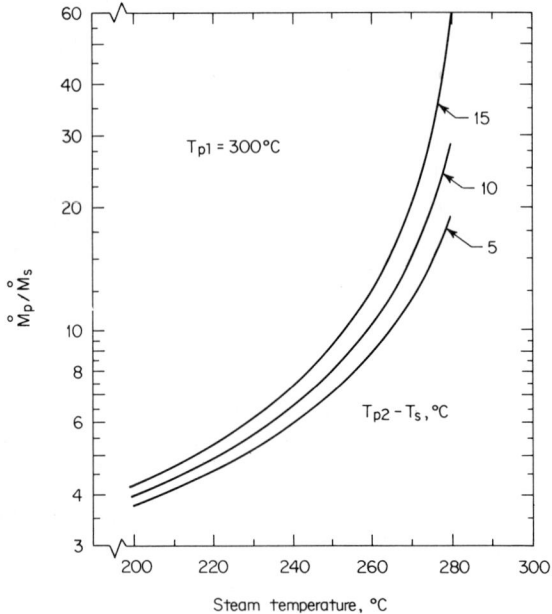

Fig. 15. Ratio of pressurized-water mean flow rate to steam mass flow versus steam temperature.

Experiments have been undertaken to obtain a basic data base to facilitate the design of a thermal storage unit in which steam is introduced into a liquid water storage medium. For this purpose, a test facility has been designed to operate at temperatures and pressures as high as 275°F and 60 bars.

REFERENCES

Research Applied to Solar Thermal Power Systems, Report NSF/RANN/SE/GI-34871. Series of reports commencing with semiannual progress report, July 1, 1972, and continuing through 1973 and 1974. These reports prepared under Grant GI-34871 from the National Science Foundation Advanced Technology Application Division of RANN, Washington, D.C.
Jordan, Richard C., Ernst R. G. Eckert, James W. Ramsey, Roger N. Schmidt, Ephraim M. Sparrow, and Gottfried K. Wehner: The Minnesota/Honeywell Terrestrial Solar Power System, The Sun in the Service of Mankind International Congress, Paris, sponsored by UNESCO and the French Government, July 2–6, 1973.
Eckert, E. R. G.: Solar Power, AIAA Paper 73–710, presented at AIAA 8th Thermophysics Conference, Palm Springs, Calif., July 16–18, 1973.
Eckert, E. R. G., R. C. Jordan, E. M. Sparrow, G. K. Wehner, J. W. Ramsey, and R. N. Schmidt, Solar-Thermal Electric Power Generation, paper presented at International Solar Energy Society, U.S. Section, Cleveland, Oct. 3–4, 1973.
Drake, R. M., Jr., and E. R. G. Eckert: "Analysis of Heat and Mass Transfer," McGraw-Hill, New York, 1972.

Beach, R. R., M. J. Bell, J. H. Hicks, and F. J. Pocock: Preoperational Water Chemistry Control for Nuclear Steam Supply Steams, Technical Paper RDTP73-3, presented to American Power Conference, 1973.

Deverall, J. E.: Mercury as a Heat Pipe Fluid, *Los Alamos Sci. Lab. Rep.* LA-4300-MS, January 1970.

Mishra, L. N.: Prospects of Power Plant Cycles Using Refrigerants, *Am. Soc. Heat., Refrig., Air Cond. Eng. J.*, January 1972.

Cunnington, G. R., and C. L. Tien: Heat Transfer in Microsphere Cryogenic Insulation, Paper C-1, Cryogenic Engineering Confreence, Boulder, Colo., August 1972.

Bajura, R. A., V. F. LeRose, and L. E. Williams: Fluid Distribution in Dividing, Combining and Reverse Flow Manifolds, ASME Paper 73-Pwr-1, Joint Power Generation Conference, New Orleans, La., Sept. 16–19, 1973.

"Steam—Its Generation and Use," 38th ed., Babcock & Wilcox Co., New York, 1972.

Tong, L. S.: "Boiling Heat Transfer and Two-Phase Flow," Wiley, New York, 1965.

Eckert, E. R. G., and R. N. Schmidt: Solar Energy-Thermal Conversion, *Proc. 1st Symp.* on RANN: Research Applied to National Needs, National Science Foundation, Washington, D.C., 1973.

High-Temperature Solar Energy

BY JESSE D. WALTON, JR.*

Clean, nonpolluting high-temperature energy can be obtained directly from the sun through the use of solar furnaces. Such devices are actually solar concentrators which use mirrors or lenses to collect and concentrate this abundant, renewable energy resource. Although high-temperature solar devices have been known throughout recorded history, it has been only within the last 25 years that solar furnaces have been developed to the point that their potential as a commercial high-temperature energy source can be determined.

DEVELOPMENT OF LARGE SOLAR FURNACES

In 1948, under the leadership of Professor F. Trombe, the Centre National de la Recherche Scientifique (CNRS) in Paris undertook the design, construction, and development of the world's first large solar furnace at Montlouis in the French Pyrenees mountains (Ref. 1). This furnace was completed in 1952, and provided 50 kilowatts of thermal energy. The Montlouis solar furnace became the prototype design for other large high-temperature solar furnaces. Basically, this design utilized a single large heliostat (array of numerous flat mirror elements), which continuously tracked the sun to direct the sun's rays onto a concentrating reflector (parabolic or spherical), consisting of many smaller mirror elements, each of which was contoured to concentrate the incident radiation at a common focal point. In the Montlouis furnace, the heliostat was 13 meters wide and 10.5 meters tall, and contained 540 flat mirrors each 50×50 centimeters. The concentrating reflector was made up of 3500 mirrors 16×16 centimeters arranged in a parabolic configuration 36 feet wide and 30 feet high with a focal length of 6 meters. Each of the 3500 flat mirror elements in the parabolic concentrator was mechanically contoured and aligned to focus the radiation received from the heliostat onto the focal point of the parabola.

The successful performance of the Montlouis solar furnace led to the use of its design as the prototype for the next three large single heliostat-concentrator solar furnaces, which were to be built during the next 20 years. All three of these furnaces were similar to the Montlouis furnace in size, operation, and thermal power level, and were constructed by (1) The U.S. Army Quartermaster Corps, Natick, Massachusetts (Ref. 2); (2) Tohoku University, Sendai, Japan (Ref. 3); and (3) the French Army's Laboratoire, Central de L'Armement, Odeillo, Font-Romeu, France (Ref. 4). In 1973, the U.S. Army's solar furnace was moved to the Nuclear Weapon Effects Laboratory, White Sands Missile Range, New Mexico, where it became operational in 1974. The characteristics of the mirror elements used in these three solar furnaces, together with those for the CNRS solar furnace at Montlouis, are presented in Table 1.

Although the Montlouis solar furnace played a major role in developing applications for high-temperature solar energy and in providing design information for the three other large solar furnaces, its most valuable contribution to the field of high-temperature solar energy was the experience and background which it provided the CNRS Solar Energy Laboratory that led them to design and construct the world's largest solar furnace, the CNRS 1000-kilowatt solar furnace.

THE CNRS 1000 KILOWATT SOLAR FURNACE

This furnace is located at Odeillo, Font-Romeu (altitude of 5900 feet), about 25 miles east of Andorra and 5 miles west of Montlouis (Ref. 5). At this location the sun shines as many as 180 days a year, and solar intensities as high as 1000 watts per square meter are

*Technical Manager, Solar Energy Programs, Applied Sciences Laboratory, Engineering Experiment Station, Georgia Institute of Technology, Atlanta, Georgia.

TABLE 1. Characteristics of Large, Single-Heliostat Solar Furnaces
35–50-kilowatt thermal capacity

	CNRS, Montlouis, France	Tohoku University, Sendai, Japan	French Army, Odeillo, Font-Romeu, France	U.S. Army, Nuclear Weapon Effects Lab.,* White Sands Missile Range, New Mexico	
Date first operated	1952	1962	1972	1974	
HELIOSTAT					
Size					
Meters	10.5 × 13	14 × 15.5	13.2 × 17.5	11 × 12.2	
Feet	34 × 43	46 × 51	43 × 57	36 × 40	
Number of mirrors	540	238	638	356	
Mirror size					
Centimeters	50 × 50 .	90 × 100	50 × 50	62 × 62	
Inches	19.7 × 19.7	35.4 × 39.4	19.7 × 19.7	24.4 × 24.4	
Total mirror area					
Square meters	135	214	159.5	137	
Square feet	1453	2306	1717	1473	
CONCENTRATOR					
Configuration	Parabolic	Parabolic	Spherical	Spherical	
Size					
Meters	9 × 11	10 (diam.)	10 × 10	8.5 × 8.5	
Feet	29.5 × 36	33 (diam.)	33 × 33	28 × 28	
Focal length					
Meters	6	3.2	10.75	10.9	
Feet	19.7	10.5	35.3	35.7	
Number of mirrors	3500	181	384	180	
Mirror size				90	90
Centimeters	16 × 16		50 × 50	62 × 62	64 × 66
Inches	6.3 × 6.3		19.7 × 19.7	24.4 × 24.4	25.2 × 26
Total mirror area					
Square meters	89.6	78.5	96	72.6	
Square feet	964	845	1033	781	
THERMAL					
PERFORMANCE†					
Total thermal power	45 (est.)	35 (est.)	42.5	32 (est.)	
Thermal efficiency (%)	55 (est.)	50 (est.)	48	50 (est.)	
Max. heat flux (W/cm²)	1200	. . .	580	400	

*Previously operated by the U.S. Army, Quartermaster Corps, Natick, Mass., 1958.
†Based on insolation = 900 to 950 W/m².

common. The solar furnace was completed October 1, 1970, after more than 10 years of construction and a cost of $2 million. The cost of the associated buildings, offices, and laboratories was approximately $4 million. A schematic diagram of the CNRS 1000-kilowatt solar furnace is given in Fig. 1. This furnace utilizes 63 heliostats to direct the sun's rays onto the surface of the giant parabolic concentrator.

The Heliostats. The 63 heliostats are each 7.5 meters wide by 6 meters high and contain 180 single flat mirror elements 50 × 50 centimeters. The total area of mirror surface in the 63 heliostats is 2835 square meters, or over one-half the playing area of a football field.

The heliostats are located directly north of the parabola and are arranged on eight terraces. Each terrace corresponds in elevation to one of the floors of the building supporting the concentrating parabola. A solar beam of constant energy is thus directed horizontally and southward from the heliostats to the mirrors which make up the concentrating parabola.

Each heliostat is designed to illuminate a specific area on the parabola and is equipped with a dual optical control system which maintains the proper orientation for each heliostat by means of a dual hydraulic system. This dual system permits each heliostat to be operated in either a "search" or a "track" mode. In both cases, the optical guidance system uses an optical tube which contains four photodiodes that control the heliostat motion in east-west and up-down direction.

When operating in the "search" mode, a short (10-centimeter) optical tube with a 40°

acceptance angle is used to activate the "fast" hydraulic system which operates in an on-off mode to quickly bring the heliostat to within the operating range of the "track" system. In the track mode, a 100-centimeter optical tube is used to control a slower-acting hydraulic system which operates in a proportional control mode. The size of the image of the sun at the base of the 100-centimeter tube is ½ inch in diameter, and the accuracy of the control is one minute of arc.

The Parabola.   The concentrating parabola has a focal length of 18 meters, is 40 meters high and 54 meters wide; and the focal axis is 13 meters above the first floor. The parabola consists of 9500 initially flat glass mirrors which were mechanically curved and adjusted to provide a solar image of minimum diameter at the focal point. Almost two years were

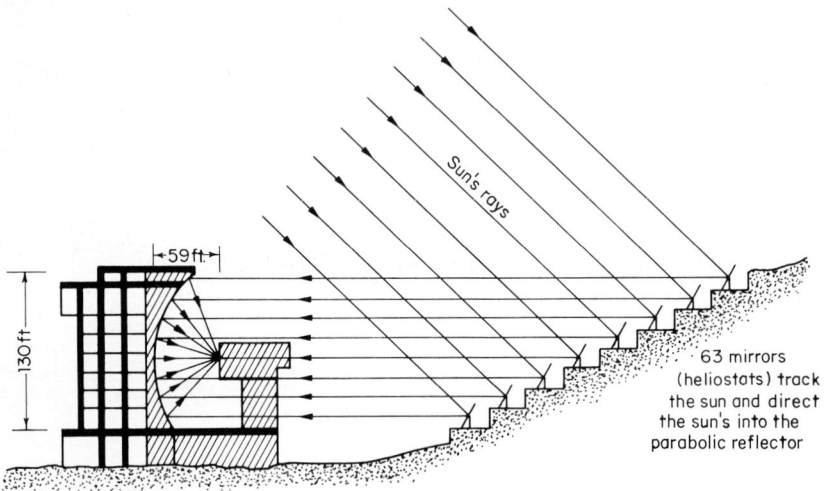

Parabolic reflector concentrates
sun's rays onto target area

Fig. 1. Schematic diagram of the CNRS 1000-kW solar furnace at Odeillo (France).

required to accomplish these two precise adjustments, which were completed October 1, 1970. The parabola and the focal building into which the concentrated solar energy is directed are shown in Fig. 2.

Thermal Characteristics.   The solar energy incident on an area of about 2000 square meters is concentrated by the parabolic reflector onto an area of less than 0.3 square meter. Sixty percent of the total thermal energy (about 600 kilowatts) is concentrated in an area of less than 0.10 square meter at the center of the focal plane of the parabola.

The diameter of the image of the sun at the focal point is 17 centimeters, and 27 percent of the thermal energy (about 270 kilowatts) is concentrated in this area. The distribution of thermal energy and the equilibrium temperature of the thermal radiation at the focal plane of the solar furnace are given in Table 2 (Ref. 6). Heat flux data in watts per square centimeter in the focal area are presented graphically in Fig. 3. Curve 0 represents the heat flux at the focal plane. Curve $d/2$ shows the heat flux and temperature at a plane removed one-half the diameter of the solar image (8.5 centimeters) behind the focal plane. Curve d presents the same data on a vertical plane removed one diameter of the solar image (17 centimeters) behind the focal plane.

An energy contour map showing the heat flux distribution in watts per square centimeter on a plane inclined at 25° (away from the parabola) from the vertical with its center at the focal point of the solar furnace is given in Fig. 4. This plane provides the most circular distribution of thermal energy at the focal point. A heat flux contour map on a horizontal plane at the focal zone and perpendicular to the optical axis of the 1000-kilowatt solar furnace is given in Fig. 5. For comparison purposes, the same data are presented for the U.S. Army solar furnace at the White Sands Missile Range. The "holes" in the center of the beams at between 20 and 30 centimeters behind the focal plane are caused by the

Fig. 2. Parabola and focal building into which concentrated solar energy is directed. CNRS 1000-kW solar furnace at Odeillo (France).

shadow of the focal building. Figure 5*a* was constructed from data presented in various CNRS reports. Figure 5*b* was reproduced from Cotton et al. (Ref. 2).

APPLICATIONS FOR SOLAR FURNACES

Historically, solar furnaces have been selected for high-temperature research and development activities in which a highly concentrated source of nonpolluting radiant energy is required. Generally such activities can be categorized as: (1) high-temperature chemistry involving the formation of very pure or otherwise unique materials; (2) high-temperature processing by which a material is fused, purified, or otherwise improved; (3) high-temperature property measurements involving the determination of the behavior of a material under conditions that require a noncontaminating environment; (4) determination of the thermal shock resistance or other behavior of materials in high temperature,

TABLE 2. Characterization of Thermal Radiation at Focal Plane of the Odeillo Solar Furnace*

Diameter of radiation:								
cm		2	6	12	16.8	20	30	40
(in.)		(0.79)	(2.36)	(4.72)	(6.61)†	(7.87)	(11.81)	(15.75)
% of total energy in area of radiation		0.5	4.50	15.5	27	35	58	75
Energy in area of radiation, kW:		5	45	155	270	350	580	750
Minimum	Watts cm^{-2}	1600	1472	1200	912	800	400	192
Heat	Cal cm^{-2} s^{-1}	383	352	287	218	191	96	46
flux‡	(Btu ft^{-2} s^{-1})	(1410)	1297)	(1057)	(804)	(705)	(352)	(169)
Average	Watts cm^{-2}	1600	1595	1370	1215	1115	820	595
Heat	Cal cm^{-2} s^{-1}	383	381	328	294	266	196	142
flux‡	(Btu ft^{-2} s^{-1})	(1410)	(1405)	(1207)	(1071)	(982)	(723)	(524)
Temperature of radiation:								
Minimum	°C	3825	3740	3540	3285	3170	2625	2140
	°F	(6915)	(6765)	(6405)	(5945)	(5740)	(4755)	(3885)
Average	°C	3825	3805	3665	3585	3465	3185	2950
	°F	(6915)	(6880)	(6630)	(6485)	(6270)	(5765)	(5340)

*For incident energy of 950 W m^{-2} ±5 percent.
†Diameter of solar image.
‡Heat flux calculated from water calorimeter and radiometry measurements.

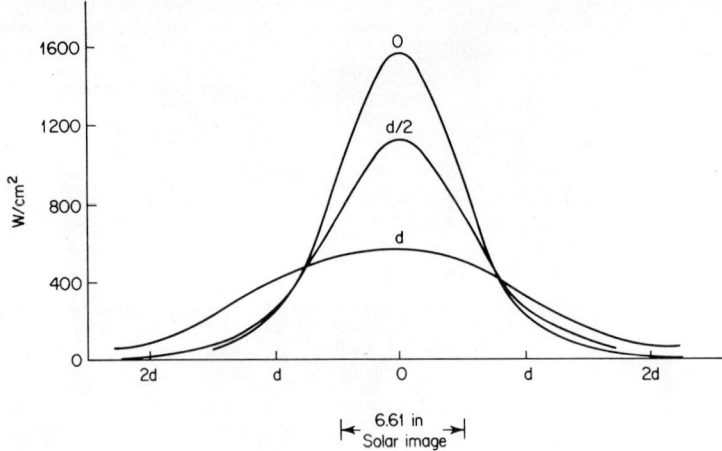

Fig. 3. Solar energy versus distance from focal point in Odeillo solar furnace. d = diameter of solar image.

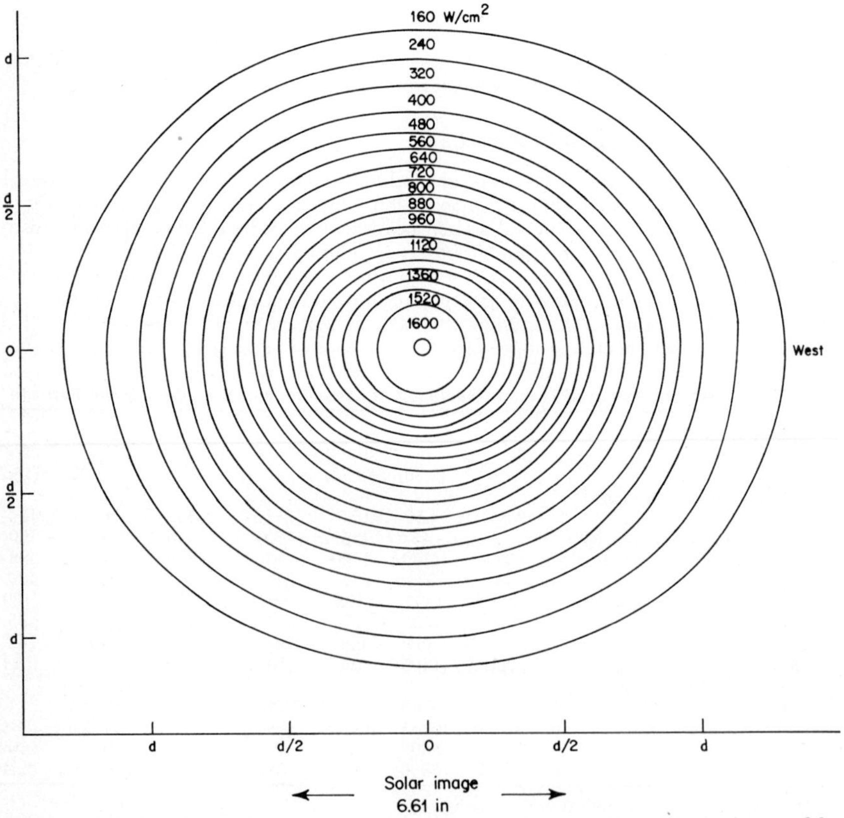

Fig. 4. Distribution of thermal energy in a plane inclined at 25° from vertical at the focal point of the Odeillo solar furnace.

high heat flux radiant energy environment; and (5) study of high-temperature solar-thermal conversion systems. Certain of these applications may be refined further by conducting the operation in an optically transparent vessel, or one containing a transparent window, such as fused quartz, through which the radiant energy may pass and in which the composition and pressure of the atmosphere can be controlled.

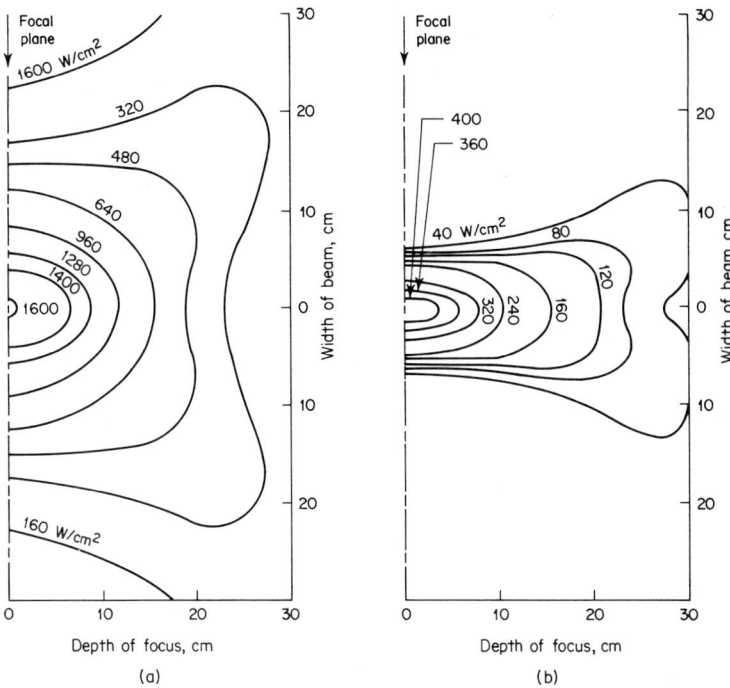

Fig. 5. Image contours in horizontal plane: (a) CNRS 1000-kW solar furnace. (b) U.S. Army 35-kW solar furnace.

A few examples of the types of high-temperature studies that have been conducted in the aforementioned categories are: (1) gas-phase reactions to form pyrolytic graphite; (2) production of very high purity fused aluminum oxide and fused silica, production of stabilized zirconia, and purification of reactive metals in a controlled atmosphere; (3) determination of microwave transmission characteristics of dielectric materials at very high temperature; (4) study of the thermal shock resistance of materials under high heat flux thermal radiation conditions simulating exposure to the thermal radiation pulse provided by a nuclear explosion; and (5) study of heat exchangers, such as boilers and superheaters, for the production of steam for electric power generation.

Processing of Fused Ceramics. The Montlouis, Japanese, and the 1000-kilowatt CNRS solar furnaces were developed primarily for use in basic and applied research in high-temperature chemistry and physics. The Montlouis and 1000-kilowatt solar furnaces have been used very effectively in the preparation of fused ceramics for the ceramic industries. By means of a large rotating cavity-type furnace, the Solar Energy Laboratory has successfully produced relatively large quantities of fused zirconia, alumina, and silica. The rates of fusion for these materials in the 1000-kilowatt solar furnace were 150 to 200 kilograms/hour, 150 kilograms/hour, and 200 kilograms/hour, respectively (Ref. 6). Because of the high melting points of these materials and the increasing cost of conventional fuels, it appears quite possible that high-temperature solar furnaces may provide an economical means of producing fused refractory oxides. In addition to simply fusing the oxides, analysis has shown that fused oxides are substantially purified by the solar furnace fusion operation. An excellent review of the uses of solar furnaces for high-temperature processing is given by Trombe (Ref. 7).

Nuclear Thermal Effects Simulation.   The French Army solar furnace at Odeillo, Font-Romeau, and the U.S. Army solar furnace at White Sands Missile Range, New Mexico, were developed primarily to provide a radiant thermal energy environment which would simulate the thermal radiation effects produced by a nuclear explosion. Through the use of pulse-shaping devices (shutters), it is possible to expose candidate materials to time-temperature-heat flux environments which can be tailored to simulate a wide range of nuclear devices. The 1000-kilowatt solar furnace has been equipped with a shutter, which allows this facility also to be used to simulate nuclear thermal radiation effects, but at much higher heat flux levels and larger exposure areas than either of the other two facilities.

Solar Thermal Conversion Systems.   In 1882 Mouchot used a large steerable parabolic solar concentrator to heat a boiler (at the focal point), which produced steam that powered a printing press at the World's Fair in Paris. In the 1950s, Baum (Ref. 8) proposed a solar power system which would use a large number of flat heliostats to direct solar radiation onto a central boiler. The heliostats would be arranged on railroad-type flatcars on several concentric circular tracks with the boiler at the common center. During a day's operation, the cars supporting the heliostats would move along the tracks to keep the image of the sun focused on the boiler. This concept was further refined by Hildebrandt (Ref. 9) by placing the boiler on a tall tower and surrounding it with flat heliostats which automatically track the sun to direct the solar radiation onto the boiler. The method used to guide the heliostats could be similar to that used by Trombe at the 1000-kilowatt solar furnace at Odeillo. *(See separate description of solar tower concept in this section of the Handbook.)*

Other methods that have been used or proposed for the production of electricity from concentrated solar energy include the linear parabolic trough or segmented mirror systems which focus solar energy onto pipes, in which either water is heated directly to steam, or some other fluid is heated which is used to produce steam through a suitable heat exchanger *(also described in this section of the Handbook.)*

However, for large-scale power production where the objective is to produce electricity at a central power station, current technology suggests that the linear parabolic trough or other type of distributed collector system may not be economical for two reasons. First, the concentrated energy provides a relatively low heat flux, which limits the practical temperature of the working fluid to the range of 150 to 300°C. Second, because of the distributed nature of such a system, long lengths of pipe are required, which must be well insulated to minimize any loss in temperature and pressure of the steam which is delivered to the turbine. Thus, the relatively low quality of the steam, coupled with a high cost for pipe and insulation, suggests that this system may be used more economically for small-scale power or thermal applications instead of for central power stations.

As of mid-1975, it is not possible to say that a central receiver (boiler-superheater) heated by a large number of heliostats can economically produce steam for central electric power plants. However, current designs for central receiver systems have the capability of generating power and heat flux levels necessary to produce the quality and quantity of steam required for central power plants using current technology. Studies related to these designs have identified two major subsystems, the cost and performance of which probably will determine the overall economics of central receiver systems. These subsystems are the heliostat and the receiver (boiler-superheater).

The heliostat, together with its guidance and control systems, has been identified as the major cost-determining subsystem. The receiver is the subsystem, which will determine the overall thermal efficiency of the system in converting the collected radiation into mechanical energy. The receiver also is the subsystem, which is the most experimental and which probably will require the most research and development effort.

Present central receiver (boiler and superheater) designs are basically either bundles of vertically oriented boiler tubes arranged on a tall tower surrounded by a field of heliostats, or a cavity-type receiver (with vertical tubes lining the interior) into which radiation from a field of heliostats is directed. Unfortunately, at present, there are no experimental data concerning the performance of such receivers at power levels sufficiently high to permit an accurate evaluation of their relative achievement, or to permit further refinement of these designs. This situation has led the National Science Foundation and the Centre

National de la Recherche Scientifique to undertake an experimental design and test program to develop a cavity-type receiver to be evaluated in the CNRS 1000-kilowatt solar furnace at Odeillo in 1976 (Ref. 10).

Although this program will provide the first high-temperature solar thermal conversion experiments at thermal power levels as high as one megawatt, one experimental program conducted in Italy should be mentioned. Under the direction of Professor G. Francia at the University of Genoa, Genoa, Italy, this program was concerned with the development of a 200-kW(th) (kilowatt-thermal) solar boiler test facility (Ref. 11). The test facility consists of a field of 271 flat mirrors, each one meter in diameter, arrayed in a hexagonal pattern. The mirrors direct the incident solar radiation into a vertically oriented cylindrical, cavity-type boiler-superheater, one meter in diameter, suspended about 9 meters above the center of the mirror field. Characteristics of the mirror elements of the solar power facility at Genoa and those of the CNRS 1000-kilowatt solar furnace are compared in Table 3.

The flat mirrors which act as heliostats in the Genoa facility are each supported on a specially designed mechanical mount called a "kinematic motion." All the kinematic motions are mechanically linked together through a common drive shaft so that they move together by a central clock-drive mechanism. This mechanism controls the motion of all

TABLE 3. Characteristics of Large, Multi-Heliostat Solar Furnaces
200–1000 kilowatts

	CNRS 1000-kilowatt solar furnace, Odeillo-Font-Romeu, France	University of Genoa, Genoa, Italy
HELIOSTAT		
Number of heliostats	63	271
Heliostat Size:		
Meters	6 × 7.5	1 (diameter)
Feet	19.7 × 24.6	3.3 (diameter)
Number of mirrors in each		
heliostat	180	1
Mirror size:		
Centimeters	50 × 50	100 (diameter)
Inches	19.7 × 19.7	39.4 (diameter)
Mirror area in each heliostat:		
Square meters	45	0.785
Square feet	484	2.69
Total number of heliostat mirrors	11,340	271
Total heliostat area:		
Square meters	2,835	212.7
Square feet	30,515	2,291
CONCENTRATOR		
Size:		
Meters	40 × 54	
Feet	131 × 177	
Number of mirrors:	9,500	No concentrator
Mirror size:		
Centimeters	45 × 45	
Inches	17.7 × 17.7	
Total mirror area:		
Square meters	1,923	
Square feet	20,707	
THERMAL PERFORMANCE*		
Total thermal power (kW)	1000	130
Average thermal power/8 hours (kW)	. . .	115
Maximum thermal efficiency (%)	58	68
Maximum heat flux (W/cm²)	1600	30

*Based on insolation = 900 to 950 W/m².

Note: The data shown for the Univ. of Genoa installation are for Unit No. 1, first operated in 1967. A second comparable unit was first operated in 1972. The mirror element of Unit No. 2 is 86 centimeters in diameter. The total thermal power is 100 kW, average thermal power per 8 hours is 88 kW, and maximum thermal efficiency is 70 percent.

the flat mirrors, so that they follow the sun through the day and keep the incident solar radiator directed into the boiler-superheater.

This system has produced steam at the rate of 150 kilograms/hour at 150 atmospheres and 500°C with an incident insolation of 900 watts/square meter, giving an overall thermal efficiency of 70 percent. Although this facility has provided significant experimental data relative to the performance of one particular central receiver design (downward-facing cavity), it is not a suitable test facility for evaluating the particular designs that are currently being considered in the United States for large-scale solar power facilities. In this respect, the 1000-kilowatt solar furnace at Odeillo remains the only large-scale [1 MW(th)] high-temperature solar test facility suitable for evaluating the thermal performance of candidate boiler-superheater designs.

This situation has resulted in the recognition of the need for a large solar energy test facility in the United States. Such a facility should be capable of evaluating central receivers at power levels above one megawatt and of evaluating advanced heliostat systems. It is presently evisioned that the test facility would be capable of providing a thermal power level of at least 5 MW(th) with some capability for expansion and that the power would be supplied by experimental as well as "standard" heliostat fields.

Also, it would be desirable to have some range of concentration factors available. For example, direct illumination of a boiler-superheater or other test volume by tracking heliostats could supply concentration factors from a few hundred to over 1000. However, for very-high-temperature research, concentration factors between 5000 and 15,000 would be required so that the test facility probably will require that a certain area of the heliostat field be used with a secondary parabolic concentrator similar to the CNRS 1000-kilowatt solar furnace. Therefore, with respect to the development of future high-temperature test facilities, the 1000-kilowatt solar furnace at Odeillo will undoubtedly play a major role in providing experience and design information in the same way that the 50-kilowatt solar furnace at Montlouis contributed to the design of the 1000 kilowatt solar furnace.

OUTLOOK FOR HIGH-TEMPERATURE SOLAR ENERGY

The future of high-temperature solar energy for electric power generation depends primarily on economics. The 1000-kilowatt solar furnace has demonstrated the technical feasibility of operating a large number (63) of heliostats to automatically track the sun to continuously concentrate the incident solar radiation onto a target area. However, as previously mentioned, the heliostat is the highest cost subsystem in current designs for central receiver solar power systems. Therefore, a major problem now facing the utilization of high-temperature solar energy is the optimization of the heliostat from the standpoint of economics as related to heliostat field design, fabrication, installation, guidance control, operation, and maintenance.

Another subsystem which must be optimized is the central receiver boiler-superheater (for steam generating systems). The future design and development of these systems depends upon experimental data which as of mid-1975 are minimal. Therefore, future research and development related to this subsystem will be directed toward experimental programs and to the development of larger solar energy test facilities upon which such programs ultimately will depend.

Finally, the optimum utilization of electric power generated from high-temperature solar energy must be determined. Problems related to overall system or plant design, energy storage, plant size, and site location must be considered; and at the present time there appears to be no "best" solution or even "best" concept. Currently, these problem areas are under study in an effort to determine the controlling parameters and to use these in developing guidelines relative to optimum plant size and mode of operation. Work by the Aerospace Corporation (Ref. 12) represents the state-of-the-art in this area.

REFERENCES

1. Trombe, F.: "Les Installations de Montlouis et le Four Solaire de 1000 kW d'Odeillo-Font-Romeu," Colloques Internationaux du Centre National de la Recherche Scientifique No. LXXXV, Applications Thermiques de l'Energie Solaire dans le Domaine de la Recherche et de l'Industrie Montlouis, June 23–28, 1958, pp. 87–128, CNRS, Paris, 1961.

2. Cotton, S., et al.: Image Quality and Use of the United States Army Quartermaster Solar Furnace, *Proc. U.N. Conf. on New Sources of Energy*, Paper S/79, Rome, Italy, 1961.
3. Sakurai, T.: Large Solar Furnace Completed in Japan, *Sun at Work*, vol. VIII, no. 2, 2d quarter 1963.
4. Meunier, R. M.: Characteristiques et Objectifs des Fours Solaires du Laboratoire Central de l'Armement, *Rev. Int. Htes, Tempér. et Réfract.*, **10**, 297–302 (1973).
5. Trombe, F., and A. Le Phat Vinh: Thousand kW Solar Furnace, Built by the National Center of Scientific Research, in Odeillo (France), *Solar Energy*, **15**, 57–61 (1973).
6. Trombe, F., L. Gion, C. Royere, and J. F. Robert: First Results Obtained with the 1000 kW Solar Furnace, *Solar Energy*, **15**, 63–66 (1973).
7. Trombe, F.: Use of Solar Energy for High Temperature Processing, *Proc. U.N. Conf. on New Sources of Energy*, vol. 6, Solar Energy, III, pp. 100–107, Rome, Italy, Aug. 21–21, 1961.
8. Baum, F. A., et al.: High Power Solar Installations, *J. Solar Energy*, **1**(1), 6 (1957).
9. Hildebrandt, A. F., G. M. Haas, W. R. Jenkins, and J. P. Colaco: Large Scale Concentration and Conversion of Solar Energy, *EOS Trans. Am. Geophys. Union*, **53**(7), 684–692 (1972).
10. Blake, F. A., and J. D. Walton: Update on the Solar Power System and Component Research Program, International Solar Energy Society meeting, Fort Collins, Colo., Aug. 21–23, 1974.
11. Francia, G.: Pilot Plants of Solar Steam Generating Stations, *Solar Energy*, **12**, 51–64 (1968).
12. Solar Thermal Conversion Mission Analysis, vol. 1, Summary Report; vol. II, Demand Analysis; vol. III, Southern California Insolation Climatology; vol. IV, Mission/System and Economic Analysis; vol. V, Area Definition and Siting Analysis; Report ATR-74 (7417-05)-1, Aerospace Corporation, El Segundo, Calif., Jan. 15, 1974.

Solar Cells and Arrays

BY HENRY OMAN* AND JOSEPH W. GELZER†

The silicon solar cell, announced to the National Academy of Sciences in April 1954 by G. L. Pearson, C. S. Fuller, and D. M. Chapin of Bell Laboratories, has been used to convert sunlight to electric power on spacecraft since 1958. The high cost of these cells, relative to alternative air-breathing power sources, had limited their use on the ground to unusual and experimental applications before 1975. Substantial U.S. Government funding of low-cost solar cell development, begun in 1975, coupled with the rising cost of alternative energy sources, will bring increasingly useful terrestrial applications for solar cells.

This Handbook section provides (1) a description of how solar cells work, (2) the electrical characteristics of solar cells, (3) methods of designing solar arrays for spacecraft, and (4) an approach to a design for a terrestrial use.

Alternatives for Space Power. Many alternatives to silicon solar cells have been developed for powering spacecraft. See Table 1. Radioisotope-thermoelectric power sources and fuel cells have actually flown on spacecraft. However, over 98 percent of the launched space-craft have been powered by silicon solar-cell arrays, the largest of which, flown on the Apollo Telescope Mount, covered 1512 ft^2 and generated 11 kilowatts of power when first deployed in space.

The term, "solar cell," generally refers to a slab of high-purity, single-crystal silicon, 0.3 to 0.5 millimeter (8 to 12 mils) thick, 2 centimeters by 1, 2, 4, or 6 centimeters in size, and doped to produce a p/n junction 0.25 to 1.0 micrometer under the illuminated surface. Alternative photovoltaic energy converters have been developed and tested. However, as of late-1975, no alternative photovoltaic energy converter has been as useful as the single-crystal silicon solar cell. See Table 2.

SILICON SOLAR CELLS

Wafers for solar cells are made of single-crystal high-purity silicon (Table 3), doped (n on p) to a prescribed volume resistivity, nominally between 1 and 10 ohm-centimeters, by the addition of minute quantities of boron or phosphorus. Diffusing in an additional dopant from the surface generates a junction within 1 micrometer of the surface. To this large flat semiconductor diode, electric contacts are added on the top and bottom surfaces. Cells having an n-doped sun-facing surface and a p-type substrate (n on p) show greater resistance to radiation degradation than p-on-n cells, and have become the predominant type. However, p-on-n cells are generally more efficient before irradiation. Electric contacts are usually metal, vacuum-deposited, and sintered; although plated and silk-screened contacts are feasible. Figure 1 shows the configurations of typical space-use solar cells. A 2 × 2 centimeter cell, 0.3 millimeter thick weighs 0.38 gram.

Cell Operation. Under equilibrium conditions, the majority carriers (electrons in n-type and holes in p-type silicon) will migrate by thermal diffusion. Some carriers migrate across the p-n junction. The electrons that migrate into the p-type silicon and the holes that migrate into the n-type silicon are called minority carriers and soon combine with the majority carriers in the new region. Each migrating electron uncovers a positive charge, and each migrating hole uncovers a negative charge. Thus, a potential difference or an electrostatic field is created across the junction.

When light strikes the cell, photons with energy more than 1.1 electron volts (the energy gap of silicon) will transfer energy to electrons, which then have enough energy to escape from their bound states, creating electron-hole pairs. Some of these photon-

*Supervisor of Electrical Power Research, Boeing Aerospace Company, Seattle, Wash.
†Manager of Product Development in the Software and Computer Systems Engineering organization of The Boeing Company, Seattle, Wash.

TABLE 1. Power Sources for Spacecraft

Power source	Feature	Principal disadvantage	Spacecraft use
Silicon solar cell	Solid state, no moving parts. Array can withstand multiple cell failures.	Over 0.1 ft² of array per watt of power. Cells degrade in radiation.	Used in nearly all spacecraft.
Radioisotope, thermoelectric	No large panels. Power available in the dark, and far from sun.	Isotopes are costly and require special handling, Isotopes and thermoelectrics degrade in power output in time. Requires hot radiators. Safety requirements impose penalties.	Mars lander; Jupiter and Saturn probes and navigational satellites.
Reactor, thermoelectric	Lots of power. No half-life problem.	Reactor is costly and heavy. Complex startup and control. Fissionable material released in launch accidents. Redundancy is costly.	One test flight.
Fuel cell	High power density.	Stored hydrogen requires much volume and weight. Fuel cell and accessories are complex and costly.	Apollo command modules.
Battery	Simple, inexpensive.	Heavier than solar panels if flight is longer than a day or two.	Gemini, Mercury spacecraft.
Monopropellant turbine	Can provide high power levels.	Fuel weight exceeds solar panels if flight is longer than a day or so. Reliability of rotating machinery is not certain.	None

TABLE 2. Photovoltaic Energy Converters

Semiconductor material	Advantages	Disadvantages	Status (1975)
Silicon	Silicon is plentiful. Efficiencies over 19% achieved.	Degrades in radiation. Efficiency drops as cell heats. Manufacturing cost is over $12/watt.	Commonly used in spacecraft. Low-cost manufacturing methods are under development.
Cadmium sulfide (copper sulfate)	Low cost, flexible. Large area cell is feasible. Low weight per square foot.	Efficiencies of over 6% are unstable.	Still under development after 10 years of intensive research.
Gallium arsenide	Operates at high temperature. Resists radiation degradation.	Efficiency generally less than 10%.	Some work being done on thin-film cells.
Aluminum-gallium arsenide	High efficiency (up to 23% with solar concentration).		New development.
Polycrystal silicon	Low cost	Efficiency less than 9%.	Under development for terrestrial use.

TABLE 3.   Physical Properties of Silicon Crystals

Property	Value
Energy gap, eV	1.107
Electron mobility, cm²/(V)(sec)	1,350 ± 100
Hole mobility, cm²/(V)(sec)	480 ± 15
Electron diffusion constant, cm²/sec	34.6
Hole diffusion constant, cm²/sec	12.3
Density, grams/cm³	2.3289
Coefficient of linear thermal expansion	2.33×10^{-6}/°C
Elastic constant (C_{11}), dynes/cm²	16.56×10^{11}
(C_{44})	7.953×10^{11}
(C_{12})	$6,386 \times 10^{11}$
Hardness, Mohs scale	7.0
Refraction index	3.420 (at 6 micrometers)
Fusion temperature, °C	1,417 ± 4
Thermal conductivity, W/(cm)(°C)	1.57
Specific heat, cal/(gram)(°C)	0.166
Latent heat of fusion, cal/gram	425 ± 6.4

released carriers will drift into the junction where the electrostatic field will displace electrons to the n side, and holes to the p side, making a voltage appear across the cell terminals. Output power is then obtained. See Fig. 2.

Spectral Response.   Because of the nonlinear relationship among absorption coefficient, wavelength, and the minimum ionization energy (1.1 eV for Si), the silicon solar cell does not respond equally to all wave-lengths of light. The energy of a photon E_p in electron volts is

$$E_p = \frac{1.2398}{\lambda}$$

where λ is the wavelength of the photon in micrometers (μm). Thus, infrared photons having a wavelength longer than 1.1 μm cannot release a hole-electron pair, and photons in the visible and ultraviolet parts of the spectrum have excess energy which merely generates heat.

Figure 3 shows the solar spectrum in space, spectrum of terrestrial sunlight, and spectral response of a typical solar cell. The spectrum and intensity of sunlight when the sun is directly overhead is referred to as "air-mass-one" sunlight. Air-mass-zero sunlight occurs in space in the vicinity of the earth. Since silicon solar cells are sensitive to the spectrum of the light source, great care must be used in interpreting the results of

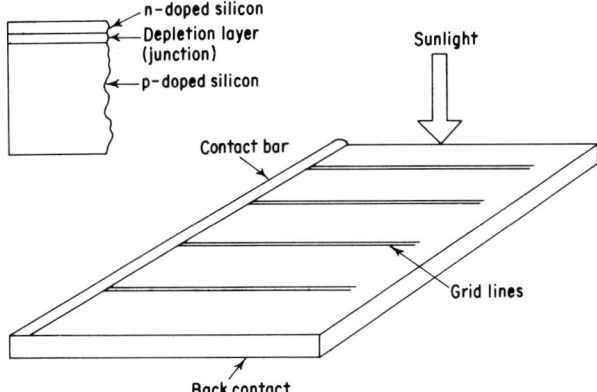

Fig. 1. Type of n-on-p silicon solar cell used for space applications.

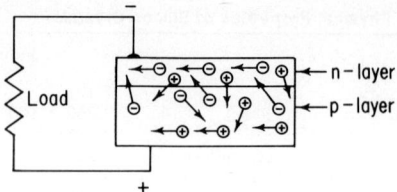

Fig. 2. Solar cell operation.

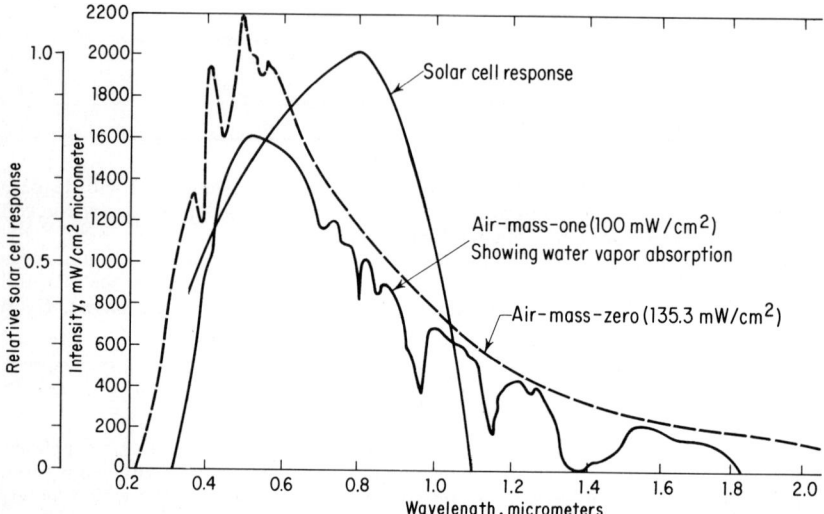

Fig. 3. Solar cell spectral response and solar spectrum at air-mass-zero and air-mass-one.

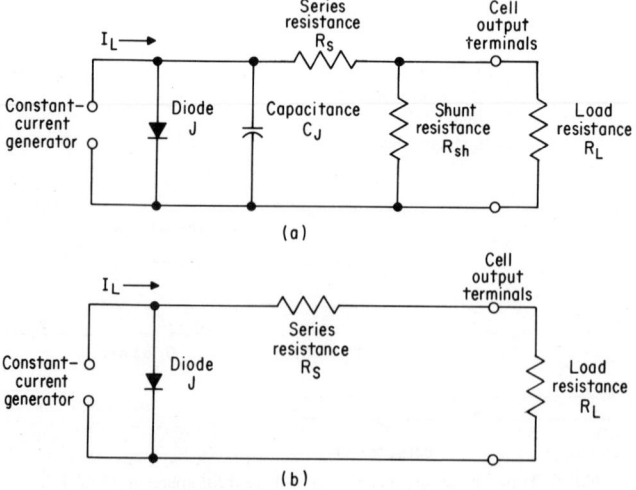

Fig. 4. Equivalent circuit of a solar cell.

terrestrial and artificial-light tests and extrapolating results from such tests to space conditions.

Equivalent Circuit.   Electrically, a solar cell consists of a current source shunted by a diode. See Fig. 4a. Constant illumination provides constant-current generation I_L. This current source is shunted by the diode characteristic of the P-N junction J and by the junction capacitance C_J. The series resistance of the cell is represented by R_S, and the shunt resistance by R_{sh}. The equivalent circuit may be simplified to that shown in Fig. 4b by ignoring the shunt resistance, which is normally quite large, and the junction capacitance, which is unimportant during steady-state power generation.

By using Kirchhoff's law, the cell output current I can be related to the light-generated current I_L, the current drained off through the junction, and the current drained through the shunt resistance.

$$I = I_L - I_0 \left\{ \exp. \left[\left(\frac{q}{AKT} \right) (V + IR_S) \right] - 1 \right\}$$

where I_0 = equivalent saturation current of the diode junction
A = a dimensionless constant between 1.0 and 5.0 for most solar cells
R_S = series resistance of the cell
q = electron charge
K = Boltzmann constant
T = absolute temperature in degrees Kelvin
$V + R_S$ = diode junction voltage
V = voltage at cell terminals

The term $I_0 \{\exp [(q/AKT) (V + IR_S)] - 1\}$ defines the diode characteristics of the junction.

Under open-circuit conditions when $I = 0$, all generated current must be conducted through the diode and the voltage will increase until this condition is satisfied. Open-circuit voltage is thus determined by the diode characteristic and is equal to

$$V_{oc} = \frac{AKT}{q} \log_e \frac{I_L + I_0}{I_0}$$

Under short-circuit conditions when $V = 0$, current through the diode is very small, and essentially all generated current is delivered to the output terminals:

$$I_{sc} = I_L$$

Thus, the short-circuit current I_{sc} of a cell can be predicted from Fig. 3, using the integration

$$I_{sc} = \int_0^\infty R(\lambda) J(\lambda) \, d\lambda$$

where $R(\lambda)$ = absolute spectral response of the solar cell
$J(\lambda)$ = spectral energy of sunlight

Parameters of Performance.   Solar cell performance depends upon temperature and illumination. The electrical parameters are:

Open-circuit voltage V_{oc}
Short-circuit current I_{sc}
Voltage at the maximum power point
Current at the maximum power point
Power at the maximum power point
Conversion efficiency

For light intensities below 300 milliwatts per square centimeter, I_{sc} varies linearly; and V_{oc} varies logarithmically with light intensity—as shown in Figs. 5 and 6. The intensity of space and sunlight in the vicinity of the earth is about 135.3 milliwatts per square centimeter.

Effects of Temperature.   Temperature affects the open-circuit voltage, short-circuit current, voltage at maximum power, and conversion efficiency, as shown in Figs. 7 and 8. As temperature increases to about 80°C, the open-circuit voltage V_{oc} decreases linearly about 2.2 millivolts per degree Celsius. This adversely affects power output. The temperature of a cell can be derived from its open-circuit voltage.

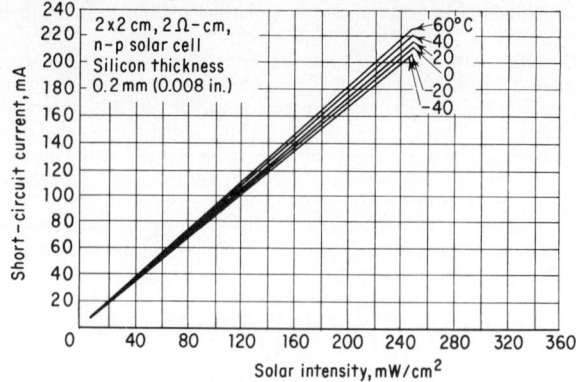

Fig. 5. Short-circuit current versus illumination.

As temperature increases, both the minority carrier lifetime and the width of the depletion layer increase. In space sunlight, this causes the short-circuit current to increase about 0.05 percent per degree Celsius. In tungsten light, the increase is much greater, about 0.1 percent per degree Celsius.

The voltage at maximum power will change in proportion to the change in open-circuit voltage. Thus, a decrease in maximum power will occur with increasing temperature. To keep cell temperature down, a transparent cover, usually of fused silica, is bonded to spacecraft cells. This replaces the low-emittance silicon surface with a glass having a higher infrared emittance, increasing heat radiation and cooling the cells better. The cover usually has an antireflection coating and a multilayer filter which removes the ultraviolet radiation that would otherwise darken the transparent adhesive used to bond the cover to the cell. Backs of solar-cell panels, if provided a view for radiating heat, are usually painted with a high-emittance coating.

DESIGN OF SPACECRAFT ARRAYS

Important to the designer of space power supplies is the current-voltage curve of the solar cell at operating conditions. This curve varies in shape and end values as cell temperature and illumination vary, and as the cell accumulates radiation damage from high-energy

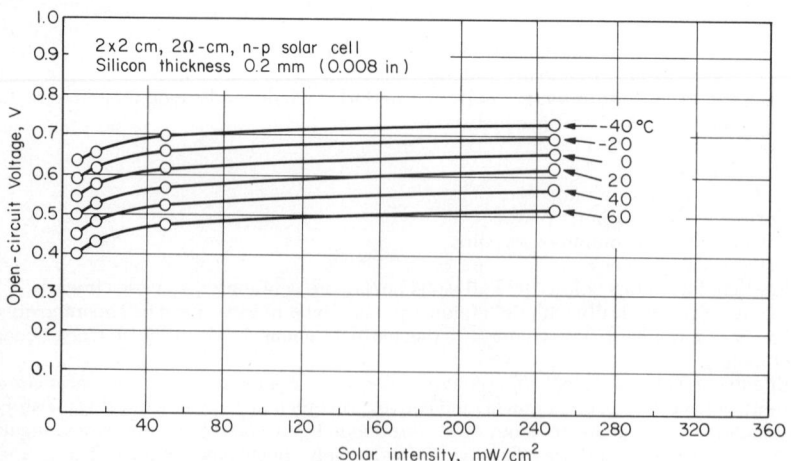

Fig. 6. Open-circuit voltage versus illumination.

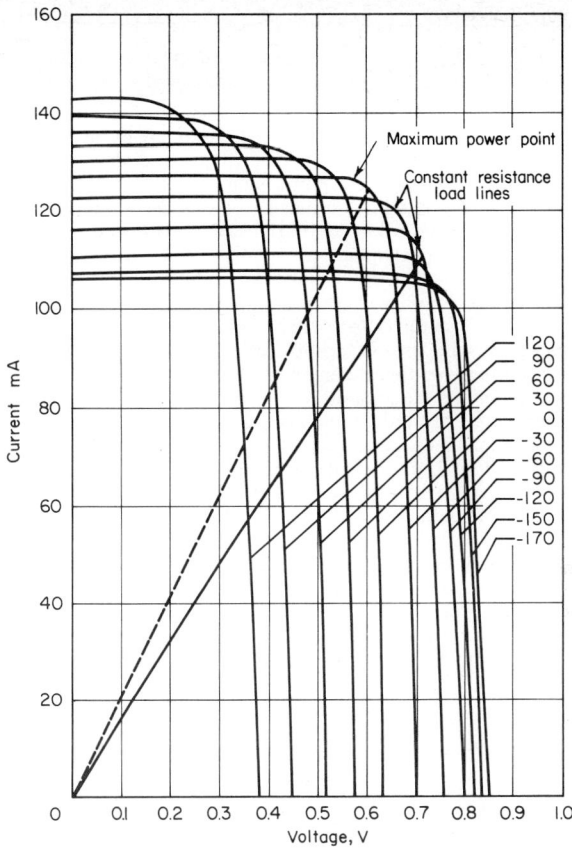

Fig. 7. Current-voltage characteristic of a silicon solar cell. *Heliotek* 2 by 2 cm cell, *n*-on-*p*, 0.012-inch thick, 140 mW per cm^2 illumination.

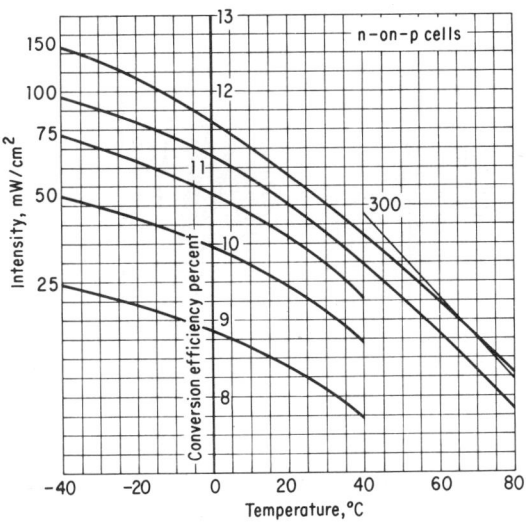

Fig. 8. Effect of temperature on conversion efficiency.

protons and electrons. The solar array must have enough cells in series to generate the minimum required voltage, and enough cells in parallel to carry the spacecraft load under worst-case end-of-mission conditions when the array has been degraded by radiation and the fused silica covers and adhesives have been darkened by ultraviolet radiation. At other times, the array could generate more power than needed.

To test a solar cell for all possible end-of-mission and other conditions is not practical. Rather, a computer is programmed to synthesize solar-cell volt-ampere curves for designated mission conditions, based upon cell performance measured under standard conditions, usually 140 milliwatts per square cm in simulated-space sunlight and 28°C cell temperature. The cell characteristics usually fed into the computer are open-circuit voltage, short-circuit current, and voltage and current at the maximum power point. The basic solar cell equation given previously can be enlarged to include second-order effects, such as shunt resistance, and greater sophistication, such as distributed resistance and capacitance. The resulting complex statements, which cannot be solved in a closed form, can be readily manipulated by a computer.

Examples of computer-generated *I-V* curves from Ref. 2 are presented in Figs. 9 and 10. The current, shown in milliamperes for a cell 1 × 2 centimeters in size, can be extrapolated for larger cells. In Fig. 9, where cell temperature is a parameter, it will be noted that short-circuit current is not significantly influenced by cell temperature. Open-circuit voltage is significantly affected by temperature. In Fig. 10, where illumination is a parameter, it can be seen that short-circuit current is directly proportional to temperature.

Manual Design of Spacecraft Solar Arrays.   Although spacecraft solar arrays are usually designed with the aid of a computer, a preliminary design engineer often must estimate the size of the array so that vehicle-level and mission trades can be made to evaluate the desirability of performance parameters which affect electric load, and hence solar array size. For example, launch vehicles come in discrete sizes, and going from a

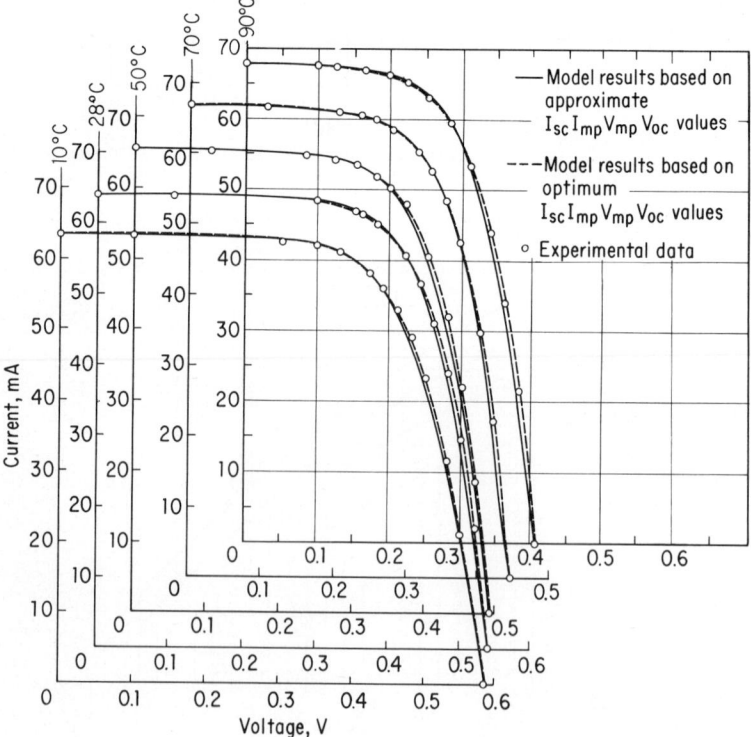

Fig. 9. Solar cell voltage-current characteristics as a function of temperature.

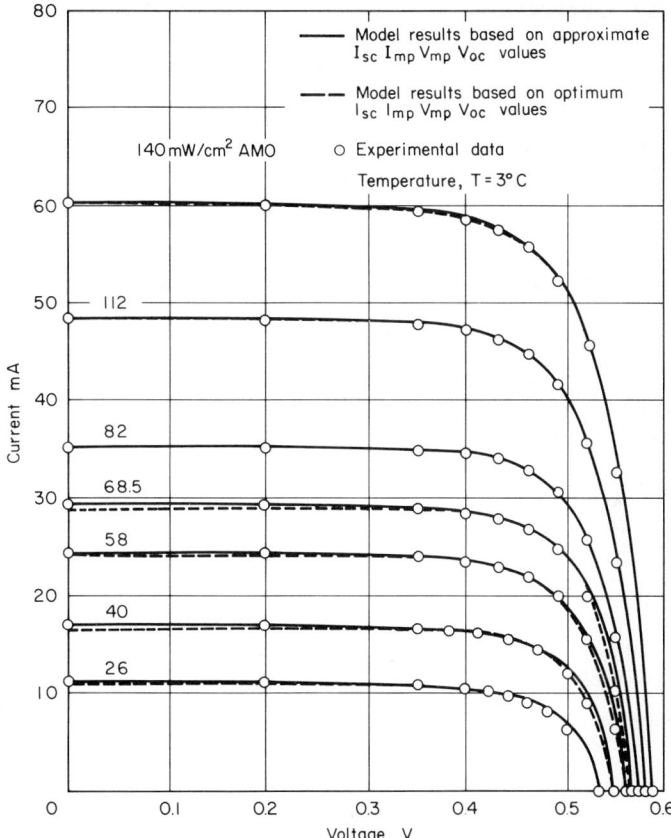

Fig. 10. Solar cell voltage-current characteristics as a function of intensity. AMO = air-mass-zero.

Titan to a Saturn means the difference between feasibility and impracticality for many spacecraft. Many trades are required in the selection of features that can be incorporated on a spacecraft without exceeding launch vehicle capability. These trades require the sizing of solar panels for many ratings, a task usually done with a desk calculator.

The preliminary design engineer starts with an estimate of solar panel temperature, which in the vicinity of Earth is around 50°C for deployed solar panels and around 75°C for solar panels mounted on the body of the spacecraft. This estimate is refined later as the spacecraft configuration is developed, to include the effects of the spacecraft and its appendages on the thermal-radiation view that the panels have. The designer next estimates the radiation influence on the solar cells. A good first assumption is that the cells are protected by 10-mil quartz covers because most trade studies have shown this to be close to optimum. The preliminary design engineer then estimates the radiation fluence interpreted into equivalent 1-MeV electrons, using tables and curves, such as published in the "Solar Cell Radiation Handbook," (Ref. 3). These radiation fluence estimates are later refined as the spacecraft orbit becomes precisely defined, by using computer programs which integrate the proton and electron flux in each zone that the spacecraft occupies.

From the solar-cell temperature and radiation fluence, the designer then proceeds to estimate the solar-cell volt-ampere curve. The designer first determines the 1-MeV fluence in the solar cells under the cover glass, using curves provided in aforementioned Ref. 3. Then, knowing the 1-MeV equivalent electron fluence, the designer can estimate the worst-case end-of-mission degraded performance of the cells from curves, such as Fig.

11, which gives the maximum power output of the cells. Unfortunately, curves such as these are plotted for a standard temperature of 28°C, and the solar cells will normally operate between 40 and 75°C. The higher temperature does not significantly affect short-circuit current, but does reduce maximum-power voltage. In preliminary design, an assumption that dV/dT of the maximum-power voltage has a temperature coefficient of -2.3 millivolts per degree Celsius produces satisfactory results, although some designers prefer to construct volt-ampere curves for the solar cell and shift them to account for temperature and radiation effects.

The solar cell commonly utilized in radiation environments, in fact in virtually all space applications, is a n-on-p cell having a nominal bulk resistivity of 10 ohm-centimeters. Cells having a 2 ohm-centimeters resistivity and p-on-n cells have higher power outputs, but are not ordinarily used in space because they degrade more rapidly than n-on-p 10-ohm-centimeter cells.

The maximum-power output of the irradiated and heated cell is further reduced by the factors shown in Table 4. To calculate the number of cells required, the resulting power output of one cell is divided into the total power required, plus I^2R losses in the array and other wiring, plus diode losses. The diode losses are typically taken to correspond with a 0.7-volt drop across the diode.

Orientation.   Important to the design of a solar array for a spacecraft is the orientation of the panels making up the array. Small spacecraft, requiring 10 watts or less, generally employ body-mounted solar panels, arrayed to give as nearly constant power as practical when the spacecraft is in any of its predicted attitudes. Larger spacecraft are often equipped with panels that are fixed with respect to the spacecraft after deployment, but again arrayed to intercept the most possible sunlight in the expected attitudes of the spacecraft. Such panels have been used in the Mariner interplanetary probes, which could most of the time be oriented to best illuminate the solar panels. Earth-orbiting spacecraft requiring hundreds of watts are generally equipped with solar panels which are continuously rotated to face the sun, unoriented panels being too inefficient.

The output of any panel not receiving normal-incidence sunlight is the product of its normal-incidence output and the cosine of the angle of incidence.

Cells in Series.   The number of cells in series must be sufficient to supply at the end-of-mission conditions the minimum allowable load voltage, plus line and diode drops. A diode is commonly installed at the end of each string of solar cells to prevent a fault within

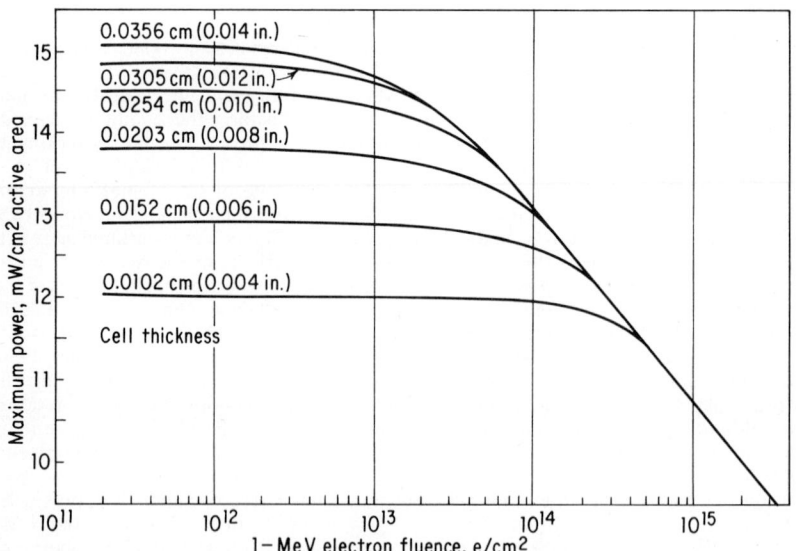

Fig. 11 Maximum power density per unit active area versus 1-MeV electron fluence for 7 to 13 ohm-centimeter n-on-p silicon cells. At 135 mW per cm², AMO illumination intensity, 28°C.

TABLE 4.   Factors Affecting Power Output of a Solar Panel

$$P_o = \eta(A)(S)(P)(K_1)(K_2)(K_3)(K_4) [1 + (T_c - 28°C) (K_5)] (K_6)$$

where

P_o	=	power output of the solar panel, watts
η	=	conversion efficiency of the solar cells at AM0 and a cell temperature of 28°C
A	=	gross area of the panel, ft^2
S	=	solar intensity in watts/ft^2
P	=	packing factor (typically 0.90) which accounts for panel area not covered by cells
K_1	=	factor to account for cover glass attenuation (0.96)
K_2	=	factor to account for process degradation and cell mismatch (0.98)
K_3	=	factor to account for standard cell calibration and test errors (0.99)
K_4	=	factor to account for solar intensity uncertainty and deviation (0.98) (used only if panels are not calibrated with balloon-flown standard cells)
T_c	=	temperature of cells in space, °C
K_5	=	temperature efficiency coefficient in the temperature range 0 to 100°C
K_6	=	factor to account for environmental degradation. This factor is a function of the mission (i.e., radiation damage, micrometeorites, and thermal cycling).

the string from disabling the entire solar array. It is also important to determine the highest array voltage, which occurs when a new panel, cold from being in a shadow, suddenly is illuminated with the brightest sunlight expected under the circumstances. This highest voltage affects the voltage regulator performance required, or voltage exposure of the load and battery if no regulator is provided.

The array designer must also establish a reserve for load growth. Between the preliminary design time and the first flight, the spacecraft loads rise and fall as the load equipment is reduced from concept to final design, but the net result is usually an increase in the power required, often around 10 to 25 percent.

Number of Cells in Parallel.   If one cell or interconnection in a series string of single solar cells breaks, the current path is interrupted and the output of the string is zero. If one cell short-circuits, the voltage and current output of the string will be reduced slightly. Short circuits to ground have not occurred on modern well-designed solar panels which have been manufactured with stringent quality-control surveillance.

If every cell in a string were paralleled with a second cell, then breakage of one cell would not reduce string current to zero. However, the good cell paralleled with the failed one would have current pushed through it, creating a voltage drop corresponding to the reversed-biased diode characteristic of the cell. As a consequence, the good cell might overheat, melting its soldered connections. For this reason, most solar arrays are designed with three or more cells in parallel, and an open-circuit failure of a single cell is hardly noticeable. In fact, many spacecraft flying today have a few broken cells because the extra increment of power obtainable by cell replacement was not worth the risk of further damage during cell replacement.

Final Solar Array Design.   In subsequent design iterations, the array output is more accurately predicted with the following refinements:

1. The spacecraft radiation environment is accurately defined and integrated.

2. Solar cell types, such as p-on-n, n-on-p, 10 ohm-centimeter, and 2 ohm-centimeter, are traded for the specific environment for greatest end-of-mission power output.

3. Cover thickness is traded, usually in the range of 0.15 to 0.5 millimeter (6 to 20 mils), to get lowest array weight and cost.

4. Solar cell efficiency is traded, usually in the 10 to 13 percent range, considering cell cost, panel weight, and overall spacecraft weight and cost as parameters.

5. Shadow degradation of array output is analyzed if the spacecraft has projecting objects, such as booms or antennas.

6. The selected solar cell is irradiated with protons and electrons having the predicted energy, and the cell volt-ampere curves are derived for the cell for pertinent points in the mission lifetime.

7. Volt-ampere curves for the selected cell are derived for all pertinent temperatures.

8. The voltage regulation method is selected.

9. The number of cells in series is calculated on the basis of supplying the required voltage, including all wiring and diode drops, under worst-case mission conditions.

10. The number of paralleled cells is determined from the current required to supply the loads, including battery charging and power subsystem losses, under the worst-case conditions. The worst case for earth-orbiting spacecraft usually is at the end of the mission and at summer aphelion.

Among alternatives available to the spacecraft designer, if the spacecraft is unable to accommodate sufficiently large solar panels, are:

1. Reduce end-of-mission load, assuming that some redundant load equipment has failed and is no longer consuming power; or simply provide less performance at end-of-mission conditions.

2. Use batteries to supplement solar panels under worst-case conditions.

TERRESTRIAL SOLAR ARRAY DESIGN

A terrestrial solar array must produce the required power at an appropriate voltage, and survive in an environment that has rain, snow, hail, high winds, dust, sand, and bird droppings. The method of calculating the solar array output for supplying a continuous load is described in the following several paragraphs.

A terrestrial array should have enough solar cells in series to produce the required voltage when the array is illuminated on the hottest expected summer day. The cells may not have to operate at their maximum power during the hottest hours, particularly if the energy storage is sized for the short days at winter solstice. Optimization of the number of cells in series requires consideration of: (1) length of time the array will be at the highest temperatures, (2) value of power during the hottest periods, (3) cost of the extra solar cells required to support array voltage at the maximum power during hot periods, (4) voltage drop in the feeders between the array and its load, (5) adaptability of the load to voltage variations, and (6) type of voltage regulation adopted.

Solar cells for terrestrial use are often made round to avoid wasting the silicon that would otherwise have to be trimmed from the disks that are sliced off a round single-crystal silicon rod, usually 3 inches in diameter. Each cell has extending across its front center a conducting bus, and current-collecting fingers perpendicular to the bus. Connections to the front surface of the cell are usually made by soldering wires to both ends of the bus, for redundancy. Redundant connections are likewise soldered to the back side of the cell.

No low-cost, proved method of protecting solar cells from the terrestrial environment was available as of late 1975. Hermetic sealing in Pyrex glass would obviously protect the cells for the 30-year lifetime normally required in power generation service, but this approach is costly. Sandwiching the cells between glass, with all voids filled with RTV 602 silicone resin, is promising, the silicone having been proved out by years of exposure in space sunlight. Life testing of various techniques of encapsulating solar cells in plastic may reveal a more economical approach to a solar cell module that can survive the terrestrial environments.

Direction of Solar Cell Development. Well-funded development programs in existence at the time of this writing (late 1975) virtually assure that higher-efficiency solar cells will be available at some time in the future at reasonable prices. Particularly promising developments include those listed in Table 5.

TERRESTRIAL SOLAR POWER GENERATION

Feasibility of decentralization and scaling down is a feature of solar cell power plants. Nuclear and fuel-burning steam power plants are economical only if built in large sizes. For example, the new Centralia coal-burning steam plant in Washington is rated 1400 megawatts, and the Trojan nuclear plant at Rainier, Oregon, is rated 1100 megawatts. Nuclear and coal-burning power plants rated a few kilowatts are not practical. On the other hand, the efficiency of solar cells in a small power plant is no different than in a large power plant, and the overall efficiency is improved by locating the plant next to the load and eliminating transmission losses. Furthermore, the cost of solar cells for a small power plant on a per-kilowatt basis need not be significantly different from the cost of solar cells for a large plant.

TABLE 5.   Representative Solar Cell Developments (1976)

Cell type	Feature	Attained efficiency, %
Violet (silicon)	Enhanced blue and ultraviolet response.	14
Helios (silicon)	Ta_2O_5 antireflecting coating and backfield doping.	14
Black (silicon)	Enhanced blue and ultraviolet absorption.	19.7
Gallium-aluminum-arsenide	Efficiency not greatly degraded by high temperatures.	23

A solar power plant that supplies continuous power has many components, each of which can be parametrically varied. For each component, there are many choices. For example, energy can be stored in batteries, hydrogen tanks, flywheels, or water pumped to a higher elevation. The amount of energy stored can vary from that required to carry minimum loads for one summer night to that required for a week of cloudy days during the winter. Each choice affects the cost of power.

In the following description, the technical data are organized to allow parametric variations of a model solar power system. The model system is a small stand-alone power generator, compatible with conventional electric utilities. It provides a decentralized energy source for homes and other small users. The system shown in Fig. 12 consists of silicon cells that convert sunlight directly to electricity, an electrolyzer that converts the electricity to hydrogen for reserve energy, a metal hydride unit for compact storage of gaseous hydrogen, and a fuel cell for conversion of hydrogen to electricity. An inverter converts the dc output from fuel cell and solar cells to standard 60 Hz ac. Several studies of similar combinations have been published (Refs. 4 and 5).

The model system has been sized to provide an average continuous power of 1.3 kilowatts, requiring 7.7-kW peak generation from the solar cell array when sunlit. The entire array output may be used to handle peak loads. At other times, most of the array output is stored for use when sunshine is not available. A peak load of 7.7 kW can also be supplied from stored energy.

All elements of this solar cell, storage, and recovery system are solid state and nonpolluting. Having no moving parts, the elements are quiet and relatively maintenance free. The system operates in a closed manner, requiring no external power or resources other

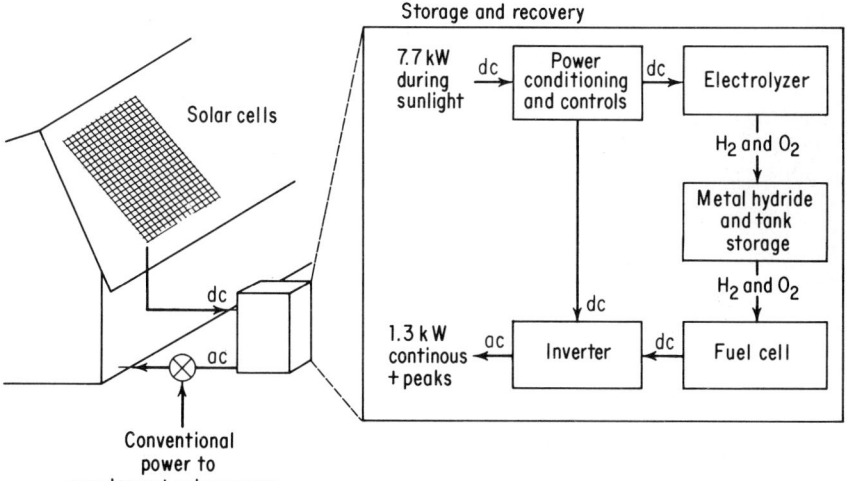

Fig. 12. Model solar power system.

TABLE 6.   Residential Electricity Utilization

Energy use	Annual consumption, kWh		Continuous average kW	Solar application
Space heating:				
Electric furnace	18,500		4.2 (or 6 months)	
Heat Pump	8,400		1.3* (or 6 months)	
				Heating & cooling by heat pump
Space cooling:		Design		
Electric conditioning	5,400	point	1.2 (or 6 months)	
Heat pump	5,700		1.3 (or 6 months)	
Water heating	6,400		0.7	All utilities
Cooking	1,300		0.2	other than
Appliances	3,200		0.4	heating &
Lighting	900		0.1	cooling
TOTALS	25,900		2.7	
	(heat pump)		(heat pump)	

*Assumes a reduction of 0.6 kW by use of the waste heat from the solar storage and recovery system.

than sunshine and occasional water. In those areas where sufficient sunshine is available, the units can be decentralized, avoiding the cost of transmission and distribution.

Residential Electricity Use.   A study (Ref. 4) of a model single-family residence in Tulsa, Oklahoma, shows the annual energy requirements for combinations of electric appliances. These data are summarized in Table 6. The residence has a single story with 1500 square feet of living space, heated to 70°F in the winter and cooled to 75°F in the summer.

Comparative data are shown for heating and cooling by heat pump. The energy economy of the heat pump in United States latitudes makes it well suited for home solar systems with electricity output. The data show that a 1.3-kW continuous electric solar power system can provide the total heating and cooling needs of this model residence if a standard commercial heat pump is used. This assumes that about two-thirds of the waste heat from the electrolyzer/fuel cell is recovered for space heating.

A 1.3-kW solar power system sized to handle the heating and cooling can also provide a backup for critical items, such as the refrigerator or freezer during utility power blackouts. By supplementing utility-furnished power, it allows the user to reduce power cost by at least 50 percent.

Cost of Silicon Solar Cells.   The cost of silicon solar cells has been decreasing rapidly. In the 1960s the cost was $200 to $600 per watt. In 1970 the cost was $100 to $200 per watt. These costs varied for different applications, most of which were for spacecraft where requirements for low weight, resistance to radiation, and rigorous quality control added to cost. Also, the cost of cells in a specific procurement is affected by the quantity purchased, and the required efficiency, cell size, quality control, and documentation.

The cost extrapolation into the future (Fig. 13) is based on the ERDA-sponsored Low Cost Solar Array Project, meeting its objective of delivering solar panels for $0.50 per watt in 1985. Procurement of terrestrial solar cells at a price of $12 per watt by the NASA Lewis Research Center in 1975 demonstrated that the cost of terrestrial cells is definitely falling.

The economic analysis that follows is based upon solar cells costing $1.00 per watt, a figure that might be attained in 1980. Total system costs can be readily adjusted for any different solar cell cost.

Solar   Array   Characteristics.   The characteristics of a solar array for supplying 1.3 kW continuous to a home are summarized by Fig. 14. The array is sized to generate 7.7 kW whenever illuminated,

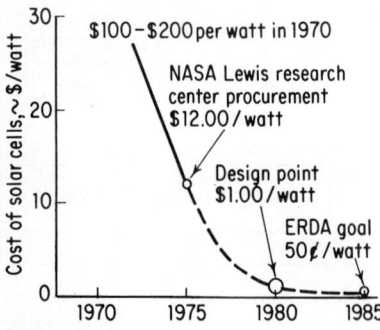

Fig. 13. Contemplated improvement in cost of silicon solar cells.

with sunlight having an intensity of 80 milliwatts per square centimeter. A cell efficiency of 15 percent is assumed. Cells having a conversion efficiency of 15 percent had been made in 1975, but were not generally available. In 1980, terrestrial-use cells having a 15 percent efficiency should be readily obtainable. The results of the analysis that follows can be adjusted for an efficiency other than 15 percent.

The selected 100-volt output of the array can be adjusted up or down, merely by changing the number of cells in series in each string. For example, having 341 cells in series and 155 cells in parallel would produce the same 7.7 kW at 150 volts. The voltage of the array will probably be established by the requirements of the electrolyzer since it will consume most of the power. The 1.3 kW of power supplied directly to the loads in the home will probably be inverted into 60 Hz ac, and the input voltage to the inverter is not critical since it does not significantly affect the inversion efficiency over the range of interest.

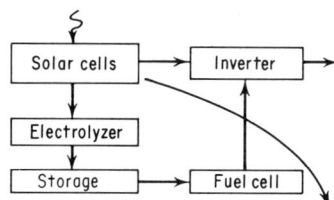

Fig. 14. Solar array characteristics.

Typical solar cell			Typical array of cells	
Efficiency	15%		Current at peak power	70A
Area	12 cm²		Voltage at peak power	110 V
Typical insolation	80 mW/cm²		Peak power	7700 W
Current at peak power	0.33 A		Cells in series (number)	250
Voltage at peak power	0.44 V		Cells in parallel (number)	212
Peak power	0.1452 W		Total array, 250 × 212	53,000
			Total area	685 ft²

Model for Hours of Sunshine. The system is sized to supply an average load of 1.3 kW continuously to a home. All the 24×1.3 kilowatthours of energy required for one day, plus losses, must be collected while the sun shines. Energy for a series of cloudy days must likewise be collected during sunlight hours.

Plotted in Fig. 15 are the hours of sunshine available at a 35°N latitude location at various times of the year. These data are approximate and are derived from measurements (Ref. 7). Table 7 summarizes the percent of possible sunshine available at selected locations where weather records are kept (Ref. 8). Large areas of California and the southwestern United States, which average 80 percent of the possible sunshine, are the most practical for initial solar power applications. An 80 percent availability of sunshine was used in the calculations that follow. The model is based on 80 percent of 9 hours, or 7.2 hours of sunshine available each day. The extra sunshine beyond 7.2 hours a day would be used to generate hydrogen which is stored for cloudy days.

Required Area for Solar Array. The area required for solar cells is shown in Fig. 16 as a function of the average continuous power output desired from the complete power system. The mathematical model for these graphs is described shortly under the heading, "Solar System Efficiency." Parameters for the variability of these functions are solar cell efficiency and average expected sunshine per day. The data assume a typical insolation of 80 percent of the solar intensity for air-mass-one conditions, or 80 percent of 100 milliwatts per square centimeter.

The design point of 1.3-kW continuous power plant output, 15 percent efficient solar cells, and 7.2 hours of average daily sunshine corresponds to 1028 square feet of required area. Not all the area is covered with solar cells. The model assumes, for the purpose of calculating output power of the array, that only 67 percent of the area is effective in terms of solar cells oriented normal to the sun's rays, or 689 square feet of effective area.

Solar System Efficiency. A simple mathematical model is used to calculate the solar cell area required as a function of the continuous power output of the solar plant. The

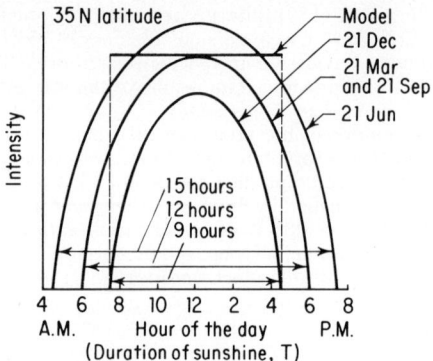

Fig. 15. Model for hours of sunshine. $e_T T = 0.8 \times 9 = 7.2$ hours.

TABLE 7.   Percent Possible Sunshine—Annual, $(100 e_7)$

City	3-year average	1973	1972	1971
Fresno, Calif.	78	78	75	82
Sacramento, Calif.	79	77	79	82
Las Vegas, Nev.	85	83	85	87
Phoenix, Ariz.	87	88	88	86
Albuquerque, N. Mex.	77	75	77	79
El Paso, Tex.	86	86	85	88
Amarillo, Tex.	75	74	73	78
Little Rock, Ark.	72	75	71	70
Average	80%			

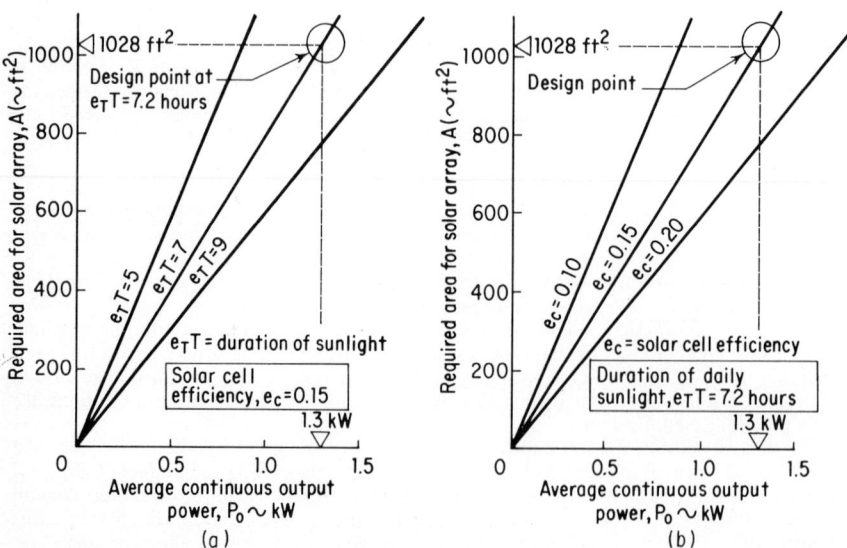

Fig. 16. Required area for solar array. *(a)* Effect of amount of sunshine on size of solar array. *(b)* Effect of solar cell efficiency on size of solar array.

nomenclature defines efficiencies and power levels at the input or output of the elements as shown in Fig. 17.

Let P_i = solar radiation for air-mass-one standard conditions at the earth's surface. This model assumes P_i = 93W/ft² (100 mW/cm²).

P_s = peak power output of the solar array, kW.

P_o = design continuous power output of the system, kW.

A = required rooftop area, ft².

T = hours of possible sunlight per day averaged over the entire year for the given geographical location.

e_T = fraction of possible sunlight, averaged over the years 1970 to 1973, reaching a given geographical location. This factor is tabulated for many cities in the United States in Ref. 8. This model assumes $e_T T$ = 7.2 hours.

e_A = efficiency of utilization, by the solar cell array, of home rooftop space. This factor accounts for (1) spacing between cells or panels of cells for cooling, wind protection, array structure, and maintenance access; (2) off-perpendicular orientation of the cells relative to the sun's rays (fixed installation); and (3) the unsuitability of a portion of the rooftop for solar arrays. This models assumes e_A = 0.67.

e_i = insolation efficiency that accounts for atmospheric opacity (e.g., haze). This model assumes e_i = 0.80.

e_c = efficiency of a silicon solar cell. This model assumes e_c = 0.15.

$e_{\text{dc-ac}}$ = efficiency of an inverter for converting dc to ac. This model assumes $e_{\text{dc-ac}}$ = 0.95.

e_{dc} = efficiency for power conditioning required at the input interface to the energy storage. This model assumes e_{dc} = 0.95.

e_{sr} = efficiency of storage and recovery, including the electrolyzer, storage device, fuel

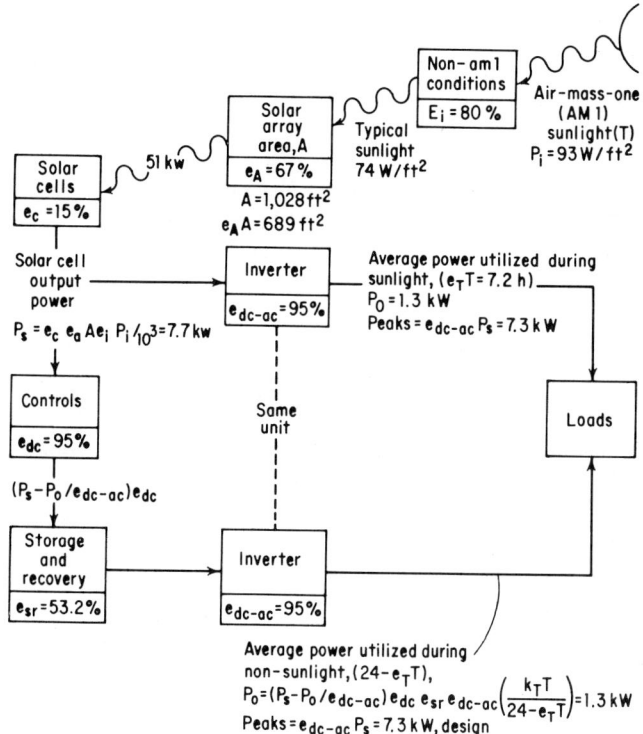

Fig. 17. Solar system efficiency.

cell, and supporting equipment. This model assumes $e_{sr} = 0.532$ (further described shortly under the heading "Storage and Recovery Efficiency").

The overall efficiency of the system is

$$\frac{P_o \times 24 \text{ hours}}{P_i \times 10^{-3} \times A \times e_T T} = 4.5\% \text{ at} \quad \begin{aligned} P_o &= 1.3 \text{ kW} \\ A &= 1{,}028 \text{ ft}^2 \\ e_T T &= 7.2 \text{ hours} \end{aligned}$$

The peak power output of the solar cell array is, for a given area A:

$$P_s = \frac{P_i e_i A e_{,1} e_c}{10^3} \tag{1}$$

The continuous power output P_o of the system is a function of area:

$$P_o = f(A) = (P_s - P_o/e_{dc-ac}) \frac{e_{dc} T e_T e_{sr} e_{dc-ac}}{24 - T e_T} \tag{2}$$

For a given required continuous power output, Eqs. (1) and (2) may be solved to determine required area.

Storage and Recovery Efficiency. Hydrogen and batteries are the only apparent practical ways of storing energy in a 1.3-kW continuous solar power system. However, at the time of writing this article (late 1976), there were no suitable batteries for the type of installation described. For example, to supply 1.3 kW for an 11-day no-sun period, 343 kWh of energy storage is required. The lowest-cost battery is the lead-acid type, used in automobiles, which costs around $50 per kWh of capacity; or $17,000 for the array of batteries required to store 345 kWh. More important, automobile batteries are not designed for long life when deeply discharged, losing one-half of their capacity within 200 charge-discharge cycles. On the other hand, an 11-day supply of hydrogen, when stored as a metal hydride, will occupy no more volume than that of an 80-gallon domestic hot-water tank.

The storage and recovery subsystems in Fig. 18 are sized to provide 1.3 kW average continuous power during nonsunlight portions of the day. The individual component efficiencies shown have been achieved in practice. They include no benefits from recycling internally generated heat. The overall efficiency will be improved by utilizing the heat assumed to be wasted from the fuel cell for hydrogen release from hydride storage. This heat replaces that lost as latent heat of fusion when storing the hydrogen as a hydride. The heat of fusion may be used by the electrolyzer to augment water distillation.

The power loss in storage and recovery, including power conditioning and inversion is

$$P_o \frac{1-e}{e} = 1.4 \text{ kW or } 4800 \text{ Btu/hour}$$

where $e = e_{dc} e_{sr} e_{dc-ac} = 0.48$

For possible home heating use, the average continuous power loss would be

$$\frac{1.4 \text{ W} \times 16.8 \text{ hours}}{24 \text{ hours}} = 0.98 \text{ kW}$$

With two-thirds of this heat loss used for space heating, the system is sized to provide the total heating and cooling needs of the model home by a combination of electrically driven heat pump and recovery of heat losses.

The 70 percent efficiency of the fuel cell suggests that loads in the home, such as water heating, might well be supplied more directly by burning hydrogen.*

Electrolyzer Characteristics. Energy storage by the conversion of electric power to hydrogen and oxygen is accomplished by commercial electrolyzers (as of 1975) with an efficiency of about 80 percent (Ref. 9). This efficiency will probably increase. A 1974 estimate (Ref. 10) for electrolyzers shows an efficiency of 85 to 90 percent and a capital cost of $70 to $100 per kilowatt.

The electrolyzer shown in Fig. 19 has a water purifier and a condenser. Water supplied

---

*See also articles on "Hydrogen," "Batteries," and "Fuel Cells" in the *Chemical Fuels Technology* section of this Handbook.

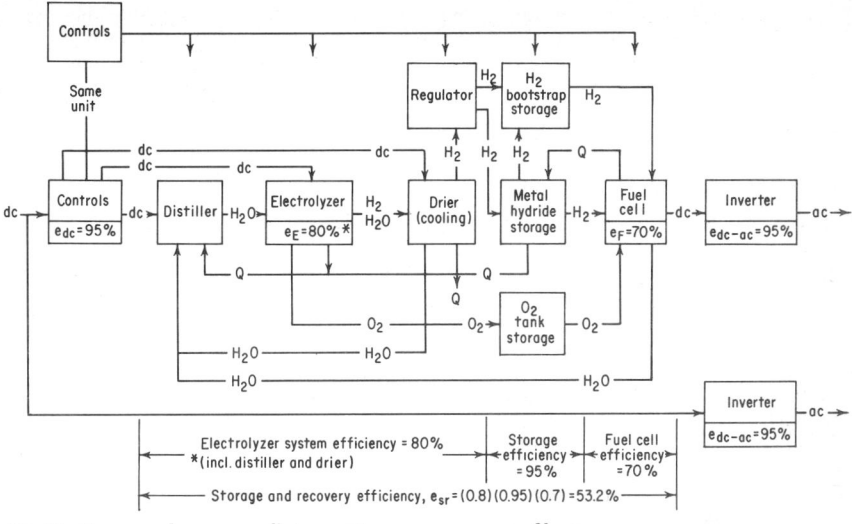

Fig. 18. Storage and recovery efficiency. Q means movement of heat.

Average unused power during sunlight, $P_s - P_o/e_{dc-ac}$ 6.318 kW

Average hours of sunlight per day, $e_T T = 80\% \times 9$ hours $\times$ 7.2 hours

                                           45.49 kWh

Controls, storage and recovery, and inversion efficiency,

$e_{dc} e_{sr} e_{dc-ac}$ $\times$ 0.48

Average energy utilization per day from storage,

$(P_s - P_o/e_{dc-ac}) e_T T e_{dc} e_{sr} e_{ac-dc} = z$ 21.84 kWh

                                    $\div (24 - 7.2)$ hours

Continuous power utilized from storage over non-sunlight portion
of day, $z/24 - e_T T$ 1.3 kW

to the electrolyzer must be of high purity (less than 10 ppm of impurities). The electrolyzer efficiency includes the energy required by the purifier. Little energy is required to purify the average 0.3 gallon/hour of water necessary for electrolysis to 0.27 pound of hydrogen per hour. The calculation: $(0.27 \text{ lb } H_2/\text{hour})(18.015/2.015) = 2.4 \text{ lb } H_2/\text{hour}$, or 0.3 gal H_2/hour. Waste heat from the electrolyzer may reduce the net energy loss for distillation. The hydrogen gas is cooled to near-ambient temperature by a condenser,

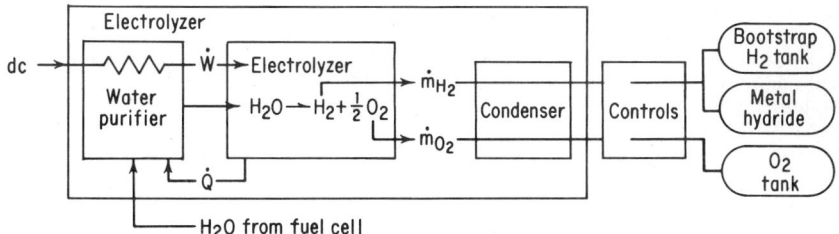

Fig. 19. Electrolyzer characteristics.

Enthalpy efficiency, e_H: $e_H = \dot{m} \Delta H/\dot{W} = 80\%$

Heat loss, $\dot{Q}$: $\dot{Q} = \dot{W} - \dot{m}\Delta H = (1 - e_H)\dot{W} = 4123$ Btu/hour

    where $\dot{W} = (P_s - P_o/e_{dc-ac})e_{dc} = 6.0$ kW

Hydrogen flow rate, $\dot{m}H_2$: $\dot{m}H_2 = e_H W/61{,}000$ Btu/lb $\times$ kWh/3413 = 0.27 lb/hour

Cell voltage, E:

    Voltage efficiency $= 0.8 e_H = 64\%$

    Theoretical minimum voltage $= 1.23$ V (reversible)

    $E = 1.23/0.64 = 1.9$V

removing water vapor from the hydrogen to improve the volume-density efficiency of the metal hydride storage unit. The small aftercooling energy losses are also included in the overall electrolyzer efficiency. Water storage is a part of the electrolyzer unit. Some of the recycled water from the fuel cell is lost, and must be replaced by makeup water.

The electrolyzer characteristics shown are based upon a mathematical model. The enthalpy efficiency is the ratio of output energy rate to input energy rate. A state-of-the-art enthalpy efficiency of 80 percent is assumed. The output energy rate is the product of the mass flow rate of water and the enthalpy of the reaction at standard conditions, or, equivalently, the product of the mass flow rate of hydrogen and the enthalpy of the oxidation of hydrogen. If the reactant water starts in a liquid state, the higher heating value is used. The input energy rate is the rate of electric work done in the cell. With this definition of input/output efficiency, the heat loss rate $\dot{Q}$ can be determined from a first-law energy balance for flow conditions. This heat is transferred to the home to augment space and water heating.

The voltage supplied to the cell, E, is related to the theoretical voltage when the cell is operated reversibly, E_{REV}, by $E = e_E E_{REV}$, where the voltage efficiency e_E is related to the enthalpy efficiency by a factor which is nearly constant over the temperature range of interest: that is, $e_E = 0.8e_H$.

Research and development in the mid-1970s is directed toward achieving voltage efficiencies greater than 80 percent, resulting in endothermic operation (enthalpy efficiency greater than 100 percent). Since the power supply to the electrolyzer must be at a relatively low voltage level, power conditioning equipment is required to interface with the solar array output.

The conductivity of the electrolyte increases with temperature. Thus high operating temperatures are desirable, and the water must be pressurized to prevent vaporization from the electrolyte solution. In many practical cell designs, overall performance increases with pressure increase. Modern cells can operate at elevated temperatures and still retain long design life. Many cells operate at 300 to 450 psia, a pressure suitable for charging the metal hydride storage unit and the oxygen storage tank.

Most electrodes are nickel, without noble metal catalysts. Current densities of electrodes range from 100 to 200 milliamperes per square centimeter.

Alternative methods of producing hydrogen from water include thermal decomposition and decomposition with hot carbon and by catalysts. Thermal decomposition of water needs temperatures of over 2000°C. Decomposition of water with hot carbon requires the use of coal which provides much of the energy. The use of catalysts permits the decomposition of water at temperatures in the 400 to 700°C range, but involves multistep processes and numerous chemicals. Only electrolysis appears practical for small installations as of late 1975.

Metal Hydride Storage Characteristics.   The hydrogen gas from the electrolyzer is stored for use during low-level or nonexistent sunlight. Candidate storage methods are compressed gas, liquefaction, and metal hydride. A prohibitive disadvantage of compressed gas is its volume. For liquefaction, the hydrogen must be cooled to −422.9°F, requiring 30 to 49 percent of the energy available from the hydrogen.

The use of metal hydride for gaseous hydrogen storage is a relatively new technology with promise. In particular, iron–titanium is an alloy with excellent qualities for a solar system application. Hydrogen at relatively low pressures can be added to or removed from the solid metal at temperatures within a few degrees of ambient. Potentially, the alloy can be cycled (charged and discharged with hydrogen) indefinitely. The alloy will absorb hydrogen up to an equilibrium pressure called the dissociation pressure of approximately 300 psi. This corresponds to the state-of-the-art pressure levels for electrolyzers. Hydrogen can be stored as a hydride at volume densities close to that of liquid hydrogen. Iron–titanium is inexpensive (about 50 cents per pound in bulk quantities, as of 1975). The formation of iron–titanium is exothermic, producing only 7000 Btu per pound of hydrogen (about 11 percent of the heat of combustion for hydrogen). During charging, this heat may be used by the water purifier of the electrolyzer in distillation. The thermal process is completely reversible, and so the hydrogen storage unit may be discharged by supplying from the fuel-cell waste heat the required 7000 Btu per pound of hydrogen.

Figure 20 shows the characteristics of an iron–titanium hydride unit developed by Brookhaven National Laboratory and operated by Public Service Electric and Gas Company at its Maplewood, New Jersey, Testing Laboratory (Refs. 4, 11, 12). Its characteristics

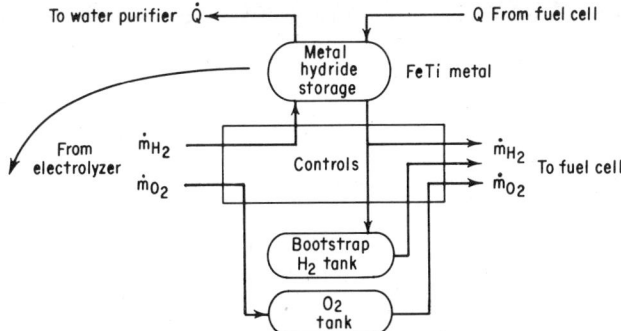

Fig. 20. Metal hydride storage characteristics. (Based on experimental unit operated by Public Service Electric and Gas Company.)
Storage efficiency, $e_S = 95\%$; charging pressure: 300 to 450 psia; metal hydride unit size: 6½ feet long by 1 foot diameter; weight of iron-titanium: 893 lb; design capacity: 12 to 14 lb hydrogen; design flowrate: 1.5 lb/hour for 8 hours; cost of iron-titanium (in bulk): 50¢/lb; reaction: $1.08\ \text{Fe}\ \text{TiH}_{0.1} + \text{H}_2 \rightleftharpoons 1.08\ \text{FeTiH}_{1.95}$.

indicate that iron hydride storage is an excellent candidate for this solar system application.

Metal hydride storage units were also developed and demonstrated in 1972 by Billings Energy Research Corporation for automotive hydrogen engines (Ref. 13).

A significant feature of metal hydride storage is safety. The hydrogen is chemically bonded with a nonreactive metal under relatively low pressure, thus lessening the probability of tank rupture. Inadvertent release of hydrogen is inherently self-limiting, since the hydrogen release rate is quickly reduced by the cooling associated with discharge. Low heat rates and temperature changes of only a few degrees are involved in charging.

Capacity of Gaseous Storage. The data plotted in Fig. 21 show that the capacity of the 1-foot-diameter by 6½-foot-long Brookhaven metal hydride tank (Ref. 4) is adequate for five consecutive cloudy days. A standard size 80-gallon (10.7 ft³) tank would provide energy for about 11 consecutive cloudy days. Hydrogen storage as a compressed gas is not practical at pressures that would be acceptable for home systems. Storage of oxygen as a

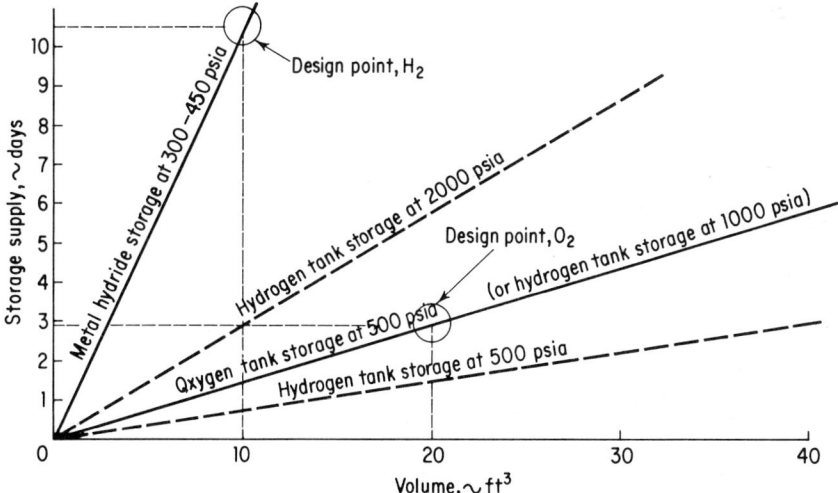

Fig. 21. Capacity of gaseous storage.

pressurized gas is feasible since a 20 ft³ tank at 500 psia will provide oxygen for almost three days. Fuel cells with flexibility to operate on either oxygen or air would reduce oxygen storage tank size, because they could be operated on air, at a slightly lower efficiency, whenever the oxygen supply was exhausted.

The data for Fig. 21 were calculated as follows:

Metal hydride storage:

$$\text{Density at nominal performance} = \frac{14 \text{ lb}}{6.5\pi/4 \text{ ft}^3} = 2.74 \text{ lb/ft}^3$$

$$\text{Maximum density} = \tfrac{3}{4} \text{ liquid hydrogen density}$$
$$= \tfrac{3}{4}\,(4.4) = 3.3 \text{ lb/ft}^3$$

$$\text{Days of supply} = \frac{\text{density} \times \text{volume}}{\text{rate of utilization}}$$

$$\text{where rate of utilization} = \frac{0.11 \text{ lb}}{\text{hour}} \times \frac{24 \text{ hours}}{\text{day}} = 2.64 \text{ lb/day}$$

Tank storage of hydrogen:

$$\text{Days of supply} = \frac{0.0056 \times \text{pressure (psia)} \times \text{volume (ft}^3)}{14.7 \times \text{rate of utilization (from above)}}$$

Tank storage of oxygen:

$$\text{Days of supply} = \frac{0.08921 \times \text{pressure (psia)} - \text{volume (ft}^3)}{14.7 \times \text{rate of utilization} \times 8}$$

Fuel Cell Characteristics. The fuel cell recovers the energy, as needed by loads, from stored hydrogen and oxygen by drawing electricity from the reaction of the gases, as shown in Fig. 22. A small bootstrap tank of hydrogen runs the fuel cell until the heat of reaction transferred to the metal hydride unit is sufficient to discharge hydrogen from the hydride. The bootstrap tank is replenished from the metal hydride storage during steady-state operation of the fuel cell.

The characteristics shown on Fig. 22 for the fuel cell are based on the same calculations at standard conditions as those used for the electrolyzer. Actual performance, as shown, is based on a 70 percent fuel cell efficiency. Reference 4 quotes an efficiency of 70 to 75 percent for advanced hydrogen-oxygen fuel cells. Pratt and Whitney's 1975 technology will produce fuel cell power plants independent of size, at a cost of $350 to $450 per kilowatt (Ref. 14).

The heat loss, computed from a first-law energy flow balance, is 1902 Btu per hour. Of this, 770 Btu per hour (7000 Btu per pound of hydrogen × 0.11 pound hydrogen per hour) is transferred to the metal hydride unit for hydrogen discharge. The remainder is available through the hot water discharge and other heat conduction for home heating.

The fuel cell will consume about 0.11 lb hydrogen per hour at the average power output level of 1.3 kilowatts. Checking the fuel cell hydrogen flow against the electrolyzer flow, the energy balance from the input of the electrolyzer to the output of the fuel cell inverter is

$$\frac{P_s - P_o}{e_{\text{dc}-\text{ac}}} \quad e_{\text{dc}}e_{\text{sr}}e_{\text{dc}-\text{ac}}e_T T = P_o(24 - e_T T)$$

which reduces to

$$\dot{m}_{\text{electrolyzer}} \times 7.2 \text{ hours} = \dot{m}_{\text{fuel cell}} \times \frac{16.8 \text{ hours}}{e_{\text{storage}}}$$

or

$$0.27 \times 7.2 = \frac{0.11 \times 16.8}{0.95}$$

which balances where e_{storage} is the storage tank efficiency, and $\dot{m}$ is the mass flow rate in pounds per hour of the electrolyzer or fuel cell shown in Figs. 19 and 22, respectively.

Solar Power System Cost. Elements of this cost are shown in Table 8. The cost to the purchaser of a 1.3-kW continuous solar power system is estimated to be $14,000 in 1975 dollars. This cost is based upon $1.00 per watt for solar cells and the other assumptions previously discussed.

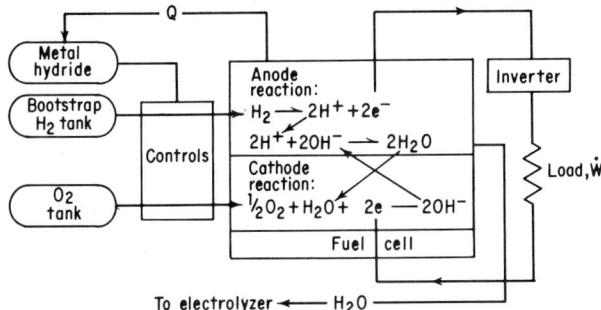

Fig. 22. Fuel cell characteristics.

Enthalpy efficiency, e_H: $e_H = \dot{W}/\dot{m}\,\Delta H = 70\%$

Heat loss, $\dot{Q}$: $\dot{Q} = \dot{m}\,\Delta H - W = (e_H - 1)\dot{m}\,\Delta H$

$$= -\dot{W}\left(\frac{1 - e_H}{e_H}\right) = 1.3\ \text{kW}\left(\frac{1 - e_H}{e_H}\right) = \frac{1902\ \text{Btu/hour}}{\text{(including inverter)}}$$

Hydrogen flow rate, $\dot{m}H_2$: $\dot{m}H_2 = \dfrac{W}{e_H \Delta H} = \dfrac{1.3}{e_{dc-ac}\,e_H \times 61{,}000\ \text{Btu/lb} \times \text{kWh}/3413\text{Btu}}$

$$= 0.11\ \text{lb/hour}$$

Cell voltage, E:

Voltage efficiency $= 0.8E_H = 56\%$

Theoretical minimum (reversible) voltage $= 1.23\text{V}$

$E = 0.56 \times 1.23 = 0.69\text{V}$

No attempt has yet been made to estimate the maintenance cost of the 1.3-kW power source and energy storage. The solar cells will be relatively maintenance free, but the electrolyzer and fuel cell may require periodic inspection and perhaps replacement of parts.

In an economic sense, the value of 1.3 kW of continuous solar power depends upon the cost of procuring it from an alternate source. Figure 23 shows the annual saving from displacing 1.3 kW of power from the public utility serving the premises, plotted as a function of price to the consumer of electricity.

An 8.8 cent per kWh rate represents an annual cost of $1000, which also is the annual interest on $14,000 investment at 7 percent. The effective interest may be reduced by the effect of tax and other government incentives which encourage the use of solar power. Thus, if power costs 8.8 cents per kWh, and a 1.3-kWh solar power plant can be procured for $14,000, then the procurement of such a plant would be economically justified. By comparison, as of the present situation, in 1975 electricity did not cost 8.8 cents a kilowatt-

TABLE 8.   Solar Power System Cost (When in Production)

Cost Item	Cost	Comment
Silicon cell	$7,700	7.7 kW output power at $1.00/watt (1980)
Solar array structure	400	
Fuel cell	2,900	7.3 kW peak power output at $400/kW
Electrolyzer	600	7.3 kW peak power input at $70–100/kW
Metal Hydride	800	800 lb FeTi at 50¢/lb + structure
Hydrogen and oxygen storage tanks	500	at $300/1,000 standard cubic feet + controls
Electric power conditioning and controls	500	
Installation	500	
Total	$14,000	

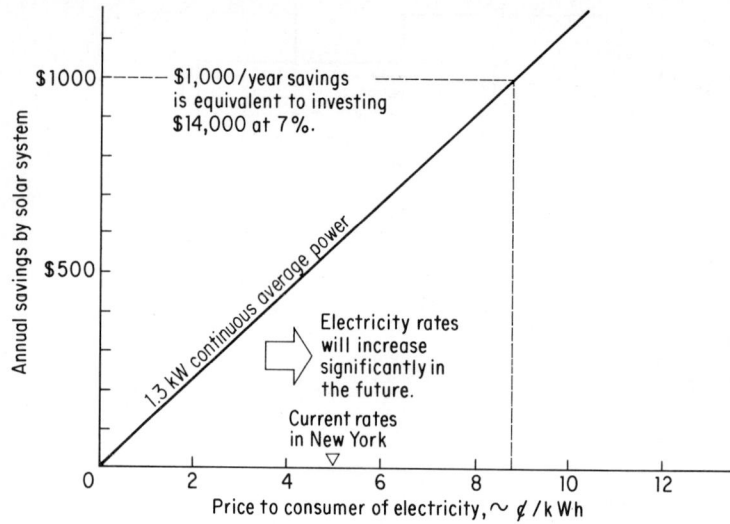

Fig. 23. Contemplated return on investment.

hour. In some parts of the United States, electricity can be obtained for less than one cent per kilowatthour. However, in some of the Eastern cities the price of electricity has risen to over 5 cents a kWh, and further increases may occur as "old" oil is decontrolled and utilities must pay full price for the oil which they consume.

REFERENCES

1. Chapin, D. M., C. S. Fuller, and G. L. Pearson: A New Silicon *p-n* Junction Photocell for Converting Solar Radiation into Electric Power, *J. Appl. Phys.*, **25**, May 1954.
2. Brown, W. D.: Computer Simulation of Solar Array Performance, *Hughes Aircraft Co. Rep.* SSD70135R.
3. Carter, J. R., Jr. and H. Y. Tada: "The Solar Cell Radiation Handbook," TRW Systems Group Publication 21945-6001-RU-00, June 28, 1973.
4. Burger, J. M., and P. A. Lewis (Public Service Electric and Gas Company); R. J. Isler, and F. J. Salzano (Brookhaven National Laboratory); and J. M. King, Jr. (Pratt and Whitney Aircraft Co.): Energy Storage for Utilities via Hydrogen Systems, *9th Intersoc. Energy Conversion Eng. Conf. Proc.*, Miami, Fla., 1974.
5. Bakus, C. E. (Arizona State University): A Solar-Electric Residential Power System, 7th Intersociety Energy Conversion Engineering Conference, 1972.
6. Calvert, F. O., and D. G. Harden (University of Oklahoma): A Comparative Study of Residential Energy Usage, 8th Intersociety Energy Conversion Engineering Conference, 1973.
7. Swet, C. J. (John Hopkins University, Applied Physics Laboratory): Heliotropic Thermal Generators, 8th Intersociety Energy Conversion Engineering Conference, 1973.
8. U.S. Department of Commerce, National Oceanic and Atmosphere Administration, Environmental Data Service: Climatological Data, National Summary, 1973, 1972, and 1971.
9. University of Washington research paper for Seattle City Light Co.: Exploratory Engineering Research in Energy Management and Utilization, Final Report, Feb. 28, 1973.
10. Fernandes, R. A. (Niagara Mohawk Power Corp.): Hydrogen Cycle Peak-Shaving for Electrical Utilities, *9th Intersoc. Energy Conversion Eng. Conf. Proc.*, Miami, Fla., 1974.
11. Strickland, G., J. J. Reilly, and R. H. Wiswahl, Jr. (Brookhaven National Laboratory): An Engineering-Scale Energy Storage Reservoir of Iron-Titanium Hydride, Hydrogen Economy Miami Energy (THEME) Conference, Miami, Fla., 1974.
12. Hoffman, K. C. (Brookhaven National Laboratory): *Hydrogen Energy Fundamentals Symp. Proc.*, Miami, Fla., 1975.
13. Billings, R. E.: Hydrogen Storage in Automobiles Using Cryogenics and Metal Hydrides, Hydrogen Economy Miami Energy (THEME) Conference, Miami, Fla., 1974.
14. Hammond, A. L., W. D. Metz, and T. H. Maugh, II: "Energy and the Future," American Association for the Advancement of Science, Washington, D.C., 1973.

Solar Climate Control*

BY JAMES C. BURKE*

The term *solar climate control* means the use of solar energy in private residences; multifamily dwellings; and commercial, industrial, and public buildings for space heating and cooling, water heating, and on-site power generation. Typically a solar climate control system will consist of the following components:

1. A *solar collector*—a surface or composite surface which by virtues of geometry or surface properties absorbs solar energy and imparts this energy to a heat-transfer fluid which circulates through the collector. Although highly concentrating collectors which track the sun have certain advantages for high-temperature collection, the less expensive, flat-plate collector and other stationary designs with little or no concentration are generally adequate for the collection temperatures required for space heating and cooling. Typically, the solar collector consists of an absorber (either blackened or selectively black to solar energy) covered by one or more transparent panes. The cover pane(s), although transparent to the incident solar radiation, greatly reduce heat loss by convection and reradiation from the absorber. Therefore, the collector serves to trap solar energy.

2. A *heat storage system*—material which has a high specific heat or experiences a change of phase and which accepts collected solar heat as available and allows it to be withdrawn as needed. To date, water and stone or gravel have been the most common thermal storage materials. However, cetain change-of-phase materials, such as eutectic salts, show promise of more compact storage if various technical and economic problems can be resolved.

3. A *source of auxiliary energy*—fuel or electricity to provide heat during extended cloudy periods, precluding the need for uneconomically large collection and storage facilities.

4. A *heat-actuated air conditioning unit*—one that can be driven by collected solar heat. A number of varieties of heat-actuated equipment are possible contenders for solar cooling. These can be divided into heat engine types, such as the Rankine cycle, absorption machines, and desiccant cooling or dehumidification systems.

5. *Ancillary equipment*—the necessary piping, controls, heat exchangers, heat-transfer fluid, valves, pumps, motors, etc., used to couple all the elements into an operating system.

6. *On-site power generation*—the use of the solar cells (incorporated into a solar collector), or thermally driven heat engines, to convert solar energy directly to electricity. (See previous article in this Handbook section on "Solar Cells and Arrays.")

A solar climate control system could include functions, such as: (1) solar hot water heating (or preheating); (2) solar space heating and hot water; (3) combined heating and cooling, using either heat-actuated air conditioning or heat pumps for cooling—any of the foregoing with thermal storage and controls specifically designed to enhance utility load management.

SOLAR COLLECTION

Devices that intercept and collect the sun's energy can be classified in terms of the degree of concentration achieved. Highly concentrated collectors, requiring tracking of the sun, are promising contenders for large-scale solar thermal power generation. Moderate to low concentrating collectors requiring only periodic seasonal adjustment have been considered for both solar thermal power generation and some climate control applications, particularly cooling. However, the flat-plate-type collector, having little or no concentration, can be adequate for the collection temperatures required for space heating and

*Arthur D. Little, Inc., Cambridge, Massachusetts.

cooling, as previously mentioned. This collector is described in the following paragraphs. Since one of the major costs involved in solar heating-cooling systems is that of a collector, much depends upon designing the least expensive practical collector.

Figure 1 is a schematic diagram of the cross section of a typical flat-plate collector. Its essential features are those of a blackened absorber surface covered by one or more panes, which are transparent to the short wave incident solar radiation. These panes greatly reduce heat loss by air conduction and convection; and in the case of glass and certain plastics, they can be opaque to the longer wave reradiated infrared energy. The working fluid used to remove the heat from the collector can be either water or some other fluid, flowing in tubes attached to the blackened plate (as shown in Fig. 1), or air flowing behind

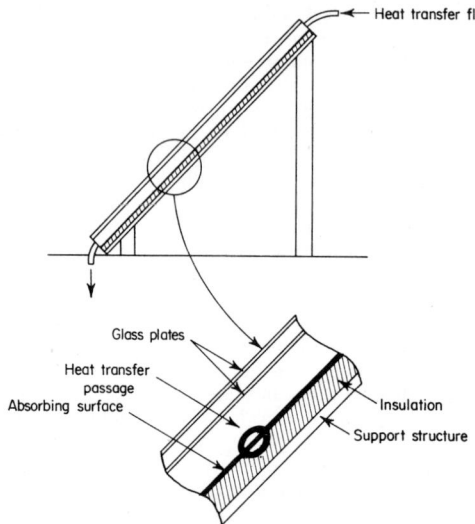

Heat transfer fluid

Glass plates

Heat transfer passage

Absorbing surface

Insulation

Support structure

Fig. 1. Flat plate solar collector.

the absorber surface or between the absorber surface and the transparent panes. Liquid heat-transfer fluids tend to have the most favorable heat-transfer pumping power characteristics. However, air eliminates many problems such as freezing, boiling, and contamination of potable water (with toxic heat-transfer fluids), and tends to minimize corrosion.

Solar collection efficiency is defined as the ratio of usable energy collected to incident solar energy. Efficiency η may be calculated as

$$\eta = \alpha\tau \frac{U\Delta T}{S}$$

where α is the solar absorptance at the absorber surface
  τ is the transmittance of the transparent pane(s)
  U is the collector heat loss coefficient
  ΔT is the temperature difference between absorber and ambient air
  S is the incident solar flux

As the above equation indicates, solar collector efficiency tends to decrease at either high collection temperature (high ΔT), or low incident solar radiation (low S)—the rapidity of this decrease being dependent on the heat loss coefficient U. An aim in high efficiency solar collector design is to maintain high solar absorptance and transmittance while minimizing the heat loss coefficient. Some of the design trade-offs involved are illustrated in Fig. 2, which shows the performance of one and two (glass) pane collectors having both flat black and selectively black surfaces. As shown in this figure, a second glass pane reduces convective and radiative heat loss—although at some transmittance penalty, as evidenced by the slight loss in efficiency at low temperature. Using a selective surface absorber (one that has a high solar absorptance but a low infrared emittance) reduces radiative heat loss and further improves high temperature performance for both one and two-pane collectors.

Further improvements in solar collector performance can be achieved by suppressing air convection and conduction between the absorber and cover pane. Suppression of convection can be achieved with a honeycomb structure which inhibits natural circulation patterns—or by reducing the air pressure in the space. Improved flat-plate collector designs employing either honeycomb or partial vacuum for convection suppression have been designed and successfully tested.

A particularly promising approach for the virtual elimination of both convection and air conduction is the evacuated tubular design. This configuration is capable of maintaining a very high vacuum for extended periods, using manufacturing techniques similar to those employed in the production of fluorescent light bulbs. Although not strictly speaking a

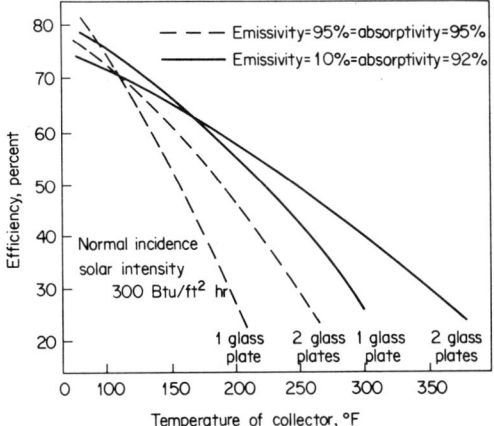

Fig. 2. Solar collector efficiency. *(Hottel and Unger, Solar Energy, October 1959, with permission.)*

flat-plate collector, this design is similar to the flat plate, in that little or no concentration is required. The evacuated tubular collector is capable of very good high temperature performance, and, when the market is large enough to justify the substantial tooling costs, collector costs may be competitive with those of the simpler flat-plate variety.

The solar collector should face south (in the Northern Hemisphere) and be tilted to maximize incident solar radiation. The optimum angle of the tilt of a flat-plate collector is latitude plus about 15 geometric degrees for winter heating and latitude minus about 10 degrees for summer cooling. For a combined heating and cooling system—or for a domestic hot water heating system—the tilt angle should approximate local latitude. The accuracy of neither the tilt angle nor the azimuthal pointing direction is very critical—departures of 10 to 15 degrees from optimum having very little effect on performance.

HEAT STORAGE

A comparison of heat storage capacity on a volume basis between various storage media shows that water can store 62.5 Btu per cubic foot per degree Fahrenheit, while rocks, bricks, and gravel can store about 36 Btu per cubic foot per degree Fahrenheit. Some salts which melt in the desired temperature range can store about 60 Btu per cubic foot per degree Fahrenheit as sensible heat, and 9500 Btu per cubic foot at the melting point as heat of fusion. For the same volume, therefore, the heat of fusion is equivalent to a temperature swing of about 150°F for water or almost 300°F for solid materials. Various change of phase materials have promise for substantial reductions in volume, particularly for applications like storing "cold" for air conditioning where the allowable temperature excursion may be only about 10°F. However, change of phase materials presents various technical/economic problems, so that in the near term either water or solids are likely to be the materials of choice. Water storage tends to be most compatible with a solar collector designed for a liquid heat-transfer fluid; solids such as rocks tend to be more compatible with an air collector

HEAT-ACTUATED REFRIGERATION

A great variety of heat-actuated refrigeration cycles have been proposed for air conditioning. These can be divided into heat engine types, such as the Rankine cycles, absorption machines, and desiccant systems.

Regardless of the type of heat-actuated refrigeration, operating temperature is a tradeoff in coupling a solar collector to a heat-actuated refrigeration machine. As discussed

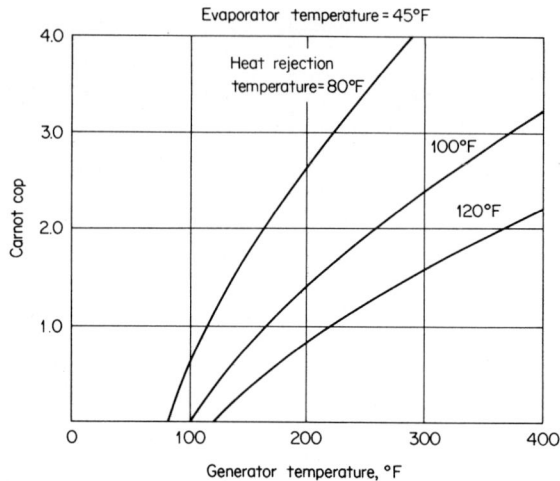

Fig. 3. Carnot coefficient of performance of heat-actuated refrigerators.

previously, and presented in Fig. 2, the efficiency of solar collectors decreases with temperatures. On the other hand, the coefficient of performance (COP) of heat-actuated refrigerators, defined as the ratio of cooling produced to heat input, increases with generator temperature. The Carnot cycle which represents the maximum theoretical performance for various temperature levels is useful for describing the effects of generator temperatures. (Actual refrigeration cycles may approach 50 to 60 percent of Carnot performance.) Figure 3 shows Carnot COP as a function of generator temperature for an evaporator temperature of 45°F, and for heat rejection temperatures of 120°F (typical of air cooling), 100°F (warm water cooling such as might be achieved with a cooling tower), and 80°F (cold water cooling such as might be achieved with well water). (For absorption machines, it is assumed that absorber and condenser operate at a common heat rejection temperature.)

The results of Figs. 2 and 3 are combined in Fig. 4 to illustrate the effects of collector/ generator temperature on overall cooling effectiveness for two collector designs. The overall cooling effectiveness, the product of solar collector efficiency and the air conditioner coefficient of performance, represents the cooling achieved per unit of solar energy incident on the collector. For this plot, temperature losses between collector and generator were neglected, and the air conditioner was assumed to achieve 50 percent of ideal Carnot performance. Figure 4 shows that the overall cooling effectiveness for the two-pane black collectors reach a maximum value at about 200°F and falls off rapidly at higher temperatures. For a more refined collector, such as a double pane with selective surface, the overall cooling effectiveness is higher and reaches a maximum at nearly 300°F. For an even more advanced design, such as an evacuated tubular collector, the maximum cooling effectiveness would be substantially higher and would occur at higher temperature.

ABSORPTION REFRIGERATION

Since absorption refrigeration and particularly the lithium bromide–water and ammonia-water cycles represent the most common technology in heat-actuated air conditioning equipment, we will at this point focus our attention on them, although the other cycles do have particular advantages. The Rankine cycle, which will become increasingly promising as high performance collectors permit higher driving temperatures, can be air cooled and can be electrically (as well as thermally) driven in the auxiliary mode and could also be used for electric power generation. Desiccant systems, used either for dehumidification alone or (with evaporative cooling) for complete air conditioning, may, in certain climatic situations, be more effective than either absorption or Rankine cycles.

Figure 5 presents minimum generator temperature (corresponding to equal concentrations in generator and absorber—and implying an infinite circulation flow in the absorber-generator circuit) as a function of heat rejection temperature (again assuming that condenser and absorber temperatures are equal) for an evaporator temperature of 45°F. This plot is equally valid in defining *minimum* generator temperatures for either lithium bromide-water or ammonia-water cycles. However, in practice, the lithium bromide-water machine is able to operate at higher circulation rates, and hence achieve generator temperatures closer to the theoretical minimum, than is normally true of ammonia-water machines. It is apparent from this plot that the minimum generator temperature is quite sensitive to heat rejection temperature and explains why most of the solar powered absorption machines operated to date with simple low temperature solar collectors have been water cooled.

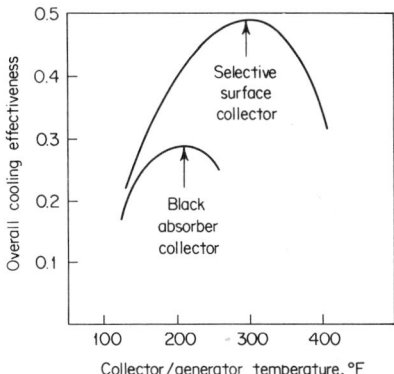

Fig. 4. Overall collector-air conditioner performance. Normal incidence solar intensity = 300 Btu/(ft^2) (hour). (*Arthur D. Little, Inc.*)

Lithium Bromide–Water Cycle.   This cycle has a number of attractions. It has fewer heat-transfer components than the ammonia-water cycle. The low (subatmospheric) cycle pressure minimizes pumping power (external pump power can be completely eliminated by the use of a heat-actuated bubble pump). However, the solubility characteristics of lithium bromide–water solutions (i.e., the tendency for crystallization at the high concentration required for air-cooled absorbers) has in practice limited these cycles to water cooling. The water-cooled LiBr cycle has been successfully used for solar air conditioning in Australia (Ref. 1), and more recently has been incorporated in a number of solar cooling demonstration projects such as those sponsored by the National Science Foundation. However, in residential units, the need for a cooling tower may be a serious penalty.

Ammonia-Water Cycle.   This cycle has no solubility limitation on absorber temperature and therefore it can be easily air cooled. It operates at high pressures (300 to 100 psi), requiring a reasonably large solution pump (½ hp in a 3-ton unit) and discouraging high circulation flows. Design point generator temperatures of existing air-cooled units are high—about 350°F.

Manufacturers indicate that they have had experience with "underfiring" their machine by 25 percent and thereby dropping the generator temperature from about 350 to 300°F with no appreciable change in design COP. It is possible that operation at 50 percent input and 250°F is feasible with only minor loss in COP. If a larger solution pump were added, further reductions in generator temperature might be achieved.

Dr. Erich Farber of the University of Florida has operated ammonia-water machines driven by solar collectors for some years (Ref. 2). He has run his machines at very low generator temperatures (design point about 175°F and part load operation as low as 120°F) by the use of water cooling.

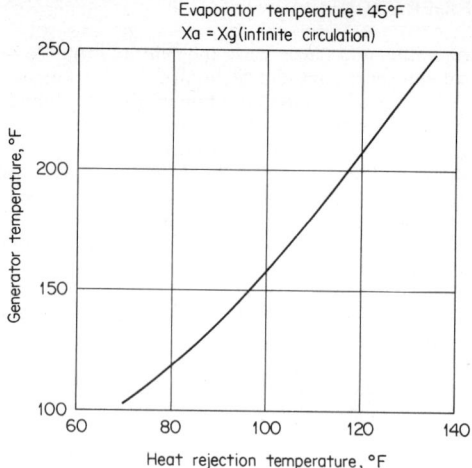

Fig. 5. Minimum generator temperatures for LiBr–H_2O and NH_3–H_2O absorption cycles.

SOLAR HEATING AND COOLING SYSTEM APPROACHES

To illustrate the operation of solar heating and cooling systems, three systems will be briefly described: (1) a heating system with hot storage, (2) a simple cooling system with cold storage, and (3) a combined heating/cooling system.

1. *Simple Heating System with Hot Storage.* As shown in Fig. 6, the basic elements of this system, exclusive of pumps, valves, controls, etc., are: (*a*) a solar collector, (*b*) an auxiliary heating device, (*c*) hot storage system, and (*d*) heater element fan and air duct system.

The solar collector absorbs heat energy from the sun and transfers it to a heat-transfer fluid which conveys the heat to a hot storage system. From the hot storage, heat is withdrawn from storage through a heater coil, where an air-circulating fan carries heat from the coils into an air duct system.

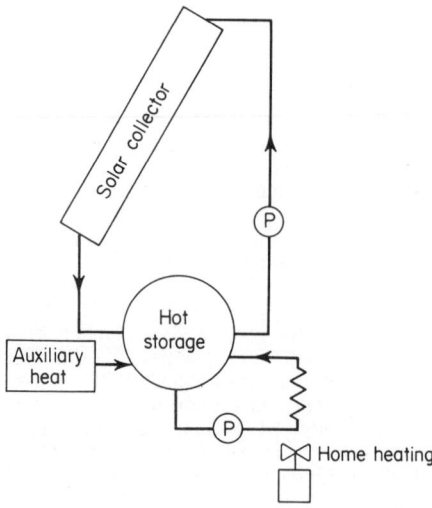

Fig. 6. Basic elements of solar-heating system.

When the solar system cannot supply an adequate amount of heat, the auxiliary heating device, such as a fuel burner, electrical resistance heating, or an electric driven heat pump, comes on. (As will be shown later, economic considerations argue against solar systems that are large enough to provide all, or almost all, of the load.) This auxiliary heat could be added to hot storage, as shown in Fig. 5, or used to directly heat the room air.

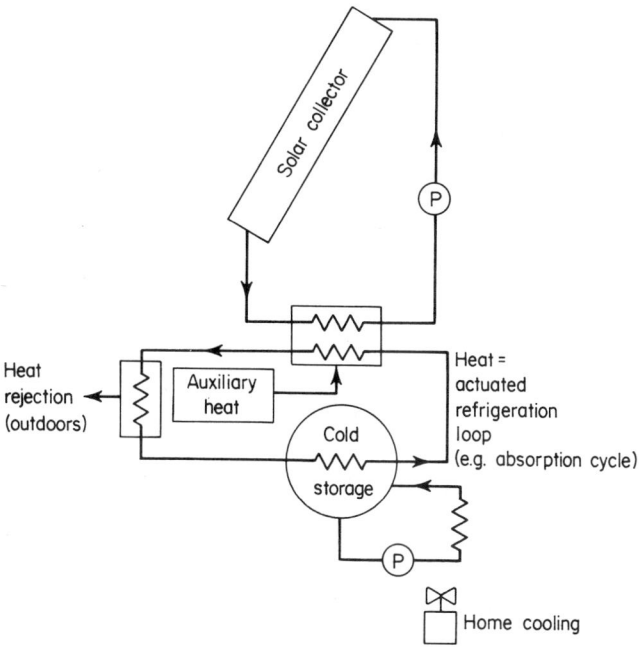

Fig. 7. Basic elements of solar cooling; cold storage only.

Applying the auxiliary heat directly to the room air results in the minimum auxiliary heat quantity, since this heat does not raise storage (and collection) temperature and thereby reduce collection efficiency. For this type of heat addition, commonly the auxiliary might be a conventional air furnace located downstream of the hot water storage heating coil which would be activated only when storage was not hot enough to supply the required heating load. On the other hand, heat addition to storage (at off-peak hours) can provide opportunities for utility load management which might more than outweigh the loss in solar system collection efficiency.

2. *Solar Cooling—Cold Storage Only.*   As shown in Fig. 7, the basic elements of this system are: (*a*) a solar collector, (*b*) an auxiliary heating device, (*c*) a cold storage system, and (*d*) a heat-actuated refrigeration loop (such as an absorption cycle).

The solar collector absorbs heat energy and transfers it to a heat-transfer fluid, which, in turn, conveys the solar heat to the generator or boiler of the heat-actuated refrigeration loop. This refrigeration loop can also be driven by auxiliary heating when the solar heat input is not adequate. If the heat-actuated refrigeration loop were of the Rankine cycle type, instead of an absorption cycle, it might also be possible to accomplish auxiliary cooling with electric drive instead of with heat—frequently the operating costs of electric air conditioning are less than those of heat-driven air conditioning.

In the system arrangement of Fig. 6, thermal storage is placed on the cold side of the refrigeration loop. The refrigeration loop cools the cold storage reservoir from which home cooling is supplied on demand. Alternatively, thermal storage could be placed on the hot side of the loop—similar to the space heating system of Fig. 5. The cold storage approach may have special advantages for load management, particularly if the auxiliary cooling is electrically driven.

3. *Combined Heating and Cooling System.*   Figure 8 illustrates a typical combined heating and cooling system, which is only one of a variety of possible systems. The major elements of this system are: (*a*) a solar collector, (*b*) an auxiliary furnace with heating coils, (*c*) a storage system (hot in winter, cold in summer), (*d*) an absorption refrigeration cycle, and (*e*) necessary valving and controls.

The system is designed to provide both heating and cooling on demand. The heat energy generated by the solar collector is directed to either the hot storage tank or the refrigeration cycle generator, according to the seasonal mode of operation.

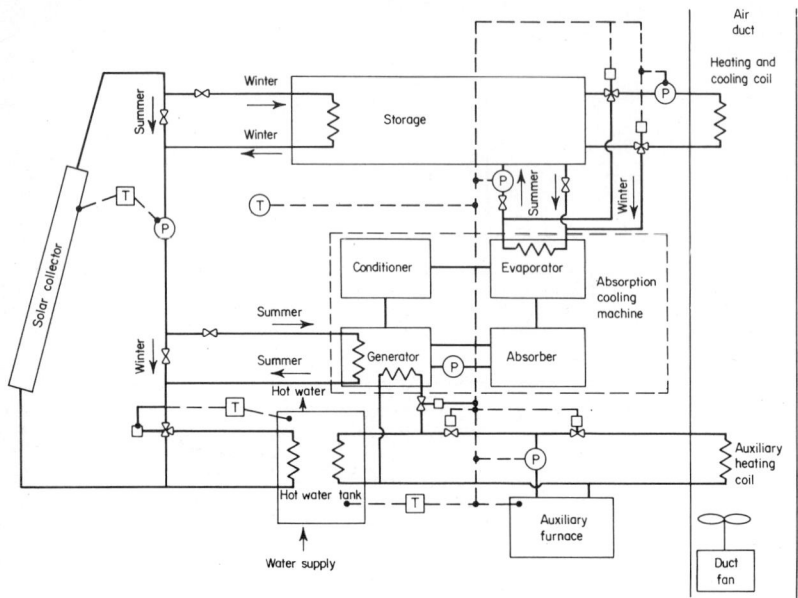

Fig. 8. Combined solar climate heating and cooling system.

When solar energy (either direct or stored) cannot supply the required heating or cooling load, auxiliary heating or cooling can be employed.

Winter Mode of Operation.   In this mode, solar energy is gathered at the collector and is pumped directly to the storage system. From the storage system, heat is extracted according to household needs through the heating/cooling coil in the main air duct. This coil is controlled by three-way valves which are open to heating and shut to cooling. Heat is then extracted from the coil by air fans and carried into the house.

At certain times, the solar collector and storage system will not be able to supply enough heat to maintain the needs of the household. In these instances, the auxiliary furnace will assume the heating load until the solar system is able to provide heating. This furnace will also furnish auxiliary heat for domestic hot water when solar energy cannot provide this function.

At times while on winter mode operation, there will be days when cooling may be required. When this condition exists, the auxiliary furnace will drive the absorption cycle. The two three-way valves on the heating/cooling coil will be actuated to permit chilled water from the absorption machine to circulate through this coil.

Summer Mode of Operation. In summer, solar energy is gathered at the collector and pumped in the form of heat directly to the absorption refrigeration machine and to the domestic hot water tank. The collected energy serves as the main driving force for the cooling system. When the collector is unable to provide the necessary energy for the cooling systems, the auxiliary furnace is activated to supplement the energy load.

Once the cooling cycle is activated, the cooling produced is directed to the same storage system used for storing heat in the winter. Cooling is extracted per household needs from the storage system through the heating/cooling coil. In this mode, the three-way valves of the coil are open to storage system and shut to direct cooling from the cooling cycle. Should the storage system not be able to provide the cooling, these valves would be closed to the storage system and open to direct cooling.

While on the summer mode, the auxiliary furnace can be utilized to heat the house on cold nights.

The sytems above are presented as typical of general concepts, and are not necessarily illustrative of final designs or optimum arrangements of components. The final detail system design will, in particular, be dependent on whether fuel or electricity is employed for auxiliary heating. The above description was based on fuel auxiliary. For the near term, fuel (natural gas or fuel oil) may be preferable to electric resistance heating in terms of economics; however, in many areas unavailability of natural gas may dictate use of electricity. Where electricity is employed the use of heat pumps for auxiliary heating and cooling may be merited, and it would be particularly desirable to use thermal storage and controls in such a way as to improve load management.

ECONOMICS OF SOLAR CLIMATE CONTROL

The economics of solar climate control vary greatly with the type of system, the climatic conditions, and the cost of conventional energy. Systems for heating domestic hot water tend to have the most favorable near-term economics because of the steady usage factor, low temperature requirements, and the fact that relatively high cost electric resistance heating is frequently the competing conventional energy. Heating systems, while promising greater overall energy savings, tend to be somewhat less cost effective because of the seasonal variation in heating load. The economics of solar cooling tends to be most complex because of the higher temperature requirements and the additional cost of the heat-actuated equipment.

Tybout and Löf (Ref. 3) have made an extensive engineering and cost study of solar house heating. In their studies, they have used the parameter solar heat costs, defined as the annual capital charges on the solar equipment (first cost times amortization rate) divided by the solar heat collected and utilized. Solar heat cost can be used as an index for comparison with the cost of conventional energy. Its numerical values of cost vary with location and system size. For a given building in a given location, minimum solar heat cost occurs at a solar system size which represents the best balance between annual solar heat collection per unit area, and capital cost per unit area. Systems larger than this optimum area are penalized by low efficiency, since the collector may be oversized for an appreciable portion of the year. On the other hand, systems smaller than the optimum area suffer from increasing cost per unit area due to the fixed (or weakly size-dependent) costs associated with items, such as controls, pumps, and sizing. Figure 9, taken from an updating and extension of the original Tybout and Löf work by Pogany, Ward, and Löf (Ref. 4), presents an example of the variation of solar heat costs with percent solar for a single-family house in Albuquerque, New Mexico. The authors present a range of parametric value of solar collector installed cost, the high value of which is taken to be indicative of present costs.

Minimum solar heat costs will also be a function of locality. Regions having abundant sunshine, and a heating load which is both appreciable and seasonally constant, will have lower solar heat costs than those typified by cloudy weather and a highly variable heating load. The Albuquerque location of Fig. 9 has excellent solar insolation. However, it is possible to achieve even lower solar heat costs in a location like Santa Maria, California, a coastal town between Santa Barbara and San Luis Obispo, which has both a relatively steady heating load and quite sunny weather. On the other hand, in a location like Boston, Massachusetts, typical of a highly variable heating load combined with reasonably cloudy weather, minimum solar heat costs will be substantially higher than those shown in Fig. 9. The cost of conventional energy is another location-dependent factor which is very important to the competitiveness of solar heat. In the near-term, locations in the northeastern United States, characterized by both high solar heat costs and higher conventional energy costs, might make for better solar economics than southwestern locations where both the solar heat cost and the cost of conventional energy are relatively lower.

In most sections of the United States, solar heat costs are likely to be competitive with electric-resistance heat costs but higher than the cost of heat provided from fossil fuels.

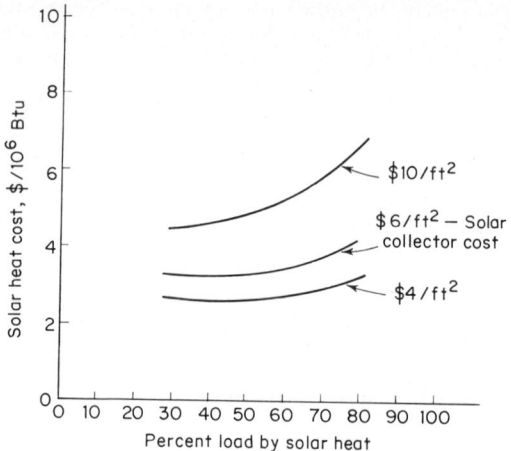

Fig. 9. Solar heating costs versus percentage of load by solar heat for Albuquerque, New Mexico. Criteria: Collector tilt = latitude plus 15°. Collector cost = $4.00, $6.00, and $10.00 per square foot. Storage costs = $0.05 per pound. Glass covers = 2. Constant cost = $375 per system. Amortization = 20 years at 8 percent. Heating demand = 25,000 Btu per degree day.

With the expectation that conventional energy costs will, on an average, increase at a faster rate than solar equipment costs, the economic attractiveness of solar climate control is expected to improve.

The average number of hours of sunshine per year in the United States is given in the map of Fig. 10.

ARCHITECTURAL AND BUILDING FACTORS

Solar climate control systems will have to be integrated with different building designs. New buildings can be designed to fit the requirement of a solar climate control system while applications to existing buildings will have to be determined on an individual basis, since the requirements for successful operation of the system were not considered when the buildings were designed. Collectors could be installed on flat roofs of commercial and industrial buildings, or designed to fit the sloping roofs of a wide variety of buildings. The solar collectors could serve either individual buildings or clusters of buildings such as apartment complexes. The requirements for maximum availability of solar radiation will determine the siting of a specific building to minimize obstruction by either man-made or natural objects.

Optimization of solar climate control will require efficient thermal control through appropriate thermal insulations, construction materials, design (roof overhang, for example), color selection, and building orientation. Climatic conditions, ranging from insolation to the temperatures encountered at the building site, will have to be known. This information is presently available only at a few specific sites and often is of questionable accuracy.

The optimization of the performance and cost of the solar climate control system will depend also on determining the optimum combination of solar energy and auxiliary energy sources. Although it is theoretically possible to design a solar climate control system which will use only solar energy, the cost-benefit ratio will not be attractive in most installations. The solar collector area required to achieve this condition would not be effectively utilized throughout the year.

The potential of solar energy applications in buildings can be greater if the solar collector is modified to include solar cells to produce electricity, while still providing heating and cooling. The electricity produced can be stored in batteries which could provide an interface with electric power supplied by utilities.

A solar climate control system introduces additional constraints which usually would not need to be considered in building design and construction. Solar energy availability

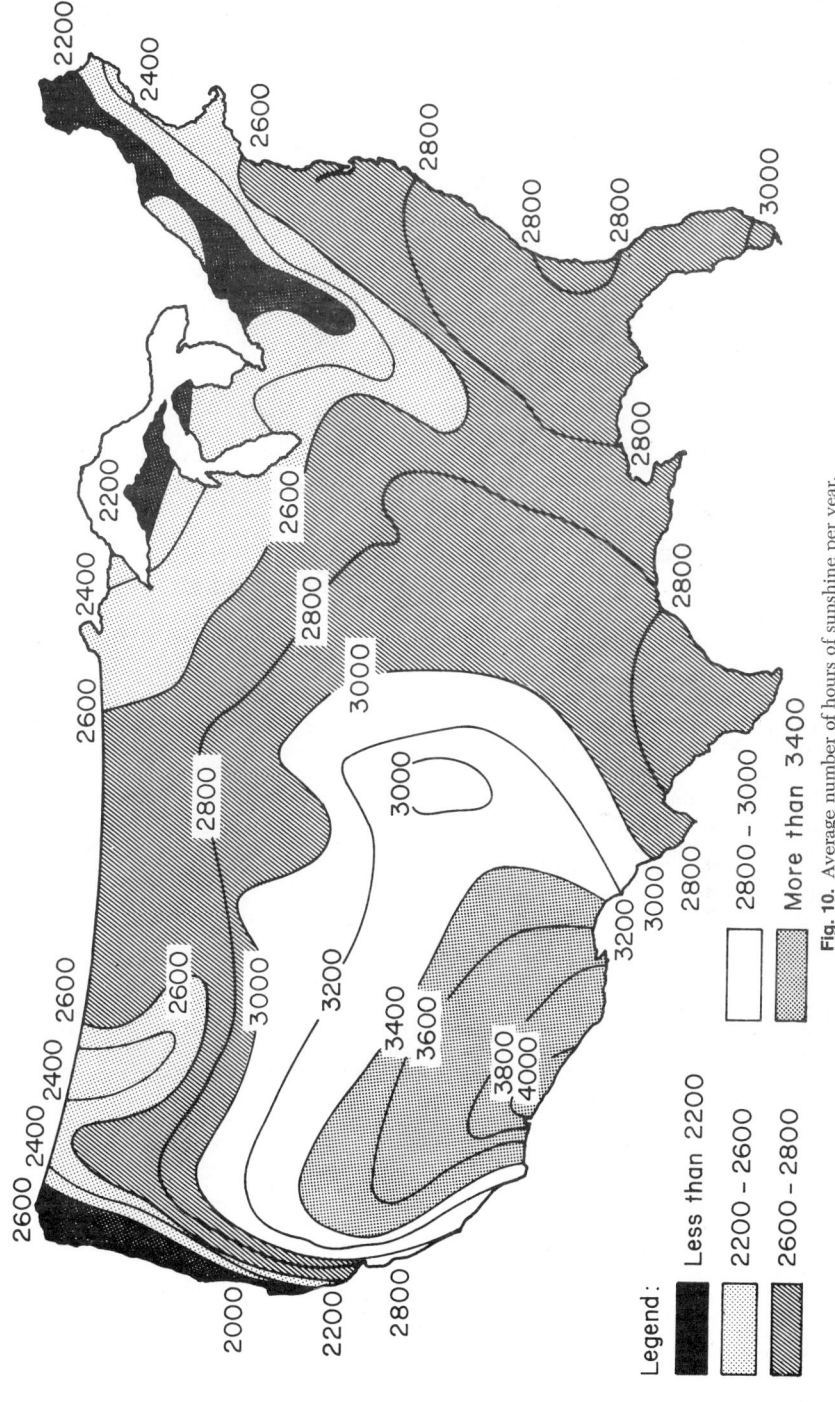

Fig. 10. Average number of hours of sunshine per year.

Legend:

Less than 2200

2200 – 2600

2600 – 2800

2800 – 3000

More than 3400

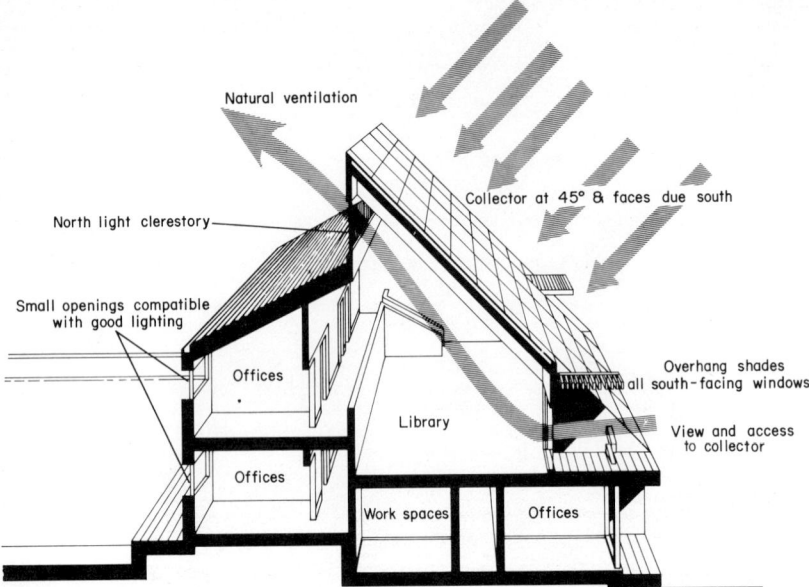

Fig. 11. Perspective section of proposed solar building. A building configuration was selected which permits a reasonably large solar collector (3500 ft²), facing south at an angle of 45°, and which accommodates approximately 8000 ft² of working space. Estimates of heat loss indicate heat demand will be in range of 40,000 to 70,000 Btu per degree day. Based on results of Tybout and Löf, it is estimated that a solar system using a two-pane 3500 ft² collector should account for between 65 and 75 percent of total seasonal heating load. *(Designed by Arthur D. Little, Inc. and Cambridge Seven Associates, Inc. for Massachusetts Audubon Society.)*

will determine the mechanical system choice for heating, ventilating, and air conditioning. Temperature controls will actuate circulation of heat-transfer fluid through the solar collector and withdrawal from thermal storage; thermal storage materials and the volume required and its placement will introduce constraints in building design.

The design of an office building (8000 ft²) is shown in Fig. 11.

See also staff-prepared Addenda A, B, and C which follow this article.

REFERENCES

1. Sheridan, N. R.: Performance of the Brisbane Solar House, Solar Energy Conference Paper, 1970.
2. Farber, E. A.: The Direct Use of Solar Energy to Operate Refrigeration and Air Conditioning Systems, *Fla. Eng. Exp. Sta., Tech. Rept. 15*, University of Florida.
3. Tybout, R. A. and G. O. G. Löf: Solar House Heating, *Natur. Res. J.*, **10** (2), (1970).
4. Pogany, D., Ward, D. S., and G. O. G. Löf: The Economics of Solar Heating and Cooling Systems, presented at the International Solar Energy Society's International Congress and Exposition, Los Angeles, California, July 1975.
5. Welles, J. G. and R. H. Waterman, Jr.: *Harvard Bus. Rev.*, **42**(4), 106(1964).
6. Wenk, E., Jr.: Priorities for Research Applicable to National Needs, Committee on Public Engineering Policy, National Academy of Engineering, Washington, D.C., 1973.
7. Arrow, K. J.: "Rate and Direction of Inventive Activity," Princeton Univ. Press, Princeton, N.J., 1962.
8. Ezra, Arthur A.: Technology Utilization: Incentives and Solar Energy, *Science*, **187**, 4178, 707–713, February 28, 1975.
9. Duffie, J. A. and W. A. Beckman: "Solar Energy Thermal Processes," Wiley, New York, 1975.
10. Duffie, J. A. and W. A. Beckman: Solar Heating and Cooling, *Science*, **191**, 4223, 143–149, January 16, 1976.
11. Kevan, P. G.: Sun-Tracking Solar Furnaces in High Arctic Flowers: Significance for Pollination and Insects, *Science*, **189** (4204), 723–726 (August 29, 1975).
12. Vodraska, K.: Determining the Availability of Solar Energy within the Contiguous United States, EI&I Associates, Newbury Park, California.

ADDENDUM A. Technological Overview

Although significant technological advances are required to make certain approaches, such as solar cells, practical for solar climate applications, much of the basic technology for the simpler direct thermal approaches is on hand as of 1976. Essentially, much of this technology has been available since the early 1960s and, of course, basic concepts date back many decades.

A major problem involved in effecting a greater utilization of the simpler thermal approaches is not all technical and not all economic. Rather, a need exists for providing incentives to commercial manufacturers to produce and market solar thermal heating and cooling systems for residences and commercial buildings and, equally, to encourage the ultimate users of such systems to give them careful consideration. Progress during the past couple of years has been made in extricating solar climate control from the academic to the commercial spheres, but nevertheless industry still remains highly reliant upon both educational institutions and government bureaus for creation of much of the fundamental technical information required. What is needed, of course, is a happy balance between essentially public-funded research and private, industrial research in this field. The essentially economic aspects of the problem are coupled with the very large task of information dissemination to homeowners and building operators, to builders and commercial property developers, to equipment manufacturers, to lending institutions, and to architects and construction engineering firms.

Unfortunately, there is no readily developed system for bringing an essentially new technology to market, particularly in the highly developed nations with sophisticated economic systems and high living standards, especially when increased energy costs tend to become lost in a generally inflated economy. By contrast, solar water heaters have been quite common in Israel, Cyprus, India, and Japan for many years in the interest of conserving wood and other fuels in short supply. A solar irrigation pump (55 horsepower) was built in Egypt as early as 1913. Pumps of this type are in operation today in Niger, Mauritania, Senegal, and Upper Volta. Of course, it must be mentioned that such areas are well suited climatically for capturing solar energy, where the ground-level solar flux may be as much as twice as high as in the case of the contiguous United States. Further, in the underdeveloped nations, conventional energy sources either are not conveniently available in many locations, or the costs are prohibitive.

Comparatively recently, serious thought has been given to developing a so-called "technology delivery system" for solar climate control systems and, in fact, for other nonconventional energy approaches. Generally, it is not the objective of this Handbook to delve into matters that are mainly in the social, marketing, financial, and political realms. Therefore, as a starter, the reader is referred to Refs. 5 through 8 at the end of the prior article.

The application of solar climate control has not quite progressed to the point where it may raise legal and environmental concerns of its own. From a legal standpoint, it is interesting to note that the "right of the sky" is not a relatively new concept, but dates back to early England. The so-called "doctrine of ancient lights" is the legal assumption whereby if light, air, and view have been enjoyed (for a given location) from the time "when the memory of man runneth not to the contrary" (taken from the time of the reign of Richard I in 1189), then there exists an enforcible right to light, air, and view. Modern law in the United States, in contrast, has established that the surface owner has a right to receive light from that area of the sky directly above the owner's property and no rights in connection with that radiation which may be received across the land of the neighboring property. Prior to the serious appearance of solar climate control, this matter has been essentially concerned with the problems of shading beaches and other recreational facilities by the erection of neighboring high-rise structures. Also, some home associations in hilly and mountainous areas have passed regulations pertaining to the blocking of attractive views. It is obvious that the future of solar energy utilization will require much further clarification and most likely revisions of present rights to the sky. It is indeed probable that sky rights may become a marketable commodity of a property as are underground mineral rights, for example, today. Thus, if one desires to protect a solar energy installation from artificially caused shadows, it may be necessary, well in advance, to procure the sky rights of neighboring property, at least those rights above a certain elevation, thus limiting the height of adjoining structures.

ADDENDUM B. Commercially Available or Announced Collectors

As of late-1976, solar collectors are being marketed by a relatively few of the large and prestigious firms in the United States, Japan, and other highly developed nations. Needless to say, the opportunistic aspects of the thermal solar collector field have attracted scores of lesser, but not necessarily less effective, smaller firms. As public awareness increases, of course, the field may attract the usual complement of technically weak and, in some cases, unscrupulous operators.

Although commercially available today, the practical application of collectors is still in a very early phase, with much experience to be gained. Remaining technical problems are not so much in the realm of solar physics and fundamentals, which have been expertly explored by academic institutions, as those problems that are posed by the day-to-day and year-to-year operating conditions arising from exposure to the rigors of local weather and atmospheric conditions. These include the need to further develop means for combating the accumulation of dirt and debris (from trees, shrubs, etc.) and atmospheric pollutants that tend to substantially lower collector efficiency, a process commencing with the day of installation. Means also must be developed for protecting installations from the lifting and destructive effects of high and unusual hurricane-type winds as well as ways to overcome corrosion from rain, dew, and other atmospheric moisture. All of these problems are accentuated because logically panels of solar collectors must be exposed to all of the elements. Thus, it is not surprising that several of the pioneering firms in this field are basically in the materials business (metals, glass, paint, etc.).

Because first costs tend to be high and because solar thermal systems are affected by numerous installation and operating factors, most of which tend to lower efficiency with time, the building and residence designer and owner must carefully weigh initial costs against long-term operating and maintenance costs, including parts replacements—rather than simply comparing first costs with projected energy savings that are based upon initial operating efficiency.

It is interesting to note the marked increase in performance of solar installations over the period of the mid-1960s to the present. Probably it is not too much to expect very significant improvements during the remainder of the 1970s. Truly mass-produced elements may not appear until the early 1980s, requiring a few additional years of practical learning experience and the time required to achieve mass public education on both the advantages and limitations of solar climate control. Further, for the mass market, additional narrowing of the margin between the cost effectiveness of solar thermal systems and that of conventional energy sources must take place.

Only a few of the commercially available solar collector units can be described briefly here. The intent of including these descriptions is to provide a reasonably representative picture of the commercial state of the art.

Example 1: One collector design* is made with integrally finned extruded aluminum tube, serpentined into the configuration of a flat-plate collector. The standard collector panel is 4×8 feet $\times 3\frac{5}{8}$ inches thick and weighs 67.8 pounds when the tubes are full of fluid. Collector area is 29 square feet. Fin tube length is 63.7 feet, with one inlet and one outlet. There are 8 tubes per panel. Inside cross-sectional area of tube is approximately 0.139 square inch and tube wall nominal thickness is 0.05 inch. The aluminum collector surface is coated with siliconized polyester flat black with approximate absorbtivity/emissivity of 0.95/0.95. The back of the collector finned-tube is mill finish aluminum. The back of the panel is insulated with 1 inch of spun glass plus 0.8 inch of closed-cell foam material. Reflective aluminum foil vapor barrier is used on the interface and painted aluminum sheet on the exterior backing. The collector frame is made from aluminum extruded shapes, including sides, ends, braces, clips, and cover glazing frames. Frame surfaces are mill-finished aluminum. Glazing is a double 4 mil *Tedlar*™ (DuPont) cover. These solar collectors are designed to operate in closed systems with inhibited water. However, they may be adapted to conventional systems of space heating, household and commercial hot water, and swimming pool heating.

Example 2. One collector design† is a flat plate type collector, $34\frac{3}{16} \times 76\frac{3}{16}$ inches $\times 1\frac{5}{16}$ inch thick and weighs approximately 6 pounds per square foot when filled with heat-transfer fluid. The cover plates are two pieces of ⅛-inch tempered glass, *Herculite*™ (PPG Industries). The absorbing surface is *Duracron*™ (PPG Industries) Super 600 L/G flat black paint coating. The collector plate is *Roll Bond*™ (Olin Brass Corp.) type 1100 aluminum solar absorber. There is a 2½-inch insulating backing of fiber glass. The edge-retaining system consists of metal channel with desiccant type spacers. These units are designed for solar-powered space heating and hot water systems, where a fluid is used to transport heat from the collector.

Example 3. One collector design‡ is an enclosed-type box collector, overall dimensions of which are approximately $44 \times 71\frac{1}{2}$ inches $\times 8\frac{1}{2}$ inches thick. When the solar heat-collecting cylinders are filled with water, the weight is approximately 545 pounds. Dry weight is approximately 104 pounds. The solar collector is comprised of 6 rigid polyethylene plastic cylinders, 6½ inches in length. The transparent cover sheet is of polycarbonate plastic. An aluminum foil reflector is used. Heat insulation is foamed plastic. Upper and lower headers and the water tank are made of rigid polyethylene. These collectors have found wide use in Japan for industrial and residential hot water heating and are now marketed in the United States.

Example 4. One collector design§ is a flat plate type collector, 36×78 inches $\times 5$ inches thick and weighs approximately 6½ pounds per square foot (unfilled and with double glass cover). The collector plate is copper as are the ½ $\times$ 1-inch rectangular tubes. A conductive sealant is used between the tubes and collector plate to maximize heat transfer. The spacing of tubes on the collector plate is adaptable to different design requirements, although 6-inch centers have proved effective in most cases. At both ends of the collector unit, the tubes are brazed to conventional ¾-inch round copper water tube headers. The header ends project through an aluminum housing for joining by conventional soldering to inlet and outlet piping. Header ends can project through either the side or bottom of the housing. Insulation housing under the collector plate is 3½-inch fiberglass and around the perimeter of the housing is 1-inch rigid insulation. Wood liners around the perimeter of the housing are optional (facilitate installation and securing to existing roof structures). Transparent coverings can be supplied

---

*Torex™ Model 14 solar collector by Reynolds Metals Co., Richmond, Virginia.
†*Baseline Solar Collector* by PPG Industries, Inc., Pittsburgh, Pennsylvania.
‡*Hi-Heater* by Hitachi Chemical Co., Ltd., Tokyo.
§*Revere Modular Solar Collector* by Revere Copper and Brass Incorporated, Rome, New York.

in different numbers of thicknesses or layers. The standard coating used is a heat-resistant, flat black paint. Glass with low iron content can be used for covers. Manufacturer suggests that glass panels be obtained locally. Developed for new and existing buildings, the principal applications for which unit has been designed are domestic water heating, space heating, swimming pool heating, and heating and air conditioning and heat pump systems.

Example 5. One collector design¶ is comprised of a flat absorber/heat exchanger and internal manifolding. The Series 2000 unit is an air-type solar collector. Air is circulated from the building to one end of the collector. As the air passes through the collector, it is heated by converted radiant energy. An optional hot water heat exchanger may be used in series with the collector to provide solar preheated water for domestic consumption. The unit is designed to work in connection with a solar heat storage system comprised of dry pebbles. Solar heated air is routed through the pebbles and returns to the collector for reheating. The cover plates of the collector are two ⅛-inch hermetically sealed low-iron, tempered glass panels. The absorber is 24 gage steel with baked-on high absorbency flat black coating.

ADDENDUM C. Daily Terrestrial Solar Energy Received versus Location

For reasons stated in the prior article (map of Fig. 10), the application of solar climate control will vary considerably from one geographical and climatic region to the next. The average daily terrestrial solar energy received on a horizontal surface at several locations in the United States is given in the accompanying table.

TABLE C.1. Daily Terrestrial Solar Energy Received on a Horizontal Surface

Figures are given in langley units* per day

	Daily average			Annual daily average
City and state	Lowest month		Highest month	
Over 500 langley units				
Inyokern, California	295	Dec	836	June	579
El Paso, Texas	313	Dec	730	June	536
Phoenix, Arizona	281	Dec	739	June	520
Tucson, Arizona	305	Dec	729	May	518
Honolulu, Hawaii	371	Dec	617	May	516
Albuquerque, New Mexico	276	Dec	726	June	512
Las Vegas, Nevada	258	Dec	748	June	509
Yuma, Arizona	271	Dec	705	June	506
Between 450 and 499 langley units				
Page, Arizona	243	Dec	707	June	495
Pearl Harbor, Hawaii	349	Dec	552	May	483
Santa Maria, California	252	Dec	694	June	481
Reno, Nevada	209	Dec	714	June	479
Fort Worth, Texas	245	Dec	651	June	477
Riverside, California	260	Dec	666	June	470
Ely, Nevada	220	Dec	708	June	467
Midland, Texas	275	Dec	617	June	466
Soda Springs, California	182	Dec	760	July	459
Miami, Florida	319	Dec	552	May	453
Tampa, Florida	300	Dec	596	May	453
Key West, Florida	292	Dec	579	May	452
Between 400 and 449 langley units				
Cape Hatteras, North Carolina	216	Dec	645	June	447
Dodge City, Kansas	234	Dec	650	June	447
Aplachicola, Florida	262	Dec	603	May	444
Lander, Wyoming	196	Dec	678	June	443
San Antonio, Texas	256	Dec	612	June	442
Pasadena, California	236	Dec	599	Aug	439
Corpus Christi, Texas	240	Dec	629	July	436
Oklahoma City, Oklahoma	237	Dec	623	June	435
Brownsville, Texas	253	Dec	619	July	435
Gainesville, Florida	254	Dec	586	May	431

¶Solaron Corporation, Denver, Colorado.

TABLE C.1.   Daily Terrestrial Solar Energy Received on a Horizontal Surface (*Continued*)

Figures are given in langley units* per day

City and state	Daily average				Annual daily average
	Lowest month		Highest month		
Flaming Gorge, Utah	215	Dec	650	June	426
Pensacola, Florida	224	Dec	568	June	416
Tallahassee, Florida	260	Dec	548	May	416
Griffin, Georgia	210	Dec	580	June	416
Grand Lake, Colorado	184	Dec	632	June	416
Lake Charles, Louisiana	232	Dec	582	June	414
Lexington, Kentucky	172	Jan	628	June	411
Dallas, Texas	221	Dec	595	June	411
Laramie, Wyoming	186	Dec	643	June	408
Charleston, South Carolina	228	Dec	556	June	406
Stillwater, Oklahoma	205	Jan	600	June	405
Jacksonville, Florida	230	Dec	556	May	404
Shreveport, Louisiana	205	Dec	578	July	400
Between 350 and 399 langley units					
Prosser, Washington	100	Dec	707	July	399
New Orleans, Louisiana	220	Dec	549	June	399
Memphis, Tennessee	184	Dec	589	June	397
Belle Isle, Florida	291	Dec	488	July	397
San Mateo, California	176	Dec	598	June	396
Boise, Idaho	124	Dec	670	July	395
Salt Lake City, Utah	146	Dec	621	June	394
Atlanta, Georgia	201	Dec	554	June	394
North Platte, Nebraska	178	Dec	599	June	393
Rapid City, South Dakota	158	Dec	590	July	392
Glasgow, Montana	116	Dec	645	July	388
Little Rock, Arkansas	187	Dec	559	June	385
Greensboro, North Carolina	197	Dec	564	June	383
Norfolk, Virginia	184	Dec	572	June	382
La Jolla, California	221	Dec	506	May	380
Columbia, Missouri	158	Dec	574	June/July	380
North Omaha, Nebraska	170	Dec	568	July	379
Pullman, Washington	96	Dec	699	July	372
Manhattan, Kansas	156	Dec	551	June	371
Topeka, Kansas	165	Dec	554	June	371
Nashville, Tennessee	161	Dec	567	June	370
Bismarck, North Dakota	124	Dec	617	July	369
Boulder, Colorado	182	Dec	525	June	367
Great Falls, Montana	112	Dec	639	July	366
Lincoln, Nebraska	159	Dec	568	July	363
Oak Ridge, Tennessee	161	Jan	551	June	363
Spokane, Washington	75	Dec	665	July	361
Louisville, Kentucky	150	Dec	560	June	360
Philadelphia, Pennsylvania	152	Dec	554	June	355
Annapolis, Maryland	155	Dec	557	June	355
Trenton, New Jersey	155	Dec	546	June	355
Moline, Illinois	134	Dec	565	July	352
Portland, Maine	138	Dec	561	July	351
Between 300 and 349 langley units					
St. Cloud, Minnesota	123	Dec	555	July	348
Indianapolis, Indiana	130	Dec	547	June	345
Ames, Iowa	143	Dec	541	June	345
Milwaukee, Wisconsin	120	Dec	565	June	345
Columbus, Ohio	128	Jan	562	June	340
Newport, Rhode Island	115	Jan	538	June	339
State College, Pennsylvania	120	Dec	544	June	335
Washington, D.C.	141	Dec	510	June	333
Sault Ste. Marie, Michigan	96	Dec	573	July	333

TABLE C.1.   Daily Terrestrial Solar Energy Received on a Horizontal Surface (*Continued*)

Figures are given in langley units* per day

City and state	Lowest month		Highest month		Annual daily average
Green Bay, Wisconsin	110	Dec	542	June	327
New York, N.Y.	128	Dec	526	June	324
Madison, Wisconsin	115	Dec	531	July	324
Cambridge, Massachusetts	124	Dec	482	July	322
Cleveland, Ohio	115	Dec	562	June/July	320
Burlington, Vermont	103	Dec	532	July	317
Caribou, Maine	107	Dec	508	July	316
Eureka, California	131	Dec	494	June	313
Summit, Montana	76	Dec	560	July	312
Boston, Massachusetts	119	Dec	499	June	311
East Lansing, Michigan	108	Dec	547	June	311
Ithaca, New York	96	Dec	515	July	302
Tacoma, Washington	64	Dec	566	July	300
Between 250 and 299 langley units					
Schenectady, New York	104	Dec	448	June	282
Medford, Oregon	77	Dec	497	July	282
Chicago, Illinois	76	Dec	473	July	273
Seattle, Washington	59	Dec	501	July	273
Annette, Alaska	40	Dec	481	July	251
Between 200 and 249 langley units					
Fairbanks, Alaska	6	Dec	504	June	224
Barrow, Alaska	6	Jan/Nov	528	June	206

*1 langley = 1 gram calorie per square centimeter = 3.687 Btu/per square foot.

Satellite Energy Systems for Earth Power

BY DANIEL L. GREGORY* AND GORDON R. WOODCOCK†

The finite limits inherent in many of the current ground-based energy systems have brought attention to the possibilities of extraterrestrial energy systems. These systems would be space-located, but provide energy for utilization on Earth. Potential advantages derive from the unattenuated solar insolation, which is nearly uninterrupted for properly selected orbits, so that storage systems and/or supplemental fuels are not required as with ground-based solar plants. The annual energy accumulation of a sun-oriented area in Earth orbit is approximately six times greater than that of an equivalent oriented area in the most favorable solar collection areas on Earth. Transmission systems for this energy are envisioned, which allow delivery to most of the inhabited regions of Earth by capitalizing on the orbital vantage point. The inherently high performance of such systems may provide economic leverage to offset the problems, including cost, involved with their orbital placement.

Symbols used in this discussion are defined in Table 1.

ORBIT CONDITIONS

All the satellite systems discussed here are assumed to be located in circular orbits in the equatorial plane of the Earth at an altitude such that the orbital period is 24 hours. Such a system remains fixed over the subsatellite point on Earth. This *geosynchronous* orbit location is currently primarily used for communication satellites, although military and weather systems are beginning operation in this region.

Due to the 23.45° inclination of the equatorial orbit plane to the ecliptic, the satellite passes "below" (south) of the shadow of the Earth for six months per year, as shown in Fig. 1, and above the shadow for six months. However, on and near the equinoxes (shadow shown on figure), the satellite will be briefly eclipsed, with a maximum duration of 70 minutes. Overall, such a satellite is illuminated 99.26 percent of the year.

In this orbit, the total direct solar insolation is 1353 W/m² (some variation occurs, primarily owing to elliptical orbit form of the Earth). The majority of the planetary radiation belts lie below this orbit; however, energetic particles from periodic solar flares add to the total exposure, which includes the electron flux of the upper belt region.

PASSIVE ORBITAL SYSTEMS

Solar Energy Reflectors. Probably the simplest conceptual utilization of geosynchronous satellite systems is that of an orbiting mirror which provides sunlight to a ground-based solar power plant. This energy might augment the direct solar radiation during daylight hours and allow operation at night. An important optical effect must be considered. Since the sun is not a point source, its angular width of 0.53°, as viewed from the region of the Earth, will be duplicated by the reflected light cone produced by any mirror, no matter what its size. (See Fig. 2.)

As a consequence, the smallest image that can be produced on Earth by any geosynchronous orbit mirror is approximately 330 km in diameter. If it is desired to have a "one sun equivalent" image strength, it is, therefore, necessary that the orbital mirror diameter be also 330 km. If we assume that the mirror has an efficiency of 0.80 (due to imperfections in reflectivity and geometry), the mirror diameter would have to be 370 km for "one sun."

The mirror must have a curved surface in order to maintain the required image. If composed of individual facets on a flat base, each facet must be oriented so that all images

*Senior Specialist Engineer, Space-Based Power Study; †Engineering Supervisor, Future Space Transportation Systems Analysis, Research and Engineering Division, Boeing Aerospace Company, Seattle, Wash.

TABLE 1.   **Symbols Used in Discussion of Satellite Energy Systems**

A_c	Wall area of cavity absorber
A_o	Area of cavity aperture
A_r	Area of receiving antenna
A_t	Area of transmitting antenna
C	Concentration ratio of solar intensity
D	Transmitter antenna diameter
e	Subscript denoting electric power
F_a	Fraction of possible energy absorption in cavity
F_r	Fraction of possible reradiation from cavity
P	Power density at point on transmitter or receiver antenna
P_o	Maximum power density on transmitter or receiver
P_{ic}	Power input to cavity
P_m	Total microwave power in receiver area
P_{oc}	Power output from cavity
P_{rad}	Power reradiated from cavity
Q	Solar insolation (1353 W/m² ± 3.5% in Earth vicinity)
r	Subscript indicating receiver element
R	Transmission range
S	Area of solar reflector
S'	Capture area of inclined reflector
t	Subscript denoting thermal power or transmitter element
T	Cavity interior wall temperature
ϵ	Emissivity
η	Efficiency
θ	Cavity view factor
λ	Microwave wavelength
ϕ	Angular location with respect to solar noon
ρ	Radial location on transmitter or receiver antenna
ρ_x	Radial location corresponding to the 63% power point
σ	Stefan-Boltzman constant (5.6697×10^{-8} W/(m²)(K⁴))

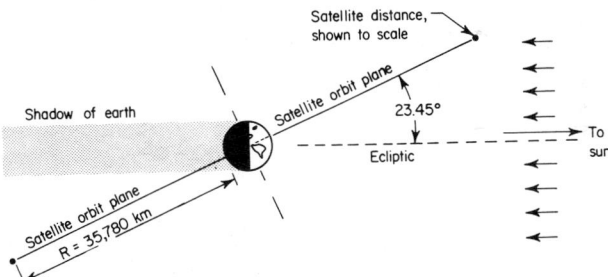

Fig. 1. Geosynchronous orbit, conditions for winter solstice.

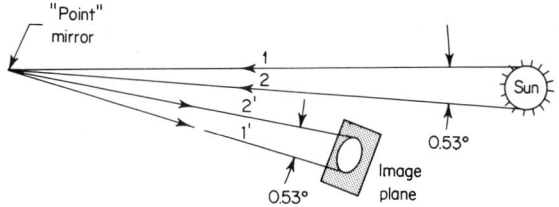

Fig. 2. Solar image size.

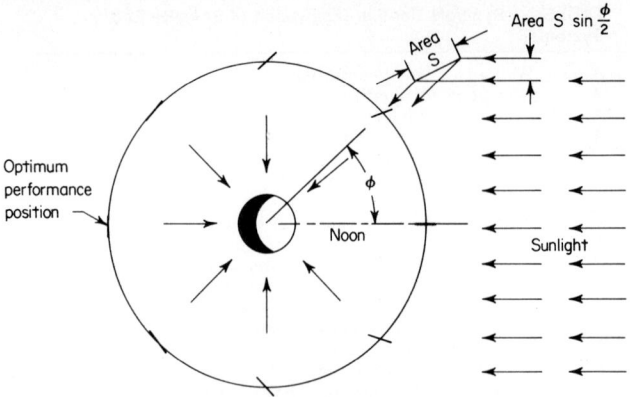

Fig. 3. Successive positions of orbital mirror (24 hours).

are superimposed. The edge facets are tilted the most, by 0.13° relative to the mirror plane. Some form of active tilt control would probably be required. An optical beacon in the center of the target area on Earth could be tracked by a sensor and servomotor system on each facet. Figure 3 shows a view of the orbit from above the North Pole of the Earth. Note that the mirror must rotate 180° in 24 hours if its image is fixed geographically, and that both sides of the mirror must be reflective or else a rather rapid additional 180° rotation must be performed at the position corresponding to noon on the Earth below. Considering the inertia of the mirror, it is probably best to use both sides of the mirror so that a continuous rotation rate will suffice. Thus, the facets must be so mounted and driven that the effective mirror curvature can be adjusted for operation in either direction.

Figure 4 provides an exaggerated indication of how the individual facets might be controlled. At solar noon, when the mirror is edge-on to the sun, the facets shift from the solid line position to the dashed line position, with the solar energy passing through the open framework. Triangular, square, or hexagonal facets could be used to provide a nearly continuous surface. When large orbital mirrors were first proposed (1929: Ref. 1), the metal sodium was suggested for the reflective facets. Now, it is logical to turn to the plastic films, with a coating such as aluminum which could provide a first surface reflectivity of more than 0.90.

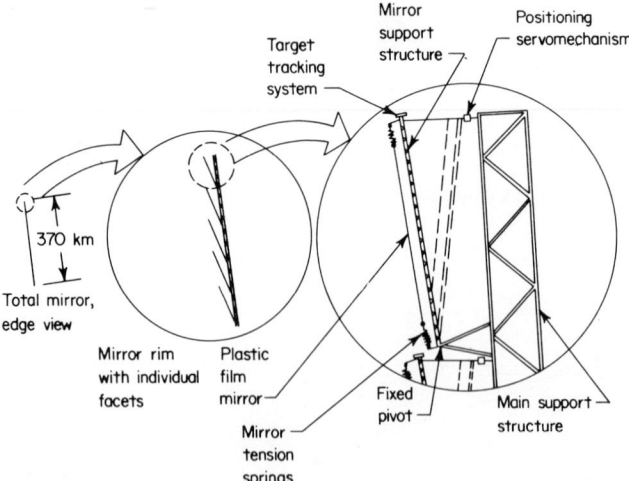

Fig. 4. Schematic representation of mirror with pointing facets. (Angular position of facet exaggerated; actually ±8 minutes maximum travel.)

TABLE 2.   Reflector Satellite Masses*

Item	Mass, 10^6 kg	Remarks
Plastic film	380	⅒ mil, aluminized both sides
Individual facet frames*	2660	Provide flatness for each facet
Facet supports, positioners	760	Pivots and servomechanisms
Main frame	1300	Base which supports facets
Thruster system	30	Attitude control, orbit trim
Electric power system	30	Operates thrusters and positioners
Miscellaneous, unknown	520	20%
Total	5680×10^6 kg	

*A description of a potential facet form is given under "Thermal Engines" later in this article.

The mirror satellite will require rocket thrusters for attitude control and orbit trim, with the thrust of sunlight as the major perturbing force. The pressure of sunlight is about 4.6 N/km² at the orbit of the Earth, yielding a force on the 370-km-diameter mirror of 5×10^5N when the mirror is normal to the incident sunlight. Maintenance of the required orbit can be maintained only by rocket thrusters. If the average sunlight pressure is 2.5×10^5N over the 24-hour orbit, even very high performance rocket thrusters will require an annual supply of at least 6×10^7 kg of reaction mass (to be exhausted as the jet). 6×10^7 kg is approximately one percent of the satellite mirror weight. A weight estimate for a conceptual "one sun" mirror satellite is given in Table 2.

This weight is approximately that of 20 of the current largest supertankers. Placement of such weights in orbit is discussed later.

Referring to Fig. 3, it can be seen that, if a mirror of area S is focused on the subsatellite point, the projected area of the mirror as seen from the sun is

$$S' = S \sin \phi/2 \tag{1}$$

since, per Snell's law, the mirror is inclined at $\phi/2$ when positioned at ϕ. If the mirror is sized to provide one sun above the atmosphere, in the optimum performance position, we may expect to achieve approximately 0.9 kw/m² through a clear atmosphere. Figure 5 shows the strength of insolation produced by such a mirror over a 24-hour period (assuming no occultations). Typical direct insolation of the sun is also shown (value appropriate to a desert region), along with the sum of the two. Note that since, in general, the true sun and the mirrored sun are not in the same place in the sky, a high-concentration ground solar plant cannot utilize both sources, but must select the stronger. Low-concentration systems may benefit from both sources.

A "hydrogen economy" variant has been proposed (Ref. 2) wherein almost all energy used within the United States is produced by solar power plants in the Southwest. Approximately 160,000 km² of plants with an efficiency of 0.25 might be required for the year 2100. The illumination area of the previously discussed satellite mirror is approximately 90,000 km². Thus, the orbital mirror could illuminate such a ground power

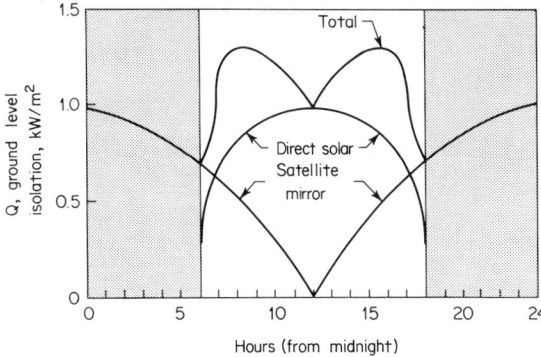

Fig. 5. Combination of orbital mirror output and direct sunlight.

installation, to fill nighttime energy requirements without a storage system (since the night demand power level would be only approximately 40 percent of that of the day).

Figure 6 shows a possible mirror configuration on orbit; note the illuminated area on the Earth. The reflected light from the mirror, of course, will not penetrate clouds. Herein lies a potential problem. A convective column of warmed air will tend to form over the illuminated area. Air from the surrounding "night" area will tend to rush into the area of "day," and clouds may be carried in, or may form when the air column reaches the height

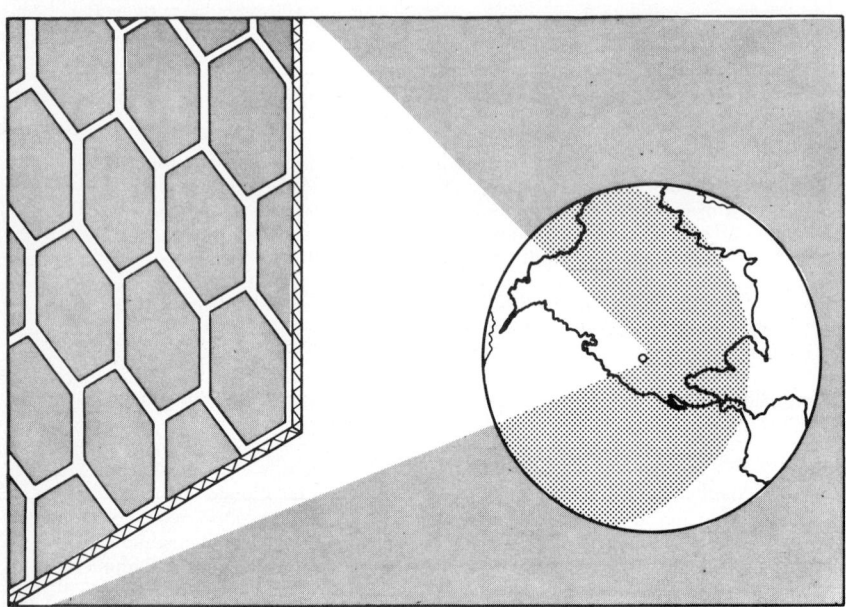

Fig. 6. Passive sunlight reflector illuminating area (shown to scale) on earth.

where condensation occurs. Thus, the passive solar reflector might be partially, or totally, self-defeating.

Microwave Energy Reflectors.   Ehricke has proposed (Ref. 3) a passive reflector to act as a relay between a microwave transmitter located near, and energized by, a ground electric power plant and a receiver located near the power utilization point. Power is transferred by the microwave beam. (See Fig. 7.) Several advantages may accrue from this concept:

1. The potential prime solar power areas of the world, such as southwestern United States and northwestern Mexico, Argentina, Australia, India, South Africa, and the Middle East, can provide power to remotely located distribution centers, even across oceans, mountain ranges, etc.

2. The satellite is relatively "light," perhaps only several hundred thousands of kilograms.

3. The microwave beam would pass, with relatively little attenuation, through clouds and rain.

4. The power-generating elements are ground-based.

5. Operation need not cease during occultations of the orbital elements.

Microwave power transmission is discussed more fully later in this article under "Microwave." The relay satellite is located beyond the near field zone of the transmitter so that a two-dimensional (flat) reflector can be used. The reflector is exactly that; no active receiving or transmitting elements are required. The surface could be a metallic sheet. However, a wire mesh will suffice if its spaces are kept appropriately small (approximately 2 mm). Such a wire mesh can weigh as little as 48,000 kg/km^2 for aluminum, or 32,000 kg/km^2 for beryllium.

Although the transmitter array for a system which delivers 10 GW(e) ground output may be 10 km on a side, beam focusing can allow the reflector to be only approximately one kilometer across. A tubular main frame structure supporting individual faceted reflectors is a possible structural approach. Occultations and varying sun angles will subject the system to changes in mesh tension. A potential saving in weight could result from maintenance of a constant mesh tension by controlling electromotors, so that the support

Fig. 7. Microwave power relay satellite.

frames may be light. Maintenance of the required flatness will be a very demanding requirement. System efficiency depends not only on beam coherence at the transmitter array (by phase control), but also on the reflector itself being planar, within $\pm\lambda/80$: i.e., approximately $\pm$ 1 mm over its 1-km² surface. The facets of the solar reflector must be tilted at the proper angles, but need not be coplanar as in this instance.

The relay satellite will require thrusters for orbit trim and attitude control. Electric power for these thrusters could be derived from the power beam by a rectifying antenna. The openings in the reflector mesh pass most of the incident sunlight, but the microwave beam is deflected because of its longer wavelength; as a consequence, the reflector satellite receives a greater thrust from the microwave beam than it does from the sunlight.

ON-ORBIT POWER GENERATION

Photovoltaic. Large [5-and 10-GW(e) ground output] orbital plants employing solar cells for energy conversion have been proposed and studied in some detail by Glaser (Refs. 4, 5, and 7). Transmission of the generated power to Earth is accomplished by microwave beam. Figure 8 (reproduced from Ref. 5) shows such a plant. The solar cell arrays are oriented by an attitude control system so as to face the sun, while the transmitting antenna constantly faces the ground receiving station. Consequently, the antenna must be mounted on gimbal joints. The frame structure joining the two large array areas is nonmetallic to allow passage of the microwave beam.

To reduce the weight and cost of the cell arrays, plane mirrors are arranged to effect

Fig. 8. Design concept for a satellite solar power station.

concentration of the incident sunlight. The concentration ratio has been limited to two in order to limit cell heating and the consequent reduction in cell efficiency (Fig. 9).

Concentrating mirrors of this type require high reflectivity, but need not be perfectly flat. Tensioned, aluminized plastic film would suffice, with a relatively light framework. The solar cell arrays and the concentrating mirrors would be mounted on a main frame assembly, which would also mount the transmitter/antenna, power switching and distribution equipment, electric thrusters, etc.

System sizing and performance are dependent upon the characteristics of the solar cells. Considerable extrapolation of cell performance has been assumed in previous sizings of such systems. Table 3 compares characteristics of the Skylab system with those assumed in Ref. 5.

Some of the improvements noted can no doubt be brought about; it is not clear that a cell can be developed which simultaneously incorporates all of them (and, in addition,

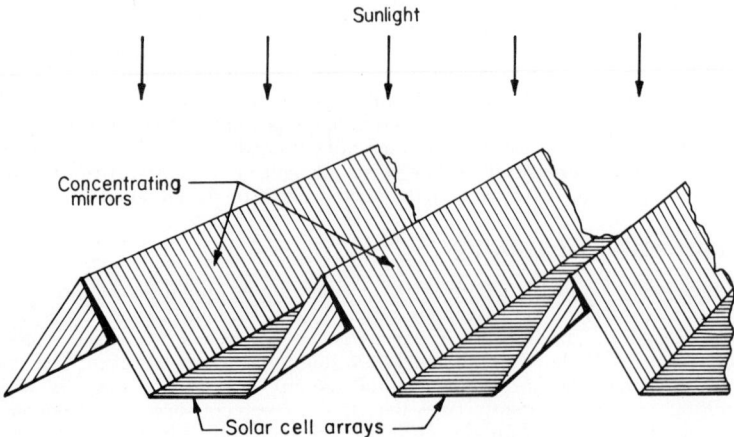

Fig. 9. Concentrating mirrors reduce system weight.

TABLE 3. Skylab and Satellite Power Plant Solar Cell Characteristics

	Skylab	SSPS (Ref. 5)
Initial efficiency	10%	18%
Performance reduction/year	~ 10.4% (of initial)	1.2% (of initial)
Thickness (cell + cover)	3 mm	0.05 mm
Panel mass/area	9.4 kg/m²	0.18 kg/m²

low cost—see discussion later in this article under "Economics"). An initial efficiency of 15 percent with a deterioration rate of 1.9 percent per year has recently been achieved in a silicon cell (Ref. 6).

A weight of 1.1×10^7 is predicted (Ref. 5) for a *powersat* (power satellite) appropriate to a 5-GW(e) ground output.

Thermal engines are proposed (Ref. 8) as an alternative energy conversion method for orbital plants; additional data were provided in Refs. 2 and 9. The basic concept is that of a faceted reflector which concentrates the received solar energy into a cavity absorber. A working fluid (inert gas or gases) transfers this energy to a thermal engine, which rotates an electric generator (Fig. 10). This is, in essence, the current "towertop" ground solar power concept, in orbit.

The collector is envisioned as consisting of individual reflecting facets, each of which is positioned so as to shine its image into the cavity absorber. Again, aluminized plastic film is base-lined, tensioned by the facet frame, which is mounted to the main frame of the powersat with pivots driven by positioning mechanisms. The main frame of the collector need be only approximately paraboloidal; if the gaps between facets are kept small, the collector efficiency can be kept high. The flat sections of the collector shown in Fig. 11 are arranged to approximate a paraboloid. The individually positioned facets reduce the requirements for main frame solar pointing accuracy, stiffness, and geometric figure.

If the plastic film is rolled for transportation, so as to avoid creasing, a tension of approximately 3 percent of yield strength produces an excellent reflecting surface. The framing members of such a facet have a weight approximately eight times that of the film itself if conventional aluminum structure is used, for an overall facet weight of 35 g/m² (¹⁄₁₀-mil film).

Transportation to the assembly orbit would be facilitated by a structure which collapses to higher density. One concept for such structure is shown in Fig. 12, along with a possible item of "orbital tooling" expansion and handling of the structure elements. A cavity absorber is selected for several reasons:

1. Reflection and reradiation losses can be low, even at very high temperatures.

2. The concentration ratio can be kept high by choice of a small aperture, while a large wall area remains available for heat transfer to the working fluid.

3. The large wall area allows fabrication of a low pressure drop heat exchanger, which promotes high Brayton cycle efficiency. See article on "Heat Engines for Solar Power," given earlier in this Handbook section.

Figure 13 shows an absorber, Brayton engine modules, and associated radiator panels at the collector focus. Figures 14 and 15 give possible overall appearances of a powersat.

The absorber has a major impact upon the total plant efficiency. Overall system optimization of a thermal engine powersat would maximize the power-to-weight or

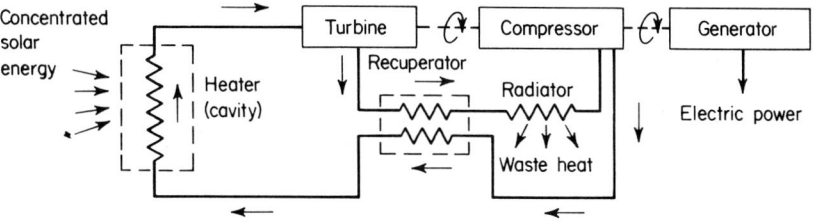

Fig. 10. Brayton cycle schematic diagram.

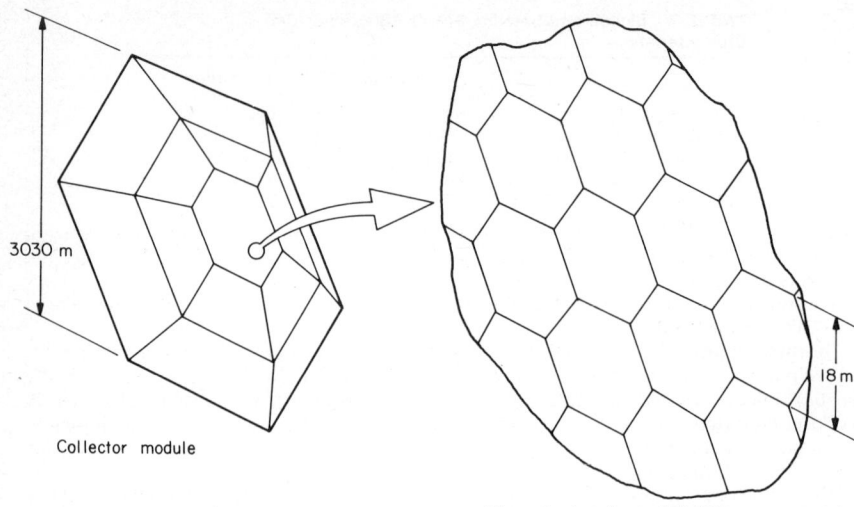

3030 m

Collector module

View showing facets (31,000 per module)

Fig. 11. Faceted solar collector module.

power-to-cost ratios. The following analysis is presented as an aid to such optimizations. The power into the cavity absorber is

$$P_{ic} = CQA_o \qquad (2)$$

θ, the cavity view factor, and interior surface emissivity ϵ determine how much of this energy is reflected back out of the opening. Stephens and Haire (Ref. 10) have determined F_a and F_r for a range of cavity geometric forms and surface emissivity, where the fraction of the possible power absorbed into the walls is F_a. Some of this power is reradiated, according to the wall area, temperature, and emissivity, with F_r representing the fraction of possible reradiation.

$$P_{rad} = F_r \sigma T^4 A_c \qquad (3)$$

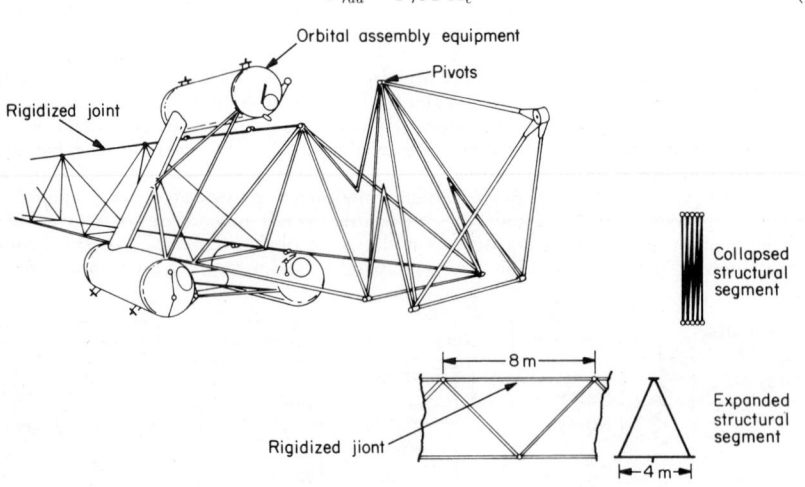

Orbital assembly equipment

Pivots

Rigidized joint

Collapsed structural segment

Expanded structural segment

├─8 m─┤

Rigidized jiont

├─4 m─┤

Major frame member
(e.g., heat absorber support arms)

Fig. 12. Orbital erection and assembly of frame elements.

The remaining power is that transferred into the flowing working fluid and lost through the cavity walls, so that

$$P_oc = F_aCQA_o - F_r\sigma T^4 A_c \tag{4}$$

The efficiency of the cavity is

$$\eta = \frac{P_oc}{P_ic} = \frac{F_aCQA_o - F_r\sigma T^4 A_c}{CQA_o} \tag{5}$$

$$= F_a - \frac{F_r\sigma T^4}{CQ}\left(\frac{A_c}{A_o}\right) \tag{6}$$

Parentheses are used to emphasize the cavity wall-area-to-aperture ratio.
The required concentration ratio can be found from

$$C = \frac{P_oc}{F_aQA_o} + \frac{F_r\sigma T^4}{F_aQ}\left(\frac{A_c}{A_o}\right) \tag{7}$$

The right-hand term yields the concentration ratio necessary to achieve the desired wall temperature if the cavity is adiabatic; the left-hand term indicates the additional energy concentration required as power is removed. Figure 16 plots Eq. (7) for representative values. The required concentration ratio for an efficient powersat is thus in the range of 1500 to 3500. Ground tower-top systems will use concentration ratios of 500 to 1500.
Figure 17 compares mid-life efficiency factors for photovoltaic and thermal engine

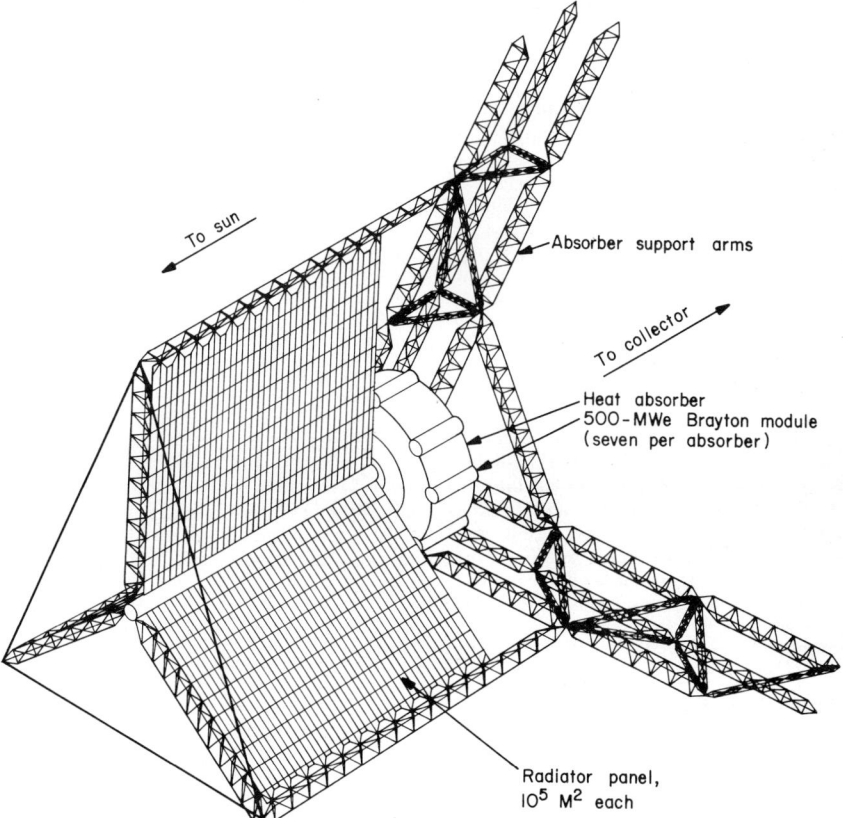

Fig. 13. Equipment at solar collector focus.

powersats; estimates are provided for near-term and eventual efficiencies. Note that the microwave transmission efficiencies are the same for the two concepts. The higher total efficiency of the thermal engine system results in it requiring less than half the solar capture area of the solar cell unit. This, of course, does not in itself imply lower cost or weights for the thermal engine system.

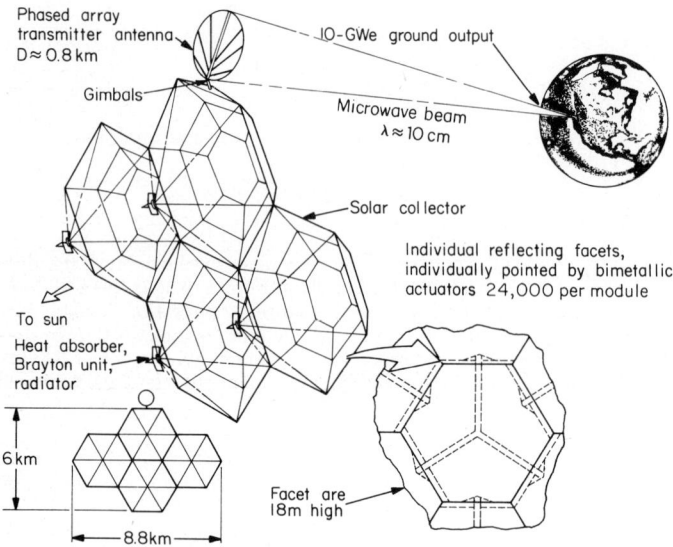

Fig. 14. Orbital power plant for continuous 10-GW ground output in geostationary orbit. Mass = 34×10^6 kg.

Item	Efficiency
Collector	0.90
Absorber	0.55
Heat engine	0.65
Alternator	0.98
Transmitter	0.90
"Space loss"	0.95
Atmosphere	0.98
Receiver	0.90
System	0.23

Nuclear power plants even in configurations considered environmentally unacceptable could conceivably be operated safely in geosynchronous orbit. Currently envisioned fusion plants do not appear to fall into this category. However, some fission reactors, including fast breeders, may be candidates. All such would involve fuels that require periodic reprocessing. Earth-surface reprocessing would involve bringing fuel elements down from orbit, as would be the fuels produced by orbital breeders for ground use. Alternatively, all fuel processing might be done in orbit.

Taylor (Ref. 11) has envisioned an orbital system wherein solar energy powers a production facility for fissionable isotopes (U-233 or Pu) for use in power reactors on the ground. The authors have not evaluated this concept.

ENERGY DELIVERY TO EARTH

Microwave.   Microwave energy transfer was initially investigated by Tesla (Ref. 12), and subsequently by Brown (Refs, 5, 13, and 14) and Goubau (Ref. 15). In the present application, the electric output of the plant is converted to microwaves which are formed into a beam by an antenna. The beam then passes through the 35,000-km space gap, and the atmosphere, and, if properly aimed, is converted to dc or ac power at a receiving station which is connected to a power distribution system. RF energy of any frequency would cross the space gap, but low attenuation atmospheric passage is dependent upon proper frequency selection. Oxygen absorption becomes significant as the wavelength λ is increased above 5 cm, while absorption and scattering losses due to rain increase rapidly

Fig. 15. Powersat, with earth in field of view. *(Boeing Aerospace Co.)*

for values of λ below 10 cm. A wavelength that is a compromise between these effects should be selected; 10 cm is a currently accepted value (frequency of 2.5 GHz).

Unfocused transmission to a far field receiver would be used by a powersat (Fig. 18) to keep the ground power level within environmentally acceptable limits by maintaining a relatively large (~ 10-km) ground beam diameter. Focusing would be used in a power relay satellite system, with a minimum beam diameter of approximately 1 km, which would occur at the satellite location.

Power transmission efficiency is a function of the total and relative areas of the transmitter and receiver antennas, the wavelength and transmission range:

$$\eta = f\left(\frac{\sqrt{A_t A_r}}{\lambda R}\right)$$

and is a maximum when the areas are equal. Since the cost of placement of the orbital antenna is higher than that of the ground antenna, minimum cost (but somewhat lower efficiency) will be obtained with a small space-located transmitter antenna and a larger

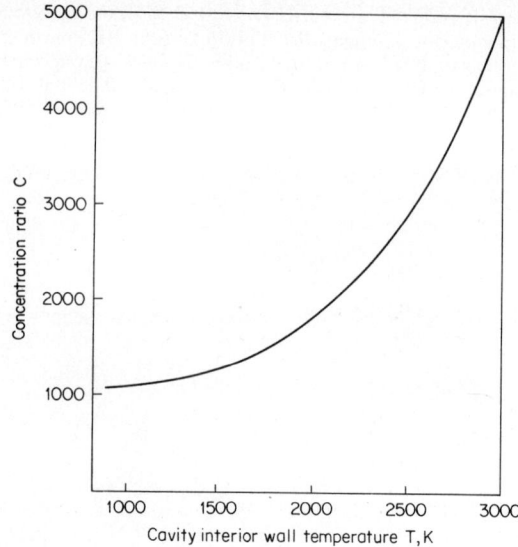

Fig. 16. Concentration ratio required to achieve cavity temperature.

$\epsilon = 0.80 \qquad P_{oc} = 5.25 \times 10^6$ GWt
$\Theta = 0.15 \qquad Q = 1.36$ 1W/m²
$F_a = 0.96 \qquad A_c = 8A_o$
$F_R = 0.14 \qquad A_o = 3.8 \times 10^3$ m²

(For heat transfer rate of 400 kW/m² through cavity walls, 43% of A_c is heat exchanger.)

ground antenna. This also reduces the peak energy density near the ground. The individual generating elements (amplitrons) of the transmitter are phase-controlled to produce a coherent beam that can be precision steered. If the transmitter could not be operated as a piloted phased array, coherence would depend on extreme geometric accuracy of the antenna (perhaps 2 cm variation over a diameter of 800 m). Phase control may be possible

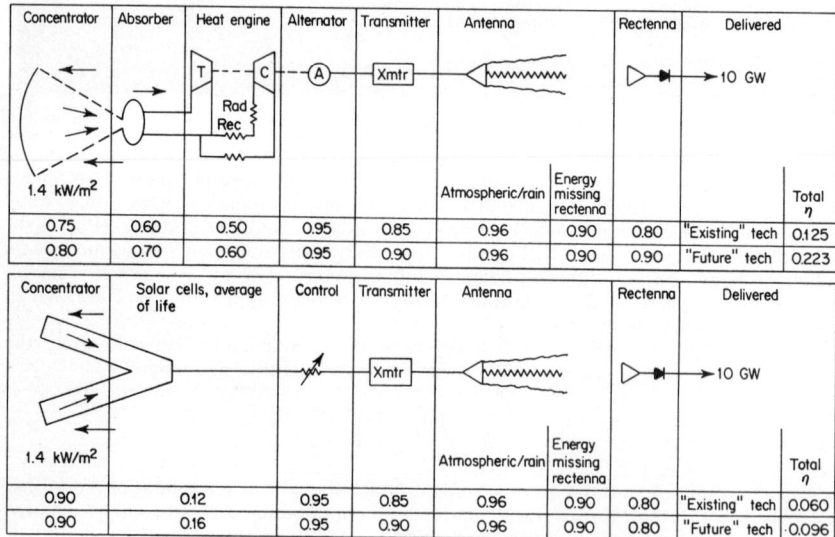

Concentrator	Absorber	Heat engine	Alternator	Transmitter	Antenna	Rectenna	Delivered	
1.4 kW/m²					Atmospheric/rain	Energy missing rectenna		Total η
0.75	0.60	0.50	0.95	0.85	0.96	0.90	0.80	"Existing" tech	0.125
0.80	0.70	0.60	0.95	0.90	0.96	0.90	0.90	"Future" tech	0.223

Concentrator	Solar cells, average of life	Control	Transmitter	Antenna	Rectenna	Delivered	
1.4 kW/m²				Atmospheric/rain	Energy missing rectenna		Total η
0.90	0.12	0.95	0.85	0.96	0.90	0.80	"Existing" tech	0.060
0.90	0.16	0.95	0.90	0.96	0.90	0.80	"Future" tech	0.096

Fig. 17. Efficiencies of thermal and solar cell Powersats compared.

by means of a system of phase shifters in the orbital antenna, which respond to the phase front from a ground-based pilot signal. This system lends itself to safety interlocking, since the power beam would become diffuse if the pilot signal is interrupted.

The ground receiving antenna which also rectifies ("rectenna") the RF energy is a physically simple device consisting primarily of a large number of dipoles, each of which is integrated with and feeds a diode rectifier. The rectenna array will be approximately 8 km in diameter if it is to lie within the region of a transmitted beam of 1.1×10^7 kW, with a transmitter diameter of 800 M and $\lambda = 12$ cm. The rectenna can be mounted on poles so that the ground below can be used for other purposes: e.g., agriculture. Only a small fraction of the incident sunlight need be interrupted.

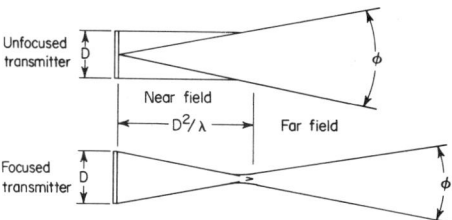

Fig. 18. Focused and unfocused collimated radio beams. $\Theta \approx 1.26\ \lambda/D$.

Per Goubau, the maximum efficiency occurs with a spherical phase front beam of radius R, and when the transmitting and receiving apertures become very large, the power distributions across the apertures become near gaussian:

$$p_t = p_{0,t} \exp\left(-\frac{\rho^2}{\rho_{xr}^2}\right) \quad \text{and} \quad p_r = p_{0,r} \exp\left(-\frac{\rho^2}{\rho_{xt}^2}\right) \tag{9, 10}$$

The required antenna areas are proportional to the characteristic power density radii:

$$A_t = 2\pi \frac{(A_t A_r)^{1/2}}{\lambda R}\rho_{xt}^2 \quad A_r = 2\pi\frac{(A_t A_r)^{1/2}}{\lambda R}\rho_{xr} \tag{11, 12}$$

The maximum power density at the receiver is

$$p_{0,r} = \frac{P_m}{\pi \rho_{xr}^2} \tag{13}$$

Specific antenna areas are thus found by

$$A_r = \frac{2P_m(A_t A_r)^{1/2}}{p_{0,r}\lambda R} \tag{14}$$

$$A_t = \frac{\lambda R p_{0,r}(A_t A_r)^{1/2}}{2P_m} \tag{15}$$

When the receiving site is nonequatorial, allowance must be made for the greater land area required for the receiving antenna, since the beam is not perpendicular to the surface. The rectenna elements should be arranged in "stadium seat" fashion: i.e., in segments set perpendicular to the beam but spaced apart to prevent shadowing. Thus, the area of the actual rectenna is independent of site location. Eventual system efficiencies are predicted (Ref. 7) to be 0.90 for microwave generation, 0.95 for beam forming/space transmission, and 0.90 for collection and rectification, for a net energy-transfer efficiency of 77 percent. For the power relay satellite, the double atmospheric pass and imperfections in the reflector itself will probably limit transmission efficiency to 0.50 or less.

Laser Energy Transmission.   Laser systems may be considered for the space-to-ground power link. Let us neglect the method and efficiency of generation of the laser beam and merely examine the overall system considerations. We can envision no practical method for reception and energy conversion of a laser beam other than to direct it upon a thermal absorber (probably a cavity unit), which acts as the heating unit for a thermal engine. Table 4 summarizes the probable resultant efficiency and some other parameters.

TABLE 4.   Comparison of Microwave and Laser Transmission Systems for 10^7 kW(e) Ground Power Output

	Microwave	Laser
Beam width at receiver	~ 10 km	~ 100 m
Conversion efficiency	0.80 → 0.9	0.30 → 0.40
Beam Power	1.2×10^7 kW	3×10^7 kW
Maximum beam power density	50 mW/cm²	0.4 kW/cm²
Wavelength	12 cm	5×10^{-5} cm
Receiver area	78 km²	7800 m²
Conversion system	Rectenna	Cavity Absorber, Brayton Engine
Loss in clouds/rain	<10%	≈ 100%
Known effect on exposed man	Rise in body temperature, discomfort	Fatal

The efficiency of reception of perhaps 35 percent is very poor compared to the 85 to 90 percent conversion efficiency of the microwave rectenna. The beam is also potentially more dangerous. In addition, clouds could totally block or at least seriously degrade transmission.

TRANSPORTATION SYSTEMS

Low Earth Orbit (Assembly Operations).   The initial satellite systems of the late fifties weighed tens of kilograms; the Apollo and Skylab boosters launched payloads weighing thousands of kilograms; while satellite systems envisioned here weigh millions of kilograms. As will be shown in the next section, a dramatic decrease in the cost of launches must also occur for powersat viability. Three factors tend to reduce launch cost per kilogram:

1. Increased launch vehicle/payload size
2. Increased launch rates
3. Repetition: i.e., identical launch operations

The partially reusable, multistage space shuttle (Fig. 19), currently under development by the NASA, may achieve launch costs of approximately \$310/kg over a projected program which involves the launching of approximately 10^7 kg. A launch rate appropriate to the placement of several powersats might reduce this cost by perhaps 40 percent. The shuttle emphasizes two-way transportation; payloads are carried in both directions within a payload bay, and can be brought down to a runway landing. A powersat transporter, tailored for the task, would emphasize one-way placement to an altitude of approximately 600 km. (The low density/high atmospheric drag characteristics of power satellites will dictate such an altitude for assembly.) See Table 5.

Total system reusability and single-stage operation may yield the required low cost. Such vehicles have been proposed, employing horizontal landing and both vertical and horizontal takeoff (Refs. 16 and 17). Again, these vehicles have emphasized two-way transportation, with payload densities approximating those of the shuttle. Some improvement in their performance could also be obtained by use of a "Separable Orbit Maneuvering System," which reduces the total inert mass that must be moved to the assembly altitude (Ref. 2).

Growth versions or derivatives of the space shuttle might provide the required cost/weight performance. A *tailored* powersat transportation system emphasizing low cost,

TABLE 5.   Space Shuttle and Powerstat Transporter Characteristics

	Shuttle	Powersat transporter
Payload to 600 km (East launch)	18,000 kg	≥30,000 kg
Down payload	≤15,000 kg	Small/infrequent
Payload density	60 kg/m³	25 kg/m³
Cost (for 10^9 kg launched)	≈ \$200/kg	≤\$40/kg

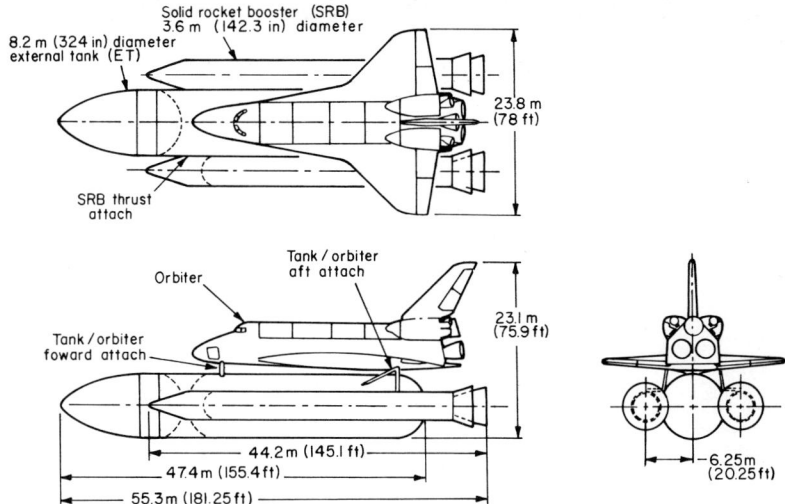

Fig. 19. Space shuttle system. Gross lift-off weight: 1889 tons (4164 klb). 104° inclination: 93 × 185 km (50 × 100 n mi). Orbiter 68 tons (150 klb) dry; SRB 1056 tons (2327 klb); ET 740 tons (1631 klb); payload 14.5 tons (32 klb).

one-way transport, and lower payload density may best employ some expendable portions; a proper balance between expendability and reusability can yield minimum cost (Ref. 18).

The majority of the inert weight of the current orbiter is involved with transport and recovery, including runway landing of the payload weight and volume. Elimination of the payload return requirement and the adaptation of a pod-type reentry and recovery system for the main engines might increase the payload by as much as 40,000 kg, so that the launch cost could be distributed over a much greater payload weight. Further development of the booster system might yield costs in the vicinity of the requirement shown in Table 5.

Geosynchronous Orbit.   Three propulsion types are candidates for the transfer of either an entire powersat or its parts from the assembly orbit to the operational orbit. See Table 6.

Electric propulsion appears to be best suited to the task, since the rocket engines (thrusters) and propellant required will add only approximately one-half to the basic powersat weight; chemical propulsion would triple it. Electric propulsion can also capitalize on the electric power production capacity of the powersat, by using this power to energize the thrusters. Depending on the fraction of the total power that is absorbed by the installed thrusters, the trip time to geosynchronous orbit may be as few as seven days, or up to several hundred, as electric rocket systems are characterized by an inherently low ratio of thrust to weight.

A modular or shuttle transportation mode can reduce the weight and cost of geosynchronous orbit placement (Ref. 2). Figure 20 shows one way that might be implemented. The trip time in each direction is approximately 50 days, so that the entire assembly operation requires a minimum of one year.

During each transfer operation, the powersat modules would be exposed to Van Allen

TABLE 6.   Geosynchronous Orbit Propulsion Candidates

Type	Typical jet velocity, kilometers/second
Chemical	4
Nuclear	8
Electric	15–150

radiation for approximately 30 days. Unless a solar cell of *extreme* radiation hardness can be developed, this means that the solar cell satellites cannot power themselves to geosynchronous orbit without significant performance degradation. An alternative approach is the use of a dedicated solar electric transportation stage to transport the power-sat modules; i.e., the transportation system is not part of the final installation.

ECONOMICS

Increasing fuel costs *and inflation* will tend to raise the national busbar costs of electricity and so provide an environment in which solar power, including space located variants, will become more and more competitive. Solar power systems which either inherently do not require storage for continuous output, or which do incorporate storage, can displace

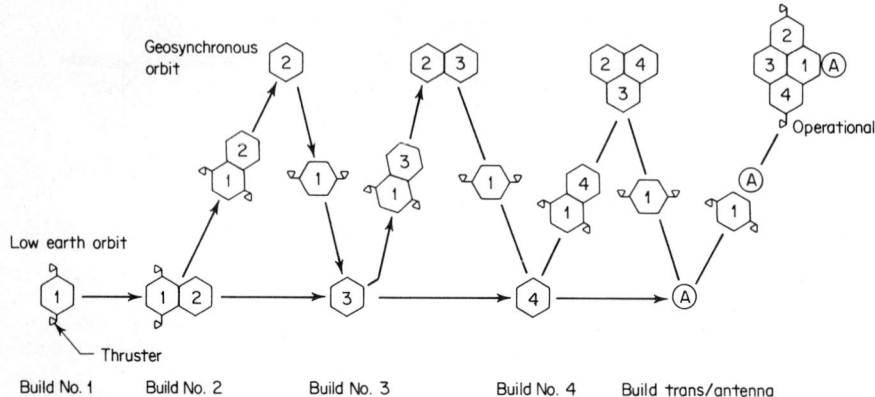

Fig. 20. Modular transportation to geosynchronous orbit.

capacity from the national power network, and their economic justification must be based on this capability. Systems that do not have storage and are highly interruptible can only be justified economically on the basis of fuel displacement, since they cannot displace capacity. Predictions of fuel cost variation with time are clouded by such factors as international politics and high inflation, at varying rates in different countries. On the assumption that inflation rates would return to a value appropriate to the previous decade, maximum permissible costs were calculated for solar power plants which would produce electricity at competitive busbar costs (Ref. 2). See Table 7.

With this guideline, we can perform a cursory evaluation of the economics of solar orbital systems. If the orbital passive solar reflector is used to augment a ground solar power plant so that storage is not required, the value added is approximately $400* per kW(e) in the year 2000. The mirror delivers approximately 6×10^{10} kW(th) to the ground plant, which would produce 6×10^9 kW(e) if an efficiency of 0.10 based on the *land* area can be achieved. The value *added* is thus $400 \times 6 \times 10^9$ or 2.4×10^{12}. The mirror weight is approximately 6×10^9 kg, for a value of $250/kg. Allowing $100/kg for boost cost permits $200/kg to be spent on the structure. This is approximately the cost of current high performance aircraft. Thus, marginal economic feasibility might be achieved, assuming that a market for the added power exists and that disabling weather changes do not occur.

Ehricke (Ref. 3) estimates that a power relay satellite (the passive microwave reflector) becomes competitive with more conventional transmission schemes for overland distances of 4000 km or more. His system is appropriate to a *global* energy solution, wherein the solar power producing areas of the world are linked to distant demand points; complex international relationships would be involved.

Estimated first unit costs for a thermal engine powersat (Fig. 15) are given on the left side of Fig. 21. The 6.5×10^9 hardware cost (which includes a ground receiving antenna) is equivalent to $650/kW. Table 7 indicates that such a plant may be "competitive" at

*1974 dollars.

TABLE 7. Estimates of Competitive Future Costs for Solar Power Plants

	Dollars/kilowatt (electric)		
	1980	1990	2000
Solar alone (capacity displacing systems)	1000	1500	2000
Solar plants displacing 60% of normal fuel consumption	610	800	1050

$1500/kW. The remaining $750/kW is available for powersat launch and assembly. The right side of Fig. 21 shows a resulting permissible launch cost of $44/kg (1974 dollars), which is approximately one-sixth of the currently estimated space shuttle cost. For similar launch costs, the photovoltaic powersat will require solar cells at less than one one-hundredth of their current costs to meet the $1500/kw goal.

ENVIRONMENTAL EFFECTS

All the satellite systems described would be placed in low-Earth orbit by rocket vehicles which exhaust particles and/or gases. Probably the highest performance engine that would be used is one burning liquid hydrogen and liquid oxygen with an exhaust of high temperature water. At least 15 kg of water would be released into the atmosphere for 1 kg to orbit. However, the total emissions from the launching of a 5×10^7-kg powersat are less than the water content of a single large thunderstorm.

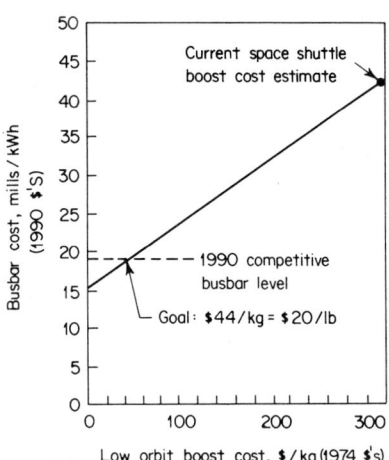

Fig. 21. Effects of boost costs (1974 dollars, except as noted). Plant costs excluding orbital transportation [10 kW(e) ground output]

	Dollars $\times 10^8$
Collector facets and orienters	6.3
Structure	4.5
Heat absorber/radiators	9.6
Rotating machinery	5.6
Transmitters/cooling	15.0
Thrusters (GSO transfer)	6.8
Attitude control	2.2
Ground antenna	15.0
Total	65.0

As previously mentioned, large orbital mirrors directing sunlight to Earth may cause cloud formations, high winds, etc., as the result of an induced convective air column. The weather impact could conceivably be felt hundreds of kilometers from the target area. Careful analysis of these effects should obviously precede any implementation of such systems. Since receivers for laser beams would probably be no more than 30 to 40 percent efficient, considerable energy would also be released on Earth by such systems, with potentially significant thermal effects.

A rectenna for reception and conversion of microwave beams may be as efficient as 90 percent so that the associated thermal energy release is small. Space generated power transmitted via microwaves may have the lowest terrestrial thermal impact of any power source other than ground-based solar power plants. However, the microwave beam has other environmental considerations. The beam is of the continuous wave type (not pulsed or otherwise modulated) and is nonionizing in tissue. A temperature rise is produced in exposed tissue (the principle of the microwave oven). The rectenna can be elevated (on poles) and fences and other security devices provided to preclude unauthorized access by men and animals. Birds will, of course, pass through the beam; tests on caged birds (Ref. 19) indicate some heating effects, but general tolerance to a field intensity of 45 mW/cm^2, a potentially economically viable signal strength. Beam areas can be "prohibited" on aeronautical charts; the only effects on metal-skinned aircraft may be on their avionics.

OUTLOOK

The reader should be cautioned that the foregoing economic analyses are speculative at best, considering the many variables involved, such as future events in international politics relative to fossil fuel availability and costs, the technical success of breeder reactor and fusion development programs, public acceptance of nuclear expansion, and numerous other factors.

Successful incorporation of these systems is generally dependent on several major technological achievements: low cost rocket transport to geosynchronous orbit, assembly of large structure in orbit, efficient/safe microwave power transmission, and either efficient thermal engine systems or very low cost lightweight solar cells. In addition, there is the problem of public acceptance of a very novel concept. In view of this, it is probable that initiation of such a project must await the precedent of successful use of solar plants for the ground generation of electric power.

REFERENCES

1. Oberth, H.: "Wege zur Raumschiffaht," Oldenbourg Verlag, Munich, 1929.
2. Woodcock, G. R., and D. L. Gregory: Economics Analyses of Solar Power, Ninth Intersociety Energy Conversion Engineering Conference (IECEC), San Francisco, Aug. 27, 1974.
3. Ehricke, K. A.: "The Power Relay Satellite," North American Space Operations, Rockwell International Corporation, March 1974.
4. Glaser, P. E.: Concept for a Satellite Solar Power System, *Chem. Technol.*, October 1971.
5. Glaser, P. E., et al.: Feasibility Study of a Satellite Solar Power Station, NASA CR-2357, February 1974.
6. *Aviation Week*, p. 47, Oct. 7, 1974.
7. Glaser, P. E.: Solar Power Via Satellite, *Astronaut. and Aeronaut.*, August 1973.
8. Woodcock, G. R.: On the Economics of Space Utilization, 23d International Astronautical Conference, Vienna, October 1972 (Raumfahrtforschung, May/June 1973).
9. Patha, J. T., and G. R. Woodcock: Feasibility of Large Scale Orbital Solar/Thermal Power Generation, 8th IECEC, 1973.
10. Stephens, C. W., and A. M. Haire: Internal Design Considerations for Cavity-Type Solar Absorbers, *ARS J.*, July 1961.
11. Taylor, T. B.: Electrical and Isotope Power from Space for Terrestrial Use, *Ann. N.Y. Acad. Sci*, 1972.
12. O'Neill, J. J. Washburn: "Prodigal Genius—The Life of Nicola Tesla," 1944.
13. Brown, W. C.: Experiments in the Transportation of Energy by Microwave Beam, *IEEE Intersoc. Conf. Rec.*, vol. 12, 1964.
14. Brown, W. C.: The Satellite Solar Power Station, *IEEE Spectrum*, March 1973.
15. Goubau, G.: Microwave Power Transmission from an Orbiting Solar Power Station, *J. Microwave Power*, 5(2), 1970.

16. Salkeld, R.: Single Stage Shuttles for Ground Launch and Air Launch, *Astronaut. and Aeronaut.*, March 1974.
17. Bangsund, E.: Feasibility Study of Single Stage to Orbit Vehicle, SAMSO TR-72-321, January 1973.
18. Gregory, D. L.: Shuttle Design Starts in Outer Orbit, *Astronaut. and Aeronaut.*, August 1971.
19. Tanner, J. A., et al.: The Effects of Microwaves on Birds: Preliminary Experiments, *J. Microwave Power*, 4(2), 1969.

Ocean Thermal Energy Conversion

BY GORDON L. DUGGER*

The idea of using the temperature difference between the warm surface water of the tropical oceans and the deeper cold layer returning from the arctic regions to run a heat engine was first proposed by d'Arsonval in 1881. The concept was first demonstrated by French scientist Georges Claude, in 1930 (Ref. 1). The development of useful power by the ocean thermal energy conversion (OTEC) may come into its own by 1981. Claude's primitive plant off Cuba demonstrated a 22-kilowatt (electric) gross power output before it was destroyed by a hurricane. French engineers demonstrated the laying of a deep-water

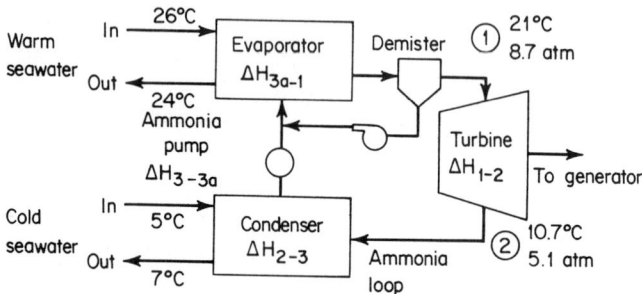

Fig. 1. Simplified loop diagram for the closed-Rankine-cycle OTEC system using ammonia as the working fluid.

pipe of a 3.5-megawatt (electric) net power output plant off the Ivory Coast of Africa in 1956 (Ref. 2), but the greater economy offered by a new hydroelectric plant nearby caused that project to be dropped.

Claude's open-Rankine cycle process depended on evacuating a chamber to the pressure (0.03 atmosphere) at which the warm seawater would flash-vaporize. The resulting low-density steam drove a turbine and then was condensed in a second evacuated chamber by direct contact with cold seawater which was falling like rain. A plant of this type, developing a significant [e.g., 100 megawatts (electric)] power level would require turbines with very large diameter rotors and ducts to match. Brown and Wechsler (Ref. 3) have recently done a preliminary analysis of an open-cycle plant that they found very encouraging. They pointed out that the forces on the turbine blades could be met by very lightweight structures similar to sailplane wings, which need not be expensive.

However, to most investigators today, use of a closed-Rankine cycle for an OTEC system looks more conventional and is certain to work after rather pedestrian engineering of some still very large components, in this case, heat exchangers. Instead of evaporating the seawater itself to drive a turbine, a working fluid, such as ammonia or propane, can be used in a closed loop, as sketched in Fig. 1. It is evaporated at 21°C (70°F) by heat exchange with 26°C (80°F) seawater. Ammonia has a vapor pressure of 8.7 atmospheres at 21°C, and so the power turbine can be two orders of magnitude smaller than in the open cycle. After driving the turbine, the ammonia is recondensed at 10.7°C by heat exchange with 5°C water drawn from 700- to 900-meter depth. The remainder of this article is devoted to this closed cycle for OTEC.

Because the overall ocean temperature difference ΔT on which this system operates is only 20°C (±4°C, depending on plant location, pipe depth, and, perhaps, the season), and the temperature differences in the evaporator and condenser (from seawater to ammonia)

*Applied Physics Laboratory, The Johns Hopkins University, Laurel, Md.

are 5°C each, only 10°C is left for the turbine ΔT. As a result, the theoretical Rankine cycle efficiency, defined as the working fluid's enthalpy change through the turbine divided by its enthalpy change through the evaporator, is only 3.3 percent. When the pumping power required for the warm seawater, the cold seawater, and the ammonia, and the efficiencies of the pumps, turbines, and generators are taken into account, this theoretical 3.3 percent

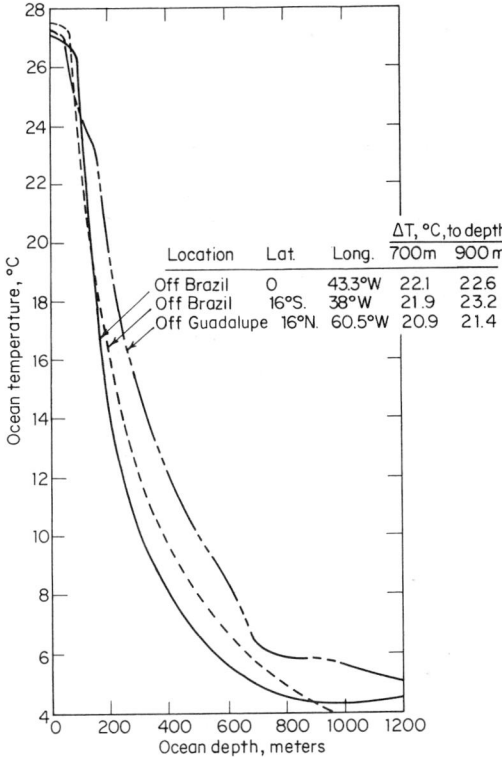

Location	Lat.	Long.	ΔT, °C, to depth	
			700 m	900 m
Off Brazil	0	43.3°W	22.1	22.6
Off Brazil	16°S.	38°W	21.9	23.2
Off Guadalupe	16°N.	60.5°W	20.9	21.4

Fig. 2. Temperature profiles at tropical sites in the Atlantic Ocean (from data in Ref. 4).

drops to a net value near 2.5 percent. Some power engineers, accustomed to a fossil-fuel power plant efficiency of 34 percent (corresponding to a 10,000 Btu/kilowatthour "heating rate") or more, have looked at this 2.5 percent efficiency and rejected OTEC as unworthy of further thought. But industrial and/or university teams that have examined and reported on this system in detail are convinced of its worth and near-term potential for contributing significantly to world energy needs. The fact that the efficiency is small does not mean that it is in danger of going to zero; it simply means that the heat exchangers will be large and that large volumes of seawater will have to be pumped through them.

THE ENERGY SOURCE

Oceans cover 71 percent of the earth's surface. They constitute a natural solar energy collection and storage system, and the resulting thermal energy can be drawn off 24 hours a day by OTEC plants. The tropical oceans are particularly attractive. Temperature differences of 20 to 23°C or more between the surface and the depths are available, as illustrated in Fig. 2. Furthermore, in the region between 10°N and 10°S latitude, an OTEC plant-ship can navigate to remain in regions where winds rarely exceed 25 knots; there are no hurricanes, and currents are less than one knot at all depths. Thus, construction and maintenance of plants designed to "graze" in this region will present minimal problems.

The available ocean area in the 10° latitude band is about 80 million square kilometers

(30 million square miles), and it receives a 24-hour average solar energy flux in excess of 215 watts per square meter, or 1.7×10^{10} megawatts total. Floating OTEC plants would need to develop electric power equal to only 0.004 percent of this incident solar energy in order to equal the projected United States electric power demand of 6.7×10^5 megawatts (electric) in the year 2000, based on an annual growth rate of 4 percent per year.

Because much of the tropical ocean belt is remote from the United States, locations in the Gulf Stream off the lower United States East Coast (Ref. 5), as well as in the Gulf of Mexico, and off Puerto Rico, Sainte Croix, and Hawaii (Ref. 6) are also being studied. A significant fraction of the power required by the Southeastern states could be provided by moored OTEC plants with undersea cables to shore. However, the available ocean temperature differences in the Gulf Stream are 20 to 30 percent smaller than in the tropics, and much sturdier construction will be needed to withstand the larger currents and frequent hurricanes. Thus, the costs of construction and maintenance of such plants are being weighed against the costs of getting the energy to shore by other means from less expensive tropical plants; or using the power at sea to produce ammonia (for fertilizer) or liquid hydrogen (for many uses, e.g., Ref. 7), or other products.

At this point, mention should be made of ocean tidal power for comparison with the OTEC approach. See Ref. 8 for example. Power can be extracted from tides by filling and emptying a dammed coastal bay or an estuary during tidal periods. A 240-megawatt (electric) plant on the Rance River in France is producing over 500 million kilowatthours per year, achieving a load factor of 0.26, in contrast to the 0.9 load factor projected for OTEC plants. Only one location in North America has aroused serious interest, namely, the Bay of Fundy–Passamaquoddy Bay area between Maine, New Brunswick, and Nova Scotia, where 15-meter tides exist. If the system designed in 1961 were built at the estimated cost of $3400 per kilowatt (electric) average power delivery (1961 dollars), it would provide only as much power as a single 230-megawatt (electric) OTEC plant. This high cost and small potential may make tidal power of relatively little interest to the United States. There is greater potential on the northern coasts of the U.S.S.R., but it is still miniscule when compared with the worldwide potential of ocean thermal power. Reference to the article on "Tidal Energy" in the section on *Hydropower Technology* in this Handbook is suggested.

Detailed comparison of OTEC potentials (Ref. 9) are beyond the scope of this article, but OTEC systems have greater economic promise than photovoltaic (solar cell) power systems in the opinion of this writer. Biomass energy systems, and possibly wind-power and direct-solar-thermal-power systems may prove competitive, but will lack the quantitative resource potential that exists for OTEC power. However, a possible synergistic payoff worth noting is the combination of a floating biomass farm (and/or possibly a fish/shellfish farm) with an OTEC plant.

CONCEPTUAL DESIGNS FOR OTEC PLANTS

In 1966, Anderson and Anderson (Refs. 10 and 11) presented a conceptual design for a 100-megawatt (electric) OTEC plant (Fig. 3) and estimated a capital cost of $167 per kilowatt (electric) of power capacity, a cost competitive then with fossil-fuel plants. Some features of continuing interest are: (1) A floating platform supports the plant, and an analogy to stable platforms for deep-hole drilling was mentioned. (2) The evaporators and condensers are submerged. (3) A working fluid with high working pressures (5 to 9 atmospheres) permits use of efficient, low-cost, single-stage turbines.

More recently, Heronemus et al. at the University of Massachusetts (Refs. 5, 6, and 12), under support from the National Science Foundation, Research Applied to National Needs (NSF/RANN) program, designed submerged catamaran configurations to be anchored in the Gulf Stream off Miami, Florida. In their "Mark II," 400-megawatt (electric) plant concept (Fig. 4), the turbines and plate-fin condensers are housed in twin 24-meter (80-foot) inside-diameter by 183-meter (600-foot) long concrete hulls. Banks of plate-fin evaporators are staggered in depth serially above the hulls to take advantage of the Gulf Stream current to reduce the pumping work. The cold-water pipe, hinged from a header between the hulls as shown in the end view, would extend to a depth of 300 meters (1000 feet) or more. The plant would be tethered from this pipe to an anchor. An undersea cable would take the power to shore.

Zener and Lavi of Carnegie-Mellon University (CMU) (Refs. 13 to 16) propose an

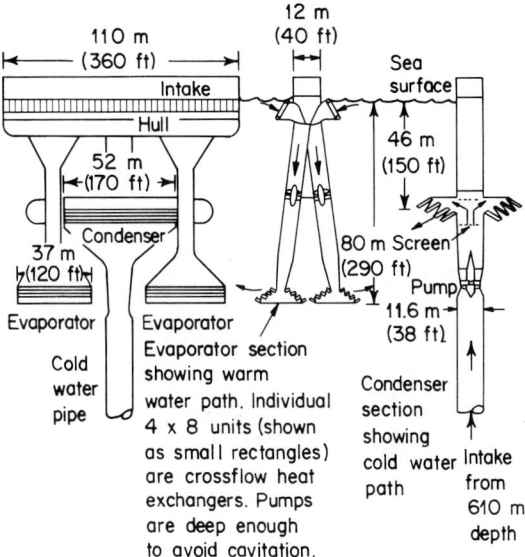

Fig. 3. The Andersons' 100-megawatt (electric) OTEC plant concept using propane as the working fluid (Ref. 10).

unmanned, automated, fully submerged plant (not shown) in which a large vertical warm-water intake pipe a few meters below sea level, and a vertical cold-water pipe reaching a depth of 500 meters (1640 feet) or more (which might, in some cases, rest on the bottom, with side inlets), to serve a symmetrical arrangement of multiple power modules in between, say, 20 to 50 meters (65 to 165 feet) depth. They have devoted more attention (under NSF/RANN support) to geometric programming to find optimum component designs, especially for fluted-tube heat exchangers (Ref. 15).

A tropical-ocean, OTEC plant-ship concept being investigated by the Applied Physics Laboratory of The Johns Hopkins University (APL/JHU) is illustrated in Fig. 5. Large banks of submerged condenser modules are located below the four deck openings nearest

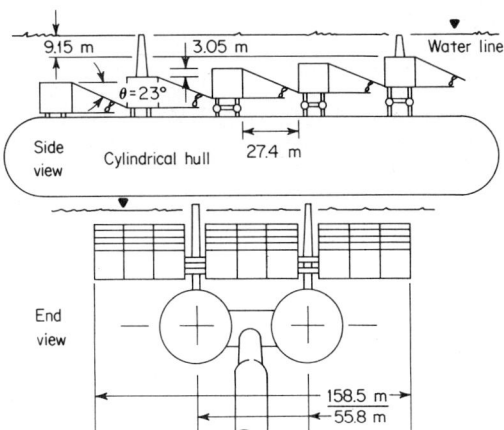

Fig. 4. Simplified sketch of the 400-megawatt (electric) submerged catamaran configuration (Mark II) OTEC plant concept developed by the University of Massachusetts for mooring in the Gulf Stream off Miami, Florida (Ref. 12).

the central cold-water pipe, which extends to 750 to 900 meters (2500 to 3000 feet) depth. Large banks of submerged evaporator modules below other deck openings are connected to the condensers in relatively small power modules [order of 5 megawatts (electric)], so that the heat exchanger units can be cleaned or serviced on a regular schedule without significantly affecting plant output. Relatively large-diameter (102- to 152-millimeter, or 4- to 6-inch) aluminum tubes are used in these two-phase-flow heat exchangers to minimize

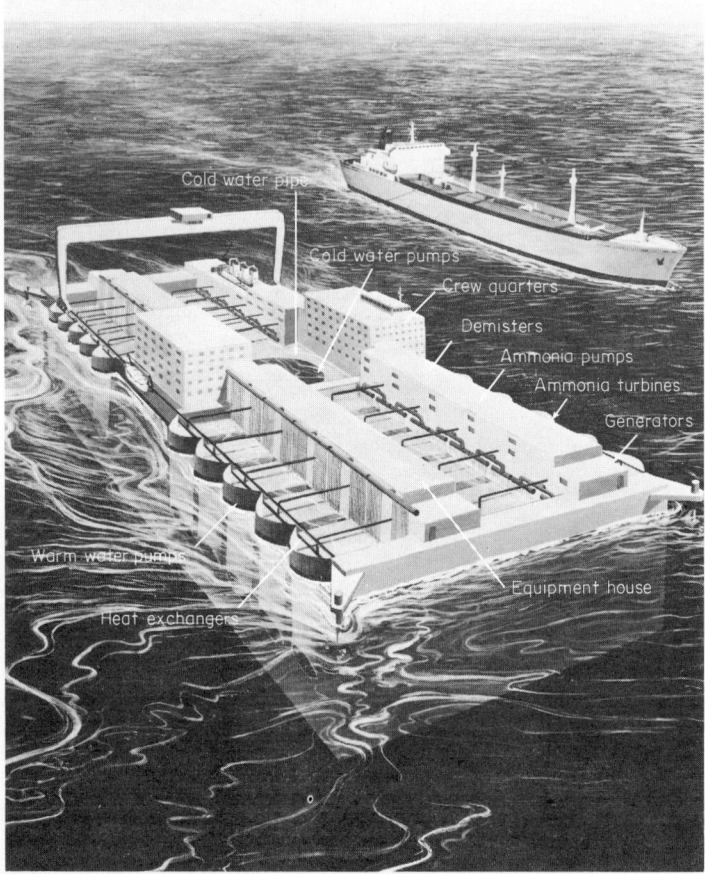

Fig. 5. Artist's concept of a dynamically positioned (or "grazing") tropical-ocean OTEC plant having submerged two-phase-flow heat exchangers, under investigation by the Applied Physics Laboratory, The Johns Hopkins University (Ref. 24).

fabrication cost, and hence plant capital cost. The exit flows from the evaporators and condensers could be directed to assist in the station-keeping needed in the mild 0.5-knot currents in which the plant is intended to operate. For 100-megawatt (electric) size, it would have a 60-meter (196 feet) beam and a length of 120 to 150 meters (395 to 490 feet). Using concrete and steel for the structure, its displacement would be on the order of 142,000 long tons, including facilities for producing and storing ammonia.

In fiscal year 1975, two industrial teams, the Lockheed Missiles and Space Co.—Ocean Systems/Bechtel Corporation/T. Y. Lin International team (hereinafter called the "Lockheed team" for brevity) and TRW Ocean and Energy Systems, TRW Inc./Global Marine Development, Inc./United Engineers and Constructors (hereinafter called the "TRW team") conducted competitive, 9-month, OTEC system studies for NSF/RANN. The very conservative base-line plant design that resulted from the Lockheed team's study (Refs. 6a and 17) is a spar-buoy configuration, primarily of reinforced concrete construction, with a telescoping, concrete cold-water pipe reaching approximately 457 meters (1500 feet) depth. The core of the vessel is 57 meters (187 feet) inside diameter, and the maximum dimension across power modules is 157 meters (515 feet); displacement is 300,000 tons.

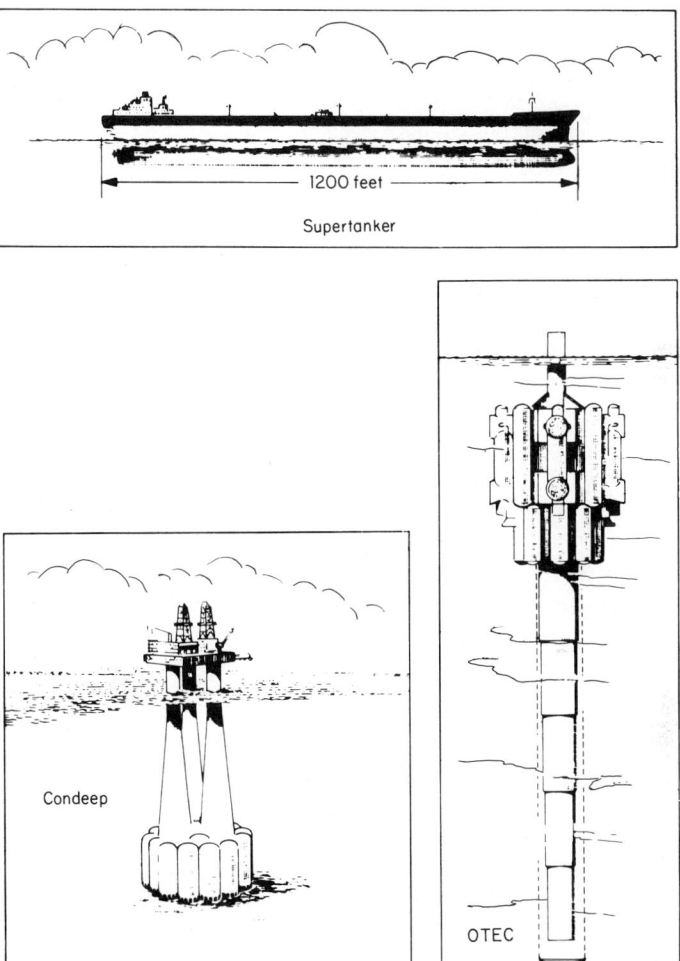

Fig. 6. Comparative sizes of a supertanker, Condeep, and the moored, spar-buoy concept for a 160-megawatt (electric) plant developed by the Lockheed Ocean Systems/Bechtel Corporation/T. Y. Lin International team (Refs. 6a and 17).

The team presented many examples of large existing steel or reinforced-concrete structures, as illustrated by comparison with two familiar examples in Fig. 6, that have demonstrated the necessary construction/deployment technology. Four detachable power modules (Fig. 7) using titanium-tubed heat exchangers, with seawater inside and ammonia outside the tubes, generate a total net output of 160 megawatts (electric). This rugged base-line plant is designed to be tethered to a deep-sea anchor (not shown) for use in virtually any appropriate location in semitropical or tropical waters with a 100-year system life expectancy, even in the hurricane belts.

The TRW team briefly investigated three spar-buoy configurations, six semisubmersible configurations, and six surface-vessel configurations before selecting the cylindrical surface vessel in Fig. 8 for their conservative, base-line concept (Refs. 6*b* and 18). It is 103 meters (314 feet) in diameter and displaces 213,000 tons. Four power modules located within the reinforced-concrete hull generate 100 megawatts (electric) net output. Design life is 40 years. As in the Lockheed base-line design, the heat exchangers use titanium

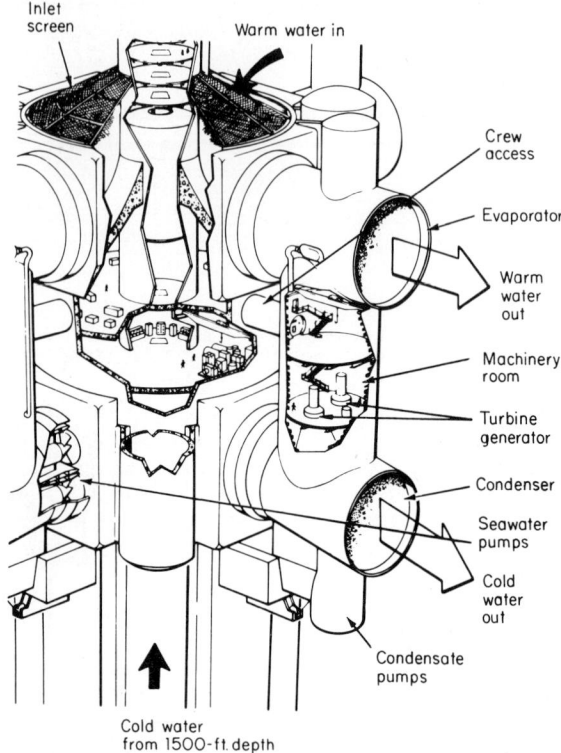

Fig. 7. One of the four detachable power modules of the Lockheed team's base-line plant concept (Ref. 6*a*).

tubes with ammonia working fluid outside the tubes. However, the TRW plant design includes a cold-water pipe made of fiber-reinforced plastic with a length of about 1220 meters (4000 feet) to achieve a higher ΔT for the plant, and they judge that a dynamic plant-positioning system (using the warm-water discharge and part of the cold-water discharge in shrouded jets) will be less expensive and more reliable than a mooring system, especially for use in deep tropical oceans.

One question often asked about OTEC plant arrangements is: Why not put the condenser at depth and eliminate the large cold-water pipe? The answer is: (1) The power required to pump the working fluid as vapor to that depth and to pump it back up as liquid exceeds that of pumping the seawater up. (2) The condenser would need to be much heavier and costlier to withstand the hydrostatic pressure at depth.

In summary, a number of approaches to the overall design of an OTEC plant may be acceptable. Design differences for application in tropical oceans versus the Gulf Stream are to be expected, just as designs for offshore drilling platforms vary for different environments. Additional studies and some experiments on platform models and heat transfer are in progress. Trade-off studies, based on additional experimental data that will be obtained in 1975 to 1976 (Ref. 6), will determine the more cost-effective configurations. The various investigators to date concur that prospects for economically competitive

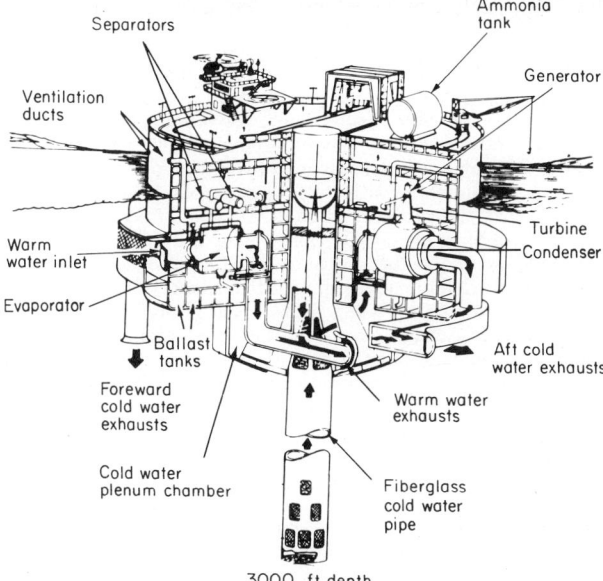

Fig. 8. The 100-megawatt (electric) cylindrical-surface-vessel baseline OTEC plant concept developed by the TRW Systems Group/Global Marine Development Corp/United Engineers and Constructors team (Refs. 6*b* and 18).

ocean power plants look very good, within the state-of-the-art or small extrapolations thereof. No "breakthrough" in any area is needed.

TURBINE TECHNOLOGY AND WORKING FLUID CONSIDERATIONS

Design of the single-stage turbines needed will be relatively straightforward, since the required characteristics, in terms of head versus theoretical mass flow per horsepower obtained, fall between those for hydroelectric water turbines and those for combustion gas turbines. Figure 9 from Ref. 19 shows variations of speed and diameter with output power rating for ammonia and propane turbines operating at OTEC plant conditions, based on a specific speed of 100 and a specific diameter of 1.3, near optimum values. Propane

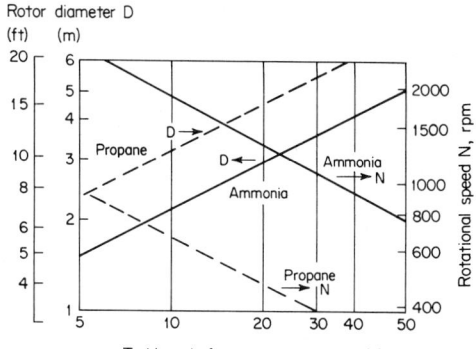

Fig. 9. Characteristics of ammonia and propane turbines for OTEC plants (from Ref. 19).

turbines are larger and slower than ammonia turbines and, therefore, cost nearly twice as much.

Materials can be rather pedestrian because the tip speeds, pressures, temperatures, and temperature variations will be moderate compared with those for conventional steam or gas turbines. Seals must be designed with care to avoid leakage. With a multiple-power-module approach, it will be possible to select a specific design point with a high efficiency (near 90 percent) without concern for operation over a broad rpm range, because load adjustment for the OTEC plant can be managed by cutting out one turbine at a time.

Table 1 summarizes the ammonia vapor turbine designs selected by TRW (for 1 turbine per 25 megawatts (electric) net-power module, Ref. 6b) and Lockheed (for 2 turbines per 40-megawatt (electric) net-power module, for a plant achieving a lower ocean ΔT, 19 versus 22°C (Ref. 6a), and shows the propane/ammonia ratios for several parameters as given by Lockheed. Table 2 summarizes the relative merits of propane, ammonia, and a halogenated hydrocarbon (or Freon type), R-12/31, based on a table presented by TRW (Ref. 6b). Halogenated hydrocarbons would require even larger turbines and heat exchangers than propane, and hence appear to be ruled out on a cost basis as well as being judged now to pose the most severe environmental problems. Thus, ammonia emerges as the choice of most investigators at this writing.

HEAT EXCHANGERS

The evaporators and condensers will be the largest and most costly components on the basic OTEC plant, because the available ΔT's are small, more similar to those in refrigeration/air conditioning equipment than in the boilers and condensers in conventional power plants. Some effects of the working fluid choice have been previously described.

In the evaporators, there may be a problem of biofouling from organisms that flourish in the warm water. This problem can be minimized by keeping the water velocity above 1.5 meters/second (5 feet/second) and, in general, by locating the plant far from shore. However, it may be necessary to provide a means for either regularly cleaning the surfaces on the seawater side or preventing the biofouling (e.g., by adding 0.1 to 0.3 part per million of chlorine to the seawater, either batchwise or continuously). Bifouling will be smaller, perhaps negligible, in the condenser, because there will be far fewer organisms in the cold water drawn from great depths (Ref. 20).

Material selection will represent a compromise. The 90/10 copper-nickel alloy commonly used in marine condensers has excellent thermal conductivity, is resistant to fouling, and could be used with propane as the working fluid; but it is expensive and is not compatible with ammonia, and the copper leached from it would have an adverse environmental impact. Titanium has excellent corrosion resistance except for galvanic corrosion, but it is very costly. Aluminum, in suitable alloys which are practically inert to seawater (Ref. 21) and are much lower in cost than Cu-Ni or Ti, has proved very

TABLE 1.   Ammonia Turbine Design Summaries and Propane/Ammonia Ratios for OTEC Plants

	Ammonia		Propane/Ammonia
	TRW (Ref. 6b)	Lockheed (Ref. 6a)	Ratio (Ref. 6a)
Shaft power, horsepower	44,236	42,846	—
Shaft power, megawatts (electric)	33	32	(same)
Shaft speed, rpm	1,800	1,800	0.37
Flow rate, pounds/second	1,933	1,657	3.25
Specific speed	—	91	—
Overall efficiency	0.896	0.902	—
Inlet total pressure, psia	122	136	—
Inlet total temperature, °F	67	73.3	—
Rotor exit static pressure, psia	87	92.7	—
Rotor diameter, tip/hub, inches	112/73	—	1.5
Rotor speed, tip/hub, feet/second	880/565	—	0.55
Turbine assembly weight, pounds	50,000	—	—
Generator (4-pole, 1800 rpm), pounds	250,000	—	—
Heat exchanger volume	—	—	2.5

TABLE 2.   Working Fluid Comparison (Ref. 6b)

	Ammonia		Propane		R-12/31	
	Liquid	Vapor	Liquid	Vapor	Liquid	Vapor
Thermal conductivity, 50–70°F, Btu/(hour)(square foot)(°F)	0.29	0.014	0.07	0.01	0.05	0.006
Heat capacity, Btu/(pound)(°F)	1.13	0.19	0.62	0.24	0.24	0.097
Heat of vaporization, Btu/pound	500		140		70	
Materials compatibility	Wet NH_3 not compatible with copper.		Excellent except some plastics.		Excellent.	
Toxicity	Severe but easily detected.		Slight— difficult to detect.		Slight—difficult to detect	
Flammability	Moderate		Explosion hazard		Not flammable	
Solubility in water	High		Low		Very low; hydrolyzes	
Effect on external environment	Slight		Undesirable local effects		Potentially severe problem	
Problem of contamination as working fluid	Moderate		Negligible		Negligible	
Availability	Good		Good		Potential problem	

satisfactory for heat exchangers for desalting plant use (Ref. 22), and it has very good thermal conductivity. Thus, an aluminum alloy, such as 5083 or 5086, would appear to be more cost effective. Carbon-filled-plastic heat exchangers are receiving some study and may prove attractive from a cost viewpoint (Ref. 6c), but would require two to four times as much surface area as metallic systems because of the low thermal conductivity of the plastic matrix and diffusivity-barrier materials required.

Fins (Ref. 12), flutes (Ref. 15), or porous surfaces (Ref. 23) may be used on the working fluid side to enhance overall heat-transfer rate in metallic heat exchangers by factors up to 2 or more, but probably will be impractical on the seawater side, particularly in the evaporator, because they would make cleaning operations more difficult. If appreciable fouling does occur, its effect will dominate the overall heat-transfer coefficient U and reduce the percentage gains in U due to surface enhancement.

Leak detection and repair (or module deactivation and replacement) must be possible and is an important factor in design of such large exchangers. For this reason, and to reduce initial assembly costs, designing to reduce the number of tube joints may be more important than striving for maximum efficiency. Thus, the APL/JHU team favors two-phase-flow heat exchangers using ammonia *inside* multipass (folded approximately 20 times), 4- to 6-inch-diameter aluminum tubes which they estimate will cost $2 to $3 per square foot (Ref. 24), compared to $9 to $12 per square foot for conventional shell-and-tube, titanium heat exchangers (Refs. 6a and 6b). With the seawater flowing single-pass, outside (and normal to) the multipass tubes, the "shells" will have seawater on both sides with very little pressure differential, so that they can have thin walls and rectangular cross sections to facilitate multimodular arrangements in a plant-ship, such as suggested by Fig. 5 that is designed to take advantage of the relatively benign, tropical-ocean environment. A multimodular, low-draft (prior to cold-water pipe installation) design is also expected to facilitate construction in existing United States shipyards. Since the heat exchangers themselves will be buoyant, the best overall structural arrangement for economy remains to be specified. The limited work done to date by APL with an integrated team including Sun Shipbuilding & Dry Dock Co., Hydronautics, Inc., The Woods Hole Oceanographic Institution, and Kaiser Aluminum and Chemical Corp. indicates that a good, economical design will emerge.

COLD WATER PIPE (CWP) AND SEAWATER PUMPS

Fabrication and deployment of cold-water pipes (CWP) for an OTEC plant will be a challenging engineering/logistics problem, just as it was for Claude in 1930 (Ref. 1) and the French team in 1956 (Ref. 2). However, the facts that (1) they did deploy pipes successfully; (2) oil, gas, and water pipelines are criss-crossing North America; and (3) complex oil-drilling, mining, and other offshore rigs are becoming commonplace; provide assurance that there will be practical solutions to this problem.

For 100 megawatts (electric) net power output, a CWP of 15 to 18 meters (50 to 60 feet) in diameter will be needed to pump approximately 2 billion kilograms per hour (8 million gallons per minute) of seawater through the condensers. These flow figures are based on an overall ΔT near 22°C with ammonia as the working fluid, and parasitic pumping power losses near 25 percent [133 megawatts (electric) gross output to net 100 megawatts (electric)]. The design of the CWP will be determined by the plant location (the ΔT versus depth trade-off) and the overall system efficiency/cost-effectiveness trade-off, which involves many parameters. For a 900-meter (3000 foot)-long CWP for a 100-megawatt (electric) plant, the loss due to friction in the CWP and the dump loss from it to the condensers will be approximately 4 megawatts (electric) for a 15-meter (50-foot)-diameter (3 meters/second or 10 feet/second water velocity), or less for a larger diameter. In comparison, pumping power required for the evaporator(s) and condenser(s) will be 6 to 10 megawatts (electric) each, and for the head loss due to seawater density change with depth in the CWP, approximately 4 megawatts (electric). Auxiliary machinery and ship and crew needs will add 3 to 6 megawatts (electric).

Among the CWP designs under investigation are a double-walled aluminum structure (Ref. 24), reinforced fiberglass (Ref. 6b), reinforced concrete (Ref. 6a), and a "stockade" design employing a ring of smaller-diameter pipes (Ref. 6c). The concrete design has not been evaluated for use at depths much greater than about 520 meters (1700 feet).

The seawater pumps will be comparable in size to the largest turbines used for hydroelectric power plants. Figure 10 and Table 3 present the Lockheed team's design, which calls for four cold-water and four warm-water pumps for each of the four 40-megawatt (electric) net-power-output modules (a total of 32 pumps) of their base-line plant (Ref. 6a). The warm-water pumps would be in the warm-water intake (hull-side)

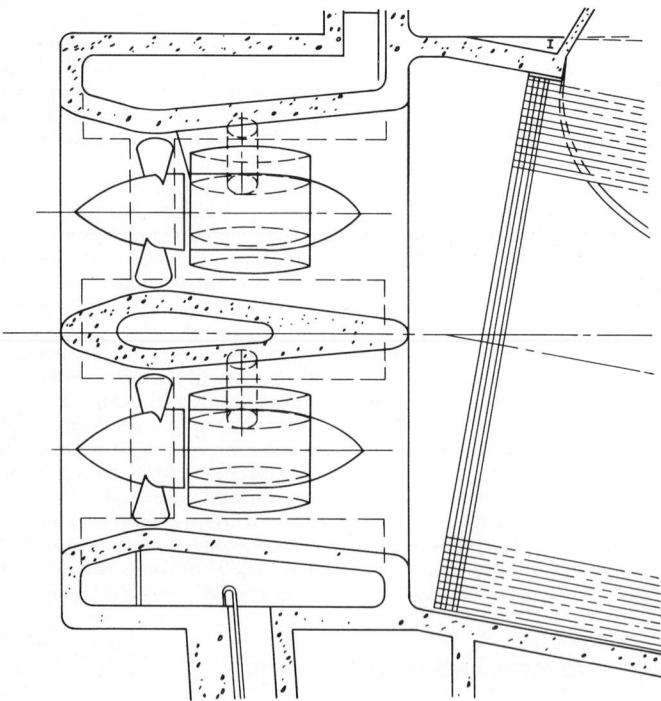

Fig. 10. Typical seawater pumps for OTEC plants as they would be installed in the Lockheed team's power module concept (Ref. 6a).

TABLE 3.   Lockheed's OTEC Seawater Pump Design Summary (Ref. 6a)

	Warm water	Cold water
Number of pumps/module	4	4
Capacity/pump, cubic meters/second (cubic feet/second)	85(3000)	113(4000)
Head requirement, meters (feet)	1.2(4.1)	2.7(8.9)
Efficiency, %	86+	86+
Revolutions per minute	35–44	54–65
Runner diameter, meters (feet)	6(20)	6(20)
Blade angle, degrees	18	16
Water velocity at runner, meters/second (feet/second)	4(13)	5(17)
Shaft horsepower	1620	4070
Bulb diameter, meters (feet)	3(10)	3(10)
Pump length, meters (feet)	12(40)	12(40)
Power supply: 4–16 kV, 60-Hz, 3-phase		

cylinder at the top of the module; the cold-water pumps in the cold-water intake (hull-side) cylinder at the bottom. Refer to Fig. 7. As another example, APL's concept for a 100-megawatt (electric) tropical-ocean plant-ship with 20 power modules of 5 megawatts (electric) each would call for a total of nineteen cold-water pumps of approximately 0.13 times the capacity per pump shown in Table 3 to send cold water to head ponds over the condenser banks, and perhaps 20 warm-water pumps of 0.17 times the capacity per pump shown in Table 3.

ENVIRONMENTAL AND SOCIAL IMPACTS

As for all other solar energy conversion systems, the basic ocean thermal energy conversion plant would be very attractive ecologically as compared with fossil-fuel or nuclear power plants, because it would be "fueled" by a portion of the inexhaustible supply of energy from the sun and would be nonpolluting. In contrast with all the other solar and nonsolar types, however, it would use no valuable land area or visible structures for power generation or energy storage to accommodate noninsolation periods. The only land use requirements would be for power transmission substations tying into a utility grid and/or the storage that will be needed in any event for products that might otherwise need to be imported from foreign land-based plants.

With respect to ecological impact in the oceans, studies are needed. However, a few points can be made here. If no changes in ocean surface layer absorptivity, currents, or mixing were induced by the power plant operation, the 2 percent or so local utilization of incident solar energy could (at worst) be compensated by the same percentage reduction in surface cooling by water evaporated from the ocean surface. Thus, if it is assumed that evaporation rate is proportional to water vapor pressure, the surface temperature within the area of influence of the plant would eventually be lowered from 26.5 to 26.1°C. But the effect will not be even this large, because the peak insolation far exceeds the average value, and a cooler layer would absorb more of this energy, which would be distributed by mixing and currents. Bathen (Ref. 6e) used local weather data (including wet-bulb temperatures) and surface water temperatures to calculate heat balances for the air-sea interface off Keahole Point, Hawaii, for average winter and summer conditions with and without the presence of a 0.8°C anomaly due to a cold-water discharge, and concluded that the net heat budgets at the interface would increase, both winter and summer. He also did a hydrodynamic analysis for cold-water-discharge plume mixing for a 20-megawatt (electric) OTEC plant off Keahole Point, and concluded that the impacts would be equivalent to temperature drops in the mixed layer (above the thermocline) of just 0.02 and 0.07°C during winter and summer conditions, respectively, and over area of only 1.1 and 2.6 square kilometers, respectively. Such small temperature changes are well below the typical diurnal changes of 0.1 and 0.3°C, respectively.

The cold water brought from the deep to cool the condenser is rich in nutrients that significantly enhance fish growth and could provide an ecological advantage, whether or not a deliberate attempt is made to pursue mariculture near OTEC plants. The landings of fishery products by the United States have lagged far behind those of other nations and have been only one-third those of Japan, the U.S.S.R., and Peru (Ref. 25). Upwelling of cold water has led to large catches of sardines off Peru, and previously off California. The possibility of reaping benefits near or downstream of OTEC plants should be included in environmental impact investigations. Bathen's analysis, again for the effects of a 20-megawatt (electric) OTEC plant off Keahole Point, Hawaii, indicated that nutrient levels in surface waters would increase by 300 percent at a 260-meter (853-foot) radius from the cold water discharge point, resulting in a 1500 percent increase in phytoplankton and a 540 percent increase in herbivores. Thus, the required strategy for a combined OTEC-mariculture operation would be to assure operation in a small current and direct the discharges downstream to the mariculture farm to avoid biofouling problems and efficiency (ΔT) effects on the plant.

Although normal operation of an OTEC plant would produce no chemical pollution of air or water, impact studies will be needed to assess the potential consequences of a spill of the working fluid, which might be ammonia or propane. With appropriate instrumentation and monitoring of plant performance, it should be possible to detect spills early and take action before they reach a significant magnitude. It would be expected that all major components of an OTEC plant would be designed for at least a 20-year life and inspected at reasonable intervals just as the components of any stationary power system are so inspected.

Another factor needing study is possible interference with shipping lanes; but operation of unmoored, grazing plants in tropical seas beyond territorial limits would be equivalent to other commercial shipping on the high seas to which the present Law of the Seas applies. Legal considerations are being studied for ERDA (Ref. 6).

Relative to the OTEC plant's ability to draw warm water from the mixed layer without drawing cold water from below the thermocline into the evaporators, Zener (Ref. 16) has estimated that for a surface current of 0.03 meter/second (0.09 foot/second), OTEC plants up to 400 megawatts (electric) size could operate without a problem. The presence of a larger current would permit an increase in plant size based on this criterion, but it seems unlikely that early plants will exceed 500 megawatts (electric), because economies of scale probably will not be significant above that size.

COST ESTIMATES FOR POWER PRODUCTION

Only approximate cost estimates for OTEC plants can be made until performance data for a pilot plant embodying the basic features of a complete system are available, a demonstration plant has been run, and costs of competitive systems have settled to some predictable pattern. Coal-fired plant costs tripled in the 1965 to 1975 decade, and nuclear fission power plant costs rose from $200 per kilowatt (electric) to projections as high as $1100 per kilowatt (electric) or more for plants that will come on line in the 1980s.

A simple comparison is presented in Table 4. An OTEC plant with an uninterruptible energy source and modular design, permitting an average use factor of 0.9, could have a capital cost, including means for getting the energy to shore, of $1000 per kilowatt (electric) and be competitive with a nuclear fission plant at approximately $500 per kilowatt (electric) and superior to a coal plant at approximately $450 per kilowatt (electric), using eastern coal at $28 to $37 per ton. The OTEC cost could go to $2500 per kilowatt (electric), if 10 percent fixed-charge rate could be attained and if it were competing with a nuclear plant at $1000 per kilowatt (electric), or an oil-fired plant using oil at $11 per barrel. The fixed charges for this comparison are taken to be 15 percent for the land-based plants, and 13 or 10 percent for tropical OTEC plants, which will be subject to no local taxes, although they may have somewhat higher insurance costs. The 10 percent rate is based upon Export-Import Bank financing at 6 percent interest rate and reduced insurance cost (Ref. 27).

Table 5 compares capital cost estimates for OTEC plants from APL/JHU, Carnegie-Mellon University (CMU), whose "medium" estimate (Ref. 14) is shown here, and the University of Massachusetts group (Ref. 5). For comparison, the estimate from the latter group for propane working fluid has been converted to use of ammonia in the last column of the table. These various estimates are reasonably consistent when the strong effects of the seawater ΔT and the working fluid are taken into account.

The cost estimates presented by the two industrial teams headed by Lockheed (Refs. 6a

TABLE 4.   Approximate Power Cost Comparisons for New Construction to Estimate Allowable Competitive Cost for an OTEC Plant
Including cost of getting power to shore

Cost element	Fossil fuel		Nuclear		Allowable OTEC range
	Oil	Coal	Low	High	
Investment, $ per kilowatt (electric)	465*	450*	500*	1000	1000–2500
Use factor*	0.75	0.75	0.6		0.9
Fixed charge rate, %	15	15	15		13–10
Costs, mills per kilowatthour:					
Fixed charges	11	10	14	29	17–32
Operating cost	1	1	1		1
Fuel cost	20†	11–14†	3		0
Power cost	32	22–25	18–33		18–33

*Values from Ref. 11. Costs include $100 per kilowatt (electric) for pollution and safety control costs, and costs for fossil fuel plants include 30-day fuel storage facilities.
†Oil at $11 per barrel and eastern coal at $28–$37 per ton. Based on heat rate of 10,000 Btu/kilowatthour.

and 17) and TRW (Refs. 6b and 18), subsequent to the initial preparation of this article, are based upon early 1975 dollars and are considerably higher than those in Table 2, which were based upon 1973 to 1974 inputs. Their conservative approach to "immediately buildable," long-life base-line designs with expensive titanium-tubed heat exchangers was consistent with the guidelines given to them for their base-line studies. The resulting base-line estimates are $1800 to $2000 per kilowatt (electric) by TRW (for 22°C ΔT); and $2500 to $2600 per kilowatt (electric) by Lockheed (for 19°C ΔT). Approximately 50 to 58 percent of these costs, respectively, are for the heat exchangers; and both teams note that these costs could be reduced by 40 to 80 percent by improved designs based on use of aluminum. The Lockheed team detailed a series of possible heat-exchanger improvements and other factors (including higher ocean ΔT) that could ultimately lead to a heat exchanger cost as low as $200 per kilowatt (electric) with sheet-metal construction. The TRW team noted that major cost reductions might be achieved in the platform as well as the heat exchangers, and they projected a plant cost reduction to $1100 per kilowatt (electric). Thus, the estimates of costs from these studies by industrial teams, which have added greatly to the creditability of near-term developments and demonstration of OTEC plants, are not nearly as far from those in Table 5 as would appear at first glance. With these comments in mind, Tables 6 and 7, based upon estimates from Table 5, are presented with the caution that some cost escalations are to be expected, but the relative cost change should not be great enough to alter substantially the foreseen attractive competitive capability of OTEC plants. Inflation will affect the competitive systems in Table 4 (via fuel costs) even more than OTEC plants, e.g., delivered fuel oil prices exceeding $11 per barrel.

TABLE 5.   Estimates of Capital Cost, $ per Kilowatt (electric), 1974 dollars, for OTEC Power Plants
Excluding power (or product) transmission to shore

Source (Ref.)	U. Mass[5]		APL/JHU[24]	CMU[14]
Plant site	Gulf Stream		±10° latitude	—
Ocean ΔT, °C(°F)	18(32)		22(39) 24(43)	20(36)
Working fluid	Propane	NH₃*	NH₃	NH₃
Costs, $/kilowatt (electric), net:				
Heat exchangers	340	254	153	280
Turbines, generators				
pumps, miscellaneous	179	100	109	282
Cold water pipe	63	63	45	58
Platform	48	48	50	36
Total, $/kilowatt (electric)	630	465	357	656

*The University of Massachusetts estimates for use of propane working fluid are from Ref. 5 and were adjusted for use of ammonia by G. L. Dugger.

TABLE 6.   Estimated Power Costs at Shore for Gulf Stream Plants
Based on University of Massachusetts costs (from Table 5)

Costs	Working fluid					
	Propane			Ammonia		
Capital costs, $/kilowatt (electric):						
Basic OTEC plant	630			465		
Power conversion/transmission						
system to shore	82			83		
Subtotal	712			548		
Add 12% for interest and escalation						
during construction	85			66		
Total capital cost, $/kilowatt (electric)	797			614		
Fixed charge rate, %	15	13	7	15	13	7
Costs in mills/kilowatthour (electric):						
Fixed charge at 0.9 load factor	15	13	7	12	10	5.4
Operating cost	1	1	1	1	1	1
Power at shore	16	14	8	13	11	6.4

Table 6 shows the University of Massachusetts estimates for total capital cost of getting the power to shore, including 12 percent for interest and escalation during construction. These capital costs for propane and ammonia working fluids are converted to power costs for three assumed fixed-charge rates, 15, 13, and 7 percent, the last being typical of public utility financing in the 1950s and 1960s. The resulting estimated power cost range of 6 to 16 mills per kilowatthour could double and still be attractive in the near term.

PRODUCTION OF AMMONIA OR OTHER PRODUCTS AT TROPICAL OCEAN PLANTS

Production of ammonia at a floating, tropical OTEC plant is attractive because: (1) ammonia is a major item of national and international commerce used in fertilizer and numerous other chemicals and products; (2) in 1974 to 1975, ammonia plants in the United States consumed 2½ percent of the natural gas supply, and, if existing plants continue to receive as much natural gas as in 1975, projections for the ammonia demand (Ref. 28) indicate a shortfall by 1985 of 10 million tons; and (3) production of ammonia at the OTEC plant would require only hydrogen from seawater and nitrogen from the air, as shown in Fig. 11.

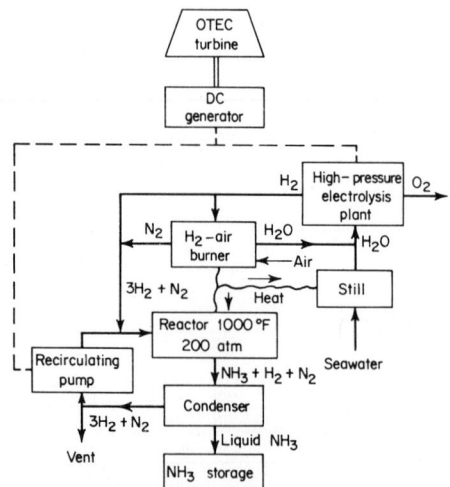

Fig. 11. Block diagram for an OTEC-ammonia plant (Ref. 24).

The nitrogen is obtained by burning the oxygen from air with one-seventh of the hydrogen from electrolysis cells to form water, leaving nitrogen plus the minor constituents of air, mainly argon and carbon dioxide. The presence of carbon dioxide and argon will require fractional venting with a resultant slight loss of product ammonia. The water vapor is condensed and returned to the electrolysis cells. The heat produced by the burner is used in part to operate the seawater still to produce the rest of the water needed for the electrolysis cells (which also provide some heat for the still), and in part to provide heat to the catalytic converter. The remaining gas from the burner is mixed with the remaining six-tenths of the hydrogen from the electrolysis cells in a molar ratio of $1N_2$ to $3H_2$ and fed to the converter, which uses a promoted-iron or other catalyst. A condenser then removes a portion of the ammonia as liquid, and the remaining gases are recirculated through the converter by a compressor. This converter and condenser would be of the same type used in existing commercial plants, but the more costly and maintenance-demanding oil- or coal-reforming portions of those plants would be eliminated. Alternative possibilities are use of a reverse osmosis system to produce the fresh water and air liquefaction to obtain the nitrogen. The latter would allow a greater ammonia production capacity at a modest increase in capital cost.

The estimates in Table 7 suggest that liquid ammonia could be delivered to United States ports from distances of about 4000 miles at a cost (before profit and taxes) of $63 to $72 per ton (1974 dollars), compared with early 1975 cost estimates of $48 to $91 per ton for a natural gas cost range of $0.40 to $1.50 per cubic foot (Ref. 29). The price quotations for ammonia at the plant gate have ranged between $145 and $165 per ton. At such prices, the value of the ammonia produced by one 500-megawatt (electric) plant (475,000 tons of ammonia per year, or 3 percent of current United States production), within three to four years, would be equal to the capital investment in the plant. Use of the oxygen produced by the electrolysis cells is not credited here, but it may prove economically attractive to liquefy the oxygen for shipment to shore for use in fuel cells, propulsion systems, or waste-treatment plants.

The electrolytic reduction of alumina (made from bauxite on shore) to aluminum on OTEC plants also is being investigated because it is an electric power-intensive process, and part of the market might be for subsequent OTEC plants. A 500-megawatt (electric)

TABLE 7.   Estimated Costs for Producing Ammonia at a Tropical OTEC Plant with 39°F ΔT*

Plant cost to busbars (Table 5), $/kilowatt (electric)	$357	
Add ammonia plant (Fig. 10)	90	
Total plant investment, $/kilowatt (electric)	$447	
Basis: 500-megawatt (electric) plant producing 475,000 tons/year at 0.9 load factor		
Plant investment (P.I.), $/ton-year	$470	
Working capital, $/ton-year	20	
Total capital required, $/ton-year	$490	
Costs in $/ton for two methods of financing*†	Conventional	Export-Import Bank†
Catalyst, chemicals, labor and overhead, $/ton	2	2
Maintenance (1% of P.I.)	5	5
Insurance (2%, 1% of P.I.)	9	5
Depreciation (20 yr, 5% of P.I.)	24	24
Interest (8%, 6% of ½ P.I.)	19	14
Production cost, subtotal	59	50
Interest on working capital	1	1
Shipment to U.S. port	12	12
Cost at port, $/ton	72	63

*Based on type of ammonia production costing used in Ref. 29. Maintenance is lower than in Ref. 29, because only a portion, which requires least maintenance, of a natural-gas-fed ammonia plant is needed here.

†The Export-Import Bank (ExIm) is currently funding about $10 billion/yr similar industrial projects at 6%, with longer repayment period loans frequently combined with private financing; in 1970, funded an ammonia plant (Ref. 27).

OTEC plant with aluminum heat exchangers and some aluminum structure will require up to 100,000 tons of aluminum, but one such plant could produce approximately 230,000 tons per year of aluminum (at a plant load factor of 0.85 and 8 kilowatthours per pound of aluminum), equal to 4 percent of the current United States production rate, for use in subsequent plants. And use of aluminum in transportation equipment, other energy systems, et al. can be expected to increase.

Since magnesium chloride is a constituent of seawater, magnesium could be produced at sea by shipping calcium oxide (from oystershells or limestone) to the platform, or in some locations, obtained from the ocean floor. The demands for magnesium today could be met by just two 500-megawatt (electric) OTEC plants. Another possibility for the late 1980s would be to ship coal or a carbonate to the platform and use the hydrogen produced from it to make synthetic oil, methane, or methanol. In the longer term, the hydrogen itself is expected to be the major product of the OTEC plants (Ref. 6).

POTENTIAL GROWTH RATE FOR TROPICAL OTEC PLANTS

The 1975 United States energy requirement is near 80×10^{15} Btu/year or 80 Q, and most projections anticipate that it will reach 160 Q by the year 2000. Tropical OTEC plants ultimately could provide many times this requirement.

Ammonia appears to be the most attractive candidate for initial production at sea. When demands for ammonia, aluminum, and other energy-intensive products have been alleviated by tropical OTEC plants, a next step could be a "hydrogen economy." (See "Hydrogen" in section on *Chemical Fuels Technology* in this Handbook.) In a broad sense, the industrial/resource/technological limitations on rate of growth of OTEC energy production will be imposed not by the basic solar energy resource, but by one or more other factors, including:

1. The ability to obtain the needed raw materials (e.g., bauxite for making aluminum; no raw materials are needed to make hydrogen and ammonia, except for periodic replacements of electrodes and potassium hydroxide in the electrolysis cells and catalyst in the ammonia converter); and/or metals for making the heat exchangers (mostly aluminum), turbines (aluminum and/or steel), generators (steel and copper), and platforms (aluminum, steel, and concrete).

2. The ability to provide the manpower and shipbuilding facilities required (but no appreciable requirement for expansion is expected before 1985).

3. The ability to provide trained, seagoing engineering/construction crews and service ships to erect the plants at sea.

4. The ability to attract operating crews and provide support facilities on shore for crew training and supply operations.

According to Ref. 26, the achievement of 200 gigawatts (electric) capacity by the year 2000 would correspond well with the expansion rate (lower curve slope in Fig. 12) for nuclear plants in the 1965 to 1980 time period derived from licensing, construction, and planning information. However, approval and construction of OTEC plants, which would be offshore and are expected to have no appreciable environmental effects or safety hazards, may proceed even faster, so that the upper curve slope in Fig. 12 may appear reasonable at this time.

REFERENCES

1. Claude, Georges: Power from Tropical Seas, *Mech. Eng.*, vol. 52, pp. 1039–1044, December 1930.
2. Massart, Georges Louis: The Tribulations of Trying to Harness Thermal Power, *MTS J.*, vol. 8, no. 9, pp. 18–21, Oct.–Nov. 1974.
3. Brown, C. E., and L. Wechsler: Engineering an Open Cycle Power Plant for Extracting Solar Energy from the Sea, Paper OTC 2254, Offshore Technology Conference, Houston, May 5–8, 1975.
4. Fuglister, F. C.: "Atlantic Ocean Atlas of Temperature and Salinity Profiles and Data from the International Geophysical Year of 1957–1958", Woods Hole Oceanographic Institute, Woods Hole, Mass., 1960.
5. McGowan, J. E., and W. E. Heronemus: Gulf Stream Based Ocean Thermal Power Plants, AIAA Paper 75-643, AIAA/AAS Solar Energy for Earth Conference, Los Angeles, Apr. 21–24, 1975.

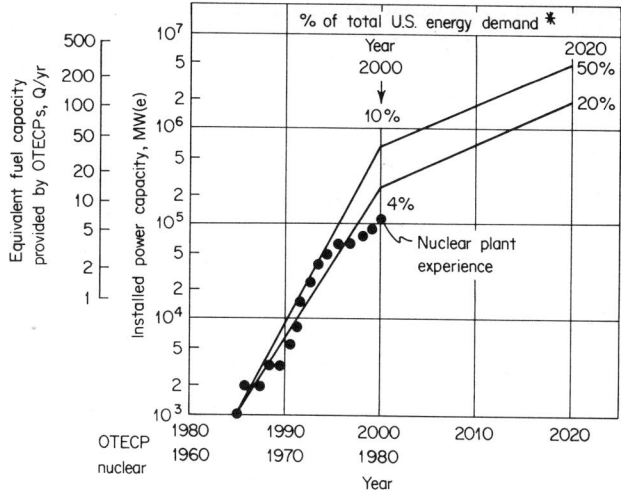

Fig. 12. Potential expansion rate for OTEC installed capacity and percentage contributions to the total United States energy demand. (Reference data for nuclear plant experience from Ref. 27.) *Based on 2% growth rate in demand, years 2000 to 2020. 20% of tropical OTECPs produce ammonia or aluminum; 80% send LH₂ to shore.

6. Dugger, G. L. (ed.): "3rd Workshop on Ocean Thermal Energy Conversion," conducted by the Applied Physics Laboratory, The Johns Hopkins University, for the Division of Solar Energy, U.S. Energy Research and Development Administration, at Houston, May 8–10, 1975:
   (a) Trimble, L. C., B. L. Messinger, and H. G. Ulbrich (Lockheed Missiles and Space Co., Sunnyvale, Calif.), and T. Y. Lin (T. Y. Lin International, San Francisco): Ocean Thermal Energy Conversion System Study Report.
   (b) Douglass, R. H. (Ocean and Energy Systems, TRW, Inc., Redondo Bench, Calif.): Ocean Thermal Energy Conversion: An Engineering Evaluation.
   (c) Surratt, W. B., G. K. Hart, and E. M. Sieder (DSS Engineers, Inc., Ft. Lauderdale, Fla.): Plastic Heat Exchangers for Ocean Thermal Energy Conversion.
   (d) Anderson, J. Hilbert, and James H. Anderson, Jr. (Sea Solar Power, Inc., York, Pa.): Cold Water Pipe: A Design Feasibility Study.
   (e) Bathen, K. T. (University of Hawaii at Manoa, Honolulu): Oceanographic and Socio-Economic Aspects of an Ocean Thermal Energy Conversion Plant in Subtropical Hawaiian Waters.
   This volume also includes reports on 28 other completed, continuing, or beginning projects supported by NSF/RANN, ERDA, the NOAA Office of Sea Grant, and the U.S. Maritime Administration.
7. Veziroglu, T. N. (ed.): The Hydrogen Economy Miami Energy (THEME) Conference, March 18–20, 1974, Miami Beach, Fla., University of Miami, Coral Gables, Fla.
8. Griffin, O. M.: The Ocean as a Renewable Source of Energy, ASME Paper 74-WA/Oct-5, ASME Winter Annual Meeting, New York, Nov. 17–22, 1974.
9. Killian, H. J., G. L. Dugger, and J. Grey (eds.): "Solar Energy for Earth, an AIAA Assessment," American Institute of Aeronautics and Astronautics, New York, Apr. 21, 1975.
10. Anderson, J. Hilbert, and James H. Anderson, Jr.: Thermal Power from Sea Water, *Mech. Eng.*, pp. 41–46, April 1966.
11. Anderson, J. H., Jr.: Economic Power and Water from Solar Energy, ASME Winter Annual Meeting, Paper 72-WA/Sol-2, November 1972.
12. McGowan, J. G., and J. W. Connell: Heat Exchanger Design for Ocean Thermal Difference Power Plants, ASME Paper 74-WA/Oct-4, ASME Winter Annual Meeting, New York, Nov. 17–22, 1974.
13. Zener, Clarence, Solar Sea Power, *Physics Today*, p. 48, January 1973.
14. Lavi, A.: Final Report, Solar Sea Power Project, Report NSF/RANN/SE/GI-39114/PR/74/6, Carnegie-Mellon University, Pittsburgh, Pa., Jan. 31, 1975.
15. Zener, C., et al.: Solar Sea Power, Third Quarterly Progress Report, Carnegie-Mellon University, Pittsburgh, Pa., Report NSF/RANN/SE/GI-39114/PR/74/3, Apr. 30, 1974.
16. Zener, Clarence: Site Limitations on Solar Sea Powerplants, AIAA Paper 75-618, AIAA/AAS Solar Energy for Earth Conference, Los Angeles, Apr. 21–24, 1975.
17. Trimble, L. C., and B. Messinger: Ocean Thermal Energy Conversion System Evaluations, AIAA Paper 75-616, AIAA/AAS Solar Energy for Earth Conference, Los Angeles, Apr. 21–24, 1975.

18. Douglass, Robert H.: Systems Aspects of Ocean Thermal Energy Conversion, AIAA Paper 75-615, AIAA/AAS Solar Energy for Earth Conference, Los Angeles, Apr. 21–24, 1975.
19. Lessard, R. D.: Technical and Economic Evaluation of Ocean Thermal Difference Powerplant Turbomachinery, Energy Program, Report NSF/RANN/SE/GI-34979/TR/73/18, University of Massachusetts, Amherst, Mass., December 1973.
20. Haderlie, E. C. (Naval Postgraduate School, Monterey, Calif.): Private communication with W. B. Shippen and E. J. Francis, APL/JHU; Mar. 4, 1975.
21. Fink, F. W., and W. K. Boyd: The Corrosion of Metals in Marine Environments, DMIC Report 245, May 1970, Defense Metals Information Center, Battelle Memorial Institute, Columbus, Ohio.
22. Verink, Ellis D., Jr.: Aluminum Alloys for Saline Waters, *Chem. Eng.*, pp. 105–110, Apr. 15, 1974.
23. Milton, R. M., and C. F. Gottzman: High Efficiency Reboilers and Condensers, *Chem. Eng. Prog.*, vol. 9, no. 9, pp. 56–61, September 1972.
24. Dugger, G. L., H. L. Olsen, W. B. Shippen, E. J. Francis, and W. H. Avery: Floating Ocean Thermal Power Plants and Potential Products, *Journal of Hydronautics*, vol. 9, no. 4, pp. 129–141, October 1975.
25. Jahnig, Charles E. (Triton Seafarms Co., Rumson, N. J., and Exxon and APL/JHU Consultant): Private communication to W. H. Avery, APL/JHU, February 1974.
26. "Solar Energy Task Force Final Report, Project Independence Blueprint," Oct. 4, 1974.
27. Export Import Bank Program Review, Senate Report S241-6, February 1974.
28. Andollent, M., and A. Hahn: International Trade in Ammonia May Rise Sharply, *Chem. Eng. News*, pp. 13–14, Aug. 15, 1974.
29. Harre, E. A., O. W. Livingston, and J. T. Shields: World Fertilizer Review and Outlook, TVA Report TA(AO)6-69, 1974, National Fertilizer Development Center, Muscle Shoals, Ala.

Photosynthesis

BY LAWRENCE F. SMALL*

Photosynthesis in the strictest sense is a set of processes whereby most plants and some bacteria convert light energy into chemical energy in the cell. In a broader sense, photosynthesis includes the subsequent utilization of the manufactured chemical energy, plus certain reduced compounds, to convert stable inorganic compounds such as carbon dioxide into the complex, energy-rich organic compounds which are the stuff of life. A basic equation can be written to explain photosynthesis in its broadest sense:

$$CO_2 + 2H_2A \xrightarrow[\text{psn pigments}]{\text{light energy}} (CH_2O) + H_2O + 2A$$

If sulfur bacteria are the photosynthetic organisms being investigated, elemental sulfur can be substituted for A above so that the equation becomes

$$CO_2 + 2H_2S \xrightarrow[\text{bacteriochlorophyll}]{\text{light energy}} (CH_2O) + H_2O + 2S$$

In green plant photosynthesis, which is by far the most prevalent type, oxygen substitutes for A to give the commonly seen expression:

$$CO_2 + 2H_2O \xrightarrow[\text{chlorophyll}]{\text{light energy}} (CH_2O) + H_2O + O_2$$

There are three main differences between bacterial and green plant photosynthesis, which can at least partly be deduced from the above equations.

1. There is no evolution of O_2 in bacterial photosynthesis, and O_2 is not required for the process. Bacterial photosynthesis thus can occur under anaerobic conditions, while green plant photosynthesis can not.

2. Bacteria will not grow photosynthetically unless they have some suitable oxidizable substance (i.e., some hydrogen-donor substrate) besides water. A great variety of compounds can serve this purpose, from simple inorganics like H_2S to organics such as simple alcohols and organic acids.

3. Bacteria have different types of photosynthetic pigments from green plants, although the pigments serve basically the same function of trapping light energy and participating in initial photochemical steps.

In green plants, including most algae, the photosynthetic apparatus is contained in rather large intracellular bodies called chloroplasts. The internal structure of chloroplasts is lamellar, and in some species the lamellae have thickened portions that resemble stacks of disks. Each stack is called a granum. Each lamellar surface in and outside the grana appears to have many smaller protrusions in which the photosynthetic pigments are located. The pigment molecules in each protrusion are thought to make up a basic photosynthetic unit. The photosynthetic unit is the smallest pigment cluster that can provide all the light-harvesting and photochemical functions to convert light energy to chemical energy.

In some algae, notably members of the *Cyanophyta* (blue-green algae), neither chloroplasts nor grana are present, but rather the lamellae exist throughout each cell. In photosynthetic bacteria, smaller pigment-containing bodies known as chromatophores exist, with bacteriochlorophyll in the photosynthetic units.

The most current formulation of the photochemistry of green-plant photosynthesis (the so-called "light" reaction) involves two photosystems. System I accepts long-wavelength light (>680 nm) and is composed of chlorophyll *a* forms and possibly some other

*Professor, School of Oceanography, Oregon State University, Corvallis, Oreg.

pigments. A single quantum of energy, absorbed anywhere in a photosynthetic unit in system I, migrates to a reaction center. The reaction center contains a specialized chlorophyll, called P700 because of its peak absorption at about 700 nm. The transfer of energy to P700 from the light-capturing pigments apparently is accomplished with high efficiency by a resonance transfer process. In photosynthetic bacteria the reaction-center chlorophyll is different, with peak absorbance at 870 or 890 nm, depending on the bacterial species. It appears that each reaction center in green plants and bacteria requires two of the proper chlorophylls in cooperation before an initial photochemical event can be carried out.

Acceptance of the energy by P700, at a potential of 0.47 volt, initiates an electron-transfer event in the reaction center. The energized electron from the reaction center is captured by a more highly reduced compound known as ferredoxin-reducing substance (FRS), at a potential of about 0.60 volt. FRS subsequently reduces the less electronegative ferredoxin, which then supplies the energy to reduce nicotinamide adenine dinucleotide diphosphate (NADP) to NADPH. Enzymes are also required at this final stage, so that conversion of NADP to NADPH is not strictly part of the photochemical reaction. Nevertheless, NADPH appears to be a universal reducing agent in green plant cells, one that can be mobilized throughout the cell to perform other reductions. As such, it can be thought of as one of the primary products of the "light" reaction of green-plant photosynthesis.

The formation of a strong reductant by system I leaves a positively charged "hole" in the reaction center. The reaction center is brought back to its ground state by an energized electron emanating from system II. A quantum of light energy below 680 nm is collected by a pigment in system II. Depending on the plant species, accessory pigments in system II can be forms of chlorophyll a, chlorophylls b or c, phycobilin pigments, or certain carotenoid pigments. The types of accessory pigments more or less determine the wavelengths most efficiently trapped. Many green plants have a complement of accessory pigments that allows the efficient collection of virtually any wavelength between about 400 and 700 nm.

The collecting pigment likely transfers the energy to a system II reaction center in a photosynthetic unit. No specific reaction center pigment, analogous to P700 in system I, has been identified, but there is reasonable evidence that system II photosynthetic units exist distinctly from system I units, and that transfer of energy to some form of chlorophyll a is accomplished. In any event, light absorbed by system II, at an approximate potential of 0.80 volt, often results in an energy-transfer event. The energized electron from system II is picked up by an electroneutral substance, Q (probably plastoquinone) if the electroneutral substance itself is in an oxidized state and therefore able to accept an electron. If the substance is already reduced, then the energized electron is obliged to "fall back" into system II and dissipate its energy as fluorescence. Once the system II electron is accepted by the electroneutral substance, it then "cascades" to the lower potential of system I through a series of cytochromes. The cytochromes thus form a link between systems I and II, being oxidized by the former and reduced by the latter.

The electron donated by system II to ultimately reduce system I to ground state leaves a positive "hole" in system II. System II is satisfied by an electron from water. By as yet unclear means, water (0.80 volt) is split to yield an electron and elemental oxygen. The oxygen is subsequently removed from the cell (its rate of evolution has been used as one index of photosynthetic rate in green plants).

The "light" reaction in green plants is thus seen to be a two-stage serial reduction from the oxidative potential of water to the high-energy potential of NADPH. The energy to "pump" the whole system against an energy gradient comes from light injected into the system at two critical stages. As part of the series involves the movement of electrons from electronegative to electropositive compounds, however, some of the energy of excitation becomes available to form high-energy phosphate bonds in a process called photophosphorylation. Thus, ATP is formed from ADP, inorganic phosphorus, and some excitation energy stripped from the electron cascade through the cytochrome chain. In addition, some ATP appears to be formed from excitation energy not required in the FRS-ferredoxin electron transfer. ATP is the universal chemical energy of the cell. The composite "light" reaction for the conversion of light energy into chemical energy, and for the development of reducing power in the form of NADPH, might be written as

$$n\,\text{NADP} + n\,\text{H}_2\text{O} + n\,\text{ADP} + n\,\text{P} \xrightarrow[\text{pigments}]{\text{light energy}} n\,\text{NADPH} + n\,\text{ATP} + n\,\text{O}_2$$

In bacterial photosynthesis, two photosystems can be used, but not together. One system is similar to system II in green plants, except that water cannot be used as a substrate, and hence oxygen is not evolved. The second system, perhaps more common, is similar to system I in green plants. In this latter system, long-wavelength light impinging on bacteriochlorophyll in a photosynthetic unit initiates an electron-transfer event involving P870 or P890. The excited electron is accepted by a strongly electronegative electron acceptor, probably FRS or ferredoxin itself. Once this strong reductant is formed, the subsequent electron transfers are exergonic but do not involve reduction of NADP. Rather, the electron "cascades downhill" through other compounds, including an electroneutral acceptor and cytochromes. The ultimate recipient of the electron is the pigment molecule from which the electron originally came. The energy of excitation is stripped from the electron as it is transferred around the circuit, and this energy is used to form ATP from ADP and inorganic phosphate. The process has been called cyclic photophosphorylation, to distinguish it from the noncyclic process undertaken by green plants.

Primary products of the "light" reaction in green-plant photosynthesis are used to reduce CO_2 to carbohydrates in a process known as the "dark" reaction of photosynthesis. The "dark" reaction is actually a series of chemical reactions in which CO_2 is incorporated into the plant through a five-carbon compound, to form a six-carbon intermediate. This intermediate is split into two three-carbon compounds, which are subsequently reduced to hexose phosphates upon input of ATP and NADPH. These energy-rich compounds are then combined with other compounds in various ways to yield different carbohydrates, or are converted back to the original five-carbon CO_2 acceptor molecule by several pathways. The "dark" reaction is slower than the "light" reaction, and thus is the rate-controlling reaction for the complete photosynthetic equation.

In some plants, notably sugarcane, sorghum, maize, and certain others, the carbon pathway is supplemented or replaced by a pathway involving immediate incorporation of CO_2 into certain organic acids through phospho-enol-pyruvate (PEP). Such a mechanism circumvents the excretion of two-carbon compounds (such as glycollate) that often occurs under high-light, high-O_2 conditions in plants not containing the alternative CO_2-incorporation scheme. Plants having the PEP pathway are known to be some of the most productive plants on earth, with respect to rapid increases in biomass. What they are able to do, in effect, is to use "light" reaction products more efficiently to fix CO_2 over a broader range of light and O_2 levels.

If the "light" reaction is combined with the basic "dark" reaction, and terms are collected, a balanced net reaction is developed which corresponds to the basic photosynthetic equation often given in textbooks:

$$2\text{H}_2\text{O} \rightarrow \text{O}_2 + 4e^- + 4\text{H}^+$$
$$2\text{NADP} + 4e^- + 2\text{H}^+ + 3\text{ADP} + 3\text{P} \rightarrow 2\text{NADPH} + 3\text{ATP}$$
$$\underline{2\text{H}^+ + 2\text{NADPH} + 3\text{ATP} + \text{CO}_2 \rightarrow 2\text{NADP} + \text{H}_2\text{O} + 3\text{ADP} + 3\text{P} + \text{CH}_2\text{O}}$$

NET: $\qquad\qquad\qquad \text{CO}_2 + \text{H}_2\text{O} \rightarrow \text{CH}_2\text{O} + \text{O}_2$

It is readily seen that each complete "light" reaction involves the net production of one electron, but that four electrons (from two water molecules) are required to reduce each CO_2 molecule. Therefore, the "light" reaction must be accomplished four times before CO_2 can be reduced to carbohydrate. Because two quanta of light are needed for each complete reaction, it follows that $4 \times 2 = 8$ quanta of light are needed, as a minimum, for every molecule of CO_2 fixed as carbohydrate (or, more correctly, eight mole quanta of light are required for each mole of CO_2 fixed). A mole quanta, or an einstein, is equivalent to 6.02×10^{23} quanta. If the "light" reaction is not 100 percent efficient, and it is not in most cases, more than eight mole quanta will be required to fix one mole of CO_2. Some experimental evidence with various plants has suggested that the average quantum requirement is nearer 10.

The approximate minimum energy required to fix one mole of CO_2 in photosynthesis can be calculated by knowing that four electrons are required, and that each electron is moved from the potential of water (0.8 volt) to the potential of NADPH and simple carbohydrate (0.4 volt), a total of 1.2 volts. A minimum of $4 \times 1.2 = 4.8$ electron volts is thus needed. Energy exchange in chemical processes is usually given in terms of kilocalories per mole. So, if we know that 23 kcal/mole = 1 electron volt, the minimal energy required to fix one mole of CO_2 can be given as $4.8 \times 23 \approx 110$ kcal/mole. The free energy of one mole of CH_2O, as sugar, is 118 kcal/mole C greater than that of a mole CO_2. Such a number is slightly greater than the minimal energy calculated to reduce CO_2 to CH_2O. Also, some energy is likely required to split water so as to produce the initial reductant and oxidant, and to keep these reactants separated in the photosynthetic unit until electron transfer is accomplished. Furthermore, photosynthetically derived energy expenditures in other than CH_2O are not accounted for in the minimal energy calculation. A quantum requirement of about 10 seems entirely feasible under these considerations.

The efficiency with which available sunlight can be used to fix CO_2 can not exceed the quantum efficiency of the photosynthetic mechanism. Presuming for the moment that 10 mole quanta of light are used, regardless of wavelength in the 400- to 675-mm range, it is easy to show that, in blue light (400 nm), the quantum efficiency is

$$\frac{118}{72 \times 10} \times 100 \approx 16\%$$

where a mole quanta of 400-nm light is equivalent to 72 kcal, and 10 mole quanta are required to produce 118 kcal/mole of CO_2 fixed as CH_2O. In red light (675 nm), with an energy equivalent of 42 kcal/mole quanta, the efficiency is

$$\frac{118}{42 \times 10} \times 100 \approx 28\%$$

An arithmetic average over all wavelengths approximates 20 percent; thus, of the light energy impinging on plant pigment systems, only about 20 percent can be utilized.

Although the average quantum requirement for many plant species over all wavelengths might be about 10, the number is not absolute and is a function of the composite absorption spectrum of all the photosynthetic pigments in a given species. If the reciprocal of the quantum requirement, called the quantum yield (the number of moles of CO_2 fixed by a mole quanta of known wavelength light), is plotted against wavelength, a reasonably uniform spectrum is derived for some plants, while for others the spectrum is nonuniform. Those plants with uniform quantum yields have accessory pigments which, as a group, can trap and transfer all wavelengths more or less equally, while the reverse is true for plants with nonuniform quantum yield spectra. If the quantum yield at each wavelength is subsequently multiplied by the actual mole quanta of light absorbed at corresponding wavelengths, action spectra (in units of moles CO_2 fixed) are developed. Action spectra show the relative light utilization efficiencies of photosynthetic cells in different wavelengths, and indicate the ability of plants to adapt to different light quality regimes (such as different depths in lakes and oceans). Even action spectra are not constant, however, with changes wrought by different nutrient and light intensity conditions.

The efficiency of utilization of the total light energy incident to the earth is of course much less than the quantum efficiency because the spacing of plants on land and sea; of leaves, fronds, and photosynthetic cells in the plants; and of chloroplasts and pigments inside the cells; and the transfer of energy from accessory pigments to reaction centers all preclude 100 percent utilization of the available sunlight. Of the 73×10^8 kcal falling on an acre of land per year in temperate latitudes, only about 0.1 to 0.5 percent is fixed as organic material. In vigorously growing croplands and certain very productive zones of the sea, the average efficiency might be on the order of 1 or 2 percent. Investigators have experimentally approached the quantum efficiency of photosynthesis in mass cultures of unicellular algae under controlled conditions, but no natural systems approach the 20 percent figure. Even though efficiency is low, the energy converted by photosynthesis annually is about 100 times greater than the heat of combustion of all the coal mined on the earth in one year, and it is about 10^4 times greater than all the energy derived by man from waterpower during one year.

REFERENCES

Bonner, J.: The Upper Limit of Crop Yield, *Science,* **137,** 11–15 (1962)

Govindjee, R.: The Absorption of Light in Photosynthesis, *Sci. American,* **231,** 68–82 (1974)

Heath, O. V. S.: "The Physiological Aspects of Photosynthesis," Stanford Univ. Press, Stanford, Calif., 1969.

Rabinowitch, E. and R. Govindjee: "Photosynthesis," Wiley, New York, 1969.

Ryther, J. H.: Potential Productivity of the Sea, *Science,* **130,** 602–608 (1959)

Whittingham, C. P.: "The Mechanism of Photosynthesis," American Elsevier, Inc., New York, 1974.

Electrical Energy from the Wind

BY E. WENDELL HEWSON*

Until comparatively recently, the assemblies of equipment used for extracting power from the wind were known as *windmills*. Additional names are now being used, including *wind turbine generators, aerogenerators,* and others.

CLASSIFICATION OF WIND MACHINES

Aerogenerators may be classified according to either capacity or configuration.

Capacity. Classification according to rated capacity tends to be arbitrary. A convenient classification is: (1) *small,* 0 to 9 kilowatts; (2) *intermediate,* 10 to 99 kilowatts; and (3) *large,* 100 to 3000 kilowatts (0.1 to 3 megawatts).

Small units are manufactured in a number of countries to supply some of the electric energy requirements of farms and homes. These small aerogenerators are being used increasingly not only in isolated areas where regular electric utility power is absent, but also in conjunction with utility sources to reduce dependence upon fossil and nuclear fueled plants. Many of these small units are described in the publication, "Spectrum," (1975), listed among the references at the end of this article. There has been relatively little activity in the design and construction of aerogenerators in the intermediate range of sizes. The remainder of this article is devoted mainly to a discussion of large aerogenerators because these provide the greatest potential for making a significant contribution to the future electric power needs of the United States and of the world.

Configuration. Aerogenerators may be classified according to the configuration of the various components making up the system. For example, aerogenerators may be categorized as being either *horizontal axis* or *vertical axis* types. Both of these types are described later.

THEORY OF WIND POWER

The most significant fact arising from the analysis of the earth's winds as a source of electric energy is that *the power produced is proportional to the cube of the wind speed.* This result is readily derived from the kinetic energy of wind. The mass of air m with speed V and density ρ flowing per unit time through area A swept by the blades of a conventional horizontal axis wind turbine is ρAV. Thus, the kinetic energy of this mass of air is given by

$$\text{Kinetic energy} = \tfrac{1}{2}mV_2 = \tfrac{1}{2}\,\rho\,AV^3 \qquad (1)$$

As shown from the theoretical analysis by Betz (1919), the maximum fraction of this kinetic energy that can be extracted from the wind is 16/27 or 0.593, so that the theoretical maximum energy output of a wind turbine is given by

$$\text{Theoretical maximum power output} = 0.297\rho AV^3 \qquad (2)$$

The process of energy conversion leads to power reductions which vary with the type of wind turbine and aerogenerator and which are roughly one-third of the theoretical maximum output. Hence, the power output actually available is

$$\text{Available power output} = (\tfrac{2}{3})0.297\rho AV^3 \approx 0.2\rho AV^3 \qquad (3)$$

If the diameter of the blades of the rotor system is D, then Eq. (3) becomes

$$\text{Available power output} \approx 0.05\pi\rho D^2 V^3 \qquad (4)$$

*Professor and Chairman, Department of Atmospheric Sciences, Oregon State University, Corvallis, Oreg.

Thus, power available for a given wind speed is *proportional to the square of the rotor diameter*.

Additional quantities need to be defined, such as the power coefficient C_p, and the overall power coefficient C_{op}, as given below:

$$C_p = \frac{\text{power output of wind turbine}}{\frac{1}{2}\rho A V^3} \tag{5}$$

$$C_{op} = \frac{\text{power output at generator}}{\frac{1}{2}\rho A V^3} \tag{6}$$

The coefficient C_{op} thus includes the inefficiencies of the transmission and generator.

The basic relations required in the analysis of the winds as a source of energy are thus given in Eqs. (1) through (4). Further aspects of the theory are developed by Wilson and Lissaman (1974). See references.

WIND SYSTEM DESIGN

Economical design requires a systems analysis of the various components of an aerogenerator, including the rotor, transmission, controls, generator, and tower in relation to the wind characteristics at the site. Some of the features of these various components are described in the following paragraphs.

Rotors. In general, rotors are designed for either horizontal axis or vertical axis operation. Horizontal axis rotors have been chosen for most wind machines.

Horizontal Axis Rotors. Such rotors come in many designs. Three well-known types and their operating characteristics are presented in Fig. 1. The modern propeller when used in a wind turbine utilizes variable pitch in order to regulate rate of shaft rotation to achieve maximum efficiency. Other types are the flexible sail-wing rotor (Sweeney and Nixon, 1973) and one consisting of a circular array of 48 thin and narrow blades stressed in tension by a surrounding circular hoop which serves as a rim drive to rotate at high speed and without step-up gearing the shaft of a generator (American Wind Turbine Co., 1974). Both these designs have been constructed in relatively small units only. The various atmospheric and other factors which cause dynamic loading of the rotors are suggested in Fig. 2. It should be noted that the blades are attached to the hub in a flexible manner which permits them to move to and fro in the direction of the wind in a vibration mode known as *flapping*. As a group, the blades of the horizontal axis rotors are driven by lifting action. As a result, the speed ratio (the ratio of blade tip speed to wind speed) is high, reaching a value as great as 12. The efficiency of a wind turbine at these large speed ratios is high. The structural aspects of horizontal axis rotors have been analyzed by Thresher (1974).

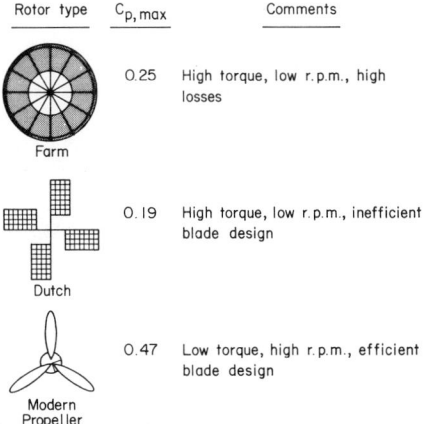

Rotor type	$C_{p,\text{max}}$	Comments
Farm	0.25	High torque, low r.p.m., high losses
Dutch	0.19	High torque, low r.p.m., inefficient blade design
Modern Propeller	0.47	Low torque, high r.p.m., efficient blade design

Fig. 1. Power extractable from three types of horizontal axis rotors as presented by Savino (1974).

Vertical Axis Rotors.   The best known vertical axis rotor is made up of two identical hemicylinders with their axes vertical. The patterns of air flow around and air pressure on two such hemicylinders are shown in Fig. 3. As described by Klemin (1925), when the configuration is as in Fig. 3*a*, there is a reduced pressure behind hemicylinder *a* as it moves against the wind, reducing the torque. If, however, there is an air passage between the hemicylinders, as in Fig. 3*b*, the air pressure behind hemicylinder *a* is increased instead of being reduced, with the result that the torque is about three times greater than without the air passage. This vertical axis rotor was developed many years ago by the Finnish engineer Savonius (1931), and is being used increasingly for small wind power

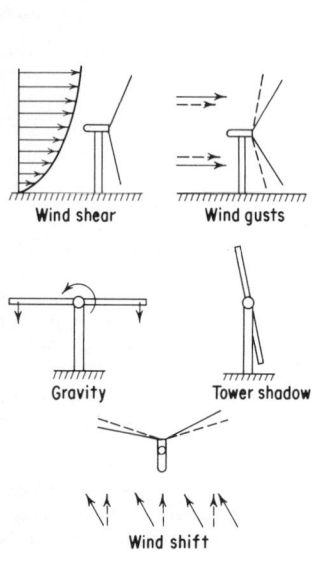

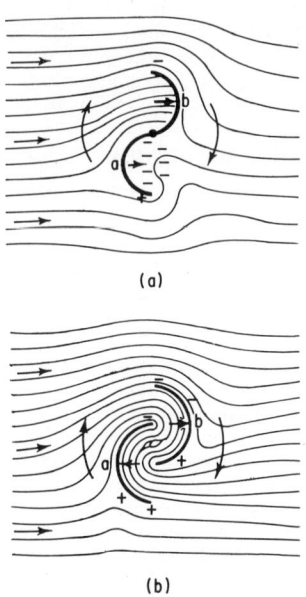

(a)

(b)

Fig. 2. Causes of cyclic rotor system loads. (Kaman Aerospace Corporation.)

Fig. 3. Principles of operation of hemispherical vertical axis rotors according to Klemin (1925): *(a)* Low-efficiency configuration. *(b)* High-efficiency configuration of Savonius (1931).

installations. Because of its vertical symmetry, the rotor does not need to be kept headed into the wind, which makes it a very simple and inexpensive rotor. On the other hand, the hemicylinders are dragged by the wind instead of being lifted, as in the horizontal axis rotors; so the speed ratio and hence the efficiency of the Savonius rotor is low.

Another type of vertical axis rotor was devised by the French engineer, G. J. M. Darrieus (1931). A modern version, developed by South and Rangi (1971, 1972), is shown in Fig. 4. The flexible metal strips take the form of a tropaskien* and act in a lifting mode as they rotate so that, for a given wind speed, the unit rotates more rapidly and is more efficient than the Savonius rotor. Unfortunately, the Darrieus rotor is not self-starting, even in high winds. These vertical axis types of aerogenerator merit further study, as recommended by Vance (1973).

Although the foregoing figures show single rotor systems only, groups of rotors will doubtless be required for outputs greater than several megawatts.

The performance characteristics of various rotor configurations, according to Wilson and Lissaman (1974), are shown in Fig. 5. This analysis suggests that the two-blade type is the most efficient form now available. Studies by General Electric Company (1975) and Kaman Aerospace Corporation (1975) reach the same conclusion. The remaining material

*Greek for "turning rope."

in this section draws heavily on the aforementioned General Electric Company document, unless otherwise indicated.

Transmissions.   The rate of rotation of large wind turbine generators operating at rated capacity or below is conveniently controlled by varying the pitch of the rotor blades, but it is low, about 40 to 50 revolutions per minute (rpm). Because optimum generator output requires much greater rates of rotation, such as 1800 rpm, it is necessary to increase greatly the low rotor rate of turning. Among the transmission options are mechanical systems involving fixed ratio gears, belts, and chains, singly or in combination, or hydraulic systems involving fluid pumps and motors. Fixed ratio gears are recommended for top-mounted equipment because of their high efficiency, known cost, and minimum system risk. For bottom-mounted equipment which requires a right-angle drive, transmission costs might be reduced substantially by using large-diameter bearings with ring gears mounted on the hub to serve as a transmission to increase rotor speed to generator speed. Such a combination offers a high degree of design flexibility as well as large potential savings.

Generators.   Either constant or variable speed generators are a possibility, but variable speed units are expensive and/or unproved. Among the constant speed generator candidates for use are synchronous, induction, and permanent magnet types. The generator of choice is the synchronous unit for large aerogenerator systems because it is very versatile and has an extensive data base. Other electrical components and systems are, however, under development as shown by the work of Reitan (1973), Allison (1973), and Hughes (1974).

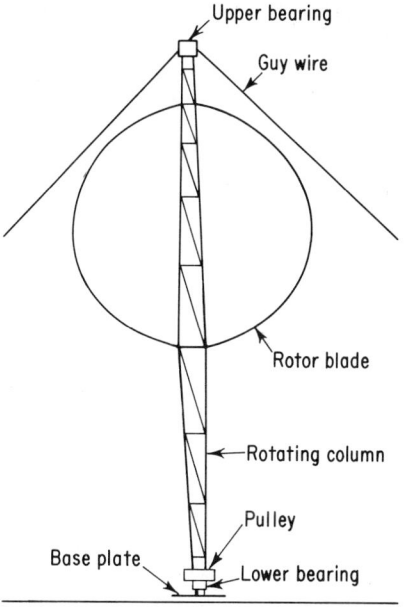

Fig. 4. The vertical axis rotor of G. J. M. Darrieus (1931) as presented by South and Rangi (1971, 1972).

Controls.   The modern large wind turbine generator requires a versatile and reliable control system to perform the following functions: (1) the orientation of the rotor into the wind (azimuth of yaw); (2) startup and cut-in of the equipment; (3) power control of the rotor by varying the pitch of the blades; (4) generator output monitoring—status, data computation, and storage; (5) shutdown and cutout owing to malfunction or very high winds; (6) protection for the generator, the utility accepting the power, and the prime mover; (7) auxiliary and/or emergency power; and (8) maintenance mode.

Many combinations are possible in terms of the control system and may involve the following components: (1) sensors—mechanical, electrical, or pneumatic; (2) decision elements—relays, logic modules, analog circuits, a microprocessor, a fluidics unit, or a mechanical unit; and (3) actuators—hydraulic, electric, or pneumatic. A recommended control system of low cost, high reliability, and flexibility and versatility is provided by the combination of electronic transducers feeding into a microprocessor which, in turn, signals electric actuators and provides protection through electronic circuits, although a pneumatic slip clutch may be required.

Towers.   Four types of supporting towers deserve consideration. These are (1) the reinforced concrete tower, (2) the pole tower, (3) the built-up shell-tube tower, and (4) the truss tower. Among these, the truss tower is favored because it is proved and widely adaptable, cost is low, parts are readily available, it is readily transported, and it is potentially stiff. Shell-tube towers also have attractive features and may prove to be competitive with truss towers. Thresher (1974) has prepared a detailed analysis of towers in relation to wind power generation.

Systems. For large aerogenerators, an attractive system configuration is the two-bladed horizontal axis propeller with variable pitch for power regulation along with a fixed ratio gearbox, microprocessor, and synchronous generator grouped aloft and rotating in the horizontal as the rotor is kept headed into the wind by a servo system. Figure 6a shows such a system in schematic form.

On the other hand, having the major components of the power train near the ground has a number of obvious advantages. Such a configuration of components is sketched in Fig. 6b. A ring gear mounted on the rotor hub as part of a right-angled power train system to increase rpm to the transmission might well serve to greatly reduce costs of transmission. Such bottom mounting of equipment has two additional attractive features: (1) the potential for reducing the natural frequency of the tower, and (2) the potential for reducing maintenance costs. ·

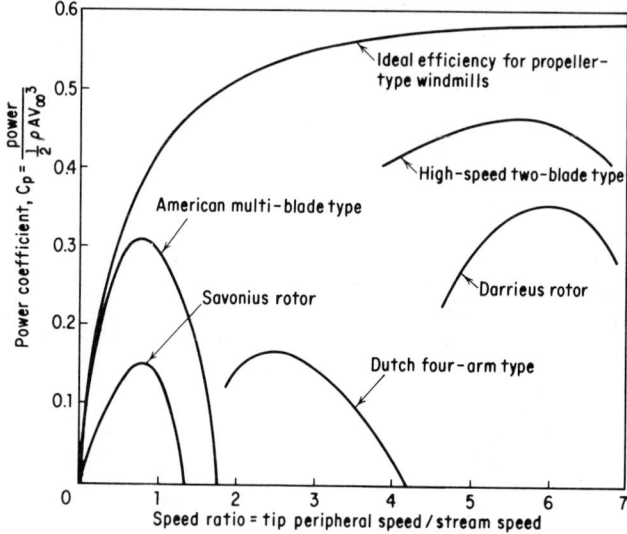

Fig. 5. Typical performance of wind power machines.

In conclusion, it should be emphasized that power outputs greater than several megawatts will require groups of wind turbine generator units, arrayed either vertically or horizontally as circumstances dictate.

THE ATMOSPHERE AS A HEAT ENGINE

The atmosphere is a great heat engine fueled by the sun. As in all such engines, only a fraction of the energy supplied can be transformed into work. The remainder must be rejected from the system, in this case by the loss of heat by long-wave radiation to space from the earth and its atmosphere. The winds and ocean currents represent the work done by the engine and may be thought of as its flywheel. The aerogenerator converts the motion of translation of the winds of this giant heat engine into a rotation capable of driving an alternator. Since the earth-atmosphere system itself disposes of the rejected heat in this great heat engine, wind power systems do not have to contend with the waste heat which presents such a problem in power plants, such as the fossil- and nuclear-fueled ones.

The power in the wind is vast. It has been estimated by Putnam (1948), based upon Brunt's (1934) calculation, that the total energy of the atmosphere is about 10^{14} megawatts. If it is assumed that one ten-millionth of this amount is available for people to use (World Meteorological Organization, 1954), then there are 10^7 megawatts of usable power in the winds. This is equivalent to the output of ten thousand typical fossil-fueled or nuclear power plants of 1000 megawatts each. Von Arx (1974) points out, however, that Putnam's value is that of a stored resource which can be drained at no more than its replenishment rate not to exceed the solar flux at the surface, 10^{10} megawatts, because the winds are driven mainly by heating from below. Von Arx's estimate of available wind power, 10^6 megawatts, is less than the previous figure by a factor of ten. Von Arx also finds the available wind power to be ten times greater than the available hydroelectric power. By way of contrast, it has been estimated that the water power potential of the whole earth is equivalent to the output of 500 such power plants (Putnam, 1948). On this basis, there is twenty times more potential for wind power than for water power.

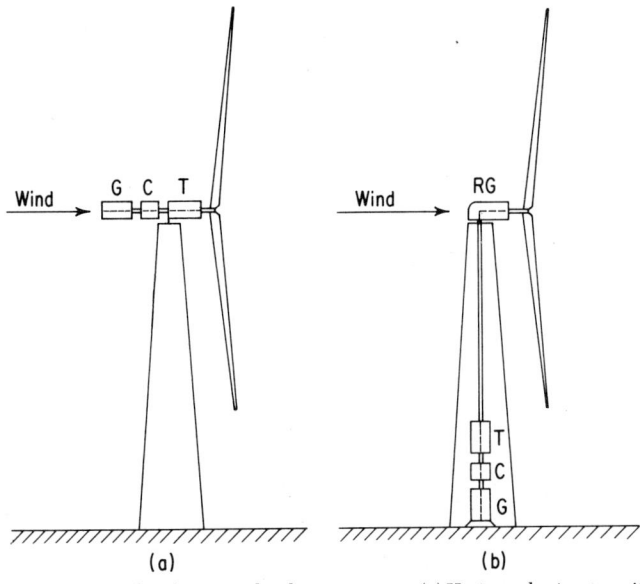

Fig. 6. Component systems for a large wind turbine generator: (*a*) Horizontal axis rotor with associated components aloft (at top of tower). (*b*) Horizontal axis rotor with associated components located below (near ground level). T = transmission; C = controls; G = generator; RG = ring gear.

An estimate of the annual average wind power available over the continental United States and coastal waters, according to Reed (1974), is presented in Fig. 7. The estimates for the mountainous western half of the country and for the Appalachians are subject to considerable uncertainty. This is true because the values given are based on measurements of standard weather stations, usually at or near airports, the locations of which are often chosen so as to avoid high winds dangerous to aircraft operation. For example, the map shows the Columbia River Gorge area lying east of Portland, Oregon, as having only 100 watts per square meter of wind power, although actual wind measurements in the Gorge show clearly a power level several times greater than this. Shoreline winds along the Oregon coast are also stronger than indicated. On the other hand, the values given for relatively level terrain are doubtless reliable. Figure 7 shows three areas of high winds: (1) along the coast and over the offshore waters of the Pacific Northwest, (2) over portions of the Great Plains, and (3) over the coastal waters off the mid-Atlantic and New England states.

HISTORY OF WIND POWER

The first windmills of which there is an historical record were developed in Persia and China many centuries B.C. According to Hodges (1975), the first mention of a windmill in western Europe comes from Yorkshire in northern England where, in 1185 A.D., a tenant at Weedle agreed to pay an annual rent of 8 shillings for the privilege of erecting and using a windmill.

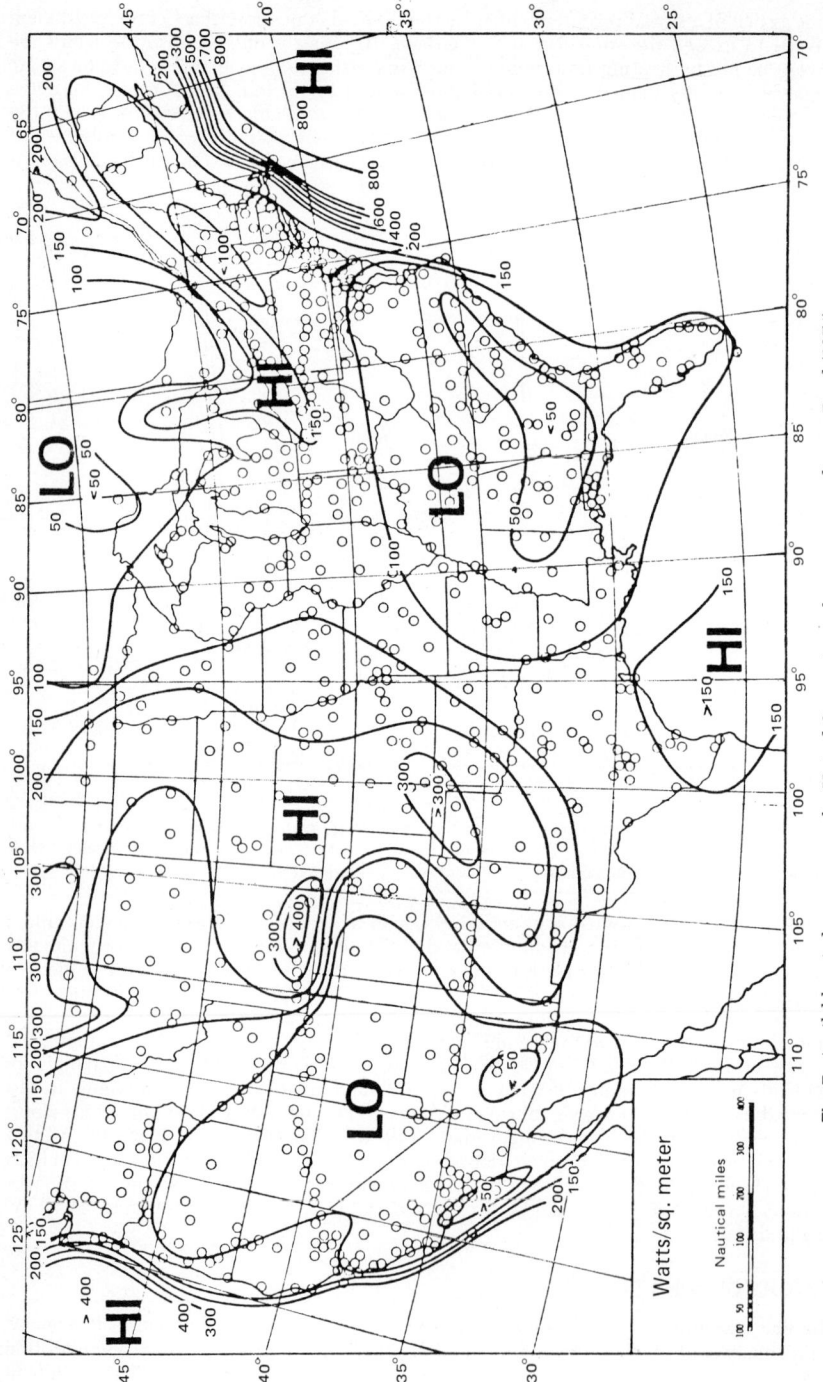

Fig. 7. Available wind power over the United States. Annual average according to Reed (1974).

Many large wind turbines with power ratings from 0.1 to 1.25 megawatts have operated successfully in Denmark, England, France, Germany, Russia, and the United States, as described by Golding (1955), Hütter (1973), and Putnam (1948). A number of these units are described briefly in the following paragraphs.

France.   An aerogenerator designed by Darrieus was built at Bourget in 1929. The tower was 20 meters (66 feet) high, and the blades 20 meters (66 feet) in diameter. The

Fig. 8. Aerogenerator of Russian design built at Yalta near the Black Sea in 1931. *(From "Power from the Wind" by P. C. Putnam, Van Nostrand Reinhold Co., New York, 1948, with permission.)*

unit was a direct-current generator rated at 0.015 megawatt at 6 meters/second (13.4 mph). A much larger unit rated at 0.8 megawatt at very high winds was built later in France and operated successfully from 1958 to 1961, according to Noel (1973).

Russia.   In 1931, at Yalta near the Black Sea, a unit rated at 0.1 megawatt was erected with a tower 23 meters (75 feet) high and blades 30.5 meters (100 feet) in diameter, as shown in Fig. 8. This was an alternating-current unit which reached its rated output at 11 meters/second (24.6 mph). The thrust of the wind is taken by the diagonal member, whose base moves on a circular rail system as the aerogenerator rotates about the central tower to head into the wind. The phrase "bold and practical" has been applied to this aerogenerator.

Germany.   In 1920, a wind turbine designed by Kumme and consisting of six blades was built in Germany. It called for a generator on the ground and a long vertical flexible shaft topped by a bevel gear to transmit torque to the ground. It was found, however, that with this arrangement, it was more expensive to have the generator on the ground than aloft.

In 1933, Honnef of Berlin proposed a design for five wind turbines, each nearly 40 meters (250 feet) in diameter supported on a single large tower 305 meters (1000 feet) or more in height. This unit, rated by the inventor at 50 megawatts, had a number of novel features, but was never built.

More recently, relatively large aerogenerators, up to 0.1 megawatt using fiberglass blades fabricated in a novel way, have been designed by U. Hütter of the University of Stuttgart (1973) and constructed in the early 1960s. Units of Hütter's design have been built and operated and have proved to possess good operating characteristics and resistance to fatigue.

United States.   The largest wind turbine ever built, an alternating-current generator

Fig. 9. The Smith-Putnam aerogenerator built near Rutland, Vermont, during World War II. *(From "Power from the Wind" by P. C. Putnam, Van Nostrand Reinhold Co., New York, 1948, with permission.)*

rated at 1.25 megawatts at 13.4 meters/second (30 mph) was constructed during World War II near Rutland, Vermont (Putnam, 1948), as shown in Fig. 9. This unit consisted of a two-bladed rotor 53 meters (175 feet) in diameter mounted on a 33-meter (110-foot) tower. The generator was located upwind from the tower and the blades downwind from it, the whole unit being kept headed into the wind by a wind vane aloft, which actuated suitable servomechanisms. The latter, in turn, controlled a yaw motor which rotated the assembly.

Tests were run at intervals between 1941 and 1945, and in March of 1945 the aerogenerator commenced continuous operation as a routine generating station on the lines of the Central Vermont Public Service Corporation. Operation was satisfactory, and initially there was no trouble, but at the end of 23 days, a defect caused by wartime shortages resulted in the loss of a blade. By then, wartime shortages were so severe that it would have taken four years to obtain a replacement blade and, because of economic considerations, it was decided to abandon the undertaking.

In addition to Putnam and his associates, P. H. Thomas (1945) of the Federal Power Commission was active for many years in developing numerous concepts of wind power generation in relation to the needs of the utilities. The validity of many of Thomas' ideas is becoming increasingly clear.

Great Britain.   In the mid-1950s, Enfield Cables Ltd., using a pneumatic transmission system devised by the French engineer J. Andreau, built an aerogenerator rated at 0.1

megawatt at 13.4 meters/second (30 mph) with blades 24 meters (80 feet) in diameter on a 30-meter (100-foot) tower, as illustrated in Fig. 10. This is described in detail by Golding (1955).

The hollow blades of the aerogenerator are open at their tips. These hollow blades are connected to an air column extending to the base of the installation. The rotating blades

Fig. 10. The Enfield-Andreau aerogenerator built in the early 1950s in Great Britain. *(From "Generation of Electricity by Wind Power" by E. W. Golding, Philosophical Library, New York, 1955.)*

act as a centrifugal air pump as they eject air from their open tips by centrifugal force. This causes reduced pressure within the blades which draws air into the machine through the circular inlet at the base, indicated in Fig. 10, up through an air turbine just above, then up the narrower vertical column, and finally out the tips of the whirling blades. Below the air inlet on the base are an alternator and a control gear.

There are several advantages to a pneumatic transmission of this type, but it is somewhat less efficient than more conventional devices. The Enfield-Andreau unit operated successfully for a number of years in Algeria.

Denmark. During World War II, Denmark constructed some 18 aerogenerators in the 70- to 90-kilowatt range, which supplied direct-current power to the dc power grids of the country. In 1957, an aerogenerator designed to produce 0.2 megawatt at a wind speed of 15 meters/second (34 mph) was built at Gedser. A sketch of this machine is shown in Fig. 11. The tower is 26 meters (85 feet) high, and the three blades sweep out an area 24 meters (79 feet) in diameter. The electric generator located on a horizontally rotating platform at the top of the tower produced 380 volts ac power. This system cost $41,000 to build, or $205 per rated kilowatt. Although shut down in 1968, this unit is the largest system still standing. A summary of the outlook for wind power in Denmark through the year 2050 is given by Sørensen (1975).

RESEARCH AND DEVELOPMENT IN THE UNITED STATES

In recent years, there has been very little research and development on wind power in the United States. In 1971, wind power research commenced at Oregon State University with funding by the four Oregon P.U.D.s (public utility districts), and has continued since that time under the same sponsorship (Hewson; *et al.*, 1973*a*, 1975; and Wilson, 1973). At the time this project started, it was the only externally sponsored wind power research undertaking in the United States. Analyses of several aspects of wind power development on a limited basis were also being conducted at Princeton University, the University of Massachusetts, and Oklahoma State University.

When it became clear that the country's fossil fuel resources and reserves were inadequate to meet both short- and long-term demands, the federal government began to take an increasingly active role in wind energy conversion research and development, at first through the Research Applied to National Needs (RANN) program of the National

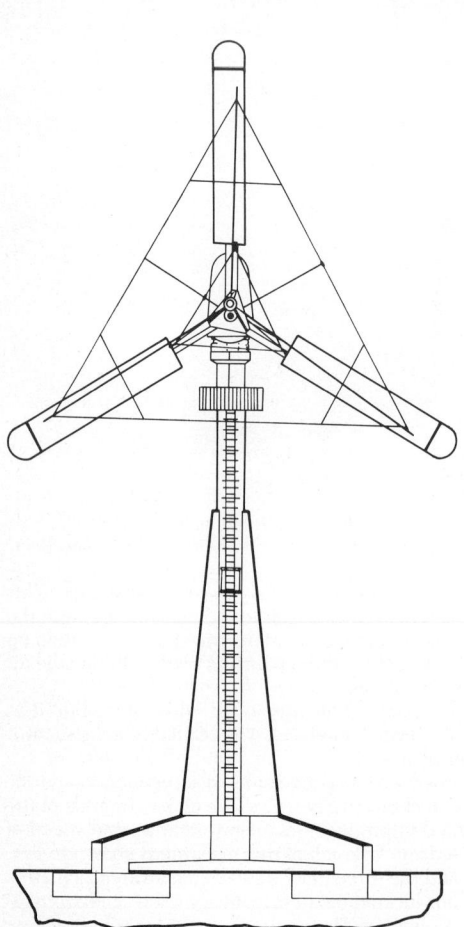

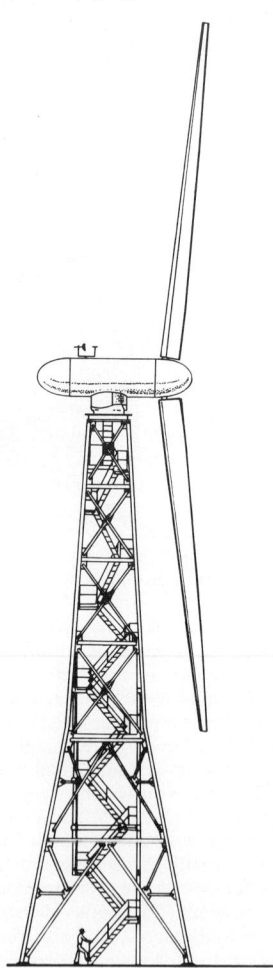

Fig. 11. Schematic representation of the 0.2-megawatt aerogenerator built in 1957 at Gedser, Denmark.

Fig. 12. Schematic representation of the 0.1-megawatt experimental wind turbine near Sandusky, Ohio.

Science Foundation (NSF), working in close cooperation with the National Aeronautics and Space Administration (NASA) through its Lewis Research Center (LeRC) at Cleveland, Ohio. With the establishment of the Energy Research and Development Administration (ERDA), this organization took over the role of NSF in the federal government's program for the development of the winds as a source of energy.

The federal government's plans have been described in broad outline by Divone (1974). The main program objectives are to advance the technology of wind energy conversion systems, and to become cost competitive and to lay the groundwork for rapid commercial expansion for producing significant amounts of energy. Some further details are described briefly here.

An important phase of ERDA's comprehensive program is the NASA/LeRC wind project as described by R. L. Thomas (1974). In 1974, NASA as part of NSF's Wind Energy Program commenced construction of an experimental wind turbine generator rated at 0.1 megawatt at a speed of 8 meters/second (18 mph). The cost of the two-year project was $865,000. A sketch of this installation is shown in Fig. 12. The tower is 30 meters (100 feet) high, and the rotor is 38 meters (125 feet) in diameter. The system has been erected and

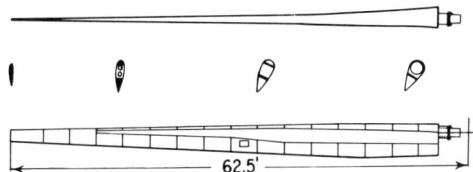

Fig. 13. Design details of the metal rotor blade of the 0.1-megawatt experimental aerogenerator at Sandusky, Ohio. (*Lockheed-California Company, Burbank, California.*) Airfoil: NACA 23000 series; total twist, 25.5 degrees (nonlinear). Size, 1.2 meters (4 feet) chord at root end, 0.46 meter (1.5 feet) at tip. Weight: approximately 900 kilograms (2000 pounds). Design features: 4130 steel roof end fitting to mate with hub; D spar main load carrying member (2024-T_4 aluminum); main ribs spaced 1.12 meters (44 inches) apart (aluminum); access door (for instrumentation and inspection).

tested at the NASA–Lewis Plum Brook test area at Sandusky, Ohio. It is expected to generate 180,000 kilowatthours per year in the form of 460-volt, 3-phase, 60-cycle ac output. Design details of the metal blades forming the rotor are shown in Fig. 13, and of the drive train assembly in Fig. 14. The associated equipment at the Plum Brook test area includes a meteorological tower 61 meters (200 feet) high, with instrumentation at 9.1 meters (30 feet), 27.4 meters (90 feet), 45.7 meters (150 feet), and 59.4 meters (195 feet), to characterize accurately the wind regime at that locality. Such a study of the wind characteristics at a site is important, as shown by Savino (1974). NASA's plans also called for substantial involvement by the power utilities in the development, according to Ragsdale (1974).

WIND POWER SURVEYS

Since the power generated by the wind is proportional to the cube of the wind speed, it is important to locate areas of persistent high wind. Thus, for example, a 10-meter/second (22 mph) wind will produce eight times as much power as a 5-meter/second (11 mph) wind will produce. Surveys to locate suitable wind power sites have been conducted in the United States, France, and Great Britain.

As part of the program for the construction of the 1.25-megawatt aerogenerator in Vermont previously described, between 1940 and 1945 twenty sites in the Green Mountains of Vermont and nearby areas were selected and instrumented with wind-measuring equipment, according to Putnam (1948). Some of the sites were studied for no more than six months, but this period turned out to be too short. It is clear from the Vermont experience that, in rough terrain especially, the wind measurements must be taken with great care.

The misreading of the wind power potential of the selected site was the key reason for the failure, economically, of the system. This fact emphasizes the necessity for a detailed meteorological study of potential sites, and especially for those that show promise of being first-class sites, including a comparison of short-term wind records with long-term climatological wind records at a nearby station in order to ascertain whether or not the short-time winds are representative. As pointed out by Hewson et al. (1973c), the minimum requirement for the meteorological evaluation of any site is a program of wind measurements at a height of 9.1 meters (30 feet); and at a site being given serious consideration, even a minimum program calls for wind measurements at 9.1 meters (30

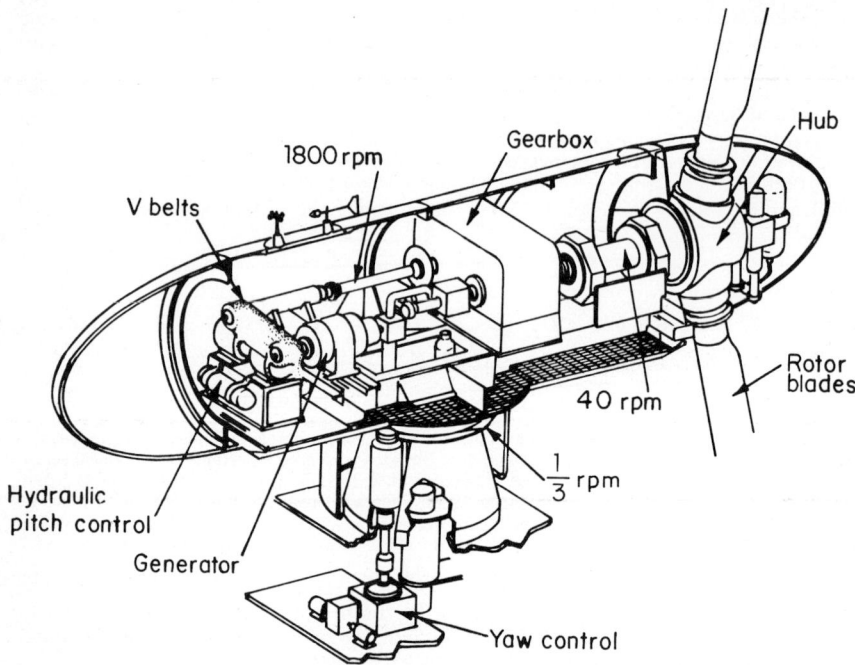

Fig. 14. Design details of the drive train assembly of the 0.1-megawatt experimental aerogenerator at Sandusky, Ohio.

feet) and at 30 meters (100 feet). In all but flat terrain, the problems of locating the best wind power sites are formidable, as the discussion in the next sections will show. There is thus a need for substantial numbers of meteorologists skilled in the methods of locating good wind-power sites, or in "wind prospecting" as this activity is sometimes described.

At the actual site chosen in Vermont, a 56-meter (185-foot) steel tower, named "The Christmas Tree" because of its branching arms and shown in Fig. 15, was fitted with anemometers at various levels. This tower with its horizontal members permitted rather complete measurements of the wind field to which the big aerogenerator would be exposed.

In France, commencing in 1946, some 150 instruments designed to give directly in kilowatthours the cumulative energy in the wind were installed on sites located in all parts of the country. The program was under the direction of a Committee on Wind Energy and was assisted in certain studies of the variation of wind speed with height by the Research Department of Electricité de France (Golding, 1955).

The Electrical Research Association in Great Britain about the same time commenced a large wind survey of various sites (Golding, 1955). Most of the better sites were located on the windier western coasts of Wales, Scotland, and Ireland. Certain similarities between the coastal terrain of western Wales and Scotland and that of western Oregon led the writer to consider the possibility that the wind might prove to be an economical supplementary source of electric power for the Pacific Northwest.

A wind power survey has been conducted in Oregon and adjacent areas since 1971 by

Oregon State University with the sponsorship of the four Oregon P.U.D.s (Hewson, 1974, 1975; Hewson *et al.*, 1973*a*, 1973*b*, and 1975). The thrust of the research has been the study of various possible wind power sites, especially those at and near the Oregon coastline and in the Columbia River Valley. These studies have involved the detailed analysis of existing wind records and the establishment of new wind-measuring stations. At the same

time, the wind tunnel at Oregon State University has been enlarged and improved to permit model studies of air flow patterns around terrain features. If a comparison of model and actual air flow patterns over and near pronounced terrain features show satisfactory agreement, then the location of desirable wind power sites will be greatly facilitated and expedited. Other projected uses of the wind tunnel will be described later in this article.

VELOCITY AND POWER DURATION CURVES

According to Golding and Stodhart (1949, 1952), the single most valuable piece of information that can be obtained about a potential wind power site is its power duration curve, which is conveniently obtained from its velocity duration curve, illustrated in Fig. 16. The horizontal axis gives the number of hours in the year, the total being 8760; and the vertical axis is wind velocity. Thus, the graph gives the number of hours in the year during which the indicated wind velocity was exceeded. A power duration curve is similar, because power is proportional to the cube of the wind speed. Such a power duration curve is shown in Fig. 17. It gives the number of hours per year that the power output exceeds the indicated values that are obtained, in part, by cubing the wind speed. This curve is for a site with relatively little time with low wind during

Fig. 15. The 56-meter (185-foot) meteorological tower and instruments at the site of the Smith-Putman aerogenerator near Rutland, Vermont. *(From "Power from the Wind" by P. C. Putnam, Van Nostrand Reinhold Co., New York, 1948, with permission.)*

the year. In the interval *ge*, the wind is too light to produce a significant amount of power. At the wind speed corresponding to *g*, say 3 meters/second (7 mph), appreciable power is being generated. The point *fg* is called the "cut-in point."

With higher wind speeds, the power output is greater, and at *c*, the aerogenerator is operating at its rated capacity. At greater wind speeds, the output is generally held constant at this value (for component cost optimization reasons with conventional components) for full load operation by adjusting the pitch of the blades, or by some other appropriate method. At some much higher wind speed *b*, called the "furling point," perhaps about 27 meters/second (60 mph), it is advisable to shut down the plant to avoid damage.

In the diagram, the hatched area *bcfgh* under the power duration curve represents

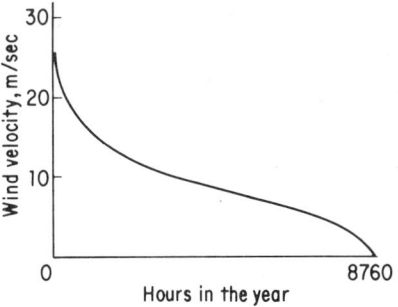

Fig. 16. A velocity duration curve for a good wind power site.

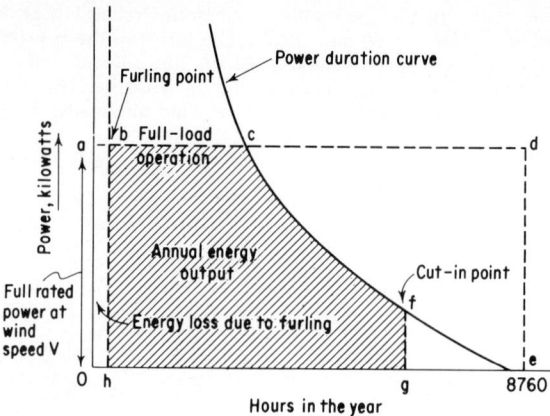

Fig. 17. A power duration curve for a good wind power site.

the actual annual output of energy to the same scale as the rectangle *adeo* represents the annual output if the plant were running at full rated power throughout the entire year. The ratio of area *bcfgh* to area *adeo* is the annual plant load factor, and multiplication of this by 8760 gives the specific output in kilowatthours per year per kilowatt. Thus, the specific output is the equivalent number of hours of full load operation.

CHARACTERISTICS OF SUITABLE WIND POWER SITES

The various characteristics of good wind power sites have been described in detail by Davidson et al. (1964) and by others. Mountainous terrain offers some attractive sites. The air flow over such terrain has been discussed by Frenkiel (1962, 1963). He defines the best hilltop terrain, provided winds are strong, as those having minimum values of the wind speed gradient in the vertical between the heights of 10 meters (33 feet) and 40 meters (131 feet). The variation of wind speed with height is conveniently expressed by the relation

$$\frac{V_{10}}{V_{40}} = \left(\frac{z_{10}}{z_{40}}\right)^{p} \tag{7}$$

where over level terrain the value of p is close to 0.14 with strong winds. It is desirable that any hilltop site should have a value of p less than 0.14, with the better ones having values less than 0.10, or even less than 0.07. This means that the hilltop area should be small with slopes both up and down wind between 1 in 6 and 1 in 3½. The low vertical wind gradient is recommended to minimize stress on the rotor blades, but wind speed and gustiness are also significant and may be the governing factors. A minimum vertical gradient of wind speed may not be so important for modern aerogenerators. Frenkiel also recommends an isolated peak in a valley whose axis lies in the direction of the prevailing wind.

The characteristics of good sites have also been described by Savino (1974), and may be summarized as follows:

1. A site should have a high annual wind speed.

2. There should be no tall obstructions for a mile or two upwind.

3. The top of a smooth, well rounded hill with gentle slopes lying on a flat plain or located on an island in a lake or sea is a good site.

4. An open plain or an open shoreline may be a good location.

5. A mountain gap which produces wind funneling is good.

Some of these ideas are incorporated in Fig. 18. Savino also lists additional aids in site selection in the form of ecological signs of persistent winds that may affect vegetation. Such ecological signs of a good wind power site include:

1. Flagging of trees, in which the branches stream downwind.

2. Throwing of trees, in which the main trunk is permanently bent downwind.

3. Wind clipping, which causes abnormally low trees with tops of a uniform height.

4. Tree or bush carpets, in which vegetation never grows taller than a low scrubby bush.

Mountainous coastal terrain provides some of the very best wind power sites, as illustrated in Fig. 19.

Site characteristics also have been discussed by Putnam (1948) and Golding (1955).

WIND PROSPECTING

A number of possible sites in the Pacific Northwest, with particular emphasis on coastal areas and the Columbia River Valley, have been evaluated by Hewson et al. (1973a and b). This evaluation is based on an analysis of past wind records from many sources and of wind records from stations established in the course of the research program, combined with wind tunnel studies.

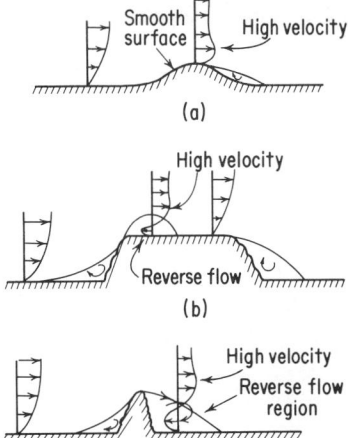

Fig. 18. A portrayal of the characteristics of wind power sites, according to Savino (1974): (a) A suitable site—well rounded hill or ridge with small top. (b) An unsuitable site—hill or ridge with abrupt sides and a large flat top. (c) A possibly suitable site—a sharp peak.

Evaluation of Wind Records. A substantial number of wind-measuring stations are or have been maintained along the Oregon coast. The locations of these and of the wind stations established, especially by the Oregon project, are shown in Fig. 20. A number of the new stations could not be located in the best possible positions because of the lack of adequate access and the difficulty in servicing the instruments when heavy snowfalls occurred in the Coast Range.

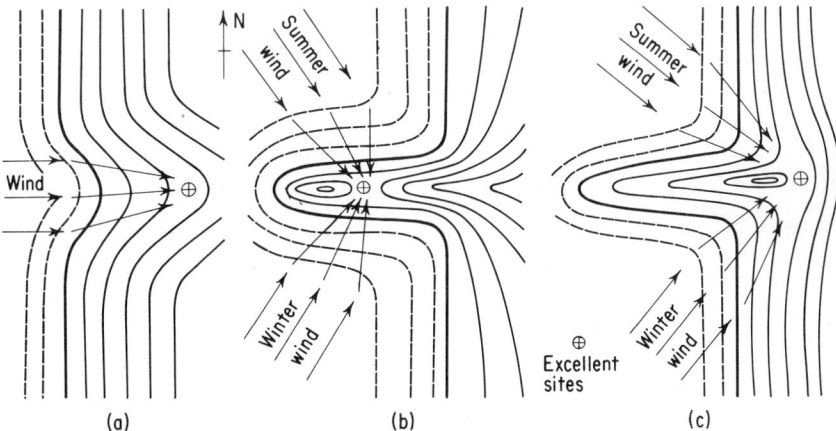

Fig. 19. Very good mountainous coastal sites: (a) Prevailing winds both summer and winter funnel as they ascend, as in portions of the coastal regions of Wales and western Scotland. (b) A saddleback in a cape provides high winds both winter and summer, as at Yaquina Head, Oregon. (c) High ground on a cape and inland from it funnels ascending strong winds through a saddleback both winter and summer, as at Cape Lookout, Oregon.

The winds in the Columbia River Valley have been analyzed less intensively. Figure 21 is a contour map of the region known locally as the Columbia River Gorge. The winds in this area are known to be relatively strong, but are also reported to be very gusty at times. Extreme gustiness, if it occurred, would place a severe strain on a conventional aerogenerator. In order to discover whether such gustiness does occur, pilot balloon observations,

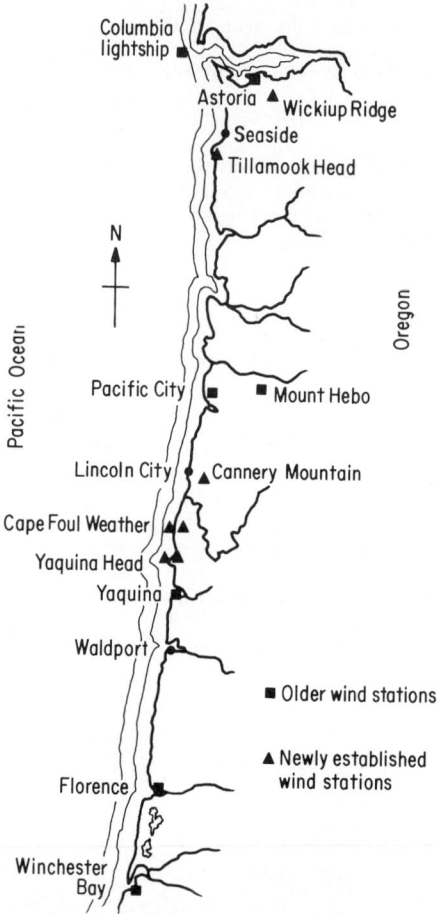

Fig. 20. Locations of previously existing and newly established wind stations along and near the Oregon coast.

initially single theodolite measurements, were made for a selected period. Figure 22 shows the occurrence of upper winds during the day in that portion of the gorge which is near Cascade Locks in terms of lines of equal wind speed component along the center line of the valley, expressed in knots. Such lines are sometimes called *isovents*. Extreme turbulence appears in such an isovent chart as extreme variations of wind speed or wind direction, or of both. Extreme gustiness or turbulence probably occurred at the times and heights indicated by the hatched area. The occurrence of such gustiness has been confirmed by later, more accurate pilot balloon observations, by means of the more reliable two-theodolite technique.

Comparison of Wind Speeds at Various Sites. A direct comparison of winds at certain locations reveals a number of interesting facts. For example, in Fig. 23 a compari-

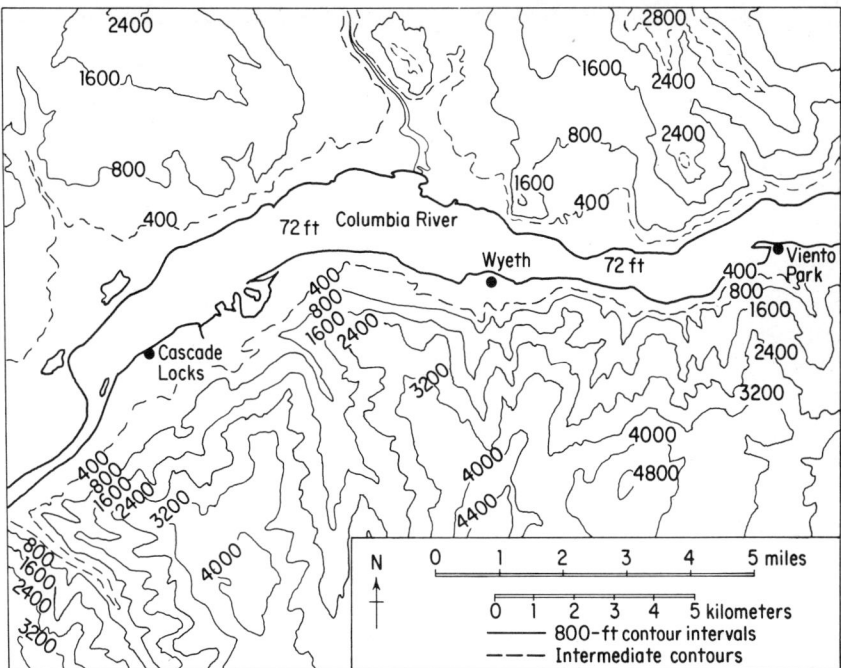

Fig. 21. Contour map of the Columbia River Gorge showing location of stations at which pilot balloon observations were made: Cascade Locks, Wyeth, and Viento Park.

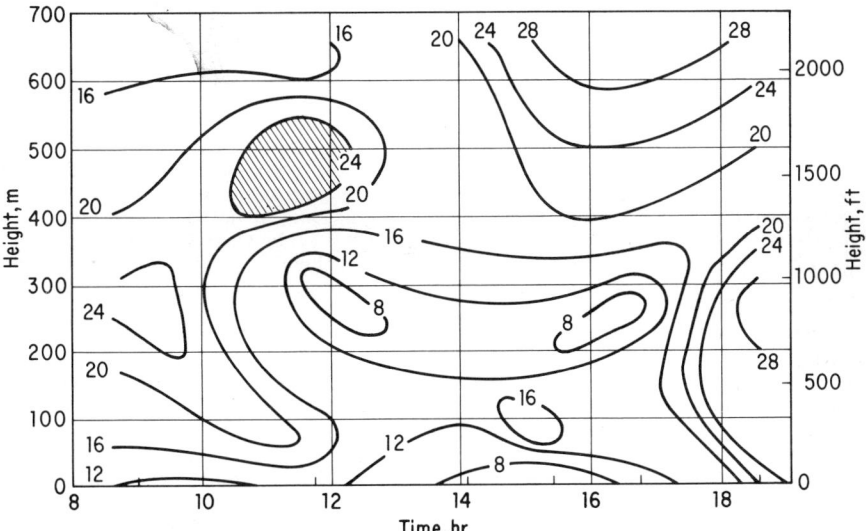

Fig. 22. Isovents of equal up-valley wind speed in knots on September 5, 1972, at Cascade Locks, Oregon. Extreme gustiness or turbulence probably occurred at the times and heights indicated by the hatched area.

son is made of the wind speeds for 1968 at the Columbia Lightship and at Astoria, which are only 25 kilometers (16 miles) apart. It would be expected that the latter, being at the mouth of the Columbia River, would be shielded to some extent by nearby terrain features. Figure 23 illustrates just how pronounced the difference is—except during the late summer and early fall of 1968, the wind at Astoria was substantially less than that at the Columbia Lightship, and it should be remembered that power output will be approximately proportional to the cube of those speeds.

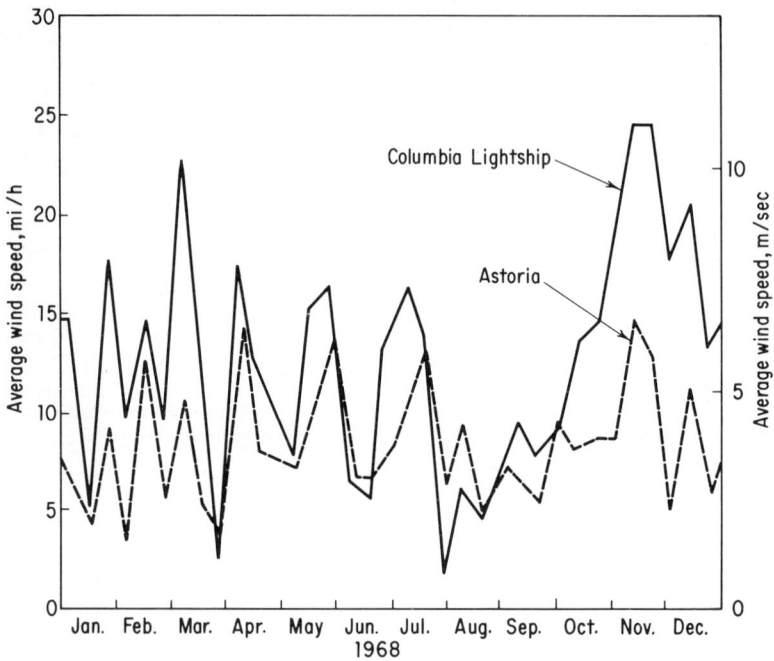

Fig. 23. A comparison of wind speeds for 1968 at Astoria, near the mouth of the Columbia River, and at the Columbia Lightship, two stations which are only about 25 kilometers (16 miles) apart.

A similar comparison is presented in Fig. 24, in which the winds for 1973 at Mount Hebo [a little less than one kilometer (3180 feet) in height] are shown with those for Astoria which lies 110 kilometers (68 miles) to the north of Mount Hebo. Although the high winds on Mount Hebo are well known locally, it is evident that such high winds are largely limited to winter storms. The longer-term annual winds for Astoria, shown by the broken curve, are presented to show that the 1973 Astoria winds are representative of long-term average weather conditions.

Velocity and Power Duration Curves for Various Sites. As previously mentioned, the wind power potential of a site is conveniently represented by a power duration curve of the type shown in Fig. 17. Such power duration curves are given in Fig. 25 by Hewson et al. (1973a and b) for three sites in Oregon and two in Great Britain. The curves for Columbia Lightship, Cascade Locks, and Astoria are based on approximately 4, 7, and 6 years of data, respectively. For comparison are shown two curves obtained from the very comprehensive wind power survey made in Great Britain by the Electrical Research Association. Rhossili Down, in western Wales, was among the best two or three sites found in Great Britain, and that marked Inland Britain was one of the worst (Golding, 1955). A comparison of the power duration curves for the Columbia Lightship and Cascade Locks shows that there is more power in the higher winds at Columbia Lightship, but less power in the lighter winds. The difference between the Astoria and

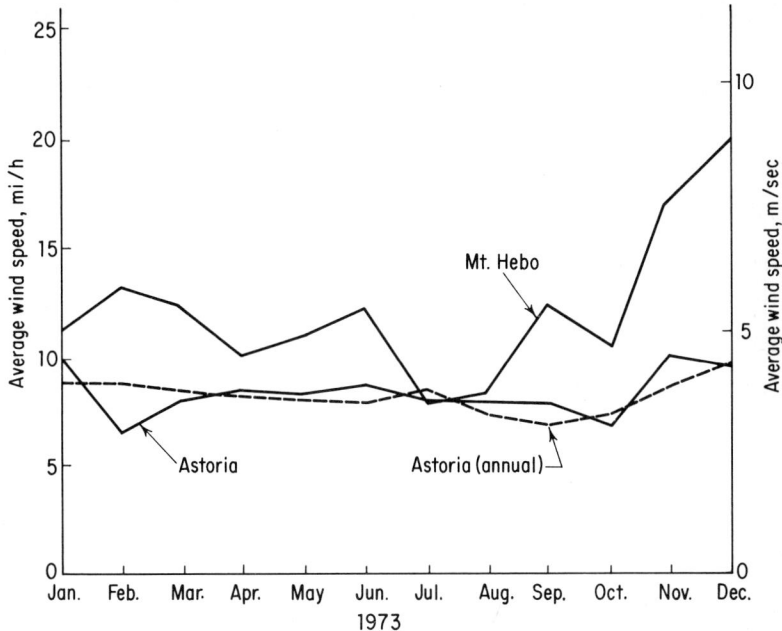

Fig. 24. A comparison of wind speeds for 1973 at Astoria, near the mouth of the Columbia River, and at Mount Hebo which lies 110 kilometers (68 miles) to the south.

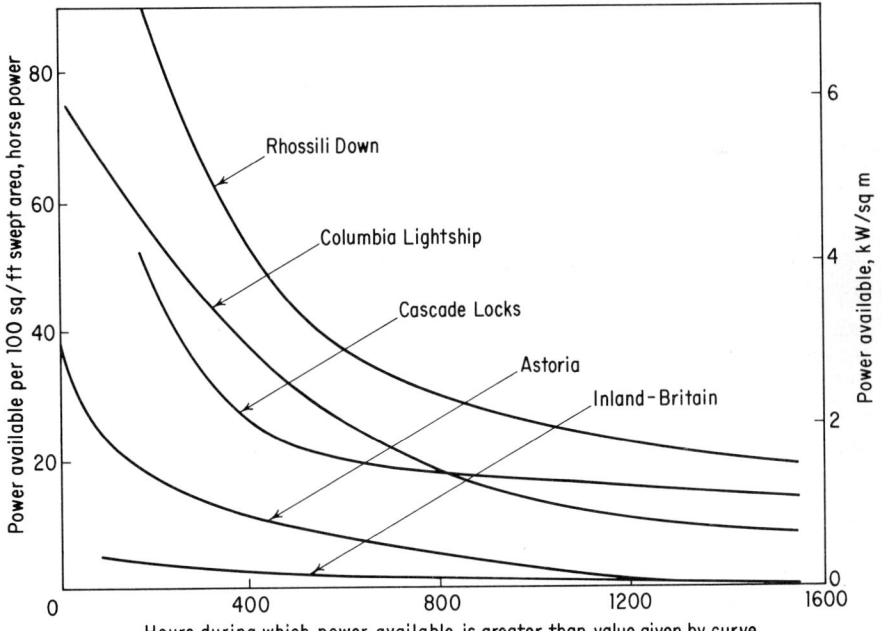

Fig. 25. Power duration curves for five sites, three in Oregon and two in Great Britain.

Columbia Lightship stations, although they are only about 25 kilometers (16 miles) apart, indicates how effective the coastal area is in reducing wind speeds.

A different and more recent set of power duration curves is presented in Fig. 26. The power in the winds at the Columbia Lightship is surprisingly greater for 1973 than it was

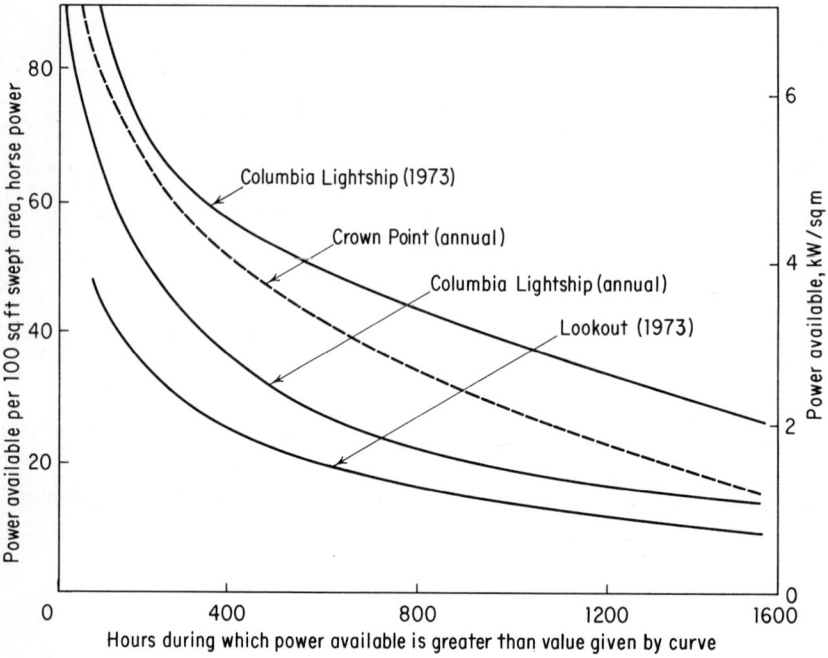

Fig. 26. Power duration curves for various Oregon sites.

for the longer-term (annual) average. This illustrates how greatly the wind speed, and hence the wind power, may vary from year to year at a given site. The average annual power in the winds at Crown Point, a promontory located at the western end of the Columbia River Gorge about 8 kilometers (5 miles) east of Corbett, is also large. The degree of turbulence in the wind at Crown Point is not known, but it may be too high for conventional aerogenerators. The lowest curve is for the lower wind station, called The Lookout, on Cape Foulweather, the location of which is shown in Fig. 20.

Three velocity duration curves are presented in Fig. 27. The winds at Cape Blanco, the location of which is shown in Fig. 33, and those at the Yaquina Communications Station are somewhat stronger than those at the Yaquina Lighthouse. The U.S. Coast Guard, which operates the Communications Station on Yaquina Head, has been most cooperative in permitting the installation of a wind instrument on one of its towers and a recorder in its building. The Coast Guard also has supplied the wind records for Cape Blanco. Figure 28 shows the calculated variation of power output with wind speed for the Yaquina Communications Station wind data given in Fig. 27. The estimate is for a wind turbine with a variable speed drive and a synchronous generator, the blades being not much different from those of the Plum Brook aerogenerator, previously shown in Fig. 12. The energy output per year is approximately 1,150,000 kilowatthours. It is clear that Yaquina Head is a fist-rate location for an aerogenerator installation.

After the first three years of the Oregon study, it was concluded, according to Hewson *et al.* (1975), that the generation of substantial amounts of electrical energy is indeed feasible in Oregon and adjacent areas. A number of conclusions were drawn: (1) Yaquina

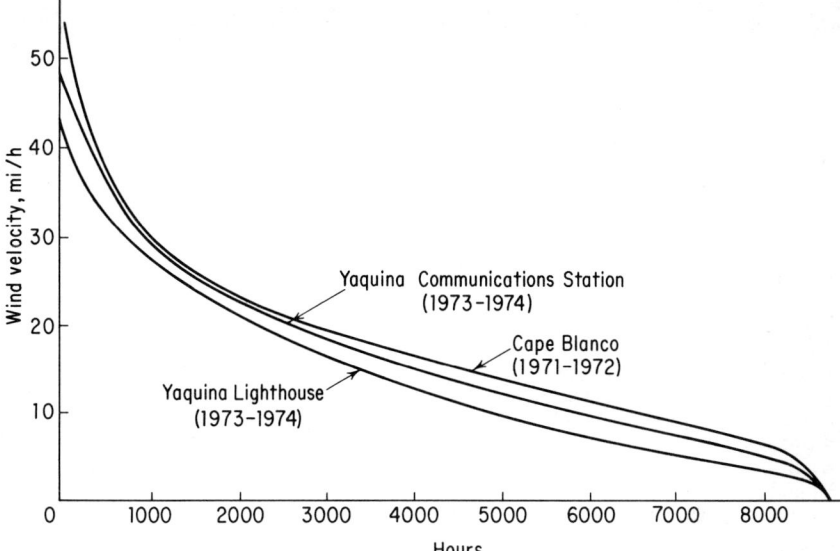

Fig. 27. Velocity duration curves for two stations on Yaquina Head and one on Cape Blanco, Oregon. The curves give the number of hours per year that the wind speeds exceed the indicated values.

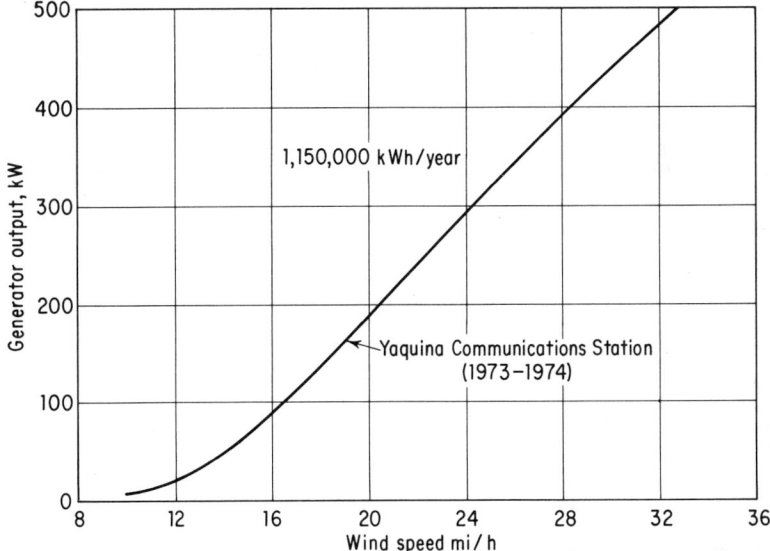

Fig. 28. The calculated variation of power output with wind speed and the yearly energy total for 1973–1974 at the U.S. Coast Guard Communications Station on Yaquina Head on the Oregon coast (see also Figs. 20 and 27) for a wind turbine with blades 38 meters (125 feet) in diameter and a power train consisting of a variable speed drive and a synchronous generator. *(Lockheed-California Company, Burbank, California.)*

Head (Fig. 20) is an excellent site for a large experimental aerogenerator. (2) As a rough estimate, 40 to 100 megawatts of capacity might be installed near Cape Blanco (Fig. 33). The area lying to the east of the Columbia River Gorge (Fig. 21) has winds that could supply as much as 500 megawatts of wind power.

Energy Storage and Site Location.   If the wind is to be thought of as a source of firm power instead of a supplementary source, then some method of storing the power generated by the wind must be devised. Pump storage is often suggested. This involves using the wind energy to pump water into a higher reservoir from which hydropower is developed as the water falls. Such a solution involves additional disturbance of the environment to create a reservoir if none is already available, as well as additional cost.

Wind Tunnel Modeling for Site Location.   Although 30 years ago, wind tunnel modeling was attempted as an aid in site selection with limited success, it is believed that modern wind tunnels and modern methods of using them have substantially greater prospects of success. One of Oregon State University's wind tunnels, therefore, was expanded and improved for this and other purposes. The tunnel now has a cross section 1.5 meters (5 feet) by 1.2 meters (4 feet), with a working section 9 meters (30 feet) long, and an adjustable ceiling. The tunnel produces winds up to 27 meters/second (60 mph).

A model of Yaquina Head has been constructed for wind tunnel tests. As Fig. 20 illustrates, two wind stations have been established on Yaquina Head to permit actual wind measurements to be made. A comparison of the actual winds as measured on Yaquina Head with those measured in the wind tunnel around the model of Yaquina Head will provide important information on the applicability of the method. If model tests

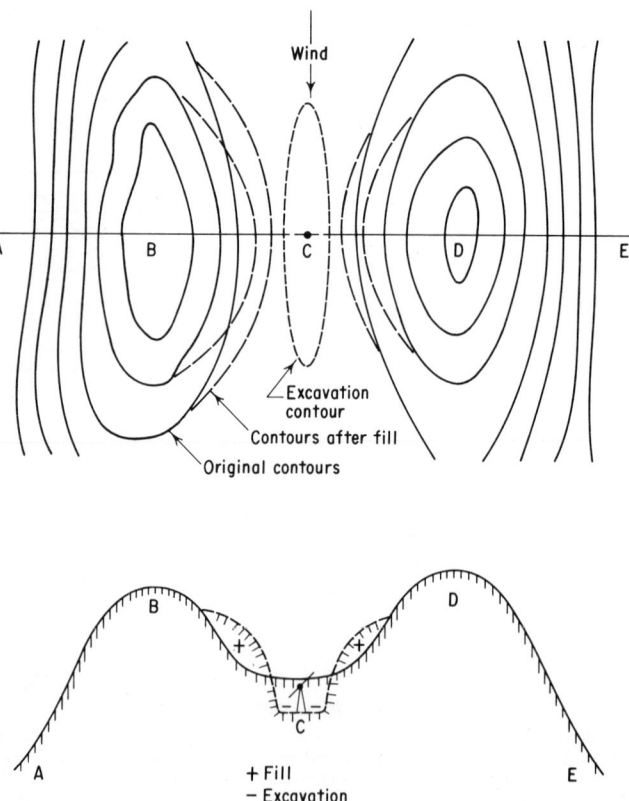

Fig. 29. An example of terrain modification designed to increase power in the wind by increased venturi action.

in the wind tunnel can be used in site selection instead of lengthy series of wind measurements, it will permit substantial savings in time and money.

Terrain Modification in Relation to Site Selections.   It has been suggested by Hewson (1973b) that, in certain cases, terrain modification may become an important factor in site selection. It is well known that a fluid entering a constriction in a channel experiences an acceleration. The higher winds of the Columbia River Gorge are an example of the operation of this principle. It is possible that minor terrain modifications of a site might increase substantially the wind power potential of the site.

Figure 29 shows a hypothetical example of how such terrain modification might increase winds. Point C is part of a saddleback system in which B and D are high ground. Excavation in the area indicated by minuses, and fill represented by pluses, may enhance the existing venturi effect caused by the saddleback. Because of the dependence of wind power on the cube of the wind speed, even a modest 10 percent increase in wind speed due to an augmented venturi effect might well prove to be worthwhile.

The effectiveness of various types of terrain modification will be tested, using models in the wind tunnel.

AEROGENERATOR "FARMS"

The wind is a low-density source of power compared with water. The energy developed is the product of the density of the air and the cube of the wind speed. The density of air near sea level is approximately one-thousandth that of water, so that about one thousand times more air than water must pass through a turbine at the same speed to generate equal amounts of power. This means that wind turbines must be much larger than water turbines to achieve equal power output. The Smith-Putnam turbine with its blades 53 meters (175 feet) in diameter and rated 1.25 megawatts at 13.4 meters/second (30 mph), shown in Fig. 8, illustrates just how large a turbine is required to achieve a relatively modest power output.

If wind power is to be considered as more than a source of electric energy for isolated homes and farms, it becomes apparent that arrays, or "farms," of wind turbines will be needed to generate enough power to justify feeding it into existing networks. A small "farm" consisting of 16 vertical rotor units of the Savonius type is sketched in Fig. 30. The units are supported in a vertical position by an inexpensive system of guy wires. The only compression members are the vertical shafts of the rotors, which are stiffened by the two hemicylinders of the Savonius rotors.

SPECIAL-DUTY AEROGENERATORS

There may be particular circumstances in which special-duty aerogenerators may fill a need. Two examples are given here.

Aerogenerators above Coastal Waters.   The possibility that offshore winds may be strong enough to provide substantial wind power deserves further investigation. The power duration curves for the Columbia Lightship shown in Figs. 25 and 26 and the analysis of wind power potential off the East Coast of the United States presented by Heronemus (1972) both support this contention. Present cable technology is such that power generated a few miles offshore can be transmitted to the shore without serious line loss of energy if oil-filled cables are used. If more distant offshore sites are chosen, such as Nantucket Shoals, New York Shoals, or Georges Bank, it might be desirable or necessary to use the offshore winds to produce hydrogen gas which, in turn, is fed into fuel cells for conversion to electricity. There is one serious uncertainty about offshore installations which requires more attention than it has received in the past. This is the distinct possibility, if not probability, that the supporting tower or mooring mechanism may be inordinately costly if it is to withstand the pounding of wind and waves of midlatitude winter storms. A detailed cost analysis of a seabed-based research tower made by Nath et al. (1973) suggests that this may indeed be so.

Two proposed three-wind-turbine aerogenerators are sketched in Figs. 31 and 32. They were designed by Heronemus (1972) for possible installation in the shallow waters off the East Coast.

The shallow waters of the Pacific Ocean lying over the Continental Shelf off the coasts of California, Oregon, and Washington may also be suitable for wind turbine installations,

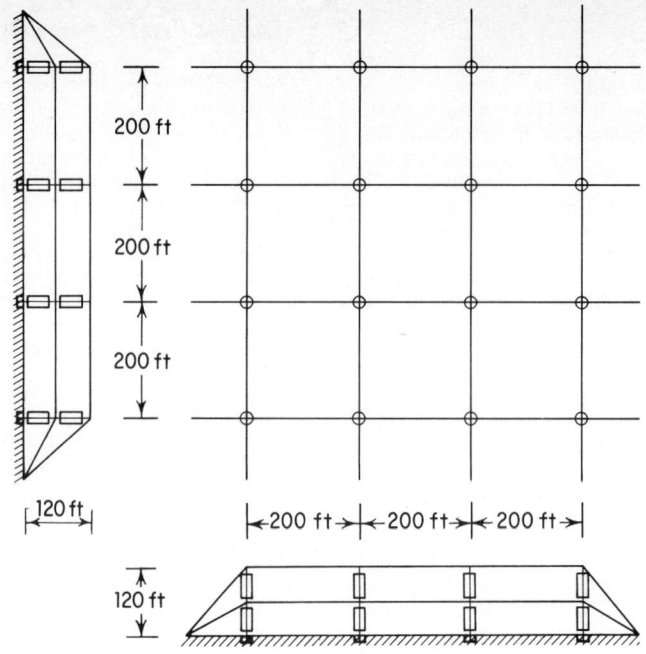

Fig. 30. A small "farm" of 16 vertical rotor aerogenerators.

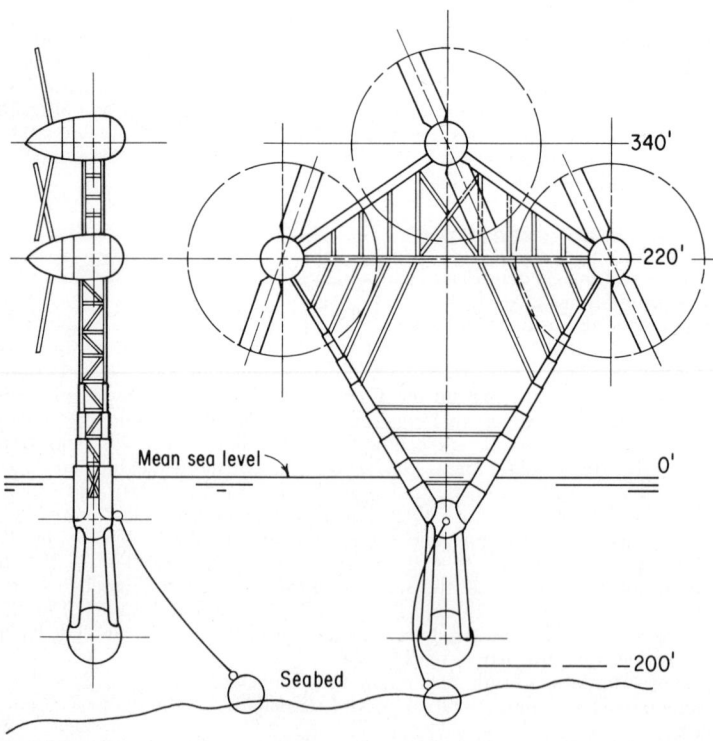

Fig. 31. Anchored three-wind-turbine installation proposed by Heronemus. The blades of each wind turbine are 60 meters (200 feet) in diameter.

provided that supporting towers which will not be swept away by winter storms can be built at reasonable cost. Figure 33 shows contour lines of ocean depth for 10, 20, and 30 fathoms (18, 37, 55 meters; 60, 120, 180 feet). The east-west scale of the contour lines is magnified five times in order to bring out details. The squares numbered 1 and 2 are wind power farms, each approximately 16 kilometers (10 miles) in the north-south direction and 3 kilometers (2 miles) in the east-west direction. They are represented as squares in the figures rather than rectangles because of the east-west magnification of the contour lines

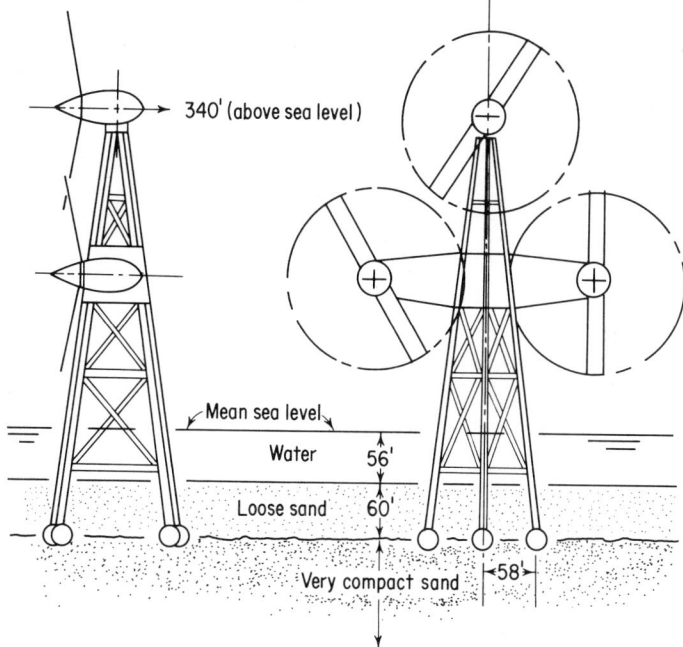

Fig. 32. Seabed-mounted three-wind-turbine installation proposed by Heronemus.

just mentioned. It is estimated that each farm would have an output of 500 megawatts, the equivalent of a small nuclear or fossil-fueled power plant.

Aerogenerators in High Mountains.   The problem at a high mountain site may be different from that of a site over coastal waters. At such a mountain station, the winds may be too strong and turbulent and the icing too severe for the rotors of conventional wind turbine generators. Another type of generator, called the wing aerogenerator, offers some possibility of surviving in such a rugged environment. This wind power generator is designed to generate electric power without any rotating parts which are directly exposed to the wind. It is an omnidirectional machine. There are no rotating blades which must be kept headed into the wind. The device is best described by reference to Fig. 34.

A number of oval-shaped hollow wings are supported horizontally, one above the other, by a central vertical column. The horizontal supports are also tubular. Figure 34 shows the upper two wings of an array. The wing aerogenerator utilizes Bernoulli's principle, which stresses that fluid pressure along a streamline varies inversely with fluid speed.

Thus, as air speed increases, the air pressure decreases, and vice versa. If this concept is applied in the circular neutral wing aerogenerator, the wind speed increases as the wind flows over and under the hollow oval-shaped wings AB and CD. Air pressure decreases just above and below the wings. A pressure differential is thereby induced between the interior and exterior of each hollow wing. As a result, air flows outward through the holes in the upper and lower surfaces of each wing, leading to pressure gradients, which cause air to flow outward through the four tubular supports and upward through the central vertical column JK. Four partitions divide each wing into four 90-degree sectors. Air is drawn from outside through the grating PQ and upward through the air turbine M, causing both it and the coaxially mounted generator or alternator N to rotate. Spring-loaded circular-plate one-way valves are located at E, F, G, and H, to prevent backflows in case atmospheric turbulence, on the topmost wing, for example, induces a momentary reverse-pressure gradient. The number of wings on one vertical column would be limited mainly by the height of the column. Twenty wings would presumably generate roughly 20 times as much power as one wing. Guy wires assist the four horizontal tubes in keeping each wing firmly in position.

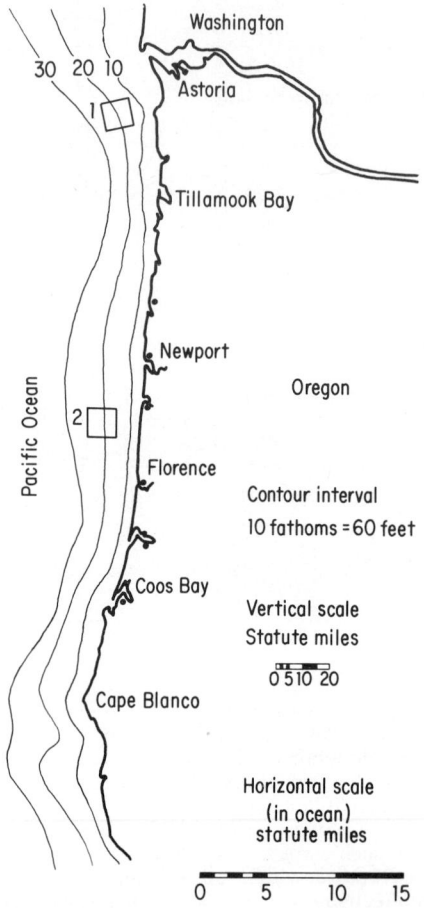

Fig. 33. The Oregon coast showing contour lines of ocean depth for 10, 20 and 30 fathoms and possible offshore wind power farm locations.

The principle of drawing air upward through an air turbine and hence through a hollow vertical column was proposed by the French engineer, J. Andreau, as pointed out in the description of Fig. 10. The advantages of the design presented here over the Enfield-Andreau unit include:

1. The wing generator operates with the wind from any direction. There is no system of blades which must be kept heading into the wind.

2. Although heavy icing may coat the wings, close the holes through which the air is aspirated, and thus stop power production, it will not damage the air turbine itself, which is well protected. Excessive icing may cause failure of the structure, a hazard to any structure exposed to severe icing conditions.

3. The buffeting of the wings by a highly turbulent wind will have little effect in inducing metal fatigue in comparison with that produced in the system of large rotating blades of the conventional aerogenerator. Turbulence-induced metal fatigue is the primary limiting factor on the life of conventional aerogenerator blades. Because such blades are one of the most expensive components of such units, blade replacement made

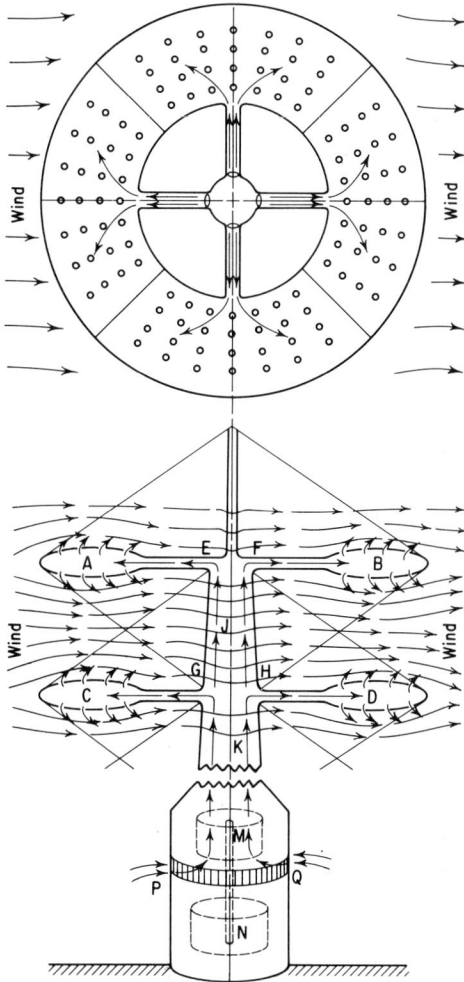

Fig. 34. The circular neutral wing aerogenerator. This aerogenerator has been described in a document submitted by E. Wendell Hewson and received on February 4, 1974, by the U.S. Patent Office, which assigned Disclosure Document No. 028124 to it.

necessary by metal fatigue may well result in high upkeep and maintenance costs in rotating blade systems. No such upkeep costs will be incurred in the wing generator.

4. Severe turbulence will also have less effect on the turbine itself. Internal viscosity effects will prevent excessive upward air flow in the vertical column and hence excess speed by the air turbine. Since air is a compressible fluid, it will serve to smooth out the effects that wind turbulence may have in causing sudden fluctuations in speed of the air turbine.

The main disadvantage of the wing aerogenerator which limits its possible use to sites with extremely high winds, such as a mountaintop, is its low efficiency. At ordinary wind speeds, the pressure differentials induced by the Bernouilli effect are not large enough to induce a strong upward air flow through the air turbine.

Two linear versions of the wing aerogenerator are compared with the circular design in Fig. 35. The linear unit shown in Fig. 35*b* is very similar to the circular design of Fig. 35*a*, the only difference being that the wing is rectangular instead of circular. It is thus

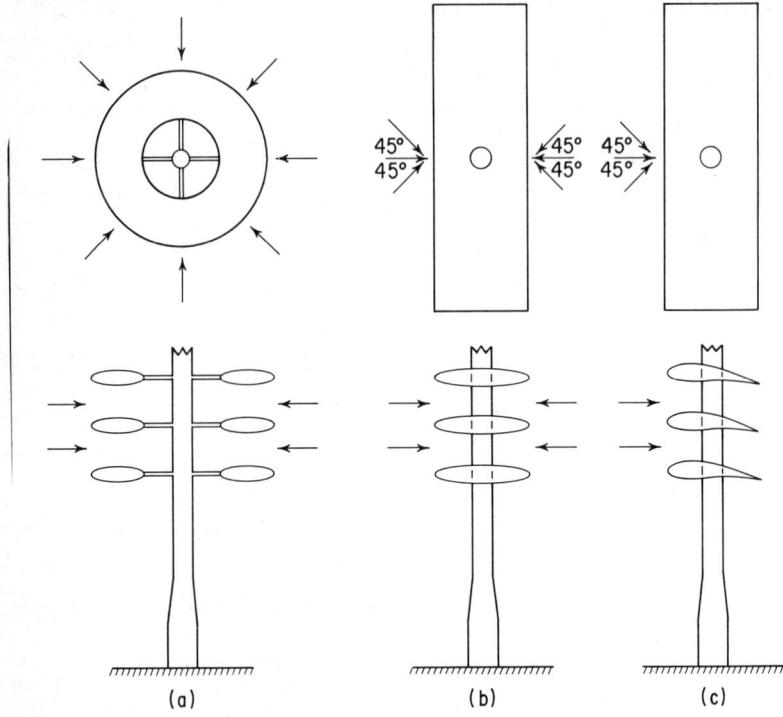

Fig. 35. Three versions of the wing aerogenerator: (*a*) The circular neutral wing aerogenerator. (*b*) The linear neutral wing aerogenerator. (*c*) The linear lift wing aerogenerator. The aspirator holes in the wings are not shown.

bidirectional instead of omnidirectional, as the arrows show, and hence suitable for special sites, such as narrow valleys with very high winds and severe turbulence. In the linear lift wing version shown in Fig. 35*c*, the wings are of the aircraft type with aspirating holes in the upper surface only, and it is thus unidirectional in its response to winds. As the arrows in the figures show, the units accept winds from 360, 180, or 90 degrees. It should be emphasized that Figs. 34 and 35 describe only the basic concept of the wing aerogenerator. There are many details to be determined through a program of optimization. Such a program would provide answers to questions, such as the following:

1. What is the optimum shape of the vertical cross section of the wings described earlier in general terms as "oval-shaped"? 2. What are the optimum dimensions (diameter, width, length, etc.) of each wing? 3. How large should the holes be, and how spaced, in each wing? How many holes should there be? What diameter should the four horizontal supporting air ducts have? 4. What should be the vertical spacing of the wings? 5. How should the diameter of the hollow vertical column vary with height? 6. What is the optimum number of wings?

ENERGY STORAGE

There are several possibilities for storing energy derived from the wind. Batteries are conveniently used for the storage of energy extracted by small aerogenerators, but their potential for storing energy from large systems requires detailed study, as pointed out by Schwartz (1973). Pump storage has already been mentioned in relation to site location. The use of wind power to compress air and store it in caverns or preferably aquifers is recommended by Szego (1973). The use of a very strong flywheel made up of radially oriented rods of high tensile strength is advocated by Rabenhorst (1973). The electricity generated by the wind may be used in electrolytic processes to dissociate water into its component hydrogen and oxygen, with the hydrogen being stored for later use as a fuel for power plants of various types, as proposed by Hughes (1973, 1974), Hausz (1973), and Titterington (1973).

Reference to articles on "Batteries," "Fuel Cells," and "Hydrogen" in the section on *Chemical Fuels Technology* in this Handbook is suggested.

Further development of these techniques holds promise of providing an economical method of storing energy. Such storage may not always be necessary, as the discussion which follows on firm power suggests.

FIRM POWER

It is sometimes assumed that the lack of wind at times at a wind power installation means that, in the absence of storage capability, the winds do not offer any potential for firm power. But this conclusion cannot be accepted without further analysis. For example, if aerogenerator farms are distributed over a large area and if these are feeding into a network, it is highly probable that, at any given time, power is being generated by at least two or three of these installations. Thus, if the total rated power of the units in the area is 500 megawatts, climatological analysis of the winds over the region might well show that the firm power capacity of the network is 100 megawatts, or as much as 200 megawatts if load factors are favorable.

PEAK POWER

Peak power demands in a service area often bring into use old and inefficient generating equipment. Wind power may prove to be especially valuable in meeting peaking needs at an attractive cost. In many areas and cities, as in Portland, Oregon, peak power demands occur in the winter when the winds are strongest. Hence, wind power will be available to help meet peaking needs.

Thus, wind power may be fed effectively into the existing grids even if satisfactory storage methods are not yet available. Wind power may thus be thought of as an important fuel saver in meeting peaking and other power needs on a large scale.

ENVIRONMENTAL IMPACTS

It is difficult to visualize a large source of electric energy, such as wind power, which has fewer serious environmental impacts. As previously mentioned, there is no rejected heat to be disposed of, there being only a conversion of a small amount of mechanical energy into heat energy as a result of friction.

Visual Pollution.   Wind turbines must, of necessity, be large structures. Thus, wind turbines, such as those shown in Figs. 8 to 12, do present a problem in visual pollution. Although more modern, attractive structures will undoubtedly be developed, the solution is probably to group such aerogenerators in "farms," as mentioned earlier, and in locations where they will be seen by comparatively few people. There would be relatively little visual pollution by such wind power farms at sites high on the sides of mountain valleys, or in offshore coastal waters.

Land Use.   Such wind power farms would, in general, occupy more land area than fossil-fueled or nuclear-fueled plants, but the difference is not great, especially for nuclear plants, if the required exclusion, mining, fuel storage, and other areas involved in the total energy package are taken into account. The nature of wind power generation is such that multiple use of the land, as for agriculture, would present few problems. For example, Vomocil (1974) has suggested that arrays of aerogenerators might function well to create shelter belts in areas where wind erosion of the soil is a serious problem.

Public Reaction to a Large Number of Wind Turbines.   A problem related to visual pollution and land use is public reaction to the idea of large numbers of aerogenerators.

People have become accustomed to high-density energy sources, such as fossil-fueled and nuclear-fueled power plants. Such plants with an output of 1000 megawatts are becoming commonplace. Although the output of these plants is high, the fact that they are high-density energy sources usually means that the environmental impacts, such as thermal pollution, are also severe.

The public has not yet adjusted its thinking to the distinct possibility that a group of one thousand aerogenerators, each with an input of one megawatt, for a total of a thousand megawatts, may be more desirable from nearly every point of view than a single fossil- or nuclear-fueled plant of comparable output. One substantial advantage is that the number of units in an aerogenerator farm may be increased in response to demonstrated need for increased output, without the danger of constructing unnecessary capacity in very large fossil fuel or nuclear units. Another is less loss of power due to equipment failure. Such failure in a nuclear or fossil fuel plant may mean a 100 percent loss of power. If 5 units in a 1000-unit aerogenerator farm are down for repairs, it means a power loss of only 0.5 percent.

Influence on Winds.   It is often asked whether or not the large-scale use of wind power might slow the earth's winds sufficiently to cause a major change in climate. Such an environmental impact appears to be most unlikely. It is improbable that any appreciable influence on climate could be detected. The possible effect on climate may be thought of as the equivalent of growing a number of groves of tall trees. In a wind, the branches of the trees extract energy from the wind as shown by their swaying, just as the rotating blades of a wind turbine do. There may be a slight slowing of the winds for a short distance downwind from an aerogenerator farm, but the winds would accelerate rapidly as a result of downward transport of momentum from the stronger winds aloft.

ECONOMIC POTENTIAL AND COST ESTIMATES

A detailed outline of the various factors that must be considered in estimating the economic potential of wind power has been presented by Dubey and Coty (1975). They list and discuss eight evaluation criteria, all of which test plausibility and economic payoff: (1) magnitude of application, (2) market potential, (3) cost goals, (4) insensitivity to system size, (5) insensitivity to output variations, (6) sensitivity to public acceptance, (7) land use, and (8) safety. Additionally, there is 1980 to 1990 availability. A thorough assessment of these factors should permit a reliable evaluation of the economic potential for wind energy conversion.

Various estimates have been made of the future electric energy demand for the United States. The projection of Tyrrell (1973) is shown in Fig. 36. Also shown is the projected wind energy capacity as given by Project Independence (1974). The indicated growth of capacity can be developed at costs competitive with fossil fuel energy sources with oil at $11/barrel. The curves show wind power providing an increasing fraction of the national

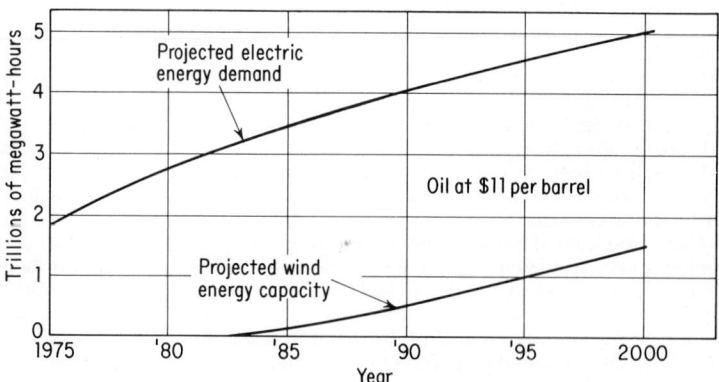

Fig. 36. Forecast of United States electric energy needs, according to Tyrrell (1973), with projected wind energy capacity made by Project Independence (1974).

TABLE 1.   Typical Wind Power System Cost Breakdown*

| | Power level, kilowatts | |
| | 100 | 1000 |
Major element of cost	Dollars per kilowatt	
Rotor	580	97
Transmission	83	57
Tower	126	36
Generator	35	34
Controls	45	19
TOTAL	870	240
Energy output, kilowatthours	405,000	4,410,000
Energy cost, cents/kilowatthour	3.6	0.9

*Capital costs only, but including tower erection costs (according to General Electric Co., 1975.)

electric energy requirements, starting in 1983, becoming 12 percent by 1990, and 30 percent by 2000.

It is relatively difficult to estimate the cost of wind power because designs must be standardized and mass production methods employed. If inflation only is allowed for, with no provision for space age technology, it was estimated by Hewson et al. (1973a) that a production run of 100 wind turbines of the Smith-Putnam design (Putnam, 1948) would cost $700 per installed kilowatt in 1970 dollars. Other cost estimates are available. General Electric Company (1975) has assembled estimates for 100-kilowatt and 1000-kilowatt aerogenerators. The general specification of these units is: Rotor, 2 blades; constant rpm, with power regulation by blade pitch change; transmission, fixed ratio gears; controls, microprocessor; generator, synchronous; and tower, truss. Equipment is top-mounted. The estimates are presented in Table 1.

It will be noted that both capital costs of equipment and the cost of energy produced decrease with increasing power. Another estimate of wind turbine generator selling prices in 1973 dollars has been presented by the Kaman Aerospace Corporation (1975), as shown by Fig. 37. Kaman's estimate for a 100-kilowatt unit is less than the General Electric figures, but the figures for a 1000-kilowatt unit are similar. It should be emphasized that the foregoing cost estimates apply to land installations only. Offshore installations in middle and high latitudes and in the hurricane belts are likely to be more

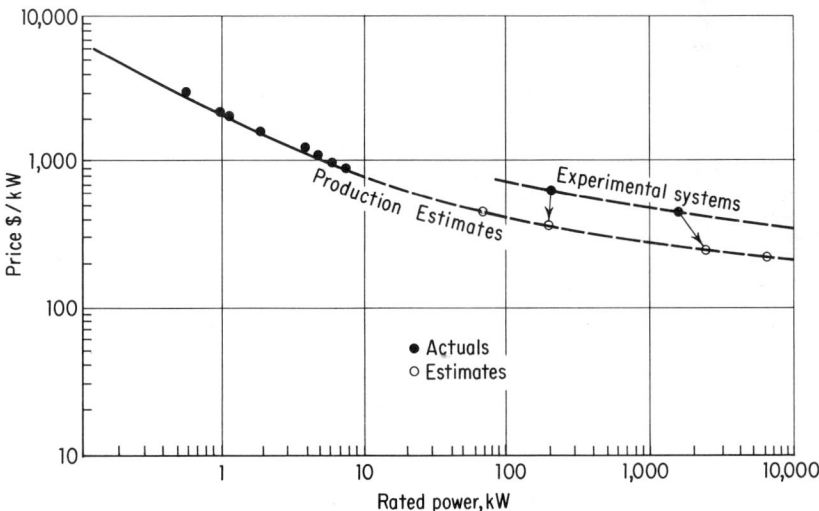

Fig. 37. Expected aerogenerator selling prices (1973 dollars). *(Kaman Aerospace Corporation, 1975.)*

expensive for obvious reasons, as shown by Nath et al. (1973). Further discussions of costs, and of many other aspects of wind energy, can be found in the report of the hearing on wind energy before the Subcommittee on Energy of the Committee on Science and Astronautics, U.S. House of Representatives, 93d Congress, 2d Session (1974).

If one is to make a realistic estimate of the comparative costs of various energy sources, it is necessary to take into account all costs involved in the production of energy. For example, for coal-fired steam plants, one should consider not only the cost of the plant itself, but also the cost of mining and transporting the coal, and of removing sulfur contained in it. Other similar examples can be given. Applying the same criteria to wind power, it is found that the significant cost is the initial cost of the aerogenerator. The energy cost in the various stages of fabrication should be evaluated carefully. If the blades can be constructed to last 25 to 30 years, then operational costs for that period will be very small. It is particularly noteworthy that wind power results in no continuing depletion of limited natural resources.

TIMING

The use of small windmills of several kilowatts capacity for farm and rural power needs and isolated locations in general has been a demonstrated concept for many years. Although advancements in technology will no doubt lead to refinements, there are a sufficient number of well-proved designs, which are commercially available to warrant the immediate installation of many such units in appropriate locations. The same cannot be said about intermediate size and large units. A number of different units in each size range must be designed, constructed, and thoroughly tested. The best can then be chosen and manufactured commercially, using the most efficient mass production methods. With adequate funding, mass production of well-tested designs might commence as early as 1982, with many thousands in operation by 1990. By the last decade of the century, it should be possible to produce as much as 10 percent of the nation's energy requirements by wind energy conversion systems.

Acknowledgments.   The research on wind power at Oregon State University referred to in this article has been sponsored since 1971 by the four Oregon P.U.D.s: those of Central Lincoln, Tillamook, Clatskanie, and Northern Wasco County. Their support, which has not been limited to funding alone, is greatly appreciated. In 1975, Oregon's Eugene Water & Electric Board joined the group of sponsors. The author and his associates have also, since 1975, been assisting the Bonneville Power Administration in the development of the meteorological aspects of its wind power program.

REFERENCES

Allison, H. J. (1973): An Electrical Generator with a Variable Speed Input–Constant Frequency Output, *Wind Energy Convers. Syst. Workshop Proc.*, Washington, D.C., June 11–13, 1973, edited by J. M. Savino, NSF/RA/W-73-006, NSF-RANN Program, Washington, D.C., and NASA–Lewis Research Center, Cleveland, pp. 115–120.

American Wind Turbine Company (1974): Lyman, Nebr.

Betz, A. (1919): Schraubenpropeller mit Geringstem Energieverlust, *Nach. der Kgl. Gesellschaft der Wiss. zu Gottingen, Math.-Phys. Klasse*, pp. 193–217; reprinted in "Vier Abhandlungen zur Hydrodynamik und Aerodynamik" by L. Prandtl and A. Betz, Gottingen, 1927. Reprint Ann Arbor: Edwards Bros., 1943, pp. 68–92.

Brunt, D. (1934): "Physical and Dynamical Meteorology," London, Cambridge University Press.

Darrieus, G. J. M. (1931): U.S. Patent 1,834,018, Dec. 8.

Davidson, B., N. Gerbier, S. D. Papagianakis, and P. J. Rijkoort (1964): Sites for Wind Power Installations, Technical Note 63; WMO 156, TP 76, World Meteorological Organization, Geneva.

Divone, L. V. (1974): The NSF National Wind Program, NSF/NASA/Utility Wind Energy Conference, Dec. 17, 1974, NASA–Lewis Research Center, Cleveland.

Dubey, M., and U. Coty (1975): The Economic Potential for Wind Energy Conversion, Greater Los Angeles Area Energy Symposium, Apr. 3, 1975, Los Angeles Council of Engineers and Scientists.

Frenkiel, J. (1962): Wind Profiles over Hills (in Relation to Wind-Power Utilization), *Quart. J. Roy. Meteorol. Soc.*, **88**, 156–169.

Frenkiel, J. (1963): Gusts over Hills (in Relation to Wind-Power Utilization). *Quart. J. Roy. Meteorol. Soc.*, **89**, 281–283.

General Electric (1975): Program for Conceptual Design, Parametric Analysis, and Preliminary Designs for Low Power (50–250 kW) and High Power (500–3000 kW) Wind Generator System, Advanced Energy Systems, Space Division, Philadelphia, GE Document 75SD542.

Golding, E. W. (1955): "The Generation of Electricity by Wind Power," New York, Philosophical Library, 1955.

Golding, E. W., and A. H. Stodhart (1949): The Potentialities of Wind Power for Electricity Generation (with Special Reference to Small-Scale Operation), Technical Report W/T 16, The British Electrical and Allied Industries Research Association, Leatherhead, Surrey, England.

Golding, E. W., and A. H. Stodhart (1952): The Selection and Characteristics of Wind-Power Sites, Technical Report C/T 108, The British Electrical and Allied Industries Research Association, Leatherhead, Surrey, England.

Hausz, W. (1973): Use of Hydrogen and Hydrogen-rich Components as a Means of Storing and Transporting Energy, *Wind Energy Convers. Syst. Workshop Proc.*, Washington, D.C., June 11–13, 1973, edited by J. M. Savino, NSF/RA/W-73-006, NSF-RANN Program, Washington, D.C., and NASA–Lewis Research Center, Cleveland, pp. 130–134.

Heronemus, W. E. (1972): Pollution-free Energy from Offshore Winds, Preprints, 8th Annual Conference and Exposition, Marine Technology Society, Sept. 11–13, 1972. Washington, D.C.

Hewson, E. W., H. F. Davis, J. H. Nath, R. W. Thresher, and R. E. Wilson (1973*a*): Wind Power Potential in Selected Areas of Oregon, Report 73-1, Oregon State University, Corvallis.

Hewson, E. W. (1973*b*): Wind Power Research at Oregon State University, *Wind Energy Conver. Syst. Proc.*, Washington, D.C., June 11–13, 1973, edited by J. M. Savino, NSF/RA/W-73-006, NSF-RANN Program, Washington, D.C., and NASA–Lewis Research Center, Cleveland, pp. 209–212.

Hewson, E. W., et al. (1973*c*): Wind Characteristics and Siting, *Wind Energy Conver. Syst. Workshop Proc.*, Washington, D.C., June 11–13, 1973, edited by J. M. Savino, NSF/RA/W-73-006, NSF-RANN Program, Washington, D.C., and NASA–Lewis Research Center, Cleveland, pp. 209–212.

Hewson, E. W. (1974): Wind Power Potential in Selected Areas of Oregon, Report PUD 74-2A, Oregon State University, Corvallis.

Hewson, E. W. (1975): Generation of Power from the Wind, *Bull. Amer. Meteorol. Soc.*, **56** (7), 660–675.

Hewson, E. W., Baker, R. W., and R. Brownlow (1975): Wind Power Potential in Selected Areas of Oregon, Report PUD 75-3, Oregon State University, Corvallis.

Hodges, H. (1975): English Post-Mills, *Brit. Hist. Illus.*, **1**, (6), 30–33.

Hughes, W. L. (1973): Energy Storage Using High-Pressure Electrolysis and Methods for Reconversion, *Wind Energy Convers. Syst. Workshop Proc.*, Washington, D.C., June 11–13, 1973, edited by J. M. Savino, NSF/RA/W-73-006, NSF-RANN Program, Washington, D.C., and NASA–Lewis Research Center, Cleveland, pp. 123–129.

Hughes, W. L. (1974): Hearing on Wind Energy before the Subcommittee on Energy of the Committee on Science and Astronautics, U.S. House of Representatives, Ninety-Third Congress, Second Session, No. 49.

Hütter, U. (1973): Past Developments of Large Wind Generators in Europe, *Wind Energy Convers. Syst. Workshop Proc.*, Washington, D.C., June 11–13, 1973, edited by J. M. Savino, NSF/RA/W-73-006, NSF-RANN Program, Washington, D.C., and NASA–Lewis Research Center, Cleveland, pp. 19–22.

Kaman Aerospace Corporation (1975): Conceptual Design, Parametric Analyses and Preliminary Designs for Low Power (50–250kw) and High Power (500–3000 kw) Wind Generator Systems, Contract NAS3-19404.

Klemin, A. (1925): The Savenius Wing Rotor, *Mech. Eng.*, **47**(11), 911–912.

Nath, J. H., J. L. Washburn, R. S. Mesecar, and D. C. Tavolacci (1973): Offshore Research Tower, School of Oceanography and Department of Mechanical Engineering, Oregon State University, Corvallis.

Noel, J. M. (1973): French Wind Generator Systems, *Wind Energy Convers. Syst. Workshop Proc.*, Washington, D.C., June 11–13, 1973, edited by J. M. Savino, NSF/RA/W-73-006, NSF-RANN Program, Washington, D.C., and NASA–Lewis Research Center, Cleveland, pp. 186–196.

Project Independence (1974): Task Force Report on Solar Energy, Federal Energy Administration, Washington, D.C., November 1974.

Putnam, P. C. (1948:) "Power from the Wind," New York, Van Nostrand.

Rabenhorst, D. W. (1973): Superflywheel Energy Storage System, *Wind Energy Convers. Syst. Workshop Proc.*, Washington, D.C., June 11–13, 1973, edited by J. M. Savino, NSF/RA/W-73-006, NSF-RANN Program, Washington, D.C., and NASA–Lewis Research Center, Cleveland, pp. 137–145.

Ragsdale, R. G. (1974): Proposed Plan for Utility Involvement, NSF/NASA/Utility Wind Energy Conference, Dec. 17, 1974. NASA–Lewis Research Center, Cleveland.

Reed, J. W. (1974): Wind Power Climatology, *Weatherwise*, **27**, 236–242.

Reitan, D. K. (1973): Wind-Powered Asynchronous AC/DC/AC Converter System, *Wind Energy Convers. Syst. Workshop Proc.*, Washington, D.C., June 11–13, 1973, edited by J. M. Savino, NSF/RA/W-73-006, NSF-RANN Program, Washington, D.C., and NASA–Lewis Research Center, Cleveland, pp. 109–114.

Savino, J. M. (1974): Energy Extraction from the Wind and Site Characteristics, NSF/NASA/Utility Wind Energy Conference, Dec. 17, 1974, NASA–Lewis Research Center, Cleveland.

Savonius, S. J. (1931): The S-Rotor and Its Applications, *Mech. Eng.*, **53**(5), 333–338.

Schwartz, H. J. (1973): Batteries for Storage of Wind-generated Electricity, *Wind Energy Convers. Syst. Workshop Proc.*, Washington, D.C., June 11–13, 1973, edited by J. M. Savino, NSF/RA/W-73-006, NSF-RANN Program, Washington, D.C., and NASA–Lewis Research Center, Cleveland, pp. 146–151.

Sørensen, B. (1975): Energy and Resources—A Plan According to Which Solar and Wind Energy Would Supply Denmark's Needs by the Year 2050, *Science,* **189**, 4199 (July 25, 1975).

South, P., and R. S. Rangi (1971): Preliminary Tests of a High Speed Vertical Axis Windmill Model, *Nat. Res. Counc.* (National Aeronautical Establishment, Ottawa, Canada) *Lab. Tech. Rep.* LTR-LA-74.

South, P., and R. S. Rangi (1972): A Wind Tunnel Investigation of a 14-Ft. Diameter Vertical Axis Windmill, *Nat. Res. Counc.* (National Aeronautical Establishment, Ottawa, Canada) *Lab. Tech. Rep.* LTR-LA-105.

SPECTRUM (1975): An Alternate Technology Equipment Directory, Alternate Sources of Energy, Inc., Route 2, Box 90A, Milaca, Minn. 56353.

Subcommittee on Energy (1974): Hearing on Wind Energy, Committee on Science and Astronautics, U.S. House of Representatives, Ninety-Third Congress, Second Session, No. 49.

Sweeney, T. E., and W. B. Nixon (1973): An Introduction to the Princeton Sailwing Windmill, *Wind Energy Convers. Syst. Workshop Proc.,* Washington, D.C., June 11–13, 1973, edited by J. M. Savino, NSF/RA/W-73-006, NSF-RANN Program, Washington D.C., and NASA–Lewis Research Center, Cleveland, pp. 70–72.

Szego, G. C. (1973): Energy Storage by Compressed Air, *Wind Energy Convers. Syst. Workshop Proc.,* Washington, D.C., June 11–13, 1973, edited by J. M. Savino, NSF/RA/W-73-006, NSF-RANN Program, Washington, D.C., and NASA–Lewis Research Center, Cleveland, pp. 152–154.

Thomas, P. H. (1945): "Electric Power from the Wind," Federal Power Commission, Washington, D.C., March 1945.

Thomas, R. L. (1974): The Lewis Research Center Wind Project, NSF/NASA/Utility Wind Energy Conference, Dec. 17, 1974. NASA–Lewis Research Center, Cleveland, Ohio.

Thresher, R. W. (1974): Structural Aspects of Wind Machines, Report No. 74-2B, Oregon State University, Corvallis.

Titterington, W. A. (1973): Status and Applicability of Solid Polymer Electrolyte Technology to Electrolytic Hydrogen and Oxygen Production, *Wind Energy Convers. Syst. Workshop Proc.,* Washington, D.C., June 11–13, 1973, edited by J. M. Savino, NSF/RA/W-73-006, NSF-RANN Program, Washington, D.C., and NASA–Lewis Research Center, Cleveland, pp. 135–136.

Tyrrell, T. J. (1973): Projections of Electricity Demand, Oak Ridge National Laboratory, Tenn., ORNL-NSF-EP50, November 1973.

Vance, W. (1973): Vertical Axis Wind Rotors—Status and Potential, *Wind Energy Convers. Syst. Workshop Proc.,* Washington, D.C., June 11–13, 1973, edited by J. M. Savino, NSF/RA/W-73-006, NSF-RANN Program, Washington, D.C., and NASA–Lewis Research Center, Cleveland, pp. 96–102.

Vomocil, J. A. (1974): Personal communication.

von Arx, W. S. (1974): Energy: Natural Limits and Abundances. *EOS Trans. Am. Geophys. Union,* **55**, 828–832.

Wilson, R. E. (1973): The Oregon State University Wind Studies, *Wind Energy Convers. Syst. Workshop Proc.,* Washington, D.C., June 11–13, 1973, edited by J. M. Savino. NSF/RA/W-73-006, NSF-RANN Program, Washington, D.C., and NASA–Lewis Research Center, Cleveland, pp. 180–185.

Wilson, R. E., and P. B. S. Lissaman (1974): "Applied Aerodynamics of Wind Power Machines," Oregon State University, Corvallis.

W.M.O. (1943): Energy from the Wind, Technical Note 4; WMO 32. TP10, World Meteorological Organization, Geneva.

Section **7**

Geothermal Energy Technology

R. S. BOLTON, B.E., Diploma Imperial College. *Chief Geothermal Engineer, Ministry of Works and Development, New Zealand, Wellington North, New Zealand. Member, New Zealand Institution of Engineers. (Geothermal Energy in New Zealand)*

R. G. BOWEN. *Consulting Geologist, Bend, Oregon. (Geothermal Energy)*

E. A. GROH. *Consulting Geologist, Portland, Oregon. (Geothermal Energy)*

BALDUR LINDAL. *Consulting Engineer, Reykjavík, Iceland. Member, The Association of Chartered Engineers in Iceland, the Association of Consulting Engineers in Iceland (Icelandic Branch of Federation International des Ingenieurs-Conseils). (Geothermal Energy for Space and Process Heating)*

Geothermal Energy

BY R. G. BOWEN* AND E. A. GROH†

Heat is conducted outward from the interior of the surface of the earth at an average rate of about 1.5 microcalories per square centimeter per second. Over a period of one year, this flux to the total surface of the earth amounts to over 10^{20} calories. Heat stored in rock beneath the United States to a depth of 10 kilometers (~ 6.2 miles) is estimated to be on the order of 8×10^{24} calories (White and Williams, 1975). Despite these large numbers, most of this heat is not usable. Heat that can presently be exploited from the earth must be concentrated in geothermal reservoirs where it is accumulated and stored through geological processes.

THE GEOLOGICAL SETTING

The earth's crust is some 5 kiometers thick under the oceans and 30 to 40 kilometers thick beneath the continents. Under the crust lies the mantle, which is 2900 kilometers thick. A distinct seismic discontinuity, most probably due to a compositional change, is present between the crust and mantle.

Quite recently, a new concept of the outer few hundred kilometers of the earth has developed. This concept is embodied in the plate tectonic model of the earth. The surface of the earth, including the sea floor, is divided into several rigid plates which are moving relative to one another. The plates are composed of lithosphere which includes oceanic or continental crust or both, veneering and combined with the uppermost part of the mantle. Oceanic lithosphere is 75 to 100 kilometers thick, and continental lithosphere is about 150 kilometers in thickness. Beneath the lithosphere lies the athenosphere, 400 to 600 kilometers thick. Its composition is not known, but seismic data indicate a zone of partial melting in the upper part and several probable density transitions in the lower part.

Along the belt of oceanic ridges the plates are moving apart at a rate of a few centimeters per year, and new mantle material—magma—is filling the gap. In the direction of plate motion away from the ridges, plates must converge, one plate sinking or subducting beneath the other. Deep oceanic trenches form at these boundaries. Beyond the trenches volcanic arcs are produced, accompanied by shallow to deep seismicity. Such boundaries are typified by Japan, Indonesia, Kamchatcka, and Aleutian peninsulas, and the Andes of South America.

Where plates are converging, both having a veneer of continental crust, the crust is less dense and cannot sink. Thrust faulting, folding, and thickening of the crust mark these boundaries, examples of which are the Himalayan and Alpine mountain regions.

At plate boundaries where neither spreading nor subduction is occurring, the plates slide past one another along great fractures called transform faults. The San Andreas fault system is a prime example, as it connects the East Pacific Ridge which enters the Gulf of California to the Gorda Ridge lying off the Oregon-California coast and marks the boundary between the North American Plate and the Pacific Plate.

Spreading of the plates is confined largely to the ocean ridges which lie in the deep ocean. The Red Sea and Gulf of California rifts probably developed only within the last few million years, and deep ocean is yet to be attained. The great East African Rift is unusual in that separation is occurring within continental lithosphere. Continued spreading of this rift may eventually split the African continent and produce new ocean floor. The driving mechanism for plate motion is not understood as yet, but appears to be connected with convective movement of the mantle. The energy is supplied, whatever the mechanism, by the internal heat of the earth.

*Consulting Geologist, State of Oregon, Department of Geology and Mineral Industries, Portland, Oreg.

†Consulting Geologist, Bend, Oreg.

It is along the spreading and converging plate boundaries that abnormal terrestrial heat flow occurs. Mass transfer of heat by magmas generated from the mantle brings heat to shallower levels of the crust. From these heat sources geothermal systems are developed. All the prospective high-enthalpy geothermal areas of the world are found within the belts of geologically young volcanism and crustal deformation produced by moving lithospheric plates.

GEOTHERMAL SYSTEMS

Geothermal systems develop in the upper few kilometers of the earth's crust from a source of heat at some greater depth. The geothermal fluid—water containing dissolved minerals and salts—is heated and becomes less dense. If the overlying rock is permeable, a convection cell or system is generated. A cover of impervious rock, to prevent escape of the fluid to the surface, must overlie this system. The thermal gradient is high in the covering rock, and decreases rapidly within the upper part of the geothermal system where convection becomes pronounced. The temperature then varies little with depth and is called the *base temperature;* this part of the system constitutes the reservoir. Leaks from the reservoir to the surface are manifested by steam vents, hot springs, geysers, and fumaroles.

Geothermal systems are divided into two general classes: one called a vapor-dominated system, and the other a liquid or water-dominated system.

A vapor-dominated system produces saturated to slightly superheated steam with temperatures around 250°C and pressures of 30 to 35 bars. The reservoir generally consists of highly fractured or porous rocks, and well flows may range from a few thousand to over 250,000 kilograms/hour from depths ranging from 1000 to 2500 meters. Noncondensable gas content of the steam varies from a few tenths percent to 5 percent or more. Noncondensable gas content may be much higher initially, but diminishes with production and indicates past accumulation in the reservoir.

The much below hydrostatic pressure in these reservoirs indicates that they are sealed from groundwater infiltration. It is believed that they developed from high-temperature liquid-dominated systems which seal their cooler margins through time by precipitation of dissolved material, mainly silica. Further slow escape of water forms a steam space and a deep liquid phase, probably a very hot brine. Heat is received from a source beneath the system, probably a magmatic intrusion.

The steam fields at The Geysers, California; Larderello, Italy; and Matsukawa, Japan are typical examples of the vapor-dominated system. Reservoir characteristics are similar for all. The Geysers reservoir rocks are indurated, highly fractured graywacke sandstone and volcanic rocks. Porous limestone and dolomite are the reservoir rocks of the Larderello region, and fractured volcanic rocks serve as the reservoir at Matsukawa.

A liquid-dominated system may be conveniently divided into two types: one having high-enthalpy fluids above 200 calories/gram, and one having low-enthalpy fluids below this point. This division tends to separate fluids useful for generating electric power from those most useful for other purposes.

An important physical difference between the liquid and the vapor-dominated systems is that the reservoir pressures in the liquid systems are near hydrostatic pressures, or around 0.1 bar per meter of depth. So at depths of 1000 to 2500 meters pressures are 100 to 250 bars, contrasting to the 30 to 35 bars in the vapor-dominated system.

High-enthalpy systems contain waters with dissolved solids ranging from around 2000 to as much as 260,000 ppm, and temperatures of 200 to as high as 388°C. The predominant anion of the dissolved solids is chloride, along with lesser amounts of sulfate and carbonate. Sodium and potassium are the main cations with a smaller amount of calcium and sometimes magnesium. Up to 800 ppm of silica may be present which, along with several ppm of fluoride and several tens of ppm of boron, are troublesome in the disposal of these high-enthalpy fluids.

Wells drilled into this type of reservoir produce a mixture of water and steam; the steam may be separated at a suitable pressure to operate a turbine. Noncondensable gas in the separated steam is usually below one percent.

The best developed high-enthalpy liquid-dominated reservoir is located at Wairakei, New Zealand, where wells are drilled into a permeable pumiceous volcanic rock capped by an impermeable sedimentary formation. Temperature of the fluid is around 260°C, and about 20 percent is flashed to steam for power production. Another such system still being developed is the Cerro Prieto reservoir in the Mexicali Valley of Mexico north of the Gulf of California. Electric power is now being produced at Cerro Prieto from a fluid having a temperature of 300°C or more and salinities of 15,000 to 25,000 ppm, in a reservoir of permeable sedimentary rock. To the north in the Imperial Valley of California, wells have been drilled to 2500 meters into reservoirs with similar characteristics, except the Salton Sea reservoir which contains a concentrated brine having as much as 260,000 ppm total solids.

Low-enthalpy liquid-dominated systems have properties more variable than those known for the high-enthalpy systems. In some the sulfate anion may be dominant, and in others carbonate-bicarbonate. The salinities tend to be lower, and some could be considered potable. Dissolved silica content, which is a function of temperature, is less; and the toxic elements fluorine and boron also are generally diminished. Temperatures in low-enthalpy systems range from about 10°C above average annual temperature to the previously mentioned arbitrary division at 200°C.

Included in this category are low-enthalpy waters found in some deep sedimentary basins where the overlying rocks have a low conductivity. Temperatures may range from 50 or 60 to 120°C, but the reservoirs are very large. The Hungarian basin and several in the U.S.S.R. are examples of this type. Along the Gulf Coast of the United States similar reservoirs exist within sands in undercompacted sediments. Temperatures above 200°C have been reported, and pressures much above hydrostatic. Deep wells, 2000 meters or more in depth, are required to tap the thermal waters of these basins. Since no connection exists with young volcanism in these basins, heat is thought to be supplied by a slightly above-normal terrestrial heat flow coupled with the insulating effect of the overlying sediments.

Iceland is perhaps best known for the many low-enthalpy reservoirs which have been discovered and are being utilized. Numerous other reservoirs are known throughout the world, among which several are being utilized in the United States, notably in Oregon, Idaho, and California. In general, the close association of these reservoirs with young volcanism suggests magmatic heat as the source.

EXPLORATION

The known higher-enthalpy geothermal systems or resources of the world are located where faulting has created uplift and subsidence of the crust, with attendant mass transfer of heat from depth by magmas and geothermal convection systems. These activities are closely associated with geologically recent movement of the lithospheric plates.

The United States has a broad region covering the western conterminous states which has been disturbed by recent interaction of the plates and changes in direction of their motions. Investigations over the last few years show large areas of this region to have above-normal heat flow with numerous hot springs and wells.

Surface displays of heat offer the simplest and easiest means of exploring for geothermal resources. Yet hot springs or geysers may be some distance from a reservoir, and drilling a deep test well at a spring may prove nonproductive. Also, some reservoirs may have little or no surface display. Therefore, geologic and geophysical methods must be used to enhance the chances of discovery.

Geologic studies can help to show the structure and stratigraphy which may outline domed areas, grabens, and calderas prospective for geothermal resources. Aerial and satellite photography and imagery are very important in the geologic investigations of such things as fault patterns and recent volcanism.

Of the many geophysical methods, measurement of the geothermal gradient and determination of heat flow from shallow drill holes are most valuable. Care must be used in extrapolating the data for greater depths, and groundwater migration can introduce serious discrepancies, but the method is direct in outlining thermal anomalies.

Gravimetric studies can indicate the presence of intrusive rock, which may be a heat source, or contrasting densities, which may define a caldera or graben. Small gravity anomalies are associated with several known thermal anomalies in the Imperial Valley, California.

Rocks that are hot and also saturated with saline waters have very low electrical resistivities. Low resistivities are characteristic of high-temperature liquid-dominated systems, and electrical and electromagnetic methods are useful in the search for and delineation of their size. Practical results have been obtained on newly discovered reservoirs in the Imperial Valley and in New Zealand.

Passive seismic surveys, including ground noise, have been performed over a number of known geothermal reservoirs. One method involves the recording of microearthquakes which many geothermal systems seem to generate. The activity probably arises from the highly faulted nature of the reservoirs and their association with regions of young tectonism. On the other hand, ground noise or geothermal noise surveys record the accoustic signals within a narrow range of amplitude and frequency. Results so far suggest that individual geothermal systems produce characteristic signals, and are related to reservoir depth and temperature gradients. If the reliability of this method can be demonstrated, it could become very important in geothermal exploration because of its simplicity and economy.

Active seismic surveys, generating and recording seismic waves produced by explosions or shock, are useful in determining subsurface structure and faults. Either the reflection or refraction method can be employed, depending upon which is most suitable for a particular location and problem. Some recent work indicates that attenuation of seismic waves may occur in geothermal systems. Further development of this procedure may increase the usefulness of active seismic investigations for geothermal resources.

Magnetic surveys involve measuring the magnetic properties of the underlying rocks. Positive magnetic anomalies often are associated with intrusive rocks, and negative anomalies occur over rocks in which the magnetic minerals have been altered by geothermal fluids. A magnetic survey would thus seem to be useful for seeking geothermal reservoirs, but so many complicating factors arise that results are generally very difficult to interpret.

Many geochemical and isotopic investigations have been performed on samples of spring waters and geothermal fluids throughout the world. As a result, certain constituents or ratios of these constituents may be used to indicate probable reservoir temperatures of liquid-dominated systems. Silica content and the sodium-potassium-calcium rates are the best indicators. High chloride content, above 50 ppm, in springs suggests that the system is liquid-dominated. Springs associated with vapor-dominated systems are said to contain less than 20 ppm chloride.

Isotopic analyses of hydrogen and oxygen in geothermal waters provide a means of determining the origin of the waters. It is now known that geothermal fluids are meteoric in origin, and any volcanic or magmatic addition is minor. By this method the hydrology of an area can be appraised concerning the recharge of water to a geothermal reservoir.

None of these exploration methods can prove the existence and size of a geothermal reservoir. Only the drilling of deep wells and testing of the product found will determine whether successful development and utilization can be obtained.

UTILIZATION

Geothermal fluids can be used in almost any manner for their heat energy, but certain problems can arise. Some fluids have corrosive properties; some precipitate constituents such as silica and calcite; and in their disposal, some fluids contain excessive amounts of toxic elements, such as boron and fluorine, or salts. Corrosion and mineral precipitation can be, and to a degree have been, overcome by use of heat exchangers and special materials, and by maintaining back pressure to prevent flashing in the pipes. Waste disposal is now being handled in new installations by returning fluids to the reservoir or other underground aquifer. Hydrogen sulfide, often present in the noncondensable gases, is both noxious and toxic; chemical absorption methods are presently being tried to eliminate this problem where dispersal in the air is unsatisfactory.

Vapor-dominated systems, such as The Geysers and Larderello, have at present electric power capacities of about 400 and 405 MW (megawatts), respectively. Units planned or being constructed will double the capacity at The Geysers over the next few years. Ultimate capacity has been variously estimated from 1200 to as much as 4800 MW. The Larderello field is now at capacity, and other discoveries being made in Italy will continue to expand that country's total capacity. Twenty megawatts are presently generated at Matsukawa, Japan; and future expansion is planned.

Steam at about 7 bars pressure is admitted to turbines at The Geysers plant. Exhaust steam is condensed by direct contact with the cooling water. Cooling towers are used, and evaporation reduces to about 20 percent the amount added by the condensed steam. This overflow is now returned to the geothermal reservoir by a disposal well. Since pressures in the reservoir are much below hydrostatic, gravity flow suffices, and no pumping is necessary. Noncondensable gases are exhausted from the condenser into the atmosphere, although steps are now being taken to remove hydrogen sulfide.

The Wairakei, New Zealand, power plant and the more recently installed plant at Cerro Prieto, Mexico, generate about 160 and 75 MW, respectively. They are the major installations using fluids from high-enthalpy, liquid-dominated systems, although 13 MW are produced at Otake, Japan; 5 MW at Pauzhetsk, Kamchatka, U.S.S.R.; and 3 MW at Namafjall, Iceland. An additional 10 MW is produced at Kawerau, New Zealand, for operation of an industrial plant.

In all these installations, steam must be separated from the steam-water mixture coming from the wells. Steam pressures at the turbine varies with the particular installation, but is in the neighborhood of 3 to 4 bars. Surface condensers are used at the Wairakei plant since river water is readily available for cooling. Both the condensate and flashed fluid from the separators are passed into the river, apparently without harm to the environment. Direct condensation and cooling towers are employed at Cerro Prieto, and the overflow and flashed fluid are presently conveyed to the Gulf of California. Long-term consequences may require ponding or disposal of the fluids underground.

Consideration must also be given to the use of steam for purposes other than electric power production. Industries, for instance, which may use large amounts of process steam in the range of pressures provided by geothermal fluids can realize economies that are not possible if this steam must be generated from other fuels. This aspect needs more attention in the utilization of geothermal resources.

Low-enthalpy geothermal fluids may in the long run prove to be the most useful to man as an alternative to other forms of energy. Much more of this geothermal energy appears to be available and, depending on initial temperature, may be conveyed long distances. Properly insulated lines should allow a fluid at 150°C to flow over a distance of 50 kilometers with perhaps a 15 to 25°C temperature loss. If gravity flow can be obtained, the economics are even more attractive. Such heat may be used for a multitude of applications. Space heating is the supreme example. Fluids with temperatures as low as 50°C at the point of use can be of service. Water heating is another important application, and, with suitable temperatures, absorption-type air conditioning and refrigeration may employ this energy.

These low-enthalpy fluids are utilized extensively in Iceland. Half the population now receives its space heating requirements from such sources. Reykjavík receives practically all its heating needs from hot water reservoirs at temperatures from about 90 to 150°C as much as 18 kilometers distant. With the numerous reservoirs located throughout the country, more of the smaller villages and farms are employing this energy. Other applications in Iceland range from greenhouse heating to the drying of diatomite.

Among the other countries of the world putting their low-enthalpy waters to similar use are Hungary and the U.S.S.R. In the United States, Klamath Falls, Oregon, and Boise, Idaho, are using low-enthalpy waters for residential, business, and school space-heating needs. Most of the wells in Klamath Falls use a heat exchanger within the well to avoid strict city ordinances on discharge of hot waters to the sewer system.

In the upper temperature range for low-enthalpy fluids, 160 to 200°C, some research and engineering has been performed, indicating that electric power may be produced. This involves exchanging heat from the geothermal fluid to a higher vapor-pressure working fluid such as isobutane or one of the freons, which then may drive a turbine to produce electricity. Thermal efficiencies are low, and large amounts of geothermal fluid must be handled for the power generated. A 680-kW freon unit operated at Paratunka, Kamchatka, in the U.S.S.R. is the only example at present.

Geothermal fluids, both low- and high-enthalpy, may also have important use for the production of fresh water. The multiple-effect or multistage-flash distillation processes are quite efficient in the use of heat. In either process the geothermal fluid may be the feed, or its heat may be exchanged with another impure water as feed. Problems concerning corrosion and scaling will determine which method is more suitable. Investigations are presently being made with geothermal fluids of Imperial Valley, California, regarding the feasibility of desalination.

HISTORY OF GEOTHERMAL ENERGY

Utilization of geothermal energy for purposes other than the heating of bathing pools began in Italy in the late eighteenth and early nineteenth century near the present site of the Larderello field. Steam from fumaroles and shallow boreholes was first used to aid the extraction of boric acid from the hot pools. This industry thrived for many years, and then in 1904, as a result of a dispute with the local electric utility, Prince Piero Conti, owner of the fields, decided to hook a generator to a steam engine driven by the natural steam. The success of this operation led to the installation of the first power plant, with a capacity of 250 kW, in 1913. Increasing exploitation has led to a progressive increase in installed capacity to the present level of 405 MW.

The development of the Larderello field has been characterized by a great deal of innovation, as numerous ideas have been tried to improve the technology of utilization, along with multipurpose utilization. Early developments in the field were directed toward the combining of electric power production with the extraction of boron and other chemicals in the geothermal fluids. By means of heat exchangers a clean fluid could be used in the turbines, but as the value of the chemicals declined and as turbines were made more resistant to corrosion and abrasion, plants employing the intermediate heat exchangers were replaced by direct intake turbines. The direct intake turbines could be constructed at lower costs, and because there were no losses at the heat exchangers, more power could be produced per unit of steam.

Another innovation practiced in Italy has been to install relatively small (1.5- to 5-MW) back-pressure turbines which exhaust directly to the atmosphere. These are used on individual wells very early in the development of a new field. There are several advantages to utilizing back-pressure turbines in this way; one is that they will handle steam containing large quantities of noncondensable gases, such as carbon dioxide, which sometimes exceed 30 percent by weight of the gases in a newly opened field. Thus, gas that has become concentrated over a long period of time in the upper portions of the reservoir is released, and the ratio of noncondensable gases to steam improved to the point where it can be used in conventional condensing turbines.

Another advantage to this practice is that the reservoir temperature-pressure-volume relationships can be determined by production testing, and reservoir life predictions can be made prior to commitment of funding for more extensive developments. Revenues obtained from the sale of electricity during this testing period, sometimes extending over two or three years, can make a significant return of exploration costs. For example, placing a 10-mil-per-kilowatt value on the electricity, a 5-MW plant operating at an 80 percent load factor would produce around $350,000/year in gross revenues. At the end of the testing period, either a condenser is added for increased efficiency, or the unit is moved to another location for further testing. Often in Italy the electricity is used locally on electric-powered drilling rigs for field development.

Another significant innovation at Larderello has been the development of the technique of using air as the drilling fluid when drilling geothermal wells. With air as the circulating fluid instead of mineral-bearing muds, it is possible to drill into low-pressured steam zones without sealing them off with a cake of mud. It has now become common practice in all vapor-dominated systems to use air instead of mud. The method greatly increases the flow per well, as more zones are able to produce steam.

As early as 1932, scientists in New Zealand, being well aware of the success of the Larderello field and knowing that the numerous thermal manifestations such as hot springs and geysers on their own North Island represented accumulated heat energy at depth, approached their government to develop this resource. However, it was not until 1948 that serious study began to appraise the geothermal resources on North Island with the idea of building a geothermal power station. By 1953, drilling in the Wairakei area showed that sufficient steam was available to construct a power station. The first Wairakei station was completed in 1960 and the second in 1963, for a combined capacity of 192 MW.

In the United States, the first attempt to harness geothermal power was in the early 1920s when several shallow wells were drilled at The Geysers and the steam was used to power a generator driven by a small reciprocating engine. The combination of lack of demand of electric power in the immediate area of the field, distance from major power-using areas, and the availablity of low-cost hydropower made The Geysers economically unattractive at that time. In 1955, development of The Geysers field was again attempted, this time leading to the building of the first power plant in 1960 with a capacity of 12.5 MW. Successive drilling and construction have led to the present steam production capacity of around 1000 MW and operating plants of 502 MW.

ECONOMICS

The capital costs of discovery, development, and installation of geothermal heating systems have tended to discourage their use for space and process heating. Exceptions are areas where the geothermal resource was found by chance or was known to exist at relatively shallow depths, or where fossil fuel imports could not be supported by the local economy. For the production of electric power from geothermal sources, however, the costs are comparable to or lower than those of all other methods except hydroelectric.

The subsidizing of fossil fuels, various tax advantages, and wellhead price control caused these depletable energy sources to displace renewable resources, such as geothermal, solar, wind, and waterpower, all of which had a history of more extensive use before the age of abundant fossil fuels.

As the costs of fossil fuels increase owing to scarcity, embargo, and concern over environmental effects of ever-increasing consumption, the higher capital costs of geothermal and other renewable resources are balanced against their lower operating costs and assured energy supplies. Where geothermal sources have been developed for heating, their costs, including amortization of the systems, range between 10 and 25 percent of the costs of fossil fuel systems. In Iceland, Bodvarsson and Zoëga (1964) quote figures ranging from 7.5 to 12 U.S. cents per million Btu for district heating in the Reykjavík area, while imported oil heating cost 102 U.S. cents per million Btu.

In the United States, Purvine (1974) reports costs for geothermal heating at Oregon Institute of Technology in Klamath Falls, Oregon. This new college campus with eight detached buildings totaling over 400,000 square feet of floor space costs $10–14,000 a year when averaged over nine years since opening. These costs included extensive modification, improvement, and experimentation during its first years of operation, but do not include amortization costs of the wells. Costs at Oregon Institute of Technology of heating an older campus using oil-fired hot water averaged $94,000 a year before the institute moved into the new buildings in 1964.

Armstead (1973), in a detailed study of geothermal energy costs, reports price of energy at point of delivery in a geothermal area of between 4 and 6 cents U.S. per million Btu, and price of energy from fossil fuels near their source of production of between 30 and 65 cents U.S. per million Btu.

Electric power production costs from geothermal sources have been reported by many, including the following: Kaufman (1964), McMillan (1970), and Armstead (1973). All show that both capital and operating costs are lower than for other base-load thermal plants. The construction costs of geothermal plants are now about one-half to two-thirds of comparable fossil-fuel plants and one-third to one-fourth those of a nuclear plant. The savings are related to the absence of a man-made boiler and the related complications of fuel handling and combustion. The operation of geothermal plants is also relatively simple, and amenable to automatic unattended operation. The Geysers normally operates unattended (but protected by automatic equipment) for 16 hours out of 24, with maintenance personnel on hand only during the day.

Economics of scale are not as important in geothermal power generation as in other types of plants. This can be an advantage. In order to achieve low power costs in nuclear, hydro, or fossil-fuel plants it is necessary to construct plants of hundreds or even thousands of megawatts. So in many instances utilities must either overbuild and wait for demand to catch up to plant capacity or sell power at wholesale to other utilities. Unit costs of geothermal plants change less than those of other thermal plants as the size increases, and so a utility can economically start small and add generating units as more steam is found or as the load increases.

The increasing dependency on electricity and on fossil fuels for many of the day-to-day functions depends critically upon a constant source. This has been illustrated dramatically in several ways in recent years as electric "brownouts" and "blackouts" have increasingly plagued several parts of the world and political manipulations have withheld fuels, thereby causing severe economic strain in some regions. The heat in geothermal systems is constant and subject to change only over a very long time cycle. Once a natural "boiler" is found, a continuous supply of energy is assured for many years. The simplicity of a system in which the geothermal fluids flow to the surface by utilizing a portion of their own energy, thence directly into the power plant, makes the operation independent of outside support; there are no railroads or mines or complex processing facilities to be put out of service by strike, natural catastrophe, or political decision. The reliability of a geothermal system is an advantage that possibly exceeds its low costs.

ENVIRONMENTAL COMPARISON

The potential environmental impacts arising from geothermal developments are similar to those from any other industrial operation. The construction of roads, drilling of wells, and installation of pipelines and power plants all contribute to the changes in land-use patterns for the particular site. The effects on the land vary, dependent upon the type of fluid and utilization.

There is less environmental impact from producing electric power from a geothermal plant than from other types of thermal power plants, and in many cases less than from a hydroelectric plant when the dislocations caused by massive construction are considered. In geothermal power production, all the steps of the fuel cycle are localized at the site. Other types of thermal power plants require considerable industrial support in the form of mines, transport facilities, and processing plants; thus the environmental impact of the fuel cycle for these operations extends far beyond the bounds of the power generating plant.

The "dry steam" or vapor-intensive type of geothermal electric power plant has a long history of production experience based on Larderello, The Geysers, and Matsukawa fields. For these areas the only continuing environment degradation has been the release of hydrogen sulfide gas. Because of the remote location and relative small size and nearby natural release of hydrogen sulfide, this was not considered to be a serious problem in the past. However, as the size and number of plants increase, the problem is of greater concern, and studies are under way to mitigate it. A main part of the hydrogen sulfide problem is that the presence of even small amounts can be detected, thus making it more obnoxious than the odor of sulfur dioxide as released by coal-fired plants, but the actual amount of sulfur released from the geothermal plant per unit of power generated is less.

The U.S. Environmental Protection Agency (EPA) limits sulfur emission from fossil-fuel plants to 1.2 pounds per million Btu. The amount of sulfur released at The Geysers is less than a quarter of the EPA limits.

At The Geysers, Pacific Gas and Electric is conducting an emission abatement program that is expected to scrub 90 percent of the hydrogen sulfide out of the noncondensable gases. The company plans to start adding this equipment to the new installations and begin a program of retrofitting on the older equipment.

The major gaseous release from geothermal plants is carbon dioxide, but here again the release per unit of power is much less than from any fossil-fuel plant. Moreover, the geothermal plant releases no oxides of nitrogen, smoke, fly ash, or other aerosols.

Some routine operations in geothermal steam fields are extremely noisy. In the past the process of well cleanout and testing generated large amounts of noise, sometimes continuing for long periods. At present the major source of noise is episodic and occurs only during the initial testing period when a productive well is first opened to clean the rock and other debris from the well bore. The noise normally lasts for only a few hours; as soon as the well stops throwing out the debris, further testing is done through silencers. Uncontrolled blowouts are also very noisy, but these are infrequent.

The "hot water" or liquid-intensive geothermal field has problems of a different nature. It takes 100 to 150 pounds of hot water (in contrast to 16 to 20 pounds of steam) to produce a kilowatthour of electricity. The handling and disposal of these large quantities of water per unit of power are what cause most of the environmental problems of the "hot water" geothermal plant.

Thermal waters carry dissolved solids ranging from a few hundred to hundreds of thousands of parts per million. The presence of these dissolved chemicals usually precludes intermingling the geothermal waters with other surface waters, and in the United

States necessitates their injection into the ground. Because of the fear of ground subsidence from the removal of large quantities of water and possible seismic effects due to reinjection, no hot-water fields have been developed in the United States. However, detailed studies have shown that, in most areas, subsidence from geothermal fields would not occur, but in cases where it presented a potential problem, it could be alleviated by reinjection, as practiced in the oil fields. Induced seismic activity relating to injection of fluids into the ground is shown to be proportional to injection pressures. Because the reinjection of geothermal fluids involves only a return of fluid to the reservoir at hydrostatic to subhydrostatic pressures, there is no reason to believe seismic activity would be induced.

The natural thermodynamic constraints placed on any steam electric power production cycle require the rejection of 60 to 70 percent of the energy produced. Because the geothermal plant operates at lower temperatures, the thermodynamic efficiency is commonly half that of modern central station practice; that is, it has an efficiency of conversion of 16 to 17 percent. It must be remembered that this represents the net efficiency of the entire power cycle from production of steam through the well bores to conversion to electricity. For the fossil fuel or the nuclear power plant the thermodynamic efficiency is only a measure of the conversion at the power plant and does not take into account the energy expended in the mining, processing, transportation, and waste handling required from these fuels. When these energy costs are added to the nuclear or fossil-fuel plant, they appear to have a net efficiency much lower than that of the geothermal plant.

Geothermal plants do not require a supplementary source of cooling water when using natural steam or the flashed cycle. The steam, after passing through the turbine, is condensed, piped to the cooling towers, and then recirculated back to cool the condenser. By this method the field at The Geysers produces about 20 percent more condensate than is evaporated. This surplus is then returned to the reservoir where it originated, thus prolonging the useful life of the field. A geothermal plant is the only type of thermal power plant that does not compete with other uses for our dwindling supplies of water.

The environmental impact of any power-production system is reflected in the number and complexity of the steps in the fuel and production cycle. Because geothermal power plants utilize naturally occurring steam, they need no complex steam-generating equipment or extensive mining, processing, storage, or transportation facilities, as do other thermal power plants. But, because all the power production steps are localized within the bounds of the geothermal field, it appears that the geothermal plant does have considerably more effect than other thermal power plants. As a practical matter, a geothermal development will displace other uses in the area being developed. But after the initial construction period, most of the area within the geothermal field can return to preexisting land-use patterns if not directly incompatible with geothermal developments. An example of this is the Larderello field in Italy where development has stabilized and other land uses such as farming and grazing take place within the field. At Larderello most of the area occupied by the geothermal field is covered with farms, orchards, and vineyards; with the wells, steam transmission lines, and power plants occupying only a small percentage of the land.

Some specific geothermal installations are described in following articles in this section of the Handbook.

Editor's Note: As of very late-1976, the major geothermal energy projects underway in the United States and as reported to the World Energy Conference Survey of Energy Resources include:

Vapor-dominated systems (T > 150°C)

The Geysers (northern California), steam turbine plants for electric power generation. Operational and privately funded. (See separate article in this Handbook section.)

Liquid dominated systems

Klamath Falls, Oregon, direct heat for residential space. Operational.

Boise, Idaho, direct heat for residential space. Operational.

Southern California Edison Co., direct flash conversion technique. Objective is to test direct flash of high saline brines. Just underway.

San Diego Gas and Electric Co., binary cycle. Objective is to test corrosion resistance problems of geothermal system. Under construction.

Raft River, Idaho, binary cycle test loop. Objective is to obtain direct heat and electrical generation. Under construction.

East Mesa, California, moderate temperature conversion system. Objective is to test conversion equipment. Just underway.

REFERENCES

1. Proceedings of the United Nations Conference on New Sources of Energy: Solar, Energy, Wind Power and Geothermal Energy, Rome, Italy, Aug. 21–31, 1961, vols. 2 and 3, "Geothermal Energy," United Nations, New York, 1964.
2. Proceedings of the United Nations Symposium on Development and Utilization of Geothermal Resources, Pisa, Italy, Sept. 22–Oct. 1, 1970, "Geothermics," Special Issue 2, 2 vols., United Nations, New York, 1971.
3. Armstead, H. C. H. (ed.): Geothermal Energy, Review of Research and Development, *UNESCO Earth Scie. Publ.* 12, Paris, 1973.
4. *Ibid.* (see Ref. 3).
5. Bodvarsson, Gunnar, and Johannes Zoëga: (see Ref. 1).
6. Cox, Allan (ed.): "Plate Tectonics and Geomagnetic Reversals," Freeman, San Francisco, 1973.
7. Kaufman, Alvin: Geothermal Power, an Economic Evaluation, *U.S. Bur. Mines Inform. Circ.* 8320, Pittsburgh, Pa., 1964.
8. Kruger, Paul, and Carrel Otte (eds.): "Geothermal Energy Resources, Production Stimulation," Stanford University Press, Stanford, Calif., 1973.
9. McMillan, D. A., Jr. (see Ref. 2).
10. Purvine, W. D.: "Utilization of Thermal Energy at Oregon Institute of Technology, Klamath Falls, Oregon," in Proceedings of the International Conference on Geothermal Energy for Industrial, Agricultural, and Commercial-Residential Uses, Klamath Falls, Oregon, October 7–9, 1974.
11. Robertson, E. C. (ed.): "The Nature of the Solid Earth," McGraw-Hill, New York, 1972.
12. White, D. E., and D. L. Williams: "Assessment of Geothermal Resources of the United States—1975," U.S. Geological Survey Circular 726, Branch of Distribution, U.S. Geological Survey, Arlington, Virginia, 1975.

Geothermal Energy in New Zealand

BY R. S. BOLTON*

New Zealand is well endowed with geothermal resources. A thermal area extends over a belt about 250 kilometers long and up to about 50 kilometers wide across the North Island between the central group of volcanic mountains (Mount Ruapehu, Mount Ngauruhoe, and Mount Tongariro) and the White Island volcano in the Bay of Plenty. See inset of Fig. 1. Within this area is to be found a diversity of thermal activity—geysers, fumaroles, hot

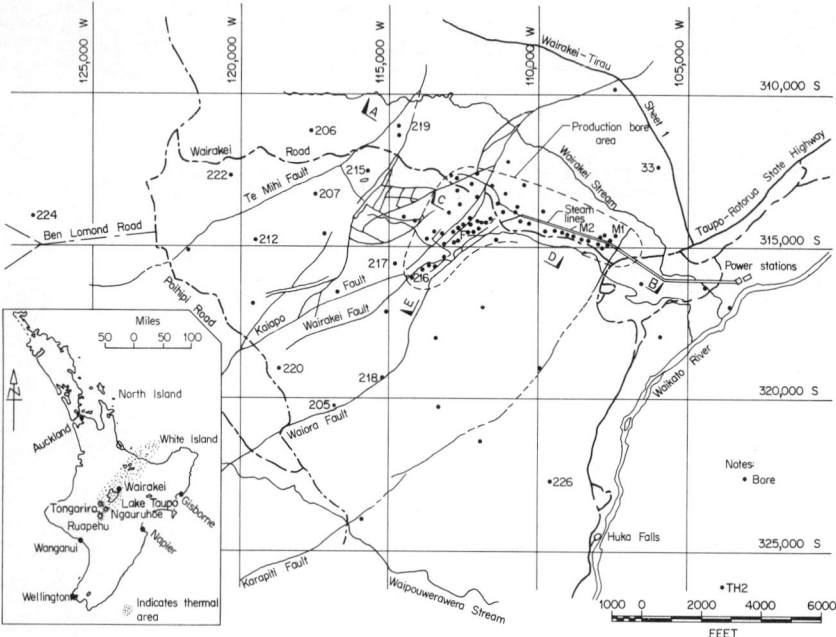

Fig. 1. Wairakei geothermal project area. Inset shows North Island, New Zealand, indicating thermal area. Geological sections are detailed on Fig. 3.

springs and pools of boiling mud. Wairakei is one of several active areas where it is known that aquifers containing water up to and exceeding a temperature of 300°C exist.

WAIRAKEI

Wairakei, the second major geothermal power station to be constructed in the world, fed its first power into New Zealand's electric system in November 1958. See Fig. 2. The New Zealand Electricity Department (NZED) controls the Wairakei Stations. This department is also responsible for all government-owned power stations and for distributing electricity through its grid system to electricity supply authorities in the various centers. The NZED also installed the turboalternators, machinery, and ancillary equipment and conducted some other works on the Wairakei project.

*Chief Geothermal Engineer, Ministry of Works and Development, New Zealand, Wellington North, New Zealand.

Fig. 2. Wairakei Valley from the air. The modern Tourist Hotel Corporation hotel is in foreground, just off State Highway No. 1.

The Ministry of Works and Development (MWD) was in charge of overall administration of construction on the project and is now responsible for all "downhole" maintenance and for the design and installation of all major modifications to the steam transmission system. MWD is responsible for investigating and developing geothermal areas in which the government is directly interested and undertakes all the drilling and associated field development. MWD also investigates alternative uses for the energy and, in conjunction with the Department of Scientific and Industrial Research (DSIR), new techniques and improvements to old ones. DSIR provides all scientific services, such as geology, geophysics, chemistry, physics, and metallurgy. A laboratory and a team of scientists and technicians are maintained at Wairakei, and fundamental research into geothermal energy and volcanism is performed.

Geology.   The geology of the Wairakei field is an important factor in its exploitation. The main features in the production area are shown in Fig. 3 and may be summarized as follows:

1. *Surface Formations*—Pumice and loosely consolidated breccias to about 125 meters.

2. *Huka Falls Formation*—Mudstone-siltstone layers, interbedded in some areas with breccias. This formation ranges from about 125 to 200 meters in thickness. It is generally impermeable and acts as a cap rock, although the natural activity indicates that fissure permeability is present.

3. *Waiora Formation*—Reasonably well-consolidated breccia with mudstone-siltstone stringer and silicified lenses. In the western part of the production area, the lower contact of this formation is about 650 meters deep, but to the east, the lower contact dips very sharply, and the formation has not been drilled to its full depth. Generally, the formation is permeable and provides a significant proportion of the production.

4. *Wairakei Ignimbrites*—The contact between the Wairakei ignimbrites and the Waiora formation is very broken and proves the main production zone. The ignimbrite itself has a very low production capability.

In the outer area, there are substantial rhyolite intrusions in the Waiora formations. In the upper levels the Huka Falls and Waiora formations thin out, and in the lower levels the contact between the Waiora formation and the Wairakei ignimbrite is displaced downward by some 500 meters.

The other important geological feature is, of course, the extensive faulting in the area. As may be seen from Fig. 1, the main faulting system runs in a southwest-northeast direction, with transverse faulting in the regions of most intensive natural activity. From

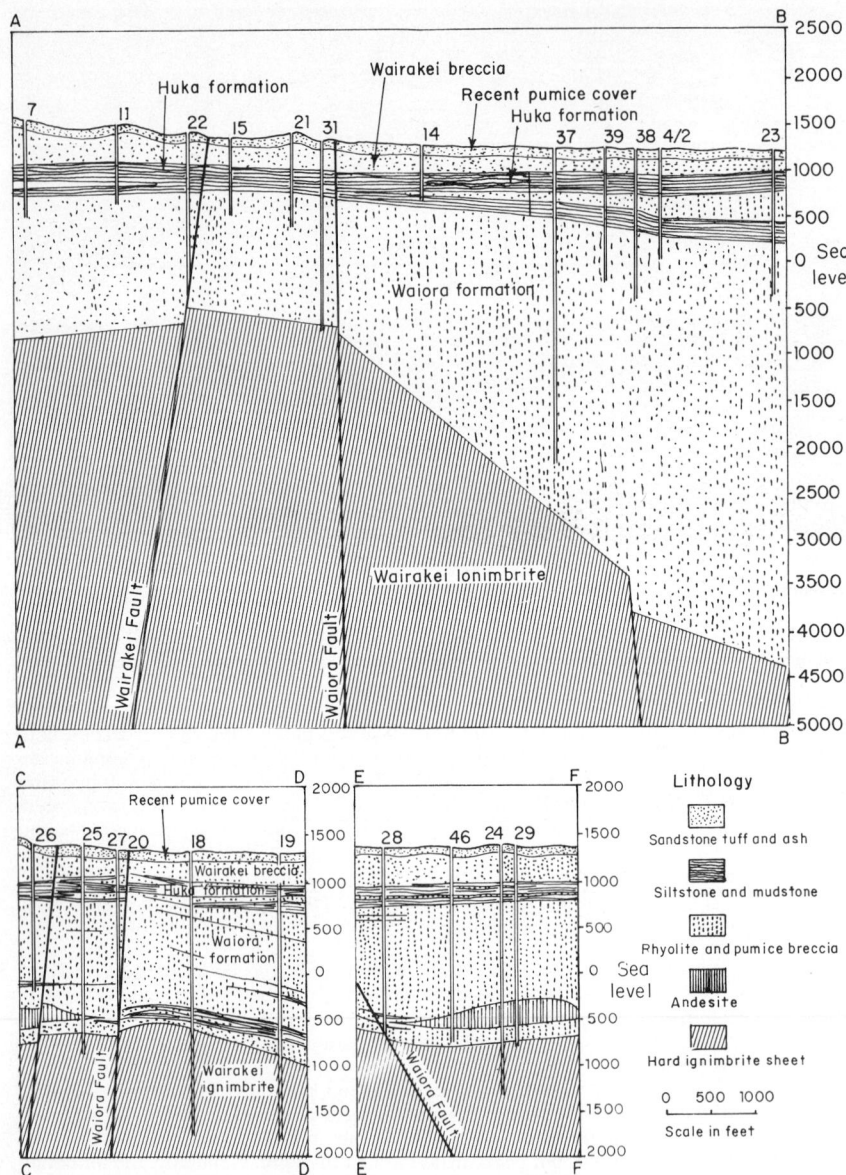

Fig. 3. Geological sections A–B, C–D, and E–F as indicated on Fig. 1.

drilling, it is evident that the faults can be open at depth, and there is no doubt that they exert a strong influence on the underground flow conditions.

Effects of Exploitation on the Field.   The production history of the field extends over nearly 20 years. In that time, substantial changes have taken place in the underground conditions. Initially, the aquifer was filled with water, with the temperature and pressure conditions following the boiling point for a depth relationship until a temperature of 260°C was reached.

Exploitation has resulted in an almost uniform pressure decline of over 21 bars, affecting an area considerably greater than the area in which the production wells are drilled. This area is defined approximately by the Wairakei Stream in the north and northeast; the Waikato River to Huka Falls in the southeast; the Waipouwerawera Stream in the south; and Poihipi Road in the west. These features may be seen in Fig. 1.

The relationship between rate of pressure decline and rate of drawoff is shown in Fig. 4. This is not a linear relationship. Although there is an immediate fall in pressures following an increase in drawoff, a prolonged period of substantially constant drawoff shows pressures tending toward stable value. The relation between pressures and drawoff suggests that conditions are being influenced by an inflow, which is supported by the rise in pressure at the beginning of 1968 in a period when the field drawoff was reduced to about one-third of normal.

Temperature trends in the upper levels reveal a similar pattern, also shown in Fig. 4. The temperature trends are for the average of the maximum temperatures in the wells in the western or main production area, and reflect the temperatures in the production area. It is important to note, however, that temperatures at greater depths have indicated no change.

Trends in individual well outputs appear in Fig. 5. The effect of the fall in pressures and temperatures in the producing zone is clearly shown. However, extrapolation of the trends in the total discharge from the field indicates that the flow tends toward a stable value sufficient to maintain an output between 125 and 140 MW (megawatts) indefinitely. This compares with an output of 150 MW in 1974.

Some wells show an increase in steam content, while others show a decrease. The enthalpy of the total discharge is, if anything, tending to decrease. Thus, although the possibility of the field going dry cannot be discounted, it would obviously be a very long-term process, of which there is no indication at this time (1975).

With the increase in the capacity of New Zealand's electric system, Wairakei's contribution as a percentage will decrease. As of 1974, the station produced 10 percent of the energy required for the North Island. It has consistently operated at an annual load factor of 89 percent since fully commissioned.

Wells.   A reinforced concrete cellar is constructed at each wellhead to carry the heavy loads imposed by the drilling rigs. The ground around the cellar is consolidated by grouting so as to render it impervious, and to give protection against the upward migration of steam and hot water to the surface around the outside of the well. This operation is considered essential for safety of workmen and equipment, and has proved worthwhile in one incident where an uncontrollable eruption occurred during drilling. In this case, the consolidation grouting deflected the eruption away from the wellhead for a sufficient length of time to enable all equipment to be safely removed.

Drilling is done with rotary-type rigs, basically following oil drilling practices. Signifi-

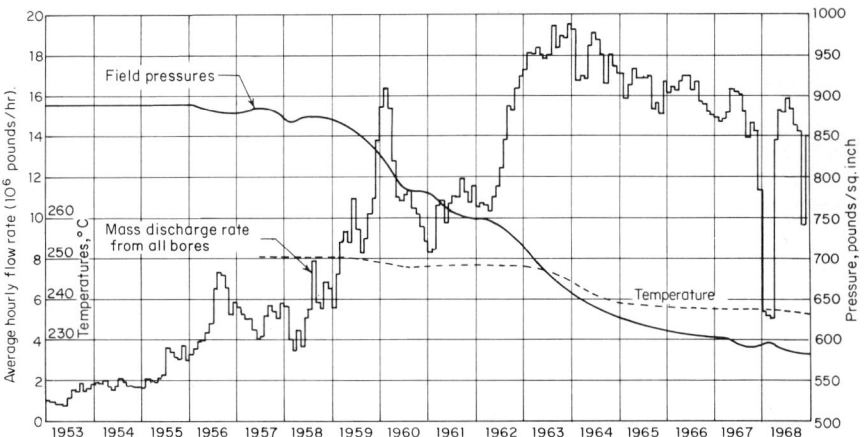

Fig. 4. Field discharge rate and pressures at Wairakei. At the end of 1975, pressures had fallen to 550 psig, and the total mass drawoff was averaging 11.5 million pounds per hour.

cant differences in geothermal drilling in New Zealand conditions have been the need to provide special cooling facilities for the drilling mud and the need to run and cement to the surface all casings except the liner. Earlier wells were 10 and 15 centimeters in diameter, drilled with rigs of 250- and 500-meter nominal capacity. The bulk of the production wells, however, are of 20 centimeters diameter, drilled with a rig having a

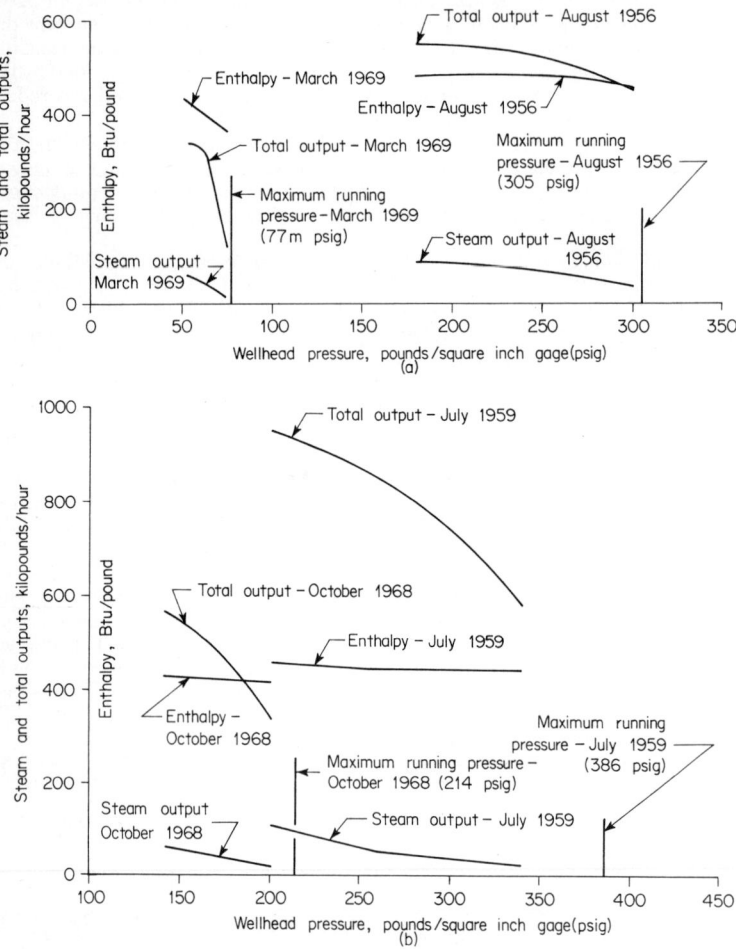

Fig. 5. Wairakei well output characteristics: *(a)* Bore 43. *(b)* Bore 81.

normal capacity of 900 meters. The depth of these wells ranges from 600 to 1400 meters. A typical arrangement of a 20-centimeter-diameter well is shown in Fig. 6.

A total of 102 wells have been drilled at Wairakei, of which 68 have supplied the powerhouse. The remainder consist of investigation wells drilled in the outer areas, and early investigation wells which were not suitable for production. They include one investigation well drilled to 2400 meters, using a 3000-meter-capacity rig.

Steam Collection and Transmission System.   The initial design of the scheme envisaged the separation of the steam and water at each wellhead, with the establishment of two separate supply systems, one operating at about 13.5 atmospheres gage, and the other at 5.5 atmospheres gage. The latter or intermediate pressure (I.P.) steam was to have supplied energy to a heavy water plant, while the high pressure (H.P.) steam was for the generation of electricity. However, the decision was made not to proceed with the production of heavy water. Because design and procurement had reached an advanced stage, and tenders had been called for construction, the two-pressure transmission system was retained.

Also, a pilot scheme was designed and commissioned to test the feasibility of transmitting the hot water to the powerhouse, where it was to be flashed to provide additional I.P. steam and also low-pressure (L.P.) pass-in steam at atmospheric pressure. The pilot hot-water scheme functioned satisfactorily on the available water. However, it was initiated during a period of accelerated drilling and development of the bore field, and largely as a

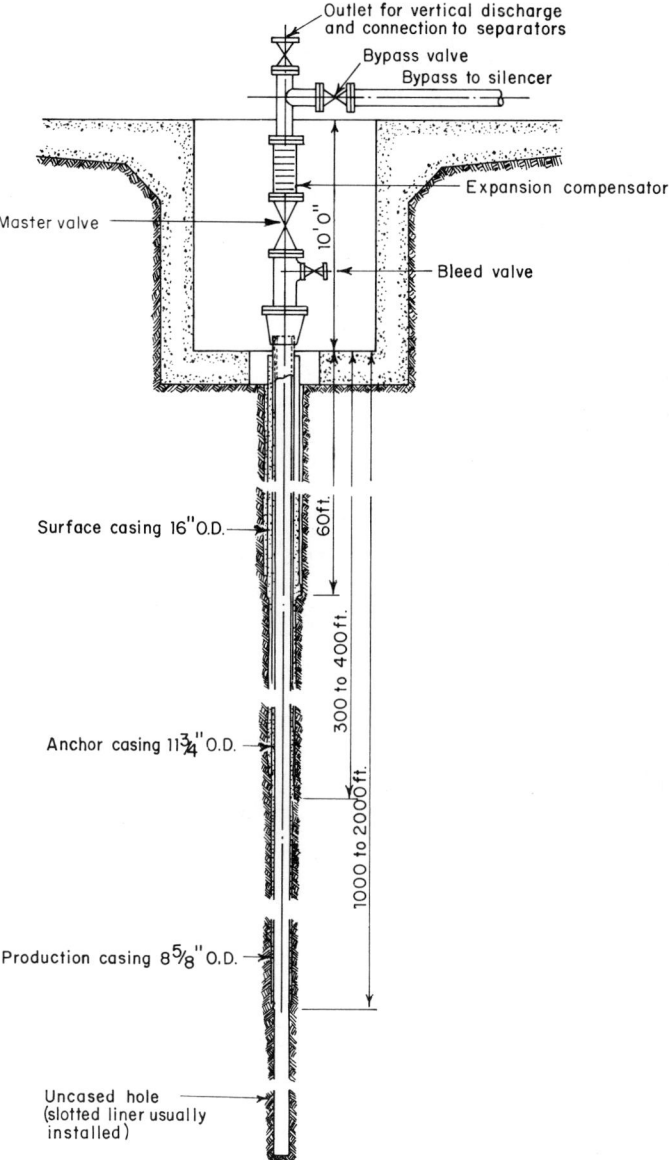

Fig. 6. Typical 20-centimeter-diameter (8-inch) bore.

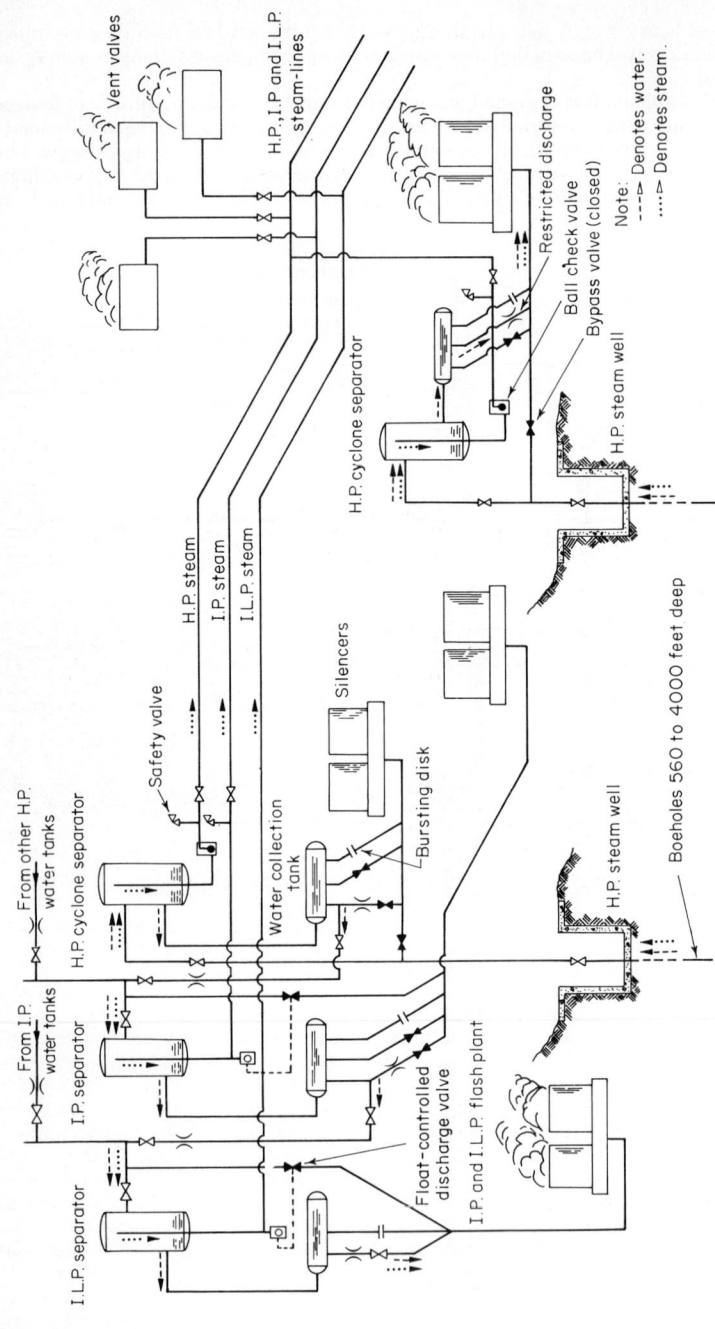

Fig. 7. Steam collection and transmission system.

Vent valves

H.P., I.P. and I.L.P. steam-lines

Restricted discharge

Ball check valve

Bypass valve (closed)

H.P. steam well

Note:
- - - → Denotes water.
...... → Denotes steam.

H.P. cyclone separator

H.P. steam

I.P. steam

I.L.P. steam

Boreholes 560 to 4000 feet deep

Safety valve

Silencers

From other H.P. water tanks

H.P. cyclone separator

Water collection tank

Bursting disk

H.P. steam well

From I.P. water tanks

I.P. separator

I.L.P. separator

Float-controlled discharge valve

I.P. and I.L.P. flash plant

7-20

result of an appreciable increase of total drawoff, the water output from the selected bores had declined considerably by the time the plant was commissioned, and the scheme was finally withdrawn from service because of lack of water.

The development of Wairakei thus continued on the basis of using the separated steam, with the energy in the hot water wasted. However, the substantial quantities of energy available in the hot water were not completely overlooked, and substantial modifications of the collection and transmission system enabling this energy to be used were completed in 1974.

In these modifications, the separated steam from the H.P. wells is collected and sent to the power station as before. The water from groups of these wells is collected and flashed to provide I.P. steam. This steam is collected and transmitted to the powerhouse, together with the separated steam from the I.P. wells. The I.P. water from both the flash vessels

Fig. 8. About 600 meters from the powerhouse, the steam lines drop down an escarpment 21 meters high. Because of the unstable pumice country, massive concrete foundations were needed. Special provision for expansion was also necessary at this point.

and the I.P. wells is collected and flashed again. This steam at 1.67 atmospheres gage is transmitted to the powerhouse, where it is let down through a pressure-reducing valve to 0.1 atmosphere gage and used as pass-in steam. The system is illustrated diagrammatically in Fig. 7. See also Fig. 8.

Choice of Working Pressures.   The original choice of pressures was largely determined by force of circumstance. The heavy-water distillation plant that was originally to have formed part of the project was to have taken geothermal steam at 3.5 atmospheres gage and to have yielded vapor at about 0.1 atmosphere gage. These requirements dictated two pressure systems within the plant area, and it was logical to install condensing turbines to absorb the L.P. steam, and topping sets to extract power from the H.P. wells before exhausting into the distillation plant. The steam from the I.P. wells could conveniently be delivered to the station at about 3.5 atmospheres gage to supplement the exhaust steam from the topping sets. The admission pressure for the topping sets required a compromise, and the original choice of 12 atmospheres gage was a matter of judgment rather than strict logic. Subsequent exploitation of the field has required this pressure to be reduced successively, and it is now reduced to 8 atmospheres gage.

As mentioned, the system has been modified from its original design to make use of energy in the water which was previously wasted. This has resulted in the introduction of a further pressure system, called the intermediate low-pressure (I.L.P.) system, operating at 1.67 atmosphere gage. Thus, the net result is that four pressure systems are in use at Wairakei:

1. *The H.P. System,* in which the feeding wells originally operated at wellhead pressures, ranging from 13.5 atmospheres gage upward, delivering steam to the H.P. back-pressure turbines at 12.0 atmospheres gage turbine stop/valve pressure. Wellhead pres-

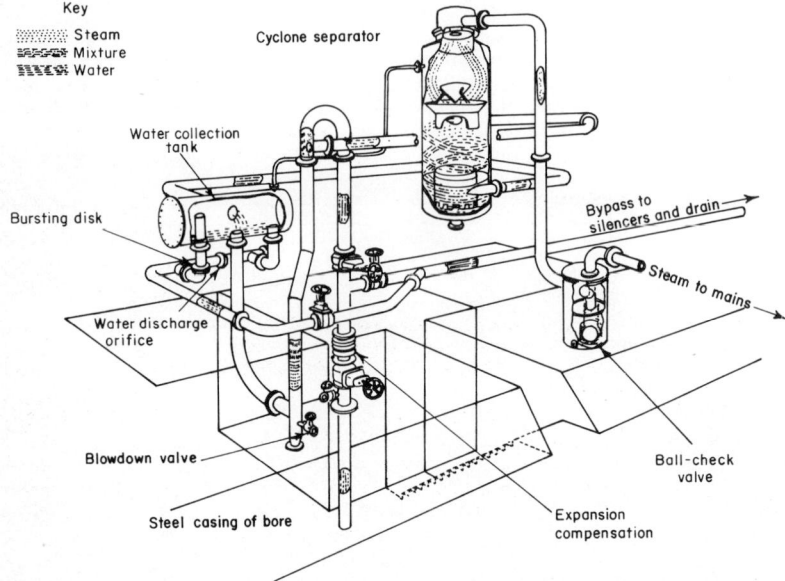

Fig. 9. Wellhead gear with U-bend and top outlet cyclone. Hot water is discarded.

sures have fallen, and the H.P. turbine stop/valve pressure is now about 8 atmospheres gage.

2. *The I.P. System* collecting steam from wells at wellhead pressures ranging from about 5.5 to 8 atmospheres gage and from the intermediate pressure section of the flash units. At the station, this is mixed with the exhaust steam from the H.P. sets and enters I.P. back-pressure turbines at a turbine stop/valve pressure of about 3.5 atmospheres gage.

3. *The I.L.P. System* collects steam from wells operating at about 2 atmospheres gage and from the intermediate low-pressure section of the flash units and delivers it to the station at about 1 atmosphere gage.

4. *The L.P. System* receives exhaust steam from the I.P. sets at about 0.1 atmosphere gage and from the letdown I.L.P. steam and delivers it to the L.P. condensing sets at a stop/valve pressure of just under 0.1 atmosphere gage.

Wellhead Equipment. The chief function of the wellhead gear is to separate the steam from the water as soon as the mixture emerges from the well. Two methods of achieving this have been adopted. The first, shown in Fig. 9, effects the separation in two stages. The mixture is first conducted around an inverted U-bend placed immediately above the well. The water is thrown by centrifugal force against the outer wall of the U-bend, and the steam is drawn off from the inner wall just after the bend. This device acts as a crude separator, removing 80 to 90 percent of the water. The outflowing wet steam is then led to a top outlet cyclone separator, from which the emerging steam is about 99.0 percent dry.

A more recent and simpler method of achieving separation is to use a simple "Webre"-type cyclone without the assistance of a U-bend, which is no longer considered necessary. With this arrangement, the steam/water mixture is led directly from the well through a spiral inlet into a vertical cylinder with a central steam takeoff pipe emerging from the bottom, as shown in Fig. 10. The water collection and pumping system installed on those wells originally connected to the pilot hot-water scheme before it was taken out of service is also shown in Fig. 10. All wells with bottom outlet cyclones now have water collection tanks similar to those shown in Fig. 9, but omitting the connection to the U-bend. See also Fig. 11.

With both types of separator, a ball float valve is fitted in the steam takeoff to prevent water from being carried over into the steam pipelines if a cyclone should be accidentally flooded. Should this ball valve close, there would be a tendency for the pressure in the wellhead gear to rise to the full shut-in pressure of the well, which may sometimes be as high as 27 atmospheres gage and which is considerably higher than the design pressure of the gear. Although this gear is protected by safety valves, a rupture disk is also provided which will burst when the ball valve closes abruptly, thus ensuring that the well output will be discharged to waste and avoid passing erosive water through the safety valves.

The separated water flows to a water drum. Excessive loss of steam through the water line or of water with the steam is controlled by making use of the ability of orifices to discharge boiling water over a wide range of flows without flooding and without loss of seal. Such orifices need not be sized with great precision, and their use has proved entirely satisfactory as self-regulating devices for controlling the discharge of the unwanted water. From time to time, resizing has been necessary owing to the water quantities falling below the range of the orifice.

At each well, there is a bypass connection to enable the well fluid to be blown to waste when not required or when the wellhead gear is out of service for inspection and maintenance. A silencer is also provided at each well to receive rejected or bypassed fluids. This silencer consists of two cylindrical concrete vessels joined tangentially to each other and open at the top. The water/steam mixture impinges onto a cusped steel erosion plate let into the walls of the concrete cylinders at the tangential point of contact. The water swirls around the cylinder walls and is led to waste after most of its kinetic energy has been dissipated through friction and turbulence, while the steam escapes from the tops of the cylinders. Thus, most of the noise is deflected skyward.

At all wellheads, steam flowmeters are installed in the steam takeoff pipes. Pressure gages are also provided for registering the wellhead pressures and the pressure drops across the equipment. By routine logging of the flowmeter and pressure readings, a continuous check is kept on the behavior of the bores.

Steam Transmission System.   The dry steam is led from the wellhead separators through branch pipes to the main steam transmission pipelines. The branch pipes vary in bore from 15 to 30 centimeters, according to length and well output. The main pipelines are 50, 75, 105, or 120 centimeters (nominal).

For the initial installation, 50-centimeter nominal bore pipe was generally the most advantageous, allowing for cost and operational flexibility. This was about the maximum size of seamless pipe readily available at that time, and it was considered desirable to keep the number of welds to a minimum.

For the second stage, it was found that larger pipes would be necessary, but sufficient experience had been obtained from the earlier installation to show that properly executed welds performed quite satisfactorily. Thus, for this stage, 75-centimeter pipe, longitudinally welded, was used.

Spiral welded pipes of 105 and 120 centimeters in diameter were used for subsequent extensions of the system, mainly as a result of the large specific volumes associated with the I.L.P. system.

For the H.P. pipelines, capacity was limited by pressure drop considerations. For a total pipeline drop of about 1.3 atmosphere, the 50-centimeter pipe will deliver 110 tons/hour, and the 75-centimeter pipe about three times that amount. The capacity of the lower-pressure mains is, however, limited by velocity considerations. For both the I.P. and I.L.P. lines, it was considered inadvisable to exceed 45 meters/second, largely because of the possible erosion from slightly wet steam. On this basis, the capacity of the 50-centimeter I.P. line is approximately 90 tons/hour; and of the 75-centimeter pipeline, 206 tons/hour. The 120-centimeter I.L.P. line, operating at approximately 1 atmosphere gage immediately upstream of the letdown valve at the power station, has a capacity of 150 tons/hour.

To take up expansion in the main pipelines, flexible loops are provided at intervals of about 300 meters or less. These loops, which impart negligible thrusts on anchor blocks, are placed in a vertical plane to permit road access beneath them. Each loop, which can absorb about 75 centimeters of pipe movement, incorporates three hinged angular bellows compensators and thus resembles a three-pin arch. Light steel structures support the loops against lateral windage and seismic forces. Between anchors, pipe movement is taken up on rollers. On branch lines, expansion is taken up by means of solid bends and loops without the assistance of flexible bellows pieces.

In transit from the bores to the station, some steam condenses. The condensate is collected in drain pots formed in the mains at intervals of about 150 meters, and is

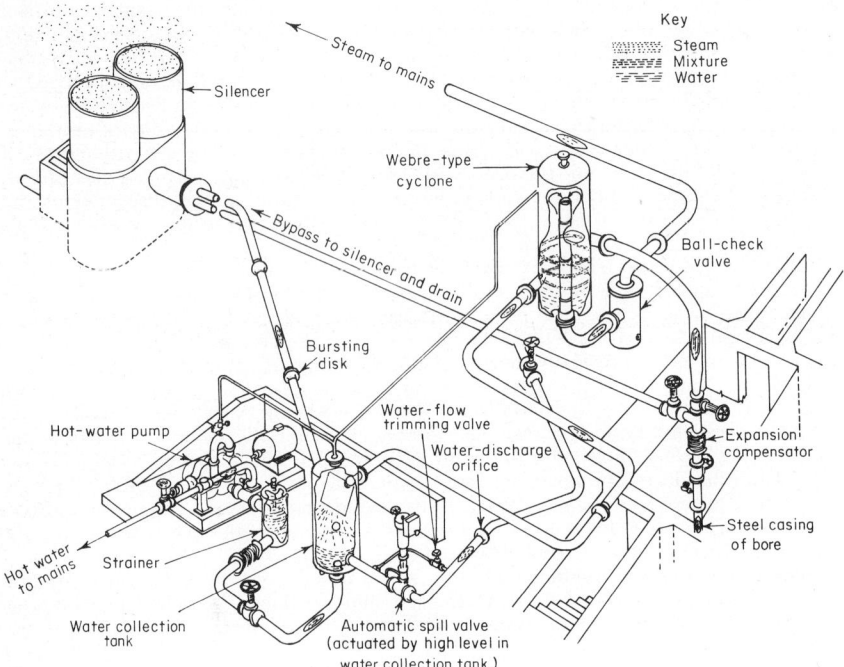

Fig. 10. Wellhead equipment with bottom outlet (Webre) cyclone. This also shows the pumping setup on wells included in the pilot hot water scheme. These have now been replaced with a simple water collection tank.

discharged to waste through traps. Despite the loss of heat entailed, a moderate amount of pipeline condensation is beneficial because, by repeated dilution and partial removal, the water phase becomes highly purified in transit. Thus, the pipeline in effect acts as an efficient scrubber.

Safety valves and bursting disks are provided at each wellhead and flash unit to protect the equipment against abnormal pressure rises, but it is considered that these should operate only as a last resort. Therefore, to protect the three transmission systems, each is provided with automatically operated vent valves. The valves are activated by the pressure at selected points in each system. They normally operate as trimming valves, balancing the supply and demand of steam, but can take the whole discharge from each system if a complete station shutdown should require it.

Hot-Water Transmission.   The major problem in the transmission of the hot water was to prevent boiling in the pipeline. The solution adopted was to raise the pressure by pumping and to lower the vapor pressure by cooling. The latter was accomplished by mixing the water from the I.P. and H.P. bores.

Metallurgical and Chemical Considerations.   Chemical impurities in the geothermal fluids require some special care in the selection of materials. Field tests on material samples initiated in 1950 gave valuable guidance on this problem. Fortunately, mild steel shows good resistance to geothermal fluids so long as no oxygen is present. This material can be used safely for wellhead gear, steam and hot-water pipes, flash vessels, and other equipment. In the initial design, it was considered advisable to add 3 millimeters extra wall thickness as a corrosion allowance. However, experience has shown this requirement to be unnecessary.

Copper and copper-based alloys, other than certain brasses, are vulnerable; and this precludes the use of brazing spelter.

Standby corrosion on stainless steel compensators caused difficulty on one occasion. Stress corrosion cracking occurred on two wellhead compensators. However, these were minor events and the installation has been remarkably free from corrosion problems.

Fig. 11. A typical wellhead setup using a bottom outlet cyclone. The twin tower silencer to the left provides complete control over the water, as well as reducing noise to acceptable levels. The well is located in the foreground and is connected to the separator by the large sweep bend. The steam and water are separated by simple centrifugal action, the steam flowing from the cyclone, to the ball check vessel located at the left of the cyclone and thence to the steam mains through the branch line running out to the right of the photo. The tangential water outlet is connected to the water drum, and thence to the silencer.

POWER PLANT

The arrangement of turbogenerators is shown in Fig. 12. In A station, high-pressure steam is received from the H.P. mains and is passed through H.P. back-pressure turbo sets which exhaust into an I.P. manifold. The exhaust steam here mixes with direct steam from I.P. bores, and the mixture is then expanded through I.P. back-pressure sets, which then exhaust into an L.P. manifold. Finally, the L.P. steam is mixed with L.P. steam from the flash units in the field, and expanded again through L.P. condensing sets.

The reasons for adopting this arrangement of three groups of turbo sets, operating in series, was covered previously in this article under "Choice of Working Pressures." It should be noted that the three mixed-pressure (M.P.) pass-in sets in B station are indicative of the design that would have been adopted had the heavy-water plant not been included in the original design.

A further extension of the B station by the addition of two 30-MW sets was originally contemplated, but this was not done, as experience showed that although the field was approaching a stable discharge rate, it was unlikely that this could be increased to an appreciable extent.

To retain as far as possible an undisturbed balance of steam flows when any set in the A station is taken out of service, bypass reducing valves are connected across the back-pressure sets, and a dump condenser is provided as a flow substitute for an L.P. condensing set. If two or more L.P. sets should be out of service simultaneously, it would be necessary to curtail the load on the L.P. sets.

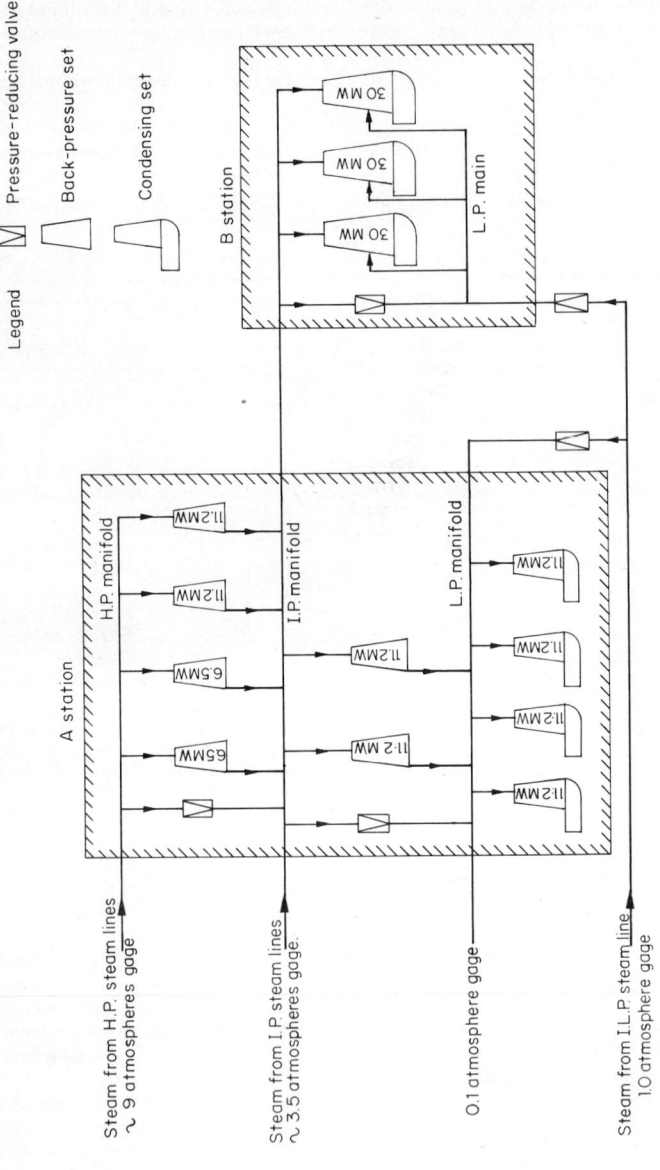

Fig. 12. Layout of turbogenerators. Installed plant: In A Station 102.6 MW
In B Station   90.0 MW

Total.   192.6 MW

Since there is no need to recover the condensate for feed purposes, jet condensers are used for the L.P. and M.P. sets, and the condensed steam passes out to the river with the cooling water. To remove the incondensable gases (bore gas, leakage air, and air released from the river water under vacuum conditions) from the condensers, each L.P. set in the A station was originally provided with a 10-stage high-speed motor-driven exhauster in four separate series units, all running at approximately 14,500 revolutions/minute together. Because of vibration problems and noise, the motor-driven exhausters were not completely satisfactory and are no longer used. Steam-operated ejectors are used in both A and B stations.

The exhaust steam from each back-pressure set would be about 6 percent wet with dry steam at the stop valve. Separators are incorporated in the pipework system to remove most of this wetness, so that steam entering the next downstream turbine may be as dry as possible. In the 30-MW M.P. sets, the wetness of the exhaust steam entering the condenser would be about 14 percent if no steps were taken to remove some of the water at an intermediate point. Before the pass-in belt, the wetness is about 6.5 percent. An integral separator is built into the turbine casing to remove about 85 percent of this moisture, so that after mixing with the pass-in steam the wetness at entry to the first stage of the L.P. cylinder is less than 1 percent. In this way, it is possible to reduce the wetness at entry to the condenser to little more than 10 percent, which is regarded as an acceptable figure from the erosion standpoint.

Choice of Generator Sets. As with the working pressures, the heavy-water plant exercised a considerable influence upon the choice of generating plant for the first stage of development. The distillation plant was to have used about 450 tons/hour of vapor, accepting it at 3.5 atmospheres gage and rejecting it at about 0.1 atmosphere gage. Design and procurement were so far advanced when the heavy-water plant was cancelled that the only reasonable course was to supply a generating plant to utilize this steam fraction. Thus, while the arrangement of three classes of turbines in the A station is cumbersome, it was under the circumstances unavoidable and, of course, was not repeated in the B station. See Fig. 13.

For the L.P. condensing sets, the blade tip speed was limited to 275 meters per second. This limitation resulted from the selection of stainless iron in the soft state, instead of the more usual hard state, as the blade material. This decision was based on the susceptibility of the hard or martensitic iron to stress corrosion cracking. The limitation on the blade tip speed, by limiting the diameter of the last row of blades and the size of the exhaust annulus, limited the capacity of the L.P. sets to under 12 MW.

Inlet and outlet conditions for the I.P. turbogenerators were fixed. With the pressure drop available, the choice was made to install two sets, using about the same steam quantity as the heavy-water plant would have used, and each set generating about 11 MW. This had the advantage of enabling alternators identical with those of the L.P. sets to be utilized.

To the H.P., or topping sets, the topping steam vapor pressure was selected initially as 12 atmospheres gage after a consideration of well characteristics and pressure drops in the mains and wellhead equipments. Two topping sets were obtained, each of 6.5 MW capacity, and rejecting the amount of steam at 3.5 atmospheres gage necessary make up the heavy-water plant steam requirements.

For Stage II, the H.P., and I.P. steam supply systems were faits accomplis, and it was decided that although further topping sets would be required to extract power from future H.P. steam winnings, the expansion from I.P. to vacuum should, in the future, be done in single machines, even though this meant introducing a fourth class of turbo set. The adoption of this longer expansion range enabled much larger turbines to be used. All the Stage I sets were to run at 3000 revolutions/minute, but by adopting a speed of 1500 revolutions/minute, at the same time making use of the greater pressure range, it was possible to design condensing sets of 30 MW rating using single sets, while still restricting the blade tip speed to the conservative figure of 275 meters/second.

Although the Stage I plant was supplied with direct steam only, it was decided to make increasing use of the hot water at future stages. The second-stage flash steam in the pilot scheme could feed the fourth L.P. set in the A station, but the problem arose of how to make use of the second-stage flash steam from future hot-water supplies in the B station. It was decided that this could be used as "pass-in" steam entering the 30-MW turbines at a stage operating slightly above atmospheric pressure.

The size of the additional topping sets was fixed at 11 MW, mainly from a desire for standardization of alternators.

The choice of 30-MW pass-in sets for the *B* station was based, in part, on the probable availability of low-pressure steam from extensions to the pilot hot-water scheme. This did not eventuate, but the pass-in capability of the *B* station machines is now being utilized by the I.L.P. steam produced by the flash units.

Cooling Water System.   To supply the necessary cooling water for the jet condensers, a riverside pumphouse has been constructed. The construction of the Aratiatia hydroelec-

Fig. 13. Interior of part of powerhouse *A*. The high-pressure turboalternators (second and third from foreground) are now supplied with steam from the mains at a pressure of 8 atmospheres gage, and exhaust into the intermediate-pressure manifold at 3.5 atmospheres gage. The intermediate-pressure turboalternators (first and fourth from foreground) are of 11,200 kilowatts each and are fed with steam at 3.5 atmospheres gage, through the H.P. turbines. They exhaust at atmospheric pressure into the low-pressure manifold which can be seen in the pipe bay. The low-pressure manifold is also fed with steam supplied through reducing valves from the intermediate low-pressure main. The fifth and sixth machines are also of 11,200 kW capacity.

tric scheme some three miles downstream on the Waikato River three years after completion of Wairakei resulted in the river being raised by 3 meters. To cater for the two conditions of river level, two points on the characteristic curve were specified when tenders for the pumps were invited.

For the *A* station, four centrifugal pumps are provided for supplying the condensers of the L.P. sets. With the river at low level, all four pumps will be required to serve three running M.P. sets; with the river at high level, one pump running set will suffice. The pump characteristics are as follows:

| *Cooling water* | *Low river level* | | *High river level* | |
pumps serving	*tons/hour*	*meters*	*tons/hour*	*meters*
A station L.P. sets	5200	22	6,900	17
B station M.P. sets	7600	25	10,000	20

Before entering the jet condensers, the cooling water is passed through rotary strainers, one strainer being provided for each condensing set. Solid matter entrained in these strainers is removed by means of scrapers and is flushed to waste. After passing through the condensers, which are of the barometric leg type, the warmed water enters an underground culvert beneath the foundation raft and is conducted to the outfall on the river bank about 200 meters downstream of the pumphouse.

The mean river-water temperature is 15°C (maximum 21°C), and the outlet from the condensers has a mean temperature of 29.5°C (maximum 34°C).

Electric Equipment.   The step-up transformers are in single phase units with one spare unit of each size; and the 220-kV substation is of the ring bus-bar type, the bars being carried on post insulators mounted on concrete stools. The outgoing 220-kV feeders are carried on single-circuit towers. Local distribution feeders are taken from a subsidiary 11-kV switchboard fed from the main 11-kV boards through 1:1 transformers which restrict the fault level and provide voltage regulating facilities.

By comparison with an orthodox thermal station, the number of sizable auxiliaries is small, these being confined mainly to the cooling water pumps. All the auxiliaries in the *A* station are fed from 3.3-kV and 400-volt station boards, but in the *B* station each alternator has a unit transformer and a unit auxiliary transformer which feed a cooling water pump at 3.3 kV and minor auxiliaries at 400 volts. There are also in the *B* station a 3.3-kV and a 400-volt general-purpose switchboard arranged for alternative feeds. A 63-kW automatic starting ac Diesel generator is installed in the *A* station to supply selected sections of main lighting under emergency conditions.

The neutral points of the 220-kV transformer windings are solidly grounded; those of the alternators are grounded through voltage transformers having alarm relays connected to their secondary windings. The 3.3-kV systems in both stations and the 11-kV distribution system in the *A* station are solidly grounded. The 400-volt supplies in both stations have a multiple-grounded neutral system.

The alternators and main step-up transformers are equipped with overcurrent and balanced current protection. All other transformers have overload and ground leakage or restricted ground leakage protection. All transformers except for 400 volts have gas relay protection. Electronic regulators control the generation voltage of each set.

Between the *A* and *B* stations, there is a small auxiliary building which accommodates the compressors for the 220-kV air-blast switchgear, the filtration plant for the transformer oil, and the 63-kW Diesel generator.

Safety Precautions.   The principal safety measures include the following:

1. *High Pressures.*   The parts of the system under pressure are adequately protected by means of spring-loaded safety valves capable of relieving the total inflow of steam from all possible sources without the design pressures of the equipment being exceeded. To avoid the use of safety valves of prodigious size, the L.P. station manifolds, which normally work at only slightly above atmospheric pressure, are designed for a pressure rise of 1.1 atmospheres. The exhaust isolating valves of the back-pressure turbines are interlocked with the stop valves to prevent the exhaust casings from being subjected to inlet pressure.

2. *Water Carry-over.*   The consequences of a slug of water reaching the turbines could be so disastrous that a strong chain of defense is needed between the bores and the station. The ball valves should normally present gross quantities of water from entering the pipelines, and the trapping arrangements along the mains should deal with small quantities of carry-over and with line condensation. However, as a further precaution, water detectors are established at strategic points in the mains to transmit distress signals to the station. These initiate sequential tripping of turbo sets.

3. *Plant Runaway.*   Wairakei power station operates in parallel with a group of hydro stations which are so governed as to allow a momentary rise in frequency up to 35 percent above normal in the event of loss of load. The Wairakei turboalternators are designed for a maximum overspeed of only 15 percent in line with normal steam practice. Hence it is necessary to take precautions against the Wairakei sets being motored by the hydro sets if at any time the whole group of stations were to lose load. Special high-speed frequency-sensitive relays are, therefore, provided which sever the electric link between Wairakei and the hydro stations by opening the step-up transformer circuit breakers when the frequency rises to 7.5 percent above normal. This should prevent the operation of the overspeed trips and leave the Wairakei sets running, ready for resynchronizing with the grid system as soon as normal conditions have been restored.

Efficiency.   The greatest value can be extracted from Wairakei if the stations are operated at the maximum possible load and at high load factor, with absolutely steady conditions. Load despatching in North Island is, therefore, manipulated so that the system load variations are as far as possible taken up by the hydro and fuel-fired stations. The control engineer at Wairakei endeavors to maintain steam pressures at the station (indicated in the control room by means of critical gages) at such values that enable operation of the sets at the highest possible outputs.

In the several years since the station was fully commissioned, it has operated at an average load factor of 89 percent. Most of the unavailable generation was due to planned outages, with unplanned outages due to malfunctioning of any part of the Wairakei system being very small and, in fact, less than unplanned outages due to system malfunctioning.

Effect on the Environment.   The main adverse effects of Wairakei on the environment beyond the bounds of the field are those due to the use of the Waikato river as a heat sink, and as a means of waste water disposal. The mean flow in the river is approximately 125 cubic meters/second, and the average temperature rise from the waste heat is 1.5°C. Wairakei is fortunate in that the total dissolved salts in the geothermal fluids is approximately 4400 ppm, and the mean-river salinity is increased by less than 40 ppm. It is, however, difficult to assess the extent to which the Wairakei, in particular, has affected the river because there have been two other major concurrent changes to the river's environment, namely, the construction of hydroelectric dams, and the reclaiming of land for farmland, necessitating the continuing use of large quantities of fertilizer.

Within the confines of the field, the main adverse effects have resulted from the gas content of the fluid, the effects of uncontrolled blowouts, and subsidence. In the field itself, gas occasionally accumulates in isolated pockets because of wellhead leaks at shut-in wells. The gas is mainly carbon dioxide, with about 5 percent hydrogen sulfide, and care must be taken when entering the cellar of any shut-in well. Other than this, it is difficult to detect any gas in the steam field. Almost all the gas comes off with the steam and is removed from the turbine condensers at the powerhouse. There, it is exhausted to atmosphere through a stack above the powerhouse. In adverse atmospheric conditions, there is some drift to ground level.

Two blowouts have occurred at Wairakei, one due to the failure of a well casing after some years of service; the other during the course of drilling a well. Environmentally, the main effects of a blowout are the waste of a resource, an increased discharge of heat and water locally, and possible danger to other structures in the vicinity. For the last-mentioned reason, one blowout was brought under control as quickly as possible. The other, however, was in an isolated area and, apart from a continuing loss of energy, was causing little trouble. It was left, therefore, unattended. After a period of 14 years, it has sealed itself. During its lifetime, it became a well-known tourist attraction.

The subsidence at Wairakei has a strong vertical and horizontal component and has formed a roughly elliptical dish-shaped depression. The maximum rate of subsidence has been about 0.4 meter per year, with a total maximum subsidence estimated to be over 3 meters. The center of the subsidence is about 450 meters from the nearest wells, and 1800 meters from the region of greatest drawoff. The movement is attributed to the bending of the mudstone cap rock, resulting from the fall in pressure and the withdrawal of the large mass of fluid. The geology of the area of maximum subsidence is not known precisely, but there is a direct correlation between the subsidence and the thickness of the breccia which underlies the mudstone in the region for which geological information is available. Precise measurements taken by tiltmeter during a partial shutdown in 1968 suggest that the subsidence may be reversible.

The ground movement has affected steam mains and drainage channels with some inconvenience, but without jeopardizing the operation of the field. However, the effect of the movements measured at Wairakei on powerhouses or similar structures would be little short of disastrous. For this reason, the possibility of ground movement must always be considered, and a comprehensive system of bench marks must be installed in sufficient time to enable potential areas of subsidence to be located before the powerhouse site is committed.

There have been, of course, some substantial benefits to the environment, and these must also be taken into account in making an overall assessment. Not the least of these benefits is the value to the economy of New Zealand of a low-cost, reliable source of indigenous energy. Others include the contribution that Wairakei has made to the development of the region, the creation of job opportunities in an expanding population, and the creation of a worthwhile tourist attraction.

OTHER GEOTHERMAL FIELDS IN NEW ZEALAND

The work at and near Wairakei has touched only a small part of the country's geothermal resource, which is estimated to approach 2000 MW. In addition to the generation of electricity at Wairakei, geothermal energy is utilized industrially at a pulp and paper mill at Kawerau, and has been employed for many years for domestic and small-scale commercial and industrial use in the city of Rotorua, and other parts of the thermal region.

There are at least another 20 areas in New Zealand showing geothermal potential. Not all will be suitable for development, but investigations are continuing. Exploration drilling has been carried out in connection with eight of these areas. One of these areas, Broadlands, has been drilled up to a proven 150 MW, and the 1975 Power Planning Report recommended the installation of a plant of this capacity for commissioning in 1981. See Fig. 14. Other areas of interest include Orakeikorako, Reporoa, Rotokawa, Tauhara,

Fig. 14. Initial discharge of well 25 at Broadlands. Estimated flow is 430 tons/hour.

Te Kopia, Waiotapu, Waikiti, and Tikitere in the thermal belt; and Ngawha Springs in the far north.

The Kawerau field supplies energy to the Tasman Pulp and Paper Company's mill. See Fig. 15.

Clean steam to supplement boiler steam is produced from two steam generators fed by geothermal steam. One produces 26.5 tons/hour of clean steam at 10 atmospheres gage and the other 22.5 tons/hour at 3.5 atmospheres gage. Feedwater for these units is pumped from the boiler plant hot well through feedwater heaters where it is heated by the geothermal condensate before entering the generator vessels.

In addition to generating clean steam required for mill processes, geothermal steam is used direct. Such applications include timber drying, log-handling equipment, and recovery boiler shatter sprays and liquor heaters.

Geothermal steam surplus to mill requirements is fed at 7 atmospheres gage to a 10-MW noncondensing turboalternator exhausting to atmosphere. Up to 5 tons/hour of the exhaust steam from the turbine is fed to a black liquor preevaporator where it is condensed. This turboalternator operates in parallel with other units fed with boiler steam and with the external power supply. If the latter fails, the geothermal turbine takes preference in the demand for steam. To generate its full-rated output, it requires 160 tons/hour of steam.

Undoubtedly, geothermal heat can be used more efficiently for heating purposes than for electric power generation and the utilization at Kawerau is a good example. The presence of geothermal energy was an important factor in selecting the location of the mill. The field is capable of producing more energy than it is producing at present; and mill expansions, coupled with the energy supply situation, have resulted in further investigations. The energy discharged from KA21, the first well completed as part of the renewed investigations, would produce nearly 30 MW if used for power generation.

SUMMARY

All New Zealand geothermal fields so far drilled are classified as hot-water fields. Formation temperatures, and consequently the enthalpy of discharge, varies from field to field. That at Wairakei is about 260°C; the highest temperature so far measured is 307°C at Broadlands. Generally, the chemistry of the fields is similar, although there are differences in detail. A common feature is a low total dissolved solid content of about 4000 parts per million. This eases a number of problems in utilization found in other fields throughout the world.

Utilization of future fields probably will be for electric power generation, although other possibilities cannot be overlooked. Future developments will be quite different as compared with Wairakei. Current work, both in New Zealand and in other countries, on

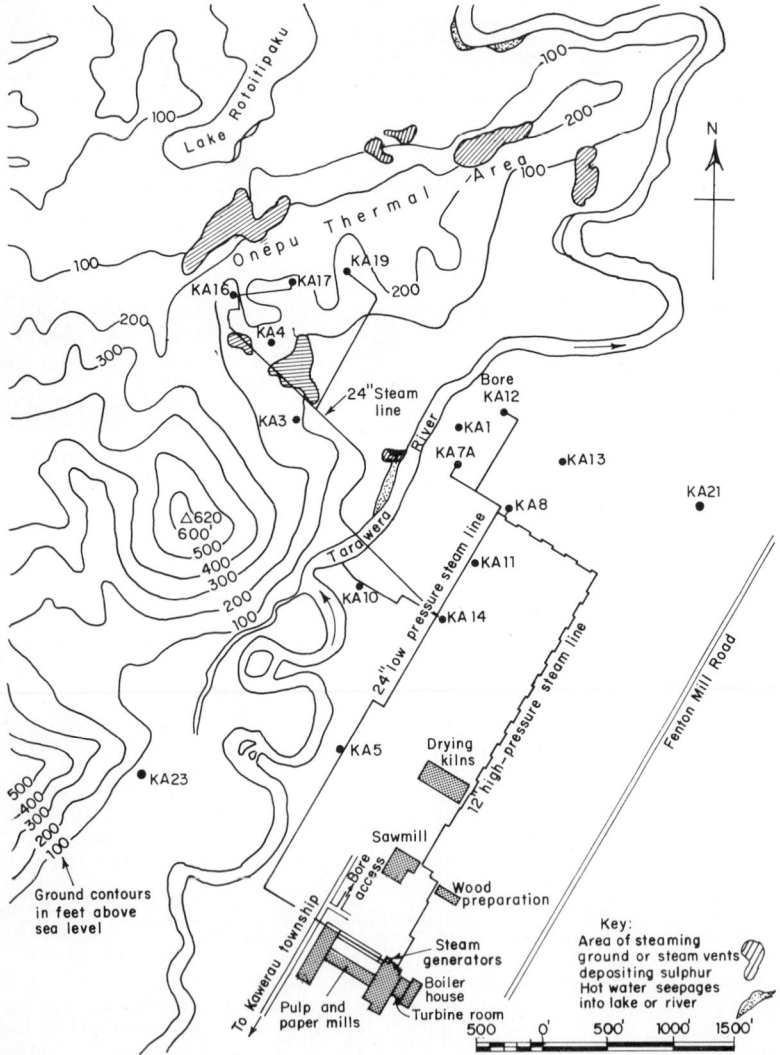

Fig. 15. Kawerau geothermal steam field layout. KA21 and KA23 are the final wells drilled under the renewed investigation program.

techniques, such as reinjection, chemical recovery, two-phase transmission, and the binary cycle, will have marked effects on the appearance and efficiency of future schemes.

REFERENCES

The following papers were presented to the United Nations Conference on New Sources of Energy, Rome, Italy, August, 1961, published by the United Nations, New York, 1961, Proceedings of the conference, vols. 2 and 3, "Geothermal Energy":

Bangma, P.: The Development and Performance of a Steam/Water Separator for Use on Geothermal Bores, Paper G13, vol. 3.

Dench, N. D.: Silencers for Geothermal Bore Discharge, Paper G18, vol. 3.

Hunt, A. M.: The Measurement of Borehole Discharges, Downhole Temperatures and Pressures, and Surface Heat Flows at Wairakei, Paper G19, vol. 3.

The following papers were presented to the United Nations Symposium on the Development and Utilization of Geothermal Resources, Pisa, Italy, Sept. 22–Oct. 1, 1970, published in "Geothermics," *Special Issue no. 2,* 1970:

Bolton, R. S.: The Behaviour of the Wairakei Geothermal Field during Exploitation.

Hatton, J.: Ground Subsidence of a Geothermal Field during Exploitation.

Smith, J. H.: Geothermal Development in New Zealand.

Smith, J. H., and G. R. McKenzie: Wairakei Power Station, New Zealand—Economic Factors of Development and Operation.

Wigley, D. M.: Recovery of Flash Steam from Hot Bore Water.

Also:

Grange, L. I., et al.: Geothermal Steam for Power in New Zealand, *New Zealand Dep. Sci. Ind. Res. Bull.* 117, Wellington North, New Zealand, 1955.

Grindley, G. W.: The Geology, Structure and Exploitation of the Wairakei Geothermal Field, Taupo, New Zealand, *New Zealand Geol. Surv., Bull. n.s.* 75, 1965.

The Geysers Geothermal Field in California

Hiking through the mountains between Cloverdale and Calistoga, California, in search of grizzly bears one day in 1847, explorer-surveyor William Bell Elliott came upon a frightening sight—steam pouring out of a canyon along a quarter mile of its length. Elliott had discovered The Geysers. Situated on the north bank of Big Sulphur Creek, 17 miles east of Cloverdale and 21 miles northeast of Geyserville, the site is in rough and remote countryside. Present geothermal facilities are about 90 highway miles north of San Francisco and located on the steep slopes of a canyon near Cobb Mountain, an extinct volcano.

The Geysers Power Plant* is the largest geothermal installation in the world and the only geothermal power plant in the United States. The units in service as of mid-1975

TABLE 1. Electric Generating Capacity of The Geysers Units

Construction date	Unit	Unit capacity, kilowatts	Cumulative total plant capacity, kilowatts
1960	1	11,000	11,000
1963	2	13,000	24,000
1967	3	27,000	51,000
1968	4	27,000	78,000
1971	5	53,000	131,000
1971	6	53,000	184,000
1972	7	53,000	237,000
1972	8	53,000	290,000
1973	9	53,000	343,000
1973	10	53,000	396,000
1974	11	106,000	502,000
1975	12	106,000	608,000
1976	13*	135,000	743,000
1976	14	110,000	853,000
1976	15†	55,000	908,000

*Steam contract with Signal Oil and Gas Company.
†Steam contract with Pacific Energy Corporation.
All other steam contracts with Union Oil Company of California.

produce > 500,000 kilowatts, using steam at about 100 psi and 350°F (177°C), piped directly from wells tapping a dry-steam reservoir.

The plant capacity at The Geysers, covering over a dozen units, through 1976 is shown in Table 1. Estimates† have been made that up to 6 million kilowatts of geothermal energy might be in operation in the United States by the year 2000. The National Petroleum Council estimated in 1972 that the geothermal generating capacity in the United States, specifically in California and Nevada, could increase to between 7 million

---

*Pacific Gas and Electric Company tested several wells in the area in 1957. These wells had been completed by the Magma and Thermal Power Companies. It was found that a small turbine generator unit could produce power at competitive cost, and thus a steam-purchase contract was negotiated. In 1967, Magma Power Company, Thermal Power Company, and the Union Oil Company of California pooled their interests in a joint venture which now holds approximately 15,000 acres of geothermal property. A surrounding area of about 165,000 acres has been identified by the U.S. Department of the Interior as a known geothermal resource area (KGRA). In 1970, Pacific Gas and Electric Company signed contracts with the joint venture for which Union Oil Company acts as the operator. Pacific Gas and Electric Company also has contracts with the Signal Oil and Gas Company and Pacific Energy Corporation covering steam from nearby holdings.

†Electric Power Research Institute, Palo Alto, Calif.

and 19 million kilowatts by 1983. A more optimistic appraisal was given by the Geothermal Resources Research Conference which indicated that, with an extensive research and development program, 132,000 megawatts of geothermal power could be developed by 1985, and 395,000 megawatts by the year 2000. This wide range of estimates indicates that considerable fundamental knowledge on the existence, exploitation, and efficient utilization of geothermal energy remains to be acquired. The Geysers installations represent, at a minimum, an excellent tangible start for the United States in tapping this energy source.

The geological situation at The Geysers is visualized about as illustrated in Fig. 1. About 20 miles below the crust of the earth, a molten mass or magma is still in the process of cooling. In some places, earth tremors of the early Cenozoic era have caused fissures to open and the magma to come quite close to the surface. This process can cause active volcanoes and, where there is surface water, hot springs and geysers. The hot magma is also responsible for steam vents, termed *fumaroles*, like those found at The Geysers. The steam thrown off by cooling magma is called *magmatic steam*. When surface water seeps down into porous rock heated by magma, the steam formed is called *metoeritic steam*, probably the biggest source of geothermal steam. Scientific investigators are still not entirely certain how the steam is formed at The Geysers.

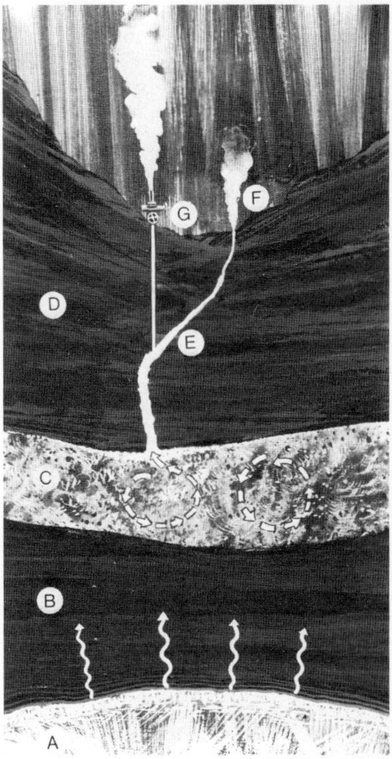

Progression of Installations. An overall map of The Geysers area is given in Fig. 2. The turbine generator used for Unit 1 had originally been installed at Sacramento, California, in 1924. Unit 2, which started up in 1963, is in the same building and has a slightly higher rating. The two units at this location have a combined capacity of 24,000 kilowatts. Units 3 and 4 are located adjacent to each other, about 1.5 miles northwest of the first two units. Unit siting is determined mainly by the location of the steam wells. Units 5 and 6 were placed in operation in 1971 and are located about midway between Units 1 and 2 and Units 3 and 4. Locations of the other units are shown on the map. Units 12 through 15 are not shown on the map, but the sites are in the overall area indicated. Units 3 and 4 are illustrated in Fig. 3. See Table 1.

Steam Supply. The early steam wells were drilled adjacent to the original natural steam vents on 200- to 500-foot (60- to 150-meter) centers to depths of 400 to 1000 feet (120 to 300 meters). These produced steam flows in the range of 40,000 to 80,000 pounds per hour. Employing improved drilling technology, the steam-supplying firms have, for a period of years, tapped a deeper and higher-pressure steam zone at depths between 2000 feet (600 meters) and 7000 feet (2100 meters). Wells up to 9000 feet (2745 meters) have been drilled. Many of these deep wells are far removed from the natural steam outcroppings, and produce considerably higher flows. One was tested at 380,000 pounds per hour flow.

Fig. 1. Cross section of geothermal field. (A) Magma (molten mass, still in the process of cooling. (B) Solid rock, conducts heat upward. (C) Porous rock, contains water that is boiled by heat from below. (D) Solid rock, prevents steam from escaping. (E) Fissure, allows steam to escape. (F) Geyser, fumarole, or hot spring. (G) Well, taps steam in fissure. (*Pacific Gas and Electric Company.*)

The steam-supplying firms provide the steam-gathering piping system, connecting the well heads with the power company's generating units. The steam supply line to a 55,000-kilowatt unit is 36-inch (outside-diameter), ⅜-inch wall carbon steel pipe. This would typically be connected to about seven steam wells. Centrifugal steam separators are installed in the steam pipes to remove any particulate matter and moisture. The steam contains about 1 percent noncondensable gases in approximately the following amounts:

Carbon dioxide	0.79%
Ammonia	0.07
Methane	0.05
Hydrogen sulfide	0.05
Nitrogen and argon	0.03
Hydrogen	0.01
	1.00%

The steam also contains powderlike dust which deposits out in protected areas of the turbines. This dust builds up on the inside of the turbine blade shrouds in the first two stages. In lower stages, the buildup appears to be washed away by water in the steam. This shroud buildup has caused blade and shroud failures. Earlier units have had heavier-duty replacement blades and shrouds installed to mitigate the problem. A turbine water-wash program also may improve the situation.

Materials of Construction.   Before starting detailed design of the first unit, extensive studies were conducted to determine the suitability of various materials for the mechanical equipment and piping. It was found that the steam, as it comes from the wells with a slight amount of superheat, is relatively noncorrosive. Carbon steel can be used successfully for piping in this area, and the turbines do not require special corrosion-resistant materials. As the steam condenses in the condenser, the noncondensable gases become more concentrated, the hydrogen sulfide partially oxidizes to very weak sulfuric acid, and the corrosiveness of the steam and condensate increases greatly. Carbon steel, copper-based alloy, zinc, and cadmium are unsuitable in this area. Austenitic stainless steels, aluminum, or epoxy-fiber glass are satisfactory. Concrete requires a coal-tar epoxy coating to prevent sulfate deterioration.

Hydrogen sulfide in the air causes serious problems in the electric equipment because it is corrosive to copper, copper alloys, and silver. Few special features were incorporated in the design of Unit 1, and thus there were serious problems with electric contacts, relay springs, wiring with exposed copper and other metal parts. Subsequently, various preventive means have been used. Tin alloy coatings have been found to resist corrosion effectively although they have not been satisfactory on current-carrying contact surfaces. Aluminum seems to be particularly impervious to attack, as are stainless steel and some of

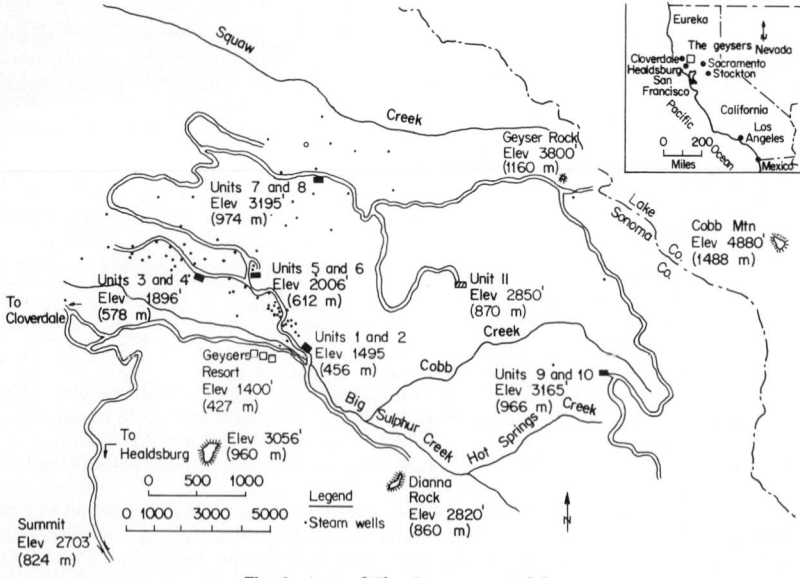

Fig. 2. Area of The Geysers in California.

Fig. 3. Units 3 and 4 at The Geysers went into operation in 1967 and 1968, respectively. These units increased total plant capacity to 82,000 kilowatts. Several additional units have been added during the intervening period. In the foreground are steam pipes with expansion loops, the latter permitting the pipe to contract when the plant has to be shut down, and to expand upon startup. The steam condensate rising from the row of five low stacks at the left marks the location of blowdown valves. When the plant must be shut down, the steam escapes through these valves. *(Pacific Gas and Electric Company.)*

the precious metals. Platinum inserts or plating appears to be a good solution to the problem with contacts, although evaluation has not been completed.

Protective relays are particularly vulnerable to attack, and special relays constructed with noncorrosive materials were procured for Units 2, 3, and 4. After Unit 4, the manufacturerer discontinued special relays. Beginning with Unit 5, the relays, communication equipment 480-volt switchgear, and generator excitation cubicle were placed in a clean-room environment. This is actually three rooms on three levels, maintained at slightly positive pressure with clean air from activated carbon filters.

OPERATION

Several factors led to the determination that The Geysers units could be operated without 24-hour attendance by shift operators. Previous experience with many unattended hydro plants on the Pacific Gas and Electric system had been satisfactory. Geothermal plants are much simpler to operate than fuel-fired steam plants because they have no boilers with all the necessary auxiliaries and controls. Since The Geysers units are basically energy producers, they can be operated at blocked load settings. All the operating and some minor maintenance is performed by roving operators. These persons perform operating duties at all plants in the area.

Protection and Alarms.   Where two units are housed in one building, they share the same high-voltage transmission line, and have a common 480-volt station service bus. Any electric faults that occur beyond either of the two generator breakers requires that both units be tripped and, in addition, that the oil circuit breakers be opened. The Geysers units do not have automatic synchronizing equipment. Therefore, any trip-out of the transmission line requires that the associated units be shut down. Any trip of the transmission-line breakers, with the accompanying unit shutdown, leaves the two units without auxiliary power, and dc emergency oil pumps must be operated to lubricate the turbine-generator bearings until rotation has stopped. Also, on loss of ac power, the hydrogen-cooled machines have their emergency seal oil pumps turned on and are left on until they are manually turned off. The generators are purged with carbon dioxide if the battery voltage is reduced to a low enough level by operation of the emergency pumps. If

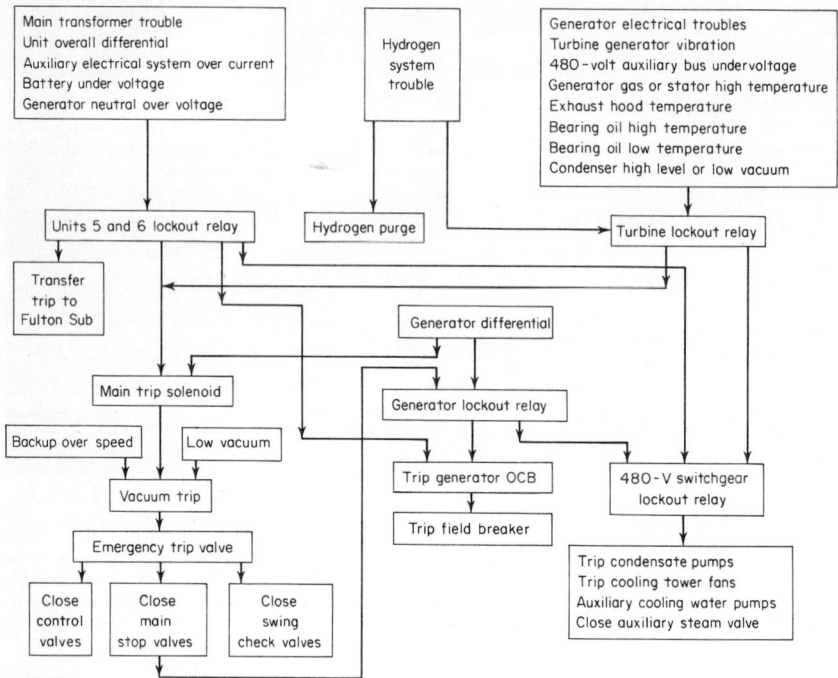

Fig. 4. Turbine-generator shutdown and lockout sequences for Unit 5 at The Geysers plant.

only one unit trips, and the transmission-line oil circuit breaker remains closed, ac auxiliary power will still be available to both units. For any individual unit trip operation that does occur, the main steam trip and check valves close, the generator breaker opens, and the major 480-volt motor loads trip out. For trips due to turbine-related troubles, the main steam trip and check valves close before the generator breaker opens. For trips due to generator electric problems, the generator breaker is tripped simultaneously with the closing of the valves. A block diagram of the major shutdown and lockout features of Unit 5 is given in Fig. 4.

Power Cycle. The power cycle for Units 5 and 6 is shown in Fig. 5. Except for the flow rates and arrangement of the equipment, it is typical of cycles employed by all the units. Steam from the wells is introduced into the turbines, which exhausts to direct-contact condensers located directly below the turbines. The combined condensed steam and cooling water are pumped by two condensate pumps to the cooling water.

The turbine back pressure on all units is about 4 inches (100 millimeters) of mercury absolute. Cooled water from the tower basin is returned to the condenser by gravity and the vacuum head developed by the condenser. Since the cooling tower evaporation rate is less than the turbine steam flow, an excess of water is developed in the cycle. This flow is dependent upon the dry-bulb temperature and relative humidity, but there is a surplus under all operating conditions. For several years, this excess water from the units has been returned to the steam suppliers for reinjection via several wells into the steam reservoir. The reinjection method was tried initially with some concern expressed by the steam suppliers to the effect that it might quench the producing steam wells. However, it has proved successful. It is believed that reinjection can extend the productive life of the steam reservoirs since it is felt that there may be more heat in the reservoir than there is vapor to extract it.

Two-stage steam-jet ejectors are used to purge the noncondensable gases from the turbine condensers. The condensers for these ejectors are also of the direct-contact design.

Turbines. The steam turbines are fabricated largely of the manufacturers' standard materials for low-pressure low-temperature service. Blades and nozzles are typically of 11 to 13 percent chrome steel. Carbon steel is used for the turbine casings. Austenitic stainless steel inserts are provided in the casings opposite the rotating blades to prevent moisture erosion of the casings. The steam inlet conditions for the turbine-generator units are shown in Table 2.

Lower steam pressures were used for the earlier units since their steam supply came from the shallow low-pressure steam reservoir. The later units are supplied from the deep higher-pressure reservoir. All turbines are steam-sealed in a conventional manner.

Cooling Towers. All cooling towers are of the induced-draft type. Structural support members of the distribution headers and basins are redwood, and the tower siding is

TABLE 2. Turbine-Generator Steam Inlet Conditions

Unit	Rating, kilowatts	Steam flow, pounds/hour	Steam pressure, psig	Temperature °F	Temperature °C
No. 1	12,500	240,000	93.9	348	175.5
No. 2	13,750	255,475	78.9	342	172.2
Nos. 3, 4	27,500	509,600	78.9	342	172.2
Nos. 5–10	55,000	907,530	113.7	355	179.4
No. 11	110,000	1,808,000	113.7	355	179.4

transite. Earlier fill material was polystyrene and polypropylene, but polyvinyl is preferred because it is a fire retardant. Cooling tower basins are reinforced concrete painted with coal-tar epoxy to prevent deterioration of the concrete. Cooling towers are designed to cool the condensate from 118.4°F(48°C) to 80°F(27°C) at 65°F(18°C) wet bulb.

Electric Generators. All The Geysers generators are typical of units of their size used elsewhere. The larger units are hydrogen-cooled and are purged automatically under certain trouble conditions. Because outdoor generator oil circuit breakers are used, no enclosed 13.8-kilovolt switchgear is associated with the generator main connections. Outdoor potential transformers are used, and the potential taps to the bus are provided by the cable-bus manufacturer. Four aluminum cables are used per phase for each generator, which gives a rating of 3000 amperes. Cables from each generator oil circuit breaker are connected directly to the single main transformer terminals. Shielded cables with cross-linked polyethylene insulation are used.

Three different types of excitation systems are used on Units 1 through 4. All units subsequent to Unit 4 use static excitation with power potential transformers and saturable current transformers. Elimination of commutators is particularly desirable in The Geysers environment.

Main Transformers. Each of the first four units had its own step-up transformer. This was important for reliability since the earlier units contributed to the local power supply. When unit size increased to 55,000 kilowatts, one main three-phase transformer of 132,000 voltamperes was used for each two units, with resultant savings in cost. Transmission voltages have ranged from 60,000 to 230,000 volts. Dual voltage step-up transformers were purchased for Units 3 through 10. Starting with Unit 11, single-voltage 230,000-volt transformers are possible because the combined plant outputs require 230,000-volt operation.

Metering. Steam purchase contracts provide for payments for power delivered to transmission. Units 1 and 2 and Units 3 and 4 have transmission voltage metering sets of 60,000 and 115,000 volts respectively. Because of the high cost of 230,000-volt metering equipment, Units 5 and 6 and subsequent units are metered at the low-voltage side of the main transformer. Transformer losses are determined from ampere-squared-hour and volt-squared-hour meters and subtracted.

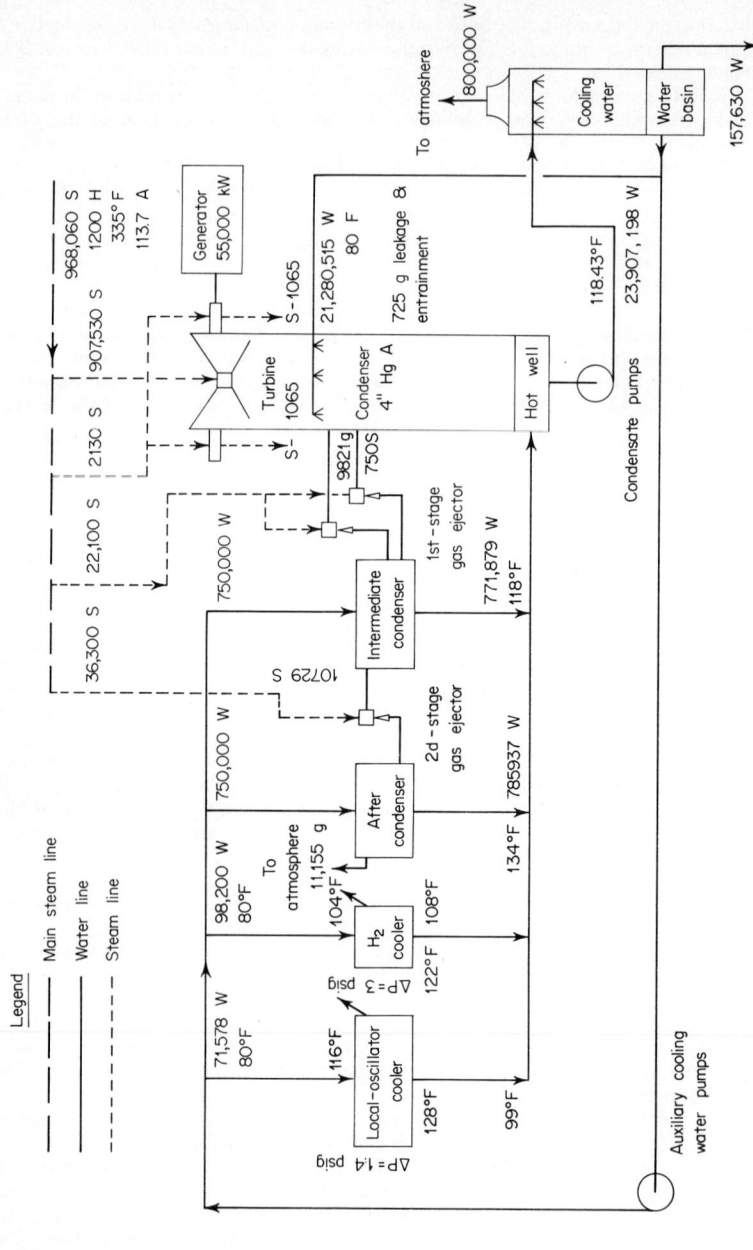

Legend

—— Main steam line

—— Water line

--- Steam line

7-40

Fig. 5. Heat-Balance diagram (designed load) for Units 5 and 6 of The Geysers plant.

S = lb/hr of steam   W = lb/hr of water

g = lb/hr of gases

H = enthalpy of steam, Btu/lb   A = absolute pressure, psia

G = gage pressure, psig

Performance:

Throttle flow .... 907,530 lb/hr

Generator electric output .... 55,000 kW

Auxiliary power, electric:

   Circulating water pumps .... 930 kW

   Cooling tower fans .... 605 kW

   Other .... 445 kW

     Total .... 1980 kW

Net unit output .... 53,020 kW

Net heat rate referred to 60°F .... 21,690 Btu/kW-hr

Conditions:

Generator power factor 0.90

Turbine exhaust back pressure .... 4 in. Hg

Dry-bulb air temperature .... 80°F

Wet-bulb air temperature .... 65°F

All steam flows include 1% noncondensable gases

REFERENCES

White, D. E., L. J. P. Muffler, and A. H. Truesdell: Vapor Dominated Hydrothermal Systems Compared to Hot Water Systems, *Econ. Geol.*, **66**, 1 (January–February 1971).

Finney, J. P., F. J. Miller, and D. B. Mills: Geothermal Power Project of Pacific Gas and Electric Company at The Geysers, California, presented at the Summer Meeting of the IEEE, July 1972, Institute of Electrical and Electronics Engineers, New York.

Bowen, Richard G., and Edward A. Groh: Geothermal—Earth's Primordial Energy: in "Energy Technology to the Year 2000," Massachusetts Institute of Technology, Cambridge, Mass., 1972.

White, D. E. and D. L. Williams: "Assessment of Geothermal Resources of the United States—1975," U.S. Geological Survey Circular 726, Branch of Distribution, U.S. Geological Survey, Arlington, Virginia, 1975.

Geothermal Energy for Space and Process Heating

BY BALDUR LINDAL*

As pointed out in a preceding article by Bowen and Groh in this Handbook section, there are geothermal sources of energy in numerous parts of the world, and there is a significant variation in the temperature, pressure, and chemical characteristics of the geothermal steam or hot water that may be produced. Perhaps generally in the literature of recent years less emphasis has been placed on the *non*power applications for geothermal energy, such as space and process heating uses, than on the use of this energy to generate electric power. Also, inadequate emphasis has been placed on the multipurpose applications of geothermal energy, where the primary purpose may be, on the one hand, to generate power, or, on the other hand, to provide commercial and residential heating, or still in another case, to provide thermal energy for processing and manufacturing, but with the opportunity to fully exploit the energy either with two or all three primary purposes and numerous secondary purposes. The flexibility with which the energy from a geothermal source can be applied varies, of course, with the quality of the output of the source, but unless the full spectrum of possibilities is thoroughly explored, numerous opportunities can be overlooked.

Application of geothermal energy to electric power generation is a reasonbly straightforward extension of existing steam technology, and the comparison of the costs of geothermal steam with the costs of fossil-fuel generated or nuclear-reactor generated steam do not involve basically new equipment and process design philosophies. In geothermal energy for electric power, it is essentially a matter of justifying an investment in a geothermal facility as compared with readily apparent alternatives. Particularly in the case of geothermal energy for processing and manufacturing, there is need for new engineering perspectives and responses. This is especially important because all geothermal energy sources are not suited to efficient electric power generation. There are, however, many sources that can provide excellent low-cost thermal energy both for space heating and for processing and manufacturing. Some of these aspects are described in this article.

GEOTHERMAL ENERGY FOR HEATING

As a rule, energy for heating is extracted from geothermal fluids which are predominantly water. These fluids are made available through drilling into hydrothermal reservoirs. The location of the sources of geothermal fluids and the feasibility of practical recovery of the fluids is a matter of local geology. Geological aspects dictate the chemical composition of the fluids and the original temperature.

The discharge of drill holes may consist of (1) water only, (2) a mixture of steam and water, or (3) dry steam. In a mixture of steam and water, the ratio of steam to water will depend on the temperature in the geothermal reservoir (which generally will have the liquid phase only) and on the back pressure applied. These relationships may be calculated according to the thermodynamic behavior of water, assuming constant enthalpy boiling. Each borehole, however, will have its own pattern for total flow, and the total flow will be governed by the back pressure in the more elevated pressure range. An example of such discharge behavior is shown in Fig. 1.

For wet steam, a discharge pressure of 6 to 10 atmospheres is common for long-time operations, but higher operating pressures are possible in some high-temperature areas. Thus, steam condensing temperatures up to around 200°C(392°F) are possible in some cases, but the range from 160 to 180°C(320 to 356°F) is the most common.

Besides water, geothermal fluids contain chemical components which may influence

---

*Consulting Engineer, Virkir Consulting Group Ltd., Reykjavík, Iceland.

the means of application. A classification of the various types of geothermal water was made by White (1957), the most important being as follows:

1. *Sodium Chloride Water.*   Most common in large underground geothermal reservoir systems. The water is usually alkaline at discharge. The most common anion is Cl⁻ (chloride); the principal cation is Na⁺ (sodium).

2. *Acid Sulfate Chloride Water.*   Relatively rare; acid at discharge.

3. *Acid Sulfate Water.*   Common in fumarole areas. Springs usually have little or no discharge.

4. *Calcium Bicarbonate Water.*   Common at low temperatures. May deposit calcium carbonate at discharge.

Within the same geothermal area, more than one type of geothermal water may occur. Sodium chloride water is of the greatest economic significance. Most practical uses of

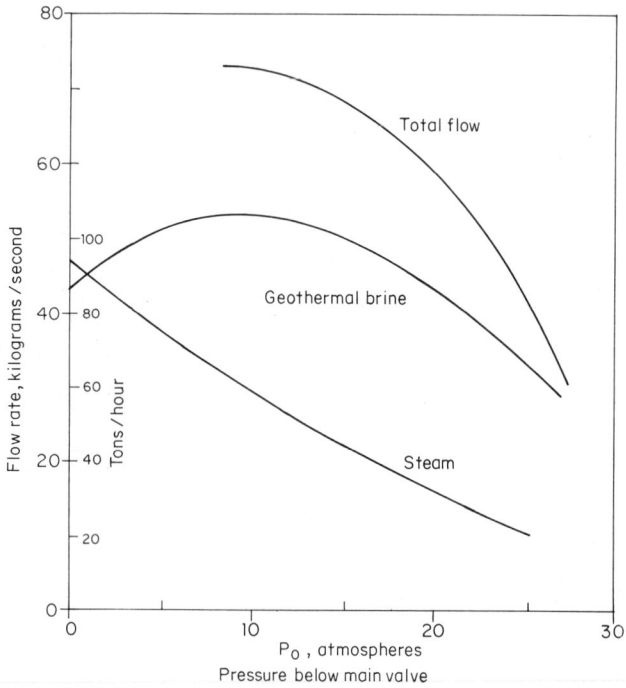

Fig. 1. Flow characteristics of Borehole 8—Reykjanes. (SOURCE: *The National Energy Authority, Reykjavik.*)

geothermal water involve this type. Unless otherwise stated, the accounts presented in this Handbook description refer to this type of geothermal water.

Geothermal fluids also contain some noncondensable gas. At low temperatures (below 100°C), this is usually mostly nitrogen. In bicarbonate water, carbon dioxide also will be present. Generally, the amount of nitrogen will be only a few milliliters per liter of water, but if carbon dioxide is also present, this may vary greatly. At more elevated temperatures in the reservoir temperature range of 200 to 300°C (392 to 572°F), carbon dioxide generally becomes predominant, and there will also be hydrogen sulfide, hydrogen, and other minor components. This noncondensable gas will be in the steam phase. When the steam condenses in a heating operation, this gas will be partly dissolved in the condensate. The amount of such gas varies widely. A common range is 500 to 2000 milliliters (at standard temperature and pressure conditions) of gas per liter of condensate. In some locations, the gas could be as much as ten times this amount.

A complete record of the chemical characteristics of the geothermal fluids under consideration should be obtained and considered carefully in each specific case of a potential application for heating purposes.

Corrosion.   Geothermal fluids may cause considerable corrosion and create fouling and other operating problems if proper preventive measures are not applied.

The geothermal fluid obtained from the wet type of reservoir is water containing, for instance, some silica, chloride, sulfate, carbonate, sodium, potassium, calcium, hydrogen sulfide, and carbon dioxide. At the more elevated temperatures, most of these elements are present in higher concentrations than at the low-temperature level.

While geothermal water below the atmospheric boiling-point temperature is used as such, steam is usually flashed from the water of more elevated temperatures before use. The volatile impurities (hydrogen sulfide and carbon dioxide) will follow the steam phase, while the other components will remain in the water phase. In a similar manner, dry steam will contain the volatile components (hydrogen sulfide and carbon dioxide). Thus, the nongaseous components are usually significant in water-phase corrosion in geothermal systems, whereas the gaseous components are of significance in the steam phase, both in the condensate and in atmospheric corrosion.

Materials of Construction.   Low carbon steels are the most significant material of construction for geothermal systems, although various other metals and alloys are used for special purposes. Regarding the higher-temperature range, a summary is made of the results of surface corrosion measurements in Table 1. The following general conclusions were drawn by Marshall and Braithwaite (1973):

1. In air-free geothermal steam and high-temperature water, corrosion rates of the common engineering alloys are usually higher (somewhat) than those encountered in clean boiler-plant steam and water under similar temperature and pressure conditions.

2. Corrosion rates of most common engineering alloys in air-free geothermal fluids, with the possible exception of copper-base alloys, are low enough for their practical use in the construction of geothermal plants.

3. Aeration of geothermal media drastically accelerates the corrosion rate of most engineering alloys, with the notable and useful exceptions of austenitic stainless steel, titanium, and chromium (plating).

4. Surface corrosion of carbon steel and galvanized steel becomes severe in atmospheres contaminated by saline spray from geothermal bores, requiring the use of aluminum, stainless steel, or protective coating if the contamination cannot be avoided. Surface tarnishing of copper and silver is also very rapid in atmospheres contaminated by hydrogen sulfide, and may become of consequence in electrical, telephonic, and building rainwater equipment. Aluminum is not tarnished by hydrogen sulfide.

The corrosion rate for mild steel in contact with low-temperature geothermal water of the sodium chloride type is usually low, provided that oxygen is kept absent. In Table 2 some specific cases are reported, according to S. Hermannson et al. (1975). There is some pitting of mild steel as a rule, and, together with the wall thickness, this will determine the useful life of the structure in question. Ordinarily, pipes of mild steel may last 30 years, or even much longer. However, as evidenced by system 9 (Table 2), major amounts of dissolved oxygen will enhance corrosion. If appreciable chloride is present, complete elimination of dissolved oxygen is essential. There are, however, exceptional cases where high rates of pitting are found without apparent explanation. Therefore, it is advisable to make corrosion rate tests in every case of a projected use of a new source of low-temperature geothermal water.

Results from sulfide stress cracking investigations have been reported (Marshall and Braithwaite, 1973) with the following general conclusions drawn:

1. Low-strength carbon and alloy steels resist sulfide stress cracking in geothermal fluids. The strength level at which susceptibility to cracking occurs is not clearly defined and probably varies with type of steel, severity of stressing, and specific environmental conditions. One investigator reports that the threshold strength level for sulfide stress cracking appears to be about 88,000 pounds per square inch for carbon and low-alloy steels, and in the range of 110,000 to 120,000 psi for high-chromium steels.

2. Medium- and high-strength carbon and alloy steels are susceptible to sulfide stress cracking under various environmental conditions in geothermal media up to at least 190°C(374°F).

3. Sulfide stress cracking in geothermal media is influenced by the physical condition of the specific environment.

TABLE 1. Surface Corrosion Rates of Metals in Geothermal Media
Values are in mils[1]

Metal	Bore water,[2] >200°C	Water,[3] ~125°C	Steam,[4] 100-200°C	Aerated steam[5] ~100°C	Condensate,[6] ~70°C	Condensate/fresh water mixture,[7] ~50°C	Highly acid thermal water[8]
Titanium	0	0	0	0	—	—	0
Chromium (plating on steel)	0	—	—	—	—	—	—
Aluminium	1	0.8-P	0-P-Ĩ	0-P	0.2	9	28
Zinc (coating on steel)	S	1	0-1-P	S	0	S	—
Austenitic stainless steels[9]	0.1	0	0	0	0	0	22
Ferritic stainless steels[10]	0-0.1	0.1-P	0-0.3-P	1-P	0.1-P	0-0.5	1000
Carbon and low alloy steels	0.3-0.4	0.3-0.5	0.3-6	20	3	30-170	—
Gray cast iron	1	0.4	1-3	10	—	90	8
High silicon cast iron	—	—	0.5	1	—	—	—
Brasses[11]	5	0.3	0.3-0.6	40	0.2	—	—
Bronze	20	—	2	9	1	—	—
Aluminium bronzes	10	—	2-3	10	—	—	—
Silicon bronze	9	—	3	20	—	—	—
Cupronickel	10	—	2	—	—	—	—
Beryllium copper	20	10	4	40	5	—	—
Copper	6	10	2	40	2	—	—
Nickel	8-10	—	1	8	—	—	—
Monel and K Monel	0.3	1	2-4	10	4	—	14
Nimonic 75	1	—	0	—	—	—	—
Inconel	—	0	0-0.3	80	—	—	20
Lead, antimonial lead	—	—	0.5	2.5-P	—	1	6

SOURCE: Marshall and Braithwaite, 1973. (Ref. 5)

1. 1 mil = 0.001 inch
2. Tests in water at bottom of a closed geothermal bore.
3. Water separated from wet geothermal steam at wellhead.
4. Steam separated from discharging geothermal bore.
5. Geothermal steam mixed with injected air.
6. Geothermal steam separated and condensed under pressure.
7. Geothermal steam condensed with fresh water to stimulate fluid in a jet condenser hot well.
8. Natural water in a volcanic crater (Tombs, 1960).
9. 18/8 CrNi, 18/8/3 CrNiMo, and 18/12/2 CrNiMo varieties.
10. 13 Cr, 17 Cr, 17/2 CrNi varieties.
11. 60/40 CuZn, arsenical 70/30 CuZn varieties.
I = internal attack with embrittlement.
P = pitting.
S = zinc coating stripped.

TABLE 2. Corrosion Rates¶ for Mild Steel in Geothermal Water of Less than 100 C and of Varying Chemical Composition of Water*

System	1 Reykjahlíð†	2 Reykir†	3 Bolholt†	4 Seltjarnarnes	5 Húsavík	6 Hvammstangi	7 Mógilsá	8 Selfoss	9 Dalvík
Temperature at testing site	86°C (186.8°F)	73°C (163.4°F)	83°C (181.4°F)	84°C (183.2°F)	80°C (176°F)	80°C (176°F)	70°C (158°F)	82°C (179.6°F)	55°C (131°F)
Corrosion rate μ/year‡	2.8	0.7	1.0	0.6	1.2	4.8	40	16	55
Max. pitting depth, μ/year‡	60	50	50	0	120	105	510	320	275
pH	9.75	9.70	9.55	8.50	9.55	9.45	8.05	8.75	10.25
Oxygen (O_2), ppm	0.0	0.0	0.0	0.0	0.0	0.0	0.0	<0.1	1.2
Sulfide (HS^-), ppm	1.4	0.4	0.3	<0.1	1.0	0.4	1.5	<0.1	<0.1
Silica (SiO_2), ppm	91	74	128	117	186	98	124	69	90
Calcium (Ca^{++}), ppm	2.0	2.8	2.4	105	9.1	27.8	10.6	25.5	1.9
Sodium (Na^+), ppm	46.4	44.0	53.6	337	47.0	160	81.0	170	42.0
Magnesium (Mg^{++}), ppm	<0.1	<0.1	0.0	0.2	0.0	0.0	0.1	0.1	0.0
Chloride (Cl^-), ppm	14.3	15.2	27.2	528	10.1	141	28.5	245	8.3
Fluoride (F^-), ppm	1.1	0.6	0.8	0.9	1.1	2.0	1.3	0.2	0.5
Sulfate (SO_4^-), ppm	20.2	15.0	17.1	180	27.8	137	50.2	51.4	13.8
Bicarbonate, (HCO_3^-), ppm	30.0	30.0	27.0	20.0	46.0	17.0	130	40.0	13.7
Carbonate (CO_3^{--}), ppm	11.0	10.0	7.0	0.6	9.6	3.0	0.9	1.0	13.7
Dissolved solids, ppm	228	194	280	1382	336	617	367	603	224

*S. Hermannson et al., 1975 (Ref. 9).
†Supply stations to Reykjavík District Heating System.
‡25.4 μ = 1 mil.
¶One-year tests.

Sulfide stress cracking is a factor of major importance in the design and operation of equipment, particularly turbines, for the utilization of geothermal steam and water. The use of medium- and high-strength steels is believed to be dangerous, on present knowledge, except in situations where the specific environment is known to be innocuous. Low-strength (less than 88,000-psi) steels are known to be safe for use in steam media under reasonable service stresses.

Stress corrosion of austenitic stainless steels in hot chloride solutions is of importance with geothermal fluids. The minimum requirements of chloride for stress corrosion are very low.

Fouling in Geothermal Systems.   Although several minor elements in geothermal water may cause fouling in heating systems and supply mains, mainly two components may cause very rapid scaling; i.e., silica and calcite. The noncondensable gas in the geothermal steam may also accumulate if it is not properly vented out of the heating system.

In a geothermal reservoir, the water will dissolve silica from the rock according to the solubility relationships of quartz at temperatures above about 180°C(356°F), but apparently equilibrium with chalcedony occurs at lower temperatures. Both these silica minerals display a rising solubility curve with temperature, the chalcedony solubility being slightly higher than that of quartz. Upon cooling of the water (coincident with any thermal utilization), concentrations of silica which are above the solubility limits of quartz and chalcedony will be reached. The silica does not readily precipitate as quartz or chalcedony upon sudden cooling at moderate temperatures, but may precipitate if the solubility limit of the much more soluble opaline form is reached and polymerization readily occurs upon supersaturation.

The solubilities of quartz and opal in pure water are shown in Fig. 2. Guiding lines in the diagram reveal how flashing of the water influences the limit of opaline saturation. High pH values of the water will increase somewhat the value for total silica required for the saturation limit, because of the dissociation of silicic acid. Arnórsson (1975) provides further details. See reference list.

When reservoir temperatures are less than 160°C(320°F), and if quartz equilibrium is involved, the maximum silica will be 150 parts per million. With reference to Fig. 2, it will be noted that opaline silica may begin to form by cooling to 25°C(77°F); or if flashing down to 100°C is involved by cooling to 34°C(93.2°F). Comparable figures for original equilibrium with chalcedony are 44°C(111.2°F) and 52°C(123.8°F), respectively. If high pH values are also involved, these limits will be somewhat lower. Since these temperatures coincide with the lower limits for most heating applications, self-induced precipitation of silica is not usually found with heating systems that use water from such low-temperature geothermal reservoirs.

However, silica may accumulate slowly in low-temperature systems. Thus, coatings of zinc, such as galvanized steel, are to be avoided—because silica will precipitate on their corrosion products as well as on the corrosion products of copper and steel. Fortunately, effective oxygen control can minimize steel corrosion. It is most often satisfactory to keep the oxygen content below 0.05 part per million in systems that operate at a maximum temperature of 100°C.

With reference to Table 2, geothermal water heating systems 1 through 6 exhibit mild scaling conditions, and the water is used directly in radiator systems except in system 5 where the silica is on the high side. A high rate of scaling is found in systems 7 and 9, in keeping with the high rate of corrosion. It is of interest to note, however, that a slow formation in such systems may be beneficial because of some retardation of oxygen corrosion.

In water systems of reservoir temperatures above 160°C(320°F), it is important to take account of the limit for opaline silica precipitation. The precipitation of silica and coincident fouling above the opal solubility limit may be extremely severe in some instances.

Calcite fouling is generally associated with the bicarbonate type of water only. When carbon dioxide is released from such a system, the pH is increased, and calcium carbonate may precipitate. This will usually occur in the supply system itself. But, if such water is used for heating purposes, a dilution with fresh water may be advisable in order to avoid fouling.

Upon the condensation of geothermal steam in a heating system, the noncondensable gas will concentrate. Since noncondensable gas will affect the heat-transfer coefficient adversely, the gas must be vented. In order to preserve a high heat-transfer coefficient, the steam is not condensed fully. At the end of the path of condensing steam, the gas, along with a small part of the original steam charge, is vented. This may coincide with the discharge of condensate, and a steam trap may not work properly except by direct bleeding of this type.

Since geothermal gas includes hydrogen sulfide, any leaks of steam or of the more concentrated noncondensables may be harmful to equipment and personnel. Such leaks commonly occur with valve stems and other equipment which may use gland packings. Since it has proved very difficult to prevent such leaks completely, any system involved in the use of primary natural steam should be compact wherever possible, and good ventilation should be provided.

TEMPERATURE RANGES OF GEOTHERMAL FLUIDS

The range of useful temperatures of geothermal fluids may extend from as low as 20°C(68°F) to as high as naturally available. At the low temperatures, the energy carrier is usually the liquid water phase; at more elevated temperatures, it is the steam phase.

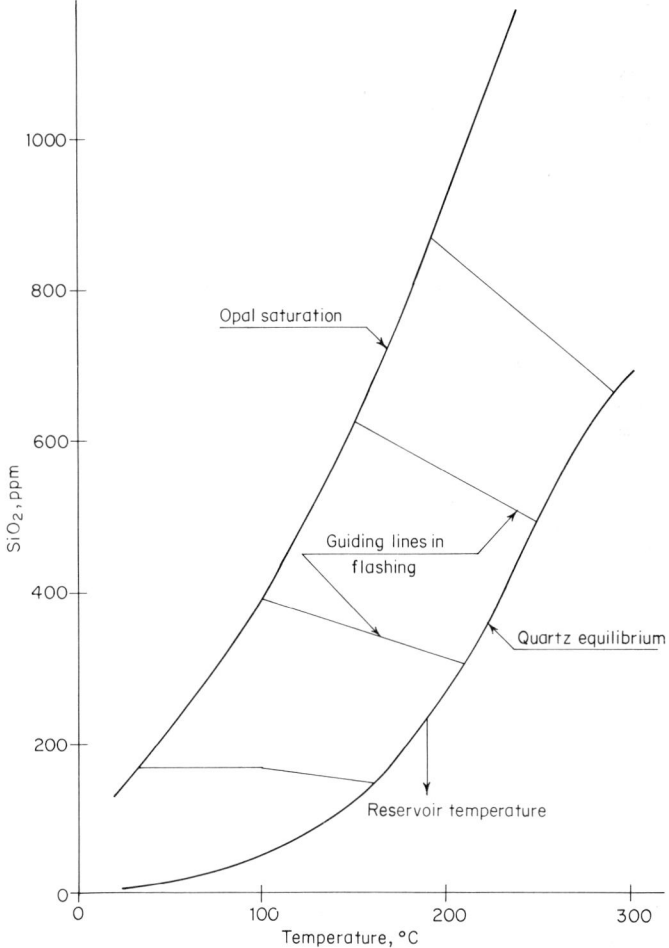

Fig. 2. Solubility of quartz and opal in water. (Based on data given in Ref. 1.)

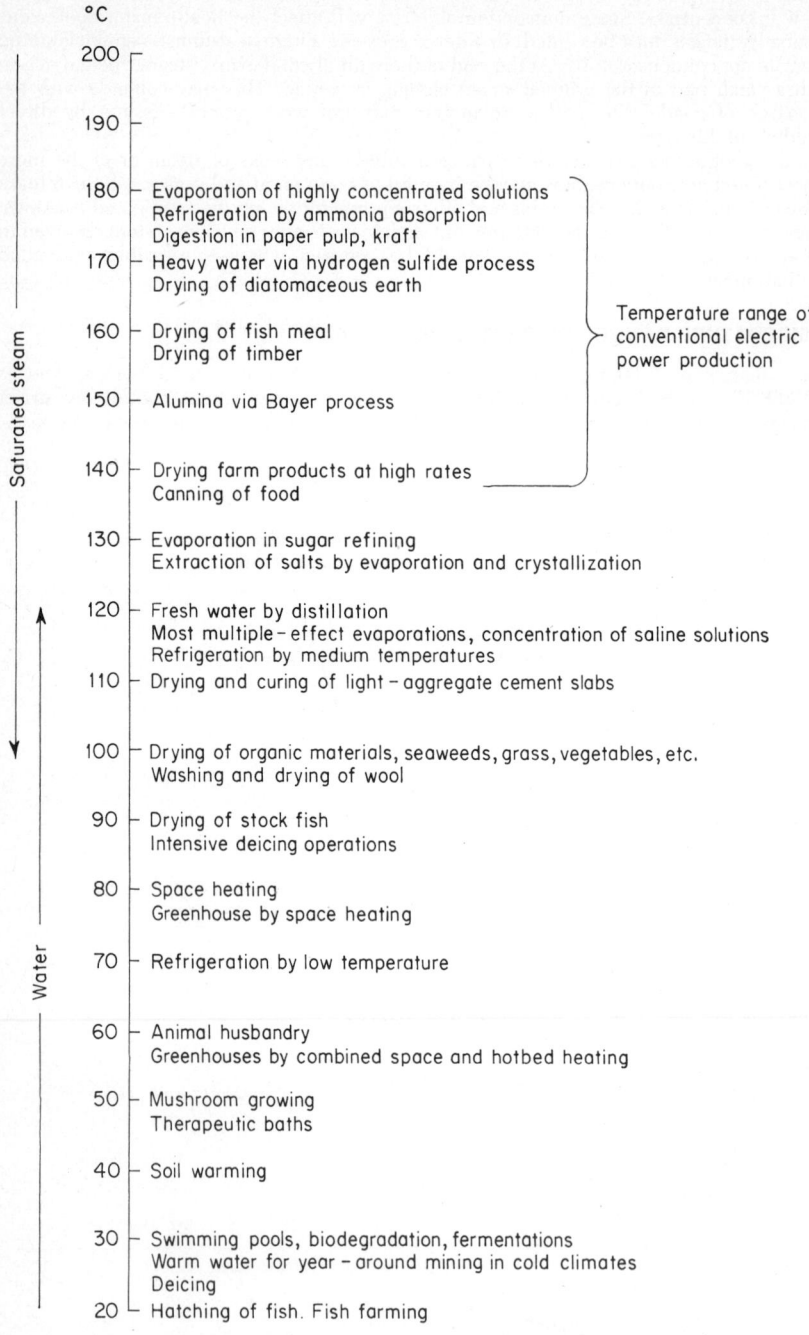

Fig. 3. Applications versus temperature range of geothermal water and steam.

Where a source of geothermal energy is available, the question arises: What use can be made of a fluid having a heating potential which is predetermined by geological factors? Considering the present pattern of applications, the dominating usefulness in the low-temperature range is for space heating and greenhouses; while the generation of electricity and, to some extent, process heating dominates in the high-temperature range.

The spectrum of temperature range for geothermal water and steam is plotted against possible applications in Fig. 3. The applications are arranged for the feed temperature to the system required. The chart is for general guidance, particularly since the useful temperature for most applications is not tightly fixed, but rather represents a range depending upon various factors. Of course, any specific application will have its lower economic temperature limit and an optimum temperature. However, a feed temperature that may be high usually can be adjusted without difficulty.

TRANSPORTING GEOTHERMAL ENERGY

Natural steam, which ordinarily will be available at pressures ranging from 5 to 20 atmospheres, is usually exploited within a few kilometers of the source. The high volume of steam makes pipelines relatively expensive, and the pressure loss resulting from long-distance pipes may decrease the usefulness as a heating agent. Therefore, piping systems for steam are usually limited to the collection of steam from a geothermal area to a suitable site for utilization, preferably quite close and if possible within the thermal area. Thus, the economics of many process heating applications is such that it may be more suitable to transport raw materials to the source of steam instead of vice versa.

Steam pipelines are made from ordinary low carbon steel, with suitable means for expansion. They are insulated as with glass wool, for example, covered with an aluminum jacket.

When the steam is wet, the water is usually separated by special cyclones located at the borehole. Hence, each phase is piped separately. However, two-phase flow in pipelines is possible.

Transportation for major distances usually takes place with the water only. Main-supply pipelines for district heating systems of up to 16 kilometers have been in use for decades in Iceland, and longer pipelines will be installed in the future. Engineering analysis has revealed that in certain instances the transportation of heat as water may be much less expensive than the transmission of electric power. This is especially true when high-energy loads are involved.

TABLE 3. Examples of Cross-Country Hot-Water Supply Pipelines for District Heating Systems

	1 Reykir–Reykjavík main supply conduit	2 Reykir–Reykjavík older supply conduit	3 Svartsengi* Keflavík supply pipeline	4 Hveravellir– Húsavík supply pipeline
Type of water	Geothermal	Geothermal	Geothermal— steam-heated fresh water	Geothermal
Temperature	86°C	86°C	95°C	98°C
	(186.8°F)	(186.8°F)	(203°F)	(208.4°F)
Length, kilometers	13	16	13	19
Pipe diameter, millimeters	700	Two 350	500	250
Capacity, tons/hour	5000	1000	1000	200
Pumping power, kilowatts	1800	700	—	Self-flowing
Material of pipe construction	Steel	Steel	Steel	Cement asbestus
Type of insulation	Rock wool	Turf	Rock wool	Earth
Temperature loss	1.5°C	5°C	2°C	18°C
	(2.7°F)	(9°F)	(3.6°F)	(32.4°F)
Means of protection	Concrete conduit	Concrete conduit	Aluminum jacket	Earth cover

*In planning stage.

The temperature of the water in pipelines is often less than 100°C, but cases up to 125°C(257°F) exist. At suitable conditions, the most economic temperatures may be 160 to 180°C(320 to 356°F) (Bödvarsson, 1964). The water in the main supply pipelines may be geothermal, heated fresh water, or a mixture of both, depending upon the available geothermal source and the heating purpose.

Hot-water-supply pipelines are generally made of ordinary steel, insulated, and the major lines are protected by a concrete conduit. The flow rate of the water usually is maintained by multistage centrifugal pumps. Examples of supply pipelines are given in Table 3, but it should be noted that system 4 is of exceptional construction.

SPACE HEATING

Geothermal energy for space heating is of great importance in some countries—in Iceland, for instance, where about 50 percent of the population enjoys such heating for its homes. Plans have been made to extend this type of heating to 70 percent of the population within a few years. The geothermal fluids for such applications usually come from geothermal reservoirs at temperatures ranging from 60 to as high as 150°C(140 to 302°F). Thermal fluids within this temperature range occur at economically acceptable depths in many parts of the world, and extensive areas are known. Thus, widespread use of this form of energy is possible.

Although geothermal space heating may serve a single house in the rural area, the most usual approach is district heating services which serve whole areas, towns, or a city. As a rule, space heating by geothermal energy causes minimal pollution problems, inasmuch as there is no smoke and the warm effluents are distributed widely to the sewage system. In many areas where such systems have been installed, the cost of energy provided is very low when compared with fossil fuels. A depreciation time for equipment of 20 to 30 years is generally used for economic evaluations.

Distribution Systems for Space Heating.   Most distribution systems where hot water is the heating medium are single-pipe systems, which involve the discharge of the water to the sewage system after use. The distribution temperature of the water is preferably in the 80 to 90°C range (176 to 194°F) and will cool down to around 40°C(104°F) upon use. The supply mains to the distribution system will ordinarily discharge into storage tanks which help in taking care of daily fluctuations in hot-water load. Booster pumping is usually necessary in order to maintain sufficient pressure in the distribution system.

The distribution network in towns or in a city is installed underground in the streets. Street mains larger than 3 inches in diameter may be placed in concrete channels and are insulated, for example, by rockwool (glass wool) or aerated concrete. The channels are embedded in gravel together with concrete drainpipes. Minimum inclination of these channels is kept at 0.5 percent. Examples of street main channels are shown in Fig. 4. At street junctions, the channels may meet in concrete chambers, where valves, fastening bolts, and expansion joints are placed. These chambers are ventilated and either drained from the bottom, or if that is not possible, they will have a pump pit. Smaller street mains and house connections from street mains may be insulated with polyurethane foam insulation, protected by a watertight jacket of high-density polyethylene (Fig. 4c). More recently, such foam-insulated pipes have sometimes been used for street mains up to 6 inches.

Distribution Loads in District Heating Systems.   A district heating system must be tailored to the local climate. The most important characteristics in this respect are the variations in daily outside temperature over the year. Since every heating arrangement for homes must ultimately have a capacity to provide comfort on the coldest day of the year, obviously there must exist some overcapacity most of the time.

The ultimate cost of geothermal energy for space heating in such systems is usually nearly proportional to the maximum capacity required. Therefore, various approaches are used in order to increase the annual *load factor*. The latter term is defined as the ratio of total energy used to the basic design capacity. Some of these methods include the following:

1. The system is designed for an outside temperature somewhat higher than that of the coldest day of the year, assuming the need for boosting from other sources for a few days each year.

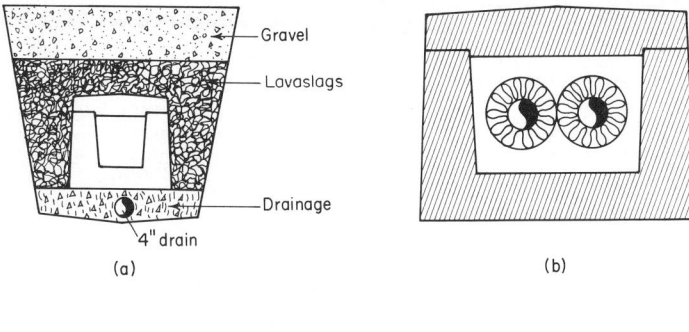

(a)   (b)

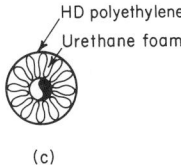

(c)

Fig. 4. Street main channels: *(a)* Buried channel. *(b)* Rock or glass-wool insulation. *(c)* House connections and minor street mains.

2. The system may include a fossil-fuel booster which is intended for raising the temperature of the water during the coldest spells.

3. The system may include a local geothermal underground reservoir, where deep well pumps are installed in the drill holes. This arrangement may yield increased production for a limited time by pumping at a draw-down of the water level.

House Heating.   Generally, central heating systems are used for houses. The hot water usually is admitted directly to these systems and discharged to sewage after use. A piping and valving arrangement is shown in Fig. 5. Hot domestic water for faucets is also supplied directly.

Inferential water meters with a magnetic coupling between the flow sensor and register mechanism are frequently used. The maximum flow of hot water is also controlled by sealed maximum-flow regulators. Sometimes only maximum-flow regulators are used. When direct supply is not advisable, as in the case of water with high mineral content that would cause much scaling, heat exchangers may be used between the hot water and the water circulating in the central heating system. All types of public buildings and commer-

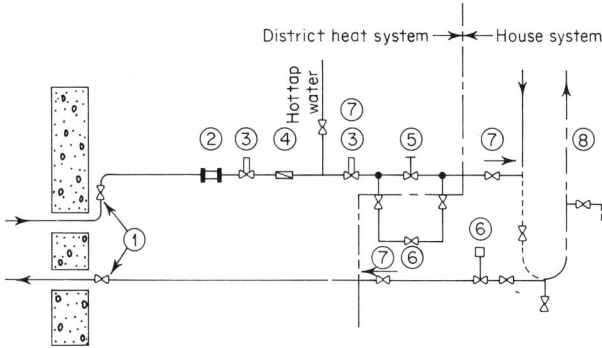

Fig. 5. Piping and valving arrangements for house connection to district heating system: (1) Main stop valves. (2) Strainer. (3) Sealed regulating valve. (4) Water meter. (5) Regulating valve. (6) Automatic valves. (7) Check valves. (8) Safety valve.

cial establishments are also heated geothermally where district heating systems are available.

Agricultural and Related Applications.   Geothermal energy for heating greenhouses is important in a number of countries. Since the temperature of the heat source will vary greatly from one location to the next, as well as variations in heating requirements, the surface area of the radiator system (often consisting of bare pipes) must be carefully tailored to local conditions. Heating fluid temperatures well above 100°C are rarely practical. Small greenhouses may take advantage of heat in the effluent from ordinary

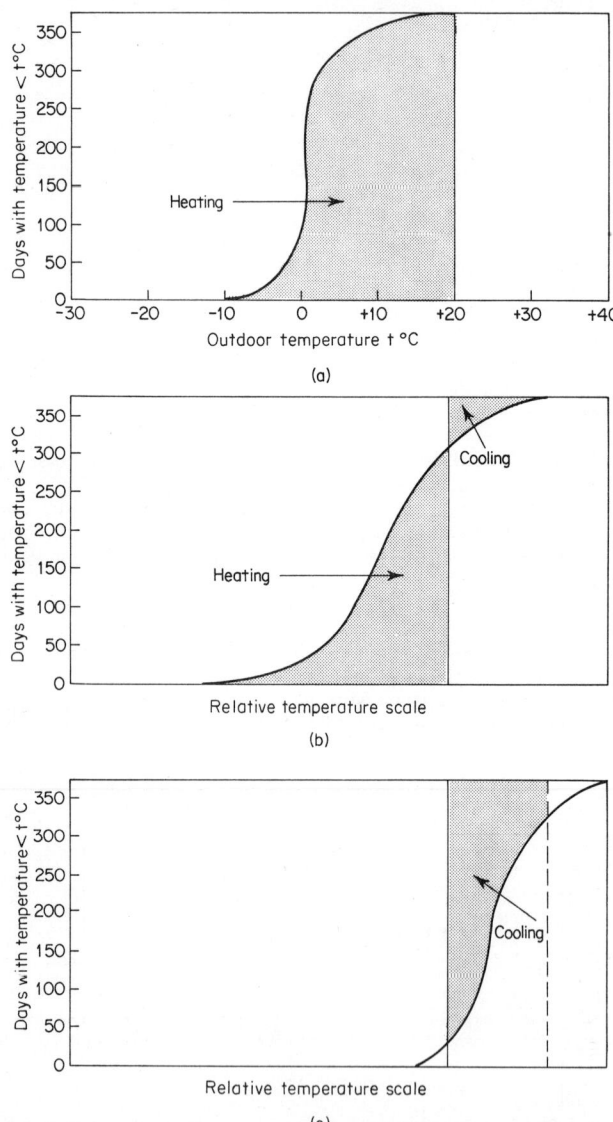

Fig. 6. Thermal characteristics of three types of climates: (*a*) Sub-Arctic island (Iceland). (*b*) Cold-temperate continental climate (qualitative). (*c*) Tropical climate (qualitative). (*Einarsson, 1973.*)

space-heating systems. The most important crops of heated greenhouses of this type include cut flowers, tomatoes, cucumbers, and seedlings of many varieties. Animal husbandry, fish farming, and hatching stations also frequently take advantage of available geothermal hot water.

Climatic Variations. Generally, geothermal energy to date has been exploited for those most obvious of applications: (1) for the generation of electric power, particularly more recently with distribution and shortage problems associated with fossil fuels; and (2) dating back many more years, for space heating and some secondary heating applications in colder climates, as has been representative of the situation in Iceland. Only modest attention has been given thus far by engineers and planners to the exploitation of geothermal energy sources which may be located in warmer climates—as for cooling and for processing and manufacturing applications. The heating and cooling requirements of three types of climates are shown in Fig. 6, according to Einarsson (1973). In certain climatic situations, since refrigeration may be achieved through heating, it is suggested that the load factor may be improved by supplying heat for refrigeration as needed, as well as for heating when needed. Refrigeration for comfort cooling by the use of geothermal energy has been described by Reynolds (1970). See reference list. District heat distribution systems which would be intended for both heating and cooling most likely would have a different optimum temperature than those intended for space heating alone. In Table 4, the load distribution and other selected physical data for the geothermal heating system in Reykjavík (1974) are summarized. The system has been expanded to serve neighboring communities.

PROCESS HEATING

Since geothermal energy resources exist in many countries, it is of interest to point out some common factors which affect the viability of exploitation. Since the applications for

TABLE 4.   Heat Distribution in Reykjavík Geothermal Heating System

Climatic Data	
Mean temperature of the year	+5°C 41°F
Mean temperature of the warmest month (July)	+11.2°C 52.2°F
Mean temperature of the coldest month (January)	−0.4°C 31.3°F
Available Heat Resources (1974)	
Reykir geothermal area, 3600 cubic meters/hour at 80°C(176°F) to the city	145 Gcal/hour
Reykjavík geothermal area, 1700 cubic meters/hour at 119°C(246.2°F) (average)	135 Gcal/hour
Own peak power boiler plants (oil-fired)	30 Gcal/hour
National Power Company peak-power boiler plant (available at electrical off-peak hours only)	20 Gcal/hour
Total	330 Gcal/hour
Heat Load	
Volume of houses connected	15×10^6 cubic meters
Number of houses connected	11,000
Heat load at −10°C(14°F) outside and 20°C(68°F) inside	285 Gcal/hour
Specific load at −10°C outside and 20°C inside	19 kcal/(hour) (cubic meter)
System Data	
Installed horsepower in pumping plants	7676
Area served by distribution system	16.8 square kilometers
Length of pipelines:	
Collecting mains	21.0 kilometers
Supply mains	46.0 kilometers
Street mains	146.5 kilometers
House connections	157.5 kilometers
Average density of population	5400 inhabitants/square kilometer
Average load density	17 Gcal/(hour) (square kilometer)
Yearly Heat Production	
Geothermal energy (1973)	1100 tcal/year
Peak power stations (1973)	40 Tcal/year
Total	1140 Tcal/year

TABLE 5. Examples of Process Design Features for Geothermal Steam and Water

Operation	Geothermal steam		Geothermal water	
	Type	Examples	Type	Examples
Drying	Indirect heating	Steam tube driers, drum driers	Indirect heating	Multideck conveyer drier
Evaporation	Primary heat exchangers accessible	Forced circulation evaporators	Countercurrent heaters	Preheaters
Distillation	Steam distillation	General equipment	—	—
Refrigeration	Freezing	Ammonia absorption	Comfort cooling	Lithium bromium absorption
Deicing	—	—	Direct application, indirect heating	Dredging aid, pavement deicing

TABLE 6. Consumption of Steam and Steam Used per $ Value in Some Established Fuel-based Processes

Product and process	Steam requirements, kilograms steam/ kilogram product	Steam per unit product value, kilograms/$ value (cost basis, 1970)
Heavy water by hydrogen sulfide process	10.000	151
Ascorbic acid	250	45
Viscose rayon	(70)	42
Lactose	40	130
Acetic acid from wood via Suida process	35	159
Ethyl alcohol from sulfite liquor	22	142
Ethyl alcohol from wood waste	19	123
Ethylene glycol via chlorohydrin	13	45
Casein	13	10
Ethylene oxide	11	33
Basic Mg carbonate	9	37
35% hydrogen peroxide	9	23
85% hydrogen peroxide from 35% H_2O_2	4¾	—
Solid caustic soda via diaphragm cells	8	121
Acetic acid from wood via solvent extraction	7½	34
Alumina via Bayers process	(7)	106
Ethyl alcohol from molasses	7	45
Beet sugar	5¾	26
Sodium chlorate	5½	28
Kraft pulp	4⅕	32
Dissolving pulp	4⅕	—
Sulfite pulp	3½	26
Aluminium sulfate	3½	79
Synthetic ethyl alcohol from ethylene	3	20
Calcium hypochloride, high test	3⅓	50
Acetic acid from wood via Othmer process	2¾	13
Ammonium chloride	2¾	21
Boric acid	2¼	20
Soda ash via Solvay process	2	60
Cottonseed oil	2	9
Natural sodium sulfate	1⅘	54
Cane sugar refining	1⅔	8
Ammonium nitrate from ammonia	1½	20
Ammonium sulfate	⅙	5
Fresh water from seawater by distillation	1½	227

heating in cold climates and the generation of electric power are obvious uses and are receiving considerable attention, it may be well to concentrate on the possiblities for process heating uses in any climate.

Perhaps the three most important questions are: (1) What products may utilize the heat in geothermal fluids? (2) What are the potential savings or advantages as compared with competitive energy approaches? (3) If there is a logistic disadvantage involved in site location, can this be offset by the lower-cost energy?

Because present technology is largely tailored to the use of fossil fuels, no conclusive answers can be sought directly from present engineering and economic practice. However, it may be helpful to begin a search by studying the conventional processes that employ fossil-fuel-generated steam. And there also will be instances in which geothermal fluids may be utilized with an advantage, even where no steam is employed in the conventional processes today. Some examples include: (1) Indirect heating in a process instead of direct contact heating. For example, a steam-tube drier may be used instead of a direct-fired drier. (2) There may exist a choice of several processes for any one specific objective. One process may permit utilization of geothermal energy with a great advantage, while another may not require any heat, but may entail other high-cost categories. (3) The availability of geothermal energy may call for a completely new process. General areas of application were given in Fig. 3. See also Table 5. For further analysis, the steam requirements and steam per unit product value for several products of the chemical and process industries are given in Table 6.

The economic importance of geothermal energy in a specific process may be judged by the share it has in the value of a product. This often can be roughly estimated in terms of the steam or the amount of fossil fuel that would otherwise be required. The effect of a different design, and hence different investment, also enters into the calculation. Numerous examples are known where the equivalent share of thermal energy may be from 5 to 20 percent of the value of a product.

Examples of existing and planned application of geothermal energy for process use are given in Table 7. See also Fig. 7.

Multipurpose Applications. The major industrial plants currently in operation have amply demonstrated that geothermal energy is a versatile source of energy. There are examples where process heating, space heating, and electric power production have been integrated into the same overall system. Because there is a large variation in geothermal energy sources, optimal utilization of that energy can be achieved only through individual analysis of each source. There are, however, a few generalizations:

1. When electric power is the main objective, there generally are ample opportunities

TABLE 7. Examples of Existing and Planned Applications of Geothermal Energy for Process Use

Product	Country	Applications	Form of geothermal energy
Pulp and paper	New Zealand	Evaporating, digesting, drying	Primary and secondary steam
Timber drying and seasoning	New Zealand, Iceland	Drying, seasoning	Steam, hot water
Diatomite processing	Iceland	Drying, heating, deicing	Steam
Hay drying*	Iceland	Drying	Hot water
Seaweed drying	Iceland	Drying	Hot water
Washing of wool	Iceland, USSR	Heating and drying	Steam
Curing and drying of building material	Iceland	Heating and drying	Steam, hot water
Stock fish drying	Iceland	Drying	Hot water
Salt recovery from seawater	Japan	Evaporation	Steam
Salts from geothermal* brine	Iceland	Evaporation	Steam
Boric acid recovery	Italy	Evaporation	Steam
Brewing and distillation	Japan	Heating and evaporation	Steam

*Planned.

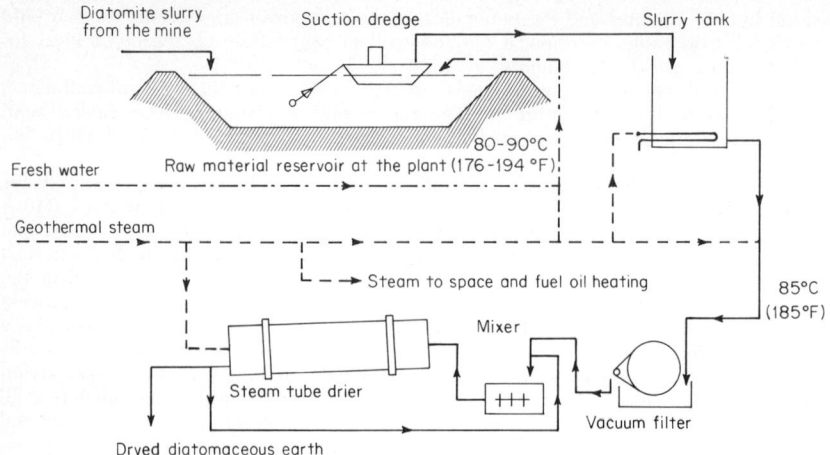

Fig. 7. Utilization of natural steam in a diatomite plant.

for use of waste heat, at least in those plants using wet steam. In such instances, geothermal water may be rejected at elevated temperatures, which subsequently can be utilized for space heating, fresh water production, and some industrial applications.

2. When space heating is the main objective, secondary electric power generation is possible in some instances. There are numerous other secondary applications (green-houses, soil warming, heating of swimming pools, etc.)

3. When process heating is the main objective, depending upon the geothermal source, some generation of needed electric power may be possible, and, as in the other cases, there usually are ample opportunities for numerous secondary heating applications.

REFERENCES

1. Arnórsson, S.: Application of the Silica Geothermometer in Low-Temperature Hydrothermal Areas in Iceland, *American Journal of Science*, vol. 25, pp. 763–784, 1975.
2. Bödvarsson, G. and J. Zoëga: Production and Distribution of Natural Heat for Domestic and Industrial Heating in Iceland, *Proc. U.N. Conf. New Sources of Energy*, vol. 3, p. 452, 1964.
3. Einarsson, S.: Geothermal District Heating, *Geothermal Energy*, p. 129, UNESCO, New York, 1973.
4. Marshall, T., and W. R. Braithwaite: Corrosion Control in Geothermal Systems, *Geothermal Energy*, pp. 151–159, UNESCO, New York, 1973.
5. Reynolds, G.: Cooling with Geothermal Heat, *U.N. Symp. Geothermal Energy*, Pisa, Italy, 1970.
6. Sigvaldason, G. E.: Geochemical Methods in Geothermal Exploration, *Geothermal Energy*, pp. 49–59, UNESCO, New York, 1973.
7. White, D. E.: Thermal Waters of Volcanic Origin, *Bull. Geol. Soc. Amer.*, vol. 68, pp. 1637–1658, 1957.
8. Zoëga, J., and G. Kristinsson: The Reykjavík District Heating System, First International District Heating Convention, London, 1970.
9. Hermannson, S., Eliasson, G., Sigurdsoon, P., and A. Einarsson: Corrosion in Geothermal Heating Systems, Industrial Research and Development Institute, Reykjavík, 1975.

Section **8**

Hydropower Technology

Hydropower

Gravitational energy available for the generation of electricity from water flowing from a higher level to a lower level is manifested on the earth in essentially two forms: (1) Descending natural watercourses, created by precipitation of rain and snow, which flow from mountains, hills, and plateaus to sea level; and (2) the changes in levels of estuaries and other ocean-associated bodies of water which occur as the result of actions of the tides. In terms of total hydrogravitational energy available, the successful exploitation of tidal energy as of the mid-1970s is but a token of what can be achieved through further geographical site selection, equipment development, and capital investment. The most successful tidal power installation is described in a subsequent article in this Handbook section.

The utilization of energy available from watercourses, on the other hand, contributes substantially to the total amount of electricity produced. As shown in Table 1, nearly 16 percent of the total electric power generated in the United States is derived from hydroelectric installations. Canada leads the world with 74.5 percent of its total electric power generated from hydroelectric facilities. Major hydroelectric plant installations throughout the world are listed in Table 2. Of course, there are numerous additional installations of lesser capacity. As will be noted from Fig. 1, hydroelectric power, as a percentage of the total installed electric generating capacity in the United States, has been falling off quite steadily for nearly fifty years. The highest percentage (41.4 percent) was achieved in 1932. It is estimated that as of the mid-1970s, only about 30 percent of the hydroelectric potential of the United States has been exploited. During a period of great concern over energy shortages, the logical expectation would be that there would be great construction activity under way; but there are numerous obstacles, as will be described briefly later.

Fig. 1. Trend of hydroelectric power as a percentage of the total installed electric generating capacity in the United States. *(Federal Power Commission.)*

TABLE 1. Hydroelectric Capabilities of Nine Leading Electric Power Producing Nations of the World (1971–1972)

Country	Installed Capacity			Production		
	Thousands of kilowatts	Percent of nation's own total	Percent of 9-nation total	Millions of kilowatt-hours	Percent of nation's own total	Percent of 9-nation total
United States	56,586	14.6	29.9	269,580	15.7	34.4
U.S.S.R.	33,448	19.1	17.7	126,099	15.8	16.1
Canada	30,601	65.5	16.1	160,984	74.5	20.6
Japan	20,176	26.4	10.6	86,849	22.5	11.1
France	15,459	37.3	8.2	48,726	32.5	6.2
Italy	15,280	42.2	8.0	40,019	32.1	5.1
Spain	11,054	58.5	5.9	32,283	52.2	4.1
Germany (West)	4,842	8.9	2.5	14,054	5.4	1.8
United Kingdom	2,158	3.0	1.1	4,311	1.7	0.6
Total	189,604		100.0	782,905		100.0

SOURCE: Federal Power Commission.

TABLE 2. Major Hydroelectric Plants

Country and name of dam or power plant	Year of dedication	Megawatts Present capacity	Megawatts Ultimate capacity
UNITED STATES			
Robert Moses-Niagara	1961	1950	1950
Castaic	1974	1250	1250
Grand Coulee	1941	2025	9771
John Day	NC	2160	2700
Chief Joseph	1956	1024	2073
St. Lawrence Power Dam (with Canada)	1958	1880	1880
The Dalles	1957	1119	1813
McNary	1953	986	1406
Hoover	1936	1345	1345
Wanapum	1963	831	1330
Priest Rapids	1959	789	1262
Rocky Reach	1961	712	1215
Dworshak	NC		1060
Northfield Mountain	NC		1000
CANADA			
Churchill Falls	NC		5225
Mica	NC		2760
W.A.C. Bennett	NC	1150	2270
Kemano	1954	835	1670
Beauharnois	1951	1586	1641
Sir Adam Beck No. 2	1954	900	1370
Daniel Johnson	NC		1353
Kettle Rapids	NC	714	1224
Manicouagan No. 3	NC		1176
Bersimis No. 1	1956	1050	1050
Manicouagan No. 2	1965		1016
U.S.S.R.			
Sayansk	NC		6400
Krasnoyarsk	1968	5080	6096
Sukhovo	NC		5225
Bratsk	1961	4500	4600
Ust-Illimsk	NC	720	4300
Nurek	NC		2700
Volga–22d Congress	1958	2543	2560
Volga–V.I. Lenin	1955	2100	2300
Cheboksary	NC		1632
Inguri	NC		1600
Saratov	1967		1359
Toktogul	NC		1200
Nizhne-Kamskaya	NC		1090
Zeya	NC		1020
Votkinsk	1961		1000
Chirkey	NC		1000
OTHER COUNTRIES			
Guri (Venezuela)	1967	524	6500
Cabora Basa (Mozambique)	NC		4000
Ilha Solteira (Brazil)	NC		3200
Marimbondo (Brazil)	NC		1400
Furnas (Brazil)	1963		1200
Jupia (Brazil)	1961		1400
Iron Gate (Rumania-Yugoslavia)	NC		2160
Saad El-Aali-High Aswan Dam (UAR)	1967	1750	2100
Tarbella (Pakistan)	NC		2100
Mangla (Pakistan)	NC	300	1000
Kariba (Rhodesia-Zambia)	1959	600	1500
Liukiahsia (China)	1963		1500
Sanmen Hsia (China)	NC		1100
Tumut-3 (Australia)	NC		1500

TABLE 2.   Major Hydroelectric Plants (Continued)

Country and name of dam or power plant	Year of dedication	Megawatts	
		Present capacity	Ultimate capacity
Talbingo (Australia)	NC		1500
Keban (Turkey)	NC	620	1240
El Chocan (Argentina)	NC		1200
Bhakra (India)	NC	450	1050
Kaniji (Nigeria)	NC		1000
Chivor (Colombia)	NC	500	1000

SOURCE: U.S. Bureau of Reclamation, Department of the Interior.
NC = Not completed.

The geographical distribution of hydroelectric power production in the United States, by regions and states, is given in Table 3. The historical record of installed hydroelectric generating capacity by type of management and ownership, since 1956, is given in Table 4. Because of the very large investment required, as well as legal problems (sometimes involving the rights and interests of more than one region or state), the majority of hydroelectric projects customarily have been essentially government-sponsored. Traditionally, investor-owned hydroelectric facilities have represented about one-third of the total.

TABLE 3.   Hydroelectric Power Production in the United States (1972)

Region and state	Hydroelectric generating capacity, megawatts	Regional percent of United States total
NEW ENGLAND	**1.480**	2.6
Maine	0.342	
New Hampshire	0.381	
Vermont	0.186	
Massachusetts	0.435	
Rhode Island	0.002	
Connecticut	0.134	
MIDDLE ATLANTIC	**5.967**	10.5
New York	3.974	
New Jersey	0.341	
Pennsylvania	1.652	
EAST NORTH CENTRAL	**0.840**	1.5
Ohio	0.002	
Indiana	0.093	
Illinois	0.032	
Michigan	0.344	
Wisconsin	0.369	
WEST NORTH CENTRAL	**3.089**	5.5
Minnesota	0.136	
Iowa	0.132	
Missouri	0.800	
North Dakota	0.400	
South Dakota	1.384	
Nebraska	0.236	
Kansas	0.001	
SOUTH ATLANTIC	**5.505**	9.7
Delaware	0.000	
Maryland	0.494	
District of Columbia	0.003	
Virginia	0.842	
West Virginia	0.101	
North Carolina	1.834	

Region and state	Hydroelectric generating capacity, megawatts	Regional percent of United States total
South Carolina	1.145	
Georgia	1.056	
Florida	0.030	
EAST SOUTH CENTRAL	**5.259**	9.3
Kentucky	0.686	
Tennessee	2.065	
Alabama	2.508	
Mississippi	0.000	
WEST SOUTH CENTRAL	**2.280**	4.0
Arkansas	0.998	
Louisiana	0.000	
Oklahoma	0.765	
Texas	0.517	
MOUNTAIN	**6.601**	11.7
Montana	1.512	
Idaho	1.255	
Wyoming	0.222	
Colorado	0.743	
New Mexico	0.024	
Arizona	1.961	
Utah	0.202	
Nevada	0.682	
PACIFIC	**25.466**	45.0
Washington	11.842	
Oregon	5.638	
California	7.986	
ALASKA AND HAWAII	**0.079**	<1.0
Alaska	0.076	
Hawaii	0.003	
TOTAL UNITED STATES	**56.566**	

SOURCE: Federal Power Commission.
Statistics are for installed capacity.

The relative advantages and limitations of hydroelectric power are given in Table 5.* Some of the obstacles that large hydroelectric projects usually encounter include: (1) High initial costs of a major dam(s), involving usually one main structure, but often requiring smaller dams to complete the reservoir complex. (2) At least to some extent, creation of a large reservoir requires relocation of highways, railroads, and power lines; and, in some instances small towns have to be moved. (3) Unless located in wilderness areas, numerous home sites must be moved. (4) Very long delivery time for major equipment, such as turbines. (5) Frequently, the abandonment of large areas of farmland. (6) Increasing pressures by special interest groups.

There are two forms of underdeveloped hydro potential, not only in the United States, but throughout the world: (1) areas of large hydropotential where no dams have yet been built; and (2) areas where dams have been constructed, but where full generating potential has not been fully exploited. This latter factor is clearly reflected in Table 2. Even though there are obstacles, the further development of hydroelectric power represents a sound approach whereby, within a period of 20 years or less, the hydroelectric generating capacity of the United States could be doubled—and essentially with no new technology required, as is the case of most other proposals for additional energy.

With reference to Table 3, it is interesting to note that although 45 percent of the present hydroelectric generating capacity in the United States is located in the Pacific

*Information in this and the four following paragraphs is based essentially on observations made in "Energy Perspectives," No. 14, September 1974, published by Battelle Memorial Institute, Columbus, Ohio.

TABLE 4.   Installed Hydroelectric Generating Capacity in the United States by Type of Management and Ownership

Thousands of kilowatts

Year	Total utility industry	Investor-owned	Coopera-tives	Municipal utilities	Federal installations	Power districts, state projects
1972	56,566	19,303	56	3330	24,333	9544
1971	55,898	19,055	56	3267	24,053	9467
1970	55,056	18,850	56	3195	23,499	9456
1969	52,753	18,444	56	3195	22,054	9004
1968	51,168	18,337	57	3177	20,905	8692
1967	48,112	18,134	57	2623	19,589	7709
1966	44,977	16,381	57	2005	19,531	7003
1965	43,782	16,096	58	2005	18,717	6906
1964	42,188	15,296	58	1979	17,947	6908
1963	40,214	14,620	58	1982	16,923	6631
1962	37,342	13,629	47	1651	16,149	5866
1961	35,481	13,522	50	1584	15,091	5234
1960	32,367	13,359	46	1521	14,637	2804
1959	31,074	13,110	43	1385	14,125	2411
1958	29,359	12,458	44	1385	13,697	1775
1957	27,036	11,327	40	1382	13,043	1244
1956	25,654	10,948	40	1384	12,135	1147

SOURCE: Federal Power Commission.

Region, there are other important hydro reservoirs in the eastern portion of the nation. The Allegheny River watershed is an example. This important stream drains much of southwestern New York and northwestern Pennsylvania; and, with the Monongahela, it forms the Ohio River. The region drained by the Allegheny lies in the snow belt at the eastern end of Lake Erie. It is in this watershed that the snow melt water, often augmented by early spring rains, produces the surges of high water which numerous times have flooded the cities of Pittsburgh, Wheeling, and Cincinnati. Part of the Alle-

TABLE 5.   Relative Advantages and Limitations of Hydroelectric Power Installations

ADVANTAGES
Continuous low-cost power production except when droughts occur.
Low maintenance costs.
No consumption of irreplaceable fossil fuel.
No air pollution.
Reservoir lakes can be used for recreation in majority but not in all cases.
Reservoirs can provide considerable but not complete flood protection to downstream areas.
Reservoirs are capable of storing large quantities of water for long periods of time, but not indefinitely.
Downstream flow can be managed to aid in water-quality control and to level out the extremes of winter versus summer stream conditions.
Ground-water reserves are increased by recharging from the reservoir

LIMITATIONS
High initial cost of construction.
Recreational facilities can be adversely affected in reservoirs where drawdown in the dry season lowers the water level.
Flood protection can best be provided by an *empty* reservoir, whereas power production is best from a *full* reservoir. A full reservoir cannot retain a major flood; an empty reservoir generates no power. The compromise then is to retain enough water in a reservoir to ensure continuous power generation, but leave a margin of free board to take the major surges out of a sudden torrential rainstorm.
Loss of land suitable for agriculture.
Power production may be curtailed or even discontinued in time of drought.
Original stream valley is inundated.
Some water is lost by evaporation from the reservoir surface.
In coastal areas, such as Oregon and Washington, the construction of dams prohibits the upstream migration of anadromous fish, such as the Pacific salmon, unless some arrangement, such as a "fish ladder" is provided.

SOURCE: Battelle Memorial Institute, Columbus, Ohio.

gheny watershed is already protected by dams; but part of it, a major tributary, the Clarion River, has a substantial, unused hydroelectric potential.

Also, the upper reaches of the Susquehanna have not been dammed. Some of the disastrous flooding following as aftereffects of East Coast hurricanes, might have been avoided had flood-control projects been in existence upstream from Harrisburg and Wilkes-Barre.

A discussion of hydroelectric projects should not overlook the water-supply situation. It is well established that water consumption is increasing and, as in Southern California, when consumption exceeds local supply, water must be imported. Projects comparable to delivery of water to Southern California from the north will inevitably be constructed in other parts of the nation and the world. Great tunnel systems will be cut through mountain ranges; and streams will be diverted, sometimes to flow in the opposite direction.

In British Columbia, for example, the Nechako River, draining the high snow fields of the Coast Range and originally flowing east, has been dammed. The water now flows west through a tunnel, then drops several hundred feet to sea level, generating thousands of kilowatts, before flowing into the Pacific Ocean. The concept of drainage reversal to gain a hydroelectric advantage has been used successfully elsewhere, but not in the United States. A possible site for such a plant could be considered for western New York, the "panhandle" of Pennsylvania, and northeastern Ohio. Through a series of interconnected reservoirs, some already in existence, water that would normally flow into the Allegheny and Ohio Rivers could be rerouted northwestward over a divide and down into Lake Erie, thereby creating a hydraulic "head" of some 700 feet. Much-needed water-storage facilities could result from such a project, plus added flood control on both the Allegheny and the Ohio Rivers.

CLASSIFICATION OF HYDROELECTRIC PLANTS

Hydroelectric plants can be classified by the following:
  1. *Extent of impounded volume:*
     (a) Storage plants
     (b) Run-of-river plants
  2. *Status of the facility in the total power system:*
     (a) Peak-load plant
     (b) Base-load plant
     (c) Isolated plant
  3. *Head:*
     (a) High-head development
     (b) Medium-head development
     (c) Low-head development

The low-head plant has a characteristic design differing in all essentials from the high-head plant. The medium-head plant may partake of the characteristics of either the high- or the low-head plant as its working head approaches either the high- or low-head range. There is no definite line of demarcation between high, medium, and low heads. However, a head of more than 500 feet can be considered a high-head development; and one lower than 50 feet, a low-head development. Briefly, the characteristics of the low-head plant are: (1) Vertical, reaction type, runners using large volume of water and requiring large water passages. (2) Substructure is both extensive and expensive, and intake works are large and complex. (3) Large-diameter generators are necessary because of the low rotational speeds. Characteristics of the high-head plant are: (1) horizontal impulse turbines, (2) small volumes of water at high pressures, (3) plant at some distance from the dam. The advantage of smaller and simpler substructure is offset by the presence of a long water conduit, or penstock, between dam and plant. The turbines are high speed and allow smaller generator diameter. The high speed is accounted for by the high heads used. Inherently, the impulse turbine has a low characteristic speed.

The possible hydroelectric development sites along the flow of a stream are of two major types, namely: (1) those suitable for run-of-the river plants (Fig. 2), and (2) those offering natural impounding basins for storage plants (Fig. 3). In general, the run-of-the-river plant is cheaper than the storage plant of equal capacity, but it suffers seasonal variation of output which is more or less proportional to the variation of stream flow. A mountain-reservoir type of plant is shown in Fig. 4.

Storage plants give a greater proportion of firm power which can be delivered day by day on a regular schedule. This firm power is in more or less direct ratio to the degree of regulation of the flow of the stream; and this, in turn, is a function of the impounded volume. Complete regulation of stream flow is rarely possible or practical, although 80 to 90 percent regulation is not infrequent.

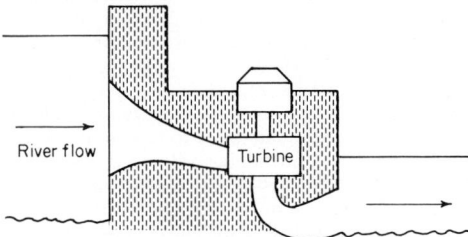

Fig. 2. River plant (low head), typified by the Bonneville and St. Lawrence projects.

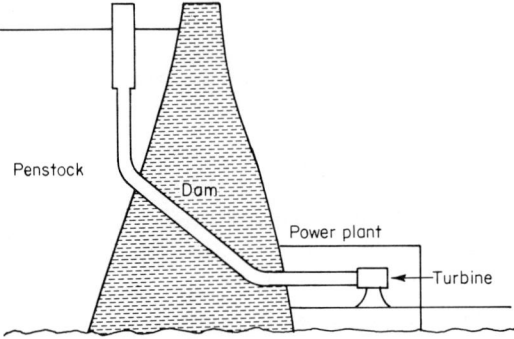

Fig. 3. River plant (high head), typified by the Hoover Dam and Niagara installations.

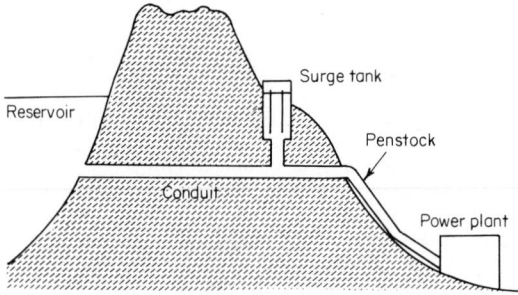

Fig. 4. Mountain reservoir plant, typified by the Appalachia project.

In any storage plant, the theoretical energy or power available over and beyond the firm power developed is known as *flash power* or *flood peak power*. Firm power commands, commercially, a considerably higher rate than flash power.

When integrated with steam generating plants, hydro plants are frequently used to give peak power outputs to take care of peak load condition and thus avoid the expensive standby service of additional steam generating equipment. Such service, of course, may still permit the delivery of a certain amount of firm power.

If all the run-of-river plants were located upstream from the storage plants, they would be operated continuously on a base-load plan, because, were they idle, their small reservoirs would quickly overflow, and water would be wasted over the crest gates. If, however, they are located between storage plants, the run of the river, as far as they are concerned, is just what the storage plants are passing on to them. Thus, located downstream from a storage plant, a run-of-river plant will produce an increase in output when the storage plant increases its output.

In the hydroelectric plant, the turbines and generators are the main items of equipment. The hydroelectric superstructure, as usually laid out, has one large building, housing the main units, and an electrical bay, or wing, of one or more stories, in which are located the switching equipment, offices, storerooms, and most of the auxiliary equipment.

Hydro sites that are developed to utilize only part of the normal stream flow are exceptions to the general rule. Very rarely is a development made in which conservation of the water and its use in the most efficient manner are not paramount features of operation. Failure to give due cognizance to this feature may wipe out the net operating profit; hence a continuous, watchful scrutiny of all natural factors which can affect the station operation is a duty of the operating personnel.

A hydraulic turbine suffers loss of efficiency at heads above or below the designed value because of shock losses. At the correct head, there will be one point of best efficiency, somewhere between 80 and 95 percent of full load. When a number of units are installed in a plant, and when steam reserve is available, it is generally possible to operate the units near the point of best efficiency. There are four faults of operation and maintenance which can reduce the maximum energy production of a hydro plant:

1. Waste of water over spillways
2. Improper distribution of the load between the station units
3. Water leakage through valves, gates, dam or flow line
4. Wear on moving parts, especially corrosion or erosion of the runner

The relative simplicity of hydroelectric equipment makes hydraulic efficiency of the turbine the prime consideration.

HYDRAULIC TURBINES

The fundamentals of the turbine were incorporated into the wheels built before the start of the nineteenth century, but its principal development has occurred since that time. Beginning with Fourenyon and his outward flow turbine, Jonval, Boyden, Swain, and Francis rapidly brought the reaction turbine to an advanced stage of development. By 1875, the inward flow turbine, as perfected by Francis and which now bears his name, had established itself in the lead, a position that it held until about 1900, when the impulse, or Pelton, type of wheel had progressed to the point of dominating the high-head field.

The inherent slow speed of the Francis-type runner on low heads was a fault that the propeller-type runner was designed to cure. During the decade 1910 to 1920, progress was made with this type of wheel, and by 1920 the propeller-type runner, often called the Nagler runner, was definitely established in the hydroelectric field. Later, it was arranged so that the blades could be adjusted and set at different angles to accommodate changes in elevation of the forebay level without undue loss of efficiency. The success of the propeller-type turbine encouraged American adoption of the Kaplan turbine, on which the blade adjustment is performed automatically, being under the same control as the turbine gates.

As between impulse and reaction types, the action in the impulse turbine is easier to understand. There is no difficulty in visualizing the transformation of pressure head into velocity head at the nozzle, or of understanding the push, or impulse, that is given to the buckets by the stream of water. The jet is directed upon the rotor tangentially, and hence this type is also called the *tangential turbine*. The velocity of the jet of water is only slightly less than the free spouting velocity under the effective head h. Impulse buckets are divided into two halves by a "splitter," and the axial thrusts which would otherwise have to be borne by special bearings are equalized.

The essential difference between the impulse and reaction types is that in the former the entire energy received by the wheel is in the velocity form, while in the latter it may be partially in the velocity form, but is also, in a large measure, still in the pressure form. The reaction of conversion of residual pressure into velocity in the runner is the source of much of the torque delivered to the reaction turbine. If the turbine were blocked stationary and had its gates opened, the water would issue from the turbine as from a nozzle. Now, by removing the blocking, let these nozzles begin to rotate, and the absolute velocity of water leaving them is found to be diminishing, the energy having been absorbed by the runner. At the best speed, the final velocity will be just sufficient to enable the water to clear the runner. At this time, the wheel may be absorbing 90 to 95 percent of the energy that the water had in the pressure form just before reaching the turbine gates.

In a Francis turbine, the water flows inward, and then downward and into the draft tube.

Classification of Hydraulic Turbines.   A convenient classification of hydraulic turbines is as follows:

1. *Reaction turbine*	Water under pressure is only partially converted into velocity before it enters the turbine runner.
(*a*) *Francis turbine*	See Fig. 5.
(*b*) *Propeller turbine*	
(1) *Fixed-blade*	See Fig. 6.
(2) *Adjustable-blade*	See Figs. 7 and 8.
(3) *Axial-flow*	See Fig. 9.
(4) *Diagonal-flow*	See Fig. 10.
2. *Impulse turbine*	Water under pressure is entirely converted into velocity before it enters the turbine runner. See Fig. 11.

Both reaction and impulse turbines have a stationary guide case and a revolving part, termed the *runner.** In the guide case of the reaction turbine, the water under pressure is only partly converted into velocity, leaving a pressure head at the entrance to the runner. This pressure head causes an acceleration of the relative velocity of flow through the runner, the discharge area of which is smaller than the entrance area. The reaction turbine thus utilizes the pressure of the water and the reactive force on the curved buckets which tend to change the direction of flow. Except in operating vented at low loads, the water passages are completely filled with water from the intake to the end of the draft tube.

In the guide case (designated the nozzle pipe, needle nozzle, and nozzle tip) of the impulse turbine, the pressure head is completely transformed into velocity before striking the runner so that air surrounds both the runner and the full jet issuing from the nozzle tip.

Reaction Turbines.   Figure 5 shows a Francis-type inward-flow reaction turbine. Water enters the spiral case from intake passages or penstocks, passes through the stay ring guided by the stationary stay-ring vanes, thence through the movable wicket gates through the runner and into the draft tube, through which it flows into the tailrace or tailwater reservoir. The movable wicket gates with axis parellel to the main shaft control the

*This and the subsequent five paragraphs are from "Standard Handbook for Electrical Engineers," 10th ed. (D. G. Fink and J. M. Carroll, eds.), McGraw-Hill, New York. Figures 6 through 12 also from same source.

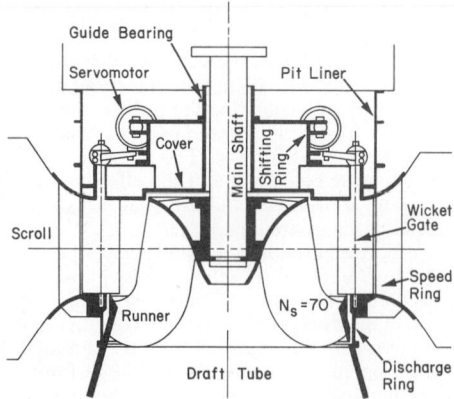

Fig. 5. Principal elements of a Francis turbine.

flow of water to the runner and thereby control the power output of the turbine. Francis turbine runners usually have the upper ends of the buckets attached to a crown and the lower ends attached to a band, thus completely enclosing the water passageway through the runner. Francis turbines are normally used for medium heads ranging from 100 to 1300 feet.

Propeller Turbine. Also of the reaction type, this turbine is differentiated from the Francis turbine in that the runner has unshrouded blades (no crown or band). The blades, three to eight in number, are either fixed or adjustable. As the names indicate, in the fixed-blade propeller runner the blades are in a permanent fixed position, while in the adjustable-blade runner the blade angle can be adjusted. For fixed-blade runners, the blade angle is usually set between 16 and 20° where maximum efficiency is obtained. For adjustable-blade runners, the blade angle may vary from 10° minimum to 32° maximum. The blades may be adjusted mechanically by hand or electric motor through a train of gears. However, this method has been largely abandoned in recent years in favor of the oil-pressure-operated blades. This

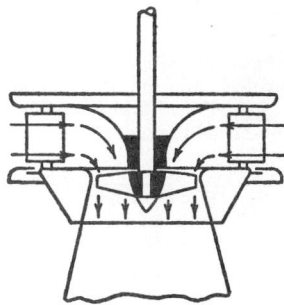

Fig. 6. Vertical-shaft reaction turbine, fixed-blade propeller type.

type of turbine is commonly called a Kaplan turbine. Figure 7 is a sectional elevation of a Kaplan turbine, and Fig. 8 is a sectional view of its rotating element. The blades are adjusted by means of an oil-operated piston located within the main shaft. The operating piston can also be located in the hub of the runner, either above or below the runner blades. The oil is admitted to and discharged from above and below the piston

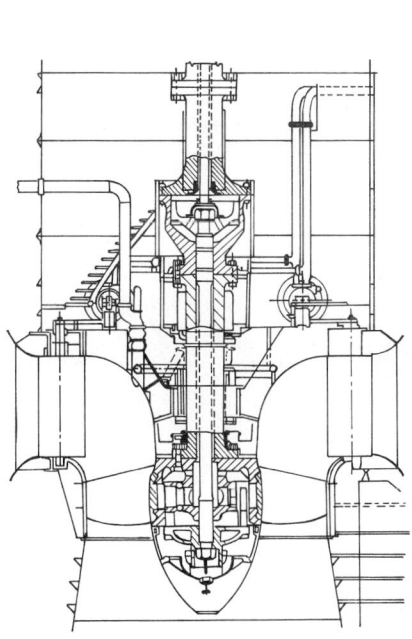

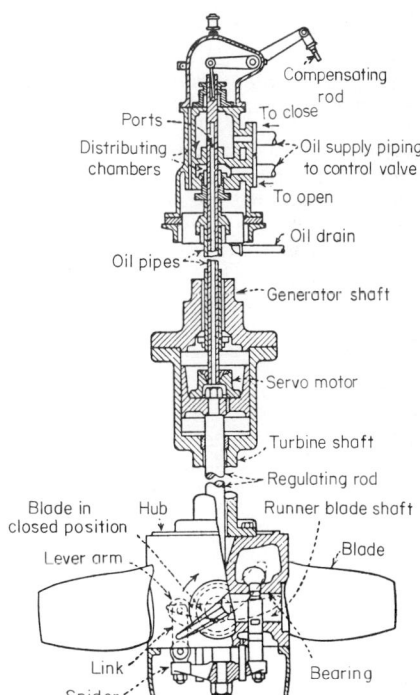

Fig. 7. Sectional elevation of a Kaplan turbine.

Fig. 8. Sectional elevation of rotating element (runner and operating mechanism) of a Kaplan turbine.

by means of an oil distributor, located either on top of the generator shaft above the generator or surrounding the main shaft below the generator. The oil pressure is supplied from the governor oil-pressure system, and the flow of oil is controlled by the governor. The control has a cam so shaped and arranged that blade tilt will vary with the wicket-gate opening so as to produce a maximum-efficiency envelope curve. The greater the wicket gate opening, the greater the angle of tilt and the greater the power output. Propeller turbines have the same arrangement of water passages as reaction turbines, and the flow of water through these passageways is the same in both types.

Axial-Flow Turbines.   These turbines use the propeller-type runner with either fixed or adjustable blades. Their characteristic feature is that the water approaches the runner

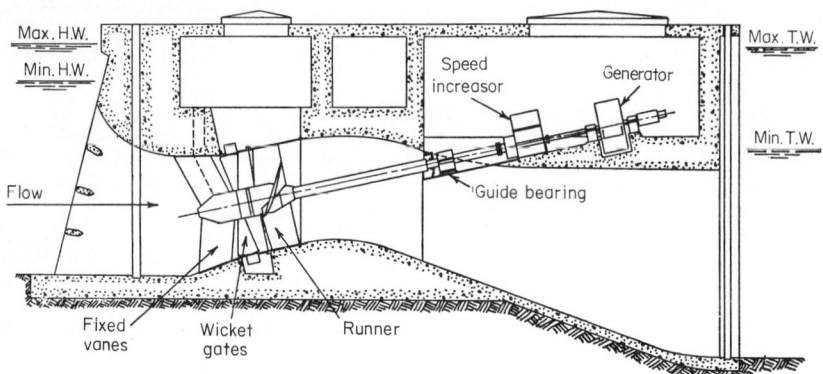

Fig. 9. Sectional elevation of an axial flow (tube) turbine.

parallel to the main shaft and thus avoids having to make a 90° turn after leaving the wicket gates in order to enter the runner, since the axes of the wicket gates extend radially outward from the main shaft, providing a straight or nearly straight-through passageway from intake to draft-tube discharge. See Fig. 9. The propeller runners previously described, with either fixed or adjustable blades, have the axis of the blades substantially at right angles to the main shaft (Figs. 6, 7, and 8). However, there is a type of propeller runner in which the axis of the blades is at about 45° with the main shaft. These are commonly known as *diagonal-flow turbines.* The blades may be either fixed or adjustable. If adjustable, they are usually known as *Dariaz turbines,* which are sometimes further characterized by having the axis of the wicket gates (either fixed or movable) set at

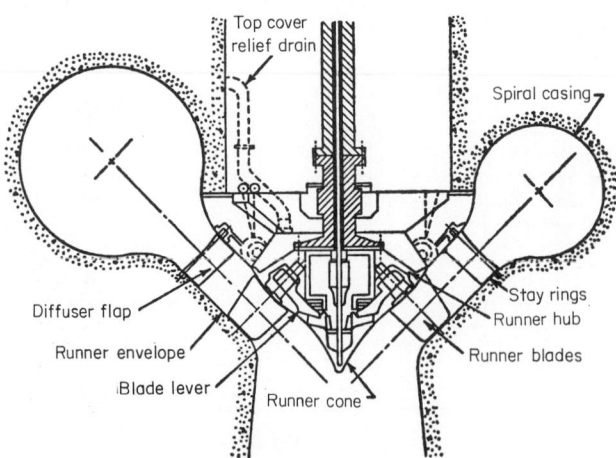

Fig. 10. Sectional elevation of a diagonal-flow (Dariaz) turbine.

a 45° angle with the main shaft and the spiral case angled accordingly (Fig. 10). Some adjustable-blade diagonal-flow runners are so designed that the blades can be closed against one another to shut off the flow of water through the runner, thus eliminating the need for adjustable wicket gates for this purpose. In general, diagonal-flow runners bridge the gap between the propeller and the Francis runners. Thus their characteristic performance lies somewhere between these two, and they are therefore used for the higher heads in the propeller range and overlap part of the Francis range. Dariaz turbines have been proposed for heads as high as 400 feet. The regular propeller turbines are used for low heads, ranging from the lowest head that is practical (one installation operates under a

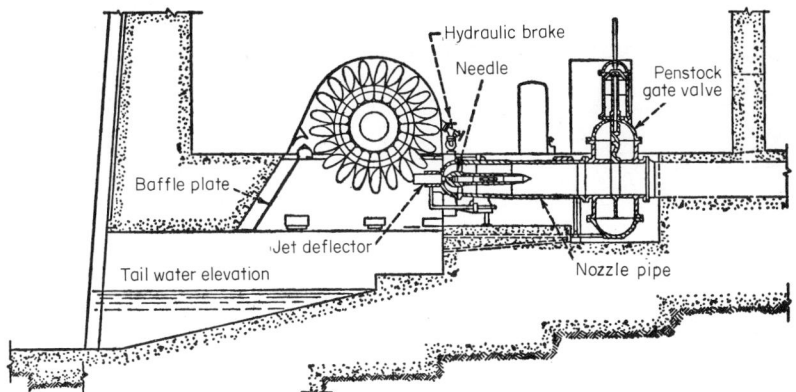

Fig. 11. Section through horizontal-impulse turbine.

7-foot head) to heads up to 200 feet, thus partly overlapping the range of heads for Francis turbines.

Impulse Turbines. The impulse turbine in its contemporaryyy configuration consists of one or more free jets of water discharging into an aerated space and impinging on a set of buckets attached around the periphery of a disk (Figs. 11 and 12). The buckets vary in some details of their construction, but in general are bowl-shaped and have a central dividing wall, or splitter, extending radially outward from the shaft. This splitter divides the stream, and the bowl-shaped portions of the bucket turn the water back, imparting the full effect of the jet to the runner. The free jet is formed by the water passing through the nozzle pipe, the needle nozzle, and thence through the nozzle tip. The size of the jet and thus the power output of the turbine are controlled by a needle in the center of the needle nozzle and needle tip. The movement of the needle is controlled by the governor. A jet deflector is located just outside the nozzle tip to deflect the jet from the buckets to effect sudden load reductions. Impulse turbines are utilized when the head is too high for practical use of Francis turbines, which is normally any head exceeding 1000 to 1300 feet.

Typical turbine efficiency curves are shown in Fig. 13. A schematic diagram of a typical hydro governor system is shown in Fig. 14.

Draft Tube. Hydraulic turbines frequently discharge the water with considerably more velocity than would be economical from the efficiency viewpoint, were it not possible to recover a great deal of that energy by the proper use of a diffusing chamber at the outlet. The diffusing chamber or tube is known as the draft tube, and there is a variety of types. However, the main objective is to convert the velocity head residing in the water leaving the turbine into pressure head. If this can be done efficiently, the turbine can be set somewhat below normal tailwater level.

The greater the specific speed of a turbine runner, the higher will be the velocity of the water discharged into the draft tube, and the more important the recovery of this velocity by draft tube design. The draft tube is to take the water from the turbine at a point where the pressure is considerably less than atmospheric, and, by efficiently reducing the velocity, convert it into pressure head so that it can emerge smoothly into the tailrace at atmospheric pressure. By "efficiently" is meant without shock or whirl loss. Not all the velocity head can be recovered, for the water must be given to the tailrace at normal

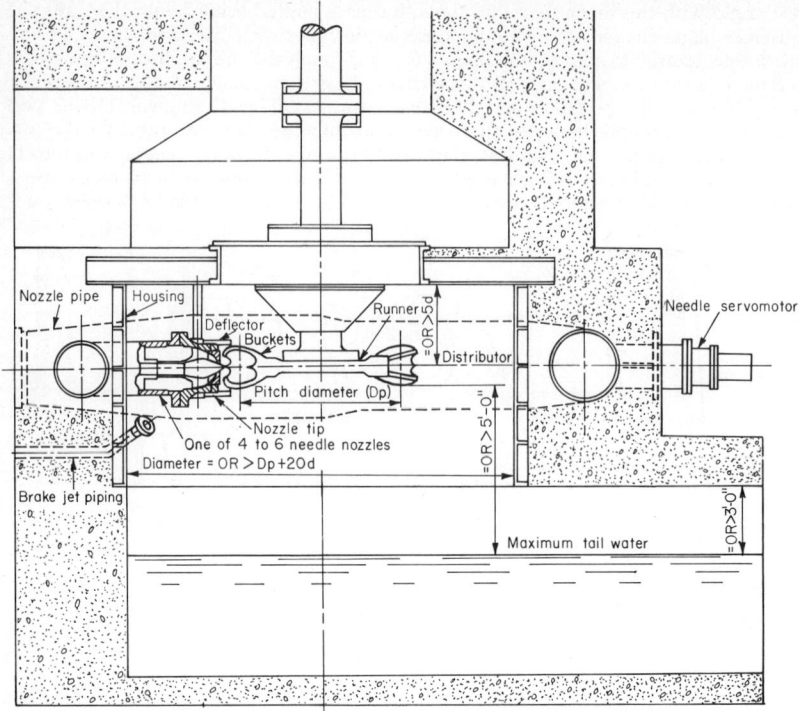

Fig. 12 Vertical-shaft multijet impulse turbine.

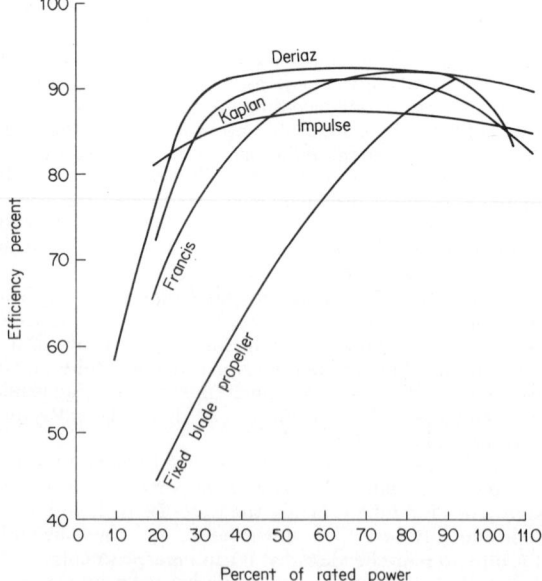

Fig. 13. Hydraulic turbine efficiency curves.

tailrace velocity to prevent its backing up into the turbine. Also, whatever friction loss occurs in the draft tube adds to this reduction of useful head.

PUMPED-STORAGE PLANTS

The history of pumped-storage plants dates back to the late 1920s when several plants were installed in Europe, and one, the Rocky River plant of Connecticut Light and Power, was built in the United States. From the point of view of plant location, there are three categories of pumped-storage installations:

1. Combined with Conventional Hydro Plant. Plants of this type are used in locations suitable for conventional hydro plants, but where rainfall or water availability and system demand are out of phase. For example, in Switzerland, demand is highest in winter when water is scarce and lowest in summer when there is an abundance of water. The available

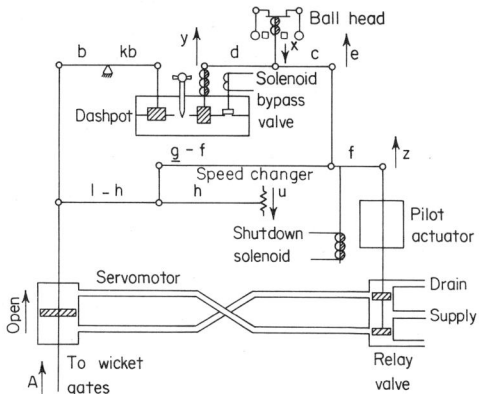

Fig. 14. Typical hydro governor system.

energy in summer can be used to fill the reservoirs and store the available water for later use in meeting the winter demand. A situation similar to this exists at Niagara where the cycle time is one day rather than one year.

2. Pure Pumped Storage. The advantages of this type of plant are its flexibility of location, in that the upper reservoir need have no source of water other than what is pumped into it, and the possibility of developing large plants with a small reservoir and high head. This type of plant is commonly used in steam-based systems which lack the many advantages of available hydro generation. As well as providing fast and reliable peaking power, pumped-storage units have the added advantage of smoothing the weekly load curve and enabling more of the efficient base-loaded steam plants to be operated continuously.

3. Pumped Storage with Diversion. This situation arises when available water must be shared between power generation and irrigation use. Water that must be pumped to a higher reservoir to feed an irrigation canal can be used as a source of peaking power if allowed to run back down through a pump turbine.

Performance of Pumped-Storage Units. From the point of view of dynamic response there are three basic types of pumped storage units:

1. *Separate Pump and Turbine on the Same Shaft.* Some thirty plants of this type (from 20 to 900 megawatts) were built in Europe during the period 1930 to 1964. Only two plants are known in the United States:

1928	Connecticut	Rocky River Plant	230-foot head	2×25 MW
1950	Texas	Buchanan Plant	120-foot head	13 MW

The Rocky River plant was converted to reversible about 1953.

The turbines in these units are Francis type for heads in the range 100 to 1000 feet, and Pelton wheels for heads up to 3000 feet. These units are usually mounted on a horizontal shaft with a clutch (hydraulic or friction) between the motor generator and the pump. The turbine is usually rigidly connected to the shaft and is dewatered, using compressed air during pumping. Sometimes small impulse turbines are installed on the shaft for starting and braking.

Typical operation for a unit with clutch and starting turbine would require two minutes for starting and loading from standstill, less than one minute for changeover from full-load pumping to full-load generating, and slightly more than one minute for the reverse operation.

As an example of the use of this fast turnaround, in an area of Germany during the 9 A.M. coffee break, 580 MW of load are dropped in about six minutes. This change is compensated by a group of pumped-storage plants going from 400 MW generating to 200 MW pumping during the same time interval.

2. *Reversible Pump-Turbines.* These units are of radial or mixed flow type [Francis *(F)* or Deriaz *(D)*] and have been designed to operate at a wide range of heads. Representative existing or planned plants using reversible pump-turbines are listed below:

Plant	Units	Head, feet	Power, MW
Flatiron	1 F	290	9
Hiwassee	1 F	190	60
Beck 2	8 D	83	8 × 34
Lewiston	12 F	75	12 × 21
Taum Sauk	2 F	790	2 × 220
Valdecanas	3 D	250	3 × 80
Smith Mountain	2 F	180	2 × 65
Yards Creek	3 F	656	3 × 113
Cruachan	4 F	1125	8 × 113
Muddy Run	8 F	355	8 × 113
San Luis	8 F	200	8 × 24
Cabin Creek	2 F	1190	2 × 166
Oroville	3 F	500	3 × 90
Thermalito	3 F	85	3 × 26
Castaic	6 F	1000	6 × 200
Kinzua	3 F	665	3 × 110
Cornwall	8 F	1050	8 × 257

F = Francis; D = Dariaz.

In connection with a study made to determine whether a pumped-storage plant could, under certain circumstances, improve the reliability of a system whose generation is based almost entirely on steam units, dynamic response data of the type indicated below were analyzed.

Hiwassee
Start pumping	Dewater, 1 minute
	Speed (reduced voltage start), 2 minutes
	Load, not less than 25 seconds
Start generating	Speed, 2 to 3 minutes
	Load, not less than 25 seconds

Lewiston
(Arbitrary 4-minute delay on all starts for auxiliaries)
Start pumping	Dewater, 58 seconds
	Speed (full-voltage start), 18 seconds
	Air release, 2 minutes
	Load, 24 seconds
Stop pumping	Unit off line after 27 seconds
	Gates close at 1 minute, 33 seconds
	Unit at standstill at 1 minute, 36 seconds
Start generating	Speed, 15 seconds
	Unit on line at 45 seconds
	Full load at 2 minutes
Stop generating	No load, unit off line at 58 seconds
	Unit at standstill at 2 minutes, 50 seconds

Taum Sauk

Start pumping	Auxiliaries and dewater, 3 minutes, 30 seconds
	Speed (starting motor), 10 minutes
	Unit on line after 16 minutes
	Full load after 33 minutes
Stop pumping	Gates closed, unit off line after 8 minutes
	Unit at standstill after 15 minutes, 30 seconds
Start generating	Open penstock valve, 3 minutes
	Speed, 1 minute, 45 seconds
	Unit on line after 8 minutes
	Full load after 20 minutes (can be started and loaded in less than 10 minutes)
Stop generating	No load, unit off line at 11 minutes
	Gates closed after 11 minutes, 30 seconds
	Unit at standstill at 26 minutes, 30 seconds

3. *Axial-Flow Units.*   These units are designed to operate at low heads (around 20 feet) and have adjustable blades similar to those of a Kaplan turbine. The bulb type is used in Europe and the tube type has been designed and built in the United States. Both can operate as turbines or as pumps in both directions of flow by reversing the pitch of the blades.

A 9-megawatt bulb unit was installed at Saint Malo, France, and in the Rance tidal project near Saint Malo there are twenty-four 10-megawatt bulb units. See next article in this Handbook section. Three 3-megawatt tube turbines were constructed in the United States and installed in Brazil in 1938. These units have been operated primarily as pumps. In 1965, five 25-megawatt tube turbines were planned for installation at Ozark dam on the Arkansas River. Although designed as turbines, they could easily be converted for pumped-storage use.

Operation during a Major Disturbance.   In 1968, IBM made a study for the Department of Public Service of the State of New York to determine the possibility of rapid turnaround of a pumped-storage plant to assist in the restoration of a system to steady 60-Hz operation immediately after a major disturbance. Such a disturbance might involve separation of a system from an interconnection from which it had been drawing power or to which it had been delivering power.

It was found at that time that most of the existing and planned major pumped-storage installations in the United States had large high-head reversible pump-turbines, which, under normal operation, require many minutes to go from full-load pumping to full-load generating, or vice-versa. It was found that under conditions in which rapid reversal is desirable, little improvement would be possible in the changeover from generating to pumping, but in the opposite direction a considerable reduction in time could be achieved. After trip-out of the motor generator it would be possible to move the wicket gates in a programmed manner so that the unit would reverse and stabilize at synchronous speed in the generating direction as quickly as possible. It could then be synchronized with the system, put back on line, and would then start to supply power. It would, however, still require a minute or two, depending on the particular unit, to go from full-load pumping to full-load generating because of the large inertia of the motor generator, and water-hammer problems associated with the long penstock.

Consider a system whose generation is entirely steam-based apart from one or two pumped-storage plants with large, high-head pump turbines. It could become separated from an interconnection at any time of day, and the following three situations will be considered:

1. *When the pumped-storage plants are pumping.*   This will usually occur at night, during the lightest load period when the system has plenty of spinning reserve because many of the base-load steam plants will be kept on line. However, it may continue some way up the slope of the daily load curve when the spinning reserve will be decreasing.

2. *When the pumped-storage plants are generating.*   This will occur during the peak-load period, and the spinning reserve of the steam units will be minimal at this time.

3. *Other times when some or all of the pumped-storage units will be idle.*   In situation 1, if the separated system has excess generation, it is possible that the pumping load could be increased, to some extent, fairly rapidly, but the question of reversal does not arise. If it

has excess load, the pumped-storage plants provide a convenient interruptible load which can be shed immediately. If the system is still lacking in generation, the frequency will decay and stabilize at some level lower than 60 Hz, depending on the generation deficit, the load-shedding schedule, the load droop characteristics (as a function of frequency), and the spinning reserve of the units on line. Because of the time required to reverse the pumped-storage plants, their generating capability would not be available in time to affect this initial transient period. However, if the resulting frequency is too low, this generation would be used to restore the system frequency to 60 Hz because of the undesirability of prolonged low-frequency operation. Only the low-head, axial-flow units previously mentioned could reverse on a time scale capable of significantly affecting the initial transient. Because of their low head, a huge plant and a very large water flow rate would be required to generate appreciable power, and this type of plant is not as economically attractive as a high-head plant.

In situation 2, if the separated system has excess load, the generation of the pumped-storage plants could be increased fairly rapidly, as with a conventional hydro unit, but the question of reversal again does not arise. If there is excess generation, the pumped-storage plants could be tripped off immediately; and, if necessary, their generation could be made available to the system shortly thereafter. If, on the other hand, a steam unit were to trip off, on overspeed, it would require longer to restart and reload than a hydro unit. If the system still has excess generation, the steam units would reduce their generation by governor action in less time than that required to reverse the pumped-storage plants to their pumping mode, and, therefore, reversal would not be feasible.

In situation 3, there are two ways in which the idle pumped-storage units could be used to advantage. A system of operating a pumped-storage plant known as "hydraulic short-circuit" has been proposed. Some of the units would be operated as turbines in order to drive others as pumps. By tripping off one group or the other, instant load or generation would be available. Since the efficiency of the complete cycle is only about 75 percent, the plant would have to be fed power in order to maintain the level of the upper reservoir. Such a scheme is economically unattractive.

A second scheme would be to operate the idle units as synchronous condensers, some spinning in the generating direction, others in the pumping direction. In this way, generation or load would be available at the maximum loading rate of the units, which would be considerably faster than the time required for startup or reversal.

The aforementioned study concluded that a pumped-storage plant could under certain circumstances improve the reliability of a system whose generation is based almost entirely on steam units. However, because the design of large pumped-storage plants is influenced mainly by economic considerations, the reversal of these units takes too long for the procedure to significantly aid a system during the initial transient period following a major disturbance.

CONVENTIONAL HYDRO PLANT PLUS PUMPED STORAGE

In 1966, the U.S. Army Corps of Engineers completed the Kinzua Dam and reservoir on the Allegheny River in northwestern Pennsylvania near Warren. See Fig. 15. The primary purposes of the project were related to flood control, low-flow augmentation, and recreation. It has been estimated that the dam prevents upward of $7 million in flood damage annually. The dam and reservoir maintain low flow of 500 cubic feet/second at the dam, plus additional flow sufficient to keep the magnitude and quality of flow within acceptable limits at points downstream on the Allegheny River. During the summer recreation season, the dam impounds a 12,000-acre lake, extending north into New York State.

Studies made in the early 1960s indicated that a conventional run-of-the-river hydroelectric power facility alone would not be economical. The studies did indicate, however, that such facilities if constructed as an increment to a pumped-storage plant would be justified. The project was continued on this basis and was completed in 1969 with a facility producing an average peaking capacity of more than 380 MW. The principal elements of the system are the Seneca Station, upper and lower reservoirs, and connecting conduits. The upper reservoir is retained within a circular dike covering an area of 139 acres. The lower reservoir is the 27-mile long lake impounded by the Kinzua Dam and previously mentioned. High-head and low-head conduits connect the Seneca Station with the upper and lower reservoirs, respectively.

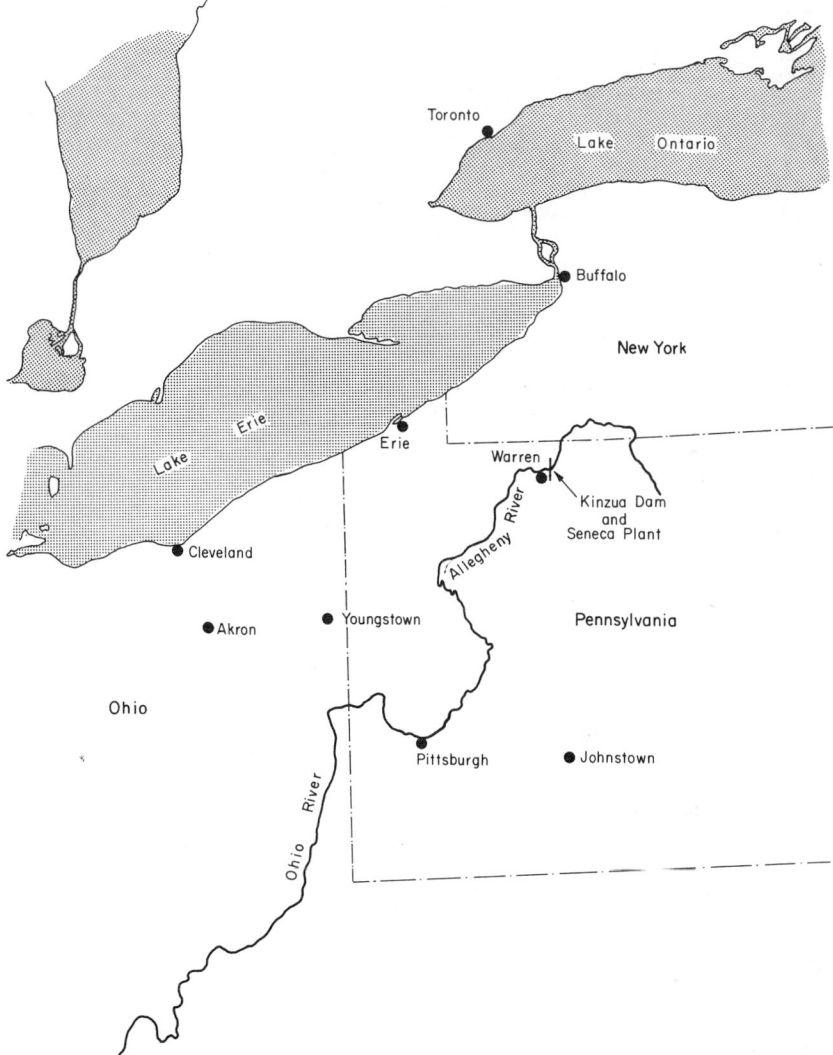

Fig. 15. Location of Kinzua Dam and Seneca Plant on Allegheny River eight miles east of Warren, Pennsylvania. The Kinzua Pumped-Storage Plant, including the Seneca Pumping and Generating Station, was built by the Cleveland Electric Illuminating Company and the Pennsylvania Electric Company. Plant ties into system, serving general areas shown.

The primary function of the Seneca Plant is to provide generating capacity for utility on-peak or emergency periods. The plant is designed so that both the pumped-storage and conventional portions are exclusively for these generating periods. With reference to Fig. 16, Unit 1, a reversible pumping/generating unit, is part of the pumped-storage portion. This unit pumps water from the Allegheny Reservoir to the upper reservoir. When generating, the unit discharges water from the upper to the lower reservoirs.

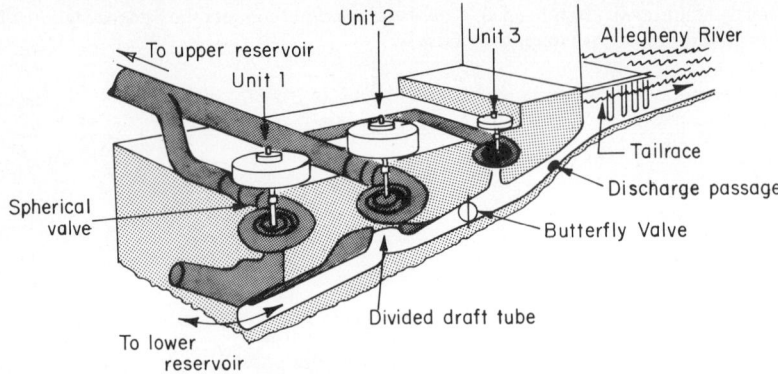

Fig. 16. The Seneca Station incorporates two pump-turbines coupled to generator-motors, one Francis turbine coupled to generator, with remote control.

Unit 2 is another reversible pumping/generating unit and can operate in one of two modes: (1) as part of the pumped-storage portion of the plant, or (2) as part of both the pumped-storage and conventional portions of the plant. Unit 2 pumps water from the lower to the upper reservoir and can generate power by discharging water from the upper to the lower reservoir—an operation that in its entirety is part of the pumped-storage portion of the hydro plant. However, Unit 2 can generate power through discharging water to Kinzua Dam tailwater through the discharge passage. This is achieved by using a specially designed, divided draft tube, and proper operation of its low-head conduit service gate in the lower intake and the butterfly valve in the discharge passage. When so operated, Unit 2 develops the combined head of the pumped-storage and conventional portions of the plant.

Unit 3, the conventional hydroelectric equipment, is part of both the pumped-storage and conventional portions of the plant. Unit 3 can operate only from water pumped into the upper reservoir by Units 1 and 2 and discharges water for generation from the upper reservoir to Kinzua Dam tailwater. Thus, during generation, Unit 3 develops the combined head of the pumped-storage and conventional portions of the plant.

The head range for the units is 644 to 813 feet between upper and lower reservoirs; and 802 to 867 feet between upper reservoir and the Kinzua Dam tailrace. Discharges through the units to the Kinzua tailrace are compensated for by reducing the discharge through the sluice control gates at the dam.

The pump/turbines and Francis turbine are equipped with double-seated, spherical-type inlet valves, selected because of the high head and velocity flow experienced by the turbine casings. The 114-inch-diameter pump/turbine valves are split horizontally; the 48-inch-diameter Francis turbine valve is split vertically. The valve bodies and plugs are constructed of cast steel. The valves are operated by high-pressure oil systems capable of supplying a full opening or closing stroke in two minutes. An automatic sequential controller programs the various events required to open or close the valves.

Water velocities in the high-head and low-head conduits are 21 to 24 feet/second (generating) and 15 to 21 feet/second (pumping). The combined water column between upper and lower reservoirs is about 3500 feet in length, with water in the column weighing 40,000 tons and possessing 60,000,000 foot-pounds of kinetic energy.

In conventional hydro plants having reaction-type turbines, pressure changes can be reduced and plant response to load changes can be enhanced by installation of a surge tank. Site considerations did not permit installation of such a tank at reasonable cost. Therefore, the equipment and conduits had to be designed for handling the very large pressure rises and resultant high flow velocities.

The generator/motors coupled to the pump/turbines are vertical-shaft machines, umbrella-type, with combined thrust and guide bearings below the rotor. These units have stronger damping properties than conventional generators. The thrust bearings have high-pressure oil-lift systems that operate only when a machine is started or stopped. The machines have closed cooling systems, with water-cooled heat exchanges for both bearing oil and air cooling. Each machine has a direct-connected, self-cooled, rotating exciter, using brushes of special design to permit operation without readjustment when the direction of rotation is reversed.

The generator coupled to the Francis turbine is a vertical-shaft machine with a combination thrust and guide bearing above the rotor and a second guide bearing below the rotor. The cooling system is the same as that used for the generator/motors. It is excited by a thyristor rectifier, which obtains power from the station service system. Because this machine is used to start Units 1 and 2 in pumping mode, it has lower-than-normal armature resistance and reactance.

The general control scheme for the Seneca Station is shown schematically in Fig. 17. In addition to radio control, an important feature of the plant is the synchronous method of

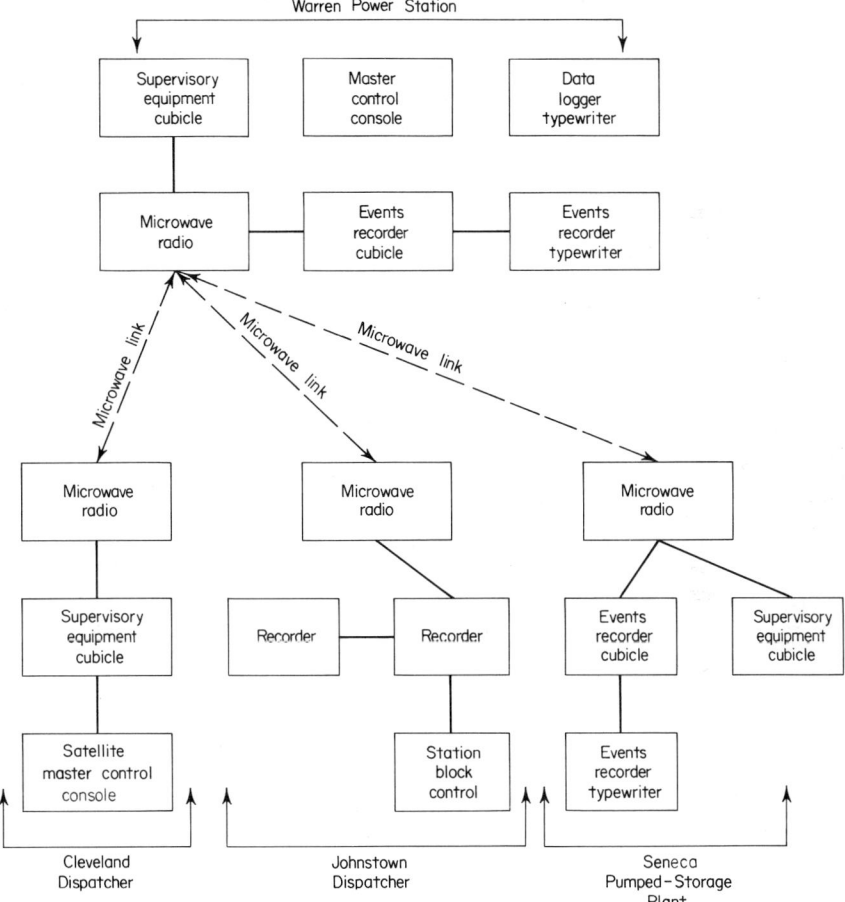

Fig. 17. General control scheme for the power plant. The Seneca Station is linked by microwave radio to the master control station at Warren, Pennsylvania, and the satellite master control station at Cleveland, Ohio. There is also a link to Pennsylvania Electric Company's main dispatching office in Johnston, Pennsylvania. The Johnstown operator is provided with watt and var data and other control features in case of emergency. The power plant is normally controlled by the Warren operator, who can transfer control to Cleveland, in which case the Warren controls are cut out. Power, energy, reservoir, and river-level quantities are inputs to both master stations at all times. The Warren station is also equipped with automatic event and data loggers. These produce printed records of such occurrences as unit starting and stopping, circuit breaker and relay operations, and periodic readings of such variables as water levels and voltages.

starting Units 1 and 2 in the pumping mode. The starting sequence for Unit 2 (similar to Unit 1) is as follows:

1. Water in the draft tube is pushed clear of runner with compressed air.
2. Field windings of Units 2 and 3 are excited, and thrust-bearing oil-lift pumps are started.
3. Wicket gates of Unit 3 are opened to 40 percent of total movement.
4. Both units accelerate to synchronous speed together, and speed is regulated by the governor for Unit 3.
5. Oil-lift pumps stop at 30 percent speed.
6. At 90 percent speed, running exciter takes over excitation of Unit 2, whose voltage regulator then operates.
7. Automatic synchronizer is placed in operation when Unit 3 terminal voltage reaches 80 percent of normal.
8. When conditions are correct, automatic synchronizer causes starting-bus circuit breaker to open, and the circuit breaker (230-kV) of Unit 2 closes by action of an auxiliary switch of the starting-bus circuit breaker.
9. Unit 3 is shut down, and water is admitted to Unit 2 runner chamber by shutting down the water depression system. After Unit 2 is fully primed, the high-head guard valve and then the wicket gates are opened.

Detailed specifications on the major equipment items comprising the Seneca Plant installation will be found in "The Electric Utility Generation Plantbook," edited by *Power* magazine (revised periodically).

REFERENCES

In addition to specific references indicated below, statistics on hydroelectric power are continuously updated and available from such organizations as: U.S. Federal Power Commission, Washington, D.C.; Edison Electric Institute, New York; Electric Power Research Institute, Palo Alto, Calif.; Bureau of Reclamation, U.S. Department of the Interior, Washington, D.C., U.S. Army Corps of Engineers.

Rheingans, W. J.: Hydroelectric Power Plants, in D. G. Fink and J. M. Carroll (eds.): "Standard Handbook for Electrical Engineers," 10 ed., McGraw-Hill, New York, 1972.
Weedy, B. M.: "Electric Power Systems," 2 ed., Wiley, New York, 1972.
"The Electric Utility Generation Plantbook," by the editors of *Power* magazine, McGraw-Hill, New York, 1972.
"Power Generation Systems," by the editors of *Power* magazine, McGraw-Hill, New York, 1967.
Creager, W. P., J. D. Justin, and J. Hinds: "Engineering for Dams," Wiley, New York, 1945.
Parker, A.: "Planning and Estimating Dam Construction," McGraw-Hill, New York, 1971.
Potter, P. J.: "Power Plant Theory and Design," Ronald Press, New York, 1959.
Cullan, A. H.: "Rivers in Harness: the Story of Dams," Chilton, Philadelphia, 1962.
"Power System Computer Feasibility Study," by IBM Research Division, San Jose, Calif., for the Department of Public Service of the State of New York, vols. II and III, 1968.
Giles, R. V.: "Fluid Dynamics and Hydraulics," 2 ed., McGraw-Hill, New York, 1967.

Tidal Energy

The tides as a source of energy were exploited as early as the Middle Ages. For example, tidal power was used in the coastal areas of Brittany in France to drive watermills from the twelth century onward, although this use progressively diminished during the present century because of the growing availability of other low-cost and more convenient energy forms. However, at the time Électricité de France* decided in 1959 to proceed with construction of what is today the world's largest tidal generating station, installed in the maritime estuary of the Rance River (Saint Malo–Dinard, Brittany), there were still a few small tidal mills in operation. See Fig. 1.

Study of methods of equipping the estuary of the Rance for production of electric energy on a large scale (240,000 kilowatts) was actually commenced early in this century, although detailed design and construction did not start until much later. Generator sets were not ordered until 1960. Electrical connection to the first group of generators was made in August 1966. Final connection of the 24 turboalternator sets to the electrical grid occurred in early December 1967.

The general location of the tidal energy installation along the English Channel (La Manche) on the northern coast of France is shown in Fig. 2. A more detailed map showing the location of the installation is given in Fig. 3.

The early research workers on the Rance estuary project did no more than take up the underlying principle of an idea which, because of the presence of the early and successful tidal mills, was not unknown to that region. The site was of fascinating interest to engineers by reason of its great tidal range, which reaches as much as 13.5 meters (44 feet). But construction and successful operation of a plant of this magnitude had to await the development of three major factors: (1) a far greater understanding of tidal phenomena; (2) development of a new type of turboalternator set, known as the *bulb unit*, which is able to use low heads to drive a turbine in either direction of flow, and also to function as a pump; and (3) cutting off the estuary of the Rance, which meant damming a river with a maximum flow of 18,000 cubic meters per second, both at flood and at ebb—treble the flow of the Rhone in flood at Avignon. The first factor, of course, involved highly analytical mathematical and theoretical investigations of tidal phenomena. The latter two factors involved solutions to difficult problems in mechanical, electrical, civil, hydraulic, and materials engineering.

Some of the fundamentals of tidal energy will be discussed later.

*Much of the information given in this summary was furnished by Électricité de France. The cooperation of EDF and of the Consulat Général de France (Los Angeles) is deeply appreciated.

Fig. 1. Very early tidal mill at Saint-Suliac (Brittany).

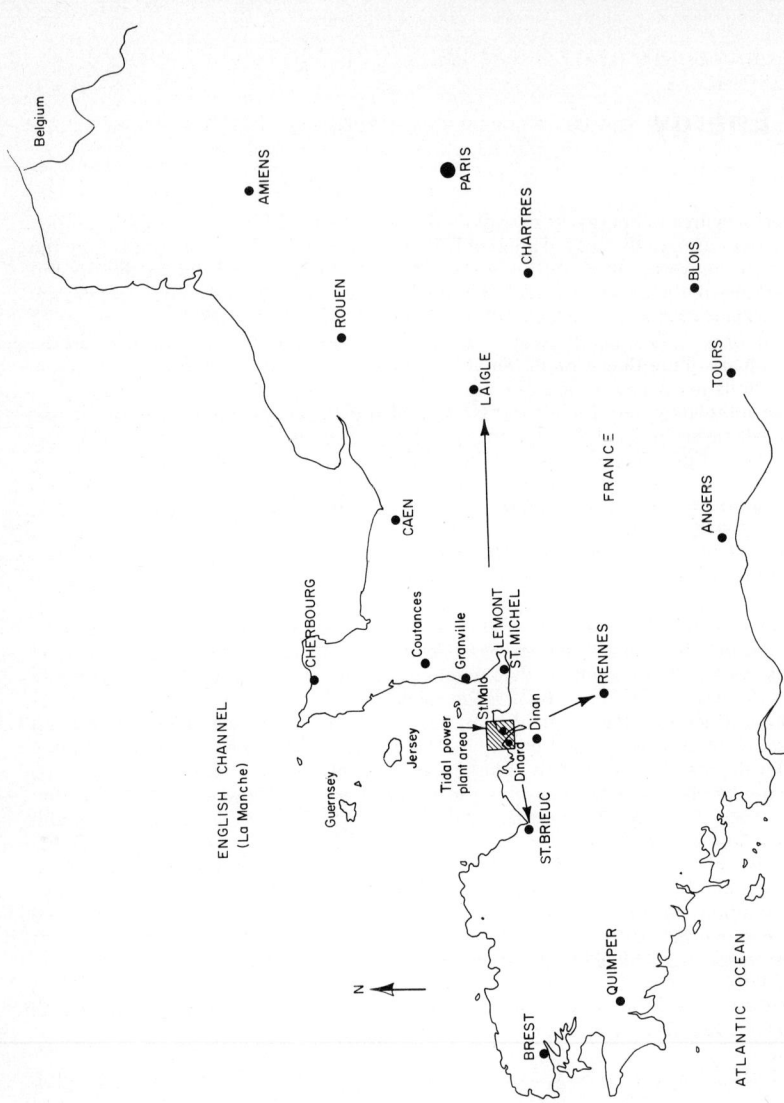

Fig. 2. Northern coast of France showing location of tidal energy power plant between Dinard and St. Servan (St. Malo) in the province of Brittany. Power lines extend toward St. Brieuc (and on the Brest), to Rennes, and to Laigle (and on to Paris).

ENGINEERING SUMMARY OF THE RANCE INSTALLATION

A plane section of the installation, extending from Pointe de la Brebis on the left bank to Pointe de la Briantais on the right bank, a distance of some 750 meters (~2460 feet), is shown in Fig. 4. The tidal generating facilities consist basically of the tidal basin constituted by the estuary of the river Rance, separated from the sea by a dam. Sluice gates built into the dam can be operated to adjust the natural rate of drainage from the basin to the sea with outgoing tides, or the rate at which the basin is filled with incoming tides. The difference in levels between the basin and the sea, therefore, can be controlled, and when it reaches an appropriate figure, the basin is filled (or emptied) through the dam. The flow of water is used to move the wheels of hydraulic turbines which, in turn, drive alternators. These turboalternator sets, of which there are 24, produce electric energy.

For optimum operating characteristics, the law defining the level as a function of time in the basin, which defines the rate of flow through the dam as a consequence, is so selected as to implement the maximum production of energy.

Apart from the three functional elements (dam, sluices, turboalternators), additional requirements had to be met, including: (1) the dam was not to be an obstacle to navigation between the sea and the estuary, and so had to incorporate a lock. (2) The general appearance of the installation had to respect the natural features of the site. (3) The construction had to serve as the basis for a road across the Rance, linking Dinard on the left bank to St. Malo and St. Servan on the right bank.

Geologically, the Rance estuary was formed by subsidence, and its banks are sharply outlined. Its general orientation, taken from the mouth, is north to south. The dam cutting the basin off from the sea runs in a straight line roughly west to east between two rocky spurs of land which cut into either end of the seaward extension of the estuary.

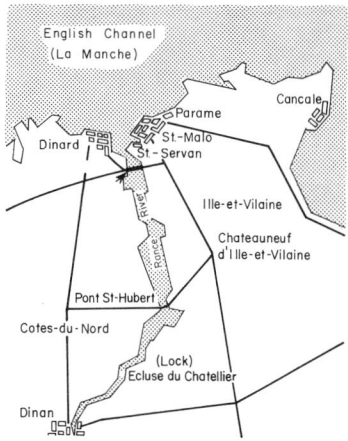

Fig. 3. Detail map of location of tidal energy plant.

The useful volume of the tidal basin is 184 million cubic meters (~6.5 billion cubic feet) between the extreme of zero elevation at ebb tide and the 13.5 meters (marine scale)* reached in exceptionally heavy spring tides. The corresponding areas of the basin are 4.3 square kilometers (~1.7 square miles) at zero elevation and 22 square kilometers (~8.5 square miles) at 13.5 meters.

The presence of a rocky island, Ilot de Chalibert, will be observed (Fig. 4) about three-fourths of the distance between the left and right extremities of the dam. The maximum

*Zero on navigational charts, which may vary along the shoreline, is defined as the lowest level reached by the sea at mean low-water spring tide at a given point. Heights given in Fig. 4 and elsewhere mentioned here are in the marine scale. In this particular location, marine zero corresponds to minus 6.85 meters.

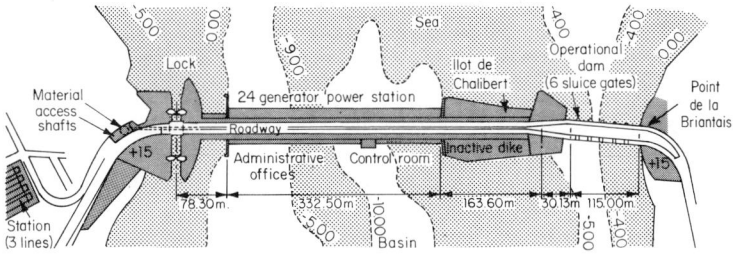

Fig. 4. Plane section of tidal energy plant on Rance estuary.

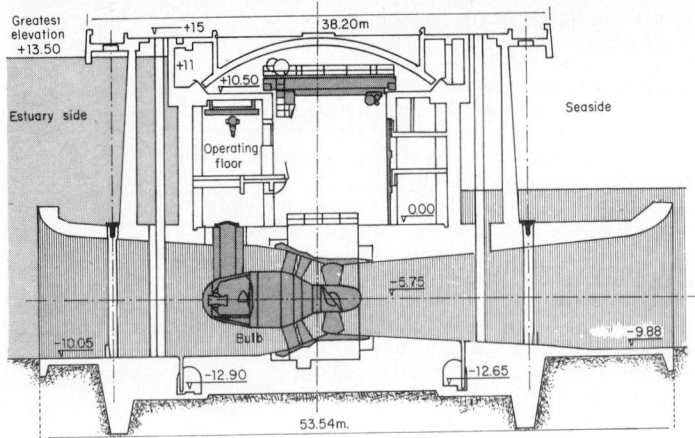

Fig. 5. Cross section of a machinery bay.

depth of the seabed between this island and the western bank of the estuary reaches an elevation of minus 11, compared with a maximum of minus 6 toward the eastern bank, In both cases, the seabed is regular and rocky. These conditions favored construction of the sluice gates between Ilot de Chalibert and Pointe de la Briantais, and it was found that a sufficient sluice-gate facility could be constructed without any need for rock blasting. Early construction of the sluice gates greatly facilitated the construction project. The required surface of sluice gates (900 square meters—about 1076 square yards) permitted a flow of about 4000 cubic meters (~141,000 cubic feet) per second per meter change in level.

With further reference to Fig. 4, the siting of the various components of the structure followed logically. The sluice gates thus occupied the eastern segment, between Ilot de Chalibert and the east bank. The lock was built in the western channel, along the left bank, far from the considerable currents generated by operation of the sluice gates, and on depths sufficiently shallow for the protective envelope used in constructing it to be put into position at low tide. The lock had to be constructed at an early stage.

The natural place for the dam structure and the generating plant was in the intervening space. In fact, the 24 sets of turboalternators in the installation did not fully absorb the available length, which would have allowed 32 sets, a solution which was envisaged at one stage of planning. Floor space had to be allowed for each set, both for initial assembly of machinery and for subsequent maintenance. Moving from the lock to Ilot de Chalibert, there is the dam-power station, consisting of 3 assembly bays, 24 machinery bays (with a control bay between two of them), and then the inactive dike which seals off the basin.

The central part of the dam structure was leveled off at elevation plus 15, i.e., 1½ meters above the greatest tidal rise, so that the roadways which top the dam could be built at that height. The aesthetic constraint which made it undesirable to erect any major structures above elevation plus 15 posed difficulties in connection with the access for heavy equipment to the interior of the plant. Thus, equipment was brought to a building which backs directly on to the left bank cliff. This building straddles a shaft through which components were lowered by an overhead traveling crane to a gallery, which runs under the lock and emerges inside the dam structure proper, below the overhead traveling cranes used to assemble the machinery sets.

Machine Bays.   Each turboalternator set occupies a bay 13.3 meters (~44 feet) long, measured along the major axis of the dam. Basically, a bay includes, as shown in Fig. 5, the hydraulic conduit for passage through the dam of the water whose current drives the propeller of the turbine, which is fully immersed in the conduit. The ends of the conduit are in the seaward and estuary sides of the concrete of the dam structure. In its central section, where it crosses the pits of the machinery bays (the floor of the pits is at elevation minus 10.5 meters), the conduit consists of a metal envelope.

Each turboalternator set is housed in a metal hull, with the propeller of the turbine wheel to be driven by the current projecting from the downstream (i.e. seaward) end of the hull. Access to the hull protecting the alternator (which is upstream from the turbine, i.e., on the estuary side) is through a flue which crosses the conduit and exits within the dam at elevation zero.

It will be noted that a double set of slots, emerging at elevation plus 15 on both the seaward and estuary sides, has been built into the hydraulic conduit. The slots nearest a turboalternator set are for the passage of gate valves which can be operated under load, and are used either to stop an uncontrollable flow resulting from damage to the set, or to isolate the conduit for later dewatering. The farther channels serve for introduction of cofferdams to enable maintenance of the fixed metal components of the nearer slots to be undertaken.

Within the dam, above the metal part of the hydraulic conduit, i.e., above the pit of the machinery set, is the free space required for movement of parts during dismantling and reassembly operations. The space required for the auxiliary machinery is located above this, over the concrete of the pit structures. In order to facilitate their use, these spaces are further divided horizontally into a number of levels. Thus, the ground floor is at elevation zero, with three intermediate levels on the seaward side, at elevations plus 4.4, plus 7, and plus 10.5, respectively, and two intermediate levels toward the estuary at plus 4 and plus 10.5.

The ceiling is an arch whose keystone is at elevation plus 14.8. The space for auxiliary equipment and dismantling operations has been found above the generating sets, but below the highest tide level. Thus, the upper surface of the structure forms a kind of seating on which the roadways have been built. These are therefore quite close to the water level.

The individual machine bays are coupled to one another, and constitute the recess in the dam corresponding to the station proper (not including the assembly and control bays)—a total length of 24 × 13.3, or 320 meters (~ 1050 feet). In building this long structure, provision had to be made for passage of personnel and equipment from end to end, for ventilation, water and oil circuits, and electric cables. These lengthwise passages had to be supplemented to allow for transversal and horizontal movement; and these channels found their natural siting in the piers between pairs of successive bays.

The upper surface of the dam at elevation plus 15 carries the highway, divided by a center island into two one-way roads, each 7 meters wide. On either side of the highway are two sets of rail tracks (4.35 meters gage) for carrying the bridge cranes used to operate the gates which shut off the hydraulic conduits.

A longitudinal gallery at elevation minus 12.65 on both the seaward and estuary sides channels drained-off products toward the sumps of the station and also contains the piping carrying water taken from the hydraulic conduits when these are emptied for maintenance.

Bulb Units.   Much study and experimental work went into the design of a satisfactory bulb unit. Comparison of the Rance scheme as visualized as early as 1951 with the finally accepted and installed design is shown in Fig. 6. In the earlier scheme, the large-diameter generator, incorporating a conventional vertical-shaft arrangement, was out of the water above the turbine; whereas the later, smaller-diameter generator is inside a bulb-shaped casing immersed in the flow.

In the earlier proposed design configuration, the power set spacing would have been such that the installed capacity of the Rance site would have been limited to 200,000

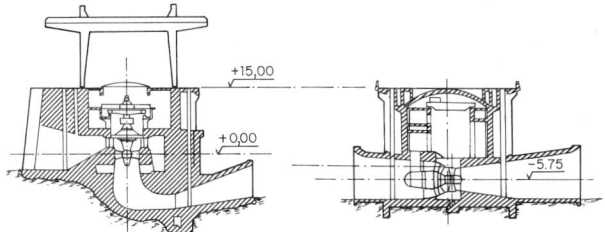

Fig. 6. At left is shown the conventional vertical-shaft unit initially contemplated in 1951. At right is shown the horizontal-bulb unit which is incorporated in the final design of the Rance installation.

kilowatts. The size of the units would have been such that it would have been necessary to provide for a gantry crane traveling on the inactive dike, while the pits accommodating the power sets would have been fitted with impermeable covers. The transformers in that case would have been installed ashore. In the final layout, the hollow dam across the estuary accommodates twenty-four 10,000-kilowatt power sets and all their accessories, the three 80,000-kilowatt transformers, medium-tension equipment, all power station ancillaries, and erection bays, leaving the top of the dam clear for the double-lane roadway.

In the early 1940s, designers were well familiar with low-head river site harnessing problems, but this knowledge could not be applied directly to the tidal plant design. One of the particular features of a tidal plant is its range of turbine operation in either direction of flow, with both head and discharge varying considerably. Power plant operational studies soon pointed to the advantage of pumping at around high tide slack water so as to *overfill* the basin; and so as to *overempty* it at low tide. This required machines capable of running as low-delivery head pumps. Design studies by Robert Gibrat clearly showed the tremendous potential scope of tidal power plant operating cycles. In order to derive maximum benefit from these cycles, it was essential to develop units capable of operating as turbines, pumps, or orifices with the flow in either direction.

Three principal designs were developed: (1) The horizontal-shaft upstream-bulb unit (2) the downstream-bulb unit and (3) the horizontal-shaft pit unit. The design shown at the right in Fig. 6* incorporated the following features: (a) A conical intake duct flaring out at the bulb accommodating the generator; (b) a metal casing containing the generator, connected to the duct wall by combined radial guide and stay rods (this assembly comprising the stay rods and the two walls they connect is known as the stay ring, or speed ring); (c) a distributor between the stay ring and the runner, consisting of a set of guide vanes conducting flow to the runner and controlled by an operating ring; (d) a propeller-type runner with variable-pitch blades; and (e) a conical draft tube after the runner. Subsequent improvements mainly concerned the flow duct at the generation section, and guide vane and runner blade design. Comprehensive electrical design research also finally resulted in a substantially reduced generator diameter, so that it would not require a flaring out of the hydraulic duct at the generator section.

After many years of research and testing at other sites (Électricité de France hydro facilities at the Wadrinau power plant on the River Moselle, at Argentat, and Cambeyrac) and careful comparison with other designs (Harza machine, Kaplan turbines, etc.), the upstream-bulb design with a single support was selected. The upstream-bulb design features a variable-pitch propeller runner and a conical distributor with adjustable guide vanes and is especially designed for operation with either direction of flow and rotation. Bulb-tupe units are now used in conventional river hydro installations in France, Italy, and the U.S.S.R.

The final hydraulic specifications are summarized in Table 1.

A large number of overall mechanical layouts were designed for an earlier experimental plant, which finally led to one featuring a single connection between the unit and the concrete in the stay-ring mounting section. See Fig. 7. The stay ring is thus the only main rigid support for the generator body and turbine guide bearing support, which are bolted to it, one on either side. Prestressed tie rods are fixed to the concrete by one end, and to a ring between the nose cap and generator body by the other. These reduce possible movement of the unit as a whole with respect to the stay ring center to negligible proportions, without, however, restricting its axial freedom of movement. The vertical access shaft to the bulb casing is not a rigid mounting, for although its top end is anchored to the concrete, its bottom end is only attached to the thin and therefore deformable cooling nose skin plating.

The main advantages of this layout are: (1) The connections between the supports and the concrete have little effect on strain in the main bulb unit components, i.e., its body and supporting cone. (2) The flanges holding the body to the stay ring and cooling nose do not transmit any high loads. This facilitates durable sealing of these flanges against seawater, which is an essential requirement for satisfactory operation. (3) Connections between supports are not readily affected by seawater corrosion. (4) The concrete intake duct has

*Patented by Service des Études Marémotrices, 'a la Direction des Études et Recherches de l'Électricité de France.

TABLE 1.   Hydraulic Specifications of Rance Installation

Flow	Heads,* meters (between)
FROM BASIN TO SEA:	
Direct turbine operation	+11 and +1
Reverse pump operation for acceleration	+2 and 0
Reverse normal pump operation	0 and −6
Direct orifice operation	+3 and 0
FROM SEA TO BASIN:	
Reverse turbine operation	−11 and −1
Direct pump operation for acceleration	−3 and 0
Direct normal pump operation	0 and +6
Reverse orifice operation	−3 and 0
PRINCIPAL DETAILS OF THE TURBINES	
Rated output	10,000 kilowatts
Runner diameter	5.35 meters (17.55 feet)
Number of runner blades	4
Range of blade angle adjustment	−5 to +35°
Number of guide vanes	24
Rate speed	93.75 revolutions per minute
Maximum overspeed condition	260 revolutions per minute

*Heads associated with basin levels above sea level are conventionally denoted by a plus sign; those with basin levels below sea level by a minus sign.

smooth simple outlines with only one small opening where joined by the vertical axis shaft.

The generator and nose cap are in the part of the hydraulic duct on the basin side. The turbine casing connects this section to the seaward side of the hydraulic duct. There is a pit by the turbine for assembly and dismantling of components of the generating set. The turbine runner and generator rotor are on a common shaft with two bearings. Vertical loads are transmitted to the concrete by: (1) stay vanes connecting the bulb casing to the duct wall; and (2) four prestressed radial tie rods. The hydraulic thrust acting on the runner is taken by a double thrust bearing between the turbine runner and generator, and transmitted to the concrete through the stay vanes.

Guide Vanes and Runner Blades.   The two-way flow axial turbines have the following control arrangements: (1) The distributor, acting as a stationary screen with adjustable vanes; and (2) a variable-pitch runner. For direct turbine operation, the direction of flow is from the distributor to the runner. In this case, the purpose of the distributor is: (1) to share

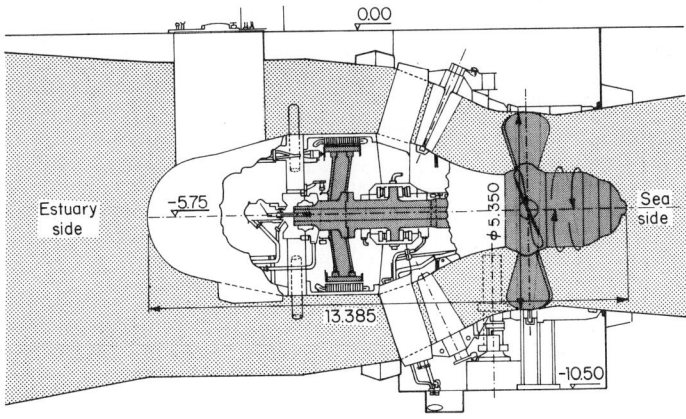

Fig. 7. Bulb unit installed in hydraulic conduit.

with the runner blades in controlling the unit for optimum efficiency, and (2) to shut the unit down in case of disconnection.

The direction of flow for reverse-turbine operation is from the runner to the distributor, and the former is liable to cause dangerous vibration of the latter in certain conditions. For this reason, the guide vanes stay fully open for this type of operation, being locked as rigidly as possible in this position. The mechanical design difficulties encountered with this machine were due mainly to the numerous different types of operation required in a tidal plant as contrasted with conventional turbines; and also to the high loads liable to very sudden variation. Changes from one set of conditions to another have necessitated an extensive study of transients in order to determine the loads on the unit so as to reduce vibration risks. Relationships governing guide-vane and runner-blade operation and combinations have been subject to prolonged research and testing. This has produced comparatively simple automatic control sequences that are easy to implement and that ensure full operational reliability of the equipment.

Conical Distributor. This comprises a set of controllable guide vanes mounted between two concentric spherical distributor rings carrying the guide-vane stem bushes, bearings, and seals. The guide vanes are connected by levers and ball-jointed connecting links to an operating ring actuated by two vertical oil servomotors controlled by the main governor system. Average loads are small compared to maximum loads in reverse-turbine synchronous operation and pump operation in either flow direction. During reverse-turbine and direct pump operation, they are liable to be pulsating, calling for a reliable means of locking the guide vanes in position.

Runner. In addition to hydraulic torque, the runner blades also must be able to resist centrifugal thrust. The maximum overall torque on each blade is +600 kNm for reverse-turbine operation under −11 meters head. The maximum centrifugal load on each blade is approximately 10,000 kN. (A newton, N, is the force imparting an acceleration of 1 m/sec² to a mass of 1 kilogram. For practical purposes: 1 N = 10^{-1} kilogram force; 1 kN = 1 kilonewton.)

The amount of runner overhang has a most important effect on the mechanical strength of the unit. The runner mounting on its shaft and the shaft seal and bearing layout are designed to reduce overhang to a minimum. For reasons of economy and ease of manufacture and assembly, the blades are bolted to carbon steel stems. The servomotor controlling the blades is accommodated inside the runner hub cap. The maximum closing force (at a pressure of 60 bars) is 6700 kN, and the opening load at 35 bars is 6000 kN. Regulations on the limitation of discharge variations impose very long operating times:

Slow opening and closure	400 seconds
Rapid closure	40
Closure at runaway speed	280

Bearings. The load-carrying bearing near the runner in big axial turbines is one of their most delicate components. Despite their low rotational speed, loads on these bearings exceed 1000 kN for a 6-meter (~19.7-foot)-diameter runner, and may be as high as 2000 kN in bigger machines. This is a difficult problem, therefore, especially as these particular components must be absolutely reliable in operation. Load-carrying bearings on the turbine side thus require very careful design and manufacture. The turbine bearing is usually the one carrying most of the load. In the Rance units, the blade control servomotor dimensions required for the various operating conditions result in a heavier runner, high loads on the turbine bearing, and lighter generator bearing loads. In order to prevent vibration due to random hydraulic load variations, the generator bearing clearance has been reduced to the absolute minimum still ensuring reliable operation and within normal manufacturing clearances. Both guide bearings and the double-thrust bearing must be designed to run in either direction of rotation.

Generator bearing load	72.6 kN	
Turbine bearing load	915	kN
Thrust on thrust bearing:		
Normal (under 11-meter head):		
Direct turbine	2,660	kN
Reverse turbine	3,080	kN
Exceptional conditions	5,000	kN

Both bearings feature a high-pressure oil injection system which automatically comes into operation when the unit starts up or stops, and also during orifice operation.

Shaft Seal. Particular care is required in sealing the bulb casing against seawater ingress. Tests and development work on prototype units produced the following arrangements: (1) a set of labyrinth seals; (2) a cylindrical carbon ring seal rubbing against a wearing piece on the shaft of the unit; and (3) two static systems operated from inside the bulb casing, enabling the seal to be dismantled without having to isolate and dewater the duct.

The sealing arrangements comprise two cylindrical seals, each with two carbon rings, made up of segments and wedges held up against the wearing piece on the shaft by a spring and the bearing water supply. A continuous supply of clean fresh water or filtered seawater is essential for satisfactory bearing operation. A fresh water supply was not practical for the Rance scheme. No supply problems arise with seawater, except that it requires very thorough filtering to prevent bacterial growth lodging and proliferating between the carbon rings. The seawater supplies are collected in tanks from which the carbon rings can be fed under pressure. The high electrolytic potential developing between carbon and aluminum bronze in salt water ruled out these materials. Synthetic rubber springs and polyester fiberglass stops are used, along with a stainless steel shaft wearing piece and other stainless steel parts throughout.

Generator. A conventional unit designed to run at 93.75 revolutions per minute would require a generator stator 9 meters (~29.5 feet) in diameter. To be able to accommodate the generator inside a casing sufficiently compact to ensure acceptable flow conditions upstream from the turbine, alternators of about half the normal diameter had to be designed. It is this essential feature which gives bulb generators their special character.

Radical changes had to be made to existing generator cooling arrangements, as the conventional cooling system with a radial air flow into hollow compartments in the machine body would have been too bulky. Axial cooling was resorted to instead, with a flow of air circulating through a number of narrow passages running parallel to the stator and rotor. Furthermore, the peripheral speed is so low that the pressure produced by a fan mounted on the shaft would be five to ten times lower than with a conventional generator. As a bulb generator offers a higher resistance to air flow than a conventional design, it was necessary to use a separate fan in order to cool the alternator economically.

If cooling air at atmospheric pressure were used, the power absorbed by the fan might amount to as much as 1 percent of the generator output, whereas air friction losses in the gap and at the ends of the machine are extremely low, i.e., about 0.05 to 0.1 percent of the generator output, as they are independent of the air flow. This is one reason why compressed air has come into general use, for analysis of temperature rise in a generator shows it to be the sum of the following three individual temperature rises: (1) temperature rises independent of cooling fluid flow and pressure, i.e., temperature differences due to the passage of losses through the insulation and magnetic circuit; (2) temperature differences between metal or insulating surfaces and the cooling fluid in contact with them, which vary as the mass flow to the 0.8th power; and (3) temperature rise of the fluid itself, which is related to specific heat, i.e., linear mass flow. Hence, mass flow fully defines the temperature rise of a given machine.

If P is the density of a fluid and V its volume flow, the mass flow is $P \times V$. It is also well known that fluid circulation power is proportional to PV^3. Comparing circulation powers required to cool a given machine to the same degree at a pressure 1 or a pressure n, it is readily seen that power consumption is n^2 times less with compressed air than with air at standard conditions. A 10,000-kilowatt generator requiring a 100-kilowatt fan to cool it at standard conditions would require only an 11-kilowatt fan in air at 3 bars. It is obvious that this possibility can be turned to advantage in seeking more effective cooling at the cost of a slight increase in cooling power in order to achieve high specific power unit output, i.e., appreciably, smaller power unit dimensions. A point to note is that as windage losses increase linearly with pressure, excessive air pressure should be avoided. There is an optimum value of n at which the sum of the forced ventilation and windage losses is a minimum. This economic operating pressure is somewhere between 2 and 4 bars. A pressure of 2 bars is used at the Rance installation.

At the foregoing pressure, the generator casing does not require any reinforcement. Being designed to take and transmit the various operating loads, it can safely withstand this pressure, much of which is in any case balanced by the water pressure outside.

Compressed air also helps to keep water out of the casing, and its availability as an adjustment parameter is advantageous. By slightly increasing the air pressure, the unit can be kept at a constant temperature when under an overload.

Casing and Windings.   The generator casing is a straightforward massive hot-rolled steel cylinder welded up in a slag bath, with two annular flanges bolting to the turbine stay ring and basin side nose cap, respectively. The stator windings are of the Roebel bar type with synthetic resin insulation. The latter has excellent mechanical and dielectric properties. Special precautions were taken for the magnetic circuit, which the specifications required to be completely watertight. Its laminations are insulated and bonded together to form a watertight block.

When the units are running under load and stopping, the active parts heat up, and the magnetic circuit expands with respect to the cold casing in contact with the seawater. This expansion is absorbed through the flexibility of the magnetic circuit assembly frame and its slide mounting on the generator body.

The rotor features a rolled-steel plate rim welded up in a slag bath and a fairly tight fit to a 6-armed rotor spider, leaving sufficient passage free between the arms for withdrawal of the turbine bearing components through the generator. The inductor pole insulation to earth, impregnated under vacuum and pressure, is an additional safeguard against damage due to accidental flooding.

The excitation system is sufficiently unique to warrant a brief description. It is entirely static. The alternators are excited in sets of four by a system comprising: (1) a transformer with its output connected to a composite diode and thyristor bridge common to four units, and (2) four transformers with their outputs connected to four diode bridges. The composite bridge supplying an excitation voltage proportional to the stator voltage and adjustable by the thyristors is connected in series with each diode bridge producing an excitation voltage proportional to the current output from each generator, and thus introducing an individual compound term into the excitation control process. The voltage regulator affords good voltage regulation by action on the grid of the composite bridge thyristors.

Coupling.   Bulb units have low inertia and take about 1 to 2 seconds to get up to speed, about one-third to one-fourth of the time required for conventional units of the same output. This low inertia could be considered a disadvantage where coupling is concerned, but with the high response available from electric governors, this is only a minor factor. The generators are coupled to the network for asynchronous operation, without a synchroscope. The turbine is started up and run up to approximately the synchronous speed in about one second while temporarily operating as an asynchronous motor. The high subtransient reactance of these machines limits the supply of current to the network to roughly 2.2 times the normal current. This is a very rapid coupling method, and the alternators are coupled to the network four at a time. Because the power plant is designed for alternative turbine and pump operation cycles, the generators are designed to run as synchronous motors during storage periods, driving the turbines as pumps. This has not involved any change in generator dimensions.

Pump-Turbine Governing System.   The distributor and runner blade controls for each unit are similar to those used with Kaplan turbines, with the following main features: (1) a hydraulic slave (feedback) system controlling the distributor guide vanes, and (2) a hydraulic slave system controlling the runner blades. Each of these systems features a "mixer," receiving electric signals either from outside sources or from the feedback circuit. Outgoing signals from the mixer are amplified and control the actuator which registers the electric signal it has received and makes a pilot valve move accordingly, which in turn causes the hydraulic motor inlet valve to open.

A speed correction signal is fed to the runner blade control mixer for each unit, bringing the latter to near the synchronous speed. This signal cancels out automatically when the unit is coupled to the network.

The runner blade settings, i.e., turbine discharge, are continuously determined for the required output by a single automatic programmer serving the whole power plant. Independently of this programmer, plant personnel are also able to control load for each group of four bulb sets constituting an operational unit. Orders from the programmer or load control system act on the mixer controlling the runner blades for each unit, and also on an electric cam, matching runner blade angle to distributor opening. For a given head, there must be a definite distributor opening (i.e., streamline heading) for each runner blade angle (i.e., turbine discharge). To achieve this, the electric cam receives one signal from the programmer and another depending on turbine head. With these two signals, it produces a command which it feeds to the mixer controlling the distributor of each power set. Eight power sets, i.e., two operational units, share the same head measurement and transmission system.

Thus, the programmer is able to control turbine discharge while allowing for any necessary limitations, for example, because of navigational requirements. The maximum permissible rate of overall discharge variation for the 24 power sets has been set at 180 cubic meters/second per minute, and, similarly, the estuary water level must not vary by more than 4 meters in any one hour.

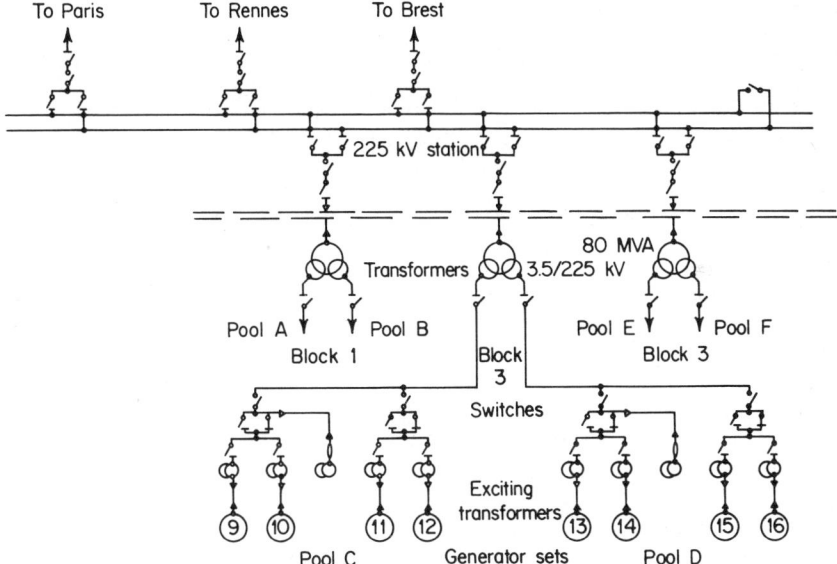

Fig. 8. Line connections of a group of eight generators.

Electrical System.   This can be broken down into three components, relating respectively to distribution of the energy produced by the machine sets, the supply of energy to the auxiliary equipment in the installation, and the command and operational control circuits.

Energy is produced by the machine sets at 3.5 kV and is delivered to the general power system at 225 kV. The relatively low unit power of each set made it necessary to connect a certain number of sets to a single 3.5 to 225-kV step-up transformer. The grouping of sets is shown in Fig. 8. There are three transformers in all, each with an 80-MVA high-tension winding and two distinct 40-MVA low tension windings. Four machine sets deliver current in parallel to each LT winding. Following conventional practice, the internal impedances of a transformer have been chosen so as to render the system equivalent, in particular for short-circuit power, to one of two distinct 40-MVA transformers delivering electricity to the busbars at 225 kV. Thus the way in which each pool of four generating sets constitutes an operating unit can be seen.

The three 80-MVA transformers are installed inside the dam structure, in three machine bays whose layout was slightly modified for the purpose. The high-tension winding is directly connected to a set of three monopolar oil-insulated 225-kV cables. The corresponding feeder line runs first through the gallery at elevation plus 7, then passes beneath the lock, ascends the shaft of the access building under whose terrace it emerges, and thence carries on to the outside 225-kV station. See Fig. 4. This station has the conventional isolated phase arrangement with two sets of bars. There are seven bays; three for arrival of current from the transformers; three departure bays toward Paris (Laigle); Rennes; and Brest (St. Brieux); and a connection bay. See Fig. 2.

Attached to the two LT windings of the transformer are the set of bars each connecting one of the two pools of machine sets. The relatively low voltage of the sets entails quite high intensities, and the usual precautions have been taken to avoid parasitic overheating of magnetizable materials, in particular concrete tie rods. This part of each set of bars and the corresponding shielding are of aluminum.

Two circuit breakers are in shunt on each set of bars, each breaker controlling a pair of machine sets. The intensity of the current at this point (3500 amperes) is such that conventional apparatus can again be used. Sectioning and reversing switchgear installed behind the circuit breakers governs the crossing of two phases for inversion of the direction of rotation of the machines, according to operating requirements. The 3.5-kV connections in shunt on this switchgear radiate toward the group of four sets in dry-insulated cables, and join the access arms of the machines whence they are connected to the stator.

For operating purposes, a pool of four machine sets is treated as a single unit (synchronization of startup and close down, load equalization). However, there is an isolating switch which can be operated manually when a set is not generating and may be used to eliminate a machine that is not functioning properly or that requires maintenance.

Current for excitation of the rotor is supplied by a static rectifier mechanism whose components are located on the 3.5-kV circuits just described. These consist of a single transformer in shunt on the circuit, common to the pool of four machine sets; and four transformers, each connected in series to the stator. The secondaries of both types of transformer are connected with the rectifiers furnishing the shunt and compound elements, respectively, of the exciting current. The rectifier components in the series-mounted transformers are simple diodes, but that part of the Graetz circuit which is fed by the shunt transformer consists of gas-filled relays whose control of input enables the voltage supplied by this rectifier to be regulated.

The use of sectioning and reversing switchgear makes it impossible to maintain phase isolation right through to the machines. A minimum impedance protective circuit is used to cover against any phase synchronization faults that might occur.

MAXIMIZING ENERGY AVAILABLE FROM THE TIDES

The output of a tidal power plant should be concentrated in the critical hours that govern the cost of expanding output facilities. There are at least two methods of doing this: (1) using several basins instead of only one, as proposed by Belidor; and (2) employing a combination of withdrawing from and pumping back into a basin so as to optimize power production with available energy and power requirements.

In terms of multiple basins, the simplest layout is an upper basin which is filled at high tide, a lower basin which is emptied at low tide, and a power plant in which the water produces power as it flows from the upper to the lower basin. There is no way, in this method, of dealing with the difference between energy availability at high and low tides. Tidal amplitude in the average neap tide is only about half the amplitude registered in the average spring tide, and the difference in amplitude as between strong spring tides and weak neap tides is considerably greater. Differences in energy output are naturally still more marked than differences in amplitudes, since the volume of water available to drive turbines—as well as the available head—is reduced by a factor which is given by the form of the basins.

By contrast, as regards the daily pattern of operation, the presence of two basins enables output to be adjusted to the hourly pattern of demand, the counterpart being some diminution in the quantity of energy produced. For example, it is always possible to run the turbines at full throughput during a limited peak-hour period, evidently under lower head during neap tides since it is out of the question for the basins' storage capacity to cover the requirements of a weekly scale of operations. However, not all sites lend themselves to this course. In particular, there is no way of introducing a two-basin system in the Rance estuary. The geography of the Rance called for a powerhouse-cum-dam installed at the mouth of the estuary, and the first choice to be made was whether it would be equipped for single cycle working, generating energy only as the basin emptied, or for a double cycle, generating both during filling and emptying of the basin.

As conventional turbines feature a spiral casing upstream from the wheel and a draft tube downstream, they cannot be reversed, and double-cycle operation is possible only with very bulky water circuits involving high civil engineering costs. The considerable interest which therefore attached to the development of axial turbines led to the study and testing of the bulb generating set.

There are other differences between single- and double-cycle operation, as regards their contribution to both capacity and production, and the hours at which they can produce most energy, given the site characteristics of the tidal power plant as defined by its "locational constant," i.e., the time interval between passage of the full moon across the meridian and the subsequent high tide. The choice between single- and double-cycle working has been under discussion since Belidor's time, but the modern view of the cost pattern of energy as being a function of the hour of the day naturally led to the investigation of possible intermediary solutions and to suggestions that a tidal power plant might be conceived which would operate alternately on a single- and a double-cycle basis. This was the question that led M. Gibrat to completely recast and extend the theory of cycles.

Gibrat introduced into the new theory the effect of the pumping operations which can easily be undertaken by virtue of the characteristics of the bulb unit. Pumping not only changes the hour of the day at which energy is produced, but by contrast with hydro pumping storage stations whose output efficiency coefficient is at best two-thirds, it actually increases the amount of energy available, the complement obviously being supplied by the tide itself. Why this should be so can be understood by considering a normal tide which is not the highest that the dam has been built to withstand. Near high tide, seawater can be pumped to increase the rate at which the estuary is filled, so raising the level upstream by one meter above what it would have been at high tide. The additional water stored in this way is kept in the estuary for 3 or 4 hours. It is then turbined to low tide elevation under a head of—say 6 or 7 meters—considerably in excess of the height by which it was raised at high tide. Hence the ratio of the additional energy produced to the energy expended in pumping can be, despite losses, considerably above unity.

The optimum operational pattern of the bulb units as between turbine functioning during filling and emptying and pump functioning to "overfill" or "overempty" is a problem that is rendered all the more complex in that it has to be resolved, not for successive identical days, but allowing for fairly substantial changes in tidal amplitude from one day to the next over the transition from spring to neap tides. It is further complicated by the fact that the user value of energy varies according to the time of day and, of course, is not the same on weekends as during the week.

The theories developed by M. Gibrat have shed considerable light on this problem. Using them as a basis, an operator with experience can obtain a higher user value from the Rance installation than would have been available with single- or double-cycle conventional equipment. Continuing computerized refinements with actual data will provide a route to ultimate determination of the absolute optimal program. See Fig. 9.

TIDES AS A RESONANCE PHENOMENON

Although Sir Isaac Newton explained the underlying causes of the tides, namely, the attraction exerted by the moon and the sun on the molecules of the oceans, and thus provided an immediate explanation for the variation of the time of high tide with the moon, for the modification of the tidal range during the half lunar month and for the spring tides at suitable conjunctions of the moon and the sun, he nevertheless did not offer explanations of why tidal variations exist from one place on earth to the next. Newton's explanations, for example, did not clarify such situations as: (1) Why is there only one high tide every 24 hours at Do-Son in Tonkin? See Fig. 10. (2) Why do the high and low tides in Tahiti occur at the same time every day, regardless of the phases of the moon? (3) Why is the tidal range at Mont Saint-Michel more than 13 meters, whereas it is measured in decimenters in the Mediterranean? See Fig. 11. (4) Why, although a double impulse causes the tide to ebb and flow along the coasts each day, simultaneous reversal of flow and the rise in water level takes place in the Atlantic, but is not observed in the English Channel?

Thus, tidal energy technology leading to an installation as significant as that of the Rance installation had to await development of new analyses and theories, including the much earlier work of Laplace who, in 1774, introduced hydrodynamics into tidal theory with the development of the concept of *resonance* between the periods of the moon and sun and the oscillatory periods of various water basins, such as the English Channel.

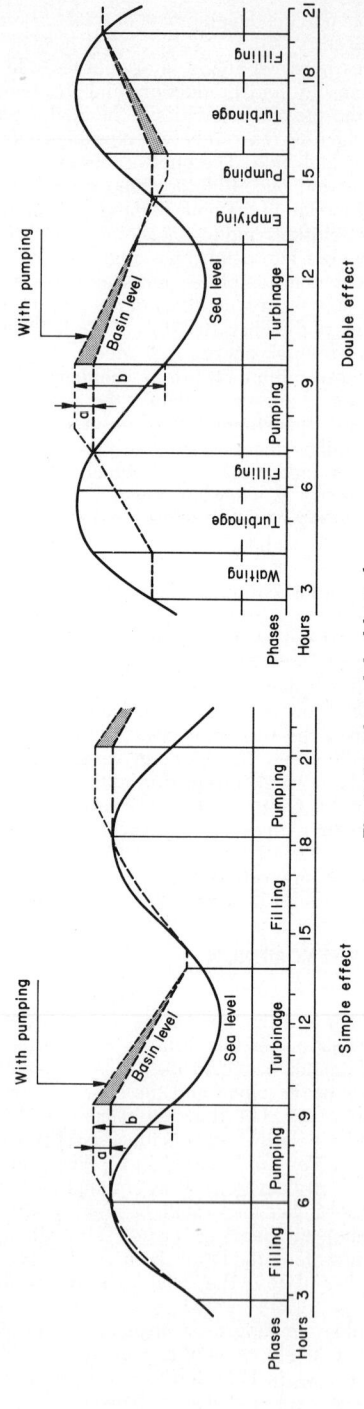

Fig. 9. Simple and double cycles.

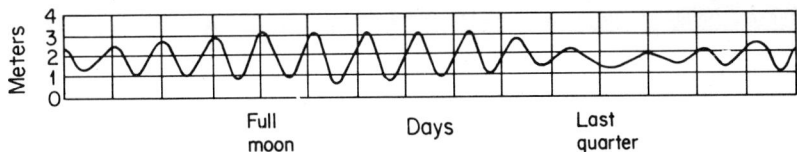

Fig. 10. One full tide daily at Do-Son (Tonkin).

While Newton earlier concentrated on basic cause, scientists who followed had to concentrate on the response characteristics of various bodies of water to the prime cause. Notably, during the first half of the twentieth century, the problem was attacked by persons including Allar, Belidor, Bonnefile, Caquot, Gibrat, Jeffreys, Lacombe, and many others. See Table 2.

The greatest variation that the moon produces in the attraction exerted by the earth on a molecule of water is six million times smaller than gravity, and the sun has a still smaller effect than the moon.* Altogether, if the free surface of the sea were to remain in equilibrium, and that is Newton's basic assumption, there would be a tide amounting to but a few decimeters as a result of the variation in the horizontal component of gravity and one amounting to 0.9 meter (2.95 feet) as a result of the variation in the vertical component. This obviously would leave the 13-meter tide at Mont Saint-Michel unexplained. Newton's theory thus is inadequate for a full understanding. As the sea's surface cannot remain in equilibrium, the water flows back and forth and never comes to rest. As programs of capturing tidal energy, such as utilization of the enormous store of energy in the English Channel, considerably more sophisticated investigations of tidal phenomena must continue. The ratio of the energy taken up to the energy actually made available by the sea can no longer be neglected. The tides will certainly be modified, and possibly in an unexpected manner since we are concerned with resonance which, by definition, is very sensitive to variations of the main parameters. So—it is necessary to know how and to what extent these will be modified.

A few calculations of an elementary nature can clarify the essential concepts and provide an introduction to the methods of developing tidal energy. Utilizing tidal energy is very simple. A dam is built in a bay or an estuary, separating a basin from the sea. Suitable generating sets, which operate in both directions, make it possible to use the water to produce electric energy as it flows back and forth. It is assumed that the tide has a range A, the difference in level between high and low tide, and that a volume V can be stored between these two levels below elevation A. What then is the upper limit, if it exists, of the energy that can be used during each tide? To simplify calculations further, let it be assumed that the machines have an efficiency of unity and that there is no limitation on the discharges that can be used.

*The following several paragraphs are based upon observations made by Robert Gibrat, consulting engineer to Électricité de France for tidal power stations.

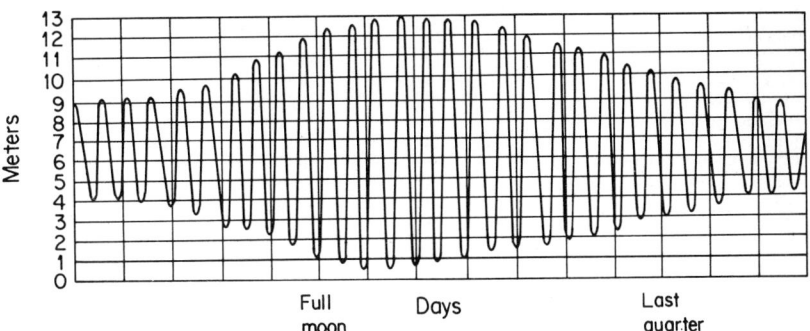

Fig. 11. Two full tides daily at Mont Saint-Michel.

TABLE 2.   Range of Tides in Major Harbors of the World

Location	Mean range		Spring range	
	Feet	Meters	Feet	Meters
Alexandria, Egypt	0.5	0.15	0.8	0.24
Anchorage, Alaska	26.7	8.14	29.6*	9.03
Antwerp, Belgium	15.7	4.79	17.8	5.43
Auckland, New Zealand	8.0	2.44	9.2	2.81
Baltimore, Maryland	1.1	0.34	1.3	0.40
Bilboa, Spain	9.0	2.75	11.8	3.60
Bombay, India	8.7	2.65	11.8	3.60
Bordeaux, France	13.9	4.24	15.7	4.79
Boston, Massachusetts	9.5	2.90	11.0	3.36
Buenos Aires, Argentina	2.2	0.67	2.4	0.73
Burntcoat Head, Nova Scotia (Bay of Fundy)	41.6	12.69	47.5	14.49
Callao, Peru	1.8	0.55	2.4	0.73
Canal Zone (Atlantic Side)	0.7	0.21	1.1*	0.34
Canal Zone (Pacific Side)	12.6	3.84	16.4	5.00
Capetown, Republic of South Africa	3.8	1.16	5.2	1.59
Charleston, South Carolina	5.1	1.56	6.0	1.83
Cherbourg, France	13.0	3.97	18.0	5.49
Dakar, Africa	3.3	1.01	4.4	1.34
Dover, England	14.5	4.42	18.6	5.67
Galveston, Texas	1.0	0.31	1.4*	0.43
Genoa, Italy	0.6	0.18	0.8	0.24
Gibraltar, Spain	2.3	0.70	3.1	0.95
Halifax, Nova Scotia	4.4	1.34	5.3	1.62
Hamburg, Germany	7.6	2.32	8.1	2.47
Havana, Cuba	1.0	0.31	1.2	0.37
Hong Kong	3.1	0.95	5.3*	1.62
Honolulu, Hawaii	1.2	0.37	1.9*	0.58
Jacksonville, Florida	2.0	0.61	2.3	0.70
Juneau, Alaska	14.0	4.27	16.6*	5.06
La Guaira, Venezuela	—	—	1.0*	0.31
Lisbon, Portugal	8.4	2.56	10.8	3.29
Liverpool, England	21.2	6.47	27.1	8.27
Manila, Philippines	—	—	3.3*	1.01
Marseilles, France	0.4	0.12	0.6	0.18
Melbourne, Australia	1.7	0.52	1.9	0.58
Miami Beach, Florida	2.5	0.76	3.0	0.92
Murmansk, U.S.S.R.	7.9	2.41	9.9	3.02
Naples, Italy	0.7	0.21	1.0	0.31
New York, New York	4.4	1.34	5.3	1.62
Norfolk, Virginia	2.8	0.85	3.4	1.04
Osaka, Japan	2.5	0.76	3.3	1.01
Oslo, Norway	1.0	0.31	1.1	0.34
Philadelphia, Pennsylvania	5.9	1.80	6.2	1.89
Portland, Maine	8.9	2.71	10.2	3.11
Portland, Oregon	1.8	0.55	2.4*	0.73
Quebec, Canada	13.7	4.18	15.5	4.73
Rangoon, Burma	13.4	4.09	17.0	5.19
Reykjavik, Iceland	9.2	2.81	12.5	3.81
Rio de Janeiro, Brazil	2.5	0.76	3.5	1.07
Rotterdam, Netherlands	5.0	1.53	5.4	1.65
St. John, New Brunswick	20.6	6.28	23.6	7.20
St. John's, Newfoundland	2.6	0.79	3.5	1.07
San Diego, California	4.2	1.28	5.8*	1.77
San Francisco, California	4.0	1.22	5.7*	1.74
San Juan, Puerto Rico	1.1	0.34	1.3	0.40
Seattle, Washington	7.6	2.32	11.3*	3.45
Shanghai, People's Republic of China	6.7	2.04	8.9	2.71
Singapore, Malaya	5.6	1.71	7.4	2.26
Southampton, England	10.0	3.05	13.6	4.15
Sydney, Australia	3.6	1.10	4.5	1.37

	Mean range		Spring range	
Location	Feet	Meters	Feet	Meters
Valparaiso, Chile	3.0	0.92	3.9	1.19
Vladivostok, U.S.S.R.	0.6	0.18	0.7	0.21
Yokohama, Japan	3.5	1.07	4.7	1.43
Zanzibar, Africa	8.8	2.68	12.4	3.78

SOURCE: U.S. Coast and Geodetic Survey.
*Diurnal range.

Under head H, a discharge q (using suitable units) will produce a power output qH. If $S(z)$ is the area of the pool at elevation z, the energy generated by emptying a section dz is $S(z)z\, dz$. The energy produced per cycle will be

$$\int_0^A S(z)z\, dz \qquad \text{(during emptying)}$$

$$\int_0^A S(z)(A-z)\, dz \quad \text{(during filling)}$$

on the assumption that horizontal sections can be emptied or filled at the turn of the tide, the sea being at elevation O or A. With the simplifying assumptions made, the total energy per cycle with double working thus will be

$$A\int_0^A S(z)\, dz = AV \tag{1}$$

or the product of the tidal range and the volume used, a formula which, it may be observed, is valid for a basin of any shape. Thus, we can call this quantity the "natural" energy of the basin for tide A. But this name can be deceptive.

Each cubic meter of water entering the basin is thus used under a head A. If all tides had the same range A, at the end of the year the 705* tides would have provided a cumulative head of $705A$ for each cubic meter of basin storage capacity. However, in fact, A varies throughout the year, together with V, which is directly dependent upon it, since the basin does not fill up or empty completely except at the maximum spring tides. The question may be asked: Is the "natural" energy the maximum amount of energy that can be obtained from a single tide? Surge-wave propagation phenomena will quickly show that this is not so.

As early as 1942, a scale-model test had demonstrated how to exceed AV. At high tide (level A), the gates are suddenly opened without using the energy as the water passes through, and a wave is propagated up the estuary and, increasing in amplitude, is reflected at the other end. The gates are closed at the moment when the wave returns and when the direction of flow at the dam is about to change. Finally, the water surface settles down to a level higher than A. On a scale model, $1.5A$ was attained without any difficulty. In a rectangular canal with a constant cross section, it can easily be seen that the level theoretically attains $2A$ in the absence of losses. In this process, the energy transformed while filling is zero; during emptying at low-tide slack water, it can be as much as

$$\tfrac{1}{2}S\,(2A)^2 = 2SA^2 = 2AV$$

The energy thus obtained is twice the natural energy and one enters into the realm of paradoxes. It might be hoped that this is a result of the surge wave and that all is worked out by nature. The answer is negative—because with a single basin, we could consider using power from the grid using a double working cycle, including pumping, a forerunner of the most complex real cycles.

---

*There are not $2 \times 365 = 730$ tides per year because the lunar day is approximately 1.035 day, i.e., 24 hours, 50 minutes, 28 seconds in length.

Consider four phases as follows:

1. At the end of a cycle when the basin has been emptied to low-tide level and the gates are closed, power from the grid is used to pump water to the sea, lowering the level in the basin to $-B$, the energy thus used is

$$E_p = \int_{-B}^{0} S(z)(-z)\, dz \quad (z \leqslant O)$$

2. The tide rises and at high tide the basin is filled during slack water, producing filling energy as follows:

$$E_r = \int_{-B}^{A} S(z)(A - z)\, dz \quad (-B \leqslant z \leqslant A)$$

3. Still at the turn of the tide, water is pumped from the sea to the basin, using power from the grid a second time, and the water level rises to C. The energy thus used is

$$E_p{}^1 = \int_{A}^{C} S(z)(z - A)\, dz \quad (A \leqslant z \leqslant C)$$

4. At low tide slack, the basin is emptied from C to zero producing the following energy:

$$E_v = \int_{C}^{0} S(z)z\, dz \quad (O \leqslant z \leqslant C)$$

After having accounted for the energy E_p and $E_p{}^1$ used for pumping, there remains a total of

$$E_r + E_v - E_p - E_p{}^1 = A \int_{-B}^{C} S\, dz = A(V + V_p + V_p{}^1)$$

The transformed energy is equal to the sum of the natural energy plus A times the total volume pumped, and can thus be as large as desired if the area of the basin for $z\, A$ can be made large enough.

A double working cycle including pumping thus seems capable of employing energy without any restrictions other than that resulting from plant limitations.

This cannot be so, and it is essential to understand why properly. If not, one would make the same mistake as an electrician who knows the voltage at a power point and wants to work out the available wattage without taking into account the voltage drop and phase change resulting from plugging in an appliance. To forget the connection imped-ance would be disastrous. It becomes necessary at this point to analyze the physical process of energy transfer by the tides and to understand that the actual propagation of the energy drawn to the power stations produces modifications to the range of the tide and the tidal currents which can appreciably alter the movement of energy. Certainly there is a natural limit for a given site, but basin capacity and tidal range are not enough to determine it.

The analogy of the energy of rivers may be helpful. In developing a site on a river, the energy produced is taken from the energy that was formerly dissipated naturally in whirlpools, eddies, and various forms of friction before the dam was built; and the transformable energy limit is obviously the total of these losses. In similar fashion, should the energy dissipated by the tide be taken as the natural limit that is being sought? The answer is negative—because a calculation shows that the mean power dissipated by friction in the Rance Estuary is at most 60,000 kilowatts while the power station has a capacity of 240,000 kilowatts and is far from taking up all the utilizable energy. It is estimated that the additional energy imparted by the heavenly bodies is less than 1000 kilowatts for the whole estuary. Thus, a further paradox. On the average, the total power dissipated in the whole English Channel is estimated at 157 million kilowatts.

Tidal energy is of a very special nature, and one cannot draw an analogy with the energy of a river without risking very great errors, The tidal phenomenon which gives rise to this energy is essentially a resonance effect, which is very exceptional. Furthermore, using such energy often causes the modifications thus made to the tidal regime to spread over a great distance and thereby considerably distort the initial phenomenon. As pointed out by Robert Gibrat, "The energy today is not where it will be used tomorrow." It should be added that the power generated by the Rance station is too small compared with that associated with the tides in the English Channel to cause such reactions. However, the same may not be true of future stations which will be considerably larger.

According to calculations made by Jeffreys, the mean power dissipated by the tide over the whole world is about 1100 million kilowatts, most being supplied by shallow narrow seas like the English Channel, the Irish Sea, the North Sea, and mainly the Bering Straits, which account for 70 percent of the total. The deep oceans, such as the Atlantic, Pacific, and Indian Ocean, hardly use a thousandth of all the energy dissipated.

At the time of Jeffreys' calculations (1952), a comparison between observations of the moon's movements and the numerical consequences of the theory led astronomers to assume the existence of a residual secular acceleration not arising from celestial mechanics. It was thus shown that the earth's rotational speed was decreasing and releasing about 1400 million kilowatts, which agrees quite satisfactorily with the order of magnitude quoted earlier. It consequently seemed very plausible to assume that the energy of the tide was derived from that of the earth's rotation by friction between the water and the seabed, thus slowing up the earth.

Some years later (early 1960s), the picture became much less clear. Observations made in the nuclear submarine *Nautilus* indicated that Jeffreys' figures should be reduced by three-fourths. Also some leading astronomers were of the opinion that the rotational speed of the earth might be constant. The basic question remains without full proof, although there are numerous scientific opinions: "Is the earth's kinetic energy, or the sun's thermal energy the source of tidal energy?"

Returning to practical considerations, Gibrat observed in 1966 that the real problem appears to be covered by the case of 27 tides (14 days), since it alone enables the Saturday and Sunday off-peak periods, when energy usually costs less, to be included and almost exactly covers a whole number of weeks. Since the number of possible cycles for one tide is 16, and 16^2 or 256 cycles are possible with two tides, there are 16^{27} or 2^{108} different cycles possible for 27 tides. This is about 3×10^{32}. Fortunately, the present detailed examinations of possible prediction by the grid control center allows one to assume that a reasonable "horizon" for such predictions would not go beyond a relatively few days, inasmuch as atmospheric conditions or plant availability hardly make it possible to aim further. With the help of modern computers, tidal power stations should, therefore, be operated within these limits, and probably should not try to go beyond four tides or fourth-order cycles. Energy production computations give 12,000 cycles out of a possible 65,536. This would seem to be within the range of a computer. The theory of waiting cycles provides useful material for the preparation of a sliding program, i.e., one that is changed with each new cycle since it would be enough to add 3282 cycles to the machine work. Undoubtedly, further simplifications will also enter the picture. It is interesting to note that if the horizon were extended to five tides, the machine work would be multiplied by approximately ten—and so on for each tide.

As pointed out by Gibrat, a study committee was formed in the early 1950s to consider basic forces at work, particularly in view of the proposed Chausey tidal power station (10,000 MW) in the Gulf of St. Malo. The work may be summarized briefly. An attempt was made to find a criterion for the Coriolis effect by setting up various representations of the tide in the Channel. In 1950, Professor Lacombe reviewed the various explanations for the tide in the channel, and arrived at the conclusion that the tide in the Gulf of St. Malo was correctly represented by the diffraction of a wave from the Atlantic by the Cotentin Peninsula. Bonnefille perfected this representation first by allowing for a wave coming from the eastern English Channel and then by introducing a limitation consisting of two parallel lines representing the northern and southern Channel coasts. Another idea that was not new at the time introduced the interference of Kelvin waves, but this raised considerable theoretical difficulties. Gibrat emphasized the energy aspect of the problem and the special parts played by the phase difference between currents and sea levels, by analogy with reactive power and the power factor in electrical engineering. Bonnefille did much work along these lines, and, using theoretical computations based on work by Allard, was able to show how much energy is dissipated, by friction and turbulence (Fig.

12), summarized these computations, and indicated the active power (in 1000 MW) and reactive power (in 1000 Mvar) which is transmitted and consumed.

In 1954, before the formation of the aforementioned study group, Électricité de France decided in principle to construct a large static scale model (horizontal scale 1/50,000, vertical scale 1/500) representing the whole of the English Channel from the Atlantic to

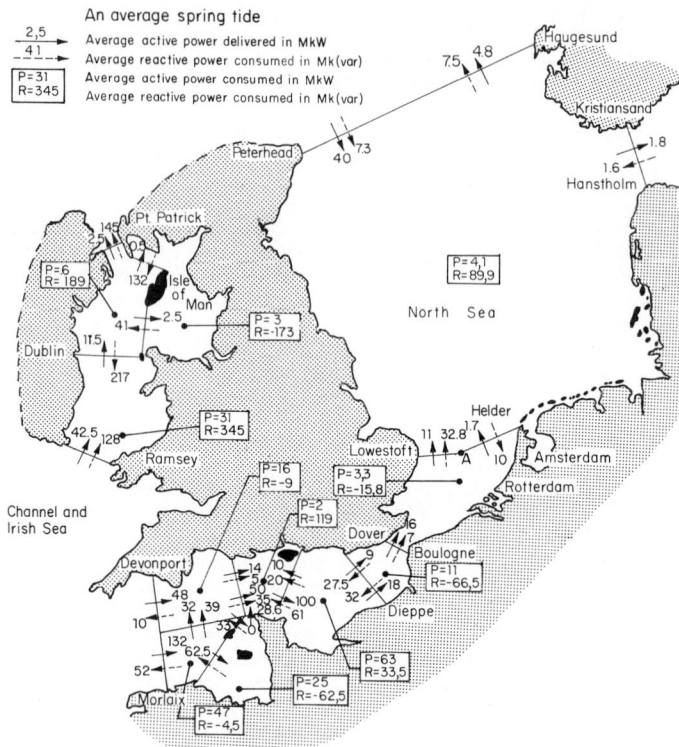

Fig. 12. Active and reactive power, delivered and consumed. (The English Channel, the Irish Sea, and the North Sea, from a study by R. Bonnefille, 1956.)

the Straits of Dover (tidal period 200). Since the model was static, it did not allow for the Coriolis acceleration (the model would have had to rotate once every 51.6 seconds to allow for this). For this reason, tidal propagation was very different from the results of observation, despite lengthy adjustments. The need for a better understanding of the influence of the Coriolis force in order to explain the tide in the English Channel and the Chausey backwater effect caused the construction of a small 4-meter-diameter rotating platform on a lighthouse-lens mounting so as to permit simple cases of tides obtained with and without friction to be compared, and in this way to determine the relative importance of the two phenomena. The platform was able to run at all speeds between one revolution in 10 seconds and one revolution in 20 seconds, and all coastal configurations were represented by partitions placed in a 10-centimeter-high plastic-walled tank with a 120-centimeter by 30-meter base. After the results of many tests, it was possible to bring the study group's discussions to a conclusion with the following observations:

1. The Coriolis force is not sufficient to explain the tide in the English Channel, and it is not the reason for the increased tidal range at St. Malo.

2. The frictional force is sufficient to explain the increased tidal range at St. Malo, but does not explain the difference between the French and English coasts.

3. The simultaneous action of both forces results in a satisfactory enough representation of nature.

4. Since the tide in the Gulf of St. Malo is mainly affected by roughness, static models can be used without imposing too many conditions, provided they are confined to the zone bounded by the Gulf, which is the most important region for a study of the Chausey power station.

The conclusion of the study group's work resulted in construction of a large rotating slab which would enable all theoretical and practical problems raised by Coriolis to be investigated. The final decision to do this was taken by the University of Grenoble in conjunction with the Laboratoire Dauphinois d'Hydraulique and Electricité de France. This tool has enabled the reproduction of tides in the whole English Channel. It has made possible the investigation of the backwater effect of Chausey power station, determining its effect on freeboard, and to gather much useful information.

In the area of tidal energy technology, the Rance installation and the associated efforts of French scientists and engineers remain by far the most outstanding achievements to date. In Canada, the earlier "Little Kodiak" project in the Bay of Fundy aroused only modest interest because it lies in a low-consumption rural area, while the St. Lawrence and the falls further north offer considerable reserves of hydropower. In the United States, an abrupt start was made with the Passamaquody project, also in the Bay of Fundy, in the early 1930s. At that time, National Recovery Administration (NRA) funds were made available to hire about 5000 workers for the project, but a subsequent and serious study showed that the scheme would not be economic at that time, and the work was halted.

WAVE ENERGY

Albert W. Stahl, USN (*Transactions ASME,* vol. 13, p. 438) states that the total energy of a series of trochoidal deep-sea waves may be expressed as follows: Hp per foot of breadth of wave = $0.0329 \times H^2 \sqrt{L} [1 - 4.935 (H^2/L^2)]$, where H = height of wave in feet, and L = length of wave between successive crests in feet. For example, with L = 25 feet and L/H = 50, hp = 0.04; with L = 100 feet and L/H = 10, hp = 31.3. Not much more than a quarter of the total energy of such waves would probably be available after reaching shallow water, and apparatus rugged enough for this purpose would doubtless be unable to utilize more than a third of this amount.

Wave motors brought out from time to time have depended for their operation largely on the lifting power of the waves. One installed at Atlantic City, New Jersey (*Power,* January 17, 1911), consisted of six 4-foot cylindrical floats 4 feet high. These, each weighing about 3100 pounds, were lifted 2 feet by the waves about 11 times per minute, and drove a horizontal shaft by means of chains and ratchets, developing but 12 hp, steadiness being obtained by the use of heavy flywheels.

A wave motor employing a hydraulic ram for raising a portion of the water to a high level was proposed by Smith (*Mechanical Engineering,* September 1927, p. 995). The waves were envisioned as entering a scoop which would be connected to the ram by a long drive pipe. The apparatus would be automatically adjusted for vertical level with tide changes.

Gravity waves may be only a few feet high, and yet develop as much as 50 kilowatts per foot of wave front. Most historical wave motors utilize: (1) the kinetic energy of the waves by a device, such as a paddle wheel, or (2) the potential energy by such a device as a series of floats. It appears that few devices proposed thus far utilize both forms of energy. Jacobs (*Power Engineering,* September 1956) analyzed the periodic fluctuations or "seiching" of the water level of harbors or basins where, with a resonant port, a 1000-foot wave front might be used to achieve a liquid-piston effect for the compression of air, the air subsequently used in an air turbine.

REFERENCES

Ailleret, Pierre: Tides in the Competition between Energy Sources, *Rev. Fr. Énerg.,* September–October 1966.
Ibid.
Gibrat, Robert: Scientific Aspects of the Use of Tidal Energy.
Olivier-Martin, Didier: The Studies of Methods for Barring the Rance.
Caquot, Albert: The Definitive Cut-off Project.
Glasser, Georges, and Francis Auroy: Research into the Development of the Bulb Unit.

Caillez, Henri, and Marc Faral: General Description of the Tidal Generating Station and Its Electrome-
    chanical Equipment.
Maubossin, Georges, and Raymond Soulas: Realization of the Project.
Cabanius, Jean, and Etienne Svilarich: The Rance Project and Its Contribution to Hydroelectric
    Technology.
Bourges, Yvon: The Rance Tidal Power Scheme and the Saint-Malo Region.

L'Usine Maremotrice de la Rance, "Électricité de France," Paris, 1973.
Gray, T. J., and O. K. Gashus (eds.): "Tidal Power," Proceedings of an international conference on the
    Utilization of Tidal Power, Nova Scotia Technical College, Halifax, Canada, 1970.
Summers, C. M.: The Conversion of Energy, *Sci. Amer.*, September 1971.

Power Technology Trends

A. N. ADDIE, B.S., M.S.M.E. *Manager, Advance Engineering, Electro-Motive Division, General Motors Corporation, LaGrange, Illinois. Member, American Society of Mechanical Engineers. Registered Professional Engineer (Illinois). (Rail Power Systems)*

M. M. ADIBI. *Industry Consultant, International Business Machines Corporation, Houston, Texas. (Trends in Technical Use of Computer Systems in the Power Industry)*

J. D. BAILIE, B.S.M.E. *Director of Technical Sales, Petroleum Chemicals Division, Ethyl Corporation, Houston, Texas. Member, American Petroleum Institute, Society of Automotive Engineers. (Methylcyclopentadienyl Manganese Tricarbonyl Combustion Improver)*

M. S. BALDWIN, B.S.E.E. *Generation Consultant, Power Generation Systems, Westinghouse Electric Corporation, Pittsburgh, Pennsylvania. Member, American Institute of Electrical Engineers, Institute of Electrical and Electronics Engineers, Registered Professional Engineer (Ohio, Pennsylvania). (Superconducting Turbine Generators)*

R. J. BUIST, B.S., M.S. *Engineering Manager, Thermoelectrics Department, Borg-Warner Corporation, Des Plaines, Illinois. Member, American Physical Society, Scientific Research Society of America. (Thermoelectric Cooling)*

JOHN O. BURCKLE. *Research Chemical Engineer, Office of Research and Development, U.S. Environmental Protection Agency, Environmental Research Center, Cincinnati, Ohio. (Particulate Emission Measurement Methods)*

L. T. CLERE, B.S.Ch.E. *Environmental Engineer, Sun Oil Company, Toledo, Ohio. Member, American Institute of*

Chemical Engineers, Ohio Society of Professional Engineers. *(Bio-Oxidation Process Reduces Water Use)*

J. H. CRONIN, B.A., B.S. *Advisory Engineer, Transmission and Distribution Systems Engineering, Westinghouse Electric Corporation, East Pittsburgh, Pennsylvania. Member, Institute of Electrical and Electronics Engineers. (Trends in Electric Power Transmission)*

W. V. DAILEY. *Technical Products Division, Mine Safety Appliances Company, Pittsburgh, Pennsylvania. (Monitoring of Carbon Monoxide Emissions) (Monitoring of Hydrocarbon Emissions)*

S. G. DUKELOW. *Manager Power Generation Marketing, Bailey Meter Company, a subsidiary of The Babcock & Wilcox Company, Wickliffe, Ohio. (Controls for Fossil Fuel/Steam Energy Conversion)*

WILLEM GROENENDAAL. *Shell Internationale Petroleum Maatschappij B. V., The Hague, The Netherlands. (Claus Sulfur Recovery Units and Claus Off-Gas Treating)*

PIET J. HALBMEYER. *Head of the Oil Gasification and Technical Back-Up, Licensing Division, Shell Internationale Petroleum Maatschappij B. V., The Hague, The Netherlands. (Gasification as Basis for Combined-Cycle Power Generation)*

VICTOR G. HERMAN, Ph.D. *New York, N.Y. (Dynamometers for Measurement of Relatively Small Forces)*

STEPHEN A. JOHNSON, B.S.Ch.E. *Senior Research Engineer, Riley Stoker Corporation, Worcester, Massachusetts. Member, American Institute of Chemical Engineers. (NO_x Control by Furnace and Burner Design)*

R. F. LAWRENCE, B.S.E.E. *Manager, Transmission and Distribution Systems Engineering, Westinghouse Electric Corporation, East Pittsburgh, Pennsylvania. Fellow, Institute of Electrical and Electronics Engineers. (Trends in Electric Power Transmission)*

DONALD R. LOHSE, B.S.C.E. *Product Specialist, Reciprocating Compressors, Air Power Division, Joy Manufacturing Company, Michigan City, Indiana. Member, National Management Association. (Modified Gas Turbine Generator) (Utilizing Waste Heat from Air Compressors)*

E. F. MOHLER, JR., M.S. Chem. *Senior Environmental Engineer, Sun Oil Company, Toledo, Ohio. Member, Air Pollution Control Association, American Chemical Society, American Institute of Chemical Engineers, American Management Association, American Society for Testing and Materials, National Association of Corrosion Engineers, National Management Association, Water Pollution Control Federation. (Bio-Oxidation Process Reduces Water Use)*

JAAP E. NABER. *Koninklijke/Shell-Laboratorium, Amsterdam, The Netherlands. (Claus Sulfur Recovery Units and Claus Off-Gas Treating)*

ROY G. NEVILLE, M.Sc., Ph.D., D.Sc., C.Chem., F.R.I.C. *Chemical and Environmental Consultant. Fellow, American Association for the Advancement of Science, American Institute of Chemists, Chemical Society of London, Royal Institute of Chemistry. Member, American Chemical Society, History of Science Society, Research Society of America, Royal Institution of Great Britain, Contributor to numerous chemical and engineering journals, textbooks, and encyclopedias. Holder of numerous patents on chemical process technology. (Wet Scrubbing Processes—SO_x and NO_x Removal Chemistry)*

W. JEFF OSBORNE, B.S. *Formerly Supervising Process Engineer, Davy Powergas Inc., Lakeland, Florida. Member, American Institute of Chemical Engineers. [Sulfur Dioxide Recovery (Wellman-Lord Process)]*

ALBERT H. RAWDON, B.S., M.S.M.E. *Director of Research and Development, Riley Stoker Corporation, Worcester, Massachusetts. Member, Air Pollution Control Association, American Society of Mechanical Engineers, Combustion Institute, Worcester Engineering Society. Registered Professional Engineer (Massachusetts). (NO_x Control by Furnace and Burner Design)*

JAMES R. SMALL, A.B., M.S. *Product Manager, Process Instruments, Instrument Products Division, E. I. du Pont de Nemours & Co., Inc., Wilmington, Delaware. Member, American Chemical Society, Instrument Society of America. (Source Monitoring of NO_x and SO_2)*

C. C. STERRETT, B.S.M.E. *Program Manager, Superconducting Electrical Machinery Department, Westinghouse Electric Corporation, Pittsburgh, Pennsylvania. Member, Institute of Electrical and Electronics Engi-*

neers, *National Electrical Manufacturers Association. Registered Professional Engineer (Pennsylvania). (Superconducting Turbine Generators)*

GORDON L. TAFT, B.S. *Product Specialist, Rotary Screw Compressors, Air Power Division, Joy Manufacturing Company, Michigan City, Indiana. Member, National Management Association. (Utilizing Waste Heat from Air Compressors)*

WALTER A. TROEGER, B.S., M.S. *Engineering Department, Weston Instruments, Inc., Newark, New Jersey. Registered Professional Engineer (New Jersey). (Electric Power and Energy Measurement)*

GODFRIED J. VAN DEN BERG, Ph.D. *Head of the Conversion Processes and Project Technology Division, Shell Internationale Petroleum Maatschappi B. V., The Hague, The Netherlands. (Gasification as Basis for Combined-Cycle Power Generation)*

PIET J. J. VAN DOORN. *Specialist, Licensing Division, Shell Internationale Petroleum Maatschappi B. V., The Hague, The Netherlands. (Gasification as Basis for Combined-Cycle Power Generation)*

JOHANNES A. WESSELINGH. *Koninklijke/Shell— Laboratorium, Amsterdam, The Netherlands. (Claus Sulfur Recovery Units and Claus Off-Gas Treating)*

JAMES A. WILLIAMSON, B.S., M.S. *Senior Applications Chemist, Instrument Products Division, E. I. du Pont de Nemours & Co., Inc., Wilmington, Delaware. Member, American Chemical Society, Instrument Society of America. (Source Monitoring of NO_x and SO_2)*

A. H. ZEITZ, Jr., B.S.M.E. *Product Application Manager, Petroleum Chemicals Division, Ethyl Corporation, Ferndale, Detroit, Michigan. Member, Society of Automotive Engineers. (Methylcyclopentadienyl Manganese Tricarbonyl Combustion Improver)*

Steam Generation*

In this brief summary, past key inventions and events pertaining to steam technology are highlighted to indicate how solutions were found to former crucial problems. Current trends in steam generation research and development also are reviewed. Several of the following articles in this handbook section present further details on various aspects of steam generation.

Except for standby and peak-load units, approximately 90 percent of the new electric capacity being installed (as of 1976) utilizes steam. The use of steam for electric power generation is expanding in most countries of the world, the principal exceptions being countries where available water power resources are greater than power needs. Steam propels most of the world's naval vessels and a high percentage of its waterborne commerce. Steam is used on a large scale in many industrial processes and is still much used for space heating.

Steam boilers currently range in size from those required to heat a small-sized home to the very large ones used in electric power-generating stations. Some single boilers deliver approximately 10 million pounds of steam and consume more than 500 tons of coal per hour. In these large units, pressures range from 2500 to about 4000 psi and the steam is usually superheated to a temperature of 1000°F (538°C) or higher. Contemporary steam boilers operate safely and dependably and remain in service for many years with cleaning and repairs usually required only at scheduled outage periods. These boilers owe their dependability and safety to more than a hundred years of experience in the design, fabrication, and operation of watertube boilers. During this period, the properties of steam and water have been accurately determined and tabulated for use by the designer. A new understanding of heat transfer, fluid flow, and boiler circulation has been developed. Means have been devised for burning large quantities of fuel economically and safely and for disposing products and by-products of combustion. Steels and alloy materials now available are stronger and more consistent in their properties, and advanced methods are used for their fabrication and inspection. Finally, industry-wide codes and standards have been adopted to regulate the design, fabrication, inspection, and operation of pressure parts.

The nuclear steam supply system is a later development, representing a union of nuclear physics and the steam boiler industry. Since the 1950s, a number of nuclear steam supply systems have been placed in operation for naval propulsion and for the generation of electricity. Some of these units for the electric power industry are comparable in capacity to the largest fossil-fueled boilers.

As industrial growth spreads and accelerates, there is growing concern for future energy shortages not only in the United States but worldwide. To meet this challenge, additional renewable energy sources and their conversion to useful work are being investigated. As reflected by other sections in this handbook, these include sources such as solar energy, nuclear fusion, fuel cells, magnetohydrodynamics, thermionics, geothermal, and tidal energy. However, until economical feasibility of these systems is established, steam will remain the primary means of electric power generation.

DEVELOPMENT OF STEAM GENERATION AND USE

Hero of Alexandria, probably in the first century A.D., described a boiler and reaction turbine, but he made no suggestions for the useful application of the device. There is no record of the practical use of steam until the seventeenth century. Commencing at that time, many conditions existed which stimulated the rapid development of steam usage in a power cycle. Mining for ores and minerals had expanded greatly and large quantities of

---

*Based on information furnished largely by Babcock & Wilcox Company, much of which is contained in "Steam: Its Generation and Use," copyrighted by The Babcock & Wilcox Company.

fuel were needed for smelting. Considerable fuel was needed for space heating and cooking. Industrial and military growth, especially in England, also demanded greater amounts of fuel. Large-scale coal mining was developed, but as mines became deeper, they often were flooded with water. The English, in particular, were faced with a serious curtailment of their growth if they could not find some economical way to pump water from the mines. The importance attached to the problem can be seen from the great number of men working on it and by the many patents issued on machines to pump water using "the expansive power of steam."

One of the first practical steam engine developments was Thomas Newcomen's atmospheric engine. Steam admitted from a boiler to a cylinder raised a piston by expansion and assistance from a counterweight on the other end of a beam actuated by the piston. The steam valve was then closed and the steam in the cylinder condensed by a jet of cold water. The vacuum formed caused the piston to be forced downward by atmospheric pressure doing work on a pump. The boiler used by Newcomen was nothing more than a plain copper brewer's kettle. In 1711, a Newcomen's engine was introduced into the mines for pumping water.

During the last half of the eighteenth century, the great inventor, James Watt, made many significant improvements to the early steam engine, which by now was completely separate from the boiler. While little is said in biographies of Watt about improvements of steam boilers, the evidence indicates that Boulton and Watt introduced the first *waggon boiler*, so named because of its appearance. This was nothing more than a closed vessel for water and steam, shaped like a covered wagon set over a fire pit.

Fire-Tube Boilers.   The next outstanding inventor and builder was Richard Trevithick, who realized the major problem of these steam systems was the manufacture of the boiler. Whereas copper was the only material heretofore available, hammered wrought iron plates could be used. In 1800, Trevithick made an engine for 65 psi pressure, having a 25-inch cylinder and a 10-foot stroke. The engine's high working pressure was possible only because of the successful construction of a high-pressure boiler. Built in 1804, the boiler had a cast iron cylindrical shell and dished end.

As the demand increased for greater amounts of power, later developments saw the single pipe flue replaced by many gas tubes which increased heating surface. This was essentially the design in widespread use up to about 1870. However, fire-tube boilers, being limited in capacity and pressure, were not destined to fulfill the requirements which developed later for higher pressure and larger unit sizes. Also, there was the ominous record of many explosions.

Watertube Boilers.   In the late 1700s, John Stevens, an American, invented a watertube boiler consisting of a group of small tubes closed at one end and connected at the other to a central reservoir. As a boiler, the design was short lived due to basic engineering problems in construction and operation. Stephen Wilcox proposed in 1856 what was to be a significant breakthrough for watertube boilers. The design incorporated inclined water tubes connecting water spaces at the front and rear, with a steam space above, allowing better water circulation and more heating surface. An added advantage was the reduced explosion hazard that was inherent in the watertube design. In 1866, George Herman Babcock became associated with Stephen Wilcox, and the first Babcock and Wilcox boiler was patented a year later.

The success and widespread acceptance of the inclined straight-tube boiler during an unprecedented period of rapid industrial growth stimulated other inventors to explore new ideas in boiler design. In 1880, Allan Stirling developed a design connecting the steam-generating tubes directly to a steam-separating drum and featuring low headroom above the furnace. These boilers featured bent tubes as contrasted to the B&W straight tubes. Merits of bent-tube boilers for special applications were soon recognized by Babcock and Wilcox, and what had become the Stirling Consolidated Boiler Company was purchased by B&W in 1906.

ELECTRIC UTILITIES

Steam was used originally to provide heat and power for local industrial use. With the advent of practical electric power generation and distribution, utility companies were formed to serve industrial and residential users over wide areas.

The plant of the Brush Electric Light Company, in Philadelphia, was the first such electric generating station in America. Four B&W boilers rated at 73 horsepower each were installed in this plant in 1881. The Fisk Street Station of the Commonwealth Edison Company, in service in 1903, was the first utility plant to use steam turbines exclusively for electric power generation. Ninety-six B&W boilers rated at 508 horsepower each were installed in this plant with turbine steam conditions of 170 psi pressure and 70°F superheat.

During the first two decades of the twentieth century, there was an increase in steam pressures and temperatures to 275 psi and 560°F (293°C) (146°F superheat), respectively. In 1921, the North Tees Station of the Newcastle Electric Supply Company (northern England) went into operation with steam at a pressure of 450 psi and a temperature of 650°F (343°C). The steam was reheated to 500°F (260°C), and regenerative feedwater heating was used to attain a boiler feedwater temperature of 300°F (149°C). Three years later (1924), the Crawford Avenue Station of the Commonwealth Edison Company and the Philo and Twin Branch stations of the present American Electric Power system were placed in service for operation with steam at 550 psi and 725°F (385°C) at the turbine throttle. The steam was reheated to 700°F (371°C).

A station designed for much higher steam pressure—the Weymouth (later named Edgar) Station of the Boston Edison Company—started up in 1925. The 3150-kilowatt high-pressure topping unit used steam at 1200 psi and 700°F, and the steam was reheated to 700°F for the main turbines.

PULVERIZED-COAL AND WATER-COOLED FURNACES

Other major changes in boiler design and construction occurred in the 1920s. Previously, as individual electric generating stations increased in capacity, the practice was simply to increase the number of boilers. This procedure eventually proved to be uneconomical; instead, the individual boilers were built larger and larger. Soon, however, the size became such that existing furnace designs and methods of coal burning, such as stokers, were no longer adequate.

Insofar as fuel burning was concerned, the development of pulverized-coal firing provided the answer. The higher volumetric combustion rates and unit sizes made possible by burning pulverized coal could not have been fully exploited without the use of water-cooled furnaces, which not only eliminated the problem of rapid deterioration of the refractory walls due to slag, but also reduced fouling of convection heating surfaces to manageable proportions by lowering the temperature of the gases leaving the furnace.

Integral-Furnace Boilers. At first, furnace water-cooling was applied to existing boiler designs, with its circulatory system essentially independent of the boiler circulation. In the early 1930s, a new concept was initiated in which the furnace water-cooled surface and the boiler surface were arranged together so that each was an integral part of a boiler unit.

PACKAGE BOILERS

The increasing need for industrial and heating boilers, combined with increasing costs of field-assembled equipment, let to the development in the late 1940s of the shop-assembled package boiler. Completely shop-assembled units are fabricated in capacities up to 350,000 pounds of steam per hour at pressures up to 1280 psi and temperatures up to 900°F (482°C).

LATER BOILER DEVELOPMENTS

In addition to reducing furnace maintenance and the fouling of convection heating surfaces, water-cooling also helped to generate more steam. Consequently, the boiler surface was reduced, since additional steam-generating surface was available in the water-cooled furnace. Increased feed and steam temperatures and increased steam pressures, for greater cycle efficiency, still further reduced boiler tube bank surface, to be replaced by additional superheater surface.

As a result of these advances, boiler units for steam pressures above 1200 psi consist essentially of furnace waterwall tubes, superheaters, and such heat recovery accessories

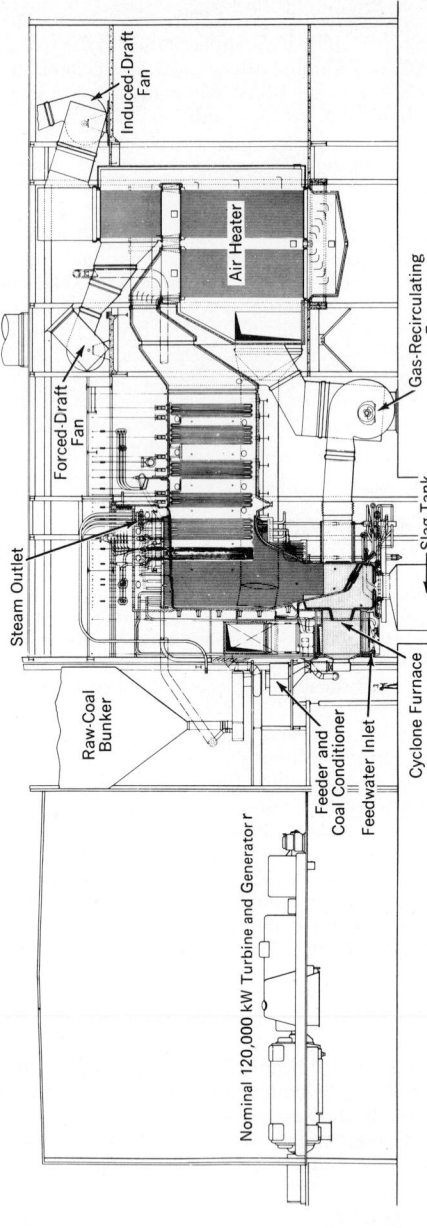

Fig. 1. Elevation view of a 120-megawatt generating unit served by a Babcock & Wilcox Universal-Pressure boiler operating at 5500 to 4500 psi, 1150°F (621°C) with reheats to 1050°F (566°C) and 1000°F (538°C).

Induced-Draft Fan

Air Heater

Gas-Recirculating Fan

Forced-Draft Fan

Slag Tank

Steam Outlet

Raw-Coal Bunker

Cyclone Furnace

Feeder and Coal Conditioner

Feedwater Inlet

Nominal 120,000 kW Turbine and Generator

as economizers and air heaters. Boiler units for lower pressures, however, have considerable steam-generating surface in tube banks in addition to the water-cooled surface in the furnace.

UNIVERSAL-PRESSURE BOILERS

An important event contributing to the production of electricity at the lowest possible cost was the successful operation in 1957 of the first commercial unit for operation at a steam pressure above the critical value (3208 psia) at the Philo Plant of the Ohio Power Company (Fig. 1). This steam generator (B&W) for the 120-megawatt unit delivers 675,000 pounds of steam per hour at 4500 psi and superheated to 1150°F (621°C), with two reheats to 1050°F (566°C) and 1000°F (538°C).

The *universal-pressure boiler,* so named because it can be designed for subcritical or supercritical operation, is capable of rapid load pickup. Increase in load at rates up to 5 percent per minute, insofar as the boiler is concerned, is easily attained. In an emergency, the boiler is capable of increasing load from 25 to 90 percent of full load in 4 minutes, and from 90 to 100 percent of full load in another 3 minutes. This boiler, with its start-up system, enables close matching of steam and turbine metal temperatures and thereby reduces thermal stresses to a minimum during either a cold start or a hot restart.

Figure 2 shows one of two universal-pressure boilers of recent design, each of which will furnish the steam for a 1300-megawatt capacity generating unit in a large utility plant.

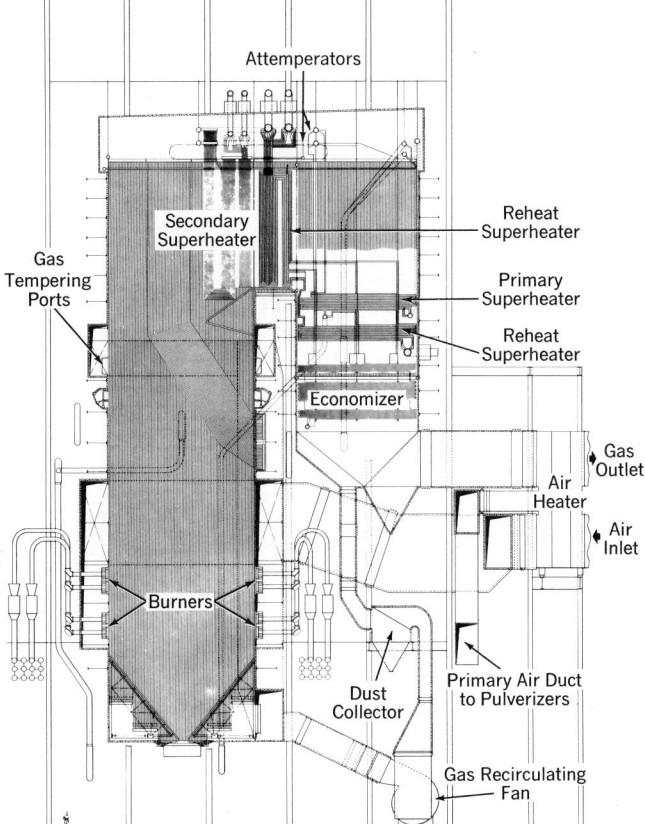

Fig. 2. Universal-Pressure boiler for generating 9,775,000 pounds of steam per hour. (*B & W*)

Each boiler will supply 9,775,000 pounds of steam per hour at 3845 psi and 1010°F (543°C), with reheat to 1000°F (538°C).

MODIFIED STEAM CYCLES

The continual quest for lower heat rates and thus higher cycle efficiency has involved modifications of the conventional steam cycle. One of these, using high-temperature low-pressure mercury vapor to top a conventional steam cycle, dates back to the binary fluid mercury-steam unit placed in service in 1928 at a New England utility. Binary fluid *topping* cycles are so named because the rejected heat of one fluid cycle is used to supply heat to another fluid operating in a lower temperature range. In the mercury-steam cycle, the mercury condenser also acts as the steam boiler.

Other high-efficiency cycles involve combinations of gas turbines and steam power and direct thermal-to-electrical energy conversion. Direct-conversion systems under study for large power sizes include a magnetohydrodynamic (MHD) unit topping the conventional steam cycle and using conventional fuel or a char by-product from coal gasification or liquefaction.

In spite of the many complex cycles devised to increase overall plant efficiency, the conventional steam cycle has, as of the mid-1970s, proven to be the most economical. The increasing use of high steam pressures and temperatures, reheaters, regenerative feedwater heaters, economizers, and air heaters has led to improved efficiency in the contemporary steam power cycle. In addition, many other developments have progressively reduced the cost of equipment.

MARINE BOILERS

The watertube boiler was successfully applied to the propulsion of naval and merchant vessels in the 1890s. An improved design, installed in 1899 in the United States cruiser *Alert* established the superiority of the watertube boiler for marine propulsion. Subsequently, the development of marine boilers for naval and merchant ship propulsion has paralleled that for stationary use.

NUCLEAR STEAM SUPPLY SYSTEMS

From 1942, when Enrico Fermi demonstrated a controlled self-sustained fission reaction, nuclear energy has been recognized as an important source of heat for power generation. The first significant application of this new source for steam generation was the U.S.S. *Nautilus* prototype, which was operated at the National Reactor Testing Station in Idaho in the early 1950s. This has been followed by a number of installations for the propulsion of naval vessels.

The first electric utility installation was the 90-megawatt unit (net capacity) at the Shippingport Atomic Power Station of the Duquesne Light Company. This plant, owned partly by Duquesne Light Company and partly by the U.S. Atomic Energy Commission, went into operation in 1957. Construction permits were issued in 1955 and 1956 to three electric utilities for nuclear units approximately twice this size. These units—for Commonwealth Edison Company's Dresden Nuclear Power Station, Yankee Atomic Electric Company's Rowe Plant, and Consolidated Edison Company's Indian Point Station—went into commercial operation in the early 1960s. Further progress in the nuclear steam-generation field is covered in the handbook section on *Nuclear Energy Technology*.

REFERENCES

See listings at the end of the next several articles in this handbook section dealing with various aspects of steam technology.

Thermodynamics of Steam Cycles*

ENERGY, WORK, AND HEAT QUANTITY BALANCES

Thermodynamic processes follow the conservation laws as applied to energy and mass exclusive of mass-energy exchanges. The latter, although important in nuclear reactions, are insignificant in combustion and heat engines. Based on this restricted mass-energy relation of the system and associated processes, a balance prevails between energy, work, and heat quantities entering and leaving the system except for short time intervals in which energy is being added or withdrawn under unsteady state (time-dependent) conditions. Because the terms heat and work refer to energy in transit, they are recognized as characteristics or properties of the process, not the system. All other energy terms represent stored energy and are the properties used in describing a particular system and its potentials for change. Expressed as an equation, the restricted energy conservation law is

$$E_2 - E_1 + E(t) = Q - W_k \tag{1}$$

$E_2 - E_1 = \Delta E$ is the change in stored energy at the boundary states 1 and 2 of the system, and $E(t)$ accounts for energy changes due to unsteady state performance [for steady state systems $E(t) = 0$]. Q is heat added to, and W_k is the work done by the system. If the system is open in the sense that mass enters and leaves, the term ΔE reflects stored energy entering and leaving with the mass.

LAWS OF THERMODYNAMICS

The first law of thermodynamics is based on the conservation law expressed in Eq. (1) and, by convention, relates the heat and work quantities of this equation to internally stored energy u. Only internal energy u and externally stored energy are directly affected by heat flow. Heat added to a system will not change the potential energy of its mass in a gravitational field nor will it affect the kinetic energy of this mass as a whole unless it is first converted to work. It is convenient to express the first law as

$$\Delta u = Q - W_k \tag{2}$$

Although the first law treats heat and work as interchangeable, it is also a matter of experience that certain qualifications apply. Briefly, all forms of energy, including the transient form *work*, can be wholly converted to heat, but the converse is not generally true. Given a source of heat coupled with a heat-work cycle, such as heat released by high-temperature combustion in a steam power plant, only a portion of this heat can be converted to work. The rest must be rejected as heat to the stored energy of a sink at lower temperature, such as the atmosphere. This is, in essence, the Kelvin-Planck statement of the second law of thermodynamics. It can also be shown that it is equivalent to the Clausius statement that heat, in the absence of some form of external assistance, can only flow from a hotter to a colder body.

Entropy. Heat flow, like work, being energy in transit, is a function of a potential difference. The potential is easily recognized as temperature. If a quantity of heat is divided by its absolute temperature, the quotient can be considered a type of distribution property or factor complementing the intensity factor of temperature. Such a property, proposed and named *entropy* by Clausius, is widely used in all branches of thermodynamics because of its close relationship to the second law.

Rather than to define entropy (symbol S) in an absolute sense, which is the problem of the third law of thermodynamics, it will suffice to explain the significance of differences in this property given by

$$S_2 - S_1 = \Delta S = \int_1^2 \frac{\delta(Q_{rev})}{T} \times \text{(total system mass)} \tag{3}$$

where ΔS = change in entropy, Btu/°R
$\quad Q_{rev}$ = reversible heat flow between thermodynamic equilibrium states 1 and 2 of the system, Btu/pound (see Fig. 2)
$\quad T$ = absolute temperature, °R

Entropy S is an extensive property. By dividing by the total mass of the system, it is converted to an intensive property s, which is also referred to as entropy; although the term *specific entropy* is more rigid.

CYCLES

Carnot Cycle. Sadi Carnot (1824) proposed the concepts of the *cycle* and *reversible processes*. The Carnot cycle (named in his honor) is still used as a comparison for heat engine performance. This cycle, on a temperature-entropy diagram, is shown in Fig. 1a for any gas and in Fig. 1b for a two-phase saturated vapor. The cycle for a nonideal gas, such as superheated steam, is shown in Fig. 1c on Mollier coordinates (entropy versus enthalpy).

The Carnot cycle consists of the following processes:

1. Heat added to the working medium at constant temperature ($dT = 0$) from an appropriate heat source resulting in expansion work and changes in enthalpy. (For an ideal gas, changes in internal energy and pv are zero and, therefore, changes in enthalpy are zero.)

2. Adiabatic isentropic expansion ($ds = 0$) with expansion work and an equivalent decrease in enthalpy.

3. Constant temperature heat rejection to the surroundings equivalent to the compression work and any changes in enthalpy.

4. Adiabatic isentropic compression back to the starting temperature with compression work and an equivalent increase in enthalpy.

This cycle has no counterpart in practice. The only way to carry out the constant temperature processes in a single-phase system would be to approximate them through a series of isentropic expansions and constant pressure reheats for heat addition and isentropic compressions with a series of intercoolers for heat rejection. Another serious disadvantage of a Carnot gas engine would be the small ratio of net work to gross work (net work refers to the difference between the work of expansion, gross work, and the work of compression). Even a two-phase cycle, such as shown by Fig. 1b, would be subject to the practical difficulties of wet compression and, to a lesser extent, of wet expansion.

Nevertheless, the Carnot cycle clearly illustrates the basic principles of thermodynamics. Since the processes are reversible, the Carnot cycle offers the maximum thermal efficiency attainable between any given temperatures of heat source and heat sink. Moreover, this thermal efficiency depends only on these temperatures:

$$\eta = \frac{T_1 - T_2}{T_1} = 1 - \frac{T_2}{T_1} \tag{4}$$

where η = thermal efficiency of heat-to-work conversion
$\quad T_1$ = absolute temperature of heat source, °R
$\quad T_2$ = absolute temperature of heat sink, °R

The efficiency statement of Eq. (4) can be extended to cover all reversible cycles, where T_1 and T_2 are defined as mean temperatures found by dividing the heats added and rejected reversibly by the Δs in each case. For this reason, it can be stated that all reversible cycles have the same efficiencies when considered between the same mean temperature limits of heat source and heat sink.

Rankine Cycle. Early thermodynamic developments were centered around the performance of steam engines and, for comparison purposes, it was natural to select a reversible cycle which more nearly approximated the processes related to its operation. The Rankine cycle, shown in Fig. 2, proposed independently by Rankine and Clausius, meets this objective. All steps are specified for the system only (working medium) and

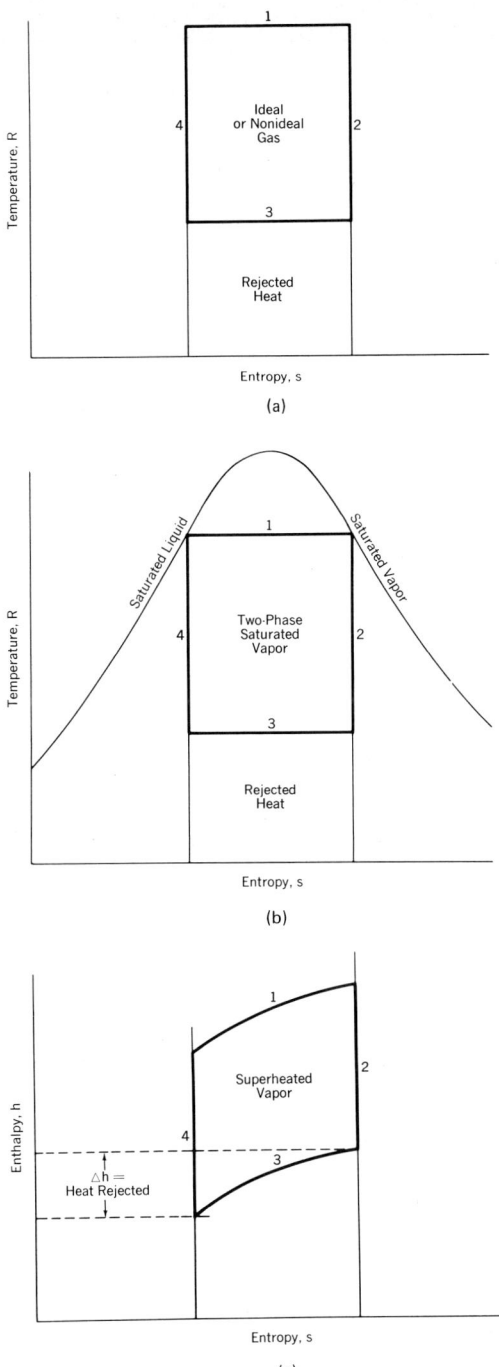

Fig. 1. Carnot cycles: (*a*) Temperature-entropy diagram—gas. (*b*) Temperature-entropy diagram—saturated vapor. (*c*) Mollier diagram—superheated vapor. (*Babcock & Wilcox Company.*)

carried out reversibly in the vapor, liquid, and two-phase states as indicated in the figure. Liquid is compressed isentropically from a to b. From b to c, heat is added reversibly in the compressed liquid, two-phase, and superheat states. Isentropic expansion with shaft work output takes place from c to d, and unavailable heat is rejected to the atmospheric sink from d to a.

The main feature of the Rankine cycle is compression confined to the liquid phase only, avoiding the high compression work and mechanical problems of a corresponding Carnot

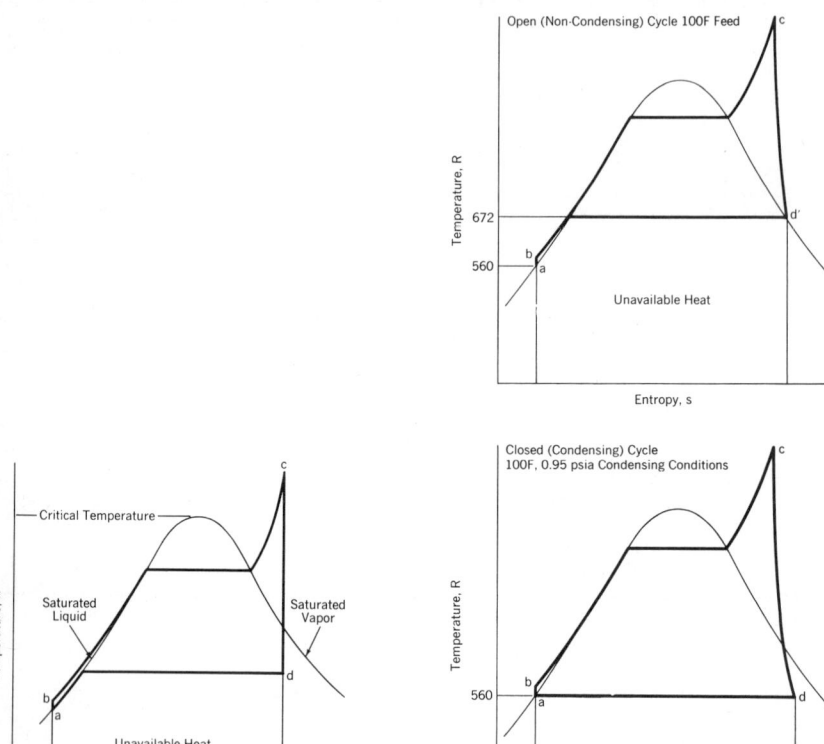

Fig. 2. Temperature-entropy diagram of the ideal Rankine cycle.

Fig. 3. Rankine cycles.

cycle with two-phase compression. This part of the cycle from a to b in Fig. 2 is greatly exaggerated since the difference between the saturated liquid line and reversible heat addition to compressed liquid is too small to show in proper scale. For example, the temperature rise with isentropic compression of water from saturation temperature of 212°F (100°C) and 1 atmosphere to 1000 psia is less than 1°.

If the Rankine cycle is closed in the sense that the same fluid repeatedly executes the various processes, it is termed a *condensing cycle*. Although the closed condensing Rankine cycle was developed to improve steam engine efficiency, a closed cycle is essential for any toxic or hazardous working fluid. Steam has the important advantage of being inherently safe in these categories. However, close control of water chemistry required in high-pressure, high-temperature power cycles also makes it desirable to recirculate steam condensate using a minimum of makeup. *Makeup* is defined as water added to the steam cycle to replace leakage and any other withdrawals. Open steam cycles are still found in small unit sizes and some special process and heating load applications coupled with power. Usually the condensate from process and heating loads is returned to the power cycle for economic reasons.

The higher efficiency of the condensing steam cycle is a result of the particular pressure-temperature relationship between water and its vapor state, steam. The lowest temperature at which an open or noncondensing steam cycle may reject heat is approximately the saturation temperature of 672°R (212°F; 100°C). This corresponds to normal atmospheric pressure of 14.7 psia. The closed or condensing cycle takes advantage of the much lower sink temperature for heat rejection available in natural bodies of water and the atmosphere. Being closed, the back pressure is no longer limited to normal atmospheric pressure, but rather to saturation pressure corresponding to a condensing temperature in the neighborhood of 100°F (38°C) and lower. Because the maximum possible pdv work per pound of a compressible fluid is directly related to a function of the pressure ratio available for expansion, as well as the initial absolute temperature, an increase in this pressure ratio means an increase in the available work. A condensing cycle with 1.5 inches mercury absolute pressure in the condenser and starting with the same initial pressure will have a pressure ratio 20 times greater than the noncondensing cycle expanding to the atmosphere.

Figure 3 illustrates the difference between an open and closed Rankine cycle (nonideal) superheated steam. Liquid compression takes place from a to b, and heat is added from b to c. The work and heat quantities involved in each of these processes are the same for both cycles. Expansion and conversion of stored energy to work is from c to d' for the open cycle and from c to d for the closed cycle. Since this process is shown for the irreversible case, there is internal heat flow and an increase in entropy. From d' to a and d to a, heat is rejected. Because this last portion of the two cycles is shown as reversible, the shaded areas are proportional to rejected heat. The larger amount of rejected heat for the open cycle is clearly indicated.

Regenerative Rankine Cycle. The reversible cycle efficiency given by Eq. (4), where T_2 and T_1 are mean absolute temperatures for rejecting and adding heat, respectively, indicates only three choices for improving ideal cycle efficiency: (1) decreasing T_2, (2) increasing T_1, or (3) both actions. Not much can be done to reduce T_2 in the Rankine cycle because of the limitations imposed by the normal temperatures of available sinks for rejected heat. There is some leeway available by selecting variable condenser pressures for the very large units with two or more exhaust hoods, since the lowest temperature in a condenser is set by the exit temperature of the cooling water. On the other hand, there are many ways to increase T_1 even though the maximum steam temperature may be limited by the materials problems of high-temperature corrosion and allowable stress at elevated temperatures.

One early improvement in the Rankine cycle was the adoption of regenerative feedwater heating. This is done by extracting steam at various stages in the turbine to heat the feedwater as it is pumped from the hot well to the boiler economizer.

Figure 4 is a cycle diagram of a widely used supercritical steam cycle showing schematically the arrangement of various components, including the feedwater heaters. This cycle also employs one stage of steam reheat which is still another method of increasing the mean T_1. Regardless of whether the cycle is high temperature, high pressure, or reheat, regeneration is used in all modern condensing steam power plants. It not only improves cycle efficiency, but has other advantages, among which are lower volume flow in the final turbine stages and a convenient means of deaerating the feed. The steam power-cycle diagram of Fig. 4 includes a heat-source system which is the combustion products of a fossil fuel. Stored energy is released as heat by the combustion of the fuel with air, and a portion is then transferred in the boiler for generating and superheating steam. The remaining heat is discharged to the surroundings as indicated. This part of the system also incorporates the principle of regeneration in an air heater to recycle low-level (low-temperature) heat from the combustion gases which would otherwise be rejected to the atmosphere as an increase in stack loss. Because of the higher feedwater temperatures entering the economizer of a boiler supplying steam to a regenerative Rankine cycle, the amount of low-level heat in combustion gases that can be transferred directly to the steam cycle is limited. In other words, these two types of regeneration are not entirely independent with respect to the overall effect on cycle efficiency.

The air heater as a regenerator uses low-level heat which otherwise would be rejected from the cycle. Feedwater heaters, on the other hand, utilize heat which could have been partially converted to work by further expansion through the turbine. Both types of regeneration will increase cycle efficiency as long as they can show a net decrease in the entropy production for the system and its surroundings.

The temperature-entropy diagram of Fig. 5 for the steam cycle of Fig. 4 illustrates how the principle of regeneration works in increasing the mean temperature level for heat addition. Instead of heat input starting at the hot well temperature of 91.7°F, the lowest temperature for heat addition from the combustion process has been raised to 550°F by the bleed heaters.

This same diagram of Fig. 5 also shows that the mean temperature level for heat addition is increased by reheating steam. Where maximum temperatures are limited by

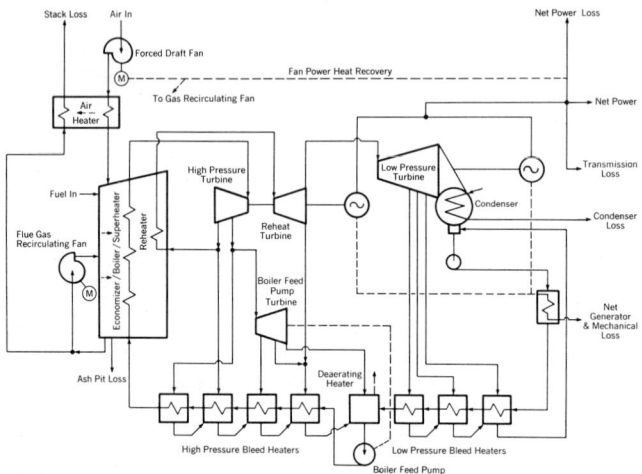

Fig. 4. Power cycle diagram using fossil fuel. Single reheat, eight-stage regenerative feed heating: 3515 psia, 1000/1000°F steam (538/538°C).

physical or economic reasons, reheating after partial expansion of the working fluid can be substituted as an effective means of raising the average T_1. The hypothetical case of an infinite number of reheat and expansion stages approaches a constant temperature heat addition, at least in the superheat region which may, in turn, be established at the maximum permissible temperature limit of the working medium or its containment. On the other hand, merely increasing T_1 may not improve efficiency. If the entropy increase accompanying reheat causes the final expansion process to terminate in superheated vapor, the mean temperature for heat rejection T_2 has also been increased unless the superheat can be extracted in a regenerator adding heat to the boiler feedwater. Such a regenerator would have to operate at the expense of the very effective bleed cycle. All these factors, plus component design problems, must be considered in any cycle analysis where the objective is to optimize the combination of physical and economic limits and fuel economy. Overall cycle characteristics, including efficiency, are illustrated more clearly by plotting on a Mollier chart. See Fig. 7 presented later.

The procedure used in preparing Fig. 5 deserves special comment since it illustrates an important function of entropy as a property. All processes on the diagram represent total entropies divided by the high-pressure steam flow rate. Total entropies at any point of the cycle are the product of the flow rate at that point and the entropy per pound corresponding to the pressure, temperature, and stage of the steam. If a point falls in the two-phase region (wet-steam zone), entropy is calculated from values in the steam tables in the same manner as enthalpy. In either case, the value for evaporation is multiplied by the steam quality (fraction of uncondensed steam) and added to the value for water at saturation conditions.

Since there are different flow rates for the various processes of the cycle, small sections of individual *T-s* diagrams are superimposed in Fig. 5 on a base diagram identified by saturated liquid and vapor parameters. This base actually applies only to those parts of the cycle representing heat addition to high-pressure steam and the expansion of this steam down to the first bleed point in the high-pressure turbine. At each bleed point, the

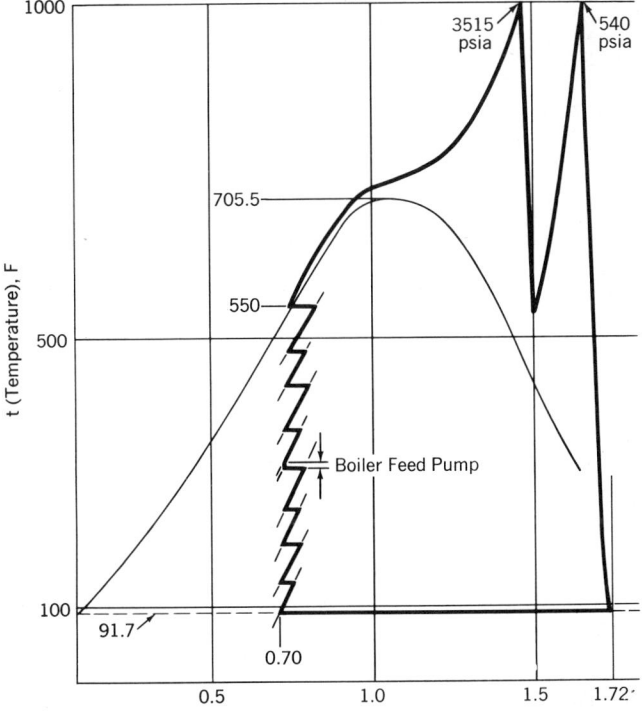

Fig. 5. Steam cycle for fossil fuel: temperature entropy diagram. Single reheat, eight-stage regenerative feed heating: 3515 psia, 1000/1000°F steam (538/538°C).

expansion line should show a decrease in entropy due to reduced flow entering the next turbine stage. However, for convenience, the individual step backs in the expansion lines have been shifted to the right in order to show the high-pressure steam expansion as one continuous process. The expansion of the reheated steam has also been treated in the same way.

Feedwater heating through the regenerators and compression by the pumps (represented by the zigzag lines in Fig. 5) result in a net increase in entropy. However, two factors are involved—an increase in entropy from the heat added to the feed and a decrease resulting from condensing and cooling the bleed steam and drain flows from higher pressure heaters.

For example, the feedwater heater just before the deaerating heater heats 5,436,519 pounds/hour feed from 188.5°F (86.9°C) to 239.9°F (115.5°C). This increases the enthalpy *h* of the feed from 156.4 to 207.1 and the entropy *s* from 0.2762 to 0.3530. The total increase in entropy per pound of high-pressure steam flow of 7,860,371 pounds/hour is

$$\frac{(0.3530 - 0.2762)\ 5,436,519}{7,860,371} = 0.0531\ \text{Btu/(pound) (°F)}$$

Feed temperature rises 51.4°F, and the total heat absorbed is

$$(207.1 - 156.4)\ 5,436,519 = 275,631,900\ \text{Btu/hour}$$

On the heat-source side of the balance, 263,238 pounds/hour of steam are bled from the low-pressure turbine at 29.1 psia, 1213.2h, and 1.7680s. This steam is desuperheated, condensed, and cooled to

$$\frac{1213.2 - 275,631,900}{263,238} = 166.1\ \text{Btu/pound} \qquad \text{at } 198.5°\text{F}$$

The corresponding entropy of the heater drain is 0.2915s. Therefore, the entropy decrease is

$$\frac{(1.7680 - 0.2915)2,632,238}{7,860,371} = 0.0494\ \text{Btu/(pound) (°F)}$$

This heater shows a net increase in entropy of $0.0531 - 0.0494 = 0.0037$ Btu/(pound) (°F), which is a measure of available heat energy loss due to pressure drop required for flow and temperature difference necessary for heat transfer. The product of this entropy increase and the absolute temperature of the sink for receiving rejected heat represents heat rendered unavailable for work by the internal heat flows of irreversible processes.

Cycles also are discussed in the article on "Heat Engines for Solar Power" in the *Solar Energy Technology* section of this Handbook.

AVAILABLE ENERGY

From the foregoing it can be stated that, associated with the property enthalpy, there is a derived quantity, formed by the product of the corresponding entropy and the absolute temperature of the available heat sink, which has the nature of a property. This is the minimum unavailable heat flow to the sink whenever work is extracted from enthalpy. The difference between h (enthalpy) and $T_0 s$ is another derived quantity called *available energy*.

$$e = h - T_0 s \tag{5}$$

where e = maximum energy available (sometimes called exergy) for h for useful work, Btu/pound

h = enthalpy, Btu/pound

T_0 = sink absolute temperature, °R

s = entropy, Btu/pound, °R

The available and unavailable energy are not properties since they are not completely defined by an equation of state but are also dependent on the sink temperature.

Available energy is useful in cycle analysis for optimizing the thermal performance of the various component pieces of equipment relative to overall cycle efficiency. In this way, small controllable changes in availability may be weighed against larger fixed unavailable heat quantities which are inherent to the cycle. As an example, an increased pressure drop through a heat exchanger may reduce surface, but it also increases capitalized fuel cost by reducing efficiency.

STEAM CYCLE FOR NUCLEAR PLANT

Figure 6 illustrates a Rankine cycle whose thermal energy source is a nuclear steam system. High-pressure cooling water (primary loop) from a pressurized water reactor is circulated through a once-through steam generator (boiler) which, in turn, supplies steam for the turbine. In this way, heat from the fission process absorbed in the primary loop is transferred through the tubes of the steam generator to generate steam at 925 psia on the low-pressure side of the steam generator and superheat it to 570°F (299°C). This steam is delivered to the high-pressure turbine at 900 psia and 566°F (296.7°C).

Pressure limitations and especially temperature limitations required for a nuclear reactor mean that expansion lines of the power cycle lie largely in the region of wet steam, in other words, a saturated or nearly saturated steam cycle. The expansion lines for the cycle of Fig. 6 are plotted on an h-s diagram in Fig. 7.

The superheated steam is delivered to the turbine at a temperature only 34°F ($\sim$ 19°C)

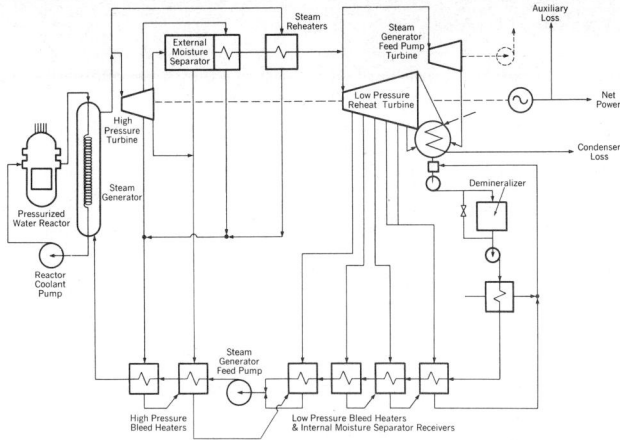

Fig. 6. Power cycle diagram: nuclear fuel. Reheat by bleed and high-pressure steam, moisture separation, and six-stage regenerative feed heating: 900 psia, 566/503°F steam (296.7/261°C).

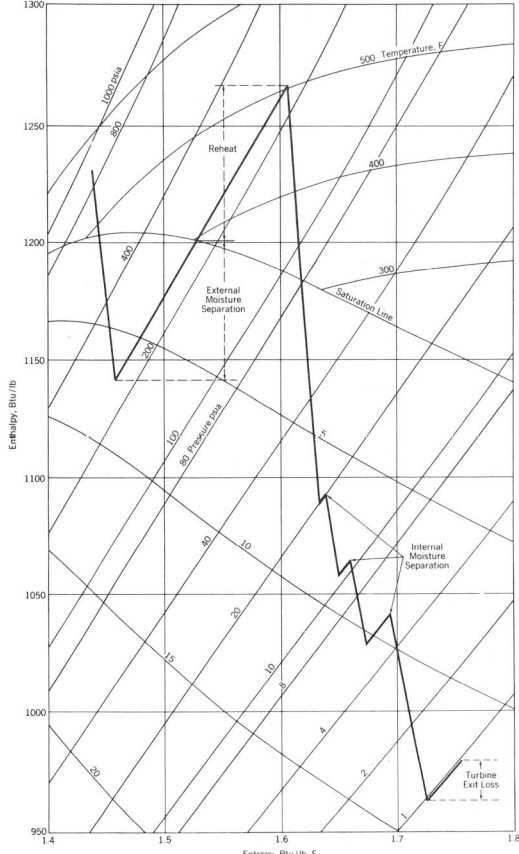

Fig. 7. Steam cycle for nuclear fuel: Mollier chart. Reheat by bleed and high-pressure steam, moisture separation, and six-stage regenerative feed heating: 900 psia, 566/503°F steam (296.7/261°C).

above saturation. Although this amount of superheat does improve cycle efficiency, it does not eliminate the problem (encountered with saturated steam cycles) of having to handle large quantities of condensed moisture in the turbine. For example, if expansion from the initial conditions shown on Fig. 7 proceeded in one step down to the back pressure of 2.0 inches mercury (approximately 1.0 psi), the moisture formed would be in excess of 20 percent. High moisture in the steam not only imposes erosion problems, especially in turbine blades, but also reduces expansion efficiency.

Although there are mechanical losses from momentum exchanges between slow-moving particles of condensate and high-velocity steam as well as moving turbine blades, there is also a strictly thermodynamic loss resulting from supersaturation. The expansion of the steam is too rapid to permit equilibrium thermodynamic properties to exist when condensation is occurring. Under this condition, the steam becomes subcooled, retaining a part of the available energy which would be released by condensation.

Figure 6 indicates the two methods of moisture removal utilized in this cycle, and Fig. 7 shows the effect of this moisture removal on the cycle. After expansion in the high-pressure turbine, the steam passes through a moisture separator, which is a low-pressure-drop separator located external to the turbine. After passing through this separator, the steam is reheated in two stages, first by bleed steam and then by high-pressure steam to 503°F (261°C), before entering the low-pressure turbine. Here, a second method is used for moisture removal, utilizing grooves on the back of the turbine blades to drain the moisture from certain stages of the low-pressure turbine. The separated moisture is then carried off with the bleed steam.

Internal moisture separation not only reduces erosion but also affords a thermodynamic advantage due to the divergence of the constant pressure lines with increasing enthalpy and entropy. This can be shown by the use of available energy e in the following manner. Consider the moisture removal stage at 10.8 psia in Fig. 7. After expansion down to 10.8 psia, the moisture content of the steam is 8.9 percent. This is reduced to 8.2 percent by internal separation. Other properties are as follows:

	End of expansion	After moisture extraction
P, psia	10.8	10.8
h, Btu/pound	1,057.9	1,064.7
s, Btu/(pound) (°F)	1.6491	1.6595
T_0 (at 2 inches mercury), °R	560.8	560.8
$T_0 s$, Btu/pound	924.8	930.7
$e = h - T_0 s$, Btu/pound	133.1	134.0

The increase in availability, Δe, due to moisture extraction is $134.0 - 133.1 = 0.9$ Btu/pound of steam.

The values of moisture and enthalpy listed are given for equilibrium conditions without considering supersaturation, which is accounted for empirically by the isentropic efficiency of the expansion line. Values of enthalpy and entropy used in the example are taken from the American Society of Mechanical Engineers' "Steam Tables," where they are listed to five significant figures. Whenever it is necessary to work with small differences between large numbers, it is also necessary to use a sufficient number of significant figures.

See articles in the *Nuclear Energy Technology* section of this Handbook, particularly "Boiling Water Reactor," and "Pressurized Water Reactor."

MULTIPURPOSE STEAM POWER PLANTS

In steam power plants operated solely for the generation of electric power, thermal efficiencies that are economically justifiable range up to about 40 percent in fossil-fuel plants and 34 percent in nuclear plants, including the optimum use of steam pressure and temperature, reheat, and bleed preheating of feedwater. This means that more than half the heat released from the fuel is wasted and must be transferred to the environment in some way. This is usually done through a condenser, resulting in the heating of some body of water. As the number of fuel plant installations increases, the heating of streams and bodies of water is approaching undesirable limits in many areas, particularly in the Northeastern part of the United States.

There is an increasing need for higher efficiency steam plants. Higher efficiency can be obtained by the use of higher steam temperatures, but, after many years of research on high-temperature metals, steam temperature is limited economically to about 1000°F (538°C).

One practical means available for improving the utilization of energy in steam is the use of multipurpose steam plants, where steam is exhausted or extracted from the turbines at a proper pressure level for use in industrial processing. With such arrangements, it is possible to obtain an overall thermal utilization between 65 and 70 percent. The use of combination power-and-process installations has been common in industry for many years, but the demand for process steam is not sufficient to permit the use of these combined cycles in many electric power generating plant installations.

REFERENCES

"Steam: Its Generation and Use," 38th ed., Babcock & Wilcox Company, New York, 1972.

Hatsopoulos, G. N., and J. H. Keenan: "Principles of General Thermodynamics," Wiley, New York, 1965.

Keenan, J. H., F. G. Keyes, P. G. Hill, and J. G. Moore (eds.): "Steam Tables: Thermodynamic Properties of Water Including Vapor, Liquid, and Solid Phases," Wiley, New York, 1969.

Holman, J. P.: "Thermodynamics," McGraw-Hill, New York, 1969.

Obert, E. F., and R. A. Gaggiolo: "Thermodynamics," 2d ed., McGraw-Hill, New York, 1963.

Reynolds, W. C.: "Thermodynamics," 2d ed., McGraw-Hill, New York, 1968.

Reynolds, W. C., and H. C. Perkins: "Engineering Thermodynamics," McGraw-Hill, New York, 1970.

Woodruff, E., and H. Lammers: "Steam Plant Operation," 3d ed., McGraw-Hill, New York, 1967.

Principles of Combustion in Steam Generation*

OBJECTIVES AND CHEMISTRY OF COMBUSTION

Combustion, for the purposes of this article, may be defined as the rapid chemical combination of oxygen with the combustible elements of a fuel. There are three combustible chemical elements of principal significance: carbon, hydrogen, and sulfur. Sulfur is usually of minor significance as a source of heat, but it can be of major concern in terms of corrosion and pollution problems.

Carbon and hydrogen when burned to completion with oxygen unite according to

$$C + O_2 \rightarrow CO_2 + 14{,}100 \text{ Btu/pound of C}$$

$$2H_2 + O_2 \rightarrow 2H_2O + 61{,}100 \text{ Btu/pound of H}_2$$

Air is the usual source of oxygen for boiler furnaces. These combustion reactions are exothermic and the heat released is indicated in the foregoing equations.

The objective of good combustion is to release all this heat while minimizing losses from combustion imperfections and superfluous air. The combination of the combustible elements and compounds of a fuel with all the oxygen requires *temperature* high enough to ignite the constituents, mixing or *turbulence*, and sufficient *time* for complete combustion. These factors are sometimes referred to as the *three T*'s of combustion.

This article details the basic chemistry necessary for understanding the phenomena of combustion in the boiler furnace. Ability to calculate the release of heat in combustion and to determine the amount and nature of the combustion products is essential for the design of a steam-generating plant and determination of its performance characteristics.

Table 1 lists the chemical elements and compounds found in fuels generally used in the commercial generation of heat, giving their molecular weights, heats of combustion, and other combustion constants. The term "100% total air" as used in Table 1 and figures and examples which appear elsewhere in this article means 100 percent of the air theoretically required for combustion without excess. Higher percentages indicate *theoretical plus excess air;* for example, 125 percent total air means 100 percent theoretical air plus 25 percent excess air.

Concept of the Mole. The mass of a substance in pounds (or other unit of mass) equal to its molecular weight is called a *pound mole* (or gram mole, etc.) of the substance. In power plant practice in the United States, the pound mole frequently is simply called a mole. For example, carbon (C) has a molecular weight of 12 (12.01115 to be highly exact). Therefore, a mole of carbon weighs 12 pounds. In the case of gases, the volume occupied by a mole is called the *molal volume* and is a constant of 394 cubic feet for "ideal gases" at 80°F and atmospheric pressure (approximately 14.7 psia or 30 inches mercury). These concepts of mass and volume are essential tools in combustion calculations.

FUNDAMENTAL LAWS

Several fundamental physical laws apply to combustion calculations. These are reviewed briefly as follows.

Conservation of Matter. This is the familiar statement that "Matter is neither destroyed nor created." There must be a weight balance between the sum of the weights entering a process and the sum leaving. In other words, A pounds of fuel combined with B pounds of air will always result in $A + B$ pounds of products. (It should be noted that when a pound of a typical coal is burned, releasing 13,500 Btu, the quantity of mass converted to energy amounts to only 3.5×10^{-10} pound, a loss too small to be measured or considered in conventional combustion calculations.) Obviously, this conversion is of significance to nuclear reactions.

Conservation of Energy. This is the familiar statement that "Energy is neither destroyed nor created." The sum of the energy (potential, kinetic, thermal, chemical, and electrical) entering a process must equal the sum of energy leaving, although the proportionate amounts of each may change. In combustion, chemical energy is exchanged for energy in the form of heat. The parenthetical observation made in the prior paragraph also applies to this relationship.

Ideal Gas Law. The volume of an ideal gas is directly proportional to its absolute temperature and inversely proportional to its absolute pressure. The proportionality constant is found to be the same for 1 mole of any ideal gas. The law may be expressed

$$v_M = \frac{RT}{p}$$

where v_M = volume, cubic feet/mole of gas
$\quad p$ = absolute pressure, pounds/square foot
$\quad T$ = absolute temperature, °R (Rankine) = °F + 460
$\quad R$ = universal gas constant, 1545 foot-pounds/(mole) (T)
This equation states that 1 mole of all ideal gases occupies the same volume for the same pressure and temperature conditions, that is, 394 cubic feet at 14.7 psia and 80°F. Experiments indicate that most gases approach this ideal.

Law of Combining Weights. All substances combine in accordance with simple definite weight relationships. These relationships are exactly proportional to the molecular weights of the constituents. For example, carbon (molecular weight = 12) combines with oxygen (molecular weight = 32) to form carbon dioxide (molecular weight = 44), so that 12 pounds of carbon plus 32 pounds of oxygen (O_2) unite to form 44 pounds of carbon dioxide (CO_2).

Avogadro's Law. Equal volumes of different gases at the same pressure and temperature contain the same number of molecules. From the concept of the mole, a pound mole of any substance contains a mass equal in pounds to the molecular weight of the substance. Thus, the ratio of mole weight to molecular weight is a constant, and a mole of a chemically pure substance contains the same number of molecules, no matter what the substance may be. Since a mole of any ideal gas occupies the same volume at a given pressure and temperature (ideal gas law), it follows that equal volumes of different gases at the same pressure and temperature contain the same number of molecules.

Dalton's Law. The total pressure of a mixture of gases is the sum of the partial pressures which would be exerted by each of the constituents if each gas were to occupy alone the same volume as the mixture. In other words, for equal volumes V of three gases (A, B, and C) all at the same temperature T but at different pressurex P_a, P_b, and P_c, when all three gases are placed in the space of the volume V, then the resulting pressure P is equal to $P_a + P_b + P_c$. For gases, each gas in a mixture fills the entire volume and exerts a pressure independent of the other gases.

Amagat's Law. The total volume occupied by a mixture of gases is equal to the sum of the volumes which would be occupied by each of the constituents when at the same pressure and temperature as the mixture. This law is related to Dalton's law, but considers the additive effects of volume instead of pressure. If all three gases are at pressure P and temperature T but at volumes V_a, V_b, and V_c, then, when combined so that T and P are unchanged, the volume of the mixture is $V = V_a + V_b + V_c$.

APPLICATION OF FUNDAMENTAL LAWS

Table 2 summarizes the molecular and weight relationships between fuel and oxygen and lists the heat of combustion for the substances commonly involved in combustion. Most of the weight and volume relationships in combustion calculations can be determined by using the information in Table 2 and the seven fundamental laws just described.

TABLE 1. Combustion Constants of Various Elements and Compounds

Element or compound	Formula	Molecular weight	Pounds per cubic foot	Cubic feet per pound	Specific gravity air = 1.0	Heat of combustion			
						Btu per cubic foot		Btu per pound	
						Gross (high)	Net (low)	Gross (high)	Net (low)
COMMON COMBUSTION SUBSTANCES									
Carbon*	C	12.01						14,093	14,093
Hydrogen	H₂	2.016	0.0053	187.723	0.0696	325	275	61,095	51,623
Oxygen	O₂	32.00	0.0846	11.819	1.1053				
Nitrogen (atm)	N₂	28.01	0.0744	13.443	0.9718				
Carbon monoxide	CO	28.01	0.0740	13.506	0.9672	321	321	4,347	4,347
Carbon dioxide	CO₂	44.01	0.1170	8.548	1.5282				
ALKANES OR PARAFFINS									
Methane	CH₄	16.04	0.0425	23.552	0.5543	1012	911	23,875	21,495
Ethane	C₂H₆	30.07	0.0803	12.455	1.0488	1773	1622	22,323	20,418
Propane	C₃H₈	44.09	0.1196	8.365	1.5617	2524	2322	21,669	19,937
n-Butane	C₄H₁₀	58.12	0.1582	6.321	2.0665	3271	3018	21,321	19,678
Isobutane	C₄H₁₀	58.12	0.1582	6.321	2.0665	3261	3009	21,271	19,628
n-Pentane	C₅H₁₂	72.15	0.1904	5.252	2.4872	4020	3717	21,095	19,507
Isopentane	C₅H₁₂	72.15	0.1904	5.252	2.4872	4011	3708	21,047	19,459
Neopentane	C₅H₁₂	72.15	0.1904	5.252	2.4872	3994	3692	20,978	19,390
n-Hexane	C₆H₁₄	86.17	0.2274	4.398	2.9704	4768	4415	20,966	19,415
ALKENES OR OLEFINS									
Ethylene	C₂H₄	28.05	0.0742	13.475	0.9740	1604	1503	21,636	20,275
Propylene	C₃H₆	42.08	0.1110	9.007	1.4504	2340	2188	21,048	19,687
n-Butene	C₄H₈	56.10	0.1480	6.756	1.9336	3084	2885	20,854	19,493
Isobutene	C₄H₈	56.10	0.1480	6.756	1.9336	3069	2868	20,737	19,376
n-Pentene	C₅H₁₀	70.13	0.1852	5.400	2.4190	3837	3585	20,720	19,359
BENZENOIDS OR AROMATICS									
Benzene	C₆H₆	78.11	0.2060	4.852	2.6920	3752	3601	18,184	17,451
Toluene	C₇H₈	92.13	0.2431	4.113	3.1760	4486	4285	18,501	17,672
Xylene	C₈H₁₀	106.16	0.2803	3.567	3.6618	5230	4980	18,650	17,760
OTHER GASES OR VAPORS									
Acetylene	C₂H₂	26.04	0.0697	14.344	0.9107	1477	1426	21,502	20,769
Naphthalene	C₁₀H₈	128.16	0.3384	2.955	4.4208	5854	5654	17,303	16,708
Methyl alcohol	CH₃OH	32.04	0.0846	11.820	1.1052	868	767	10,258	9,066
Ethyl alcohol	C₂H₅OH	46.07	0.1216	8.221	1.5890	1600	1449	13,161	11,917
Ammonia	NH₃	17.03	0.0456	21.914	0.5961	441	364	9,667	7,985
Sulfur*	S	32.06						3,980	3,980
Hydrogen sulfide	H₂S	34.08	0.0911	10.979	1.1898	646	595	7,097	6,537
Sulfur dioxide	SO₂	64.06	0.1733	5.770	2.2640				
Water vapor	H₂O	18.02	0.0476	21.017	0.6215				
Air			0.0766	13.063	1.0000				

For 100% total air

Element or compound	Moles per mole of combustible or cubic feet per cubic foot of combustible						Pounds per pound of combustible					
	Required for combustion			Flue products			Required for combustion			Flue products		
	O₂	N₂	Air	CO₂	H₂O	N₂	O₂	N₂	Air	CO₂	H₂O	N₂
Carbon*	1.0	3.76	4.76	1.0		3.76	2.66	8.86	11.53	3.66		8.86
Hydrogen	0.5	1.88	2.38		1.0	1.88	7.94	26.41	34.34		8.94	26.41
Oxygen												
Nitrogen (atm)												
Carbon monoxide	0.5	1.88	2.38	1.0		1.88	0.57	1.90	2.47	1.57		1.90
Carbon dioxide												
Methane	2.0	7.53	9.53	1.0	2.0	7.53	3.99	13.28	17.27	2.74	2.25	13.28
Ethane	3.5	13.18	16.68	2.0	3.0	13.18	3.73	12.39	16.12	2.93	1.80	12.39
Propane	5.0	18.82	23.82	3.0	4.0	18.82	3.63	12.07	15.70	2.99	1.63	12.07
n-Butane	6.5	24.47	30.97	4.0	5.0	24.47	3.58	11.91	15.49	3.03	1.55	11.91
Isobutane	6.5	24.47	30.97	4.0	5.0	24.47	3.58	11.91	15.49	3.03	1.55	11.91
n-Pentane	8.0	30.11	38.11	5.0	6.0	30.11	3.55	11.81	15.35	3.05	1.50	11.81
Isopentane	8.0	30.11	38.11	5.0	6.0	30.11	3.55	11.81	15.35	3.05	1.50	11.81
Neopentane	8.0	30.11	38.11	5.0	6.0	30.11	3.55	11.81	15.35	3.05	1.50	11.81
n-Hexane	9.5	35.76	45.26	6.0	7.0	35.76	3.53	11.74	15.27	3.06	1.46	11.74
Ethylene	3.0	11.29	14.29	2.0	2.0	11.29	3.42	11.39	14.81	3.14	1.29	11.39
Propylene	4.5	16.94	21.44	3.0	3.0	16.94	3.42	11.39	14.81	3.14	1.29	11.39
n-Butene	6.0	22.59	28.59	4.0	4.0	22.59	3.42	11.39	14.81	3.14	1.29	11.39
Isobutene	6.0	22.59	28.59	4.0	4.0	22.59	3.42	11.39	14.81	3.14	1.29	11.39
n-Pentene	7.5	28.23	35.73	5.0	5.0	28.23	3.42	11.39	14.81	3.14	1.29	11.39
Benzene	7.5	28.23	35.73	6.0	3.0	28.23	3.07	10.22	13.30	3.38	0.69	10.22
Toluene	9.0	33.88	42.88	7.0	4.0	33.88	3.13	10.40	13.53	3.34	0.78	10.40
Xylene	10.5	39.52	50.02	8.0	5.0	39.52	3.17	10.53	13.70	3.32	0.85	10.53
Acetylene	2.5	9.41	11.91	2.0	1.0	9.41	3.07	10.22	13.30	3.38	0.69	10.22
Naphthalene	12.0	45.17	57.17	10.0	4.0	45.17	3.00	9.97	12.96	3.43	0.56	9.97
Methyl alcohol	1.5	5.65	7.15	1.0	2.0	5.65	1.50	4.98	6.48	1.37	1.13	4.98
Ethyl alcohol	3.0	11.29	14.29	2.0	3.0	11.29	2.08	6.93	9.02	1.92	1.17	6.93
Ammonia	0.75	2.82	3.57	—	1.5	3.32	1.41	4.69	6.10	—	1.59	5.51
				SO₂								
Sulfur*	1.0	3.76	4.76	1.0		3.76	1.00	3.29	4.29	2.00		3.29
Hydrogen sulfide	1.5	5.65	7.15	1.0	1.0	5.65	1.41	4.69	6.10	1.88	0.53	4.69

Note: All gas volumes corrected to 60°F and 30 inches mercury (dry).
*Carbon and sulfur are considered as gases for molal calculations only.
SOURCE: Data from "Steam: Its Generation and Use," Babcock & Wilcox Company, New York, 1972.

TABLE 2.   Common Chemical Reactions of Combustion

Combustible	Reaction	Moles	Pounds	Heat of combustion (high), Btu/pound fuel
Carbon (to CO)	$2C + O_2 = 2CO$	$2 + 1 = 2$	$24 + 32 = 56$	4,000
Carbon (to CO_2)	$C + O_2 = CO_2$	$1 + 1 = 1$	$12 + 32 = 44$	14,100
Carbon monoxide	$2CO + O_2 = 2CO_2$	$2 + 1 = 2$	$56 + 32 = 88$	4,345
Hydrogen	$2H_2 + O_2 = 2H_2O$	$2 + 1 = 2$	$4 + 32 = 36$	61,100
Sulfur (to SO_2)	$S + O_2 = SO_2$	$1 + 1 = 1$	$32 + 32 = 64$	3,980
Methane	$CH_4 + 2O_2 = CO_2 + 2H_2O$	$1 + 2 = 1 + 2$	$16 + 64 = 80$	23,875
Acetylene	$2C_2H_2 + 5O_2 = 4CO_2 + 2H_2O)$	$2 + 5 = 4 + 2$	$52 + 160 = 212$	21,500
Ethylene	$C_2H_4 + 3O_2 = 2CO_2 + 2H_2O$	$1 + 3 = 2 + 2$	$28 + 96 = 124$	21,635
Ethane	$2C_2H_6 + 7O_2 = 4CO_2 + 6H_2O$	$2 + 7 = 4 + 6$	$60 + 224 = 284$	22,325
Hydrogen sulfide	$2H_2S + 3O_2 = 2SO_2 + 2H_2O$	$2 + 3 = 2 + 2$	$68 + 96 = 164$	7,100

The data for C and H_2 can be expressed as follows:

C	+	O_2	=	CO_2
1 molecule	+ 1 molecule		→ 1 molecule	
1 mole	+ 1 mole		= 1 mole	
*	+ 1 cubic foot		→ 1 cubic foot	
12 pounds	+ 32 pounds		= 44 pounds	

*When 1 cubic foot of oxygen (O_2) combines with carbon (C), it forms 1 cubic foot of carbon dioxide (CO_2). If carbon were an ideal gas instead of a solid, 1 cubic foot of carbon would be required.

$2H_2$	+	O_2	=	$2H_2O$
2 molecules	+ 1 molecule		→ 2 molecules	
2 moles	+ 1 mole		= 2 moles	
2 cubic feet	+ 1 cubic foot		→ 2 cubic feet	
4 pounds	+ 32 pounds		= 36 pounds	

While there is a weight balance in these equations, there is not a molecular or volume balance. For example, 2 cubic feet of H_2 unite with 1 cubic foot of O_2 to form only 2 cubic feet of H_2O. This relationship is based on Avogadro's law and the law of combining weights.

The Mole in Combustion Calculations. Combustion calculations involving gaseous mixtures can be simplified by the use of the mole. Since equal volumes of gases at any given pressure and temperature contain the same number of molecules (Avogadro's law), the weights of equal volumes of gases are proportional to their molecular weights. If M is the molecular weight of the gas, 1 mole equals M pounds. Actual values are available from Table 1, for example:

$$1 \text{ mole of } O_2 = 32 \text{ pounds oxygen}$$
$$1 \text{ mole of } H_2 = 2 \text{ pounds hydrogen}$$
$$1 \text{ mole of } CH_4 = 16 \text{ pounds of methane}$$

Data from Table 1 can be used to demonstrate that the volume of 1 mole at a given pressure and temperature is approximately fixed and independent of the kind of gas.

At 60°F and atmospheric pressure (30 inches mercury), the specific volume of oxygen is 11.819 cubic feet/pound. Therefore, 1 mole of oxygen has a volume of $32 \times 11.819 = 378.21$ cubic feet. Similarly, under the same conditions, the specific volume of hydrogen is 187.723 cubic feet/pound, and 1 mole has a volume of $2.016 \times 187.723 = 378.45$ cubic feet. This volume, usually taken as 379 cubic feet, therefore approximates the volume of 1 mole of any gas at 60°F and atmospheric pressure.

The mole faction of a component of a mixture is the number of moles of the component divided by the sum of the number of moles of all the components of the mixture. As a mole of every ideal gas occupies the same volume, it follows by Avogadro's law that in a mixture of ideal gases the mole fraction of a component will exactly equal the volume fraction:

$$\frac{\text{Moles of component}}{\text{Total moles}} = \frac{\text{volume of component}}{\text{volume of total mixture}}$$

This is a valuable concept, since the volumetric analysis of a mixture of gases automatically gives the mole fractions of the different components.

In power plant practice, the practical source of oxygen is primarily air, which includes, along with the oxygen, a mixture of nitrogen, water vapor, and small amounts of inert gases, such as argon, neon, and helium. Data on the composition of air are given in Table 3.

The information in Table 2 can be used for air instead of O_2 if 3.76 moles of nitrogen (N_2) are added to both the left and right sides of each equation for every mole of O_2 involved. For example, the burning of CO in air becomes

$$2CO + O_2 + 3.76N_2 = 2CO_2 + 3.76N_2$$

or for methane (CH_4):

$$CH_4 + 2O_2 + 2(3.76)N_2 = CO_2 + 2H_2O + 7.52N_2$$

As indicated by the following example for a fuel gas, molal calculations have a simple and direct application to gaseous fuels, where the analyses are usually reported in percent on a volume basis.

Fuel gas analysis,
% by volume

CH_4	85.3
C_2H_6	12.6
CO_2	0.1
N_2	1.7
O_2	0.3
Total	100.0

This analysis may also be expressed as 85.3 moles CH_4 per 100 moles of fuel; 12.6 moles C_2H_6 per 100 moles of fuel; and so on.

The elemental breakdown of each constituent may also be designated in moles per 100 moles of fuels, as follows:

$$\text{C in } CH_4 = 85.3 \times 1 = 85.3 \text{ moles}$$
$$\text{C in } C_2H_6 = 12.6 \times 2 = 25.2 \text{ moles}$$
$$\text{C in } CO_2 = 0.1 \times 1 = 0.1 \text{ moles}$$
$$\text{Total C per 100 moles fuel} = 110.6 \text{ moles}$$

$$H_2 \text{ in } CH_4 = 85.3 \times 2 = 170.6 \text{ moles}$$
$$H_2 \text{ in } C_2H_6 = 12.6 \times 3 = 37.8 \text{ moles}$$
$$\text{Total } H_2 \text{ per 100 moles fuel} = 208.4 \text{ moles}$$

$$O_2 \text{ in } CO_2 = 0.1 \times 1 = 0.1 \text{ moles}$$
$$O_2 = 0.3 \times 1 = 0.3 \text{ moles}$$
$$\text{Total } O_2 \text{ per 100 moles fuel} = 0.4 \text{ moles}$$

$$N_2 \text{ per 100 moles fuel} = 1.7 \text{ moles}$$

An analysis of the flue gas produced by burning a gas fuel of the composition just described could be

Constituent	% by volume
CO_2	10.4
O_2	2.8
N_2	86.8
Total	100.0

TABLE 3. Composition of Air

Component	Composition of dry air	
	Percent by volume	Percent by weight
Oxygen (O_2)	20.99	23.15
Nitrogen (N_2)	78.03	76.85*
Inerts*	0.98	
Equivalent molecular weight of air	= 29.0*	
Percent moisture	= 1.3 percent by weight	
(Standard for the boiler industry—ABMA)†		
Moles air/mole oxygen ⎱ Cubic feet air/cubic feet oxygen ⎰	$= \dfrac{100}{20.99} = 4.76$	
Moles nitrogen/mole oxygen	$= \dfrac{79.01}{20.99} = 3.76$	
Pounds air (dry)/pound oxygen	$= \dfrac{100}{23.15} = 4.32$	
Pounds nitrogen/pound oxygen	$= \dfrac{76.85}{23.15} = 3.32$	

*It is convenient in combustion calculations to account for inerts as equivalent nitrogen. The equivalent weight percentage of 76.85 and the equivalent molecular weight of 29.0 have been corrected to account for the extra weight of the inerts.

†Air containing 0.013 pound water/pound dry air is often referred to as *standard air.*

Analyses of flue gases are reported on a volume basis, *dry,* when an Orsat or other type of gas analysis is used. Flue gases are cooled to room temperature and bubbled through water in most gas analyses, so that the gas becomes saturated with water vapor. This would occur even if no water vapor were formed during combustion. Proportionate parts of the water vapor content of the gas will be absorbed with the different constituents of the gas so that the resulting analysis may be safely assumed to be that of dry gas. These percentages may also be expressed as 10.4 moles CO_2, 2.8 moles O_2, and 86.8 moles N_2—each per 100 moles of dry flue gas.

For each mole of C burned, 1 mole of CO_2 is formed. From the fuel analysis used, there are 110.6 moles C per 100 moles of fuel, and there are also 110.6 moles CO_2 formed from the 110.6 moles C in the fuel. From the flue gas analysis, there are $^{100}/_{10.4} = 9.62$ moles of dry flue gas per mole of CO_2. The 100 moles of fuel will then yield $110.6 \times 9.62 = 1064$ moles of dry flue gas. By the application of the mole method, an important value has been quickly determined through knowing only the flue gas analysis and the fuel analysis.

From the flue gas analysis, the molecular weight of the dry flue gas can be easily determined, as

$$
\begin{aligned}
10.4 \text{ moles } CO_2 \text{ weigh } 10.4 \times 44 \text{ pounds} &= 457.6 \text{ pounds} \\
2.8 \text{ moles } O_2 \text{ weigh } 2.8 \times 32 \text{ pounds} &= 89.6 \\
\underline{86.8 \text{ moles } N_2 \text{ weigh } 86.8 \times 28 \text{ pounds}} &= \underline{2430.4} \\
100.0 \text{ moles dry flue gas} &= 2977.6 \text{ pounds}
\end{aligned}
$$

Therefore, one mole equivalent of dry flue gas = 29.8 pounds, or the equivalent molecular weight of the dry flue gas = 29.8. Hence, the weight of 1064 moles of dry flue gas is 1064 × 29.8 = 31,700 pounds, or 100 moles of fuel yields 31,700 pounds of dry flue gas.

The weight of 100 moles of fuel is the sum of the products of each constituent in the fuel and its molecular weight:

$$
\begin{aligned}
CH_4 \quad 85.3 \times 16 &= 1365 \text{ pounds} \\
C_2H_6 \quad 12.6 \times 30 &= 378 \\
CO_2 \quad 0.1 \times 44 &= 4.4 \\
N_2 \quad 1.7 \times 28 &= 47.6 \\
O_2 \quad 0.3 \times 32 &= \underline{9.6} \\
100.0 \text{ moles} &= 1804.6 \text{ pounds}
\end{aligned}
$$

Thus, 1804.6 pounds of gas fuel yields 31,700 pounds of dry flue gas, and each pound of gas fuel yields $^{31,700}\!/_{1805} = 17.6$ pounds dry flue gas.

HEAT OF COMBUSTION

In a boiler furnace (where no mechanical work is done), the heat energy evolved from the union of combustible elements with oxygen depends upon the ultimate products of combustion and not on any intermediate combinations that may occur in reaching the final result.

A simple demonstration of this law is the union of 1 pound of carbon with oxygen to produce a specific amount of heat (about 14,100 Btu—Table 2). The union may be in one step to form the gaseous production of combustion, CO_2; or under certain conditions, the union may be in two steps: first to form carbon monoxide (CO), producing a much smaller amount of heat (4345 Btu), and second, the union of the CO so obtained to form CO_2, releasing an additional 9755 Btu. However, the sum of the heats released in the two steps equals the 14,100 Btu evolved when carbon is burned in a single step to form CO_2 as the final product.

That carbon may enter into these two combinations with oxygen is of the utmost importance in the design of combustion equipment. Firing methods must assure complete mixture of fuel and oxygen to be certain that all the carbon burns to CO_2 and not to CO. Failure to meet this requirement will result in appreciable losses in combustion efficiency and in the amount of heat released by the fuel, since only about 28 percent of the available heat in the carbon is released if CO is formed instead of CO_2.

Measuring Heat of Combustion. In boiler practice, the heat of combustion of a fuel is the amount of heat, expressed in Btu, generated by the complete combustion (or oxidation) of a unit weight (1 pound in the United States) of fuel. Calorific value or fuel Btu value are terms also used.

The amount of heat generated by complete combustion is a constant for any given combination of combustible elements and compounds and is not affected by the manner in which the combustion takes place, provided it is complete.

The heat of combustion of a fuel is usually determined by direct measurement in a calorimeter of the heat evolved during combustion. See "Bomb Calorimetry for Testing Solid and Liquid Fuels" in the Handbook section on *Coal Technology* and "Calorimetric Measurement of Gaseous Fuels" in the Handbook section on *Gas Technology*.

The heat of combustion of most gases encountered in boiler practice is given in Table 1. If the content of any gas mixture is known, its heat of combustion can be accurately determined by adding the products of the volume percentage of each constituent times its heat of combustion.

For accurate heat values of solid and liquid fuels, calorimeter determinations are required. However, approximate heat values may be determined for most coals if the ultimate chemical analysis is known. Dulong's formula gives reasonably accurate results (within 2 to 3 percent) for most coals and is often used as a routine check of values determined by calorimeter:

$$\text{Btu/pound} = 14,544C + 62,028 (H_2 - O_2/8) + 4050S \qquad (1)$$

See also "Coal—Properties and Statistics" in the handbook section on *Coal Technology*.

In formula (1), the symbols represent the proportionate parts by weight of the constituents of the fuel—carbon, hydrogen, oxygen, and sulfur—as determined by an ultimate analysis. The coefficients represent the approximate heating values of the constituents in Btu per pound. The term $O_2/8$ is a correction applied to the hydrogen in the fuel to account for the hydrogen already combined with the oxygen in the form of moisture. This formula is not generally suitable for calculating the Btu values of gaseous fuels.

High and Low Heat Values. Water vapor is one of the products of combustion for all fuels that contain hydrogen. The heat content of a fuel depends on whether this water vapor is allowed to remain in the vapor state or is condensed to liquid. In the bomb calorimeter, the products of combustion are cooled to the initial temperature and all the water vapor formed during combustion is condensed to liquid. This gives the high, or gross, heat content of the fuel with the heat of vaporization included in the reported value. For the low, or net heat of combustion, it is assumed that all products of combustion remain in the gaseous state.

While the high, or gross, heat of combustion can be accurately determined by established (ASTM) procedures, direct determination of the low heat of combustion is difficult. Therefore, it is usually calculated by using

$$Q_L = Q_H - 1040W \qquad (2)$$

where Q_L = low heat of combustion of fuel, Btu/pound
$\quad Q_H$ = high heat of combustion of fuel, Btu/pound
$\quad W$ = pounds of water formed per pound of fuel
$\quad 1040$ = factor to reduce high heat of combustion at constant volume to low heat of combustion at constant pressure

In the United States, the practice is to use the high heat of combustion in boiler combustion calculations. In Europe, the low heat value is used.

IGNITION TEMPERATURES

Ignition temperature may be defined as the temperature at which more heat is generated by combustion than is lost to the surroundings so that the combustion process becomes self-sustaining. The term is usually applied to rapid combustion in air at atmospheric pressure.

Ignition temperatures of combustible substances vary greatly as shown in Table 4. This table lists minimum temperatures and temperature ranges in air for fuels and for the combustible constituents of fuels commonly used in the commercial generation of heat. Many factors influence ignition temperature so that any tabulations can be used only as a guide. Pressure, velocity, enclosure configuration, catalytic materials, air-fuel mixture uniformity, and ignition source are only some of the variables. Ignition temperature usually decreases with rising pressure and increases with increasing moisture content in the air.

The ignition temperatures of the gases of a coal vary considerably and are appreciably higher than the ignition temperatures of the fixed carbon of the coal. However, the ignition temperature of coal may be considered as the ignition temperature of its fixed carbon content, since the gaseous constituents are usually distilled off but not ignited before this temperature is attained.

ADIABATIC FLAME TEMPERATURE

The adiabatic flame temperature is the maximum theoretical temperature which can be reached by the products of combustion of a specific fuel and air (or oxygen) combination, assuming no loss of heat to the surroundings until combustion is complete. This theoreti-

TABLE 4.   Ignition Temperatures of Fuels in Air
(Approximate values and ranges at atmospheric pressure)

Combustible	Formula	Temperature °F	Temperature °C
Sulfur	S	470	243
Charcoal	C	650	343
Fixed carbon (bituminous coal)	C	765	407
Fixed carbon (semibituminous coal)	C	870	466
Fixed carbon (anthracite)	C	840–1115	449–602
Acetylene	C_2H_2	580–825	304–441
Ethane	C_2H_6	880–1165	471–630
Ethylene	C_2H_4	900–1020	482–549
Hydrogen	H_2	1065–1095	574–591
Methane	CH_4	1170–1380	632–749
Carbon monoxide	CO	1130–1215	610–657
Kerosine		490–560	254–293
Gasoline		500–800	260–427

cal temperature also assumes no dissociation, a phenomenon described shortly. The heat of combustion of the fuel is the major factor in flame temperature, but increasing the temperature of the air or of the fuel will also have the effect of raising the flame temperature. As may be expected, this adiabatic temperature is a maximum with zero excess air (only enough air chemically required to combine with the fuel, i.e., stoichiometric quantities), since any excess is not involved in the combustion process and only diminishes the temperature of the products of combustion.

The adiabatic temperature is determined from the adiabatic enthalpy of the flue gas:

$$h_g = \frac{\left(\begin{array}{c}\text{heat of}\\\text{combustion}\end{array}\right) + \left(\begin{array}{c}\text{sensible heat}\\\text{in fuel}\end{array}\right) + \left(\begin{array}{c}\text{sensible heat}\\\text{in air}\end{array}\right)}{\text{weight of products of combustion}}$$

where h_g = adiabatic enthalpy (adiabatic heat content of the products of combustion), Btu/pound

Knowing the moisture content of the products of combustion and its enthalpy, the theoretical flame or gas temperature can be obtained from Fig. 1.

The adiabatic temperature is a fictitiously high temperature that does not exist in fact. Actual flame temperatures are lower for two main reasons:

1. Combustion is not instantaneous. Some heat is lost to the surroundings as combustion takes place. The faster the combustion occurs, the less heat is lost before combustion is complete. If combustion is slow enough, the gases may be cooled sufficiently for combustion to be incomplete with some of the fuel unburned. This is related to the time factor in the three T's of combustion mentioned previously.

2. At temperatures above 3000°F (1649°C), some of the CO_2 and H_2O in the flue gases dissociates, absorbing heat in the process. At 3500°F (1926°C), about 10 percent of the CO_2 in a typical flue gas dissociates to CO and O_2 with a heat absorption of 4345 Btu/pound of CO formed, and about 3 percent of the H_2O dissociates to H_2 and O_2 with a heat absorption of 61,100 Btu/pound of H_2 formed. As the gas cools, the CO and H_2 dissociated recombine with the O_2 and liberate the heat absorbed in dissociation, so the heat is not lost. However, the effect is to lower the maximum actual flame temperature.

COMBUSTION CALCULATIONS

Combustion calculations are the starting point for all design and performance determinations for boilers and their related component parts. They establish (1) the quantities of the constituents involved in the chemistry of combustion, (2) the quantity of heat released, and (3) the efficiency of the combustion process under both ideal and actual conditions.

Combustion Air. Since carbon, hydrogen, and sulfur are the only combustible elements found in the common fuels used for commercial steam generation, the air (pounds) theoretically required for the complete combustion of 1 pound of fuel is

$$11.53C + 34.34(H_2 - O_2/8) + 4.29S \tag{3}$$

where C, H_2, O_2, and S represent the fraction by weight (percent/100) of carbon, hydrogen, oxygen, and sulfur, and the constants are those given in Table 1. The factor $O_2/8$ has been previously explained.

With gaseous fuels, instead of breaking down the hydrocarbons into their constituent elements, it is simpler to use the amount of air for the various compounds directly as given in Table 1. For instance, for a gaseous fuel containing the combustible gases indicated in the following expression, the theoretical air required for complete combustion (pounds air/pounds fuel) is

$$2.47CO + 34.34H_2 + 17.27CH_4 + 13.30C_2H_2 + 14.81C_2H_4 + 16.12C_2H_6 + 6.10H_2S \\ - 4.32O_2 \tag{4}$$

The molecular symbols represent the fraction by weight of the gaseous compounds and elements.

If, as is the usual custom, the analyses of gaseous fuels are given on a volumetric basis, the cubic feet of combustion air required, as given in Table 1, should be used. Thus, for a

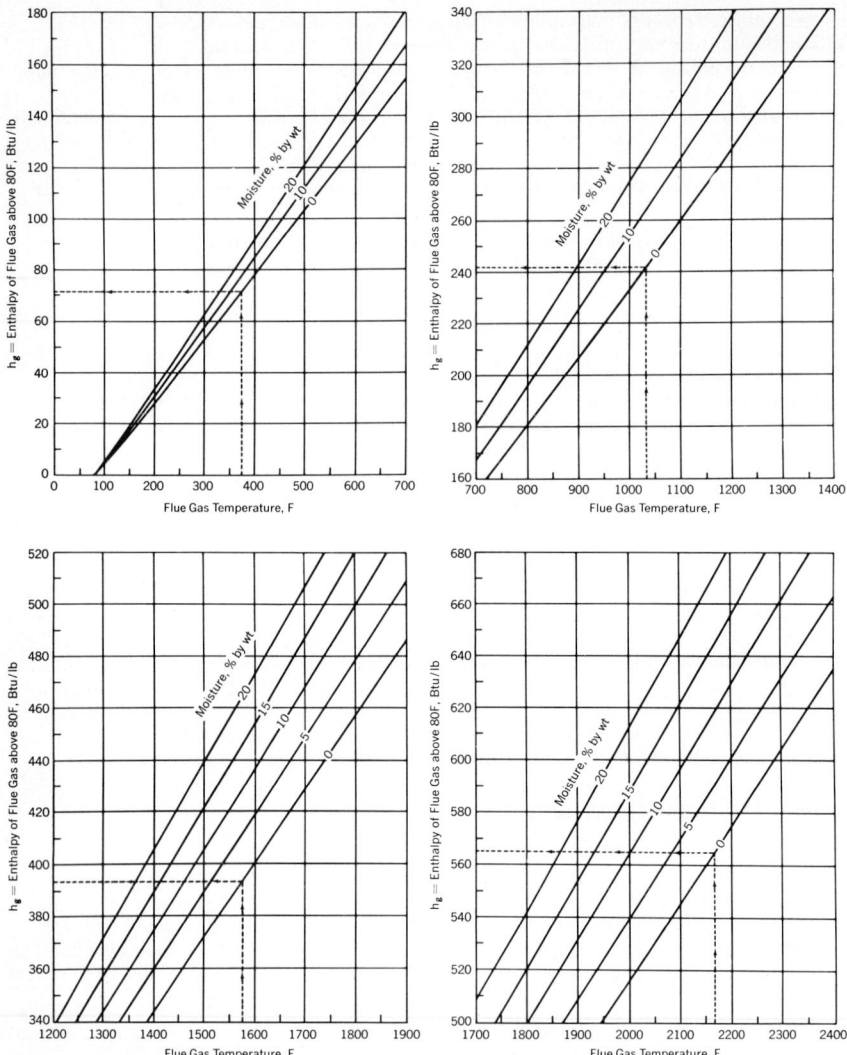

Fig. 1. Enthalpy of flue gases above 80°F at 30 inches of mercury, expressed in Btu per pound. Series of charts covers range of flue gas temperature from 0 to 4000°F. (*Babcock & Wilcox Company.*)

gaseous fuel containing the combustible gases indicated in the following expression, the cubic feet of theoretical air required per cubic foot of fuel for complete combustion is

$$2.38(CO + H_2) + 9.53CH_4 + 11.91C_2H_2 + 14.29C_2H_4 + 16.68C_2H_6 + 7.15H_2S \\ - 4.76O_2 \quad (5)$$

where the molecular symbols represent the fraction by volume of the gaseous compounds and elements. Note that the total air requirement is reduced if oxygen is one of the constituents of the fuel.

The products of combustion can also be determined from the data given in Table 1. Assuming complete combustion with theoretical air of the fuels ordinarily used for commercial steam generation, the products of combustion in pounds (including the nitrogen carried with the combustion air) per pound of fuel are

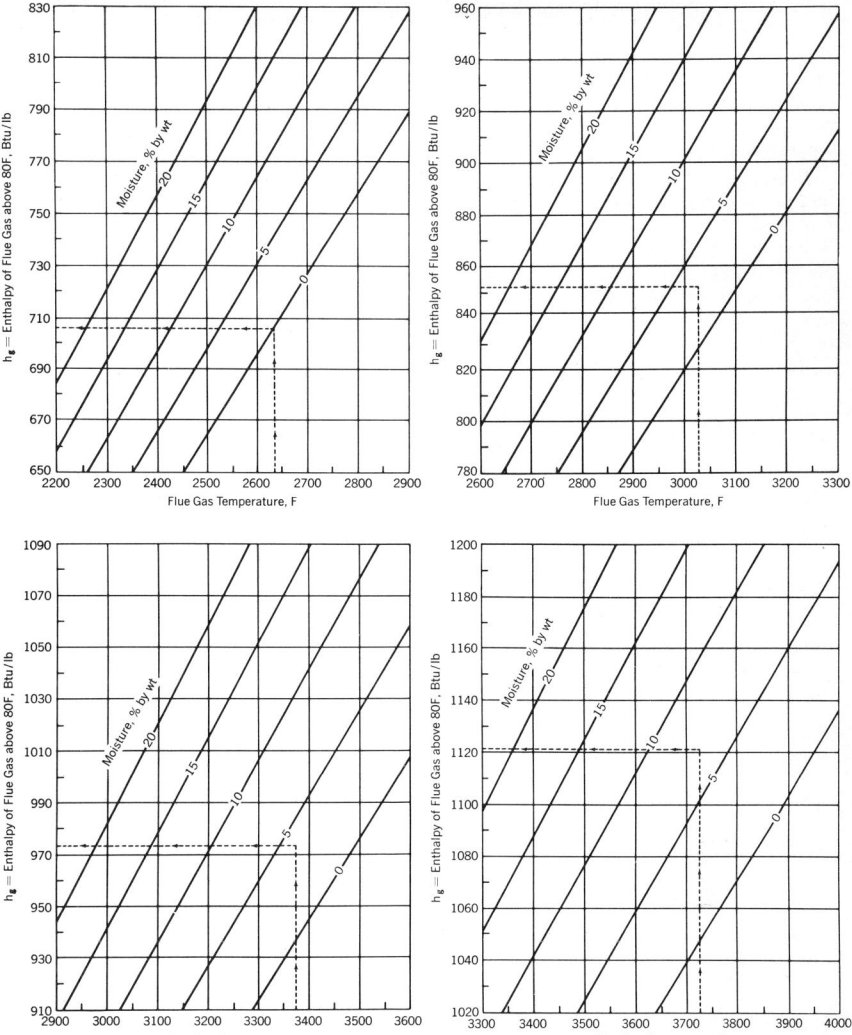

Fig. 1. (*continued*)

$$CO_2 = 3.66C$$
$$H_2O = 8.94H_2 + H_2O*$$
$$SO_2 = 2.00S$$
$$N_2 = 8.86C + 26.41(H_2 - O_2/8) + 3.29S + N_2\dagger$$

The moisture introduced with the combustion air must be added to this theoretical quantity to obtain the total weight of combustion products. The molecular symbols represent the fraction by weight of the constituents in the fuel.

Energy Losses.   Not all the Btu content of the fuel is converted to heat and absorbed by the steam generation equipment. Some of the fuel may be unburned, leaving carbon in

*Fraction by weight of H_2O (percent/100) in the fuel as moisture.
†Fraction by weight of N_2 in the fuel as nitrogen.

the ash, or carbon may be burned incompletely to form some CO instead of all CO_2. Usually all the H_2 in the fuel is burned. By far the greatest heat loss is the loss that goes up the stack. Since the heat in the fuel is determined from a base of ambient temperature, all the products of combustion must be cooled to the same temperature if all the heat is to be utilized. Higher temperatures then represent a loss, which is the sum of four items: (1) the sensible heat in dry flue gas, (2) the sensible heat in the moisture in the air, (3) the sensible heat in the H_2O in the fuel, and (4) the latent heat of the moisture in the fuel.

It is necessary from a practical standpoint to use more than the theoretical air requirements to assure sufficient oxygen for complete combustion. Excess air would not be required if it were possible to have an ideally perfect union of air and fuel. It is necessary, however, to keep the excess at a minimum in order to hold down stack losses. The excess air that is not used in the combustion of the fuel leaves the unit at stack temperature. The heat required to heat this air from room temperature to stack temperature serves no purpose and is lost heat. Table 5 gives realistic values of excess air for the fuel-burning equipment which experience has shown is required to assure complete combustion for various fuels and methods of firing. See also the article later in this handbook section on "NO_x Control by Furnace and Burner Design."

In most furnaces operating under suction there is also some leakage of air into the setting, and, consequently, the excess air leaving the furnace and the unit is greater than that at the fuel-burning equipment. Another loss which must be considered is the radiation loss from the unit setting.

In summary, there are certain inherent heat losses over which there is no control and others which are subject to some control. The inherent losses are the result of (1) the discharge of the products of combustion at a temperature higher than ambient and (2) the moisture content of the fuel plus the combination of some of the hydrogen with the

TABLE 5. Usual Amount of Excess Air Supplied to Fuel-Burning Equipment

Fuel	Type of furnace or burners	Excess air % by weight
Pulverized coal	Completely water-cooled furnace for slag-tap or dry-ash removal	15–20
	Partially water-cooled furnace for dry-ash removal	15–40
Crushed coal	Cyclone furnace—pressure or suction	10–15
Coal	Spreader stoker	30–60
	Water-cooled vibrating-grate stoker	30–60
	Chain-grate and traveling-grate stokers	15–50
	Underfeed stoker	20–50
Fuel oil	Oil burners, register-type	5–10
	Multifuel burners and flat-flame	10–20
Acid sludge	Cone and flat-flame type burners, steam-atomized	10–15
Natural, coke-oven, and refinery gas	Register-type burners	5–10
	Multifuel burners	7–12
Blast-furnace gas	Intertube nozzle-type burners	15–18
Wood	Dutch oven (10–23% through grates) and Hofft-type	20–25
Bagasse	All furnaces	25–35
Black liquor	Recovery furnaces for kraft and soda-pulping processes	5–7

oxygen in the fuel. The avoidable heat losses, or those which can be controlled by good design and careful operation, can be minimized by (1) careful control of excess air, (2) tolerating virtually no unburned solid combustible matter in ash or refuse, (3) permitting no unburned gaseous combustibles in the exit gases, and (4) a well-insulated setting for the steam-generating unit to reduce radiation loss.

The efficiency of combustion in a heat exchanger or boiler is 100 minus the sum of the heat losses expressed in percent.

Example of Combustion Calculation.   The detailed steps in the solution of combustion problems are best illustrated by an example. In the following example, the amount of moisture in the combustion air is taken as 0.013 pound per pound of dry air, corresponding to conditions of 80°F dry bulb temperature and 60% relative humidity. This value is often used as standard in combustion calculations.

Use of the weight method is illustrated. The values in the following example are generally developed from the data in Table 1, and the procedure followed is apparent from the execution.

Example: Weight Method–Solid Fuel–Standard Air*
(Complete combustion assumed) See Table 1 for multipliers

Ultimate analysis, pound/pound fuel as fired			Required for combustion, pound/pound fuel	
			O_2	Dry air
			@ 100% total air	
C	0.728	× 2.66 and × 11.53	1.936	8.394
H_2	0.048	× 7.94 and × 34.34	0.381	1.648
O_2	0.062			
N_2	0.015			
S	0.022	× 1.00 and × 4.29	0.022	0.094
H_2O	0.035			
Ash	0.090			
Sum	1.000		2.339	10.136
Less O_2 in fuel (deduct)			−0.062	−0.268†
Required (at 100% total air)			2.277	9.868
			@ 125% total air	
O_2 and air × 125/100, total			2.846	12.335
Excess air = 12.335 − 9.868				2.467
Excess O_2 = 2.846 − 2.277			0.569	

PRODUCTS OF COMBUSTION

			pound/pound fuel @ 125% total air
CO_2	0.728	× 3.66	2.664
H_2O	0.048	× 8.94 + 0.035	
		+ 0.013 × 12.335	0.624
SO_2	0.022	× 2.00	0.044
O_2	(excess)		0.569
N_2	12.335	× 0.7685 + 0.015	9.629
Weight, wet			13.530
Weight, dry = 13.530 − 0.624			12.906

*Air at 60% relative humidity and 80°F dry bulb, or 0.013 pound of moisture per pound dry air.
†Air equivalent of O_2 in fuel.

FLUE GAS ANALYSIS

In the continuously recording or indicating flue gas analyzer, a gas is continuously drawn from a selected location, and samples are analyzed at intervals of 1 minute or longer. Both the analysis and the indication or recording of the results are commonly automatic. The analysis may or may not include all the constituents of the products of combustion. Many continuously recording instruments give the percentage of CO_2 only; others only give the

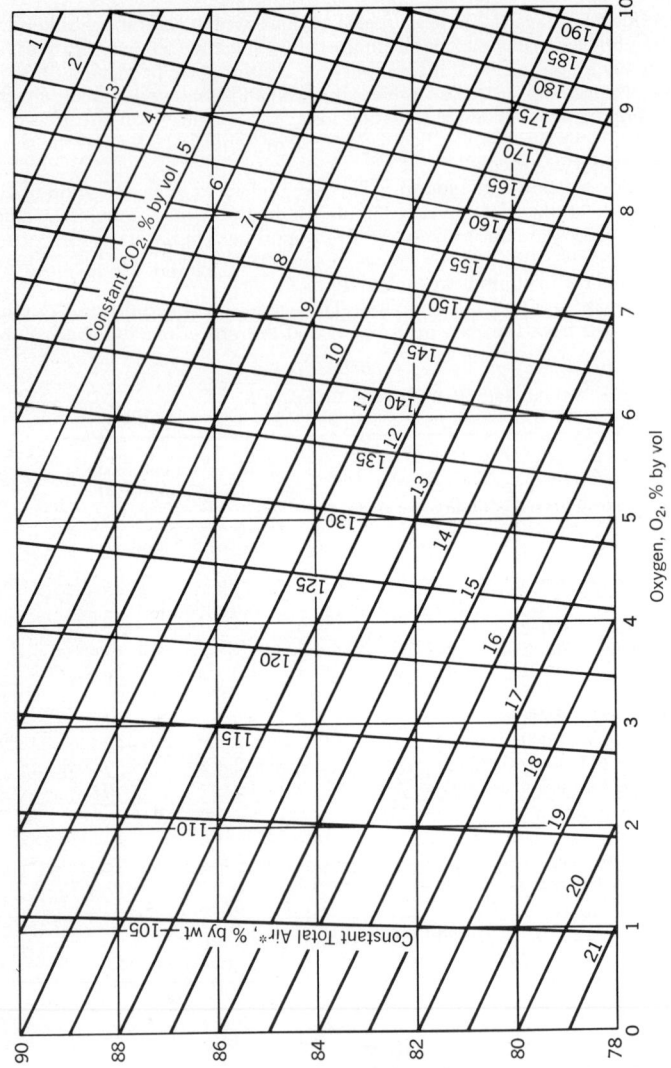

Fig. 2. Chart for approximating total air from flue gas analysis.

*Total air = 100 + % excess air by wt

percentage of O_2; still others indicate the percent of unburned gases. Unburned gases, however, should not be present with proper combustion in the field of heat production.

The amount of O_2 in the flue gases is significant in defining the status of the combustion process. Its presence always means that more oxygen (excess air) is being introduced than is being used. Assuming complete combustion, low values of O_2 in the flue gases reflect moderate (nearly correct) excess air and reduced heat losses to the stack, while higher values of O_2 mean needlessly higher stack losses. The quantitative determination of total air (total air = 100 percent excess air) admitted to an actual combustion process requires a complete flue gas analysis for CO_2, O_2, CO, and N_2 (by difference) or the direct measurement of the air supplied by a suitable fluid meter.

The approximate percent total air from flue gas analyses may be determined from the curves of Fig. 2. A formula that has long been used for approximating the percent excess air from an Orsat analysis is

$$\% \text{ excess air} = 100 \times \frac{O_2 - CO/2}{0.264 N_2 - (O_2 - CO/2)} \tag{6}$$

The results are reasonably close for the flue gases from hydrocarbon fuels of low nitrogen content. The error increases with increase in nitrogen content in the fuel.

Flue Gas Sampling.   Great care should be taken to secure truly representative samples of the gas for analysis. The usual practice is to take successive samples from a number of points laid out in checkerboard fashion over a cross section of the flue or area traversed by the gas. Gas samples should be drawn at regular intervals over a relatively long period (during the entire period of a formal test).

When the temperatures exceed 900°F (482°C), gas samples should be drawn through a water-cooled pipe to avoid the loss of O_2 from oxidation of pipe material at such temperature. Uncooled pipes of special steels have been used with fair success, but there is still the uncertainty of the loss of some O_2 to the metal. A properly designed water-cooled sampler can be safely used to obtain gas samples from the hottest regions of a furnace (3000°F; 1649°C and above).

Metering Combustion Air.   Metering the flow of air or the flow of combustion gases can be done to establish the amount of total air used in a boiler or other similar combustion heat exchanger. For a given total airflow, the flow so measured has a nearly straight-line relationship to the Btu input. In a steam boiler, the steam output (and, consequently, the Btu output) bears a nearly constant relation to the Btu input. Metered steam is, therefore, a suitable index of Btu input, and the operator may thus proportion the combustion air to the fuel at any rate of steam output. See also "Controls for Fossil Fuel/ Steam Energy Conversion." later in this Handbook section.

REFERENCES

"Steam: Its Generation and Use," 38th ed., Babcock & Wilcox Company, New York, 1972.

Beer, J., and N. Chigier: "Combustion Aerodynamics (Fuel & Energy Science Monographs Series)," Wiley, New York, 1972.

Monroe, E. S., Jr.: Heat Generation, in R. H. Perry and C. H. Chilton (eds.), "Chemical Engineers' Handbook," 5th ed., sec. 9, McGraw-Hill, New York, 1973.

Hawkins, G. A.: Combustion, in T. Baumeister (ed.), "Marks' Standard Handbook for Mechanical Engineers," 7th ed., McGraw-Hill, New York, 1967.

von Elbe, Lewis: "Combustion, Flames and Explosions of Gases," Academic, New York, 1961.

Oil and Gas Burners*

EFFECTIVENESS OF BURNERS

The burner is the principal equipment component for the firing of oil and gas. Burners are normally located in the vertical walls of the furnace. The burners introduce the fuel and air into the furnace to sustain the exothermic chemical reactions for the most effective release of heat. That effectiveness is judged by the following factors:

1. The rate of feed of the fuel and air shall comply with the load demand on the boiler over a predetermined operating range.

2. The efficiency of the combustion process shall be as high as possible with the minimum of unburned combustibles and minimum excess air in the products.

3. The physical size and complexity of the furnace and burners shall be as small as possible to minimize the required investment and to meet the limitations on space, weight, and flexibility imposed by the service conditions.

4. The design of the burners, including the materials used, shall provide reliable operation under specified service conditions, and shall assure meeting accepted standards of maintenance for the burners and the furnaces in which they are installed.

5. Safety shall be paramount under all conditions of operation of burners, furnace, and boiler, including starting, stopping, load changes, and variations in the fuel.

The normal use of a steam generator requires operation at different outputs to meet varying load demands. The specified operating range or *load range* for a burner is the ratio of full load on the burner to the minimum load at which the burner must be capable of reliable operation. For example, with a boiler of 100,000 pounds/hour capacity (steam delivered), a load range of 4:1 on the burners means that the unit can be operated from 100,000 pounds/hour down to 25,000 pounds/hour without changing the number of burners in operation and with complete combustion.

Combustion air is generally delivered to the burners by fans. It is necessary to supply more than the theoretical air quantity to assure complete combustion of the fuel in the combustion chamber (furnace). The amount of excess air provided should be just enough to burn the fuel completely in order to minimize the sensible heat loss in the stack gases. See also prior article on "Principles of Combustion in Steam Generation" and subsequent article on "NO_x Control by Furnace and Burner Design" in this Handbook section.

Continuity of service is enhanced by designing the furnace and arranging the burners to minimize slagging and fouling of heat-absorbing surfaces for the normal range of fuels burned.

Maintenance costs of the burner are minimized by (1) the least exposure to furnace heat and (2) provision for replacement or repair of vulnerable parts while the unit continues in operation.

CONTEMPORARY BURNER TYPES

The most frequently used burners are the *circular burner* and the *cell burner*. Figure 1 shows a single circular register burner for oil firing. Figure 2 shows a cell burner for natural gas firing. Burners of this type are also used for pulverized coal or for firing any one of these three principal fuels singly or in combination. Coal-burning systems are described in the next article in this Handbook section.

The maximum capability of the individual circular burner ranges up to 165 million Btu/hour. Cell burners have maximum capability up to 495 million Btu/hour. In both circular and cell burners, the tangentially disposed "doors" built into the air register provide the turbulence necessary to mix the fuel and air and produce short, compact flames.

While the fuel is introduced to the burner in a fairly dense mixture in the center, the

*The principal portion of this description is extracted with permission from "Steam: Its Generation and Use," copyrighted by The Babcock & Wilcox Company.

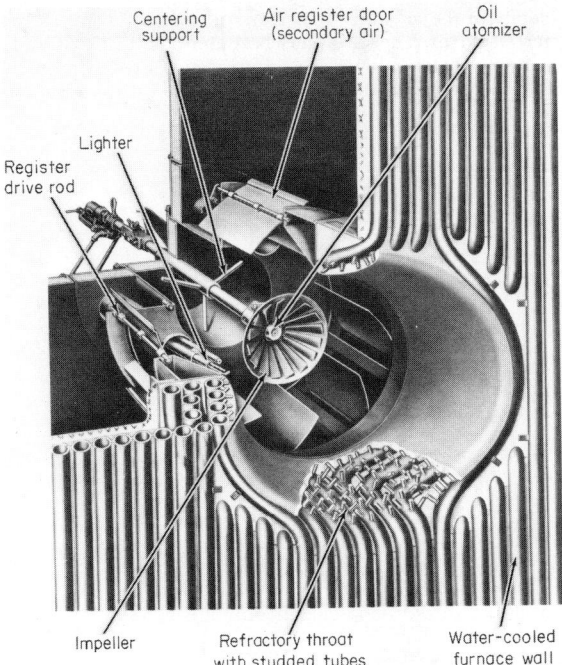

Centering support · Air register door (secondary air) · Oil atomizer · Lighter · Register drive rod · Impeller · Refractory throat with studded tubes · Water-cooled furnace wall

Fig. 1. Circular register burner with water-cooled throat for oil firing. *(Babcock & Wilcox Company.)*

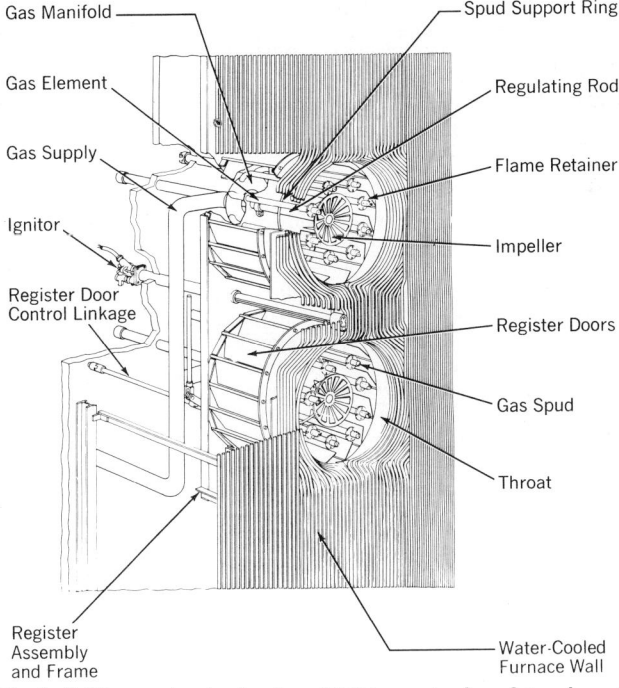

Gas Manifold · Spud Support Ring · Gas Element · Regulating Rod · Gas Supply · Flame Retainer · Ignitor · Impeller · Register Door Control Linkage · Register Doors · Gas Spud · Throat · Register Assembly and Frame · Water-Cooled Furnace Wall

Fig. 2. Cell burner showing location of C-B type natural-gas-firing elements.

direction and velocity of the air, plus dispersion of the fuel, completely and thoroughly mixes it with combustion air. These burners best meet the five requirements previously listed. The only burner part subject to serious maintenance is the impeller, which promotes mixing of the fuel with the secondary air. Normally, an impeller should last a year or more unless it is accidentally damaged.

OIL BURNERS

Atomization. To burn fuel oil at the high rates demanded of modern boiler units, the oil must be "atomized," i.e., dispersed into the furnace as a fine mist, somewhat like a heavy fog. This exposes a large amount of oil particle surface for contact with the combustion air to assure prompt ignition and rapid combustion. There are many ways of atomizing fuel oil, but the description here is limited to the two most popular ways: (1) steam or air, and (2) mechanical atomizers.

For proper atomization, oil of grades heavier than No. 2 must be heated to reduce the viscosity to 135–150 Saybolt Seconds Universal. Steam or electric heaters are required to raise the oil temperature to the required degree, i.e., approximately to:

135°F (57°C)	No. 4 oil
185°F (85°C)	No. 5 oil
200–220°F (93–104°C)	No. 6 oil

With certain oils, better combustion is obtained at somewhat higher temperatures than required for atomization. However, oil temperature must not be raised to the point where vapor binding occurs in the pump supplying the oil, since this could cause flow interruptions, followed by loss of ignition. It is also important, of course, that oil be free from acid, grit, and other foreign matter to avoid clogging or damage to burners and control valves. Fuel oil properties are described in the article on "Petroleum Fuels Characteristics" in the Handbook section on *Petroleum Technology.*

Steam or Air Atomizers. The steam (or air) atomizer (Fig. 3) is the most widely used of burners. In general, it operates on the principle of producing a steam-fuel (or air-fuel) emulsion which, when released into the furnace, atomizes the oil through the rapid expansion of the steam. The atomizing steam must be dry because moisture causes pulsations which can lead to loss of ignition. Where steam is not available, moisture-free compressed air can be substituted.

Steam atomizers are available in sizes up to 165 million Btu/hour input—about 9200 pounds of oil per hour. Oil pressure is much lower than for mechanical atomizers. The steam and oil pressure required are dependent on the design of the steam atomizer. Maximum oil pressure is usually about 100 psi; steam pressure is generally about 20 to 40 psi higher than oil pressure.

The steam atomizer performs more efficiently over a wider load range than other types. It normally atomizes the fuel properly down to 20 percent of rated capacity, and in some instances steam atomizers have been successfully operated at 5 percent capacity. Frequently, these extremes in range cannot be fully utilized because the temperature of the combustion space falls off to such a degree that, despite the excellent quality of atomization, there is not sufficient temperature to complete the combustion process adequately.

A disadvantage of the steam atomizer is its consumption of steam. A good steam atomizer uses not over 0.1 pound of steam per pound of fuel oil at its maximum capacity. For a large unit over a long period of time, this amounts to a substantial quantity of steam and consequent heat loss to the stack. Where the boiler unit supplies a substantial amount of steam for a process where condensate recovery is small, the additional makeup for the steam atomizer is inconsequential. However, in a large utility boiler where turbine losses are low and where there is very little makeup, the use of atomizing steam can have an important effect on the total makeup requirements. The cost of employing a compressed air supply as a substitute for steam may or may not be justified, depending upon an overall study of steam system costs.

Mechanical Atomizers. In the mechanical atomizer, the pressure of the fuel itself is used as the means for atomization. Many forms have been developed. Those with moving parts close to the furnace have lost favor because of the excessive maintenance required to keep them operational.

The return-flow atomizer (Fig. 4) is used for many marine installations and some

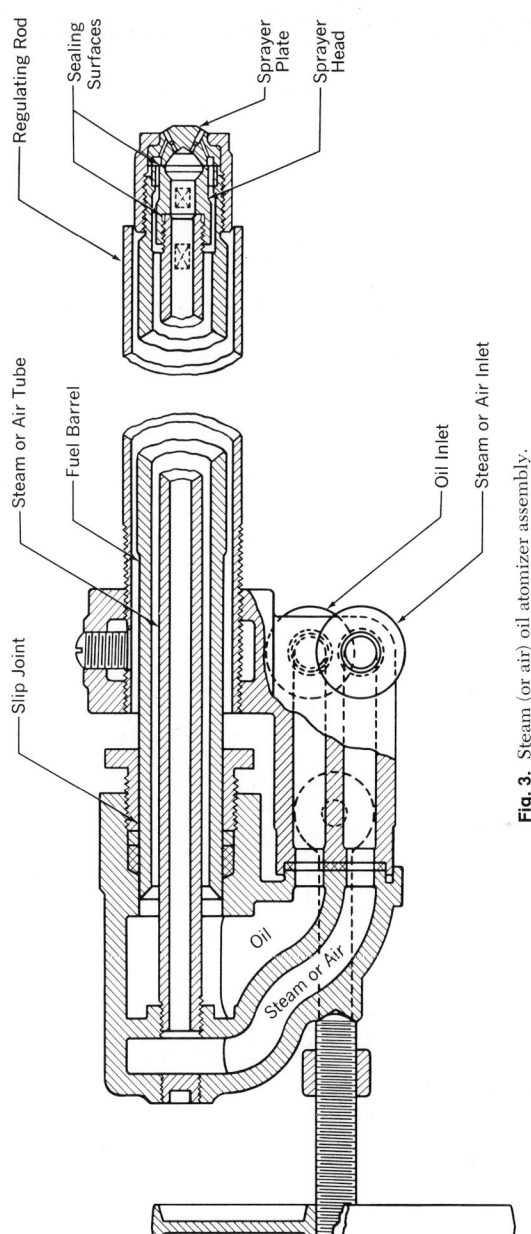

Regulating Rod

Sealing Surfaces

Sprayer Plate

Sprayer Head

Steam or Air Tube

Fuel Barrel

Oil Inlet

Steam or Air Inlet

Slip Joint

Oil

Steam or Air

Fig. 3. Steam (or air) oil atomizer assembly.

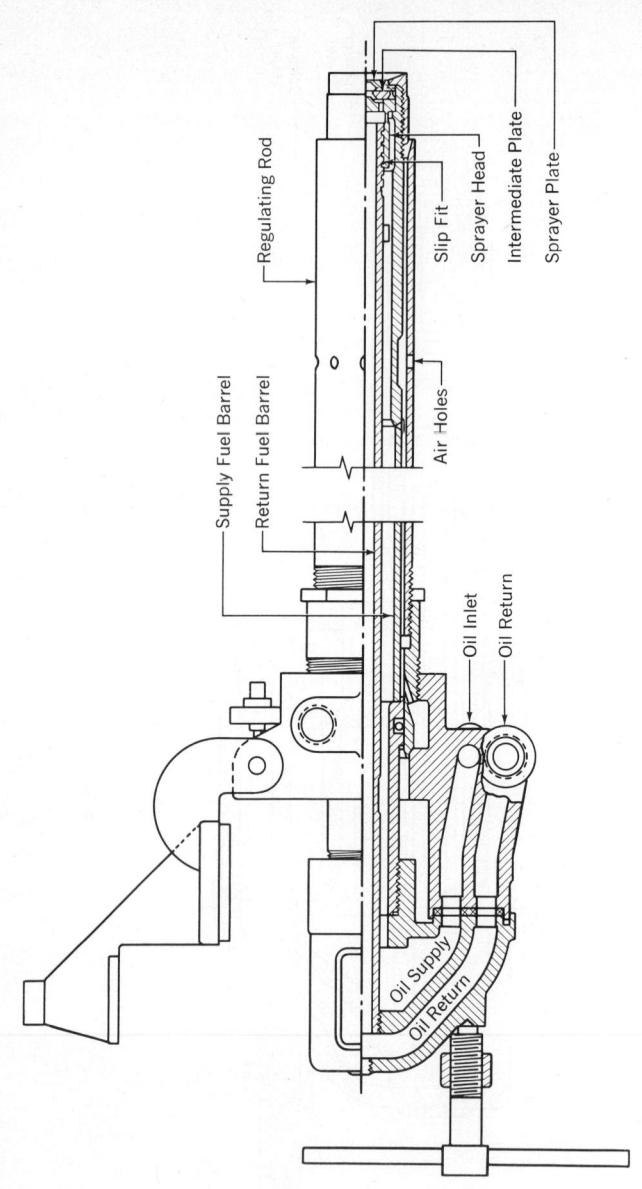

Fig. 4. Mechanical return-flow oil atomizer assembly.

Regulating Rod

Slip Fit
Sprayer Head
Intermediate Plate
Sprayer Plate

Supply Fuel Barrel
Return Fuel Barrel

Air Holes

Oil Inlet
Oil Return

Oil Supply
Oil Return

stationary units where the use of atomizing steam is objectionable or impractical. The oil pressure required at the atomizer for maximum capacity ranges from 600 to 1000 psi, depending on capacity, load range, and fuel. The fuel flows through tangentially disposed slots in a "sprayer plate" (Fig. 5) into a "whirl" chamber, from which it issues through an orifice of the sprayer plate as a fine conical mist or spray. After the oil passes through the

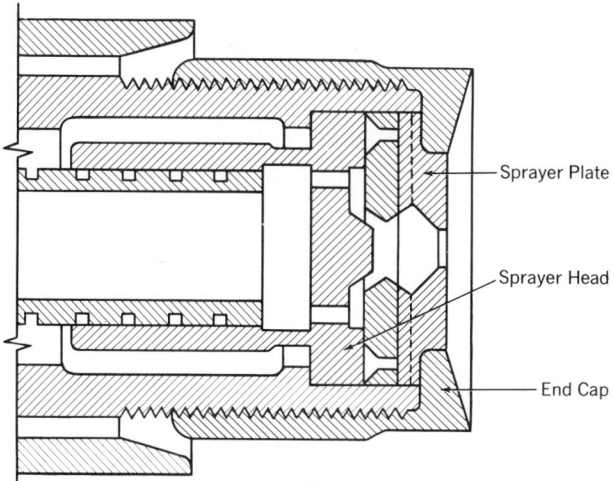

Fig. 5. Mechanical return-flow oil atomizer detail at furnace end of atomizer assembly, showing sprayer head, sprayer plate, and end cap.

tangentially disposed slots, that which exceeds the boiler input requirements is returned from the base of the whirl chamber to the fuel oil system.

Mechanical atomizers are available in sizes up to 180 million Btu/hour input—about 10,000 pounds of oil per hour. The acceptable operating range may be as much as 10:1 or as little as 3:1, depending on the maximum oil pressure used for the system, the furnace configuration, air temperature, and burner throat velocity. Return-flow atomizers are ideally suited for standard grades of fuel oil where it is desired to meet load variations without changing sprayer plates or cutting burners in and out of service.

For good performance over an operating range of 10:1, the combustion chamber should be of relatively small cross section at the burner zone, the oil pressure at the burners for full load should be 1000 psi, the temperature of air for combustion should be significantly above ambient throughout the load range, and the air resistance across the burner should be 8 to 12 inches of water at full load, depending upon the size of the particular burner. Departures downward from any of these values markedly affect the satisfactory load range obtainable from the burner. In a properly designed and operated system, the high-pressure return-flow mechanical atomizer will provide combustion efficiency comparable to that obtainable with a good steam atomizer.

NATURAL GAS BURNERS

The C-B type gas element (Fig. 2) was developed for the cell burner. Distinctive features of this gas element include:

1. A large part of the gas discharges in front of the impeller, which serves as an ignition stabilizer at high loads.

2. Each gas burner comprises several "spuds," each of which is a gas pipe with multiple holes at the end to discharge gas for ignition at the burner throat. Each spud is fitted with a 4-inch-diameter flame holder to stabilize ignition at low inputs.

3. The fuel ports are relatively large to minimize plugging in service.

4. The gas spuds are located so that oil will not impinge on them when provision is made for firing this fuel.

Because of its unusually good ignition stability, even under severe variations in airflow, the C-B type element is expected eventually to replace most natural gas element designs now used circular-type burners.

With the proper selection of control equipment, a multifuel-fired burner with a C-B type gas element is capable of changing from one fuel to another without a drop in load or boiler pressure. Simultaneous firing of natural gas and oil in the same burner is acceptable on burners equipped with C-B type gas elements.

This type element is designed for use with natural gas or other gaseous fuels containing by volume, at least

70% methane (CH_4) or
70% propane (C_3H_8) or
25% hydrogen (H_2)

This element is designed for a maximum input of 173 million Btu/(hour) (burner). In certain respects, natural gas is an ideal fuel for a burner since it requires no preparation to be suitable for rapid and intimate mixing with the combustion air flowing through the burner throat. However, this characteristic of easy ignition, under most operating conditions, has (in some cases) led to operator carelessness with resulting damaging explosions.

To provide safe operation, ignition of a gas burner should remain close to the burner wall throughout the full range of allowable gas pressures, not only with normal airflows, but also with much more airflow through the burner than is theoretically required. Ideally, it should be possible at the minimum load to pass full-load airflow through the burner, and at full load as much as 25 percent in excess of theoretical air without loss of ignition. With this latitude in airflow, it is not likely that ignition can be lost, even momentarily, during some upset in airflow due to improper operation or error.

BURNERS FOR OTHER GASES

Many industrial applications utilize coke oven gas, blast furnace gas, refinery gas, or other industrial by-product gases. With these gases, the heat release per unit volume of fuel gas may be very different from that of natural gas. Hence, gas elements must be designed to accommodate the particular characteristics of the gas to be burned. Also, burners must be designed with reference to ignition stability and load-range factors which govern in each case. Other special problems may be introduced by the presence of impurities in industrial gases, such as sulfur in coke oven gas, and entrained dust in blast furnace gas. As substitute natural gas (SNG) becomes more widely available as the result of coal gasification and other processes, careful attention to burner design and use becomes paramount.

LIGHTERS (IGNITORS) AND PILOTS

Equipment is available, for boiler units ranging from the smallest to the largest, that allows the boiler operator to ignite the main fuel by simply pressing a button. This equipment ranges from spark devices that ignite fuel oil directly, to gas or light oil equipment, in itself spark-ignited, which is used for ignition of the main streams of gas and fuel oil. These devices are available with control equipment that ranges from the simplest push button requiring observation of ignition by the operator at the burner, to a fully programmed starting sequence, complete with interlocks and flame-sensing equipment, all remotely operated from the boiler control room. More details are presented in the article on "Controls for Fossil Fuel/Steam Energy Conversion" later in this Handbook section.

Usually the ignition device is energized only long enough to assure that the main flame is self-sustaining. With the fuel normally used in oil or natural gas burners, ignition should be self-sustaining within 1 or 2 seconds after the fuel reaches the combustion air. On a fully automated burner, it is customary to allow 10 to 15 seconds "trial for ignition" so that the fuel can reach the burner after the fuel shutoff valve on the burner is operated.

In general, lighters should be used sparingly. Where special fuel is used for the lighters, it is usually more expensive than the fuel used in the regular burners, and thus unnecessary use adds to operating costs. However, there are applications where a continuously burning lighter or pilot is needed. This is particularly true in the use of a by-product fuel, such as gas from a chemical process. In most installations, such fuels are piped directly to the boiler house without an intervening accumulator. The quantity and quality of fuel supplied to the boiler house are subject to any malfunction in the chemical process. Supply pressure may vary beyond the range of main burner stability, or the combustible content of the gas may change. Such variations require continuous operation of a pilot. It may be necessary also to provide supplementary fuel to each burner reliably and continuously to keep the process gas burning during abnormal conditions in the process. In these instances, the boiler operators have an awareness of the vagaries of the process fuel that does not exist in the operation of conventional oil- and natural-gas-fired boilers. As a consequence, they know the safety of the boiler operation is highly dependent upon the reliability of the ignitor and are alert to any malfunction that may occur.

EXCESS AIR

When firing oil or natural gas at high loads, there normally is no fuel distribution problem, but distribution of combustion air in the burner wind box is not perfect, and this dictates the use of some excess air to assure complete combustion from all burners. On most units, it is possible to operate with as little as 5 to 7 percent excess air at the furnace outlet at full load, and with extra care in design and operation, some large utility-type boilers have operated with as little as 2.5 percent excess air without excessive unburned combustible loss.

At partial loads on all units regardless of the particular fuel fired, it is necessary to increase the excess air as the load is reduced. If all burners are kept in service, the furnace temperature and the air velocity, which provides burner turbulence, decrease with decrease in load. Raising the excess air tends to compensate for the deterioration of these items as the load drops. Also, it may be desirable to operate with higher-than-normal excess air to maintain steam temperature or minimize corrosion of the "cold" end of the unit. Increased excess air is desirable even when the number of burners in use is reduced at lower loads. Burner dampers are designed not to close tightly to permit the air to protect the idle burner(s) from overheating by radiant heat from nearby operating burners.

BURNER PULSATION

One of the most mystifying problems associated with gas burners and, to a much lesser degree, with oil burners is that of burner pulsation. It appears to result from certain combinations of combustion chamber size and configuration, coupled with some characteristic of the burners, perhaps "too perfect" mixing of fuel and air at the burner. When one or more burners on a large unit starts to pulsate, it may become alarmingly violent, at times shaking the whole boiler. Making an adjustment of only one burner may start or stop pulsation. At times, only minor burner adjustments eliminate the pulsation. In other instances, it is necessary to alter the burners. This may involve modifying the gas ports, impinging gas streams on one another, or using some other device that effectively alters the mixing of the gas with the air.

To avoid pulsations, burner and boiler manufacturers incorporate the latest information available into the designs of burners and furnaces. However, a better understanding of this problem remains to be developed.

OPERATING RANGE OR LOAD RANGE

Various claims have been made over the years for the operating range or load range characteristics of specific burners. Erroneous interpretation of test results has led to performance predictions which could not be substantiated. There have been misunderstandings between suppliers and users because of interpretation of results. The most persistent differences in opinion are in the acceptability of flame appearance, the length of time the unit is operated at minimum load, and the evaluation of potential hazard from deposits that may occur during the low-load tests. The following considerations are basic to understanding the difficulties arising from low-load operation.

1. Large boilers with air heaters have a sizable "heat inertia" that requires more than 2 hours at full load for temperatures throughout the unit to stabilize. Conversely, a significant time period is necessary for them to retrograde on dropping load.

2. A brief drop in load, e.g., for an hour or two, may cause no problems with ignition

stability or the deposit of combustible products. However, if the low load is held for a longer period, 4 hours or more, operation may become intolerable toward the end of the period because of a drop in temperature of the furnace and the combustion air. In rare instances, a load drop for less than an hour may cause trouble. Some boiler installations require carrying low loads consistently over weekends. All such service demands must be taken into account in the design and selection of burners.

REFERENCES

"Steam: Its Generation and Use," 38th ed., Babcock & Wilcox Company, New York, 1972.

Smith, R. B.: Heat Generation, in R. H. Perry and C. H. Chilton (eds.), "Chemical Engineers' Handbook," 5th ed., McGraw-Hill, New York, 1973.

Kessler, G. W.: Steam Boilers, in T. Baumeister (ed.), "Marks' Standard Handbook for Mechanical Engineers," 7th ed., McGraw-Hill, New York, 1967.

Burkhardt, C. H.: "Domestic and Commercial Oil Burners," 3d. ed., McGraw-Hill, New York, 1969.

Coal-Burning Systems*

Consumption of bituminous coal and lignite in the United States approximates 500 million tons annually. More than three-fourths of this tonnage is used to generate steam for the electrical and other industries. A high percentage of the coal used for the generation of steam is burned in pulverized form.

SELECTION OF COAL-BURNING EQUIPMENT

Equipment selection for a particular installation consists of balancing the investment, operating characteristics, efficiency, and type of coal to be used with the objective of achieving the most economical installation. Almost any coal can be burned successfully in pulverized form or on some type of stoker. The capacity limitations imposed by stokers have been overcome by the development of pulverized-coal and cyclone-furnace firing. These improved methods also provide (1) ability to use any size of coal available, (2) improved response to load changes, (3) increase in thermal efficiency because of lower excess air for combustion and lower carbon loss than with stoker firing, (4) a reduction in personnel required for operation, and (5) improved ability to burn coal in combination with oil and gas.

Experience shows that stoker firing is more economical for steam-generating units of capacity less than 100,000 pounds of steam per hour, where the lower efficiency of a stoker can be tolerated. In larger plants, where fuel cost is a larger fraction of the operating cost, pulverized-coal or cyclone-furnace firing is more economical except in special cases. Operating characteristics may be of controlling significance in the choice of firing method. For example, where unit size is suitable for pulverized-coal, cyclone-furnace, or stoker firing, the extremely wide load range of stoker firing may make it preferable. Where rotary kilns and industrial furnaces are fired by coal, pulverized-coal firing is generally used.

The type of coal also influences the choice of the method of firing for boiler furnaces. Primary considerations include:
1. *Pulverized-coal firing:* grindability, rank, moisture, volatile matter, and ash
2. *Stoker firing:* rank of coal, volatile matter, ash, and ash-softening temperature
3. *Cyclone-furnace firing:* volatile matter, ash, and ash viscosity

Convenient approximations for the selection of bituminous coals for the firing of boilers are given in Table 1.

TABLE 1.   Coal Characteristics and Method of Firing

Characteristic	Stoker	Pulverized-coal	Cyclone-furnace
Maximum total moisture (as fired), %*	15–20	15	20
Minimum volatile matter (dry basis), %	15	15	15
Maximum total ash (dry basis), %	20	20	25
Maximum sulfur (as fired), %	5		

*These limits may be exceeded for lower rank, higher inherent-moisture-content coals, i.e., subbituminous and lignite.

PULVERIZED-COAL SYSTEMS

A pulverized-coal system pulverizes the coal, delivers it to the fuel-burning equipment, and accomplishes complete combustion in the furnace with a minimum of excess air. The system must operate as a continuous process, and, within specified design limitations, the coal supply or feed must be varied as rapidly and as widely as required by the combustion

*The principal portion of this description is extracted with permission from "Steam: Its Generation and Use," copyrighted by The Babcock & Wilcox Company.

process. A small portion of the air required for combustion (15 to 20 percent in contemporary installations) is used to transport the coal to the burner. This is known as *primary air*. In the direct-firing system, primary air is also used to dry the coal in the pulverizer. The remainder of the combustion air (80 to 85 percent) is introduced at the burner and is known as *secondary air*.

Basic Components.   The two basic equipment components of a pulverized-coal system are:

1. The pulverizer, which pulverizes the coal to the fineness required
2. The burner, which accomplishes the mixing of the pulverized-coal–primary-air mixture with secondary air in the right proportions and delivers the mixture to the furnace for combustion

Other necessary requirements are:

3. Hot air for drying the coal for effective pulverization
4. Fan(s) to supply air to the pulverizer and deliver the coal-air mixture to the burner(s)
5. Coal feeder to control the rate of coal feed to each pulverizer
6. Coal and air conveying lines

Two principal systems—the *bin system* and the *direct-firing system*—have been used for processing, distributing, and burning pulverized coal. The direct-firing system is being installed almost exclusively in current projects.

Bin System.   This system is primarily of historical interest, although a large number of units of this type remain in operation. The system was used before pulverizing equipment had reached the stage of development where it could be relied upon for uninterrupted operation, flexibility, and consistent performance.

In the bin system, the coal is processed at a location apart from the furnace, and the end product is pneumatically conveyed to cyclone collectors which recover the fines and clean the moisture-laden air before returning it to the atmosphere. The pulverized coal is discharged into storage bins and later conveyed by pneumatic transport through pipelines to utilization bins which may be as far as 5000 feet from the point of preparation. A bin system is shown in Fig. 1, a transport system in Fig. 2.

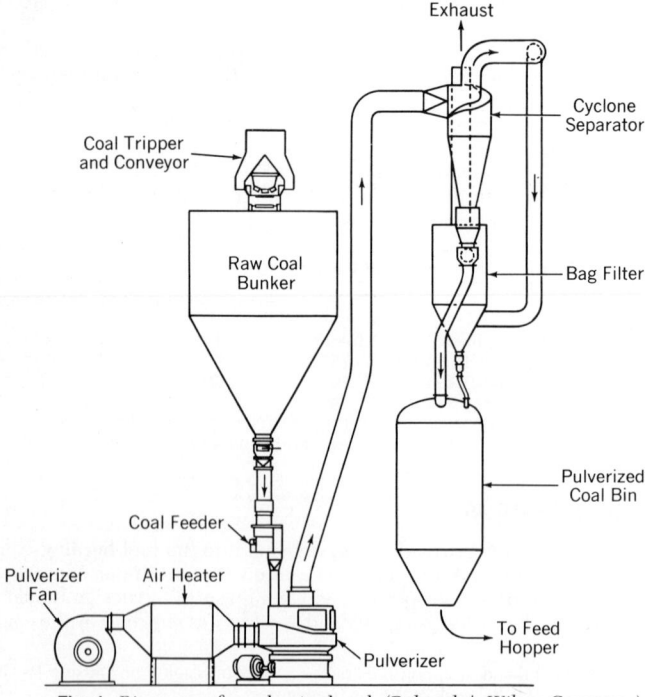

Fig. 1. Bin system for pulverized coal. (*Babcock & Wilcox Company.*)

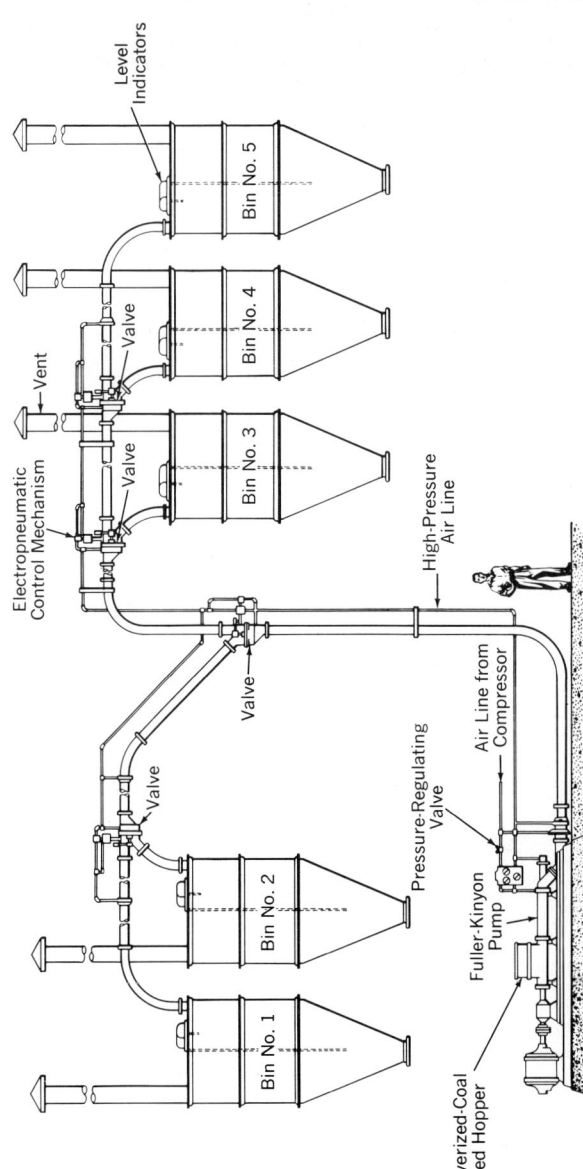

Fig. 2. Pneumatic transport system for conveying pulverized coal. Capacity, 1 to 100 tons per hour.

For the coal-air transport system, a differential-pitch screw pump is provided to feed pulverized coal continuously into a pipeline, where coal is aerated or fluidized at the entrance so that it flows through the pipe somewhat like a viscous fluid. Through a system of two-way valves (Fig. 2), the coal may be distributed to any number of bins. The system may be arranged for manual, remote, or automatic control. These air-transport systems are built in sizes from 1 to 100 tons of coal per hour. For successful operation, the surface moisture in the pulverized coal must not exceed 3 percent, and the fineness should not be less than 90 percent through a 50-mesh sieve. Although bin systems installed in older plants are still operating successfully, this system is no longer competitive with the direct-firing system. Furthermore, the drying, transportation, and storage of pulverized coal, other than anthracite, involves a fire hazard from spontaneous combustion.

Direct-Firing System.   This system is characterized by much greater safety, plant cleanliness, greater simplicity, lower initial investment, lower operating cost, and less space requirement. The pulverizing equipment developed for the direct-firing system permits continuous utilization of raw coal directly from the bunkers where coal is stored in the condition in which it is received at the plant. This is accomplished by feeding the raw coal directly into the pulverizer, where it is dried as well as pulverized, and then delivering it to the burners in a single continuous operation.

Components of the direct-firing system (Figs. 3 and 4) include:

1. Raw-coal feeder
2. Source (steam or gas air heater) to supply hot primary air to the pulverizer for drying the coal
3. Pulverizer fan, also known as the primary-air fan, arranged as a blower (or exhauster)
4. Pulverizer arranged to operate under pressure (or suction)
5. Coal-and-air conveying lines
6. Burners

Two direct-firing methods are in use: (1) the pressure type, which is the more commonly used, and (2) the suction type. The principal differences are summarized in Table 2.

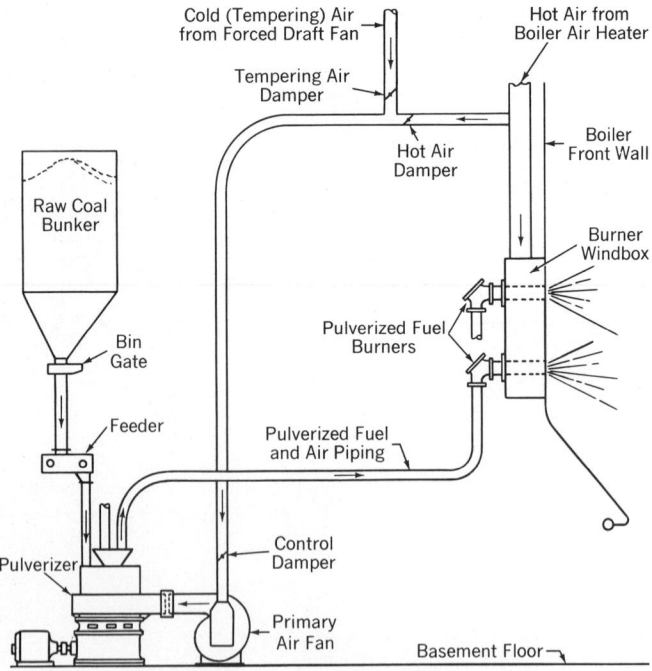

Fig. 3. Direct-firing system for pulverized coal.

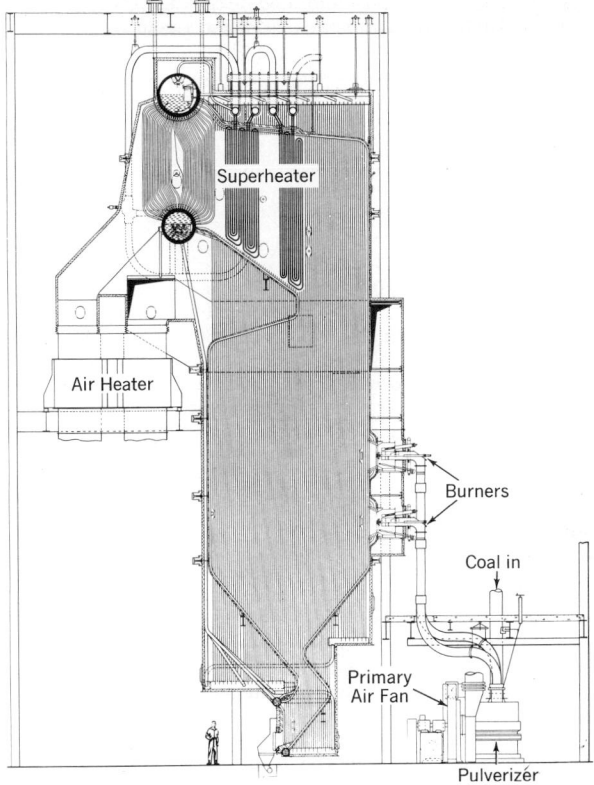

Fig. 4. Two-drum boiler direct-fired with pulverized coal.

In the pressure method, the primary-air fan, located on the inlet side of the pulverizer, forces the hot primary air through the pulverizer where it picks up the pulverized coal and delivers the proper coal-air mixture to the burners. Where a separate air heater is provided, the fan operates on cold air, forcing the air first through the air heater and then the pulverizer. In either event, the coal is delivered to the burners by a fan operating entirely on air, so that no entrained dust passes through the fan. One pulverizer generally furnishes the coal for several burners. With the pressure method, it is usual to supply each burner with a single conveying line direct from the pulverizer, thus eliminating the expense of a distributor.

In the suction method, the air and entrained coal are drawn through the pulverizer

TABLE 2. Comparative Features of Direct-Firing Pressure and Suction Systems

System	Pressure	Suction
Type of fan	Blower	Exhauster
Location of fan	Pulverizer inlet	Pulverizer outlet
Fan construction	Standard	Explosion-proof
Fan handles	Air only	Pulverized coal and air
Relative fan efficiency	High	Low
Fan wear	Low to none	High
Pulverized-coal distribution to burners	Good	Distributor required

under negative pressure by an exhauster located on the outlet side of the pulverizer. With this arrangement, the fan handles a mixture of coal and air, and distribution of the mixture to more than one burner must be obtained by a distributor beyond the fan discharge.

The feeding of coal and air to the pulverizer is controlled by either of two methods: (1) The coal feed is proportioned to the load demand, and the primary-air supply is adjusted to the rate of coal feed; or (2) the primary air through the pulverizer is proportioned to the load demand, and the coal feed is adjusted to the rate of airflow. In either case, a predetermined air-coal ratio is maintained for any given load.

The direct-firing system, in addition to eliminating separately fired dryers and storage facilities for pulverized coal, permits the use of inlet air temperatures to the pulverizer up to 650°F (343°C) and higher for drying high-moisture coals (total moisture 20 percent; surface moisture 15 percent) or high-moisture lignites (20 to 40 percent total moisture) in the pulverizer. The direct-firing system has one minor disadvantage. The operating range of a pulverizer is usually not more than 3:1 (without change in the number of burners in service) because the air velocities in lines and other parts of the system must be maintained above the minimum values to keep the coal in suspension. In practice, most boiler units are provided with more than one pulverizer, each feeding multiple burners. Load variations beyond 3:1 are generally accommodated by shutting down (or starting up) a pulverizer and the burners it supplies.

PULVERIZERS

The design fundamentals of pulverizers include:

1. Feeding. In direct-firing systems, the fuel rate must be capable of automatic control by the boiler load demand. Pulverizer airflow must be proportioned to fuel rate to provide the air required for drying, the correct primary-air–coal ratio, and the velocity required for transporting the fuel to the burners. One design (B&W) uses a control that maintains a predetermined variable air-fuel ratio over the entire operating range of the pulverizer. Airflow is measured by an orifice or a pitot tube in the duct supplying air to the pulverizer, and fuel flow is measured by feeder speed, a gravimetric feeder, or by the static pressure differential across the pulverizer grinding zone.

2. Drying. In order to pulverize and circulate fuel pneumatically within a pulverizer, enough of the moisture must be removed to leave the fuel dry and dusty. For most commercially available coal, preheated air to the pulverizer is required. Drying is accomplished quickly as the coal is being circulated and ground. The use of preheated air permits control of the temperature of the fuel-air mixture to the burners for the most stable ignition.

3. Grinding. The pulverizer must do adequate work on each passage of the material through the grinding zone and without the production of excessive superfines. This is best accomplished by internal recirculation of coarse material. The pulverizer should maintain its grinding ability over the life cycle of the grinding elements and be able to reject foreign matter that enters with the feed. These objectives should be attained without excessive wear and power consumption.

4. Circulating. Circulation of coal within a pulverizer is required (*a*) to promote rapid drying by mixing the incoming feed with dry material in the pulverizer, (*b*) to keep the grinding elements loaded at all times, and (*c*) to remove pulverized material from the grinding elements.

5. Classifying. It is not feasible to grind all the coal to the desired fineness in a single passage through the grinding elements. Therefore, a device called a *classifier* is provided. Coal pulverized to the proper fineness leaves the pulverizer and goes to the burners. Oversize coal is separated out by the classifier and returned to the grinding zone. If separation is not discriminative, oversize particles will go to the furnace and cause unburned combustible loss. Separation must be effective over the entire operating range, and the classifier must be adjustable for product size, since the required fineness is not the same for all applications.

6. Transporting. The velocity in pulverizer discharge lines must be sufficiently high to prevent settling and drifting of coal. At the burners, the air-coal mixture must be uniform and the velocity suitable.

Pulverizer requirements may be summarized by:

a. Rapid response to load change and adaptability to automatic control.

b. Continuous service for long operating periods.

c. Maintenance of prescribed performance throughout the life of pulverizer grinding elements.

d. A wide variety of coals should be acceptable.

e. Ease of maintenance with the minimum number and variety of parts, and space for adequate access.

f. Minimum building volume required.

The rank of coal and its end use govern the fineness to which coal must be ground. The data of Table 3 are helpful in this specification.

Grindability Index.   Some coals are harder and hence more difficult to pulverize than others. The grindability of a coal is expressed as an index showing the relative hardness of that coal compared with a standard coal chosen as 100 (unity) grindability. Thus, a coal is

TABLE 3.   Required Pulverized Fuel Fineness
(Percent through 200 U.S. Sieve*)

	ASTM classification of coals by rank					
	Fixed carbon, %			Fixed carbon below 69%		
					Btu/pound	
Type of furnace	97.9–86 (petroleum coke)	85.9–78	77.9–69	above 13,000	12,900–11,000	below 11,000
Marine boiler		85	80	80	75	
Water-cooled	80	75	70	70	65†	60†
Cement kiln	90	85	80	80	80	
Metallurgical	(As determined by process, generally from 80 to 90%)					

*The 200-mesh screen (sieve) has 200 openings per linear inch or 40,000 openings per square inch. For U.S. and ASTM sieve series, the nominal aperture for 200 mesh is 0.0029 inch, or 0.074 millimeters. The ASTM designation for 200 mesh is 74 micrometers.

†Extremely high ash content coals will require higher fineness than indicated.

harder or easier to grind if its grindability index is less or greater, respectively, than 100. The capacity of a pulverizer is related to the grindability index of the coal. A method of testing to determine the grindability index of coal is described in ASTM Standard D 409. See also the article on "Coal Testing" in the Handbook section on *Coal Technology.*

TYPES OF PULVERIZERS

The reduction of materials to a fine-particle size for countless uses is a very old art. Coal-pulverizing equipment is based, generally, on rock and mineral-ore grinding machinery. The principles involved in all pulverizing machinery are grinding (1) by impact, (2) by attrition, (3) by compression, or (4) by a combination of two or more of these methods. The principal types of coal pulverizers may be classified under high, slow, and medium speed:

Medium speed (75–225 revolutions per minute):
  Ball-and-race pulverizer
  Roll-and-race pulverizer
  Standard roll-and-race
  Bowl mill

Slow speed (below 75 revolutions per minute):
  Ball-and-race pulverizer
  Roll-and-race pulverizer
  Tube mill

High speed (above 225 revolutions per minute):
  Impact mill

Medium-Speed Pulverizers.   The same principle of pulverizing by a combination of crushing under pressure, impact, and attrition between grinding surfaces and material is used in both the ball-and-race and the roll-and-race pulverizers, but the method is different. In both groups, air is the predominant means for circulating the material through the grinding zone and conveying the finished product to the burners or cyclone collectors. Ball-and-race pulverizers may be operated under pressure or suction; roll-and-race pulverizers are normally operated under suction.

Medium-Speed Ball-and-Race Pulverizers. This type of pulverizer operates on the ball-bearing principle. The grinding elements constitute the distinguishing feature. In the ball-and-race pulverizer, these elements consist of a row of balls with one race below them and another above. High load circulation through the grinding zone, very desirable for effective drying and classification of the finished product, is a feature of this type of pulverizer. One pulverizer design of this type (B&W) is shown in Fig. 5. This design has one stationary top ring, one rotating bottom ring, and one set of balls that compose the

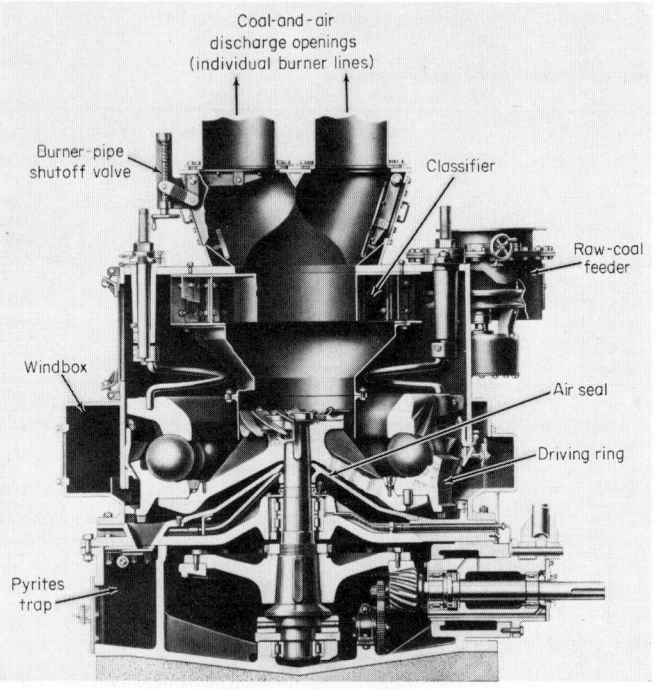

Fig. 5. Single-row ball-and-race pulverizer. *(B & W Type EL.)*

grinding elements. The pressure required for efficient grinding is obtained from externally adjusting dual-purpose springs. The bottom ring is driven by the yoke which is attached to the vertical main shaft of the pulverizer. The top ring is held stationary by the dual-purpose springs.

Raw coal is fed into the grinding zone where it mixes with partially ground coal that forms the circulating load. Pulverizer air causes the coal to circulate through the grinding elements where some of it is pulverized in each pass through the row of balls. As the coal becomes fine enough to be picked up by the air, it is carried to the classifier where coal of the desired fineness is separated from the stream entering the classifier and is carried out with the air. Oversize material is returned to the grinding zone. The classifier is a multiple-inlet cyclone with adjustable inlet vanes to permit varying inlet velocity as required to obtain the desired fineness of the pulverized coal.

This type of pulverizer (Type EL) is made in 18 sizes from a model with balls located on a 17-inch pitch circle to a model with a 76-inch pitch circle of balls. Capacities range from 1½ to 20 tons per hour of 50 grindability (ASTM) coal with a product fineness of 70 percent through a 200-mesh sieve.

The ball-and-race pulverizer has the advantage of operation under positive pressure, with drying and conveying air supplied by the primary-air fan, and its consequent avoidance of wear on the fan rotor and housing. Thus, there is wide latitude in the selection of a high-efficiency fan. A further advantage is better and more uniform coal-and-air distribution to a plurality of burners in a direct-firing system, since each burner is supplied from a common source. A comparison of ball-and-race and roll-and-race pulverizers is given in Table 4.

Medium-Speed Roll-and-Race Pulverizers.   There are two classes of this type of pulverizer: (1) standard roll-and-race pulverizer, where the ring is stationary and the rolls rotate, and (2) the bowl mill, where the rolls are stationary and the ring rotates.

In the standard roll-and-race pulverizer, the grinding elements of four or more rolls, suspended from driving arms, revolve in a horizontally positioned race. The pressure exerted by the rolls against the race is proportional to the centrifugal force resulting from the speed of the driving shaft. Roll-and-race pulverizers of this type have been used extensively in processing coal for the bin system and for grinding various materials in the ceramic and chemical industries. However, this type of pulverizer is not usually considered suitable for direct-firing conditions, as it has to be taken out of service to lubricate the internal roll journals.

The principal components of the bowl mill are a rotating bowl equipped with a replaceable grinding ring, two or more tapered rolls in stationary journals, an automatically controlled feeder, a classifier, and a main drive. Instead of depending on centrifugal force for the grinding action, the rolls in the bowl mill are held in position relative to the grinding ring by mechanical springs, and centrifugal force is used only to feed the coal between the race and the rolls. The bowl mill is suitable for the direct-firing system since the roller journals can be lubricated and the rolls adjusted without shutting down. The classifier is an internal cone type, housed in a high-sided chamber, and provided with adjustable vanes and a reject discharge trap at the bottom. Coal fineness is externally controlled by adjusting the entrance vanes in the classifier and by adjusting the compression springs to control the pressure of the rolls on the material.

The bowl mill is usually designed to operate under suction, and the pulverizer fan is placed on the outlet side of the classifier. Heavy scroll liners and rotor blades are generally provided to withstand the abrasive action of the material on the fan. These mills are used extensively in direct-firing systems, for steam boilers, and for cement kilns.

Slow-Speed Pulverizers.   One type (B&W MPS) of pulverizer of this class is shown in Fig. 6. This is a roll-and-race type utilizing three large-diameter rolls equally spaced around the mill. The rolls are mounted on axles to assure their rolling in a true path

TABLE 4.   Characteristics of Ball-and-Race and Roll-and-Race Pulverizers

Characteristic	Ball and race*	Roll and race†
Output, maximum tons/hour	1½–20	55–60
Speed	Medium	Slow
Range of revolutions per minute of main shaft	231–90	23.5
Effect of moisture on output	None up to 15%, if air is hot enough	None up to 10%. Slight reduction above this
Effect of wear on fineness	None	Slight
Method of control	Control primary air, which controls coal feed	Control coal feed, which controls primary air
Maintenance (material only)	Low, 2 to 7¢ per ton	Low, 2 to 6¢ per ton
Noise level	Medium	Low

Common characteristics:
  Operates under pressure
  Fan location is at pulverizer inlet
  Load circulation is high
  Classification is internal
  Classification control is by mechanical adjustment
  Drying range of moisture: up to 20 percent total; 15 percent surface
  Response to load change: fast
  Power input per unit output, including fan: low, 14 kilowatthours/ton

  *B&W, Type EL.
  †B&W, Type MPS.

around the mill. Roll assemblies are attached (by a unique pivoted connection) to a stationary frame above them to keep them in position. Grinding pressure is supplied by springs that apply force to the axles of the rolls. The grinding ring rotates at low speed. It is shaped to form a race in which the rolls run.

Circulation of coal is similar to that of the previously described (Type EL) pulverizer. Raw coal is fed into the mill either inside or outside the grinding race. It immediately

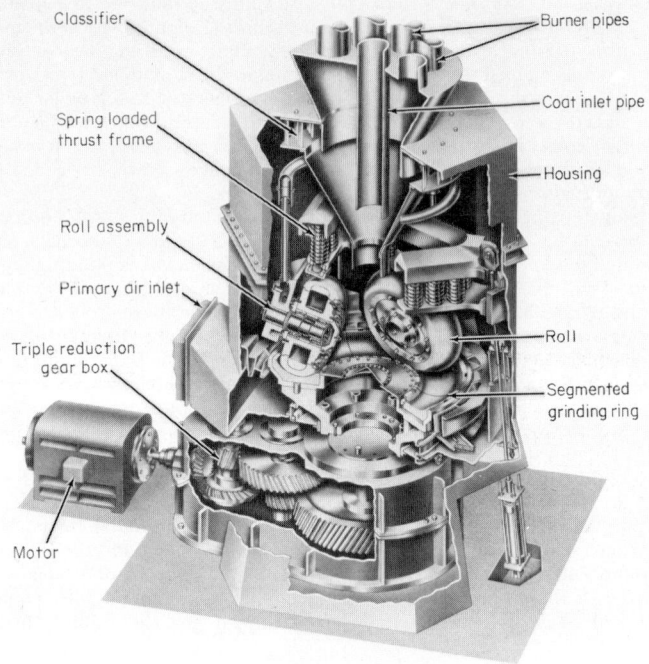

Fig. 6. Type MPS mill. *(B & W.)*

mixes with the partially ground coal that is circulated within the grinding zone by the airflow through the pulverizer. As the coal is reduced in size, the air carries it to the classifier (also similar to Type EL). The fine coal, along with the air, leaves the pulverizer through the outlet pipes. The oversize coal is returned to the grinding zone through the classifier-discharge seal. This unit (size 89—89-inch pitch circle of the rolls where they contact the ring) weighs 150 tons, is 22½ feet high, and is about 12 feet across. Each roll assembly weighs 10 tons. The pulverizer is driven by a 700-horsepower motor. The unit operates under pressure, thus eliminating erosion of the primary-air fan by pulverized coal. This permits selecting primary-air fans of high efficiency and readily allows using one fan for two or more pulverizers. Good distribution of coal to the various burners is obtained by taking individual burner lines directly from the mill outlet.

Disk-and-Roller Mill. This machine is a roll-and-race type, similar to the bowl mill in construction and operation. The principal differences are that the bowl has been flattened into a disk and the replaceable ring driven by the rotating disk is essentially flat. There are two or more rollers loaded by external springs or air cylinders to produce the desired pressure between the rollers and the race. These pulverizers are equipped with an integral classifier of the cyclone type, or a rotating member driven by a separate motor-gear reducer unit.

Tube Mill. This is one of the oldest practical pulverizers. A charge of mixed-size forged-steel balls in a horizontally supported grinding cylinder is activated by gravity as the cylinder is rotated. The coal is pulverized by attrition and impact as the ball charge ascends and falls within the coal. Air is circulated over the charge to convey the finished product to the classifiers. The coal rejected in the classifiers returns to the grinding zone by gravity. This machine is used for direct-firing systems by positioning the pulverizer fan so that it draws from the classifier outlets and discharges to the burner lines. The conical-end feature of the grinding cylinder causes size segregation in both ball charge and material within the grinding zone. See Fig. 7.

The outstanding features of the tube mill are dependability and low maintenance as a result of the simple arrangement of liners and ball charge in the grinding zone. No provision is required for removing tramp iron from the raw coal. Because of its simple and sturdy construction, low maintenance costs at rated capacities, and ability to grind very

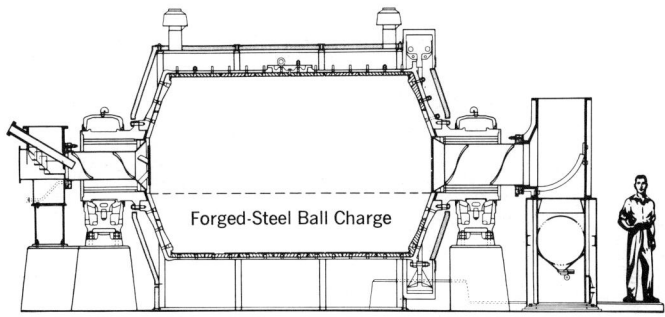

Fig. 7. Tube mill pulverizer.

abrasive materials, the tube mill is still being used almost exclusively for wet and dry grinding for various process materials.

For any given capacity, the tube mill is larger in size and heavier in construction than high- or medium-speed pulverizers using different principles of grinding. The tube mill requires more power per unit output than other types. Because of the absence of load circulation within the mill, the excessive amount of fine product within the ball charge, and circulation of preheated air over the top of the charge, it is not as efficient in drying the coal as medium-speed pulverizers with high air circulation through the grinding zone. This results in capacity reduction when handling wet coal.

The tube mill is still successfully applied to the grinding of coal where there is ample room for installation and where power cost is not the governing portion of the operating cost. For the most part, however, the tube mill has been replaced by more efficient machines for grinding coal.

High-Speed Pulverizers. In high-speed impact pulverizers, the grinding elements consist of hammerlike beaters revolving in a chamber lined with wear-resistant plates. These machines are used in Europe with subbituminous and brown coals of high moisture content (up to 50 percent). Flue gas from the furnace is introduced into the pulverizer to dry the coal. The product obtained is more closely comparable to the crushed coal used for the cyclone furnace than to pulverized coal as used in the United States.

SELECTION OF PULVERIZER EQUIPMENT

If selection anticipates the use of a variety of coals, the pulverizer should be sized for the coal that gives the highest *base capacity*. Base capacity is the desired capacity divided by the capacity factor. The latter is a function of the grindability of the coal and the fineness required. See Fig. 8.

The extent of drying in a given pulverizer depends upon its design and the method used to introduce preheated air into the grinding zone. Raw coal with very high surface moisture (over 15 percent) can be efficiently dried when fed into the grinding zone of a pulverizer designed for a high internal circulating load, i.e., a high ratio of coal recirculated to coal feed. As the recirculated material is dry, the more of it there is, the less effect the wet feed has on the performance of the mill.

As a practical matter, temperature is the only variable for controlling the heat input for drying, since the weight of primary air is usually a fixed quantity at any given output.

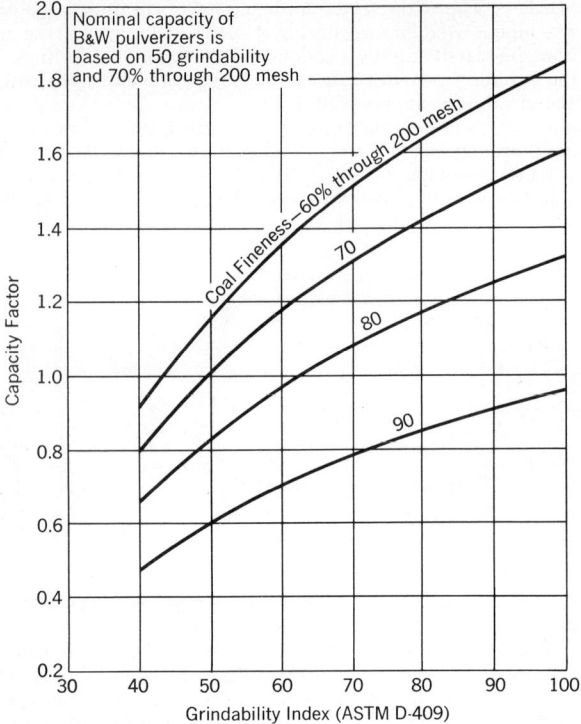

Fig. 8. Effect of grindability and fineness on pulverizer capacity.

The percentage of volatile matter in the fuel has a direct bearing on the probability of premature ignition of the primary-air–fuel mixture at the burners. The generally accepted safe values for exit fuel-air temperatures are given in Table 5. The temperature of the primary air entering the pulverizer may run 650°F (343°C) or more, depending on the amount of surface moisture and the type of pulverizer.

Fine grinding of coal is necessary to assure complete combustion of the carbon for maximum efficiency and to minimize the deposit of ash and carbon on the heat-absorbing surfaces. This applies not only in the firing of steam boilers but also in other applications where close temperature control and the avoidance of carbon contamination are important. Chemical and metallurgical processes using pulverized coal as a source of thermal or chemical energy generally require very finely ground coal to assure the optimum reaction in a limited combustion zone and often under difficult firing conditions.

Fineness is expressed as the percentage of the product passing through various sizes of sieves, graded from No. 16 to No. 325 in the ASTM designation. Coal classification by rank and the end use of the product determine the fineness to which coal should be ground

TABLE 5. Prevalent Pulverizer Exit Primary-Air–Fuel Temperature

Fuel	Exit temperature	
	°F	°C
Lignite	120–140	49–60
High-volatile bituminous	150	66
Low-volatile bituminous	150–175	66–80
Anthracite	200	93
Petroleum coke	200–250	93–121

(Table 3). The effect of fineness on capacity is shown in Fig. 8, where the capacity factor is given as a function of the fineness factor and the grindability index.

The range through which the equipment will operate must be considered in selecting pulverizing equipment for direct firing. The range through which a single pulverizer can operate is an inherent feature of the pulverizer. The range for safe operation depends on the type and number of burners, type of fuel, and whether the furnace is "hot" or "cold." A range of about 3:1 is common.

Pulverizer Control Equipment.   In one type of pulverizer (B&W), control is effected by proportioning coal feed and pulverizer airflow to load demand. Full automatic control of the output is obtained by varying the airflow through the pulverizer. Airflow indication is obtained from modified pitot tubes, orifice, or venturi installed in the air duct ahead of the pulverizer. The air-metering device is calibrated on the job and adjusted so that all pulverizers on a unit show the same coal flow for the same airflow. Coal-feed rate is obtained from the feeder speed or from differential pressure across the pulverizer. The control equipment converts the primary-air differential from a square root to a linear value and balances it against the coal-feed signal. The control is adjustable so that the desired air-fuel ratio can be maintained over the complete operating range of the pulverizer.

Sampling Equipment.   There is no universally adopted procedure for sampling pulverized coal, but the procedures given in ASME PTC 4.2 (Performance Test Code for Coal Pulverizers) and ASTM Standard D 197 may be considered as good practice. Two general arrangements of the equipment required for sampling pulverized coal in a direct-firing system are shown in Fig. 9, one using a bag collector, the other using a cyclone. The results of sieve analyses of pulverized-coal samples should be plotted on a Rosin and Rammler chart (Fig. 10). This graphical form of presenting size distribution is explained in the U.S. Bureau of Mines Information Circular 7346 (1946). If the points fall on a relatively straight line, the sample can be considered as representative. If there is a sharp or abrupt break in the curve, the sample should be discarded.

Exhausters and Blowers.   Primary air is required for conveying the pulverized coal to the burners. In the direct-firing system, the primary air is supplied through the pulverizers. With a pressure system, the primary-air fan handles clean air and is not subjected to abrasion by the pulverized coal. In this case, a high-efficiency fan can be used since the conditions permit an efficient rotor design and high tip speed.

With a suction system, the fan or exhauster must handle pulverized-coal-laden air. To comply with regulations (National Fire Protection Association et al.), the exhauster housing must be designed to withstand a pressure of 200 psi in case of an explosion within the fan. Furthermore, since the exhauster is subject to excessive wear, the design is limited to a paddle-wheel type of heavy construction and hard-metal or other protective-

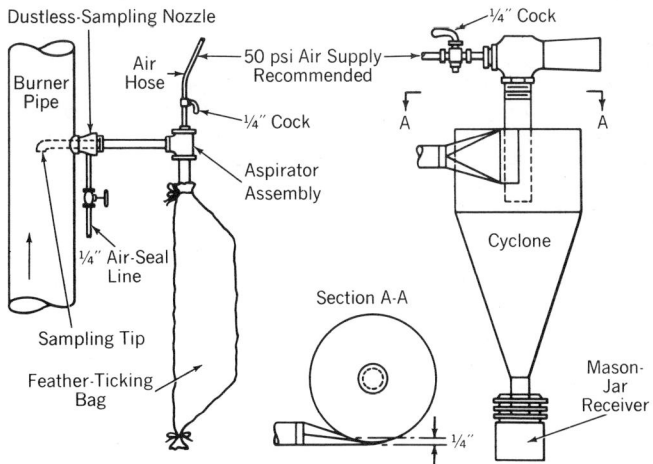

Fig. 9. Suggested arrangement for sampling pulverized coal in a direct-firing system using aspirator and bag or cyclone collector.

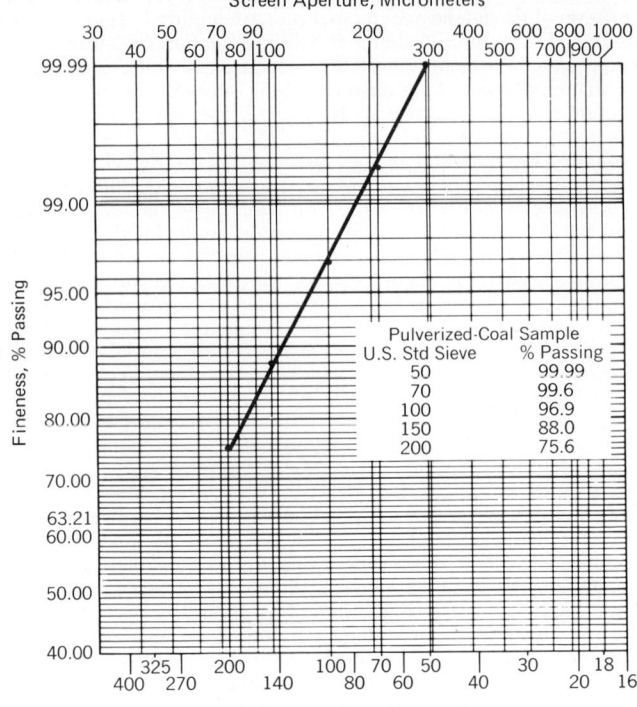

Fig. 10. Rosin and Rammler chart for plotting pulverized-coal sample sieve analyses.

surface coatings. All these construction features are detrimental to the mechanical efficiency of the fan.

The speed of the pulverizer fan depends on the pressure requirements of the system, but is generally about 1800 revolutions per minute. The fans are usually independently driven, but in small sizes, one motor may be used to drive both the pulverizer and the fan.

Power consumption of the pulverizer fan is an important item in the operating cost, and the selection of the fan should receive careful consideration. For a direct-firing installation, the fan must overcome the flow resistances through the pulverizers and burner lines as well as provide the pressure required at the burner. The resistance through the pulverizer depends on the inherent characteristics of the pulverizer and the amount of drying accomplished in it. The resistance through the discharge lines is a function of their length and the air-coal velocity that must be maintained to prevent settling. See Table 6.

Figure 11 shows average relative power consumption, under given conditions, for a standard commercial blower handling clean air and a fan used as an exhauster handling coal-laden air.

COST OF PULVERIZING COAL

The operating costs per unit of output are affected by (1) size of the installation, (2) characteristics of the coal, and (3) fineness of the finished product. The life of the grinding

TABLE 6.   Average Velocity of Primary Air and Coal Mixture through Conveying Pipes

(Air-coal ratio, 2 pounds air/pound coal)

Pulverized-coal fineness, % through 200-mesh sieve	70	80	85
Normal velocity, feet per minute	5000	4500	4000
Minimum velocity, feet per minute	3000	3000	3000

elements of the ball-and-race type of pulverizer varies from 6000 to 32,000 hours of operation. For bituminous coal, the rate of wear is generally a function of the volatile matter and the sulfur in the coal. The operating cost per unit of output is a function of the total coal pulverized during the life of the grinding elements. The maintenance cost is usually expressed in cost per ton of coal pulverized. For direct-firing systems, the power

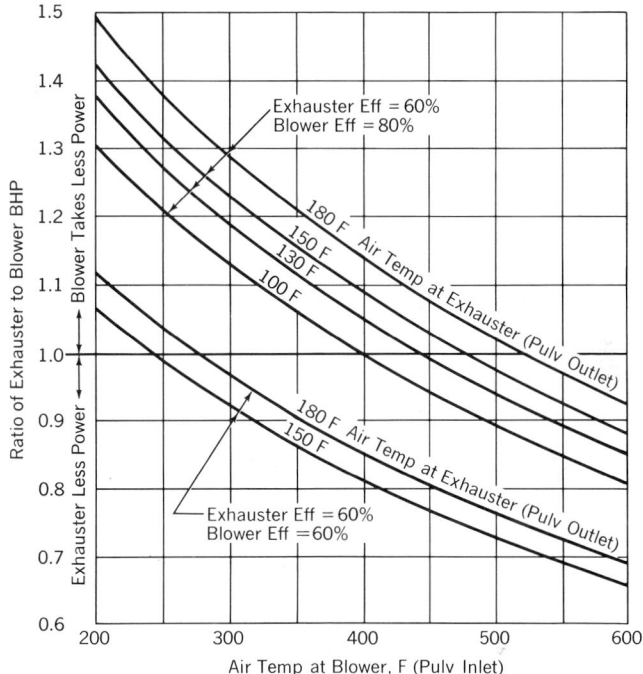

Fig. 11. Comparison between power consumption of blower and exhauster for given set of conditions. BHP = brake horsepower.

consumption per ton of coal delivered to the burners is the sum of the power required to drive (1) the auxiliaries delivering coal to the pulverizer, (2) the pulverizer, and (3) the fan delivering air to the pulverizer and coal to the burners.

PULVERIZED-COAL BURNING EQUIPMENT

As for oil and gas, the burner is the principal equipment component for the firing of pulverized coal. Much of the description relating to the burning of oil and gas in the article on "Oil and Gas Burners," given earlier in this Handbook section, is basically applicable to pulverized coal. However, the use of solid fuel in pulverized form presents additional problems in the design of boilers and furnaces.

As oil must be atomized to expose a large amount of oil particle surface to combustion air, so coal must be pulverized to the point where particles are small enough, i.e., surface is sufficiently large per unit of mass to assure proper combustion. See Table 3.

In addition, the surface moisture must be removed from the coal. In the direct-firing system, the coal is delivered to the burner in suspension in the primary air, and this mixture must be adequately mixed with the secondary air at the burner.

Piping and Nozzle Sizing Requirements.   Size selection of pulverized-coal piping and burner nozzles requires flow velocities that are high enough to keep the coal particles in suspension in the primary airstream. The airflow characteristics must be completely understood for the installation at hand.

Contemporary pulverizers generally use 25 to 50 percent of their full-load airflow requirement at zero output. Horizontal sections of burner pipe should be sized for no less than 3000 feet per minute velocity at the minimum expected capacity of the pulverizer. Horizontally arranged burner nozzles should be sized for no less than 3000 feet per minute at the minimum pulverizer capacity. Vertically disposed coal nozzles may be sized for velocities as low as 2200 feet per minute, the actual value depending on the burner configuration.

Stability of Ignition.   To assure stability of ignition, the temperature of the primary air and coal leaving the pulverizer must be at least 130°F (54°C) for units burning coal with 30+ percent volatile matter, going up to temperatures as high as 180°F (82°C) where the volatile matter of the coal is as low as 22 percent. The maximum temperature of the primary-air–coal mixture leaving the pulverizer, for coals with 25+ percent volatile matter, is approximately 150°F (66°C) because higher temperatures increase the tendency for coking on the burner parts.

Standards of Burner Performance.   Operators of pulverized-coal equipment should expect burner performance to meet the following conditions:

1. The rate of feed of the coal and air must comply with the load demand over a predetermined operating range. For most contemporary applications, ignition of the pulverized coal should be stable, without the use of supporting fuel, over a load range of approximately 3:1. Most steam boilers are equipped with a multiplicity of burners so that a wider capacity range is readily obtained by varying the number of burners and pulverizers in use.

2. Unburned combustible loss should be less than 2 percent. With most well-designed and coordinated installations, it is possible to keep the unburned combustible loss under 1 percent with excess air in the range of 15 to 22 percent, measured at the furnace outlet.

3. The burner should not require adjustments to maintain flame shape. The design should avoid the formation of deposits that may interfere with the continued efficient and reliable performance of the burner over the operating range.

4. Only minor repairs should be necessary during the annual overhaul. Burner parts subject to abrasion may require replacement at more frequent intervals. Alloy steel should be used for parts that cannot be protected by cooling or other means, to avoid damage from high temperatures.

5. Safety must be paramount under all conditions of operation.

Contemporary Burner Types.   As with oil and gas, the most frequently used burners are the circular and cell types. Figure 12 shows a circular-type burner equipped for firing pulverized coal only. It can be used singly or in multiples. Figure 13 shows a cell burner equipped to fire pulverized coal, oil, or gas. Either of these burner types can be equipped to fire any combination of the three principal fuels. However, combination pulverized-coal firing with oil in the same burner should be restricted to short emergency periods. It is not recommended for long operating periods due to possible coke formation on the pulverized-coal element.

The maximum nozzle input per individual burner is 165 million Btu per hour. The secondary-air port velocity at full load on the boiler ranges from 4000 feet per minute for small boilers where unheated secondary air is used to 6000 feet per minute for a dry-ash-removal furnace with 600°F (316°C) air. When circular burners are installed in slag-tap furnaces, velocities of 7500 feet per minute are common. Figure 14 shows the arrangement of pulverized-coal-firing equipment for an electric utility boiler.

Lighters (Ignitors) and Pilots.   Ignition and control equipment available for pulverized-coal firing is similar to that for oil and gas, although there are differences in the way it is used. In the case of pulverized coal, it may be necessary to keep the ignitors operating for hours until the temperature in the combustion zone becomes high enough to assure self-sustaining ignition of the main fuel.

The self-igniting characteristics of pulverized coal vary from one fuel to another, but on most coals it should be possible to maintain ignition without auxiliary fuel down to one-third capacity of the boiler. In some instances, completely reliable ignition is obtained down to a quarter load. When firing pulverized coal with volatile matter less than 25 percent, it may be necessary to activate the ignitors even at high loads. This can occur with any coal if it is excessively wet, frozen, or feeding sporadically into the pulverizers. If the ignitor is not activated during the intervals when coal is not reaching the pulverizer, or is reaching it in small amounts, ignition of the burner may be lost momentarily; then upon reestablishing coal flow, the burner may ignite with explosive force from an adjacent burner.

Excess Air.   Pulverized coal requires more excess air for satisfactory combustion than either oil or natural gas. One reason for this is the inherent maldistribution of coal both to

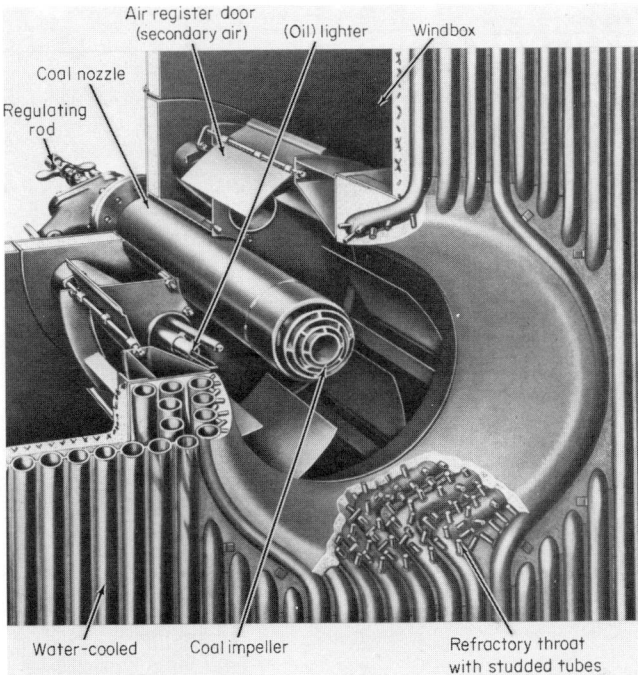

Air register door
(secondary air)   (Oil) lighter   Windbox

Coal nozzle

Regulating
rod

Water-cooled   Coal impeller   Refractory throat
with studded tubes

Fig. 12.   Circular register burner for pulverized-coal firing. *(B & W.)*

individual burner pipes and to the fuel discharge nozzles. The minimum acceptable quantity of unburned combustible is usually obtained with 15 percent excess air as measured at the furnace outlet at high loads. This allows for the normal maldistribution of both primary-air coal and secondary air. Higher excess-air values may be necessary to avoid slagging or fouling of the heat-absorption equipment.

In the design of the burner and furnace of a pulverized-coal-fired unit, consideration must be given to the burner arrangement and furnace configuration to minimize slagging or fouling from coal ash. Increasing excess air will permit most designs to perform satisfactorily, but this can be an uneconomical long-term substitute for good basic design.

Operating Range or Load Range.   In general, the pulverizer-burner combination can operate satisfactorily from full load to approximately 40 percent of full load with all pulverizers and burners in service. In some installations, a pulverizer and set of burners, in addition to the number actually required, are provided to assure availability of the boiler unit in case of unscheduled outage of a pulverizer. Where spares are provided, it is generally most economical to operate with the greatest number of burners and pulverizers in service consistent with the capacity demand on the unit. Although the use of this excess equipment raises the minimum load which can be obtained without cutting out pulverizers and burners, other benefits offset this disadvantage.

It is easier to pick up load with an operating pulverizer than to bring an idle unit into service. Also, at high loads on a boiler unit, the burner elements in idle burners deteriorate quickly because of radiant heat. Air that is admitted through idle burners to reduce overheating does not enter into the combustion reaction but is excess air which lowers the boiler efficiency.

As the output of the boiler unit is reduced from full load, it is desirable to reduce the number of pulverizers and burners in service to avoid settling of pulverized coal in the burner piping or burner nozzles. It may also be necessary to activate the ignition fuel to

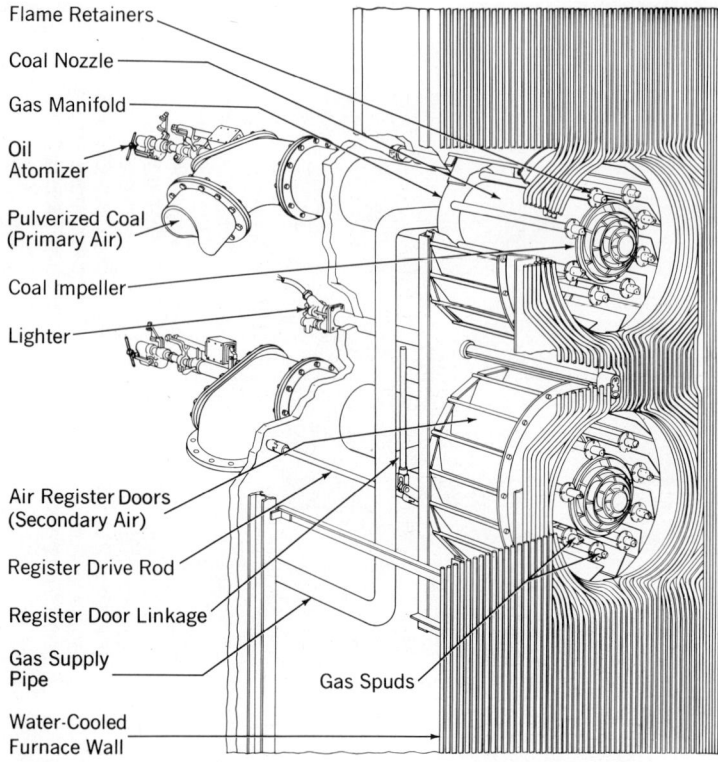

Flame Retainers

Coal Nozzle

Gas Manifold

Oil Atomizer

Pulverized Coal (Primary Air)

Coal Impeller

Lighter

Air Register Doors (Secondary Air)

Register Drive Rod

Register Door Linkage

Gas Supply Pipe

Gas Spuds

Water-Cooled Furnace Wall

Fig. 13. Cell burner for pulverized-coal, oil, and natural-gas firing. (*B & W.*)

sustain ignition of the coal fire because of the reduced furnace temperature that prevails at the lower load.

Where a unit has no more than two pulverizers, it is usually satisfactory to take individual burners out of service as the boiler load is reduced and thus maintain adequate burner line and nozzle velocities. For units with three or more pulverizers, it is better to take a pulverizer and its full complement of burners out of service or bring it back into service, on a change in load.

Idle burners are subject to considerable radiant heat from the furnace and can attain temperatures above the coking temperatures of the coals. The use of alloy metals provides longer life for burner parts. However, if they are not adequately cooled below the coking temperature before being placed in service, coke may form and cause severe damage to the parts. The easiest way to cool the nozzle is to run cold primary air through the pulverizer and burners for 5 to 10 minutes and then immediately feed coal before the nozzles can reheat. It is for this reason that a pulverizer and its group of burners should be operated as a unit rather than as individual burners. There is no simple way to air cool individual burners before bringing them into service.

Large boilers with air heaters have a sizable "heat inertia" which requires upward of 6 hours at full load for temperatures throughout the unit to stabilize. Conversely, a significant time period is necessary for them to retrograde on dropping load. When a fuel containing ash—either oil or pulverized coal—is burned, the length of time for temperatures to stabilize increases.

Starting Cold Boilers and Operating at Low Loads.   Starting with pulverized coal probably is next safest to gas. While coal is difficult to ignite and probably will require continuous operation of the ignitors, the unburned fuel that escapes is dry dust with a high-ignition temperature. It does not cling readily to surfaces and is carried out of the unit with the products of combustion. The only potential problem is dust that accumulates in hoppers under boiler components or in dust collectors. These should be emptied

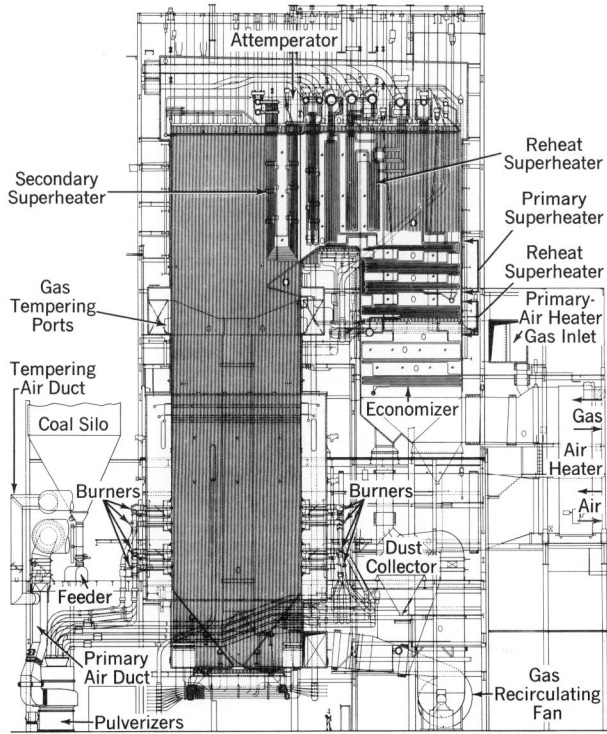

Fig. 14. Pulverized-coal-firing equipment for an electric utility boiler. *(B & W.)*

frequently so the unburned material cannot build up to a point where it may ignite and burn with damaging effect.

Oil burners with mechanical atomizers can be used for sustaining ignition of pulverized coal during start-ups with little risk of air-heater fires. Only a small amount of oil is used, and the resulting deposits are inconsequential.

Coordination of Pulverizers and Burners.   To design a system for pulverized coal, components must be chosen to function together over the load range desired. In a storage system, all auxiliary equipment must be selected to operate at its most efficient point when the pulverizer is at maximum rating. In a direct-firing system, the pulverizer and its burners have to be considered as a unit and selected not only to meet maximum load requirements but also to provide a stable operating minimum load. Because of variations in coal, job requirements, and equipment ratings, each application must be considered separately.

Pulverizers and burner coordination curves (Fig. 15) can be plotted to show the maximum capability, operating range, and permissible limits of both the pulverizer and its burners. The curves shown in Fig. 15 were developed for B&W equipment and design parameters.

1. *Curve A.*   The maximum steam flow must be known. In this example, 4 million pounds per hour is used. The heat input to the furnace divided by the Btu per pound of coal, as fired, gives the coal rate (pounds per hour) required for the boiler. The number of pulverizers is then determined. The choice is made on the basis of operating load range or

permissible loss of load, if one pulverizer is out of service, or on available pulverizer size. Four pulverizers are selected here as providing the necessary flexibility. The coal required per pulverizer is 93,675 pounds per hour.

Some overcapacity is required in the pulverizers to maintain maximum boiler output continuously. In this example, 10 percent is used so that the coal per pulverizer is raised

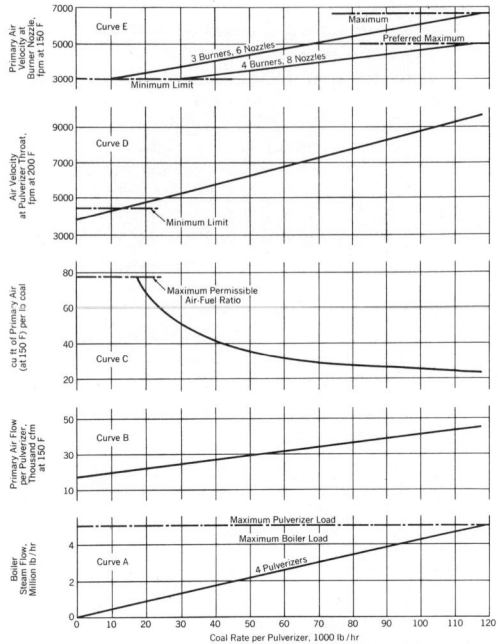

Fig. 15. Pulverizer-burner coordination curves.

to 103,000 pounds per hour. The pulverizer capability is fixed by the fineness of the pulverized coal (Table 3) and its grindability. A fineness of 70 percent through a 200-mesh screen and a grindability of 60 are used in the example. Figure 8 gives a capacity factor of 1.17 for these values. Dividing the coal required by the capacity factor gives the base capacity required for the pulverizer, that is, 88,000 pounds per hour. A pulverizer with 50 tons per hour base capacity has the required capacity. For the conditions given, it will grind 117,000 pounds per hour. Curve A can now be plotted as a straight line by drawing a line from zero through the intersection of 4 million pounds of steam and 93,675 pounds of coal, and extending this line to 117,000 pounds of coal per hour. The horizontal line through this point is the limit line for pulverizer capability and represents the boiler load that four pulverizers can theoretically carry.

2. *Curve B.* Each size of pulverizer is designed to operate at a specific airflow. The full-load airflow for the 50 tons per hour pulverizer is 45,000 cubic feet per minute at 150°F (66°C), and airflow at zero output is 18,000 cubic feet per minute at the same temperature. Curve B is a straight line between these points. This curve shows the primary air going through each pulverizer and the burners it serves, under any combination of load and number of pulverizers in service.

3. *Curve C.* This air-fuel ratio curve is obtained by dividing airflow rate by coal per unit of time. This curve is helpful in showing when the mixture gets too "lean" for stable ignition. The limit is a function of the volatile matter in the coal. For the coal used in this example, 78 cubic feet of 150°F air per pound of coal is as lean as is possible with stable ignition. The limit line can be drawn at this point.

4. *Curve D.* The passage where air enters the pulverizer grinding zone is called the *throat.* The velocity of the air at this point prevents coal from falling down through the

throat and being rejected from the bottom of the pulverizer. At maximum pulverizer load, the throat velocity should be 9750 feet per minute of 200°F (93°C) air, and at zero load it would be 3900 feet per minute. Curve D is a straight line joining these points. With the pulverizer loaded along Curve B, the minimum throat velocity is 4500 feet per minute of 200°F air, as shown by the limit line.

5. *Curve E.* Before these curves can be drawn the number and size of burners must be determined. At maximum pulverizer airflow, the preferred primary-air velocity through the burner is 5000 feet per minute of 150°F air. For short periods, it is acceptable to operate as high as 6500 feet per minute, but at this velocity burner erosion is too high for continuous use. In this example, each pulverizer is assumed to supply four cell burners, each having two nozzles of 14¼-inch diameter. This gives a nozzle velocity of 5080 feet per minute at maximum pulverizer load, with 2032 feet per minute at zero pulverizer load. On this curve, a horizontal line is drawn and marked *minimum limit* at 3000 feet per minute velocity. A second horizontal line is drawn at 5000 feet per minute velocity and labeled *preferred maximum.* A third horizontal line is drawn at 6500 feet per minute and marked *maximum.* As the minimum limit of the burner is reached (with decreasing load) considerably before the other limits, the burner range can be extended by cutting out burners. Therefore, another burner velocity curve can be drawn using three burners, six nozzles, as shown.

The coordination curves thus developed show that the pulverized-coal equipment will permit operation of the boiler with four pulverizers and sixteen burners in service, at the maximum of 4 million pounds of steam per hour (limited by the boiler) and down to a minimum load of 1.3 million pounds per hour. This range should be entirely possible on automatic control with no attention from the operator. Under starting conditions, with extremely low firing rates, some burners may be shut off to permit operation of a pulverizer at a lower rating.

Where coals of different characteristics must be fired, as is frequently the case, pulverizer and burner performance on each coal should be plotted on the coordination curves. When the coals vary widely, flexibility of operation of the boiler may be severely curtailed.

CYCLONE FURNACE

The introduction of pulverized-coal firing in the 1920s was a major advance in methods for burning coal. Currently, pulverized-coal firing is highly developed and is the best way to burn many types of coal, particularly the higher grades and ranks.

Since about 1940, the cyclone furnace has been developed and is now widely used. Cyclone-furnace firing represents a significant advance in coal firing since the introduction of pulverized coal. The cyclone furnace is applicable to coals having a slag viscosity of 250 poise at 2600°F (1,427°C) or lower, provided the ash analysis does not indicate excessive formation of iron or iron pyrites. With these coals, cyclone-furnace firing provides the benefits obtainable with pulverized-coal firing plus the advantages of (1) reduction in fly ash content in the flue gas, (2) saving in the cost of fuel preparation, since only crushing is required instead of pulverization, and (3) reduction in furnace size.

When coal is burned in boiler furnaces, the combustion of hydrogen is accomplished without difficulty, but successful combustion of carbon to CO_2 requires special measures to assure a continuing supply of oxygen in contact with carbon particles as long as they remain unburned. Not only must there be intimate mixing of the coal particles and air, but there must also be sufficient turbulence to remove the combustion products as they form at the surface of the fuel and provide fresh air at the fuel surface to continue combustion. The greater the turbulence, the more rapid the process. Hence, less time is required for combustion.

With pulverized-coal firing, the coal is reduced to a powder, so fine that approximately 70 percent will pass a 200-mesh screen. The finely pulverized coal is then very intimately mixed with combustion air in the burner. However, after this initial mixing, the tiny coal particles are merely carried along in the airstream. Very little additional scrubbing by the air occurs. Thus, further contact of oxygen with the coal must be largely by diffusion. The furnace consequently has to be relatively large to give the necessary retention time for oxygen to diffuse through the blanketing CO_2 layer to reach the coal particles, and, at the same time, temperatures must be sufficiently high to complete combustion. After combustion, since the residual ash particles are much smaller than even the original tiny pulverized-coal particles, the former are easily carried along with the flue gases from the furnace and through the boiler setting. At the same time, the pulverized-coal-fired boiler furnace also has the function of cooling the combustion gases, so that when they enter the convection surfaces they are below the temperature at which slagging occurs. This function conflicts with that of maintaining the high temperatures necessary to complete combustion. It would, therefore, be advantageous to separate these functions by providing a separate small combustion chamber where high turbulence and temperature may be maintained, and by using the main boiler furnace primarily to cool the combustion gases.

For many years, engineers recognized this need and actively explored basic changes in the design of furnaces and fuel-burning equipment to improve combustion and furnace performance. In addition, significant changes in availability and use of coal further increased the need for new designs; e.g., demands for higher grades of coal have depleted many seams, and others have been reserved for metallurgical and other uses. Mechanization in coal mining has increased the ash content of mined coal. Washing is widely used to lower ash and sulfur contents, but at added expense. The industrial growth of the western portion of the United States, rich in reserves of subbituminous and lignitic coal, has caused greater attention to and consumption of these lower ranks of coal. This has furthered the need for equipment fully suitable for the lower grades and ranks of high-ash, low-fusion-temperature coal. The cyclone furnace thus is an outgrowth of efforts to meet these needs and to overcome difficulties encountered with other firing methods.

Cyclone-Furnace Operating Principle.   Figure 16 shows a water-cooled horizontal cylinder in which fuel is fired, heat is released at extremely high rates, and combustion is completed. This is a cyclone furnace. Its water-cooled surfaces are studded and covered with refractory over most of their area. Coal crushed in a simple crusher, so that approximately 95 percent will pass a 4-mesh screen, is introduced into the burner end of the cyclone. About 20 percent of the combustion air, termed primary air, also enters the burner tangentially and imparts a whirling motion to the incoming coal. Secondary air is

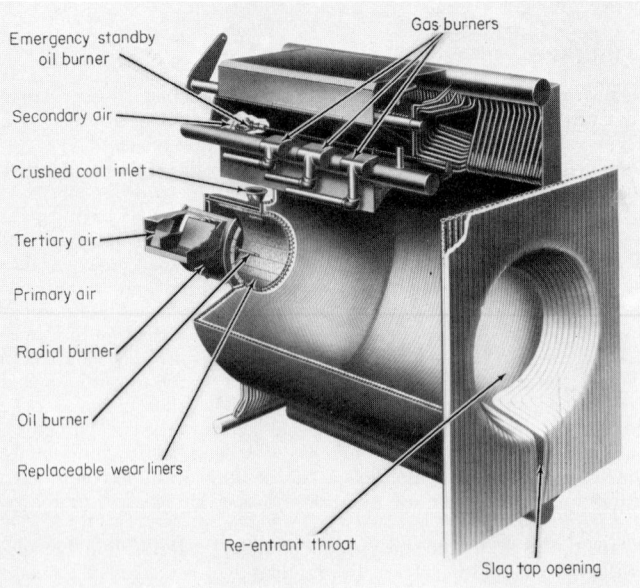

Fig. 16. Cyclone furnace, in the form of a horizontal cylinder, is completely water-cooled by connection to the main boiler circulation. All combustion gases leave through the reentrant throat at the rear. Molten slag drains from the bottom at the rear through a small opening into the adjacent boiler furnace. *(B & W.)*

admitted in the same direction tangentially at the roof of the main barrel of the cyclone and imparts a further whirling or centrifugal action to the coal particles. A small amount of air is admitted at the center of the burner.

The combustible is burned from the fuel at heat release rates of 450,000 to 800,000 Btu/ (cubic foot) (hour) and gas temperatures exceeding 3000°F (1,649°C) are developed. These temperatures are sufficiently high to melt the ash into a liquid slag, which forms a layer on the walls of the cyclone. The incoming coal particles are thrown to the walls by

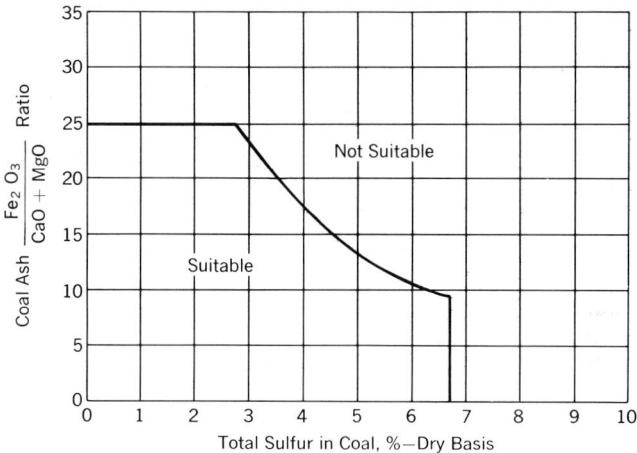

Fig. 17. Coal suitability for cyclone furnaces based on tendency to form iron and iron sulfide.

centrifugal force, held in the slag, and scrubbed by the high-velocity tangential secondary air. Thus the air required to burn the coal is quickly supplied, and the products of combustion are rapidly removed.

The gaseous products of combustion are discharged through the water-cooled reentrant throat of the cyclone (Fig. 16) into the gas-cooling boiler furnace. Molten slag in excess of the thin layer retained on the walls continually drains away from the burner end and discharges through the slag tap opening to the boiler furnace, from which it is tapped into a slag tank, solidified, and disintegrated for further disposition.

Suitability of Fuels for Cyclone Furnace.   The cyclone furnace is capable of burning successfully a large variety of fuels. A wide range of coals varying in rank from low volatile bituminous to lignite may be burned and, in addition, other solid fuels, such as woodbark, coal chars, and petroleum coke, may be satisfactorily fired in combinations with other fossil fuels. Fuel oils and gases are also suitable for firing.

The suitability of coals is dependent on the moisture, ash, and volatile contents, together with the chemical composition of the ash. The volatile matter should be higher than 15 percent, on a dry basis, to obtain the required high combustion rate. The ash content should be a minimum of about 6 percent to provide a proper slag coating in the cyclone and can be as high as 25 percent on a dry basis. A wide range of moisture content is permissible depending on coal rank, secondary-air temperature, and fuel-preparation equipment that may include capability for predrying the fuel.

One of the two important criteria for coal suitability is the total amount of sulfur compared to the ratio of iron to calcium and magnesium (Fig. 17). This comparison gives an indication of the tendency of the coal to form iron and iron sulfide, both of which are very undesirable in the cyclone furnace. Coals with too high sulfur content and/or a high iron ratio are not considered suitable. The other important criterion for establishing the suitability of coal for firing in the cyclone is the viscosity of the slag formed from the ash.

Slag will just flow on a horizontal surface at a viscosity of 250 poise. The temperature at which this viscosity occurs (T_{250}) is used as the criterion to determine the suitability of a coal from this point of view. The T_{250} is calculated from a chemical analysis of the coal ash, and a value of 2600°F (1427°C) is considered maximum. Somewhat lower temperatures may be desirable for fuels with high moisture contents and low heating values.

The suitability of other solid fuels, such as woodbark, petroleum coke, or chars, must be considered on an individual basis and the amount of supplementary fuel required carefully calculated.

Cyclone-Furnace Design Features.   The two general firing arrangements used for the cyclone furnace are (1) one-wall firing and (2) opposed firing. These configurations are shown in Fig. 18. For smaller units, sufficient firing capacity is usually attained with cyclone furnaces located in only one wall. For large units, furnace width can often be reduced by using opposed firing.

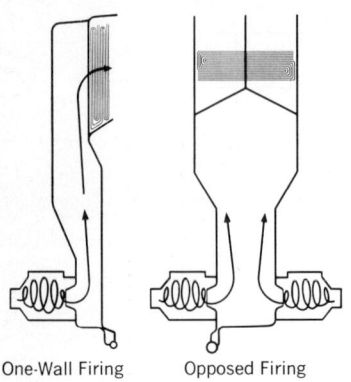

One-Wall Firing   Opposed Firing

Fig. 18. Firing arrangements used with cyclone furnaces. *(B & W.)*

There are two general types of coal preparation and feeding systems used: (1) the bin system and (2) the direct-firing system. See Fig. 19. The former is preferred for most bituminous coals when the plant layout permits. The range of sizing of crushed coal required with either system is given in Fig. 20.

The coal feeders normally used are of the belt type. A rotating distributor is provided at the coal discharge from the feeder to assure a continuous and uniform rate of feed. This is necessary because the coal is burned almost instantaneously when it reaches the cyclone furnace, and fluctuations in feed are reflected in combustion conditions. The rapidity of combustion makes the cyclone furnace very responsive to load demands, and it has been demonstrated that boiler output can be made to respond very quickly to demand by changing coal-feeder speed. Continuous weighing devices can be applied to the belt feeder so that it can serve the dual function of coal scale and feeder.

Oil and gas also are satisfactory cyclone fuels. These fuels can be burned at ratings and with performance equal to those with coal firing. Oil may be injected either into the secondary-air stream or through the center of the front coal burner (Fig. 21), where the oil is picked up and atomized by the high-velocity airstream. Gas is fired through flat ports located in the secondary-air entrance to the cyclone. Slag-handling equipment for a cyclone-furnace boiler unit is similar to that for a pulverized-coal slag-tap unit. The capacity of the slag-handling equipment must be greater since the percentage of ash recovered in the cyclone furnace is higher. The batch-removal system, shown in Fig. 22, is the system generally used.

Combustion Controls.   Automatic combustion controls for cyclone-furnace boilers are generally based on maintaining equal coal weights and equal total airflows in the proper proportion to each cyclone furnace. Where volumetric-type feeders are used, equal coal weights are obtained by maintaining equal feeder speeds. Where gravimetric-type feeders are used, they measure and control the coal weights to the cyclone furnaces.

Combustion airflow is measured separately to each cyclone, and the controls maintain equal coal rates and airflows to each cyclone furnace. The overall excess air is controlled in the usual manner with a boiler meter based on steam flow and airflow.

Performance Factors.   The first commercial cyclone-furnace boiler was designed to burn Central Illinois coal and was installed at the Calumet Station of the Commonwealth Edison Company, Chicago, Ill., in 1944. Since then, over 600 cyclone-furnace units have been installed in boilers throughout the United States and Europe. In the United States, coals of the following constituent range have been burned in commercial cyclone-furnace boilers:

Moisture, %	2 to 40
Volatile matter (dry), %	18 to 45
Fixed carbon (dry), %	35 to 75
Ash (dry), %	4 to 25

Since the only coal preparation is crushing, the power required is low compared with that for pulverizing coal. To offset this, the forced-draft fan power required is relatively high. Cyclone-furnace air pressure drop is in the range of 20 to 40 inches water compared

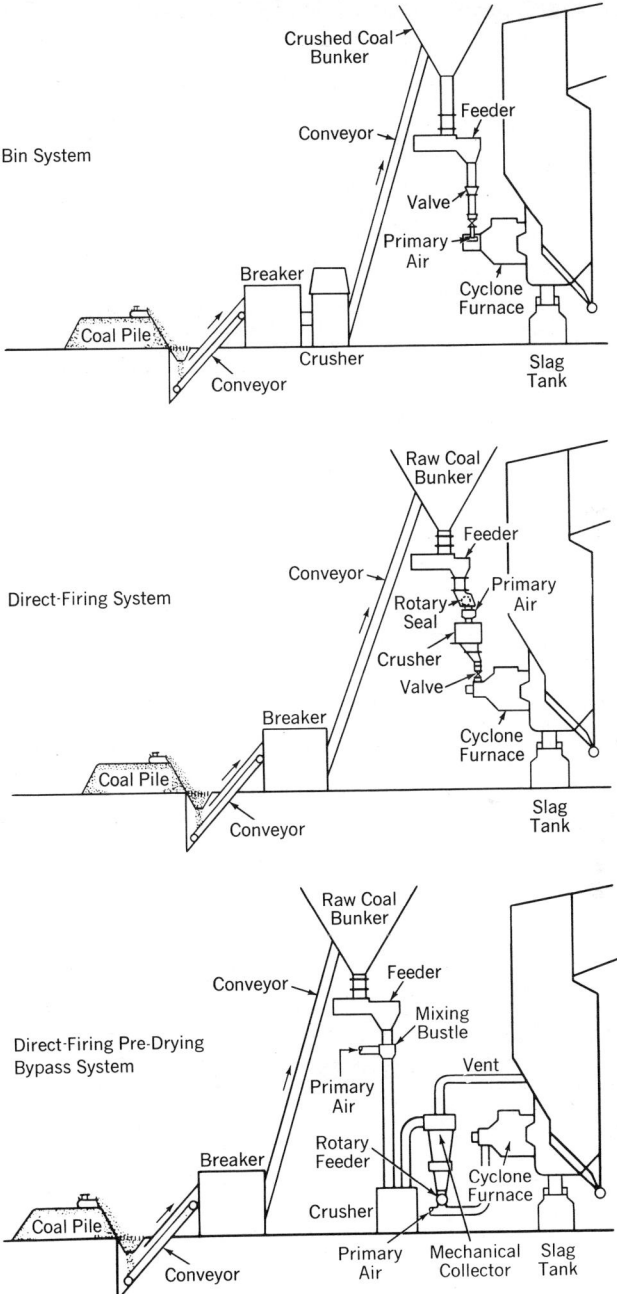

Fig. 19. Bin, direct-firing, and direct-firing predrying bypass systems for coal preparation and feeding to a cyclone furnace.

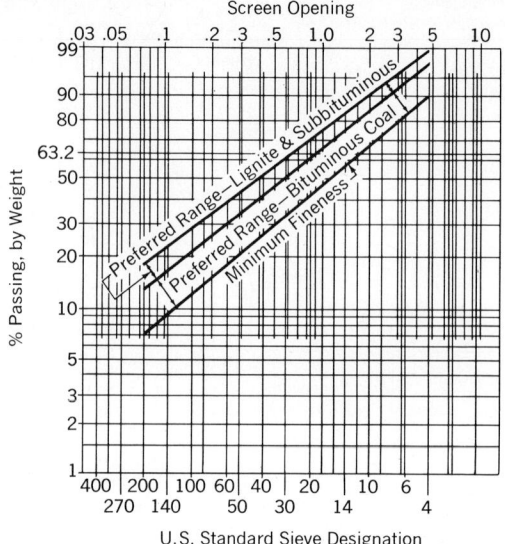

Fig. 20. Sizing of crushed coal fired in a cyclone furnace.

with 2 to 10 inches water for pulverized-coal burners. Figure 23 shows that the comparative power requirements vary considerably for different types of bituminous coals and lignite. The excess air required for satisfactory combustion of an individual cyclone furnace is less than 10 percent. However, where automatic controls are used and particularly where there are several cyclone furnaces for one boiler, excess air is usually maintained between 10 and 15 percent to assure that no individual cyclone is operating with insufficient air.

The dust loading of the flue gas from coal-fired cyclone units is in the range of 20 to 30 percent of the ash in the coal, compared with about 80 percent for a dry-ash pulverized-

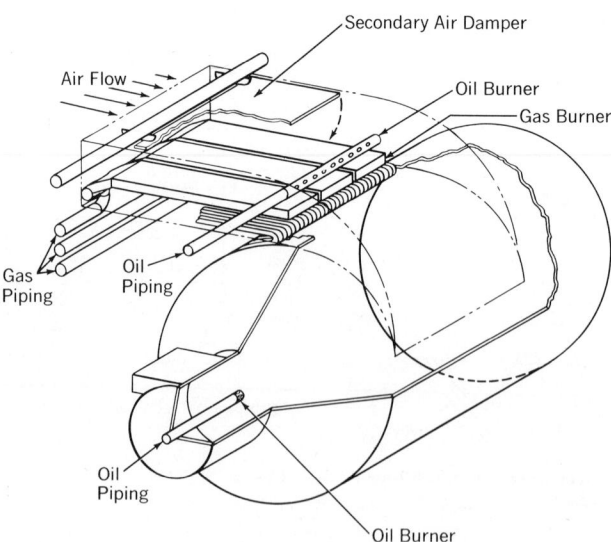

Fig. 21. Arrangement of gas and oil burner in cyclone furnace.

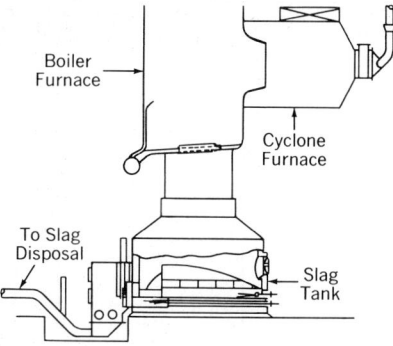

Fig. 22. Batch-removal slag-handling system for cyclone-furnace boiler.

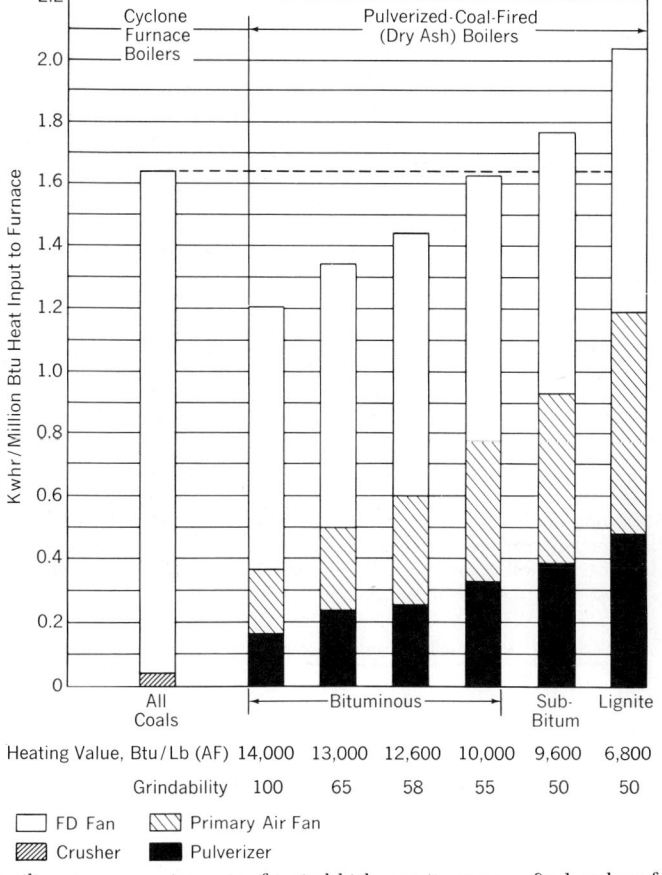

Fig. 23. Auxiliary power requirements of typical high-capacity pressure-fired cyclone furnace and pulverized-coal units. (*B & W.*)

coal-fired unit. This means that, if both units are equipped with 95 percent efficient precipitators or dust collectors, the ash discharged from the stack of a cyclone-fired unit will be less than half that from the stack of a dry-ash pulverized-coal-fired unit. This comparison is shown in Fig. 24 for a large utility cyclone-furnace unit and a pulverized-coal unit arranged for dry ash removal.

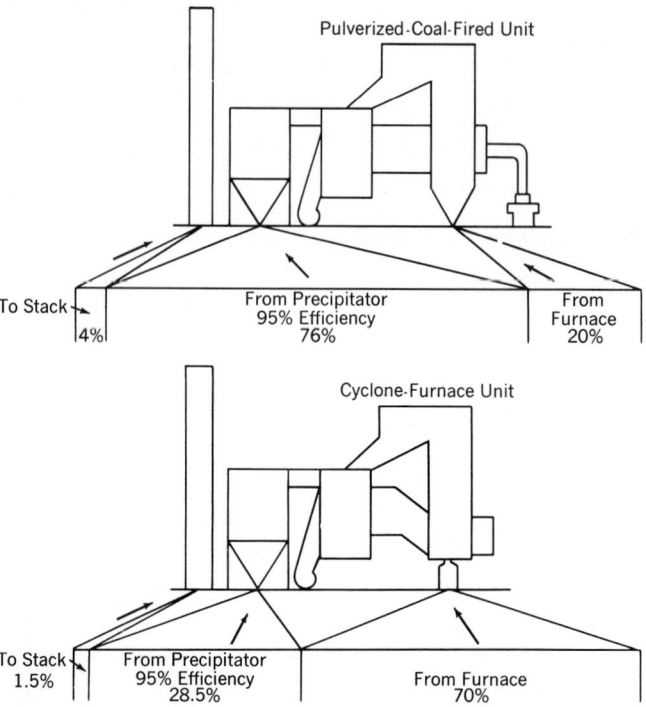

Fig. 24. Comparison of fly-ash emission from typical large dry-ash-removal pulverized-coal-fired unit and cyclone-furnace unit.

STOKERS

A successful stoker installation requires the selection of the correct type and size for the fuel to be used and the desired capacity. Also, the associated boiler unit should have the necessary instruments for the proper control of the stoker. The grate area required for a given stoker type and capacity is determined from allowable rates established by experience. Table 7 lists allowable fuel-burning rates [Btu/(square foot) (hour)] for various types of stokers, based on using coals suited to the stoker type in each case.

TABLE 7.   Maximum Allowable Fuel-Burning Rates for Stokers

Type of stoker	Btu/(square foot) (hour)
Spreader:	
Stationary and dumping grate	450,000
Traveling grate	750,000
Vibrating grate	400,000
Underfeed:	
Single or double retort	425,000
Multiple retort	600,000
Water-cooled vibrating grate	400,000
Chain grate and traveling grate	500,000

For a boiler of a given steam capacity, these maximum fuel-burning rates determine the plan area for a stoker-fired furnace. As boiler unit size is increased, practical considerations limit stoker size and, consequently, the maximum rate of steam generation with this method of firing. Because of the greater flexibility in furnace design with pulverized-coal and cyclone-furnace firing and the trend toward larger boiler units, the demand for stokers is less than in former years. The practical steam output limit of boilers equipped with mechanical stokers is about 400,000 pounds per hour, although many engineers limit the application of stokers to lower steam capacities. However, within their capacity range, mechanical stokers are an important and valued element of contemporary equipment for the production of steam or hot water. When applicable, stokers are often preferred over pulverizers because of their greater operating range, capability of burning a wide range of solid fuels, and their lower power requirements. Almost any coal can be burned successfully on some type of stoker. In addition, many by-products and waste fuels, such as coke breeze, wood wastes, pulpwood bark, and bagasse, can be used either as base or auxiliary fuel.

Mechanical stokers can be classified in four main groups, based on the method of introducing fuel to the furnace: (1) spreader stokers, (2) underfeed stokers, (3) water-cooled vibrating-grate stokers, and (4) chain-grate and traveling-grate stokers. Among these several types, the spreader stoker is most generally used in the capacity range from 75,000 to 400,000 pounds of steam per hour, because it responds rapidly to load swings and can burn a wide range of fuels.

Underfeed stokers of the single-retort, ram-feed, side-ash-discharge type are used principally for heating and for small industrial units of less than 30,000 pounds of steam per hour capacity. Larger size underfeed stokers of multiple-retort, rear-ash-discharge type have been largely displaced by spreader stokers and by the water-cooled vibrating-grate stokers in the intermediate range. Chain- and traveling-grate stokers, while still used in some areas, are gradually being displaced by the spreader and vibrating-grate types.

Spreader Stokers.   The spreader stoker is capable of burning a wide range of coals, from high-rank eastern bituminous to lignite or brown coal and a variety of by-product waste fuels. As the name implies, the spreader stoker projects fuel into the furnace over the fire with a uniform spreading action, permitting suspension burning of the fine fuel particles. See Fig. 25. The heavier pieces that cannot be supported in the gas flow fall to the grate for combustion in a thin fast-burning bed. This method of firing provides excellent sensitivity to load fluctuations as ignition is almost instantaneous on increase of firing rate and the thin fuel bed can be burned out rapidly when desired.

Figure 26 illustrates a fuel feeder-distributor unit of the variable stroke, reciprocating-feed-plate type. The reciprocating-feed plate moves coal from the supply hopper over an adjustable spill plate to fall onto an overthrow rotor. This rotor is equipped with curved blades for uniform coal distribution over the furnace area.

Introduction of the continuous-ash-discharge traveling grate of the air-metering design in the late 1930s brought the spreader stoker into immediate and widespread popularity. Since there are no interruptions for removing ashes and because of the thin, fast-burning fuel bed, average burning rates were increased approximately 70 percent over the stationary- and dumping-grate types. This type of stoker is generally competitive in sizes up to about 525 square feet of grate area, corresponding to steam capacity somewhat over 400,000 pounds of steam per hour. The furnace width required for stokers above this size usually results in increased boiler costs as compared with pulverized-coal- or cyclone-furnace-fired units with narrower and higher furnaces.

Although continuous-cleaning grates of reciprocating and vibrating designs have also been developed, the continuous-ash-discharge traveling-grate stoker is preferred for large boilers because of its higher burning rates.

Furnace Design.   An example of good furnace design is shown in Fig. 27, which also illustrates the fly-carbon reinjection system. This unit has water-cooled walls that are actually a necessity for traveling or continuous-cleaning spreader-stoker grates, where slag or clinker formation adjacent to the stoker would interfere with movement of the fuel bed. Furnaces with refractory walls are sometimes installed with stationary or intermittent-dumping-grate spreader stokers, but the high maintenance cost of such refractories makes this application questionable. The waterwalls are usually vertical, or nearly so, as arches are not desirable.

An overfire air system, with pressures from 27 to 30 inches water (gage), is essential to

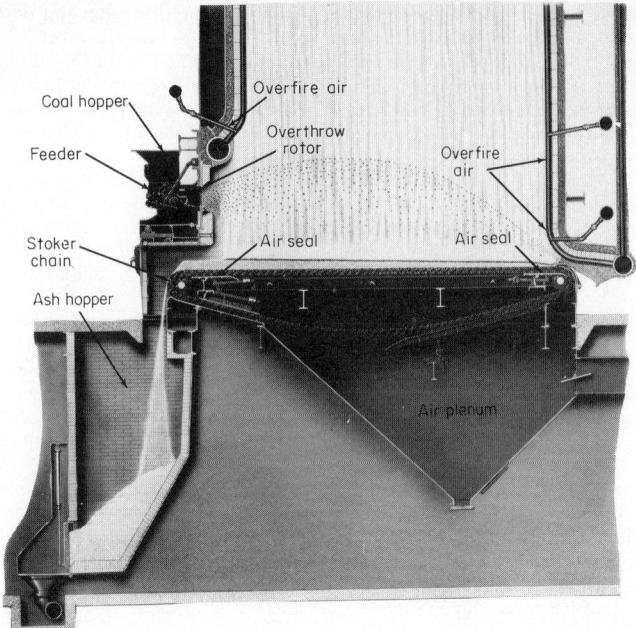

Fig. 25. Traveling-grate spreader stoker with front ash discharge. *(B & W.)*

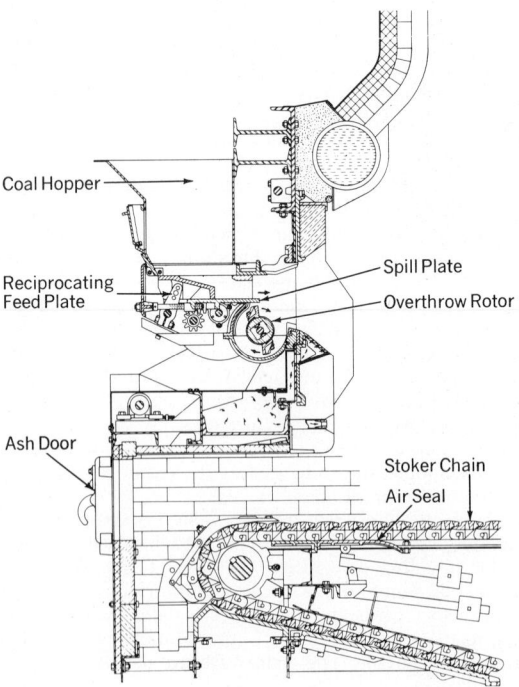

Fig. 26. Reciprocating-feeder distributor and overthrow rotor for spreader stokers.

successful suspension burning. It is customary to provide at least two rows of evenly spaced high-pressure air jets in the furnace rear wall and one in the front wall (Fig. 25). This air mixes with the furnace gases and creates the turbulence required to complete combustion.

Fly-Carbon Collection and Reinjection Systems.   Partial suspension burning results in a greater carry-over of particulate matter in the flue gas than with other types of stokers. Dust collectors are consequently required with spreader stokers. The collector generally used is a cyclone-type precipitator with a selective feature so that the fines are deposited

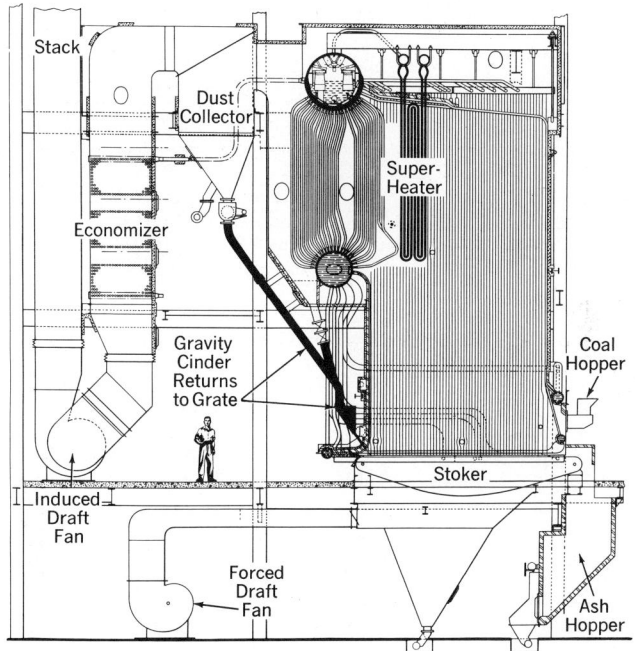

Fig. 27. Spreader-stoker installation with gravity fly-carbon return. *(B & W.)*

in a hopper for discharge to the ash disposal system, and the coarse carbon-bearing particles are skimmed off and returned to the furnace for further burning.

Fuels and Fuel Bed.   All spreader stokers, and in particular the traveling-grate spreader type, have an extraordinary ability to burn fuels with a wide range of burning characteristics, including coals with caking tendencies. High-moisture, free-burning bituminous and lignite coals are commonly used, and some low volatile fuels, such as coke breeze, have been burned in a mixture with higher volatile coal. Anthracite coal, however, is not a satisfactory fuel for spreader-stoker firing.

The ideal fuel and ash bed for a coal-fired spreader stoker is evenly distributed and from 2 to 4 inches thick. Maximum heat release rates are from 450,000 Btu/(square foot) (hour) on stationary or dumping rates to 750,000 Btu/(square foot) (hour) on traveling-grate spreader stokers. Higher releases are practical with certain of the waste fuels, such as pulpwood bark.

Residue Handling.   With stationary grates, the ash is removed from the grates one section at a time by hoe or rake through suitable doors at the grate level. Intermittent-dumping-grate stokers discharge the ash to pits. These may be shallow for firing-floor cleanout in the absence of basement space. The practical maximum net length or length open to airflow for a stationary grate is about 9 feet. With a dump grate arranged for floor cleaning, the net length should be held under 12 feet. In the case of an arrangement with a basement ash pit, this dimension may be 15 feet. The recommended maximum net length of reciprocating and vibrating grates for spreader stokers is 15 feet, while for traveling grates it is about 18 feet with current designs.

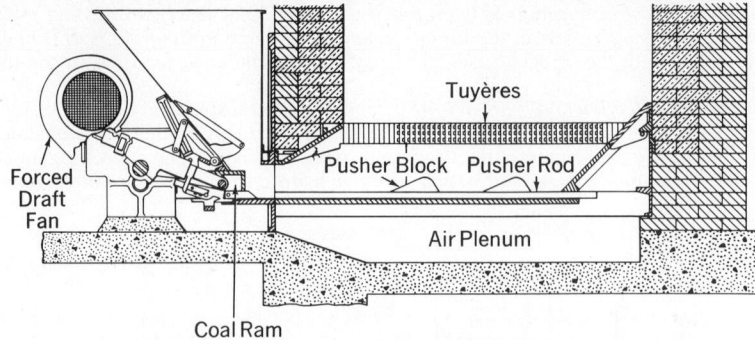

Fig. 28. Single-retort, horizontal-feed, side-ash-discharge underfeed stoker.

Control. With its sensitivity to adjustment for varying loads, the spreader stoker requires close control of fuel and air supply to achieve best results. There are many types of automatic combustion controls available, from simple positioning types sometimes used on relatively small installations to the more elaborate airflow–steam-flow equipment found in large plants.

Underfeed Stokers. These are generally of two types: (1) horizontal-feed, side-ash-discharge type (Fig. 28) and (2) the gravity-feed, rear-ash-discharge type (Fig. 29). In the side-ash-discharge underfeed stoker, fuel is fed from the hopper by means of a reciprocating ram to a central trough called the *retort*. On very small heating stokers, a screw conveys the coal from the hopper to the retort. A series of small auxiliary pushers in the bottom of the retort assist in moving the fuel rearward, and, as the retort is filled, the fuel is moved upward to spread to each side over the air-admitting tuyères and side grates.

As the fuel rises in the retort and is subjected to heat from the burning fuel above, volatile gases are distilled off and mixed with air supplied through the tuyères above each side of the retort and through the side grates. The volatile mixture burns as it passes upward through the incandescent zone, sustaining ignition of the rising fuel. Burning continues as the incoming raw coal continually forces the fuel bed to each side. Combustion is completed by the time the bed reaches the side-dumping grates. The ash is intermittently discharged to shallow pits, quenched, and removed through doors at the front of the stoker.

The single (and double) retort, horizontal-type stokers are generally limited to 25,000 to 30,000 pounds of steam per hour with burning rates of 425,000 Btu/(square foot) (hour) in furnaces with water-cooled walls. For refractory-walled furnaces, the maximum rate

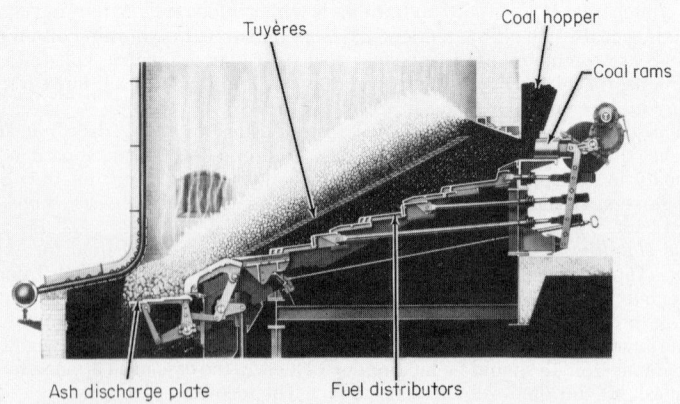

Fig. 29. Multiple-retort, gravity-feed type, rear-ash-discharge underfeed stoker. *(B & W.)*

should be reduced to about 300,000 Btu/(square foot)(hour). The multiple retort, rear-end-cleaning type (Fig. 29) has a retort and grate inclination of 20 to 25°. This type of stoker can be designed for boiler units generating up to 500,000 pounds of steam per hour. Burning rates up to 600,000 Btu/(square foot) (hour) are practical.

In general, underfeed stokers are able to burn caking coals. The range of agitation imparted to the fuel bed in different stoker designs permits the use of coals with varying degrees of caking properties. The only certain way to select suitable coal is by actual field tests for a given unit. The ash-fusion temperature is an important factor in the selection of coals. Usually, the lower the ash-fusion temperature, the greater the possibility of clinker trouble.

Water-Cooled Vibrating-Grate Stokers.   Several manufacturers build spreader stokers with air-cooled vibrating or oscillating grates. An entirely different design of stoker is the water-cooled vibrating-grate hopper-feed type (Fig. 30). The water-cooled vibrating-grate

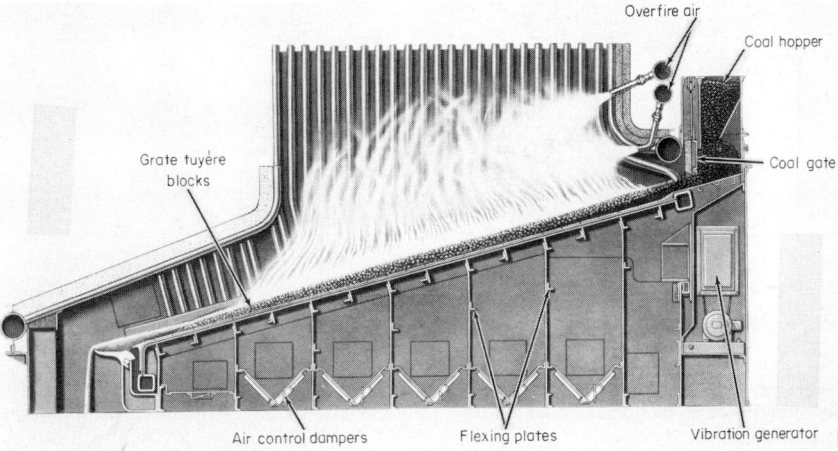

Fig. 30. Water-cooled vibrating-grate stoker.

stoker consists of a tuyère grate surface mounted on, and in intimate contact with, a grid of water tubes interconnected with the boiler circulation system for positive cooling. The entire structure is supported by a number of flexing plates allowing the grid and its grate to move freely in a vibrating action that conveys coal from the feeding hopper onto the grate and gradually to the rear of the stoker. Ashes are automatically discharged to either a shallow or basement ash pit.

Vibration of the grates is intermittent, and the frequency of the vibration periods is regulated by a timing device. Timing is regulated by the automatic combustion control system to conform to load variations, synchronizing the fuel feeding rate with the air supply.

Furnace Design.   The water-cooled vibrating-grate stoker is suitable for burning a wide range of bituminous and lignite coals. Even with coals having a high free-swelling index, the gentle agitation and compaction of the fuel bed tends to keep the bed porous without the formation of large clinkers generally associated with low ash-fusion coals. A well-distributed, uniform fuel bed is maintained without blow holes or thin spots.

The furnace design for this stoker should include water-cooled walls to prevent slag formation adjacent to the stoker. A rear arch extending over approximately one-third of the stoker length directs the gases forward to mix with the rich volatile gases released in the ignition zone. A short front arch is adequate for most bituminous fuels. The use of high-pressure air jets—from 27 to 30 inches water (gage)—through the front arch provides turbulent gas mixing and promotes combustion. In rare cases, with extremely low-volatile fuels, some refractory facing of the front water-cooled arch may be desirable to increase the temperature over the ignition section.

Burning rates of these stokers vary with different fuels but, in general, the maximum heat release rate should not exceed 400,000 Btu/(square foot) (hour). In this range, fly-carbon carry-over is held to a minimum.

Chain-Grate and Traveling-Grate Stokers.   Traveling-grate stokers, including the specific type known as the chain-grate stoker, have assembled links, grates, or keys joined together in endless belt arrangements that pass over the sprockets or return bends located at the front and rear of the furnace. Coal, fed from the hopper (Fig. 31) onto the moving assembly, enters the furnace after passing under an adjustable gate to regulate the

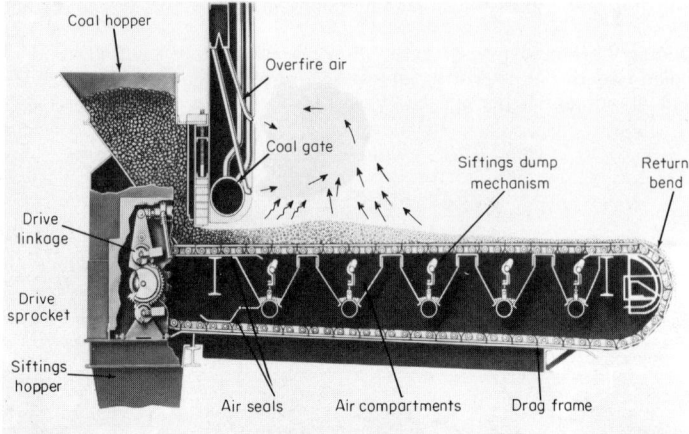

Fig. 31. Chain-grate stoker. *(Laclede Stoker Co.)*

thickness of the fuel bed. The layer of coal on the grate, as it enters the furnace, is heated by radiation from the furnace gases and is ignited together with the hydrocarbon and other combustible gases driven off by distillation. The fuel bed continues to burn as it moves along with the bed, growing progressively thinner as combustion continues. At the far end of the travel, ash is discharged from the end of the grate into the ash pit. Although there are structural differences, the operation of the chain-grate and other traveling-grate types are quite similar. Generally, these stokers use furnace arches (front and/or rear) to improve combustion by reflecting heat onto the fuel bed. The front arch also serves to break up and mix rich streams of volatile gases that might otherwise travel through the unit unburned.

Chain- and traveling-grate stokers can burn a wide variety of fuels. Almost any solid fuel—peat, lignite, subbituminous, free-burning bituminous, anthracite, and coke breeze—of suitable size can be burned on these stokers. Chain- and traveling-grate stokers are offered for a maximum continuous burning rate of 425,000 Btu/(square foot)(hour), with high-moisture (20 percent), high-ash (20 percent) bituminous coal, and 500,000 Btu/(square foot)(hour), with lower-moisture (10 percent) lower-ash (8 to 12 percent) bituminous coal. For anthracite, the corresponding burning rate is 350,000 Btu/(square foot)(hour).

REFERENCES

"Steam: Its Generation and Use," 38th ed., Babcock & Wilcox Company, New York, 1972.
Smith, R. B.: Heat Generation, in R. H. Perry and C. H. Chilton (eds.), "Chemical Engineers' Handbook," 5th ed., McGraw-Hill, New York, 1973.
Kessler, G. W.: Steam Boilers, in T. Baumeister (ed.), "Marks' Standard Handbook for Mechanical Engineers," 7th ed., McGraw-Hill, New York, 1967.

Boilers and Associated Equipment*

In the effective contemporary steam generator, various components are arranged to absorb heat efficiently from the products of combustion. These components are generally described as (1) boiler, (2) superheater, (3) reheater, (4) economizer, and (5) air heater.

CLASSIFICATION OF BOILERS

Boiler surface may be defined as those parts of tubes, drums, and shells that are part of the boiler circulatory system and that are in contact with the hot gases on one side and water, or a mixture of water and steam, on the other side. Although the term boiler may refer to the overall steam-generating unit, the term *boiler surface* does not include the economizer or any component other than the boiler itself. Boilers may be broadly classified as shell, fire-tube, and watertube types.

Advantage of Watertube Boiler. Modern boilers are of the watertube type. The safety and dependability of operation that characterize the boilers currently used had their beginning in the introduction of this type of boiler. In the watertube boiler, the water and steam are inside the tubes and the hot gases are in contact with the outer tube surfaces. The boiler is constructed of a number of sections of tubes, headers, and drums joined together in such a way that circulation of water is provided for adequate cooling of all parts, and the large indeterminate stresses of the fire-tube boilers are eliminated. With watertube designs it is possible to protect thick drums from the hot gases and resultant high thermal stresses. With correct operation, explosive failures have been essentially eliminated with watertube boilers. Further, the water space is divided into sections so arranged that, should any section fail, no general explosion occurs and the destructive effects are limited.

The watertube construction facilitates obtaining greater boiler capacity, and the use of higher pressure. In addition, the watertube boiler offers greater versatility in arrangement, and this permits the most efficient use of the furnace, superheater, reheater, and other heat-recovery components.

Watertube boilers may be classified as straight-tube and bent-tube. Straight-tube boilers have been supplanted by updated designs of bent-tube boilers, which are more economical and serviceable than the straight-tube designs.

BOILING PROCESS

The majority of fossil-fuel steam generators and all current commercial nuclear steam supply systems operate at subcritical pressures. A comprehension of the boiling process is essential in the design of these units.

Circulation. In contemporary boilers, as much as 1 to 3 cubic feet of steam per minute may be formed in a 5-foot length of 2½-inch tubing in the furnace area. Water must be continuously supplied for this steam generation and, in many designs, excess water is provided to protect the tubes from overheating. The movement of water and steam (within the boiler) which provides this water supply and removes the steam generated is known as *circulation*. In some boiler designs, circulation is accomplished by a pump. In all supercritical-pressure boilers, a pump must be provided to maintain circulation. However, at pressures below the critical, "natural" circulation occurs inherently in a watertube boiler if the heating surface is disposed so that the steam is free to rise in inclined or vertical tubes as it forms.

One of the important criteria in the design of a boiler or nuclear reactor is the ability to

*The principal portion of this description is extracted with permission from "Steam: Its Generation and Use," copyrighted by The Babcock & Wilcox Company.

remove the maximum amount of heat from a tube or flow channel at a rate equal to the maximum supplied by the source.

In order to obtain information on the nature of boiling, a number of investigators have experimented with electrically heated wires in a pool of water. Other investigators have performed the experiment of heating a tube or other type of flow channel cooled by a flow of water at a pressure below critical, and subjecting the tube to various levels of heat input.

Figure 1 is a generalized curve that summarizes the results of these investigations. This curve can be regarded as a general correlation of test results at a number of different heat

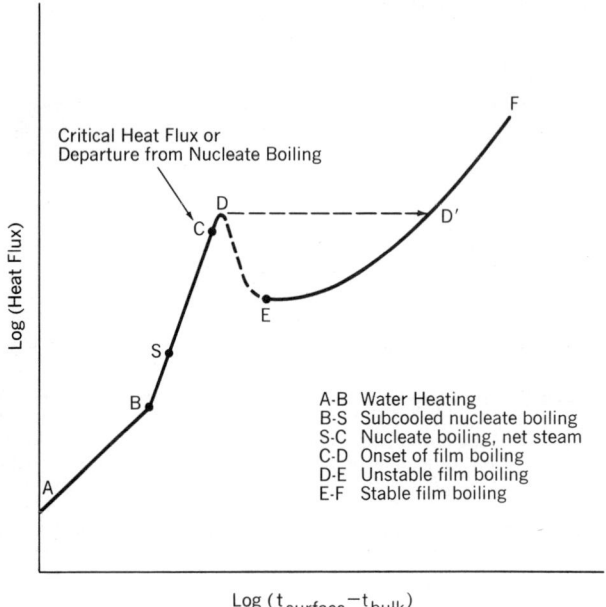

Fig. 1. Heat transfer to water and steam in a heated flow channel. Relation of heat flux to temperature difference between channel-wall and bulk-water or steam temperature. *(Babcock & Wilcox Company.)*

inputs to heated wires in a pool of water or to heated tubes or flow channels. It can also be regarded as a series of different heat inputs to a single flow channel. In this case, the points on the curve represent a series of temperature differences (surface temperature minus bulk water or steam temperature) corresponding to the water and steam conditions existing at a single location on the flow channel for different levels of heat flux or heat input. If the channel is evenly heated along its length, the location represented is the outlet end of the heated section of the channel. Absolute values on the curve are dependent on many factors, including pressure, flow-channel geometry, mass velocity, flux patterns, and degree of water subcooling.

For all heat input conditions (points on Fig. 1), water pressure and temperature at the inlet to the channel remain constant. Hence, the amount of subcooling (saturation temperature minus water temperature) at the inlet also remains constant. Ideally, water flow through the tube is maintained at a fixed rate.

The initial heat flux at point A is shown increasing on a logarithmic scale for points to the right of A. Until point B is reached, the heat input is not sufficient to produce boiling.

At B, the local heat flux is sufficient to raise the water temperature adjacent to the heated surface to saturation temperature, or slightly above, and a change from the liquid to the vapor state occurs locally. This change is characterized by the coexistence of both phases at essentially the same temperature locally, differing only in a few degrees of liquid superheat necessary for heat transfer, and by heat absorption required to overcome the molecular binding forces of the liquid phase. Here the change of state is accompanied by ebullition of the vapor as opposed to evaporation at a free surface, and the term *boiling* is used to describe the process. Also, the ebullition takes place at an interface other than that of the liquid and its vapor, actually at a solid-liquid interface. Hence, this boiling is described as *nucleate boiling*.

Nucleate Boiling.   The bulk of the water does not reach saturation temperature until the heat flux of point S is reached. Between B and S, the steam bubbles formed at the heated surface condense quickly in the main stream, giving up their latent heat to raise the temperature of the water. This condition is known as *subcooled-nucleate* or *local* boiling. Nucleate boiling occurs at all points up to C; beyond S, the bubbles do not collapse, since this part of the curve represents boiling with the water bulk temperature at saturation.

Both nucleate-boiling regimes, subcooled and saturated, are characterized by very high heat-transfer coefficients. These are ascribed to the high secondary velocities of water caused by the liberation of surface tension energies available in the liquid-vapor-solid interfaces at the instant of bubble release from the heating surface. This is a convection-type transfer coefficient based on bubble kinetics and is also affected to some extent by bulk mass velocity, depending on the velocity range. As the result of these high heat-transfer coefficients, tube- or flow-channel surface temperatures do not greatly exceed the saturation temperature.

Film Boiling.   Beyond the nucleate boiling region (B-C in Fig. 1), the bubbles of steam forming on the hot tube surface begin to interfere with the flow of water to the surface and eventually coalesce to form a film of superheated steam over part or all of the heating surface. This condition is known as *film boiling*. From D to E, film boiling is unstable; beyond point E, film boiling becomes stable.

Point of Departure from Nucleate Boiling.   In a fossil-fuel-fired boiler or in a nuclear reactor, when the local heat flux exceeds that corresponding to point D, the surface temperature may rise very quickly, along the horizontal dotted line in Fig. 1, to the point D'. If the temperature at D' is sufficiently high, heating surface burns out or melts. Hence D is known as the *burnout point*, and C, which may be very close to it, is known as the *point of departure from nucleate boiling (DNB)*, or the *critical heat flux*.

Stable and even unstable film boiling is acceptable in certain types of heat-transfer equipment where the temperature of the heat source is within the safe operating range of the equipment, or where the boiling film heat-transfer coefficient is the controlling resistance to heat flux. Steam generators for pressurized-water reactor systems, which are actually water-to-boiling-water heat exchangers, and certain types of process heat-exchange equipment are in this category. Film boiling conditions are also encountered in once-through boilers.

Factors Affecting DNB.   The point of departure from nucleate boiling *(DNB)* is defined in Fig. 1. The effect of steam quality and heat flux on the location of the *DNB* point are demonstrated by Fig. 2. This figure shows the various heat-transfer regimes taking place along the length of a uniformly heated vertical tube cooled by water flowing upward. On this figure, the inner wall temperatures are plotted as functions of enthalpy and steam quality, starting with hot water, passing through the region where steam is being generated (0 to 100 percent quality), and finally into the superheated region.

By following the line for moderate heat flux, it is seen that the metal temperature in the subcooled region is parallel to the water temperature, and only slightly above it. When boiling starts, the heat-transfer coefficient increases and the metal temperature remains just above saturation temperature. Finally, at high steam quality, the *DNB* point is reached where the nucleate boiling process breaks down. The metal temperature increases at this point but decreases again as steam quality approaches 100 percent. In the superheat region, the wall temperature again increases with, and approximately parallel to, the superheated steam temperature.

For the curve marked *high heat flux*, the *DNB* point is reached at a lower steam quality, and the peak metal temperature is higher. At very high heat fluxes, the *DNB* can occur in subcooled water. Avoidance of this last type of *DNB* is an important criterion in the design of nuclear reactors of the pressurized-water type.

Figures 1 and 2 present the *DNB* phenomenon from the standpoint of a heated flow channel in which flow, pressure, and inlet temperature (inlet subcooling) remain constant. *DNB* is also affected by variations in mass velocity, pressure, subcooling, and channel

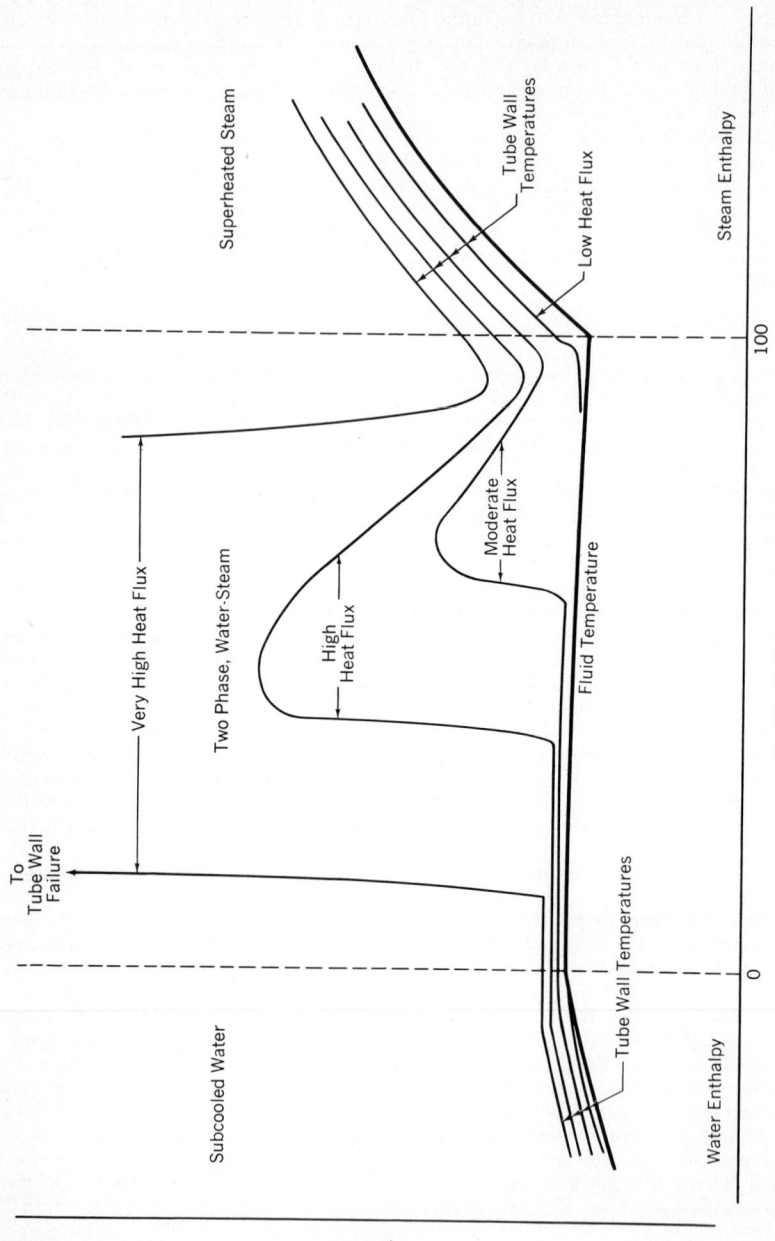

Fig. 2. Fluid and tube wall temperatures under conditions of water heating, nucleate boiling, film boiling, and superheating steam.

dimensions. See also the articles on "Boiling Water Reactor" and "Pressurized Water Reactor" in the Handbook section on *Nuclear Energy Technology.*

Many fossil-fuel boilers are designed to operate in the range between 2000 psi and the critical pressure. In this range, pressure has an important effect, shown in Fig. 3, in that the steam quality limit for nucleate boiling falls rapidly near the critical pressure; i.e., at constant heat flux, the *DNB* point occurs at a decreasingly low steam quality as pressure rises. Many correlations of critical heat flux or *DNB* have been proposed and are satisfactory within certain limits of pressure, mass velocity, and heat flux. Figure 4 is an example of a correlation which is useful in the design of fossil-fuel natural-circulation boilers. This correlation defines safe and unsafe regimes for two heat-flux levels at a given pressure in terms of steam quality and mass velocity. Additional factors must be introduced when tubes are used in membrane or tangent walls or in any position other than vertical. Such factors include inside diameter of tubes and surface condition. The last of these factors, where the character of the inside tube surface is purposely altered, will be discussed shortly.

The above discussion applies only to subcritical pressures. As the operating pressure is increased, the various flow and boiling regimes gradually disappear. However, there are tube metal temperature excursions in low-velocity supercritical-pressure operations simi-

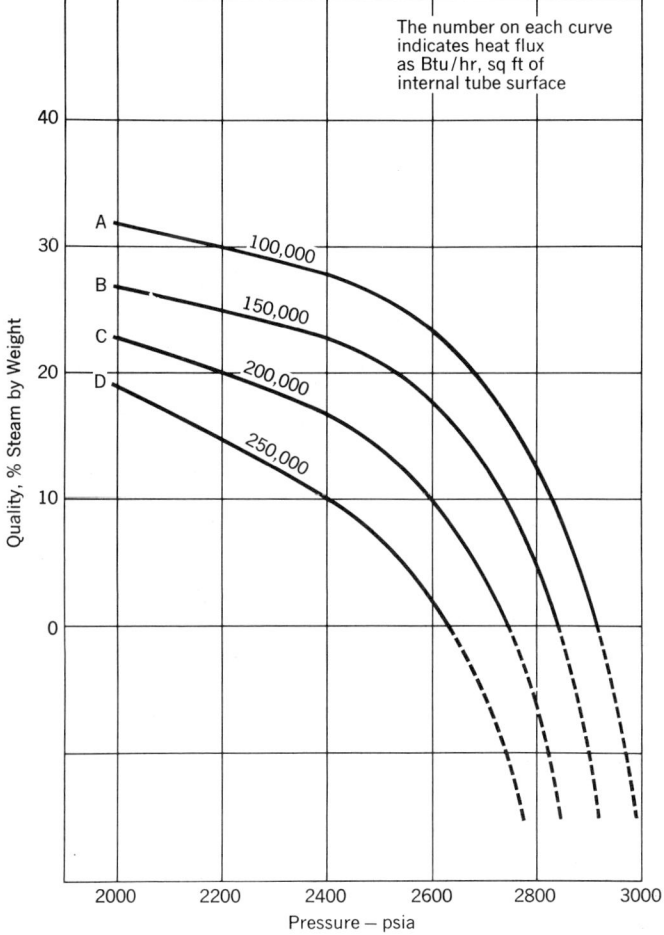

Fig. 3. Steam quality limit for nucleate boiling as a function of pressure.

lar to those found in subcritical boiling. This phenomenon, known as *pseudofilm boiling*, is under intensive experimental investigation.

Ribbed Tubes.   To inhibit or delay the onset of *DNB*, a large number of devices, including internal twisters, springs, and various grooved, ribbed, and corrugated tubes, have been tested.* The most satisfactory overall performance was obtained with tubes

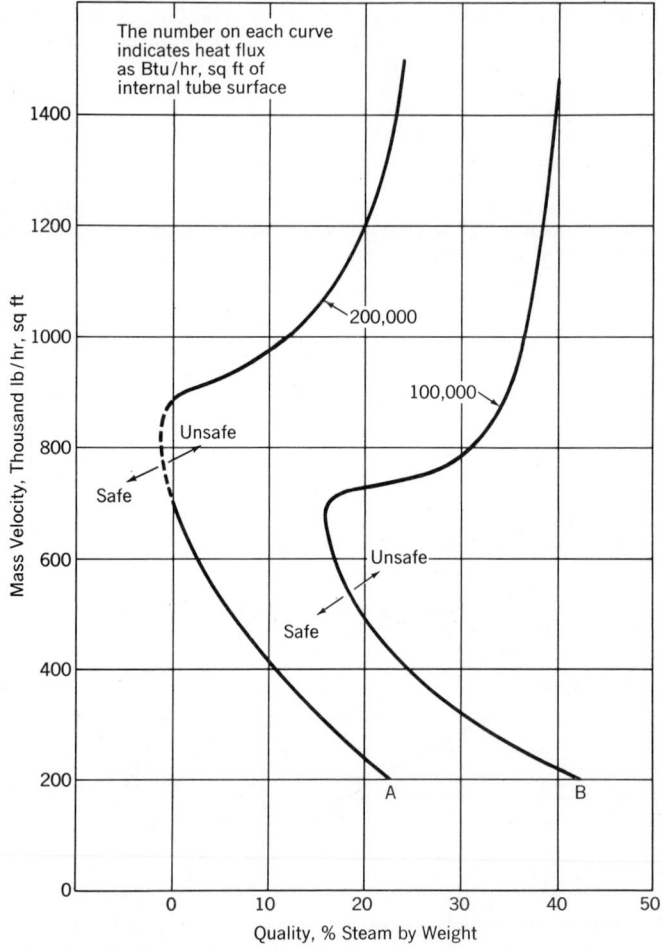

Fig. 4. Steam quality limit for nucleate boiling at 2700 psia, as a function of mass velocity.

having helical ribs on the inside surface. Two types of rib configuration were developed: (1) single-lead ribbed tube (Fig. 5a) for small internal diameters used in once-through subcritical-pressure boilers, and (2) multiple-lead ribbed tubes (Fig. 5b) for larger internal diameters used in natural-circulation boilers. Both these tubes have shown an effective ability to delay the breakdown of nucleate boiling.

Since the ribbed tube is more expensive than a smooth-bore tube, its utilization in design involves an economic balance of several design factors. In most instances, there is no incentive to use ribbed tubes below 2200 psi.

*Including work by The Babcock & Wilcox Company.

STEAM SEPARATION AND PURITY

Boilers operating below the critical point, except for once-through types, are customarily provided with a steam drum in which saturated steam is separated from the steam-water mixture discharged by the boiler tubes. Saturated steam leaves, and feedwater enters this drum through their respective nozzles (with some exceptions in multidrum boilers). However, the primary functions of this drum are to provide a free controllable surface for separation of saturated steam from water and a housing for any mechanical separating devices. Steam drums are designed to provide the volume necessary, in combination with the controls and firing equipment, to prevent excessive rise of water into the steam separators, resulting in carry-over of water with the steam.

Solids in Boiler Water.   Boiler water contains solid materials, principally in solution. Steam contamination (solid particles in the superheated steam) comes from the boiler water, largely in the carry-over of water droplets. Therefore, in general, as boiler-water

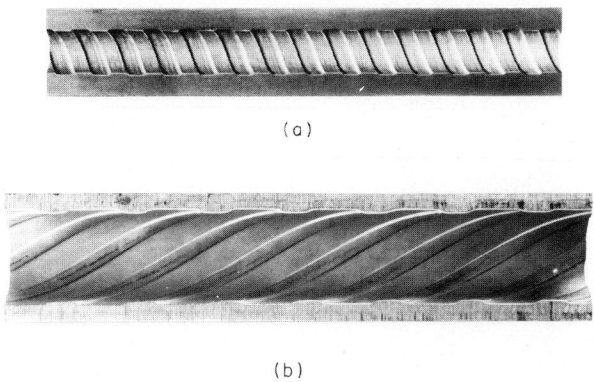

(a)

(b)

Fig. 5. Examples of ribbed tubes: (*a*) Single lead. (*b*) Multiple lead. (*Babcock & Wilcox Company.*)

concentration increases, steam contamination may be expected to increase. Historically, the carry-over of water into superheater tubes resulted in deposit of entrained solids in the superheater tubes. This caused increased tube temperatures and distortion and burnout of tubes. Therefore, it was necessary to develop devices to remove water from the steam. The need for extreme purity of steam for use in contemporary high-pressure turbines has provided additional incentive for reducing the carry-over of solids in steam. Troublesome deposits on turbine blades may occur with surprisingly low (0.6 ppm) total solids contamination of steam.

Factors Affecting Steam Separation.   Separation of steam from the mixture discharged into the drum from steam-water risers is related to both design and operating factors, which include:

Design factors:
  Design pressure
  Drum size, length, and diameter
  Rate of steam generation
  Circulation ratio—water circulated to heated tubes divided by steam generated
  Type and arrangement of mechanical separators
  Feedwater supply and steam discharge equipment and arrangement
  Arrangement of downcomer and riser circuits in the steam drum

Operating factors:
  Operating pressure
  Boiler load (steam flow)
  Type of steam load
  Chemical analysis of boiler water
  Water level carried

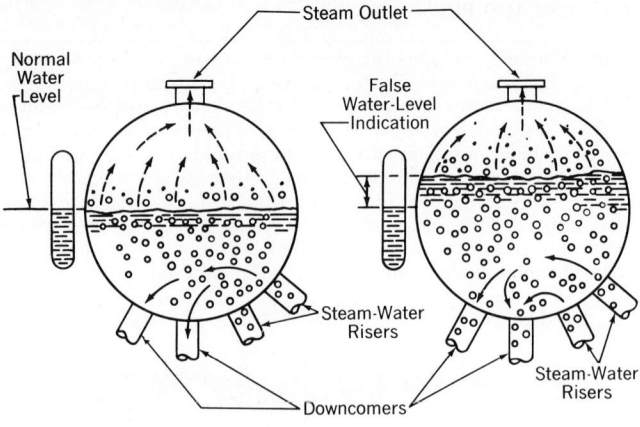

(a) Low Steaming Rate   (b) High Steaming Rate

Fig. 6. Effect of rate of steam generation on steam separation in a boiler drum without separation devices.

In steam drums without separation devices, where separation is by gravity only, the manner in which some of the above items affect separation is indicated in simplified form in Figs. 6 and 7.

Mechanical Steam Separators for Drums.   Gravity steam separation alone is generally unsatisfactory for boilers of the usual sizes and operating requirements. Most steam drums, therefore, are fitted with some form of primary separator. Simple types of primary separators are shown in Fig. 8. These facilitate or supplement gravity separation. The extent and arrangement of the various baffles and deflectors should always allow for access to the drums.

In the cyclone steam separator (B&W), shown in Fig. 9, centrifugal force many times the force of gravity is used to separate the steam from the water. Cyclones, essentially cylindrical in form, and corrugated scrubbers are the basic components of this type of separator.

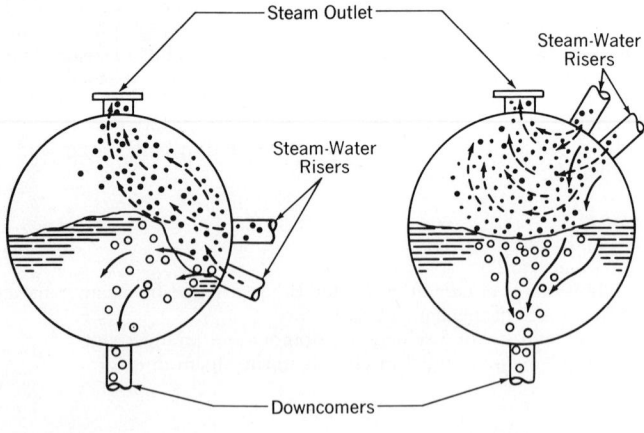

(a) Discharge Tubes Near   (b) Discharge Tubes Above
    Drum Center Line   Drum Center Line

Fig. 7. Effect of location of discharge from risers on steam separation in a boiler drum without separation devices.

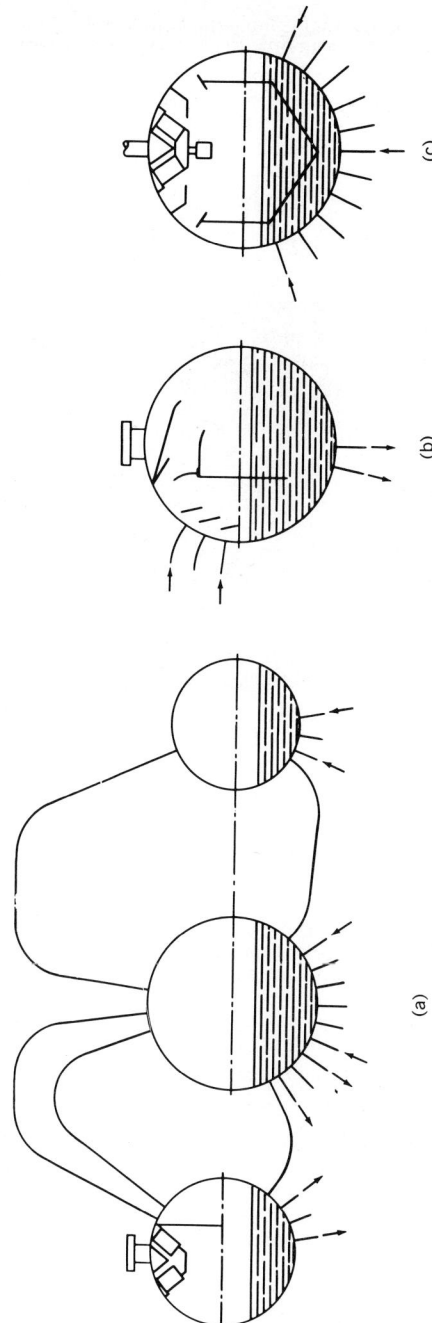

Fig. 8. Simple types of primary steam separators in boiler drums: (*a*) Deflector baffle. (*b*) Another type of deflector baffle. (*c*) Compartment baffle.

The cyclones are arranged internally along the length of the drum, and the steam-water mixture is admitted tangentially. The water forms a layer against the cylinder walls, and the steam (of less density) moves to the core of the cylinder and then upward. The water flows downward in the cylinder and is discharged through an annulus at the bottom, below the drum water level. Thus, with the water returning from drum storage to the

Fig. 9. Double-row arrangement of cyclone-type primary steam separators, with scrubber elements at top of drum for secondary separation. (*Babcock & Wilcox Company.*)

downcomers virtually free of steam bubbles, maximum net head is available for producing flow in the circuits, which is the important factor in the successful use of natural circulation. The steam moving upward from the cylinder passes through a small primary corrugated scrubber at the top of the cyclone for additional separation. Under many conditions of operation, no further refinement in separation is required, although the cyclone separator is considered only as a primary separator.

When wide load fluctuations and variations in water analyses are expected, large corrugated secondary scrubbers may be installed at the top of the drum to provide nearly perfect steam separation. These scrubbers may be termed *secondary separators.*

The combination of cyclone separators and scrubbers just described provides the means for obtaining steam purity corresponding to less than 1.0 ppm solids content under a wide variation of operating conditions. This purity is generally adequate in commercial practice. However, further refinement in steam purification is required where it is necessary to remove boiler-water salts, such as silica, which are entrained in steam by vaporization or solution mechanism. Washing the steam with condensate or feedwater of acceptable purity may be used for this purpose.

Steam-Washing for Silica-Laden Steam.   It is often impractical to maintain boiler-water concentrations of silica sufficiently low to prevent turbine fouling, and other measures, such as steam-washing, are used to control this type of steam contamination. In steam-washing, silica-laden steam is brought into intimate contact with relatively pure wash water, such as condensate or feedwater, and silica is absorbed from the steam by the wash water.

A steam-drum arrangement employing steam-washing is shown in Fig. 10. The drum is equipped with primary mechanical separators of the centrifugal type and corrugated scrubbers. Steam leaving the primary separators flows to a steam-washer arranged in the top of the steam drum. The washer consists of a rectangular column approximately the

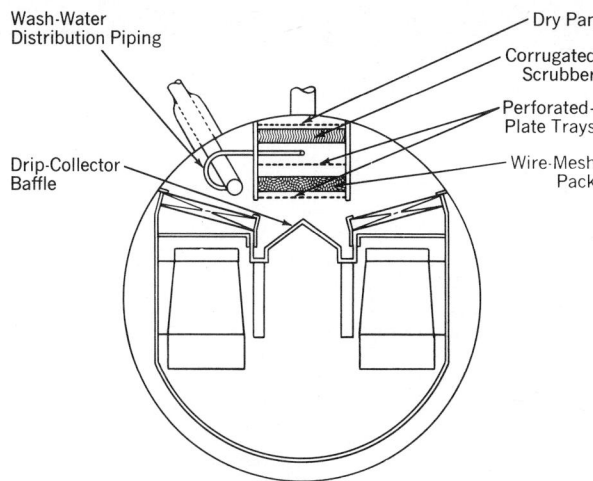

Wash-Water
Distribution Piping

Dry Pan

Corrugated
Scrubber

Perforated-
Plate Trays

Drip-Collector
Baffle

Wire-Mesh
Pack

Fig. 10. Arrangement of steam-drum internals for washing silica-laden steam.

length of the steam drum. Steam passes vertically upward through a perforated plate, a pack of stainless steel wire mesh, a second perforated plate, and finally a corrugated scrubber element. Wash water enters the drum through a nozzle and flows downward through the washer, counterflow to the steam. The steam velocity through the tray perforations maintains, above each tray, a layer of wash water which is kept in violent agitation by the steam. The wire mesh provides a large surface area for achieving intimate contact between the steam and the wash water.

BENT-TUBE BOILERS

Many important contemporary designs of boilers, such as the two-drum Stirling, the integral-furnace, the radiant, and the universal-pressure, are included in the *bent-tube* classification. All bent-tube boilers, with the exception of those with stoker or flat refractory floors, have water-cooled walls and floors or hoppers. The principal designs are illustrated in Figs. 11 to 16.

Fig. 11. Integral-furnace boiler (Type FM). Shop-assembled unit, complete and ready to operate. (*B & W.*)

Integral-Furnace Boiler.   This is a two-drum boiler which, in the smaller capacities, is adaptable to shop assembly and shipment as a package. Figure 11 shows a low-capacity (Type FM integral-furnace) boiler designed for shop assembly. This package boiler is shipped complete with support steel, casing, forced-draft fan (unmounted in larger sizes), firing equipment, and controls—ready for operation when water, fuel, and electrical

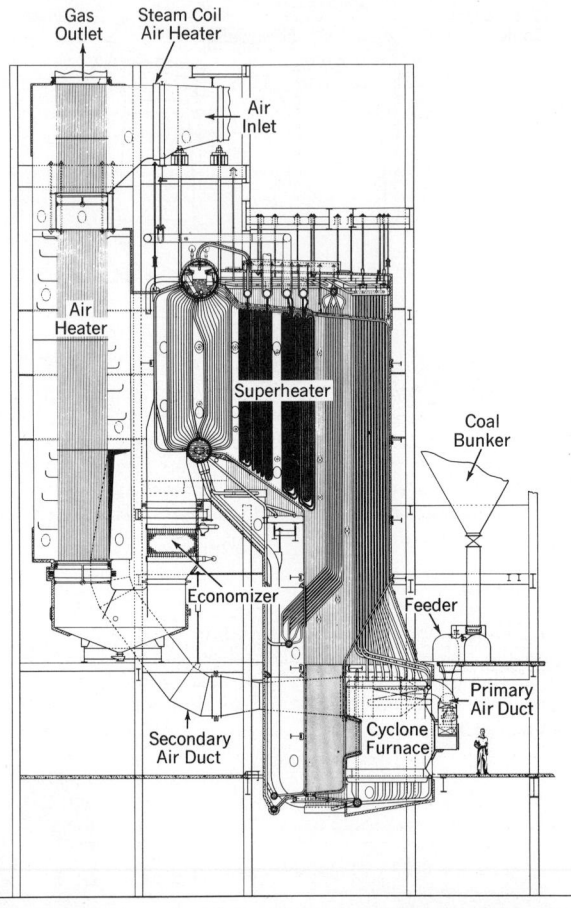

Fig. 12.   Two-drum Stirling boiler for cyclone-furnace firing. Design pressure 1575 psi; steam temperature 900°F (482°C); maximum continuous steam output 550,000 pounds/hour. *(B & W.)*

connections are made. Only a stub stack is required. It is built for outputs from 8000 to 160,000 pounds of steam per hour. Steam pressures range to 925 psi and temperatures to 825°F (441°C). Units can be fired with oil, gas, or a combination of the two fuels. Only a forced-draft fan is required, as the casing is airtight (welded) and the combustion gases are under pressure. Size of the unit is varied principally by changes in setting depth and drum length. Two combinations of width and height facilitate standardization of parts and assembly.

Two-Drum Stirling Boiler.   The simple arrangement possible for the connecting tubes, with one upper steam drum directly over one lower drum, led to the development of a series of designs known as the two-drum Stirling boiler (Fig. 12). These designs are standardized over a wide range of capacities and pressures, with steam flows varying from 200,000 to 1.2 million pounds per hour, design pressures to 1750 psi, and steam temperatures up to 1000°F (538°C). The firing may be by cyclone furnace, pulverized coal, oil, or gas. The two-drum Stirling boiler is furnished for industrial and utility applications. It may be considered as a transition unit, covering an intermediate size range. Because of its versatility and economy, the design enjoys worldwide acceptance.

High-Pressure and High-Temperature Boilers.   In the rapid development of power-plant economy, the single-boiler, single-turbine combination has been adopted for central station use, where electric power is the end product of heat transformation. There is an incentive to use very large electrical generators, since the heat rate, investment, and labor costs decrease as size increases. In the design of large boiler units for this application, the important factors are (1) high steam pressure, (2) high steam temperature, (3) bleed feedwater heating, and (4) reheat. High steam pressure means high saturation temperature and low temperature difference between steam and exit gas. High steam temperature

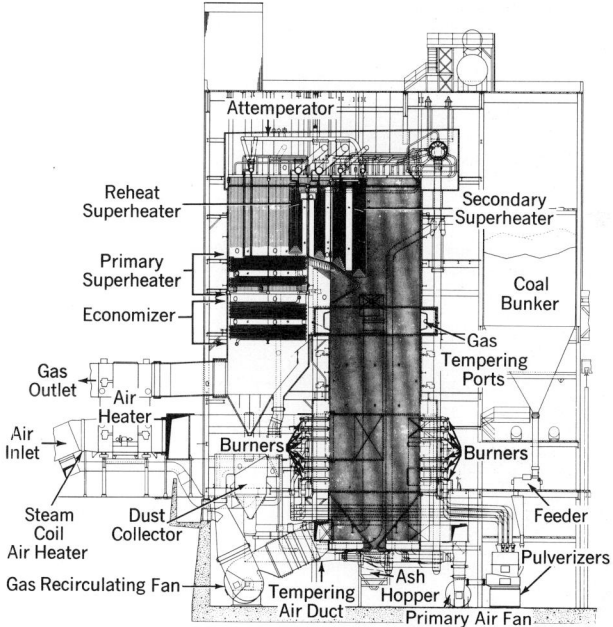

Fig. 13. Radiant boiler for pulverized coal firing. Design pressure 2875 psi; primary and reheat steam temperatures of 1000°F (538°C); maximum continuous steam output 1,750,000 pounds/hour. *(B & W.)*

means high initial temperature and, usually, reheating to high temperature for reuse of the steam. Bleed feedwater heating lowers the mean temperature difference in an economizer and increases the gas temperature leaving the economizer. An air heater is then required to lower the exit-gas temperature. These factors and, above all, the economic need for continuity of operation to realize an optimum return on large investment involved have combined to produce boiler units different in many respects from earlier concepts. Thus, the principle of the integrated boiler unit is firmly established for very large boilers, as well as for boilers of smaller outputs.

As steam pressures have increased, steam temperatures also have increased. This necessitates proportionally more superheating surface and less boiler surface. When pressures exceed 1500 psi in a drum-type boiler, the heat absorbed in furnace and boiler-screen tubes is normally almost enough to generate the steam. Thus, it is usually more economical to use economizer surface for any additional evaporation required as well as to raise the feedwater to saturation. All the steam is then generated in the furnace, water-cooled wall enclosures of superheater and economizer, boiler screen, division walls, and, in some cases, the outlet end of a steaming economizer.

Radiant Boiler.   This boiler, shown in Figs. 13 and 14, is a high-pressure, high-temperature, high-capacity boiler of the drum type. It is adaptable to pulverized-coal or cyclone-furnace firing, and also to natural gas and oil firing. Boiler convection surface is a minimum in these units. The unit shown in Fig. 13 is a pulverized-coal-fired unit with hopper-bottom construction for dry-ash removal. It has an output of 1.75 million pounds of steam per hour for continuous operation. Design pressure is 2875 psi, and primary and reheat steam temperatures are 1000°F (538°C). Preengineered boilers of this type are available in capacities from 300,000 to 7 million pounds of steam per hour, corresponding to approximately 40 to 1000 megawatts of electric-power-generating capacity. Standard components, such as furnaces, superheaters, reheaters, economizers, and air heaters, are integrated to coordinate the fuel fired with the turbine throttle requirements. Standard

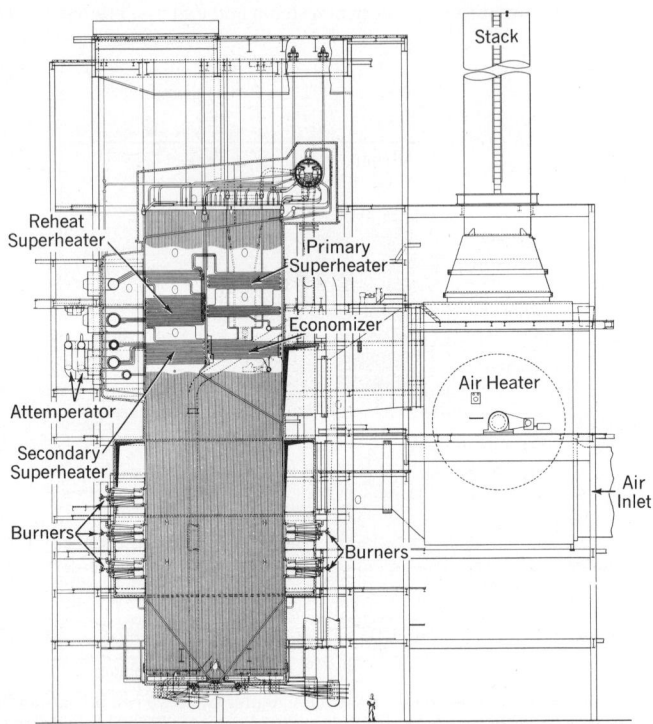

Fig. 14. El Paso type radiant boiler for natural gas and oil firing. Superheater outlet pressure 2625 psi; primary and reheat steam temperature 1005°F (541°C); capacity 3,777,000 pounds of steam per hour. (*B & W.*)

sizes are available in reasonable increments of width and height to permit selection of an economical unit for the required steam conditions and capacity.

Figure 14 illustrates the El Paso-type radiant boiler, a standardized unit developed for natural gas and oil firing. This compact and economical design is suitable for these fuels because of the cleanliness of natural gas and the relatively minor ash problems encountered with oil, as compared with coal.

Universal-Pressure Boiler.   This is a high-capacity, high-temperature boiler of the once-through or Benson type. See Figs. 15 and 16. Functionally applicable at any boiler pressure, it is applicable economically in the pressure range from 2000 to 4000 psi. Firing may be by coal, either pulverized or cyclone-furnace-fired, or by natural gas or oil.

The working fluid is pumped into the unit as liquid, passes sequentially through all the pressure-part heating surfaces, where it is converted to steam as it absorbs heat, and leaves as steam at the desired temperature. There is no recirculation of water within the unit, and, for this reason, a drum is not required to separate water from steam. The universal-pressure boiler may be designed to operate at either subcritical or supercritical pressures. The basic design and concept of this boiler permits very large installations. Figure 15 illustrates a pulverized-coal-fired unit for a 1300-megawatt-capacity electric generating unit. Figure 16 shows a natural-gas-fired installation, supplying steam for a 750-megawatt generator.

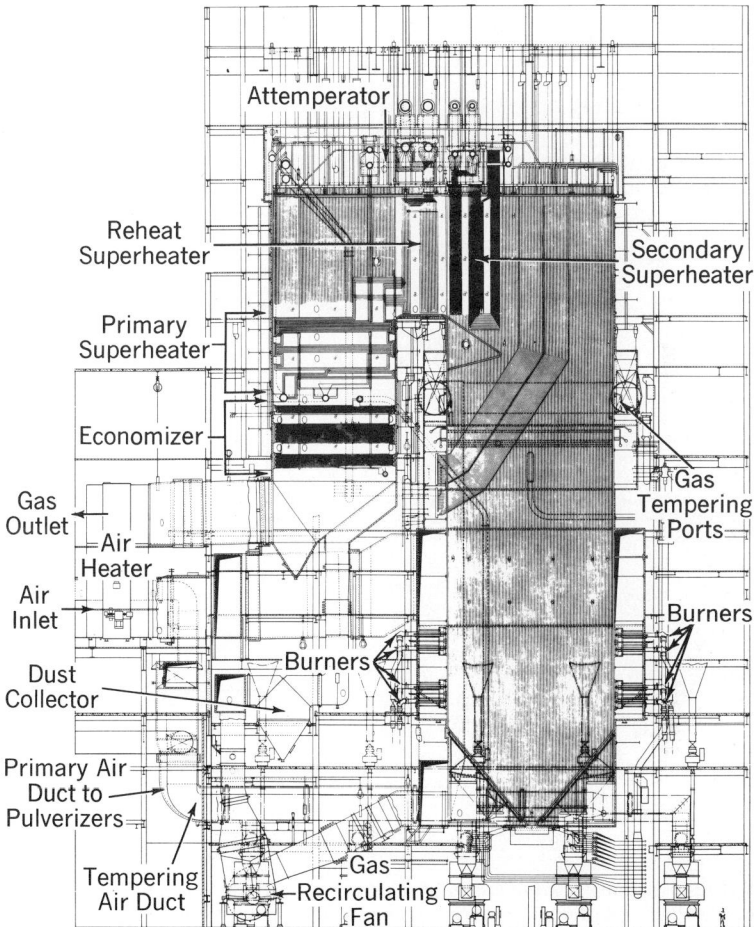

Fig. 15. Universal-Pressure boiler for pulverized-coal firing. Superheater outlet pressure 3650 psi; primary and reheat steam temperature 1003° F (~ 540°C); capacity 9,300,000 pounds of steam per hour. (*B & W.*)

BOILER DESIGN

A boiler may be a unit complete in itself without auxiliary heat-absorbing equipment, or it may constitute a rather small part of a large steam-generating complex in which the steam is generated primarily in the furnace tubes, and the convection surface consists of a superheater, reheater, steaming economizer, and air heater. In the latter case, it is possible to consider that a drum-type boiler comprises only the steam drum and the screen tubes between the furnace and the superheater. However, the furnace waterwall tubes, and usually a number of sidewall and support tubes in the convection portion of the unit, discharge steam into the drum and, therefore, effectively form a part of the boiler.

Major Factors in Boiler Design.   In the case of the universal-pressure boiler, there is no steam drum but rather an arrangement of tubes in which steam is generated and superheated. Whether the boiler is a drum or once-through type, whether it is an

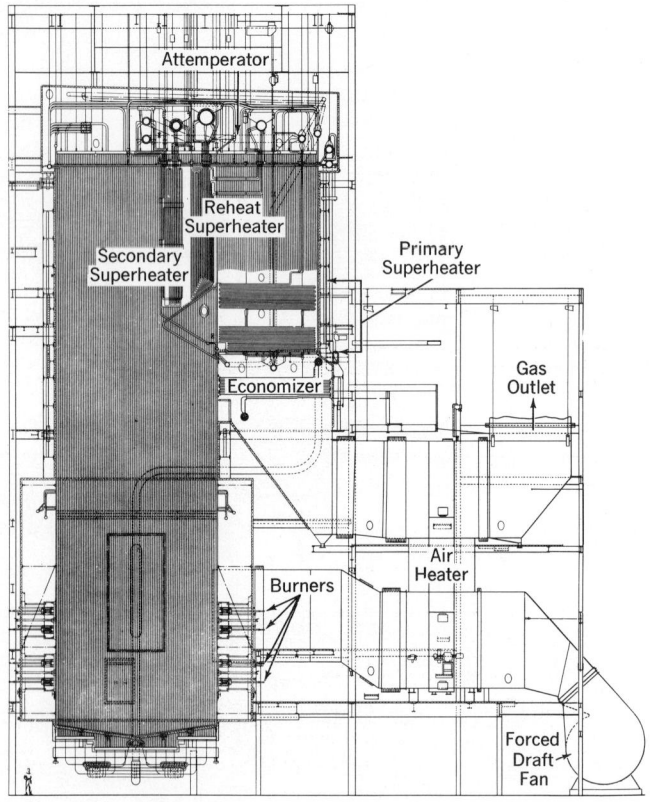

Fig. 16. Universal-Pressure boiler for natural gas firing. Superheater outlet pressure 3850 psi; primary and reheat steam temperature 1005°F (~ 540°C); capacity 5,455,000 pounds of steam per hour. (*B & W.*)

individual unit or a small part of a large complex, it is necessary in design to give proper consideration to the performance required from the total complex of the steam generating unit. Within this framework, the important items which must be accomplished in boiler design include:

1. Determine the heat to be absorbed in the boiler and other heat-transfer equipment, the optimum efficiency to use, and the type of fuel or fuels for which the unit is to be designed. When a particular fuel is selected, determine the amount of fuel required, the necessary or preferred preheated-air temperature, and the quantities of air required and flue gas to be generated.

2. Determine the size and shape required for the furnace, giving consideration to location, the space requirements of burners or fuel bed, and incorporating sufficient furnace volume to accomplish complete combustion. Provision must also be made for proper handling of the ash contained in the fuel, and water-cooled surface must be provided in the furnace walls to reduce the gas temperature leaving the furnace to the desired value.

3. The general disposition of convection heating surfaces must be so planned that the superheater and reheater, when provided, are located at the optimum temperature zone

where the gas temperature is high enough to afford good heat transfer from the gas to the steam, yet not so high as to result in excessive tube temperatures or excessive fouling from ash in the fuel.

While there is flexibility in the location of saturation or boiler surface, there must be enough total convection surface either before or after the superheater to transfer the heat required to heat the feedwater to saturation temperature and to generate the remainder of the steam required which is not generated in the furnace. This can be accomplished without an economizer, or an economizer can be provided to heat the feedwater to saturation temperature or even to generate up to 20 percent of the full-load steam requirement.

The foregoing must be accomplished in a design that provides for proper cleanliness of heating surfaces without buildup of slag or ash deposits and without corrosion of pressure parts.

4. Pressure parts must be designed in accordance with applicable codes using approved materials with stresses not exceeding those allowable at the temperatures experienced during operation.

5. A tight boiler setting or enclosure must be constructed around the furnace, boiler, superheater, reheater and air heater, and gas-tight flues or ducts must be provided to convey the gases of combustion to the stack.

6. Supports for pressure parts and setting must be designed with adequate consideration for expansion and local requirements, including wind and earthquake loading.

Combustion Data. The basis for the designer's selection of equipment includes factors involved in the selection of fuels. In some areas, there are several fuels available, and their availability and cost may be expected to change during the lifetime of the plant, with the result that the unit must be designed to burn more than one fuel. It is usually possible to determine which fuel is the most difficult from the standpoint of combustion and ash handling, and the unit is, therefore, designed for the most difficult fuel which will be used.

After steam requirements—steam flow, steam pressure, and temperature—and boiler feedwater temperature are determined, the required rate of heat absorption, q, is determined from

$$q = w'(h'_2 - h'_1) + w''(h''_2 - h''_1) \tag{1}$$

where q = rate of heat absorption, Btu/hour
w' = primary steam or feedwater flow, pounds/hour
w'' = reheat steam flow, pounds/hour
h'_1 = enthalpy of feedwater entering, Btu/pound
h'_2 = enthalpy of primary steam leaving superheater, Btu/pound
h''_1 = enthalpy of steam entering reheater, Btu/pound
h''_2 = enthalpy of steam leaving reheater, Btu/pound

To determine unit efficiency, it is necessary to know the temperature of the flue gas leaving the unit. This temperature may be set at the point where further addition of heating surface to reduce gas temperature would not be justified by the increased economy obtained. In the case of sulfur-bearing fuels, flue gas temperature is usually kept above the dew point to avoid sulfur corrosion of economizer or air heater surfaces.

The efficiency of combustion is 100 minus the sum of the heat losses expressed in percent. For a fuel with known characteristics and a given flue gas temperature, heat losses are evaluated by methods described under the prior article "Principles of Combustion in Steam Generation" in this Handbook section. The fuel input rate is then determined from Eq. (1) and the following equation:

$$w_F = \frac{q}{Q_H \times \text{eff}} \tag{2}$$

where w_F = fuel input rate, pounds/hour
Q_H = high heat value of fuel, Btu/pound
eff = efficiency

From the quantity of fuel to be burned per hour, the corresponding weight of air required and the weight of combustion gases produced are determined.

FURNACE DESIGN

When pulverized-coal or cyclone-furnace firing is used, the wall(s) in which the burners or cyclones are located must be designed to accommodate them and the necessary fuel- and air-supply lines. Minimum clearances, established by experience, must be maintained between burners to avoid interference of the fuel streams from the various burners with each other. Minimum clearances must also be provided between burners and sidewalls and between each burner and the opposite wall to avoid flame impingement on furnace walls with consequent possible overheating of wall tubes or excessive deposits of ash or slag.

Where fuel is burned on stokers or hearths, the size of the furnace is usually set by providing a plan area based on a specified release of heat per square foot of grate area per hour. These fuel-burning rates are based on experience and vary for different types of stokers.

The furnace must also be proportioned so that combustion is completed with due regard to the factors of temperature, turbulence, and time. Adequate combustion temperature is partly a function of burner (or other fuel-burning equipment) design, i.e., the burner equipment must be designed to provide the proper mixing of air and gas, and it must be sized for the job to be done so that the burner is not over- or underloaded. Likewise, there must not be too much cooling surface in the furnace in proportion to the fuel to be burned. Preheated air is beneficial in obtaining adequate combustion temperature and is required for pulverized-coal and cyclone-furnace firing, as well as for residual or heavy oils.

Turbulence is primarily a function of the fuel-burning equipment, and its importance lies in supplying air, not only to individual fuel particles, but also to any unburned or partially burned gases until combustion is completed. The time factor is fulfilled primarily by providing sufficient furnace volume so that the combustion gases remain in the furnace long enough to assure completeness of combustion.

Water-Cooled Walls   Most contemporary boiler furnaces have all walls water-cooled. This not only reduces maintenance on the furnace walls but also serves to reduce the gas temperature entering the convection bank to the point where slag deposit and superheater corrosion can be controlled by sootblowers. Furnace wall tubes are spaced on close centers to obtain maximum heat absorption. Tangent tube construction, used on earlier units, has been replaced with "membrane walls" in which a steel bar or membrane is welded between adjacent tubes. This construction, used on both natural- and forced-circulation units for all types of firing, consists of flat wall sections composed of panels of single rows of tubes on centers wider than a tube diameter connected by means of a membrane bar securely welded to the tube on its center line. This results in a continuous wall surface of rugged, pressure-tight construction capable of transferring a maximum amount of heat to the tube. The individual panels are of a width and length suitable for economical manufacture and assembly, with bottom and top headers attached in the shop prior to shipment for field assembly. Membrane walls with refractory lining are used in the lower furnace walls of cyclone-fired units.

Handling of Ash.   In the case of coal and, to a lesser extent, with oil, a very important consideration is the presence of ash in the fuel. If this ash is not properly considered in the design and operation, it can and does deposit not only on furnace walls and floor but through the convection banks. This not only reduces the heat absorbed by the unit but also increases draft loss, corrodes pressure parts, and eventually can cause shutdown of the unit for cleaning and repairs. With certain fuels, the proper handling of ash constitutes an overriding consideration in the design of furnaces, boilers, and other heat-transfer equipment. In coal-fired furnaces, the ash problems are more severe. There are two approaches to the handling of ash, namely, (1) the dry-ash furnace and (2) the slag-tap furnace.

1. *Dry-Ash Furnace.*   In this type of furnace, which is particularly applicable to coals with high ash fusion temperatures, the furnace is provided with a hopper bottom and with sufficient cooling surface so that the ash impinging on furnace walls or hopper bottom is solid and dry and can be removed essentially as dry particles. When pulverized coal is burned in a dry-ash furnace, about 80 percent of the ash is carried through the convection banks. Most of this fly ash is normally removed by particulate-removal equipment located

just ahead of the stack. See "Particulate Emission Measurement Methods" later in this Handbook section.

2. *Slag-Tap Furnace.* With many coals having low ash fusion temperatures, it is difficult to utilize a dry-bottom furnace because the slag is either molten or sticky and tends to cling and build up on furnace walls and bottom. The slag-tap furnace has been developed to handle coals of these types. The most successful form of the slag-tap furnace is that used in conjunction with cyclone-furnace firing. The furnace comprises a two-stage arrangement. In the lower part of the furnace, gas temperature is maintained high enough so that the slag drops in liquid form onto a floor where a pool of liquid slag is maintained and tapped into a slag tank containing water. In the upper part of the furnace, the gases are cooled below the ash fusion point so that ash carried over into the convection banks is dry and does not adhere.

Convection Boiler Surface.   The gas temperature leaving the furnace or entering the boiler depends mainly on the ratio of the heat released to the amount of furnace-wall cooling surface installed. Because the cost of furnace-wall cooling surface is relatively higher than that of boiler surface, the furnace size and surface are limited to the amount required to lower sufficiently the gas temperature entering the convection tube banks to avoid ash deposits.

The first few rows of tubes in the convection bank may be boiler tubes widely spaced to provide gas lanes wide enough to prevent plugging with ash and slag and to facilitate cleaning. These widely spaced boiler tubes are known as the *slag screen* or *boiler screen.* In many large units, they are used to support the furnace rear wall tubes. These screen tubes receive heat by radiation from the furnace, and by radiation and convection from the combustion gases passing through them.

In the larger contemporary units, the superheater generally replaces the boiler screen or, if not, is located immediately beyond it (Fig. 17). The gas temperature entering the superheater must be high enough to give the superheat desired with a reasonable amount

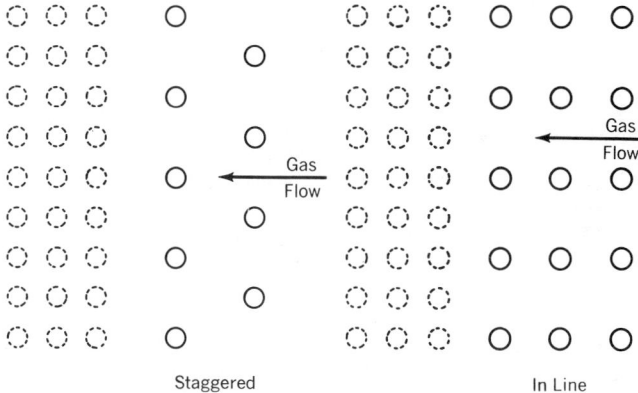

Staggered   In Line

Fig. 17. Arrangement of slag screens ahead of tube banks.

of heating surface and the use of economical materials. The arrangement shown in Fig. 17 shows two or three rows of boiler-screen tubes ahead of superheater tubes on close spacing. However, it may also be necessary to locate several rows of superheater tubes on wide spacing.

Design of boiler surface after the superheater will depend on the particular type of unit selected, desired gas temperature drop, and acceptable gas pressure drop (draft loss) through the boiler surface. Typical arrangements of boiler surface for various types of boilers have been illustrated. The object in the design of convection heating surfaces is to establish the combination of tube diameter, tube spacing, length of tubes, number of tubes wide and deep, and gas baffling that will give the desired gas temperature drop with the pressure drop permissible.

Heating surface and pressure drop are directly interrelated since both are primarily dependent on gas mass velocity. If either heating surface or pressure drop is increased, the other must decrease in order to maintain the desired gas temperature drop (heat transfer). Hence, there is an optimum gas mass velocity which results in the optimum combination of heating surface and gas pressure drop.

For a given gas mass velocity (pounds of gas per hour per square foot of gas flow channel), or for a given gas velocity, a considerably higher gas film conductance, heat absorption, and draft loss result when the gases flow at right angles to the tubes (crossflow) than when they flow parallel to the tubes (longitudinal flow). Gas turns between tube banks generally add draft loss with little or no benefit to heat absorption and should be designed for easy flow.

From a long record of experience, given sets of conditions for each fuel to be burned have been effectively established as the conditions of economic practice. While these conditions vary as improvements occur over a period of years, at any particular time competitive economics act to hold most of the variables involved within a fairly limited range.

Design of Pressure Parts.   Boilers have achieved the safety and reliability which they now have through the use of sound materials and safe practices for determining acceptable stresses in tubes, drums, and other pressure parts. Boilers are always designed to applicable codes. Most stationary boilers in the United States are designed to the ASME Code. In each case, the stress allowable depends on the maximum temperature to which the part is subjected and, therefore, it is important that pressure parts be so designed that design temperatures are known and are not exceeded in operation. In boilers, the material temperatures are normally designed to be only a few degrees above the saturation temperature corresponding to the boiler pressure. In boiler tubes, this is accomplished by providing sufficient water to avoid the occurrence of departure from nucleate boiling (DNB). This means that an adequate supply of water must be provided for each tube, and this is particularly important in furnace and screen tubes where heat input is high.

Because steam drums have thick walls, it is necessary to limit the heat flow through them to avoid excessively high thermal gradients. Where the drum is penetrated by a number of tube holes, the flow of water through these holes serves to cool the ligaments between. Where the heat input through a drum would be too high, because of high gas temperature or velocity, insulation may be provided on the outside of the drum. This is particularly necessary where there are no tube penetrations to provide cooling.

In a drum-type boiler, equipment is provided in the steam drum for the reduction of moisture and solids in the steam to acceptable values. In once-through boilers, all moisture is evaporated in the tubes where boiling and superheating occur sequentially. In boilers of this type, steam purity depends on maintaining adequate purity of the feedwater.

The boiler safety valve constitutes a very important item in the safety of modern boilers. By law, the boiler design pressure for which the pressure parts must be designed cannot be less than the safety-valve relief pressure. As a practical matter, to avoid unnecessary losses and maintenance from frequent action of the safety valves, the first safety valve should be set to relieve at not less than the desired boiler operating pressure plus about 5 percent. The operating pressure in the boiler steam drum, in turn, depends on the pressure required at the point of use and the intervening pressure drop. As an example, where the steam is used in a turbine, the boiler-operating pressure is determined by adding to the turbine throttle pressure and pressure drop through the steam piping, nonreturn valve, superheater, and drum internals at maximum steam flow.

BOILER SUPPORTS

Boiler and furnace-wall tubes are usually supported by the drums or headers to which they are connected. In the design of proper supports, the following considerations are important:

1. The tubes must be so arranged that they will not be subject to excessive bending-moment stresses in carrying the weight of the tubes, drums, other parts which they support, and the contained water. When the unit is bottom-supported, the tubes must satisfy column requirements.

2. The holding strength of the tube seats must not be exceeded.

3. Provision must be made to accommodate the required expansion of the pressure

parts. For a top-supported unit, the hanger rods must be designed to swing at the proper angle, and they must be long enough to take the movement without excessive stresses in either the rods or the pressure parts. Bottom-supported boilers should be anchored only at one point, guided along one line, and allowed to expand freely in all other directions. To reduce the frictional forces and resultant stresses in the pressure parts, roller saddles or mountings are desirable for bottom-supported heavy loads.

SUPERHEATERS AND REHEATERS

Early in the nineteenth century, it was demonstrated that substantial savings in fuel could be experienced when steam engines were run with some superheat in the steam. In the late 1800s, lubrication problems were encountered with reciprocating engines, but once these were overcome, development of superheaters continued. Commercial development of the steam turbine hastened the general use of superheat. By 1920, steam temperatures of 650°F (343°C), representing superheats of 250°F (140°C), were generally accepted. In the early 1920s, the regenerative cycle, using steam bled from turbines for feedwater heating, was developed to improve station economy without going to higher steam temperatures. At the same time, superheater development permitted raising the steam temperature to 725°F (385°C). A further gain in economy by still higher temperature was at that time limited by allowable superheater tube-metal temperature. This led to the commercial use of reheat, where the steam leaving the high-pressure stage of the turbine was reheated in a separate reheat superheater and returned at higher temperature and enthalpy to the low-pressure stage.

The first reheat unit for a central station was proposed in 1922 and went into service in September 1924. It was designed for 650 psi and operated at 550 psi and 725°F. Exhaust steam from the high-pressure turbine was reheated to 725°F at 135 psi. A much higher-pressure reheat unit, designed in 1924 for 1200 psi and 700°F (371°C), went into service in December 1925.

When saturated steam is utilized in a steam turbine, the work done results in a loss of energy by the steam and consequent condensation of a portion of the steam, even though there is a drop in pressure. The amount of work that can be done by the turbine is limited by the amount of moisture which can be handled by the turbine without excessive wear on the turbine blades. This is normally somewhere between 10 and 15 percent. It is possible to increase the amount of work done by moisture separation between turbine stages, but this is economical only in special cases. Even with moisture separation, the total energy that can be transformed to work in the turbine is small compared to the amount of heat required to raise the water from feedwater temperature to saturation and then evaporate it. Thus, moisture constitutes the basic limitation in turbine design.

Because a turbine generally transforms the heat of superheat into work without forming moisture, the heat of superheat is essentially all recoverable in the turbine. This is illustrated in the temperature-entropy diagram of the ideal Rankine cycle where the heat added to the right of the saturated vapor line is shown as 100 percent recoverable. While this is not always entirely correct, the Rankine cycle diagrams in Fig. 18 indicate that this is essentially true in practical cycles. See also "Thermodynamics of Steam Cycles" given earlier in this Handbook section. Figure 2 of that article shows this same benefit from superheat in a regenerative cycle plus an additional substantial benefit from reheat, which also embodies the principle of high utilization of heat added above the saturated vapor line.

The foregoing discussion is not specifically applicable, however, at steam pressures in the vicinity of the critical point. The term *superheat* is not quite appropriate in defining the temperature of the working fluid at or above the critical point. However, even at pressures exceeding 3208 psia, heat added at temperatures above 705°F (374°C) is essentially all recoverable in a turbine.

Superheater Types.   The original and somewhat basic type of superheater and reheater was the convection unit, for gas temperatures where heat transfer by radiation was very small. With a unit of this type, the steam temperature leaving the superheater increases with boiler output because of the decreasing percentage of heat input that is absorbed in the furnace, leaving more heat available for superheater absorption. Since convection heat-transfer rates are almost a direct function of output, the total absorption in the superheater per pound of steam increases with increase in boiler output (Fig. 19). This

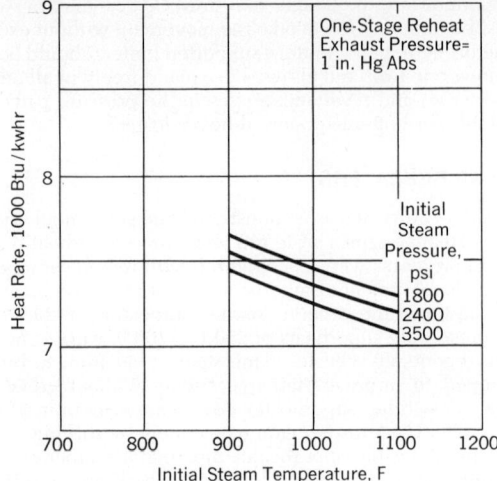

Fig. 18. Effects of changes in steam temperature and pressure on performance of ideal Rankine cycle with one-stage reheat.

effect is increasingly pronounced the further the superheater is removed from the furnace, i.e., the lower the gas temperature entering the superheater.

On the other hand, the radiant superheater receives its heat through radiation and practically none from convection. Because the heat absorption of furnace surfaces does not increase in direct proportion to boiler output, but at a considerably lesser rate, the curve of radiant superheat is a function of load slopes downward with increase in boiler output.

In certain cases, the two opposite-sloping curves have been coordinated by the combi-

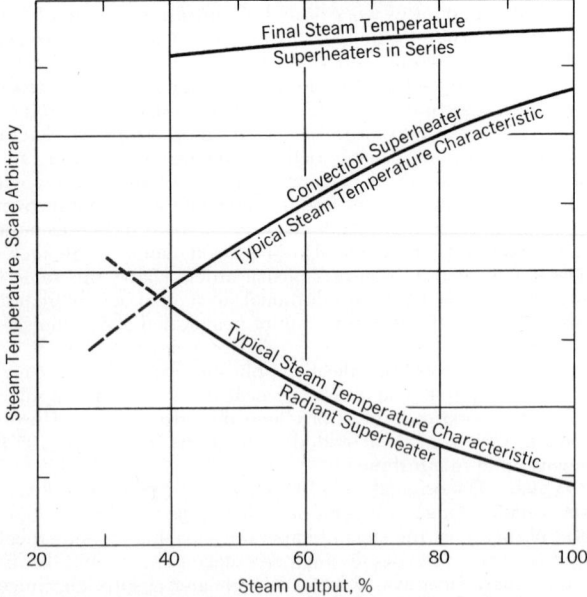

Fig. 19. A substantially uniform final steam temperature over a range of output can be attained by a series arrangement of radiant and convection superheater components.

nation of radiant and convection superheaters to give flat superheat curves over wide ranges in load, as typically indicated in Fig. 19. A separately fired superheater has the characteristic that it can be fired to produce a flat superheat curve.

Development of Superheaters.   The early convection superheaters were placed above or behind a deep bank of boiler tubes to shield them from the fire or from the higher-temperature gases. The greater heat absorption required in the superheater for higher steam temperatures made it necessary to move the superheater closer to the fire. This new location brought with it problems which were not apparent with the superheaters located in the original lower-gas-temperature zone. Steam- and gas-distribution difficulties and instances of general overheating of tube metal were ultimately resolved by improved superheater design, including higher mass velocity of the steam. This increased the heat conductance through the steam film, resulting in lower tube-metal temperatures, and also improved steam distribution by increasing pressure drop through the tubes.

Steam mass velocity in modern superheaters ranges from as low as 100,000 to 1 million pounds/(square foot)(hour) or higher, depending on pressure, steam, and gas temperatures, and the tolerable pressure drop in the superheater.

The fundamental considerations governing superheater design apply also to reheater design. However, the pressure drop in reheaters is critical because the gain in heat rate with the reheat cycle can be completely nullified by too much pressure drop through the reheater system. Hence, steam mass flows are generally somewhat lower in the reheater.

Tube Sizes.   Plain cylindrical tubes of 2- or 2½-inch outside diameter predominate in superheaters and reheaters in stationary practice. Smaller diameters (1- or 1¼-inch) are used to conserve weight and space in marine units. Steam pressure drop is higher and alignment is more difficult with the smaller diameters. Larger diameters bring about higher pressure stresses.

More recent designs have called for greater spans between supports for horizontal superheater tubes and for wider tube spacing or fewer tubes per row to avoid slag accumulation. The 2½-inch tube has met these new conditions with a minimum sacrifice of the smaller-tube advantages, and 3-inch tubes are used to advantage in some cases. When steam temperatures increase, the allowable stresses may force a return to the smaller-diameter, thinner-wall tube.

Plain tubes are used almost exclusively in contemporary superheater practice. Extended surface on superheater tubes in the form of fins, rings, or studs not only makes gas-side cleaning difficult, but the added thickness increases metal temperature and thermal stress beyond tolerable limits.

Relationships in Superheater Design.   Effective superheater design calls for the resolution of several factors. The outstanding considerations include:

   1. The steam temperature desired
   2. The superheater surface required to give this steam temperature
   3. The gas temperature zone in which the surface is to be located
   4. The type of steel, alloy, or other material best suited to make up the surface and the supports
   5. The rate of steam flow through the tubes (mass velocity), which is limited by the permissible steam pressure drop, but which, in turn, exerts a dominant control over tube-metal temperatures
   6. The arrangement of surface to meet the characteristics of the fuels anticipated, with particular reference to the spacing of the tubes to prevent accumulations of ash and slag, or to provide for the removal of such formations in their early stages
   7. The physical design and type of superheater as a structure

A change in any one of the first six items will call for a counterbalancing change in all other items.

The steam temperature desired in advanced power station design is the maximum for which the superheater designer and manufacturer can produce an economical structure. Economics in this case requires the resolution of two interrelated factors: (1) the first, or investment, cost, and (2) the later cost of upkeep for minimum operating troubles, outages, and replacements. A higher first cost is warranted if the upkeep cost is thereby reduced sufficiently to cover, within a reasonable time, the extra initial cost. The steam temperature desired is, therefore, based on the complete coordinated knowledge available for the optimum evaluation of the combination of the other five items and the necessities of the particular project. Operating experience in recent years has resulted in the use of approximately 1000°F (538°C) steam temperature both for primary superheat and reheat in nearly all large units procured for installation in the United States.

After the steam temperature desired is established or specified, the next consideration is the amount of surface necessary to give this superheat. The amount of superheater surface required is dependent on the four remaining items, and, since there is no single correlation, the amount of surface must be determined by trial, locating it in a zone of gas temperature that is likely to be satisfactory. In the so-called *standard boilers*, the zone is fairly well established by the physical arrangements and by the space preempted for superheater surface.

Steam mass velocity, steam pressure drop, and superheater tube-metal temperatures are calculated after the amount of surface is established for the trial location and the trial tube spacing. The proper type of material is then selected for the component tubes, headers, and other parts. It may be necessary to compare several arrangements to obtain an optimum combination that will:

1. Require an alloy of lesser cost
2. Give a more reasonable steam pressure drop without jeopardizing the tube temperatures
3. Give a higher steam mass velocity in order to lower the tube temperatures
4. Give a different spacing of tubes that will provide more protection against the ash accumulations with uncertain types of fuel
5. Permit closer spacing of the tubes, thereby making a more economical arrangement for a fuel supply that is known to be favorable
6. Give an arrangement of tubes that will reduce the draft loss for an installation where draft loss evaluation is crucial
7. Permit the superheater surface to be located in a zone of a higher gas temperature, with a consequent saving in surface, that will compensate for deviation from a standard arrangement

It is possible to achieve a practical design with optimum economic and operational characteristics and with all criteria reasonably satisfied, but a large measure of experience and the application of sound physical principles are required for satisfactory results.

Relationships in Reheater Design. The same general similarity exists between superheater and reheater considerations, but the reheater is limited to ruggedness of design by the permissible steam pressure drop. Steam mass velocities in reheater tubes should be sufficient to keep the steam-film temperature drop below 150°F (83°C). Ordinarily this may be done with less than 5 percent pressure drop through the reheater tubes. This allows another 5 percent pressure drop for the reheater piping and valves without exceeding the usual 10 percent total allowable.

Metals for Tubes. Oxidation resistance, maximum allowable stress, and economics determine the choice of materials for superheater and reheater tubes. The use of carbon steel should be extended as far as these considerations permit. Beyond this point, carefully selected alloy steels should be used. The steels commonly used for this application are shown in Table 1, which also lists the maximum allowable stress values.

Supports for Superheaters and Reheaters. Because superheaters and reheaters are located in zones of relatively high gas temperature, it is preferable to have the major support loads carried by the tubes themselves. In horizontal superheaters, the support load is usually transferred to the boiler or wall tubes by means of lugs, one welded to the boiler tubes and the other to the superheater tubes (Fig. 20). In many cases, these lugs are made of high-chromium-nickel alloy. They must slide on one another to provide relative movement between the boiler tubes and the superheater tubes. Supports of the saddle type, also shown in Fig. 20, provide for relative movement between adjacent components of the superheater itself.

As units grow in size, the span of the superheater tubes may become so great that it is inadequate to support them only at the ends. Most of the larger-size units utilize "stringer" tubes, generally from the economizer outlet, to support the superheater tubes, as indicated in Fig. 21. Tube spacing within the section is maintained by the use of saddle-type supports.

With pendant superheaters, the major support points are located outside the gas stream with the pendant loops supporting themselves in simple tension. Figure 22 shows a standard support arrangement for a pendant-superheater outlet section with major section supports above the roof line. Where adequate side spacing is available, steam-cooled

TABLE 1. Superheater and Reheater Tubes*

Material	ASME Specification number	Grade	Maximum allowable design stress, pounds/square inch, for metal temperatures not exceeding							
			°C 482 / °F 900	510 / 950	538 / 1000	566 / 1050	593 / 1100	621 / 1150	649 / 1200	704 / 1300
Carbon steel	SA210	A1	5,000	3,000						
Carbon moly	SA209	T1a	12,500	8,500						
Croloy ½	SA213	T2	12,500	10,000	6,250					
Croloy 1¼	SA213	T11	13,100	11,000	6,550	4,050				
Croloy 2¼	SA213	T22	13,100	11,000	7,800	5,800	4200			
Croloy 5	SA213	T5			5,600	4,150	3050	2000		
Croloy 9M	SA213	T9			8,500	5,500	3300	2200		
Croloy 18-8	SA213	TP304H†			13,750	12,150	9750	7700	6050	3700
Croloy 18-8	SA213	TP304H			9,750	9,500	8850	7700	6050	3700
Croloy 18-8	SA213	TP321H†			14,000	11,700	9050	6900	5350	3150
Croloy 18-8	SA213	TP321H			10,450	10,100	8800	6900	5350	3150

*The figures listed are the maximum allowable stress values in accordance with the "ASME Boiler and Pressure Vessel Code," sec. I, Power Boilers, June 30, 1970 addenda.

†Due to relatively low yield strength of these materials, these higher stress values were established at temperatures where the short-time tensile properties govern to permit the use of these alloys where slightly greater deformation is acceptable. The stress values in this range exceed 62½ percent but do not exceed 90 percent of the yield strength at temperature. Use of these stresses may result in dimensional changes due to permanent strain. These stress values are not recommended for the flanges of gasketed joints or other applications where slight amounts of distortion can cause leakage or malfunction.

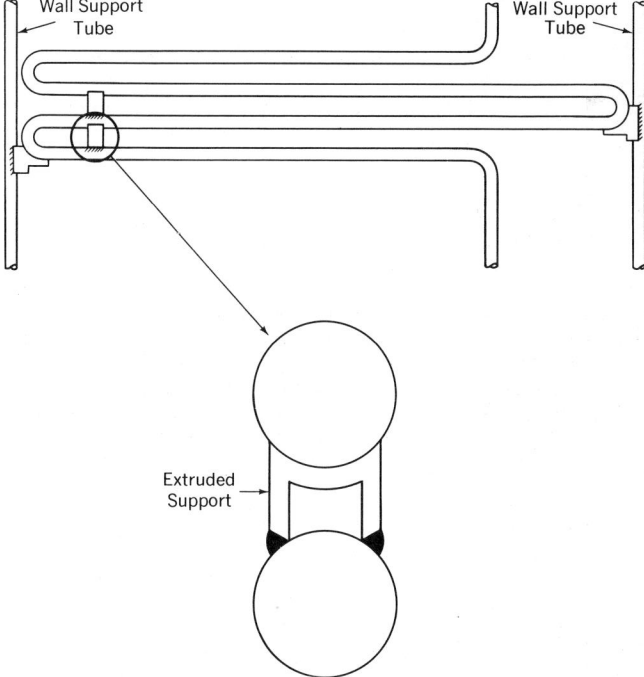

Fig. 20. Horizontal superheater section end-supported on walls. *(B & W.)*

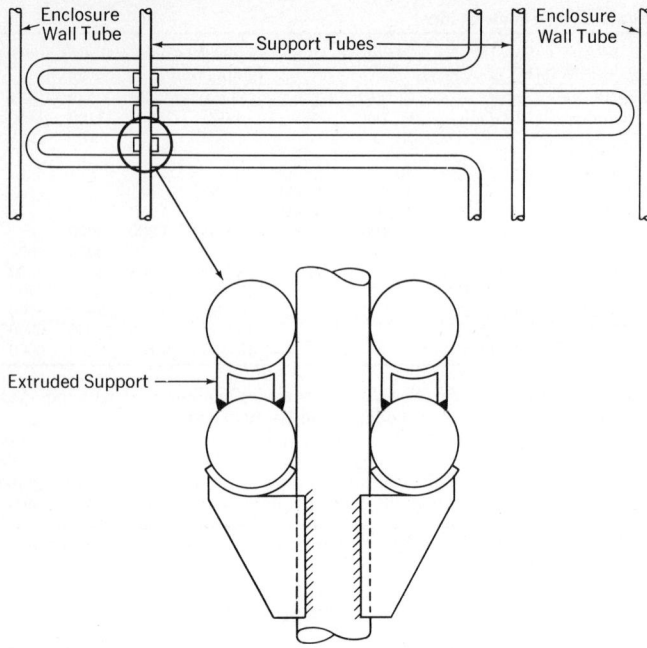

Fig. 21. Horizontal superheater section with intermediate stringer supports. *(B & W.)*

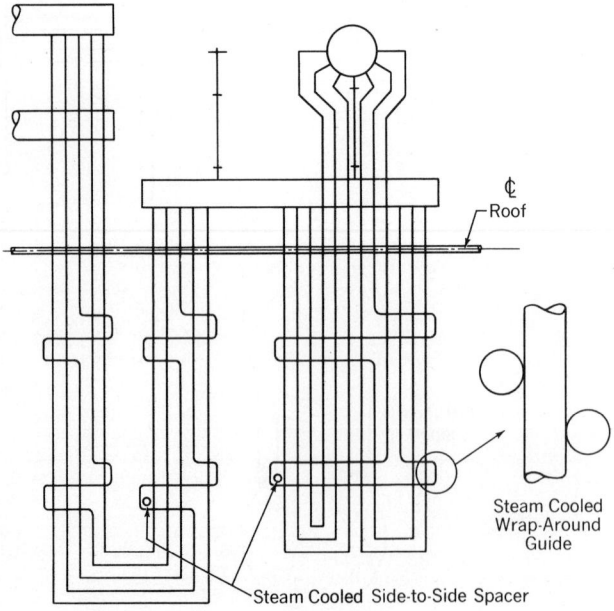

Fig. 22. Secondary superheater section. *(B & W.)*

wraparound ties are used, thus radically reducing the amount of attachment to tubes and assuring maximum flexibility for accommodation of differential thermal expansion of individual loops within a section. In addition, in the higher-gas-temperature zones, steam-cooled side-to-side ties are utilized to maintain side spacings. For closer side-space elements, steam-cooled wraparounds are not practical, and mechanical ties are used to maintain alignment (Fig. 23). In this case, the clear back spacing between tubes and the size of attachments have been kept to a minimum. This serves to reduce the operating temperatures of the attachment and also to reduce the thermal stresses imposed on the

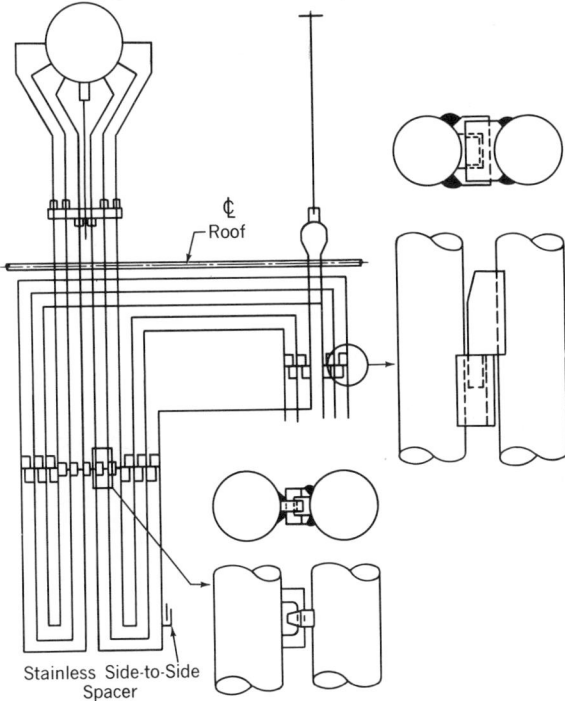

Fig. 23. Pendant reheater section. *(B & W.)*

tube wall. Figure 23 shows a typical arrangement of a pendant reheater section and illustrates the support of a separated bank by a special loop of reheater which permits all major supports to be kept above the roof and out of the gas stream.

Internal Cleaning. In the early designs of superheaters and reheaters, access to tube ends was provided, and the tubes were so arranged that they could be turbined for internal cleaning. Contemporary designs provide for internal cleaning when such provision does not adversely affect other important features. Access for internal cleaning of the tubes is now subordinated to permit greater freedom in design, so necessary for good performance. The infrequent cleaning required can normally be accomplished by washing with water.

External Cleaning. In the early designs of superheaters and reheaters, the importance of maintaining clean surfaces was not stressed as it is now. The units designed currently are furnished for continuous operation, in some cases for 18 to 24 months between outages. Usually, 1 year between outages is considered acceptable.

The superheater sections (platens) of modern standard utility units are spaced according to the zone of gas temperature and the fuel fired. For pulverized-coal- and cyclone-furnace-fired units, a transverse spacing of 18 inches has been found satisfactory in the high-temperature-gas zone at the furnace exit. With progressive gas cooling, the spacing is decreased, usually varying from 18 to 9 inches and from 9 to 6 inches, and, as the horizontal surface is crossed, the tubes are spaced 4½ and possibly 4 inches across the gas flow. The back spacing in the direction of gas flow is usually set to allow ½- to ¾-inch clear space between tubes in the high-temperature zones with an increase allowable in the 1500°F (816°C) gas-temperature zone of the horizontal surface. These spacings are empirical, based on tube-fouling and erosion experience, and on manufacturing requirements.

STEAM-TEMPERATURE ADJUSTMENT AND CONTROL

Improvement in the heat rate of the modern boiler unit and turbine results in large part from the high cycle efficiency possible with high steam temperatures. The importance of regulating steam temperature within narrow limits is evident from Fig. 18, which shows that a change of 35 to 40°F (~ 20°C) corresponds to a change of about 1 percent in heat rate at pressures from 1800 to 3500 psi.

Other important reasons for accurate regulation of steam temperature are to prevent failures from overheating parts of the superheater, reheater, or turbine, to prevent thermal expansion from reducing turbine clearances to the danger point, and to avoid erosion from excessive moisture in the last stages of the turbine.

The control of fluctuations in temperature from uncertainties of operation, such as slag or ash accumulation, is important. However, superheat and reheat steam temperatures in steam generation are mainly affected by variations in steam output (Fig. 19).

With drum-type boilers, steam output and pressure are maintained constant by firing rate, while the resulting superheat and reheat steam temperatures depend on basic design and other important operating variables, such as the ratio of convection to radiant heat-absorbing surface, excess air, feedwater temperature, changes in fuel that affect turbine characteristics and ash deposits on the heating surfaces, and the specific burner combination in service. In the universal-pressure once-through boiler, which has a variable transition zone, steam output and pressure are controlled by the boiler feed pump and steam temperature by the firing rate, leaving reheat steam temperature as a dependent variable.

Standard performance practice for steam-generating equipment permits a tolerance of ±10°F (5.5°C) in a specified steam temperature.

Some of the terms used in describing superheater and reheater control problems include:

Adjustment: A change in the arrangement of equipment which affects steam temperature, but cannot be used to vary steam temperature during operation. Example: Removal of superheater tubes.

Control: Regulation of steam temperature during operation without changing the arrangement of equipment. Example: Operation of an attemperator.

Attemperator: Apparatus for reducing and controlling the temperature of a superheated vapor or of a fluid passing through it. Example: A bank of tubes, submerged in the boiler water, through which all or a part of the superheated steam is diverted to give up some of its heat, thereby regulating the final steam temperature.

Radiant superheater or reheater: Heating surface containing steam and receiving heat primarily by radiation. Example: Tubes which form part of a furnace enclosure and within which steam is flowing.

Convection superheater or reheater: Heating surface containing steam and receiving heat primarily by convection. Example: A bank of tubes within which steam is flowing and across or along which hot gases pass.

Effect of Operating Variables.   Many operating variables affect steam temperatures in drum- or separator-type units. To maintain constant steam temperature, means must be provided to compensate for the effect of such variables, the most important of which are:

Load: As the load increases, the quantity and temperature of combustion gases increase. In a convection superheater, steam temperature increases with load, the slope of the curve being less steep as superheater location is brought closer to the furnace. In a radiant superheater, steam temperature decreases as load increases. Sometimes a convection and a radiant superheater of proper proportions are installed in series in a single steam-generating unit to maintain substantially constant steam temperature over a considerable range of load (Fig. 19).

Excess air: For a change in the amount of excess air entering at the burners, there is a corresponding change in the quantity of gas flowing over a convection superheater and, therefore, an increase in excess air tends to raise the steam temperature.

Feedwater temperature: Increase in feedwater temperature causes a reduction in superheat since, for a given steam flow, less fuel is fired and less gas passes over the superheater.

Heating-surface cleanliness: Removal of ash or slag deposits from heat-absorbing surfaces ahead of the superheater will reduce gas temperature and steam temperature. Removal of deposits from superheater surface will increase superheater absorption and raise steam temperature.

Use of saturated steam: If saturated steam from the boiler is used for sootblowers or for auxiliaries, such as pumps and fans, an increased firing rate is required to maintain constant main steam output, and this raises the steam temperature.

Blowdown: The effect of blowdown is similar to the use of saturated steam, but in lesser degree because of the low enthalpy of water as compared with steam.

Burner operation: The distribution of heat input among burners at different positions or a change in the adjustment of a burner usually has an effect on steam temperature through changes in furnace heat-absorption rate.

Fuel: Variations in steam temperature may result from changing the type of fuel burned or from changes in the characteristics of a given fuel from time to time.

REFERENCES

"Steam: Its Generation and Use," 38th ed., Babcock & Wilcox Company, New York, 1972.

Smith, R. B.: Heat Generation, in R. H. Perry and C. H. Chilton (eds.), "Chemical Engineers' Handbook," 5th ed., McGraw-Hill, New York, 1973.

Kessler, G. W.: Steam Boilers, in T. Baumeister (ed.), "Marks' Standard Handbook for Mechanical Engineers," 7th ed., McGraw-Hill, New York, 1967.

Tong, L. S.: "Boiling Heat Transfer and Two-Phase Flow," Wiley, New York, 1965.

Keenan, J. H., F. G. Keyes, P. G. Hill, and J. G. Moore (eds.): "Steam Tables: Thermodynamic Properties of Water Including Vapor, Liquid, and Solid Phases," Wiley, New York, 1969.

Schack, A.: "Industrial Heat Transfer: Practical and Theoretical with Basic Numerical Examples," Wiley, New York, 1965.

Shields, C.: "Boilers," McGraw-Hill, New York, 1961.

Higgins, A.: "Boiler Room Questions and Answers," McGraw-Hill, New York, 1945 (historical interest).

Spring, H.: "Boiler Operator's Guide," McGraw-Hill, New York, 1940 (historical interest).

Woodruff, E., and H. Lammers: "Steam-Plant Operation," 3d ed., McGraw-Hill, New York, 1967.

Eckert, E. R. G., and R. M. Drake, Jr.: "Heat and Mass Transfer," 2d ed., McGraw-Hill, New York, 1959.

Gebhart, B.: "Heat Transfer," 2d ed., McGraw-Hill, New York, 1970.

Holman, J. P.: "Heat Transfer," 3d ed., McGraw-Hill, New York, 1972.

Kays, W. M.: "Convective Heat and Mass Transfer," McGraw-Hill, New York, 1966.

Myers, G. E.: "Analytical Methods in Conduction Heat Transfer," McGraw-Hill, New York, 1971.

Pincus, L.: "Practical Boiler Water Treatment," McGraw-Hill, New York, 1962.

"Power Generation Systems," McGraw-Hill, New York, 1967.

Simonson, J.: "Introduction to Engineering Heat Transfer," McGraw-Hill, New York, 1968.

Controls for Fossil Fuel/Steam Energy Conversion

By S. G. DUKELOW*

OBJECTIVES OF STEAM POWER PLANT INSTRUMENTATION

Measurement and control equipment serves three basic purposes in the steam generation or boiler portion of the steam power plant: (1) *safety*—interlocks assure proper sequence during start-up, shutdown, and emergency operations while continually monitoring potential hazards and taking appropriate actions; (2) *automatic operation*—continuously regulates the flows of fuel, air, and feedwater to meet load demands while maintaining the operating pressures, levels, and temperatures at their desired conditions; and (3) *efficiency*—by proper system design with competent and regular maintenance and tuning so that optimum safe and automatic operation is continuously attained.

MAJOR FOSSIL-FUEL BOILER CONTROL SYSTEMS

Controls for boilers are made up of several systems that appear at first glance to be independent systems. These systems, however, are highly interdependent so that they should be designed as a single coordinated system to minimize the effect of interactions between systems. The various control systems for fossil-fuel boilers include:
1. Combustion control systems
2. Steam temperature control systems
3. Feedwater control systems
4. Burner-sequence and safety systems
5. Integrated control systems to coordinate all the above systems with the turbogenerator control

BASIC BOILER CONTROL

Boiler control is the regulating of the boiler outlet conditions of steam flow, pressure, and temperature to their desired values. In control terminology, the boiler outlet steam conditions are called the *output* or *controlled variables,* and the desired values of the outlet conditions are the *set points* or *input-demand signals.* The quantities of fuel, air, and water are adjusted to obtain the desired outlet steam conditions and are called the *manipulated* or *controlled variables.*

The boiler is referred to as the *system, plant,* or *process,* and disruptive influences on the boiler, both internal and external, such as variations in heat content of fuel or cycle efficiency are the *disturbance inputs.* The controller or control system has the function of "looking at" the desired (set points) and actual values (output variables) of the outlet steam conditions and adjusting the amounts of fuel, air, and water (manipulated variables) to make the outlet conditions match their desired values. The controller can be *manual,* with an operator making the adjustments, or it can be *automatic,* with a pneumatic or electronic analog computer or a digital computer making the adjustments.

While it is theoretically possible to operate a boiler satisfactorily with manual control, the operator must maintain a tedious, constant watch for the occurrence of a disturbance. Time is required for the boiler to respond to a correction, and this can lead to overcorrection with further "upset" to the boiler. An automatic controller, on the other hand, does not experience tedium and, once properly adjusted, will always make the proper adjust-

*Manager Power Generation Marketing, Bailey Meter Company, a subsidiary of The Babcock & Wilcox Company, Wickliffe, Ohio.

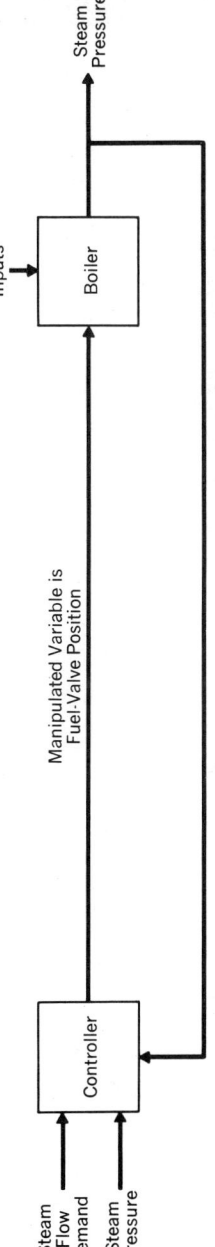

Fig. 1. Simplified block diagram of a boiler control function. *(Babcock & Wilcox Company.)*

ment to minimize the upsets to the boiler and, therefore, will control the boiler more efficiently and reliably.

The various types of automatic control can be illustrated by use of the block diagram in Fig. 1, which represents the control of one output variable, namely, steam pressure, in a hypothetical boiler. The set points for steam pressure and steam flow will be used as reference values, and the manipulated variable is the fuel-valve position. Disturbance inputs come from the boiler, which is the system to be controlled.

Open-Loop Control   The simplest control mode is the open-loop, feedforward, or nonfeedback control, where the manipulated variables of fuel, air, and water are adjusted only from the input-demand signals without monitoring the outlet conditions or output variables. As an example, Fig. 2 illustrates an open-loop control system to accomplish the control function illustrated by Fig. 1. Figure 2a is a block diagram, representing the action to be taken, which is "feedforward" only. Figure 2b is a calibration curve, expressing fuel-valve position as $f(D)$, a function of steam flow. This calibration curve, established by manually determining the fuel-valve position required to maintain the desired steam flow with a constant steam pressure, is entered into the controller. The response of this open-loop control is very fast and depends only on the accuracy of the calibration curve.

One of the difficulties with open-loop control is that the calibration curve is accurate only as long as the boiler conditions remain as they were when the calibration was established.

Another problem with open-loop control is that the output changes as the load changes on the output signal (the fuel valve in Fig. 2a). For example, a change in friction on the valve stem will cause a variation in valve position in response to a given input-demand

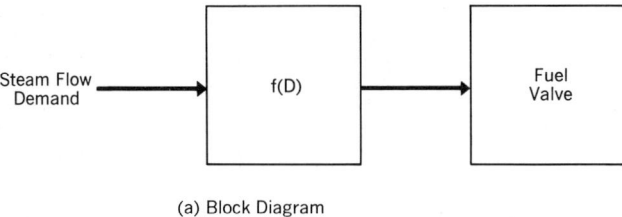

Steam Flow Demand → $f(D)$ → Fuel Valve

(a) Block Diagram

(b) Calibration Curve,
Expressing Fuel-Valve Position
as a Function of Steam Demand

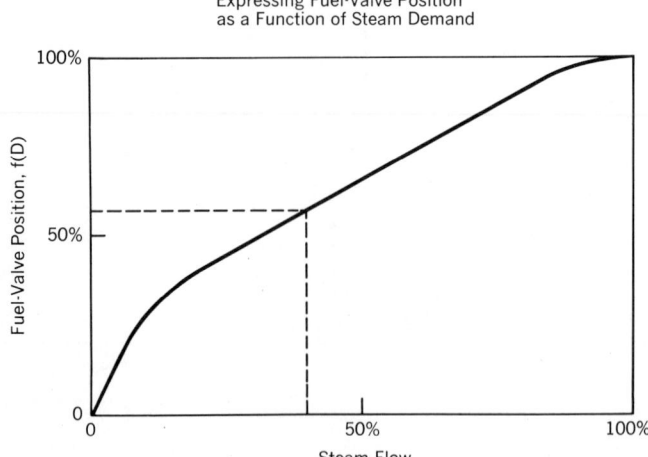

Fig. 2. Open-loop mode to control steam pressure by varying fuel input.

signal, thus introducing an error into the performance of the open-loop control. Such disadvantages normally outweigh the open-loop advantages of stability and simplicity.

An open loop of this sort cannot be used without some sort of recorrection. It thus cannot be used as a complete system since there must be feedback and the loop closed for a system to be of practical use.

Closed-Loop Control   Since the system requirements cannot be met by an open-loop control, then a closed-loop or feedback control must be used. In the closed-loop control mode, the actual output of the system is measured and compared to the input-demand signal, with the difference between the two signals (the error signal) used to reduce the difference between the demand and output signals to zero.

The open-loop system of Fig. 2 could be closed by having an operator observe the measured steam pressure and then manually adjust the fuel-valve position to obtain the fuel flow necessary to maintain the desired steam pressure at the required steam flow. To avoid the disadvantages of manual control, an automatic controller can be provided in a closed-loop or feedback system.

Proportional Control.   The simplest type of closed-loop system is proportional control, where the manipulated variable or controller output is proportional to the deviation of the controlled variable from its set or set point value. The deviation of the controlled variable from its set point is called the *error signal*. Depending on the arrangement of the controller, the output signal of a proportional control system will always be either directly or inversely proportional to the controlled variable.

Figure 3 illustrates the application of proportional control to the control function of Fig. 1. (See Table 1 for explanation of Δ and K in Fig. 1, as well as other symbols in subsequent figures in this article.)

In the system of Fig. 3 for every deviation of steam pressure from its set point, the fuel-valve signal will be changed to a specific value, as shown in Fig. 4 for three values of proportional gain. In Fig. 4 the proportional gain is said to have a value of 1.0 if the full 100 percent input range of the pressure input initiates a 100 percent change in the fuel-valve control signal. The gain is 2.0 when 50 percent of steam pressure input signal range produces 100 percent range in the fuel control signal and 4.0 when 25 percent of range produces 100 percent change in the fuel control signal. In actual practice, the range of

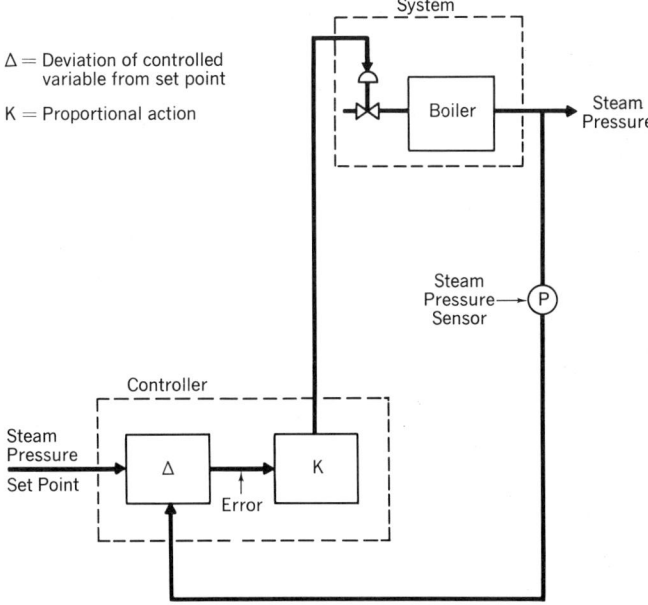

Fig. 3. Proportional-control system.

TABLE 1. Symbols for Control Terms

O	—Transmitter
K	—Proportional action (gain)
∫	—Integral action
Σ	—Summing action
Δ	—Difference or subtracting action
<	—Low select auctioneer
>	—High select auctioneer
≮	—Low limiting
≯	—High limiting
d/dt	—Derivative (rate)
Σ/n	—Averaging
⬦(A/T)	—Hand-automatic selector station (analog control)
⬦(A/T/A)	—Hand-automatic selector station (analog control) with bias
H/A	—Hand-automatic selector station (digital control)
T	—Transfer
±	—Bias action
f(x)	—Power device (valves, drives, etc.)

From "Steam," 38th ed., The Babcock & Wilcox Company, New York, 1972.

pressure input is 1.5 to 2.0 times the operating pressure and the proportional gain is in the order of 6.0, requiring a change of approximately 20 percent of the operating pressure to move the fuel valve from closed to open.

The response of steam pressure to step increases in load on the boiler is shown in Fig. 5 for two values of proportional gain on the controller. The application of a step increase in load or steam flow immediately decreases the steam pressure which, through the proportional controller, causes the fuel valve to open by a proportional amount. The additional heat input due to the increased fuel flow causes the steam pressure to return toward its set point, which in turn reduces the fuel-valve opening.

There may be a certain amount of cycling or hunting, depending on the proportional gain of the controller, before the steam pressure stabilizes. It can be noted in Fig. 5 that

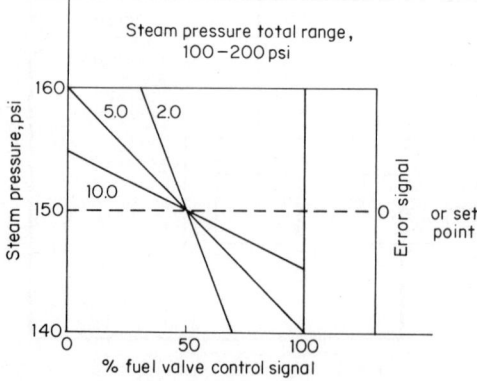

Fig. 4. Fuel valve signal for various pressure deviations in a proportional-control system.

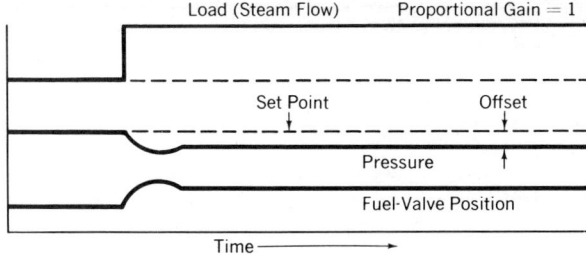

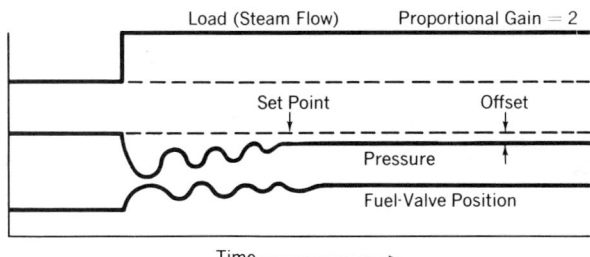

Fig. 5. Response of steam pressure to step increases of load in a proportional control system. (*Babcock & Wilcox Company.*)

the steam pressure does not stabilize at its set point but is offset to a value below the set point. A characteristic of "proportional-only" control is that an error or offset is necessary to provide a steady-state fuel-valve opening which will support the desired load. For every different load there is thus a different steady-state steam pressure and fuel-valve opening.

An increase in the relative proportional gain or sensitivity will reduce the final offset in the steam pressure, as shown in Fig. 5. An increase in proportional gain may increase the time required for the system to reach a steady-state condition. This offset is always present with proportional control, and if the gain is increased even more to reduce the offset, undamped oscillations will result.

Integral or Reset Control. This offset may be eliminated by the addition of the integral or reset mode of control to the proportional-control system, as illustrated in Fig. 6. The response of the hypothetical boiler to a step increase in load when using a proportional-plus-integral control system is shown in Fig. 7. The steam pressure is returned to its set point without the offset which is present with the proportional-only control. However, the system is sensitive to the matching of the controller-integrating time constant to the response time of the process and must be carefully tuned for stable operation.

Integral control, as its name implies, is based on the integration of the deviation between the controller variable and its set point over the time the deviation occurs.

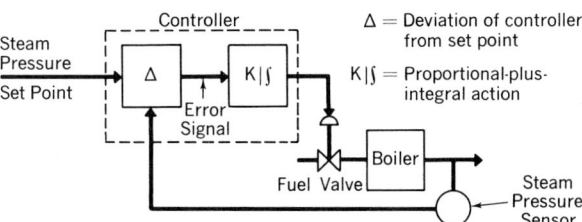

Fig. 6. Proportional-plus-integral control.

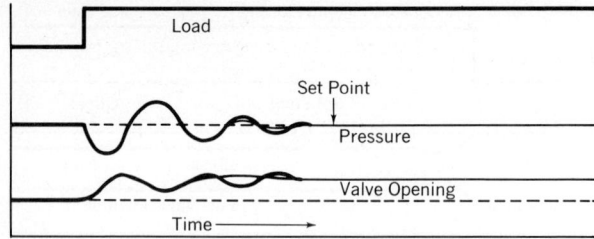

Fig. 7. Response of steam pressure to step increase of load in a proportional-plus-integral-control system.

Integral control is also referred to as *reset control* since the band of proportional action is shifted or reset so that the manipulated variable operates about a new base point.

Derivative Control. The stability and response of the system may be improved still further by adding a third mode of control action to the controller called *derivative* or *rate control*. Derivative control is a function of the rate of change of the controlled variable from its set point, as shown in Fig. 8. The addition of the derivative control mode to the controller-boiler is shown in Fig. 9. As soon as the step change is made, the pressure starts to drop and the proportional mode begins to open the fuel valve. The derivative mode will also open the fuel valve further as a function of the rate at which the pressure is changing, providing anticipation of where the fuel valve should be positioned. When the rate of change of the steam pressure decreases, the derivative control has less effect and the proportional and integral modes do the final positioning of the fuel valve.

Feedforward-Feedback Control In a simple feedback closed-loop control system, the controlled variable always has to deviate from its set point before any corrective action is initiated by the controller. In this respect the open-loop or feedforward system has a faster response since it takes corrective action before the controlled variable starts to change. When the feedforward system is combined with the closed-loop or feedback system, the result is a system with fast response that is able to compensate for changes in the calibration curve of Fig. 2b. Figure 10 represents a feedforward-feedback control system to accomplish the control function of Fig. 1.

As the step change in load occurs, the feedforward signal immediately positions the fuel valve to meet the requirements of the calibration curve. If this curve is exact, no error develops and the feedback loop has no work to do. If the calibration curve is in error, the feedback loop readjusts the fuel-valve position to eliminate any pressure error which may

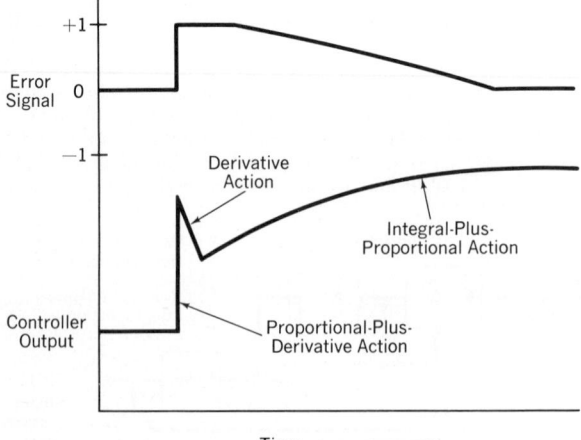

Fig. 8. Proportional-plus-integral-plus-derivative action.

Δ = Deviation of controller from set point

$K|\int|\frac{d}{dt}$ = Proportional-plus-integral-plus-derivative action

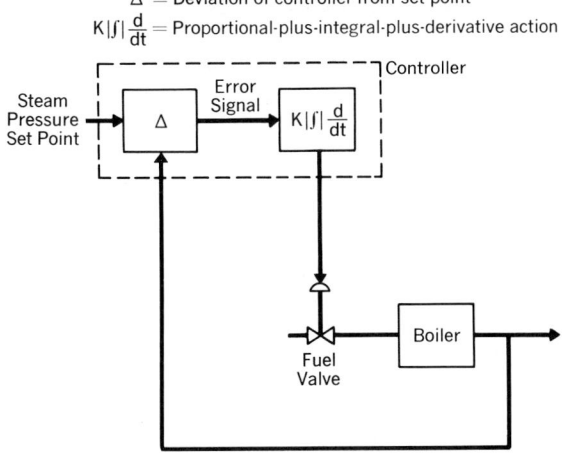

Fig. 9. Proportional-plus-integral-plus-derivative control.

develop due to shifts in the calibration. However, the response will be faster and the magnitude of the system upset will be smaller since the feedforward signal positions the valve near its final steady-state position, leaving a smaller range of action to be accomplished by the feedback loop.

Requirements of the Controller   The requirements to be met with a controller are that the overall controlled boiler system be stable and responsive. This means that the values of the controlled variables or outlet steam conditions are maintained close to their desired values under equilibrium or steady-state conditions with no cycling or self-sustained oscillations, and that the system recovers quickly without excessive hunting, overshoot or oscillations from system disturbances and changes in the reference values or set points.

The discussion illustrated by Figs. 3 to 10 considers only one of the subloop controls on the boiler. However, each of the other boiler-output variables will be controlled by one of these same modes in its control subloop, depending on the accuracy and speed of response required for the overall controlled system to be stable and have fast response.

FEEDWATER CONTROL SYSTEMS

The purpose of the feedwater control is to regulate the flow of water to a drum-type boiler so as to maintain the level in the boiler drum between the desired limits. The control

f(D) = Calibration for feedforward action

Σ = Summing action

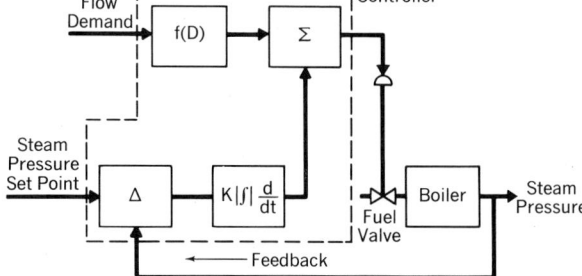

Fig. 10. Feedforward-feedback control.

system will vary with the type and capacity of the boiler as well as the characteristics of the load.

Most shop-assembled boilers in the lower capacity range and the lower operating-pressure range are equipped with self-contained feedwater control systems of the thermohydraulic or thermostatic types. The thermohydraulic type is generally applied to boilers having an operating pressure in the range between 60 and 600 psi and capacities up to 40,000 pounds/hour under steady load conditions.

The self-operated feedwater regulator, illustrated in Fig. 11, is actuated by a closed hydraulic system consisting of the annular space between the two concentric tubes of a "generator," the connecting copper tubing, the metal bellows of the regulating valve, and the water necessary to fill the system. The inner tube of the generator is connected to the steam drum of the boiler, the lower end to the water space, and the upper end to the steam space.

The water level in the inner tube of the generator follows the actual level in the drum. When the water level in the drum decreases, the heat from the steam in that portion of the tube vacated by the drop in level causes water in the outer tube to flash into steam, displacing water from the outer tube through the connecting tubing into the bellows. This causes the bellows to expand, increasing the regulating-valve opening to admit more water to the drum.

In the thermohydraulic regulator, water input is controlled in proportion to drum level, not to load. As a result, the level maintained at higher loads will be somewhat lower than the level maintained at relatively lower loads. The amount of this regulated-level varia-

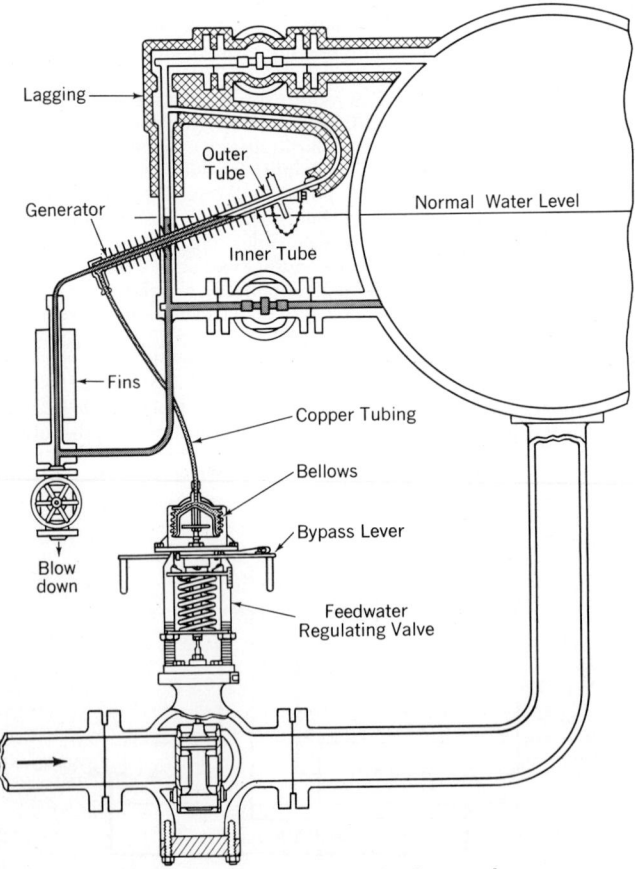

Fig. 11. Self-operated thermohydraulic control.

tion will depend on the extent of the load variation, the sizing of the regulating valve, and the slope of the "generator" (Fig. 11). The unregulated level variation will depend on the extent of "swell and shrink" effects, drum pressure changes, and supply water variations.

For higher capacity boilers, those subject to variable loads and those operating at higher pressures, a pneumatic or electrically operated feedwater control system is applied. Controls for these types use varying degrees of control action to suit the particular application. They are classified as single-, two-, or three-element feedwater control systems. The details of these systems are illustrated in Figs. 12 to 14, using the symbols given in Table 2.

Single-Element Feedwater Control   In single-element feedwater control (Fig. 12), the water in the drum is at the desired level when the signal from the level transmitter equals its set point.

If an undesirable water level exists, the controller applies proportional-plus-integral action to the difference between the drum-level and set point signals to change the position of the regulating valve. A hand-automatic station gives the operator complete control over the valve. A valve positioner can be included in the control-valve assembly to match the valve characteristics to the individual requirements of the system.

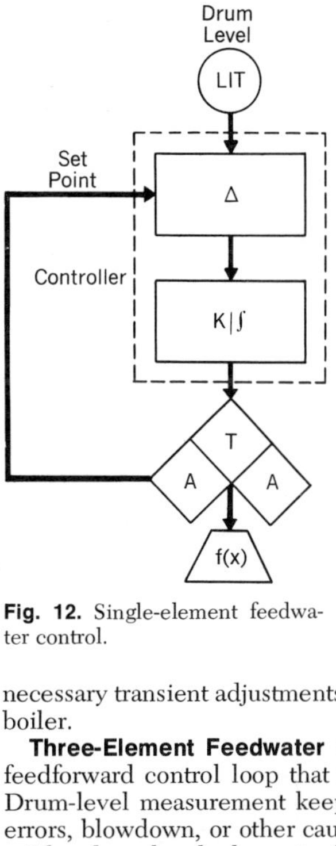

Single-element control will maintain a constant drum level for slow changes in load, steam pressure, or feedwater pressure. However, since the control signal satisfies the requirements of drum level only, excessive "swell or shrink" effects will result in wider drum-level variations and a longer time for restoring drum level to set point following a load change with single-element control than with two- or three-element control.

Two-Element Feedwater Control   Two-element control (Fig. 13) comprises a feedforward control loop that uses steam-flow measurement to control feedwater input, with level measurement assuring correct drum level.

The drum-level element of the controller applies proportional action to the difference between the drum-level signal and its set point. The sum of the drum-level error signal and the flow signal determines the valve position. Thus, the steam-flow measurement maintains feedwater flow proportional to steam flow; the drum-level measurement corrects for any imbalance in water input versus steam output caused by deviations in the valve position/water-flow relationship, and provides the necessary transient adjustments to cope with the "swell and shrink" characteristics of the boiler.

Fig. 12. Single-element feedwater control.

Three-Element Feedwater Control   Three-element control (Fig. 14) is a cascaded-feedforward control loop that maintains water-flow input equal to feedwater demand. Drum-level measurement keeps the level in the drum from drifting due to flowmeter errors, blowdown, or other causes.

The drum-level element of the controller applies proportional action to the error between the drum-level signal and its set point. The sum of the drum-level error signal and the steam-flow signal is the feedwater-demand signal. This is the output of the "summer." The feedwater-demand signal is compared with the water-flow input, and the difference is the combined output of the controller. Proportional-plus-integral action is incorporated to provide a feedwater correction signal for valve regulation or pump speed control.

In single boiler-turbine units, the turbine first-stage shell pressure can be used as a substitute variable for steam flow.

Three-element feedwater control systems can be adjusted to restore a predetermined drum level at all loads; or in boilers with severely fluctuating loads, the system can be adjusted to permit water level in the drum to vary with the boiler loading to compensate for "swell and shrink" effects. If the drum level is allowed to vary in this manner, a nearly constant inventory of water, as opposed to a constant level, is maintained.

TABLE 2. Symbols for Feedwater Control

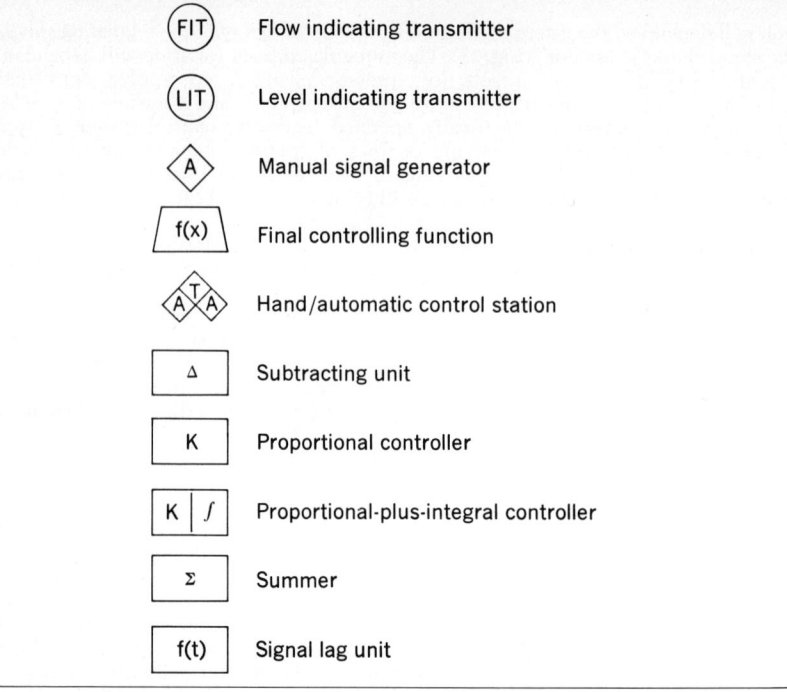

Symbol	Description
FIT	Flow indicating transmitter
LIT	Level indicating transmitter
A	Manual signal generator
f(x)	Final controlling function
A T A	Hand/automatic control station
Δ	Subtracting unit
K	Proportional controller
K ∫	Proportional-plus-integral controller
Σ	Summer
f(t)	Signal lag unit

From "Steam," 38th ed., The Babcock & Wilcox Company, New York, 1972.

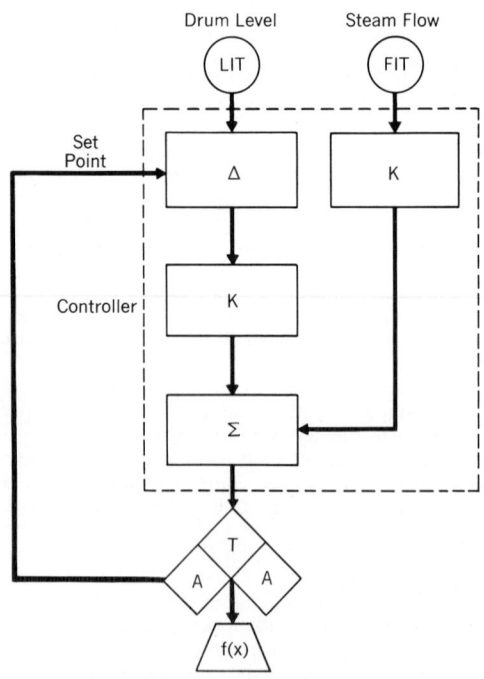

Fig. 13. Two-element feedwater control.

Three-element feedwater control systems of this type can be precalibrated so that only a few simple adjustments are needed to match the system to the individual requirements of the boiler. Control characteristics can be easily changed by direct adjustment.

Basically the choice of feedwater control systems is determined by capacity and operating conditions. The thermohydraulic regulator is used for low capacities and steady load conditions. As capacities increase and loads are more rapid and variable, more

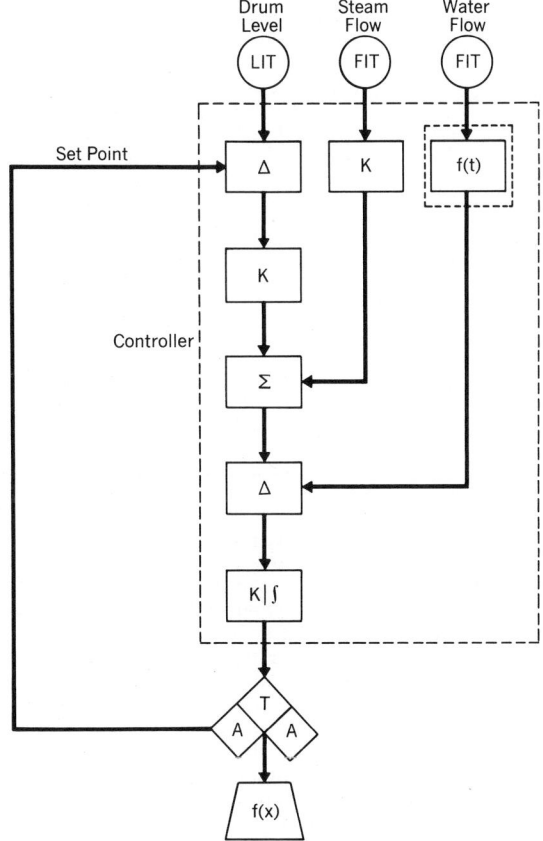

Fig. 14. Three-element feedwater control.

sophisticated systems such as two or three element are required. The technical choice between two and three element generally relates to the ability to predict a repeatable valve position versus flow relationship of the feedwater control valve so that the steam-flow versus valve-position relationship is meaningful and can be calibrated in the system. Where this relationship is not sufficiently repeatable, three element is the choice. Three element is often used on boilers 100,000 pounds per hour and above simply because it is more easily tuned and maintained than two element.

COMBUSTION CONTROL SYSTEMS

The function of a combustion control system is to control the fuel and air input or firing rate to the furnace in response to a load index representing a demand for the level of fuel input. The demand for firing rate is, therefore, a demand for energy input into the system to match a withdrawal of energy at some point in the cycle. For boiler operation and

Table 3.   Differences in Performance of Control Systems

SOURCE OF EXCESS AIR VARIATIONS AND ALLOWANCES	JACKSHAFT PARALLEL CONTROL (SINGLE DRIVE) MIN.	MAX.	PARALLEL CONTROL (MULTIPLE DRIVE) MIN.	MAX.	PRESSURE RATIO CONTROL MIN.	MAX.	CROSS-LIMITED METERING MIN.	MAX.	CROSS-LIMITED METERING WITH O2 CORRECTION MIN.	MAX.	CORRECTIVE MECHANISM
FUEL											
1 FUEL VARIATIONS (PRESS., TEMP., GRAV., BTU)	2	4	2	4	2	4	2	4	–	–	O2 CORRECTION
FUEL SYSTEM VARIATIONS (P.H. SET, VALVE, PIPING, BURNER WEAR, INTERACTION)	4	10	4	10	4	10	–	–	–	–	FUEL FLOW MEASUREMENT
COMBUSTION AIR											
2 COMBUSTION AIR VARIATIONS (PRESS., TEMP., HUMIDITY)	7	14	7	14	4	9	4	9	–	–	O2 CORRECTION
COMBUSTION AIR SYSTEM VARIATIONS 3 (FAN PERFORMANCE, DAMPERS, DUCTING, INTERACTION)	2	5	2	5	1	3	–	–	–	..	AIR FLOW MEASUREMENT
EXHAUST EXHAUST GAS AND SYSTEM VARIATIONS (BOILER FOULING, STACK EFFECT, INTERACTION)	7	14	7	14	3	7	–	–	..	–	AIR FLOW MEASUREMENT
CONTROL											
FUEL POSITIONING ACCURACY	1	4	1	4	–	–	–	–	–	–	
AIR POSITIONING ACCURACY	2	6	2	6	–	–	–	–	–	–	
CONTROL SYSTEM ACCURACY	1	2	1	2	1	2	1	2	1	2	HARDWARE QUALITY AND SYSTEM DESIGN
CONTROL ALIGNMENT	6	10	3	5	3	5	1	2	1	2	
CONTROL SYNCHRONIZATION	3	7	2	4	2	4	–	–	1	–	
SUM OTHER UNACCOUNTED FOR ERRORS	1	2	1	2	1	2	1	2	1	2	ALL THE EXCESS AIR VALUES SHOWN INCLUDE THE PRACTICAL VALUE REQUIRED FOR COMPLETE COMBUSTION
4 SUM OF ERRORS	36%	78%	32%	70%	19%	46%	9%	19%	3%	6%	
5 RMS OF ERRORS	4%	8%	3.5%	7.5%	2.6%	5.5%	1.5%	3%	5%	1%	
6 AVG ERROR % (SUM + RMS)	±31		±26		±19		±8		0 + 2.5		
7 TOTAL AIR	143		137		130		118		110		109 FOR BURNER
LOSS % FUEL LOST DUE TO 8 CONTROL ERROR BAND	5.0		3.7		3.0		1.3		0%		

control systems, variations in the boiler outlet pressure are often used as an index of an unbalance between fuel-energy input and energy withdrawal in the output steam.

An additional function of a combustion control system is to control the fuel and air input in a manner that will minimize fuel consumption while matching the steam system energy withdrawal. This is accomplished through system design to minimize the combustion airflow for a given fuel input while avoiding the presence of combustible gases, smoke, particulate matter, or other pollutants in the flue gases leaving the boiler.

The presence of combustible gas in the flue gases is not only wasteful of fuel but potentially hazardous, hence system design should also include provision for minimum airflow. Fuel should also be limited to available combustion airflow and airflow should not be allowed to exceed that required to properly burn the fuel.

The precision with which combustion can potentially be controlled is dependent on the precision of the measurement and controlled devices used. The more imprecise the control action the greater cushion of excess air that must be allowed to avoid combustible gas loss. Table 3 demonstrates the additional excess-air cushions required due to relative adequacies of the basic types of combustion control systems shown in Figs. 15 to 19.

This analysis is based on analyzing whether and to what degree additional inputs to the combustion control system can compensate for all the normal day-to-day or month-to-month variations in the various inputs to the boiler or total steam generation system.

The economic impact of operating a boiler at total air (100 + percent excess air) levels higher than necessary is shown on the chart Fig. 20. In this demonstration, 15 percent excess or 115 percent total air is the base as compared to higher levels of excess air and the higher flue gas temperatures which result. As a result of sharply increased fuel prices, more complete combustion control systems, such as Figs. 18 and 19, are economically attractive on boilers as small as 25,000 pounds of steam per hour. Past practice has been to use very simple systems, such as shown in Fig. 15, on boilers of this capacity. As fuel prices increase, the economic benefit of more complete systems increase, and these systems can be justified on smaller boilers.

A variety of combustion control arrangements have been developed over the years to fit

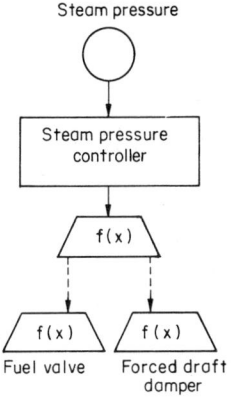

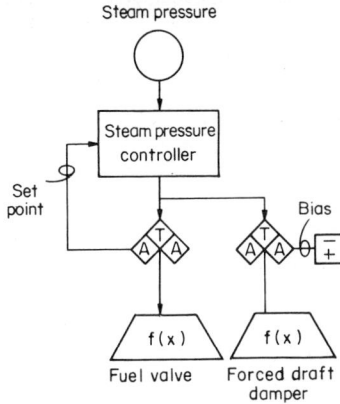

Fig. 15. Simple parallel positioning control system.

Fig. 16. Simple parallel positioning system with provision for operator adjustment.

the needs of specific applications. Load demands, operating philosophy, plant layout, types of firing, and fuel costs must all be considered before the final selection is made.

The principle of automatic readjustment of fuel/air ratio from flue gas percent oxygen content often is economically beneficial due to the capability of operating the boiler consistently with a lower excess-air cushion and the fuel saving that results. Such a system is shown in Fig. 19.

Examples of specific applications to different firing methods and fuels are shown in the following. There are many variations of these systems from simple systems, such as that in Fig. 15, that will operate the boiler automatically to the most complete that account for the minimum fuel consumption and maximum control stability, such as shown in Fig. 19.

Stoker-Fired Boilers   Stoker-fired boilers are regulated by positioning fuel and combustion air from changes in steam pressure. A change in steam demand initiates a signal from the steam-pressure controller, through the boiler master controller, to increase or decrease both fuel and air simultaneously and in parallel to satisfy the demand. As long as an error in pressure exists (a departure of actual pressure from the set point value), the steam-pressure controller will continue to integrate the fuel and air until the pressure has returned to its set point. See Fig. 21.

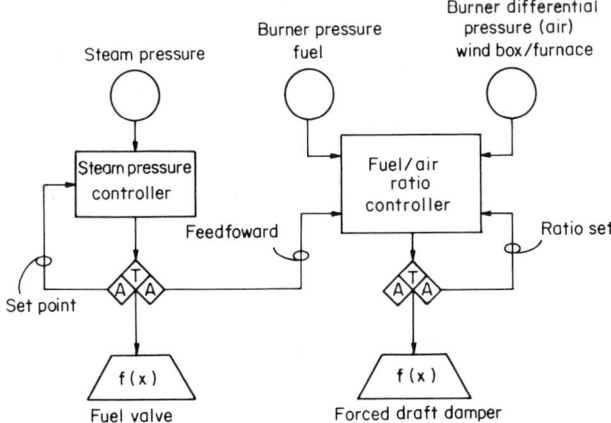

Fig. 17. Parallel feedforward fuel/air ratio system with readjustment from burner fuel/air pressure relationship.

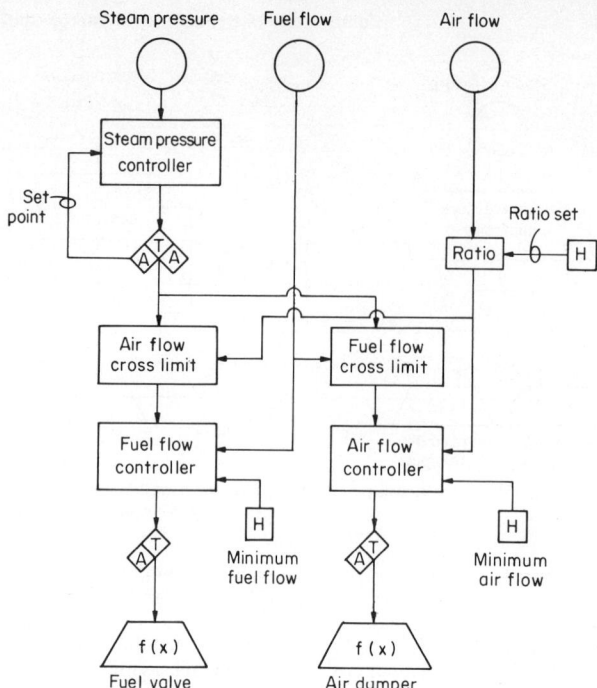

Fig. 18. Fully metered fuel-flow/air-flow system with fuel and air cross limits and manual fuel/air ratio adjustment.

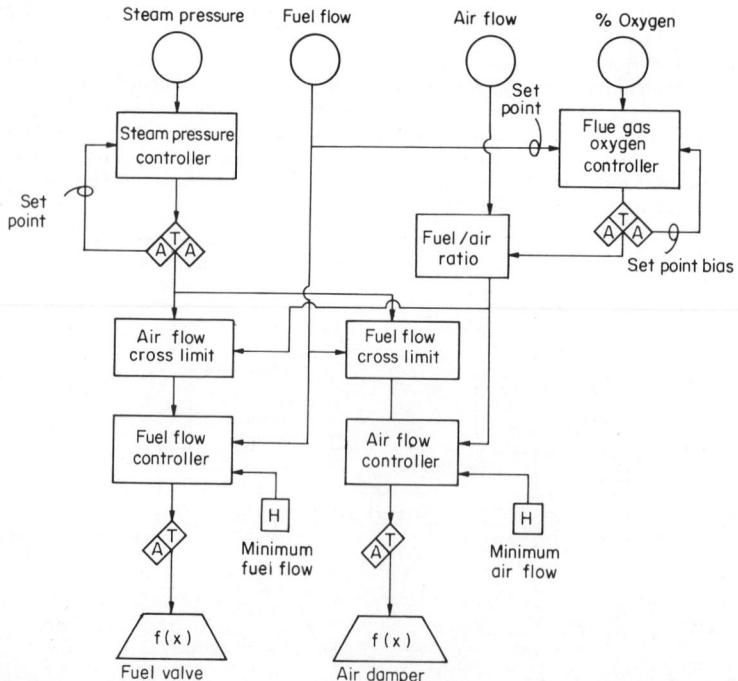

Fig. 19. Fully-metered fuel-flow/air-flow system with fuel and air cross limits and automatic fuel/air ratio adjustment from flue gas percent oxygen.

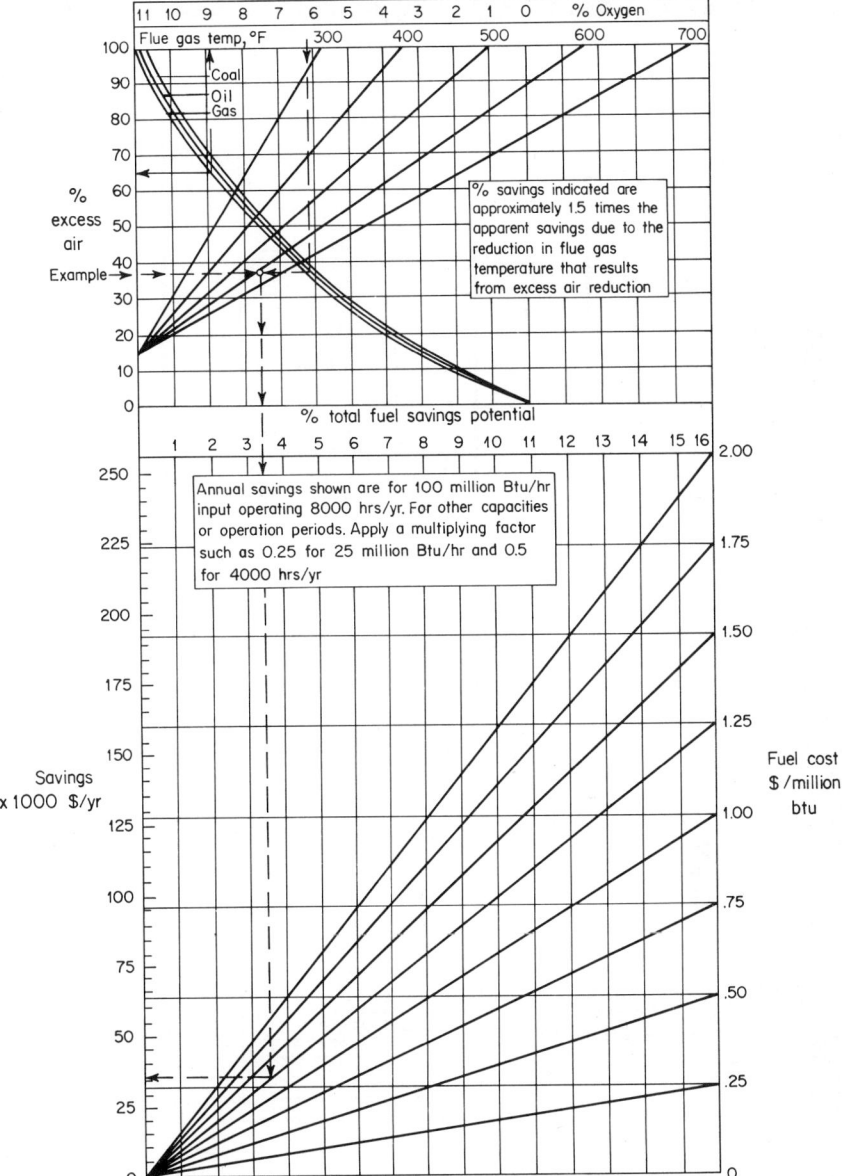

Fig. 20. Chart for determination of fuel savings. (*S. G. Dukelow, Bailey Meter Company.*)

A combustion guide, in this case steam flow/airflow, compares steam flow, an index of fuel being burned, and a calibrated relative airflow measurement. Should any of the various disturbance inputs cause this relationship to change, the error signal produced then modifies the stoker feedforward signal to produce a corrected airflow demand. In this way, the desired fuel-air ratio is continuously maintained. Just as in the control system of Fig. 19, Percent Flue Gas Oxygen can be used as a final trim of the fuel-air ratio for more precise control.

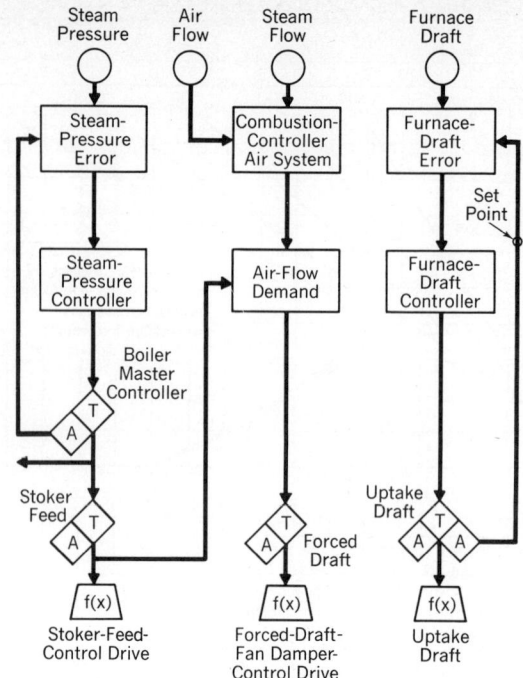

Fig. 21. Combustion control for a spreader-stoker-fired boiler. *(Babcock & Wilcox Company.)*

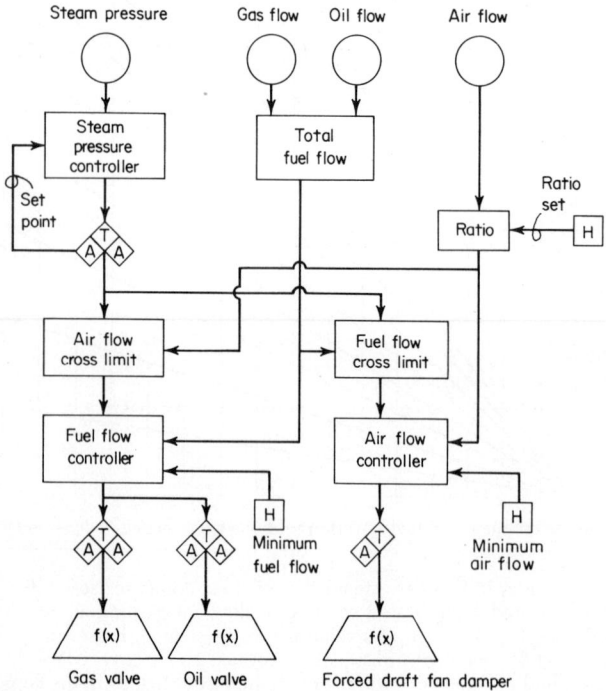

Fig. 22. Metered cross-limited combustion control system for gas and/or oil.

Furnace draft is regulated separately through the use of a furnace-draft controller and a power operator that positions the uptake damper.

Gas- and Oil-Fired Boilers Figure 22 illustrates a system applicable to the burning of gas and oil, separately or together. The fuel flows and airflows are controlled from steam pressure which provides a demand signal for fuel flow and airflow in parallel. Fuel is

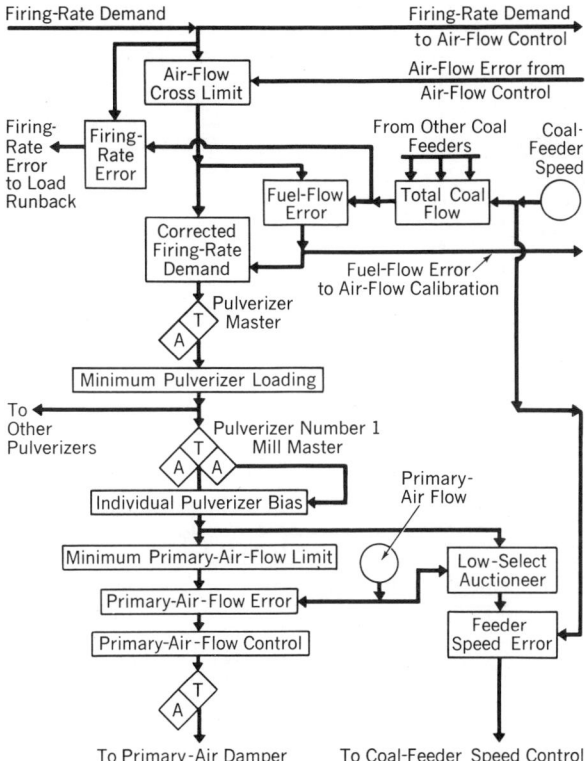

Fig. 23. Combustion control for a pulverized-coal-fired boiler.

limited to available airflow, and air is limited to fuel available. This system is very similar to the basic system shown in Fig. 18 with dual fuel provision being made in this modification. This demonstrates how basic concepts and system design can be modified to accept differences in fuel or other factors.

Pulverized-Coal-Fired Boilers Figure 23 illustrates a more sophisticated combustion control system that would be used on larger boilers having several pulverizers, each supplying several burners. It is not only important to maintain the correct fuel-flow/airflow ratio at all times, but the fuel flow from each pulverizer to its associated burners should be properly proportioned and distributed for good boiler operation.

The firing rate is compared to the total measured fuel flow (summation of all feeders in service delivering coal) to develop the demand to the pulverizer master. The pulverizer master demand signal is then applied in parallel to all operating pulverizers. The controls below the master are for one pulverizer. Other pulverizers have duplicate controls that are not shown in this diagram.

Should an upset occur in the fuel-flow/airflow ratio so that the total airflow is low, an error signal from airflow control will reduce firing rate demand to the pulverizer master to restore the proper fuel-flow/airflow ratio.

The coal flow which is a function of the amount of primary airflow through the pulverizer is varied directly with airflow, since the primary-air damper and coal-feeder speed control receive the same demand signals. When an error develops between demand and measured primary airflow or coal-feeder speed, proportional-plus-integral action will be applied through the controllers to adjust the damper position or feeder speed to reduce the error to zero. A low primary-airflow cutback is applied in the individual pulverizer control. If measured primary airflow is low relative to coal rate or feeder speed demand, this condition is recognized in the "low-select auctioneer" in the

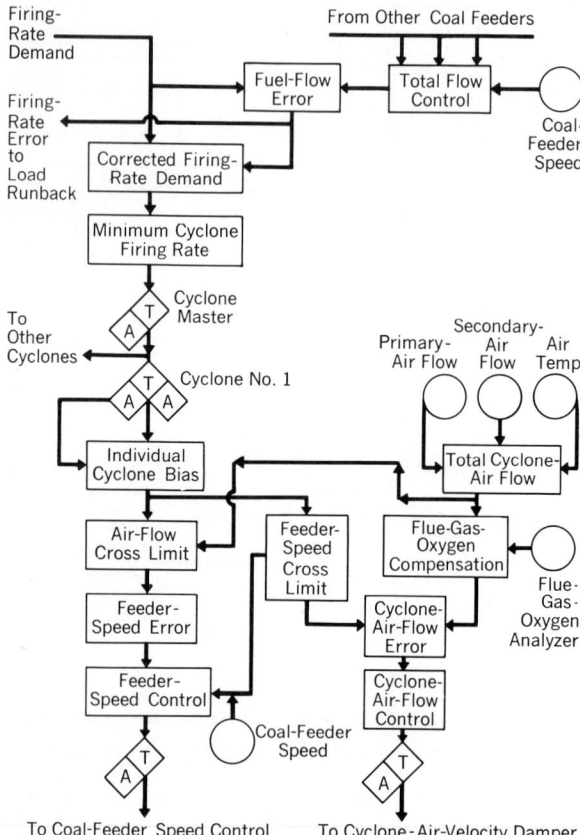

Fig. 24. Combustion control for a cyclone-fired boiler.

coal feeder demand to reduce the demand to that equivalent to the measured primary airflow. A minimum pulverizer loading plus a minimum primary-airflow limit are applied to keep the pulverizers above their minimum safe operation load, to maintain sufficient burner nozzle velocities at all times, and to maintain the primary-air/fuel ratio above a prescribed level at all pulverizer loadings.

Cyclone-Furnace-Fired Boilers   The cyclone controls shown in Fig. 24 are similar to those described for pulverized-coal-fired units. Although the cyclone functions as an individual furnace, the principles of combustion control are the same.

On multicyclone installations, the feeder drives are calibrated so that all feeders run at the same speed for the same master signal. The total airflow is controlled by the velocity damper in each cyclone to maintain the proper fuel-air relationship. This airflow is automatically temperature-compensated to provide the correct amount of air under all boiler loads. The total airflow to the cyclone is controlled by the wind box-to-furnace differential pressure, which is varied as a function of load to increase or decrease the forced-draft-fan output.

Automatic compensation for the number of cyclones in service has been incorporated along with the additional feature of an oxygen analyzer as a final excess-air compensation.

Combustion Guides   In the preceding diagrams of control application, combustion guides have been shown in some instances as automatically applied. On simpler systems, combustion guides are necessary so that an operator may have their results as periodic manual adjustments to the combustion control systems are made. As stated previously, any excess air not actually required constitutes a loss, resulting in lower boiler efficiency and higher fuel bills.

On the other hand, excess air at too low a value results in fuel losses due to combustible gases leaving the boiler, and thus higher fuel costs. Excess air for any fuel is a function of

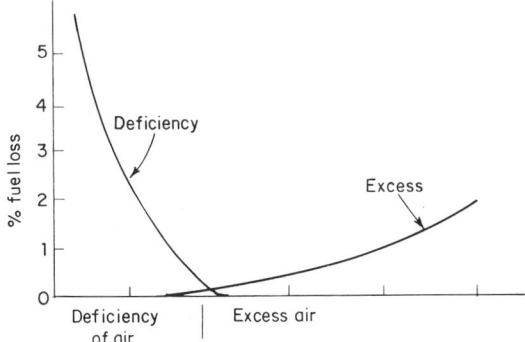

Fig. 25.   Relationship of excess air and deficiency of air-to-fuel loss at constant flue gas temperature.

percent net oxygen concentration in the flue gases. Figure 25 shows generally the relationship between operation with excess air as compared to combustible as it relates to fuel consumption.

Three basic types of combustion guides are applied to steam-generating units. They are (1) gas analysis of the flue gases, (2) steam-flow/airflow ratio, and (3) fuel-flow/airflow ratio. Each has distinct advantages. A combination is often incorporated in the instrumentation system.

Gas Analysis.   Gas analyzers are used to ascertain correct fuel-air relationships. Representative flue gas samples are continuously analyzed either chemically or electronically to produce chart records of the oxygen and combustibles present in the products of combustion. Since there is direct correlation between the percentage of oxygen in the flue gases and the quantity of excess air supplied to the combustion zone, the operator is furnished with a continuous and direct reading indicative of combustion efficiency. The indicated or recorded oxygen signal can be used for control purposes in adjusting the total airflow.

Figure 26 diagrams a modern percent net oxygen analyzer of the fuel cell type, mounted on a flue gas duct. This type of analyzer and sampling system has eliminated the maintenance of the extensive sampling systems previously required. Because of the maintenance reduction and the higher cost of fuel, such an analyzer may be easily justified on small boilers.

Steam-Flow/Airflow.   The steam-flow/airflow concept (Fig. 27) can be applied to obtain an instantaneous indication of the efficiency of fuel consumption. Because boiler efficiency at a specific load remains essentially constant for a given excess air, fuel consumption can be determined from steam flow, which is established by the use of standard flow-measuring devices. Steam flow is used as an index of heat absorption.

The airflow indication or index can be established by primary elements located on either the air side or the flue-gas side of the unit.

The measurement of combustion air supplied to a boiler is made difficult and costly by several factors, including:

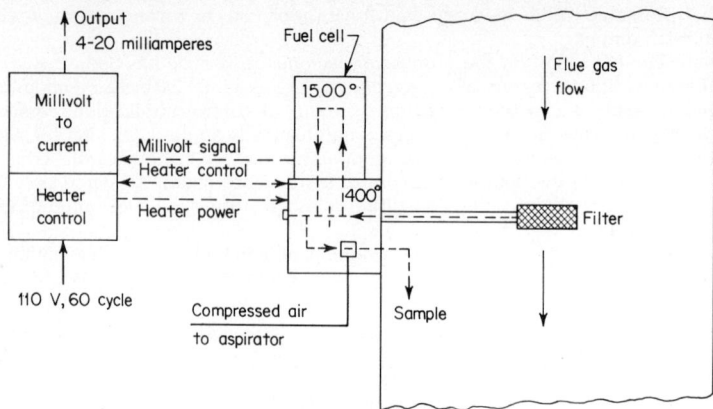

Fig. 26. Fuel cell oxygen analyzer.

1. The air-measuring device is a restrictive type, similar to the primary elements used in other flow measurements. The resulting unrestored head loss means added fan power.

2. On large and complex units, it is usually impossible to find one duct through which all the combustion air passes. Metering equipment capable of accurately totaling several flows must then be employed.

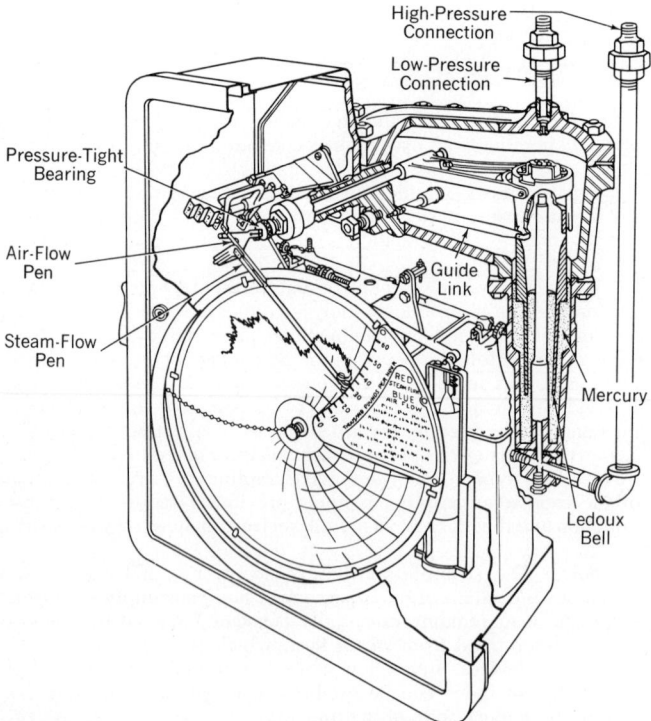

Fig. 27. Steam-flow/air-flow combustion guide for maintaining desired excess air. Fuel-flow/air-flow device is similar, but may have different flow-measuring mechanism. *(Babcock & Wilcox Company.)*

3. Air density must be considered in both the calibration and operation of any airflow meter. Either manual or automatic temperature compensation must be applied to the airflow indication for it to reflect the mass rate of airflow.

The steam-flow/airflow meter is calibrated after the unit is placed in operation. The calibration is based on results obtained from a number of combustion tests. Proper calibration makes possible a visual presentation of the relative correctness of the fuel-air proportioning. The operator is able to achieve optimum firing conditions by maintaining the correct relative positions of the two measurement pens.

Fuel-Flow/Airflow.   Where it is practicable to measure the rates of admission of fuel and air prior to combustion, a fuel-flow/airflow guide can be installed. This may be for direct control, or for backup or setting up limits of fuel flow or airflow.

Fuel-Flow Measurement.   The measurement of gaseous fuels requires consideration of the physical properties of the gas, particularly temperature, pressure, and specific gravity. Where widely changing properties are encountered, automatic compensation should be provided. Such meters normally use an orifice in the main gas header to produce a differential pressure indicative of flow rate.

With liquid fuels, various types of flowmeters can be installed directly in the supply header to indicate and record fuel flow.

The measurement of solid-fuel flow can be accomplished through the use of weight scales or feeder speeds. The use of feeder speed, however, is more applicable to the control and indication of fuel flow. The feeder can be calibrated in pounds of fuel flow per feeder revolution to produce a continuous flow rate. On large units several feeder speeds can be totalized to give the total fuel flow to the unit.

Comparison of Combustion Guides.   Each combustion guide has its particular merits; none is infallible. The informed combustion engineer utilizes the advantages of one to overcome the disadvantages of another.

The fuel-flow/airflow ratio guide proportions fuel and air continuously during severe load swings. It is, therefore, popular on gas- and/or oil-fired units. It has the disadvantage of being in error for wide variations in heating value of the fuel. When the fuel heating value changes, the calibration of the correct fuel-air relationship also changes. Hence, to maintain a given excess air, the proportions of air and fuel must also be changed. Often the fuel-flow/airflow ratio guide is used for coal-fired units having a variable-Btu fuel but with a feedback calibration of airflow based on oxygen measurement or a Btu-input index such as steam flow or megawatt generation.

On major load changes, the steam-flow/airflow device is temporarily in error, i.e., the fuel consumption is normally higher or lower than indicated by the steam-flow load index, because of the overfiring and underfiring necessary for steam pressure control. This guide is also affected by changing feedwater or steam temperatures, since a variation of either temperature demands more or less transfer of heat to each pound of steam produced at the specified pressure and temperature. These errors are inherently minimized where megawatt generation is used as the index of Btu absorption. It is necessary, when substantial changes may occur in steam or feedwater temperature, to provide manual or automatic compensation to the steam-flow/airflow calibration.

A gas analyzer gives true excess-air determinations. Some delay is introduced, since combustion must be completed before a correct sample can be obtained. If the sample is withdrawn close to the combustion zone, this delay is not usually objectionable. Considerable study may be required to locate sample points that will give correct average excess-air values. The dirty, hot, and corrosive conditions of the flue gases at these points may make periodic maintenance necessary for continuous dependable sampling. The gas analyzer is an accurate index for feedback control.

It should be pointed out that at least one of these combustion guides is needed for proper control of the combustion process. Each of these can be applied individually or in combination for automatic adjustment of fuel-air ratio.

BURNER CONTROLS

Burner-control systems of various types are now applied to almost all boilers to prevent continued operation of a boiler where a hazardous furnace condition could exist and to assist the operator in starting and stopping of burners and fuel equipment.

The most important burner-control function is to prevent furnace or pulverizer explosions which could threaten the safety of operating personnel and damage the boiler. The control system must also prevent damage to burners and fuel equipment from faulty operation while avoiding false trips of fuel equipment when a truly unsafe condition does not exist.

Other important factors in the design of the burner-control system are the method and location to be used in the start-up, shutdown, operation, and control of the fuel equipment. These factors must be understood before purchase, design, or application of a burner-control system.

The different categories of burner control are summarized in the following discussion.

Manual Control.   With complete manual control (Fig. 28) it is necessary to operate the burner equipment at the burner platform. Positioning of the burner components is

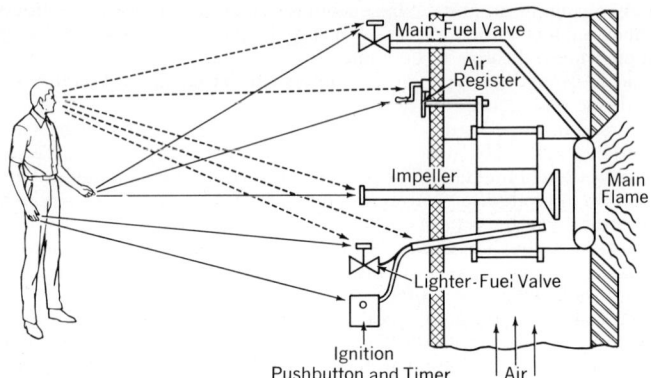

Fig. 28. Fuel-burner manual control.

performed manually. Checks on the existing conditions are dependent on observation and evaluation by the local operator. Good communication between the local operator and the control room is required to coordinate the start-up and shutdown of burner equipment. Manual supervision and control are of interest principally on older boilers, since some form of automatic burner control is provided with almost all boilers now being purchased.

Manual Control with Lighter-Flame-Proving Systems.   The recommended initial step is to apply a semiautomatic lighter control including a flame-proving and interlock system (Fig. 29). From local-panelboard initiation, assuming first that various prefiring and purge interlocks are satisfied, the lighter will be started and proved continuously, and thus provide a permissive signal for introduction of main fuel to that burner.

Flame detectors or other means to "prove" lighters in service are recommended by the Boiler-Furnace Explosions Committee of the National Fire Protection Association (NFPA). This is the recognized industry committee for establishing standards for boiler-furnace safety.

For this type of limited system the fuel equipment should still be operated only from the burner platform as with complete manual control.

Manual Control with Lighter- and Main-Flame Proving.   Many industrial boilers with gas and/or oil firing by multiple burners are protected by a burner interlock and trip system using individual-burner supervisory cocks and trip valves from main-flame detection (Fig. 30). The lighter is initiated from a local panel with manual operation of the air registers and supervisory cocks. After a short ignition-time delay, the individual burner valve will trip on loss of main flame. Normally no other monitoring is provided at each burner except by the operator's intelligence. This is a local manual system and should not be operated from a remote location. Such systems have limited burner interlocks and trips, or none, except from lighter- and main-flame detectors. Cross interlocks between burners are also nonexistent or limited. This relatively inexpensive burner-control system has proved satisfactory where only a few burners are operated from the burner platform. NFPA now requires a system of this scope as a minimum for oil- and gas-fired boilers.

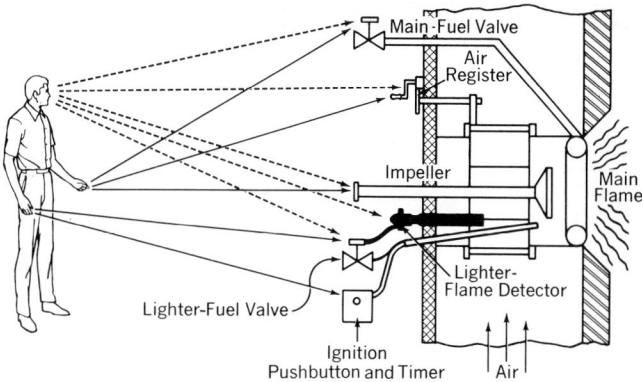

Fig. 29. Manual control with lighter-flame-proving system.

Remote Manual Sequence Control. Illustrated in Fig. 31, this system represents a major step, permitting remote manual operation, and using instrumentation systems and position switches in the control room for intelligence. This system is not recommended for remote operation where safety is dependent only on flame detection. Various burner permissives, interlocks, and trips should also be provided to account for the position of fuel valves and air registers in addition to main-flame detection. With this system the operator participates in the operation of the fuel equipment. He controls each sequence of the burner-operating procedure from the control room and no steps are taken except by his initiation.

Automatic Sequence Control. The next logical step (Fig. 32) is to automate the sequence control to permit start-up of burner equipment from a single push button or switch control. Automation then replaces the operator in control of the operating sequences. Since the operator initiates the demand for each fuel-utilization unit, it is expected that he participates, or that he at least monitors with his intelligence the operating sequence, as indicated by signal lights and instrumentation signals, as the start-up process proceeds to completion with the fuel-utilization unit in service on automatic control. This category has been widely applied to gas burners but also to oil- and coal-burning systems as well.

Fuel Management. Finally a degree of fuel system automation can be achieved that will permit fuel equipment to be placed in service without supervision by the operator (Fig. 33). A fuel-management system can be applied that will recognize the level of fuel demand to the boiler, will know the operating range of the fuel equipment in service, will

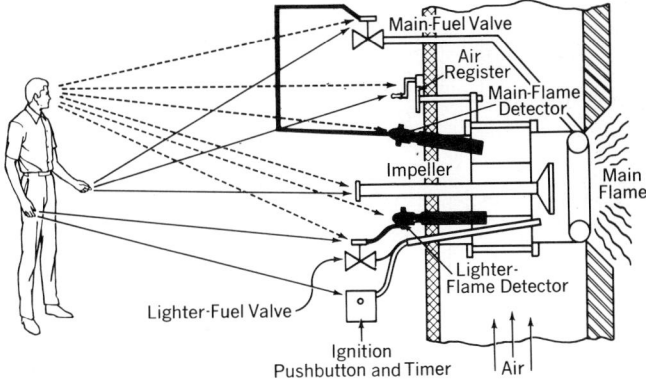

Fig. 30. Manual control with lighter- and main-flame-proving systems.

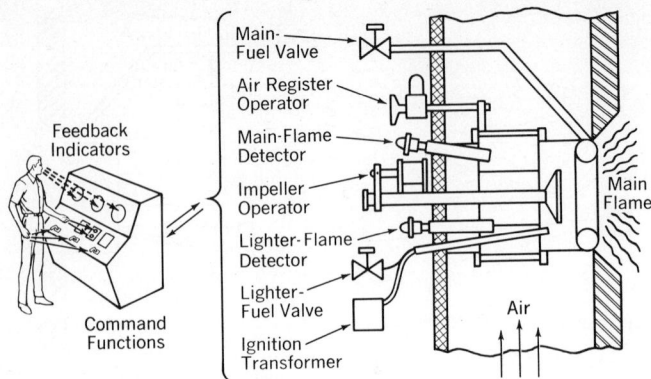

Fig. 31. Remote manual sequence control.

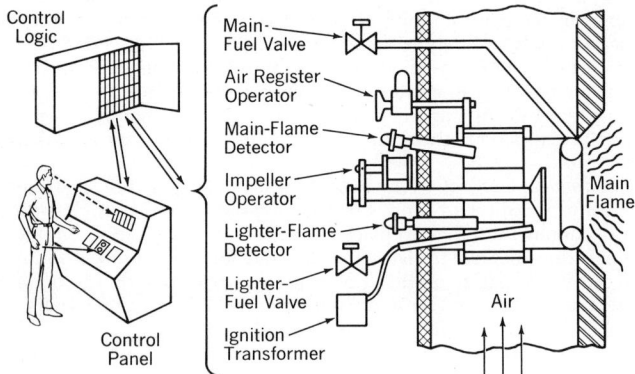

Fig. 32. Automatic sequence control.

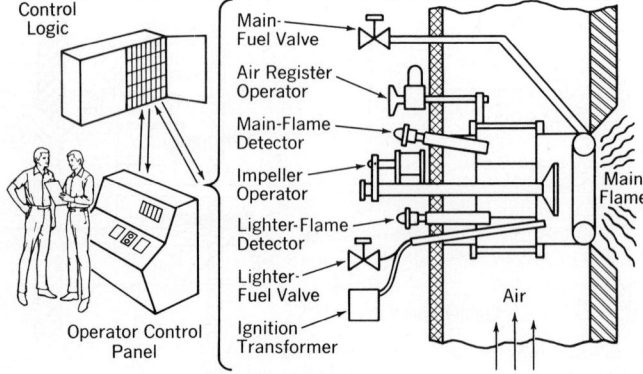

Fig. 33. Fuel management.

reach a decision concerning the need for starting up or shutting down the next increment of fuel equipment, and will select the next increment based on the firing pattern of burners in service. Such demands for the start-up or shutdown of fuel preparation and burning equipment can be initiated by the management system without the immediate knowledge of the operator and, in fact, his attention may be diverted from the firing and fuel preparation equipment at this time.

The degree of operating flexibility allowed with a burner-control system is closely related to the degree of operator participation. A higher level of automation reduces the

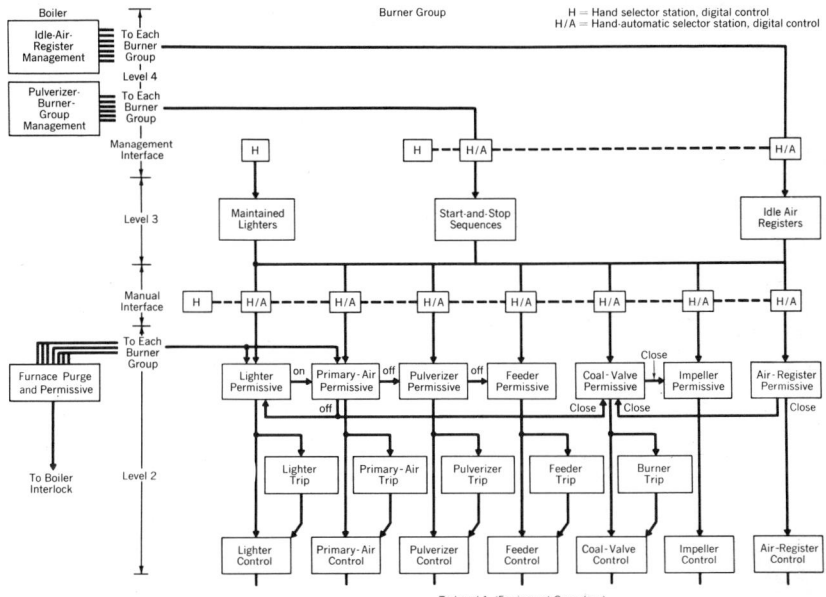

Fig. 34. Pulverizer-burner control system.

flexibility of the operator in handling situations where a piece of equipment fails to perform as expected. One method used to provide more operating flexibility is to allow operator participation at two or more levels. Increased flexibility is also obtained by careful grouping of equipment so that a fault anywhere in the system will affect a limited amount of equipment. An example of this is the grouping of burners on a pulverizer so that the failure of a piece of hardware on a burner will only affect the failed burner, and not the pulverizer and its remaining burners.

Figure 34 illustrates provision for manual intervention in a fuel-management-type system. In this figure, functions are detailed at four levels of a pulverizer-burner-control system. Manual intervention is provided at level 2 to permit remote manual operation with interlock protection features.

The cost of the burner-control system is also important, but it varies significantly with the functional requirements and degree of operating flexibility, as well as the type of logic equipment (solid state or relay) and packaging desired for the system, so that a true evaluation of cost cannot be reached without a complete understanding between the customer and vendor of the factors involved.

A very common error in applying burner controls is attempting to use a local manual-control system as a limited-scope remote-manual-control system.

Flame Detection.   One of the most important items required for a burner-control system is the means of assuring the presence or absence of flame on each individual burner regardless of the type of fuel being burned.

Ultraviolet (UV) flame detectors have been applied to all types of fuels with varying degrees of success. The UV detector is especially well suited to natural-gas firing due to the abundance of ultraviolet radiation produced by the combustion of the hydrogen in natural gas. A detector now coming into general use has recently been developed using the high-frequency dynamic flickering of the primary combustion process in combination with the intensity of the visible radiation of the flame. Application of this dual-signal detector indicates that the Flicker* detector provides greater discrimination and reliability than the UV detector on oil and coal firing. The two detectors can be used in combination on a burner where it is necessary to monitor both gas and oil or coal firing on the same burner, such as on a coal burner with a gas lighter. Such a combination detector is shown in Fig. 35.

The present philosophy in applying flame detectors to any type of fuel requires either

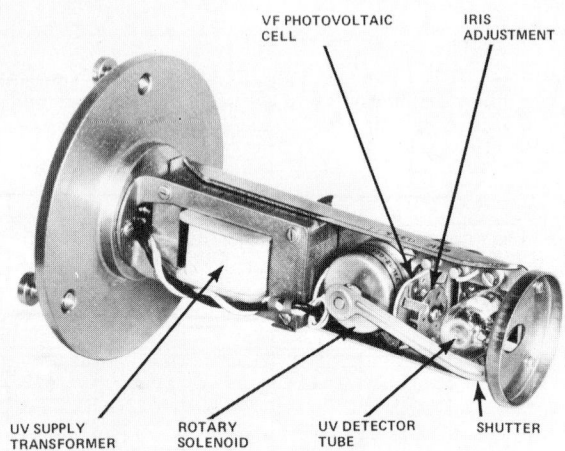

Fig. 35. Combination ultraviolet (UV) and visible flicker (VF) flame detector. (*S. G. Dukelow, Bailey Meter Company.*)

the use of self-checking detectors or the use of redundant detectors on each burner in order to provide the reliability necessary for operation with a burner-control system.

The proper location of the flame detectors on a burner is dependent on many factors, including burner-control-system design and, therefore, should be determined by the burner-control manufacturer as an integral part of burner-control-system design.

Flame-monitoring devices in current use are designed for an on-off type of operation based on the presence or absence of flame. Although the basic detector units provide a variable analog output signal which can be read on a meter, there is no direct claim to an established relationship between this relative output signal and the flame quality. Therefore, the analog signals should never be relied upon as flame-quality indicators. They should be used only to provide helpful information for initial setup adjustments and continuous "on-line" observance.

Many power plants today are using closed-circuit television to provide a means of continuous furnace observation by the operator from the control room. Closed-circuit television has been successfully applied to all fuels, and more applications are expected in the future.

CONTROL SYSTEM DESIGN FOR ONCE-THROUGH BOILERS

The introduction and increased use of the once-through boiler design in the United States stems from the continuing efforts of the utility industry and its suppliers to reduce the cost

*Registered trademark of Bailey Meter Company.

of generating electric power. The once-through boiler makes possible the increase in cycle efficiency obtainable from operation at full steam temperature over a wider load range and operation at supercritical pressures. The once-through boiler also reduces capital costs since elimination of the steam drum and use of smaller-diameter tubes result in considerable saving in material.

Operation of the Once-Through Boiler The once-through concept can best be described by thinking of the boiler as a single tube. Feedwater is pumped into one end, heat is applied along the length of the tube, and steam flows out the other end. The output from the tube is a function of the feedwater flow and the amount of heat supplied. The output fluid enthalpy, or heat content, depends only on the ratio of the heat input to the feedwater flow. The presence of a valve at the outlet of this one-tube boiler provides a means of varying the pressure level. When the pressure level is maintained constant, the outlet steam temperature also is dependent only on the ratio of the heat input to the feedwater flow.

At steady-state conditions the feedwater flow-in equals the steam flow-out. The pressure level will be influenced not only by the valve restriction at the outlet but also by the density of the fluid throughout the system. Therefore, it is important to note that a change in heat input will influence both pressure and temperature. It is possible to vary the flow and pressure at the outlet by changing the amount of valve restriction or at the inlet by changing both the feedwater flow and the heat input. It is important to note that a change in feedwater flow without a corresponding change in the heat input will result in a change in the outlet steam temperature. During transient conditions other factors, such as the fluid and energy storage requirements which change with load, must be considered since they influence the feedwater flow and heat inputs.

Characteristics of Control Systems Experience with once-through boiler operation has indicated the importance of a control system which fully exploits the capability of the once-through concept. The three most basic control system modes, which have been used with the once-through boiler, are the (1) turbine-following, (2) boiler-following, and (3) integrated boiler-turbine-generator control. The latter combines the advantages of the first two. These three operating modes are essentially the master portion of the system which sets up the basic demand for fuel, air, and feedwater.

Turbine-Following Control. A conventional turbine-following control system is shown in Fig. 36. With a turbine-following system the turbine generator is assigned the responsibility of throttle-pressure control, while megawatt load control is the responsibility of the boiler. When an increase in load is demanded, the boiler control increases the pumping and firing rates, which in turn starts raising the throttle pressure. In order to maintain a constant throttle pressure, the turbine control valves open and accept the additional boiler output, thus increasing the load. When a decrease in load is demanded, the same procedure occurs in the reverse direction. Load response with this type of system is rather slow since the turbine generator must wait for the boiler to change its output before repositioning the turbine control valves to change load.

Boiler-Following Control. A conventional boiler-following control system (used also in most drum boilers) is illustrated in Fig. 37. With a boiler-following system the boiler is assigned the responsibility of throttle-pressure control while megawatt-load control is the responsibility of the turbine generator. The demand for a load change goes directly to the turbine control valves to reposition them to achieve the desired load. Following a load change, the boiler control modifies the pumping and firing rates to reach the new load level and restore throttle pressure to its normal operating value. Load response with this type of system is very rapid, since the stored energy in the boiler is used to provide the initial change in load. The fast load response is obtained at the expense of less stable throttle-pressure control.

Integrated Control. While both these systems can provide satisfactory control of the unit, each has certain inherent disadvantages and neither exploits fully the capabilities of both the combined boiler and turbine generator. The turbine-following and boiler-following systems have been combined into an integrated-control system which provides the advantages of both systems while minimizing the disadvantages.

The integrated boiler-turbine-generator control system is shown in Fig. 38. Megawatt-load control and throttle-pressure control are the responsibility of both the boiler and the turbine generator. One type of an integrated system assigns the responsibility of throttle-

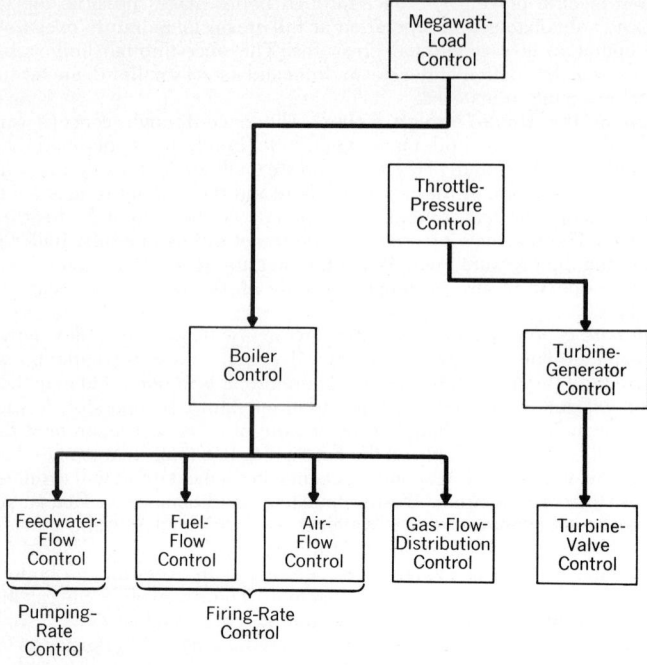

Fig. 36. Turbine-following control system. *(Babcock & Wilcox Company.)*

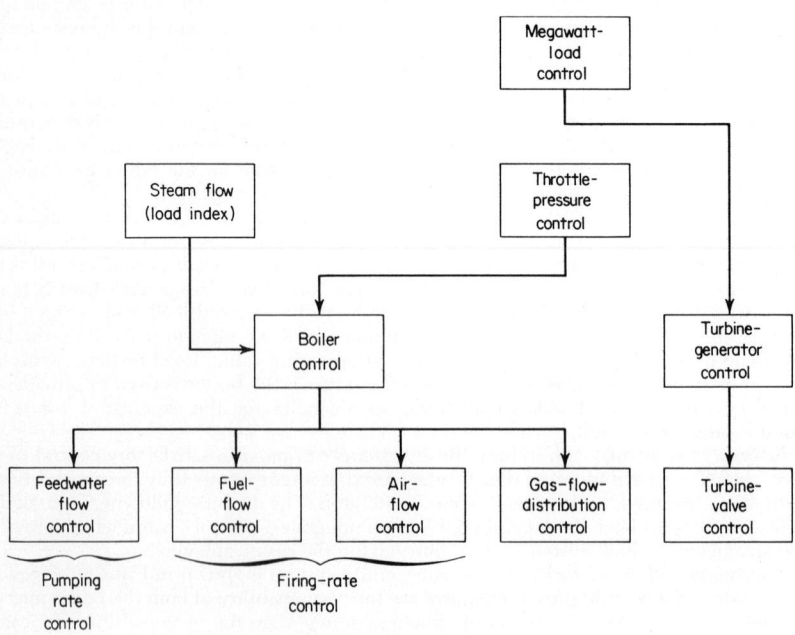

Fig. 37. Boiler-following control system.

pressure control to the turbine generator, thus taking advantage of the stability of a turbine-following system. In addition the system utilizes the stored energy in the boiler, thus taking advantage of the fast load response of a boiler-following system.

Since the boiler is not capable of producing rapid changes in steam generation at constant pressure, the turbine is used to provide the initial load response. When a change in load is demanded, the throttle-pressure set point is modified using megawatt error

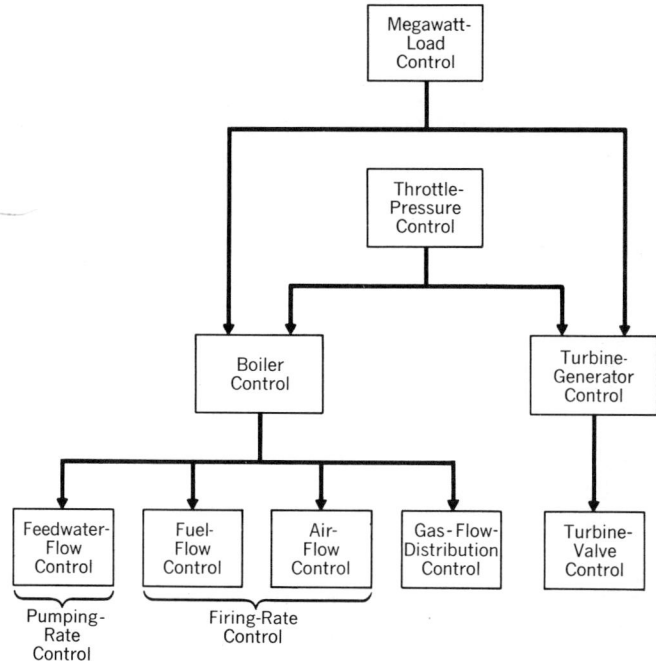

Fig. 38. Integrated-boiler-turbine-generator control system. See also Fig. 41.

(difference between the actual load and the load demand), and the turbine control valves respond to the change in set point to give the new load level rather quickly. As the boiler modifies the pumping and firing rate to reach the new load and restore throttle pressure, the throttle-pressure set point returns to its normal value. The result is a fast and efficient production of electric power through proper coordination of the boiler and turbine generator. The speed of response is much faster than a turbine-following system, as shown in Fig. 39. However, it is not as fast as a boiler-following system because the effect of megawatt error is limited to maintain a balance between boiler response and boiler stability. Wider limits would provide more rapid response, while narrower limits would provide more stability.

Development of the integrated-control system received added impetus from the introduction by the electric power generating companies of the wide-area economic-load-dispatch system. This system, through a means involving incremental cost control, requires precise load control. Each unit under control is required to produce a specific level of megawatt generation, and the area-load-control system continues to readjust the turbine control valves until this level is reached. Now, even relatively minor interactions between steam flow and throttle pressure can create a continuous interaction between the boiler, the turbine generator, and the area-load control. The conventional boiler-following system, which is subject to such interactions between steam flow and throttle pressure, does not lend itself to use with this type of an area-load control. This is not due to any incapability of the boiler-following system as such but is the result of combining two highly responsive systems.

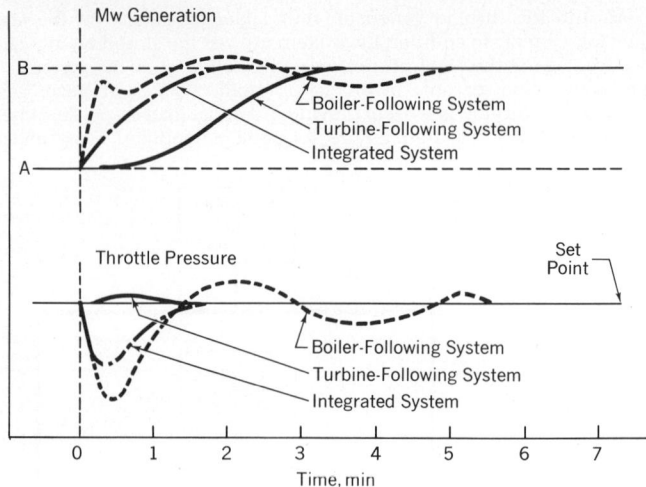

Fig. 39. Comparison of response characteristics of various arrangements of boiler-turbine-generator control systems for a large change in load demand.

Since wide-area economic-load-dispatch systems are being utilized increasingly, greater emphasis is placed on controlling the boiler and turbine generator together in an integrated system to achieve automatic control of megawatt generation over a wide operating-load range.

While the integrated-control system was developed primarily for use with the once-through boiler, it can also be applied to drum boilers especially where wide-area economic-load-dispatch systems are in service. A description of the integrated boiler-turbine-generator control system is presented in the following section.

Basic Once-Through Boiler Control in Integrated Mode   The control system for fuel, air, and feedwater in its basic form consists of a number of ratio controls which compare related pairs of controlled inputs, such as:

1. Boiler energy input to generator energy output.
2. Superheater-spray-water flow to feedwater flow.
3. Fuel flow to feedwater flow.
4. Fuel flow to airflow.
5. Recirculated-gas flow to airflow. This in effect is a ratio of reheater absorption to absorption in primary water and steam.
6. Fuel flow to primary airflow for pulverized-coal-fired units.

Figure 40 is a simplified diagram showing all system inputs controlled in a parallel relationship from the megawatt-load demand by means of ratio controls.

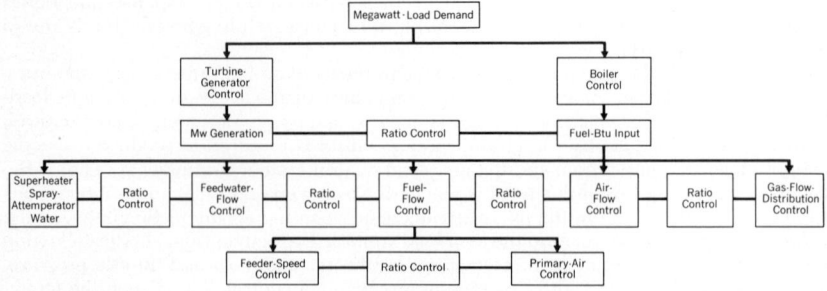

Fig. 40. Ratio controls.

The integrated-control system coordinates the boiler and turbine generator for fast and efficient production of electrical power in response to load demands initiated by the automatic load-dispatch system. A functional description of the integrated-control system, shown in Fig. 41, is presented in the sections following.

Megawatt-Load Control.   Most electric utility companies have an area-load-control system which introduces a demand on the boiler–turbine generator unit. The megawatt-

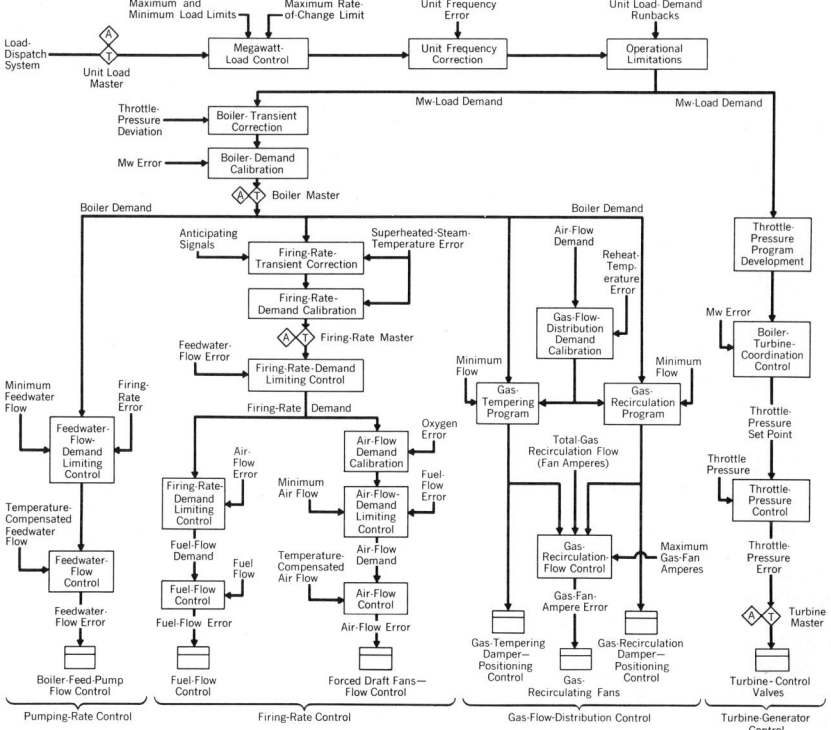

Fig. 41. Integrated boiler-turbine-generator control system.

load control conditions these demands to make them compatible with the state of the unit and its ability to change load. Also considered are such factors as maximum and minimum load limits and maximum permissible rate of load change. The megawatt-load demand is modified for any frequency error, recognizing that, whenever a frequency error occurs, the turbine-speed governor will change the turbine-control-valve position, thus changing load.

A control feature is included which will "run back" (i.e., reduce) the megawatt-load demand whenever unit capability is limited for any reason such as the loss of a boiler feed pump. The unit-load master permits the unit to be separated from the load-dispatch system and placed on manual control with the operator establishing the unit's output. The final megawatt-load demand signal is applied in parallel to the boiler and turbine generator.

Throttle-Pressure Control.   Throttle-pressure control is the responsibility of both the boiler and the turbine generator. During steady-state operation, the turbine-control valves maintain a constant pressure at the throttle. During load transients the throttle pressure is controlled within plus or minus approximately 100 psi of its normal set point. The amount of pressure variation from the normal set point depends on the size of the megawatt-load error, i.e., the size of the load change on the unit. This is discussed below under Turbine-Generator Control.

Whenever throttle pressure deviates from its normal set point, a signal proportional to the size of the error modifies the basic boiler demand to return pressure to its normal operating value. This corrective action provides a means of over- or under-pumping and firing during load changes to accommodate the changes required in the boiler-fluid and energy-storage levels.

Turbine-Generator Control.   The megawatt-load demand is used to develop a steady-state throttle-pressure program. The normal program for a unit operating a supercritical pressure requires maintaining a throttle-pressure set point of 1000 psi up to approximately 8 percent of load. This is "ramped" up to 3500 psi as the load is increased to 25 percent. Above this load, throttle pressure is maintained at 3500 psi.

The normal steady-state throttle-pressure set point is compared with the actual measurement of throttle pressure. Any error causes the throttle-pressure control to adjust the position of the turbine-control valves to maintain the desired throttle pressure. When a change in load is demanded, the boiler-turbine-coordination control uses megawatt error to modify the steady-state throttle-pressure set point and program the opening and closing of the turbine-control valves in a controlled manner in order to provide a fast response to the megawatt-load demand while maintaining stable throttle-pressure control. As the megawatt generation changes to satisfy the megawatt-load demand, the throttle-pressure set point returns to its normal value. The turbine master permits the operator to change the turbine-control valves manually to establish the desired unit load, in which case the boiler is controlled in a "boiler-following" mode.

Boiler Control.   The megawatt-load demand is used by the boiler control to establish demands for feedwater, fuel flows and airflows, and gas-flow distribution. The megawatt-load demand is modified for any deviation in throttle pressure as discussed earlier in Throttle-Pressure Control in order to change the fluid- and energy-storage levels of the boiler. This correction assists in maintaining boiler output consistent with turbine steam requirements.

The boiler and turbine must be maintained in the proper relationship at all times to produce the load demanded by the system. Any permanent change in the normal relationship between the boiler and turbine, such as a change in the turbine efficiency, will show up under steady-state conditions as a megawatt error. Consequently, megawatt error is used to calibrate the boiler demand to keep the boiler and turbine in the correct relationship. The boiler master permits the operator to establish the boiler demand manually. The boiler demand is applied in parallel to feedwater-flow, firing-rate, and gas-flow-distribution controls.

Feedwater-Flow Control.   The boiler demand is used to establish a demand for feedwater flow. A minimum feedwater-flow demand (usually 25 or 33 percent of full load) assures that an adequate feedwater flow is maintained at all loads to protect the furnace tubes from overheating. Whenever the firing rate is unable to satisfy its demand, a firing-rate error will modify the demand for feedwater flow. This assists in maintaining the feedwater flow and the firing rate in the proper relationship, thus minimizing steam-temperature upsets.

The feedwater-flow demand is calibrated using feedwater-temperature error since any deviation from the expected feedwater temperature requires more or less feedwater flow to compensate for the change in extraction flow to the feedwater heaters.

Firing-Rate Control.   Boiler demand is used to establish the base-firing-rate demand. Various control signals are used to correct the firing-rate demand during transient conditions in anticipation of and in response to steam-temperature or gas-temperature changes. This minimizes the steam-temperature upset associated with a load change.

Gas temperature is used at low loads to improve the response of the steam-temperature control.

The proper calibration of firing rate to feedwater flow is required in order to maintain the desired superheat temperature. Continuous calibration of the firing-rate demand is provided to correct for steam-temperature errors which result from changes in boiler cleanliness, and changes in the fuel Btu content.

Spray attemperation is used above half-load to achieve a rapid, temporary correction of steam temperature while the firing-rate demand is being modified to produce a slower, permanent change. Whenever the feedwater flow is unable to satisfy its demand, a feedwater-flow error will modify the demand for firing rate. This assists in maintaining feedwater flow and firing rate in the proper relationship, thus minimizing steam-temperature upsets.

The final firing-rate demand is used to establish a demand for airflows and fuel flows. This arrangement produces fast firing-rate response to load changes by moving the fuel and air simultaneously. The firing-rate master permits manual operation of the fuel-flow and airflow controls.

The fuel-flow demand is compared with the measured fuel flow. Any error will cause the fuel-flow control to modulate the control valve to satisfy the fuel-flow demand. Whenever the airflow demand cannot be satisfied, an airflow error will modify the demand for fuel flow. This assists in maintaining fuel flow and airflow in the proper relationship to assure safe and efficient boiler operation.

An airflow demand calibration is included in the control system to account for any disturbance caused by fuel-flow and/or airflow metering errors or by changes in the Btu content of the fuel. This assists in maintaining the correct amount of excess air required for the combustion process. Airflow demand is compared with the measurement of airflow which has been temperature-compensated for accuracy. Any error will cause the airflow control to reposition the forced-draft-fan controls to satisfy the airflow requirements.

A minimum airflow demand (usually 25 percent of full load airflow) assures safe boiler operation at low loads. Whenever the fuel-flow demand cannot be satisfied, fuel-flow error will modify the demand for fuel flow. This assists in maintaining the proper relationship between the fuel flows and airflows.

Gas-Flow-Distribution Control.   Combustion gases leaving the economizer are recirculated and introduced through the bottom of the furnace as gas recirculation for the dual purpose of controlling reheat-steam temperature at high loads and for furnace protection throughout the load range, particularly at low loads. Gas recirculation alters the heat-absorption throughout the unit. On coal-fired units combustion gases leaving the economizer are introduced near the furnace outlet as gas tempering to control the furnace-exit-gas temperature.

The airflow demand is used to develop the high load program for gas-flow distribution. This program is continuously calibrated to control the reheater absorption in order to maintain the desired reheat-steam temperature. The boiler demand is used to develop the low-load program for gas recirculation and gas tempering. Whenever the reheat steam temperature is higher than desired, the amount of gas-recirculation flow is reduced. When gas recirculation has been reduced to its minimum value and reheat temperature is still high, reheat spray attemperation is used to control reheat temperature.

The sum of the gas-recirculation and the gas-tempering demands is the total flow required from the gas recirculation fans. The gas-fan amperes, which represent an accurate measurement of flow, are used to control the gas-recirculation flow. The total gas-recirculation-flow demand is a demand for gas-fan amperes and is compared with the actual measurement of gas-fan amperes. Any error will cause the gas-recirculation-flow control to reposition the fan-inlet damper to satisfy the demand for gas-fan amperes. Fan overload protection is provided through a maximum gas-fan-amperes limit. On some units excess air may be used instead of gas recirculation to control reheat-steam temperature and provide furnace protection.

Start-up and Low-Load Control.   During start-up and operation at loads below the minimum feedwater flow, some or all of the flow is diverted through a bypass system to the flash tank. This enables the boiler to provide the desired steam flow to the turbine while maintaining a minimum feedwater flow for protection of the furnace tubes from overheating. Operation of various valves in the bypass system provides load and pressure control.

The flash-tank control automatically maintains control of both flash-tank pressure and flash-tank level while the unit is operating on the bypass system. The controls selectively distribute the flash-tank drains and steam flow to the condenser, deaerator, and the high-pressure heaters, initially to clean up the water and then to recover heat from both the steam and drain flows.

The preceding description for once-through boilers covers the control application for units manufactured by one of the four once-through boiler suppliers in the United States. Due to the difference in external piping, valving and flash-tank arrangement plus differences in boiler design and operating philosophy, the application is different though similar for boilers by other manufacturers.

INSTALLATION AND SERVICE REQUIREMENTS

It is important to make certain that instrumentation and control equipment are properly installed and serviced. Supervision of installation and the required calibration adjustments should be performed by qualified and experienced personnel. Adequate time for tune-up of controls should be provided during initial boiler operation.

Experienced field engineers, employed by control manufacturers, are available to assist customers. The field adjustment of a sophisticated control system is a job for an expert. Once done, maintenance can be performed by trained plant personnel. The plant personnel who will be responsible for this maintenance should be available to observe and assist the control manufacturer's representatives in the installation and calibration of the control system. Most control manufacturers offer training programs to familiarize customer personnel with the control equipment. A planned program of preventive maintenance should be developed for the control system.

FUTURE DIRECTIONS OF BOILER-CONTROL SYSTEMS

As described in the preceding pages boiler control is made up of a number of separate systems which should each be designed with the total system in mind.

The modulating portion of the total system has generally been accomplished with an analog control system. The use of solid state digital computers as a power plant tool for data logging, alarm monitoring, performance computation, etc., has greatly expanded in the last decade. Since 1958 when the initial installation was made until today such application now is made in a very large percentage of new power plant units.

Digital sequence logic burner-control applications, started in the 1950s with relay logic, are applied today on almost all new units and use solid state logic elements. During this period, the analog controls switched from pneumatic to electronic in the early 1960s on power plant boilers, though a majority of smaller boilers today are still pneumatic.

Figures 42 and 43 are representative of much of the current practice in control center design and the arrangement of the cabinets that house the electronic systems involved.

The following lists areas of expected change in the control of boilers in the relatively near-term:

1. Smaller industrial boilers move from predominantly pneumatic to predominantly electronic.

2. Burner-control systems will move from wired logic systems to programmable systems using microprocessor computers.

3. Boiler-modulating control will move from analog control to direct digital control.

4. Expanded use of computers to analyze plant operating data to a significantly greater level, providing the operator with levels of information not now available.

5. Improved control center human communications by combining digital and analog function presentation to ease the operator's task in operating complex equipment, such as large pulverizers.

6. Move toward computer-driven control centers where expanded computer applications analyze plant data and present meaningful information to the operator for optimizing plant operation.

Industrial Boiler Control   Future changes for smaller boilers will likely be in the direction of more implementation with electronic rather than penumatic control. The motive force for this change will be cost as electronic equipment costs rise more slowly than pneumatic equipment costs. Counteracting this force will be the requirement to retrain personnel in electronic maintenance.

Direct Digital Control   A greater use of the digital computer in performing all the control functions in addition to data logging, alarm monitoring, turbine start-up, etc., which have been the prime computer functions in the past. These added control functions are called DDC (direct digital control). Complete systems of this type have been in operation successfully since 1972.

Fig. 42. Control center of a large steam-electric generating station. *(S. G. Dukelow, Bailey Meter Company.)*

The use of a digital computer with DDC does not in itself require a change in control center design. Figures 44 and 45 are DDC installations with two different approaches to control center design. Figure 44 is conventional and to an operator appears very similar to earlier systems. Similarly, Fig. 45, while using a different design criteria, appears only slightly different to the operator than if the installation had been implemented with analog control.

DDC applications to date have illustrated the following distinct advantages and benefits not previously available with conventional analog approaches:

Fig. 43. Arrangement of electronic control systems cabinets.

Fig. 44. Direct digital control (DDC) control center with conventional operator interface design.

Fig. 45. DDC control center with condensed operator interface design.

1. Input transducer signals can be tested in many ways before being used to develop a control signal. When an input signal fails in an analog control system, it generally results in an immediate faulty control action.

2. The computer being used for direct digital control is capable of diagnosing itself for the majority of its troubles and can place the control loops on manual or backup so that no faulty control action will occur. This is not possible with analog hardware. If a component failure occurs, a faulty control action generally results.

3. Often, peculiarities are discovered about each process, and control strategies must be changed to accommodate these peculiarities. The computer has virtually unlimited arithmetic capability. The control engineer need not be concerned about cost or time delays when the need arises to add extra control operations, such as proportional, reset, function generation, as well as more advanced forms of control.

4. Direct digital control offers unlimited flexibility. Control calculations can simply duplicate well-known analog actions or can be expanded in many ways to achieve better control through the creation of specialized modules. DDC will contribute to advancing the state of the art, which can only be reflected as an advantage to the user.

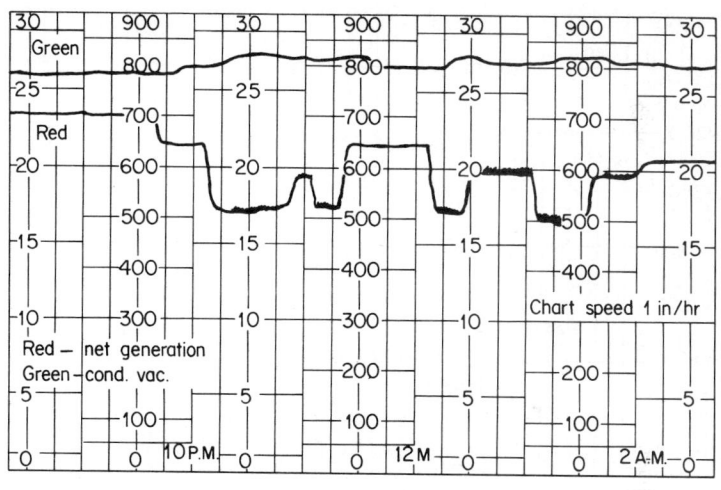

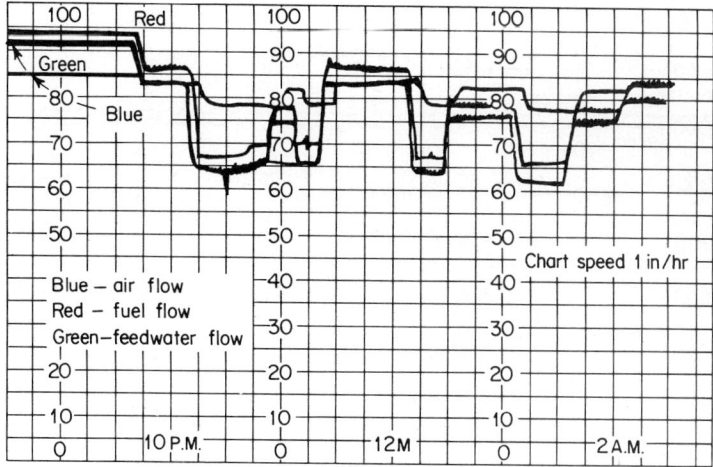

Fig. 46. Load tests for documentation. See also Fig. 47.

5. Direct digital control has the capability of providing pure delay in its control calculations. This is not achievable with analog control. With pure delay, control action can be delayed any desired length of time in order to duplicate pure delay in the process. Also, an analysis of why a measurement has changed can be made before control action is taken.

6. Because digital computer calculations are error- and drift-free, direct digital control settings are exact, reproducible, and noninteracting. For example, the setting of the gain constant and reset rate have no effect on each other whatsoever. Control settings developed by simulation can be duplicated precisely in the field.

The ease by which the serviceman can obtain a record of control settings and the system configuration greatly assists him in doing his job more easily and quickly.

7. A high degree of reliability is obtained because of the relatively few components involved. Note that the integrated-boiler and burner-control systems presently represented by 30 to 40 cabinets of hardware on a large conventional system are reduced to 10 to 20 cabinets for DDC. The user, in the future, can obtain a considerable economic savings by reflecting this reduction in his space requirements for control and relay rooms.

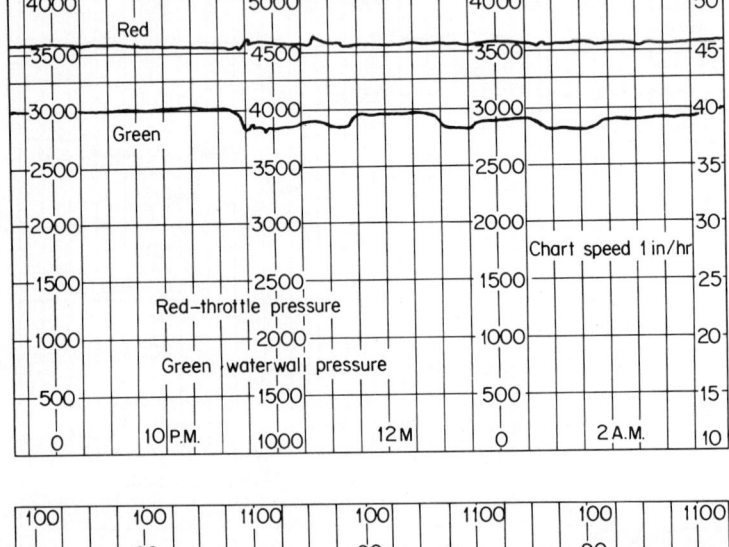

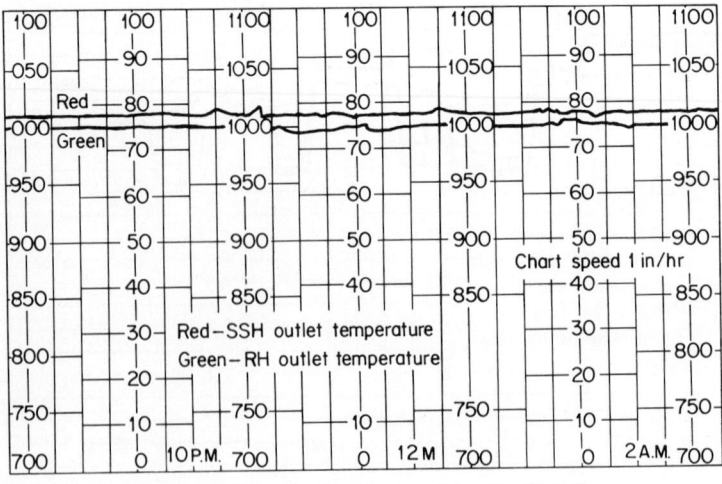

Fig. 47. Load tests for documentation. See also Fig. 46.

8. A computer control system permits control engineers and maintenance personnel to readily and continually monitor the status within the system, thus simplifying trouble-shooting. The computer provides the flexibility equivalent to having a digital voltmeter on every wire in an analog system.

9. Some additional unanticipated benefits which experience to date have revealed are:

a. All calibration, tuning, and testing work has been performed in the presence of the unit operator at the BTG board using the operator's control console. As a result of the continuous dialog between the control engineer and operators, smoother, better-coordinated control commissionings have occurred. Also, the participation of the operator in this process speeds his training and confidence in the system.

b. The merging of control and information systems has resulted in a symbiosis that provides access to complete information on the status of the control system (permissives, timers, transfers, logic action) through cathode-ray-tube (CRT) messages, alarms, and status summaries. These programs provide printed records that automatically document the progress and state of the control system throughout the tuning and calibration period. Also, the diagnostic programs, trend recording, logging, and CRT displays provide powerful control debugging tools heretofore not available, all of which greatly assist the serviceman in doing his job more accurately, easily, and quickly.

c. Generating unit performance has been noticeably improved; i.e., unit response, availability and controllability exceed that which we have experienced with conventional analog implementation.

Chart records showing the superior performance of a system of this type are shown in Figs. 46 and 47.

Expanded Use of Digital Computer   Greater use of the digital computer also provides the ability to analyze data and supply it to the operator in meaningful terms rather than as

Fig. 48. Advanced design for a computer-driven boiler-turbine-generator control center. *(Ebasco Services Inc.)*

a mass of data he must absorb and analyze before taking action. By such expansion of the operator's brain potential, greater plant availability and reliability can be realized with a higher quality of operator action during emergencies and improved actions during normal operation to forestall emergencies occurring.

Computer-Driven Control Center   An advanced view of a boiler-turbine-generator control center that takes full advantage of computer capability is shown in Fig. 48. Control centers such as this or similar designs for the increasingly larger and more complex generating units can be expected as the need for the expansion of the capability of the operator and the solution to this problem is fully exploited.

Steam Turbine Design Trends*

For fossil fuel applications in the 500 to 600 megawatt range, a turbine consisting of a combined high-pressure/intermediate-pressure turbine element and two double flow low-pressure turbine elements is commonly provided. The low-pressure turbine elements may have 25-, 28½-, or 31- inch long last row blades, depending upon the economic optimization with available cooling water, fuel costs, and other cycle parameters. The world's first 500 megawatt tandem-compound steam turbine was a unit of this type, placed into service late in 1965. Combined elements, with the high- and intermediate-pressure blading in a single casing, offer significant advantages. The installed length of such a unit is about 18 feet less than that required when the high-pressure and intermediate-pressure elements are contained in separate turbine casings. With one less turbine element, the savings in capital, erection, and maintenance costs are appreciable. A similar arrangement is provided for units in the 150 to 400 megawatt size class, consisting of a combined high-pressure/intermediate-pressure turbine coupled in tandem with one double-flow low-pressure turbine element. The last row blades of the low-pressure element may be 23, 25, 28½, or 31 inches long, depending upon the unit rating cycle, fuel costs, and expected load factor. Many of the improvements developed for the higher-rated units are also available in units rated less than 150 megawatts. An example of this is the single-case, single-exhaust-flow unit for nonreheat applications, such as the 100 megawatt turbines designed for some combined-cycle plants.

For larger fossil fuel applications, above 650 megawatts, the complete unit may consist of a separate double-flow high-pressure turbine element and a separate double-flow intermediate-pressure turbine element coupled in tandem with either two or three double-flow low-pressure turbine elements. This type unit may utilize either 28½- or 31-inch long last row blades. Units of both the four-casing four-flow type and five-casing six-flow type are in service. Maximum turbine-generator capabilities of units in service are slightly over 800 megawatts and, before the end of the 1970s, six-flow types are projected in the 1200 megawatt size range.

Prior to the great increases in cost of raw fuel, the history of how the economy of scale in turbine-generator unit sizes drove down the cost of electric power generation is well known. Larger units provide better thermal performance, lower capital costs, and fewer parts to maintain than multiple sets of smaller ratings for the same total capacity. The sharply accelerating trend to even larger unit sizes, which began in the 1950s, continues to move upward in the mid-1970s. The largest unit in service as of 1973 was slightly over 1,000 megawatts, while the largest unit scheduled for delivery in 1976 is about 1300 megawatts.

TECHNOLOGICAL ADVANCEMENTS

Some of the technological advances of the 1970s which have contributed to improved steam turbines include:

1. Development of an anchored throttle valve-steam chest system which eliminates the concern for possible turbine casing distortion from main steam inlet pipe reactions. The steam chest, anchored to the foundation, isolates the reactions to the main steam station piping so they are not imposed on the turbine casing.

2. Improvements in the design of internal stationary parts have consisted of innovations to minimize the effects of thermal gradients and thermal cycling. Fossil units in particular are designed with the expectation that they will be called upon to support intermediate peaking or cyclic duty operation. Separate nozzle chambers are used to eliminate locked-in thermal stresses when only a portion of the admission arc is active while operating at reduced load. Inner casings are mounted in the outer casings so they are free to expand radially and axially. The alignment of the inner casing is maintained by

*Basic information for this summary provided by Westinghouse Electric Corporation.

supports at the horizontal joint and keys on the vertical center line. All blade rings, dummy rings, and gland rings are separately mounted in the inner or outer casing with their alignment maintained by supports on the horizontal joint and dowel pins at the vertical center line. Each component is free to expand independently, without being restricted or restrained by surrounding parts. Thermal distortion due to temperature gradients and load cycling is practically eliminated. Thermal cracking of casings and nozzle chambers, once a major industry problem, is solved where these features are applied.

3. The first stage in a high-pressure turbine, the control stage, is subjected to the shock load associated with partial arc admission and the flow excitation produced by the nozzle vane wakes. Research has been carried out regarding the operating characteristics of control-stage blading. Troublesome modes of vibration have been identified and their frequencies established. Nozzle vane sections have been developed to provide lower levels of excitation and better control of exciting frequencies.

4. In fossil fuel turbine units, steam temperatures are high enough to materially affect the useful creep life of a turbine rotor at the operating stress level. Considerable design effort has concentrated on reducing the exposed rotor temperature at the two inlet zones of the unit by blanketing these rotor areas with cooling steam. Steam, which has been cooled by expansion through the control stage, blankets the zone between the nozzle chambers and the rotor. For 1,000°F (538°C) inlet temperature, the maximum rotor temperature is limited to about 925°F (496°C). At this reduced temperature, the creep life of the rotor is nearly doubled.

5. Many innovations have been applied to rotating parts to improve thermal performance and reliability. Improved mechanical design in the form of a side-entry blade root has simplified rotor design. This route allows the use of a full row of blades, thereby eliminating the loss in performance and the additional blade loading which exists due to the flow disturbance created when the closing blade of the row must be omitted. This type of blade root was developed nearly 40 years ago (Westinghouse).

6. One of the most significant innovations during the last few years has been the application of sophisticated control techniques made possible by the development of solid-state technology. A particularly valuable feature of digital electrohydraulic control is the ability to switch control between single valve and sequential valve operation. This valve management gives the operator the flexibility to more favorably match temperatures in the first-stage zone of the high-pressure turbine. This minimizes starting and loading periods required for intermediate or cyclic duty application. Automatic startup programs also have been developed.

REFERENCES

Silvestri, J. G., Jr., Aanstad, O. A., and J. T. Ballantyne: "A Review of Sliding Throttle Pressure for Fossil Fueled Steam-Turbine Generators," American Power Conference, Chicago, Illinois, April 18–20, 1972.

Jaegtnes, K. O., McDonald, M. P., and W. T. F. Broer: "Turbine-Generator Control Systems for Dual Units in Nuclear Applications," American Power Conference, Chicago, Illinois, April 29–May 1, 1974.

Donato, V., and S. P. Davis: "Radio Telemetry for Strain Measurements in Turbines," Sound and Vibration (April 1973).

Bannister, R. L., and P. M. Niskode: "Analysis and Control of Steam Turbine-Generator Noise," Noise Control Engineering (Winter 1974).

Gasification as Basis for Combined-Cycle Power Generation

BY GODFRIED J. VAN DEN BERG,* PIET J. J. VAN DOORN,† AND PIET J. HALBMEYER‡

Fuel gas, produced by partial oxidation of high-sulfur residual feedstocks and followed by effective desulfurization processing, can be used in a combined-cycle power station where electricity is generated partly in a gas turbine and partly in a steam turbine cycle. The potential high efficiency of such a combined-cycle power station enables the generation of electricity under nonpolluting conditions at a cost that may compare favorably with alternative routes, considering both pollution abatement and limited supplies of low-sulfur fuels.

COMBINED-CYCLE VERSUS CONVENTIONAL POWER STATION

In a conventional power station (CPS), the fuel oil is burned with air at atmospheric pressure in a boiler where the heat of combustion is used to produce superheated high-pressure steam. In a combined-cycle power station (CCPS), the fuel oil is partially oxidized with air at elevated pressure, whereby the fuel oil is converted into a raw fuel gas. This gas, after removal of contaminants, such as ash and sulfur, is subsequently burned in a combustor and expanded in a gas turbine.

In a CPS, all electricity is produced by the expansion of steam in turbogenerators. In a CCPS, electricity is partly produced by expansion of gas in gas turbogenerators and partly by expansion of steam in steam turbogenerators. In addition to high efficiency, the CCPS permits (1) recovery of up to 95 percent of the sulfur in fuel oil as elemental sulfur, (2) operation with no emission of particulate matter, (3) operation with low emission of nitrogen oxides because of low-flame temperature, (4) lower demand for cooling water than required by a CPS, because only part of the electricity is generated via the steam expansion (and subsequent steam condensation) cycle, and (5) operation at elevated pressure which results in the use of compact, shop-fabricated equipment.

A comparison of the CPS and CCPS approaches is given graphically in Fig. 1. The CCPS schemes discussed here are all based upon the use of residual fuel oil as fuel to the power station. The gasification portion of the process is the SGP (Shell Gasification Process), which has been developed with special emphasis on the use of heavy residual fuel oil as feedstock. Commercial operation of the SGP units has shown that the reliability and on-stream efficiency of the process is high, even in cases where high-ash fuels are being processed. An on-stream efficiency of 95 percent can be taken as a realistic figure. A coal-gasification-based CCPS has been built in West Germany (Ref. 1).

MAJOR ELEMENTS OF CCPS APPROACH

With reference to Fig. 1, the CCPS approach, compared with a CPS, incorporates three new elements:

Fuel Gas Preparation Step.　This is the SGP portion of the total process. In this step, the fuel oil is first partially oxidized in a reactor at elevated pressure (15 to 25 atmospheres) with air, whereby the oil is converted into a gas with carbon monoxide and hydrogen as the main components. Details are described in the article on the SGP.* The

---

*Head of the Conversion Processes and Project Technology Division, †Specialist, Licensing Division, and ‡Head of the Oil Gasification and Technical Back-up, Licensing Division, Shell Internationale Petroleum Maatschappij B. V., The Hague, the Netherlands.
*For description of the SGP, see "Noncatalytic, Partial-Oxidation Gasification Process" in the *Gas Technology* section of this handbook.

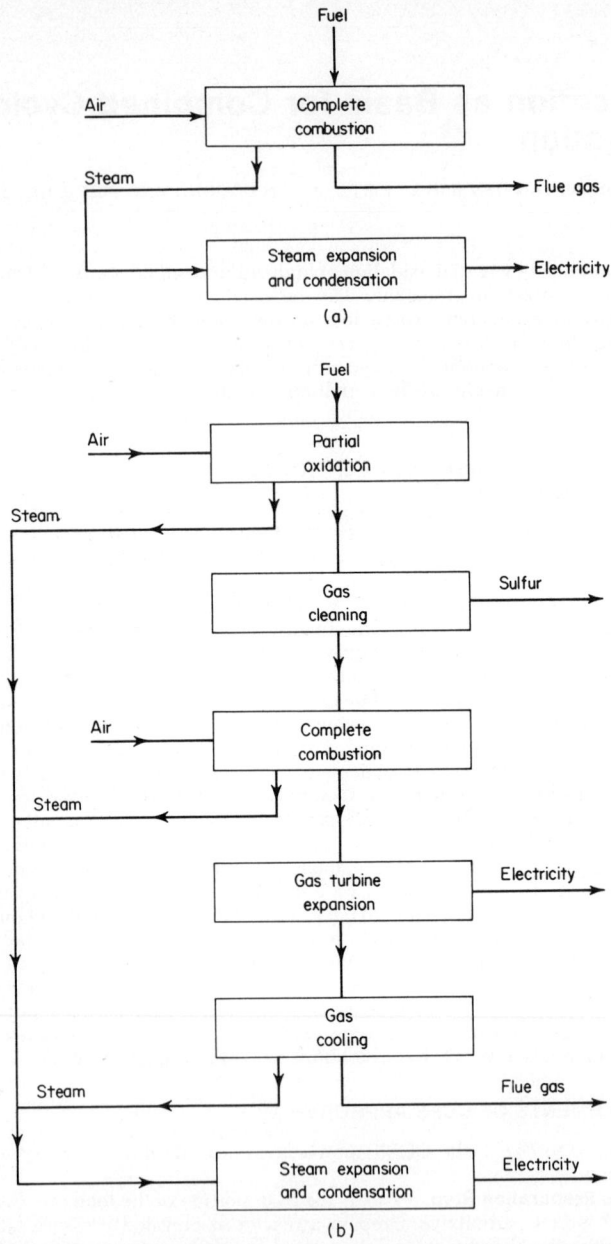

Fig. 1. Comparison of power generation schemes: (*a*) Conventional power station. (*b*) Combined-cycle power station. (*Shell Internationale Petroleum Maatschappij B. V.*)

sulfur of the fuel oil is mainly converted into hydrogen sulfide which, along with carbonyl sulfide and carbon dioxide, can be removed by a selective treating process. See "Absorption of Acidic Gases from Natural Gas and SNG" in the *Gas Technology* section of this handbook.

Supercharged Boiler. The clean fuel gas, as produced from the foregoing steps, is burned in a supercharged boiler at about 10 to 20 atmospheres pressure. In this boiler, the high-pressure saturated steam produced in the waste heat boilers of the gasification unit is superheated. By use of the supercharged boiler, the steam conditions are made independent of the gas turbine outlet temperature. This means that the steam superheat temperature can be about 1000°F ($\sim$ 540°C) instead of 660 to 750°F ($\sim$ 340 to 400°C) if the steam is superheated in a nonfired gas turbine exhaust boiler installed downstream of a gas turbine with an inlet temperature of 1560 to 1740°C ($\sim$ 850 to 950°C). The latter represents current technology for industrial gas turbines. The higher steam superheat temperature results in a higher net efficiency for the power station. A fired-exhaust boiler, although superior to a nonfired boiler, would show higher stack losses as compared with a supercharged boiler. The high gas pressure and the high heat transfer rates result in a compact boiler, which is fully shop fabricated. The NO_x emissions may be expected to be lower than in direct combustion of the gas in the gas turbine combustion chamber.

By controlling both the combustion air dosage and the amount of steam superheated in the boiler, the temperature of the gas leaving the supercharged boiler can be regulated. This gas then is sent to the gas expansion turbine.

Gas Expansion Turbine. The incorporation of gas turbines in natural-gas- (or light-distillate-fuel) fired combined-cycle power stations has found increasing application because of (1) the high efficiencies that can be obtained, and/or (2) the capital cost for such power stations is relatively low (Ref. 9). An important aspect of using a gas expansion turbine is that the inlet temperature of such a turbine can be considerably higher [at present, 1560 to 1740°F ($\sim$ 850 to 950°C)] than the temperature at which a steam turbine can be operated, 1020°F ($\sim$ 550°C). The handling of high-pressure–high-temperature steam presents difficult technical problems.

The combination of a gas expansion turbine cycle with a steam expansion cycle, therefore, enables the conversion of heat into electricity, starting at a very high-temperature level, which favorably affects the conversion efficiency.

The reliability and availability of suitable gas turbines also are important factors. The use of gas turbines in power stations generally has been confined to those stations that are operated for peak-shaving purposes, for which duty the low capital costs are of advantage and availability is of lesser importance. Reports indicate that the availability of the gas turbine cycle can be better than that of the steam turbine cycle (Ref. 9), and also that long periods between maintenance are being obtained (Ref. 10). An example of the increasing confidence in the reliability and availability of gas turbines is their use in high-capital natural gas liquefaction plants (Ref. 11). As contrasted with steam turbine, boiler, and superheater design problems for high-temperature operation, there are promising indications that, through a combination of blade-cooling techniques and blade material developments, the allowable inlet temperature of gas turbines will be continuously increased. Thus, the efficiency of converting heat into electricity can be expected to gradually increase for power stations incorporating gas turbines (Ref. 12). See Fig. 2.

EFFICIENCY OF SGP-BASED CCPS INSTALLATIONS

The combination of the various elements in a CCPS installation, as previously described, together with a conventional steam cycle, leads to a power station where the efficiency loss caused by the clean fuel gas preparation step is compensated to a large extent by the high heat-to-electricity conversion efficiency obtained through the use of the gas turbine. See Fig. 3. In Table 1, the effect of the gas turbine inlet temperature on the overall efficiency of the CCPS system is shown. From this table it can be concluded that at a gas turbine inlet temperature of around 1650°F ($\sim$ 900°C), the efficiency of a CCPS system is equal to that of a conventional oil-fired power station. This means that at 1650°F, the favorable effect of this high-temperature level on the overall plant efficiency has fully compensated for the efficiency losses caused by the fuel gas preparation step.

Inasmuch as in the CCPS system electricity is generated both by a gas expansion cycle and by a steam expansion cycle, considerable freedom exists in optimizing toward alternative aspects, such as efficiency, capital outlay, and cooling water requirements. If, for instance, thermal pollution is an overriding consideration, the cooling water requirement can be reduced by diverting part of the steam into the gas expansion cycle. In this

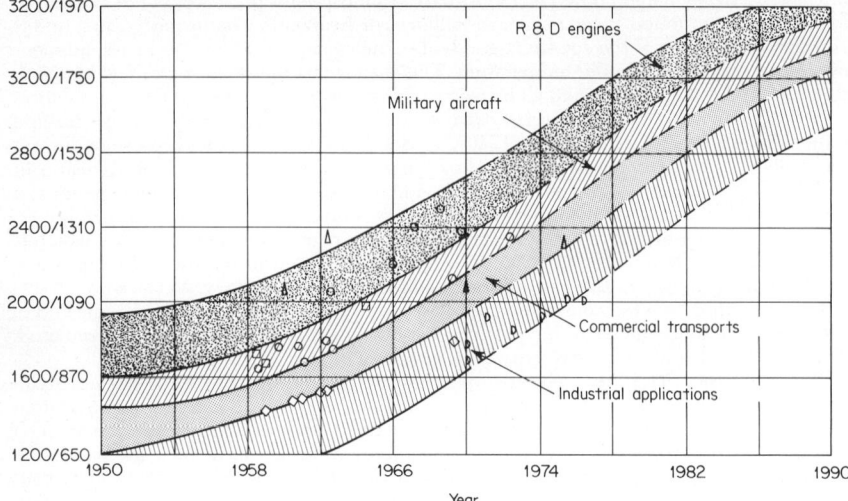

Turbine inlet temperature, °F/°C

Fig. 2. Estimated turbine inlet temperature progression. (*Advanced Nonthermally Polluting Gas Turbines in Utility Applications, United Technology Res. Lab. Rep.* J-970978-8, *prepared for U.S. Environmental Protection Agency, Water Quality Office, Contract No.* 14-12-593; *March* 1971.)

way, electricity generation via the gas expansion cycle is increased; the cooling water requirement for steam condensation is decreased. Of course, this scheme would, at the same time, decrease electricity generation via the steam cycle and would result in consumption of boiler feedwater.

As an example, it has been calculated that at 1560°F (~ 850°C) turbine inlet temperature, a steam injection into the gas turbine inlet stream rate at a rate of 2.5 lb/lb (2.5 kg/kg) power station oil feed would have the following effects (compare with Table 1): Percentage of power of the expansion gas turbine cycle would increase from 24 to 36 percent. Steam to be condensed would be reduced from 87 to 60 percent (pounds per kilowatthour as percent of conventional power station). Plant efficiency would be reduced from 38.5 to 37.7 percent.

COMPARATIVE ECONOMICS OF SGP-BASED CCPS SYSTEMS

In Table 2, the economics of a CCPS system are compared with those of a CPS system.* The additional costs incurred in the CCPS, as compared with the costs of a CPS system,

TABLE 1.   Efficiencies of Combined-Cycle Power Generation Stations*
(Based upon use of SGP† fuel gas preparation step)‡

Gas turbine inlet temperature:				
°F	1560	1830	2190	2550
~ °C	850	1000	1200	1400
Plant efficiency, %	38.5	40.8	43.0	44.7
Percentage power: expanded gas turbine cycle, %	24	29	35	40
Steam to be condensed (lb/kWh) — % of conventional power station	87	81	74	69

*Reproduced by permission of source: Shell Internationale Petroleum Maatschappij B. V.
†Shell Gasification Process.
‡The efficiency of a conventional power station (CPS) comprising steam turbines with an efficiency equal to those used in the SGP-based systems shown here was calculated to be 39.5 percent.

*The figures carry some uncertainty concerning capital cost figures for the various CCPS schemes.

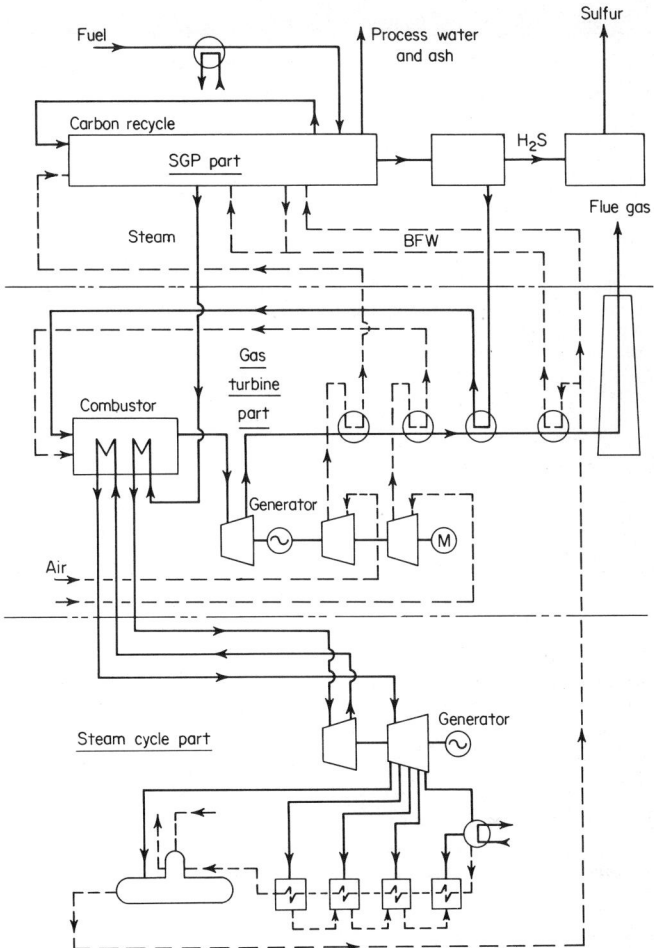

Fig. 3. Shell gasification process-based (SGP) combined-cycle power station (CCPS). BFW = boiler feedwater; M = motor. *(Shell Internationale Petroleum Maatschappij B. V.)*

are charged in this table as a "sulfur removal cost" against the fuel oil used. In this way, the operation of a CCPS system can be compared with alternative ways of removing sulfur from fuel oil. An alternative, for example, might be the hydrodesulfurization of residual fuel (also termed direct hydrodesulfurization process). This process currently projects desulfurization costs ranging from $200 to $360/ton sulfur removed, depending upon origin of the crude (Ref. 13). Pending further development, hydrodesulfurization for residual oils of certain high-ash content crudes are not considered here. From Table 2, it can be concluded that a power station utilizing the SGP and using currently available gas turbines with an inlet temperature of 1560°F (~ 850°C) results in sulfur removal costs that are on the order of 60 to 70 percent of the costs of alternative desulfurization techniques. A further increase in gas turbine inlet temperature would result in further reduction of the sulfur removal cost of a system using the SGP.

REFERENCES

1. Bund, K., K. A. Henney, and K. H. Krieb: Kombiniertes Gas/Dampf-Turbine-Kraftwerk mit Steinkohle-Druckvergasungsanlage im Kraftwerk Kellermann in Lünen, *Brennst-Waerme-Kraft*, **23**(6):258–262 (1971).

TABLE 2.   Relative Economics of Combined-Cycle Power Generation Station (CCPS) versus Conventional Power Station (CPS)[a,b]

Factor					
Turbine inlet temperature:					
°F		1560	1830	2190	2550
~°C		850	1000	1200	1400
Plant efficiency, %	39.5	38.5	40.8	43.0	44.7
Capital,[d] U.S. $ × 10⁶	40	48–46.5	48–46	48–45	48–44.5
Cost of sulfur removal:[e]					
U.S. $/ton sulfur	250[f]	165–134	130–85	96–26	70–(−14)
U.S. ¢/(barrel)·(% sulfur)	40[f]	26–21	20–13	15–4	11–(−2)

[a]Reproduced by permission of source: Shell Internationale Petroleum Maatschappij B. V.

[b]Based upon: 200-MW unit; 6000 hours annual service period; fuel with 4 percent (wt) sulfur; 90 percent desulfurization. Operating costs plus a capital charge taken as 20.5 percent on capital (5 percent for operating, maintenance, and overhead; 0.5 percent for catalysts and chemicals; 15 percent for repayment of capital, tax, and return on capital). Sulfur credit, $20/ton.

[c]Based upon system using SGP (Shell Gasification Process).

[d]Capital figures taken from a 1970 Shell/Sulzer Study (Refs. 3 and 7) comparing a 200-MW CPS with a two-stage expansion CCPS and escalated for 1972 economic conditions. For case I through IV, two assumptions were made: (1) capital remains $48 × 10⁶; and (2) capital is reduced proportionally with the increase in the gas turbine contribution to power generation. Costs have risen proportionately during the 1970–1976 period.

[e]Calculated on the basis of the difference in price between high-sulfur fuel (for the CCPS) and the clean fuel (for the CPS) at a constant electricity price.

[f]Hydrodesulfurization of long residue. Cost can be as high as $360/ton sulfur, or 57¢/(barrel)·(% sulfur). See Ref. (13).

2. Dautzenberg, F. M., J. E. Naber, and A. J. J. van Ginneken: The Shell Flue Gas Desulfurization Process, *Chem. Eng. Prog.*, **67**(8):86–91 (August 1971).

3. Van den Berg, G. J., P. J. J. van Doorn, and W. A. Aeberli: "Residual Fuel Oil Based Power Station Using Partial Oxidation." *Note:* This paper (E.3) and ref. 4 to 6 and 13 were presented at the Seminar of the Desulphurization of Fuels and Combustion Gases, United Nations Commission for Europe, Geneva (Nov. 16–20, 1970).

4. Rudolph R., and G. Kempf: "Gasification of Fuels with Gas Desulphurization for Thermal Power Plants," paper E.4. See note under ref. 3.

5. Robson, F. L., and A. J. Giramonti: "An Advanced Cycle Power Station System Burning Gasified and Desulphurized Coal," paper E.1. See note under ref. 3.

6. Khristianovich, S. A., and V. M. Mailennikov: "Method of Desulphurization of High Sulphurous Fuel Oil at Power Station by Preliminary Gasification under Pressure with Wet Gas Cleaning," paper E.2. See note under ref. 3.

7. Van den Berg, G. J., P. J. J. van Doorn, and W. A. Aeberli: *Verfahrenstechnik*, **5**(10):406–408 (1971). *Note:* This is an abstract of ref. 3.

8. Klein, J. P.: Developments in Sulfinol and Adip Processes Increases Uses, *Oil Gas J.*, **10**(9):109–111 (1970).

9. Neuzeitliche Dampf-Kraftwerke, "Bericht über die 1971 Steam Plant Convenin's-Gravenhage, *Brennst-Waerme-Kraft*, **23**(8):367–371 (1971).

10. Davis, C. Dale: How Gas Turbine Driven Centrifugal and Power Train Work for Shell, *Oil Gas J.*, **66**(13):132–135 (1968).

11. Dyer, A. F.: LNG from Alaska to Japan, *Chem. Eng. Prog.*, **65**(4):53–57 (1969).

12. Biancardi, F. R., G. T. Peters, and A. M. Lauterman: Advanced Nonthermally Polluting Gas Turbines in Utility Applications, *United Aircr. Lab. Rep.* J-970978-8, prepared for the U.S. Environmental Protection Agency, Water Quality Office, Contract No. 14-12-593, March 1971.

13. Van Ginneken, A. J. J.: "The Desulphurization of Residual Fuel Oils from Middle East Crudes," paper A.5. See note under ref. 3.

Coal Gasification for Electric Power Generation

BACKGROUND

A multistage coal-gasification process* proposed by Westinghouse Electric Corporation consists of three process units: (1) a dryer, (2) a devolatilizer/desulfurizer, and (3) a gasifier/combustor, as shown in the flowsheet (Fig. 1). An artist's conception of the fluidized-bed coal-gasification system as it will appear in a 240-megawatt combined-cycle generating station is shown in Fig. 2. The process arrangement capitalizes on the potential advantages of the fluidized-bed reactor, but at the same time realizes the thermal efficiency inherent in the fixed-bed gasifier's counterflow of gases and solids.

COAL PREPARATION

Crushed coal is fed to a fluidized-bed dryer where the coal's water content is reduced so that particles will flow freely and thus be more readily transported and introduced into the devolatilizer/desulfurizer unit. There the devolatilization, desulfurization, and partial hydrogasification functions are combined in a single recirculating fluidized-bed reactor operating at 1300 to 1700°F (704 to 927°C). Dry coal is introduced through a central draft tube in this reactor. Inside this tube, the raw coal and large quantities of recycled solids—char and lime sorbent—are carried upward by gases flowing at velocities greater than 15 feet/second. The recycled solids that continually dilute the coal feed and temper the hot inlet gases descend in an annular downcomer—a fluidized bed surrounding the draft tube. The recirculating solids, which flow at weight rates up to 100 times the coal feed rate, effectively prevent or control agglomeration of the coal feed as it devolatilizes and passes through a phase in which it becomes sticky. Volatile products are driven off the coal in an atmosphere containing hydrogen, which reacts with the coal and char to form methane and higher hydrocarbons and release heat. The dry char that results from devolatilization has a particle size and density that cause it to concentrate in the top section of the fluidized bed where it can be withdrawn.

Lime sorbent is added near the top of the bed and mixes with the recirculating solids, removing hydrogen sulfide from the fuel gases. Spent (sulfided) sorbent is withdrawn from the reactor after stripping out the char, either in the char transfer line or in a separator of special design. Although some heat is provided by the hydrogasification and devolatilization reactions, much of the heat required is supplied by the high-temperature fuel gas fed from the gasifier/combustor unit. Some heat is also transported by solids carried over in the gases from the gasifier.

The gasification of low-sulfur char is conducted in the upper section of the fluidized bed in the gasifier/combustor unit. Steam reacts with char, absorbing heat and forming fuel gases (hydrogen and carbon monoxide). The gasification section operates at temperatures of 1800 to 2000°F (982 to 1093°C). The lower leg of the unit serves as the combustor. Char fed from the devolatilizer/desulfurizer unit is burned with air, forming normal combustion products (steam and carbon dioxide) and releasing the major portion of the heat required by the total process. Combustor temperature is 2100°F (1149°C). Heat is transported from the combustor to the upper gasifier section by combustion gases and by solids that flow between the combustor and gasifier sections.

---

*The ultimate goal of the coal-gasification power system project described here is the demonstration on a commercial scale of an economic and environmentally acceptable electric generating plant that links a coal-gasification system with a combined-cycle plant adapted for burning low-Btu fuel gas. The project is being carried out by a six-member industry/government partnership.

The work is progressing in phases, beginning with the development of a small-scale fluidized-bed gasifier to demonstrate the process described here. From this and other government-funded development work, a conceptual gasification process will be chosen for final scale-up to prototype size. Once the scale-up procedures have been verified, a commercial-size gasifier plant will be constructed for operation by Public Service Indiana at its Dresser Station.

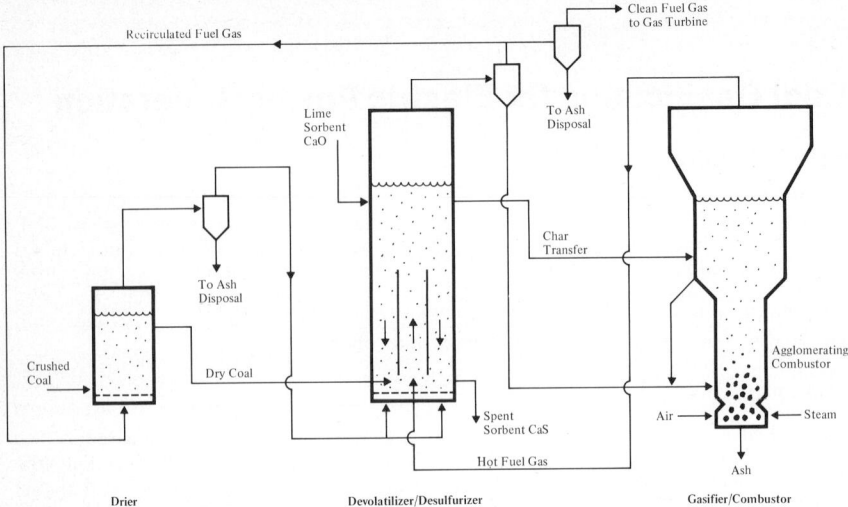

Fig. 1. General flowsheet for coal gasification for electric power generation. (*Westinghouse Electric Corporation.*)

At the 2100° combustion temperature, ash agglomerates and segregates in the lower bed leg for removal

CHARACTERISTICS OF PROCESS

This multiple fluidized-bed concept has the potential for overcoming the inherent limitations of some gasification processes and for providing an economic gasification system for power plants.

Fig. 2. Artist's concept of fluidized-bed coal gasification system as it will appear in a 240-megawatt combined-cycle generating station. (*Westinghouse Electric Corporation.*)

A wide variation in fuels including caking coals and high-ash coals can be used without costly and inefficient pretreatment. This flexibility will allow power plants to utilize local coal resources and minimize coal transportation costs.

A wide variation in coal particle size, ranging from ¼-inch diameter down to fine dust, can be accommodated. Since mechanically mined coal may contain as much as 20 percent fine particles, the ability of a gasifier to use coal as crushed without extraction of fines will provide a further economic benefit.

Good process heat economy is realized through the countercurrent movement of gases and solids between stages. The multistage arrangement provides the long residence time required for high carbon conversion with good temperature control in both stages of gasification. The proposed design also minimizes loss of carbon in ash. Fluidized-bed agglomeration of coal ash with low carbon loss (1 to 2 percent) has been demonstrated on both small- and large-scale equipment.

Clean fuel gas can be produced without the heat loss that would occur if the fuel gases had to be passed through scrubbers, because the fluidized-bed concept permits fuel gas desulfurization by limestone or dolomite at process temperature. Since this temperature is above the range in which tars condense, these should pass through the system with no harmful effects.

Although this advanced gasifier concept is unique, it is composed of subsystems that have been successfully demonstrated in other processing systems. The development task thus becomes one of testing these processes in combination and of learning to design and build economic and reliable apparatus.

The major development steps, ranging through mid-1981, are (1) development and operation of a multiple fluidized-bed process development unit; (2) select gasifier concept for further development; (3) scale-up concept and, if necessary, build and operate a 5 ton/hour gasifier pilot plant; (4) complete design of generating pilot plant for the Dresser station; (5) complete construction of generating pilot plant; and (6) operate combined-cycle plant with coal gasifier.

GASIFICATION PROCESS

Four basic sequential reactions occur in the gasification process. The principal heat-producing reaction is *oxidation*, which results when oxygen reacts with the fuel to form carbon dioxide and water (steam):

$$C + O_2 \rightarrow CO_2$$
$$H_2 + \tfrac{1}{2}O_2 \rightarrow H_2O$$

The *gasification* reaction is the greatest absorber of heat; this reaction occurs when unburned carbon from the fuel reacts with steam and carbon dioxide to form hydrogen and carbon monoxide:

$$C + \binom{H_2O}{CO_2} \rightarrow CO + \binom{H_2}{CO}$$

A third reaction, *hydrogasification*, occurs when hydrogen reacts with carbon from the fuel to form methane:

$$C + 2H_2 \rightarrow CH_4$$

This reaction is moderately exothermic; i.e., some heat is released. And, finally, the fuel when heated also undergoes *devolatilization:*

$$\text{Coal} + \text{heat} \rightarrow C + CH_4 + HC$$

where C is carbon in the form of char and HC consists of higher hydrocarbons and tars. This reaction may be thermally neutral or it may yield heat depending on coal type and process conditions.

In the overall gasification process, these four basic reactions can occur simultaneously throughout a reactor, or each reaction may be localized in a region of the reactor or in a separate reaction vessel. In any case, most gasification processes are designed so that the heat released by oxidation, hydrogasification, and devolatilization matches the heat required by the gasification reaction (plus the sensible heat in the reaction products). The overall heat balance is achieved by adjusting the amount of air (or oxygen) and steam added to the process. If the reactions are carried out in separate regions or reactors, some means are required to transfer heat between regions.

The fuel gas product composition produced by a gasification process depends primarily on the nature of the fuel and on the temperature, pressure, and gas composition in the regions where gasification, hydrogasification, and devolatilization occur. These quantities determine the kinetic rates and the thermodynamic limits of the reactions. When water, carbon monoxide, hydrogen, and carbon dioxide coexist at high temperature, they can also undergo the shift reaction.

$$H_2 + CO_2 \rightarrow H_2O + CO$$

However, this reaction has a negligible effect on process heat balance. While the reaction's equilibrium does affect product gas composition it has little effect on the gas-heating value.

The basic gasification process that has been described up to this point generates a product gas whose heating value comes primarily from the carbon monoxide and hydrogen provided by the gasification reaction, because the hydrogasification and devolatilization reactions contribute only small quantities of methane. To upgrade the mixture into methane (CH_4) to match the heating value of natural gas would require further relatively expensive process steps—especially, additional hydrogen must be produced and reacted chemically with carbon monoxide to form methane. Furthermore, the oxidation reaction should be fed pure oxygen rather than air so that the product gas will not be diluted with inert nitrogen.

On the other hand, if the purpose of gasifying coal is solely to remove ash and sulfur so that the fuel gas is nonpolluting to the environment and nonerosive to gas turbines, the heating value need only be sufficiently high to maintain a stable flame in the gas turbine combustor. It happens that air-blown gasifiers performing the four basic process reactions described can produce a gas mixture with a heating value of 120 to 160 Btu/standard cubic foot, which is adequate for a gas turbine fuel. Eliminating the needs for pure oxygen and for methane production makes such gasification systems less expensive than those systems designed to produce a natural gas substitute. Thus, the plants being developed for the production of pipeline gas and for power fuel gas differ markedly in complexity. In general, pipeline gas processes employ either pure oxygen or hydrogen together with steam at pressures ranging from 30 to 60 atmospheres to produce a product gas high in methane. Processes for power fuel gas use air and steam at 10 to 20 atmospheres to produce a much lower-cost, lower-Btu-content product. The basic differences in the characteristic properties required of these two product gases are summarized in Table 1.

Fluidized-Bed Reactor.   Because of problems inherent in fixed-bed gasifiers, this coal-gasification power system project has been oriented toward fluidized-bed technology. In fluidized-bed reactors, gases flow through a mass of particles at a sufficiently high velocity to support the fuel particles but not high enough to carry them out of the bed. Thus, the

TABLE 1.   Fuel Gas Properties Required for Pipeline Gas and for Combined-Cycle Power Plant Fuel

Characteristic	Pipeline gas	Power plant fuel
Heat content, Btu/standard cubic foot	~ 1000	>120
Pressure, atmospheres	>30	10–20
Temperature:		
°F	~ 70	70–1800*
°C	21	21–982
Composition	Primarily CH_4	CO, H_2, N_2, CO_2, H_2O, CH_4
Cleanliness:		
Sulfur	<1 ppm	1.2 pounds SO_2/10^6 Btu (~550 ppm)†
Particulates	<<0.01 pound/10^6 Btu†	0.1 pound/10^6 Btu‡

*A high temperature is advantageous and may be necessary if the heating value of the gas is low.
†Limits established by process requirements.
‡Limits established by either gas turbine materials or laws regulation emissions (both subject to change with time).

fluidized bed intensifies the interactions of the fuel particles with air and steam and prevents agglomeration and clinkering through its rolling motion. Higher gas velocity and more uniform gas flow enables the fluidized-bed reactor to handle the fine particles that tend to plug fixed-bed reactors.

The other major anticipated advantage of the fluidized-bed reactor is its potential capability to remove sulfur within the gasification process. This capability stems from the ease and versatility a fluidized bed provides in handling solids. Solid materials can be readily added or selectively removed because particles can be segregated by their relative size and density. Gas velocities can be chosen to promote particle mixing in the bed or to cause separation between particles of different size and density. This versatility in control of solids flow permits a controlled stream of limestone particles to be introduced and to flow through the bed. As sulfur in the coal is converted to hydrogen sulfide, the limestone sorbent [a mixture of high-calcia limestone ($CaCO_3$) or dolomite ($MgCO_3 \cdot CaCO_3$) and lime (CaO)] removes this pollutant from the fuel gases, adding another major reaction to the process:

$$\begin{matrix} CaCO_3 & & CO_2 \\ (\quad) + H_2S \rightarrow CaS + H_2O + (\quad) \\ CaO & & \end{matrix}$$

The relatively dense calcium sulfide particles that result can be tapped off. The fuel gas bled from the top of the bed contains very little hydrogen sulfide and requires further cleaning only to remove particulates and any high-temperature tars that remain.

Although fluidized-bed gasification reactors have not been applied commercially, at least five fluidized-bed units are currently under development to produce pipeline gas and/or liquid fuels from coal. Other fluidized-bed processing units have been developed to produce pipeline gas from oil, and fluidized-bed reactors are now used commercially in the catalytic cracking of oil, roasting of sulfide ores, incineration of oil wastes and sludges, production of organic chemical monomers, making of cement, and conversion of nuclear materials for fuel elements.

Gas and Expansion Turbines*

BRAYTON OR JOULE CYCLE

Basically, the gas turbine operates on the concept of the Brayton or Joule cycle (constant-pressure cycle) which was originally used to describe the operation of an air engine, a compressor, and a combustion chamber. In the air engine, air entered the compressor where the pressure was increased. Fuel burning in the combustion chamber raised the temperature of the compressed air under constant-pressure conditions. The resulting high-temperature gases were then introduced to the engine where they expanded and performed work. The excess work of the engine over that required to compress the air was available for operating other devices, such as a generator. The cycle is illustrated and briefly defined in Fig. 1.

In the gas turbine, the air compressor and engine are replaced by an axial-flow compressor and gas turbine. Although the turbine is only part of the whole assembly, in modern terminology the complete assembly is frequently referred to simply as a *gas turbine*. Air is compressed in the compressor, after which it enters a combustion chamber where the temperature is increased while the pressure remains constant. The resulting high-temperature air then enters the turbine, thereby performing work.

GAS-TURBINE DESIGN CONSIDERATIONS

Gas turbines often are rated according to power output (sea level and 80°F; 26.7°C). Some American and most European designs are rated at 59°F (15°C). The power output and efficiency are larger for those fuels which produce larger volumes of products of combustion, since the compressor does not do any work on additional volume. Gas turbines are classified by the physical arrangements of the component parts, and categories include (1)

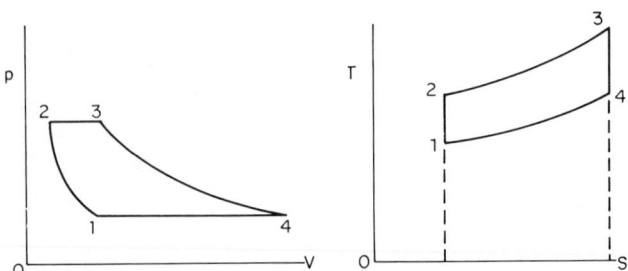

$$V_3/V_2 = V_4/V_1 = T_3/T_2 = T_4/T_1$$

where V = total volume; T = t + 459.69 = absolute temperature = deg R

$$\frac{T_2}{T_1} = \frac{T_3}{T_4} = \left(\frac{V_1}{V_2}\right)^{k-1} = \left(\frac{V_4}{V_3}\right)^{k-1} = \left(\frac{p_2}{p_1}\right)^{k-1/k}$$

where $k = c_p/c_v$; c_p = specific heat at constant pressure; c_v = specific heat at constant volume; p = absolute pressure, pounds / square foot

$$W = Jmc_p(T_3 - T_2 - T_4 + T_1)$$

where W = external work performed on surroundings during change of state, foot-pounds; J = mechanical equivalent of heat = 778.26 foot-pounds per BTU = 4.1861 joules per cal; m = mass of substance under consideration, lb$_m$

Efficiency = $W/JQ_{23} = 1(T_1/T_2)$

where Q = quantity of heat absorbed by the system from the surroundings, Btu

Fig. 1. Brayton or Joule cycle.

*Much of the basic information for this article was furnished by the Solar Division, International Harvester Company, San Diego, California.

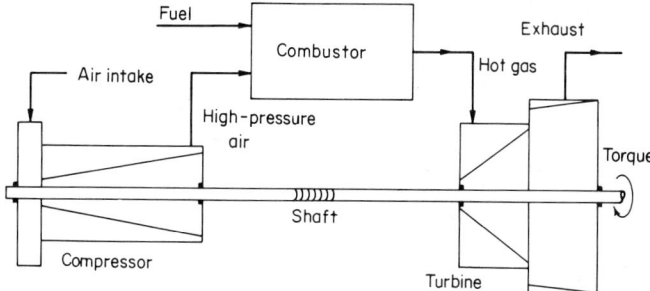

Fig. 2. Gas-turbine configuration exhibiting basic Brayton or Joule cycle.

single-shaft, (2) two-shaft, (3) regenerative (heat exchanger is used to recover exhaust losses and heat air to the combustor(s), (4) intercooled (heat removed between compressors), and (5) reheat (heat added between turbines). Various configurations of gas-turbine systems are shown in Figs. 2 to 8.

Efficiency.   The overall efficiency of a gas turbine is a function of the compressor and turbine efficiencies, ambient air temperature, nozzle inlet temperature, and the type of cycle used. The compressor and turbine are designed for high efficiency. The first-stage gas temperature establishes material and stress conditions for the first set of rotating blades. To the gas temperature at these blades is added the temperature drop across the first-stage nozzles to determine the inlet temperature of the turbine. This may vary from 1300 to 1500°F (704 to 816°C) for industrial turbines and usually will be higher for aviation gas turbines. The higher values are usually used in impulse turbines.

In a simple-cycle turbine, there is (for each turbine inlet temperature) an optimum pressure ratio producing the highest possible efficiency. The efficiency and optimum pressure ratio increase with increasing turbine inlet temperatures. These pressure ratios vary from 4 (at 130°F) up to 6 (at 1500°F).

Regenerative cycles favor lower pressure ratios which result in low compressor discharge temperatures, thus allowing greater recovery of heat from the turbine exhaust gases. High-ratio regenerative plants use intercoolers in the compressor circuit to lower the compressor discharge air temperature.

Although any type of efficient compressor can be used, such as positive displacement (Lysholm), centrifugal, and axial flow, most industrial gas turbines use axial-flow compressors. The turbine may have impulse or reaction blading. To minimize losses, air from the compressor discharge flows through the combustor directly into the turbine nozzle.

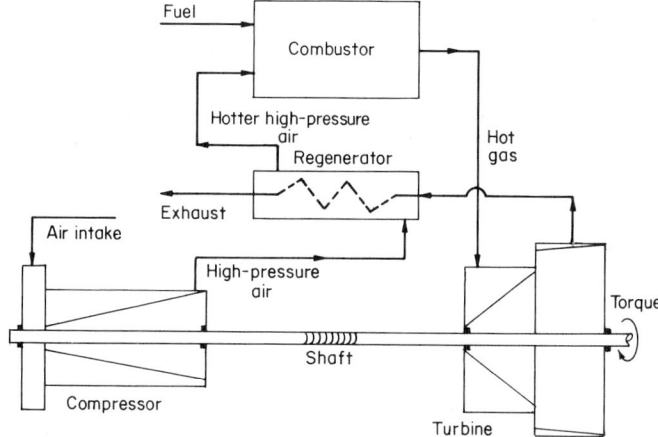

Fig. 3. Gas-turbine configuration with regeneration.

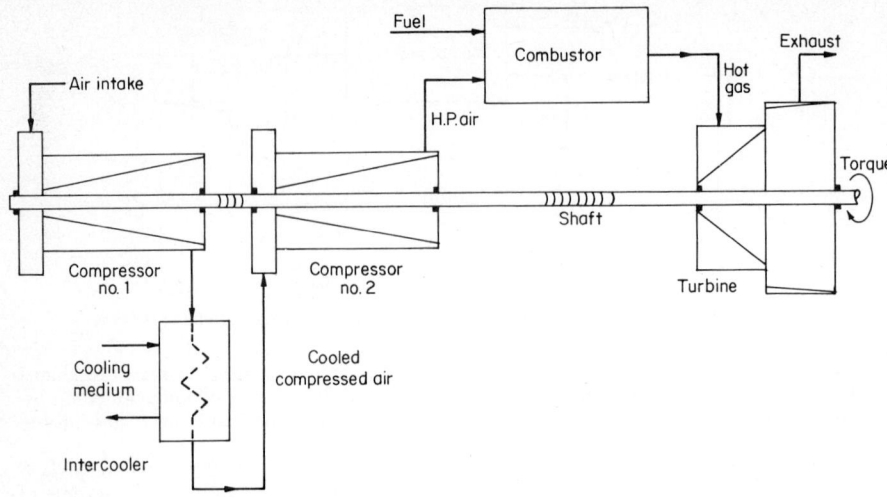

Fig. 4. Gas-turbine configuration with intercooling.

Throttle valves are not used because the resulting pressure drop decreases overall efficiency.

A gas turbine has a large amount of excess air. The combustor is designed with an inner portion burning only part of the air to achieve high combustion temperatures and efficiency. Products of combustion are effectively mixed with the remainder of the air to minimize temperature stratification. Each turbine may have one large combustor or several smaller combustors operating in parallel.

Open- and Closed-Cycle Types.   Most gas-turbine installations are of the open-cycle type, using atmospheric air as the working medium and burning relatively clean fuels. Where dirty fuels are used, it is possible to locate the burner in the gas-turbine discharge, using a heat exchanger to heat the air discharged by the compressor. In closed-cycle installations (Figs. 7 and 8), it may be desirable to use other gases, since efficiency increases as the specific heat ratio (c_p/c_r) decreases. Optimum plant efficiency occurs at increasingly higher pressure ratios with decreasing values of c_p/c_r. However, for convenience most closed systems use air.

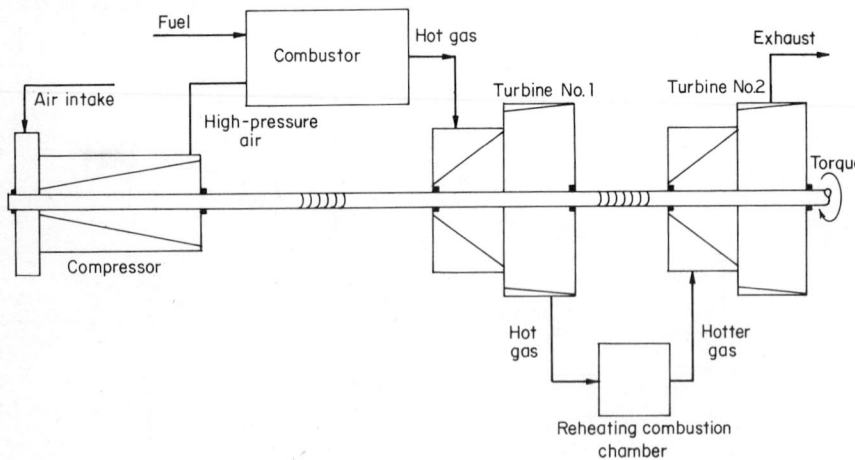

Fig. 5. Gas-turbine configuration with reheating.

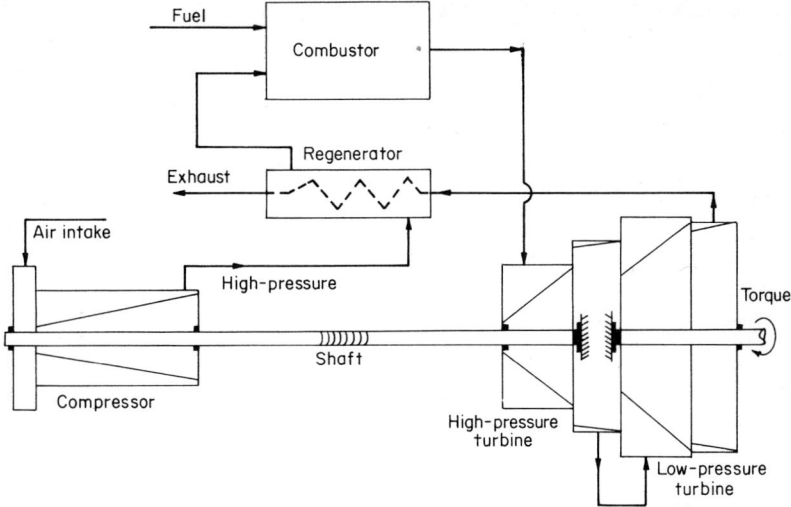

Fig. 6. Regenerative-cycle gas turbine. Two-shaft arrangement with separate power turbines in series.

Closed systems can provide a high plant efficiency over a power range from 25 to 100 percent by varying the turbine exhaust and compressor inlet pressure from atmospheric to about 60 psig. These installations require costly heaters, located between compressor discharge and turbine inlet, and large coolers, located between the turbine exhaust and the compressor suction. Usually, combustion of a fuel provides the heat source and cooling water the coolant medium.

Overloads.   Even if only temporary in nature, a large overload can cause a single-shaft gas turbine to shut down, since its fuel input is limited by the inlet overtemperature protective system. If the torque requirements of the driven machine do not decrease sufficiently with speed reduction, then the gas turbine will continue to slow down. This

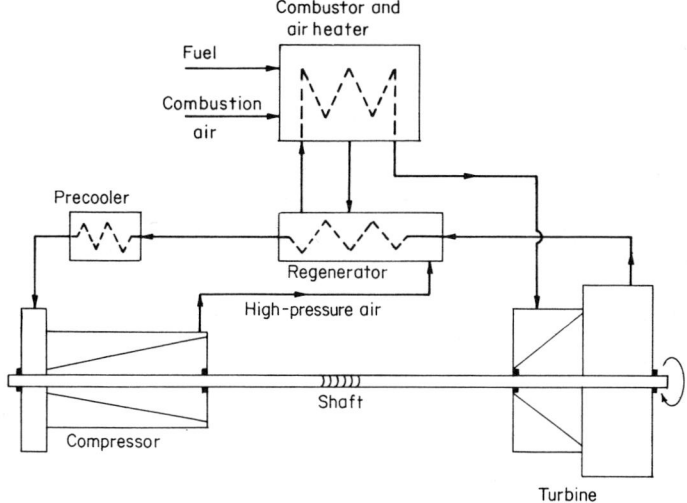

Fig. 7. Closed-cycle gas turbine.

results in higher exhaust temperatures. The exhaust-temperature control system will either shut off the fuel valve or further reduce fuel input, causing the turbine to decrease its speed and finally shut down. Carefully matching the load characteristics of the driven equipment with those of the driver can prevent such occurrences.

Single-Shaft Gas Turbines. The wide acceptance of the single-shaft turbine arises from its low cost and compactness in terms of power output per cubic foot of machinery space. Disadvantages include a relatively low operating speed range and sensitivity to atmospheric temperature. The low operating speed range arises from (1) the quantity of

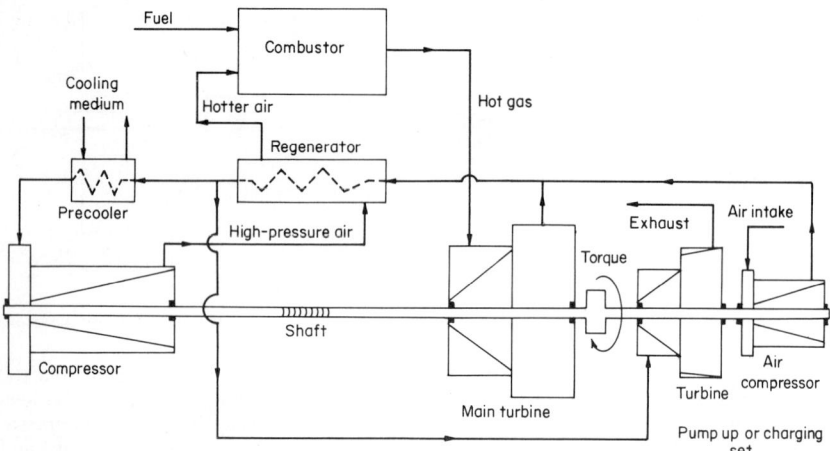

Fig. 8. Semiclosed internally-fired gas-turbine cycle.

airflow induced by the compressor is proportional to its speed and (2) the back pressure produced by the turbine nozzles is proportional to airflow. At low speeds, the turbine power is decreased by low airflows and secondarily by the effect of low pressures on allowable inlet-air temperatures. At low flows, the decreased pressure at the turbine inlet may require a reduction of turbine inlet temperature to maintain the exhaust temperature within design limitations. This results in a further reduction in power. In most applications it is necessary to unload the turbine during start-up.

Two-Shaft Gas Turbines. A wider operating speed range is provided by the more costly two-shaft machine, which consists of a high-pressure turbine driving the air compressor and a low-pressure turbine on a separate shaft to provide output power. See Fig. 9. A variable-area nozzle can be used in the low-pressure turbine to increase the operating speed range. Change in the fuel input to the high-pressure turbine causes the speed and quality of airflow to change. The low-pressure turbine power output is changed by varying the quantity of airflow and the nozzle area of the power turbine.

Air/Temperature Relationships. The airflow to a gas turbine is inversely proportional to the absolute air temperature at the compressor inlet. Since the compressor discharge pressure is set by the turbine nozzles (proportional to flow), this results in decreased turbine power output during hot weather and increased power during cold weather. In hot, dry areas, hot incoming air can be cooled by evaporation using water injection. In locations where the summer season is short, it may be possible to obtain rated power by increasing the turbine temperature for a short period without appreciably shortening the life of the equipment. In extremely cold temperatures, high air pressures will exist at the turbine inlet and should be considered in the design of the gas turbine.

Since the power required by the compressor is approximately twice as great as the shaft output, a 1 percent change in compressor efficiency will result in a 2 percent change in shaft power. A 1 percent change in turbine efficiency will produce a 3 percent change in shaft power. Therefore, it is important that all losses be minimized and that sufficiently large inlet and exhaust piping or ducts be used.

START-UP OF GAS TURBINES

A gas turbine is started by bringing it up to starting speed and maintaining this speed for several minutes in order to purge the casing. Some machines require that the casing or rotor be heated slowly by burning a nominal amount of fuel in the combustors for several minutes. The turbine inlet temperature is then increased rapidly to a value above the design temperature, thus producing sufficient power in the turbine to bring the set up to full speed. Some installations will require a blowoff valve to prevent surging during start-

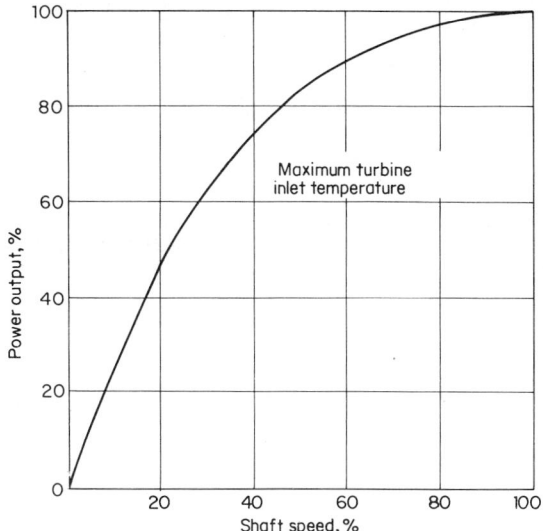

Fig. 9. Operating range for various speeds and loads for two-shaft gas turbine.

up. The starting power requirements of an unloaded gas turbine will range between 5 and 10 percent of the full-load speed. Two-shaft turbines will require slightly more starting power than single-shaft machines. By opening the nozzles of the low-pressure turbine, the load is not driven during start-up.

FUELS

A wide variety of fuels can be used in gas turbines. The major fuel requirements are that (1) the fuel does not form ashes which will deposit on the blades and interfere with operation, (2) the fuel does not contain dust which will erode the blades, and (3) the fuel does not contain uninhibited vanadium. Commonly used fuels include natural and refinery gas, blast-furnace gas, fuel oils (including heavy residuals), and the growing application of gas turbines in cycles involving gases derived from coal and other previously nontraditional sources. The latter kinds of uses are described elsewhere in this Handbook and reference should be made to the alphabetical index.

The simple-cycle gas turbine is relatively inefficient with almost all its losses in the hot exhaust gases. When exhaust gases can be used in a boiler or for process heating, the combination of turbine and heat-recovery apparatus results in a high-efficiency plant. Integration of the gas turbine with process requirements also can result in high efficiency. In particular, see "The Total Energy Concept in Refinery and Process Plant Planning" in the section on *Petroleum Technology* in this Handbook.

GAS-TURBINE STANDBY GENERATOR SETS

Over the last few years, with growing awareness of the possibility of more frequent utility power outages, thousands of gas-turbine standby generator sets have been installed in industrial plants, hospitals, apartment and office buildings, hotels, motels, schools, and other public facilities. Further, to ensure the continued reliable operation of communications systems and computer centers in the event of brownouts and blackouts, gas-turbine standby generator sets have been widely accepted as emergency substitutes. As described later, generator sets also find extensive application in offshore exploration and drilling activity. Mobile generator sets are used on construction projects, in military operations,

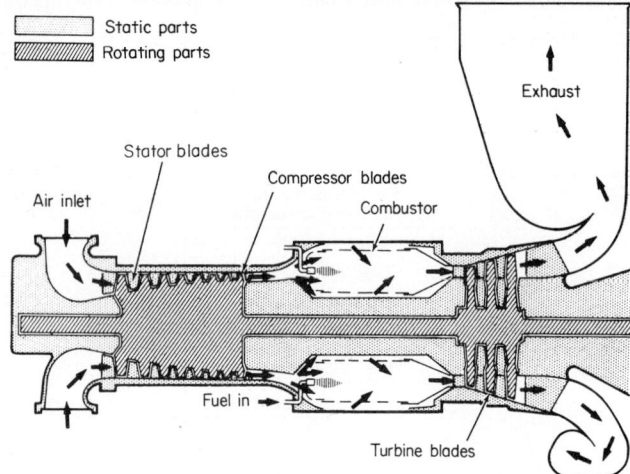

Fig. 10. Cross section of Solar *Saturn* gas turbine (*Solar Division, International Harvester Co.*)

and various emergency operations. Such sets have been used to furnish portable power for pumps in connection with fighting forest fires, for example.

Popular standby electric generator sets are designed for 225, 900, and 2800 kilowatts. Although gas-turbine sets have a slightly higher initial cost, they offer favorable long-term economy when compared with reciprocating-engine generator sets. The cross section of a gas turbine used in standby generator sets* is shown in Fig. 10. This is a single-shaft industrial gas-turbine engine with antifriction bearings and integral speed-reducing gearbox to 1800 r/min. The nominal engine rating is 1200 hp, based on standard conditions of 59°F (15°C) at sea level, with zero inlet and exhaust losses. A flexible drive coupling connects the output shaft of the gearbox to the generator shaft. A cutaway view of the turbine is shown in Fig. 11; and of the reduction drive in Fig. 12.

An eight-stage axial compressor fully annular combustion chamber and three-stage axial-flow turbine are arranged with a straight-through flow path. Advantages of this configuration include minimization of flow losses and a cylindrical structure of considerable rigidity and stability. The accessory drive gearbox with starter mounting pad is located behind the air inlet. The air inlet housing, compressor case, compressor diffuser, combustion chamber case, turbine case, exhaust diffuser, and turbine output housing are interconnected to form the main structural shell of the engine. The reduction drive assembly is composed of a two-stage planetary gear train which reduces the turbine output shaft speed from 22,300 to 6000 r/min in the first stage and to 1800 (1500 for 50-Hz sets) r/min in the second stage. The primary stage consists of the turbine-connected sun gear, stationary planets, and a rotating ring gear. The second stage is a true planetary system, composed of a sun gear, rotating planet gears, and a stationary ring gear. The final output is transmitted through the rotating gear carrier to the output shaft.

Air is drawn into the air inlet and compressed by the eight-stage axial-flow compressor. The compressed air is directed into the combustion chamber in a steady flow. Within the annular combustion chamber, fuel is injected into the pressurized air. During the engine starting cycle, the fuel-air mixture is ignited by a high-voltage spark from the ignitor plug

*Similar units are available as continuous-duty generator sets.

and continuous burning is maintained as long as there is an adequate flow of pressurized air and fuel. The hot, pressurized gas from the combustion chamber expands through the turbine, dropping in pressure and temperature as it drives the turbine. In this way, the energy of the fuel is transformed into the rotating power of the turbine output shaft.

The turbine drives the compressor and the external load (generator) through a common

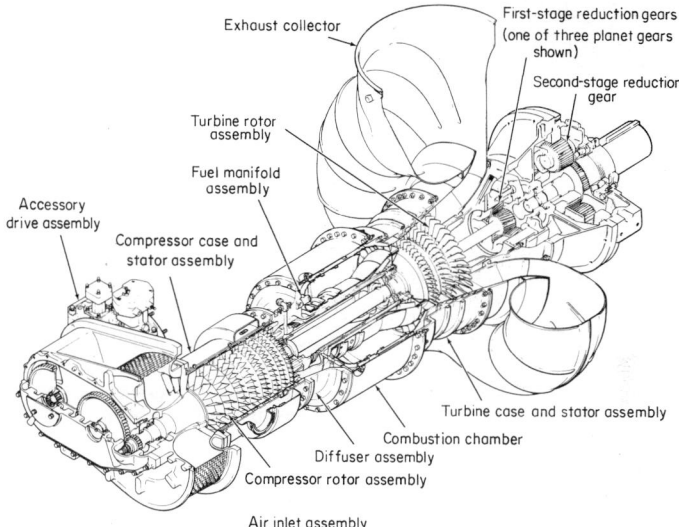

Fig. 11. Solar *Saturn* single-shaft gas turbine.

shaft. During the starting cycle, the starter supplies the driving power for the gas turbine up to approximately 15 percent rated speed, at which time the gas turbine "lights off" and accelerates to 100 percent rated speed. The starter remains engaged after light-off and assists in the acceleration from 15 to 60 percent speed. Approximately 20 percent of the available air mass is required for combustion; the remainder is used to cool the combustion chamber and reduce the temperature of the combustion gases to less than 1550°F (843°C) at the inlet to the first turbine stage. The exhaust temperature varies with the generator load and has a maximum value of 935°F (~ 502°C). Exhaust products have a 17 to 19 percent oxygen content.

The compression ratio of the compressor is 6.2:1; flow is 12.8 pounds/second. There are nine fuel nozzles in the combustion chamber. Materials of construction are:

Air intake housing	Aluminum	Turbine case	422 AISI Stainless steel
Compressor case	17-4 PH Stainless steel	Turbine disks	A286 Stainless steel
Compressor blades	410 Cast stainless steel	Turbine blades (1st and 2nd stage)	S816 Alloy steel
Combustor liner	N155 High-temperature alloy	Turbine blades (3rd stage)	713 Cast alloy steel
Combustor case	N155 High-temperature alloy	Exhaust diffuser	17-4 PH Stainless steel
Turbine nozzles (1st and 2nd stage)	N155 High-temperature alloy	Exhaust collector	321 Stainless steel
Turbine nozzles (3rd stage)	347 Stainless steel	Accessory gear housing	Aluminum

Voltage Regulator. Shown in Fig. 13, the voltage regulator is completely static, of the silicon-controlled rectifier type, employing a Zener reference, and arranged for single-phase sensing. All rectifiers are silicon and hermetically sealed. Regulators include cross-current-compensation for parallel operation. The regulator has a ±5 percent voltage

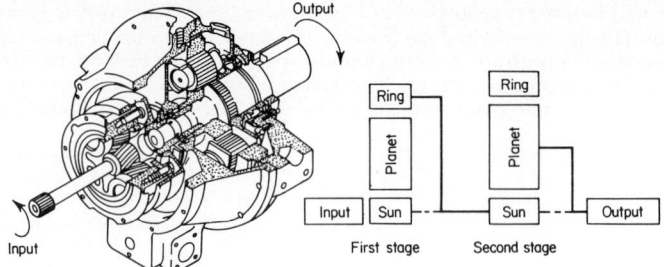

Fig. 12. Reduction drive used with turbine illustrated in Fig. 11.

adjustment range by means of a rheostat which can be mounted up to 100 feet from the generator.

Exciter. As shown in Fig. 14, the exciter unit consists essentially of two separate assemblies: (1) a three-phase, rotating armature type, alternating current generator, and (2) a three-phase, full-wave bridge rectifier. The rotating armature and the rotating rectifier assembly are mounted on the generator rotor shaft and are electrically interconnected with each other and with the generator field windings. The stator for the exciter consists of a wound laminated core installed in a flange ring which forms an integral part of the generator front bearing bracket. The exciter system has sufficient capacity to provide excitation when the generator is carrying 150 percent rated current for 1 minute.

Liquid-Fuel System. Although operable with natural gas, liquid fuel is normally used for standby power because of the ease in storing quantities of fuel at low cost. Grades 1 and 2 fuel oil, Grades 1 and 2 diesel fuel, and kerosine (JP-4 or JP-5) in the temperature range of $-65°F$ ($-54°C$) to $+130°F$ ($+54°C$) permit gas-turbine operation providing the viscosity of the fuel is maintained at 12 centistokes (or less) kinematic viscosity and does not contain excessive amounts of damaging elements. The system functions properly with fuels having a lower heating value (LHV) of 18,000 to 20,000 Btu/pound.

The turbine liquid-fuel system is completely integrated with the generator package. As

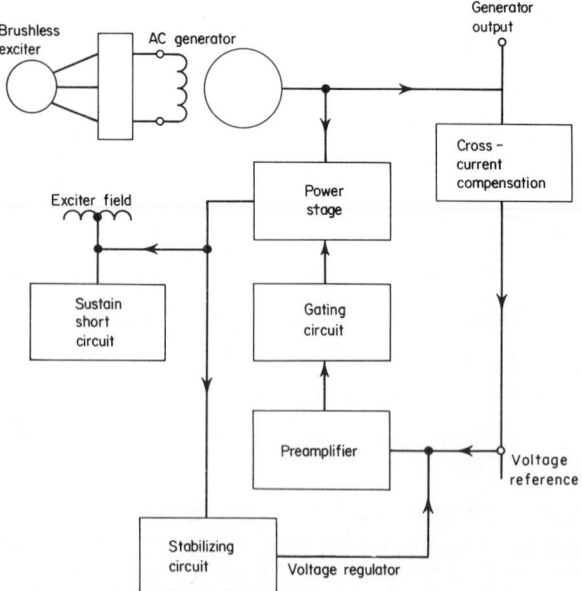

Fig. 13. Voltage regulator schematic diagram.

shown in Fig. 15, the order in which the fuel passes from the supply connection and through the major part of the fuel system is a 5-micrometer, low-pressure filter (replaceable-element type); main turbine-driven fuel pump; high-pressure filter (10 micrometers); main fuel control (combines function of an acceleration control fuel-metering valve, fuel shutoff valve, and flow divider); and into the annular fuel manifold to the nine fuel nozzles equally spaced in the tubular manifold.

The fuel pressure required at the 5-micrometer fuel filter inlet is 20 psig. A 24-volt dc motor-driven fuel boost pump with a 75-micrometer suction strain is sometimes used. The pump supplies fuel at 2.5 gallons/minute and has a 20-foot equivalent lift capacity. A duplex, low-pressure, 5-micrometer filter arrangement is sometimes used to facilitate filter servicing during operation.

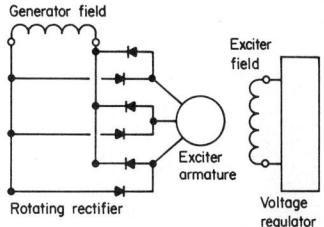

Fig. 14. Exciter system.

Governors. Three types of governing systems are used: a mechanical governor, an electric governor, and a combination mechanical/electrical system. Advantages claimed for the latter type include (1) isochronous control in multiunit operation, (2) automatic load-sharing capability in multiunit operation, and (3) improved frequency response characteristics under transient conditions. The system consists of an electrical load sensing governor for primary frequency control and a mechanical speed-sensing governor for backup. The system is always used when the automatic paralleling system is employed. Characteristics of the electric governor: steady state ±0.25 percent; full-load deviation, 3 percent; full-load recovery time, 3 seconds. Characteristics of the mechanical governor; steady state ±0.25 percent; full-load deviation, 4 percent; full-load recovery time, 4 seconds.

Starting Systems. Two basic starting systems are available for the generator set just described: (1) dc electric starting system and (2) air/gas starting system. The electric starting system shown schematically in Fig. 16 incorporates a 24-volt dc shunt-wound starting motor mounted on the turbine accessory drive pad. The start contactor for this motor is mounted on a panel with the dc power control circuit breakers. Standard battery systems are available for starting and control service.

The air/gas-turbine starting system shown schematically in Fig. 17 is especially suited for gas-turbine requirements and can use either compressed air or natural gas as the power

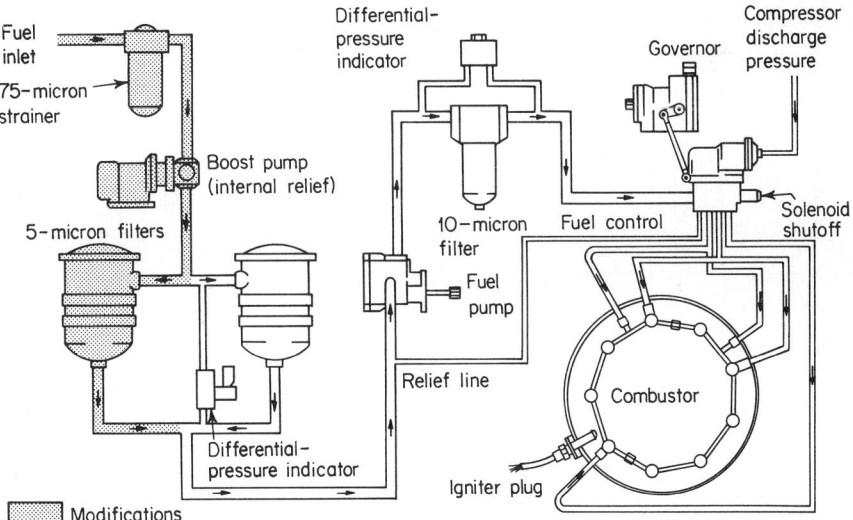

Fig. 15. Liquid fuel system schematic diagram. Shaded equipment may be added to handle more severe fuel situations, and for convenience in servicing.

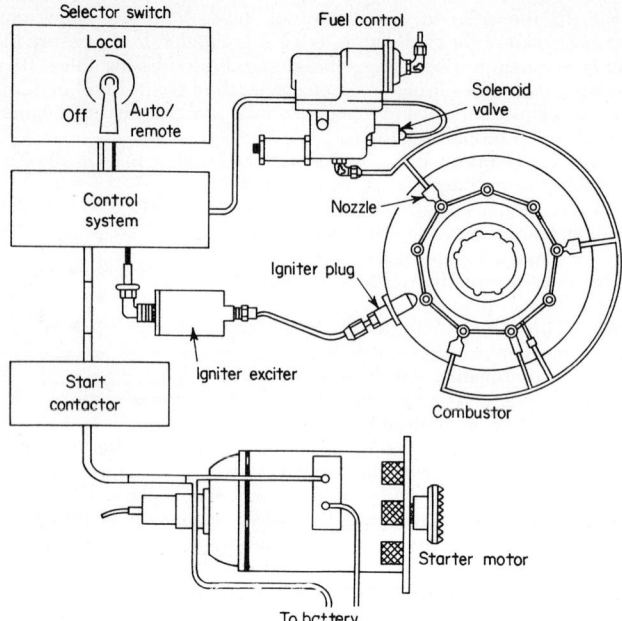

Fig. 16. Dc electric starting system.

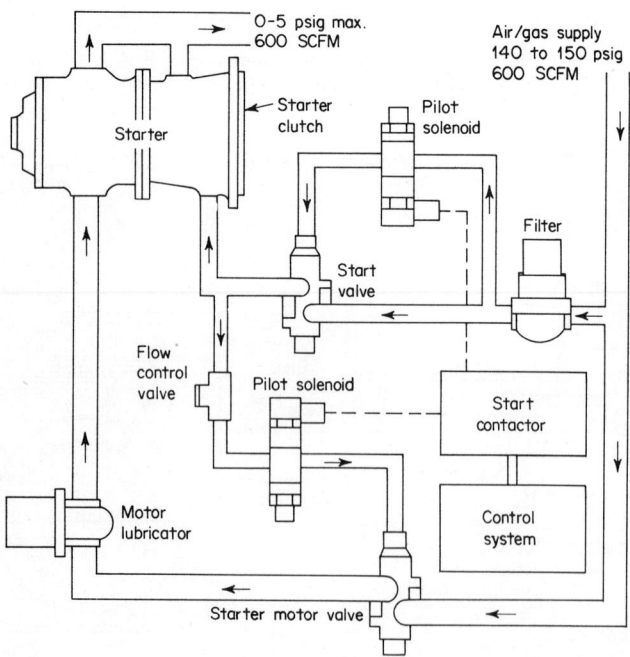

Fig. 17. Air/gas starting system schematic diagram. SCFM = standard cubic feet per minute.

source. The starter motor is mounted on a drive adapter attached to the rear face of the accessory drive assembly. Starter motor torque is transmitted to the engine through a pneumatically operated jaw-type clutch. When a start is initiated, pneumatic pressure causes the starter jaw to engage the mating jaw on the accessory drive shaft. Once engaged, pneumatic pressure is admitted to the starter motor. At starter dropout speed, the penumatic valves close, causing the starter to stop motoring and the clutch jaws to disengage. The starter requires a supply of natural gas or compressed air to 140 psig (minimum) through a 1.5-inch diameter line. The total quantity of gas used by the starter during a start sequence is approximately 400 standard cubic feet, with a maximum demand rate of about 600 standard cubic feet per minute. The discharge should be suitably vented to atmosphere.

Running Sequence. During normal operation, turbine speed is automatically controlled by the governor and the generator is automatically controlled by the voltage regulator. Speed and voltage may be manually adjusted as required.

Stop Sequence. Unit shutdown can be accomplished by local or remote push button control, by automatic contact opening from a remote device, or by unit malfunction. The shutdown sequence is as follows:

1. When all engine-run circuits are deenergized, the main fuel valve closes and fuel flow to the engine ceases. A signal is initiated to trip the main power circuit breaker and the generator field is removed as the engine begins to decelerate.

2. Engine speed decreases below 60 percent and the hourmeter is deenergized. Engine speed drops below 13 percent and the restart time delay relay is deenergized and begins to time out, ensuring that the engine will coast to a complete stop before a subsequent start can be initiated.

3. After engine rotation has stopped, the restart relay times out to arm the electrical system for restart. When shutdown is initiated by a malfunction, the start sequence is blocked until the reset button is depressed.

During operation, the unit is automatically protected for turbine overspeed, high turbine temperature, low oil pressure, high oil temperature, and turbine underspeed.

Output power versus engine inlet air temperature for the gas-turbine standby generator set just described is given in Fig. 18. Approximate exhaust gas composition curve is given in Fig. 19.

NATURAL GAS COMPRESSOR SETS

There has been a rapidly growing trend since about mid-1969 (the concept was first introduced in 1961) to apply gas-turbine-powered centrifugal compressors in the oil and

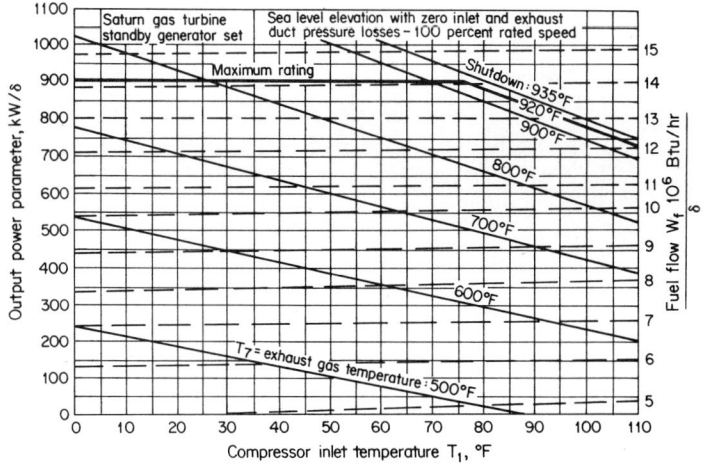

Fig. 18. Output power versus engine inlet air temperature.

gas industry. The sets consist of a two-shaft, axial-flow gas turbine driving a centrifugal gas compressor. One firm* offers two systems, one designed to operate on gas taken from the pipeline, filtered and regulated to 185 psig (13 kg/cm²), and another for gas at 150 psig (10.5 kg/cm²). Such systems are used for a variety of purposes in the natural gas industry, both onshore and offshore. They are available with from one to eight stages to meet various flow and pressure conditions. The systems can compress gas at ratios of over 5:1

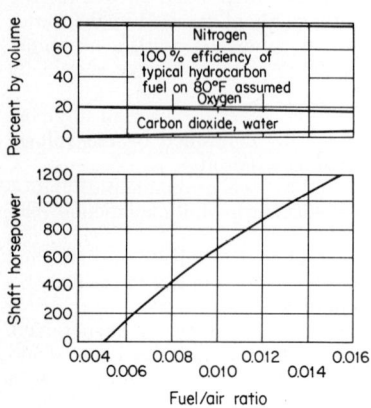

Fig. 19. Approximate exhaust gas composition curve.

for a wide range of flows. Tandem units, with a turbine driving two or three compressors, can achieve ratios up to 35:1. Systems are operating at an altitude of more than 7000 feet in New Mexico where they are subject to temperatures ranging from below 0°F (-17.8°C) to 100°F (37.8°C). Systems also are operating in desert areas (Libya, Iran, West Pakistan, etc.), in jungle climates (South America, etc.), and on drilling platforms in various seas of the world. There is nearly a 10:1 weight reduction in the gas-turbine–centrifugal compressor approach as compared with reciprocating-compressor units. In one example, for comparable performance, a gas-turbine system will weigh about 33,500 pounds (16,344 kg) as compared with a reciprocating-compressor unit that will weigh more than 300,000 pounds (136,080 kg). The gas-turbine–centrifugal compressor sets are used in connection with gas transmission, gathering, production, boosting, and refrigeration.

The cross section of a two-shaft gas-turbine engine† is shown in Fig. 20. The gas-turbine engine has an output-load shaft (power turbine) which is mechanically independent of the gas-producer section and permits a wide range of speeds at full power. See power transfer diagram of Fig. 21. This design contributes to better part-load fuel economy. In this engine configuration, called a two-shaft (or split-shaft) design, the first two turbine stages drive the engine compressor only; gases leaving the gas-producer turbine flow through the free power turbine, forming a fluid coupling. The remaining energy is absorbed by the power turbine and is transferred to the output shaft. The speed of the gas producer is closely associated with the power level of the engine. Conse-

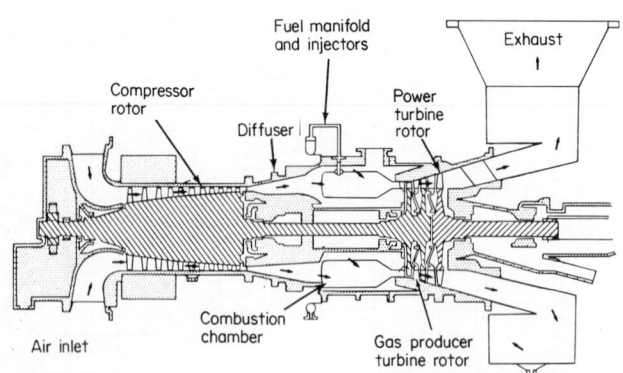

Fig. 20. Sectional view of Solar *Centaur* gas turbine engine, indicating air flow.

*Solar Division, International Harvester Company (*Saturn* and *Centaur* systems).
†Solar *Centaur* engine.

quently, a speed governor on this shaft is used to control the power level. The power turbine runs at a speed which is dependent only upon the load and the engine is protected against overspeed in the event the load is removed.

The gas-turbine engine requires, for stoichiometric combustion, approximately one-quarter of the total air handled. The excess air is used to cool the combustion chamber and mixes with the products of combustion to reduce the gas temperature at the inlet to the first turbine stage. The cooling employed at the combustion chamber of the engine keeps metal temperatures in the turbine section at acceptable levels. Cooling is accomplished by air, thus eliminating requirement for cooling water.

Recuperator. The addition of a recuperator to a system of this type will lower fuel consumption by about 20 percent, or from 9800 Btu/(hp) (hour) to 7800 Btu/(hp)(hour). A system with recuperator installed is shown in Fig. 22.

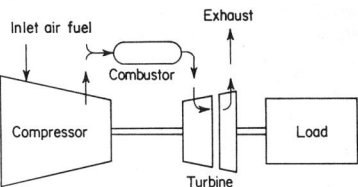

Fig. 21. Power transfer in two-shaft turbine engine.

GAS TURBINES IN OFFSHORE ACTIVITY

The compactness and relatively light weight of gas-turbine power packages has been particularly attractive to offshore installations in the petroleum and gas field. Representa-

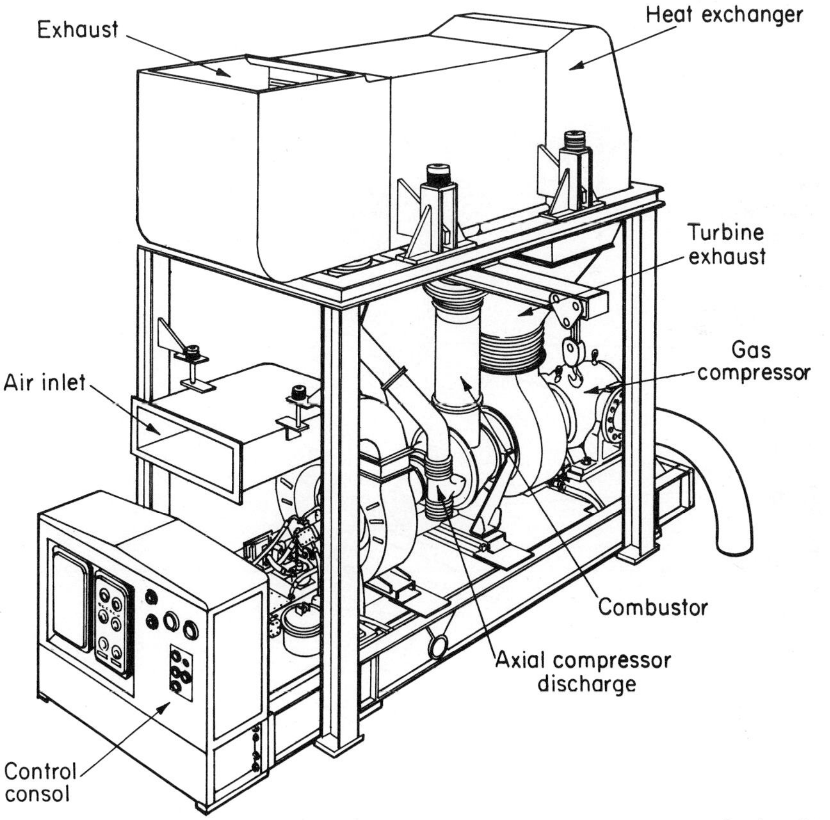

Fig. 22. Solar *Centaur* gas turbine natural gas compressor set equipped with new retrofitted combustor and recuperator.

tive application of gas-turbine power installations is indicated by the diagram of the Grayling Platform, Cook Inlet, Alaska, shown in Fig. 23.

GAS TURBINES IN THE UTILITY POWER FIELD

In addition to the wide use of gas turbines for standby and relatively low kilowatt production for continuous service as previously described, gas turbines have made marked inroads in the utility power field as greater emphasis has been placed on combined-cycle plants. In mid-1970s, installations of gas turbines in the utility field in the United States were being made at the rate of about 8000 megawatts/year. Some of the

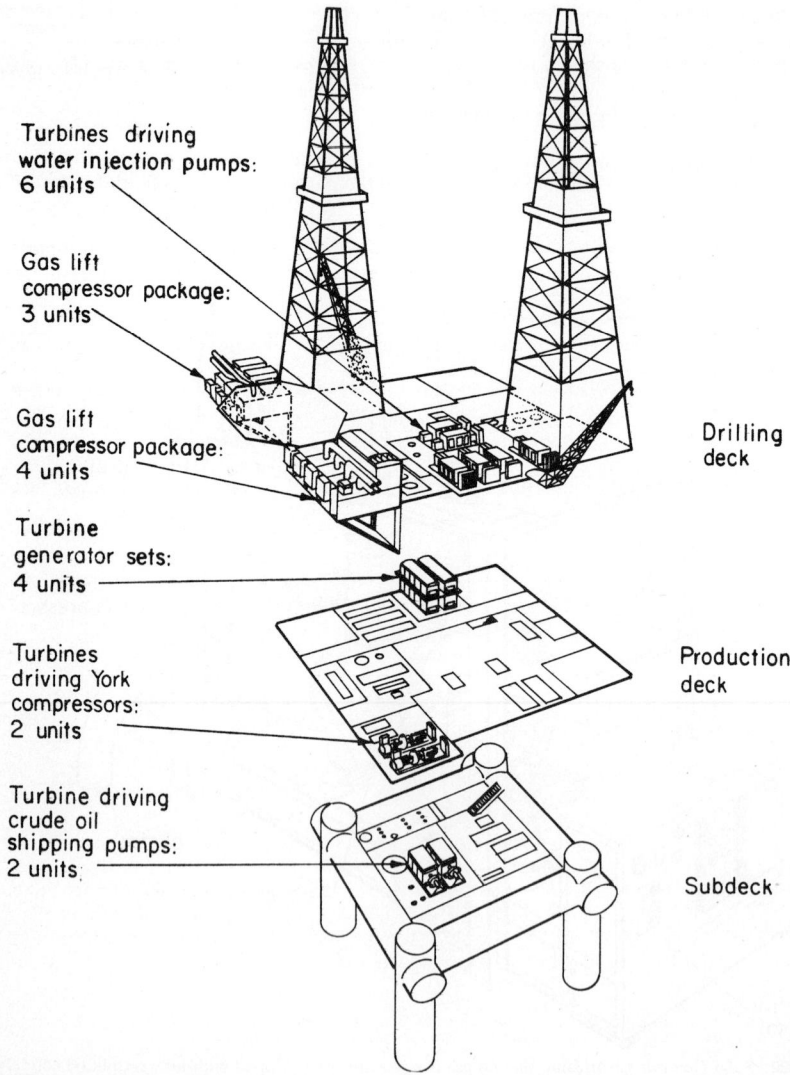

Fig. 23. Use of gas turbine-powered systems on Grayling Platform, Cook Inlet, Alaska. (*Solar Division, International Harvester Company.*)

reasons for acceptance of gas turbines in the utility power field include the following. (1) High availability of gas turbines can be achieved when these units are operated at relatively constant load for extended periods and when they are maintained in accordance with manufacturers' instructions. Outstanding performance of this equipment in pipeline service has been proved. (2) In some cases, it is more economical to order a combined-cycle plant than to invest capital in pollution-control equipment for an older station that might not withstand the rigors of cycling operation. (3) Environmental impact of combined-cycle systems should be less than that of peaking/cycling units because of reduced cooling water requirements and the utilization of higher quality fuels. (4) Construction costs of combined-cycle plants are relatively low because most components are assembled at the factory and shipped in modules to the plant site. A major problem remaining for preengineered combined-cycle systems is the reduction of nitrogen oxides from gas turbines.

A sampling of gas turbines installed or on order by gas and electric power utilities is summarized in Table 1.

Combined-Cycle Plants.* Over the past decade, electrical energy shortages have resulted in the use of many peaking units at operating levels in excess of their optimum economic levels. Those peaking units are employed to supply the intermediate load band, along with former base-load units that have had to be shifted to intermediate service. The latter are usually older units that lack modern control systems and therefore lack the control flexibility that intermediate generation requires. In short, the intermediate load band often is not being supplied as efficiently and as flexibly as it could be.

In addition, large interconnecting ties require increasingly close control of area generation to meet the security requirements of the grid, i.e., to keep the system from breaking up. Such control can best be provided by reserve power supplied by peaking and intermediate plants that are able to start up rapidly and alter their power outputs rapidly in response to the grid's dynamic needs.

Well-engineered combined-cycle plants† can meet those needs for increased efficiency and flexibility because they blend the best features of peaking and base-load generation by combining one steam and two gas turbines. Gas turbines are fast in starting up and in responding to changes in power demands, but they are relatively inefficient by themselves; steam turbines are slower in start-up and response but are more efficient. A good engineering design will combine them in such a way as to use waste heat from the gas turbines to generate steam for the steam turbine. See system diagram of Fig. 24.

The result is a plant that has an excellent heat rate when the steam turbine is operated with one or both gas turbines and yet has fast start-up and control flexibility. In the plant shown, heat rate is 8000 Btu/kilowatthour with the steam turbine and both gas turbines operating. For comparison, the heat rates of typical modern steam base-load plants and gas-turbine peaking plants are, respectively, about 9000 and 12,000 Btu/kilowatthour. The gas turbines can be operated independently of the steam turbine, and of each other, for flexible plant operation. Start-up of the entire plant from hot standby requires about one-fourth the time required by typical steam plants.

Thus, although a plant of this type may be designed primarily for intermediate generation, it can provide service ranging from peaking to base-load generation—as needed and as influenced by the relative costs of fuels.

Plant operation includes both transient and steady-state modes, which are defined in the control room by the state of bistable or multistable switches, breakers, or valves, and by the state of closed-loop controls, such as on/off switches. The transients are long-term modes, such as acceleration and the building up of steam pressure. Steady-state operating modes are:

1. *Hot standby, ready to start.* Steam-cycle components are kept hot by electric heaters. The plant can reach full load from this mode in approximately 1 hour.

2. *Running mode.* The turbine-generator units selected for operation are running at 3600 r/min and all associated auxiliaries are in normal operation. Excitation has been

---

*Basic data obtained with permission from the *Westinghouse Eng.*, April 1974 (PACE Power-Plant Control System Provides Operating Flexibility and High Plant Availability, by T. C. Giras and P. A. Berman).

†Combined-cycle plants are also described elsewhere in this Handbook. Consult the alphabetical index.

TABLE 1. Abridged List of Representative Gas Turbines Installed in Electric and Gas Utility Industries*

No. of Units	Location	Primary Use†	Black plant installation	Method of remote operation‡	Normal power, megawatts	Time to reach normal power, minutes	Turbine speed, r/min	Fuel§
6	Southeastern U.S.	PP	No	Dw, Spv	64.4	30	3600	Various
4	Minnesota	PP	No	Dw	48.4	20	3600	DsO
4	New Jersey	BL	No	Dw	44.2	20	3600	DsO, CrO
1	Wisconsin	PP	Yes	Dw, Spv	36–63	8	6200–8800	G, DsO
1	South Africa	PP	Yes	Dw	40.0	3		DsO
12	Florida	PP	Yes	Dw	38.0	5	8800–6500	DF
1	New Jersey	PP	Yes	Dw, Spv	31.1	20	4894	Various
16	New York	PP	Yes	Spv	20.1	10	5100	DsO
1	Alaska	PP	No	Dw	27.4	8	5100	G
1	New Jersey	MD	No	Dw	17.5		8600–6350	K
2	Saudi Arabia	MD	No		11.0	10	5100	DF
2	Michigan	PP	Yes			8	7650	DF
1	Louisiana	PP	Yes		10.2	7	8600–6350	DsO
2	Gulf of Mexico	GH	No	Spv		10	6900	G
5	Ontario, Canada	GH	No	Dw			7500	G
4	West Germany	PP	Yes		3.5	1	18,000	DsO
2	Illinois	GH	Yes	Dw	3.6	1	13,820	G
8	Libya	GH	Yes			1	14,300	G
2	Alberta, Canada	SB	Yes	Dw	2.5	1	14,300	DsO
3	Persian Gulf	BL	Yes		2.4	1	14,300	G
1	Illinois	TE	No			1	17,100	PG
8	Indonesia	TE	Yes	Dw	1.0	1	17,100	DsO
3	Persian Gulf	BL	No	Dw	1.0	1	17,100	PG
1	Maine	SB	Yes	Dw	0.9	1	22,300	DsO
8	California	SB	Yes		0.8	1	22,300	DsO
2	Alberta, Canada	TE	Yes	Dw	0.8	1	22,300	G

*Installed or on order (mid-1970s).

†PP = peaking power and standby; BL = base-load; MD = mechanical drive; GH = gas handling; SB = standby power; TE = total energy.

‡Dw = direct-wire; Spv = supervisor (telemetry).

§DsO = distillate oil; CrO = crude oil; G = natural gas; DF = diesel fuel; K = kerosine jet A; PG = process gas.

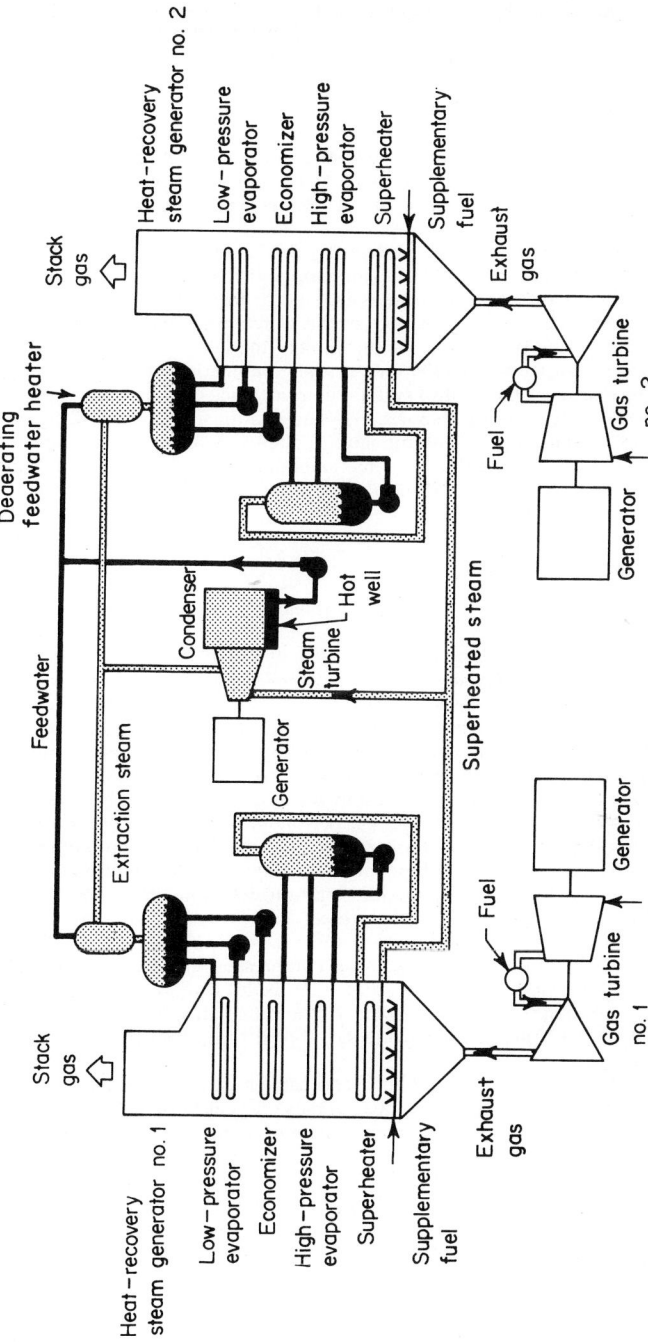

Fig. 24. Combined-cycle power plant. (*PACE 260 system engineered by Westinghouse.*) Exhaust energy from two gas turbine-generator units is used to develop steam for a conventional steam turbine-generator unit. Plant burns either gas or oil. Because the steam cycle provides only about one-half the plant output, the need for cooling water is approximately one-half that of a conventional all-steam power plant of same size. Each gas turbine exhausts into its own heat-recovery steam generator, where additional fuel is burned. The hot gases generate steam, which is supplied to a common header at 1200 psia and 950°F (510°C). From there it goes to the single-cylinder axial-exhaust turbine. All three generators are hydrogen-cooled.

applied to the generators and all equipment is ready for synchronization with the line. The running turbines are operating primarily on speed control, and the afterburners (which are auxiliary firing units in the heat-recovery steam generators) are on. The steam turbine bypass valve acts as a back-pressure control, opening and closing as needed to maintain constant pressure in the main steam header.

3. *Power generation, gas turbines only.* When the steam turbine is not needed for power generation or is shut down for routine maintenance, the gas turbines can be used alone to generate power. Start-up to base load takes about 35 minutes. The steam generated can be bypassed to the condenser, or the heat-recovery steam generators can be drained and vented.

4. *Power generation, combined cycle.* Ordinarily, the two gas turbines are synchronized first and then the steam turbine. The steam temperature required to roll the steam turbine (determined by the rotor temperature) is attained by loading the gas turbines and/ or firing the afterburners. It is also possible to start the plant with one gas turbine and the steam turbine and then bring the second gas turbine on line later.

Regardless of the plant-operating configuration (the combination of turbine-generator units selected by the operator), the control system offers a choice of four control-operating levels. From highest to lowest level, they are (1) plant-coordinated control, (2) operator automatic control, (3) operator analog control, and (4) manual control. It is not necessary for all generating units to be operated at the same level of control. Consult the alphabetical index for further descriptions in this Handbook pertaining to power plant controls and instrumentation.

MODIFIED GAS TURBINE GENERATOR*

Somewhat analogous to the pumped storage principle used in some hydroelectric plants for storing energy to be available for peak system loads is a system under test in the Federal Republic of Germany in which large quantities of air are compressed and stored. (Pumped storage is described in article on "Hydropower—Conventional Systems" in the Handbook section on *Hydropower Technology.*)

In review, it should be recalled that about two-thirds of the total power generated by a gas turbine is used to supply air for combustion and cooling, with the remaining one-third of the power available for external work. Thus, for every horsepower delivered to the output shaft, two horsepower go into driving the axial flow compressor. Another characteristic is that most gas turbines require about 10 cubic feet per minute of air, compressed to about 75 psig (6 atmospheres) for each brake horsepower (bhp) delivered to the output shaft. Considering the foregoing fundamentals, how can they be applied to a gas-turbine/electric power peaking station consisting of (1) a modified gas-turbine generator; (2) an electric-motor-driven, multistage reciprocating air compressor; (3) a very large underground air storage volume capable of containing high pressure?

If it were not necessary to drive the usual axial compressor with the turbine wheel, almost all of the power developed could be utilized for useful work. In this concept, the modified turbine has no compressor as a part of it. The required air comes from the underground receiver that has been pumped up during the off-peak time of day by the reciprocating compressor(s).

How large must the auxiliary equipment be in comparison with the turbine generator? It appears that the ratio will not vary if the peak-versus-off-peak time ratios do not vary much. Consider a peaking situation of 4.6 hours. This will require about 275 compressor hp for each 1000 hp of turbine output.

A 7100 kVA power plant requires about 10,000 bhp. In addition to the fuel, the turbine will require about 33,300 cubic feet per minute of air at 6 atmospheres pressure; or for 4.6 hours, this amounts to $4.6 \times 60 \times 33,300 = 9,190,800$ cubic feet. Since air is storable and compressible, let it be assumed that it can be stored at a pressure of 1500 psig, or about 100 atmospheres. Applying Boyle's law, $P_1V_1 = P_2V_2$, theoretically it takes one volume of air to increase the pressure by one atmosphere.

Thus, availability of air would be $100 - 6 = 94$ atmospheres of air. Dividing 9,190,800 by 94 yields 97,774 cubic feet; or a cube 46 feet on a side would be an equivalent volume.

---

*Information furnished by D. R. Lohse, Product Specialist, Reciprocating Compressors, Joy Manufacturing Company, Michigan City, Indiana.

In terms of compressor capacity, the off-peak period of 19.4 hours is equivalent to 1,164 minutes. The compressor system will have to deliver an average rate of 7,896 cubic feet per minute during the pump-up period; pumping initially from atmospheric to 75 psig with about 1345 bhp and ending the period at 1500 psig (from atmospheric) with 3432 bhp, for an average hp of 2389. Thus,

Supply side (storage): 2389 hp × 19.4 hours = 46,346 hp hours

Demand side: 10,000 hp × 4.6 hours = 46,000 hp hours

The foregoing comparison of power utilized to compress the air into storage to the power derived from the turbine during its 4.6 hours of operation results in a small difference. This amounts to less than 1 percent, and the larger number shows that there is some loss in filling an air receiver, and then expanding it back out again. The justification to propose this type of system then can come from (1) the increased cost of a larger basic power system that would be utilized for only 4 or 5 hours per day, and (2) that same power system operating at less than full capacity would not be as efficient as if it were designed for the lower capacity.

EMISSIONS ASPECTS OF GAS TURBINES*

To get the best performance from heat engines, engineers are concerned with obtaining high combustion efficiency. It is unfortunate that the easiest way to achieve high efficiency is to operate at maximum combustion temperatures. While this produces low levels of unburned hydrocarbons and carbon monoxide emissions, it also results in the highest levels of nitrogen oxide (NO_x) emissions. As shown by Fig. 25, the generation of

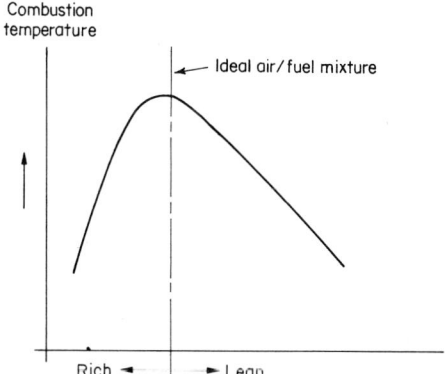

Fig. 25. Temperature-mixture relationships.

NO_x is directly related to high combustion temperatures. There are several ways by which combustion temperatures can be reduced; however, the most effective way is to burn fuel at lean mixtures. Combustion temperatures are related to air/fuel ratios with peak temperatures occurring at the ideal air/fuel ratio of about 15 pounds of air to 1 pound of fuel. From Fig. 26, it is noted that only by combustion at high air/fuel ratios (lean mixtures) will emissions of all components be at a reduced level. It is evident that the key to controlling combustion emissions is the development of successful methods of efficiently burning fuel at very lean mixtures, i.e., low combustion temperatures. This requires modification of existing combustion systems. Relatively speaking, such physical modifications are less difficult with the gas turbine than with the piston engine.

The relative levels of NO_x emission for gasoline engines, diesel engines, large gas engines, and several gas turbines are shown in Fig. 27. The NO_x data on piston engines have been taken from publications and reports from various sources and represent a fairly wide cross section of designs. The fact that engine design characteristics, such as rated r/min, fuel injection method, turbocharged or unturbocharged, have an effect on NO_x levels

*Additional data can be found in "Exhaust Emissions from Stationary Powerplants," Solar Division, International Harvester Company, San Diego, Calif., 1973.

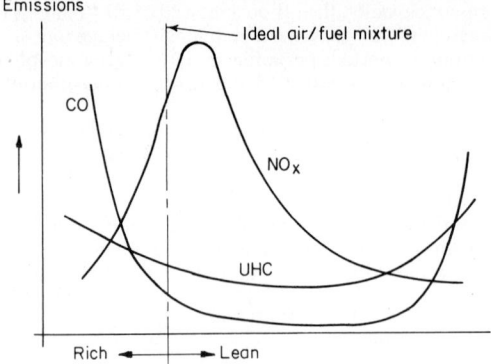

Fig. 26. Emissions-mixture relationship. CO = carbon monoxide, NO = nitrogen oxides, UHC = unburned hydrocarbons.

accounts for the wide range of NO_x level in the piston engine. It is evident from Fig. 27 that gas turbines, even without an attempt to control their emissions, have lower levels of NO_x than controlled large gas engines. Although data are not included here, this is also true of carbon monoxide and unburned hydrocarbons.

There are two factors inherent to the conventional gas turbine that account for low emissions of the three major pollutants:

1. Combustion temperatures in gas turbines are low since they use about three times the air needed to burn the fuel consumed. Typically, gas turbines have an available air/fuel ratio of 50 to 60:1 which results in an oxygen-rich environment. In comparison, the gasoline engine (spark ignition) and diesel engines (including gas engines) require an operating air/fuel ratio of about 15:1 and 25:1, respectively. The air/fuel-ratio operating

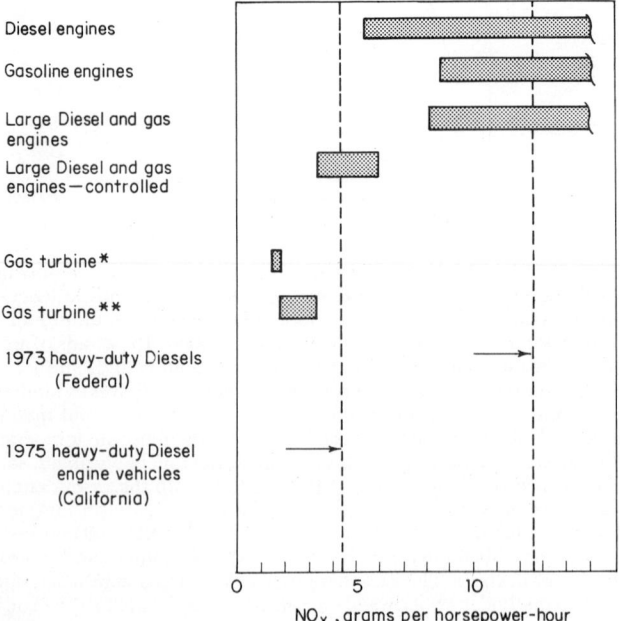

Fig. 27. Interassociation of emissions and type of engine. *Solar *Saturn* turbine; **Solar *Centaur* turbine.

ranges of piston engines and gas turbines and the effect on NO_x formation are shown in Fig. 28.

2. Since the gas turbine is a continuous-flow machine, the time available for combustion of the fuel and the attainment of thermal equilibrium is considerably greater than in a piston engine.

A significant difference influencing the control of exhaust pollutants from piston engines and gas-turbine engines lies in their mechanical concept. With the piston engine,

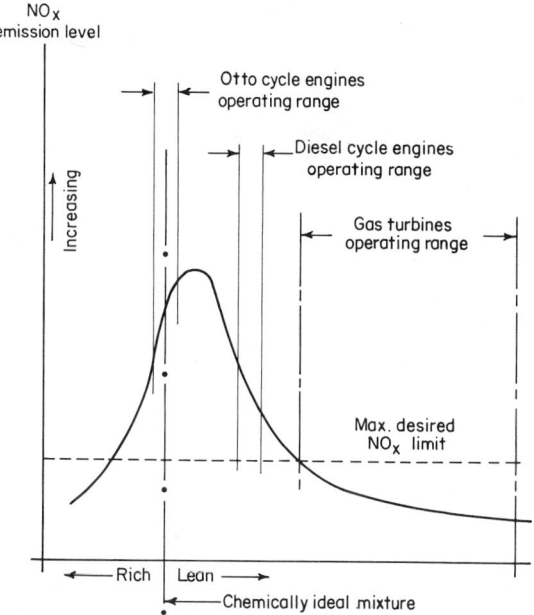

Fig. 28. Correlation of emission level and engine-type operating range.

all the thermodynamic processes (compression of air, combustion of fuel, and expansion of gases to produce power) take place within the confines of a single element, the cylinder. The gas turbine, on the other hand, achieves its compression, combustion, and expansion in three separate and distinct components, which permit a great deal of latitude in the design of each component. This is of special significance since, with the gas turbine, there is freedom to design pollutant-reducing combustion systems techniques that cannot be used in the piston engine. The turbine combustor can be varied in size and shape and can utilize various arrangements for injecting the fuel, premixing the air, and controlling combustion temperatures (air/fuel ratio) to minimize the generation of NO_x, CO, and unburned hydrocarbons. Further, these techniques generally can be applied without penalty to engine performance, power, or fuel consumption.

The piston engine does not have such flexibility. Within the more or less rigid confines of the cylinder combustion chamber, there is little latitude for design innovations. The problem is aggravated by the very short time available to burn fuel within the cylinder, a characteristic unavoidable with the intermittent, explosionlike combustion of the piston engine. The short combustion time interval makes it essential that the Otto-cycle (spark ignition) engine operate closely to a chemically ideal ratio of air and fuel, about 15:1, if the engine is to run smoothly and efficiently. The explosion of this mixture generates pressures that push combustion temperatures to a peak of about 4500°F (~ 2482°C), a condition that produces nearly maximum NO_x. Although diesel engines (including natural gas engines) operate at somewhat higher air/fuel ratios of 20 to 25:1, they also produce relatively high levels of NO_x emissions because they operate at even higher combustion pressures, which lead to high combustion temperatures.

Although the piston and cylinder arrangement of the piston engine offers limited opportunities to reduce emissions through design innovation, the combustion process itself can be modified or influenced to reduce the levels of NO_x, CO, and unburned hydrocarbons. Some of the techniques are:

1. Retard the ignition timing
2. Operate at the leanest possible air/fuel mixture (sacrificing efficiency)
3. Reduce compression ratio
4. Controlled burning of fuel (stratified charge)
5. Dilute combustion mixture by recirculating exhaust gases
6. Increase the operating temperature of the cylinder and head

It is significant to note that experience to date indicates that to make substantial reductions to NO_x, more than one of the techniques is required. Whatever combination is used, the objective is to lower peak combustion temperatures and, therefore, cylinder gas pressure. In the long run, this penalizes the performance of the engine in power and/or fuel consumption. The bar entitled "Large Diesel and gas engines—controlled" in Fig. 27, illustrates the level to which NO_x emissions from large engines of this type can be controlled—accompanied by varying degrees of power and fuel consumption penalities.

Another approach to lowering emissions from piston engines is that of treating the exhaust gases before they are ejected into the atmosphere. This consists of adding an afterburner (thermal reactor) to complete the combustion of CO and unburned hydrocarbons and a catalytic reactor to break down NO_x $(NO + CO \rightarrow N + CO_2)$. For the catalyst to reduce a nitric oxide by a significant amount, the exhaust must contain a high level of carbon monoxide. This, in turn, requires that the piston engine be operated with a rich fuel/air mixture which unavoidably increases the fuel consumption. See Fig. 28.

Emission Techniques for Gas Turbines.   Research indicates that combustion temperatures in gas turbines must be held within a range of 2200 to 3100°F (1204 to 1705°C) to hold all emissions to an acceptable level. The gas-turbine combustor is relatively insensitive to shape and size in regard to its ability to operate efficiently. This gives the designer much greater freedom to modify and arrange the combustor so that it can operate at low combustion temperatures than is available to the designer of the piston engine. Burning fuel efficiently at lean mixtures and low temperatures requires that:

1. The fuel be thoroughly evaporated and uniformly mixed with the air before burning
2. Burning the mixture takes place uniformly and as rapidly as possible
3. Additional (secondary) air be added downstream of the combustion zone to assure complete burning of CO and unburned hydrocarbons
4. The combustor liner wall temperature be high enough to eliminate fuel quenching, reducing the tendency to form CO, unburned hydrocarbons, and smoke

A comparison of the traditional combustion design concept with that for achieving lower emissions is given in Fig. 29. In the traditional combustor design (a), the primary concern is stable combustion over all engine operating conditions with good combustion efficiency. This is accomplished by establishing a central combustion zone that will burn at near stoichiometric conditions. This approach accomplishes stability and efficiency objectives, but the resulting high combustion temperatures produce NO_x levels that are higher than attained otherwise.

By modifying the combustor, as shown in (b), the combustion process can be controlled to a lower temperature. An abundance of primary air is premixed with the fuel to produce a homogeneous mixture of air and fuel vapor. This lean mixture, which ranges between a 30:1 air/fuel ratio at full engine load to 45:1 at no load, is burned in the primary combustion zone to maintain a combustion temperature within the range of 3000 to 2200°F (1649 to 1204°C). Secondary air brought in through the combustor liner downstream from the combustion zone completes the burning process, controls the combustion liner metal temperature, and dilutes the combustion gas to the proper operating temperature level.

EXPANSION TURBINES

An expansion turbine converts the energy of a gas or vapor stream into mechanical work as the gas or vapor expands through the turbine. The expansion process occurs rapidly, and heat transferred to or from the gas is usually very small. Consequently, in accordance with the first law of thermodynamics, the internal energy of the gas decreases as work is done, and the resultant temperature of the gas may be quite low, thus giving the expander the ability to act as a refrigerator as well as a work-producing device. As a result turboexpanders have been widely used in the cryogenic field to produce the refrigeration required for the separation and liquefaction of gases. By common usage, the terms *turboexpanders* and *expansion turbines* specifically exclude steam turbines and combustion gas turbines.

Turboexpanders may be classed into two broad categories: (1) axial-flow and (2) radial-flow. Axial-flow turbines are those in which the gas flow is essentially parallel to the axis

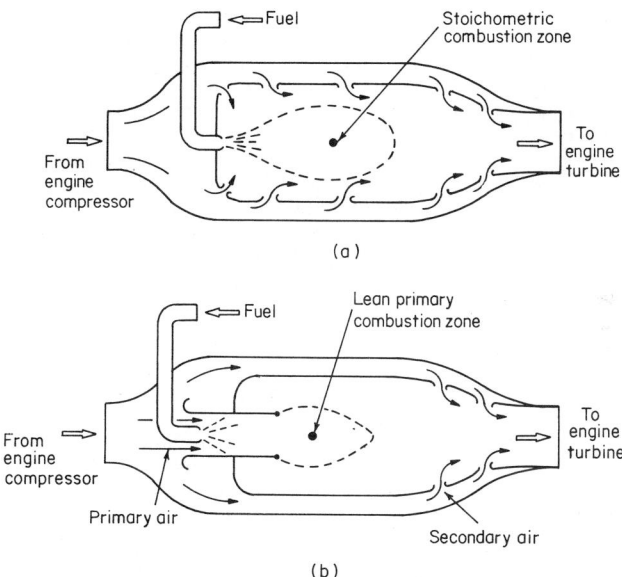

Fig. 29. Gas turbine combustion design principles: (*a*) Traditional. (*b*) Low-emissions design approaches.

or shaft of the turbine. Turbines of this type resemble a conventional steam turbine and may be single-stage or multistage with impulse or reaction blading, or some combination of impulse and reaction blading. Turbines of this type are not usually used for producing low temperatures, but are basically power-recovery devices and find application where flow rates, inlet temperatures, or total energy drops are quite high. Radial-flow turbines are those in which the gas flow is essentially at right angles to the turbine shaft. Flow may be radially inward or outward, but commercially available turbines are usually the radial-inward-flow type. Radial-flow turbines are usually single-stage and have combination impulse-reaction blades and a rotor that resembles a centrifugal-pump impeller. The gas is jetted tangentially into the outer periphery of the rotor and flows radially inward to the "eye," from which the gas is jetted backward by the angle of the blades so that it leaves the rotor without spin and flows axially away.

These machines usually have an efficiency of from 75 to 88 percent, usually operate at very low temperature, operate often on small or moderate streams, dictating a comparatively high rotating speed, and incorporate effective shaft seals to conserve the process stream. Commonly established operating limitations for turboexpanders are an enthalpy drop of 40 to 50 Btu/pound per stage of expansion and a rotor-tip speed of 1000 feet/second. Commercial turboexpanders are available up to 2500 psig inlet pressure and inlet temperatures of over 1000°F (538°C). The permissible liquid production in the expanding stream varies with discharge pressure; it may be as high as 20 percent (wt) in the discharge, provided the turboexpander has been specifically designed to handle liquids.

Power Recovery. A potential application for the turboexpander exists whenever a large flow of gas is reduced from a high pressure to some lower pressure, or when high-temperature process streams (waste heat) are available at moderate pressures. When such conditions exist, they should be examined to determine if the use of a turboexpander is justified. In such cases, a turbine can be used to drive a pump, compressor, or electric generator, thus recovering a large portion of the otherwise wasted energy. In applications of this type, careful consideration should be given to the temperature drop which will occur in the expander. Sometimes, it may be necessary to heat or dry the inlet gas to avoid low exhaust temperatures, or the formation of liquids.

Refrigeration. Turboexpanders used as components of refrigeration systems offer many possibilities to the designer of refrigeration cycles. They may be used in closed cycles with a pure gas, such as nitrogen, which is alternately compressed and expanded to provide the required refrigeration through a heat exchanger. Various types of open cycles also can be devised so that the process stream to be cooled passes through the expander, thus eliminating the need for the low-temperature heat exchanger. Liquid products can be produced directly from the turboexpander in this manner provided that the expander is specifically designed for this type of service.

The first turboexpander designs took advantage of a "free" pressure drop that was available at particular locations. Since then, the technique has been refined and enlarged to embrace practically every situation encountered in extracting hydrocarbons from a mixed gas stream, even where "free" pressure drop is not available. Designs were improved through better utilization of heat exchangers, improved optimization of expansion pressure levels, improved utilization of construction materials, design of the control system by simulating operations in a computer, and the development of interlocking instrumentation.

The foregoing refinements enabled the utilization of the turboexpander economically in plants where full pressure restoration is required. The turboexpander system also has been used for recovery of propane alone and in some cases has been found to be more economical than the oil absorption process. Other turboexpander processes include dehydration and dew point control of wellhead gas streams, utilizing pressure reduction of 2000 to 10,000 psig and nitrogen rejection together with heavier hydrocarbon recovery. Turboexpanders were first used a number of years ago in air separation plants and ethylene purification systems.

Some inherent advantages of turboexpander systems include (1) the final cryogenic operating temperature is obtained from the turboexpander and not achieved by costly low-level external refrigeration; (2) product separation pressure is set to give the most desirable equilibrium conditions; (3) plants are compact and inherently simple; (4) capital investment and operating costs are usually low, as much as 40 percent savings over conventional ethane recovery method; and (5) maintenance requirements are low.

With reference to Fig. 30, the turboexpander system operates as follows: Feed gas is first dehydrated and sometimes alcohol is injected at stratetic points to protect further against the formation of ice or hydrates. Next, the feed is chilled by heat exchange with the residue gas. Condensed liquids are then separated and the vapors delivered to the expander. A direct-connected compressor recovers the expander energy, boosting the pressure of the condensate stripper overhead gas. Residue gas can be further compressed to any delivery pressure. Alternatively, the compressor can be used to boost the pressure of the feed gas to obtain additional refrigeration. By using a turboexpander to remove energy from the gas, refrigeration is materially increased and the temperature is lowered below those conditions obtainable from simple adiabatic expansion.

Liquids condensed at the expander outlet and in the feed-chilling step are fed to a fractionation system. The bottoms from fractionation represent the desired product mixture, which can have an ethane content equivalent to 90 percent or more of the ethane in the feed gas. Virtually 100 percent of propane and heavier hydrocarbons in the feed gas can be recovered. Flexibility in product composition is obtained either by adjusting the expander outlet pressure, or by stripping undesirable components overhead in the condensate stripper. Pressure difference across the expander is the principal energy source. Thus, minimal amounts of electric power and fuel are required for pumps, dehydrator regeneration heat, and stripper reboiler heat.

Because of the small size of turboexpander rotating elements, the forces and stresses at high speed (7500 to 45,000 r/min) are equal to or less than those encountered in lower-speed rotating machinery. Any part of a turboexpander usually can be replaced in about 3 hours downtime.

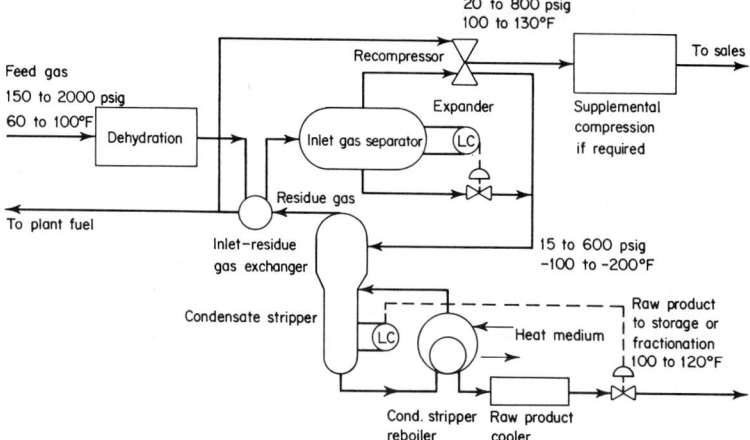

Fig. 30. Expander system for gas processing. LC = level controller *(Fluor Corporation.)*

REFERENCES

Csanady, G. T.: "Theory of Turbomachines," McGraw-Hill, New York, 1964.

Lee, J. F.: "The Theory and Design of Steam and Gas Turbines," McGraw-Hill, New York, 1954.

Treager, I. W.: "Aircraft Gas Turbine Engine Technology," McGraw-Hill, New York, 1970.

Morrison, R.: "Gas Turbines," in T. Baumeister and L. Marks (eds.), "Standard Handbook for Mechanical Engineers," 7th ed., McGraw-Hill, New York, 1967.

Journal of Engineering for Power, *Trans. ASME.*, **A81**, American Society of Mechanical Engineers, New York, 1959.

"Bibliography on Gas Turbines," ASME Gas Turbine Power Division, American Society of Mechanical Engineers, New York, periodically updated.

"The Gas Turbine Catalog," Gas Turbine Publications, Inc., New York, issued annually.

Dusinberre, G. M., and J. C. Lester: "Gas Turbine Power," 2d ed., International Textbook, Scranton, Pa., 1958.

Jennings, B. H., and W. L. Rogers: "Gas Turbines," McGraw-Hill, New York, 1953.

Baumeister, T.: Gas Turbines, in D. G. Fink and J. M. Carroll (eds.), "Standard Handbook for Electrical Engineers," 10th ed., McGraw-Hill, New York, 1968.

Shepherd, H.: "Principles of Turbomachinery," Macmillan, New York, 1956.

Abadie, V. H.: Expansion Turbines, in R. H. Perry and C. H. Chilton (eds.), "Chemical Engineers' Handbook," 5th ed., McGraw-Hill, New York, 1973.

Li, K. W.: "Compressed Air Storage in Gas Turbine Systems," *Trans. of the ASME*, Paper No. 75 Pwr F, October 1975.

Evans, R. W., and C. J. Otterholm: "Use of Cold Accumulations in the Air Conditioning Field," ASHVE Research Paper 1197, *American Society of Heating and Ventilating Engineers, Transactions*, **48**, 123 (1942).

Boester, C. F.: "The Application of Storage Refrigeration to Air Conditioning," ASHVE Research Paper 1142, *American Society of Heating and Ventilating Engineers, Transactions*, **45**, 675 (1939).

Dudley, V. C.: "Thermal Energy Storage in Air Conditioning: A Means for Reducing Peak Electrical Power Demand," PhD dissertation, University of Pennsylvania, Philadelphia (1973).

Shepherd, D. G.: "A Low-Pollution On-Site Energy System for Peak-Power Supply," ASME Paper No. 74-GT-134, 1974.

Superconducting Turbine Generators

BY M. S. BALDWIN* AND C. C. STERRETT†

PROGRESS IN GENERATOR COOLING SYSTEMS

Turbine generator ratings have increased since the formation of the electric utility industry early in this century. Presently, single generators rated 1200 MVA (megavoltamperes) are in service and higher rated units will be installed within a few years. Generators have never been the limiting factor in larger power-generating-unit designs and there is no reason to believe that they will be in the foreseeable future.

Earlier increases in rating per unit of size or weight have been achieved by improving the cooling of the heat-generating components of the generator. Cooling system advances which have allowed increased generator ratings may be listed progressively as:

1. Air cooled
2. 0.5 psig hydrogen
3. 15.0 psig hydrogen
4. 30.0 psig hydrogen
5. 45.0 psig hydrogen (inner cooled)
6. 60.0 psig hydrogen (inner cooled)
7. 75.0 psig hydrogen (inner cooled)
8. Water-cooled stator winding and 75.0 psig hydrogen rotor

These cooling system advances are illustrated in Fig. 1. Note that an all-water-cooled system does not appreciably extend the megavoltamperes per cubic foot.

Concurrent with the cooling system advances and better utilization of materials, there has been an increase in physical sizes, in rotor diameters, lengths, and weights, and in the use of more nearly optimum designs.

Shipping weight and size limitations have both been increased by the construction of special generator shipment cars which are capable of handling generators 14 feet in diameter and weighing 600 tons. See Fig. 2. These cars may be used to carry either a complete generator stator or a wound inner frame. The latter construction is used when it is necessary to ship the maximum possible active stator volume and weight. The outer frame is then shipped in sections. The generator and car together weigh a total of 900 tons. The articulated pivots on each end of the center section permit shifting the load to negotiate bridges, curves, and tunnels on the railroad right-of-way and sidings.

Manufacturing plants have been expanded in size and capacity to handle, machine, and test the large components required. Special machines, such as an eight-axis computer-controlled machining center, weighing 860 tons, can handle generator frames up to 16 feet in diameter by 38 feet long. The tape-controlled machine can position a 200-ton frame to within 0.001 inch of the command position. Rotating machine tools can accomodate a 67-inch-diameter forging weighing up to 200 tons. See Fig. 3.

It is believed that current technology will provide for ratings up to the order of 2500 MVA. Such ratings will require high-cost materials, complex shipping procedures, and expensive designs. Thus, superconducting turbine generators may be first applied at ratings beyond 2500 MVA to raise the rating limit. They appear to offer economic benefits in the larger ratings, but opinions vary as to how low a rating will someday be practical and economical.

FUTURE GENERATION REQUIREMENTS

It is prudent to examine the projected growth rates of the large power pools and large individual companies to determine the time schedule for development of large generating

*Generation Consultant, Power Generation Systems, and †Program Manager, Superconducting Electrical Machinery Department, Westinghouse Electric Corporation, Pittsburgh, Pa.

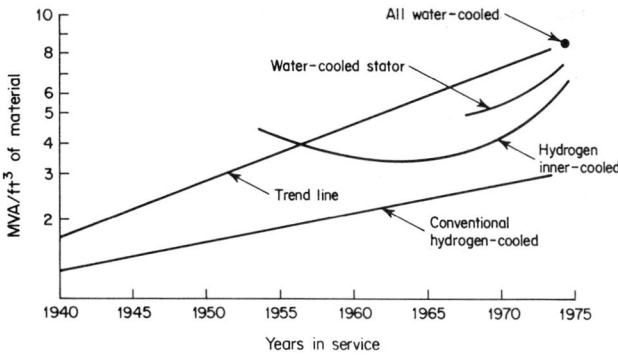

Fig. 1. MVA per ft³ versus years in service for 3600-rpm turbine generators.

units. This subject has long been of interest to power system planners. Since the mid-1950s, numerous studies have been made in an attempt to determine the optimum rating of the largest units on a power system. These studies, which include economics, reserve requirements, system stability, and transmission requirements, nearly all result in establishing the optimum rating between 4 and 10 percent of the peak load or total capacity of the system or pool considered. An investigation was made of the practices being followed by the power pools, planning groups, and larger electric utility companies in the United

Fig. 2. Generator shipment by railcar. *(Westinghouse Electric Corporation.)*

Fig. 3. Rotor machining tool. *(Westinghouse Electric Corporation.)*

TABLE 1.   Range of Size of Largest Unit Installed or Scheduled for Installation from 1971 to 1980 as a Percentage of the System or Pool Required Capacity for the In-Service Year

Pool or utility	Maximum	Minimum
Pennsylvania–New Jersey–Maryland	3.2	1.8
Virginia-Carolina Group	3.9	2.7
South Central Electrical Companies	3.8	1.9
West Group	3.5	2.3
New York Pool	4.5	2.0
Tennessee Valley Authority	6.2	3.8
Southern Co.	5.0	3.3
Northwest Pool	4.9	1.9
New England Pool	6.1	2.5
Commonwealth Edison	7.8	4.4
Northern California	7.4	4.2
American Electric Power	8.9	5.3
	6.3	5.1

States for the 1970s. The results shown in Table 1 reveal that the large units actually being installed and scheduled for installation through the early 1980s are approximately between 2 to 9 percent of the required capacity of the systems on which they will operate. As might be expected, the lower percentage number is associated with the larger pools and the higher percentage with the large individual companies or smaller power pools.

Relating these figures to the projected growth rates of the various power pools and large utility companies provides an estimate of the rate at which the unit ratings should increase to meet the future requirements of the industry. This rate is shown in Fig. 4. The growth is shown as a band rather than a single line for two reasons: (1) The top of the band represents the requirements of the largest pools, and (2) the bottom of the band represents the requirements of the smaller pools or individual companies. Accordingly, the band width shown will accommodate what is believed to represent the requirements of over 70 percent of the utility industry in the United States.

It is anticipated that the future ratings will not increase as a continuous curve but rather as a series of steps or plateaus, dictated largely by the capability of the nuclear steam supply system. Such steps are indicated as dashed lines in Fig. 4.

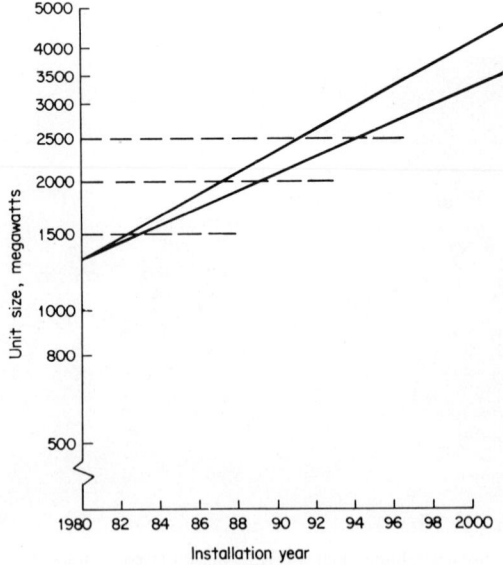

Fig. 4. Growth of generator ratings.

The slope of this band represents a more conservative practice than is presently being used and a much lower rate of increase than in the past. For example, between 1944 and 1970, the cumulative annual growth rate of unit sizes was over 10 percent, while the growth rate of the peak load was 7.5 percent for the same period. A maximum unit size of approximately 4000 MW is indicated in Fig. 4, whereas a simple extrapolation of past and current practices would suggest a maximum unit rating in the year 2000 between 7000 and

Fig. 5. U.S. Air Force 5000-kVA, 12,000-rpm rotor on test. *(Westinghouse Electric Corporation.)*

11,000 MW. An additional consideration is the sheer number of units that will be required to provide the projected generation additions if superconducting generators are not applied.

SUPERCONDUCTING AC GENERATOR TECHNOLOGY

Small superconducting generators have been built and tested, followed by the publication of reports and technical papers. Research in the field of superconductivity and cryogenics has been carried out by some organizations* since the very early 1950s. The first 60-Hz machine rated 45 kVA was built and tested by the Massachusetts Institute of Technology in 1969.† MIT then proceeded to build and test a 200-kVA model.‡ The work at MIT is supported by EEI-EPRI (Edison Electric Institute-Electric PowResearch Institute).

A 5000-kVA, 3-phase, 60Hz, 4160-volt, 3600 r/min generator with a superconducting rotor was built and successfully tested by Westinghouse in September 1972.¶ As of the

*A. Wexler, Evaporation Rate of Liquid Helium I, *J. Appl. Phys.*, **22**(12):1463–1470 (1951).

†P. Thullen, J. L. Smith, Jr., J. C. Dudley, H. H. Woodson, and D. L. Greene, An Experimental Alternator with a Superconducting Rotating Field Winding, *IEEE Trans. Power Appar. Syst.*, **PAS-90**(2):611–619, (March/April 1971).

‡J. L. Kirtley, Jr., J. L. Smith, Jr., P. Thullen, and H. H. Woodson, MIT-EEI Program on Large Superconducting Machines, *Trans. Paper* T73 137-7, presented at the IEEE Winter Power Meeting, New York, January 1973.

¶D. C. Litz, A. Patterson, C. K. Jones, S. A. Karpathy, M. S. Walker, and Y. W. Chang, Development of a 5 MVA Superconducting Generator, Testing, and Evaluation. *IEEE Trans. Power Appar. Syst.*, **PAS-93**(2):496–499 (March/ April 1974); C. J. Mole and C. C. Sterrett, A Superconducting Machine for Central Station Power Generation, *Proc. Am. Power Conf.*, **35**:1035–1047 (1973).

mid-1970s, this is the world's largest superconducting generator. See Fig. 5. A sectional view of the generator is shown in Fig. 6. In January 1974, Westinghouse (under contract †to the U.S. Air Force) successfully tested a 5000-kVA, 12,000 r/min superconducting rotor and demonstrated the feasibility of supplying 400-Hz power for aircraft with high-speed, lightweight generators.

Further Study and Development.   Certain areas which require further research and study have been identified. These can be put in terms of objectives which must be met in order to carry out meaningful designs of futher machines: (1) Develop long-life seals for transferring helium in and out of the rotor; (2) select and characterize cryogenic structural materials; (3) develop optimum configurations for superconducting field windings; (4)

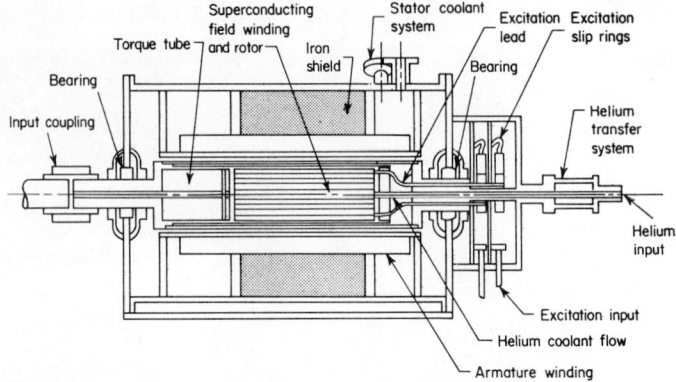

Fig. 6. Cross section of Westinghouse 5000-kVA superconducting generator.

develop stator windings; (5) develop rotor damper shields and structures; and (6) develop refrigeration systems.

For example, 10,000 hours life for rotating face seal systems has been demonstrated. Also noteworthy is the purchase in 1973 of a refrigeration system using rotary compressors and expanders. This is being adapted to the Westinghouse 5000-kVA superconducting generator and will be evaluated. This is a first step in determining the refrigeration system which should be developed for electric utility use.

In addition to the programs on seals, structural materials, refrigeration, superconductor performance, damper shields, cryogenic insulation, stator windings, and supports, a number of other programs are being conducted for the United States government in superconducting electrical machinery for both ac and dc applications. The machinery development is of interest for ship propulsion and aircraft where the small size and light weight of superconducting electrical machines offer many potential advantages.

The NBS (National Bureau of Standards, Cryogenic Division, Boulder, Col.) earlier supported studies under its CEI (Cryogenics for the Electric Industry) program and presently Westinghouse is performing basic research work for NBS for cryogenic structural materials. Involved is the study of materials, metallurgy, processing, fabricating, and performance of materials for cryogenic applications.

POTENTIAL ADVANTAGES OF SUPERCONDUCTING TURBINE-DRIVEN GENERATORS

The potential advantages of superconducting turbine-driven generators have been generally identified worldwide as follows:

1. Size and weight reduced 1 to 3
2. Potential for higher ratings
3. Higher efficiency
4. Capital plant cost reduced
5. Capitalized cost of losses

†Work supported by USAF, Contract F33615-7-1C-1591, Aero Propulsion Laboratory, WPAFB.

TABLE 2.   Cost Benefits for a 2000-MVA Generator

Generator (price)	$ 4,000,000	$2.0/kVA
Efficiency:		
Incremental plant	5,000,000	2.5
operation	1,400,000	0.7
Exciter	500,000	0.25
System stability	100,000	0.05
Lower inertia	−(300,000)	−0.15
Transportation	20,000	0.01
Higher voltage	100,000	0.05
Refrigeration redundancy	−(500,000)	−0.25
Evaluated incremental benefit for a 2000-MVA superconducting generator	$10,320,000	$5.16/kVA

6. Improved power system stability
7. Simpler excitation system
8. Reduced foundation and building cost
9. Easier shipment

A superconducting generator will have approximately twice the steady state stability limit of a conventional generator on the same power system. At the same time, it appears to offer better transient and dynamic stability performance with a relatively uncomplicated excitation system. It also promises to have better capability for withstanding unbalanced faults.

The listed advantages are of interest to electric utility companies because they result in economic benefits. All the potential advantages listed have a financial value so it is interesting to contemplate the potential savings to the electric utility industry. Because 2000-MVA generators are predicted for commercial operation by the late 1980s, a cost comparison on this basis can be made, as indicated in Table 2. The cost benefits shown in Table 2 are reasonable and conservative for a superconducting generator as compared with a conventional unit.

The forecast for generator additions by the electric utility industry for the years 1990–2000 may be given as shown in Table 3. From the forecast, it may be concluded that the nuclear and fossil additions will be base-load turbine generators totaling 790,000 MW. This represents the installation of three hundred ninety-five 2200-MVA turbine generators (0.9 power factor assumed). Thus, the potential savings to the electric utility industry for the years 1990–2000 can be

$11 million (per generator at $5/kVA) × 395 generators = $4.345 billion total savings

This is an extremely large savings—over $4 billion.

This can be translated to a single electric utility with a 1970 installed capacity of 4000 MW. This utility will require the following capacity to follow the predicted growth rate based on doubling every 10 years:

Year	Megawatts
1970	4,000
1980	8,000
1990	16,000
2000	32,000

Note that 16,000 MW will be added between the years 1990 and 2000.
This represents the installation of eight 2200-MVA turbine generators. The savings to

TABLE 3.   Electric Utility Generation Additions, Years 1990 to 2000

Nuclear	675,000 MW
Fossil	115,000
Combined cycle	192,000
Hydro	42,000
Gas turbine	95,000
Total additions	1,119,000 MW

an electric utility with a 1970 installed capacity of 4000 MW thus would be $11 million $\times$ 8 generators = $88 million total savings.

Development and Demonstrations.   The development of large superconducting turbine generators will require continued close cooperation between the manufacturer and the electric utility industry in order to secure the overwhelming technical and economic benefits at the earliest possible time. The work accomplished to date has been achieved through joint efforts resulting in successful machine models and paper studies for larger generators. Large generators have been shown to be very attractive to the electric utilities.*

The next step is expected to be the construction of a 100-MW turbine generator demonstration unit which will be installed on an electric utility power system and generate power under actual system conditions. A 7-year 100-MW machine test program is suggested which would be completed in early 1982. The overall time scale includes a logical progression of machine sizes escalating up to 2500 MVA in steps of 600, 1600, 2000, and 2500 MVA. The start of the 600-MVA program is tentatively scheduled for early 1979, and completion of the 2500-MVA program is expected to be 1995.

*A. D. Little, INC., "Study of the Markets for Large Rotating Electrical Machinery, 1970–2000," report to U.S. Department of Commerce, National Bureau of Standards, February 1973; General Electric Company, "Market Appraisal for Superconductive Generators," report to U.S. Department of Commerce, National Bureau of Standards, March 1973; Westinghouse Electric Corp., "Problems and Possible Solutions for Power Generation Using Superconducting Machinery," report to Cryogenics Division, Institute for Basic Standards, U.S. Department of Commerce, National Bureau of Standards, March 1973.

Magnetohydrodynamic Generators and Thermionic Converters

The use of magnetohydrodynamic (MHD) generation in combination with conventional steam generation in base-loaded plants has been under investigation since the mid-1960s (Refs. 1, 2). Although encouraging progress is reported, the inefficiency of MHD conversion with the present state of technology remains a serious difficulty. However, the simplicity of the device leads to a low capital cost. Thus, it has been proposed to develop and install MHD generators for emergency and peaking use in situations where annual operating time would be perhaps less than 100 hours per year (Ref. 3).

ADVANTAGES AS AN EMERGENCY GENERATOR

An emergency MHD generator has several potential advantages: (1) It can be brought to full power from a standby condition in 5 seconds or less; (2) it has an anticipated low capital cost; (3) the output can be changed from no-load to full-load in fractions of a second; and (4) it has no moving parts; the main items requiring maintenance can be replaced at low cost within 6 to 8 hours. These advantages have yet to be demonstrated on a large scale. One of the largest MHD generators currently produced is installed in a wind tunnel and has a rating of 20 megawatts. Development efforts are proceeding to produce the advantages claimed for units in excess of 200 megawatts.

FUNDAMENTAL OPERATING PRINCIPLES

An MHD generator is a device that directly converts the kinetic and thermal energy of an ionized gas to electrical energy* by forcing the gas to move in a duct or channel through a magnetic field. The motion induces an electomotive force which can then be used to power an external load. A primitive MHD generator is shown in Fig. 1.

The steady-state operation of this unit is governed approximately by the following one-dimensional flow equations (Ref. 4):

Force/unit volume:

$$\rho \frac{dv}{dx} + \frac{dp}{dx} - (\mathbf{J} \times \mathbf{B})_x + F = 0 \tag{1}$$

Power/unit volume:

$$\rho v \left(\frac{dH}{dx} + v \frac{dv}{dx} \right) - \mathbf{J} \cdot \mathbf{E} + G = 0 \tag{2}$$

Mass flow:

$$\rho v A = \text{constant} \tag{3}$$

and by Ohm's law, modified to account for magnetic fields and Hall effect:

$$\mathbf{J} + \mu(\mathbf{J} \times \mathbf{B}) = \sigma[\mathbf{E} + (v \times \mathbf{B})] \tag{4}$$

where ρ = gas density
$\mathbf{v}$ = gas velocity vector
p = gas pressure
$\mathbf{J}$ = current density vector
$\mathbf{B}$ = applied magnetic field
F = wall friction force/unit volume

*Much of the basic information reported here is from Ref. 14.

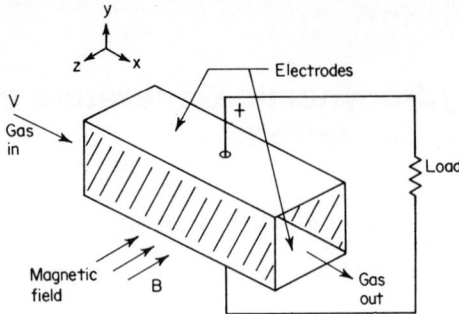

Fig. 1. Highly simplified diagram of MHD generator.

H = gas enthalpy
E = electric field vector
G = heat loss through walls/unit volume
A = cross-sectional area of duct
μ = electron mobility in gas
σ = electrical conductivity in gas

Equation (4) is of the most interest in working with the electrical load, since it greatly influences the way in which power is drawn from the unit. The axial component of $(\mathbf{J} \times \mathbf{B})$ produces an axial field, called the *Hall field,* which may be several thousand volts per

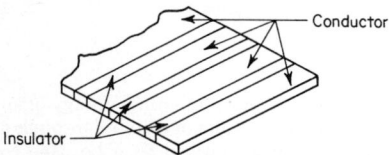

Fig. 2. Section of segmented electrode.

meter (Ref. 2). This causes currents to flow along the electrodes, creating high ohmic losses. To minimize these losses, the electrodes are segmented and separated by strips of insulating material, as shown in Fig. 2. A large MHD unit would typically have over 100 segments.

For a duct with segmented electrodes, there are several ways to connect the load. Two of these connections are illustrated schematically in Fig. 3.

Faraday Connection.   In the Faraday connection (Fig. 3a), the total load is divided among the electrodes; each electrode segment then behaves like a separate MHD generator. The electrical properties can be derived from Eq. (4) by setting J_x (the axial component of current) $= 0$ because of the segmenting, and $J_y = -\alpha E_y$, where α represents the load admittance.

The Faraday connection makes most efficient use of the thermal energy of the gas (Ref. 4), but the requirement of multiple load connections is a drawback. This shortcoming can be overcome, however, by paralleling inverter-transformers at the ac line.

Hall Connection.   The Hall connection (Fig. 3b) uses the induced Hall potential to drive the load. The top and bottom of each electrode segment are joined, so that no transverse electric field exists. This configuration can be analyzed by setting $E_y = 0$ in Eq. (4). J_x is not zero, and is, in fact, the load current density $(J_x = -\alpha E_x)$. Successful application of this configuration depends on obtaining very strong magnetic fields and highly ionized gases. These are not readily available.

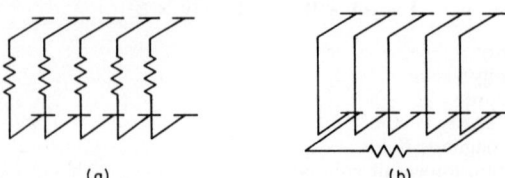

(a)   (b)

Fig. 3. Load configurations for segmented electrode generator: (a) Faraday connection. (b) Hall connection.

Other interconnections of the electrodes have been proposed (Ref. 5), but their superiority over the two configurations mentioned remains to be demonstrated on a practical scale.

DC/AC Inverters.   The dc electrical equivalent circuit for a Faraday-connected MHD generator may be derived from the modified Ohm's law, Eq. (4). Since $J_x = 0$, the equation becomes

$$J_y = \sigma[E_y + Bv] \tag{5}$$

$$-\mu B J_y = \sigma E_x \tag{6}$$

Assume the MHD channel is of square cross section with height h and length L. Then the total current I_y, which is the sum of the currents at all electrodes, is given by

$$I_y = J_y h L \tag{7}$$

and the terminal voltage V_y by

$$V_y = -E_y h \tag{8}$$

Using Eqs. (7) and (8), Eq. (5) may be written

$$V_y = hBv - \frac{1}{\sigma L} I_y \tag{9}$$

Maximum power is delivered if the load resistance R is equal to the internal resistance $1/(\sigma L)$, and

$$P_{max} = \tfrac{1}{4}\sigma B^2 V^2(h^2 L) \tag{10}$$

Typical averaged values might be

$L = 10$ meters
$h = 1.5$ meters
$B = 2.2$ webers/square meter (Ref. 3)
Fuel: JP $- 4/0_2$ (Ref. 1)
Combustion temperature $= 3430$ K
Average velocity $= 1,250$ meters/second (Ref. 6)
$\sigma = 5$ mho/meter (based on average gas temperature of 2500 K (Ref. 1)
$\mu = 0.5$ square meters/volt-second (Ref. 7)

For these values,

$$V_y = 4120 - 0.02 I_y$$

and

$$P_{max} = 213 \text{ MW}$$

The numerical values calculated using these figures agree surprisingly well with the electrical parameters of a proposed 2000-V, 200-MW magnetohydrodynamic emergency generator (Ref. 3). Note, however, that the velocity, conductivity, and temperature of the gas vary drastically along the duct, so that the foregoing numbers are only representative.

The dc equivalent circuit for the Hall connection can be derived in a similar manner. Equation (4) becomes

$$J_x - \mu B J_y = \sigma E_x \tag{11}$$

$$J_y = \sigma B v - \mu B J_x \tag{12}$$

The total current I_x is

$$I_x = J_x \cdot h^2 \tag{13}$$

and the axial voltage is

$$V_x = -E_x \cdot L \tag{14}$$

These equations yield the formulas

$$V_x = \mu B (LBv) - \frac{L(1 + \mu^2 B^2)}{\sigma h^2} I_x \tag{15}$$

and

$$P_{\text{max}} = \frac{1}{4} \frac{\mu^2 B^2}{1 + \mu^2 B^2} \sigma B^2 V^2 (h^2 L) \tag{16}$$

For the same physical data, these equations become

$$V_x = 30,000 - 1.96 I_x$$

and

$$P_{\text{max}} = 117 \text{ MW}$$

Reference 3 describes a Hall-connected generator, but the electrical characteristics more closely resemble a Faraday-connected unit. The relative inefficiency of the Hall generator outweighs the advantage of the single-load connection. Hence, the Faraday connection would probably be chosen.

The time required to generate full power from an initially unexcited MHD generator is dependent largely on how fast the magnetic field can be built up (Ref. 8). For an emergency generator, a bootstrap operation is envisioned, whereby part of the power generated is delivered to the load; the rest is used to build up the magnetic field.

An approximate equation for the bootstrap start (Ref. 8) is

$$\frac{1}{P} \frac{dP}{dt} = \left(\frac{z - 1}{z}\right) \frac{P_0}{W_0} - \frac{1}{z} \frac{dz}{dt} \tag{17}$$

$$P(t = 0) = P_{gi}$$

where
P = power delivered to the load
P_0 = maximum rated power output
W_0 = maximum field energy
P_{gi} = power generated due to residual or standby excitation
$\frac{1}{z}(0 \leq \frac{1}{z} \leq 1)$ = the fraction of generated power delivered to the load

For constant z, Eq. (17) implies an exponential buildup of power until $P = P_0/z$, at which time the magnet is fully energized and the load power can be increased to P_0 less

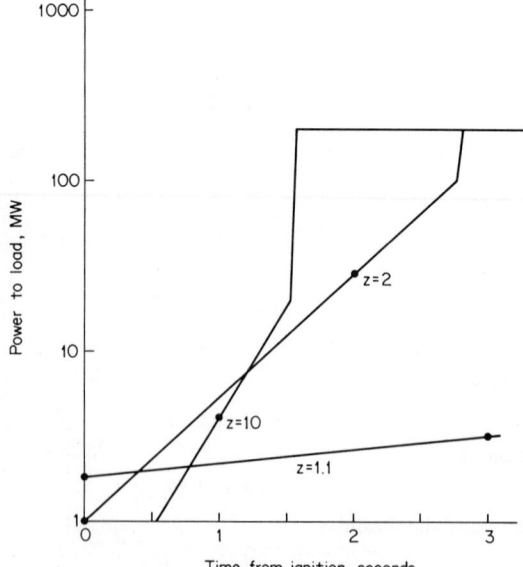

Fig. 4. MHD power delivered during startup.

magnet losses. Equation (9) shows that the start-up process may be modeled by a time-varying voltage in series with a constant resistor. The voltage is proportional to the magnetic field B and has a small initial value determined by residual magnetism or standby excitation.

Typical responses for various values of z are shown in Fig. 4. The plots indicate that the process of switching the power from the magnet to the load is not accomplished instantaneously, but has a time constant associated with it. This time constant is modeled by the equation

$$T_M P_L + P_L = P \qquad P_L(0) = P(0) \tag{18}$$

where P_L = actual load power
$\quad\ P$ = solution of Eq. (17)
The effect of Eq. (18) is negligible except at the switching instant. For convenience, then, in simulation, P_L is considered to be the load power at all times.

For the 200-MW emergency generator previously mentioned, the representative values (Ref. 8) are

$$
\begin{aligned}
P_0 &= 215 \text{ MW} \\
W_0 &= 65 \text{ MJ} \\
P_{gi} &= 2 \text{ MW (for residual magnetism)} \\
M &= \text{magnet losses} = 15 \text{ MW} \\
T_M &= 0.04 \text{ second}
\end{aligned}
$$

Thus, for constant z,

$$P = \frac{2}{z} \exp\left(3.36 \frac{z-1}{z} t\right) \text{MW} \qquad (0 \le t < t_s)$$

$$= 200 \text{ MW} \qquad\qquad\qquad (t_s < t) \tag{19}$$

$$215 = 2 \exp\left(3.36 \frac{z-1}{z} t_s\right)$$

The value of z can be selected to satisfy certain performance requirements. For example, to maximize the integrated power (energy) delivered to the load, one should make z infinite, investing all the power in building up the magnet. On the other hand, if the electrical network cannot withstand a large surge of power, z should be smaller to allow a more gradual buildup of power delivered to the system.

The output of an MHD generator appears as a dc potential. To make this power available to the electric network, it must be converted to alternating current. Devices performing this job are called inverters, and may be analyzed as rectifiers operating in reverse. Inverter circuits are already in use in high-voltage dc transmission links and have been studied in some detail (Refs. 9–12).

A simple inverter circuit is shown in Fig. 5. Practical inverter circuits employ different transformer connections and/or bridge-connected diodes, but their operation is fundamentally the same (Ref. 13). The dc current I is synchronously switched between the transformer legs by the controlled diodes (thyristors or mercury-arc rectifiers), thereby

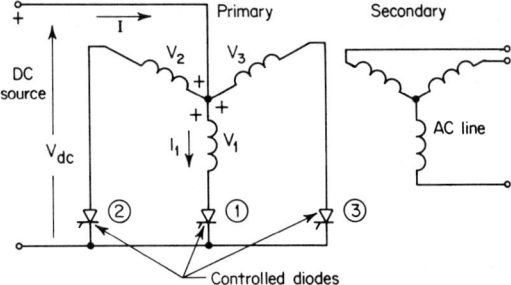

Fig. 5. Basic three-phase inverter circuit.

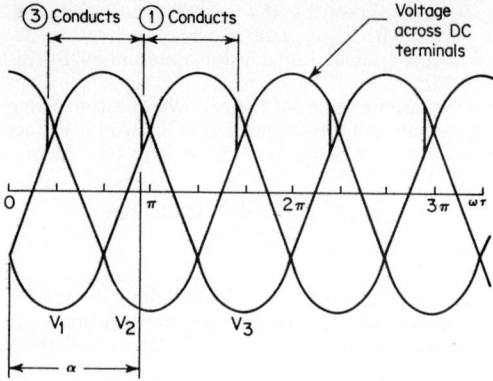

Fig. 6. Typical inverter waveforms.

producing three-phase currents at the ac line. The operation is controlled by the relative phase of the firing signal and the transformer voltage. Typical inverter waveforms are sketched in Fig. 6.

In practice, the current does not instantaneously switch between legs, and a description of the commutation process is necessary in order to model the inverter's electrical characteristics.

Let the primary voltages be

$$V_1 = E \sin\left(\omega t - \frac{5\pi}{6}\right)$$

$$V_2 = E \sin\left(\omega t - \frac{3\pi}{2}\right) \tag{20}$$

$$V_3 = E \sin\left(\omega t - \frac{\pi}{6}\right)$$

and assume that diode 3 is conducting. The voltage D_1 across diode 1 is given by

$$D_1 = V_3 - V_1 = 2E \sin\frac{\pi}{3}\ \sin \omega t$$

Diode 1 will conduct if it is fired when the voltage across it is positive. Because of the transformer inductive reactance, the current does not instantaneously switch from leg 3 to leg 1, but changes continuously. The equivalent circuit is shown in Fig. 7.

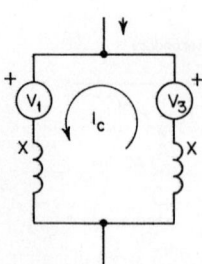

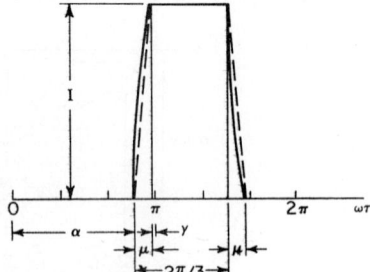

Fig. 7. Equivalent circuit for commutation analysis.

Fig. 8. Current waveform in transformer winding.

The loop current I_c goes from 0 at the instant diode 1 is fired to $I_c = I$ (the dc current) at the instant diode 3 is extinguished. The loop equation is

$$\dot{I}_c = \frac{\omega}{2X}(V_3 - V_1) = \frac{\omega E}{X}\sin\frac{\pi}{3}\sin\omega t \tag{21}$$

where X = transformer reactance at frequencey ω. Let
   $\omega t = \alpha$ at the instant diode 1 is fired
   $\omega t = \pi - \gamma$ at the instant diode 3 is extinguished
   $u = \pi - \gamma - \alpha$ be the overlap angle
Note that $\omega t = \pi$ when $V_3 = V_1$. Then

$$I = \frac{E}{X}\sin\frac{\pi}{3}(\cos\alpha + \cos\gamma)$$

which may be rewritten

$$\cos\gamma = \frac{XI}{E\sin\pi/3} + \cos(u + \gamma) \tag{22}$$

The dc voltage across the inverter may be calculated from

$$V_{dc} = \frac{1}{2\pi/3}\left[\int_\alpha^{\alpha + 2\pi/3} V_1\, d(\omega t) + \int_\alpha^{\alpha + u}\frac{X}{\omega}\dot{I}_c\, d(\omega t)\right]$$

since the commutation process is repeated every 120 electrical degrees. The integration yields

$$V_{dc} = \frac{E\sin\pi/3}{\pi/3}\cos(\pi - \alpha) + \frac{3}{2\pi}XI \tag{23}$$

so that there is a natural load droop if the firing angle is held constant.

Now consider the ac aspects of the inverter. The current waveform through leg 1 of the transformer primary is shown in Fig. 8. For simplicity, assume the shape is trapezoidal as indicated by the dotted lines. The fundamental component is

$$I_1 = \frac{2}{\pi}I\sin\frac{\pi}{3}\frac{\sin u/2}{u/2}\sin\left(\omega t - \frac{5\pi}{6} + \gamma + \frac{u}{2}\right)$$

Recall that

$$V_1 = E\sin\left(\omega t - \frac{5\pi}{6}\right)$$

Hence the real and reactive power supplied to the ac network are

$$P = EI\frac{\sin\pi/3}{\pi/3}\frac{\sin u/2}{u/2}\cos\left(\gamma + \frac{u}{2}\right) \tag{24}$$

$$Q = -EI\frac{\sin\pi/3}{\pi/3}\frac{\sin u/2}{u/2}\sin\left(\gamma + \frac{u}{2}\right) \tag{25}$$

Note that inverters cannot supply reactive power. Thus shunt capacitors or other sources of reactive power are also required. Note also that the ac system must supply the voltage E. Hence inverters cannot be used to supply ac power to a "dead" system.

To complete the description of the inverter, it is necessary to describe how the extinction and overlap angles vary with the terminal voltage and current. Define a regulation factor ρ as the ratio of commutation voltage drop at full load to nominal no-load voltage. In other words,

$$\rho = \frac{(3/2\pi)\,XI_f}{(3/\pi)\,E_N \sin \pi/3} = \frac{XI_f}{2E_N \sin \pi/3} \tag{26}$$

where I_f = rated full-load current and
$\quad E_N$ = nominal primary voltage (proportional to bus node voltage).
From Eqs. (22) and (23), one can write

$$\cos \gamma = 2\rho\,\frac{I}{I_f}\,\frac{E_N}{N} + \cos(u + \gamma) \tag{27}$$

$$V_{dc} = E_N\,\frac{\sin \pi/3}{\pi/3}\left[\frac{E}{E_N}\cos(u+\gamma) + \rho\,\frac{I}{I_f}\right] \tag{28}$$

and

$$V_{dc} = E_N\,\frac{\sin \pi/3}{\pi/3}\left(\frac{E}{E_N}\cos \gamma - \rho\,\frac{I}{I_f}\right) \tag{29}$$

Note that E may not equal E_N if the bus node voltage departs from its nominal value.
   It is convenient at this point to normalize all the equations by introducing the quantities

$$V^* = \frac{V_{dc}}{E_N\,\dfrac{\sin \pi/3}{\pi/3}} \qquad I^* = \frac{I}{I_f} \qquad E^* = \frac{E}{E_N}$$

$$P^* = \frac{P}{I_f E_N\,\dfrac{\sin \pi/3}{\pi/3}} \qquad Q^* = \frac{Q}{I_f E_N\,\dfrac{\sin \pi/3}{\pi/3}} \tag{30}$$

Then the relations of interest become

$$P^* = E^* I^*\,\frac{\sin u/2}{u/2}\cos\left(\gamma + \frac{u}{2}\right) = V^* I^* \tag{31}$$

$$Q^* = -E^* I^*\,\frac{\sin u/2}{u/2}\sin\left(\gamma + \frac{u}{2}\right) \tag{32}$$

$$V^* = E^* \cos(u + \gamma) + \rho I^* \tag{33}$$

$$V^* = E^* \cos \gamma - \rho I^* \tag{34}$$

The manner in which power is transferred from dc source to ac load depends largely on the way the firing angle is controlled. Many options are possible. Among these are:
1. Constant firing angle
2. Constant extinction angle
3. Constant current
4. Constant impedance
Constant extinction angle control is preferred for operation at full load. There are two factors involved. A study of the inverter equations shows that operation with a small extinction angle gives least reactive power consumption. On the other hand, the extinction angle must be kept larger than some small value (~ 3 to $5°$) to allow time for the extinguished diode to de-ionize and come under control of the firing grid once again. Constant extinction angle control ensures that the extinction angle is kept as small as possible consistent with de-ionization.
If the dc source voltage is not near its nominal value, constant extinction angle control is not appropriate, and another control policy must be used. Recall that the model of a Faraday-connected MHD generator consists of a time-varying voltage in series with a constant resistance. To transfer maximum power during the start-up, it is desirable to control the firing angle so that the inverter presents a constant impedance to the MHD

unit. The value of the impedance should be approximately the internal impedance of the generator.

Let the firing angle be controlled so that

$$I^* = kV^*$$

and select k such that $I^* = 1.0$ when $\gamma = 0°$. Using Eq. (34),

$$k = \frac{1}{1 - \rho} \tag{35}$$

Let the MHD unit be represented by

$$V_{dc} = V_{MHD} - RI$$
$$V_{dc}^* = V_{MHD}^* - \sigma I^* \tag{36}$$

where

$$\sigma = \frac{R}{E_N \dfrac{\sin \pi/3}{\pi/3}}$$
$$\quad I_f$$

To simulate the emergency generator mentioned previously, the following nominal parameter values were used:

$$\rho = 0.05 \quad \text{(Ref. 14)}$$
$$I_f = 100 \text{ kA}$$

$$E_N \frac{\sin \pi/3}{\pi/3} = 2000 \text{ volts}$$
$$R = 0.02 \ \Omega$$
$$\sigma = 1.0$$
$$V_M^*(\text{max}) = 2.0$$

The variation of P^*, Q^*, γ, and u as a function of the internal MHD voltage V_M^* is shown in Figs. 9 and 10. Control is for constant impedance if $\gamma > 3°$, and for constant extinction angle of 3° otherwise.

Summary of Equations for Simulation of MHD/Inverter Operation. Generated power during start-up ($t = 0$ when unit is activated):

$$P_G = \frac{P_{gi}}{z} \exp\left(\frac{P_0}{W_0} \frac{z-1}{z} t\right) \quad \text{MW} \quad 0 \leq t < t_s$$
$$= P_0 - M \quad \text{MW} \quad t_s < t \tag{19}$$

$$P_0 = P_{gi} \exp\left(\frac{P_0}{W_0} \frac{z-1}{z} t_s\right) \quad \text{MW}$$

Power available at the inverter:

$$T_M P_L + P_L = P_G \tag{18}$$

Normalized power available at the inverter:

$$P_L^* = \frac{P_L}{I_f E_N \dfrac{\sin \pi/3}{\pi/3}} \tag{30}$$

Normalized internal voltage:

$$V_{MHD}^* = \sqrt{\frac{P_L^*}{1 - \rho}} \ (\sigma + 1 - \rho) \tag{37}$$

Normalized dc current:

$$I^* = \frac{V^*_{\text{MHD}}}{\sigma + 1 - \rho} \qquad \text{if } \frac{V^*_{\text{MHD}}}{\sigma + 1 - \rho} < E^* \cos \gamma_0 \qquad (38)$$

$$I^* = \frac{V^*_{\text{MHD}} - E^* \cos \gamma_0}{\sigma - \rho} \qquad \text{if } \frac{V^*_{\text{MHD}}}{\sigma + 1 - \rho} \geq E^* \cos \gamma_0 \qquad (39)$$

Extinction angle:

$$\cos \gamma = \frac{I^*}{E^*} \qquad \text{if } \frac{V^*_{\text{MHD}}}{\sigma + 1 - \rho} < E^* \cos \gamma_0 \qquad (40)$$

$$\gamma = \gamma_0 \qquad \text{if } \frac{V^*_{\text{MHD}}}{\sigma + 1 - \rho} \geq E^* \cos \gamma_0 \qquad (41)$$

Inverter terminal voltage:

$$V^* = V^*_M - \sigma I^* \qquad (36)$$

Overlap angle:

$$\cos (u + \gamma) = \frac{V^* - \rho I^*}{E^*} \qquad (33)$$

Power actually delivered to network:

$$P^* = V^* I^* \qquad (31)$$

Reactive power

$$Q^* = -E^* I^* \frac{\sin u/2}{u/2} \sin \left(\gamma + \frac{u}{2} \right) \qquad (32)$$

APPLICATIONS OF MHD UNITS

Simulation and other investigations and developments to date would indicate the following potential areas of application for MHD units:

1. Supplement for System Load Shedding.   Particularly in connection with the investigation of the effectiveness of MHD units to supplement or reduce required system load shedding in a situation where an island is formed with insufficient generation and with resulting declining frequency, it is believed that the use of quick-responding MHD units can significantly reduce the amount of required underfrequency load shedding, mainly in situations where the sum of the rating of the MHD units plus fast-acting system reserve is approximately equal to the load/generation imbalance at the time of separation. The benefits appear to be less pronounced, but are still significant even if this sum is only one-half the imbalance.

Providing continuous power from the system to the MHD units (say either 1.5 MW or 15.0 MW per unit) to maintain a level of magnetism above the residual magnetism and hence improve response time does not always add significant improvement and thus may not warrant the power drain cost, even though it may only be provided during peak load hours (where it could possibly be better used as additional system reserve). Significant improvement will not be realized unless the ratings of the MHD units total to at least one-half or more of the load/generation imbalance.

The MHD unit seems most effective when all the bootstrap power output of the unit is used to build up the unit magnetization to rate level before providing power to the system. Bus voltage oscillations seem less severe for this procedure than for the approach where the MHD magnet is preenergized, or where the unit magnetization is allowed to build up more slowly (by using only part of the bootstrap start-up power for energizing purposes, with the remainder allowed to flow into the power system).

Studies also indicate that significant reduction in the amount of load shed would only be realized if the MHD unit(s) can be brought up to full power in less than 2 to 3 seconds,

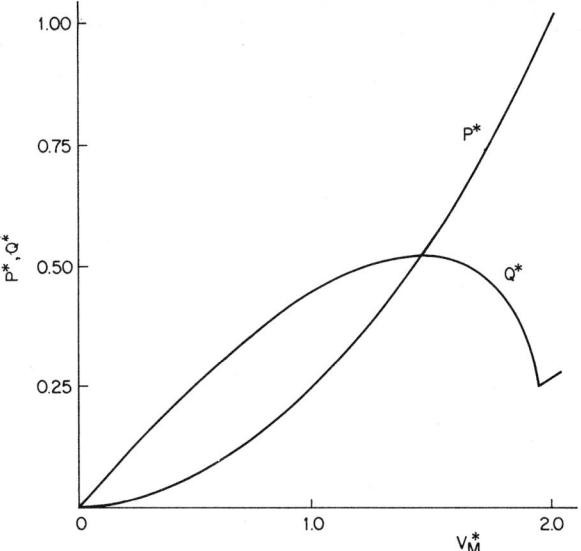

Fig. 9. Normalized powers for constant impedance plus constant extinction angle control.

depending on the percentage of load imbalance relative to the rating of the unit. A 5-second response would appear to be completely inadequate for this application. Figure 11 shows the effect of the time required for one 200-MW unit (approximately 5 percent of system load) to reach rated power. Considerable liberty has been taken in drawing curves between the few available data points.

2. Subduing Short-term Transients. MHD units (as proposed for quick-start, short-service devices) do not seem particularly appropriate for subduing short-term (0 to 3 seconds) power system swings because of several factors: (1) In order to provide the response time required to be effective, 15 MW of continuous excitation would be required to obtain quick availability (20 to 40 milliseconds) of 200 MW for each unit (1.5 MW of

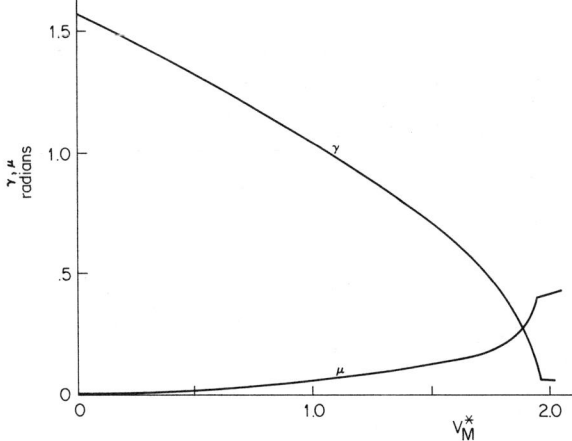

Fig. 10. Extinction angle and overlap angle for constant impedance plus constant extinction angle control.

continuous excitation power would yield a time response of about 0.7 second, which is much too slow to be effective for this purpose; and (2) a fair number of these units would have to be scattered throughout the system to provide coverage for even a fairly small number of possible contingencies. It is obvious that the costs associated with the required continuous excitation power and the number of units needed for reasonable system coverage would not make this application of MHD units attractive.

3. Use in Place of Underfrequency Load Shedding. It is generally concluded that

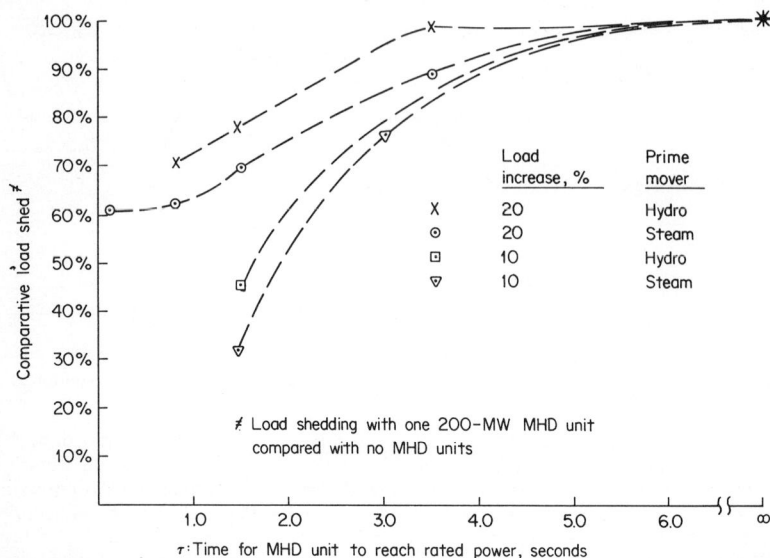

Fig. 11. Effect of time required for one 200-MW unit (approximately 5% of system load) to reach rated power.

quick-start MHD units should not be considered as an alternative to underfrequency load-shedding practices, but rather as a supplement to these practices. Underfrequency load shedding is attractive for combating rapidly dropping frequency because it is effective, has comparatively low installation costs, can be implemented independently at many scattered substations throughout the system with resulting high reliability, and can be tailored rather easily to the power system's individual requirements.

However, if a particular power system did not choose to incorporate underfrequency load shedding, a sufficient number of quick-responding MHD units could be used to boost drooping frequency during the time period (up to several minutes) when the reserve capability of the other prime movers is being realized and hence would potentially allow time for centrally controlled manual load shedding, if required. In this situation, the rating of the MHD units would be more effective than the spinning reserve of conventional prime movers because the MHD units are much faster to respond and do not suffer from such problems as the steam depletion problem inherent to many steam units after about 1 minute of significantly increased output.

4. Other Uses. In emergency situations where the power system does not split into islands, but where fast-responding local generation may be required to relieve overloaded line flows, quick-acting MHD units may be economically attractive for reducing the spinning reserve required to respond to this type of situation, particularly if installation costs of $50/kW can be realized. With such comparatively fast-acting units, line overloads could be relieved sooner, and hence the chances of encountering other system problems associated with extended line overloads would be reduced.

The potential use of such MHD units as infrequently used peaking units may be considered, but other types of peaking units, such as gas turbines and diesel units, become very competitive for usages exceeding 100 hours per year.

THERMIONIC CONVERTERS

The comparative simplicity and absence of moving parts, characteristic of MHD generators, also are present in the thermionic converter. This is based upon a different physical principle and for larger power installations at an even earlier stage of development than the MHD generator.

When subjected to a high temperature, a properly selected cathode material will emit electrons from its surface. As shown by Fig. 12, a current of J amperes will be generated

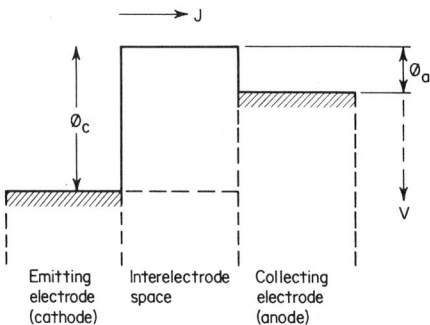

Fig. 12. Schematic diagram of operating principle for thermionic converter.

when electrons within the cathode have been elevated above their normal energy level by a minimum of ϕ_c volts, where ϕ_c represents the value of the *cathode work function*. This quantity varies with the material used and is of critical importance to thermionic power devices because it relates directly to the temperature at which the cathode must be operated. The electrons, after crossing an evacuated interelectrode gap, give up energy in the amount of ϕ_a, this value being referred to as the *anode work function*. Proper selection of electrode materials thus results in a flow of electrons at the remaining voltage V through an external load, such as a motor. In some designs, a vapor-filled space rather than a vacuum can be used. Thus, in essence, the thermionic converter is a high-temperature static heat engine that obeys the Carnot cycle. Electrons emitted from the cathode are the source of current; the heat input to the emitter is the driving force that provides that potential.

There is a tendency for an electron cloud to develop a space charge barrier in the interelectrode gap, tending to prevent the flow of electrons. This space charge effect can be reduced if the space is kept extremely small, in terms of thousandths of an inch, but obviously such designs carry serious size limitations, low power densities, and difficulties in fabrication. Some of these disadvantages can be overcome by spacing of mils or tens of mils and neutralizing the resulting space charges by introduction of positive ions in the electron cloud. The ionization of cesium atoms on the hot cathode surface of a thermionic power device is one method of doing this, or a third electrode can be introduced into wider-space devices to ionize the cesium.

Conventionally, thermionic power devices have been classified as (1) vacuum diodes, (2) cesium diodes, and (3) cesium triodes. Each type has different requirements for materials, temperature, and fabrication, and each type has different operating and output characteristics. Normally, vacuum diodes are best operated at a maximum cathode temperature in the order of 1200°C and yield about ½ watt of power per square centimeter of cathode surface at overall efficiencies in the range of 3 to 6 percent or more. Cesium diodes may produce between 10 to 20 watts per square centimeter when operated at temperatures between 1300 and 2300°C, with corresponding efficiencies approaching 10 and 30 percent. Triode devices generally occupy an intermediate position in terms of temperature, efficiency, and output.

Provided that sufficient energy can be supplied, the source of heat for the cathode is not critical. Chemical fuels, such as the fossil fuels, nuclear, and solar energy sources have entered into the development work of thermionic power devices over a number of years. Because of its basic characteristics, the thermionic power device has figured in space-vehicle power supply developments. With an input temperature of about 2000 K and an output temperature of about 1000 K, the high heat-rejection temperature theoretically makes for an ideal heat engine for use in outer space. To operate a heat engine, it is just as important to remove the rejected heat as to put heat in at a high temperature. Since the only way to remove heat in outer space is by radiation and since radiation varies as the fourth power of the temperature, it is apparent that a high reject temperature results in very lightweight and comparatively small radiator areas. Development work over the years also has concentrated on use of thermionic power devices in connection with nuclear fuel systems, particularly fission reactors designed for space application. Cathode heating as high as 2300°C is possible with thermionic nuclear fuel elements, resulting in very high density power output.

Much of the development work has concentrated on the cesium diode concept. The fuel material surface may be the cathode, or cladding around the fuel may serve the purpose. Other development work has concentrated on gas- or oil-fired thermionic converters for remote power supplies. The high emitter temperatures dictate the use of specially designed burners. Since most flames contain hydrogen, which diffuses through refractory metals, there has been the problem of obtaining a protective coating for the emitter to prevent hydrogen diffusion. Concepts for larger applications of thermionic power devices have included their possible use in connection with residence heating. A small thermionic converter may operate on a pilot light. When the large flame came on as the result of the system calling for heat, the flame would heat larger thermionic converters whose electrical output would be used to operate the blowers or pumps as may be required.

Applications suggested over the years for application of fully developed thermionic power devices have included small power plants for "peaking" in commercial power generation systems; power generation from high-temperature industrial process heat recovery; auxiliary power plants for "topping" in commercial power generation; remote power units for irrigation pumping; portable power units for lawn mowers, boats, and power tools; portable power units for recreational camp sites; remote power units for navigation aids, oceanographic data transmittal, and weather reporting; power units for automobiles; remote power units for pipeline operations; power units for worldwide space communications systems; remote power units for microwave relay stations; and, for military use, small power plants for combat area use; portable power units for communications and radar; and units for underwater operations.

In terms of electrode materials, extensive investigations have been made of tungsten, tungsten impregnated with barium, molybdenum, zirconium carbide-uranium carbide, thorium carbide, tantalum, and rhenium for cathodes; and of nickel with barium, silver oxide, tungsten oxide, stainless steel, copper, molybdenum, tantalum, and niobium for anodes.

REFERENCES

1. Brogan, T. R.: MHD Power Generation, *IEEE Spectrum*, 1:58–65 (February 1964).
2. Tsu, T. C.: MHD Power Generators in Central Stations, *IEEE Spectrum*, 4:59–65 (June 1967).
3. Avco Everett Research Laboratory, "Magneto-Hydrodynamic Emergency and Peaking Power Generation," June 1968.
4. Carter, C., and J. B. Heywood: Optimization Studies on Open-Cycle MHD Generators, *Proc. IEEE*, 56:1409–1419 (September 1968).
5. deMontardy, A., and J. Pericart: Electrical Performance of a Duct with Segmented Electrodes Under Various Conditions, *Proc. IEEE*, 56:1547–1555 (September 1968).
6. TRW Space Technology Laboratories, *Space Data*, 62–63 (1965)
7. Freck, D. V.: On the Electrical Conductivity of Seeded Air Combustion Products, *B. J. Appl. Phys.*, 15:301–310 (1964).
8. Rosa, J. J.: informal communication, July 29, 1968.
9. Hingorani, N. G., et al.: Dynamic Simulation of HVDC Power Transmission Systems on Digital Computers—Generalized Mesh Analysis Approach, *IEEE Trans. Power Appar. Syst.*, PAS-87:989–996 (April 1968).

10. Krause, P. C., and D. P. Carroll: A Hybrid Computer Study of a DC Power System: I, *IEEE Trans. Power Appar. Syst.*, **PAS-87**:970–979 (April 1968).
11. Ainsworth, J. D.: The Phase-Locked Oscillator—A New Control System for Controlled Static Converters, *IEEE Trans. Power Appar. Syst.* **PAS-87**:859–865 (March 1968).
12. Hingorani, N. G., and P. Chadwick: A New Constant Extinction Angle Control for AC/DC/AC Static Converters, *IEEE Trans. Power Appar. Syst.* **PAS-87**:866–872 (March 1968).
13. Fink, D. G., and J. M. Carroll (eds.): "Standard Handbook for Electrical Engineers," 10th ed., pp. 12-158–12-174, McGraw-Hill, New York, 1968.
14. "Power System Computer Feasibility Study," prepared by IBM Research Division, San Jose, Calif., for the Department of Public Service of the State of New York, vols. II and III, December 1968.

Rail Power Systems

BY A. N. ADDIE*

GENERAL BACKGROUND OF RAIL POWER

Systems of propulsion for rail locomotion may be divided into two categories: (1) on-board energy conversion systems and (2) stationary energy conversion supply systems. Included in the first category are vehicles employing steam engines, diesel engines, and gas turbines in combination with electric, hydraulic, and mechanical transmissions. Included in the second category are electric vehicles deriving their power from overhead wires supplied by alternating or direct current, or third rails supplied by direct current, both from a stationary electric generating plant.

Data from Ref. 1 show that in 1970 the total railroad route mileage in the world totaled about 746,160 miles, of which about 10 percent, or 74,900 miles are electrified. Table 1 shows the distribution of the various forms of locomotive propulsion by number of units employed by the principal countries of the world.

The steam locomotive was the principal energy conversion system used by railroads in the United States until the late 1940s, when the diesel-electric system began to replace the steam system. In the United States today, the trend to dieselization is complete, as is also the case for many other countries. Of the approximately 111,900 locomotives in operation in the world in 1970, 41.8 percent were diesel electric, 34.1 percent were steam, 7.4 percent were diesel hydraulic, 2.8 percent were diesel mechanical, and the remaining 13.9 percent were straight electric. The United States is, by far, the major user of diesel-electric locomotives, with 26,830 units. India has the greatest number of steam locomotives in operation, with approximately 19,590 units. West Germany leads the free world in the use of both diesel-hydraulic (3466 units) and straight electric locomotives (1945 units).

The rail mode is one of the most efficient modes for the transport of goods. Typical transport efficiencies obtained from Ref. 2 for several modes of transport are listed in Table 2.

Railroads in the United States (1971) used 3.8 billion gallons of diesel oil, accounting for about 1.65 percent of the total petroleum usage, and 3.2 percent of the petroleum used for transportation. For this expenditure of fuel energy, the railroads hauled 746,000 million net tons of freight, amounting to 38.5 percent of the total intercity freight and 9031 million passenger miles (5.7 percent of nonautomotive passenger travel). See Ref. 3.

The basic reasons for the high level of efficiency in the movement of goods in the railroad mode using modern motive power are found in two areas: (1) the inherently low resistance of the train-type vehicle using the steel wheel on the steel rail and (2) the high energy conversion efficiency of the diesel engine as a prime mover for the motive power.

RESISTANCE OF LOCOMOTIVES AND ROLLING STOCK

In the United States, it is current practice to use the resistance of locomotives and cars at speeds up to 50 miles per hour, as determined by the Davis formula (Ref. 4), which for locomotives and cars is as follows:

Lead unit:
$$R_L = 1.3 + \frac{29}{W} + .03V + \frac{.0024AV^2}{Wn}$$

Trailing:
$$R_T = 1.3 + \frac{29}{W} + .03V + \frac{.00034AV^2}{Wn}$$

*Manager, Advance Engineering, Electro-Motive Division, General Motors Corporation, LaGrange, Ill.

Freight car:
$$R_F = 1.3 + \frac{29}{W} + .045V + \frac{.0005AV^2}{Wn}$$

Passenger car:
$$R_P = 1.3 + \frac{29}{W} + .03V + \frac{.00034AV^2}{Wn}$$

where R = rolling resistance, pounds/ton
W = weight per axle, tons
V = speed, miles per hour
n = number of axles per car
A = cross-sectional area of car body and trucks, square feet

The first two terms represent those effects which are related to speed, such as wheel flange friction. The fourth term represents resistance in still air, which increases with the square of the speed. The coefficient of the air resistance term is affected by the profile of the nose of the locomotive.

At the relatively low speeds at which freight trains operate in the United States, the effect of streamlining is relatively unimportant, and the truncated wedge profile having a Davis air resistance coefficient of .0024, corresponding to the drag coefficient Cd = .938, adequately describes the air resistance of freight locomotives. At high speeds, profile drag is of predominant importance, and the drag coefficient can be reduced to about one-third of the freight locomotive by streamlining (Cd = .315). Using the Totten corrections to the Davis formula for high speed (Ref. 6), a streamlined lightweight train having a 65-ton locomotive and four 42-ton cars has resistance and power-required characteristics shown in Fig. 1. The contribution of profile drag of a lead locomotive having a cross-sectional area of 115 square feet and high degree of streamlining (Cd = .315) amounts to 2371 pounds at 160 miles per hour, or 23.8 percent of the total.

By far the most significant factor in train resistance is the effect of grade. Railroads in the United States have attempted to limit main-line maximum grades to about 2.2 percent for this reason. The resistance due to grade is the force required to lift the train vertically 1

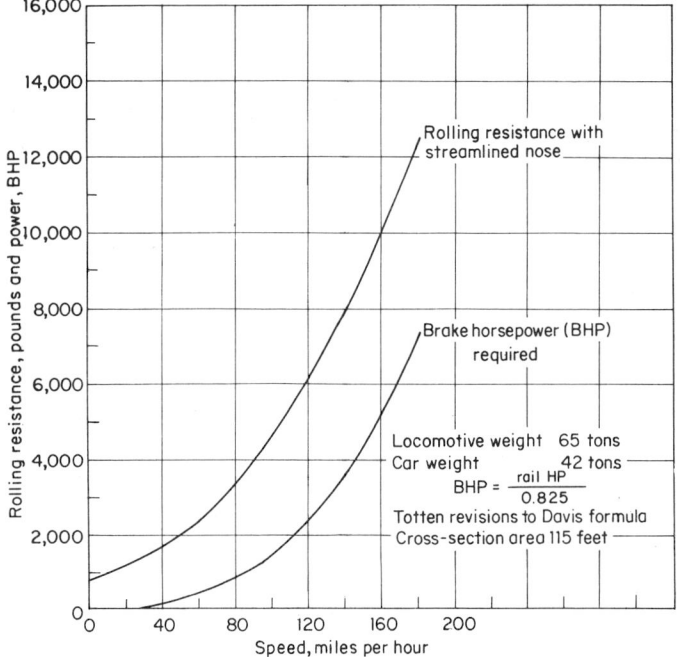

Fig. 1. Resistance and power required for streamlined lightweight train (1 locomotive plus 4 cars).

TABLE 1. World Railways Mileage and Motive Power

		Total route miles	Electric route miles	Steam	Diesel electric	Electric	Diesel hydraulic	Diesel mechanical
1.	United States	204,000	1,100	13	26,830	212		
2.	U.S.S.R.	84,000	21,040					
3.	Canada	40,458	27		3,615		11	18
4.	India	37,086	2,208	19,593	1,985	1,104	46	244
5.	Australia	26,853	532	342	1,073	76	167	57
6.	Argentina	25,122	76	2,350	357	10		861
7.	China	23,900	200					
8.	France	22,318	5,755		1,350	1,517	195	
9.	West Germany	18,861	5,435			1,945	3466	
10.	Poland	16,506	2,406					
11.	Brazil	15,269	1,601	430	738	66	52	474
12.	Japan	14,774	5,913	1,893	152	1,872	1104	18
13.	South Africa	13,302	2,706	2,448	406	1,175	33	12
14.	Czechoslovakia	13,053	521	2,700		1,150		720
15.	United Kingdom	11,799	1,965		1,680	322	409	
16.	Italy	9,950	4,919		196	1,935	159	
17.	East Germany	9,108	843					
18.	Spain	8,493	1,724	980	480	423	38	
19.	Mexico	8,436	64		964	9		
20.	Jugoslavia	7,367	938		347	403		
21.	Pakistan	7,110	178	1,048	480	29		
22.	Sweden	7,009	4,324		113	838	382	
23.	Romania	6,424	247		600+	30		
24.	Chile	5,163	476	492	282	142		6
25.	Hungary	5,029	520					
26.	Turkey	4,943	89	849	44		65	30
27.	Indonesia	4,515	68	1,057		13		271
28.	East Africa	3,663		326	55		49	18
29.	Finland	3,600	42		114	27	261	
30.	Korea	3,378		115	282			
31.	Austria	3,373	1,408		38	495	295	
32.	Bulgaria	3,350	620					
33.	Egypt	3,185	15	27	407		154	4
34.	Congo (Dem. Rep.)	3,158	529	198	65	51	77	28
35.	New Zealand	3,012	62	18	369	14	65	37
36.	Mozambique	3,102		234	42		14	
37.	Sudan	2,955		127	107		55	
38.	Norway	2,636	1,516		31	175	39	
39.	Algeria	2,634	186		145	39		30
40.	Belgium	2,588	756		516	226	457	
41.	Iran	2,464			217			
42.	Thailand	2,339		239	137		14	
43.	Nigeria	2,178		212	91		10	
44.	Bolivia	2,199	6	121	20	3	9	2
45.	Colombia	2,083		147	147			
46.	Rhodesia	2,040		238	79			
47.	Netherlands	1,955	1,023		484	146		
48.	Burma	1,910		287	33		41	
49.	Uruguay	1,866		161	103			2
50.	Switzerland	1,810	1,800		103	1,008		
51.	Angola	1,765		153	42		25	11
52.	Portugal	1,760	466		84	15	36	
53.	Cuba	1,749	11					
54.	Denmark	1,461	51		147		183	
55.	Ireland	1,361						
56.	Ecuador	1,296		54				6
57.	Morocco	1,274	454		31	54		39
58.	Tunisia	1,243			96			35

TABLE 1.   World Railways Mileage and Motive Power (*Continued*)

		Total route miles	Electric route miles	Steam	Diesel electric	Electric	Diesel hydraulic	Diesel mechanical
59.	Peru	1,122		125	64		8	10
60.	Malaysia	1,055	?	109		49	50	4
61.	Greece	969						
62.	Ceylon	954		16	48		96	
63.	Mongolia	868						
64.	Vietnam	777		12	53		10	2
65.	Ivory Coast	730						41
66.	Iraq	714		123	35			3
67.	Philippines	713			79		5	
68.	Zambia	649		92	20		18	
69.	Senegal	641			37		21	
70.	Taiwan	621		257	52			
71.	Ghana	592		110	84		10	
72.	Camaroon	590			43			
73.	Malagasy	549			33		14	4
74.	Syria	528		76				
75.	Guinea	510			7			29
76.	Guatamala	509		136	8			21
77.	Congo (Rep. Pop.)	497			57		16	18
78.	Ethiopia	485			25		10	
79.	Israel	430			24		22	
80.	Fiji	420						
81.	Mauretania	404			31			
82.	Mali	399			22			
83.	Costa Rica	386	79	3	57			3
84.	El Salvador	385		53				2
85.	Cambodia	383		24	6			10
86.	Saudi Arabia	380			25			
87.	Dahomey	359			11		6	4
88.	Malawi	343		5	10		12	
89.	Paraguay	312		29				
90.	Jordan	310		26				
91.	Venezuela	293		4	43			3
92.	Togo	275			7		4	
93.	Lebanon	259		32	6			
94.	Luxembourg	210						
95.	Dominican Rep.	207		1	11			17
96.	Jamaica	204		1	27		2	
97.	Northern Ireland	203						
98.	Nicaragua	197		1	6		1	
99.	Eritrea	191		17				8
100.	Panama	185			30		2	16
101.	Liberia	166			23		3	3
102.	Swaziland	136						
103.	Albania	135						
104.	Haiti	112		3		2		5
105.	Libya	108						
106.	Honduras	106		20	26	16		10
107.	Sabah	96		5			6	3
108.	Surinam	81		11				
109.	Sierra Leone	58					32	
110.	Nepal	64		13				
111.	Guyana	37		4	2		2	44
112.	Hong Kong	22			9			
	Total	746,164	74,899	38,160	46,698	15,591	8261	3183

Source: Ref. 1.

TABLE 2. Efficiencies of Transportation Modes in the United States Based on National Averages

Transportation mode	Fuel, Btu/net ton mile
Rail (line haul)*	680
Barge	780
Highway truck*	2,800
Cargo plane	27,000
Helicopter	187,000

*SOURCE: Ref. 2.

foot for every 100 horizontal feet traveled (1 percent grade) and is found to be 20 pounds/ (ton)(percent grade).

In addition to rolling resistance and grade resistance, the effect of track curvature must be accounted for where present. Curvature is measured in terms of the central angle subtended by a chord of 100 feet. The resistance due to curvature is approximately 0.8 pounds/(ton)(degree).

Table 3 compares the rolling resistance on level tangent track of a train composed of a six-axle, 195-ton, 3000-horsepower locomotive with 2000 trailing tons (1150 tons of cargo) with a highway 180-horsepower tractor trailer vehicle of 50 tons gross weight (28 tons of cargo).

It can be seen that the specific resistance for the highway tractor trailer is about four times that of the train. The specific resistance of the locomotive unit by itself exceeds that of the train due to the air drag resulting from its frontal area.

FACTORS AFFECTING THE PERFORMANCE OF LOCOMOTIVES

The capability of a locomotive to exert a force to overcome the rolling resistance of a train is dependent on the frictional properties of the wheel-rail interface. Because of the damage caused by the static and dynamic loads imposed on the rail by the locomotive, it becomes necessary to limit the allowable axle loading. In the United States, a customary axle load of 65,000 pounds is used, while in Europe 20 metric tons (44,000 pounds) is frequently used. Thus the maximum tractive force per axle that may be exerted is limited to a fraction of the axle load. While the adhesion (friction) which exists at the wheel-rail interface varies widely with the surface conditions of the rail, with optimum conditions, adhesion coefficients in excess of 33 percent can be obtained for starting trains. Transmissions for locomotives are frequently designed to permit the development of tractive force for a limited period of time sufficient to utilize this adhesion coefficient under starting conditions. To increase the capability of a locomotive to exert tractive force, the number of axles may be increased. Thus locomotives may be classified with regard to the number of driving axles. Small shunting or switching locomotives have as few as two axles, and large heavy road locomotives may have as many as eight driving axles.

In the United States, couplers in widespread use on freight cars have usable force limits of 250,000 pounds. Locomotives commonly are designed to utilize starting adhesion coefficients of 33⅓ percent. The most common number of axles employed per unit in heavy drag service is six, which results in a starting tractive effort of 130,000 pounds. Thus two of these units would stress the coupler to its usable force limit if full tractive effort capability of the locomotive were employed and sufficient wheel-rail adhesion were

TABLE 3. Specific Rolling Resistance Comparison

Speed, miles/hour	Tractor plus trailer (50 tons), pounds/ton*	Locomotive plus 40 cars (2195 tons), pounds/ton	Locomotive alone (195 tons), pounds/ton
20	20.2	4.6	3.1
40	25.3	6.3	6.1
60	33.5	8.99	10.5

*Computed from tractor-trailer rolling resistance relations presented in Ref. 3.

obtained. Train tonnage is normally determined on the basis of an 18 percent adhesion coefficient; hence, three 6-axle units are permissable at the head end of a freight train unless special couplers are used.

PRIME MOVER LOCOMOTIVE TRANSMISSIONS

Except in the case of small mechanical shunters, locomotives employing on-board energy conversion systems generally require a transmission to permit the prime mover to operate at its optimum rated power level throughout the locomotive speed range. In the steam locomotive, this function was performed by varying the cylinder cutoff to obtain an approximately linearly decreasing relationship of tractive effort to speed. In locomotives utilizing internal combustion prime movers, the transmission function is provided through electric or hydraulic means to obtain a relationship approximately the ideal hyperbolic-type curve of tractive effort versus speed representing constant rail power. The departure of the actual rail power curve from that which would be obtained if the rated prime mover power were available at the rail is a measure of the efficiency of the transmission and the magnitude of losses due to locomotive auxiliaries. A typical tractive effort speed curve for a modern diesel-electric locomotive showing part-load performance is shown in Fig. 2. Representative figures for the approximate best efficiency of energy conversion for a typical United States four-axle diesel-electric locomotive can be obtained from a tabulation of the component losses and efficiencies. See Table 4.

Using a typical thermal efficiency for a modern diesel engine of 38 percent at 60°F (15.6°C) and at sea level, the best full-load thermal efficiency at the rail is, therefore, 31.4 percent. The variation of transmission efficiency as a function of locomotive speed and at full power is shown in Fig. 3.

Typical duty cycles for freight locomotives in the United States are given in Table 5.

To obtain realistic values of the operating efficiency of the modern diesel-electric locomotive, it is necessary to consider not only the performance characteristics of the motive power and the resistance values for the trailing load but also the speed limits existing on a particular railroad profile. This task can be carried out most efficiently by using a digital computer to simulate the operation. To illustrate the efficiency of energy conversion in a modern diesel-electric locomotive, two railroad profiles, one typical of those existing in the Western United States (Fig. 4) and another typical of the relatively flat profiles existing in the Midwestern United States (Fig. 5), were selected for computer simulation. A three-unit consist of 3600-horsepower freight locomotives geared for a top speed of 72 miles per hour was selected to pull tonnages of 4536 tons at the minimum continuous speed of 11 miles per hour on the ruling grade of 2.2 percent over the 462 miles of mountainous profile. The same motive power was selected to pull 9648 tons at a minimum continuous speed of 11 miles per hour on the ruling grade of 1 percent over the 450 miles of a relatively flat profile. The results of the simulation for freight trains are presented in Table 6.

The Btu per gross ton mile figures resulting from the simulation represent the performance of the most modern diesel-electric motive power with no allowance for unavoida-

TABLE 4. Typical Distribution of Power Losses for Diesel-Electric Locomotive at 33.8 Miles per Hour

Factor and component	Horsepower	%
Rated diesel engine power	3244.5	100
Accessory power cooling fans, air compressor (unloaded), generator blowers, traction motor blowers, auxiliary generator power	244.5	7.54
Generator loss (95.87% efficiency)	123.9	3.81
Traction motor loss (94% efficiency)	172.6	5.32
Gear loss (99% efficiency)	27	.83
Total losses	568	17.50
Net rail horsepower	2676.5	
Overall efficiency (engine to wheels)		82.5

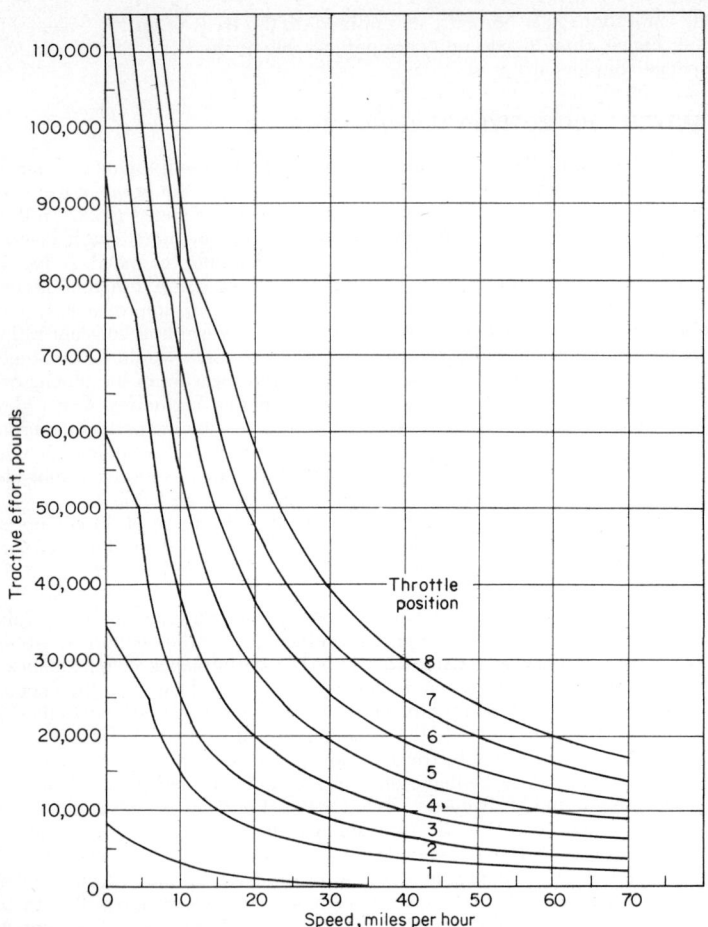

Fig. 2. Speed-tractive effort curves for SD-45 3600-horsepower freight locomotive. One AR10 genera-tor. Six D77 motors. 40-inch wheels, 62/15 gear ratio. *(Electro-Motive Division, General Motors Corporation.)*

TABLE 5.   Typical Duty Cycle for Diesel-Electric Locomotive

Throttle position	% rated power	Heavy	Medium	Light
			% of time	
8	100	30	17	9
7	86	3	4	3
6	66	3	4	3
5	51	3	4	3
4	35	3	4	3
3	23	3	4	3
2	12	3	4	3
1	5	3	4	3
Idle	0.75*	41	46	66
Dynamic brake	3*	8	9	5

*Locomotive auxiliary power only.

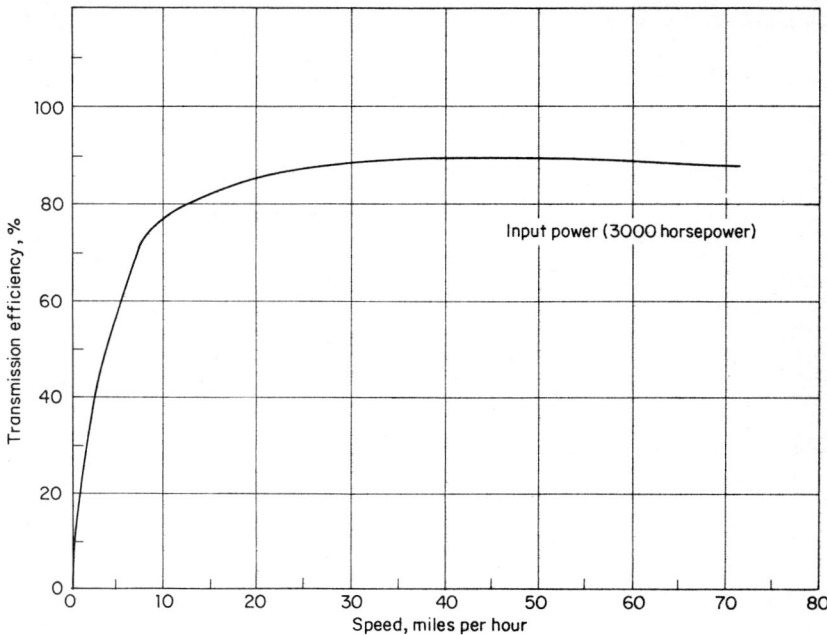

Fig. 3. Full-load transmission efficiency of a four-axle 3000-horsepower freight locomotive.

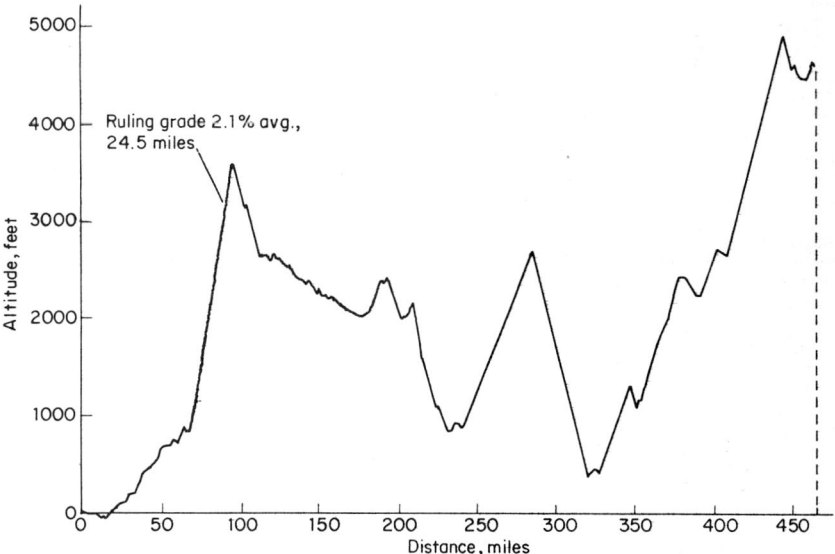

Fig. 4. Typical mountainous profile of American western railroad.

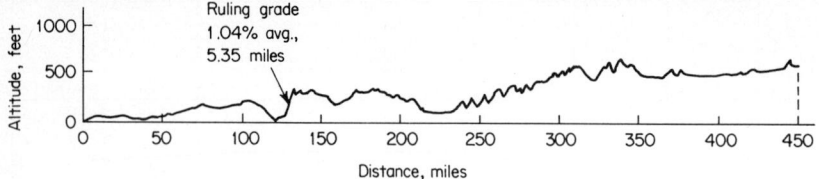

Fig. 5. Typical profile of an American midwestern railroad.

ble delays experienced in actual railroad service. A true measure of the efficiency of transport is obtained by computing the energy expended per net ton mile. Using the ratio of average loaded weight to unloaded weights of freight cars in the United States as 2.35:1, the Btu per net ton miles are 401 and 829.8, respectively, for the two conditions of Table 3. These figures compare well with the average figure of 680 Btu/net ton mile given in Table 2.

DIESEL-ELECTRIC LOCOMOTIVE TYPES

Diesel-electric locomotives have received the greatest amount of technological development in the United States and, except for a small number of electric locomotives (see Table 1), are the principal form of motive power. The most modern types of diesel-electric locomotives available in the United States (mid-1970s) are listed in Table 7.

Photographs of the locomotives are given in Figs. 6 to 11.

Although both two- and four-stroke cycle diesel engines are used in diesel-electric locomotives, the majority of diesel locomotives operating in the United States utilize naturally aspirated or turbocharged two-stroke cycle diesel engines (see Fig. 12).* These engines drive directly into an alternator (Fig. 13) equipped with a full-wave rectifier bridge which supplies rectified direct current to series dc motors (Fig. 14) mounted in the trucks. The Sprague drive, consisting of an axle-suspended traction motor (Fig. 15) with a nose support on the truck frame, and pinion located on the armature shaft meshing with a final drive gear pressed on the axle, is universally used on locomotives in the United States.

The successful application of the diesel locomotive in the United States is largely attributable to the high degree of standardization for mass production of major components including engines, generators, traction motors, and trucks, which has resulted in low cost per horsepower with excellent reliability.

TABLE 6.   Computer Simulation Data Results for Freight-Train Performance Over Contrasting Profiles

Performance factors	Flat profile	Mountainous profile
Trailing tonnage	9648	4536
Distance traveled, miles	450	462
Average speed	43.4	38.4
Elapsed time, hours	10.4	12.0
Nominal horsepower/trailing ton	1.12	2.38
Gross ton miles/trip	4,341,600	2,095,632
Total fuel consumed, gallons	5686	5679
Equivalent fuel energy, kWh	217,664	217,396
Rail energy expended, kWh	69,305	64,356
Average conversion efficiency, %	31.8	29.6
Gross ton miles/gallons of fuel	763.6	369
Btu/gross ton mile	170.6	353.1

*Built by Electro-Motive Division, General Motors Corporation.

TABLE 7.   Basic Types of Diesel-Electric Locomotives Available in the United States

Type	Rated power, horsepower	Axles, no.	Weight, pounds
Industrial switcher	1000	4	240,000
General-purpose locomotives	2000	4	260,000
General-purpose locomotives	3000	4	260,000
Heavy-duty freight locomotives	3000	6	390,000
Heavy-duty, high-speed freight locomotives	3600	6	390,000
Heavy-duty, high-speed, special-purpose freight locomotives	6600	8	550,000
Passenger locomotives	3000	6	260,000

DIESEL-HYDRAULIC LOCOMOTIVE TYPES

Diesel-hydraulic locomotives have received the greatest amount of technological development in West Germany. The invention of the hydraulic coupling and torque converter was made by Foettinger in 1905 for propeller drives on ships. In 1933, the J. M. Voith Company began to apply the hydraulic transmission to rail vehicles, and in 1939, Maybach Motorenbau collaborated with Voith to produce hydraulic transmissions for diesel-

Fig. 6. 1000-horsepower industrial switcher. (*Electro-Motive Division, General Motors Corporation.*)

Fig. 7. 2000-horsepower general-purpose locomotive. (*Electro-Motive Division, General Motors Corporation.*)

Fig. 8. 3000-horsepower general-purpose locomotive. *(Electro-Motive Division, General Motors Corporation.)*

Fig. 9. 3600-horsepower heavy-duty high-speed freight locomotive. *(Electro-Motive Division, General Motors Corporation.)*

Fig. 10. 6600-horsepower heavy-duty special-purpose freight locomotive. *(Electro-Motive Division, General Motors Corporation.)*

Fig. 11. 3000-horsepower passenger locomotive. *(Electro-Motive Division, General Motors Corporation.)*

propelled railcars and locomotives. Development continued after World War II, and the German Federal Railways decided to concentrate on electrification of high traffic density lines and to use the diesel-hydraulic locomotive for shunting, secondary lines, and less dense freight and passenger service. The majority of applications have been made using the Voith multiple hydraulic circuit torque converter principle. In diesel-hydraulic road locomotives the diesel engine generally drives the hydraulic torque converter through step-up gearing. The output of the converter is reduced in speed by step-down gearing and torque is transmitted to truck-axle-mounted bevel gear reductions through Cardan shafts. Thus the engine is permitted to run at nearly optimum speed and load while the locomotive runs through a variable speed range. An illustration of the largest and most modern of locomotives utilizing the diesel-hydraulic principle is the NY7 locomotive produced by Henschel. See Fig. 16. The NY7 locomotive is an 82-metric-ton (180,000 pounds) unit having a total of 5400 engine brake horsepower. It is capable of speeds up to 160 kilometers per hour (99 miles per hour) with peak transmission efficiency of about 82 percent.

Fig. 12. 3300-brake-horsepower turbocharged two-stroke-cycle. Diesel engine. (16-645E3, *Electro-Motive Division, General Motors Corporation.*)

Fig. 13. Alternator and rectifier. (AR10, *Electro-Motive Division, General Motors Corporation.*)

In contrast to the relatively few types of diesel-electric locomotives used in the United States, there are a large number of diesel-hydraulic locomotive types applied in Germany and throughout the world.

ELECTRIC LOCOMOTIVE TYPES

As shown in Table 1, about 10 percent of the world's railroad route miles are electrified. Electrification of railroad routes began in Europe and the United States in the late 1800s and early 1900s. Early locomotives were built to run on various types of current, including 600-volt dc, 11,000-volt ac, 25-cycle, and 3000-volt dc. The only remaining freight and passenger electrified operations in the United States at the present time is on the Penn Central, which currently operates GG-1 locomotives built by Baldwin-Westinghouse

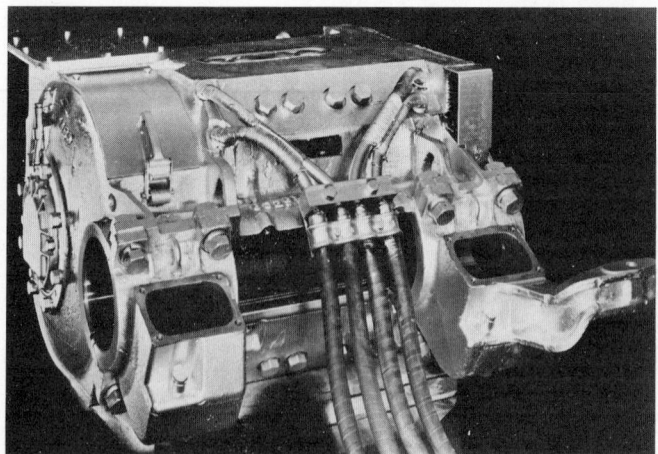

Fig. 14. Direct-current traction motor. (D77, *Electro-Motive Division, General Motors Corporation.*)

Fig. 15. Three-axle truck with direct-current traction motors. (*Electro-Motive Division, General Motors Corporation.*)

Electric Company and General Electric Company and E44 rectifier locomotives using dc series motors built by General Electric Company.

Modern electric locomotives utilize the silicon-controlled rectifier (thyristor) as the power-rectifying and control device to transmit power from the transformer to the dc traction motor.

The ASEA (Sweden) RC_2 locomotive (Fig. 17), which embodies the most advanced technology in electric locomotives, operates on 25-kV catenary with a full rectification and control of the current supply to separately excited, dc, truck-frame-mounted motors using thyristor bridges.

Electric Locomotive Transmissions. In contrast to the diesel-electric locomotive, which transports its own energy conversion equipment, the electric locomotive merely receives electrical energy from the generating station and transmits it in the appropriate form to the propulsion motors. The first electrification on the main line began on the London, England, underground railways in 1890 and used 500-volt dc third-rail supply to supply 100-horsepower locomotives for passenger service. In the United States in 1894, the Baltimore and Ohio Railroad employed a 675-volt dc overhead trolley to supply current to 1440-horsepower locomotives. In 1899, the Burgdorf-Thun Railway in Switzerland utilized 750-volt, three-phase alternating current to power 300-horsepower locomo-

Fig. 16. Diesel hydraulic locomotive, 5400 brake horsepower. (*NY7, Henschel Lokomotiven.*)

tives. The first extensive use of electrification for freight and passenger operations began in 1902 when 15-cycle, 3000-volt, three-phase alternating current was used to power 1000-horsepower locomotives on the Simplon Line in Switzerland.

The obvious difficulty with three-phase power distribution on a railroad is the problem of current collection. This problem is much simplified when only a single phase is utilized, but the problem then becomes one of producing a series traction motor with safe commutation. To obviate this difficulty, motor generator sets were used to convert the single-phase alternating current to 700 volts for the dc series traction motors. Somewhat later a satisfactory low-frequency (15 hertz) single-phase commutator-type ac motor was developed which enabled the motor generator converter to be eliminated. Locomotives

Fig. 17. RC_2 Electric locomotive. *(ASEA, Sweden.)*

operating on 11 kV, 25 hertz utilizing a tap-changing transformer to feed single-phase series traction motors were widely used from 1915 to the present on what is now the Penn Central Railroad.

A major advance in the technology of electric locomotives occurred when means became available to rectify the output of the transformer and to supply rectified alternating current to dc series traction motors similar in design to those used in diesel-electric locomotives. In the United States this was first accomplished on the Pennsylvania Railroad in 1951 using the ignitron (mercury arc) rectifier. Further technical advances have been made in recent years, particularly in Sweden by the ASEA Company, which in 1967 introduced the first thyristor (silicon-controlled rectifier) locomotive.

Thus, the most modern electric locomotive receives power from a single-phase, high-voltage (11, 25 or 50 kV) catenary system at commercial frequency (50 or 60 hertz) and transforms the voltage to a lower level. The current at this lower voltage is rectified in one or more rectifier bridges using combinations of diodes and thyristors to control the duration of current flow to dc traction motors. Series or separately excited field motors may be used in modern thyristor locomotives; however, there are significant advantages to be gained in locomotive adhesion by separately exciting the field and controlling each motor individually.

Typical of the characteristics of a modern electric locomotive are those of the ASEA RC_2 locomotive shown in Fig. 18. In contrast to diesel-electric and diesel-hydraulic locomotives, which have tractive effort curves reflecting nearly constant power output from the prime mover, the electric locomotive operates at constant tractive effort (constant current) up to a base speed (40 miles per hour for the RC_2), after which the line voltage is reached. As speed increases, the tractive effort decreases with motor field weakening as required by the constant line voltage.

Maximum full-load efficiency of electric locomotives of the RC_2 type, including all auxiliaries, is approximately 85 percent.

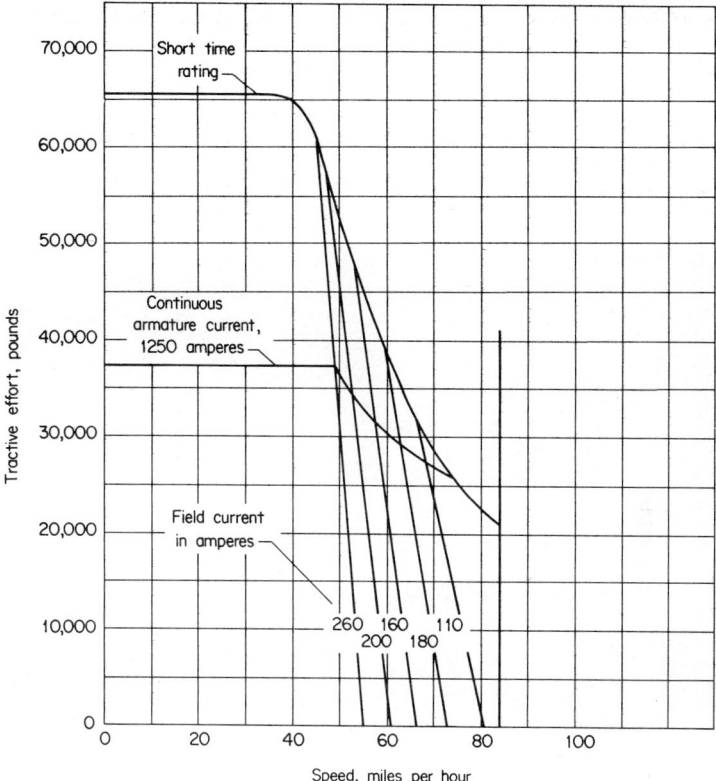

Fig. 18. Speed-tractive effort curves for the ASEA RC_2 electric locomotive.

In the United States, the most recent application of thyristor technology to electric freight locomotive propulsion is found in the GM-6-C locomotive shown in Fig. 19. This unit, built by Electro-Motive Division of General Motors Corporation, is a six-axle Diesel equivalent horsepower unit and is intended for heavy freight operations under 12 kV, 25 Hz, or 50 kV, 60 Hz catenaries. See Fig. 19.

The energy conversion efficiency of a diesel-electric locomotive from fuel to rail is about 31.4 percent. The corresponding energy conversion efficiency of an electrified system must include the electrical power generation plant, substation, catenary, and locomotive. A typical full-load conversion efficiency for an electrification system, assuming no transmission line losses, is indicated in Table 8. It will be noted that the overall peak energy conversion efficiency of an electrified system closely approximates the corresponding value for a diesel-electric locomotive.

RAIL POWER SYSTEMS OUTLOOK

Railroad motive power evolved from inefficient steam engine propulsion to efficient diesel-electric and diesel-hydraulic propulsion. There have been several attempts to

TABLE 8. Electric Railroad Efficiency

Steam power plant	0.38
Substation	0.99
Catenary	0.95
Locomotive	0.85
Overall conversion	0.304

apply the gas turbine prime mover to locomotives, both with direct drive and electric transmissions. In the field of freight operations, the gas turbine has met with only limited success, largely as a result of its inferior fuel consumption, and has yielded to the diesel-electric system. Specific propulsion systems generally find application in those areas where they afford economic advantage. With the current concern for the conservation of liquid petroleum fuels and the rapidly escalating costs of this energy source, only those propulsion systems which excel in efficiency, utilize alternate sources of energy, or offer a specific advantage not capable of being matched by competive systems will survive. The diesel-electric propulsion system is the most efficient system yet devised for locomotive

Fig. 19. 6000 Diesel-equivalent horsepower electric freight locomotive. *(Electro-Motive Division, General Motors Corporation.)*

propulsion. Though extremely reliable, it is a relatively bulky and heavy system. It has proven to be ideally suited to yard switching and secondary-line operations as well as main-line drag freight operations. Up to about 120 miles per hour, the diesel electric is a strong competitor for passenger train service. It may be anticipated that the diesel-electric system will be predominant in the United States for some time to come. As a long-range possibility, electrification is a viable alternative which must be assessed comparatively against the diesel-electric system on an economic basis, including energy costs and fuel availability.

In the field of short distance (about 300 miles) intercity passenger travel, the high-speed passenger train may become a viable alternative to the airplane, providing that investments are made to upgrade the unsatisfactory conditions of much of the main-line trackage in the United States. If the railroad track is maintained in sufficiently good alignment to permit speeds up to 200 miles per hour, the gas-turbine electric locomotive may be a strong contender for this market because it has the attributes of light weight, high power, compactness and freedom from the difficult problems of current collection at high speeds, which limits electric locomotive operation.

The RTG gas turbine train developed by ANF Frangico and the SNCF is in operation in the United States under a lease agreement with AMTRAK. It is capable of a top speed of 125 miles per hour. This five-car train weighs 250 tons including 300 passengers, which is about half the equivalent weight of a diesel-electric-powered conventional train. Total propulsion power is 4560 horsepower supplied by four engines, two of which are installed in each power car. Power is transmitted via a Voith hydraulic transmission to one truck on each power car. Advanced forms of this type train are being evaluated at speeds up to 185 miles per hour by the French SNCF.

At speeds above about 200 miles per hour, it is generally conceded that propulsion via the steel wheel on the steel rail becomes impractical because of dynamic loading of the rail and danger of derailment. For propulsion above this speed, systems employing means for support of the vehicle independent of the means for propulsion may be necessary. Vehicle support may take the form of magnetic levitation, both attractive and repulsive arrangements, or by means of air cushions. Propulsion can be provided by a reaction system employing gas-turbine-driven propeller, such as has been employed on the Bertin Aerotrain or by jet engines as in the Grumman TACV. Alternatively a linear induction motor can be employed whereby on-board ac electrical energy is supplied to a primary coil and an induced field is set up in the stationary track system. Interaction of the stationary and moving magnetic fields result in propulsive forces. Experimental development of these systems are under way under the title Tracked Air Cushion Vehicle (TACV) financed by the U.S. Department of Transportation.

The magnetic levitation (MAGLEV) approach to vehicle support is being developed in West Germany by Messerschmitt-Bolkow-Blohn (MBB) and also by Krauss-Maffei. Both groups are using the attraction principle which requires a control system with rapid response to regulate the clearance gap (10 to 20 millimeters). Lift power is 1 to 2 kW per metric ton compared to about 20 kW per metric ton for air cushion vehicles.

In order to operate with large clearance gaps, repulsive systems are used which require powerful magnetic fields obtained with superconducting, cryogenically cooled systems. The Japanese National Railways has achieved clearance gaps up to 100 millimeters using this system, and Siemens-AEG-BBC in Europe is currently developing this system.

Success of the hover train principle for intercity trains appears to be keyed to the need for high vehicle speeds since the expenditure of lift energy does not permit them to be competitive with wheeled vehicles in the low speed range.

REFERENCES

1. "Jane's World Railways, 1971–1972," McGraw-Hill, New York, 1972.
2. "Battelle Topical Report on Energy Requirements for the Movement of Intercity Freight," for Association of American Railroads, Battelle Memorial Institute, Columbus, Ohio, Dec. 15, 1972.
3. "Yearbook of Railroad Facts," Association of American Railroads, Washington, D.C., 1973.
4. Davis, W. J.: The Tractive Resistance of Electric Locomotives and Cars, *Gen. Electr. Rev.*, **29** (October 1926).
5. Hay, William W.: "Transportation Engineering," Wiley, New York, 1961.
6. Hay, William W.: "Railroad Engineering," vol. 1, Wiley, New York, 1953.

Thermoelectric Cooling

BY R. J. BUIST*

Thermoelectric cooling has much in common with conventional refrigeration methods. In a conventional refrigeration system, the main working parts are the freezer, condenser, and compressor. The freezer surface is where the liquid refrigerant boils, changes to vapor, and absorbs heat energy. The compressor circulates the refrigerant above ambient level. The condenser helps discharge the absorbed heat into surrounding ambient.

In thermoelectric refrigeration, the freezer surface becomes cold through absorption of energy by the electrons as they pass from one semiconductor to another instead of energy absorption by the refrigerant as it changes from liquid to vapor. The compressor is replaced by a direct-current power source which pumps the electrons from one semiconductor to another. A heat sink replaces the conventional condenser fins, discharging the accumulated heat energy from the system. A thermoelectric cooling system refrigerates without the use of mechanical devices, except perhaps in the auxiliary sense, and without refrigerant.

ELEMENTS OF THERMOELECTRIC COOLERS

The components of a thermoelectric cooler can be shown best by way of a cross section of a typical unit, as shown in Fig. 1. Thermoelectric coolers such as this are actually small heat pumps which operate on physical principles established over a century ago. In a thermoelectric cooler, semiconductor materials with dissimilar characteristics are connected electrically in series and thermally in parallel so that two junctions are created.

The semiconductor materials are N- and P-type.† Heat absorbed at the cold junction is pumped to the hot junction at a rate proportional to carrier current passing through the circuit and the number of couples. Good thermoelectric semiconductor materials, such as bismuth telluride, greatly impede conventional heat conduction from hot to cold areas yet provide an easy flow for the carriers. In addition these materials have carriers with a capacity for carrying more heat. Only since refinement of semiconductor materials in the early 1950s has thermoelectric refrigeration been considered practical for many applications.

CHARACTERISTICS AND OPERATIVE RANGES

In practical use, couples similar to the single couple shown in Fig. 1 are combined in a module where they are connected in series electrically and parallel thermally, as evident from Fig. 2. Normally, a module is the smallest component available. The user can tailor quantity, size, or capacity of the module to fit exact requirements without procuring more total capacity than actually is needed. Modules are available in a wide variety of sizes, shapes, operating currents, operating voltages, number of couples, and ranges of heat-pumping levels. The present trend is toward a larger number of couples operating at a low current.

Uses for thermoelectric coolers can be grouped into three categories: (1) electronic components, (2) temperature control units, and (3) medical and laboratory instruments, with numerous examples of use in each category. Thermoelectric coolers are capable of operating from +100°C to −125°C. Special units can be fabricated to withstand tempera-

---

*Engineering Manager, Thermoelectrics Department, Borg-Warner Corporation, Des Plaines, Ill.

†Semiconductor materials that have more electrons than necessary to complete a perfect molecular lattice structure are known as N-type. Where there are not enough electrons to complete a lattice structure, they are known as P-type. The extra electrons in the N-type material and the holes left in the P-type material are called *carriers;* they are the agents that move the heat energy from the cold to the hot junction.

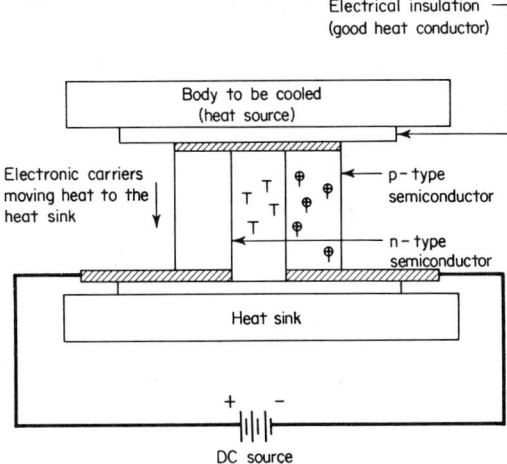

Fig. 1. Cross section of a typical thermoelectric cooler.

tures in excess of 150°C. Simple electronic control schemes allow control within a fraction of a degree of desired load temperatures above or below ambient.

Modules normally contain from 2 to 71 couples with ceramic-metal laminate plates (Fig. 3) at both the hot and cold junctions to provide good thermal conduction and good electrical insulation. A module has a single pair of connecting leads.

If modules are to be used in cooling chambers or large components, a total surface area of virtually any size can be made by placing the appropriate number of modules side by side. The interfaces at the cold junction and the hot junction must be constructed to transfer heat in and out of the module with little difference in temperature. This is accomplished with metal-ceramic laminate plates that give strength and permit good thermal bonding between the two interfaces. The outer plate surface is usually tinned to facilitate soldering to heat sinks. Where soldering is not practical, as in the case of thermal expansion differences, heat transfer grease is recommended. Epoxy bonding agents are available where a more permanent solderless bond is required.

CASCADING

The single-stage module is capable of pumping heat where the difference in temperature of the cold junction and the hot junction is 70°C or less; however, in those applications

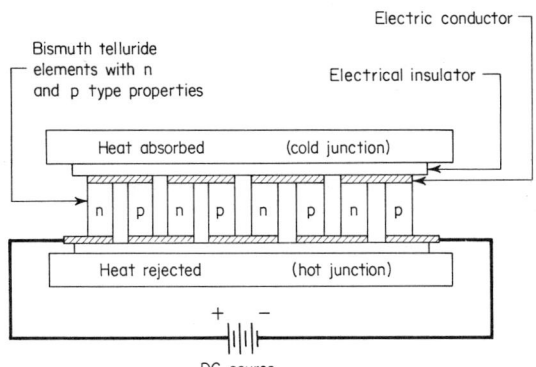

Fig. 2. Typical module assembly. Elements are electrically in series and thermally in parallel.

which require higher delta T's, the modules can be cascaded. Cascading is a mechanical stacking of the modules so that the cold junction of one module becomes the heat sink for a smaller module placed on top.

In addition to the heat pumped by any given stage, the next lower stage must also pump the heat resulting from the input power to that upper stage. Consequently, each succeeding stage must be larger—from the top of the cascade downward.

With any given set of heat sink and cold spot temperatures, there exists an optimum heat-pumping capacity or "size" ratio between each adjacent pair or stages. The optimum size ratio increases as the overall delta T increases but decreases as the number of stages (N) increases. It is not necessarily a constant from stage to stage even with delta T and '

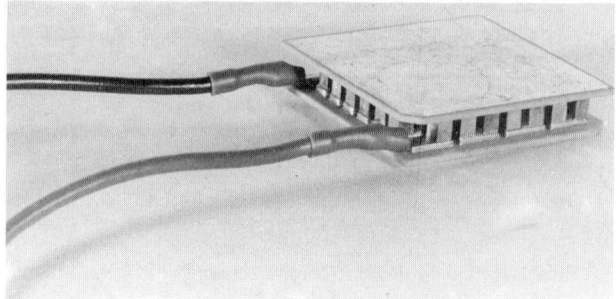

Fig. 3. Modules contain from 2 to 71 couples with ceramic-metal laminate plates at both the hot and cold junctions.

fixed. Optimization of a cascade design requires accurate temperature-dependent data on the thermoelectric materials in combination with a computerized numerical design theory. Examples of optimized units are shown in Fig. 4.

Standard Cascades. Applications requiring low-temperature thermoelectric coolers usually have very strict limitations on available power. Therefore, it is not practical to fabricate and stock numerous different cascades which can be optimized for only one set of conditions. On the other hand, fully optimized prototypes involve engineering and manufacturing costs which may prove uneconomical for some applications. An alternative approach* has been developed for responding to such requirements. Standard cascades are fabricated by assembling partials of standard modules. The number of different standard cascades is virtually unlimited due to "free" variables, such as number of stages, couple distribution, and the basic building block module.

To determine the best standard cascade for a given application the desired hot side temperature, cold side temperature, and thermal load are entered into a computer. The result is a listing of numerous standard coolers which meet these specifications with various combinations of input power and cost. Generally, the lowest input power devices are of higher cost, and vice versa. Thus, cost/performance trade-offs are immediately discernible. Examples of some typical standard cascades are shown in Fig. 5.

POWER SUPPLY

These range from the simple open-loop direct current supply with a switch to feedback systems with close temperature regulation and fast response. Portable power supplies are made that will operate from both ac and dc power. The only limitation on the supply is that ripple be maintained at a point lower than 10 to 15 percent. Open-loop systems will generally contain a transformer, rectifiers, choke, and chassis with heat sink for the rectifiers. In feedback systems, a thermistor is used to sense temperature at the cold junction, and this signal is compared with the desired temperature setting to obtain an

*Borg-Warner design system.

Fig. 4. Examples of optimized thermoelectric cooling units.

Fig. 5 Typical standard cascade.

error signal. The amplified error signal determines the power applied to the thermoelectric cooler.

EXTRANEOUS HEAT SOURCES

The load to be cooled should be isolated from other sources of heat to obtain maximum efficiency. Other than the load, there are three losses to consider when applying thermoelectric coolers: (1) conduction losses, (2) convection losses, and (3) radiation losses.

Conduction loss is directly proportional to the temperature difference between the hot and cold junctions and to the thermal conductivity of the materials in between. Thus, at large delta T's, conduction losses increase in importance. Convection losses are greatly reduced when the cold plate is protected from the gaseous environment by some form of insulation. Roughly, convection losses are equivalent to 1 milliwatt multiplied by the area of the cold plate in square centimeters and the temperature difference in degrees Celsius between the cold plate and the ambient [1 mW/(cm²) (°C)]. Radiation losses are approximately equivalent to 50 milliwatts per square centimeter, at cold plate temperatures near −75°C in an ambient of +27°C.

When devices operate in an evacuated enclosure, convection losses are virtually eliminated. If shielding is used, radiation losses are also reduced. To keep convection losses at a low level, a vacuum better than 10^{-3} torr is required.

Electric Power Production and Requirements

STATISTICAL BACKGROUND

Where statistics are included in this Handbook, the editors believe that stress should be given to the fact that rarely are statistics for any given segment of energy and power technology the full domain of a central collecting agency. Rather, there are several private and governmental organizations involved in both the collection and interpretation of data. In terms of installed physical facilities, particularly in the United States, data in the electric power field are quite reliable and reasonably current. This type of information is not subject to widely divergent opinions as in the case of estimating coal, petroleum, or natural gas reserves. However, in terms of forecasting future installations and requirements, this may be even more tenuous because of the large number of factors, such as technological advancements, raw energy availability, environmental restrictions, and even national and world political decisions, which in recent years have been subject to considerable oscillation. The nuclear power situation in the United States is covered in more detail under "Nuclear Fission Reactors" in the Handbook section *Nuclear Energy Technology.*

WORLDWIDE ELECTRIC POWER

As shown by Fig. 1, nine nations generate and consume over three-fourths of total world electric power. As of 1971, the United States led the world in terms of installed generating capacity, with a total of 386.7 million kilowatts. This is more than double the second leading nation, the U.S.S.R., with an installed capacity of 175.4 million kilowatts. Not shown on the chart with the United States, the U.S.S.R., Japan, the United Kingdom, the Federal Republic of Germany (West), Canada, France, Italy, and Spain, are the German Democratic Republic (East), with 1.3 percent of the 1970 electric power production of 58.8 billion kilowatthours; Norway, about 1.3 percent, 57.2 billion kilowatthours; Sweden, about 1.3 percent, 57 billion kilowatthours; Poland, about 1.3 percent, 55.7 billion kilowatthours; and, in terms of generating capacity, the People's Republic of China, estimated at 17 million kilowatts; and India, estimated at 15.5 million kilowatts. Significantly, the installed generating capacity (name plate) of the United States had increased to 495.4 million kilowatts by the end of 1974. It is important in considering generating capacity figures to note that capacity is measured in terms of *kilowatts*, whereas electric power production is measured in terms of *kilowatthours*. Since all generating equipment is not on line 100 percent of the time, this varying with locale and climate, there is not always necessarily a direct relationship between capacity (ability to produce) and the actual amount produced.

Probably more significant in terms of industrial and residential life style is the kilowatthours per capita figure, in which Canada leads the world with 9936 kilowatthours produced annually per Canadian citizen. With exception of the last three of the nine leading electric-power-producing nations, shown in Fig. 1, the ranking per capita, as shown in Fig. 2, differs from the ranking for total productive capacity. Unfortunately, detailed requirements or electric power use data are not readily available for the nations of the world, as they are given later in this description for the United States, but there is no question that electric power production and consumption reflect the degree of industrialization, including power-consuming mechanization and automation, the degree of residential conveniences—appliances, lighting, communications, heating—the extent of certain specialized high power-consuming industries, such as the mining and metallurgical fields (aluminum, magnesium, steel, etc.), and obviously to some degree, a wasteful attitude toward electric power.

Annual use of electricity by the worker in the United States averaged an estimated 45,751 kilowatthours in 1971, an increase of 3739 kilowatthours over 1970. A steady increase (data available since 1961) in this figure is indicated by Table 1. The total amount

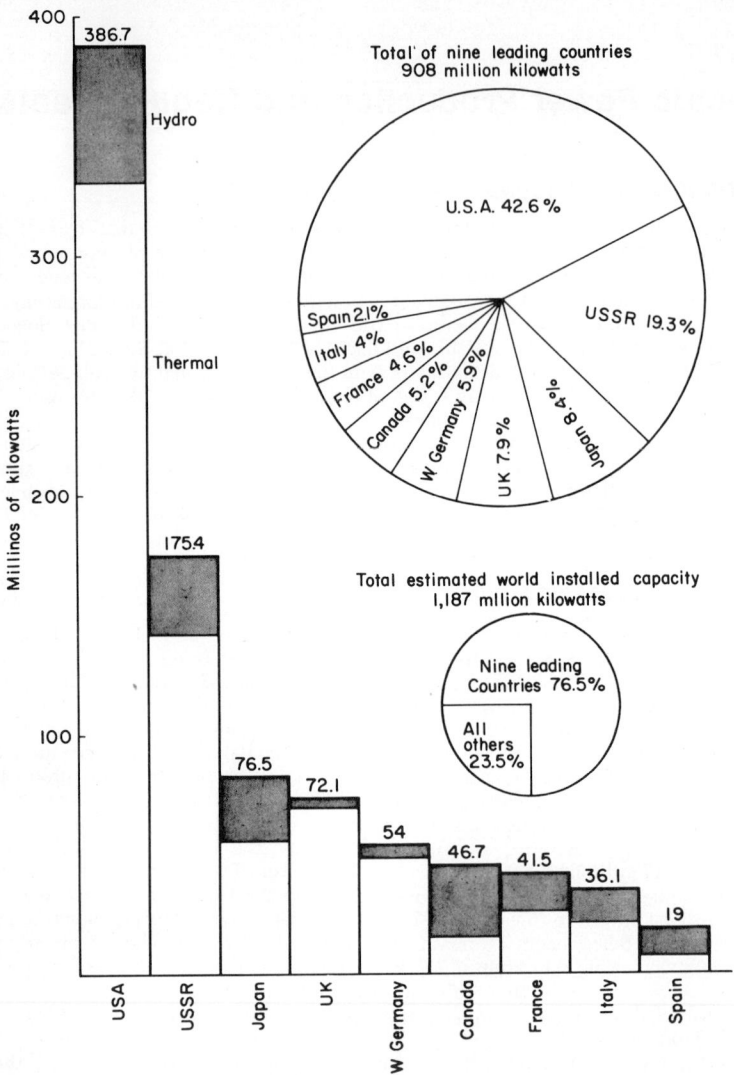

Fig. 1. World electric power generating capacity. The nine countries with the greatest installed capacity (Early 1970s) are shown. The top, shaded portion of each bar indicates that portion of the total electric power generated from hydro sources. The lower portion of each bar indicates all thermal systems, including nuclear fuels. The figures for the United States do not include nonutility plants, such as captive industrial power plants, captive mine power, and railway electric power plants. Also, the net imports for the United States are not reflected by these figures. (*U.S. Federal Power Commission, Washington.*)

of electricity used by United States manufacturing industries has gone up approximate 68 percent since 1961. No worker, by sheer muscle power, can produce in a day th energy represented by just 1 kilowatthour of electricity. The average power which a person can exert is calculated at about 35 watts. Thus, a person averaging 250 eight-hour days of manual work a year is estimated to expend energy equivalent to about 67 kilowatthours. In 1971, therefore, a factory worker using 45,751 kilowatthours annually had the equivalent energy of 683 people helping on the job all year long.

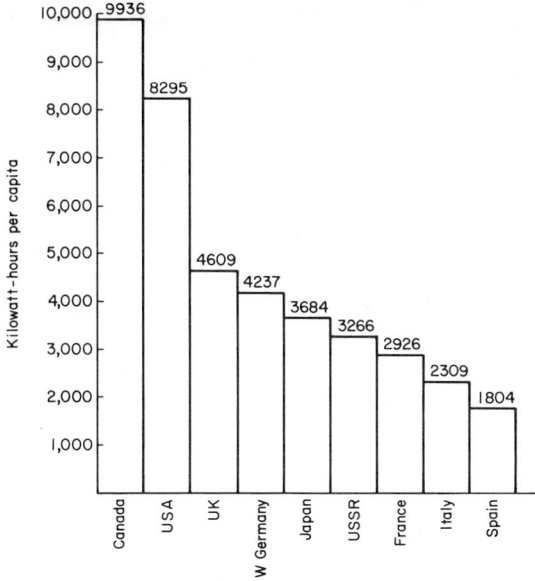

Fig. 2. Kilowatt-hours of installed capacity per capita for the nine leading electric power producing nations. (*U.S. Federal Power Commission, Washington.*)

ELECTRIC POWER GENERATION IN THE UNITED STATES

The growth of electric power generation in the United States is shown in Fig. 3, with an indicated production in 1975 of 1906 billion kilowatthours.* This is equivalent, of course, to 1.906 trillion kilowatthours. In determining fuels requirements for the electric power

TABLE 1.   Electric Energy Used by Average Production Worker in the United States

Year	Estimated production workers in the manufacturing industries, thousands	Electricity consumed by manufacturing industries, millions of kilowatthours	Average annual kilowatthours per worker
1971	13,487	617,046	45,751
1970	14,033	589,560	42,012
1969	14,767	571,688	38,714
1968	14,514	533,803	36,778
1967	14,308	505,855	35,355
1966	14,297	482,599	33,755
1965	13,434	453,881	33,786
1964	12,769	437,132	34,234
1963	12,558	406,461	32,367
1962	12,488	388,385	31,101
1961	12,085	367,751	30,430

Basic data source: U.S. Census information.

Kilowatts measure the production capacity or capability of electric generators and also the power requirement of electrical equipment and appliances. Electric generators now operate in sizes from less than 100 kilowatts to 1,150,000 kilowatts. The average size of units under construction is about 500,000 kilowatts, but the trend is increasing. *Kilowatthours* measure the amount of electricity generated and also the amount consumed. A 100,000-kilowatt generator running at full capacity for 1 hour will generate 100,000 kilowatthours of electricity. Since there are 8760 hours in a year, such a generator operated continuously at full capacity for the whole year will generate 100,000 × 8760 or 876,000,000 kilowatthours in a year.

generation industry, the National Petroleum Council's "Electricity Task Group" has estimated that the utility generation of electric power will total nearly 10 trillion kilowatthours by the year 2000. Forecasts of this type are generally based upon on estimate of growing requirements and the extrapolation of past requirements curves. Numerous factors contribute to what is no less than an amazing growth rate, such factors including a continuation of industrialization (extensions of mechanization and automation), large

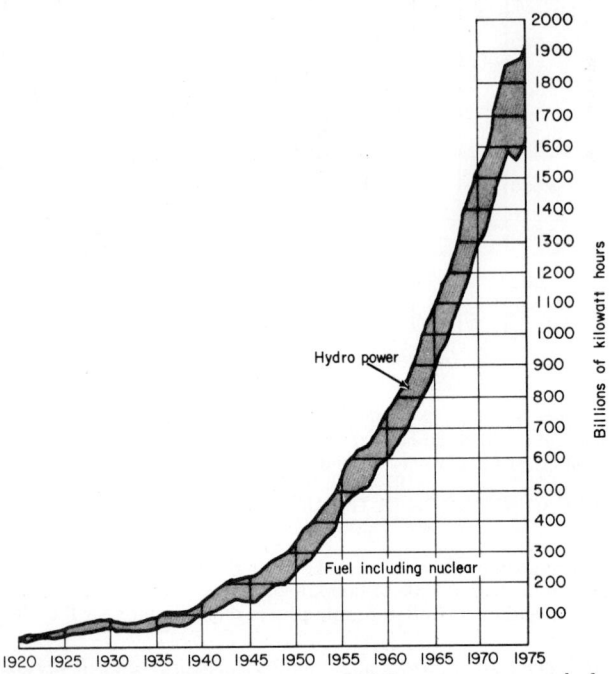

Fig. 3. Electricity generated in the United States. Shaded portion represents hydro power. Lower portion of chart includes all thermal energy (coal, fuel oil, natural gas, SNG, nuclear). *(Source: U.S. Federal Power Commission, Washington.)*

quantities of electrical energy required for environmental protective measures, population increase, a continuation of residential utilization of electrical energy. Well-constructed forecasts also take into consideration as much as may be practical the development timetable of raw energy sources, possible restrictions on energy consumption, and certainly not least among such factors, the availability of capital to expand electric-power-generating and distributing facilities. Exemplary of the complexities which enter into long-range electric power forecasts is the fact that the aforementioned task group established six possible and differing conditions, any one of which may arise in actuality—or quite possibly a new set of conditions not cranked into the forecast. Thus, the user of this Handbook who is concerned with long-range forecasting in this field is urged to contact all major data sources several times each year in order to keep up with the continuing changes (and even surprises) that are bound to occur during the next couple of decades. See the list of references at the end of this description.

The ubiquitous nature of electric power in the United States is dramatized by the backbone transmission map showing the power supply regions as defined by the U.S. Federal Power Commission. See Fig. 4.

The Electric Power Survey Committee of the Edison Electric Institute, in cooperation with representatives of the electric power systems or power areas throughout the contiguous United States and the nation's principal manufacturers of heavy electric power equipment, issues an "Electric Power Survey" twice each year. Contained in this data-packed survey are charts of peak capabilities, peak loads, and gross margins. Summer and

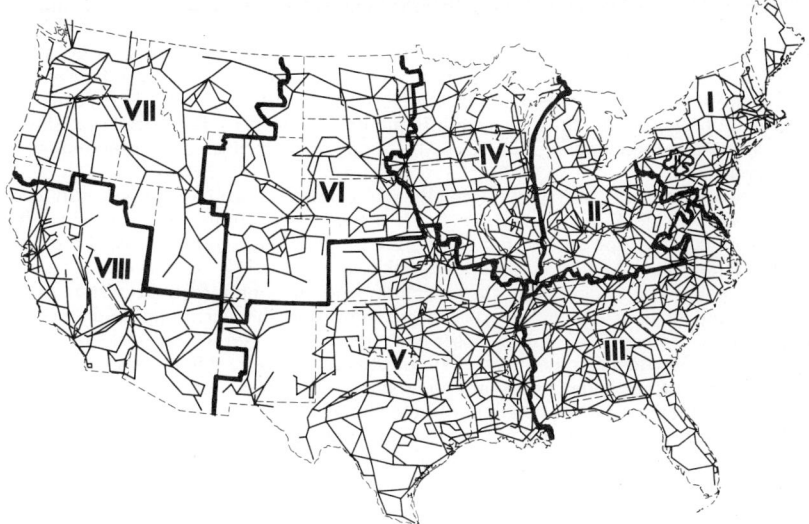

Fig. 4. Backbone transmission map showing the power supply regions as defined by the U.S. Federal Power Commission, Washington.

winter conditions are reported separately. The historical development of December or winter peak capabilities and gross margins for the total electric utility industry of the United States, excluding Alaska and Hawaii, is indicated in Fig. 5. Definitions of terms used in these surveys include:

Capability.   The capability of the power systems is defined as the maximum kilowatt capability of the systems with all power sources available, with no allowance for outages,

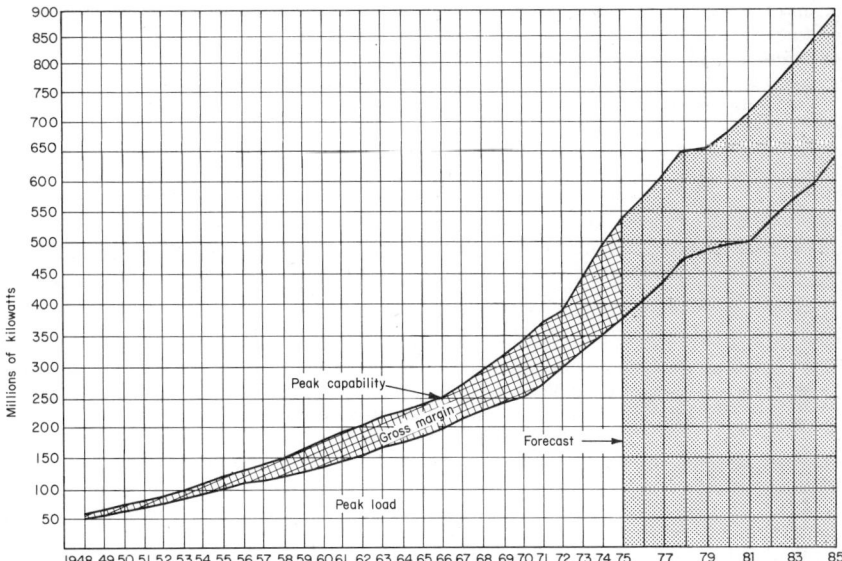

Fig. 5. Historical development of December peak capabilities and gross margins for total electric utility industry of the United States, excluding Alaska and Hawaii. (*Source: Edison Electric Institute, New York.*)

and (for systems including hydroinstallations) with sufficient kilowatthours to supply the energy requirements of the system.

The capability must provide for scheduled maintenance, emergency outages, and system operating requirements, in addition to the estimated load and any unforeseen load. The capability of existing installations is determined by the dependable capacity at the time of the peak load and the energy requirements of the system. Estimated capabilities of those systems served wholly or in part by hydropower sources are determined for median hydroconditions.

The capabilities of all systems normally interconnected are based upon fully coordinated operation. Peak capability refers to the maximum capability, as previously defined, at the approximate time and for the duration of the load referred to, that is, at the time of the December peak load or the summer peak load.

Peak Load.   This is defined as the maximum load encountered on the system during a given period of time. December peak loads represent the maximum load during the winter period. Generally, there is closer coincidence between winter peaks on different systems than between summer peaks. Winter peaks are governed in part by events related to the calendar (preholiday and year-end activity), whereas summer peaks are governed by local weather conditions which may vary somewhat widely with locations.

Gross Margin (or Capability Margin).   This represents the difference between capability, as previously defined, and peak load. Gross margin, therefore, should be sufficient to provide for scheduled maintenance, emergency outages, and system operating requirements, if practical operating conditions are to be maintained. Any excess in the gross margin over and above the provision for these items is available for unforeseen loads.

The 1974 year-end capability of the industry was 466.5 million kilowatts, an increase of 34.0 million kilowatts, or 7.9 percent above the capability at the end of 1973. The December capability for 1975 for the entire electric utility industry approximated 504.3 million kilowatts (under median hydro conditions). This represented an increase of 37.8 million kilowatts, or 8.1 percent over the corresponding 1974 capability. Forecasts for 1976 through 1980 show slightly reduced values, as much as 1.9 percent below those expected for these years when estimated as recently as mid-1975.

For December 1985, the total capability of the industry under median hydro conditions is forecast to reach 875.7 million kilowatts. This gives an average annual increase for the 11-year period from the end of 1974 of 5.9 percent. The forecast also considers the following changes in the mix of generating equipment that will likely occur during the 1974–1985 period:

   Conventional steam-turbine generators, down to 53.6 percent from 63.8 percent
   Nuclear steam-turbine generators, up to 24.2 percent from 7.5 percent
   Conventional hydrogenerators, down to 10.2 percent from 15.5 percent
   Pumped-storage generators, up to 2.9 percent from 1.9 percent
   Gas-turbine generators, down to 7.4 percent from 10.4 percent
   Combined-cycle generators up to 1.2 percent from 0.4 percent
   Diesel-engine generators and others remaining at about 0.5 percent

Weekly output for the electric utility industry in the United States, excluding Alaska and Hawaii, for 1972 through 1974, is shown in Fig. 6. Seasonal factors are based on 52 weeks ending on Saturday.

The total amount of electricity generated by the electric utility industry during the mid-1970s is broken down by the 50 states in Fig. 7. Three states, Texas, California, and New York exceeded 100 billion kilowatthours. Tennessee, Kentucky, Indiana, Alabama, North Carolina, Michigan, Florida, Washington, Illinois, Ohio, and Pennsylvania, in increasing order, exceeded 50 billion kilowatthours. The average for all 50 states is approximately 37 billion kilowatthours annually. Seventeen states exceed the average; the remaining states generate less than the average, with Rhode Island at the low end of the group with a figure of 1231 million kilowatthours. It should be observed, of course, that there is no direct correlation between electricity generation and consumption because heavy-consuming states, in some instances, will import power from nearby states. The state of Washington, for example, with its heavy hydropower installations, has exportable power. The shaded portions of the bars in Fig. 7 indicate the portion of hydropower.

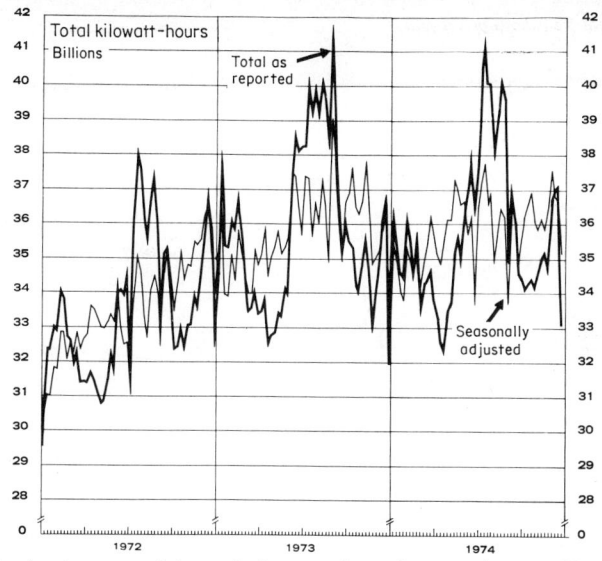

Fig. 6. Weekly electric output of the total electric utility industry in the United States, excluding Alaska and Hawaii. (*Source: Edison Electric Institute, New York.*)

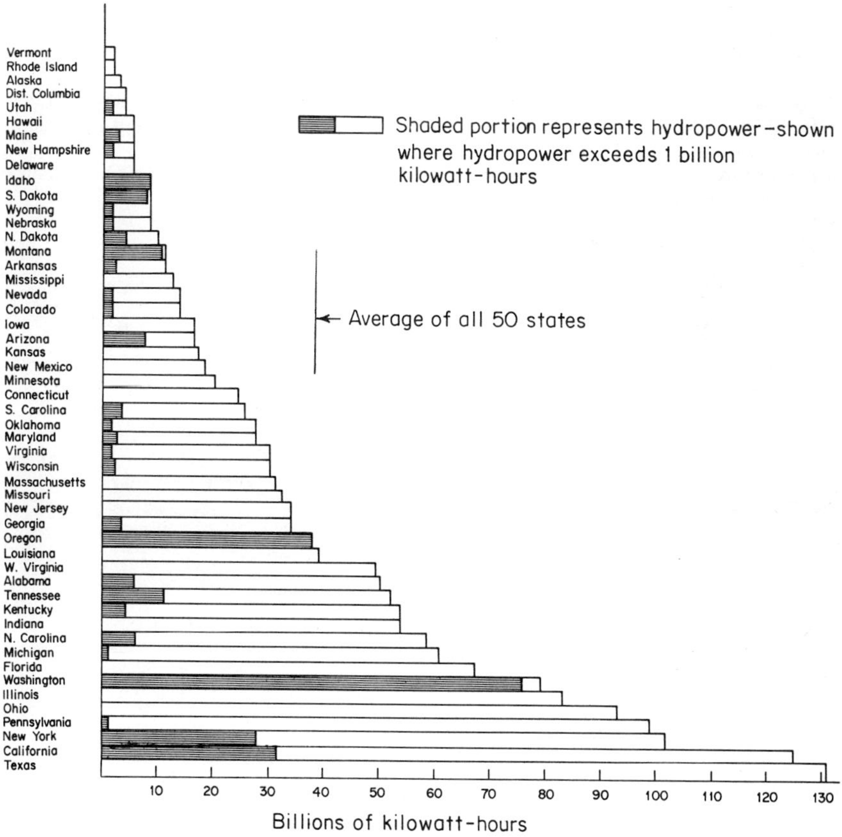

Shaded portion represents hydropower—shown where hydropower exceeds 1 billion kilowatt-hours

← Average of all 50 states

Billions of kilowatt-hours

Fig. 7. Production of electricity by states.

Fig. 8. Major geographic divisions as reported by the Edison Electric Institute in weekly publication, *Electric Output.* Percentage figures within regions are for net generation in each region, adjusted for net imports.

Rather than by state, statisticians in the electric power field generally prefer analysis of the total situation on a regional basis. In some cases, state lines are used in determining the borders of regions; in other cases, other criteria are used. For example, major geographic divisions as reported by the Edison Electric Institute in its weekly publication, *Electric Output,* are shown in Fig. 8. The percentages indicated within the regions are for net generation in each region (adjusted for net imports) as of Jan. 1, 1973. However, for other reporting, these figures are translated into the geographic divisions used by the U.S. Bureau of the Census and as indicated by the map of Fig. 9.

Fig. 9. Geographic divisions used by the U.S. Bureau of the Census. The data given in Table 2 correlate with these regions.

TABLE 2. Geographic Aspects of the Electric Utility Industry in the United States

Region	% of total U.S. generating capacity	% of total U.S. electric power delivered	% of total U.S. overhead electric lines, 22,000 volts and above	Installed generating capacity, thousands of kilowatts	Electricity delivered, millions of kilowatt hours	Circuit miles of overhead electric lines, 22,000 volts and above
			Comparison of region with remainder of the United States			
New England	3.9	3.7	3.1	18,528	69,237	14,562
Middle Atlantic	13.9	12.7	8.5	66,041	235,594	40,406
East North Central	18.2	18.3	15.6	86,642	340,517	73,845
West North Central	7.4	6.5	12.6	35,186	121,042	59,936
South Atlantic	18.1	18.0	16.2	86425	336,145	76,822
East South Central	8.9	9.4	9.1	42,102	175,756	42,965
West South Central	12.6	12.2	11.9	60,121	227,070	56,738
Mountain	5.2	5.7	10.0	24,689	105,622	47,524
Pacific	11.4	13.2	12.7	54,441	246,819	60,623
Alaska	0.1	0.1	(0.01)	592	1,871	52
Hawaii	< 0.3	< 0.3	0.3	1,221	5,288	1,116
Total United States	100.0	100.0	100.0	475,988	1,864,961	474,589

Notes: (1) Installed generator capacity as of Jan. 1, 1975. (2) Electricity delivered for year 1974. (3) Fuel data as of Jan. 1, 1975. (4) Circuit miles of overhead electric line do not include 53,500 miles of line installed by the Rural Electrification Administration cooperatives (as of Jan. 1, 1973).

Composition of characteristics within region

Region	Mix of prime movers within region: type of prime mover, % of generator capacity				Mix of fuels used within region: type of fuel used, % of total			
	Hydro	Steam Conventional	Nuclear	Internal combustion	Coal	Oil	Gas	Nuclear
New England	6.9	68.2	24.5	0.4	8.1	64.4	1.3	26.2
Middle Atlantic	12.6	78.9	8.4	0.1	49.2	39.2	2.3	9.3
East North Central	0.9	90.3	8.4	0.4	82.7	5.1	3.8	8.4
West North Central	10.8	79.8	7.7	1.7	60.9	2.4	28.1	8.6
South Atlantic	5.2	87.3	7.4	0.1	58.0	27.7	6.5	7.8
East South Central	14.5	81.9	3.6		89.4	3.5	2.9	4.2
West South Central	4.2	95.2	0.2	0.4	3.2	5.6	91.1	< 0.1
Mountain	30.5	69.3		0.2	66.7	8.4	24.9	
Pacific	66.9	30.3	2.7	< 0.1	5.3	49.9	36.3	8.5
Alaska	17.4	73.9		8.7	19.2	12.3	68.5	
Hawaii	< 0.1	97.0		2.9		100.0		
Total United States	16.1	77.5	6.1	0.3	53.3	19.2	20.2	7.3

GEOGRAPHICAL STATISTICS AND DIFFERENCES

The regions indicated on Table 2 correspond with those shown on map of Fig. 9. At the left of the table, the principal statistics for each region are given, along with an indication of the percentage of that region in terms of the remainder of the United States. In the right portion of the table, the mix of prime movers and the mix of fuels used within each region are given. Study of the table shows a number of significant contrasts. The differences between percent of generating capacity and percent of electric power delivered indicates to a degree whether or not a given region is an importer or exporter of power. The number of circuit miles of overhead electric lines corresponds with what would be expected in terms of concentrated areas of population versus a lot of open country. One of the most striking differences is in the area of hydropower, with a high of 66.9 percent in the Pacific Region (and 30.3 percent in the Mountain Region) as compared with a low of 0.9 percent in the East North Central Region (and 4.2 percent in the West South Central Region). With the exception of the Pacific Region (30.3 percent) and to a lesser extent the New England Region (68.2 percent) and the Mountain Region (69.3 percent), the other regions hover closely to the national average of 77.5 percent as conventional steam prime mover users. The table clearly shows the pioneering use of nuclear-fueled steam generators in the New England Region.

On the fuel side, to be discussed in more detail shortly, the Pacific Region stands out with little use of coal as of 1975 because of tight environmental restrictions and also the prior economic availability of natural gas (36.3 percent in 1975; 78.2 percent in 1973). The New England Region is also a low user of coal (8.1 percent), mainly turning to low-sulfur imported fuel oils (64.4 percent). Logically, the West South Central Region is a heavy user (91.7 percent) of natural gas as a fuel.

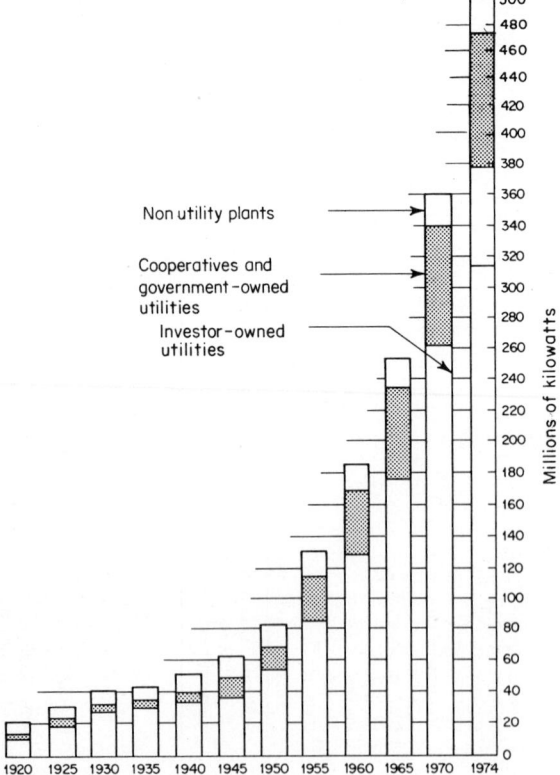

Fig. 10. Growth of the electric utility industry in the United States, indicating principal segments of ownership and operating responsibility.

OWNERSHIP OF ELECTRIC UTILITIES

The growth of the utility industry in the United States in terms of ownership and operating responsibility is illustrated by Fig. 10. Percentagewise, all ownership segments have been expanding essentially in a proportion with that of overall growth, with exception of a slowing down in the percentage of growth of nonutility (essentially captive) plants.

The principal segments of utility costs are plotted in Fig. 11 for the period commencing

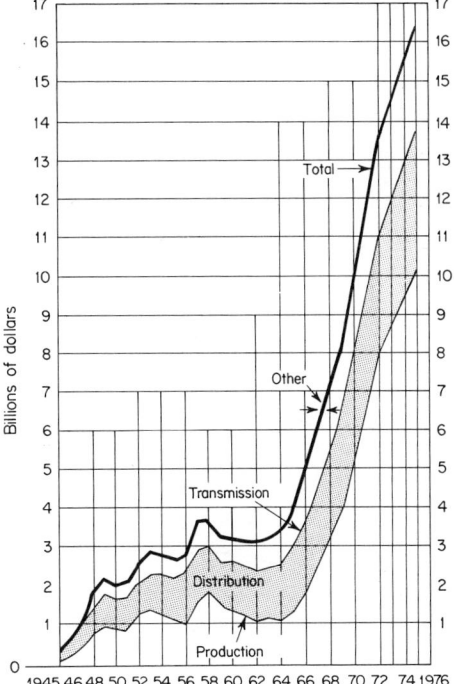

Fig. 11. Principal elements of electric utility costs. (*Source: Edison Electric Institute, New York.*)

with 1945. Rising costs give emphasis to the problems of the private utilities in obtaining capital for the vast expansions required during the next few decades.

ELECTRIC POWER REQUIREMENTS

For statistical and analytical purposes, the electric power markets have been traditionally broken down into the major segments indicated by Figs. 12 and 13. A major part of the classification termed *Large Light and Power* is composed of the manufacturing industries. An interesting comparison of the consumption of this market for the years 1939 and 1967 is given in Table 3. Kilowatthour use per man-hour for 1971 was over 16 times that of 1920, increasing from 1.21 to 19.67 kilowatthours. Production output of manufacturers in 1971, according to the U.S. Federal Reserve Board Index of Manufactures, was about seven times the annual production in 1920. During the same period, power use per point of the FRB Index was three times that in 1920, with a 17 percent gain per Index point between 1961 and 1971. This comparison is shown in Table 4. The second-largest category of electric power use is *Residential*. The majority of homes and apartments in the United States now have at least 13 different electrical appliances, namely, the first 13 items appearing in the list of Table 5. The average number of kilowatthours used by residential consumers of electric power has risen from 1329 kilowatthours in 1946, to 4016 kilowatthours in 1961, to 7379 kilowatthours in 1971.

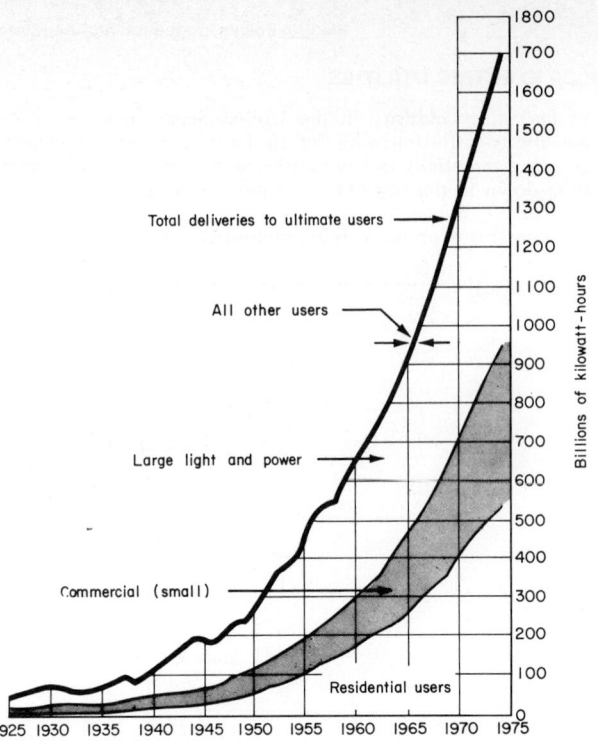

Fig. 12. Major traditional classifications of electric power users. (*Source: Edison Electric Institute, New York.*)

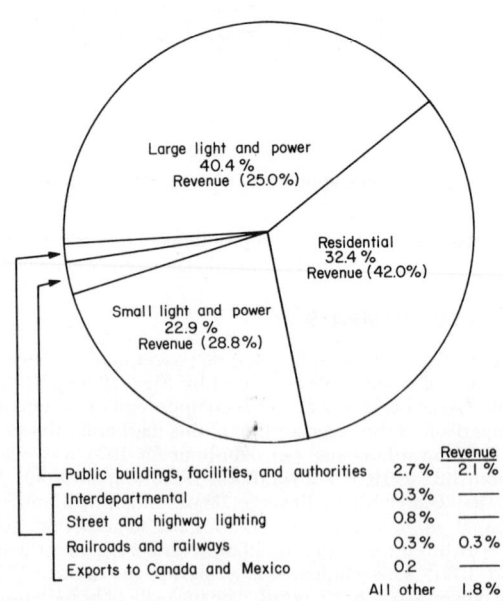

		Revenue
Public buildings, facilities, and authorities	2.7%	2.1%
Interdepartmental	0.3%	—
Street and highway lighting	0.8%	—
Railroads and railways	0.3%	0.3%
Exports to Canada and Mexico	0.2	—
All other		1.8%

Fig. 13. Distribution of electric power requirements in the United States. Note substantial differences between deliveries and revenues for various segments of electric power users. (*Source of data: Edison Electric Institute, New York.*)

TABLE 3.   Electricity Requirements by Manufacturing Industries in the United States

Industry	Electricity consumption, millions of kilowatthours		Cost of purchased power, percent of product value	
	1939	1967	1939	1967
Primary metal industries and fabricated metal products	18,191	146,730	1.8	2.4*
Chemicals and allied products	9,811	116,828	2.0	1.7
Paper and allied products	9,394	49,069	3.9	1.9
Food and kindred products	6,388	26,788	1.1	0.4
Transportation equipment	2,949	23,555	0.7	0.4
Petroleum and coal products	3,440	22,280	1.2	0.8
Stone, clay, and glass products	4,852	20,811	3.2	1.5
Textile mill products	6,805	20,796	2.3	0.9
Electrical machinery	1,432	19,205	1.1	0.5
Machinery, except electrical	1,985	17,255	0.9	0.5
Rubber products	1,584	10,768	1.6	1.0
Lumber and wood products	1,238	7,972	1.8	0.9
Printing and publishing	859	5,817	0.8	0.4
Apparel and related products	353	3,611	0.4	0.3
Instruments and related products	†	3,083	†	0.4
Furniture and fixtures	605	2,520	1.0	0.5
Leather and leather products	402	1,334	0.6	0.4
Tobacco manufacturers	115	850	0.2	0.2
Miscellaneous manufacturers ‡	466	6,583	0.9	0.4
All manufacturing	70,869	505,855	1.41	0.79

*1.8 for primary metal industries; 0.6 for fabricated metal products.
†Not available.
‡Includes ordnance and accessories.
SOURCE: U.S. Census of Manufacturers.

TABLE 4.   Relationship of Electricity Consumption and Manufacturing

Year	Federal Reserve Board Index of Manufactures*	Kilowatthours per point of FRB Index, millions†	Kilowatthours per man-hour†
1971	105.2	5031	19.67
1970	105.2	4791	18.05
1969	110.5	4416	16.28
1968	105.7	4303	15.40
1967	100.0	4218	14.52
1966	98.3	4063	13.52
1965	89.1	4147	13.35
1964	81.2	4286	13.39
1963	75.8	4174	12.54
1962	71.4	4218	11.94
1961	65.6	4288	11.74
1950	45.0	3650	6.61
1940	25.4	3111	4.62
1930	18.7	2761	3.21
1920	15.3	1669	1.21

*Manufacturing industries only. FRB Index uses base year of 1967 equaling 100.
†Power used for reduction of aluminum and magnesium and by the Atomic Energy Commission not included, since these processes consume large quantities of electric power with very few man-hours.
SOURCES: "Handbook of Labor Statistics," U.S. Department of Labor; "Monthly Labor Review"; "Historical Statistics of the United States"; U.S. Bureau of the Census; Federal Reserve Board; and the Edison Electric Institute.

TABLE 5.   Principal Appliances Used in Homes in the United States
(As of Dec. 31, 1971)

Appliance	Homes with one or more of listed appliances
Radio	65.4 million
Television (black and white)	65.3
Television (color)	33.5
Refrigerator	65.2
Vacuum cleaner	61.9
Clothes washer	61.8
Toaster	61.7
Coffeemaker	59.7
Steam iron (with steam spray)	59.1
Electric iron (no steam)	6.4
Mixer	55.3
Range	38.2
Frypan	38.0
Electric blanket	33.5
Can opener	31.5
Room air conditioner	29.2
Blender	26.2
Water heater	22.2
Freezer	21.4
Dishwasher	19.4
Food waste disposer	18.6
Hotplate and/or buffet range	16.3
Fully electrically heated home	5.6

FUELS FOR ELECTRIC POWER GENERATION*

Historically, the consumption of fossil fuels for electric generation in the United States, commencing with the year 1951, is given in Table 6. This table demonstrates the steady growth of the use of coal despite environmental problems and the rapdily increasing rates of use of both fuel oil and gas. Increases in efficiency of the use of fuels are also indicated, with 13,641 Btu required per kilowatthour in 1951 as contrasted with 10,479 Btu per kilowatthour in 1972. Expressed in terms of coal, this indicates that in 1951, 1.142 pounds of coal were required per kilowatthour as contrasted with 0.911 pound of coal per kilowatthour in 1972.

Trends in the use of various fuels for electric power generation are illustrated in Fig. 14, along with a lowering of the heat rate. The effect of fuel price and efficiency of use upon cost of fuel per kilowatthour generated is illustrated in Fig. 15.

Regional Analyses of Future Fuels Requirements.   In the work of the "Electricity Task Group" of the National Petroleum Council, six geographic regions of the United States, as depicted by the map of Fig. 16, were studied. One of these regions is that served by the TVA (Tennessee Valley Authority).

New England Region.   Traditionally, the reliability of fuel supply was the prime criterion for procurement of fuels. Because of the economic unavailability of a fuel that will meet critical air pollution requirements from nearby United States sources, New England electric power producers turned to other nations for low-sulfur fuel oil. International sources, of course, are subject to political factors that involve both import and export policies. It appears that planning for the New England Region assumes an increasing reliance on nuclear base-load plants to meet new growth, particularly after 1978. An estimate of peak loads through 1985 indicates the growth anticipated:

1974	16,279 MWe†
1980	25,903
1985	38,100

*MW e = megawatts electric (the electric output in megawatts at the generator terminals of a nuclear power plant).

†Permission to use much of the basic information on fuel supplies for electric power generation over the period 1975–1985 from the National Petroleum Council task force report is gratefully acknowledged. Much additional information is available from the full report.

.**TABLE 6.   Consumption of Fossil Fuels for Electric Generation in the United States***

Year	Coal, thousand short tons	Fuel oil, thousand 42-gallon barrels	Gas, million cubic feet	Total fossil fuel in coal equivalent,† thousand short tons	Net generation by fuels,† kWh in millions	Heat rate,‡ Btu per kWh	Lb of Coal per kWh
1972	351 045	493 927	3 978 673	646 308	1 418 781	10 479	0.911
1971	327 926	396 238	3 992 981	600 496	1 308 857	10 536	0.918
1970	320 818	335 504	3 931 996	573 152	1 261 474	10 508	0.909
1969	310 641	251 027	3 487 642	517 937	1 177 127	10 457	0.880
1968	297 779	188 642	3 147 909	475 475	1 093 614	10 371	0.870
1967	274 185	161 278	2 746 352	428 165	984 560	10 396	0.870
1966	266 477	140 949	2 609 949	409 854	943 552	10 399	0.869
1965	244 788	115 203	2 321 101	367 567	857 286	10 384	0.858
1964	225 425	101 141	2 322 896	344 083	803 222	10 407	0.857
1963	211 332	93 314	2 144 473	319 840	747 530	10 438	0.856
1962	193 238	85 768	1 965 974	292 501	681 533	10 493	0.858
1961	182 121	85 736	1 825 117	275 544	638 277	10 552	0.863
1960	176 634	85 340	1 724 762	266 064	607 142	10 701	0.876
1959	168 423	88 263	1 628 509	254 525	571 883	10 879	0.890
1958	155 724	77 668	1 372 853	228 136	504 497	11 090	0.904
1957	160 769	79 693	1 336 141	232 576	501 098	11 365	0.928
1956	158 279	72 711	1 239 311	233 733	478 487	11 456	0.935
1955	143 759	75 274	1 153 280	206 929	433 786	11 699	0.954
1954	118 385	66 745	1 165 498	180 367	364 354	12 180	0.990
1953	115 897	82 238	1 034 272	178 491	337 042	12 889	1.059
1952	107 071	67 218	910 117	160 872	293 640	13 361	1.096
1951	105 768	63 945	763 898	154 498	270 531	13 641	1.142

*Alaska and Hawaii included since 1963.
†Excludes geothermal, wood, waste, and nuclear fuels.
‡Edison Electric Institute estimate.
SOURCE: Federal Power Commission.

Even with the heavy nuclear commitment indicated, the fossil-fuel requirements in the New England Region will experience continued growth. Inasmuch as it is expected that this increment of fossil growth will be oil-fueled, the reliability of supplies outside the United States will remain an important concern in the future.

East Region.   This region has been heavily reliant on Appalachian coal as a utility fuel. However, in recent years, the utilities serving the coastal areas commenced a shift to oil, mainly for economic reasons. However, because of strict sulfur oxide emission regulations, several inland utilities also shifted to low-sulfur oil. Although serious efforts are underway to increase the supply of acceptable fuel oil, any event which would drive requirements above projected levels, such as delays in nuclear power plant schedules, could prove disruptive. In the longer term, electric power producers in the East Region are counting heavily on the development of low-Btu gasification and/or stack-gas scrubbing techniques that would enable them to return to the vast Appalachian coal fields as a ready source of acceptable energy.

Other alternatives which may exist for some utilities in the East Region include the possible importation of liquefied natural gas (LNG) where extremely severe sulfur emission controls are in force. Long-term commitments for some quantities of Canadian hydropower are also a possibility. However, it is not likely that much hydroelectric power will be available from Canada. Some sites from which interim power might be available in the late 1970s and early 1980s include Churchill Falls, Labrador; Nelson River, Manitoba; Peace and Upper Columbia, British Columbia; and Rupert River, James Bay, Quebec.

Power interties would permit that some power, ranging from 300 to 2000 MWe, would be available at several locations. The Rupert River development in Quebec has promise of providing electric power for the Northeast United States on an interim basis commencing in the late 1970s. Estimates indicate that the available capacity would be no more than about 2000 MWe for an interim period commencing about 1980, that is, less than the capacity of a single large nuclear power plant. Canadian hydroelectric power thus would provide only a temporary availability of limited energy.

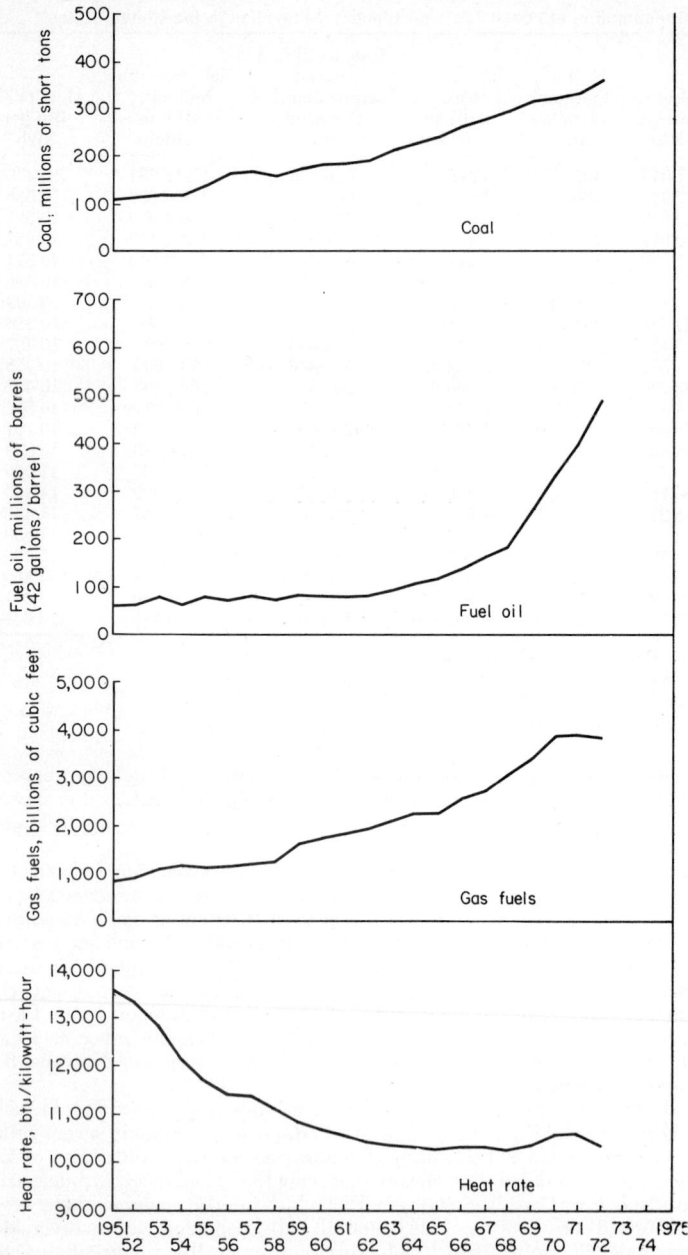

Fig. 14. Growth in use of various fuels for electric power generation; and lowering of heat rate. (*Source: U.S. Federal Power Commission, Washington.*)

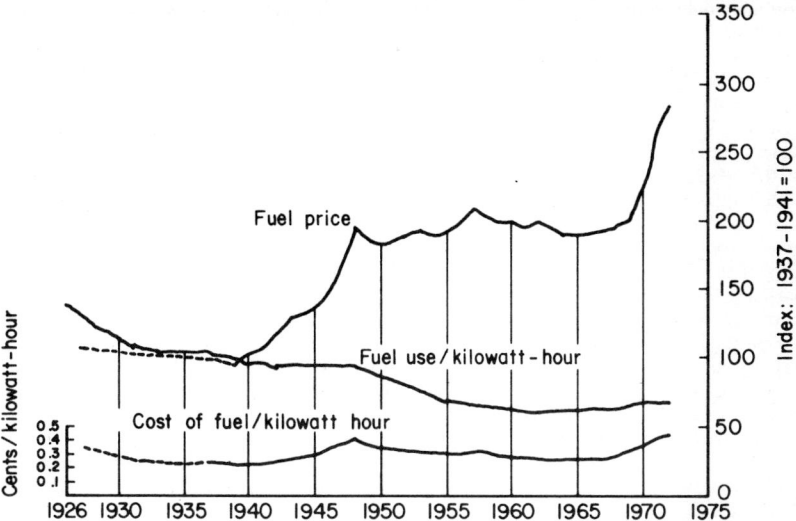

Fig. 15. Effect on fuel price and efficiency of use on cost of fuel per kilowatt-hour generated. Figures are based on the total electric utility industry in the United States, including Alaska and Hawaii, since 1963. The data are based on all fuel used in electric generation, and expressed in units of equivalent coal. (*Sources: 1927–1932, U.S. Census of Central Electric Stations; 1937–1958, U.S. Federal Power Commission; 1959–1975, U.S. Federal Power Commission and Edison Electric Institute.*)

Also, utilities in the East Region that are dependent on imported low-sulfur fuel may have additional options by the late 1970s. Western European refineries operating on North Sea crude may be a source of some extra supplies. However, such supplies would be limited because of the large requirement for European consumption. Also, the development of refining capacity in the Maritime Provinces of Canada may provide an additional source of products to supplement those coming from the Caribbean.

Midwest Region. Traditionally, electric power utilities in this region have relied on

Fig. 16. Regions selected by Electricity Task Group of the National Petroleum Council in its study of future requirements for fuels for electric power generation.

regional coal resources. Restrictions on sulfur dioxide emissions, however, have rendered much midwestern coal unsuitable for power production, at least until gasification and stack-gas scrubbing technologies are further improved. Other sources of low-sulfur coal exist in the West, but these fuels represent a limited alternative because of their physical and chemical characteristics. Their relatively low heat content results in greater freight costs per Btu delivered, and problems with unburned carbon in both boilers and precipitators make the use of these coals in existing installations difficult. As a result, the utilities in the Midwest Region are confronted with an acute, near-term fuel supply problem which appears to dictate a much greater reliance on fuel oil. A switch to fuel oil, however, is complicated by possible import and allocation regulations and also by logistic problems. Pipeline capacity would be overtaxed if even only a small portion of the consumption of the Midwest Region were converted to oil. Water transportation offers only a partial solution because many generating plants are not situated on navigable waterways. Power stations depending on Great Lakes shipping would be plagued by the annual winter halt in navigation. In view of these complexities, utility companies in this region will be able to use oil only if they can solve the specific supply and transportation problems for specific locations under consideration.

Over the longer term, it appears that the Midwest Region will look to nuclear power for a large part of its electric energy supply. However, there is increasing concern over thermal pollution of the Great Lakes, which are ideal as heat sinks for power production. This may require location of nuclear plants elsewhere at greater cost. If low-Btu gasification of coal and stack-gas sulfur dioxide removal technologies develop at an accelerated rate, coal will return as the prime fuel. Of course, if exploration for oil in the Canadian Arctic locates large low-sulfur resources, the resulting supplies could be moved by pipeline to midcontinent areas.

For the short term, oil from areas outside North America appears to be the obvious swing fuel.

South Central Region.   Generally, this is a gas-producing area where long-term gas supply contracts have been obtainable. Until the early 1970s, construction had been designed around natural gas as a primary fuel, with provision to operate on oil in emergencies. Continuous operation of these units on oil presents serious problems insofar as availability is concerned because of excessive boiler-tube failures occasioned by hot spots encountered while burning oil. Such conditions are not encountered when burning gas for which the boilers are mainly designed. Further, units designed to burn gas as the primary fuel lose approximately 10 percent of capacity immediately upon switching to oil. They lose another approximately 10 percent of capacity when they are subjected to continuous oil firing because of soot buildup on heat-transfer surfaces.

Conversion from natural gas to oil as the primary fuel would involve a costly and long program because utilities would necessarily first have to install new generating capacity equal to approximately 20 percent of the existing and under-construction capacity simply to offset capacity losses just described.

The South Central Region of investor-owned electric utilities burned approximately 50 percent of the 3900 quadrillion Btu of natural gas consumed in the United States for electric generation in 1970. Gas-fired generating capacity in the South Central Region is estimated at 44 million kilowatts. This capacity would have to be supplemented by approximately 8.8 million kilowatts of capacity, costing an estimated $1.5 billion, before a full switchover to fuel oil could be effected. The program would require about 8 years to effect an orderly conversion.

The fuel oil equivalent of gas consumed in the South Central Region in the early 1970s was nearly 300 million barrels/year, obviously a large imposition on the fuel oil supply capability that now exists worldwide. This, coupled with other user demands for fuel oil, would require an orderly program of refinery and tanker construction in order to satisfy demand in the event that gas is not available.

Generally, capacity under construction and committed for operation subsequent to the mid-1970s is being designed to utilize fuels other than natural gas. Nuclear capacity is predominant in plants for long-range base-load requirements. Oil-fired units are being planned for some base-load use as well as for intermediate-load operation as gas supply fades out and nuclear capacity is brought on line to cover base loads. Oil will also play an important part in meeting peak loads, since it can be transported and stored at atmospheric pressure and requires only moderate investment in transport and storage facilities.

It is believed that coal may play but a minor role in the base-load electric generating picture in the South Central Region because of transportation costs coupled with environmental problems, making it uncompetitive with nuclear costs. However, beyond 1985, coal may play an important role in fueling peaking capacity if technology for liquefaction of coal results in products which are competitive with fuel oil.

For the South Central Region, through 1985, it is usually assumed that (1) generation requirements will double about each 7 years, commencing with 1970, and (2) in effect, the nuclear program will result in 8.5 percent of all energy being supplied from nuclear plants in 1980 and 50 percent by 1985. The change in fuel mix is indicated by Table 7. Since the nuclear estimates given in Table 7 are believed to be the maximum practical for the area, and since swings in the period must be substantially covered by a combination of oil and gas, the National Petroleum Council task group believes that the oil figures for 1980 and 1985 could be as high as 5150 and 4100 trillion Btu, respectively, if all gas were taken out of power-plant usage.

Tennessee Valley Authority (TVA) Region. Hydroelectric power was the initiating force behind the TVA system. When TVA built its first thermal unit in the early 1940s, coal was selected because large deposits of economical, utility-grade coal were available in or near the area. Alternate fuels were either unavailable or significantly more costly. The use of coal continued from 1942 until 1966 when the 3400-MWe Browns Ferry Nuclear Plant was announced. Since 1966, plans have been made for adding a 2600-MWe coal-fired plant, three additional nuclear plants, 28 gas turbine units, and a pumped-storage plant. At the end of 1972, the installed capacity of the TVA system totaled nearly 22,000 MWe, consisting of 16,400 MWe of coal-fired capacity, 1100 MWe of gas-turbine capacity, and 4300 MWe of hydroelectric capacity. One plant with 900 MWe of coal-fired capacity is designed for gas and coal, and gas is supplied to the plant on a seasonal basis. The gas-turbine installations are designed to use gas or oil, and gas is used during the summer months when available.

Between the mid-1970s and 1985, approximately 23,000 MWe of capacity will be added to supply the expected increase in system load. Nuclear capacity could account for as much as 18,000 MWe, with fossil-fueled capacity or pumped-storage peaking capacity accounting for the remainder. Generation requirements in 1985 are expected to exceed 200 billion kilowatthours. Hydroelectric generation will then supply less than 10 percent of the requirements. It is expected that, in addition to hydroelectric, gas, oil, and nuclear fuel, over 25 million tons of coal per year will be used to generate the required power.

In 1985, the present gas turbines and three steam units that can use gas will represent about 5 percent of the total system capacity, and the estimated output from these units will be about 2 percent of the total generation requirements. There is a critical need for this capacity in the TVA system, and the curtailment or reduction of gas and petroleum usage would be serious. Further, the gas-turbine units and the three gas-fired steam units are located on or near the perimeter of the TVA system where the transmission of substitute power would be costly.

West Region. This region is composed of distinct subregions, each unique with respect to resources for electric generation. The northwestern part has relatively abundant hydroelectric resources in conjunction with nuclear and coal generation. In the mountain subregion (east of the coastal states), coal is the predominant source, supplemented with oil and gas-fueled generation. In California, oil, natural gas, nuclear and hydroelectric energy compose the fuel mix. Alaska has abundant oil, gas, and hydroelectric resources, and Hawaii is heavily dependent on offshore oil for its electric generation requirements.

TABLE 7. Possible Changing Fuel Mix for Electric Generating Plants in the South Central Region

(Trillion Btu)

Year	Nuclear	Gas	Oil	Total
		Fuel		
1970	0	1800	60	1860
1975	60	1640	1420	3120
1980	470	1340	3810	5620
1985	4000	900	3200	8100

SOURCE: National Petroleum Council, Washington, D.C.

Although the aggregate raw energy reserves of the West Region are over 20 times the projected total requirements in this century, more than half these reserves are in the form of coal. Thus, in the near-term prior to perfection of coal gasification and liquefaction technology, the West Region will be increasingly dependent on extraregional fuel supplies, particularly oil.

To satisfy the enormous projected oil requirements for this region, it is anticipated that a combination of available alternatives will be used, including (1) conversion of select electric-generating plants to burn low-sulfur crude oil, (2) topping plants to produce low-sulfur residuals and naphthas, (3) importation of low-sulfur residual oils, and (4) desulfurization of imported or domestic crude oils.

NATIONAL ELECTRIC RELIABILITY COUNCIL

Data for the semi-annual electric power surveys made by the Edison Electric Institute include power system capabilities, loads, capability margins, kilowatt-hour requirements, and fuel consumption. Survey results are presented for each of the nine reliability councils of the National Electric Reliability Council (NERC). This council encompasses essentially all utility systems in the United States and four Canadian provinces. The nine councils in the United States are:

ECAR—East Central Area Reliability Coordination Agreement
ERCOT—Electric Reliability Council of Texas
MAAC—Mid-Atlantic Area Council
MAIN—Mid-America Interpool Network
MARCA—Mid-Continent Area Reliability Coordination Agreement
NPCC—Northeast Power Coordinating Council
SERC—Southeastern Electric Reliability Council
SPP—Southwest Power Pool
WSCC—Western Systems Coordinating Council

REFERENCES

Electricity Task Group: "U.S. Energy Outlook—Fuels for Electricity," National Petroleum Council, Washington, D.C., 1973.

Staff: "Statistical Year Book of the Electric Utility Industry," Edison Electric Institute, New York, published annually.

Staff: "Semi-Annual Electric Power Survey," Edison Electric Institute, New York, published semiannually.

Multiple authors: Energy Technology to the Year 2000, *Technol. Rev.* Massachusetts Institute of Technology, (1972).

Additional statistical reference material is available from American Public Power Association, Washington, D.C.; National Rural Electric Cooperative Association, Washington, D.C.; Federal Power Commission, Washington, D.C.; Rural Electrification Administration, Washington, D.C.; National Coal Association, Washington, D.C.; American Petroleum Institute, Washington, D.C.; American Gas Association, Arlington, Va.; National Electrical Manufacturers Association, New York; The Electric Power Research Institute, Palo Alto, Calif.

Trends in Electric Power Transmission

BY R. F. LAWRENCE* AND J. H. CRONIN†

In 1960, there were only 2500 miles of extra-high voltage (EHV) transmission in the United States and only at the 345-kilovolt level. By 1970, there was a tenfold increase in EHV to a total circuit mileage of over 25,000 miles. The patterns for this increase are shown in Fig. 1. The prime reasons for this growth are (1) the needs for increased power transfer between generation and load points and (2) the increased interconnections for achieving more reliable power supply systems. By 1980, it is projected that there will be over 68,000 miles of EHV lines in the United States.

GROWTH IN VOLTAGE

A plot of the maximum operating voltage for the electric utility industry in the United States is shown in Fig. 2. From its inception, the increase in voltage was based on technology and economics. Early changes in transmission voltage reveal a doubling of the previous level when a change was made. One exception is the single, unique use of 287 kilovolts from Hoover Dam to Los Angeles. Generally, transmission remained at a level of 220 kilovolts from 1922 until 1953 when the first EHV, 345-kilovolt transmission was placed into service. By 1972, some 1000 circuit miles of 765 kilovolts were placed in operation.

Most of the EHV lines through the mid-1970s have been alternating current. However, in 1969 a 750-kilovolt direct current line was placed into operation. The line transmits power from the Pacific Northwest to the southern California areas on a unique (for transmissions systems in the United States) point-to-point basis over a long distance.

If technical problems can be solved in time to permit availability of economic lines and equipment, application for 1200-kilovolt lines can be foreseen in the mid-1980s. Steps to higher voltages are always preceded by extensive research and field investigations to make certain that both the economics and technology of a system can be achieved. A major reason for the step to higher voltage is the improvement in economics that is possible. The cost per megawatt at 765 kilovolts is only about 8 percent of the cost for 138-kilovolt transmission. The cost savings is further demonstrated by the fact that in going from 345 to 765 kilovolts, the line cost doubles but the rating increases by four times.

GROWTH IN CIRCUIT MILES

Annual additions in circuit miles are shown in Fig. 3. The chart indicates that installations from 1966 through 1973 were approximately 50 percent greater than for the pre-1966 period.

GROWTH IN MEGAWATT-MILES

Simply viewing transmission in terms of circuit miles does not present a true picture of transmission growth. A more meaningful way to measure this growth is in terms of capability of the transmission systems to transmit power. Thus, the term *megawatt-miles* is an excellent approach and is analogous to passenger-miles as used to express the capability to transport people; or of ton-miles in terms of freight.

Typical circuit capabilities for various line voltages are shown in Fig. 4. This capability

*Manager, Transmission and Distribution Systems Engineering, Westinghouse Electric Corporation, East Pittsburgh, Pa. †Advisory Engineer, Transmission and Distribution Systems Engineering, Westinghouse Electric Corporation, East Pittsburgh, Pa.

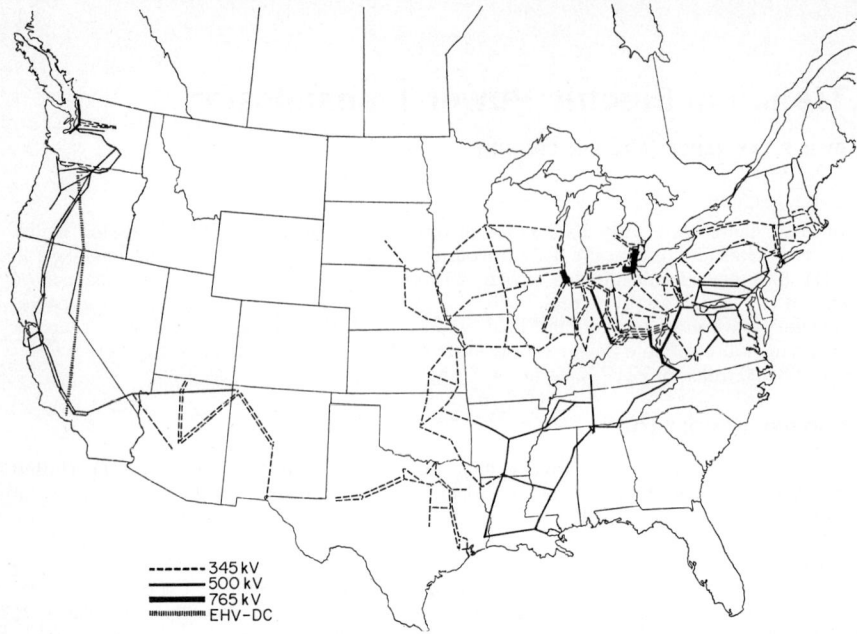

Fig. 1. Extra-high voltage (EHV) transmission lines in the United States (1970).

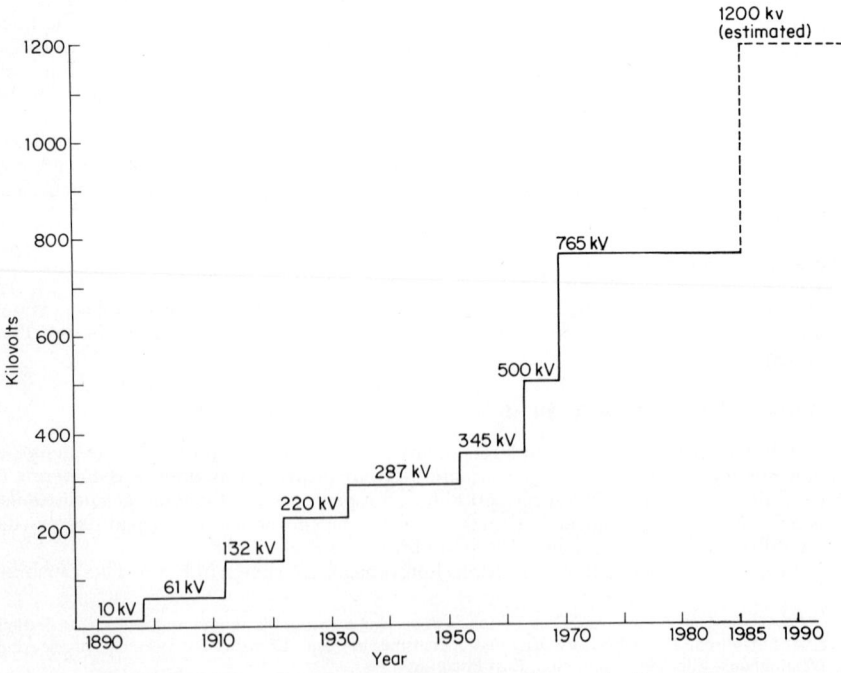

Fig. 2. Record of increased voltage of transmission lines in the United States.

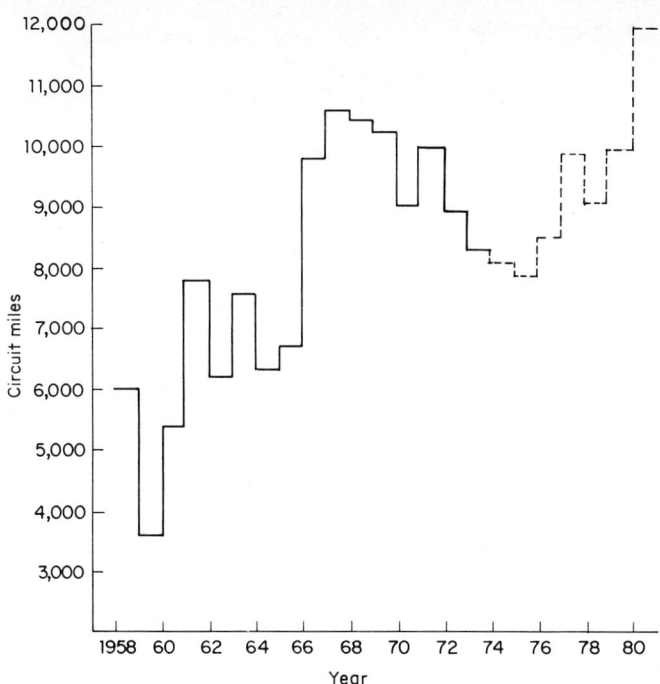

Fig. 3. Annual additions of circuit miles of transmission in the United States (115 kilovolts and above).

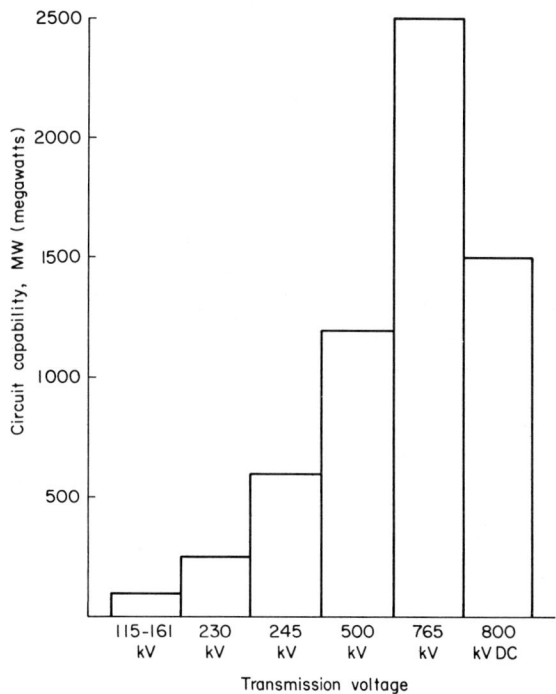

Fig. 4. Typical circuit capabilities.

varies approximately as the square of the voltage, where, as examples, 138 kilovolts is 100 megawatts, 345 kilovolts is 600 megawatts, and 765 kilovolts is 2500 megawatts.

Transmission capability using these conversion factors applied to circuit miles at various voltage levels for lines installed and projected is shown in Fig. 5. This chart also illustrates the impact of EHV on transmission capability. Starting in 1953 with the first 345-kilovolt system, a steady increase in capability occurred even with the circuit miles being relatively constant. As 345 kilovolts continued to grow in the early 1960s, this capability increased and in the mid-1960s advanced very rapidly with the introduction of the 500-kilovolt systems. With the additional lines installed in the late 1960s, the impact of extensive 500- and 765-kilovolt as 750-kilovolt dc systems is clearly seen. The annual

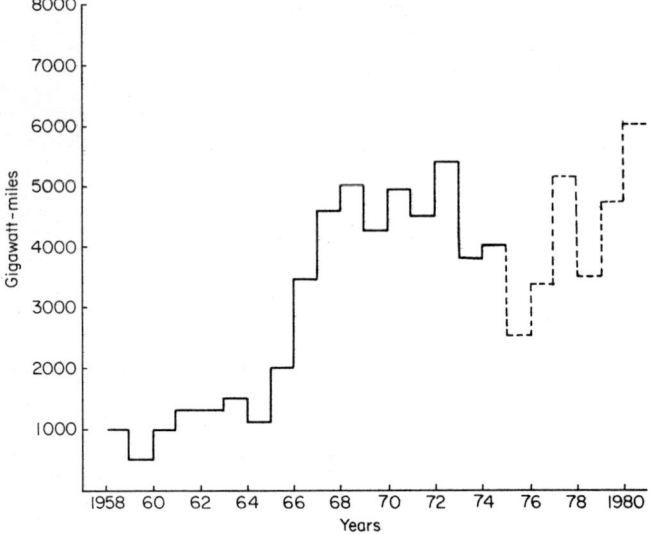

Fig. 5. Annual addition of gigawatt-miles of transmission in the United States (115 kilovolts and above).

additions declined in the mid-1970s due to conservation and the general business recession, but it is expected to recover in the early 1980s and exceed those additions of the 1970s. The wider use of EHV in interconnections and overlaying of existing systems with EHV are the major reasons for this high level of capacity additions.

How does transmission capability growth compare to that of generation? The average generation and circuit mile additions for various time periods are shown in Fig. 6. Note that for the period 1961–1965 generation installed capacity and transmission were about even. For the period 1966–1970 the transmission capacity was double that for generation. For the later years, the generation capacity installed will be higher and the transmission will continue to exceed that of generation.

The formation of pools with heavy interconnection between member companies is responsible for much of the growth in transmission lines in recent years. This growth is illustrated for one of the larger pools in the United States. When this pool was formed in 1963, there were no 345-kilovolt lines. There were 400 miles of 230-kilovolt lines, with the largest unit in the individual companies being 250 megawatts. In 1972, nine years later, this pool had 2250 miles of 345 kilovolts, 1900 miles of 230 kilovolts, and maximum unit size of 800 megawatts.

ULTRAHIGH VOLTAGE ECONOMIC STUDY

The U.S. Federal Power Commission Regional Advisory Committee reports provide projections of transmission growth through the periods 1970–1980 and 1980–1990. Estimated numbers of circuits, voltage levels, and other factors have been consolidated from the FPC reports onto a map of the United States.* By using the aforementioned reports and map, a simplification of the transmission situation in the United States during the period 1970–1980 can be made. The single-line transmission equivalents for the transmission additions in this period are shown in Fig. 7. On this map, substations with many line terminals would exist between transmission intersections. The number of major substations is 329 and is somewhat arbitrary to the extent that it represents the large generation

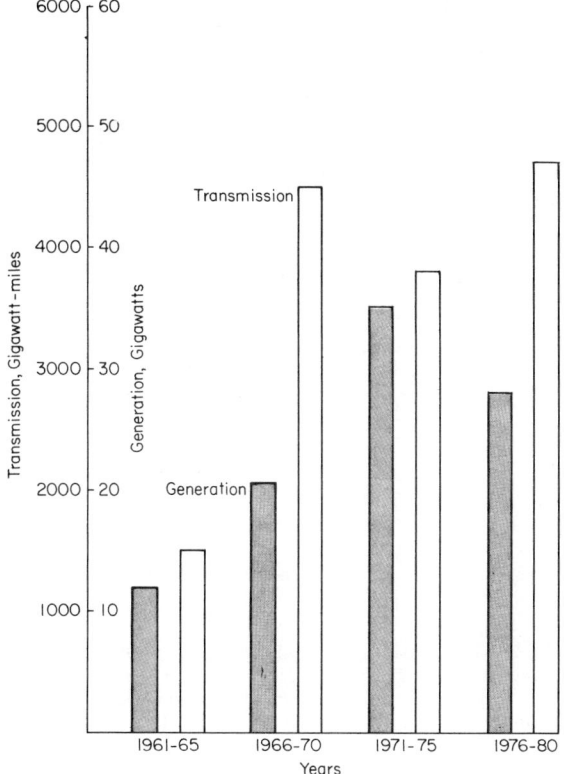

Fig. 6. Average annual additions of generation and transmission capacity in the United States.

and load points. Each of the 329 substations, however, consists of step-up and step-down transformations. Although there would be a much larger number of substations than indicated on the map, the figure does represent a system expansion plan on which an economic study with 1100 kilovolts can be made.

In performing this study, all the transmission line additions between each pair of substations were assigned a capability associated with its voltage level and line length. The total capability of transmission between each pair of substations was thus determined by summing up all the individual capabilities for each line addition between substations. Using this total transmission capacity and the capacity of 1100-kilovolt transmission, the number of 1100-kilovolt transmission lines required to match or to exceed this capability was determined. The costs of the two systems are compared, and 1100-kilovolt transmission is assumed to be used where it was the economic choice. In places where 1100 kilovolts was not the economic choice, the planned transmission lines remained.

System Representation.

1. *A 329-Substation Transmission System* The transmission additions from FPC reports and the map resulted in the base 329-substation transmission system for the

Electr. Light Power (March 1970).

United States for the period 1970–1990. These 329 substations represent major load and/or generation areas and are located at major transmission intersections shown in Fig. 8. The transmission additions connecting the substations on the map are equivalent single lines.

2. *A 144-Substation Transmission System*   For this system, each power supply area (PSA) from the National Power Survey shown was individually analyzed. Each PSA was reduced to three major load substation areas. In effect, there are three electrical centers of

Fig. 7. 1970–1980 transmission additions with high-voltage (HV) and extra-high voltage (EHV) 329-substation system.

gravity in each PSA. The transmission connecting the substations are the projected additions for the 20-year period (1970–1990).

3. *A 76-Substation Transmission System*   For this system, each PSA was analyzed and reduced to a minimum number of major load-generation terminals.

4. *A 48-Substation Transmission System*   In this system, each PSA was reduced to one major load area. Consequently, an electrical center was established in each PSA resulting in 48 locations.

Economic Substitution.   Using transmission line and terminal costs along with the line addition distance, a system cost was derived for each of the four transmission models. These system costs were calculated during three periods: 1970–1980, 1980–1990, and 1970–1990.

The same process used to obtain the base and substitute system cost for the 329-substation model was also used for the 144-, 76-, and 48-substation models.

The base costs for each model are considerably less than the 329-substation transmission system, since all internal transmission additions within a consolidated substation area are neglected. As the number of terminals for the United States becomes less, the basic system cost decreases proportionally since more internal transmission in a PSA is neglected.

Results of Study.   The economics of transmission additions at 1200 kilovolts compared to presently planned transmission is summarized in Fig. 8. This is a plot of percent cost difference (savings) versus the number of substations for the substitute voltage levels during the time period 1970–1980 and 1980–1990. It is important here to think in terms of percent savings instead of actual dollars, since the transmission *does not include* all voltage levels and every substation in the United States. From this curve, there is an optimum number of substations for the United States which offers the greatest savings. Of the four transmission models studied, the 76-substation transmission system gives the greatest savings with 1100-kilovolts. These curves also show that the cost of transmission addition with ultrahigh voltage was not the best economic choice with the 329-substation network. On the other hand, the cost of transmission additions with 1100 kilovolts was the most economic in the 144-, 76-, and 48-substation networks. This approaches the interregional interconnection concept.

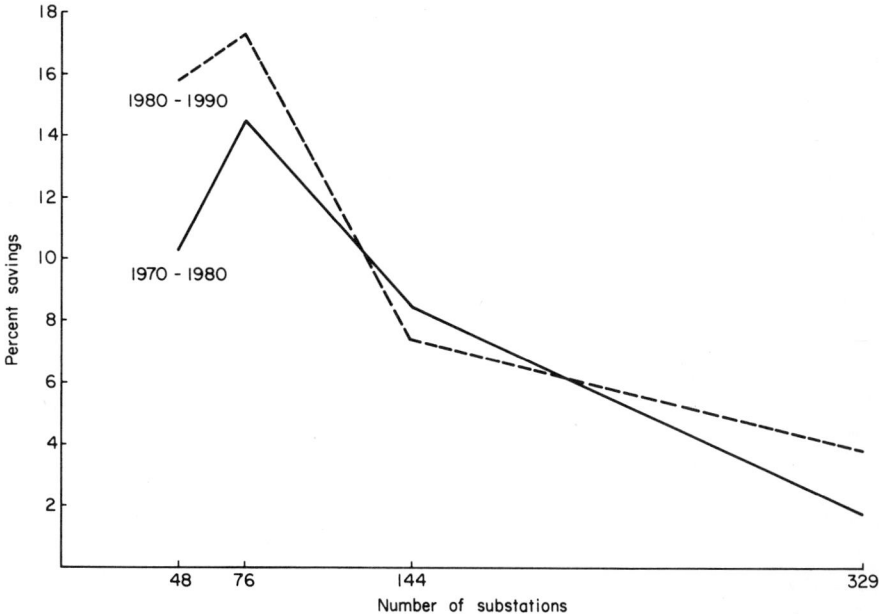

Fig. 8. Transmission savings from proposed capital expenditures with ultrahigh-voltage (UHV) transmission (1200 kilovolts).

The locations where 1100 kilovolts was the economic choice with the 329-substation system are shown in Fig. 9. The use of 1100 kilovolts in this model is only 2356 miles during the 1970–1990 period. In this transmission expansion pattern, the use of 1100 kilovolts is fragmented. There appears to be no interconnection pattern of ultrahigh voltage system that might suggest improved power interchange capacity between pools and regions. This is extremely important for the interregional interconnection, extraterritorial generation, and system security concepts.

As the number of substations for the United States decreases, the locations where 1100 kilovolts is the economic choice increase. The 1100-kilovolt transmission also is more interconnected and not as fragmented as in the 329-substation system. These locations in the 76-substation system are indicated in Fig. 10. The total use of 1100 kilovolts for this system in the period 1970–1990 is 8274 miles. The total use of 1100 kilovolts for the 144- and 48-substation models are 5582 and 8886 miles, respectively.

A longer time period in the range of 20 years should be considered in planning transmission systems since it can be seen that the potential industry savings during the 20-year period 1970–1990 is greater than two 10-year periods 1970–1980 and 1980–1990. The longer-range planning period will become even more important in the future in view of the increased restrictions on the use of land for transmission right-of-way.

UHV Terminal Equipment Requirements.   The exact amount of terminal equipment required for 1100 kilovolts is difficult to determine since none has been installed. Using the existing voltage level of 765 kilovolts, a reasonable projection of 1100 kilovolts can be extrapolated to give an indication of the size of the requirements.

The results of the economic study indicated that for the period 1970–1990, fifty-five 1100-kilovolt lines could be economically installed. Since 1100 kilovolts is not presently available, the study was made in two periods. The first was 1970–1980 and for this period,

Fig. 9. System of 329 substation locations with installed ultrahigh-voltage (UHV), 1200 kilovolts, 1970–1990 period.

seventeen 1100-kilovolt lines could be justified. The second period was 1980–1990 and twenty-five 1100-kilovolt lines could be justified.

The twenty-five 1100-kilovolt lines represent a minimum number for the 1980–1990 period and, in fact, if a UHV development program were started in the mid-1970s, many utilities might delay presently planned transmission lines to achieve the economic

Fig. 10. System of 76 substation locations with installed ultrahigh-voltage (UHV), 1200 kilovolts, 1970–1990 period.

benefits of future 1100-kilovolt systems. For this reason, the number of lines economically justified for 1980–1990 was assumed to be 25 plus 17, or 42 lines with 84 terminals.

The UHV requirements for the 1980–1990 period were calculated. From these, the UHV terminal equipment for the period would represent a cost of approximately $400 million (1973 dollars).

UNDERGROUND TECHNICAL DEVELOPMENTS

With the large cost differential of as high as 20:1 existing between overhead (OH) and underground (UG) systems, it is evident that new technology must emerge to meet the

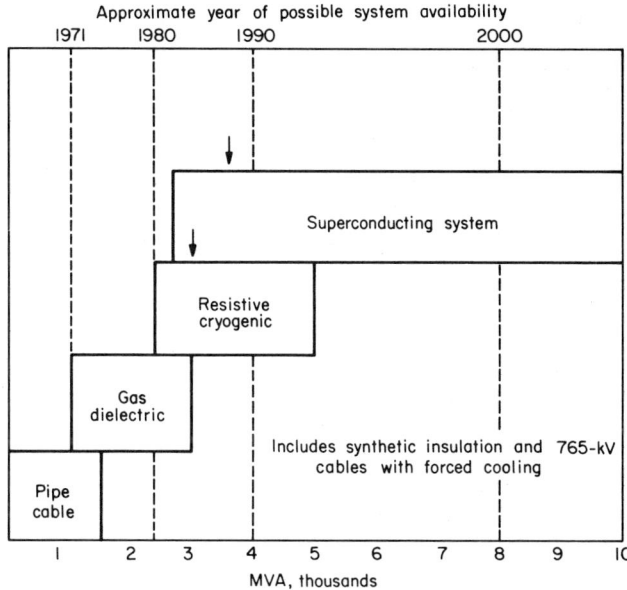

Fig. 11. Comparative capabilities of underground transmission systems utilizing new technology.

future UG power transmission requirements. This technology is being investigated, and research and development is underway on many new systems. Such developments, of course, require much time and money and must proceed in an orderly, systematic program to achieve optimum output.

Substations. Land requirements for conventional, open-air substations are large, but, in general, are not extremely critical since these stations can be located in the more remote areas. These remote locations enable OH transmission lines to enter and exit the stations. As population densities increase, suitable locations for these stations are becoming more difficult to find. For this reason, new station designs are being studied that will use gas insulation to make a compact, integrated metal-enclosed equipment package. The most common gas used to date for this application is SF_6 (sulfur hexafluoride). With these designs, the land area is about one-tenth that required for open air and only about one-fiftieth the volume of a conventional station is required.

Transmission Systems. The new types of underground transmission systems, with a projected time schedule for their commercial application, are shown in Fig. 11. As technology is developed and as larger power transfers are required, each of these systems should serve as a means of transmitting electrical energy underground.

1. *Pipe* These systems utilize existing oil-impregnated paper-insulated cables contained in metal pipes with oil under high pressures. Systems are in service at voltage levels up to 345 kilovolts. Testing is being done on systems up to 500 kilovolts. High-quality insulation systems are being investigated in both paper and synthetic tapes for use in 500- and 765-kilovolt systems.

2. *Gas Dielectrics*   As of the mid-1970s, these systems are beginning to be used for special applications involving short runs. The highest voltage being used is 345 kilovolts and is in a single-phase design. In this system, SF_6 gas is used for the insulating and cooling medium. Research is being done on enclosing all three phases in a common enclosure to reduce the overall system size and installation cost. Research and development also is being applied to enclosures that are flexible, another factor to contribute to ease of installation.

3. *Resistive Cryogenics*   Since electrical resistance is a function of the conductor temperature, it is possible to reduce the heat losses in a system when it can be operated at a low temperature. As an example, aluminum has an electrical resistance at liquid-nitrogen temperature ($-320°F$; $-196°C$) or one-tenth that for normal ambient temperatures. With these lower losses, higher transmission capacities can be achieved, and projected costs indicate that a system of this type might be competitive to conventional systems in the range of 2000 to 5000 MVA (megavoltamperes).

4. *Superconducting Systems*   These systems operate at even lower temperatures, down into the liquid-helium range of $-452°F$ ($-266.9°C$). Certain metals, such as niobium, lose all electrical resistance at these temperatures with the result that extremely high currents can be transmitted with low losses. The complexity of these systems, together with the difficult operational problems, indicate that it will require extensive research to achieve commercial equipment. Preliminary estimates are that these might be competitive for capacities of 5000 to 10,000 MVA.

DIRECT CURRENT TRANSMISSION

The extent of dc lines in the United States in 1970 totaled about 900 miles and was less than 0.5 percent of the total transmission installed. An accurate projection of the amount of direct current to be installed in the future is difficult to make, but its application should increase. The use of direct current will be confined to limited applications where special problems exist. These general areas of use are:

1. Special stability problems involving remote generation
2. Long-distance point-to-point power transfers with overhead systems
3. Asynchronous ties for control of power flow
4. Dc cable systems for power supply to high-density load areas

In early 1973, the Electric Research Council issued a request to manufacturers for a proposal to install a 100-MW dc link on an electric utility system. The objectives of this prototype are to demonstrate advances in the state of the art for direct current in control and operation and in reducing the size of the terminal stations. These terminals are to be installed on the Consolidated Edison Company system in New York. Two other direct current systems are under construction to deliver power from coal fields in North Dakota to major load areas in Minnesota.

Trends in Technical Use of Computer Systems in the Power Industry

BY M. M. ADIBI*

BACKGROUND

In scheduling its day-to-day operation and in planning for its future growth, the power industry has made extensive use of analytical tools and mathematical modes which, through optimization and simulation, help in the decision-making process. A schematic description of the ingredients of all such models is given in Fig. 1. The industry has long been one of the largest users of computers and among the most sophisticated in its modeling and computational techniques. This is quite understandable when one considers the great cost of power system equipment, the complexity of power systems, and the severe operational, reliability, and environmental requirements on the electricity supply. Computer applications have assisted the industry in achieving its objectives of reducing cost of energy delivered to consumers, improving quality of service, and enhancing the quality of the environment. These objectives have been achieved to a large degree in the following manner:

1. Since the industry is one in which capital investment is usually high (over 10 percent of total spending by the nation's industries), unit costs have been reduced by operating facilities closer to their design limits, allowing better utilization of equipment.

2. Unit costs have also been reduced by automation, allowing operation with fewer personnel, and by optimization, lowering fuel consumption per kilowatthour of delivery.

3. Electricity cannot readily be stored. Therefore, production and consumption must be simultaneous. Hence, enough capacity is required to meet the coincident demand, or peak load, of all customers. Since the peak loads of various customers occur at different times, interconnections between power systems can provide important economies arising from different time patterns or diversity of use of the component systems in the network.

4. Quality of service has been improved by reducing the number, extent, and duration of service interruptions.

5. Quality of environment has been maintained by operating facilities within acceptable bounds of emission, thermal discharge, and waste disposal.

As of the mid-1970s, the industry has reached a stage where computer systems are no longer simply engineering tools. The effectiveness of computer applications is one of the key elements in achieving the basic functions associated with the planning, designing, construction, operation, and maintenance of the power system. In fact, engineering and computers have been integrated. This integration may be viewed as tending toward the construction of a utility industry information system. Such a system is shown in Fig. 2. This figure graphically illustrates a typical information system which may be viewed as a combination and integration of eight functional information systems.

Such an information system can extend management capabilities through information accessibility by providing the ability (1) to economically maintain complete central information files which are immediately accessible from any location in the company or system for information entry and retrieval, (2) to give status information on any major maintenance or construction project, (3) to summarize all revenue and expense for closer cost control, or (4) to furnish daily operating statements. The information system can

*Industry Consultant, International Business Machines Corporation, Houston, Tex. The author and editor are deeply grateful to the following individuals who reviewed this summary prior to publication: Vince Converti, Manager, Systems Engineering, Arizona Public Service Company and Secretary, Power System Engineering Committee, PES, IEEE; Lloyd W. Coombe, Director, Technical Services, Detroit Edison Company; Denos C. Gazis, Consultant, General Sciences, T. J. Watson Research Center, IBM Corporation; Lester H. Fink, Program Manager, Energy Research and Development Administration; and Richard J. Schulte, Program Coordinator, Electric Power Research Institute.

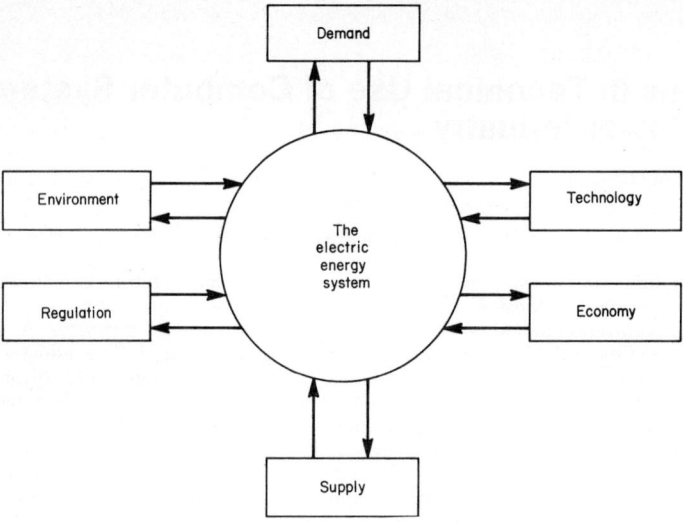

Fig. 1. Electric energy model.

provide meaningful data regarding all facets of utility operations, at the proper time and location, to assist every level of management in making decisions which will result in close control over operations. In this summary, only a part of such an information system can be described briefly. No attempt is made to deal with business data processing, which is a very extensive field of computer use and includes a great variety of computer applications. In the future, the division between business and technical computing will become less distinct as comprehensive data bases and data communications paths are formed to serve both purposes.

GOALS OF THE POWER INDUSTRY

The industry's dedicated purpose is to provide adequate, reliable, environmentally compatible, reasonable-cost electricity with the ultimate goal of improving the net earnings per common share. In spite of the apparent differences between publicly and

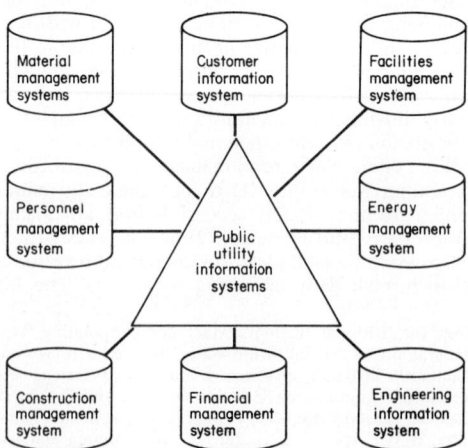

Fig. 2. Public utility information systems.

privately owned utilities, the foregoing goal is applicable to both situations in different forms.

The prime goal is reached by pursuing a number of objectives as shown in Fig. 3 and described briefly as follows.

IMPROVE FINANCIAL MANAGEMENT

1. Raising New Capital.   Demand for electric power is growing at an annual rate of about 7 percent so that the industry must double its size in less than 10 years. To finance

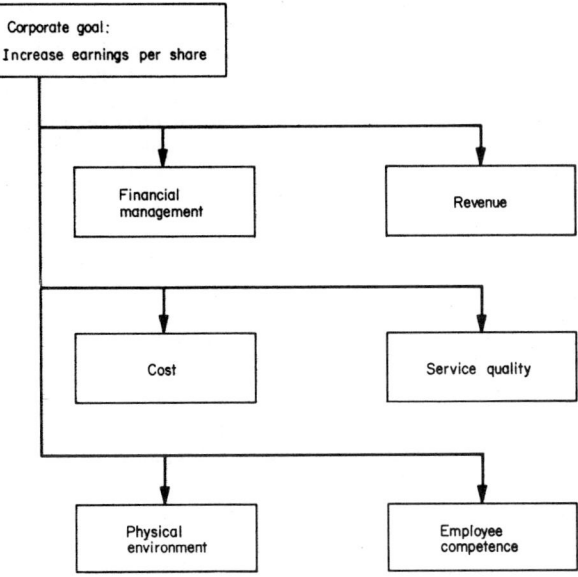

Fig. 3. Goals of the power industry.

new plants, the industry has to raise more new capital over the next 5 years than in the last 10 years. The projected construction for the industry in the next 10 years runs into hundreds of billions of dollars. This, in view of the interest and inflation rates of recent years, is a most difficult financial task, but nevertheless an important goal to reach.

2. Plant Investment.   In the next 10 years, utilities will be spending considerable sums in generating plants and transmission facilities. Cost of a modern power plant and associated transmission lines is enormous. Present-day decisions on such additions, together with proper selection of plant sites and acquisition of transmission right-of-ways, have long-range financial implications affecting earnings. The industry's goal here is to make the right decision based on long-range plans.

3. Long-Term Contracts.   Fuel constitutes about 35 percent of the industry's total annual operating expenditures. A modern power plant consumes about 500 tons of coal each hour. Average life of such a power plant is about 30 years. A nuclear power plant of similar size requires an initial nuclear core expense consisting of about $100 million and then a significant annual expenditure for the next 30 years. The goal is to procure these fossil and fissile fuels through long-term contracts, providing a continuous supply of reasonable-cost fuel throughout the lifetime of the plant.

4. Growth Through Affiliation.   There has been a significant number of corporate mergers between large and small utilities. The goal in these affiliations is to meet the growth in demand for energy by taking advantage of economy of scale, consolidation of administration, engineering research and development, and increasing reliability of bulk power supply.

INCREASE REVENUE

For 30 years (1935–1965), utilities were in a spiral of lowering costs, lowering rates, and increasing sales. During this period of falling rates and due to regulatory lag (in rate adjustment), utilities enjoyed a higher revenue and, therefore, were motivated to be efficient. The costs were reduced by installation of larger generators, higher transmission voltages (economy of scale), lower fuel costs, and shifts to the available gas and oil from coal.

In the past decade (since 1965), the utilities have gone through a period of rising costs due to (1) rise in fuel cost, (2) environmental impact, (3) reaching a diminishing return in the increase in size and improvement in efficiency of units, and (4) investing in a new technology (the nuclear energy plants).

During this rising-cost period, the regulatory lag in rate adjustments has had an adverse economic impact. Utility executives are seeking, and regulatory commissioners are just now beginning to project and apply, a regulatory lead. This approach demands detailed analysis of past and present operations and projection of future requirements by financial modeling, optimization schemes, and simulation techniques.

REDUCE COST

Cost reduction can be achieved (1) by reducing investment per kilowatt of installation capacity for generation transmission and distribution and (2) by reducing operating cost per kilowatthour of energy delivered.

1. Reduced plant investment can be achieved by (a) installing larger generating plants (3000 megawatts) and (b) higher transmission (750 kV) and distribution (34.5 kV) voltages. Power pooling, interconnection planning, and coordination can be used to gain further advantages of scale.

Improving production and distribution facility utilization (i.e., capacity factor and load factor) through peak sharing, reserve sharing, load diversity, and distribution load balance are important considerations. Other factors include designing facilities with more precision and reducing the factor of safety—and operating the facilities closer to their design limits.

2. Reduced operating expenditure can be achieved by (a) adopting new technology (such as breeder reactors), requiring lower fuel costs; and (b) by improving conventional and established methods of higher energy conversion efficiencies.

Other factors include reducing energy losses in transmission and distribution facilities, interchanging energy with different time zones in different seasons to take advantage of diversity, and, also, producing and distributing electricity with fewer personnel. Minimizing labor force and material inventory required for maintenance, repairs and restoration of generation, transmission, and distribution facilities are associated factors. Added to these are means to reduce customer accounting, general accounting, and administrative expense.

IMPROVE QUALITY OF SERVICE

Factors include reducing frequency, duration, and extent of outages in power supply, plus reducing voltage and frequency discontinuities and sudden excursions (power line disturbances) to sensitive electronic loads and digital equipment.

Another factor is improving customer service through prompt response to inquiries, requests, or complaints. This includes maintaining power supply within prescribed ranges and specifications, thus avoiding over- and under-voltages and frequencies, and restoring interruptions to service quickly.

ENHANCE ENVIRONMENT

Reducing thermal discharge to natural bodies of water through the use of artificial lakes, cooling towers, and desalination processes are desirable techniques, as well as advancing direct conversion of heat energy to electrical energy, such as via magnetohydrodynamics, thermionics, and fuel cells. Other factors include reduction of release of combustion products (sulfur dioxide, nitrogen oxides, and particulate matters) into the atmosphere and reducing the frequency, duration, and intensity of pollution concentrations in urban areas. Providing more productive uses for fly ash and safer storage of nuclear waste are important considerations.

Major problems include the selection of remote or underground sites for generating stations and improving the aesthetics by increased use of underground distribution facilities. Beautifying transmission towers and lines in harmony with the countryside is another goal.

IMPROVE EMPLOYEE COMPETENCE

1. Labor. The industry has a labor force of about half a million employees, which is not very large considering the tremendous output of the industry. In the past 10 years, while the generating capacity doubled, the number of employees remained substantially the same. This has been achieved through (a) operating large installations with fewer personnel per output, (b) centralized control of generation and transmission, (c) unattended substations, and (d) minimizing maintenance and repair crews by automating dispatch procedures. Such trends will continue in the future.

2. Professional. The design and construction of large installations, such as generating stations or extra-high voltage lines, are often contracted out and are engineered and supervised by consulting firms. Thus, in effect, the consultants provide a common professional pool for all utilities. The electrical equipment manufacturers have been primarily responsible for research and development of the industry; the practice of accepting turn-key contracts is common, and thus manufacturers also provide a common pool of professional personnel.

However, the advent of nuclear power, extra-high voltage, and environmental limitations requires significant changes in utility systems and calls for an increase in both quality and quantity of professional skills. The industry recognizes the need for this rapid increase of "in-house" know-how and competence. This goal could be reached by (a) improving the productivity and effectiveness of employees, (b) merging and affiliating with the neighboring companies forming regional groups and taking advantage of "critical mass," and (c) supporting institutions of higher learning and research organizations by sponsoring research and development efforts.

SPECTRUM OF COMPUTER USE

A general review of the engineering and operating computer application indicates that these tasks may fall within six broad categories, as shown in Figs. 4 and 5, and as described briefly as follows:

1. Long-Range Planning. These applications are related to 20-, 10-, 5-, 3-, and 1-year "construction programs" and embrace planning, design, and construction of new facilities. These functions are performed at least once per year and use long-range load forecasts and other predictions as input data. Batch data processing using large-scale computers, bulk auxiliary storage, and fast peripheral equipment, such as printers and card-read-punch, can effectively meet the computational requirements of this category.

2. Short-Range Planning. These applications deal with annual construction of new facilities and the economic and reliable operation of these additions in conjunction with other interconnected power systems. Nuclear fuel management, annual hydrothermal coordination, and coordination for firm transmission and generation planning are among these functions. Because these programs are more frequently called upon, they could reside in large random access storage devices. Thus, only changes in data and programs need be entered when using specific programs.

3. Long-Range Scheduling. These applications are related to monthly and daily operation of the power system. In this category, are transmission and generation maintenance scheduling, unit commitment and withdrawal, and other functions dealing both with reliability and economy of operation. Remote data processing consisting of central computer system, associated peripheral equipment, random access files, and remote operators' terminal meet the requirements of reliability and economy of operation. The operator enters data and inquiries for new schedules through an appropriate operator-machine communication terminal. These terminals can be located at all engineering and operating centers.

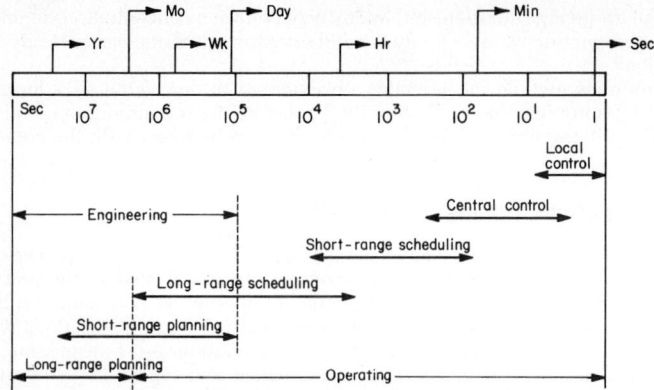

Fig. 4. Computational requirements in terms of time availability.

4. Short-Range Scheduling. These applications are related to determination of reserve indices, hourly data recording, and like data. These schedules are performed at least once an hour. They are based on historical data, but also need current power system data, such as facilities in and out of operation, generation outputs, and line flows. Therefore, they require direct data flow into the computer. The results, however, are presented to the operator for consideration and execution.

Because of the scheduling nature of these applications, very fast data acquisition is not a prerequisite. However, accuracy and timeliness of schedules are related to the extent that they include direct data acquisition.

5. Central Control. These regulating functions are carried out to meet the changing demand of the power system. Power system monitoring and display, rescheduling, and control of system frequency, tie-line flows, voltage conditions, and transmission flows are examples of this category. These functions are performed in the few-seconds to several-minutes range and, therefore, require not only direct data flows into the central computer,

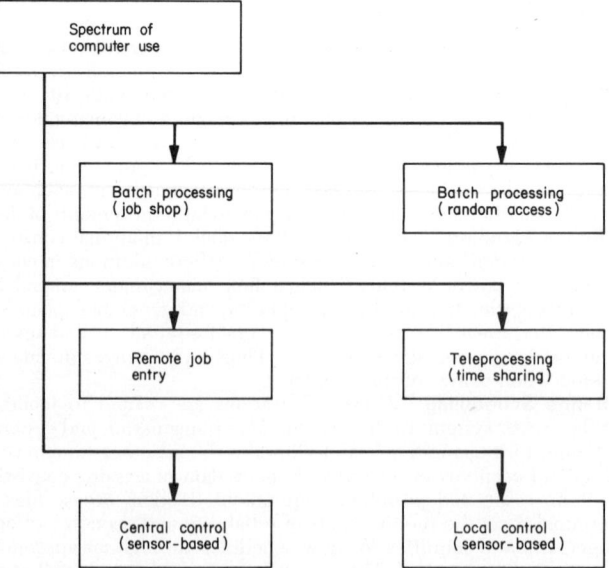

Fig. 5. Spectrum of computer use in electric utility industry.

but, in addition, require signals from the computer to the various remote controllers and actuators.

6. Local Control. These applications require a response speed beyond the capability of central computer control. Most of these functions are initiated immediately after a fault develops or a variable exceeds certain limits. Their objective is to react quickly and correct the situation, or to isolate and contain a disturbance. These functions are performed in the few-milliseconds to several-seconds range and can best be handled by local computers (*a*) by directly sensing variables and controlling through actuators (examples of this type of operation are direct digital control of boilers or digital relaying of the substations); and (*b*) by superimposing the computer on the local controllers or protective relays in order to reset their operating points. (These are in the 1- to 10-second range).

The computational requirements shown in Fig. 4 cover both engineering and operating functions. These two areas of computer activities are interrelated. From the foregoing discussion, it is clear that the power system operating functions do not all necessarily have to be performed on a real-time basis.

CONTEMPORARY COMPUTER APPLICATIONS

The major technical computer applications can be grouped into (1) power system planning, (2) operation, (3) plant monitoring and control, (4) design, (5) management science, and (6) construction.

POWER SYSTEM PLANNING

As the electric utility industry has grown in size and complexity, modifications and additions to existing electric power networks have become increasingly costly. Therefore, it is vital that different design possibilities for additions and modifications to the network be studied in detail to determine their effect on the network, their effect during abnormal operating conditions, and their applicability as a solution to current and future power demands.

To assist electric utility planning engineers in evaluating various network configurations, a group of highly specialized digital computer programs is being used by the electric utility industry. These programs simulate the steady-state operation of an electric network (power flow), bus voltages and line flows in an electric network under fault conditions (fault studies), and synchronous machine swings during network disturbance (transient stability). See Fig. 6.

With the growth in size of electric power networks, the large volume of data to be handled becomes a limiting factor in the number of studies or alternative design plans that can be considered for any one network in a reasonable period of time. Network data must be collected, classified, prepared for computer input, and filed for convenient and easy retrieval.

Types of programs in use include:

1. *The power flow program* is one of the major tools of the system planning responsibility and is utilized extensively. This includes the smallest subtransmission (69 kV) region to be studied up to a very large (4000 bus), highly interconnected network.

Important to the system planner is that input data errors be minimized and that there be an easy and rapid turnaround for answers when the program frequency of use is high. To accomplish this, an inexpensive terminal capability is provided with the ability to store base cases or numerous power system models on the computer's direct access storage files.

The storage capability provides many different cases that the planning engineer can access for analysis for studying or varying a particular system condition.

The simulation programs associated with the power flow program comprise a system of linked programs that have the following capability:

a. Basic programs involving calculation of voltages, power flows, angles, and interchanges between areas of a power system

b. A network reduction program to represent large networks as equivalents in conjunction with the specified area to be studied

c. A current distribution program that indicates the sensitivity of response of the various circuits to outages of specified transmission lines

  d. The series of programs associated with the stored power flow files which permit accessing a particular case, deleting a case, adding a case, and changing a case

  e. The var allocation program that selects the minimum amount of kilovars of compensation necessary to maintain bus voltages within specified limits under normal and/or emergency conditions

2. *Other major simulation programs* used in the power system include (*a*) the *tran-*

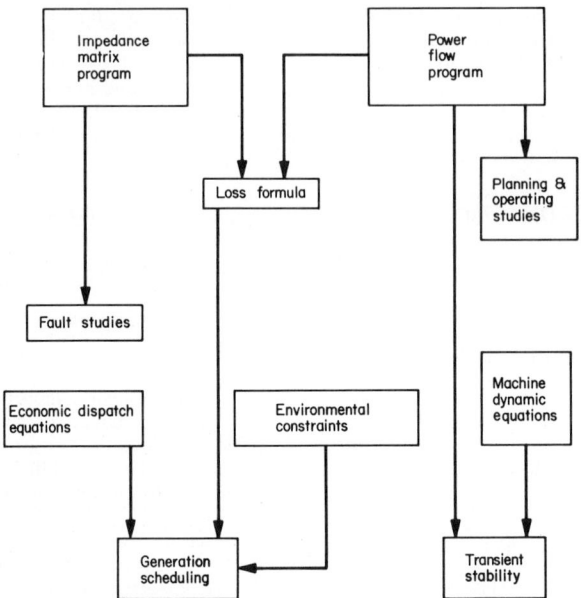

Fig. 6. Power system planning programs in frequent use.

sient stability program, which is used for determining the effects of large power system disturbances; (*b*) the *short circuit program*, which is employed to calculate three-phase and line-to-ground fault currents; (*c*) the *relay coordination program* used for determining relay settings; and (*d*) the *lightning and switching surge program*, which calculates the voltage and current transients resulting from switching surges and lightning strikes.

3. *The distribution circuit analysis program* calculates for a radial distribution circuit, load currents, voltages, losses, impedances, and fault currents. The program is used to assess the performance of the many distribution circuits in operation on the power system and to plan future distribution facilities. A distribution circuit file is maintained in a similar manner to that for power flow base cases, and remote terminal input/output capability is provided and used to serve the distribution engineers who are located remotely from the computer center.

4. *Other simulation programs* used in power production include (*a*) power plant production cost for generation planning, (*b*) generalized heat balance for steam power plants, and (*c*) nuclear plant fuel management.

  Major recent work involving simulation is associated with the development of a corporate model. The corporate model is designed as a management analytical tool. It will permit examining, over a 5-year period, the effects of capital deferrals and advancements associated with construction, changes in electric rates or fuel costs, the impacts of various load growth or marketing patterns, alternative financing methods, various changes in government taxes, and other factors.

  Other engineering computer activities include the development of a power plant unit maintenance scheduling program to handle the constraints associated with scheduling maintenance of the many generating units on typical systems.

POWER SYSTEM OPERATION

The prime concern of the electric utility industry is to meet the consumer's power demand at all times and under all conditions. Electric utilities are continually seeking and have been most receptive to every available technique which would (1) reduce capital investment per kilowatt of installation, (2) reduce the operating expenditures per kilowatthour of energy delivered, and (3) improve quality of service to the consumer.

Digital computers can be found in use at all levels of operation in power systems today. At the generating-plant level, for example, they are used to control unit start-up and to monitor operating conditions. In bulk power substations, they serve such functions as monitoring, event recording, and switching. At the system operating center level, digital computers help (1) to improve economy of operation, (2) to improve quality of service, and (3) to simplify system operation. In this descriptive summary, it is possible only to consider computer applications at the system operating center level; and only a few topics can be described in any depth. To give a better perspective of the topics that will be covered, it is worthwhile to list some of the major applications in each of the three foregoing categories.

Under economy of operation, there are generation scheduling, interchange scheduling, energy accounting, unit commitment, maintenance scheduling, and hydrothermal coordination. In the category of quality of service are included security assessment, reserve requirements, load relief strategy, contingency checks. equipment and relay coordination, load frequencey control, and voltage scheduling and control. In the category of system operation, there are system monitoring and display, dispatching, maintenance and repair of equipment, logging of power system conditions and events, and aids to dispatchers for making decisions.

1. Computer, Display, Data Acquisition, and Control Systems. The basic requirements in the operation of a power system are (a) measurement and recording (storage of power system variables), (b) correlation and reduction of the measurements into meaningful data indicative of power system status and performance, (c) comparison of the data with the established information and past record to arrive at appropriate decisions, and (d) execution of these decisions to control the power system for reliable and efficient operation.

The primary purpose of a computer system is to improve quality of service and economy of operation. Service quality is measured by service availability, voltage regulation, and frequency regulation. Economy of operation is measured by cost of delivered energy. Accordingly, the required functions of a computer system can conveniently be divided into two broad categories: (1) those that deal more directly with improving service quality and (2) those functions that are related more closely to economy of operation. The two categories are, of course, interdependent.

a. *Service Quality Programs.*   This category of programs includes:
(System Security)
    (1) Frequency rescheduling and control
    (2) Transmission flow rescheduling and control (active and reactive)
    (3) Tie-line flow rescheduling and control
    (4) Voltage profile rescheduling and control
    (5) Indices of system reserve and security
    (6) Area (island) isolation planning
    (7) Load-shedding schedules
    (8) First contingency check
    (9) Equipment coordination check
    (10) Relay coordination check
    (11) System state monitor and display
    (12) Prearranged clearance requests
    (13) Unscheduled clearance requests
    (14) Work permit "program"preparation
    (15) Simulator for dispatcher's exercise and training
(Training)
A special case of the first and third programs is the conventional load-frequency control. The voltage profile rescheduling can also be viewed as a reactive power-dispatching program.

b. Economy Operation Programs.   The major programs in this category include:
(1) Generation scheduling (economic dispatch)
(2) Interchange scheduling (inter- and intra-)
(3) Interchange energy accounting
(4) Unit commitment
(5) Maintenance scheduling
(6) Hydrofossil nuclear coordination
(7) Pumped-storage scheduling

The foregoing programs used generating plants' cost characteristics, transmission line loss coefficients, water and equipment availability, and short- and long-range load forecasts in order to determine an operating schedule which will result in minimum fuel and labor costs, while improving equipment utilization.

The generation scheduling program is based on daily load forecast and, usually is run at least once an hour. Because the generating plants' cost characteristics usually have been established several months before scheduling, they do not completely reflect the current operating conditions of the plant.

The interchange scheduling and energy accounting programs are normally used for hourly tansactions with the neighboring systems.

There are some advantages in performing the first three functions of the preceding list on a sensor-based computer, because the current data flow from the power system allows calculation of current line losses. However, such incremental benefits must be compared with the added cost of the sensor-based computer requirements.

Pumped-storage scheduling is a daily program. Hydrothermal coordination is performed both on a short-range (daily) and a long-range (annually) basis. Water value coefficients are generally determined in the long-range studies and are then utilized in the daily and hourly use of hydrothermal coordination equations.

Selection of generating units to be operated or withdrawn from operation (unit commitment) is based on the daily load curve, production cost characteristics, start-up and shutdown cost and time, and duration of operation. This program can be executed as often as every 6 hours. Maintenance scheduling also depends on the foregoing factors as well as maintenance costs. It is typically performed weekly or monthly. However, because of a forced outage, it may be necessary to have additional "on demand" runs with this program and revise the schedule.

2. System Description.   Before describing the computer, display, data acquisition, and control systems, it is in order to describe briefly the power system operating facilities.

Load Dispatching.   Most power systems are under control of automatic load regulators which are installed in the system operating center (SOC). These analog load regulators receive signals representing frequency, plant generation, tie-line flows, and nonconforming loads over telemetering equipment via communication channels, such as leased telephone lines, microwave systems, or power-line carriers. They adjust power system generation by transmitting raise and lower signals to the plant in order to maintain system frequency, meet the interchange commitment with neighboring power systems, and allocate generation among the various plants on an economic basis.

The load dispatcher collects power system data from indicating and recording instruments and uses these data together with load forecasts to estimate the next day's load. The dispatcher then determines how much generation is required and which generators should carry the load. The dispatcher also determines schedules of preventive maintenance for generators or transmission lines. In addition, the load dispatcher governs bulk power substation and transmission switching for safe and economical system operation.

Power systems serving extensive areas tend not to centralize all operations in one system operating center, but divide the system into several regional operating centers (ROC). Each region has jurisdiction over local operations. The functions delegated to the regions vary with the utility's policy and requirements.

The load dispatchers and regional dispatchers supervise three main functions: (1) reliability of service, (2) quality of service (i.e., voltage and frequency), and (3) economy of production. For these operations, both SOCs and ROCs would be equipped with appropriate facilities, such as communication, metering, pilot boards, and load regulators.

System Dispatching.   The transmission system is normally supervised and operated from a central office. Generally, this is the same office (SOC) where load-dispatching facilities are located. Here a one-line diagram representing generating plants, transmission circuits, and bulk power substations is usually provided on a curved display wall-board. The system dispatchers use this diagram of the power system as a guide for clearing equipment for scheduled and unscheduled maintenance.

In order to prepare clearance instructions for field personnel, the system dispatcher studies the board and determines what steps should be taken to maintain service while the requested equipment is removed from service. As these steps are followed by the field personnel, information is relayed back by telephone and the system diagram is manually updated. For scheduled outages, it is estimated that in 1 to 2 percent of the outages the dispatcher cannot forsee what may actually happen as a result of a decision. Under these circumstances, alternative operations may be tried in order to find the correct procedure. Sometimes the whole operation may be postponed until periods of very light loads (weekends, for example), but this procedure is undesirable since it could result in higher labor cost due to overtime and in interruption of service. The percentage of errors is understandably higher for unscheduled outages when the system dispatcher is under duress.

Switching Operation.   Many bulk power substations are equipped with substation monitoring and control systems (SMAC). This equipment monitors positions of all the substation circuit breakers and notifies the dispatcher at the regional operating center by audio and visual alarms when circuit breakers trip out. The dispatcher is also provided with a means for selecting circuit breakers for close or open operation. In addition, the dispatcher can select certain lines for measurements of currents, voltages, or other variables. The dispatcher normally keeps a log of all switching operations and trip-outs.

In some metropolitan power systems, the bulk power substations are attended and act as receiving stations for all the distribution substations. The distribution substations are monitored and remotely controlled from the bulk power substation. The most significant information is relayed directly to the system dispatcher.

Power Pool Operation.   Most utilities are members of an interconnection or a power pool. These pools are organized to coordinate generation and transmission expansion and operation within and between pools. Some also coordinate interpool operation.

A power pool takes advantage of load diversity, generation and transmission outage diversity, and economy of size in order to reduce operating expenditure and installation cost and to improve reliability. The operation of a power pool is directed from a pool operating center (POC) that is typically equipped with facilities similar to a system operating center (SOC), but which performs some different functions.

The POC is linked to all SOCs within the pool by telemetering and communication facilities. In addition, it directly monitors inter- and intra-pool tie-line flows, major transmission lines, and system frequency. The POC primarily performs scheduling functions based on short-and long-range load forecasts. These functions include, but are not limited to, generation scheduling, unit commitment, generation and transmission maintenance scheduling, pool-to-pool interchange scheduling, and inter- and intra-pool energy accounting.

Energy Management System (EMS).   Regardless of the type of functions performed or the frequency of execution, the SOC, ROC, and SMAC areas just described all deal with the same physical and operational data related to a given power system and all help to improve the power system performance. It is then economical and desirable to have a computer system which will encompass the entire power system by using common data banks. This type of system may be called an *energy management system.*

The EMS may have a general configuration along the lines of Fig. 7, which consists of the following subsystems.

1. *A computer system at the SOC* performs load dispatching, system security, and system-dispatching functions related to the high-tension transmission network. It includes central processing unit(s), large random access storage, card input/output unit, and printers. It provides an interface to the operator's console(s), which include cathode-ray tubes with graphic capability, the large instrument board, and the wall display boards. The communication control unit provides communications paths between the SOC system and the POC, the ROCs, and the terminals at the generating plants.

2. *A relatively smaller computer system* is required at each ROC to perform system dispatching for distribution and switching operation for distribution and transmission for the entire region. The ROC system functionally overlaps the SOC system as well as adjacent ROC systems. It consists of central processing unit(s) with peripheral equipment,

such as card input/output unit, and output printer, auxiliary storage and display units. It is capable of interfacing with the operator's console and communication control unit. The latter links this center with bulk power substations, maintenance crew trucks, and the SOC.

3. *The substation monitoring and control system (SMAC)* includes a master terminal (or small computer) with appropriate input/output signal capability at every bulk power substation (and step-up substation), and facilities at the ROC for proper interface to the computer and/or master terminal. The SMAC system monitors substation parameters, provides for remote switching operation through the regional operation, and functions as

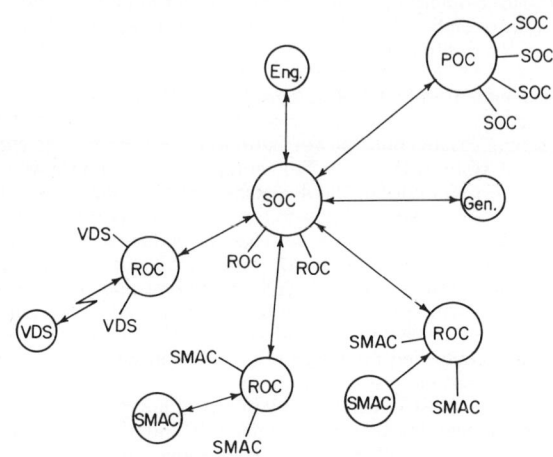

Fig. 7. Energy management system (EMS).
POC   Pool operating center
SOC   System operating center
ROC   Regional operating center
SMAC   Substation monitoring and control
VDS   Vehicle dispatch system
Gen.   Generating plants
Eng.   Engineering department

the data acquisition medium for pool, system, and regional operating centers. The SMAC system is the interface between the power system and the EMS.

4. *The vehicle dispatch system (VDS)* includes a mobile terminal in each maintenance crew truck and facilities at the region for interfacing with the regional operator and the ROC system. The mobile terminal may include a typewriter to give hard copy output in the truck and a keyboard to enter data into the ROC system. The VDS system provides an efficient means of communication between regional operators and trouble crews in the field. This speeds up restoration of service, avoids misunderstanding in times of emergency, and saves labor and material through efficient service dispatching.

5. *The SOC system has a direct link to computers* installed in major generating stations. Both load dispatching and system security functions are improved by having current information on the status of generating units. In the absence of plant computers, the SMAC terminals placed in the step-up substations can be expanded to handle the required data flow between plant and system operating center via the ROC system.

6. *The SOC system may also have a number of terminals* in the engineering department to provide engineers with current data on power system operation and a means of modifying the application programs in the SOC and ROC systems. These terminals may have typewriters, cathode-ray-tube display units, and input card units.

7. *The SOC system is linked with the POC* to transmit and receive information required for pool operation. The POC system is very similar to the SOC system except that it communicates only with the SOC system of pool members. At every member's system operating center, there is a terminal, pool console, and interface to its SOC system.

POWER PLANT MONITORING AND CONTROL

Fossil Fuel Plants.   Computers are used at power plants to record (log) measurements of steam pressures, temperatures, flows, and electrical loads and to check each measurement for proper operating range. In some cases, when unusual conditions occur, these data are automatically printed on the system printers as an alarm report. In other cases, outputs are printed only upon request of the system operators (demand reports) or on a periodic basis. In all cases, the information is used for subsequent review and analysis.

The computer is also used to assist the operator in controlling the start-up to near-synchronization speed.

The installation and operation of the data acquisition and control system has enabled utilities to realize significant operational and monitoring improvements in the daily performance of the fossil fuel unit. Examples include:

1. *Consolidation of Operations.*   Using the operator's control panel, a single operator and assistant perform all the control and monitoring functions for the unit or plant. Without installation of a centralized control room and a computer system, operators have to be stationed throughout the plant at boilers, pumps, switches, and turbines. With a computer system, each of the turbogenerating systems is controlled as a unit by one person in one location with a minimal amount of manual intervention.

2. *Speed of Data Collection.*   Operators have more information about the functioning of the equipment and have it faster than they ever did previously. Input points are continuously alarm-checked by the system without operator intervention or observation. Rapid changes in pressure and flow measurements can be instantaneously observed and, when required, remedial manual action can be initiated immediately.

3. *Safety Standards.*   The computer system provides additional knowledge about operating conditions and possible malfunctions of plant equipment. As a result of this expanded intelligence, equipment can more easily remain within safe operational limits during each day's processing.

Nuclear-Fueled Plants.   In a nuclear generating plant of large capacity, the data-logging function assumes major proportions. Initially, data loggers were considered for the purpose of data acquisition only. Later, it became apparant that the analytical procedures related to nuclear power generation would require the greater capabilities of a digital computer. In addition, the ability to analyze various plant parameters assists in optimizing plant operation. Daily accumulation of reactor plant data provides reactor engineering personnel with information to indicate important operational trends and to perform more detailed simulation studies on a larger computer. Some of the advantages of these computer applications include:

1. Data as accumulated by the computer serve the needs of the total plant.

2. The process-monitoring capabilities help the plant operators in minute-to-minute operations.

3. Data that are gathered in systematic form assist in various equipment maintenance considerations.

4. Instrument personnel can use the system's capabilities to troubleshoot instrumentation problems.

5. Reactor engineering personnel can obtain analytical data that help in the optimization of reactor output. Adjustment of the predicted burnup rate of the nuclear fuel is done on a continuing basis to provide management with information necessary for the scheduling of the plant-refueling operation.

Plant Management System (PMS).   Each plant has a wide spectrum of data-processing requirements in areas that are presently satisfied mostly by manual means. Some of these requirements are in areas where computers have been conventionally applied, such as in maintenance management, parts and material control, and personnel management. In addition to these areas, each plant is required to collect operations data for use by such groups as system dispatching, design, planning, engineering, and results engineering.

In earlier times, it was not practical to consider installing computer equipment at existing plants because the cost for equipment to provide the necessary data-processing capabilities was prohibitive. The advent of computers having teleprocessing capabilities and other allied developments warrants a renewed investigation into providing computer capabilities for power plants.

The concept of the plant management system (PMS) applies a centrally located computer to the operations of a utility's remote generating plants. It provides data-processing services for all power plants through a communication-based computer system, requiring only terminal equipment at the power plant sites. The terminals communicate with the central computer by means of conventional communication facilities. All data storage, computation, and related activities are provided by the central computer.

The application requirements will differ, of course, for coal or nuclear steam-generating stations and for gas/oil-fired steam plants. It is also recognized that specific procedures among electric utilities can vary significantly.

Some application areas are described briefly as follows:

1. *Plant Data Collection*   Plant data are collected automatically by reading instruments. Data are stored and retrieved as required to prepare reports, perform analyses, and related functions.

2. *Report Preparation*   Reports and logs are automatically generated. These reports include those required by the plant management, load dispatchers, and planning and engineering groups.

3. *Operator Guide Control*   Collects and analyzes operating data. Plant supervisors are informed when an inefficient or abnormal condition exists. Periodic reports are prepared to reflect plant operation.

4. *Determination of Unit Incremental Generation Cost*   Periodically updates unit cost data by collecting data that continuously reflect actual operating conditions. This information is transmitted to the load dispatcher for use in loading dispatching.

5. *Maintenance Management*   Stores pertinent plant maintenance information. This information is later retrieved as maintenance history for analysis of maintenance costs and evaluation of equipment performance.

6. *Parts and Material Control*   Maintains inventory records of parts and materials stored at each plant. Many of the functions of normal inventory control are performed, such as stock count, reorder-at-minimum-inventory cost, and charge for parts of material issued.

7. *Personnel Management*   Performs the tasks of timekeeping and maintenance of personnel records. Entries are made on an exception basis and PMS stores all information required by the payroll group. In addition, information, such as overtime hours, vacation time, and sick time, is stored and continually updated for quick retrieval.

DESIGN APPLICATIONS

The design of planned facilities involves the efforts of a large engineering staff and a substantial investment in facilities. To provide support in these activities, computer programs have been developed for analysis of specified designs. The application of these programs contributes to the installation of reliable and economic facilities. An appreciable percentage of the total computer production hours and use is devoted to runs associated with design activities. Major programs developed for the design of power plants and associated facilities include:

1. Structural steel framing program which is used to design the beams, columns, girders, and base plates of power plant structures.

2. The foundation slab analysis program which is used to design large, complex foundation mats. The results provided include evaluation of various slab thicknesses, soil-bearing pressures, shears and bending moments, and reinforcement areas.

3. The concrete stack analysis program which is used to analyze proposed stacks by determining loadings, resulting stresses, and required steel reinforcement. This program is used extensively in the design of very tall concrete stacks selected for new power plants. The analysis includes stack weight, wind load, earthquake, and temperature parameters.

4. The nuclear containment structure program performs an evaluation of the structure with regard to seismic-loading conditions.

5. The power plant piping program analyzes the flexibility of a piping system under the influence of temperature changes.

6. In the area of transmission line design, several additional programs have been developed to aid the transmission engineer, including the tower analysis program which provides a design bar force summary, giving the maximum tension and maximum

compression for each member of a three-dimensional structure over the entire load range specified.

MANAGEMENT SCIENCE APPLICATIONS

In the early stages of computer applications, independent programs were developed to meet specific computation requirements of engineering. These programs were important to the solution of problems that had increased in complexity and scale such that alternative methods of solution were no longer economic or tenable. This provided a sound base for moving ahead to approach engineering problems as a system of engineering and to direct attention to an associated facet of engineering, namely, management science. This functional area is informational in nature and, as such, requires the building of data base systems. The field encompasses such broad areas as (1) material control record systems for construction projects, (2) cost control systems for construction, (3) activity scheduling of engineering jobs, (4) maintenance scheduling, (5) equipment performance statistics, (6) equipment load management, and (7) equipment basic record systems.

Management science is directed toward controlling and reducing costs in the operation, maintenance, and construction of power system facilities. Engineers and technical managers have always maintained manual information systems to accomplish their job tasks. The computer permits much improved methods of cost control and requires that the source information only be captured once instead of redundant manual manipulation. Examples of projects associated with management science include:

1. Transformer Load Monitoring. In service on the lines of a large power company are hundreds of thousands of distribution transformers rated at millions of kilovoltamperes and representing an installed cost of hundreds of millions of dollars. This investment is increasing at a rate of approximately 7 percent per year. Since distribution transformers represent such a substantial proportion of system investment, it is imperative that they be utilized to their fullest economic capability.

A large percentage of distribution transformers are nominally underloaded, i.e., they are oversized for the load being served and hence waste money through overinvestment and excessive core losses. Overloaded transformers also waste money in terms of copper losses, loss of life, fuse burnouts and replacements, transformer burnouts and replacements, and the investigation of low-voltage complaints. Records indicate that about 50 percent of customer voltage complaints are directly associated with heavily loaded transformers.

Studies indicate that if one could achieve a loading pattern which attempts to maintain transformers within their economic loading range of 80 to 160 percent of nameplate capacity, one could obtain an annual savings of millions of dollars.

2. Power System Equipment Performance. The purpose of establishing an equipment operation data base is to record and summarize the specific causes of service interruption to generation, transmission, distribution, and communication system as well as to customer causes. The data provide the basis for (a) designing new systems to specific reliability levels, (b) monitoring equipment and manufacurer adherence to desired availability standards, and (c) carrying out maintenance scheduling activities.

The data base consists of a main file for each major equipment category and supplementary files which supply input to a family of programs designed to provide the engineers with periodic statistical reports. The engineer also has the ability to retrieve from this data base any combination of data that may be desired.

CONSTRUCTION MANAGEMENT SYSTEM

Public utility companies on the average spend more for new construction each year than any other industry. Although their primary objective is the usual production and sale of a product, they must be concerned with a large capital investment program. Efficient planning, scheduling, and control of people and material resources are necessary if customer demand is to be met and, at the same time, a fair rate of return is to be provided on the investment. The construction management system, utilizing a central computer with teleprocessing and direct access file capabilities, is an application designed to fullfill these requirements. The features of the construction management system include:

1. Improved Estimating Procedures Estimates of work standards are continuously updated for changes in engineering technology and time standard evaluation. Standard labor rates are automatically updated with current overhead, as it is affected by lost time, vacation, and supervision. Manual computation, data manipulation, and clerical filing are eliminated, thus providing the estimator with more time for design. With these improvements, project estimates are more accurate and require fewer modifications.

2. Automatic Personnel Planning and Scheduling Integration of individual project schedule requirements into a single schedule for all projects provides maximum utilization and control of personnel. Personnel schedules are available to all responsible groups and reflect the most recent status. High-speed data handling allows rapid appraisal of schedules. Construction project management can have multiple personnel schedules prepared to show the comparative effect of different decisions.

3. Rapid Cost Reporting Uniform methods of cost accumulation are applied to standard units of work. The result is an accurate comparison of all labor trades and work crews. Units of work are established to conform to property record accounting classifications and permit the automatic recording of property in the asset accounts. Description, location, and cost of property are available in the computer for tax and insurance reporting.

4. Better Cost Control To effect cost control, actual cost of the construction project is continuously compared with the original estimate. The percentage of project completion is factored against the estimate and compared against the actual cost to date at any time to identify cost deviation. This approach permits corrective measures early in the project schedule to prevent excessive construction costs in later stages.

5. Reduced Clerical Work The need to manually compile data, maintain information files, perform certain computations, and prepare reports is substantially reduced and, in some instances, completely eliminated in the construction management system.

6. Dynamic Management Control Current expenditures with forecasted cost deviations are available upon demand by individual construction projects as well as for the entire construction program. Personnel projections and material delivery schedules are monitored and reported immediately for evaluation. Thus, the effectiveness of management control for a single project, or for the entire construction program, can be evaluated frequently.

TRENDS IN COMPUTER USE

It is projected that between 1975 and 1985 the *technical use* of computers in the electrical utility industry will develop more rapidly than the *business use*. This will be made possible by the development in hardware and software systems. Figure 8 shows major computer application opportunities. The following brief descriptions are not intended to be complete, but are used to indicate some trends.

1. Boiler-Turbine Generator Response. The large investments required for plants and present (1975) cutbacks in construction plans will force utilities to operate with smaller margins. This, among other corrective actions, dictates operation with less reserve, which in turn calls for improvement of plant (steam boiler-turbine) response to sudden loss of load if blackouts are to be avoided. This represents a significant opportunity for off-line simulation of power plants.

2. Nuclear Power Plant Load Factor. Much of the industry's commitment to large nuclear generation facilities presents a significant opportunity for further computer use. Through system engineering and off-line analysis of nuclear reactors, the load factors on these plants could be improved from the present 56.1 percent average load factor on over 60 contemporary plants to about 75 percent.

3. Fuel and Environmental Dispatching. Such dispatching would allow use of the national coal resources while maintaining acceptable air quality standards. This daily and weekly off-line analysis and simulation could also relieve the industry from pressure to commit to large investment for certain desulfurization processes.

4. Automatic Meter Reading. As the cost of electricity and labor (for meter reading) increases and the price of integrated circuit electronic kilowatthour meters reduces, automatic meter reading can become a more economical proposition. Interest in automatic meter reading and associated equipment is also increasing because of its possible use to regulate customers' thermal inertia loads for diversity benefits and system security purposes.

5. Corporate Computer Concept.   Most computer applications in planning, design, construction, and operation are performed at such a frequency that they need not be computed on a real-time basis. This fact, together with the availability of new operator-machine terminals, promotes the joint use of large-scale corporate computers by all utility departments. These terminals or consoles provide engineering, operating, and production

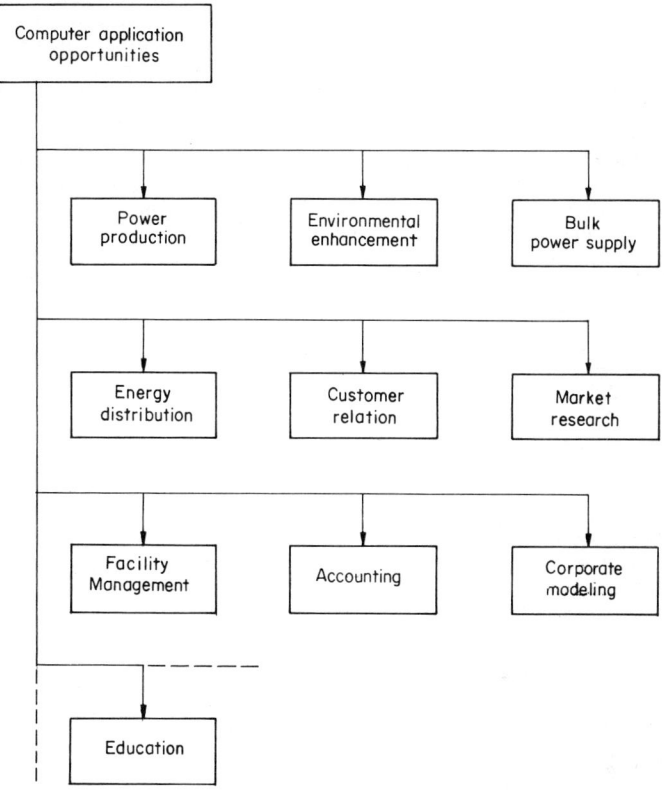

Fig. 8. Trends in technical computer use in electric utility industry.

"work stations," making available the high response of the corporate computer and the large corporate data base for interactive use.

6. Interactive Loadflow.   An engineering computer application has been developed that will enable the power system planning engineer to simulate and solve various power system expansion alternatives in an interactive mode, utilizing a display console, presenting the results in graphic form. The system utilizes a limited graphics color cathode-ray-tube terminal specified with a special set of graphic characters that enable the presentation of loadflow results in the form of system one-line diagrams. The multicolor feature of the cathode-ray-tube console is utilized to indicated heavily loaded lines, bus voltages outside of normal limits, open circuit breakers, and allied information. The engineer working at such a console may, with a light pen and alphanumeric keyboard, remove, add, or change elements of the system being studied and request a solution from the host computer.

7. Corporate Models.   Financial models have been in use in some companies for several years. One of the prime developmental areas of the present appears to be in modularizing the design and development of models. Modules related to various corporate functions having a large impact on future financial forecasts, such as operations costs, marketing and revenue forecasting, and construction expenditures, are included.

Models are being developed to provide short-range monthly financial forecasts as well as annual forecasts for much longer periods of time—up to 20 years into the future. Data base capabilities allowing storage of many base cases for immediate retrieval, modification, and study are being implemented.

8. Automated Mapping.   Geographical information generally related to a grid coordinate system is obtained by aerial photography as well as digitization from existing map documents. These geographic data, along with facilities data transcribed from existing map documents or other sources, are combined into a computerized data base for use with appropriate systems software and hardware, including digital plotters. Hard copy maps are produced under computer control. Facilities are generally categorized in layers in the data base, thereby making it possible to produce maps showing only those facilities desired for a specific purpose.

9. Distribution Facilities Information System.   The objectives of a computer-aided distribution facilities management system are (a) to create a data base containing information about installed and planned facilities that is needed to plan, engineer, and provide gas and electric services to new and existing customers; (b) to maintain the data base in a timely and economical manner; (c) to provide up-to-date information about installed and planned facilities to the utility employees in the most convenient manner and form that can be cost-justified; and (d) to improve the management and utilization of distribution facilities through the use of justified computer applications. A system which meets these objectives can help meet the larger objective of providing better service at lower cost.

10. Nuclear Fuel Analysis.   Computer programs simulating nuclear core physics in various levels of detail are being utilized for analysis purposes. Continual improvement and enhancement is proceeding. These programs are used not only in the design of cores for nuclear reactors, but also for the planning and decision making regarding the fuel burnup and associated electrical generation for nuclear units. In such cases, core-refueling schedules are of vital concern. Safety analysis for emergency situations is another area that utilizes the simulation program.

11. Data Base Management.   The increasing necessity for accurate, up-to-date, easily accessible, well-managed data for use in management decision-making, as well as for processing computer application systems, points to the need for utilization of generalized data base management systems. In addition, as more electric utility application systems are developed, it is becoming increasingly apparent that there is a considerable interrelationship and overlap of the data requirements for the various application systems. Several electric utilities have developed and are in the process of continual development of generalized data base management systems. Utilization of such systems is relatively new in the electric utility field. However, there is a strong trend in this direction.

12. Time Sharing.   Many utility companies are using time-sharing computer systems for engineering and technical computations. As in the case of data base management systems, considerable developmental and application effort is under way in this area. Several companies are developing computer application systems that utilize large data bases created and updated in the corporate computer by a batch process in such a way that the data bases are available to engineering and operating personnel in various offices and locations through time-sharing terminals. These personnel contribute to the updating of the data bases and also have the capability of utilizing the information stored for computation and reporting purposes associated with their corporate functions. The availability of time-sharing capabilities has prompted a large increase in development work associated with on-line systems applications. The advantages of interactive capability and the ability to perform calculations, analyses, and output reporting with rapid responses is, with justification, becoming well accepted by the electric utility industry.

REFERENCES

"Technical Computer Systems for Electric Utilities Today and Tomorrow," presented by EEI at the UNIPED, Madrid, Spain (October 1974).

Gabrielle, A. F.: "The Integration of the Computer with Engineering, Design, and Construction," presented at IBM Executive Conference, London (September 1972).

Coombe, L. W., et al.: "Interactive Loadflow System, *PICA Conf. Proc.* (1975).

Nagel, T. J.: "Reliability and Interconnected Power System Planning," American Power Conference, Chicago, Ill., April 1968.

Patton, A. D., et al.: Reliability of Generation Supply, *IEEE Trans.* (February 1968).

Savas, E.: "Computer Control of Industrial Processes," McGraw-Hill, New York, 1965.

Stagg, G. W.: "Expanding Role of Computers at American Electric Power, IBM Executive Seminar, October 1968.

Stevenson, W. D., Jr.: "Elements of Power System Analysis," McGraw-Hill, New York, 1962.

Stagg, G. W., and A. H. El-Abiad: "Computer Methods in Power System Analysis," McGraw-Hill, New York, 1968.

Elgerd, O. I.: "Electrical Energy Systems Theory: An Introduction," McGraw-Hill, New York, 1971.

Electric Power and Energy Measurement

BY WALTER A. TROEGER*

TERMINOLOGY

Over the years, the term *power* in association with electricity has tended to lose its true meaning. Thus, power is often found used in nontechnical literature where actually the correct term *energy* should be used. By definition, power is the rate at which energy is transformed or made available and is measured in *watts*. Energy may be defined as the time integral of power, or is the total energy supplied and is measured in *watthours*.

From an economic viewpoint, the most important of all electrical measurements is the measurement of energy. The watthour meter in various forms can be found in nearly every home, factory, highway billboard, and other locations where electric energy is being purchased. Metering, installation, and wiring have been governed by national, industry, and local codes for so many years that at least in the United States a particular type of installation is nearly identical everywhere.

For homes or small stores where energy demand is low and fractional horsepower motors are used, the common supply is single-phase, 3-wire. Two voltage levels are available, 120 and 240 volts, depending upon which pair of wires is selected. Electric ranges and heavy-duty home air conditioner motors can take advantage of the high voltage to reduce line currents which reduce losses and permit smaller-size wiring.

Measurement of energy is almost always by means of fixed-installation metering. This provides safety through grounding of the meter enclosure and ease of reading through proper location and mounting. Tamper-proof housings, which are also weatherproof where necessary, are common practice to ensure the integrity of readings.

On the other hand, the measurement of power (watts) follows no such set of rigid rules. Very often considerable planning must go into a watt measurement to properly use existing metering or to purchase new equipment so that the test will be valid and results will be within the expected accuracy limits.

Whereas the measurement of energy is almost entirely restricted to 60-Hz power, power measurements range from direct current to alternating current, including distorted waves, chopped waves, and missing pulses. A variety of circuits for connecting wattmeters to single-phase and polyphase systems have been developed over the years. Basic connection diagrams appear in many electrical engineering textbooks. However, the user of a wattmeter is hard pressed to find diagrams covering practical or unusual situations. Wattmeter manufacturers usually offer an instruction book or bound set of connection diagrams with their instruments. Basic wattmeter connection diagrams appear later in this article.

POWER THEORY

Since energy is simply the total power over a time period, an understanding of the power equations will provide some background into both power and energy terminology. A dc circuit under steady-state conditions will produce power, computed as the product of the voltage across the circuit and the current in amperes in the circuit. This will also apply to ac circuits as long as instantaneous values of volts and amperes are used. The product of volts and amperes at any instant will give the instantaneous power in watts. But, such a measurement is unusual, difficult to make, and the resulting information is of limited use. Instantaneous power is of interest, of course, in the study of transient phenomena.

Average power in an ac circuit is of far more interest since it is equivalent to dc power and is a measure of mechanical work being done or heat liberated. Wattage or average power has an exact mathematical relation to horsepower or Btu.

*Engineering Department, Weston Instruments, Newark, N.J.

The most basic equation for power, relating voltage, current, power, and the phase angle between the voltage and current, is derived as follows. If both voltage and current are sinusoidal, the average power over a cycle is

$$P = \frac{1}{T} \int_0^T ei \, dt$$

$$= \frac{1}{2\pi} \int_0^{2\pi} E_m \sin \theta \, I_m \sin (\theta - \phi) \, d\theta$$

where E_m and I_m are maximum values and ϕ is the phase angle by which the current lags behind the voltage.

From $\sin (\theta - \phi) = \sin \theta \cos \phi - \cos \theta \sin \phi$,

$$P = \frac{E_m I_m}{2\pi} \left(\int_0^{2\pi} \sin^2 \theta \cos \phi \, d\theta - \int_0^{2\pi} \sin \theta \cos \theta \sin \phi \, d\theta \right)$$

$$= \frac{E_m I_m}{2\pi} \left\{ \left[\frac{\theta}{2} - \frac{\sin 2\theta}{4} \right]_0^{2\pi} \cdot \cos \phi - \left[\frac{1}{2} \sin^2 \theta \right]_0^{2\pi} \sin \phi \right\}$$

$$= \frac{E_m I_m}{2} \cos \phi$$

The rms values of sinusoidal voltage and current are

$$E = \frac{E_m}{\sqrt{2}} \quad \text{and} \quad I = \frac{I_m}{\sqrt{2}}$$

Substitution in the previous equation yields

$$P = EI \cos \phi$$

An immediate concern is what happens to the indications of a wattmeter if the voltage or current or both are not sinusoidal. Since it is possible to synthesize an odd wave shape with higher harmonics of the fundametal frequency, a wattmeter will give correct indications if it is frequency-compensated over the span of harmonics. It is to be noted that if a particular harmonic is present in either the current or the voltage, but not the other, it does not contribute to the average or active power. Frequency compensation of the wattmeter is still necessary for an accurate measurement.

WATTMETER CONSTRUCTION

Dynamometer.   All wattmeters of this type contain a fixed coil (usually divided into two coils) which carries the current and a moving coil having series resistance connected for voltage, turning within the fixed coil. The torque on the moving system is proportional to the product of the currents in the fixed and moving coils.

$$\text{Torque} = K_1 i_m i_f \frac{dM}{d\theta}$$

$$\text{Deflection} = \frac{K_2}{T} \int_0^T ei \, dt$$

which is identical to the average power equation

$$P = \frac{1}{T} \int_0^T ei \, dt$$

multiplied by a constant.

It follows then that a dynamometer wattmeter is a *true rms wattmeter* and will take into account the magnitudes of voltage and current as well as the phase angle between them. Furthermore, meter indication will reverse when the flow of power reverses.

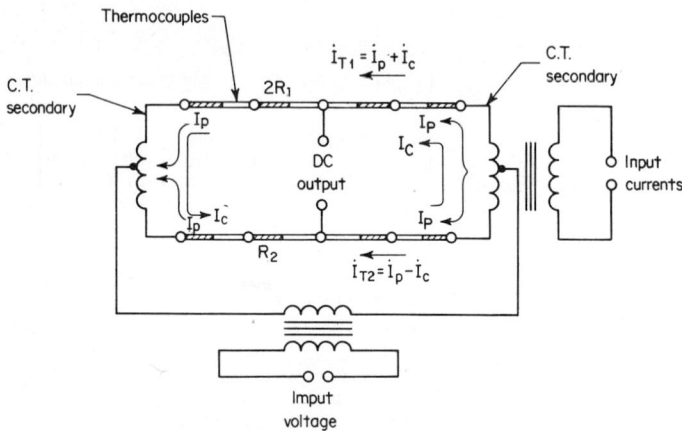

Fig. 1. Basic circuit of single-element thermal watt converter.
$\dot{I}_P$ = current due to potential transformer (P.T.)
$\dot{I}_c$ = current due to current transformer (C.T.)
$\dot{I}_{T1}$ = vectorial sum of currents in R_1
$\dot{I}_{T2}$ = vectorial difference of currents in R_2
$R_1 = R_2$

Thermal Watt Converter.   As with the dynamometer, this type of instrument dates back to the early days of electricity. Heat produced by the voltage and current directly heats thermocouples which are arranged in a network to provide a dc output directly proportional to the wattage input. Figure 1 shows the essential parts of a thermal watt converter, namely, the potential transformer which is connected to monitor voltage, the current transformer which has a double-wound, center-tapped secondary, and the two sets of thermocouples.

The quantity of heat in R_1 is proportional to $R_1 (I_{T1})^2$ and in R_2 is equal to $R_2 (I_{T2})^2$. A vector diagram can be drawn for the sum and difference of I_p and I_c. From this diagram, equations can be developed which yield the difference in wattage in resistance R_1 and R_2 as $4RE1 \cos \phi$. The variable part of this term is $EI \cos \phi$ which is the expression for ac power.

Some early thermal wattmeters used bimetallic elements or liquid-filled thermometers to measure the difference in heat between resistors R_1 and R_2. Thermocouples were used in more recent designs. In one design of thermal wattmeter (Weston), the resistors and the thermocouples are one and the same. They act both as the heating resistors and as the temperature-sensing elements and show a very fast response to power changes. Also because the impedances of the several parts of the circuit are inherently balanced, there is little tendency for interchange of currents or potentials between the ac and the dc portions of the network.

Used mostly by the power industry, especially in totalizing of electric system loads, the watt converter has also found widespread use for measuring the power taken by very large motors where a remote readout is needed. Classed as a true rms wattmeter, this device will respond to magnitude of voltage and current as well as the phase angle between them. Converter output will also reverse if the flow of power in the system reverses.

Specific Design Considerations.   For clarity, it is believed best to comment pertaining to design factors pertaining to a few specific commercial instruments, realizing of course that other designs are also available commercially. The Weston Model 868 single-element and the Model 878 two-element watt converters have a response to 99 percent in 0.7 second. Due to this rapid response, some protection of the thermocouples is necessary during overload conditions. Converters made by Weston use transformer cores which will saturate at moderate overload and minimize thermocouple burnout or other damage. If repeated overloads occur, the current circuit is usually operated below the 5-ampere normal level. Output is 50 millivolts open circuit per element when connected to a 500-watt load. The Model 878 two-element watt converter has an output of 100 millivolts for 1000 watts of circuit load. Provision is made for adjusting resistors in the output network so that the output can be reduced to achieve a particular ratio between input wattage and output millivolts.

Since a thermal watt converter has no moving parts to wear, it has been made rugged by encapsulating the entire circuit in a steel case. Polyphase wattmeters are easily made by assembling two or more single-element wattmeters and providing for a common output signal. The Weston Model 866 three-element wattmeter was designed for use on military motor-generator sets having 3-phase, 4-wire distribution systems. Such systems can be expected to operate under unbalanced conditions and to correctly read system power; a three-element wattmeter must be used.

If the internal potential and current transformers are of good quality, the thermal watt converter can be used on frequencies extending to 20,000 Hz. The Weston Model 865 high-frequency type has a working frequency span of 180 to 20,000 Hz. In general, all thermal watt converters use internal transformers, which excludes their use on direct currents.

The Low Power Factor Wattmeter.   All the so-called true rms wattmeters will operate over the full range of power factor from zero to unity. Many low power factor examples can be found in the laboratory, such as motor or transformer testing, core loss tests, and power supply circuits. At zero power factor, a wattmeter will indicate zero watts even with rated voltage and current flowing. It quickly becomes apparent that the major difficulty in making a low power factor measurement is the low indication obtained on the meter. For example, if a wattmeter indicates full scale with a unity power factor load, it will indicate half scale with a 50 percent power factor load for the same level of voltage and current. At 20 percent power factor, the pointer will move only one-fifth the distance up-scale.

Special wattmeters are available for use on low power factor circuits. They are commonly called 20 percent power factor meters since the full-scale wattage is equal to the maximum voltage times the maximum current times 0.2. Both the accuracy and readability are improved through the use of this type of instrument. Since the 20 percent power factor wattmeter is designed to develop five times the torque of a unit-power-factor-type instrument, care must be taken not to apply voltage or current above the maximum values shown on the instrument rating. Large overloads will soon burn out the resistors, fixed coils, or moving coils. Small overloads continuously applied will cause deterioration of the overheated insulation. A wattmeter designed for low power factor can be used on higher power factor circuits provided either the voltage or the current is sufficiently reduced to keep the pointer on the scale. Likewise, normal unity power factor meters may be used at low circuit power factors provided maximums are not exceeded.

POWER MEASUREMENT

Power in an ac or dc circuit may be determined indirectly by making appropriate measurements of voltage, current, and, where necessary, power factor. Power factor is usually expressed as a decimal value ranging between 0 and 1 and is derived from the cosine of the angle between the voltage and current. Power factor is further designated lead or lag, depending upon whether the current vector is ahead of or behind the voltage vector based on counterclockwise rotation of the vectors. When making power calculations, it is not necessary to know if power factor is lead or lag.

Therefore, calculated power is a valid procedure, but with some reservations. Results are accurate only if all quantities of voltage, current, and power factor are correctly measured. The most elusive quantity is power factor. It is rare that a single-phase power factor meter can be used on other than 60 Hz. Even the 3-phase power factor meter has the requirement that the load must be balanced, although some designs have covered several thousand Hz. Modern electronic phase-angle voltmeters give excellent results on good sine waves over a wide frequency span. However, when distorted waves are encountered, results are questionable since many instruments of this type operate on the zero crossing principle. Further, the power factor of a distorted wave has little meaning since by definition it is based on a sine wave. The conclusion is soon reached that the only way to measure power accurately is through the use of a wattmeter.

In recent years, the phrase "true rms wattmeter" has been reserved for the description of the ultimate in wattmeters. This is because some of the types of wattmeters available today will be accurate at only one frequency, must operate over a narrow voltage span, will not operate on direct current, or must be worked at a high power factor. Although more descriptive than technically correct, the phrase will probably remain in the literature.

The original, basic, true rms wattmeter was the dynamometer. Even until recently, this type of instrument was used as the standard wattmeter at the U.S. Bureau of Standards. A dynamometer wattmeter can be calibrated very accurately on direct current and then used on alternating current. It is often the standard used to check other wattmeter devices because it can be made to a high accuracy, is a passive device, and will retain its accuracy for many years.

Blondel Theorem. Probably the most important theorem in electric power measurement is that proposed by Blondel. In essence, it states that to correctly measure total system power, it is permissible to use one less wattmeter than current-carrying conductors. Also, the common point for the potential circuits is the conductor without a wattmeter current connection. The circuit being so measured may be operated at any power factor or condition of current or voltage unbalance. Strict adherence to Blondel's theorem would require the use of three wattmeters or a three-element-type meter to correctly measure a 3-phase, 4-wire system. Since large commercial systems strive to maintain good voltage balance, a less expensive wattmeter of the 2½-element design may be used and still achieve good accuracy.

Many questions often arise as to the proper wattmeter connections for various types of loads. For example, in a 3-phase, 3-wire circuit, the load can be delta, wye, or some other configuration. The wattmeter is only concerned with the three wires. This leads to a simple pictorial concept for the connection of a wattmeter. Visualize a laboratory bench with an unknown power supply on the left, the connecting wiring across the bench, and an unknown load to the right. Without knowledge of source or load, a true measurement of total system power can be made by following Blondel's theorem. Assume four wires are present and that all may be carrying current. Provide three wattmeters, making the wire without a meter the potential common. It is to be noted that one terminal of the meter current circuit and potential circuit carries an instantaneous polarity marking (usually plus or minus). That means that if (+) of a direct current supply is applied to each of these terminals, the meter will deflect up-scale. Likewise if (−) were so applied, the meter will still go up-scale. If one terminal is made the opposite polarity, the meter will move down-scale. In a multimeter connection, the (±) current terminals should all be toward the source. Even so, due to load reactance one wattmeter may produce a reversed indication. Total power then will be the algebraic sum of all meter readings. The measurement has been made without any knowledge of the source or the load. Voltage and current levels must be within the range of the meter to avoid overheating damage.

If more knowledge of the load can be obtained, the immediate benefit would be reduced metering costs. If we still have a 3-phase, 4-wire system, but know that the neutral wire carries no current, then only two wattmeters are needed to give a true reading.

If we further know that both voltage and currents are balanced around the phases, then metering costs can be further reduced by using a single wattmeter and a wye box. When combined with the meter, the two arms of resistance in the wye box form a wye having a neutral point. The wattmeter will then measure a phase power which is a known fraction of the total power. The scales of switchboard meters are usually direct reading, but the

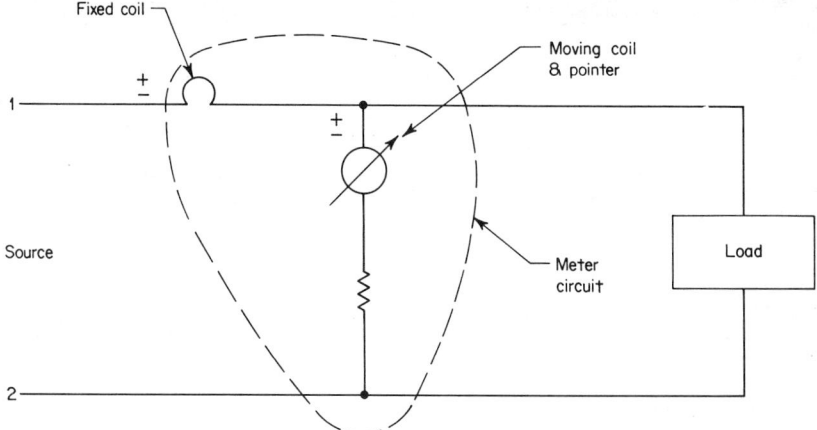

Fig. 2. Single-element wattmeter with potential circuit connected on load side. This connection is most often used.

indication from a portable wattmeter must be multiplied by the wye box multiplying factor.

If only voltages in this 3-phase, 4-wire example are known to be balanced, then a so-called 2½-element wattmeter is satisfactory.

Adherence to Blondel's theorem will always provide a true power measurement regardless of circuit conditions. A knowledge of the circuit will often point the way to less expensive metering which can provide adequate results.

For the measurement of power in a 2-wire circuit, Figs. 2 and 3 show the possible connections. The connection shown in Fig. 2 indicating the wattmeter potential circuit connected on the load side of the current coils is most often used. Readings taken on small wattage loads may easily be corrected for meter loss by opening the load and reading the wattmeter. Although not truly correct under varying loads, this "tare" reading will much improve the accuracy of the measurement.

If the wattmeter current coils are wound for low currents, such as 0.1 ampere, there is sufficient voltage drop across them under load to make the connection shown in Fig. 3 and thus yield a more accurate result.

Figure 4 clearly demonstrates Blondel's theorem of two wattmeters in a 3-wire circuit where the common potential circuit connection is made in the line without a current coil.

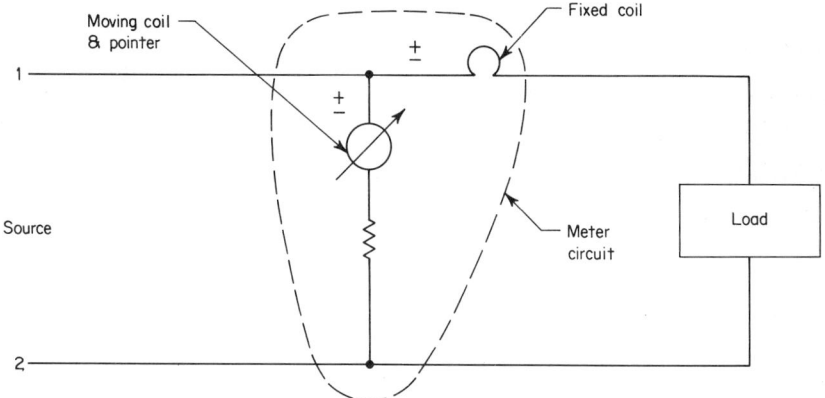

Fig. 3. Single-element wattmeter with potential circuit connected on source side. Generally used when fixed coil has a low current rating and a high voltage drop.

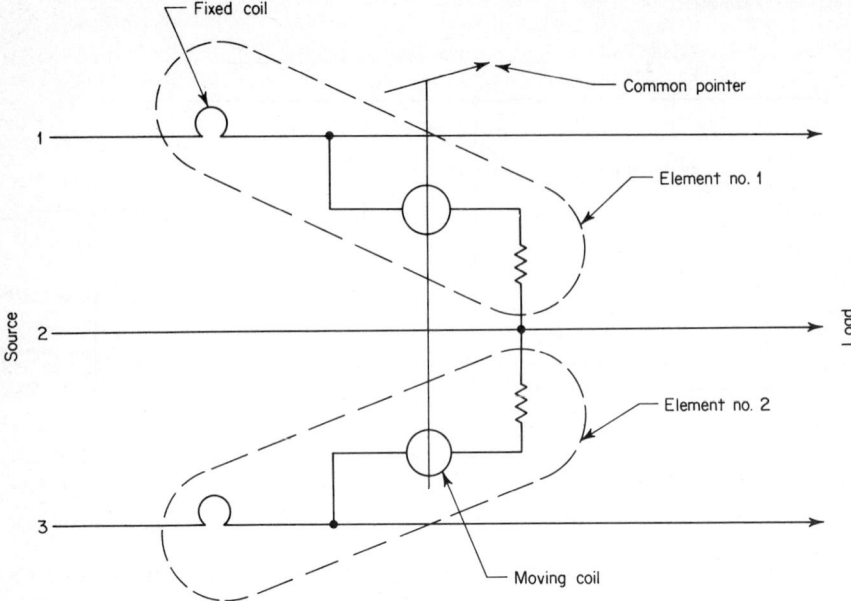

Fig. 4. Two-element wattmeter connected to a three-phase three-wire system.

There are some two-element wattmeters available which connect the two moving coils into line 2. Such an arrangement is adequate for moderate voltages and accuracy, but the mechanical force set up between the fixed and moving coil due to the electrostatic effect precludes the use of this connection where high accuracy is needed.

The 2½-element wattmeter of Fig. 5 will monitor all the current flowing and so is capable of a correct wattage measurement where current unbalance exists. Line-to-line voltages should be nearly balanced and it is assumed that they are 120° from one another in vector rotation.

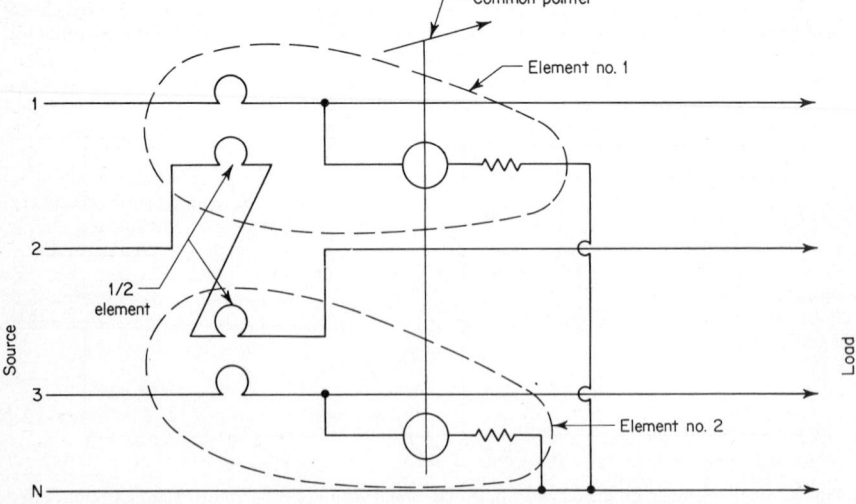

Fig. 5. A 2½-element wattmeter connected to a three-phase, four-wire system.

POWER FACTOR

Whenever the voltage and current are not in phase, a third term called *power factor* must be introduced. The formula is expressed by

$$\text{Power factor} = \cos\left(\tan^{-1} \frac{Q}{P}\right)$$

where Q = reactive power
P = active power

This basic formula may be used on simple, 2-wire circuits as well as a 3-phase system as long as the polyphase system is balanced in both voltage and current.

When a polyphase system is unbalanced in any manner, the system power factor will no longer have any specific physical meaning. A numerical ratio can be obtained and is defined as the interval power factor. This is not, in general, equal to the average value of the power factor during the interval.

Power factor is not usually one of the measured quantities when testing low-power circuits under laboratory conditions. Commercial users of bulk power very often monitor power factor so as to be able to make adjustments of condenser banks or synchronous motors to keep power factor as high as possible. This, in turn, usually reduces the cost of power purchased.

Power Factor Meter.   Since the power factor meter can show at a glance the operating condition of a power system, it is most often found on the switchboards of both consumer and supplier of power.

Ratio-type movements are commonly found in both single-phase and polyphase power factor meters. The single-phase power factor meter measures the ratio of vars to watts which corresponds to the tangent of the power factor angle when the voltage and current are sinusoidal. A scale can then be drawn for power factor which is the cosine of the angle.

The polyphase power factor meter also uses the same basic ratio mechanism as the single-phase meter and is connected 3-phase, 3-wire. This instrument will indicate vector power factor by measuring the angle between line current 2 and line voltage 3-2 and 1-2. The indication is the polyphase power factor only for balanced voltages and currents when both are sinusoidal. Some years ago, Weston offered a polyphase power factor meter for use on unbalanced systems. However, due to limited acceptance because of the high cost resulting from the complexity of construction, it was discontinued. It is doubtful that such a meter can be purchased today from any manufacturer.

Single-phase instruments can be scaled in a variety of combinations, such as 0–1 power factor lag or lead, or 0.3–1–0.3 power factor. Due to the principle of the instrument, not every range combination is possible in the polyphase power factor meter.

Varmeters.   When compared to the other electrical quantities, the *var* is a relatively new term, having been recognized by international agreement in 1930. The letters were taken from volt-ampere-reactive and represent power incapable of producing work. Voltages used in this form of metering are always 90° in vector rotation from that used in wattage measurement. In any ac system having sinusoidal voltages and currents operating at other than unity power factor, the real power is less than the voltampere product and is related by the familiar right triangle. See Fig. 6.

Since more iron and copper are required to deliver a given amount of power at a low than at a high power factor, design allowances must be made for any reactive voltamperes in addition to the designed load power. Line losses are higher and voltage regulation poorer when power factor is low. Any customer contributing to low power factor must be expected to pay for this added loss in addition to the actual energy consumed. Var metering is generally used to obtain an estimate of the average power factor of a fluctuating load over a period of time. A recording-type meter would be used for this measurement.

$$\text{Var} = EI \sin \phi = EI \cos (90° - \phi)$$

A wattmeter can be converted to a varmeter if the voltage is shifted into quadrature with the line voltage at the load. In single-phase varmeters, this voltage shift is done by means of added reactance. In polyphase systems, the voltage shift is most easily done by reconnection if all necessary points are available. On 3-phase, 3-wire systems, a var connection may be made to wye-connected resistors forming an artificial neutral. A special polyphase, phase-shifting autotransformer, also called a *potential converter* or *phasing transformer,* is available for var service. When a varmeter is purchased from the manufacturer, the proper scaling, instrument adjustment (watts), and connection diagrams are provided. If it is desired to convert an existing wattmeter to a varmeter, several problems must be avoided (or at least understood).

Varmeter indication can be either up- or down-scale, depending upon whether clockwise or counterclockwise rotation is selected for the new var voltages. Either is correct, but it should be determined that up-scale indication will occur for leading or lagging power factor. Another more serious problem is that of voltage level. If the new var voltages differ from those used by the basic wattmeter, then for the same power supplied to the load the full-scale value and, therefore, all scale points will be in error. In order to correct this, the meter series resistance would have to be changed (a task for a meter shop with wattmeter calibration equipment).

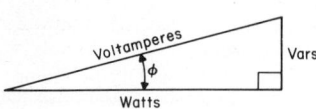

Fig. 6. Real power is less than volts × amperes. Φ = power factor angle.

Varmeters are usually arranged to deflect to the right (up-scale) on leading power factor circuits with phase rotation 1-2-3. Zero center varmeters also have the words IN at the left and OUT on the right. The OUT refers to the flow of power from a source, considered to be located at the left of a diagram, to a load somewhere to the diagram right.

INSTRUMENT TRANSFORMERS

For reasons of personnel safety and good instrument design, it becomes impractical to connect meters directly to circuits having voltages above 1000 volts and currents above 50 amperes. Furthermore, instruments tend to become inaccurate when directly connected to high voltage because of electrostatic forces that act on the moving system. Shunts are often considered for large ac currents but have the disadvantages of large power loss at high current ratings and no voltage isolation. Unless a shunt is especially designed for ac use, its inductance, nonuniform split of current in the blocks, skin effect, proximity effect, and nonuniformity of the material all contribute to ac shunt error.

By contrast, potential and current transformers offer a practical and safe means of reducing voltage and current by an exact ratio for instrument use. All instrument transformers have the following basic design objectives: (1) careful attention to an exact primary-to-secondary ratio and (2) as small a phase angle as possible. Instrument transformers by design have small load (burden) capability because meters do not need large amounts of power, and low burden will enhance the design for best possible accuracy.

In a power station, the instrument transformers that are used for station metering, relay operation, and control services are often several feet tall, resulting from a design to withstand very high voltages. For laboratory and shop testing at low distribution voltages, both potential transformers and current transformers are very small in comparison, and with a weight of perhaps 15 pounds can be hand-carried. Quality potential transformers for laboratory use will support a 25-voltampere burden, have a ratio accuracy from 0.1 to 0.5 percent, and a phase angle of 10 minutes. A core-type design in which the winding surrounds the iron provides the necessary insulation for a high range of 2300/1150 to 115 volts. The iron is usually grain-oriented silicon steel from 0.012 to 0.025 inch thick, depending upon the quality of the transformer and frequency range. Current transformers of the toroidal type, using tape-wound, high nickel-iron cores, have phase angles of less than 2 minutes and a ratio error of plus or minus 0.02 percent. Burden capability is up to 25 VA. Primary-to-secondary insulation is difficult to provide in a toroidal transformer. A rating of 2500 volts is common for stock transformers, with 5000 working volts pushing the practical limit for this construction in custom designs.

Dynamometers and Torque Measurements

In connection with the design and use of apparatus and machines, ranging from the very large to the very small, dynamometers and torque meters provide useful information relating to the efficient conversion and application of energy. Although the dynamometer sometimes is thought of as an instrument for measuring power, actually it measures only the force factor F in the equation hp $= FD/33,000$, where 1 hp $= 33,000$ ft-lb/min, $F =$ force, and $D =$ distance.

The SI unit of force is the *newton* (N). A force of one newton, when applied for one second, will give to a kilogram mass a speed of one meter per second (an acceleration of one meter per second per second). One newton equals approximately two-tenths of a pound of force.

Generally, dynamometers are of two types: (1) *absorption dynamometers,* in which power is absorbed by friction and dissipated as heat, and (2) *transmission dynamometers* in which power measured is passed on and very little wasted on friction. The general ranges of application accuracies of dynamometers are given in Table 1.

Prony Brake. A rather crude form of absorption dynamometer for measurements generally ranging from 1 to 100 hp and for machine shaft speeds of 1000 r/min or less is the prony brake configuration. Because of difficulties in maintaining adjustment for constant power and of heat dissipation, the device is normally confined to laboratory and instructional uses. As shown in Fig. 1, a prony brake is composed of (1) a hollow drum that is attached to the motor or engine shaft under test and (2) an arm that is attached to some form of band with friction lining that passes around the drum. The free end of the arm either is attached to a hanging scale or rests on the platform of a bench scale.

Load is applied by increasing the tension of the band. The drum usually contains water to dissipate heat developed. Sometimes the braking force is applied by means of two lined wooden blocks which are screw-tightened against the rotating member. In other cases the braking may be accomplished by a series of lined wooden cleats placed radially about the rotating member.

A closely associated device is the rope brake, in which the braking action simply results

TABLE 1. Characteristics of Dynamometers

Type of dynamometer*	Approximate speed limit, r/min†	Usual hp limit, hp‡	Probable error, %§
Prony brakes:			
Block	2000	10	1–5
Band	1000	5	1
Wooden cleats on bands	1000	200	½–1
Rope (¾-inch) with cleats	1000	50	1
Fluid-friction dynamometers:			
Froude (ordinary water brake)	10,000	25,000	⅛–½
Turbine	4,000	5,000	1⁄10–⅛
Alden	1,000	5,000	½–1
Fan brake	2,000	200	1–5
Electric dynamometers:			
Eddy-current brake	6000	300	⅛–½
Electric generator	750–4000	30,000	1⁄10–⅛
Transmission dynamometers:			
Torsion	1000–3000	50,000	1–5
Kenerson	1500	100	2

*Dynamometers for testing small apparatus and equipment not included.
†Speed limits are approximately the highest available commercially.
‡Hp refers to largest sizes obtainable.
§Probable error is very approximate and refers to apparatus in fair adjustment.

from tightening a rope wrapped around the rotating member. Rope and band brakes may be used for the range of 0.1 to 50 hp and for speeds up to 4000 r/min.

$$\text{bhp} = 2 \ln \frac{W - W_0}{33,000}$$

where bhp = brake horsepower
$\quad l$ = arm length, ft
$\quad n$ = r/min
W and W_0 = scale loads, lb, with brake operating and brake free, respectively
By using a suitable counterweight, W_0 can be made to be zero, thus simplifying computations.

Fan-Brake Dynamometer.   This is a device for determining the approximate horsepower of high-speed engines. Two plates, one attached to each end of an arm, are affixed to the rotating shaft of the machine under test such that a "fan" configuration is obtained. Power is absorbed by the fan action of the plates on the surrounding air, the amount being dependent upon the size of the plates, their distance from the center of rotation, and the cube of the r/min. A typical configuration would be a "fan" with rectangular blades about 10 inches wide and in a radial direction, 14 inches wide axially, and ⅛ inch thick. Calibration is empirical. Compensation must be made for changes in pressure, temperature, and location of nearby objects that would disturb the measurement process.

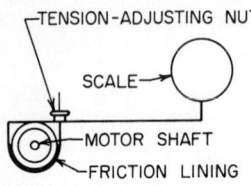

Fig. 1. Arrangement of a Prony brake.

Fluid-Friction Dynamometer.   This device is similar in construction to a centrifugal pump. The dynamometer casing is supported on antifriction bearings and tends to revolve with the rotor. An attached brake arm, supported on scales, measures the turning moment, and the horsepower absorbed is calculated in the same way as for a prony brake. One or more smooth rotors may be used in a water brake, or additional friction may be obtained by vanes and recesses on the casing and rotor. The horsepower absorbed by a water brake varies approximately as the cube of the r/min and as the fifth power of the rotor diameter.

Electric Dynamometer.*   In one form, the electric dynamometer essentially is an electric generator that is driven by the engine or motor under test. The power developed is dissipated into heat or may be channeled into the power line. Another form, the eddy-current dynamometer, consists of one or more conducting disks rotating in a magnetic field.

The power developed by the engine or motor under test is determined by measuring the torque applied to the dynamometer and multiplying by the speed. Since 1 hp = 33,000 ft-lb/min,

$$\text{hp} = \frac{2\pi L \times F \times \text{r/min}}{33,000}$$

where hp = horsepower
$\quad L$ = radius (length of arm), ft
$\quad F$ = force, lb
$\quad$ r/min = revolutions per minute
Since most dynamometer arms are measured in inches, for L in inches,

$$\text{hp} = \frac{F \times L \times \text{r/min}}{63,025}$$

To simplify calculations, L generally is established as 63,025 divided by some round number, such as 2000 ($L = 31.513$ inches); 3000 ($L = 21.008$ inches); or 6000 ($L = 10.504$ inches). The round number then is the *dynamometer constant*, and the formula becomes

*Information furnished by Donald B. Kendall, formerly Chief Engineer, Toledo Scale Division, Reliance Electric Company, Toledo, Ohio.

$$\text{hp} = \frac{F \times \text{r/min}}{\text{dynamometer constant}}$$

To determine the rating for a dynamometer scale (or other force-measuring device), the formula may be rewritten as

$$F = \frac{\text{hp} \times \text{dynamometer constant}}{\text{r/min}}$$

Assume a dynamometer with a continuous rating of 200 hp with 150 percent (300 hp) rating for 15 min, 2500 to 6000 r/min, with a torque arm of 21.008 inches (dynamometer constant = 3000):

$$F = \frac{300 \times 3000}{2500} = 360 \text{ lb}$$

A further simplification, common in laboratories with a number of dynamometers of different sizes, is to calibrate the scales in arbitrary units so that for each dynamometer,

$$\text{hp} = \frac{\text{scale reading in arbitrary units} \times \text{r/min}}{1000}$$

In each case, the pounds pull per arbitrary unit is 63.025 inches divided by the actual length of the torque arm in inches.

In most cases, the dynamometer will be called upon to measure torque in either direction of rotation. When mechanical scales, either spring or pendulum type, are used, a *reversing linkage*, as shown in Fig. 2, is required. An actuating bar with two knife-edges in opposite directions is bolted to the dynamometer frame. If the direction of rotation is clockwise, the left linkage, in tension, applies an upward force to its pivot in the lever and a downward force is exerted on the scale. The right linkage is free. If the direction is counterclockwise, the right linkage, in compression, applies a downward force to its pivot in the lever and a downward force is exerted on the scale. The left linkage is free. The "push-pull" fulcrum pivot is located midway between the tension and compression pivots in the lever so that the scale reads the same in either direction.

Use of Strain Gages. Torque from the dynamometer also can be measured by coupling a "universal" (calibrated in tension and compression) strain-gage load cell to the arm, or by coupling two compression-type hydraulic or pneumatic load cells, one above and one below the arm. Or the torque can be measured by mounting four strain gages, as shown in Fig. 3, on a calibrated section of shaft and inserting this unit between the engine or motor and the dynamometer. Slip rings are used to carry the leads from the conventional bridge circuit.

To minimize errors in torque measurement, the electrical leads to the dynamometer should be very flexible in the direction of rotation, and the shaft bearings should have the least friction possible. On some large precision dynamometers, the outer races of the shaft bearings are rotated so that the bearing friction is always "rolling" and "breakaway" friction is never encountered.

Torque Transducer (Coupling Type). Where applications permit breaking into a coupling, an electronic transducer of the kind shown in Fig. 4 may be used for making both static and dynamic measurements of torque. The sensing element comprises a portion of the power transmission shaft and there is no contact, such as by slip rings, between the sensing element (shaft) and the stationary housing (coil). The device utilizes a linear-variable differential transformer to provide a proportional ac output signal. The output signal is linear with and proportional to the amount of torque applied and reversed phase 180° through null when the torque is reversed. The transducer can be operated into conventional recorders and amplifiers that employ ac bridge circuitry. Units are available for measurement of torque, in increments, from as low as 2.5 to as high as 60,000 inch-pounds. Temperature range is from −65 to +250°F and, with special construction, up to 700°F. Input voltage is 5 V rms, 400 to 1000 Hz; output sensitivity is 40 to 50 mV full scale per volt input; nonlinearity is ¾ percent over half range—1 percent over full range; hysteresis is less than 0.25 percent; repeatability is 0.15 percent of full scale.

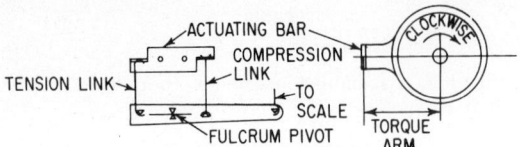

Fig. 2. Dynamometer connection to a scale.

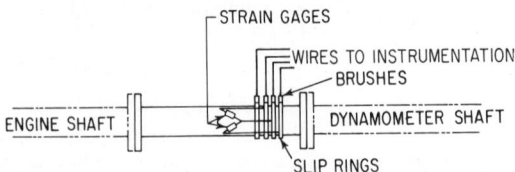

Fig. 3. Strain gages used for torque measurement.

Dynamometers for Measurement of Relatively Small Forces.* The forces and
torques involved in instruments, controls, small apparatus, appliance devices, etc., are
usually quite critical to their successful and reliable operation. Added emphasis is also
given to these measurements in the design of equipment to achieve maximum efficiency
in the use of energy. In the automotive field, such components as distributors, circuit
breakers, fuel gages, horns, and light commutators require dynamometer measurements
during their manufacture and quality control. In the aircraft field, there are servo controls,
panel instruments, such as artificial horizons, temperature and pressure gages, and automatic pilot systems. Other fields where dynamometer measurements are critical to checking, adjusting, regulating, and testing components and systems include such products as
relays, contactors, office machines, cameras, clockwork toys, watches and clocks, electronic components, springs, and in arms manufacturing (firing devices, firearms, and
torpedo equipment). Often, the dynamometers and torque-measuring devices in these
fields take the form of portable, often hand-held, configurations.

The general principle of a small transmission dynamometer is shown in Fig. 5. The

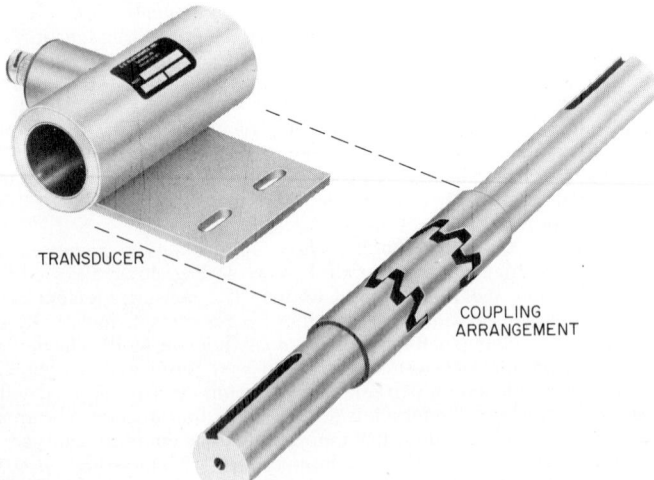

Fig. 4. Coupling-type electronic torque transducer. *(C-E Electronics/AMETEK.)*

*The assistance of Dr. Victor George Herman, New York, in the assembly of this information is
gratefully acknowledged.

principle is that of establishing a comparison with a rated force. The force developed by draw spring R_1 (object of measurement) is measured by the angular displacement of blade spring R_2 (rated force in the instrument). The force can be read on an indexed dial. So-called fan-type dynamometers, although relatively low in cost, can be used only for rough measurements. Particularly, if the tongue (feeler) is bent, the accuracy will suffer.

The principle of operation of a much more sophisticated instrument* is illustrated in Fig. 6. Various models of this configuration permit measurement of forces from as small as 0.3 gram up to 2000 grams (from 0 to 20 newtons). Calibration may be in grams, grams/newtons, or ounces. The feeler (1), on whose tip the force to be measured is applied, carries on to the other end of the toothed segment (2) and pressure wheel (5). The pressure wheel presses against one or the other (clockwise or counterclockwise) of the two rated blade springs (3) (4), depending upon which side of the feeler the force is applied. The rotating movement of the feeler is amplified by a set of gears (2) (6), the indicator needle being firmly attached to pinion (6). Hair spring (7) takes up any play that may be present.

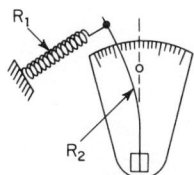

Fig. 5. Operating principle of small transmission dynamometer.

Checking of a dynamometer can be carried out easily by hanging a known weight on the tip of the feeler, making certain that the feeler remains in a horizontal position. The indicator needle then should give a reading corresponding to within 2 percent of the known weight. Some dynamometers† gages have a slide indicator which enables unskilled operators to make tests without, at the same time, looking at the face of the instrument. The slide indicator stops at the highest point reached. Use of cup and jewel bearings in the higher-quality instruments minimize friction; springs of beryllium bronze are used in the more accurate instruments. Carpo gages are equipped with movable arrows which can be set for "maximum" and "minimum" positions, thus making it convenient for unskilled operators to check parts for being within prescribed tolerances. The longer the feeler, the easier it is to reach hidden locations and also the more sensitive the gage. Carpo gages, for example, have long, heavily nickel plated feelers.

Applications of precision dynamometers are shown in Fig. 7a and b.

Torquemeters for measuring small forces are of two types: (1) *static* types that measure the torque necessary to start rotation of a given body (or torque necessary to stop rotation)

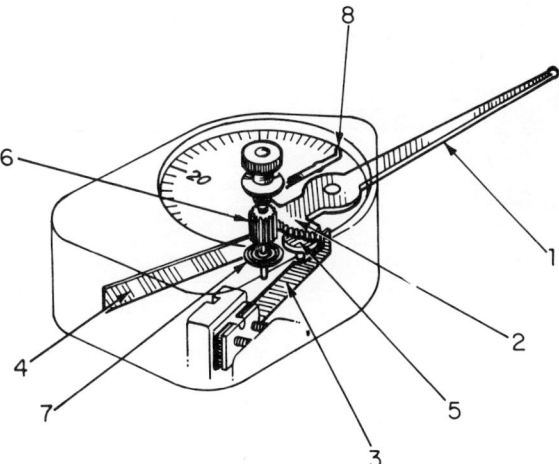

Fig. 6. Mechanical principle of precision-type dynamometer: (1) Feeler. (2) Toothed segment. (3, 4) Blade springs of known ratings. (5) Pressure wheel. (6) Pinion. (7) Spiral or hairspring for taking up play. (8) Indicator needle. (*CARPO.*)

*Such as the CARPO, CORREX, and HALDEX dynamometers.
†Such as the CARPO and CORREX dynamometers.

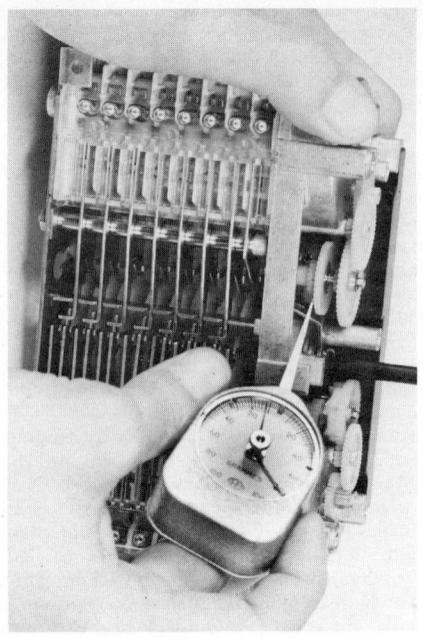

Fig. 7. Applications of precision dynamometers: (*a*) Testing clutch spring in washing machine timer. (*b*) Testing a telephone relay. (*CARPO.*)

and (2) dynamic types for measuring the torque of a body during rotation, without slowing or stopping the rotation.

Torque is frequently measured in kilogram-meters (multiples and submultiples), or in accordance with the international metric system in meters-newton (mN), also in multiples and submultiples. A kilogram-meter equals 9.81 meters newton. One meter-newton is the torque provided by a force of one newton applied at the end of a one-meter lever arm.

The principle of one type* of static torquemeter is shown in Fig. 8. The end of the drive shaft (1) is connected to the arbor whose torque is to be measured. A calibrated spring (3) fixed to the anchor (2) firmly attached to the axle, is connected at the other end to a cone

*CARPO.

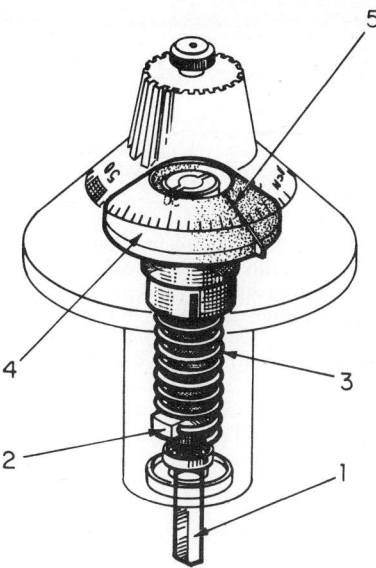

Fig. 8. Mechanical principle of operation of a static torquemeter: (1) Drive shaft. (2) Spring anchor. (3) Calibrated spring. (4) Needle holding cone. (5) Indicator needle. *(CARPO.)*

(4) on which the needle (5) is fixed. The conical shape of the dial allows easy reading from all directions. Torques may be measured over the range from 10 cm-cN to 16 cm-dN. Small torquemeters find wide application in connection with measurements made on small motors, servomechanisms, brakes, clutches, gears, resolvers, gyroscopes, potentiometers, commutators, switching dials, locks, record players, movie cameras, and calculators.

Wet Scrubbing Process—SO_x AND NO_x Removal Chemistry

BY ROY G. NEVILLE*

The combustion of fossil fuels (e.g., coal, oil, natural gas) liberates three major air pollutants: (1) particulates; (2) sulfur dioxide (SO_2), and (3) oxides of nitrogen (NO_x). Fly ash particulates can be removed satisfactorily by electrostatic precipitators, wet scrubbers, and occasionally by bag houses. Sulfur oxides (SO_2, SO_3, that is, SO_x) and nitrogen oxides (NO, N_2O_3, NO_2, that is, NO_x) cannot be removed by mechanical filters or electrostatic precipitators. Therefore, increasing attention has been given to wet scrubbers as a means of absorbing SO_x and NO_x from waste gases. Although water is partially effective as a scrubbing medium for SO_2 absorption, with removal efficiencies up to 20 percent, solutions containing appropriate chemical absorbents have been shown to be particularly effective, and sulfur dioxide removal efficiencies ranging from 80 to 99.8 percent have been achieved by treating waste gases with a variety of chemical reactants. In recent years, several developments in NO_x wet scrubbing have been made. However, the removal of NO_x presents a much more difficult problem due to the very low solubility of NO_x in water as compared with that of SO_x.

MAGNITUDE OF THE PROBLEM

Although the concentration of SO_2 in stack gases emitted by steam generation plants is usually in the range of 400 to 2000 parts per million (ppm), the *volume* of gases produced by the utility industry in the United States and worldwide results in the liberation of large tonnages of sulfur dioxide into the atmosphere. Estimates have been made, and the figure for the emission of oxides of sulfur by power plants in the United States alone exceeded 35 million tons of SO_x in 1970. Other heavily industrialized countries also contribute significantly to the total SO_x pollution of the atmosphere. Unless checked, by the year 2000 it is estimated that the annual emission of SO_x into the atmosphere will exceed 200 million tons!

WET SCRUBBING PROCESSES FOR SO_2 ABSORPTION

Over 60 different processes for the removal of SO_x and NO_x have been disclosed in the patent literature. While technically ingenious, many of the processes contain limiting features which should be recognized. On the basis of practical, economic, and industrial considerations, only the most promising chemical processes will be reviewed.

Developmental work has concentrated mainly on scrubbing systems adaptable to the utility industry, involving the combustion of such fossil fuels as coal, oil, or natural gas to generate steam for the production of electricity. Fossil fuels typically contain between 1 and 6 percent sulfur in the form of pyrites and organic sulfur compounds.

Chemical scrubbing systems for sulfur dioxide absorption fall into two broad categories: (1) disposable ("throwaway") systems and (2) regenerable ("product recovery") systems. Companies using or recommending individual processes will not be mentioned in this survey, except in cases in which the process carries the name of the company. At present, processes are classified according to their state of development and the scale on which they are applied. First-generation processes are defined as those currently in use with a proven record of reliability and typically refer to systems employed in power stations of 100 megawatts or more. Second-generation systems involve processes which are in a later

*Chemical and Environmental Consultant, Linus Pauling Institute of Science and Medicine, Menlo Park, California.

stage of development, or have been used industrially in smaller power stations (for example, 1–100 megawatts). Third-generation systems involve processes still in the pilot stage, or those which have been used to a limited extent in small industrial facilities (e.g., less than a megawatt).

Calcium Systems. Typical of first-generation systems are those which employ an aqueous slurry of an insoluble calcium compound, which can be discarded after use. Disposable SO$_2$-removal systems employ aqueous slurries of finely ground materials such as lime, limestone, or dolomite to produce a mixture of insoluble sulfites and sulfates. On passing through the scrubber, SO$_2$ from the waste gas dissolves to form sulfurous acid:

$$SO_2 + H_2O \leftrightharpoons H_2SO_3$$

The dissolved SO$_2$ reacts with the lime [calcium hydroxide, Ca(OH)$_2$], or limestone (calcium carbonate, CaCO$_3$) slurry to form insoluble calcium sulfite, CaSO$_3$:

$$Ca\,(OH)_2 + H_2SO_3 = CaSO_3 + 2H_2O$$
$$CaCO_3 \quad + H_2SO_3 = CaSO_3 + H_2O + CO_2$$

In warm (50–60°C) solutions, calcium sulfite is less soluble (0.001 percent) than either calcium hydroxide (0.12 percent) or calcium carbonate (0.002 percent). This results in the encapsulation of each particle of Ca(OH)$_2$ or CaCO$_3$ with CaSO$_3$, and the lime or limestone particles are thus inactivated toward reaction with further SO$_2$ at alkaline pH values. This encapsulation phenomenon is wasteful and costly. When lime or limestone slurries are used in alkaline (> pH 7) solution, it is found experimentally that 1.5 to 2.5 times the theoretical amounts of slurry chemicals are required to remove each mole of SO$_2$, even though (theoretically) a 1:1 molar ratio of calcium chemical/SO$_2$ is required. If the pH value of the slurry is maintained in the range of approximately pH 5–6 (i.e., slightly acid), less calcium chemicals are required. At pH 5–6, the calcium sulfite initially formed on the surface of the lime or limestone particles reacts with additional incoming SO$_2$ to produce soluble calcium bisulfite:

$$CaSO_3 + SO_2 + H_2O = Ca(HSO_3)_2$$

Unfortunately, SO$_2$ is less soluble (and hence less easily removed by scrubbing) in slightly acid solutions, so that it is extremely difficult in practice to operate a calcium-based (e.g., lime or limestone) system in such a manner that SO$_2$ removal is maximized while the quantities of calcium chemicals are minimized in order to approach stoichiometric conditions.

As calcium-based slurry systems are usually operated at pH 6–10, disposal of the very large masses of used slurry (after removal of most of the water) presents a major problem. A typical power station using a calcium-based SO$_2$-removal slurry system produces several hundred tons of spent slurry per day. The economic disposal of such large masses of material is difficult or impossible in many parts of the United States because there are insufficient sites available for dumping. Although environmentally innocuous (consisting largely of CaSO$_3$·2H$_2$O, CaSO$_4$·2H$_2$O, and unreacted CaCO$_3$), objections to dumping this material in landfills have been made. Some companies have attempted to offer a partial solution to the disposal problem by processing the waste sludge into building materials, but considerable developmental work remains.

A further disadvantage of lime or limestone systems is their marked tendency to precipitate insoluble calcium salts *inside* the scrubber. In almost all calcium-based scrubbing systems it has been found that the internal components (e.g., spray nozzles, demister sections) of wet scrubbers, *regardless of scrubber design*, become heavily coated with a hard "scale" after a relatively short period, which reduces scrubber efficiency. Unless the scale is removed, the scrubber shortly becomes inoperable. Various types of scrubbers have been proposed (e.g., those having rotating heavy chains which knock the scale off the scrubber walls) to minimize scaling but there is no fully satisfactory solution. The scaling problem is a consequence of using a *calcium*-based system. It is for this reason that some users are converting to sodium or ammonium systems, which are not subject to the scaling problem as sodium and ammonium salts are completely soluble in water. Better efficiencies of SO$_2$ removal can be achieved also with sodium or ammonium systems with less tonnages of scrubbing chemicals. The use of completely soluble scrubbing chemicals (e.g., sodium, ammonium) eliminates the need to shut down the scrubber for descaling.

The comparatively low cost of limestone for preparing slurries is offset by the low efficiency of SO_2 removal (in usual practice, 50–75 percent, although claims as high as 85 percent have been made) and the downtime required for cleaning scale from the scrubbers. Thus, in considering a calcium system, operating efficiency and downtime must be weighed along with material costs.

During the past several years viable calcium-based SO_2-removal systems using carbide sludge have been installed in power stations in the United States and Japan. Carbide sludge, the waste residue after calcium carbide has been treated with water to produce acetylene, consists primarily of finely divided calcium hydroxide. Sulfur dioxide removal efficiencies of 85 percent and higher have been achieved in United States and Japanese power plants. A guarantee level of 80 percent SO_2 removal has been achieved with a total liquid/gas (L/G) ratio of 40 gallons of scrubbing liquor per 1000 cubic feet of gas. While these operations to date have been promising, more operating time will be required before the carbide sludge system can be considered reliable. Also its use is confined largely to power plants which are within reasonable distance of calcium carbide and acetylene generating plants.

Magnesium Systems.   Although chemically analogous to calcium-based systems, first-generation magnesium-based scrubbing systems possess several advantages over their calcium analogs. A slurry of finely divided magnesium hydroxide, $Mg(OH)_2$, is pumped through the scrubber to remove SO_2 from stack gases. Insoluble magnesium sulfite, $MgSO_3$, is formed:

$$Mg(OH)_2 + SO_2 = MgSO_3 + H_2O$$

Hydrated magnesium sulfite ($MgSO_3 \cdot 6H_2O$) can be disposed of as such; however, in practice, it is usually strongly heated to produce a rich stream of SO_2 and regenerate MgO. The SO_2 is compressed, liquefied, and stored in tanks for future use; or it is catalytically oxidized to sulfur trioxide, SO_3, and treated with water to produce sulfuric acid. Alternatively, the SO_2 is mixed with hydrogen sulfide, H_2S, to produce elemental sulfur by the Claus process:

$$SO_2 + 2H_2S = 3S + 2H_2O$$

Economics determine which alternative will be selected. The magnesium oxide is treated with water, and the slurry of magnesium hydroxide is recycled to the scrubber:

$$MgSO_3 + heat = MgO + SO_2$$
$$MgO \quad + H_2O = Mg(OH)_2$$

The magnesium system is versatile as insoluble $MgSO_3$ can absorb another molecule of SO_2 to form soluble magnesium bisulfite, $Mg(HSO_3)_2$:

$$MgSO_3 + SO_2 + H_2O = Mg(HSO_3)_2$$

The use of magnesium bisulfite solution as a solvent for lignin in the pulp-and-paper industry could provide an outlet for the $Mg(HSO_3)_2$ produced. Wood chips are heated with $Mg(HSO_3)_2$ solution to dissolve the lignin and leave a pulp of undissolved cellulose fibers.

Absorption efficiency of SO_2 attainable in an $Mg(OH)_2$ system is good, and removal efficiencies from 90 to 95 percent have been claimed without difficulty at reasonable liquor recirculation and MgO feed rates.

As with the calcium system, serious scaling occurs inside the scrubber due to the buildup of insoluble $MgSO_3$. As previously indicated, however, if the scrubbing liquor is allowed to become slightly acid (e.g., pH 4.5–5.0), by adjusting the rate of addition of $Mg(OH)_2$ slurry, soluble magnesium bisulfite, $Mg(HSO_3)_2$, is formed and the $MgSO_3$ scale redissolves. An advantage of the magnesium over the calcium system is that, on atmospheric oxidation, $MgSO_3$ is converted to soluble magnesium sulfate, $MgSO_4$ (compare with $CaSO_4$, which is insoluble). Formation of $MgSO_4$ represents a loss of magnesium from the system, however, as the sulfate is thermally stable and cannot be as readily decomposed to MgO. Magnesium sulfate can be treated with alkalies and the hydroxide regenerated:

$$MgSO_4 + 2NH_4OH = Mg(OH)_2 + (NH_4)_2SO_4$$
$$MgSO_4 + Na_2CO_3 = MgCO_3 + Na_2SO_4$$
$$MgCO_3 + heat = MgO + CO_2$$

Sodium Systems. Scrubbing solutions containing sodium (or other alkali metal) compounds have been extensively studied for the removal of SO$_2$ from stack gases, and they are among the most successful of the large-scale first-generation processes. Scientific and economic justification for using sodium compounds include:

1. Complete solubility in water with no formation of scale.
2. Reactions with SO$_2$ are straightforward and can be represented by balanced chemical equations.

Environmental regulations prohibit, however, the disposal of high concentrations of sodium compounds into natural waters, and prospective users of sodium-based scrubbing solutions should check local water-pollution regulations.

Such compounds as sodium carbonate, sodium bicarbonate, sodium sesquicarbonate, and sodium hydroxide (caustic soda) or mixtures of these compounds are effective for SO$_2$ absorption. Each of these compounds reacts with SO$_2$ to produce sodium sulfite, Na$_2$SO$_3$, according to the following equations:

$$Na_2CO_3 + SO_2 = Na_2SO_3 + CO_2$$
$$2NaHCO_3 + SO_2 = Na_2SO_3 + 2CO_2 + H_2O$$
$$2NaOH + SO_2 = Na_2SO_3 + H_2O$$

In one commercially used process* a scrubbing solution of sodium sulfite is employed, which readily absorbs SO$_2$ to form the bisulfite:

$$Na_2SO_3 + H_2O + SO_2 = 2NaHSO_3$$

In practice, only a portion of the Na$_2$SO$_3$ is converted to NaHSO$_3$ because the SO$_2$ absorption efficiency diminishes as the bisulfite concentration increases. The resulting solution is heated to decompose the bisulfite and thermally regenerate the sulfite:

$$2NaHSO_3 = Na_2SO_3 + H_2O + SO_2$$

The gaseous SO$_2$ is compressed, liquefied, and stored; oxidized catalytically and converted to sulfuric acid; or treated with hydrogen sulfide to produce elemental sulfur by the Claus process, previously described.

Disadvantages of the sodium sulfite process include the progressive and irreversible atmospheric oxidation of Na$_2$SO$_3$ to Na$_2$SO$_4$, which acts as an inert diluent'and does not react chemically with SO$_2$. The rate of oxidation is increased by the presence of certain ions (e.g., copper, iron, manganese, chromium) which function as "oxygen-carriers." Unless deionized water is used, these ions are frequently present in the water available to industrial facilities. Due to the buildup of Na$_2$SO$_4$ in the scrubbing solution, the number of times that the process Na$_2$SO$_3$ → NaHSO$_3$ → Na$_2$SO$_3$ + SO$_2$ can occur in practice is limited, unless fresh NA$_2$SO$_3$ is added to the solution as system makeup and a portion of the scrubbing solution is removed from the process as system blowdown.

In a modification of the sodium sulfite process, a portion of the scrubbing solution is treated with a slurry of calcium hydroxide to produce sodium hydroxide and insoluble calcium sulfite and sulfate by double decomposition:

$$Na_2SO_3 + Ca(OH)_2 = 2NaOH + CaSO_3$$
$$Na_2SO_4 + Ca(OH)_2 = 2NaOH + CaSO_4$$

Precipitated calcium salts are separated from the solution, and the regenerated NaOH is returned to the scrubbing solution to absorb incoming SO$_2$. This modification is commonly described as the *double-alkali* process and is generally recognized as a viable system.

One possible disadvantage to the process in its present form may be the gradual precipitation of calcium sulfite and sulfate inside the scrubber, due to the reaction of the SO$_2$ with residual calcium salts in the recirculated NaOH stream. Traces of calcium salts can be removed from the NaOH solution by carbonation followed by polishing by the addition of small quantities of oxalic acid solution. Oxalic acid reacts immediately with residual calcium ions in solution to precipitate calcium oxalate, CaC$_2$O$_4$, one of the most insoluble calcium salts known (0.0006% at 20°C).

Scrubbing solutions containing approximately 2 to 20% of sodium hydroxide, carbonate, bicarbonate, sesquicarbonate or sulfite, or any mixture of these compounds, consistently

*Wellman-Lord process.

remove 90 to 99.8 percent of the SO_2 typically present in waste gases (for example, 300–4000 ppm), depending upon scrubbing solution temperature, gas-to-liquid contact time, number of stages in the scrubber, efficiency of gas-to-liquid contact, and the spray pressure employed.

In a novel second-generation process,* a 10% solution of sodium dihydrogen phosphate is employed. Flue gas is first prehumidified to adiabatic saturation temperature in a spray chamber, prescrubbed to reduce dust loadings to 0.04 to 0.05 grains/cubic foot, and then contacted in a wet scrubber with a 10% aqueous solution of NaH_2PO_4. The NaH_2PO_4 acts as a buffer, which has a pH value of approximately 4.0. Sulfur dioxide dissolves readily to form sulfurous acid, and the warm (54–71°C) dilute acid is passed to a stirred tank in which it reacts with H_2S, to produce a precipitate of elemental sulfur by the Claus reaction. The sulfur slurry (2–3% corresponding to 2800–3000 ppm SO_2) is filtered, and the filtrate (containing NaH_2PO_4) is recycled to the scrubber. The finely divided sulfur is melted in an autoclave to 99.9 percent purity. One-third of the sulfur is sold as a by-product, while the other two-thirds is used to produce H_2S by reaction with a gaseous hydrocarbon (e.g., methane, CH_4). Liquid hydrocarbon oils or coal have also been employed on the small scale. The phosphate process has proven to have on-stream reliabilities of 98 to 99+ percent when used for several months cleaning a 100 cubic feet per minute slipstream from a 165-megawatt boiler. It is claimed that this process cuts SO_2 emissions to 50 ppm or below from an input gas stream containing 2000 ppm SO_2.

Possible disadvantages of this process are the formation of *small* quantities of sodium sulfate (Na_2SO_4), which may be separated and removed from the process by crystallization, and the inevitable loss of NaH_2PO_4. If NaH_2PO_4 is allowed to pass in quantity into natural waters, it provides nutrition for the rampant growth of blue-green algae, which leads to the eutrophication and eventual death of flora and fauna in lakes and rivers.

Ammonia-Based Systems.   First-generation scrubbing solutions containing ammonia-based chemicals have important advantages, as well as disadvantages, over sodium systems for SO_2 absorption. Ammonia-based systems are competitive with the alkali metal systems. Regeneration by conventional means can be accomplished, and the by-product, ammonium sulfate, is a marketable commodity in the fertilizer and chemical industry.

Solutions containing ammonium sulfite, with or without the addition of ammonium hydroxide, have been widely used commercially. A disadvantage of an ammonium system, when compared with a sodium-based system, is that an ammonium system can operate effectively only within a pH range of 4.0 to 7.0. As the pH value of the scrubbing solution increases above 7.0, progressively more *gaseous* ammonia is liberated, and this *reacts in the gas phase* with water vapor and SO_2 in the waste gas stream to produce a dense aerosol (i.e., visible white "plume") of ammonium salts which is difficult for many scrubbers to remove. Work is now progressing and reports have been made substantiating a fumeless ammoniacal scrubbing process by carefully adjusting the conditions of the gas and scrubbing liquor.

A "plume" of colloidally dispersed salt particles cannot form when a sodium system is used. Moreover, a sodium-based scrubbing system can operate over a much wider pH range.

In order to regenerate the scrubbing solution, the ammonium bisulfite and sulfite mixture is heated to drive off gaseous SO_2:

$$2NH_4HSO_3 = (NH_4)_2SO_3 + H_2O + SO_2$$

Alternatively, the ammonium bisulfite/sulfite mixture can be treated with calcium hydroxide ["milk of lime," $Ca(OH)_2$]. Gaseous NH_3 is copiously evolved and is trapped in water, which is then recirculated to the scrubber:

$$(NH_4)_2SO_3 + Ca(OH)_2 = CaSO_3 + 2H_2O + 2NH_3$$
$$NH_3 + H_2O = NH_4OH$$

Excellent SO_2 removal (>95 percent) and low loss of NH_3 or ammonium salts can be attained by employing a properly designed three- or four- stage scrubber.

Details on the relative merits of thermal stripping of NH_3, versus lime treatment of

*Stauffer sodium phosphate process.

spent ammonium-based scrubbing solutions, are presented in the comprehensive work of Slack (see references).

Organic Systems.   Two systems involving organic compounds are described here.

Citrate System   Sodium citrate is used in an interesting third-generation process for SO$_2$ removal from stack gases. The solution is buffered at pH 3.0 to 3.7 by the citrate ion, sulfur dioxide is absorbed, and an equilibrium mixture of sodium bisulfite and citric acid is produced:

$$\begin{array}{ccc}
\text{CH}_2\text{COONa} & & \text{CH}_2\text{COOH} \\
| & & | \\
\text{HO-C-COONa} + 3\text{SO}_2 + 3\text{H}_2\text{O} = 3\text{NaHSO}_3 + & \text{HO-C-COOH} \\
| & & | \\
\text{CH}_2\text{COONa} & & \text{CH}_2\text{COOH}
\end{array}$$

The bisulfite leaving the scrubber is then reduced with gaseous hydrogen sulfide (H$_2$S), which precipitates elemental sulfur by a modified Claus reaction:

$$\begin{array}{ccc}
\text{CH}_2\text{COOH} & & \text{CH}_2\text{COONa} \\
| & & | \\
3\text{NaHSO}_3 + \text{HO-C-COOH} + 6\text{H}_2\text{S} = & \text{HO-C-COONa} + 9\text{H}_2\text{O} + 9\text{S} \\
| & & | \\
\text{CH}_2\text{COOH} & & \text{CH}_2\text{COONa}
\end{array}$$

The scrubbing solution can be sodium citrate only or a mixture of sodium sulfite and sodium citrate. Sodium citrate is regenerated when the spent scrubbing solution is treated with H$_2$S, as shown in the above equation.

This appears chemically to be an elegant system, but it has several limitations. First, the low pH of the citric and acid-sodium citrate buffer (3.0–3.7) reduces the quantity and efficiency of SO$_2$ removal, and if the system is operated at higher pH values, complicated side reactions occur in which sulfur is lost by the irreversible formation of sodium sulfate and salts of higher oxyacids of sulfur (e.g., tetrathionate, pentathionate). Several of the higher oxyacid sodium salts react with H$_2$S, but there is a net loss of sulfur by irreversible formation of sulfate. Some of the reactions occurring in the sodium sulfite/bisulfite/citrate solution, when operated in the pH range in which SO$_2$ is more efficiently absorbed (e.g., pH 4–7), are as follows:

$$\begin{array}{rl}
2\text{NaHSO}_3 + \text{S} & \rightleftharpoons \text{Na}_2\text{S}_2\text{O}_3 + \text{H}_2\text{O} + \text{SO}_2 \\
6\text{Na}_2\text{S}_2\text{O}_3 + 9\text{SO}_2 & \rightleftharpoons 4\text{Na}_2\text{S}_3\text{O}_6 + \text{Na}_2\text{S}_4\text{O}_6 + \text{Na}_2\text{S}_5\text{O}_6 \\
3\text{Na}_2\text{S}_3\text{O}_6 & \rightleftharpoons 2\text{Na}_2\text{SO}_4 + \text{Na}_2\text{S}_5\text{O}_6 + 2\text{SO}_2
\end{array}$$

When, for example, the tetrathionate, Na$_2$S$_4$O$_6$, is treated with H$_2$S, elemental sulfur is precipitated and the pH value rises due to the formation of sodium hydroxide (which immediately reacts with SO$_2$ present to produce sodium sulfite):

$$\begin{array}{rl}
\text{Na}_2\text{S}_4\text{O}_6 + 5\text{H}_2\text{S} & = 2\text{NaOH} + 4\text{H}_2\text{O} + 9\text{S} \\
2\text{NaOH} + \text{SO}_2 & = \text{Na}_2\text{SO}_3 + \text{H}_2\text{O}
\end{array}$$

Overall, the reaction is

$$\text{Na}_2\text{S}_4\text{O}_6 + 5\text{H}_2\text{S} + \text{SO}_2 = \text{Na}_2\text{SO}_3 + 5\text{H}_2\text{O} + 9\text{S}$$

Further potential problems with the relatively high-cost citrate system appear to be the very large tonnages of citric acid required, the difficulty of separating the finely divided (colloidal) sulfur from the citrate, and the formation of sodium sulfate as one of the by-products.

Despite these difficulties, this process is receiving increasing attention as it is relatively uncomplicated and has the desirable feature of producing elemental sulfur that is saleable and is stored easily.

Formate System   This SO$_2$ removal process* employs two reactions involving potassium formate, HCOOK, which is regenerated after recovery of elemental sulfur.

In pilot plant studies, SO$_2$ (~ 4000 ppm) in a simulated stack gas is contacted in a wet scrubber with a concentrated (~ 85%) solution of HCOOK at about 200°F (93°C). Under these conditions SO$_2$ reacts rapidly with formate to produce potassium thiosulfate, K$_2$S$_2$O$_3$:

$$2\text{HCOOK} + 2\text{SO}_2 = \text{K}_2\text{S}_2\text{O}_3 + 2\text{CO}_2 + \text{H}_2\text{O}$$

*Consolidation Coal Company.

The mixture of $K_2S_2O_3$ and unused- HCOOK leaving the scrubber is treated with additional formate at about 540°F (282°C), under pressure, to produce potassium hydrosulfide (KSH) and potassium carbonate (K_2CO_3):

$$4HCOOK + K_2S_2O_3 = 2K_2CO_3 + 2KSH + 2CO_2 + H_2O$$

The KSH is treated with carbon dioxide (CO_2) and steam, to produce gaseous hydrogen sulfide (H_2S) and K_2CO_3:

$$2KSH + CO_2 + H_2O = K_2CO_3 + 2H_2S$$

The combined streams of concentrated K_2CO_3 are then treated at 540°F and 1000 psia with carbon monoxide (CO) to regenerate the formate:

$$K_2CO_3 + 2CO + H_2O = 2HCOOK + CO_2$$

The H_2S and SO_2 are combined to produce elemental sulfur, by the Claus reaction:

$$2H_2S + SO_2 = 3S + 2H_2O$$

The potassium formate method has the advantage over some wet scrubbing methods for SO_2 removal that no precipitation of insoluble intermediates occurs at any stage of the process. Homogeneous aqueous solutions can be handled easily throughout every step of the recycled scrubbing and regeneration processes.

Other alkali metal and ammonium formates may be substituted for HCOOK, but potassium formate is best to use as it is nonvolatile, stable, and very soluble. Ammonium formate is thermally unstable, and other alkali metal formates are less desirable alternatives as they are much less soluble.

Disadvantages of the formate process are the necessity of heating the K_2CO_3 solution at 540°F (282°C), at a pressure of approximately 67 atmospheres (that is, 985 psia), with carbon monoxide to regenerate potassium formate; and the energy-consuming requirement of high pressure and temperature (540°F) to produce potassium hydrosulfide and carbonate.

WET SCRUBBING PROCESS FOR SO_2 AND NO_x REMOVAL

While it has been demonstrated repeatedly that solutions of NaOH, $NaHCO_3$, and Na_2CO_3 are effective for the removal of SO_2 from waste gas streams, these solutions are *not* effective for the removal of mixtures of NO and NO_2 (that is, NO_x), particularly when the gas stream is flowing at linear velocities of several hundred to several thousand feet per minute. Even when highly efficient scrubbers are employed, industrial waste gas streams pass at such high velocities through the scrubber that there is insufficient time available for complete dissolution of the NO_x. For example, at 500 to 1000 feet per minute, 95 to 99 percent SO_2 removal from a flue gas stream containing 2000 ppm SO_2 can be achieved in a two- or three-stage scrubber, employing 10 to 20% NaOH solution. However, under the *same* conditions, with a concentration of 2000 ppm of NO, NO_2, or mixtures of these gases, only 5 to 15 percent removal of NO_x can be achieved.

For situations in which SO_2 and NO_x must be removed simultaneously, it is necessary to employ scrubbing solutions specially formulated to react chemically with SO_2 and NO_x. The principle of *reducing* the NO_x in waste gases to an innocuous mixture of N_2 and a small amount of N_2O has recently been developed in a process* to a point where its application is commercially viable.

This process uses sodium compounds containing small concentrations of catalysts that have the dual function of fixing and reducing the NO_x gases as fast as they are exposed to the scrubbing solution and of oxidizing the chemicals used to a saleable product.

The process can be adapted to remove SO_2 and NO_x simultaneously, or NO_x only if SO_2 is not present in the gas stream. It possesses the advantages that (1) no scale-producing insoluble compounds can form as reaction products in the scrubber; (2) the method is simple and inexpensive to operate using proven and conventional gas scrubbing equipment; (3) readily obtainable chemicals are used and (if desired) the main scrubbing chemical can be continuously regenerated for reuse with minimal loss to the system; and

*Neville-Krebs process.

(4) the scrubbing solution operates at pH 6 to 8 and is therefore not corrosive to scrubber components.

For simultaneously removing both SO$_2$ and NO$_x$ from a waste gas stream, the process employs a solution of a suitably formulated and buffered mixture containing sodium sulfite and other compounds, plus small concentrations of inexpensive, proprietary, water-soluble catalysts. Scrubbing solutions containing a total concentration of approximately 5 to 20 percent solids are normally used.

The NO$_x$ gases in the waste stream react with the catalysts in the scrubbing solution to form labile nonvolatile compounds which react immediately with the Na$_2$SO$_3$ to produce gaseous nitrogen (containing a trace of nitrous oxide).

Depending on the waste gas flow rate through the scrubber, the number of scrubber stages used, the temperature and composition of the scrubbing solution, etc., removal efficiencies of 95 to 99.8 percent from a stream containing 2000 to 4000 ppm SO$_2$, and efficiencies of 60 to 90 percent from a stream containing 1000 to 2000 ppm NO$_x$ have been achieved.

In the case of NO$_x$ removal only up to six scrubbing stages are normally recommended in order to allow sufficient residence time for the NO$_x$ to dissolve in the scrubbing solution as the gas passes through the scrubber. The process achieves the objectives required of an industrial gas scrubbing process, namely, low-cost operation, and the use of nontoxic, noncorrosive, completely soluble, nonscaling chemicals, which operate at roughly neutral pH.

WET SCRUBBING PROCESSES FOR NO$_x$ ABSORPTION

The fundamental difference between SO$_2$ and NO$_x$ removal from waste gases is that NO$_x$ gases (that is, NO, N$_2$O$_3$, NO$_2$) are approximately 1000 to 2000 times less soluble in water than SO$_2$ at a given temperature. Consequently, in order to remove NO$_x$ *effectively* from a gas stream, four essential conditions must be satisfied; namely; (1) the waste gas must be exposed to the *maximum surface area* of scrubbing solution; (2) the gas must be exposed to sufficient volume of water to dissolve the NO$_x$; (3) the scrubbing solution must contain chemicals that instantly react with the dissolved NO$_x$, either to "fix" it as nonvolatile nitrite and nitrate ions or to reduce the NO$_x$ to environmentally innocuous gases, e.g., nitrogen (N$_2$) or nitrous oxide (N$_2$O); and (4) the scrubbing solution should operate at the lowest practical temperature to optimize the solubility of the NO$_x$.

It has been found experimentally that, with *few* exceptions, conventionally designed wet scrubbers (e.g., moving bed, packed tower, venturi-type) do not provide sufficient liquid-to-gas contact surface areas, or residence times, to allow the NO$_x$ to dissolve in the scrubbing solution and thereby react with it. To attain these objectives, a scrubber requires a rapidly moving, finely dispersed liquid spray, in three (but preferably up to six or more) separate spray stages. Scrubbers having from three to six stages or more employing solution spray pressures of 100 to 200 psig, with the solution being as cool as practicable [for example, 60 to 122°F (15.6° to 50°C) to maximize NO$_x$ solubility] have been shown to be capable of reducing the NO$_x$ concentration in a gas (or air) stream from 1500 to 2000 ppm to less than 100 ppm (i.e., less than No. 1 Ringelmann).*

The number of gas-scrubbing stages required depends largely on the amount of NO$_x$ to be removed. Concentrations of NO$_x$ in the range 20,000 to 40,000 ppm require 6 to 12 stages, or more, depending on such variables as gas flow rate, residence time of the gas stream in the scrubber, temperature and gallonage of the scrubbing solution, spray pressure, concentration of chemicals in the solution, and other factors.

NO$_x$ Scrubbing: Chemical Basis. In combustion processes above 1000°C, atmospheric oxygen and nitrogen chemically react to form the colorless gas nitric oxide (NO):

$$N_2 + O_2 \rightleftharpoons 2NO$$

In the presence of excess oxygen (on cooling) nitrogen dioxide (NO$_2$) is formed comparatively slowly by a termolecular reaction:

$$2NO + O_2 = 2NO_2$$

In most industrial combustion reactions (e.g., power stations), oxygen is *not* in excess,

*R. G. Neville, experimental data, not yet published.

so that NO is frequently present in comparatively large amounts, and may be 70 to 95 percent of the nitrogen oxides in the waste gas. Unless nitric oxide is absorbed during its passage through the scrubber, it passes to the atmosphere where it reacts with atmospheric oxygen to produce NO_2, which can be observed as a brownish haze in the vicinity of the stack and often for considerable distances downstream.

Due to the very low solubility of nitric oxide (NO) in water (i.e., only 0.0062% by weight at 20°C), adequate residence time must be allowed within the wet scrubber for the NO to dissolve in the water of the solution (for example, 1–5 seconds or more, depending on the concentration of NO in the gas stream). If insufficient *residence time* is allowed while the NO-bearing gas is passing through the scrubber, only a *portion* of the NO present will dissolve, regardless of the *volume* of solution passing through the scrubber. Conversely, if sufficient residence time is allowed for the NO to dissolve, but insufficient *gallonage* of water is provided for the amount of NO present in the gas stream, then only a portion of the NO will be dissolved. Thus, it is of the greatest importance to provide sufficient residence time and sufficient volume of solution to dissolve the NO present. Two further requirements are that the solution should be as cool as practicable (NO has a negative solubility coefficient) and it should be passed through the scrubber as a *very finely dispersed* spray, to obtain maximum liquid-to-gas contact by the creation of maximum dynamic liquid surface areas.

The same requirements apply for the effective dissolution of nitrogen dioxide (NO_2). On *standing*, dissolved NO_2 reacts with H_2O to give a mixture of nitrous and nitric acids. However, this reaction is comparatively *slow* (i.e., it takes several seconds to go to completion) compared with the time of *dissolution* of NO_2 in H_2O.

Unless nitrous acid (HNO_2) is fixed as a nonvolatile salt by reaction with a base, it decomposes in warm solutions, forming a mixture of dinitrogen trioxide (N_2O_3), nitric acid, and *gaseous* nitric oxide (which is undesirable when the object of scrubbing is the removal of NO):

$$2HNO_2 \rightleftharpoons N_2O_3 + H_2O$$

$$3HNO_2 = HNO_3 + H_2O + 2NO$$

Unlike dinitrogen trioxide (N_2O_3) and nitrogen dioxide (NO_2), NO is *not* the anhydride of a stable acid, and therefore cannot be fixed as a salt. The reaction of NO with water is *very slow* and occurs to only a *minute* extent:

$$4NO + H_2O = N_2O + 2HNO_2$$

Similarly, the reaction of NO with water containing an alkali (for example, NaOH), mentioned in some textbooks, is *extremely slow* at atmospheric pressure and temperature, and can therefore be ignored for industrial applications:

$$4NO + 2NaOH = 2NaNO_2 + H_2O + N_2O$$

The comparatively few economically feasible and commercially practical systems available for the removal of NO_x will now be described.

Calcium- and Magnesium-Based Systems. It is commonly believed that nitrogen oxides in waste gases can be scrubbed by solutions of alkali metal hydroxides or carbonates, or suspensions of alkaline earth hydroxides [e.g., milk of lime, $Ca(OH)_2$; magnesia, $Mg(OH)_2$]. However, this is an oversimplification. As indicated earlier, *no* removal of NO_x can occur until the NO_x has dissolved in the water of the scrubbing solution. Therefore, sufficient *residence time* within the scrubber must be allowed to accomplish the dissolution of the NO_x. Further, the oxides of nitrogen present in most industrial waste gases comprise two main compounds: NO and NO_2. If the NO/NO_2 ratio is 1:1, this mixture behaves as if it were dinitrogen trioxide (N_2O_3) and reacts with water to form nitrous acid (HNO_2):

$$NO + NO_2 = N_2O_3$$
$$N_2O_3 + H_2O = 2HNO_2$$

If the NO_2 is in excess, it reacts with water to form a mixture of nitrous and nitric acids:

$$2NO_2 + H_2O = HNO_2 + HNO_3$$

These reactions take from several seconds to several minutes to go to completion, depending upon physical conditions.

In many industrial processes the degree of oxidation of NO$_x$ is *less* than 1:1; that is, NO is in excess. Since NO is not the anhydride of a stable acid, no salts can be formed by using alkaline scrubbing solutions. Thus, the excess NO is *not* removed by scrubbing with alkaline solutions.

Numerous experiments have demonstrated that slurries of calcium hydroxide, Ca(OH)$_2$, or calcium carbonate, CaCO$_3$, are almost totally ineffective for removing NO and NO$_2$ from waste gas streams that are flowing at *practical industrial* velocities. The literature contains numerous references to laboratory-scale and small pilot-plant scale experiments that "prove" that slurries or solutions of Ca(OH)$_2$, CaCO$_3$, NaOH, Na$_2$CO$_3$, etc., can be used to remove NO$_x$ from gas streams, particularly if the NO$_2$/NO ratio is greater than 1:1. However, study of the patent literature on such processes reveals that concentrated slurries or solutions (>10–20%) are employed and that the gas flow-rates are very slow (1– 10 liters/minute). Removal efficiencies of 45 to 85 percent are claimed for such slow-flowing, high liquid-to-gas systems, which are not suited to large-scale industrial applications.

Ammonia-Based Systems.   The use of ammonium sulfite, (NH$_4$)$_2$SO$_3$, and bisulfite, NH$_4$HSO$_3$, mixtures to reduce the concentration of NO$_x$ in a waste gas stream has been demonstrated.* The ratio of bisulfite/sulfite employed ranges from 0.1 to 0.4:1, and the procedure resembles the sodium-sulfite† process, except that ammonium salts are used. The reactions involved are as follows, in which the NO and NO$_2$ are reduced to a mixture of nitrous oxide (N$_2$O) and nitrogen (N$_2$):

$$(NH_4)_2SO_3 + 2NO = (NH_4)_2SO_4 + N_2O$$
$$4(NH_4)_2SO_3 + 2NO_2 = 4(NH_4)_2SO_4 + N_2$$
$$2NH_4HSO_3 + 2NO = 2NH_4HSO_4 + N_2$$

Both ammonium sulfite and bisulfite reduce NO and NO$_2$ to N$_2$O and N$_2$, and the foregoing equations are merely illustrative. It is claimed that a solution containing 2.3 moles of (NH$_4$)$_2$SO$_3$ and 0.8 mole of NH$_4$HSO$_3$ (the bisulfite/sulfite ratio being 0.35), can remove up to 70 percent of the NO$_x$ from the tail gases of a nitric acid plant.

This process appears to offer a practical means for the removal of NO$_x$ from some industrial waste gases. However, the method suffers from the disadvantage that the solution cannot be used above pH 7 (NH$_3$ gas is liberated), and the sulfite/bisulfite mixture is rapidly oxidized by the NO$_x$ and atmospheric oxygen to sulfate and bisulfate. On recrystallization, the solid product contains a significant amount of sulfite, which makes it unsuitable as an agricultural fertilizer. Nevertheless, this and similar processes represent a viable approach toward handling the NO$_x$-containing waste gases from many industrial facilities.

Sodium-Based Systems.   If no SO$_2$ is present, a sodium-based process may be employed to remove NO$_x$ efficiently and effectively from a waste gas stream. In one proprietary third-generation process,‡ removal efficiencies of 60 to 90 percent have been achieved from gas streams containing up to 1500 to 2000 ppm of NO$_x$ passing through a three-stage scrubber (1 cubic foot per stage) at 150 to 500 cubic feet per minute.

Organic Systems.   Of the organic systems, the urea system and the electron-donor system are described here.

Urea System   A third-generation scrubbing solution for the removal of NO$_x$ from low-volume, slow-flowing waste gas streams uses an aqueous, slightly acid solution of urea, CO(NH$_2$)$_2$. Due to the high cost of urea, compared with most inorganic scrubbing chemicals, this system is likely to be useful *only* in the removal of NO$_x$ from *slow-moving* gas streams from *small* industrial plants.

The urea process¶ employs an aqueous slightly acid solution in which the NO$_x$ dissolves. The nitrous acid formed then reacts with urea to form nitrogen, carbon dioxide, and water:

$$NO + NO_2 = N_2O_3$$
$$N_2O_3 + H_2O = 2HNO_2$$
$$2HNO_2 + CO(NH_2)_2 = CO_2 + 2N_2 + 3H_2O$$

*R. Garlet, U.S. Patent 3,329,478 (1967).
†Wellman-Lord process.
‡Neville-Krebs process (patents applied for).
¶A. Warshaw, U.S. Patent 3,565,575 (1971).

Typically, a 1 to 30 percent solution of urea is used, containing up to 10 percent by volume of an acid (for example, HCl, HNO_3, H_2SO_4, CH_3COOH). The solution temperature is usually between 30 and 90°C, as the rate of reaction of NO_x increases at higher temperatures. The solubility of NO_x decreases rapidly at temperatures much above ambient, however, so that the advantage of increased reaction rate at elevated temperatures is largely offset by the reduced solubility of NO_x. For this reason, the use of urea as a scrubbing chemical for NO_x is limited to slow-flowing, relatively cool gas streams; and the cost of urea suggests that its use would only be economically feasible in small industrial plants.

Using an acidified 10 percent solution of urea at 60°C, laboratory-scale and small pilot-plant tests have shown that up to 98 percent removal of the NO_x gases from an experimental nitric acid plant can be achieved providing the NO_x-containing gas is passed at the ratio of only 100 milliliters per minute per 800 milliliters of urea solution. This very slow gas flow-rate and high liquid/gas ratio allows sufficient time for the NO_x to dissolve in the water of the scrubbing solution and react with the urea to form nitrogen. For some industrial applications in which large volumes of rapidly flowing NO_x-containing gases must be treated, the urea scrubbing method is inappropriate.

Electron-Donor Systems   Several third-generation pilot scale processes employing electron-donor compounds have been claimed to remove NO_x effectively from waste gas streams. These compounds include tri-n-butyl phosphate, dimethylformamide, triethyleneglycol dimethyl ether, dimethylsulfoxide, hexamethylphosphoramide, diethyleneglycol dimethyl ether, tricresyl phosphate, dioxane, etc. Most of these compounds are expensive compared with the inorganic compounds used in other scrubbing systems.

As in other NO_x absorption systems, sufficient residence time must be allowed within the scrubber for the NO_x to dissolve in, and react with, the electron-donor solution. In a typical case, a gas stream containing 1200 ppm of NO and 2100 ppm of NO_2 (a by-product from the nitration of cellulose) was sparged through a two-phase system of water (16 percent) and tributyl phosphate (84 percent) at a rate of 40 cubic feet per minute per cubic foot of scrubbing solution at ambient temperature. The concentration of NO_x in the effluent gas stream was reduced by 90 percent. However, when the rate of passage of the NO_x-containing gas was increased tenfold (i.e., to 400 cubic feet per minute), the NO_x absorption fell markedly, and absorption efficiencies in the range of only 10 to 30 percent were achieved. As in the case of urea, these data suggest that organic systems are appropriate only for relatively *small-scale* operations involving limited volumes of waste gases produced at *slow* rates, which allow sufficient residence time in the scrubber for the NO_x to dissolve and react with the chemicals in the scrubbing solution.

Miscellaneous Systems.   Nitrogen oxides (NO_x) have been removed from waste gases by a variety of exotic organic and inorganic solutions, some of which use a high-pressure step (for example, 10–100 atmospheres). As systems involving pressurization are of limited general application, they will not be considered further. Information on these specialized applications will be found in Lawrence (Ref. 2).

Organic solvents for NO_x include solutions of dienes, particularly cyclopentadiene, in nonreactive solvents (e.g., kerosene). The NO and NO_2 add chemically to the double bonds of the diene and are fixed as a nongaseous organic compound. High efficiency (>90 percent) of NO_x removal is claimed, but only from slow-flowing gas streams.

Inorganic systems for NO_x removal that appear to show promise include solutions of chromous salts (e.g., chromous sulfate, $CrSO_4$), or sodium chlorite ($NaClO_2$):

$$NaClO_2 + 2NO + 2NaOH = NaCl + NaNO_2 + NaNO_3 + H_2O$$

No economic or industrially meaningful information has yet appeared on these NO_x-removal processes, which have been demonstrated only on the laboratory or pilot plant scale. For details on such processes the reader is referred to the account given by Lawrence (Ref. 2).

REFERENCES

1. Slack, A. V.: "Sulfur Dioxide Removal from Waste Gases, 1971," Noyes Data Corporation, Noyes Building, Park Ridge, N.J.
2. Lawrence, A. A.: "Nitrogen Oxides Emission Control, 1972," Noyes Data Corporation, Noyes Building, Park Ridge, N.J.

3. Slack, A. V., G. G. McGlamery, and H. L. Falkenberry: Economic Factors in the Recovery of Sulfur Dioxide from Power Plant Stack Gas, *J. Air Pollut. Control Assoc.*, **21:** 9–15(1971).
4. Dennis, C. S.: Potential Solutions to Utilities SO$_2$ Problems in the '70's, *Combustion*, **42:**12–21(1970).
5. Dayton, S.: Wet Scrubbing of Weak SO$_2$ Gets a Trial at New McGill Pilot Plant, *Eng. Min. J.*, **172:**66–68(December, 1971).
6. Slack, A. V., H. L. Falkenberry, and R. E. Harrington: Sulfur Oxide Removal from Waste Gases: Lime-Limestone Scrubbing Technology, *J. Air Pollut. Control Assoc.*, **22:**159–166(1972).
7. Meisel, G. M.: Sulfur Recovery, *J. Met.*, 31–39(May 1972).
8. New Pilot Plant will Test Ammonia Scrubbing Process for Copper Smelter Gas, *Eng. Min. J.*, **175:**24,32(1974).
9. Gilbert, W.: Selecting Materials for Wet Scrubbing Systems, *Pollut. Eng.*, **5:**28–29(1973).
10. SO$_2$ Removal Technology Enters Growth Phase—Tighter Emission Standards Spur Commercialization, *Environ. Sci. Technol.*, **6:**687–691(1972).
11. Wiedersum, G. C.: Control of Power Plant Emissions, *Chem. Eng. Prog.*, **66:**49–55(1970).
12. Tieman, J. W.: Controlling SO$_2$ Emissions from Coal-Burning Boilers, *Min. Eng.*, **24:**47–55(1972).
13. Olds, F. C.: SO$_2$ & NO$_x$. A Critical Look at a Confused Situation. *Power Eng.*, **77:**32–39(1973).
14. New SO$_2$-Cleanup Contenders, *Chem. Eng.*, **81:**46–47(1974).
15. Yavorsky, P. M., N. J. Mazzocco, G. D. Rutledge, and E. Gorin: Potassium Formate Process for Removing SO$_2$ from Stack Gas, *Environ. Sci. Technol.*, **4:**757–765(1970).
16. Dunham, J. T., C. Rampacek, and T. A. Henrie: High-Sulfur Coal for Generating Electricity, *Science*, **184:**346–351(1974).
17. Train, R. E.: Engineering Environmental Rehabilitation, *Consult. Eng.*, **43:**60–64(1974).

Sulfur Dioxide Recovery (Wellman-Lord Process)

BY W. JEFF OSBORNE*

The Wellman-Lord Sulfur Dioxide Recovery Process is designed to produce a concentrated sulfur dioxide (SO_2) gas from lean off-gas streams of any type of plant producing a stack gas containing SO_2. That includes practically any plant burning fuel oil or coal for its power. The most suited application is in treating gas containing 0.15 to 3.0 percent SO_2 by volume.

The process allows considerable flexibility in end-product choice. The concentrated SO_2 gas can be fed to a conventional sulfuric acid plant, it can be converted to elemental sulfur by a reduction process, or it can be converted into liquid SO_2. Any of these products can be sold to help defray the recovery system's operating costs.

As of late 1975, 19 plants were currently in operation and 13 other plants were in the design and construction stage. The total overall capacity currently in operation or in design is about 3 million standard cubic feet per minute of SO_2, more than 70 percent of this total capacity for use in treating boiler stack gas.

The Wellman-Lord process is based on a sodium sulfite/bisulfite system. The flue gas from a typical fossil-fuel-fired boiler requires saturation and particulate removal prior to absorption of the SO_2. After appropriate pretreatment, the flue gas enters the SO_2 absorber, which is a simple two- or three-stage contacting device that can reduce the sulfur concentration to the required level and can accommodate a wide range of turndown conditions.

The absorption system can be located a considerable distance from the regenerative system. Also, it is practical to treat gases from more than one unit by installing separate absorbers for each SO_2 source, with all the absorbers being supplied by a common regeneration system.

SO_2-rich gas is contacted countercurrently in the absorber by the sodium sulfite solution and passes out the top after being stripped of SO_2. Now rich in bisulfite, the solution leaving the bottom of the tower is discharged to a surge tank. See flowsheet (Fig. 1).

From the surge tank, the bisulfite solution flows at a steady rate into a forced-circulation evaporator-crystallizer, the heart of the regeneration system. The solution can be heated by low-pressure exhaust steam that normally would be discharged into the atmosphere. Plants currently are utilizing steam at pressures as low as 15 psi. Even lower pressures are practical.

In large plants, such as power plants, it is practical and economical to operate the regeneration system as a double-effect evaporator. This will reduce steam consumption by a further 40 to 45 percent.

In the evaporator, clearly shown in the upper portion of the total installation illustrated by Fig. 2, the bisulfite solution is thermally decomposed into sodium sulfite and sulfur dioxide. Overhead vapor leaving the evaporator is subjected to one or more stages of partial condensation to remove water and to achieve the desired product gas quality.

The sodium sulfite (Na_2SO_3) formed precipitates out of solution and builds a dense slurry of crystals in the evaporator. A portion of this slurry is withdrawn from the evaporator and sent to a dissolving tank, where condensate from the overhead system is added to dissolve the sulfite crystals. The resulting solution is sent to another surge tank and is then fed back to the absorber to complete the process loop.

Feeding solutions from the absorber system and from the regeneration system through surge tanks enables the entire recovery process to operate smoothly and reliably despite frequent gas flow and concentration fluctuations.

In addition, the surge tanks allow a shutdown of the regeneration system for up to 3 days without interfering with the SO_2 removal in the absorption section. This allows time

*Formerly supervising Process Engineer, Davy Powergas Inc., Lakeland, Fla.

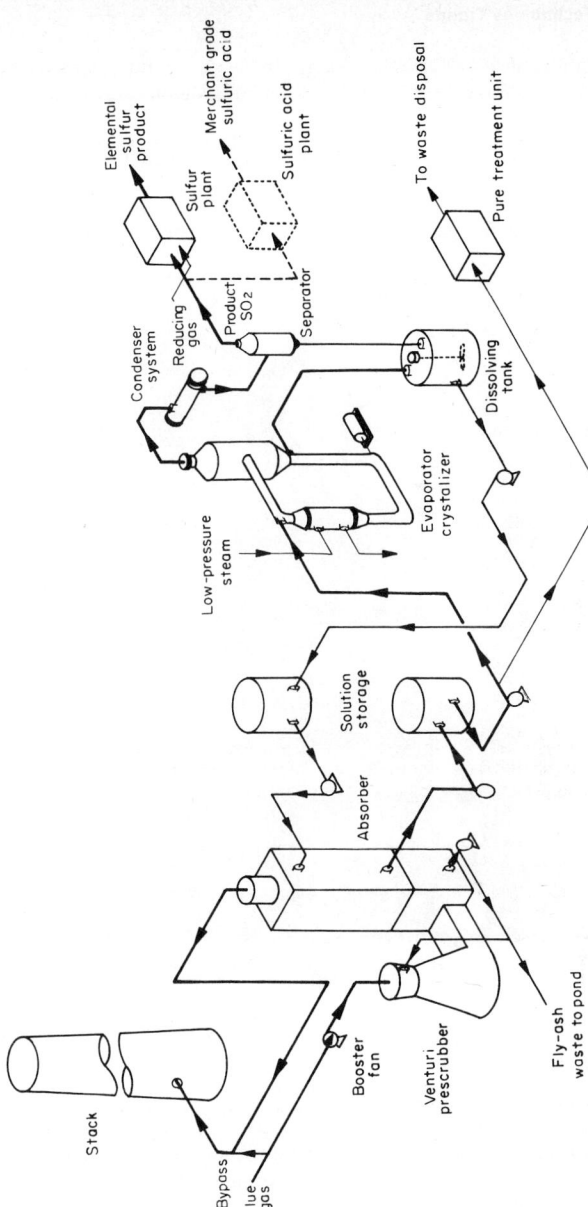

Fig. 1. Essential elements and flow of Wellman-Lord sulfur dioxide recovery process.

for scheduled maintenance or necessary repairs and assists in increasing the reliability of the system as well as reducing the need for spare equipment.

A moderate amount of the circulating solution is oxidized to nonregenerable sulfate (Na_2SO_4). To control the sulfate level, a small stream of the solution is purged from the system. This purge stream can be dried for sale or for disposal or it can be treated to permits its discharge as an innocuous effluent.

Operating Experience.   Two boilers at an industrial plant in Japan with evaporation rates of 130 million tons per hour, equivalent to a 75-megawatt power plant, equipped

Fig. 2. Wellman-Lord sulfur dioxide recovery process installed at petroleum refinery (Standard Oil Company, El Segundo, California). Evaporator-crystallizer is shown in immediate foreground at left; top of evaporator can be seen in upper portion of view; dissolving tank is shown at right.

with the Wellman-Lord process, began operation in mid-1971. At this installation, SO_2 concentration varies from 400 to 2000 ppm, due to the varying quantities of waste process gas from the plant. During the first 10 months of operation, the operating costs for this system were approximately 50 percent for interest and carrying charges, 15 percent for steam, 10 percent for electric power, 10 percent for caustic soda consumption, and 15 percent for labor and miscellaneous costs.

Sulfur dioxide removal efficiencies are consistently greater than 90 percent, and outlet concentrations are consistently less than 200 ppm. Only 0.5 man per shift is required to operate the system, including waste disposal and production. During the first 3 years of operation of this system, reliability has been essentially 100 percent.

Another system of this type is installed on a power-generating unit of a major power company in Japan. The power plant is a 220-megawatt, peak-shaving oil-fired unit with a daily load variation of from 100 to 33 percent capacity, a variation which can take place in 22 minutes. A high-efficiency electrostatic precipitator is installed ahead of the scrubber to reduce the particulate matter to 0.015 grains per standard cubic foot of stack gas. During the start-up phase of this installation in May 1973, the operators inadvertently switched from 0.7 percent sulfur fuel to 4.0 percent sulfur fuel. The control system, designed to give a constant outlet SO_2 concentration, detected the change in inlet sulfur load and automatically reset the controlling conditions. At no time did the SO_2 emissions exceed 150 ppm. The emission level obtained at this facility when burning as high as 4 percent sulfur residual fuel has consistently remained at 130 to 135 ppm SO_2. The product SO_2 is fed to an 80 ton per day sulfuric acid plant.

Capital costs at the aforementioned plant were the equivalent of $28 per kilowatt. This breaks down to 30 percent for absorption, 35 percent for SO_2 recovery, 10 percent for sulfuric acid plant, and 25 percent for water treatment.

A demonstration installation partially funded by the U.S. Environmental Protection Agency began operation in 1975 at the Dean H. Mitchell Station of Northern Indiana Public Service Company in Gary, Indiana. This station burns midwestern United States bituminous coal with 3.0 to 3.5 percent sulfur and a heat content of about 11,000 Btu per pound. This fuel produces a flue gas with 2100 ppm SO_2(vol.) The Wellman-Lord process being installed on a 115-megawatt steam/electric unit is designed to reduce SO_2 emission levels by 90 percent, i.e., to 200 ppm SO_2(vol.) or less. The end product will be bright yellow sulfur with a purity of 99.5 percent. Another process is undergoing installation at a coal-burning power plant in the Southwestern United States, with a capacity of more than 670 megawatts. The SO_2 recovered will be converted into marketabel elemental sulfur.

Claus Sulfur Recovery Units and Claus Off-Gas Treating

BY JAAP E. NABER,* JOHANNES A. WESSELINGH,† AND WILLEM GROENENDAAL‡

Sulfur recovery units of the Claus type, with a thermal and one or more catalytic stages, have been applied for more than 20 years in oil refineries and natural gas treating plants in those instances where large quantities of hydrogen sulfide are produced. In modern Claus units, with 94 to 96 percent sulfur recovery efficiency, the theoretically possible yield of sulfur is closely approached. The remaining 4 to 6 percent sulfur is present in the Claus unit off-gas and is converted into sulfur dioxide in a catalytic or thermal incinerator and discharged to the atmosphere.

In the thermal stage of a Claus unit, part of the hydrogen sulfide is burned with air to form sulfur dioxide, such that the H_2S/SO_2 mole ratio in the combustion gases theoretically would be 2. However, under the conditions prevailing (temperature, 1100–1500°C; and pressure slightly above atmospheric), H_2S and SO_2 react to form S_2 with a 50 to 70 percent yield. After cooling of the process gas and condensation and removal of the sulfur in a steam-raising waste heat boiler, the formation of sulfur (S_6-S_8) from the remaining H_2S and SO_2 is further promoted in one or more catalytic stages. Each reactor is preceded by a process gas reheat facility and followed by a sulfur condenser. The reactors operate at temperatures slightly above the sulfur dew point (300–200°C), since, otherwise, sulfur condensation would take place in the pores of the bauxite or γ-alumina Claus catalyst and the catalyst would lose its activity.

Single-train Claus units ranging in capacity from a few tons per stream-day H_2S intake up to 1800 tons per stream-day intake are in operation.

Although the total quantity of sulfur dioxide originating from Claus units is small as compared with the emissions originating from fossil-fuel-fired installations, the desulfurization of Claus off-gases (Ref. 1,2) has received special attention because (1) the increased demand for natural gas and oil products and a deeper desulfurization of crude oil by hydrodesulfurization and hydrocracking techniques have caused an increase in the capacities of Claus units, with the consequence that Claus units are becoming an important source of sulfur dioxide emission in oil refineries and natural gas processing plants; and (2) the concentration of sulfur compounds in Claus off-gas is large as compared with that in flue gases. Furthermore, the gas is free of particulate matter and its desulfurization is, in principle, less complex than the desulfurization of flue gases originating from fossil-fuel-fired installations.

Criteria for an effective Claus off-gas desulfurization process would include (1) easy and flexible operation, (2) familiar technology and easy adaptability to existing Claus units, (3) no secondary-air or water pollution, or waste disposal, and (4) a high degree of desulfurization over a wide range of operating conditions. To meet air-pollution requirements, the typical sulfur dioxide content of the desulfurized stack gas should be about 300 ppm(vol.) as compared with about 9000 ppm(vol.) sulfur dioxide in the stack gas of a Claus unit with 94 percent sulfur recovery efficiency. The total sulfur recovery efficiency of a Claus unit, followed by a desulfurization unit, should be about 99.8 percent on sulfur intake.

CLAUS OFF-GAS TREATING PROCESSES

Processes for the desulfurization of Claus off-gas can be classified into two groups (Ref. 3).

1. Low-temperature Claus Processes. The formation of sulfur from hydrogen sulfide and sulfur dioxide is further enhanced by operating a catalytic system at a temperature at

*Head, Gas Manufacturing and Treating Section, Oil Process Development Department, Koninklijke/Shell-Laboratorium, Amsterdam, The Netherlands.
†Koninklijke/Shell-Laboratorium, Amsterdam, The Netherlands.
‡Shell Internationale Petroleum Maatschappij B.V., The Hague, The Netherlands.

which the thermodynamic equilibrium is more favorable than under normal Claus conditions. These systems are usually operated below the sulfur dew point but above the sulfur solidification point. With a low-temperature Claus process, such as the IFP* and the Sulfreen processes, an improvement of the overall sulfur recovery up to 98.5 percent can be achieved. These types of processes have found widespread commercial application in those locations where the achievable improvement of the overall sulfur recovery efficiency is sufficient to meet local air-pollution abatement requirements.

2. Conversion/Concentration Processes. In the conversion step, the various sulfur compounds, including elemental sulfur, are converted into hydrogen sulfide or sulfur dioxide. This single sulfur species is subsequently isolated and converted into elemental sulfur, sulfuric acid, or gypsum.

Processes which have found commercial application are: Wellman-Lord, Chiyoda's Thoroughbred 101, Parsons' Beavon, and SCOT.† A number of processes could be classified as hybrids: First sulfur is removed in a low-temperature Claus process and then the remaining sulfur compounds are recovered in a conversion/concentration process. The IFP-2 process belongs to this group.

The SCOT process contains a reduction section and an amine absorption section, as illustrated in Fig. 1. In the reduction section, all sulfur compounds and any free sulfur in the Claus off-gas are completely converted into hydrogen sulfide with hydrogen, or a mixture of hydrogen and carbon monoxide over a cobalt/molybdenum-on-alumina catalyst at a temperature of about 300°C. As the reducing gas, hydrogen from a hydrogen plant or catalytic reformer make-gas can be applied. If such sources are not available, a suitable reducing gas can be generated by substoichiometric combustion of a light hydrocarbon in the direct heater preceding the reduction reactor.

The reactor effluent gas is cooled and the water is condensed. The cooled gas, which normally contains up to 3 percent (vol.) hydrogen sulfide and up to 40 percent (vol.) carbon dioxide, is countercurrently scrubbed with an alkanolamine solution in the absorption column. The absorber top gas, which now contains only about 300 ppm (vol.) hydrogen sulfide, is incinerated in the Claus incinerator. Typical compositions of the relevant gas streams in Claus and SCOT units are given in Table 1.

The design of the absorption column is such that hydrogen sulfide is essentially completely removed, whereas only a fraction of the carbon dioxide is coabsorbed; that is, 10 to 40 percent of the carbon dioxide present is in the absorber feed gas, depending upon contactor design and type of amine used.

The fat amine solution is freed from absorbed acid gases in a conventional stripper (Ref.

TABLE 1. Typical Compositions of Gas Streams in Claus and SCOT Units
(Basis: Sulfur recovery in Claus unit is 94 percent)

Composition, % vol.	Claus intake	Claus off-gas	SCOT off-gas	Incinerated SCOT off-gas
H_2S	89.9	0.85	0.03	<10 ppm vol
SO_2		0.42		0.02
S_8 vapor and mist		0.05		
COS		0.05	10 ppm vol.	
CS_2		0.04	1 ppm vol.	
CO		0.22		
CO_2	4.6	2.37	3.05	4.42
HC (MW:30)	0.5			
H_2		1.60	0.96	
H_2O	5.0	33.10	7.00	9.84
N_2		61.30	88.96	83.94
O_2				1.78
Total	100.0	100.00	100.00	100.00
Temperature, °C	40	140	40	650
Pressure, kg/cm² abs.	1.5	1.3	1.0	1.0
Gas quantity, mole relative	1	3.0	2.2	3.5

*IFP (Institut Francais du Pétrole).
†SCOT (Shell Claus Off-Gas Treating).

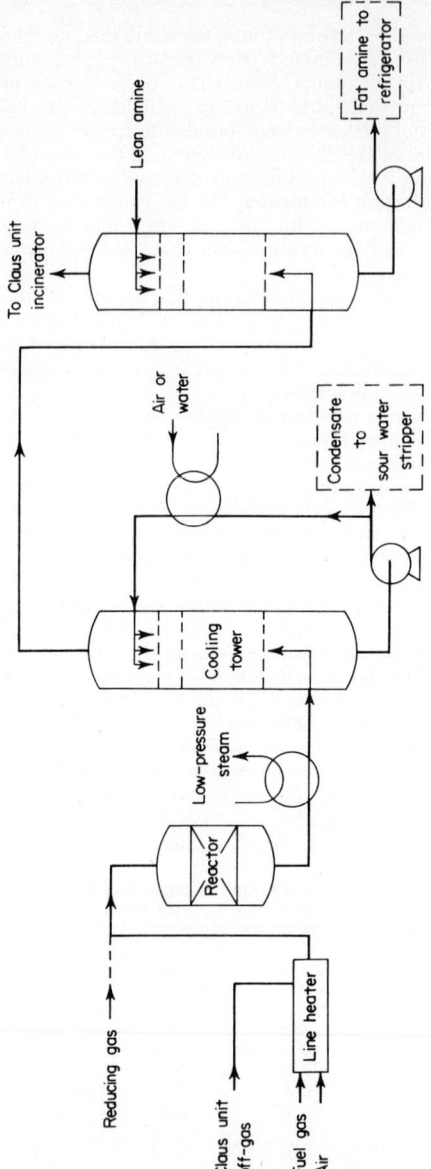

Fig. 1. Shell Claus off-gas treating process.

4). The top gas, consisting of hydrogen sulfide and carbon dioxide, is recycled to the feed of the Claus unit. The recycle stream has only a marginal effect on the operation of the Claus unit. In a typical case (Table 1) at 94 percent sulfur recovery in the Claus unit and 30 percent coabsorption of the carbon dioxide in the alkanolamine solution, the total feed flow to the Claus unit increased by only 7 percent, while the hydrogen sulfide concentration in the acid gas feed stream dropped from 90 to 86 percent (vol.).

The main feed stream to the Claus complex is usually obtained from gas purification units using an alkanolamine absorption/regeneration cycle. Therefore, especially for new installations, the possibility exists of integrating the regeneration of the fat amine solution coming from the SCOT unit with that of the gas purification unit.

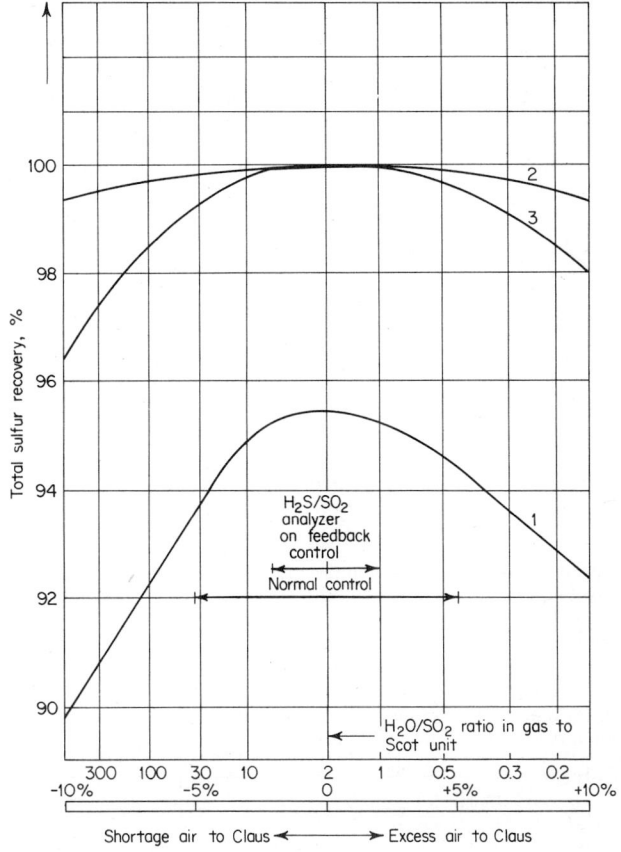

Fig. 2. Relation between total sulfur recovery and air to the main burner of the Claus unit. Theoretical sulfur recovery of Claus unit: (1) With two catalytic stages. (2) With two catalytic stages followed by the Scot process (with extra flexibility in Scot design). (3) With two catalytic stages followed by the Scot process. (Scot unit designed for optimum Claus operation: no extra flexibility.)

Composition of Sour Gas to Claus Unit

	% vol.
H_2S	93.0
CO_2	0.0
HC	0.5*
H_2O	6.5
	100.0

*Molecular weight 30.

The SCOT process has been designed for minimum pressure drop so that it can be easily added to an existing Claus unit. If insufficient pressure is available, a gas booster can be installed, preferably between cooling tower and amine absorption tower. Familiar, easily operable equipment is used and no secondary waste streams are produced.

With this process, the total sulfur recovery can be increased to above 99.8 percent of sulfur intake to the Claus unit. An operating range from 20 to 100 percent on intake can be obtained, and changes in the SCOT feed gas composition have only a small effect on the overall sulfur recovery. In Fig. 2, the relation between total sulfur recovery and air to the main burner of the Claus unit is shown. The figure demonstrates that the total sulfur recovery is only marginally affected by normal variations in the air supply rate.

PROCESS CHEMISTRY

There are three major phases of the process: (1) reduction, (2) amine absorption, and (3) amine regeneration.

Reduction.   Sulfur dioxide and elemental sulfur are reduced by hydrogen as follows:

$$SO_2 + 3H_2 \rightarrow H_2S + 2H_2O \qquad \Delta H_{560K} = -51 \text{ kcal/mole} \qquad (1)$$
$$S_8 + 8H_2 \rightarrow 8H_2S \qquad \Delta H_{560K} = -70 \text{ kcal/mole} \qquad (2)$$

With an excess of H_2, virtually complete conversion of elemental sulfur and sulfur dioxide into H_2S is obtained [i.e., residual SO_2 contents below 10 ppm (vol.)]. The reduction of SO_2 can be described as a first-order reaction in H_2 and a zero-order reaction in SO_2. The activation energy is about 20 kcal/mole.

When carbon monoxide is also present as a reducing agent, it is equivalent to H_2 mole per mole.

The presence of CO results in an increased reaction rate for the reduction of sulfur compounds in Claus off-gases as compared with the reduction with H_2 alone.

Carbonyl sulfide (COS) and carbon disulfide (CS_2), which together make up 5 to 30 percent of the sulfur present in the Claus tail gas, are mainly converted into H_2S by hydrolysis. The contribution of the reduction reaction to the conversion into H_2S is minor only:

$$COS + H_2O \rightarrow H_2S + CO_2 \qquad (3)$$
$$CS_2 + 2H_2O \rightarrow 2H_2S + CO_2 \qquad (4)$$

The concentrations of COS and CS_2 approach thermodynamic equilibrium values, i.e., 10 ppm (vol.) COS and 1 ppm (vol.) CS_2 in the reactor effluent for gas compositions similar to those given in Table 1.

Methanization of CO and CO_2 takes place to an almost negligible extent, resulting in a methane content of the off-gas of less than 20 ppm (vol.) methane at temperatures in the reduction reactor up to about 450°C.

Amine Absorption.   The H_2S and CO_2 present in the cooling tower top gas can react with the secondary amine solution as follows:

$$H_2S + \overset{R}{\underset{R}{>}}NH \rightleftharpoons HS^- + \overset{R}{\underset{R}{>}}N^+H_2 \qquad \Delta H_{313K} = -10 \text{ kcal/mole} \qquad (5)$$

$$CO_2 + 2\overset{R}{\underset{R}{>}}NH \rightleftharpoons \overset{R}{\underset{R}{>}}NCOO^- + \overset{R}{\underset{R}{>}}NH^+_2 \qquad \Delta H_{313K} = -24 \text{ kcal/mole} \qquad (6)$$

The equilibrium solubility for CO_2 under absorber conditions is usually higher than that of H_2S, and the solubility of H_2S decreases in the presence of CO_2 (Refs. 4,5). Thus, a selective H_2S absorption cannot be expected on the basis of the solubility data alone. However, by making use of the much larger absorption rate of H_2S as compared with that of CO_2, selective H_2S absorption is possible.

Absorption rates of H_2S and CO_2 are shown as a function of the partial pressure in Fig. 3. The data were derived from laboratory experiments in a stirred cell with a well-defined gas/liquid interface. A 2 molar dipropylamine solution was used.

Hydrogen sulfide reacts with amines almost instantaneously and the reaction of CO_2 is moderately fast. Therefore, the overall absorption rates are primarily determined by mass

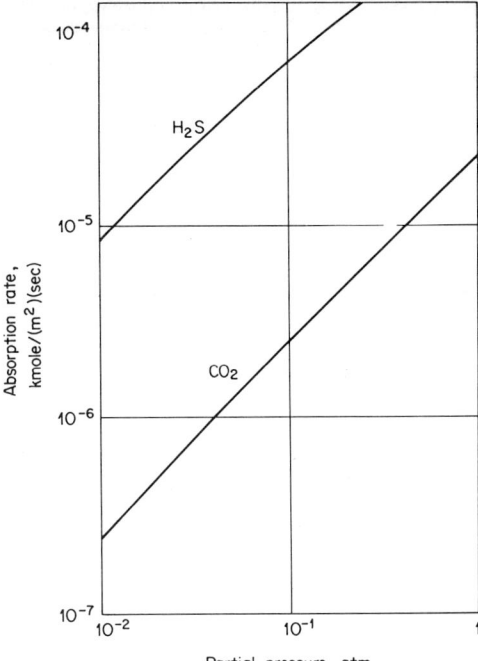

Fig. 3. Typical absorption rates of hydrogen sulfide and carbon dioxide through the interface of an amine solution.

transfer. The absorption of H_2S is controlled by diffusion in the gas phase and that of CO_2 by mass transfer through the liquid film at the gas/liquid interface, which is the more severe limitation. Accordingly, the absorption of H_2S in the amines is much faster than that of CO_2, resulting under SCOT absorber conditions in almost complete absorption of the H_2S and a limited coabsorption of the CO_2.

TABLE 2.   SCOT Process Cost Data for Add-On and Integrated Units

| | Capacity of Claus unit, tons of sulfur intake/stream day | | | | | |
| | Add-on SCOT unit† | | | Integrated SCOT unit‡ | | |
Item of cost	100	200	1000	100	200	1000
Total capital investment, US $ × 10⁶*	0.9	1.6	3.6	0.7	1.2	2.8
Operating costs, $ per stream day (333 stream days per annum):						
Direct costs	270	460	1880	270	460	1880
Capital charge (17% on equip. capital)	450	770	1680	370	570	1280
Total	720	1230	3560	640	1030	3160

Note: The capital investment for the add-on SCOT unit corresponds to about 100 percent of the capital investment of the preceding Claus unit. For the integrated SCOT, the investment is about 75 percent.

*Basis is Western Europe (1973). For the United States, these figures should be increased by 10 percent.

†Add-on unit (SCOT unit with gas blower, separate alkanolamine regeneration facilities and separate sour water stripper).

‡Integrated unit (SCOT unit fully integrated, but bearing a share of the costs for combined amine regeneration facilities and sour water stripper. There is no gas blower, but the costs of pressure increase in upstream units has been added).

With tertiary amines, only the slower reaction to bicarbonate occurs:

$$CO_2 + H_2O + \diagdown N \rightleftharpoons HCO_3^- + NH^+ \qquad (7)$$

The rate of CO_2 absorption is about half that for dipropylamine. In spite of this, triethanolamine is not attractive, since it is too weak a base to give acceptable H_2S loadings under SCOT absorber conditions. However, methyl diethanolamine becomes attractive for special cases with a high CO_2 percentage in SCOT absorber feed.

Amine Regeneration.   Under the conditions prevailing in the regenerator, reactions (5) and (6) are reversed and the H_2S and CO_2 are liberated.

CAPITAL INVESTMENT

The capital investment and operating cost estimates given in Table 2 for a unit treating the tail gas of a Claus unit with one thermal and two catalytic stages have been based on erection in Western Europe in 1973, with an indication of costs for the United States. They include inside and outside plot costs, tie-in and first fill of catalyst and chemicals.

REFERENCES

1. Barry, B.: Reduce Claus Sulfur Emission, *Hydrocarbon Process.*, 102–106 (April 1972).
2. Hyne, J. B.: Methods for Desulfurization, *Oil Gas J.*, 64–78 (Aug. 28, 1972).
3. Groenendaal, W., and H. C. A. van Meurs: Shell Launches Its Claus Off-gas Desulphurization Process, *Pet. Petrochem. Int.*, **12**(9): 54–58 (September 1972).
4. Kohl, A. L., and F. C. Riesenfeld: "Gas Purification," McGraw-Hill, New York, 1960.
5. Perry, J. H., et al. (eds.): "Chemical Engineers' Handbook," 4th ed., pp. 14-10–14-11, McGraw-Hill, New York, 1963.

NO$_x$ Control by Furnace and Burner Design

BY STEPHEN A. JOHNSON* AND ALBERT H. RAWDON†

Since the early 1950s, when it was learned that oxides of nitrogen (NO$_x$) react with hydrocarbons in the presence of oxygen and sunlight to form photochemical smog (Ref. 1), researchers have been attempting to minimize the emission of these gases from stationary combustion sources. In the meantime, governmental legislation and subsequent promulgation of emission regulations (Ref. 2) have made NO$_x$ control one of the primary design parameters for utility and industrial power boilers. Boiler manufacturers have had to adapt their designs to control oxides of nitrogen emissions from utility power boilers without reducing boiler performance and reliability. With the cooperation of the utility power companies, major design innovations have been made. However, NO$_x$ control techniques are continually being refined as new data are obtained, and will continue until the state of the art is reduced to standard operating procedure.

SPECIES OF NO$_x$

The primary oxide of nitrogen produced in a boiler furnace is nitric oxide (NO), although some nitrogen dioxide (NO$_2$) is formed as the NO is further oxidized downstream. These two species are normally grouped together and called NO$_x$. NO$_x$ is measured in parts per million (ppm) by volume by either wet chemical methods (usually the phenoldisulfonic acid method) (Ref. 2) or by a variety of nonprime instruments. These results are converted to "pounds of NO$_x$ (expressed as NO$_2$) per million Btu fired" for the purpose of comparison with emission regulations. Figure 1 presents graphs which facilitate this conversion for representative fuels.

NO$_x$ FORMATION

To fully understand NO$_x$ control methods utilized by boiler manufacturers, one must first know how NO$_x$ is formed during the combustion process. The following section will briefly summarize combustion as it occurs in a boiler.

In theory, flames may be generally classified into two groups: *premixed flames,* in which the rate of combustion is controlled by reactant concentrations and the rate at which they are injected into the combustion chamber, and *diffusion flames,* in which the rate of combustion is controlled by the diffusion of oxidant to the fuel and diffusion of combustion products to the bulk gases. In practice, however, premixed flames are seldom found outside the laboratory, and pure diffusion flames are difficult to attain. Utility boiler flames are vaguely referred to as "turbulent diffusion" (Ref. 3) flames; that is, fuel and air are introduced as two separate streams which are artificially swirled and mixed by diffusion.

The combustion mechanism varies with different fuels (Refs. 4, 5), but basically there is a region early in the flame development where gaseous hydrocarbons break down and rapidly disappear, while products of partial oxidation (CO, OH, etc.) are formed. This is followed by a second region within the flame where combustion is completed at a slower rate (Ref. 6). Coal combustion introduces a third region in which oxygen diffuses to the carbon residue left after devolatilization to complete combustion.

Oxides of nitrogen are formed by two distinct mechanisms in two separate regions of the flame. Thermal NO$_x$ is formed in the second region of the flame during the same time span in which gaseous hydrocarbon combustion is completed. This mechanism of NO$_x$ formation, which has been fully characterized by Zeldovich (Ref. 7), is summarized below:

*Senior Research Engineer, and †Director of Research and Development, Riley Stoker Corporation, Worcester, Mass.

$$O_2 \rightleftharpoons 2O \qquad \text{initiation and termination} \qquad (1)$$
$$N_2 + O \rightleftharpoons NO + N \qquad \text{propagation, } T > 2800°F \ (1538°C) \qquad (2)$$
$$O_2 + N \rightleftharpoons NO + O \qquad \text{propagation, } T > 1500°F \ (816°C) \qquad (3)$$

Oxygen molecules are first dissociated in the high-temperature furnace environment early in the flame to form oxygen atoms in superequilibrium amounts. These atoms are preferentially consumed by carbon, but when the temperature rises to a certain level, these oxygen atoms contain enough energy to split the very stable nitrogen-nitrogen bond.

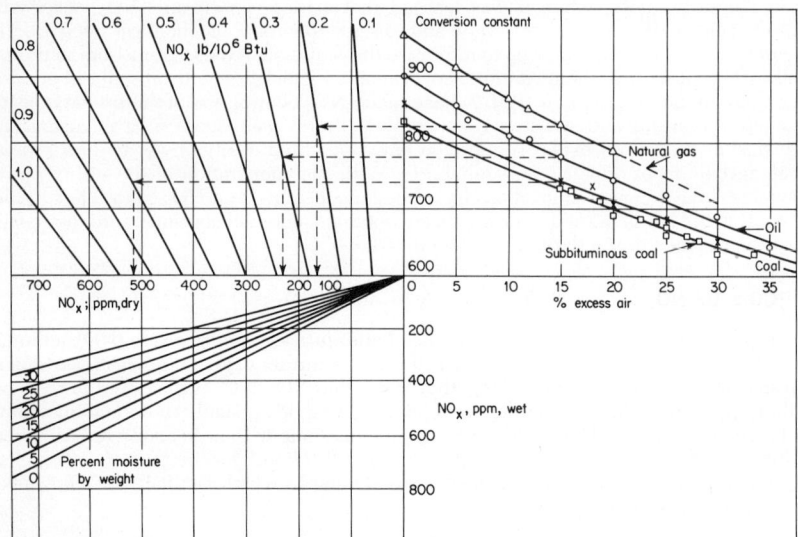

Fig. 1. Conversion of NO_x, ppm, to NO_x, pounds/million Btu, for various fuels and excess airs:

$$NO_x \times \frac{lb}{10^6 \, Btu} = \frac{MW \, NO_2}{MW \, Flue \, gas} - \left(\frac{(lb \, Flue \, gas/lb \, Fuel) \times NO_x \, ppm}{H.H.V.} \right)$$

where MW = molecular weight (lb/lb mole); H.H.V = higher heating value (Btu/lb).

When this occurs, NO is formed and a nitrogen atom is released to further react with oxygen to produce NO. The whole chain reaction terminates when the temperature is cool enough so that no more free radicals are produced.

It can be seen from the above discussion that thermal NO_x is a function of two things: furnace temperature and oxygen concentration. If the local oxygen concentration is decreased by lowering the overall excess air, there are fewer oxygen atoms made available by reaction (1) to react with the ample supply of nitrogen in the combustion air, and less thermal NO_x is formed. If the temperature falls below 2800°F, the number of oxygen atoms that contain enough energy to react with N_2 via reaction (2) is statistically negligible, but this number increases almost exponentially as furnace temperatures increase. It is also worthy of note that if a gas stream already containing NO (such as turbine exhaust gas) is injected into a furnace environment where temperatures are below 2800°F, reaction (2) cannot proceed, thus preventing reaction (3) from taking place. But reaction (3) reverse can occur, thus reducing the NO_x concentration of that stream. In fact, this phenomenon has been observed in some combined cycle utility boilers (Ref. 8) and laboratory experiments (Ref. 13).

The amount of thermal NO_x formed is also a function of residence time of the combustion gases in the flame zone (Ref. 9); however, this has been shown to be extremely short in power boilers. In a matter of seconds, the gases pass through the high-temperature zone into an atmosphere where free radicals recombine and NO_x formation is "quenched." The residence time varies according to the turbulence inherent in each particular burner-furnace configuration.

A second mechanism of oxides of nitrogen formation is oxidation of the chemically bound nitrogen in the fuel. At present this mechanism is only partially understood, but a great deal of research by a number of investigators (Refs. 10–13) has resulted in a tentative theory.

It has been shown that fuel NO$_x$ is formed relatively early in the flame from nitrogen compounds in the volatile gas surrounding the burning fuel particle. However, only a portion of the total amount of nitrogen in the fuel is converted to NO$_x$. Various researchers have presented the following observations concerning the degree of fuel nitrogen conversion to NO$_x$:

1. Fuel nitrogen conversion efficiency increases as the amount of nitrogen in the fuel decreases (Refs. 10–12).

2. Fuel nitrogen conversion efficiency decreases as the oxygen concentration in the burner zone decreases, particularly during substoichiometric combustion (Refs. 10–13).

3. Fuel nitrogen conversion is relatively insensitive to furnace temperature (Refs. 6, 11, 14, 16).

4. Fuel nitrogen conversion decreases with departure from well-mixed conditions. Actual boiler tests have shown nitrogen conversion rates to decrease by 50 percent during low-load firing (No. 6 oil doped) where flame turbulence is very low. In another utility boiler, total NO$_x$ emissions more than doubled (375 to 800 ppm) when the same No. 6 oil fuel was fired first with low-turbulence oil tips and then with high-turbulence oil tips, though in this case it was impossible to separate the thermal NO$_x$ from the fuel NO$_x$ increase (Refs. 6, 8, 14). The oil was doped with organic nitrogen compounds.

5. Fuel nitrogen conversion is less for diffusion flames than it is for premixed flames (Ref. 13) as shown in Fig. 2.

6. If NO is injected into a flame, it is quickly converted to an intermediate which reconverts completely to NO in premixed flames and partially to NO in diffusion flames (Refs. 8, 13, 15).

As a result of the above evidence, the following model is presented. The fuel is pyrolized upon entering the furnace, releasing nitrogen which is probably in the form of N, NH, or CN. These free radicals can then be oxidized to form NO or can recombine to form N$_2$, depending upon the localized oxygen concentrations which are determined by turbulence in the diffusion flame. The portion of the nitrogen that forms NO can still be reduced by the following reaction scheme:

$$RN + RN \overset{K_4}{\rightleftharpoons} N_2 + 2R \tag{4}$$

$$RN + O \overset{K_5}{\rightleftharpoons} NO + R \tag{5}$$

$$NO + RN \overset{K_6}{\rightleftharpoons} N_2 + OR \tag{6}$$

The specie R can be either hydrogen or carbon. It can be seen that if we only consider the forward reactions, the fractional conversion (f_N) of RN to NO is

$$f_N \rightarrow NO = \frac{K_5[O]}{K_4[RN] + K_5[O] + K_6[NO]}$$

It is also possible that NO formed by the above mechanism can interact with thermal NO$_x$ via the Zeldovich mechanism later in the flame. The presence of fuel NO would tend to drive Eqs. (2) and (3) in the reverse direction, even though temperature would favor the forward reaction. Therefore, a certain amount of thermal NO$_x$ is prevented from forming. Tests performed by Pershing et al. (Ref. 16), in which coal and oil were burned in N$_2$/O$_2$ to determine total NO$_x$, and argon/O$_2$ to determine fuel NO$_x$, show a much higher apparent fuel nitrogen conversion when the fuel and thermal NO$_x$ are separated than one would expect from previous results.

The effect of fuel nitrogen on NO$_x$ emitted during the combustion of oil can be found in Fig. 3. Similar tests performed using coal fuel have yielded indeterminate results, presumably due to the interaction of other variables (coal moisture, percent volatiles, slagging, etc.).

NO$_x$ EMISSION DATA

From the preceding discussion, it can be concluded that NO$_x$ emissions from power boilers are mainly a function of furnace temperature for thermal NO$_x$, and localized air/ fuel ratio for thermal and fuel NO$_x$. Many boiler manufacturers (Refs. 17, 18) and some researchers (Ref. 19) utilize this fact to predict NO$_x$ emissions. In general, furnace temperatures increase with a heat input increase (from increased fuel flow or sensible heat in air, flue gas, and fuel) and decrease with an increase in waterwall cooling surface in the flame zone. Therefore, seat-of-the-pants logic tells us that NO$_x$ emissions can be

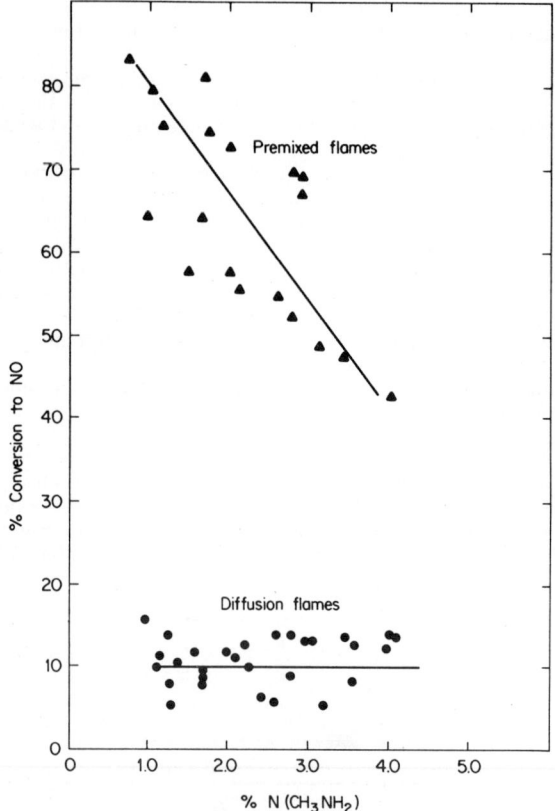

Fig. 2. Conversion of methylamine to nitric oxide.

correlated by the ratio of heat input to cooling surface area. It only remains to refine the correlation by determining what surface area is most effective in cooling the flame and by separating out the effects of other variables (Ref. 21).

Another way to correlate NO$_x$ data is to plot them against the quantity "megawatts/ furnace firing wall" (Ref. 19). Megawatt output of a boiler is proportional to heat released in the furnace (assuming that each furnace has the same furnace heat transfer efficiency), and the number of firing walls is roughly proportional to the furnace cooling surface if the effect of water-cooled platens in the burning zone is considered. Using this empirical method, NO$_x$ emissions have been predicted to within a least-squares standard deviation of 50 to 70 ppm (Ref. 21). Others have utilized Zeldovich kinetics in combination with the heat transfer profile of the boiler to predict NO$_x$ emission as a function of time and temperature (Refs. 22, 23). This method, though more rigorous than the previous method, depends on the reliability of the rate coefficients employed. There is considerable controversy surrounding the universal applicability of these rate coefficients to every furnace condition. They seem to work for tangential boilers firing gas and oil but fall short for other types of units. The kinetic analysis also suffers if it assumes that oxygen atoms are in equilibrium with their molecules.

Typically, for a given fuel, NO$_x$ emissions from utility boilers can be arranged in order of increasing NO$_x$ emission as follows:

Tangential < turbo < wall-fired (both 1 wall and opposed fired units) < cyclone

However, it should be noted that all types of boilers (with the possible exception of

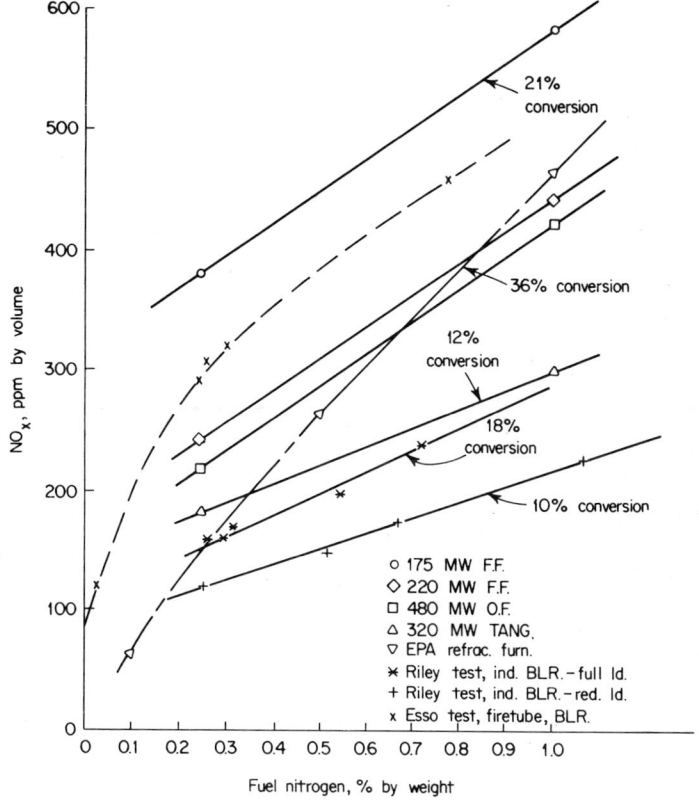

Fig. 3. Effect of fuel nitrogen on NO$_x$ emissions.

cyclone-type units) are amenable to control by the methods mentioned in the next section (Ref. 24). Tangential and turbo units are low NO$_x$ emitters due to slow mixing of fuel and air along with high utilization of furnace cooling surface area. Wall-fired furnaces conventionally contain a highly turbulent fire in a relatively confined volume. The resultant high temperatures produce high NO$_x$ emissions. It is possible to lengthen the flame by burner modification in this type of unit as long as sufficient time is allocated in the furnace for complete combustion to take place (Ref. 25). Typical flame patterns for utility furnaces are illustrated in Fig. 4.

Generally, coal-fired units emit the most NO$_x$ (500–1000 ppm), while gas-fired units (100–1000 ppm) and oil-fired units (100–500 ppm) generally produce less NO$_x$ in a given boiler. Typical NO$_x$ emissions from utility boilers can be found in Refs. 18, 21–33.

It is much more difficult to compare the NO$_x$-emitting characteristics of industrial boilers. Front-fired boilers with a single burner tend to be low NO$_x$ emitters when firing natural gas and distillate oil with cold (80°F; 26.7°C) air. The emission becomes considerable if preheated air and/or residual oil is utilized in the furnace. Cross drum-type industrial boilers can be either high or low NO_x emitters depending on the cooling surface area incorporated into the design. Stoker firing of coal or solid wastes produces very little NO_x because the slow, diffusion-controlled burning results in low temperatures and limited fuel nitrogen conversion.

Combined cycle units in which turbine exhaust gas is utilized in the steam cycle combine high cycle efficiency with reduced NO_x emission. The NO_x produced in a gas turbine is amenable to control by such techniques as water injection and by combustor

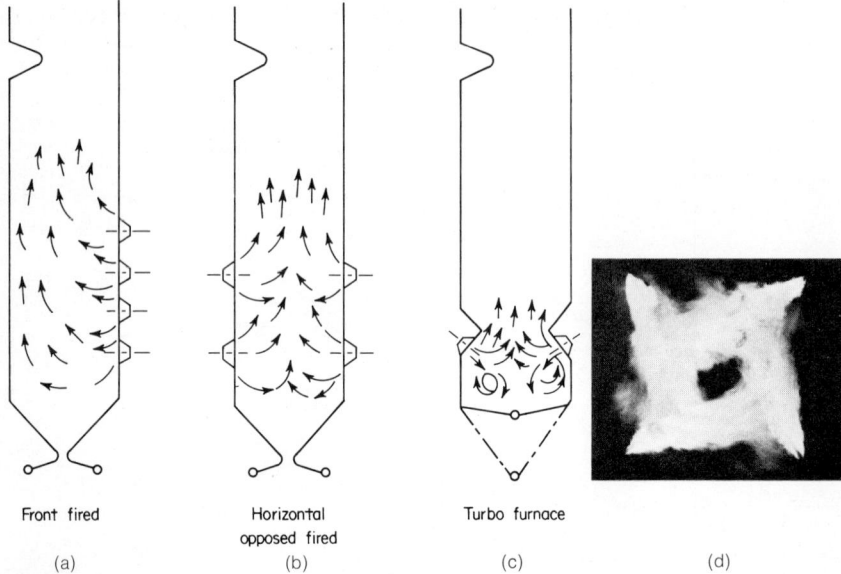

Front fired

Horizontal
opposed fired

Turbo furnace

(a)   (b)   (c)   (d)

Fig. 4. Utility boiler flame patterns: (*a*) Front-fired. (*b*) Horizontal opposed-fired. (*c*) Turbo furnace. (*d*) Tangential flame pattern viewed from top of furnace. (*Ref.22. Based on data used with permission from Combustion Engineering, Inc.*)

modifications (Refs. 26, 27). However, by utilizing the turbine exhaust gas [typically about 900°F (482°C) and 15 percent O_2] as a source of O_2 in the steam boiler wind box, NO_x emissions are minimized because the resultant low flame temperatures not only limit NO_x formation but can reduce NO_x formed in the turbine by reversing Eq. (2) (Ref. 8).

NOₓ CONTROL BY FURNACE DESIGN

The main objective in designing a furnace for minimum NO_x emission is to increase the furnace cooling surface in the flame zone. This can be done by increasing the horizontal and vertical burner spacing. Some boiler manufacturers also specify that a large number of relatively small burners be utilized so that the heat from the fuel can be released more uniformly over the largest possible area (Ref. 25). Another way to increase furnace cooling surface is to provide water-cooled division walls or platens in the burners zone, although this may complicate the boiler circulation (Ref. 21).

The furnace designer must be very careful in sizing a furnace for NO_x control, especially when controlled combustion (discussed later) is used. If the furnace is too deep, excess air must be increased to complete the combustion. If the furnace is too shallow or the burners are too close to a waterwall, flames may impinge on the walls causing incomplete combustion, high tube-metal temperatures, tube corrosion on either the steam or water side, or slagging.

A furnace must also be designed for maximum utilization of furnace cooling surface. This can be done by choosing a furnace type (as mentioned before) in which the flame is directed along a wall. Another way to utilize cooling surface is to maximize heat removal rates by providing adequate wall blowers to keep the walls clean. An accumulation of ash

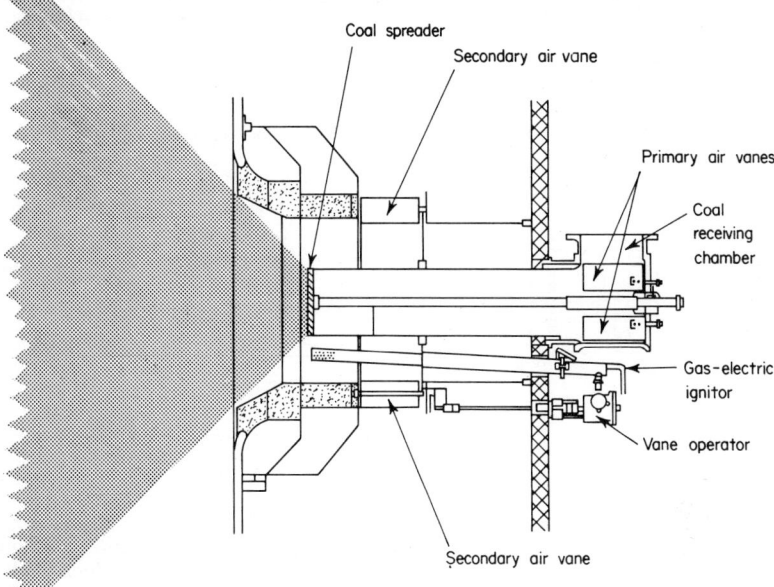

Fig. 5. Flare-type burner.

on the furnace walls acts as a barrier to radiative heat transfer, leading to an increase in furnace temperatures and NO$_x$ (Refs. 28, 29).

NO$_x$ CONTROL BY BURNER DESIGN

There are three different philosophies utilized by burner designers to control NO$_x$. Each one is briefly discussed below.

The first type of low NO$_x$ burner design delays the mixing of fuel and air to produce a low-temperature, diffusion-controlled flame. This type of burning also has the potential to minimize fuel nitrogen conversion by providing a localized reducing atmosphere early in the flame.

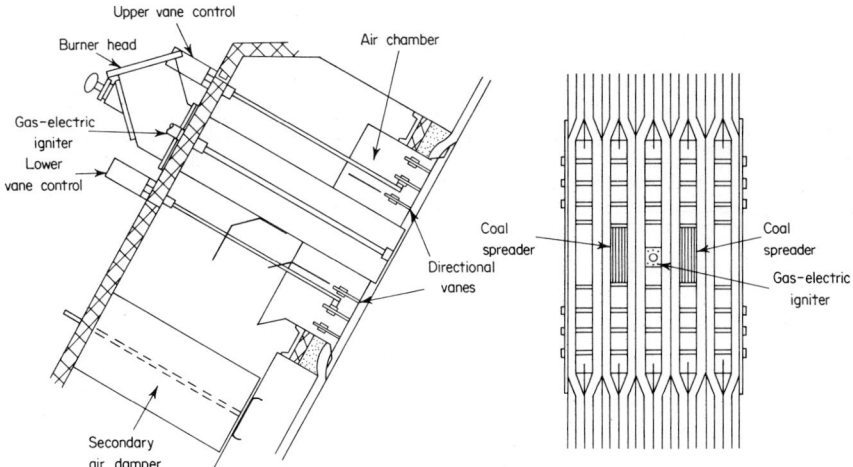

Fig. 6. Directional flame burner.

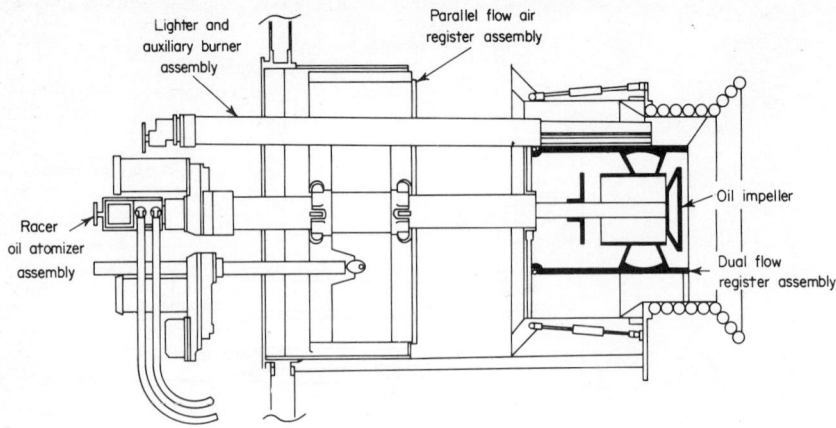

Fig. 7. Entrainment-type burner.

Three types of fuel-burning equipment to produce low NO_x diffusion flames are diagrammed in Figs. 5 to 8. Slot-type burners are generally used to burn coal in tangential or turbo furnaces. Double-register burners generally burn oil or gas. They have the flexibility of operating with high swirl for stable off-stoichiometric firing or with low swirl for diffusion burning. Entrainment burners are also low-swirl burners in which the fuel-primary airstream enters the furnace with sufficient velocity to entrain the secondary airstream and complete combustion.

Turbulent flame burners are inherently high NO_x producers, though it has been found that injection of fuel perpendicular to the secondary airstream rather than parallel to the airstream can slightly reduce the emission (Fig. 9). However, this type of burner can be a powerful tool for NO_x control when used in conjunction with flue gas recirculation or two-stage combustion because it maintains efficient combustion and prevents smoke, even when burning velocities are very low.

Burners designed to complete combustion with a minimum amount of excess air also lower NO_x emissions by limiting the oxygen available to combine with free nitrogen radicals. Tight control of individual burners is necessary to utilize this technique. How-

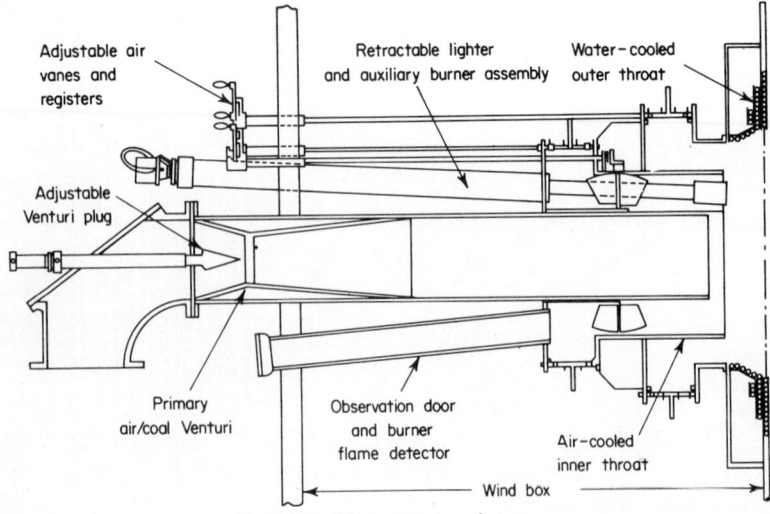

Fig. 8. Double-register-type burner.

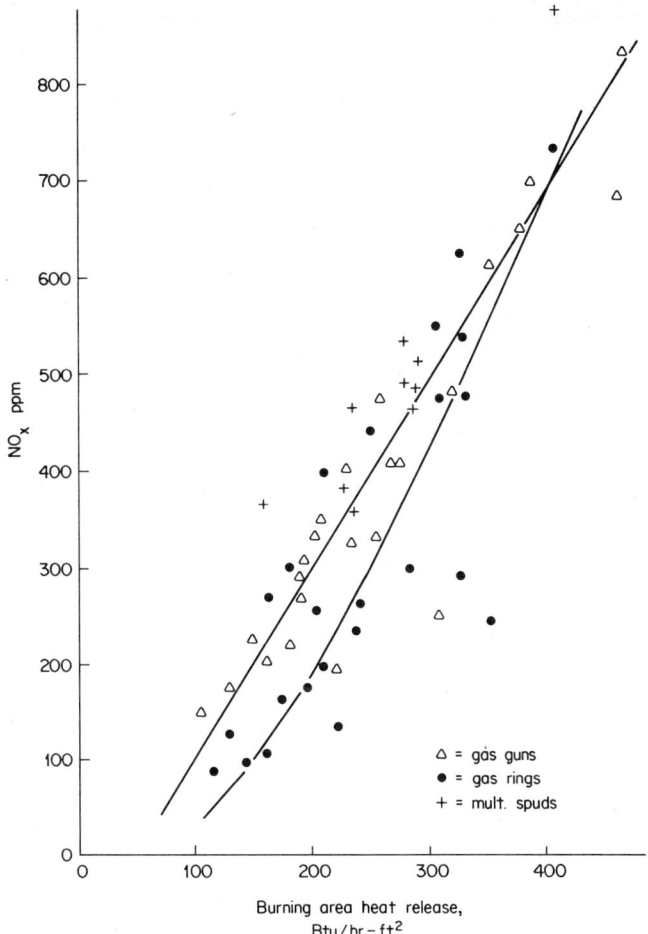

Fig. 9. Effect of burner types on NO$_x$. Front-fired units burning natural gas fuel.

ever, only a 20 to 30 percent decrease in NO$_x$ is feasible by this method, and the reduction could be less if the low excess-air conditions increase furnace wall deposits or if increased turbulence is necessary for complete combustion.

NO$_x$ CONTROL BY COMBUSTION MODIFICATION

The two most common and most effective methods of combustion modification are two-stage combustion and flue gas recirculation. These NO$_x$ controls are incorporated into the design of new power boilers over and above the NO$_x$ controls already mentioned if it is necessary to meet strict NO$_x$-emission regulations. They are especially effective in minimizing the NO$_x$ emissions from existing power boilers that were not originally designed to control this air pollutant.

Two-stage combustion (Ref. 30) is achieved by either of two methods: (1) by firing selected burners fuel-rich while the remaining burners operate air-rich or on air alone (called *biased firing*) or (2) by utilizing overfire air ports whereby the entire array of burners would be firing fuel-rich with the remaining air necessary to complete the combustion being added above through special openings or ports. In this way the bulk of the burning is done in an oxygen-starved atmosphere where both thermal and fuel NO$_x$

are minimized. Combustion is later completed at lower temperatures when the products of partial combustion contact the air-rich stream.

The amount of NO_x reduction obtained with two-stage combustion is directly related to the stoichiometric ratio of air to fuel at the fuel-rich burners. Figures 10 to 12 show data obtained from actual utility power boilers using both biased firing and overfire air (Refs.

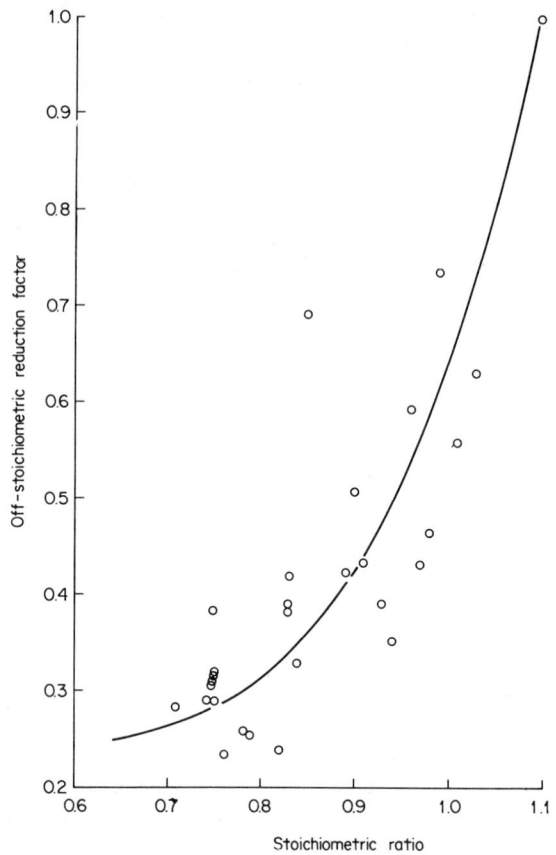

Fig. 10. NO_x reductions during two-stage combustion. Natural gas fuel used.

21, 31, 32). Normally, NO_x reductions of 60 to 70 percent can be achieved when firing natural gas, and 40 to 50 percent reductions are possible when firing coal or oil.

The degree of off-stoichiometry achieved (and thus, the NO_x reduction achieved) is limited by the emission of excess carbon monoxide when firing natural gas, smoke when firing residual oil, and unburned carbon when firing coal. Therefore, it is sometimes necessary to slightly increase excess air to the furnace to prevent the emission of secondary pollutants and the decrease of combustion efficiency.

The long-term effects of off-stoichiometric firing are currently under investigation. A growing number of power companies are accepting this technique as standard operating procedure. One problem with two-stage combustion is that it sometimes shifts the heat transfer characteristics of the boiler. In most cases when a change occurs furnace heat transfer is increased, although the opposite has happened in some short-term tests (Ref. 8). A more important question must be answered concerning the two-stage combustion of coal. It is possible that the reducing atmosphere during two-stage combustion may cause a slagging condition and subsequent tube corrosion due to the reduction of the protective oxide layer by CO or H$_2$S. Tests by Esso Research (Ref. 33) using specially designed corrosion probes have not revealed significant weight loss of tube specimens. However furnaces are being designed with increased plan areas and with air bleed streams along the lower furnace walls to decrease the probability of a slagging problem occurring.

Flue gas recirculation, during which a portion of the flue gas is rerouted from before the

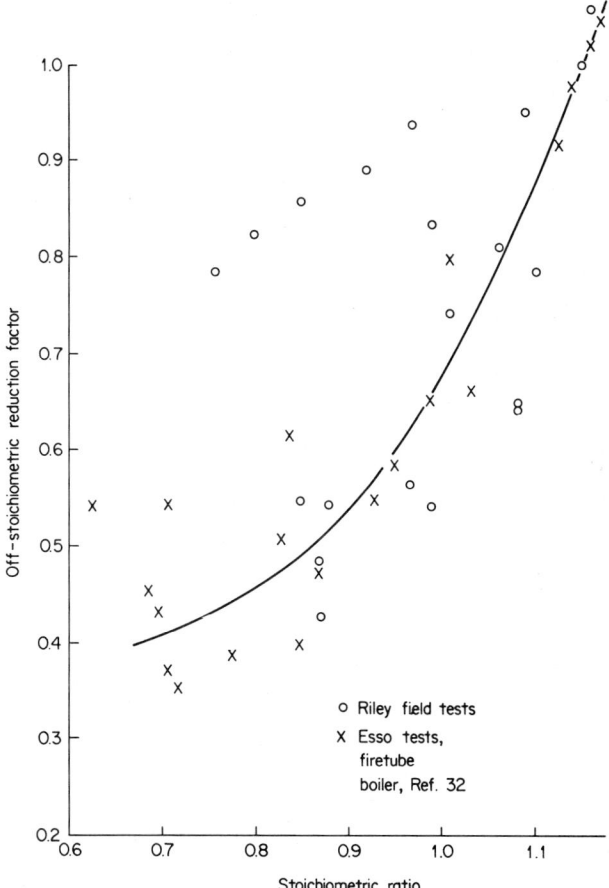

Fig. 11. NO$_x$ reductions during two-stage combustion. Fuel oil used.

air heater through a fan to the wind box, is used by boiler designers to control NO$_x$ emissions and boiler steam temperature. In the past, flue gas has been recirculated to the furnace hopper for steam temperature control. This has very little effect on NO$_x$ emissions because the inert gases tend to hug the walls and never mix with the main flame. Flue gas must also be intimately mixed with combustion air to prevent flame instability and ensure optimum NO$_x$ reductions.

Recycle of relatively cool (800°F; 427°C) combustion products lowers NO$_x$ by cooling the bulk flame temperature and slightly diluting the flame oxygen concentration. The amount of flue gas recycle is limited by flame stability and diminishing returns after 20 percent recirculation (Fig. 13). About a 75 percent reduction in thermal NO$_x$ can be expected with 20 percent flue gas recirculation (Ref. 24).

Although it is relatively easy to operate a boiler with flue gas recirculation, the cost of installation is very high. This is especially true when retrofitting an existing power boiler

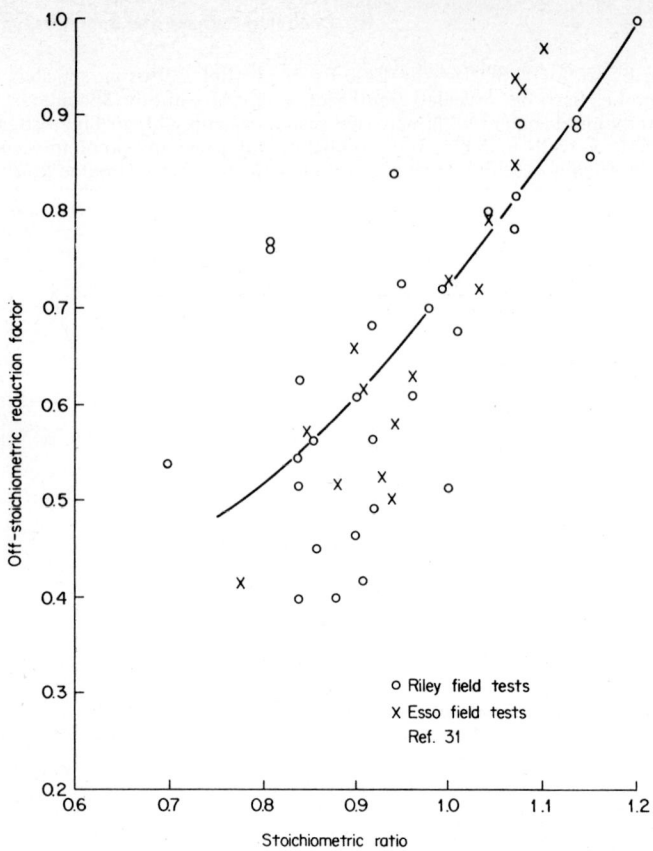

Fig. 12. NO$_x$ reductions during two-stage combustion. Coal used.

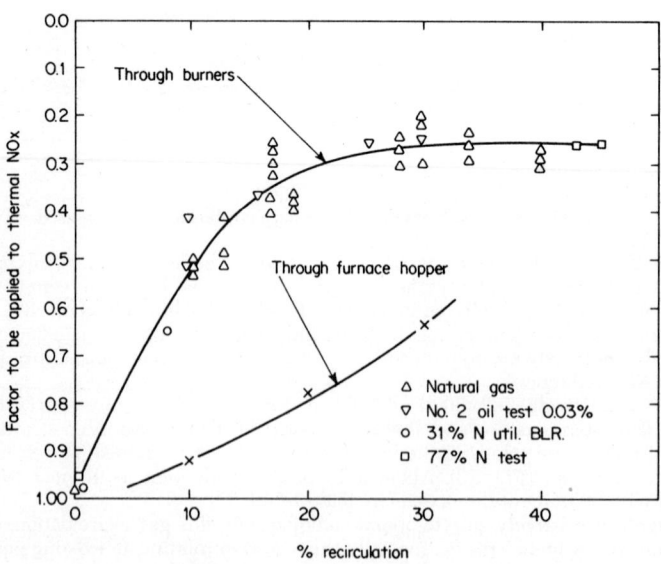

Fig. 13. Flue gas recycle NO$_x$ reduction.

because it is necessary to enlarge the boiler back passes to handle the extra volume of hot gases without causing erosion or acoustic resonance (Ref. 25). In addition, it is usually necessary to add economizer surface and remove superheater surface to prevent excessive steam temperatures and tube-metal temperatures. Flame stability is only a problem when the wind box oxygen concentration is diluted below a critical value, at which time the flame may wander away from the burner. However, experience has shown that the increase in mass flow through the burner usually increases turbulent mixing so that flame stability may even increase with the addition of flue gas recirculation, especially when it is utilized in combination with two-stage combustion (Ref. 34). When firing oil or coal, care must also be taken to eliminate particulate matter from the recirculated gas because this dust could erode or plug burner parts or obstruct the flame scanners (Ref. 21).

RECENT OBSERVATIONS

Research work of the type described here is a continuing endeavor. Important observations revealed at the NO$_x$ Control Technology Seminar sponsored by the Electric Power Research Institute, Palo Alto, California, in February 1976 (see Ref. 35), included:

1. Design of fluidized bed boilers to fire coal has reached the pilot stage. Laboratory results indicate that this type of combustor would produce about one-half the NO$_x$ that is emitted by conventional boilers firing pulverized coal.

2. Field tests of coal-fired boilers utilizing new low-NO$_x$ delayed mixing-type burners have shown that NO$_x$ can be reduced from 30 to 50 percent from the NO$_x$ produced in the same furnace using conventional flare burners.

3. Prolonged boiler operation using two-staged combustion has not resulted in significant long-term corrosion, except where the flame has impinged on furnace water walls.

4. Flue gas recirculation has achieved marginal success in reducing NO$_x$ from coal and residual oil fired boilers. Thermal NO$_x$ is reduced, but in some cases the added turbulence caused by the flue gas recirculation may increase the fuel nitrogen conversion. In one such test, 10 percent flue gas recirculation resulted in a slight total NO$_x$ increase.

REFERENCES

1. Haagen-Smit, A. J.: *Ind. Eng. Chem.*, **44**: 1342 (1952).
2. *Fed. Regist.*, **36**: 247 (Dec. 23, 1971).
3. Essenhigh, R. H.: A New Application of Perfectly Stirred Reactor Theory to Design of Combustion Chambers, *Tech. Rep.* FS67-1(u), Department of the Navy, Office of Naval Research, Power Branch, March 1967.
4. Heap, M. P., T. M. Lowes, R. Walmsley, and H. Bartelda: "Burner Design Principles for Minimum NO$_x$ Emissions," presented at EPA Coal Combustion Seminar, Research Triangle Park, N.C., June 1973.
5. Keston, A. S., J. J. Sangiovanni, and C. T. Bowman: "NO Formation in Fuel Droplet Combustion," presented at 1st American Flame Days, sponsored by American Flame Research Committee, Chicago, Ill., September 1972.
6. Sarofim, A. F., and W. Bartok: "Methods for Control of Nitrogen Oxide Emissions from Stationary Sources," advanced seminar, pp. III-1 to III-36, sponsored by American Institute of Chemical Engineers, New York, 1970.
7. Zeldovich, J.: The Oxidation of Nitrogen in Combustion and Explosions, *Acta Physiochim. URSS*, **21**: 577 (1946).
8. Riley Stoker Corporation test data.
9. Breen, B. P.: A General Understanding of Low Nitric Oxide Operation of Stationary Combustors, prepared for *Fundamentals of Pollutant Formation in Combustion Processes*, sponsored by the University of California, Los Angeles, August 1970.
10. Fenimore, C. P.: Formation of Nitric Oxide from Fuel Nitrogen in Ethylene Flames, *Combust. Flame*, **19** (2): 289–296 (1972)
11. Martin, G. B., and E. E. Berkau: "An Investigation of the Conversion of Various Fuel Nitrogen Compounds to NO in Oil Combustion," presented at 70th AIChE National Meeting, Atlantic City, N.J., American Institute of Chemical Engineers, New York, August 1971.
12. Turner, D. W., R. L. Andrews, and C. W. Siegmund: "Influence of Combustion Modification and Fuel Nitrogen Content on Nitrogen Oxides Emissions from Fuel Oil Combustion," presented at 64th Annual AIChE Meeting, San Francisco, Calif., American Institute of Chemical Engineers, New York, 1971.
13. Sarofim, A. F., G. C. Williams, M. Modell, and S. M. Slater: "Conversion of Fuel Nitrogen to Nitric Oxide in Premixed and Diffusion Flames," 66th Annual AIChE Meeting, Philadelphia, Pa., American Institute of Chemical Engineers, New York, November 1973.

14. Jonke, A. A.: Argonne National Laboratory Rep. Contract No. ANL-ES-CEN-F009 (1970).
15. Merryman, E. L., H. R. Hazard, R. E. Barrett, and A. Levy: "Recent Studies of the Conversion of Fuel Nitrogen to NO_x," presented to Central State Section, Combustion Institute, University of Wisconsin, Madison, Wis., March 1974.
16. Pershing, D. W., G. B. Martin, and E. E. Berkau: "Influence of Design Variables on the Production of Thermal and Fuel NO from Residual Oil and Coal Combustion," presented at 66th Annual AIChE Meeting, Philadelphia, Pa., American Institute of Chemical Engineers, New York, November 1973.
17. Rawdon, A. H., and R. S. Sadowski: "An Experimental Correlation of Oxides of Nitrogen from Power Boilers Based on Field Data," presented at ASME Winter Annual Meeting, American Society of Mechanical Engineers, New York, Nov. 26–30, 1972.
18. Sommerlad, R. E., R. P. Weldon, and R. H. Pai: "Nitrogen Oxides Emission—an Analytical Evaluation of Test Data," presented at 33d American Power Conference, sponsored by Illinois Institute of Technology, Chicago, Ill., April 1971.
19. Bartok, W., A. R. Crawford, and G. J. Piegari: Systematic Field Study of NO_x Emission Control Methods for Utility Boilers, *Esso Res. Final Rep. No.* GRU 4G, 71, Contract No. CPA 70-90, Esso Research & Engineering, Florham Park, N.J., December 1971.
20. Bartok, W., A. R. Crawford, A. R. Cunningham, H. J. Hall, E. H. Manny, and A. Skopp: A Systematic Study of Nitrogen Oxide Control Methods for Stationary Sources, *Esso Res. Final Rep. No.* GR-2-NOS-69, EPA Contract PH 22-68-55, Esso Research & Engineering, Florham Park, N.J., November 1969.
21. Rawdon, A. H., and S. A. Johnson: "Application of NO_x Control Technology to Power Boilers," presented at the 35th Annual Power Conference, sponsored by Illinois Institute of Technology, Chicago, Ill., May 10, 1973.
22. Bueters, K. S., W. W. Habelt, C. E. Blakeslee, and H. E. Burback: "NO_x Emissions from Tangentially Fired Utility Boilers," presented at 66th Annual AIChE Meeting, Philadelphia, Pa., American Institute of Chemical Engineers, New York, Nov. 11–15, 1973.
23. Bagwell, F. A., K. E. Rosenthal, B. P. Breen, N. Bayard de Volo, and A. W. Bell: "Oxides of Nitrogen Emission Reduction Program for Oil and Gas Fired Utility Boilers," presented at 32d American Power Conference, sponsored by Illinois Institute of Technology, Chicago, Ill., April 1970.
24. Bagwell, F. A., K. E. Rosenthal, D. P. Teixiera, B. P. Breen, N. Bayard de Volo, and S. Kehro: "Utility Boiler Operating Modes for Reduced Nitric Oxide Emissions," presented at 64th Annual Meeting, Air Pollution Control Association, Atlantic City, N.J., 1971.
25. Krippene, B. C.: "Nitric Oxide Reduction through Controlled Combustion—A Challenge to the Boiler Designer," presented at 16th National Power Instrumentation Symposium, Chicago, Ill., Instrument Society of America, Pittsburgh, Pa., May 1973.
26. Murad, Richard J.: "Emission and Control of NO_x in Industrial Gas Turbine Combustors: Experimental Results," presented at 66th Annual AIChE Meeting, Philadelphia, Pa., American Institute of Chemical Engineers, New York, Nov. 11–15, 1973.
27. Klapatch, R. D., and T. R. Koblish: Nitrogen Oxide Control with Water Injection in Gas Turbines, *ASME Paper* 71-WA/GT-9, presented at the ASME Winter Annual Meeting, Washington, D.C., American Society of Mechanical Engineers, New York, November 1971.
28. Balakrishnan, A., and D. K. Edwards: Radiative Flame Cooling for Reduction of Nitric Oxide Emissions, *ASME Paper* 73-HT-32, presented at Atlanta, Ga., meeting, American Society of Mechanical Engineers, New York Aug. 5–8, 1973.
29. Matthews, K. J.: Natural Gas Firing on a Small Scale C.E.G.B. Standard Oil Burner, *J. Ins. Fuel*, **193** (June 1970).
30. Barnhart, D. H., and E. K. Diehl: "Control of Nitrogen Oxide in Boiler Flue Gases by Two-Staged Combustion," presented at 52d Annual Meeting of Air Pollution Control Association, Los Angeles, June 1959.
31. Bartok, W., A. R. Crawford, E. H. Manny, L. Berkowitz, and R. E. Hall: Field Testing: Application of Combustion Modifications to Control NO_x Emissions from Large Utility Boilers, *EPA Contract No.* 68-02-2227, Environmental Protection Agency, Research Triangle Park, N.C., June 1974.
32. Turner, D. W., and C. W. Siegmund: "Staged Combustion and Flue Gas R_e Cycle: Potential for Minimizing NO_x from Fuel Oil Combustion," presented at 1st American Flame Days, sponsored by American Flame Research Committee, Chicago, Ill., September 1972.
33. Crawford, A. R., E. H. Manny, and W. Bartok: "NO_x Emission Control for Coal-fired Utility Boilers," presented at Pulverized Coal Combustion Seminar, sponsored by Environmental Protection Agency, Research Triangle Park, N.C., June 19–20, 1973.
34. Armento, W. J., and W. L. Sage: "The Effect of Design and Operation Variables on NO_x Formation in Coal-fired Furnaces," presented at Pulverized Coal Combustion Seminar, sponsored by Environmental Protection Agency, Research Triangle Park, N.C., June 19–20, 1973.
35. *Proceedings of the NO_x Control Technology Seminar* (D. P. Teixeira, Chairman), sponsored by the Electric Power Research Institute, Palo Alto, California, February 1976.

Methylcyclopentadienyl Manganese Tricarbonyl Combustion Improver

BY J. D. BAILIE* AND A. H. ZEITZ, JR.†

Concentrations of Use. The versatile metal-organic compound, methylcyclopentadienyl manganese tricarbonyl, has been used as a fuel oil additive to reduce deposits and SO_3 formation in boilers since 1961 and in gas turbine fuels since 1965 to reduce exhaust smoke and particulates. The additive is 24.7 percent (weight) manganese and weighs 11.5 pounds/gallon. Use concentrations range from 0.025 gram manganese/gallon of fuel oil for boilers (9.4 pounds CI 2/1000 barrels) to 0.31 gram manganese/gallon of fuel for gas turbines to reduce smoke (about 100 ppm of manganese). This compound is known commercially as "Ethyl" Combustion Improver 2, or simply CI 2.

Smoke Reduction. In distillate fuel oils and heavier residual fuels, CI 2 reduces smoke as a function of concentration, fuel composition, and operating conditions. Such reductions may be attractive from an air pollution view alone. However, gains in combustion efficiency are possible where combustion air is reduced with CI 2 to produce the same smoke as with the untreated fuel. In practice, these gains range from 2 to 7 percent. Considering current high fuel costs, a modest 4 percent gain is attractive. As an example, at $0.25/gallon for No. 6 heating oil, the net savings are $9400 per 1 million gallons burned, including the cost of the CI 2.

SO_3 Reduction. Belyea's work (Ref. 1) shows that CI 2 gave a 45 percent reduction in sulfur trioxide (SO_3) across a range of excess air in a large utility boiler. The basis for this CI 2 effect lies in the conditioning of high-temperature boiler surfaces, thus reducing the catalytic effect of Fe_2O_3 (ferric oxide) on converting SO_2 to SO_3. A similar reduction (65 percent) also occurred in coal-fired units where 20 grams of CI 2 per ton were used.

A reduction in SO_3 content in exhaust gas reduces the dew point of the gas. Frequently, this reduction in dew point with CI 2 is sufficient to alter air heater deposits from sticky, highly corrosive types to drier, less acid, easier-to-remove deposits. This avoids troublesome maintenance operations.

Examinations of fireside deposits following operation with fuel treated with CI 2 have shown reduced amounts and easier removal of deposits. Analyses of deposits from three package boilers (6000 to 20,000 pounds of steam/hour) showed about 25 percent reduction in deposits and large reductions in Fe_2O_3 and $Fe_2(SO_4)_3$ in deposits after operation on CI 2. Since the oxide is an active catalyst and the sulfate an active corroding agent, these reductions are attractive.

The technical paper, "Reduction of Gas Turbine Smoke and Particulate Emissions by a Manganese Fuel Additive," by Plonsker, Rifkin, Gluckstein, and Bailie, presented at the March 1974 meeting of The Combustion Institute provides current data on the effectiveness of CI 2 in turbine fuels. The work shows that CI 2 in the range of 20 to 100 ppm manganese generally reduces smoke, particulate, and carbon emissions by 50 to 90 percent. A continuation of this work is presented in technical paper, "Effect of CI 2 on Emissions from a Large Power-Generating Gas Turbine," which shows, at 25 ppm manganese, CI 2 in the fuel reduced the Bacharach smoke number from a baseline value of 5.3 to 3.0, where the smoke is invisible to the eye. At the same time, total particulate weight was reduced by 35 to 50 percent and carbon was reduced by 75 percent. At 50 ppm manganese, CI 2 reduced total particulate an additional 20 percent (55 to 70 percent reduction) and essentially eliminated carbon.

The recent work confirms earlier CI 2 work by turbine manufacturers, presented to the American Society of Mechanical Engineers in 1967 (Refs. 2–4). CI 2 has been widely used

*Director of Technical Sales, and †Product Application Manager, Petroleum Chemicals Division, Ethyl Corporation, Houston, Tex.

in turbines since 1965 without any significant side effects on blade deposits, filter operation, or turbine durability.

REFERENCES

1. Belyea, A. R.: Manganese Additive Reduces SO_3, *Power* (November 1966).
2. Taylor, W. G.: Smoke Elimination from Gas Turbines Burning Distillate Oil, *ASME Paper* 67-PWR-3 (1967).
3. Davis, F. D., Jr.: Smoke Abatement in Gas Turbines in Industrial Use, *ASME Paper* 67-PWR-4 (1967).
4. DeCorso, M., C. E. Hussey, and M. J. Ambose: Smokeless Combustion in Oil Burning Gas Turbines, *ASME Paper* 67-PWR-5 (1967).

Utilizing Waste Heat from Air Compressors

BY GORDON L. TAFT* AND DONALD R. LOHSE†

HEAT AVAILABILITY FROM A COMPRESSOR

One may ask the question, "How much heat is available from a compressor?" The answer is approximately all that is put into it—heat in the form of brake horsepower (bhp), of course. To calculate the amount of heat energy available, multiply the horsepower of the compressor by 42 to get the Btu per minute. This is true of all types of compressors: reciprocating, centrifugal, rotary sliding vane, and screw; air-cooled and water-cooled; single-stage or multistage. For example, a 50-horsepower compressor will generate 50 × 42 or 2100 Btu/minute or 126,000 Btu/hour. This much heat is sufficient to heat two 6-room houses.

Air-cooled units are well adapted for using waste heat for space heating inasmuch as little or no additional equipment is necessary. Air-cooled compressors have fans that blow ambient air over the radiator type of heat exchangers, making this warmed air readily available to heat any large space. Estimating the amount of heating that would be realized from an air-cooled compressor installed in a presently unheated building is relatively easy to calculate.

CALCULATIONS FOR SPACE HEATING WITH AIR-COOLED COMPRESSOR

Assume a warehouse 100 feet long, 50 feet wide, and 20 feet high. The walls are 12-inch plain cinder block, the roof is 2-inch wood uninsulated construction, and the floor is 6-inch bare concrete. There are two 10- by 10-foot overhead doors to the outside. No windows, roof vents, or exhaust fans are in the building. The warehouse is located adjacent to a manufacturing facility so that zero heat loss through the separating wall is assumed. Thus, only three walls are considered in the following calculations.

Further assume that a 125 horsepower air-cooled compressor with an aftercooler and air receiver is installed in the warehouse near an outside wall. Further assume that the compressor will run at full load. Heat rejected by the compressor into the surrounding air will be

$$125 \text{ bhp} \times 42 \text{ Btu/(bhp)(minute)} \times 60 \text{ minutes/hour} = 315,000 \text{ Btu/hour}$$

Next, determine the temperature rise in the room with an estimate of heat loss through the walls, roof, and floor. The following basic formula may be used:

$$\text{Heat loss in Btu/hour} = K \times A \times \Delta t$$

where K = coefficient of heat transfer
A = number of square feet of exposed area
Δt = temperature difference (indoor versus outdoor)

For convenience, set up the formula for solving for Δt:

$$\Delta t = \frac{\text{Btu/hour input}}{\text{Btu/hour loss}}$$

Based upon a 15 miles/hour wind velocity, K factor values of Btu loss/(square foot)(hour)(°F) are

12-inch plain cinder block	0.37
2-inch built-up wood roof	0.32

*Product Specialist, Rotary Screw Compressors, and †Product Specialist, Reciprocating Compressors, Air Power Division, Joy Manufacturing Company, Michigan City, Ind.

6-inch bare concrete slab floor	0.10
3-inch solid wood door	0.33

Exposed areas are:

Walls: $(100 + 50 + 50)(20)$−doors $(2)(10 \times 10)$	= 3800 square feet
Doors: $(2)(10 \times 10)$	= 200 square feet
Roof: 100×50	= 5000 square feet
Floor: 100×50	= 5000 square feet

An air change, due to leakage, of one change per hour is considered a minimum (atmospheric air occupies 14 cubic feet/pound and will absorb 0.23 Btu/(pound)(°F) temperature rise), so this is equal to

$$\text{Volume} \times 0.075 \times 0.23 \times \Delta t$$

Thus, the following formula emerges:

$$\Delta t = \frac{315{,}000}{(3800 \times 0.37) + (200 \times 0.33) + (5000 \times 0.32) + (5000 \times 0.10) + (100{,}000 \times 0.075 \times 0.23)}$$

$$= \frac{315{,}000}{1406 + 66 + 1600 + 500 + 1725} = \frac{315{,}000}{5297}$$

$$= 59°F \text{ temperature rise}$$

Thus, if the outside temperature were 10°F, the compressor could conceivably maintain an indoor temperature in the warehouse of 69°F. On an air-cooled unit, ventilating air flowing over it at the rate of 100 cubic feet/(minute)(hp) will increase in temperature by 26°F (double the ventilating air flow rate and the temperature increase will be 13°F).

PIPING AND DUCTING ARRANGEMENTS

Distribution of heat from air-cooled compressors is usually simpler than that from a water-cooled unit, that is, if the machine is relatively close to the area to be heated. In any case, the machine should be placed near an outside wall, principally so that the heat may be ejected to the outside in hot weather.

In recent years, there has been a tendency to locate smaller packaged air compressor units that can operate unattended close to where the air is needed. One Canadian compressor user devised a rather elaborate heat-distribution system from several rotary screw compressors* in an equipment room and commented that the next time it will look closely at silenced machines for placement right out in the plant.

A midwestern stamping plant placed its equipment room in a new addition. Two 50-horsepower air-cooled rotary screw compressors* were placed in the room. The outside wall has louvers to supply outside air. Elbows (90°) were fabricated to fit over the grill in the top of the machine where cooling air is exhausted. Ductwork from the elbows goes through interior walls into the plant area. A deflector directs the air upward. Currently, in warm weather, the elbow is rotated 90° and the hot air exhausts to the outside. Consideration is being given to replacing this with a simple system utilizing a diverter damper, perhaps thermostatically controlled. A third compressor now in the same equipment room with the other two units will be moved to a new equipment room where, along with any future compressors to be added, it will help heat the original building. A comparison of gas utility bills has shown that the doubled space has been heated with the same amount of gas as the original space during winters with comparable degree days.

A Canadian heavy equipment manufacturing firm with six 40-horsepower rotary screw machines uses its equipment room as a plenum. Outside cooling air is pulled into the room by a two-speed roof fan. Another fan discharges heat rejected by all compressors into a factory area. A third fan on the outside wall discharges the heat outdoors in warm weather.

Another manufacturer ran ducting from the outlet grill of its three rotary screw compressors to a common duct hung from the ceiling—a single outlet from this duct ran down the length of a work area, distributing heated air through regularly located deflectors. Since a

*Joy Twistair.

considerable length of duct was involved, a fan was installed in the distribution duct at the point it leaves the common duct. A booster fan, such as this, is often necessary to convey the air to the extremities of the area to be heated. Should this fan stop at any time, heat may build up at the compressor, causing high temperatures at both the inlet and discharge. To prevent this occurrence, the booster fan may be electrically connected through a relay to the compressor and shut down the compressor should the fan stop.

CALCULATION FOR SPACE HEATING WITH WATER-COOLED COMPRESSOR

The 42 Btu/bhp minimum heat-rejection figure can be related to a cooling water temperature increase by using the following formula based upon the fundamental relationships: One Btu raises the temperature of one pound of water one degree Fahrenheit; one gallon of water weights 8.33 pounds:

$$\frac{(\text{Compressor bhp}) \, [\text{Btu}/(\text{bhp})(\text{minute})]}{(120°F - \text{cooling water temperature, }°F) \, (8.33)} = \text{gallons/minute}$$

Note that the recommended discharge water temperature from a reciprocating compressor is from 110 to 120°F.

Assuming 60°F cooling water and a 100-horsepower compressor, the formula then indicates a water flow rate of 8.4 gallons/minute.

By examination, it will be noted that when the water temperature difference for the water is reduced, the water flow rate will increase. Water flow will be proportional to the brake horsepower. A 200-horsepower machine will have twice the water flow for a given temperature difference as compared with a 100-horsepower compressor.

Perhaps one of the most universal applications for use of this available heat is for space heating. Hot water can be readily piped to a point of use (in an insulated pipe if necessary), circulating through a commercial water-to-air, fan-cooled heat exchanger or radiator(s) and returned as cool water to the compressor. This water is not wasted. No additional water is needed, eliminating water and sewage charges.

WARM-WEATHER USES

Space heating of the type described generally is not required in most areas during the summer months. Rather than to dump the hot water into sewage, there are some alternatives:

1. Use the water in a process. For example, a cutlery manufacturer uses this as a warm water rinse in manufacturing operations.

2. Divert this water to an outdoor fan-cooled radiator, saving the water by recirculation, but not putting the heat into the building.

3. If water is needed for a lawn, use the warm water in this fashion during the summer season. There is no extra charge for the water and there is no sewage charge. When the lawn has had enough water, revert back to recirculating through the fan-cooled radiator.

4. As was done over 30 years ago for the first time in a Belgium coal mine, use the warm water for shower and washroom facilities year-round.

Bio-Oxidation Process Reduces Water Use

BY E. F. MOHLER, JR., AND L. T. CLERE*

BACKGROUND AND PERFORMANCE

The reuse–bio-oxidation process (Refs. 1–6) successfully handles refinery waste water to reduce pollution, conserve water, and cut costs, while maintaining a satisfactory water quality through the plant. Twenty-one years of continuous operation have proved the success of the process.

Phenolic-type compound removal exceeds 99.9 percent, which is better than standard commercial processes such as activated sludge or trickling filters can provide. COD (chemical oxygen demand), BOD (biochemical oxygen demand), and TOC (total organic carbon) removal is consistently above 90 percent, with a residual TOC average of 19 pounds per 1000 barrels of crude oil. Established low-pollutant loadings to the dry ditch are now being reduced by effective use of upflow sand filters and use of the clean water as cooling tower makeup.

Exceptionally low water consumption rates are expected to be reduced from the current level of 28 gallons per barrel of crude oil to less than 10 gallons per barrel of crude oil. Further reuse of water for cooling tower makeup in a typical 120,000 barrel per day refinery is expected to conserve in excess of $150,000 per year with the improved filtration.

The reuse–bio-oxidation system requires equipment for which capital investment is less than for other standard terminal waste treatment processes and operating costs considerably lower per increment of treating than other processes. Maintenance requirements are low.

Although the reuse water has a high dissolved solids loading, it is satisfactory for cooling tower makeups without special conditioning treatment. Continued monitoring confirms a trouble-free operation with respect to corrosion, heat transfer, and cooling tower wood deterioration.

PROCESS DESCRIPTION AND FLOWSHEET

The reuse–bio-oxidation process has evolved from experiments in the Toledo refinery of Sun Oil Company, based on the use of cooling towers to provide bio-oxidation of phenolic materials while using the waters for conventional equipment process cooling. The presently installed system (Fig. 1) consists of five stages of reuse bio-treatment incorporated into the overall refinery waste water system.

1. *Secondary separator.* Significant quantities of phenol-type compounds and oxygen-demanding constituents are removed during residence in this separator.

2. *Cooling tower.* The first pass down the cooling tower (second stage) removes 90 percent of the phenol loading.

3. *Plant heat exchange system.* Essentially all the remaining phenol is bio-oxidized as water recirculates through the operating heat exchange system (third stage).

4. *Final cooling and aeration.* Passing the water over an induced-draft cooling tower gives additional removal of trace amounts of phenolic-type compounds and adds sufficient oxygen to establish effluent requirements.

5. *Clean-up.* Final removal of solids is accomplished, using filters, after which the reclaimed water is used for makeup to general cooling towers in the refinery.

Successful operation of a reuse–bio-oxidation process system requires removal of oil and sulfur compounds from refinery effluent water prior to entering the system. Conven-

*Environmental Engineering, Sun Oil Company, Toledo, Ohio. This paper is an updated adaptation of paper presented to the National Conference on Complete Water Reuse, Washington, D.C., April 1973, and an article in *Hydrocarbon Processing*, October 1973.

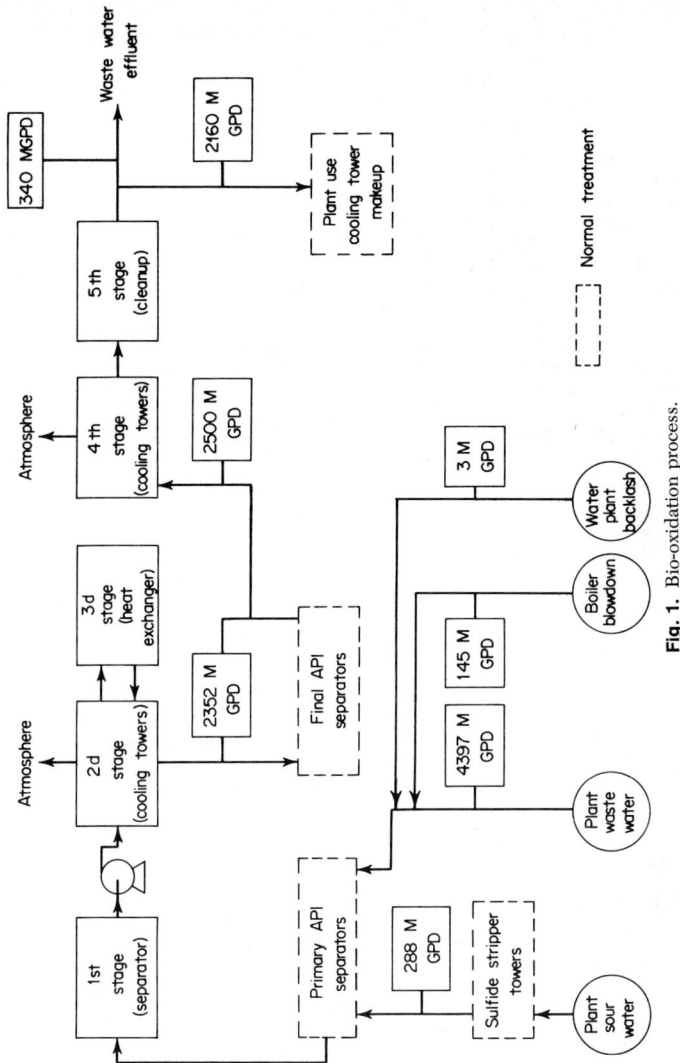

Fig. 1. Bio-oxidation process.

tional processing is used for this purpose to produce an effluent water containing no hydrogen sulfide and only trace quantities of oil. Water in the API separator is approximately 8 pH before entering the first stage, an impounding basin large enough for surge requirements. The API separator removes 99.78 percent of the incoming oil.

Bio-Oxidation Mechanism. No formal bacterial seeding was needed to initiate bio-oxidation. Within hours after the reuse water was introduced to the cooling tower an excellent biological floc was generated by phenol-oxidizing bacteria of both air and water types. Achromobacter and micrococcus have been identified as good phenol oxidizers (Ref. 7). Electron microscope examination of biological floc from the cooling tower water shows presence of rodlike and spherical forms at 21,700 magnification.

Examination of circulating cooling water shows diverse life forms, including bacteria, single-cell algae, protozoa, diatoms, and nematodes. The bacterial colonies include many different species of bacilli and cocci similar to those found in river water. Five of the species utilized 200 parts per million (ppm) of phenol and two of the five grown in 1000 ppm phenol/mineral salts medium (Ref. 8). These bacteria are unicellular plants without a well-defined nucleus, do not contain the green pigment possessed by higher plants, and reproduce by fission. A single spherical bacterium is only 0.0005 millimeter in diameter. Surface area/volume ratio accounts for the high biological activity of bacteria contrasted to other forms (Ref. 9). The bacteria are aerobic and heterotrophic, depending on organic compounds for their food supply, and mesophilic because their optimum activity is in the range of 60 to 100°F (15.6 to 38°C).

OPERATIONS

Bio-oxidation of phenolic-type and other organic compounds is a function of type and concentration of the compounds and presence of sufficient nutrients. Ammonium salts and some phosphate are present in the reuse makeup water.

Design and operating data for the induced-draft cooling tower used in bio-oxidation are shown in Table 1. These towers offer an ideal environment for efficient bio-oxidation since hydraulic loading, air/water weight ratio, and temperature are conducive to maximum performance. Furthermore, the most important advantage is the dilution ratio which ranges from 16 to 60, thus preventing disturbance of the bio-oxidation equilibrium due to sudden shift in phenol-loading or other chemical changes.

Typical chemical compositions of makeup and discharge waters to the reuse towers (second and third stage) are shown in Table 2.

Trends in phenolic-type compound removal for forced- and induced-draft cooling

TABLE 1   Reuse Cooling Tower (Second Stage) Design and Operating Data

	Counterflow induced draft	Counterflow induced draft
Number of cells	3	11
Cell dimensions, ft:		
Width	54	24
Length	30	36
Height	18	38' 9.5"
Fans:		
Number per cell	2	1
Size	16'−4 blade	14'−4 blade
Water rates, gal/min:		
Circulating	16,000	32,000
Makeup	1000–2000	500–1000
Blowdown	400–1500	200–500
Hydraulic loading, gal/(ft²)(min)	3.30	3.37
Air rate, cu ft/min	2,200,000	2,838,000
Air/water weight ratio	1.26	0.78
Temperatures, °F:		
Water inlet	70–120	70–120
Water outlet	64–90	64–90

TABLE 2.   Chemical Composition, Reuse Cooling Water (Second Stage)

	Makeup			Blowdown		
	Max	Min	Av	Max	Min	Av
pH	9.05	7.80	8.50	8.00	7.40	7.60
Total alkalinity, ppm	175	104	144	157	95	133
Chloride, ppm	204	98	150	240	103	146
Sulfates, ppm	226	106	149	255	120	170
Iron (total), ppm	0.290	0.105	0.198	0.370	0.203	0.261
Copper (total), ppm	0.107	0.056	0.072	0.166	0.072	0.118
Total hardness	149	119	132	159	122	132
Suspended solids, ppm	106	54	72	156	59	85
Dissolved solids, ppm	700	460	583	725	500	557
Silica, ppm	5.4	0.6	3.9	9.4	6.7	7.8
Total phosphorus, ppm	4.8	3.5	3.9	5.2	3.7	4.4
Metophosphate, ppm				6.9	1.2	3.1
Orthophosphate, ppm				4.1	0.8	1.6
Cycles of concentration				1.2*	1.0*	1.1*
Total organic carbon, ppm	151	84	123	146	64	94
Total nitrogen, ppm	37.0	4.3	25.6	31.4	7.1	20.3
Basic nitrogen, ppm	30.0	3.5	21.2	31.1	0.63	17.6
Oil (total), ppm	121	59	80	75	54	66
Oil (nonvolatile), ppm	118	40	66	69	40	65
Phenolic type compounds, ppm	31.4	12.3	20.1	0.131	0.444	0.083

*Based on chloride ion concentration.

towers are illustrated in Table 3. Although the individual tower phenol compound loadings range from 102 to 840 pounds per day, performance is consistently satisfactory at all rates.

Fundamental Data.   Samples taken at key locations in the circulating system (Table 4) indicate initiation of bio-oxidation when water contacts the air in the cooling tower. Furthermore, over 90 percent of the reaction occurs during the fall through the tower structure. The balance of bio-oxidation is carried out in the cooling tower basin before the water is pumped into the circulating system. Overall removal exceeds 99.9 percent.

Induced-draft tower performance is less effective since a trace of phenolic compounds

TABLE 3.   Phenolic Compound Removal, Forced- and Induced-Draft Cooling Towers (Second and Third Stages)

Sample location	Phenolic-type compounds				Flow, gal/min	Type of cooling tower
	ppm	lb/day	lb/day removed	% removal		
Makeup	48.0	346			600	Forced draft
Blowdown	0.082	0.099	345.9	99.9+	100	
Makeup	70.0	840			1000	Forced draft
Blowdown	0.41	2.46	837.5	99.7	500	
Makeup	15.7	133			710	Forced Draft
Blowdown	0.194	0.23	132.8	99.8	100	
Makeup	19.6	141			600	Forced draft
Blowdown	0.800	0.96	140	99.4	100	
Makeup	17.2	165			800	Induced draft
Blowdown	0.115	0.31	164.69	99.8	225	
Makeup	10.0	102			850	Induced draft
Blowdown	0.033	0.109	101.89	99.9	275	
Makeup	13.2	146			925	Induced draft
Blowdown	0.033	0.133	145.87	99.9	350	
Makeup	33.2	364			910	Induced draft
Blowdown	0.043	0.196	363.804	99.9+	370	

TABLE 4. Forced-Draft Cooling Tower Phenol Bio-Oxidation Data (Second Stage)

Sampling location	Flow rate, gal/min	Phenolic compounds		% of total load of phenolic compounds*
		ppm	lb/day	
Circulating water	17,500	0.082	17.3	3.81
Cooling tower makeup water	800	48.0	454	
Cooling tower blowdown water	300	0.082	0.30	0.07
Water distributor (at top of tower)		1.71	360	79.30
Upper section of cooling tower		0.95	200	43.1
Midway down cooling tower		0.35	73.6	16.2
Bottom of cooling tower		0.168	35.5	7.82

*99.93 percent removal.

is present in the bacterial sludge in the blowdown water from the cooling tower (Table 5). Total phenol removal is over 99.9 percent, however, which again demonstrates essentially complete reaction in the cooling tower and basin.

System Monitoring.   The system is monitored on a continuous, automatic, composite sampling basis using two 24-hour samples. In addition, spot samples are obtained as necessary.* Parameters required in regular monthly reports include flow rate, temperature, dissolved oxygen, pH, BOD, COD, TOC, oil and grease, phenols, ammonia, phosphorus, suspended solids, cyanide (total), and chromium (total). Zinc, copper, cadmium, and lead are reported quarterly. Flow rate, temperature, and dissolved oxygen of the final effluent are continuously monitored and recorded. The receiving dry ditch is also checked for flow rate, temperature, pH, dissolved oxygen, COD, TOC, oil and grease, and suspended solids at locations before and after the final treated effluent is introduced.

Alkalinity Balance.   A useful indicator of the efficiency of bio-oxidation is the alkalinity balance. Although the systems operate at two cycles of concentration, the alkalinity of the circulating stream falls below the makeup water, since carbon dioxide is the end product of phenolic compound destruction.

Continuous composite samples of reuse makeup water and refinery final effluent are used to determine performance. The phenolic-type compounds are chemically fixed during sampling in order to prevent oxidation before analysis. The low-range, 4-aminoantipyrine solvent extraction procedure is used for determining phenols in treated water from the bio-oxidation cooling towers. Phenol removal surveys are normally conducted on bio-oxidation cooling towers once per week.

Total organic carbon (TOC) is a more reliable index of bio-oxidation system efficiency than COD or BOD. Data for a complete TOC removal survey (Table 6) show a total loading of 9295 pounds reduced to 1050 pounds for an overall removal of 88.8 percent.

COD determination reflects the effect of inorganic constituents, while BOD is subject to interference as well as requiring extended analysis time. Correlations have been

TABLE 5. Induced-Draft Cooling Tower Phenol Bio-Oxidation Data (Second Stage)

Sampling location	Flow rates	Phenolic compounds		% of total load of phenolic compounds*
		ppm	lb/day	
Circulating water	23,850 gal/min			
Cooling tower makeup water	800 gal/min	22.3	214	
Cooling tower blowdown water	333 gal/min	0.043	0.172	0.08
Bacterial sludge from cooling tower blowdown	0.228 lb/min	8.08	0.0027	0.0013
Air leaving the cooling tower (induced-draft air)	2,320,000 cu ft/min	0.026	7.18	3.36

*99.92 percent removal.

*The refinery is operating under a 5-year NPDES permit (March 15, 1973 to March 15, 1978).

TABLE 6. Total Organic Carbon Removal, Bio-Oxidation Process

Stage	In	Out	Removed	%
1	9295	3643	5652	60.6
2	3643			
3		1132	2511	68.8
4	1132	1053	79	7.0
5	1053	1050	3	0.3
Overall	9295	1050	8245	88.8

established for effluent water to show a BOD/TOC ratio and a COD/TOC ratio of 1.3 and 3.21, respectively.

Chemical control of the reuse water requires close control of the pH to prevent upsetting the bio-oxidation equilibrium. When neutralization is necessary, it is conducted by field personnel.

Continuous acid addition is not required. Alkalinity is naturally controlled in the process by the formation of carbon dioxide as a buffering system.

Hydrocarbon Leakage. This type of leakage into the cooling water from the process condensers, coolers, and exchangers is undesirable because of its effect on heat transfer and bio-oxidation. Remote alarm hydrocarbon monitors are installed on all cooling water systems. These devices utilize instrument air to strip hydrocarbon from the water, thereby forming an explosive mixture which actuates an alarm in the control room. Immediate steps are necessary to locate the leak and remove the hydrocarbon from the system.

MAINTENANCE

Heat Transfer. Bio-oxidation intentionally forms appreciable amounts of suspended solids in circulating cooling water (typically 110 ppm). Use of this water does not require chlorination for maintenance of heat transfer equipment. Condensers and exchangers are backwashed (Ref. 10) on a routine basis to maintain proper heat transfer levels.

Corrosion. Corrosion rates have been satisfactory throughout the years of waste water reuse bio-oxidation. See Table 7. The oxygen-demanding nature of reuse water accounts for the virtual absence of dissolved oxygen in the circulating cooling water. Low corrosion rates for steel are a result of the low dissolved oxygen concentration. Addition of corrosion inhibitors is not necessary.

Long-term corrosion rates for admiralty tubes and brass tube sheets average less than 0.5 millimeters per year. Absence of hydrogen sulfide and mercaptans from the water preclude pitting of the admiralty and brass.

Wood Deterioration. Cooling tower structures are protected by an adequate program of inspection and testing. Wood conditions for towers using fresh makeup and reuse

TABLE 7. Corrosion Rate Data for Reuse Cooling Water System

Metal	Test period exposure in days	Corrosion rate, mm/year
Steel	374	3.3
	372	3.6
	380	1.9
	49	1.6
	27	0.7
	136	1.0
	40	2.5
	140	3.1
	55	1.9
	56	6.7
Brass	260	0.2
	182	1.1
	189	0.8
	366	0.1
	230	0.8

TABLE 8.   California Redwood Association Inspection Data; Sun Oil Company Toledo Refinery Cooling Towers

| Makeup water | Age when inspected | Deterioration rate, % per year | | | |
| | | Surface | | Internal | |
		Fill	Eliminators	Frame	Casing
Fresh	4 yr, 11 mo	0	0	0	0
Fresh	7 yr, 10 mo	0	0	0	0
Reuse	4 yr, 11 mo	0	0	0	0
Reuse	7 yr, 10 mo	1.3	0	0	0

makeup water are compared for control purposes. Typical data for a series of inspections are shown in Table 8. Three sets of wood specimens representing use in fresh water makeup towers and reuse towers show no evidence of soft rot or typical decay attack. Several types of fungus hyphae and bacteria were found after a 5-year period of observation.

WATER CLEANUP

Suspended solids in the form of biological floc constitute a source of turbidity, carry traces of absorbed oil, and account for most of the remaining oxygen demand in the effluent from the fourth stage of bio-oxidation. The suspended solids are colloidal dispersion from the multistage bio-oxidation operation and recirculation pumps. Over 50 percent of the suspended solids are smaller than 5 micrometers in diameter and present a difficult challenge for removal. Several approaches have been evaluated to remove this material, including use of three types of filters, two air-flotation systems, and two microstrainers. The microstrainers and induced air-flotation equipment were unsatisfactory. Study of one upflow and two downflow filters narrowed ultimate selection to the use of sand filters, either upflow or downflow. Upflow filtration allowed the establishment of a high-solids loading which extends the operating cycle, and makes addition of chemicals easier. Excellent agglomeration of suspended solids is observed throughout the entire sand bed in upflow filtration.

Upflow filtration test runs on a pilot system operated in the range of 4.5 to 6.4 gallons/ (minute)(square foot) with an air scour showed an average removal of 88 percent and 84 percent at the two rates, respectively. See Table 9. COD reduction is in the 50 percent region, as is TOC removal. Limited data also reflect good removal of trace oil and phosphorus. Capacity is in the order of 4 to 5 pounds of dry solids per square foot of filter area, which permits operating cycles up to 24 hours. Preliminary evaluation of the addition of alum and/or polyelectrolytes shows promise.

Operation of the fully automated upflow sand filters, however, has still not shown the removal efficiency equivalent to the pilot unit. A number of design problems were found and corrected and additional modifications are either in progress or planned.

TABLE 9.   Suspended Solids Removal in Upflow Pilot Sand Filter (Stage 5) at Different Flow Rates and Inlet Concentrations

| Flow rate, gal/(min)(ft²) | Suspended solids, ppm | | Removal efficiency, % |
	Inlet	Outlet	
4.5	48	10	79
4.5	46	7	85
4.5	36	2	94
4.5	32	2	94
6.4	42	7	83
6.4	52	14	73
6.4	38	2	95

WATER CONSERVATION

Reuse bio-oxidation successfully conserves fresh water. The average industry demand for fresh water is 2150 gallons per barrel of crude oil refined, compared to 55 gallons per barrel of crude using this system. Water consumption is 214 gallons per barrel of crude on the national average, compared to 28 gallons with this system. Water consumption has been reduced to as little as 17 gallons per barrel of crude.

Additional reuse from the operation of the sand filter system will reduce water consumption below 10 gallons per barrel. For a refinery of 120,000 barrels per day, the savings amounts to about $174,000 per year, plus that required for chemically treating the fresh water. Additional conservation and use of the sand filter system will result in another savings in the order of $150,000 per year.

REFERENCES

1. Elkin, H. F.: Biological Oxidation and Reuse of Refinery Waste Water for Pollution Control," *Proc. Am. Pet. Inst.* (May 1956).
2. Elkin, H. F., E. F. Mohler, Jr., and L. R. Kumnick: Biological Oxidation of Oil Refinery Wastes in Cooling Tower Systems, *Sewage Ind. Wastes*, **28** (12): 1475 (December 1956).
3. Mohler, E. F., Jr.: "Purify While They Cool," Industrial Pollution Control Symposium, American Institute of Chemical Engineers, Akron, Ohio, May 10, 1957.
4. Mohler, E. F., Jr., H. F. Elkin, and L. R. Kumnick: Experience with Reuse and Bio-oxidation of Refinery Wastewater in Cooling Tower Systems, *J. Water Pollut. Control Fed.*, **36** (11): 1380 (November 1964).
5. Mohler, E. F., Jr.: Extended Experience in Biological Treatment and Reuse of Refinery Waste Water in Cooling Towers, *Proc. 26th Ann. Int. Water Conf.* (October 1965).
6. Mohler, E. F., Jr.: "The Key Role of Monitoring in Optimizing Petroleum Refinery Water Pollution Abatement and Conservation," Air-Water-Noise Pollution Conference, University of Michigan, Ann Arbor, Mich., Nov. 22, 1969.
7. Lynn, G. E., and T. J. Powers: Bacterial Studies in Oxidation of Phenolic Wastes, *Sewage Ind. Wastes* (January 1955).
8. Tallon, G. R.: Prepared discussion of paper in ref. 5.
9. Thimann, K. V.: "The Life of Bacteria," Macmillan, New York, 1955.
10. Hitzeman, T. E., and H. W. Feist: To Clean Your Condenser—Backwash is the Cheapest, *Power Eng.* (July 1958).
11. Mohler, E. F., Jr., and L. T. Clere: Bio-oxidation Process Saves H_2O, *Hydrocarbon Process.* (October 1973).

Source Monitoring of NO_x and SO_2

BY JAMES R. SMALL* AND JAMES A. WILLIAMSON†

Sulfur dioxide (SO_2) and nitrogen oxides (NO_x) measurement in stack emissions is becoming increasingly important to such major industries as petroleum refining and gas processing, sulfuric acid and nitric acid manufacturing, fossil-fuel-fired power plants, smelters, and pulp and paper mills. SO_2 and NO_x measurement in off-gases from these processes, in many instances, is required by local, state, and federal regulations that are either proposed or in effect. Consequently, there is a real need for continuous SO_2 and NO_x monitors that are accurate and reliable.

Stack-monitoring requirements for SO_2 and NO_x have brought about the design of special monitoring systems for this use, rather than the wholesale adaptation of ambient monitors. This is because (1) stack samples contain extremely high concentrations of SO_2 and/or NO_x compared to ambient air, and ambient monitors usually are not capable of handling these much higher levels of pollutants; and (2) in addition to high concentrations of SO_2 and NO_x, stack gases often contain water vapor, particulates, corrosive compounds, and other chemical species that may interfere either directly with the measurement or with the mechanical operation of the analyzer.

SAMPLING TECHNIQUES

Whether the concern is with reference methods of analysis or continuous monitoring with instruments, sampling techniques are of the utmost importance.

Materials of Construction. Some typical materials from which various sampling components may be made, commencing with the best at the top and following in order of decreasing usefulness with regard to preserving the integrity of the sample, are listed as follows:

1. Teflon ® TFE and FEP fluorocarbon resins (and similar fluorocarbons)
2. Glass
3. Carpenter 20 (alloy 20) stainless steel
4. 316 stainless steel
5. 304 stainless steel

There are, obviously, other factors to consider in some sampling cases. If one must have maximum structural strength in some component, it cannot be built of fluorocarbon or glass. With temperatures over 300°F (149°C), most fluorocarbons are generally unacceptable. In some compliance tests, glass is specified, thus ruling out fluorocarbons for that use. But, given a choice, the higher the material is on the above list, the less likely it is to absorb part of the sample, or react with part of the sample, or allow any sample components to stick to tubing walls and other exposed parts. Fluorocarbons are particularly advantageous in this respect. An all-fluorocarbon system will be very unlikely to exhibit plugging problems.

Generally forbidden (in the interest of obtaining good results) in the category of sample tubing or other components are materials, such as gum rubber, polyethylene or polypropylene, polyvinyl chloride tubing, or other materials containing plasticizers. Plasticizers will "wash out" over a period of time, causing interference in the detection of low concentrations of NO_x and SO_2. Also, some gases permeate through these materials rather readily, again leading to errors.

Copper or brass will react with components of sample streams, and the reaction products can adsorb or desorb SO_2 and NO_x. Even 316 stainless used in a critical valve can corrode sufficiently to allow leakage. Sample valves should be constructed with critical parts of Carpenter 20 alloy if possible.

*Product Manager, Process Instruments, and †Senior Applications Chemist, Instrument Products Division, E. I. du Pont de Nemours & Co., Inc., Wilmington, Del.

Sampling System Components.   These will vary depending on the analytical method used, the source, and ambient conditions. There are instruments for stack monitoring which require no sampling system but monitor gases in the stack. These are discussed later. A categorization of sampling systems might appear as follows:

1. Cooled, water removed versus "isothermal" sampling
2. Pumped versus aspirated
3. Completely filtered versus partial filtration

Water Vapor Removal.   Some instruments, such as the chemiluminescence monitors, the electromonitors, and some of the nondispersive infrared (NDIR) monitors, show interference at various levels from water vapor in the sample. Others, such as (ultraviolet/visible) photometric analyzers, are unaffected by moisture. Consequently, the need for removal of water vapor by cooling and trapping is determined by the kind of analyzer in the system. Several instrument manufacturers offer refrigeration units to accomplish this when required. If water vapor is removed, then some means must be provided for disposing of the condensate.

If water vapor is not to be removed, then generally a heated and insulated sample line (and analyzer cabinet) will be used. This tends to maintain sample integrity to a greater extent than any system where portions of the sample are removed, and a heated system will reduce corrosion by elimination of corrosive condensate. The system must be kept above the dew point of the sample stream, however, and this can be difficult if sampling is done just downstream of a scrubber. Scrubbers pose some of the more difficult sample-handling problems. Liquids frequently splash onto and are drawn into the sample line inlet unless an efficient shield is used. Scrubbing media, such as magnesium oxide, are often deposited on the in-stack filter by evaporation of the carrier water, eventually plugging the filter. An effective shield can prevent most of this plugging.

Aspirators and Pumps.   Several manufacturers offer systems in which the sample is moved by means of a vacuum aspirator, operated by air pressure. If this aspirator is properly designed and made of Teflon ® TFE flourocarbon, it can give very long, trouble-free service. A good aspirator can pull 15 to 18 inches of mercury vacuum, dead headed, which is more than adequate for sampling in most circumstances.

By contrast, pumps, even when well designed and made of stainless steel, will have a more limited lifetime between overhauls. The very corrosive nature of a typical stack gas is very tough on finely machined parts of which pumps are made. In either case, pressure regulation is usually necessary.

Filtration.   Filtration can be of many types. Some systems require very little filtration, since system design makes the system almost insensitive to particulates. In this case, an in-stack filter of fine stainless steel screen, retaining particles to about 20 micrometers, has proven effective for most installations. Where this is not adequate, in-line filters which can be removed and cleaned or replaced are available with retentions below 1 micrometer. These have limited capacity, however, and should be preceded by another filter with more capacity and less retention.

A *blowback* cycle, in which compressed air is used to blow out the sample line, back through the in-stack filter, has proven very effective in many installations. For photometric analyzers, this also provides an opportunity for automatically checking and resetting the instrument zero at some time interval.

Figure 1 illustrates a generalized sample system design that has been successfully deployed with analyzers utilizing visible or ultraviolet photometric measurement techniques.

The basic design of an NO$_x$ analyzer's sample system is essentially the same as for SO$_2$. This is illustrated in Fig. 2. However, provision is made for the automatic introduction of oxygen under pressure for oxidizing NO to NO$_2$ in the sample cell in the analyzer shown. Fig. 3 shows the system enclosed in a cabinet for weather protection.

MEASURING INSTRUMENTATION

After the sample has been extracted from the sample point and brought to the analyzer by the sample train, the final step in this sequence takes place. One or more of the components in the sample are measured to determine their relative concentrations. If the sample train has done its job properly, the measured sample is representative of the material in the stack at the sample point, so that the analyzer has at least a chance of

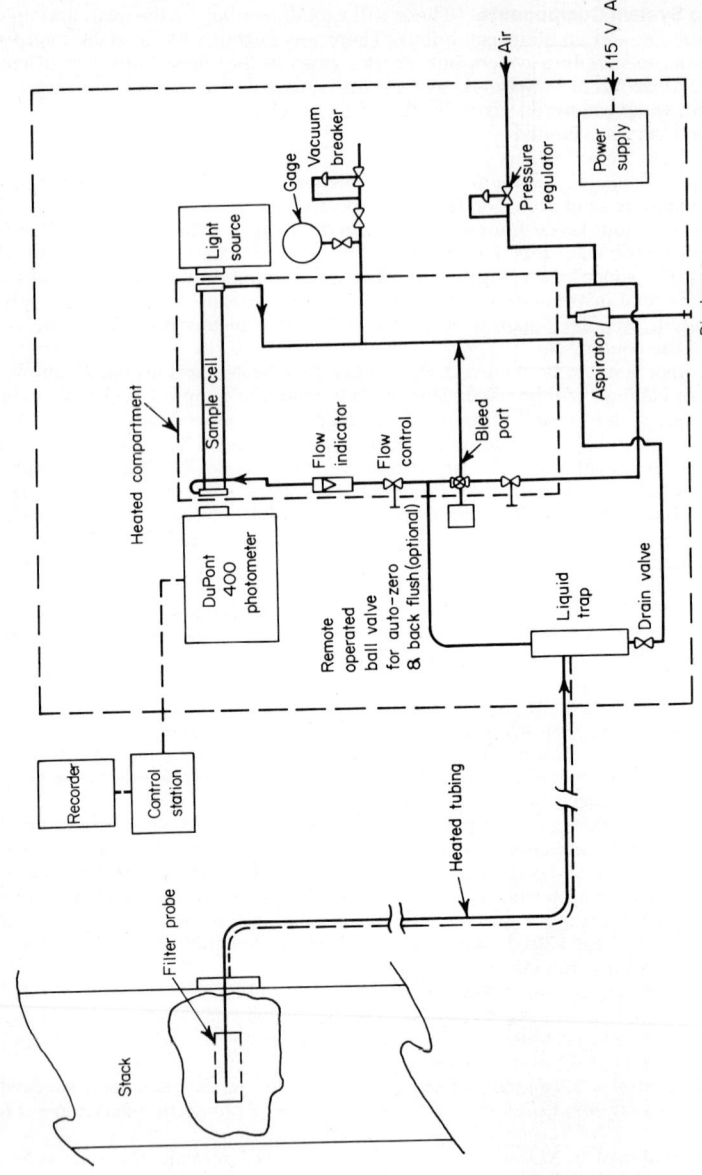

Fig. 1. Generalized sampling system for analyzers utilizing visible or ultraviolet photometric measurement techniques. (*DuPont* 460 SO₂ Analyzer System.)

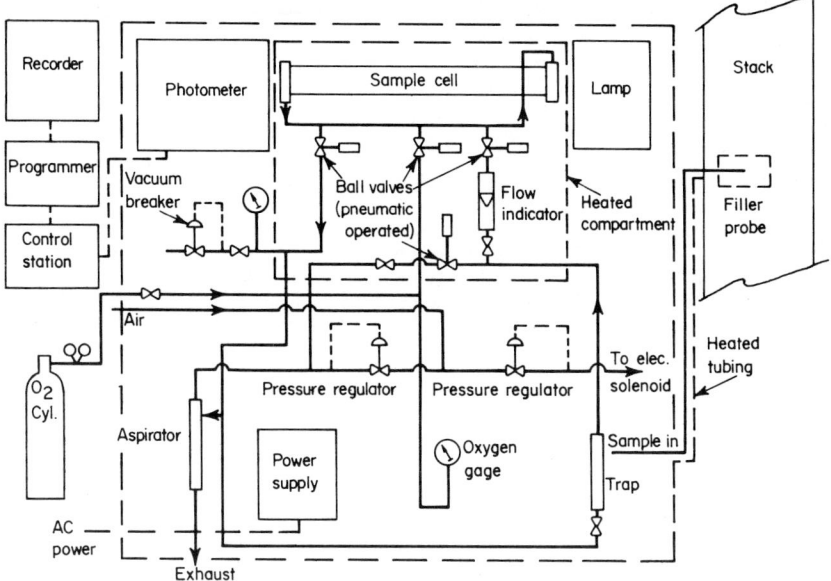

Fig. 2. Basic design of an NO$_x$ analyzer sampling system.

Fig. 3. NO$_x$ analyzer enclosed in cabinet for weather protection. *(DuPont Instrument Products Division.)*

making a valid measurement. How valid that measurement is depends in part on whether any sample components interfere with the measurement of others. It is here that the correct choice of analyzers can prevent many problems.

The following types of measuring instruments will be discussed, and represent the major portion of pollution-monitoring instruments in use today:
1. Infrared analyzers
2. Ultraviolet-visible photometric analyzers
3. Chemiluminescence analyzers
4. Electrochemical analyzers
5. Fluorescence analyzers

Energy Absorption in Ultraviolet-Visible-Infrared Regions.   In four of the five types of analyzers listed, the detection process begins with the absorption of energy by the species being detected. Therefore, a very brief discussion of energy absorption in the UV-visible-IR regions is probably in order. More detailed discussions of this subject are available in many publications, several of which are listed in the references (Refs. 3.5,6).

The absorption of electromagnetic radiation in the visible, UV, and IR regions of the spectrum occurs in basically two different ways: (1) absorption of UV-visible energy with excitation of an electron to a higher energy level and (2) absorption of infrared energy with energy added to a molecule as bond vibration, rotation, etc.

The amount of energy absorbed is generally expressed as the *extinction coefficient* or the *molar absorptivity*, and this amount of absorbed energy is a constant at a fixed wavelength for a particular chemical compound, so that it can be used in determining the concentration of that compound in a gaseous mixture. This is accomplished by using Beer's law, a mathematical relationship between a measured quantity, light transmission (measured as *absorbance*), and the concentration of the compound to be determined. Beer's law can be expressed as

$$A = abc$$

where A = quantity measured on a readout device
a = molar absorptivity defined above
b = path length through which the light and sample interact
c = concentration of the material being measured

A can further be defined as

$$A = \log \frac{I}{I_0}$$

where I/I_0 = fraction of light transmitted. Rewriting the first equation, the exponential nature of the absorption of light becomes evident:

$$\log \frac{I}{I_0} = -abc \quad \text{or} \quad \frac{I}{I_0} = e^{-abc}$$

Examining these equations, one can see that A will vary directly with concentration, and if the output of the phototube is processed in a logarithmic amplifier, the variation will be linear. Otherwise, a nonlinear relationship will exist. This then is the fundamental basis of any analyzer using the absorption of energy as its measuring principle.

The greater sensitivity available for analysis of nitrogen oxides and sulfur dioxide by UV-visible analysis as opposed to infrared is in part due to the molar absorptivity in the UV-visible range exceeding that in the infrared by about a factor of 3, and in part due to more stable sources and detection systems for UV and visible energy, so that UV-visible analyzers can frequently measure to a full-scale concentration of 100 ppm for NO_x and SO_2. Infrared analyzers are usually limited to higher full-scale levels.

It is interesting to note here that the energy absorbed in the chemiluminescent process is derived not from incident photons (electromagnetic radiation) but from a chemical reaction occurring in the sample. This is an energy absorption and reemission process, even though the input energy differs quantitatively from that in a straight photometric absorption system. Chemiluminescent reactions are quite common and are present in all combustion processes which emit light.

Ultraviolet-Visible Photometric Instruments.   The ultraviolet analyzers are of two fundamental types with regard to stack sampling: extractive and in-stack. In the extractive technique, a sample is withdrawn from a stack or other source and carried through a sample line to the analyzer. In the in-stack technique, the analyzer is mounted on the stack and looks directly at the sample. There are advantages and disadvantages to each technique.

A typical extractive type of UV analyzer will be discussed first. The analyzer has a light source, a sample system, and a means of detecting the light once it has passed through the sample. The light source typically is a medium-pressure mercury vapor lamp, which gives a series of discrete emission lines and is very useful in pollution monitoring. The photometer may be of a split-beam type as shown in Fig. 4. In this photometer, the entire

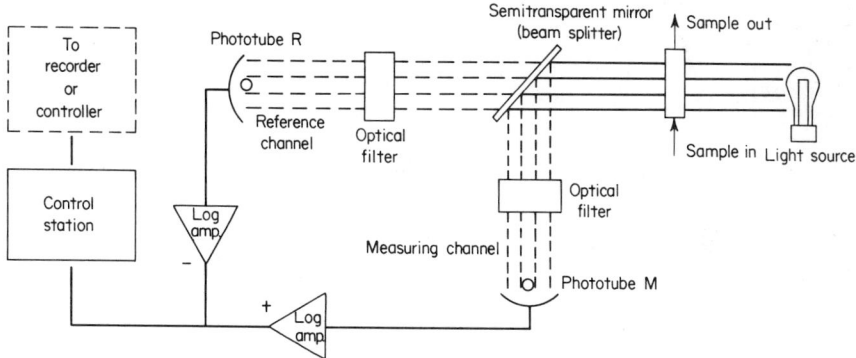

Fig. 4. Split-beam type photometer.

light path is first passed through the sample and then is divided by a semitransparent mirror into two beams, one of which is the measuring beam and the second is the reference beam. The measuring beam is passed through a series of filters, which isolate a particular wavelength to be used for the measuring wavelength, and then strikes a phototube which creates a current proportional to the light energy passing through the phototube. The second beam, the reference beam, passes through a second series of filters which isolate the reference wavelength from the metallic vapor emission spectra, and this too strikes a phototube generating a current proportional to the reference wavelength light. The currents from these two phototubes are amplified in logarithmic amplifiers, and the difference in the amplified signal is then taken and is proportional to the concentration of the sample.

The use of logarithmic amplifiers allows one to achieve a signal which is proportional in a linear fashion to the concentration of the material being analyzed in the sample. Furthermore, the use of metallic vapor emission spectra as the source further enhances the linearity by eliminating some of the disadvantages of using prisms or gratings with continuous sources to isolate a wavelength for analysis.

The split-beam analyzer shown in the diagram virtually eliminates the interference normally seen from particulate matter in the sample stream and requires in most cases only coarse filtration (20 micrometers). A light loss of as much as 25 percent has only a minimal effect on the accuracy of SO$_2$ or NO$_x$ measurements (as little as 0.5 percent error is introduced).

Mercury vapor discharge lamps are used as light sources for both SO$_2$ and NO$_x$ measurements. The 280-nanometer (nm) mercury emission line in the ultraviolet region, when isolated by narrow bandpass filters, has proven useful in SO$_2$-monitoring applications. Likewise, the 436-nm mercury emission line is useful for NO$_2$ measurements.

A reference wavelength of 578 nm was chosen for the following reason. Fig. 5 illustrates the superimposed absorption spectra of SO$_2$ and NO$_2$. The spectra of these compounds overlap. By choosing a measuring wavelength of 280 nm for SO$_2$ and a reference wavelength of 578 nm, interference of NO$_2$ is eliminated in the SO$_2$ measurement. As shown in Fig. 5, when 436 nm is used for NO$_2$ measurement, SO$_2$ does not interfere. Also in this optical region, there is no interference in either measurement from CO, CO_2, H_2O, or other common stack gas components.

NO is essentially transparent in the wavelength region of interest, and the measurement of this compound is accomplished in a unique manner. Williamson (Ref. 8) developed a technique for oxidizing NO to NO_2 with 99+ percent conversion in a matter of minutes applicable for emission source concentrations. This technique was subsequently

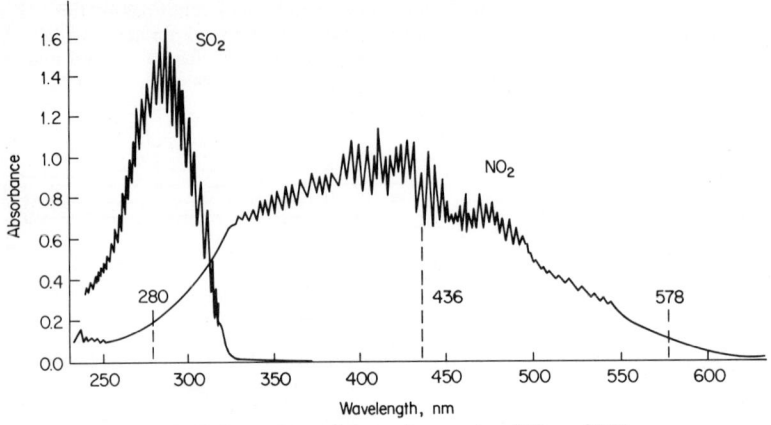

Fig. 5. Superimposed absorption spectra of SO_2 and NO_x.

incorporated into a viable analytical device to determine both NO and NO_2, or NO_x, employing visible photometric principles.

This device automatically oxidizes NO to NO_2 in the photometer cell by introducing oxygen under pressure. Consequently, in a single 10-minute analysis cycle, NO_2, NO, and NO_x levels can be determined.

The obvious disadvantage of the extractive method described above is that a sample must be removed from the stack and carried in a sample system to the analyzer which is often located some distance away. Extractive systems have been refined over the past 5 to 6 years, however, and most of the problem areas have been eliminated from the systems so that such problems as corrosion and line plugging are eliminated to a large extent in some present-day systems.

In an ultraviolet-visible-type analyzer, the removal of water vapor from the sample is unnecessary, and consequently a further simplification is achieved with this type of analyzer in contrast to infrared analyzers and certain other types of analyzers requiring the removal of water vapor and essentially all the particulate matter.

The extractive technique has several distinct advantages over the in-stack-type monitor. One of these is its use in multiple sample point monitoring. A single analyzer can be used to monitor three or four stacks at somewhat less expense than placing in-stack monitors in each of the three or four stacks as would be required with that type analyzer. Second, calibration and zeroing of the analyzer are considerably easier with extractive techniques where a sample of calibration gas or zero gas can be introduced into the sample line at the source and allowed to pass through the sample line to the analyzer. Zeroing in particular is difficult with in-stack techniques and must be achieved by some method other than using a zero gas. Further, when one considers the maintenance which is always required at some frequency in analyzers of any type, it is quite obvious that an analyzer on the ground is going to be more readily accessible for maintenance than an analyzer 100 to 200 feet up the side of a stack.

A typical in-stack-type analyzer will consist of a light source and light detector of some sort mounted external to the stack, with a probe extending into the stack as shown in Fig. 6 and 7. In a typical analyzer, a mirror at the far end of the probe reflects the transmitted portion of the ultraviolet light back into the spectrometer head, and this reflected spectrum is compared electronically with a photograph of a characteristic spectral fingerprint of SO_2 absorption. An electronic signal proportional to the sulfur dioxide content results

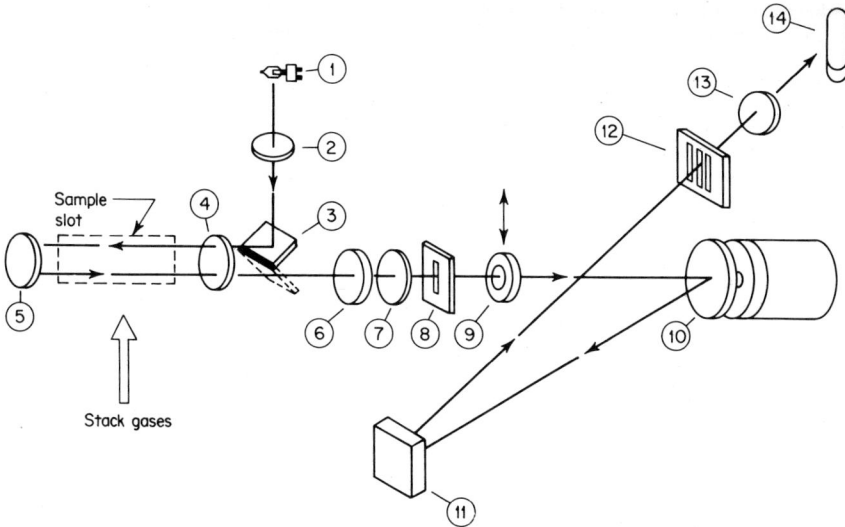

Fig. 6. Typical in-stack type analyzer consisting of light source and light detector mounted external to stack with probe extending into stack: (1) Tungsten halogen lamp. (2) Collimating lens. (3) Zero/read mirror. (4) Lamp focusing lens—primary. (5) Probe mirror. (6) Lamp focusing lens—secondary. (7) Filter. (8) Entrance slit. (9) Motorized calibration cell. (10) Modulator mirror. (11) Diffraction grating. (12) Exit correlation mask. (13) Exit mask lens. (14) Photomultiplier.

without interference from other chemical substances that may be present. This type system is not yet available for NO$_x$ monitoring. Ranges for sulfur dioxide for this system generally extend from 0 to 1000 to 0 to 5000 ppm. In this type analyzer, it is necessary to prevent the incursion of flue gases into the analyzer itself. An air curtain prevents flue gases from entering this spectrometer via the optical path.

Another analyzer of this general type used a second derivative measurement technique to obtain the sulfur dioxide measurement.

While the in-stack analyzers do have significant advantages in that they do not have to

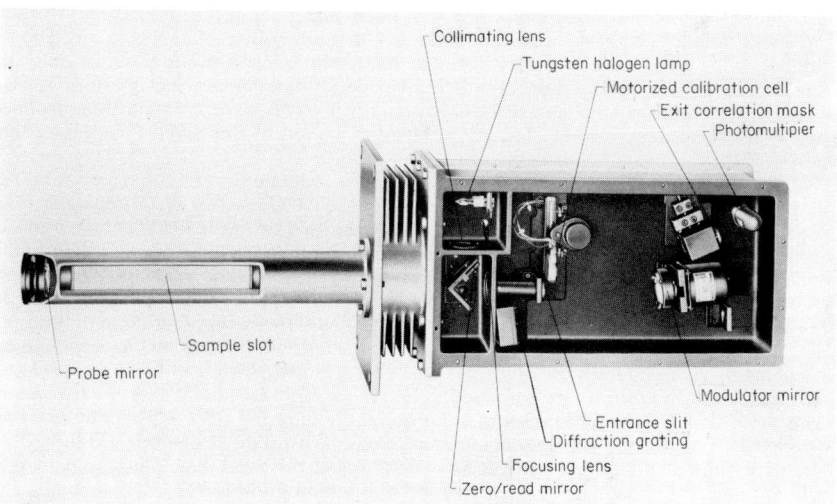

Fig. 7. Correlation spectrophotometer essentially of format shown in Fig. 6.

extract a sample, treat the sample, and convey it to the analyzer itself, they do have disadvantages in being more difficult to calibrate, more difficult to zero, more difficult to maintain, and more subject to weather interferences such as wide temperature swings.

In some cases, the electronics are also more complicated; that is, they have more moving parts and more precisely tuned components than the extractive techniques generally require.

Nondispersive Infrared (NDIR).   The typical nondispersive infrared analyzer is a dual beam design and generally uses positive filtering and a microphone-type detector as shown in Fig. 8. The detector contains a gas which absorbs energy at the wavelength being used for analysis. The two beams from similar nichrome wire sources pass through

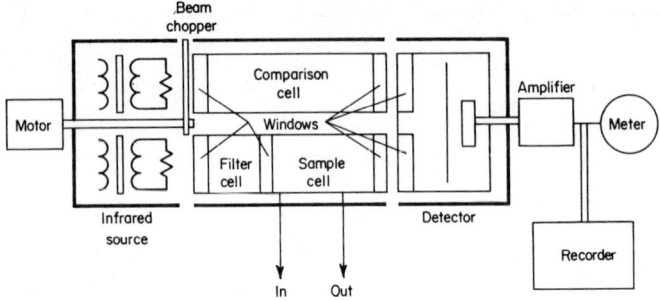

Fig. 8. Typical nondispersive infrared analyzer of dual-beam design, using positive filtering and a microphone-type detector. (*Mine Safety Appliances Company.*)

two cells, a sample cell and a reference cell filled with a known gas mixture. As the gas to be detected passes through the sample cell, it absorbs some of the energy from the source, causing less energy to reach the detector gas. As a result, the detector gas cools and contracts, causing the membrane of the condenser microphone to move. This movement is converted to an output electrical signal. The chopper causes energy to pass alternately through the two beams, and the oscillation of the membrane in the detector resulting from alternate heating and cooling of the gas generates an alternating output at the frequency of the chopper rotation. If identical gases are in both cells, no heating and cooling occurs, and the output is zero.

Problems occur in nondispersive infrared analysis when the absorption band of the material being measured lies close to that of other materials in the sample. Both water vapor and carbon dioxide absorb rather strongly in the infrared spectral region and tend to interfere with many measurements made by this technique. Such interferences can be partially eliminated by filtering, in which a sealed cell containing the interfering substance is placed in the IR beam, and absorbs energy at its characteristic wavelength. This essentially eliminates further changes in absorption due to this same material in the sample cell.

Chemiluminescent Analyzers.   The chemiluminescent analyzer offered by a number of manufacturers is based on the reaction of nitric oxide (NO) and ozone (O_3) in a reaction chamber. A photomultiplier tube detects the light emitted by the reaction between nitric oxide and ozone. An optical filter located between the reaction chamber and the photocathode is selected to eliminate light energy from reactions that might possibly occur simultaneously with the NO plus O_3 reaction. This results in a very sensitive interference-free instrument for the quantitative measurement of NO. However, in order to determine NO_x, the NO_2 content of the sample must first be converted back to NO by a reduction mechanism. This generally involves passage of the sample over a reduction catalyst to convert NO_2 to NO prior to its introduction into the sample chamber. This NO is reacted along with the NO initially present in the sample, giving the total NO_x content of the sample. This type analyzer provides a continuous measurement of nitric oxide and is essentially linear and quite sensitive. Its sensitivity extends well down into the parts per million range, allowing it to be used also for ambient monitoring.

The chemiluminescent analyzer, while it has some advantages over other systems, does require careful sample preparation. The necessity for the removal of essentially all water vapor requires a refrigeration unit, and the chemiluminescence technique requires extremely good sample filtration as particles would create problems in the chemiluminescent reaction. Further, while not all the chemiluminscent analyzers require a high-vacuum chamber, some still utilize a very-low-pressure chamber in which to accomplish the reaction. The determination of nitrogen dioxide is slightly less certain in that the chemiluminescent analyzer must convert NO$_2$ back to NO and the conversion is not entirely quantitative in most cases. A conversion test should be run where this analyzer is used for the determination of a significant quantity of nitrogen dioxide to determine the efficiency of the converter in reducing nitrogen dioxide back to nitric oxide. The chemiluminescent analyzer is not capable of determining sulfur dioxide.

Fluorescence SO$_2$ Monitors.   A relatively new type analyzer has been introduced into the market in the past 2 years involving the generating of fluorescence in sulfur dioxide and the detection of this fluorescence by photomultiplier tube. This analyzer has a UV lamp and an interference filter to isolate a certain portion of the UV spectra, a sample cell through which the sample gas is passed, and a second filter placed at 90° to the incoming light path, which allows separation of the fluorescent spectra prior to its introduction to a photomultiplier tube. The current from this photomultiplier tube is then transmitted to an amplifier and then to a meter reading out the sulfur dioxide content. This technique has the advantage of being very specific to sulfur dioxide and is relatively free of interferences, although there is some possibility that water vapor in large amounts might create a significant interference. It might be necessary to remove water vapor by use of a refrigeration unit prior to determination of sulfur dioxide by fluorescence. Water vapor causes quenching or reduction of the fluorescent signal. The fluorescence SO$_2$ monitor has a range of 0 to 5000 ppm. One version was initially developed by the National Bureau of Standards, who then licensed a manufacturer to build the analyzer for stack monitoring. A second manufacturer has a similar analyzer based on a pulsed signal to generate fluorescence in SO$_2$ available.

Electrochemical Analyzers.   The last category to be discussed is the electrochemical analyzer. This is offered by at least two manufacturers and involves the passage of a portion of the sample through a membrane to a thin-film electrolyte, then to a sensing electrode. Here the material being measured is oxidized (or reduced) and the resulting cell current is amplified and read out as a signal proportional to the concentration of the material in the sample. The sampling system for this type analyzer will typically contain a refrigeration unit, a filtration unit, a sample pump, and if NO$_x$ is being monitored, an SO$_2$ scrubber. Sulfur dioxide must be removed from the portion of the sample being monitored for NO$_2$ or NO$_x$ as sulfur dioxide would constitute a severe interference in the measurement of NO$_x$. The electrochemical monitor is essentially linear over its entire range. Ranges extend from 0 to 500 up to 0 to 5000 ppm. The transducers, which are the sensing component in the electrochemical analyzer, are available for either NO$_x$ or SO$_2$ monitoring and are replaceable units so that when one ceases to work it can be conveniently replaced. Electrochemical monitors have been observed to be temperature sensitive in some evaluations, and temperature compensation should be provided for the most accurate results.

CALIBRATION

Photometric analyzers may be calibrated by one of three methods: (1) internal optical standards, (2) chemical tests, and (3) calibrated gas mixtures. Other analyzers can use only the last two methods.

Optical standards (filters) that simulate the light-absorbing properties of NO$_2$ or SO$_2$ are provided as a part of some instrument packages. One needs only to purge the sample cell with clean air or nitrogen, zero the analyzer, position the calibration filter in the light beam, and reset the span (if necessary) to give a reading corresponding to the calibration filter value. The entire task requires only a few minutes. The inherent stability and subsequent reproducibility of optical standards makes them ideal for field calibration and rapid span checks.

The Reference Methods.   Reference test methods have been adopted by the U.S. Environmental Protection Agency (EPA).* The method for sulfur dioxide is identified as

Fed. Regist., **36:**24890–24893 (Dec. 23, 1971).

Method 6, and for nitrogen oxides, Method 7. Since continuous-source-monitoring-instruments performance is frequently compared with test results by these methods, a brief discussion follows.

The method for SO_2 is rather simple and straightforward. The gas to be analyzed for SO_2 is bubbled through a solution of aqueous hydrogen peroxide, which converts SO_2 to sulfur trioxide (SO_3). This readily reacts with the water present to form sulfuric acid. The sulfate ion (in sulfuric acid) is titrated with barium perchlorate, precipitating barium sulfate. Excess barium is detected by "Thorin" indicator solution at the point where all sulfate has been precipitated. The procedure is reasonably simple, so that with a little practice, those competent in laboratory analytical techniques can achieve good results on standard gases. Stack samples present more problems but still can be analyzed with good results if reasonable care is taken to use good sampling technique. The analysis can be completed in 1 to 2 hours.

The method for NO_x is far from straightforward, and though capable of yielding good results when run with rigorous attention to detail by an experienced analyst, it has many potential sources of error for those not familiar with the technique. The gas to be analyzed for NO_x is pulled into an evacuated flask containing a solution of sulfuric acid and hydrogen peroxide, which slowly oxidizes the nitric oxide (NO) present to nitrogen dioxide (NO_2). This requires about 16 hours to complete once the flask is filled with sample.

This solution is made basic and evaporated, then taken up in phenol disulfonic acid (from which the method gets its name—PDS). After several more steps, the color of the solution resulting from the reaction of the nitrate ion (formed from NO_2 in water) with PDS is read on a spectrophotometer at 420 nm. This color is proportional to the total NO_x content.

Some of the problems encountered in this method, each of which can lead to erroneous results, are (1) incomplete conversion of NO to NO_2, (2) adsorption (and loss) of NO_2 on walls of flask, (3) loss of sample in evaporating to dryness, and (4) errors resulting from incorrect pH adjustment.In addition, the method requires nearly 24 hours from start to finish.

Considering the difficulty of getting good results from the PDS method, it is probably wise for anyone requiring this test on their facility to consider contracting the job to an experienced analytical laboratory or to a consultant making use of such a laboratory.

Calibration Gases. Calibrated gas mixtures (span gases) may be used to calibrate emission-monitoring systems. If the gases are carefully prepared, accurately analyzed, stored under satisfactory conditions, and not contaminated during use, they can provide a valid means of calibration.

FIELD PERFORMANCE

Analysis systems based on measurement of the strong ultraviolet light absorption of SO_2 and the strong visible light absorption of NO_x have been described by Saltzman and Williamson (Ref.4). These are complete emission-monitoring systems including photometer (detector) and sample transport system.

Instruments employing ultraviolet or visible photometric measurement principles and sample systems similar to those described have been successfully used in a number of plants to measure SO_2 and NO_x in stack gases.

Table 1 lists pertinent performance data with respect to process type, length of instrument operation, malfunctions, and required maintenance time.

Maintenance data in Table 1 are particularly impressive since the figures listed represent the best estimate of system users on the total amount of time spent per week to correct malfunctions, calibrate, and perform preventive maintenance. Very few instrument problems occurred and none of these caused significant downtime. From the data it can easily be seen that with careful design of sample systems, coupled with unique photometric techniques, analysis systems can be provided to reliably measure SO_2 and NO_x in off-gases from combustion or chemical processes. Low maintenance requirements can be attributed to ruggedized construction, careful selection of corrosion-resistant components, and unique photometric design. Figure 3 illustrates a typical system housed in a weather-resistant cabinet.

TABLE 1.   Typical Analyzer System Performance Data

Process	Measurement	Installation date	Est. downtime by user	System failures description	Est. maintenance, hours/week
Sulfuric acid plant	SO$_2$	July 1971	<1%	Sample line temperature controller*	~ 2
Steam-generating plant	SO$_2$	Feb. 1972	<1%	Sample line plugging†	~1
Steam-generating plant	NO$_x$	Feb. 1972	<1%	Sample line leak	~1
Claus sulfur recovery plant	H$_2$S/SO$_2$ ratio	March 1970	<1%	None	2.5
Black liquor recovery boiler	SO$_2$	March 1972	<1%	Probe blocking‡	~ 2
Nitric acid plant	NO	Nov. 1971	1%	None	~ 2

*Printed circuit board failure.
†Increased filter probe blowback frequency from 60 minutes and this corrected the problem.
‡Use of probe shield corrected the problem.

EMISSION LIMITS

Table 2 gives the SO$_2$ and NO$_x$ emission limits for all states, and the federal limits, for new installations except as noted. These limitations were in effect in mid-1974, and obviously are subject to subsequent change. Thus this table is intended as a guide only. Those readers concerned with the state limits should check with their local environmental officials for up-to-date regulations.

The limits are generally given in terms of pounds per million Btu heat input, but it should be noted that some are as percent sulfur in the fuel, some are expressed as concentration in the emitted gas, and some are absolute limits in terms of total pounds allowed per hour, regardless of plant size. Where limits are arrived at by complex formulas, no attempt was made here to include such information in the table.

In the United States, the EPA is continuing to promulgate new rules covering emissions testing and monitoring. New rules may include further definition of initial testing and evaluation when one purchases a monitoring system. Also, the fundamental calibration techniques may be changed considerably within the near future, as EPA requirements become more carefully defined as to calibration requirements for a monitoring system.

There is a wide variety of monitoring equipment available commercially. In procuring such equipment, in summary, a number of points must be considered:

1. Extract the sample or measure in the stack?
2. Which type of monitor is best, that is, UV, IR, etc.?
3. Single-point sampling or multiple-point sampling?
4. Initial expense versus the expense of maintaining an initially expensive system.

Other questions which must be considered include accuracy. One must consider the need for obtaining a system which will give accurate results. To achieve this objective, it is suggested that one might choose an analyzer which already has been evaluated and which has shown satisfactory results in EPA tests. Several of the analyzers available fall into this category (Refs.1,2,7).

REFERENCES*

1. Homolya, J. B.: Coupling Continuous Gas Monitors to Emission Sources, *Chem. Technol.*, 426 (July 1974).
2. Jaye, F. G.: "Monitoring Instrumentation for the Measurement of Sulfur Dioxide in Stationary Source Emissions," EPA-R2-73-163, Office of Research and Monitoring, Environmental Protection Agency, Washington, D.C. (1973).

*Including additional reading suggestions.

TABLE 2. Emission Limits
(Unless otherwise noted, all figures are pounds/million Btu)

	SO₂		NOₓ		
	Oil fired	Coal fired	Oil fired	Coal fired	Gas fired
EPA (Federal)	0.8	1.2	0.3	0.7	0.2
Alabama	1.2	1.2	0.3	0.7	0.2
Alaska		500 ppm			
Arizona	0.8	0.8	0.3	0.7	0.2
Arkansas		0.2 ppm beyond plant			
California		Locally regulated		Locally regulated	
Colorado	0.8	1.2	0.3	0.7	0.2
		500 ppm/source			
Connecticut	0.55	0.55	0.3	0.7	0.2
Delaware		1% (w/w) fuel oil, 0.3% for	0.3	0.3	0.2
		distillate (N.C. County)			
Florida	0.8	1.2	0.3	0.7	0.2
Georgia		(Varies with source)*	0.3	0.7	0.2
Hawaii		0.5% (w/w)			
Idaho		1.75% (w/w) 1% (w/w)			
		0.3% (w/w) (distillate)			
Illinois		(Varies with source)*		(Varies with source)*	
Indiana		(Varies with source)*			
Iowa	1.5	5	0.3		0.2
Kansas	0.8	1.2	0.3	0.7	0.2
Kentucky		(Varies with region)	0.3	0.7	0.2
Louisiana		2000 ppm @ source			
		0.14 ppm off plant 24 hr. max.			
Maine		1.5% (w/w) (Portland AQCR)			
		2.5% (w/w) (Other AQCR's)			
Maryland		1% (w/w) max. S all fuel	0.2	0.5	0.2
Massachusetts	0.28†		0.3	0.3	0.3
	0.55				
	0.17 (distillate)				
Michigan	2% (w/w) (<500,000 lb				
	steam/hr)				
	1.5% (w/w) (others)				
Minnesota	2%(w/w)		0.3		0.2
	1.5% (w/w) (Source			(Priority I Regions)	
	>250 × 10⁶Btu/hr				
	in Minneapolis/				
	St. Paul AQCR)				
Mississippi	4.8	4.8			
	2.4 (modified,				
	<250 × 10⁶Btu)				
Missouri		1000 lb/hr	0.3	0.7	0.2
Montana	1.0	1.0			
Nebraska	0.8	1.2	0.3	0.7	0.2
Nevada		0.105 (>250 × 10⁶Btu)			
		0.7 (<250 × 10⁶Btu)			
New Hampshire	0.4% (No. 2)‡	1.5%			
	1.0% (No. 4)				
	1.0% (No. 5, 6)				
New Jersey	0.2% (No. 2) or				
	0.2 lb/10⁶Btu				
	0.3% (No. 4)				
New Mexico		0.34 (new)	0.3	0.45 (new)	0.2 (new)
		1.00 (existing)		0.7 (existing)	0.3 (existing)
New York	1.1 (existing)	1.8 (existing)	0.3	0.7	0.2
	0.4 (new)	0.6 (new)			
N.Y. City	0.3				
Metropolitan	0.1 (distillate)				
area					
North Carolina	1.6 (new) (any fuel)				
	2.3 (existing) (any fuel)				
North Dakota	3.0 (any fuel)				
Ohio	0.1 (any fuel, by 7/1/75)		0.3	0.9	0.2
Oklahoma	0.8 (until 2.0 7/1/75,		0.3	0.7	0.2
	then 0.3) (0.2 for gas)				
Oregon	1.75% (w/w)	1% (w/w)			
	0.3% (distillate)				
	0.5% (No. 2)				
	1.4	1.6			
		(< 250 × 10⁶Btu)			

(Unless otherwise noted, all figures are pounds/million Btu)

	SO₂		NO_x		
	Oil fired	Coal fired	Oil fired	Coal fired	Gas fired
	0.8	1.2 ($> 250 \times 10^6$Btu)			
Pennsylvania	(Varies with source and area)*		0.3	0.7	0.2
Rhode Island	0.8	1.2	0.3	0.7	0.2
South Carolina	2.3 (all sources)				
South Dakota	0.8	1.2	0.3	0.7	0.2
	3.0 (<250 $\times 10^6$Btu)				
Tennessee	620 ppm/10^6Btu (7/1/75)		−.3	0.7	C.2
	440 ppm/10^6Btu (new)	620 ppm/10^6Btu (new)			
Texas	0.8	1.2	0.3	0.7	0.2
Utah	20% of input sulfur or less				
	1.5% (w/w)	1% (w/w)			
Vermont	1% (w/w) all fuels		0.3	0.3	0.3
Virginia	2.64 $\times$ Btu/hr cap. in 10^6		0.7	0.9	0.4
	1.06 $\times$ Btu/hr cap. in 10^6 (for AQC Region 7)				
Washington	1000 ppm (7/1/75)				
West Virginia	(Varies with source and area)*				
Wisconsin	0.8	1.2	0.3	0.7	0.2
Wyoming	No source emission stds., ambient only.		0.2		0.3

*Complex formulas for calculating limits.

†For Arlington, Belmont, Boston, Brookline, Cambridge, Chelsea, Everett, Malden, Medford, Newtòn, Somerville, Waltham, Watertown.

‡(No. 2), etc., refers to fuel oil grades.

3. Orchin, M., and H. H. Jaffe: "Symmetry, Orbitals, and Spectra," chap. 9, Wiley, New York, 1974.
4. Saltzman, R. S:, and J. A. Williamson: Monitoring Stationary Source Emissions for Air Pollutants with Photometric Analyzer Systems, *Anal. Instrum.,* 9 (1971).
5. Schenk, G. H.: "Absorption of Light and Ultraviolet Radiation: Fluorescence and Phosphorescence Emission," Allyn and Bacon, Boston, 1973.
6. Scott, A. I.: "Interpretation of the Ultraviolet Spectra of Natural Products," introduction and chap. 1, Macmillan, New York, 1964.
7. Snyder, A. D., P. L. Sherman, E. C. Eimutis, F. C. Jaye, and M. G. Konicek: Laboratory and Field Evaluation of Stationary Source Instrumentation for Oxices of Nitrogen Emissions, *Anal. Instrum.,* 10(1972).
8. Williamson, J. A.: "Oxidation of Nitric Oxide to Nitrogen Dioxide for Photometric Measurement of NO_x in Emission Source Monitoring," *Proc. 27th Annu. Conf. Exhibit,* Instrument Society of America, Pittsburgh, Pennsylvania (1972).
9. Wolf, P. C.: Systems for Continuous Stack-Gas Monitoring, *Mech. Eng.,* pt. I (April 1974); pt. II (May 1974).

Monitoring of Carbon Monoxide Emissions

BY W. V. DAILEY*

CARBON MONOXIDE AS HEALTH HAZARD

Approximately 90 percent of the carbon monoxide (CO) in the air is from natural sources, such as the decay of organic matter (Ref.1) and forest fires. The remainder results from incomplete burning of fossil fuels, such as oil, coal, and gas—from automobiles and other vehicles, chemical, metallurgical, and other combustion processes. Carbon monoxide can be found, of course, within vehicular tunnels and enclosed parking garages and occurs more frequently in industrial environments than any other gas classified as dangerous. Carbon monoxide also is one of the most insidious of all toxic gases, being odorless, colorless, imperceptible to the senses, thus giving no warning of its presence. For these reasons, carbon monoxide is considered a serious health hazard. See Fig. 1.

When inhaled, CO is quickly absorbed by the hemoglobin, and as the hemoglobin becomes saturated with the gas, the blood loses its ability to carry oxygen. Headache, nausea, unconsciousness, and death may soon follow upon continued exposure to the gas. In the absence of an automatic method of detection, toxic concentrations of CO can build up before exposed individuals have any indication of danger. Small concentrations of CO have been reported to affect driver response, and the *Federal Register* (Ref.2) indicates health hazard occurs at 9 ppm, the primary air standard for CO, and has thus mandated reduction (Ref.3) of CO emissions from the prime source, the internal combustion engine.

Sampling Systems. The importance of gas-sampling systems to the accuracy and reliability of associated instrumentation is stressed in the article on "Monitoring of Hydrocarbon Emissions" in this section of the handbook. Sampling systems also are described in some detail in the article on "Source Monitoring of NO_x and SO_2."

INSTRUMENTATION

Nondispersive Infrared Analyzers. Ambient air measurements (Ref.4) for carbon monoxide have been made since the early 1950s on a continuous basis with nondispersive infrared (NDIR) analyzers, calibrated 0 to 100 ppm CO full scale. Most of the ambient measurements in the United States are presently done with NDIR instruments by local, state, and federal agencies. Performance specifications for such equipment are documented in the *Federal Register*. The NDIR analyzer, in which CO is measured at 4.6 micrometers, is described in some detail in Appendix B of the article on "Monitoring of Hydrocarbon Emissions" and in the article on "Source Monitoring of NO_x and SO_2."

Levels of CO within urban areas have been determined with NDIR instrements, but to accurately measure background (natural) levels of CO, the typical NDIR instrument lacks the required sensitivity of 0.1 ppm or less. Recently developed is a modification of the automatic process gas chromatograph (Ref. 5) that provides an ambient air analysis of CO in full-scale ranges of 1 to 5 ppm and also provides a measurement of methane and nonmethane hydrocarbons. Such instruments are used in many field studies and have proven that the ocean (plankton) is a natural source of CO emissions. (Ref. 6) This instrument also is described in some detail in Appendix A of the article on "Monitoring of Hydrocarbon Emissions."

Catalytic-Oxidation-Type Analyzers. Another recently available highly sensitive air quality instrument for CO is based upon the catalytic oxidation principle. This method continuously drys the sample, catalytically oxidizes the CO to CO_2, and measures the heat of combustion, using sensitive heat-detection circuits operating in a temperature-controlled (to $\pm0.5°C$) chromatograph oven. Full-scale ranges as low as 0 to 10 ppm are possible. It is interesting to note that early instruments in the 1920s for CO detection used a catalytic oxidation principle.

*Technical Products Division, Mine Safety Appliances Company, Pittsburgh, Pa.

In modern CO analyzers, samples of air to be analyzed are drawn continuously through sampling lines to a centrally located instrument cabinet. Here, the air passes through an insulated heated detector cell where the CO content is determined. See Fig. 2. When the air sample passes through the hopcalite (a low-temperature catalyst), any CO present is immediately oxidized. Heat from the oxidation creases an imbalance in the cell and is indicated on a direct-reading meter, recording potentiometer, contact-making meters, or a combination of these devices.

CO analyzers are calibrated 0 to 400 ppm (or lower ranges) with warning alarms set at 50 ppm and high alarms at approximately 200 ppm, depending upon the particular require-

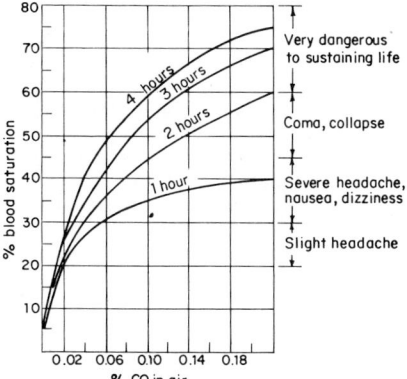

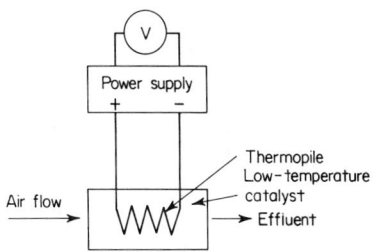

Fig. 1. Effects of carbon monoxide on human beings. This chart can be considered only as a general guide because the percent of CO blood saturation will vary with exertion, excitement, fear, depth of respiration, anemia, and general physical condition of the individual.

Fig. 2. Operating principle of CO detector utilizing the catalytic oxidation of CO.

ments of the installation. The CO analyzer can be time-shared between a number of sample points, and the monitoring system (Ref.10) can be supplied with local or remote, visual or audible alarms and recorders (often essential for proving compliance with regulatory requirements), and with switches to actuate ventilation fans automatically. CO monitoring for OSHA (Occupational Safety and Health Administration) (Ref.9) purposes and personnel protection is necessary in metallurgical processing, operation of cupolas, blast furnaces, coke ovens, heat-treating shops, open hearths, tunnels, underground parking garages, engine test cells, roll-on/roll-off ships, and chemical plants and other facilities where carbon monoxide may be present as an air pollutant.

Threshold limit values are described in Appendix G of the article on "Monitoring of Hydrocarbon Emissions."

Process Applications. Specific applications in the process industries include the determination of ppm CO in ammonia synthesis gas streams for the protection of expensive catalysts (Ref.11); in blast furnace operation, analysis of CO in top gases can indicate changes in wind rate, burden, slippage, and other important operating factors. Other process applications include ethylene oxide production (Ref.12), catalyst regeneration, CO/CO_2 ratio determinations of several processes, including some coal gasification processes, and in the monitoring of heat-treating atmospheres. For example, inert atmospheres are sometimes monitored for total combustibles (CO and H_2) by using a modified catalytic-combustion-type instrument.

Total Combustibles Analyzers. A combustible gas detector operates on the principle that a burning gas releases heat proportional to the concentration of the combustible gas (or fuel) in the sample. By measuring the heat evolved, the concentraction of the gas can be determined. A catalytic platinum filament, functioning as one arm of a wheatstone bridge, is heated to a preselected temperature by the passage of a controlled electric current from a power source. The bridge is balanced on fresh air. Then, a sample of the atmosphere to be tested is drawn into the chamber surrounding the filament. Heat

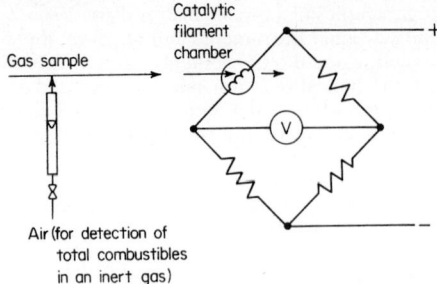

Fig. 3. Operating principle of total combustibles gas analyzer.

produced by burning of any combustible gas present raises the filament temperature, increasing its resistance and unbalancing the bridge. The bridge imbalance is indicated by a galvanometer which is calibrated in terms of percentage of combustible gases, or in terms of concentration by volume of the compound(s) of interest.

In connection with the analysis of inert gas atmospheres for combustibles, such as hydrogen and carbon monoxide, a standard combustible gas analyzer cannot be used because inert gases do not contain oxygen for combustion. In such cases, air is mixed with the sample prior to entry into the instrument, as indicated in Fig. 3.

Portable Instruments.   Portable instruments, utilizing fuel cells or various chemical methods for CO detection, are available for spot checks and survey studies. In the carbon monoxide detector tube, for example, the air sample is drawn through a tube in which silicone gel is placed that is impregnated with a chemical that reacts with CO to produce either a stain or color change. The operator estimates the presence and concentration of CO in the atmosphere by comparing the color or length of the stain with a reference chart supplied by the manufacturer. Although such methods are accurate only to within ± 15 to ± 25 percent, they are sometimes suitable for determination of potential trouble spots (Ref. 13). Fuel cells, which in actual usage are encapsulated redox cells with the appropriate electrodes, electrolytes, and potentials, provide meter readouts directly in volume concentration.

COMBUSTION CONTROL

Energy conservation measures have revived interest in the use of gas analyzers in combustion control. Particularly in combustion situations where it is desired to maintain

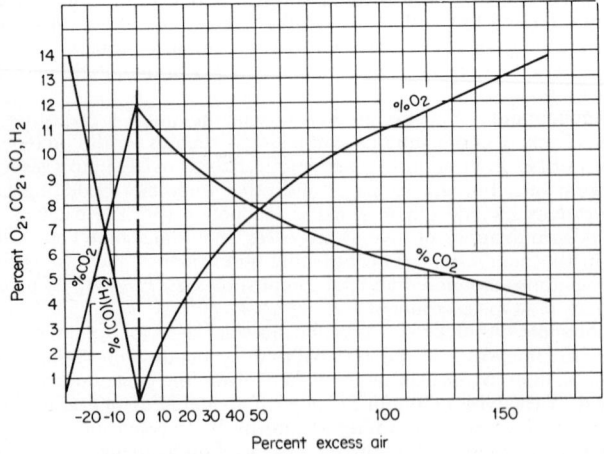

Fig. 4. Products of combustion versus excess air.

near-stoichiometric oxygen-fuel mixes, a carbon monoxide analyzer can be useful for immediately sensing any condition where there is no excess air. From Fig. 4, note the steep rise of the CO line as less-than-stoichiometric oxygen-fuel conditions are encountered.

REFERENCES

1. Instrumentation for Environmental Monitoring, Lawrence Berkeley Laboratory, University of California, Berkeley, Calif., December 1973.
2. *Fed. Reg.*, 36(228): 22369–22572 (Nov. 25, 1971).
3. Federal Clean Air Act of 1963.
4. Jacobs, M. B., et al.: Continuous Determination of Carbon Monoxide and Hydrocarbons in Air by a Modified Infrared Analyzer, *J. Air Pollut. Control Assoc.*, 9(2):110–114 (August 1959).
5. Poli, A. A.: Air Monitoring for Carbon Monoxide and Methane with a Single Analyzer, *17th Anal. Instrum. Div. Symp.*, Instrument Society of America, Pittsburgh, Pa., May 1972.
6. Swinnerton, J. W.: The Ocean as a Natural Source of CO, *Science,* 167::984–986 (1970).
7. Source of information, New Jersey State Department of Health, Trenton, N.J.
8. Cline, E. L., and L. Tinkham: "A Realistic Vehicle Emission Inspection System," paper no. 68-152, Air Pollution Control Association (1968).
9. Occupational Safety and Health Standards: National Concensus Standards and Established Federal Standards, *Fed. Reg.*, pt. II, Vol. 36(105)(May 29, 1971).
10. Dailey, W. V.: Monitoring Work Areas for Toxic and Flammable Hazards, *ISA Ann. Meeting,* Instrument Society of America, Pittsburgh, Pa. (1972).
11. A Guide to Process Analytical Instrumentation, *Tech. Bull.* 4068, Beckman Instruments, Inc., Fullerton, Calif. (1966).
12. Dailey, W. V.: "Infrared Analyzers for On-Stream Analysis of Ethylene Oxide Process Streams," 64th Annual Meeting of AIChE., American Institute of Chemical Engineers, New York (Nov. 29–Dec. 2, 1971).
13. Air Quality Criteria for Carbon Monixide, AP-62, U.S. Dept. of Health, Education, and Welfare, Washington, D.C. (March 1970).
14. Verdin, A.: "Gas Analysis Instrumentation," Wiley, New York, 1973.
15. Siggia, S.: "Continuous Analysis of Chemical Process Streams," Wiley, New York, 1959.

Monitoring of Hydrocarbon Emissions

BY W. V. DAILEY*

Before describing methods for the continuous analysis of hydrocarbon emissions, it should be pointed out that there are various classes of hydrocarbons (Ref. 1) which greatly affect the nature of the pollutants and the type of instrumentation used. Two major classes of hydrocarbons are (1) those hydrocarbons consisting only of hydrogen and carbon atoms (aliphatic, olefinic, and aromatic compounds) and (2) hydrocarbon derivatives, such as ketones, alcohols, halogenated hydrocarbons, naphthas, and solvent vapors. Wide ranges of calibrations are involved, from fractional parts per million (ppm) to volume percent, creating many analytical problems and requiring different measurement methodologies.

Some methods for hydrocarbon determinations are essentially of a laboratory nature, requiring special skills ranging from expert technicans through chemists and spectroscopists, and may involve computers and quite complex chemical analysis schemes. The descriptions here will concentrate on continuous and semicontinuous automatic-industrial-type instruments which, once installed and calibrated, can furnish practical operating information when carefully and routinely maintained. Industrial analyzers in this category include chromatography, nondispersive infrared spectroscopy, hydrogen flame ionization detectors (sometimes called total hydrocarbon analyzers), and catalytic combustion methods. See also Appendixes A to H.

SAMPLING SYSTEMS

Such instruments require automatic sampling systems, *always* the most important part of an analysis system. The cliche "The analyzer is no better than the sampling system" is well founded. This is further confirmed by the following statement which appears in the American Petroleum Institute's "Manual on Installation of Refinery Instruments and Control Systems," API RP 550, 2d edition, May (1965):

> There is no single simple solution to the sampling problem and no universal system suitable for most applications. Each installation has its own problems and requires individual study before accurate analysis can be expected. Experience has shown that the study will prove valuable, providing reliable data and in reducing maintenance.

Analyzers *must be* provided with a relatively clean sample, condensate-free, and at a relatively constant pressure and flow rate. The system must be designed to ensure sample integrity. For example, selection of the improper filtering media may result in loss of hydrocarbons due to absorption on the filter. Careful calibration is also important for reliable data. Certified standards should be obtained from reliable suppliers and spares should be readily available. Total hydrocarbon analyzers require hydrocarbon-free combustion air and hydrogen for proper operation.

HYDROCARBONS AND AIR QUALITY

The average concentrations of the most abundant hydrocarbons in air have been measured (Ref. 1) as follows (expressed in ppm as carbon): methane, 3.22; toluene, 0.7; *n*-butane, 0.26; isopentane, 0.21; ethane, 0.20; benzene, 0.19; *n*-pentane, 0.18; propane, 0.15; ethylene, 0.12; as well as the presence of other aromatics, olefins, acetylenes, and so on.

The above compounds, and partially or noncombusted gasoline vapors, are the source of hydrocarbon pollution and components in the formation (Ref. 1) of peroxyactylnitrate (PAN) in smog. Historically (Ref. 2), this analysis initally involved the use of infrared

---

*Technical Products Division, Mine Safety Appliances Company, Pittsburgh, Pa.

spectroscopy with nondispersive infrared (NDIR) analyzers sensitized at 3.4 micrometers, where all hydrocarbons absorb infrared energy. Such instruments were calibrated 0 to 3 ppm full-scale range in terms of *n*-hexane.

The analysis system, operating at pressures of approximately 10 atmospheres, had high sample system maintenance and, although the instruments exhibited high response to saturated hydrocarbons, NDIR analyzers have been found to have relatively low response to the olefinic/aromatic materials, e.g., the so-called *reactive* types of smog ingredients. See Appendix F.

To simplify the system and to provide better data more indicative of the reactive hydrocarbons present, the total hydrocarbon analyzers (T/HC) were used, responding at ranges as low as 0 to 4 ppm hydrocarbons (as methane). However, methane is a naturally occurring material that is considered nonreactive. Thus, corrections of the T/HC analysis signal were necessary to compensate for methane concentrations.

One method of performing this correction is to use an activated charcoal scrubber (Refs. 4, 5) to remove all hydrocarbons but methane and alternate the air sample through it to a T/HC. Thus, the recorded signals will be alternately methane and total hydrocarbons. With some further calculation, the data can be presented as methane and nonmethane hydrocarbons. Sometimes the nonmethane hydrocarbons are referred to as reactive hydrocarbons, but this is a misnomer because nonmethane hydrocarbons also may be ethane, hexane, benzene, and others—all considered nonreactive photochemically. See Appendix F.

In some cases, two total hydrocarbon instruments were used with the activated charcoal with each analyzer, providing a continuous signal for one of the two parameters. An obvious limitation of this system is that the charcoal scrubber could pass all hydrocarbons (as the result of channeling) and the methane analysis data would be unreliable.

Another method of analysis, developed by the Environmental Protection Agency (EPA) and others (Ref. 3), utilized the high sensitivity of the hydrogen flame ionization detector (HFID) and the separation technique of chromatography. By selection of the proper column, methane and nonmethane hydrocarbons and carbon monoxide could be separated and measured by an HFID. Carbon monoxide was measured by first separation, then conversion to methane by hydrogen and a nickel catalyst (a methanator), and subsequently measured with the methane and nonmethane hydrocarbons by an HFID. Such instruments are on many air-quality monitoring programs.

More recently developed is a method (Ref. 6) to continously measure methane and nonmethane hydrocarbons by comparing signals from two HFIDs: one, a total hydrocarbon signal from the air sample; the other, a partially oxidized air sample (for removal of all hydrocarbons but methane). A simple electronic subtraction system then provides the methane and nonmethane hydrocarbon signals.

To date, it would appear that none of the foregoing analysis systems provides a satisfactory measurement of the *reactive* hydrocarbons present in the air. But, all these systems do provide a reliable index of air-quality contamination by hydrocarbons (as methane and nonmethane compounds). In order to perform an analysis of the reactive hydrocarbons, it is necessary to utilize laboratory gas chromatographic analytical techniques. Unfortunately, such methods are not easily adapted to continuous monitoring.

Engine Exhaust.   Hydrocarbons and carbon monoxide have proven to be precursors of smog, with, of course, automobile exhaust gas as the prime source of hydrocarbons for the photochemical reactions which produce smog. The growing use of supplemental diesel engine and gas turbine power must be considered along with vehicular engine exhausts as sources of air pollutants. The Clean Air Act of 1963 mandated the reduction of hydrocarbons, carbon monoxide, and nitrogen oxides from engine exhaust gases. To accomplish these goals, it was necessary to study the automobile engine and to determine emissions when operating under various loads, speeds, air/fuel ratios, and other factors.

Considerable progress was made in using NDIR analyzers in these studies. (Ref. 8) Specifically with regard to hydrocarbons, the instruments were sensitized at 3.4 micrometers and calibrated in terms of *n*-hexane. Two instruments were initially used at ranges of 0 to 0.1 and 0 to 1 percent. To correlate data between industrial and regulatory laboratories, the sample systems were standardized and a specified fuel was used to compensate for the low response of the NDIR to olefins and aromatics.

As the internal combustion engine was improved to produce lower hydrocarbon emissions, the test procedures were also altered. Total hydrocarbon analyzers (T/HC) are now used for this measurement (Ref. 9) because the T/HC responds better to the olefins and aromatics. Also, the instrument is more sensitive and is calibrated in terms of methane.

However, it should be pointed out that NDIR analyzers are not obsolete for the measurement of hydrocarbons from engine exhaust gases. Several states (Ref. 10) use such sensors for car inspection (for carbon monoxide and hydrocarbon emissions) and surveillance studies. Also, NDIR analyzers are used in many garages for diagnostic work and engine tune-ups.

Stack (Solvent) Analysis. Stack emissions from chemical processes, paint-drying ovens, solvent recovery plants, rotogravure printing operations, and the like are common sources of emissions. These emissions can be analyzed on a continuous basis by NDIR and T/HC analyzers, depending upon the analysis range required, components to be monitored, and background compounds present.

Such monitoring and resulting analytical data are not only useful for air pollution purposes, but also can be used to detect process upsets, for alarm or control purposes, for the detection of solvent breakthrough from solvent recovery beds to conserve steam and heat in bed regeneration, and to detect possible emissions of toxic materials (Ref. 11).

PROCESS ANALYSIS FOR HYDROCARBONS

Selecting an analyzer for a process application obviously depends upon numerous factors. Most petrochemical/refinery analysis applications use the gas chromatograph. No other analytical technique is currently capable of measuring solids, liquids, and gases for hydrocarbon content over a range from a few ppm to volume percentage. NDIR, T/HC, and other sensors are also used on processes, such as air separation (Ref. 12), ethylene production (Ref. 13), and ethylene oxide production (Ref. 14). The satisfactory solution of such analytical problems requires careful cooperation between prospective user and instrument supplier.

Flammable Hydrocarbons. Two important instrument applications involve the monitoring of air and gases for flammable hydrocarbons: (1) detection of LEL (lower explosive limit) and (2) detection of TLV (threshold limit value) for toxic compounds. These terms are defined in Appendix G. See also Refs. 15 and 16.

The need for measuring toxic and flammable gases is well established. However, there are some processes, such as drying ovens (Ref. 17), that normally may not be considered dangerous but that operate near the LEL. Accumulation of toxic gases within work areas obviously represents a potential health hazard and there are increasing governmental regulations in this area (Ref. 11). Some of the analytical techniques and systems used to monitor air quality for pollution purposes also can be used for TLV and LEL measurements. Because hazardous concentrations range from 200 ppm for the TLV of methyl alcohol, for example, to 6.7 percent for the LEL, different analytical methods may have to be used for monitoring the same substance. In this case, an NDIR may be used for the LEL, a T/HC for the TLV.

The conventional "hot wire" detector is probably the simplest instrument for measuring the LEL of hydrocarbons, such as gasoline vapors and natural gas. The instrument operates on the principle that a burning gas releases heat that is proportional to the concentration of the combustible gas (or fuel) in the sample. Catalytic combustion is used and, by measuring the heat evolved, the concentration of the gas can be determined. A typical instrument incorporates a wheatstone bridge circuit, with one arm being a catalytic filament. This filament is heated to a preselected temperature by passage of a controlled electrical current from a power source. The bridge is balanced on fresh air and then the atmosphere to be tested is drawn into the chamber surrounding the filament. Heat produced by burning the combustible gas raises the temperature of the filament, increasing its resistance, and unbalancing the bridge. The instrument can be calibrated in terms of 0 to 100 percent LEL. The method responds to all combustible gases and thus can be used as a universal LEL detector for hydrocarbons, but with the instrument calibrated for a particular gas of interest. Limitations involve vapor pressures of the compounds and, if silicone vapors are present, the catalytic filaments can become inactive. Where silicones are present, a reliable substitute is the NDIR (Ref. 20).

The design of such combustible gas detectors ranges from portable battery-powered types for spot checks to continuous monitors with remote sensors, readouts, and alarms. Applications (Refs. 19, 20) include the measurement of combustible gases in petrochemical plants, offshore drilling rigs, control rooms, compressor rooms, and flammable storage areas. The instruments can be calibrated for alcohols, ethers, ketones, gasoline, and other hydrocarbons.

Other methods for monitoring combustible gases include thermal conductivity, optical interferometry, gas chromatography, gas-sensing semiconductors, and total hydrocarbon analyzers.

TOXIC GAS MONITORING

Threshold level value ranges are from a fraction of a ppm to a few thousand ppm. The TLVs for airborne contaminants are outlined in Ref. 16. Generally, TLVs of 5 ppm or less can be monitored by a T/HC-type instrument. For values greater than 50 ppm, NDIR analyzers are applicable. Usually the sensor is time-shared between a number of sampling points, with an alarm circuit to actuate remote alarms and ventilation equipment (Ref. 19). Recorders or data loggers can be used in the system for record-keeping requirements of regulatory agencies.

OTHER ANALYSIS METHODS

A recently developed instrument (Ref. 23) for TLV analyzers uses a principle of photo-ionization where the air sample containing organic components is ionized by an ultraviolet light. The degree of ionization is related to concentration; and although such a method is somewhat nonselective, it is sensitive from a few ppm to fractional percentage ranges. This instrument is particularly attractive in that no reagents are employed, the device is nonresponsive to methane, water vapor, or CO_2 commonly found in air samples.

The next generation (Ref. 24) of TLV (and LEL) sensors will involve semiconductors that are constructed of a metal oxide with the proper doping. The sensor is heated to a specific temperature and at that temperature reducing or oxidizing gases produce a valency change of the metal ion on the surface. This surface conductance change is detected by the electronic circuits for meter readout, alarms, etc.

Such methods are sensitive to fullscale ranges of about 10 to 100 ppm, but lack some specificity, being responsive to water vapor and other compounds. Once the technology is developed to overcome these obstacles, such sensors will find wide use for detection of hazardous gases in TLV and LEL ranges.

REFERENCES

See appendixes which follow for diagrams and further details on some of the instruments mentioned in this article.

1. "Air Quality Criteria for Hydrocarbons," AP-64, Department of Health, Education and Welfare, Washington, D.C., March 1970.
2. Ortman, G. C., and V. L. Thompson: Performance of Hydrocarbon Monitoring Instrumentation, *ASTM Symp. Instrum. Monitoring Air Quality*, American Society for Testing and Materials, Philadelphia, Pa., Aug. 14–16, 1973.
3. Poli, A. A.: Air Monitoring for Carbon Monoxide and Methane with a Single Analyzer, *17th Anal. Instrum. Div. Symp.*, Instrument Society of America, Pittsburgh, Pa., May 1972.
4. Altshuller, A. P., G. C. Ortman, and B. E. Saltzman: Reactive Hydrocarbons and Nonmethane Hydrocarbons, *Proc. 7th Conf. Methods in Air Pollut. Studies*, Los Angeles, Calif.- Jan. 25–26, 1965.
5. Altshuller, A. P., G. C. Ortman, B. E. Saltzman, and R. E. Neligan: *J. Air Pollut. Control*, **16**: 87-91 (February 1966).
6. Poli, A. A., and T. L. Zinn: A New Continuous Monitor for Methane and Nonmethane Total Hydrocarbons, *19th Anal. Instrum. Div. Symp.* Instrument Society of America, Pittsburgh, Pa., May 1973.
7. Federal Clean Air Act of 1963, U.S. Government Printing Office, Washington, D.C.
8. *Fed. Reg.*, **33**(108): 8304-8324 (June 4, 1968).
9. *Fed. Reg.*, **35**(136): 11334-11359 (July 15, 1970).
10. Appropriate regulatory agencies: State of New Jersey (Trenton); State of California (Sacramento).
11. Occupational Safety and Health Standards; National Concensus Standards and Established Federal Standards, *Fed. Reg., Pt. II,* **36**(105) (May 29, 1971).
12. Houser, E. A.: The Application of On-Stream Analytical Instrumentation in Air Separation Plants Tech. Bull. 5036, Beckman Instruments, Inc., Fullerton, Calif. (periodically revised).

13. A Guide to Process Analytical Instrumentation, Tech. Bull. 4068, Beckman Instruments, Inc., Fullerton, Calif. (periodically revised).
14. Dailey, W. V.: Infrared Analyzers for On-Stream Analysis of Ethylene Oxide Process Streams," 64th Annual AIChE Meeting, Nov. 29–Dec. 2, 1971, American Institute of Chemical Engineers, New York.
15. Zabetakis, M. G.: "Flammable Characteristics of Combustible Gases and Vapors," Bul. 627, U.S. Bureau of Mines, Pittsburgh, Pa., 1965
16. "Threshold Limit Values of Airborne Contaminants and Physical Agents, with Intended Changes Adopted by ACGIH for 1975." American Conference of Governmental Industrial Hygienists, Cincinnati, Ohio.
17. "Standard of the National Board of Fire Underwriters for Ovens and Furnaces as Recommended by the NFPA, NBFU," No. 86A, National Board of Fire Underwriters, Chicago, Ill.
18. Strange, J. P.: Sensors for Automatic Analysis, *Instrum. Control Syst.* (November 1962).
19. Dailey, W. V.: "Monitoring Work Areas for Toxic and Flammable Hazards," Annual Meeting, 1972, Instrument Society of America, Pittsburgh, Pa.
20. Bossert, C. J.: Monitoring and Control of Combustible Gas Concentrations Below the Lower Explosive Limit, 20th National Symposium, Analysis Instrumentation Division, May, 1974, Instrument Society of America, Pittsburgh, Pa.
21. Verdin, A.: "Gas Analysis Instrumentation," Wiley, New York, 1973.
22. Siggia, S.: "Continuous Analysis of Chemical Process Systems," Wiley, New York, 1959.
23. Driscoll, J. M., and F. F. Spazian: A New Instrument for Continuous Monitoring of Odorous Sulphur Compounds, ISA Conference, Instrument Society of America, Pittsburgh, Pa., October 1974.
24. Fredericks, G. E., and R. E. Scott: Design of a Combustible Gas Measuring Circuit, *Rev. of Sci. Instruments* **46**, 6 (June 1975).

APPENDIX A   Gas Chromatograph for Methane, Nonmethane, and Carbon Monoxide Air-Quality Analysis

The block diagram of Fig. A-1 shows the complete chromatographic system. The sample valve loop introduces ambient air at atmospheric pressure on command from the programmer. Chromatographic column 1 is used with the backflush valve for hydrocarbons other than methane analysis from the sample. Column 2 is the analytical column that splits the sample into components. These three components (air, methane, and carbon monoxide) are then passed through the methanator or converter where only the carbon monoxide is affected (i.e., converted to methane with hydrogen enrichment over a nickel catalyst) and then transported by the helium carrier gas to the flame ionization detector as separate detectable peaks.

APPENDIX B   Nondispersive Infrared Analyzer (NDIR)

A nondispersive infrared analyzer is an optical instrument designed to monitor the concentration of a single component (at a selected wavelength) in a multicomponent stream. The optics consist of a source of polychromatic infrared energy, an absorption (or sample) cell, a comparison (or reference) cell, and an infrared detector (monochromatic), sensitized at a wavelength where the component of interest absorbs infrared energy and the background components are transparent.

The dual-chamber, Luft-type, selective pneumatic detector is usually used because it is easily sensitized to various infrared absorption bands. The sensitization of a Luft-type detector involves selecting a detector gas (usually the component of interest) that exhibits infrared absorption at the same wavelength as the component(s) of interest. This gas is used to provide response of the detector at the desired wavelength. By changing the detector gas, the wavelength of response also can be changed. Thus, responses at various wavelengths can be "built in" to the detector as required by the analyst.

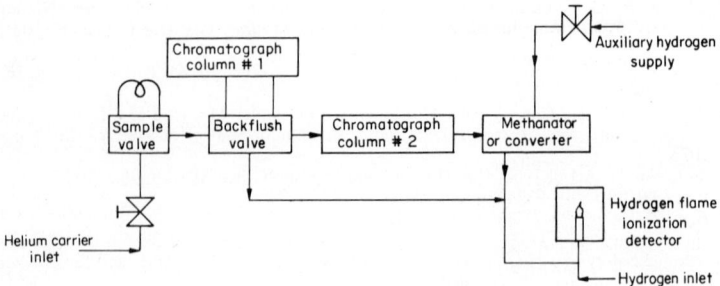

Fig. A-1. Block diagram of chromatographic system.

The range of the instrument is controlled by the sample cell length. As shown in Fig. B-1, two similar Nichrome filaments are used as sources of infrared radiation. Beams from these filaments pass through parallel gold-plated stainless steel cells. One beam traverses the sample cell and the other the comparison cell. The emergent radiation is directed into the single detector cell. As the gas in the detector absorbs radiation, its temperature and pressure increase. An expansion of the detector gas causes a condenser microphone membrane to move. This movement is converted and electrically amplified to produce an output signal.

Between the source and the cells, a semicircular beam chopper alternately blocks the radiation to the

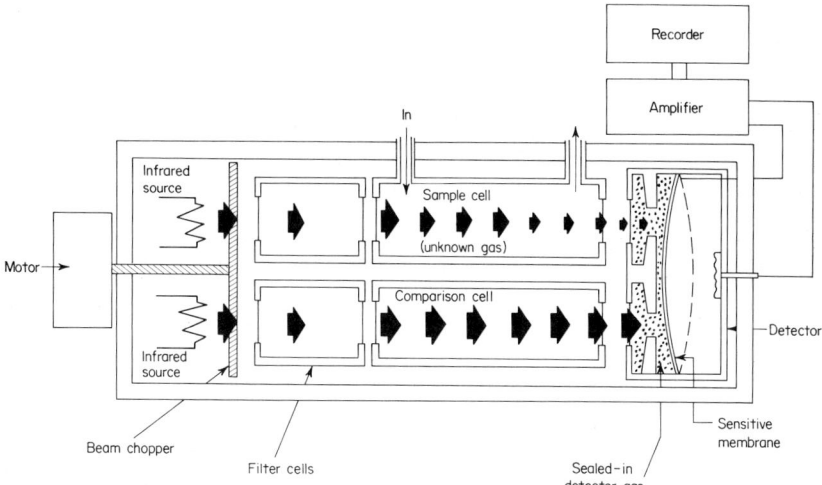

Fig. B-1. Schematic diagram of nondispersive infrared analyzer (NDIR).

sample cell and to the comparison cell. When the beams are equal, an equal amount of radiation enters the detector from each beam. The amplifier is turned so that only variations in light intensity occurring at the chopped frequency produce an output signal. Therefore, when the beams are equal, the output is zero.

When the gas to be analyzed is introduced into the sample cell, it absorbs infrared energy and thus reduces the radiation reaching the detector from the sample beam. As a result, the beams become unequal and the radiation entering the detector pulses as the beams are alternated. The detector gas expands and contracts in accordance with the pulse. In this way, the detector directly indicates the difference between the two beams. The expansion and contraction of the gas causes the membrane to move in response. The membrane movement varies the condenser microphone capacity; the variation in capacity generates an electric signal which is proportional to the difference between the two radiation beams. The signal is then amplified and fed to an indicating meter and/or recording potentiometer.

An NDIR analyzer can be used to monitor SO_2 or NO in stack gases by replacing the optics and field recalibration. Optics are stable and do not require resensitization in the field. More detail can be found in the paper by N. W. Hartz and J. L. Waters, "An Improved Luft Type Infrared Analyzer," Instrument Society of America National Conference, Houston, Tex., 1951.

APPENDIX C   Hydrogen Flame Ionization Detector (HFID)—Total Hydrocarbon Analyzer (T/HC)

A hydrogen flame ionization or total hydrocarbon analyzer is composed of four elements: (1) sample flow system, (2) combustion gases system, (3) burner assembly, and (4) electrometer and power supplies. See Fig. C-1.

The operation of the system is based upon the ionization of carbon atoms in a hydrogen flame. Normally, a flame of pure hydrogen contains an almost negligible number of ions. Adding organic compounds, even in traces, results in a large number of ions in the flame.

In the analyzer, the sample to be analyzed is mixed with a hydrogen fuel and passed through a small jet. Air supplied to the annular space around the jet supports combustion. Any hydrocarbon carried into the flame results in the formation of carbon ions. An electrical potential across the flame jet and an "ion collector" electrode suspended above the flame produces an ion current that is proportional to the hydrocarbon count. This is measured by an electrometer circuit whose output then provides an analysis signal for the direct-reading meter or for an optional potentiometric recorder.

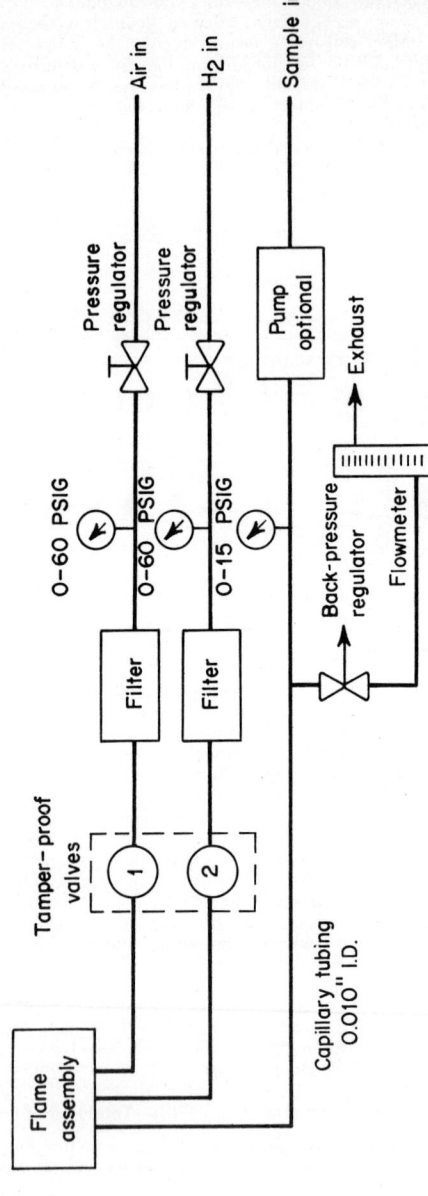

Fig. C-1. Schematic diagram of total hydrocarbon analyzer (T/HC).

APPENDIX D   Combustible Gas Detector (Catalytic Combustion)

A combustible gas detection system depends upon the heat developed by the actual combustion of the flammable portion of a sampled atmosphere over two heated catalytic elements that form one-half of a balanced electrical circuit. Two inactive elements form the other half. When a mixture of flammable gas in air is brought into contact with these catalytic elements, a rapid combustion of the combustible gas takes place on the surfaces. See Fig. D-1.

This burning of the combustible gas increases the temperature of the element, creating a propor-

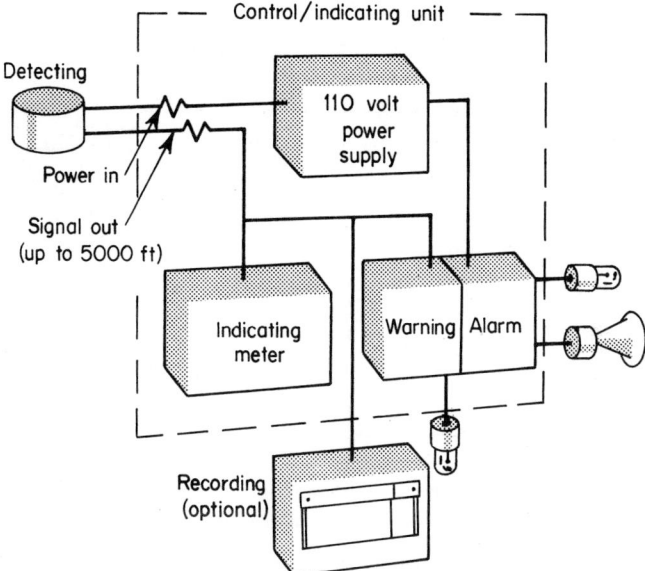

Fig. D-1. Schematic diagram of a combustible gas detector using catalytic combustion.

tional change in its resistance. The change in resistance creates an imbalance in the bridge circuitry, the result of which is calibrated on a meter indicating the concentration of combustible gases in the sample. Thus, a reading on the meter between 0 and 100 percent shows how closely the monitored atmosphere approaches the minimum concentration required for a flammable mixture.

APPENDIX E   Methane/Nonmethane Dual Flame Analyzer

The methane/nonmethane dual hydrogen flame ionization detector, as shown schematically in Fig. E-1, is designed to continuously and simultaneously analyze ambient air for methane and for total hydrocarbons minus methane. A dual flame head is closely coupled with individual electrometers whose output is directed to an electronic system to obtain continuous individual signals for the two measurements. An equal-output signal is set on each electrometer output. The sample containing hydrocarbons, including methane, is fed by a self-contained pump to a parallel inlet system. Burner 1 is fed the unaltered sample. In burner 2, the sample is passed through a cutting catalyst bed. The catalyst oxidizes all hydrocarbons except methane. More than 99 percent of all other hydrocarbons are removed, while methane passes through unaltered. Thus, one burner detects all hydrocarbons including methane, while the other burner detects only methane. By electronic subtraction, the output signals then indicate total hydrocarbons less methane through burner 1, and methane only through burner 2. The oxidizer provides clean air for both combustion air and zero air.

APPENDIX F   Reactive Hydrocarbons

The following statements are taken from the Air Pollution Control District Rules and Regulations, Chapters 2 and 3, Division 20, California State Health and Safety Code, Rule 66 (page 25d):

> For the purpose of this rule, a photochemically reactive solvent is any solvent with an aggregate of more than 20% of its total volume composed of the chemical compounds classified below or which exceeds any of the following individual percentage composition limitations, referred to the total volume of solvent:

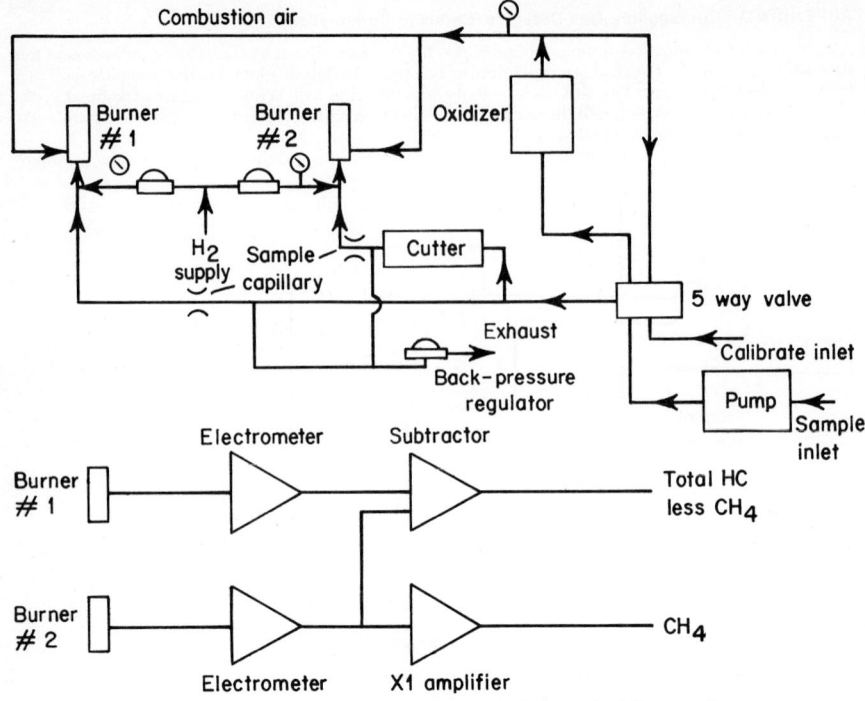

Fig. E-1. Schematic diagram of methane/nonmethane dual flame analyzer.

1. A combination of hydrocarbons, alcohols, aldehydes, esters, ethers, or ketones having an olefinic or cycloolefinic type of unsaturation: 5%

2. A combination of aromatic compounds with eight or more carbon atoms to the molecule except ehtylbenzene: 8%

3. A combination of ethylbenzene, ketones having branched hydrocarbon structures, trichloroethylene or toluene: 20%

Whenever any organic solvent or any constituent of an organic solvent may be classified from its chemical structure into more than one of the above groups of organic compounds, it shall be considered as a member of the most reactive chemical group, that is, that group having the least allowable percent of the total volume of solvents.

APPENDIX G   Hydrocarbon Limits

*Flammable (explosive) limits.** In the case of gases or vapors that form flammable mixtures with air or oxygen, there is a minimum concentration of vapors in air or oxygen below which propagation of flame does not occur on contact with a source of ignition. There is also a maximum proportion of vapor or gas in boundary-line mixtures of vapor or gas with air, which if ignited will just propagate flame. These are known as the *lower and upper flammable or explosive limits* and are usually expressed in terms of percentage by volume of gas or vapor in air.

Threshold Limit Values of Airborne Contaminants and Physical Agents.† The following statements are from the referenced document (important reading for anyone involved with TLV monitoring):

Threshold limit values (TLV) refer to airborne concentrations of substances and represent conditions under which it is believed that nearly all workers may be repeatedly exposed day after day without adverse effect. Because of wide variation in individual susceptability, however, a small percentage of workers may experience discomfort from some substances at concentrations at or below the threshold limit, and a smaller percentage may be affected more seriously by aggravation of a preexisting

---

*Fire Hazards of Flammable Liquids, Gases, Volatile Solids, NFPA No. 325M, National Fire Protection Agency, 1969.

†Threshold Limit Values of Airborne Contaminents and Physical Agents with Intended Changes Adopted by ACGIH for 1973.

TABLE H-1. Comparison of Combustible Detection Principles Used in Hydrocarbon Monitors

Parameter	Gas chromatograph	Hydrogen flame ionization detector	Thermal conductivity	Interferometer	Infrared	Semiconductor*	Catalytic combustion*
Sensitivity	Excellent	Best	Poor to good	Good	Very good	Excellent	Very good
Reliability	Good	Good	Good	Good	Good	Poor	Very good
Selectivity	Excellent	Very good‡	Poor to good	Poor‡§	Very good†	Poor§	Very good
Response time	Poor	Good	Good	Good	Good	Excellent	Excellent (1 sec)
Stability	Good	Very good	Poor to good	Good	Very good	Good	Good
Simplicity	Very complex	Medium complexity	Simple	Medium complexity	Medium complexity	Very simple	Very simple
Sample system required	Yes	Yes	Yes	Yes	Yes	No	No
Relative cost	Very high	Medium to high	Medium	Medium	Medium	Low	Low
Rangeability	ppm to 100%	ppm to low %	Wide	Wide	Very wide	Very wide	Below UEL¶
Maintenance	High	Medium	Medium	Medium	Medium	Low	Low
Auxiliary gas supplies	Yes	Yes	Sometimes	No	No	No	No

*Diffusion head type of sensor.
†Will not see hydrogen.
‡Hydrocarbons and hydrogen of opposite polarity signals.
§Responds to products of combustion.
¶UEL = upper explosive limit.

condition, or by development of an occupational illness. Threshold limit values refer to time-weighted concentrations (TWA) for a 7- or 8-hour workday and 40-hour workweek. They should be used as guides in the control of health hazards and should not be used as fine lines between safe and dangerous concentrations. TWA = time-weighted average.

A listed value bearing a "C" designation refers to a "ceiling" value that should not be exceeded; all values should fluctuate below the listed value. This, in effect, makes the "C" designation a maximal allowable concentration (MAC). In general, the basis for assigning or not assigning a "C" value rests on whether excursions of concentration above a proposed limit *for periods up to 15 minutes* may result in (1) intolerable irritation, (2) chronic, or irreversible tissue change, or (3) narcosis of sufficient degree to increase accident proneness, impair self-rescue, or materially reduce work efficiency.

APPENDIX H   Comparison of Combustible Detection Principles

In Table H-1, seven instruments used in the monitoring of hydrocarbons are compared.

Particulate Emission Measurement Methods

BY JOHN O. BURCKLE*

The measurement and characterization of particulate emissions from energy production systems are vital links to successfully controlling process emissions. Instrumentation is needed for support of research and development activities, process monitoring, and acceptance and compliance testing. Each application will have performance criteria reflecting the end use of the data and the conditions under which the instrument must operate. While specifications for an actual source require a definitive study of that source, general guidelines can be derived from a consideration of the ranges of particulate and source characteristics, the problems relating to sampling low concentrations in moving gas streams, and a review of the operating principles of available measuring techniques.

SOURCES AND TYPES OF PARTICULATES

Atmospheric particulate matter is known to arise from several origins: as particles emitted from both natural and anthropogenic sources, particles formed by condensation upon cooling of hot gases in the ambient air, particles formed by evaporation of liquid from droplets containing dissolved matter, and particles formed in the ambient air through reactions converting gases to particulate matter (secondary particles).

Multimodal Character of Atmospheric Aerosols. Recent investigations of the particle size distribution of atmospheric aerosols (Ref. 1) have revealed a multimodal character (Fig. 1), usually with a bimodal mass, volume, or surface area distribution and frequently trimodal surface area distribution near sources of fresh combustion aerosols. These modes are attributed to the following factors:

1. The course mode, 2 micrometers (μm) and greater, is formed by relatively large particles generated mechanically or by evaporation of liquid from droplets containing dissolved substances.

2. The nuclei mode, 0.03 μm and smaller, is formed by condensation of vapors from high-temperature processes or by gaseous reactions products.

3. The intermediate, or "accumulation" mode, 0.1 to 1.0 μm, is formed by coagulation of nuclei.

This evidence indicates that atmospheric particles tend to form a stable aerosol having a size distribution ranging from about 0.1 to 1.0 μm. The larger settleable particles (greater than 1.0 μm in size) fall out, and the very fine particles (smaller than 0.1 μm) tend to agglomerate to form larger particles which remain suspended. The nuclei mode tends to be highly transient and is concentration-limited by coagulation with both other nuclei and also particles in the accumulation mode. It appears that further growth of particle size from the accumulation mode to the coarse mode is limited to 5 percent or less (by mass). Therefore, the particulate content of a source emission and the ambient air can be viewed as composed of two portions, i.e., settleable and suspended.

Control Efficiency/Particle Size Relationship. Control of emissions in both size ranges is required because both settleable and suspended atmospheric particulates have deleterious effects upon the environment. Significantly, it is the suspended particles from an upper level of about 2 to 5 μm and smaller that health experts consider most harmful to human health because particles of this size range have been found to penetrate the body's natural defense mechanisms and reach most deeply into the lungs. Efforts to control particulate emissions to the atmosphere have historically been geared to maximizing the efficiency of control (by weight) of the overall particulate loading emanating from the generating process. This work has lead to the empirical understanding that present systems can perform with high control efficiencies down to a particle size of about 2 to 3

*Research Chemical Engineer, Office of Research and Development, U.S. Environmental Protection Agency, Environmental Research Center, Cincinnati, Ohio.

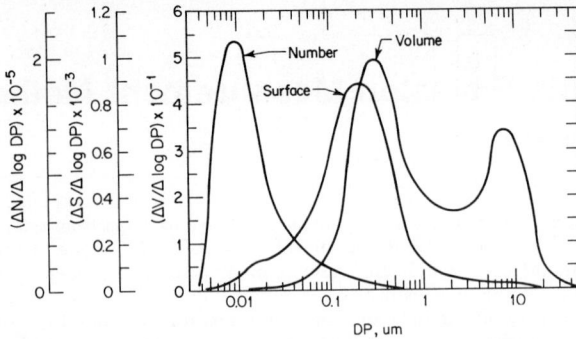

Fig. 1. Normalized frequency plots of number, surface, and volume distributions for the 1969 grand average Los Angeles smog aerosol. Note the bimodal distribution of volume and the fact that each weighting shows features not indicated by other curves. From Ref. 1.

μm, but, below this size, the control efficiency appears to decrease with decreasing particle size to a minimum between 1.0 and 0.1 μm and then increases again (Ref. 2). See Fig. 2.

This relationship of control efficiency and particle size is highly significant to any strategy for controlling particulate air pollution, and serves to underscore the need to adequately measure and evaluate both ambient particulate air pollution and source emissions.

MEASUREMENT METHODS

Particulate measurement methods are used for the measurement of total aerosol concentration, aerosol particle size distribution, and chemical and other physical particle characteristics. The accuracy of aerosol measurements is governed by these major factors: (1) sampling, (2) quantitative collection and assay (manual), or (3) analysis with assay (instrumental).

The most significant factors influencing accurate measurements of particulate emissions are the physical properties of the particles composing the aerosol, their aerodynamic behavior in the effluent gas, the spatial and temporal variations in the aerosol flux and particle size distribution, and the method and equipment applied for sampling, collection or analysis, and assay. The aspects of sampling and analysis to determine total mass emissions and size distribution are discussed in the following sections. Methods for chemical characterization are briefly summarized (Ref. 3) in Fig. 3. Detailed methods for chemical and physical characterization are available in the open literature.

Sampling. The volumetric flow rate of the effluent from a stationary source is much too large to allow collection of the total gas stream for measurement, and some techniques must be used to obtain a representative fraction of the total stream for analysis. Sampling is the action of obtaining a small portion of the aerosol emission which truly represents the total emission over the sampling cross section and sampling period. Successful sampling

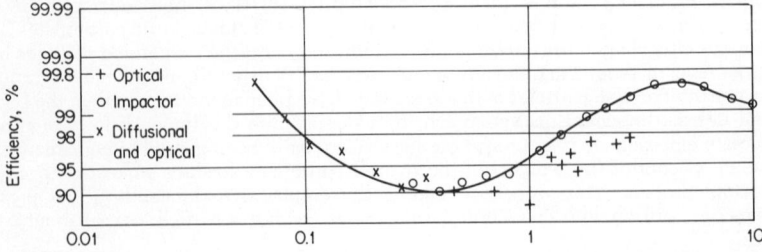

Fig. 2. Typical curve for the fractional efficiency of an electrostatic precipitator. From Ref. 2.

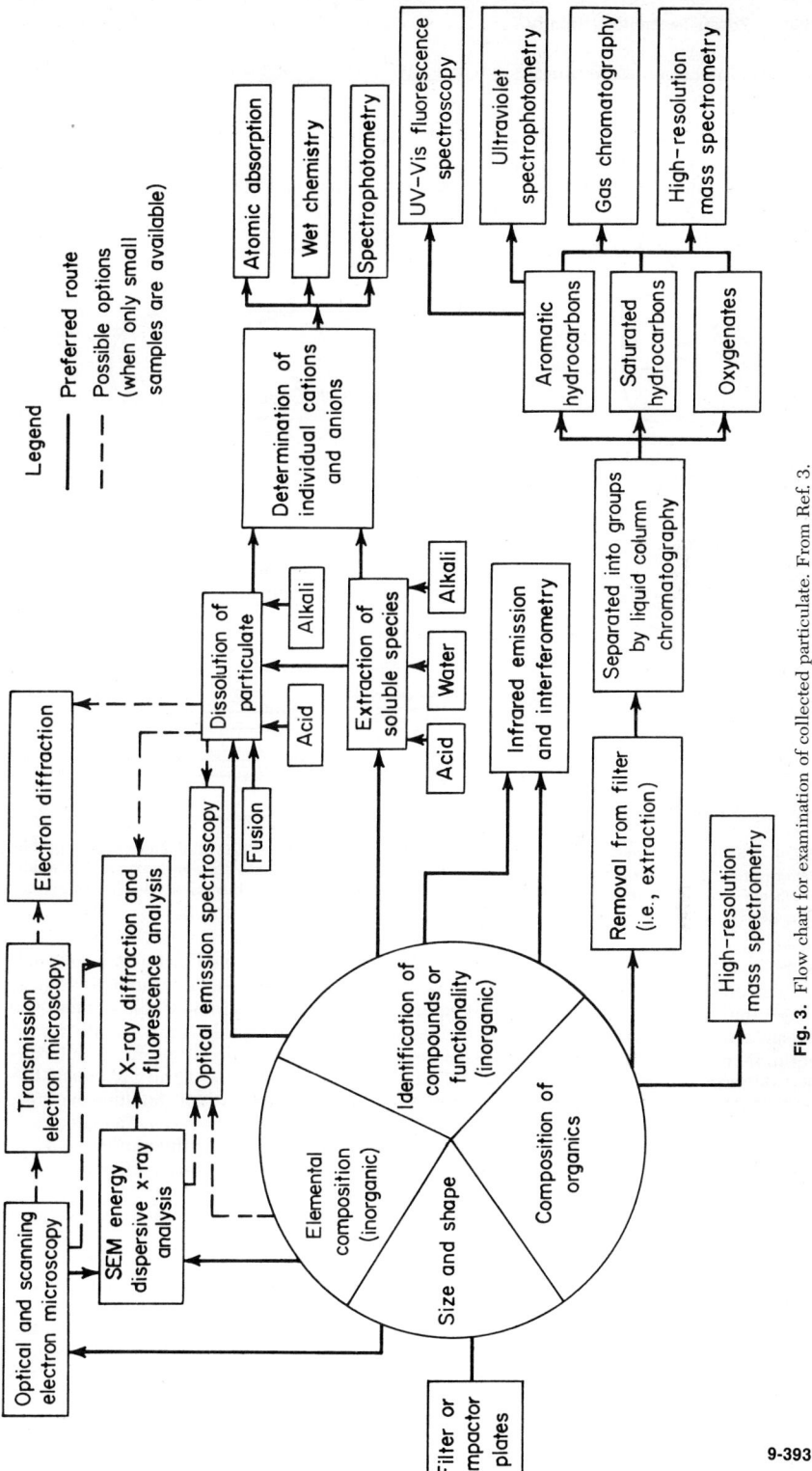

Fig. 3. Flow chart for examination of collected particulate. From Ref. 3.

9-393

must take into account the process dynamics and flow characteristics of the aerosol at the sampling point.

If a representative sample is to be obtained, flow behavior of the gross aerosol in the source must be carefully considered. Actual particle flow in turbulent source streams has not been extensively studied for the dilute case encountered in stack emissions, but some conclusions can be drawn from existing data for pneumatic transport and studies of single-

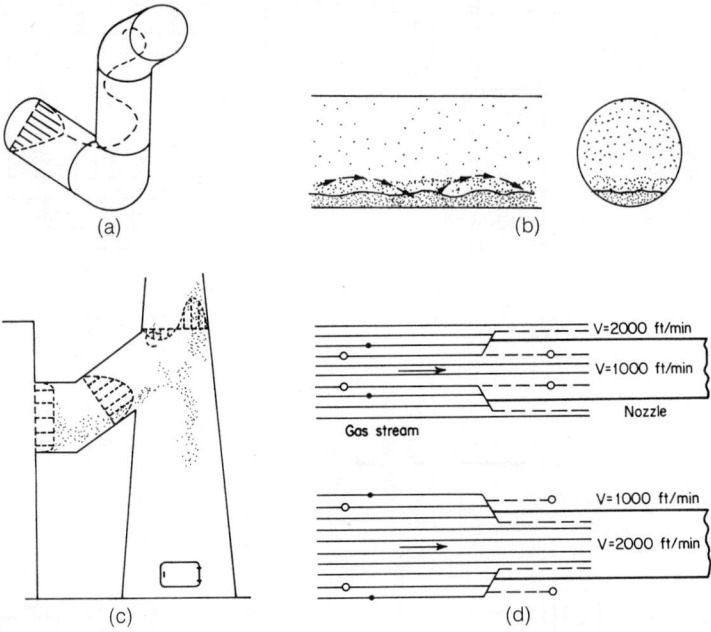

Fig. 4. Illustration of aerosol flow behavior. From Ref. 5. (*a*) Particle segregation and velocity profile distortion in a typical boiler exhaust duct and stack. (*b*) Two views of settling segregation and saltation of particles in a long horizontal duct. (*c*) Example of angular flow induced by multiple change of duct direction. (*d*) Effect of non-isokinetic sampling.

particle motion (Ref. 4). The motion of a single particle is determined by the gravitational, inertial, and aerodynamic forces exerted upon it and the characteristics of the particle size, shape, and density. In the normal configuration of industrial equipment, the effluent gas stream encounters many directional changes in passing through the process and control equipment, and, as a result of these changes, the gas stream velocity profile is substantially skewed (Ref. 5) as shown in Fig. 4. During directional changes of the gas stream, the particles encounter a variety of forces of varying magnitude which, on the average, accelerate them toward the outside of any turns. The trajectory of a given particle depends upon the point of entry of the particle into the turn, the velocities of the particle and the gas stream, and the aerodynamic properties of the particle. Those particles having diffusivities of the same order of magnitude as the gas eddies will continue to follow the gas flow lines. Particles with smaller diffusivities, however, will tend to leave the main gas profile and become concentrated toward the outside of the turn.

An additional factor that enters into the problem of particulate maldistribution is nonuniformity of total gas flow. For "steady-state" boiler operation, changes in operating conditions cause gas temperature and volume to cycle around a mean value with periods ranging from 5 to 30 minutes, and larger changes result from changes in process throughput (Ref. 6). These variations in total gas flow have the effect of shifting particulate flow profiles within the duct by altering the gas velocity and hence particle distribution at the measurement point. Because of the complex interactions, particle flow distribution cannot be accurately predicted for practical situations, but it is clear that the aerosol is not well mixed and homogeneous at or near a directional change. The homogeneity will tend to become reestablished downstream of directional changes by the action of turbulent transport, but the mixing length required is not well established. Many manual sampling procedures recommend 8 to 10 duct diameters as a minimum distance. Few large ducts run such distances without further directional change or obstructions, and the gas velocity profiles obtained during many source-sampling studies over the years would indicate that well-mixed streams should not be anticipated in stationary sources.

One of two basic sampling approaches can be applied, depending upon the measure-

Extraction sampling:

  1. Manual measurement:

      – Point sampling

      – Gross composite sample for total mass or size distribution with filter

      – Real-time size classification with inertial sizing device

  2. Continuous monitoring:

      – Point sampling

      – Total mass with sensor only

      – Combination of inertial sizing and sensing sensing devices gives size distribution monitoring

      – Single-particle analyzer gives size distribution

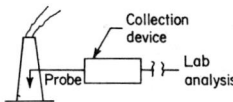

In-situ sampling:

  1. Point technique

      – Continuous with on-site analysis

      – Discrete with laboratory analysis

  2. Integrated-path technique:

      – Continuous with on-site analysis

      – Discrete with laboratory analysis

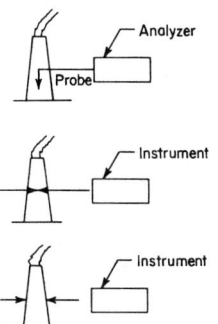

Fig. 5. Sampling approaches for stationary sources.

ment to be performed: (1) direct measurements in the gas stream (in-situ) or (2) extraction of a small portion of the effluent stream followed by external analysis. See Fig. 5. Both techniques have certain advantages and disadvantages when applied to particulate mass measurements. Extraction of a sample from a flowing stream has been utilized extensively for air pollution source emission measurements. Basically, the sample is aspirated from the emission stream through a probe and subsequently analyzed for the particulate and gas components. This technique normally used for manual, time-average measurements can also be utilized for monitoring by incorporating a sensor capable of instantaneous measurement of total particulate loading or size distribution.

Extraction sampling involves several factors which limit its usefulness. First, nonrepresentative sampling bias is introduced when the distribution of particulate flux across the duct is not uniform in time or space, or both. (Ref. 7) In addition, nonrepresentative sampling bias (sample modification) is introduced when the sample entering the probe nozzle is not the same composition (particle/gas ratio) as that in the bulk emission stream. To prevent the occurrence of such a bias, it is necessary to sample isokinetically (Refs. 8–10) when the particle properties are such that the particles do not follow the gas flow lines. See Fig. 4d. Sampling bias may also be introduced by the configuration and orientation of the nozzle. Such effects are documented to varying extents in the literature, although no conclusive data have been presented pertaining to nozzle size. Modification of the composition of the sample by chemical interactions of the gas, particulate, and sample train components can occur, but little is known of the magnitude or mechanisms (Refs. 3, 11). Another aspect of sample bias which has an influence on continuous particulate instruments and size distribution measurement is deposition and reentrainment of particulate in the probe (Ref. 12). Investigation of these phenomena has begun only recently since they had little significance in manual measurements for total mass concentration (where probe cleaning was practiced).

In-situ techniques utilize an intrinsic property of the gas/particulate stream, which is related to the total particulate concentration. Such systems require only an energy source which transmits a signal into the duct and a sensor reponsive to changes (reflected or transmitted) in intensity of the incident energy. One of the major advantages of the transmission approach is that it provides an integrated measurement over the entire path length. While these techniques avoid many of the problems associated with extracted sampling, they are not fully satisfactory. The problems of particle flux distribution is not

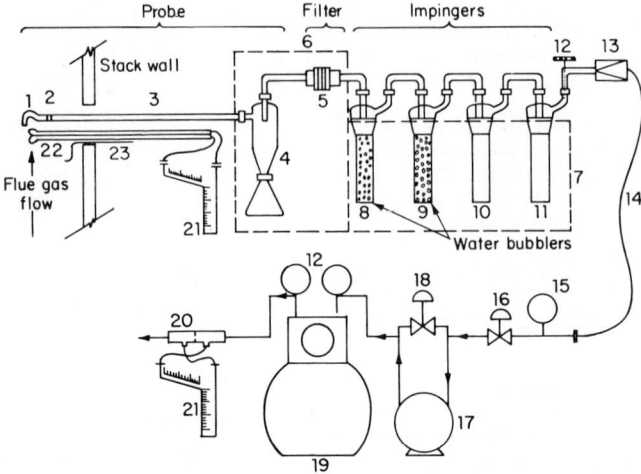

Fig. 6. EPA particulate sampling train. From Ref. 11. Principal parts include: (1) Stainless steel, buttonhook-type probe tip. (2) Stainless steel coupling. (3) Probe body (⅝-inch O.D., medium-wall Pyrex tube logarithmically wound with 25 feet of 26-gage nickel-chromium wire). (4) Cyclone and flask (optional). (5) Fritted-glass filter holder. (6) Electrically heated enclosed box. (7) Ice bath containing four impingers connected in series. (8) The Greenburg-Smith-type impinger with tip removed. (9) Second impinger with tip. (10) Third impinger with tip removed. (11) Fourth impinger with tip removed and containing approximately 175 grams of accurately weighed dry silica gel. (12) Pressure gage. (13) Check valve. (14) Flexible rubber vacuum tubing. (15) Vacuum gage. (16) Needle valve. (17) Leakless vacuum pump. (18) Bypass valve. (19) Dry-gas meter. (20) Calibrated orifice. (21) Draft gage. (22) S-type pitot tube. (23) Thermocouple or other device for stack temperature measurement.

entirely solved by the application of an across-the-stack measurement because the sample size is still small. For example, a 4-inch energy beam projected across a 10-foot duct will "see" only 4 percent of the area in the measurement plane. Maldistribution of particulate flow could easily distort the measurement, especially if the energy beam were poorly oriented with respect to the maldistribution.

Manual Methods. The first documented method to measure particulate emissions from stationary sources was developed by Maximillian Ringelmann in the 1890s. While this approach still enjoys a certain popularity because of its low cost and simplicity, it is not sufficiently accurate to permit evaluation of the actual mass of particulate in the gas stream. The need for more quantitative measurements to determine the efficiency of combustion and dust control equipment led to the development, during the 1930s and 1940s, of several gravimetric sampling procedures. The most familiar of these include the ASME PTC-27 by the American Society of Mechanical Engineers, the IGCI Publication No. 101 by the Industrial Gas Cleaning Institute, and Bulletin WP-50 by the Western Precipitation Company. These procedures are based upon the aspiration of a gas sample from the effluent stream, filtration, weighing of the filter catch, and measurement or calculation of the total sample gas volume. A more recent adaptation of these methods was promulgated by the U.S. Environmental Protection Agency as the test method for measuring particulate emissions from various stationary sources subject to federal regulation (Refs. 11, 13). See Fig. 6.

Increasing efforts to define the extent of air pollution and develop control technology for particulate emissions have made evident the many disadvantages of applying manual methods to routine measurement on a major scale. The most significant of these are the high cost of performing a series of tests and the inability of the techniques to provide a continuous record of instantaneous emission rates. Since procedures are not continuous or instantaneous, instrumentation capable of monitoring particulate mass flow and particle size distribution in an effluent gas stream is needed.

PARTICULATE INSTRUMENTATION

The accuracy of a particulate mass monitor is ultimately a function of the representativeness of the sample and the degree to which the sensor responds to true mass. The most significant factors influencing both areas are the properties of the particulate matter or, more precisely, the net effect of the properties of many minute, discrete particles. For purposes of measurement, there are six basic particle properties of importance as listed in Table 1. These properties may affect the measurement in any one or all of three major areas: particulate flow, sampling, and analysis. The significance of the "entries" is the number of particle properties that can influence the accuracy of measurements and the occurrence of effects on measurement parameters.

While the individual particle properties are important, it is the wide variations in the ranges of various properties that are of most concern in the application of continuous monitors. If all particles had the same set of properties—that is, if the particulate were nearly homogeneous—appropriate compensations for the effects could be accomplished through calibrations against actual mass measurements, and the response of the system would be an accurate measure of mass thereafter. Unfortunately, there are few cases where the particulate in a stationary source effluent is homogeneous and stable as exemplified by the data (Table 2) for size and density (Refs. 6, 14, 15), two very important properties which affect mass measurements. It is apparent from the data that these properties vary considerably, even within the same type of process. The implication may be drawn, then, that the net properties of a mixture of particles are the result of the interrelationships between various parameters derived from the combination of the individual particles. A rigorous proof of the exact effect of any given property for a mixture of different particles is not usually possible. Overall, variations in particle properties can be related to the process feedstock, design, and operation and the location of the "measurement" or "sampling point."

A difference in the process raw material has an obvious effect on the chemical composition and color of particulate emissions. The properties of the feed material can also have an effect on the particulate even when firing a given type of fuel, as demonstrated by the close correlation of the chemical composition of the particulate with the mineral components of coal under conditions of complete combustion (Ref. 14). The mineral constituents in coal vary considerably from mine to mine, and even within the same mine, and particulate properties vary similarly when the coal is burned. Other studies have shown that the particulate emissions vary with the size distribution of the coal feed. For example, the finer particles of coal fed by spreader stokers are entrained and carried out by the combustion gases. This particulate is relatively high in carbon content, a factor which affects not only color and chemical composition but also particle density.

The effects of process design (firing method) on the size distributions from coal-fired

TABLE 1.   Measurement Parameters Affected by Particle Properties*

Particle property	Measurement parameter		
	Particulate flow	Sampling	Analysis
Size	+	+	+
Shape	+	+	+
Density	+	+	+
Color	O	O	+
Resistivity	O	O	+
Composition	O	O	+

*Particle properties that may have a significant effect on measurement accuracy; + indicates a major influence is to be expected, while O is indicative of little or no effect.

TABLE 2. Particulate Properties of Selected Sources

Source	Concentration (uncontrolled), grains/standard cubic foot (Ref. 6)	Size distribution: Weight percent less than* (Ref. 14)				Density, grams/cubic centimeter (Ref. 15)
		10 μm	50 μm	100 μm		
Coal-fired power plants:						
Pulverized	2–15	5–70 (30)	45–90 (75)	70–98 (90)		0.6 carbon
Cyclone	0.4–4	60–90 (75)	80–99 (91)	88–99 (95)		2.5 fused ash
Stoker, Spreader	0.2–10	8–11 (10)	20–60 (40)	25–80 (60)		5.0 magnetite
Oil-fired power plants	0.02–1	(90 < 1)				2.5

Densities of particle-size fractions for pulverized-coal-fired emissions (Ref. 13)

Particle-size fraction, μm	Weight percent in fraction	Density, grams/cubic centimeter
<44	78	1.78
44 to 74	10	1.70
74 to 149	8.3	1.60
Total sample	100	1.75

*Mean values are listed in parentheses.

units (Table 2) are the result of particle size of the coal fed to the boiler, different gas velocities, temperatures, and residence times in the furnace. The effects also extend to density, color, and chemical composition because the time-temperature relationships influence the burnout of carbon and the fusion of the mineral components. As noted previously, for example, the carbon content of particulate from spreader-stoker units is relatively high, a situation which also results in significant differences in the properties of the particulate from this type unit compared to fly ash resulting from other firing methods (Ref. 15).

The effects of process operation are perhaps of greater significance to monitor design than feed or process design because they occur in a single source, and any variations in particulate properties must be handled by a single measurement technique. The most obvious example of such variations is apparent to anyone who has observed stack emissions from a combustion source during boiler load changes or "soot blowing." Particles emitted from a fuel-rich oil-burning source are shiny black spheres of oil and partially oxidized fuel. During soot blowing, a large number of "lacy," brownish-black, glassy particles are present, whereas during normal operations, the particulate is a mixture of grayish ash, black carbon, and acid mist (Ref. 15).

Changes in particulate properties as a function of location in the processing equipment can be caused by many factors, including temperature or pressure changes, continuing reactions, ductwork configuration, and control equipment. Condensables in the effluent stream can result in additional particulate loading as across an induced-draft fan. The effect of control equipment on reducing the size distribution range is evident in the efficiency curves for the various types of devices utilized for stationary particulate control. However, it is also possible for control devices to alter other properties of the particulate. Tests of a cyclone collector on a stoker-fired power plant showed that the unit was 85 percent efficient for total particulate but only 47 percent efficient for collection of the trace metals (Ref. 16).

Because of the wide range of physical and chemical characteristics of the particulate in stationary source effluent gases, the most desirable principles for application are those which respond directly to mass. As can be seen from the display in Table 3, there is no sensing principle that can be used for obtaining indirect measures of mass which are not subject to responses caused by variations in particle properties other than mass. A brief discussion of each principle and the problems involved in its application to in-situ or extraction-type sampling follows.

For in-situ measurements, the attentuation of visible light, gamma rays, and acoustical energy has been investigated. The most widely attempted techniques have employed the attenuation of electromagnetic energy in the visible portion of the spectrum (Refs. 17–19). See Fig. 7. Instrument response is a function of particle concentration, path length, and the attenuation factor for the particle. Two problems associated with this type device are the attenuation factor—which is a function of particle size, shape, retractive index, surface characteristics and orientation, and the wavelength of the illuminating source—and the measurement of a concentration related to the number of particles rather than the mass (Ref. 6). The low cost, simplicity of operation, and reliability of the components are

TABLE 3.   Sensor Principles Affected by Particle Properties*

Particle property	Mass	Light	Gamma	Beta	Electrical	Acoustical
Size	+†	+	○	+†	+	+
Shape	○	+	○	○	+	+
Density	○	+	○	○	+	+
Color	○	+	○	○	○	○
Resistivity	○	○	○	○	+	+
Composition	○	○	+	○	○	○

*Particle properties that may have a significant effect on mass measurement accuracy; + indicates a major influence is to be expected, while ○ is indicative of little or no effect.

†Size affects such measurements only in that the quantitative collection on a filter is necessary, that is, 100 percent filterability; note that particles between about 0.5 and 0.05 μm are not 100 percent filterable, and some errors may result from filter penetration.

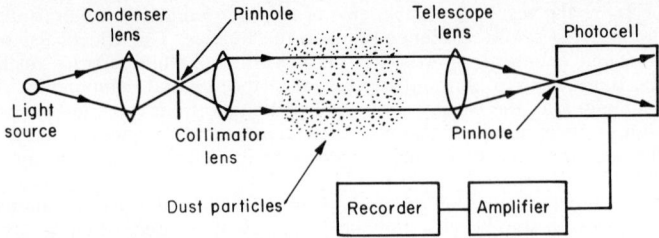

Fig. 7. Light-transmission method. From Ref. 22.

distinct advantages, however, and this approach can be useful where the particulate emission has constant characteristics.

A similar approach utilized gamma radiation from a nuclear source and a Geiger-tube detector (Ref. 20). Absorption of the energy is again a function of path length, concentration, and absorption factor. Interaction of the energy with the particle is on the atomic level, and the absorption is related to the number of atoms in the beam and the chemical composition of the particles. The equipment required for this type instrumentation is somewhat complex and expensive, and the sensitivity to the particulate loadings of effluent streams is not high. However, insensitivity to many of the variables in effluent streams makes the approach seem attractive where chemical composition is reasonably uniform.

A third technique which has been investigated for in-situ measurements is the absorption of acoustical energy (Refs. 6, 21). The technique has many of the same problems as visible radiation absorption since the interaction is with the particle and is influenced by individual particle characteristics. Transmission of the energy also involves the gas medium, and changes in gas density will influence overall response. Sensitivity to source concentrations is not good, and the approach has not been exploited.

A number of approaches have been used for the continuous analysis of particulate extracted from the main gas stream (Refs. 6, 22). Several of these are a direct measure of the mass collected, and the primary problems affecting the measurement are related to representative sampling and quantitative collection upon a sensor surface. A system based on an electrobalance has been used as a direct measure of mass. The extracted particulate is deposited on a collection media which is then automatically cut off, transferred to the balance, and weighed. A second technique is based on collection of the extracted particulate on an oscillating quartz crystal. See Fig. 8. The frequency of oscillation is a function of the mass of material deposited on the surface. The most widely investigated extraction approach is based on collection of the particulate on a filter media and detection

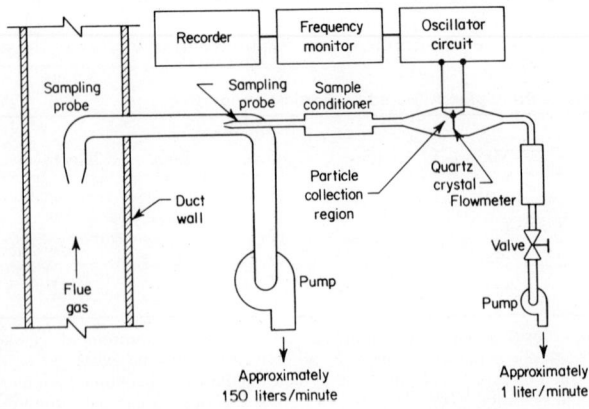

Fig. 8. Piezoelectric microbalance. From Ref. 22.

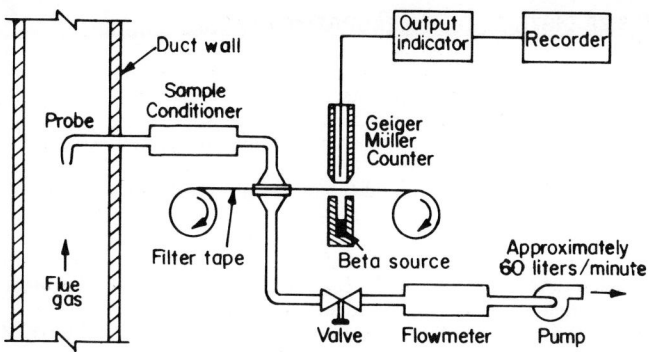

Fig. 9. Beta monitor. From Ref. 22.

of the mass through absorption of low-energy beta radiation (Refs. 6, 22 to 24). See Fig. 9. Absorption of the energy is a function of the total mass collected and the chemical composition of the sample. Theory predicts a very small effect due to composition, and in practice this has been reasonably demonstrated.

Indirect measurements based on electrical devices, light absorbance, and light reflectance have been employed. The electrical devices measure current transferred by the particles from a charging zone to discharge collection plates (Refs. 6, 25, 26). Sensors of this type assume that the charge placed on a particle is a function of particle mass. Significant errors can result since the total charge acquired by each particle is related to its size and electrical properties (resistivity). Light scattering, absorbance, and reflectance have also been applied to measurements of extracted samples (Refs. 6, 26). For absorbance or reflectance, the particulate matter is deposited on a filter tape and is read after deposition is completed. For scattering, the aerosol is "collimated" so that it passes one particle at a time through an optical view volume where the scattered light is measured. See Fig. 10. The scattering patterns are related through light-scattering theory to the particle size and mass calculated therefrom; these operations are normally performed automatically by the more sophisticated instruments. These optical approaches have the same dependence upon the optical properties of the particle as the in-situ devices; that is, response is a function of size, shape, and color, and additionally a uniform density must be assumed to calculate mass from size.

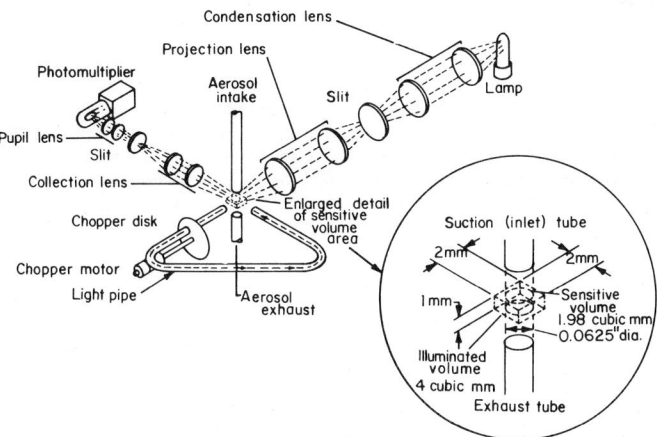

Fig. 10. Operating principle of optical particle counter which senses light scattered at 90 degrees from the incident beam. From Ref. 6. (*Royco Instruments, Inc.*)

PARTICLE SIZE DISTRIBUTION MEASUREMENTS

The acquisition of information regarding the nature of a particle size distribution is dependent upon three principal elements:
1. Analytical model used to display the distribution
2. Analytical method of expressing, or "weighting," the particle concentration within each size interval
3. Definition of the particle size and the technique used to measure the particle size characteristic

Particle size distributions are displayed graphically as a relation of particle concentration to particle size. There are a number of analytical models for graphic representation which have been derived, and each has certain useful properties and provides different types of information regarding the size distribution. In addition, there are a number of ways to express the quantity of particulate matter within each size interval. Commonly used concentration weightings include number, surface area, volume, or mass. It is not always clear which way of expressing size distribution is most advantageous in any given situation. Each has its limitations. A mass distribution generally emphasizes large particles and deemphasizes or ignores small particles which are important but do not contain significant mass compared to the large particles. A number distribution, on the other hand, emphasizes small particles which are present in large numbers. A number distribution ignores the one or two large particles in a cloud of millions of small ones, even though the one or two large particles may weigh more than the millions of small particles. A concentration distribution shows the actual particle concentration level, an important factor determining particle interaction rates and an essential factor for pollution control monitoring (Ref. 27).

The concentration weighting can be expressed as a percentage of the total which is contained within each size category rather than in terms of the actual particle concentration in numbers, milligrams, square centimeters, or cubic centimeters of particles per unit air volume. For example, in the case of a particle mass percentage weighting, the result would be a graph showing the percentage of the mass of particles contained within each size category, or more commonly, the percentage of the mass of particle contained by particles larger than (or smaller than) a given size. The percentage distribution, as well as several other distributions, shows quickly, clearly, and without ambiguity exactly what size range contains the majority of the particles. Although one can usually convert mathematically from one distribution to another, any measurement error becomes magnified with each conversion. One cannot convert from a percentage distribution to a concentration distribution without knowing the total concentration of the particle sample (Ref. 27).

Additionally, there are a number of different measures of particle size, such as inertia, terminal-settling velocity, a characteristic physical dimension, surface area, or volume, each of which is associated with particular measurement techniques. The various methods of analysis measure a "particle size" based on one of these size-discriminating techniques. As a result, the same sample, when subjected to a number of analytical methods, will yield a unique distribution, some of which will be significantly different, for each. Therefore when comparability of test results between studies is an objective, the method of analysis becomes a significant factor which can prevent useful comparisons of data.

Therefore there are many ways to display a particle size distribution, each resulting in a curve with a unique shape depending upon which combination of the three principal elements is used. The investigator who desires to measure the particle size distribution of an aerosol must decide first what type of analytical model and concentration measurement is desired. This decision fixes the choice of the particle size characteristic to be measured. The final step is, then, selection of the appropriate sampling/analytical technique and equipment most suitable to the satisfaction of the measurement problem. The selection of the method and equipment should be based upon obtaining the most direct measurement of the selected size characteristic and must also take into account the effective range of the method, the operating environment (temperature, pressures, concentration, corrosiveness, etc.), and the sampling factors discussed previously.

There are a number of techniques for performing particle size measurements which have been reduced to practice. These are reviewed by Lapple (1971) (Ref. 28), Carver

(Ref. 29), and Sem et al. (1972) (Ref. 27), among others. The accuracy of the measurement is limited by the sampling operation, i.e., the degree to which the sample obtained represents the aerosol as it exists in its dispersed form; classification, i.e., the degree to which each particle sampled is accurately classed into its appropriate size interval; and assay, i.e., the method used to quantify the amount of particles in each size interval. Selection of a particular technique fixes the measurement strategy which comprises one of the combinations of sampling, classification, and assay described below.

1. A gross composite sample is obtained by filtration. The sample is removed to the laboratory, redispersed, and analyzed by one of the numerous techniques available. The

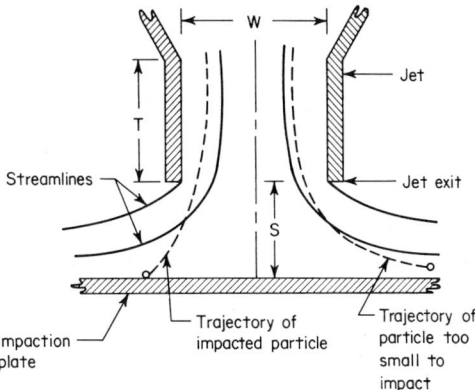

Fig. 11. Impactor stage: W = jet diameter. T = jet length. S = jet/plate clearance. From Ref. 27.

greatest shortcoming of this strategy is the inability to ensure that the redispersed particles are those present in the original aerosol since the size distribution is highly subject to modification in the bulk sampling and redispersion operations.

2. Real-time classification. The particles in the aerosol sample are classified into the appropriate size interval, in real-time, as the sample is obtained. This strategy yields a separate composite sample for each size interval which is assayed with a suitable laboratory method not requiring redispersion. The principal measurement system applicable to this strategy is the impactor (Fig. 11) or cyclone. The classification takes place so that the particles are deposited directly from the aerosol onto the proper stage, thus eliminating errors of redispersing a bulk sample.

3. Sample collection with real-time classification and assay. The sample is analyzed (classified and assayed) in a continuous operation to periodically provide an output of the size distribution. At this time the only instruments of this type available are based on the optical single-particle counter. See Fig. 10. Adaptation of the suitable sensing system to an aerodynamic classifier, shown conceptually in Fig. 12, would provide a much needed instrument for continuous measurement of aerodynamic particle size (Refs. 27, 30, 31). A system for measurement of particle size below 0.1 μm has been devised (Fig. 13) and applied to stack emission measurements to provide size distribution information for both collector inlet and outlet aerosol conditions (Ref. 32).

4. In-situ sampling with real-time classification followed by laboratory assay. A signal providing a record of the particles in the aerosol is obtained by the interaction of energy with the aerosol. This record, containing information on a large number of particles in the aerosol field of view, is subsequently analyzed to extract the desired information regarding the particle size distribution, number density, and degree of dispersion. Laser holography has been successfully applied (with limitations) to make such measurements (Refs. 6, 27, 33, 34). However laser holography is limited by the same operating principles as the more conventional optical techniques.

5. In-situ sampling with real-time classification and assay. In a manner similar to that employed in the previous strategy, the signal resulting from the interaction of energy with the particles in the aerosol are obtained and processed electronically in real time. Laser instrument systems (the laser doppler velocimeter) have been used for this type of

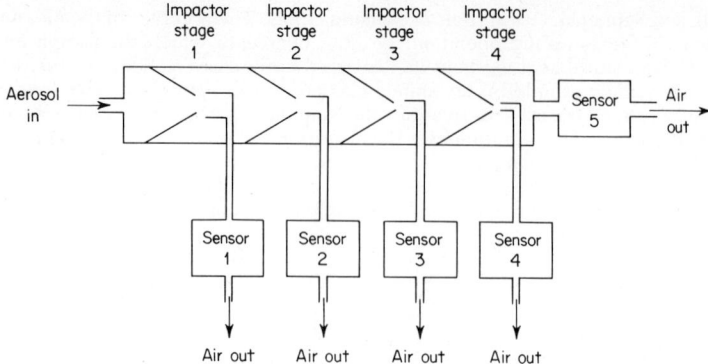

Fig. 12. Apparent-surface cascade impactor. From Ref. 27.

measurement (Refs. 35, 36). However, the LDV sets up a small view volume in space (i.e., remotely within the aerosol) so that the signal is generated from individual particles rather than an entire portion of the aerosol. This method thus is analogous to the optical single-particle counter and is limited by the same principles of operation.

A review of several conceptual approaches for an automatic, continuous particle size distribution monitor (Ref. 27) is presented in Table 4. These approaches are based upon the combination of size classification or discrimination techniques with appropriate methods for quantifying the amount of particles within each size range.

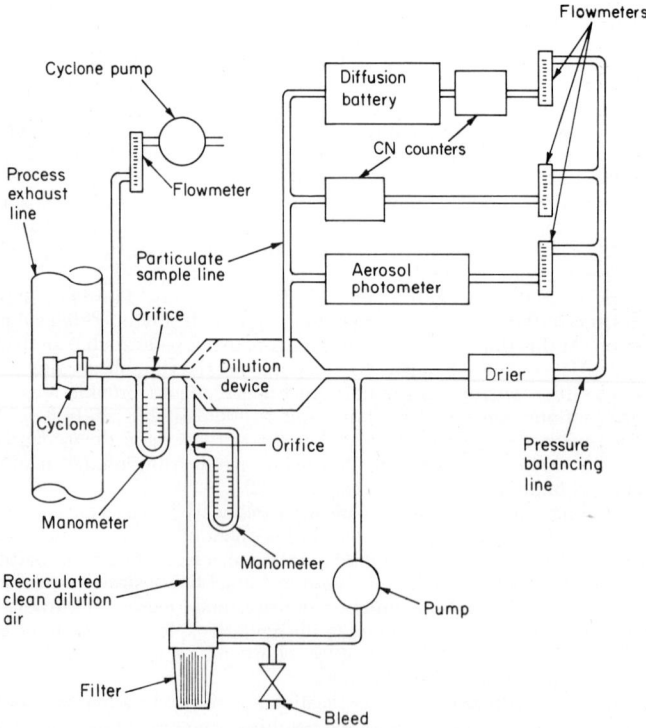

Fig. 13. Optical and diffusion sizing system. From Ref. 32.

TABLE 4. Operable Size Ranges of Practical Combinations of Size Classifiers and Sensors

[All numbers are the diameter (μm) of a spherical particle with a density of 1 gram/cubic centimeter. Data from Ref. 6.]

Particle size classifier	Concentration-sensing technique						
	Mass		Optical		Soiling potential 0.2–50	Electrostatic ion capture and attenuation 0.005–100	Nuclei counter 0.001–0.05
	Beta radiation attenuation 0.01–100	Piezoelectric quartz crystal 0.01–20	Light transmission 0.2–100	Photometry 0.2–50			
Aerodynamic:							
Impactor 0.2*–30	0.2†–30	0.2†–20	0.2–30	0.2–30	0.2–30	0.2–30	
Cyclone 0.5–30	0.5†–30	0.5†–20	0.5–30	0.5–30	0.5–30	0.5–30	
Gravitational elutriator 1.0–100	1.0†–100	1.0†–20	1.0–100	1.0–50	1.0–50	1.0–100	
Gravitational sedimentation 1.0–50	1.0†–50	1.0†–20	1.0–50	1.0–50	1.0–50	1.0–50	
Electrostatic 0.005–0.6						0.005–0.6	0.005–0.05
Brownian diffusion 0.001–0.05							0.001–0.05

*Low-pressure impactor may prove useful down to 0.05 μm for some applications where volatile particles are not present.

†Although this is the lowest size cutoff of the classifier, the sensor can detect smaller particles below this size cutoff, lumping them into one size range.

REFERENCES

1. Whitby, K. T.: On the Multimodal Nature of Atmospheric Aerosol Size Distributions, *Proc. VIII Int. Conf. Nucleation*, Leningrad, USSR, Sept. 28, 1973.
2. Nichols, G. B.: Theoretical and Practical Aspects of Fine Particulate Collection by Electrostatic Precipitators, *Proc. Symp. Control Fine Particulate Emissions Ind. Sources*, Joint US–USSR Working Group, San Francisco, January 1974 (USEPA).
3. Funkhouser, J. T., et al.: "Manual Methods for Sampling and Analysis of Particulate Emissions from Municipal Incinerators," September 1973 (USEPA-650/2-73-023).
4. Soo, S. L., et al.: *Ind. Eng. Chem. Fundam.*, 3: 97 (1963).
5. Engdahl, R. B.: Measurement of Fine Particles Suspended in Exhaust Gases, *Proc. Symp. Source Sampling Atmos. Contaminants*, H. G. McAdie (ed.), Toronto Section of the Chemical Institute of Canada, Feb. 23, 1971.
6. Sem, G. J., et al.: State of the Art: 1971—Instrumentation for Measurement of Particulate Emissions from Combustion Sources, Vol. II; Particulate Mass-Detail Report; USEPA/APTD 0734; NTIS/PB 202-666.
7. Hawksley, P. G. W., S. Badzioch, and J. H. Blackett: "Measurement of Solids in Flue Gases," British Coal Utilization Research Association, Leatherhead, Surrey, England (1961).
8. Vitols, V.: Ph.D. thesis, University of Michigan (1964).
9. Badzioch, J.: *Jl. Inst. Fuels*, 33: 106 (1960).
10. Hemeon, W. C., and G. G. Haines: *Air Repair*, 3: 159 (1954).
11. Hillenbrand, L. J., R. B. Engdahl, and R. E. Barrett: Chemical Composition of Particulate Air Pollutants from Fossil-Fuel Combustion Sources, March 1973, USEPA: T2-73-216; NTIS: PB219-009.
12. Sehmel, G. H.: *Proc. Int. At. Energy Symp.*, Vienna (July 1967).
13. Smith, W. S., et al.: Stack Gas Sampling Improved and Simplified with New Equipment, *APCA Paper No. 67-119, 60th Annual Meeting of the APCA*, Cleveland, Ohio, June 11–16, 1967.
14. Smith, W. S. and C. W. Gruber: Atmospheric Emissions from Coal Combustion—An Inventory Guide, USEPA/AP-024; NTIS PB 170-851, April 1966.
15. McCrone, W. C., R. G. Draftz, and J. G. Delly: "The Particle Atlas," Ann Arbor Science Publishing, Ann Arbor, Mich., 1967.
16. Cuffe, S. T., and R. W. Gerstle: Emissions from Coal-Fired Power Plants: A Comprehensive Summary, USEPA/AP-035; NTIS: PB 174-798; USGPO: FS2-300, AP-35, (1967).
17. Dorizin, V. G., and V. K. LaMer: *Colloid Sci.*, 14: 74 (1959).
18. Kerker, M., and E. Matijevic: *Jl. Air Pollut. Control Assoc.*, 18:665-(1968).
19. Wolber, W. G.: *Research/Development*, 12:18 (1968).
20. Holzhe, J., and H. Demmrich: *Neue Huette*, 14:198 (1969).
21. Mitchell, R. I., and R. B. Engdahl: *Jl. Air Pollut. Control Assoc.*, 13, 558 (1963).
22. Sem, G. J., and J. A. Borgos: State of the Art: 1971—Instrumentation for Measurement of Particulate Emissions from Combustion Sources, Vol. IV; Experiments and Final Report, USEPA/6502-73-022; NTIS: PB 231-919/AS, September 1973.
23. Nader, J. S., and B. R. Allen: A Mass Leading and Radioactivity Analyzer for Atmospheric Particulate, *AIHA Jl.*, 21:300-307 (1960).
24. McShane, W. P., and E. Bulba: *Jl. Air Pollut. Control Assoc.*, 18:216 (1968).
25. Grindell, D. H.: *Proc. Institute of Electrical Engineers*, 34 (1960).
26. Konig, W., and H. Rock: *Staub*, 20:212 (1960).
27. Sem, G. J., et al.: State of the Art: 1971—Instrumentation for Measurement of Particulate Emissions from Combustion Sources, Vol. III: Particulate Size, USEPA/APTD-1524; NTIS: PB 233-393/AS, July 1972.
28. Lapple, C. E.: Particle Size Analysis and Analyzers, *Chem. Eng.*, 149-156 (May 20, 1968).
29. Carver, L. D.: Particle Size Analysis, *Ind. Res.*, 40-43 (August 1971).
30. Conner, W. D.: An Inertial-type Particle Separator for Collecting Large Samples, *Jl. Air Pollut. Control Assoc.*, 16:35 (1966).
31. Wagman, J., and C. M. Peterson: A Continuous Monitor for Size Distribution of Airborne Particulate Emissions, *Proc. 3rd Int. Clean Air Congr.*, Dusseldorf, October 1973.
32. Smith, W. B., K. M. Cushing, and J. D. McCain: Particle Sizing Techniques for Control Device Evaluation, USEPA 650/2-74-102, October 1974.
33. Allen, J. B., et al.: Velocity of Particulate in Laminar and Turbulent Gas Flow by Holographic Techniques, USEPA-APTD-0918; NTIS: PB 206-950, October 1971.
34. Matkin, J. H.: Determination of Aerosol Size and Velocity by Holography and Steam-Water Critical Flow, Ph.D. thesis, University of Washington, University Microfilms, Seattle, 1968.
35. Andrews, D. G., and H. S. Seifert: Investigations of Particle-Size Determination from Optical Response of a Laser-Doppler Velocimeter, *Project Squid Tech. Rep. SU-1-PU*, Project Squid Headquarters, Purdue University, Lafayette, Ind., February 1972.
36. Farmer, W. M.: On the Measurement of Particle Size, Number Density, and Velocity Using a Laser Interferometer, Contract No. F40600-72-C-0003; AEDC; Arnold Air Force Station, Tenn. (1972).

Index

Adenosine diphosphate, 6-139
Adenosine triphosphate, 4-5 to 4-6, 6-139
Adhesion coefficient, railway locomotive,
    9-228 to 9-229
Adiabatic isentropic compression, 9-24
Adiabatic isentropic expansion, 9-24
Adiabatic system, calorimeter, 1-64
Adiabatic wall temperature, solar absorber,
    6-12
Adjustable-blade hydraulic turbine, 8-12
Admissible draft, 3-213
Aerial survey, pipeline, 2-72
Aeroboic organisms in biosynthesis, 4-6
Aerogenerator farm, 6-165
Aerogenerators, 6-165 to 6-170
Aerosol flow behavior, 9-394
Aerosols, atmospheric, 9-391 to 9-392
Aerospace Corporation, 6-35, 6-54
Aerospace power, solar cell for, 6-56, 6-64
    to 6-68
Afghanistan:
    natural gas production, 2-39
    natural gas reserves, 2-42, 2-44
    oil wells producing, number of, 3-88
    production and exploratory oil wells
        completed, 3-86
    ultimate recovery potential for crude oil,
        3-92
Africa:
    crude oil production, annual, 3-80
    crude oil reserves, recoverable, 3-82
    East African Rift separation, 7-4
    liquid natural gas exports to the United
        Kingdom, 2-103
    natural gas production, 2-39
    natural gas reserves, 2-42, 2-44
    petroleum pipeline mileage, 2-74
    seaborne imports of petroleum, 3-199
    ultimate recoverable crude oil reserves,
        3-92
    (See also specific African nations)
After perpendicular, tanker and product
    carrier, 3-203
AGA (see American Gas Association)
Age, oil reservoir rocks, 3-51
Age ranking, coals, 1-16
Agglomerating coals, 1-18
Agglutinating coals, 1-17
Agricultural chemicals (see
    Petrochemicals)
Agricultural limestone, land reclamation
    use of, 1-105
Agricultural processes, geothermal energy
    for, 7-54 to 7-55
Agricultural wastes, energy conversion of,
    4-5 to 4-6, 4-12 to 4-25
Agriculture in spoiled-land areas, 1-115
Air:
    composition of, 9-40
    versus fuel loss in boiler combustion
        system, 9-141
    physical constants of, 9-36 to 9-37
    preheating for combustion, 3-279 to 3-
        283
    versus total sulfur recovery, Claus unit,
        9-331
Air atomizer, oil burner, 9-52

Air-blown gasifier, Lurgi, 1-188
Air cargo transportation efficiency, 9-228
Air classification of solid wastes, 4-13
Air cleaners, coal-preparation, 1-122
Air compressor, waste heat utilization of, 9-
    351 to 9-353
Air compressor drives, energy conservation
    in, 3-294 to 3-296
Air conditioners, number in United States,
    9-260
Air conditioning, solar (see Solar climate
    control)
Air-cooled turbine generators, 9-202
Air drilling, oil well, 3-118 to 3-119
Air heater regenerator, 9-26
Air leakage, energy losses from, 3-296, 9-46
Air pollution, scope of problem, 9-312
    (See also various pollutants, such as
        Nitrogen oxides; Sulfur oxides;
        etc.)
Air Preheater Company, 3-283
Air pressure versus output, pneumatic tool,
    3-297
Air quality versus hydrocarbon emissions,
    9-380
Air requirements, oil shale retorting (see
    Oil shale, retorting of)
Air separation plant, turboexpanders used
    in, 9-200
Air-temperature relationship, gas turbine,
    9-180
Air-to-oil ratio, combustion optimum, 3-285
Air washing, coal-preparation, 1-126
Aircraft:
    electric generators for, 9-206
    engines, 4-43
    fuels for (see Aviation fuels)
    turbine inlet temperature of, 9-168
Aircraft starters, propellants for, 4-77
Alabama:
    coal characteristics and reserves, 1-27, 1-
        29 to 1-32, 1-34, 1-37, 1-92 to 1-93,
        1-243
    crude oil productive capacity, 3-56, 3-60
    crude oil reserves, 3-46 to 3-47
    electric power generating capacity, 9-
        253
    emission limits for SO_2 and NO_x, 9-374
    hydroelectric power production, 8-7
    natural gas reserves and production, 2-
        10 to 2-13, 2-16 to 2-18
    nuclear power installations, 5-13
    oil field, largest, 3-62
    tar sand deposits, 3-159 to 3-161
    wood-to-oil process development, 3-168
        to 3-173
    [See also Southeastern District (API
        Survey)]
Alabama Bituminous Coal District, 1-169
Alabama By-Products Corporation, 1-31
Alabama Power Company, 5-12
Alarm, geothermal system safety, 7-29, 7-37
    to 7-38
Alaska:
    American Gas Association area of, 2-9
    American Petroleum Institute Survey
        District of, 3-50

Aluminum poisoning, land spoils, 1-103
Aluminum powder, propellant use of, 4-83
Amagat's law, 9-35
Amarillo, Texas, percent sunshine possible, 6-72
Amax Coal Company, 1-31, 1-42, 1-118 to 1-120
Amblygonite, 5-183
American Association of Petroleum Geologists, 3-45
classification of wells, 2-23
American Bureau of Shipping, 3-178, 3-196
American Cyanamid Company, 4-65, 4-68
American Electric Power Company, 9-19, 9-204
American Gas Association, 1-201, 1-211, 1-217, 1-219
major natural gas producing areas and associated pipelines reported by, 2-61
natural gas potential supply estimates of, 2-20 to 2-33
natural gas reserves estimates of, 2-9 to 2-19
undersea drilling platform standards of, 3-109
American Geological Institute, 1-15
American Motors, 4-47
American Oil Company, 1-248, 1-259
American Petroleum Institute, 3-103
API gravity: of Athabasca bitumen, 3-162 to 3-163
aviation fuels, 3-37
crude oils: natural, 3-13 to 3-17, 3-253
synthetic: from Athabasca bitumen, 3-165
from wood-to-oil process, 3-173
crude shale oils: from N-T-U process, 3-149
from Paraho process, 3-155
from Tosco II process, 3-156
from Union Oil process, 3-151
from U.S. Bureau of Mines process, 3-151
definition of, 3-13
Diesel fuel oils, 3-33 to 3-37
gasoline, 3-25
versus heat of combustion for fuels, 3-281
hydrofluoric acid alkylate, 3-238
crude oil reserves estimates of, 3-45 to 3-72
natural gas potential supply estimates of, 2-20
natural gas reserves estimates of, 2-9 to 2-19, 2-33
petroleum products surveys of, 3-22, 3-44
production division of, 3-107 to 3-111
tankcar registry of, 3-176, 3-195
American Potash Company, 5-184
American Society of Mechanical Engineers (ASME), 9-71
Boiler and Pressure Vessel Code, 9-117
American Society for Testing and Materials, testing procedures of:
aviation and jet fuels, 3-38

American Society for Testing and Materials, testing procedures of (*Cont.*):
coal: ash analysis of, 1-51
ash determination of, 1-49
ash fusibility of, 1-51
bomb calorimetry tests of, 1-79 to 1-80
calorific value of, 1-50, 1-69
carbon: fixed, 1-49 to 1-50
total, 1-50
characterization tests of, 1-48
chlorine content of, 1-50
classification system for, 1-16 to 1-17, 1-20, 1-22, 1-53 to 1-55
fineness of, 9-70
fixed carbon, 1-49 to 1-50
free swelling index of, 1-51
Gieseler plasticity, 1-52
grindability of, 1-26, 1-51
gross heating value of, 1-50
Hardgrove grindability, 1-51
heating value of, 1-50, 1-69
hydrogen, total, 1-50
mechanical sampling of, 1-120
moisture content of, 1-49
petrographic techniques, 1-52
plasticity of, 1-52
proximate analysis of, 1-48 to 1-49
pulverized coal, 9-71
quality control of, 1-61
roof and rock bolting standards, 1-75
sampling techniques, 1-59 to 1-60, 1-120
size testing of, 1-27
Sole heated-oven test, 1-53
sulfur content of, 1-50 to 1-51
total carbon content of, 1-50
total hydrogen content of, 1-50
trace elements in, 1-51
ultimate analysis of, 1-69
volatile matter in, 1-49
Diesel fuels, 3-33 to 3-36
gaseous fuels, 2-91 to 2-93
gasoline, 3-25
petroleum products, 3-22
tankcar testing procedures, 3-195
American Thermogen process, 4-19 to 4-20
Ames, Iowa, daily terrestrial solar energy received at, 6-96
Amine absorption, Claus process, 9-332 to 9-333
Ammonia:
combustion constants of, 9-36 to 9-37
fuel cell use of, 4-70 to 4-72
geothermal fluids and gases content of. 7-36
Lurgi process removal of, 1-197, 1-199
natural gas as raw material for, 3-267
ocean thermal energy conversion plant working fluid use of, 6-124 to 6-127
petrochemical production importance of, 3-267
production of: biological processes, 4-8
clean coke process, 1-249
Sasol complex, 1-295
propellant use of, 4-80 to 4-82
Ammonia-based system:
nitrogen oxides absorption by, 9-321
sulfur dioxide absorption by, 9-316 to 9-317

Ammonia dissociation process, 4-36
Ammonia-LPG carrier, 3-204 to 3-205
Ammonia production:
  by ocean thermal energy conversion
    plants, 6-132 to 6-134
  relative to methanol production, 2-123
Ammonia recovery at Sasol complex, 1-287
Ammonium chloride electrolyte, battery
  use of, 4-52
Ammonium nitrate, propellant use of, 4-83
Ammonium perchlorate, propellant use of,
  4-79, 4-83 to 4-85
Ammonium sulfite and bisulfite, gas-
  processing system use of, 9-321
Amoco Production Company, 3-120 to 3-129
Anaerobic conditions, biochemical
  synthesis, 4-6
Anaerobic digestion, methane from, 4-8 to
  4-9
Anahuac oil field, 3-64
Analysis:
  of coal: ash content, 1-49, 1-51, 1-57
    ash fusibility, 1-51, 1-57
    carbon: fixed, 1-49 to 1-50
      total, 1-50
    chlorine content, 1-50, 1-58
    coking properties, 1-53, 1-58
    dilation, 1-53
    grindability, 1-51 to 1-52, 1-58
    heating value, 1-51, 1-63 to 1-70
    moisture content, 1-49, 1-58
    plasticity, 1-52 to 1-53
    proximate analysis, 1-48 to 1-50
    quality control, 1-61
    specific gravity, 1-56
    sulfur content, 1-50 to 1-51, 1-58
    swelling, 1-51
    trace elements, 1-51
    volatile matter, 1-49
    washability, 1-55 to 1-56
  of combustion processes, 3-286 to 3-287
  of design stress for aluminum alloys, 2-
    109
  of gases and air pollutants: carbon
    monoxide, 9-376 to 9-379
    catalytic rich gas feedstock, 2-139
    catalytic rich gas product, 2-139, 2-141
    catalytic rich gas Series A town gas, 2-
      144
    catalytic rich gas synthetic gas: by
      double methanation, 2-144
      by hydrogasification, 2-146
    flue gas, 9-39, 9-47 to 9-49, 9-141
    heating value, calorimetric
      measurements, 2-90 to 2-98
    hydrocarbons, 9-380 to 9-390
    methane rich gas methanation section
      product gas, 2-156
    methane rich gas second-stage
      methanator product gas, 2-157
    natural gas, Panhandle sample, 2-129
    nitrogen oxides, 9-362 to 9-375
    partial-oxidation gasification process
      residual feed, 2-173
    particulates, 9-391 to 9-406
    Shell gasification process product gas,
      2-174

Analysis, of gases and air pollutants (Cont.):
    Sulfinol process effluent, 2-134
    sulfur oxides, 9-362 to 9-375
  of geothermal fluids and gases, 7-11 to 7-
    12, 7-24 to 7-25, 7-30, 7-35 to 7-36,
    7-43 to 7-49
  of petroleum and petroleum products:
    alkylates and alkylate feedstocks, 3-
      238
    Athabasca bitumen, 3-163
    crude oil, 3-13 to 3-17
    oil shale, 3-147
    petroleum fuels, 3-22 to 3-44
    synthetic crude oil from oil shale
      retorting, 3-155 to 3-156
  of solid wastes, 4-22
  of stress and strain in coal mine roofs, 1-
    81 to 1-82
Anchor lines, pipelaying vessels, 2-81 to 2-
  83
Anchorage, Alaska, tidal range of, 8-40
Anchorage of coal-mine roof bolts, 1-76
Anchored throttle valve-steam chest
  system, 9-163
"Ancient lights, doctrine of," 6-93
Andector oil field, 3-64
Anderson ocean thermal energy conver-
  sion plant concept, 6-120 to 6-
  121
Andes Mountains, geothermal resources of,
  7-4
Andreades operating group, 3-207
ANF Frangico gas turbine train, 9-241
Angle of repose, coal, 1-147
Angola:
  crude oil: annual production of, 3-80
    production and exploratory wells, 3-
      86, 3-88
    ultimate recovery potential for, 3-92
  metric tons to barrels conversion factor,
    3-84
  natural gas: production of, 2-39
    reserves of, 2-42, 2-44
  railway mileage and motive power, 9-
    226
Anhydrous ammonia, fuel-cell use of, 4-70
  to 4-72
Aniline by-product, clean coke process, 1-
  252 to 1-253
Aniline-gravity constant number, aviation
  turbine fuels, 3-37
Aniline point:
  Diesel fuel oils, 3-33 to 3-36
  turbine fuels, 3-37
Animal manure, energy conversion of, 4-
  13
Annapolis, Maryland, daily terrestrial solar
  energy received at, 6-96
Annette, Alaska, daily terrestrial solar
  energy received at, 6-97
Annihilation products, 4-89
Anthracene oil solvent, coal purification
  use of, 1-282
Anthracite coal, 1-15 to 1-27
  Lurgi process gasification of, 1-195
  pneumoconiosis from, 1-93
  pulverizer exit temperature of, 9-70

Arkansas (*Cont.*):
  sunshine, percent possible at Little
    Rock, 6-72
  tar sand deposits in, 3-159, 3-161
Arkansas Power and Light Company, 5-12,
  5-22
Arkansas River hydro reserve, 8-19
Arkose, 3-51
Arkwright No. 1 coal mine, 1-29
Arnold nuclear facility, 5-21
Aromatic alkylation, 3-241
Aromatic content, aviation turbine fuel, 3-
  37
Aromatic crude oils, 3-12
Aromatic hydrocarbons:
  catalytic reforming production of, 3-228
  crude oil sources of, 3-19
  heat of combustion of, 9-36 to 9-37
Arsenic, oil well treating acid use of, 3-134
Articulated pontoon, 2-81 to 2-82
Artificial lift, oil well, 3-143
Arzew, Algeria, liquefied natural gas
  facility, 2-102 to 2-103
ASEA (Sweden) RC$_2$ locomotive, 9-237 to
  9-238
Ash:
  coal content of, 1-17, 1-49
  coal-fired equipment, importance of, 9-
    59
  crude shale oil content of: N-T-U
    process, 3-149
    U.S. Bureau of Mines process, 3-155
  Diesel fuel oil content of, 3-33 to 3-36
  versus natural gamma radiation of coal,
    1-45
  solid-waste processing system produc-
    tion of, 4-14, 4-17, 4-20, 4-22,
    4-24
  versus total sulfur for cyclone-furnace
    burning, 9-81
Ash-free basis, coal analysis, 1-17
Ash handling:
  boiler-furnace, 9-110 to 9-111
  stoker, 9-86 to 9-92
Ash lock, Lurgi process, 1-189 to 1-190
Ashtabula, Ohio, coal transfer station, 1-
  170
Asia:
  crude oil: annual production of, 3-80
    production and exploratory oil wells,
    3-86
    recoverable crude oil reserves, 3-82 to
    3-83
    ultimate recoverable crude oil
    reserves, 3-92
  natural gas: production of, 2-39
    reserves of, 2-42, 2-44 to 2-45
  (*See also specific Asian nations*)
ASME (*see* American Society of
  Mechanical Engineers)
Asphalt:
  consumption of, United States, 3-96, 3-
    98
  crude oil content of, 3-13 to 3-15
Asphalt Ridge, Utah, tar sand deposits at, 3-
  159

Asphalt rock, 3-159
Asphaltenes, Athabasca bitumen content
  of, 3-163
Asphaltic-base crude oils, 3-12
Asphaltic oils (*see* Tar sands)
As-received moisture, coal, definition of, 1-
  49
Assembly:
  of marine pipeline, 2-78
  nuclear reactor fuel (*see* Fuel elements,
    rods and bundles, nuclear reactor)
  solar energy satellite, 6-106, 6-112 to 6-
    115
ASSEMBLY drilling assistance computer
  program, 3-128
Assistance programs, drilling, 3-126 to 3-
  128
Associated-dissolved natural gas:
  definition of, 2-16
  probable reserves of, in Canada, 2-34
  proved reserves of: in Canada, 2-34
    in United States, 2-10
Association of American Railroads:
  interchange rules of, 1-161
  tankcar standards of, 3-178, 3-191, 3-195
Aster, spoiled-land area use of, 1-110
ASTM (*see* American Society for Testing
  and Materials)
Astoria, Oregon, wind power site at, 6-158,
  6-160 to 6-162, 6-168
Aswan Dam hydroelectric plant, 8-5
Atchison, Topeka and Santa Fe Railway
  Company, 1-153 to 1-167
Athabasca bitumen, 3-162
  (*See also* Tar sands)
Athabasca tar sand deposit, 3-159 to 3-167
Athenosphere, 7-4
Atlanta, Georgia, daily terrestrial solar
  energy received at, 6-96
Atlantic City wave-motor project, 8-45
Atlantic Coast NPC petroleum province, 2-62
Atlantic nuclear plant, 5-15, 5-22
Atlantic Ocean NPC petroleum province,
  2-62
Atlantic-Richfield Corporation, 1-259, 4-48
Atlantic Seaboard Corporation, 2-64
Atlantic Seacoke process, 1-253 to 1-254
Atmosphere:
  energy absorption by, 6-12
  heat engine role of, 6-146 to 6-147
  pollution of, 9-312
    (*See also various pollutants, such as*
    Nitrogen oxides; Sulfur oxides; *etc.*)
  total energy of, 6-146
Atmospheric distillation, crude oil (*see*
  Distillation, crude oil)
Atmospheric residue, 3-226
Atom, 5-39
Atomic disintegration, 5-39
Atomic hydrogen, 4-27
Atomic mass, 5-39
Atomic mass unit, 5-39
Atomic nucleus, 5-137
Atomic number, 5-39
  versus energy level for propellant
    oxides, 4-76

Atomic structure, hydrogen, 4-27
Atomic weight, 5-39
Atomization, oil burner, 9-52 to 9-55
ATP (*see* Adenosine triphosphate)
Attemperator, boiler, 9-107, 9-120
Auckland, New Zealand, tidal range at, 8-40
Auctioneer control system, boiler, 9-126
Audibert-Arnu coal test, 1-17, 1-21, 1-23
Audible sensor, coal mine roof analysis use of, 1-82
Auger, pipeline, 2-73
Auger mining, coal, 1-29 to 1-33, 1-97
Australia:
    coal reserves of, 1-36
    crude oil: annual production of, 3-80
        production and exploratory wells, 3-86, 3-88
        ultimate recovery potential for, 3-93
    hydroelectric plants in, 8-5 to 8-6
    lithium resources of, 5-183
    metric tons to barrels conversion factor, 3-84
    natural gas: production of, 2-40
        reserves of, 2-42, 2-45
    nuclear energy facilities in, 5-25, 5-37 to 5-38
    railway mileage and motive power, 9-226
    seaborne imports of petroleum, 3-199
    tidal range at Melbourne and Sydney, 8-40
    uranium resources of, 5-176
Austria:
    crude oil: annual production of, 3-81
        production and exploratory wells, 3-86
        ultimate recovery potential for, 3-93
    metric tons to barrels conversion factor, 3-84
    natural gas: production of, 2-40
        reserves of, 2-42, 2-45
    nuclear energy facilities in, 5-25, 5-36 to 5-38
    railway mileage and motive power, 9-226
Automated mapping, electric utility use of, 9-294
Automatic control:
    boiler and combustion processes, 9-122 to 9-162
    nuclear facilities (*see* Nuclear energy technology)
    petroleum plants (*see* Petroleum)
    (*See also* Instrumentation; Monitoring)
Automatic meter reading, electric utility, 9-293
Automatic sampler, coal, 1-120
Automatic sequence control, boiler burner, 9-145
Automobiles:
    demand for, in United States, 3-101
    electric, 4-55 to 4-58
    engines (*see* Engines)
    pollution from (*see* Engines)
Automotive chemicals, petroleum-derived (*see* Petrochemicals)

Auxiliary systems, reactor (*see* Reactors)
Availability:
    energy, xxv–xxvii
    solar electric power, 6-6 to 6-7
Available energy, 9-30
Average power, alternating current circuit, 9-296 to 9-297
Avgas (*see* Aviation fuels)
Aviation fuels:
    additives for, 3-38
    annual United States survey of, 3-37 to 3-39
    aviation gasoline, 3-37
    consumption of, United States, 3-97
    hydrotreatment improvement of, 3-235
    jet fuels, 3-37 to 3-39
    physical properties of, 3-181
    pipeline transportation of, 3-174
    tankcar transportation of, 3-181
    turbine fuels, 3-37 to 3-39
Avignon, France, flooding of, 8-25
Avogadro's law, 9-35
AVR pebble-bed, gas-cooled reactor, 5-89 to 5-90
Axial compressor, liquefaction process use of, 2-107 to 2-108
Axial-flow hydraulic turbine, 8-12, 8-14, 8-19
Axial peaking, boiling water reactor, 5-71
Axle classification, railway locomotive, 9-228
Axle loading, railway equipment, 9-228 to 9-229, 9-233
Axle weight, railway rolling stock, 9-225
Ayrcoe coal mine, 1-30
Ayrgem coal mine, 1-29

Babcock, George Herman, 9-18
Babcock and Wilcox Company, 1-16, 9-17 to 9-18, 9-23, 9-34, 9-50, 9-59
Back-pressure turbine, geothermal system, 7-9
Bacon, F. T., 4-59
Bacterial photosynthesis, 6-137
Bacterial seeding, bio-oxidation process, 9-356
Baden, E. R., oil well depth record, 3-129
Badgett Terminal Corporation, 1-141
Baffle, boiler, 9-100 to 9-102
Bagasse:
    fuel use of, 4-13
    normal excess air required for burning, 4-13
Bahamas:
    crude oil: production and exploratory wells, 3-87 to 3-88
        ultimate recovery potential for, 3-954
    liquefied natural gas production, 2-102
    natural gas, ultimate recovery potential for, 2-46
Bahrain:
    crude oil: production and exploratory wells, 3-80, 3-86, 3-88
        ultimate recovery potential for, 3-92
    metric tons to barrels conversion factor, 3-84

Bahrain (*Cont.*):
  natural gas: production of, 2-39
    reserves of, 2-42, 2-44
  oil field discovery, 3-90
Bailly nuclear plant, 5-13, 5-22
Baldwin-Westinghouse GG-1 locomotive,
    9-236
Ball-and-race coal pulverizer, 9-65 to 9-68
Ball heater:
  oil shale retorting process use of, 3-155
  Toscoal process use of, 1-273 to 1-275
Ballast, definition of, 3-203
Ballast tank, definition of, 3-206
Ballistic properties, propellant, 4-87
Baltimore, Maryland:
  coal pipeline terminal at, 1-180
  tidal range at, 8-40
  waste-conversion facility at, 4-19
Baltimore and Ohio Railroad, 1-154, 9-237
Baltimore Gas and Electric Company, 4-
    19, 5-14
Band-type prony brake, 9-305 to 9-306
Bangladesh:
  crude oil: producing wells in, 3-88
    reserves of, 3-82
    ultimate potential for recovery of, 3-92
  natural gas: production of, 2-39
    reserves of, 2-42, 2-44
  nuclear power facilities planned, 5-25, 5-
    37 to 5-38
Bantry Bay deep crude terminal, 3-207
Bar screen, coal (*see* Screening, coal)
Barbados:
  crude oil: producing wells in, 3-87 to 3-
    88
    proved recoverable reserves of, 3-83
    ultimate recovery potential for, 2-46
  natural gas, ultimate recovery potential
    for, 2-46
Bareboat charter, 3-203
"Barefoot" oil well completion, 3-131
Barge:
  coal, loading systems for, 1-141 to 1-144,
    1-174 to 1-176
    design of, 1-176
  marine pipeline laying, 2-77 to 2-89
  petroleum products, 3-203
Barge-shape hull, 2-79
Barge transportation efficiency, 9-228
Barium additives, antismoking use of, 3-31
Barium anode, thermionic converter use of,
    9-222
Barium-137, fragment of nuclear fission, 5-7
Barrels to metric tons conversion factors for
    various nations, 3-84
Barrow, Alaska, daily terrestrial solar
    energy received at, 6-97
Barry, Theodore, and Associates, 1-85
Bartley, West Virginia, mine disaster, 1-92
Barton, Alan R., nuclear plant, 5-12
Barvøys vessel, 1-128 to 1-129
Base-load electric generating plant,
    hydroelectric, 8-9
Base-load plant installations, gas turbine,
    9-192
Base-load solar plant, 6-6, 6-24 to 6-25, 6-
    35, 6-69 to 6-80

Base metal oxide catalysts, fuel-cell, 4-62
Base temperature, geothermal, 7-5
Baseball II fusion device, 5-165, 5-167
Baseline solar collector, 6-94
Basin, tidal power, 8-27 to 8-28, 8-36 to 8-39
Battelle Memorial Institute, 1-223, 2-12, 4-
    56, 8-8 to 8-9
Battery power, 4-50 to 4-54
  electric automobile, 4-55 to 4-58
  spacecraft, 6-57
Baum jig, 1-120, 1-127
Bay de Chene oil field, 3-65
Bay of Fundy:
  tidal energy project at, 8-45
  tidal range at, 8-40
Bay of Plenty, 7-14
Bay Marchand oil field, 3-62
Bay Saint Elaine oil field, 3-64
Bayer process, geothermal energy for, 7-50,
    7-56
Beam, tanker and product carrier, 3-199, 3-
    203, 3-207 to 3-214, 3-219 to 3-220
Bearing, two-way flow axial turbine, 8-32 to
    8-33
Bearing friction, railway locomotive and
    rolling stock, 9-225
Beatrice coal mine, 1-30
Beatrice Pocahontas Company, 1-31
Beaufort Sea, oil exploration in, 3-94
Beauharnois hydroelectric plant, 8-5
Beaulah coal mine, 1-30
Beaver Lodge oil field, 3-65
Beaver Valley nuclear plants, 5-17, 5-22
Beck, Sir Adam, No. 2 hydroelectric plant,
    8-5
Beck 2 pumped-storage plant, 8-18
Bed, coal, 1-37 to 1-47
Bee Veer coal mine, 1-31
Beer Canal, 3-221, 3-223
Beer's law, 9-366
Beets, fermentation of, 4-7
Belgium:
  coal reserves of, 1-36
  crude oil: producing wells in, 3-88
    ultimate recovery potential for, 3-93
  gas-cooled fast breeder reactor compo-
    nents, development of, 5-135
  natural gas, ultimate recovery potential
    for, 2-45
  nuclear power facilities in, 5-25, 5-36 to
    5-38
  nuclear fuel study, 5-179
  radium, early use of, 5-173
  railway mileage and motive power, 9-226
  tidal range at Antwerp, 8-40
Belidor, tidal studies of, 8-39
Belize:
  producing oil wells in, 3-88
  ultimate recovery potential: for crude oil
    in, 3-94
    for natural gas in, 2-46
Bell Creek oil field, 3-64
Belle Isle, Florida, daily terrestrial solar
    energy received at, 6-96
Bellefonte nuclear plant, 5-12, 5-22
Belmont County, Ohio, coals, 1-20
Belridge South oil field, 3-63

Belt conveyor, coal, 1-118 to 1-119, 1-146 to
    1-150
Belt scale, coal, 1-119
Belt-tripper coal loading system, 1-139 to
    1-141
Beltman accidents, coal mine, 1-87 to 1-91
Bender, coal mine roof support bolt, 1-81
Bending, pipeline, 2-73
Bennett, W. A. C., hydroelectric plant, 8-5
Bent-tube boiler, 9-93, 9-103 to 9-107
Bentonite clay, drilling mud use of, 3-123
Benwood, West Virginia, mine disaster, 1-
    92
Benzene:
    alkylation of, 3-241
    clean coke process by-product source of,
        1-252
    combustion constants of, 9-36 to 9-37
    crude oil sources of, 3-19
    fuel properties of, 4-45
    hydrogenation of, 3-267
    production of, 1-253 to 1-254
Benzenoids, heat of combustion of, 9-36 to
    9-37
Benzoic acid, calorimeter use of, 1-65
Benzonaphthothiophenes, 3-252
Benzothiophenes, crude oil content of, 3-
    254
Bering Sea, oil exploration in, 3-94
Berkeley gas-cooled reactor, 5-89
Bernic Lake lepidolite resources, 5-183
Bersimis No. 1 hydroelectric plant, 8-5
Bertha Rogers, the, oil well depth record,
    3-129
Bertin Aerotrain, 9-241
Beryllium, propellant use of, 4-83, 4-85
Beryllium-antimony start-up sources,
    boiling water reactor, 5-69 to 5-70
Beryllium fluoride, fusion reactor blanket
    use of, 5-156
Bessemer and Lake Erie Railroad, 1-154
Bessemer City, North Carolina, lithium
    plant at, 5-184
Beta-alumina solid electrolyte battery, 4-52
Beta monitor, 9-401
Beta particle, 5-39
Bethlehem Mines Corporation, 1-31
Bharkra hydroelectric plant, 8-6
Bhutan:
    number of producing oil wells in, 3-88
    ultimate recovery potential for crude oil
        in, 3-92
Biased firing, nitrogen oxides control by, 9-
    343 to 9-344
BI-GAS coal gasification process, 1-212 to
    1-218
    design and testing program, 1-215 to 1-
        217
    gas processing in, 1-212 to 1-214
    two-stage gasifier, 1-212
Big Horn Coal Company, 1-31
Big Horn coal mine, 1-30
"Big Inch" pipeline, 3-174
Big Rock Point nuclear facility, 5-14, 5-21
Big Sky coal mine, 1-29
Big Sulphur Creek, California, 7-34
Bikita lepidolite resources, 5-183

Bilboa, Spain, tidal range at, 8-40
Bin, coal, 1-118 to 1-119, 1-122
Bin-fired cyclone furnace, 9-83
Bin system, pulverized-coal, 9-60 to 9-62
Binary fluid topping cycle, 9-22
Binding energy, 5-39
Binding ingredients, propellant, 4-85
Biochemical fuel cell, 4-9
Biochemical fuel sources, 4-5 to 4-11
Biochemical oxygen demand (BOD), 9-354
Biochemical process, coal formed by, 1-15
Biodegradable synthetic detergents, 3-20
Biological methane production, 4-16 to 4-
    18
Biological shield, 5-34, 5-171
Bio-oxidation process, water use reduction
    by, 9-354 to 9-361
Biosynthetic hydrogen, 4-6
Bipolar electrode-type electrolyzer, 4-34
Bipropellants, 4-77, 4-82
Bishop, Virginia, mine disaster, 1-92
Bishop Coal Company, 1-31
Bishop coal mine, 1-30
Bismarck, North Dakota, daily terrestrial
    solar energy received at, 6-96
Bismuth-lead glass, superconductor use of,
    5-162
Bismuth telluride, thermoelectric cooling
    system use of, 9-243
Bit:
    coal mine roof drilling, 1-78
    oil well drilling, 3-113 to 3-118, 3-121 to
        3-128
BITOP drilling assistance computer
    program, 3-128
Bituminous coal:
    barge transportation of, 1-141 to 1-144, 1-
        170 to 1-177
    Lurgi process gasification of, 1-188 to 1-
        200, 1-285 to 1-296
    pneumonoconiosis from, 1-93
    production of, 1-25 to 1-33
        auger mining, 1-30 to 1-33
        contour (strip) mining, 1-30 to 1-33
        underground mining, 1-29 to 1-32
    pulverizer exit temperature of, 9-70
    ranking of, 1-54
    sizes of, 1-25
    slurry pipeline transportation of, 1-178 to
        1-187
    testing of, 1-48 to 1-70
Bituminous Coal Act of 1937, 1-167 to 1-
    169
Bituminous Coal Research, Inc., 1-212 to 1-
    218
Bituminous fermentation, 1-15
Bituminous sands (see Tar sands)
Bjorne Formation, tar sand deposits in, 3-
    159 to 3-162
Black coal, 1-15
Black Fox nuclear plants, 5-16
Black lignite, 1-15
Black liquor fuel, excess air normally
    supplied for, 9-46
Black liquor recovery plant, analyzer for, 9-
    373
Black lung, 1-92 to 1-93

British-Dutch-West German Tripartite nuclear fuel enrichment agreement, 5-179

British energy statistics (see United Kingdom)

British Gas Corporation, 2-139 to 2-140, 2-142 to 2-147, 2-160 to 2-163

British Gas Council, 1-238
  catalytic rich gas process development, 2-137
  hydrocracking and hydrogasification process development, 2-164
  liquefied natural gas developments, 2-102

British inland wind power sites, 6-161

British Methane, Ltd., 2-102

British Railways, tankcar operations of, 3-194 to 3-195

British thermal unit, definition of, xxiv

Brittany, tidal power plant located in, 8-25 to 8-27
  (See also Rance Estuary tidal power plant)

Brittle materials, compressive forces in, 1-75

Broadlands geothermal field, 7-31 to 7-33

Broken anthracite, 1-27

Brome grass, spoiled-land area use of, 1-107 to 1-108

Brookhaven National Laboratory, 4-38, 4-43

Brown & Root International Ltd., 2-77 to 2-89

Brown-Boveri & Cie, 4-62

Brown coal, 1-16 to 1-17
  international classification of, 1-24

Browns Ferry nuclear facilities, 5-12, 5-21 to 5-22

Brownsville, Texas, daily terrestrial solar energy received at, 6-95

Bruceton, Pennsylvania, Synthane coal processing pilot plant at, 1-232

Brunei:
  crude oil: producing wells in, 3-88
    ultimate recovery potential for, 3-92
  liquefied natural gas facility in, 2-102 to 2-103
  metric tons to barrels conversion factor, 3-84
  natural gas, ultimate recovery potential for, 2-44

Brunswick nuclear plants, 5-16, 5-22

Brush Electric Light Company, 9-18

Btu content:
  coal, 3-280
  compounds and elements, 9-36 to 9-37
  fuel oil, 3-280
  liquefied gases, 3-280
  natural gas, 3-280
  (See also Calorific value)

Btu control system, gas-mixture, 2-96 to 2-99

Bubble, steam, 9-94 to 9-97

Buchanan hydroelectric plant, 8-17

Bucket:
  calorimeter, 1-63 to 1-70
  hydraulic turbine, 8-11, 8-13

Bucket elevator, coal, 1-148

Buckheart No. 17 coal mine, 1-29

Buckwheat, spoiled-land area use of, 1-109

Buckwheat coal, 1-27

Buenos Aires, Argentina, tidal range at, 8-40

Buffalo, New York, early natural gas storage experiment at, 2-75

Buffering, pressurized water reactor control element assembly, 5-84

Bugey I gas-cooled reactor, 5-89

Building construction factors, solar heating, 6-90 to 6-92

Building heating:
  geothermal energy for, 7-8, 7-52 to 7-58
  solar systems for (see Solar climate control)

Bulb unit, hydropower, 8-19, 8-29 to 8-34
  governing system for, 8-34 to 8-36

Bulgaria:
  crude oil: producing wells in, 3-88
    proved recoverable reserves of, 3-83
    ultimate recovery potential for, 3-93
  metric tons to barrels conversion factor, 3-84
  natural gas: production of, 2-40
    reserves of, 2-42, 2-45
  nuclear power facilities in, 5-25, 5-36 to 5-38
  railway mileage and motive power, 9-226

Bulk cargo, tanker and product carrier, 3-203

Bump, coal mine roof, 1-73

Bunker C fuel oil, 3-40

Buoy, pipeline location, 2-85

Burbank oil field, 3-105

Burgdorf-Thun Railway, 9-237

Burial, pipeline: barges for, 2-88
  offshore, 2-77 to 2-89
  onshore, 2-70 to 2-76

Burlington, Vermont, daily terrestrial solar energy received at, 6-97

Burlington Generating Station, 1-144 to 1-146

Burlington Northern Railroad, 1-154

Burma:
  crude oil: annual production of, 3-80
    producing wells in, 3-88
    ultimate recovery potential for, 3-92
  metric tons to barrels conversion factor, 3-84
  natural gas: production of, 2-39
    reserves of, 2-42, 2-44
  railway mileage and motive power, 9-226
  tidal range at Rangoon, 8-40

Burn parameters, fusion reactor reference design, 5-169

Burnable poison, nuclear reactor, 5-68, 5-71

Burner control systems, boiler, 9-143 to 9-147

Burner fuel oils, 3-39 to 3-43
  annual survey, United States, 3-39
  characteristics of, 3-43
  consumption, United States, 3-97 to 3-99, 3-244

Calorific value (*Cont.*):
 methanol, 4-7, 2-125
 natural gas, 4-7
 refuse, 4-12
 urban wastes, 4-12
 waste materials, 4-12
Calorific value control system, gas-mixture, 2-96 to 2-99
Calorimeter, bomb-type, 1-63 to 1-70
Calorimetry:
 definition of, 1-63
 gaseous fuels, 2-90 to 2-98
Calormixer gas calorimeter, 2-92 to 2-93
Caloroptic-type gas calorimeter, 2-92 to 2-93
Calvert Cliffs nuclear plant, 5-14, 5-21 to 5-22
Cambodia, railway mileage and motive power in, 9-227
Cambria County, Pennsylvania, coals of, 1-20
Cambrian Geologic System, oil and gas formation during, 3-51
 in Canada, 3-76 to 3-77
 in United States, 2-14, 3-59
Cambridge, Massachusetts, daily terrestrial solar energy received at, 6-97
Cambridge Seven Associates, Inc., 6-92
Cameroon:
 crude oil: production and exploratory wells in, 3-86, 3-88
 ultimate recovery potential for, 3-92
 natural gas, ultimate recovery potential for, 2-44
 railway mileage and motive power, 9-227
Campbell County, Wyoming, coals in, 1-20
Canada:
 coal reserves, 1-36
 crude oil: production and exploration of, 3-81, 3-87 to 3-88
 recoverable reserves of, 3-85
 ultimate recovery potential for, 3-94
 electric power generating capacity, 9-248 to 9-249
 fuel cell installations, 4-70
 hydroelectric power capacity, 8-3 to 8-6, 9-261
 hydrogen-fuel air mixture tests by Captain King, 4-38
 Maritime Provinces, petroleum refining developments in, 9-263
 metric tons to barrels conversion factor, 3-84
 natural gas: reserves of, 2-34 to 2-37, 2-40, 2-43, 2-46
 underground storage experiment, Welland County, Ontario, 2-75
 United States imports of, 2-66
 nuclear power facilities, 5-25 to 5-26, 5-36 to 5-38
 Chalk River experimental plant, 5-10
 fuel enrichment, 5-179
 heavy water reactor, 5-40
 lithium resources, 5-183 to 5-185
 uranium resources, 5-174 to 5-176
 pipeline systems, 2-64 to 2-65, 2-74

Canada (*Cont.*):
 radium, early use of, 5-173
 railway mileage and motive power, 9-226
 seaborne exports, petroleum, 3-199
 tar sand deposits, 3-159 to 3-167
 tidal energy projects: Bay of Fundy, 8-45
 Little Kodiak, 8-45
 tidal range at Burntcoat Head, Halifax, Quebec, Saint John (New Brunswick), St. John's (Newfoundland), 8-40
Canadian Arctic Gas Study Limited, 2-35
Canadian Export Development Corporation, 3-197
Canadian Frontier pipeline system, 2-64
Canadian Petroleum Association:
 natural gas reserves estimates of, 2-34 to 2-38
 oil reserves, 3-73 to 3-77
Canal Zone, tidal range at Atlantic and Pacific sides, 8-40
Candelia wax, propellant use of, 4-83
Candida pseudotropicalis, 4-7
Candida utilis, 4-7
Candle coal, 1-16
Cannel coal, 1-16
Canopy protection, coal mine roof bolting, 1-79, 1-82
Canvey Island liquefied gas facility, 2-102
Cap rock, hydrocarbon entrapment, 3-104
Capability, electric power system, definition of, 9-251 to 9-252
Capability margin, electric utility industry, 9-252
Capacity:
 battery, 4-52 to 4-53, 4-55 to 4-58
 coal barge, 1-171 to 1-172, 1-176
 compressed gas tankcars, 3-180 to 3-181
 electric generating (*see* Electric generating capacity)
 HYGAS process, 1-208
 rail coal cars, 1-160 to 1-161
Capacity classification, wind machine, 6-142
Capacity displacement system, 6-6
Cape Blanco, Oregon, wind power site, 6-163
Cape Foul Weather, Oregon, wind power site, 6-158
Cape Hatteras, North Carolina, daily terrestrial solar energy received at, 6-95
Cape Lookout, Oregon, wind power site, 6-157
Capetown, Republic of South Africa, tidal range at, 8-40
Capital costs (*see* Costs and economics)
Capital management, electric utility, 9-279
Captain coal mine, 1-29
Caquot, tidal studies of, 8-39
Car haulers, 1-122, 1-139
Carbide-sludge gas absorption system, 9-314
Carbohydrates, fermentation of, 4-7 to 4-9
Carbon:
 in coals, 1-15 to 1-33, 1-41 to 1-70
 combustion of, 9-34
 air ignition temperature, 9-42
 combustion constants, 9-36 to 9-38
 fixed, 1-49
 total, 1-50

I-17

Carbon black:
  by-product of clean coke process, 1-252
  propellant use of, 4-83
Carbon content:
  Athabasca bitumen, 3-163
  wood-to-oil process synthetic crude oil,
    3-173
Carbon dioxide:
  content of: Claus off-gas, 9-329
    geothermal fluids and gases, 7-11 to 7-
      12, 7-36, 7-44 to 7-45
    natural gas, 2-129
    synthetic pipeline gases, 2-118
    vent gas, 2-118
  formation of: anaerobic digestion
      process, 4-8
    BI-GAS process, 1-214
    biological processes, 4-6 to 4-7
      methane process, 4-17 to 4-18
    carbon dioxide acceptor process, 1-
      219
    COGAS process, 1-264
    gas turbine combustion, 9-188
    Hygas process, 1-203
    Lurgi process, 1-191
    noncatalytic partial oxidation of
      hydrocarbons, 4-36
    steam reforming of hydrocarbons, 4-37
    Union Carbide Purox process, 4-16
    Westinghouse coal gasification
      process, 9-173
  versus hydrogen sulfide absorption,
    amine unit, 9-333
  oil well treatment use of, 3-135
  photochemical reduction of, 4-5
  photosynthesis acceptor role of, 6-139
  physical constants of, 9-36 to 9-37
  reactor coolant use of, 5-10, 5-89
Carbon dioxide acceptor coal gasification
    process, 1-219 to 1-222
  advantages of process, 1-221
  feedstock and acceptor characteristics, 1-
    221 to 1-222
  fundamentals of, 1-219 to 1-221
  gas cleanup and methanation, 1-221
Carbon dioxide removal:
  methane-rich gas process, 2-158
  Sulfinol process, 2-133 to 2-136
Carbon disulfide content, Claus off-gas, 9-
    329
Carbon-filled plastic heat exchangers,
    ocean thermal energy conversion
    plant use of, 6-127
Carbon-hydrogen ratio:
  Athabasca bitumen, 3-163
  catalytic rich gas process, 2-146
  Tosco II shale oil process, 3-156
Carbon moly superheater and reheater
    tubes, 9-117
Carbon monoxide:
  combustion constants of, 9-36 to 9-38
    air ignition temperature, 9-42
  content of: Claus off-gas, 9-329
    SGP process product gas, 2-174
  emissions of, 9-196
  feedstock for methanol production, 2-
    120

Carbon monoxide (*Cont.*):
  formation of: BI-GAS process, 1-214
    COGAS process, 1-264
    gas turbine combustion, 9-196 to 9-198
    gasoline combustion, 3-22
    Hygas process, 1-203
    Lurgi process, 1-191
    Westinghouse coal gasification
      process, 9-173 to 9-175
    wood-to-oil process, 3-168 to 3-169
  hydrogen burning, absence of, 4-31
  monitoring of, 9-376 to 9-379
Carbon monoxide boiler furnace, 2-174
Carbon monoxide-carbon dioxide ratio,
    determination of, 9-377
Carbon monoxide-hydrogen ratio:
  Arge process, 1-294
  COGAS process, 1-266
  Lurgi process, 1-194 to 1-195
  Synthol process, 1-294
Carbon residue:
  burner fuel oil, 3-40
  Diesel fuel oil, 3-33 to 3-36
  Khafji crude oil, 3-253
Carbon-steam equilibrium, COGAS
    process, 1-265
Carbonate rock:
  gas reservoir, 2-11 to 2-12
  oil reservoir, 3-51, 3-58
Carboniferous age, coal formed during, 1-
    15, 1-37, 1-48
Carbonization (*see* Char; Gasification, of
    coal)
Carbonization-hydrogenation method,
    metallurgical coke production by, 1-
    249 to 1-254
Carbonyl sulfide content, Claus process, 9-
    329, 9-332
Carbonyl sulfide removal, Sulfinol process,
    2-133 to 2-136
Carburetted water gas, calorific value of, 2-
    90
Cardan shaft, railway locomotive, 9-235
Cargo, bulk, definition of, 3-203
Cargo plane efficiency, 9-228
Caribou, Maine, daily terrestrial solar
    energy received at, 6-97
Carling, France, coal slurry pipeline
    terminus at, 1-180
Carload coal movement, 1-154
Carmichaels, Pennsylvania, mine disaster
    at, 1-92
Carnegie-Mellon University, 4-38
  ocean thermal energy conversion plant
    concept of, 6-120 to 6-121
Carnot cycle, 4-59, 9-24, 9-221
Carnot cycle efficiency:
  heat engine, 6-7 to 6-9
  liquefaction processes, 2-107 to 2-109
Carnot gas engine, 9-24
Carolina Power and Light Company, 5-16
    to 5-17
Carpo dynamometers, 9-309 to 9-311
Carrier, seagoing petroleum (*see* Seagoing
    vessels, petroleum carrying)
Carrier fluids, oil well sand control use of,
    3-133

Cascade Locks, Oregon, wind power site at, 6-159, 6-161
Cascade Pipeline Limited, 1-180
Cascade refrigeration cycle, 2-102 to 2-109
Cascaded cycles, solar engine, 6-19 to 6-22
Cascading, thermoelectric cooling system module, 9-243 to 9-244
Case Western Reserve University, 4-33
Casing, oil well, 3-117, 3-119, 3-121 to 3-122, 3-128 to 3-132
CASING drilling assistance computer program, 3-128
Casinghead gas, 2-17
Castaic hydroelectric plant, 8-5, 8-18
Castle Gate, Utah, mine disaster at, 1-92
Catalysts:
  catalytic rich gas process use of, 2-139
  Claus process use of, 9-329
  cobalt, fuel cell use of, 4-64
  cobalt-molybdenum, Synthoil process use of, 1-270
  Comox, Lurgi gasification process use of, 1-198
  Consol (CSF) process use of, 1-276
  copper, methanol process use of, 2-121
  costs of, 1-210
  ethanol synthesis use of, 4-8
  ethylene hydration use of, 4-8
  fluid catalytic cracking process use of, 3-231 to 3-232
  fuel cell use of, 4-59 to 4-68
  Hydrane process use of, 1-224
  hydrocracking process use of, 2-234 to 2-235
  hydrodesulfurization use of, 3-259
  hydrogen liquefaction, 4-27
  hydrogenation use of, 4-29
  methanation process use of, 1-204, 1-213, 1-217, 1-241, 1-266 to 1-267
  methane-rich gas process use of, 2-149 to 2-153
  methanol production use of, 2-121
  nickel: COGAS methanation use of, 1-267
    Lurgi methanation use of, 1-204, 4-29
  palladium, 4-29
  partial oxidation of hydrocarbons process use of, 4-36
  platinum, 4-29
  poison removal from, 3-19
  poisoning by crude oils, 3-13
  Raney nickel, Synthane process use of, 1-241
  Sasol complex use of, 1-289, 1-291 to 1-293
  selenium, 4-29
  SELEXOL process use of, 1-213
  silicon dioxide-aluminum oxide, Synthoil process use of, 1-270
  Synthoil process use of, 1-269 to 1-270
  wood-to-oil process use of, 3-170
  zinc chloride process use of, 1-281
Catalytic combustion gas detector, 9-387, 9-389
Catalytic cracking:
  effect on fuel oil quality, 3-40
  petroleum, 3-18 to 3-21, 3-227

Catalytic-oxidation carbon monoxide analyzer, 9-376 to 9-377
Catalytic reformate, 3-225
Catalytic reforming, 3-19, 3-226 to 3-229, 3-249, 3-278
Catalytic rich gas (CRG) process, 2-137 to 2-148
Catawba nuclear plants, 5-17, 5-22
Catenary, electric railway, 9-236 to 9-237, 9-239
Caterpillar Tractor Company, 1-94 to 1-101
Cathode material, battery, 4-51 to 4-53
Caustic washing, burner fuel oil, 3-40
Cavendish, Henry, 4-27
Cavern, compressed air storage in, 9-194 to 9-195
Caving, coal mine roof, 1-73
Cavity absorber, solar, 6-11 to 6-14
C-B Diesel fuel oils, 3-28, 3-33
C-B type natural-gas firing elements, 9-51, 9-56
CEGB nuclear plant, 5-33 to 5-35
Cell(s):
  battery, 4-50 to 4-53
  fuel cell, 4-61 to 4-62
  solar (see Solar cells and arrays)
Cell burner, 9-50 to 9-52
  for pulverized coal, 9-76
Cellulose, synthetic crude oil from, 3-168 to 3-173
Cellulose degradation, 4-9
Cement kiln furnace, fuel fineness requirements of, 9-65
Cement mills, coal requirements of, 1-28
Cenozoic Era, gas and oil formed during, 3-51
  in Canada, 3-76 to 3-77
  in United States, 2-14, 3-59
Center tank, definition of, 3-206
Centistoke, definition of, 3-42
Central African Republic:
  crude oil: production of, 3-88
    ultimate recovery potential for, 3-92
  natural gas, ultimate recovery potential for, 2-44
  uranium resources of, 5-176
Central America (see specific country)
Central Appalachian Coal Company, 1-31
Central control systems, electric utility, 9-282 to 9-283
Central District, API Survey, 3-50
Central Ohio Coal Company, 1-31
Central Power and Light Company, 5-18
Central Region, Diesel fuel survey of, 3-32 to 3-36
Central Region Petroleum Products Survey, 3-32 to 3-36
Central Vermont Public Service Corporation, 6-150
Centralia, Illinois, mine disaster at, 1-92
Centralia coal mine, 1-29
Centre National de la Recherche Scientifique, 6-46 to 6-55
Centrifugal compressor:
  liquefaction process use of, 2-107
  methanol process use of, 2-121
Centrifugal separators, evaluation of, 3-128

Centrifuge:
coal separation use of, 1-118 to 1-119, 1-126, 1-130 to 1-135
tar sand process use of, 3-165
Ceramic research, solar furnace used for, 6-51 to 6-52
Cereals, fermentation of, 4-7
Cermet, 5-43
Cerro Prieto geothermal reservoir, 7-6, 7-8
Certification, coal mining job, 1-91
Cesium diodes and triodes, 9-221 to 9-222
Cetane number:
definition of, 3-44
of Diesel fuels, 3-28 to 3-36
Ceylon, railway mileage and motive power in, 9-227
Chad:
crude oil: production of, 3-88
ultimate recovery potential for, 3-92
natural gas, ultimate recovery potential for, 2-44
Chain-grate stokers, 9-92
Chain reaction, 5-7, 5-43
Chalk River nuclear experimental plant, 5-10
Channel:
magnetohydrodynamic, 9-211
nuclear reactor fuel (see Fuel elements, rods and bundles, nuclear reactor)
superport: curved, 3-216
depth of access lanes, 3-213 to 3-215
maneuvering areas, 3-216 to 3-218
width of access channels, 3-215 to 3-216
Channel dredged level, definition of, 3-215
Channel marking, superport, 3-220
Chantiers de l'Atlantique, 3-210
Chapapote (see Tar sands)
Chapelcross gas-cooled reactor, 5-89
Char, formation of: BI-GAS process, 1-212
carbon dioxide acceptor process, 1-219 to 1-220
clean coke process, 1-249
COED process, 1-256 to 1-260
COGAS process, 1-261 to 1-265
Hydrane process, 1-224 to 1-225, 1-229 to 1-230
Monsanto Landgard process, 4-18
pyrolysis process (Garrett), 4-14
Synthane process, 1-239 to 1-240
Toscoal process, 1-272 to 1-274
Char-oil-energy development (COED) process, 1-255 to 1-260
char, 1-260
coal pyrolysis, 1-257
gas product, 1-259 to 1-260
hydrotreating, 1-259
major results to date, 1-256 to 1-257
major technological factors, 1-257 to 1-259
oil filtration, 1-257 to 1-259
process flow, 1-255 to 1-256
product usage, 1-259
synthetic crude oil product, 1-259
Charging, process (see Feeding and charging systems)

Charleston, South Carolina:
daily terrestrial solar energy received at, 6-96
tidal range at, 8-40
Charleston, West Virginia, waste-conversion facility at, 4-16
Charleston nuclear plants, 5-22
Charter party, tanker and product carrier, 3-203
Chausey, France, tidal power station at, 8-43 to 8-45
Cheborksary hydroelectric plant, 8-5
Chelating agent, jet fuel use of, 3-38
Chemical analysis (see Analysis)
Chemical energy storage:
by batteries, 4-50 to 4-58
by photosynthesis processes, 4-5
Chemical feedstocks, natural gas source of, 2-129 to 2-132
Chemical fuels:
batteries, 4-50 to 4-58
for electric automobile, 4-55 to 4-58
biochemical, 4-5 to 4-11
butanol, 4-8
cellulose degradation, 4-9
conversion economics, 4-9
ethanol, 4-7 to 4-8
fermentation, 4-7 to 4-8
fuel cells, 4-9, 4-58 to 4-73
hydrogen, 4-26 to 4-35
characteristics of, 4-27 to 4-28
chemical applications of, 4-35
fermentation source of, 4-9
fuel applications of, 4-29 to 4-33
photosynthesis source of, 4-5 to 4-6
sources of, 4-33 to 4-37
hydrogen economy, 4-26 to 4-27
isopropanol, 4-8
methane, 4-8 to 4-9, 4-16 to 4-18
methanol, 4-44 to 4-49
oxidation, biochemical, 4-6 to 4-7
photosynthesis, 4-5 to 4-6
plant cultivation, 4-5 to 4-7
propellants, 4-74 to 4-90
pyrolysis, 4-13 to 4-23
shale oil extraction, 4-9
solid wastes, 4-12 to 4-25
Chemical industry:
electricity consumption of, 9-259
gas requirements of, 2-52 to 2-56
Chemical plants, energy conservation in, 3-301
Chemical potential energy, xxiv
Chemical reactions, battery, 4-52 to 4-53
Chemical rocket propellants (see Propellants)
Chemical shim, 5-43
Chemical treatment:
oil well, 3-133
spoiled-land areas, 1-105 to 1-106
Chemiluminescent analyzer, nitrogen oxides, 9-370 to 9-371
Chemistry:
combustion, 9-34 to 9-41
wet-scrubbing, 9-312 to 9-327
ChemShare Process Systems Ltd., 2-116 to 2-119

Cherbourg, France, tidal range at, 8-40
Cherokee nuclear plants, 5-17, 5-22
Cherry, Illinois, mine disaster at, 1-92
Chesapeake and Ohio Railroad, 1-154
Chest x-rays, coal miners', 1-92 to 1-93
Cheswick, Pennsylvania, mine disaster at, 1-92
Chevron Research Company, 4-45
Chicago, Illinois:
  barge shipments of LNG to, 2-102
  coal transfer station at, 1-170
  daily terrestrial solar energy received at, 6-97
Chicago and Northwestern Railroad, 1-154
Chief Joseph hydroelectric plant, 8-5
Chile:
  crude oil: annual production, 3-81
    production and exploratory wells completed, 3-87 to 3-88
    ultimate recovery potential for, 3-95
  lithium resources, 5-185
  metric tons to barrels conversion factor, 3-84
  natural gas: production of, 2-41
    reserves of, 2-43, 2-46
  nuclear power facilities in, 5-26, 5-37 to 5-38
  railway mileage and motive power, 9-226
  tidal range at Valparaiso, 8-41
China [see People's Republic of China; Republic of China (Taiwan)]
Chinon gas-cooled reactor, 5-89
Chinook coal mine, 1-30
Chirkey hydroelectric plant, 8-5
Chivor hydroelectric plant, 8-6
Chlorine:
  bomb calorimetry determination of, 1-69
  in coal, 1-50
  removal from catalytic rich gas process, 2-140
Chlorine trifluoride, propellant use of, 4-80, 4-82
Chlorophylls, 6-138
Chloroplast, 4-6
Chromatographic gas analyzer, 9-384, 9-389
Chromium-zinc catalyst, methanol process use of, 2-121
Chromous salts in gas-processing systems, 9-322
Chrysler Corporation, 3-26
Chubu EPCO nuclear plants, 5-29 to 5-30
Chukchi Sea, oil exploration in, 3-94
Churchill Falls hydroelectric unit, 8-5, 9-261
C &I/Girdler Corporation, 4-37
Cimarron Coal Corporation, 1-31
Cincinnati, Ohio, flooding of, 8-8
Cincinnati Gas and Electric Company, 5-16
Circuit:
  magnetohydrodynamic system, 9-213 to 9-216
  silicon solar cells and arrays, 6-61
  wattmeter, 9-300 to 9-302

Circuit capability versus transmission voltage, electric power, 9-269
Circuit miles:
  electric power transmission, 9-267 to 9-276
  by regions in United States, 9-255
Circular burner, 9-50 to 9-52
Circular neutral wing aerogenerator, 6-169
Circular register burner, pulverized-coal, 9-74 to 9-75
Circulation:
  boiler, 9-93 to 9-95
  pulverized coal, 9-64
Circulation ratio, boiler, 9-99
Circulator, helium, 5-47 to 5-48, 5-91 to 5-94, 5-101
Cities Service Gas Company, 2-64
Cities Service Research and Development Company, 3-251 to 3-264
Citrate gas-processing system, 9-317
City gate, pipeline, 2-74
City water costs, 3-274
C & K Coal Company, 1-31
Claddant (see Fuel elements, rods and bundles, nuclear reactor)
Clamshell bucket coal barge loader, 1-142 to 1-143, 1-175
Clarion River hydro reserve, 8-9
Classes of coal, definitions of, 1-17
Classification:
  air, solid waste, 4-13
  electric power requirements, 9-257 to 9-258
  gas consumers, 2-47 to 2-51
  natural gas wells, 2-23
Classifying cyclone, coal, 1-119
Clastic carbonate oil-reservoir rocks, 3-51
Claude's ocean thermal energy conversion plant, 6-118, 6-127
Claus off-gas treating, 2-174
Claus sulfur recovery process, 1-198, 1-200, 1-204, 1-213, 1-225, 1-250, 1-252, 1-254 to 1-255, 2-170, 9-314 to 9-316, 9-318, 9-328 to 9-334
  analyzer for, 9-373
Clausium, 9-23
Clay:
  definition of, 1-71
  drilling mud use of, 3-123
  incorporated in coal, 1-37
  oil shale content of, 3-147
  process catalysts of, 3-231
Clay products manufacture:
  gas requirements of, 2-52 to 2-56
  power requirements of, 9-259
Clayton Valley lithium resources, 5-183
Clean ballast, definition of, 3-203
Clean coke process, United States Steel Corporation, 1-249 to 1-254
Clean fuel process:
  Lurgi, 1-200
  Union Carbide Purox, 4-16
Cleaning:
  boiler tube, 9-119 to 9-120
  coal, 1-117 to 1-120, 1-125 to 1-126

Cleanup systems:
  boiling water reactor, 5-72
  industrial water systems, 9-360
  oil well (see Treating, oil well)
Clearance, underkeel, 3-213 to 3-215
Clearfield County, Pennsylvania, coals in,
  1-24
Cleavage, bond, 3-251
Cleavage plane, coal, 1-120
Clementine reactor, 5-111
Cleveland, Ohio:
  coal pipeline terminal at, 1-178, 1-180
  daily terrestrial solar energy received at,
    6-97
  hydroelectric power control station at, 8-
    23 to 8-24
  liquefied natural gas storage disaster at,
    2-100
Cleveland Electric Illuminating Company,
  5-16, 8-21
Climate control, solar (see Solar climate
  control)
Climate versus geothermal energy use, 7-
  54 to 7-55
Climatic effects, gas turbine operation, 9-180
Clinch River nuclear plant, 5-17, 5-22, 5-
  112
Clinchfield Railroad, 1-154
Clinker formation, oil shale processing, 3-
  149, 3-153
Clinton nuclear plant, 5-13
Closed-cycle gas turbine, 9-178 to 9-179
Closed-loop control, boiler, 9-125 to 9-128
Closed-Rankine-cycle ocean thermal
  energy conversion system, 6-118
Clostridium butylicum, 4-8, 4-10
Clothes washers, 9-260
Cloud point, Diesel fuel oils, 3-33 to 3-36,
  3-42
Clover, spoiled-land areas use of, 1-109
Cloverdale, California, 7-34
CMDB propellant, 4-83
CNTRFUGE drilling assistance computer
  program, 3-128
Coal:
  analysis of, 1-16 to 1-33, 1-41 to 1-42, 1-
    48 to 1-70
  auger mining of, 1-29 to 1-31
  anthracite, 1-15 to 1-27, 1-34 to 1-36, 1-
    54, 1-93, 1-195, 9-70
  ASTM classification of, 1-17 to 1-20
  bituminous, 1-15, 1-25 to 1-33, 1-93, 1-
    141 to 1-144, 1-170 to 1-177, 1-188
    to 1-200, 1-285 to 1-296
  breaking and crushing of, 1-15, 1-120 to
    1-124
  brown, 1-24
  cannel, 1-16
  characteristics of, 1-15 to 1-25, 1-38 to 1-
    40, 1-48 to 1-62
  classification of, 1-16 to 1-25, 1-53 to 1-55
  cleaning of, 1-125 to 1-126
  coking of, 1-58, 1-249 to 1-254
  consumption of, 1-25 to 1-33, 5-43
  contour mining of, 1-94 to 1-101
    seeding and planting of spoiled lands,
      1-102 to 1-116

Coal (Cont.):
  conversion (see extraction of, gasifica-
    tion of, liquefaction of, and pyrol-
    ysis of below)
  crushing and breaking of, 1-120 to 1-124
  dewatering of, 1-134 to 1-136
  drying of, 1-134 to 1-136
  excess air normally supplied for burning
    of, 9-46
  exploration for, 1-37 to 1-47
  extraction of, 1-276 to 1-284
  gasification of: Arge process, 1-293 to 1-
    295
    BI-GAS process, 1-212 to 1-218
    carbon dioxide acceptor process, 1-
      219 to 1-222
    char-oil-energy process, 1-255 to 1-
      260
    COED process, 1-255 to 1-260
    COGAS process, 1-260 to 1-268
    Fischer-Tropsch progress, 1-285 to 1-
      288
    Hydrane process, 1-223 to 1-231
    Hygas process, 1-201 to 1-211
    in-situ processes, 1-242 to 1-248
    Lurgi process, 1-188 to 1-200, 1-285 to
      1-296
    pyrolysis processes, 1-257, 1-272 to 1-
      275
    Rectisol process, 1-293 to 1-296
    Synthane process, 1-232 to 1-241
    Synthol process, 1-288 to 1-296
  geology of, 1-15 to 1-16, 1-37 to 1-41
  grades of, 1-16 to 1-33
  handling systems for, 1-137 to 1-150
  international classification of, 1-24
  lignite, 1-15 to 1-16, 1-25, 1-36, 1-54, 1-
    189, 1-195 to 1-196, 1-212, 1-227
  liquefaction of, 1-269 to 1-271
  mining of, 1-25 to 1-33, 1-71 to 1-93
  origin of, 1-15 to 1-16
  peat, 1-16
  preparation plants for, 1-117 to 1-136
  production of, 1-25 to 1-33
  pyrolysis of, 1-257, 1-272 to 1-275
  quality control of, 1-61
  ranks of, 1-16 to 1-33
  reserves of, 1-34 to 1-36, 1-170, 1-246 to
    1-247
  sampling of, 1-59 to 1-60, 1-120
  screening of, 1-124 to 1-125
  semianthracite, 1-15 to 1-27, 1-54, 1-196
  semibituminous, 1-15, 1-54
  sizes of, 1-25
  sizing for cyclone furnace firing, 9-84
  solvent refining of, 1-262 to 1-284
  storage of, 1-150 to 1-152
  strip mining of, 1-94 to 1-101, 1-103 to 1-
    116
  subbituminous, 1-15 to 1-27, 1-35, 1-54,
    1-195 to 1-196, 1-219 to 1-222
  testing of, 1-48 to 1-70
  transportation of, 1-137 to 1-187
    barge, 1-141 to 1-144, 1-170 to 1-177
    conveyor, 1-146 to 1-152
    pipeline slurry, 1-178 to 1-187
    railroad, 1-137 to 1-140, 1-153 to 1-169

Colorado (*Cont.*):
hydroelectric power production of, 8-7
natural gas: production and reserves of, 2-10 to 2-13, 2-16 to 2-18
underground gas storage capacity, 2-76
nuclear power installations in, 5-12
Fort Saint Vrain Nuclear Generating Station, 5-47 to 5-56
oil shale resources of, 3-147, 3-149 to 3-156
solar energy received at Boulder and Grand Lake, 6-96
Colorado Bituminous Coal District, 1-169
Colorado Interstate Corporation, 2-64
Colorado School of Mines Foundation, Inc., 2-20
Colstrip coal mine, 1-29
Columbia, Missouri, daily terrestrial solar energy received at, 6-96
Columbia Gulf Transmission Company, 2-64
Columbia Lightship, Oregon, wind power site at, 6-158, 6-160 to 6-162
Columbia River Gorge, wind power sites at, 6-147
Columbia River Valley, wind power sites at, 6-158
Columbium hydride, 4-43
Columbus, Ohio, daily terrestrial solar energy received at, 6-96
Combination carrier, crude oil and petroleum products, 3-203
Combined-cycle generating station, growth forecast for, 9-207
Combined-cycle power generation, 9-165 to 9-175, 9-191 to 9-194, 9-261
Combining weights, law of, 9-35
Combustible detection principles, 9-389
Combustion, 9-34 to 9-49
calculations for, 9-35, 9-38 to 9-40, 9-43 to 9-47
constants for various fuels, 9-36 to 9-37
definition of, 9-34
flue gas analysis, 9-47 to 9-49
fundamental laws of, 9-34 to 9-35
heats of, 9-41 to 9-42
Combustion air:
boiler system control of, 9-134
definition and role of, 9-43 to 9-47
metering of, 9-49
preheating of, 3-279 to 3-283
Combustion analysis, energy savings from, 3-286 to 3-287
Combustion chamber deposit modifiers, 3-27
Combustion control, boiler-furnace, 9-133 to 9-143
Combustion Engineering, Inc., 1-232 to 1-241, 5-75 to 5-86
Combustion-gasification operations, COGAS process, 1-261 to 1-266
Combustion guides, control system, 9-141 to 9-143
Combustion improvement chemicals, 9-349 to 9-350

Combustion modification, nitrogen oxides control by, 9-343 to 9-347
Combustion Power Company, Inc., 4-24
Combustion products versus excess air, 9-378
Combustion temperature:
gas turbines, 9-196
magnetohydrodynamic generator, 9-211
(*See also* Combustion)
Combustion-type engines (*see* Engines)
Combustion zone variables, Lurgi process, 1-193 to 1-194
Combustor, gas turbine, 9-177 to 9-181
Cominco Limited, 4-34
Commanche nuclear plants, 5-17 to 5-18, 5-22
Commercial aircraft, turbine inlet temperature of, 9-168
Commercial building heating, solar (*see* Solar climate control)
Commercial consumption, natural gas, 2-47 to 2-49
Commercial Testing & Engineering Company, 1-48 to 1-70
Commonwealth Edison Company, 5-13, 5-22, 9-19, 9-204
Communication:
coal mine, 1-84
pipeline system, 2-73
Compact tension fracture toughness, nickel steel, 2-112
Compagnie Nationale de Navigation, S.A., 3-210
Compartment conveyor, coal, 1-148
Compensation, gas calorimeter, 2-93
Competent rock, coal bed, 1-71
Composition determination (*see* Analysis)
Compoundable ship, petroleum-carrying, 3-203
Compressed Gas Association, 3-176
Compressed gas tankcars, 3-179 to 3-184
Compression of plasma, magnetic, 5-153, 5-158 to 5-172
Compression ratio:
engine, 4-42
gas turbine versus reciprocating compressor, 9-188
Compressive forces, brittle materials, 1-75
Compressive loading, coal mine roof, 1-75
Compressor:
centrifugal, 2-121
gas turbine, 9-176 to 9-183
liquefied natural gas use of, 2-102 to 2-109
waste heat utilization from, 9-351 to 9-353
Computational model:
solar energy absorber tube, 6-40
solar power system, 6-69 to 6-70
Computational requirements, electric utility, 9-282
Computer control, boiler systems, 9-157 to 9-162
Computer systems, power industry use of, 9-277 to 9-295
employee management, 9-281

Computer systems, power industry use of (*Cont.*):
    financial management, 9-279 to 9-281
    management science applications, 9-291 to 9-292
    new construction management, 9-291 to 9-292
    power plant monitoring and control, 9-289 to 9-291
    power system operation, 9-285 to 8-289
    power system planning, 9-283 to 9-284
Concentrating table, coal, 1-122
Concentration:
    Claus process, 9-329 to 9-332
    coal slurry, 1-179
Concentrator, solar optical, 6-10, 6-36 to 6-39, 6-46 to 6-54, 6-98 to 6-108
Concord coal mine, 1-29
Concrete coating, pipeline, 2-83
Concrete containment structure (*see* Containment structure, nuclear reactor)
Concrete lining, coal mine roof, 1-74
Concrete products manufacture, gas requirements of, 2-52 to 2-56
Concretion, coal bed, 1-71
Concretions, coal-associated, 1-39
Condeep supertanker, 6-123
Condensate costs, 3-274
Condenser:
    heat-transfer coefficients of, 3-285
    ocean thermal energy conversion system use of, 6-118 to 6-129
Condensing cycle, 9-26
Condoluminescence, 4-31
Conduction loss, thermoelectric cooling system, 9-246
Cone separator, coal, 1-130
Confinement, plasma, 5-141 to 5-147
Conglomerate rocks, oil reservoir, 3-51
Congo (Brazzaville):
    crude oil: exploratory and production wells in, 3-86, 3-88
        proved recoverable reserves of, 3-82
        ultimate recovery potential for, 3-92
    natural gas, ultimate recovery potential for, 2-44
    railway mileage and motive power, 9-226
Conical distributor, two-way flow-axial turbine, 8-32
Conical stockpile, coal, 1-137 to 1-138
Coning, oil well, 3-131
Conneaut, Ohio, coal transfer station at, 1-170
Connecticut:
    electric power generating capacity of, 9-253
    emission limits for SO_2 and NO_x, 9-374
    hydroelectric power production of, 8-6
    nuclear power installations in, 5-12 to 5-13
Connecticut Light and Power Company, 8-17
Connecticut Yankee Atomic Power Company, 5-12

Connections, magnetohydrodynamic generator, 9-210 to 9-211
Conoco Coal Development Company, 1-219 to 1-222, 1-276 to 1-281
Conradson carbon conversion, Shell SNG process, 2-165
Conradson carbon residue:
    crude oil, 3-15
    crude shale oil, 3-149, 3-151
Conroe oil field, 3-62, 3-105
Conservation, refinery and process plant energy, 3-272 to 3-303
    air system losses, reduction of, 3-294 to 3-296
    air-to-oil ratio, optimization of, 3-285
    checklists for conservation, 3-300 to 3-303
    combustion analyzers, savings from, 3-287
    fluid catalytic cracking process, 3-275
    fuel and products conservation, 3-297 to 3-300
        tank design, 3-297
        vapor recovery systems, 3-297 to 3-300
    gas turbines with heat recovery, 3-287 to 3-288
    heat exchangers, 3-284
    hydrogen unit, 3-277
    industrial water system, 9-361
    insulation, 3-289 to 3-290
    lighting system optimization, 3-291
    partial-oxidation unit, 3-277
    power factor optimization, 3-296 to 3-297
    waste heat recovery from process streams, 3-283 to 3-284, 3-290 to 3-294
Conservation of energy, law of, xxv, 9-35
Conservation of matter, law of, 9-34 to 9-35
Consistency (size), coal slurry, 1-179 to 1-183
Consol synthetic fuel (CSF) process, 1-276 to 1-281
    coal extraction, 1-276
    Cresap, West Virginia, pilot plant, 1-277 to 1-281
    economic potential, 1-277
    extract hydrogenation, 1-276
    zinc chloride hydrocracking, 1-276 to 1-277, 1-281
Consolidated Edison Copany, 5-15, 9-22, 9-276
Consolidated Gas Supply Corporation, 2-64
Consolidated Natural Gas Service Company, 1-261
Consolidation Coal Company, 1-31, 1-156, 1-178 to 1-180, 1-281
Consolidation Coal Pipeline Company, 1-178, 1-180
Constant extinction angle, 9-216 to 9-218
Constant firing angle, 9-216
Constant tension coil, fusion reactor, 5-164
Construction:
    pipeline: marine, 2-77 to 2-89
        overland, 2-65, 2-70 to 2-74
    Port of Rotterdam, 3-220 to 3-224

Costs and economics, electric power systems (*Cont.*):
  superconducting turbine generators, 9-206 to 9-208
  superheater, 9-115 to 9-116
  thermoelectric cooling systems, 9-244
  environmental systems: Claus process, 9-333 to 9-334
    limestone gas absorption system, 9-314
    magnesium-based scrubbing system, 9-314
    Wellman-Lord process, 9-326
  geothermal systems, 7-10 to 7-11, 7-29, 7-55 to 7-58
  hydrogen systems: electrolysis of water, 4-34
    as energy transport, 4-31 to 4-33
    as heating fuel, 4-30 to 4-31
    for motive power, 4-29, 4-40
  nuclear power systems: fusion reactor reference designs, 5-170
    uranium mining versus coal mining costs, 5-118
    water reactor versus fast reactors, 5-11
  petroleum production and refining: gas supply potential forecasting, 2-24 to 2-25
    oil well drilling, 3-121 to 3-129
    oil well economic recovery, 3-61 to 3-71
    refining and processing energy costs, 3-272 to 3-303
  petroleum transportation: pipeline movement, 3-174
    overland, 2-63 to 2-70
    tanker movement, 3-210
  solar energy systems: electric power generation, 6-6 to 6-7, 6-78 to 6-80
    ocean thermal energy conversion plants, 6-130 to 6-135
    satellite systems, 6-114 to 6-115
    silicon cells, 6-70
    solar climate control, 6-89 to 6-90
    storage of energy, 6-22 to 6-23, 6-35, 6-43 to 6-44
    tower energy collector, 6-27 to 6-28, 6-32 to 6-33
    wind power, 6-172 to 6-174
  solid-waste conversion systems: American Thermogen process, 4-20
    biological methane process, 4-17 to 4-18
    CPU-400 system, 4-25
    Garrett pyrolysis process, 4-15
    Monsanto Landgard process, 4-18 to 4-19
    supplemental fuel system, 4-23
    Union Carbide Purox process, 4-16
    waterwall incineration process, 4-21
    wood waste-to-oil process, 3-168
Cote Blanche Bay oil field, 3-63
Cote Blanche Island oil field, 3-64
Cotentin Peninsula, tidal range at, 8-43
Couple, thermoelectric cooling system, 9-242 to 9-246

Coupling, bulb-type turbine generator, 8-34
Coupling agents, oil well sand control use of, 3-133
Coupling-type torque transducer, 9-307
Cowden (North) oil field, 3-62
CPU-400 waste conversion system, 4-23 to 4-25
Cracking phase, partial-oxidation gasification process, 2-171
Cracking processes, petroleum refining, 3-18 to 3-21, 3-227, 3-231 to 3-235, 3-248, 3-251 to 3-263, 3-275 to 3-276
Crash bags, propellants for, 4-77
Crash stop, petroleum carrier, 3-219
Crawler, contour-mining use of, 1-94 to 1-101
Creep, steam turbine rotor, 9-164
Cresap, West Virginia, CSF process pilot plant located at, 1-277 to 1-281
Cresols, clean coke process production of, 1-253
Cretaceous Geologic System, fossil fuels formed during, 3-51
  coal, 1-37
  crude oil: in Canada, 3-76 to 3-77
    in United States, 3-59
  natural gas, in United States, 2-14
Critical field, superconducting material, 5-159, 5-161
Critical heat flux, boiler, 9-94 to 9-95
Critical mass, 5-8, 5-44
Critical size, 5-44
Critical temperature, 9-26
  superconducting materials, 5-158 to 5-165
Criticality, secondary, liquid-metal fast breeder reactor, 5-105 to 5-106
Croloy superheater and reheater tubes, 9-117
Crooked hole, drilling of, 3-112 to 3-115
Crop wastes, energy conversion of, 4-5 to 4-6, 4-13
Crops for spoiled-land areas, 1-115
Cross-drum boilers, emission characteristics of, 9-340
Cross section:
  magnetohydrodynamic, 9-211
  nuclear (*see* Nuclear energy technology)
Crown Point, Oregon, wind power site at, 6-162
Crownvetch, spoiled-land area use of, 1-109
Cruacham pumped-storage plant, 8-18
Crucible-swelling number, hard coals, 1-21 to 1-23
Crude gas shift conversion, Lurgi process, 1-198
Crude oil:
  basic classes of, 3-12 to 3-13
  composition of, 3-12 to 3-17, 3-252
  definition of, 3-12, 3-55
  processing of, 3-21
    (*See also* Petroleum)
  production of (*see* Production, crude oil)
  products of, 3-18
    (*See also* Petroleum)

Crude oil (*Cont.*):
  reserves of (*see* Reserves, crude oil)
  synthetic crude oil: from Athabasca
    bitumen, 3-165
    from wood-to-oil process, 3-168 to 3-173
    (*See also* Synthoil coal conversion
    process)
Crude petroleum carriers (*see* Seagoing
  vessels, petroleum carrying)
Crude shale oil, properties of: from N-T-U
  process, 3-149
  Paraho process, 3-155
  Tosco II process, 3-156
  Union Oil process, 3-151
  U.S. Bureau of Mines process, 3-151,
    3-155
Crushed coal:
  excess air normally supplied for burning,
    9-46
  sizing for cyclone furnace, 9-84
Crusher, coal, 1-118 to 1-124, 1-181 to 1-183
Crust, earth's, 7-4
Crust formation, spoiled-land area, 1-105
Cryogenic aircraft fuels, 3-39
Cryogenic cooling, superconducting
  magnet, 5-163 to 5-165
Cryogenic electric power transmission
  systems, 9-276
Cryogenic liquid propellants, 4-77
Cryogenic recovery of hydrocarbons from
  natural gas, 2-131 to 2-132
Cryogenic service alloys, 2-109 to 2-113
Cryogenic system:
  expansion turbine use of, 9-200 to 9-201
  fusion reactor reference design of, 5-169
Cryogenic Systems for the Electric
  Industry program, 9-207
Cryogenic upgrading of low-Btu gases, 2-116 to 2-119
Cryogenically cooled turbine generators,
  9-205 to 9-208
Crystal River nuclear plant, 5-13, 5-22
Crystalline oxidizers, propellant use of, 4-79 to 4-85
Cuba:
  crude oil: production of, 3-88
    ultimate recovery potential for, 3-94
  metric tons to barrels conversion factor,
    3-84
  natural gas, ultimate recovery potential
    for, 2-46
  railway mileage and motive power, 9-226
  tidal range at Havana, 8-40
Cultivation, plant, 4-5 to 4-6
Cumulative production:
  of crude oil: API definition of, 3-58
    in Canada, 3-74
    in United States, 3-57
    worldwide, 3-79 to 3-82
  of natural gas: in Canada, 2-35
    in United States, 2-31
Curing processes, geothermal energy for,
  7-50, 7-55 to 7-57

Curley cannel, 1-16
Current, silicon solar cell, 6-61
  (*See also* Battery power)
Current effects, marine pipeline, 2-77,
  2-85 to 2-89
Current forces, drilling structure, 3-110 to
  3-111
Curvature, railway track, 9-228
Customer information system, electric
  utility, 9-278, 9-280, 9-285, 9-292
Cutler-Hammer, Inc., 2-90 to 2-99
Cutler-Hammer type of recording gas
  calorimeter, 2-91 to 2-94
Cutter operator accidents, coal mine, 1-86
  to 1-91
Cyanophyta, 6-137
Cycle:
  liquefied natural gas refrigeration, 2-102
    to 2-109
  steam, 9-24 to 9-30
Cycle efficiency, liquefaction processes, 2-106 to 2-109
Cycle oil, 3-225, 3-231 to 3-232
Cycle time, contour-mining, 1-99
Cyclic operation, tidal power plant, 8-37 to
  8-39, 8-41 to 8-45
Cyclic photophosphorylation, 6-139
Cyclic rotor system loads, 6-144
Cycling, oil reservoir, 3-66
Cyclohexane, synthesis of, 3-20
Cyclone:
  coal separation by, 1-118 to 1-119, 1-122,
    1-129 to 1-131
  fluid catalytic cracking process use of, 3-232
  oil shale retorting use of, 3-152
  Weber, 7-25
CYCLONE drilling assistance computer
  program, 3-128
Cyclone-fired power plants, particulate
  properties of, 9-398
Cyclone furnace, 9-79 to 9-86
  controls and performance, 9-82 to 9-86
  design features of, 9-82
  fuel suitability for, 9-81 to 9-82
  operating principle of, 9-80 to 9-81
Cyclone-furnace-fired boiler control
  system, 9-140 to 9-141
Cyclone steam separator, 9-100 to 9-102
Cyclopentadiene, 9-322
Cyclopentane content, natural gas, 2-129
Cyclotetramethylene tetranitramine,
  propellant use of, 4-83
Cyclotrimethylene trinitramine, propellant
  use of, 4-83
Cylinder, compressed gas, 3-178
Cylindrical concentrator, solar energy, 6-10
Cyprus:
  crude oil: production of, 3-88
    ultimate recovery potential for, 3-92
  natural gas, ultimate recovery potential
    for, 2-44
  solar water heaters in, 6-93
Cytochromes, 6-138 to 6-139
Czechoslovakia:
  coal reserves of, 1-36

Czechoslovakia (*Cont.*):
  crude oil: production of, 3-88
    proved recoverable reserves of, 3-83
    ultimate recovery potential for, 3-93
  natural gas: production of, 2-40
    reserves of, 2-43, 2-45
  nuclear power facilities in, 5-26, 5-36, 5-38
  radium, early use of, 5-173
  railway mileage and motive power, 9-226

Dahomey:
  crude oil: production of, 3-88
    ultimate recovery potential for, 3-92
  natural gas, ultimate recovery potential for, 2-44
  railway mileage and motive power, 9-227
Daily terrestrial solar energy received on a horizontal surface, 6-95 to 6-97
Dairy wastes, energy conversion of, 4-13
Dairyland Power Cooperative, 5-18
Dakar, tidal range at, 8-40
Dallas, Texas, daily terrestrial solar energy received at, 6-96
Dallas Power & Light Company, 1-144 to 1-145
Dalles, The, hydroelectric plant, 8-5
Dalton's law, 9-35
Dams:
  Europoort use of, 3-221
  hydropower, 8-5 to 8-11, 8-18 to 8-22
  tidal power, 8-27 to 8-28, 8-36 to 8-39
Dandelion, spoiled-land area use of, 1-110
Daniel Johnson hydroelectric plant, 8-5
Dariaz turbine, 8-15 to 8-16
Dark reaction, photosynthesis, 6-139
Darrieus vertical axis rotor, 6-144
Data compilation and processing, coal-exploration programs, 1-44 to 1-47
Dave Johnston coal mine, 1-29
Davis air resistance coefficient, 9-225
Davis-Besse nuclear plant, 5-16, 5-22
Davy Powergas Ltd., 2-120 to 2-128, 9-324 to 9-327
Dawson, New Mexico, mine disaster at, 1-92
DC/AC inverters, magnetohydrodynamic generator systems, 9-211 to 9-218
Dead load, drilling structure, 3-107
Dead storage, coal, 1-138
Deadweight, tanker and product carrier, 3-203
Deadweight tonnage:
  seagoing petroleum carrying vessels, 3-198 to 3-200, 3-207 to 3-214, 3-219 to 3-220
  versus stopping distance of ship, 3-218 to 3-219
Dealkylation, clean coke process use of, 1-253
Deasphalting, solvent (*see* Solvent deasphalting)
De Beque, Colorado, oil shale test site at, 3-150

Debutanization, natural gas, 2-129 to 2-132
Debutanizer:
  economic reflux ratio for, 3-295
  hydrocracking process use of, 3-235
Decay probability, 5-44
Decca port navigational aid, 3-219, 3-222
Deceleration (ion), electrostatic direct conversion of plasma energy use of, 5-151
Dechlorination, catalytic rich gas process, 2-140
Decomposed rock, contour-mining handling of, 1-101
Deep mining, coal, 1-30 to 1-35
Deepwater pileing (*see* Marine pipelines)
Deflection drilling, oil well, 3-115 to 3-118
Deflector, boiler, 9-100 to 9-102
Deicing, geothermal energy for, 7-56
Deisohexanizer, 3-242
Deisopentanizer, 3-243
Delaware:
  electric power generating capacity of, 9-253
  emission limits for SO_2 and NO_x, 9-374
  nuclear power installations, 5-13
Delaware basin, 2-23
Delayed coking:
  petroleum, 3-18, 3-230, 3-246 to 3-248
  tar sand bitumen upgrading use of, 3-164 to 3-166
Delayed-mixing burners, nitrogen oxides control with, 9-347
Delayed neutron, 5-44
Delhi oil field, 3-63
Delmarva Power and Light Company, 5-13
Delta coal mine, 1-30
Delta plains, coal formation in, 1-37
Delta T, seawater, 6-118
Delta West oil fields, 3-62 to 3-63
Demand, petroleum products, 3-96 to 3-102
  (*See also* Consumption, United States)
Demand Regions, Petroleum Administration for Defense, 2-62
Demetallization, residuums, 3-262
Demethanization, natural gas, 2-129 to 2-132
Demister, oil shale retorting use of, 3-152
Denmark:
  crude oil: annual production, 3-88
    exploratory and production wells completed, 3-86
    proved recoverable reserves of, 3-83
    ultimate recovery potential for, 3-93
    metric tons to barrels conversion factor, 3-84
  natural gas, reserves of, 2-42, 2-45
  railway mileage and motive power, 9-226
  uranium resources of, 5-176
  wind power experience and research, 6-151
Dense media separation, coal, 1-117 to 1-120, 1-126 to 1-133
Dense-medium cyclone, 1-131
Dense phase, fluid catalytic cracking process, 3-232

Density, coal handling importance of, 1-147
Densmore tankcar development, 3-176
Denver, University of, 2-20 to 2-21, 2-47 to 2-59
Denver and Rio Grande Western Railroad, 1-154
Denver Research Institute, oil shale pilot plant of, 3-154
Department of Public Services, State of New York, 8-19
Departure from nucleate boiling (DNB), 9-94 to 9-98
Depentanizer, 3-242
Deposits, coal, 1-37 to 1-40
  (*See also* Reserves)
Depreciation, tankcar, 3-194
Depressurization system, boiling water reactor, 5-73
Depropanization, natural gas, 2-129 to 2-132
Depth:
  coal mine, 1-35 to 1-36
  versus drilling time, oil well, 3-122
  versus mud weight, geopressured wells, 3-121
  oil and gas wells, 2-24 to 2-25, 2-27
  oil well records, 3-129
  tanker and product carrier, 3-203
Derichment, liquefied natural gas, 2-141
Derivative control system, boiler, 9-128
Derrick:
  oil well, 3-107
  pipeline, 2-73
Desalination, geothermal energy for, 7-50, 7-55 to 7-57
Desalting, crude oil, 3-225
Desert operation, gas turbine, 9-188
Desert Southwest, solar energy available in, 6-26
Design:
  boiler, 9-108 to 9-110
  electric utility, computer aids for, 9-290 to 9-291
  furnace and burner, 9-340 to 9-348
  gas and expansion turbines, 9-176 to 9-177
  petroleum carrying ships (*see* Seagoing vessels, petroleum carrying)
  superports (*see* Superports)
Desulfurization:
  BI-GAS process, 1-213 to 1-214
  carbon dioxide acceptor process, 1-221
  carbonization/hydrogenation metallurgical coke process, 1-249 to 1-251
  catalytic rich gas process, 2-140
  char-oil-energy development (COED) process, 1-257, 1-259
  clean fuel gas process, 1-200
  coal gasification process, 9-171, 9-174
  COGAS process, 1-262 to 1-264, 1-266 to 1-267
  Consol (CSF) process, 1-276
  crude oil, 3-228
  fluidized-bed hydrogenation process, 2-161

Desulfurization (*Cont.*):
  fuel gas, 2-174
  fuel oil, 3-40
  Hydrane process, 1-225 to 1-228
  Hygas process, 1-203 to 1-204
  methane rich gas process, 2-152
  Rectisol process, 1-198 to 1-199, 1-287, 1-294 to 1-295
  SELEXOL process, 1-213
  solvent refining process, 1-283 to 1-284
  steam reforming of hydrocarbons, 4-37
  Synthane process, 1-232 to 1-237, 1-240
  Synthoil process, 1-269 to 1-271
  Toscoal process, 1-272 to 1-274
  (*See also* Hydrogen sulfide; Sulfur; Sulfur dioxide)
Detector, boiler flame, 9-147 to 9-149
  (*See also* Instrumentation)
Detergent additives:
  Diesel fuels, 3-30
  gasoline, 3-27
Detergents, petroleum sources for, 3-20
Detonation hazard, propellant, 4-87
Detroit Edison Company, 1-156, 5-14, 9-277
Deuterium, 4-27, 4-29, 5-44
  reaction chain, 5-137 to 5-138, 5-150, 5-182
  tritium fuel pellet, 5-141
Deuteron, 5-44, 5-138
Development Engineering, Inc., oil shale vertical retorting kiln of, 3-153
Deviation drilling, oil well, 3-115 to 3-118
Deviation surveying, 3-112
Devolatilization, coal, 1-192, 1-202, 1-219, 1-250, 1-255
Devolatilizer, coal gasification facility, 9-171, 9-174
Devonian Geologic System, gas and oil formed during, 3-51
  in Canada, 3-76 to 3-77
  in United States, 2-14, 3-51, 3-59
DEV PLAN drilling assistance computer program, 3-128
Dewar system, Princeton Reactor, 5-166
Dewatering, coal, 1-118 to 1-120, 1-134 to 1-136
D. EXP drilling assistance computer program, 3-128
D-exponent, oil well drilling, 3-128
DIABIT drilling assistance computer program, 3-128
Diablo Canyon nuclear plant, 5-12, 5-22
Diagonal-flow hydraulic turbine, 8-12
Diameter, gas pipelines, 2-63 to 2-66, 2-73
Diamond M oil field, 3-63
Diatomite processing, geothermal energy for, 7-57 to 7-58
Dibenzothiophenes, crude oil content of, 3-254
Dielectrics (gas), electric power transmission use of, 9-276
Diesel auxiliary, nuclear plant use of, 5-99
Diesel-electric locomotives, 9-224, 9-226 to 9-227, 9-232 to 9-233
Diesel-electric power plants, thermal efficiency of, 4-71
Diesel engine emission levels, 9-196 to 9-197
  (*See also* Engines)

Diesel fuel components, crude oil, 3-15
Diesel fuels:
    additives for, 3-29 to 3-31
    annual United States survey of, 3-27 to 3-36
    API gravity of, 3-33 to 3-37
    cetane number of, 3-28 to 3-36
    characteristics of, 3-27 to 3-36
    consumption of, United States, 3-97 to 3-98
    distillation range end point of, 3-28 to 3-36
    hydrotreating for improvement of, 3-235
    sulfur content of, 3-28 to 3-36
    viscosity of, 3-28 to 3-36
Diesel-hydraulic locomotives, 9-224, 9-226 to 9-227, 9-233 to 9-236
Diesel-mechanical locomotives, 9-223, 9-226 to 9-227
Diesel oil feed, synthetic natural gas process, 2-167
Diethylphthalate, propellant use of, 4-83
Differential etching, oil well treatment use of, 3-135
Diffusion:
    gaseous, 5-44
    hydrogen through palladium film, 4-37
Diffusion flame, 9-335
Digester, sludge, 4-16 to 4-17
Digital computer:
    boiler system use of, 9-161 to 9-162
    electric utility industry use of, 9-277 to 9-295
Digital sequence logic burner control system, 9-156
Diisopropanolamine solvent, 2-133
Dike, liquefied natural gas facility, 2-100
Dilatometer, Audibert-Arnu, 1-21 to 1-23
Dilute phase, fluid catalytic cracking process, 3-232
Dilute phase reactor, Hydrane process, 1-226 to 1-229
Dimensional trends, seagoing petroleum carrying vessels, 3-199, 3-207 to 3-214, 3-219 to 3-220
Dimensions, tankcar, 3-181 to 3-182, 3-187 to 3-190
Dimethylbutane, 3-241 to 3-242
Dimethylformamide, 9-322
Dimethylhexane, alkylate content of, 3-238
Dimethylhydrazine (unsymmetrical), propellant use of, 4-81
Dinard, France, tidal power plant located at, 8-25 to 8-27
Dinitrogen trioxide, 9-230
Diode, 9-221
Direct absorber, solar, 6-11 to 6-13
Direct current electric power transmission, 9-276
Direct digital control, boiler, 9-157 to 9-161
Direct energy conversion fusion reactor, 5-150 to 5-154
Direct-fired cyclone furnace, 9-83
Direct-firing system, pulverized coal, 9-264
Directional drilling, oil well, 3-115 to 3-118
Directional flame burner, 9-341
Disasters, coal mine, 1-92

Discharge characteristics, battery, 4-53
Discharging, tankcar, 3-183 to 3-193
Discovery, crude oil: API definition of, 3-61, 3-66
    estimates of, 3-66 to 3-71
    (See also Offshore gas and oil drilling and production; Reserves)
Dishwashers, automatic, 9-260
Disk-and-roller mill coal pulverizer, 9-68
Dislocations, coal deposit, 1-40
Dispatcher training computer program, electric utility, 9-285
Dispatching:
    Seneca pumped-storage plant, 8-23
    tidal power plant, 8-34 to 8-36
Dispatching computer programs, electric utility, 9-286 to 9-287
Displacement, tanker and product carrier (see Draft, tanker and product carrier)
Dissociation, hydrogen, 4-27
Dissolved-gas drive hydrocarbon reservoir, 3-105 to 3-106
Dissolved natural gas, Canadian reserves of, 2-34
Dissolved solids, geothermal fluid content of, 7-11 to 7-12, 7-35 to 7-36, 7-43 to 7-49
Dissolver, coal solvent refining use of, 1-282 to 1-283
Distance factors, contour-mining, 1-98
Distillate fuels:
    consumption of, in United States, 3-96 to 3-98
    pipeline transportation of, 3-174
    production of, 3-225
Distillate heating oils (see Burner fuel oils)
Distillation, crude oil, 3-18, 3-225 to 3-226, 3-245 to 3-246, 3-259
Distillation range:
    aviation fuels, 3-37
    burner fuel oils, 3-41 to 3-43
    crude oil, 3-13 to 3-17
    crude shale oil, 3-149, 3-151, 3-155 to 3-156
    Diesel fuels, 3-28 to 3-36
    gasoline, 3-23 to 3-25
    hydrofluoric acid alkylate, 3-238
    synthetic crude oil from Athabasca bitumen, 3-165
Distilling, geothermal energy for, 7-56 to 7-57
Distribution of petroleum and petroleum products (see Transportation, petroleum and petroleum products)
Distribution circuit analysis computer program, electric utility, 9-284
Distribution cost, electric utility, 9-257
Distributor, two-way flow-axial turbine, 8-32
District of Columbia:
    daily terrestrial solar energy received at, 6-96
    electric power generating capacity of, 9-253
    hydroelectric power production of, 8-6
Disturbance, electric power pumped-storage plant, 8-19

Disulfides, crude oil content of, 3-254
Diurnal variation, solar power, 6-30
Divers, marine pipelaying use of, 2-87
Diverting agents, oil well acidizing use of, 3-134
Dixie pipeline, 3-175
DNB (see Departure from nucleate boiling)
Dodge City, Kansas, daily terrestrial solar energy received at, 6-95
Dogleg, drilling hole, 3-113
Dogwood, spoiled-land area use of, 1-112
Dole, West Virginia, mine disaster at, 1-92
Dolomite:
    hydrogen sulfide removal with, 9-175
    oil shale content of, 3-147
Dolomite acceptor, 1-219
Dolomite systems, sulfur dioxide removal with, 9-313 to 9-314
Dome-shaped oil and gas structures, 3-104
Dome-type tankcars, 3-183, 3-191 to 3-193, 3-197
Domestic heating oil (see Burner fuel oils)
Dominican Republic:
    crude oil: production of, 3-88
        ultimate recovery potential for, 3-94
    natural gas, ultimate recovery potential for, 2-46
    railway mileage and motive power, 9-227
Doppler effect, fast reactor, 5-106 to 5-107
Doppler radar, superport use of, 3-219 to 3-220
Dos Cuadras oil field, 3-63
Do-Son, tides at, 8-37, 8-39
Dotiki coal mine, 1-30
Double-alkali gas absorption process, 9-315
Double-cycle operation, tidal power plant, 8-39, 8-41 to 8-45
Double-methanzation route, catalytic rich gas process, 2-143 to 2-144
Double-register burner, 9-342
Double-roll crusher, coal, 1-123 to 1-125
Doubling time, 5-117, 5-155, 5-170
Douglas Point nuclear plants, 5-14, 5-22
Dounreay factor, 5-111
Dover, England, tidal range at, 8-40
Dow Chemical Company, 3-133
Down-hole pack-off devices, 3-139
Downflow fixed-bed hydrodesulfurization reactor, 3-255 to 3-257
Draft:
    admissible, 3-213
    pipeline vessel, 2-79
    tanker and product carrier, 3-199, 3-203, 3-207 to 3-214, 3-219 to 3-220
Draft tube, hydraulic turbine, 8-15 to 8-17
Drag coefficient, railway rolling stock, 9-225
Dragon reactor, 5-88 to 5-90
Drainage and erosion control contour-mining, 1-94 to 1-115
Dravo Corporation, 1-143, 1-170 to 1-177
Draw slate, 1-75
Drawoff versus pressure, geothermal field, 7-17 to 7-18

Drawworks, oil derrick, 3-107
Dredge, pipeline, 2-73
Dresden nuclear facilities, 5-13, 5-21
Dresser Station, Public Service of Indiana, 9-171
Drill bench, contour-mining, 1-94
Drill-off test, 3-128
Drill pipe design, 3-128
Drill string, 3-113 to 3-118
Drillability determination, coal mine roof, 1-77 to 1-78
Drilling:
    coal exploration, 1-40 to 1-41
    crude oil exploration, 3-61 to 3-71, 3-86 to 3-87, 3-90 to 3-91, 3-94
        (See also Production, crude oil)
    geothermal system wells, 7-7, 7-17 to 7-18, 7-22, 7-31 to 7-36, 7-51 to 7-52
Drilling assistance programs, 3-126 to 3-128
Drilling depth versus potential for natural gas, 2-27
Drilling depth records, oil well, 3-129
Drilling objectives, natural gas well, 2-23
Drilling platform, gas turbines for, 9-188 to 9-190
Drilling rate versus hydraulic horsepower, oil well, 3-124
Drilling structures, oil and gas, 3-107 to 3-112
Drilling time versus depth, oil well, 3-122
DRILLOFF drilling assistance computer program, 3-128
DRILPIPE drilling assistance computer program, 3-128
Drive, control rod (see Fuel elements, rods and bundles, nuclear reactor)
Drop-shatter test, coking coal, 1-58
Drum, boiler, 9-99 to 9-109, 9-111 to 9-115, 9-120 to 9-121
Dry-ash furnace, boiler, 9-110 to 9-111
Dry basis, flue gas analysis, 9-40
Dry-bulk carriers, 3-212
Dry-habitat welding, marine pipeline, 2-86
Dry mineral-matter-free fixed carbon, definition of, 1-16
Drying:
    coal, 1-118 to 1-120, 1-134 to 1-138
    pulverized coal, 9-64
    wood wastes, 3-169
Drying operations:
    geothermal energy for, 7-50, 7-55 to 7-57
    waste energy conversion system use of, 4-13
Dual refrigerant cascade, 2-106
Duane Arnold Energy Center, 5-13
Dugger coal mine, 1-30
Duke Power Company, 5-16 to 5-17
Dumping fee, solid waste, 4-15 to 4-18, 4-20 to 4-21, 4-23, 4-25
Dumping operations, contour-mining, 1-94 to 1-101
Dungeness gas-cooled reactor, 5-89
Du Pont de Nemours, E. I., & Co., Inc., 1-248, 3-26, 9-362 to 9-375
Duquesne Light Company, 1-31, 5-17, 9-22
Dust, coal mine, 1-84 to 1-85, 1-92 to 1-93

Dust collection, coal mine roof drilling, 1-78
Dust loading, cyclone furnace, 9-84
Dustiness index, coking-coal, 1-58
Dutch Government, Department of Waterways, 3-210
Duty cycle, Diesel-electric locomotive, 9-230
Dworshak hydroelectric plant, 8-5
Dyes, gasoline, 3-27
Dynamic loading conditions, floating rig, 3-109
Dynamometer and torque measurements, 9-305 to 9-311
Dynamometer constant, 9-306 to 9-307
Dynamometer-type wattmeter, 9-297 to 9-298

Eagle No. 2 coal mine, 1-30
Earth, geothermal gradient of, 7-4
Earth-storable liquid propellants, 4-77
Earthmoving equipment, contour-mining, 1-94 to 1-101
Earthquake load, drilling structure, 3-107
Easement grant, pipeline, 2-72
East Africa, railway mileage and motive power of, 9-226
East African Rift, 7-4
East Central Area Reliability Coordination Agreement (ECAR), 9-266
East Coast of Canada, offshore drilling in, 3-73
East Lansing, Michigan, daily terrestrial solar energy received at, 6-97
East Mesa, California, geothermal installation at, 7-13
East North Central Region:
  electric utility industry in, 9-254 to 9-256
  hydroelectric power production of, 8-6
  petroleum products survey of, 3-32 to 3-36
  [See also Northeastern District (API Survey)]
East Pacific Ridge, 7-4
East Region, fuel requirements for electric power generation, 9-261 to 9-264
East South Central Region:
  electric utility industry in, 9-254 to 9-255
  hydroelectric power production of, 8-7
  petroleum products survey of, 3-32 to 3-36
  [See also Southeastern District (API Survey)]
East Texas crude oil, characteristics of, 3-14 to 3-15
East Texas oil field, 3-62
Eastern Associated Coal Corporation, 1-31
Eastern Canada, natural gas reserves in, 2-34 to 2-36
Eastern forbs, surface-mine revegetation use of, 1-109
Eastern grass species, surface-mine revegetation use of, 1-108
Eastern Interior NPC petroleum province, 2-62

Eastern Interior Region, coal formations in, 1-37, 1-40
Eastern Region:
  coal transporation in, 1-165 to 1-166
  petroleum products survey of, 3-32 to 3-36
  [See also Northeastern District (API Survey); Southeastern District (API Survey)]
Eastern Rocky Mountain NPC petroleum province, 2-62
Eastern Transvaal Highveld, coal fields of, 1-285
Eastern trees and shrubs, surface-mine revegetation use of, 1-113
Ebasco Services, Inc., 5-47, 9-161
Ebullating-bed reactor:
  hydrocracking process use of, 3-234 to 3-235
  hydrodesulfurization process use of, 3-257 to 3-259
Eccles, West Virginia, mine disaster at, 1-92
Economics and costs (see Costs and economics)
Economizer, pressurized water reactor, 5-77
Economy:
  boiler fuel, 9-133 to 9-138
  gasification-based combined-cycle power station, 9-168 to 9-170
  superconducting turbine generators, 9-207 to 9-208
Ecuador:
  crude oil: annual production of, 3-81
    percent of recoverable reserves of, 3-85
    production and exploratory wells completed, 3-87
    ultimate recovery potential for, 3-95
    metric tons to barrels conversion factor, 3-84
  natural gas: production of, 2-41
    reserves of, 2-43, 2-46
  railway mileage and motive power, 9-226
  tar sand deposits in, 3-160 to 3-161
Eddy-current brake dynamometer, 9-305
Edison Electric Institute, 9-205, 9-251, 9-253 to 9-254, 9-257 to 9-259, 9-261, 9-263
Edmonton, Alberta, pipeline terminal at, 2-65
Efficiency:
  Carnot cycle, 9-24
  coal gasification plant, 1-223
  combined-cycle power station, 9-168 to 9-170
  electric railroad, 9-239
  electrostatic precipitator, 9-392
  fired heater, 3-278 to 3-283
    air preheating system, 3-280
    steam generation, 3-288 to 3-289
  fuel cell, 4-71
  fuel nitrogen conversion, 9-337
  gas turbine, 9-177 to 9-178, 9-180

Electricity in United States, forecast to year 2000, 5-42
Electricity Task Group, National Petroleum Council, 9-260 to 9-266
Electrification, railway, 9-224, 9-226 to 9-227, 9-235 to 9-240
Electrochemical analyzer, nitrogen oxides and sulfur dioxide determined by, 9-371
Electrochemical system, battery, 4-52 to 4-53
Electrode:
  battery, 4-50 to 4-53
  fuel cell, 4-59 to 4-72
  magnetohydrodynamic generator, 9-210 to 9-211
  thermionic converter, 9-221
Electrode polarization loss, fuel cell, 4-61
Electrolysis, hydrogen production by, 4-32, 4-34
Electrolyte:
  battery, 4-50 to 4-53
  fuel cell, 4-59 to 4-69
Electrolyzer, solar array system use of, 6-69, 6-74
Electromagnetic pump, 5-44
Electromagnetic radiation, definition of, xxiv
Electro-Motive Division, General Motors Corporation, 9-224 to 9-241
Electron, 5-44
Electron-donor gas-processing system, 9-322
Electron flow, thermionic converter, 9-221 to 9-222
Electron separation, electrostatic direct conversion of plasma energy use of, 5-150 to 5-151
Electron transport system, biochemical oxidation, 4-6
Electron volt, definition of, xxv, 5-44
Electronic components, thermoelectric cooling of, 9-242
Electronic theory, solar cell, 6-56 to 6-59
Electronics industry, gas requirements of, 2-52 to 2-56
Electrostatic direct energy conversion in fusion reactors, 5-150 to 5-152
Electrostatic hazards, jet fuel, 3-38 to 3-39
Electrostatic precipitator, 4-15, 4-20, 9-392
Electrothermal hydrogen source, Hygas process, 1-207
Elevating-boom barge loader, 1-141 to 1-142
Elevating scraper, 1-99
Elevator, coal, 1-148
Elk Basin oil field, 3-64
Elk Hills oil field, 3-62
Ellef Ringnes Island pipeline system, 2-65
Elliott, William Bell, 7-34
Ellsworth No. 51 coal mine, 1-30
El Paso, Texas:
  daily terrestrial solar energy received at, 6-95
  sunshine, percent possible, 6-72
El Paso Natural Gas Company, 2-64
El Paso radiant boiler, 9-106

El Salvador:
  crude oil: number of wells producing, 3-88
    ultimate recovery potential for, 3-94
  natural gas, ultimate recovery potential for, 2-46
  railway mileage and motive power, 9-227
Ely, Nevada, daily terrestrial solar energy received at, 6-95
Embden-Meyerof-Parnas glycolytic pathway, 4-6
Emergency-generator, magnetohydrodynamic, 9-209, 9-219 to 9-221
Emerson, Manitoba, pipeline terminal at, 2-65
Emery County, Utah, coals of, 1-20
Emissions:
  automobile mileage, effect of control of, 3-26
  Diesel engines, 9-196 to 9-197
  electric power generating plants, 9-260 to 9-261, 9-264 to 9-266
  gas turbines, 9-195 to 9-198
  gasoline engines, 9-196 to 9-197
  geothermal energy installations, 7-11 to 7-13, 7-30, 7-36 to 7-37, 7-43 to 7-45
  interface with energy programs, 9-312 to 9-406
  internal combustion engines, 4-38 to 4-43, 4-48 to 4-49
  limits of, in various states, 9-374 to 9-375
  monitoring of: carbon monoxide, 9-376 to 9-379
    hydrocarbons, 9-381 to 9-390
    nitrogen oxides, 9-362 to 9-375
    particulates, 9-391 to 9-406
    sulfur dioxide, 9-362 to 9-375
  nuclear reactors (see Nuclear energy technology)
  Otto cycle engines, 9-196 to 9-197
  particulate, measurement of, 9-391 to 9-406
  from petroleum fuels, 3-22 to 3-45
  scrubbing and other treatment systems:
    Claus process, 9-328 to 9-335
    combustion improving chemicals, 9-349 to 9-350
    furnace and burner redesign, 9-335 to 9-348
    nitrogen oxides, 9-312 to 9-323
    sulfur oxides, 9-312 to 9-327
    Wellman-Lord process, 9-324 to 9-327
    wet-scrubbing processes, 9-312 to 9-327
  (See also Carbon dioxide; Carbon monoxide; Nitrogen oxides; Sulfur oxides)
Emissivity, solar, 6-12
Emitting electrode, thermionic converter, 9-221
Empire District Electric Company, 1-146
Empire oil field, 3-63
Empire Transportation Company, 3-176
Employee management system, electric utility, 9-281

Excess air (*Cont.*):
    definition of, 9-34
    firing, normal percentage requirements
        of, 9-46
    fuel loss as function of, 3-286
Exchanger, cold-train, 2-102 to 2-103
Exchanger heat losses (*see* Conservation,
    refinery and process plant energy)
Excitation current, generator, 8-36
Excitation system, superconducting
    turbine generator, 9-207
Exciter, gas turbine, 9-184
Exhaust analysis, engine, 9-381 to 9-382
Exhaust pollutants, various engines, 9-196
    to 9-197
    (*See also* Emissions)
Exhauster, pulverized coal system, 9-71 to
    9-73
Expanded-bed hydrodesulfurization
    reactor, 3-257 to 3-259
Expansion, plasma, 5-150
Expansion dome, tankcar, 3-183, 3-187, 3-
    191 to 3-193
Expansion-shell type coal mine roof bolt,
    1-75
Expansion turbine:
    for combined-cycle power generation, 9-
        167
    definition of, 9-198 to 9-199
    gas processing applications of, 9-201
    power recovery applications of, 9-199 to
        9-200
    refrigeration applications of, 9-200 to 9-
        201
Experience factor, coal mining safety, 1-89
    to 1-92
Exploration:
    coal, 1-37 to 1-47, 1-55
    crude oil, 3-61 to 3-71, 3-86 to 3-87, 3-90
        to 3-91, 3-94
    geothermal energy, 7-6 to 7-7
Explosions:
    boiler, 9-93
    coal mine, 1-84
    liquefied natural gas, 2-100 to 2-102
    of propellants, 4-87
Export-Import Bank, The, 6-133
Extension, oil field, API definition of, 3-66
Extent, oil field, API definition of, 3-70
External cleaning, boiler tube, 9-119 to 9-120
Extinction angle, magnetohydrodynamic
    system, 9-216 to 9-219
Extinction coefficient, energy absorption,
    9-366
Extra-high voltage transmission systems, 9-
    267 to 9-273
Extract hydrogenation, CSF process, 1-276
Extraction, coal, 1-276
Extraction loss volume, natural gas, 2-59
Extraction processes, geothermal energy
    for, 7-50, 7-55 to 7-57
Extraction sampling, particulate, 9-395
Extractive distillation, toluene, 3-19
Exxon, 2-103
    tanker study of, 3-210
Exxon Research and Engineering
    Company, 4-46

Fabius coal mine, 1-30
Faceted solar collector module, 6-106
Facies change, coal bed, 1-71
Facilities management computer system,
    electric utility, 9-278, 9-282 to 9-283,
    9-285 to 9-295
Factory building heating, solar (*see* Solar
    climate control)
Fairbanks, Alaska, daily terrestrial solar
    energy received at, 6-97
Fairway oil field, 3-63
Fairweather gas calorimeter, 2-91 to 2-93
Fall, coal mine, definition of, 1-73
Fan-brake dynamometer, 9-305
Far East:
    petroleum pipeline mileage in, 3-74
    seaborne imports of petroleum by, 3-
        199
    (*See also* specific country)
Faraday connection, magnetohydro-
    dynamic generator, 9-210
Farley, Joseph M., nuclear plants, 5-12, 5-
    22
Farm, aerogenerator, 6-165
Farm tap, pipeline, 2-72
Farmington, West Virginia, mine disaster
    at, 1-92
Fast Flux Test facility, 5-110
Fast neutron, 5-44
Fast reactors, 5-9, 5-44
    gas-cooled fast breeder reactor, 5-123 to
        5-136
    liquid-metal fast breeder reactor, 5-102
        to 5-123
Fatality analysis coal mine, 1-85 to 1-89
Fault:
    coal deposit, 1-40, 1-71
    San Andreas, 7-4
    Wairakei, 7-16
Faulting-related hydrocarbon entrapment,
    3-104
Federal Coal Mine Health and Safety Act
    of 1969, 1-84
Federal Interagency Task Force Report, oil
    shale retorting, 3-147
Federal No. 1 coal mine, 1-30
Federal Oil Conservation Board, 3-45
Federal Power Commission (*see* United
    States of America, Federal Power
    Commission)
Federal Register, 1-76
Federal Reserve Board Index, 3-100 to 3-
    101, 9-257, 9-259
Federally-owned power systems, 8-8
Feedforward-feedback control, boiler, 9-
    128 to 9-129
Feeding and charging systems:
    coal processes: carbon dioxide acceptor
        process, 1-219
    char-oil-energy development (COED)
        process, 1-256
    clean coke process, 1-250
    coal-handling facility, 1-138 to 1-141
    coal-preparation plant, 1-118 to 1-119
    coal-using facility, 1-144 to 1-152
    flood-loading, 1-139 to 1-141, 1-162
    Hygas process, 1-202

Fitzpatrick, James A., nuclear plant, 5-15, 5-22
Fixed-bed catalytic systems:
  catalytic reforming, 3-228 to 3-229
  hydrocracking, 3-234
  hydrodesulfurization, 3-255 to 3-257
Fixed-blade hydraulic turbine, 8-12, 8-16
Fixed carbon, coal, definition of, 1-49
Fixed carbon criterion, coal, 1-16 to 1-19
Fixed-end beam analogy, coal mine roof, 1-72
Fixed platform, offshore drilling, 3-108
Flame, hydrogen, 4-31, 4-39
  theory of, 9-335
Flame detection systems, boiler, 9-147 to 9-148
Flame patterns, utility boiler, 9-340
Flame speed:
  acetylene, ethane, ethylene, gasoline, hydrogen, and methane, 4-39
  methanol, 4-47
Flame temperature, adiabatic, 9-42 to 9-43
Flame temperature type of gas calorimeter, 2-92 to 2-93
Flameless hydrogen catalytic heater, 4-28, 4-31
Flaming Gorge, Utah, daily terrestrial energy received at, 6-96
Flammable explosive limits, 9-388
Flammable hydrocarbons analysis, 9-382 to 9-383
Flange, insulation of, 3-290
Flapping, wind power rotor blade, 6-143
Flare system design, maximum liquid and gas recovery by, 3-299
Flare-type furnace burner, 9-341
Flash point:
  definition of, 3-42
  gasoline, kerosine, and propane, 3-180
Flash power, 8-10
Flash tank control, boiler, 9-155
Flat plate absorber, 6-9
Flatiron pumped-storage plant, 8-18
Flaw detection, pipeline, 2-73, 2-79 to 2-80, 2-83 to 2-85
Fletcher, J. H., & Company, 1-71 to 1-83
Float-and-sink test, coal, 1-56
Floating ocean thermal energy conversion plants, 6-120 to 6-129
Floating rig, 3-109
Flocculation unit, coal-cleaning, 1-122
Flood loading, coal, 1-139 to 1-141, 1-162
Flood peak power, 8-10
Flood protection, 8-8 to 8-9
Florence, Oregon, wind power site at, 6-168
Florence Mining Company, 1-31
Florida:
  AGA area of, 3-9
  API Survey district of, 3-50
  crude oil: largest oil fields of, 3-62
    offshore oil jurisdictional boundary of, 3-91
    production of, 3-53, 3-56, 3-60
    reserves of, 3-46 to 3-47, 3-57 to 3-59
  daily terrestrial solar energy received: at Aplachicola, Gainesville, Key West, Miami, and Tampa, 6-95

Florida, daily terrestrial solar energy received (Cont.):
  at Belle Isle, Jacksonville, Pensacola, and Tallahassee, 6-96
  electric power generating capacity of, 9-252 to 9-253
  emission limits for SO_2 and NO_x, 9-374
  gas turbine installations in, 9-192
  hydroelectric power production of, 8-7
  Miami ocean thermal energy conversion plant site, 6-120 to 6-121
  natural gas, production and reserves of, 2-10 to 2-13, 2-16 to 2-18
  nuclear power installations in, 5-13
  tidal range at Jacksonville and Miami Beach, 8-40
Florida, University of, 6-85
Florida Everglades, 1-37
Florida Gas Transmission Company, 2-64
Florida Power and Light Company, 5-13
Flotation separators, coal, 1-122, 1-126 to 1-128, 1-133
Flow, boiler gases and fluids, 9-109 to 9-112
Flow-axial turbine (see Bulb unit, hydropower)
Flow channel, boiler, 9-94
Flow conditions, coal slurry pipeline, 1-179
Flow measurement, boiler fuel, 9-143
  (See also Instrumentation)
Flow sheets and system diagrams:
  coal energy systems: Arge process, 1-295
    BI-GAS process, 1-213, 1-217
    carbon dioxide acceptor process, 1-220
    char-oil-energy development (COED) process, 1-256, 1-258
    clean coke process, 1-249 to 1-253
    COGAS process, 1-262 to 1-263, 1-265
    Consol (CSF) process, 1-278 to 1-279
    drying processes, 1-134
    handling system, Martin Lake Steam Electric Station (Texas), 1-145
    Hydrane process, 1-225 to 1-226
    Hygas process, 1-205 to 1-207
    in-situ gasification, 1-242
    Lurgi process, 1-189, 1-198 to 1-199
    preparation plant: Amax Coal Company, Leahy Mine, 1-119
      Island Creek Coal Company, Virginia Pocahontas No. 3 mine, 1-118
    Sasol complex, 1-286, 1-290
    slurry pipeline, Black Mesa Pipeline Company, 1-198 to 1-199
    solvent-refining process, 1-283
    Synthoil process, 1-270
    Synthol process, 1-295
    Toscoal process, 1-272
  fuel cell systems: ammonia-air fuel cell system, 4-72
    fuel cell power plant, 4-71, 6-79
  gas energy systems: absorption-oil process for recovery of ethane and heavier hydrocarbons from natural gas, 2-130

Free pressure drop, 9-200
Free radicals, 4-89
Free Swelling Index, coal, 1-17, 1-21 to 1-23, 1-51
Freeboard, pipeline vessel, 2-79
Freeman Coal Mining Corporation, 1-31
Freeze-wall technique, 5-50
Freezer, electric, 9-260
Freezing point, aviation fuels, 3-37 to 3-39
Freight car, frictional resistance of, 9-225
Freight locomotives (see Locomotives, rail system)
Freight miles, railway, 9-224
Freight rate minimum, compressed gas tankcar, 3-180 to 3-181
Freight train, coal, 1-155
French Army Laboratoire, Central de L'Armement solar furnace, 6-46 to 6-47
Freon type working fluid, ocean thermal energy conversion plant use of, 6-127
Frequency rescheduling and control computer program, electric utility, 9-285
Fresno, California, percent possible sunshine at, 6-72
Friction, tidal force aspects of, 8-43
Frictional resistance, locomotives and rolling stock, 9-224 to 9-225
Front-end loader, coal, 1-161 to 1-162
Front-fired boiler, emission characteristics of, 9-339
Froth-flotation separation, coal-preparation use of, 1-122, 1-133
Froude fluid-friction dynamometer, 9-305
Fruit juice, fermentation of, 4-7
Fruit tree prunings, energy conversion of, 4-13
Frypan, electric, 9-260
Fuel and containment pools, boiling water reactor, 5-72
Fuel cells:
    aqueous electrolyte used in, 4-63 to 4-68
    biochemical-type, 4-9
    depolarization-type, 4-9
    fuels for, 4-59 to 4-72
    high-temperature designs of, 4-62 to 4-63
    mass and energy balance of, 4-68 to 4-69
    oxygen analyzer use of, 9-142
    product type, 4-9, 4-59 to 4-72
    redox type, 4-9
    single cell type, 4-59 to 4-60
    solar array system use of, 6-74 to 6-80
    spacecraft use of, 6-57
    voltage of, 4-60 to 4-61
Fuel conversion factor, 5-44
Fuel cycle, 5-44
Fuel displacement system, 6-6 to 6-7
Fuel economy, methanol-gasoline, 4-46
Fuel elements, rods and bundles, nuclear reactor: boiling water reactor, 5-67 to 5-69, 5-71 to 5-72
    gas-cooled fast breeder reactor, 5-126 to 5-128
    high-temperature gas-cooled reactor, 5-54, 5-87, 5-91, 5-94 to 5-97

Fuel elements, rods and bundles, nuclear reactor (Cont.):
    liquid-metal fast breeder reactor, 5-114 to 5-116
    pressurized water reactor, 5-76, 5-80 to 5-85
Fuel injection nozzle, cleanliness of, 3-30
Fuel loading, nuclear reactor: boiling water reactor, 5-67 to 5-69
    gas-cooled fast breeder reactor, 5-128 to 5-130
    high-temperature, gas-cooled reactor, 5-53, 5-94 to 5-95
    liquid-metal fast breeder reactor, 5-106, 5-112, 5-114 to 5-115
    pressurized water reactor, 5-80 to 5-85
Fuel loss versus air, boiler combustion system, 9-141
Fuel management boiler, 9-145 to 9-147
    nuclear (see Fuel elements, rods and bundles, nuclear reactor)
Fuel measurement, boiler, 9-143
Fuel nitrogen versus nitrogen oxide formation, 9-339
Fuel oil (see Burner fuel oils; Fuels)
Fuel reaction, fusion reactor reference design, 5-170
Fuel requirements:
    electric power generation, 9-207, 9-260 to 9-266
    transportation, 9-228
Fuel saving, boiler control system, 9-133
    (See also Thermal energy, conservation of)
Fuels:
    analysis of (see Analysis)
    biochemical, 4-5 to 4-11
    butanol, 4-1 to 4-8
    calorific value of (see Calorific value)
    coal, 1-1 to 1-296
    costs of (see Costs and economics)
    desulfurization of (see Desulfurization)
    ethanol, 4-7 to 4-8
    fossil (see Coal; Natural gas; Petroleum)
    fuel gas, 3-225
        costs of, 3-274
        desulfurization, 2-174
        Lurgi process for making, 1-200
        preparation of, 9-165 to 9-166
        synthetic, 2-173
        (See also Substitute natural gas)
    fuel oil: consumption of, 9-261 to 9-262
        costs of, 3-274
        excess air normally supplied for burning, 9-46
        synthetic (see Synthetic fuel oil)
        waste conversion system use of, 4-13
        (See also Burner fuel oils)
    gas-turbine, 9-181
        (See also Gas turbines)
    geothermal fluids, 7-1 to 7-58
    heating value of (see Calorific value)
    hydrogen: direct uses of, 4-29 to 4-31
        heating uses of, 4-30 to 4-31
        motive power uses of, 4-29 to 4-33, 4-38 to 4-43
    isopropanol, 4-7 to 4-8

Germany, Federal Republic of
(*Cont.*):
  natural gas: production of, 2-40
    reserves of, 2-43, 2-45
  nuclear power facilities in, 5-27 to 5-
    28, 5-36 to 5-38
    gas-cooled fast breeder reactor
      component developments of, 5-
      135
    high-temperature gas-cooled
      reactor developments of, 5-88
    nuclear fuels study of, 5-179
    Nuclear Research Center at Julich,
      4-35
    sodium-cooled fast breeder reactor
      developments of, 5-110 to 5-113
  pumped-air energy storage
    experiment of, 9-194
  railway mileage and motive power of,
    9-224, 9-226
  Siemens AG fuel cell, 4-64 to 4-65, 4-
    69
  tar sand deposits in, 3-160
  tidal range at Hamburg, 8-40
  Varta AG fuel cell, 4-64 to 4-65, 4-69
  wind power experience and research
    of, 6-148 to 6-150
  (*See also* German Democratic Republic)
Germination, reclamation vegetation, 1-
  109
Geysers, The, geothermal field, 7-5, 7-7 to
  7-8. 7-11 to 7-12, 7-34 to 7-42
Geyserville, California, 7-34
Ghana:
  crude oil: annual production of, 3-86
    proved recoverable reserves of, 3-82
    ultimate recovery potential for, 3-92
  natural gas, ultimate recovery potential
    for, 2-44
  railway mileage and motive power of, 9-
    227
Gibraltar, Spain, tidal range at, 8-40
Gibraltar Coal Corporation, 1-31, 1-142
Gibraltar coal mine, 1-30
Gibrat, Robert, tidal energy studies of, 8-37
  to 8-39, 8-41 to 8-45
Gieseler plasticity test, coal, 1-52 to 1-53
Gigawatt-mile, electric-power
  transmission, 9-270 to 9-276
Gilsonite, 3-159
Gin pole, oil derrick, 3-107
Ginna, R. E., nuclear facility, 5-15, 5-21
Girbitol hydrogen production process, 4-36
  to 4-37
Glance pitch, 3-159
Glasgow, Montana, daily terrestrial solar
  energy received at, 6-96
Glass, solid-waste content of, 4-22
Glass industry:
  electricity consumption of, 9-259
  gas requirements of, 2-52 to 2-56
Glenharold coal mine, 1-30
Global Marine Development, Inc., 6-122
Globtik Tokyo oil tanker, 3-211 to 3-212
Glucose, degradation of, 4-6
Glycerol, jet fuel use of, 3-38
Glycols, extraction use of, 3-19

Glycolysis, 4-6
Golden Trend oil field, 3-64
Goldsmith oil field, 3-63
Gondola car, coal, 1-158
Gorda Ridge, 7-4
Gorgas, Alabama, in-situ coal gasification
  experiments at, 1-243
Gotaverken shipyard, 3-213
Governing system:
  gas turbine, 9-185
  hydraulic turbine, 8-15, 8-17
  pump turbine, 8-34 to 8-35
Government-owned electric power
  utilities, 9-256 to 9-257
Gradation (size consist), coal slurry, 1-179
Grade, railway, 9-225, 9-231 to 9-232
Grade factors, contour mining, 1-99
Gradient, geothermal, 7-4
Grading, coal, 1-15 to 1-33
Graetz circuit, 8-36
Grahamite, 3-159
Grain, fermentation of, 4-8
Grain characteristics, propellant, 4-87 to 4-
  88
Gram-calorie, definition of, xxiv, 1-63
Gram-mole, definition of, 9-34
Grama grass, spoiled-land area use of, 1-
  107
Grand Coulee hydroelectric plant, 8-5
Grand Gulf nuclear plant, 5-15, 5-22 to 5-23
Grand Isle oil fields, 3-63 to 3-64
Grand Lake, Colorado, daily terrestrial
  solar energy received at, 6-96
Grande Valley, Colorado, oil shales at, 3-
  148 to 3-150
Granite Point oil field, 3-64
Granite wash, 3-51
Graphite:
  gas-cooled fast breeder reactor use of, 5-
    126
  high-temperature gas-cooled reactor use
    of, 5-47, 5-88, 5-90, 5-95
  nuclear reactor use of, 5-8
Graphitic ring structure, char, 1-223
Grass roots versus modernization,
  petroleum refinery, 3-273
Gravel, contour mining handling of, 1-100
Gravel-packed liner, oil well, 3-129, 3-131
Gravel storage system, solar heat, 6-83, 6-
  95
Gravimetric exploration, geothermal field,
  7-6
Gravitational potential energy, xxiii
Gravity, API (*see* API gravity)
Gravity cleaning system, coal, 1-125 to 1-
  126, 1-131 to 1-132
Gravity steam separator, 9-100
Gravity waves, energy potential of, 8-45
Gray-King coking test, 1-17
Grayling Platform, Cook Inlet, Alaska
  location of, 9-190
Graywacke, 3-51
Grazing-type tropical ocean thermal
  energy conversion plant, 6-122
Great Britain Royal Airship Works, 4-38
Great Falls, Montana, daily terrestrial solar
  energy received at, 6-96

Great Lakes:
coal boats for, 1-174
thermal pollution of, 9-264
Great Lakes coal transport system, 1-154
Great Lakes Gas Transmission Company, 2-64
Great Lakes region, gas consumption of, 2-51, 2-53
Great Lakes/Saint Lawrence Waterway, coal transportation in, 1-172 to 1-174
Great Plains Provinces, coal formation in, 1-38
Great Salt Lake, lithium resources of, 5-183
Greece:
crude oil: annual production of, 3-88
exploratory and production wells completed, 3-86
ultimate recovery potential for, 3-93
natural gas, ultimate recovery potential for, 2-45
nuclear installations in, 5-28, 5-37 to 5-38
railway mileage and motive power of, 9-227
Green Bay, Wisconsin, daily terrestrial solar energy received at, 6-97
Green County nuclear plant, 5-15
Green Mountains, wind power survey of, 6-153
Green plant photosynthesis, 6-137
Green River System, coal movement in, 1-172
Greenhouse effect, 6-27
Greenhouse heating, geothermal energy for, 7-50, 7-52, 7-54
Greenland, uranium resources of, 5-176
Greensboro, North Carolina, daily terrestrial solar energy received at, 6-96
Greenville, Kentucky, mine disaster at, 1-92
Greenwood nuclear plants, 5-14, 5-23
Gremlin engine, 4-47
Grenoble, University of, 8-45
Grid (drilling), coal exploration, 1-46
Griffin, Georgia, daily terrestrial solar energy received at, 6-96
Grindability, pulverized coal firing importance of, 9-59
Grindability index:
coal, definition of, 9-65, 9-70
coal-handling importance of, 1-147
test of, 1-51 to 1-52
Grinder, coal slurry, 1-181 to 1-183
Grinding:
pulverized coal, 9-64
wood wastes, 3-169
Groningen, the Netherlands, gas discovery at, 3-91
Gross heating value, coal, definition of, 1-50
Gross margin, electric utility, 9-251 to 9-252
Gross national product, petroleum demand relationship to, 3-96, 3-100 to 3-101
Ground control, mine safety, 1-84
Ground water, reserves of, 8-8

Grounding, tankcar, 3-191
Grouting, coal mine roof, 1-74, 1-76
Growth:
of electric power generation, 9-271
of electric power transmission systems, 9-267 to 9-275
of electric utility industry in United States, 9-207, 9-256, 9-279
of superconducting turbine generator use, 9-207
Grumman TACV advanced railway developments, 9-241
Guatemala:
crude oil: annual production of, 3-88
ultimate recovery potential for, 3-94
natural gas, ultimate recovery potential for, 2-46
railway mileage and motive power of, 9-227
Guide vane, two-way flow-axial turbine, 8-31 to 8-32
Guinea:
crude oil: annual production of, 3-88
ultimate recovery potential for, 3-92
natural gas, ultimate recovery potential for, 2-44
railway mileage and motive power of, 9-227
Gulf of Alaska, oil exploration in, 3-94
Gulf of California, 7-4, 7-6, 7-8
Gulf of Mexico:
natural gas reserves and production of, 2-10
NPC petroleum province of, 2-62
ocean thermal energy conversion sites of, 6-120
offshore fields, gas turbine installations in, 9-192
oil resource developments of, 3-90 to 3-91, 3-94
(See also Offshore gas and oil drilling and production)
pipe laying costs in, 2-87 to 2-89
Gulf of Nigeria, characteristics of crude oil in, 3-16 to 3-17
Gulf of Saint Malo, tides of, 8-43
(See also Rance Estuary tidal power plant)
Gulf of Tonkin, tides at, 8-37, 8-39
Gulf Coast geosyncline, 2-23
Gulf Coast naphthenic crude oils, 3-12 to 3-15
Gulf Coast region:
gas consumption of, 2-51, 2-53
NPC petroleum province of, 2-62
Gulf Intercoastal Waterway, coal movement in, 1-171 to 1-174
Gulf Research and Development Company, 1-243 to 1-244
Gulf Resources and Chemical Corporation, 5-184
Gulf States Utilities Company, 5-14, 5-18
Gulf Stream, ocean thermal energy conversion plant sites in, 6-120
Gulf terminal, Bantry Bay, 3-211
Guliev studies of ship squat versus speed, 3-213

Heat balance:
  electric utility computer program for, 9-284
  geothermal power plant, 7-40 to 7-41, 7-54 to 7-55
  (*See also* Material and energy balance)
Heat conduction, interior-to-earth surface, 7-4
Heat engine:
  atmosphere as, 6-146 to 6-147
  solar power, 6-6 to 6-25
Heat exchange:
  energy conservation in, 3-284 to 3-285
  heat-transfer coefficients of, 3-285
  in nuclear reactors: boiling water reactor, 5-57 to 5-70
    gas-cooled fast breeder reactor, 5-124 to 5-128, 5-131 to 5-134
    high-temperature gas-cooled reactor, 5-47 to 5-49, 5-54 to 5-55, 5-87, 5-90 to 5-96, 5-100 to 5-101
    liquid-metal fast breeder reactor, 5-102 to 5-114
    pressurized water reactor, 5-75 to 5-81
  in ocean thermal energy conversion plant, 6-126 to 6-127
Heat flow, definition of, 9-23
Heat generation rate, boiling water reactor, 5-70
Heat pipe, solar energy system, 6-39 to 6-40
Heat pump, thermoelectric, 9-242
Heat rate, electric power generation, 9-261 to 9-262
Heat recovery:
  COGAS process, 1-266
  gas turbines, 3-287 to 3-288
  [*See also* Boiler(s)]
Heat release, nuclear fission reaction, 5-7 to 5-8
Heat sink:
  geothermal system (*see* Acoustic problems, geothermal systems)
  thermoelectric cooling system, 9-242 to 9-246
Heat storage (*see* Storage)
Heat transfer:
  in calorimeters, 1-64 to 1-65
  coefficients of, 3-285
  in pressurized water reactor, 5-77
  solar collector, 6-12 to 6-14
  solar heat pipe, 6-40 to 6-42
  water-to-steam, boiler, 9-94
Heat-transport fluid, solar tower energy system, 5-32
Heat value, definition of, 9-41 to 9-42
  (*See also* Calorific value)
Heat-work cycle, 9-23
Heater, fired, efficiency of, 3-278 to 3-282
Heating:
  solar (*see* Solar climate control)
  tankcar, 3-189
Heating fuel, hydrogen as, 4-30 to 4-31
  (*See also* Fuels)
Heating oils (*see* Burner fuel oils)
Heating value:
  aviation turbine fuels, 3-37

Heating value (*Cont.*):
  gas mixing control system for, 2-96 to 2-99
  low-Btu gas, 2-118
  (*See also* Calorific value)
Heavy-duty freight locomotives, 9-233
Heavy-media cycloid, 1-131
Heavy residuum feedstock, partial-oxidation gasification process, 2-172
Heavy water, 5-8 to 5-10, 5-40, 5-44, 5-137
  geothermal energy for production of, 7-18 to 7-19, 7-27, 7-50
Helicopter delivery, pipeline use of, 2-73
Helicopter transportation, efficiency of, 9-228
Helides, 4-89
Helios (silicon) solar cell, 6-69
Heliostat, solar energy system, 6-12, 6-26 to 6-27, 6-46 to 6-48
Helium:
  boron-neutron capture reaction release of, 5-72
  cryogenic gas upgrading recovery of, 2-118
  fusion reactor shielding use of, 5-171
  high-Btu gas content of, 2-118
  in metastable propellants, 4-89
  natural gas content of, 2-129
  nuclear reactor circulator use of, 5-47 to 5-48, 5-91 to 5-94, 5-101
  nuclear reactor cooling use of, 5-47 to 5-49, 5-55, 5-87 to 5-96
  turbine-generator cooling use of, 9-205 to 9-208
  vent gas content of, 2-118
Helvetia Coal Company, 1-31
Hemispherical vertical axis rotors, 6-144
Henschel NY7 locomotive, 9-235
Herbicides, petroleum-derived (*see* Petrochemicals)
Hero of Alexandria, 9-17
Heronemus three-wing turbine installation, 6-167
Hewitt-Robins coal boat unloading system, 1-174
Hewson circular neutral wing aerogenerator, 6-169
Hexafluoroarsenate electrolyte, battery use of, 4-51
Hexafluorophosphate electrolyte, battery use of, 4-51
n-Hexane, 3-241 to 3-242
  combustion constants of, 9-36 to 9-37
  natural gas content of, 2-129
Hexose sugars, fermentation of, 4-7
Heysham gas-cooled reactor, 5-89
High-enthalpy geothermal systems, 7-5 to 7-6, 7-14 to 7-33
High-head hydroelectric plant, 8-9
High heat value, definition of, 9-41 to 9-42
High-pressure, high-temperature boiler, 9-105 to 9-107
High-pressure gas drive as fluid injection system for oil well, 3-142
High-pressure methanol process, 2-121
High-speed coal pulverizer, 9-69

High-temperature gas-cooled reactor, 5-11, 5-40, 5-47 to 5-56, 5-87 to 5-101
  AVR reactor, 5-90
  characteristics of, 5-54 to 5-55, 5-100 to 5-101
  construction and startup of, 5-47 to 5-56
  Dragon reactor, 5-89 to 5-90
  early development of, 5-88 to 5-89
  Fort St. Vrain Generation Station use of, 5-47 to 5-56, 5-90
  instrumentation of, 5-97 to 5-98
  Peach Bottom 1 reactor, 5-90
  radioactive waste management of, 5-97
  safeguard systems for, 5-95 to 5-97
High-temperature research, solar furnace used for, 6-51 to 6-54
High-temperature solar energy, 6-46 to 6-55
High-volatile bituminous coal, 1-17, 1-54
Highway and street lighting, power requirements for, 9-258
Highway trucks, efficiency of, 9-228
Hi-Heater solar energy collector, 6-94
Hilbig oil field, 3-105
Hillsboro coal mine, 1-29
Himalayan Mountains, geothermal resources of, 7-4
Hinkley Point gas-cooled reactors, 5-89
Hit on the Euphrates, oil sands at, 3-160
Hiwassee pumped-storage plant, 8-18
HMX (cyclotetramethylene tetranitramine), propellant use of, 4-83
Hobbs oil field, 3-65
Hokkoh, Japan, town gas plant at, 2-149
Hole drilling, oil well, 3-112 to 3-119
Holiday detector, 2-73
Holten, Germany, Lurgi process pilot plant at, 1-190
Home cooking and heating, hydrogen for, 4-30
Home heating oil (see Burner fuel oils)
Homestead coal mine, 1-29
Homogeneous flow, coal slurry, 1-179
Hondo oil field, 3-63
Honduras:
  crude oil: annual production of, 3-88
    exploratory and production wells completed, 3-87
    ultimate recovery potential for, 3-95
  natural gas, ultimate recovery potential for, 2-46
  railway mileage and motive power of, 9-227
Honeysuckle, spoiled-land area use of, 1-113
Honeywell, Inc., 6-37
Hong Kong:
  nuclear facilities planned for, 5-28, 5-37 to 5-38
  railway mileage and motive power of, 9-227
  tidal range at, 8-40
Honolulu, Hawaii, daily terrestrial solar energy received at, 6-95
Hook of Holland, 3-223

Hoover Dam:
  hydroelectric plant at, 8-5 to 8-10
  and Los Angeles electric power transmission line, 9-267
Hope Creek nuclear plants, 5-15, 5-23
Hopper, coal, 1-118 to 1-119, 1-122
Hopper car, coal, 1-157 to 1-158
Horizontal-axis rotors, 6-143
Horizontal impulse turbine, 8-15
Horizontal-shaft bulb unit, 8-29
Horsepower:
  railway locomotive, 9-224 to 9-241
  seagoing petroleum-carrying vessels, 3-199
Horticultural crops, spoiled-land area use of, 1-115
Hot air storage system, solar heat, 6-95
Hot junction, thermoelectric cooling system, 9-242 to 9-246
Hot sodium-sulfur battery, 4-52 to 4-53, 4-55, 4-57
Hot springs, geothermal, 7-5
Hot Springs Creek, California, 7-36
Hot standby, ready-to-start mode, gas turbine, 9-191
Hot-water separation plant, tar sand, 3-163 to 3-165
Hot water storage systems, solar heat, 6-43 to 6-44, 6-94 to 6-95
Hot-water transmission, geothermal, 7-24, 7-51 to 7-55
Hot wire detector, 9-382
Hotplate, electric, 9-260
Houillieres du Bassin du Lorraine coal slurry pipeline, 1-180
House heating:
  geothermal energy for, 7-53 to 7-54
  solar energy for (see Solar climate control)
Household appliances, electric, 9-260
Household statistics, petroleum-demand-related, 3-101
Houston, University of, 6-10, 6-26 to 6-33
Houston Lighting and Power Company, 5-18
Houston nuclear plant, 5-23
Hover train principle, 9-241
Howaldtswerke-Deutsche Werft, 3-210
Howard Glasscock oil field, 3-63
Huka Falls geothermal formation, 7-15 to 7-18
Hull shape, pipeline vessel, 2-79
Humble Gas Transmission Company, 2-64
Humble Oil and Refining Company, 3-152
Humboldt Bay nuclear plant, 5-12, 5-21
Humphrey No. 7 coal mine, 1-29
Hungarian Basin, geothermal resources of, 7-6
Hungary:
  crude oil: annual production of, 3-81, 3-88
    ultimate recovery potential for, 3-93
    metric tons to barrels conversion factor, 3-84
  natural gas: production of, 2-40
    reserves of, 2-42, 2-45

Hydrogen (*Cont.*):
 sources of: ammonia dissociation, 4-36
  biochemical conversion, 4-27
   generation: by *Escherichia coli* and
    Clostridium spp, 4-9
   by purple bacteria, 4-6
  catalytic reforming, 3-229
  electrolysis, 4-34
  electrothermal process, 1-207
  geothermal gases, 7-36
  hydrocarbons: catalytic partial-
   oxidation of, 4-3 to 4-6
   Shell gasification process, 2-174
   steam-iron system, 1-205 to 1-207,
    4-36
   steam-methanol process, 4-36
   steam reforming, 4-36
   substitute natural gas process, 2-168
  petroleum refining processes, 3-250
  photolysis, 4-26
  thermal dissociation, 4-36
  thermochemical splitting, 4-35
 storage of: cryogenic systems, 4-26
  as hydrides, 4-28, 4-43
  for solar systems, 6-23 to 6-25
  underground, 4-27
  vehicular, 4-42
 transmission costs of, 4-33
 uses of: aircraft, 3-39, 4-43
  automotive, 4-38 to 4-43
  chemical production, 4-35
  energy transport, 4-31 to 4-33
  heating, 4-30 to 4-31
  hydrodesulfurization, 3-251 to 3-264
  hydrogasification and hydrogenation
   (*see* Hydrogenation)
  motive power, 4-29 to 4-30, 4-38 to 4-
   43
  ore reduction, 4-26
  rocket propulsion, 4-27
  welding, 4-27
 wood-to-oil process synthetic crude oil
  content of, 3-173
Hydrogen-air fuel cell, 4-59 to 4-61
Hydrogen-air mixtures, 4-43
Hydrogen-carbon monoxide cycle, energy
 transmission by, 6-32
Hydrogen-carbon monoxide ratio, 1-194 to
 1-195, 1-266, 1-294, 9-173 to 9-174
Hydrogen chloride, zinc chloride process
 recovery of, 1-281
Hydrogen-cooled turbine generators, 9-202
Hydrogen economy, 4-26 to 4-27, 6-16
Hydrogen feedstock, methanol process, 2-
 120
Hydrogen flame ionization detector, 9-381,
 9-385 to 9-386, 9-389
Hydrogen induction technique (HIT), 4-40
Hydrogen ion concentration, spoiled-land
 soil, 1-103 to 1-104
Hydrogen nuclide, fusion of, 5-7
Hydrogen-oil ratio, substitute natural gas
 process, 2-166
Hydrogen peroxide, propellant use of, 4-
 80, 4-82 to 4-83
Hydrogen production unit, energy
 summary of, 3-277

Hydrogen sulfide:
 BI-GAS process product gas, content of,
  1-214
 versus carbon dioxide absorption, amine
  unit, 9-333
 catalytic rich gas process, removal of, 2-
  140
 Claus processing of, 9-328 to 9-334
 coal gasification process, removal of, 9-
  171
 COED process char, content of, 1-266
 COGAS process, removal of, 1-262 to 1-
  263
 COGAS process dry synthesis gas,
  content of, 1-266
 COGAS process pyrolysis gas, content
  of, 1-264
 combustion constants of, 9-36 to 9-38
 fluidized-bed hydrogenation process,
  removal of, 2-161
 geothermal fluids and gases, content of,
  7-11 to 7-12, 7-36 to 7-37, 7-44 to 7-
  49
 Hydrane process, removal of, 1-225
 Hydrane process pipeline gas, content
  of, 1-228
 Hygas process gasifier gas, content of, 1-
  203
 Lurgi process, removal of, 1-198 to 1-
  199, 1-290
 methane-rich gas process, removal of, 2-
  150
 oil well content of, 3-134
 petroleum products, content of, 3-252
 scrubber for removal of, 3-255
 SELEXOL process, removal of, 1-213
 solvent-refining-of-coal process, removal
  of, 1-282 to 1-284
 Sulfinol process, removal of, 2-133 to 2-
  136
 Synthane process shift-conversion gas:
  content of, 1-236
  removal of, 1-233, 1-237, 1-240
 Synthoil process, removal of, 1-270
 tar sand process gas, removal of, 3-166
 Westinghouse coal gasification process
  gas, content of, 9-174 to 9-175
Hydrogen sulfide cracking, oil well pipe, 3-
 135
Hydrogenation:
 BI-GAS process use of, 1-214
 burner fuel oil production use of, 3-40
 carbonization/hydrogenation metallurgical coke process use of,
  1-251
 char-oil-energy development (COED)
  process use of, 1-257 to 1-259
 COGAS process use of, 1-262 to 1-266
 Consol (CSF) process use of, 1-277 to 1-
  279
 fluidized-bed substitute natural gas
  process use of, 2-160 to 2-164
 Hydrane process use of, 1-223 to 1-230
 hydrotreating process use of, 3-235 to 3-236
 Hygas process use of, 1-203 to 1-207
 Lurgi process use of, 1-194 to 1-195, 1-
  198 to 1-199

Hydrogenation (*Cont.*):
   of petroleum products, 3-19 to 3-30
   products of, 4-29
   solvent-refining process use of, 1-282 to
       1-283
   Synthane process use of, 1-236 to 1-241
   Synthoil process use of, 1-269 to 1-270
Hydrogenolysis, definition of, 3-251
Hydrogovernor system, 8-17
Hydrogravitational energy, 8-3
Hydrolysis:
   ethyl sulfate, 4-8
   starches, 4-7
Hydronautics, Inc., 6-127
Hydrophobic electrode substrate, fuel cell,
       4-63
Hydropower technology, 8-3 to 8-56
   advantages of hydropower, 8-8 to 8-9
   availability of hydropower, xxvi
   classification of hydroelectric plants:
       base-load plants, 8-9 to 8-10
       mountain reservoir plants, 8-10 to 8-11
       peak-load plants, 8-9 to 8-10
       pumped-storage plants, 8-17 to 8-24
       run-of-river plants, 8-9 to 8-11
       storage plants, 8-9 to 8-11
   draft tubes, 8-15 to 8-17
   growth and potential of hydropower, 8-9,
       9-207
   limitations of hydropower, 8-8 to 8-9
   management of hydro plants in United
       States, 8-8
   ownership of hydro plants, United
       States, 8-8
   production of hydropower: in Canada, 9-
       261
       in United States, 8-3 to 8-8, 9-255
       worldwide, 8-3 to 8-7, 9-248
   tidal power plants, 8-25 to 8-45
       Rance Estuary installation, 8-25 to 8-
           36
       tidal energy phenomena, 8-36 to 8-45
   trends as percent of total power, United
       States, 8-3
   turbines for (*see* Hydraulic turbines)
   wave energy projects, 8-45
Hydro Quebec nuclear plants, 5-26
Hydroseeding, reclaimed-area, 1-113
Hydroseparators, coal, 1-122
Hydrotreating processes, 3-19 to 3-20, 3-
       228, 3-235 to 3-236, 3-250, 3-262
   Athabasca bitumen, upgrading by, 3-166
   COED process use of, 1-257 to 1-259
Hygas coal gasification process, 1-201 to 1-
       211
   chemistry of, 1-201 to 1-202
   coal preparation and pretreatment for, 1-
       202, 1-208
   economics of, 1-207 to 1-208
   gas purification, 1-203 to 1-204
   hydrogasifier reactor, 1-203
   hydrogen supply for, 1-204 to 1-207
   methanation, 1-204
   steam-oxygen and steam-iron systems, 1-
       204 to 1-207
   water requirements of, 1-211
Hygiene, coal mine, 1-85

Ice forces, drilling structure reactions to, 3-
       111
Iceland:
   crude oil: production of, 3-88
       ultimate recovery potential for, 3-93
   geothermal resources of, 7-6, 7-10, 7-43
       to 7-58
   natural gas, ultimate recovery potential
       for, 2-45
   tidal range at Reykjavik, 8-40
Icing suppressants:
   aviation fuels, 3-38
   gasoline, 3-27
Idaho:
   AGA area of, 2-9
   API Survey district of, 3-50
   daily terrestrial solar energy received at
       Boise, 6-96
   electric power generating capacity of, 9-
       253
   emission limits for SO_2 and NO_x, 9-374
   geothermal resources of, 7-6, 7-12 to 7-13
   hydroelectric power production of, 8-7
Ideal air-fuel mixture, 9-196
Ideal gas law, 9-35
Igneous rocks, oil reservoir, 3-51
Ignimbrites, Wairakei, 7-15
Ignition stability, pulverized-coal, 9-74
Ignition system:
   bomb calorimeter, 1-63 to 1-70
   internal combustion engine, 4-40
Ignition temperature, definition of, 9-42
Ignition time delay, 9-144
Ignitor, boiler burner, 9-56 to 9-57, 9-75
Ilha Solteira hydroelectric plant, 8-5
Illinois:
   AGA area of, 2-9
   [*See also* Central District (API
       Survey)]
   Bituminous Coal District of, 1-16, 1-169
   coals in, 1-20, 1-24, 1-27, 1-29 to 1-30, 1-
       32, 1-34 to 1-35
   crude oil: production of, 3-53, 3-56, 3-60
       reserves of, 3-46 to 3-47, 3-57 to 3-59
   daily terrestrial solar energy received: at
       Chicago, 6-97
       at Moline, 6-96
   electric power generating capacity of, 9-
       252 to 9-253
   emission limits for SO_2 and NO_x, 9-374
   gas turbine installations in, 9-192
   natural gas: production and reserves of,
       2-10 to 2-13, 2-16 to 2-18
   underground gas storage capacity, 2-
       74 to 2-75
   Norris City terminus of "Big Inch"
       pipeline, 3-174
   nuclear power installations in, 5-13
   State Geological Survey of, 1-41, 1-51
Illinois Central Gulf Railroad, 1-154
Illinois Power Company, 5-13
Illinois River system, coal movement in, 1-
       172
Illumination, coal mine, 1-84
Ilot de Chalibert, France, tidal power plant
       located at, 8-27 to 8-28
IMP quadrupole fusion device, 5-165, 5-167

Impact mill coal pulverizer, 9-65, 9-69
Impact pipeline coating method, 2-83
Impactor stage, 9-403
Imperial Chemical Industries, Ltd., 2-121
  to 2-124
Imperial Valley, California, geothermal
  resources of, 7-6
Imports:
  liquefied natural gas, 2-66 to 2-69
  natural gas, 2-66 to 2-69
  petroleum products, 3-99
Impsonite, 3-159
Impulse blading, gas turbine, 9-177
Impulse bucket, hydraulic turbine, 8-11, 8-
  13
Impulse turbine, hydraulic, 8-11, 8-15 to 8-
  16
Incinerator, waterwall, 4-20 to 4-21
Inclined drilling (see Directional drilling)
In-core instrumentation, nuclear reactor, 5-
  85 to 5-86
Independent Natural Gas Association, 2-20
Independent tank, definition of, 3-206
India:
  coal reserves of, 1-36
  crude oil: annual production of, 3-80, 3-
    88
    ultimate recovery potential for, 3-92
  electric power production of, 9-247
  hydroelectric plants in, 8-6
  metric tons to barrels conversion factor,
    3-84
  natural gas: production of, 2-39
    reserves of, 2-42, 2-44
  nuclear power plants in, 5-29, 5-36 to 5-
    38
  railway mileage and motive power of, 9-
    224, 9-226
  solar water heaters, use of, 6-93
  tidal range at Bombay, 8-40
  uranium resources of, 5-176
India Department of Atomic Energy,
  nuclear plant of, 5-28
Indian Head coal mine, 1-30
Indian Ocean, tidal range of, 8-43
Indian Point nuclear facility, 5-15, 5-21, 5-
  23
Indiana:
  AGA area of, 2-9
  API Survey district of, 3-50
  Bituminous Coal District of, 1-169
  coals in, 1-29 to 1-30, 1-32, 1-34 to 1-35,
    1-37, 1-92 to 1-93
  crude oil: productive capacity of, 3-56, 3-
    60
    reserves of, 3-46 to 3-47, 3-57 to 3-59
  daily terrestrial solar energy received at
    Indianapolis, 6-96
  electric power generating capacity of, 9-
    252 to 9-253
  emission limits for SO_2 and NO_x, 9-374
  hydroelectric power production of, 8-6
  natural gas: production and reserves of,
    2-10 to 2-13, 2-16 to 2-18
    underground gas storage capacity, 2-
      74 to 2-75
  nuclear power installations in, 5-13

Indiana and Michigan Electric Company,
  5-14
Indiana County, Pennsylvania, coals in, 1-
  20
Indianapolis, Indiana, daily terrestrial solar
  energy received at, 6-96
Indicated additional reserves of crude oil,
  API definition of, 3-61
  (See also Reserves)
Indonesia:
  crude oil: annual production of, 3-80
    exploratory and production wells
      completed, 3-86, 3-88
    percent of recoverable reserves of, 3-
      85
    ultimate recovery potential for, 3-92
  gas turbine installations in, 9-192
  geothermal resources of, 7-4
  metric tons to barrels conversion factor,
    3-84
  natural gas: production of, 2-39
    reserves of, 2-42, 2-44
  railway mileage and motive power of, 9-
    226
  tar sand deposits of, 3-160
Induced Hall potential, magnetohydro-
  dynamic generator, 9-210
Industrial switching locomotives, 9-233
Inerts, combustion, definition of, 9-40
Inexperience, coal mine accident factor of,
  1-89 to 1-92
Inferential-type gas calorimeter, 2-90
Infrared analyzer, nondispersive, 9-370
Infrared hydrocarbon monitor, 9-389
Infrared region, energy absorption in, 9-
  366
Infrared scanning, coal mine roof analysis
  use of, 1-82
Inguri hydroelectric plant, 8-5
Inhibitors, oil well acidizing use of, 3-
  134
In-hole screening device, oil well, 3-132 to
  3-133
Initial core, nuclear fission reactor, 5-40
Injection wells, in-situ coal gasification, 1-
  243
Inland coal mine, 1-29
Inland Steel Company, 1-31
Inland Waterways System, United States,
  1-170
Insert bit, drilling, 3-125
Inserter, coal mine roof support bolt, 1-81
In-situ burning, oil recovery improvement
  by, 3-143
In-situ coal gasification, 1-242 to 1-248
  Gorgas, Alabama, experiments in, 1-244
  Hanna, Wyoming, experiments near, 1-
    244 to 1-246
  Gary Higgins, gasification concept of, 1-
    246 to 1-248
  United Kingdom, early work of, 1-242 to
    1-244
  U.S.S.R., early work of, 1-242 to 1-244
Inspection, pipeline, 2-73, 2-79 to 2-80, 2-
  83 to 2-85
In-stack ultraviolet gas analyzer, 9-369
Installation, boiler control system, 9-156

Instantaneous power, definition of, xxiii
Institute of Gas Technology, 1-201 to 1-211, 1-238, 2-164 to 2-170, 4-32, 4-34, 4-63
Instrument industry:
  electricity consumption of, 9-259
  gas requirements of, 2-52 to 2-56
Instrument transformers, 9-304
Instrumentation:
  boiler, 9-122 to 9-162
  electric power generation, 9-296 to 9-304
  flue gas analysis, 9-47 to 9-49
  nuclear reactor: boiling water reactor, 5-63 to 5-65, 5-72 to 5-73
    gas-cooled fast breeder reactor, 5-132 to 5-134
    high-temperature gas-cooled reactor, 5-51 to 5-53, 5-97 to 5-98
    pressurized water reactor, 5-85 to 5-86
  polluting gas and particulate emissions, 9-362 to 9-406
  Seneca pumped-storage plant, 8-23 to 8-24
  thermoelectric cooling system, 9-242
  tidal power plant, 8-34 to 8-36
  wind power systems, 6-153 to 6-154
Insulated tankcars, 3-185 to 3-190
Insulation:
  cryogenic upgrading process use of, 2-119
  energy conservation by use of, 3-289 to 3-290
Integral control system, boiler, 9-127 to 9-128
Integral furnace-boiler, 9-19, 9-103 to 9-104
Integral-plus-proportional action, boiler control system, 9-128
Integral tank, definition of, 3-206
Integrated boiler-turbine generator control system, 9-151
Integrated hydrogen energy system, 6-16
Integrated process plants, conservation of energy in, 3-272 to 3-303
Intentional caving, coal mine roof, 1-73
Interactive loadflow computer programs, electric utility use of, 6-293
Interchangeability, heat and work, 9-23
Interconnected tank train, 3-179
Intercooling, gas turbine, 9-178
Interdependence, petrochemical complex, 3-270
Interelectrode gap, thermionic converter, 9-221
Intergovernmental Marine Consultative Organization of the United Nations, 3-196, 3-210, 3-223
Intermediate, definition of, 3-265
Intermediate load solar plant, 6-6, 6-69 to 6-80
Intermediate neutron, 5-44
Intermediate reactor, 5-44
Intermediate string, oil well, 3-129
Intermetallic compounds, superconductor use of, 5-161
Intermountain Forest and Range Experiment Station, 1-107 to 1-110
Internal cleaning, boiler tube, 9-119

Internal energy, 9-23
Internal heat, earth, 7-4
International Atomic Energy Agency, 5-174
International Business Machines Corporation, 8-19, 9-209, 9-277 to 9-295
International classification of coals, 1-16
International Harvester Company, Solar Division, 9-176 to 9-198
International Nickel Company, Inc., 2-113 to 2-114
International regulations, shipping, 3-210
Iowa:
  AGA area of, 2-9
  Bituminous Coal District of, 1-169
  coals in, 1-27, 1-34 to 1-35, 1-37, 1-93
  daily terrestrial solar energy received at Ames, 6-96
  electric power generating capacity of, 9-252 to 9-253
  emission limits for SO_2 and NO_x, 9-374
  hydroelectric power production of, 8-6
  natural gas, production and reserves of, 2-10 to 2-13, 2-16 to 2-18
  nuclear power installations in, 5-13
Iowa Electric Light and Power Company, 5-13
Iowa-Illinois Gas and Electric Company, 5-13
Iowa Southern Utilities, 1-144 to 1-146
Iran:
  ancient windmills in, 6-147
  crude oil: annual production of, 3-80
    characteristics of, 3-16
    exploratory and production wells completed, 3-86
    percent of recoverable reserves, 3-85
    ultimate recovery potential for, 3-92
  gas turbines installed in, 9-188
  metric tons to barrels conversion factor, 3-84
  natural gas: production of, 2-39
    reserves of, 2-42, 2-44
  nuclear power facilities planned, 5-29, 5-37 to 5-38
  railway mileage and motive power of, 9-226
Iraq:
  crude oil: annual production of, 3-80, 3-88
    percent of recoverable reserves, 3-85
    ultimate recovery potential for, 3-92
  metric tons to barrels conversion factor, 3-84
  natural gas: production of, 2-39
    reserves of, 2-42, 2-44
  railway mileage and motive power of, 9-227
Ireland:
  Bantry Bay deep crude terminal, 3-207
  crude oil: production of, 3-88
    production and exploratory wells completed, 3-86
    ultimate recovery potential for, 3-93
  natural gas, ultimate recovery potential for, 2-45

Japan (*Cont.*):
  Japan Gasoline Company, Ltd., 2-149 to 2-159
  Keiyo Gas Company, Ltd., 2-149, 2-152, 2-155 to 2-156
  liquefied natural gas imports from Brunei, 2-103
  metric tons to barrels conversion factor, 3-84
  natural gas: production of, 2-39
    reserves of, 2-42, 2-45
  Nikki Chemical Company, Ltd., 2-149 to 2-150
  nuclear power installations in, 5-29 to 5-31, 5-36 to 5-38
    advanced thermal reactor design, 5-40
    sodium-cooled fast breeder reactor development, 5-110 to 5-113
    uranium resources of, 5-176
  Osaka Gas Company, Ltd., 2-149
  petroleum: seaborne imports of, 3-199
    tanker service for, 3-210
      Itachi shipyard, 3-211
  railway mileage and motive power of, 9-226
  railway system advancements by Japanese National Railways, 9-241
  solar furnace developed by Tohoku University, 6-46 to 6-47
  solar water heaters, use of, 6-93 to 6-94
  tar sand deposits in, 3-160
  tidal range: at Osaka, 8-40
    at Yokohama, 8-41
JAPCO nuclear plants, 5-29, 5-31
Jay oil field, 3-62
Jeffrey Manufacturing Company, 1-124
Jeffreys, tidal studies of, 8-39, 8-43
Jersey Central Power and Light Company, 5-15, 5-17
Jet deflection bit, 3-118
Jet fuels, United States consumption of, 3-97 to 3-98
  (*See also* Aviation fuels)
Jet pump recirculation system, boiling water reactor, 5-63 to 5-64
Jet sled, marine pipeline, 2-84, 2-88
Jig, coal cleaning, 1-122, 1-127
Job mobility, coal mine safety factor of, 1-90
John Day hydroelectric plant, 8-5
Johns Hopkins University, The, 6-118 to 6-135
Johnstown, Pennsylvania, mine disaster at, 1-92
Joint, coal bed, 1-71
Jordan:
  crude oil: annual production of, 3-89
    ultimate recovery potential for, 3-93
  natural gas, ultimate recovery potential for, 2-45
  railway mileage and motive power of, 9-227
Joule cycle, gas turbine, 9-176
Joy Manufacturing Company, 9-194 to 9-195, 9-351 to 9-353
Joyo reactor, 5-112
JP military jet fuels, 3-38

JPN propellant, 4-83
J-tube pipelaying method, 2-86 to 2-87
Juneau, Alaska, tidal range at, 8-40
Junegrass, spoiled-land area use of, 1-107
Jungle climate, gas turbine operation in, 9-188
Junipers, spoiled-land area use of, 1-112
Jupia hydroelectric plant, 8-5
Jupiter Corporation, The, 2-64
Jurassic Geologic System, gas and oil formed during: in Canada, 3-76 to 3-77
  in United States, 2-14, 3-59
Jurisdictional boundary, offshore oil drilling, 3-91

Kaiser Aluminum and Chemical Corporation, 6-127
Kaiser Steel Corporation, 1-139, 1-159
Kaman Aerospace Corporation, 6-144
Kamchatka Peninsula, geothermal resources of, 7-4, 7-8
Kansas:
  AGA area of, 2-9
  API Survey district of, 3-50
  coals in, 1-27, 1-32, 1-34 to 1-35, 1-37
  crude oil, production and reserves of, 3-46 to 3-47, 3-53, 3-56, 3-60
  daily terrestrial solar energy received: at Dodge City, 6-95
    at Manhattan and Topeka, 6-96
  electric power generating capacity of, 9-253
  emission limits for SO_2 and NO_x, 9-374
  hydroelectric power production of, 8-6
  natural gas: production and reserves of, 2-10 to 2-13, 2-16 to 2-18
    underground storage capacity, 2-74 to 2-75
  nuclear power installations in, 5-14
  tar sand deposits in, 3-159, 3-161
Kansas Gas and Electric Company, 5-14
Kaplan turbine, 8-11, 8-13, 8-16, 8-30, 8-34
Kariba hydroelectric plant, 8-5
Karibib lepidolite resources, 5-183
Karlsruhe Nuclear Research Center, 5-124
Kawerau geothermal installation, 7-8, 7-31 to 7-32
Keahole Point, Hawaii, ocean thermal energy conversion plant site at, 6-129
Keban hydroelectric plant, 8-6
Keiyo Gas Company, Ltd., 2-149, 2-152, 2-155 to 2-156
Kellerman coal mine, 1-31
Kellerman Mining Company, 1-31
Kellogg, M. W., Company, 2-130 to 2-131
Kelly Snyder oil field, 3-62
Kelvin-Planck statement, 9-23
Kelvin waves, tidal effects of, 8-43
Kemano hydroelectric plant, 8-5
Kemmerer Coal Company, 1-31
Ken coal mine, 1-29
Kenai, Alaska, liquefied natural gas facility at, 2-102 to 2-103
Kenerson dynamometer, 9-305
Kennametal, Inc., 1-78

Lynch Winifrede coal mine, 1-30
Lynnville coal mine, 1-29

McArthur River oil field, 3-62
McComb, Mississippi crude oil,
  characteristics of, 3-14 to 3-15
McDonnell Douglas Astronautics, 6-26
McDowell County, West Virginia, coals in,
  1-20, 1-24, 1-34
McElroy coal mine, 1-29
McElroy oil field, 3-65
McGuire, William B., nuclear plants, 5-16,
  5-23
Machinery industry:
  electricity consumption of, 9-259
  gas requirements of, 2-52 to 2-56
Mackenzie Delta, oil reserves of, 3-73 to 3-
  74
Mackenzie Delta pipeline system, 2-65
McKittrick oil field, 3-65
McMurray Formation, tar sands in, 3-159
McNally Norton coal washer, 1-129
McNally-Pittsburg Manufacturing
  Corporation, 1-117 to 1-152
McNally stacker/reclaimer, 1-151 to 1-152
McNally Trom three-product vessel, 1-127
McNary hydroelectric plant, 8-5
Madison, Wisconsin, daily terrestrial solar
  energy received at, 6-97
Madisonville, Kentucky, in-situ coal
  gasification experiments at, 1-244
Magma Power Company, 7-34
Magmatic steam, 7-35
Magnesia gas-processing system, 9-320 to
  9-321
Magnesium bisulfite, utilization of, 9-314
Magnesium electrode, battery, 4-52
Magnesium hydride, 4-28, 4-43
Magnesium oxide acceptor, 1-221
Magnesium perchlorate electrolyte,
  battery, 4-52
Magnesium sulfite disposal and treatment,
  9-314
Magnesium system:
  nitrogen oxides absorption with, 9-320 to
    9-321
  sulfur dioxide absorption with, 9-314
Magnetic compression-expansion direct
  energy conversion in fusion reactors,
  5-152 to 5-153
Magnetic expander, direct plasma energy
  conversion use of, 5-151
Magnetic jack, 5-44
Magnetic levitation vehicle support, 9-241
Magnetic potential energy, definition of,
  xxiv
Magnetic surveys:
  geothermal system, 7-7
  oil drilling, 3-116 to 3-117
Magnetization curve:
  ideal Type-I superconducting cylinder,
    5-160
  ideal Type-II superconducting cylinder,
    5-160 to 5-161
Magnetohydrodynamic generators, 9-209
  to 9-221

Magnetohydrodynamic solar energy
  converters, 6-15 to 6-17
Magnets:
  superconducting, fusion reactor use of
    (see Superconducting magnets,
    fusion reactor use of)
  tramp iron, 1-118 to 1-119, 1-122
Magnox, 5-44
Main Pass oil field, 3-63 to 3-65
Maine:
  daily terrestrial solar energy received: at
    Caribou, 6-97
    at Portland, 6-96
  electric power generating capacity of, 9-
    253
  emission limits for SO_2 and NO_x, 9-374
  gas turbine installations in, 9-192
  hydroelectric power production, 8-6
  nuclear power installations in, 5-14
  tidal range at Portland, 8-40
Maine Yankee nuclear facility, 5-14, 5-21
Maintenance:
  of boiler control system, 9-156
  of magnetohydrodynamic generator, 9-
    209
Maize, 6-139
Makeup water, definition of, 9-26
Malagasy Republic (Madagascar):
  crude oil: producing wells for, 3-89
    production and exploratory wells
      completed, 3-86
    ultimate recovery potential for, 3-92
  natural gas, ultimate recovery potential
    for, 2-44
  railway mileage and motive power of, 9-
    227
  tar sand deposits in, 3-160
Malawi:
  crude oil: producing wells for, 3-89
    ultimate recovery potential for, 3-92
  natural gas, ultimate recovery potential
    for, 2-44
  railway mileage and motive power of, 9-
    227
Malaya:
  nuclear installations planned, 5-37 to 5-
    38
  tidal range at Singapore, 8-40
Malaysia:
  crude oil: producing wells for, 3-89
    production and exploratory wells
      completed, 3-86
    proved recoverable reserves of, 3-82
    ultimate recovery potential for, 3-93
    metric tons to barrels conversion factor,
      3-84
  railway mileage and motive power of, 9-
    227
Maldives:
  crude oil: producing oil wells for, 3-89
    ultimate recovery potential for, 3-92
  natural gas, ultimate recovery potential
    for, 2-45
Mali:
  crude oil: producing oil wells for, 3-89
    ultimate recovery potential for, 3-
      92

Mali (Cont.):
    natural gas, ultimate recovery potential for, 2-44
    railway mileage and motive power of, 9-227
Malta:
    crude oil: producing oil wells for, 3-89
        production and exploratory wells completed, 3-87
        ultimate recovery potential for, 3-93
    natural gas, ultimate recovery potential for, 2-45
Maltha (see Tar sands)
Mammoth, Pennsylvania, mine disaster at, 1-92
Management:
    of boiler fuel system, 9-145 to 9-147
    of mined-out areas, 1-102 to 1-115
Maneuverability, ship, 3-216 to 3-218
Manganese dioxide electrode, battery, 4-52
Manganese poisoning, spoiled-land area, 1-103
Mangla hydroelectric plant, 8-5
Manhattan, Kansas, daily terrestrial solar energy received at, 6-96
Manicouagan hydroelectric plant, 8-5
Manila, Philippines, tidal range at, 8-40
Manitoba:
    crude oil reserves of, 3-74 to 3-77
    hydroelectric facilities in, 9-261
    pipeline terminus at Emerson, 2-65
    tar sand deposits of, 3-159 to 3-161
Manned platform, offshore drilling, 3-108
Mantle, heat generation in, 7-5
Manual control, boiler burner system, 9-144 to 9-145
Manufacturers Light and Heat Company, 2-64
Manufacturing industries:
    coal requirements of, 1-28
    electricity consumption of, 9-259
    gas requirements of, 2-52 to 2-56
Manure:
    energy conversion of, 4-13
    synthetic oil from, 3-168
Manway, tanker, 3-183
Mapco pipeline, 3-175
Maple Creek coal mine, 1-29
Maples, spoiled-land area use of, 1-112 to 1-113
Mapping:
    coal resource, 1-38
    electric utility use of, 9-294
    spoiled-land area, 1-105
Maracaibo sedimentary basin, tar sands in, 3-160
Marble Hill nuclear plant, 5-13
Marcasitic sulfur, coal content of, 1-22
Marcoule gas-cooled reactor, 5-89
Marianna, Pennsylvania, mine disaster at, 1-92
Marianna 58 coal mine, 1-30
Marimbondo hydroelectric plant, 8-5
Marine boilers, 9-22, 9-65
Marine coal traffic, 1-173
Marine Diesel fuel oils, 3-31, 3-36
Marine fouling, drilling structure, 3-111

Marine pipelines, 2-77 to 2-89
Marine scale, tide measurement by, 8-27
Marine Transport Lines, Inc., 3-196 to 3-207
Maritime Provinces, petroleum refining developments of, 9-263
Marketable natural gas:
    in Canada, 2-35
    Canadian Petroleum Association definition of, 2-37
Marsa el Brega, Libya, liquefied natural gas facility at, 2-102 to 2-103
Marseilles, France:
    tanker terminal at, 3-210
    tidal range at, 8-40
Martin Incinerator Systems, 4-21
Martin Lake Steam Electric Station, 1-144 to 1-145
Marty coal mine, 1-30
Marty Corporation, 1-31
Maryland:
    AGA area of, 2-9
    coals in, 1-27, 1-30, 1-32, 1-34 to 1-35
    daily terrestrial solar energy received at Annapolis, 6-96
    electric power generating capacity of, 9-253
    emission limits for SO_2 and NO_x, 9-374
    hydroelectric power production of, 8-6
    natural gas, production and reserves of, 2-10 to 2-13, 2-16 to 2-18
    nuclear power installations in, 5-14
    tidal range at Baltimore, 8-40
    waste conversion plant at Baltimore, 4-19
Masonry support, coal mine roof, 1-74
Mass balance, fuel cell, 4-68 to 4-69
Mass decrement, 5-44
Mass-energy equation, xxii
Mass spectrometric gas calorimeter, 2-92 to 2-93
Mass velocity versus steam quality, 9-98
Massachusetts:
    daily terrestrial solar energy received at Boston and Cambridge, 6-97
    electric power generating capacity of, 9-253
    emission limits for SO_2 and NO_x, 9-374
    hydroelectric power production of, 8-6
    nuclear power installations in, 5-14
    solar-heated building in Cambridge, 6-92
    tidal range at Boston, 8-40
Massachusetts, University of, ocean thermal energy conversion plant concept of, 6-120
Massachusetts Audubon Society, solar-heated building of, 6-92
Massachusetts Institute of Technology, 5-108, 5-137 to 5-149, 9-205 to 9-206
Mast feed, coal mine roof drilling, 1-79
Material, superconducting, 5-161 to 5-163
Material and energy balance:
    catalytic reformer heater, 3-278
    COGAS process, 1-267
    gas treating system, 3-296
    Hydrane process, 1-226

Material and energy balance (*Cont.*):
hydrocracking/hydrogasification process, 2-169
fuel cell, 4-68 to 4-69
Lurgi process gasifier, 1-192
reformer heater, steam-methane process, 3-279
steam generation, process heat, 3-288
steam power plant: high-pressure, 3-289
low-pressure, 3-288
Synthane process: acid gas removal and sulfur recovery, 1-237
methanation system, 1-238
pretreatment and gasification system, 1-233
raw gas scrubbing system, 1-233
shift conversion system, 1-236
Tosco II oil retorting process, 3-157
Material management system, electric utility, 9-278
Material recovery system, solid-waste processing, 4-14 to 4-15, 4-17 to 4-18, 4-20, 4-22 to 4-24
Materials selection, liquefied natural gas equipment, 2-100 to 2-101, 2-109 to 2-113
Mathematical model:
oil well drilling, 3-126
solar energy absorber tube, 6-40
solar power system, 6-69 to 6-70
Mather, Pennsylvania, mine disaster at, 1-92
Mathies Coal Company, 1-31
Mathies coal mine, 1-29
Matrix-type fuel cell, 4-67 to 4-70
Matsukawa geothermal field, 7-5, 7-11
Matter, conservation law of, 9-34
Matthews coal mine, 1-30
Mauritania:
crude oil: production of, 3-89
ultimate recovery potential for, 3-92
natural gas, ultimate recovery potential for, 2-44
railway mileage and motive power of, 9-227
solar water heaters used in, 6-93
Mauritius:
crude oil: production of, 3-89
ultimate recovery potential for, 3-92
natural gas, ultimate recovery potential for, 2-44
Maxine coal mine, 1-30
Maybach Motorenbau, 9-233
Mean sea level, 3-221
Measurement:
electric power, 9-296 to 9-304
heat of combustion, 9-41
(*See also* Analysis; Instrumentation)
Mecco coal mine, 1-30
Mechanical atomizer, oil burner, 9-52 to 9-55
Mechanical-sampler bias testing, coal, 1-60
Mechanical steam separator, 9-100 to 9-103
Mechanics, coal mine roof support, 1-71 to 1-83
Medford, Oregon, daily terrestrial solar energy received at, 6-97

Mediterranean Sea, tidal range of, 8-37
Medium-head hydroelectric plant, 8-9
Medium-volatile bituminous coal, 1-17, 1-54
Megapascal unit, definition of, 5-171
Megawatt electric unit, definition of, 9-260
Megawatt-load control, boiler, 9-153
Megawatt-mile unit, definition of, 9-267
Megawatt output, boiler, 9-338
Meinel optical concentrator, 6-10, 6-27
Melbourne, Australia, tidal range at, 8-40
Meltdown, fast breeder reactor, 5-105
Melting point:
of butane, 3-180
of propane, 3-180
Melville Island, tar sand deposits at, 3-159 to 3-162
Membrane, fuel cell, 4-63 to 4-72
Memphis, Tennessee, daily terrestrial solar energy received at, 6-96
Mendocino nuclear facilities, 5-23
Mercaptan removal, Sulfinol process, 2-133 to 2-136
Mercaptans, crude oil content of, 3-254
Mercer County, North Dakota, coals in, 1-20, 1-34
Mercuric oxide electrode, battery, 4-52
Mercury, resistance versus temperature, 5-158
Mercury battery, 4-52 to 4-53
Mercury coolant, fast breeder reactor, 5-111
Mercury-steam systems, 6-19, 9-22
Mesopotamian-Persian Gulf sedimentary basin, tar sand deposits in, 3-160
Mesozoic Era, fossil fuels formed during, 3-51
coal, 1-15
crude oil: in Canada, 3-76 to 3-77
in United States, 3-59
natural gas, in United States, 2-14
Messerschmitt-Bolkow-Blohn (MBB), magnetic levitation developments of, 9-241
Metabituminous coal, 1-15
Metabolic processes, plant-to-energy conversion by, 4-5 to 4-6, 4-8, 4-10
Metal:
Athabasca bitumen content of, 3-163
crude oil content of, 3-13 to 3-17
solid-waste content of, 4-22
Metal deactivators, gasoline, 3-27
Metal hydride storage system, 6-74 to 6-80
Metallurgical coke:
clean coke process for producing, 1-249 to 1-254
testing of coal for, 1-58
Metallurgical furnaces, fuel fineness requirements for, 9-65
Metals industry:
electricity consumption of, 9-259
energy conservation in, 3-301
gas requirements of, 2-52 to 2-56
Metamorphic rocks, oil reservoir, 3-51
Metamorphism, coal formation by, 1-15
Metastable propellant ingredients, 4-88 to 4-89

Mirrors and reflectors, solar energy system (*Cont.*):
  solar furnaces, 6-46 to 6-50, 6-53 to 6-54
  solar tower energy configuration, 6-26 to 6-32
Miscible flooding for oil well treatment, 3-142
Missiles, propellants for, 4-76
Mississippi:
  AGA area of, 2-9
  API Survey district of, 3-50
  crude oil: characteristics of, 3-13 to 3-14
    productive capacity for, 3-53, 3-56, 3-60
    reserves of, 3-46 to 3-47, 3-57 to 3-59
  electric power generating capacity of, 9-253
  emission limits for SO_2 and NO_x, 9-374
  natural gas: production and reserves of, 2-10 to 2-13, 2-16 to 2-18
    underground gas storage capacity, 2-75 to 2-76
  nuclear power installations in, 5-15
Mississippi embayment, 2-23
Mississippi-Ohio River system, coal movement in, 1-170 to 1-172
Mississippi Power and Light Company, 5-15
Mississippi River Transmission Corporation, 2-64
Mississippian Geologic System, fossil fuels formed in, 3-51
  crude oil: in Canada, 3-76 to 3-77
    in United States, 3-59
  natural gas, in United States, 2-14
Missouri:
  AGA area of, 2-9
  coals in, 1-27, 1-31 to 1-32, 1-34 to 1-35, 1-37
  daily terrestrial solar energy received at Columbia, 6-96
  electric power generating capacity of, 9-253
  emission limits for SO_2 and NO_x, 9-374
  hydroelectric power production of, 8-6
  nuclear power installations in, 5-15
  tar sand deposits in, 3-159, 3-161
Mixed-refrigerant cascade cycle, 2-102 to 2-108
Mixer, electric, 9-260
Mixing systems, gas, 2-96 to 2-99
Mixture calculations, gas, 2-96
Mobil Oil Corporation, 3-152
Model:
  electric power system, 9-277 to 9-278, 9-284 to 9-286, 9-293 to 9-294
  English Channel tidal phenomena, 8-43 to 8-45
  Europoort superport scheme, 3-222
  solar energy absorber tube, 6-40
  solar power system, 6-69 to 6-70
Moderator:
  beryllium oxide as, 5-8
  boron-carbide control rods as, 5-71
  burnable gadolinia poison as, 5-71

Moderator (*Cont.*):
  definition of, 5-8 to 5-9, 5-45
  graphite as, 5-8
  heavy water as, 5-8 to 5-10, 5-44
Modernization, petroleum refinery, 3-272 to 3-273
Modular transportation, solar energy satellite, 6-114
Module, liquefaction, 2-103
Modulus of elasticity, aluminum cryogenic alloy, 2-107
Mogul washbox, 1-119
Mohave Generating Station, 1-178 to 1-182
Moist, mineral-matter-free basis, definition of, 1-16
Moisture:
  coal content of, 1-15 to 1-26, 1-48 to 1-62
  coal-fired equipment importance of, 9-59
  coal handling importance of, 1-147
  solid-waste content of, 4-22
Molal volume, definition of, 9-34
Molar absorptivity, energy, 9-366
Molasses, fermentation of, 4-7
Molded beam, tanker and product carrier, 3-203
Molded depth, tanker and product carrier, 3-203
Mole concept, combustion calculations use of, 9-38 to 9-40
Mole fraction, definition of, 9-38
Molecular sieve, water removal with, 2-116
Molex process, 3-269
Moline, Illinois, daily terrestrial solar energy received at, 6-96
Mollier diagram, 2-106, 9-25, 9-31
Molten-salt electrolyte, battery, 4-51
Molybdenum anode, thermionic-converter, 9-222
Molybdenum-nickel catalyst, catalytic rich gas process use of, 2-140
Momentum storage, solar system, 6-23 to 6-25
Mongolia:
  crude oil: production of, 3-89
    ultimate recovery potential for, 3-93
    metric tons to barrels conversion factor, 3-84
  natural gas, ultimate recovery potential for, 2-45
  railway mileage and motive power of, 9-227
Monitoring:
  bio-oxidation process, 9-358
  nuclear plant (*see* Nuclear energy technology)
  polluting gases and particulates, 9-362 to 9-406
  power plant, 9-289 to 9-290
  (*See also* Analysis; Instrumentation)
Monju reactor, 5-112
Monoethanolamine, carbon dioxide removal with, 2-116
Monogah, West Virginia, mine disaster at, 1-92
Monomethylhydrazine, propellant use of, 4-81

Nebraska (*Cont.*):
natural gas, production and reserves of, 2-10 to 2-13, 2-16 to 2-18
nuclear power installations in, 5-15
Nebraska Public Power District, 5-15
Nechako River hydropower reserves, 8-9
Neches oil field, 3-64
Needlegrass, spoiled-land area use of, 1-107
Nelson River hydroelectric facility, 9-261
Nemacolin mine fire, 1-84
Neopentane, combustion constants of, 9-36 to 9-37
Nepal:
crude oil: production of; 3-89
ultimate recovery potential for, 3-93
natural gas, ultimate recovery potential for, 2-45
railway mileage and motive power of, 9-227
Net calorific value gas calorimeter, 2-90
Net heat of combustion, calorimetry, 1-67 to 1-68
Netherlands, the:
coal reserves of, 1-36
crude oil: annual production of, 3-81
exploratory and production wells completed, 3-87, 3-89
ultimate recovery potential for, 3-93
fuel cell developments of, 4-63
metric tons to barrels conversion factor, 3-84
natural gas: discovery at Groningen, 3-91
production of, 2-40
reserves of, 2-42 to 2-45
nuclear installations in, 5-31, 5-36 to 5-38
fuel study of, 5-179
Port of Rotterdam and Europoort, 3-220 to 3-224
railway mileage and motive power of, 9-226
Royal Netherlands Meteorological Institute, 3-222
tidal range at Rotterdam, 8-40
Neutralization, spoiled-land area, 1-105 to 1-106
Neutron, 5-7, 5-45
Neutron absorption, 5-8, 5-10
Neutron cross section, 5-45
Neutron density, 5-9
Neutron flux, 5-8 to 5-9, 5-45, 5-68
Neutron lifetime, liquid-metal fast breeder reactor, 5-105
Neutron multiplication, 5-155
Neutron sources, boiling water reactor, 5-69 to 5-70
Nevada:
AGA area of, 2-9
Black Mesa pipeline terminus, 1-178
Clayton valley lithium resources, 5-183
daily terrestrial solar energy received at Ely, Las Vegas, and Reno, 6-95
electric power generating capacity of, 9-253
emission limits for SO_2 and NO_x, 9-374
hydroelectric power production of, 8-7

Nevada-Texas-Utah oil shale retorting process, 3-148 to 3-149
New Brunswick, tidal range at St. John, 8-40
New England:
electric utilities in, 9-254 to 9-256
fuel requirements of, 9-260 to 9-261
gas consumption of, 2-51 to 2-52
hydroelectric power production of, 8-6
power pool of, 9-204
[*See also* Northeastern District (API Survey)]
New England Electric System, 5-19
New Guinea, metric tons to barrels conversion factor of, 3-84
New Hampshire:
electric power generating capacity of, 9-253
emission limits for SO_2 and NO_x, 9-374
hydroelectric power production of, 8-6
nuclear power installations in, 5-15
New Jersey:
daily terrestrial solar energy received at Trenton, 6-96
electric power generating capacity of, 9-253
emission limits for SO_2 and NO_x, 9-374
gas turbine installations in, 9-192
hydroelectric power production of, 8-6
Linden terminus, "Big Inch" pipeline, 3-174
nuclear power installations in, 5-15
synthetic gas plant at Harrison, 2-147
wave motor project in Atlantic City, 8-45
New Mexico:
AGA area of, 2-9
API Survey district of, 3-50
Bituminous Coal District of, 1-169
coals in, 1-27, 1-29, 1-32, 1-34 to 1-35, 1-92 to 1-93
crude oil: characteristics of, 3-13 to 3-14
production of, 3-53 to 3-54, 3-56, 3-60, 3-63 to 3-65
reserves of, 3-48, 3-57 to 3-59
daily terrestrial solar energy received at Albuquerque, 6-95
electric power generating capacity of, 9-253
emission limits for SO_2 and NO_x, 9-374
gas transmission systems in, 2-63
hydroelectric power production of, 8-6
natural gas, production and reserves of, 2-10 to 2-13, 2-16 to 2-18
sunshine, percent possible at Albuquerque, 6-72
tar sand deposits of, 3-159, 3-161
New oil fields, API classification of, 3-70 to 3-72
New Orleans, Louisiana, coal transfer port at, 1-171 to 1-172
New pool discoveries, crude oil, 3-61 to 3-71
(*See also* Exploration, crude oil)
New York:
AGA area of, 2-9
API Survey district of, 3-50

Nitrogen (*Cont.*):
 coal, content of, 1-22, 1-50
 COGAS process: moderate-Btu gas,
   content of, 1-266
  pipeline gas, content of, 1-266
  pyrolysis gas, content of, 1-265
 freeze-wall technique use of, 5-50
 geothermal gases, content of, 7-36, 7-44
 high-Btu gas, content of, 2-118
 Hydrane process: coal content of, 1-227
  feed gas content of, 1-228
 Kjeldahl-Gunning determination of, 1-50
 low-Btu gas, content of, 2-118
 Lurgi process, raw coal content of, 1-192,
   1-286
 natural gas, content of, 2-129
 oil-well treatment use of, 3-135
 oxides of (*see* Nitrogen oxides)
 physical constants of, 9-36 to 9-37
 refrigeration use of, 9-200
 Shell gasification process product gas,
   content of, 2-174
 spoiled-land enrichment with, 1-106
 Synthane process: raw coal content of, 1-
   233
  raw gas content of, 1-233
  shift conversion, presence of, 1-236
 Toscoal process: char content of, 1-273
  process oil content of, 1-274
 vent gas, content of, 2-118
Nitrogen dioxide, 9-319 to 9-321
Nitrogen-methane mixtures, upgrading of,
  2-116
Nitrogen oxides:
 emission limits of, 9-374 to 9-375
 engine horsepower related to, 9-196
 furnace and burner design for control of,
   9-335 to 9-348
 removal of, 9-312 to 9-323
 source monitoring of, 9-362 to 9-375
 wavelength versus absorbance of, 9-368
Nitrogen-16, reactor steam content of, 5-57
Nitrogen tetroxide, propellant use of, 4-81
Nitroglycerin, propellant use of, 4-83
Nitronium perchlorate, propellant use of,
  4-85
Nizhne-Kamskaya hydroelectric plant, 8-5
Noise:
 coal mine, 1-84 to 1-85
 geothermal system, 7-11, 7-19, 7-24
Nominal channel bed level, definition of,
  3-215
Nonagglomerating coals, 1-18
Nonaqueous electrolytes, battery, 4-51 to 4-53
Nonassociated natural gas:
 definition of, 2-16
 probable reserves of, in Canada, 2-34
 proved reserves of: in Canada, 2-34
  in United States, 2-10
Noncaking coals, 1-17
Noncatalytic partial-oxidation:
 gasification process, 2-171 to 2-174
 of hydrocarbons, hydrogen produced by,
   4-36
Nonconventional reserves, hydrocarbon,
  Canadian Petroleum Association
  definition of, 3-73

Nondestructive testing, pipeline, 2-73, 2-
  79 to 2-80, 2-83 to 2-85
Nondispersive infrared analyzer, 9-370, 9-
  376, 9-384 to 9-385
Nonferrous metals, industry, gas
  requirements of, 2-52 to 2-56
Nonproductive formation, definition of, 2-
  24
Norco nuclear plant, 5-19
Norfolk, Virginia:
 daily terrestrial solar energy received at,
   6-96
 tidal range at, 8-40
Norfolk and Western Railroad, 1-154
Norris coal mine, 1-30
Norske Veritas, Det, 3-196
North America:
 crude oil: annual production of, 3-81
  imports of, 3-199
  production and exploratory oil wells
    completed, 3-87
  recoverable reserves of, 3-83
  ultimate recoverable reserves of, 3-
    94
 natural gas: production of, 2-40
  reserves of, 2-43, 2-46
North American Coal Corporation, 1-31
North Anna nuclear plants, 5-18, 5-23
North Carolina:
 daily terrestrial solar energy received: at
   Cape Hatteras, 6-95
  at Greensboro, 6-96
 electric power generating capacity of, 9-
   252 to 9-253
 emission limits for SO_2 and NO_x, 9-374
 hydroelectric power production of, 8-6
 lithium resources of, 5-184
 nuclear power installations in, 5-16
North Central District (API Survey), 3-50
North Dakota:
 AGA area of, 2-9
 Bituminous Coal District of, 1-169
 coals in, 1-20, 1-27, 1-30, 1-32, 1-34 to 1-
   35, 1-37, 1-212, 1-221, 1-227
 crude oil: productive capacity of, 3-53, 3-
   56, 3-60, 3-65
  reserves of, 3-46 to 3-47, 3-57 to 3-59
 daily terrestrial solar energy received at
   Bismarck, 6-96
 direct current power transmission
   systems in, 9-276
 electric power generating capacity of, 9-
   253
 emission limits for SO_2 and NO_x, 9-374
 hydroelectric power production of, 8-6
 natural gas, production and reserves of,
   2-10 to 2-13, 2-16 to 2-18
North Island, New Zealand, geothermal
  fields on, 7-14 to 7-33
North Kenai Peninsula, crude oil
  characteristics of, 3-14 to 3-15
North Omaha, Nebraska, daily terrestrial
  solar energy received at, 6-96
North Platte, Nebraska, daily terrestrial
  solar energy received at, 6-96
North Rocky Mountain District (API
  Survey), 3-50

North Sea:
  crude oil found in, 9-263
  navigation in, 3-223
  petroleum resources of, 3-91
  tidal range of, 8-43 to 8-44
North Slope pipeline system, 2-65 to 2-69
Northeast Nuclear Energy Company, 5-12
  to 5-13
Northeast Power Coordinating Council
  (NPCC), 9-266
Northeast Texas AGA area, 2-9
Northeast Utilities, 5-14
Northeastern District (API Survey), 3-50
Northeastern Forest Experiment Station, 1-
  102 to 1-115
Northern California power pool, 9-204
Northern Indiana Public Service
  Company, 5-13
Northern Natural Gas Company, 2-64
Northern Plains region, gas consumption
  of, 2-51, 2-53
Northern States Power Company, 5-15, 5-
  19
Northfield Mountain hydroelectric plant,
  8-5
Northwest power pool, 9-204
Norway:
  crude oil: production of, 3-89
    production and exploratory wells
      completed, 3-87
    proved recoverable reserves of, 3-83
    ultimate recovery potential for, 3-93
  electrolytic hydrogen production, 4-34
  metric tons to barrels conversion factor,
    3-84
  nuclear facilities planned, 5-31, 5-37 to
    5-38
  power production, electric, 9-247
  railway mileage and motive power, 9-
    226
  tidal range at Oslo, 8-40
Nova Scotia:
  tidal energy project of, 8-45
  tidal range at Burntcoat Head and
    Halifax, 8-40
Novovolynskaya mine (U.S.S.R.), coal
  slurry pipeline from, 1-180
Nozzle, pulverized coal burner, 9-74 to 9-
  77
Nozzle chamber, steam turbine, 9-163
N-T-U (Nevada-Texas-Utah) oil shale
  retorting process, 3-148 to 3-149
Nuclear cross section, 5-71
Nuclear energy, availability of, xxvii
Nuclear Energy Agency of the Organization for Economic Cooperation
  and Development, 5-174
Nuclear energy technology, 5-1 to 5-185
  fission reactors, 5-7 to 5-136, 5-173 to 5-
    181
    boiling water reactor, 5-57 to 5-74
    construction and start-up, 5-47 to 5-56
    gas-cooled fast breeder reactor, 5-124
      to 5-136
    glossary of nuclear terms, 5-39 to 5-47
    high-temperature gas-cooled reactor,
      5-87 to 5-101

Nuclear energy technology, fission
  reactors (Cont.):
    installations of nuclear power plants:
      in United States, 5-12 to 5-24
      worldwide, 5-25 to 5-39
    liquid-metal fast breeder reactor, 5-
      102 to 5-123
    pressurized water reactor, 5-75 to 5-86
    uranium resources and processing, 5-
      173 to 5-181
  forecast of nuclear power in United
    States, 5-21, 5-37 to 5-39, 9-207
  fusion power, 5-137 to 5-172, 5-182 to 5-
    185
    direct energy conversion in, 5-150 to
      5-154
    lithium for thermonuclear fusion, 5-
      182 to 5-185
    superconducting magnets for fusion
      reactors, 5-158 to 5-172
    tritium breeder requirements in
      fusion reactor blankets, 5-155 to 5-
      157
  uranium (see Uranium)
  monitoring of plants, 9-289
  power plant load factor computer
    programs, 9-292
  reactor containment structures, 9-290
  regional distribution of plants, 9-255
  solar energy satellite, nuclear-powered,
    6-108
  steam-nuclear plants, 9-261 to 9-266
  steam supply systems, 9-23
Nuclear potential energy, xxiii
Nuclear Research Center in West
  Germany, 4-35
Nuclear stimulation, gas field, 2-66 to 2-69
Nuclear thermal effects studies, solar
  furnace used for, 6-52
Nuclear Weapon Effects Laboratory, solar
  furnace at, 6-46 to 6-47
Nucleate boiling, 9-95
Nuclei, particulate, 3-391
Nucleon, 5-45
Nuclide, 5-7, 5-45
Nurek hydroelectric plant, 8-5
Nusselt's number, 6-12
Nut coal, 1-25, 1-27
Nylon, cyclohexane for, 3-20

Oak, spoiled-land area use of, 1-112 to 1-113
Oak Ridge, Tennessee, daily terrestrial
  solar energy received at, 6-96
Oak Ridge National Laboratory:
  fusion reactor blankets, study of, 5-147
  IMP quadrupole fusion device,
    development of, 5-165, 5-167
  lithium isotope plant at, 5-184
  tritium breeder requirements in fusion
    reactor blankets, study of, 5-155 to
    5-157
Oak Ridge Research reactor, 5-135
Oats and oatgrass, spoiled-land area use of,
  1-108
O-B-O (ore, bulk, or oil cargoes)
  combination carrier, 3-206

Oil-fired generating plant, regional distribution in United States of, **9-255**
Oil-in-place estimates:
API definition of, **3-67**
CPA definition of, **3-73**
(*See also* Reserves)
Oil Makers Company, **3-133**
Oil-ore carriers, **3-212**
Oil sands (*see* Tar sands)
Oil shale, retorting of, **3-147** to **3-159**
    crude oil obtained from, properties of, **3-155**
    horizontal configuration, **3-154** to **3-158**
        Tosco II process, **3-154** to **3-156**
        material balance of, **3-157**
        process gas, analysis of, **3-156**
        shale resources for, **3-147**
    vertical configuration, **3-129** to **3-154**
        gas combustion process (U.S. Bureau of Mines), **3-150** to **3-153**
        N-T-U process, **3-148**
        Occidental Petroleum process, **3-150**
        Paraho retort, **3-153** to **3-155**
        Union Oil process, **3-149** to **3-150**
Oil Shale Corporation, The, **1-272** to **1-275**, **3-147** to **3-158**
Okefenokee Swamp, coal formation in, **1-37**
Okinawa, deep crude terminal on, **3-207**
Oklahoma:
AGA area of, **2-9**
API Survey district of, **3-50**
coals in, **1-27**, **1-32**, **1-34** to **1-35**, **1-37**, **1-92** to **1-93**
crude oil: Burbank oil field, **3-105**
    characteristics of, **3-14**
    largest oil fields, **3-62**, **3-64** to **3-65**
    oil well depth record, **3-129**
    productive capacity of, **3-53**, **3-56**, **3-60**
    reserves of, **3-46** to **3-47**, **3-57** to **3-59**
daily terrestrial solar energy received: at Oklahoma City, **6-95**
    at Stillwater, **6-96**
electric power generating capacity of, **9-253**
emission limits for SO_2 and NO_x, **9-374**
hydroelectric power production of, **8-7**
natural gas: production and reserves of, **2-10** to **2-13**, **2-16** to **2-18**, **2-60**
    underground gas storage capacity, **2-75**
nuclear power installations in, **5-16**
Oklahoma City, Oklahoma, daily terrestrial solar energy received at, **6-95**
Oklahoma State University, **4-39**
Old Ben Coal Corporation, **1-31**
Old Ocean gas field, **3-105**
Oldbury gas-cooled reactor, **5-89**
Olefins:
aviation turbine fuel content of, **3-37**
clean coke process production of, **1-254**
defection of, **9-380**
heat of combustion of, **9-36** to **9-37**
Sasol complex production of, **1-295**
Olenek anticline, tar sands in, **3-160**

Olex process, **3-269**
Olga Coal Company, **1-31**
Oligocene Period, fossil fuels formed during: crude oil, **3-51**
    natural gas, **2-14**
    tar sands, **3-160**
Oliphant, Marcus, **4-27**
Olive trees, spoiled-land area use of, **1-112**
Omaha (Nebraska) Public Power District, **5-15**
Oman:
crude oil: annual production of, **3-80**
    production and exploratory wells completed, **3-86**, **3-89**
    recoverable reserves of, **3-85**
    ultimate recovery potential for, **3-93**
    metric tons to barrels conversion factor, **3-84**
natural gas: production of, **2-39**
    reserves of, **2-42**, **2-45**
On-board energy conversion supply system, railway locomotive, **9-224**
On-orbit solar power generation, **6-103** to **6-108**
Onassis operating group, **3-207**
Once-through boiler, control system for, **9-148** to **9-158**
One-wall firing, cyclone furnace use of, **9-82**
Onshore potential, worldwide natural gas, **2-44** to **2-46**
Ontario:
crude oil reserves of, **3-74** to **3-77**
gas turbine installations in, **9-192**
natural gas: first underground storage facility at Welland County for, **2-75**
    reserves of, **3-74** to **3-77**
Ontario hydroelectric-nuclear plants, **5-25** to **5-26**
O-O (ore-oil) combination carrier, **3-206**
Opal, geothermal fluid content of, **7-48** to **7-49**
OP DRILLING computer assistance drilling program, **3-126**
Open-circuit voltage, hydrogen-air cell, **4-61**
Open-cycle gas turbine, **9-178** to **9-179**
Open-hopper coal barge, **1-176**
Open-loop control, boiler, **9-124** to **9-125**
Open-pit mining, tar sands, **3-163** to **3-167**
Open-Rankine-cycle ocean thermal energy conversion plant, **6-118**
Open-top hopper coal, **1-157** to **1-158**
Operating costs (*see* Costs and economics)
Operating range, definition of burner, **9-57**, **9-75**
Operating variables, boiler, **9-120** to **9-162**
Opposed firing, cyclone furnace use of, **9-82**
Optical and diffusion sizing system, particulate measurement with, **9-404**
Optical concentrator, solar, **6-10**
Optical particle counter, **9-401**
Optimization:
energy use: refinery and process plant, **3-276**
    heat exchangers, **3-285**

Oxygen, requirements for (*Cont.*):
  solid-waste processing, 4-15
  steel production, 4-26
  Synthoil process, 1-271
  synthetic crude oil, content of, 3-173
Oxygen-bomb calorimeter, 1-63 to 1-70
Oxygen electrode, battery, 4-52
Oxygen handling, 1-69
Oxygen-steam ratio, Lurgi process, 1-193
Oxygenation, Sasol complex use of, 1-295
Oyster Bayou oil field, 3-64
Oyster Creek nuclear facility, 5-15, 5-21
Ozark Dam hydroelectric plant, 8-19
Ozokerite, 3-159
Ozone, propellant use of, 4-81

Pacific Coast AGA area, 2-9
Pacific Coast District (API Survey), 3-50
Pacific Energy Corporation, 7-34
Pacific Gas and Electric Company, 1-248, 5-12, 5-57, 7-34 to 7-42
Pacific Gas Transmission Company, 2-64
Pacific Northwest-California electri power transmission line, 9-267
Pacific Northwest region, gas consumption of, 2-51, 2-55
Pacific Ocean NPC petroleum province, 2-62
Pacific Power and Light Company, 1-31
Pacific region:
  electric utility industry in, 9-254 to 9-255
  gas consumption of, 2-51
  hydroelectric power production of, 8-7
Pacific Southwest region, gas consumption of, 2-51, 2-54
Pack sand, 3-133
Package boilers, 9-19
Packaged power system, offshore, 9-189 to 9-190
Packer, oil well, 3-132
Pacol process, 3-259
Page, Arizona, daily terrestrial solar energy received at, 6-95
Paint chemicals, petroleum-derived (*see* Petrochemicals)
Pakistan:
  crude oil: production of, 3-86, 3-89
    proved recoverable reserves of, 3-82
    ultimate recovery potential for, 3-93
  gas turbine installations in, 9-188
  hydroelectric power production of, 8-5
  natural gas: production of, 2-39
    reserves of, 2-42, 2-45
  nuclear power installations in, 5-31, 5-37 to 5-38
  railway mileage and motive power of, 9-226
Paleocene Period:
  crude oil formed during, 3-51
  natural gas formed during, 2-14
Paleozoic Age, natural gas formed during, 2-14
Palisades nuclear facility, 5-14, 5-21
Palladium film by hydrogen diffuser, 4-37
Palo Verde nuclear plants, 5-12, 5-23

Panama:
  crude oil: production of, 3-87, 3-89
    ultimate recovery potential for, 3-95
  natural gas, ultimate recovery potential for, 2-46
  railway mileage and motive power of, 3-227
Panama Canal, tidal ranges at Atlantic and Pacific sides, 8-40
Pan American Petroleum Corporation, 3-152
Panhandle Eastern Pipe Line Company, 1-261, 2-64
Panhandle oil field, 3-62
Paper:
  energy conversion of, 4-13
  solid-wastes content of, 4-22
Paper manufacture:
  electric power consumption for, 9-259
  energy conservation in, 3-301
  gas requirements for, 2-52 to 2-56
  geothermal energy for, 7-31, 7-50, 7-55 to 7-57
Parabolic concentrator, 6-46 to 6-54
Paradise coal mine, 1-30
Paraffin isomers, 3-241
Paraffin-isoparaffin separation, 3-20
Paraffin plugging, 3-138
Paraffinic crude oils, 3-12
Paraffins, heat of combustion of, 9-36 to 9-37
Paraguay:
  crude oil: production of, 3-89
    ultimate recovery potential for, 3-94
  natural gas, ultimate recovery potential for, 2-46
  railway mileage and motive power of, 9-227
Paraho Oil Shale Demonstration, Inc., oil shale retorting process of, 3-153 to 3-154
Parahydrogen, 4-27
Parallel connected solar cells, 6-67
Parallel feedforward fuel-air ratio control system, boiler use of, 9-135
Parallel positioning control system, boiler use of, 9-135
Paratunka geothermal installation, 7-8
Parex process, 3-269
Paris, France, tidal power for, 8-26, 8-35
Parr, S. W., 1-64
Parr formulas, coal analysis, 1-19
Parr Instrument Company, 1-63 to 1-70
Parr-type gas calorimeter, 2-92 to 2-93
Parrot coal, 1-16
Partial-oxidation gasification process, 2-171 to 2-174
Partial-oxidation unit, energy system for, 3-277
Particle counter, 9-401
Particle size:
  of coal slurries, 1-179
  distribution measurement of, 9-402 to 9-405
  of drilling muds, 3-123
  of particulate emissions, 9-391 to 9-392
Particulate emissions, measurement of, 9-391 to 9-406

Parting, coal deposit, 1-38
Pasadena, California, daily terrestrial solar energy received at, 6-95
Passamaquody tidal energy project, 8-45
Passenger railway car, frictional resistance of, 9-225
Passenger railway locomotives (*see* Locomotives, rail system)
Passive orbital system, solar energy satellite, 6-98 to 6-103
Passive sunlight reflector, 6-102
Pasture crops, spoiled-land area use of, 1-115
Pathological study, coal miner, 1-93
Pauzhetsk geothermal installation, 7-8
PBAN propellant, 4-83
Pea coal, 1-27
Peabody Coal Company, 1-31, 1-180 to 1-181
Peace River hydroelectric facilities, 9-261
Peace River tar sand deposits, 3-159 to 3-161
Peach Bottom nuclear facility, 5-16 to 5-17, 5-21, 5-23, 5-47, 5-88, 5-90
Peak capability, electric utility industry, 9-251
Peak load, electric power system:
    definition of, 9-252
    hydroelectric plant, 8-9
    solar plant, 6-6, 6-35, 6-69 to 6-80
Peak power, wind generated, 6-171
Peaking power:
    boiling water reactor, 5-71
    gas turbine, 9-192
    magnetohydrodynamic generator, 9-209, 9-218 to 9-221
Pearl Harbor, Hawaii, daily terrestrial solar energy received at, 6-95
Peat, 1-16
Peat-to-anthracite theory, 1-15
Peatification, 1-16
Pebble Bed gas-cooled reactor, 5-89 to 5-90
Pechelbronn tar sand deposits, 3-160
Pellets:
    coke, 1-253
    nuclear fuel (*see* Fuel elements, rods and bundles, nuclear reactor; Uranium)
Pelton waterwheel drive, 5-48, 5-53
Pendant reheater, boiler, 9-119
Pendulum effect, oil well drilling, 3-114
Penetration rate versus solids concentration, oil well drilling, 3-123
Penex unit, 3-242
Pennsylvania:
    AGA area of, 2-9
    API Survey district of, 3-50
    Bituminous Coal District of, 1-167 to 1-168
    coals in, 1-20, 1-24, 1-29 to 1-30, 1-32, 1-34 to 1-35, 1-37, 1-92 to 1-93, 1-209 to 1-210, 1-227
    crude oil: paraffinic, 3-13
        production of, 3-56, 3-60
        reserves of, 3-46 to 3-47, 3-57 to 3-59
    daily terrestrial solar energy received at Philadelphia and State College, 6-96

Pennsylvania (*Cont.*):
    electric power generating capacity of, 9-252 to 9-253
    emission limits for SO_2 and NO_x, 9-375
    first oil well treating process carried out in, 3-133
    first rail shipment of oil made from, 3-176
    hydroelectric power: Kinzua Dam project, 8-20 to 8-24
        potential of, 8-8 to 8-9
        production of, 8-6
    natural gas: production and reserves of, 2-10 to 2-13, 2-16 to 2-18
        underground gas storage capacity, 2-74 to 2-75
    nuclear power installations in, 5-16 to 5-17
    tidal range at Philadelphia, 8-40
Pennsylvania, University of, 4-5 to 4-11
Pennsylvania Electric Company, 8-21
Pennsylvania-New Jersey-Maryland Power Pool, 9-204
Pennsylvania Power and Light Company, 5-17
Pennsylvania Railroad, electrification pioneering of, 9-238
Pennsylvanian Geologic System, fossil fuels formed during, 3-51
    crude oil: in Canada, 3-76 to 3-77
        paraffinic, 3-12
        in United States, 3-59
    natural gas, in United States, 2-14
Pennsylvanian Period, coal formed during, 1-37
Pensacola, Florida, daily terrestrial solar energy received at, 6-96
Penstock, 8-15, 8-19
Pentaborane, propellant use of, 4-81
Pentane:
    characteristics of, 3-241 to 3-242
    combustion constants of, 9-36 to 9-37
    detection of, 9-380
    natural gas content of, 2-129
Pentanol, Sasol complex production of, 1-295
Pentene, combustion constants of, 9-36 to 9-37
Pentoses, fermentation of, 4-7
People's Republic of China:
    coal reserves of, 1-36
        mine disasters, 1-92
    crude oil: production of, 3-80, 3-88
        recoverable reserves of, 3-85
    electric power production of, 9-247
    hydroelectric plants in, 8-5
    metric tons to barrels conversion factor, 3-84
    natural gas: production of, 2-39
        reserves of, 2-42, 2-44
    railway mileage and motive power of, 9-226
    tidal range at Shanghai, 8-40
    uranium resources of, 5-174
    wind mills in, 6-147
Per capita electric power production, various nations, 9-247

Photophosphorylation, 4-5, 6-139
Photopotentiometer sun angle sensor, 6-39
Photosynthesis, 4-5, 6-137 to 6-141
Photovoltaic energy converters (*see* Solar
    cells and arrays)
Photovoltaic systems, solar energy
    satellite, 6-103 to 6-105
Phycobilin pigments, 6-138
Piezoelectric microbalance, 9-400
Pigments, photosynthesis, 6-138
Pike County, Kentucky, coals in, 1-20
Pilgrim nuclear facilities, 5-14, 5-21, 5-23
Piling, offshore drilling platform, 3-112
Pillar, coal bed, 1-71, 1-73
Pillar mining, coal, 1-71, 1-73
Pilot, boiler burner, 9-56 to 9-57, 9-75
Pilot plant:
    COED process, 1-259
    CSF process, 1-277 to 1-281
    oil shale retorting process, 3-147 to 3-159
    tar sands process, 3-163 to 3-167
    wood-to-oil process, 3-168 to 3-173
Pinch devices, plasma confinement, 5-140
    (*See also* Theta pinch plasma
    experiment)
Pine trees, spoiled-land area use of, 1-112
    to 1-113
Pinegrass, spoiled-land area use of, 1-107
Pipe, insulation of, 3-290
Pipe-lay vessel, 2-79 to 2-86
Pipe systems, electric power transmission
    use of, 9-275
Pipeline gas, synthetic (*see* Substitute
    natural gas)
Pipelines:
    coal slurry, 1-178 to 1-187
    crude oil (*see* petroleum and petroleum
        products *below*)
    geothermal system, 7-18 to 7-24, 7-35 to
        7-36, 7-51 to 7-55
    marine, gas and oil, 2-77 to 2-89
    natural gas, 2-60 to 2-74
        gas compressor sets for, 9-187 to 9-189
    oil (*see* petroleum and petroleum
        products *below*)
    petroleum and petroleum products, 3-
        174 to 3-175
Piping, pulverized coal, 9-73 to 9-74
Piping design computer program, electric
    utility use of, 9-290
Piston engines (*see* Engines)
Pitch, 3-159
Pitch coke, 1-250
Pitching, weather effects on, 3-214
Pittsburg and Midway Coal Mining
    Company, 1-31
Pittsburgh, Pennsylvania, flooding of, 8-8
Pittsburgh seam coal, 1-209 to 1-210, 1-227,
    1-276
Pittston Company, 1-31
Planck's constant and equation, xxii
Planning, mined-out area, 1-102 to 1-115
Planning systems, electric utility, 9-278, 9-
    283 to 9-284, 9-295
Plant cultivation, 4-5 to 4-6
Plant investment, electric utility, 9-279
Plasma, fusion power, 5-140 to 5-172

Plasma configuration, fusion reactor
    reference program, 5-169
Plasma Physics Laboratory (*see* Princeton
    Plasma Physics Laboratory)
Plastic film reflector, 6-101
Plastics:
    natural gas requirements for producing,
        2-52 to 2-56
    petroleum-derived (*see* Petrochemicals)
    solid waste content of, 4-22
Plastometer, Gieseler, 1-52 to 1-53
Plate boundary, 7-4
Plate sampler, coal, 1-120 to 1-121
Platform:
    drilling, 3-108 to 3-112
    marine pipeline laying, 2-77 to 2-89
Platforming catalytic reforming process, 3-
    229
Platinum catalyst, fuel cell use of, 4-62, 4-
    64
Platteville, Colorado nuclear facility (*see*
    Fort St. Vrain Nuclear Generating
    Station)
Pleistocene Period, oil formed during, 3-51
Pliocene Period, fossil fuels formed
    during, 3-51
    crude oil, 3-51
    natural gas, 2-14
    tar sands, 3-160
Plotter, pipeline location, 2-85
Plum Brook wind power installation, 6-153
Plutonium:
    characteristics of, 5-45
    gas-cooled fast breeder reactor,
        recycling of, 5-131 to 5-132
    liquid-metal fast breeder reactor fuel of,
        5-111, 5-114 to 5-117
    plutonium metal versus plutonium
        oxide in, 5-106
    open-market recycle of, 5-85
    recycling and recovery in various types
        of reactors, 5-40 to 5-41
Plutonium-239 fissile nuclide, 5-7 to 5-8, 5-
    173
Plymouth, Pennsylvania, mine disaster at,
    1-92
Pneumatic dust collection, coal mine roof
    drilling, 1-78 to 1-79
Pneumatic system, energy storage (*see*
    Storage)
Pneumatic transmission, wind power
    system, 6-150 to 6-151
Pneumoconiosis, 1-92 to 1-93
Pocahontas, Virginia, mine disaster at, 1-
    92
Point Beach nuclear plants, 5-18, 5-21
Point de la Briantais, France, tidal power
    plant located at, 8-28
Point-of-custody transfer, API definition of,
    3-68, 3-70
Point-to-point electric power transmission,
    9-267
Poison, nuclear reactor, 5-68, 5-71
Poisoning, catalyst (*see* Catalysts)
Poland:
    coal reserves of, 1-36
    coal slurry pipeline in, 1-180

Poland (*Cont.*):
 crude oil: production of, 3-89
  proved recoverable reserves of, 3-83
  ultimate recovery potential for, 3-93
 cryogenic upgrading of natural gas in, 2-117
 electric power production of, 9-247
 metric tons to barrels conversion factor, 3-84
 natural gas: production of, 2-40
  reserves of, 2-42, 2-45
 nuclear facilities planned for, 5-32, 5-37 to 5-38
 railway mileage and motive power of, 9-226
Polar additives, jet fuel use of, 3-39
Polarization, fuel cell, 4-61
Pollution:
 energy-environmental interface, 9-312 to 9-406
 hydrodesulfurization for reducing, 3-251 to 3-264
 nuclear system, 5-97, 5-103 to 5-122, 5-139
 (*See also* Wastes; *and various pollutants, such as* Nitrogen oxides; Sulfur oxides)
Polycrystal silicon photovoltaic energy converter, 6-58
Polycyclic hydrocarbons, aircraft fuel use of, 3-39
Polymer resin lining, coal mine roof, 1-74
Polymer-type binders, propellant, 4-85
Polymer well drilling muds, 3-124 to 3-125
Polymerization:
 char, 1-223
 light olefins, 3-18
Polyphase power factor meter, 9-303
Pontiac engine, 4-42
Pontoon, pipelaying, 2-78, 2-80 to 2-81
Pool:
 electric power, 9-201 to 9-205, 9-287
 refinery (*see* Refinery gasoline pool)
Pool arrangement, liquid-metal fast breeder reactor, 5-114
Pool operation computer programs, electric utility use of, 9-287
Poplar trees, spoiled-land area use of, 1-112
Population relationship:
 to petroleum demand, 3-96, 3-100 to 3-101
 to solid waste generation, 3-168
Porosity test, coking-coal, 1-58
Porous membrane, fuel cell, 4-68
Port, petroleum handling (*see* Superports)
Port of Rotterdam, 3-220 to 3-224
Port radar, 3-219
Portage, Pennsylvania, mine disaster at, 1-92
Portland, Maine:
 daily terrestrial solar energy received at, 6-96
 tidal range at, 8-40
Portland, Oregon, tidal range at, 8-40
Portland General Electric Company, 5-16

Portugal:
 crude oil: production of, 3-89
  ultimate recovery potential for, 3-93
 natural gas, ultimate recovery potential for, 2-45
 nuclear facility planned, 5-32, 5-37 to 5-38
 railway mileage and motive power, 9-226
 tidal range at Lisbon, 8-40
 uranium resources of, 5-176
Portuguese Guinea:
 crude oil production of, 3-89, 3-92
 natural gas, 2-44
Positioning, marine pipeline, 2-82 to 2-83
Possible potential gas supply:
 definition of, 2-22
 source map of, 2-29
Post supports, coal mine roof, 1-74
Postdisaster survival and rescue, coal mine, 1-84
Postle oil field, 3-65
Potassium carbonate, gases removed by, 1-240
Potassium formate, gas processing system use of, 9-317 to 9-318
Potassium hydroxide electrolyte:
 battery, 4-52
 fuel cell, 4-62 to 4-64
Potassium metal heat-transport fluid, solar system, 6-10
Potassium nitrate, propellant use of, 4-85
Potassium perchlorate, propellant use of, 4-85
Potassium sulfate, propellant use of, 4-83
Potatoes, fermentation of, 4-7
Potential converter, 9-304
Potential energy, definition of, xxiii, 9-23
Potential gas, PGC definition of, 2-21
Potential Gas Committee, report of, 2-20 to 2-23
Potential supply of natural gas in United States, 2-20 to 2-39
Potomac Electric Power Company, 5-14
Pound mole, definition of, 9-34
Pour point:
 Athabasca bitumen, 3-163
 crude oil, 3-13 to 3-17
 definition of, 3-42
 Diesel fuel oils, 3-33 to 3-36
 synthetic crude oil from shale: N-T-U process, 3-151
  Paraho process, 3-155
  Tosco II process, 3-156
  Union oil process, 3-151
  U.S. Bureau of Mines process, 3-151
Powder River Basic coal reserves, 1-246 to 1-247
Powdered aluminum, propellant use of, 4-83
Power:
 definition of, xxii, 9-296
 energy conservation in production of, 3-289
 fusion, reactor reference design, 5-169
 (*See also* Nuclear energy technology)
 generation of (*see* production of *below*)

Preheating:
    combustion air, 3-279 to 3-283
    process gas, 3-290
Premixed flame, 9-335
Prentice oil field, 3-65
Preparation plant, coal, 1-117 to 1-136
Pressure:
    boiler system, 9-17 to 9-22
    geothermal fluids, 7-4 to 7-10, 7-17 to 7-18,
        7-28 to 7-31, 7-39, 7-41, 7-45 to 7-58
    oil burner fuel, 9-52
    pipeline, 2-73
    versus steam quality, 9-97
Pressure-arch theory, coal mine roof
    support, 1-72
Pressure gasifier, Lurgi process, 1-189 to 1-
    192, 1-197 to 1-198
Pressure limits, superheater and reheater
    tubes, 9-117
Pressure maintenance, oil reservoir, 3-66
    in Canadian fields, 3-75
Pressure profile, oil well, 3-119, 3-121
Pressure-type direct-firing pulverized coal
    system, 9-63
Pressure-type tankcars, 3-179 to 3-184
Pressurized water, solar heat-pipe system
    use of, 6-40 to 6-42
Pressurized water reactor, 5-9, 5-11, 5-40,
    5-75 to 5-86
    control element assemblies, 5-83 to 5-85
    fuel and control rods, 5-80 to 5-83
    instrumentation and control of, 5-85 to 5-
      86
    pressurizer, 5-79
    reactor coolant pumps, 5-77 to 5-79
    reactor vessel, 5-79 to 5-85
    steam generator, 5-75 to 5-77
Prestressed concrete structure, 5-48
    (See also Containment structure, nuclear
      reactor)
Pretreating:
    low-Btu gases, 2-117
    Synthane process, 1-237
Priest Rapids hydroelectric plant, 8-5
Prilling, solvent-refined coal, 1-283
Primary battery, 4-50, 4-52 to 4-53
Primary metals industries, electricity
    consumption of, 9-259
Primary separator, steam, 9-100 to 9-102
Primary system, nuclear power plant, 5-9,
    5-45
Prime mover, rail power, 9-224 to 9-241
Princeton, New Jersey, COED coal
    processing pilot plant located at, 1-255
    to 1-260
Princeton Plasma Physics Laboratory:
    fusion reactor reference design of, 5-166
      to 5-171
    superconducting magnets as applied to
      fusion reactor, development of, 5-
      158 to 5-172
    TCT experiment of, 5-145
Princeton Reactor Studies Group, 5-164
Printing and publishing industry:
    electricity consumption of, 9-259
    gas requirements of, 2-52 to 2-56

Probable potential gas supply:
    definition of, 2-22
    source map of, 2-29
Probable reserves:
    of crude oil, CPA definition of, 3-73
    (See also Reserves)
    of natural gas, CPA definition of, 2-34
PROBO (Product-Oil-Bulk-Ore)
    combination carrier, 3-206
Process heating, geothermal energy for, 7-
    8, 7-31, 7-43 to 7-58
Process plants, energy conservation
    programs for, 3-272 to 3-303
Process streams, heat recovery from, 3-283
    to 3-285
Process variable (see Instrumentation)
Processing, petroleum (see Petroleum)
Procor Limited (U.K.), 3-195
Procurement, tankcar, 3-193 to 3-194
Producer gas, calorific value of, 2-90
Producing wells, crude oil, worldwide
    statistics of, 3-86 to 3-89
Product carrier, seagoing, 3-202
Production:
    coal (see Coal)
    crude oil: definition of, 3-57
      drilling and well servicing structures,
        3-107 to 3-112
        API standards for, 3-107 to 3-109
        offshore structures, 3-108 to 3-112
        platforms. 3-108
      geology of oil occurrence, 3-48 to 3-53,
        3-103 to 3-107
      production processes and techniques:
        acidizing, 3-133 to 3-136
        air and gas drilling, 3-118 to 3-119
        artificial lift, 3-143
        computer assisted drilling, 3-125 to
          3-129
        directional drilling, 3-115 to 3-118
        dissolved-gas drive, 3-105 to 3-106
        fluid injection, 3-141 to 3-143
        gas-cap drive, 3-105
        geometry of drilling, 3-112 to 3-115
        geopressured drilling, 3-119
        hydraulic fracturing, 3-137 to 3-
          138
        optimization of drilling, 3-120 to 3-
          129
        sand control, 3-132 to 3-133
        water control, 3-131 to 3-132
        water drive, 3-105 to 3-107
        well completion, 3-129 to 3-130
        well control, 3-119
        work-overs, 3-138 to 3-141
      statistics, 3-57 to 3-60
      in Canada, 3-73 to 3-77
      offshore, 3-46, 3-48, 3-57 to 3-60, 3-
        62 to 3-65, 3-88 to 3-95
      in United States, 3-45 to 3-72
    natural gas: in Canada, 2-34 to 2-37
      in United States, 2-9 to 2-19
      worldwide, 2-38 to 2-46
Production cost, electric utility, 9-257
Production wells, in-situ coal gasification,
    1-243

Productive capacity, crude oil: API
definition of, 3-67
maximum daily crude rate, 3-68
Productive formation, definition of, 2-24
Productivity, coal, comparison of American
and European, 1-91
Profile, railway, 9-231 to 9-232
Profile drag, railway rolling stock, 9-225
Programming:
electric utility computer use for, 9-277 to
9-295
of tidal power plants, 8-35
Prompt critical, 5-45
Prony brake, 9-305 to 9-306
Propane:
aircraft fuel use of, 3-39
alkylation feedstock use of, 3-238
calorific value of, 2-90
combustion constants of, 9-36 to 9-37
cryogenic separation of, 9-200
detection of, 9-380
filling density of, 3-180
high heating value of, 3-280
natural gas content of, 2-129
physical properties of, 3-180
tankcar shipment of, 3-180
Propane gas drive oil well fluid injection
system, 3-142
Propane-MRC refrigeration cycle, 2-102 to
2-108
Propane-propylene product, Sasol
complex, 1-295
Propanol, Sasol complex production of, 1-
295
Propellants, 4-74 to 4-90
antimatter rockets, 4-89
burning rate, 4-85 to 4-87
classification of, 4-74
function of, 4-74
hybrid rocket, 4-88
liquid, 4-77 to 4-78, 4-80 to 4-81
metastable, 4-88 to 4-89
nonrocket uses of, 4-76 to 4-77
grain of, 4-87 to 4-88
solid, 4-78 to 4-79
beryllium and aluminum hydride, 4-
83, 4-85
binding ingredients of, 4-85
boron, 4-85
HMX and RDX, 4-83
nitrates, 4-83
perchlorates, 4-79 to 4-83
powdered aluminum, 4-83
specific impulse of, 4-75 to 4-76
Propeller-type hydraulic turbine, 8-11 to 8-
14, 8-16
Proportional action, boiler control system,
9-125 to 9-127
Proportional-plus-derivative action, boiler
control system, 9-128
Propping agent, oil well treatment use of,
3-137, 3-140 to 3-141
Propulsion:
pipeline vessel, 2-79
rocket (see Propellants)
solar energy satellite, 6-113
Propulsion power, tanker, 3-200

Propylene:
alkylate feedstock use of, 3-238
alkylation of, 3-19, 3-228
(See also Alkylation)
catalytic cracking process, product of, 3-
227
Propylene carbonate electrolyte, battery, 4-
51
Prosser, Washington, daily terrestrial solar
energy received at, 6-96
Protection, geothermal system safety, 7-29,
7-37 to 7-38
Protective devices:
for coal mining roof support, 1-71 to 1-83
for mining equipment, 1-90 to 1-92
Protective string, oil well, 3-129
Protective system, nuclear reactor (see
Safety and emergency factors, nuclear
reactor)
Protium, 4-27, 4-29, 5-45
Proved acreage, oil field, API definition of,
3-66
Proved category, natural gas: definition of,
2-23
source map of, 2-30
Proved developed reserves, crude oil, API
definition of, 3-61
(See also Reserves)
Proved drilled area, oil field, definition of,
3-72
Proved recoverable reserves, worldwide
crude oil, 3-82 to 3-84
Proved reserves:
of crude oil: API definition of, 3-58
CPA definition of, 3-73
of natural gas, CPA definition of, 2-34
Proved underdeveloped reserves of crude
oil, API definition of, 3-61
(See also Reserves)
Providence No. 1 coal mine, 1-30
Proximate analysis, coal, 1-48 to 1-49
Prudhoe Bay, oil exploration in, 3-94
Prudhoe Bay oil field, 3-62
Prudhoe Bay pipeline system, 2-74
Public Service of Indiana, 5-13, 9-171
Public Service Company of Colorado, 5-12,
5-47 to 5-56, 5-90
Public Service Company of New
Hampshire, 5-15
Public Service Company of Oklahoma, 5-
16
Public Service Electric and Gas Company
of New Jersey, 4-34, 5-15
Public utilities (see Power; Power
technology)
Puerto Rico:
nuclear power plans of, 5-19
ocean thermal energy conversion plant
site in, 6-120
tidal range at San Juan, 8-40
Puerto Rico Water Resources Authority, 5-
19
Puget Sound Power and Light Company, 5-18
Pullman, Washington, daily terrestrial solar
energy received at, 6-96
Pullman-Standard Corporation, 1-157, 1-
159 to 1-160

Recuperated Brayton cycle, **6**-18 to **6**-20
Recuperator, gas turbine, **9**-189
Recycle gas, oil shale retorting process, 3-148, 3-152
Red Lodge, Montana, mine disaster at, 1-92
Red Sea, **7**-4
Red top, spoiled-land area use of, 1-108
Redstone, Colorado, mine disaster at, 1-92
Reduced crude, 3-21
Reduction drive, gas turbine, **9**-184
Reduction-oxidation reactor, 1-207
Redwood viscosity, 3-44
Reel method, marine pipeline, 2-78
Reference material, calorimeter, 1-65
Refinery:
  conservation of energy in, 3-272 to 3-303
  gasoline pools in, 3-25 to 3-26, 3-225, 3-234
  size trends of, 3-244 to 3-245
  (*See also* Petroleum)
Refinery gas:
  calorific value of, **2**-90
  excess air normally supplied for burning of, **9**-46
Reflector, solar energy (*see* Mirrors and reflectors, solar energy system)
Reflux ratio economics, debutanizer, 3-295
Reformer, methanol process, **2**-121 to **2**-123
Reformer heater, heat recovery from, 3-279
Reforming, 1-286, 1-290, **2**-149 to **2**-152, 3-40, 3-141, 4-36
Refract treatment, oil well, 3-141
Refrigerated electric power transmission system, **9**-276
Refrigeration:
  geothermal energy for, 7-50 to 7-56
  thermoelectric, **9**-242 to **9**-246
Refrigeration cycle, liquefied natural gas, **2**-102 to **2**-109
Refrigeration system:
  expansion turbines used in, **9**-200 to **9**-201
  natural gas processing use of, **2**-129 to **2**-131
  turbine generator cooling use of, **9**-205 to **9**-208
Refueling, fusion reactor, 5-152 to 5-153
  (*See also* Nuclear energy technology)
Refuse:
  coal-preparation plant, 1-117 to 1-119
  energy conversion of, 4-12 to 4-25
  synthetic oil from, 3-168
Regeneration system, gas processing (*see* Wet scrubbing processes)
Regenerative cycle, gas turbine, **9**-177 to **9**-180
Regenerative feedwater heating, **9**-27, **9**-29
Regenerative Rankine cycle, **9**-27 to **9**-30
Regenerative reactor, 5-45
Regenerative-reheat steam Rankine cycle, **6**-17
Regenerator:
  carbon dioxide acceptor process, 1-220
  fluid catalytic cracking process, 3-232
Regional operating center (ROC), electric utility, **9**-286 to **9**-288

Registry, ship, definition of, 3-206
Regular price gasoline, 3-23
  (*See also* Gasoline)
Regulating rod (*see* Fuel elements, rods and bundles, nuclear reactor)
Regulations:
  shipping, 3-210
  steam temperature, **9**-120 to **9**-122
Regulators:
  feedwater, **9**-130
  gas turbine voltage, **9**-183 to **9**-184
  pipeline, **2**-73
Reheater, steam, **9**-113 to **9**-119
Reheating, gas turbine, **9**-178
Reid vapor pressure:
  aviation fuels, 3-37
  crude oil, 3-14 to 3-17
  definition of, 3-41
  gasoline, 3-23 to 3-25
Reineke gas calorimeter, **2**-91 to **2**-93
Reinforced concrete lining, coal mine roof, 1-74
Relative biological effectiveness (RBE), 5-45
Relay, geothermal system use of, 7-29, 7-38 to 7-39
Relay coordination, electric utility computer program for, **9**-284
Relay testing, dynamometer, **9**-310
Reliability, nuclear plant (*see* Safety and emergency factors, nuclear reactor)
Reliance Electric Company, **9**-306
Relief valve, boiling water reactor, 5-63
Remaining reserves:
  of crude oil, CPA definition of, 3-73
  of natural gas: in Canada, **2**-36
  CPA definition of, **2**-37
Remote control, coal mine roof bolting, 1-81
Remote manual sequence control, boiler burner, **9**-145
Removal, polluting gas (*see* Pollution)
Rennes, France, tidal power for, 8-26, 8-35
Reno, Nevada, daily terrestrial solar energy received at, **6**-95
Rensselaer Polytechnic Institute, 4-58
Repairman accidents, coal mine, 1-85 to 1-91
Replacement loading, nuclear fission reactor, 5-40
Reporoa geothermal area, 7-31
Republic of China (Taiwan):
  crude oil: production and exploratory wells completed, 3-86, 3-88
    proved recoverable reserves of, 3-82
    ultimate recovery potential for, 3-92
    metric tons to barrels conversion factor, 3-84
  natural gas: production of, **2**-39
    reserves of, **2**-42, **2**-44
  nuclear facilities planned, 5-33, 5-36, 5-38
  railway mileage and motive power of, **9**-227
Republic of South Africa:
  coal reserves of, 1-36

Republic of South Africa (*Cont.*):
crude oil: production and exploratory
wells completed, 3-86, 3-89
ultimate recovery potential for, 3-92
gas turbine installations in, 9-192
natural gas, ultimate recovery potential
for, 2-44
nuclear facilities planned, 5-32, 5-37 to
5-38
railway mileage and motive power of, 9-
226
Sasol complex coal gasification and
petrochemical production of, 1-285
to 1-296
tidal range at Capetown, 8-40
uranium resources of, 5-174 to 5-176
Requirements:
electric power, 9-257 to 9-267
fuels for electric power generation, 9-
260 to 9-266
gas, 2-47 to 2-59
petroleum products, 3-96 to 3-102
Reradiation spectrum, solar absorber, 6-9
Rescue, coal mine disaster, 1-84
Research, oil well drilling, 3-126 to 3-128
Research octane number (*see* Octane
number)
Reserves:
coal, 1-34 to 1-36
crude oil: in Canada, 3-73 to 3-77
in United States, 3-45 to 3-72
worldwide, 3-78 to 3-95
in geothermal reservoirs, 7-3 to 7-7, 7-16
to 7-18, 7-31 to 7-33, 7-35 to 7-36, 7-
55, 7-57
groundwater, 8-8
in hydropower reservoirs, 8-3, 8-7 to 8-9
lithium, 5-184 to 5-185
natural gas: in Canada, 2-34 to 2-37
in United States, 2-9 to 2-19, 2-21 to 2-
33
worldwide, 2-38 to 2-46
oil shale, 3-147
tar sands, 3-73, 3-159 to 3-167
tidal energy (potential), 8-25, 8-40 to 8-
41, 8-43 to 8-45
uranium, 5-173 to 5-181
wave energy (potential), 8-45
Reservoirs:
geothermal, 7-7, 7-17 to 7-18, 7-22, 7-31
to 7-33, 7-36, 7-51 to 7-52
(*See also* Reserves, in geothermal
reservoirs)
hydrocarbon: crude oil, 3-51 to 3-53
geologic structure of, 3-103 to 3-105
natural gas, 2-10 to 2-13
hydropower, 8-5 to 8-11, 8-18 to 8-22
reserves in, 8-3, 8-7 to 8-9
Reset control system, boiler, 9-127 to 9-128
Residence heating, solar (*see* Solar climate
control)
Residential power consumption:
electric, 9-257
gas, 2-47 to 2-49
Residual fuels, 3-40, 3-96, 3-225, 3-244
Residual magnetism, magnetohydrodynamic system, 9-218

Residue handling, spreader stoker, 9-89
Residuum feedstock, partial-oxidation
gasification process use of, 2-172
Residuums:
crude oil, 3-18
demetallization of, 3-262
Resin lining, coal mine roof, 1-74
Resins:
Athabasca bitumen content of, 3-163
oil well sand control use of, 3-133
petroleum-derived (*see* Petrochemicals)
Resistance, frictional, locomotives and
rolling stock, 9-224 to 9-225
Resistive cryogenics, electric power
transmission use of, 9-276
Resource Sciences Company, 1-248
Respiration, biochemical process, 4-6
Response:
boiler control system, 9-123 to 9-140
boiler-turbine-generator control system,
9-152
pumped-storage plant, 8-17
Restricted draft tanker, 3-207, 3-211 to 3-
212
Retarded acids, oil well treatment use of, 3-
135
Retort, underfeed stoker, 9-90
Retorting, oil shale (*see* Oil shale, retorting
of)
Return-flow atomizer, oil burner, 9-52 to 9-
55
Reuse, bio-oxidation process for water, 9-
354 to 9-361
Revegetation, mined-out area, 1-102 to 1-
115
Revenue management, electric utility, 9-
280
Revere modular solar collector, 6-94
Reverse pump operation, Rance Estuary
tidal power plant, 8-31
Reversible pump-turbine, 8-18
Reversible processes, 9-24
Reversible work, liquefaction process, 2-
106 to 2-109
Reversing linkage, electric dynamometer,
9-307
Revision estimate, oil field, API definition
of, 3-66
Revolving grate, gasifier, 1-188
Reykir-Reykjavik geothermal water line, 7-
51
Reykjavik, Iceland:
geothermal installations at, 7-10, 7-43 to
7-58
tidal range at, 8-40
Reynolds number, coal particle, 1-228
Rhaetic tar sand deposits, 3-160
Rhine graben, tar sand deposits in, 3-160
Rhizobium bacteria, 1-111
Rhode Island:
daily terrestrial solar energy received at
Newport, 6-96
electric power generating capacity of, 9-
252 to 9-253
emission limits for SO_2 and NO_x, 9-
375
hydroelectric power production of, 8-6

Rhodesia:
  hydroelectric plants in, 8-5
  lithium resources of, 5-183
  railway mileage and motive power of, 9-226
Rhone River, flooding of, 8-25
Rhossili Down, Wales, wind power site at, 6-160 to 6-161
Ribbed boiler tube, 9-98
Rice coal, 1-27
Ricegrass, spoiled-land area use of, 1-107
Rich air-fuel mixtures, 9-196
Rich gas composition, catalytic rich gas process, 2-139
Rifle, Colorado, oil shale test site at, 3-150, 3-153
Rig, floating, 3-109
Right of the sky doctrine, 6-93
Right of way, pipeline, 2-72
RIGSEL drilling assistance computer program, 3-128
Riley Stoker Corporation, 9-335 to 9-348
Rio Algom Mines Limited, 5-173 to 5-181
Rio de Janeiro, Brazil, tidal range at, 8-40
Ripped rock, contour mining handling of, 1-100
Riser:
  fluid catalytic cracking process, 3-232
  marine pipeline, 2-84, 2-86 to 2-87
River Bend nuclear plants, 5-14, 5-24
River crossing, pipeline, 2-73
River King coal mine, 1-29
River Queen coal mine, 1-29
River-rail system, coal transport by, 1-154
Riverside, California, daily terrestrial solar energy received at, 6-95
Road conditions, contour mining, 1-98 to 1-100
Road oil, consumption of, 3-99
Robena coal mine, 1-29
Robert Moses-Niagara hydroelectric plant, 8-5, 8-10
Robinson, H. B., nuclear plant, 5-17, 5-21
Robinson Run No. 95 coal mine, 1-29
Rochester and Pittsburgh Coal Company, 1-31
Rochester Gas and Electric Company, 5-15
Rock:
  coal-associated, 1-37, 1-71
  contour mining handling of, 1-94 to 1-101
Rock drillability determination, coal mine roof, 1-77 to 1-79
Rock oil, 3-12
Rock spoils, 1-103
Rock storage systems, solar heat, 6-83, 6-95
Rocket propellants (see Propellants)
Rocky Mountain Energy Company, 1-248
Rocky Mountain region:
  AGA area of, 2-9
  API Survey district of, 3-50
  coal formation in, 1-38
  Diesel fuel survey of, 3-32 to 3-36
  gas consumption of, 2-51, 2-54
  hydroelectric power production of, 8-7
  NPC petroleum province of, 2-62
  petroleum products survey of, 3-32 to 3-36

Rocky Reach hydroelectric plant, 8-5
Rocky River hydroelectric plant, 8-17
Rod, fuel (see Fuel elements, rods and bundles, nuclear reactor)
Rod mill, coal slurry, 1-181
Roentgen, 5-45
Roga index, 1-23
Roll-and-race coal pulverizer, 9-65 to 9-68
Roll crusher, coal, 1-23
Rolling, weather effects on ship, 3-214
Rolling resistance factors:
  contour mining, 1-99
  railway operation, 9-224 to 9-228
Romania:
  crude oil: annual production of, 3-81, 3-89
    ultimate recovery potential for, 3-93
  hydroelectric plants in, 8-5
  metric tons to barrels conversion factor, 3-84
  natural gas: production of, 2-40
    reserves of, 2-42, 2-45
  nuclear facilities planned, 5-32, 5-37 to 5-38
  railway mileage and motive power of, 9-226
  tar sand deposits in, 3-160
Roof, coal deposit, 1-39
Roof bolting, coal mine, 1-74 to 1-83
Roof bolting accidents, 1-85 to 1-91
Roof control, underground mining, 1-71 to 1-83
Rope-type prony brake, 9-305 to 9-306
Rose Valley No. 6 coal mine, 1-30
Rosebud coal mine, 1-30
Rosebud Coal Sales Company, 1-31
Rosin and Ramler, coal size distribution chart of, 9-71 to 9-72
Rotary breaker, coal, 1-118 to 1-119, 1-123 to 1-124
Rotary bucket stacker-reclaimer, 1-152
Rotary drilling:
  coal exploration use of, 1-41 to 1-42
  coal mine roof testing, 1-79
Rotary drilling optimization, oil well, 3-122 to 3-128
Rotary-drum sampler, coal, 1-120
Rotary-dump gondola car, coal, 1-158
Rotary sampler, coal, 1-120 to 1-121
Rotating equipment, energy recovery from, 3-292
Rotokawa geothermal area, 7-31
Rotor:
  machining of, 9-203
  superconducting (see Superconducting turbine generators)
  wind-machine, 6-143 to 6-145
Rotorua geothermal installations, 7-31
Rotterdam, the Netherlands:
  port of, 3-220 to 3-224
  tidal range at, 8-40
Rotterdam Waterway, 3-222
Royal Netherlands Meteorological Institute, 3-222
R-R Diesel fuel oils, 3-30, 3-35
RTG gas turbine train, 9-240
Ruanda, lithium resources of, 5-183

Screen, boiler, 9-111
Screening:
  coal, 1-117 to 1-120, 1-124 to 1-125
  solid wastes, 4-14
Screening device, oil well in-hole, 3-132 to 3-133
Screw conveyor, coal, 1-148
Scrubbing processes, gas, 9-312 to 9-327
  Garret pyrolysis process, 1-190 to 1-191
  Lurgi process gas, 1-190 to 1-191
  nitrogen oxides and sulfur dioxide, 9-312 to 9-327
  oil shale retorting process, 3-148 to 3-152
  steam, 9-101 to 9-102
  Synthane process raw gas, 1-233, 1-240
Scylla IV plasma, 5-140, 5-143 to 5-146
Sea:
  tidal energy of, 8-27 to 8-46
  wave energy of, 8-46
Sea Wolf reactor, 5-111
Seaboard Coast Line Railroad, 1-154
Seabrook nuclear plants, 5-15, 5-24
Seagoing vessels, petroleum carrying, 3-196 to 3-206
  barges, 3-203
  classification of, 3-198, 3-207 to 3-212
  combination carriers, 3-200, 3-202, 3-212
  crude petroleum carriers, 3-198 to 3-200, 3-207 to 3-223
  definition of terms, 3-203, 3-206
  dry bulk carriers, 3-212 to 3-213
  gas carriers, 3-202 to 3-203
  liquefied petroleum gas carriers, 3-202, 3-204 to 3-205
  procurement of, 3-197
  product carriers, 3-202
  registry of, 3-196 to 3-197
  regulatory agencies, 3-196
  size and number of, 3-196, 3-207 to 3-223
  superports for, 3-207 to 3-224
  tankers, 3-198 to 3-199, 3-201, 3-207 to 3-208, 3-210 to 3-214
Seal:
  tankcar enclosure, 3-191
  two-way flow axial turbine shaft, 8-33
Sealant, oil well water control, 3-131
Seam, coal, 1-34 to 1-36
Searles Lake, lithium resources of, 5-183
Seasonal variation, solar mirror field, 6-28 to 6-29
Seat rock, coal deposit, 1-40
Seattle, Washington:
  daily terrestrial solar energy received at, 6-97
  tidal range at, 8-40
Seawater, deuterium content of, 5-138
Seawater pumps, ocean thermal energy conversion plant, 6-127 to 6-129
Seaweed drying, geothermal energy for, 7-57
Secondary criticality, liquid-metal fast breeder reactor, 5-105 to 5-106

Secondary recovery, oil reservoir, 3-66
Secondary system, nuclear power plant, 5-9, 5-45
Sediment:
  crude oil content of, 3-13 to 3-17
  gasoline content of, 3-22
Sedimentary materials, coal-associated, 1-38 to 1-40, 1-71
Sedimentary rock spoils, 1-103
Seeding, mined-out area, 1-102 to 1-115
SEFOR reactor experimental program, 5-110 to 5-111
Segco No. 1 coal mine, 1-30
Segmented electrode, magnetohydro-dynamic generator, 9-210
Segregated ballast, 3-203
Seismic data, geothermal system, 7-4
Seismic survey, geothermal system, 7-7
Selection, coal burning equipment, 9-59
Self-cleaning coal car, 1-157
Self-sustaining nuclear reaction, 5-7, 5-9, 5-43
Self-unloading lake boat, coal, 1-174
Semianthracite coal, 1-15 to 1-27
  Lurgi process reaction temperature of, 1-196
  ranking of, 1-54
Semibituminous coal, 1-15, 1-54
Semicarbonization, 1-16
Semiclosed internally-fired gas-turbine cycle, 9-180
Semiconductor gas detectors, 9-383, 9-389
Semiconductors, thermoelectric cooling system use of, 9-242 to 9-246
Seminole oil field, 3-62
Semipermeable membrane, fuel cell use of, 4-62
Semisubmersible hull, pipelaying vessel, 2-79
Semiwildcat well, 3-121 to 3-122
Seneca Pumping and Generating Station, 8-21 to 8-24
Senegal:
  crude oil: production of, 3-86, 3-89
    ultimate recovery potential for, 3-92
    metric tons to barrels conversion factor, 3-84
  natural gas, ultimate recovery potential for, 2-44
  railway mileage and motive power of, 9-227
  solar water heaters used in, 6-93
Senonian tar sand deposits, 3-160
Sensible heat, definition of, 9-43
Separation:
  of crude oil components, 3-226 to 3-227
  electron, electrostatic direct energy conversion of plasma energy use of, 5-150 to 5-151
  of noncombustibles in solid waste processing, 4-14 to 4-24
  steam, 9-99 to 9-103
  p-xylene, 3-267
Separator:
  geothermal system use of (see Steam, geothermal)
  magnetic, 1-118

I-98

Separator cone, coal, 1-118 to 1-119, 1-130
Sequence control system, burner, 9-143 to
    9-147
Sequential operation, gas turbine, 9-187
Sequoyah nuclear plants, 5-17, 5-24
Series-connected solar cells, 6-66 to 6-67
Serpentine plug, hydrocarbon entrapment,
    3-104
Service, boiler control system, 9-156
Service quality computer programs,
    electric utility, 9-285
Servo system, sun-tracking, 6-37 to 6-39
Sewage:
    energy conversion of, 4-12 to 4-25
    methane production from, 4-16 to 4-18
Shaft furnace, 4-15, 4-19
Shaft seal, two-way flow axial turbine, 8-33
SHAKER drilling assistance computer
    program, 3-128
Shale:
    coal-associated, 1-37
    definition of, 1-71
Shale oil, biological extraction of, 4-9
    (See also Oil shale, retorting of)
Shale spoils, 1-103
Shanghai, tidal range at, 8-40
Shear modulus, cryogenic aluminum alloy,
    2-107
Shear stress, coal mine roof support, 1-71 to
    1-83
Shearon Harris nuclear plants, 5-16
Shelf life, battery, 4-53
Shell boiler, 9-93
Shell capacity, tankcar, 3-191
Shell gasification process, 2-171 to 2-174, 9-
    165 to 9-170
Shell hydrogen production process, 4-36
Shell innage, tankcar, 3-191
Shell Internationale Petroleum
    Maatschappij, B.V., 9-165 to 9-170, 9-
    328
Shell Oil Company, 2-133 to 2-136, 2-171
    to 2-174, 3-288
Shell outage, tankcar, 3-191
Shell Pipe Line Corporation, 2-102
Shell tankers, 3-212
Sheltering devices, coal mine roof bolting,
    1-80 to 1-82
Sheridan County, Wyoming, coals in, 1-
    -20
Shield:
    fusion reactor reference design, 5-171
    nuclear reactor, 5-9, 5-45
Shift conversion:
    COGAS process, 1-262 to 1-264
    Lurgi process, 1-189, 1-198
    Synthane process, 1-236, 1-240
    Synthoil process, 1-269 to 1-270
Shim rod (see Fuel elements, rods and
    bundles, nuclear reactor)
Ship:
    petroleum carrying (see Seagoing
        vessels, petroleum carrying)
    pipeline laying, 2-77 to 2-89
Ship-shape hull, 2-79
Ship Shoal oil fields, 3-63, 3-65, 3-90
Shipowners' equipment, tanker, 3-220

Shipping, methanol versus liquefied
    natural gas, 2-125 to 2-127
Shippingport Atomic Power Station, 9-22
Shippingport nuclear facility, 5-17, 5-21
SHOEFRAC drilling assistance computer
    program, 3-128
Shoemaker coal mine, 1-29
Shoreham nuclear plant, 5-15, 5-24
Shoreline pipeline laying, 2-88 to 2-89
Short circuit computer program, electric
    utility, 9-284
Short-pole tankcar gaging, 3-191 to 3-192
Short-range planning and scheduling
    systems, electric utility, 9-281 to 9-
    282
Short-term transient, magnetohydro-
    dynamic units for, 9-219
Shot-firer accidents, coal mine, 1-85 to 1-91
Shot rock, contour mining handling of, 1-
    100
Sho-Vel-Tum oil field, 3-62
Shredding, waste energy system, 4-13
Shreveport, Louisiana, daily terrestrial
    solar energy received at, 6-96
Shunting locomotives, 9-228, 9-233
Shutdown sequence, geothermal system,
    7-38
Shuttle operator accidents, coal mine, 1-85
    to 1-91
Siberian tar sand deposits, 3-160
Side-boom tractor, 2-73
Siderite, coal-associated, 1-39
Siemens-AEG-BBC, advanced railway
    developments of, 9-241
Siemens AG fuel cell, 4-64 to 4-65, 4-69
Sierra Leone:
    crude oil: production of, 3-89
        ultimate recovery potential for, 3-92
    natural gas, ultimate recovery potential
        for, 2-44
    railway mileage and motive power of, 9-
        227
Sigma Colliery, 1-289 to 1-291
Sigma gas calorimeter, 2-91 to 2-93
Signal Oil and Gas Company, 7-34
Silane resin-to-sand coupling agents, 3-133
Silencer, geothermal system, 7-19 to 7-20,
    7-23, 7-25
Silica, geothermal fluid content of, 7-48
Silica production, solar furnace used for, 6-51
Silica removal, steam, 9-102 to 9-103
Silicon crystals, physical properties of, 6-59
Silicon solar cells and arrays, 6-56 to 6-80
Silicon-vanadium superconductor, 5-162
Silo, coal storage, 1-137
Siltstone, 1-71, 3-51
Silurian Geologic System, fossil fuels
    formed during, 3-51
    crude oil: in Canada, 3-76 to 3-77
        in United States, 3-59
    natural gas, in United States, 2-14
Silver catalyst, fuel cell use of, 4-62, 4-64
Silver oxide anode, thermionic converter
    use of, 9-222
Silver oxide electrode, battery use of, 4-52
Silver Peak, Nevada, lithium resource at, 5-
    184

Silver potassium iodide cyanide solid electrolyte, battery use of, 4-53
Silver rubidium iodide solid electrolyte, battery use of, 4-53
Silver-zinc battery, 4-52 to 4-53, 4-55
Simco-Peabody coal mine, 1-30
Simulation programs:
  electric utility, 9-284 to 9-286, 9-293 to 9-294
  freight train performance for contrasting railway profiles, 9-232
  Hydrane process performance, 1-229 to 1-230
  magnetohydrodynamic system, 9-216 to 9-218
Sinclair coal mine, 1-29
Sinclair Research, Incorporated, 3-152
Singapore:
  crude oil: production of, 3-89
    ultimate recovery potential for, 3-93
  natural gas, ultimate recovery potential for, 2-45
  nuclear facility planned for, 5-37 to 5-38
  tidal range at, 8-40
Single carload coal movements, 1-154
Single-cycle operation, tidal power plant, 8-39, 8-41 to 8-45
Single-element boiler feedwater control, 9-131
Single-heliostat solar furnaces, 6-47 to 6-50
Single-roll crusher, coal, 1-123 to 1-125
Single-shaft gas turbine, 9-180
Sir Adam Beck No. 2 hydroelectric plant, 8-5
Site evaluation, mined-out areas, 1-103 to 1-105
Site location:
  liquefied natural gas facilities, 2-100 to 2-101
  ocean thermal energy conversion plants (see Ocean thermal energy conversion)
Site preparation, mined-out areas, 1-105 to 1-106
Size:
  coal handling factor of, 1-147
  of rail coal cars, 1-160
  of turbine generators, 9-204
Size classification, coal, 1-25
Size consist, coal slurry, 1-179
Size grading, coal, 1-25, 1-27, 1-117 to 1-118, 1-124
Size trends, refinery, 3-244 to 3-245
Sizewall gas-cooled reactor, 5-89
Sizing of coal, cyclone furnace firing requirements for, 9-84
Skagit nuclear plants, 5-18, 5-24
Skikda, Algeria, liquefied natural gas facility at, 2-102 to 2-103
Sky rights, solar system, 6-93
Sky temperature, 6-41
Slack, coal, 1-25
Slag:
  American Thermogen process production of, 4-19
  cyclone furnace handling of, 9-81 to 9-82

Slag screen, boiler, 9-111
Slag-tap furnace, boiler, 9-111
Slate, definition of, 1-71
Slaughter oil field, 3-62
Slide, submarine, 3-112
Slide valve, Synthol reactor, 1-292
Slip-tube gaging device, tankcar, 3-184
Slop tank, definition of, 3-206
Slotted liner, oil well, 3-131
Slow neutron, 5-7, 5-45
Sluice gate, tidal power plant, 8-27 to 8-28
Slurry feed system, Hygas process, 1-202
Slurry handling:
  fluid catalytic cracking process, 2-232
  wood-to-oil process, 3-169 to 3-170
Slurry pipeline, coal, 1-179 to 1-187
Slurry production, solvent refining of coal, 1-282 to 1-284
Slurry sealants, oil well drilling use of, 3-119
Slurry systems, gas absorption (see Scrubbing processes, gas)
S-M Diesel fuel oils, 3-31, 3-36
Smith Mountain pumped-storage plant, 8-18
Smith-Putman aerogenerator, 6-150, 6-153
Smog detector (see Instrumentation, polluting gas and particulate emissions)
Smoke point:
  aviation fuels, 3-37
  definition of, 3-42
Smoke production, Diesel-fuel, 3-31
Smoke reduction, combustion-improving chemicals for, 9-349
Smoke volatility index, aviation fuels, 3-37
Smokeless coal, 1-15
SNCF gas turbine train development of, 9-240
Snell's law, 6-28
SNG (see Substitute natural gas)
Soaking phase, partial-oxidation gasification process, 2-171
Sociéte Maritime Shell, 3-210
Society of Automotive Engineers, 3-26
Soda Springs, California, daily terrestrial solar energy received at, 6-95
Sodium:
  fusion reactor use of, 5-171
  liquid-metal fast breeder reactor use of, 5-102 to 5-103, 5-111
Sodium carbonate, wood-to-oil process use of, 3-168
Sodium chloride, crude oil content of, 3-13 to 3-17
Sodium chloride-sodium nitrate storage system, solar energy system, 6-43 to 6-44
Sodium chloride water, geothermal, 7-44
Sodium chlorite, gas-processing system occurrence of, 9-322
Sodium compounds, disposal of, 9-315
Sodium dihydrogen phosphate, gas-processing system occurrence of, 9-316
Sodium formate, wood-to-oil process formation of, 3-168

Southern California, water delivery system for, 8-9
Southern California Edison Company, 5-12, 5-24, 7-12
Southern Company, The, 1-284, 9-204
Southern Electric Generating Company, 1-31
Southern Natural Gas Company, 2-64
Southern Pacific Pipelines, Incorporated, 1-180
Southern Railroad, 1-154, 1-156
Southern region:
    Diesel fuel survey of, 3-32 to 3-36
    petroleum products survey of, 3-32
Southwest Africa:
    crude oil, ultimate recovery potential for, 3-92
    lithium ores of, 5-183
    natural gas, ultimate recovery potential for, 2-44
Southwest Experimental Fast Oxide Reactor, 5-110 to 5-111
Southwest Gas Corporation, 2-64
Southwest Power Pool (SPP), 9-266
Southwest Texas crude oil, characteristics of, 3-14 to 3-15
Southwestern Illinois Coal Corporation, 1-31
Soviet Union (see Union of Soviet Socialist Republics)
Soybean, spoiled-land area use of, 1-109
Space heating:
    geothermal energy for, 7-8, 7-43 to 7-58
    waste heat for, 9-351 to 9-353
    (See also Solar climate control)
Space heating oils (see Burner fuel oils)
Space launch vehicles, propellants for, 4-76
Space power, solar cell, 6-56, 6-64 to 6-68
    (See also Fuel cells)
Space shuttle, 6-112 to 6-113
Spacecraft, propellants for, 4-76
    (See also Fuel cells)
Spacer, boiler tube, 9-116 to 9-118
Spain:
    crude oil: production of, 3-87, 3-89
        ultimate recovery potential for, 3-93
    electric power generating capacity of, 9-247 to 9-249
    hydroelectric power production of, 8-4 to 8-5
    metric tons to barrels conversion factor, 3-84
    natural gas, reserves of, 2-43, 2-45
    nuclear installations in, 5-32, 5-36 to 5-38
    railway mileage and motive power of, 9-226
    tidal range at Bilboa and Gibraltar, 8-40
    uranium resources of, 5-176
Spangler, Pennsylvania, mine disaster at, 1-92
Spans, pipeline, 2-73
Specific entropy, definition of, 9-24
Specific gravity:
    API (see API gravity)
    Athabasca bitumen, 3-162 to 3-163
    petroleum fuels, 3-42
    various elements and compounds, 9-36 to 9-37

Specific gravity test, coal, 1-56
Specific gravity type of gas calorimeter, 2-92 to 2-93
Specific heat:
    Athabasca bitumen, 3-163
    cryogenic alloys, 2-112, 2-114
Specific impulse, rocket-engine performance, 4-75 to 4-76, 4-82 to 4-83, 4-85
Specific power, nuclear fission reactor, 5-40
Spectral response, silicon solar cell, 6-59 to 6-61
Spectrophotometric analyzer, nitrogen oxides and sulfur dioxide determined by, 9-363 to 9-370
Speculative potential gas supply:
    definition of, 2-22
    source map of, 2-30
Speed:
    aircraft, versus fuel requirements, 3-39
    electric locomotive, speed-tractive effort curves, ASEA RC$_2$ locomotive, 9-239
    freight locomotive: speed-tractive effort curves, 9-230
    versus tractive effort and throttle position, 9-230
    versus transmission efficiency, 9-231
    railway rolling stock, 9-225, 9-230 to 9-239
    tanker and supertanker: versus squat calculations, 3-213 to 3-215
    versus stopping distance, 3-218 to 3-219
Spencer Chemical Company, 1-282
Spent-air power recovery system, 3-294
Spent shale handling, oil shale retorting, 3-148, 3-151 to 3-153, 3-155
Spherical concentrator, solar, 6-47
Spillage, liquefied natural gas, 2-102
Spinach chloroplast, 4-6
Spiral conveyor, coal, 1-148
Split-beam photometer for gas detection, 9-367 to 9-370
Splitter, hydraulic turbine, 8-11
Splitter column, hydrocracking, 3-235
Spodumene, 5-183
Spoil, contour mining, 1-94 to 1-101
Spokane, Washington, daily terrestrial solar energy received at, 6-96
Spray attemperation, 9-154
Sprayberry Trend oil field, 3-65
Spreader stoker-fired boiler, control system for, 9-138
Spreader stokers, 9-87 to 9-89
Springs, dynamometer testing of, 9-310
Spud, gas burner, 9-55
Spud bit, 3-118
Square collar, oil well drilling use of, 3-113
Squat, seagoing vessel, 3-213 to 3-215
Squaw Creek, California, 7-36
Squaw Creek coal mine, 1-29
Squeezing, coal mine roof definition of, 1-73
Squirreltail, spoiled-land area use of, 1-107

Sri Lanka:
    crude oil: production of, 3-89
        ultimate recovery potential for, 3-93
    natural gas, ultimate recovery potential
        for, 2-45
Stability:
    of coal mine roof, 1-71 to 1-83
    of marine pipeline, 2-77 to 2-84
    of pipeline laying vessel, 2-79
Stabilization, superconducting magnet, 5-
    162
Stabilization control, gas-mixture, 2-97 to
    2-98
Stabilized mirror plasma experiment, 5-141
    to 5-143, 5-145
Stabilizer, oil well drilling, 3-118
    hydrotreating process, 3-236
    platforming process, 3-229
Stack analysis, 9-382
    electric utility computer program for, 9-
        290
    (See also Instrumentation, polluting gas
        and particulate emission)
Stacker, coal, 1-151 to 1-152
Stahl, Albert W., 8-45
Stainless steel, gas turbine use of, 9-183
Stainless steel anode, thermionic
    converter, 9-222
Standard air, definition of, 9-40
Standard boiler, definition of, 9-116
Standard cascade refrigeration cycle, 2-102
    to 2-104
Standard cubic foot, definition of, 1-208
Standard Oil Company (Ohio), 3-153 to 3-
    154
Standardization, bomb calorimeter, 1-69 to
    1-70
Standby generator sets, 9-181 to 9-182
Standby liquid control system, boiling
    water reactor, 5-73
Star coal mine, 1-30
Star feeder, 3-152
Starches, fermentation of, 4-7
Starting system, gas turbine, 9-185 to
    9-187
Start-up:
    boiler control system, 9-155 to 9-156
    gas turbine, 9-181
    nuclear reactor, 5-9, 5-53 to 5-55
    pulverized coal system, 9-77
Start-up power versus power to load,
    magnetohydrodynamic generator, 9-
    212, 9-220
State College, Pennsylvania, daily
    terrestrial solar energy received at, 6-
    96
Stationary energy conversion supply
    system, railway locomotive, 9-224
Stationary power plant, exhaust emissions
    from, 9-195
Steam:
    combustion for generation of, 9-34 to 9-
        49
    adiabatic flame temperature, 9-42 to 9-
        43
    application of fundamental laws, 9-35
        to 9-41

Steam, combustion for generation
    of (Cont.):
    calculations for combustion reactions,
        9-43 to 9-47, 9-109
    combustion constants for various
        fuels, 9-36 to 9-37
    composition of combustion air, 9-
        40
    flue gas analysis, 9-47 to 9-49
    heat of combustion for various fuels,
        9-41 to 9-42
    ignition temperatures, 9-42
costs of, 3-274
generating systems for, 9-17 to 9-22, 9-93
    to 9-121
    bent-tube boilers, 9-103 to 9-107
    boiler developments and trends, 9-19
        to 9-21, 9-93, 9-103 to 9-109, 9-119
        to 9-129
    boiler output versus temperature, 9-
        114
    boiler systems, 9-17 to 9-22, 9-93 to 9-
        121
    coal-fired systems, 9-19, 9-59 to 9-92
    control systems, 9-120 to 9-162
    fire-tube boilers, 9-18
    installations in United States, by
        region, 9-255
    integral-furnace boilers, 9-19, 9-104
    marine boilers, 9-22, 9-115
    modified steam cycles, 9-22
    multipurpose steam power plants, 9-
        32 to 9-33
    nuclear plants, 9-22
        boiling water reactor, 5-58 to 5-64
        gas-cooled fast breeder reactor, 5-
            124 to 5-126, 5-130 to 5-131
        high-temperature gas-cooled
            reactor, 5-47 to 5-48, 5-55, 5-92 to
            5-95
        liquid-metal fast breeder reactor, 5-
            124 to 5-126, 5-130 to 5-131
        pressurized water reactor, 5-75 to 5-
            77
    oil- and gas-fired systems, 9-50 to 9-58
    package boilers, 9-19
    quality of steam produced, 9-97 to 9-
        98
    separation and purification systems, 9-
        99 to 9-103
    Stirling boilers, 9-104 to 9-105
    superheaters and reheaters, 9-113 to
        9-119
    waste heat systems, 3-278 to 3-279
    waste-solid combustion: American
        Thermogen process, 4-19 to 4-20
        Monsanto Landgard process, 4-18
            to 4-19
        supplementary fuel system, 4-21 to
            4-22
    water-cooled furnaces, 9-19, 9-110
    watertube boilers, 9-18, 9-93
geothermal, 7-5 to 7-6, 7-11, 7-16 to 7-24,
    7-35 to 7-36, 7-44
Steam atomizer, oil burner, 9-52
Steam-carbon equilibrium, COGAS
    process, 1-265

Straight-hole, oil well drilling, 3-112 to 3-115

Straight-run gasoline, 3-19, 3-22, 3-225 to 3-226

(*See also* Gasoline)

Straight-tube boiler, 9-93

Straightener, coal mine roof support bolt, 1-81

Strain gage dynamometers, 9-307 to 9-308

Strain measurement, coal mine roof, 1-80 to 1-82

Straits of Dover, tidal range of, 8-44

Stratification, coal bed, 1-71

Stratigraphic trap:
  crude oil, 3-52 to 3-53, 3-59
  natural gas, 2-11 to 2-12
  tar sands, 3-161

Streamline coal mine, 1-29

Streamlining, railway equipment, 9-225 to 9-227

Street and highway lighting, power requirements for, 9-258

Stress corrosion, geothermal system, 7-45 to 7-48

Stress distribution, coal mine roof supports, 1-72

Stress values, superheater and reheater tubes, 9-117

Stretford sulfur removal plant, 2-140

Stretford unit, solvent refining of coal use of, 1-240, 1-283

String, liquefaction, 2-103

Stringer tube, superheater, 9-116 to 9-118

Strip mining, 1-102 to 1-115

Structural engineering, coal mine roof support, 1-71 to 1-83

Structural entrapment:
  crude oil, 3-51 to 3-52, 3-59
  natural gas, 2-11

Structural framing computer program, electric utility, 9-290

Stub-dead-end track coal terminal, 1-164

Stuttgart, University of, 6-150

Styrene, ethylbenzene for, 3-20

Subbituminous coal, 1-15 to 1-27
  carbon dioxide acceptor process gasification of, 1-219 to 1-222
  Lurgi process gasification of, 1-196, 1-219 to 1-222
  ranking of, 1-54
  reserves of, 1-35

Subcooled-nucleate boiling, 9-95

Subcritical operation, boiler, 9-21

Submarine, fast breeder reactor for, 5-111

Submarine slides, 3-112

Subsidence of land, geothermal field, 7-12, 7-30

Substation monitoring and control system computer program, electric utility, 9-288

Substation transmission systems, 9-271 to 9-275

Substitute natural gas (SNG):
  consumption forecast for electric power generation, 5-43
  cost comparison with other fuels, 4-7

Substitute natural gas (SNG) (*Cont.*):
  gas supply versus need for SNG, 2-66 to 2-69
  (*See also* Natural gas)
  heat of combustion comparison with other fuels, 4-7
  pipeline attachment costs of, 2-63
  production processes for: BI-GAS process, 1-212 to 1-218
    biological methane process, 4-16 to 4-18
    carbon dioxide acceptor process, 1-219 to 1-222
    catalytic rich gas process, 2-137 to 2-148
    COGAS process, 1-261 to 1-268
    fluidized-bed hydrogenation process, 2-160 to 2-163
    Hydrane process, 1-223 to 1-231
    hydrocracking-hydrogasification process, 2-164 to 2-170
    Hygas process, 1-201 to 1-211
    in-situ coal gasification process, 1-242 to 1-248
    Lurgi process, 1-188 to 1-200, 1-285 to 1-296
    methane rich gas process, 2-149 to 2-159
    noncatalytic, partial-oxidation process, 2-171 to 2-174
    Toscoal pyrolysis process, 1-272 to 1-275
    Union Carbide Purox process, 4-15 to 4-16
  as supplement to natural gas supplies, 2-51, 2-57

Suction pressure, liquefied natural gas refrigerant system, 2-108

Suction-type, direct-firing pulverized coal system, 9-63

Suction volume, liquefied natural gas refrigerant system, 2-108

Sudan:
  crude oil: production of, 3-89
    ultimate recovery potential for, 3-92
  natural gas, ultimate recovery potential for, 2-44
  railway mileage and motive power of, 9-226

Sudangrass, spoiled-land area use of, 1-108

Suez Canal, tanker accommodation of, 3-210

Sugar beets, fermentation of, 4-7

Sugarcane, 6-139 to 6-140

Sugarland oil field, 3-105

Sukhovo hydroelectric plant, 8-5

Sulfate sulfur, coal content of, 1-51

Sulfide stress cracking, geothermal system, 7-45

Sulfinol process, 2-133 to 2-136

Sulfolane process, 3-368
  solvent used for, 2-133

Sulfur:
  acid soils in mining spoils caused by, 1-103
  air ignition temperature of, 9-42

Supervisor accidents, coal mine, 1-85 to 1-91
Supervisory control:
  of Seneca pumped-storage plant, 8-23 to 8-24
  of tidal power plant, 8-34 to 8-36
Supplemental gas sources, Gas Requirements Committee definition of, 2-51, 2-57
Supply-Technical Advisory Task Force, 1-208
Supplyman accidents, coal mine, 1-85 to 1-91
Support:
  boiler, 9-112 to 9-113
  coal mine roof, 1-71 to 1-83
  superheater and reheater, 9-116 to 9-119
Supported electrode, fuel cell, 4-65
Surface, boiler, 9-93
Surface area, gas-processing system, 9-319
Surface mining, coal, 1-34 to 1-36
Surfactants:
  jet fuel use of, 3-39
  oil well acidizing use of, 3-134
Surge bin, coal, 1-139
SURGE drilling assistance computer program, 3-128
Surge effects, magnetohydrodynamic generators, 9-219 to 9-220
Surinam:
  lithium resources of, 5-183
  railway mileage and motive power of, 9-227
Surry nuclear plants, 5-18, 5-21, 5-24
Surveying, deviation, 3-112
Surveys:
  annual United States petroleum fuel, 3-22 to 3-44
  coal exploration, 1-42 to 1-47
  pipeline, 2-72
  wind power, 6-153 to 6-155
Survival, coal mine disaster, 1-84
Susquehanna nuclear plants, 5-17, 5-24
Svartsengi-Keflavik geothermal water line, 7-51
Swamp, coal formation in, 1-37
Swan Hunter shipyard, 3-213
Swanson River oil field, 3-64
Swaziland:
  crude oil: production of, 3-89
    ultimate recovery potential for, 3-92
  natural gas, ultimate recovery potential for, 2-44
  railway mileage and motive power of, 9-227
Sweden:
  crude oil: production of, 3-87
    ultimate recovery potential for, 3-93
  electric power production of, 9-247
  lithium resources of, 5-183
  locomotive (ASEA RC₂), development of, 9-237 to 9-238
  natural gas, ultimate recovery potential for, 2-45
  nuclear installations in, 5-32 to 5-33, 5-36 to 5-38

Sweden (*Cont.*):
  railway mileage and motive power of, 9-226
  shipyards of, 3-213
    study of ship squat versus speed, 3-213
Sweden State Power Board, 5-32 to 5-33
Sweet crude oils, 3-12
Swell, vessel movement effects of, 3-214
Swelling index, coal, 1-17
Swimming pool, geothermal energy for, 7-50
Switchgear, geothermal system, 7-29, 7-37 to 7-39
Switching locomotives, 9-228, 9-233
Switching operation computer program, electric utility, 9-287
Switching surge computer program, electric utility, 9-284
Switchover problems, gas-heating, 4-31
Switzerland:
  Burgdorf-Thun railway, 9-237 to 9-238
  crude oil: production of, 3-89
    ultimate recovery potential for, 3-95
  gas-cooled fast breeder reactor components, development of, 5-135
  hydroelectric power plants in, 8-17
  nuclear installations in, 5-33, 5-36 to 5-38
  railway mileage and motive power of, 9-226
  Simplon Line, 9-238
Sydney, Australia, tidal range at, 8-40
Symbols:
  for boiler control system, 9-126
  for feedwater control system, 9-132
Synchronization:
  geothermal system, 7-29, 7-37 to 7-39
  pumped-storage plant, 8-19 to 8-24
  tidal power plant, 8-34 to 8-36
Syngas (*see* Substitute natural gas)
Synthane coal gasification process, 1-232 to 1-241
Synthesis gas:
  catalytic rich gas process, 2-144 to 2-147
  methanol process, 2-120
  Sulfinol process for purification of, 2-133 to 2-136
Synthetic chemicals, petroleum-derived (*see* Petrochemicals)
Synthetic crude oil:
  oil shale source of, 3-147 to 3-159
  tar sand source of, 3-159 to 3-167
  wood waste source of, 3-168 to 3-173
Synthetic detergents, petroleum sources for, 3-20
Synthetic ethanol, 4-7 to 4-8
Synthetic fuel oil, 1-269 to 1-271
Synthetic gas, forecast for electric power generation requirements of, 5-43
Synthetic isopropanol, 4-7 to 4-8
Synthetic methane, 4-1 to 4-8
  (*See also* Gasification, coal)
Synthetic natural gas (*see* Substitute natural gas)
Synthetic polymers, 3-266

Texas (*Cont.*):
  NPC petroleum provinces of, **2**-62
  nuclear power installations in, **5**-17 to **5**-18
  sunshine, percent possible at Amarillo and El Paso, **6**-72
  tar sand deposits in, **3**-159, **3**-161
  tidal range at Galveston, **8**-40
  (*See also* Solar energy technology, tower energy collector)
Texas, University of, **3**-116
Texas Eastern Transmission Corporation, **1**-180, **2**-64
Texas Electric Service Company, **1**-144 to **1**-145
Texas Gulf Coast AGA area, **2**-9
Texas Instruments Incorporated, **4**-63
Texas Mid-Continent Oil and Gas Association, **3**-106
Texas Power and Light Company, **1**-144 to **1**-145
Texas Utilities Services, Incorporated, **1**-243, **5**-17 to **5**-18
Textile industry:
  electricity consumption of, **9**-259
  gas requirements of, **2**-52 to **2**-56
Thailand:
  crude oil: production of, **3**-90
    recoverable reserves of, **3**-83
    ultimate recovery potential for, **3**-93
  natural gas, ultimate recovery potential for, **2**-45
  nuclear facilities planned, **5**-33, **5**-36 to **5**-38
  railway mileage and motive power of, **9**-226
Thames Estuary, liquefied gas facility at, **2**-102
Theoretical air, definition of, **9**-34
Theory, wind power, **6**-142 to **6**-143
Thermal coking, **3**-227, **3**-229 to **3**-230
Thermal conductivity:
  ammonia, **6**-127
  low-temperature alloys: aluminum, **2**-107
    nickel, **2**-114
    propane, **6**-127
Thermal conductivity type of hydrocarbon monitor, **9**-389
Thermal cracking, petroleum, **3**-18, **3**-40, **3**-227 to **3**-231, **3**-246
Thermal efficiency:
  Carnot cycle, **9**-24
  coal gasification process, **1**-223
  fuel cell power plants, **4**-71
  nuclear fission reactor, **5**-40
  Rankine cycle, **9**-26 to **9**-27
  regenerative Rankine cycle, **9**-27 to **9**-30
Thermal energy:
  air conditioning system storage of, **9**-194 to **9**-195
  conservation of, **3**-278 to **3**-296, **3**-302
Thermal engines, solar energy satellite, **6**-105 to **6**-108
Thermal expansion, cryogenic nickel alloys, **2**-113
Thermal gasoline, **3**-225, **3**-229 to **3**-230

Thermal liquid metal storage, solar system use of, **6**-23 to **6**-25
Thermal neutron, **5**-8, **5**-46
Thermal nitrogen oxides, **9**-335
Thermal oxidizer, preheat system for, **3**-292
Thermal performance, solar furnace, **6**-47, **6**-53
Thermal pollution:
  comparison of various fuel causes of, **5**-122
  geothermal system, **7**-9, **7**-11 to **7**-12, **7**-28 to **7**-29
Thermal Power Company, **7**-34
Thermal processes, oil recovery improvement with, **3**-143
Thermal reactor, **5**-40, **5**-46
Thermal stability, aviation fuels, **3**-37
Thermal stabilization, superconducting magnet, **5**-162
Thermal watt converter, **9**-298 to **9**-299
Thermalito pumped-storage plant, **8**-18
Thermalization, neutron (*see* Moderator)
Thermeter gas calorimeter, **2**-91 to **2**-93
Thermionic converters, **9**-221 to **9**-223
Thermionic solar energy converters, **6**-14
Thermochemical solar energy converters, **6**-14 to **6**-15
Thermodynamic efficiency, liquefaction process, **2**-106 to **2**-109
Thermodynamics, steam cycle, **9**-23 to **9**-33
  available energy, **9**-30
  Carnot cycle, **9**-24
  cycles, **9**-24 to **9**-30
  energy, work, and heat quantity balances, **9**-23
  entropy concept, **9**-23 to **9**-24
  laws of, **9**-23 to **9**-24
  nuclear plant steam cycle, **9**-30 to **9**-32
  Rankine cycle, **9**-24
  regenerative Rankine cycle, **9**-27 to **9**-30
Thermoelectric cooling, **9**-242 to **9**-246
Thermoelectric power system, spacecraft, **6**-57
Thermoelectric solar energy converters, **6**-15
Thermometer, calorimeter, **1**-63 to **1**-70
Thermonuclear reaction, **5**-46
  (*See also* Fusion power)
Theta pinch plasma experiment, **5**-141, **5**-143, **5**-145
Thickening, coal slurry, **1**-117
Thickness:
  of coal bed, **1**-38 to **1**-41
  of pipeline wall, **2**-73
Thiols, crude oil content of, **3**-254
Thionyl chloride electrolyte, battery use of, **4**-51
Thiophenes, crude oil content of, **3**-254
Third rail, **9**-224, **9**-236 to **9**-237
Thompson oil field, **3**-63
Thomsen type of gas calorimeter, **2**-92 to **2**-93
Thorium, definition of, **5**-46
Thorium-232 fissionable nuclide, **5**-8
Thorium-uranium fuel cycle, high-temperature gas-cooled reactor use of, **5**-88, **5**-96

Three-element boiler feedwater control, 9-131 to 9-133

Three Mile Island nuclear plants, 5-17, 5-24

Three-phase inverter circuit, magnetohydrodynamic system, 9-213

Threshold curves, superconductor, 5-159

Threshold limit value, 9-382, 9-388

Throttle position:
  versus percent rated power, Diesel-electric locomotive, 9-230
  versus tractive effort and speed, freight locomotive, 9-230

Throttle pressure control, boiler, 9-153 to 9-154

Tidal energy, availability of, xxvi–xxvii

Tidal mill, 8-25

Tidal power, source of, 6-120

Tidal power technology, 8-25 to 8-45

Tide forces, drilling structure, 3-110

Tides:
  energy in, 8-36 to 8-37
  in English Channel, 8-37 to 8-45
  Marine Scale for measurement of, 8-27
  number of, 8-41
  resonance phenomena of, 8-37 to 8-45
  spring, 8-27

Tidewater coal transport system, 1-154

Tie-line flow rescheduling and control computer program, electric utility, 9-285

Tiger Leasing Group, 3-194

Tikitere geothermal area, 7-31

Tillamook Bay, Oregon, wind power site at, 6-168

Timbalier Bay oil field, 3-64

Timber drying, geothermal energy for, 7-57

Timbering, coal mine roof, 1-73 to 1-74

Timberman accidents, coal mine, 1-86 to 1-91

Time:
  combustion importance of, 9-34
  magnetohydrodynamic generator power rise, 9-212, 9-220

Time sharing, electric utility computer, 9-294

Timothy, spoiled-land area use of, 1-107 to 1-108

Tin-niobium binary alloy, superconducting magnet use of, 5-161, 5-163, 5-167

Titanium alloys, ocean thermal energy conversion plant heat exchanger use of, 6-126

Titanium-niobium binary alloy, superconducting magnet use of, 5-161, 5-167

TNT, toluene for, 3-19

Toaster, 9-260

Tobacco industry:
  electricity consumption of, 9-259
  gas requirements of, 2-52 to 2-56

Tobago (see Trinidad and Tobago)

Togo:
  crude oil: production of, 3-90
    production and exploratory wells completed, 3-86
    ultimate recovery potential for, 3-92

Togo (Cont.):
  natural gas, ultimate recovery potential for, 2-44
  railway mileage and motive power of, 9-227

Tohuku University, solar furnace at, 6-46 to 6-47

Tokai-Mura gas-cooled reactor, 5-89

Tokamaks, 5-140 to 5-142, 5-145, 5-169

Toktogul hydroelectric plant, 8-5

Tokyo Tankers, 3-210

Toledo, Ohio, coal transfer station at, 1-170

Toledo Edison Company, 5-16

Toluene:
  combustion constants of, 9-36 to 9-37
  crude oil processing as source of, 3-19
  demethylation of, 3-267
  detection of, 9-380

Tom O'Connor oil field, 3-62

Tonga:
  crude oil: production of, 3-90
    ultimate recovery potential for, 3-93
  natural gas, ultimate recovery potential for, 2-45

Tonkin, tides at, 8-37, 8-39

Tonnage, tanker and crude carrier (see Deadweight, tanker and product carrier)

Top tower focused solar energy system, 6-10, 6-26 to 6-33

Topeka, Kansas, daily terrestrial solar energy received at, 6-96

Topping use, thermionic converters for, 9-222

Torex solar energy collector, 6-94

Toroidal devices, plasma confinement, 5-141 to 5-146

Toroidal magnetic coil, 5-164 to 5-165

Torque converter, railway locomotive, 9-233

Torque measurement, 9-307 to 9-311

Torsion-type dynamometer, 9-305 to 9-308

Toscoal process, pyrolysis of coal by, 1-272 to 1-275

Total air approximation chart, flue gas analysis use of, 9-48

Total calorific value gas calorimeter, 2-90

Total carbon, definition of, as element of coal, 1-50

Total combustibles analyzer, 9-377 to 9-378

Total hydrocarbon detectors, 9-381, 9-385 to 9-386

Total hydrogen, definition of, as element of coal, 1-50

Total moisture, definition of, 1-49

Total nitrogen, definition of, as element of coal, 1-50

Total organic carbon (TOC), 9-354

Total proved area, oil field, definition of, 3-72

Total sulfur, definition of, as element of coal, 1-50

Totten correction, rolling frictional resistance, 9-225

Towboat, coal-barge, 1-171, 1-173, 1-176

Tower, aerogenerator, 6-145

Tower platform, offshore drilling, 3-112

United Kingdom (*Cont.*):
National Coal Board, 1-91 to 1-92
nuclear power installations in, 5-33 to 5-34, 5-36 to 5-38
advanced gas reactor, development of, 5-40
Culham Laboratory, 5-146
gas-cooled fast breeder reactor, development of, 5-133
high-temperature gas-cooled nuclear reactor, development of, 5-88 to 5-89
sodium-cooled fast breeder reactor, development of, 5-110 to 5-113
rail tank wagon operations of, 2-194 to 2-195
railway mileage and motive power of, 9-226
tidal range at Dover, Liverpool, and Southampton, 8-40
town gas production in, 2-140
wind power: developments of, 6-150 to 6-151
windmills of the 12th century, 6-147
wind power surveys of Ireland, Scotland, and Wales, 6-154
United Nations, Inter-Governmental Maritime Consultative Organization of, 3-196
United States of America:
Air Force, 3-39, 4-74, 4-83, 9-205 to 9-206
Army Corps of Engineers, 8-20
Army Quartermaster Corps, 6-46 to 6-47
Bureau of the Census, 3-101, 9-249, 9-254, 9-256, 9-259
Bureau of Mines, 1-23 to 1-24, 1-28, 1-49, 1-51, 1-55, 1-74, 1-76, 1-79, 1-85 to 1-92, 1-223 to 1-241, 1-244 to 1-246, 1-269 to 1-271, 2-9 to 2-19, 2-59, 2-100 to 2-102, 3-22, 3-33 to 3-37, 3-39, 3-57, 3-78, 3-84, 3-98 to 3-99, 3-168 to 3-173, 4-24, 9-71
Bureau of Reclamation, 8-6
Coast and Geodetic Survey, 8-40 to 8-41
Coast Guard, 3-178
Department of Agriculture, 1-102 to 1-115
Department of the Interior, 1-201, 1-211, 1-217
Department of Labor, 9-259
Department of Transportation, 2-72, 3-178 to 3-192, 4-58, 9-241
Energy Research and Development Administration, 1-201, 1-211, 1-232 to 1-241, 1-255, 1-276, 1-284, 5-12 to 5-20, 5-24 to 5-38, 5-174
Environmental Protection Agency, 1-225, 2-57, 4-12 to 4-25, 7-11, 9-374 to 9-375, 9-391 to 9-406
Federal Power Commission, 8-3, 8-7 to 8-8, 9-248 to 9-251, 9-261 to 9-263, 9-270
Federal Reserve Board, 9-257, 9-259
Geological Survey, 1-34, 2-9 to 2-19, 2-38 to 2-46, 3-45, 3-78 to 3-79, 3-90
Interstate Commerce Commission, 3-178, 4-70

United States of America (*Cont.*):
National Bureau of Standards, 1-61, 3-41, 9-206
Navy, 3-38
Nuclear Regulatory Commission, 1-242, 5-12 to 5-20, 5-24 to 5-38, 5-40 to 5-41, 5-47, 5-49, 5-108, 5-134, 5-144, 5-174, 5-184
United States Census of Central Electric Stations, 9-263
United States Steel Corporation, 1-31, 1-249 to 1-254
United Technology Corporation, 4-70 to 4-71, 4-77
Universal, Pennsylvania, coal mine at, 1-29
Universal Oil Products Company, 3-12, 3-14 to 3-20, 3-228 to 3-232, 3-235, 3-240, 3-243, 3-250, 3-267 to 3-269, 3-271
Universal-pressure boiler, 9-20 to 9-22, 9-106 to 9-107
Unloading, tankcar, 3-183 to 3-193
Unloading systems, coal, 1-162 to 1-163
Unmanned platform, offshore drilling, 3-108
UOP characterization factor, Athabasca bitumen, 3-163
Upflow expanded-bed hydrodesulfurization factor, 3-257 to 3-259
Upgrading:
Athabasca bitumen, 3-164
low-Btu gases, 2-116 to 2-119
Upper Columbia hydroelectric facilities, 9-261
Upper cylinder lubricants, gasoline content of, 3-27
Upper Volta:
crude oil: production of, 3-90
ultimate recovery potential for, 3-92
natural gas, ultimate recovery potential for, 2-44
solar water heaters, use of, 6-93
Uranium:
absorption properties of uranium-238, 5-7
boiling water reactor fuel use of, 5-67 to 5-69
cost of, 5-11, 5-118, 5-134
definition of, 5-46
enrichment of, 5-44, 5-178 to 5-179
exploration for, 5-179 to 5-180
fission of uranium-235, 5-7
gas-cooled fast breeder reactor fuel use of, 5-126
high-temperature gas-cooled reactor fuel use of, 5-87 to 5-88, 5-94 to 5-96
liquid-metal fast breeder reactor fuel use of, 5-106 to 5-107, 5-110 to 5-111, 5-114 to 5-117, 5-122
packaging and shipment of, 5-179
pressurized water reactor fuel use of, 5-81 to 5-82
processing of, 5-175 to 5-179
reserves of, 5-173 to 5-175
Uranium dioxide, 5-177 to 5-178
(*See also* Fuel elements, rods and bundles, nuclear reactor)

Uranium tetrafluoride, 5-178
Urban Vehicle Design Competition, 4-39
Urban wastes, energy conversion of, 4-12
    to 4-25
Urea gas-processing system, 9-321 to 9-
    322
Urey, 4-27
Uruguay:
    crude oil: production of, 3-90
        ultimate recovery potential for, 3-95
    natural gas, ultimate recovery potential
        for, 2-46
    railway mileage and motive power of, 9-
        226
Ust-Illimsk hydroelectric plant, 8-5
Utah:
    AGA area of, 2-9
    API Survey district of, 3-50
    Bituminous Coal District of, 1-169
    coals in, 1-20, 1-27, 1-32, 1-34 to 1-35, 1-
        92 to 1-93
    crude oil: productive capacity of, 3-53, 3-
        56, 3-60, 3-64
        reserves of, 3-46 to 3-47, 3-57 to 3-59
    daily terrestrial solar energy received at
        Flaming Gorge and Salt Lake City,
        6-96
    electric power generating capacity of, 9-
        253
    emission limits for SO_2 and NO_x, 9-375
    hydroelectric power production of, 8-7
    lithium reserves in Great Salt Lake, 5-
        183
    natural gas: production and reserves of,
        2-10 to 2-13, 2-16 to 2-18
        underground gas storage capacity, 2-
            76
    oil shale, reserves of, 3-147
    tar sand deposits of, 3-159, 3-161, 3-
        163
Utah, University of, 4-58
Utah International Incorporated, 1-31
Utilities, coal requirements of, 1-28
Utility applications, gas turbine, 9-190 to 9-
    191
Utility costs, refinery and process plants, 3-
    274
Utility industry (see Electric utility
    industry)
U-tube steam generator, pressurized water
    reactor use of, 5-75 to 5-77
UWMAK-I fusion power reactor design, 5-
    169 to 5-171

Vacuum chamber, fusion reactor reference
    design, 5-170 to 5-171
Vacuum cleaner, 9-260
Vacuum diode, 9-221
Vacuum distillation, crude oil (see
    Distillation, crude oil)
Vaccum oil field, 3-63
Valdecanas pumped-storage plant, 8-18
Vallecitos boiling water reactor, 5-57
Valley Camp Coal Company, 1-31
Valley fill method, contour mining use of,
    1-96 to 1-97
Valparaiso, Chile, tidal range at, 8-41

Valve(s):
    boiler system control, 9-124 to 9-162
    boiling water reactor, 5-59, 5-63 to 5-66,
        5-72 to 5-73
    gas-cooled fast breeder reactor, 5-128, 5-
        133
    high-temperature gas-cooled reactor, 5-
        47, 5-49, 5-91 to 5-93, 5-95 to 5-99
    hydraulic turbine, 8-22
    insulation of, 3-290
    liquid-metal fast breeder reactor, 5-119,
        5-121
    pressurized water reactor, 5-77
    slide, 1-292
    tankcar, 3-183 to 3-184, 3-186, 3-190
Valve management, steam turbine, 9-164
Valving, pipeline, 2-73
Van oil field, 3-63
Vanadium:
    Athabasca bitumen content of, 3-163
    crude oil content of, 3-13 to 3-17, 3-253
        removal of, 3-262
Vanadium-gallium binary alloy,
    superconducting magnet use of, 5-161
Vanadium-hafnium-zirconium alloy,
    superconducting magnet use of, 5-162
Vanadium hydrides, 4-43
Vanadium-silicon superconductor, 5-162
Vancouver, British Columbia, coal
    pipeline terminus at, 1-180
Vapor-dominated geothermal system, 7-5
    to 7-6, 7-11 to 7-12, 7-34 to 7-42
Vapor explosion, liquefied natural gas, 2-
    102
Vapor-liquid ratio, gasoline, 3-25
Vapor lock:
    jet fuel, 3-37 to 3-38
    methanol-gasoline mixture, 4-46
Vapor pressure:
    aviation gasoline, 3-181
    butane, 3-180
    fuel oil, 3-181
    gasoline, 3-181
    kerosine, 3-180
    liquid propellants, 4-80 to 4-81
    propane, 3-180
Vapor recovery systems, 3-297 to 3-300
Vapor valve, tankcar, 3-183 to 3-184
Variable, process: boiler operation, 9-120
    to 9-162
    Lurgi gasification, 1-193 to 1-194
    pumped-storage plant, 8-23 to 8-24
    Sasol complex, 1-268 to 1-269
Varmeter, 9-303 to 9-304
Varta AG fuel cell, 4-64 to 4-65, 4-69
V. C. No. 1 coal mine, 1-30
Vector analysis, oil well drilling use of, 3-
    115
Vegetable matter, carbonization of, 1-15 to
    1-16, 1-37 to 1-41
Vegetation, spoiled-land area use of, 1-105
    to 1-115
Vehicle dispatch system computer
    program, electric utility, 9-288
Velma, Oklahoma, characteristics of crude
    oil from, 3-14 to 3-15
Velocity:
    coal slurry, 1-179

West Virginia-New York City coal pipeline, 1-180
Wester, A. G., shipyard, 3-211
Western coals, pipeline movement of, 1-183 to 1-185
Western Energy Company, 1-31
Western forms, spoiled-land area use of, 1-110
Western grass species, spoiled-land area use of, 1-107
Western Gulf Basin NPC petroleum province, 2-62
Western Interior Region, coal formation in, 1-37 to 1-40
Western railroad, mountainous profile of, 9-231
Western region:
  coal transportation in, 1-166 to 1-167
  fuel requirements for electric power generation, 9-265 to 9-266
  petroleum products survey of, 3-32 to 3-36
  [See also Pacific Coast District (API Survey)]
Western Rocky Mountain NPC petroleum province, 2-62
Western Samoa:
  crude oil: production of, 3-88
    ultimate recovery potential for, 3-93
  natural gas, ultimate recovery potential for, 2-45
Western Systems Coordinating Council (WSCC), 9-266
Western trees and shrubs, spoiled-land area use of, 1-112
Westinghouse Electric Corporation, 9-163 to 9-164, 9-171 to 9-175, 9-202 to 9-208
Westmoreland County, Pennsylvania, coals in, 1-20
Weston Instruments, 9-296 to 9-304
Wet scrubbing processes:
  nitrogen oxides absorption, 9-318 to 9-323
  sulfur dioxide absorption, 9-312 to 9-318
  sulfur dioxide recovery, 9-324 to 9-327
  sulfur recovery, Claus, 9-328 to 9-334
Wharton coal mine, 1-30
Wheat, spoiled-land area use of, 1-107 to 1-108
Wheatgrass, spoiled-land area use of, 1-107
Wheel flange friction, railway rolling stock, 9-225
Wheel-tractor scrapers, 1-98 to 1-99
Wheelabrator-Frye Incorporated, 1-282 to 1-284
Wheeling, West Virginia, flooding of, 8-8
Whipstock deflection tool, 3-118
White Island volcano, 7-14
White River Oil Shale Group, 3-154
Wicket gate, hydroelectric plant, 8-12, 8-14, 8-24
Width, ship channel, 3-215 to 3-217
Wietze tar sand deposits, 3-160
Wilcox, Stephen, 9-18
Wildlife habitat, reclaimed areas for, 1-114
Wildrye, spoiled-land area use of, 1-107
Wilkes-Barre, Pennsylvania, flooding of, 8-9
Williams No. 98 coal mine, 1-30

Williamson County, Illinois, coals in, 1-20, 1-24
Wilmington oil field, 3-62
Wind energy, electric power from, 6-142 to 6-176
  aerogenerator farms, 6-165
  aerogenerators, 6-165 to 6-170
  atmosphere as a heat engine, 6-146 to 6-147
  classification of wind machines, 6-142
  controls for, 6-145
  economics and cost factors of, 6-172 to 6-174
  energy availability in United States, 6-148
  environmental impact of, 6-171 to 6-172
  generators for, 6-145
  history of, 6-147 to 6-149
  installations of: Denmark, 6-151
    France, 6-149
    Germany, 6-149 to 6-150
    United Kingdom, 6-150 to 6-151
    United States, 6-150, 6-152 to 6-153
    U.S.S.R., 6-149
  peak power handling, 6-171
  rotors for, 6-143 to 6-145
  storage of wind energy, 6-164, 6-170 to 6-171
  system design of, 6-143 to 6-146
  terrain modification of wind power sites, 6-165
  theory of, 6-142 to 6-143
  transmission for, 6-145
  velocity and power duration curves, 6-155 to 6-156, 6-160 to 6-164
  wind power site selection, 6-156 to 6-157
  wind prospecting, 6-157 to 6-165
  wind speeds, 6-158 to 6-160
  wind surveys, 6-153 to 6-155
  wind tunnel modeling, 6-164
Wind loads, drilling structure, 3-107 to 3-110
Wind records, evaluation of, 6-157 to 6-158
Windings, bulb-type generator, 8-34
Windmills (see Wind energy, electric power from)
Windscale AGR gas-cooled reactor, 5-89
Wing aerogenerator, 6-169 to 6-170
Wing tank, definition of, 3-206
Winkler gasifier, 1-188
Winnipeg, Manitoba, pipeline terminal at, 2-65
Wisconsin:
  daily terrestrial solar energy received: at Green Bay and Madison, 6-97
    at Milwaukee, 6-96
  electric power generating capacity of, 9-253
  emission limits for SO_2 and NO_x, 9-375
  gas turbine installations in, 9-192
  hydroelectric power generation of, 8-6
  nuclear power installations in, 5-18, 5-24
Wisconsin, University of:
  fusion blanket studies of, 5-146 to 5-147, 5-165
  UWMAK-I fusion power plant design, 5-169 to 5-171